HANDBOOK

of tables for

APPLIED

ENGINEERING

SCIENCE

SECOND EDITION

EDITORS

RAY E. BOLZ, D.Eng.
Dean of Engineering,
Case Western Reserve University

GEORGE L. TUVE, Sc.D.
Formerly Professor of Engineering,
Case Institute of Technology

A DIVISION OF
THE CHEMICAL RUBBER CO.
18901 Cranwood Parkway · Cleveland, Ohio 44128

Second Printing, 1976
By CRC Press, Inc.

International Standard Book Number 0-8493-0252-8

Former International Standard Book Number 0-87819-252-2

Library of Congress Catalog Card Number 75-117044

HANDBOOK SERIES

Handbook of Chemistry and Physics, 53rd edition
Standard Mathematical Tables, 20th edition
Handbook of tables for Mathematics, 4th edition
Handbook of tables for Organic Compound Identification, 3rd edition
Handbook of Biochemistry, selected data for Molecular Biology, 2nd edition
Handbook of Clinical Laboratory Data, 2nd edition
Manual for Clinical Laboratory Procedures, 2nd edition
Manual of Nuclear Medicine Procedures, 1st edition
Handbook of Laboratory Safety, 2nd edition
Handbook of tables for Probability and Statistics, 2nd edition
Fenaroli's Handbook of Flavor Ingredients, 1st edition
Handbook of Analytical Toxicology, 1st edition
Manual of Laboratory Procedures in Toxicology, 1st edition
Handbook of tables for Applied Engineering Science, 2nd edition
Handbook of Chromatography, 1st edition
Handbook of Environmental Control, 1st edition
Handbook of Food Additives, 2nd edition
Handbook of Radioactive Nuclides, 1st edition
Handbook of Lasers, 1st edition
Handbook of Microbiology, 1st edition
Atlas of Spectral Data and Physical Constants for Organic Compounds, 1st edition
*Handbook of Engineering Medicine and Biology, 1st edition
*Handbook of Marine Sciences, 1st edition
*Handbook of Material Science, 1st edition
*Handbook of Spectroscopy, 1st edition
*Handbook of Laboratory Animal Science

*Currently in preparation.

Division of THE CHEMICAL RUBBER CO.

Editor-in-Chief
Robert C. Weast, Ph.D.
Vice President, Research, Consolidated Gas Service Company, Inc.

Preface to the Second Edition

Desktop access to numerical data as provided by this handbook is often required for preliminary study of a subject, and for planning, reviewing, estimating, and evaluating the feasibility of engineering projects. The convenience of a one-volume databook gives it a uniquely useful position as the "first source" to be consulted in preference to a specialized library. The engineer needs quick access to minimum basic data, even before he reaches the stage of using a computer or consulting specialists. This handbook is designed to serve such needs by providing the practicing engineer and the engineering student with a wide spectrum of data covering many fields of modern engineering with references to more complete sources.

In recognition of the growing importance of the metric SI system of units, special attention has been paid throughout this edition to the presentation of data in metric as well as in conventional units. This has added importance because of the scarcity of tabular data in the international metric units. Comparisons of parallel tables and the frequent appearance of the necessary conversion factors will enable the user to adapt quickly to the new system of units, which will come into common usage during the next decade.

Tables that are new in this edition cover such subjects as lasers, radiation, cryogenics, ultrasonics, semiconductors, high-vacuum techniques, eutectic alloys, and organic and inorganic surface coatings. Another major addition is the expansion of the sections on engineering materials and composites, with more detailed indexing by name, class and usage. The special Index of Properties, the first item in Section 1, allows ready comparisons with respect to a single property, whether physical, chemical, electrical, radiant, mechanical or thermal.

The sections providing data on the environment and its control have been greatly expanded. Human safety, environmental hazards, and the protection of the individual are given major attention. Pollution of air and water, the problems of solid wastes, and protection against the hazards of toxicity, ionizing radiation, fire, noise and odors are all treated in some detail. Data are included on the outdoor and the indoor environment, including temperature, humidity, solar radiation, acoustics and illumination.

The readiness with which a single item of data may be located is most important to the usefulness of any reference work. The user of this handbook is assisted not only by a comprehensive index but also by cross references and by numerically keyed subject headings at the top of each page. Each table is self-explanatory, with units, abbreviations and symbols clearly defined, and tabular material subdivided for easy reading.

This handbook is condensed from thousands of pages of original data and the objective is to quote accepted and critically evaluated data, to indicate ranges and trends, and to provide specific references. In the many sources available, there are often two or more values reported for the same item. When recognized authorities do not agree, an effort is made to favor the latest and most authentic source, or to use the values selected by engineering standards or codes or those published by the National Standard Reference Data System (NSRDS).

In attempting to cover the broad engineering field in a single volume, the editors are well aware of wide gaps and omissions. Most of the common mathematical tables have been intentionally omitted, as have the practical and empirical data on industrial and shop processes. The book is designed as a supplement to the excellent specialized literature already available covering the theory, the mathematical formulations, the applications, and the descriptions of equipment for use in the various fields.

Ideas for the contents of this handbook have come from many sources and the editors are concerned that there may be inadvertent omissions of recognition and credit. Furthermore, data of importance to engineers that appear in other volumes of the CRC Handbook Series, and particularly in the *Handbook of Chemistry and Physics,* have been freely quoted. Appreciation is expressed to the CRC Handbook editors and contributors. Colleagues, librarians, authors, compilers, and publishers have generously cooperated and sincere thanks are offered to each. Every effort has been made to determine and quote the original sources of specific data. Permissions to reprint from these sources have been freely granted by authors and copyright owners and are acknowledged with thanks.

The best reference book is actually designed by its users, and the editors urgently invite suggestions and corrections.

Cleveland, Ohio
December 1972

RAY E. BOLZ
GEORGE L. TUVE

ADVISORY BOARD

CONTRIBUTORS

Edmond E. Bisson
Fluid System Components Division
Lewis Research Center, NASA
Cleveland, Ohio 44135

Roland Breitwieser
Thermionics Branch
Lewis Research Center, NASA
Cleveland, Ohio 44135

William L. Bryan
School of Engineering
Case Western Reserve University
Cleveland, Ohio 44106

Donald H. Buckley
Lubrication Fundamentals Section
Lewis Research Center, NASA
Cleveland, Ohio 44135

Henry Burlage, Jr.
Propulsion Division
California Institute of Technology
Pasadena, California 91103

Robert T. Carpenter
Space Nuclear Systems Division
U.S. Atomic Energy Commission
Washington, D.C. 20545

Donald C. Dilley
Consulting Metallurgical Engineer
Cleveland Heights, Ohio 44118

Lowell C. Domholdt
Fenn College of Engineering
Cleveland State University
Cleveland, Ohio 44115

John C. Evvard
Chief Scientist
Lewis Research Center, NASA
Cleveland, Ohio 44135

Arthur P. Fraas
Reactor Division
Oak Ridge National Laboratory, AEC
Oak Ridge, Tennessee 37830

Donald F. Gibbons
Center for the Study of Materials
Case Western Reserve University
Cleveland, Ohio 44106

Frank J. Hendel
Technical Director, U.S. Air Force
SAMTEC
Vandenberg Air Force Base, California
93437

Ralph Hultgren
Department of Metallurgy
University of California
Berkeley, California 94720

Duane P. Jordan
Department of Mechanical Engineering
Texas Tech University
Lubbock, Texas 79409

Louis Meites
Department of Chemistry
Clarkson College of Technology
Potsdam, New York 13676

Harry W. Mergler
School of Engineering
Case Western Reserve University
Cleveland, Ohio 44106

Floro Miraldi
School of Engineering
Case Western Reserve University
Cleveland, Ohio 44106

J. H. Mulligan
National Academy of Engineering
Washington, D.C. 20418

Glenn Murphy
Department of Nuclear Engineering
Iowa State University
Ames, Iowa 50010

David H. Navon
Department of Electrical Engineering
University of Massachusetts
Amherst, Massachusetts 01002

CONTRIBUTORS *(Continued)*

T. D. Northwood
Building Physics Section
National Research Council of Canada
Ottawa, Ontario, Canada, KIA-OR6

Yoh-Han Pao
School of Engineering
Case Western Reserve University
Cleveland, Ohio 44106

Charles J. Parker
Research and Development Laboratory
Corning Glass Works
Corning, New York 14830

Marcia L. Parsons
Sears Library
Case Western Reserve University
Cleveland, Ohio 44106

William M. Ritchey
Millis Science Center
Case Western Reserve University
Cleveland, Ohio 44106

B. W. Roberts
Research and Development Center
General Electric Company
Schenectady, New York 12301

Glenn T. Seaborg
Lawrence Berkeley Laboratory
University of California
Berkeley, California 94720

Robert S. Shankland
Department of Physics
Case Western Reserve University
Cleveland, Ohio 44106

Gail P. Smith
Technical Staff Services
Corning Glass Works
Corning, New York 14830

Lauriston S. Taylor
National Council on Radiation Protection
and Measurements
Washington, D.C. 20014

William A. Tomazic
Chemical Propulsion Division
Lewis Research Center, NASA
Cleveland, Ohio 44135

Y. S. Touloukian
Thermophysical Properties Research Center
Purdue University
West Lafayette, Indiana 47906

Richard L. Tuve
Consultant on Fire Technology
Johns Hopkins Applied Physics Laboratory
Silver Spring, Maryland 20910

Donald J. Vild
Architectural Services
Libbey-Owens-Ford Company
Toledo, Ohio 43605

L. L. Winter
Materials Research Division
Union Carbide Corporation
Cleveland, Ohio 44101

ACKNOWLEDGMENTS

Assistance and suggestions from the following individuals are acknowledged with thanks.

Abraham Abramowitz
City University of New York
New York, New York 10031

Edward G. Bobalek
University of Maine
Orono, Maine 04473

Byron L. Bondurant
Ohio State University
Columbus, Ohio 43210

LeRoy W. Davis
Nevada Engineering & Technology Corporation
Long Beach, California 90802

Charles M. Harman
Duke University
Durham, North Carolina 27706

Charles A. Harper
Westinghouse Electric Corporation
Baltimore, Maryland 21203

Herbert S. Isbin
University of Minnesota
Minneapolis, Minnesota 55455

John E. Kaufman
Illuminating Engineering Society
New York, New York 10017

William F. Kerka
Cleveland State University
Cleveland, Ohio 44115

Irving Lefkowitz
Case Western Reserve University
Cleveland, Ohio 44106

Terry McGowan
General Electric Company
Cleveland, Ohio 44112

Douglas Muster
University of Houston
Houston, Texas 77004

Herbert B. Nottage
University of California
Los Angeles, California 90024

C. O. Peterson, Jr.
PPG Industries, Inc.
Pittsburgh, Pennsylvania 15222

Albert B. Pincince
Camp Dresser & McKee
Pasadena, California 91101

Eli Reshotko
Case Western Reserve University
Cleveland, Ohio 44106

Adel S. Sanda
Case Western Reserve University
Cleveland, Ohio 44106

Ronald G. Schultz
Cleveland State University
Cleveland, Ohio 44115

L. G. Seigel
United States Steel Corporation
Pittsburgh, Pennsylvania 15221

Walter W. Soroka
University of California
Berkeley, California 94720

William L. Wolfe
University of Arizona
Tucson, Arizona 85721

Daniel K. Wright, Jr.
Case Western Reserve University
Cleveland, Ohio 44106

TABLE OF CONTENTS

TABLE OF CONTENTS *(Continued)*

Engineering Materials and Their Properties

INDEX OF PROPERTIES

In listing the properties of engineering materials in this handbook, an effort has been made to show all the important properties of a single material in the same table. The materials that are useful for a variety of purposes are listed in Section 1. Each is classified according to its usual physical state, as a gas, liquid, or solid.

The chemical elements and many chemical compounds appear in Section 3. Certain materials for distinctive uses, e.g., fuels, solvents, nuclides, and magnetic materials, are treated in that section of the handbook that is devoted to the special application.

The following index of properties is provided for quick reference to specific properties. The index items contain the common name of the property and the page location of the tables listing the property in question. Data on a specific material may also be located in the main index.

Table 1-1. PROPERTIES OF DRY AIR
AT ATMOSPHERIC PRESSURE—ENGLISH UNITS*

For properties of dry air in SI units, see Table 1-2.

SYMBOLS AND UNITS:

$°K$ = degrees Kelvin

$°R$ = degrees Rankine

$°F$ = degrees Fahrenheit

ρ = density, lb_m/ft^3

c_p = specific heat capacity,
      $Btu/lb_m°R = cal/g°K$

c_p/c_v = specific heat capacity, ratio dimensionless

μ = viscosity. For $lb_m/sec\,ft$ multiply by 10^{-4}.

k = thermal conductivity, $Btu/hr\,ft\,°R$

Pr = Prandtl number, dimensionless

h = enthalpy, Btu/lb_m. For cal/g multiply by 0.5555.

V_s = sound velocity, ft/sec

Temperature			Properties							
$°K$	$°R$	$°F$	ρ	c_p	c_p/c_v	μ	k	Pr	h	V_s
100	180	−280	.2247	.2456		.0466	.00534	.770	42.3	651
110	198	−262	.2033	.2440	1.4202	.0513	.00586	.768	46.7	685
120	216	−244	.1858	.2430	1.4166	.0559	.00638	.766	51.1	717
130	234	−226	.1711	.2423	1.4139	.0604	.00690	.763	55.5	747
140	252	−208	.1586	.2418	1.4119	.0648	.00742	.761	59.8	776
150	270	−190	.1478	.2414	1.4102	.0691	.00793	.758	64.2	804
160	288	−172	.1384	.2411	1.4089	.0733	.00844	.754	68.5	831
170	306	−154	.1301	.2408	1.4079	.0774	.00895	.750	72.9	856
180	324	−136	.1228	.2406	1.4071	.0815	.00946	.746	77.2	882
190	342	−118	.1163	.2405	1.4064	.0854	.00996	.743	81.5	906
200	360	−100	.1104	.2404	1.4057	.0892	.01045	.739	85.8	930
205	369	−91	.1078	.2403	1.4055	.0911	.01070	.738	88.0	941
210	378	−82	.1051	.2403	1.4053	.0930	.01095	.736	90.2	953
215	387	−73	.1027	.2403	1.4050	.0949	.01119	.734	92.3	964
220	396	−64	.1003	.2402	1.4048	.0967	.01143	.732	94.5	976
225	405	−55	.0981	.2402	1.4046	.0986	.01168	.731	96.7	987
230	414	−46	.0959	.2402	1.4044	.1004	.01191	.729	98.8	998
235	423	−37	.0939	.2402	1.4042	.1022	.01215	.727	101.0	1008
240	432	−28	.0919	.2401	1.4040	.1039	.01239	.725	103.1	1019
245	441	−19	.0901	.2401	1.4038	.1057	.01263	.724	105.3	1030
250	450	−10	.0882	.2401	1.4036	.1074	.01287	.722	107.5	1040
255.4	459.7	0	.0865	.2401	1.4034	.1092	.01310	.721	109.6	1051
260	468	8	.0848	.2401	1.4032	.1109	.01334	.719	111.8	1061
265	477	17	.0832	.2402	1.4030	.1126	.01357	.717	114.0	1071
270	486	26	.0817	.2402	1.4029	.1143	.01380	.716	116.1	1081
275	495	35	.0802	.2402	1.4026	.1160	.01403	.715	118.3	1091
280	504	44	.0787	.2402	1.4024	.1176	.01426	.713	120.4	1101
285	513	53	.0774	.2402	1.4022	.1192	.01448	.711	122.6	1111
290	522	62	.0760	.2403	1.4020	.1208	.01472	.710	124.8	1120
295	531	71	.0747	.2403	1.4018	.1224	.01494	.709	126.9	1130
300	540	80	.0735	.2404	1.4017	.1241	.01516	.708	129.1	1140
305	549	89	.0723	.2404	1.4015	.1257	.01539	.707	131.3	1149
310	558	98	.0711	.2405	1.4013	.1272	.01561	.705	133.4	1158
315	567	107	.0700	.2405	1.4010	.1287	.01583	.704	135.6	1167
320	576	116	.0689	.2406	1.4008	.1303	.01606	.703	137.7	1177
325	585	125	.0678	.2407	1.4006	.1319	.01627	.702	139.9	1186
330	594	134	.0668	.2407	1.4004	.1334	.01649	.701	142.1	1195
335	603	143	.0658	.2408	1.4001	.1349	.01670	.700	144.2	1204
340	612	152	.0648	.2409	1.3999	.1364	.01692	.699	146.4	1213
345	621	161	.0639	.2410	1.3996	.1379	.01713	.698	148.6	1221

Table 1-1. PROPERTIES OF DRY AIR
AT ATMOSPHERIC PRESSURE—ENGLISH UNITS (Continued)

Temperature			Properties							
°K	°R	°F	ρ	c_p	c_p/c_v	μ	k	Pr	h	V_s
350	630	170	.0630	.2411	1.3993	.1394	.01735	.697	150.7	1230
355	639	179	.0621	.2411	1.3990	.1409	.01758	.696	152.9	1239
360	648	188	.0612	.2412	1.3987	.1423	.01779	.695	155.1	1247
365	657	197	.0604	.2413	1.3984	.1437	.01800	.694	157.3	1256
370	666	206	.0595	.2415	1.3981	.1452	.01820	.693	159.4	1264
375	675	215	.0588	.2416	1.3978	.1465	.01841	.692	161.6	1273
380	684	224	.0580	.2417	1.3975	.1479	.01862	.691	163.8	1281
385	693	233	.0572	.2418	1.3971	.1494	.01883	.690	166.0	1289
390	702	242	.0565	.2420	1.3968	.1508	.01904	.690	168.1	1298
395	711	251	.0559	.2421	1.3964	.1522	.01925	.689	170.3	1306
400	720	260	.0551	.2422	1.3961	.1536	.01945	.689	172.5	1314
410	738	278	.0537	.2425	1.3953	.1563	.01985	.688	176.9	1330
420	756	296	.0525	.2428	1.3946	.1590	.02026	.687	181.2	1346
430	774	314	.0512	.2432	1.3938	.1617	.02066	.686	185.6	1361
440	792	332	.0501	.2435	1.3929	.1643	.02106	.684	190.0	1377
450	810	350	.0490	.2439	1.3920	.1670	.02144	.684	194.4	1392
460	828	368	.0479	.2443	1.3911	.1695	.02183	.683	198.8	1407
470	846	386	.0469	.2447	1.3901	.1719	.02222	.682	203.2	1421
480	864	404	.0459	.2451	1.3892	.1744	.02260	.681	207.6	1436
490	882	422	.0450	.2456	1.3881	.1769	.02298	.680	212.0	1450
500	900	440	.0441	.2460	1.3871	.1794	.02335	.680	216.4	1464
510	918	458	.0432	.2465	1.3861	.1818	.02373	.680	220.8	1478
520	936	476	.0424	.2469	1.3851	.1842	.02409	.680	225.3	1492
530	954	494	.0416	.2474	1.3840	.1867	.02445	.680	229.7	1506
540	972	512	.0408	.2479	1.3829	.1891	.02482	.680	234.2	1520
550	990	530	.0400	.2484	1.3818	.1914	.02518	.680	238.7	1533
560	1008	548	.0393	.2490	1.3806	.1937	.02554	.680	243.1	1546
570	1026	566	.0386	.2495	1.3795	.1960	.02589	.680	247.6	1559
580	1044	584	.0380	.2500	1.3783	.1983	.02624	.680	252.1	1572
590	1062	602	.0373	.2506	1.3772	.2005	.02659	.680	256.6	1585
600	1080	620	.0367	.2511	1.3760	.2027	.02694	.680	261.1	1597
620	1116	656	.0355	.2522	1.3737	.2071	.02762	.681	270.2	1622
640	1152	692	.0344	.2533	1.3714	.2115	.02829	.682	279.3	1647
660	1188	728	.0334	.2545	1.3691	.2156	.02896	.682	288.4	1671
680	1224	764	.0324	.2556	1.3668	.2198	.02962	.683	297.6	1695
700	1260	800	.0315	.2568	1.3646	.2239	.03026	.684	306.8	1718
720	1296	836	.0306	.2579	1.3623	.2279	.03089	.685	316.1	1742
740	1332	872	.0298	.2591	1.3601	.2320	.03151	.686	325.4	1764
760	1368	908	.0290	.2602	1.3580	.2359	.03214	.687	334.8	1787
780	1404	944	.0282	.2613	1.3559	.2398	.03275	.688	344.1	1808
800	1440	980	.0275	.2624	1.354	.2435	.03337	.689	353.6	1830
850	1530	1070	.0259	.2653	1.349	.2529	.03485	.693	377.3	1883
900	1620	1160	.0245	.2678	1.345	.2618	.03627	.696	401.3	1934
950	1710	1250	.0231	.2704	1.340	.2705	.03768	.699	425.5	1985
1000	1800	1340	.0220	.2728	1.336	.2790	.03903	.702	450.0	2032
1100	1980	1520	.0200	.2774	1.329	.2954			499.5	2126
1200	2160	1700	.0184	.2817	1.322	.3108			549.8	2215
1300	2340	1880	.0169	.2860	1.316	.3256			600.9	2300
1400	2520	2060	.0157	.2900	1.310	.3398			652.7	2381
1500	2700	2240	.0147	.2940	1.304	.3535			705.3	2459
1600	2880	2420	.0138	.2984	1.299	.3667			758.6	2535
1800	3240	2780	.0122	.3076	1.288	.3917			867.7	2676
2000	3600	3140	.0110	.3196	1.274				980.5	2807
2400	4320	3860	.0092	.3760	1.238				1226.8	3033
2800	5040	4580	.0078	.5396	1.196				1547.3	3225

*Condensed and computed from: "Tables of Thermal Properties of Gases", National Bureau of Standards Circular 564, U.S. Government Printing Office, November 1955.

Table 1-2. PROPERTIES OF DRY AIR AT ATMOSPHERIC PRESSURE—SI UNITS*

For properties of dry air in English units, see Table 1-1.

SYMBOLS AND UNITS:

K = absolute temperature, degrees Kelvin
deg C = temperature, degrees Celsius
deg F = temperature, degrees Fahrenheit
ρ = density, kg/m^3
c_p = specific heat capacity, $kJ/kg \cdot K$
c_p/c_v = specific heat capacity ratio, dimensionless
μ = viscosity. For $N \cdot s/m^2$ ($= kg/m \cdot s$) multiply tabulated values by 10^{-6}
k = thermal conductivity, $MW/m \cdot K$
Pr = Prandtl number, dimensionless
h = enthalpy, kJ/kg
V_s = sound velocity, m/s

Temperature			Properties							
K	deg C	deg F	ρ	c_p	c_p/c_v	μ	k	Pr	h	V_s
100	−173.15	−280	3.598	1.028		6.929	9.248	.770	98.42	198.4
110	−163.15	−262	3.256	1.022	1.420 2	7.633	10.15	.768	108.7	208.7
120	−153.15	−244	2.975	1.017	1.416 6	8.319	11.05	.766	118.8	218.4
130	−143.15	−226	2.740	1.014	1.413 9	8.990	11.94	.763	129.0	227.6
140	−133.15	−208	2.540	1.012	1.411 9	9.646	12.84	.761	139.1	236.4
150	−123.15	−190	2.367	1.010	1.410 2	10.28	13.73	.758	149.2	245.0
160	−113.15	−172	2.217	1.009	1.408 9	10.91	14.61	.754	159.4	253.2
170	−103.15	−154	2.085	1.008	1.407 9	11.52	15.49	.750	169.4	261.0
180	−93.15	−136	1.968	1.007	1.407 1	12.12	16.37	.746	179.5	268.7
190	−83.15	−118	1.863	1.007	1.406 4	12.71	17.23	.743	189.6	276.2
200	−73.15	−100	1.769	1.006	1.405 7	13.28	18.09	.739	199.7	283.4
205	−68.15	−91	1.726	1.006	1.405 5	13.56	18.52	.738	204.7	286.9
210	−63.15	−82	1.684	1.006	1.405 3	13.85	18.94	.736	209.7	290.5
215	−58.15	−73	1.646	1.006	1.405 0	14.12	19.36	.734	214.8	293.9
220	−53.15	−64	1.607	1.006	1.404 8	14.40	19.78	.732	219.8	297.4
225	−48.15	−55	1.572	1.006	1.404 6	14.67	20.20	.731	224.8	300.8
230	−43.15	−46	1.537	1.006	1.404 4	14.94	20.62	.729	229.8	304.1
235	−38.15	−37	1.505	1.006	1.404 2	15.20	21.04	.727	234.9	307.4
240	−33.15	−28	1.473	1.005	1.404 0	15.47	21.45	.725	239.9	310.6
245	−28.15	−19	1.443	1.005	1.403 8	15.73	21.86	.724	244.9	313.8
250	−23.15	−10	1.413	1.005	1.403 6	15.99	22.27	.722	250.0	317.1
255	−18.15	−1	1.386	1.005	1.403 4	16.25	22.68	.721	255.0	320.2
260	−13.15	8	1.359	1.005	1.403 2	16.50	23.08	.719	260.0	323.4
265	−8.15	17	1.333	1.005	1.403 0	16.75	23.48	.717	265.0	326.5
270	−3.15	26	1.308	1.006	1.402 9	17.00	23.88	.716	270.1	329.6
275	+1.85	35	1.285	1.006	1.402 6	17.26	24.28	.715	275.1	332.6
280	6.85	44	1.261	1.006	1.402 4	17.50	24.67	.713	280.1	335.6
285	11.85	53	1.240	1.006	1.402 2	17.74	25.06	.711	285.1	338.5
290	16.85	62	1.218	1.006	1.402 0	17.98	25.47	.710	290.2	341.5
295	21.85	71	1.197	1.006	1.401 8	18.22	25.85	.709	295.2	344.4
300	26.85	80	1.177	1.006	1.401 7	18.46	26.24	.708	300.2	347.3
305	31.85	89	1.158	1.006	1.401 5	18.70	26.63	.707	305.3	350.2
310	36.85	98	1.139	1.007	1.401 3	18.93	27.01	.705	310.3	353.1
315	41.85	107	1.121	1.007	1.401 0	19.15	27.40	.704	315.3	355.8
320	46.85	116	1.103	1.007	1.400 8	19.39	27.78	.703	320.4	358.7

*Condensed and computed from: "Tables of Thermal Properties of Gases", National Bureau of Standards Circular 564, U.S. Government Printing Office, November 1955.

Table 1-2. PROPERTIES OF DRY AIR AT ATMOSPHERIC PRESSURE—SI UNITS *(Continued)*

Temperature			Properties							
K	deg C	deg F	ρ	c_p	c_p/c_v	μ	k	Pr	h	V_s
325	51.85	125	1.086	1.008	1.400 6	19.63	28.15	.702	325.4	361.4
330	56.85	134	1.070	1.008	1.400 4	19.85	28.53	.701	330.4	364.2
335	61.85	143	1.054	1.008	1.400 1	20.08	28.90	.700	335.5	366.9
340	66.85	152	1.038	1.008	1.399 9	20.30	29.28	.699	340.5	369.6
345	71.85	161	1.023	1.009	1.399 6	20.52	29.64	.698	345.6	372.3
350	76.85	170	1.008	1.009	1.399 3	20.75	30.03	.697	350.6	375.0
355	81.85	179	0.994 5	1.010	1.399 0	20.97	30.39	.696	355.7	377.6
360	86.85	188	0.980 5	1.010	1.398 7	21.18	30.78	.695	360.7	380.2
365	91.85	197	0.967 2	1.010	1.398 4	21.38	31.14	.694	365.8	382.8
370	96.85	206	0.953 9	1.011	1.398 1	21.60	31.50	.693	370.8	385.4
375	101.85	215	0.941 3	1.011	1.397 8	21.81	31.86	.692	375.9	388.0
380	106.85	224	0.928 8	1.012	1.397 5	22.02	32.23	.691	380.9	390.5
385	111.85	233	0.916 9	1.012	1.397 1	22.24	32.59	.690	386.0	393.0
390	116.85	242	0.905 0	1.013	1.396 8	22.44	32.95	.690	391.0	395.5
395	121.85	251	0.893 6	1.014	1.396 4	22.65	33.31	.689	396.1	398.0
400	126.85	260	0.882 2	1.014	1.396 1	22.86	33.65	.689	401.2	400.4
410	136.85	278	0.860 8	1.015	1.395 3	23.27	34.35	.688	411.3	405.3
420	146.85	296	0.840 2	1.017	1.394 6	23.66	35.05	.687	421.5	410.2
430	156.85	314	0.820 7	1.018	1.393 8	24.06	35.75	.686	431.7	414.9
440	166.85	332	0.802 1	1.020	1.392 9	24.45	36.43	.684	441.9	419.6
450	176.85	350	0.784 2	1.021	1.392 0	24.85	37.10	.684	452.1	424.2
460	186.85	368	0.767 7	1.023	1.391 1	25.22	37.78	.683	462.3	428.7
470	196.85	386	0.750 9	1.024	1.390 1	25.58	38.46	.682	472.5	433.2
480	206.85	404	0.735 1	1.026	1.389 2	25.96	39.11	.681	482.8	437.6
490	216.85	422	0.720 1	1.028	1.388 1	26.32	39.76	.680	493.0	442.0
500	226.85	440	0.705 7	1.030	1.387 1	26.70	40.41	.680	503.3	446.4
510	236.85	458	0.691 9	1.032	1.386 1	27.06	41.06	.680	513.6	450.6
520	246.85	476	0.678 6	1.034	1.385 1	27.42	41.69	.680	524.0	454.9
530	256.85	494	0.665 8	1.036	1.384 0	27.78	42.32	.680	534.3	459.0
540	266.85	512	0.653 5	1.038	1.382 9	28.14	42.94	.680	544.7	463.2
550	276.85	530	0.641 6	1.040	1.381 8	28.48	43.57	.680	555.1	467.3
560	286.85	548	0.630 1	1.042	1.380 6	28.83	44.20	.680	565.5	471.3
570	296.85	566	0.619 0	1.044	1.379 5	29.17	44.80	.680	575.9	475.3
580	306.85	584	0.608 4	1.047	1.378 3	29.52	45.41	.680	586.4	479.2
590	316.85	602	0.598 0	1.049	1.377 2	29.84	46.01	.680	596.9	483.2
600	326.85	620	0.588 1	1.051	1.376 0	30.17	46.61	.680	607.4	486.9
620	346.85	656	0.569 1	1.056	1.373 7	30.82	47.80	.681	628.4	494.5
640	366.85	692	0.551 4	1.061	1.371 4	31.47	48.96	.682	649.6	502.1
660	386.85	728	0.534 7	1.065	1.369 1	32.09	50.12	.682	670.9	509.4
680	406.85	764	0.518 9	1.070	1.366 8	32.71	51.25	.683	692.2	516.7
700	426.85	800	0.504 0	1.075	1.364 6	33.32	52.36	.684	713.7	523.7
720	446.85	836	0.490 1	1.080	1.362 3	33.92	53.45	.685	735.2	531.0
740	466.85	872	0.476 9	1.085	1.360 1	34.52	54.53	.686	756.9	537.6
760	486.85	908	0.464 3	1.089	1.358 0	35.11	55.62	.687	778.6	544.6
780	506.85	944	0.452 4	1.094	1.355 9	35.69	56.68	.688	800.5	551.2
800	526.85	980	0.441 0	1.099	1.354	36.24	57.74	.689	822.4	557.8
850	576.85	1 070	0.415 2	1.110	1.349	37.63	60.30	.693	877.5	574.1
900	626.85	1 160	0.392 0	1.121	1.345	38.97	62.76	.696	933.4	589.6
950	676.85	1 250	0.371 4	1.132	1.340	40.26	65.20	.699	989.7	604.9
1 000	726.85	1 340	0.352 9	1.142	1.336	41.53	67.54	.702	1 046	619.5
1 100	826.85	1 520	0.320 8	1.161	1.329	43.96			1 162	648.0
1 200	926.85	1 700	0.294 1	1.179	1.322	46.26			1 279	675.2
1 300	1 026.85	1 880	0.271 4	1.197	1.316	48.46			1 398	701.0
1 400	1 126.85	2 060	0.252 1	1.214	1.310	50.57			1 518	725.9
1 500	1 220.85	2 240	0.235 3	1.231	1.304	52.61			1 640	749.4
1 600	1 326.85	2 420	0.220 6	1.249	1.299	54.57			1 764	772.6
1 800	1 526.85	2 780	0.196 0	1.288	1.288	58.29			2 018	815.7
2 000	1 726.85	3 140	0.176 4	1.338	1.274				2 280	855.5
2 400	2 126.85	3 860	0.146 7	1.574	1.238				2 853	924.4
2 800	2 526.85	4 580	0.124 5	2.259	1.196				3 599	983.1

Table 1-3.　PSYCHROMETRIC TABLES*

Properties of Moist Air at 29.92 in. Hg (101 325 N/m²)

SYMBOLS AND UNITS:[a]

P_s = pressure of steam at saturation, psia. For N/m² multiply by 6894.8.

W_s = humidity ratio at saturation, mass of water vapor associated with unit mass of dry air.

V_a = specific volume of dry air, cu ft per lb. For m³/kg multiply by 0.062420.

V_s = specific volume of saturated mixture, cu ft per lb dry air. For m³/kg dry air multiply by 0.062420.

[b]h_a = specific enthalpy of dry air, Btu/lb. For J/kg multiply by 2326.

h_s = specific enthalpy of saturated mixture, Btu per lb dry air. For J/kg dry air multiply by 2326.3.

s_s = specific entropy of saturated mixture, Btu/°R per lb dry air. For J/K·kg dry air multiply by 4186.8.

Temperature		Properties						
°F	°C	P_s	W_s	V_a	V_s	h_a	h_s	s_s
− 100	− 73.33	.00002	—	9.046	9.046	− 24.04	− 24.04	− .0590
− 90	− 67.78	.00005	—	9.300	9.300	− 21.63	− 21.63	− .0524
− 80	− 62.22	.00012	—	9.553	9.553	− 19.23	− 19.22	− .0459
− 70	− 56.67	.00024	.00001	9.806	9.806	− 16.82	− 16.81	− .0397
− 60	− 51.11	.00050	.00002	10.059	10.059	− 14.42	− 14.39	− .0335
− 50	− 45.56	.00098	.00004	10.313	10.314	− 12.01	− 11.97	− .0275
− 45	− 42.78	.00135	.00006	10.440	10.441	− 10.81	− 10.76	− .0246
− 40	− 40	.00186	.00008	10.566	10.567	− 9.609	− 9.526	− .0217
− 38	− 38.89	.00211	.00009	10.617	10.619	− 9.129	− 9.035	− .0205
− 36	− 37.78	.00239	.00010	10.668	10.670	− 8.648	− 8.542	− .0193
− 34	− 36.67	.00270	.00011	10.718	10.720	− 8.168	− 8.047	− .0182
− 32	− 35.56	.00305	.00013	10.769	10.771	− 7.687	− 7.551	− .0170
− 30	− 34.44	.00344	.00015	10.820	10.822	− 7.207	− 7.053	− .0158
− 28	− 33.33	.00388	.00017	10.870	10.873	− 6.726	− 6.553	− .0147
− 26	− 32.22	.00436	.00019	10.921	10.924	− 6.246	− 6.050	− .0135
− 24	− 31.11	.00490	.00021	10.972	10.976	− 5.765	− 5.546	− .0123
− 22	− 30	.00551	.00023	11.022	11.026	− 5.285	− 5.039	− .0112
− 20	− 28.89	.00618	.00026	11.073	11.078	− 4.804	− 4.527	− .0100
− 18	− 27.78	.00693	.00029	11.124	11.129	− 4.324	− 4.014	− .0088
− 16	− 26.67	.00776	.00033	11.174	11.180	− 3.843	− 3.495	− .0077
− 14	− 25.56	.00868	.00037	11.225	11.232	− 3.363	− 2.974	− .0065
− 12	− 24.44	.00969	.00041	11.275	11.283	− 2.882	− 2.446	− .0053
− 10	− 23.33	.01082	.00046	11.326	11.334	− 2.402	− 1.915	− .0041
8	− 22.22	.01207	.00051	11.376	11.385	− 1.922	− 1.379	− .0029
− 6	− 21.11	.01344	.00057	11.427	11.437	− 1.441	− 0.835	− .0017
− 4	− 20	.01496	.00064	11.477	11.489	− 0.961	− 0.286	− .0005
− 2	− 18.89	.01664	.00071	11.528	11.541	− 0.480	0.271	.0007
0	− 17.78	.01849	.00079	11.578	11.593	0.000	0.835	.0019
1	− 17.22	.01948	.00083	11.604	11.619	0.240	1.120	.0025
2	− 16.67	.02052	.00087	11.629	11.645	0.480	1.408	.0032
3	− 16.11	.02161	.00092	11.654	11.671	0.721	1.698	.0038
4	− 15.56	.02276	.00097	11.679	11.697	0.961	1.991	.0044
5	− 15	.02396	.00102	11.705	11.724	1.201	2.286	.0051
6	− 14.44	.02521	.00107	11.730	11.750	1.441	2.583	.0057
7	− 13.89	.02653	.00113	11.756	11.777	1.681	2.883	.0064
8	− 13.33	.02791	.00119	11.781	11.803	1.922	3.188	.0070
9	− 12.78	.02936	.00125	11.806	11.830	2.162	3.494	.0077
10	− 12.22	.03087	.00132	11.831	11.856	2.402	3.803	.0083
11	− 11.67	.03246	.00138	11.857	11.883	2.642	4.116	.0090
12	− 11.11	.03412	.00145	11.882	11.910	2.882	4.432	.0097
13	− 10.56	.03585	.00153	11.907	11.936	3.123	4.753	.0103
14	− 10	.03767	.00161	11.933	11.963	3.363	5.076	.0110
15	− 9.44	.03957	.00169	11.958	11.990	3.603	5.403	.0117
16	− 8.89	.04156	.00177	11.983	12.017	3.843	5.735	.0124
17	− 8.33	.04363	.00186	12.009	12.044	4.083	6.071	.0131
18	− 7.78	.04581	.00195	12.034	12.072	4.324	6.412	.0138
19	− 7.22	.04808	.00205	12.059	12.099	4.564	6.756	.0145
20	− 6.67	.05045	.00215	12.084	12.126	4.804	7.106	.0153
21	− 6.11	.05293	.00226	12.110	12.154	5.044	7.460	.0160
22	− 5.56	.05552	.00237	12.135	12.181	5.284	7.820	.0168

Table 1-3.   PSYCHROMETRIC TABLES *(Continued)*

Temperature		Properties						
°F	°C	P_s	W_s	V_a	V_s	h_a	h_s	s_s
23	−5	.05823	.00249	12.160	12.209	5.525	8.186	.0175
24	−4.44	.06105	.00261	12.186	12.237	5.765	8.557	.0183
25	−3.89	.06400	.00273	12.211	12.265	6.005	8.934	.0191
26	−3.33	.06708	.00287	12.236	12.293	6.245	9.317	.0199
27	−2.78	.07030	.00300	12.262	12.321	6.485	9.706	.0207
28	−2.22	.07365	.00315	12.287	12.349	6.726	10.10	.0215
29	−1.67	.07715	.00330	12.312	12.377	6.966	10.51	.0223
30	−1.11	.08080	.00345	12.338	12.406	7.206	10.92	.0232
31	−.56	.08461	.00362	12.363	12.434	7.446	11.33	.0240
32	0	.08858	.00379	12.388	12.463	7.686	11.76	.0249
33	+.56	.092227	.00394	12.413	12.492	7.927	12.17	.0257
34	1.11	.095999	.00411	12.438	12.520	8.167	12.59	.0266
35	1.67	.099908	.00428	12.464	12.549	8.407	13.01	.0274
36	2.22	.10396	.00445	12.489	12.578	8.647	13.44	.0283
37	2.78	.10815	.00463	12.514	12.607	8.887	13.87	.0292
38	3.33	.11249	.00482	12.540	12.637	9.128	14.32	.0301
39	3.89	.11699	.00501	12.565	12.666	9.368	14.77	.0310
40	4.44	.12164	.00521	12.590	12.695	9.608	15.23	.0319
41	5	.12646	.00542	12.616	12.725	9.848	15.70	.0328
42	5.56	.13145	.00564	12.641	12.755	10.09	16.17	.0338
43	6.11	.13660	.00586	12.666	12.785	10.33	16.66	.0347
44	6.67	.14194	.00609	12.691	12.815	10.57	17.15	.0357
45	7.22	.14746	.00633	12.717	12.846	10.81	17.65	.0367
46	7.78	.15317	.00658	12.742	12.876	11.05	18.16	.0377
47	8.33	.15907	.00684	12.767	12.907	11.29	18.68	.0387
48	8.89	.16517	.00710	12.792	12.938	11.53	19.21	.0398
49	9.44	.17148	.00737	12.818	12.969	11.77	19.75	.0408
50	10	.17799	.00766	12.843	13.001	12.01	20.30	.0419
51	10.56	.18473	.00795	12.868	13.032	12.25	20.86	.0430
52	11.11	.19169	.00826	12.894	13.064	12.49	21.44	.0441
53	11.67	.19888	.00857	12.919	13.097	12.73	22.02	.0453
54	12.22	.20630	.00889	12.944	13.129	12.97	22.62	.0465
55	12.78	.21397	.00923	12.970	13.162	13.21	23.22	.0476
56	13.33	.22188	.00958	12.995	13.195	13.45	23.84	.0488
57	13.89	.23006	.00993	13.020	13.228	13.69	24.48	.0501
58	14.44	.23849	.01030	13.045	13.261	13.93	25.12	.0513
59	15	.24720	.01069	13.071	13.295	14.17	25.78	.0526
60	15.56	.25618	.01108	13.096	13.329	14.41	26.46	.0539
61	16.11	.26545	.01149	13.121	13.363	14.65	27.15	.0552
62	16.67	.27502	.01191	13.147	13.398	14.89	27.85	.0566
63	17.22	.28488	.01235	13.172	13.433	15.13	28.57	.0579
64	17.78	.29505	.01280	13.197	13.468	15.37	29.31	.0594
65	18.33	.30554	.01326	13.222	13.504	15.61	30.06	.0608
66	18.89	.31636	.01374	13.247	13.539	15.86	30.83	.0623
67	19.44	.32750	.01424	13.273	13.576	16.10	31.62	.0638
68	20	.33900	.01475	13.298	13.613	16.34	32.42	.0653
69	20.56	.35084	.01528	13.323	13.650	16.58	33.25	.0668
70	21.11	.36304	.01582	13.348	13.687	16.82	34.09	.0684
71	21.67	.37561	.01639	13.373	13.724	17.06	34.95	.0700
72	22.22	.38856	.01697	13.398	13.762	17.30	35.83	.0717
73	22.78	.40190	.01757	13.424	13.801	17.54	36.74	.0734
74	23.33	.41564	.01819	13.449	13.841	17.78	37.66	.0751
75	23.89	.42979	.01882	13.474	13.881	18.02	38.61	.0769
76	24.44	.44435	.01948	13.499	13.921	18.26	39.57	.0787
77	25	.45935	.02016	13.525	13.962	18.50	40.57	.0806
78	25.56	.47478	.02086	13.550	14.003	18.74	41.58	.0825
79	26.11	.49066	.02158	13.575	14.045	18.98	42.62	.0844
80	26.67	.50701	.02233	13.601	14.087	19.22	43.69	.0864
81	27.22	.52382	.02310	13.626	14.130	19.46	44.78	.0884
82	27.78	.54112	.02389	13.651	14.174	19.70	45.90	.0905

Table 1-3. PSYCHROMETRIC TABLES *(Continued)*

Temperature		Properties						
°F	°C	P_s	W_s	V_a	V_s	h_a	h_s	s_s
83	28.33	.55892	.02471	13.676	14.218	19.94	47.04	.0926
84	28.89	.57722	.02555	13.702	14.262	20.18	48.22	.0948
85	29.44	.59604	.02642	13.727	14.308	20.42	49.43	.0970
86	30	.61540	.02731	13.752	14.354	20.66	50.66	.0993
87	30.56	.63530	.02824	13.777	14.401	20.90	51.93	.1016
88	31.11	.65575	.02919	13.803	14.448	21.14	53.23	.1040
89	31.67	.67678	.03017	13.828	14.496	21.39	54.56	.1064
90	32.22	.69838	.03118	13.853	14.545	21.63	55.93	.1089
91	32.78	.72059	.03223	13.879	14.595	21.87	57.33	.1115
92	33.33	.74340	.03330	13.904	14.645	22.11	58.78	.1141
93	33.89	.76684	.03441	13.929	14.697	22.35	60.25	.1168
94	34.44	.79091	.03556	13.954	14.749	22.59	61.77	.1195
95	35	.81564	.03673	13.980	14.802	22.83	63.32	.1223
96	35.56	.84103	.03795	14.005	14.856	23.07	64.92	.1252
97	36.11	.86711	.03920	14.030	14.911	23.31	66.55	.1282
98	36.67	.89388	.04049	14.056	14.967	23.55	68.23	.1312
99	37.22	.92137	.04182	14.081	15.023	23.79	69.96	.1343
100	37.78	.94959	.04319	14.106	15.081	24.03	71.73	.1375
102	38.89	1.0083	.04606	14.137	15.200	24.51	75.42	.1441
104	40	1.0700	.04911	14.207	15.324	24.99	79.31	.1510
106	41.11	1.1351	.05234	14.258	15.452	25.47	83.42	.1584
108	42.22	1.2035	.05578	14.308	15.586	25.95	87.76	.1660
110	43.33	1.2754	.05944	14.359	15.724	26.43	92.34	.1742
112	44.44	1.3510	.06333	14.409	15.869	26.92	97.18	.1826
114	45.56	1.4305	.06746	14.460	16.020	27.40	102.3	.1916
116	46.67	1.5139	.07185	14.510	16.178	27.88	107.7	.2011
118	47.78	1.6014	.07652	14.561	16.343	28.36	113.5	.2111
120	48.89	1.6933	.08149	14.611	16.516	28.84	119.5	.2216
122	50	1.7897	.08678	14.662	16.696	29.32	126.0	.2329
124	51.11	1.8907	.09242	14.712	16.886	29.80	132.8	.2446
126	52.22	1.9966	.09841	14.763	17.086	30.29	140.1	.2571
128	53.33	2.1075	.1048	14.813	17.295	30.77	147.8	.2703
130	54.44	2.2237	.1116	14.864	17.516	31.25	155.9	.2844
132	55.56	2.3452	.1189	14.915	17.749	31.73	164.7	.2993
134	56.67	2.4725	.1267	14.965	17.994	32.21	174.0	.3151
136	57.78	2.6055	.1350	15.016	18.253	32.69	183.9	.3318
138	58.89	2.7446	.1439	15.066	18.528	33.17	194.4	.3496
140	60	2.8900	.1534	15.117	18.819	33.66	205.7	.3686
142	61.11	3.0419	.1636	15.167	19.128	34.14	217.7	.3888
144	62.22	3.2006	.1745	15.218	19.457	34.62	230.6	.4104
146	63.33	3.3662	.1862	15.268	19.807	35.10	244.4	.4335
148	64.44	3.5390	.1989	15.319	20.181	35.58	259.3	.4583
150	65.56	3.7194	.2125	15.369	20.580	36.06	275.3	.4848
155	68.33	4.2046	.2514	15.496	21.709	37.27	320.8	.5599
160	71.11	4.7424	.2990	15.622	23.068	38.47	376.3	.6511
165	73.89	5.3372	.3581	15.748	24.733	39.68	445.0	.7629
170	76.67	5.9936	.4327	15.874	26.812	40.88	531.5	.9030
175	79.44	6.7168	.5292	16.001	29.476	42.09	643.2	1.083
180	82.22	7.5119	.6578	16.127	32.997	43.29	791.8	1.319
185	85	8.3845	.8363	16.253	37.854	44.50	997.7	1.646
190	87.78	9.3403	1.099	16.379	44.959	45.70	1301	2.122
195	90.56	10.386	1.519	16.506	56.291	46.91	1784	2.879
200	93.33	11.526	2.295	16.632	77.142	48.12	2677	4.266

Note: The P_s column in this table gives the vapor pressure of pure water at temperature intervals of one degree Fahrenheit. For the latest data on vapor pressures at intervals of 0.1°C, from 0–100°C, see "Vapor Pressure Equation for Water", A. Wexler and L. Greenspan, *J. Res. Nat. Bur. Stand.*, 75A(3): 213–229, May–June 1971.

[a]The Btu in this table is defined as 778.3 ft lbf in the source. The conversion factors listed are consistent with this definition.

[b]For very low barometric pressures and high wet-bulb temperatures, the values of h_a given in this table are somewhat low. For corrections see "ASHRAE Handbook of Fundamentals".

*Condensed with permission from: "ASHRAE Handbook of Fundamentals", American Society of Heating, Refrigerating and Air-Conditioning Engineers, 1972.

Table 1-4.   PSYCHROMETRIC TABLE—SI UNITS*

Properties of Moist Air at 101 325 N/m²

For properties of moist air in English units, see Table 1-3.

SYMBOLS AND UNITS:

P_s = pressure of water vapor at saturation, N/m²
W_s = humidity ratio at saturation, mass of water vapor associated with unit mass of dry air
V_a = specific volume of dry air, m³/kg
V_s = specific volume of saturated mixture, m³/kg dry air
$h_a{}^a$ = specific enthalpy of dry air, kJ/kg
h_s = specific enthalpy of saturated mixture, kJ/kg dry air
s_s = specific entropy of saturated mixture, J/K·kg dry air

Temperature			Properties						
C	K	F	P_s	W_s	V_a	V_s	h_a	h_s	s_s
−40	233.15	−40	12.838	0.000 079 25	0.659 61	0.659 68	−22.35	−22.16	−90.659
−30	243.15	−22	37.992	0.000 234 4	0.688 08	0.688 33	−12.29	−11.72	−46.732
−25	248.15	−13	63.248	0.000 390 3	0.702 32	0.702 75	−7.265	−6.306	−24.706
−20	253.15	−4	103.19	0.000 637 1	0.716 49	0.717 24	−2.236	−0.6653	−2.2194
−15	258.15	+5	165.18	0.001 020	0.730 72	0.731 91	+2.794	5.318	21.189
−10	263.15	14	259.72	0.001 606	0.744 95	0.746 83	7.823	11.81	46.104
−5	268.15	23	401.49	0.002 485	0.759 12	0.762 18	12.85	19.04	73.365
0	273.15	32	610.80	0.003 788	0.773 36	0.778 04	17.88	27.35	104.14
5	278.15	41	871.93	0.005 421	0.787 59	0.794 40	22.91	36.52	137.39
10	283.15	50	1 227.2	0.007 658	0.801 76	0.811 63	27.94	47.23	175.54
15	288.15	59	1 704.4	0.010 69	0.816 00	0.829 98	32.97	59.97	220.22
20	293.15	68	2 337.2	0.014 75	0.830 17	0.849 83	38.00	75.42	273.32
25	298.15	77	3 167.0	0.020 16	0.844 34	0.871 62	43.03	94.38	337.39
30	303.15	86	4 242.8	0.027 31	0.858 51	0.896 09	48.07	117.8	415.65
35	308.15	95	5 623.4	0.036 73	0.872 74	0.924 06	53.10	147.3	512.17
40	313.15	104	7 377.6	0.049 11	0.886 92	0.956 65	58.14	184.5	532.31
45	318.15	113	9 584.8	0.065 36	0.901 15	0.995 35	63.17	232.0	783.06
50	323.15	122	12 339	0.086 78	0.915 32	1.042 3	68.21	293.1	975.27
55	328.15	131	15 745	0.115 2	0.929 49	1.100 7	73.25	372.9	1 221.5
60	333.15	140	19 925	0.153 4	0.943 72	1.174 8	78.29	478.5	1 543.5
65	338.15	149	25 014	0.205 5	0.957 90	1.272 1	83.33	621.4	1 973.6
70	343.15	158	31 167	0.278 8	0.972 07	1.404 2	88.38	820.5	2 564.8
75	348.15	167	38 554	0.385 8	0.986 30	1.592 4	93.42	1 110	3 412.8
80	353.15	176	47 365	0.551 9	1.000 5	1.879 1	98.47	1 557	4 710.9
85	358.15	185	57 809	0.836 3	1.014 6	2.363 2	103.5	2 321	6 892.6
90	363.15	194	70 112	1.416	1.028 8	3.340 9	108.6	3 876	11 281

Note: The P_s column in this table gives the vapor pressure of pure water at temperature intervals of five degrees Celsius. For the latest data on vapor pressures at intervals of 0.1 deg C, from 0–100 deg C, see "Vapor Pressure Equation for Water", A. Wexler and L. Greenspan, *J. Res. Nat. Bur. Stand.*, 75A(3):213–229, May–June 1971.

[a]For very low barometric pressures and high wet-bulb temperatures, the values of h_a in this table are somewhat low; for corrections see "ASHRAE Handbook of Fundamentals".

*Computed from: Psychrometric Tables, in "ASHRAE Handbook of Fundamentals", American Society of Heating, Refrigerating and Air-Conditioning Engineers, 1972.

Table 1-5.   WATER VAPOR AT LOW PRESSURES—SI UNITS*

Perfect Gas Behavior $pv/T = R = 0.461\,51$ kJ/kg·K

SYMBOLS AND UNITS:

t = thermodynamic temperature, deg C

T = thermodynamic temperature, K

$pv = RT$, kJ/kg

u_o = specific internal energy at zero pressure, kJ/kg

h_o = specific enthalpy at zero pressure, kJ/kg

s_l = specific entropy of semiperfect vapor at 0.1 MN/m², kJ/kg·K

ψ_l = specific Helmholtz free energy of semiperfect vapor at 0.1 MN/m², kJ/kg

ζ_l = specific Gibbs free energy of semiperfect vapor at 0.1 MN/m², kJ/kg

p_r = relative pressure, pressure of semiperfect vapor at zero entropy, TN/m²

v_r = relative specific volume, specific volume of semiperfect vapor at zero entropy, mm³/kg

c_{po} = specific heat capacity at constant pressure for zero pressure, kJ/kg·K

c_{vo} = specific heat capacity at constant volume for zero pressure, kJ/kg·K

$k = c_{po}/c_{vo}$ = isentropic exponent, $-(\partial \log p/\partial \log v)_s$

t	T	pv	u_o	h_o	s_l	ψ_l	ζ_l	p_r	v_r	c_{po}	c_{vo}	k
0	273.15	126.06	2 375.5	2 501.5	6.804 2	516.9	643.0	.252 9	498.4	1.858 4	1.396 9	1.330 4
10	283.15	130.68	2 389.4	2 520.1	6.871 1	443.9	574.6	.292 3	447.0	1.860 1	1.398 6	1.330 0
20	293.15	135.29	2 403.4	2 538.7	6.935 7	370.2	505.5	.336 3	402.4	1.862 2	1.400 7	1.329 5
30	303.15	139.91	2 417.5	2 557.4	6.998 2	296.0	435.9	.385 0	363.4	1.864 7	1.403 1	1.328 9
40	313.15	144.52	2 431.5	2 576.0	7.058 7	221.1	365.6	.439 0	329.2	1.867 4	1.405 9	1.328 3
50	323.15	149.14	2 445.6	2 594.7	7.117 5	145.6	294.7	.498 6	299.1	1.870 5	1.409 0	1.327 5
60	333.15	153.75	2 459.7	2 613.4	7.174 5	69.5	223.2	.564 2	272.5	1.873 8	1.412 3	1.326 8
70	343.15	158.37	2 473.8	2 632.2	7.230 0	−7.2	151.2	.636 3	248.9	1.877 4	1.415 9	1.325 9
80	353.15	162.98	2 488.0	2 651.0	7.284 0	−84.3	78.6	.715 2	227.9	1.881 2	1.419 7	1.325 1
90	363.15	167.60	2 502.2	2 669.8	7.336 6	−162.1	5.5	.801 5	209.1	1.885 2	1.423 7	1.324 2
100	373.15	172.21	2 516.5	2 688.7	7.387 8	−240.3	−68.1	.895 7	192.26	1.889 4	1.427 9	1.323 2
120	393.15	181.44	2 545 1	2 726.6	7.486 7	−398.3	−216.8	1.109 7	163.50	1.898 3	1.436 7	1.321 2
140	413.15	190.67	2 573.9	2 764.6	7.581 1	−558.2	−367.5	1.361 7	140.03	1.907 7	1 446 2	1.319 1
160	433.15	199.90	2 603.0	2 802.9	7.671 5	−720.0	−520.1	1.656 4	120.69	1.917 7	1.456 2	1.316 9
180	453.15	209.13	2 632.2	2 841.3	7.758 3	−883.5	−674.4	1.999 1	104.61	1.928 1	1.466 6	1.314 7
200	473.15	218.4	2 661.6	2 880.0	7.841 8	−1 048.7	−830.4	2.396	91.15	1.938 9	1.477 4	1.312 4
300	573.15	264.5	2 812.3	3 076.8	8.218 9	−1 898.4	−1 633.9	5.423	48.77	1.997 5	1.536 0	1.300 5
400	673.15	310.7	2 969.0	3 279.7	8.545 1	−2 783.1	−2 472.5	10.996	28.25	2.061 4	1.599 9	1.288 5
500	773.15	356.8	3 132.4	3 489.2	8.835 2	−3 699	−3 342	20.61	17.310	2.128 7	1.667 2	1.276 8
600	873.15	403.0	3 302.5	3 705.5	9.098 2	−4 642	−4 239	36.45	11.056	2.198 0	1.736 5	1.265 8
700	973.15	449.1	3 479.7	3 928.8	9.340 3	−5 610	−5 161	61.58	7.293	2.268 3	1.806 8	1.255 4
800	1 073.15	495.3	3 663.9	4 159.2	9.565 5	−6 601	−6 106	100.34	4.936	2.338 7	1.877 1	1.245 9
900	1 173.15	541.4	3 855.1	4 396.5	9.776 9	−7 615	−7 073	158.63	3.413	2.407 8	1.946 2	1.237 1
1 000	1 273.15	587.6	4 053.1	4 640.6	9.976 6	−8 649	−8 061	244.5	2.403	2.474 4	2.012 8	1.299 3
1 100	1 373.15	633.7	4 257.5	4 891.2	10.166 1	−9 702	−9 068	368.6	1.719	2.536 9	2.075 4	1.222 4
1 200	1 473.15	679.9	4 467.9	5 147.8	10.346 4	−10 774	−10 094	544.9	1.248	2.593 8	2.132 3	1.216 4
1 300	1 573.15	726.0	4 683.7	5 409.7	10.518 4	−11 863	−11 137	791.0	.918	2.643 1	2.181 6	1.211 5

*Adapted from: "Steam Tables", J.H. Keenan, F.G. Keyes, P.G. Hill, and J.G. Moore, John Wiley & Sons, Inc., 1969 (International Edition—Metric Units).

REFERENCE

For other steam tables in metric units, see "Steam Tables in SI Units", Ministry of Technology, London, 1970.

Table 1-6.　SATURATED STEAM AND SATURATED WATER (TEMPERATURE)**

"1967 ASME Steam Tables"

For steam properties in SI units, see Tables 1-5 and 1-7.

Temp, °F	Press, psia	Volume, ft³/lb_m Water, v_f	Evap, v_{fg}	Steam, v_g	Enthalpy, Btu/lb_m Water, h_f	Evap, h_{fg}	Steam, h_g	Entropy, Btu/lb_m °F Water, s_f	Evap, s_{fg}	Steam, s_g	Temp, °F
705.47	3208.2	0.05078	0.00000	0.05078	906.0	0.0	906.0	1.0612	0.0000	1.0612	705.47
705.0	3198.3	0.04427	0.01304	0.05730	873.0	61.4	934.4	1.0329	0.0527	1.0856	705.0
704.5	3187.8	0.04233	0.01822	0.06055	861.9	85.3	947.2	1.0234	0.0732	1.0967	704.5
704.0	3177.2	0.04108	0.02192	0.06300	854.2	102.0	956.2	1.0169	0.0876	1.1046	704.0
703.5	3166.8	0.04015	0.02489	0.06504	848.2	115.2	963.5	1.0118	0.0991	1.1109	703.5
703.0	3156.3	0.03940	0.02744	0.06684	843.2	126.4	969.6	1.0076	0.1087	1.1163	703.0
702.5	3145.9	0.03878	0.02969	0.06847	838.9	136.1	974.9	1.0039	0.1171	1.1210	702.5
702.0	3135.5	0.03824	0.03173	0.06997	835.0	144.7	979.7	1.0006	0.1246	1.1252	702.0
701.5	3125.2	0.03777	0.03361	0.07138	831.5	152.6	984.0	0.9977	0.1314	1.1291	701.5
701.0	3114.9	0.03735	0.03536	0.07271	828.2	159.8	988.0	0.9949	0.1377	1.1326	701.0
700.5	3104.6	0.03697	0.03701	0.07397	825.2	166.5	991.7	0.9924	0.1435	1.1359	700.5
700.0	3094.3	0.03662	0.03857	0.07519	822.4	172.7	995.2	0.9901	0.1490	1.1390	700.0
699.0	3073.9	0.03600	0.04149	0.07749	817.3	184.2	1001.5	0.9858	0.1590	1.1447	699.0
698.0	3053.6	0.03546	0.04420	0.07966	812.6	194.6	1007.2	0.9818	0.1681	1.1499	698.0
697.0	3033.5	0.03498	0.04674	0.08172	808.4	204.0	1012.4	0.9783	0.1764	1.1547	697.0
696.0	3013.4	0.03455	0.04916	0.08371	804.4	212.8	1017.2	0.9749	0.1841	1.1591	696.0
695.0	2993.5	0.03415	0.05147	0.08563	800.6	221.0	1021.7	0.9718	0.1914	1.1632	695.0
694.0	2973.7	0.03379	0.05370	0.08749	797.1	228.8	1025.9	0.9689	0.1983	1.1671	694.0
693.0	2954.0	0.03345	0.05587	0.08931	793.8	236.1	1029.9	0.9660	0.2048	1.1708	693.0
692.0	2934.5	0.03313	0.05797	0.09110	790.5	243.1	1033.6	0.9634	0.2110	1.1744	692.0
690.0	2895.7	0.03256	0.06203	0.09459	784.5	256.1	1040.6	0.9583	0.2227	1.1810	690.0
688.0	2857.4	0.03204	0.06595	0.09799	778.8	268.2	1047.0	0.9535	0.2337	1.1872	688.0
686.0	2819.5	0.03157	0.06976	0.10133	773.4	279.5	1052.9	0.9490	0.2439	1.1930	686.0
684.0	2782.1	0.03114	0.07349	0.10463	768.2	290.2	1058.4	0.9447	0.2537	1.1984	684.0
682.0	2745.1	0.03074	0.07716	0.10790	763.3	300.4	1063.6	0.9406	0.2631	1.2036	682.0
680.0	2708.6	0.03037	0.08080	0.11117	758.5	310.1	1068.5	0.9365	0.2720	1.2086	680.0
678.0	2672.5	0.03002	0.08440	0.11442	753.8	319.4	1073.2	0.9326	0.2807	1.2133	678.0
676.0	2636.8	0.02970	0.08799	0.11769	749.2	328.5	1077.6	0.9287	0.2892	1.2179	676.0
674.0	2601.5	0.02939	0.09156	0.12096	744.7	337.2	1081.9	0.9249	0.2974	1.2223	674.0
672.0	2566.6	0.02911	0.09514	0.12424	740.2	345.7	1085.9	0.9212	0.3054	1.2266	672.0
670.0	2532.2	0.02884	0.09871	0.12755	735.8	354.0	1089.8	0.9174	0.3133	1.2307	670.0
668.0	2498.1	0.02858	0.10229	0.13087	731.5	362.1	1093.5	0.9137	0.3210	1.2347	668.0
666.0	2464.4	0.02834	0.10588	0.13421	727.1	370.0	1097.1	0.9100	0.3286	1.2387	666.0
664.0	2431.1	0.02811	1.10947	0.13757	722.9	377.7	1100.6	0.9064	0.3361	1.2425	664.0
662.0	2398.2	0.02789	0.11306	0.14095	718.8	385.1	1103.9	0.9028	0.3434	1.2462	662.0
660.0	2365.7	0.02768	0.11663	0.14431	714.9	392.1	1107.0	0.8995	0.3502	1.2498	660.0
658.0	2333.5	0.02748	0.12023	0.14771	711.1	399.0	1110.1	0.8963	0.3570	1.2533	658.0
656.0	2301.7	0.02728	0.12387	0.15115	707.4	405.7	1113.1	0.8931	0.3637	1.2567	656.0
654.0	2270.3	0.02709	0.12754	0.15463	703.7	412.2	1115.9	0.8899	0.3702	1.2601	654.0
652.0	2239.2	0.02691	0.13124	0.15816	700.0	418.7	1118.7	0.8868	0.3767	1.2634	652.0
650.0	2208.4	0.02674	0.13499	0.16173	696.4	425.0	1121.4	0.8837	0.3830	1.2667	650.0
648.0	2178.1	0.02657	0.13876	0.16534	692.9	431.1	1124.0	0.8806	0.3893	1.2699	648.0
646.0	2148.0	0.02641	0.14258	0.16899	689.4	437.2	1126.6	0.8776	0.3954	1.2730	646.0
644.0	2118.3	0.02625	0.14644	0.17269	685.9	443.1	1129.0	0.8746	0.4015	1.2761	644.0
642.0	2088.9	0.02610	0.15033	0.17643	682.5	448.9	1131.4	0.8716	0.4075	1.2791	642.0
640.0	2059.9	0.02595	0.15427	0.18021	679.1	454.6	1133.7	0.8686	0.4134	1.2821	640.0
638.0	2031.2	0.02580	0.15824	0.18405	675.8	460.2	1136.0	0.8657	0.4193	1.2850	638.0
636.0	2002.8	0.02566	0.16226	0.18792	672.4	465.7	1138.1	0.8628	0.4251	1.2879	636.0
634.0	1974.7	0.02553	0.16633	0.19185	669.1	471.1	1140.2	0.8599	0.4307	1.2907	634.0
632.0	1947.0	0.02539	0.17044	0.19583	665.9	476.4	1142.2	0.8571	0.4364	1.2934	632.0
630.0	1919.5	0.02526	0.17459	0.19986	662.7	481.6	1144.2	0.8542	0.4419	1.2962	630.0
628.0	1892.4	0.02514	0.17880	0.20394	659.5	486.7	1146.1	0.8514	0.4474	1.2988	628.0
626.0	1865.6	0.02501	0.18306	0.20807	656.3	491.7	1148.0	0.8486	0.4529	1.3015	626.0
624.0	1839.0	0.02489	0.18737	0.21226	653.1	496.6	1149.8	0.8458	0.4583	1.3041	624.0
622.0	1812.8	0.02477	0.19173	0.21650	650.0	501.5	1151.5	0.8430	0.4636	1.3066	622.0
620.0	1786.9	0.02466	0.19615	0.22081	646.9	506.3	1153.2	0.8403	0.4689	1.3092	620.0
618.0	1761.2	0.02455	0.20063	0.22517	643.8	511.0	1154.8	0.8375	0.4742	1.3117	618.0
616.0	1735.9	0.02444	0.20516	0.22960	640.8	515.6	1156.4	0.8348	0.4794	1.3141	616.0
614.0	1710.8	0.02433	0.20976	0.23409	637.8	520.2	1158.0	0.8321	0.4845	1.3166	614.0
612.0	1686.1	0.02422	0.21442	0.23865	634.8	524.7	1159.5	0.8294	0.4896	1.3190	612.0
610.0	1661.6	0.02412	0.21915	0.24327	631.8	529.2	1160.9	0.8267	0.4947	1.3214	610.0
608.0	1637.3	0.02402	0.22394	0.24796	628.8	533.6	1162.4	0.8240	0.4997	1.3238	608.0
606.0	1613.4	0.02392	0.22881	0.25273	625.9	537.9	1163.8	0.8214	0.5048	1.3261	606.0
604.0	1589.7	0.02382	0.23374	0.25757	622.9	542.2	1165.1	0.8187	0.5097	1.3284	604.0
602.0	1566.3	0.02373	0.23875	0.26248	620.0	546.4	1166.4	0.8161	0.5147	1.3307	602.0
600.0	1543.2	0.02364	0.24384	0.26747	617.1	550.6	1167.7	0.8134	0.5196	1.3330	600.0
598.0	1520.4	0.02354	0.24900	0.27255	614.3	554.7	1169.0	0.8108	0.5245	1.3353	598.0
596.0	1497.8	0.02345	0.25425	0.27770	611.4	558.8	1170.2	0.8082	0.5293	1.3375	596.0
594.0	1475.4	0.02337	0.25958	0.28294	608.6	562.8	1171.4	0.8056	0.5342	1.3398	594.0
592.0	1453.3	0.02328	0.26499	0.28827	605.7	566.8	1172.6	0.8030	0.5390	1.3420	592.0
590.0	1431.5	0.02319	0.27049	0.29368	602.9	570.8	1173.7	0.8004	0.5437	1.3442	590.0
588.0	1410.0	0.02311	0.27608	0.29919	600.1	574.7	1174.8	0.7978	0.5485	1.3464	588.0
586.0	1388.6	0.02303	0.28176	0.30478	597.3	578.5	1175.9	0.7953	0.5532	1.3485	586.0
584.0	1367.6	0.02295	0.28753	0.31048	594.6	582.4	1176.9	0.7927	0.5580	1.3507	584.0
582.0	1346.7	0.02287	0.29340	0.31627	591.8	586.1	1178.0	0.7902	0.5627	1.3528	582.0
580.0	1326.2	0.02279	0.29937	0.32216	589.1	589.9	1179.0	0.7876	0.5673	1.3550	580.0

Table 1-6.   SATURATED STEAM AND SATURATED WATER (TEMPERATURE) *(Continued)*

Temp, °F	Press, psia	Volume, ft³/lbₘ			Enthalpy, Btu/lbₘ			Entropy, Btu/lbₘ °F			Temp, °F
		Water, v_f	Evap, v_{fg}	Steam, v_g	Water, h_f	Evap, h_{fg}	Steam, h_g	Water, s_f	Evap, s_{fg}	Steam, s_g	
580.0	1326.17	0.02279	0.29937	0.32216	589.1	589.9	1179.0	0.7876	0.5673	1.3550	580.0
578.0	1305.84	0.02271	0.30544	0.32816	586.4	593.6	1179.9	0.7851	0.5720	1.3571	578.0
576.0	1285.74	0.02264	0.31162	0.33426	583.7	597.2	1180.9	0.7825	0.5766	1.3592	576.0
571.0	1265.89	0.02256	0.31790	0.34046	581.0	600.9	1181.8	0.7800	0.5813	1.3613	574.0
572.0	1246.26	0.02249	0.32429	0.34678	578.3	604.5	1182.7	0.7775	0.5859	1.3634	572.0
570.0	1226.88	0.02242	0.33079	0.35321	575.6	608.0	1183.6	0.7750	0.5905	1.3654	570.0
568.0	1207.72	0.02235	0.33741	0.35975	572.9	611.5	1184.5	0.7725	0.5950	1.3675	568.0
566.0	1188.80	0.02228	0.34414	0.36642	570.3	615.0	1185.3	0.7699	0.5996	1.3696	566.0
564.0	1170.10	0.02221	0.35099	0.37320	567.6	618.5	1186.1	0.7674	0.6041	1.3716	564.0
562.0	1151.63	0.02214	0.35797	0.38011	565.0	621.9	1186.9	0.7650	0.6087	1.3736	562.0
560.0	1133.38	0.02207	0.36507	0.38714	562.4	625.3	1187.7	0.7625	0.6132	1.3757	560.0
558.0	1115.36	0.02201	0.37230	0.39431	559.8	628.6	1188.4	0.7600	0.6177	1.3777	558.0
556.0	1097.55	0.02194	0.37966	0.40160	557.2	632.0	1189.2	0.7575	0.6222	1.3797	556.0
554.0	1079.96	0.02188	0.38715	0.40903	554.6	635.3	1189.9	0.7550	0.6267	1.3817	554.0
552.0	1062.59	0.02182	0.39479	0.41660	552.0	638.5	1190.6	0.7525	0.6311	1.3837	552.0
550.0	1045.43	0.02176	0.40256	0.42432	549.5	641.8	1191.2	0.7501	0.6356	1.3856	550.0
548.0	1028.49	0.02169	0.41048	0.43217	546.9	645.0	1191.9	0.7476	0.6400	1.3876	548.0
546.0	1011.75	0.02163	0.41855	0.44018	544.4	648.1	1192.5	0.7451	0.6445	1.3896	546.0
544.0	995.22	0.02157	0.42677	0.44834	541.8	651.3	1193.1	0.7427	0.6489	1.3915	544.0
542.0	978.90	0.02151	0.43514	0.45665	539.3	654.4	1193.7	0.7402	0.6533	1.3935	542.0
540.0	962.79	0.02146	0.44367	0.46513	536.8	657.5	1194.3	0.7378	0.6577	1.3954	540.0
538.0	946.88	0.02140	0.45237	0.47377	534.2	660.6	1194.8	0.7353	0.6621	1.3974	538.0
536.0	931.17	0.02134	0.46123	0.48257	531.7	663.6	1195.4	0.7329	0.6665	1.3993	536.0
534.0	915.66	0.02129	0.47026	0.49155	529.2	666.6	1195.9	0.7304	0.6708	1.4012	534.0
532.0	900.34	0.02123	0.47947	0.50070	526.8	669.6	1196.4	0.7280	0.6752	1.4032	532.0
530.0	885.23	0.02118	0.48886	0.51004	524.3	672.6	1196.9	0.7255	0.6796	1.4051	530.0
528.0	870.31	0.02112	0.49843	0.51955	521.8	675.5	1197.3	0.7231	0.6839	1.4070	528.0
526.0	855.58	0.02107	0.50819	0.52926	519.3	678.4	1197.8	0.7206	0.6883	1.4089	526.0
524.0	841.04	0.02102	0.51814	0.53916	516.9	681.3	1198.2	0.7182	0.6926	1.4108	524.0
522.0	826.69	0.02097	0.52829	0.54926	514.4	684.2	1198.6	0.7158	0.6969	1.4127	522.0
520.0	812.53	0.02091	0.53864	0.55956	512.0	687.0	1199.0	0.7133	0.7013	1.4146	520.0
518.0	798.55	0.02086	0.54920	0.57006	509.6	689.9	1199.4	0.7109	0.7056	1.4165	518.0
516.0	784.76	0.02081	0.55997	0.58079	507.1	692.7	1199.8	0.7085	0.7099	1.4183	516.0
514.0	771.15	0.02076	0.57096	0.59173	504.7	695.4	1200.2	0.7060	0.7142	1.4202	514.0
512.0	757.72	0.02072	0.58218	0.60289	502.3	698.2	1200.5	0.7036	0.7185	1.4221	512.0
510.0	744.47	0.02067	0.59362	0.61429	499.9	700.9	1200.8	0.7012	0.7228	1.4240	510.0
508.0	731.40	0.02062	0.60530	0.62592	497.5	703.7	1201.1	0.6987	0.7271	1.4258	508.0
506.0	718.50	0.02057	0.61722	0.63779	495.1	706.3	1201.4	0.6963	0.7314	1.4277	506.0
504.0	705.78	0.02053	0.62938	0.64991	492.7	709.0	1201.7	0.6939	0.7357	1.4296	504.0
502.0	693.23	0.02048	0.64180	0.66228	490.3	711.7	1202.0	0.6915	0.7400	1.4314	502.0
500.0	680.86	0.02043	0.65448	0.67492	487.9	714.3	1202.2	0.6890	0.7443	1.4333	500.0
498.0	668.65	0.02039	0.66743	0.68782	485.6	716.9	1202.5	0.6866	0.7486	1.4352	498.0
496.0	656.61	0.02034	0.68065	0.70100	483.2	719.5	1202.7	0.6842	0.7528	1.4370	496.0
494.0	644.73	0.02030	0.69415	0.71445	480.8	722.1	1202.9	0.6818	0.7571	1.4389	494.0
492.0	633.03	0.02026	0.70794	0.72820	478.5	724.6	1203.1	0.6793	0.7614	1.4407	492.0
490.0	621.48	0.02021	0.72203	0.74224	476.1	727.2	1203.3	0.6769	0.7657	1.4426	490.0
488.0	610.10	0.02017	0.73641	0.75658	473.8	729.7	1203.5	0.6745	0.7700	1.4444	488.0
486.0	598.87	0.02013	0.75111	0.77124	471.5	732.2	1203.7	0.6721	0.7742	1.4463	486.0
484.0	587.81	0.02009	0.76613	0.78622	469.1	734.7	1203.8	0.6696	0.7785	1.4481	484.0
482.0	576.90	0.02004	0.78148	0.80152	466.8	737.2	1204.0	0.6672	0.7828	1.4500	482.0
480.0	566.15	0.02000	0.79716	0.81717	464.5	739.6	1204.1	0.6648	0.7871	1.4518	480.0
478.0	555.55	0.01996	0.81319	0.83315	462.2	742.1	1204.2	0.6624	0.7913	1.4537	478.0
476.0	545.11	0.01992	0.82958	0.84950	459.9	744.5	1204.3	0.6599	0.7956	1.4555	476.0
474.0	534.81	0.01988	0.84632	0.86621	457.5	746.9	1204.4	0.6575	0.7999	1.4574	474.0
472.0	524.67	0.01984	0.86345	0.88329	455.2	749.3	1204.5	0.6551	0.8042	1.4592	472.0
470.0	514.67	0.01980	0.88095	0.90076	452.9	751.6	1204.6	0.6527	0.8084	1.4611	470.0
468.0	504.83	0.01976	0.89885	0.91862	450.7	754.0	1204.6	0.6502	0.8127	1.4629	468.0
466.0	495.12	0.01973	0.91716	0.93689	448.4	756.3	1204.7	0.6478	0.8170	1.4648	466.0
464.0	485.56	0.01969	0.93588	0.95557	446.1	758.6	1204.7	0.6454	0.8213	1.4667	464.0
462.0	476.14	0.01965	0.95504	0.97469	443.8	761.0	1204.8	0.6429	0.8256	1.4685	462.0
460.0	466.87	0.01961	0.97463	0.99424	441.5	763.2	1204.8	0.6405	0.8299	1.4704	460.0
458.0	457.73	0.01958	0.99467	1.01425	439.3	765.5	1204.8	0.6381	0.8342	1.4722	458.0
456.0	448.73	0.01954	1.01518	1.03472	437.0	767.8	1204.8	0.6356	0.8385	1.4741	456.0
454.0	439.87	0.01950	1.03616	1.05567	434.7	770.0	1204.8	0.6332	0.8428	1.4759	454.0
452.0	431.14	0.01947	1.05764	1.07711	432.5	772.3	1204.8	0.6308	0.8471	1.4778	452.0
450.0	422.55	0.01943	1.07962	1.09905	430.2	774.5	1204.7	0.6283	0.8514	1.4797	450.0
448.0	414.09	0.01940	1.10212	1.12152	428.0	776.7	1204.7	0.6259	0.8557	1.4815	448.0
446.0	405.76	0.01936	1.12515	1.14452	425.7	778.9	1204.6	0.6234	0.8600	1.4834	446.0
444.0	397.56	0.01933	1.14874	1.16806	423.5	781.1	1204.6	0.6210	0.8643	1.4853	444.0
442.0	389.49	0.01929	1.17288	1.19217	421.3	783.2	1204.5	0.6185	0.8686	1.4872	442.0
440.0	381.54	0.01926	1.19761	1.21687	419.0	785.4	1204.4	0.6161	0.8729	1.4890	440.0
438.0	373.72	0.01923	1.22293	1.24216	416.8	787.5	1204.3	0.6136	0.8773	1.4909	438.0
436.0	366.03	0.01919	1.24887	1.26806	414.6	789.7	1204.2	0.6112	0.8816	1.4928	436.0
434.0	358.46	0.01916	1.27544	1.29460	412.4	791.8	1204.1	0.6087	0.8859	1.4947	434.0
432.0	351.00	0.01913	1.30266	1.32179	410.1	793.9	1204.0	0.6063	0.8903	1.4966	432.0
430.0	343.67	0.01909	1.33055	1.34965	407.9	796.0	1203.9	0.6038	0.8946	1.4985	430.0

Table 1-6. SATURATED STEAM AND SATURATED WATER (TEMPERATURE) *(Continued)*

Temp, °F	Press, psia	Volume, ft³/lb_m Water, v_f	Evap, v_{fg}	Steam, v_g	Enthalpy, Btu/lb_m Water, h_f	Evap, h_{fg}	Steam, h_g	Entropy, Btu/lb_m °F Water, s_f	Evap, s_{fg}	Steam, s_g	Temp, °F
430.0	343.674	0.01909	1.3306	1.3496	407.9	796.0	1203.9	0.6038	0.8946	1.4985	430.0
428.0	336.463	0.01906	1.3591	1.3782	405.7	798.0	1203.7	0.6014	0.8990	1.5004	428.0
426.0	329.369	0.01903	1.3884	1.4075	403.5	800.1	1203.6	0.5989	0.9034	1.5023	426.0
424.0	322.391	0.01900	1.4184	1.4374	401.3	802.2	1203.5	0.5964	0.9077	1.5042	424.0
422.0	315.529	0.01897	1.4492	1.4682	399.1	804.2	1203.3	0.5940	0.9121	1.5061	422.0
420.0	308.780	0.01894	1.4808	1.4997	396.9	806.2	1203.1	0.5915	0.9165	1.5080	420.0
418.0	302.143	0.01890	1.5131	1.5320	394.7	808.2	1202.9	0.5890	0.9209	1.5099	418.0
416.0	295.617	0.01887	1.5463	1.5651	392.5	810.2	1202.8	0.5866	0.9253	1.5118	416.0
414.0	289.201	0.01884	1.5803	1.5991	390.3	812.2	1202.6	0.5841	0.9297	1.5137	414.0
412.0	282.894	0.01881	1.6152	1.6340	388.1	814.2	1202.4	0.5816	0.9341	1.5157	412.0
410.0	276.694	0.01878	1.6510	1.6697	386.0	816.2	1202.1	0.5791	0.9385	1.5176	410.0
408.0	270.600	0.01875	1.6877	1.7064	383.8	818.2	1201.9	0.5766	0.9429	1.5195	408.0
406.0	264.611	0.01872	1.7253	1.7441	381.6	820.1	1201.7	0.5742	0.9473	1.5215	406.0
404.0	258.725	0.01870	1.7640	1.7827	379.4	822.0	1201.5	0.5717	0.9518	1.5234	404.0
402.0	252.942	0.01867	1.8037	1.8223	377.3	824.0	1201.2	0.5692	0.9562	1.5254	402.0
400.0	247.259	0.01864	1.8444	1.8630	375.1	825.9	1201.0	0.5667	0.9607	1.5274	400.0
398.0	241.677	0.01861	1.8862	1.9048	372.9	827.8	1200.7	0.5642	0.9651	1.5293	398.0
396.0	236.193	0.01858	1.9291	1.9477	370.8	829.7	1200.4	0.5617	0.9696	1.5313	396.0
394.0	230.807	0.01855	1.9731	1.9917	368.6	831.6	1200.2	0.5592	0.9741	1.5333	394.0
392.0	225.516	0.01853	2.0184	2.0369	366.5	833.4	1199.9	0.5567	0.9786	1.5352	392.0
390.0	220.321	0.01850	2.0649	2.0833	364.3	835.3	1199.6	0.5542	0.9831	1.5372	390.0
388.0	215.220	0.01847	2.1126	2.1311	362.2	837.2	1199.3	0.5516	0.9876	1.5392	388.0
386.0	210.211	0.01844	2.1616	2.1801	360.0	839.0	1199.0	0.5491	0.9921	1.5412	386.0
384.0	205.294	0.01842	2.2120	2.2304	357.9	840.8	1198.7	0.5466	0.9966	1.5432	384.0
382.0	200.467	0.01839	2.2638	2.2821	355.7	842.7	1198.4	0.5441	1.0012	1.5452	382.0
380.0	195.729	0.01836	2.3170	2.3353	353.6	844.5	1198.0	0.5416	1.0057	1.5473	380.0
378.0	191.080	0.01834	2.3716	2.3900	351.4	846.3	1197.7	0.5390	1.0103	1.5493	378.0
376.0	186.517	0.01831	2.4279	2.4462	349.3	848.1	1197.4	0.5365	1.0148	1.5513	376.0
374.0	182.040	0.01829	2.4857	2.5039	347.2	849.8	1197.0	0.5340	1.0194	1.5534	374.0
372.0	177.648	0.01826	2.5451	2.5633	345.0	851.6	1196.7	0.5314	1.0240	1.5554	372.0
370.0	173.339	0.01823	2.6062	2.6244	342.9	853.4	1196.3	0.5289	1.0286	1.5575	370.0
368.0	169.113	0.01821	2.6691	2.6873	340.8	855.1	1195.9	0.5263	1.0332	1.5595	368.0
366.0	164.968	0.01818	2.7337	2.7519	338.7	856.9	1195.6	0.5238	1.0378	1.5616	366.0
364.0	160.903	0.01816	2.8002	2.8184	336.5	858.6	1195.2	0.5212	1.0424	1.5637	364.0
362.0	156.917	0.01813	2.8687	2.8868	334.4	860.4	1194.8	0.5187	1.0471	1.5658	362.0
360.0	153.010	0.01811	2.9392	2.9573	332.3	862.1	1194.4	0.5161	1.0517	1.5678	360.0
358.0	149.179	0.01809	3.0117	3.0298	330.2	863.8	1194.0	0.5135	1.0564	1.5699	358.0
356.0	145.424	0.01806	3.0863	3.1044	328.1	865.5	1193.6	0.5110	1.0611	1.5721	356.0
354.0	141.744	0.01804	3.1632	3.1812	326.0	867.2	1193.2	0.5084	1.0658	1.5742	354.0
352.0	138.138	0.01801	3.2423	3.2603	323.9	868.9	1192.7	0.5058	1.0705	1.5763	352.0
350.0	134.604	0.01799	3.3238	3.3418	321.8	870.6	1192.3	0.5032	1.0752	1.5784	350.0
348.0	131.142	0.01797	3.4078	3.4258	319.7	872.2	1191.9	0.5006	1.0799	1.5806	348.0
346.0	127.751	0.01794	3.4943	3.5122	317.6	873.9	1191.4	0.4980	1.0847	1.5827	346.0
344.0	124.430	0.01792	3.5834	3.6013	315.5	875.5	1191.0	0.4954	1.0894	1.5849	344.0
342.0	121.177	0.01790	3.6752	3.6931	313.4	877.2	1190.5	0.4928	1.0942	1.5871	342.0
340.0	117.992	0.01787	3.7699	3.7878	311.3	878.8	1190.1	0.4902	1.0990	1.5892	340.0
338.0	114.873	0.01785	3.8675	3.8853	309.2	880.5	1189.6	0.4876	1.1038	1.5914	338.0
336.0	111.820	0.01783	3.9681	3.9859	307.1	882.1	1189.1	0.4850	1.1086	1.5936	336.0
334.0	108.832	0.01781	4.0718	4.0896	305.0	883.7	1188.7	0.4824	1.1134	1.5958	334.0
332.0	105.907	0.01779	4.1788	4.1966	302.9	885.3	1188.2	0.4798	1.1183	1.5981	332.0
330.0	103.045	0.01776	4.2892	4.3069	300.8	886.9	1187.7	0.4772	1.1231	1.6003	330.0
328.0	100.245	0.01774	4.4030	4.4208	298.7	888.5	1187.2	0.4745	1.1280	1.6025	328.0
326.0	97.506	0.01772	4.5205	4.5382	296.6	890.1	1186.7	0.4719	1.1329	1.6048	326.0
324.0	94.826	0.01770	4.6418	4.6595	294.6	891.6	1186.2	0.4692	1.1378	1.6071	324.0
322.0	92.205	0.01768	4.7669	4.7846	292.5	893.2	1185.7	0.4666	1.1427	1.6093	322.0
320.0	89.643	0.01766	4.8961	4.9138	290.4	894.8	1185.2	0.4640	1.1477	1.6116	320.0
318.0	87.137	0.01764	5.0295	5.0471	288.3	896.3	1184.7	0.4613	1.1526	1.6139	318.0
316.0	84.688	0.01761	5.1673	5.1849	286.3	897.9	1184.1	0.4586	1.1576	1.6162	316.0
314.0	82.293	0.01759	5.3096	5.3272	284.2	899.4	1183.6	0.4560	1.1626	1.6185	314.0
312.0	79.953	0.01757	5.4566	5.4742	282.1	901.0	1183.1	0.4533	1.1676	1.6209	312.0
310.0	77.667	0.01755	5.6085	5.6260	280.0	902.5	1182.5	0.4506	1.1726	1.6232	310.0
308.0	75.433	0.01753	5.7655	5.7830	278.0	904.0	1182.0	0.4479	1.1776	1.6256	308.0
306.0	73.251	0.01751	5.9277	5.9452	275.9	905.5	1181.4	0.4453	1.1827	1.6279	306.0
304.0	71.119	0.01749	6.0955	6.1130	273.8	907.0	1180.9	0.4426	1.1877	1.6303	304.0
302.0	69.038	0.01747	6.2689	6.2864	271.8	908.5	1180.3	0.4399	1.1928	1.6327	302.0
300.0	67.005	0.01745	6.4483	6.4658	269.7	910.0	1179.7	0.4372	1.1979	1.6351	300.0
298.0	65.021	0.01743	6.6339	6.6513	267.7	911.5	1179.2	0.4345	1.2031	1.6375	298.0
296.0	63.084	0.01741	6.8259	6.8433	265.6	913.0	1178.6	0.4317	1.2082	1.6400	296.0
294.0	61.194	0.01739	7.0245	7.0419	263.5	914.5	1178.0	0.4290	1.2134	1.6424	294.0
292.0	59.350	0.01738	7.2301	7.2475	261.5	915.9	1177.4	0.4263	1.2186	1.6449	292.0
290.0	57.550	0.01736	7.4430	7.4603	259.4	917.4	1176.8	0.4236	1.2238	1.6473	290.0
288.0	55.795	0.01734	7.6634	7.6807	257.4	918.8	1176.2	0.4208	1.2290	1.6498	288.0
286.0	54.083	0.01732	7.8916	7.9089	255.3	920.3	1175.6	0.4181	1.2342	1.6523	286.0
284.0	52.414	0.01730	8.1280	8.1453	253.3	921.7	1175.0	0.4154	1.2395	1.6548	284.0
282.0	50.786	0.01728	8.3729	8.3902	251.2	923.2	1174.4	0.4126	1.2448	1.6574	282.0
280.0	49.200	0.01726	8.6267	8.6439	249.2	924.6	1173.8	0.4098	1.2501	1.6599	280.0

Table 1-6.   SATURATED STEAM AND SATURATED WATER
(TEMPERATURE) *(Continued)*

Temp, °F	Press, psia	Volume, ft³/lbₘ Water, v_f	Evap, v_{fg}	Steam, v_g	Enthalpy, Btu/lbₘ Water, h_f	Evap, h_{fg}	Steam, h_g	Entropy, Btu/lbₘ °F Water, s_f	Evap, s_{fg}	Steam, s_g	Temp, °F
280.0	49.200	0.017264	8.627	8.644	249.17	924.6	1173.8	0.4098	1.2501	1.6599	280.0
278.0	47.653	0.017246	8.890	8.907	247.13	926.0	1173.2	0.4071	1.2554	1.6625	278.0
276.0	46.147	0.017228	9.162	9.180	245.08	927.5	1172.5	0.4043	1.2607	1.6650	276.0
274.0	44.678	0.017210	9.445	9.462	243.03	928.9	1171.9	0.4015	1.2661	1.6676	274.0
272.0	43.249	0.017193	9.738	9.755	240.99	930.3	1171.3	0.3987	1.2715	1.6702	272.0
270.0	41.856	0.017175	10.042	10.060	238.95	931.7	1170.6	0.3960	1.2769	1.6729	270.0
268.0	40.500	0.017157	10.358	10.375	236.91	933.1	1170.0	0.3932	1.2823	1.6755	268.0
266.0	39.179	0.017140	10.685	10.703	234.87	934.5	1169.3	0.3904	1.2878	1.6781	266.0
264.0	37.894	0.017123	11.025	11.042	232.83	935.9	1168.7	0.3876	1.2933	1.6808	264.0
262.0	36.644	0.017106	11.378	11.395	230.79	937.3	1168.0	0.3847	1.2988	1.6835	262.0
260.0	35.427	0.017089	11.745	11.762	228.76	938.6	1167.4	0.3819	1.3043	1.6862	260.0
258.0	34.243	0.017072	12.125	12.142	226.72	940.0	1166.7	0.3791	1.3098	1.6889	258.0
256.0	33.091	0.017055	12.520	12.538	224.69	941.4	1166.1	0.3763	1.3154	1.6917	256.0
254.0	31.972	0.017039	12.931	12.948	222.65	942.7	1165.4	0.3734	1.3210	1.6944	254.0
252.0	30.883	0.017022	13.358	13.375	220.62	944.1	1164.7	0.3706	1.3266	1.6972	252.0
250.0	29.825	0.017006	13.802	13.819	218.59	945.4	1164.0	0.3677	1.3323	1.7000	250.0
248.0	28.796	0.016990	14.264	14.281	216.56	946.8	1163.4	0.3649	1.3379	1.7028	248.0
246.0	27.797	0.016974	14.744	14.761	214.53	948.1	1162.7	0.3620	1.3436	1.7056	246.0
244.0	26.826	0.016958	15.243	15.260	212.50	949.5	1162.0	0.3591	1.3494	1.7085	244.0
242.0	25.883	0.016942	15.763	15.780	210.48	950.8	1161.3	0.3562	1.3551	1.7113	242.0
240.0	24.968	0.016926	16.304	16.321	208.45	952.1	1160.6	0.3533	1.3609	1.7142	240.0
238.0	24.079	0.016910	16.867	16.884	206.42	953.5	1159.9	0.3505	1.3667	1.7171	238.0
236.0	23.216	0.016895	17.454	17.471	204.40	954.8	1159.2	0.3476	1.3725	1.7201	236.0
234.0	22.379	0.016880	18.065	18.082	202.38	956.1	1158.5	0.3446	1.3784	1.7230	234.0
232.0	21.567	0.016864	18.701	18.718	200.35	957.4	1157.8	0.3417	1.3842	1.7260	232.0
230.0	20.779	0.016849	19.364	19.381	198.33	958.7	1157.1	0.3388	1.3902	1.7290	230.0
229.0	20.394	0.016842	19.707	19.723	197.32	959.4	1156.7	0.3373	1.3931	1.7305	229.0
228.0	20.015	0.016834	20.056	20.073	196.31	960.0	1156.3	0.3359	1.3961	1.7320	228.0
227.0	19.642	0.016827	20.413	20.429	195.30	960.7	1156.0	0.3344	1.3991	1.7335	227.0
226.0	19.274	0.016819	20.777	20.794	194.29	961.3	1155.6	0.3329	1.4021	1.7350	226.0
225.0	18.912	0.016812	21.140	21.166	193.29	962.0	1155.3	0.3315	1.4051	1.7365	225.0
224.0	18.556	0.016805	21.529	21.545	192.27	962.6	1154.9	0.3300	1.4081	1.7380	224.0
223.0	18.206	0.016797	21.917	21.933	191.26	963.3	1154.5	0.3285	1.4111	1.7396	223.0
222.0	17.860	0.016790	22.313	22.330	190.25	963.9	1154.2	0.3270	1.4141	1.7411	222.0
221.0	17.521	0.016783	22.718	22.735	189.24	964.6	1153.8	0.3255	1.4171	1.7427	221.0
220.0	17.186	0.016775	23.131	23.148	188.23	965.2	1153.4	0.3241	1.4201	1.7442	220.0
219.0	16.857	0.016768	23.554	23.571	187.22	965.8	1153.1	0.3226	1.4232	1.7458	219.0
218.0	16.533	0.016761	23.986	24.002	186.21	966.5	1152.7	0.3211	1.4262	1.7473	218.0
217.0	16.214	0.016754	24.427	24.444	185.21	967.1	1152.3	0.3196	1.4293	1.7489	217.0
216.0	15.901	0.016747	24.878	24.894	184.20	967.8	1152.0	0.3181	1.4323	1.7505	216.0
215.0	15.592	0.016740	25.338	25.355	183.19	968.4	1151.6	0.3166	1.4354	1.7520	215.0
214.0	15.289	0.016733	25.809	25.826	182.18	969.0	1151.2	0.3151	1.4385	1.7536	214.0
213.0	14.990	0.016726	26.290	26.307	181.17	969.7	1150.8	0.3136	1.4416	1.7552	213.0
212.0	14.696	0.016719	26.782	26.799	180.17	970.3	1150.5	0.3121	1.4447	1.7568	212.0
211.0	14.407	0.016712	27.285	27.302	179.16	970.9	1150.1	0.3106	1.4478	1.7584	211.0
210.0	14.123	0.016705	27.799	27.816	178.15	971.6	1149.7	0.3091	1.4509	1.7600	210.0
209.0	13.843	0.016698	28.324	28.341	177.14	972.2	1149.4	0.3076	1.4540	1.7616	209.0
208.0	13.568	0.016691	28.862	28.878	176.14	972.8	1149.0	0.3061	1.4571	1.7632	208.0
207.0	13.297	0.016684	29.411	29.428	175.13	973.5	1148.6	0.3046	1.4602	1.7649	207.0
206.0	13.031	0.016677	29.973	29.989	174.12	974.1	1148.2	0.3031	1.4634	1.7665	206.0
205.0	12.770	0.016670	30.547	30.564	173.12	974.7	1147.9	0.3016	1.4665	1.7681	205.0
204.0	12.512	0.016664	31.135	31.151	172.11	975.4	1147.5	0.3001	1.4697	1.7698	204.0
203.0	12.259	0.016657	31.736	31.752	171.10	976.0	1147.1	0.2986	1.4728	1.7714	203.0
202.0	12.011	0.016650	32.350	32.367	170.10	976.6	1146.7	0.2971	1.4760	1.7731	202.0
201.0	11.766	0.016643	32.979	32.996	169.09	977.2	1146.3	0.2955	1.4792	1.7747	201.0
200.0	11.526	0.016637	33.622	33.639	168.09	977.9	1146.0	0.2940	1.4824	1.7764	200.0
199.0	11.290	0.016630	34.280	34.297	167.08	978.5	1145.6	0.2925	1.4856	1.7781	199.0
198.0	11.058	0.016624	34.954	34.970	166.08	979.1	1145.2	0.2910	1.4888	1.7798	198.0
197.0	10.830	0.016617	35.643	35.659	165.07	979.7	1144.8	0.2894	1.4920	1.7814	197.0
196.0	10.605	0.016611	36.348	36.364	164.06	980.4	1144.4	0.2879	1.4952	1.7831	196.0
195.0	10.385	0.016604	37.069	37.086	163.06	981.0	1144.0	0.2864	1.4985	1.7848	195.0
194.0	10.168	0.016598	37.808	37.824	162.05	981.6	1143.7	0.2848	1.5017	1.7865	194.0
193.0	9.956	0.016591	38.564	38.580	161.05	982.2	1143.3	0.2833	1.5050	1.7882	193.0
192.0	9.747	0.016585	39.337	39.354	160.05	982.8	1142.9	0.2818	1.5082	1.7900	192.0
191.0	9.541	0.016578	40.130	40.146	159.04	983.5	1142.5	0.2802	1.5115	1.7917	191.0
190.0	9.340	0.016572	40.941	40.957	158.04	984.1	1142.1	0.2787	1.5148	1.7934	190.0
189.0	9.141	0.016566	41.771	41.787	157.03	984.7	1141.7	0.2771	1.5180	1.7952	189.0
188.0	8.947	0.016559	42.621	42.638	156.03	985.3	1141.3	0.2756	1.5213	1.7969	188.0
187.0	8.756	0.016553	43.492	43.508	155.02	985.9	1140.9	0.2740	1.5246	1.7987	187.0
186.0	8.568	0.016547	44.383	44.400	154.02	986.5	1140.5	0.2725	1.5279	1.8004	186.0
185.0	8.384	0.016541	45.297	45.313	153.02	987.1	1140.2	0.2709	1.5313	1.8022	185.0
184.0	8.203	0.016534	46.232	46.249	152.01	987.8	1139.8	0.2694	1.5346	1.8040	184.0
183.0	8.025	0.016528	47.190	47.207	151.01	988.4	1139.4	0.2678	1.5379	1.8057	183.0
182.0	7.850	0.016522	48.172	48.189	150.01	989.0	1139.0	0.2662	1.5413	1.8075	182.0
181.0	7.679	0.016516	49.178	49.194	149.00	989.6	1138.6	0.2647	1.5446	1.8093	181.0
180.0	7.511	0.016510	50.208	50.225	148.00	990.2	1138.2	0.2631	1.5480	1.8111	180.0

Table 1-6. SATURATED STEAM AND SATURATED WATER
(TEMPERATURE) *(Continued)*

Temp, °F	Press, psia	Volume, ft³/lbm			Enthalpy, Btu/lbm			Entropy, Btu/lbm °F			Temp, °F
		Water, v_f	Evap, v_{fg}	Steam, v_g	Water, h_f	Evap, h_{fg}	Steam, h_g	Water, s_f	Evap, s_{fg}	Steam, s_g	
180.0	7.5110	0.016510	50.21	50.22	148.00	990.2	1138.2	0.2631	1.5480	1.8111	180.0
179.0	7.3460	0.016504	51.26	51.28	147.00	990.8	1137.8	0.2615	1.5514	1.8129	179.0
178.0	7.1840	0.016498	52.35	52.36	145.99	991.4	1137.4	0.2600	1.5548	1.8147	178.0
177.0	7.0250	0.016492	53.46	53.47	144.99	992.0	1137.0	0.2584	1.5582	1.8166	177.0
176.0	6.8690	0.016486	54.59	54.61	143.99	992.6	1136.6	0.2568	1.5616	1.8184	176.0
175.0	6.7159	0.016480	55.76	55.77	142.99	993.2	1136.2	0.2552	1.5650	1.8202	175.0
174.0	6.5656	0.016474	56.95	56.97	141.98	993.8	1135.8	0.2537	1.5684	1.8221	174.0
173.0	6.4182	0.016468	58.18	58.19	140.98	994.4	1135.4	0.2521	1.5718	1.8239	173.0
172.0	6.2736	0.016463	59.43	59.45	139.98	995.0	1135.0	0.2505	1.5753	1.8258	172.0
171.0	6.1318	0.016457	60.72	60.74	138.98	995.6	1134.6	0.2489	1.5787	1.8276	171.0
170.0	5.9926	0.016451	62.04	62.06	137.97	996.2	1134.2	0.2473	1.5822	1.8295	170.0
169.0	5.8562	0.016445	63.39	63.41	136.97	996.8	1133.8	0.2457	1.5857	1.8314	169.0
168.0	5.7223	0.016440	64.78	64.80	135.97	997.4	1133.4	0.2441	1.5892	1.8333	168.0
167.0	5.5911	0.016434	66.21	66.22	134.97	998.0	1133.0	0.2425	1.5926	1.8352	167.0
166.0	5.4623	0.016428	67.67	67.68	133.97	998.6	1132.6	0.2409	1.5961	1.8371	166.0
165.0	5.3361	0.016423	69.17	69.18	132.96	999.2	1132.2	0.2393	1.5997	1.8390	165.0
164.0	5.2124	0.016417	70.70	70.72	131.96	999.8	1131.8	0.2377	1.6032	1.8409	164.0
163.0	5.0911	0.016412	72.28	72.30	130.96	1000.4	1131.4	0.2361	1.6067	1.8428	163.0
162.0	4.9722	0.016406	73.90	73.92	129.96	1001.0	1131.0	0.2345	1.6103	1.8448	162.0
161.0	4.8556	0.016401	75.56	75.58	128.96	1001.6	1130.6	0.2329	1.6138	1.8467	161.0
160.0	4.7414	0.016395	77.27	77.29	127.96	1002.2	1130.2	0.2313	1.6174	1.8487	160.0
159.0	4.6294	0.016390	79.02	79.04	126.96	1002.8	1129.8	0.2297	1.6210	1.8506	159.0
158.0	4.5197	0.016384	80.82	80.83	125.96	1003.4	1129.4	0.2281	1.6245	1.8526	158.0
157.0	4.4122	0.016379	82.66	82.68	124.95	1004.0	1129.0	0.2264	1.6281	1.8546	157.0
156.0	4.3068	0.016374	84.56	84.57	123.95	1004.6	1128.6	0.2248	1.6318	1.8566	156.0
155.0	4.2036	0.016369	86.50	86.52	122.95	1005.2	1128.2	0.2232	1.6354	1.8586	155.0
154.0	4.1025	0.016363	88.50	88.52	121.95	1005.8	1127.7	0.2216	1.6390	1.8606	154.0
153.0	4.0035	0.016358	90.55	90.57	120.95	1006.4	1127.3	0.2199	1.6426	1.8626	153.0
152.0	3.9065	0.016353	92.66	92.68	119.95	1007.0	1126.9	0.2183	1.6463	1.8646	152.0
151.0	3.8114	0.016348	94.83	94.84	118.95	1007.6	1126.5	0.2167	1.6500	1.8666	151.0
150.0	3.7184	0.016343	97.05	97.07	117.95	1008.2	1126.1	0.2150	1.6536	1.8686	150.0
149.0	3.6273	0.016337	99.33	99.35	116.95	1008.7	1125.7	0.2134	1.6573	1.8707	149.0
148.0	3.5381	0.016332	101.68	101.70	115.95	1009.3	1125.3	0.2117	1.6610	1.8727	148.0
147.0	3.4508	0.016327	104.10	104.11	114.95	1009.9	1124.9	0.2101	1.6647	1.8748	147.0
146.0	3.3653	0.016322	106.58	106.59	113.95	1010.5	1124.5	0.2084	1.6684	1.8769	146.0
145.0	3.2816	0.016317	109.12	109.14	112.95	1011.1	1124.0	0.2068	1.6722	1.8789	145.0
144.0	3.1997	0.016312	111.74	111.76	111.95	1011.7	1123.6	0.2051	1.6759	1.8810	144.0
143.0	3.1195	0.016308	114.44	114.45	110.95	1012.3	1123.2	0.2035	1.6797	1.8831	143.0
142.0	3.0411	0.016303	117.21	117.22	109.95	1012.9	1122.8	0.2018	1.6834	1.8852	142.0
141.0	2.9643	0.016298	120.05	120.07	108.95	1013.4	1122.4	0.2001	1.6872	1.8873	141.0
140.0	2.8892	0.016293	122.98	123.00	107.95	1014.0	1122.0	0.1985	1.6910	1.8895	140.0
139.0	2.8157	0.016288	125.99	126.01	106.95	1014.6	1121.6	0.1968	1.6948	1.8916	139.0
138.0	2.7438	0.016284	129.09	129.11	105.95	1015.2	1121.1	0.1951	1.6986	1.8937	138.0
137.0	2.6735	0.016279	132.28	132.29	104.95	1015.8	1120.7	0.1935	1.7024	1.8959	137.0
136.0	2.6047	0.016274	135.55	135.57	103.95	1016.4	1120.3	0.1918	1.7063	1.8980	136.0
135.0	2.5375	0.016270	138.93	138.94	102.95	1016.9	1119.9	0.1901	1.7101	1.9002	135.0
134.0	2.4717	0.016265	142.40	142.41	101.95	1017.5	1119.5	0.1884	1.7140	1.9024	134.0
133.0	2.4074	0.016260	145.97	145.98	100.95	1018.1	1119.1	0.1867	1.7178	1.9046	133.0
132.0	2.3445	0.016256	149.64	149.66	99.95	1018.7	1118.6	0.1851	1.7217	1.9068	132.0
131.0	2.2830	0.016251	153.42	153.44	98.95	1019.3	1118.2	0.1834	1.7256	1.9090	131.0
130.0	2.2230	0.016247	157.32	157.33	97.96	1019.8	1117.8	0.1817	1.7295	1.9112	130.0
129.0	2.1642	0.016243	161.32	161.34	96.96	1020.4	1117.4	0.1800	1.7335	1.9134	129.0
128.0	2.1068	0.016238	165.45	165.47	95.96	1021.0	1117.0	0.1783	1.7374	1.9157	128.0
127.0	2.0507	0.016234	169.70	169.72	94.96	1021.6	1116.5	0.1766	1.7413	1.9179	127.0
126.0	1.9959	0.016229	174.08	174.09	93.96	1022.2	1116.1	0.1749	1.7453	1.9202	126.0
125.0	1.9424	0.016225	178.58	178.60	92.96	1022.7	1115.7	0.1732	1.7493	1.9224	125.0
124.0	1.8901	0.016221	183.23	183.24	91.96	1023.3	1115.3	0.1715	1.7533	1.9247	124.0
123.0	1.8390	0.016217	188.01	188.03	90.96	1023.9	1114.9	0.1697	1.7573	1.9270	123.0
122.0	1.7891	0.016213	192.94	192.95	89.96	1024.5	1114.4	0.1680	1.7613	1.9293	122.0
121.0	1.7403	0.016208	198.01	198.03	88.96	1025.0	1114.0	0.1663	1.7653	1.9316	121.0
120.0	1.6927	0.016204	203.25	203.26	87.97	1025.6	1113.6	0.1646	1.7693	1.9339	120.0
119.0	1.6463	0.016200	208.64	208.66	86.97	1026.2	1113.2	0.1629	1.7734	1.9362	119.0
118.0	1.6009	0.016196	214.20	214.21	85.97	1026.8	1112.7	0.1611	1.7774	1.9386	118.0
117.0	1.5566	0.016192	219.93	219.94	84.97	1027.3	1112.3	0.1594	1.7815	1.9409	117.0
116.0	1.5133	0.016188	225.84	225.85	83.97	1027.9	1111.9	0.1577	1.7856	1.9433	116.0
115.0	1.4711	0.016184	231.93	231.94	82.97	1028.5	1111.5	0.1559	1.7897	1.9457	115.0
114.0	1.4299	0.016180	238.21	238.22	81.97	1029.1	1111.0	0.1542	1.7938	1.9480	114.0
113.0	1.3898	0.016177	244.69	244.70	80.98	1029.6	1110.6	0.1525	1.7980	1.9504	113.0
112.0	1.3505	0.016173	251.37	251.38	79.98	1030.2	1110.2	0.1507	1.8021	1.9528	112.0
111.0	1.3123	0.016169	258.26	258.28	78.98	1030.8	1109.8	0.1490	1.8063	1.9552	111.0
110.0	1.2750	0.016165	265.37	265.39	77.98	1031.4	1109.3	0.1472	1.8105	1.9577	110.0
109.0	1.2385	0.016162	272.71	272.72	76.98	1031.9	1108.9	0.1455	1.8146	1.9601	109.0
108.0	1.2030	0.016158	280.28	280.30	75.98	1032.5	1108.5	0.1437	1.8188	1.9626	108.0
107.0	1.1684	0.016154	288.09	288.11	74.99	1033.1	1108.1	0.1419	1.8231	1.9650	107.0
106.0	1.1347	0.016151	296.16	296.18	73.99	1033.6	1107.6	0.1402	1.8273	1.9675	106.0
105.0	1.1017	0.016147	304.49	304.50	72.99	1034.2	1107.2	0.1384	1.8315	1.9700	105.0

Table 1-6. SATURATED STEAM AND SATURATED WATER (TEMPERATURE) *(Continued)*

Temp, °F	Press, psia	Volume, ft³/lb_m			Enthalpy, Btu/lb_m			Entropy, Btu/lb_m °F			Temp, °F
		Water, v_f	Evap, v_{fg}	Steam, v_g	Water, h_f	Evap, h_{fg}	Steam, h_g	Water, s_f	Evap, s_{fg}	Steam, s_g	
105.0	1.10174	0.016147	304.5	304.5	72.990	1034.2	1107.2	0.1384	1.8315	1.9700	105.0
104.0	1.06965	0.016144	313.1	313.1	71.992	1034.8	1106.8	0.1366	1.8358	1.9725	104.0
103.0	1.03838	0.016140	322.0	322.0	70.993	1035.4	1106.3	0.1349	1.8401	1.9750	103.0
102.0	1.00789	0.016137	331.1	331.1	69.995	1035.9	1105.9	0.1331	1.8444	1.9775	102.0
101.0	0.97818	0.016133	340.6	340.6	68.997	1036.5	1105.5	0.1313	1.8487	1.9800	101.0
100.0	0.94924	0.016130	350.4	350.4	67.999	1037.1	1105.1	0.1295	1.8530	1.9825	100.0
99.0	0.92103	0.016127	360.5	360.5	67.001	1037.6	1104.6	0.1278	1.8573	1.9851	99.0
98.0	0.89356	0.016123	370.9	370.9	66.003	1038.2	1104.2	0.1260	1.8617	1.9876	98.0
97.0	0.86679	0.016120	381.7	381.7	65.005	1038.8	1103.8	0.1242	1.8660	1.9902	97.0
96.0	0.84072	0.016117	392.8	392.9	64.006	1039.3	1103.3	0.1224	1.8704	1.9928	96.0
95.0	0.81534	0.016114	404.4	404.4	63.008	1039.9	1102.9	0.1206	1.8748	1.9954	95.0
94.0	0.79062	0.016111	416.3	416.3	62.010	1040.5	1102.5	0.1188	1.8792	1.9980	94.0
93.0	0.76655	0.016108	428.6	428.6	61.012	1041.0	1102.1	0.1170	1.8837	2.0006	93.0
92.0	0.74313	0.016105	441.3	441.3	60.014	1041.6	1101.6	0.1152	1.8881	2.0033	92.0
91.0	0.72032	0.016102	454.5	454.5	59.016	1042.2	1101.2	0.1134	1.8926	2.0059	91.0
90.0	0.69813	0.016099	468.1	468.1	58.018	1042.7	1100.8	0.1115	1.8970	2.0086	90.0
89.0	0.67653	0.016096	482.2	482.2	57.020	1043.3	1100.3	0.1097	1.9015	2.0112	89.0
88.0	0.65551	0.016093	496.8	496.8	56.022	1043.9	1099.9	0.1079	1.9060	2.0139	88.0
87.0	0.63507	0.016090	511.9	511.9	55.024	1044.4	1099.5	0.1061	1.9105	2.0166	87.0
86.0	0.61518	0.016087	527.5	527.5	54.026	1045.0	1099.0	0.1043	1.9151	2.0193	86.0
85.0	0.59583	0.016085	543.6	543.6	53.027	1045.6	1098.6	0.1024	1.9196	2.0221	85.0
84.0	0.57702	0.016082	560.3	560.3	52.029	1046.1	1098.2	0.1006	1.9242	2.0248	84.0
83.0	0.55872	0.016079	577.6	577.6	51.031	1046.7	1097.7	0.0988	1.9288	2.0275	83.0
82.0	0.54093	0.016077	595.5	595.6	50.033	1047.3	1097.3	0.0969	1.9334	2.0303	82.0
81.0	0.52364	0.016074	614.1	614.1	49.035	1047.8	1096.9	0.0951	1.9380	2.0331	81.0
80.0	0.50683	0.016072	633.3	633.3	48.037	1048.4	1096.4	0.0932	1.9426	2.0359	80.0
79.0	0.49049	0.016070	653.2	653.2	47.038	1049.0	1096.0	0.0914	1.9473	2.0387	79.0
78.0	0.47461	0.016067	673.8	673.9	46.040	1049.5	1095.6	0.0895	1.9520	2.0415	78.0
77.0	0.45919	0.016065	695.2	695.2	45.042	1050.1	1095.1	0.0877	1.9567	2.0443	77.0
76.0	0.44420	0.016063	717.4	717.4	44.043	1050.7	1094.7	0.0858	1.9614	2.0472	76.0
75.0	0.42964	0.016060	740.3	740.3	43.045	1051.2	1094.3	0.0839	1.9661	2.0500	75.0
74.0	0.41550	0.016058	764.1	764.1	42.046	1051.8	1093.8	0.0821	1.9708	2.0529	74.0
73.0	0.40177	0.016056	788.8	788.8	41.048	1052.4	1093.4	0.0802	1.9756	2.0558	73.0
72.0	0.38844	0.016054	814.3	814.3	40.049	1052.9	1093.0	0.0783	1.9804	2.0587	72.0
71.0	0.37549	0.016052	840.8	840.9	39.050	1053.5	1092.5	0.0764	1.9852	2.0616	71.0
70.0	0.36292	0.016050	868.3	868.4	38.052	1054.0	1092.1	0.0745	1.9900	2.0645	70.0
69.0	0.35073	0.016048	896.9	896.9	37.053	1054.6	1091.7	0.0727	1.9948	2.0675	69.0
68.0	0.33889	0.016046	926.5	926.5	36.054	1055.2	1091.2	0.0708	1.9996	2.0704	68.0
67.0	0.32740	0.016044	957.2	957.2	35.055	1055.7	1090.8	0.0689	2.0045	2.0734	67.0
66.0	0.31626	0.016043	989.0	989.1	34.056	1056.3	1090.4	0.0670	2.0094	2.0764	66.0
65.0	0.30545	0.016041	1022.1	1022.1	33.057	1056.9	1089.9	0.0651	2.0143	2.0794	65.0
64.0	0.29497	0.016039	1056.5	1056.5	32.058	1057.4	1089.5	0.0632	2.0192	2.0824	64.0
63.0	0.28480	0.016038	1092.1	1092.1	31.058	1058.0	1089.0	0.0613	2.0242	2.0854	63.0
62.0	0.27494	0.016036	1129.2	1129.2	30.059	1058.6	1088.6	0.0593	2.0291	2.0885	62.0
61.0	0.26538	0.016035	1167.6	1167.6	29.059	1059.1	1088.2	0.0574	2.0341	2.0915	61.0
60.0	0.25011	0.016033	1207.6	1207.6	28.060	1059.7	1087.7	0.0555	2.0391	2.0946	60.0
59.0	0.24713	0.016032	1249.1	1249.1	27.060	1060.2	1087.3	0.0536	2.0441	2.0977	59.0
58.0	0.23843	0.016031	1292.2	1292.2	26.060	1060.8	1086.9	0.0516	2.0491	2.1008	58.0
57.0	0.23000	0.016029	1337.0	1337.0	25.060	1061.4	1086.4	0.0497	2.0542	2.1039	57.0
56.0	0.22183	0.016028	1383.6	1383.6	24.059	1061.9	1086.0	0.0478	2.0593	2.1070	56.0
55.0	0.21392	0.016027	1432.0	1432.0	23.059	1062.5	1085.6	0.0458	2.0644	2.1102	55.0
54.0	0.20625	0.016026	1482.4	1482.4	22.058	1063.1	1085.1	0.0439	2.0695	2.1134	54.0
53.0	0.19883	0.016025	1534.7	1534.8	21.058	1063.6	1084.7	0.0419	2.0746	2.1165	53.0
52.0	0.19165	0.016024	1589.2	1589.2	20.057	1064.2	1084.2	0.0400	2.0798	2.1197	52.0
51.0	0.18469	0.016023	1645.9	1645.9	19.056	1064.7	1083.8	0.0380	2.0849	2.1230	51.0
50.0	0.17796	0.016023	1704.8	1704.8	18.054	1065.3	1083.4	0.0361	2.0901	2.1262	50.0
49.0	0.17144	0.016022	1766.2	1766.2	17.053	1065.9	1082.9	0.0341	2.0953	2.1294	49.0
48.0	0.16514	0.016021	1830.0	1830.0	16.051	1066.4	1082.5	0.0321	2.1006	2.1327	48.0
47.0	0.15904	0.016021	1896.5	1896.5	15.049	1067.0	1082.1	0.0301	2.1058	2.1360	47.0
46.0	0.15314	0.016020	1965.7	1965.7	14.047	1067.6	1081.6	0.0282	2.1111	2.1393	46.0
45.0	0.14744	0.016020	2037.7	2037.8	13.044	1068.1	1081.2	0.0262	2.1164	2.1426	45.0
44.0	0.14192	0.016019	2112.8	2112.8	12.041	1068.7	1080.7	0.0242	2.1217	2.1459	44.0
43.0	0.13659	0.016019	2191.0	2191.0	11.038	1069.3	1080.3	0.0222	2.1271	2.1493	43.0
42.0	0.13143	0.016019	2272.4	2272.4	10.035	1069.8	1079.9	0.0202	2.1325	2.1527	42.0
41.0	0.12645	0.016019	2357.3	2357.3	9.031	1070.4	1079.4	0.0182	2.1378	2.1560	41.0
40.0	0.12163	0.016019	2445.8	2445.8	8.027	1071.0	1079.0	0.0162	2.1432	2.1594	40.0
39.0	0.11698	0.016019	2538.0	2538.0	7.023	1071.5	1078.5	0.0142	2.1487	2.1629	39.0
38.0	0.11249	0.016019	2634.1	2634.2	6.018	1072.1	1078.1	0.0122	2.1541	2.1663	38.0
37.0	0.10815	0.016019	2734.4	2734.4	5.013	1072.7	1077.7	0.0101	2.1596	2.1697	37.0
36.0	0.10395	0.016020	2839.0	2839.0	4.008	1073.2	1077.2	0.0081	2.1651	2.1732	36.0
35.0	0.09991	0.016020	2948.1	2948.1	3.002	1073.8	1076.8	0.0061	2.1706	2.1767	35.0
34.0	0.09600	0.016021	3061.9	3061.9	1.996	1074.4	1076.4	0.0041	2.1762	2.1802	34.0
33.0	0.09223	0.016021	3180.7	3180.7	0.989	1074.9	1075.9	0.0020	2.1817	2.1837	33.0
32.018	0.08865	0.016022	3302.4	3302.4	0.0003	1075.5	1075.5	0.0000	2.1872	2.1872	32.018
*32.0	0.08859	0.016022	3304.7	3304.7	−0.0179	1075.5	1075.5	−0.0000	2.1873	2.1873	32.0

*The states here shown are metastable.

**Reproduced by permission of the publishers and copyright owners of the "1967 ASME Steam Tables". Further data and information on the thermodynamic and transport properties of steam and water are contained in the above ASME publication. It is obtainable from The American Society of Mechanical Engineers, United Engineering Center, 345 East 47th Street, New York, New York 10017.

For further information on the "ASME Steam Tables", see note at end of Table 1-42.

Table 1-7. SATURATED STEAM, WATER, AND ICE—SI UNITS*

For English units see Table 1-6.

SUBSCRIPTS:

- f refers to a property of liquid in equilibrium with vapor
- g refers to a property of vapor in equilibrium with liquid
- i refers to a property of solid in equilibrium with vapor
- fg refers to a change by evaporation
- ig refers to a change by sublimation

Temperature		Pressure, MN/m^2	Specific volume, m^3/kg		Specific internal energy, kJ/kg		Specific enthalpy, kJ/kg			Specific entropy, kJ/kg·K	
C	K										

SOLID—VAPOR

C	K		v_i	v_g	u_i	u_g	h_i	h_{ig}	h_g	s_i	s_g
−40	233.15	0.000 012 9	0.001 084 1	83.54	−411.70	2 319.6	−411.70	2 838.9	2 427.2	−1.532	10.644
−30	243.15	0.000 038 1	0.001 085 8	29.43	−393.23	2 333.6	−393.23	2 839.0	2 445.8	−1.455	10.221
−20	253.15	0.000 103 5	0.001 087 4	11.286	−374.03	2 347.5	−374.03	2 838.4	2 464.3	−1.377	9.835
−10	263.15	0.000 260 2	0.001 089 1	4.667	−354.09	2 361.4	−354.09	2 837.0	2 482.9	−1.299	9.481
0	273.15	0.000 610 8	0.001 090 8	2.063	−333.43	2 375.3	−333.43	2 834.8	2 501.3	−1.221	9.157
0.01	273.16	0.000 611 3	0.001 090 8	2.061	−333.40	2 375.3	−333.40	2 834.8	2 501.4	−1.221	9.156

LIQUID—VAPOR

C	K		v_f	v_g	u_f	u_g	h_f	h_{fg}	h_g	s_f	s_g
0	273.15	0.000 610 9	0.001 000 2	206.278	−0.03	2 375.3	−0.02	2 501.4	2 501.3	−0.000 1	9.156 5
0.01	273.16	0.000 611 3	0.001 000 2	206.136	0	2 375.3	+0.01	2 501.3	2 501.4	0	9.156 2
5.00	278.15	0.000 872 1	0.001 000 1	147.120	+20.97	2 382.3	20.98	2 489.6	2 510.6	+0.076 1	9.025 7
6.98	280.13	0.001 000 0	0.001 000 2	129.208	29.30	2 385.0	29.30	2 484.9	2 514.2	0.105 9	8.975
10.00	283.15	0.001 227 6	0.001 000 4	106.379	42.00	2 389.2	42.01	2 477.7	2 519.8	0.151 0	8.900 8
13.03	286.18	0.001 500 0	0.001 000 7	87.980	54.71	2 393.3	54.71	2 470.6	2 525.3	0.195 7	8.827 9
15.00	288.15	0.001 705 1	0.001 000 9	77.926	62.99	2 396.1	62.99	2 465.9	2 528.9	0.224 5	8.781 4
17.50	290.65	0.002 000 0	0.001 001 3	67.004	73.48	2 399.5	73.48	2 460.0	2 533.5	0.260 7	8.723 7
20.00	293.15	0.002 339	0.001 001 8	57.791	83.95	2 402.9	83.96	2 454.1	2 538.1	0.296 6	8.667 2
24.08	297.23	0.003 000 0	0.001 002 7	45.665	101.04	2 408.5	101.05	2 444.5	2 545.5	0.354 5	8.577 6
25.00	298.15	0.003 169	0.001 002 9	43.360	104.88	2 409.8	104.89	2 442.3	2 547.2	0.367 4	8.558 0
28.96	302.11	0.004 000	0.001 004 0	34.800	121.45	2 415.2	121.46	2 432.9	2 554.4	0.422 6	8.474 6
30.00	303.15	0.004 246	0.001 004 3	32.894	125.78	2 416.6	125.79	2 430.5	2 556.3	0.436 9	8.453 3
32.88	306.03	0.005 000	0.001 005 3	28.192	137.81	2 420.5	137.82	2 423.7	2 561.5	0.476 4	8.395 1
35.00	308.15	0.005 628	0.001 006 0	25.216	146.67	2 423.4	146.68	2 418.6	2 565.3	0.505 3	8.353 1
36.16	309.31	0.006 000	0.001 006 4	23.739	151.53	2 425.0	151.53	2 415.9	2 567.4	0.521 0	8.330 4
39.00	312.15	0.007 000	0.001 007 4	20.530	163.39	2 428.8	163.40	2 409.1	2 572.5	0.559 2	8.275 8
40.00	313.15	0.007 384	0.001 007 8	19.523	167.56	2 430.1	167.57	2 406.7	2 574.3	0.572 5	8.257 0
41.51	314.66	0.008 000	0.001 008 4	18.103	173.87	2 432.2	173.88	2 403.1	2 577.0	0.592 6	8.228 7
43.76	316.91	0.009 000	0.001 009 4	16.203	183.27	2 435.2	183.29	2 397.7	2 581.0	0.622 4	8.187 2
45.00	318.15	0.009 593	0.001 009 9	15.258	188.44	2 436.8	188.45	2 394.8	2 583.2	0.638 7	8.164 8
45.81	318.96	0.010 000	0.001 010 2	14.674	191.82	2 437.9	191.83	2 392.8	2 584.7	0.649 3	8.150 2
50.00	323.15	0.012 349	0.001 012 1	12.032	209.32	2 443.5	209.33	2 382.7	2 592.1	0.703 8	8.076 3
53.97	327.12	0.015 000	0.001 014 1	10.022	225.92	2 448.7	225.94	2 373.1	2 599.1	0.754 9	8.008 5
55.00	328.15	0.015 758	0.001 014 6	9.568	230.21	2 450.1	230.23	2 370.7	2 600.9	0.767 9	7.991 3
60.00	333.15	0.019 940	0.001 017 2	7.671	251.11	2 456.6	251.13	2 358.5	2 609.6	0.831 2	7.909 6
60.06	333.21	0.020 000	0.001 017 2	7.649	251.38	2 456.7	251.40	2 358.3	2 609.7	0.832 0	7.908 5
65.00	338.15	0.025 030	0.001 019 9	6.197	272.02	2 463.1	272.06	2 346.2	2 618.3	0.893 5	7.831 0
69.10	342.25	0.030 000	0.001 022 3	5.229	289.20	2 468.4	289.23	2 336.1	2 625.3	0.943 9	7.768 6
70.00	343.15	0.031 190	0.001 022 8	5.042	292.95	2 469.6	292.98	2 333.8	2 626.8	0.954 9	7.755 3
75.00	348.15	0.038 580	0.001 025 9	4.131	313.90	2 475.9	313.93	2 321.4	2 635.3	1.015 5	7.682 4
75.87	349.02	0.040 000	0.001 026 5	3.993	317.53	2 477.0	317.58	2 319.2	2 636.8	1.025 9	7.670 0
80.00	353.15	0.047 390	0.001 029 1	3.407	334.86	2 482.2	334.91	2 308.8	2 643.7	1.075 3	7.612 2
81.33	354.48	0.050 000	0.001 030 0	3.240	340.44	2 483.9	340.49	2 305.4	2 645.9	1.091 0	7.593 9
85.00	358.15	0.057 830	0.001 032 5	2.828	355.84	2 488.4	355.90	2 296.0	2 651.9	1.134 3	7.544 5
85.94	359.09	0.060 000	0.001 033 1	2.732	359.79	2 489.6	359.86	2 293.6	2 653.5	1.145 3	7.532 0
89.95	363.10	0.070 000	0.001 036 0	2.365	376.63	2 494.5	376.70	2 283.3	2 660.0	1.191 9	7.479 7
90.00	363.15	0.070 140	0.001 036 0	2.361	376.85	2 494.5	376.92	2 283.2	2 660.1	1.192 5	7.479 1
93.50	366.65	0.080 000	0.001 038 6	2.087	391.58	2 498.8	391.66	2 274.1	2 665.8	1.232 9	7.434 6
95.00	368.15	0.084 550	0.001 039 7	1.981 9	397.88	2 500.6	397.96	2 270.2	2 668.1	1.250 0	7.415 9

*Condensed from: J.H. Keenan, F.G. Keyes, P.G. Hill, and J.G. Moore, "Steam Tables: Thermodynamic Properties of Water Including Vapor, Liquid, and Solid Phases". Copyright © 1969, by John Wiley & Sons, Inc. Reprinted by permission of the authors and publisher.

Table 1-7.   SATURATED STEAM, WATER, AND ICE—SI UNITS *(Continued)*

| Temperature | | Pressure, MN/m^2 | Specific volume, m^3/kg | | Specific internal energy, kJ/kg | | Specific enthalpy, kJ/kg | | | Specific entropy, kJ/kg·K | |
C	K		v_f	v_g	u_f	u_g	h_f	h_{fg}	h_g	s_f	s_g
LIQUID—VAPOR											
96.71	369.86	0.090 000	0.001 041 0	1.869	405.06	2 502.6	405.15	2 265.7	2 670.9	1.269 5	7.394 9
99.63	372.78	0.100 000	0.001 043 2	1.694 0	417.36	2 506.1	417.46	2 258.0	2 675.5	1.302 6	7.359 4
100.00	373.15	0.101 350	0.001 043 5	1.672 9	418.94	2 506.5	419.04	2 257.0	2 676.1	1.306 9	7.354 9
110.00	383.15	0.143 270	0.001 051 6	1.210 2	461.14	2 518.1	461.30	2 230.2	2 691.5	1.418 5	7.238 7
111.37	384.52	0.150 000	0.001 052 8	1.159 3	466.94	2 519.7	467.11	2 226.5	2 693.6	1.433 6	7.223 3
120.00	393.15	0.198 530	0.001 060 3	0.891 9	503.50	2 529.3	503.71	2 202.6	2 706.3	1.527 6	7.129 6
120.23	393.38	0.200 000	0.001 060 5	0.885 7	504.49	2 529.5	504.70	2 201.9	2 706.7	1.530 1	7.127 1
130.00	403.15	0.270 100	0.001 069 7	0.668 5	546.02	2 539.9	546.31	2 174.2	2 720.5	1.634 4	7.026 9
133.55	406.70	0.300 000	0.001 073 2	0.605 8	561.15	2 543.6	561.47	2 163.8	2 725.3	1.671 8	6.991 9
140.00	413.15	0.361 300	0.001 079 7	0.508 9	588.74	2 550.0	589.13	2 144.7	2 733.9	1.739 1	6.929 9
143.63	416.78	0.400 000	0.001 083 6	0.462 5	604.31	2 553.6	604.74	2 133.8	2 738.6	1.776 6	6.895 9
150.00	423.15	0.475 800	0.001 090 5	0.392 8	631.68	2 559.5	632.20	2 114.3	2 746.5	1.841 8	6.837 9
151.86	425.01	0.500 000	0.001 092 6	0.374 9	639.68	2 561.2	640.23	2 108.5	2 748.7	1.860 7	6.821 3
160.00	433.15	0.617 800	0.001 102 0	0.307 1	674.87	2 568.4	675.55	2 082.6	2 758.1	1.942 7	6.750 2
170.00	443.15	0.791 700	0.001 114 3	0.242 8	718.33	2 576.5	719.21	2 049.5	2 768.7	2.041 9	6.666 3
179.91	453.06	1.000 000	0.001 127 3	0.194 44	761.68	2 583.6	762.81	2 015.3	2 778.1	2.138 7	6.586 5
180.00	453.15	1.002 100	0.001 127 4	0.194 05	762.09	2 583.7	763.22	2 015.0	2 778.2	2.139 6	6.585 7
190.00	463.15	1.254 400	0.001 141 4	0.156 54	806.19	2 590.0	807.62	1 978.8	2 786.4	2.235 9	6.507 9
198.32	471.47	1.500 000	0.001 153 9	0.131 77	843.16	2 594.5	844.89	1 947.3	2 792.2	2.315 0	6.444 8
200.00	473.15	1.553 800	0.001 156 5	0.127 36	850.65	2 595.3	852.45	1 940.7	2 793.2	2.330 9	6.432 3
210.00	483.15	1.906 200	0.001 172 6	0.104 41	895.53	2 599.5	897.76	1 900.7	2 798.5	2.424 8	6.358 5
212.42	485.57	2.000 000	0.001 176 7	0.099 63	906.44	2 600.3	908.79	1 890.7	2 799.5	2.447 4	6.340 9
220.00	493.15	2.318 000	0.001 190 0	0.086 19	940.87	2 602.4	943.62	1 858.5	2 802.1	2.517 8	6.286 1
223.99	497.14	2.500 000	0.001 197 3	0.079 98	959.11	2 603.1	962.11	1 841.0	2 803.1	2.554 7	6.257 5
230.00	503.15	2.795 000	0.001 208 8	0.071 58	986.74	2 603.9	990.12	1 813.8	2 804.0	2.609 9	6.214 6
233.90	507.05	3.000 000	0.001 216 5	0.066 68	1 004.78	2 604.1	1 008.42	1 795.7	2 804.2	2.645 7	6.186 9
240.00	513.15	3.344 000	0.001 229 1	0.059 76	1 033.21	2 604.0	1 037.32	1 766.5	2 803.8	2.701 5	6.143 7
242.60	515.75	3.500 000	0.001 234 7	0.057 07	1 045.43	2 603.7	1 049.75	1 753.7	2 803.4	2.725 3	6.125 3
250.00	523.15	3.973 000	0.001 251 2	0.050 13	1 080.39	2 602.4	1 085.36	1 716.2	2 801.5	2.792 7	6.073 0
250.40	523.55	4.000 000	0.001 252 2	0.049 78	1 082.31	2 602.3	1 087.31	1 714.1	2 801.4	2.796 4	6.070 1
260.00	533.15	4.688 000	0.001 275 5	0.042 21	1 128.39	2 599.0	1 134.37	1 662.5	2 796.9	2.883 8	6.001 9
263.99	537.14	5.000 000	0.001 285 9	0.039 44	1 147.81	2 597.1	1 154.23	1 640.1	2 794.3	2.920 2	5.973 4
270.00	543.15	5.499 000	0.001 302 3	0.035 64	1 177.36	2 593.7	1 184.51	1 605.2	2 789.7	2.975 1	5.930 1
275.64	548.79	6.000 000	0.001 318 7	0.032 44	1 205.44	2 589.7	1 213.35	1 571.0	2 784.3	3.026 7	5.889 2
280.00	553.15	6.412 000	0.001 332 1	0.030 17	1 227.46	2 586.1	1 235.99	1 543.6	2 779.6	3.066 8	5.857 1
285.88	559.03	7.000 000	0.001 351 3	0.027 37	1 257.55	2 580.5	1 267.00	1 505.1	2 772.1	3.121 1	5.813 3
290.00	563.15	7.436 000	0.001 365 6	0.025 57	1 278.92	2 576.0	1 289.07	1 477.1	2 766.2	3.159 4	5.782 1
295.06	568.21	8.000 000	0.001 384 2	0.023 52	1 305.57	2 569.8	1 316.64	1 441.3	2 758.0	3.206 8	5.743 2
300.00	573.15	8.581 000	0.001 403 6	0.021 67	1 332.0	2 563.0	1 344.0	1 404.9	2 749.0	3.253 4	5.704 5
303.40	576.55	9.000 000	0.001 417 8	0.020 48	1 350.51	2 557.8	1 363.26	1 378.9	2 742.1	3.285 8	5.677 2
310.00	583.15	9.856 000	0.001 447 4	0.018 350	1 387.1	2 546.4	1 401.3	1 326 0	2 727.3	3.349 3	5.623 0
311.06	584.21	10.000 000	0.001 452 4	0.018 026	1 393.04	2 544.4	1 407.56	1 317.1	2 724.7	3.359 6	5.614 1
320.00	593.15	11.274 000	0.001 498 8	0.015 488	1 444.6	2 525.5	1 461.5	1 238.6	2 700.1	3.448 0	5.536 2
324.75	597.90	12.000 000	0.001 526 7	0.014 263	1 473.0	2 513.7	1 491.3	1 193.6	2 684.9	3.496 2	5.492 4
330.00	603.15	12.845 000	0.001 560 7	0.012 996	1 505.3	2 498.9	1 525.3	1 140.6	2 665.9	3.550 7	5.441 7
336.75	609.90	14.000 000	0.001 610 7	0.011 485	1 548.6	2 476.8	1 571.1	1 066.5	2 637.6	3.623 2	5.371 7
340.00	613.15	14.586 000	0.001 637 9	0.001 079 7	1 570.3	2 464.6	1 594.2	1 027.9	2 622.0	3.659 4	5.335 7
347.44	620.59	16.000 000	0.001 710 7	0.009 306	1 622.7	2 431.7	1 650.1	930.6	2 580.6	3.746 1	5.245 5
350.00	623.15	16.513 000	0.001 740 3	0.008 813	1 641.9	2 418.4	1 670.6	893.4	2 563.9	3.777 7	5.211 2
357.06	630.21	18.000 000	0.001 839 7	0.007 489	1 698.9	2 374.3	1 732.0	777.1	2 509.1	3.871 5	5.104 4
360.00	633.15	18.651 000	0.001 892 5	0.006 945	1 725.2	2 351.5	1 760.5	720.5	2 481.0	3.914 7	5.052 6
365.81	638.96	20.000 000	0.002 036	0.005 834	1 785.6	2 293.0	1 826.3	583.4	2 409.7	4.013 9	4.926 9
370.00	643.15	21.030 000	0.002 213	0.004 925	1 844.0	2 228.5	1 890.5	441.6	2 332.1	4.110 6	4.797 1
373.80	646.95	22.000 000	0.002 742	0.003 568	1 961.9	2 087.1	2 022.2	143.4	2 165.6	4.311 0	4.532 7
374.136	647.286	22.090 000	0.003 155	0.003 155	2 029.6	2 029.6	2 099.3	0	2 099.3	4.429 8	4.429 8

Table 1-8. SPECIFIC HEAT AT CONSTANT PRESSURE OF STEAM AND OF WATER*

For specific heat in J/kg·K, multiply values in Btu/lb$_m$ °F by 4186.8.

c_p, Btu/lb$_m$ °F — Pressure, psia

Temp, °F	1	1.5	2	3	4	6	8	10	15	20	30	40	60	80	100	Temp, °F
1500	0.559	0.559	0.559	0.559	0.559	0.559	0.559	0.559	0.559	0.559	0.560	0.560	0.560	0.561	0.561	1500
1480	0.557	0.557	0.557	0.557	0.557	0.557	0.557	0.558	0.558	0.558	0.558	0.558	0.559	0.559	0.559	1480
1460	0.556	0.556	0.556	0.556	0.556	0.556	0.556	0.556	0.556	0.556	0.556	0.557	0.557	0.557	0.558	1460
1440	0.554	0.554	0.554	0.554	0.554	0.554	0.554	0.554	0.554	0.554	0.555	0.555	0.555	0.556	0.556	1440
1420	0.552	0.552	0.552	0.552	0.552	0.552	0.553	0.553	0.553	0.553	0.553	0.553	0.554	0.554	0.555	1420
1400	0.551	0.551	0.551	0.551	0.551	0.551	0.551	0.551	0.551	0.551	0.551	0.552	0.552	0.553	0.553	1400
1380	0.549	0.549	0.549	0.549	0.549	0.549	0.549	0.549	0.549	0.549	0.550	0.550	0.550	0.551	0.551	1380
1360	0.547	0.547	0.547	0.547	0.547	0.547	0.547	0.547	0.548	0.548	0.548	0.548	0.549	0.549	0.550	1360
1340	0.546	0.546	0.546	0.546	0.546	0.546	0.546	0.546	0.546	0.546	0.546	0.547	0.547	0.548	0.548	1340
1320	0.544	0.544	0.544	0.544	0.544	0.544	0.544	0.544	0.544	0.544	0.545	0.545	0.545	0.546	0.546	1320
1300	0.542	0.542	0.542	0.542	0.542	0.542	0.542	0.542	0.542	0.543	0.543	0.543	0.544	0.544	0.545	1300
1280	0.540	0.540	0.540	0.540	0.540	0.540	0.540	0.541	0.541	0.541	0.541	0.541	0.542	0.543	0.543	1280
1260	0.538	0.539	0.539	0.539	0.539	0.539	0.539	0.539	0.539	0.539	0.539	0.540	0.540	0.541	0.541	1260
1240	0.537	0.537	0.537	0.537	0.537	0.537	0.537	0.537	0.537	0.537	0.538	0.538	0.539	0.539	0.540	1240
1220	0.535	0.535	0.535	0.535	0.535	0.535	0.535	0.535	0.535	0.535	0.536	0.536	0.536	0.537	0.538	1220
1200	0.533	0.533	0.533	0.533	0.533	0.533	0.533	0.533	0.534	0.534	0.534	0.534	0.535	0.536	0.536	1200
1180	0.531	0.531	0.531	0.531	0.531	0.531	0.532	0.532	0.532	0.532	0.532	0.533	0.533	0.534	0.535	1180
1160	0.529	0.529	0.530	0.530	0.530	0.530	0.530	0.530	0.530	0.530	0.530	0.531	0.532	0.532	0.533	1160
1140	0.528	0.528	0.528	0.528	0.528	0.528	0.528	0.528	0.528	0.528	0.529	0.529	0.530	0.531	0.531	1140
1120	0.526	0.526	0.526	0.526	0.526	0.526	0.526	0.526	0.526	0.527	0.527	0.527	0.528	0.529	0.530	1120
1100	0.524	0.524	0.524	0.524	0.524	0.524	0.524	0.524	0.525	0.525	0.525	0.525	0.526	0.527	0.528	1100
1080	0.522	0.522	0.522	0.522	0.522	0.522	0.522	0.522	0.523	0.523	0.523	0.523	0.524	0.525	0.526	1080
1060	0.520	0.520	0.520	0.520	0.520	0.521	0.521	0.521	0.521	0.521	0.522	0.522	0.523	0.524	0.524	1060
1040	0.518	0.519	0.519	0.519	0.519	0.519	0.519	0.519	0.519	0.519	0.520	0.520	0.521	0.522	0.523	1040
1020	0.517	0.517	0.517	0.517	0.517	0.517	0.517	0.517	0.517	0.518	0.518	0.518	0.519	0.520	0.521	1020
1000	0.515	0.515	0.515	0.515	0.515	0.515	0.515	0.515	0.515	0.516	0.516	0.517	0.518	0.519	0.519	1000
980	0.513	0.513	0.513	0.513	0.513	0.513	0.513	0.513	0.514	0.514	0.514	0.515	0.516	0.517	0.518	980
960	0.511	0.511	0.511	0.511	0.511	0.511	0.512	0.512	0.512	0.512	0.513	0.513	0.514	0.515	0.516	960
940	0.509	0.509	0.509	0.509	0.509	0.510	0.510	0.510	0.510	0.510	0.511	0.511	0.512	0.514	0.515	940
920	0.507	0.508	0.508	0.508	0.508	0.508	0.508	0.508	0.508	0.509	0.509	0.510	0.511	0.512	0.513	920
900	0.506	0.506	0.506	0.506	0.506	0.506	0.506	0.506	0.506	0.507	0.507	0.508	0.509	0.510	0.512	900
880	0.504	0.504	0.504	0.504	0.504	0.504	0.504	0.504	0.505	0.505	0.506	0.506	0.508	0.509	0.510	880
860	0.502	0.502	0.502	0.502	0.502	0.502	0.503	0.503	0.503	0.503	0.504	0.505	0.506	0.507	0.509	860
840	0.500	0.500	0.500	0.500	0.500	0.501	0.501	0.501	0.501	0.502	0.502	0.503	0.504	0.506	0.507	840
820	0.498	0.498	0.499	0.499	0.499	0.499	0.499	0.499	0.499	0.500	0.501	0.501	0.503	0.504	0.506	820
800	0.497	0.497	0.497	0.497	0.497	0.497	0.497	0.497	0.498	0.498	0.499	0.500	0.501	0.503	0.505	800
780	0.495	0.495	0.495	0.495	0.495	0.495	0.495	0.496	0.496	0.496	0.497	0.498	0.500	0.502	0.503	780
760	0.493	0.493	0.493	0.493	0.493	0.494	0.494	0.494	0.494	0.495	0.496	0.497	0.499	0.500	0.502	760
740	0.491	0.491	0.491	0.492	0.492	0.492	0.492	0.492	0.493	0.493	0.494	0.495	0.497	0.499	0.501	740
720	0.490	0.490	0.490	0.490	0.490	0.490	0.490	0.491	0.491	0.492	0.493	0.493	0.496	0.498	0.500	720
700	0.488	0.488	0.488	0.488	0.488	0.488	0.489	0.489	0.490	0.490	0.491	0.492	0.495	0.497	0.500	700
680	0.486	0.486	0.486	0.486	0.487	0.487	0.487	0.487	0.488	0.489	0.490	0.491	0.494	0.496	0.499	680
660	0.484	0.485	0.485	0.485	0.485	0.485	0.485	0.486	0.486	0.487	0.489	0.490	0.493	0.496	0.499	660
640	0.483	0.483	0.483	0.483	0.483	0.484	0.484	0.484	0.485	0.486	0.487	0.489	0.492	0.495	0.499	640
620	0.481	0.481	0.481	0.481	0.482	0.482	0.482	0.483	0.483	0.484	0.486	0.488	0.491	0.495	0.499	620
600	0.479	0.480	0.480	0.480	0.480	0.480	0.481	0.481	0.482	0.483	0.485	0.487	0.491	0.495	0.499	600
580	0.478	0.478	0.478	0.478	0.478	0.479	0.479	0.480	0.481	0.482	0.484	0.486	0.491	0.495	0.500	580
560	0.476	0.476	0.476	0.477	0.477	0.477	0.478	0.478	0.479	0.481	0.483	0.485	0.490	0.496	0.501	560
540	0.475	0.475	0.475	0.475	0.475	0.476	0.476	0.477	0.478	0.480	0.482	0.485	0.491	0.497	0.503	540
520	0.473	0.473	0.473	0.474	0.474	0.475	0.475	0.476	0.477	0.479	0.482	0.485	0.491	0.498	0.505	520
500	0.472	0.472	0.472	0.472	0.473	0.473	0.474	0.475	0.476	0.478	0.481	0.485	0.492	0.500	0.508	500
480	0.470	0.470	0.470	0.471	0.471	0.472	0.473	0.473	0.475	0.477	0.481	0.485	0.493	0.502	0.511	480
460	0.469	0.469	0.469	0.469	0.470	0.471	0.472	0.472	0.475	0.477	0.481	0.486	0.495	0.505	0.516	460
440	0.467	0.467	0.468	0.468	0.469	0.470	0.470	0.471	0.474	0.476	0.481	0.487	0.498	0.509	0.522	440
420	0.466	0.466	0.466	0.467	0.467	0.468	0.470	0.471	0.473	0.476	0.482	0.488	0.501	0.514	0.528	420
400	0.464	0.465	0.465	0.466	0.466	0.467	0.469	0.470	0.473	0.476	0.483	0.490	0.504	0.520	0.536	400
380	0.463	0.463	0.464	0.464	0.465	0.466	0.468	0.469	0.473	0.477	0.484	0.492	0.509	0.527	0.546	380
360	0.462	0.462	0.462	0.463	0.464	0.466	0.467	0.469	0.473	0.477	0.486	0.495	0.515	0.536	0.558	360
340	0.460	0.461	0.461	0.462	0.463	0.465	0.467	0.469	0.473	0.478	0.488	0.499	0.521	0.546	_0.572_	340
320	0.459	0.460	0.460	0.461	0.462	0.464	0.467	0.469	0.474	0.480	0.491	0.504	0.530	_0.558_	1.036	320
300	0.458	0.459	0.459	0.460	0.462	0.464	0.466	0.469	0.475	0.482	0.495	0.509	_0.539_	1.029	1.029	300
280	0.457	0.458	0.458	0.460	0.461	0.464	0.467	0.469	0.477	0.484	0.500	_0.516_	1.022	1.022	1.022	280
260	0.456	0.457	0.457	0.459	0.461	0.464	0.467	0.470	0.478	0.487	_0.505_	1.017	1.017	1.017	1.016	260
240	0.455	0.456	0.457	0.458	0.460	0.464	0.468	0.471	0.481	_0.491_	1.012	1.012	1.012	1.012	1.012	240
220	0.454	0.455	0.456	0.458	0.460	0.464	0.468	0.473	_0.484_	1.008	1.008	1.008	1.008	1.008	1.008	220
200	0.453	0.454	0.455	0.458	0.460	0.465	_0.470_	0.475	1.005	1.005	1.005	1.005	1.005	1.005	1.005	200
180	0.452	0.454	0.455	0.458	0.460	_0.466_	1.003	1.003	1.003	1.003	1.003	1.003	1.003	1.002	1.002	180
160	0.451	0.453	0.455	_0.458_	_0.461_	1.001	1.001	1.001	1.001	1.001	1.001	1.001	1.001	1.001	1.001	160
140	0.451	0.453	_0.454_	1.000	1.000	1.000	1.000	1.000	0.999	0.999	0.999	0.999	0.999	0.999	0.999	140
120	_0.450_	_0.452_	0.999	0.999	0.999	0.999	0.999	0.999	0.999	0.999	0.998	0.998	0.998	0.998	0.998	120
100	0.998	0.998	0.998	0.998	0.998	0.998	0.998	0.998	0.998	0.998	0.998	0.998	0.998	0.998	0.998	100
80	0.998	0.998	0.998	0.998	0.998	0.998	0.998	0.998	0.998	0.998	0.998	0.998	0.998	0.998	0.998	80
60	1.000	1.000	1.000	1.000	1.000	1.000	1.000	1.000	1.000	1.000	1.000	1.000	0.999	0.999	0.999	60
40	1.004	1.004	1.004	1.004	1.004	1.004	1.004	1.004	1.004	1.004	1.004	1.004	1.004	1.004	1.003	40
32	1.007	1.007	1.007	1.007	1.007	1.007	1.007	1.007	1.007	1.007	1.007	1.007	1.007	1.007	1.006	32
Sat. water	0.998	0.998	0.999	1.000	1.000	1.002	1.003	1.004	1.007	1.010	1.014	1.019	1.026	1.033	1.039	Sat. water
Sat. steam	0.450	0.452	0.454	0.458	0.461	0.466	0.471	0.475	0.485	0.493	0.508	0.521	0.543	0.564	0.582	Sat. steam

Table 1-8. SPECIFIC HEAT AT CONSTANT PRESSURE OF STEAM AND OF WATER (Continued)

Temp, °F	c_p, Btu/lb$_m$ °F Pressure, psia															Temp, °F
	150	200	300	400	600	800	1000	1500	2000	3000	4000	6000	8000	10000	15000	
1500	0.562	0.563	0.565	0.567	0.571	0.576	0.580	0.590	0.601	0.623	0.645	0.691	0.737	0.780	0.868	1500
1480	0.561	0.562	0.564	0.566	0.570	0.575	0.579	0.590	0.601	0.623	0.647	0.694	0.742	0.786	0.878	1480
1460	0.559	0.560	0.562	0.565	0.569	0.573	0.578	0.589	0.601	0.624	0.648	0.698	0.747	0.793	0.888	1460
1440	0.557	0.559	0.561	0.563	0.568	0.572	0.577	0.589	0.600	0.625	0.650	0.701	0.753	0.800	0.900	1440
1420	0.556	0.557	0.559	0.562	0.566	0.571	0.576	0.588	0.600	0.625	0.651	0.705	0.759	0.808	0.909	1420
1400	0.554	0.555	0.558	0.560	0.565	0.570	0.575	0.587	0.600	0.626	0.653	0.709	0.765	0.817	0.926	1400
1380	0.553	0.554	0.556	0.559	0.564	0.569	0.574	0.587	0.600	0.627	0.655	0.714	0.773	0.827	0.939	1380
1360	0.551	0.552	0.555	0.558	0.563	0.568	0.573	0.586	0.600	0.628	0.657	0.719	0.781	0.838	0.953	1360
1340	0.549	0.551	0.553	0.556	0.561	0.567	0.572	0.586	0.600	0.629	0.660	0.725	0.790	0.850	0.968	1340
1320	0.548	0.549	0.552	0.555	0.560	0.566	0.571	0.585	0.600	0.630	0.663	0.731	0.800	0.864	0.983	1320
1300	0.546	0.548	0.550	0.553	0.559	0.565	0.570	0.585	0.600	0.632	0.666	0.738	0.811	0.879	0.998	1300
1280	0.545	0.546	0.549	0.552	0.558	0.564	0.570	0.585	0.600	0.634	0.669	0.746	0.824	0.897	1.014	1280
1260	0.543	0.544	0.547	0.550	0.556	0.563	0.569	0.585	0.601	0.636	0.673	0.755	0.838	0.918	1.033	1260
1240	0.541	0.543	0.546	0.549	0.555	0.562	0.568	0.584	0.601	0.638	0.678	0.765	0.855	0.942	1.053	1240
1220	0.540	0.541	0.544	0.548	0.554	0.561	0.567	0.584	0.602	0.641	0.683	0.777	0.875	0.969	1.072	1220
1200	0.538	0.540	0.543	0.546	0.553	0.560	0.567	0.584	0.603	0.644	0.689	0.790	0.897	1.000	1.095	1200
1180	0.536	0.538	0.541	0.545	0.552	0.559	0.566	0.584	0.604	0.647	0.696	0.805	0.922	1.033	1.117	1180
1160	0.535	0.536	0.540	0.544	0.551	0.558	0.565	0.585	0.606	0.652	0.704	0.823	0.952	1.070	1.143	1160
1140	0.533	0.535	0.539	0.542	0.550	0.557	0.565	0.585	0.607	0.656	0.713	0.843	0.986	1.107	1.167	1140
1120	0.531	0.533	0.537	0.541	0.549	0.557	0.565	0.586	0.609	0.662	0.723	0.866	1.025	1.149	1.190	1120
1100	0.530	0.532	0.536	0.540	0.548	0.556	0.564	0.587	0.612	0.668	0.735	0.893	1.070	1.180	1.200	1100
1080	0.528	0.530	0.534	0.538	0.547	0.555	0.564	0.588	0.615	0.676	0.749	0.924	1.120	1.242	1.240	1080
1060	0.527	0.529	0.533	0.537	0.546	0.555	0.564	0.590	0.618	0.685	0.765	0.960	1.176	1.295	1.260	1060
1040	0.525	0.527	0.532	0.536	0.545	0.555	0.565	0.592	0.622	0.695	0.783	1.002	1.238	1.351	1.282	1040
1020	0.523	0.526	0.530	0.535	0.545	0.555	0.565	0.594	0.627	0.707	0.804	1.051	1.306	1.399	1.298	1020
1000	0.522	0.524	0.529	0.534	0.544	0.555	0.566	0.597	0.633	0.721	0.829	1.110	1.382	1.471	1.306	1000
980	0.520	0.523	0.528	0.533	0.544	0.555	0.567	0.601	0.640	0.737	0.858	1.180	1.475	1.531	1.312	980
960	0.519	0.521	0.527	0.532	0.543	0.556	0.568	0.605	0.648	0.756	0.893	1.267	1.598	1.595	1.310	960
940	0.517	0.520	0.526	0.531	0.543	0.556	0.570	0.610	0.658	0.778	0.934	1.376	1.708	1.639	1.299	940
920	0.516	0.519	0.525	0.531	0.544	0.558	0.573	0.617	0.669	0.803	0.984	1.520	1.819	1.667	1.281	920
900	0.515	0.518	0.524	0.530	0.544	0.559	0.576	0.624	0.683	0.834	1.048	1.716	1.932	1.660	1.259	900
880	0.513	0.516	0.523	0.530	0.545	0.561	0.580	0.633	0.699	0.870	1.100	1.990	2.000	1.633	1.232	880
860	0.512	0.515	0.523	0.530	0.546	0.564	0.584	0.644	0.718	0.918	1.240	2.316	2.019	1.593	1.212	860
840	0.511	0.514	0.522	0.530	0.548	0.568	0.590	0.657	0.740	0.977	1.395	2.653	1.978	1.547	1.192	840
820	0.510	0.514	0.522	0.531	0.550	0.572	0.597	0.672	0.767	1.054	1.620	2.886	1.888	1.503	1.175	820
800	0.509	0.513	0.522	0.532	0.553	0.577	0.605	0.690	0.800	1.160	1.967	2.872	1.768	1.459	1.157	800
780	0.508	0.513	0.522	0.533	0.557	0.584	0.615	0.712	0.840	1.312	2.550	2.547	1.670	1.416	1.142	780
760	0.507	0.512	0.523	0.535	0.561	0.592	0.628	0.738	0.892	1.542	4.462	2.156	1.576	1.370	1.126	760
740	0.507	0.512	0.524	0.537	0.567	0.602	0.642	0.770	0.960	1.913	8.119	1.886	1.493	1.332	1.114	740
720	0.506	0.512	0.525	0.540	0.574	0.613	0.660	0.811	1.052	2.584	3.458	1.696	1.421	1.290	1.100	720
700	0.506	0.513	0.528	0.544	0.582	0.627	0.681	0.861	1.181	6.145[⊙]	2.237	1.557	1.358	1.250	1.089	700
680	0.506	0.514	0.530	0.549	0.592	0.644	0.707	0.927	1.365	2.469	1.789	1.450	1.303	1.217	1.079	680
660	0.507	0.515	0.534	0.555	0.604	0.665	0.738	1.015	1.639	1.851	1.587	1.369	1.256	1.187	1.071	660
640	0.507	0.517	0.538	0.562	0.619	0.690	0.777	1.135	2.219	1.601	1.454	1.303	1.216	1.157	1.063	640
620	0.509	0.519	0.543	0.571	0.637	0.720	0.826	1.308	1.614	1.455	1.362	1.252	1.184	1.136	1.056	620
600	0.510	0.522	0.550	0.582	0.659	0.757	0.888	1.580	1.453	1.358	1.295	1.211	1.157	1.118	1.052	600
580	0.513	0.526	0.558	0.595	0.685	0.804	0.969	1.393	1.351	1.289	1.243	1.178	1.134	1.102	1.046	580
560	0.516	0.531	0.568	0.611	0.717	0.862	1.079	1.309	1.281	1.237	1.202	1.151	1.115	1.087	1.039	560
540	0.519	0.538	0.580	0.630	0.756	0.937	1.272	1.249	1.229	1.196	1.169	1.128	1.098	1.074	1.031	540
520	0.524	0.545	0.594	0.653	0.804	1.035	1.221	1.204	1.180	1.164	1.142	1.109	1.083	1.002	1.024	520
500	0.530	0.554	0.611	0.680	0.865	1.187	1.181	1.169	1.157	1.137	1.120	1.092	1.069	1.051	1.017	500
480	0.537	0.565	0.632	0.714	1.159	1.154	1.150	1.140	1.131	1.115	1.101	1.077	1.057	1.041	1.010	480
460	0.545	0.578	0.657	0.755	1.132	1.128	1.125	1.117	1.110	1.096	1.084	1.064	1.047	1.033	1.004	460
440	0.556	0.594	0.687	1.113	1.110	1.107	1.104	1.098	1.092	1.080	1.070	1.052	1.038	1.025	0.999	440
420	0.568	0.614	0.724	1.094	1.091	1.089	1.087	1.081	1.076	1.067	1.058	1.042	1.029	1.018	0.994	420
400	0.583	0.636	1.079	1.078	1.076	1.074	1.072	1.067	1.063	1.055	1.047	1.034	1.022	1.011	0.990	400
380	0.601	1.066	1.065	1.065	1.063	1.061	1.059	1.056	1.052	1.044	1.038	1.026	1.015	1.006	0.986	380
360	0.622	1.054	1.054	1.053	1.052	1.050	1.049	1.045	1.042	1.036	1.030	1.019	1.009	1.001	0.982	360
340	1.045	1.044	1.044	1.043	1.042	1.040	1.039	1.036	1.033	1.028	1.022	1.013	1.004	0.996	0.979	340
320	1.036	1.036	1.035	1.034	1.033	1.032	1.031	1.028	1.026	1.021	1.016	1.007	0.999	0.992	0.976	320
300	1.028	1.028	1.028	1.027	1.026	1.025	1.024	1.022	1.019	1.015	1.010	1.002	0.995	0.988	0.973	300
280	1.022	1.022	1.021	1.021	1.020	1.019	1.018	1.016	1.014	1.009	1.005	0.998	0.991	0.985	0.971	280
260	1.016	1.016	1.016	1.015	1.014	1.013	1.013	1.011	1.009	1.005	1.001	0.994	0.988	0.982	0.968	260
240	1.012	1.011	1.011	1.011	1.010	1.009	1.008	1.006	1.004	1.001	0.997	0.991	0.985	0.979	0.966	240
220	1.008	1.008	1.007	1.007	1.006	1.005	1.005	1.003	1.001	0.998	0.994	0.988	0.982	0.977	0.964	220
200	1.005	1.004	1.004	1.004	1.003	1.002	1.002	1.000	0.998	0.995	0.992	0.986	0.980	0.975	0.963	200
180	1.002	1.002	1.002	1.001	1.001	1.000	0.999	0.998	0.996	0.993	0.989	0.983	0.978	0.973	0.961	180
160	1.000	1.000	1.000	0.999	0.999	0.998	0.997	0.996	0.994	0.991	0.987	0.981	0.976	0.971	0.959	160
140	0.999	0.999	0.998	0.998	0.997	0.997	0.996	0.994	0.992	0.989	0.986	0.980	0.974	0.969	0.958	140
120	0.998	0.998	0.997	0.997	0.996	0.996	0.995	0.993	0.991	0.988	0.984	0.978	0.972	0.967	0.957	120
100	0.997	0.997	0.997	0.996	0.996	0.995	0.994	0.992	0.990	0.986	0.983	0.976	0.970	0.965	0.955	100
80	0.998	0.997	0.997	0.996	0.995	0.994	0.994	0.991	0.989	0.985	0.981	0.974	0.968	0.962	0.951	80
60	0.999	0.999	0.998	0.997	0.996	0.995	0.994	0.991	0.989	0.984	0.979	0.970	0.963	0.956	0.942	60
40	1.003	1.003	1.002	1.001	1.000	0.998	0.997	0.993	0.989	0.983	0.976	0.965	0.954	0.945	0.920	40
32	1.006	1.006	1.005	1.004	1.002	1.000	0.999	0.994	0.990	0.983	0.975	0.962	0.949	0.937	0.904	32
Sat. water	1.054	1.067	1.093	1.118	1.168	1.224	1.286	1.492	1.841	7.646	—	—	—	—	—	Sat. water
Sat. steam	0.624	0.661	0.729	0.792	0.915	1.046	1.191	1.667	2.557	13.66	—	—	—	—	—	Sat. steam

⊙Critical point.

*From: "1967 ASME Steam Tables", The American Society of Mechanical Engineers.

Table 1-9. THERMAL CONDUCTIVITY OF STEAM AND WATER*

For viscosity and thermal conductivity of steam and water in SI units, see Table 1-13.

For conductivity in W/m·K, multiply values in Btu/hr ft °F by 1.7296.

Temp, °F	k, (Btu/hr ft °F) 10^3 — Pressure, psia												
	1	2	5	10	20	50	100	200	500	1000	2000	5000	7500
1500	63.7	63.7	63.7	63.7	63.7	63.8	64.0	64.3	65.4	67.1	70.7	82.0	92.2
1450	61.4	61.4	61.5	61.5	61.5	61.6	61.8	62.1	63.2	64.9	68.5	80.1	90.6
1400	59.2	59.2	59.2	59.2	59.3	59.4	59.6	59.9	60.9	62.7	66.3	78.2	89.2
1350	57.0	57.0	57.0	57.0	57.1	57.2	57.3	57.7	58.7	60.5	64.2	76.3	87.9
1300	54.8	54.8	54.8	54.8	54.8	54.9	55.1	55.5	56.5	58.3	62.0	74.6	86.9
1250	52.6	52.6	52.6	52.6	52.6	52.7	52.9	53.2	54.3	56.1	59.9	73.0	86.3
1200	50.4	50.4	50.4	50.4	50.4	50.5	50.7	51.0	52.1	53.9	57.8	71.6	86.2
1150	48.2	48.2	48.2	48.2	48.2	48.3	48.5	48.9	49.9	51.8	55.7	70.5	87.0
1100	46.0	46.0	46.0	46.0	46.1	46.2	46.3	46.7	47.8	49.6	53.7	69.8	89.0
1050	43.9	43.9	43.9	43.9	43.9	44.0	44.2	44.6	45.6	47.5	51.8	69.7	93.4
1000	41.7	41.7	41.8	41.8	41.8	41.9	42.1	42.4	43.5	45.5	50.0	70.7	102.9
950	39.6	39.6	39.7	39.7	39.7	39.8	40.0	40.3	41.4	43.5	48.3	73.5	115.5
900	37.6	37.6	37.6	37.6	37.6	37.7	37.9	38.3	39.4	41.5	46.8	80.2	138.7
850	35.5	35.6	35.6	35.6	35.6	35.7	35.9	36.3	37.4	39.7	45.6	96.7	178.8
800	33.6	33.6	33.6	33.6	33.6	33.7	33.9	34.3	35.5	37.9	44.9	129.6	223.2
750	31.6	31.6	31.6	31.6	31.7	31.8	32.0	32.3	33.6	36.3	45.2	202.5	258.3
700	29.7	29.7	29.7	29.7	29.8	29.9	30.1	30.4	31.8	35.0	47.5⊛	262.8	295.1
650	27.8	27.8	27.9	27.9	27.9	28.0	28.2	28.6	30.1	34.1	55.7	304.3	326.7
600	26.0	26.0	26.1	26.1	26.1	26.2	26.4	26.9	28.7	34.1	301.9	333.7	349.3
550	24.3	24.3	24.3	24.3	24.4	24.5	24.7	25.2	27.5	36.1	333.7	356.1	368.0
500	22.6	22.6	22.6	22.6	22.7	22.8	23.0	23.6	26.9	350.8	357.4	373.8	383.6
450	21.0	21.0	21.0	21.0	21.0	21.2	21.4	22.3	368.1	370.6	375.3	387.9	396.5
400	19.4	19.4	19.4	19.4	19.5	19.6	20.0	21.3	383.0	384.9	388.5	398.6	406.4
350	17.9	17.9	17.9	17.9	18.0	18.2	18.8	392.0	392.9	394.4	397.4	406.1	413.2
300	16.5	16.5	16.5	16.5	16.6	16.9	396.9	397.2	398.0	399.3	402.0	409.9	416.4
250	15.1	15.1	15.1	15.2	15.3	396.9	397.0	397.3	398.1	399.4	402.1	409.7	415.8
200	13.8	13.8	13.9	14.0	391.6	391.6	391.8	392.1	393.0	394.4	397.2	404.9	410.6
150	12.7	12.7	380.5	380.5	380.6	380.7	380.8	381.1	382.1	383.7	386.7	394.7	400.3
100	363.3	363.3	363.3	363.3	363.3	363.4	363.6	363.9	365.0	366.6	369.8	378.3	384.1
50	339.1	339.1	339.1	339.1	339.2	339.3	339.4	339.8	340.8	342.5	345.7	354.6	361.0
32	328.6	328.6	328.6	328.6	328.6	328.7	328.9	329.2	330.3	331.9	335.1	344.1	350.8
Sat. water	364.0	373.1	383.8	390.4	395.2	397.4	394.7	386.2	361.7	327.6	271.8	—	—
Sat. steam	11.6	12.2	13.0	13.8	14.8	16.6	18.4	21.1	27.2	36.5	61.3	—	—

⊛Critical point.
*From: "1967 ASME Steam Tables", The American Society of Mechanical Engineers.

Table 1-10. SUPERHEATED STEAM AND COMPRESSED WATER*

For properties in SI units, see Table 1-11.

$$v = \text{cu ft/lb}_m \qquad h = \text{Btu/lb}_m \qquad s = \text{Btu/lb}_m \, °R$$

Pressure, psia (sat temp, °F)		Temperature, °F									
		100°	200°	300°	400°	500°	600°	700°	800°	900°	1000°
0.5 (79.59)	v	665.9	785.5	904.8	1024.0	1143.2	1262.3	1381.5			
	h	1105.4	1150.3	1195.8	1241.9	1288.6	1336.2	1384.5			
	s	2.0537	2.1276	2.1917	2.2487	2.3001	2.3472	2.3908			
1 (101.74)	v	0.01613	392.5	452.3	511.9	571.5	631.1	690.7			
	h	68.00	1150.2	1195.7	1241.8	1288.6	1336.1	1384.5			
	s	0.1295	2.0509	2.1152	2.1722	2.2237	2.2708	2.3144			
2 (126.07)	v	0.01613	196.03	225.99	255.87	285.70	315.52	345.32	375.12	404.91	434.71
	h	68.00	1149.8	1195.5	1241.7	1288.5	1336.1	1384.4	1433.7	1483.8	1534.8
	s	0.1295	1.9740	2.0385	2.0957	2.1472	2.1943	2.2380	2.2787	2.3170	2.3532
3 (141.47)	v	0.01613	130.54	150.57	170.52	190.43	210.31	230.19	250.06	269.93	289.79
	h	68.00	1149.4	1195.2	1241.5	1288.4	1336.0	1384.4	1433.6	1483.8	1534.8
	s	0.1295	1.9289	1.9936	2.0509	2.1024	2.1496	2.1932	2.2340	2.2723	2.3085
5 (162.24)	v	0.01613	78.14	90.24	102.24	114.21	126.15	138.08	150.01	161.94	173.86
	h	68.01	1148.6	1194.8	1241.3	1288.2	1335.9	1384.3	1433.6	1483.7	1534.7
	s	0.1295	1.8716	1.9369	1.9943	2.0460	2.0932	2.1369	2.1776	2.2159	2.2521
10 (193.21)	v	0.01613	38.84	44.98	51.03	57.04	63.03	69.00	74.98	80.94	86.91
	h	68.02	1146.6	1193.7	1240.6	1287.8	1335.5	1384.0	1433.4	1483.5	1534.6
	s	0.1295	1.7928	1.8593	1.9173	1.9692	2.0166	2.0603	2.1011	2.1394	2.1757
15 (213.03)	v	0.01613	0.01664	29.899	33.963	37.985	41.986	45.978	49.964	53.946	57.926
	h	68.04	168.09	1192.5	1239.9	1287.3	1335.2	1383.8	1433.2	1483.4	1534.5
	s	0.1295	0.2940	1.8134	1.8720	1.9242	1.9717	2.0155	2.0563	2.0946	2.1309
20 (227.96)	v	0.01613	0.01664	22.356	25.428	28.457	31.466	34.465	37.458	40.447	43.435
	h	68.05	168.11	1191.4	1239.2	1286.9	1334.9	1383.5	1432.9	1483.2	1534.3
	s	0.1295	0.2940	1.7805	1.8397	1.8921	1.9397	1.9836	2.0244	2.0628	2.0991
25 (240.07)	v	0.01613	0.01664	17.829	20.307	22.740	25.153	27.557	29.954	32.348	34.740
	h	68.06	168.12	1190.2	1238.5	1286.4	1334.6	1383.3	1432.7	1483.0	1534.2
	s	0.1295	0.2940	1.7547	1.8145	1.8672	1.9149	1.9588	1.9997	2.0381	2.0744
30 (250.34)	v	0.01613	0.01664	14.810	16.892	18.929	20.945	22.951	24.952	26.949	28.943
	h	68.08	168.13	1189.0	1237.8	1286.0	1334.2	1383.0	1432.5	1482.8	1534.0
	s	0.1295	0.2940	1.7334	1.7937	1.8467	1.8946	1.9386	1.9795	2.0179	2.0543
35 (259.29)	v	0.01613	0.01664	12.654	14.453	16.207	17.939	19.662	21.379	23.092	24.803
	h	68.09	168.14	1187.8	12.371	1285.5	1333.9	1382.8	1432.3	1482.7	1533.9
	s	0.1295	0.2940	1.7152	1.7761	1.8294	1.8774	1.9214	1.9624	2.0009	2.0372
40 (267.25)	v	0.01613	0.01664	11.036	12.624	14.165	15.685	17.195	18.699	20.199	21.697
	h	68.10	168.15	1186.6	1236.4	1285.0	1333.6	1382.5	1432.1	1482.5	1533.7
	s	0.1295	0.2940	1.6992	1.7608	1.8143	1.8624	1.9065	1.9476	1.9860	2.0224
50 (281.02)	v	0.01613	0.01663	8.769	10.062	11.306	12.529	13.741	14.947	16.150	17.350
	h	68.13	168.17	1184.1	1234.9	1284.1	1332.9	1382.0	1431.7	1482.2	1533.4
	s	0.1295	0.2940	1.6720	1.7349	1.7890	1.8374	1.8816	1.9227	1.9613	1.9977
60 (292.71)	v	0.01613	0.01663	7.257	8.354	9.400	10.425	11.438	12.446	13.450	14.452
	h	68.15	168.20	1181.6	1233.5	1283.2	1332.3	1381.5	1431.3	1481.8	1533.2
	s	0.1295	0.2939	1.6492	1.7134	1.7681	1.8168	1.8612	1.9024	1.9410	1.9774
70 (302.93)	v	0.01613	0.01663	0.01745	7.133	8.039	8.922	9.793	10.659	11.522	12.382
	h	68.18	168.22	269.72	1232.0	1282.2	1331.6	1381.0	1430.9	1481.5	1532.9
	s	0.1295	0.2939	0.4372	1.6951	1.7504	1.7993	1.8439	1.8852	1.9238	1.9603
80 (312.04)	v	0.01613	0.01663	0.01745	6.218	7.018	7.794	8.560	9.319	10.075	10.829
	h	68.21	168.24	269.74	1230.5	1281.3	1330.9	1380.5	1430.5	1481.1	1532.6
	s	0.1295	0.2939	0.4371	1.6790	1.7349	1.7842	1.8289	1.8702	1.9089	1.9454
90 (320.28)	v	0.01613	0.01663	0.01745	5.505	6.223	6.917	7.600	8.277	8.950	9.621
	h	68.23	168.26	269.76	1228.9	1280.3	1330.2	1380.0	1430.1	1480.8	1532.3
	s	0.1295	0.2939	0.4371	1.6646	1.7212	1.7707	1.8156	1.8570	1.8957	1.9323
100 (327.82)	v	0.01613	0.01663	0.01745	4.935	5.588	6.216	6.833	7.443	8.050	8.655
	h	68.26	168.29	269.77	1227.4	1279.3	1329.6	1379.5	1429.7	1480.4	1532.0
	s	0.1295	0.2939	0.4371	1.6516	1.7088	1.7586	1.8036	1.8451	1.8839	1.9205
110 (334.79)	v	0.01612	0.01663	0.01745	4.468	5.068	5.642	6.205	6.761	7.314	7.865
	h	68.29	168.31	269.79	1225.8	1278.3	1328.9	1379.0	1429.2	1480.1	1531.7
	s	0.1295	0.2939	0.4371	1.6396	1.6975	1.7476	1.7928	1.8344	1.8732	1.9099
120 (341.27)	v	0.01612	0.01663	0.01745	4.0786	4.6341	5.1637	5.6813	6.1928	6.7006	7.2060
	h	68.31	168.33	269.81	1224.1	1277.4	1328.2	1378.4	1428.8	1479.8	1531.4
	s	0.1295	0.2939	0.4371	1.6286	1.6872	1.7376	1.7829	1.8246	1.8635	1.9001

*Condensed from: "1967 ASME Steam Tables", The American Society of Mechanical Engineers, Table 3 (112 pages).

Table 1-10. SUPERHEATED STEAM AND COMPRESSED WATER
(Continued)

Pressure, psia (sat temp, °F)		400°	500°	600°	700°	800°	900°	1000°	1100°	1200°	1300°
130 (347.33)	v	3.7489	4.2672	4.7589	5.2384	5.7118	6.1814	6.6486	7.1140	7.5781	8.0411
	h	1222.5	1276.4	1327.5	1377.9	1428.4	1479.4	1531.1	1583.6	1636.9	1691.1
	s	1.6182	1.6775	1.7283	1.7737	1.8155	1.8545	1.8911	1.9259	1.9591	1.9907
140 (353.04)	v	3.4661	3.9526	4.4119	4.8588	5.2995	5.7364	6.1709	6.6036	7.0349	7.4652
	h	1220.8	1275.3	1326.8	1377.4	1428.0	1479.1	1530.8	1583.4	1636.7	1690.9
	s	1.6085	1.6686	1.7196	1.7652	1.8071	1.8461	1.8828	1.9176	1.9508	1.9825
150 (358.43)	v	3.2208	3.6799	4.1112	4.5298	4.9421	5.3507	5.7568	6.1612	6.5642	6.9661
	h	1219.1	1274.3	1326.1	1376.9	1427.6	1478.7	1530.5	1583.1	1636.5	1690.7
	s	1.5993	1.6602	1.7115	1.7573	1.7992	1.8383	1.8751	1.9099	1.9431	1.9748
160 (363.55)	v	3.0060	3.4413	3.8480	4.2420	4.6295	5.0132	5.3945	5.7741	6.1522	6.5293
	h	1217.4	1273.3	1325.4	1376.4	1427.2	1478.4	1530.3	1582.9	1636.3	1690.5
	s	1.5906	1.6522	1.7039	1.7499	1.7919	1.8310	1.8678	1.9027	1.9359	1.9676
180 (373.08)	v	2.6474	3.0433	3.4093	3.7621	4.1084	4.4508	4.7907	5.1289	5.4657	5.8014
	h	1213.8	1271.2	1324.0	1375.3	1426.3	1477.7	1529.7	1582.4	1635.9	1690.2
	s	1.5743	1.6376	1.6900	1.7362	1.7784	1.8176	1.8545	1.8894	1.9227	1.9545
200 (381.80)	v	2.3598	2.7247	3.0583	3.3783	3.6915	4.0008	4.3077	4.6128	4.9165	5.2191
	h	1210.1	1269.0	1322.6	1374.3	1425.5	1477.0	1529.1	1581.9	1635.4	1689.8
	s	1.5593	1.6242	1.6773	1.7239	1.7663	1.8057	1.8426	1.8776	1.9109	1.9427
220 (389.88)	v	2.1240	2.4638	2.7710	3.0642	3.3504	3.6327	3.9125	4.1905	4.4671	4.7426
	h	1206.3	1266.9	1321.2	1373.2	1424.7	1476.3	1528.5	1581.4	1635.0	1689.4
	s	1.5453	1.6120	1.6658	1.7128	1.7553	1.7948	1.8318	1.8668	1.9002	1.9320
240 (397.39)	v	1.9268	2.2462	2.5316	2.8024	3.0661	3.3259	3.5831	3.8385	4.0926	4.3456
	h	1202.4	1264.6	1319.7	1372.1	1423.8	1475.6	1527.9	1580.9	1634.6	1689.1
	s	1.5320	1.6006	1.6552	1.7025	1.7452	1.7848	1.8219	1.8570	1.8904	1.9223
260 (404.44)	v	0.01864	2.0619	2.3289	2.5808	2.8256	3.0663	3.3044	3.5408	3.7758	4.0097
	h	375.11	1262.4	1318.2	1371.1	1423.0	1474.9	1527.3	1580.4	1634.2	1688.7
	s	0.5666	1.5899	1.6453	1.6930	1.7359	1.7756	1.8128	1.8480	1.8814	1.9133
280 (411.07)	v	0.01863	1.9037	2.1551	2.3909	2.6194	2.8437	3.0655	3.2855	3.5042	3.7217
	h	375.13	1260.0	1316.8	1370.0	1422.1	1474.2	1526.8	1579.9	1633.8	1688.4
	s	0.5666	1.5798	1.6361	1.6841	1.7273	1.7671	1.8043	1.8395	1.8730	1.9050
300 (417.35)	v	0.01863	1.7665	2.0044	2.2263	2.4407	2.6509	2.8585	3.0643	3.2688	3.4721
	h	375.15	1257.7	1315.2	1368.9	1421.3	1473.6	1526.2	1579.4	1633.3	1688.0
	s	0.5665	1.5703	1.6274	1.6758	1.7192	1.7591	1.7964	1.8317	1.8652	1.8972
320 (423.31)	v	0.01863	1.6462	1.8725	2.0823	2.2843	2.4821	2.6774	2.8708	3.0628	3.2538
	h	375.17	1255.2	1313.7	1367.8	1420.5	1472.9	1525.6	1578.9	1632.9	1687.6
	s	0.5665	1.5612	1.6192	1.6680	1.7116	1.7516	1.7890	1.8243	1.8579	1.8899
340 (428.99)	v	0.01863	1.5399	1.7561	1.9552	2.1463	2.3333	2.5175	2.7000	2.8811	3.0611
	h	375.20	1252.8	1312.2	1366.7	1419.6	1472.2	1525.0	1578.4	1632.5	1687.3
	s	0.5664	1.5525	1.6114	1.6606	1.7044	1.7445	1.7820	1.8174	1.8510	1.8831
360 (434.41)	v	0.01862	1.4454	1.6525	1.8421	2.0237	2.2009	2.3755	2.5482	2.7196	2.8898
	h	375.22	1250.3	1310.6	1365.6	1418.7	1471.5	1524.4	1577.9	1632.1	1686.9
	s	0.5664	1.5441	1.6040	1.6536	1.6976	1.7379	1.7754	1.8109	1.8445	1.8766
380 (439.61)	v	0.01862	1.3606	1.5598	1.7410	1.9139	2.0825	2.2484	2.4124	2.5750	2.7366
	h	375.24	1247.7	1309.0	1364.5	1417.9	1470.8	1523.8	1577.4	1631.6	1686.5
	s	0.5663	1.5360	1.5969	1.6470	1.6911	1.7315	1.7692	1.8047	1.8384	1.8705
400 (444.60)	v	0.01862	1.2841	1.4763	1.6499	1.8151	1.9759	2.1339	2.2901	2.4450	2.5987
	h	375.27	1245.1	1307.4	1363.4	1417.0	1470.1	1523.3	1576.9	1631.2	1686.2
	s	0.5663	1.5282	1.5901	1.6406	1.6850	1.7255	1.7632	1.7988	1.8325	1.8647
440 (454.03)	v	0.01861	1.1517	1.3319	1.4926	1.6445	1.7918	1.9363	2.0790	2.2203	2.3605
	h	375.31	1239.7	1304.2	1361.1	1415.3	1468.7	1522.1	1575.9	1630.4	1685.5
	s	0.5662	1.5132	1.5772	1.6286	1.6734	1.7142	1.7521	1.7878	1.8216	1.8538
480 (462.82)	v	0.01861	1.0409	1.2115	1.3615	1.5023	1.6384	1.7716	1.9030	2.0330	2.1619
	h	375.36	1234.1	1300.8	1358.8	1413.6	1467.3	1520.9	1574.9	1629.5	1684.7
	s	0.5661	1.4990	1.5652	1.6176	1.6628	1.7038	1.7419	˙ 1.7777	1.8116	1.8439
520 (471.07)	v	0.01861	0.9466	1.1094	1.2504	1.3819	1.5085	1.6323	1.7542	1.8746	1.9940
	h	375.40	1228.3	1297.4	1356.5	1411.8	1465.9	1519.7	1573.9	1628.7	1684.0
	s	0.5659	1.4853	1.5539	1.6072	1.6530	1.6943	1.7325	1.7684	1.8024	1.8348
560 (478.84)	v	0.01860	0.8653	1.0217	1.1552	1.2787	1.3972	1.5129	1.6266	1.7388	1.8500
	h	375.45	1222.2	1293.9	1354.2	1410.0	1464.4	1518.6	1572.9	1627.8	1683.3
	s	0.5658	1.4720	1.5431	1.5975	1.6438	1.6853	1.7237	1.7598	1.7939	1.8263

Table 1-10. SUPERHEATED STEAM AND COMPRESSED WATER
(Continued)

Pressure, psia (sat temp, °F)		500°	600°	700°	800°	900°	1000°	1100°	1200°	1300°	1400°
						Temperature, °F					
600 (486.20)	v	0.7944	0.9456	1.0726	1.1892	1.3008	1.4093	1.5160	1.6211	1.7252	1.8284
	h	1215.9	1290.3	1351.8	1408.3	1463.0	1517.4	1571.9	1627.0	1682.6	1738.8
	s	1.4590	1.5329	1.5884	1.6351	1.6769	1.7155	1.7517	1.7859	1.8184	1.8494
650 (494.89)	v	0.7173	0.8634	0.9835	1.0929	1.1969	1.2979	1.3969	1.4944	1.5909	1.6864
	h	1207.6	1285.7	1348.7	1406.0	1461.2	1515.9	1570.7	1625.9	1681.6	1738.0
	s	1.4430	1.5207	1.5775	1.6249	1.6671	1.7059	1.7422	1.7765	1.8092	1.8403
700 (503.08)	v	0.0243	0.7928	0.9072	1.0102	1.1078	1.2023	1.2948	1.3858	1.4757	1.5647
	h	487.93	1281.0	1345.6	1403.7	1459.4	1514.4	1569.4	1624.8	1680.7	1737.2
	s	0.6889	1.5090	1.5673	1.6154	1.6580	1.6970	1.7335	1.7679	1.8006	1.8318
750 (510.84)	v	0.0242	0.7313	0.8409	0.9386	1.0306	1.1195	1.2063	1.2916	1.3759	1.4592
	h	487.90	1276.1	1342.5	1401.5	1457.6	1512.9	1568.2	1623.8	1679.8	1736.4
	s	0.6887	1.4977	1.5577	1.6065	1.6494	1.6886	1.7252	1.7598	1.7926	1.8239
800 (518.21)	v	0.02041	0.6774	0.7828	0.8759	0.9631	1.0470	1.1289	1.2093	1.2885	1.3669
	h	487.88	1271.1	1339.3	1399.1	1455.8	1511.4	1566.9	1622.7	1678.9	1735.7
	s	0.6885	1.4869	1.5484	1.5980	1.6413	1.6807	1.7175	1.7522	1.7851	1.8164
850 (525.24)	v	0.02039	0.6296	0.7315	0.8205	0.9034	0.9830	1.0606	1.1366	1.2115	1.2855
	h	487.86	1265.9	1336.0	1396.8	1454.0	1510.0	1565.7	1621.6	1678.0	1734.9
	s	0.6883	1.4763	1.5396	1.5899	1.6336	1.6733	1.7102	1.7450	1.7780	1.8094
900 (531.95)	v	0.02038	0.5869	0.6858	0.7713	0.8504	0.9262	0.9998	1.0720	1.1430	1.2131
	h	487.83	1260.6	1332.7	1394.4	1452.2	1508.5	1564.4	1620.6	1677.1	1734.1
	s	0.6881	1.4659	1.5311	1.5822	1.6263	1.6662	1.7033	1.7382	1.7713	1.8028
950 (538.39)	v	0.02037	0.5485	0.6449	0.7272	0.8030	0.8753	0.9455	1.0142	1.0817	1.1484
	h	487.81	1255.1	1329.3	1392.0	1450.3	1507.0	1563.2	1619.5	1676.2	1733.3
	s	0.6878	1.4557	1.5228	1.5748	1.6193	1.6595	1.6967	1.7317	1.7649	1.7965
1000 (544.58)	v	0.02036	0.5137	0.6080	0.6875	0.7603	0.8295	0.8966	0.9622	1.0266	1.0901
	h	487.79	1249.3	1325.9	1389.6	1448.5	1505.4	1561.9	1618.4	1675.3	1732.5
	s	0.6876	1.4457	1.5149	1.5677	1.6126	1.6530	1.6905	1.7256	1.7589	1.7905
1100 (556.28)	v	0.02034	0.4531	0.5440	0.6188	0.6865	0.7505	0.8121	0.8723	0.9313	0.9894
	h	487.75	1237.3	1318.8	1384.7	1444.7	1502.4	1559.4	1616.3	1673.5	1731.0
	s	0.6872	1.4259	1.4996	1.5542	1.6000	1.6410	1.6787	1.7141	1.7475	1.7793
1200 (567.19)	v	0.02031	0.4016	0.4905	0.5615	0.6250	0.6845	0.7418	0.7974	0.8519	0.9055
	h	487.72	1224.2	1311.5	1379.7	1440.9	1499.4	1556.9	1614.2	1671.6	1729.4
	s	0.6868	1.4061	1.4851	1.5415	1.5883	1.6298	1.6679	1.7035	1.7371	1.7691
1300 (577.42)	v	0.02089	0.3570	0.4451	0.5129	0.5729	0.6287	0.6822	0.7341	0.7847	0.8345
	h	487.68	1209.9	1303.9	1374.6	1437.1	1496.3	1554.3	1612.0	1669.8	1727.9
	s	0.6863	1.3860	1.4711	1.5296	1.5773	1.6194	1.6578	1.6937	1.7275	1.7596
1400 (587.07)	v	0.02027	0.3176	0.4059	0.4712	0.5282	0.5809	0.6311	0.6798	0.7272	0.7737
	h	487.65	1194.1	1296.1	1369.3	1433.2	1493.2	1551.8	1609.9	1668.0	1726.3
	s	0.6859	1.3652	1.4575	1.5182	1.5670	1.6096	1.6484	1.6845	1.7185	1.7508
1500 (596.20)	v	0.02025	0.2820	0.3717	0.4350	0.4894	0.5394	0.5869	0.6327	0.6773	0.7210
	h	487.63	1176.3	1287.9	1364.0	1429.2	1490.1	1549.2	1607.7	1666.2	1724.8
	s	0.6855	1.3431	1.4443	1.5073	1.5572	1.6004	1.6395	1.6759	1.7101	1.7425
2000 (635.80)	v	0.02014	0.02332	0.2488	0.3072	0.3534	0.3942	0.4320	0.4680	0.5027	0.5365
	h	487.53	614.48	1240.9	1335.4	1408.7	1474.1	1536.2	1596.9	1657.0	1717.0
	s	0.6834	0.8091	1.3794	1.4578	1.5138	1.5603	1.6014	1.6391	1.6743	1.7075
2500 (668.11)	v	0.02004	0.02302	0.1681	0.2293	0.2712	0.3068	0.3390	0.3692	0.3980	0.4259
	h	487.50	612.08	1176.7	1303.4	1386.7	1457.5	1522.9	1585.9	1647.8	1709.2
	s	0.6815	0.8048	1.3076	1.4129	1.4766	1.5269	1.5703	1.6094	1.6456	1.6796
3000 (695.33)	v	0.01995	0.02276	0.0982	0.1759	0.2161	0.2484	0.2770	0.3033	0.3282	0.3522
	h	487.52	610.08	1060.5	1267.0	1363.2	1440.2	1509.4	1574.8	1638.5	1701.4
	s	0.6796	0.8009	1.1966	1.3692	1.4429	1.4976	1.5434	1.5841	1.6214	1.6561
3200 (705.08)	v	0.0199	0.0227	0.0335	0.1588	0.1987	0.2301	0.2576	0.2827	0.3065	0.3291
	h	487.5	609.4	800.8	1250.9	1353.4	1433.1	1503.8	1570.3	1634.8	1698.3
	s	0.6788	0.7994	0.9708	1.3515	1.4300	1.4866	1.5335	1.5749	1.6126	1.6477
4000 (—)	v	0.0198	0.0223	0.0287	0.1052	0.1463	0.1752	0.1994	0.2210	0.2411	0.2601
	h	487.7	606.9	763.0	1174.3	1311.6	1403.6	1481.3	1552.2	1619.8	1685.7
	s	0.6760	0.7940	0.9343	1.2754	1.3807	1.4461	1.4976	1.5417	1.5812	1.6177
5000 (—)	v	0.0196	0.0219	0.0268	0.0591	0.1038	0.1312	0.1529	0.1718	0.1890	0.2050
	h	488.1	604.6	746.0	1042.9	1252.9	1364.6	1452.1	1529.1	1600.9	1670.0
	s	0.6726	0.7880	0.9153	1.1593	1.3207	1.4001	1.4582	1.5061	1.5481	1.5863

Table 1-11. SUPERHEATED STEAM—SI UNITS*

SYMBOLS: v = specific volume, m³/kg h = specific enthalpy, kJ/kg
u = specific internal energy, kJ/kg s = specific entropy, kJ/K·kg

Pressure, MN/m² (saturation temperature)		50 C 323.15 K	100 C 373.15 K	150 C 423.15 K	200 C 473.15 K	300 C 573.15 K	400 C 673.15 K	500 C 773.15 K	700 C 973.15 K	1 000 C 1 273.15 K	1 300 C 1 573.15 K
0.001 (6.98 C) (280.13 K)	v	149.093	172.187	195.272	218.352	264.508	310.661	356.814	449.117	587.571	726.025
	u	2 445.4	2 516.4	2 588.4	2 661.6	2 812.2	2 969.0	3 132.4	3 479.6	4 053.0	4 683.7
	h	2 594.5	2 688.6	2 783.6	2 880.0	3 076.8	3 279.7	3 489.2	3 928.7	4 640.6	5 409.7
	s	9.242 3	9.512 9	9.752 0	9.967 1	10.344 3	10.670 5	10.960 5	11.465 5	12.101 9	12.643 8
0.002 (17.50 C) (290.65 K)	v	74.524	86.081	97.628	109.170	132.251	155.329	178.405	224.558	293.785	363.012
	u	2 445.2	2 516.3	2 588.3	2 661.6	2 812.2	2 969.0	3 132.4	3 479.6	4 053.0	4 683.7
	h	2 594.3	2 688.4	2 783.6	2 879.9	3 076.7	3 279.7	3 489.2	3 928.7	4 640.6	5 409.7
	s	8.921 9	9.192 8	9.432 0	9.647 1	10.024 3	10.350 6	10.640 6	11.145 6	11.782 0	12.323 9
0.004 (28.96 C) (302.11 K)	v	37.240	43.028	48.806	54.580	66.122	77.662	89.201	112.278	146.892	181.506
	u	2 444.9	2 516.1	2 588.2	2 661.5	2 812.2	2 969.0	3 132.3	3 479.6	4 053.0	4 683.7
	h	2 593.9	2 688.2	2 783.4	2 879.8	3 076.7	3 279.6	3 489.2	3 928.7	4 640.6	5 409.7
	s	8.600 9	8.872 4	9.111 8	9.327 1	9.704 4	10.030 7	10.320 7	10.825 7	11.462 1	12.004 0
0.006 (36.16 C) (309.31 K)	v	24.812	28.676	32.532	36.383	44.079	51.774	59.467	74.852	97.928	121.004
	u	2 444.6	2 515.9	2 588.1	2 661.4	2 812.2	2 969.0	3 132.3	3 479.6	4 053.0	4 683.7
	h	2 593.4	2 688.0	2 783.3	2 879.7	3 076.6	3 279.6	3 489.1	3 928.7	4 640.6	5 409.7
	s	8.412 8	8.684 7	8.924 4	9.139 8	9.517 2	9.843 5	10.133 6	10.638 6	11.275 0	11.816 8
0.008 (41.51 C) (314.66 K)	v	18.598	21.501	24.395	27.284	33.058	38.829	44.599	56.138	73.446	90.753
	u	2 444.2	2 515.7	2 588.0	2 661.4	2 812.1	2 969.0	3 132.3	3 479.6	4 053.0	4 683.7
	h	2 593.0	2 687.7	2 783.1	2 879.6	3 076.6	3 279.6	3 489.1	3 928.7	4 640.6	5 409.7
	s	8.279 0	8.551 4	8.791 4	9.006 9	9.384 4	9.710 7	10.000 8	10.505 8	11.142 2	11.684 1
0.010 (45.81 C) (318.96 K)	v	14.869	17.196	19.512	21.825	26.445	31.063	35.679	44.911	58.757	72.602
	u	2 443.9	2 515.5	2 587.9	2 661.3	2 812.1	2 968.9	3 132.3	3 479.6	4 053.0	4 683.7
	h	2 592.6	2 687.5	2 783.0	2 879.5	3 076.5	3 279.6	3 489.1	3 928.7	4 640.6	5 409.7
	s	8.174 9	8.447 9	8.688 2	8.903 8	9.281 3	9.607 7	9.897 8	10.402 8	11.039 3	11.581 1
0.020 (60.06 C) (333.21 K)	v	7.412	8.585	9.748	10.907	13.219	15.529	17.838	22.455	29.378	36.301
	u	2 442.2	2 514.6	2 587.3	2 660.9	2 811.9	2 968.8	3 132.2	3 479.5	4 053.0	4 683.7
	h	2 590.4	2 686.2	2 782.3	2 879.1	3 076.3	3 279.4	3 489.0	3 928.6	4 640.6	5 409.7
	s	7.849 8	8.125 5	8.366 9	8.583 1	8.961 1	9.287 6	9.577 8	10.082 9	10.719 3	11.261 2
0.040 (75.87 C) (349.02 K)	v	3.683	4.279	4.866	5.448	6.606	7.763	8.918	11.227	14.689	18.151
	u	2 438.8	2 512.6	2 586.2	2 660.2	2 811.5	2 968.6	3 132.1	3 479.4	4 052.9	4 683.6
	h	2 586.1	2 683.8	2 780.8	2 878.1	3 075.8	3 279.1	3 488.8	3 928.5	4 640.5	5 409.6
	s	7.519 2	7.800 3	8.044 4	8.261 7	8.640 6	8.967 4	9.257 7	9.762 9	10.399 4	10.941 2
0.060 (85.94 C) (359.09 K)	v	2.440	2.844	3.238	3.628	4.402	5.174	5.944	7.484	9.792	12.100
	u	2 435.3	2 510.6	2 585.1	2 659.5	2 811.2	2 968.4	3 131.9	3 479.4	4 052.9	4 683.6
	h	2 581.7	2 681.3	2 779.4	2 877.2	3 075.3	3 278.8	3 488.6	3 928.4	4 640.4	5 409.6
	s	7.321 2	7.607 9	7.854 6	8.073 1	8.452 8	8.779 9	9.070 4	9.575 7	10.212 2	10.754 1
0.080 (93.50 C) (366.65 K)	v	1.818 3	2.127	2.425	2.718	3.300	3.879	4.458	5.613	7.344	9.075
	u	2 431.7	2 508.7	2 583.9	2 658.8	2 810.8	2 968.1	3 131.7	3 479.3	4 052.8	4 683.5
	h	2 577.2	2 678.8	2 777.9	2 876.2	3 074.8	3 278.5	3 488.3	3 928.3	4 640.4	5 409.5
	s	7.177 5	7.469 8	7.719 1	7.938 8	8.319 4	8.646 8	8.937 4	9.442 8	10.079 4	10.621 3
0.100 (99.63 C) (372.78 K)	v	1.445 0	1.695 8	1.936 4	2.172	2.639	3.103	3.565	4.490	5.875	7.260
	u	2 428.2	2 506.7	2 582.8	2 658.1	2 810.4	2 967.9	3 131.6	3 479.2	4 052.8	4 683.5
	h	2 572.7	2 676.2	2 776.4	2 875.3	3 074.3	3 278.2	3 488.1	3 928.2	4 640.3	5 409.5
	s	7.063 3	7.361 4	7.613 4	7.834 3	8.215 8	8.543 5	8.834 2	9.339 8	9.976 4	10.518 3
0.200 (120.23 C) (393.38 K)	v	.696 9	.834 0	.959 6	1.080 3	1.316 2	1.549 3	1.781 4	2.244	2.937	3.630
	u	2 409.5	2 496.3	2 576.9	2 654.4	2 808.6	2 966.7	3 130.8	3 478.8	4 052.5	4 683.2
	h	2 548.9	2 663.1	2 768.8	2 870.5	3 071.8	3 276.6	3 487.1	3 927.6	4 640.0	5 409.3
	s	6.684 4	7.013 5	7.279 5	7.506 6	7.892 6	8.221 8	8.513 3	9.019 4	9.656 3	10.198 2
0.300 (133.55 C) (406.70 K)	v	.445 5	.546 1	.633 9	.716 3	.875 3	1.031 5	1.186 7	1.495 7	1.958 1	2.420 1
	u	2 389.1	2 485.4	2 570.8	2 650.7	2 806.7	2 965.6	3 130.0	3 478.4	4 052.3	4 683.0
	h	2 522.7	2 649.2	2 761.0	2 865.6	3 069.3	3 275.0	3 486.0	3 927.1	4 639.7	5 409.0
	s	6.431 9	6.796 5	7.077 8	7.311 5	7.702 2	8.033 0	8.325 1	8.831 9	9.469 0	10.011 0
0.400 (143.63 C) (416.78 K)	v	.317 7	.401 7	.470 8	.534 2	.654 8	.772 6	.889 3	1.121 5	1.468 5	1.815 1
	u	2 366.3	2 473.8	2 564.5	2 646.8	2 804.8	2 964.4	3 129.2	3 477.9	4 052.0	4 682.8
	h	2 493.4	2 634.5	2 752.8	2 860.5	3 066.8	3 273.4	3 484.9	3 926.5	4 639.4	5 408.8
	s	6.224 8	6.631 9	6.929 9	7.170 6	7.566 2	7.898 5	8.191 3	8.698 7	9.336 0	9.878 0
0.500 (151.86 C) (425.01 K)	v		.314 6	.372 9	.424 9	.522 6	.617 3	.710 9	.896 9	1.174 7	1.452 1
	u		2 461.5	2 557.9	2 642.9	2 802.9	2 963.2	3 128.4	3 477.5	4 051.8	4 682.5
	h		2 618.7	2 744.4	2 855.4	3 064.2	3 271.9	3 483.9	3 925.9	4 639.1	5 408.6
	s		6.494 5	6.811 1	7.059 2	7.459 9	7.793 8	8.087 3	8.595 2	9.232 8	9.774 9

*Condensed from: J.H. Keenan, F.G. Keyes, P.G. Hill, and J.G. Moore, "Steam Tables: Thermodynamic Properties of Water Including Vapor, Liquid, and Solid Phases". Copyright © 1969 by John Wiley & Sons, Inc. Reprinted by permission of the authors and publisher. (International Edition—Metric Units).

Table 1-11.   SUPERHEATED STEAM—SI UNITS *(Continued)*

Pressure, MN/m² (saturation temperature)	Temperature									
	150 C 423.15 K	200 C 473.15 K	250 C 523.15 K	300 C 573.15 K	350 C 623.15 K	400 C 673.15 K	500 C 773.15 K	700 C 973.15 K	1 000 C 1 273.15 K	1 300 C 1 573.15 K
0.60 v	.307 5	.352 0	.393 8	.434 4	.474 2	.513 7	.592 0	.747 2	.978 8	1.210 1
(158.85 C) u	2 551.1	2 638.9	2 720.9	2 801.0	2 881.2	2 962.1	3 127.6	3 477.0	4 051.5	4 682.3
(432.00 K) h	2 735.6	2 850.1	2 957.2	3 061.6	3 165.7	3 270.3	3 482.8	3 925.3	4 638.8	5 408.3
s	6.710 5	6.966 5	7.181 6	7.372 4	7.546 4	7.707 9	8.002 1	8.510 7	9.148 5	9.690 6
0.70 v	.260 6	.299 9	.336 3	.371 4	.405 8	.439 7	.507 0	.640 3	.838 9	1.037 2
(164.97 C) u	2 544.0	2 634.8	2 718.2	2 799.1	2 879.7	2 960.9	3 126.8	3 476.6	4 051.3	4 682.1
(438.12 K) h	2 726.5	2 844.8	2 953.6	3 059.1	3 163.7	3 268.7	3 481.7	3 924.8	4 638.5	5 408.1
s	6.622 0	6.886 5	7.105 3	7.297 9	7.472 9	7.635 0	7.929 9	8.439 1	9.077 1	9.619 3
0.80 v	.225 4	.260 8	.293 1	.324 1	.354 4	.384 3	.443 3	.560 1	.734 0	.907 6
(170.43 C) u	2 536.7	2 630.6	2 715.5	2 797.2	2 878.2	2 959.7	3 126.0	3 476.2	4 051.0	4 681.8
(443.58 K) h	2 717.0	2 839.0	2 950.0	3 056.5	3 161.7	3 267.1	3 480.6	3 924.2	4 638.2	5 407.9
s	6.542 3	6.815 8	7.038 4	7.232 8	7.408 9	7.571 6	7.867 3	8.377 0	9.015 3	9.557 5
0.90 v	.197 88	.230 3	.259 6	.287 4	.314 4	.341 1	.393 8	.497 7	.652 4	.806 7
(175.38 C) u	2 528.9	2 626.3	2 712.7	2 795.2	2 876.7	2 958.5	3 125.2	3 475.8	4 050.8	4 681.6
(448.53 K) h	2 707.0	2 833.6	2 946.3	3 053.8	3 159.7	3 265.5	3 479.6	3 923.7	4 637.9	5 407.6
s	6.468 9	6.752 2	6.978 7	7.175 0	7.352 1	7.515 5	7.811 9	8.322 2	8.960 8	9.503 0
1.00 v		.206 0	.232 7	.257 9	.282 5	.306 6	.354 1	.447 8	.587 1	.726 1
(179.91 C) u		2 621.9	2 709.9	2 793.2	2 875.2	2 957.3	3 124.4	3 475.3	4 050.5	4 681.3
(453.06 K) h		2 827.9	2 942.6	3 051.2	3 157.7	3 263.9	3 478.5	3 923.1	4 637.6	5 407.4
s		6.694 0	6.924 7	7.122 9	7.301 1	7.465 1	7.762 2	8.273 1	8.911 9	9.454 3
1.50 v		.132 48	.151 95	.169 66	.186 56	.203 0	.235 2	.298 1	.391 3	.484 1
(198.32 C) u		2 598.1	2 695.3	2 783.1	2 867.6	2 951.3	3 120.3	3 473.1	4 049.2	4 680.2
(471.47 K) h		2 796.8	2 923.3	3 037.6	3 147.5	3 255.8	3 473.1	3 920.2	4 636.1	5 406.3
s		6.454 6	6.709 0	6.917 9	7.101 7	7.269 0	7.569 8	8.083 7	8.723 8	9.266 4
2.0 v		.095 27	.111 44	.125 47	.138 57	.151 20	.175 68	.223 2	.293 3	.363 1
(212.42 C) u		2 570.6	2 679.6	2 772.6	2 859.8	2 945.2	3 116.2	3 470.9	4 048.0	4 679.0
(485.57 K) h		2 761.1	2 902.5	3 023.5	3 137.0	3 247.6	3 467.6	3 917.4	4 634.6	5 405.1
s		6.260 8	6.545 3	6.766 4	6.956 3	7.127 1	7.431 7	7.948 7	8.590 1	9.132 9
2.50 v			.087 00	.098 90	.109 76	.120 10	.139 98	.178 32	.234 6	.290 5
(223.99 C) u			2 662.6	2 761.6	2 851.9	2 939.1	3 112.1	3 468.7	4 046.7	4 677.8
(497.14 K) h			2 880.1	3 008.8	3 126.3	3 239.3	3 462.1	3 914.5	4 633.1	5 404.0
s			6.408 5	6.643 8	6.840 3	7.014 8	7.323 4	7.843 5	8.486 1	9.029 1
3.0 v			.070 58	.081 14	.090 53	.099 36	.116 19	.148 38	.195 41	.242 06
(233.90 C) u			2 644.0	2 750.1	2 843.7	2 932.8	3 108.0	3 466.5	4 045.4	4 676.6
(507.05 K) h			2 855.8	2 993.5	3 115.3	3 230.9	3 456.5	3 911.7	4 631.6	5 402.8
s			6.287 2	6.539 0	6.742 8	6.921 2	7.233 8	7.757 1	8.400 9	8.944 2
3.5 v			.058 72	.068 42	.076 78	.084 53	.099 18	.126 99	.167 43	.207 49
(242.60 C) u			2 623.7	2 738.0	2 835.3	2 926.4	3 103.8	3 464.3	4 044.1	4 675.5
(515.75 K) h			2 829.2	2 977.5	3 104.0	3 222.3	3 450.9	3 908.8	4 630.1	5 401.7
s			6.174 9	6.446 1	6.657 9	6.840 5	7.157 2	7.683 7	8.328 8	8.872 3
4.0 v			.049 69	.058 84	.066 45	.073 41	.086 43	.110 95	.146 45	.181 56
(250.40 C) u			2 601.1	2 725.3	2 826.7	2 919.9	3 099.5	3 462.1	4 042.9	4 674.3
(523.55 K) h			2 799.9	2 960.7	3 092.5	3 213.6	3 445.3	3 905.9	4 628.7	5 400.5
s			6.067 2	6.361 5	6.582 1	6.769 0	7.090 1	7.619 8	8.266 2	8.810 0
4.5 v				.051 35	.058 40	.064 75	.076 51	.098 47	.130 13	.161 39
(257.49 C) u				2 712.0	2 817.8	2 913.3	3 095.3	3 459.9	4 041.6	4 673.1
(530.64 K) h				2 943.1	3 080.6	3 204.7	3 439.6	3 903.0	4 627.2	5 399.4
s				6.282 8	6.513 1	6.704 7	7.030 1	7.563 1	8.210 8	8.754 9
5.0 v				.045 32	.051 94	.057 81	.068 57	.088 49	.117 07	.145 26
(263.99 C) u				2 698.0	2 808.7	2 906.6	3 091.0	3 457.6	4 040.4	4 672.0
(537.14 K) h				2 924.5	3 068.4	3 195.7	3 433.8	3 900.1	4 625.7	5 398.2
s				6.208 4	6.449 3	6.645 9	6.975 9	7.512 2	8.161 2	8.705 5
5.5 v				.040 34	.046 65	.052 13	.062 07	.080 33	.106 39	.132 06
(270.02 C) u				2 683.1	2 799.3	2 899.8	3 086.6	3 455.4	4 039.1	4 670.8
(543.17 K) h				2 905.0	3 055.9	3 186.5	3 428.0	3 897.2	4 624.2	5 397.1
s				6.136 9	6.389 7	6.591 5	6.926 2	7.465 9	8.116 3	8.660 8
6.0 v				.036 16	.042 23	.047 39	.056 65	.073 52	.097 49	.121 06
(275.64 C) u				2 667.2	2 789.6	2 892.9	3 082.2	3 453.1	4 037.8	4 669.6
(548.79 K) h				2 884.2	3 043.0	3 177.2	3 422.2	3 894.2	4 622.7	5 396.0
s				6.067 4	6.333 5	6.540 8	6.880 3	7.423 4	8.075 1	8.619 9

Table 1-11. SUPERHEATED STEAM—SI UNITS *(Continued)*

Pressure, MN/m² (saturation temperature)		Temperature									
		300 C 573.15 K	350 C 623.15 K	400 C 673.15 K	450 C 723.15 K	500 C 773.15 K	600 C 873.15 K	700 C 973.15 K	800 C 1073.15 K	1000 C 1273.15 K	1300 C 1573.15 K
6.5 (280.91 C) (554.06 K)	v	.032 58	.038 47	.043 38	.047 85	.052 07	.060 08	.067 76	.075 25	.089 95	.111 75
	u	2 650.3	2 779.7	2 885.8	2 983.5	3 077.8	3 263.8	3 450.8	3 641.3	4 036.6	4 668.5
	h	2 862.1	3 029.7	3 167.7	3 294.5	3 416.3	3 654.3	3 891.3	4 130.5	4 621.3	5 394.8
	s	5.998 9	6.279 8	6.493 0	6.674 7	6.837 6	7.127 2	7.384 1	7.618 0	8.037 2	8.582 2
7.0 (285.88 C) (559.03 K)	v	.029 47	.035 24	.039 93	.044 16	.048 14	.055 65	.062 83	.069 81	.083 50	.103 77
	u	2 632.2	2 769.4	2 878.6	2 978.0	3 073.4	3 260.7	3 448.5	3 639.5	4 035.3	4 667.3
	h	2 838.4	3 016.0	3 158.1	3 287.1	3 410.3	3 650.3	3 888.3	4 128.2	4 619.8	5 393.7
	s	5.930 5	6.228 3	6.447 8	6.632 7	6.797 5	7.089 4	7.347 6	7.582 2	8.002 0	8.547 3
7.5 (290.59 C) (563.73 K)	v	.026 72	.032 43	.036 94	.040 96	.044 73	.051 81	.058 55	.065 10	.077 90	.096 85
	u	2 612.5	2 758.7	2 871.3	2 972.4	3 068.9	3 257.6	3 446.2	3 637.8	4 034.1	4 666.2
	h	2 812.9	3 001.9	3 148.3	3 279.6	3 404.3	3 646.2	3 885.4	4 126.0	4 618.3	5 392.6
	s	5.861 4	6.178 5	6.404 7	6.592 9	6.759 8	7.054 0	7.313 4	7.548 7	7.969 2	8.514 7
8.0 (295.06 C) (568.21 K)	v	.024 26	.029 95	.034 32	.038 17	.041 75	.048 45	.054 81	.060 97	.073 01	.090 80
	u	2 590.9	2 747.7	2 863.8	2 966.7	3 064.3	3 254.4	3 443.9	3 636.0	4 032.8	4 665.0
	h	2 785.0	2 987.3	3 138.3	3 272.0	3 398.3	3 642.0	3 882.4	4 123.8	4 616.9	5 391.5
	s	5.790 6	6.130 1	6.363 4	6.555 1	6.724 0	7.020 6	7.281 2	7.517 3	7.938 4	8.484 2
8.5 (299.33 C) (572.48 K)	v	.022 02	.027 76	.032 00	.035 69	.039 11	.045 48	.051 51	.057 33	.068 69	.085 47
	u	2 567.1	2 736.3	2 856.2	2 961.0	3 059.8	3 251.3	3 441.6	3 634.2	4 031.6	4 663.9
	h	2 754.8	2 972.2	3 128.2	3 264.4	3 392.2	3 637.9	3 879.4	4 121.5	4 615.4	5 390.3
	s	5.716 8	6.082 7	6.323 7	6.519 0	6.690 0	6.988 9	7.250 9	7.487 7	7.909 5	8.455 5
9.0 (303.40 C) (576.55 K)	v	.019 948	.025 80	.029 93	.033 50	.036 77	.042 85	.048 57	.054 09	.064 85	.080 72
	u	2 540.2	2 724.4	2 848.4	2 955.2	3 055.2	3 248.1	3 439.3	3 632.5	4 030.3	4 662.7
	h	2 719.7	2 956.6	3 117.8	3 256.6	3 386.1	3 633.7	3 876.5	4 119.3	4 614.0	5 389.2
	s	5.638 2	6.036 1	6.285 4	6.484 4	6.657 6	6.958 9	7.222 1	7.459 6	7.882 1	8.428 4
9.5 (307.31 C) (580.46 K)	v		.024 03	.028 08	.031 53	.034 67	.040 49	.045 94	.051 20	.061 41	.076 48
	u		2 712.1	2 840.5	2 949.3	3 050.5	3 244.9	3 437.0	3 630.7	4 029.1	4 661.6
	h		2 940.3	3 107.3	3 248.8	3 379.9	3 629.5	3 873.5	4 117.1	4 612.5	5 388.1
	s		5.990 1	6.248 2	6.451 1	6.626 5	6.930 3	7.194 8	7.433 0	7.856 2	8.402 7
10.0 (311.06 C) (584.21 K)	v		.022 42	.026 41	.029 75	.032 79	.038 37	.043 58	.048 59	.058 32	.072 65
	u		2 699.2	2 832.4	2 943.4	3 045.8	3 241.7	3 434.7	3 628.9	4 027.8	4 660.5
	h		2 923.4	3 096.5	3 240.9	3 373.7	3 625.3	3 870.5	4 114.8	4 611.0	5 387.0
	s		5.944 3	6.212 0	6.419 0	6.596 6	6.902 9	7.168 7	7.407 7	7.831 5	8.378 3
11.0 (318.15 C) (591.30 K)	v		.019 611	.023 51	.026 68	.029 52	.034 70	.039 50	.044 09	.052 98	.066 05
	u		2 671.6	2 815.7	2 931.2	3 036.3	3 235.2	3 430.0	3 625.4	4 025.3	4 658.2
	h		2 887.3	3 074.3	3 224.7	3 361.0	3 616.9	3 864.5	4 110.4	4 608.1	5 384.8
	s		5.852 7	6.142 0	6.357 7	6.540 0	6.851 4	7.119 9	7.360 5	7.785 6	8.332 9
12.0 (324.75 C) (597.90 K)	v		.017 209	.021 08	.024 12	.026 80	.031 64	.036 10	.040 34	.048 53	.060 55
	u		2 641.2	2 798.3	2 918.8	3 026.6	3 228.7	3 425.2	3 621.8	4 022.8	4 655.9
	h		2 847.7	3 051.3	3 208.2	3 348.2	3 608.3	3 858.4	4 105.9	4 605.3	5 382.6
	s		5.759 6	6.074 7	6.299 8	6.487 1	6.803 7	7.074 9	7.317 0	7.743 5	8.291 3
13.0 (330.93 C) (604.08 K)	v		.015 108	.019 007	.021 94	.024 50	.029 05	.033 22	.037 17	.044 77	.055 90
	u		2 606.9	2 780.1	2 906.1	3 016.8	3 222.1	3 420.5	3 618.2	4 020.4	4 653.6
	h		2 803.3	3 027.2	3 191.3	3 335.2	3 599.7	3 852.3	4 101.4	4 602.4	5 380.3
	s		5.662 6	6.009 1	6.244 6	6.437 1	6.759 1	7.033 1	7.276 7	7.704 7	8.252 9
14.0 (336.75 C) (609.90 K)	v		.013 221	.017 216	.020 07	.022 52	.026 83	.030 75	.034 45	.041 54	.051 91
	u		2 567.5	2 760.9	2 893.0	3 006.8	3 215.4	3 415.7	3 614.5	4 017.9	4 651.4
	h		2 752.6	3 001.9	3 174.0	3 322.0	3 591.1	3 846.2	4 096.9	4 599.5	5 378.1
	s		5.558 5	5.944 8	6.191 6	6.389 7	6.717 2	6.993 9	7.239 2	7.668 5	8.217 3
15.0 (342.24 C) (615.39 K)	v		.011 470	.015 649	.018 445	.020 80	.024 91	.028 61	.032 10	.038 75	.048 45
	u		2 520.4	2 740.7	2 879.5	2 996.6	3 208.6	3 410.9	3 610.9	4 015.4	4 649.1
	h		2 692.4	2 975.5	3 156.2	3 308.6	3 582.3	3 840.1	4 092.4	4 596.6	5 376.0
	s		5.442 1	5.881 1	6.140 4	6.344 3	6.677 6	6.957 2	7.204 0	7.634 8	8.184 0
16.0 (347.44 C) (620.59 K)	v		.009 749	.014 262	.017 018	.019 296	.023 23	.026 74	.030 03	.036 30	.045 43
	u		2 459.7	2 719.4	2 865.7	2 986.2	3 201.8	3 406.0	3 607.3	4 013.0	4 646.9
	h		2 615.7	2 947.6	3 138.0	3 294.9	3 573.5	3 833.9	4 087.8	4 593.8	5 373.8
	s		5.302 0	5.817 5	6.090 7	6.300 7	6.639 9	6.922 4	7.170 8	7.603 1	8.152 8
17.0 (352.37 C) (625.52 K)	v		.007 787	.013 021	.015 754	.017 967	.021 74	.025 09	.028 22	.034 14	.042 76
	u		2 364.2	2 696.9	2 851.5	2 975.6	3 195.0	3 401.2	3 603.7	4 010.5	4 644.6
	h		2 496.6	2 918.2	3 119.3	3 281.1	3 564.6	3 827.7	4 083.3	4 590.9	5 371.6
	s		5.096 6	5.753 6	6.042 2	6.258 7	6.604 0	6.889 4	7.139 5	7.573 2	8.123 4

Table 1-11.   SUPERHEATED STEAM—SI UNITS *(Continued)*

Pressure, MN/m² (saturation temperature)		400 C 673.15 K	450 C 723.15 K	500 C 773.15 K	550 C 823.15 K	600 C 873.15 K	650 C 923.15 K	700 C 973.15 K	800 C 1073.15 K	1000 C 1273.15 K	1300 C 1573.15 K
						Temperature					
18.0 (357.06 C) (630.21 K)	v	.011901	.014625	.016784	.018681	.02042	.02206	.02362	.02660	.03223	.04039
	u	2672.8	2836.8	2964.9	3079.6	3188.0	3293.1	3396.3	3600.0	4008.0	4642.4
	h	2887.0	3100.1	3267.0	3415.9	3555.6	3690.2	3821.5	4078.8	4588.1	5369.4
	s	5.6887	5.9947	6.2181	6.4048	6.5696	6.7195	6.8580	7.1098	7.5449	8.0956
19.0 (361.54 C) (634.69 K)	v	.010881	.013612	.015724	.017562	.019241	.02081	.02231	.02515	.03051	.03827
	u	2647.0	2821.7	2954.0	3071.0	3181.1	3287.3	3391.4	3596.3	4005.6	4640.2
	h	2853.8	3080.4	3252.7	3404.7	3546.6	3682.8	3815.3	4074.3	4585.3	5367.3
	s	5.6224	5.9479	6.1786	6.3692	6.5366	6.6882	6.8281	7.0814	7.5181	8.0692
20.0 (365.81 C) (638.96 K)	v	.009942	.012695	.014768	.016555	.018178	.019693	.02113	.02385	.02897	.03636
	u	2619.3	2806.2	2942.9	3062.4	3174.0	3281.4	3386.4	3592.7	4003.1	4638.0
	h	2818.1	3060.1	3238.2	3393.5	3537.6	3675.3	3809.0	4069.7	4582.5	5365.1
	s	5.5540	5.9017	6.1401	6.3348	6.5048	6.6582	6.7993	7.0544	7.4925	8.0442
22.0 (373.80 C) (646.95 K)	v	.008253	.011101	.013115	.014815	.016341	.017756	.019092	.02161	.02630	.03306
	u	2556.0	2773.6	2920.0	3044.7	3159.7	3269.6	3376.4	3585.3	3998.2	4633.5
	h	2737.6	3017.9	3208.6	3370.6	3519.2	3660.2	3796.5	4060.6	4576.8	5360.8
	s	5.4074	5.8105	6.0658	6.2691	6.4444	6.6014	6.7451	7.0036	7.4448	7.9974
25.0	v	.006004	.009162	.011123	.012724	.014137	.015433	.016646	.018912	.02310	.02910
	u	2430.1	2720.7	2884.3	3017.5	3137.9	3251.6	3361.3	3574.3	3990.9	4626.9
	h	2580.2	2949.7	3162.4	3335.6	3491.4	3637.4	3777.5	4047.1	4568.5	5354.4
	s	5.1418	5.6744	5.9592	6.1765	6.3602	6.5229	6.6707	6.9345	7.3802	7.9342
30.0	v	.002790	.006735	.008678	.010168	.011446	.012596	.013661	.015623	.019196	.024266
	u	2067.4	2619.3	2820.7	2970.3	3100.5	3221.0	3335.8	3555.5	3978.8	4616.0
	h	2151.1	2821.4	3081.1	3275.4	3443.9	3598.9	3745.6	4024.2	4554.7	5344.0
	s	4.4728	5.4424	5.7905	6.0342	6.2331	6.4058	6.5606	6.8332	7.2867	7.8432
35.0	v	.002100	.004961	.006927	.008345	9.527	.010575	.011533	.013278	.016410	.020815
	u	1914.1	2498.7	2751.9	2921.0	3062.0	3189.8	3309.8	3536.7	3966.7	4605.1
	h	1987.6	2672.4	2994.4	3213.0	3395.5	3559.9	3713.5	4001.5	4541.1	5333.6
	s	4.2126	5.1962	5.6282	5.9026	6.1179	6.3010	6.4631	6.7450	7.2064	7.7653
40.0	v	.0019077	.003693	.005622	.006984	.008094	.009063	.009941	.011523	.014324	.018229
	u	1854.6	2365.1	2678.4	2869.7	3022.6	3158.0	3283.6	3517.8	3954.6	4594.3
	h	1930.9	2512.8	2903.3	3149.1	3346.4	3520.6	3681.2	3978.7	4527.6	5323.5
	s	4.1135	4.9459	5.4700	5.7785	6.0114	6.2054	6.3750	6.6662	7.1356	7.6969
50.0	v	.0017309	.002486	.003892	.005118	.006112	.006966	.007727	.009076	.011411	.014616
	u	1788.1	2159.6	2525.5	2763.6	2942.0	3093.5	3230.5	3479.8	3930.5	4572.8
	h	1874.6	2284.0	2720.1	3019.5	3247.6	3441.8	3616.8	3933.6	4501.1	5303.6
	s	4.0031	4.5884	5.1726	5.5485	5.8178	6.0342	6.2189	6.5290	7.0146	7.5808
60.0	v	.0016335	.002085	.002956	.003956	.004834	.005595	.006272	.007459	.009480	.012215
	u	1745.4	2053.9	2390.6	2658.8	2861.1	3028.8	3177.2	3441.5	3906.4	4551.4
	h	1843.5	2179.0	2567.9	2896.2	3151.2	3364.5	3553.5	3889.1	4475.2	5284.5
	s	3.9318	4.4121	4.9321	5.3441	5.6452	5.8829	6.0824	6.4109	6.9127	7.4837
70.0	v	.0015664	.0018931	.002466	.003227	.003976	.004650	.005256	.006318	.008110	.010509
	u	1713.2	1990.2	2290.6	2563.9	2783.4	2965.3	3124.5	3403.4	3882.1	4530.0
	h	1822.8	2122.7	2463.2	2789.7	3061.7	3290.8	3492.4	3845.7	4449.8	5265.6
	s	3.8775	4.3068	4.7619	5.1715	5.4925	5.7479	5.9606	6.3066	6.8240	7.3999
80.0	v	.0015154	.0017746	.002188	.002763	.003386	.003976	.004518	.005477	.007093	.009236
	u	1687.1	1945.0	2218.9	2483.9	2711.8	2904.6	3073.2	3365.6	3857.8	4508.8
	h	1808.3	2086.9	2394.0	2704.9	2982.7	3222.7	3434.6	3803.8	4425.2	5247.6
	s	3.8330	4.2321	4.6425	5.0323	5.3601	5.6276	5.8512	6.2128	6.7451	7.3259
90.0	v	.0014743	.0016908	.002013	.002458	.002971	.003483	.003967	.004837	.006309	.008252
	u	1665.1	1909.9	2165.6	2418.6	2648.2	2848.0	3024.0	3328.5	3833.5	4487.6
	h	1797.7	2062.0	2346.7	2639.8	2915.6	3161.6	3381.1	3763.8	4401.3	5230.3
	s	3.7952	4.1737	4.5542	4.9216	5.2469	5.5211	5.7528	6.1277	6.6738	7.2594
100.0	v	.0014399	.0016266	.0018903	.002246	.002671	.003115	.003546	.004338	.005690	.007469
	u	1646.0	1881.1	2123.8	2365.7	2592.7	2796.2	2977.7	3292.3	3809.2	4466.6
	h	1790.0	2043.8	2312.8	2590.3	2859.8	3107.7	3332.3	3726.1	4378.2	5213.6
	s	3.7620	4.1256	4.4852	4.8329	5.1509	5.4271	5.6642	6.0499	6.6087	7.1989

Table 1-12. VISCOSITY OF STEAM AND WATER*

For other conversion factors see Table 9-30. For viscosity in SI units, see Table 1-13.

UNITS: Viscosity $= \mu = (\text{lbf·sec/ft}^2) \times 10^7$. (Units of lbf·sec/ft^2 are identical with slugs/ft·sec.) For centipoises multiply the printed numbers by 0.00479. For N·s/m^2 multiply lbf·sec/ft^2 by 47.880.

Temp, °F	Pressure, psia														
	1	2	5	10	20	50	100	200	500	1000	2000	5000	7500	10000	12000
1500	8.6	8.6	8.6	8.6	8.6	8.6	8.6	8.6	8.7	8.7	8.8	9.2	9.6	10.1	10.5
1450	8.4	8.4	8.4	8.4	8.4	8.4	8.4	8.4	8.4	8.5	8.6	9.0	9.4	9.9	10.3
1400	8.1	8.1	8.1	8.1	8.1	8.1	8.2	8.2	8.2	8.3	8.4	8.8	9.2	9.8	10.2
1350	7.9	7.9	7.9	7.9	7.9	7.9	7.9	7.9	8.0	8.0	8.1	8.6	9.1	9.6	10.2
1300	7.7	7.7	7.7	7.7	7.7	7.7	7.7	7.7	7.7	7.8	7.9	8.4	8.9	9.5	10.1
1250	7.4	7.4	7.4	7.4	7.4	7.4	7.4	7.5	7.5	7.6	7.7	8.2	8.8	9.5	10.1
1200	7.2	7.2	7.2	7.2	7.2	7.2	7.2	7.2	7.3	7.3	7.5	8.0	8.6	9.4	10.1
1150	7.0	7.0	7.0	7.0	7.0	7.0	7.0	7.0	7.0	7.1	7.2	7.8	8.5	9.4	10.3
1100	6.7	6.7	6.7	6.7	6.7	6.7	6.7	6.7	6.8	6.9	7.0	7.7	8.4	9.5	10.5
1050	6.5	6.5	6.5	6.5	6.5	6.5	6.5	6.5	6.6	6.6	6.8	7.5	8.4	9.8	10.9
1000	6.3	6.3	6.3	6.3	6.3	6.3	6.3	6.3	6.3	6.4	6.6	7.4	8.5	10.2	11.5
950	6.0	6.0	6.0	6.0	6.0	6.0	6.0	6.0	6.1	6.2	6.4	7.3	8.8	10.9	12.4
900	5.8	5.8	5.8	5.8	5.8	5.8	5.8	5.8	5.9	5.9	6.1	7.3	9.5	12.0	13.4
850	5.5	5.5	5.5	5.5	5.5	5.5	5.6	5.6	5.6	5.7	5.9	7.4	10.9	13.3	14.6
800	5.3	5.3	5.3	5.3	5.3	5.3	5.3	5.3	5.4	5.5	5.7	8.4	12.9	14.8	15.7
750	5.1	5.1	5.1	5.1	5.1	5.1	5.1	5.1	5.2	5.3	5.5	12.0	14.8	16.2	17.0
700	4.8	4.8	4.8	4.8	4.8	4.8	4.9	4.9	4.9	5.0	5.4[⊗]	14.8	16.5	17.7	18.5
650	4.6	4.6	4.6	4.6	4.6	4.6	4.6	4.6	4.7	4.7	<u>4.9</u>	17.2	18.3	19.3	20.0
600	4.4	4.4	4.4	4.4	4.4	4.4	4.4	4.4	4.4	4.4	18.1	19.1	20.0	21.0	21.7
550	4.1	4.1	4.1	4.1	4.1	4.1	4.1	4.1	4.1	<u>4.0</u>	19.9	21.0	21.9	22.8	23.5
500	3.9	3.9	3.9	3.9	3.9	3.9	3.9	3.8	<u>3.8</u>	21.6	22.0	23.1	23.9	24.8	25.5
450	3.7	3.7	3.7	3.7	3.6	3.6	3.6	3.6	24.1	24.2	24.6	25.6	26.5	27.4	28.1
400	3.4	3.4	3.4	3.4	3.4	3.4	3.4	<u>3.3</u>	27.4	27.6	27.9	28.9	29.8	30.6	31.3
350	3.2	3.2	3.2	3.2	3.2	3.2	<u>3.1</u>	31.8	31.9	32.1	32.4	33.4	34.2	35.1	35.7
300	2.9	2.9	2.9	2.9	2.9	<u>2.9</u>	38.1	38.2	38.3	38.4	38.7	39.7	40.5	41.3	42.0
250	2.7	2.7	2.7	2.7	<u>2.7</u>	47.6	47.6	47.6	47.7	47.8	48.2	49.1	49.8	50.6	51.2
200	2.5	2.5	<u>2.5</u>	<u>2.5</u>	62.6	62.6	62.6	62.7	62.8	62.9	63.2	64.0	64.7	65.4	65.9
150	<u>2.2</u>	<u>2.2</u>	89.1	89.1	89.1	89.1	89.1	89.1	89.2	89.3	89.5	90.1	90.7	91.2	91.7
100	142.0	142.0	142.0	142.0	142.0	142.0	142.0	142.0	142.0	142.1	142.1	142.3	142.5	142.6	142.7
50	271.4	271.4	271.4	271.4	271.4	271.4	271.4	271.3	271.2	271.0	270.6	269.3	268.2	267.1	266.3
32	366.1	366.1	366.1	366.1	366.1	366.1	366.1	366.0	365.7	365.3	364.5	361.9	359.8	357.7	356.0
Sat. water	139.4	109.4	81.0	65.3	53.3	41.2	34.3	28.8	23.1	19.7	16.3	—	—	—	—
Sat. steam	2.0	2.1	2.3	2.4	2.6	2.8	3.0	3.2	3.6	4.0	4.7	—	—	—	—

[⊗]Critical point.
*From: "1967 ASME Steam Tables", The American Society of Mechanical Engineers.

Table 1-13.   VISCOSITY AND THERMAL CONDUCTIVITY OF STEAM AND WATER—SI UNITS*

For English units see Tables 1-9 and 1-12.

SYMBOLS AND UNITS:

   μ = dynamic viscosity. For N·s/m^2 (=kg/m·s) multiply tabulated values by 10^{-6}

   v = kinematic viscosity. For m^2/s multiply tabulated values by 10^{-6}

   k = thermal conductivity in MW/m·K

| Temperature | | Pressure | | | | | | | | |
| | | 0.1 MN/m^2 | | | 0.5 MN/m^2 | | | 1.0 MN/m^2 | | |
C	K	μ	v	k	μ	v	k	μ	v	k
0	273.15	1 750	1.75	569	1 750	1.75	569	1 750	1.75	570
50	323.15	544	0.551	643	544	0.551	644	544	0.550	644
100	373.15	12.11	20.54	24.8	279	0.291	681	279	0.291	681
150	423.15	14.15	27.39	28.7	181	0.198	687	181	0.198	687
200	473.15	16.18	35.14	33.2	16.02	6.81	33.8	15.85	0.327	35.1
250	523.15	18.22	43.83	38.2	18.14	8.61	38.6	18.06	0.420	39.3
300	573.15	20.25	53.44	43.4	20.23	10.57	43.8	20.22	0.522	44.4
350	623.15	22.3	64.02	49.0	—	—	49.4	—	—	49.9
400	673.15	24.3	75.40	54.9	24.4	15.06	55.3	24.4	7.48	55.7
450	723.15	26.4	88.0	61.1	26.4	17.5	61.4	26.5	0.88	61.8
500	773.15	28.4	101.3	67.4	28.4	20.2	67.7	28.5	0.101	68.2
550	823.15	30.4	115.4	73.9	30.5	23.1	74.3	30.5	0.115	74.7
600	873.15	32.5	130.9	80.6	32.5	26.1	80.9	32.6	0.131	81.4
650	923.15	34.5	146.9	87.4	34.5	29.3	87.7	34.6	0.147	88.2
700	973.15	36.5	163.9	94.3	36.6	32.8	94.6	36.6	0.164	95.0

| Temperature | | Pressure | | | | | | | | |
| | | 5.0 MN/m^2 | | | 10 MN/m^2 | | | 20 MN/m^2 | | |
C	K	μ	v	k	μ	v	k	μ	v	k
0	273.15	1 750	1.75	573	1 750	1.74	577	1 740	1.72	585
50	323.15	545	0.550	647	545	0.549	651	546	0.548	659
100	373.15	280	0.291	684	281	0.292	688	283	0.293	695
150	423.15	182	0.198	690	183	0.199	693	186	0.200	700
200	473.15	135	0.155	668	136	0.156	672	138	0.158	681
250	523.15	107	0.134	618	108	0.134	625	111	0.136	639
300	573.15	20.06	0.909	52.5	90.5	0.126	545	93	0.127	571
350	623.15	—	—	55.4	—	—	68.8	73.5	0.122	454
400	673.15	25.0	1.45	60.2	25.8	0.682	68.6	28.6	0.285	107
450	723.15	26.9	1.70	65.9	27.6	0.821	72.4	29.6	0.376	93
500	773.15	28.9	1.98	72.0	29.5	0.967	77.6	31.1	0.459	93
550	823.15	30.9	2.28	78.4	31.5	1.123	83.5	32.8	0.543	96
600	873.15	32.9	2.59	85.0	33.4	1.282	89.8	34.6	0.629	101
650	923.15	34.9	2.92	91.7	35.4	1.452	96	36.5	0.719	107
700	973.15	36.9	3.27	98.6	37.4	1.630	103	38.4	0.812	113

| Temperature | | Pressure | | | | | | | | |
| | | 30 MN/m^2 | | | 40 MN/m^2 | | | 50 MN/m^2 | | |
C	K	μ	v	k	μ	v	k	μ	v	k
0	273.15	1 740	1.71	592	1 730	1.70	599	1 720	1.68	606
50	323.15	547	0.547	666	548	0.545	672	549	0.544	678
100	373.15	285	0.293	701	287	0.294	707	289	0.295	713
150	423.15	188	0.201	706	190	0.203	713	192	0.204	720
200	473.15	140	0.159	689	143	0.161	697	145	0.162	704
250	523.15	113	0.137	652	116	0.139	662	118	0.140	671
300	573.15	95.5	0.127	592	98.1	0.128	609	101	0.130	622
350	623.15	78.5	0.122	496	82.5	0.122	529	85	0.123	552
400	673.15	45.8	0.128	264	62.8	0.120	390	28.6	0.120	436
450	723.15	33.1	0.223	138	41.1	0.152	220	29.6	0.130	301
500	773.15	33.4	0.290	116	36.9	0.208	153	31.1	0.164	206
550	823.15	34.6	0.352	112	36.9	0.258	134	32.8	0.205	163
600	873.15	36.1	0.413	114	37.9	0.307	130	34.6	0.245	149
650	923.15	37.7	0.475	118	39.2	0.355	132	36.5	0.286	147
700	973.15	39.5	0.540	124	40.8	0.406	135	38.4	0.327	148

*Adapted from: J.H. Keenan, F.G. Keyes, P.G. Hill, and J.G. Moore, "Steam Tables: Thermodynamic Properties of Water Including Vapor, Liquid, and Solid Phases". Copyright © 1969 by John Wiley & Sons, Inc. Reprinted by permission of the authors and publisher. (International Edition—Metric Units).

Table 1-14. PROPERTIES OF GASES

Gases and Vapors, Including Fuels and Refrigerants, English and Metric Units

For effects of temperature, see Tables, 1-15, 1-19, 1-20, and 1-23. For effects of pressure, see Tables 1-20 through 1-23.

The properties of pure gases are given at 25 deg C (77 deg F, 298 K) and atmospheric pressure (except as stated).

Common name(s)	Acetylene (Ethyne)	Air [mixture]	Ammonia, anhyd.	Argon
Chemical formula	C_2H_2		NH_3	Ar
Refrigerant number	—	729	717	740
CHEMICAL AND PHYSICAL PROPERTIES				
Molecular weight	26.04	28.966	17.02	39.948
Specific gravity, air = 1	0.90	1.00	0.59	1.38
Specific volume, ft^3/lb	14.9	13.5	23.0	9.80
Specific volume, m^3/kg	0.93	0.842	1.43	0.622
Density of liquid (at atm bp), lb/ft^3	43.0	54.6	42.6	87.0
Density of liquid (at atm bp), kg/m^3	693.	879.	686.	1 400.
Vapor pressure at 25 deg C, psia			145.4	
Vapor pressure at 25 deg C, MN/m^2			1.00	
Viscosity (abs), lbm/ft·sec	6.72×10^{-6}	12.1×10^{-6}	6.72×10^{-6}	13.4×10^{-6}
Viscosity (abs), centipoises[a]	0.01	0.018	0.010	0.02
Sound velocity in gas, m/sec	343	346	415	322
THERMAL AND THERMO-DYNAMIC PROPERTIES				
Specific heat, c_p, Btu/lb·deg F or cal/g·deg C	0.40	0.240 3	0.52	0.125
Specific heat, c_p, J/kg·K	1 674.	1 005.	2 175.	523.
Specific heat ratio, c_p/c_v	1.25	1.40	1.3	1.67
Gas constant R, ft-lb/lb·deg F	59.3	53.3	90.8	38.7
Gas constant R, J/kg·deg C	319	286.8	488.	208.
Thermal conductivity, Btu/hr·ft·deg F	0.014	0.015 1	0.015	0.010 2
Thermal conductivity, W/m·deg C	0.024	0.026	0.026	0.017 2
Boiling point (sat 14.7 psia), deg F	− 103	− 320	− 28.	− 303.
Boiling point (sat 760 mm), deg C	− 75	− 195	− 33.3	− 186
Latent heat of evap (at bp), Btu/lb	264	88.2	589.3	70.
Latent heat of evap (at bp), J/kg	614 000	205 000.	1 373 000	163 000
Freezing (melting) point, deg F (1 atm)	− 116	− 357.2	− 107.9	− 308.5
Freezing (melting) point, deg C (1 atm)	− 82.2	− 216.2	− 77.7	− 189.2
Latent heat of fusion, Btu/lb	23.	10.0	143.0	
Latent heat of fusion, J/kg	53 500	23 200	332 300	
Critical temperature, deg F	97.1	− 220.5	271.4	− 187.6
Critical temperature, deg C	36.2	− 140.3	132.5	− 122
Critical pressure, psia	907.	550.	1 650.	707.
Critical pressure, MN/m^2	6.25	3.8	11.4	4.87
Critical volume, ft^3/lb		0.050	0.068	0.029 9
Critical volume, m^3/kg		0.003	0.004 24	0.001 86
Flammable (yes or no)	Yes	No	No	No
Heat of combustion, Btu/ft^3	1 450	—	—	—
Heat of combustion, Btu/lb	21 600	—	—	—
Heat of combustion, kJ/kg	50 200	—	—	—

[a]For N·sec/m^2 divide by 1 000.

Table 1-14.   PROPERTIES OF GASES (*Continued*)

Common name(s)	Butadiene	n-Butane	Isobutane (2-Methyl propane)	1-Butene (Butylene)
Chemical formula	C_4H_6	C_4H_{10}	C_4H_{10}	C_4H_8
Refrigerant number	—	600	600a	—
CHEMICAL AND PHYSICAL PROPERTIES				
Molecular weight	54.09	58.12	58.12	56.108
Specific gravity, air = 1	1.87	2.07	2.07	1.94
Specific volume, ft³/lb	7.1	6.5	6.5	6.7
Specific volume, m³/kg	0.44	0.405	0.418	0.42
Density of liquid (at atm bp), lb/ft³		37.5	37.2	
Density of liquid (at atm bp), kg/m³		604.	599.	
Vapor pressure at 25 deg C, psia		35.4	50.4	
Vapor pressure at 25 deg C, MN/m²		0.024 4	0.347	
Viscosity (abs), lbm/ft·sec		4.8×10^{-6}		
Viscosity (abs), centipoises[a]		0.007		
Sound velocity in gas, m/sec	226	216	216	222
THERMAL AND THERMO-DYNAMIC PROPERTIES				
Specific heat, c_p, Btu/lb·deg F or cal/g·deg C	0.341	0.39	0.39	0.36
Specific heat, c_p, J/kg·K	1 427.	1 675.	1 630.	1 505.
Specific heat ratio, c_p/c_v	1.12	1.096	1.10	1.112
Gas constant R, ft-lb/lb·deg F	28.55	26.56	26.56	27.52
Gas constant R, J/kg·deg C	154.	143.	143.	148.
Thermal conductivity, Btu/hr·ft·deg F		0.01	0.01	
Thermal conductivity, W/m·deg C		0.017	0.017	
Boiling point (sat 14.7 psia), deg F	24.1	31.2	10.8	20.6
Boiling point (sat 760 mm), deg C	−4.5	−0.4	−11.8	−6.3
Latent heat of evap (at bp), Btu/lb		165.6	157.5	167.9
Latent heat of evap (at bp), J/kg		386 000	366 000	391 000
Freezing (melting) point, deg F (1 atm)	−164.	−217.	−229	−301.6
Freezing (melting) point, deg C (1 atm)	−109.	−138	−145	−185.3
Latent heat of fusion, Btu/lb		19.2		16.4
Latent heat of fusion, J/kg		44 700		38 100
Critical temperature, deg F		306	273.	291.
Critical temperature, deg C	171.	152.	134.	144.
Critical pressure, psia	652.	550.	537.	621.
Critical pressure, MN/m²		3.8	3.7	4.28
Critical volume, ft³/lb		0.070		0.068
Critical volume, m³/kg		0.004 3		0.004 2
Flammable (yes or no)	Yes	Yes	Yes	Yes
Heat of combustion, Btu/ft³	2 950	3 300	3 300	3 150
Heat of combustion, Btu/lb	20 900	21 400	21 400	21 000
Heat of combustion, kJ/kg	48 600	49 700	49 700	48 800

[a]For N·sec/m² divide by 1 000.

Table 1-14. PROPERTIES OF GASES *(Continued)*

	cis-2- *Butene* C_4H_8	*trans-2-* *Butene* C_4H_8	*Isobutene* C_4H_8	*Carbon* *dioxide* CO_2
Common name(s)				
Chemical formula				
Refrigerant number	—	—	—	*744*
CHEMICAL AND PHYSICAL PROPERTIES				
Molecular weight	56.108	56.108	56.108	44.01
Specific gravity, air = 1	1.94	1.94	1.94	1.52
Specific volume, ft^3/lb	6.7	6.7	6.7	8.8
Specific volume, m^3/kg	0.42	0.42	0.42	0.55
Density of liquid (at atm bp), lb/ft^3				—
Density of liquid (at atm bp), kg/m^3				—
Vapor pressure at 25 deg C, psia				931.
Vapor pressure at 25 deg C, MN/m^2				6.42
Viscosity (abs), lbm/ft·sec				9.4×10^{-6}
Viscosity (abs), centipoises^a				0.014
Sound velocity in gas, m/sec	223.	221.	221.	270.
THERMAL AND THERMO-DYNAMIC PROPERTIES				
Specific heat, c_p, Btu/lb·deg F or cal/g·deg C	0.327	0.365	0.37	0.205
Specific heat, c_p, J/kg·K	1 368.	1 527.	1 548.	876.
Specific heat ratio, c_p/c_v	1.121	1.107	1.10	1.30
Gas constant R, ft-lb/lb·deg F				35.1
Gas constant R, J/kg·deg C				189.
Thermal conductivity, Btu/hr·ft·deg F				0.01
Thermal conductivity, W/m·deg C				0.017
Boiling point (sat 14.7 psia), deg F	38.6	33.6	19.2	-109.4^b
Boiling point (sat 760 mm), deg C	3.7	0.9	−7.1	−78.5
Latent heat of evap (at bp), Btu/lb	178.9	174.4	169.	246.
Latent heat of evap (at bp), J/kg	416 000.	406 000.	393 000.	572 000.
Freezing (melting) point, deg F (1 atm)	−218.	−158.		
Freezing (melting) point, deg C (1 atm)	−138.9	−105.5		
Latent heat of fusion, Btu/lb	31.2	41.6	25.3	—
Latent heat of fusion, J/kg	72 600.	96 800.	58 800.	—
Critical temperature, deg F				88.
Critical temperature, deg C	160.	155.		31.
Critical pressure, psia	595.	610.		1 072.
Critical pressure, MN/m^2	4.10	4.20		7.4
Critical volume, ft^3/lb				
Critical volume, m^3/kg				
Flammable (yes or no)	Yes	Yes	Yes	No
Heat of combustion, Btu/ft^3	3 150.	3 150.	3 150.	—
Heat of combustion, Btu/lb	21 000.	21 000.	21 000.	—
Heat of combustion, kJ/kg	48 800.	48 800.	48 800.	—

^{a}For N·sec/m^2 divide by 1 000.
^{b}Sublimes.

Table 1-14.   PROPERTIES OF GASES *(Continued)*

Common name(s)	Carbon monoxide	Chlorine	Deuterium	Ethane
Chemical formula	CO	Cl_2	D_2	C_2H_6
Refrigerant number	—	—	—	170
CHEMICAL AND PHYSICAL PROPERTIES				
Molecular weight	28.011	70.906	2.014	30.070
Specific gravity, air = 1	0.967	2.45	0.070	1.04
Specific volume, ft^3/lb	14.0	5.52	194.5	13.025
Specific volume, m^3/kg	0.874	0.344	12.12	0.815
Density of liquid (at atm bp), lb/ft^3		97.3		28.
Density of liquid (at atm bp), kg/m^3		1 559.		449.
Vapor pressure at 25 deg C, psia			0.756	
Vapor pressure at 25 deg C, MN/m^2			0.005 2	
Viscosity (abs), lbm/ft·sec	12.1×10^{-6}	9.4×10^{-6}	8.75×10^{-6}	$64. \times 10^{-6}$
Viscosity (abs), centipoises[a]	0.018	0.014	0.013	0.095
Sound velocity in gas, m/sec	352.	215.	930.	316.
THERMAL AND THERMO-DYNAMIC PROPERTIES				
Specific heat, c_p, Btu/lb·deg F or cal/g·deg C	0.25	0.114	1.73	0.41
Specific heat, c_p, J/kg·K	1 046.	477.	7 238.	1 715.
Specific heat ratio, c_p/c_v	1.40	1.35	1.40	1.20
Gas constant R, ft-lb/lb·deg F	55.2	21.8	384.	51.4
Gas constant R, J/kg·deg C	297.	117.	2 066.	276.
Thermal conductivity, Btu/hr·ft·deg F	0.014	0.005	0.081	0.010
Thermal conductivity, W/m·deg C	0.024	0.008 7	0.140	0.017
Boiling point (sat 14.7 psia), deg F	−312.7	−29.2		−127.
Boiling point (sat 760 mm), deg C	−191.5	−34.		−88.3
Latent heat of evap (at bp), Btu/lb	92.8	123.7		210.
Latent heat of evap (at bp), J/kg	216 000.	288 000.		488 000.
Freezing (melting) point, deg F (1 atm)	−337.	−150.		−278.
Freezing (melting) point, deg C (1 atm)	−205.	−101.		−172.2
Latent heat of fusion, Btu/lb	12.8	41.0		41.
Latent heat of fusion, J/kg		95 400.		95 300.
Critical temperature, deg F	−220.	291.	−390.6	90.1
Critical temperature, deg C	−140.	144.	−234.8	32.2
Critical pressure, psia	507.	1 120.	241.	709.
Critical pressure, MN/m^2	3.49	7.72	1.66	4.89
Critical volume, ft^3/lb	0.053	0.028	0.239	0.076
Critical volume, m^3/kg	0.003 3	0.001 75	0.014 9	0.004 7
Flammable (yes or no)	Yes	No		Yes
Heat of combustion, Btu/ft^3	310.	—		
Heat of combustion, Btu/lb	4 340.	—		22 300.
Heat of combustion, kJ/kg	10 100.	—		51 800.

[a]For N·sec/m^2 divide by 1 000.

Table 1-14. PROPERTIES OF GASES *(Continued)*

Common name(s)	Ethyl chloride	Ethylene (Ethene)	Fluorine
Chemical formula	C_2H_5Cl	C_2H_4	F_2
Refrigerant number	160	1 150	—
CHEMICAL AND PHYSICAL PROPERTIES			
Molecular weight	64.515	28.054	37.996
Specific gravity, air = 1	2.23	0.969	1.31
Specific volume, ft^3/lb	6.07	13.9	10.31
Specific volume, m^3/kg	0.378	0.87	0.706
Density of liquid (at atm bp), lb/ft^3	56.5	35.5	
Density of liquid (at atm bp), kg/m^3	905.	569.	
Vapor pressure at 25 deg C, psia			
Vapor pressure at 25 deg C, MN/m^2			
Viscosity (abs), lbm/ft·sec		6.72×10^{-6}	16.1×10^{-6}
Viscosity (abs), centipoises[a]		0.010	0.024
Sound velocity in gas, m/sec	204.	331.	290.
THERMAL AND THERMO-DYNAMIC PROPERTIES			
Specific heat, c_p, Btu/lb·deg F or cal/g·deg C	0.27	0.37	0.198
Specific heat, c_p, J/kg·K	1 130.	1 548.	828.
Specific heat ratio, c_p/c_v	1.13	1.24	1.35
Gas constant R, ft-lb/lb·deg F	24.0	55.1	40.7
Gas constant R, J/kg·deg C	129.	296.	219.
Thermal conductivity, Btu/hr·ft·deg F		0.010	0.016
Thermal conductivity, W/m·deg C		0.017	0.028
Boiling point (sat 14.7 psia), deg F	54.	−155.	−306.4
Boiling point (sat 760 mm), deg C	12.2	−103.8	−188.
Latent heat of evap (at bp), Btu/lb	166.	208.	74.
Latent heat of evap (at bp), J/kg	386 000.	484 000.	172 000.
Freezing (melting) point, deg F (1 atm)	−218.	−272.	−364.
Freezing (melting) point, deg C (1 atm)	−138.9	−169.	−220.
Latent heat of fusion, Btu/lb	29.3	51.5	11.
Latent heat of fusion, J/kg	68 100.	120 000.	25 600.
Critical temperature, deg F	368.6	49.	−200
Critical temperature, deg C	187.	9.5	−129.
Critical pressure, psia	764.	741.	810.
Critical pressure, MN/m^2	5.27	5.11	5.58
Critical volume, ft^3/lb	0.049	0.073	
Critical volume, m^3/kg	0.003 06	0.004 6	
Flammable (yes or no)	No	Yes	
Heat of combustion, Btu/ft^3	—	1 480.	
Heat of combustion, Btu/lb	—	20 600.	
Heat of combustion, kJ/kg	—	47 800.	

[a]For N·sec/m^2 divide by 1 000.

Table 1-14. PROPERTIES OF GASES (*Continued*)

Common name(s)	Fluorocarbons			
Chemical formula	CCl_3F	CCl_2F_2	$CClF_3$	$CBrF_3$
Refrigerant number	*11*	*12*	*13*	*13B1*
CHEMICAL AND PHYSICAL PROPERTIES				
Molecular weight	137.37	120.91	104.46	148.91
Specific gravity, air = 1	4.74	4.17	3.61	5.14
Specific volume, ft^3/lb	2.74	3.12	3.58	2.50
Specific volume, m^3/kg	0.171	0.195	0.224	0.975
Density of liquid (at atm bp), lb/ft^3	92.1	93.0	95.0	124.4
Density of liquid (at atm bp), kg/m^3	1 475.	1 490.	1 522.	1 993.
Vapor pressure at 25 deg C, psia		94.51	516.	234.8
Vapor pressure at 25 deg C, MN/m^2		0.652	3.56	1.619
Viscosity (abs), lbm/ft·sec	7.39×10^{-6}	8.74×10^{-6}		
Viscosity (abs), centipoises[a]	0.011	0.013		
Sound velocity in gas, m/sec				
THERMAL AND THERMO-DYNAMIC PROPERTIES				
Specific heat, c_p, Btu/lb·deg F or cal/g·deg C	0.14	0.146	0.154	
Specific heat, c_p, J/kg·K	586.	611.	644.	
Specific heat ratio, c_p/c_v	1.14	1.14	1.145	
Gas constant R, ft-lb/lb·deg F				
Gas constant R, J/kg·deg C				
Thermal conductivity, Btu/hr·ft·deg F	0.005	0.006		
Thermal conductivity, W/m·deg C	0.008 7	0.010 4		
Boiling point (sat 14.7 psia), deg F	74.9	−21.8	−114.6	−72.
Boiling point (sat 760 mm), deg C	23.8	−29.9	−81.4	−57.8
Latent heat of evap (at bp), Btu/lb	77.5	71.1	63.0	51.1
Latent heat of evap (at bp), J/kg	180 000.	165 000.	147 000.	119 000.
Freezing (melting) point, deg F (1 atm)	−168.	−252.	−294.	−270.
Freezing (melting) point, deg C (1 atm)	−111.	−157.8	−181.1	−167.8
Latent heat of fusion, Btu/lb				
Latent heat of fusion, J/kg				
Critical temperature, deg F	388.4	233.	83.9	152.
Critical temperature, deg C	198.	111.7	28.8	66.7
Critical pressure, psia	635.	582.	559.	573.
Critical pressure, MN/m^2	4.38	4.01	3.85	3.95
Critical volume, ft^3/lb	0.028 9	0.287	0.027 7	0.021 5
Critical volume, m^3/kg	0.001 80	0.018	0.001 73	0.001 34
Flammable (yes or no)	No	No	No	No
Heat of combustion, Btu/ft^3	—	—	—	—
Heat of combustion, Btu/lb	—	—	—	—
Heat of combustion, kJ/kg	—	—	—	—

[a] For N·sec/m^2 divide by 1 000.

Table 1-14.   PROPERTIES OF GASES *(Continued)*

Common name(s)	Fluorocarbons			
Chemical formula	CF_4	$CHCl_2F$	$CHClF_2$	$C_2Cl_2F_4$
Refrigerant number	14	21	22	114
CHEMICAL AND PHYSICAL PROPERTIES				
Molecular weight	88.00	102.92	86.468	170.92
Specific gravity, air = 1	3.04	3.55	2.99	5.90
Specific volume, ft³/lb	4.34	3.7	4.35	2.6
Specific volume, m³/kg	0.271	0.231	0.271	0.162
Density of liquid (at atm bp), lb/ft³	102.0	87.7	88.2	94.8
Density of liquid (at atm bp), kg/m³	1 634.	1 405.	1 413.	1 519.
Vapor pressure at 25 deg C, psia		26.4	151.4	30.9
Vapor pressure at 25 deg C, MN/m²		0.182	1.044	0.213
Viscosity (abs), lbm/ft·sec		8.06×10^{-6}	8.74×10^{-6}	8.06×10^{-6}
Viscosity (abs), centipoises[a]		0.012	0.013	0.012
Sound velocity in gas, m/sec				
THERMAL AND THERMO-DYNAMIC PROPERTIES				
Specific heat, c_p, Btu/lb·deg F or cal/g·deg C		0.139	0.157	0.158
Specific heat, c_p, J/kg·K		582.	657.	661.
Specific heat ratio, c_p/c_v		1.18	1.185	1.09
Gas constant R, ft-lb/lb·deg F				
Gas constant R, J/kg·deg C				
Thermal conductivity, Btu/hr·ft·deg F			0.007	0.006
Thermal conductivity, W/m·deg C			0.012	0.010
Boiling point (sat 14.7 psia), deg F	−198.2	48.1	−41.3	38.4
Boiling point (sat 760 mm), deg C	−127.9	9.0	−40.7	3.55
Latent heat of evap (at bp), Btu/lb	58.5	104.1	100.4	58.4
Latent heat of evap (at bp), J/kg	136 000.	242 000.	234 000.	136 000.
Freezing (melting) point, deg F (1 atm)	−299.	−211.	−256.	−137.
Freezing (melting) point, deg C (1 atm)	−183.8	−135.	−160.	−93.8
Latent heat of fusion, Btu/lb	2.53			
Latent heat of fusion, J/kg	5 880.			
Critical temperature, deg F	−49.9	353.3	204.8	294.
Critical temperature, deg C	−45.5	178.5	96.5	
Critical pressure, psia	610.	750.	715.	475.
Critical pressure, MN/m²	4.21	5.17	4.93	3.28
Critical volume, ft³/lb	0.025	0.030 7	0.030 5	0.027 5
Critical volume, m³/kg	0.001 6	0.001 91	0.001 90	0.001 71
Flammable (yes or no)	No	No	No	No
Heat of combustion, Btu/ft³	—	—	—	—
Heat of combustion, Btu/lb	—	—	—	—
Heat of combustion, kJ/kg	—	—	—	—

[a] For N·sec/m² divide by 1 000.

Table 1-14. PROPERTIES OF GASES *(Continued)*

Common name(s)	Fluorocarbons			Helium
Chemical formula	C_2ClF_5	$C_2H_3ClF_2$	$C_2H_4F_2$	He
Refrigerant number	115	142b	152a	704
CHEMICAL AND PHYSICAL PROPERTIES				
Molecular weight	154.47	100.50	66.05	4.002 6
Specific gravity, air = 1	5.33	3.47	2.28	0.138
Specific volume, ft³/lb	2.44	3.7	5.9	97.86
Specific volume, m³/kg	0.152	0.231	0.368	6.11
Density of liquid (at atm bp), lb/ft³	96.5	74.6	62.8	7.80
Density of liquid (at atm bp), kg/m³	1 546.	1 195.	1 006.	125.
Vapor pressure at 25 deg C, psia	132.1	49.1	86.8	
Vapor pressure at 25 deg C, MN/m²	0.911	0.338 5	0.596	
Viscosity (abs), lbm/ft·sec				13.4×10^{-6}
Viscosity (abs), centipoises[a]				0.02
Sound velocity in gas, m/sec				1 015.
THERMAL AND THERMO-DYNAMIC PROPERTIES				
Specific heat, c_p, Btu/lb·deg F or cal/g·deg C	0.161			1.24
Specific heat, c_p, J/kg·K	674.			5 188.
Specific heat ratio, c_p/c_v	1.091			1.66
Gas constant R, ft-lb/lb·deg F				386.
Gas constant R, J/kg·deg C				2 077.
Thermal conductivity, Btu/hr·ft·deg F				0.086
Thermal conductivity, W/m·deg C				0.149
Boiling point (sat 14.7 psia), deg F	−38.0	14.	−13.	−452.
Boiling point (sat 760 mm), deg C	−38.9	−10.0	−25.0	4.22 K
Latent heat of evap (at bp), Btu/lb	53.4	92.5	137.1	10.0
Latent heat of evap (at bp), J/kg	124 000.	215 000.	319 000.	23 300.
Freezing (melting) point, deg F (1 atm)	−149.			[b]
Freezing (melting) point, deg C (1 atm)	−100.6			—
Latent heat of fusion, Btu/lb				—
Latent heat of fusion, J/kg				—
Critical temperature, deg F	176.		387.	−450.3
Critical temperature, deg C				5.2 K
Critical pressure, psia	457.6			33.22
Critical pressure, MN/m²	3.155			
Critical volume, ft³/lb	0.026 1			0.231
Critical volume, m³/kg	0.001 63			0.014 4
Flammable (yes or no)	No	No	No	No
Heat of combustion, Btu/ft³	—	—	—	—
Heat of combustion, Btu/lb	—	—	—	—
Heat of combustion, kJ/kg	—	—	—	—

[a] For N·sec/m² divide by 1 000.
[b] Helium cannot be solidified at atmospheric pressure.

Table 1-14.   PROPERTIES OF GASES *(Continued)*

Common name(s)	Hydrogen	Hydrogen chloride	Hydrogen sulfide	Krypton
Chemical formula	H_2	HCl	H_2S	Kr
Refrigerant number	702	—	—	—
CHEMICAL AND PHYSICAL PROPERTIES				
Molecular weight	2.016	36.461	34.076	83.80
Specific gravity, air = 1	0.070	1.26	1.18	2.89
Specific volume, ft^3/lb	194.	10.74	11.5	4.67
Specific volume, m^3/kg	12.1	0.670	0.093·0	0.291
Density of liquid (at atm bp), lb/ft^3	4.43	74.4	62.	150.6
Density of liquid (at atm bp), kg/m^3	71.0	1 192.	993.	2 413.
Vapor pressure at 25 deg C, psia				
Vapor pressure at 25 deg C, MN/m^2				
Viscosity (abs), lbm/ft·sec	6.05×10^{-6}	10.1×10^{-6}	8.74×10^{-6}	16.8×10^{-6}
Viscosity (abs), centipoises[a]	0.009	0.015	0.013	0.025
Sound velocity in gas, m/sec	1 315.	310.	302.	223.
THERMAL AND THERMO-DYNAMIC PROPERTIES				
Specific heat, c_p, Btu/lb·deg F or cal/g·deg C	3.42	0.194	0.23	0.059
Specific heat, c_p, J/kg·K	14 310.	812.	962.	247.
Specific heat ratio, c_p/c_v	1.405	1.39	1.33	1.68
Gas constant R, ft-lb/lb·deg F	767.	42.4	45.3	18.4
Gas constant R, J/kg·deg C	4 126.	228.	244.	99.0
Thermal conductivity, Btu/hr·ft·deg F	0.105	0.008	0.008	0.005 4
Thermal conductivity, W/m·deg C	0.018 2	0.014	0.014	0.009 3
Boiling point (sat 14.7 psia), deg F	− 423.	− 121.	− 76.	− 244.
Boiling point (sat 760 mm), deg C	20.4 K	− 85.	− 60.	− 153.
Latent heat of evap (at bp), Btu/lb	192.	190.5	234.	46.4
Latent heat of evap (at bp), J/kg	447 000.	443 000.	544 000.	108 000.
Freezing (melting) point, deg F (1 atm)	− 434.6	− 169.6	− 119.2	− 272.
Freezing (melting) point, deg C (1 atm)	− 259.1	− 112.	− 84.	− 169.
Latent heat of fusion, Btu/lb	25.0	23.4	30.2	4.7
Latent heat of fusion, J/kg	58 000.	54 400.	70 200.	10 900.
Critical temperature, deg F	− 399.8	124.	213.	
Critical temperature, deg C	− 240.0	51.2	100.4	− 63.8
Critical pressure, psia	189.	1 201.	1 309.	800.
Critical pressure, MN/m^2	1.30	8.28	9.02	5.52
Critical volume, ft^3/lb	0.53	0.038	0.046	0.017 7
Critical volume, m^3/kg	0.033	0.002 4	0.002 9	0.001 1
Flammable (yes or no)	Yes	No	Yes	No
Heat of combustion, Btu/ft^3	320.	—	700.	—
Heat of combustion, Btu/lb	62 050.	—	8 000.	—
Heat of combustion, kJ/kg	144 000.	—	18 600.	—

[a]For N·sec/m² divide by 1 000.

Table 1-14.   PROPERTIES OF GASES (Continued)

Common name(s)	Methane	Methyl chloride	Neon	Nitric oxide
Chemical formula	CH_4	CH_3Cl	Ne	NO
Refrigerant number	50	40	720	—
CHEMICAL AND PHYSICAL PROPERTIES				
Molecular weight	16.044	50.488	20.179	30.006
Specific gravity, air = 1	0.554	1.74	0.697	1.04
Specific volume, ft^3/lb	24.2	7.4	19.41	13.05
Specific volume, m^3/kg	1.51	0.462	1.211	0.814
Density of liquid (at atm bp), lb/ft^3	26.3	62.7	75.35	
Density of liquid (at atm bp), kg/m^3	421.	1 004.	1 207.	
Vapor pressure at 25 deg C, psia		82.2		
Vapor pressure at 25 deg C, MN/m^2		0.567		
Viscosity (abs), lbm/ft·sec	7.39×10^{-6}	7.39×10^{-6}	21.5×10^{-6}	12.8×10^{-6}
Viscosity (abs), centipoises[a]	0.011	0.011	0.032	0.019
Sound velocity in gas, m/sec	446.	251.	454.	341.
THERMAL AND THERMO-DYNAMIC PROPERTIES				
Specific heat, c_p, Btu/lb·deg F or cal/g·deg C	0.54	0.20	0.246	0.235
Specific heat, c_p, J/kg·K	2 260.	837.	1 030.	983.
Specific heat ratio, c_p/c_v	1.31	1.28	1.64	1.40
Gas constant R, ft-lb/lb·deg F	96.	30.6	76.6	51.5
Gas constant R, J/kg·deg C	518.	165.	412.	277.
Thermal conductivity, Btu/hr·ft·deg F	0.02	0.006	0.028	0.015
Thermal conductivity, W/m·deg C	0.035	0.010	0.048	0.026
Boiling point (sat 14.7 psia), deg F	− 259.	− 10.7	− 410.9	− 240.
Boiling point (sat 760 mm), deg C	− 434.2	− 23.7	− 246.	− 151.5
Latent heat of evap (at bp), Btu/lb	219.2	184.1	37.	
Latent heat of evap (at bp), J/kg	510 000.	428 000.	86 100.	
Freezing (melting) point, deg F (1 atm)	− 296.6	− 144.	− 415.6	− 258.
Freezing (melting) point, deg C (1 atm)	− 182.6	− 97.8	− 248.7	− 161.
Latent heat of fusion, Btu/lb	14.	56.	6.8	32.9
Latent heat of fusion, J/kg	32 600.	130 000.	15 800.	76 500.
Critical temperature, deg F	− 116.	289.4	− 379.8	− 136.
Critical temperature, deg C	− 82.3	143.	− 228.8	− 93.3
Critical pressure, psia	673.	968.	396.	945.
Critical pressure, MN/m^2	4.64	6.67	2.73	6.52
Critical volume, ft^3/lb	0.099	0.043	0.033	0.033 2
Critical volume, m^3/kg	0.006 2	0.002 7	0.002 0	0.002 07
Flammable (yes or no)	Yes	Yes	No	No
Heat of combustion, Btu/ft^3	985.		—	—
Heat of combustion, Btu/lb	2 290.		—	—
Heat of combustion, kJ/kg			—	—

[a]For N·sec/m^2 divide by 1 000.

Table 1-14.   PROPERTIES OF GASES *(Continued)*

Common name(s)	Nitrogen	Nitrous oxide	Oxygen	Ozone
Chemical formula	N_2	N_2O	O_2	O_3
Refrigerant number	728	744A	732	—
CHEMICAL AND PHYSICAL PROPERTIES				
Molecular weight	28.013 4	44.012	31.998 8	47.998
Specific gravity, air = 1	0.967	1.52	1.105	1.66
Specific volume, ft^3/lb	13.98	8.90	12.24	8.16
Specific volume, m^3/kg	0.872	0.555	0.764	0.509
Density of liquid (at atm bp), lb/ft^3	50.46	76.6	71.27	
Density of liquid (at atm bp), kg/m^3	808.4	1 227.	1 142.	
Vapor pressure at 25 deg C, psia				
Vapor pressure at 25 deg C, MN/m^2				
Viscosity (abs), lbm/ft·sec	12.1×10^{-6}	10.1×10^{-6}	13.4×10^{-6}	8.74×10^{-6}
Viscosity (abs), centipoises[a]	0.018	0.015	0.020	0.013
Sound velocity in gas, m/sec	353.	268.	329.	
THERMAL AND THERMO-DYNAMIC PROPERTIES				
Specific heat, c_p, Btu/lb·deg F or cal/g·deg C	0.249	0.21	0.220	0.196
Specific heat, c_p, J/kg·K	1 040.	879.	920.	820.
Specific heat ratio, c_p/c_v	1.40	1.31	1.40	
Gas constant R, ft-lb/lb·deg F	55.2	35.1	48.3	32.2
Gas constant R, J/kg·deg C	297.	189.	260.	173.
Thermal conductivity, Btu/hr·ft·deg F	0.015	0.010	0.015	0.019
Thermal conductivity, W/m·deg C	0.026	0.017	0.026	0.033
Boiling point (sat 14.7 psia), deg F	− 320.4	− 127.3	− 297.3	− 170.
Boiling point (sat 760 mm), deg C	− 195.8	− 88.5	− 182.97	− 112.
Latent heat of evap (at bp), Btu/lb	85.5	161.8	91.7	
Latent heat of evap (at bp), J/kg	199 000.	376 000.	213 000.	
Freezing (melting) point, deg F (1 atm)	− 346.	− 131.5	− 361.1	− 315.5
Freezing (melting) point, deg C (1 atm)	− 210.	− 90.8	− 218.4	− 193.
Latent heat of fusion, Btu/lb	11.1	63.9	5.9	97.2
Latent heat of fusion, J/kg	25 800.	149 000.	13 700.	226 000.
Critical temperature, deg F	− 232.6	97.7	− 181.5	16.
Critical temperature, deg C	− 147.	36.5	− 118.6	− 9.
Critical pressure, psia	493.	1 052.	726.	800.
Critical pressure, MN/m^2	3.40	7.25	5.01	5.52
Critical volume, ft^3/lb	0.051	0.036	0.040	0.029 8
Critical volume, m^3/kg	0.003 18	0.002 2	0.002 5	0.001 86
Flammable (yes or no)	No	No	No	No
Heat of combustion, Btu/ft^3	—	—	—	—
Heat of combustion, Btu/lb	—	—	—	—
Heat of combustion, kJ/kg	—	—	—	—

[a]For N·sec/m^2 divide by 1 000.

Table 1-14. **PROPERTIES OF GASES** (Continued)

Common name(s)	Propane	Propylene (Propene)	Sulfur dioxide	Xenon
Chemical formula	C_3H_8	C_3H_6	SO_2	Xe
Refrigerant number	290	1 270	764	—
CHEMICAL AND PHYSICAL PROPERTIES				
Molecular weight	44.097	42.08	64.06	131.30
Specific gravity, air = 1	1.52	1.45	2.21	4.53
Specific volume, ft^3/lb	8.84	9.3	6.11	2.98
Specific volume, m^3/kg	0.552	0.58		
Density of liquid (at atm bp), lb/ft^3	36.2	37.5	42.8	190.8
Density of liquid (at atm bp), kg/m^3	580.	601.	585.	3 060.
Vapor pressure at 25 deg C, psia	135.7	166.4	56.6	
Vapor pressure at 25 deg C, MN/m^2	0.936	1.147	0.390	
Viscosity (abs), lbm/ft·sec	53.8×10^{-6}	57.1×10^{-6}	8.74×10^{-6}	15.5×10^{-6}
Viscosity (abs), centipoises^a	0.080	0.085	0.013	0.023
Sound velocity in gas, m/sec	253.	261.	220.	177.
THERMAL AND THERMO-DYNAMIC PROPERTIES				
Specific heat, c_p, Btu/lb·deg F or cal/g·deg C	0.39	0.36	0.11	0.115
Specific heat, c_p, J/kg·K	1 630.	1 506.	460.	481.
Specific heat ratio, c_p/c_v	1.2	1.16	1.29	1.67
Gas constant R, ft-lb/lb·deg F	35.0	36.7	24.1	11.8
Gas constant R, J/kg·deg C	188.	197.	130.	63.5
Thermal conductivity, Btu/hr·ft·deg F	0.010	0.010	0.006	0.003
Thermal conductivity, W/m·deg C	0.017	0.017	0.010	0.005 2
Boiling point (sat 14.7 psia), deg F	−44.	−54.	14.0	−162.5
Boiling point (sat 760 mm), deg C	−42.2	−48.3	−10.	−108.
Latent heat of evap (at bp), Btu/lb	184.	188.2	155.5	41.4
Latent heat of evap (at bp), J/kg	428 000.	438 000.	362 000.	96 000.
Freezing (melting) point, deg F (1 atm)	−309.8	−301.	−104.	−220.
Freezing (melting) point, deg C (1 atm)	−189.9	−185.	−75.5	−140.
Latent heat of fusion, Btu/lb	19.1		58.0	10.
Latent heat of fusion, J/kg	44 400.		135 000.	23 300.
Critical temperature, deg F	205.	197.	315.5	61.9
Critical temperature, deg C	96.	91.7	157.6	16.6
Critical pressure, psia	618.	668.	1 141.	852.
Critical pressure, MN/m^2	4.26	4.61	7.87	5.87
Critical volume, ft^3/lb	0.073	0.069	0.03	0.014 5
Critical volume, m^3/kg	0.004 5	0.004 3	0.001 9	0.000 90
Flammable (yes or no)	Yes	Yes	No	No
Heat of combustion, Btu/ft^3	2 450.	2 310.	—	—
Heat of combustion, Btu/lb	21 660.	21 500.	—	—
Heat of combustion, kJ/kg	50 340.	50 000.	—	—

^{a}For N·sec/m^2 divide by 1 000.

Table 1-15. THERMAL CONDUCTIVITY OF GASES AT VARIOUS TEMPERATURES*

The values in this table are given as $\dfrac{\text{cal}}{\text{sec cm }^{\circ}\text{C}} \times 10^{6}$. For W/m·K multiply by 0.0004184. For Btu/hr·ft·deg F multiply by 0.0002419.

Gas	°F −400 °C −240	−300 −184.4	−200 −128.9	−100 −73.3	−40 −40	−20 −28.9	0 −17.8	20 −6.7	40 4.4	60 15.6	80 26.7	100 37.8	120 48.9	200 93.3
Acetylene				28.10	34.71	37.19	39.67	42.15	45.04	47.94	50.83	53.72	56.62	69.43
Air					50.09	52.15	54.22	56.24	58.31	60.34	62.20	64.22	66.04	
Ammonia					43.39	45.87	48.35	50.83	53.31	55.79	58.68	61.58	64.47	
Argon					34.30	35.95	37.19	38.85	40.09	41.33	42.57	44.22	45.46	
Bromine							9.09					11.57		
n-Butane					27.90	29.75	31.70	30.99	33.06	35.54	38.02	40.91	43.39	54.14
i-Butane								32.65	33.89	36.37	38.85	41.74	44.22	55.79
Carbon dioxide								33.68	35.62	37.61	39.67	41.74	43.81	
Carbon disulfide											19.01	19.84		
Carbon monoxide					47.94	50.00	51.95	53.85	55.87	57.86	59.92	61.99	63.89	
Chlorine					15.29	16.53	17.36	18.18	16.53	17.77	21.08	21.90	23.14	
Deuterium					274.82	285.15	295.07	305.81	309.95	322.34	334.74	343.01	355.40	
Ethane				23.97	32.65	35.54	38.43	41.33	44.63	47.94	51.24	54.55	58.27	74.39
Ethanol								29.34	30.99	32.65	34.71	36.78		
Ethylamine								31.41	33.47	35.54	37.61	39.67	42.15	
Ethylene				26.86	33.06	35.54	38.02	40.50	43.39	46.29	49.18	52.07	54.96	68.19
Fluorine		18.18	30.58	43.39	50.83	52.90	55.38	57.86	59.92	61.99	64.06	66.12	68.19	76.04
Helium	84.31	163.24	221.51	274.8	304.99	314.49	324.00	333.50	343.42	352.10	360.36	368.63	376.07	
Hydrogen	59.92	142.57	227.29	308.7	357.47	371.93	388.46	405.00	417.39	433.92	446.32	458.72	471.11	
Hydrogen bromide					15.29	16.11	16.49	17.77	18.60	19.84	20.66	21.49		
Hydrogen chloride					25.62	26.86	28.51	29.75	30.99	32.23	33.89	35.12		
Hydrogen cyanide							23.97	25.62	26.86	28.10	29.75	30.99	32.65	
Hydrogen sulfide							28.10	29.75	31.41	33.47		36.78		
Krypton							19.84					23.56		
Methane		22.32	36.86	52.07	61.37	64.55	67.86	71.08	74.39	78.11	81.83	85.54	89.26	106.62
Neon					97.94	100.84	104.14	107.03	109.93	112.82	115.71	118.19	121.09	
Nitric oxide			30.91	42.40	49.01	51.24	53.39	55.54	57.65	59.76		64.06	66.12	
Nitrogen		20.25	33.06	44.22	50.42	52.48	54.55	56.20	58.27	60.34	62.40	64.06	65.71	74.39
Nitrous oxide					28.93	30.91	32.90	35.04	37.15	39.30	41.45	43.81	46.08	
Oxygen		18.84	31.66	43.72	50.54	52.81	54.96	57.24	59.43	61.58	63.64	65.91	68.19	76.87
n-Propane					27.69	29.75	32.23	34.71	37.19	39.67	42.47	45.46	48.35	60.75
R-11(CCl₃F)						12.81	13.64	14.88	15.70	16.53	17.77	18.60		
R-12(CCl₂F₂)						17.36	18.60	19.42	20.66	21.49	22.73	23.56		
R-21(CHCl₂F)							21.90	22.32	22.73	23.14	23.56	23.97		
R-22(CHClF₂)							24.80	25.62	26.45	27.28	28.10	28.93		
Water							34.71	36.78	38.85	40.50	42.57	44.63	46.70	54.96

*From: "CRC Handbook of Chemistry and Physics", 53rd ed., R.C. Weast, Ed., The Chemical Rubber Co., 1972.

Table 1-16.   CRITICAL TEMPERATURES, PRESSURES, AND VOLUMES*

For pressure in N/m, multiply by 10 132. For volume in m^3/kg, divide by molecular weight and multiply by 10^{-3}.

Name	Formula	Critical temperature, T_c, °C	Critical pressure, P_c, atm	Critical volume, cm^3/mole
Acetaldehyde	C_2H_4O	187.8	54.7	168
Acetic acid	$C_2H_4O_2$	321.6	57.1	171
Acetone	C_3H_6O	235.9	47	211
Acetylene (Ethyne)	C_2H_2	36.3	61.6	113
Air (R–729)	—	−140.7	37.2	90
Ammonia (R–717)	NH_3	132.5	112.5	88
Aniline	C_6H_7N	425.6	52.3	274
Argon	Ar	−122	48	75
Benzene	C_6H_6	289	48.6	260
Bromine	Br	311	102	144
n-Butane (R–600)	C_4H_{10}	152	37.5	255
Carbon dioxide	CO_2	31	72.9	94
Carbon disulfide	CS_2	279	79	170
Carbon monoxide	CO	−140	34.5	93
Carbon tetrachloride (R–10)	CCl_4	283	45	270
Carbon tetrafluoride (R–14)	CF	−45.5	41.4	153
Chlorine	Cl_2	144	76.1	124
Chloroform (R–20)	$CHCl_3$	263	54	240
Cyanogen	C_2N_2	127	50	200
Cyclohexane	C_6H_{12}	280	40	308
n-Decane	$C_{10}H_{22}$	344.5	20.8	602
Deuterium	D_2	−234.8	16.4	—
Ethane	C_2H_6	32.2	48.2	148
Ethane, 1,1-dichloro- (R–150a)	$C_2H_4Cl_2$	249.8	50	244
Ethane, 1,2-dichlorotetrafluoro- (R–114)	$C_2Cl_2F_4$	145.8	32.3	292
Ethane, tetrachlorodifluoro- (R–112)	$C_2Cl_4F_2$	278	32.9	354
Ethane, trichlorotrifluoro- (R–113)	$C_2Cl_2F_3$	214.2	33.7	325
Ethanol (Ethyl alcohol)	C_2H_6O	243	63	167
Ethene (Ethylene)	C_2H_4	9.9	50.5	124
Ether, diethyl-	$C_4H_{10}O$	192.6	35.6	274
Ether, dimethyl-	C_2H_6O	127	52.6	178
Ethylamine	C_2H_7N	183.3	55.5	185
Ethyl chloride	C_2H_5Cl	187	52	199
Fluorine	F_2	−129	55	—
Formic acid, methyl- (Methyl formate)	$C_2H_4O_2$	214	59.2	172
Helium[4]	He	−268	2.26	56
n-Heptane	C_7H_{16}	267.1	27	426
n-Hexane	C_6H_{14}	234.2	29.9	368
Hydrazine	N_2H_2	380	14.5	96
Hydrogen	H_2	−239.9	12.8	65
Hydrogen bromide	HBr	90	84.5	105
Hydrogen chloride	HCl	51.4	82	87.6
Hydrogen cyanide	HCN	183.5	50	139
Hydrogen sulfide	H_2S	100.4	89	95
Iodine	I_2	512	116	—
Isobutane (R–600a)	C_4H_{10}	135	36	263
Isobutyl alcohol	$C_4H_{10}O$	275	42.4	272
Ketone, diethyl-	$C_5H_{10}O$	287.8	36.9	336
Krypton	Kr	−63.8	54.3	92
Methane (R–50)	CH_4	−82.4	45.8	99
Methane, bromotrifluoro- (R–13B1)	$CBrF_3$	66.5	39	205
Methane, chlorotrifluoro- (R–13)	$CClF_3$	28.85	38	180
Methane, chlorodifluoro- (R–22)	$CHClF_2$	96.5	48.5	165
Methane, dichlorodifluoro- (R–12)	CCl_2F_2	111.5	39.6	218
Methane, dichloromonofluoro- (R–21)	$CHCl_2F$	178.5	51	197
Methane, tetrafluoro-	CF_4	−45.7	41.4	—

Table 1-16. CRITICAL TEMPERATURES, PRESSURES, AND VOLUMES (Continued)

Name	Formula	Critical temperature, T_c, °C	Critical pressure, P_c, atm	Critical volume, $cm^3/mole$
Methane, trichlorofluoro- (R–11)	CCl_3F	198	43.2	248
Methanol (Methyl alcohol)	CH_4O	240	78.5	118
Methyl acetate	$C_3H_6O_2$	233.7	46.3	228
Methyl chloride (R–40)	CH_3Cl	143	66	143
Methyl fluoride (R–41)	CH_3F	44.6	58	113
Methylene chloride (R–30)	CH_2Cl_2	237	60	193
Naphthalene	$C_{10}H_8$	474.8	40.6	—
Neon	Ne	−228.7	26.9	42
Nitric oxide	NO	−93	64	58
Nitrogen	N_2	−147	33.5	90
Nitrogen dioxide	NO_2	157.8	100	82
Nitrous oxide	N_2O	36.5	71.7	96
n-Nonane	C_9H_{20}	321	22.5	543
n-Octane	C_8H_{18}	296	24.8	486
Oxygen	O_2	−118.4	50.1	78
Ozone	O_3	−5.16	67	89.4
n-Pentane	C_5H_{12}	196.6	33.3	311
Propane (R–290)	C_3H_8	96.8	42	200
Propyne	C_3H_4	127.8	52.8	167
Silane	SiH_4	−3.45	47.8	—
Styrene	C_8H_8	374.4	39.4	—
Sulfur dioxide	SO_2	157.8	77.7	122.
Sulfur trioxide	SO_3	218	83	126
Toluene	C_7H_8	320.8	41.6	320
Water	H_2O	374.2	218.3	56
Xenon	Xe	16.6	58	119

Note: R numbers in parentheses are ASHRAE Standard Refrigerant numbers.
*Table compiled from several sources.

REFERENCE

"CRC Handbook of Chemistry and Physics", 53rd ed., R.C. Weast, Ed., The Chemical Rubber Co., 1972.

Table 1-17. VAN DER WAALS' CONSTANTS FOR GASES*

Calculated from "Landolt-Bornstein Physical Chemical Tables"

Van der Waals' equation is an equation of state for real gases. It may be written

$$\left(P + \frac{a}{V^2}\right)(V - b) = RT \text{ for one mole or } \left(P + \frac{n^2 a}{V^2}\right)(V - nb) = nRT \text{ for } n \text{ moles.}$$

The term a is a measure of the attractive force between the molecules. The term b is due to the finite volume of the molecules and to their general incompressibility. It is known that a and b vary to some extent with temperature. The values for a and b in the following table are those to be used when the pressure is in atmospheres and the volume is in liters. Thus R in the above equation will be 0.08206 liter atmospheres per mole per degree. T is degrees Kelvin.

Name	Formula	a $\frac{(liters)^2 \times atm}{(mole)^2}$	b $\frac{liters}{mole}$	Name	Formula	a $\frac{(liters)^2 \times atm}{(mole)^2}$	b $\frac{liters}{mole}$
Acetic acid	CH_3CO_2H	17.59	0.1068	Ammonia	NH_3	4.170	0.03707
Acetic anhydride	$(CH_3CO)_2O$	19.90	0.1263	Amyl formate	$HCO_2C_5H_{11}$	27.58	0.1730
Acetone	$(CH_3)_2CO$	13.91	0.0994	Amylene	C_5H_{10}	15.90	0.1207
Acetonitrile	CH_3CN	17.58	0.1168	Isoamylene	C_5H_{10}	18.08	0.1405
Acetylene	C_2H_2	4.390	0.05136	Aniline	$C_6H_5NH_2$	26.50	0.1369

Table 1-17. VAN DER WAALS' CONSTANTS FOR GASES *(Continued)*

Name	Formula	a $\dfrac{(liters)^2 \times atm}{(mole)^2}$	b $\dfrac{liters}{mole}$	Name	Formula	a $\dfrac{(liters)^2 \times atm}{(mole)^2}$	b $\dfrac{liters}{mole}$
Argon	Ar	1.345	0.03219	Hydrogen sulfide	H_2S	4.431	0.04287
Benzene	C_6H_6	18.00	0.1154	Iodobenzene	C_6H_5I	33.08	0.1656
Benzonitrile	C_6H_5CN	33.39	0.1724	Krypton	Kr	2.318	0.03978
Bromobenzene	C_6H_5Br	28.56	0.1539	Mercury	Hg	8.093	0.01696
n-Butane	C_4H_{10}	14.47	0.1226	Mesitylene	$(CH_3)_3C_6H_3$	34.32	0.1979
iso-Butane	C_4H_{10}	12.87	0.1142	Methane	CH_4	2.253	0.04278
iso-Butyl acetate	$CH_3CO_2C_4H_9$	28.50	0.1833	Methyl acetate	$CH_3CO_2CH_3$	15.29	0.1091
iso-Butyl alcohol	C_4H_9OH	17.03	0.1143	Methyl alcohol	CH_3OH	9.523	0.06702
iso-Butyl benzene	$C_6H_5C_4H_9$	38.59	0.2144	Methylamine	CH_3NH_2	7.130	0.05992
iso-Butyl formate	$HCO_2C_4H_9$	22.54	0.1476	Methyl butyrate	$C_3H_7CO_2CH_3$	23.94	0.1569
Butyronitrile	C_3H_7CN	25.72	0.1596	Methyl isobutyrate	$C_3H_7CO_2CH_3$	24.50	0.1637
Capronitrile	$C_5H_{11}CN$	34.16	0.1984	Methyl chloride	CH_3Cl	7.471	0.06483
Carbon dioxide	CO_2	3.592	0.04267	Methyl ether	$(CH_3)_2O$	8.073	0.07246
Carbon disulfide	CS_2	11.62	0.07685	Methyl ethyl ether	$CH_3OC_2H_5$	11.95	0.09775
Carbon monoxide	CO	1.485	0.03985	Methyl ethyl sulfide	$CH_3SC_2H_5$	19.23	0.1304
Carbon oxysulfide	COS	3.933	0.05817	Methyl fluoride	CH_3F	4.631	0.05264
Carbon tetrachloride	CCl_4	20.39	0.1383	Methyl formate	HCO_2CH_3	10.84	0.08068
Chlorine	Cl_2	6.493	0.05622	Methyl propionate	$C_2H_5CO_2CH_3$	19.91	0.1360
Chlorobenzene	C_6H_5Cl	25.43	0.1453	Methyl sulfide	$(CH_3)_2S$	12.87	0.09213
Chloroform	$CHCl_3$	15.17	0.1022	Methyl valerate	$C_4H_9CO_2CH_3$	28.96	0.1845
m-Cresol	$CH_3C_6H_4OH$	31.38	0.1607	Naphthalene	$C_{10}H_8$	39.74	0.1937
Cyanogen	C_2N_2	7.667	0.06901	Neon	Ne	0.2107	0.01709
Cyclohexane	C_6H_{12}	22.81	0.1424	Nitric oxide	NO	1.340	0.02789
Cymene	$C_{10}H_{14}$	42.16	0.2336	Nitrogen	N_2	1.390	0.03913
Decane	$C_{10}H_{12}$	48.55	0.2905	Nitrogen dioxide	NO_2	5.284	0.04424
Di-isobutyl	C_8H_{18}	34.97	0.2296	Nitrous oxide	N_2O	3.782	0.04415
Diethylamine	$(C_2H_5)_2NH$	19.15	0.1392	n-Octane	C_8H_{18}	37.32	0.2368
Dimethylamine	$(CH_3)_2NH$	10.38	0.08570	Oxygen	O_2	1.360	0.03183
Dimethylaniline	$C_6H_5N(CH_3)_2$	37.49	0.1970	n-Pentane	C_5H_{12}	19.01	0.1460
Diphenyl	$(C_6H_5)_2$	52.79	0.2480	iso-Pentane	C_5H_{12}	18.05	0.1417
Diphenyl methane	$(C_6H_5)_2CH_2$	38.20	0.2240	Phenetole	$C_6H_5OC_2H_5$	35.16	0.1963
Dipropylamine	$(C_3H_7)_2NH$	27.72	0.1820	Phosphine	PH_3	4.631	0.05156
Di-isopropyl	$(C_3H_7)_2$	23.13	0.1669	Phosphonium chloride	PH_4Cl	4.054	0.04545
Durene	$C_{10}H_{14}$	45.32	0.2424	Phosphorus	P	52.94	0.1566
Ethane	C_2H_6	5.489	0.06380	Propane	C_3H_8	8.664	0.08445
Ethyl acetate	$CH_3CO_2C_2H_5$	20.45	0.1412	Propionic acid	$C_2H_5CO_2H$	20.11	0.1187
Ethyl alcohol	C_2H_5OH	12.02	0.08407	Propionitrile	C_2H_5CN	16.44	0.1064
Ethylamine	$C_2H_5NH_2$	10.60	0.08409	Propyl acetate	$CH_3CO_2C_3H_7$	24.63	0.1619
Ethyl benzene	$C_2H_5C_6H_5$	28.60	0.1667	Propyl alcohol	C_3H_7OH	14.92	0.1019
Ethyl butyrate	$C_3H_7CO_2C_2H_5$	30.07	0.1919	Propylamine	$C_3H_7NH_2$	14.99	0.1090
Ethyl isobutyrate	$C_3H_7CO_2C_2H_5$	28.87	0.1994	Propyl benzene	$C_6H_5C_3H_7$	35.85	0.2028
Ethyl chloride	C_2H_5Cl	10.91	0.08651	iso-Propyl benzene	$C_6H_5C_3H_7$	35.64	0.2025
Ethyl ether	$(C_2H_5)_2O$	17.38	0.1344	Propyl chloride	C_3H_7Cl	15.91	0.1141
Ethyl formate	$HCO_2C_2H_5$	14.80	0.1056	Propyl formate	$HCO_2C_3H_7$	18.95	0.1280
Ethyl mercaptan	C_2H_5SH	11.24	0.08098	Propylene	C_2H_6	8.379	0.08272
Ethyl propionate	$C_2H_5CO_2C_2H_5$	24.39	0.1615	Pseudo-cumene	$C_6H_3(CH_3)_3$	36.61	0.2021
Ethyl sulfide	$(C_2H_5)_2S$	18.75	0.1214	Silicon fluoride	SiF_4	4.195	0.05571
Ethylene	C_2H_4	4.471	0.05714	Silicon tetrahydride	SiH_4	4.320	0.05786
Ethylene bromide	$(CH_2Br)_2$	13.98	0.08664	Stannic chloride	$SnCl_4$	26.91	0.1642
Ethylene chloride	$(CH_2Cl)_2$	16.91	0.1086	Sulfur dioxide	SO_2	6.714	0.05636
Ethylidene chloride	CH_3CHCl_2	15.50	0.1073	Thiophene	C_4H_4S	20.72	0.1270
Fluorobenzene	C_6H_5F	19.93	0.1286	Toluene	$C_6H_5CH_3$	24.06	0.1463
Germanium tetrachloride	$GeCl_4$	22.60	0.1485	Triethylamine	$(C_2H_5)_3N$	27.17	0.1831
Helium	He	0.03412	0.02370	Trimethylamine	$(CH_3)_3N$	13.02	0.1084
n-Heptane	C_7H_{16}	31.51	0.2065	Xenon	Xe	4.194	0.05105
n-Hexane	C_6H_{14}	24.39	0.1735	m-Xylene	$C_6H_4(CH_3)_2$	30.36	0.1772
Hydrogen	H_2	0.2444	0.02661	o-Xylene	$C_6H_4(CH_3)_2$	29.98	0.1755
Hydrogen bromide	HBr	4.451	0.04431	p-Xylene	$C_6H_4(CH_3)_2$	30.93	0.1809
Hydrogen chloride	HCl	3.667	0.04081	Water	H_2O	5.464	0.03049
Hydrogen selenide	H_2Se	5.268	0.04637				

*From: "CRC Handbook of Chemistry and Physics", 53rd ed., R.C. Weast, Ed., The Chemical Rubber Co., 1972.

Table 1-18. PRESSURE-TEMPERATURE RELATIONS FOR SATURATED VAPORS AND REFRIGERANTS*

Boiling Points in °F at Various Pressures

For composition and properties of refrigerants, see Tables 1-28 through 1-36.

Vapor	Press, atm → 20	13.6	10	6.8	5.0	3.4	2.0	1.0	.68	.50	.10	.068	.05	.01
Press, psia	294	200	147	100	73.5	50	29.4	14.7	10	7.35	1.47	1.0	.735	.147
Air	−244	−257	−265	−276	−282	−291	−306	−314						
Ammonia	122	96	78	56	40	22	−1	−28.0	−41	−51	−102	−104.5		
Argon	−222.5	−240	−247	−261	−268	−279	−290	−302.6	−309					
Azeotrope R–500	150	119	96	70	51	29	1	−28.3	−43	−55				
Azeotrope R–502	120	90	68	43	25	4	−21	−50	−64	−75				
Butane	238.1	2.2	175.6	145.5	123.4	97.9	65.9	31.0	8.3	0				
Carbon dioxide	−1	−24	−40	−58	−71	−80	−93	−109.3	−119	−128				
Carbon disulfide	348	302.9	277.5	241	220.6	186.4	156.5	116						
Carbon monoxide	−237.5	−250.6	−258	−269	−275	−287.5	−298.4	−312.4						
Carbon tetrachloride	418	369	347	311	290	241	214	172	150	136	64	42	30	−13
Chloroform	370	327	303	268	249	214	183	142						
Ethane	20	−6	−25	−46.7	−67.6	−80	−102.6	−127.5	−140	−149.2	−189	−197	−202.9	
Ethanol	361.4	325	305	276	259	231.5	207.5	173	156	143	86	72	62	16
Ethylene	−18	−41.8	−60.7	−80	−95	−111.5	−130.5	−155	−166	−174.4	−210.8	−218.1	−223.6	−249.1
Ethyl ether	304	267	246	215	198	165	136	95.3						
Fluorine refrigerants														
R–11	297	259	230	198	174	148	113	74.9	56	42	−19	−31	−40	−82
R–12	164	132	108	81	61	38	10	−21.6	−37.5	−49	−100	−109	−117	−151
R–13	34	8	−11	−33	−49	−67	−90	−114.6	−127	−137	−177	−185	−190	
R–13B1	94	65	44	20	2	−18	−44	−72.0	−86	−96	−141	−150	−157	
R–14	−86	−106	−120	−137	−148	−162	−179	−198.4	−207.7	−214				
R–21				160	138	113	83	48.1	31	18	−38			
R–22	126	90	75	51	33	12	−12	−41.4	−56	−66	−114	−122	−129	
R–23	21	0	−20	−35	−48	−67	−92	−116	−127	−136	−174	−181	−187.5	
R–113	354	311	280	249	222.4	195	160	117.6	98	83	18	6	−4.4	
R–114	247	210	184	153	131	106	74	38.4	21	8	−48	−59	−68	−107.5
R–115		107	84	58	40	18	−8	−38.4	−53	−65	−114			
R–142b	208	175	150	122.5	102	79	48.4	14	−2.5	−15	−71.9	−83.4	−92	
R–152a	163	133	112	87	68	47	18	−12	−29	−40	−93	−103	−111.6	−148.5
R–216	323	284	255	222	197	170	135	97	77.5	63	2	−10	−19	
R–C 318		177.5	153	125	105	82	53	21	5	−7				
Glycerol			709	676.8	658.4	620	586.4	547	527.6	508	418.5		385.6	320
Helium							−451	−452.1	−452.8	−453				
Hexane	409	358	330	290	267	230	198	155.4						
Hydrogen			−403	−408	−411	−414	−419	−423	−423	−427	−434	−435		

Table 1-18.   PRESSURE-TEMPERATURE RELATIONS FOR SATURATED VAPORS AND REFRIGERANTS
(Continued)

Vapor	Press, atm	.01	.05	.068	.10	.50	.68	1.0	2.0	3.4	5.0	6.8	10	13.6	20
	Press, psia	.147	.735	1.0	1.47	7.35	10	14.7	29.4	50	73.5	100	147	200	294
Isobutane								10.3	45.3	76.9	101.5	123.6	152.6	178.3	213.6
Mercury			438.1	458.1	473.5	609.1	637	674	748.5	804.8	869	903	963	1005	1078
Methane						-272.8	-267	-259	-242.6	-227.9	-215.8	-205	-190.7	-171.6	-160
Methanol								150	172.6	195.6	222	236	260	293	351
Methyl chloride (R-40)						-38	-26.5	10.8	20	48	70	90	116	140	
Neon						-415	-413.2	-411	-407	-402	-400	-396	-393	-386	-383
Nitrogen						-331	-326.5	-320.4	-309	-298	-290.5	-283	-277	-264	-251
Octane								258.8	307	343	385	411	457	483	545
Oxygen			-354	-334	-329	-309	-304	-297.3	-285	-272	-264	-256	-244	-234	-220
Pentane								97	136.8	167.3	200	217	247.5	277.9	332
Propane						-70	-59.2	-44.2	-12.5	15	36.6	55.7	81.6	105	135.9
Propylene								-53.9	-24.2	3	20.4	42.2	67.6	90	120.8
Sulfur dioxide								14.0	43.4	65.4	90	105	132	152	189
Water		45	92	102	115	179	193	212	249	281	306	328	357	382	416

* Compiled from several sources.

Table 1-19. EFFECT OF TEMPERATURE ON PROPERTIES OF GASES AND VAPORS AT ATMOSPHERIC PRESSURE*

SYMBOLS AND CONVERSION FACTORS:

ρ = density in g/cm^3. For lb/ft^3 multiply by 62.428. For kg/m^3 multiply by 1000.

c_p = specific heat in cal/g °K. For Btu/lb·deg R multiply by 1. For J/kg·K multiply by 4184.0.

k = thermal conductivity in cal/sec cm °K. For W/m·K multiply by 418.4. For Btu/hr·ft·deg R multiply by 241.9.

μ = absolute viscosity in centipoises. For lb/ft·hr multiply by 2.419. For N·s/m^2 multiply by 0.001.

| Substance | Temperature | | ρ, g/cm^3 | c_p, cal/g °K | k, cal/sec cm °K | μ, centipoise |
	°C	°F				
Ammonia	0	32	9.56×10^{-4}	.52	5.23×10^{-5}	9.18×10^{-3}
	20	68	8.94×10^{-4}	.52	5.69×10^{-5}	9.82×10^{-3}
	50	122	8.11×10^{-4}	.52	6.48×10^{-5}	1.09×10^{-2}
	100	212	7.02×10^{-4}	.53		1.28×10^{-2}
	200	392	6.20×10^{-4}			1.64×10^{-2}
	300	572	5.12×10^{-4}			1.99×10^{-2}
Argon	−13	9	1.87×10^{-3}	.125	3.74×10^{-5}	2.04×10^{-2}
	−3	37	1.81×10^{-3}	.125	3.87×10^{-5}	2.11×10^{-2}
	7	45	1.74×10^{-3}	.125	3.99×10^{-5}	2.17×10^{-2}
	27	81	1.62×10^{-3}	.125	4.22×10^{-5}	2.30×10^{-2}
	77	171	1.39×10^{-3}	.124	4.79×10^{-5}	2.59×10^{-2}
	227	441	9.74×10^{-4}	.124	6.31×10^{-5}	3.37×10^{-2}
	727	1,341	4.87×10^{-4}	.124	1.02×10^{-4}	5.42×10^{-2}
	1,227	2,241	3.25×10^{-4}	.124	1.31×10^{-4}	7.08×10^{-2}
	1,727	3,141	2.43×10^{-4}	.124		
Butane	0	32	2.59×10^{-3}	.3802	3.16×10^{-5}	6.84×10^{-3}
	100	212	1.90×10^{-3}	.4842	5.60×10^{-5}	9.26×10^{-3}
	200	392	1.50×10^{-3}	.5865	8.70×10^{-5}	1.17×10^{-2}
	300	572	1.24×10^{-3}	.6721	1.24×10^{-4}	1.40×10^{-2}
	400	752	1.05×10^{-3}	.7474	1.66×10^{-4}	1.64×10^{-2}
	500	932	9.16×10^{-4}	.8131	2.15×10^{-4}	1.87×10^{-2}
	600	1,112	8.12×10^{-4}	.8704	2.69×10^{-4}	2.11×10^{-2}
Carbon dioxide	−13	9	2.08×10^{-3}	.1944	3.25×10^{-5}	1.31×10^{-2}
	−3	27	2.00×10^{-3}	.1967	3.42×10^{-5}	1.36×10^{-2}
	7	45	1.93×10^{-3}	.1989	3.60×10^{-5}	1.40×10^{-2}
	17	63	1.86×10^{-3}	.2012	3.78×10^{-5}	1.45×10^{-2}
	27	81	1.80×10^{-3}	.2035	3.96×10^{-5}	1.49×10^{-2}
	77	171	1.54×10^{-3}	.2146	4.89×10^{-5}	1.72×10^{-2}
	227	441	1.07×10^{-3}	.2424	8.01×10^{-5}	2.32×10^{-2}
	727	1,341	5.36×10^{-4}	.2946		3.89×10^{-2}
	1,227	2,241	3.57×10^{-4}	.3166		
Carbon monoxide	−13	9	1.31×10^{-3}	.2489	5.30×10^{-5}	1.59×10^{-2}
	−3	27	1.27×10^{-3}	.2489	5.49×10^{-5}	1.64×10^{-2}
	7	45	1.22×10^{-3}	.2489	5.67×10^{-5}	1.69×10^{-2}
	17	63	1.18×10^{-3}	.2489	5.85×10^{-5}	1.74×10^{-2}
	27	81	1.14×10^{-3}	.2489	6.03×10^{-5}	1.79×10^{-2}
	77	171	9.75×10^{-4}	.2493	6.89×10^{-5}	2.01×10^{-2}
	227	441	6.82×10^{-4}	.2542	9.23×10^{-5}	2.61×10^{-2}
	727	1,341	3.41×10^{-4}			4.17×10^{-2}
	1,227	2,241	2.27×10^{-4}			5.44×10^{-2}
Ethane	0	32	1.342×10^{-3}	.3934	4.52×10^{-5}	8.60×10^{-3}
	100	212	9.83×10^{-4}	.4938	7.59×10^{-5}	1.14×10^{-2}
	200	392	7.76×10^{-4}	.5947	1.13×10^{-4}	1.41×10^{-2}
	300	572	6.40×10^{-4}	.6854	1.55×10^{-4}	1.68×10^{-2}
	400	752	5.45×10^{-4}	.7676	2.04×10^{-4}	1.93×10^{-2}
	500	932	4.74×10^{-4}	.8405	2.57×10^{-4}	2.20×10^{-2}
	600	1,112	4.20×10^{-4}	.9045	3.15×10^{-4}	2.45×10^{-2}
Ethanol	100	212	1.49×10^{-3}	.403	5.50×10^{-5}	1.08×10^{-2}
	200	392	1.18×10^{-3}	.480	8.39×10^{-5}	1.37×10^{-2}
	300	572	9.74×10^{-4}	.554	1.19×10^{-4}	1.67×10^{-2}
	400	752	8.28×10^{-4}	.624	1.59×10^{-4}	1.97×10^{-2}
	500	932	7.20×10^{-4}	.691	2.05×10^{-4}	2.26×10^{-2}
Helium	−240	−400	1.463×10^{-3}		8.43×10^{-5}	3.74×10^{-3}
	−129	−200	3.38×10^{-4}		2.22×10^{-4}	1.19×10^{-2}
	0	32	3.68×10^{-4}	1.23	3.40×10^{-4}	1.86×10^{-2}
	20	68	1.67×10^{-4}	1.24	3.55×10^{-4}	1.94×10^{-2}
	40	104	1.56×10^{-4}	1.24	3.70×10^{-4}	2.03×10^{-2}
	49	120		1.24	3.76×10^{-4}	2.06×10^{-2}

Table 1-19. EFFECT OF TEMPERATURE ON PROPERTIES OF GASES AND VAPORS AT ATMOSPHERIC PRESSURE *(Continued)*

Substance	Temperature °C	Temperature °F	ρ, g/cm³	c_p, cal/g °K	k, cal/sec cm °K	μ, centipoise
Hydrogen	−13	9	9.44×10^{-5}	3.373	3.86×10^{-4}	8.14×10^{-3}
	−3	37	9.10×10^{-5}	3.388	3.98×10^{-4}	8.35×10^{-3}
	7	45	8.77×10^{-5}	3.400	4.11×10^{-4}	8.55×10^{-3}
	27	81	8.47×10^{-5}	3.410	4.22×10^{-4}	8.76×10^{-3}
	77	171	8.19×10^{-5}	3.418	4.34×10^{-4}	8.96×10^{-3}
	227	441	7.02×10^{-5}		4.91×10^{-4}	9.94×10^{-3}
	727	1,341	4.912×10^{-5}	3.467	6.50×10^{-4}	1.26×10^{-2}
Methane	0	32	7.16×10^{-4}	.5172	7.33×10^{-5}	1.04×10^{-2}
	100	212	5.25×10^{-4}	.5848	1.11×10^{-4}	1.32×10^{-2}
	200	392	4.14×10^{-4}	.6704	1.52×10^{-4}	1.59×10^{-2}
	300	572	3.42×10^{-4}	.7584	1.96×10^{-4}	1.83×10^{-2}
	400	752	2.91×10^{-4}	.8430	2.43×10^{-4}	2.07×10^{-2}
	500	932	2.53×10^{-4}	.9210	2.91×10^{-4}	2.29×10^{-2}
	600	1,112	2.24×10^{-4}	.9919	3.43×10^{-4}	2.52×10^{-2}
Nitrogen	−13	9	1.31×10^{-3}	.2488	5.52×10^{-5}	
	−3	37	1.27×10^{-3}	.2487	5.71×10^{-5}	
	7	45	1.22×10^{-3}	.2487	5.89×10^{-5}	
	27	81	1.18×10^{-3}	.2487	6.06×10^{-5}	
	77	171	1.14×10^{-3}	.2487	6.24×10^{-5}	1.79×10^{-2}
	227	441	9.75×10^{-3}	.2490	7.11×10^{-5}	2.00×10^{-2}
	727	1,341	6.82×10^{-3}	.2524	9.49×10^{-5}	2.57×10^{-2}
Oxygen	−13	9	1.50×10^{-3}	.2188	5.60×10^{-5}	1.85×10^{-2}
	3	37	1.45×10^{-3}	.2190	5.80×10^{-5}	1.90×10^{-2}
	7	45	1.39×10^{-3}	.2193	5.98×10^{-5}	1.96×10^{-2}
	27	81	1.35×10^{-3}	.2195	6.22×10^{-5}	2.01×10^{-2}
	77	171	1.30×10^{-3}	.2198	6.40×10^{-5}	2.06×10^{-2}
	227	441	1.11×10^{-3}	.2221	7.33×10^{-5}	2.32×10^{-2}
	727	1,341	7.80×10^{-3}	.2324	9.97×10^{-5}	2.99×10^{-2}
Propane	0	32	1.97×10^{-3}	.3701	3.62×10^{-5}	7.50×10^{-3}
	100	212	1.44×10^{-3}	.4817	6.26×10^{-5}	1.00×10^{-2}
	200	392	1.14×10^{-3}	.5871	9.56×10^{-5}	1.25×10^{-2}
	300	572	9.39×10^{-4}	.6770	1.34×10^{-4}	1.40×10^{-2}
	400	752	7.99×10^{-4}	.7550	1.78×10^{-4}	1.72×10^{-2}
	500	932	6.94×10^{-4}	.8237	2.28×10^{-4}	1.94×10^{-2}
	600	1,112	6.16×10^{-4}	.8831	2.83×10^{-4}	2.18×10^{-2}

*Compiled from several sources.

Table 1-20. EFFECT OF HIGHER PRESSURES ON DENSITY AND COMPRESSIBILITY OF GASES*

SYMBOLS: ρ = density, lb$_m$/ft³. For g/cm³ multiply by 0.016018. For kg/m³ multiply by 16.018.

Z = compressibility factor, dimensionless

For pressure in N/m², multiply values in psia by 6894.

Pressure atm	Pressure psia	He† ρ	He† Z	H₂‡ ρ	H₂‡ Z	N₂‡ ρ	N₂‡ Z
136	2,000	1.32	1.076	0.63	1.086	9.5	1.024
204	3,000	1.90	1.106	0.90	1.130	13.7	1.064
272	4,000	2.46	1.136	1.17	1.174	17.3	1.119
340	5,000	3.00	1.166	1.42	1.218	20.4	1.189
408	6,000	3.54	1.194	1.65	1.262	22.9	1.264
476	7,000	4.00	1.222	1.87	1.306	25.2	1.348
544	8,000	4.46	1.250	2.07	1.350	27.0	1.432
612	9,000	4.92	1.278	2.25	1.395	28.6	1.516
680	10,000	5.38	1.306	2.43	1.440	30.2	1.600
749	11,000	—	—	2.59	1.484	31.6	1.684
817	12,000	—	—	—	—	32.9	1.768
885	13,000	—	—	—	—	34.0	1.852

†At 70°F.
‡At 80.6°F.
*Based largely on: NASA Tech Brief B67–10610, John F. Kennedy Space Center, 1967.

Table 1-21. EFFECT OF TEMPERATURE ON PROPERTIES OF GASES

Table A gives the density, viscosity, and thermal conductivity of seven common gases, in both metric and English units. Table B lists the dimensionless ratios of each of these properties at three selected pressures and a range of temperatures; from these ratios and the values in Table A, the actual properties can be readily calculated.

The tables were condensed from "Tables of the Thermal Properties of Gases", National Bureau of Standards Circular 564, U.S. Government Printing Office, November 1955 (488 pages). Tables in the original source cover the same gases (plus steam) but are much more extensive with respect to temperature and pressure.

SYMBOLS FOR DIMENSIONLESS VALUES IN TABLES:

ρ/ρ_o = relative density

c_p/R = specific heat

c_p/c_v = specific heat ratio

μ/μ_o = relative viscosity

k/k_o = relative thermal conductivity

$Pr = c_p \, \mu/k$ = Prandtl number

$Z = \dfrac{PV}{RT}$ = compressibility factor

Note: Subscript "o" denotes values at standard conditions: $T = 273.16$ K and $P = 1$ atmosphere, or 0.1013 MN/m². Viscosity equivalents: g/s·cm = poise = dyne·s/cm²; N·s/m² = kg/m·s.

Table A.   PROPERTIES OF GASES AT STANDARD CONDITIONS

Properties	Units	Air	Argon	CO_2	CO	H_2	N_2	O_2
ρ_o, density	kg/m³	1.293	1.784	1.977	1.250	8.989×10^{-2}	1.250	1.429
	g/cm³	1.293×10^{-3}	1.784×10^{-3}	1.977×10^{-3}	1.250×10^{-3}	8.989×10^{-5}	1.250×10^{-3}	1.429×10^{-3}
	lb/ft³	8.072×10^{-2}	0.1114	0.1234	7.807×10^{-2}	5.611×10^{-3}	7.806×10^{-2}	8.921×10^{-2}
μ_o, absolute viscosity	N·s/m²	1.716×10^{-5}	2.125×10^{-5}	1.370×10^{-5}	1.657×10^{-5}	8.411×10^{-6}	1.663×10^{-5}	1.919×10^{-5}
	g/s·cm	1.716×10^{-4}	2.125×10^{-4}	1.370×10^{-4}	1.657×10^{-4}	8.411×10^{-5}	1.663×10^{-4}	1.919×10^{-4}
	lb/s·ft	1.153×10^{-5}	1.428×10^{-5}	9.207×10^{-6}	1.113×10^{-5}	5.652×10^{-6}	1.117×10^{-5}	1.290×10^{-5}
k_o, thermal conductivity	W/m·K	2.414×10^{-2}	1.634×10^{-2}	1.455×10^{-2}	2.322×10^{-2}	1.682×10^{-1}	2.41×10^{-2}	2.455×10^{-2}
	cal/cm·s·K	5.770×10^{-5}	3.905×10^{-5}	3.477×10^{-5}	5.549×10^{-5}	4.021×10^{-4}	5.77×10^{-5}	5.867×10^{-5}
	Btu/ft·hr·deg R	1.395×10^{-2}	9.444×10^{-3}	8.407×10^{-3}	1.342×10^{-2}	9.724×10^{-2}	1.40×10^{-2}	1.419×10^{-2}
To convert c_p/R to c_p, multiply by the following:								
For cal/g·K or Btu/lb·deg R		0.0686	0.0497	0.0451	0.0709	0.9851	0.0709	0.0621
For J/kg·K		287.0	208.2	188.9	296.8	4124.	296.8	259.8

Table 1-21. EFFECT OF TEMPERATURE ON PROPERTIES OF GASES (Continued)

Table B. RATIOS OF PROPERTIES

Temperature		Pressure: 1 atm, 14.7 psia (or 0.1013 MN/m²)							Pressure: 10 atm, 1470 psia				Pressure: 0.01 atm, 0.147 psia		
K	R	ρ/ρ_o	c_p/R	c_p/c_v	μ/μ_o	k/k_o	P_r	Z	Z	ρ/ρ_o	c_p/R	c_p/c_v	ρ/ρ_o	c_p/R	c_p/c_v
AIR															
260	468.0	1.051	3.503	1.403	.9617	.9561	.719	.99923	.9557	109.9	4.344	1.663	.01050	3.495	1.401
270	486.0	1.012	3.503	1.403	.9909	.9894	—	.99937	.9675	104.5	4.248	1.634	.01011	3.496	1.401
280	504.0	.9755	3.504	1.402	1.020	1.022	.713	.99949	.9775	99.74	4.167	1.609	.009750	3.498	1.400
290	522.0	.9417	3.505	1.402	1.048	1.055	.708	.99960	.9860	95.47	4.101	1.589	.009413	3.499	1.400
300	540.0	.9102	3.506	1.402	1.076	1.087	—	.99970	.9933	91.61	4.046	1.571	.009100	3.501	1.400
350	630.0	.7800	3.516	1.399	1.209	1.244	.580	1.0000	1.0176	76.65	3.872	1.511	.007800	3.512	1.398
500	900.0	.5458	3.588	1.387	1.556	1.674	.702	1.0003	1.0393	52.53	3.735	1.4321	.005460	3.587	1.387
1,000	1,800	.2729	3.979	1.336	2.420	2.798	—	1.0003	1.0333	26.42	4.012	1.342	.002730	3.979	1.336
1,500	2,700	.1820	4.289	1.304	3.066	—	—	1.0002	1.0244	17.77	4.302	1.306	.001820	4.301	1.304
2,000	3,600	.1365	4.662	1.274	—	—	—	1.0004	1.0188	13.40	4.605	1.281	.001363	5.326	1.243
ARGON															
260	468.0	1.051	2.508	1.671	.9600	.958	.679	.99882	.9055	115.9	3.40	2.17	.01050	2.500	1.667 (Constant)
270	486.0	1.012	2.507	—	.9906	.991	.678	.99899	.9208	109.8	3.30	—	.01011	2.500	
280	504.0	.9755	2.507	1.670	1.021	1.022	.677	.99913	.9340	104.4	3.23	2.06	.009746	2.500	
290	522.0	.9417	2.506	—	1.050	1.052	.677	.99926	.9454	99.54	3.17	—	.009410	2.500	
300	540.0	.9102	2.506	1.670	1.080	1.081	.677	.99937	.9553	95.22	3.12	1.96	.009097	2.500	
350	630.0	.7799	2.504	1.668	1.219	1.227	.673	.99976	.9879	78.93	2.92	1.753	.007797	2.500	
500	900.0	.5457	2.502	1.667	1.588	1.616	.666	1.0000	1.0224	53.38	2.670	1.680	.005458	2.500	
1,000	1,800	.2728	2.500	—	2.551	2.611	.661	1.0003	1.0265	26.59	2.536	1.671	.002729	2.500	
1,500	2,700	.1819	2.500	—	3.331	3.352	.671	1.0002	1.0203	17.83	2.513	1.667	.001819	2.500	
2,000	3,600	.1364	2.500	—	—	—	—	1.0002	1.0159	13.43	2.507	—	.001365	2.500	
CARBON DIOXIDE															
260	468.0	1.052	4.311	1.317	.9542	.935	.788	.99197	—	—	—	—	.01044	4.249	1.308
270	486.0	1.012	4.361	1.311	.9891	.984	.784	.99291	—	—	—	—	.01005	4.307	1.303
280	504.0	.9750	4.411	1.304	1.024	1.035	.779	.99372	—	—	—	—	.009690	4.365	1.297
290	522.0	.9407	4.462	1.299	1.058	1.087	.775	.99441	—	—	—	—	.009356	4.421	1.293
300	540.0	.9088	4.513	1.293	1.091	1.139	.770	.99501	—	—	—	—	.009044	4.477	1.288
350	630.0	.7774	4.758	1.271	1.256	1.406	.755	.99705	.6507	119.1	8.67	2.40	.007752	4.737	1.268
500	900.0	.5429	5.429	1.230	1.697	2.304	.702	.99927	.9365	57.94	6.014	1.387	.005426	5.367	1.229
1,000	1,800	.2712	6.533	1.181	2.840	—	—	1.0002	1.0248	26.47	6.640	1.196	.002713	6.532	1.181
1,500	2,700	.1808	7.021	1.166	3.737	—	—	1.0002	1.0253	17.64	7.061	1.172	.001809	7.021	1.166
2,000	3,600	—	—	—	—	—	—	—	—	—	—	—	—	—	—
CARBON MONOXIDE															
260	468.0	1.051	3.511	1.402	.9605	.956	.747	.99911	—	—	—	—	.01050	3.502	1.400
270	486.0	1.012	3.511	1.402	.9906	.989	.745	.99928	—	—	—	—	.01011	3.503	1.400
280	504.0	.9755	3.510	1.402	1.020	1.022	.742	.99942	.9752	99.97	4.091	1.698	.009749	3.503	1.400
290	522.0	.9417	3.511	1.402	1.049	1.054	.740	.99955	.9851	95.55	4.049	1.663	.009413	3.504	1.399
300	540.0	.9102	3.511	1.401	1.078	1.087	.737	.99965	.9935	91.59	4.016	1.633	.009099	3.505	1.399

Table 1-21. EFFECT OF TEMPERATURE ON PROPERTIES OF GASES *(Continued)*
Table B. RATIOS OF PROPERTIES *(Continued)*

K	R	Pressure: 1 atm, 14.7 psia (or 0.1013 MN/m²)							Pressure: 100 atm, 1470 psia				Pressure: 0.01 atm, 0.147 psia		
		ρ/ρ_o	c_p/R	c_p/c_v	μ/μ_o	k/k_o	Pr	Z	ρ/ρ_o	c_p/R	c_p/c_v	Z	ρ/ρ_o	c_p/R	c_p/c_v
CARBON MONOXIDE (Cont.)															
350	630.0	.7799	3.517	1.399	1.213	1.242	.728	1.0000	76.34	3.897	1.542	1.0216	.007799	3.513	1.398
500	900.0	.5457	3.585	1.387	1.574	1.664	.718	1.0004	52.15	3.763	1.441	1.0469	.005459	3.583	1.387
1,000	1,800	.2729	—	1.334	2.519	—	—	1.0004	26.27	4.029	1.342	1.0391	.002730	3.991	1.334
1,500	2,700	.1819	—	1.309	3.285	—	—	1.0003	17.69	4.251	1.311	1.0286	.001820	4.236	1.309
2,000	3,600	.1365	—	1.298	—	—	—	1.0002	13.35	4.366	1.298	1.0221	.001365	4.359	1.298
HYDROGEN															
260	468.0	1.051	3.425	1.413	.9672	.959	.712	1.0006	98.55	3.504	1.431	1.0667	.01051	—	—
270	486.0	1.012	3.439	1.410	.9922	.990	—	1.0006	95.05	3.510	1.427	1.0651	.01012	—	—
280	504.0	.9756	3.451	1.408	1.017	1.021	.709	1.0006	91.78	3.516	1.423	1.0636	.009762	—	—
290	522.0	.9420	3.462	1.406	1.041	1.051	—	1.0006	88.74	3.521	1.420	1.0621	.009425	—	—
300	540.0	.9106	3.470	1.405	1.065	1.080	.706	1.0006	85.90	3.526	1.417	1.0607	.009111	—	—
350	630.0	.7806	—	—	1.182	1.222	—	1.0005	74.09	—	—	1.0541	.007809	—	—
500	900.0	.5464	3.519	1.397	1.503	1.616	.675	1.0004	52.56	3.536	1.398	1.0400	.005467	—	—
1,000	1,800	—	—	—	2.393	—	—	—	—	—	—	—	—	—	—
1,500	2,700	—	—	—	—	—	—	—	—	—	—	—	—	—	—
2,000	3,600	—	—	—	—	—	—	—	—	—	—	—	—	—	—
NITROGEN															
260	468.0	1.051	3.509	1.402	—	.957	.725	.99937	108.1	4.28	1.65	.9716	.01050	3.502	1.400
270	486.0	1.012	3.508	1.402	—	.990	—	.99951	103.0	4.20	1.62	.9822	.01011	3.502	1.400
280	504.0	.9755	3.508	1.402	—	1.021	.719	.99963	98.38	4.198	1.602	.9911	.00975	3.502	1.400
290	522.0	.9418	3.508	1.401	—	1.051	—	.99973	94.28	4.070	1.582	.9986	.00941	3.503	1.400
300	540.0	.9103	3.508	1.401	1.074	1.081	.713	.99982	90.52	4.021	1.566	1.0054	.00910	3.503	1.400
350	630.0	.7800	3.512	—	1.203	1.232	—	1.0001	75.93	3.858	—	1.0274	.00780	3.508	1.391
500	900.0	.5459	3.560	1.391	1.546	1.645	.684	1.0004	52.20	3.710	1.437	1.0461	.00546	3.558	1.341
1,000	1,800	.2729	3.933	1.341	2.406	2.673	.724	1.0004	26.34	3.965	1.347	1.0365	.00273	3.933	1.313
1,500	2,700	.1820	4.191	1.313	3.040	—	—	1.0003	17.73	4.203	1.315	1.0264	.00182	4.191	1.313
2,000	3,600	.1365	4.327	1.301	—	—	—	1.0002	13.38	4.333	1.301	1.0202	.00137	4.327	1.301
OXYGEN															
260	468.0	1.051	3.524	1.400	.9617	.954	.722	.99883	116.2	4.537	1.721	.9033	.01050	3.516	1.398
270	486.0	1.012	3.527	—	.9909	.989	.718	.99900	110.0	4.407	—	.9191	.01011	3.519	—
280	504.0	.9755	3.531	1.398	1.019	1.02	.717	.99915	104.5	4.307	1.648	.9326	.00975	3.524	1.396
290	522.0	.9417	3.535	—	1.048	1.06	.710	.99928	99.67	4.229	—	.9441	.00941	3.529	—
300	540.0	.9102	3.540	1.396	1.075	1.09	.709	.99939	95.34	4.165	1.599	.9541	.00910	3.535	1.395
350	630.0	.7799	3.576	1.366	1.208	1.25	.702	.99979	78.86	3.986	—	.9887	.00780	3.572	1.365
500	900.0	.5457	3.742	—	1.560	1.70	.697	1.0002	53.21	3.920	1.420	1.0256	.00546	3.740	1.313
1,000	1,800	.2728	4.195	1.313	2.485	—	—	1.0003	26.51	4.232	1.321	1.0296	.00273	4.195	1.294
1,500	2,700	.1819	4.398	1.294	3.237	—	—	1.0002	17.79	4.412	1.297	1.0224	.00182	4.398	—
2,000	3,600	.1364	4.544	1.282	3.897	—	—	1.0002	13.41	4.551	1.283	1.0175	.00136	4.544	1.282

Table 1-22. EFFECT OF PRESSURE ON SPECIFIC HEAT AND COMPRESSIBILITY OF GASES*

At 68°F (20°C)

SYMBOLS: Z = compressibility factor, dimensionless
c_p/c_v = specific heat ratio, dimensionless
c_p = specific heat, Btu/lb_m °F = cal/g °C

For pressure in MN/m^2, multiply values in atm by 0.1013.

Gas	Properties	.01 atm / .147 psia	.1 atm / 1.47 psia	.4 atm / 5.9 psia	1 atm / 14.7 psia	7 atm / 103 psia	10 atm / 147 psia	40 atm / 590 psia	70 atm / 1030 psia	100 atm / 1470 psia
Air	Z	1.0000	.9999	.9998	.9996	.9975	.9965	.9889	.9861	.9884
	c_p/c_v	1.4002	1.4003	1.4008	1.4019	1.4131	1.4188	1.4763	1.5323	1.5828
	c_p	.2399	.2400	.2401	.2403	.2427	.2440	.2569	.2700	.2824
Argon	Z	1.0000	.9999	.9997	.9993	.9951	.9931	.9744	.9594	.9487
	c_p/c_v	1.667	1.667	1.667	1.670	1.690	1.699	1.80	1.90	2.00
	c_p	.1243	.1243	.1244	.1246	.1264	.1276	.1372	.1472	.1566
CO₂	Z	.9999	.9995	.99784	.99461	.9615	.9443	—	—	—
	c_p/c_v	1.2909	1.2914	1.2933	1.297	1.338	1.361	—	—	—
	c_p	.2003	.2005	.2010	.2021	.2132	.2146	—	—	—
CO	Z	1.0000	.9999	.9998	.9996	.9962	.9961	.9880	.9853	.9879
	c_p/c_v	1.399	1.399	—	1.402	—	1.422	1.491	1.567	1.653
	c_p	.2484	.2485	—	.2493	—	.2530	.2657	.2750	.2863
H₂	Z	1.0000	1.0001	—	1.0006	1.0042	1.0060	1.0242	1.0426	1.0616
	c_p/c_v	1.400	—	—	1.406	—	1.408	—	—	1.419
	c_p	—	—	—	3.412	—	3.418	—	—	3.469
N₂	Z	1.0000	.9999	.9999	.9998	.9984	.9978	.9940	.9947	1.0009
	c_p/c_v	1.400	1.400	1.400	1.401	1.413	1.419	1.475	1.529	1.577
	c_p	.2483	.2483	.2484	.2487	.2511	.2523	.2637	.2782	.2874
O₂	Z	1.0000	.9999	.9997	.9993	.9953	.9933	.9742	.9586	.9474
	c_p/c_v	1.396	1.396	—	1.397	1.410	1.416	1.490	1.546	1.610
	c_p	.2190	.2190	.2191	.2195	.2219	.2231	.2301	.2487	.2611

Pressure

*From: "Tables of the Thermal Properties of Gases", National Bureau of Standards Circular 564, U.S. Government Printing Office, November 1955.

Table 1-23.   EFFECTS OF PRESSURE AND TEMPERATURE ON PROPERTIES OF GASES—METRIC UNITS*

For additional data on effects of pressure and temperature, see Tables 1-21 and 1-22. For saturation tables and other properties of gases, consult the Index.

For pressure in MN/m^2, multiply values in atm by 0.1013. For density in kg/m^3, multiply by 1000. For specific heat in J/kg·K, multiply by 4184. For enthalpy in J/kg, multiply by 4184.

AIR

Temp, °K	Pressure, atmospheres						
	1	4	7	10	40	70	100
COMPRESSIBILITY FACTOR							
150	0.9941	0.9759	0.9572	0.9378	0.6832		
250	0.9991	0.9962	0.9935	0.9908	0.9668	0.9498	0.9417
300	0.9997	0.9988	0.9980	0.9972	0.9914	0.9900	0.9933
500	1.0003	1.0014	1.0024	1.0035	1.0145	1.0265	1.0393
1,000	1.0003	1.0013	1.0023	1.0033	1.0132	1.0233	1.0333
1,500	1.0002	1.0010	1.0017	1.0025	1.0097	1.0171	1.0244
DENSITY, g cm^{-3}							
150	0.0024	0.0096	0.0172	0.0251	0.1378		
250	0.0014	0.0057	0.0099	0.0143	0.0584	0.1041	0.1499
300	0.0012	0.0047	0.0083	0.0118	0.0474	0.0832	0.1185
500	0.0007	0.0028	0.0049	0.0070	0.0278	0.0481	0.0679
1,000	0.0004	0.0014	0.0025	0.0035	0.0140	0.0241	0.0342
1,500	0.0002	0.0009	0.0016	0.0023	0.0093	0.0162	0.0230
SPECIFIC HEAT, c_p, cal g^{-1} K^{-1}							
150	0.2415	0.2480	0.2553	0.2637			
250	0.2403	0.2420	0.2438	0.2455	0.2648	0.2858	0.3062
300	0.2405	0.2416	0.2427	0.2439	0.2553	0.2668	0.2776
500	0.2462	0.2465	0.2469	0.2472	0.2505	0.2536	0.2565
1,000	0.2730	0.2730	0.2732	0.2733	0.2739	0.2746	0.2752
1,500	0.2942	0.2943	0.2943	0.2943	0.2946	0.2949	0.2951
ENTHALPY, cal g^{-1}							
150	35.67	35.11	34.52	33.94			
250	59.741	59.507	59.276	59.046	56.73	54.49	52.42
300	71.759	71.594	71.433	71.275	69.701	68.234	66.90
500	120.30	120.27	120.22	120.19	119.84	119.55	119.32
1,000	250.14	250.18	250.22	250.25	250.68	251.11	251.56
1,500	392.08	392.11	392.21	392.26	392.94	393.59	394.21
ENTROPY, cal g^{-1} K^{-1}							
150	1.4740	1.3763	1.3354	1.3082			
250	1.5970	1.5011	1.4618	1.4368	1.334	1.289	1.257
300	1.6408	1.5452	1.5064	1.4814	1.3816	1.3387	1.3100
500	1.7647	1.6695	1.6309	1.6063	1.5097	1.4701	1.4445
1,000	1.9436	1.8485	1.8101	1.7854	1.6900	1.6514	1.6265
1,500	2.0585	1.9633	1.9249	1.9002	1.8047	1.7659	1.7412

Table 1-23. EFFECTS OF PRESSURE AND TEMPERATURE ON PROPERTIES OF GASES—METRIC UNITS *(Continued)*

ARGON

Temp, °K	Pressure, atmospheres						
	1	*4*	*7*	*10*	*40*	*70*	*100*
COMPRESSIBILITY FACTOR							
100	0.9782	0.9079					
200	0.9971	0.9882	0.9792	0.9702	0.8778	0.7838	0.6917
300	0.9994	0.9975	0.9957	0.9938	0.9773	0.9643	0.9553
500	1.0002	1.0007	1.0013	1.0018	1.0079	1.0147	1.0224
1,000	1.0003	1.0010	1.0018	1.0026	1.0105	1.0185	1.0265
1,500	1.0002	1.0008	1.0014	1.0002	1.0081	1.0142	1.0203
DENSITY, g cm^{-3}							
100	0.0050	0.0214					
200	0.0024	0.0099	0.0174	0.0251	0.1109	0.2174	0.3519
300	0.0016	0.0065	0.0114	0.0163	0.0664	0.1179	0.1699
500	0.0010	0.0039	0.0068	0.0097	0.0386	0.0672	0.0952
1,000	0.0005	0.0019	0.0034	0.0049	0.0193	0.0335	0.0474
1,500	0.0003	0.0012	0.0023	0.0033	0.0129	0.0224	0.0318
SPECIFIC HEAT, c_p, cal g^{-1} K^{-1}							
100	0.1297	0.150	0.177				
200	0.1251	0.1275	0.1299	0.1325	0.1647	0.2089	0.2590
300	0.1247	0.1255	0.1264	0.1279	0.1363	0.1458	0.1552
500	0.1245	0.1247	0.1250	0.1252	0.1279	0.1304	0.1328
1,000	0.1244	0.1244	0.1245	0.1246	0.1251	0.1256	0.1262
1,500	0.1244	0.1244	0.1244	0.1244	0.1247	0.1249	0.1250
ENTHALPY, cal g^{-1}							
100	12.142	11.171	10.074				
200	24.782	24.501	24.215	23.926	20.792	17.666	
300	37.265	37.125	36.987	36.847	35.469	34.137	32.887
500	62.172	62.130	62.085	62.042	61.629	61.235	60.381
1,000	124.38	124.40	124.41	124.43	124.62	124.81	125.00
1,500	186.57	186.61	186.64	186.68	187.03	187.37	187.74
ENTROPY, cal g^{-1} K^{-1}							
100	0.7882	0.7128	0.6776				
200	0.8759	0.8060	0.7772	0.7584	0.6786	0.6383	0.6069
300	0.9266	0.8572	0.8290	0.8109	0.7382	0.7068	0.6856
500	0.9902	0.9211	0.8932	0.8753	0.8052	0.7763	0.7575
1,000	1.0764	1.0074	0.9796	0.9618	0.8926	0.8646	0.8466
1,500	1.1268	1.0579	1.0300	1.0123	0.9432	0.9153	0.8975

Table 1-23. EFFECTS OF PRESSURE AND TEMPERATURE ON PROPERTIES OF GASES—METRIC UNITS *(Continued)*

CARBON DIOXIDE

Temp, °K	Pressure, atmospheres						
	1	*4*	*7*	*10*	*40*	*70*	*100*
COMPRESSIBILITY FACTOR							
250	0.9909	0.9629	0.9337	0.9022			
300	0.9950	0.9798	0.9644	0.9486	0.7611		
500	0.9993	0.9971	0.9950	0.9928	0.9721	0.9531	0.9365
1,000	1.0002	1.0008	1.0015	1.0022	1.0092	1.0167	1.0248
1,500	1.0002	1.0010	1.0017	1.0025	1.0100	1.0176	1.0253
DENSITY, g cm^{-3}							
250	0.0022	0.0089	0.0161	0.0238			
300	0.0018	0.0073	0.0130	0.0188	0.0939		
500	0.0011	0.0043	0.0075	0.0108	0.0441	0.0788	0.1145
1,000	0.0005	0.0021	0.0037	0.0054	0.0213	0.0369	0.0523
1,500	0.0004	0.0014	0.0025	0.0036	0.0142	0.0246	0.0349
SPECIFIC HEAT, c_p, cal g^{-1} K^{-1}							
250	0.1924	0.2043	0.2245	0.2605			
300	0.2038	0.2088	0.2140	0.2195	0.3364		
500	0.2427	0.2436	0.2446	0.2456	0.2560	0.2653	0.2716
1,000	0.2950	0.2951	0.2953	0.2954	0.2970	0.2984	0.2998
1,500	0.3170	0.3171	0.3172	0.3172	0.3178	0.3183	0.3188
ENTHALPY, cal g^{-1}							
250	41.097	40.049	38.926	37.668			
300	51.001	50.310	49.620	48.904	40.086		
500	95.885	95.650	95.428	95.181	92.776	90.223	87.337
1,000	232.22	232.18	232.15	232.11	231.76	231.32	230.60
1,500	385.97	385.98	385.99	386.00	386.13	386.12	385.80
ENTROPY, cal g^{-1} K^{-1}							
250	1.1251	1.0595	1.0310	1.0112			
300	1.1611	1.0969	1.0700	1.0523	0.9676		
500	1.2749	1.2119	1.1863	1.1698	1.1034	1.0739	1.0528
1,000	1.4618	1.3992	1.3738	1.3577	1.2944	1.2682	1.2505
1,500	1.5862	1.5234	1.4983	1.4822	1.4192	1.3933	1.3760

Table 1-23.   EFFECTS OF PRESSURE AND TEMPERATURE ON PROPERTIES OF GASES—METRIC UNITS *(Continued)*

HYDROGEN

Temp, °K	Pressure, atmospheres						
	1	4	7	10	40	70	100
COMPRESSIBILITY FACTOR							
100	0.9998	0.9992	0.9987	0.09983	1.0029	1.0222	1.0560
200	1.0007	1.0028	1.0048	1.0068	1.0283	1.0513	1.0760
300	1.0006	1.0024	1.0042	1.0059	1.0238	1.0420	1.0607
400	1.0005	1.0020	1.0034	1.0048	1.0193	1.0339	1.0486
500	1.0004	1.0016	1.0028	1.0040	1.0160	1.0280	1.0400
600	1.0003	1.0012	1.0023	1.0034	1.0136	1.0237	1.0337
DENSITY, g cm^{-3}							
100	0.0002	0.0009	0.0017	0.0025	0.0100	0.0168	0.0232
200	0.0001	0.0005	0.0009	0.0012	0.0048	0.0082	0.0114
300	0.0001	0.0003	0.0006	0.0008	0.0032	0.0055	0.0077
400	0.0001	0.0002	0.0004	0.0006	0.0024	0.0042	0.0059
500	0.0000	0.0002	0.0003	0.0005	0.0019	0.0033	0.0047
600	0.0000	0.0002	0.0003	0.0004	0.0016	0.0028	0.0040
SPECIFIC HEAT, c_p, cal g^{-1} K^{-1}							
100	2.683	—	—	2.750	—	—	3.248
200	3.235	—	—	3.249	—	—	3.364
300	3.420	—	—	3.426	—	—	3.476
400	3.461	—	—	3.464	—	—	3.488
500	3.469	—	—	3.471	—	—	3.485
600	3.477	—	—	3.479	—	—	3.486
ENTHALPY, cal g^{-1}							
100	376.10	—	—	372.97	—	—	351.62
200	675.40	—	—	675.37	—	—	677.93
300	1010.3	—	—	1011.2	—	—	1021.6
400	1354.8	—	—	1356.2	—	—	1370.2
500	1701.3	—	—	1702.9	—	—	1718.9
600	2048.6	—	—	2050.3	—	—	2067.5
ENTROPY, cal g^{-1} K^{-1}							
100	12.094	—	—	9.795	—	—	7.294
200	14.147	—	—	11.871	—	—	9.550
300	15.501	—	—	13.229	—	—	10.941
400	16.492	—	—	14.222	—	—	11.944
500	17.265	—	—	14.996	—	—	12.722
600	17.899	—	—	15.629	—	—	13.359

Table 1-23. EFFECTS OF PRESSURE AND TEMPERATURE ON PROPERTIES OF GASES—METRIC UNITS *(Continued)*

NITROGEN

Temp, °K	Pressure, atmospheres						
	1	4	7	10	40	70	100
COMPRESSIBILITY FACTOR							
150	0.9944	0.9773	0.9597	0.9416	0.736		
250	0.9992	0.9969	0.9946	0.9924	0.9731	0.9613	0.9893
300	0.9998	0.9993	0.9988	0.9984	0.9962	0.9984	1.0054
500	1.0004	1.0016	1.0029	1.0041	1.0173	1.0313	1.0461
1,000	1.0004	1.0014	1.0025	1.0036	1.0145	1.0255	1.0365
1,500	1.0003	1.0011	1.0018	1.0026	1.0105	1.0185	1.0264
DENSITY, g cm^{-3}							
150	0.0023	0.0093	0.0166	0.0242	0.1237		
250	0.0014	0.0055	0.0096	0.0138	0.0561	0.0994	0.1424
300	0.0011	0.0046	0.0080	0.0114	0.0457	0.0800	0.1132
500	0.0007	0.0027	0.0048	0.0068	0.0268	0.0463	0.0653
1,000	0.0003	0.0014	0.0024	0.0034	0.0134	0.0233	0.0329
1,500	0.0002	0.0009	0.0016	0.0023	0.0090	0.0156	0.0222
SPECIFIC HEAT, c_p, cal g^{-1} K^{-1}							
150	0.2503	0.2568	0.2642	0.2727			
250	0.2489	0.2507	0.2524	0.2542	0.2731	0.2925	0.3107
300	0.2488	0.2500	0.2511	0.2523	0.2638	0.2751	0.2852
500	0.2525	0.2528	0.2532	0.2535	0.2569	0.2602	0.2632
1,000	0.2790	0.2790	0.2791	0.2792	0.2799	0.2806	0.2812
1,500	0.2973	0.2973	0.2973	0.2974	0.2976	0.2979	0.2981
ENTHALPY, cal g^{-1}							
150	36.990	36.428	35.841	35.238			
250	61.930	61.699	61.470	61.240	58.977	56.809	54.832
300	74.372	74.212	74.055	73.898	72.369	70.953	69.674
500	124.38	124.35	124.32	124.29	124.01	123.79	123.62
1,000	257.08	257.13	257.18	257.23	257.74	258.27	258.80
1,500	401.63	401.70	401.77	401.85	402.58	403.32	404.06
ENTROPY, cal g^{-1} K^{-1}							
150	1.4620	1.3611	1.3188	1.2907	1.1547		
250	1.8894	1.4904	1.4499	1.4239	1.3180	1.2710	1.2390
300	1.6348	1.5360	1.4958	1.4700	1.3669	1.3226	1.2931
500	1.7625	1.6640	1.6241	1.5987	1.4989	1.4578	1.4312
1,000	1.9453	1.8470	1.8073	1.7819	1.6833	1.6434	1.6179
1,500	2.0623	1.9640	1.9243	1.8990	1.8006	1.7608	1.7354

Table 1-23. EFFECTS OF PRESSURE AND TEMPERATURE ON PROPERTIES OF GASES—METRIC UNITS (Continued)

OXYGEN

Temp, °K	Pressure, atmospheres						
	1	4	7	10	40	70	100
COMPRESSIBILITY FACTOR							
150	0.9928	0.9708	0.9477	0.9236			
250	0.9986	0.9945	0.9904	0.9863	0.9462	0.9108	0.8845
300	0.9994	0.9976	0.9958	0.9940	0.9773	0.9636	0.9541
500	1.0002	1.0009	1.0015	1.0022	1.0094	1.0173	1.0256
1,000	1.0003	1.0012	1.0020	1.0029	1.0117	1.0206	1.0296
1,500	1.0002	1.0009	1.0016	1.0022	1.0089	1.0156	1.0224
DENSITY, g cm^{-3}							
150	0.0026	0.0107	0.0192	0.0281	0.1775		
250	0.0016	0.0063	0.0110	0.0158	0.0659	0.1199	0.1764
300	0.0013	0.0052	0.0091	0.0131	0.0532	0.0944	0.1362
500	0.0008	0.0031	0.0055	0.0078	0.0309	0.0537	0.0760
1,000	0.0004	0.0016	0.0027	0.0039	0.0154	0.0267	0.0379
1,500	0.0003	0.0010	0.0018	0.0026	0.0103	0.0179	0.0254
SPECIFIC HEAT, c_p, cal g^{-1} K^{-1}							
150	0.2195	0.2264	0.2348	0.2454			
250	0.2187	0.2204	0.2222	0.2240	0.2440	0.2671	0.2925
300	0.2199	0.2210	0.2221	0.2233	0.2351	0.2472	0.2586
500	0.2323	0.2327	0.2330	0.2334	0.2368	0.2402	0.2434
1,000	0.2605	0.2606	0.2607	0.2607	0.2614	0.2621	0.2628
1,500	0.2731	0.2731	0.2731	0.2732	0.2734	0.2737	0.2740
ENTHALPY, cal g^{-1}							
150	32.349	31.757	31.134	30.471			
250	54.219	53.972	53.721	53.471	50.907	48.328	45.852
300	65.179	65.001	64.821	64.643	62.849	61.101	59.456
500	110.26	110.21	110.15	110.10	109.58	109.09	108.62
1,000	234.42	234.45	234.48	234.50	234.77	235.06	235.35
1,500	368.14	368.19	368.24	368.29	368.78	369.27	369.78
ENTROPY, cal g^{-1} K^{-1}							
150	1.3812	1.2925	1.2550	1.2298			
250	1.4929	1.4061	1.3706	1.3477	1.2539	1.2113	1.1814
300	1.5329	1.4463	1.4111	1.3884	1.2974	1.2579	1.2311
500	1.6477	1.5615	1.5266	1.5043	1.4167	1.3805	1.3570
1,000	1.8187	1.7326	1.6978	1.6757	1.5893	1.5543	1.5319
1,500	1.9270	1.8409	1.8061	1.7840	1.6978	1.6630	1.6407

*Condensed from: "Tables of the Thermal Properties of Gases", National Bureau of Standards Circular 564, U.S. Government Printing Office, November 1955.

REFERENCE

"ASHRAE Handbook of Fundamentals", American Society of Heating, Refrigerating and Air-Conditioning Engineers, 1967.

Table 1-24. GENERAL PROPERTIES OF REFRIGERANTS*

For properties of saturated and superheated refrigerants at usual operating pressures, see Tables 1-25 to 1-36.

	R–11	R–12	R–13	R–22	R–113	R–114	R–500	R–502	R–717
Chemical formula	CCl_3F	CCl_2F_2	$CClF_3$	$CHClF_2$	CCl_2F- $CClF_2$	$C_2Cl_2F_4$	‡	**	NH_3
Molecular weight	137.38	120.93	104.47	86.48	187.39	170.94	99.31	111.6	17.03
Boiling temperature at 14.7 psia, °F	74.9	−21.6	−114.6	−41.4	117.6	38.8	−28.3	−50.1	−28.0
Freezing temperature at 14.7 psia, °F	−168	−252	−294	−256	−31	−137	−254	—	−108
Critical temperature, °F	388.4	233.6	83.9	204.8	417.4	294.3	221.9	194	271.4
Critical pressure, psia	640	597	561	721.9	498.9	473	641.9	619	1 657
Critical pressure, MN/m²	4.41	4.12	3.87	4.98	3.44	3.26	4.43	4.27	11.4
Critical density, lb/cu ft	34.6	34.84	36.1	32.8	36.0	36.3	31.0	34.91	14.6
Critical density, kg/m³	554	558	578	525	577	581	496	559	234
Density of liquid, 86°F, lb/cu ft	91.39	80.67	81.05^{-22}	73.28	96.96	89.95	71.06	76.13	37.16
Density of liquid, 303.15 K, kg/m³	1 464	1 292	$1\ 298^{-22}$	1 174	1 553	1 441	1 138	1 219	595.2
Sp vol of sat vapor, 5°F, cu ft/lb	12.205	1.458	0.304	1.243	27.04	4.226	1.501	0.825	8.150
Sp vol of sat vapor, 258.15 K, m³/kg	0.761 9	0.091 02	0.018 98	0.077 60	1.688	0.263 8	0.093 7	0.051 50	0.508 8
Sp heat of liquid, 86°F, Btu/lb °F	0.21	0.235	0.247	0.305	0.218	0.246	0.290	0.305	1.14
Sp heat of liquid, 303.15 K, kJ/kg·K	0.878	0.983	1.03	1.28	0.912	1.03	1.21	1.28	4.77
Sp heat ratio (c_p/c_v); vapor at 86°F and 14.7 psia	1.13	1.139	1.17	1.18	1.12	1.09	1.14	1.135	1.29
Thermal conductivity									
Sat liquid, 5°F	0.058	0.052	0.06^{-95}	0.069	0.044	0.041		0.052	0.29
Sat liquid, 258.15 K	100	90	100^{-95}	120	76	71		90	500
Sat liquid, 86°F	0.049	0.040		0.050	0.037	0.033		0.037	0.29
Sat liquid, 303.15 K	85	69		86	64	57		64	500
Vapor at sat press, 5°F	0.003 4	0.004 7		0.005 1	0.003 5	0.004 7		0.005 4	0.012
Vapor at sat press, 258.15 K	5.9	8.1		8.8	6.0	8.1		9.3	21
Vapor at 14.7 psia, 86°F	0.004 5	0.005 9		0.006 5	0.004 5	0.006 2		0.006 9	0.014
Vapor at 0.101 3 MN/m², 303.15 K	7.8	10		11	7.8	11		12	24
Viscosity, N·s/m²									
Sat liquid, 5°F	0.630	0.335	$.037^{-95}$	0.298	1.28	0.614	0.292	0.334	0.250
Sat liquid, 258.15 K	0.000 630	0.000 335	$0.000\ 037^{-95}$	0.000 298	0.001 28	0.000 614	0.000 292	0.000 334	0.000 250
Sat liquid, 86°F	0.404	0.254		0.230	0.638	0.356	0.220	0.240	0.207
Sat liquid, 303.15 K	0.000 404	0.000 254		0.000 230	0.000 638	0.000 356	0.000 220	0.000 240	0.000 207
Vapor at sat press, 5°F	0.008 7	0.010 8		0.011 2	0.007 9	0.009 6		0.011 2	0.008 5
Vapor at sat press, 258.15 K	0.000 008 7	0.000 010 8		0.000 011 2	0.000 007 9	0.000 009 6		0.000 011 2	0.000 008 5
Vapor at 14.7 psia, 86°F	0.010 8	0.012 7		0.013 2	0.009 6	0.011 4		0.013 1	0.010 2
Vapor at 0.101 3 MN/m², 303.15 K	0.000 010 8	0.000 012 7		0.000 012 3	0.000 009 6	0.000 011 4		0.000 013 1	0.000 010 2
Relative dielectric strength of vapor at 73°F and 14.7 psia (nitrogen = 1)	3.1	2.4	1.4	1.3	3.9				0.82 (84°F)
Toxicity (Underwriters' Laboratories Classification)†	Group 5a	Group 6	Group 6+	Group 5a	Group 4½	Group 6	Group 5a	Group 5a	Group 2

†See explanation at end of table.
‡R–500 is azeotrope 73.8% (by wt) CCl_2F_2 and 26.2% (by wt) CH_3-CHF_2.
**R–502 is azeotrope $CHClF_2 = 48.8\%$ and $CClF_2CF_3 = 51.2\%$.

Table 1-24.   GENERAL PROPERTIES OF REFRIGERANTS *(Continued)*

Property	R–13B1	R–14	R–40, Methyl chloride	R–50, Methane	R–170, Ethane	R–290, Propane	R–600, n-Butane	R–744, Carbon dioxide
	Fluorocarbons							
Chemical formula	$CBrF_3$	CF_4	CH_3Cl	CH_4	C_2H_6	C_3H_8	C_4H_{10}	CO_2
Molecular weight	148.9	88.01	50.48	16.03	30.04	44.09	58.12	44.01
Boiling point at 14.7 psia, °F	−72.0	−198.4	−10.8	−258.9	−127.5	−44.2	31.3	−109.3 subl.
Freezing point at 14.7 psia, °F	−270	−299	−144	−297	−278	−309.8	−217	−69.9[a]
Critical temperature, °F	152.6	−50	289.4	−115.8	90.1	206	306	87.8
Critical pressure, psia	575	543	968.7	673.1	708.3	617.4	550.1	1 057.4
Critical pressure, MN/m^2	3.96	3.74	6.68	4.64	4.88	4.26	3.79	7.29
Critical density, lb/cu ft	46.5	39	23.3	10.1	13.2	13.7	14.2	28.6
Critical density, kg/m^3	745	625	373	162	211	219	227	458
Density of liquid, 86°F, lb/cu ft	93.58	82.2[b]	56.24		16.57	36.2	35.62	
Density of liquid, 303.15 K, kg/m^3	1 499	1 317[b]	900.9		265.4	579.9	570.6	
Sp vol of sat vapor, 5°F, cu ft/lb	0.379 6		4.471		0.531 3	2.509	9.98	0.266 1
Sp vol of sat vapor, 258.15 K, m^3/kg	0.023 70		0.279 1		0.033 17	0.156 6	0.623 0	0.016 61
Toxicity (Underwriters' Laboratories Classification)[d]	Group 6	Group 6[c]	Group 4	Group 5[a]	Group 5[a]	Group 5[a]	Group 5	Group 5

[a] At 76.4 psia.
[b] At −112°F (317.59 K).
[c] Unofficial.
[d] The Underwriters' Laboratories Classification of toxicity is as follows:
    Group 1: Lethal concentration 0.5 to 1.0 percent for durations of 5 minutes.
    Group 2: Lethal concentration 0.5 to 1.0 percent for durations of 30 minutes.
    Group 3: Lethal concentration 2.0 to 2.5 percent for durations of 1 hour.
    Group 4: Lethal concentration 2.0 to 2.5 percent for durations of 2 hours.
    Group 5a: Less toxic than group 4, more toxic than group 6.
    Group 5b: Available data would classify these as 5a or 6.
    Group 6: Concentrations up to about 20 percent for 2 hours do not appear to produce injury.

*Based largely on: "ASHRAE Handbook of Fundamentals", American Society of Heating, Refrigerating and Air-Conditioning Engineers, 1972.

REFERENCE
"Properties of Commonly-Used Refrigerants", Air-Conditioning Refrigeration Institute, 1967.

Table 1-25. REFRIGERANT 11—LIQUID AND SATURATED VAPOR*

For conversions to SI units, see Table 1-41.

Temp, °F	Pressure, psia	Volume, ft³/lb Vapor, v_g	Density, lb/ft³ Liquid, $1/v_f$	Enthalpy, Btu/lb† Liquid, h_f	Latent, h_{fg}	Vapor, h_g	Entropy, Btu/lb °R† Liquid, s_f	Vapor, s_g
−150	0.003272	7393.8	109.18	−22.345	97.280	74.935	−0.061775	0.25237
−125	.01625	1608.3	107.43	−17.217	94.779	77.562	−.045851	.23735
−100	.06241	449.88	105.65	−12.128	92.420	80.293	−.031184	.22577
−90	.10075	268.35	104.93	−10.101	91.512	81.412	−.025625	.22193
−80	.15791	187.57	104.21	−8.077	90.622	82.544	−.020224	.21846
−70	.24091	126.12	103.49	−6.056	89.746	83.690	−.014971	.21534
−60	.35855	86.857	102.76	−4.037	88.885	84.847	−.009855	.21254
−50	.52163	61.144	102.02	−2.019	88.034	86.015	−.004868	.21002
−45	.62419	51.695	101.65	−1.010	87.612	86.603	−.002419	.20886
−40	.74317	43.917	101.28	0.000	87.193	87.193	.000000	.20776
−35	.88059	37.483	100.91	1.010	86.775	87.784	.002392	.20673
−30	1.0386	32.133	100.54	2.020	86.358	88.378	.004756	.20574
−25	1.2195	27.664	100.16	3.031	85.942	88.974	.007095	.20481
−20	1.4258	23.914	99.788	4.043	85.528	89.570	.009408	.20393
−15	1.6602	20.754	99.410	5.055	85.114	90.169	.011696	.20310
−10	1.9254	18.079	99.031	6.068	84.700	90.768	.013961	.20232
−5	2.2245	15.805	98.650	7.082	84.287	91.369	.016202	.20158
0	2.5607	13.866	98.268	8.098	83.873	91.970	.018421	.20088
5	2.9373	12.205	97.884	9.114	83.459	92.572	.020619	.20023
10	3.3578	10.778	97.498	10.131	83.044	93.175	.022795	.19961
15	3.8259	9.5465	97.111	11.150	82.628	93.778	.024951	.19903
20	4.3456	8.4810	96.722	12.170	82.212	94.382	.027086	.19848
25	4.9207	7.5560	96.331	13.191	81.794	94.985	.029202	.19796
30	5.5556	6.7503	95.938	14.214	81.374	95.588	.031299	.19748
35	6.2546	6.0465	95.544	15.238	80.953	96.192	.033377	.19703
40	7.0223	5.4298	95.147	16.264	80.530	96.794	.035437	.19660
45	7.8633	4.8879	94.748	17.291	80.105	97.396	.037480	.19621
50	8.7825	4.4105	94.347	18.320	79.678	97.998	.039505	.19584
55	9.7850	3.9887	93.944	19.350	79.248	98.598	.041512	.19549
60	10.876	3.6151	93.539	20.382	78.816	99.198	.043504	.19517
65	12.061	3.2834	93.132	21.416	78.380	99.796	.045478	.19487
70	13.345	2.9882	92.722	22.451	77.942	100.393	.047437	.19459
75	14.733	2.7248	92.310	23.488	77.501	100.989	.049381	.19433
80	16.233	2.4893	91.895	24.527	77.056	101.583	.051309	.19409
85	17.848	2.2783	91.477	25.567	76.608	102.175	.053222	.19387
90	19.587	2.0888	91.057	26.610	76.156	102.765	.055121	.19367
95	21.454	1.9183	90.635	27.654	75.700	103.354	.057005	.19348
100	23.456	1.7645	90.209	28.700	75.240	103.940	.058875	.19331
110	27.890	1.4999	89.349	30.798	74.307	105.105	.062575	.19301
120	32.943	1.2824	88.476	32.905	73.355	106.260	.066223	.19277
130	38.668	1.1024	87.589	35.021	72.383	107.404	.069822	.19257
140	45.123	0.95232	86.689	37.147	71.388	108.535	.073374	.19242
150	52.364	0.82650	85.772	39.284	70.368	109.653	.076883	.19230
160	60.451	0.70237	84.839	41.433	69.322	110.755	.080351	.19222
170	69.447	0.63035	83.887	43.596	68.245	111.841	.083781	.19216
180	79.414	0.55357	82.916	45.774	67.135	112.909	.087178	.19213
190	90.416	0.48776	81.924	47.969	65.989	113.958	.090544	.19212
200	105.52	0.43108	80.908	50.183	64.802	114.985	.093884	.19212
300	300.21	0.13860	68.696	74.205	49.183	123.387	.12703	.19177
388.4	639.50	0.028927	34.570	112.080	0.000	112.080	.17219	.17219

†Based on zero for the saturated liquid at −40°F.
*From published data of E.I. du Pont de Nemours & Co., Inc., 1965. Used by permission.

REFERENCES
"ASHRAE Handbook of Fundamentals", American Society of Heating, Refrigerating and Air-Conditioning Engineers, 1967.
"Properties of Commonly-Used Refrigerants", Air-Conditioning Refrigeration Institute, 1967.

Table 1-26. SUPERHEATED REFRIGERANT 11*

For conversions to SI units, see Table 1-41.

SYMBOLS: v = volume, ft³/lb h = enthalpy, Btu/lb s = entropy, Btu/lb °R

Temp, °F	1.0 psia			5.0 psia			10 psia			20 psia		
	v	h	s	v	h	s	v	h	s	v	h	s
0	35.75	92.1	.2147									
50	39.69	98.6	.2281	7.84	98.3	.2044						
100	43.63	105.4	.2408	8.65	105.1	.2172	4.275	104.8	.2068	2.087	104.2	.1959
120	45.20	108.2	.2457	8.97	108.0	.2221	4.440	107.7	.2118	2.174	107.1	.2010
140	46.77	111.0	.2505	9.29	110.8	.2270	4.604	110.5	.2166	2.260	110.0	.2060
160	48.33	113.9	.2552	9.61	113.7	.2317	4.766	113.4	.2214	2.345	112.9	.2108
180	49.90	116.7	.2598	9.93	116.6	.2363	4.928	116.4	.2260	2.429	115.9	.2155
200	51.47	119.7	.2643	10.24	119.5	.2408	5.090	119.3	.2306	2.512	118.9	.2201
220	53.04	122.6	.2687	10.56	122.5	.2453	5.250	122.3	.2350	2.595	121.9	.2246
240	54.60	125.6	.2730	10.88	125.5	.2496	5.411	125.3	.2394	2.678	124.9	.2290
260	56.17	128.6	.2773	11.19	128.5	.2539	5.571	128.3	.2437	2.759	128.0	.2333
280				11.51	131.6	.2580	5.730	131.4	.2479	2.841	131.1	.2376
300				11.82	134.6	.2622	5.889	134.5	.2520	2.922	134.2	.2417
350							6.286	142.3	.2620	3.124	142.1	.2517
400										3.324	150.1	.2613

Temp, °F	30 psia			40 psia			60 psia			80 psia		
	v	h	s	v	h	s	v	h	s	v	h	s
140	1.478	109.4	.1994	1.086	108.8	.1945						
160	1.537	112.4	.2043	1.133	111.9	.1995	.7265	110.8	.1924			
180	1.595	115.4	.2091	1.178	115.0	.2044	.7595	114.0	.1974			
200	1.653	118.5	.2138	1.223	118.0	.2091	.7915	117.1	.2023	.5750	116.2	.1970
220	1.710	121.5	.2183	1.267	121.1	.2138	.8228	120.3	.2070	.6001	119.4	.2019
240	1.766	124.6	.2228	1.310	124.2	.2183	.8535	123.5	.2116	.6246	122.7	.2066
260	1.822	127.7	.2271	1.353	127.3	.2226	.8837	126.6	.2161	.6485	125.9	.2112
280	1.878	130.8	.2314	1.396	130.5	.2269	.9134	129.8	.2205	.6718	129.2	.2156
300	1.933	133.9	.2356	1.438	133.6	.2312	.9427	133.0	.2247	.6948	132.4	.2200
320	1.988	137.1	.2397	1.480	136.8	.2353	.9717	136.2	.2289	.7174	135.7	.2242
340	2.042	140.2	.2437	1.521	140.0	.2393	1.00	139.5	.2330	.7397	138.9	.2283
360	2.097	143.4	.2476	1.563	143.2	.2433	1.03	142.7	.2370	.7617	142.2	.2324
380	2.151	146.6	.2515	1.604	146.4	.2471	1.06	145.9	.2409	.7835	145.5	.2363
400	2.205	149.8	.2553	1.645	149.6	.2509	1.09	149.2	.2447	.8051	148.8	.2402
450							1.16	157.4	.2540	.8583	157.0	.2496

Temp, °F	100 psia			120 psia			160 psia			200 psia		
	v	h	s	v	h	s	v	h	s	v	h	s
200	.4440	115.1	.1926									
220	.4658	118.5	.1977	.3754	117.5	.1939						
240	.4867	121.8	.2025	.3941	120.9	.1989	.2769	119.0	.1927			
260	.5069	125.2	.2072	.4120	124.4	.2037	.2923	122.6	.1978	.2188	120.7	.1925
280	.5265	128.5	.2117	.4293	127.7	.2084	.3069	126.2	.2027	.2322	124.5	.1977
300	.5457	131.8	.2161	.4461	131.1	.2129	.3208	129.7	.2073	.2447	128.2	.2026
320	.5645	135.1	.2204	.4624	134.5	.2172	.3342	133.2	.2119	.2566	131.8	.2073
340	.5830	138.4	.2246	.4784	137.8	.2215	.3472	136.6	.2162	.2679	135.4	.2119
360	.6012	141.7	.2287	.4941	141.2	.2256	.3598	140.1	.2205	.2788	138.9	.2162
380	.6192	145.0	.2327	.5095	144.5	.2297	.3721	143.5	.2246	.2893	142.5	.2205
400	.6369	148.3	.2366	.5247	147.9	.2336	.3842	146.9	.2287	.2996	145.9	.2246
450	.6805	156.7	.2460	.5619	156.3	.2431	.4135	155.5	.2383	.3243	154.6	.2344
500	.7233	165.0	.2550	.5982	164.7	.2521	.4418	164.0	.2474	.3479	163.3	.2437

*From published data of E.I. du Pont de Nemours & Co., Inc., 1965. Used by permission.
See Table 1-25 for References.

Table 1-27. REFRIGERANT 12—LIQUID AND SATURATED VAPOR*

For conversions to SI units, see Table 1-41.

Temp, °F	Pressure, psia	Volume, ft³/lb Vapor, v_g	Density, lb/ft³ Liquid, $1/v_f$	Enthalpy, Btu/lb† Liquid, h_f	Latent, h_{fg}	Vapor, h_g	Entropy, Btu/lb °R† Liquid, s_f	Vapor, s_g
−150	.15359	178.65	104.36	−22.697	83.534	60.837	−.062619	.20711
−125	.51641	57.283	102.29	−17.587	81.096	63.509	−.046754	.19554
−100	1.4280	22.164	100.15	−12.466	78.714	66.248	−.032005	.18683
−90	2.0509	15.821	99.274	−10.409	77.764	67.355	−.026367	.18398
−80	2.8807	11.533	98.382	−8.3451	76.812	68.467	−.020862	.18143
−70	3.9651	8.5687	97.475	−6.2730	75.853	69.580	−.015481	.17916
−60	5.3575	6.4774	96.553	−4.1919	74.885	70.693	−.010214	.17714
−50	7.1168	4.9742	95.616	−2.1011	73.906	71.805	−.005056	.17533
−45	8.1540	4.3828	95.141	−1.0519	73.411	72.359	−.002516	.17451
−40	9.3076	3.8750	94.661	0.0000	72.913	72.913	.000000	.17373
−35	10.586	3.4373	94.178	1.0546	72.409	73.464	.002492	.17299
−30	11.999	3.0585	93.690	2.1120	71.903	74.015	.004961	.17229
−25	13.556	2.7295	93.197	3.1724	71.391	74.563	.007407	.17164
−20	15.267	2.4429	92.699	4.2357	70.784	75.110	.009831	.17102
−15	17.141	2.1924	92.197	5.3020	70.352	75.654	.012234	.17043
−10	19.189	1.9727	91.689	6.3716	69.824	76.196	.014617	.16989
−5	21.422	1.7794	91.177	7.4444	69.291	76.735	.016979	.16937
0	23.849	1.6089	90.659	8.5207	68.750	77.271	.019323	.16888
5	26.483	1.4580	90.135	9.6005	68.204	77.805	.021647	.16842
10	29.335	1.3241	89.606	10.684	67.651	78.335	.023954	.16798
15	32.415	1.2050	89.070	11.771	67.090	78.861	.026243	.16758
20	35.736	1.0988	88.529	12.863	66.522	79.385	.028515	.16719
25	39.310	1.0039	87.981	13.958	65.946	79.904	.030772	.16683
30	43.148	.91880	87.426	15.058	65.361	80.419	.033013	.16648
35	47.263	.84237	86.865	16.163	64.767	80.930	.035240	.16616
40	51.667	.77357	86.296	17.273	64.163	81.436	.037453	.16586
45	56.373	.71149	85.720	18.387	63.550	81.937	.039652	.16557
50	61.394	.65537	85.136	19.507	62.926	82.433	.041839	.16530
55	66.743	.60453	84.544	20.634	62.290	82.924	.044015	.16504
60	72.433	.55839	83.944	21.766	61.643	83.409	.046180	.16479
65	78.477	.51642	83.335	22.905	60.982	83.887	.048336	.16456
70	84.888	.47818	82.717	24.050	60.309	84.359	.050482	.16434
75	91.682	.44327	82.089	25.204	59.621	84.825	.052620	.16412
80	98.870	.41135	81.450	26.365	58.917	85.282	.054751	.16392
85	106.47	.38212	80.802	27.534	58.198	85.732	.056877	.16372
90	114.49	.35529	80.142	28.713	57.461	86.174	.058997	.16353
95	122.95	.33063	79.470	29.901	56.705	86.606	.061113	.16334
100	131.86	.30794	78.785	31.100	55.929	87.029	.063227	.16315
110	151.11	.26769	77.376	33.531	54.313	87.844	.067451	.16279
120	172.35	.23326	75.906	36.013	52.597	88.610	.071680	.16241
130	195.71	.20364	74.367	38.553	50.768	89.321	.075927	.16202
140	221.32	.17799	72.748	41.162	48.805	89.967	.080205	.16159
150	249.31	.15564	71.035	43.850	46.684	90.534	.084531	.16110
160	279.82	.13604	69.209	46.633	44.373	91.006	.088927	.16053
170	313.00	.11873	67.244	49.529	41.830	91.359	.903418	.15985
180	349.00	.10330	65.102	52.562	38.999	91.561	.098039	.15900
190	387.98	.089418	62.728	55.769	35.792	91.561	.10284	.15793
200	430.09	.076728	60.026	59.203	32.075	91.278	.10789	.15651
·225	550.26	.046900	49.868	69.763	17.888	87.651	.12298	.14911
233.6	596.9	.02870	34.84	78.86	0	78.86	.1359	.1359

†Based on zero for the saturated liquid at −40°F.
*From published data of E.I. du Pont de Nemours & Co., Inc., 1955 and 1956. Used by permission.
  See Table 1-25 for References.

Table 1-28. SUPERHEATED REFRIGERANT 12*

For conversions to SI units, see Table 1-41.

SYMBOLS: v = volume, ft³/lb h = enthalpy, Btu/lb s = entropy, Btu/lb °R

Temp, °F	3.0 psia			10 psia			20 psia			40 psia		
	v	h	s	v	h	s	v	h	s	v	h	s
−40	12.29	73.46	.1933									
−20	12.90	76.06	.1993	3.791	75.5	.1787						
0	13.50	78.71	.2052	3.981	78.2	.1847	1.939	77.6	.1722			
20	14.10	81.43	.2110	4.169	81.0	.1906	2.039	80.4	.1783			
40	14.70	84.19	.2167	4.356	83.8	.1964	2.137	83.3	.1842	1.026	82.1	.1711
60	15.30	87.02	.2222	4.541	86.7	.2020	2.234	86.2	.1899	1.079	85.2	.1771
80	15.90	89.89	.2276	4.725	89.6	.2075	2.330	89.2	.1955	1.131	88.3	.1829
100	16.50	92.81	.2329	4.908	92.5	.2128	2.424	92.2	.2010	1.181	91.4	.1885
120	17.09	95.79	.2382	5.090	95.5	.2181	2.518	95.2	.2063	1.231	94.5	.1940
140	17.69	98.81	.2433	5.272	98.6	.2233	2.611	98.3	.2115	1.280	97.6	.1993
160	18.28	101.9	.2483	5.453	101.7	.2283	2.704	101.4	.2166	1.328	100.8	.2045
180	18.88	105.0	.2532	5.634	104.8	.2333	2.796	104.5	.2216	1.376	104.0	.2096
200	19.47	108.1	.2581	5.815	108.0	.2381	2.887	107.7	.2265	1.424	107.2	.2146
250				6.264	116.0	.2499	3.116	115.8	.2384	1.541	115.4	.2266
300										1.657	123.8	.2380

Temp, °F	60 psia			80 psia			100 psia			120 psia		
	v	h	s	v	h	s	v	h	s	v	h	s
60	.6921	84.1	.1689									
80	.7296	87.3	.1750	.5280	86.3	.1689						
100	.7659	90.5	.1808	.5573	89.6	.1749	.4314	88.7	.1700	.3466	87.7	.1656
120	.8011	93.7	.1864	.5856	92.9	.1807	.4556	92.1	.1760	.3684	91.2	.1718
140	.8355	96.9	.1919	.6129	96.2	.1863	.4788	95.5	.1817	.3890	94.7	.1778
160	.8693	100.2	.1972	.6394	99.5	.1917	.5012	98.9	.1873	.4087	98.2	.1835
180	.9025	103.4	.2023	.6654	102.9	.1970	.5229	102.3	.1926	.4277	101.6	.1889
200	.9353	106.7	.2074	.6910	106.2	.2021	.5441	105.6	.1978	.4461	105.1	.1942
220	.9678	110.0	.2123	.7161	109.5	.2071	.5649	109.0	.2029	.4640	108.5	.1993
240	.9999	113.3	.2171	.7409	112.9	.2119	.5854	112.4	.2078	.4816	111.9	.2043
260	1.032	116.7	.2218	.7654	116.3	.2167	.6055	115.8	.2126	.4989	115.4	.2092
280	1.063	120.0	.2264	.7898	119.7	.2214	.6255	119.3	.2173	.5159	118.9	.2139
300	1.095	123.4	.2310	.8139	123.1	.2259	.6452	122.7	.2219	.5327	122.3	.2186
350				.8735	131.7	.2370	.6938	131.4	.2330	.5739	131.1	.2297

Temp, °F	160 psia			200 psia			300 psia			400 psia		
	v	h	s	v	h	s	v	h	s	v	h	s
120	.2576	89.3	.1645									
140	.2756	93.1	.1709	.2058	91.1	.1648						
160	.2922	96.7	.1770	.2212	95.1	.1713						
180	.3080	100.3	.1827	.2354	98.9	.1774	.1348	94.6	.1654			
200	.3230	103.9	.1882	.2486	102.7	.1831	.1470	99.0	.1722	.0910	93.7	.1609
220	.3375	107.5	.1935	.2612	106.3	.1886	.1577	103.1	.1784	.1032	99.0	.1689
240	.3516	111.0	.1986	.2732	110.0	.1939	.1676	107.1	.1842	.1130	103.7	.1757
260	.3653	114.5	.2036	.2849	113.6	.1990	.1769	111.0	.1897	.1216	108.1	.1818
280	.3787	118.0	.2084	.2962	117.2	.2039	.1856	114.9	.1950	.1295	112.3	.1876
300	.3919	121.6	.2131	.3073	120.8	.2087	.1940	118.7	.2000	.1368	116.3	.1930
350	.4241	130.4	.2244	.3341	129.8	.2202	.2139	128.0	.2120	.1535	126.2	.2055
400	.4554	139.4	.2352	.3599	138.8	.2310	.2326	137.3	.2231	.1689	135.8	.2170
450							.2506	146.7	.2336	.1833	145.3	.2278

*From published data of E.I. du Pont de Nemours & Co., Inc., 1955 and 1956. Used by permission.
See Table 1-25 for References.

Table 1-29. REFRIGERANTS 13 AND 113

For conversions to SI units, see Table 1-41.

REFRIGERANT 13, SATURATED†

Temp, °F	Pressure, psia	Volume, ft³/lb Vapor, v_g	Density, lb/ft³ Liquid, $1/v_f$	Enthalpy, Btu/lb‡ Liquid, h_f	Latent, h_{fg}	Vapor, h_g	Entropy, Btu/lb °R‡ Liquid, s_f	Vapor, s_g
−200	.4329	61.33	105.6	−34.551	73.096	38.545	−.10081	.18066
−180	1.238	22.99	103.2	−30.298	70.970	40.672	−.08575	.16800
−160	3.104	9.750	100.8	−26.083	68.808	42.725	−.07213	.15747
−140	6.455	4.950	98.33	−21.902	66.696	44.794	−.05844	.15019
−120	12.48	2.681	95.69	−17.671	64.473	46.802	−.04590	.14390
−100	22.23	1.5642	93.02	−13.387	62.138	48.751	−.03286	.13889
−80	36.98	.9689	90.17	−9.052	59.672	50.620	−.02230	.13486
−60	58.19	.6289	87.26	−4.604	56.993	52.389	−.01106	.13153
−40	87.43	.4234	84.10	0.000	54.023	54.023	.00000	.12872
−20	126.4	.2930	80.71	4.809	50.668	55.477	.01096	.12620
0	176.8	.2066	76.98	10.052	46.638	56.690	.02234	.12380
20	240.4	.14732	72.73	15.443	42.100	57.543	.03351	.12128
40	319.6	.10455	67.70	21.370	36.450	57.820	.04516	.11811
60	417.0	.07189	61.09	28.310	28.677	56.987	.05824	.11342
80	535.5	.04131	48.85	38.527	13.565	52.092	.07672	.10185

REFRIGERANT 13, SUPERHEATED†

Temp, °F	2.0 psia v	h	s	10 psia v	h	s	100 psia v	h	s	300 psia v	h	s
−100	18.38	50.19	.1876	3.602	49.64	.1560						
0	23.56	63.45	.2200	4.670	63.20	.1889	.4180	60.11	.1403			
100	28.71	78.34	.2492	5.719	78.19	.2184	.5425	76.41	.1724	.1562	71.64	.1453
200	33.86	.94.76	.2763 *	6.755	94.64	.2453	.6559	93.41	.2003	.2040	90.46	.1761
300	38.99	112.46	.3011	7.787	112.37	.2705	.7658	111.44	.2258	.2455	109.27	.2027
400	44.14	131.2	.3244	8.819	131.2	.2937	.8732	130.4	.2492	.2847	128.7	.2268

REFRIGERANT 113, SATURATED**

Temp, °F	Pressure, psia	Volume, ft³/lb Vapor, v_g	Density, lb/ft³ Liquid, $1/v_f$	Enthalpy, Btu/lb‡ Liquid, h_f	Latent, h_{fg}	Vapor, h_g	Entropy, Btu/lb °R‡ Liquid, s_f	Vapor, s_g
−20	.4288	58.61	104.96	3.96	72.09	76.05	.0092	.1732
0	.8377	31.31	103.56	7.98	70.92	78.89	.0182	.1725
20	1.534	17.81	102.10	12.03	69.72	81.75	.0268	.1722
40	2.655	10.68	100.60	16.16	68.50	84.65	.0352	.1723
60	4.374	6.713	99.05	20.35	67.22	87.57	.0434	.1728
80	6.902	4.392	97.45	24.63	65.88	90.51	.0515	.1736
100	10.48	2.976	95.79	28.99	64.46	93.45	.0594	.1746
120	15.40	2.078	94.09	33.48	62.93	96.41	.0673	.1758
140	21.93	1.491	92.33	38.05	61.31	99.36	.0750	.1773
160	30.44	1.094	90.53	42.74	59.55	102.29	.0827	.1788
180	41.22	.8193	88.67	47.53	57.66	105.19	.0903	.1804
200	54.66	.6241	86.76	52.45	55.62	108.07	.0978	.1821
300	175.6	.190	75.63	75.82	44.33	120.15	.1297	.1881
400	434.3	.049	55.02	105.78	16.84	122.62	.1659	.1855

REFRIGERANT 113, SUPERHEATED**

Temp, °F	1.0 psia v	h	s	2.0 psia v	h	s	10 psia v	h	s	50 psia v	h	s
0												
50	29.10	86.20	.1857	14.50	86.16	.1783						
100	31.97	93.81	.2000	15.95	93.77	.1926	3.124	93.48	.1751			
200	37.73	109.91	.2264	18.83	109.88	.2190	3.716	109.63	.2017	.6899	108.3	.1833
300	43.46	127.16	.2507	21.71	127.15	.2433	4.302	126.94	.2261	.8207	125.8	.2081
400							4.887	145.4	.2489	.9480	144.5	.2311

‡Based on zero for the saturated liquid at −40°F.

†Reprinted from *Ind. Eng. Chem.*, 44:188, January, 1952. Copyright 1952 by the American Chemical Society. Reprinted by permission of the copyright owner.

**From published data of E.I. du Pont de Nemours & Co., Inc., 1938. Used by permission.

See Table 1-25 for References.

Table 1-30. REFRIGERANTS 13B1, 14, AND 23

For conversions to SI units, see Table 1-41.

Temp, °F	Pressure, psia	Volume, ft³/lb	Density, lb/ft³	Enthalpy, Btu/lb†		Entropy, Btu/lb °R†	
		Vapor, v_g	Liquid, $1/v_f$	Liquid, h_f	Vapor, h_g	Liquid, s_f	Vapor, s_g

R–13B1 (BROMOTRIFLUOROMETHANE), SATURATED‡

Temp, °F	Pressure, psia	Vapor, v_g	Liquid, $1/v_f$	Liquid, h_f	Vapor, h_g	Liquid, s_f	Vapor, s_g
−160	0.6380	33.748	136.73	−18.51	38.60	−.0516	.1389
−140	1.551	14.769	134.07	−15.58	40.22	−.0422	.1324
−120	3.344	7.2407	131.33	−12.61	41.87	−.0332	.1272
−100	6.542	3.8894	128.51	−9.58	43.54	−.0245	.1232
−80	11.81	2.2495	125.58	−6.48	45.20	−.0161	.1200
−60	19.95	1.3816	122.53	−3.29	46.86	−.0080	.1175
−40	31.88	.8911	119.35	0.00	48.47	.0000	.1155
−20	48.58	.5979	116.01	3.38	50.03	.0078	.1139
0	71.16	.4143	112.48	6.87	51.51	.0155	.1126
20	100.8	.2944	108.71	10.47	52.88	.0230	.1115
40	138.6	.2132	104.65	14.19	54.11	.0305	.1104
60	186.0	.1565	100.20	18.08	55.15	.0380	.1093
80	244.4	.1156	95.22	22.15	55.92	.0454	.1080
100	315.2	.0852	89.42	26.52	56.32	.0531	.1064
150	559.2	.0324	60.71	41.77	52.51	.0781	.0957

R–14 (CARBON TETRAFLUORIDE), SATURATED**

Temp, °F	Pressure, psia	Vapor, v_g	Liquid, $1/v_f$	Liquid, h_f	Vapor, h_g	Liquid, s_f	Vapor, s_g
−220	5.7264	5.0230	105.996	44.112	105.150	.3638	.6185
−200	13.748	2.2311	102.268	48.394	107.083	.3809	.6069
−180	28.690	1.1232	98.352	53.007	108.920	.3979	.5978
−160	53.665	.6209	94.135	57.952	110.592	.4148	.5905
−140	92.070	.3674	89.549	63.224	112.014	.4316	.5842
−120	147.49	.2280	84.449	68.801	113.075	.4482	.5785
−100	223.81	.1451	78.569	74.715	113.589	.4646	.5727
−80	325.74	.0920	71.296	81.154	113.163	.4813	.5656
−60	460.61	.0534	60.307	89.236	110.394	.5011	.5540

R–23 (TRIFLUOROMETHANE), SATURATED§

Temp, °F	Pressure, psia	Vapor, v_g	Liquid, $1/v_f$	Liquid, h_f	Vapor, h_g	Liquid, s_f	Vapor, s_g
−200	0.344	115.5	97.68	−118.84	0.03	.5539	1.0114
−180	1.069	39.97	96.22	−112.02	2.44	.5793	.9886
−160	2.776	16.43	94.42	−106.41	4.92	.5986	.9701
−140	6.284	7.687	92.46	−100.96	7.38	.6162	.9551
−120	12.76	3.980	90.32	−95.53	9.78	.6327	.9428
−100	23.72	2.232	87.98	−89.94	12.07	.6487	.9323
−80	41.00	1.334	85.42	−83.96	14.21	.6648	.9233
−60	66.70	.8392	82.59	−77.59	16.14	.6809	.9154
−40	103.1	.5498	79.47	−70.77	17.79	.6974	.9084
−20	152.5	.3715	75.98	−63.55	19.09	.7139	.9019
0	217.7	.2563	72.03	−55.90	19.91	.7305	.8954
20	301.7	.1782	67.45	−47.66	20.03	.7475	.8886
40	408.3	.1220	61.93	−38.22	18.93	.7661	.8805
60	543.2	.0750	54.54	−24.97	14.30	.7910	.8666

†R–13B1: Based on zero for the saturated liquid at −40°F.
R–14: Based on zero for the saturated solid at 0°R.
R–23: Based on zero at −200°F for the ideal gas state.
‡From published data of E.I. du Pont de Nemours & Co., Inc., 1963. Used by permission.
**From published data, N.C.S. Chari, Ph.D. Dissertation, University of Michigan, 1960. Copyright by E.I. du Pont de Nemours & Co., Inc., 1961. Reprinted by permission.
§From unpublished data of General Chemical Division, Allied Chemical Corp., 1963. Used by permission.
See Table 1-25 for References.

Table 1-31. REFRIGERANT 22—LIQUID AND SATURATED VAPOR*

For conversions to SI units, see Table 1-41.

Temp, °F	Pressure, psia	Volume, ft³/lb	Density, lb/ft³	Enthalpy, Btu/lb†			Entropy, Btu/lb °R†	
		Vapor, v_g	Liquid, $1/v_f$	Liquid, h_f	Latent, h_{fg}	Vapor, h_g	Liquid, s_f	Vapor, s_g
−150	.27163	141.23	98.236	−25.974	113.495	87.521	−.07147	.29501
−125	.88551	46.692	96.035	−20.326	110.760	90.433	−.05394	.27700
−100	2.3983	18.433	93.770	−14.564	107.935	93.371	−.03734	.26274
−90	3.4229	13.235	92.843	−12.216	106.759	95.544	−.03091	.25787
−80	4.7822	9.6949	91.905	−9.838	105.548	95.710	−.02457	.25342
−70	6.5522	7.2318	90.952	−7.429	104.297	96.868	−.01832	.24932
−60	8.8180	5.4844	89.986	−4.987	103.001	98.014	−.01214	.24556
−50	11.674	4.2224	89.004	−2.511	101.656	99.144	−.00604	.24209
−45	13.354	3.7243	88.507	−1.260	100.963	99.703	−.00301	.24046
−40	15.222	3.2957	88.006	0.000	100.257	100.257	.00000	.23888
−35	17.290	2.9256	87.501	1.269	99.536	100.805	.00300	.23737
−30	19.573	2.6049	86.991	2.547	98.801	101.348	.00598	.23591
−25	22.086	2.3260	86.476	3.834	98.051	101.885	.00894	.23451
−20	24.845	2.0826	85.956	5.131	97.285	102.415	.01189	.23315
−15	27.865	1.8695	85.431	6.436	96.502	102.939	.01483	.23184
−10	31.162	1.6825	84.901	7.751	95.704	103.455	.01776	.23058
−5	34.754	1.5177	84.366	9.075	94.889	103.964	.02067	.22936
0	38.657	1.3723	83.825	10.409	94.056	104.465	.02357	.22817
5	42.888	1.2434	83.277	11.752	93.206	104.958	.02645	.22703
10	47.464	1.1290	82.724	13.104	92.338	105.442	.02932	.22592
15	52.405	1.0272	82.164	14.466	91.451	105.917	.03218	.22484
20	57.727	.93631	81.597	15.837	90.545	106.383	.03503	.22379
25	63.450	.85500	81.023	17.219	89.620	106.839	.03787	.22277
30	69.591	.78208	80.441	18.609	88.674	107.284	.04070	.22178
35	76.170	.71655	79.852	20.010	87.708	107.719	.04351	.22081
40	83.206	.65753	79.255	21.422	86.720	108.142	.04632	.21986
45	90.719	.40625	78.648	22.843	85.710	108.553	.04911	.21894
50	98.727	.55606	78.033	24.275	84.678	108.953	.05190	.21803
55	107.25	.51238	77.408	25.718	83.621	109.339	.05468	.21714
60	116.31	.47272	76.773	27.172	82.540	109.712	.05745	.21627
65	125.93	.43663	76.126	28.638	81.432	110.070	.06021	.21541
70	136.12	.40373	75.469	30.116	80.298	110.414	.06296	.21456
75	146.91	.37369	74.799	31.606	79.135	110.741	.06571	.21372
80	158.33	.34621	74.116	33.109	77.943	111.052	.06846	.21288
85	170.38	.32101	73.420	34.626	76.719	111.345	.07120	.21205
90	183.09	.29789	72.708	36.158	75.461	111.619	.07394	.21122
95	196.50	.27662	71.980	37.704	74.168	111.873	.07668	.21039
100	210.60	.25702	71.236	39.267	72.838	112.105	.07942	.20956
110	241.04	.22222	69.689	42.446	70.052	112.498	.08491	.20787
120	274.60	.19238	68.054	45.705	67.077	112.782	.09042	.20613
130	311.50	.166661	66.312	49.059	63.877	112.936	.09598	.20431
140	351.94	.14418	64.440	52.528	60.403	112.931	.10163	.20235
150	396.19	.12448	62.402	56.143	56.585	112.728	.10739	.20020
160	444.53	.10701	60.145	59.948	52.316	112.263	.11334	.19776
170	497.26	.091279	57.581	64.019	47.419.	111.438	.11959	.19490
180	554.78	.076790	54.549	68.498	41.570	110.068	.12635	.19133
190	617.59	.062837	50.677	73.711	34.023	107.734	.13409	.18646
200	686.36	.0347438	44.571	80.862	21.990	102.853	.14460	.17794
204.81	721.91	.030525	32.760	91.329	0.000	91.329	.16016	.16016

†Based on zero for the saturated liquid at −40°F.
*From published data of E.I. du Pont de Nemours & Co., Inc., 1964. Used by permission.
  See Table 1-25 for References.

Table 1-32. SUPERHEATED REFRIGERANT 22*

For conversions to SI units, see Table 1-41.

SYMBOLS: v = volume, ft³/lb h = enthalpy, Btu/lb s = entropy, Btu/lb °R

Temp, °F	3.0 psia			10 psia			20 psia			40 psia		
	v	h	s	v	h	s	v	h	s	v	h	s
−60	16.40	98.54	.2713									
−40	17.24	101.3	.2779	5.084	100.7	.2493						
−20	18.08	104.0	.2843	5.346	103.5	.2559	2.616	102.8	.2387			
0	18.92	106.9	.2906	5.606	106.4	.2623	2.752	105.8	.2454			
20	19.76	109.8	.2968	5.864	109.4	.2686	2.887	108.8	.2518	1.396	107.5	.2340
40	20.59	112.7	.3029	6.121	112.4	.2747	3.020	111.8	.2580	1.468	110.7	.2405
60	21.43	115.7	.3088	6.377	115.4	.2807	3.152	114.9	.2641	1.538	114.0	.2468
80	22.26	118.8	.3146	6.632	118.5	.2865	3.282	118.1	.2701	1.607	117.2	.2530
100	23.09	121.9	.3203	6.886	121.7	.2923	3.412	121.3	.2759	1.675	120.5	.2589
120	23.93	125.1	.3259	7.139	124.9	.2979	3.541	124.5	.2816	1.742	123.8	.2648
140	24.76	128.4	.3314	7.391	128.2	.3035	3.670	127.8	.2872	1.809	127.2	.2705
160	25.59	131.7	.3368	7.643	131.5	.3090	3.798	131.2	.2927	1.875	130.6	.2761
180	26.42	135.1	.3422	7.895	134.9	.3143	3.926	134.6	.2981	1.941	134.0	.2816
200	27.25	138.5	.3475	8.146	138.3	.3196	4.053	138.1	.3034	2.006	137.5	.2869
250				8.773	147.1	.3325	4.370	146.9	.3164	2.168	146.5	.3000
300										2.328	155.7	.3126

Temp, °F	60 psia			80 psia			100 psia			120 psia		
	v	h	s	v	h	s	v	h	s	v	h	s
40	.9486	109.6	.2295	.6878	108.3	.2211						
60	.9987	112.9	.2361	.7282	111.8	.2279	.5650	110.7	.2212			
80	1.048	116.3	.2424	.7671	115.3	.2345	.5982	114.3	.2280	.4849	113.3	.2224
100	1.095	119.7	.2486	.8048	118.8	.2408	.6300	117.9	.2345	.5131	117.0	.2291
120	1.142	123.1	.2545	.8415	122.3	.2470	.6608	121.5	.2408	.5401	120.7	.2356
140	1.188	126.5	.2604	.8775	125.8	.2529	.6908	125.1	.2469	.5661	124.3	.2418
160	1.234	130.0	.2660	.9129	129.3	.2587	.7201	128.7	.2528	.5914	128.0	.2478
180	1.279	133.5	.2716	.9477	132.9	.2644	.7489	132.3	.2586	.6161	131.7	.2537
200	1.324	137.0	.2771	.9821	136.5	.2699	.7771	135.9	.2642	.6404	135.4	.2594
220	1.368	140.6	.2824	1.016	140.1	.2753	.8051	139.6	.2696	.6642	139.1	.2649
240	1.412	144.2	.2877	1.050	143.8	.2806	.8326	143.3	.2750	.6877	142.8	.2703
260	1.456	147.9	.2928	1.083	147.5	.2858	.8599	147.0	.2803	.7109	146.6	.2756
280	1.499	151.6	.2979	1.117	151.2	.2909	.8869	150.8	.2854	.7338	150.4	.2808
300	1.542	155.4	.3029	1.150	155.0	.2960	.9137	154.6	.2905	.7565	154.2	.2860
350							.9801	164.3	.3028	.8126	164.0	.2984

Temp, °F	160 psia			200 psia			300 psia			400 psia		
	v	h	s	v	h	s	v	h	s	v	h	s
100	.3657	115.0	.2198	.2755	112.8	.2117						
120	.3882	118.9	.2267	.2960	117.0	.2191						
140	.4096	122.8	.2333	.3149	121.1	.2261	.1852	116.2	.2104			
160	.4301	126.6	.2396	.3327	125.1	.2327	.2006	120.9	.2182	.1305	115.5	.2046
180	.4499	130.4	.2456	.3497	129.1	.2390	.2146	125.4	.2253	.1446	121.0	.2133
200	.4691	134.2	.2515	.3661	133.0	.2450	.2276	129.8	.2320	.1567	126.0	.2210
220	.4879	138.0	.2572	.3820	137.0	.2509	.2398	134.0	.2384	.1677	130.8	.2281
240	.5064	141.9	.2627	.3974	140.9	.2566	.2515	138.2	.2445	.1779	135.3	.2347
260	.5245	145.7	.2681	.4125	144.8	.2621	.2628	142.4	.2503	.1874	139.8	.2410
280	.5423	149.6	.2734	.4273	148.7	.2675	.2737	146.5	.2560	.1965	144.1	.2470
300	.5599	153.5	.2786	.4419	152.7	.2727	.2843	150.6	.2615	.2052	148.5	.2527
320	.5774	157.4	.2837	.4563	156.6	.2779	.2947	154.7	.2668	.2137	152.7	.2583
340	.5946	161.3	.2887	.4705	160.6	.2830	.3048	158.9	.2720	.2219	157.0	.2637
360	.6117	165.3	.2936	.4845	164.7	.2879	.3148	163.0	.2771	.2299	161.3	.2690
380	.6287	169.3	.2985	.4984	168.7	.2928	.3247	167.1	.2821	.2378	165.5	.2741
400				.5122	172.8	.2976	.3344	171.3	.2871	.2455	169.8	.2791

*From published data of E.I. du Pont de Nemours & Co., Inc., 1964. Used by permission.
See Table 1-25 for References.

Table 1-33. REFRIGERANTS 50, 290, AND 600

For conversions to SI units, see Table 1-41.

REFRIGERANT 50 (METHANE), SATURATED†

Temp, °F	Pressure, psia	Volume, ft³/lb	Density, lb/ft³	Enthalpy, Btu/lb‡		Entropy, Btu/lb °R‡	
		Vapor, v_g	Liquid, $1/v_f$	Liquid, h_f	Vapor, h_g	Liquid, s_f	Vapor, s_g
−280	4.90	24.04	27.51	.00	228.20	.0000	1.2699
−270	8.44	14.61	27.04	8.20	232.30	.0423	1.2236
−260	13.80	9.31	26.55	16.60	236.40	.0823	1.1830
−250	21.71	6.13	26.05	25.00	240.30	.1201	1.1468
−240	32.40	4.24	25.54	33.30	243.90	.1578	1.1164
−230	46.40	3.04	25.01	42.00	247.30	.1962	1.0900
−220	64.50	2.23	24.44	50.60	250.20	.2333	1.0660
−210	87.60	1.67	23.85	59.50	252.80	.2693	1.0434
−200	115.70	1.28	23.22	68.80	254.80	.3062	1.0224
−190	150.00	.990	22.57	78.20	256.20	.3419	1.0019
−180	191.50	.773	21.86	87.20	257.00	.3767	.9816
−170	204.00	.610	21.07	98.00	257.20	.4127	.9622
−160	297.00	.483	20.23	108.70	256.50	.4476	.9411
−150	364.00	.381	19.24	120.30	254.50	.4839	.9169
−140	440.00	.301	19.14	133.20	251.20	.5214	.8905
−130	527.00	.232	16.67	148.10	245.90	.5656	.8622
−120	627.00	.161	14.37	171.80	231.40	.6329	.8083

REFRIGERANT 290 (PROPANE), SATURATED**

Temp, °F	Pressure, psia	Vapor, v_g	Liquid, $1/v_f$	Liquid, h_f	Vapor, h_g	Liquid, s_f	Vapor, s_g
−80	5.65	16.200	37.74	59.20	250.60	.1854	.6892
−60	9.78	9.770	37.00	69.30	256.60	.2120	.6800
−40	16.00	6.160	36.19	79.60	262.30	.2375	.6730
−20	25.05	4.060	35.39	90.40	268.10	.2628	.6670
0	37.81	2.740	34.57	101.60	273.80	.2872	.6615
20	55.00	1.930	33.67	113.20	279.20	.3110	.6570
40	77.80	1.330	32.73	124.50	284.10	.3343	.6533
60	106.90	.984	31.75	136.20	288.80	.3571	.6501
80	143.60	.745	30.59	148.50	293.00	.3797	.6473
100	188.70	.558	29.50	161.20	296.80	.4023	.6448
120	243.40	.426	28.31	174.60	300.40	.4255	.6428
140	308.40	.320	27.01	189.30	303.60	.4490	.6407
160	385.00	.240	25.24	205.00	305.40	.4740	.6363
180	473.20	.180	22.90	224.10	304.20	.5030	.6283
200	575.00	.113	19.19	250.10	294.90	.5420	.6100

REFRIGERANT 600 (BUTANE), SATURATED§

Temp, °F	Pressure, psia	Vapor, v_g	Liquid, $1/v_f$	Liquid, h_f	Vapor, h_g	Liquid, s_f	Vapor, s_g
0	7.3	11.1	38.59	−57.7	113.8	.864	1.240
20	11.6	7.23	37.89	−47.5	120.1	.886	1.239
40	17.7	4.87	37.19	−37.3	126.5	.908	1.239
60	26.3	3.40	36.45	−26.7	132.7	.931	1.238
80	37.2	2.4663	35.87	−17.62	138.87	.9513	1.2386
100	51.6	1.8003	35.00	−5.54	145.27	.9699	1.2397
120	69.9	1.3396	34.18	6.71	151.58	.9910	1.2414
140	92.9	1.0114	33.31	19.40	157.83	1.0131	1.2436
160	120.9	.7750	32.35	32.95	164.08	1.0349	1.2464
180	155.0	.5993	31.26	47.21	170.27	1.0568	1.2494
200	195.6	.4652	30.05	61.22	175.90	1.0785	1.2522
220	243.5	.3604	28.75	75.93	180.58	1.0999	1.2537
240	299.4	.2769	27.26	91.33	184.06	1.1220	1.2542

†Reprinted from *Transactions of AIChE*, 42(1):55, 1946.
‡R–50: Based on zero for saturated liquid at −280°F.
  R–600: Base point adjusted to zero at 0°R and ideal gas state.
**Based on data from *Ind. Eng. Chem.*, 35:602, 1943, but with base point for enthalpy and entropy adjusted to zero for saturated liquid at −200°F.
§Based on material from Dana *et al.*, originally published in *Refrigerating Engineering*, p. 402, June, 1962, for zero through 70°F, and based on data in *Ind. Eng. Chem.*, 29:1188, 1937 for the temperature range of 70 through 240°F.
  See Table 1-25 for References.

Table 1-34.   REFRIGERANTS 114 AND 717

For conversions to SI units, see Table 1-41.

REFRIGERANT 114 (DICHLOROTETRAFLUOROETHANE), SATURATED*

Temp, °F	Pressure, psia	Volume, ft³/lb Vapor, v_g	Density, lb/ft³ Liquid, $1/v_f$	Enthalpy, Btu/lb† Liquid, h_f	Latent, h_{fg}	Vapor, h_g	Entropy, Btu/lb °R† Liquid, s_f	Vapor, s_g
−100	.198	113.847	107.112	−11.740	67.523	55.783	−.03015	.15758
−50	1.378	18.549	102.945	−2.026	64.668	62.642	−.00488	.15296
0	5.949	4.7536	98.502	8.420	61.455	69.875	.01914	.15283
20	9.686	3.0185	96.630	12.830	60.004	72.834	.02851	.15360
40	15.078	1.99586	94.694	17.375	58.435	75.810	.03777	.15471
60	22.574	1.36601	92.686	22.055	56.735	78.790	.04693	.15610
80	32.659	.96283	90.593	26.865	54.895	81.760	.05597	.15768
100	45.851	.695777	88.4049	31.801	52.905	84.706	.06490	.15942
120	62.695	.513434	86.1018	36.855	50.754	87.610	.07371	.16126
140	83.757	.385479	83.6622	42.023	48.430	90.453	.08239	.16315
160	109.632	.293419	81.0553	47.300	45.913	93.212	.09095	.16504
180	140.952	.225618	78.2377	52.687	43.170	95.857	.09939	.16688
200	178.417	.174539	75.1446	58.194	40.148	98.342	.10773	.16859
250	304.872	.090480	65.2865	72.781	30.473	103.254	.12855	.17148

REFRIGERANT 114 (DICHLOROTETRAFLUOROETHANE), SUPERHEATED*

Temp, °F	2.0 psia v	h	s	10 psia v	h	s	100 psia v	h	s	300 psia v	h	s
0	14.33	70.09	.1658									
50	15.92	78.06	.1822	3.120	77.70	.1631						
100	17.50	86.47	.1980	3.448	86.17	.1789						
150	19.08	95.28	.2131	3.773	95.03	.1941						
200	20.66	104.45	.2275	4.095	104.24	.2086	.3632	101.5	.1789			
300				4.736	123.6	.2358	.4415	121.6	.2073	.1181	115.9	.1889
400							.5141	142.3	.2329	.1529	138.5	.2168

REFRIGERANT 717 (AMMONIA), SATURATED‡

Temp, °F	Pressure, psia	Volume, ft³/lb Vapor, v_g	Density, lb/ft³ Liquid, $1/v_f$	Enthalpy, Btu/lb† Liquid, h_f	Latent, h_{fg}	Vapor, h_g	Entropy, Btu/lb °R† Liquid, s_f	Vapor, s_g
−100	1.24	182.4	45.52	−63.3	635.8	572.5	−.1626	1.6055
−80	2.74	86.50	44.72	−42.2	623.5	581.2	−.1057	1.5368
−60	5.55	44.73	43.91	−21.2	610.8	589.6	−.0517	1.4769
−40	10.41	24.86	43.07	0.0	597.6	597.6	.0000	1.4242
−20	18.30	14.68	42.22	21.4	583.6	605.0	.0497	1.3774
0	30.42	9.116	41.34	42.9	568.9	611.8	.0975	1.3352
20	48.21	5.910	40.43	64.7	553.1	617.8	.1437	1.2969
40	73.32	3.971	39.49	86.8	536.2	623.0	.1885	1.2618
60	107.6	2.751	38.50	109.2	518.1	627.3	.2322	1.2294
80	153.0	1.955	37.48	132.0	498.7	630.7	.2749	1.1991
100	211.9	1.419	36.40	155.2	477.8	633.0	.3166	1.1705
120	286.4	1.047	35.26	179.0	455.0	634.0	.3576	1.1427

REFRIGERANT 717 (AMMONIA), SUPERHEATED‡

Temp, °F	10 psia v	h	s	50 psia v	h	s	100 psia v	h	s	200 psia v	h	s
−20	27.26	608.5	1.454									
0	28.58	618.9	1.477									
50	31.85	644.4	1.530	6.135	635.4	1.329						
100	35.07	670.0	1.578	6.843	663.7	1.382	3.304	655.2	1.289	1.520	635.6	1.181
150	38.26	695.8	1.622	7.521	691.1	1.429	3.672	685.0	1.340	1.740	671.8	1.243
200	41.45	722.2	1.664	8.185	718.5	1.472	4.021	713.7	1.385	1.935	703.9	1.294
300				9.489	774.0	1.550	4.695	770.8	1.466	2.295	764.5	1.379

†Based on zero for the saturated liquid at −40°F.
‡From National Bureau of Standards Circulars No. 142, 1945, and No. 472, 1948.
*From published data of E.I. du Pont de Nemours & Co., Inc., 1966. Used by permission.
  See Table 1-25 for References.

Table 1-35. REFRIGERANTS 170, 1150, AND 40

For conversions to SI units, see Table 1-41.

REFRIGERANT 170 (ETHANE), SATURATED*

Temp, °F	Pressure, psia	Volume, ft³/lb	Density, lb/ft³	Enthalpy, Btu/lb†		Entropy, Btu/lb °R†	
		Vapor, v_g	Liquid, $1/v_f$	Liquid, h_f	Vapor, h_g	Liquid, s_f	Vapor, s_g
−220	.27	310.5000	37.47	117.60	353.90	.8249	1.8107
−200	.85	107.8000	36.76	128.70	359.70	.8691	1.7587
−180	2.20	44.9000	36.05	139.90	365.50	.9133	1.7201
−160	4.97	21.1100	35.32	151.10	371.20	.9523	1.6865
−140	9.97	11.0500	34.57	162.30	376.60	.9891	1.6593
−120	18.33	6.3160	33.82	174.30	381.80	1.0240	1.6346
−100	31.32	3.8300	33.01	186.20	386.80	1.0570	1.6143
−80	50.34	2.4510	32.18	198.20	391.20	1.0890	1.5974
−60	77.02	1.6380	31.26	210.50	395.20	1.1205	1.5825
−40	113.10	1.1270	30.28	223.20	398.60	1.1519	1.5697
−20	159.90	.7983	29.22	236.30	401.70	1.1824	1.5583
0	219.70	.5754	28.01	250.30	403.90	1.2132	1.5476
20	294.00	.4198	26.64	265.10	404.90	1.2445	1.5361
40	385.00	.3062	25.06	281.00	404.50	1.2762	1.5234
60	494.20	.2164	22.95	299.30	401.30	1.3100	1.5064
80	630.70	.1411	19.75	323.70	391.40	1.3505	1.4751

REFRIGERANT 1150 (ETHYLENE), SATURATED‡

Temp, °F	Pressure, psia	Volume, ft³/lb	Density, lb/ft³	Enthalpy, Btu/lb†		Entropy, Btu/lb °R†	
		Vapor, v_g	Liquid, $1/v_f$	Liquid, h_f	Vapor, h_g	Liquid, s_f	Vapor, s_g
−260	.0544	1405.000		752.00	992.90	.7745	1.9816
−240	.2601	328.600		763.60	998.50	.8209	1.8896
−220	.8906	103.100		775.20	1004.20	.8714	1.8266
−200	2.4836	39.740		786.80	1009.90	.9174	1.7766
−180	5.9078	17.920		798.20	1015.40	.9598	1.7366
−160	12.3005	9.047		809.70	1020.00	.9993	1.7016
−140	23.1828	5.005	34.76	821.40	1023.30	1.0345	1.6659
−120	40.2315	2.987	33.74	833.60	1027.00	1.0744	1.6436
−100	65.5672	1.879	32.03	845.60	1030.00	1.1091	1.6217
−80	100.9564	1.732	31.46	858.00	1032.70	1.1470	1.6070
−60	148.4139	.857	30.23	870.70	1034.70	1.1762	1.5865
−40	206.3128	.593	28.84	883.60	1035.60	1.2031	1.5652
−20	289.8325	.419	27.31	897.50	1035.50	1.2341	1.5479
0	387.9277	.301	25.56	913.10	1033.80	1.2673	1.5298
20	507.7433	.212	23.30	932.40	1029.40	1.3064	1.5087
40	654.5554	.139	19.86	961.50	1019.00	1.3629	1.4779

REFRIGERANT 40 (METHYL CHLORIDE), SATURATED**

Temp, °F	Pressure, psia	Volume, ft³/lb	Density, lb/ft³	Enthalpy, Btu/lb†		Entropy, Btu/lb °R†	
		Vapor, v_g	Liquid, $1/v_f$	Liquid, h_f	Vapor, h_g	Liquid, s_f	Vapor, s_g
−80	1.953	41.08	66.98	−13.988	184.75	−.0351	.4882
−60	3.799	22.09	65.66	−7.039	187.74	−.0172	.4703
−40	6.878	12.72	64.39	0.000	190.66	.0000	.4544
−20	11.71	7.761	63.17	7.146	193.49	.0166	.4405
0	18.90	4.969	62.00	14.39	196.23	.0327	.4284
20	29.16	3.312	60.72	21.73	198.84	.0484	.4177
40	43.25	2.286	59.38	29.17	201.17	.0636	.4079
60	62.00	1.624	58.00	36.71	203.33	.0784	.3991
80	86.26	1.183	56.69	44.36	205.27	.0928	.3910
100	116.7	.8814	55.31	52.09	206.94	.1069	.3836
120	154.2	.6710	53.79	59.93	208.39	.1206	.3768
140	199.6	.5189	52.22	67.87	209.58	.1341	.3705
160	253.5	.4070	50.56	75.90	210.56	.1473	.3646

*Based on data from *Transactions of AIChE*, 43(1):25, 1947.

†R–170: Based on zero for the solid at 0°R.
  R–40: Based on zero for the saturated liquid at −40°F.

‡Based on data from *Transactions of AIChE*, 40:227, October, 1944, but with base point for enthalpy and entropy adjusted to zero at 0°R for the ideal gas state.

**From published data of E.I. du Pont de Nemours & Co., Inc., 1947. Used by permission.
  See Table 1-25 for References.

Table 1-36.   REFRIGERANTS 500 AND 502 (AZEOTROPES)

For conversions to SI units, see Table 1-41.

REFRIGERANT 500, SATURATED*

Temp, °F	Pressure, psia	Volume, ft³/lb Vapor, v_g	Density, lb/ft³ Liquid, $1/v_f$	Enthalpy, Btu/lb† Liquid, h_f	Latent, h_{fg}	Vapor, h_g	Entropy, Btu/lb °R† Liquid, s_f	Vapor, s_g
−100	1.735	22.18	89.64	−13.34	93.11	79.78	−.03410	.22476
−50	8.395	5.115	85.20	−2.30	88.74	86.44	−.00553	.21108
0	27.98	1.657	80.46	9.71	83.10	92.81	.02197	.20274
20	41.96	1.129	78.44	14.79	80.43	95.22	.03270	.20037
40	60.75	.7916	76.34	20.05	77.47	97.53	.04335	.19839
60	85.33	.5687	74.14	25.48	74.23	99.71	.05387	.19671
80	116.7	.4167	71.79	31.12	70.63	101.75	.06435	.19523
100	155.9	.3100	69.28	36.97	66.63	103.60	.07481	.19386
120	204.1	.2332	66.55	43.10	62.11	105.22	.08534	.19249
140	262.4	.1765	63.51	49.58	56.94	106.51	.09603	.19098
160	332.1	.1336	60.03	56.51	50.85	107.36	.10706	.18911
180	414.6	.1000	55.81	64.12	43.38	107.49	.11871	.18652
200	511.1	.0726	50.16	72.87	33.47	106.34	.13167	.18240

REFRIGERANT 500, SUPERHEATED*

Temp, °F	5.0 psia v	h	s	50 psia v	h	s	200 psia v	h	s	400 psia v	h	s
0	9.82	94.7	.2403									
50	10.93	103.0	.2573	1.012	100.2	.2072						
100	12.03	111.6	.2734	1.141	109.6	.2247						
150	13.12	120.6	.2888	1.264	119.0	.2409	.2683	112.7	.2054			
200	14.21	129.9	.3035	1.382	128.6	.2560	.3090	123.9	.2230	.1227	115.2	.1989
250	15.30	139.6	.3176	1.497	138.5	.2705	.3450	134.8	.2389	.1499	128.7	.2186
300				1.611	148.7	.2843	.3785	145.6	.2536	.1717	140.9	.2352
400							.4415	167.5	.2808	.2090	164.3	.2641

REFRIGERANT 502, SATURATED‡

Temp, °F	Pressure, psia	Volume, ft³/lb Vapor, v_g	Density, lb/ft³ Liquid, $1/v_f$	Enthalpy, Btu/lb† Liquid, h_f	Latent, h_{fg}	Vapor, h_g	Entropy, Btu/lb °R† Liquid, s_f	Vapor, s_g
−100	3.230	10.837	98.49	−15.15	82.52	67.37	−.0388	.1906
−50	14.74	2.6428	93.47	−2.57	76.44	73.87	−.0062	.1804
0	45.94	.9061	87.84	10.54	69.61	80.15	.0239	.1753
20	67.14	.6283	85.39	15.99	66.50	82.49	.0354	.1740
40	94.90	.4466	82.80	21.57	63.09	84.66	.0466	.1729
60	130.3	.3234	80.04	27.32	59.30	86.62	.0578	.1719
80	174.6	.2372	77.07	33.24	55.02	88.26	.0687	.1707
100	229.1	.1750	73.80	39.37	50.10	89.47	.0796	.1692
120	295.0	.1288	70.08	45.71	44.31	90.02	.0905	.1669
140	373.8	.09368	65.59	52.17	37.38	89.55	.1011	.1634
160	467.3	.06764	59.49	58.21	29.53	87.74	.1105	.1582

REFRIGERANT 502, SUPERHEATED‡

Temp, °F	10 psia v	h	s	40 psia v	h	s	100 psia v	h	s	300 psia v	h	s
0	4.445	81.54	.2050	1.053	80.39	.1781						
50	4.948	89.42	.2212	1.189	88.44	.1947	.4326	86.17	.1751			
100				1.323	96.84	.2105	.4951	94.95	.1915			
150							.5543	103.97	.2069	.1473	96.85	.1782
200										.1765	107.70	.1953
250										.2016	118.2	.2107

†Based on zero for the saturated liquid at −40°F.
‡From published data of E.I. du Pont de Nemours & Co., Inc., 1963. Used by permission.
*From unpublished data of General Chemicals Division, Allied Chemical Corp. Used by permission.
  See Table 1-25 for References.

Table 1-37. THERMODYNAMIC PROPERTIES OF ARGON*

Refrigerant 740: Density, Enthalpy, and Entropy; NSRDS Data

UNITS AND CONVERSION FACTORS:

K = degrees Kelvin. For degrees Rankine multiply by 1.8.

ρ_f = density of liquid in mol/liter. For lb/ft^3 multiply by 2.493.

ρ_g = density of saturated vapor in mol/liter

h = enthalpy in J/mol, above 0 K. For cal/g multiply by 0.006 0. For Btu/lb, multiply by 0.010 8.

s = entropy in J/mol·K. For cal/g·K multiply by 0.006 0. For Btu/lb·deg R, multiply by 0.010 8.

Table A. SATURATED VAPOR AND SATURATED LIQUID

Tempera-ture, K	Pressure		Density		Enthalpy		Entropy	
	atm	psia	ρ_f	ρ_g	h_f	h_g	s_f	s_g
83.8[a]	0.68	9.99	35.41	.101 3	2 814.	9 388.	52.77	131.2
85	0.78	11.46	35.24	.114 9	2 869.	9 407.	53.41	130.3
90	1.32	19.40	34.47	.186 4	3 099.	9 481.	56.02	126.9
95	2.11	31.01	33.67	.287 0	3 335.	9 544.	58.55	123.9
100	3.21	47.18	32.83	.423 5	3 575.	9 595.	60.98	121.2
105	4.68	68.78	31.96	.603 4	3 820.	9 631.	63.32	118.7
110	6.59	96.85	31.04	.835 8	4 069.	9 652.	65.58	116.3
115	9.01	132.4	30.06	1.13	4 322.	9 653.	67.76	114.1
120	12.00	176.4	29.03	1.51	4 583.	9 633.	69.89	112.0
125	15.64	229.9	27.91	1.98	4 853.	9 587.	71.99	109.9
130	20.01	294.1	26.67	2.59	5 140.	9 508.	74.11	107.7
135	25.19	370.2	25.26	3.39	5 453.	9 383.	76.33	105.4
140	31.30	460.0	23.58	4.49	5 807.	9 190.	78.71	102.9
145	38.44	565.0	21.36	6.18	6 238.	8 871.	81.51	99.7
150	46.77	687.4	17.04	10.20	6 978.	8 109.	86.23	93.8

[a]Triple point.

Table B. PROPERTIES OF SUPERHEATED VAPOR (GAS) AND COMPRESSED LIQUID

Temp, K	0.10 atm (1.47 psia)			1.0 atm (14.7 psia)			2 atm (29.4 psia)			5 atm (73.5 psia)		
	ρ	h	s	ρ	h	s	ρ	h	s	ρ	h	s
90	.013 58	9 556.	149.0	.139 6	9 501.	129.4	34.48	3 100.	56.0	34.51	3 106.	56.0
100	.012 21	9 764.	151.1	.124 7	9 718.	131.7	.255 5	9 664.	125.6	32.85	3 578.	61.0
120	.010 17	10 181.	154.9	.102 9	10 147.	135.6	.208 8	10 109.	129.6	.546 4	9 985.	121.3
140	.008 71	10 597.	158.1	.087 8	10 572.	138.9	.177 2	10 542.	133.0	.455 3	10 451.	124.9
160	.007 62	11 014.	160.9	.076 6	10 993.	141.7	.154 1	10 970.	135.8	.392 4	10 899.	127.9
180	.006 77	11 430.	163.4	.068 0	11 413.	144.2	.136 5	11 395.	138.3	.345 5	11 337.	130.5
200	.006 10	11 846.	165.6	.061 1	11 832.	146.4	.122 6	11 817.	140.5	.309 1	11 770.	132.8
220	.005 54	12 262.	167.5	.055 5	12 251.	148.4	.111 3	12 237.	142.6	.279 9	12 198.	134.8
240	.005 08	12 678.	169.3	.050 9	12 668.	150.2	.101 9	12 657.	144.4	.255 9	12 623.	136.7
260	.004 69	13 094.	171.0	.046 9	13 085.	151.8	.094 0	13 076.	146.1	.235 7	13 047.	138.4
280	.004 35	13 510.	172.6	.043 6	13 502.	153.4	.087 2	13 494.	147.6	.218 5	13 469.	139.9
300	.004 06	13 926.	174.0	.040 6	13 919.	154.8	.081 3	13 912.	149.0	.203 7	13 889.	141.4

Temp, K	10 atm (147 psia)			50 atm (735 psia)			100 atm (1 470 psia)			1 000 atm (14 700 psia)		
	ρ	h	s	ρ	h	s	ρ	h	s	ρ	h	s
90	34.55	3 114.	55.9	34.86	3 185.	55.4	35.23	3 276.	54.8	—	—	—
100	32.91	3 585.	60.9	33.31	3 646.	60.3	33.76	3 727.	59.6	—	—	—
120	1.20	9 746.	114.2	29.77	4 600.	69.0	30.54	4 645.	67.9	36.54	6 243.	58.9
140	.958 6	10 287.	118.4	24.82	5 704.	77.4	26.66	5 615.	75.4	34.88	6 984.	64.6
160	.810 3	10 776.	121.6	6.29	9 298.	101.2	20.95	6 858.	83.7	33.31	7 698.	69.4
180	.706 1	11 240.	124.4	4.38	10 302.	107.2	12.51	8 685.	94.4	31.81	8 407.	73.6
200	.627 6	11 690.	126.7	3.59	10 984.	110.8	8.60	9 958.	101.2	30.39	9 101.	77.3
220	.565 8	12 131.	128.8	3.09	11 567.	113.6	6.89	10 814.	105.2	29.02	9 780.	80.5
240	.515 7	12 566.	130.7	2.74	12 100.	115.9	5.87	11 506.	108.3	27.73	10 444.	83.4
260	.474 1	12 998.	132.4	2.48	12 604.	117.9	5.18	12 117.	111.0	26.50	11 095.	86.0
280	.438 9	13 426.	134.0	2.26	13 088.	119.7	4.66	12 679.	112.8	25.33	11 733.	88.3
300	.408 7	13 853.	136.0	2.09	13 560.	121.3	4.26	13 210.	114.6	24.24	12 359.	90.5

Note: In the original source the increments for Table A are 1.0 K from 84.0 K to 150.0 K. The superheat tables start at 0.01 atm and comprise 114 pages of tables, each with increments of 1.0 K.

*From: NSRDS-NBS-27, "Thermodynamic Properties of Argon from the Triple Point to 300 K at Pressures to 1 000 Atmospheres", A.L. Gosman, R.D. McCarty, and J.G. Hust, National Standard Reference Data Series—National Bureau of Standards, 27 March, 1969.

Table 1-38. PROPERTIES OF FLUORINE*

From the Triple Point to 300 K

Triple point—53.48 K; 0.000 252 MN/m² or 0.002 49 atm
Critical point—144.3 K; 15.10 g-mol/l; 5.215 MN/m² or 51.47 atm

MELTING POINT OF SOLID (MEASURED)

Temp, K	MN/m²	atm	Temp, K	MN/m²	atm
53.49	0.101	1.00	54.31	8.532	84.20
53.61	1.314	12.97	54.51	10.606	104.67
53.81	3.365	33.21	54.71	12.698	125.32
53.91	4.390	43.32	54.81	13.739	135.59
54.11	6.449	63.65			

SATURATED LIQUID—BOILING POINT

Temp, K	Pressure[a] MN/m²	Pressure[a] atm	Density[b] g-mol/l	Density[b] lb/ft³	Enthalpy[c] J/mol	Enthalpy[c] Btu/lb	Specific[d] heat, c_p (cal/g·deg C or Btu/lb·deg F)
53.481 1	.000 25	0.002 5	44.86	106.4	−6 025.	−68.23	.345 8
60	.001 6	0.015	43.83	103.9	−5 667.	−64.18	.343 6
70	.012 3	0.121	42.18	100.0	−5 108.	−57.85	.346 6
80	.055 5	0.547	40.44	95.9	−4 540.	−51.42	.352 3
84.95	.101 3	1.000	39.54	93.8	−4 256.	−48.20	.358 8
90	.174 0	1.718	38.59	91.5	−3 963.	−44.88	.367 0
100	.428 0	4.224	36.59	86.8	−3 368.	−38.14	.378 8
110	.888 5	8.769	34.36	81.5	−2 744.	−31.08	.404 4
120	1.633 3	16.119	31.77	75.3	−2 071.	−23.45	.448 2
130	2.747 5	27.116	28.53	67.7	−1 309.	−14.82	.556 7
140	4.337 1	42.804	23.52	55.8	−320.	−3.62	1.186

SATURATED VAPOR—CONDENSING POINT

Temp, K	Pressure[a] MN/m²	Pressure[a] atm	Density[b] g-mol/l	Density[b] lb/ft³	Enthalpy[c] J/mol	Enthalpy[c] Btu/lb	Specific[d] heat, c_p (cal/g·deg C or Btu/lb·deg F)
53.481 1	.000 25	0.002 5	.000 57	0.001	1 553.	17.59	.183 2
60	.001 6	0.015	.003 1	0.007	1 741.	19.72	.183 5
70	.012 3	0.121	.021 3	0.029	2 022.	22.90	.185 2
80	.055 5	0.547	.085 2	0.352	2 285.	25.88	.189 8
84.95	.101 3	1.000	.148 2	0.579	2 402.	27.20	.193 8
90	.174 0	1.718	.244 1	0.579	2 512.	28.45	.199 5
100	.428 0	4.224	.564 9	1.34	2 688.	30.44	.217 2
110	.888 5	8.769	1.139 0	2.70	2 795.	31.65	.248 0
120	1.633 3	16.119	2.111 1	5.01	2 806.	31.78	.305 2
130	2.747 5	27.116	3.784 8	8.98	2 668.	30.22	.441 6
140	4.337 1	42.804	7.322 6	17.36	2 161.	24.47	1.198

SUPERHEATED VAPOR—GAS

Temp, K	At 0.101 33 MN/m² = 1.00 atm lb/ft³	h, Btu/lb	c_p	At 5 MN/m² = 49.35 atm lb/ft³	h, Btu/lb	c_p
84.95[e]	.351	27.20	.193 8			
100	.295	32.39	.189 2			
120	.244	39.14	.186 6			
140	.208	45.84	.185 6			
143.32[e]				24.56	18.82	—
160	.181	52.53	.185 6	12.33	38.84	.345 1
180	.161	59.22	.186 3	9.55	49.39	.260 7
200	.145	65.95	.187 5	8.04	58.23	.234 3
220	.136	72.73	.189 2	7.02	66.43	.222 4
240	.121	79.57	.191 1	6.28	74.33	.216 5
260	.111	86.49	.193 1	5.69	82.06	.213 5
280	.103	93.48	.195 3	5.22	89.72	.212 2
300	.096	100.55	.197 6	4.82	97.35	.212 0

Note: Velocity of sound is tabulated in the original source. This varies in the range of 130–170 m/sec in the saturated vapor and up to 1 030 m/sec in the saturated liquid.

[a]To convert meganewtons/square meter (MN/m²) to psia, multiply by 145.0. To convert MN/m² to kg/m², multiply by 101 970.
[b]The molecular weight of fluorine is 37.996. To convert gram-mol/l to kg/m³, divide by 37.996.
[c]For J/kg multiply J/g-mol by 26.3. For cal/g multiply J/mol by 110.
[d]For J/kg·K multiply cal/g·deg C by 4 184.
[e]Saturation.

*Abridged from the extensive tables in "The Thermodynamic Properties of Compressed Gaseous and Liquid Fluorine", NBS Technical Note 392, National Bureau of Standards, 1970 (82 pages).

Table 1-39. LOW-TEMPERATURE PROPERTIES OF HELIUM*

Composition and atomic weight. Naturally occurring helium is a mixture of 99.99987 percent He[4] and 0.00013 percent He[3]. The atomic weight of helium is 4.0026.

Liquefaction of helium.

Table A. CRITICAL VALUES

Isotope	Temp, °K	Press, atm	Density, g/cm³
He⁴	5.2	2.26	0.0693
He³	3.35		

Table B. PROPERTIES AT THE NORMAL BOILING POINT

Isotope	Temp, °K	Press, atm	Density, g/cm³		Weight of 1 gal liquid	Gas/liquid volume ratio	Heat of vaporization, cal/g
			Vapor	Liquid			
He⁴	4.215	1	0.0170	0.125	1.043 lb	7.35:1	4.9
He³	3.20	1		0.057			

Liquid helium. Liquid helium-4 exists in two forms: He-I (above the λ-transition temperature) and He-II (below it). He-I is a normal liquid, but He-II is unlike any other known substance. Its peculiar properties include the fact that it expands on cooling and its conductivity for heat is enormous; neither its heat conduction nor viscosity obey normal rules.

Table C. λ-TRANSITION LIQUID He-I TO He-II

Temp, °K	Press, atm	Liquid density, g/cm³	Heat of transition, cal/g
2.174	0.050	0.146	0.0

Some measurements of the thermal conductivity of He-II at 1.4 and 1.75°K gave values of about 190 cal cm⁻¹ sec⁻¹ °K⁻¹, or 46,000 Btu ft⁻¹ hr⁻¹ °R⁻¹. (This value is about 200 times that of copper at ordinary temperatures.) The thermal conductivity of He-II as a function of temperature has a very pronounced maximum near 1.92°K; under some conditions a thermal conductivity as high as 810 cal cm⁻¹ sec⁻¹ °K⁻¹ was observed. Hence liquid He-II is by far the *best heat-conducting substance known.* Observed thermal conductivity of He-II depends markedly on the test conditions, and the values obtained cannot be treated in the same way as for other liquids.

The conductivity of helium-I is much lower than that of helium-II. In fact the ratio is about one to four million. Conductivities for liquid helium-I at a pressure of one atmosphere are shown in Table D.

Table D. THERMAL CONDUCTIVITY OF LIQUID HELIUM-I**
Helium-I or Naturally Occurring Helium
Milliwatt cm⁻¹ °K⁻¹

Temp, °K	Thermal cond.†	Temp, °K	Thermal cond.†	Temp, °K	Thermal cond.†
2.4	0.192	3.4	0.227	4.4	(0.335)‡
2.6	0.193	3.6	0.241	4.6	(0.366)‡
2.8	0.197	3.8	0.260	4.8	(0.400)‡
3.0	0.204	4.0	0.282	5.0	(0.437)‡
3.2	0.214	4.2	0.307	5.2	(0.477)‡

†Under saturation vapor pressure.
‡Normal boiling point 4.215°K.

Table 1-39. LOW-TEMPERATURE PROPERTIES OF HELIUM
(Continued)

Solid helium. Helium is the only liquid that cannot be solidified by lowering the temperature. It remains liquid down to absolute zero at ordinary pressures, but it can readily be solidified by increasing the pressure. Pressures necessary for solidification are given in Table E.

Solid helium can be compressed substantially by increasing the pressure. Values for the specific volume at 0°K are as shown in Table F.

Table E. SOLIDIFICATION OF HELIUM

Temp, °K	Press. atm	Solid density, g/cm³	Heat of fusion, cal/g
3.5	100		1.25
2.174	40	0.200	
1.0	26		

Table F. SPECIFIC VOLUME OF SOLID HELIUM AT 0°K

Press. atm	Specific volume, cm³/g
50	4.74
200	4.00
460	3.50
1070	3.00

*Further information on the technology and low-temperature properties of helium may be found in "American Institute of Physics Handbook", 2nd ed., D.E. Gray, Ed., McGraw-Hill Book Co., 1963, and "Technology of Liquid Helium", R.H. Kropschot, B.W. Birmingham, and D.B. Mann, Eds., National Bureau of Standards Monograph 111, 1968.

**From: "Thermal Conductivity of Selected Materials", R.W. Powell, C.Y. Ho, and P.E. Liley, National Standard Reference Data Series—National Bureau of Standards—8, 1966.

Table 1-40.   PROPERTIES OF HELIUM*

Two-phase Liquid-vapor Region

For conversion to SI units, see Table 1-41.

Temperature, °K	Pressure, atm	Saturated vapor				Saturated liquid			
		Specific volume, cm³/g	Enthalpy, J/g	Internal energy, J/g	Entropy, J/g-K	Specific volume, cm³/g	Enthalpy, J/g	Internal energy, J/g	Entropy, J/g-K
3.00	0.241	224.11	28.96	23.49	10.247	7.0852	5.30	5.13	2.356
3.20	0.320	174.83	29.48	23.81	9.911	7.1849	5.85	5.62	2.527
3.40	0.416	138.70	29.93	24.08	9.606	7.3025	6.47	6.16	2.690
3.60	0.529	111.48	30.34	24.36	9.322	7.4388	7.18	6.78	2.883
3.80	0.661	90.50	30.59	24.53	9.021	7.5965	7.93	7.40	3.059
4.00	0.814	73.86	30.72	24.63	8.736	7.7785	8.81	8.17	3.255
4.20	0.990	60.61	30.79	24.71	8.443	7.9885	9.85	9.05	3.460
4.40	1.190	49.80	30.65	24.65	8.150	8.2305	10.96	9.97	3.674
4.60	1.417	40.72	30.32	24.47	7.845	8.5470	12.33	11.10	3.920
4.80	1.672	32.87	29.49	23.92	7.473	9.0334	13.88	12.35	4.188
5.00	1.959	25.67	28.02	22.92	7.000	9.9404	16.00	14.03	4.598

*From: "Technology of Liquid Helium", R.H. Kropschot, B.W. Birmingham, and D.B. Mann, Eds., National Bureau of Standards Monograph 111, October 1968.

Table 1-41.   THERMODYNAMIC PROPERTIES OF SATURATED MERCURY*

Enthalpy and Entropy Measured from 32°F

For pressures in MN/m², multiply value in lbf/in.² by 0.006 894 8. For temperature in K, add 459.67 to value in deg F and multiply the result by 5/9. For enthalpy in J/kg, multiply value in Btu/lb by 2 324.4. For entropy in J/kg·K, multiply value in Btu/lb·deg F by 4 186.8. For specific volume in m³/kg, multiply value in ft³/lb$_m$ by 0.062 420.

Pressure lb$_f$/in.²	Temper- ature, °F	Enthalpy, Btu/lb$_m$			Entropy, Btu/lb$_m$ °R			Specific volume, sat vapor, ft³/lb$_m$
		Saturated liquid	Evap- oration	Saturated vapor	Saturated liquid	Evap- oration	Saturated vapor	
0.020	259.9	7.532	127.614	135.146	0.01259	0.17735	0.18994	1893
0.040	288.3	8.463	127.486	135.949	0.01386	0.17044	0.18430	986
0.075	316.2	9.373	127.361	136.734	0.01504	0.16415	0.17919	545
0.100	329.7	9.814	127.300	137.114	0.01561	0.16126	0.17687	416
0.200	364.3	10.936	127.144	138.080	0.01699	0.15432	0.17131	217.3
0.400	402.0	12.159	126.975	139.134	0.01844	0.14736	0.16580	113.7
0.600	425.8	12.929	126.868	139.797	0.01932	0.14328	0.16260	77.84
0.800	443.5	13.500	126.788	140.288	0.01994	0.14038	0.16032	59.58
1.00	457.7	13.959	126.724	140.683	0.02045	0.13814	0.15859	48.42
2.00	504.9	15.476	126.512	141.988	0.02205	0.13116	0.15321	25.39
4.00	557.9	17.161	126.275	143.436	0.02373	0.12434	0.14787	13.38
6.00	591.2	18.233	126.124	144.357	0.02477	0.12002	0.14479	9.26
8.00	616.5	19.035	126.011	145.046	0.02551	0.11712	0.14264	7.12
10	637.0	19.685	125.919	145.604	0.02610	0.11483	0.14093	5.81
20	706.0	21.864	125.609	147.473	0.02800	0.10779	0.13579	3.09
40	784.4	24.345	125.255	149.600	0.03004	0.10068	0.13072	1.648
60	835.7	25.940	125.024	150.964	0.03127	0.19652	0.12779	1.144
80	874.8	27.159	124.849	152.008	0.03218	0.09356	0.12574	0.885
100	906.8	28.152	124.706	152.858	0.03290	0.09127	0.12417	0.725
120	934.3	29.005	124.582	153.587	0.03350	0.08938	0.12288	0.617
140	958.3	29.748	124.474	154.222	0.03401	0.08778	0.12179	0.538
160	979.9	30.415	124.376	154.791	0.03447	0.08640	0.12087	0.478
180	999.5	31.018	124.288	155.306	0.03488	0.08518	0.12006	0.431
200	1017.2	31.560	124.209	155.769	0.03523	0.08411	0.11934	0.392
225	1038.0	32.204	124.115	156.319	0.03565	0.08287	0.11852	0.354
250	1057.2	32.784	124.029	156.813	0.03603	0.08178	0.11871	0.322
275	1074.8	33.322	123.950	157.272	0.03637	0.08079	0.11716	0.297
300	1091.2	33.824	123.876	157.700	0.03669	0.07989	0.11658	0.276
350	1121.4	34.747	123.740	158.487	0.03725	0.07828	0.11553	0.241
400	1148.4	35.565	123.620	159.185	0.03775	0.07688	0.11463	0.215
500	1196.0	37.006	123.406	160.412	0.03861	0.07455	0.11316	0.177
600	1236.8	38.245	123.221	161.466	0.03932	0.07264	0.11196	0.151
800	1306.1	40.324	122.910	163.234	0.04047	0.06961	0.11008	0.118
1,000	1364.0	42.056	122.649	164.705	0.04139	0.06726	0.10865	0.098
1,100	1390.0	42.828	122.533	165.361	0.04179	0.06625	0.10804	0.090

*Abridged from: "Thermodynamic Properties of Mercury Vapor", by L.A. Sheldon. Courtesy of General Electric Company.

Table 1-42. PROPERTIES OF LIQUID WATER*

See also Tables 1-6, 1-8, 1-9, and 1-10.

SYMBOLS AND UNITS:

ρ = density, lbm/ft^3. For g/cm^3 multiply by 0.016018. For kg/m^3 multiply by 16.018.

c_p = specific heat, Btu/lbm·deg R = cal/g·K. For J/kg·K multiply by 4186.8.

μ = viscosity. For lbf·sec/ft^2 = slugs/sec·ft, multiply by 10^{-7}. For lbm/sec·ft multiply by 10^{-7} and by 32.174. For g/sec·cm (poises) multiply by 10^{-7} and by 478.80. For N·sec/m^2 multiply by 10^{-7} and by 47.880.

k = thermal conductivity, Btu/hr·ft·deg R. For W/m·K multiply by 1.7307.

Temp, °F	At 1 atm or 14.7 psia				At 1,000 psia				At 10,000 psia			
	ρ	c_p	μ	k	ρ	c_p	μ	k	ρ	c_p	μ	k†
32	62.42	1.007	366	0.3286	62.62	0.999	365	0.3319	64.5	0.937	357	0.3508
40	62.42	1.004	323	0.334	62.62	0.997	323	0.337	64.5	0.945	315	0.356
50	62.42	1.002	272	0.3392	62.62	0.995	272	0.3425	64.5	0.951	267	0.3610
60	62.38	1.000	235	0.345	62.58	0.994	235	0.348	64.1	0.956	233	0.366
70	62.31	0.999	204	0.350	62.50	0.994	204	0.353	64.1	0.960	203	0.371
80	62.23	0.998	177	0.354	62.42	0.994	177	0.358	64.1	0.962	176	0.376
90	62.11	0.998	160	0.359	62.31	0.994	160	0.362	63.7	0.964	159	0.380
100	62.00	0.998	142	0.3633	62.19	0.994	142	0.3666	63.7	0.965	142	0.3841
110	61.88	0.999	126	0.367	62.03	0.994	126	0.371	63.7	0.966	126	0.388
120	61.73	0.999	114	0.371	61.88	0.995	114	0.374	63.3	0.967	114	0.391
130	61.54	0.999	105	0.374	61.73	0.995	105	0.378	63.3	0.968	105	0.395
140	61.39	0.999	96	0.378	61.58	0.996	96	0.381	63.3	0.969	98	0.398
150	61.20	1.000	89	0.3806	61.39	0.996	89	0.3837	63.0	0.970	91	0.4003
160	61.01	1.001	83	0.383	61.20	0.997	83	0.386	62.9	0.971	85	0.403
170	60.79	1.002	77	0.386	60.98	0.998	77	0.389	62.5	0.972	79	0.405
180	60.57	1.003	72	0.388	60.75	0.999	72	0.391	62.5	0.973	74	0.407
190	60.35	1.004	68	0.390	60.53	1.001	68	0.393	62.1	0.974	70	0.409
200	60.10	1.005	62.5	0.3916	60.31	1.002	62.9	0.3944	62.1	0.975	65.4	0.4106
250	boiling point 212°F				59.03	1.001	47.8	0.3994	60.6	0.981	50.6	0.4158
300					57.54	1.024	38.4	0.3993	59.5	0.988	41.3	0.4164
350					55.83	1.044	32.1	0.3944	58.1	0.999	35.1	0.4132
400					53.91	1.072	27.6	0.3849	56.5	1.011	30.6	0.4064
500					49.11	1.181	21.6	0.3508	52.9	1.051	24.8	0.3836
600					boiling point 544.58°F				48.3	1.118	21.0	0.3493

†At 7,500 psia.

*From: "1967 ASME Steam Tables", American Society of Mechanical Engineers, Tables 9, 10, and 11 and Figures 6, 7, 8, and 9.

The ASME compilation is a 330-page book of tables and charts, including a 2½ × 3½-ft Mollier chart. All values have been computed in accordance with the 1967 specifications of the International Formulation Committee (IFC) and are in conformity with the 1963 International Skeleton Tables. This standardization of tables began in 1921 and was extended through the International Conferences in London (1929), Berlin (1930), Washington (1934), Philadelphia (1954), London (1956), New York (1963) and Glasgow (1966). Based on these world-wide standard data, the 1967 ASME volume represents detailed computer output in both tabular and graphic form. Included are density and volume, enthalpy, entropy, specific heat, viscosity, thermal conductivity, Prandtl number, isentropic exponent, choking velocity, p-v product, etc., over the entire range (to 1500 psia 1500°F). English units are used, but all conversion factors are given.

Table 1-43. DENSITY AND VISCOSITY OF WATER*
At Atmospheric Pressure

See also Tables 1-6, 1-10, and 1-12.

For temperature in K, add 273.15 to the values in deg C. For absolute viscosity in $N \cdot s/m^2$ ($= kg/m \cdot s$), multiply values in centipoises by 10^{-3}. For kinematic viscosity in m^2/s, multiply values in centistokes by 10^{-6}.

Temperature		Specific gravity	Absolute or dynamic viscosity			Kinematic viscosity	
°C	°F		Centi-poises	$lb_m/ft\ sec$	$slugs/ft\ sec$	Centi-stokes	ft^2/sec
0	32.0	0.9999	1.787	1.200×10^{-3}	3.720×10^{-5}	1.787	1.922×10^{-5}
2	35.6	1.0000	1.671	1.123×10^{-3}	3.490×10^{-5}	1.671	1.798×10^{-5}
4	39.2	1.0000	1.567	1.053×10^{-3}	3.273×10^{-5}	1.567	1.686×10^{-5}
6	42.8	1.0000	1.472	9.891×10^{-4}	3.074×10^{-5}	1.472	1.584×10^{-5}
8	46.4	.9999	1.386	9.314×10^{-4}	2.895×10^{-5}	1.386	1.491×10^{-5}
10	50.0	.9997	1.307	8.783×10^{-4}	2.730×10^{-5}	1.307	1.406×10^{-5}
12	53.6	.9995	1.235	8.299×10^{-4}	2.579×10^{-5}	1.236	1.330×10^{-5}
14	57.2	.9993	1.169	7.855×10^{-4}	2.441×10^{-5}	1.170	1.259×10^{-5}
16	60.8	.9990	1.109	7.452×10^{-4}	2.316×10^{-5}	1.110	1.194×10^{-5}
18	64.4	.9986	1.053	7.076×10^{-4}	2.199×10^{-5}	1.054	1.134×10^{-5}
20	68.0	.9982	1.002	6.733×10^{-4}	2.093×10^{-5}	1.004	1.080×10^{-5}
22	71.6	.9978	.9548	6.416×10^{-5}	1.994×10^{-5}	.9569	1.030×10^{-5}
24	75.2	.9973	.9111	6.122×10^{-4}	1.903×10^{-5}	.9135	9.829×10^{-6}
26	78.8	.9968	.8705	5.849×10^{-4}	1.818×10^{-5}	.8732	9.396×10^{-6}
28	82.4	.9963	.8327	5.595×10^{-4}	1.739×10^{-5}	.8358	8.993×10^{-6}
30	86.0	.9957	.7975	5.358×10^{-4}	1.666×10^{-5}	.8009	8.618×10^{-6}
32	89.6	.9951	.7647	5.139×10^{-4}	1.597×10^{-5}	.7685	8.269×10^{-6}
34	93.3	.9944	.7340	4.932×10^{-4}	1.533×10^{-5}	.7381	7.942×10^{-6}
36	96.8	.9937	.7052	4.739×10^{-4}	1.473×10^{-5}	.7097	7.636×10^{-6}
38	100	.9930	.6783	4.558×10^{-4}	1.417×10^{-5}	.6831	7.350×10^{-6}
40	104	.9922	.6529	4.387×10^{-4}	1.364×10^{-5}	.6580	7.080×10^{-6}
42	108	.9915	.6291	4.227×10^{-4}	1.314×10^{-5}	.6345	6.827×10^{-6}
44	111	.9907	.6067	4.077×10^{-4}	1.267×10^{-5}	.6124	6.589×10^{-6}
46	115	.9898	.5856	3.935×10^{-4}	1.223×10^{-5}	.5916	6.366×10^{-6}
48	118	.9890	.5656	3.801×10^{-4}	1.181×10^{-5}	.5719	6.154×10^{-6}
50	122	.9881	.5468	3.674×10^{-4}	1.142×10^{-5}	.5534	5.955×10^{-6}
52	126	.9871	.5290	3.555×10^{-4}	1.105×10^{-5}	.5359	5.766×10^{-6}
54	129	.9862	.5121	3.441×10^{-4}	1.070×10^{-5}	.5193	5.588×10^{-6}
56	133	.9852	.4961	3.334×10^{-4}	1.036×10^{-5}	.5036	5.419×10^{-6}
58	136	.9842	.4809	3.232×10^{-4}	1.004×10^{-5}	.4886	5.257×10^{-6}
60	140	.9832	.4665	3.135×10^{-4}	9.743×10^{-6}	.4745	5.106×10^{-6}
62	144	.9822	.4528	3.043×10^{-4}	9.456×10^{-6}	.4610	4.960×10^{-6}
64	147	.9811	.4398	2.955×10^{-4}	9.185×10^{-6}	.4483	4.824×10^{-6}
66	151	.9800	.4273	2.871×10^{-4}	8.924×10^{-6}	.4360	4.691×10^{-6}
68	154	.9789	.4155	2.792×10^{-4}	8.678×10^{-6}	.4245	4.567×10^{-6}
70	158	.9778	.4042	2.716×10^{-4}	8.442×10^{-6}	.4134	4.448×10^{-6}
72	162	.9766	.3934	2.644×10^{-4}	8.216×10^{-6}	.4028	4.334×10^{-6}
74	165	.9755	.3831	2.574×10^{-4}	8.001×10^{-6}	.3927	4.225×10^{-6}
76	169	.9743	.3732	2.508×10^{-4}	7.794×10^{-6}	.3830	4.121×10^{-6}
78	172	.9731	.3638	2.445×10^{-4}	7.598×10^{-6}	.3738	4.022×10^{-6}
80	176	.9718	.3537	2.377×10^{-4}	7.387×10^{-6}	.3640	3.917×10^{-6}
82	180	.9706	.3460	2.325×10^{-4}	7.226×10^{-6}	.3565	3.836×10^{-6}
84	183	.9693	.3377	2.269×10^{-4}	7.053×10^{-6}	.3484	3.749×10^{-6}
86	187	.9680	.3297	2.215×10^{-4}	6.886×10^{-6}	.3406	3.665×10^{-6}
88	190	.9667	.3221	2.164×10^{-4}	6.727×10^{-6}	.3332	3.588×10^{-6}
90	194	.9653	.3147	2.115×10^{-4}	6.573×10^{-6}	.3260	3.508×10^{-6}
92	198	.9640	.3076	2.067×10^{-4}	6.424×10^{-6}	.3191	3.434×10^{-6}
94	201	.9626	.3008	2.021×10^{-4}	6.282×10^{-6}	.3125	3.363×10^{-6}
96	205	.9612	.2942	1.977×10^{-4}	6.144×10^{-6}	.3061	3.294×10^{-6}
98	208	.9598	.2879	1.934×10^{-4}	6.013×10^{-6}	.3000	3.228×10^{-6}
100	212	.9584	.2818	1.894×10^{-4}	5.885×10^{-6}	.2940	3.163×10^{-6}

*Compiled from several sources.

Table 1-44. PROPERTIES OF COMMON LIQUIDS
At 1.0 Atm Pressure, 77°F (25°C), Except as Noted

For thermal properties see Table 1-45. For properties of liquids in SI units, see Table 1-46.

For viscosity in $N \cdot s/m^2$ ($= kg/m \cdot s$), multiply values in centipoises by 0.001. For surface tension in N/m, multiply values in dyne/cm by 0.001.

Common name	Chemical formula	Molecular weight	Density, $\frac{lb}{ft^3}$	Specific gravity	Viscosity $lb_m/ft\ sec \times 10^4$	Viscosity cp	Sound velocity, meters sec	Surface tension, dynes cm	Dielectric constant	Refractive index
Acetic acid	$C_2H_4O_2$	60.0537	65.493	1.049	7.76	1.155	1584[50]	27.3	6.15	1.37
Acetone	C_3H_6O	58.081	48.98	.787	2.12	0.316	1174	23.1	20.7	1.36
Alcohol, ethyl	C_2H_5OH	46.070	49.01	.787	7.36	1.095	1144	22.33	24.3	1.36
Alcohol, methyl	CH_3OH	32.043	49.10	.789	3.76	0.56	1103	22.2	32.6	1.33
Alcohol, propyl	C_3H_8O	60.098	49.94	.802	12.9	1.92	1205	23.5	20.1	1.38
Ammonia (aqua)	—	17.698	51.411	.826	—	—	—	—	16.9	—
Benzene	C_6H_6	78.117	54.55	.876	4.04	0.601	1298	28.18	2.2	1.50
Bromine	Br_2	159.818	—	—	6.38	0.95	—	41.5	3.20	—
Carbon disulfide	CS_2	76.140	78.72	1.265	2.42	0.36	1149	32.33	2.64	1.63
Carbon tetrachloride	CCl_4	153.824	98.91	1.59	6.11	0.91	924	26.3	2.23	1.46
Castor oil	—	—	59.69	0.960	—	650	1474	—	4.7	—
Chloroform	$CHCl_3$	119.378	91.44	1.47	3.56	0.53	995	27.14	4.8	1.44
Decane	$C_{10}H_{22}$	142.290	45.34	.728	5.77	0.859	—	23.43	2.0	1.41
Dodecane	$C_{12}H_{26}$	170.345	47.11	—	9.23	1.374	—	—	—	1.41
Ether	$C_4H_{10}O$	74.125	44.54	0.715	1.50	0.223	985	16.42	4.3	1.35
Ethylene glycol	$C_2H_6O_2$	62.070	68.47	1.100	109	16.2	1644	48.2	37.7	1.43
Fluorine refrigerant R–11	CCl_3F	137.369	92.14	1.480	2.82	0.42	—	18.3	2.0	1.37
Fluorine refrigerant R–12	CCl_2F_2	120.914	81.84	1.315	—	—	—	8.87	2.0	1.29
Fluorine refrigerant R–22	CHF_2Cl	86.469	74.53	1.197	—	—	—	8.35	2.0	1.26
Glycerine	$C_3H_8O_3$	92.096	78.62	1.263	6380	950	1909	63.0	40	1.47
Heptane	C_7H_{16}	100.208	42.42	.681	2.53	0.376	1138	19.9	1.92	1.38
Hexane	C_6H_4	86.181	40.88	.657	2.00	0.297	1203	18.0	—	1.37
Iodine	I_2	253.809	—	—	—	—	—	—	11	—
Kerosene	—	—	51.2	0.823	11.0	1.64	1320	—	—	—
Linseed oil	—	—	58.0	0.93	222	33.1	—	—	3.3	—
Mercury	Hg	200.59	—	13.633	10.3	1.53	1450	484		—
Octane	C_8H_{18}	114.235	43.61	.701	3.43	0.51	1171	21.14	—	1.40
Phenol	C_6H_6O	94.116	66.94	1.071	54	8.0	1274[100]	40.4	9.8	—
Propane	C_3H_8	44.098	30.81	.495	0.74	0.11	—	6.6	1.27	1.34
Propylene	C_3H_6	42.082	32.11	.516	0.60	0.09	—	7.0	—	1.36
Propylene glycol	$C_3H_8O_2$	76.097	60.26	.968	—	42	—	36.3	—	1.43
Sea water	—	18.52	64.0	1.03	—	—	1535	—	—	—
Toluene	C_7H_8	92.144	53.83	0.865	3.70	0.550	1275[30]	27.3	2.4	1.49
Turpentine	$C_{10}H_{16}$	136.242	54.2	0.87	9.24	1.375	1240	—	—	1.47
Water	H_2O	18.0153	62.247	1.00	6.0	0.89	1498	71.97	78.54[a]	1.33

[a]The dielectric constant of water near the freezing point is 87.8; it decreases with increase in temperature to about 55.6 near the boiling point.

Table 1-45. THERMAL PROPERTIES OF COMMON LIQUIDS*

At 1.0 Atm Pressure, 77°F (25°C), Except as Noted

For other properties of liquids, see Table 1-44. For properties of liquids in SI units, see Table 1-46.

Common name	Specific heat, Btu/lbm °F	Thermal conductivity, Btu/ft hr °F	Freezing point, °F	Latent heat of fusion, Btu/lb	Boiling point, °F	Latent heat of evaporation, Btu/lb	Coefficient of cubical expansion per °F
Acetic acid	0.522	0.099	62	77.7	245	173	0.0006
Acetone	0.514	0.093	−137.4	42.3	133	223	0.00082
Alcohol, ethyl	0.584	0.099	−174.2	46.4	172.96	364	0.0006
Alcohol, methyl	0.606	0.117	−143.7	42.5	148.4	474	0.00075
Alcohol, propyl	0.567	0.093	−197	37.2	208	335	—
Ammonia (aqua)	1.047	0.204	—	—	—	—	—
Benzene	0.414	0.083	41.96	54.4	176.2	168	0.0007
Bromine	0.113		17.15	28.7	137.3	83	0.00065
Carbon disulfide	0.237	0.093	−169.5	24.80	115.26	151.2	0.0007
Carbon tetrachloride	0.207	0.060	−9.04	74.8	169.7	83.5	0.0007
Castor oil	0.47	0.104	14.1	—	—	—	—
Chloroform	0.25	0.068	−82.3	33.14	142.2	106.4	0.00073
Decane	0.528	0.085	−21.4	86.6	345.3	113	—
Dodecane	0.528	0.081	−14.74	93.0	421.3	110	—
Ether	0.529	0.075	−177	41.4	94.2	160	0.0009
Ethylene glycol	0.565	0.149	8.6	77.9	387	344	—
Fluorine refrigerant R–11	0.208†	0.054†	−168	—	74.9	77.58‡	—
Fluorine refrigerant R–12	0.232†	0.041†	−252	14.8	−21.6	71.04‡	—
Fluorine refrigerant R–22	0.300†	0.050†	−256	78.7	−41.4	100.05‡	—
Glycerine	0.627	0.166	17.0	86	554.5	419	0.0003
Heptane	0.536	0.074	−131.1	60.2	209.1	137	—
Hexane	0.541	0.072	−139.3	65.3	155.65	157	—
Iodine	0.513	—	236.3	26.74	363.8	70.71	—
Kerosene	0.5	0.084	—	—	—	108	—
Linseed oil	0.44	—	−4	—	549	—	—
Mercury	0.0333	—	−38.0	5.0	674	126.9	0.0001
Octane	0.514	0.076	−70.2	78.0	257	128	0.0004
Phenol	0.342	0.11	109.4	52.2	360	—	0.0005
Propane, R–290	0.576†	—	−305.8	34.38	−43.73	184‡	—
Propylene	0.682	—	−301.5	30.70	−53.86	147	—
Propylene glycol	0.598	—	−76	—	369	393	—
Sea water	0.90–.98	—	27.5	—	—	—	—
Toluene	0.41	0.077	−139	30.90	230.8	156	—
Turpentine	0.425	0.070	−75	—	320	126	0.00055
Water	0.998	0.352	32	143.3	212	970.3	0.00011

†At 75°F, liquid.
‡At 14.7 psia, saturation temperature.
*Compiled from several sources.

Table 1-46. PROPERTIES OF COMMON LIQUIDS—SI UNITS

At 1.0 Atm Pressure (0.101 325 MN/m²), 300 K, Except as Noted

For other data on these liquids, see Tables 1-44 and 1-45.

Common name	Density, kg/m³	Specific heat, kJ/kg·K	Viscosity, N·s/m²	Thermal conductivity, W/m·K	Freezing point, K	Latent heat of fusion, kJ/kg	Boiling point, K	Latent heat of evapora-tion, kJ/kg	Coefficient of cubical expansion per K
Acetic acid	1 049	2.18	.001 155	0.171	290	181	391	402	0.001 1
Acetone	784.6	2.15	.000 316	0.161	179.0	98.3	329	518	0.001 5
Alcohol, ethyl	785.1	2.44	.001 095	0.171	158.6	108	351.46	846	0.001 1
Alcohol, methyl	786.5	2.54	.000 56	0.202	175.5	98.8	337.8	1 100	0.001 4
Alcohol, propyl	800.0	2.37	.001 92	0.161	146	86.5	371	779	
Ammonia (aqua)	823.5	4.38		0.353					
Benzene	873.8	1.73	.000 601	0.144	278.68	126	353.3	390	0.001 3
Bromine		.473	.000 95		245.84	66.7	331.6	193	0.001 2
Carbon disulfide	1 261	.992	.000 36	0.161	161.2	57.6	319.40	351	0.001 3
Carbon tetrachloride	1 584	.866	.000 91	0.104	250.35	174	349.6	194	0.001 3
Castor oil	956.1	1.97	.650	0.180	263.2				
Chloroform	1 465	1.05	.000 53	0.118	209.6	77.0	334.4	247	0.001 3
Decane	726.3	2.21	.000 859	0.147	243.5	201	447.2	263	
Dodecane	754.6	2.21	.001 374	0.140	247.18	216	489.4	256	
Ether	713.5	2.21	.000 223	0.130	157	96.2	307.7	372	0.001 6
Ethylene glycol	1 097	2.36	.016 2	0.258	260.2	181	470	800	
Fluorine refrigerant R-11	1 476	.870[a]	.000 42	0.093[a]	162		297.0	180[b]	
Fluorine refrigerant R-12	1 311	.971[a]		0.071[a]	115	34.4	243.4	165[b]	
Fluorine refrigerant R-22	1 194	1.26[a]		0.086[a]	113	183	232.4	232[b]	
Glycerine	1 259	2.62	.950	0.287	264.8	200	563.4	974	0.000 54
Heptane	679.5	2.24	.000 376	0.128	182.54	140	371.5	318	
Hexane	654.8	2.26	.000 297	0.124	178.0	152	341.84	365	
Iodine		2.15			386.6	62.2	457.5	164	
Kerosene	820.1	2.09	.001 64	0.145				251	
Linseed oil	929.1	1.84	.033 1		253		560		
Mercury		.139	.001 53		234.3	11.6	630	295	0.000 18
Octane	698.6	2.15	.000 51	0.131	216.4	181	398	298	0.000 72
Phenol	1 072	1.43	.008 0	0.190	316.2	121	455		0.000 90
Propane	493.5	2.41[a]	.000 11		85.5	79.9	231.08	428[b]	
Propylene	514.4	2.85	.000 09		87.9	71.4	225.45	342	
Propylene glycol	965.3	2.50	.042		213		460	914	
Sea water	1 025	3.76–4.10			270.6				
Toluene	862.3	1.72	.000 550	0.133	178	71.8	383.6	363	
Turpentine	868.2	1.78	.001 375	0.121	214		433	293	0.000 99
Water	997.1	4.18	.000 89	0.609	273	333	373	2 260	0.000 20

[a] At 297 K, liquid.
[b] At .101 325 meganewtons, saturation temperature.

Table 1-47. SURFACE TENSION OF LIQUIDS*
At Atmospheric Pressure and Room Temperature

For surface tension of other common liquids, see Table 1-44. For surface tension of water and of alcohol solutions, see Table 1-49.

For surface tension in N/m, multiply values in dynes/cm by 0.001.

Name	In contact with	Surface tension, dynes/cm	Name	In contact with	Surface tension, dynes/cm
Acetaldehyde	Vapor	21.2	Glycol	Air or vapor	47.7
Acetic acid	Vapor	27.3	n-Hexane	Air	18.4
Acetone	Air or vapor	23.1	Isobutyl alcohol	Vapor	23.0
Allyl alcohol	Air or vapor	25.8	Isopropyl alcohol	Air or vapor	21.7
Aniline	Vapor	42.9	Mercury	Air	484.
Benzene	Air	28.2	Methyl acetate	Air or vapor	24.6
Bromine	Air or vapor	41.5	Methyl alcohol	Air	22.2
n-Butyl alcohol	Air or vapor	24.6	Nitrobenzene	Air or vapor	43.9
Carbon bisulfide	Vapor	32.3	Nitromethane	Vapor	36.8
Carbon tetrachloride	Vapor	26.3	n-Octane	Vapor	21.8
Chlorobenzene	Vapor	33.6	Phenol	Air or vapor	40.9
Chloroform	Air	27.1	n-Propyl acetate	Air or vapor	24.3
Diethylaniline	Vapor	34.2	n-Propyl alcohol	Vapor	23.8
Ethyl acetate	Air	23.9	Styrene	Air	32.1
Ethyl alcohol	Vapor	22.4	Sulfuric acid (98.5%)	Air or vapor	55.1
Ethyl ether	Vapor	16.5	Tetrachloroethylene	Vapor	31.7
Formic acid	Air	37.6	Toluene	Vapor	28.5
Glycerol	Air	63.0	Water	Air	72.

*Condensed from: "CRC Handbook of Chemistry and Physics", 53rd ed., R.C. Weast, Ed., The Chemical Rubber Co., 1972. For surface tension of other liquids, liquid elements, and metals, see this source.

Table 1-48. INDEX OF REFRACTION OF LIQUIDS AND SOLUTIONS
At Room Temperature

For the refractive index of other common liquids, see Table 1-44.

The refractive index varies somewhat with wavelength. Most determinations are made at 0.589 μ, the sodium D-line.

Liquid	Refractive index, relative to air, n_D	Liquid	Refractive index, relative to air, n_D
Acetaldehyde	1.332	n-Hexane	1.375
Acetic acid	1.372	Hydrochloric acid, 10% sol.	1.356
Acetic acid, 20% sol.	1.348	Magnesium chloride, 20% sol.	1.385
Acetone	1.359	Methanol	1.329
Allyl alcohol	1.414	Methanol, 20% sol.	1.338
Ammonium chloride, 20% sol.	1.371	Methyl acetate	1.360
Ammonium hydroxide, 20% sol.	1.338	Nitric acid, 20% sol.	1.360
Amyl acetate	1.400	Nitrobenzene	1.555
Amyl alcohol	1.410	Octane	1.398
Aniline	1.586	Phosphoric acid, 20% sol.	1.352
Benzene	1.501	Potassium carbonate, 20% sol.	1.367
Butyl acetate	1.396	Potassium chloride, 20% sol.	1.361
Butyl alcohol	1.440	Potassium chromate, 20% sol.	1.380
Calcium chloride, 20% sol.	1.384	Potassium iodide, 20% sol.	1.363
Carbon bisulfide	1.630	Potassium nitrate, 20% sol.	1.352
Carbon tetrachloride	1.465	Propylene glycol	1.432
Chlorobenzene	1.525	Silver nitrate, 20% sol.	1.357
Chloroform	1.445	Sodium acetate, 20% sol.	1.361
Citric acid, 20% sol.	1.360	Sodium chloride, 20% sol.	1.368
n-Decane	1.412	Sodium hydroxide, 20% sol.	1.384
Ethanol	1.361	Sodium nitrate, 20% sol.	1.356
Ethanol, 20% sol.	1.347	Sodium sulfate, 20% sol.	1.362
Ferric chloride, 10% sol.	1.361	Sodium thiosulfate, 20% sol.	1.376
Formic acid	1.371	Sucrose, 20% sol.	1.364
Formic acid, 20% sol.	1.344	Sulfuric acid, 20% sol.	1.358
Glycerol	1.474	Toluene	1.495
Glycerol, 20% sol.	1.357		

REFERENCE
"CRC Handbook of Chemistry and Physics", 53rd ed., R.C. Weast, Ed., The Chemical Rubber Co., 1972; gives data on refractive index and other properties of inorganic compounds, organic compounds, and aqueous solutions (at 1% intervals).

Table 1-49. SURFACE TENSION OF WATER
AND OF ALCOHOL SOLUTIONS*

In Contact with Air

For surface tension in N/m, multiply values in dynes/cm by 0.001 000.

PURE WATER

Temperature		Surface tension, dynes/cm	Temperature		Surface tension, dynes/cm
°C	°F		°C	°F	
−8	17.6	77.0	25	77.	72.0
−5	23.	76.4	30	86.	71.2
0	32.	75.6	40	104.	69.6
5	41.	74.9	50	122.	67.9
10	50.	74.2	60	140.	66.2
15	59.	73.5	70	158.	64,4
18	64.4	73.1	80	176.	62.6
20	68.	72.8	100	212.	58.9

ALCOHOL SOLUTIONS

Temperature		Surface tension, dynes/cm			
		Ethyl alcohol		Methyl alcohol	
°C	°F	% by volume	Surface tension	% by volume	Surface tension
20	68	10		10	59.0
30	86	10		10	57.3
40	104	10	48.3	10	
50	122	10	46.8	10	55.0
20	68	25		25	46.4
30	86	25		25	45.3
40	104	25	35.1	25	
50	122	25	34.0	25	43.2
20	68	60	27.6	60	33.0
30	86	60		60	32.3
40	104	60	26.2	60	
50	122	60	25.5	60	30.8
20	68	80	24.9	80	27.3
30	86	80		80	26.5
40	104	80	23.4	80	
50	122	80	22.6	80	25.0
20	68	100	22.3	100	22.6
30	86	100	21.5	100	21.6
40	104	100		100	
50	122	100		100	19.5

*Adapted from: "CRC Handbook of Chemistry and Physics", 53rd ed., R.C. Weast, Ed., The Chemical Rubber Co., 1972.

REFERENCES
"American Institute of Physics Handbook", 2nd ed., D.E. Gray, Ed., McGraw-Hill Book Company, 1963.

"Handbook of Chemistry", 10th ed., N.A. Lange, Ed., McGraw-Hill Book Company, 1961.

Table 1-50. ISOTHERMAL COMPRESSIBILITY OF LIQUIDS*

The following table lists the approximate compressibility (change in volume/unit volume) per atmosphere increase in pressure at constant temperature. Unit compressibility is less at the higher pressures, but no large amount of data is available. In the usual range of 0–100°C, compressibility is not greatly affected by temperature (see References). For water and many common liquids, a rough approximation is that the volume is reduced about 5 percent by a pressure of 1,000 atmospheres. The reciprocal of the compressibility varies almost linearly with the pressure. Adiabatic compressibility can hardly be determined experimentally but may be computed, using the velocity of sound in the liquid. The resulting values are generally slightly lower than those given here for the isothermal compressibility. The compressibility of concentrated water solutions is ordinarily lower than that of pure water. Compressibility of liquified gases is of about the same order of magnitude as that of other liquids.

Liquid	Compressibility at room temperature $\Delta V/V$, in % for each atmosphere change in pressure at or around pressure of			
	1 atm†	250 atm	1,000 atm	5,000 atm
Acetic acid	.009	—	—	—
Acetone	.0125	—	.0055	.002
Aniline	.0045	—	.003	—
Benzene	.0095	—	.005	—
n-Butyl alcohol	.0095	—		
Carbon disulfide	.0095	—	.005	.002
Carbon tetrachloride	.0106	—	.0052	—
Chloroform	.0100	.008	.0052	—
Dodecane	.009	—	.005	
Ethanol	.0114	.009	.005	.002
Ethyl ether	.019	—	.006	—
Glycerol	.0025	—	.002	.0012
n-Heptane	.014	—	.006	
n-Hexane	.016	.011	.0064	—
Kerosene	—	—	.0045	—
Mercury	.0004	—	.0004	—
Methanol	.012	—	.005	.002
n-Octane	.012	—	.0056	.002
Oils, petroleum	.007	—	—	—
Oils, vegetable	.005	—	—	—
Phenol	.005	—	—	—
n-Propyl alcohol	.009	—	.004	—
Toluene	.009	—	—	—
Water	.0046	.004	.0035	.0016
m-Xylene	.008	—	—	—

†1 atm = 14.696 psi = 1.013 bars = 1.033 kg/cm² = 1.013 × 10⁶ dynes/cm² = 0.101 3 MN/m².
*Compiled from several sources.

REFERENCES

"American Institute of Physics Handbook", 2nd ed., D.E. Gray, Ed., McGraw-Hill Book Company, 1963, pp. 2–162 to 2–179 (gives extensive data and references).

"CRC Handbook of Chemistry and Physics", 53rd ed., R.C. Weast, Ed., The Chemical Rubber Co., 1972 (lists 45 liquids).
"International Critical Tables of Numerical Data", National Research Council, Vol. I, McGraw-Hill Book Company, 1926.
"The Physics of High Pressure", P.W. Bridgman, G. Bell & Sons, 1949.
"Smithsonian Physical Tables", 9th ed., W.E. Forsythe, Ed., The Smithsonian Institution, 1956.

Table 1-51. ANTIFREEZE SOLUTIONS*

Specific gravity is given at 60 deg F. Specific heat is in Btu/lbm·deg R = cal/g·K; for kJ/kg·K multiply by 4.184. Viscosity is in centipoises; for N·s/m² (= kg/m·s) divide by 1 000; for lbm/s·ft multiply by 0.000 672. Heat conductivity is in Btu/hr·ft·deg F; for W/m·K multiply by 1.729 6.

Solution and property	Percentage by weight—pure antifreeze agent†								
	5	10	15	20	25	30	35	40	50
ETHANOL									
Specific gravity	.991	.984	.977	.970	.963	.956	.947	.937	.915
Freezing point, °F	28.2	23.9	18.8	12.3	4.2	−4.9			
Specific heat, 68°F	1.00	1.00	1.00	0.99	0.98	0.96	0.92	0.90	
Specific heat, 32°F									
Viscosity, cp, 68°F									
Viscosity, cp, 32°F									
Heat conductivity, 68°F									
Heat conductivity, 32°F									
METHANOL									
Specific gravity	.991	.983	.976	.968	.960	.953	.945	.937	.916
Freezing point, °F	26.6	20.2	13.0	4.9	−4.2	−13.5	−25.3		
Specific heat, 68°F	1.00	1.00	0.99	0.98	0.97	0.95	0.93	0.91	
Specific heat, 32°F									
Viscosity, cp, 68°F									
Viscosity, cp, 32°F									
Heat conductivity, 68°F									
Heat conductivity, 32°F									
ETHYLENE GLYCOL									
Specific gravity		1.012		1.025		1.04		1.055	1.065
Freezing point, °F		24		15		4		−12	−32
Specific heat, 68°F		.97		.94		.89		.84	.79
Specific heat, 32°F		.96		.93		.87		.81	.76
Viscosity, cp, 68°F		1.4		1.9		2.4		3.1	4.1
Viscosity, cp, 32°F		2.5		3.0		4.0		5.3	8.
Heat conductivity, 68°F		.33		.31		.28		.26	.24
Heat conductivity, 32°F		.32		.30		.28		.26	.24
PROPYLENE GLYCOL									
Specific gravity		1.006		1.016		1.026		1.035	1.042
Freezing point, °F		27		18		8		−7	−25
Specific heat, 68°F		.99		.96		.94		.89	.86
Specific heat, 32°F		.99		.96		.94		.89	.85
Viscosity, cp, 68°F		1.5		2.0		3.0		6.1	9.
Viscosity, cp, 32°F		2.8		4.2		6.5		11.	30.
Heat conductivity, 68°F		.32		.29		.27		.24	.23
Heat conductivity, 32°F		.31		.28		.26		.24	.23
GLYCEROL									
Specific gravity	1.012	1.023	1.036	1.048	1.060	1.074	1.87	1.10	1.13
Freezing point, °F	31.	29.5	26.5	23.5	19.5	15.	10.	4.5	−9.5
Specific heat, 68°F	0.97	0.96	0.94	0.93	0.91	0.90	0.88	0.86	0.83
Specific heat, 32°F	.96	0.94	0.92	0.90	0.88	0.87	0.86	0.84	0.80
Viscosity, cp, 68°F		1.31		1.76		2.50		3.72	6.0
Viscosity, cp, 32°F		2.44		3.44		5.14		8.25	14.6
Heat conductivity, 68°F									
Heat conductivity, 32°F									
SODIUM CHLORIDE									
Specific gravity	1.035	1.073	1.110	1.151	1.190				
Freezing point, °F	26.7	20.2	12.3	2.4	+16				
Specific heat, 68°F	.94	.89	.85	.81	.79				
Specific heat, 32°F	.93	.88	.84	.80	.78				
Viscosity, cp, 68°F		1.4	1.6	1.9	2.3				
Viscosity, cp, 32°F		2.0	2.3	2.7	3.4				
Heat conductivity, 68°F	.32	.31	.29	.28	.27				
Heat conductivity, 32°F	.31	.29	.28	.26	.25				
CALCIUM CHLORIDE									
Specific gravity	1.043	1.088	1.138	1.183	1.230	1.287			
Freezing point, °F	27.7	22.	13.5	−1.0	−21.	−48.			
Specific heat, 68°F	.91	.86	.80	.74	.69	.66			
Specific heat, 32°F	.90	.85	.79	.73	.68	.65			
Viscosity, cp, 68°F		1.3	1.6	2.0	3.0	3.8			
Viscosity, cp, 32°F		2.4	2.7	3.1	4.1	5.9			
Heat conductivity, 68°F	.34	.34	.33	.33	.33	.32			
Heat conductivity, 32°F	.32	.32	.31	.31	.31	.30			

†For commercial grades divide by decimal purity.
*Compiled from several sources.

Table 1-52. CRYOGENIC AND REFRIGERATING LIQUIDS*

Fluid	Boiling point °K	°R	°C	°F	Density, lb/ft³	Volume ratio (to room temp), gas/liq.	Latent heat of vaporization, Btu/lbm	Specific heat, c_p liquid, Btu/lbm °R	gas, Btu/lbm °R	Viscosity liquid, lbm/ft hr	gas, lbm/ft hr	Thermal conductivity liquid, Btu/hr ft °F	gas, Btu/hr ft °F	Dielectric constant
Air	79	142	−194	−318	54.6	740:1	88.2	.470	.245	.195	.016		.0043	
Argon	87	157	−186	−303	87.6	840:1	69.5	.272	.127	.610	.0200	.0712	.00350	1.52
Carbon dioxide[a]	195	350	−79	−110	97.6	730:1	246	.318	.190				.0085[c]	1.59[c]
Ethane	185	334	−88	−126	34.2	420:1	210	.600	.250				.0145[d]	
Fluorine	85	154	−188	−306	94.0	880:1	71.6	.37	.154	.592	.0178	.078	.00416	1.43
Helium 3	3.20	5.76	−270	−454	3.68	600:1	3.65	1.10		.00392		.0099		
Helium 4	4.215	7.59	−269	−452	7.80	600:1	8.92	1.09	1.63	.00864	.00309	.0156	.00560	1.0492
Hydrogen	20	36.7	−253	−423	4.43	800:1	192	2.34	2.85	.0316	.00254	.0683	.0090	1.226
Methane	111	201	−162	−259	25.5	550:1	219	.825	.398	.287	.0110	.0642		1.68
Neon	27	48.8	−246	−411	75.2	1400:1	37.3	0.44	.280	.30		.075	.0057	
Nitrogen	77.4	139.2	−196	−320	50.6	700:1	85.3	.487	.259	.382	.0134	.0804	.00415	1.434
Propane	231	416	−42	−44	35.3	310:1	183	.526	.290					1.27
Propylene	225	406	−48	−54	33.3	330:1	188	.619	.363					
Refrigerant 12	243	438	−30	−22	92.9	280:1	71.1	.213	.109	.897	.0264	.035		2.13
Refrigerant 13	192	346	−81	−114	95.0	350:1	64	.214	.0987					
Refrigerant 13B1	215	388	−58	−72	124	310:1	51.1	.405	.203					
Refrigerant 22	232	419	−41	−41	83.2	380:1	100.5	.252	.111	.849	.0254	.0635[b]	.00450	2.44

[a] Sublimes.
[b] At −4°F.
[c] At 32°F.
[d] At 125°F.
*Compiled from several sources.

Table 1-53. PROPERTIES OF LIQUID METALS*
At Atmospheric or Pumping Pressures

For properties of sodium, potassium, and sodium-potassium mixtures and for conversion factors, see Table 1-77. For other data see Tables 1-60, 3-1, and 3-2.

Metal (Melting point, °F)	Temperature		Specific gravity	Specific heat	Thermal conductivity		Absolute viscosity	
	°F	°C			$\dfrac{Btu}{hr\ ft\ °F}$	$\dfrac{cal}{sec\ cm\ °C†}$	$\dfrac{lb_m}{ft\ sec}$	centipoises
Aluminum (1220)	1250	677	2.38	.259				
	1300	704	2.37	.259	60.2	.249	1.88×10^{-3}	2.8
	1350	732	2.36	.259	63.4	.262	1.61×10^{-3}	2.4
	1400	760	2.35	.259	64.3	.266	1.34×10^{-3}	2.0
	1450	788	2.34	.259	69.9	.289	1.08×10^{-3}	1.6
Bismuth (520)	600	316	10.0	.0345	9.5	.039	1.09×10^{-3}	1.62
	800	427	9.87	.0357	9.0	.037	9.0×10^{-4}	1.34
	1000	538	9.74	.0369	9.0	.037	7.4×10^{-4}	1.10
	1200	649	9.61	.0381	9.0	.037	6.2×10^{-4}	.923
Cesium (83)	83	28	1.84	.060	10.6	.044		
	150	66					3.84×10^{-4}	.571
	250	121					2.95×10^{-4}	.439
	350	177					2.47×10^{-4}	.368
	400	204					2.30×10^{-4}	.343
Lead (621)	700	371	10.5	.038	9.3	.038	1.61×10^{-3}	2.39
	850	454	10.4	.037	9.0	.037	1.38×10^{-3}	2.05
	1000	538	10.4	.037	8.9	.036	1.17×10^{-3}	1.74
	1150	621	10.2	.037	8.7	.036	1.02×10^{-3}	1.52
	1300	704	10.1		8.6	.035	9.20×10^{-4}	1.37
Lithium (355)	400	204	.506	1.0	24.	.10	4.0×10^{-4}	.595
	600	316	.497	1.0	23.	.095	3.4×10^{-4}	.506
	800	427	.489	1.0	22.	.090	3.7×10^{-4}	.551
	1200	649	.471				2.9×10^{-4}	.432
	1800	942	.442				2.8×10^{-4}	.417
Magnesium (1203)	1250	677	1.55	.318				
	1301	705	1.53	.320				
	1350	732	1.49	.322				
Mercury (−38)	50	10	13.6	.033	4.8	.020	1.07×10^{-3}	1.59
	200	93	13.4	.033	6.0	.025	8.4×10^{-4}	1.25
	300	149	13.2	.033	6.7	.028	7.4×10^{-4}	1.10
	400	204	13.1	.032	7.2	.030	6.7×10^{-4}	.997
	600	316	12.8	.032	8.1	.033	5.8×10^{-4}	.863
Tin (449)	500	260	6.94	.058	19	.079	1.22×10^{-3}	1.82
	700	371	6.86	.060	19.4	.080	9.8×10^{-4}	1.46
	850	454	6.81	.062	19	.079	8.5×10^{-4}	1.26
	1000	538	6.74	.064	19	.079	7.6×10^{-4}	1.13
	1200	649	6.68	.066	19	.079	6.7×10^{-4}	.997
Zinc (787)	600	316	6.97	.123	35.4	.146		
	850	454	6.90	.119	33.7	.139	2.10×10^{-3}	3.12
	1000	538	6.86	.116	33.2	.137	1.72×10^{-3}	2.56
	1200	649	6.76	.113	32.8	.136	1.39×10^{-3}	2.07
	1500	816	6.74	.107	32.6	.135	9.83×10^{-4}	1.46

†For watt/cm °C, multiply by 418.4.
*Based largely on: "Liquid Metals Handbook", 3rd ed., Office of Naval Research, U.S. Government Printing Office, 1955.

Table 1-54. BINARY AZEOTROPES*

The following compounds form a binary azeotrope, a liquid of constant boiling point, when mixed in the given proportions.

No.	Component A	Per-cent of A	Component B	Per-cent of B	Boiling points, °C A	B	Azeotrope
1	Acetic acid	3.0	Water	97.0	118.1	100.0	76.6
2	Acetic acid	2.0	Benzene	98.0	118.1	80.1	80.1
3	Acetic acid	58.5	Chlorobenzene	41.5	118.1	132.0	114.7
4	Acetone	88.5	Water	11.5	56.2	100.0	56.1
5	Acetone	33.0	Carbon disulfide	67.0	56.2	46.3	39.3
6	Acetone	59.0	Hexane	41.0	56.2	69.0	49.8
7	Acetone	88.0	Methanol	12.0	56.2	64.7	55.7
8	Allyl alcohol	72.9	Water	27.1	97.1	100.0	88.2
9	Allyl alcohol	17.4	Benzene	82.6	97.1	80.1	76.8
10	Allyl alcohol	4.5	Hexane	95.5	97.1	69.0	65.5
11	Benzene	91.1	Water	8.9	80.1	100.0	69.4
12	Benzene	67.6	Ethanol	32.4	80.1	78.5	67.8
13	Benzene	60.5	Methanol	39.5	80.1	64.7	58.3
14	l-Butanol	55.5	Water	44.5	117.7	100.0	93.0
15	Butyl acetate	72.9	Water	27.1	126.5	100.0	90.7
16	Butyl ether	66.6	Water	33.4	142.0	100.0	94.1
17	Butyl ether	93.6	Ethylene glycol	6.4	142.0	197.2	139.5
18	Carbon disulfide	98.0	Water	2.0	46.3	100.0	43.6
19	Carbon disulfide	1.0	Ethyl ether	99.0	46.3	34.6	34.4
20	Carbon tetra-chloride	95.9	Water	4.1	76.8	100.0	66.8
21	Carbon tetra-chloride	57.0	Ethyl acetate	43.0	76.8	77.1	74.8
22	Carbon tetra-chloride	79.4	Methanol	20.6	76.8	64.7	55.7
23	Chloroform	97.0	Water	3.0	61.2	100.0	56.3
24	Chloroform	93.0	Ethanol	7.0	61.2	78.5	59.4
25	Chloroform	87.0	Methanol	13.0	61.2	64.7	53.5
26	Ethanol	95.6	Water	4.4	78.5	100.0	78.2
27	Ethanol	31.0	Ethyl acetate	69.0	78.5	77.1	71.8
28	Ethanol	48.0	Heptane	52.0	78.5	98.4	72.0
29	Ethanol	21.0	Hexane	79.0	78.5	69.0	58.7
30	Ethanol	68.0	Toluene	32.0	78.5	110.6	76.7
31	Ethyl acetate	91.9	Water	8.1	77.1	100.0	70.4
32	Ethyl ether	98.8	Water	1.2	34.6	100.0	34.2
33	Formic acid	77.5	Water	22.5	100.7	100.0	107.1
34	Heptane	87.1	Water	12.9	98.4	100.0	79.2
35	Heptane	48.5	Methanol	51.5	98.4	64.7	59.1
36	Hexane	94.4	Water	5.6	69.0	100.0	61.6
37	Hexane	73.1	Methanol	26.9	69.0	64.7	50.0
38	Hexane	96.0	Propanol	4.0	69.0	97.2	65.7
39	Hydrogen chloride	20.2	Water	79.8	−83.7	100.0	108.6
40	Isobutyl alcohol	70.0	Water	30.0	108.4	100.0	89.7
41	Isopropyl ether	95.4	Water	4.6	67.5	100.0	62.2
42	Methanol	72.0	Octane	28.0	64.7	125.8	63.0
43	Methanol	72.4	Toluene	27.6	64.7	110.6	63.7
44	Nonane	60.2	Water	39.8	150.8	100.0	95.0
45	Pentane	98.6	Water	1.4	36.1	100.0	34.6
46	Phenol	9.2	Water	90.8	182.0	100.0	99.5
47	Propanol	71.8	Water	28.2	97.2	100.0	88.1
48	Propanol	49.0	Toluene	51.0	97.2	110.6	92.6
49	Styrene	59.1	Water	40.9	145.2	100.0	93.9
50	Toluene	79.8	Water	20.2	110.6	100.0	85.0
51	m-Xylene	60.0	Water	40.0	139.1	100.0	94.5

*From: "CRC Handbook of Chemistry and Physics", 53rd ed., R.C. Weast, Ed., The Chemical Rubber Co., 1972; this source gives data on 1,000 azeotropes (binary and ternary).

Table 1-55. CORRECTION OF BOILING POINTS
TO STANDARD PRESSURE*

H.B. HASS AND R.F. NEWTON

This correction may be made by using the equation

$$\Delta t = \frac{(273.1 + t)(2.8808 - \log p)}{\phi + .15(2.8808 - \log p)}$$

where Δt = degrees C to be added to the observed boiling point
t = the observed boiling point
$\log p$ = the logarithm of the observed pressure in millimeters of mercury
ϕ = the entropy of vaporization at 760 mm.

The value of ϕ may be estimated from the graph and the table. Substances not included in the table may be classified by grouping them with compounds which bear a close physical or structural resemblance to them.

Compound	Group	Compound	Group	Compound	Group
Acetaldehyde	3	Cyanogen	4	Methyl amine	5
Acetic acid	4	Cyanogen chloride	3	Methyl benzoate	3
Acetic anhydride	6	Dibenzyl ketone	2	Methyl ether	3
Acetone	3	Dimethyl amine	4	Methyl ethyl ether	3
Acetophenone	4	Dimethyl oxalate	4	Methyl ethyl ketone	2
Amines	3	Dimethyl silicane	2	Methyl fluoride	3
n-Amyl alcohol	8	Esters	3	Methyl formate	4
Anthracene	1	Ethanol	8	Methyl salicylate	2
Anthraquinone	1	Ethers	2	Methyl silicane	1
Benzaldehyde	2	Ethylamine	4	α-, β-Naphthols	3
Benzoic acid	5	Ethylene glycol	7	Nitrobenzene	3
Benzonitrile	2	Ethylene oxide	3	Nitromethane	3
Benzophenone	2	Formic acid	3	o-, m-, p-Nitrotoluenes	2
Benzyl alcohol	5	Glycol diacetate	4	o-, m-, p-Nitrotoluidines	2
Butylethylene	1	Halogen derivatives	Same group as	Phenanthrene	1
Butyric acid	7		though halogen	Phenol	5
Camphor	2		were hydrogen	Phosgene	2
Carbon monoxide	1	Heptylic acid	7	Phthalic anhydride	2
Carbon oxysulfide	2	Hydrocarbons	2	Propionic acid	5
Carbon suboxide	2	Hydrogen cyanide	3	n-Propyl alcohol	8
Carbon sulfoselenide	2	Isoamyl alcohol	7	Quinoline	2
m-, p-Chloroanilines	3	Isobutyl alcohol	8	Sulfides	2
Chlorinated derivatives	Same group as	Isobutyric acid	6	Tetranitromethane	3
	though Cl	Isocaproic acid	7	Trichloroethylene	1
	were H	Methane	1	Valeric acid	7
o-, m-, p-Cresols	4	Methanol	7	Water	6

Table 1-55. CORRECTION OF BOILING POINTS
TO STANDARD PRESSURE *(Continued)*

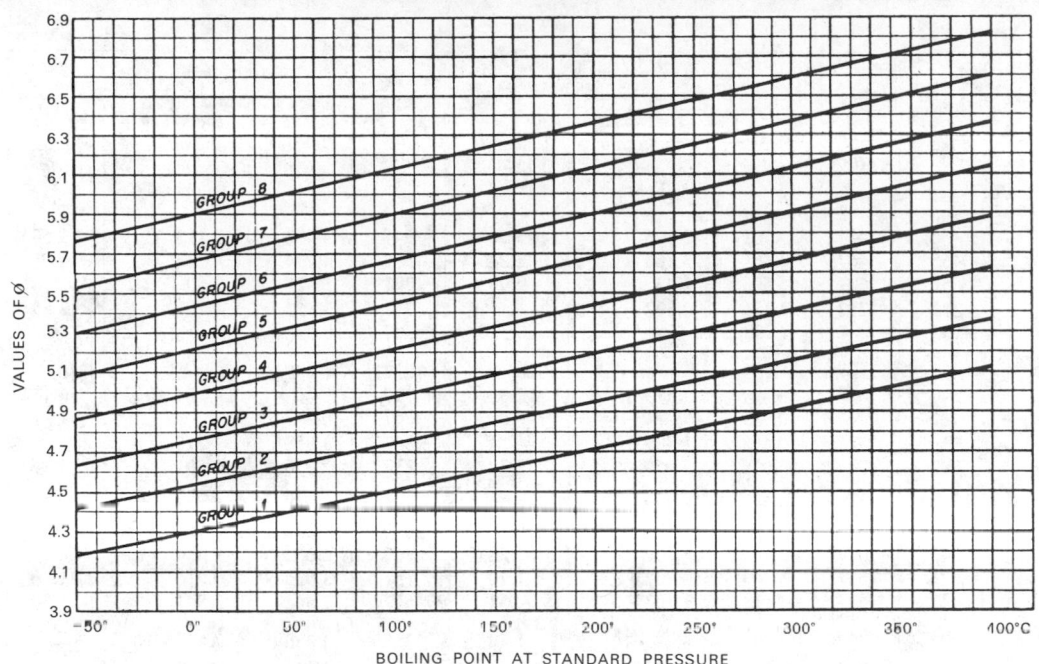

BOILING POINT AT STANDARD PRESSURE

EXAMPLES:

Example 1. Benzene boils at 20°C at 75 mm pressure. What is its normal boiling point? We do not find benzene in the table but we find hydrocarbons in group 2, and a group 2 compound with a boiling point of 20° has a ϕ of 4.6.

Substitute in the equation

$$\Delta t = \frac{(273.1 + 20)(2.8808 - 1.8751)}{4.60 + .15(2.8808 - 1.8751)} = 62°.$$

Adding this to 20° gives 82° as a first approximation.

The graph shows that the ϕ for a compound of group 2 boiling at 82° is 4.72 instead of 4.60, which we originally used. Since ϕ is in the denominator, this increase will lower out Δt by the ratio 4.60/4.72, or the corrected Δt is $62 \times 4.60/4.72 = 60.4$. Adding Δt to t gives 80.4° as a second approximation.

The formula can best be used in a slightly different form when the reverse calculation is desired, i.e., when one calculates the vapor pressure at a given temperature, lower than the normal boiling point.

$$2.8808 - \log p = \frac{\phi \Delta t}{273.1 + t - .15 \Delta t}.$$

Example 2. Alcohol boils at 78.4°C. What is its vapor pressure at 20°C? Substitute in equation 2

$$2.8808 - \log p = \frac{6.06 \times 58.4}{293.1 - (.15 \times 58.4)} = 1.245$$
$$\log p = 2.8808 - 1.245 = 1.6358$$
$$p = 43.2 \text{ mm}.$$

Here no second approximation is necessary, since the correct value of ϕ was taken immediately, the normal boiling point having been known.

*From: "CRC Handbook of Chemistry and Physics", 53rd ed., R.C. Weast, Ed., The Chemical Rubber Co., 1972.

Table 1-56. PROPERTIES OF MOLTEN SALTS*

For density in kg/m^3, multiply values in g/cm^3 by 1000. For viscosity in N·s/m^2, multiply values in centipoise by 0.001.

Salts	Melting point, °C	Near melting point				+100° (approx.)			
		Temperature, °C	Density, g/cm^3	Electrical conductivity per ohm-cm	Viscosity, centipoise	Temperature, °C	Density, g/cm^3	Electrical conductivity per ohm-cm	Viscosity, centipoise
AgBr	430	447	5.562	2.896	3.30	547	5.458	3.113	2.53
AgCl	455	467	4.861	3.868	2.24	567	4.774	4.225	1.78
AgI	558	607	5.526	2.43	2.95	707	5.425	2.60	2.28
AgNO$_3$	210	217	3.954	0.692	—	317	3.852	1.122	2.55
AlCl$_3$	192	207	1.209	—	0.320	307	0.938	—	—
B$_2$O$_3$	450	1137	1.508	—	5020.	1237	1.499	—	3580.
BeF	540	697	—	0.61×10^{-5}	2.62	797	—	13.77×10^{-5}	0.154
Bu$_4$NBF$_4$	162	167	0.935	—	9.00	267	0.877	—	1.91
CaBr$_2$	730	747	3.108	1.409	—	847	3.058	1.716	—
CaCl$_2$	782	787	2.078	2.059	3.34	887	2.036	2.501	1.96
CdBr$_2$	568	587	4.054	1.097	2.73	687	3.946	1.301	—
CdCl$_2$	568	577	3.381	1.884	—	677	3.299	2.150	1.91
CsCl	645	667	2.768	1.167	1.28	767	2.662	1.474	0.94
CsI	621	657	3.140	0.707	1.72	757	3.022	0.916	1.26
CuCl	422	527	3.618	3.703	2.54	627	3.542	3.841	1.80
HgCl$_2$	277	287	4.336	3.49×10^{-5}	1.74	387	4.050	7.447×10^{-5}	—
HgI$_2$	257	277	5.164	0.0282	2.53	377	4.841	0.0180	—
KBr	735	747	2.116	1.639	1.18	847	2.034	1.886	0.92
KCl	770	787	1.518	2.203	1.15	887	1.460	2.439	0.86
K$_2$CO$_3$	896	907	1.892	2.053	—	1007	1.848	2.342	—
K$_2$Cr$_2$O$_7$	398	417	2.273	0.247	11.87	517	2.204	0.507	—
KI	685	727	2.404	1.369	1.45	827	2.308	1.569	1.13
KNO$_3$	337	357	1.856	0.691	2.63	457	1.783	0.986	1.61
KOH	360	407	1.714	2.56	2.21	507	1.670	3.14	1.25
LaCl$_3$	870	877	3.196	1.521	—	977	3.118	1.810	3.94
LiBr	550	597	2.499	4.951	1.52	697	2.433	5.470	1.10
LiCl	610	637	1.490	5.864	1.59	737	1.447	6.354	1.09
Li$_2$CO$_3$	618	737	1.826	4.097	—	837	1.789	4.892	2 91
LiI	449	487	3.093	3.967	2.17	587	3.001	4.427	1.53
LiNO$_3$	254	277	1.768	0.928	5.85	377	1.713	—	2.95
NaBr	750	787	2.309	3.018	1.28	887	2.227	3.319	1.02
NaCl	800	817	1.547	3.629	1.38	917	1.493	3.926	0.95
Na$_2$CO$_3$	854	867	1.968	2.900	—	967	1.923	3.288	1.63
NaF	980	997	1.944	4.937	—	1097	1.888	5.211	—
NaI	662	677	2.726	2.292	1.45	777	2.631	2.603	1.09
Na$_2$MoO$_4$	687	747	2.765	1.243	—	847	2.703	1.575	—
NaNO$_2$	285	297	1.801	1.329	3.04	397	1.726	1.872	—
NaNO$_3$	310	317	1.898	1.015	2.86	417	1.827	1.453	1.77
NaOH	318	357	1.767	2.44	3.79	457	1.719	3.34	2.11
Na$_2$WO$_4$	698	777	3.792	1.145	—	877	3.712	1.35	—
PbCl$_2$	498	507	4.942	1.486	4.41	607	4.792	1.957	2.69
RbBr	680	697	2.699	1.125	1.45	797	2.592	1.359	1.13
RbCl	715	737	2.229	1.549	1.29	837	2.141	1.818	0.98
RbI	640	657	2.886	0.879	1.39	757	2.772	1.056	1.06
RbNO$_3$	316	327	2.466	0.457	3.68	427	2.369	0.673	—
SnCl$_2$	245	267	3.339	0.884	—	367	3.214	1.493	—
SnCl$_4$	−33	37	2.186	—	0.73	137	1.917	—	0.34
SrCl$_2$	875	897	2.713	2.082	3.18	997	2.655	2.466	—
TlCl	429	447	5.597	1.154	—	547	5.417	1.515	—
ZnCl$_2$	275	327	2.514	0.00268	2900.	427	2.469	0.0323	—

*From: "Molten Salts: Electrical Conductance, Density, and Viscosity Data", G.J. Janz, F.W. Dampier, G.R. Lakshminarayanan, P.K. Lorenz, and R.P.T. Tomkins, Vol. I, National Standard Reference Data Series—National Bureau of Standards 15, October 1968. Tables in the original source are in 10-degree increments, and several other salts are included. "Best equations" for each property are listed, and over 250 keyed references are included.

Table 1-57.   MECHANICAL PROPERTIES OF METALS AND ALLOYS*

Typical Composition, Properties, and Uses of Common Materials

See also Tables 1-59, 1-69, 3-1, and 3-2.
For MN/m² multiply strength in thousands of psi by 6.895.

FERROUS ALLOYS

Ferrous alloys comprise the largest volume of metal alloys used in engineering. The actual range of mechanical properties in any particular grade of alloy steel depends on the particular history and heat treatment. The steels listed in this table are intended to give some idea of the range of properties readily obtainable. Many hundreds of steels are available. Cost is frequently an important criterion in the choice of material; in general the greater the percentage of alloying elements present in the alloy, the greater will be the cost.

No.	Material	Nominal composition	Form and condition	Yield strength (0.2% offset), 1000 lb/sq. in.	Tensile strength, 1000 lb/sq in.	Elongation, in 2 in., %	Hardness, Brinell	Comments
1	*IRON* Ingot iron (Included for comparison)	Fe 99.9	Hot-rolled Annealed	29 19	45 38	26 45	90 67	
	PLAIN CARBON STEELS							
2	AISI-SAE 1020	C 0.20 Mn 0.45 Si 0.25 Fe bal.	Hot-rolled Hardened (water-quenched, 1000°F-tempered)	30 52	55 90	25 25	111 179	Bolts, crankshafts, gears, connecting rods; easily weldable
3	AISI 1025	C 0.25 Fe bal. Mn 0.45	Bar stock Hot-rolled Cold-drawn	32 54	58 64	25 15	116 126	
4	AISI-SAE 1035	C 0.35 Mn 0.75 Fe bal.	Hot-rolled Cold-rolled	39 67	72 80	18 12	143 163	Medium-strength, engineering steel
5	AISI-SAE 1045	C 0.45 Fe bal. Mn 0.75	Bar stock Annealed Hot-rolled Cold-drawn	73 45 77	80 82 91	12 16 12	170 163 179	
6	AISI-SAE 1078	C 0.78 Fe bal. Mn 0.45	Bar stock Hot-rolled; spheroidized Annealed	55 72	100 94	12 10	207 192	
7	AISI-SAE 1095	C 0.95 Fe bal. Mn 0.40						
8	AISI-SAE 1120	C 0.2 Mn 0.8 S 0.1	Cold-drawn	58	69	—	137	Free-cutting, leaded, resulphurized steel; high-speed, automatic machining
9	*ALLOY STEELS* ASTM A202/56	C 0.17 Mn 1.2 Cr 0.5 Si 0.75	Stress-relieved	45	75	18	—	Low alloy; boilers, pressure vessels

Table 1-57. MECHANICAL PROPERTIES OF METALS AND ALLOYS (Continued)

No.	Material	Nominal composition	Form and condition	Yield strength (0.2% offset), 1000 lb/sq in.	Tensile strength, 1000 lb/sq in.	Elongation, in 2 in., %	Hardness, Brinell	Comments
10	AISI 4140	C 0.40, Si 0.3 Cr 1.0, Mo 0.2 Mn 0.9	Fully-tempered Optimum properties	95 132	108 150	22 18	240	High strength; gears, shafts
11	12% Manganese steel	12% Mn C	Tempered 600°F Rolled and heat-treated stock	200 44	220 160	10 40	— 170	Machine tool parts; wear, abrasion-resistant
12	VASCO 300	Ni 18.5, Ti 0.6 Co 9.0, C 0.03 Mo 4.8	Solution treatment 1500°F; aged 900°F	110	150	18	—	Very high strength, maraging, good machining properties in annealed state
13	T1 (AISI)	W 18.0, V 1.0 Cr 4.0, C 0.7	Quenched; tempered				R(c)	High speed tool steel, cutting tools, punches, etc.
14	M2 (AISI)	W 6.5, Mo 5.0 Cr 4.0, C 0.85 V 2.0	Quenched; tempered				65–66	M-grade, cheaper, tougher
15	Stainless steel type 304	Ni 9.0, C 0.08 max Cr 19.0	Annealed; cold-rolled	35 to 160	85 to 185	60 8	160 to 400	General purpose, weldable; nonmagnetic austenitic steel
16	Stainless steel type 316	Cr 18.0, C 0.10 max Ni 11.0, Fe bal. Mo 2.5	Annealed	30 to 120	90 to 150	50 8	165 275	For severe corrosive media, under stress; nonmagnetic austenitic steel
17	Stainless steel type 431	Cr 16.0, Si 1.0 Ni 2.0, C 0.20 Mn 1.0, Fe bal.	Annealed Heat-treated	85 150	120 195	25 20	250 400	Heat-treated stainless steel, with good mechanical strength; magnetic
18	Stainless steel 17–4 PH	Cr 17.0, Co 0.35 Ni 4.0, C 0.07 Cu 4.0, Fe bal.	Annealed	110	150	10	363	Precipitation hardening; heat-resisting type; retains strength up to approx. 600°F

Table 1-57. MECHANICAL PROPERTIES OF METALS AND ALLOYS (Continued)

CAST IRONS AND CAST STEELS

These alloys are used where large and/or intricate-shaped articles are required or where over-all dimensional tolerances are not critical. Thus the article can be produced with the fabrication and machining costs held to a minimum. Except for a few heat-treatable cast steels, this class of alloys does not demonstrate high-strength qualities.

No.	Material	Nominal composition	Form and condition	Typical mechanical properties				Comments
				Yield strength (0.2% offset), 1000 lb/sq in.	Tensile strength, 1000 lb/sq in.	Elongation, in 2 in., %	Hardness, Brinell	
CAST IRONS								
19	Cast gray iron ASTM A48-48, Class 25	C 3.4, Mn 0.5, Si 1.8	Cast (as cast)	—	25 min	0.5 max	180	Engine blocks, fly-wheels, gears, machine-tool bases
20	White	C 3.4, Mn 0.6, Si 0.7	Cast	—	25	0	450	
21	Malleable iron ASTM A47	C 2.5, Mn 0.55 max, Si 1.0	Cast (annealed)	33	52	12	130	Automotives, axle bearings, track wheels, crankshafts
22	Ductile or nodular iron (Mg-containing) ASTM A339 ASTM A395	C 3.4, Mn 0.40, Ni 1%, Si 2.5, P 0.1 max, Mg 0.06, Fe bal.	Cast / Cast (as cast) / Cast (quenched, tempered)	53 / 68 / 108	70 / 90 / 135	18 / 7 / 5	170 / 235 / 310	Heavy-duty machines, gears, cams, crankshafts
23	Ni-hard type 2	C 2.7, Mn 0.5, Cr 2.0, Si 0.6, Ni 4.5, Fe bal.	Sand-cast / Chill-cast (tempered)	— / —	55 / 75	— / —	550 / 625	Strength, with heat- and corrosion-resistance
24	Ni-resist type 2	C 3.0, Mn 1.0, Cr 2.5, Si 2.0, Ni 20.0, Fe bal.	Cast (as cast)	—	27	2	140	
CAST STEELS								
25	ASTM A27-62 (60-30)	C 0.3, Si 0.8, Cr 0.4, Mn 0.6, Ni 0.5, Mo 0.2		30	60	24	—	Low alloy, medium strength, general application
26	ASTM A148-60 (105-85)			85	105	17	—	High strength; structural application

Table 1-57. MECHANICAL PROPERTIES OF METALS AND ALLOYS (Continued)

No.	Material	Nominal composition	Form and condition	Typical mechanical properties				Comments
				Yield strength (0.2% offset), 1000 lb/sq in.	Tensile strength, 1000 lb/sq in.	Elongation, in 2 in., %	Hardness, Brinell	
27	Cast 12 Cr alloy (CA–15)	C 0.15 max　Mn 1.00 max Si 1.50 max　Cr 11.5–14 Ni 1.00 max　Fe bal.	Air-cooled from 1800°F; tempered at 600°F Air-cooled from 1800°F; tempered at 1400°F	150 75	200 100	7 30	390 185	Stainless, corrosion-resistant to mildly corrosive alkalis and acids
28	Cast 29–9 alloy (CE–30) ASTM A296 63T	C 0.30 max　Mn 1.50 max Si 2.00 max　Cr 26–30 Ni 8–11　Fe bal.	As cast	60	95	15	170	Greater corrosion resistance, especially for oxidizing condition
29	Cast 28–7 alloy (HD) ASTM A297–63T	C 0.50 max　Mn 1.50 max Si 2.00 max　Cr 26–30 Ni 4–7　Fe bal.	As cast	48	85	16	190	Heat-resistant

SUPER ALLOYS

The advent of engineering applications requiring high temperature and high strength, as in jet engines and rocket motors, has lead to the development of a range of alloys collectively called super alloys. These alloys require excellent resistance to oxidation together with strength at high temperatures, typically 1800°F in existing engines. These alloys are continually being modified to develop better specific properties, and therefore entries in this group of alloys should be considered "fluid". Both wrought and casting-type alloys are represented. As the high temperature properties of cast materials improve, these alloys become more attractive, since great dimensional precision is now attainable in investment castings.

No.	Material	Nominal composition	Form and condition	Yield strength (0.2% offset), 1000 lb/sq in.	Tensile strength, 1000 lb/sq in.	Elongation, in 2 in., %	Hardness, Brinell	Comments
	NICKEL BASE							
30	Hastelloy X	Co 1.5 max　Fe 18.5 Cr 22.0　Mo 9.0 W 0.6 max　C 0.15 max (wrought) C 0.20 max (cast)　Ni bal.	Wrought sheet Mill-annealed As investment cast	52 — 46.5	113.2 67 —	43 17 —	194 172 —	
31	Hastelloy C	Cr 16.0　Fe 6.0 W 4.0　C 0.15 max Mo 17.0　Ni bal.	Sand-cast (annealed) Rolled (annealed) Investment cast	50 71 50	78 130 80	5 45 10	199 204 215	

Table 1-57. MECHANICAL PROPERTIES OF METALS AND ALLOYS (Continued)

No.	Material	Nominal composition	Form and condition	Yield strength (0.2% offset), 1000 lb/sq in.	Tensile strength, 1000 lb/sq in.	Elongation, in 2 in., %	Hardness, Brinell	Comments
	NICKEL BASE (Cont.)							
32	Inconel 713C	Ni (+Co) bal. Cr 13.0 Cb 2.0 Ti 0.6 Mo 4.5 Al 6.0	Investment cast	102	120	6	—	
33	In 100	C 18.0 Cr 10.0 Mo 3.0 Ti 4.7 Al 55.0 Co 15.0 V 1.0	Cast					
34	Taz 8	C 125.0 Cr 5.0 Mo 4.0 Al 5.0 W 4.0 Zr 1.0 Ta 8.0 V 2.5	Cast					
35	Nimonic 90	Ni (+Co) 57.00 C 0.05 Mn 0.50 Fe 0.45 S 0.007 Si 0.20 Cu 0.05 Cr 20.55 Al 1.65 Ti 2.60 Co 16.90	Annealed; wrought	90	155	—	260	General elevated temperature applications
36	Inconel X	Ni (+Co) 72.85 C 0.04 Mn 0.65 Fe 6.80 S 0.007 Si 0.30 Cu 0.05 Cr 15.0 Al 0.75 Ti 2.50 Cb (+Ta) 0.85	Annealed / Annealed; age-hardened	50 / 115	115 / 175	50 / 25	150 / 300	
37	Waspaloy	C 0.08 Cr 19.5 Mo 4.3 Ti 3.0 Co 13.5	Cold-rolled	270	275	8	Rc 51	
38	Rene 41	C 0.09 Cr 19.0 Mo 10.0 Ti 3.1 Al 1.5 Co 11.0	Wrought	100	145	—	—	

Table 1-57. MECHANICAL PROPERTIES OF METALS AND ALLOYS (Continued)

No.	Material	Nominal composition	Form and condition	Yield strength (0.2% offset), 1000 lb/sq in.	Tensile strength, 1000 lb/sq in.	Elongation, in 2 in., %	Hardness, Brinell	Comments
39	Udimet 700	C 0.08, Cr 15.0, Mo 5.0, Ti 3.5, Al 4.3, Co 18.5	Cold-rolled	280	285	6	Rc 53	High temperature; jet engine parts
40	T.D. Nickel	ThO$_2$ 2.4, Ni 97.5	Extended and cold-worked	85	100	13	—	Wrought products
	COBALT BASE							
41	Haynes Stellite alloy 25 (L605)	C 0.15 max, Cr 20.0, W 15.0, Co bal., Ni 10.0, Mn 1.5	Wrought sheet; mill annealed	63	140	60	244	
42	Haynes Stellite alloy 21 AMS 5385 (cast)	C 0.25, Mo 5.5, Co bal., Ni 2.5, Cr 28.5	As investment cast	82	103	8	313 max	For castings

ALUMINUM ALLOYS

Although the strength of aluminum alloys is in general less than that attainable in ferrous alloys or copper-base alloys, their major advantage lies in their high strength-to-weight ratio due to the low density of aluminum. Aluminum alloys have good corrosion resistance for most applications except in alkaline solutions.

No.	Material	Nominal composition	Form and condition	Yield strength (0.2% offset), 1000 lb/sq in.	Tensile strength, 1000 lb/sq in.	Elongation, in 2 in., %	Hardness, Brinell	Comments
43	3003 ASTM B221	Cu 0.12, Al bal., Mn 1.2	Annealed-O / Cold-rolled-H14 / Cold-rolled-H18	6 / 21 / 27	16 / 22 / 29	40 / 16 / 10	28 / 40 / 55	Good formability, weldable, medium strength; chemical equipment
44	2017 ASTM B221	Mn 0.5, Mg 0.5, Cu 4.0, Al bal.	Annealed-O / Heat-treated-T4	10 / 40	26 / 62	22 / 22	45 / 105	High strength; structural parts, aircraft, heavy forgings
45	2024 ASTM B211	Cu 4.5, Mg 1.5, Mn 0.6, Al bal.	Heat-treated-T4	47	68	19	120	
46	5052 ASTM B211	Cr 0.25, Mg 2.5, Al bal.	Annealed-O / Cold-rolled and stabilized-H34	13 / 31	28 / 38	30 / 14	47 / 68	Medium strength, good fatigue properties; street-light standards
47	ASTM B209		Cold-rolled and stabilized-H38	37	42	8	77	

Table 1-57. MECHANICAL PROPERTIES OF METALS AND ALLOYS (Continued)

No.	Material	Nominal composition	Form and condition	Yield strength (0.2% offset), 1000 lb/sq in.	Tensile strength, 1000 lb/sq in.	Elongation, in 2 in., %	Hardness, Brinell	Comments
48	7075 ASTM B211	Cu 1.6 Mg 2.5 Cr 0.3 Al bal. Zn 5.6	Annealed-O Heat-treated and artificially aged-T6	15 73	33 83	17 11	60 150	High strength, good corrosion resistance
49	380 ASTM SC84B	Si 9.0 Al bal. Cu 3.5	Die-cast	24	48	3	—	General purpose die casting
50	195 ASTM C4A	Si 0.8 Al bal. Cu 4.5	Sand-cast; heat-treated-T4 Sand-cast; heat-treated and artificially aged-T6	16 24	32 36	8.5 5	60 75	Structural elements, aircraft, and machines
51	214 ASTM G4A	Mg 3.8 Al bal.	Sand-cast-F	12	25	9	50	Chemical equipment, marine hardware, architectural
52	220 ASTM G10A	Mg 10.0 Al bal.	Sand-cast; heat-treated-T4	26	48	16	75	Strength with shock resistance; aircraft

COPPER ALLOYS

Because of their corrosion resistance and the fact that copper alloys have been used for many thousands of years, the number of copper alloys available is second only to the ferrous alloys. In general copper alloys do not have the high-strength qualities of the ferrous alloys, while their density is comparable The cost per strength-weight ratio is high; however, they have the advantage of ease of joining by soldering, which is not shared by other metals that have reasonable corrosion resistance

No.	Material	Nominal composition	Form and condition	Yield strength (0.2% offset), 1000 lb/sq in.	Tensile strength, 1000 lb/sq in.	Elongation, in 2 in., %	Hardness, Brinell	Comments
53	Copper ASTM B152 ASTM B124, B133 ASTM B1, B2, B3	Cu 99.9 plus	Annealed Cold-drawn Cold-rolled	10 40 40	32 45 46	45 15 5	42 90 100	Bus-bars, switches, architectural, roofing, screens
54	Gilding metal ASTM B36	Cu 95.0 Zn 5.0	Cold-rolled	50	56	5	114	Coinage, ammunition
55	Cartridge 70–30 brass ASTM B14 ASTM B19 ASTM B36 ASTM B134 ASTM B135	Cu 70.0 Zn 30.0	Cold-rolled	63	76	8	155	Good cold-working properties; radiator covers, hardware, electrical
56	Phosphor bronze 10% ASTM B103 ASTM B139 ASTM B159	Cu 90.0 Sn 10.0 P 0.25	Spring temper	—	122	4	241	Good spring qualities, high-fatigue strength

Table 1-57. MECHANICAL PROPERTIES OF METALS AND ALLOYS (Continued)

No.	Material	Nominal composition	Form and condition	Yield strength (0.2% offset), 1000 lb/sq in.	Tensile strength, 1000 lb/sq in.	Elongation, in 2 in., %	Hardness, Brinell	Comments
57	Yellow brass (high brass) ASTM B36 ASTM B134 ASTM B135	Cu 65.0 Zn 35.0	Annealed Cold-drawn Cold-rolled (HT)	18 55 60	48 70 74	60 15 10	55 115 180	Good corrosion resistance; plumbing, architectural
58	Manganese bronze ASTM B138	Cu 58.5 Zn 39.2 Fe 1.0 Sn 1.0 Mn 0.3	Annealed Cold-drawn	30 50	60 80	30 20	95 180	Forgings
59	Naval brass ASTM B21	Cu 60.0 Zn 39.25 Sn 0.75	Annealed Cold-drawn	22 40	56 65	40 35	90 150	Condenser tubing; high resistance to salt-water corrosion
60	Muntz metal ASTM B111	Cu 60.0 Zn 40.0	Annealed	20	54	45	80	Condenser tubes; valve stress
61	Aluminum bronze ASTM B169, alloy A ASTM B124 ASTM B150	Cu 92.0 Al 8.0	Annealed Hard	25 65	70 105	60 7	80 210	
62	Beryllium copper 25 ASTM B194 ASTM B197 ASTM B196	Be 1.9 Cu bal. Co or Ni 0.25	Annealed, solution-treated Cold-rolled Cold-rolled	32 104 70	70 110 190	45 5 3	B60 (Rockwell) B81 C40	Bellows, fuse clips, electrical relay parts, valves, pumps
63	Free-cutting brass	Cu 62.0 Zn 35.5 Pb 2.5	Cold-drawn	44	70	18	B80 (Rockwell)	Screws, nuts, gears, keys
64	Nickel silver 18% Alloy A (wrought) ASTM B122, No. 2	Cu 65.0 Zn 17.0 Ni 18.0	Annealed Cold-rolled Cold-drawn wire	25 70 —	58 85 105	40 4 —	70 170 —	Hardware, optical goods, camera parts
65	Nickel silver 13% (cast) 10A ASTM B149, No. 10A	Ni 12.5 Pb 9.0 Sn 2.0 Cu bal. Zn 20.0	Cast	18	35	15	55	Ornamental castings, plumbing; good machining qualities
66	Cupronickel 10% ASTM B111 ASTM B171	Cu 88.35 Ni 10.0 Fe 1.25 Mn 0.4	Annealed Cold-drawn tube	22 57	44 60	45 15	— —	Condenser, salt-water piping

Table 1-57. MECHANICAL PROPERTIES OF METALS AND ALLOYS (Continued)

No.	Material	Nominal composition	Form and condition	Yield strength (0.2% offset), 1000 lb/sq in.	Tensile strength, 1000 lb/sq in.	Elongation, in 2 in., %	Hardness, Brinell	Comments
67	Cupronickel	Cu 70.0 Ni 30.0	Wrought					Heat-exchanger process equipment, valves
68	Red brass (cast) ASTM B30, No. 4A	Cu 85.0 Zn 5.0 Pb 5.0 Sn 5.0	As-cast	17	35	25	60	
69	Silicon bronze ASTM B30, alloy 12A	Si 4.0 Fe 2.0 Zn 4.0 Al 1.0 Mn 1.0	Castings					Cheaper substitute for tin bronze
70	Tin bronze ASTM B30, alloy 1B	Sn 8% Zn 4.0	Castings					Bearings, high-pressure bushings, pump impellers
71	Navy bronze		Cast					

TIN AND LEAD-BASE ALLOYS

Major uses for these alloys are as "white"-metal bearing alloys, extruded cable sheathing, and solders. Tin forms the basis of pewter used for culinary applications.

No.	Material	Nominal composition	Form and condition	Yield strength (0.2% offset), 1000 lb/sq in.	Tensile strength, 1000 lb/sq in.	Elongation, in 2 in., %	Hardness, Brinell	Comments
72	Lead-base Babbitt ASTM B23, alloy 19	Pb 85.0 Sn 5.0 Sb 10.0 As 0.6 Cu 0.5	Chill cast	—	10	5	19	Bearings, light loads and low speeds
73	Arsenical-lead Babbitt ASTM B23, alloy 15	Pb 83.0 Sn 1.0 Sb 16.0 As 1.1 Cu 0.6	Chill cast	—	10.3	2	20	Bearings, high loads and speeds, diesel engines, steel mills
74	Chemical lead	Pb 99.9 Cu 0.06 Bi 0.005 max	Rolled 95%	1.5	2.5	50	5	
75	Antimonial lead (hard lead)	Pb 94.0 Sb 6.0	Chill cast Rolled 95%	— —	6.8 4.1	22 47	(500 kg) 9	Good corrosion resistance and strength
76	Calcium lead	Pb 99.9 Ca 0.025 Cu 0.10	Extruded and aged	—	4.5	25	—	Cable sheathing, creep-resistant pipe
77	Tin Babbitt alloy ASTM B23-61, grade 1	Sb 4.5 Cu 4.5 Sn bal.	Chill cast	—	9.3	2	17	General bearings and die casting
78	Tin die-casting alloy ASTM B102-52	Sb 13.0 Cu 5.0 Sn bal.	Die-cast	—	10	1	29	Die-casting alloy

Table 1-57. MECHANICAL PROPERTIES OF METALS AND ALLOYS (Continued)

No.	Material	Nominal composition	Form and condition	Yield strength (0.2% offset), 1000 lb/sq in.	Tensile strength, 1000 lb/sq in.	Elongation, in 2 in., %	Hardness, Brinell	Comments
79	Pewter	Sn 91.0 Cu 2.0 Sb 7.0	Rolled sheet, annealed	—	8.6	40	9.5	Ornamental and household items
80	Solder 50–50	Sn 50.0 Pb 50.0	Cast	4.8	6.1	60	14	General-purpose solder
81	Solder	Sn 20.0 Pb 80.0	Cast	3.6	5.8	16	11	Coating and joining, filling seams on automobile bodies

MAGNESIUM ALLOYS

Because of their low density these alloys are attractive for use where weight is at a premium. The major drawback to the use of these alloys is their ability to ignite in air (this can be a problem in machining); they are also costly. Magnesium alloys are used in both the wrought and die-cast forms, the latter being the most frequently used form.

No.	Material	Nominal composition	Form and condition	Yield strength (0.2% offset), 1000 lb/sq in.	Tensile strength, 1000 lb/sq in.	Elongation, in 2 in., %	Hardness, Brinell	Comments
82	Magnesium alloy AZ31B	Zn 1.0 Al 3.0 Mn 0.20 min Mg bal.	Rolled-plate (strain-hardened, then partially annealed)	24	37	18	—	Structural applications of medium strength
			Rolled-sheet (strain-hardened, then partially annealed)	32	42	15	73	
			Annealed	22	37	21	56	
			Extruded	28	38	14	—	
83	Magnesium alloy AZ80A	Zn 0.5 Al 8.5 Mn 0.15 min Mg bal.	Extruded	36	49	11	60	General extruded and forged products
			Extruded (age-hardened)	39	53	6	82	
			Forged (age-hardened)	34	50	6	72	
84	Magnesium alloy AZ92A	Zn 2.0 Al 9.0 Mn 0.10 min Mg bal.	Sand-cast (as cast)	14	24	6	50	
			Sand-cast (solution heat-treated)	14	40	12	55	
			Sand-cast (solution heat-treated and aged)	19	40	5	83	
			Sand-cast (age-hardened)	16	30	18	—	Pressure-tight sand and permanent mold castings; high UTS and good yield strength
			Sand-cast and tempered	22	40	3	81	
85	Magnesium alloy ZK60A	Zn 5.7 Zr 0.55 Mg bal.	Extruded	43	52	12	82	

Table 1-57. MECHANICAL PROPERTIES OF METALS AND ALLOYS (Continued)

No.	Material	Nominal composition	Form and condition	Yield strength (0.2% offset), 1000 lb/sq in.	Tensile strength, 1000 lb/sq in	Elongation, in 2 in, %	Hardness, Brinell	Comments
86	Magnesium alloy AZ91A and AZ91B	Zn 0.6, Al 9.0, Mn 0.13 min, Mg bal.	Die-cast (as cast)	22	33	3	67	General die-casting applications

BERYLLIUM

No.	Material	Nominal composition	Form and condition	Yield strength (0.2% offset), 1000 lb/sq in.	Tensile strength, 1000 lb/sq in	Elongation, in 2 in, %	Hardness, Brinell	Comments
87	Beryllium		Hot-pressed	27 / 38	33 / 51	1–3	—	Windows, X-ray tubes
			Cross-rolled	40 / 60	60 / 90	10–40	—	Moderator- and reflector-cladding nuclear reactors; heat-shield and structural-member missiles

NICKEL ALLOYS

Nickel and its alloys are expensive and used mainly either for their high-corrosion resistance in many environments or for high-temperature and strength applications. (See Super Alloys, above.)

No.	Material	Nominal composition	Form and condition	Yield strength (0.2% offset), 1000 lb/sq in.	Tensile strength, 1000 lb/sq in	Elongation, in 2 in, %	Hardness, Brinell	Comments
88	Nickel (cast)	Ni 95.6, Fe 0.5, Si 1.5, Cu 0.5, Mn 0.8, C 0.8	As cast	25	57	22	110	Good corrosion-resistance applications
89	K Monel	Ni (+Co) 65.25, Mn 0.60, S 0.005, Cu 29.60, Ti 0.45, C 0.15, Fe 1.00, Si 0.15, Al 2.75	Annealed	45	100	40	155	High strength and corrosion resistance; aircraft parts, valve stems, pumps
			Annealed, age-hardened	100	155	25	270	
			Spring	140	150	5	300	
			Spring, age-hardened	160	185	10	335	
90	A nickel ASTM B160 ASTM B161 ASTM B162	Ni (+Co) 99.40, Mn 0.25, S 0.005, Cu 0.05, C 0.06, Fe 0.15, Si 0.05	Annealed	20	70	40	100	Chemical industry for resistance to strong alkalis, plating nickel
			Hot-rolled	25	75	40	110	
			Cold-drawn	70	95	25	170	
			Cold-rolled	95	105	5	210	
91	Duranickel	Ni (+Co) 93.90, Mn 0.25, S 0.005, Cu 0.05, Ti 0.45, C 0.15, Fe 0.15, Si 0.55, Al 4.50	Annealed	45	100	40	160	High strength and corrosion resistance; pump rods, shafts, springs
			Annealed, age-hardened	125	170	25	330	
			Spring	—	175	5	320	
			Spring, age-hardened	—	205	10	370	

Table 1-57. MECHANICAL PROPERTIES OF METALS AND ALLOYS (Continued)

No.	Material	Nominal composition	Form and condition	Typical mechanical properties				Comments
				Yield strength (0.2% offset), 1000 lb/sq in.	Tensile strength, 1000 lb/sq in.	Elongation, in 2 in., %	Hardness, Brinell	
92	Cupronickel 55–45 (Constantan)	Cu 55.0 Ni 45.0	Annealed Cold-drawn Cold-rolled	30 50 65	60 65 85	45 30 20	— — —	Electrical-resistance wire; low temperature coefficient, high resistivity
93	Nichrome	Ni 80.0 Cr 20.0						Heating elements for furnaces

94	"S" Monel	Ni 60.0 Cu 29.0
		Fe 2.50 Mn 1.5
		     max max
		Si 4.0 Al 0.5
		            max
			Sand-casting	80–115	110–145	2	270–350	High-strength casting alloy; good bearing properties for valve seats

TITANIUM ALLOYS

The main application for these alloys is in the aerospace industry. Because of the low density and high strength of titanium alloys, they present excellent strength-to-weight ratios.

No.	Material	Nominal composition	Form and condition	Yield strength (0.2% offset), 1000 lb/sq in.	Tensile strength, 1000 lb/sq in.	Elongation, in 2 in., %	Hardness, Brinell	Comments
95	Commercial titanium ASTM B265–58T	Ti 99.4	Annealed at 1100 to 1350°F (593 to 732°C)	70	80	20	—	Moderate strength, excellent fabricability; chemical industry pipes
96	Titanium alloy ASTM B265–58T–5 Ti–6 Al–4V		Water-quenched from 1750°F (954°C); aged at 1000°F (538°C) for 2 hr	160	170	13	—	High-temperature strength needed in gas-turbine compressor blades
97	Titanium alloy Ti–4 Al–4Mn		Water-quenched from 1450°F (788°C); aged at 900°F (482°C) for 8 hr	170	185	13	—	Aircraft forgings and compressor parts
98	Ti–Mn alloy ASTM B265–58T–7	Fe 0.5 Ti bal. Mn 7.0–8.0	Sheet	140	150	18		Good formability, moderate high-temperature strength; aircraft skin

ZINC ALLOYS

A major use for these alloys is for low-cost die-cast products, such as household fixtures, automotive parts, and trim.

No.	Material	Nominal composition	Form and condition	Yield strength (0.2% offset), 1000 lb/sq in.	Tensile strength, 1000 lb/sq in.	Elongation, in 2 in., %	Hardness, Brinell	Comments
99	Zinc ASTM B69	Cd 0.35 Zn bal. Pb 0.08	Hot-rolled	—	19.5	65	38	Battery cans, grommets, lithographer's sheet

Table 1-57. MECHANICAL PROPERTIES OF METALS AND ALLOYS (Continued)

No.	Material	Nominal composition	Form and condition	Typical mechanical properties				Comments
				Yield strength (0.2% offset), 1000 lb/sq in.	Tensile strength, 1000 lb/sq in.	Elongation, in 2 in., %	Hardness, Brinell	
100	Zilloy-15	Cu 1.00 Zn bal. Mg 0.010	Hot-rolled Cold-rolled	— —	29 36	20 25	61 80	Corrugated roofs, articles with maximum stiffness
101	Zilloy-40	Cu 1.00 Zn bal.	Hot-rolled Cold-rolled	— —	24 31	50 40	52 60	Weatherstrip, spun articles
102	Zamac-5 ASTM 25	Zn (99.99% pure remainder) Al 3.5–4.3 Cu 0.75–1.25 Mg 0.03–0.08	Die-cast	—	47.6	7	91	Die casting for automobile parts, padlocks; used also for die material

ZIRCONIUM ALLOYS

These alloys have good corrosion resistance but are easily oxidized at elevated temperatures in air. The major application is for use in nuclear reactors.

No.	Material	Nominal composition	Form and condition	Yield strength (0.2% offset), 1000 lb/sq in.	Tensile strength, 1000 lb/sq in.	Elongation, in 2 in., %	Hardness, Brinell	Comments
103	Zirconium, commercial	O₂ 0.07 C 0.15 Hf 1.90 Zr bal.	Annealed	40	65	27	B80 (Rockwell)	Nuclear power-reactor cores at elevated temperatures
104	Zircaloy 2	Hf 0.02 Ni 0.05 Fe 0.15 Other 0.25 Sn 1.46 Zr bal.	Annealed	50	75	22	B90 (Rockwell)	

*Compiled from various sources.

Table 1-58.   TRADE NAMES OF ALLOYS

Trademark	Owner
ALUMEL	Hoskins Manufacturing Company
CARPENTER STAINLESS NO. 20	The Carpenter Steel Company
CARPENTER 426	The Carpenter Steel Company
CHROMEL	Hoskins Manufacturing Company
COBENIUM	Wilbur B. Driver Company
CONPERNIK	Westinghouse Electric Corporation
COR-TEN	United States Steel Corporation
CUFENLOY	Phelps Dodge Corporation
DISCALOY	Westinghouse Electric Corporation
DURANICKEL	The International Nickel Company, Inc.
DYNALLOY	Alan Wood Steel Corporation
DYNAVAR	Precision Metals Division
ELGILOY	Elgin National Watch Company
ELINVAR	Hamilton Watch Company
GEMINOL	Driver-Harris Company
HASTELLOY	Union Carbide Corporation
HI-STEEL	Inland Steel Corporation
HIPERNIK	Westinghouse Electric Corporation
HP	Republic Steel Company
HY-TUF	Crucible Steel Company
ILLIUM	Stainless Foundry & Engineering Inc.
INCOLOY	The International Nickel Company, Inc.
INCONEL	The International Nickel Company, Inc.
INVAR	Soc. Anon. de Commentry-Fourchambault et Decaziville (Acieries d'Imphy)
KANTHAL	The Kanthal Corporation
KOVAR	Westinghouse Electric Corporation
MAGARI-R	Bethlehem Steel Corporation
MANGANIN	Driver-Harris Company
MINOVAR	The International Nickel Company, Inc.
MONEL	The International Nickel Company, Inc.
MONIMAX	Allegheny Ludlum Steel Corporation
NICROTUNG	Westinghouse Electric Corporation
NIMOCAST	The International Nickel Company, Inc.
NIMONIC	The International Nickel Company, Inc.
NISILOY	The International Nickel Company, Inc.
NI-SPAN-C	The International Nickel Company, Inc.
PERMALLOY	Allegheny Ludlum Steel Corporation
PERMANICKEL	The International Nickel Company, Inc.
REFRACTALOY	Westinghouse Electric Corporation
RENE 41	Allvac Metals Corporation (Division of Teledyne)
RODAR	Wilbur B. Harris Company
SD	The International Nickel Company, Inc.
SIMINEX	Allegheny Ludlum Steel Corporation
SEALMET	Allegheny Ludlum Steel Corporation
STAINLESS STEEL W	United States Steel Corporation
STAINLESS STEEL 17-4PH	Armco Steel Corporation
SUPERMALLOY	Allegheny Ludlum Steel Corporation
T-1	United States Steel Corporation
TRI-TEN	United States Steel Corporation
TRW	TRW, Inc.
UDIMET	Special Metals Corporation
UNITEMP	Universal Cyclops Speciality Steel Division, Cyclops Corporation
USS STRUX	United States Steel Corporation
WASPALLOY	Pratt and Whitney Aircraft
WELCON	Japanese Steel Works, Ltd.
WEL-TEN	Yawata Iron & Steel Company, Ltd.
YOLOY	Youngstown Sheet & Tube Company

Table 1-59. METALS AND ALLOYS—MISCELLANEOUS PROPERTIES*

Table A. PURE METALS

At Room Temperature

| Common name | PROPERTIES (TYPICAL ONLY) | | | | | | |
	Thermal conductivity, Btu/hr ft °F	Specific gravity	Coeff. of linear expansion, μ in./ in. °F	Electrical resistivity, microhm-cm	Poisson's ratio	Modulus of elasticity, millions of psi	Approximate melting point, °F
Aluminum	137	2.70	14	2.655	0.33	10.0	1220
Antimony	10.7	6.69	5	41.8		11.3	1170
Beryllium	126	1.85	6.7	4.0	0.024–.030	42	2345
Bismuth	4.9	9.75	7.2	115		4.6	521
Cadmium	54	8.65	17	7.4		8	610
Chromium	52	7.2	3.3	13		36	3380
Cobalt	40	8.9	6.7	9		30	2723
Copper	230	8.96	9.2	1.673	0.36	17	1983
Gold	182	19.32	7.9	2.35	0.42	10.8	1945
Iridium	85.0	22.42	3.3	5.3		75	4440
Iron	46.4	7.87	6.7	9.7		28.5	2797
Lead	20.0	11.35	16	20.6	0.40–.45	2.0	621
Magnesium	91.9	1.74	14	4.45	0.35	6.4	1200
Manganese		7.21–7.44	12	185		23	2271
Mercury	4.85	13.546		98.4			− 38
Molybdenum	81	10.22	3.0	5.2	0.32	40	4750
Nickel	52.0	8.90	7.4	6.85	0.31	31	2647
Niobium (Columbium)	30	8.57	3.9	13		15	4473
Osmium	35	22.57	2.8	9		80	5477
Platinum	42	21.45	5	10.5	0.39	21.3	3220
Plutonium	4.6	19.84	30	141.4	0.15–.21	14	1180
Potassium	57.8	0.86	46	7.01			146
Rhodium	86.7	12.41	4.4	4.6		42	3569
Selenium	0.3	4.8	21	12.0		8.4	423
Silicon	48.3	2.33	2.8	1×10^5		16	2572
Silver	247	10.50	11	1.59	0.37	10.5	1760
Sodium	77.5	0.97	39	4.2			208
Tantalum	31	16.6	3.6	12.4	0.35	27	5400
Thorium	24	11.7	6.7	18	0.27	8.5	3180
Tin	37	7.31	11	11.0	0.33	6	450
Titanium	12	4.54	4.7	43	0.3	16	3040
Tungsten	103	19.3	2.5	5.65	0.28	50	6150
Uranium	14	18.8	7.4	30	0.21	24	2070
Vanadium	35	6.1	4.4	25		19	3450
Zinc	66.5	7	19	5.92	0.25	12	787

Table 1-59. METALS AND ALLOYS—MISCELLANEOUS PROPERTIES
(Continued)

Table B. COMMERCIAL METALS AND ALLOYS

CLASSIFICATION AND DESIGNATION		PROPERTIES (TYPICAL ONLY)					
Material No. (from Table 1-57)	Common name and classification	Thermal conductivity, Btu/hr ft °F	Specific gravity	Coeff. of linear expansion, μ in./ in. °F	Electrical resistivity, microhm-cm	Modulus of elasticity, millions of psi	Approximate melting point, °F
1	Ingot iron (included for comparison)	42.	7.86	6.8	9.	30	2800
2	Plain carbon steel						
	AISI–SAE 1020	30.	7.86	6.7	10.	30	2760
15	Stainless steel type 304	10.	8.02	9.6	72.	28	2600
19	Cast gray iron						
	ASTM A48–48, Class 25	26.	7.2	6.7	67.	13	2150
21	Malleable iron						
	ASTM A47	—	7.32	6.6	30.	25	2250
22	Ductile cast iron						
	ASTM A339, A395	19	7.2	7.5	60.	25	2100
24	Ni-resist cast iron, type 2	23	7.3	9.6	170.	15.6	2250
29	Cast 28–7 alloy (HD)						
	ASTM A297–63T	1.5	7.6	9.2	41.	27	2700
31	Hastelloy C	5	3.94	6.3	139.	30	2350
36	Inconel X, annealed	9	8.25	6.7	122.	31	2550
41	Haynes Stellite alloy 25 (L605)	5.5	9.15	7.61	88.	34	2500
43	Aluminum alloy 3003, rolled						
	ASTM B221	90	2.73	12.9	4.	10	1200
44	Aluminum alloy 2017, annealed						
	ASTM B221	95	2.8	12.7	4.	10.5	1185
49	Aluminum alloy 380						
	ASTM SC84B	56	2.7	11.6	7.5	10.3	1050
53	Copper						
	ASTM B152, B124, B133,						
	B1, B2, B3	225	8.91	9.3	1.7	17	1980
57	Yellow brass (high brass)						
	ASTM B36, B134, B135	69	8.47	10.5	7.	15	1710
61	Aluminum bronze						
	ASTM B169, alloy A;						
	ASTM B124, B150	41	7.8	9.2	12.	17	1900
62	Beryllium copper 25						
	ASTM B194	7	8.25	9.3	—	19	1700
64	Nickel silver 18% alloy A (wrought)						
	ASTM B122, No. 2	19	8.8	9.0	29.	18	2030
67	Cupronickel 30%	17	8.95	8.5	35.	22	2240
68	Red brass (cast)						
	ASTM B30, No. 4A	42	8.7	10.	11.	13	1825
74	Chemical lead	20	11.35	16.4	21.	2	621
75	Antimonial lead (hard lead)	17	10.9	15.1	23.	3	554
80	Solder 50–50	26	8.89	13.1	15.	—	420
82	Magnesium alloy AZ31B	45	1.77	14.5	9.	6.5	1160
89	K Monel	11	8.47	7.4	58.	26	2430
90	Nickel						
	ASTM B160, B161, B162	35	8.89	6.6	10.	30	2625
92	Cupronickel 55–45 (Constantan)	13	8.9	8.1	49.	24	2300
95	Commercial titanium	10	5.	4.9	80.	16.5	3300
99	Zinc						
	ASTM B69	62	7.14	18	6.	—	785
103	Zirconium, commercial	10	6.5	2.9	41.	12	3350

*Compiled from several sources.

Table 1-60. THERMAL PROPERTIES OF PURE METALS—METRIC UNITS

Columns under **AT ATMOSPHERIC PRESSURE**: Latent heat of fusion; At 100°K (Thermal conductivity, Specific heat); At 25°C (77°F) (Specific heat, Coeff. of linear expansion, Thermal conductivity); Specific heat (liquid) at 2000°K. Columns under **LIQUID METAL**: Vapor pressure (Boiling point temperatures, °K).

Metal	Melting point, °C	Boiling point, °C	Latent heat of fusion, cal/g	Thermal conductivity, watts/cm °C (100°K)	Specific heat, cal/g °C (100°K)	Specific heat, cal/g °C (25°C)	Coeff. of linear expansion ($\times 10^6$) $({}^{\circ}C)^{-1}$ (25°C)	Thermal conductivity, watt/cm °C (25°C)	Specific heat (liquid) at 2000°K, cal/g °C	Vapor pressure 10^{-3} atm	10^{-6} atm	10^{-9} atm
Aluminum	660.	2441.	95	3.00*	.115	0.215	25	2.37	.26	1,782	1,333	1,063
Antimony	630.	1440.	38.5	—	.040	.050	9	.185	.062	1,007	741	612
Beryllium	1285.	2475.	324.	—	.049	.436	12	2.18	.78	1,793	1,347	1,085
Bismuth	271.4	1660.	12.4	—	.026	.030	13	.084	.036	1,155	851	677
Cadmium	321.	767.	13.2	1.03	.047	.055	30	.93	.063	655	486	388
Chromium	1860.	2670.	79	1.58	.046	.110	6	.91	.224	1,992	1,530	1,247
Cobalt	1495.	2925.	66	—	.057	.10	12	.69	.164	2,167	1,652	1,345
Copper	1084.	2575.	49	4.85*	.061	.092	16.6	3.98	.118	1,862	1,391	1,120
Gold	1063.	2800.	15	3.45*	.026	.031	14.2	3.15	.0355	2,023	1,510	1,211
Iridium	2450.	4390.	33	—	.022	.031	6	1.47	.0434	3,253	2,515	2,062
Iron	1536.	2870.	65	1.52*	.052	.108	12	.803	.197	2,093	1,594	1,297
Lead	327.5	1750.	5.5	0.396	.028	.031	29	.346	.033	1,230	889	698
Magnesium	650.	1090.	88.0	1.69	.016	.243	25	1.59	.32	857	638	509
Manganese	1244.	2060.	64	—	.064	.114	22	—	.20	1,495	1,131	913
Mercury	−38.86	356.55	2.7	—	.029	.033	—	.0839	.089	393	287	227
Molybdenum	2620.	4651.	69	1.79	.033	.060	5	1.4	.089	3,344	2,558	2,079
Nickel	1453.	2800.	71	1.58	.055	.106	13	.899	.175	2,156	1,646	1,343
Niobium (Columbium)	2470.	4740.	68	0.552	.045	.064	7	.52	.083	3,523	2,721	2,232
Osmium	3025.	4225.	34	—	.024	.031	5	.61	.039	—	—	—
Platinum	1770.	3825.	24	0.75*	.024	.032	9	.73	.043	2,817	2,155	1,757
Plutonium	640.	3230.	3	—	.019	.032	54	.08	.041	2,200	1,596	1,252
Potassium	63.3	760.	14.5	—	.150	.180	83	.99	.092	606	430	335
Rhodium	1965.	3700.	50	—	.058	.058	8	1.50	—	—	—	—
Selenium	217.	700.	16	—	—	.077	37	.005	.217	—	—	—
Silicon	1411.	3280.	430	4.50*	.062	.17	3	.835	.068	2,340	1,749	1,427
Silver	961.	2212.	26.5	—	.045	.057	19	4.27	—	1,582	1,179	952
Sodium	97.83	884.	27	0.592	.234	.293	70	1.34	.040	701	504	394
Tantalum	2980.	5365.	41	—	.026	.034	6.5	.54	.047	3,959	3,052	2,495
Thorium	1750.	4800.	17	—	.024	.03	12	.41	.058	3,251	2,407	1,919
Tin	232.	2600.	14.1	0.85	.039	.054	20	.64	.188	1,857	1,366	1,080
Titanium	1670.	3290.	100	0.312	.072	.125	8.5	.2	.040	2,405	1,827	1,484
Tungsten	3400.	5550.	46	2.35*	.021	.032	4.5	1.78	.048	4,139	3,228	2,656
Uranium	1132.	4140.	12	—	.022	.028	13.4	.25	.207	2,861	2,128	1,699
Vanadium	1900.	3400.	98	—	.061	.116	8	.60	—	2,525	1,948	1,591
Zinc	419.5	910.	27	1.32	.063	.093	35	1.15	.207	752	559	449

*Temperatures of maximum thermal conductivity (conductivity values in watts/cm °C): Aluminum 13°K, cond. = 71.5; copper 10°K, cond. = 196; gold 10°K, cond. = 28.2; iron 20°K, cond. = 28.2; platinum 8°K, cond. = 12.9; silicon 7°K, cond. = 193; tungsten 8°K, cond. = 85.3.

Table 1-61. PROPERTIES OF HIGH-TEMPERATURE METALS

Approximate or Typical Values

See Tables 1-57, 1-60, 3-1, and 3-2 for other properties.
Metals are listed in order of their melting points. Actual properties depend on purity, condition, form, and treatment.

Melting point		Metal	Properties at room temperatures						At 1000°C (1830°F)		
°C	°F		Specific gravity	Specific heat	Thermal conductivity, $\frac{Btu}{hr\,ft\,°F}$	Electrical resistivity, microhm-cm	Coef of linear expansion, $(\times 10^6)$ $(°C)^{-1}$	Tensile strength, kpsi	Specific heat, $\frac{Btu}{lb\,°F}$	Thermal conductivity, $\frac{Btu}{hr\,ft\,°F}$	Tensile strength, kpsi
3400	6150	Tungsten (Wolfram)	19.3	0.032	103	5.65	4.5	430	0.037	66	100
3180	5760	Rhenium	21.0	0.033	41.1	19	7	337	0.039		120
3025	5477	Osmium	22.6	0.031	35	9	5		0.036		
2980	5400	Tantalum	16.6	0.034	31.2	12.4	6.5	55	0.037	35	30
2620	4750	Molybdenum	10.2	0.060	81	5.2	5	85	0.075	60	
2470	4470	Niobium (Columbium)	8.57	0.064	30.1	13	7	40	0.075	40	8
2450	4440	Iridium	22.4	0.031	85	5.3	6	90	0.038		48
2400	4350	Ruthenium	12.4	0.057	—	7.5	9	78	0.074		
2220	4030	Hafnium	13.3	0.035	12.7	35	6	65	0.044		
1965	3569	Rhodium	12.4	0.058	86.7	4.6	8	138	0.080		
1900	3450	Vanadium	6.1	0.116	34.7	25	8	78	0.158		
1860	3380	Chromium	7.20	0.11	52	13	6	12	0.169	36	
1850	3370	Zirconium	6.53	0.067	12.2	41	5.5	49	0.082	15	
1770	3220	Platinum	21.5	0.032	42.2	10.5	9	22	0.038	45.5	
1750	3180	Thorium	11.7	0.03	23.7	18	12	32	0.041		
1670	3040	Titanium	4.54	0.125	11.6	43	8.5	35	0.152	13	
1550	2820	Palladium	12.0	0.058	41.1	10.8	12	25	0.071		
1536	2797	Iron	7.87	0.108	46.4	9.7	12	40	0.148	17.0	
1495	2723	Cobalt	8.9	0.10	39.9	9	12	137	0.192		
1453	2647	Nickel	8.9	0.106	52	6.85	13	46	0.145	45	

Table 1-62.   PROPERTIES OF HIGH-TEMPERATURE METALS—SI UNITS

Approximate or Typical Values

Metals are listed in order of their melting points. Actual properties depend on purity, condition, form, and treatment.

Melting point deg C	Melting point deg F	Melting point K	Metal	Specific gravity	Specific heat, J/kg·K	Thermal conductivity, W/m·K	Electrical resistivity, μΩ·m	Coefficient of linear expansion, μm/m·K	Tensile strength, MN/m²	Specific heat, J/kg·K (1000 °C)	Thermal conductivity, W/m·K (1000 °C)	Tensile strength, MN/m² (1000 °C)
3 400	6 150	3 675	Tungsten (Wolfram)	19.3	134	178	0.056 5	4.5	3 000	155	114	690
3 180	5 760	3 455	Rhenium	21.0	138	71.1	0.19	7	2 300	163		830
3 025	5 477	3 300	Osmium	22.6	130	60.5	0.09	5		151		
2 980	5 400	3 255	Tantalum	16.6	142	54.0	0.124	6.5	380	155	60.5	210
2 620	4 750	2 895	Molybdenum	10.2	251	140	0.052	5	590	314	104	
2 470	4 470	2 745	Niobium (Columbium)	8.57	268	52.1	0.13	7	280	314	69.2	55
2 450	4 440	2 725	Iridium	22.4	130	147	0.053	6	620	159		330
2 400	4 350	2 675	Ruthenium	12.4	238		0.075	9	540	310		
2 220	4 030	2 493	Hafnium	13.3	146	22.0	0.35	6	450	184		
1 965	3 569	2 240	Rhodium	12.4	243	150	0.046	8	950	335		
1 900	3 450	2 175	Vanadium	6.1	485	60.0	0.25	8	540	661	62.3	
1 860	3 380	2 135	Chromium	7.20	460	89.9	0.13	6	83	707	25.9	
1 850	3 370	2 125	Zirconium	6.53	280	21.1	0.41	5.5	340	343	78.7	
1 770	3 220	2 045	Platinum	21.5	134	73.0	0.105	9	150	159		
1 750	3 180	2 025	Thorium	11.7	126	41.0	0.18	12	220	172		
1 670	3 040	1 945	Titanium	4.54	523	20.1	0.43	8.5	240	636	22.5	
1 550	2 820	1 825	Palladium	12.0	243	71.1	0.108	12	170	297		
1 536	2 797	1 810	Iron	7.87	452	80.2	0.097	12	280	619	29.4	
1 495	2 723	1 770	Cobalt	8.9	418	69.0	0.09	12	940	803		
1 453	2 647	1 725	Nickel	8.9	444	89.9	0.068 5	13	320	607	77.8	

Table 1-63.   HEAT CAPACITIES OF COPPER, SILVER, AND GOLD AT LOW TEMPERATURES*

See Table 5-29 for thermal conductivity of metals.

Specific heat values are in cal/mol·K; for J/mol·K multiply by 4.184. Enthalpy values are in cal/mol; for J/mol multiply by 4.184. Gram atomic weights are Cu = 63.540; Ag = 107.87; Au = 196.97. One Btu = 252 calories = 1 054.4 J.

Temperature		Copper		Silver		Gold	
°K	°R	c_p†	H	c_p†	H	c_p†	H
1.0	1.8	0.000177	0.000086	0.000196	0.000088	0.000282	0.000114
2.0	3.6	0.000423	0.000378	0.000633	0.000472	0.00121	0.000779
3.0	5.4	0.000805	0.000978	0.00155	0.00151	0.00337	0.00294
4.0	7.2	0.00139	0.00206	0.00320	0.00382	0.00732	0.00811
5.0	9.0	0.00225	0.00385	0.00581	0.00823	0.0136	0.0184
6.0	10.8	0.00346	0.00668	0.00964	0.0158	0.0228	0.0363
7.0	12.6	0.00508	0.0109	0.0150	0.0280	0.0356	0.0652
8.0	14.4	0.00720	0.0170	0.0222	0.0464	0.0526	0.109
9.0	16.2	0.00990	0.0255	0.0316	0.0731	0.0748	0.172
10.0	18.0	0.0133	0.0370	0.0437	0.110	0.103	0.261
11.0	19.8	0.0174	0.0523	0.0590	0.162	0.138	0.380
12.0	21.6	0.0224	0.0721	0.0778	0.230	0.180	0.539
13.0	23.4	0.0284	0.0974	0.101	0.318	0.230	0.744
14.0	25.2	0.0355	0.129	0.128	0.432	0.288	1.002
15.0	27.0	0.0439	0.169	0.160	0.576	0.352	1.321
16.0	28.8	0.0538	0.218	0.197	0.754	0.424	1.709
17.0	30.6	0.0651	0.277	0.240	0.972	0.501	2.170
18.0	32.4	0.0783	0.348	0.287	1.235	0.584	2.712
19.0	34.2	0.0933	0.434	0.338	1.547	0.671	3.339
20.0	36.0	0.110	0.536	0.394	1.912	0.762	4.055
25.0	45.0	0.230	1.363	0.733	4.689	1.254	9.074
30.0	54.0	0.405	2.928	1.141	9.354	1.763	16.62
35.0	63.0	0.631	5.496	1.580	16.15	2.246	26.66
40.0	72.0	0.894	9.296	2.012	25.14	2.682	38.99
45.0	81.0	1.178	14.47	2.417	36.23	3.073	53.40
50.0	90.0	1.471	21.09	2.786	49.25	3.415	69.64
55.0	99.0	1.765	29.18	3.117	64.03	3.709	87.47
60.0	108.0	2.054	38.73	3.411	80.36	3.964	106.7
65.0	117.0	2.332	49.70	3.669	98.08	4.185	127.1
70.0	126.0	2.596	62.03	3.896	117.0	4.377	148.5
75.0	135.0	2.843	75.63	4.096	137.0	4.544	170.8
80.0	144.0	3.072	90.43	4.272	157.9	4.691	193.9
85.0	153.0	3.285	106.3	4.429	179.7	4.820	217.7
90.0	162.0	3.480	123.2	4.568	202.2	4.934	242.1
95.0	171.0	3.660	141.1	4.692	225.3	5.034	267.0
100.0	180.0	3.825	159.8	4.804	249.1	5.123	292.4
120.0	216.0	4.362	242.0	5.148	348.8	5.393	397.7
140.0	252.0	4.749	333.3	5.382	454.3	5.571	507.5
160.0	288.0	5.032	431.3	5.550	563.7	5.691	620.2
180.0	324.0	5.244	534.1	5.676	676.0	5.772	734.9
200.0	360.0	5.408	640.7	5.773	790.5	5.835	850.9
220.0	396.0	5.537	750.2	5.852	906.8	5.892	968.2
240.0	432.0	5.640	862.0	5.917	1025.	5.944	1087.
260.0	468.0	5.722	975.7	5.973	1143.	5.992	1206.
280.0	504.0	5.788	1091.	6.020	1263.	6.037	1326.
273.15	491.7	5.767	1051.	6.005	1222.	6.022	1285.
298.15	536.7	5.840	1196.	6.059	1373.	6.075	1436.

†This is the instantaneous specific heat at the given temperature. Mean specific heat over any range may be computed from the enthalpies.
*From: "Critical Analysis of the Heat-Capacity Data of the Literature and Evaluation of Thermodynamic Properties of Copper, Silver, and Gold from 0 to 300 K", G.T. Furukawa, W.G. Saba, and M.L. Reilly, NSRDS–NBS–18, National Bureau of Standards, 1968.

Table 1-64. SILVER AND SILVER ALLOYS*

For properties of pure silver, see the index.

TYPICAL SILVER ALLOYS

Typical composition, %	Names and uses	Important property[a]
Ag 99.9+	Fine silver; contacts, chemical uses	Conductivity; ductility
Ag 99+, Mg 0.25, Ni 0.2+	High temperature contacts and spring contacts	Hardness; high conductivity
Ag 99+ (layer, 0.025 in.)	High-service machine bearings	Fatigue strength; thermal conductivity
Ag 92.5, Cu 7.5	Sterling silver; tableware	Appearance; value
Ag 90, Cu 10	Coin silver (to 1966)	Value; corrosion resistance
Ag 90 (outside layer)	Pressure-bonded laminates; coins	Appearance; conductivity
Ag 90, Pd 10	Contacts; brazing alloys	High conductivity
Ag 90, CdO 10 (sintered)	Non-sticking contacts	Heat resistance; hardness
Ag 85, Cd 15	Arc-quenching contacts	Wear resistance; ductility
Ag 80, Cu 20	Laminated coins; tableware	Appearance; value
Ag 80, In 15, Cd 5	Nuclear reactor control	Neutron absorption
Ag 77, Cd 22.6, Ni 0.4	Spring contacts	Wear resistance; ductility
Ag 72, Cu 28	Brazing alloy	Eutectic alloy; hardness
Ag 60, Ni 40 (sintered)	Circuit-breaker contacts	Hardness; heat resistance
Ag 49, Au 41.7, Cu 9, Zn 0.3	10-karat green gold; jewelry	Appearance
Ag 40, Mo or W 60 (sintered)	Switching contacts	Hardness; wear resistance
Ag 40, Au 30, Pd 30	Corrosive-chemical apparatus (e.g., for halogens and nitric acid)	Corrosion resistance; strength
Ag 35, Cu 26, Zn 21, Cd 18	Thin-joint brazing	Wetting ability; ductility
Ag 33, Hg 52, Sn 12.5, Cu 2, Zn 0.5	Dental amalgam fillings	Amalgam hardening
Ag 30.5, Au 50, Cu 17.5, Zn 2	Gold solder; jewelry	Appearance; strength
Ag 15, Au 60, Cu 10, Pd 10, Pt 4, Zn 1	White gold denture metal	Strength; wear resistance
Ag 15, Au 60, Cu 15	Precious metal solder	Appearance; strength
Ag 10, Au 58.3, Cu 29.7, Zn 2	14-karat yellow gold	Appearance; wear
Ag 2.5, Pb 97.5	Soft solder	Eutectic alloy

[a]Properties of pure silver and the high-silver alloys vary with temperature; they are subject to work hardening, and the room-temperature properties depend on the annealing temperature.

CHEMICAL USES OF SILVER

Type of use	Form of silver	Processes and remarks
Batteries	Powder; oxide; chloride	Silver oxide with zinc or cadmium; silver chloride with magnesium
Catalysts	Shapes, screens, powder, salts	In dehydrogenation, oxidation, desulfurization
Coatings	Powder; vapor	Air-drying paints and fired-glass conductive coatings
Electroplating	Cyanide; anode shapes	Usual plate less than 0.001 5 in. thickness
Medicine	Nitrate; metal, colloidal	Sterilization by metal, nitrate, or colloidal suspension
Photography	Nitrate; halide colloidal solutions	Metallic silver from salt by radiation and "developer"
Reflectors (not ultraviolet)	Nitrate; metal vapor	Silver nitrate reacts with reducing solution; vacuum evaporation

*Compiled from various sources.

REFERENCES
"Metals Handbook", Vol. 1, American Society for Metals, 1961, pp. 1174–1186.
"Silver, Economics, Metallurgy and Use", A. Butts and C.D. Coxe, Eds., D. Van Nostrand Co., 1967.
"Silver in Industry", L. Addicks, Reinhold Publishing Corp., 1940.
ASTM Standards B253, B260, B413, and E56.

Table 1-65. EQUIVALENT HARDNESS NUMBERS FOR STEEL*

Comparison of Methods for Hardness Testing

For tensile strength in MN/m², multiply value in thousands of psi by 6.895.

Brinell indentation, diam, mm†	Brinell 10-mm ball, 3000-kg load		Diamond Pyramid hardness (Vickers)	Rockwell, Brale penetrator			Rockwell, Superficial Brale penetrator		Shore Sclero-scope	Tensile strength, 1000 psi (approx.)
	Standard ball	Carbide ball		A 60-kg load	C 150-kg load	D 100-kg load	15–N 15–kg	45–N 45–kg		
2.25	—	745	840	84.1	65.3	74.8	92.3	72.2	91	—
—	—	733	820	83.8	64.7	74.3	92.1	71.8	90	—
—	—	722	800	83.4	64.0	73.8	91.8	71.0	88	—
2.30	—	712	—	—	—	—	—	—	—	—
—	—	710	780	83.0	63.3	73.3	91.5	70.2	87	—
—	—	698	760	82.6	62.5	72.6	91.2	69.4	86	—
—	—	684	740	82.2	61.8	72.1	91.0	68.6	—	—
2.35	—	682	737	82.2	61.7	72.0	91.0	68.5	84	—
—	—	670	720	81.8	61.0	71.5	90.7	67.7	83	—
—	—	656	700	81.3	60.1	70.8	90.3	66.7	—	—
2.40	—	653	697	81.2	60.0	70.7	90.2	66.5	81	—
—	—	647	690	81.1	59.7	70.5	90.1	66.2	—	—
—	—	638	680	80.8	59.2	70.1	89.8	65.7	80	—
—	—	630	670	80.6	58.8	69.8	89.7	65.3	—	—
2.45	—	627	667	80.5	58.7	69.7	89.6	65.1	79	—
2.50	—	601	640	79.8	57.3	68.7	89.0	63.5	77	—
2.55	—	578	615	79.1	56.0	67.7	88.4	62.1	75	—
2.60	—	555	591	78.4	54.7	66.7	87.8	60.6	73	298
2.65	—	534	569	77.8	53.5	65.8	87.2	59.2	71	288
2.70	—	514	547	76.9	52.1	64.7	86.5	57.6	70	274
2.75	{ 495	—	539	76.7	51.6	64.3	86.3	56.9	—	269
	{ —	495	528	76.3	51.0	63.8	85.9	56.1	68	264
2.80	{ 477	—	516	75.9	50.3	63.2	85.6	55.2	—	258
	{ —	477	508	75.6	49.6	62.7	85.3	54.5	66	252
2.85	{ 461	—	495	75.1	48.8	61.9	84.9	53.5	—	244
	{ —	461	491	74.9	48.5	61.7	84.7	53.2	65	242
2.90	{ 444	—	474	74.3	47.2	61.0	84.1	51.7	—	231
	{ —	444	472	74.2	47.1	60.8	84.0	51.5	63	230
2.95	429	429	455	73.4	45.7	59.7	83.4	49.9	61	219
3.00	415	415	440	72.8	44.5	58.8	82.8	48.4	59	212
3.05	401	401	425	72.0	43.1	57.8	82.0	46.9	58	202
3.10	388	388	410	71.4	41.8	56.8	81.4	45.3	56	193
3.15	375	375	396	70.6	40.4	55.7	80.6	43.6	54	184
3.20	363	363	383	70.0	39.1	54.6	80.0	42.0	52	177
3.25	352	352	372	69.3	37.9	53.8	79.3	40.5	51	171
3.30	341	341	360	68.7	36.6	52.8	78.6	39.1	50	164
3.35	331	331	350	68.1	35.5	51.9	78.0	37.8	48	159
3.40	321	321	339	67.5	34.3	51.0	77.3	36.4	47	154
3.45	311	311	328	66.9	33.1	50.0	76.7	34.4	46	149
3.50	302	302	319	66.3	32.1	49.3	76.1	33.8	45	146
3.55	293	293	309	65.7	30.9	48.3	75.5	32.4	43	141
3.60	285	285	301	65.3	29.9	47.6	75.0	31.2	—	138
3.65	277	277	292	64.6	28.8	46.7	74.4	29.9	41	134
3.70	269	269	284	64.1	27.6	45.9	73.7	28.5	40	130
3.75	262	262	276	63.6	26.6	45.0	73.1	27.3	39	127
3.80	255	255	269	63.0	25.4	44.2	72.5	26.0	38	123
3.85	248	248	261	62.5	24.2	43.2	71.7	24.5	37	120
3.90	241	241	253	61.8	22.8	42.0	70.9	22.8	36	116
3.95	235	235	247	61.4	21.7	41.4	70.3	21.5	35	114
4.00	229	229	241	60.8	20.5	40.5	69.7	20.1	34	111
4.25	201	201	212	—	13.8	—	—	—	31	98
4.50	179	179	188	—	8.0	—	—	—	27	87
5.00	143	143	150	—	—	—	—	—	22	71
5.50	116	116	122	—	—	—	—	—	18	58

†The values corresponding to Brinell indentations of 2.35, 2.40 and 2.45 through 4.00 correspond to the values shown in the corresponding joint SAE–ASM–ASTM Committee on Hardness Conversions as printed in ASTM E 140, American Society for Testing and Materials, Table 3.

*Abridged from: "SAE Handbook", Society of Automotive Engineers, 1967.

Table 1-66.   COMPOSITION AND MELTING POINTS OF BINARY EUTECTIC ALLOYS*

Binary Alloys and Solid Solutions of Metallic Components

This table represents most of the common binary combinations of metals. For many pairs no eutectic exists; for many others the information is uncertain or unavailable. In a fair number of cases, there is complete mutual solubility in all proportions; hence, there is a smooth temperature vs. composition curve, with no point of inflection from the melting point of one constituent to that of the other. For purposes of comparison, all values must be considered approximate in view of the experimental difficulties and the many sources of data.

Those pairs for which the liquidus curve exhibits more than one cusp are designated by a superscript a. In a few cases the cusp selected for this table does not represent the lowest possible melting point for the binary mixture.

Constituents		Composition		Melting point		Constituents		Composition		Melting point	
A	B	Mol % B	Weight % B	K	deg F	A	B	Mol % B	Weight % B	K	deg F
Ag	Al	57	25	835	1 044	Au	Bi	86.8	85	514	466
Ag	As	24	18	813	1 004	Au	Cd	70	57.1	773	932
Ag	Ca^a	37	18	820	1 017	Au	Ce^a	86	81	793	968
Ag	Ce^a	80	84	798	977	Au	Ge	27	12	629	673
Ag	Cu	40	28	1 050	1 431	Au	La^a	83	78	834	1 042
Ag	Ge	25	18	924	1 204	Au	Mg	93	62	848	1 067
Ag	La^a	72	77	791	964	Au	Mn^a	32	12	1 233	1 760
Ag	Li	99	89	418	293	Au	Na	17	2.3	1 149	1 609
Ag	Mg^a	83	52	745	882	Au	Pb	84	85	488	419
Ag	Pb	95.3	97.5	577	579	Au	Sb	34.8	24.8	633	680
Ag	Pd	25.9	25.6	924	1 204	Au	Si	18.6	3.15	636	685
Ag	Sb	41	44	758	905	Au	Sn^a	29.3	19.9	551	532
Ag	Si	10.5	2.96	1 110	1 539	Au	Te	88	83	689	781
Ag	Sr^a	77	73	709	817	Au	Tl	72	73	404	268
Ag	Te	65	69	623	662	Au	U	14	16	1 128	1 571
Ag	Th	7.6	15	1 167	1 641	B	Hf	13	71	2 130	3 375
Ag	Zr	97	93	1 100	1 521	B	Ni	57	88	1 263	1 814
Al	Au^a	59.5	90.0	842	1 056	B	Ti	7	25	1 700	2 601
Al	Ca^a	65	73	818	1 013	B	Zr	88	98	1 920	2 997
Al	Cd	81	90	1 650	2 511	Ba	Mg	97	87	891	1 144
Al	Ce	69	92	928	1 211	Be	Ni	33	76	1 468	2 183
Al	Cu^a	17.3	33.0	821	1 018	Be	Pu	97	99	910	1 179
Al	Fe	32	49.34	1 426	2 107	Be	Si	33	61	1 363	1 994
Al	Ge	29	55	700	801	Be	Ti	75	94	1 300	2 061
Al	In	5	18	910	1 179	Be	Y	61	94	1 390	2 043
Al	Mg	70	67.0	710	819	Be	Zr	65	95	1 250	1 791
Al	Ni^a	76	87	1 658	2 525	Bi	Ca	88	58.5	1 059	1 447
Al	Pt^a	57	90	1 533	2 300	Bi	Cd	56	40	420	297
Al	Si	13	13	850	1 071	Bi	In^a	78	66	340	153
Al	Th	80	97	1 510	2 259	Bi	K	50	16	615	648
Al	Zn	88.7	95.0	655	720	Bi	Mg	85	40	820	1 017
As	Co	75	70	1 189	1 681	Bi	Na	22	3.0	500	441
As	Cu^a	81.6	78.0	958	1 265	Bi	Pb	44	44	397	255
As	Fe	75	69	1 103	2 017	Bi	Sn	57	43	415	288
As	In	13	18	1 004	1 348	Bi	Te	90	84	686	775
As	Mn	57	49	1 143	1 598	Bi	Tl^a	53	52	465	378
As	Ni^a	63	57	1 077	1 479	C	Cr	87	96	1 775	2 736
As	Sb	80	87	878	1 121	C	Hf	35	88	3 450	5 751
As	Sn^a	40	51	852	1 074	C	Mo	17	45	2 480	4 005
As	Zn^a	20	18	996	1 333	C	Nb	40	84	3 580	5 985

*Compiled from several sources.

Table 1-66.   COMPOSITION AND MELTING POINTS OF BINARY EUTECTIC ALLOYS *(Continued)*

| Constituents | | Composition | | Melting point | | Constituents | | Composition | | Melting point | |
A	B	Mol % B	Weight % B	K	deg F	A	B	Mol % B	Weight % B	K	deg F
C	Ti	36	69	3 050	5 031	Gd	Ni[a]	32	15	943	1 238
C	V	84	96	1 900	2 961	Ge	Mg	38	17	953	1 256
C	W	59	96	2 980	4 905	Ge	Mn[a]	48	41	970	1 287
Ca	Cu	51	62	833	1 040	Hf	Ta	24	24	1 300	1 881
Ca	Mg[a]	32	22	718	833	In	Ni	30	17.97	1 143	1 598
Ca	Na	22	14	983	1 310	In	Sb	68	69	780	945
Ca	Ni	16	22	878	1 121	In	Sn	47	48	390	243
Ca	Sn	19	41	1 010	1 359	Ir	Mo	68	52	2 350	3 771
Cd	Cu	52	38	810	999	Ir	Nb	55	23	2 110	3 339
Cd	In	74	74	400	261	Ir	W	22	12	2 590	4 203
Cd	Pb	71	82	540	513	K	Na	32	21.67	260	−8.6
Cd	Pu	40	59	1 170	1 647	K	Rb	70	84	307	93
Cd	Sb	7.4	8	563	554	K	Sb[a]	68	84	680	765
Cd	Sn	68	69	450	351	K	Tl	84	96	440	333
Cd	Tl	73	83	475	396	La	Mg[a]	38	9.7	970	1 287
Cd	Zn	27	18	540	513	La	Pb[a]	11	15	1 049	1 429
Ce	Cu[a]	28	15	688	779	La	Sn[a]	10	9	993	1 328
Ce	Ru	33	26	923	1 202	La	Tl	16	22	913	1 184
Co	Gd	65	83	913	1 184	Mg	Ni	11	22.98	780	945
Co	Mo	27	38	1 610	2 439	Mg	Pr	4.9	23	858	1 085
Co	Nb	15	22	1 500	2 241	Mg	Pu	15	63	815	1 008
Co	Si[a]	71	54	1 486	2 215	Mg	Sb[a]	86	97	855	1 080
Co	Sn	21	35	1 380	2 025	Mg	Si	53	57	1 223	1 742
Co	Ti[a]	22	19	1 430	2 115	Mg	Sr[a]	70	89	699	799
Co	V	41	38	1 521	2 278	Mg	Th	7	42	855	1 080
Cr	Mo	14	23	1 973	3 092	Mg	Zn	30	53	615	648
Cr	Ni	46	47	1 610	2 439	Mn	Ni	40	42	1 300	1 881
Cr	Ta	13	34	1 950	3 051	Mn	Pd	26	41	1 398	2 057
Cr	Ti	86	85	950	1 251	Mn	Sb	82	91	843	1 058
Cr	V	33	32	2 050	3 231	Mn	Ti[a]	9	7.9	1 460	2 169
Cs	K	50	23	235	−36	Mn	U[a]	75	93	988	1 319
Cs	Na	20.9	4.37	241	−26	Mn	Y[a]	65	75	1 163	1 634
Cs	Rb	50	39	282	48	Mo	Nb	66	65	2 570	4 167
Cu	Ge	34	37	913	1 184	Mo	Ni	64	52	1 590	2 403
Cu	Mg[a]	85.5	69.3	758	905	Mo	Os	21	34	2 650	4 311
Cu	Mn	37	34	1 143	1 598	Mo	Pd	54	57	2 020	3 177
Cu	Pb	15	36	1 230	1 755	Mo	Re	48	64	2 780	4 545
Cu	Pr[a]	69	83	745	882	Mo	Ru	41	42	2 200	3 501
Cu	Sb[a]	63	76	800	981	Mo	Si[a]	17	5.7	2 350	3 771
Cu	Si	30	16	1 075	1 476	Na	Rb	82.1	94.5	269	25
Cu	Te	69	82	617	207	Na	Sb	60	89	678	761
Cu	Ti[a]	27	22	1 133	1 580	Na	Sn	37	75	718	833
Cu	Tl	14.5	35.3	1 357	1 983	Na	Te	55	87	592	606
Cu	U	8.2	25	1 213	1 724	Nb	Ni	58	47	1 450	2 151
Cu	Zr	9.4	13	1 253	1 796	Nb	Pt	54	71	1 970	3 087
Fe	Gd	69	86	1 123	1 562	Nb	Rh	45	31	1 770	2 727
Fe	Mo	21	31	1 725	2 646	Nb	Ru[a]	64	49	2 050	3 231
Fe	Nb	12	18.49	1 643	2 498	Nb	Zr	77	77	2 010	3 159
Fe	Sb	88	94.10	1 021	1 378	Ni	Sb	22	36.90	1 375	2 016
Fe	Si[a]	35	21	1 475	2 196	Ni	Sn	19	32.16	1 403	2 066
Fe	Sn	31	49	1 400	2 061	Ni	Th[a]	35	68	1 303	1 886
Fe	Y	65	75	1 173	1 652	Ni	Ti[a]	39	34	1 390	2 043
Fe	Zr[a]	11	17	1 600	2 421	Ni	V	52	48	1 473	2 192
Ga	Mg[a]	80	58	698	797	Ni	W	20.7	45	1 773	2 732
Ga	Ni	70	66	1 477	2 199	Ni	Zn	69	71	1 148	1 607

Table 1-66. COMPOSITION AND MELTING POINTS OF BINARY EUTECTIC ALLOYS *(Continued)*

Constituents		Composition		Melting point		Constituents		Composition		Melting point	
A	*B*	Mol % B	Weight % B	K	deg F	*A*	*B*	Mol % B	Weight % B	K	deg F
Pb	Pr	40	31	1 315	1 908	Si	Th[a]	88	98	1 710	2 619
Pb	Pt	5.3	5.0	563	554	Si	Ti[a]	86	91	1 600	2 421
Pb	Sb	18	11	520	477	Si	Zr[a]	9	24	1 570	2 367
Pb	Sn	73	61	460	369	Sn	Te	84	85	678	761
Pb	Te	85	78	680	765	Sn	Tl	31	44	440	333
Pb	Ti	92	74	998	1 337	Sn	Zn	16	9.5	465	378
Pd	Sb	89	90	868	1 103	Te	Tl	30	41	483	410
Pt	Sn	40	29	1 345	1 962	Th	Ti	40	12	1 463	2 174
Pu	Zn	73	42	1 100	1 521	Th	Zn[a]	49	21	1 220	1 737
Re	W	26	26	3 100	5 121	Ti	U	17	51	933	1 220
Sb	Tl	70	80	468	383	Ti	Y	6.8	12	1 593	2 408
Sb	Zn	33	21	780	945	Ti	Zr	50	66	790	963
Sb	Zr	82	77	1 700	2 601	U	Zr	70	47	879	1 123
Se	Sn	39	49	913	1 184						
Se	Tl	26	48	424	304						

REFERENCES

"Selected Values of Thermodynamic Properties of Metals and Alloys", R. Hultgren, R.L. Orr, P.D. Anderson, K.K. Kelley, John Wiley & Sons, Inc., 1963; a supplement to this publication has been issued periodically by the University of California, 1964–1971.

"Constitution of Binary Alloys", 2nd ed., M. Hansen, McGraw-Hill Book Company, 1958.

"Metals Reference Book", 4th ed., C.J. Smithells, Vol. 2, Butterworth & Co., London, 1967.

"Handbook of Binary Metallic Systems", 2 volumes; translated from Russian, Israel Program for Scientific Translations, Jerusalem. Available from Clearinghouse for Federal Scientific and Technical Information, Springfield, Virginia 22151.

See also *Trans. AIME, J. Inst. Metals,* and *Z. Metallkunde,* by indexes.

Table 1-67. STANDARD GRADES OF BOLTS

As Specified by SAE and ASTM

For strength in MN/m², multiply value in psi by 0.006895.

SAE grade	ASTM desig- nation	Tensile strength,* lb/in.²	Proof load or yield strength,* lb/in.²	Material
0				Low carbon
1	A307	55,000		Low carbon
2		69,000	55,000	Low carbon, stress-relieved
3		110,000	85,000	Medium carbon, cold-worked
5		120,000	85,000	Medium carbon, quenched and tempered
	A325	125,000	90,000	Medium carbon, quenched and tempered
6		140,000	110,000	Special medium carbon, quenched and tempered
7		130,000	105,000	Alloy steel, quenched and tempered
8		150,000	120,000	Alloy steel, quenched and tempered

*Properties for sizes ½ in. diameter and smaller.

For complete data and for properties of larger bolts, see SAE Handbook, ASME Handbook, and ASTM Specifications.

Table 1-68.   ALLOYS OF LOW MELTING POINTS*

Melting point °C	Melting point °F	Name	Composition, wt %				Melting point °C	Melting point °F	Name	Composition, wt %				
−48	−54	Binary eutectic	Cs 77.0	K 23.0			98	208.5	D'Arcet's metal	Bi 50.0	Sn 25.0	Pb 25.0		
−40	−40	Binary eutectic	Cs 87.0	Rb 13.0			100	212	Onion's or Lichtenberg's metal	Bi 50.0	Sn 20.0	Pb 30.0		
−30	−22	Binary eutectic	Cs 95.0	Na 5.0			102.5	216	Ternary eutectic	Bi 54.0	Sn 26.0	Cd 20.0		
−11	+12	Binary eutectic	K 78.0	Na 22.0			109	228	Rose's metal	Bi 50.0	Pb 28.0	Sn 22.0		
−8	17.5	Binary eutectic	Rb 92.0	Na 8.0			117	242.5	Binary eutectic	In 52.0	Sn 48.0			
10.7	51	Ternary eutectic	Ga 62.5	In 21.5	Sn 16.0		120	248	Binary eutectic	In 75.0	Cd 25.0			
17	62.5	Ternary eutectic	Ga 82.0	Sn 12.0	Zn 6.0		123	253	Malotte's metal	Bi 46.1	Sn 34.2	Pb 19.7		
33	91.5	Binary eutectic	Rb 68.0	K 32.0			124	255	Binary eutectic	Bi 55.5	Pb 44.5			
46.5	115	Quinternary eutectic	Sn 10.65	Bi 40.63	Pb 22.11	In 18.1 · Cd 8.2	130	266	Ternary eutectic	Bi 56.0	Sn 40.0	Zn 4.0		
58.2	137	Quaternary eutectic	Bi 49.5	Pb 17.6	Sn 11.6	In 21.3	140	284	Binary eutectic	Bi 58.0	Sn 42.0			
60.5	141	Ternary eutectic	In 51.0	Bi 32.5	Sn 16.5		183	361.5	Eutectic solder	Sn 63.0	Pb 37.0			
70	158	Wood's metal	Bi 50.0	Pb 25.0	Sn 12.5	Cd 12.5	185	365	Binary eutectic	Tl 52.0	Bi 48.0			
70	158	Lipowitz's metal	Bi 50.0	Pb 26.7	Sn 13.3	Cd 10.0	192	377.5	Soft solder	Sn 70.0	Pb 30.0			
70	158	Binary eutectic	In 67.0	Bi 33.0			198	388.5	Binary eutectic	Sn 91.0	Zn 9.0			
91.5	197	Ternary eutectic	Bi 51.6	Pb 40.2	Cd 8.2		199	390	Tin foil or white metal	Sn 92.0	Zn 8.0			
95	203	Ternary eutectic	Bi 52.5	Pb 32.0	Sn 15.5		221	430	Binary eutectic	Sn 96.5	Ag 3.5			
97	206.5	Newton's metal	Bi 50.0	Sn 18.8	Pb 31.2		226	439	Matrix	Bi 48.0	Pb 28.5	Sn 14.5	Sb 9.0	
							227	440.5	Binary eutectic	Sn 99.25	Cu 0.75			
							240	464	Antimonial tin solder	Sn 95.0	Sb 5.0			
							245	473	Tin-silver solder	Sn 95.0	Ag 5.0			

*From: "Handbook of Chemistry and Physics", 53rd ed., R.C. Weast, Ed., The Chemical Rubber Co., 1972.

Table 1-69.   MELTING POINTS OF MIXTURES OF METALS**

Melting Points, °C

Metals	Percentage of metal in second column 0%	10%	20%	30%	40%	50%	60%	70%	80%	90%	100%	Metals	Percentage of metal in second column 0%	10%	20%	30%	40%	50%	60%	70%	80%	90%	100%
Pb. Sn.	* 326	295	276	262	240	220	190	185	200	216	232	Ni. Sn.	* 1455	1380	1290	1200	1235	1290	1305	1230	1060	800	232
Bi.	322	290	...	...	179	145	126	168	205	...	268	Na. Bi.	96	425	520	590	645	690	720	730	715	570	268
Te.	322	710	790	880	917	760	600	480	410	425	446	Cd.	96	125	185	245	285	325	330	340	360	390	322
Ag.	328	460	545	590	620	650	705	775	840	905	959	Cd. Ag.	322	420	520	610	700	760	805	850	895	940	954
Na.	...	360	420	400	370	330	290	250	200	130	96	Tl.	321	300	285	270	262	258	245	230	210	235	302
Cu.	326	870	920	925	945	950	955	985	1005	1020	1084	Zn.	322	280	270	295	313	327	340	355	370	390	419
Sb.	326	250	275	330	395	440	490	525	560	600	632	Au. Cu.	1063	910	890	895	905	925	975	1000	1025	1060	1084
Al. Sb.	650	750	840	925	945	950	970	1000	1040	1010	632	Ag.	1064	1062	1061	1058	1054	1049	1039	1025	1006	982	963
Cu.	650	630	600	560	540	580	610	755	930	1055	1084	Pt.	1075	1125	1190	1250	1320	1380	1455	1530	1610	1685	1775
Au.	655	675	740	800	855	915	970	1025	1055	675	1062	K. Na.	62	17.5	−10	−3.5	5	11	26	41	58	77	97.5
Ag.	650	625	615	600	590	580	575	570	650	750	954	Hg.	...	...	...	...	90	110	135	162	265	...	
Zn.	654	640	620	600	580	560	530	510	475	425	419	Tl.	62.5	133	165	188	205	215	220	240	280	305	301
Fe.	653	860	1015	1110	1145	1145	1220	1315	1425	1500	1515	Cu. Ni.	1080	1180	1240	1290	1320	1355	1380	1410	1430	1440	1455
Sn.	650	645	635	625	620	605	590	570	560	540	232	Sn.	1082	1035	990	945	910	870	830	788	814	875	960
Sb. Bi.	632	610	590	575	555	540	520	470	405	330	268	Ag.	1084	1005	890	755	725	680	630	580	530	440	232
Ag.	630	595	570	545	520	500	505	545	680	850	959	Zn.	1084	1040	995	930	900	880	820	780	700	580	419
Sn.	622	600	570	525	480	430	395	350	310	255	232	Ag. Zn.	959	850	755	705	690	660	630	610	570	505	419
Zn.	632	555	510	540	570	565	540	525	510	470	419	Sn.	959	870	750	630	550	495	450	420	375	300	232
												Na. Hg.	96.5	90	80	70	60	45	22	55	95	215	...

*The data in this table are compiled from various sources—hence the variations in the melting point of the metals as shown in this column.
**Based largely on: "Smithsonian Physical Tables", 9th rev. ed., W.E. Forsythe, Ed., The Smithsonian Institution, 1956.

Table 1-70.   PROPERTIES OF RARE-EARTH METALS*

F.H. Spedding

To convert density from g/cm³ to kg/m³, multiply by 1000. To convert Young's modulus from kg/cm² to N/m², multiply by 98,067. Values in parentheses are estimates.

Element	Melting point, °C	Boiling point, °C	Heat of sublimation, kcal/mole ΔH 298°K	Density, g/cm³ 298°K	Atomic volume, cm³/mole	Metallic radius, Å	Electrical resistivity at 298°K, microhm-cm	Residual resistivity at 4.2°K, microhm-cm	Compressibility, cm²/kg†	Young's modulus kg/cm², millions	Poisson's ratio
Scandium	1539	2832	91.0	2.989	15.04	1.641	50.9	3.7	2.26	0.809	(0.269)
Yttrium	1523	3337	99.6	4.457	19.95	1.801	59.6	3.2	2.68	0.663	0.265
Lanthanum	920	3454	103.0	6.166	22.53	1.871	79.8	S.C.‡	4.04	0.384	0.288
Cerium	798	3257	111.60	6.771	20.69	1.824	75.3	0.7	4.10	0.306	0.248
Praseodymium	931	3212	89.09	6.772	20.81	1.828	68.0		3.21	0.332	0.305
Neodymium	1010	3127	77.3	7.003	20.60	1.822	64.3	6.8	3.0	0.387	0.306
Promethium	1080	(2460)	(64)	—	—	—	—	—	(2.8)	(0.430)	(0.278)
Samarium	1072	1778	49.3	7.537	19.95	1.802	105.0	6.2	3.34	0.348	0.352
Europium	822	1597	42.5	5.253	28.93	1.983	91.0	0.6	8.29	0.150	(0.286)
Gadolinium	1311	3233	95.75	7.898	19.91	1.801	131.0	4.4	2.56	0.573	0.259
Terbium	1360	3041	93.96	8.234	19.30	1.783	114.5	3.5	2.45	0.586	0.261
Dysprosium	1409	2335	71.2	8.540	19.03	1.775	92.6	2.4	2.55	0.644	0.243
Holmium	1470	2720	71.7	8.781	18.78	1.767	81.4	7.0	2.47	0.684	0.255
Erbium	1522	2510	74.5	9.045	18.49	1.758	86.0	4.7	2.39	0.748	0.238
Thulium	1545	1727	58.3	9.314	18.14	1.747	67.6	5.6	2.47	(0.770)	(0.235)
Ytterbium	824	1193	38.2	6.972	24.82	1.939	25.1	0.29	7.39	0.182	0.284
Lutetium	1656	3315	102.16	9.835	17.79	1.735	58.2	4.5	2.38	(0.860)	(0.233)

† All values in this column should be divided by 10⁶.

‡ S.C.—Superconductor.

*Condensed from: "Handbook of Chemistry and Physics", 53rd ed., R.C. Weast, Ed., The Chemical Rubber Co., 1972.

Table 1-71. SPECIFIC STIFFNESS OF METALS, ALLOYS, AND CERTAIN NON-METALLICS*

Specific stiffness is usually expressed as the modulus of elasticity (in tension) per unit weight-density, i.e., E/ρ, in units of pounds and inches. While the stiffness of similar alloys varies considerably, there are definite ranges and groups to be recognized. Since the specific stiffness of steel is about 100 million, the values in the following table are also approximately the percentage stiffness, referred to steel.

Material	Specific stiffness, millions
Beryllium	650
Silicon carbide	600
Alumina ceramics	400
Mica	350
Titanium carbide cermet	250
Alumina cermet	200
Molybdenum and alloys; silica glass	130
Titanium and alloys; cobalt superalloys; soda-lime glass	110
Carbon and low-alloy steel; wrought iron	105
Stainless steel; nodular cast iron; magnesium and alloys; aluminum and alloys	100
Nickel and alloys; malleable iron	95
Iron silicon alloys (cast); iridium; vanadium	90
Monel alloys; tungsten	80
Gray cast iron; columbium alloys	70
Aluminum bronze; beryllium copper	65
Nickel silver; cupronickel; zirconium	55
Yellow brass; nickel cast iron; bronze; Muntz metal; antimony	50
Copper; red brass; tantalum	45
Silver and alloys; pewter; platinum and alloys; white gold	30
Tin; thorium	25
Gold	20
Tin-lead alloy	10
Lead	5

*Compiled from several sources.

Table 1-72. APPROXIMATE EFFECT OF TEMPERATURE ON STRENGTH-WEIGHT RATIO OF METALS

Material	Temperature, °F					
	200°	400°	600°	800°	1000°	1200°
Titanium MST6A1–4V	773	696	627	550	454	
Titanium RC–130	727	636	554	455	236	
Stainless 17–7PH	654	632	582	496	250	
Magnesium AZ80X	491	196	77.3			
Inconel X	435	426	418	400	368	322
Aluminum SAP	350	306	225	155		
Molybdenum	468	400	350	309	273	245
Aluminum 7075–T6	650	127	61.8			
L–605	196	159	127	109	109	109
Stainless Type 321	112	100	91.0	81.7	70.0	59.0

REFERENCE

For tensile strengths of 40 superalloys at 1200–2000°F, see "Guide to Selection of Superalloys", W.F. Simmons, *Metal Progress* (American Society for Metals), 91:86–86C, June 1967.

Table 1-73. METAL POWDERS

Typical Metal Powders and Granules

Varieties and Uses. The largest use of metal powders is for production of parts, shapes, and electrodes by powder metallurgy. Other major uses are in metal coatings, applied by flame spraying or as paints, lacquers or inks, chemical catalysts, reducing agents, fuels, and explosives. Many composite materials contain metal powder fillers. Annual U.S. consumption of iron powder and alloys is well over 200 million pounds. Some 200 kinds are available. Copper and its alloys represent almost as great a variety, but the total consumption is less than one-third that of iron. Aluminum, nickel, chromium, tungsten, and silver powders are used in quantity, but there are very important uses for some of the other metals also. The following table gives only a few typical examples.

Particle Size and Surface. Actual dimensions and size-distribution are not yet independent of the methods of measurement. For instance, fine particles that pass a 100-mesh sieve can be dimensionally analyzed by three or four methods with divergent results (see Table 10-12). The apparent density of a powder is not independent of the method of handling, and the constant-volume "tap density" is affected by the shape of particles. Even a single method, such as a sieve analysis, depends on the condition and quality of the sieves. Particle sizes above 40 microns in the table are based on standard ASTM sieve analyses and their metric equivalents in microns ($25\,400\mu = 1$ in.). Particle size distribution and surface area of particles are not given in this table, but they are important for applications such as those involving either surface coverage or gas adsorption by the particles. Surface areas greater than one square meter per gram are attainable for metals. Integration of the area under the size distribution curve gives a low value of surface area, since it is only practical to assume spherical particles. Other methods depend on air permeability, liquid surface spread, liquid mixture turbidity, or gas adsorption. Comparable results from the same sample are very difficult to obtain if more than one test method is used.

Purity and Analysis. Metal purity to 99.9999% is often quoted, but the purity should be specified according to the use. Metals of purity 99% or less are much cheaper. A metal powder can be produced from an alloy of almost any composition. Although most of the common and proprietary alloys are commercially available in powder form, a special order may be required if the particle-size specification is unusual.

For suppliers of metal powders, consult the Metal Powder Industries Federation, 201 E. 42nd Street, New York, NY 10017.

Metal or alloy	Purity, %	Largest impurity	Particle size		Specific gravity		Designation
			microns (av. or range)	% passing-sieve No.	Powder	Solid	
Aluminum	99.9+	Fe	75–150	100–100	2.5	2.7	Pure
Aluminum	99.5	Fe or Si	<60	95–325	2.5	2.7	Atomized, fine
Aluminum	99.3	Oxygen	30–75	90–325	1.7	2.7	Atomized, spherical
Aluminum		Oxygen	6–600	95–40	1.2	2.7	Granular
Aluminum	95	Oxygen	0.03	—	0.2	2.7	Fine, spheroidal
Antimony	99.5		<100	95–325	2.42	6.7	Fine
Beryllium	99.0	Oxygen	<80	98–200		1.85	Structural
Bismuth	99.6	Sb	40–150	80–325	4.29	9.75	Commercial
Cadmium	99.9	Pb	40–150	95–325		8.65	Fine
Chromium	99.5	Fe	<45	100–325		7.2	Fine
Cobalt	99.8	Ni	1.2	99.9–400	0.7	8.9	Very fine
Copper	99.9	Sn	>1 000	—	5.3	8.95	Granular
Copper	99.8	Sn	15	95–325	4.55	8.95	Atomized shot
Copper	99.5	Sn	40–850	60–325	2.7	8.95	Electrolytic, coarse
Brass	90 Cu	Pb	40–250	55–325	3.0	8.5	10% Zn, atomized

Table 1-73.   METAL POWDERS *(Continued)*

Metal or alloy	Purity, %	Largest impurity	Particle size		Specific gravity		Designation
			microns (av. or range)	% passing-sieve No.	Powder	Solid	
Bronze	90 Cu		40–150	55–325	2.7	8.5	10% Sn
Gold	99.9+		<35	100–400	—	19.3	Pure, reduced
Gold	99.+		0.5 × 10	100–325	2.0	19.3	Flake 0.5 × 10 microns
Iron	99.8	Oxygen	7–10	—	3.2	7.9	Very fine
Iron	98.5	SiO$_2$	40–150	30–325	2.9	7.9	Reduced, annealed
Iron	91.	C	100–900	95–100	2.3	7.9	Ground cast iron
Stainless steel	71.	Si	20–150	40–325	3.0	8.0	Stainless steel 18-8
Lead	99.8	Bi	10–45	100–325	5.0	11.35	Very fine
Magnesium	99.8	Mn	<75	100–200	0.6	1.74	Commercial
Manganese	99.8	Oxygen	<100	80–325	2.8	7.3	Commercial
Ferromanganese	80 Mn	C	40–300	30–325			Medium carbon
Molybdenum	99.9	Oxygen	3–6	—	1.4	10.2	Hydrogen reduced
Nickel	99.9+	Co	10–100	80–250	2.0	8.9	High purity
Nickel	99.5	Co	<150	50–325	3.4	8.9	Electrolytic
Niobium	99.6	Oxygen	<75	100–200		8.57	High purity
Palladium	99.9+	Pt	<45	100–325		12.0	Pure
Platinum	99.0	Ag	<420	100–40		21.4	Sponge
Silver	99.9+	Si	<150	90–325	1.5	10.5	Precipitated, spongy
Silver braze	70.	Optional	<175	100–80			28% Cu brazing alloy
Tin	99.5	Oxygen	<150	30–250	3.4	7.3	Commercial
Titanium	99.		8	98–325		4.54	Commercial
Tungsten	99.9	Mo	<175	100–80	9.8	19.3	Reduced
Vanadium	99.5	Fe	<800	100–20		6.1	Commercial
Zinc	99.	Pb	<50	98–325	7.0	7.0	Reduced
Zirconium	99.+	Oxygen	<175	100–80	3.5	6.53	Reactor

REFERENCES

"Handbook of Metal Powders", A.R. Poster, Reinhold Publishing Corp., 1966.

"Powder Metallurgy", R.L. Sands and C.R. Shakespeare, Newnes, London, 1966 (The Chemical Rubber Co., U.S. distributor).

"Treatise on Powder Metallurgy", C.G. Goetzel, Interscience Books, Inc., 1949; extensive literature and patent surveys in four volumes.

"Powder Metallurgy", W. Leszynski, Ed., Interscience Books Inc., 1961; proceedings of an international conference.

"Modern Developments in Powder Metallurgy", H.H. Hausner, Ed., Plenum Publishing Corp., 1966, 3 volumes.

"Tungsten Powder Metallurgy", V.D. Barth and H.O. McIntire, NASA SP-5035, National Aeronautics and Space Administration, 1965.

See also publications of the Metal Powder Industries Federation, the American Powder Metallurgy Institute, and the Metal Powder Producers Association.

Table 1-74. PRODUCTS OF POWDER METALLURGY

For other data and references on powder metallurgy, see Table 1-73.

Powder metallurgy refers to the production of parts by a process of molding metal powders and agglomerating the form by heat. The powder mixture is often hot-molded under pressure (10 000–100 000 psi) and is sintered in an inert or a reducing atmosphere, at a temperature between 400–2 000 deg F, depending on the metal mixture. For the refractory metals higher temperatures are necessary. The methods of powder metallurgy provide a close control of the composition and allow use of mixtures that could not be fabricated by any other process. As dimensions are determined by the mold, finish machining or grinding is often eliminated, thereby reducing cost and handling, especially for large lots. Special properties of the finished product, such as porosity, friction coefficient, and electrical conductivity, can be varied somewhat by changing the proportions of the powder components.

Class	Composition or constituents	Applications and uses	Desirable properties and advantages
Small, finished parts	Various ferrous, copper, and nickel alloys	Complex shapes; small parts not requiring high strength or ductility; plain bearings	Control of dimensions and finish; two-phase bearing metals; low cost in large production lots
Refractory metals	Pure W, Mo, Ta, Nb, Re, Ti alloys	Production of high-purity tungsten, molybdenum, tantalum, niobium, etc.; beryllium; cobalt alloys	Metals used in high-temperature service; electrical, electronic, and nuclear applications
Porous metals	Copper; copper-lead; bronze; stainless steel	Porous bearings, oil-impregnated, or with graphite or plastic; friction materials; metal filters; porous electrodes; catalysts; throttle plates	Interconnected pores in the size range 5–50 microns; porosity about 20–30%
Composite metals	Al, Cu, etc. with W, Mo, Co, or stainless steel reinforcing; reactor fuel elements	Services requiring high strength with lightness, high electrical and thermal conductivities; nuclear reactor components	High-strength materials from common metals, durability of nuclear materials
Metal-nonmetal composites	Filament-reinforced ceramics; dispersion strengthening by oxides	Ceramics with good structural properties; lightweight materials for high temperature (*e.g.*, SAP)	Strengthened ceramics; heat-resistant aluminum
Magnetic materials	Nickel-iron; cobalt mixtures; ferrites	High-permeability materials; permanent magnets; ferrite cores; magnetic storage	Very high magnetic properties and close control of magnetic properties
Cermets, oxide	Al_2O_3-Cr; Al_2O_3-Cr-W; Al_2O_3-Cr-Mo; ThO_2-W	Combustion and rocket nozzles; furnace muffler, tubes, seals, extrusion dies; power-tube cathodes	High-temperature strength (2 000 deg F and above); resistance to thermal shock; high thermal conductivity; corrosion resistance
Cermets, carbide	TiC-Ni; TiC-Fe-Cr; TiC-Co-Cr-W; Cr_3C_2-Ni-W	High-temperature bearings, seals, and dies; gage blocks	Strength, toughness, and corrosion resistance at high temperatures (to 1 700 deg F); hardness
Cemented carbides	WC-Co; WC-TaC-Co; TiC-Ni; Cr_3C_2-WC-Ni	Tips for cutting tools, lathe centers, gages; wire-drawing dies; rock drills; crushers; blast nozzles	Very high hardness, compressive strength, and elastic modulus; wear and corrosion resistance; high conductivity; high-temperature strength

Table 1-75. PARTICLE-REINFORCED METALS*

For data on fiber-reinforced metals, see Table 1-76.

Low-percentage Dispersions. Non-metallic particles are dispersed in a metal matrix to impede stress-induced dislocations. Diameter and volume-spacing of the particles must be kept small to preserve certain properties of the matrix material. The objective of the dispersion is to increase the strength and the creep resistance at the higher temperatures. Insoluble oxides, borides, and carbides are used, in particle sizes usually smaller than one-tenth micron and in volume concentration less than 15 percent. Dispersion-strengthened materials are intermediate between the superalloys and the cermets.

Examples of test results reported for low-percentage dispersion strengthening of metals are given in the following table.

Table A. METALS STRENGTHENED BY DISPERSED POWDERS

| Matrix metals | Strengthener | | Stress, kpsi | |
	Component	% vol	Matrix only, no reinforcement	Composite material
Pure iron	Al_2O_3	4	2.2[a]	10.[a]
Pure iron	Al_2O_3	10	2.2[a]	21.[a]
Pure copper	Al_2O_3	10	2.2[b]	20.[b]
Platinum	ThO_2	12.5	0.6[c]	6.1[c]
Uranium	Al_2O_3	3.5	6.[d]	14.[d]
Uranium	Al_2O_3	7.5	6.[d]	20.[d]
Copper	W (powder)	60.	$(E = 18)$[e]	$(E = 34)$[e]

[a]Stress for 100-hr rupture life, 650 deg C.
[b]Stress for 100-hr rupture life, 450 deg C.
[c]Stress for 100-hr rupture life, 1 100 deg C.
[d]Stress for 10^{-4} in./in./hr creep rate.
[e]Modulus of elasticity, E = stress/strain.

Table B. STRENGTH RATIOS AT HIGH TEMPERATURES

| Matrix metal | Dispersed particles | | Strength by test | |
	Material	% by volume (range)	Range of test temperatures, deg F	Strength ratio, (referred to matrix at same temperature)
Aluminum	Al_2O_3	10–15	400–800	2–4[a]
Copper	Al_2O_3	3–10	800–1 100	5–10
Iron	Al_2O_3	8–10	1 100–1 300	5–10
Nickel	ThO_2	3	1 600–2 100	Strength higher than that of many super-alloys
Platinum	ThO_2	12	2 000–2 400	8–11
Uranium	Al_2O_3	4–7	900–1 100	2–3

[a]The tensile strength of the final product is two to four times the strength of the aluminum.

Metal-ceramic Dispersions. In a metal matrix the ceramic particles do not deform under load, and high hardness is attained. Tungsten and titanium carbides are combined with steel and with nickel, cobalt, and their alloys. These materials are widely used in cutting tools and for drawing and blanking dies. The modulus of elasticity of WC is about 102 million psi, and the compressive strength of the Co-WC composites is in the range 500–900 kpsi. Aluminum oxide is used with chromium and molybdenum. Similar materials for atomic energy applications may use uranium or plutonium oxides. Some properties of typical materials are given in Table 1-76.

Metal-metal Dispersions. Dispersed phases of two or more metals are produced by powder metallurgy (see Table 1-74). Strength, hardness, wear resistance, and improved magnetic qualities are among the properties attained. Metal combinations include copper-iron, copper-tungsten, copper-nickel, iron-nickel-tungsten, copper-nickel-tungsten, and aluminum-nickel-cobalt.

*Compiled from various sources.

Table 1-76.   FIBER-REINFORCED METALS

Ductile and low-strength metals have been reinforced with various fibers. Fiber bundles or mats in molten metals, powder mixtures pressed or extruded, and electroplating are some of the fabrication methods. Copper, aluminum, silver, nickel and titanium are among the matrix materials, with reinforcement by steel, tungsten, boron, molybdenum, silica, glass, oxides, and carbides. The ratio of fiber-strength/matrix-strength determines a certain minimum fiber volume for effective reinforcement, but the fiber-matrix bond and fiber-to-stress alignment are also critical. Increase of strength is almost linear with fiber volume. Short fibers are not fully effective so that the strength is increased much less for a given fiber-volume fraction. Typical test results for fiber-reinforced metals are included in the following table.

TEST RESULTS ON COMPOSITE METALS*

Matrix metals	Strengthener		Stress, kpsi	
	Component	% vol	Matrix only, no reinforcement	Composite material
METALS STRENGTHENED BY FIBERS				
Copper	W fibers	60	20	200[a]
Silver	Al_2O_3 whiskers	35	10[b]	75[b]
Aluminum	Glass fibers	50	(23%)[c]	(94%)[c]
Aluminum	Al_2O_3	35	25[d]	161[d]
Aluminum	Steel	25	25[d]	173[d]
Nickel	B	8	70[d]	384[d]
Iron	Al_2O_3	36	40[d]	237[d]
Titanium	Mo	20	80[d]	96[d]
METALS STRENGTHENED BY SINTERED CARBIDES				
Cobalt	WC	90	(E = 30)[e]	(E = 85)[e]
Nickel	TiC	75	(E = 31)[f]	(E = 55)[f]

[a] Tensile strength with continuous fibers.
[b] Tensile strength at 350 deg C; modulus of elasticity: Cu = 17, composite 42 (millions of psi).
[c] Percentage of tensile strength at room temperature retained when tested at 300 deg C.
[d] Tensile strength, room temperature.
[e] Modulus of elasticity, E, measured in compression; hardness, 90 R-A; compressive strength, about 600 000 psi.
[f] Modulus of elasticity, E, measured in compression; hardness about 85 R-A.

*Compiled from various sources.

REFERENCES
"Metals Handbook: Properties and Selection", Vol. 1, American Society for Metals, 1961.
"Modern Composite Materials", L.J. Broutman and R.H. Krock, Addison-Wesley Publishing Company, 1967.

Table 1-77. PHYSICAL PROPERTIES
OF SODIUM POTASSIUM MIXTURES*

SYMBOLS AND UNITS:

M.P. = melting point, deg C. For K add 273.15.

ρ = density, lbm/ft^3. For specific gravity multiply by 0.016018. For kg/m^3 multiply by 16.018.

c_p = specific heat, Btu/lbm·deg R or cal/g·deg C. For J/kg·K multiply by 4184.0.

k = thermal conductivity, Btu/hr·ft·deg R. For W/m·K multiply by 1.7296.

μ = absolute viscosity, lbm/hr·ft. For centipoises multiply by 0.41338. For N·s/m^2 multiply value in lbm/hr·ft by 0.00041338 or value in lbm/sec·ft by 1.4882.

% potassium	M.P.	Property	Temperature						
			°C 100	200	300	400	500	600	700
			°F 212	392	572	752	932	1112	1292
0	98	ρ	57.9	56.4	55.1	53.6	52.1	50.5	48.9
		c_p	.331	.320	.312	.306	.302	.300	.300
		k	—	46.7	43.7	41.1	38.9	36.9	35.3
		μ	1.71	1.09	0.83	0.69	0.59	0.51	0.45
44	17	ρ	55.6	54.1	52.6	51.1	49.6	48.1	46.5
		c_p	.269	.261	.255	.251	.249	.248	.250
		k	—	—	—	—	—	—	—
		μ	1.31	0.92	0.72	0.59	0.50	0.43	0.62
56‡	7	ρ	53.8	52.6	51.4	49.2	48.0	46.8	45.6
		c_p	.254	.248	.241	.237	.233	.233	.236
		k	13.5	14.5	15.0	15.6	15.9	—	—
		μ	—	—	—	—	—	—	—
78	−10	ρ	53.1	51.6	50.1	48.6	47.1	45.7	44.0
		c_p	.225	.217	.212	.210	.209	.209	.211
		k	13.8	14.3	15.0	15.1	15.1	15.0	14.7
		μ†	1.28	0.86	0.67	0.55	0.47	0.41	0.35
100‡	64	ρ	51.1	49.5	48.2	46.8	45.5	43.6	42.1
		c_p	.193	.190	.188	.184	.182	.183	.186
		k	—	25.8	24.6	23.4	22.2	21.0	19.8
		μ	1.07	0.74	0.55	0.44	0.40	0.35	0.33

Note: The surface heat-transfer coefficient for liquid metals in long tubes is approximately as follows:

$$h = k/D \left[7 + 0.025 \, (RePr)^{0.8} \right].$$

For details see "Liquid Metals Handbook".

†Viscosity at 66.9% potassium.
‡Data approximated from graph.

*Adapted from: "Liquid Metals Handbook", 3rd ed., Office of Naval Research, U.S. Government Printing Office, 1955.

Table 1-78. SELECTION AND APPLICATION OF PLASTICS*

Application or service	Properties required	Suitable plastics
Chemical and thermal equipment	Resistance to temperature extremes and to wide range of chemicals. Minimum moisture absorption. Fair to good strength.	Fluorocarbons, chlorinated polyether, polyvinylidene fluoride, polypropylene, high-density polyethylene, and epoxy-glass.
Heavily stressed mechanical components	High-tensile plus high-impact strength. Good fatigue resistance and stability at elevated temperatures. Machinable or moldable to close tolerance.	Nylons, TFE-filled acetals, polycarbonates, and fabric-filled phenolics.
Electrostructural parts	Excellent electrical resistance in low to medium frequencies. High-strength and -impact properties; good fatigue and heat resistance. Good dimensional stability at elevated temperatures.	Allylics, alkyds, aminos, epoxies, phenolics, polycarbonates, polyesters, polyphenylene oxides, and silicones.
Low-friction applications	Low coefficient of friction, even when non-lubricated. High resistance to abrasion. Fair to good form stability and heat and corrosion resistance.	Fluorocarbons (TFE and FEP), filled fluorocarbons (TFE), TFE fabrics, nylons, acetals, TFE-filled acetals, and high-density polyethylenes.
Light-transmission components, glazing	Good light transmission in transparent or translucent colors. Good to excellent formability and moldability. Shatter resistance. Fair to good tensile strength.	Acrylics, polystyrenes, cellulose acetates, cellulose butyrates, ionomers, rigid vinyls, polycarbonates, and medium-impact styrenes.
Housings, containers, ducts	Good to excellent impact strength and stiffness. Good formability and moldability. Moderate cost. Good environmental resistance. Fair to good tensile strength and dimensional stability.	ABS, high-impact styrene, polypropylene, high-density polyethylene, cellulose acetate butyrate, modified acrylics, polyester-glass and epoxy-glass combinations.

*Based on: "Selecting Plastics", *Machine Design*, 40 (29):4–11, Dec. 12, 1968. For complete tables with properties, see this source.

Table 1-79. PROPERTIES OF COMMERCIAL PLASTICS†

Of the many plastics commercially available in each chemical class, only one or a very few examples have been selected for this table as typical of the class. In some cases the range of properties has been expanded to include several grades or types. It is impractical to include a comprehensive list of materials or known properties of these materials in a table of convenient size. Properties vary widely with amount and kind of modifier, such as filler and plasticizer. Within any type of thermoplastic resins, molecular weight is an important variable. This property is controlled to afford the best physical properties available consistent with economical processing properties.

The information shown refers in all cases, except for "Forms available" and "Fabrication," to material in the fabricated form, which in the case of thermosetting materials means commercially cured. Physical and electrical properties will vary, to a greater or lesser degree, with different materials, with humidity conditioning environment and with orientation. Strength values are quoted on the basis of short-time tests at normal room temperature and are not suitable for engineering design purposes for load-bearing applications. Maximum continuous service temperature refers to unloaded structures. The user of this table is referred to the specifications and test procedures of the American Society for Testing Materials.

For data on corresponding synthetic fibers, see Tables 1-106 to 1-108. For data on synthetic resin coatings, see Tables 3-17 and 3-20. For data on adhesives, see Table 3-27. For data on plastics composites, see Tables 1-81, 1-82, 1-85, and 1-87. For other data on plastics, see Tables 1-83 through 1-88.

†From: "Handbook of Chemistry and Physics", 53rd ed., R.C. Weast, Ed., The Chemical Rubber Co., 1972.

Table 1-79.   PROPERTIES OF COMMERCIAL PLASTICS (Continued)

To convert psi to N/m², multiply by 6,895. For specific heat in J/kg·K, multiply by 4,184.

Properties	Chemical class	Cellulose acetate	Cellulose acetate	Cellulose acetate butyrate
	Resin type	Thermoplastic	Thermoplastic	Thermoplastic
	Subclass or modification	Soft	Hard	Soft
ELECTRICAL PROPERTIES				
D.C. resistivity, ohm-cm		$10^{10}-10^{13}$	$10^{10}-10^{13}$	$10^{10}-10^{12}$
Dielectric constant, 60 cps		3.5–7.5	3.5–7.5	3.5–6.4
Dielectric constant, 10^6 cps		3.2–7.0	3.2–7.0	3.2–6.2
Dissipation factor, 60 cps		0.01–0.06	0.01–0.06	0.01–0.04
Dissipation factor, 10^6 cps		0.01–0.10	0.01–0.10	0.01–0.04
MECHANICAL PROPERTIES				
Modulus of elasticity, 10^3 psi		86–250	190–400	74–126
Tensile strength, psi		1,900–4,700	4,600–8,500	1,900–3,800
Ultimate elongation, %		32–50	6–40	60–74
Yield stress, psi		2,200–4,200	4,100–7,600	1,200–2,600
Yield strain, %				
Rockwell hardness		R 49–R 103	R 101–R 123	R 59–R 95
Notched Izod impact strength, ft lb/in.		2.0–5.2	0.4–2.7	2.5–5.4
Specific gravity		1.27–1.34	1.27–1.34	1.15–1.22
THERMAL PROPERTIES				
Burning rate		Medium	Medium	Medium
Heat distortion, 264 psi, °C		44–57	60–113	49–58
Specific heat, cal/g		0.3–0.42	0.3–0.42	0.3–0.4
Linear thermal expansion coefficient, 10^{-5}, °C		8–16	8–16	11–17
Maximum continuous service temperature, °C				
CHEMICAL RESISTANCE				
Mineral acids, weak		Fair to good	Fair to good	Good
Mineral acids, strong		Poor	Poor	Fair to good
Oxidizing acids, concentrated		Very poor	Very poor	
Alkalies, weak		Poor	Poor	Good
Alkalies, strong		Very poor	Very poor	Poor
Alcohols		Poor	Poor	Poor
Ketones		Poor	Poor	Poor
Esters		Poor	Poor	Poor
Hydrocarbons, aliphatic		Fair to good	Fair to good	Fair to good
Hydrocarbons, aromatic		Poor to fair	Poor to fair	Poor
Oils: vegetable, animal, mineral		Fair to good	Fair to good	Good
MISCELLANEOUS PROPERTIES				
Clarity		Excellent	Excellent	Good to excellent
Color		Pale to colorless	Pale to colorless	Pale to colorless
Refractive index, n_D		1.46–1.50	1.46–1.50	1.46–1.49
Applicable ASTM specifications and test methods		D786, D706, D257, D150, D638, D785, D256, D792, D648, D696, D543, D542	D786, D706, D257, D150, D638, D785, D256, D792, D648, D696, D543, D542	D707, D257, D150, D638, D785, D256, D792, D648, D696
FORMS AVAILABLE Cs—castings, F—film, Fb—fibers, I—impregnants, L—laminations, Lq—lacquers, Mf—monofilaments, P—powder, pellet, or granules, R—rods, tubes, or other extruded forms, S—sheets.		F, Lq, P, R, S	F, Lq, P, R, S	F, Lq, P, R, S
FABRICATION Cl—calendering, Cs—casting, E—extrusion, F—hot forming or drawing, I—impregnation, MB—blow molding, MC—compression molding, MI—injection molding, S—spreading.		Cs, E, F, MB, MC, MI, S	Cs, E, F, MB, MC MI, S	Cs, E, F, MB, MC, MI, S

Table 1-79. PROPERTIES OF COMMERCIAL PLASTICS (Continued)

Properties	Chemical class	Cellulose acetate butyrate	Nylon	Polycarbonates
	Resin type	Thermoplastic	Thermoplastic	Thermoplastic
	Subclass or modification	Hard	6/6	Unfilled
ELECTRICAL PROPERTIES				
D.C. resistivity, ohm-cm		10^{10}–10^{12}		2×10^{16}
Dielectric constant, 60 cps		3.5–6.4	4.0–4.6	3.17
Dielectric constant, 10^6 cps		3.2–6.2	3.4–3.6	2.96
Dissipation factor, 60 cps		0.01–0.04	0.014–0.04	0.0009
Dissipation factor, 10^6 cps		0.01–0.04	0.04	0.01
MECHANICAL PROPERTIES				
Modulus of elasticity, 10^3 psi		150–200		290–325
Tensile strength, psi		5,000–6,800	9,000–12,000	8,000–9,500
Ultimate elongation, %		38–54	60–300	20–100
Yield stress, psi		3,600–6,100		8,000–10,000
Yield strain, %				
Rockwell hardness		R 108–R 117	R 108–R 120	M 70–M 180
Notched Izod impact strength, ft lb/in.		0.7–2.4	1.0–2.0	8–16
Specific gravity		1.19 1.25	1.13–1.15	1.2
THERMAL PROPERTIES				
Burning rate		Medium	Self-extinguishing	Self-extinguishing
Heat distortion, 264 psi, °C		70–99		135–145
Specific heat, cal/g		0.3–0.4	0.4	0.3
Linear thermal expansion coefficient, 10^{-5}, °C		11–17	8.0	6.6
Maximum continuous service temperature, °C			80–150	138–143
CHEMICAL RESISTANCE				
Mineral acids, weak		Good	Very good	Excellent
Mineral acids, strong		Fair to good	Poor	Fair
Oxidizing acids, concentrated			Poor	
Alkalies, weak		Good	No effect	Poor
Alkalies, strong		Poor	No effect	Poor
Alcohols		Poor	Good	Poor
Ketones		Poor	Good	Poor
Esters		Poor	Good	Poor
Hydrocarbons, aliphatic		Fair to good	Very good	Poor
Hydrocarbons, aromatic		Poor	Fair to good	Poor
Oils: vegetable, animal, mineral		Good	Good	Poor
MISCELLANEOUS PROPERTIES				
Clarity		Good to excellent	Clear	Clear
Color		Pale to colorless	Pale amber to colorless	Colorless
Refractive index, n_D		1.46–1.49	1.53	1.60
Applicable ASTM specifications and test methods		D707, D257, D150, D256, D792, D648, D542, D638, D785, D696, D543	D257, D150, D638, D792, D648, D696, D785, D256, D542, D543	D257, D150, D638, D792, D648, D696, D785, D256, D542, D543
FORMS AVAILABLE Cs—castings, F—film, Fb—fibers, I—impregnants, L—laminations, Lq—lacquers, Mf—monofilaments, P—powder, pellet, or granules, R—rods, tubes, or other extruded forms, S—sheets.		F, Lq, P, R, S	F, Fb, Mf, P, R, S	F, Fb, Mf, P, R, S
FABRICATION Cl—calendering, Cs—casting, E—extrusion, F—hot forming or drawing, I—impregnation, MB—blow molding, MC—compression molding, MI—injection molding, S—spreading.		Cs, E, F, MB, MC, MI, S	E, F, MB, MC, MI	Cs, E, F, MB, MC, MI

Table 1-79. PROPERTIES OF COMMERCIAL PLASTICS *(Continued)*

To convert psi to N/m², multiply by 6,895. For specific heat in J/kg·K, multiply by 4,184.

Properties	Chemical class	Polyethylene	Polyethylene	Polyethylene
	Resin type	Thermoplastic	Thermoplastic	Thermoplastic
	Subclass or modification	Low density	Medium density	High density
ELECTRICAL PROPERTIES				
D.C. resistivity, ohm-cm		$>10^{15}$	$>10^{15}$	$>10^{15}$
Dielectric constant, 60 cps		2.3–2.35	2.3	2.3–2.35
Dielectric constant, 10^6 cps		2.3–2.35	2.3	2.3–2.35
Dissipation factor, 60 cps		<0.0005	<0.0005	<0.0005
Dissipation factor, 10^6 cps		<0.0005	<0.0005	<0.0005
MECHANICAL PROPERTIES				
Modulus of elasticity, 10^3 psi		14–38	35–90	85–160
Tensile strength, psi		1,000–1,400	1,200–3,500	3,100–5,500
Ultimate elongation, %		400–700	50–600	15–100
Yield stress, psi		1,100–1,700	1,500–2,600	2,400–5,000
Yield strain, %		20–40	10–20	5–10
Rockwell hardness				R 30–R 50
Notched Izod impact strength, ft lb/in.		No break	0.5–>16	1.5–20
Specific gravity		0.91–0.925	0.926–0.941	0.941–0.965
THERMAL PROPERTIES				
Burning rate		Very slow	Slow	Slow
Heat distortion, 264 psi, °C				
Specific heat, cal/g		0.55	0.55	0.55
Linear thermal expansion coefficient, 10^{-5}, °C		10–20	14–16	11–13
Maximum continuous service temperature, °C		60–77	71–93	92–200
CHEMICAL RESISTANCE				
Mineral acids, weak		Good	Excellent	Excellent
Mineral acids, strong		Good	Excellent	Excellent
Oxidizing acids, concentrated		Good to poor	Good to poor	Good to poor
Alkalies, weak		Good	Excellent	Excellent
Alkalies, strong		Good	Excellent	Excellent
Alcohols		Excellent to poor	Excellent to poor	Excellent to poor
Ketones		Excellent to poor	Excellent to poor	Excellent to poor
Esters		Excellent to poor	Excellent to poor	Excellent to poor
Hydrocarbons, aliphatic		Fair	Fair	Fair
Hydrocarbons, aromatic		Fair	Good	Fair
Oils: vegetable, animal, mineral		Good	Excellent	Good
MISCELLANEOUS PROPERTIES				
Clarity		Translucent	Translucent	Translucent
Color		Colorless	Colorless	Colorless
Refractive index, n_D		1.50–1.54	1.52–1.54	1.54
Applicable ASTM specifications and test methods		D702, D788, D257, D638, D696, D543, D150, D412, D1248, D542	D257, D150, D412, D256, D696, D543, D638, D785, D1248, D542	D257, D150, D412, D256, D696, D543, D638, D785, D1248, D542
FORMS AVAILABLE Cs—castings, F—film, Fb—fibers, I—impregnants, L—laminations, Lq—lacquers, Mf—monofilaments, P—powder, pellet, or granules, R—rods, tubes, or other extruded forms, S—sheets.		F, Mf, P, R, S	F, Mf, P, R, S	F, Fb, Mf, P, R, S
FABRICATION Cl—calendering, Cs—casting, E—extrusion, F—hot forming or drawing, I—impregnation, MB—blow molding, MC—compression molding, MI—injection molding, S—spreading.		Cl, E, F, MB, MC, MI	Cl, E, F, MB, MC, MI	Cl, E, F, MB, MC, MI

Table 1-79. PROPERTIES OF COMMERCIAL PLASTICS *(Continued)*

Properties	Chemical class	Methylmethacrylate	Polypropylene	Polypropylene
	Resin type	Thermoplastic	Thermoplastic	Thermoplastic
	Subclass or modification	Unmodified	Unmodified	Copolymer
ELECTRICAL PROPERTIES				
D.C. resistivity, ohm-cm		$>10^{14}$	$>10^{15}$	10^{17}
Dielectric constant, 60 cps		3.5–4.5	2.2–2.6	2.3
Dielectric constant, 10^6 cps		3.0–3.5	2.2–2.6	2.3
Dissipation factor, 60 cps		0.04–0.06	<0.0005	0.0001–0.0005
Dissipation factor, 10^6 cps		0.02–0.03	0.0005–0.002	0.0001–0.002
MECHANICAL PROPERTIES				
Modulus of elasticity, 10^3 psi		350–500	1.4–1.7	
Tensile strength, psi		7,000–11,000	4,300–5,500	2,900–4,500
Ultimate elongation, %		2.0–10	>220	200–700
Yield stress, psi			4,900	
Yield strain, %			15	
Rockwell hardness		M 80–M 105	93	R 50–R 96
Notched Izod impact strength, ft lb/in.		0.3–0.6	1.0	1.1–12
Specific gravity		1.18–1.20	0.90	0.90
THERMAL PROPERTIES				
Burning rate		Slow·	Medium	Medium
Heat distortion, 264 psi, °C		66–99		
Specific heat, cal/g		0.35	0.5	0.5
Linear thermal expansion coefficient, 10^{-5}, °C		5.0–9.0	5.8–10	8–10
Maximum continuous service temperature, °C		60–93		190–240
CHEMICAL RESISTANCE				
Mineral acids, weak		Good	Excellent	Excellent
Mineral acids, strong		Fair to poor	Excellent	Excellent
Oxidizing acids, concentrated		Attacked	Good to poor	Poor
Alkalies, weak		Good	Excellent to good	Excellent
Alkalies, strong		Poor	Excellent to good	Good
Alcohols			Excellent to good	Good below 80°C
Ketones		Dissolves	Excellent to good	Good below 80°C
Esters		Dissolves	Excellent to good	Good below 80°C
Hydrocarbons, aliphatic		Good	Good to fair	Good below 80°C
Hydrocarbons, aromatic		Softens	Good to fair	Good below 80°C
Oils: vegetable, animal, mineral		Good	Good	
MISCELLANEOUS PROPERTIES				
Clarity		Excellent	Transparent	Transparent
Color		Colorless	Colorless to sl. yellow	Colorless to sl. yellow
Refractive index, n_D		1.48–1.50	1.49	
Applicable ASTM specifications and test methods		D257, D150, D638, D792, D648, D696, D785, D256, D543, D542	D257, D150, D412, D256, D648, D543, D638, D785, D542	D257, D150, D412, D256, D648, D543, D638, D785, D542
FORMS AVAILABLE Cs—castings, F—film, Fb—fibers, I—impregnants, L—laminations, Lq—lacquers, Mf—monofilaments, P—powder, pellet, or granules, R—rods, tubes, or other extruded forms, S—sheets.		Cs, P, R, S	F, Fb, Mf, P, R, S	F, Fb, Mf, P, R, S
FABRICATION Cl—calendering, Cs—casting, E—extrusion, F—hot forming or drawing, I—impregnation, MB—blow molding, MC—compression molding, MI—injection molding, S—spreading.		Cs, E, F, Lq, MB, MC, MI	Cl, E, F, MB, MC, MI	Cl, E, F, MB, MC, MI

Table 1-79.   PROPERTIES OF COMMERCIAL PLASTICS *(Continued)*

To convert psi to N/m², multiply by 6,895. For specific heat in J/kg·K, multiply by 4,184.

Properties	Chemical class	Polystyrene	Polystyrene-acrylonitrile	Polytetrafluoro-ethylene
	Resin type	Thermoplastic	Thermoplastic	Thermoplastic
	Subclass or modification	Unmodified	Unmodified	Unmodified
ELECTRICAL PROPERTIES				
D.C. resistivity, ohm-cm		$>10^{16}$	$10^{13}-10^{17}$	10^{18}
Dielectric constant, 60 cps		2.5–2.65	2.6–3.4	2.
Dielectric constant, 10^6 cps		2.5–2.65	2.5–3.1	2.
Dissipation factor, 60 cps		0.0001–0.0003	0.006–0.008	0.0002
Dissipation factor, 10^6 cps		0.0001–0.0004	0.007–0.01	0.0002
MECHANICAL PROPERTIES				
Modulus of elasticity, 10^3 psi		400–600	$>10^{16}$	33–65
Tensile strength, psi		5,000–10,000	9,000–12,000	2,000–4,500
Ultimate elongation, %		1.0–2.5	1.0–2.5	200–400
Yield stress, psi				1,600–2,000
Yield strain, %				50–75
Rockwell hardness		M 65–M 85	M 75–M 90	D 50–D 65
Notched Izod impact strength, ft lb/in.		0.25–0.60	0.3–0.6	2.5–4.0
Specific gravity		1.04–1.08	1.05–1.1	2.1–2.3
THERMAL PROPERTIES				
Burning rate		Medium to slow	Slow	Self-extinguishing
Heat distortion, 264 psi, °C			91–104	60
Specific heat, cal/g		0.32–0.35	0.32–0.35	0.25
Linear thermal expansion coefficient, 10^{-5}, °C		6.0–8.0	3.6–3.8	10
Maximum continuous service temperature, °C		66–82	77–88	260
CHEMICAL RESISTANCE				
Mineral acids, weak		Excellent	Excellent	Excellent
Mineral acids, strong		Excellent	Good to excellent	Excellent
Oxidizing acids, concentrated		Poor	Poor	Excellent
Alkalies, weak		Excellent	Excellent	Excellent
Alkalies, strong		Excellent	Good to excellent	Excellent
Alcohols		Excellent	Good to excellent	Excellent
Ketones		Dissolves	Dissolves	Excellent
Esters		Poor	Dissolves	Excellent
Hydrocarbons, aliphatic		Poor	Good	Excellent
Hydrocarbons, aromatic		Dissolves	Fair to good	Excellent
Oils: vegetable, animal, mineral		Fair to poor	Good to excellent	Excellent
MISCELLANEOUS PROPERTIES				
Clarity		Transparent	Transparent	Translucent
Color		Colorless	Colorless to amber	Colorless to gray
Refractive index, n_D		1.59–1.60	1.56–1.57	1.30–1.40
Applicable ASTM specifications and test methods		D257, D150, D638, D792, D648, D696, D785, D256, D543, D542	D257, D150, D638, D792, D648, D696, D785, D256, D543, D542	
FORMS AVAILABLE Cs—castings, F—film, Fb—fibers, I—impregnants, L—laminations, Lq—lacquers, Mf—monofilaments, P—powder, pellet, or granules, R—rods, tubes, or other extruded forms, S—sheets.		F, Fb, Mf, P, R, S	F, Mf, P, R, S	F, L, P, R, S
FABRICATION Cl—calendering, Cs—casting, E—extrusion, F—hot forming or drawing, I—impregnation, MB—blow molding, MC—compression molding, MI—injection molding, S—spreading.		E, F, MB, MC, MI	Cl, E, F, MB, MC, MI	E, F, MC, MI

Table 1-79. PROPERTIES OF COMMERCIAL PLASTICS *(Continued)*

Properties	Chemical class	Polytrifluorochloro-ethylene	Polyvinylchloride and vinylchloride acetate	Polyvinylchloride and vinylchloride acetate
	Resin type	Thermoplastic	Thermoplastic	Thermoplastic
	Subclass or modification	Unmodified	Unmodified, rigid	Plasticized, non-rigid
ELECTRICAL PROPERTIES				
D.C. resistivity, ohm-cm		10^{18}	10^{12}–10^{16}	10^{11}–10^{14}
Dielectric constant, 60 cps		2.2–2.8	3.2–4.0	5.0–9.0
Dielectric constant, 10^6 cps		2.3–2.5	3.0–4.0	3.0–4.0
Dissipation factor, 60 cps		0.001	0.01–0.02	0.03–0.05
Dissipation factor, 10^6 cps		0.005	0.006–0.02	0.06–0.1
MECHANICAL PROPERTIES				
Modulus of elasticity, 10^3 psi		150	200–600	
Tensile strength, psi		4,500–6,000	5,000–9,000	1,500–3,000
Ultimate elongation, %		250	2.0–40	200–400
Yield stress, psi		4,200		
Yield strain, %		10	1.0–5.0	
Rockwell hardness		J 75–J 95	R 110–R 120	
Notched Izod impact strength, ft lb/in.		2.5–4.0	0.4–2.0	
Specific gravity		2.1–2.3	1.36–1.4	1.15–1.35
THERMAL PROPERTIES				
Burning rate		Self-extinguishing	Self-extinguishing	Slow to self-extinguishing
Heat distortion, 264 psi, °C			60–80	
Specific heat, cal/g		0.22	0.2–0.28	0.36–0.5
Linear thermal expansion coefficient, 10^{-5}, °C		7.0	5.0–18	7.0–25
Maximum continuous service temperature, °C		200	70–74	80–105
CHEMICAL RESISTANCE				
Mineral acids, weak		Excellent	Excellent	Fair to good
Mineral acids, strong		Excellent	Good to excellent	Fair to good
Oxidizing acids, concentrated		Excellent	Fair to good	Poor to fair
Alkalies, weak		Excellent	Excellent	Fair to good
Alkalies, strong		Excellent	Good	Fair to good
Alcohols		Excellent	Excellent	Fair
Ketones		Excellent	Poor	Poor
Esters		Excellent	Poor	Poor
Hydrocarbons, aliphatic		Excellent	Excellent	Poor
Hydrocarbons, aromatic		Excellent	Poor	Poor
Oils: vegetable, animal, mineral		Excellent	Excellent	Poor
MISCELLANEOUS PROPERTIES				
Clarity		Transparent	Transparent	Transparent
Color		Colorless to pale	Colorless to amber	Colorless to amber
Refractive index, n_D		1.43	1.54	1.50–1.55
Applicable ASTM specifications and test methods		D1430, D257, D150, D256, D792, D648, D542, D638, D785, D696, D543	D708, D728, D257, D256, D792, D648, D542, D150, D638, D696, D543	D1432, D257, D150, D543, D542
FORMS AVAILABLE Cs—castings, F—film, Fb—fibers, I—impregnants, L—laminations, Lq—lacquers, Mf—monofilaments, P—powder, pellet, or granules, R—rods, tubes, or other extruded forms, S—sheets.		F, Mf, P, R, S	F, Fb, I, Lq, Mf, P, R, S	F, L, P, R, S
FABRICATION Cl—calendering, Cs—casting, E—extrusion, F—hot forming or drawing, I—impregnation, MB—blow molding, MC—compression molding, MI—injection molding, S—spreading.		Cs, E, F, I, MC, MI, S	Cl, Cs, E, F, I, MB, MC, MI, S	Cl, Cs, E, MB, MC, MI, S

Table 1-79. PROPERTIES OF COMMERCIAL PLASTICS *(Continued)*

To convert psi to N/m^2, multiply by 6,895. For specific heat in J/kg·K, multiply by 4,184.

Properties	Chemical class	Epoxy	Melamine-formaldehyde	Melamine-formaldehyde
	Resin type	Thermosetting	Thermosetting	Thermosetting
	Subclass or modification	Unfilled	α-Cellulose filled	Mineral filled (electrical)
ELECTRICAL PROPERTIES				
D.C. resistivity, ohm-cm		10^{12}–10^{14}	10^{12}–10^{14}	10^{13}–10^{14}
Dielectric constant, 60 cps		3.5–5.0	7.9–9.4	10.2
Dielectric constant, 10^6 cps		3.4–4.4	7.2–8.4	6.1
Dissipation factor, 60 cps		0.001–0.005	0.03–0.08	0.10
Dissipation factor, 10^6 cps		0.03–0.05	0.03–0.043	0.051
MECHANICAL PROPERTIES				
Modulus of elasticity, 10^3 psi		>300	1,300	1,950
Tensile strength, psi		4,000–13,000	7,000–13,000	5,500–6,500
Ultimate elongation, %		2.0–6.0	0.6–0.9	
Yield stress, psi				
Yield strain, %				
Rockwell hardness		M 75–M 110	M 110–M 124	E 90
Notched Izod impact strength, ft lb/in.		0.2–1.0	0.24–0.35	0.3–0.4
Specific gravity		1.115	1.47–1.52	1.78
THERMAL PROPERTIES				
Burning rate		Slow	Self-extinguishing	Self-extinguishing
Heat distortion, 264 psi, °C		Up to 120	204	130
Specific heat, cal/g		0.25–0.4	0.4	
Linear thermal expansion coefficient, 10^{-5}, °C		4.5–9.0	2.0–5.7	2.1–4.3
Maximum continuous service temperature, °C		80	99.0	149
CHEMICAL RESISTANCE				
Mineral acids, weak		Excellent	Good	Fair
Mineral acids, strong		Fair to good	Poor	Poor
Oxidizing acids, concentrated		Poor	Poor	Poor
Alkalies, weak		Excellent	Good	Fair
Alkalies, strong		Excellent	Poor	Poor
Alcohols		Excellent	Good	Good
Ketones		Poor	Good	Good
Esters			Good	Good
Hydrocarbons, aliphatic		Excellent	Good	Good
Hydrocarbons, aromatic		Excellent	Good	Good
Oils: vegetable, animal, mineral		Excellent	Good	Good
MISCELLANEOUS PROPERTIES				
Clarity		Transparent	Translucent	Opaque
Color		Colorless	Colorless	Dark
Refractive index, n_D		1.58		
Applicable ASTM specifications and test methods		D257, D150, D651, D792, D648, D696, D785, D256, D543	D704, D257, D150, D256, D792, D648, D638, D785, D543, D696	D704, D257, D150, D256, D792, D648, D638, D785, D543, D696
FORMS AVAILABLE		Cs, Lq	P, R, S	P, R, S
Cs—castings, F—film, Fb—fibers, I—impregnants, L—laminations, Lq—lacquers, Mf—monofilaments, P—powder, pellet, or granules, R—rods, tubes, or other extruded forms, S—sheets.				
FABRICATION		Cs, I, S	MC	MC
Cl—calendering, Cs—casting, E—extrusion, F—hot forming or drawing, I—impregnation, MB—blow molding, MC—compression molding, MI—injection molding, S—spreading.				

Table 1-79. PROPERTIES OF COMMERCIAL PLASTICS (Continued)

Properties	Chemical class	Phenol-formaldehyde	Phenol-formaldehyde	Phenol-formaldehyde
	Resin type	Thermosetting	Thermosetting	Thermosetting
	Subclass or modification	Cord filled	Cellulose filled	Unfilled cast phenolic, mechanical and chemical grade
ELECTRICAL PROPERTIES				
D.C. resistivity, ohm-cm		10^{11}–10^{12}	10^{11}–10^{13}	1.0–7.0×10^{12}
Dielectric constant, 60 cps		7.0–10.0	5.0–9.0	6.5–7.5
Dielectric constant, 10^6 cps		5.0–6.0	4.0–7.0	4.0–5.5
Dissipation factor, 60 cps		0.1–0.3	0.04–0.3	0.10–0.15
Dissipation factor, 10^6 cps		0.04–0.09	0.03–0.07	0.04–0.05
MECHANICAL PROPERTIES				
Modulus of elasticity, 10^3 psi		900–1,300	800–1,200	4.0–5.0
Tensile strength, psi		6,000–9,000	6,500–8,500	6,000–9,000
Ultimate elongation, %		0.5–1.0	0.6–1.0	1.5–2.0
Yield stress, psi				
Yield strain, %				
Rockwell hardness			M 110–M 120	M 93–M 120
Notched Izod impact strength, ft lb/in.		4.0–8.0	0.24–0.34	0.25–0.4
Specific gravity		1.36–1.43	1.32–1.55	1.307–1.318
THERMAL PROPERTIES				
Burning rate		Self-extinguishing	Self-extinguishing	Self-extinguishing
Heat distortion, 264 psi, °C		121–127	143–171	74–80
Specific heat, cal/g			0.35–0.40	
Linear thermal expansion coefficient, 10^{-5}, °C			3.0–4.5	6.0–8.0
Maximum continuous service temperature, °C		121	149–177	
CHEMICAL RESISTANCE				
Mineral acids, weak		Variable	Variable	Fair to good
Mineral acids, strong		Poor	Poor	Poor to good
Oxidizing acids, concentrated		Poor	Poor	Poor
Alkalies, weak		Variable	Variable	Poor to good
Alkalies, strong		Poor	Poor	Poor
Alcohols		Good	Good to excellent	Good to excellent
Ketones		Poor to fair	Fair	Fair
Esters		Fair to good	Fair to good	Fair to good
Hydrocarbons, aliphatic		Fair to good	Excellent	Good to excellent
Hydrocarbons, aromatic		Fair to good	Excellent	Good
Oils: vegetable, animal, mineral		Good	Excellent	Excellent
MISCELLANEOUS PROPERTIES				
Clarity		Opaque	Opaque	Clear
Color				Colorless to amber
Refractive index, n_D				
Applicable ASTM specifications and test methods		D700, D257, D150, D785, D256, D792, D638, D651, D543, D648	D700, D257, D150, D785, D256, D792, D543, D638, D651, D648, D696	D257, D150, D638, D792, D648, D696, D785, D256, D543
FORMS AVAILABLE Cs—castings, F—film, Fb—fibers, I—impregnants, L—laminations, Lq—lacquers, Mf—monofilaments, P—powder, pellet, or granules, R—rods, tubes, or other extruded forms, S—sheets		L, P, S	L, P, S	Cs, R, S
FABRICATION Cl—calendering, Cs—casting, E—extrusion, F—hot forming or drawing, I—impregnation, MB—blow molding, MC—compression molding, MI—injection molding, S—spreading.		MC	MC	Cs, F

Table 1-79.　PROPERTIES OF COMMERCIAL PLASTICS *(Continued)*

To convert psi to N/m², multiply by 6,895. For specific heat in J/kg·K, multiply by 4,184.

Properties	Chemical class	Polyester (styrene-alkyd)	Silicones	Urea formaldehyde
	Resin type	Thermosetting	Thermosetting	Thermosetting
	Subclass or modification	Glassfiber mat reinforced	Mineral filled	α-Cellulose filled
ELECTRICAL PROPERTIES				
D.C. resistivity, ohm-cm		10^{11}	$>10^{12}$	0.5–5.0
Dielectric constant, 60 cps		4.0–5.5	3.5–3.6	7.7–9.5
Dielectric constant, 10^6 cps		4.0–5.5	3.4–3.6	6.7–8.0
Dissipation factor, 60 cps		0.01–0.04	0.004	0.036–0.043
Dissipation factor, 10^6 cps		0.01–0.06	0.005–0.007	0.025–0.035
MECHANICAL PROPERTIES				
Modulus of elasticity, 10^3 psi		500–1,500		1,300–1,400
Tensile strength, psi		30,000–50,000	3,000–4,000	5,500–13,000
Ultimate elongation, %		0.5–1.5		0.6
Yield stress, psi				
Yield strain, %				
Rockwell hardness		M 80–M 120	M 85–M 95	E 94–E 97
Notched Izod impact strength, ft lb/in.		7.0–30	0.25–0.35	0.24–0.40
Specific gravity		1.5–2.1	1.8–2.8	1.47–1.52
THERMAL PROPERTIES				
Burning rate		Self-extinguishing	Self-extinguishing	Self-extinguishing
Heat distortion, 264 psi, °C		93–288	>260	130
Specific heat, cal/g		0.2–0.4	0.2–0.3	0.6
Linear thermal expansion coefficient, 10^{-5}, °C		1.8–3.0	2.0–4.0	2.2–3.6
Maximum continuous service temperature, °C		121–204	288	77
CHEMICAL RESISTANCE				
Mineral acids, weak		Good	Fair to good	Poor
Mineral acids, strong		Poor	Poor to good	Poor
Oxidizing acids, concentrated		Poor		Poor
Alkalies, weak		Good	Fair	Fair
Alkalies, strong		Poor	Poor	Poor
Alcohols		Good	Poor	Good
Ketones		Poor	Poor	Good
Esters		Good		Good
Hydrocarbons, aliphatic		Good	Fair to good	Good
Hydrocarbons, aromatic		Poor to fair	Poor	Good
Oils: vegetable, animal, mineral		Good	Good	
MISCELLANEOUS PROPERTIES				
Clarity		Translucent	Opaque	Translucent
Color		Colorless	Pale to dark	Colorless
Refractive index, n_D				1.54–1.56
Applicable ASTM specifications and test methods		D257, D150, D638, D792, D648, D696, D785, D256, D543	D257, D150, D785, D648, D696, D543, D256, D792	D705, D257, D150, D256, D792, D648, D638, D785
FORMS AVAILABLE		L, S	P	P, R, S
Cs—castings, F—film, Fb—fibers, I—impregnants, L—laminations, Lq—lacquers, Mf—monofilaments, P—powder, pellet, or granules, R—rods, tubes, or other extruded forms, S—sheets.				
FABRICATION		I	MC	MC
Cl—calendering, Cs—casting, E—extrusion, F—hot forming or drawing, I—impregnation, MB—blow molding, MC—compression molding, MI—injection molding, S—spreading.				

Table 1-79. PROPERTIES OF COMMERCIAL PLASTICS *(Continued)*

Properties	Chemical class	Acrylonitrile-butadiene-styrene (ABS)	Acetal	Alkyd resins
	Resin type	Thermoplastic	Thermoplastic	Thermosetting
	Subclass or modification	High-heat resistant	Homopolymer	Synthetic-fiber filled
ELECTRICAL PROPERTIES				
D.C. resistivity, ohm-cm				
Dielectric constant, 60 cps		2.4–5.0		3.8–5.0
Dielectric constant, 10^6 cps		2.4–3.8	3.7	3.6–4.7
Dissipation factor, 60 cps		0.003–0.008		0.012–0.026
Dissipation factor, 10^6 cps		0.007–0.015	0.004	0.01–0.016
MECHANICAL PROPERTIES				
Modulus of elasticity, 10^3 psi				
Tensile strength, psi		7,000–8,000	10,000–12,000	4,500–6,500
Ultimate elongation, %		1.0–20	15–75	
Yield stress, psi		4,000–9,000		10,000–13,000
Yield strain, %				
Rockwell hardness		R 110–R115	M 94, R 120	E 76
Notched Izod impact strength, ft lb/in.		2.0–4.0	1.4–2.3	0.50–4.5
Specific gravity		1.06–1.08	1.43	1.24–2.6
THERMAL PROPERTIES				
Burning rate		Slow	Slow	Self-extinguishing
Heat distortion, 264 psi, °C		115–118		
Specific heat, cal/g		0.3–0.4	0.35	
Linear thermal expansion coefficient, 10^{-5}, °C		6.0 6.5	8.1	4.0–5.5
Maximum continuous service temperature, °C		88–110	84	149–220
CHEMICAL RESISTANCE				
Mineral acids, weak		Good	Fair	Good
Mineral acids, strong		Good	Poor	Fair
Oxidizing acids, concentrated		Poor	Poor	
Alkalies, weak		Good	Poor	Good
Alkalies, strong		Good	Poor	Fair
Alcohols		Good	Good	Fair to good
Ketones		Poor	Good	Fair to good
Esters		Poor	Good	Fair to good
Hydrocarbons, aliphatic		Fair	Good	Fair to good
Hydrocarbons, aromatic		Fair	Good	Fair to good
Oils: vegetable, animal, mineral		Good	Good	
MISCELLANEOUS PROPERTIES				
Clarity		Translucent to opaque	Translucent to opaque	Opaque
Color		Colorless	Colorless	Colorless
Refractive index, n_D			1.48	
Applicable ASTM specifications and test methods		D638, D150, D792, D256, D758, D696, D651, D648, D543	D638, D150, D792, D256, D758, D696, D651, D648, D543	D638, D150, D792, D256, D758, D543, D651, D648
FORMS AVAILABLE Cs—castings, F—film, Fb—fibers, I—impregnants, L—laminations, Lq—lacquers, Mf—monofilaments, P—powder, pellet, or granules, R—rods, tubes, or other extruded forms, S—sheets.		P, S, L, R	C, R	P
FABRICATION Cl—calendering, Cs—casting, E—extrusion, F—hot forming or drawing, I—impregnation, MB—blow molding, MC—compression molding, MI—injection molding, S—spreading.		Cl, E, MB, MI	MI, E	Cs, MC, MI

REFERENCES

"Handbook of Common Polymers", W.J. Roff, J.R. Scott, J. Pacitti, Butterworth & Co., Ltd. (London), 1971; The Chemical Rubber Co., U.S. distributor.

ASTM Standards, American Society for Testing and Materials, 1972, Part 26.

"Modern Plastics Encyclopedia", Vol. 44, No. 1A, McGraw-Hill, Inc., 1967.

Table 1-80.　THERMAL PROPERTIES OF PLASTICS*

Typical Values Within Usual Ranges

Name	Fire-proof rating[a]	Max. temp., continuous service	Temperature, deg C			
			Deflection or softening	Melting or decomposition	Ignition	
					Flash, flame	Self
Acetal	3–4	100	160			
Acrylonitrile-butadiene-styrene	3–4	85	110	150	405	
Acrylic	4	80	100	175	390	
Alkyd	2–3	165	200	250		
Allyl	2	150	150			
Cellulose acetate	3	75	140	230	340	450
Cellulose acetate butyrate	3–4	85	110	250		
Cellulose nitrate	6	60	85	125		
Epoxy	2–3	180	200			
Ethyl cellulose	2–3	70	90		290	296
Fluorocarbon	1	225	225	325	600	
Melamine	2	120	150			
Methyl methacrylate	3–4	80	100		300	460
Phenolics	2–3	140	175		480	
Polyamide	2	140	200	250		
Polycarbonate	2	125	155			
Polyester	2–3	150	220	290		
Polyethylene	3	100	110	140	340	350
Polypropylene	4	140	145	165	345	390
Polystyrene	3–4	75	90	110		
Polyurethane	3–4	100	185	250		
Polyvinyl chloride	2	85	85	150		
Silicone	1–2	275	300			
Styrene-butadiene	3–4	60	80			
Vinyl	2–3	75	75			

[a]Rated as follows:　1—non-combustible　　4—slow burning
　　　　　　　　　　　2—self-extinguishing　5—moderate burning rate
　　　　　　　　　　　3—very slow burning　　6—highly flammable

*Compiled from several sources.

Table 1-81.　FILLERS FOR PLASTICS AND ELASTOMERS*

Non-Directional Strengthening by Powdered or Ground Materials

Elastomers. In both natural rubber and synthetic elastomers, the most common reinforcing filler is carbon black. With 25–40 percent carbon by weight, the strength (or the stress at 300 percent extension) is increased 3 to 10 times. Usually, the finer the carbon black, the greater the increase in strength.

Plastics. Both organic and mineral fillers are used in powder or ground form for modifying the properties of plastics. Most fillers increase the hardness, stiffness, and dimensional stability as compared with the unfilled plastics. In many cases the strength is increased (see Table 1-85), but directional reinforcement by fibers, cloth, glass, and metals is much more effective for high-strength plastic composites. Some of the common plastic fillers and their contributions to plastics properties and uses are indicated in the following table.

ISOTROPIC FILLERS FOR PLASTICS

Filler material	Examples of properties affected	Typical uses of filler material
Metal powder: aluminum, copper, bronze, lead	Electrical and thermal conductivity increased	Conductive sheets; patching fillers for metal; radiation shielding
Mineral powders: clay, limestone, alumina, talc, silica, mica	Opacity; hardness; resistance to heat, chemicals, and wear increased	Low-cost fillers for smooth, hard surface of tile, molded products, and parts
Organic powders: wood flour, nutshell flour, ground bark	Lightness, strength, dimensional stability increased	Smooth-surfaced and lightweight molded products of high impact strength
Flake and porous materials: chopped paper, sawdust, cellulose, vermiculite, soybean meal	Thermal conductivity, density	Light structural and heat-insulating materials and molded parts
Carbon: graphite and carbon blacks	Electrical conductivity, color	Black molded products; conductive materials
Nondirectional fibers: glass, asbestos, chopped cloth	Strength, toughness, wear resistance, heat resistance increased	Floor and wall coverings; heat and chemical-resistant molded products; containers
Woven fabrics: glass cloth, canvas, mats	Toughness, impact resistance, wear increased	Machine parts and gears; electrical sheet; aircraft structures; boats

*Compiled from various sources.

REFERENCE

"Fillers for Plastics", W.C. Wake, Ed., Iliffe Books, Ltd., 1968.

Table 1-82. FIBER-REINFORCED PLASTICS*

Fiber reinforcement in plastics varies from a composite in which a few directional fibers, comprising less than 10 percent by volume, are used to increase the tensile strength to a composite that is 90 percent fibrous, with the resin serving as a binder. Impact strength, as well as tensile and compressive strengths, are increased. Other intentional improvements for which fibers and fillers are used may include increases in hardness and stiffness, chemical resistance, electrical conductivity, thermal conductivity, or heat resistance. Additives may also be used as plasticizers, antioxidants, flame retardants, and ultraviolet absorbers.

The most common directional reinforcement of plastics is by glass fibers. Reinforcement is hand- or machine-placed or it is wound from continuous filament. Molding pressures are applied by contact, vacuum bag, or positive pressure with bag or autoclave; the glass content in these cases is usually less than 50 percent. For higher pressures and for glass content over 50 percent, matched-die, closed molds are used. Rods and tubing are extrusion molded, with continuous strands drawn through a resin bath. Resin content is then in the range of 20–40 percent.

The resins most widely used for reinforced plastics are the polyesters, especially for marine, automotive, and architectural uses. Epoxies are favored for high-performance military and space uses. For glass-wound products the strand may be preimpregnated with semi-cured epoxy, then final-cured at high temperature after assembly. Phenolics, melamines, and silicones are other thermosets used for certain purposes with glass reinforcement. Thermoplastic (melting) resins have also been used to some extent with reinforcement, nylon being the most common.

Other materials for reinforcement include metal wires, filaments, and whiskers. Among these materials are high-strength and stainless steels, boron, beryllium, titanium, carbon, carbides, and silica. Epoxy resins are favored. The following table gives typical test results for fiber-reinforced plastics.

MECHANICAL PROPERTIES OF FIBER-REINFORCED PLASTICS

| Plastic resin matrix | Fiber reinforcement | | Specific gravity | Composite | | | | Modulus of elasticity, millions of psi |
| | Type | Percent v = vol w = wt | | Ultimate strength, kpsi | | | | |
				Tension	Compression	Shear	Flexure	
Polyester	Glass cloth	65 w	1.8	70	50		90	
Epoxy	Glass cloth	65 w	2.0	60	70		100	
Phenolic	Glass cloth	65 w	1.9	60	40		95	
Polystyrene	Glass, $\frac{1}{4}$-in. fibers	30 w	1.28	14[a]				1.2
Polycarbonate	Glass, $\frac{1}{4}$-in. fibers	40 w	1.44	19[b]				1.5
Nylon 6/6	Glass, $\frac{1}{4}$-in. fibers	40 w	1.41	29[c]				1.6
Epoxy	Glass (E), directional	73 w	2.17	238				8.1
Epoxy	Glass (E), directional	56 w	1.97	149				6.2
Polyester	Glass, chopped strand, random	45 w	1.6	25	26			1.8
Epoxy	Al_2O_3 whiskers	14 v	1.64	113[d]				6.
Epoxy	Boron filaments	70 v			330	12.	300	40 (flex.)
Epoxy	Carbon filaments	38 v	1.42	80.	134.	11.3	107.	3.1

[a] Without reinforcement, tensile strength = 6 500 psi.
[b] Without reinforcement, tensile strength = 9 500 psi.
[c] Without reinforcement, tensile strength = 10 000 psi.
[d] Specimen only 0.02 × 1.25 in. size.

*Compiled from several sources.

REFERENCES

"Modern Plastics Encyclopedia", Vol. 44, No. 1A, McGraw-Hill, Inc., 1967.
"Modern Composite Materials", L.J. Broutman and R.H. Krock, Addison-Wesley Publishing Company, 1967.

Table 1-83.   TRADE NAMES, COMPOSITION, AND MANUFACTURERS OF VARIOUS PLASTICS

Trade name	Composition	Manufacturer	Trade name	Composition	Manufacturer
Abson	Acrylonitrile-butadiene, ABS polymers	B. F. Goodrich Chemical Co.	Forticel	Cellulose propionate sheet films, molding powders	Celanese Plastics Co.
Alathon	Polyethylene resins	E. I. du Pont de Nemours & Co., Inc.	Fortiflex	Polyethylene resins	Celanese Plastics Co.
Alkor	Furane resin cement	Atlas Minerals & Chemicals Div., The Electric Storage Battery Co.	Fosta-Tuf-Flex	Polystyrene, high-impact	Foster-Grant, Inc.
Amres	Phenolics, urea, and melamine resins	American Marietta Co., Pacific Resins & Chemicals, Inc.	Furnane	Furanes	Atlas Minerals & Chemicals Div., The Electric Storage Battery Co.
Araldite	Epoxy resins	CIBA Products Co., Div. CIBA Corp.	GenEpoxy	Epoxy resins for adhesives, coatings, etc.	General Mills, Inc., Chemical Div.
Atlac	Polyester resins	Atlas Chemical Industries, Inc.	Genetron	Fluorinated hydrocarbons, monomers, and polymers	Allied Chemical Corp., General Chemical Div.
Bakelite	Acrylics, epoxies, phenolics, polyethylenes, copolymers	Union Carbide Corp., Chemicals and Plastics Div.	Geon	Polyvinyl chloride materials	B. F. Goodrich Chemical Co.
Bavick-11	Methyl methacrylate and methylstyrene copolymer	J. T. Baker Chemical Co.	Grex	High-density polyethylenes	Allied Chemical Corp., Plastics Div.
Boltaflex	Supported and unsupported flexible vinyl sheeting	The General Tire & Rubber Co.	Halon	Fluorohalocarbon resins	Allied Chemical Corp.
Boltaron	Rigid polyvinyl chloride sheet	The General Tire & Rubber Co., Chemical & Plastics Div.	Hetron	Fire-retardant polyester resin	Hooker Chemical Corp., Durez Plastics Div.
Butacite	Polyvinyl butyral resins	E. I. du Pont de Nemours & Co., Inc.	Isothane	Polyurethane foam, ester, and ether	Bernel Foam Products Co., Inc.
Conolite	Polyester resins and laminates	Shellmar-Betner, Div. Continental Can Co. Woodall Industries, Inc., Conolite Div.	Kel-F	Chlorotrifluoroethylene, molding resins, and dispersions	3M Company
Corvel	Epoxies, vinyls Fusion-bond finishes	The Polymer Corp., Export-Polypenco Div.	Kralac	High-styrene resins, styrene-butadiene copolymers	Uniroyal Chemical, Div. of Uniroyal Inc.
Cumar	Paracoumarone-indene resins	Allied Chemical Corp., Plastics Div.	Kralastic	ABS polymers, copolymers	Uniroyal Chemical, Div. of Uniroyal Inc.
Cycolac	ABS polymers, acrylonitrile-butadiene-styrene copolymers	Marbon Chemical Div., Borg-Warner Corporation	Kynar	Polyvinylidene fluoride	Pennsalt Chemical Corp.
Dacovin	Polyvinyl chlorides	Diamond Shamrock Corp.	Lexan	Polycarbonate resin, film, and sheet	General Electric Company, Plastics Dept.
Dapon	Diallyl phthalate resins	FMC Corp., Organic Div.	Lucite	Acrylic resin and syrup	E. I. du Pont de Nemours & Co., Inc.
Delrin	Acetal resin and pipe	E. I. du Pont de Nemours & Co., Inc.	Lustran	ABS polymers	Monsanto Co.
Dylan	Polyethylene	Sinclair-Koppers Co.	Lustrex	Styrene molding and extrusion resins	Monsanto Co.
Dylene	Polystyrene	Sinclair-Koppers Co.	Lytron	Styrene molding and extrusion resins	Monsanto Co.
Dylite	Expandable polystyrene	Sinclair-Koppers Co.	Madurit	Melamine resins and compounds	Cassella Farbwerke Mainkur, A.G.
Epi-Rez	Epoxy resins	Celanese Coatings Co., Celanese Resin Div.	Maraglas	Epoxy-casting resin	The Marblette Corporation, Div. of Allied Products
Epolene	Low molecular-weight polyethylene resins	Eastman Chemical Products, Inc., Sub. Eastman Kodak Company	Marlex	Polyethylenes, polypropylenes, copolymers	Phillips Petroleum Co.
Epoxical	Epoxy resins	United States Gypsum Co.	Marvinol	Vinyl chloride resins and compounds	Uniroyal Chemical, Div. of Uniroyal Inc.
Epon	Epoxy resins and curing agents	The Shell Chemical Company, Plastics and Resins Div.	Merlon	Polycarbonate resins	Mobay Chemical Co.
Escon	Polypropylene resins	Enjay Chemical Co., Div. Humble Oil & Refining Company	Micarta	Melamines, phenolics, polyesters	Westinghouse Electric Co., Industrial Micarta Div.
Estane	Polyurethane materials	B. F. Goodrich Chemical Company	Microthene	Polyethylenes, polyolefins	U.S. Industrial Chemicals Co.
Fluorogreen	Teflon with glass and ceramic fibers, fluorocarbons	John L. Dore Co.	Multrathane	Urethane elastomers	Mobay Chemical Company
			Nopcofoam	Polyurethane plastics	Nopco Chemical Co., Plastics Div.
			Novodur	Polyacrylonitrile-butadiene-styrene	Farbenfabriken Bayer, A. G.
Fluororay	Ceramic-filled fluorocarbons	Raybestos-Manhattan, Inc., Plastic Products Div.	Opalon	Vinyl chloride resins and compounds	Monsanto Co.
Formica	Melamines	Formica Corp. of American Cyanamid	Paraplex	Polyester resins, acrylic-modified polyester resins	Rohm & Haas Company

Table 1-83. TRADE NAMES, COMPOSITION, AND MANUFACTURERS
OF VARIOUS PLASTICS *(Continued)*

Trade name	Composition	Manufacturer	Trade name	Composition	Manufacturer
Permelite	Melamines	Melamine Plastics, Inc., Div. of Fiberlite Corp.	Super Dylan	Polyethylene	Sinclair-Koppers Co.
Petrothene	Polyethylene resins, polypropylene resins	U.S. Industrial Chemicals Co.	Supreme	Polyethylenes	Johns-Manville Company
			Sylplast	Urea-formaldehyde compounds	FMC Corp., Organic Chemicals Div.
Piccoflex	Styrene-copolymer resins	Pennsylvania Industrial Chemical Corp.	Teflon	Fluorocarbon resins	E. I. du Pont de Nemours & Co., Inc.
Piccolastic	Styrene-polymer resins	Pennsylvania Industrial Chemical Corp.	Tenite	Cellulose acetate, cellulose-acetate-polyethylene, poly-propylenes, urethane elastomers, copolymers	Eastman Chemical Products, Inc., Sub. Eastman Kodak Co.
Plaskon	Nylons, melamines, phenolics, polyesters	Allied Chemical Corp.			
Pleogen	Alkyds, polyesters, copolymers	Mol-Rez Div., American Petrochemical Corp.	Tetran	Fluorocarbons	Pennsalt Chemicals Corp.
Plexiglas	Acrylics	Rohm & Haas Company	Texin	Urethane elastomers	Mobay Chemical Company
Pliovic	Polyvinyl chlorides	The Goodyear Tire & Rubber Co., Chemical Div.	Thioment	Polyisoprenes	Atlas Minerals & Chemicals Div., The Electric Storage Battery Co.
Plyophen	Phenolic resins	Reichhold Chemicals, Inc.			
Poly-Eth	Polyethylene resins	Gulf Oil Corp., U.S. Div. of Gulf Oil Corp.	Ultrapas	Melamine resins	Dynamit Nobel, A. G.
			Ultrathene	Ethylene-vinyl acetates	U.S. Industrial Chemicals Co
Polylite	Polyester resins	Reichhold Chemicals, Inc.			
Polypenco	Acrylics, chlorinated polyethers, fluoro-carbons, nylons, polycarbonates	Polymer Corp.	Ultron	Polyvinyl chlorides	Monsanto Co.
			Vibrathane	Urethane elastomers	Uniroyal Chemical, Div. of Uniroyal Inc.
			Vibrin	Polyester resins	Uniroyal Chemical, Div. of Uniroyal Inc.
Resimene	Urea and melamine resins	Monsanto Co.	Vitel	Polyesters	The Goodyear Tire & Rubber Co., Chemical Div.
Resinox	Phenolic resins and compounds	Monsanto Co.			
Rhonite	Urea resins	Rohm & Haas Company	Viton	Synthetic rubbers	E. I. du Pont de Nemours & Co., Inc.
Roylar	Polyurethanes	Uniroyal Chemical, Div. of Uniroyal Inc.	Vitroplast	Polyester cements	Atlas Minerals & Chemicals Div., The Electric Storage Battery Co.
Ryertex	Laminated phenolics and rigid polyvinyl chloride extrusions	Joseph T. Ryerson & Son, Inc., Industrial Plastics and Bearings Sales Div.	Vyron	Polyvinyl chlorides	Industrial Vinyls, Inc.

Table 1-84. PROPERTIES OF PLASTIC ELECTRICAL INSULATIONS*

As Used in Flat-Cable Construction

For data on standard flat conductor cables, see Table 2-26.
For density in kg/m^3, multiply specific gravity by 1000. For elastic modulus or strength multiply value in psi by 6895 to obtain N/m^2.

Characteristic	TFE Teflon[x]	TFE-Teflon glass cloth	FEP Teflon[x]	FEP-Teflon glass cloth	Kapton polyimide[x]	Kel-F[y]	PVF Tedlar[x]	Poly-propylene	Mylar[x] polyester	Polyvinyl chloride	Poly-ethylene
Appearance	Translucent	Tan	Clear bluish	Tan	Amber	White and opaque	Clear	Clear	Clear	Translucent	Clear
Specific gravity	2.15	2.2	2.15	2.2	1.42	2.10	1.38	0.905	1.395	1.25	0.93
Service temperature Minimum, °C	−60	−60	−60	−60	−60	—	−70	−55	−60	−40	−20
Maximum, °C	280	250	250	200	400	400+	105	125	150	85	80
Modulus of elasticity, psi	58,000	—	50,000	—	510,000	190,000	280,000	170,000	550,000	—	50,000
Tensile strength, psi at 25°C	3,000	20,000[b]	3,000	20,000[b]	25,000	4,500	13,000	5,700	20,000	3,000	2,000
Thermal expansion coefficient, °C	.00010	Low[b]	.00009	Low[b]	.00002	.00008	.00005	.00011	.00003	—	.00018
Dielectric constant, 10^2–10^8 hz	2.2	2.5/5[b]	2.1	2.5/5[b]	3.5	2.5	7.0	2.0	2.8–3.7	3.6–4.0	2.2
Volume resistivity, ohm-cm	>10^{18}	10^{16}	>$2/10^{18}$	10^{16}	10^{16}	$3.1/10^{16}$	$3/10^{13}$	10^{16}	10^{18}	10^{10}	10^{16}
Chemical resistance	Excellent	Excellent	Excellent	Excellent	Excellent	Excellent	Good	Good	Excellent	Good	Good
Sunlight resistance	Excellent	Excellent	Excellent	Excellent	Excellent	Excellent	Excellent	Low	Fair	Fair	Low
Water absorption, %	<0.01/24 hr	0.10–0.68[b]	<0.01/24 hr	0.18–0.30[b]	3/24 hr	0	0.5/2 hr	0.01	0.8/24 hr	0.10	0.01/24 hr
Bondability with adhesives	Good[a]	Good[a]	Good[a]	Good[a]	Good	Good[a]	Good[a]	Poor	Good	Good	Poor
Bondability to itself	Good	Poor	Good	Good	Poor	Good	Good	Good	Poor	Good	Good
Flammable	No	No	No	No	Self extinguishing	No	Yes	Slow-burning	Yes	Self extinguishing	Yes

[a] Must be treated.
[b] Depends on % glass cloth.
[x] Trademark, E. I. du Pont de Nemours & Co., Inc., Wilmington, Delaware.
[y] Trademark, 3M Company, St. Paul, Minnesota.

*Based on: "Flat Conductor Cable Technology", NASA–SP5043, 1968, Table 1.

Table 1-85. COMPARATIVE PROPERTIES OF REINFORCED PLASTICS*

SYMBOLS: U = unreinforced   R = reinforced

Multiply tabular values in psi by 6 895 to obtain N/m^2.

Property	Polyamide U	R	Polystyrene** U	R	Polycarbonate U	R	Styrene acrylonitrile† U	R	Polypropylene U	R	Acetal U	R	Linear polyethylene U	R
Tensile strength, 1000 psi	11.8	30.0	8.5	14.0	9.0	20.0	11.0	18.0	5.0	6.6	10.0	12.5	3.3	11.0
Impact strength, notched, ft-lb/in.														
At 73°F	0.9	3.8	0.3	2.5	2.0§	4.0§	0.45	3.0	1.3-2.1	2.4	1.4	3.0	—	4.5
At −40°F	0.6	4.2	0.2	3.2	1.5§	4.0§	—	4.0	—	2.2	—	3.0	—	5.0
Tensile modulus, 10^5 psi	4.0	—	4.0	12.1	3.2	17.0	5.2	15.0	2.0	4.5	4.0	8.1	1.2	9.0
Shear strength, 1000 psi	9.6	14.0	—	9.0	9.2	12.0	—	12.5	4.6	4.7	9.5	9.1	—	5.5
Flexural strength, 1000 psi	11.5	37.0	11.0	20.0	12.0	26.0	17.0	26.0	6-8	7.0	14.0	16.0	—	12.0
Compressive strength, 1000 psi	4.9††	24.0	14.0	17.0	11.0	19	17.0	22.0	8.5	6.0	5.2	13.0	2.7-3.6	6.0
Deformation, 4000-psi load, %	2.5	0.4	1.6	0.6	0.3	0.1	—	0.3	—	6.0	—	1.0	—	0.4‡
Elongation, %	60.0	2.2	2.0	1.1	60-100	1.7	3.2	1.4	>200	3.6	9-15	1.5	60.0	3.5
Water absorption, in 24 hr, %	1.5	0.6	0.03	0.07	0.3	0.09	0.2	1.15	0.01	0.05	0.20	1.1	0.01	0.04
Hardness, Rockwell	M79	E75 to 80	M70	E53	M70	E57	M83	E55	R101	M50	M94	M90	R64	R60
Specific gravity	1.14	1.52	1.05	1.28	1.2	1.52	1.07	1.36	0.90	1.05	1.43	1.7	0.96	1.30
Heat distortion temperature, at 264 psi, °F	150	502	190	220	280	300	200	225	155	280	212	335	126	260
Coeff. of thermal expansion, per F × 10^{-5}	5.5	0.9	4.0	2.2	3.9	0.9	4.0	1.9	4.7	2.7	4.5	1.9	9.0	1.7
Dielectric strength, short time, v/mil	385	480	500	396	400	482	450	515	750	—	500	—	—	600
Volume resistivity, ohm-cm × 10^{15}	450	2.6	10.0	36.0	20.0	1.4	10^{16}	4.5	17.0	15.0	0.6	38.0	10^{15}	29.0
Dielectric constant, at 60 hz	4.1	4.5	2.6	3.1	3.1	3.8	3.0	3.6	2.3	—	—	—	2.3	2.9
Power factor, at 60 hz	0.0140	0.009	0.0030	0.0048	0.0009	0.0030	0.0085	0.005	—	—	—	—	—	0.001
Approximate cost, ¢/in.³	3.0	8.0	0.5	2.5	3.6	6.5	0.9	2.5	0.6	2.1	3.3	7.8	0.7	3.1

† Heat-resistant grade.
‡ 1000-psi load.
** Medium-flow, general purpose grade.
§ Impact values for polycarbonates are a function of thickness.
†† At 1% deformation.
* From: "Reinforced Thermoplastics", W. Lachowecki, *Machine Design*, 40(29):34, Dec. 12, 1968.

Table 1-86. COMPARISON OF FIVE HIGH-TEMPERATURE WIRE-INSULATING MATERIALS*

Characteristic	Modified Teflon	Teflon	Silicone enamel DC1360	Formvar (vinyl acetal)	Plain enamel
Upper temp limit	+250°C	+250°C	+180°C	+105°C	+80°C
Lower temp limit	−100°C	−100°C	−40°C	−40°C	−40°C
Dielectric strength	Excellent	Very good	Very good	Good	Good
Dielectric constant (60 hz–30,000 Mhz)	2.0–2.05†	2.0–2.05†	Inferior	Inferior	Inferior
Power factor (60 hz–10,000 Mhz)	0.0002†	0.0002†	Inferior, about 0.006–0.007	Inferior	Inferior
Space factor	Excellent	Excellent	Excellent	Excellent	Excellent
Solvent resistance	Excellent	Excellent	Fair	Fair	Poor
Abrasion resistance	Good	Fair	Very good	Excellent	Good
Thermoplastic flow	Good	Fair	Excellent	Excellent	Good
Crazing resistance	Excellent	Very good	Fair	Fair	Fair
Flame resistance	Excellent	Excellent	Fair	Poor	Poor
Fungus resistance	Excellent	Excellent	Good	Good	Poor
Moisture resistance	Excellent	Excellent	Good	Good	Good
Continuity of insulation	Excellent	Excellent	Good	Good	Good
Arc resistance	Excellent	Excellent	Good	Good	Good
Flexibility	Excellent	Very good	Good	Good	Good

†Stable at temperatures up to 250°C.
*From: "Choosing Wire Insulation for High Temperatures", J. Holland, *Electronic Design*, 2:14, July 1954.

Table 1-87. TYPICAL PROPERTIES OF GLASS-FIBER-REINFORCED RESINS*

For conversion factors see Table 1-79.

Property	Base resin Polyester	Phenolic	Epoxy	Melamine	Polyurethane
Molding quality	Excellent	Good	Excellent	Good	Good
Compression molding Temperature, °F	170 to 320	280 to 350	300 to 330	280 to 340	300 to 400
Pressure, psi	250 to 2000	2000 to 4000	300 to 5000	2000 to 8000	100 to 5000
Mold shrinkage, in./in.	0.0 to 0.002	0.0001 to 0.001	0.001 to 0.002	0.001 to 0.004	0.009 to 0.03
Specific gravity	1.35 to 2.3	1.75 to 1.95	1.8 to 2.0	1.8 to 2.0	1.11 to 1.25
Tensile strength, 1000 psi	25 to 30	5 to 10	14 to 30	5 to 10	4.5 to 8
Elongation, %	0.5 to 5.0	0.02	4		10 to 650
Modulus of elasticity, 10^{-5} psi	8 to 20	33	30.4	24	
Compression strength, 1000 psi	15 to 30	17 to 26	30 to 38	20 to 35	20
Flexural strength, 1000 psi	10 to 40	10 to 60	20 to 26	15 to 23	7 to 9
Impact. Izod, ft-lb/in. or notch	2 to 10	10 to 50	8 to 15	4 to 6	No break
Hardness, Rockwell	M70 to M120	M95 to M100	M100 to M108	M100 to M108	M28 to R60
Thermal expansion, per °C	2 to 5×10^{-5}	1.6×10^{-5}	1.1 to 3.0×10^{-5}	1.5×10^{-5}	10 to 20×10^{-5}
Volume resistivity at 50% RH, 23°C, ohm-cm	1×10^{14}	7×10^{12}	3.8×10^{15}	2×10^{11}	2×10^{11} to 10^{14}
Dieletric strength, $\frac{1}{8}$ in. thickness, v/mil	350 to 500	140 to 370	360	170 to 300	330 to 900
Dielectric constant At 60 hz	3.8 to 6.0	7.1	5.5	9.7 to 11.1	5.4 to 7.6
At 1 khz	4.0 to 6.0	6.9			5.6 to 7.6
Dissipation factor At 60 hz	0.01 to 0.04	0.05	0.087	0.14 to 0.23	0.015 to 0.048
At 1 khz	0.01 to 0.05	0.02			0.043 to 0.060
Water absorption, %	0.01 to 1.0	0.1 to 1.2	0.05 to 0.095	0.09 to 0.21	0.7 to 0.9
Sunlight (change)	Slight	Darkens	Slight	Slight	None to slight
Chemical resistance	Fair**	Fair**	Excellent	Very good†	Fair
Machining qualities	Good	Good	Good	Good	Good

Note: Filament-wound components with high glass content, highly oriented, have higher strengths. The decreasing order of tensile strength is: roving, glass cloth, continuous mat, and chopped-strand mat.

**Attacked by strong acids or alkalies.
†Attacked by strong acids.
*From: "Reinforced Thermosets", C.A. Spang and G.I. Davis, *Machine Design*, 40(29): 32, Dec. 12, 1968.

Table 1-88.   PROPERTIES OF COMMERCIAL NYLON RESINS*

For conversion factors see Table 1-79.

Property	Type 6/6	Type 6	Type 6/10	Type 11	Glass-reinforced Type 6/6, 40%	MoS$_2$-filled, 2½%	Direct polymerized, castable
Mechanical							
Tensile strength, psi	11,800	11,800	8200	8500	30,000	10,000 to 14,000	11,000 to 14,000
Elongation, %	60	200	240	120	1.9	5 to 150	10 to 50
Tensile yield stress, psi	11,800	11,800	8500		30,000		
Flexural modulus, psi	410,000	395,000	280,000	151,000	1,800,000	450,000	
Tensile modulus, psi	420,000	380,000	280,000	178,000		450,000 to 600,000	350,000 to 450,000
Hardness, Rockwell	118R	119R	111R	55A	75E–80E	110R–125R	112R–120R
Impact strength, tensile, ft-lb/sq in.	76		160			50–180	80–100
Impact strength, Izod, ft-lb/in. of notch	0.9	1.0	1.2	3.3	3.7**	0.6	0.9
Deformation under load, 2000 psi, 122°F, %	1.4	1.8	4.2	2.02†	0.4§	0.5 to 2.5	0.5 to 1
Thermal							
Heat-deflection temp, °F							
At 66 psi	360	365	300	154	509	400 to 490	400 to 425
At 264 psi	150	152	135	118	502	200 to 470	300 to 425
Coefficient of thermal expansion, per °F	4.5×10^{-5}	4.6×10^{-5}	5×10^{-5}	10×10^{-5}	0.9×10^{-5}	3.5×10^{-5}	5.0×10^{-5}
Coefficient of thermal conductivity, Btu in./hr ft³ °F	1.7	1.7	1.5				
Specific heat	0.3–0.5	0.4	0.3–0.5	0.58			
Brittleness temp, °F	−112		−166				
Electrical							
Dielectric strength, short time, v/mil	385	420	470	425	480	300 to 400	500 to 600‡
Dielectric constant							
At 60 hz	4.0	3.8	3.9		4.45		3.7
At 10³ hz	3.9	3.7	3.6	3.3	4.40		3.7
At 10⁶ hz	3.6	3.4	3.5		4.10		3.7
Power factor							
At 60 hz	0.014	0.010	0.04	0.03	0.009		0.02
At 10³ hz	0.02	0.016	0.04	0.03	0.011		0.02
At 10⁶ hz	0.04	0.020	0.03	0.02	0.018		0.02
Volume resistivity, ohm-cm	10^{14} to 10^{15}	3×10^{15}	10^{14} to 10^{15}	2×10^{13}	2.6×10^{15}	2.5×10^{13}	
General							
Water absorption, 24 hr, %	1.5	1.6	0.4	0.4	0.6	0.5 to 1.4	0.9
Specific gravity	1.13 to 1.15	1.13	1.07 to 1.09	1.04	1.52	1.14 to 1.18	1.15 to 1.17
Melting point, °F	482 to 500	420 to 435	405 to 430	367	480 to 490	496±9	430±10
Flammability	self ext	self ext	self ext	self ext	self ext	self ext	self ext
Chemical resistance to							
Strong acids	Poor	Poor	Poor	Poor	Poor	Poor	Poor
Strong bases	Good	Good	Good	Fair	Good	Good	Good
Hydrocarbons	Excellent	Excellent	Excellent	Good	Excellent	Excellent	Excellent
Chlorinated hydrocarbons	Good	Good	Good	Fair	Good	Good	Good
Aromatic alcohols	Good	Good	Good	Good	Good	Good	Good
Aliphatic alcohols	Good	Good	Good	Fair	Good	Good	Good

Notes:

Most nylon resins listed in this table are used for injection molding, and test values are determined from standard injection-molded specimens. In these cases a single typical value is listed. Exceptions are MoS$_2$-filled nylon and direct-polymerized (castable) nylon, which are sold principally in semifinished stock shapes. Ranges of values listed are based on tests on various forms and sizes produced under varying processing conditions.

Because single values apply only to standard molded specimens, and properties vary in finished parts of different sizes and forms produced by various processes, these values should be used for comparison and preliminary design considerations only. For final design purposes the manufacturer should be consulted for test experience with the form being considered. Listed values should not be used for specification purposes.

†2000 psi, 73°F.

‡0.040-in. thick.

**½ × ¼-in. bar.

§4000 psi, 122°F.

*From: "Nylons", D.D. Carswell, *Machine Design*, 40(29):62, Dec. 12, 1968.

Table 1-89. RUBBERS AND ELASTOMERS*

Elastomers cannot be classified in any brief and simple manner, nor are they well characterized by the usual mechanical tests. The terms *rubber* and *synthetic rubber* are loosely applied to a great variety of elastic materials, from pure gum natural rubber and pure synthetics to cured, compounded, filled, and even reinforced products.

ASTM designations (D1418) by chemical polymer description are used in the following table; yet within each class the properties can vary widely, depending on the exact composition, heat treatment service temperature, and application. Typical uses, such as rubber springs and cushioning, permit an almost unlimited number of combinations of design variables.

Mechanically, rubbers may be expected to lose strength rapidly with increase in temperature, to show a large hysteresis in stress-strain behavior, to exhibit marked creep and set, and to be greatly affected by rates of load application or frequency of repeated stress. "Heat build-up", i.e., increase in temperature in service, as well as deterioration from environment (sunlight, oils, ozone, etc.) will reduce the valuable properties of many rubbers, both natural and synthetic.

The following data apply to typical samples of commercial elastomers for common uses.

KEY:

A—Acetone	J—Alkalies	S—Salts
B—Benzene	K—Ketones	T—Heat or high temperature
C—Carbon tetrachloride	L—Alcohols	U—Ultraviolet
D—Carbon disulfide	M—Ammonia	V—Vegetable oils
E—Phenol	N—Turpentine	W—Weathering
F—Sulfur compounds	O—Coal derivatives; bitumens	X—Oxidation
G—Glycerol or glycol	P—Petroleum products	Y—Aging
H—Hexane	R—Aromatics	Z—Ozone
I—Acids		

Chemical name	Polyisoprene	Butadiene	Styrene-butadiene	Acrylonitrile butadiene
Other names	Natural (or synthetic) rubber NR (IR)	BR Cis 4	Buna S Styrene SBR, GR-S	Nitrile, Buna N Hycar NBR, GR-A
CHEMICAL AND PHYSICAL				
Specific gravity	0.93	1.0	1.0	1.0
Specific heat	0.40	0.45	0.40	0.47
Thermal conductivity				
W/cm·K	0.001 7	0.002 5	0.002 6	0.002 5
Btu/hr·ft·deg F	0.10	0.14	0.15	0.14
Service temperature, deg C				
min	−25	−40	−20	−20
max	90	90	75	110
Solvents, softeners	D,K,P,V	D,H,N,P	K,P,R,V	C,K,O,R
Resistant to	A,I,J,L	G,I,J,W,Y	G,I,L,S,X	G,I,K,L,P,S, T,V,W
Swelled by	D,P,V	A,P,V	P,V	A,E,N
MECHANICAL AND ELECTRICAL				
Tensile strength				
kg/cm² (max)	300.	210.	210.	295.
kpsi (max)	4.3	3.0	3.0	4.2
Elongation at break, %	600.	700.	600.	600.
Vol. resistivity, ohm-cm	10^{15}	10^{15}	10^{14}	10^{10}
Dielectric strength				
kV/cm	235		235	185
V/mil	600.		600.	475.
Dielectric constant	3.0	2.3	2.8	3.0
Power factor (50–100 Hz)	0.003	0.005	0.005	0.007
Rebound	Good	Good	Fair	Good
COMPARATIVE RATINGS—RESISTANCE TO				
Abrasion	Good	Excellent	Good	Excellent
Cold flow (set)	Excellent		Good	Good
Tearing	Good		Poor	Fair
Air permeability	Fair	Good	Fair	Excellent
Oxidation	Fair	Fair	Fair	Fair
Flame	Poor		Poor	Poor

*Compiled from several sources.

Table 1-89.   RUBBERS AND ELASTOMERS *(Continued)*

Chemical name	Polychloro-prene	Isobutylene-isoprene	Polysulfide	Polymethane
Other names	Neoprene[a] CR, GR-M	Butyl IIR, GR-I	Thiokol[a] PS, GR-P	Adiprene[a] PU
CHEMICAL AND PHYSICAL				
Specific gravity	1.25	0.95	1.4	1.2
Specific heat	0.5	0.45	0.31	0.45
Thermal conductivity				
W/cm·K	0.002 1	0.001 3	0.003	0.001 3
Btu/hr·ft·deg F	0.12	0.075	0.17	0.075
Service temperature, deg C				
min	-20	-40	-15	-35
max	100	120	90	120
Solvents, softeners	A,B,C,D,I,N,R	D,P	C	
Resistant to	G,L,P,S,T,U,V, W,Y,Z	E,G,J,S,U,V, W,X,Y,Z	L,P,U,Z	P,V,X,Z
Swelled by	C,D,N,R	D,H,P	C,R	B,C,K,R
MECHANICAL AND ELECTRICAL				
Tensile strength				
kg/cm^2 (max)	240.	175.	90.	350.
kpsi (max)	3.5	2.5	1.3	5.0
Elongation at break, %	800.	700.	500.	550.
Vol. resistivity, ohm-cm	10^{11}	10^{17}	10^8	10^{11}
Dielectric strength				
kV/cm	195	295	125	195
V/mil	500	750	325	500
Dielectric constant	7.	2.4	8.	7.
Power factor (50–100 Hz)	.04	0.004	0.02	0.04
Rebound	Good	Poor	Poor	
COMPARATIVE RATINGS—RESISTANCE TO				
Abrasion	Excellent	Fair	Poor	Excellent
Cold flow (set)	Excellent	Fair	Poor	Poor
Tearing	Good	Good	Poor	Excellent
Air permeability	Good	Excellent	Good	Excellent
Oxidation	Good	Good	Good	Good
Flame	Excellent	Poor	Poor	Poor

[a]Proprietary.

Table 1-89.   RUBBERS AND ELASTOMERS *(Continued)*

Chemical name	Chloro-sulfonated polyethylene	Ethylene propylene	Fluorocarbon	Polysiloxane
Other names	Hypalon[a] CSM, HYP	EP, EPR, R, EPDM, EPT	Viton[a] Kel-F[a] FLU, FPM	Silastic Silicone SIL
CHEMICAL AND PHYSICAL				
Specific gravity	1.2	0.86	1.85	1.0
Specific heat		0.55		0.37
Thermal conductivity				
W/cm·K		0.003 6		0.002 9
Btu/hr·ft·deg F		0.208		0.168
Service temperature, deg C				
min		−50	−30	−60
max	250	150	275	300
Solvents, softeners				C,P,R
Resistant to	P,T,W,Y,Z	G,J,K,L,T,W, X,Z	I,J,P,T	J,T,V,W,X, Y,Z
Swelled by		C,N,P	B,E,R	C,D,M
MECHANICAL AND ELECTRICAL				
Tensile strength				
kg/cm² (max)	280	220	175	125
kpsi (max)	4.0	3.1	2.5	1.8
Elongation at break, %	500.	700.	250.	350.
Vol. resistivity, ohm-cm	10^{14}	10^{16}	10^{12}	5×10^{14}
Dielectric strength				
kV/cm	235	315	195	275
V/mil	600	800	500	700
Dielectric constant	6.0	2.6	2.5	3.0
Power factor (50–100 Hz)	0.03		0.015	0.005
Rebound	Good	Fair	Good	Good
COMPARATIVE RATINGS—RESISTANCE TO				
Abrasion	Good	Good	Good	Poor
Cold flow (set)	Poor	Fair	Fair	Fair
Tearing	Good	Good	Fair	Fair
Air permeability	Good	Fair	Excellent	Poor
Oxidation			Excellent	Good
Flame	Excellent		Good	Fair

[a]Proprietary.

REFERENCES

"Rubber Technology and Manufacture", C.M. Blow, Ed., Butterworth & Co. Ltd., 1971; published in the United States by The Chemical Rubber Co., 1971.

"Handbook of Common Polymers", W.J. Roff and J.R. Scott, Butterworth & Co. Ltd., 1971; published in the United States by The Chemical Rubber Co., 1971.

ASTM Standards, American Society for Testing and Materials, 1972, Part 28.

Table 1-90. WAXES AND RELATED MATERIALS*

Most waxes are commercially available in several grades, depending on source, refinement, bleaching, or processing. Thus the properties of a single type of wax may vary over a rather wide range, and the values given in this table should be considered only as typical.

Name, source, type, color	Specific gravity	Melting point, deg C	Flash point, deg C	Acid value[a]	Iodine value[b]	Saponifi-cation value[a]	Dielectric constant, at 10^6 Hz
VEGETABLE WAXES							
Bayberry (myrtle)	.93	50		3.5	3	212	
Candelilla (brown) from Mexico	.98	68	241	16	24	57	2.5
Carnauba (palm) from Brazil	1.00	85	300	6	12	83	2.9
Castor oil (hydrogenated)	.98	86	315	2	4	180	
Cotton (yellow)	.96	80		30	24	70	
Esparto (grass)	.99	74	255	25	16	68	
Japan (sumac)	.98	51	200	10	9	220	3.1
Ouricuri (palm) from Brazil	1.02	82	280	12	8	80	
Sugar cane (brown)	.97	80		15	20	60	2.9
ANIMAL WAXES							
Beeswax, yellow	.96	64	245	20	10	93	2.8
Chinese insect	.96	82		10	2	90	3.7
Lanolin (sheepswool) refined	.94	40		8	25	107	
Shellac (insect)	.97	77		15	5	110	
Spermaceti (whale)	.93	45	245	2	4	123	8.
MINERAL, SYNTHETIC, WAXLIKE							
Ceresin	.90	70		0	8	0	
Microcrystalline[c]	.93	70	260				2.3
Moutan (lignite)	1.02	85		40		100	
Ozocerite	.90	75		0	8	0	2.4
Paraffin[c]	.92	60	205				2.2
Polyethylene	.93	100	315	15		25	
Polyethylene glycol	1.1	45	250				

Note: The *index of refraction* of most waxes is between 1.4 and 1.5.

[a] *Acid value* is mg of KOH to neutralize the free fatty acids. *Saponification value* is mg of KOH for complete saponification.
[b] *Iodine value* is a measure of unsaturated linkages present and is expressed as g iodine absorbed per 100 g of sample.
[c] There are so many grades of paraffin wax and microcrystalline petroleum wax that the properties of any specific wax should be checked.

*Compiled from several sources.

REFERENCES
"Industrial Waxes", H. Bennett, Vol. 1, Chemical Publishing Company, 1963.
"Handbook of Chemistry and Physics", 52nd ed., R.C. Weast, Ed., The Chemical Rubber Co., 1971.
"Handbook of Chemistry", 9th ed., N.A. Lange, Ed., Handbook Publishers, Inc., 1956.
"Polishes", A. Davidsohn and B.M. Milwidsky, CRC Press, 1968.
"Adventures in Man's First Plastic", N.S. Knaggs, Reinhold Publishing Corp., 1947.

Table 1-91.   PROPERTIES OF WINDOW GLASS*

PROPERTIES OF CLEAR AND TINTED WINDOW SHEET AND PLATE GLASS

Applicable Federal Specification Standard DD-G-451c

Specific gravity	2.5	Tensile strength,	
Specific heat	0.21	thousands of psi	6–6.5
Hardness (Moh's)	5–6	Poisson's ratio	0.23
Softening point, deg F	1345	Coefficient of linear	
Refractive index,		expansion/deg F	49×10^{-7}
sodium D-line	1.52	Dielectric constant, 1 Mhz	7.1
Modulus of elasticity,			
millions of psi	10.0		

APPROXIMATE RADIATION TRANSMITTANCE OF SHEET AND PLATE GLASS

Type or tint	Nominal thickness, in.	Weight, lb/ft²	Transmittance	
			Total visible daylight, %	Direct 90° solar energy, %
Sheet	$\frac{1}{16}$	0.81	91	89
Sheet	$\frac{5}{64}$	1.00	91	88
Sheet	$\frac{3}{32}$	1.22	90	87
Sheet	$\frac{1}{8}$	1.64	90	86
Sheet	$\frac{3}{16}$	2.47	89	84
Sheet	$\frac{7}{32}$	2.85	89	82
Plate or float	$\frac{1}{8}$	1.64	90	86
Plate or float	$\frac{1}{4}$	3.28	88	79
Plate or float	$\frac{5}{16}$	4.09	88	77
Plate or float	$\frac{3}{8}$	4.91	87	74
Plate or float	$\frac{1}{2}$	6.55	86	70
Plate or float	$\frac{5}{8}$	8.18	85	65
Plate or float	$\frac{3}{4}$	9.83	83	60
Plate or float	$\frac{7}{8}$	11.45	81	55
Plate or float	1	13.13	79	49
Gray[a]	$\frac{1}{4}$	3.28	43	46
Bronze[a]	$\frac{1}{4}$	3.28	49	45
Green[a]	$\frac{1}{4}$	3.28	75	46
Double[b]	$\frac{1}{4}$ each	6.56	78	—

Note: Many types of glass are available, including tempered heat-strengthened glass, laminated shatter-proof glass, conductive-coated glass, reflective-coated glass. Several double-pane combinations are offered.

Direct 90° transmittance of solar ultraviolet radiation through non-tinted window glass is about 85 percent as high as the values for total solar energy transmittance. Ultraviolet transmittance of gray or bronze glass is lower.

Infrared transmittance is considerably lower than visual transmittance. This is significant in view of the large percentage of infrared radiation from most sources.

Visible reflectance of untinted glass is about 8 percent.

Approximate shading coefficients, ASHRAE, $\frac{1}{4}$-in. glass only: clear, 0.93; gray, 0.67; bronze, 0.65; green, 0.67.

Overall heat transfer coefficient of window area (air to air) is usually assumed to be 1.0 Btu/ft² hr, but it is lower if there is no wind.

For other data on shading coefficients, spectral transmittance, coated glass, special glasses, etc., see Tables 1-91, 2-19, and 7-19.

[a]Transmittance of tinted glass depends on depth of tint.
[b]Two $\frac{1}{4}$-in. panes with $\frac{1}{2}$-in. air space, sealed.

*Tables compiled from several sources.

Table 1-92. PROPERTIES OF SILICATE GLASSES*

Most of the commercially produced glass is for windows, bottles, and inexpensive containers; it is a soda-lime-silica glass of fairly uniform composition, similar to glass No. 0080 in the table below and in Table 1-93. The following tables on glasses (Tables 1-92 through 1-103) deal largely with that one-tenth of the glass output for which special properties are required. All data are subject to normal manufacturing variations.

Silica glass is inherently high in viscosity and melting point. These are reduced by fluxes such as Na_2O, K_2O, and B_2O_3. Soda and potash glasses have a high expansion coefficient (column 7), while that of fused silica is very low. Because the borosilicate glasses are intermediate, and their thermal shock resistance is high (e.g., Corning Code 7740 glass), they are widely used for laboratory and kitchen glassware. Aluminosilicate glasses are hard, heat-resisting, and of high chemical durability. Glass hardness (indentation) correlates closely with the elastic modulus (column 14). Lead oxide is also used as flux, with a result of reduced softening point and high refractive index: hence its uses for optical glass and art glass.

Sealing of glass with metal calls for close control of the coefficient of expansion (column 7 and Figure 1-99).

EXPLANATION OF COLUMNS:

Column 5:

B—blown ware	P—pressed ware	S—plate glass
M—multiform	R—rolled sheet	T—tubing and rod
U—panels	LC—large castings	

Column 6:
[2]Since weathering is determined primarily by clouding, which changes transmission, a rating for the opal glasses is omitted.
[3]These borosilicate glasses may rate differently if subjected to excessive heat treatment.

Column 8:
Normal service: No breakage from excessive thermal shock is assumed.
Extreme limits: Glass will be very vulnerable to thermal shock. Recommendations in this range are based on mechanical stability considerations only. Tests should be made before adopting final designs. These data are approximate only.

Column 9:
Based on plunging sample into cold water after oven heating. Resistance of 100°C means no breakage if heated to 110°C and plunged into water at 10°C. Tempered samples have over twice the resistance of annealed glass. These data are approximate only.

Column 10:
[4]These data are estimated.
Resistance in °C is the temperature differential between the two surfaces of a tube or a constrained plate that will cause a tensile stress of 1000 psi on the cooler surface.

Column 11:
Viscosity is given in poises. At the strain point the stresses are significantly reduced in a matter of hours, while at the annealing point there is adequate stress reduction in minutes.

Column 12:
Data show relative resistance to sandblasting.

Column 15:
Data at 25°C are extrapolated from high temperature readings and are approximate only.

---

*From: "Properties of Selected Commercial Glasses", Publication B–83, Corning Glass Works.

Table 1-92. PROPERTIES OF SILICATE GLASSES (Continued)

1	2	3	4	5	6			7		8				9		
					Corrosion Resistance			Thermal Expansion 10^{-7} in./in./°C.		Upper Working Temperatures (Mechanical Considerations Only)				Thermal Shock Res. Plates 6" × 6"		
									Room	Annealed		Tempered		Annealed		
Glass Code†	Type	Color	Principal Use	Forms Usually Available	Weathering	Water	Acid	0–300°C 32–572°F	Temp.-Setting Point	Normal Service °C.	Extreme Limit °C.	Normal Service °C.	Extreme Limit °C.	⅛" Thk. °C.	¼" Thk. °C.	½" Thk. °C.
0010	Potash Soda Lead	Clear	Lamp Tubing	T	2	2	2	93	100	110	380	—	—	65	50	35
0080	Soda Lime	Clear	Lamp Bulbs	B M T	3	2	2	92	103	110	460	220	250	65	50	35
0120	Potash Soda Lead	Clear	Lamp Tubing	T M	2	2	2	89	98	110	380	—	—	65	50	35
1720	Aluminosilicate	Clear	Ignition Tube	B T	1	1	3	42	52	200	650	400	450	135	115	75
1723	Aluminosilicate	Clear	Electron Tube	B T	1	1	3	46	54	200	650	400	450	125	100	70
1990	Potash Soda Lead	Clear	Iron Sealing	—	3	3	4	124	136	100	310	—	—	45	35	25
2405	Borosilicate	Red	General	B P U	—	—	—	43	51	200	480	—	—	135	115	75
2475	Soda Zinc	Red	Neon Signs	T	3	2	2	93	—	110	440	—	—	65	50	35
3320	Borosilicate	Canary	Tungsten Sealing	—	³1	³1	³2	40	43	200	480	—	—	145	110	80
6720	Soda Zinc	Opal	General	P	²–	1	2	80	92	110	480	220	275	70	60	40
6750	Soda Barium	Opal	Lighting Ware	B P R	²–	2	2	88	—	110	420	220	220	65	50	35
6810	Soda Zinc	Opal	Lighting Ware	B P R	²–	1	2	69	—	120	470	240	270	85	70	45
7040	Borosilicate	Clear	Kovar Sealing	B T	³3	³3	³4	48	54	200	430	—	—	—	—	—
7050	Borosilicate	Clear	Series Sealing	T	³3	³3	³4	46	51	200	440	235	235	125	100	70
7052	Borosilicate	Clear	Kovar Sealing	B M P T	³2	³2	³4	46	53	200	420	210	210	125	100	70
7056	Borosilicate	Clear	Kovar Sealing	B T P	2	2	4	51	57	200	460	—	—	—	—	—
7070	Borosilicate	Clear	Low Loss Electrical	B M P T	³2	³2	³2	32	39	230	430	230	230	180	150	100
7250	Borosilicate	Clear	Seal Beam Lamps	P	³1	³2	³2	36	38	230	460	260	260	160	130	90
7570	High Lead	Clear	Solder Sealing	—	1	1	4	84	92	100	300	—	—	—	—	—
7720	Borosilicate	Clear	Tungsten Sealing	B P T	³2	³2	³2	36	43	230	460	260	260	160	130	90
7740	Borosilicate	Clear	General	B P S T U	³1	³1	³1	33	35	230	490	260	290	180	150	100
7760	Borosilicate	Clear	General	B P	2	2	2	34	37	230	450	250	250	160	130	90
7900¹	96% Silica	Clear	High Temp.	B P T U M	1	1	1	8	7	800	1100	—	—	1250	1000	750
7913	96% Silica	Clear	High Temp.	B P R S T	1	1	1	8	7	900	1200	—	—	—	—	—
7940	Fused Silica	Clear	Ultrasonic	U	1	1	1	5.5	7	900	1100	—	—	1250	1000	750
8160	Potash Soda Lead	Clear	Electron Tubes	P T	2	2	3	91	100	110	380	—	—	65	50	35
8161	Potash Lead	Clear	Electron Tubes	P T	2	1	4	90	97	110	390	—	—	—	—	—
8363	High Lead	Clear	Radiation Shielding	L C	3	1	4	104	112	100	200	—	—	—	—	—
8871	Potash Lead	Clear	Capacitors	—	2	1	4	102	113	125	300	—	—	55	45	35
9010	Potash Soda Barium	Grey	TV Bulbs	P	2	2	2	89	102	110	380	—	—	—	—	—
9700	Borosilicate	Clear	u v Transmission	T U	³1	³1	³2	39	39	220	500	—	—	150	120	80
9741	Borosilicate	Clear	u v Transmission	B U T	³3	³3	³4	39	49	200	390	—	—	150	120	80

†Corning Glass Works code numbers are used in Tables 1-92 and 1-93.

Table 1-92. PROPERTIES OF SILICATE GLASSES (Continued)

10	11				12	13	14			15			16			17	18
Thermal Stress Resistance °C.	Viscosity Data†				Impact Abrasion Resistance	Density grams per C.C.	Young's Modulus		Poisson's Ratio	Log₁₀ of Volume Resistivity			Dielectric Properties at 1 Mc and 20°C			Refractive Index Sod. D Line (.5893 Microns)	Glass Code
	Strain Point °C.	Anneal-ing Point °C.	Softening Point °C.	Working Point °C.			$(10^6 lb/sq. in)$	$(10^6 kg/cm^2)$		25°C. 77°F	250°C. 482°F	350°C. 662°F	Power Factor	Dielectric Const.	Loss Factor		
19	395	435	625	985	0.8	2.86	8.9	0.63	.21	17.+	8.9	7.0	.16%	6.7	1.%	1.539	0010
17	470	510	695	1005	1.2	2.47	10.0	0.70	.24	12.4	6.4	5.1	.9	7.2	6.5	1.512	0080
20	395	435	630	980	0.8	3.05	8.6	0.60	.22	17.+	10.1	8.0	.12	6.7	.8	1.560	0120
28	670	715	915	1190	2.0	2.52	12.7	0.89	0.25	—	11.4	9.5	.38	7.2	2.7	1.530	1720
25	670	710	910	1175	2.0	2.64	12.5	0.88	0.25	—	13.5	11.3	.16%	6.3	1.0%	1.547	1723
14	330	360	500	755	—	3.47	8.4	0.59	.25	—	10.1	7.7	.04	8.3	.33	—	1990
⁴37	500	530	770	1085	—	2.50	9.9	0.70	0.21	—	—	—	—	—	—	1.507	2405
⁴17	440	480	690	1040	—	2.59	10.0	0.70	—	—	7.8	6.2	—	—	—	1.511	2475
⁴40	500	540	780	1155	—	2.27	9.4	0.66	0.19	—	8.6	7.1	.30	4.9	1.5	1.481	3320
19	510	550	775	1010	—	2.58	10.2	0.72	.21	—	—	—	—	—	—	1.507	6720
⁴18	445	485	670	1040	—	2.59										1.513	6750
⁴23	490	530	770	1010	—	2.65				—	—	—	—	—	—	1.508	6810
37	450	490	700	1080	—	2.24	8.6	0.60	.23	—	9.6	7.8	.20	4.8	1.0	1.480	7040
39	460	500	705	1025	—	2.24	8.7	0.61	.22	16.	8.8	7.2	.33	4.9	1.6	1.479	7050
41	435	480	710	1115	—	2.28	8.2	0.58	.22	17.	9.2	7.4	.26	4.9	1.3	1.484	7052
34	470	510	720	1045	—	2.29	9.2	0.65	.21	—	10.2	8.3	.27	5.7	1.5	1.487	7056
66	455	495	—	1070	4.1	2.13	7.4	0.52	.22	17.+	11.2	9.1	.06	4.1	.25	1.469	7070
48	490	540	780	1190	3.2	2.24	9.2	0.65	.20	15.	8.2	6.7	.27	4.7	1.3	1.475	7250
21	340	365	440	560	—	5.42	8.0	0.56	.28	—	10.6	8.7	.22	15.	3.3	—	7570
49	485	525	755	1140	3.2	2.35	9.1	0.64	.20	16.	8.8	7.2	.27	4.7	1.3	1.487	7720
53	515	565	820	1245	3.1	2.23	9.1	0.64	.20	15.	8.1	6.6	.50	4.6	2.6	1.474	7740
52	480	525	780	1210	—	2.23	9.1	0.64		17.	9.4	7.7	.18	4.5	.79	1.473	7760
202	820	910	1500	—	3.5	2.18	10.0	0.70	.19	17.	9.7	8.1	.05	3.8	.19	1.458	7900¹
211	820	910	1500	—	3.5	2.18	9.6	0.67	.19	—	9.7	8.1	.04	3.8	0.15	1.458	7913
290	990	1050	1580	—	3.6	2.20	10.5	0.74	.16	—	11.8	10.2	.001	3.8	.0038	1.459	7940
⁴18	395	435	630	975	—	2.98	—	—	—	—	10.6	8.4	.09	7.0	.63	1.553	8160
22	400	435	600	860	—	4.00	7.8	0.55	.24	—	12.0	9.9	.06	8.3	0.50	1.659	8161
19	300	315	380	460	—	6.22	7.4	0.52	.27	—	9.2	7.5	.19	17.0	3.2	1.97	8363
17	350	385	525	785	—	3.84	8.4	0.59	.26	—	11.1	8.8	.05	8.4	.42	—	8871
18	405	445	650	1010	—	2.64	9.8	0.69	.21	—	8.9	7.0	.17	6.3	1.1	1.507	9010
45	520	565	805	1200	—	2.26	9.6	0.67	.20	15.	8.0	6.5	—	—	—	1.478	9700
55	410	450	705	—	—	2.16	7.2	0.51	.23	17.+	9.4	7.6	—	—	—	1.468	9741

†Viscosities at these four temperatures are approximately as follows: $10^{14.5}$ poises at the strain point, 10^{13} poises at the annealing point, $10^{7.8}$ poises at the softening point, at 10^4 poises at the working point.

Table 1-93. COMPOSITIONS OF SILICATE GLASSES*

Approximate Percentages by Weight

Glass No. (See Table 1-92)	SiO_2, silica	Na_2O, soda	K_2O, potash	PbO, lead	CaO, lime	B_2O_3, boric oxide	Al_2O_3, aluminum oxide	Other
0010	63	7	7	22	—	—	1	—
0080	73	17	—	—	5	—	1	4% MgO
0120	56	4	9	29	—	—	2	—
1720	62	1	—	—	8	5	17	7% MgO
1723	57	—	—	—	10	5	15	6% BaO, 7 MgO
1990	41	5	12	40	—	—	—	2% Li_2O
2405	70	5	—	—	—	12	1	11% ZnO + CdS, Se
2475	67	10	7	—	—	—	—	12% ZnO, 2% CdO + F^-
3320	76	4	2	—	—	14	3	1% U_3O_8
6720	60	9	2	—	5	1	10	9% ZnO + 4% F^-
6750	61	15	—	—	—	1	11	9% BaO + 3% F^-
6810	56	7	1	3	4	1	10	12% ZnO + 6% F^-
7040	67	4	3	—	—	23	3	—
7050	67	7	—	—	—	24	2	—
7052	65	2	3	—	—	18	7	3% BaO + F^-, 1% Li_2O
7056	70	1	8	—	—	17	3	1% Li_2O
7070	71	0.5	1	—	—	26	1	0.5% Li_2O
7250	78	5	—	—	—	15	2	—
7570	3	—	—	75	—	11	11	—
7720	73	4	—	6	—	15	2	—
7740	81	4	—	—*	—	13	2	—
7760	79	2	2	—	—	15	2	—
7900	96	—	—	—	—	3	0.3	—
7913	96.5	—	—	—	—	3	0.5	—
7940	99.9	—	—	—	—	—	—	0.1% H_2O
8160	56	3	10	23	1	—	2	5% BaO + F^-
8161	40	—	5	51	—	—	—	2% BaO + 2% Rb_2O
8363	5	—	—	82	—	10	3	—
8871	42	2	6	49	—	—	—	1% Li_2O
9010	67	7	7	2	—	—	4	12% BaO + Co_3O_4 + NiO + F^-, 1% Li_2O
9606	56	—	—	—	—	—	20	9% TiO_2, 15% MgO
9700	80	5	—	—	—	13	2	—
9741	66	2	—	—	—	24	6	1% F^-, 1% Li_2O

*From: "Glass", J.R. Hutchins III and R.V. Harrington, *Kirk-Othmer Encyclopedia of Chemical Technology*, Vol. 10, p. 542. Copyright © 1966 by John Wiley & Sons, Inc. Reprinted by permission.

Table 1-94. COMPOSITIONS OF SOME NON-SILICATE GLASSES*

Approximate Percentages by Weight

	Glass No.						
	1	*2*	*3*	*4*	*5*	*6*	*7*
Name:	Lindeman	Morey (24)†	Jena 3061 III	Morey (25)†	Stanworth (26)†		Blau (27)†
Use:	X-ray transmitting	Optical glass	Sodium-vapor lamp tubing	Borate crown	Heat absorbing	HF resistant	Infrared transmission
	B_2O_3 83.3 Li_2O 14.2 BeO 2.4	GeO_2 78.5 B_2O_3 7.9 K_2O 5.7 Na_2O 5.6 BaO 2.1	BaO 33 Al_2O_3 25.0 B_2O_3 22.5 CaO 18.0 K_2O 1.5	B_2O_3 63.8 Al_2O_3 18.0 Na_2O 8.0 K_2O 3.5 BaO 3.5 PbO 3.0	P_2O_5 70.7 Al_2O_3 10.0 K_2O 10.0 B_2O_3 4.0 MgO 4.0 FeO 2.5 ZnO 1.0	P_2O_5 72 Al_2O_3 18 ZnO 10	PbO 78 GeO_2 18 Al_2O_3 4

Notes:
Rare-earth borate glasses have a high refractive index and low dispersion.
Phosphate glasses are resistant to hydrofluoric acid and have a low index of refraction and low dispersion.
The phosphate glass with FeO absorbs strongly in the infrared and shows almost no visible color. See Tables 2-19 and 2-20 for infrared and ultraviolet windows.
Silicate glasses discolor in sodium vapor lamps.

†See numbered references in original source.
*From "Glass", J.R. Hutchins III and R.V. Harrington, *Kirk-Othmer Encyclopedia of Chemical Technology*, Vol. 10, p. 544. Copyright © 1966 by John Wiley & Sons, Inc. Reprinted by permission.

Table 1-95.   SPECIFIC GLASS PROPERTIES
AND COMMERCIAL USES*

ABBREVIATIONS:

AO—American Optical Company
B&L—Bausch & Lomb, Inc.
Beckman—Beckman Instruments Inc.
Bell Labs—Bell Telephone Laboratories
CGW—Corning Glass Works
EK—Eastman Kodak Company

3M—Minnesota Mining & Manufacturing Company
Mosaic—Mosaic Fabrications Inc.
Narumi—Narumi, Japan
OC—Owens-Corning Fiberglas Corporation
OI—Owens-Illinois Glass Company
PPG—Pittsburgh Plate Glass Company

Strength

1. Flexible automobile window (CGW, PPG).
2. Glass-ceramic tableware (three times the strength of china) (CGW).
3. Filament-wound fiber glass-reinforced plastic (OC).
4. Hollow glass fibers for better strength-to-weight ratio (PPG).
5. High-modulus S-glass fibers (OC).
6. Chemically strengthened pipettes and centrifuge tubes (CGW).
7. Deep-submergence vehicles for instrument housing and flotation (to 20,000 ft). Benthos is designed and built for possible use at depths of 37,000 ft (CGW).

Formability

8. Glass beads, for oil-well propping (CGW).
9. Chemically machined fluid amplifiers (CGW).
10. Beta yarn—glass fiber 0.00012 in. diameter produces a supple, soft fabric; can be knitted; resistent to abrasion (OC).
11. Multilead, 1-mil wires on 4-mil centers in glass for Videograph that copies information at rates of 20,000 characters/second (CGW).
12. 62-in. fused-silica telescope mirror blank with minimum thermal distortion due to low expansion of silica (CGW).
13. 36-in. light-weight (post and plate construction) fused silica telescope mirror blank (CGW).
14. Short radius 180° curvature in high temperature windows (CGW).
15. Fiber optics (light pipes) (Mosaic, AO, CGW).
16. Photosensitive glass molds to produce rubber plates for the flexographic printing industry (CGW).

Electrical

17. Low-resistivity glass target for image orthicon tubes in television camera (CGW).
18. Substrate for high-reliability film resistors (Minuteman resistor load life test gave no failures in 23 million part hours) (CGW).
19. Glass electrodes specific for the alkali and alkaline earth metals (Beckman, CGW).
20. Conductive roving for ignition cables (OC).
21. High-reliability capacitors (CGW).

Thermal

22. Glass-ceramic "zero expansion" strong cookware (freezer to oven to table) (CGW, Narumi) and heating surfaces (CGW).
23. Ceramic honeycomb heat exchangers for gas turbines (CGW).
24. Ceramic honeycomb burners (CGW).
25. Fluffed glass insulation, prepared on job by expanding through compressed air nozzle (PPG).
26. Improved solar-shielding windows (PPG).
27. Space-vehicle windows (CGW).
28. Frostfree windshields and windows (PPG, CGW).
29. High-temperature filters (to 1100°C) (CGW).
30. Solder-sealing glass (CGW, OI).

Table 1-95. SPECIFIC GLASS PROPERTIES
AND COMMERCIAL USES *(Continued)*

Radiation

31. Glass lasers (AO, EK, CGW).
32. Fiber optics (AO, Moasic, CGW).
33. Glass-ceramic microwave transmitting radomes (CGW).
34. Dosimeter glasses for x-ray (CGW, B&L).
35. Short ultrasonic delay line glasses (CGW).
36. IR domes (infrared transmitting glasses) (CGW, EK).
37. Glass free from radioactivity for photomultipliers (CGW).
38. Breakdown-proof radiation windows (CGW).
39. Laser beam reflectors (CGW).
40. Photochromic glass (CGW).
41. High-index (up to 2.7) oxide-based glass beads; added to road paints to increase reflectivity (3M).

Chemical

42. Transistor and solid-state circuit encapsulation for protection from environment (Bell Labs, CGW).
43. Porous glass getters for use in tube envelopes (CGW).
44. Impact-resistant, plastic-clad glass pipe (CGW, OI).

Economic

45. Throw-away beverage containers (OI).

*From: "Glass", J.R. Hutchins III and R.V. Harrington, *Kirk-Othmer Encyclopedia of Chemical Technology*, Vol. 10, p. 600. Copyright © 1966 by John Wiley & Sons, Inc. Reprinted by permission.

Table 1-96.   ELECTRICAL PROPERTIES OF VARIOUS
KINDS OF GLASS*

Values are for room temperature. In general the volume resistivity is reduced at the higher temperatures, but the dissipation factor increases rapidly above 100–200°C.

Type of glass	Volume resistivity, ohm-cm	Dielectric constant, 1 Mhz	Dissipation factor, 1 Mhz
Fused silica	10^{12}	3.8	0.0002
96% silica (7900, 7910–11–12)†	10^{10}	3.8	0.0005
Soda lime			
General-purpose	10^6–10^7	7.0–7.6	0.004–0.011
Lamp bulb (0080)	10^7	7.2	0.009
Lead alkali silicate			
Electrical (0010)	10^9	6.6	0.0016
High lead (8870)	10^{12}	9.5	0.0009
Alumino borosilicate			
(Kimble N51a)	10^7	5.6	0.010
Borosilicate			
Low expansion (7740)	10^8	4.6	0.0046
Low electrical loss (7070)	10^{11}	4.0	0.0006
Tungsten sealing (7050)	10^9	4.9	0.0033
Aluminosilicate (1710–20)	10^{11}	6.3	0.0037

†Numbers in parentheses indicate equivalent Corning glass code numbers.
*From: "Electrical Insulating Materials", *Machine Design*, 39:161, Sept. 28, 1967.

Table 1-97. THERMAL PROPERTIES OF SEVERAL GLASS PRODUCTS*

For specific heat in J/kg·K, multiply values by 4184. For thermal conductivity in W/m·K, multiply values in cal/cm·sec·deg C by 418.4.

Material	Specific heat			Thermal conductivity, cal/cm sec °C × 10⁴			
	$25°C$	$500°C$	$1000°C$	$-100°C$	$0°C$	$100°C$	$400°C$
Fused silica	0.173	0.268	0.292	25.0	31.5	35.4	
7900	0.18	0.24	0.29	24	30	34	
7740	0.17	0.28		21	26	30	
1723	0.18	0.26			29	33	
0311 (chemically strengthened)	0.21	0.28			27	29	35
Soda-lime window glass	0.190	0.300	0.333	19	24	27	
Heavy flint, 80% PbO, 20% SiO_2				10	12	14	
Foamglass insulation	0.20			(0.97)	1.3	1.73	(2.81)
Fibrous glass					(0.8)		
9606 glass-ceramic	0.185	0.267	0.311		90	86	75
9608 low-expansion glass-ceramic	0.195	0.286			48	51	55

Notes:

Parentheses indicate extrapolated values.

Specific heat increases with temperature and approaches zero at 0°K. There are no critical temperatures or phase changes. Thermal conductivity increases with temperature and is very high for glass ceramics.

*From: "Glass", J.R. Hutchins III and R.V. Harrington, *Kirk-Othmer Encyclopedia of Chemical Technology*, Vol. 10, p. 598. Copyright © 1966 by John Wiley & Sons, Inc. Reprinted by permission.

Table 1-98. SOME COMMERCIAL GLASS-CERAMICS*

Commercial identification	Crystal phases	Properties	Application
Corning 8603	$Li_2O—2SiO_2$, SiO_2	Photochemically machineable	Fotoceram printed-circuit boards
Corning 9606	$2MgO·2Al_2O_3·5SiO_2$, SiO_2, TiO_2	Low-expansion, transparent to radar	Pyroceram brand radomes for ballistic missiles
Corning 9608	β-spodumene ss,[a] TiO_2	Low expansion, good chemical durability	Corning Ware cooking utensils
Neoceram (Japan)	β-spodumene ss[a]	Low expansion	Cookmaster cooking ware
Corning 0303	$Na_2O·Al_2O_3·2SiO_2$, $BaO·Al_2O_3·2SiO_2$	High strength	Centura tableware and Pyroceram brand dinnerware
Owens-Illinois Cervit	β-quartz ss[a]	Zero expansion	Telescope mirror blanks
Anchor Hocking Cookware	β-quartz ss[a]	Low expansion	Cooking ware

[a] ss = solid solution.

*From: *Ind. Eng. Chem.*, 50:3, 1966. Copyright 1966 by the American Chemical Society. Reprinted by permission of the copyright owner.

REFERENCE

"Glass-Ceramics", P.W. McMillan, Academic Press, Inc., 1964.

Figure 1-99. THERMAL EXPANSION AND SEALING OF GLASS*

For other properties of glass, see Table 1-92.

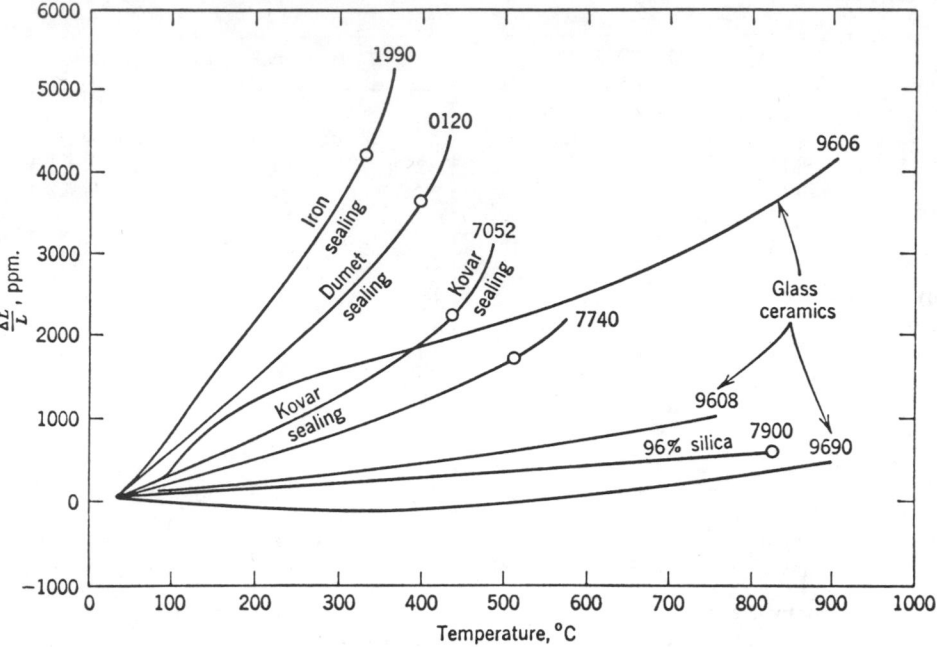

Open circles show the "setting point", below which strain cannot be effectively removed when cooling a seal. Here it is defined as the strain point plus 5°C.

*Courtesy Corning Glass Works.

Figure 1-100. SPECTRAL TRANSMISSION FOR SOME COMMERCIAL GLASSES*

See also Table 2-19.

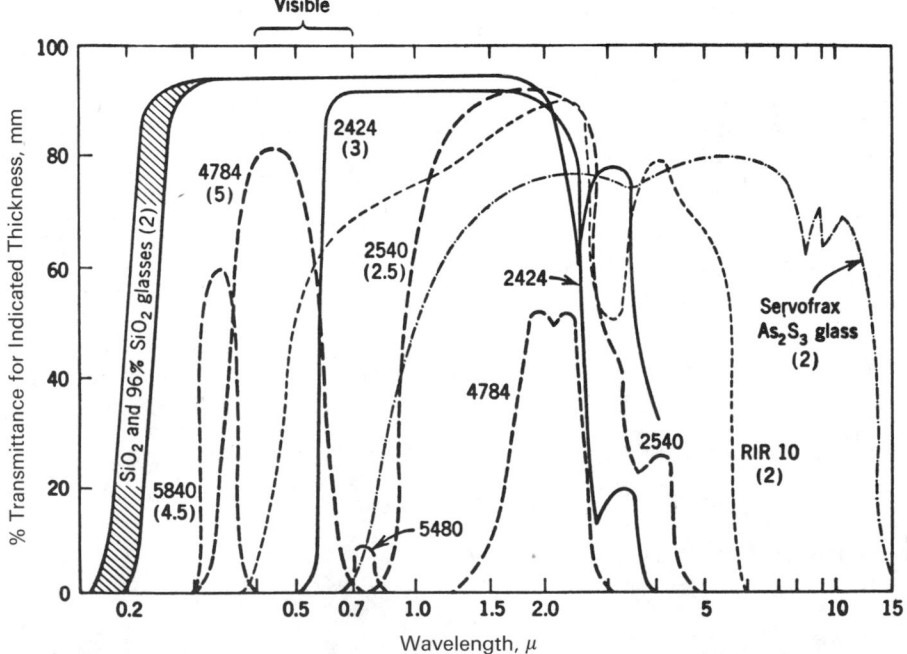

*From: "Glass", J.R. Hutchins III and R.V. Harrington, *Kirk-Othmer Encyclopedia of Chemical Technology*, Vol. 10, p. 595. Copyright © 1966 by John Wiley & Sons, Inc. Reprinted by permission.

Table 1-101. PURPOSE OF GLASS ADDITIVES*
Minor Constituents, Usually Less Than Five Percent

Purpose	Additives
Red color	Colloidal Au or Cu, CdSe (S, Te)
Pink color	MnO_2–CeO_2, Se^{2-}
Orange color	CdS (Se)
Amber color	FeS_x, Fe_2O_3–TiO_2
Yellow color	UO_2, CeO_2–TiO_2, CdS
Green color	Cr_2O_3, Fe_2O_3, CuO, U_2O_3
Blue color	CoO, FeO, CuO
Violet color	NiO, Mn_2O_3
Gray color	Co_3O_4–NiO
Black color	Mn_2O_3–Cr_2O_3, PbS, FeS, $CoSe_x$
Ultraviolet absorption	CeO_2, TiO_2, Fe_2O_3, V_2O_5, CrO_3
Infrared absorption	FeO, CuO
Decolorization, i.e., mask Fe_2O_3 color	MnO, Se^{2-}, NiO, Co_3O_4
Opacification, i.e., white opals	CaF_2, NaF, ZnS, $Ca_2(PO_4)_3$
Photosensitivity	Cu^+, Ce^{3+}, Au^+
Fluorescence	Rare earth, Mn, U, Cu, Tl, Sn, Pb, and V ions
Protection against radiation discoloration	CeO_2
Fining	SO_4^{2-}, F^-, Cl^-, Br^-, I^-, Sb_2O_3, As_2O_3

Note: There are also unintentional impurities that result in undesirable properties. These are often introduced through the use of cheaper raw materials. Yellow-green (observed through an edge) results from iron and chromium oxides. Darkening by sunlight results from cerium-arsenic constituents.

*Courtesy Corning Glass Works.

Table 1-102. REFRACTIVE INDEX FOR TYPICAL OPTICAL GLASSES*
At the Four Colors Represented by the Fraunhofer Lines C, D, F, and G′

The basic index of refraction for ray tracing and specification of focal lengths is the sodium D-line, n_D. This is close to the peak of the visual sensory brightness curve. For *achromatization* the index n_C at the red end and n_F or $n_{G'}$ at the blue end are usually used.

Medium	Symbol	I.C.T. type	v	n_C	n_D	n_F	$n_{G'}$
Borosilicate crown	BSC	500/665	66.5	1.49776	1.50000	1.50529	1.50937
Borosilicate crown	BSC-2	517/645	64.5	1.51462	1.51700	1.52264	1.52708
Spectacle crown	SPC-1	523/586	58.8	1.52042	1.52300	1.52933	1.53435
Light barium crown	LBC-1	541/599	59.7	1.53828	1.54100	1.54735	1.55249
Telescope flint	TF	530/516	51.6	1.52762	1.53050	1.53790	1.54379
Dense barium flint	DBF	670/475	47.5	1.66650	1.67050	1.68059	1.68882
Light flint	LF	576/412	41.2	1.57208	1.57600	1.58606	1.59441
Dense flint	DF-2	617/366	36.6	1.61216	1.61700	1.62901	1.63923
Dense flint	DF-4	649/338	33.9	1.64357	1.64900	1.66270	1.67456
Extra dense flint	EDF-3	720/291	29.1	1.71303	1.72000	1.73780	1.75324
Fused quartz	SiO_2		67.9		1.4585		
Crystal quartz (O ray)	SiO_2		70.0		1.5443		
Fluorite	CaF_2		95.4		1.4338		

*From: "Fundamentals of Optics", F. A. Jenkins and H. F. White, McGraw-Hill Book Company, 1957, p. 150.

Table 1-103. IDEAL AND PRACTICAL STRENGTHS OF GLASS, GLASS FIBERS, AND GLASS-CERAMICS*

The strength of individual samples of the same glass shows a large variation. Failure usually occurs at a surface flaw, and the breaking strength is much lower after prolonged application of the load, especially in air environment. As long as glass is solid, its strength is increased by an increase in temperature. For structural purposes the strength of common glass is usually assumed to be about 2,000 psi.

For tensile strength in N/m^2, multiply value in psi by 6,895.

Type of glass	Tensile strength, psi	Strength/weight ratio, psi/lb per cu in.
UNTREATED GLASS		
Theoretical strength	1,000,000–4,000,000	45,000,000
Fibers, protected in vacuum	Up to 2,000,000	—
Fibers, in air, commercially available	250,000 average	6,000,000 average
Fibers, effective strength in plastic	150,000 average	1,800,000–4,500,000
Bulk glass, protected in vacuum	Up to 500,000	—
Blown ware, unabraded	Up to 100,000	90,000
Pressed ware, unabraded	8,000 average	55,000
Bulk glass, abraded	4,000–8,000 average	30,000
Bulk glass, abraded, 1,000-hr stress	2,000 minimum	—
Bulk glass, abraded, design strength	500–1,500	—
TEMPERED GLASS		
Bulk glass, abraded	15,000–35,000	—
Normal design strength	1,500–6,000	—
CHEMICALLY STRENGTHENED GLASS		
Bulk glass, abraded	100,000 and more	—
GLASS-CERAMICS		
Bulk material, unabraded	20,000–35,000	27,000
Bulk material, abraded	10,000–24,000	—
Design strength	3,000–6,000	—
CHEMICALLY STRENGTHENED GLASS-CERAMICS		
Bulk material, abraded	200,000 and more	—

*Compiled from several sources.

Table 1-104.   CLASSIFICATION OF NATURAL FIBERS*

Name	Source	Principal producers	Approximate production, $10^6 lb/yr$	Typical uses
ANIMAL ORIGIN				
Wool (pure, dry)	Sheep	Australia, Argentina	3 000	Warm clothing, carpet, felt
Silk	Silkworm cater-pillar	Japan, China	70	Fine fabrics
Cashmere	Goat	India, Tibet	70	Quality clothes
Mohair	Goat	U.S.A., Turkey	30	Upholstery, linings, suitings, rugs
Camel hair	Camel	China, Mongolia	2	Overcoats, knits
VEGETABLE ORIGIN				
Cotton	Seed	U.S.A., U.S.S.R.	25 000	Almost all textile uses
Jute	Bast	India, Pakistan	4 500	Sacking, bale wrapping, curtains, bags, oakum
Sisal	Leaf	Mexico, Brazil	1 300	Rope, twine, rugs
Flax	Bast	U.S.S.R., Poland	1 200	Strong fabrics, paper
Kenaf	Bast	India, Pakistan	1 100	Rope, twine, carpet, canvas, bags
Hemp	Bast	U.S.S.R., Yugoslavia	500	Rope, sacking, canvas
Henequen	Leaf	Mexico, Cuba	300	Rope, twine, canvas, bags
Abaca (Manila)	Leaf	Phillipines, Guatemala	200	Rope, marine cable
Sunn	Bast	India, Pakistan	150	Rope, twine, carpet, paper
Ramie	Bast	China, Japan	25	Canvas
MINERAL ORIGIN				
Asbestos	Ore	Canada, U.S.S.R.	2 000	Brakes and clutches, building materials, packings, fire-proofing
Glass[a]	Sand	U.S.A.	—	Composites, insulation, draperies, tire cord, filters
Aluminum silicate[a]	Ore	U.S.A.	—	Packings and insulation for high temperatures

[a] Here classified as natural fibers for convenience, although they are man-made by processing.

*Compiled from several sources.

REFERENCE

"Matthews' Textile Fibers", 6th ed., H.R. Mauersberger, Ed., John Wiley & Sons, Inc., 1954.

Table 1-105. PROPERTIES OF NATURAL FIBERS*

Because there are great variations within a given fiber class, average properties may be misleading. The following typical values are only a rough comparative guide.

Name	Specific gravity	Tenacity, g/denier	Tensile strength, 10^3 psi	Elongation at break (dry), %	Standard regain, % of dry[b]	Fiber diameter, microns	Fiber length, in.	Fiber shape and kind	Resistant to
ANIMAL ORIGIN									
Wool	1.32	1.0–1.7	17–29	23–35	15–18	17–40	1.5–5	Oval, crimped, scales	Age, weak acids, solvents
Silk	1.25	3.5–5	90	20–25	10	10–13		Flexible, soft, smooth	Heat, solvents, weak acids, wear
Cashmere						15–16	1–4	Round, scales, soft	
Mohair	1.32	1.2–1.5		30	13	24–50	6–12	Round, silky	Wear, age, solvents, weak acids
Camel hair	1.32	1.8		40	13	10–40	1–6	Oval, striated	Age, solvents
VEGETABLE ORIGIN									
Cotton	1.54	2–5	30–120	5–11	7.5–8.5	10–20	0.5–2	Flat, convoluted, ribbon	Age, heat, washing, wear, solvents, alkalies, insects
Jute (bast)	1.5		50	1–1.5	14	15–20		Woody, rough, polygon	
Sisal (leaf)	1.49	2.2	75	2–2.5	13	10–30	Strand 30–40	Stiff, straight	
Flax (bast)	1.52	4–7		2–3	12	15–18	Strand 40–50	Soft, fine	Age, solvents, washing, insects, weak acids, and alkalies
Kenaf (bast)			45			15–30		Polygon or oval	
Hemp (bast)	1.48			2		18–25	Strand 30–70	Polygon or oval, irregular	
Henequen (leaf)			60				Strand 30–60	Finer than sisal	
Abaca (leaf) (Manila)	1.48	2.3–2.9	100	2–3	13		Strand 30–120		
MINERAL ORIGIN									
Asbestos	2.5		40–200			Various	0.5–10	Smooth, straight	Heat to 400 deg C, acids, chemicals, organisms
Glass[a]	2.5	7–12	200–500	3–4.5	0	Various		Circular, smooth	Chemicals, insects
Silicate[a] (Ca, Al, Mg)	2.85				0				Heat to 900 deg C, most chemicals, insects, rot

Note: Wide variations may be expected, especially for different grades of cotton. Wet strength is lower (for rayon, very much lower), but it depends on the duration of soaking. The strength of yarn is only a fraction of the cumulative strength of all individual fibers.

Most fibers exhibit relaxation of stress at constant strain and also increase in elongation at constant load (creep). The stress-strain curve is greatly affected by the rate of extension. When the stress is removed, there is a quick elastic recovery, a delayed recovery, and a permanent set. Hence the elastic behavior of any fiber depends on its stress-strain history. The elastic recoveries of nylon and wool are high; those of cotton, flax, and rayon are much lower.

The heat capacity (specific heat) of most fibers is about one-third that of water.

Other fibers: Fur hair is slightly coarser than silk fibers. Camel and llama hairs are almost as coarse as wool but only about one-third the size of human hair. Horse hair is over 100 microns; hog bristles, over 200 microns. Jute, sisal, and hemp are intermediate between cotton and wool. These are rough average sizes, and many natural fibers range 50% above or below such averages.

[a] Here classified as natural fibers for convenience, although they are man-made by processing.
[b] Expected equilibrium moisture regain of dry fiber, in percent of dry weight, when exposed in air at 70 deg F, 65% relative humidity.

*Compiled from several sources.

Table 1-106. CLASSIFICATION OF MAN-MADE FIBERS AND FABRICS

KEY TO U.S. MANUFACTURERS:

A—American Viscose Corp.
B—Beaunit Fibers, Div. of Beaunit Corp.
C—Celanese Fibers Co., Div. of Celanese Corp.
E—Tennessee Eastman Co., Div. of Eastman Kodak Co.
F—Fiber Industries, Inc.

G—W.R. Grace & Co., Dawbarn Div.
H—Hercules Powder Co.
K—American Enka Corp.
M—Monsanto Co., Textiles Div. (Chemstrand)
N—Vectra Co., Div. of National Plastics Products Co., Inc.

P—E.I. du Pont de Nemours & Co., Inc.
S—Firestone Synthetic Fibers Co.
U—Uniroyal, Inc., Textile Div.
UC—Union Carbide Corp.
V—FMC Corp., American Viscose Div.

Chemical class; common name (sources)	Typical proprietary names (and manufacturer)	Resistant to	Damaged by[a]	Typical uses
CELLULOSE FIBERS (NATURAL)				
Acetate	Esteron, etc. (E) Celacloud, etc. (C)	Petroleum chemicals, dilute acids, weak alkalies, mildew, moths	Oxidizers, many solvents, strong alkalies, heat (above 140 deg C)	Clothing, satins, drapes, linings, knits
Triacetate	Arnel (C)	Petroleum solvents, bleaches, insects, mildew, heat	Strong acids, strong alkalies, most solvents	Pleated garments, knits, drip-dry wear, table covers
Viscose rayon	Enka, etc. (K) Avisco, etc. (A)	Dilute alkalies, insects, solvents	Acids, strong alkalies, heat (above 150 deg C), moisture, mildew	Clothing, carpets, curtains, upholstery, linings
High-tenacity viscose	Tenasco (V)	Moisture, solvents	Strong acids, strong alkalies, heat (above 150 deg C)	Tire cord, belting, hose
Polynosic viscose	Avril (A)	Moisture, solvents, insects	Strong acids, strong alkalies, heat (above 150 deg C)	Dress fabrics, knits, drapes
Cuprammonium rayon (cupro)	Bemberg, etc. (B)	Bleaches, weak alkalies	Strong oxidizers, mildews, some insects, strong acids, heat (above 230 deg C)	Sheer fabrics, drapes, upholstery, satins
PROTEIN FIBERS (NATURAL)				
Animal: casein (milk)	Now little used	Solvents	Strong acids, alkalies, mildew, heat (above 100 deg C)	Largely for blending with wool, cotton, rayon, etc.
Vegetable—seed: soybeans, peanuts, corn	Now little used	Acids, solvents, moths, mildew	Alkalies, heat above 150 deg C	Blends with wool, cotton, etc.
Vegetable—latex: rubber (vulcanized)	Lastex (U)	Moisture, insects	Heat (above 110 deg C), oxidation, ozone, oils, hydrocarbons, fats, solvents	Corsetry, swimwear, footwear, supports

Table 1-106.   CLASSIFICATION OF MAN-MADE FIBERS AND FABRICS (Continued)

Chemical class; common name (sources)	Typical proprietary names (and manufacturer)	Resistant to	Damaged by[a]	Typical uses
SYNTHETIC FIBERS				
Polyacrylonitrile (acrylic)	Orlon (P) Acrilan (M) Creslan (AC) Cantrece (P) Verel-copolymer (E) Dynel-copolymer (P)	Dilute acids and alkalies, solvents, insects, mildew, weather	Strong alkalies and acids, heat (above 180 deg C), acetone, ketones	Outdoor fabrics, carpets, knits, fur-like fabrics, blankets
Polyamide	Nylon[b] (P) 6, 6.6, etc. Chemstrand nylon (M)	Alkalies, molds, solvents, moths	Strong acids, phenol, bleaches, heat (above 170 deg C)	Tire cord, carpet, upholstery, apparel, belting, hose, tents
Polyester	Dacron (P) Kodel (E) Fortrel (F)	Weak acids and alkalies, solvents, oils, mildew, moths	Phenol; heat (above 200 deg C)	Apparel, curtains, rope, twine, sailcloth, belting, fiberfill
Polyethylene (olefin, low density)	DLP (G)	Alkalies, acids (except nitric), insects, mildew	Oil and grease; heat (above 90 deg C) oxidizers	Outdoor fabrics; filter fabrics; decorative coverings
Polyethylene (olefin, high density)	DLP (G)	Alkalies, acids (except nitric), insects, mildew	Oil and grease; heat (above 100 deg C) oxidizers	Rope, twine, fishnets
Polypropylene (olefin)	Herculon (H) Polycrest (U)	Alkalies, acids, solvents, insects, mildew	Heat (above 110 deg C)	Rope, twine, outdoor fabrics, carpets, upholstery
Polyurethane; spandex[b]	Lycra (P) Spandelle (S)	Solvents, oils, alkalies, insects, oxidation	Heat (above 140 deg C); strong acids	Elastic garments, swimwear, hosiery, tricot fabrics, knits
Polyvinyl chloride (PVC)	Vinyon[b] HH (V) Dynel-copolymer (U)	Acids and alkalies, insects, mildew, alcohol, oils	Ethers, esters, aromatic hydrocarbons, ketones; hot acids; heat (above 70 deg C)	Non-woven materials; felts; filters; blends with other fibers
Polyvinyl alcohol (PVA)	(Foreign manufacture)	Acids, alkalies, insects, mildew, oils	Heat (above 160 deg C); phenol, cresol, formic acid	Wide range of industrial and apparel uses; rope, work clothes; fish nets
Polyvinylidene chloride	Saran[b] by Vectra (N)	Acids, most alkalies, alcohol, bleaches, insects, mildew, weather	Heat (above 90 deg C); many solvents	Outdoor fabrics; insect screen; curtains, upholstery; carpet; work clothes
Polytetrafluoroethylene (PTFE)	Teflon (P)	Almost all chemicals, solvents, insects, mildew	Heat (above 250 deg C); fluorine at high temperature	Corrosion-resistant packings, etc., tapes, filters, bearings

[a] Fabrics are often damaged by heat in ironing. The synthetics will not withstand the usual 400 deg F that is used for cotton and linen. For acetate and olefins the ironing temperature should be below 250 deg F, as for silk and wool. For most other synthetics a 300 deg F ironing temperature is recommended, but triacetate will tolerate higher temperature without damage.

[b] The names nylon, vinyon, saran, and spandex are recognized as generic terms rather than proprietary names.

Table 1-107.   PROPERTIES OF MAN-MADE FIBERS*

For additional properties of fibers, see Tables 1-105 and 1-115.

Chemical class; common name (sources)	Specific gravity	Tenacity, g/denier	Tensile strength, 10^3 psi	Elongation at break, %	Regain (standard)	Softening point, deg C	Melting point, deg C	Flammability	Brittleness temp, deg C
CELLULOSE FIBERS (NATURAL)									
Acetate	1.30	1.–1.3	18–25	20–30	6.5	140	230	Melts and burns	
Triacetate	1.32	1.2–1.4	20–28	25–30	3–4.5	225	300	Melts and burns	
Viscose rayon	1.51	2–2.6	30–46	17–25	13.		200[a]	Burns readily	
High-tenacity viscose	1.53	3–5	60–80	10–12	10		200[a]	Burns readily	< -114
Polynosic viscose	1.53	3–5	60–80	8–20	7		200[a]	Burns readily	
Cuprammonium rayon (cupro)	11.52	1.7–2.3	30–45	10–17	12.5		250[a]	Burns readily	
PROTEIN FIBERS (NATURAL)									
Animal: casein (milk)	1.3	1.0	15	60–70	14	100	150	Slow	
Vegetable—seed: soybeans, peanuts, corn	1.3	0.7–0.9	11–14	40–60	11–15	150	250	Slow	
Vegetable latex: rubber (vulcanized)	1.0	0.4–0.6	4–7	700–900	0	300		Burns	−60
SYNTHETIC FIBERS									
Polyacrylonitrile (acrylic)	1.17	2–5	50–75	25–40	2	190	260	Burns	
Polyamide (nylon)	1.14	4–9	70–120	20–40	4	200	215–250	Slow	< -100
Polyester (PET dacron)	1.38	4–8	70–120	10–50	0.4	225	250–290	Low	
Polyethylene (olefin, low density)	0.92	3–6	40–70	25–40	0.15	90–120	120	Slow	−114
Polyethylene (olefin, high density)	0.95	5–7	60–80	10–20	0.01	120–130	140	Slow	−114
Polypropylene (olefin)	0.91	4.5–8	45–80	15–30	0–0.5	145	160–170	Self-ext. low	−70
Polyurethane (spandex)	1.1	0.5–1.0	7–16	500–700	1.0	190	250	Burns	
Polyvinyl chloride (PVC)	1.38	0.7–2	12–17	100–125	0.1	70	140[a]	No; chars	< -100
Polyvinyl alcohol (PVA)	1.3	3–7	60–90	15–28	5	230	240	Slow	
Polyvinylidene chloride (saran)	1.7	2	40	20–30	0.1	115–135	170	No	
Polytetrafluoroethylene (PTFE)	2.1	1.2–1.4	33	15–30	0	225	300[a]	No	

Note: Mechanical properties are for room temperature and humidity and based on unstressed cross section.

[a]Decomposition; does not melt.

*Compiled from several sources.

Table 1-108. FIBERS FOR SPECIAL USES*

Desired characteristics	Fibers that are superior for these requirements	Desired characteristics	Fibers that are superior for these requirements
Moisture resistant Non-absorbent, fast drying, high wet strength, non-swelling, non-shrinking	Glass, Teflon, saran, PVC, rubber, polyethylene, poly-propylene saran, spandex, acrylic polyester	**Fire resistant** Non-flammable or very slow burning; flame resistant	Asbestos, Teflon, PVC, saran, polypropylene, polyvinyl alcohol
Climate resistant Minimum deterioration from sun, rain, sea-water, insects, mildew, and other environmental factors	Glass, Teflon, saran, PVC, rubber, acrylic, poly-propylene, spandex, poly-ester, polyvinyl alcohol	**Light weight** Low density and high strength-weight ratio	Polypropylene, polyethylene, polyurethane, nylon
Chemical resistant Unharmed by acids, alkalies, salts, oils, common solvents	Asbestos, Teflon, saran, polypropylene, polyvinyl alcohol, polyester	**High tenacity** High breaking strength for given diameter; high ulti-mate tensile strength	Glass, nylon, polyester, ramie, polypropylene, poly-vinyl alcohol, flax, silk, high-tensile viscose
Temperature resistant Retains properties at high and low temperatures; heat resistant	Asbestos, Teflon, nylon, poly-ester, cupro, triacetate	**Hard wearing** High resistance to friction and abrasion	Flax, silk, cotton, kenef, sunn, PVC, polyvinyl alcohol, polyester, polypropylene, nylon, acrylic
		Elastic and resilient No permanent set after large deformation; springy	Rubber, spandex, polyester, wool, PVC, nylon, acrylic, silk

*Compiled from several sources.

Table 1-109.   TEXTILE YARN AND FIBER SIZES

The *fineness* of silk, cotton, rayon, and other man-made yarns is usually expressed in terms of the length of yarn per unit weight, or the weight of a given length. The *denier* is the weight in grams of 9 000 meters of yarn. (Sometimes the *international denier* is used; it is defined as the weight in grams of 10 000 meters of yarn.) A similar unit is the *count*, which is the number of hanks or skeins per pound. The hank, for silk or cotton, is 840 yards, but for wool, worsted, linen, or man-made fiber, it may be different. In the metric system the *count* is the number of meters per gram of yarn.

CONVERSION TABLE

To convert from	To	Multiply by
Gram/meter	denier	.000 111 1
Milligram/kilometer	denier	111.11
Gram/meter	international denier	.000 1
Denier	ounce/1 000 yards	0.003 584
Denier	pound/10 000 yards	0.002 240
Count (cotton, silk)	yard/lb	840

Note: There is no simple conversion from these units of fineness to the diameter in microns, nor between breaking strength per denier (tenacity) and tensile strength per unit area.

Table 1-110.   PROPERTIES OF COMMON SOLID MATERIALS*

For the density of various solids, see Table 1-111.

Material	Specific gravity	Specific heat			Thermal conductivity		
		$\dfrac{Btu}{lbm \cdot deg\ R} = \dfrac{cal}{g \cdot K}$		$\dfrac{kJ}{kg \cdot K}$	$\dfrac{Btu}{hr \cdot ft \cdot deg\ F}$	$\dfrac{cal}{sec \cdot cm \cdot deg\ C}$	$\dfrac{W}{m \cdot K}$
Asbestos cement board	1.4	0.2		.837	0.35	.001 45	0.607
Asbestos millboard	1.0	0.2		.837	0.08	.000 33	0.14
Asphalt	1.1	0.4		1.67			
Beeswax	0.95	0.82		3.43			
Brick, common	1.75	0.22		.920	0.42	.001 7	0.71
Brick, hard	2.0	0.24		1.00	0.75	.003 1	1.3
Chalk	2.0	0.215		.900	0.48	.002 0	0.84
Charcoal, wood	0.4	0.24		1.00	0.05	.000 21	0.088
Coal, anthracite	1.5	0.3		1.26			
Coal, bituminous	1.2	0.33		1.38			
Concrete, light	1.4	0.23		.962	0.25	.001 0	0.42
Concrete, stone	2.2	0.18		.753	1.0	.004 1	1.7
Corkboard	0.2	0.45		1.88	0.025	.000 1	0.04
Earth, dry	1.4	0.3		1.26	0.85	.003 5	1.5
Fiberboard, light	0.24	0.6		2.51	0.035	.000 14	0.058
Fiber hardboard	1.1	0.5		2.09	0.12	.000 5	0.2
Firebrick	2.1	0.25		1.05	0.8	.003 3	1.4
Glass, window	2.5	0.2		.837	0.55	.002 3	0.96
Gypsum board	0.8	0.26		1.09	0.1	.000 41	0.17
Hairfelt	0.1	0.5		2.09	0.03	.000 12	0.050
Ice (32°)	0.9	0.5		2.09	1.25	.005 2	2.2
Leather, dry	0.9	0.36		1.51	0.09	.000 4	0.2
Limestone	2.5	0.217		.908	1.1	.004 5	1.9
Magnesia (85%)	0.25	0.2		.837	0.04	.000 17	0.071
Marble	2.6	0.21		.879	1.5	.006 2	2.6
Mica	2.7	0.12		.502	0.4	.001 7	0.71
Mineral wool blanket	0.1	0.2		.837	0.025	.000 1	0.04
Paper	0.9	0.33		1.38	0.07	.000 3	0.1
Paraffin wax	0.9	0.69		2.89	0.15	.000 6	0.2
Plaster, light	0.7	0.24		1.00	0.15	.000 6	0.2
Plaster, sand	1.8	0.22		.920	0.42	.001 7	0.71
Plastics, foamed	0.2	0.3		1.26	0.02	.000 08	0.03
Plastics, solid	1.2	0.4		1.67	0.11	.000 45	0.19
Porcelain	2.5	0.22		.920	0.9	.003 7	1.5
Sandstone	2.3	0.22		.920	1.0	.004 1	1.7
Sawdust	0.15	0.21		.879	0.05	.000 2	0.08
Silica aerogel	0.11	0.2		.837	0.015	.000 06	0.02
Vermiculite	0.13	0.2		.837	0.035	.000 14	0.058
Wood, balsa	0.16	0.7		2.93	0.03	.000 12	0.050
Wood, oak	0.7	0.5		2.09	0.10	.000 41	0.17
Wood, white pine	0.5	0.6		2.51	0.07	.000 29	0.12
Wool, felt	0.3	0.33		1.38	0.04	.000 17	0.071
Wool, loose	0.1	0.3		1.26	0.02	.000 8	0.3

*Compiled from several sources.

Table 1-111. DENSITY OF VARIOUS SOLIDS*

Approximate Density of Solids at Ordinary Atmospheric Temperature

In the case of substances with voids, such as paper or leather, the bulk density is indicated rather than the density of the solid portion. For density in kg/m³, multiply values in g/cm³ by 1 000.

Substance	Grams per cu cm	Pounds per cu ft	Substance	Grams per cu cm	Pounds per cu ft	Substance	Grams per cu cm	Pounds per cu ft
Agate	2.5–2.7	156–168	Glass			Tallow		
Alabaster			Common	2.4–2.8	150–175	Beef	0.94	59
Carbonate	2.69–2.78	168–173	Flint	2.9–5.9	180–370	Mutton	0.94	59
Sulfate	2.26–2.32	141–145	Glue	1.27	79	Tar	1.02	66
Albite	2.62–2.65	163–165	Granite	2.64–2.76	165–172	Topaz	3.5–3.6	219–223
Amber	1.06–1.11	66–69	Graphite†	2.30–2.72	144–170	Tourmaline	3.0–3.2	190–200
Amphiboles	2.9–3.2	180–200	Gum arabic	1.3–1.4	81–87	Wax, sealing	1.8	112
Anorthite	2.74–2.76	171–172	Gypsum	2.31–2.33	144–145	Wood (seasoned)		
Asbestos	2.0–2.8	125–175	Hematite	4.9–5.3	306–330	Alder	0.42–0.68	26–42
Asbestos slate	1.8	112	Hornblende	3.0	187	Apple	0.66–0.84	41–52
Asphalt	1.1–1.5	69–94	Ice	0.917	57.2	Ash	0.65–0.85	40–53
Basalt	2.4–3.1	150–190	Ivory	1.83–1.92	114–120	Balsa	0.11–0.14	7–9
Beeswax	0.96–0.97	60–61	Leather, dry	0.86	54	Bamboo	0.31–0.40	19–25
Beryl	2.69–2.7	168–169	Lime, slaked	1.3–1.4	81–87	Basswood	0.32–0.59	20–37
Biotite	2.7–3.1	170–190	Limestone	2.68–2.76	167–171	Beech	0.70–0.90	32–56
Bone	1.7–2.0	106–125	Linoleum	1.18	74	Birch	0.51–0.77	32–48
Brick	1.4–2.2	87–137	Magnetite	4.9–5.2	306–324	Blue gum	1.00	62
Butter	0.86–0.87	53–54	Malachite	3.7–4.1	231–256	Box	0.95–1.16	59–72
Calamine	4.1–4.5	255–280	Marble	2.6–2.84	160–177	Butternut	0.38	24
Calcspar	2.6–2.8	162–175	Meerschaum	0.99–1.28	62–80	Cedar	0.49–0.57	30–35
Camphor	0.99	62	Mica	2.6–3.2	165–200	Cherry	0.70–0.90	43–56
Caoutchouc	0.92–0.99	57–62	Muscovite	2.76–3.00	172–187	Dogwood	0.76	47
Cardboard	0.69	43	Ochre	3.5	218	Ebony	1.11–1.33	69–83
Celluloid	1.4	87	Opal	2.2	137	Elm	0.54–0.60	34–37
Cement, set	2.7–3.0	170–190	Paper	0.7–1.15	44–72	Hickory	0.60–0.93	37–58
Chalk	1.9–2.8	118–175	Paraffin	0.87–0.91	54–57	Holly	0.76	47
Charcoal			Peat blocks	0.84	52	Juniper	0.56	35
Oak	0.57	35	Pitch	1.07	67	Larch	0.50–0.56	31–35
Pine	0.28–0.44	18–28	Porcelain	2.3–2.5	143–156	Lignum vitae	1.17–1.33	73–83
Cinnabar	8.12	507	Porphyry	2.6–2.9	162–181	Locust	0.67–0.71	42–44
Clay	1.8–2.6	112–162	Pressed wood			Logwood	0.91	57
Coal			pulp board	0.19	12	Mahogany		
Anthracite	1.4–1.8	87–112	Pyrite	4.95–5.1	309–318	Honduras	0.66	41
Bituminous	1.2–1.5	75–94	Quartz	2.65	165	Spanish	0.85	53
Cocoa butter	0.89–0.91	56–57	Resin	1.07	67	Maple	0.62–0.75	39–47
Coke	1.0–1.7	62–105	Rock salt	2.18	136	Oak	0.60–0.90	37–56
Copal	1.04–1.14	65–71	Rubber, hard	1.19	74	Pear	0.61–0.73	38–45
Cork	0.22–0.26	14–16	Rubber, soft			Pine		
Cork linoleum	0.54	34	Commercial	1.1	69	Pitch	0.83–0.85	52–53
Corundum	3.9–4.0	245–250	Pure gum	0.91–0.93	57–58	White	0.35–0.50	22–31
Diamond	3.01–3.52	188–220	Sandstone	2.14–2.36	134–147	Yellow	0.37–0.60	23–37
Dolomite	2.84	177	Serpentine	2.50–2.65	156–165	Plum	0.66–0.78	41–49
Ebonite	1.15	72	Silica			Poplar	0.35–0.5	22–31
Emery	4.0	250	Fused trans-			Satinwood	0.95	59
Epidote	3.25–3.50	203–218	parent	2.21	138	Spruce	0.48–0.70	30–44
Feldspar	2.55–2.75	159–172	Translucent	2.07	129	Sycamore	0.40–0.60	24–37
Flint	2.63	164	Slag	2.0–3.9	125–240	Teak		
Fluorite	3.18	198	Slate	2.6–3.3	162–205	Indian	0.66–0.88	41–55
Galena	7.3–7.6	460–470	Soapstone	2.6–2.8	162–175	African	0.98	61
Gamboge	1.2	75	Spermaceti	0.95	59	Walnut	0.64–0.70	40–43
Garnet	3.15–4.3	197–268	Starch	1.53	95	Water gum	1.00	62
Gas carbon	1.88	117	Sugar	1.59	99	Willow	0.40–0.60	24–37
Gelatin	1.27	79	Talc	2.7–2.8	168–174			

†Some values reported as low as 1.6
*Based largely on: "Smithsonian Physical Tables", 9th rev. ed., W.E. Forsythe, Ed., The Smithsonian Institution, 1956, p. 292.

Table 1-112.   MECHANICAL AND PHYSICAL PROPERTIES OF BRITTLE MATERIALS*

Property	Factors on which indicated property depends
Elastic moduli: Young's modulus, shear modulus, compressibility, Poisson's ratio Thermal-expansion coefficient Melting temperature Theoretical fracture strength Surface tension	**Group A.** Properties related to the *strength of the bonds between atoms and the nature of the force-versus-interatomic spacing curve.* Strong bonds promote high elastic moduli, low thermal-expansion coefficients, high melting temperatures, high theoretical fracture strengths, and high surface tension. In polycrystalline and/or polyphase material, these properties tend toward weighted averages of the crystals or phases present and are virtually independent of the microstructural arrangement. They are also little influenced by dislocations or stress concentrations.
Electrical conductivity Thermal conductivity Optical properties	**Group B.** Properties related to the *type of bonding between atoms.* Metallic bonding promotes good electrical and thermal conductivity and absorption of visible wavelengths of light; ionic bonding and covalent bonding result in poor electrical and thermal conductivity and allow visible light to be transmitted. These properties are also little influenced by microstructural arrangement or by stress concentrations. Defects on an atomic scale, however, such as dislocations and vacancies, may have an important effect.
Yield strength	**Group C.** This property is related to the *arrangement* of atoms and to the *presence of mobile dislocations.* Simple atom arrangements with much symmetry favor glide of adjacent planes of atoms under the action of a shear stress. Mobile dislocations lower drastically the shear stress required to produce such glide. In polycrystalline and/or polyphase materials, the yield strength depends strongly on the microstructural arrangement; fine grain size and finely dispersed mixtures of phases promote high yield strength.
Ductile-fracture strength Hardness Ductility Toughness Fatigue strength	**Group D.** Properties related to the yield strength and to the *way in which moving dislocations interact.* High yield strength is accompanied by high ductile-fracture strength, high hardness, and high fatigue strength. The ductility and toughness will depend on the ability of dislocations to generate additional dislocations and to avoid being stopped by barriers, such as grain boundaries, phase boundaries, or other dislocations; thus, they are strongly dependent on microstructural arrangement.
Brittle-fracture strength	**Group E.** This property is related to the *strength of the bonds between atoms* and to the *presence of flaws* that act as powerful stress concentrators. Strong bonds and the absence of flaws promote high fracture strengths.
Creep and relaxation behavior Diffusion coefficient Internal friction	**Group F.** Properties related to the *nonelastic, nonglide movement of atoms* with respect to their neighbors. They depend strongly on the presence of atomic defects, such as dislocations and vacancies, and on the microstructural arrangement.

*From: "The Evaluation and Interpretation of Mechanical Properties of Brittle Materials", A. Rudnik, C.W. Marschall, W.H. Duckworth, and B.R. Emrich, Defense Ceramic Information Center, Report 68-3, Battelle Memorial Institute, 1968. See this reference for bibliography.
See also "Bibliography on Ceramic Fibers and Fibrous Composite Materials", Part 1, DCIC Report 69-4, August 1969, 1760 references.

Table 1-113.   PROPERTIES OF CARBON AND GRAPHITE

Manufactured carbon and graphite should not be viewed as single specific materials, but rather as families of materials. Each member of the family is essentially pure carbon, but each varies from the other in such characteristics as orientation of the crystallites, the size and number of pore spaces, grain size, degree of crystallization, and apparent density.

In the carbon industry the term *carbon* is used to refer to materials in which the small crystallites have low orientation. The term *graphite* is used to refer to material that has highly ordered structure. Specific grades, having controlled characteristics, are produced by selecting raw materials and by varying processing techniques. A broad familiarity with the properties and characteristics of the various grades is important when selecting materials for a specific application.

Information regarding characteristics and recommended applications for various types of industrial graphite is available in the "Industrial Graphite Engineering Handbook", published by Union Carbide Corporation, Carbon Products Division.

Table 1-113.   PROPERTIES OF CARBON AND GRAPHITE (Continued)

Table A.   TYPICAL PROPERTIES AT ROOM TEMPERATURE*

| Type of product | Bulk density | Porosity, percent | Strength, psi | | Elastic modulus, 10^6 psi | Specific resistance, 10^{-5} ohm-in. | Thermal conductivity, $\frac{Btu \cdot ft}{hr \cdot ft^2 \cdot deg\ F}$ | Coefficient of thermal expansion, $10^{-7}/deg\ F$ |
			Compres- sive	Flex- ural				
CARBON								
Porous carbon, grade 45	1.04	47	900	500	0.4	700	1	16
Carbon furnace lining (24″ × 30″ cross section)	1.60	21	2 500	600	0.8	160	9	13
Carbon refractory brick	1.63	17	4 400	1 200	1.2	195	16	18
Carbon chemical brick	1.56	22	8 800	2 600	1.9	160	4	13
Carbon pipe	1.55	22	9 000	2 600	1.9	150	3	13
Carbon electrodes, 8-in. diam. and equivalent rectangular	1.57	21	2 400	1 100	1.2	110	9	13
Carbon electrodes, 17- to 45-in. diam. and equivalent rectangular	1.60	21	1 700	400	0.7	170	9	13
GRAPHITE								
Porous graphite, grade 45	1.04	53	500	300	0.3	130	45	11
Nuclear graphite see Table 4-28								
Fine-grain, premium graphite	1.73	23	8 300	4 000	1.5	43	68	13
High-density, premium graphite	1.84	19	8 400	3 700	1.7	47	63	11
Recrystallized graphite	1.95	13	7 200	5 400	2.7	28	104	3
Graphite brick	1.56	31	3 100	1 650	1.4	34	86	10
Graphite pipe	1.67	26	5 000	2 800	1.7	34	86	10
Medium-grain, dense graphite, cylinders, and plates to 2¾-in. diam. and to ¾-in. thick	1.70	25	5 600	4 000	1.8	27	100	7
3 to 11-in. diam. and 2 to 12-in. thick	1.70	24	5 600	2 700	1.7	30	95	7
20 to 24-in. diam. and 20 × 20 cross section	1.75	24	5 500	2 200	1.2	35	83	12
Graphite electrodes, anodes, cylinders, and plates to 2¾-in. diam. and to ¾-in. thick	1.58	30	4 000	2 000	1.5	33	88	6
6 to 12-in. diam. and 6 to 12-in. thick	1.57	30	2 900	1 300	1.5	33	88	7
14 to 35-in. diam. and 20 to 24-in. thick	1.58	32	2 000	1 000	0.8	33	88	5

*Courtesy of Union Carbide Corporation, Carbon Products Division.

Table B.   VARIATION OF PROPERTIES WITH TEMPERATURE†

Property	500 deg C / 932 deg F	1 000 deg C / 1 832 deg F	1 500 deg C / 2 732 deg F	2 000 deg C / 3 632 deg F	2 500 deg C / 4 532 deg F
Thermal expansion[a] as percent elongation from room temperature					
Anthracite carbon	0.14	0.38	0.63	—	—
Graphite	0.12	0.32	0.55	—	—
Thermal conductivity as percent of room temperature value					
Fabricated anthracite carbon	100	103	123	—	—
Fabricated graphite	60	40	30	25	—
Instantaneous specific heat[b] as percent of room temperature value: graphite	225	262	282	294	302
Short-time breaking strength[c] as percent of room temperature value: graphite	107	120	135	153	181

[a]These are longitudinal expansion; transverse expansion is 10–60 percent greater.
[b]The specific heat at room temperature is about 0.17.
[c]Strength increases up to 2 500 deg C, then decreases rapidly; above 2 200 deg C appreciable creep will occur at high-stress levels.

†From: "Carbon Products Pocket Handbook", Union Carbide Corporation, Carbon Products Division, 1964.

Table 1-114. TYPICAL ENGINEERING APPLICATIONS OF CARBON PRODUCTS*

KEY TO ADVANTAGEOUS PROPERTIES OF CARBON PRODUCTS:[a]

1. Stability and strength at high temperatures (to 4 500 deg F in non-oxidizing atmospheres)
2. High resistance to thermal shock
3. High thermal conductivity of solid; low conductivity of porous foam, cloth, and tape
4. Low coefficient of thermal expansion
5. High radiation emissivity
6. Good electrical conductivity
7. High compressive strength
8. Stiffness of solid; flexibility of filament, cloth, or tape
9. High resistance to erosion
10. Good machinability
11. Low friction; self-lubrication
12. High resistance to chemical attack and corrosion
13. High adsorption of gases and vapors
14. High moderating ratio, *i.e.*, ratio of fast neutron slowing-down power to bulk neutron absorption coefficient
15. High ratio of thermal neutron scattering to absorption cross section

Typical application	Type and form of carbon product	Desirable properties[b]
Electrodes for arc furnaces, welding, and lighting arcs	Extruded or molded carbon or graphite	1, 2, 3, 4, 5, 6, 7, 8, 9, 10, 12
Elements for electric-resistance furnaces	Resistors fabricated from round or rectangular stock, usually graphite	1, 2, 3, 4, 5, 8, 10, 12
Metallurgical crucibles, boats, trays	Castings from carbon or graphite	1, 2, 4, 7, 8, 9, 10, 12
Foundry molds, chills, cores, risers, cupola linings	Castings, rods, bricks, shapes	1, 2, 4, 7, 8, 9, 10, 12
Brazing jigs; extrusion guides; dies	Blocks; machined shapes	1, 2, 7, 8, 9, 10, 12
High-temperature refractories and insulations	Solid or foamed bricks and shapes; tapes and laminates	1, 2, 3, 4, 7, 8, 9, 12
Rocket and missile nozzles, vanes, cones, shields	Castings; built-up shapes from graphite fabric, tape, and resins	1, 2, 3, 4, 5, 7, 8, 9, 10
Chemical reactor vessels; heat exchangers; pipes and fittings	Cast, machined, or extruded graphite or carbon, pure or impregnated with resins	1, 2, 3, 4, 7, 8, 9, 10, 12
Packing and sealing rings; bushings, joint packings	Graphite blocks, sheet, cloth, and yarn	1, 2, 3, 4, 7, 8, 9, 11, 12
Electrical contacts, brushes, resistors	Blocks, rods, machined parts	1, 3, 5, 6, 7, 10, 11
Anodes, electrodes, battery components	Extruded and molded rod, granular carbon	6, 12
Air purification, gas separations, odor, and vapor removal	Activated carbon granules	13
Nuclear reactor—moderator and reflector	Purified "nuclear graphite" bars and machined shapes	1, 2, 3, 4, 14, 15

[a]*Note:* Protection from oxidizing atmosphere is necessary in all high-temperature uses of carbon or graphite, *e.g.*, above 700 deg F.
[b]Numbers refer to items listed above.

*Compiled from various sources.

Table 1-115.   STRENGTH OF FILAMENTS, WHISKERS, FIBERS, WIRES, AND BARS*

For other properties of fibers, see Tables 1-105 and 1-107.

The following table gives typical tensile properties of individual strands, tension elements, and reinforcement materials at room temperature.

Material description	Approximate ultimate strength, thousands of psi[a]	Modulus of elasticity in tension, millions of psi	Specific gravity	Material description	Approximate ultimate strength, thousands of psi[a]	Modulus of elasticity in tension, millions of psi	Specific gravity
Graphite whisker[b]	3,000	100.	1.7	Glass fiber[i]	200	11	2.5
Quartz whisker[c]	3,000	10	2.65	Stainless steel,			
Silicon carbide				heat-treated,			
whisker[d]	3,000	70	3.2	Cr 0.18 bar	180	30	7.8
Sapphire or alun-				Flax	150	15	1.5
dum (Al_2O_3)				Hardwood	150	—	—
whisker[e]	2,600	70.	4.0	Hemp	140	8.2	1.5
Iron whisker[f]	1,800	28.	7.8	Wood, kraft	130	10.5	1.5
Silica glass[c]	650	11.	2.55	Steel wire, low-			
Steel (high carbon)				carbon	100	30	7.8
fine wire	600	30.	7.8	Nylon	80	0.7	1.07
Tungsten filament	500	50	19.3	Copper wire, hard	70	17	8.9
Boron fiber[g]	350	51.	2.3	Aluminum wire	60	10	2.7
Steel piano wire	350	30.	7.8	Catgut	60	—	—
Molybdenum wire	300	53	10.2	Cotton	60	—	—
Titanium wire	300	15.	—	Mild steel, rolled			
"Superalloy" (Cr,				bar	60	30	7.8
Co, Mo),				Bamboo fiber	50	4.9	—
wrought	275	—	—	Copper wire, soft	35	17	8.9
Steel wire, bridge				Aluminum, rolled	30	10	2.7
or spring	225	30	7.8	Viscose	12	1.2	1.2
Beryllium wire	220	42	1.85	Acetate	8	0.4	1.26
Asbestos whisker[h]	215	26.5	2.4				

[a] Unit strength (psi) is given for small diameters. Larger sizes often show much lower unit strength.

[b] Graphite and carbon fibers and whiskers: textiles and laminates with high strength, high conductivity, and refractory properties; reinforced plastics.

[c] Quartz and silica: reinforced phenolics for high temperatures; reinforced ceramics.

[d] Carbides and nitrides: refractory cloth; high-temperature electrical and thermal insulations; flame-resistant composites with metals and plastics; high-temperature filters.

[e] Metal oxide and silicate fibers: high-temperature yarn and rope; plastic and metal composites; high-temperature electrical insulation.

[f] Iron and ferrous alloy: wear-resistant composites; reinforced ceramics.

[g] Boron: filament-wound, lightweight plastic composites; nuclear shielding.

[h] Asbestos: fire-resistant paper, tape, textiles; composites with metal, plastic, glass, or ceramic; oblative materials.

[i] Glass: most widely used plastic reinforcements, chopped, woven, or wound.

*Compiled from several sources.

REFERENCES

"Advanced Materials: Refractory Fibres, Fibrous Metals, Composites", C.Z. Carroll-Porczynski, Chemical Publishing Company, 1969.

"Modern Composite Materials", L.J. Broutman and R.H. Krock, Addison-Wesley Publishing Company, Inc., 1967.

"Fibre Reinforced Materials", G.S. Holister and C. Thomas, American Elsevier Publishing Company, Inc., 1966.

Mechanical Properties of Metal Matrix Composites, C.T. Lynch, J.P. Kershaw, and B.R. Collins, *CRC Critical Reviews in Solid State Sciences*, 1(4):481, 1970.

"Whisker Technology", A.P. Levitt, Ed., John Wiley & Sons, 1970.

Table 1-116.   PROPERTIES OF CARBIDES*

For other data on heat-resistant materials, see Tables 1-61, 1-122, and 3-6.

Most carbides are industrially useful because of their hardness and heat resistance, e.g., carbides of boron, chromium, molybdenum, silicon, tantalum, titanium, vanadium, tungsten, zirconium. A few carbides have special chemical or radiative properties, e.g., carbides of beryllium, calcium, plutonium and uranium.

Pure carbides are very difficult and expensive to produce, and most available products are mixtures containing free carbon, metal, oxides and impurities. Many of the phase diagrams are very complex. The following tables attempt to give properties of the pure materials, but since experimental values are obtained under very difficult conditions, the extent of disagreement between sources is considerable.

Table A.   PHYSICAL PROPERTIES OF CARBIDES

| Carbide | Molecular weight | Specific gravity | Weight % carbon | Specific heat[a] | Melting point[b] | | Hardness,[c] kg/mm^2 (max.) |
					deg C	deg F	
Al_4C_3	143.96	2.36	25.0		1 980.	3 600.	
B_4C	55.36	2.5	21.7		2 450.	4 440.	2 800.
Be_2C	30.04	2.0	29.7	0.33	2 100.[d]	3 810.	2 700.
CaC_2	64.10	2.22	37.4				
Cr_3C_2	180.02	6.7	13.3	0.044	1 880.	3 415.	1 300.
Cr_7C_3	400.01	6.92	9.0	0.018	1 700.	3 090.	1 600.
Fe_3C	179.55	7.69	6.7		1 837.	3 340.	
HfC	190.52	12.5	6.3	0.047	3 850.	6 960.	2 500.
Mn_3C	176.83	6.9	6.8				
MoC	107.95	8.2	11.1		2 695.	4 880.	
Mo_2C	203.89	9.0	5.9		2 205.	4 000.	
Na_2C_2	70.00	1.58	34.4				
NbC	104.92	7.7	11.5	0.056	3 550.	6 420.	2 400.
Ni_3C	188.14	7.8	6.4				
SeC_2	102.98	2.68	23.4				
SiC	40.10	3.22	30.0				
SrC_2	111.64	3.2	21.6		>1 700.	>3 100.	
TaC	192.96	14.1	6.3	0.045	4 000.	7 230.	1 800.
Ta_2C	373.91		3.2		3 300.	5 970.	
ThC	244.05	10.7	4.9				850.
ThC_2	256.06	8.96	9.4	0.055	2 650.	4 800.	600.
TiC	59.91	4.9	20.0	0.135	3 100.	5 610.	2 700.
UC	250.04	13.6	4.8	0.048	2 500.	4 530.	2 800.
UC_2	262.05	11.3	9.2	0.055	2 400.	4 350.	
VC	62.95	5.8	19.1	0.123	2 800.	5 070.	2 900.
WC	195.86	15.5	6.2		2 800.	5 070.	2 400.
W_2C	379.71	17.3	3.2		2 700.	4 890.	3 000.
ZrC	103.23	6.7	11.6	0.089	3 450.	6 240.	2 700.

[a]These are specific heats at room temperature, but the value increases rapidly with temperature. At about 700 deg C or above the specific heat is 50% higher. At or above 2 500 deg C the specific heat is double that given in this column. See Reference 4 for thermal properties tables.
[b]Melting points are affected by impurities and by departure from the exact formula composition, and they are very difficult to determine.
[c]This numerical value depends on the size and shape of the indenter and on the method of test. It is also affected by excess carbon or impurities.
[d]Decomposes or sublimes above this temperature.
  Other recognized carbides include: B_6C, $Cr_{23}C_6$, Nb_2C, Nb_4C_3, Ta_2C, ThC, U_2C_3.

*Compiled from several sources.

Table 1-116.   PROPERTIES OF CARBIDES *(Continued)*

Table B.   MISCELLANEOUS PROPERTIES

Carbide	Modulus of elasticity, kpsi	Compressive strength, kpsi	Electrical resistivity, μohm-cm	Thermal conductivity, W/cm·deg C
B_4C	42 000	414		0.27
Be_2C	32 000	105		0.90
SiC				0.25
TaC	41 500		30.	0.22
ThC				0.56
TiC		400	170.	0.17
UC				0.84
UC_2				0.33
VC	39 000	90	150	
WC	100 000	650	53	0.90
W_2C	61 000		80	
ZrC		238	70	0.21

CHEMICAL REACTIONS

The following carbides are decomposed by water or water vapor: Be_2C, CaC_2, Mn_3C, Na_2C_2, ThC, ThC_2, UC, and UC_2.

The following carbides will burn in air: NbC, TaC, ThC. Other carbides that oxidize are Cr_3C_2, MoC, TiC, VC, ZrC.

Almost all carbides are attacked by HNO_3 and more rapidly by HNO_3 plus HF. Sulfuric acid attacks the carbides of thorium and vanadium. Hydrochloric acid attacks ThC, ThC_2, and UC.

ThC and UC are attacked by fluorides; VC and UC_2 react with chlorine; Be_2C reacts with halogens at higher temperatures.

The following carbides react with nitrogen (at higher temperatures): TiC, VC (gradual), ZrC. The following carbides react with CO or CO_2 (at higher temperatures), TiC, and ZrC.

REFERENCES

"Handbook of Chemistry and Physics", 53rd ed., R.C. Weast, Ed., The Chemical Rubber Co., 1972.
"Handbook of Chemistry", 9th ed., N.A. Lange, Ed., Handbook Publishers, Inc., 1956.
"Metals Handbook", 8th ed., T. Lyman, Ed., Vol. 1, American Society for Metals, 1961.
"The Refractory Carbides", E.K. Storms, Academic Press, 1967.
"Reactor Handbook", 2nd ed., C.R. Tipton, Jr., Ed., Vol. 1: Materials, Interscience Books (a division of John Wiley & Sons), 1960.

Table 1-117. PROPERTIES OF ICE*

The nature and properties of ordinary water ice are reasonably stable and constant at or below atmospheric pressure, but they are definitely affected by slight impurities and may change markedly at very low temperatures. The melting point of ice first decreases as pressure is applied, up to about 2 000 atm. Several forms of ice exist at successively higher pressures, and here the melting point increases with pressure. (It is above 100 deg C at pressures in excess of 25 000 atm.)

Table A. PROPERTIES OF ORDINARY ICE

Property	At 1 atm and near 0 deg C	At other conditions
Melting point, deg C	0.0	Lower at high pressure, to about -22 deg C at 2 000 atm
Specific gravity	0.91	High-pressure forms of ice are heavier than water
Specific heat	0.5	Lower at low temperatures; about 0.11 at 50 K
Thermal conductivity		
w/cm·K	0.021 ⎫	Higher at lower temperatures; about 0.075 w/cm·K
Btu/hr·ft·K	1.2 ⎬	at 100 K
Coefficient of linear thermal expansion/deg C	5×10^{-5}	Lower at low temperatures; 2×10^{-5} at 150 K
Ultimate compressive strength, kpsi		Increases rapidly as temperature is reduced
Modulus of elasticity, kpsi	1 290–1 430	Ice subject to creep
Poisson's ratio	0.31–0.36	
Total radiation emissivity	0.9	Lower for solar absorption
Dielectric constant	100.	Increases as temperature is reduced

Note: The crystal form of ordinary ice is hexagonal, but below 150 K it may be cubic.

Table B. ICE AT HIGH PRESSURES

Approximate Ranges of Properties[a]

Properties	Usual phase-diagram designation				
	III	V	VI	VII	VIII
Melting-point curve					None
Minimum pressure, atm	2 100	3 450	6 250	22 000	—
Corresponding temperature, K	250	255	273	355	—
Maximum pressure, atm	3 450	6 250	22 000	200 000	—
Corresponding temperature, K	255	273	355	710[b]	—
Crystal form	Tetragonal	Monoclinic	Tetragonal	Cubic	Cubic
Typical specific gravity	1.14	1.23	1.31	1.5	1.5
Dielectric constant	117.	144.	193.	150.	5.

*Compiled from various sources.

[a] For a phase diagram, see Reference 1 below, p. 51.
[b] Ice form VII does not exist below 275 K. Ice form VIII does not exist above 275 K.

REFERENCES

"The Chemical Physics of Ice", N.H. Fletcher, Cambridge University Press, 1970.
"Properties and Structure of Ice", N. Bjerrum, *Science*, 115:385, 1952.
"Properties of Ordinary Water Substance", N.E. Dorsey, Reinhold Publishing Corp., 1940.
"The Structure and Properties of Water", D. Eisenberg and W. Kauzmann, Oxford University Press, 1969.

Table 1-118.   GENERAL CLASSIFICATION OF COMMON ROCKS*

IGNEOUS ROCKS

Solidified from a Molten State

Coarse-grained crystalline	Fine-grained crystalline (or crystals and glass)	Fragmental (crystalline or glassy)
Origin: deep intrusion, slowly cooled	Origin: quickly-cooled, volcanic or shallow intrusive	Origin: explosive volcanic fragments deposited as sediments
Granite	Rhyolite	Ash and pumice (volcanic dust or cinders)
Diorite	Andesite	Tuff (consolidated volcanic ash)
Gabbro	Basalt	Agglomerate (coarse and
	Obsidian and pitchstone (essentially glass—suddenly chilled, few or no crystals)	fine volcanic debris)

MINERAL CONSTITUENTS OF IGNEOUS ROCKS

MINERALS KEY:

Q = quartz (SiO_2): hard, shiny, no true cleavage
O = orthoclase feldspar: silicates; regular cleavage
P = plagioclase feldspar: nearly white, good cleavage
A = amphibole (magnesium, iron, calcium) and/or biotite (black mica)
B = pyroxene; nearly black

Coarsely crystalline	Principal constituent minerals	Finely crystalline or porphyritic
Granite	Q+O+A (+P)†	Rhyolite
Syenite	O+A (+P)†	Trachyte
Quartz monzonite	Q+O+P+A	Dellenite
Monzonite	O+P+A	Latite
Quartz diorite	Q+P+A or B (+O)†	Dacite
Diorite	P+A or B (+O)†	Andesite
Gabbro	P+B	Basalt

†Small amount.

SEDIMENTARY ROCKS

Sediments Transported by Water, Air, Ice, Gravity

Mechanically deposited	Chemically or bio-chemically deposited
A—Unconsolidated Clay Silt Sand According to particle size Gravel Cobbles	*A—Calcareous* Limestone ($CaCO_3$) Dolomite ($CaCO_3 \cdot MgCO_3$) Marl (calcareous shale) Caliche (calcareous soil) Coquina (shell limestone)
B—Consolidated Shale (consolidated clay) Siltstone (consolidated silt) Sandstone (consolidated sand) Conglomerate (consolidated gravel or cobbles—rounded) Breccia (angular fragments)	*B—Siliceous* Chert Flint Agate Spring deposit, Opal vein or Chalcedony cavity filling *C—Others* Coal, phosphate, salines, etc.

METAMORPHIC ROCKS

Igneous or Sedimentary Rocks Changed by Heat, Pressure

A—Foliated
Slate: dense, dark, splits into thin plates (metamorphosed shale)
Schist: predominantly micaceous, semi-parallel lamellae
Gneiss: granular, banded, subordinately micaceous

B—Massive
Marble: coarsely crystalline, calcareous (metamorphosed limestone)
Quartzite: dense, very hard, quartzose (metamorphosed sandstone)

*From: "Concrete Manual", 7th ed., U.S. Bureau of Reclamation, 1966, Tables 10 and 11.

Table 1-119.   CHEMICAL COMPOSITION OF ROCKS*

Element	Average igneous rock	Average shale	Average sandstone	Average limestone	Average sediment
SiO_2	59.14	58.10	78.33	5.19	57.95
TiO_2	1.05	0.65	0.25	0.06	0.57
Al_2O_3	15.34	15.40	4.77	0.81	13.39
Fe_2O_3	3.08	4.02	1.07	0.54	3.47
FeO	3.80	2.45	0.30		2.08
MgO	3.49	2.44	1.16	7.89	2.65
CaO	5.08	3.11	5.50	42.57	5.89
Na_2O	3.84	1.30	0.45	0.05	1.13
K_2O	3.13	3.24	1.31	0.33	2.86
H_2O	1.15	5.00	1.63	0.77	3.23
P_2O_5	0.30	0.17	0.08	0.04	0.13
CO_2	0.10	2.63	5.03	41.54	5.38
SO_3		0.64	0.07	0.05	0.54
BaO	0.06	0.05	0.05		
C		0.80			0.66
	99.56	100.00	100.00	99.84	99.93

*From: "Sedimentary Rocks", F.J. Pettijohn, Harper Brothers, 1948, p. 82.

Table 1-120.   PROPERTIES OF CLEAR FUSED QUARTZ*

Property	English or metric system value	International system of units (SI) value	Property	English or metric system value	International system of units (SI) value
Density	2.2 g/cm^3	2.2×10^3 kg/m^3	Annealing point	1140°C	1410°K
Hardness	4.9 Mohs' scale		Strain point	1070°C	1340°K
Tensile strength	7,000 psi	4.8×10^7 N/m^2	Electrical resistivity	$10^{9.5}$ ohm cm	
Compressive strength	>160,000 psi	$>1.1 \times 10^9$ N/m^2		(350°C)	
Bulk modulus	5.3×10^6 psi	3.7×10^{10} N/m^2	Dielectric properties	(20°C and 1 Mc)	(293°K and 1 Mhz)
Rigidity modulus	4.5×10^6 psi	3.1×10^{10} N/m^2	Constant	3.75	3.75
Young's modulus	10.4×10^6 psi	7.17×10^{10} N/m^2	Strength	410 volts/mil	1.6×10^7 V/m
Poisson's ratio	.16	.16	Loss factor	$<4 \times 10^{-4}$	$<4 \times 10^{-4}$
Coefficient of thermal expansion	5.5×10^{-7} cm/cm °C (20°C–320°C)	5.5×10^{-7} m/m °K (293°K–593°K)	Dissipation factor	$<1 \times 10^{-4}$	$<1 \times 10^{-4}$
			Index of refraction	1.4585	1.4585
Thermal conductivity	3.3×10^{-3} g cal cm/cm^2 sec °C	1.4 W m/m^2 °K	Velocity of sound-shear wave	3.75×10^5 cm/sec	3.75×10^3 m/s
Specific heat	.18 g cal/g	750 J/kg	Velocity of sound-compressive wave	5.90×10^5 cm/sec	5.90×10^3 m/s
Fusion temperature	1800°C	2070°K			
Softening point	1670°C	1940°K	Sonic attenuation	<.033 db/ft Mc	<.11 db/m Mhz

Note: These data apply to a specific grade of commercially available material· The term "fused silica" is often used to include the entire group of materials made by fusing of silica (SiO_2). All such products contain small amounts of impurities, but the clear varieties can have a purity of 99.98 percent. Alumina (Al_2O_3) is the major impurity. Fused silica has an extremely low coefficient of expansion and does not react with most acids, metals, chlorine, or bromine at ordinary temperatures. It has good mechanical and electrical properties and is almost perfectly elastic. Its radiant transmission is high in the ultraviolet as well as in the visible region. Fused quartz or silica products are available in rod, ribbon, and other solid forms, such as tubing, chemical glassware, fiber, yarn, matt, and woven products.

*Properties data from: "Fused Quartz Catalog", courtesy of The General Electric Company.

Table 1-121.   PHYSICAL CONSTANTS OF MINERALS*

RALPH KRETZ

The following table presents data for many of the more common minerals.

In order to avoid duplication and save space, very few cross references are given in the body of the table. If the name sought is not found in the table, consult the **synonym index** given below.

Specific gravities are given at normal atmospheric temperatures, a more precise statement being valueless considering the large variations in natural minerals.

Hardness is given in terms of Mohs' scale.

Indices of refraction for the sodium line, $\lambda = 5893$ Å, unless otherwise indicated. Li, $\lambda = 6708$ Å. Indices will invariably be given in the order ω, ε or α, β, γ. Uniaxial crystals are considered positive if $\varepsilon > \omega$, negative if $\omega > \varepsilon$. Biaxial crystals are considered positive if β is nearer α in value than it is γ, and negative if β is nearer γ than α.

ABBREVIATIONS

Abbreviation	Meaning of abbreviation	Abbreviation	Meaning of abbreviation	Abbreviation	Meaning of abbreviation
bl	blue	grn	green	rhbdr	rhombohedral
blk	black	grnsh	greenish	rhomb	rhombic
blksh	blackish	hex	hexagonal	somet	sometimes
blsh	bluish	iridesc	iridescent	tarn	tarnishes
br	brown	monocl	monoclinic	tetr	tetragonal
brnsh	brownish	oft	often	tricl	triclinic
col	colorless	pa	pale	vlt	violet
cub	cubic	purp	purple	wh	white
dk	dark	(R)	radioactive	yel	yellow
Fe	Fe, ferrous iron	redsh	redish	yelsh	yellowish
Fe^{+3}	Fe, ferric iron				

SYNONYM INDEX

Compound sought	Listed	Compound sought	Listed
Acmite	Aegirine	Lead sulfate	Anglesite
Agate	Quartz (impure)	Lead sulfide	Galena
Aluminum hydroxide	Boehmite, Diaspore, Gibbsite	Limonite	Goethite (impure)
Amphibole	Actinolite, Anthophyllite, Cummingtonite, Glaucophane, Hornblende, Riebeckite, Tremolite	Lithiophyllite	Triphylite
		Lithium mica	Lepidolite
		Lodestone	Magnetite
Antimony oxide	Senarmontite, Valentinite	Magnesium carbonate	Magnesite
Antimony sulfide	Stibnite	Magnesium hydroxide	Brucite
Arsenic oxide	Arsenolite, Claudetite	Magnesium oxide	Periclase
Arsenic sulfide	Orpiment, Realgar	Magnesium sulfate	Kieserite
Barium carbonate	Witherite	Manganese carbonate	Rhodochrosite
Barium sulfate	Barite	Manganese hydroxide	Pyrochroite
Barytes	Barite	Manganese oxide	Hausmannite, Manganosite, Pyrolusite
Bauxite	Gibbsite, Boehmite, Diaspore		
Brimstone	Sulfur	Manganese sulfide	Alabandite
Bronzite	Orthopyroxene	Meerschaum	Serpentine
Cadmium sulfide	Greenockite	Mica	Muscovite, Paragonite, Phlogopite, Biotite, Lepidolite
Calamine	Hemimorphite		
Calcium carbonate	Aragonite, Calcite, Vaterite	Native copper	Copper
Calcium sulfate	Anhydrite, Gypsum	Native gold	Gold
Calcium sulfide	Oldhamite	Nickel oxide	Bunsenite
Carborundum	Moissanite	Nickel sulfide	Millerite
Chalcedony	Quartz (impure, fibrous)	Orthite	Allanite
Chinaclay	Kaolinite	Penninite	Chlorite
Chloanthite	Skutterodite	Peridote	Olivine
Chromespinel	Chromite	Pistacite	Epidote
Chrysolite	Serpentine	Pitchblende	Uraninite
Clinoptolite	Heulandite	Plagioclase	Albite, Oligoclase, Andesine, Anorthite
Clayminerals	Illite, Kaolinite, Montmorillonite		
Clinochlore	Chlorite	Potassium chloride	Sylvite
Cobaltbloom	Erythrite	Potassium sulfate	Arcanite
Copper chloride	Nantokite	Pyroxene	Diopsite, Angite, Aegirine, Jadeite, Pigeonite, Eustatite, Orthopyroxene
Copper oxide	Cuprite		
Copper sulfide	Chalcocite, Covellite, Digenite		
Emerald	Beryl	Rocksalt	Halite
Emery	Mixture of Corundum, Magnetite and other minerals	Ruby	Corundum
		Sapphire	Corundum
Epsom salt	Epsomite	Silica	Christobalite, Quartz, Tridymite
Feldspar	Orthoclase, Microcline, Anorthoclase, Albite, Oligoclase, Andesine, Anorthite	Silver chloride	Cerargyrite
		Silver iodide	Jodyrite, Miersite
		Silver sulfide	Acanthite, Argentite
Fibrolite	Sillimanite	Smalltite	Skutterodite
Flint	Quartz (impure)	Soapstone	Mixture of Talc and other minerals
Fluorapatite	Apatite	Sodium chloride	Halite
Fluorspar	Fluorite	Sodium sulfate	Thenardite
Garnet	Almandine, Pyrope, Spessartite, Andradite, Grossularite, Uvarovite, Hydrogrossularite	Strontium carbonate	Strontianite
		Strontium sulfate	Celestite
		Thorium oxide	Thorianite
Garnierite	Serpentine (Ni-bearing)	Tin oxide	Cassiterite
Glauber salt	Mirabilite	Titanite	Sphene
Hyacinth	Zircon	Titanium oxide	Anatase, Brookite, Rutile
Iceland spar	Calcite	Uranium oxide	Uraninite
Idocrase	Vesuvianite	Zeolite	Natrolite, Mesolite, Scolecite, Thomasonite, Harmatome, Eddingtonite, Heulandite, Stilbite, Phillipsite, Chabazite, Gmelinite, Levyn, Laumontite, Mordenite
Iron carbonate	Siderite		
Iron hydroxide	Goethite, Lepidocrocite		
Iron oxide	Hematite, Magnetite		
Iron spinel	Hercynite		
Iron sulfide	Marcasite, Pyrite, Pyrrhotite	Zincblende	Sphalerite
Lapis lazuli	Lazurite	Zinc carbonate	Smithsonite
Lead carbonate	Cerussite	Zinc oxide	Zincite
Lead chloride	Cotunnite	Zinc spinel	Gahnite
Lead chromate	Crocoite	Zinc sulfide	Sphalerite, Wurtzite
Lead oxide	Litharge, Minium	Zirconium oxide	Baddeleyite

Table 1-121. PHYSICAL CONSTANTS OF MINERALS (Continued)

Name	Formula	Sp. gr.	Hardness	Crystalline form and color	Index of refraction (Na) η; ω ϵ / α β γ
Acanthite	AgS	7.2–7.3	2–2.5	rhomb.(?), iron-blk.	
Actinolite	$Ca_2((Mg,Fe)_5Si_8O_{22}(OH,F)_2$	3.02–3.44	5–6	monocl., pa. to dk. grn.	1.599–1.688, 1.612–1.697, 1.622–1.705
Aegirine	$NaFe^{+3}Si_2O_6$	3.55–3.60	6	monocl., dk. grn. to grnsh. blk.	1.750–1.776, 1.780–1.820, 1.800–1.836
Åkermanite	$Ca_2MgSi_2O_7$	2.944	5–6	tetr., col., gray-grn., br.	1.632, 1.640
Alabandite	MnS	4.050	3.5–4	cub., iron-blk., tarn., br.	
Albite	$NaAlSi_3O_8$	2.63	6–6.5	tricl., col., wh., somet. yel., pink, grn.	1.527, 1.531, 1.538
Allanite	$(Ca,Mn,Ce,La,Y,Th)_2(Fe,Fe^{+3},Ti)(Al,Fe^{+3})_2Si_3O_{12}(OH)$	3.4–4.2	5–6.5	monocl., pa. br. to blk.	1.690–1.791, 1.700–1.815, 1.706–1.828
Allemontite	$AsSb$	5.8–6.2	3–4	hex., tin-wh. to redsh., gray, tarn. gray–brnsh. blk.	
Almandine	$Fe_3Al_2Si_3O_{12}$	4.318	6–7.5	cub., red, dk. red, blk.	1.830
Altaite	$PbTe$	8.15	3	cub., tin-wh., tarn. bronze-yel.	
Aluminite	$Al_2(SO_4)(OH)_4.7H_2O$	1.66–1.82	1–2	monocl.(?), wh.	1.459, 1.464, 1.470
Alunite	$(K,Na)Al_3(SO_4)_2(OH)_6$	2.6–2.9	3.5–4	rhbdr., wh., gray, yel., redsh., br.	1.572, 1.592
Alunogen	$Al_2(SO_4)_3.18H_2O$	1.77	1.5–2	tricl., col., wh., yelsh. wh., redsh. wh.	1.459–1.475, 1.461–1.478, 1.470–1.485
Amblygonite	$(Li,Na)Al(PO_4)(F,OH)$	3.0–3.1	5.5–6	tricl., wh., yelsh. wh., grnsh. wh., blsh. wh., gray	1.591, 1.604, 1.613
Analcite	$NaAlSi_2O_6.H_2O$	2.24–2.29	5.5	cub., wh., pink, gray	1.479–1.493
Anatase	TiO_2	3.90	5.5–6	tetr., br., yelsh. br., redsh. br., bl., blk., grn., gray	2.5612, 2.4880
Andalusite	Al_2OSiO_4	3.13–3.16	6.5–7.5	rhomb., pink, wh., red	1.629–1.640, 1.633–1.644, 1.638–1.650
Andesine	$([NaSi]_{0.7-0.5}[CaAl]_{0.3-0.5})AlSi_2O_8$	2.65–2.68	6–6.5	tricl., wh., gray, grn.	1.544–1.555, 1.548–1.558, 1.551–1.563
Andorite	$PbAgSb_3S_6$	5.33–5.37	3–3.5	rhomb., dk. steel gray, somet. tarn. yel. or iridesc.	
Andradite	$Ca_3Fe_2^{+3}Si_3O_{12}$	3.859	6–7.5	cub., brnsh. red, blk., somet. yel., grn.	1.887
Anglesite	$PbSO_4$	6.27–6.39	2.5–3	rhomb., col., wh., somet. gray, yelsh., grn. tinge	1.8771, 1.8826, 1.8937
Anhydrite	$CaSO_4$	2.98	3.5	rhomb., col., blsh. wh., vlt.	1.5698, 1.5754, 1.6136
Ankerite	$Ca(Fe,Mg,Mn)(CO_3)_2$	2.8–3.1	3.5–4	rhbdr., br., yelsh. br., brnsh. br., pink	1.690–1.750, 1.510–1.548
Anorthite	$CaAl_2Si_2O_8$	2.76	6–6.5	tricl., wh., yel., grn., blk.	1.577, 1.585, 1.590
Anorthoclase	$(Na,K)AlSi_3O_8$	2.56–2.60	6	tricl., col., wh.	1.523, 1.528, 1.529
Anthophyllite	$(Mg,Fe)_7Si_8O_{22}(OH,F)_2$	2.85–3.57	5.5–6	rhomb., wh., gray, grn., br., yelsh. br., dk. br.	1.596–1.694, 1.605–1.710, 1.615–1.722
Antimony	Sb	6.61–6.72	3–3.5	hex., tin-wh.	
Apatite	$Ca_5(PO_4)_3(OH,F,Cl)$	3.1–3.35	5	hex., grn., wh., yel., br. red, bl.	1.629–1.667, 1.624–1.666
Apophyllite	$KFCa_4Si_8(O)_{20}.8H_2O$	2.33–2.37	4.5–5	tetr., col., wh., pink, pa. yel., pa. grn.	1.534–1.535, 1.535–1.537
Aragonite	$CaCO_3$	2.94–2.95	3.5–4	rhomb., col., wh.	1.530–1.531, 1.680–1.681, 1.685–1.686
Arcanite	K_2SO_4	2.663		rhom., col., wh.	1.4935, 1.4947, 1.4973
Argentite	Ag_2S	7.2–7.4	2–2.5	cub., blksh. lead gray	
Arsenic	As	5.63–5.78	3.5	hex., tin-wh., tarn. dk. gray	
Arsenolite	As_2O_3	3.86–3.88	1.5	cub., wh., somet. blsh., yelsh., redsh. tinge	1.755
Arsenopyrite	$FeAsS$	5.9–6.2	5.5–6	monocl., silver-wh., to steel gray	
Atacamite	$Cu_2(OH)_3Cl$	3.74–3.78	3–3.5	rhomb., grn., dk. grn., blksh. grn.	1.831, 1.861, 1.880
Augelite	$Al_2(PO_4)(OH)_3$	2.696	4.5–5	monocl., col., wh., yelsh. wh., rose	1.5736, 1.5759, 1.5877
Augite	$(Ca,Mg,Fe,Fe^{+3},Ti,Al)_2(Si,Al)_2O_6$	3.23–3.52	5.5–6	monocl., pa. br., br., purp. br., grn., blk.	1.671–1.735, 1.672–1.741, 1.703–1.761
Autunite	$Ca(UO_2)_2(PO_4)_2.10–12H_2O$	3.1–3.2	2–2.5	tetr., yel., somet. grnsh. yel. to pa. grn.	1.577, 1.553
Axinite	$(Ca,Mn,Fe)_3Al_2BO_3Si_4O_{12}(OH)$	3.26–3.36	6.5–7	tricl., br., yelsh.	1.674–1.693, 1.681–1.701, 1.684–1.704
Azurite	$Cu_3(OH)_2(CO_3)_2$	3.77	3.5–4	monocl., azure bl., dk. bl., pa. bl.	1.730, 1.758, 1.838
Baddeleyite	ZrO_2	5.4–6.02	6.5	monocl., col., wh., gr., redsh. br., br., blk.	2.13, 2.19, 2.20
Barite	$BaSO_4$	4.50	3–3.5	rhomb., col., wh., somet. br., dk. br., gray	1.6362, 1.6373, 1.6482
Benitoite	$BaTi(SiO_3)_3$	3.65	6–6.5	rhbdr., bl., purp., col.	1.757, 1.804
Bertrandite	$Be_4Si_2O_7(OH)_2$	2.6	6	rhomb., col.	1.589, 1.602, 1.613
Beryl	$Be_3Al_2Si_6O_{18}$	2.66–2.83	7.5–8	hex., col., wh., blsh. grn., grnsh. yel., yel., bl., bl.	1.565–1.590, 1.567–1.598
Beryllonite	$NaBe(PO_4)$	2.81	5.5–6	monocl., col., wh., pa. yel.	1.5520, 1.5579, 1.561
Biotite	$K(Mg,Fe)_3AlSi_3O_{10}(OH,F)_2$	2.7–3.3	2.5–3	monocl., blk., br., redsh. br.	1.565–1.625, 1.605–1.696, 1.605–1.696
Bismuth	Bi	9.70–9.83	2–2.5	rhbdr., silver-wh. to redsh wh.	
Bismuthinite	Bi_2S_3	6.75–6.81	2	rhomb., lead gray to tin-wh., tarn. yel. or iridesc.	
Bixbyite	$(Mn,Fe)_2O_3$	4.945	6–6.5	cub., blk.	
Bloedite	$Na_2Mg(SO_4)_2.4N_2O$	2.22–2.28	2.5–3	monocl., col., somet. blsh.-grn. or redsh.	1.483, 1.486, 1.487
Boehmite	$AlO(OH)$	3.01–3.06	3.5–4	rhomb., wh.	1.64–1.65, 1.65–1.66, 1.65–1.67
Boracite	$Mg_3B_7O_{13}Cl$	2.91–2.97	7–7.5	rhomb., col., wh., gray, yel., blsh.-grn., grn.	1.66, 1.66, 1.67
Borax	$Na_2B_4O_7.10H_2O$	1.715	2–2.5	monocl., col., wh., gray, blsh. or grnsh-wh.	1.4466, 1.4687, 1.4717
Bornite	Cu_5FeS_4	5.06–5.08	3	cub., copper red to pinchbeck br., tarn. purp., iridesc.	
Boulangerite	$Pb_5Sb_4S_{11}$	6.0–6.2	2.5–3	monocl., blsh. lead gray, oft. with yel. spots	
Bournonite	$PbCuSbS_3$	5.80–5.86	2.5–3	rhomb., steel gray to blk.	
Braggite	PtS	10.0		tetr., steel gray	
Braunite	$(Mn,Si)_2O_3$	4.72–4.83	6–6.5	tetr., brns. blk. to steel gray	
Bravoite	$(Ni,Fe)S_2$	4.62	5.5–6	cub., steel gray	
Breithauptite	$NiSb$	8.23	5.5	hex., pa. copper red to vlt., tarn.	
Brochantite	$Cu_4(SO_4)(OH)_6$	3.79	3.5–4	monocl., emerald-grn. to blksh. grn., pa. grn.	1.728, 1.771, 1.800
Bromyrite	$AgBr$	6.47	2.5	cub., col., gray, yelsh., grnsh.-br.	2.253
Brookite	TiO_2	4.08–4.20	5.5–6	rhomb., br., yelsh. br., redsh. br., blk.	2.5831, 2.5843, 2.7004
Brucite	$Mg(OH)_2$	2.38–3.40	2.5	hex., wh., pa. grn., gray, bl., yel., br.	1.560–1.590, 1.580–1.600
Bunsenite	NiO	6.898	5.5	cub., dk. pistachio-grn.	(Li) 2.37
Cacoxenite	$Fe_4(PO_4)_3(OH)_3.12H_2O$	2.2–2.4	3–4	hex., yel. to brnsh.-yel., redsh. yel., somet. grnsh.	1.575–1.585, 1.635–1.656
Calcite	$CaCO_3$	2.715–2.94	3	rhbdr., col., wh., somet. gray, red., pink, bl.	1.658–1.740, 1.486–1.550
Caledonite	$Cu_2Pb_5(SO_4)_3(CO_3)(OH)_6$	5.75–5.77	2.5–3	rhomb., dk. grn., blsh. grn.	1.815–1.821, 1.863–1.869, 1.906–1.912
Calomel	$HgCl$	7.15	1.5	tetr., col., wh., gray, yelsh. wh., br.	1.973, 2.656
Cancrinite	$(Na,Ca)_{7-8}Al_6Si_6O_{24}(CO_3SO_4Cl)_{1.5-2}.1–5H_2O$	2.51–2.42	5–6	hex., col., wh., pa. bl., pa. grn., yel., redsh.	1.528–1.507, 1.503–1.495
Carnallite	$KMgCl_3.6H_2O$	1.602	2.5	rhomb., col., wh., oft. redsh., somet. yel., bl.	1.466, 1.475, 1.494
Carnotite	$K_2(UO_2)_2(VO_4)_2.3H_2O$		1–2	rhomb. or monocl., bright yel., yel., grnsh. yel.	1.75, 1.92, 1.95
Cassiterite	SnO_2	6.99	6–7	tetr., yelsh. or redsh. br., brnsh.-blk.	2.006, 2.0972
Celestite	$SrSO_4$	3.96	3–3.5	rhomb., col., wh., pa. bl., redsh., grnsh., brnsh.	1.621–1.622, 1.623–1.624, 1.630–1.631
Celsian	$BaAl_2Si_2O_8$	3.10–3.39	6–6.5	monocl., col., wh., yel.	1.579–1.587, 1.583–1.593, 1.588–1.600
Cervantite	$Sb_2O_4(?)$	6.64	4–5	rhomb.(?), yel., wh., somet. redsh.-wh.	
Cerargyrite	$AgCl$	5.55	2.5	cub., col., gray, grnsh.-br., tarn. purp., yelsh.	2.071
Cerussite	$PbCO_3$	6.53–6.57	3–3.5	rhomb., col., wh., gray, somet. bl., blk., grn.	1.8036, 2.0765, 2.0786
Chabazite	$(Ca,Na_2)Al_2Si_4O_{12}.6H_2O$	2.05–2.10	4.5	rhbdr., redsh. wh., wh., yelsh., grnsh.	1.470–1.494
Chalcocite	Cu_2S	5.5–5.8	2.5–3	rhomb., blksh., lead gray	
Chalcanthite	$CuSO_4.5H_2O$	2.28	2.5	tricl., dk. bl. to sky bl., somet. grnsh.	1.514, 1.537, 1.543
Chalcopyrite	$CuFeS_2$	4.1–4.3	3.5–4	tetr., brass-yel., tarn., iridisc.	
Chiolite	$Na_5Al_3F_{14}$	3.00	3.5–4	tetr., wh. to col.	1.349, 1.342
Chlorite	$(Mg,Al,Fe)_{12}(Si,Al)_8O_{20}(OH)_{16}$	2.6–3.3	2–3	monocl., grn., wh., yel., pink, br., red	1.57–1.66, 1.57–1.67, 1.57–1.67
Chloritoid	$(Fe,Mg,Mn)_2(AlFe^{+3})Al_3O_2SiO_4(OH)_4$	3.51–3.80	6.5	monocl., tricl., dk. grn.	1.713–1.730, 1.719–1.734, 1.723–1.740

Table 1-121. PHYSICAL CONSTANTS OF MINERALS *(Continued)*

Name	Formula	Sp. gr.	Hardness	Crystalline form and color	Index of refraction (Na) $\eta;\ \omega\ \epsilon$ $\alpha\ \beta\ \gamma$
Chondrodite	$Mg(OH,F)_2.2Mg_2SiO_4$	3.16–3.26	6.5	monocl., yel., br., red	1.592–1.615, 1.602–1.627, 1.621–1.646
Chromite	$FeCr_2O_4$	4.5–5.1	5.5	cub., blk.	2.16
Chrysoberyl	$BeAl_2O_4$	3.65–3.85	8.5	rhomb., (?), grn., yel.	1.746, 1.748, 1.756
Chrysocolla	$CuSiO_3.2H_2O$	~2.4	2	rhomb., (?.), grn., bl., br., blk.	1.575, 1.597, 1.598
Cinnabar	HgS	8.090	2–2.5	hex., red, brnsh. red, gray	(Li) 2.814, 3.143
Claudetite	As_2O_3	4.15	2.5	monocl., col. to wh.	1.87, 1.92, 2.01
Clinozoisite	$Ca_2Al_3Si_3O_{12}(OH)$	3.21–3.38	6.5	monocl., col., pa. gray, grn.	1.670–1.715, 1.674–1.725, 1.690–1.734
Cobaltite	$CoAsS$	6.33	5.5	cub., silver wh., redsh., steel gray, blk.	....
Colemanite	$Ca_2B_6O_{11}.5H_2O$	2.42–2.43	4.5	monocl., col., wh., yelsh. wh., gray	1.586, 1.592, 1.614
Columbite	$(Fe,Mn)(Cb,Ta)_2O_6$	5.15–5.25	6	rhomb., iron blk. to br. blk.	....
Connellite	$Cu_{19}(SO_4)Cl_4(OH)_{32}.3H_2O(?)$	3.36	3	hex., azure bl.	1.724–1.738, 1.746–1.758
Copiapite	$(Fe,Mg)Fe_4^{+3}(SO_4)_6(OH)_2.20H_2O$	2.08–2.17	2.5–3	tricl., yel., grnsh. yel.	1.51–1.53, 1.53–1.55, 1.58–1.60
Copper	Cu	8.95	2.5–3	cub., red	....
Coquimbite	$Fe_2(SO_4)_3.9H_2O$	2.10–2.12	2.5	hex., pa. vlt. to dk. amethystine, yelsh., grnsh.	1.53–1.55, 1.55–1.57
Cordierite	$Al_3(Mg,Fe)_2Si_5AlO_{18}$	2.53–2.78	7	rhomb., gray-bl., bl., dk. bl.	1.522–1.558, 1.524–1.574, 1.527–1.578
Corundum	Al_2O_3	4.022	9	hex., col., bl., yel., purp., grn., pink, red	1.767–1.772, 1.759–1.763
Cotunnite	$PbCl_2$	5.80	2.5	rhomb., col. to wh., somet. yelsh., grnsh.	2.199, 2.217, 2.260
Covellite	CuS	4.6–4.76	1.5–2	hex., indigo bl., dk. bl., iridesc. brass yel. to red	....
Cristobalite	SiO_2	2.33	6–7	tetr.(?.), col., wh., yel.	1.487, 1.484
Crocoite	$PbCrO_4$	5.96–6.02	2.5–3	monocl., red, orange red, orange yel.	2.29, 2.36, 2.66
Cryolite	Na_3AlF_6	2.96–2.98	2.5	monocl., col. to wh., brnsh., redsh., blk.	1.338, 1.338, 1.339
Cryolithionite	$Na_3Li_3Al_2F_{12}$	2.77	2.5–3	cub., col. to wh.	1.3395
Cubanite	$CuFe_2S_3$	4.03–4.18	3.5	rhomb., brass to bronze yel.	....
Cummingtonite	$(Mg,Fe)_7Si_8O_{22}(OH)_2$	3.2–3.5	5–6	monocl., dk. grn., br.	1.635–1.665, 1.644–1.675, 1.655–1.698
Cuprite	Cu_2O	6.14	3.5–4	cub., red, somet. blk.	....
Danburite	$CaSi_2B_2O_8$	3.0	7	rhomb., pa. yel., col., dk. yel., yelsh. br.	1.63, 1.63–1.64, 1.63–1.64
Datolite	$CaBSiO_4(OH)$	2.96–3.00	5–5.5	monocl., col., wh., yelsh., grnsh., pinksh.	1.622–1.626, 1.649–1.654, 1.666–1.670
Daubreelite	Cr_2FeS_4	3.80–3.82	?	cub., blk.	....
Derbylite	$FeSTi_3Sb_2O_{23}(?)$	4.53	5	rhomb., pitch blk.	2.45, 2.45, 2.51
Diamond	C	3.50–3.53	10	cub., col., pa. yel. to dk. yel., pa. br. to dk. br., wh., blsh. wh.	2.4175
Diaspore	$AlO(OH)$	3.3–3.5	6.5–7	rhomb., wh., graysh. wh., col.	1.682–1.706, 1.705–1.725, 1.730–1.752
Digenite	Cu_2S	5.546	2.5–3	cub., bl. to blk.	....
Diopside	$CaMgSi_2O_6$	3.22–3.38	5.5–6.5	monocl., wh., pa. grn., dk. grn.	1.664–1.695, 1.672–1.701, 1.695–1.721
Dioptase	$CuSi_6O_{18}.6H_2O$	3.5	5	rhbdr., emerald grn.	1.64–1.66, 1.70–1.71
Dolomite	$CaMg(CO_3)_2$	2.86	3.5–4	rhbdr., wh., oft. yel. or br. tinge, col.	1.679, 1.500
Douglasite	$K_2FeCl_4.2H_2O(?)$	2.16		pa. grn., tarn. brnsh. red	1.485–1.491, 1.497–1.503
Dyscrasite	Ag_3Sb	9.67–9.81	3.5–4	rhomb., silver wh., tarn. gray, yelsh. or blksh.	....
Eddingtonite	$BaAl_2Si_3O_{10}.4H_2O$	2.7–2.8		rhomb. or monocl., col., pink, br. wh.	1.541, 1.553, 1.557
Eglestonite	Hg_4OCl_2	8.4	2.5	cub., yel., orange-yel. to dk. brnsh., tarn. bl.	2.47–2.51
Emplectite	$CuBiS_2$	6.38	2	rhomb., gray to tin wh.	....
Empressite	$AgTe$	7.510	3–3.5	pa. bronze	....
Enargite	Cu_3AsS_4	4.4–4.5	3	rhomb., gray-blk. to iron-blk.	....
Enstatite	$MgSiO_3$	3.209	5–6	rhomb., col., gray, grn., yel., brn.	1.650–1.662, 1.653–1.671, 1.658–1.680
Epidote	$Ca_2Fe^{+3}Al_2Si_3O_{12}(OH)$	3.38–3.49	6	monocl., grn., yel., gray	1.715–1.751, 1.725–1.784, 1.734–1.797
Epsomite	$MgSO_4.7H_2O$	1.675–1.679	2–2.5	rhomb., col., wh. pink, grn.	1.4325, 1.4554, 1.4609
Erythrite	$(Co,Ni)_3(AsO_4)_2.8H_2O$	3.06	1.5–2.5	monocl., crimson-red, red, pa. pink	1.626, 1.661, 1.699
Eucairite	$CuAgSe$	7.6–7.8	2.5	silver wh. to lead gray	....
Euclasite	$BeAlSiO_4(OH)$	3.0–3.1	7.5	monocl., col., pa. grn., bl.	1.651, 1.655, 1.671
Eudialyte	$(Na,Ca,Fe)_6ZrSi_6O_{18}(OH,Cl)(?)$	2.8–3.1	5–6	hex., pa. pink, red, br.	1.59–1.61, 1.59–1.61
Eulytite	$Bi_4Si_3O_{12}$	6.6	4.5	cub., br., yel., gray	2.05
Euxenite	$(Y,Ca,Ce,U,Th)(Cb,Ta,Ti)_2O_6$	5.0–5.9	5.5–6.5	rhomb., blk., grnsh. or brnsh. tint.	~2.2
Fayalite	Fe_2SiO_4	4.392	6.5	rhomb., grnsh., yelsh.	1.827, 1.869, 1.879
Ferberite	$FeWO_4$	7.51	4–4.5	monocl., br. to blk.	(Li) 2.37–2.43
Fergussonite	$(Y,Er,Ce,Fe)(Cb,Ta,Ti)O_4$	5.6–5.8	5.5–6.5	tetr., gray, yel., br., dk. br.	2.1
Fluorite	CaF_2	3.18	4	cub., bl., purp., wh., col., yel., grn.	1.433–1.435
Forsterite	Mg_2SiO_4	3.222	7	rhomb., wh., grnsh., yelsh.	1.635, 1.651, 1.670
Franklinite	$ZnFe_2^{+3}O_4$	5.07–5.34	5.5–6.5	Cub., blk. to br.-blk.	(Li) ~2.36
Gahnite	$ZnAl_2O_4$	4.62	7.5–8	cub., dk. bl.-grn., somet. yelsh. or brnsh.	1.79–1.81
Galena	PbS	7.57–7.59	2.5–2.75	cub. lead gray	....
Galenabismuthite	$PbBi_2S_4$	7.04	2.5–3.5	rhomb., pa. gray to tin-wh., lead gray, somet. tarn., yel. or irid.	....
Ganomalite	$(Ca,Pb)_{10}(OH,Cl)_2(Si_2O_7)_3$	5.4–5.7	3–4	hex., col., gray	1.910, 1.945
Gaylussite	$Na_2Ca(CO_3)_2.5H_2O$	1.991	2.5–3	monocl., col. to yelsh. wh., graysh. wh., wh.	1.4435, 1.5156, 1.5233
Gehlenite	$Ca_2Al_2SiO_7$	3.038	5–6	tetr., col., gray-grn., br.	1.669, 1.658
Geikielite	$MgTiO_3$	4.05	5–6	rhbdr., brnsh blk., blsh.	2.31, 1.95
Gibbsite	$Al(OH)_3$	2.38–2.42	2.5–3.5	monocl., wh., graysh., grnsh. or redsh.-wh.	1.56–1.58, 1.56–1.58, 1.58–1.60
Glauberite	$Na_2Ca(SO_4)_2$	2.75–2.85	2.5–3	monocl., col., wh., somet. col., redsh.	1.515, 1.535, 1.536
Glauconite	$(K,Na,Ca)_{1.2-2}(Fe^{+3},Al,Fe,Mg)_4Si_{7-7.6}$ $Al_{1-0.4}O_{20}(OH)_4.nH_2O$	2.4–2.95	2	monocl., col., ye.lsh. grn., grn., blsh. gray	1.592–1.610, 1.614–1.641, 1.614–1.641
Glaucophane	$Na_2Mg_3Al_2Si_8O_{22}(OH)_2$	3.08–3.30	6	monocl., gray, lavender bl.	1.606–1.661, 1.622–1.667, 1.627–1.670
Gmelinite	$(Ca,Na_2)Al_2Si_4O_{12}.6H_2O$	~2.1	4.5	rhbdr., wh., redsh.-wh., yelsh., grnsh.	1.476–1.494, 1.474–1.480
Goethite	$FeO(OH)$	3.3–4.3	5–5.5	rhomb., blksh.-br., yelsh. or redsh.-br., yel.	2.260–2.275, 2.393–2.409, 2.398–2.515
Gold	Au	19.3	2.5–3	cub., yel.	....
Goslarite	$ZnSO_4.7H_2O$	1.978	2–2.5	rhomb., col., wh., somet. br., grn., bl.	1.4568, 1.4801, 1.4844
Graphite	C	2.09–2.23	1–2	hex., iron-blk. to steel gray	....
Greenockite	CdS	4.9	3–3.5	hex., yel. to orange	2.506, 2.529
Grossularite	$Ca_3Al_2Si_3O_{12}$	3.594	6–7.5	cub., wh., yel., grn., br., red	1.734
Gummite (R)	$UO_3.H_2O$	3.9–6.4	2.5–5	yel., orange, redsh.-yel., red, br. blk.	....
Gypsum	$CaSO_4.2H_2O$	2.30–2.37	2	monocl., wh., col., somet. gray, red, yel., br.	1.519–1.521, 1.523–1.526, 1.529–1.531
Halite	$NaCl$	2.16–2.17	2.5	cub., col., wh., orange, red	1.544
Hambergite	$Be_2(OH)(BO_3)$	2.36	7.5	rhomb., col. to gray, wh.	1.56, 1.59, 1.63
Hanksite	$Na_{22}K(SO_4)_9(CO_3)_2Cl$	2.562	3–3.5	hex., col., somet. pa.-yelsh. or gray	1.481, 1.461
Harmotome	$BaAl_2Si_6O_{16}.6H_2O$	2.41–2.47	4.5	monocl., or rhomb., col., wh., pink, gray, yel.	1.503–1.508, 1.505–1.509, 1.508–1.514
Hausmannite	Mn_3O_4	4.83–4.85	5.5	tetr. brnsh.-blk.	(Li) 2.46, 2.15
Haüyne	$(Na,Ca)_{4-8}Al_6Si_6O_{24}(SO_4S)_{1-2}$	2.44–2.50	5–6	cub., wh., gray, grn., bl.	1.496–1.505
Hedenbergite	$CaFeSi_2O_6$	3.50–3.56	6	monocl., brnsh.-grn., dk. grn., blk.	1.716–1.726, 1.723–1.730, 1.741–1.751
Helvite	$Mn_4Be_3Si_3O_{12}S$	3.20–3.44	6	cub., yel., br., redsh.-brn.	1.728–1.749
Hematite	Fe_2O_3	5.26	5–6	rhbdr., steel gray, dull red to bright red	3.22, 2.94
Hemimorphite	$Zn_4Si_2O_7(OH)_2.H_2O$	3.45	5	rhomb., col., wh., pa. bl., pa. grn., br.	1.614, 1.617, 1.636
Hercynite	$FeAl_2O_4$	4.40	7.5	cub., blk.	1.835
Herderite	$CaBe(PO_4)(Fe,OH)$	2.95–3.01	5–5.5	monocl., col. to pa. yel. or grnsh.-wh.	1.592, 1.612, 1.621
Hessite	Ag_2Te	8.24–8.45	2–3	monocl., (<149.5°), cub. (>149.5°), gray	....
Heulandite	$(Ca,Na_2)Al_2Si_7O_{18}.6H_2O$	2.1–2.2	3.5–4	pseudo-monocl., col., wh., yel., pink, red, gray, br.	1.491–1.505, 1.493–1.503, 1.500–1.512

Table 1-121. PHYSICAL CONSTANTS OF MINERALS *(Continued)*

Name	Formula	Sp. gr.	Hardness	Crystalline form and color	Index of refraction (Na) η; ω ϵ / ω β γ
Hopeite	$Zn_2(PO_4)_2.4H_2O$	3.0–3.1	3.25	rhomb., col. to grayish-wh., pa. yel.	1.57–1.59, 1.58–1.60, 1.58–1.60
Hornblende	$(Ca,Na,K)_{2-3}(Mg,Fe,Fe^{+3}Al)_5Si_6(Si,Al)_2$ $O_{22}(OH,F)_2$	3.02–3.45	5–6	monocl., grn., dk. grn., blk.	1.615–1.705, 1.618–1.714, 1.632–1.730
Huebnerite	$MnWO_4$	7.12	4–4.5	monocl., yel.-br. to red br., somet. br., blk.	2.17, 2.22, 2.32
Humite	$Mg(OH,F)_2.3MgSiO_4$	3.2–3.32	6	rhomb., yel., orange	1.607–1.643, 1.619–1.653, 1.639–1.675
Huntite	$Mg_3Ca(CO_3)_4$	2.696		rhomb.(?)., wh.	
Hydrogrossularite	$Ca_3Al_2Si_2O_8(SiO_4)_{1-m}(OH)_{4m}$	3.594–3.13	6–7.5	cub., wh., buff, pa. grn., gray, pink	1.734–1.675
Hydromagnesite	$Mg_4(OH)_2(CO_3.3H_2O$	2.236	3.5	monocl., col. to wh.	1.520–1.526, 1.524–1.530, 1.544–1.546
Illite	$K_{1-1.5}Al_4Si_{7-6.5}Al_{1-1.5}O_{20}(OH)_4$	2.6–2.9	1–2	monocl., wh.	1.54–1.57, 1.57–1.61, 1.57–1.61
Ilmenite	$FeTiO_3$	4.68–4.76	5–6	rhbdr., iron-blk.	
Iodyrite	AgI	5.69	1.5	hex., col. on exposure to light, yel., br.	2.21, 2.22
Jadeite	$NaAlSi_2O_6$	3.24–3.43	6	monocl., col., wh., grn., grnsh. blk.	1.640–1.658, 1.645–1.663, 1.652–1.673
Jamesonite	$Pb_4FeSb_6S_{14}$	5.63	2.5	monocl., gray-blk., somet. tarn. iridesc.	
Jarosite	$KFe_3(SO_4)_2(OH)_6$	2.91–3.26	2.5–3.5	rhbdr., ocherous, amber yel. to dk. br.	1.820, 1.715
Kainite	$KMg(SO_4)Cl.3H_2O$	2.15	2.5–3	monocl., col., gray bl., vlt., yelsh., redsh.	1.494, 1.505, 1.516
Kaliophyllite	$KAlSiO_4$	2.61	6	hex., col.	1.532, 1.537
Kaolinite	$Al_4Si_4O_{10}(OH)_8$	2.61–2.68	2–2.5	tricl. or monocl., wh., redsh.-wh., grnsh.-wh.	1.533–1.565, 1.559–1.569, 1.560–1.570
Kernite	$Na_2B_4O_7.4H_2O$	1.908	2.5	monocl., col., wh.	1.454, 1.472, 1.488
Kieserite	$MgSO_4.H_2O$	2.571	3.5	monocl., col., gray, wh., yelsh.	1.520, 1.533, 1.584
Kyanite	Al_2OSiO_4	3.53–3.65	5–7	tricl., bl., wh., gray, grn., yel., pink	1.712–1.718, 1.721–1.723, 1.727–1.734
Lanarkite	$Pb_2(SO_4)O$	6.92	2–2.5	monocl., gray to grnsh. wh., pa. yel.	1.925–1.931, 2.004–2.010, 2.033–2.039
Lanthanite	$(La,Ce)_2(CO_3)_3.8H_2O$	2.69–2.74	2.5–3	rhomb. to wh., pink, yelsh.	1.51–1.53, 1.584–1.590, 1.610–1.616
Laumontite	$CaAl_2Si_4O_{12}.4–3.5H_2O$	2.2–2.3	3–3.5	monocl., col., wh., red, yel., brn.	1.502–1.514, 1.512–1.522, 1.514–1.525
Laurionite	$Pb(OH)Cl$	6.24	3–3.5	rhomb., col. to wh.	2.08, 2.12, 2.16
Lawsonite	$CaAl_2(OH)_2Si_2O_7.H_2O$	3.05–3.10	6	rhomb., col., wh.	1.655, 1.674–1.675, 1.684–1.686
Lazulite	$(Mg,Fe)Al_2(PO_4)_2(OH)_2$	3.02–7.78	5.5–6	monocl., bl., blsh. wh., dk. bl., blsh. grn	1.604–1.626, 1.626–1.654, 1.637–1.663
Lazurite	$Na_4SSi_3Al_3O_{12}$	2.38–2.45	5–5.5	cub., berlin bl., azure bl., grnsh. bl., vlt.	1.500
Leadhillite	$Pb_4(SO_4)(CO_3)_2(OH)_2$	6.55	2.5–3	monocl., col. to wh., gray, pa. grn., pa. bl., yelsh.	1.87, 2.00, 2.01
Lepidocrocite	$FeO(OH)$	4.05–4.31	5	rhomb., ruby-red to red-br.	1.94, 2.20, 2.51
Lepidolite	$K_2(Li,Al)_{5-6}Si_{6-7}Al_{2-1}O_{20}(OH,F)_4$	2.80–2.90	2.5–4	monocl., col., pink, pa. purp.	1.525–1.548, 1.551–1.585, 1.554–1.587
Leucite	$KAlSi_2O_6$	2.47–2.50	5.5–6	tetr., (pseudo-cub.) wh., gray	1.508–1.511
Levyne	$(Ca,Na_2)Al_2Si_4O_{12}.6H_2O$	~2.1	4,5	rhbdr., wh., redsh. wh., yelsh., grnsh.	1–496–1.505, 1.491–1.500
Litharge	PbO	9.14	2	tetr., red	(Li) 2.665, 2.535
Loellingite	$FeAs_2$	7.39–7.41	5–5.5	rhomb., silver wh. to steel-gray	
Magnesite	$MgCO_3$	2.98–3.44	3.5–4.5	rhbdr., wh., col., somet. yel., br.	1.700–1.782, 1.509–1.563
Magnetite	Fe_3O_4	5.175	5.5–6.5	cub., blk. to br.-blk.	2.42
Malachite	$Cu_2(OH)_2(CO_3)$	4.03–4.07	3.5–4	monocl., bright grn. to dk. grn., blksh. grn.	1.652–1.658, 1.872–1.878, 1.906–1.912
Manganite	$MnO(OH)$	4.32–4.43	4	monocl., dk. steel-gray to iron-blk.	(Li) 2.25, 2.25, 2.53
Manganosite	MnO	5.364	5.5	cub., emerald grn., tarn. bl.	
Marcasite	FeS_2	4.887	6–6.5	rhomb., pa. bronze-yel., tin-wh.	
Marialite	$Na_4Al_3Si_9O_{24}Cl$	2.50–2.62	5–6	tetr., col., wh., pa. grnsh. yel., gray, br.	1.546–1.550, 1.540–1.541
Marshite	CuI	5.68	2.5	cub., col. to pa. yel., on exposure to light, red	2.346
Mascagnite	$(NH_4)_2SO_4$	1.768	2–2.5	rhomb., col., gray, yelsh.	1.5202, 1.5230, 1.5330
Matlockite	$PbFCl$	7.12	2.5–3	tetr., col. or yel. to pa. amber, grnsh.	2.145, 2.006
Meionite	$Ca_4Al_6Si_6O_{24}CO_3$	2.78	5–6	tetr., col., wh., pa. grnsh. yel., gray, br.	1.500–1.600, 1.556–1.562
Melanterite	$FeSO_4.7H_2O$	1.898	2	monocl., grn., grnsh. bl., grnsh. wh.	1.47, 1.48, 1.49
Melilite	$(Ca,Na,K)_2(Mg,Fe,Fe^{+3},Al,Si)_3O_7$	2.95–3.05	5–6	tetr., yelsh., br., grn.-br.	1.624–1.666, 1.616–1.661
Mellite	$Al_2C_{12}O_{12}.18H_2O$	1.64	2–2.5	tetr., yel., redsh., brnsh., somet. wh.	1.5393, 1.5110
Mendipite	$Pb_3O_2Cl_2$	7.24	2.5	rhomb., col. to wh., gray, oft. yel., red, bl. tinge	2.22–2.26, 2.25–2.29, 2.29–2.33
Mesolite	$Na_2Ca_2(Al_3Si_3O_{10})_3.8H_2O$	~2.26	5	monocl., col., wh., yel., pink, red	$\beta = 1.504–1.508$
Metacinnabar	HgS	7.65	3	cub., graysh.-blk.	
Microcline	$KAlSi_3O_8$	2.56–2.63	6–6.5	tricl., col., wh., pink, red, yel., grn.	1.514–1.529, 1.518–1.533, 1.521–1.539
Microlite	$(Na,Ca)_2Ta_2O_6(O,OH,F)$	4.2–6.4	5–5.5	cub., pa. yel. to br., somet. red, grn	~2.0
Miersite	AgI	5.64–5.68	2.5	cub., canary-yel.	2.18–2.22
Millerite	NiS	5.3–5.7	3–3.5	hex., pa. brass-yel. to bronze-yel., gray, tarn. iridesc.	
Mimetite	$Pb_5(AsO_4,PO_4)_3Cl$	7.24	3.5–4	hex., pa. yel. to yelsh. br., orange-yel., gray	2.147, 2.128
Minium	Pb_3O_4	8.9–9.2	2.5	scarlet red, bl. red, somet. yel. tint.	(Li) 2.40–2.44
Mirabilite	$Na_2SO_4.10H_2O$	1.490	1.5–2	monocl., col. to wh.	1.391–1.397, 1.393–1.399, 1.395–1.401
Moissanite	SiC	3.218	9.5	hex., grn. to blk., somet. blsh., red	2.647–2.649, 2.689–2.693
Molybdenite	MoS_2	4.62–4.73	1–1.5	hex., lead-gray	
Monazite	$(Ce,La,Th)PO_4$	5.0–5.3	5	monocl., yel., br., redsh. br.	1.774–1.800, 1.777–1.801, 1.828–1.851
Monetite	$CaH(PO_4)$	2.929	3.5	tricl., wh., pa. yelsh.-wh.	1.587, ~1.615, 1.640
Monticellite	$CaMgSiO_4$	3.08–3.27	5.5	rhomb., col.	1.630–1.654, 1.646–1.664, 1.653–1.674
Montmorillonite	$(0.5Ca,Na)_{0.7}(Al,Mg,Fe)_4(Si,Al)_8O_{20}(OH)_4.nH_2O$	2–3	1–2	monocl., wh., yel., grn.	1.48–1.61, 1.50–1.64, 1.50–1.64
Montroydite	HgO	11.23	2.5	rhomb., dk. red to brnsh. red, br.	(Li) 2.37, 2.5, 2.65
Mordenite	$(Na_2,K_2Ca)Al_2Si_{10}O_{24}.7H_2O$	2.12–2.15	3–4	rhomb., col., wh., red, yel., br.	1.472–1.483, 1.475–1.485, 1.477–1.487
Muscovite	$KAl_2Si_3AlO_{10}(OH,F)_2$	2.77–2.88	2.5–3	monocl., col., pa. grn., pa. red, pa. br.	1.552–1.574, 1.582–1.610, 1.587–1.616
Nantokite	$CuCl$	4.136	2.5	cub., col. to wh., grayish, grn.	1.925–1.935
Natrolite	$Na_2Al_2Si_3O_{10}.2H_2O$	2.20–2.26	5	rhomb., col., wh., gray, yel., pink, red	1.473–1.483, 1.476–1.486, 1.485–1.496
Nepheline	$Na_3KAl_4Si_4O_{16}$	2.56–2.665	5.5–6	hex., col., wh., gray	1.529–1.546, 1.526–1.542
Newberyite	$MgH(PO_4).3H_2O$	2.10	3.0–3.5	rhomb., col.	1.511–1.517, 1.514–1.520, 1.530–1.536
Niccolite	$NiAs$	7.784	5	hex., pa. copper-red, tarn. gray to blk.	
Nosean	$Na_8Al_6Si_6O_{24}SO_4$	2.30–2.40	5.5	cub., gray, bl., br.	1.495
Oldhamite	CaS	2.58	4	cub., pa. chestnut-br.	2.137
Oligoclase	$([NaSi]_{0.9-0.7}[CaAl]_{0.1-0.3})AlSi_2O_8$	2.63–2.65	6–6.5	tricl., col., wh., gray, grnsh., pink	1.533–1.544, 1.537–1.548, 1.543–1.552
Olivenite	$Cu_2(AsO_4)(OH)$	3.9–4.5	3	rhomb., olive grn., grnsh.-br., br., gray	1.75–1.78, 1.79–1.82, 1.83–1.87
Olivine	$(Mg,Fe)SiO_4$	3.22–4.39	6.5–7	rhomb., olive grn., grayish grn. to yelsh. br.	1.63–1.83, 1.65–1.87, 1.67–1.88
Opal	$SiO_2.nH_2O$	1.73–2.16	~6	col., wh., yel., br., red, grn., bl., blk.	1.41–1.46
Orpiment	As_2S_3	3.49	1.5–2	monocl., lemon-yel.	(Li) 2.4, 2.81, 3.02
Orthoclase	$KAlSi_3O_8$	2.55–2.63	6–6.5	monocl., col., wh., pink, red, yel., grn.	1.518–1.529, 1.522–1.533, 1.522–1.539
Orthopyroxene	$(Mg,Fe)SiO_3$	3.209–3.96	5–6	rhomb., col., gray, grn., yel., dk. brn.	1.650–1.768, 1.653–1.770, 1.658–1.788
Paragonite	$NaAl_2Si_3AlO_{10}(OH)_2$	2.85	2.5	monocl., col., pa. yel.	1.564–1.580, 1.594–1.609, 1.600–1.609
Parisite	$(Ce,La,Na)FCO_3.CaCO_3$	4.42	4.5	hex., brnsh., yel.	1.672, 1.771
Pectolite	$Ca_2NaH(SiO_3)_3$	2.86–2.90	4.5–5	tricl., col., wh.	1.595–1.610, 1.605–1.615, 1.632–1.645
Penfieldite	$Pb_2Cl_3(OH)$	6.6		hex., wh.	2.13, 2.21
Pentlandite	$(Fe,Ni)_9S_8$	4.6–5.0	3.5–4	cub., pa. bronze-yel.	
Percylite	$PbCuCl_2(OH)_2(?)$	?	2.5	cub(?), sky bl.	2.04–2.06
Periclase	MgO	3.55–3.68	5.5	cub., col. to gray-wh., yel., brnsh. yel., grn., bl.	1.7350
Pekovskite	$CaTiO_3$	3.97–4.26	5.5	pseudo cub., blk., gray-blk., brnsh. bl., redsh. br., bl.	2.30–2.38
Petalite	$LiAlSi_4O_{10}$	2.412–2.422	6.5	monocl., wh., gray, somet. pink, grn.	1.504–1.507, 1.510–1.513, 1.516–1.523

Table 1-121. PHYSICAL CONSTANTS OF MINERALS (*Continued*)

Name	Formula	Sp. gr.	Hardness	Crystalline form and color	Index of refraction (Na) $\eta; \ \omega \ \epsilon$ $\alpha \ \beta \ \gamma$
Pharmacosiderite...	$Fe_3(AsO_4)_2(OH)_3.5H_2O$..............	2.797	2.5	cub., olive-grn. to yel., br., redsh.	1.676–1.704
Phenakite...	Be_2SiO_4.	2.98	7.5	rhbder., col., rose, yel., br.	1.654, 1.670
Phillipsite...	$(0.5Ca,Na,K)_3Al_3Si_5O_{16}.6H_2O$.	2.2	4–4.5	monocl. or rhomb., col., wh., pink, gray, yel.	1.483–1.504, 1.484–1.509, 1.496–1.514
Phlogopite...	$KMg_3AlSi_3O_{10}(OH,F)_2$.	2.76–2.90	2–2.5	monocl., col., yelsh.-br., grn., redsh.-br., br.	1.530–1.590, 1.557–1.637, 1.558–1.637
Phosgenite......	$Pb_2(CO_3)Cl_2$.	6.133	2–3	tetr., yelsh. wh. to yelsh. br., br., somet. wh., rose, gray	2.1181, 2.1446
Piemontite.........	$Ca_2(Mn,Fe^{+3},Al)_2AlSi_3O_{12}(OH)$.	3.45–3.52	6	monocl., redsh. brn., blk.	1.732–1.794, 1.750–1.807, 1.762–1.829
Pigeonite...	$(Mg,Fe,Ca)(Mg,Fe)Si_2O_6$.	3.30–3.46	6	monocl., br., br., br., blk.	1.682–1.722, 1.684–1.722, 1.705–1.751
Platinum...	Pt.	14–19	4–4.5	cub., whitish, steel gray to dk. gray	
Pollucite...	$CsAlSi_2O_6$.	2.9	6.5	tetr., (pseudo-cub.) col.	1.507–1.527
Polybasite...	$(Ag,Cu)_{16}Sb_2S_{11}$.	6.0–6.2	2–3	monocl., iron-blk.	
Powellite...	$Ca(Mo,W)O_4$.	4.21–4.25	3.5–4	tetr., straw-yel., br., grnsh., somet. gray, bl., blk.	1.959–1.982, 1.967–1.993
Prehnite...	$Ca_2Al_2Si_3O_{10}(OH)_2$.	2.90–2.95	6–6.5	rhomb., pa. grn., yel., gray, wh.	1.611–1.632, 1.615–1.642, 1.632–1.665
Proustite...	Ag_3AsS_3.	5.57	2–2.5	rhbdr., scarlet-vermillion	3.0877, 2.7924
Pseudobrookite...	Fe_2TiO_5.	4.33–4.39	6	rhomb., dk. red-br. to brnsh. blk. and blk.	(Li) 2.38, 2.39, 2.42
Psilomelane...	$BaMn^{+2}Mn_8^{+4}O_{16}(OH)_4$.	4.71	5–6	rhomb., iron-blk. to steel-gray	
Pumpellyite...	$Ca_4(Mg,Fe,Mn)(Al,Fe^{+3},Ti)_5(OH)_3Si_6O_{21}.2H_2O$.	3.18–3.23	6	monocl., grn., blsh. grn., br.	1.674–1.702, 1.675–1.715, 1.688–1.722
Pyrargyrite......	Ag_3SbS_3.	5.85	2.5	rhbdr., deep red	(Li) 3.084, 2.881
Pyrite...	FeS_2.	5.018	6–6.5	cub., pa. brass-yel., tarn. iridesc.	
Pyrochlore...	$NaCaCb_2O_6F$.	4.2–6.4	5–5.5	cub., br. to blk., yelsh., redsh. or blksh. br.	
Pyrochroite...	$Mn(OH)_2$.	3.23–3.27	2.5	hex., col. to pa. grn. or bl., tarn. br. to blk.	1.72, 1.68
Pyrolusite...	MnO_2.	5.04–5.08	6–6.5	tetr., pa. steel-gray, iron-gray, blk., blsh.	
Pyromorphite...	$Pb_5(PO_4,AsO_4)_3Cl$.	7.00–7.08	3.5–4	hex., grn., br., orange, brnsh. red., gray	2.058, 2.048
Pyrope...	$Mg_3Al_2Si_3O_{12}$.	3.582	6–7.5	cub., red, pink	1.714
Pyrophyllite...	$Al_2Si_4O_{10}(OH)_2$.	2.65–2.90	1–2	monocl., wh., yel., pa. bl., gray-grn., brnsh.-grn.	1.534–1.556, 1.568–1.589, 1.596–1.601
Pyrrhotite...	$Fe_{1-x}S$.	4.58–4.65	3.5–4.5	hex., bronze-yel. to br., tarn., somet. iridesc.	
Quartz...	SiO_2.	2.65	7	rhbdr., col., wh., blk., purp., grn., bl., rose	1.544, 1.553
Rammelsbergite...	$NiAs_2$.	7.0–7.2	5.5–6	tin. wh., redsh. tinge	
Raspite...	$PbWO_4$.	8.46	2.5–3	monocl., yelsh. br., pa. yel., gray	1.25–1.29, 1.25–1.29, 1.28–1.32
Realgar...	AsS.	3.56	1.5–2	monocl., aurora-red to orange-yel.	2.538, 2.684, 2.704
Riebeckite...	$Na_2Fe_3Fe_2^{+3}Si_8O_{22}(OH,F)_2$.	3.02–3.42	5	monocl., dk. bl., bl.	1.654–1.701, 1.662–1.711, 1.668–1.717
Rhodochrosite...	$MnCO_3$.	3.70	3.5–4	rhbdr., pink, red, br., brnsh.-yel.	1.816, 1.597
Rhodonite...	$(Mn,Fe,Ca)SiO_3$.	3.57–3.76	5.5–6.5	tricl., pink to brnsh. red	1.711–1.738, 1.716–1.741, 1.724–1.751
Rutile...	TiO_2.	4.23–5.5	6–6.5	tetr., redsh. brn. to red, somet. yelsh., blsh.	2.605–2.613, 2.899–2.901
Safflorite...	$(Co,Fe)As_2$.	7.0–7.5	4.5–5	rhomb., tin-wh., tarn. dk. gray	
Samarskite...	$(Y,Er,Ce,U,Ca,Fe,Pb,Th)(Cb,Ta,Ti,Sn)_2O_6$.	5.69	5–6	rhomb., velvet blk., somet. brnsh. tint	~2.20
Sapphirine...	$(Mg,Fe)_2Al_4O_6SiO_4$.	3.40–3.58	7.5	monocl., pa. bl., pa. grn.	1.701–1.717, 1.703–1.720, 1.705–1.724
Scapolite...	$(Na,Ca)_4Al_3(Al,Si)_3Si_6O_{24}(Cl,F,OH,CO_3,SO_4)$	2.50–2.78	5–6	tetr., col., wh., pa. grnsh. yel., gray, bl.	1.546–1.600, 1.540–1.562
Scheelite...	$CaWO_4$.	6.08–6.12	4.5–5	tetr., yelsh. wh., pa. yel., brnsh., col., wh., gray	1.920, 1.936
Scolecite...	$CaAl_2Si_3O_{10}.3H_2O$.	2.25–2.29	5	monocl., col., wh., gray, yel., pink, red	1.507–1.513, 1.516–1.520, 1.517–1.521
Scorodite...	$Fe^{+3}(AsO_4).2H_2O$.	3.28	3.5–4	rhomb., pa. grn., gray grn., br. somet. col., blsh., yel.	1.784, 1.795, 1.814
Sellaite...	MgF_2.	3.15	5	tetr., col. to wh.	1.378, 1.390
Senarmontite...	Sb_2O_3.	5.50	2–2.5	pseudo-cub., col., gray-wh.	2.087
Serpentine...	$Mg_3Si_2O_5(OH)_4$.	~2.55	2.5–3.5	monocl., wh., yel., gray, grn., blsh. grn.	1.53–1.57, 1.56, 1.54–1.57
Siderite...	$FeCO_3$.	3.96	4–4.5	rhbdr., yelsh. br., br., dk. br.	1.875, 1.635
Sillimanite...	Al_2OSiO_4.	3.23–3.27	6.5–7.5	rhomb., col., wh., yelsh., br., grnsh.	1.654–1.661, 1.658–1.662, 1.637–1.683
Silver...	Ag.	10.1–11.1	2.5–3	cub., wh., tarn. gray or blk.	
Skutterudite...	$(Co,Ni)As_3$.	6.1–6.9	5.5–6	cub., between tin-wh. and silver-gray, tarn. gray or iridesc.	
Smithsonite...	$ZnCO_3$.	4.42–4.44	4–4.5	rhbdr., grayish wh. to dk. gray, grnsh., brnsh. wh.	1.848, 1.621
Sodalite...	$Na_8Al_6Si_6O_{24}Cl_2$.	2.27–2.33	5.5–6	cub., bl., grn., yel., gray, pink	1.483–1.487
Sperrylite...	$PtAs_2$.	10.58	6–7	cub., tin-wh.	
Spessartite...	$Mn_3Al_2Si_3O_{12}$.	4.190	6–7.5	cub., blk., dk. red, brnsh. red., bl., yelsh. orange	1.800
Sphalerite...	ZnS.	3.9–4.1	3.5–4	cub., br., blk., yel., red, wh.	2.369
Sphene...	$CaTiSiO_4(O,OH,F)$.	3.45–3.55	5	monocl., col., yel., grn., br., blk.	1.843–1.950, 1.870–2.034, 1.943–2.110
Spinel...	$MgAl_2O_4$.	3.55	7.5–8	cub., grn., red, bl., br. to col.	1.719
Spodumene...	$LiAlSi_2O_6$.	3.03–3.22	6.5–7	monocl., col., gray-wh., pa. bl., pa. grn., yelsh.	1.648–1.663, 1.655–1.669, 1.662–1.679
Stannite...	Cu_2FeSn_4.	4.3–4.5	4	tetr., steel gray to iron blk.	
Staurolite...	$(Fe,Mg)_2(AlFe^{+3})_9O_6SiO_4(O,OH)_2$.	3.74–3.83	7.5	monocl., brn., redsh., yelsh.	1.739–1.747, 1.745–1.753, 1.752–1.761
Stercorite...	$Na(NH_4)H(PO_4).4H_2O$.	1.615	2	tricl., wh., yelsh., brnsh.	1.439, 1.442, 1.469
Stibiotantalite...	$Sb(Ta,Cb)O_4$.	5.7–7.5	5.5	rhomb., dk. br. to pa. yel.-br., red-br., grnsh.-yel.	2.38, 2.41, 2.46
Stibnite...	Sb_2S_3.	4.61–4.65	2	rhomb., lead-gray to steel-gray	
Stilbite...	$(Ca,Na_2K_2)Al_2Si_7O_{18}.7H_2O$.	2.1–2.2	3.5–4	monocl., wh., yel., pink, red, gray, br.	1.484–1.500, 1.492–1.507, 1.494–1.513
Stilpnomelane...	$(K,Na,Ca)_{0-1.4}(Fe^{+3}Fe,Mg,Al)_{6-8}Si_8O_{20}(OH)_4(O,OH,H_2O)_{4-8}$	2.59–2.96	3–4	monocl., br., dk. br., redsh. br., blk., dk. grn.	1.543–1.634, 1.576–1.745, 1.576–1.745
Stolzite...	$PbWO_4$.	7.9–8.4	2.5–3	tetr., redsh. br., yelsh. gray, straw-yel., grnsh.	2.26–2.28, 2.18–2.20
Strengite...	$Fe^{+3}(PO_4).2H_2O$.	2.90	3.5–4.5	rhomb., red, carmine, vlt., near col.	1.707, 1.719, 1.741
Strontianite...	$SrCO_3$.	3.72	3.5	rhomb., col., wh., yel., grnsh., brnsh.	1.516–1.520, 1.664–1.667, 1.666–1.669
Struvite...	$Mg(NH_4)(PO_4).6H_2O$.	1.71	2	rhomb., col., somet. yelsh., brnsh.	1.495, 1.496, 1.504
Sulfur...	S.	2.07	1.5–2.5	rhomb., yel., brnsh., grnsh., redsh., gray	1.9579, 2.0377, 2.2452
Sylvanite...	$(Ag,Au)Te_2$.	8.161	1.5–2	monocl., steel-gray to silver-wh.	
Sylvite...	KCl.	1.99	2	cub., col., wh., somet. grayish, blsh., yelsh., red	1.49031
Talc...	$Mg_3Si_4O_{10}(OH)_2$.	2.58–2.83	1	monocl., col., wh., pa. grn., dk. grn., br.	1.539–1.550, 1.589–1.594, 1.589–1.600
Tantalite...	$(Fe,Mn)(Ta,Cb)_2O_6$.	7.90–8.00	6.5	rhomb., iron-bl. to br.-blk.	2.26, 2.32, 2.43
Tapiolite...	$FeTa_2O_6$.	7.9	6–6.5	tetr., blk.	(Li) 2.27, 2.42
Tellurobismuthite...	Bi_2Te_3.	7.800–7.830	1.5–2	rhbdr., pa. lead-gray	
Terlinguaite...	Hg_2OCl.	8.725	2.5	monocl., yel. to grnsh.-yel., somet. br.	(Li) 2.33–2.37, 2.62–2.66, 2.64–2.68
Tetrahedrite...	$(Cu,Fe)_{12}Sb_4S_{13}$.	4.6–5.1	3–4.5	cub., flint-gray to iron-blk. to dull-blk.	
Thenardite...	Na_2SO_4.	2.664	2.5–3	rhomb., col., grayish-wh., yelsh., yelsh. br., redsh.	1.464–1.471, 1.473–1.477, 1.481–1.485
Thermonatrite...	$Na_2CO_3.H_2O$.	2.255	1–1.5	rhomb., col. to wh., grayish, yelsh.	1.420, 1.506, 1.524
Thomsenolite...	$NaCaAlF_6.H_2O$.	2.981	2	monocl., col. to wh., somet. brnsh., redsh.	1.4072, 1.4136, 1.4150
Thomsonite...	$NaCa_2([Al,Si]_5O_{10})_2.6H_2O$.	2.10–2.39	5–5.5	rhomb., col., wh., pink, br.	1.497–1.530, 1.513–1.533, 1.518–1.544
Thorianite (R)...	ThO_2.	9.7	6.5	cub., dk. gray to brnsh.-blk., blk.	~2.20
Thorite (R)...	$ThSiO_4$.	5.2–5.4	4.5–5	tetr., orange-yel., brnsh. to blk.	~1.8
Topaz...	$Al_2SiO_4(OH,F)_2$.	3.49–3.57	8	rhomb., col., wh., yel., gray, grn., red, bl.	1.606–1.629, 1.609–1.631, 1.616–1.638

Table 1-121. PHYSICAL CONSTANTS OF MINERALS (Continued)

Name	Formula	Sp. gr.	Hard-ness	Crystalline form and color	Index of refraction (Na) $\eta;\ \omega\ \epsilon$ $\omega\ \beta\ \gamma$
Torbernite (R)....	$Cu(UO_2)_2(PO_4)_2.8-12H_2O$	3.22	2-2.5	tetr., various shades of grn.	1.592, 1.582
Tourmaline........	$Na(Mg,Fe,Mn,Li,Al)_3Al_6Si_6O_{18}(BO_3)_3$ $(OH,F)_4$	3.03-3.25	7	rhbdr., blk., bl., grn., yel., red, col., br.	1.635-1.675, 1.610-1.650
Tremolite.........	$Ca_2Mg_5Si_8O_{22}(OH,F)_2$.	3.0	5-6	monocl., col., gray, wh.	1.599, 1.612, 1.622
Tridymite.........	SiO_2.	2.27	7	rhomb., col., wh.	1.471-1.479, 1.472-1.480, 1.474-1.483
Triphyllite-Lithiophyllite........	$Li(Fe,Mn)PO_4$.	3.34-3.58	4-5	rhomb., blsh. or grnsh. gray to yelsh. br., br.	1.66-1.70, 1.67-1.70, 1.68-1.71
Troegerite (R).....	$(UO_2)_3(AsO_4)_2.12H_2O$.		2-3	tetr., lemon-yel.	1.58-1.59, 1.625-1.635
Trona............	$Na_3H(CO_3)_2.2H_2O$.	2.14	2.5-3	monocl.. gray or yelsh. wh., col.	1.412, 1.492, 1.540
Turquois.........	$Cu(Al,Fe^{+3})_6(PO_4)_4(OH)_8.4H_2O$.	2.6-3.2	4.5-6	tricl., bl., grn., grnsh.-gray	1.61-1.78, 1.62-1.84, 1.65-1.84
Ullmannite.......	$NiSbS$.	6.61-6.69	5-5.5	cub., steel-gray to silver-wh.	
Uraninite (R).....	UO_2.	8.0-11	5-6	cub., steel-blk., brnsh.-blk., grayish, grn.	....
Uvarovite........	$Ca_3Cr_2Si_3O_{12}$.	3.90	6-7.5	cub., emerald-grn.	1.86
Valentinite.......	Sb_2O_3.	5.76	2.5-3	rhomb., col. to wh., somet. yelsh., redsh., gray, br.	2.18, 2.35, 2.35
Vanadinite........	$Pb_5(VO_4)_3Cl$.	6.5-7.1	2.75-3	hex., orange-red, red, brnsh.-red, br., brnsh.-yel., yel.	2.416, 2.350
Variscite-Strengite..	$(Al,Fe^{+3})(PO_4).2H_2O$.	2.57-2.87	3.5-4.5	rhomb., pa. grn., grn., blsh.-grn., red, vlt., col.	1.563-1.707, 1.588-1.719, 1.594-1.741
Vaterite.........	$CaCO_3$.	2.645	....	hex., col.	1.550, 1.640-1.650
Vermiculite.......	$(Mg,Ca)_{0.7}(Mg,Fe^{+3}Al)_6(Al,Si)_8O_{20}(OH)_4.$ $8H_2O$	~2.3	~1.5	monocl., col., yel., grn., br.	1.525-1.564, 1.545-1.583, 1.545-1.583
Vesuvianite.......	$Ca_{10}(Mg,Fe)_2Al_4(Si_2O_7)_2(SiO_4)_5(OH,F)_4$	3.33-3.43	6-7	tetr., yel., grn., br.	1.700-1.746, 1.703-1.752
Villiaumite.......	NaF.	2.79	2-2.5	cub., carmine, (nat.), col. (artif.)	1.327
Vivianite.........	$Fe_3(PO_4)_2.8H_2O$.	2.67-2.69	1.5-2	monocl., col., tarn. pa. bl., grnsh. bl., dk. bl., blsh. blk.	1.579-1.616, 1.602-1.656, 1.629-1.675
Wagnerite........	$Mg_2(PO_4)F$.	0.15	5.5	monocl., yel., gray, somet. red, grn.	1.569, 1.579, 1.582
Wavellite........	$Al_3(OH)_3(PO_4)_2.5H_2O$.	2.36	3.25-4	rhomb., grnsh. wh., grn. to yel., somet. br., bl., wh.	1.520-1.535, 1.526-1.543, 1.545-1.561
Whewellite.......	$Ca(C_2O_4).H_2O$.	2.23	2.5-3	monocl., col., somet. yelsh., brnsh.	1.491, 1.554, 1.650
Willemite........	Zn_2SiO_4.	3.9-4.1	5.5	rhbdr., wh., yel., grn., red, gray, br.	1.691, 1.719
Witherite........	$BaCO_3$.	4.29-4.30	3.5	rhomb., col., wh., gray, yelsh. br.	1.529, 1.676, 1.677
Wolframite.......	$(Fe,Mn)WO_4$.	7.12-7.51	4-4.5	monocl., dk. gray, brnsh. blk. to iron blk.	(Li) ~2.26, 2.32, 2.42
Wollastonite......	$CaSiO_3$	2.87-3.09	4.5-5	tricl., wh., col., gray, pa. grn.	1.616-1.640, 1.628-1.650, 1.631-1.653
Wulfenite........	$PbMoO_4$.	6.5-7.0	2.75-3	tetr., orange-yel. to yel., gray, grn., br., red	2.403, 2.283
Wurtzite.........	ZnS.	3.98	3.5-4	hex., brnsh. blk.	2.356, 2.378
Xenotime........	$Y(PO_4)$.	4.4-5.1	4-5	tetr., yelsh. br. to redsh. br., somet. gray, wh., pa. yel., grnsh.	1.721, 1.816
Zeunerite (R).....	$Cu(UO_2)_2(AsO_4)_2.10-16H_2O$.			tetr.	1.602-1.610
Zincite...........	ZnO.	5.64-5.68	4	hex., orange-yel. to dk. red, somet. yel.	2.013, 2.029
Zircon...........	$ZrSiO_4$.	4.6-4.7	7.5	tetr., redsh. br., yel., gray, grn., col.	1.923-1.960, 1.968-2.015
Zoisite...........	$Ca_2Al_3Si_3O_{12}(OH)$.	3.15-3.365	6	rhomb., gray, grnsh., brnsh.	1.685-1.705, 1.688-1.710, 1.697-1.725

*From: "Handbook of Chemistry and Physics", 53rd ed., R.C. Weast, Ed., The Chemical Rubber Co., 1972.

Table 1-122. COMMON REFRACTORY MATERIALS

For density in kg/m^3, multiply value in lb/ft^3 by 16.02. For specific heat in $J/kg \cdot K$, multiply value in $Btu/lbm \cdot deg\ F$ by 4,184. For thermal conductivity in $W/m \cdot K$, multiply value in $Btu/ft \cdot hr \cdot deg\ F$ by 1.73.

Name	Formula	Density, lb_m/ft^3	Specific heat, Btu/lb_m	Mean thermal conductivity, $Btu/ft\ hr, °F$ (to 1000°)	Coefficient of cubical expansion per °F	Maximum use temperature, °F	Melting point, °F
Alumina	Al_2O_3	230	.24	2.0	8×10^{-6}	3,300	3,700
Beryllium oxide	BeO	190	.24	—	9×10^{-6}	4,000	4,600
Calcium oxide	CaO	200	.18	4.5	13×10^{-6}	4,200	4,700
Carbon, graphite	C	120	.36	7	3×10^{-6}	4,000	6,500[a]
Chrome	40% Cr_2O_3	200	.20	1.0	8×10^{-6}	3,200	3,800
Corundum	90% Al_2O_3	200	.22	1.5	7×10^{-6}	3,200	—
Forsterite	$2\ MgO \cdot SiO_2$	160	.23	1.2	10×10^{-6}	3,000	3,300
Magnesia	MgO	210	.25	2.3	11×10^{-6}	4,000	5,000
Magnesium oxide	MgO	175	.25	2.0	10×10^{-6}	3,500	5,000
Mullite	$3\ Al_2O_3 \cdot SiO_2$	160	.23	1.2	5×10^{-6}	3,000	3,350
Silica	SiO_2	110	.24	1.0	7×10^{-6}	2,800	3,100
Silicon carbide	SiC	170	.23	8	3×10^{-6}	3,000	4,000[b]
Spinel	$MgO \cdot Al_2O_3$	220	.23	5	7×10^{-6}	3,300	—
Titanium oxide	TiO_2	260	.17	2.2	8×10^{-6}	3,000	3,300
Zircon	$ZrO_2 \cdot SiO_2$	220	.15	1.3	5×10^{-6}	3,500	4,500
Zirconium oxide	ZrO_2	360	.13	1.3	4×10^{-6}	4,400	4,800

Most of these materials are available commercially as refractory tile, brick, and mortar. Properties will depend on form, purity, and mixture. Temperatures given should be considered as high limits.

[a] Sublimes.
[b] Dissociates.

Table 1-123.   AMERICAN WOODS—PROPERTIES AND USES*

For weight-density in kg/m³, multiply value in lb/ft³ by 16.02.

Species	Specific gravity		Characteristics	Uses	Weight		
	Green	Dry			lb/cu ft, green	lb/cu ft, air-dry 12%	lb/1000 board ft, air-dry 12%
Alder, red	0.37	0.41	Low shrinkage; moderate in strength, shock resistance, hardness, and weight†	Furniture; sash; doors; millwork	46	28	2330
Ash, black	0.45	0.49	Light in weight†	Cabinets; veneer; cooperage, containers	52	34	2830
Ash, Oregon	0.50	0.55	Similar to but lighter than white ash†	Similar to white ash	46	38	3160
Ash, white	0.54	0.58	Heavy; hard; stiff; strong; high shock resistance†	Handles; ladder rungs; baseball bats; farm implements; car parts	48	41	3420
Bald cypress (Southern cypress)			Moderate in strength, weight, hardness, and shrinkage**	Building construction; beams; posts; ties; tanks; ships; paneling	51	32	2670
Beech, American	0.56	0.64	Heavy; high strength, shock resistance, and shrinkage; uniform texture†	Flooring; furniture; handles; kitchenwear; ties (treated)	54	45	3750
Birch	0.57	0.63	Heavy; high strength, shock resistance, and shrinkage; uniform texture†	Interior finish; dowels; ties (treated); veneer; musical instruments	57	44	3670
Cottonwood	0.37	0.40	Uniform texture; does not split readily; moderate in weight, strength, hardness, and shrinkage	Crates; trunks; car parts; farm implements	49	28	2330
Douglas fir	0.41	0.44	Moderate in strength, weight, shock resistance, and shrinkage‡	Building and construction; poles; veneer; plywood; ships; furniture; boxes	38	34	2830
Elm	0.57	0.63	Moderate in strength, weight, and hardness; high in shock resistance and shrinkage; good in bending†	Cooperage; baskets; crates; veneer; vehicle parts	54	34	2920
Hemlock, Eastern	0.38	0.40	Moderate in weight, strength, and hardness†	Building and construction; boxes	50	28	2330
Hemlock, Western	0.38	0.42	Moderate in weight, strength, and hardness†	Sash; doors; posts; piles; building and construction	41	29	2420
Hickory, true	0.65	0.73	High toughness, hardness, shock resistance, strength, and shrinkage†	Dowels; spokes; poles; shafts; gymnasium equipment	63	51	4250
Incense cedar	0.35		Uniform texture; easy to season; low shrinkage; shock resistance, weight, and stiffness**	Lumber; fence posts; ties; poles; shingles	45		
Larch, Western	0.48	0.52	Moderate in strength, weight, shock resistance, hardness, and shrinkage‡	Doors; sash; posts; pilings; building and construction	48	36	3000

Table 1-123.   AMERICAN WOODS—PROPERTIES AND USES (Continued)

| Species | Specific gravity | | Characteristics | Uses | Weight | | |
	Green	Dry			lb/cu ft, green	lb/cu ft, air-dry 12%	lb/1000 board ft, air-dry 12%
Locust, black	0.66	0.69	High in shock resistance, weight, and hardness; very high strength; moderate shrinkage**	Mine timbers; posts; poles; ties	58	48	4000
Maple	0.44	0.48	High in hardness, weight, strength, shock resistance, and shrinkage; uniform texture†	Flooring; furniture; trim; spools; farm implements	54	40	3330
Oak, red and white	0.57	0.63	High in hardness, weight, strength, shock resistance, and shrinkage; red†, white†	Trim; ships; flooring; ties; furniture; cooperage; piles	64	44	3670
Pine, jack			Coarse texture; low strength, stiffness, shock resistance, and shrinkage	Box lumber; fuel; mine timber; ties; poles; posts			
Pine, lodgepole	0.38	0.41	Moderate in weight, hardness, strength, shock resistance, and shrinkage; easy to work‡	Poles; mine timber; ties; construction	39	29	2420
Pine			High shrinkage; moderate strength, stiffness, hardness, and shock resistance	General construction; ties; poles; posts			
Pine, Ponderosa	0.38	0.40	Moderate in weight, shock resistance, shrinkage, and hardness; easy to work†	Building; paneling; sash; frames	45	28	2330
Pine, S. yellow	0.47	0.51	Moderate in shock resistance, shrinkage, and hardness; high in strength‡	Building and construction; poles; pilings; boxes	55	41	3420
Pine, sugar	0.35	0.36	Low shock resistance; easy to work; moderate strength†	Sash; counters; blinds; patterns	52	25	2080
Pine, Western white	0.36	0.38	Moderate in strength, shock resistance, shrinkage, and hardness; easy to work†	Building and construction; patterns; boxes	35	27	2250
Red cedar, Eastern and Western	0.44	0.47	High shock resistance; low stiffness and shrinkage; moderate in strength and hardness**	Fence posts; closet liners; chests; flooring	37	37	2750
Redwood	0.38	0.40	Low shrinkage; medium in weight, strength, hardness, and shock resistance**	Posts; doors; interiors; cooling towers	50	28	2330
Spruce, Eastern	0.38	0.40	Moderate in hardness, shock resistance, weight, shrinkage, and strength†	Building; millwork; boxes; ladders	34	28	2330

Table 1-123. AMERICAN WOODS—PROPERTIES AND USES *(Continued)*

Species	Specific gravity		Characteristics	Uses	Weight		
	Green	Dry			lb/cu ft, green	lb/cu ft, air-dry 12%	lb/1000 board ft, air-dry 12%
Spruce, Engelmann	0.31	0.33	Generally straight grained; light in weight; low strength as a beam or post; low shock resistance; moderate shrinkage	Mine timber; ties; poles; flooring; studding; paper	39	23	1920
Spruce, Sitka	0.37	0.40	Moderate in weight, hardness, strength, shock resistance, and shrinkage†	Important in boat and plane construction; sash; doors; boxes; siding	33	28	2330
Sycamore	0.46	0.49	High shrinkage; moderate in weight, strength, hardness, and shock resistance†	Boxes; ties; posts; veneer; flooring; butcher blocks	52	34	2830
Tamarack	0.49	0.53	Coarse texture; moderate in strength, hardness, shrinkage, and shock resistance	Ties; mine timber; posts; poles; tanks; scaffolding	47	37	3080
Tupelo			Uniform texture; moderate in strength, hardness, shock resistance; high shrinkage; interlocked grain makes splitting difficult†	Flooring; planking; crates; furniture			
Walnut, black	0.51	0.55	Moderate shrinkage; high weight, strength, hardness, and shock resistance; easily worked and glued**	Gun stocks; cabinets; plywood; furniture; veneer	58	38	3170
White cedar	0.31	0.32	Low shrinkage, weight, shock resistance, and strength; soft; easily worked**	Poles; posts; ties; tanks; ships	24	23	1920
Willow, black			High strength and shock resistance; low beam strength and weight; interlocked grain	Lumber; veneer; charcoal; furniture; sub-flooring; studding			

†Decay resistance low.
‡Decay resistance medium.
**Decay resistance high.
*From: "Materials Data Book", E.R. Parker, McGraw–Hill Book Company, 1967, pp. 252–255.

Table 1-124. ALLOWABLE UNIT STRESSES FOR LUMBER
Grading and Specifications of Softwood Lumber

American Softwood Lumber Standard. A voluntary standard for softwood lumber has been developing since 1922. Five editions of Simplified Practice Recommendation R16 were issued from 1924–53 by the Department of Commerce; the present NBS voluntary Product Standard PS 20-70, "American Softwood Lumber Standard", was issued in 1970. It was supported by the American Lumber. Standards Committee, which functions through a widely representative National Grading Rule Committee.

Table A. NOMINAL AND MINIMUM-DRESSED SIZES*

This table applies to boards, dimensional lumber, and timbers. The thicknesses apply to all widths and all widths to all thicknesses.

Item	Thicknesses			Face widths		
		Minimum-dressed			Minimum-dressed	
	Nominal	$Dry,^a$ inches	Green, inches	Nominal	$Dry,^a$ inches	Green, inches
Boards[b]				2	$1\frac{1}{2}$	$1\frac{9}{16}$
				3	$2\frac{1}{2}$	$2\frac{9}{16}$
				4	$3\frac{1}{2}$	$3\frac{9}{16}$
				5	$4\frac{1}{2}$	$4\frac{5}{8}$
	1	$\frac{3}{4}$	$\frac{25}{32}$	6	$5\frac{1}{2}$	$5\frac{5}{8}$
				7	$6\frac{1}{2}$	$6\frac{5}{8}$
	$1\frac{1}{4}$	1	$1\frac{1}{32}$	8	$7\frac{1}{4}$	$7\frac{1}{2}$
				9	$8\frac{1}{4}$	$8\frac{1}{4}$
	$1\frac{1}{2}$	$1\frac{1}{4}$	$1\frac{9}{32}$	10	$9\frac{1}{4}$	$9\frac{1}{4}$
				11	$10\frac{1}{4}$	$10\frac{1}{2}$
				12	$11\frac{1}{4}$	$11\frac{1}{2}$
				14	$13\frac{1}{4}$	$13\frac{1}{2}$
				16	$15\frac{1}{4}$	$15\frac{1}{2}$
Dimension				2	$1\frac{1}{2}$	$1\frac{9}{16}$
				3	$2\frac{1}{2}$	$2\frac{9}{16}$
				4	$3\frac{1}{2}$	$3\frac{9}{16}$
	2	$1\frac{1}{2}$	$1\frac{9}{16}$	5	$4\frac{1}{2}$	$4\frac{5}{8}$
	$2\frac{1}{2}$	2	$2\frac{1}{16}$	6	$5\frac{1}{2}$	$5\frac{5}{8}$
	3	$2\frac{1}{2}$	$2\frac{9}{16}$	8	$7\frac{1}{4}$	$7\frac{1}{2}$
	$3\frac{1}{2}$	3	$3\frac{1}{16}$	10	$9\frac{1}{4}$	$9\frac{1}{2}$
				12	$11\frac{1}{4}$	$11\frac{1}{2}$
				14	$13\frac{1}{4}$	$13\frac{1}{2}$
				16	$15\frac{1}{4}$	$15\frac{1}{2}$
Dimension				2	$1\frac{1}{2}$	$1\frac{9}{16}$
				3	$2\frac{1}{2}$	$2\frac{9}{16}$
				4	$3\frac{1}{2}$	$3\frac{9}{16}$
				5	$4\frac{1}{2}$	$4\frac{5}{8}$
	4	$3\frac{1}{2}$	$3\frac{9}{16}$	6	$5\frac{1}{2}$	$5\frac{5}{8}$
	$4\frac{1}{2}$	4	$4\frac{1}{16}$	8	$7\frac{1}{4}$	$7\frac{1}{2}$
				10	$9\frac{1}{4}$	$9\frac{1}{2}$
				12	$11\frac{1}{4}$	$11\frac{1}{2}$
				14		$13\frac{1}{2}$
				16		$15\frac{1}{2}$
Timbers	5 and thicker		$\frac{1}{2}$ off	5 and wider		$\frac{1}{2}$ off

[a] Maximum moisture content of 19% or less.

[b] Boards less than the minimum thickness for 1 in. nominal but $\frac{5}{8}$ in. or greater thickness dry ($\frac{11}{16}$ in. green) may be regarded as American Standard Lumber, but such boards shall be marked to show the size and condition of seasoning at the time of dressing. They shall also be distinguished from 1-in. boards on invoices and certificates.

*Reprinted from: "American Softwood Lumber Standard", NBS PS 20-70, National Bureau of Standards, 1970; available from Superintendent of Documents.

Table 1-124. ALLOWABLE UNIT STRESSES FOR LUMBER (Continued)

The "American Softwood Lumber Standard", PS 20-70, gives the size and grade provisions for American Standard lumber and describes the organization and procedures for compliance enforcement and review. It lists commercial name classifications and complete definitions of terms and abbreviations.

Eleven softwood species are listed in PS 20-70, viz., cedar, cypress, fir, hemlock, juniper, larch, pine, redwood, spruce, tamarack, and yew. Five dimensional tables show the standard dressed (surface planed) sizes for almost all types of lumber, including matched tongue-and-grooved and shiplapped flooring, decking, siding, etc. Dry or seasoned lumber must have 19% or less moisture content, with an allowance for shrinkage of 0.7–1.0% for each four points of moisture content below this maximum. Green lumber has more than 19% moisture. Table A illustrates the relation between nominal size and dressed or green sizes.

National Design Specification.   Table B is condensed from the 1971 edition of "National Design Specification for Stress-Grade Lumber and Its Fastenings", as recommended and published by the National Forest Products Association, Washington, D.C. This specification was first issued by the National Lumber Manufacturers Association in 1944; subsequent editions have been issued as recommended by the Technical Advisory Committee. The 1971 edition is a 65-page bulletin with a 20-page supplement giving "Allowable Unit Stresses, Structural Lumber", from which Table B has been condensed. The data on working stresses in this Supplement have been determined in accordance with the corresponding ASTM Standards, D245-70 and D2555-70.

Table B. SPECIES, SIZES, ALLOWABLE STRESSES, AND MODULUS OF ELASTICITY

Normal Loading Conditions: Moisture Content Not Over 19 Percent, No. 1 grade, Visual Grading

To convert psi to N/m^2, multiply by 6 895.

Species[a]	Sizes, nominal	Typical grading agency, 1971[b]	Allowable unit stresses, psi[d]				Modulus of elasticity, psi
			Extreme fiber in bending[c]	Tension parallel to grain	Compression perpendicular	Compression parallel	
CEDAR							
Northern white	2 × 4	NL, NH	1 100	600	205	675	800 000
	2 or 4 × 6+	NL, NH	1 000	575	205	675	800 000
Western	2 × 4	NC	1 450	725	285	975	1 100 000
	2 or 4 × 6+	NC, WW	1 250	725	285	975	1 100 000
FIR							
Balsam	2 × 4	NL, NH	1 300	675	170	825	1 200 000
	2 or 4 × 6+	NL, NH	1 150	650	170	825	1 200 000
Douglas (larch)	2 × 4	WC, NC	2 400	1 200	385	1 250	1 800 000
	2 or 4 × 6+	WC, NC	1 750	1 000	385	1 250	1 800 000
HEMLOCK							
Eastern (tamarack)	2 × 4	NL, NH	1 750	900	365	1 050	1 300 000
	2 or 4 × 6+	NL, NH	1 500	875	365	1 050	1 300 000
Hem-fir	2 × 4	WC, NC	1 600	825	245	1 000	1 500 000
	2 or 4 × 6+	WC, NC	1 400	800	245	1 000	1 500 000
Mountain	2 × 4	WC, WW	1 700	850	370	1 000	1 300 000
	2 or 4 × 6+	WC, WW	1 450	850	370	1 000	1 300 000
PINE							
Idaho white	2 × 4	WW	1 400	725	240	925	1 400 000
	2 or 4 × 6+	WW	1 200	700	240	925	1 400 000
Lodgepole	2 × 4	WW	1 500	750	250	900	1 300 000
	2 or 4 × 6+	WW	1 300	750	250	900	1 300 000

Table 1-124.   ALLOWABLE UNIT STRESSES FOR LUMBER (Continued)

Species[a]	Sizes, nominal	Typical grading agency, 1971[b]	Allowable unit stresses, psi[d]				Modulus of elasticity, psi
			Extreme fiber in bending[c]	Tension parallel to grain	Compression perpendicular	Compression parallel	
PINE (continued)							
Northern	2 × 4	NL, NH	1 600	825	280	975	1 400 000
	2 or 4 × 6+	NL, NH	1 400	800	280	975	1 400 000
Ponderosa (sugar)	2 × 4	WW, NC	1 400	700	250	850	1 200 000
	2 or 4 × 6+	WW, NC	1 200	700	250	850	1 200 000
Red	2 × 4	NC	1 350	700	280	825	1 300 000
	2 or 4 × 6+	NC	1 150	675	280	825	1 300 000
Southern	2 × 4	SP	2 000	1 000	405	1 250	1 800 000
	2 or 4 × 6+	SP	1 750	1 000	405	1 250	1 800 000
REDWOOD							
California	2 or 4 × 2 or 4	RI	1 950	1 000	425	1 250	1 400 000
	2 or 4 × 6 to 12	RI	1 700	1 000	425	1 250	1 400 000
SPRUCE							
Eastern	2 × 4	NL, NH	1 300	750	255	900	1 100 000
	2 or 4 × 6+	NL, NH	1 250	750	255	900	1 400 000
Engelmann	2 × 4	WW	1 300	675	195	725	1 200 000
	2 or 4 × 6+	WW	1 150	650	195	725	1 200 000
Sitka	2 × 4	WC	1 550	775	280	925	1 500 000
	2 or 4 × 6+	WC	1 300	775	280	925	1 500 000

Note: Allowable unit stresses in horizontal shear are in the range of 60–100 psi for No. 1 grade.

[a] Grade designations are not entirely uniform. Values in the table apply approximately to "No. 1." There is seldom more than one better grade than No. 1, and this may be designated as select, select structural, dense, or heavy. In addition to lower grades 2 and 3, there may be other lower grades, designated as construction, standard, stud, and utility. In bending and tension the allowable unit stresses in the lowest recognized grade (utility) are of the order of $\frac{1}{8}$ to $\frac{1}{6}$ of the allowable stresses for grade No. 1. The tabular values for allowable bending stress are for the extreme fiber in "repetitive member uses," and edgewise use. The original tables give correction factors, which are less than unity for moist locations and for short-time loading; they are greater than unity if the moisture content of the wood in service is 15% or less. In general, all data apply to uses within covered structures. From the extensive tables, only the No. 1 grade in nominal 2 × 4 size and 2-in. or 4-in. planks, 6 in., and wider have been selected for illustration.

In a few cases the allowable stresses specified for the Canadian products will vary slightly from those given here for the same species by the U.S. agencies.

[b] Grading agencies represented by letters in this column are as follows:
   NC = National Lumber Grades Authority (a Canadian agency)
   NH = Northern Hardwood and Pine Manufacturers Association
   NL = Northern Lumber Manufacturers Association
   RI = Redwood Inspection Service
   SP = Southern Pine Inspection Bureau
   WC = West Coast Lumber Inspection Bureau
   WW = Western Wood Products Association

[c] It is assumed that all members are so framed, anchored, tied, and braced that they have the necessary rigidity.

[d] For short term loads, these values may be increased: add 15% for 2-month snow load; add 33% for wind or earthquake; add 100% for impact load.

REFERENCES

"Wood Handbook", Handbook No. 72, U.S. Department of Agriculture, 1955.

"Timber Construction Manual", American Institute of Timber Construction, John Wiley & Sons, Inc., 1966.

"National Design Specification for Stress-Grade Lumber and Its Fastenings", National Forest Products Association, Washington D.C., 1971.

Table 1-125. STRESS-GRADE LUMBER—MAXIMUM END LOADS*

Allowable Unit Stresses of Wood for End Grain in Bearing Parallel to Grain

These allowable unit stresses apply to the net area in bearing. When the stress in end-grain bearing exceeds 75 percent of the adjusted allowable unit stresses, bearing shall be on a metal plate, strap, or other durable, rigid, homogeneous material of adequate strength.

To convert stresses to N/m^2, multiply by 6 895.

Species	Unseasoned, psi	Seasoned, psi	Species	Unseasoned, psi	Seasoned, psi
Ash, commercial white	1 510	2 060	Maple, black and sugar	1 260	1 710
Balsam fir	980	1 330	Mountain hemlock	1 170	1 600
Beech	1 310	1 780	Northern pine	1 150	1 570
Birch, sweet and yellow	1 260	1 720	Northern white cedar	810	1 110
California redwood (close grain)	1 710	2 340	Oak, red and white	1 160	1 590
California redwood (open grain)	1 270	1 730	Ponderosa pine—sugar pine	1 000	1 360
Cottonwood, Eastern	840	1 150	Red pine	970	1 320
Douglas fir—larch (dense)	1 730	2 360	Sitka spruce	1 090	1 480
Douglas fir—larch	1 480	2 020	Southern cypress	1 460	1 990
Douglas fir, South	1 340	1 820	Southern pine (dense)	1 730	2 360
Eastern hemlock—tamarack	1 270	1 730	Southern pine (med. grain)	1 480	2 020
Eastern spruce	1 060	1 450	Southern pine (open grain)	1 260	1 720
Eastern white pine	990	1 360	Spruce—pine—fir	1 040	1 410
Engelmann spruce	860	1 170	Subalpine fir	840	1 140
Hem-fir	1 230	1 680	Sweetgum and tupelo	1 120	1 530
Hickory and pecan	1 510	2 050	Western cedars	1 150	1 570
Idaho white pine	1 080	1 470	Western white pine	1 030	1 400
Lodgepole pine	1 060	1 450	Yellow poplar	980	1 340

*Reprinted from: "National Design Specification for Stress-Grade Lumber and Its Fastenings", National Forest Products Association, 1971, p. II-3.

Table 1-126. ELECTRICAL RESISTANCE OF VARIOUS SPECIES OF WOOD*

This table gives the average of measurements made along the grain between two pairs of needle electrodes $1\frac{1}{4}$ inches apart and driven to a depth of $\frac{5}{16}$ inch, measured at 80°F.

Species	Electrical resistance, megohms, for various moisture contents						
	7%	8%	9%	10%	12%	16%	20%
SOFTWOODS							
Cypress, Southern	12,600	3,980	1,410	630	120	11.2	1.78
Douglas fir (coast region)	22,400	4,780	1,660	630	120	11.2	2.14
Fir, white	57,600	15,850	3,980	1,120	180	16.6	3.02
Hemlock, Western	22,900	5,620	2,040	850	185	16.2	2.52
Larch, Western	39,800	11,200	3,980	1,445	250	19.9	3.39
Pine							
Eastern white	20,900	5,620	2,090	850	200	19.9	3.31
Ponderosa	39,800	8,910	3,310	1,410	300	25.1	3.55
Southern longleaf	25,000	8,700	3,160	1,320	270	24.0	3.72
Southern shortleaf	43,600	11,750	3,720	1,350	255	22.4	3.80
Sugar	22,900	5,250	1,660	645	140	15.9	3.02
Redwood	22,400	4,680	1,550	615	100	7.2	1.74
Spruce, Sitka	22,400	5,890	2,140	830	165	15.5	3.02
HARDWOODS							
Ash, commercial white	12,000	2,190	690	250	55	5.0	0.89
Birch	87,000	19,950	4,470	1,290	200	18.2	3.55
Gum, red	38,000	6,460	2,090	815	160	15.1	2.63
Hickory, true		31,600	2,190	340	50	3.7	0.71
Maple, sugar	72,400	13,800	3,160	690	105	10.2	2.24
Oak							
Commercial red	14,400	4,790	1,590	630	125	11.3	2.09
Commercial white	17,400	3,550	1,100	415	80	7.2	1.15

*From: "Wood Handbook", U.S. Department of Agriculture, Handbook No. 72, 1955.

Table 1-127. INDUSTRIAL WOOL FELT SPECIFICATIONS AND DATA*

NTA class	Corresponding to		Wool content fiber basis, minimum %	Other fibers, maximum %	Wool content chemical basis, minimum %	Tensile strength, minimum psi[a]	Density, lb/cu ft	Color	Compressional resistance, %[b]	Tear strength[f]	Durometer hardness[d]
	SAE number	CF 206 ASTM CS 185									
14R1	—	—	45	55	40	—	8.7	White	35	—	8
17R1	—	—	50	50	45	—	10.6	White	45	—	12
17R2	F26	8R5	45	55	40	—	10.6	Gray	40	—	12
18R1	F10	9R1	100	—	95	240	11.2	White	50	15	15
18R2	F11	9R2	95	5	92	225	11.2	Gray	50	10	15
18R3	F13	9R4	80	20	75	150	11.2	Gray	50	10	15
26R1	F5	12R1	100	—	95	425	16.2	White	65	20	23
26R2	F6	12R2	95	5	92	300	16.2	Gray	65	18	23
26R3	F7 & F55	12R3	85	15	80	275	16.2	Gray	65	15	23
34R1	F1 & F50	16R1	100	—	95	600	21.2	White	75	25	27
34R2	F2	16R2	100	—	95	500	21.2	Blk/Gry	75	20	27
34R3	F3 & F51	16R3	95	5	90	400	21.2	Gray	75	15	27
38R1	—	18R1	100	—	95	650	23.7	White	80	25	37
38R2	—	—	100	—	95	550	23.7	Gray	80	20	37
43S1	—	20S1	100	—	95	700	26.8	White	85	35	45
56S1	—	26S1	100	—	95	800	35.0	White	90	40	62
68S1	—	32S1	100	—	95	900	42.5	White	92	45	73

APPLICATIONS: Vibration isolation
Liquid and gas filtration
Sealing
Fluid transfer and retention (wicking and lubrication)
Spacing, padding, and shock absorption
Polishing
Cushioning and packaging
Sound absorption and attenuation
Thermal insulation
Frictional material

[a] ASTM D-461 method.
[b] NTA test method.
[f] ASTM D-2262 method.
[d] Shore durometer.

*From: Northern Textile Association—Felt Manufacturers Council.

Section **2**

Electrical Science and Radiation

Table 2-1. ELECTROMAGNETIC RADIATION AND STABLE ELEMENTARY PARTICLES

	Charge	Mass	Examples or sources
Alpha particle	+2	4	Alpha "rays" emitted by heavy radiosotopes; cosmic rays
Electron	−1	1/1836	Ionosphere; atoms of matter; beta rays from radioactive elements
Gamma ray	0	0	Radioactive decay; nuclear transitions; nuclear reactors; cosmic rays
Neutrino	0	0	Emitted by sun, stars, nuclear reactors. Accompanies radioactive emission (beta decay)
Neutron†	0	1	Vicinity of planets and sun; atomic nuclei; nuclear reactors
Photon	0	0	All light flux from sun, stars, etc.; radiation belts
Positron	+1	1/1836	Fast anti-electrons emitted from radioactive materials
Proton	+1	1	Cosmic rays; radiation belts; atomic nuclei
X-ray	0	0	Radiation belts; solar radiation; high-voltage vacuum tubes

†Secondary particle; not stable; life about 1,000 seconds.

Electrons are negatively-charged "atoms of electricity"; in ordinary matter they form an ordered "cloud" surrounding the heavy, positively-charged atomic nuclei.

Photons are electromagnetic waves; they carry energy in discrete quantity, proportional to the frequency of the associated wave.

Beta decay involves the emission of an electron or positron. (The terms *beta-ray* and *beta-particle* are sometimes used.)

Gamma rays consist of high-energy photons (electromagnetic waves); they are emitted in radioactive decay.

X-rays consist of photons emitted in the acceleration (deceleration) of charged particles, as when high-speed electrons strike a heavy, metal target.

Atomic nucleus is the heavy core of the atom, consisting of protons and neutrons. The number of protons is called the *atomic number*. The number of neutrons plus protons is called the *mass number*. The energy required to separate all of the neutrons and protons of the nucleus is called the *binding energy*.

Radioactive nucleus is one that spontaneously changes by radioactive decay, electron capture, or fission. It becomes ultimately transformed into a different kind of nucleus.

Isotopes of an element contain the same number of protons but slightly different numbers of neutrons. They are chemically indistinguishable, except by very much refined procedures and in some biological reactions.

Ions are electrically-charged atoms. If the negative electron charges just balance the total positive charge of the nucleus, the atom is neutral; with more electrons the atom becomes a negative ion, and with fewer electrons it becomes a positive ion.

Fission is the breakup of nuclei into fragments that are themselves nuclei. Mass is usually lost; hence energy is released.

Fusion is the coalescing of two nuclei to form a heavier one.

Table 2-2. TERMS AND UNITS FOR RADIANT ENERGY AND ILLUMINATION

Note: Any of the following quantities may be restricted to a narrow wavelength interval by addition of the word *spectral.*

Measure of	Terms in use	Meaning or definition	Usual units
Quantity	Radiant energy Luminous energy	Total quantity of radiant energy	Erg, joule, calorie, kilowatt-hour, Btu
Rate	Radiant flux Luminous flux	Time rate of flow of radiant energy (power)	Erg/sec, watt, Btu/hr, lumen
Intensity	Radiant intensity Luminous intensity	Radiant flux per unit solid angle (point source)	Watts per steradian, candela* = lumens/steradian
Density at surface	Radiant emittance Radiant excitance Irradiance Illumination Illuminance (Emittance)	Density of radiant flux incident upon (or emitted from) a surface	Watts/sq cm, foot-candle = lumens/ sq ft, lux = lumens/sq m, phot = lumens sq cm, Btu/hr × sq ft
Density of beam (at surface)	Radiance	Unit intensity normal to the beam per unit of projected area in that direction	Watts/sr × sq cm, $\dfrac{Btu/hr}{sr \times sq\ ft}$
Effectiveness (radiating)	Emissivity (Absorptivity)	Ratio of radiant emittance (or absorptance) to that of a perfect blackbody	Dimensionless
Brightness	Luminance	Photometric brightness per unit area	Candela/ sq ft, stilb = cd/sq cm, nit = cd/sq m, foot-lambert = cd/π sq ft, lambert = cd/π sq cm, apostilb = cd/π sq m

*The candela (cd) was formerly called "candlepower". (One international candle will illuminate a sphere at one foot distance with 4π lumens, or one sq ft of the sphere with one lumen.)

Table 2-3. ELECTROMAGNETIC FREQUENCY SPECTRA

Ranges and Applications

For electronic communication ranges see Table 2-13.

Application or common name	Typical frequency, cps†	Typical wavelength‡	Approximate frequency range, cps†
Electric a-c power	60	5×10^6 m	25–60
Eddy-current heating (metals)	60	5×10^6 m	50–1,000
Servo and instrument power	400	7.5×10^5 m	100–1,000
Audio frequency standard	440	6.8×10^5 m	440 and 600
Induction furnace power	2,000	1.5×10^5 m	500–3,000
R-F heating of metals	10 kc	3×10^4 m	1kc–1mc
Power-line communication	30 kc	10^4 m	wide
Maritime and radio beacon	400 kc	750 m	20–550 kc
Radio broadcasting	1,000 kc	300 m	550–1,600 kc
Shortwave radio	20 mc	15 m	3–300 mc
Microwave diathermy	27 mc	11 m	—
Dielectric heating and drying	40 mc	7.5 m	10–200 mc
F-M radio	100 mc	3 m	91–108 mc
Television (channels 2–13)	180 mc	1.67 m	54–216 mc
Radar	500 mc	60.0 cm	200–1,200 mc
Television (channels 14–83)	800 mc	37.5 cm	470–890 mc
Tracking stations	960 mc	31.3 cm	440–5,600 mc
Intercity relay	2,000 mc	15.0 cm	1,200–20,000 mc
Radar	10,000 mc	3.0 cm	1,200–20,000 mc
Super high frequency	20,000 mc	1.5 cm	3,000–30,000 mc
Far infrafred (germanium detector)	3×10^{13}	10 μm	
Infrared (PbS detector)	1.25×10^{14}	2.4 μm	
Infrared heaters	1.5×10^{14}	2. μm	
Night infrared searchlight	3×10^{14}	1. μm	
Near infrared photography	3.75×10^{14}	0.8 μm	
Cadmium red line	4.65×10^{14}	.64385 μm	
Yellow (max visual)	5.3×10^{14}	.56 μm	
Solar max intensity	7.1×10^{14}	.42 μm	
Germicidal lamps—ultraviolet	10^{15}	.3 μm	
Soft X-rays	10^{18}	3 Å	
Hard X-rays	10^{20}	.03 Å	
Gamma rays	10^{21}	.003 Å	
Cosmic rays	3×10^{23}	10^{-5} Å	

†The name *hertz* is widely used by electrical engineers for cycles per second.
‡**Units:** 1 meter = 100 cm = 39.37 in. = 10^6 micrometers (μm) = 10^{10} angstrom units (Å).
 Velocity = 186,290 mi/s = 2.99793×10^8 m/s = frequency × wavelength.

VISIBLE SPECTRUM—REPRESENTATIVE COLORS

Color	Frequency	Wavelength
Violet	7.3×10^{14}	0.41
Blue	6.38×10^{14}	0.47
Green	5.75×10^{14}	0.52
Yellow	5.17×10^{14}	0.58
Orange	5.0×10^{14}	0.60
Red	4.6×10^{14}	0.65

Table 2-4. SPECTRAL SOLAR IRRADIANCE*
At Normal Incidence and Mean Solar Distance

Wave-length, milli-microns	Irradiance in watts/sq meter		Wave-length, milli-microns	Irradiance in watts/sq meter	
	Outside earth's atmosphere	At sea level and air mass = 2		Outside earth's atmosphere	At sea level and air mass = 2
300	450	.08	540	1,894	1,198
320	726	54	560	1,861	1,182
340	856	151	580	1,819	1,168
360	976	233	600	1,762	1,167
380	1,121	336	620	1,690	1,165
400	1,304	470	640	1,616	1,175
420	1,766	733	660	1,543	1,166
440	1,939	911	680	1,473	1,149
460	2,096	1,080	700	1,405	1,108
480	2,127	1,183	720	1,337	832
500	2,061	1,215	740	1,270	1,041
520	1,954	1,199	760	1,205	566

*Based largely on: "Proposed Standard Solar Radiation Curves for Engineering Use", P. Moon, *J. Frank. Inst.*, 230:583, Nov. 1940. These values have been adopted by the Committee on Natural Lighting of the Illuminating Engineering Society (IES) for use in dealing with the transmittance of glass. Later refinements indicate that the corresponding value of 1,322 w/m² for the total solar radiation outside the earth's atmosphere should be nearly 1,400 w/m², 444 Btu/hr ft²; this later value has been used for the data in Table 7-18.

REFERENCE
"Survey of the Literature on the Solar Constant and the Spectral Distribution of Solar Radiant Flux", M.P. Thekaekara, NASA SP-74, 1965. This paper reviews all available data and recommends a slight revision of the above table. Detailed tables are given over the range from 220–7,000 millimicrons.

Table 2-5. BLACKBODY RADIATION

Temperature		Wavelength of maximum intensity, microns, μ	Maximum normal intensity†		Total maximum hemispherical radiation†	
°K	°R		$W/cm^2\,\mu$	$Btu/hr\,ft^2\,\mu$	W/cm^2	$Btu/hr\,ft^2$
10	18	290	1.290×10^{-10}	4.092×10^{-7}	5.679×10^{-8}	1.801×10^{-4}
50	90	58.0	4.030×10^{-7}	1.278×10^{-3}	3.549×10^{-5}	1.126×10^{-1}
100	180	29.0	1.290×10^{-5}	4.092×10^{-2}	5.679×10^{-4}	1.801
200	360	14.5	4.127×10^{-4}	1.309	9.086×10^{-3}	2.882×10
300	540	9.66	3.134×10^{-3}	9.941	4.600×10^{-2}	1.459×10^2
350	630	8.28	6.774×10^{-3}	2.149×10	8.522×10^{-2}	2.703×10^2
400	720	7.25	1.321×10^{-2}	4.190×10	1.454×10^{-1}	4.612×10^2
450	810	6.44	2.380×10^{-2}	7.550×10	2.328×10^{-1}	7.385×10^2
500	900	5.80	4.030×10^{-2}	1.278×10^2	3.549×10^{-1}	1.126×10^3
550	990	5.27	6.484×10^{-2}	2.057×10^2	5.207×10^{-1}	1.652×10^3
600	1,080	4.83	1.003×10^{-1}	3.181×10^2	7.360×10^{-1}	2.335×10^3
700	1,260	4.14	2.168×10^{-1}	6.877×10^2	1.364	4.327×10^3
800	1,440	3.63	4.226×10^{-1}	1.341×10^3	2.326	7.378×10^3
900	1,620	3.22	7.616×10^{-1}	2.416×10^3	3.726	1.182×10^4
1,000	1,800	2.90	1.290	4.092×10^3	5.679	1.801×10^4
1,200	2,160	2.42	3.209	1.018×10^4	1.178×10	3.737×10^4
1,400	2,520	2.07	6.936	2.200×10^4	2.181×10	6.918×10^4
1,600	2,880	1.81	1.352×10	4.289×10^4	3.722×10	1.181×10^5
1,800	3,240	1.61	2.437×10	7.730×10^4	5.961×10	1.891×10^5
2,000	3,600	1.49	4.127×10	1.309×10^5	9.086×10	2.882×10^5
2,500	4,500	1.156	1.260×10^2	3.997×10^5	2.218×10^2	7.036×10^5
3,000	5,400	0.966	3.134×10^2	9.941×10^5	4.600×10^2	1.459×10^6
4,000	7,200	0.725	1.321×10^3	4.190×10^6	1.454×10^3	4.612×10^6
6,000	10,800	0.483	1.003×10^4	3.181×10^7	7.360×10^3	2.335×10^7
8,000	14,400	0.363	4.226×10^4	1.340×10^8	2.326×10^4	7.378×10^7

Notes: One half of the blackbody radiation lies on either side of the wavelength computed from $\lambda = 4107/T$, where λ is in microns and T is °K.

1 cm = 0.3937 in. = 10,000 microns = 10^8 Angstrom units. To convert Btu/hr·ft² to W/m², multiply by 3.1525.

†Zero temperature receiver; no reradiation.

Figure 2-6. BLACKBODY SPECTRAL INTENSITIES

For Source Temperatures between 273°K and 2600°K

For maximum emittances (to 8000°K), see Table 2-5. For tabular values, visual region, see Table 2-7.

The dashed line indicates the position of radiation peaks. The shaded area is the region of visible wavelengths. One micron = 1,000 nanometers. One watt/cm² micron = 3,170 Btu/hr ft² micron.

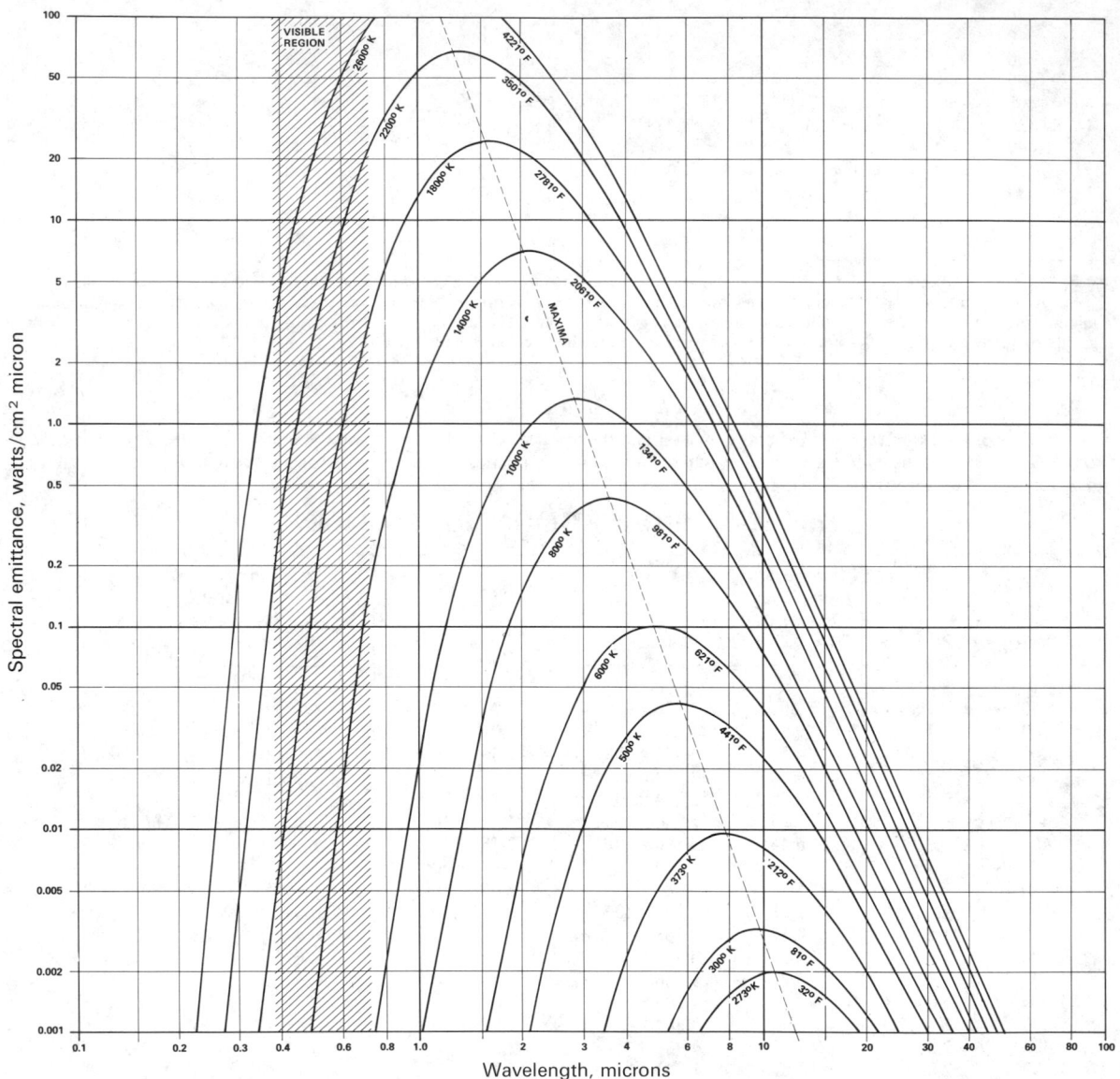

Wavelength, microns

Table 2-7.　SPECTRAL RADIANT EMITTANCE OF A BLACKBODY*

For emittance in $W/cm^2 \times$ micrometers, multiply the selected factor in the column by the value on the bottom line. To convert to $W/m^2 \cdot \mu m$, multiply the result by 10,000.

λ, milli-microns†	Temperature, °K							
	1,000°	1,200°	1,400°	1,600°	1,800°	2,000°	2,200°	2,400°
300								.2129
310							.1571	.3442
320						.1079	.2590	.5372
330						.1828	.4124	.8124
340					.1188	.2989	.6361	1.194
350					.2010	.4731	.9531	1.708
360				.1222	.3292	.7272	1.391	2.387
370				.2093	.5229	1.088	1.981	3.264
380			.1172	.3471	.8077	1.588	2.760	4.375
390			.2058	.5590	1.216	2.265	3.767	5.756
400		.1031	.3502	.8762	1.788	3.164	5.046	7.447
410		.1892	.5791	1.339	2.572	4.335	6.645	9.484
420		.3364	.9321	2.002	3.627	5.835	8.610	11.91
430	.1592	.5807	1.463	2.927	5.018	7.723	10.99	14.75
440	.3035	.9753	2.245	4.195	6.823	10.07	13.84	18.05
450	.5608	1.597	3.371	5.903	9.129	12.94	17.21	21.83
460	1.006	2.552	4.960	8.164	12.03	16.41	21.14	26.12
470	1.758	3.989	7.163	11.11	15.63	20.54	25.69	30.95
480	2.993	6.107	10.16	14.90	20.05	25.43	30.89	36.33
490	4.975	9.169	14.19	19.69	25.40	31.14	36.80	42.28
500	8.088	13.52	19.51	25.69	31.81	37.75	43.43	48.81
510	12.88	19.59	26.43	33.09	39.42	45.33	50.83	55.92
520	20.09	27.93	35.33	42.15	48.34	53.95	59.02	63.60
530	30.78	39.22	46.63	53.09	58.73	63.67	68.02	71.87
540	46.34	54.30	60.81	66.19	70.71	74.55	77.84	80.70
550	68.61	74.16	78.39	81.73	84.43	86.64	88.50	90.08
560‡	100.0	100.0	100.0	100.0	100.0	100.0	100.0	100.0
570	143.6	133.2	126.3	121.3	117.6	114.6	112.3	110.4
580	203.4	175.5	157.9	145.9	137.2	130.6	125.5	121.3
590	284.3	228.7	195.8	174.2	159.1	148.0	139.5	132.7
600	392.3	294.9	240.6	206.5	183.3	166.7	154.2	144.5
610	535.1	376.7	293.2	243.0	209.9	186.8	169.8	156.7
620	721.5	476.8	354.7	284.1	239.1	208.3	186.0	169.3
630	962.4	598.2	425.9	330.1	270.8	231.1	203.0	182.2
640	1,271.	744.2	507.9	381.3	305.1	255.3	220.6	195.4
650	1,661.	918.7	601.6	438.0	342.2	280.8	238.9	208.8
660	2,152.	1,125.	708.2	500.4	381.9	307.7	257.8	222.5
670	2,764.	1,369.	828.7	568.7	424.4	335.8	277.2	236.3
680	3,519.	1,654.	964.1	643.3	469.6	365.1	297.2	250.3
690	4,444.	1,984.	1,116.	724.3	517.6	395.6	317.5	264.4
700	5,570.	2,367.	1,284.	811.8	568.3	427.2	338.3	278.5
710	6,930.	2,806.	1,471.	906.1	621.7	459.9	359.4	292.7
720	8,561.	3,307.	1,677.	1,007.	677.8	493.6	380.9	306.9
730	10,505.	3,877.	1,903.	1,115.	736.5	528.3	402.6	321.0
740	12,807.	4,522.	2,150.	1,231.	797.7	563.8	424.5	335.1
750	15,518.	5,248.	2,419.	1,353.	861.4	600.2	446.6	349.1
760	18,691.	6,060.	2,711.	1,483.	927.5	637.2	468.7	362.9
770	22,385.	6,967.	3,027.	1,620.	995.9	674.9	491.0	376.7
780	26,664.	7,974.	3,367.	1,764.	1,067.	713.2	513.2	390.2
790	31,594.	9,088.	3,732.	1,915.	1,139.	752.1	535.5	403.5
(560)‡	.4785 $\times 10^{-7}$	.3456 $\times 10^{-5}$	.7349 $\times 10^{-4}$	.7277 $\times 10^{-3}$	.4329 $\times 10^{-2}$	.01802	.05792	.1532

Table 2-7. SPECTRAL RADIANT EMITTANCE OF A BLACKBODY
(Continued)

λ, milli-microns†	Temperature, °K							
	2,600°	2,800°	3,000°	4,000°	5,000°	6,000°	8,000°	10,000°
300	.4345	.8007	1.360	8.677	26.29	54.77	135.0	227.9
310	.6684	1.181	1.933	10.84	30.40	60.16	139.1	226.1
320	.9959	1.690	2.674	13.29	34.66	65.37	142.4	223.4
330	1.442	2.357	3.609	16.01	39.03	70.34	144.8	219.8
340	2.033	3.209	4.766	19.00	43.44	75.02	146.5	215.6
350	2.799	4.274	6.167	22.24	47.85	79.40	147.5	210.9
360	3.771	5.581	7.837	25.70	52.23	83.43	147.9	205.7
370	4.981	7.155	9.793	29.35	56.52	87.11	147.8	200.3
380	6.460	9.023	12.05	33.17	60.70	90.43	147.1	194.6
390	8.240	11.21	14.63	37.13	64.73	93.39	146.0	188.7
400	10.35	13.72	17.52	41.19	68.59	95.98	144.6	182.8
410	12.82	16.59	20.75	45.33	72.25	98.22	142.8	176.8
420	15.67	19.82	24.29	49.51	75.71	100.1	140.7	170.9
430	18.92	23.41	28.16	53.72	78.93	101.7	138.3	164.9
440	22.59	27.38	32.34	57.91	81.93	102.9	135.8	159.1
450	26.69	31.71	36.82	62.06	84.69	103.9	133.1	153.4
460	31.24	36.41	41.59	66.15	87.21	104.6	130.3	147.8
470	36.23	41.47	46.62	70.16	89.48	105.0	127.4	142.3
480	41.67	46.87	51.89	74.06	91.52	105.2	124.4	136.9
490	47.56	52.60	57.39	77.85	93.32	105.1	121.4	131.7
500	53.87	58.63	63.09	81.50	94.90	104.9	118.3	126.7
510	60.61	64.95	68.96	85.00	96.25	104.4	115.2	121.8
520	67.76	71.54	74.98	88.35	97.39	103.8	112.1	117.1
530	75.29	78.36	81.12	91.52	98.33	103.1	109.1	112.6
540	83.20	85.40	87.35	94.53	99.07	102.2	106.0	108.2
550	91.44	92.62	93.65	97.36	99.62	101.1	103.0	104.0
560	100.0	100.0	100.0	100.0	100.0	100.0	100.0	100.0
570	108.9	107.5	106.4	102.5	100.2	98.78	97.07	96.11
580	117.9	115.1	112.7	104.7	100.3	97.47	94.12	92.39
590	127.3	122.8	119.1	106.9	100.2	96.09	91.37	88.81
600	136.8	130.5	125.3	108.8	99.98	94.65	88.61	85.37
610	146.5	138.3	131.6	110.5	99.65	93.16	85.92	82.08
620	156.4	146.0	137.7	112.1	99.19	91.62	83.29	78.92
630	166.3	153.8	143.7	113.4	98.64	90.05	80.73	75.90
640	176.3	161.4	149.6	114.7	97.99	88.45	78.24	73.00
650	186.4	169.0	155.3	115.7	97.25	86.83	75.81	70.22
660	196.4	176.5	160.9	116.6	96.43	85.20	73.46	67.56
670	206.5	183.9	166.4	117.4	95.54	83.56	71.17	65.01
680	216.5	191.2	171.7	118.0	94.58	81.91	68.95	62.57
690	226.4	198.3	176.8	118.5	93.57	80.26	66.80	60.24
700	236.3	205.2	181.7	118.8	92.51	78.62	64.72	58.00
710	246.0	212.0	186.4	119.1	91.40	76.98	62.70	55.85
720	255.6	218.6	190.9	119.2	90.25	75.36	60.74	53.80
730	265.1	225.0	195.2	119.1	89.07	73.74	58.85	51.83
740	274.4	231.2	199.3	119.0	87.85	72.15	57.02	49.95
750	283.5	237.2	203.2	118.8	86.62	70.57	55.25	48.14
760	292.4	242.9	206.9	118.5	85.36	69.01	53.54	46.42
770	301.0	248.5	210.4	118.1	84.08	67.47	51.88	44.76
780	309.5	253.8	213.7	117.6	82.80	65.95	50.28	43.17
790	317.7	258.8	216.8	117.1	81.41	64.46	48.74	41.65
(560)‡	.3489	.7066	1.302	11.08	40.19	95.36	285.7	564.2

†Millimicrons are the same as nanometers.

‡All values are normalized so that the spectral emittance at λ = 560 millimicrons equals 100.0. The absolute value at 560 millimicrons is given at the bottom of each column, and here the unit is watts per square centimeter per 10 millimicrons wavelength interval [w/cm²/(10 nm)].

*From: "Color Science", G. Wyszecki and W.S. Stiles. Copyright © 1967 by John Wiley & Sons, Inc. Reprinted by permission.

Table 2-8.   TOTAL RADIATION EMISSIVITIES

Radiation at All Wavelengths Included

For other data on emissivities and reflectivities, see Tables 2-9, 2-10, 2-12, 2-15, 2-16, 2-17, 5-32, and 5-39.

Class	Material	32–100°F[a]	500–1000°F[b]
	METALLIC MATERIALS (CLEAN; DRY)		
1	Highly polished aluminum, silver, gold, brass, tin	.02–.04	.03–.10
2	Polished brass, copper, steel, nickel, chromium, platinum, clean mercury	.03–.08	.06–.2
3	Dull, smooth, clean aluminum and alloys, copper, brass, nickel, stainless steel, iron, lead, zinc	.08–.20	.15–.45
4	Rough-ground or smooth-machined castings; steel-mill products, sprayed metal, molten metal	.15–.25	.3 –.65
5	Smooth, slightly oxidized aluminum, copper, brass, lead, zinc	.2 –.4	.3 –.7
6	Bright aluminum, gilt, or bronze paints	.3 –.55	.4 –.7
7	Heavily oxidized and rough iron, steel, copper, aluminum	.6 –.85	.7 –.9
	NON-METALLIC MATERIALS[c]		
8	White or light colored paint, plaster, brick, tile, porcelain, paper, plastics, asbestos	.80–.95	.85–.60
9	Medium red, brown, green, buff, and other colors of paint, brick, tile, inks, clays, stone, concrete, wood, water	.85–.95	.85–.70
10	Glass and translucent plastics, oil, varnish, ice, crystals	.85–.95	.95–.75
11	Carbon black, tar, asphalt, matte-black paints	.90–.97	.90 .97
	SIMULATED BLACKBODY		
12	A small opening in a large box, sphere, furnace, or enclosure with high-emissivity walls	.97–.99	.97–.99

Notes:

Factors apply to either radiation or absorption.

For a small body in a large enclosure, use the emissivity of the small body only: $F_\varepsilon = \varepsilon_1$.

For rectangles or disks, either parallel or perpendicular and with a common side, use the product of the emissivities: $F_\varepsilon = \varepsilon_1 \times \varepsilon_2$.

For infinite parallel planes, concentric cylinders, or large enclosed bodies, use both emissivities in the equation

$$F_\varepsilon = \frac{1}{1/\varepsilon_1 + 1/\varepsilon_2 - 1}$$

It makes little difference whether the values in this table are regarded as total hemispherical emissivity or total normal emissivity. For non-metals the difference is very small. For metals the hemispherical emittance of a given sample may be 20 or 30 percent above the normal emittance, but in any case the difference between a measured emittance and a typical emissivity may be of the same order as the uncertainty of the description of the material. Moreover, a tolerance for the accuracy of the experimental determination is appreciable. For a thorough discussion see NASA SP-31, "Measurement of Thermal Radiation Properties of Solids", 1963 (585 pages) and "Thermal Radiation Properties Survey", G. G. Gubareff, J. E. Janssen, and R. H. Torborg, Honeywell Research Center, Minneapolis, Minn., 1960.

[a]32–100°F = 0–38°C = 273–311 K; 500–1 000°F = 260–538°C = 533–811 K.

[b]These values apply only to metals with higher melting points and to heat-resistant surfaces. Any contamination greatly increases the emissivity of a metal surface. Absorptivities of metals at cryogenic temperatures are almost the same as the corresponding emissivities at atmospheric temperatures.

[c]Emissivities of very rough surfaces are usually higher than those for smooth surfaces.

Table 2-9. SOLAR ABSORPTION COEFFICIENTS

Total absorptivity is given in the range from 0.3 to 2.5 microns for surfaces at atmospheric temperature.

Class	Surface material	Absorptivity
1	White surfaces: paint, paper, plaster, plastics, white oxides, fresh snow	0.1 –0.3
2	Light-colored surfaces: paint, paper, plastics, textiles, concrete, wood, sand, bricks, stone, dry grass	0.25–0.5
3	Darker colors: paint, inks, brick, tile, slate, leaves, rusted iron, soil	0.4 –0.8
4	Black asphalt, tar, slate, carbon, rubber, water	0.85–0.95
5	Clean, dark metals: iron and steel, lead, zinc; metallic paints	0.2 –0.5
6	Polished, bright metals: aluminum, silver, tin, magnesium, copper, chromium, nickel	0.07–0.3

Notes:

For a non-metallic surface the long-wave emissivity is higher than the solar absorptivity. The opposite is true for polished metals.

For metals the absorptivity increases as the temperature of the absorber increases.

With solar simulators the absorptivities may be different from those measured in the sun.

Vapors and air contaminants affect solar absorptivity.

Table 2-10. RATIOS OF EMISSIVITY TO ABSORPTIVITY

For a given surface it is often assumed that emissivity and absorptivity are equal, and thus total radiation or total absorption is dependent entirely on the temperature difference. For many materials, however, these radiation properties vary greatly with wavelength. The same is true for transmissivity, and this is the basis for optical and radiation filters.

If a surface is heated by a high-temperature source but radiates at a low temperature, the net heat gain may be positive or negative, depending on the material and the temperature differences. The most common case is that of surfaces exposed to sun heat. In general for clean and polished metals the solar absorptance is five to fifteen times as great as the low-temperature emissivity. On the other hand, titanium white or snow is "black" at low temperatures (long wavelengths). Visual appearance and color are very poor guides. These phenomena are of major importance in such cases as roofs, tank walls, car tops, and surfaces of satellites, solar heaters, and solar batteries.

Surface conditions vary so greatly that tabular data are of little use. A thin film of oxide or oil on a metal surface greatly increases the low temperature emissivity. A small amount of dark pigment mixed into "white paint" greatly increases the solar absorptance. Where quantitative results are needed, comparative tests should be made. The following data may be used as a general guide.

Surface material	Emissivity at atmospheric temperatures	Solar absorptivity
Highly polished "white" metals, gold, yellow brass	0.02–0.08	0.1–0.4
Clean, "dark" metals	0.1–0.35	0.3–0.6
Metallic-pigment paints	0.35–0.55	0.4–0.6
White, non-metal surfaces	0.7–0.9	0.1–0.35
Dark-colored non-metals	0.7–0.9	0.45–0.8
Black paint, asphalt, carbon, water	0.85–0.95	0.7–0.9

Table 2-11. RADIATION OF CARBON DIOXIDE AND WATER VAPOR

Approximations for the Calculation of Total Gaseous Radiation

Gases radiate and absorb selectively in narrow wavelength bands. Nevertheless, the Stefan-Boltzmann law is usually used for computing total energy radiated or absorbed. Emissivities depend on the product of partial pressure of the radiating gas P and beam length L. The effective total emissivity of a gas mixture depends on the constitutents and on the shape of the gas volume (see Notes below).

$P \times L$, ft × atm[a] (Total pressure 1.0 atm)	TOTAL EMISSIVITY					
	Gas temperature					
°R: 500°	1000°	1500°	2000°	2500°	3000°	
°F: 40°	540°	1040°	1540°	2040°	2540°	
°C: 4°	282°	560°	838°	1116°	1393°	
PURE CARBON DIOXIDE						
.01	.03	.03	.03	.03	.02	.01
.02	.04	.04	.04	.04	.03	.02
.05	.06	.06	.06	.06	.05	.04
.10	.08	.07	.07	.07	.06	.05
.20	.10	.09	.09	.09	.08	.07
.50	.13	.12	.12	.12	.11	.09
1.0	.15	.14	.14	.14	.13	.12
2.0	.17	.16	.16	.16	.16	.15
5.0	.20	.19	.19	.19	.19	.18
PURE WATER VAPOR						
.01	.03	.02	.01	—	—	—
.02	.06	.04	.02	.02	.01	—
.05	.10	.07	.06	.04	.03	.02
.10	.15	.11	.08	.06	.05	.04
.20	.20	.16	.12	.10	.08	.06
.50	.30	.24	.20	.16	.14	.12
1.0	.37	.32	.27	.23	.19	.16
2.0	.44	.39	.35	.30	.26	.22
5.0	.55	.50	.47	.41	.36	.31

Notes:

For emissivities at total pressures other than atmospheric, see References below.

The above tabular values for pure CO_2 may also be used with fair accuracy for mixtures of CO_2 and non-radiating gases.

The emissivity of a mixture of water vapor and air or other gases is considerably lower than given by the above table (see References). The following correction factors are roughly approximate: emissivities from the table should be multiplied by 0.90 if the partial pressure of the water vapor is 0.5 atm, by 0.75 if it is 0.2 atm, and by 0.55 if it is 0.05 atm.

For mixtures containing both carbon dioxide and water vapor, the total emissivity is somewhat less than the sum of the individual emissivities. The maximum correction (equal percentages of CO_2 and H_2O) is about 10 percent.

The beam length L for a spherical or cubical gas volume is $\frac{2}{3}$ D; for a long cylinder it is D; and for a thin space between two walls, it is 2 D. For tube banks the effective beam length is greater than the minimum clearance between tubes.

When the gas radiates to surrounding gray walls of low temperature and low emissivity, the net heat exchange is somewhat less than that computed from the table. (See References for methods of correction.)

[a]1.0 ft = 0.3048 m; 1.0 atm = 101 325 N/m².

REFERENCES

"Fundamentals of Heat Transfer", 3rd ed., H. Gröber and S. Erk, rev. by U. Grigull (trans. by J. Moszynski), McGraw-Hill Book Company, 1961.

"Heat Transmission", 3rd ed., W.H. McAdams, McGraw-Hill Book Company, 1954.

"Standard Handbook for Mechanical Engineers", 7th ed., T. Baumeister, Ed., McGraw-Hill Book Company, 1967.

Table 2-12.   TOTAL EMISSIVITY OF GASES

For explanation and references, see Table 2-11.

Radiation of a gas envelope to its enclosure is dependent on the partial pressure of the gas (P, atmospheres) and the thickness of the gas envelope (L, feet).

Tempera-ture, deg R	$P \times L$, atm $\times$ ft					$P \times L$, atm $\times$ ft					Tempera-ture, K
	.01	.05	.10	.50	1.0	.01	.05	.10	.50	1.0	
	WATER VAPOR					**CARBON DIOXIDE**					
1 000	.02	.07	.11	.24	.32	.03	.06	.07	.12	.14	538
1 500	.01	.06	.08	.20	.27	.03	.06	.07	.12	.14	816
2 000	—	.04	.06	.16	.23	.03	.06	.07	.12	.14	1 093
2 500	—	.03	.05	.14	.19	.02	.05	.06	.11	.13	1 371
	METHANE					**AMMONIA**					
1 000	.02	.04	.06	.12	.17	.05	.14	.20	.50	.60	538
1 500	.02	.05	.07	.15	.19	.02	.08	.13	.34	.47	816
2 000	.02	.05	.07	.15	.19	.01	.04	.07	.20	.30	1 093
2 500	.02	.05	.06	.14	.18	—	—	—	—	—	1 371
	CARBON MONOXIDE					**SULFUR DIOXIDE**					
1 000	.01	.02	.03	.05	.06	.02	.08	.13	.28	.35	538
1 500	.02	.04	.05	.08	.10	.01	.06	.10	.24	.32	816
2 000	.02	.04	.05	.05	.09	.01	.04	.07	.20	.28	1 093
2 500	—	—	—	—	—	.01	.03	.05	.15	.23	1 371

Note:

While the data on water vapor and carbon dioxide have been checked by various investigators, there are few reported results on the emissivities of other gases.

Principal emissivity bands (micrometers):
Water vapor: 2.55–2.84, 5.6–7.6, 12–25
Carbon dioxide: 2.64–2.84, 4.13–4.5, 13–17
Methane: 2.37, 3.31, 7.65

For methods of calculation for gas mixtures, consult references listed in Table 2-11.

To account for "mean effective" beam lengths, the values of PL should be about 10 percent lower than the usual computed values. A fair approximation of mean beam length is

$$L_m = 3.6 \, V/A,$$

where V is the total volume of the enclosure and A is the surface area of the walls.

Although radiation from non-luminous or blue flames is little different from that computed for the constituent gases, radiation from luminous flames is much higher. Emissivity of high-temperature luminous flames is usually in the range 0.6–0.9.

Total gas pressure has only a small effect on the total radiation from a gas, except at high pressures, such as 50 atm or more.

Table 2-13. RADIOCOMMUNICATION FREQUENCIES

Table A. FREQUENCY BAND DESIGNATIONS

Frequency range,† hz	Band No.	Wavelength designation	Adjective designation
30–300	2	Megametric	ELF Extremely low frequency
300–3 khz	3	—	VF Voice frequency
3 khz–30 khz	4	Myriametric	VLF Very low frequency
30 khz–300 khz	5	Kilometric	LF Low frequency
300 khz–3 Mhz	6	Hectometric	MF Medium frequency
3 Mhz–30 Mhz	7	Decametric	HF High frequency
30 Mhz–300 Mhz	8	Metric	VHF Very high frequency
300 Mhz–3 Ghz	9	Decimetric	UHF Ultra high frequency
3 Ghz–30 Ghz	10	Centimetric	SHF Super high frequency
30 Ghz–300 Ghz	11	Millimetric	EHF Extremely high frequency
300 Ghz–3 000 Ghz	12	Decimillimetric	—

The major portions of bands 9, 10, and 11 are designated as microwave bands with assigned letters, e.g., the X band is from 2.75 cm and 10.9 Ghz to 5.77 cm and 5.2 Ghz.

†Lower limit is excluded from each band.

Table B. EXAMPLES OF U.S. FREQUENCY ALLOCATIONS

Frequencies,[a] hz	Uses	Exclusive or shared
535–1 605 khz	AM broadcast	E
54–72 Mhz	TV channels 2–4	E
76–88 Mhz	TV channels 5–6	E
124–216 Mhz	TV channels 7–13	E
470–890 Mhz	TV channels 14–83	E
88–92 Mhz	FM, non-commercial	E
92–108 Mhz	FM, commercial	E
152–170 Mhz	Remote pickup broadcast	S
6.87–7.05 Ghz	Intercity relay	E
10.55–10.7 Ghz	Intercity relay	S
26–30 Ghz	Intercity relay	E
200–285, 325–415	Aeronautical mobile	
1.75–1.8 Mhz	Disaster	
2.80–2.81 Mhz	Interzone police	
1.6–173 Mhz[b]	Industrial	
1.6–47 Mhz[b]	Public safety (e.g., police and fire)	
20 Mhz–32 Ghz[b]	Earth-space research	
3.7–8.4 Ghz[b]	Communication satellites	

[a]FCC regulations stipulate frequency tolerances (departure from center or assigned frequency, usually not over 100 ppm or 0.01%).
[b]Several narrow bands within this range.

Table C. RADIO SERVICE ALLOCATIONS
As Defined by the International Telecommunications Union (ITU), Geneva

Allocations

Fixed: between specified fixed points
Mobile: station mobile; may be moving
Aeronautical Mobile: between aircraft or land and aircraft
Maritime Mobile: between ships or land and ships
Land Mobile: between vehicles or base station and vehicle
Radio Navigation: for navigation position determination
Aeronautical Radio Navigation: RN for aircraft (e.g., VOR)
Maritime Radio Navigation: RN for ships, including beacons and radar
Radio Location: non-navigational (e.g., radar and tracking)
Broadcasting: for reception by public

Table 2-14. SPECTRAL LUMINOUS INTENSITIES*

These relative values for luminance (photometric brightness) of the blackbody at different temperatures hold for measurements made with a field brightness above about 1 millilambert but do not hold for measurements made for low field brightness.

Wave-length, microns	Temperature, °K											
	2000°	2042.16°†	2100°	2200°	2300°	2400°	2500°	2600°	2700°	2800°	2900°	3000°
.42	.00002	.00002	.00002	.00003	.00004	.00004	.00005	.00006	.00007	.00007	.00008	.00009
.44	.00019	.00021	.00023	.00028	.00032	.00037	.00042	.00047	.00053	.00058	.00064	.00069
.46	.00083	.00088	.00096	.00111	.00125	.00140	.00155	.00171	.00186	.00202	.00217	.00233
.48	.00297	.00313	.00336	.00374	.00413	.00452	.00490	.00528	.00565	.00602	.00638	.00673
.50	.01024	.01067	.01125	.01223	.01318	.01411	.01501	.01587	.01670	.01750	.01827	.01901
.52	.03217	.03313	.03442	.03654	.03853	.04042	.04220	.04387	.04545	.04694	.04834	.04967
.54	.05972	.06086	.06236	.06475	.06692	.06890	.07072	.07238	.07390	.07530	.07659	.07776
.56	.08356	.08432	.08528	.08675	.08800	.08905	.08996	.09073	.09139	.09198	.09243	.09284
.58	.09545	.09544	.09539	.09518	.09488	.09449	.09405	.09358	.09307	.09256	.09203	.09150
.60	.08833	.08757	.08654	.08483	.08319	.08163	.08013	.07873	.07739	.07611	.07491	.07379
.62	.06663	.06554	.06409	.06178	.05966	.05774	.05595	.05432	.05281	.05141	.05012	.04893
.64	.03752	.03663	.03547	.03366	.03204	.03061	.02930	.02813	.02708	.02610	.02522	.02442
.66	.01576	.01528	.01466	.01371	.01287	.01215	.01150	.01092	.01041	.00995	.00953	.00916
.68	.00521	.00502	.00477	.00440	.00408	.00381	.00357	.00335	.00317	.00300	.00285	.00272
.70	.00147	.00141	.00133	.00121	.00111	.00102	.00095	.00088	.00083	.00078	.00073	.00069
.72	.00044	.00041	.00039	.00035	.00032	.00029	.00026	.00024	.00023	.00021	.00020	.00019
.74	.00012	.00011	.00010	.00009	.00008	.00007	.00007	.00006	.00006	.00005	.00005	.00005
.76	.00003	.00003	.00003	.00002	.00002	.00002	.00002	.00002	.00001	.00001	.00001	.00001
Relative light output:	.775	1.000	1.399	2.398	3.927	6.178	9.383	13.810	19.765	27.594	37.661	50.372
λ max:	.5825	.5820	.5805	.5785	.5770	.5755	.5745	.5730	.5715	.5705	.5695	.5685

†Platinum point.
*From: "Smithsonian Physical Tables", 9th ed., W.E. Forsythe, Ed., The Smithsonian Institution, 1956.

Table 2-15. RADIATION PROPERTIES OF LIGHT REFLECTORS AND DIFFUSERS*

Measurements are by spectrophotometer. Reflectance in infrared is relative to evaporative aluminum on glass.

SYMBOLS: R = reflectance, %; T = transmittance, %; μ = wavelength, microns (or micrometers)

REFLECTORS
Zero Transmittance

Materials	Visible range		Heat range			
	μ	R	μ	R	μ	R
Specular aluminum	.4– .6	85	1–7	90	10–15	19
Diffuse aluminum	.4– .6	79	1–7	87	10–15	54
White porcelain	.4–1	75	2.0	38	4–15	9
White enamel	.4–1	77	2.0	45	4–15	5

DIFFUSERS AND ENCLOSURES

Materials	Visible range			Heat range					
	μ	R	T	μ	R	T	μ	R	T
Clear glass, .125 in.	.4– .6	8	92	1.0	5	92	4–15	7	0
Opal glass, .155 in.	.4– .6	26	39	1–2	14	65	4–15	7	0
Clear acrylic, .12 in.	.4–1	6	92	2.0	8	53	4–15	3	0
Clear polystyrene, .12 in.	.4–1	8	89	2.0	11	61	4–15	4	0
White acrylic, .125 in.	.4– .6	27	27	1–2	10	50	4–15	3	0
White polystyrene, .12 in.	.4– .6	29	26	1–2	16	42	4–15	3	0
White vinyl, .30 in.	.4– .6	8	75	1–2	12	80	4–15	3	0

*Based on data in: "Lighting and Air Conditioning", Report CP-28, Illuminating Engineering Society, 1966, p. 8.

Table 2-16.   LIGHT REFLECTION AND TRANSMISSION

At Room Temperature

For other data concerning emissivities and reflectivities, see Tables 2-8, 2-9, 2-15, and 2-20.

REFLECTORS (Zero transmittance)	Percent reflectance†	DIFFUSERS OR ENCLOSURES	Percent transmittance†
Polished silver, clean	95	Thin quartz or silica	90
Aluminized glass, front surface	92	Clear glass or plastic, $\frac{1}{8}$-in.	90
Silvered mirror, back surface	88	Ground or frosted glass	75
Polished aluminum, specular	83	Opal-white glass	50
White porcelain enamel	78	Heat-absorbing plate glass, $\frac{1}{4}$-in.	60
White plastic	78		
Smooth aluminum, diffuse	76		
White paint, gloss	75		
Chrome plate, specular	65		
Stainless steel, specular	60		
Bright aluminum paint	60		

†Approximate figures subject to an uncertainty of at least ±5%.

Table 2-17.   RADIATION PROPERTIES OF GLASS*

Total Normal Emissivity of Glass at Elevated Temperatures

See also Tables 2-8 and 2-19.

Type of glass	Total normal emissivity†							
	$\frac{1}{4}$-in. thick (6.35 mm)				$\frac{1}{2}$-in. thick (12.7 mm)			
	50°C	253°C	451°C	693°C	50°C	253°C	451°C	693°C
Borosilicate, low-expansion	0.89	0.90	0.88	0.78	0.89	0.90	0.89	0.81
96% silica	0.87	0.81	0.72	0.56	0.87	0.83	0.76	0.62
Soda-lime plate	0.91	0.91	0.88	0.71	0.91	0.92	0.90	0.83

The transmissivity of glass is closely related to its emissivity and absorptivity. Some control of the spectral distribution curves is possible, as is evidenced by the colored glasses, infrared windows, and ultraviolet windows (see Tables 2-19 and 2-20 and Figure 1-100); the common glasses are largely opaque and absorbent in the long infrared and short ultraviolet ranges and transmit in the visible range only.

A recent development is "photochromic" glass, which, by means of silver halides, can be made to reduce its transmission on exposure to light. The changes are not instantaneous, but the glass will protect against sun glare. Glass of this type has been produced that will respond by a reduction of visual transmissivity to one-half or one-third of its normal value when exposed to sunlight, and the transmissivity is restored when the light intensity is again reduced.

Fixed-color, non-glare glass has long been used in automobiles and office windows, and certain combinations offer a considerable reduction in heat transmission and air-conditioning load in summer (see Shading Coefficients, Table 7-18).

Glass is subject to radiation damage from X-rays, gamma rays, and ion bombardment. In neutron fields glass should be boron free. Lead glasses are discolored by gamma rays.

†The total hemispherical emissivity is approximately 94% of the total normal emissivity.
*From: C.J. Parker, Corning Glass Works (private communication).

REFERENCES

"ASHRAE Handbook of Fundamentals", American Society of Heating, Refrigerating and Air-Conditioning Engineers, Inc., 1967, pp. 476–483.

"Glass in Building Design and Construction", *Building Research*, 4(3): 5–64, May-June 1967.

Table 2-18.   APPLICATIONS OF INFRARED TECHNIQUES*

For additional infrared data, consult the Index.

Functions. Five kinds of functional objectives are indicated by single-word designations in the following table.

Measure.   Measure total radiation or spectral band; analyze or identify by infrared spectra; measure infrared reflection.

Heat.   Process or treat by increased temperature; dry by radiation.

Detect.   Locate; identify; signal.

Image.   Display and/or record thermal picture.

Search.   Determine range; track; control moving object.

Range of Wavelengths. The wavelengths used in a particular infrared application depend largely on the optical materials used in the equipment (see Tables 2-20 and 2-21) and the radiation source (see Figure 2-6 and Table 7-33). In the following table the term *total* refers to the total energy at all wavelengths, while the term *all* refers to the spectral curve or spectral effects. The term *bands* indicates selective bands (two or more). Numerical values indicate typical values only.

Function	Range of wavelengths	Description
LABORATORY AND SCIENTIFIC APPLICATIONS		
Detect	All or band	Infrared spectral chemical analysis
Measure	Total	Thermal insulation studies; radiation pyrometry
Detect	Total	Railroad hotbox detection; space communication
Measure		Temperature measurements in astronomy
Search		Space vehicle and satellite tracking and flight control
INDUSTRIAL, LEGAL, AND CONSUMER APPLICATIONS		
Search		Aircraft flight control and collision warning
Detect	Total	Forest-fire detection
Measure	Band	Industrial gas analysis
Heat	Total	Radiant heating and drying
Image	0.75–1.25 μm	Detection of diseased crops
Measure	All or band	Detection of impurities
Image	0.8–1.3 μm	Surveying, mapping, geology
Image	0.75–1.3 μm	Examination of art works and documents
Image	0.8–2.0 μm	Nondestructive testing of glass, crystals, plastics, semiconductors
Detect	2–12 μm	Railway car "hotbox" detector
Measure	Total	Radiation pyrometry
MEDICAL APPLICATIONS		
Measure		Measuring eye movements and pupil diameter by reflection
Measure	Total	Skin temperature measurement; monitoring healing processes
Heat	Total	Heat therapy
Detect		Obstacle detection for the blind
Image	0.8–1.2 μm	Diagnostic and microscopic photography
Image	0.7–1.2 μm	Thermal imaging for monitoring
MILITARY APPLICATIONS		
Search		Navigation and flight control; fire control
Measure	All	Analysis of terrain
Search	Total	Missile guidance; night driving and flying, docking
Detect	Total	Detection of vehicles, personnel, etc.
Image	0.8–1.2 μm	Night photography
Image	0.75–1.25 μm	Infrared searchlights
Detect	0.8–3 μm	Infrared "telephones"

*Compiled from several sources.

REFERENCES

   "Infrared System Engineering", R.D. Hudson, Jr., John Wiley & Sons, Inc., 1969; this reference gives extensive literature abstracts.

   "Infra-red Radiation", A. Vasko, Iliffe Books Ltd., 1968.

Table 2-19. SPECTRAL TRANSMISSION: UV, IR, AND COLORED GLASSES

Range	Color	Function	Transmission wavelengths, μm		TYPICAL COMMERCIAL GLASSES		
			Range†	Max.	Corning	Bausch and Lomb	Jena
UV	Black or blue	Transmits UV (opaque to light)	.27– .38	.32	7–54, 9863	Blue 3	UG series
UV	Clear or blue	Transmits UV and visible light	.24–3.2	.40–2.0	9–54, 7910	Blue 1	WG8
Visible	Blue	Signal lens	.35– .52	.42	5–56, 5031	Blue 8	BG7
Visible	Blue-green	Signal lens	.36– .55	.46	4–70, 4308	IR–2	BG18
Visible	Green	(Transmits some IR)	.42– .53	.49	4–74, 4445	BG1	VG10
Visible	Yellow	Bright color (No UV, some IR)	.52–2.7	.60	3–70, 3384	Y–10	GG10
Visible	Amber	Signal lens	.55–2.5	.66	3–77, 3307	Y–4	FG–8
Visible	Red	Sharp cutoff. Signals; color separations	.64–2.5	.70–.75	2–59, 2404	R–3	RG–5
Visible	Clear	Common window or plate glass	.34–2.5	.5	—	—	—
Visible	Clear	Heat-absorbing glass	.30– .87	.55	1–69, 4600		KG–1
IR	Clear	Transmits IR and visible light	.40–4.0	2.0	3–138, 9–30	IR–2	—
IR	Dark	Transmits IR	1.6	1–2.7	7–57, 2550	IR–10	RG–9

†Over 50% transmission.

Table 2-20. INFRARED WINDOW MATERIALS*

Maximum Wavelength Limits for Transmission Through Window of Practical Thickness

See also Table 2-21.

SYMBOLS: H = hygroscopic; not water resistant   S = soft; easily scratched

Class or group	Approx. transmission limits, microns	Material or major constituent	Class or group	Approx. transmission limits, microns	Material or major constituent
1	2–3	Common glass	5	13–20	Sodium fluoride, NaF (H)
2	4–5	Quartz; silica, SiO_2			Zinc selenide, ZnSe
3	6–8	Lithium fluoride, LiF			Zinc sulfide, ZnS
		Magnesium fluoride, MgF_2	6	20–50	Cesium bromide, CsBr (H, S)
		Sapphire, Al_2O_3			Cesium iodide, CsI (H)
4	9–12	Arsenic trisulfide, As_2S_3 (H)			Potassium bromide, KBr (H, S)
		Barium fluoride, BaF_2			Potassium chloride, KCl (H)
		Calcium fluoride, CaF_2			Potassium iodide, KI (H)
		Sodium chloride, NaCl			Silver chloride, AgCl (S)
					Tellurium bromide, $TeBr_2$

*Compiled from several sources.

Table 2-21.   INFRARED OPTICAL MATERIALS*

Materials for Prisms, Lenses, Filters, and Mirrors in the Spectral Range of 1–50 μm

For infrared window materials see Table 2-20.

Material	Wavelength transmission limit, μm	Index of refraction (mid-range)
Arsenic trisulfide, As_2S_3	12	2.4
Cadmium telluride, CdTe	30	2.7
Calcium aluminate, $CaAl_2O_4$	6	1.65
Calcium fluoride, CaF_2	12	1.4
Cesium bromide, CsBr	37	1.65
Cesium iodide, CsI	50	1.5
Crystalline quartz or fused silica, SiO_2	4	1.45
Germanium, Ge	25	4.0
Lithium fluoride, LiF	6	1.35
Magnesium fluoride, MgF_2	7–9	1.35
Magnesium oxide, MgO	9	1.65
Optical glass; quartz glass	2–3	1.5
Potassium bromide, KBr	25	1.5
Sapphire, Al_2O_3	5–7	1.7
Silicon, Si	16	3.4
Sodium chloride, NaCl	14	1.5
Strontium titanate, $SrTiO_3$	7	2.2
Zinc selenide, ZnSe	21	2.4
Zinc sulfide, ZnS	15	2.2

Lenses and Prisms

In the use of these materials, mounting and protection are important, and thermal stress and thermal expansion must be considered. Resistance to moisture, abrasion, and other surface attack must be investigated. Some materials are available only in small diameters (e.g., strontium titanate). Spectral transmittance curves should be obtained, and variation of transmission and refraction with temperature may be important.

Filters

Dyes and plastics absorb certain wavelengths and are used as "pass filters," passing short wavelengths, or passing long wavelengths. Bandpass filters transmit only specific bands of wavelengths, with sharp boundaries.

Many proprietary filters are available, with sharp optical edges as specified, in the range of 0.6 to 1.0 μm. These are colored glass or plastic, with good thermal endurance, and their appearance is dark red-black.

Reflectors

Films of evaporated metal are widely used. Silver films have the highest reflectance (0.99) but tarnish quickly. Aluminum and copper reflect >97.5% if untarnished. Rhodium has been used (95% reflectance).

Antireflection Coatings

The following have been used for the spectral range up to about 7 microns:

Low index of refraction, 1.35–1.6	*High index of refraction, >2*
Fluorides of Ce, Mg, Th	Dioxides of Ce, Ti, Zr
Cryolite	Zinc sulfide

Atmospheric Transmission

When infrared optical instruments are used for long-distance transmission through the atmosphere, the absorption properties of the air mixture must be taken into account. Water vapor and carbon dioxide are the main absorbers, but other gases, such as CH_4, N_2O, and O_3 may interfere if present. Atmospheric absorption is selective and is high at 2.6–2.8 μm and 4.2–4.4 μm. Atmospheric humidity increases the absorption and scattering, but total transmission is fair, even through haze, fog, cloud, smog, or even light rain. Extensive tables of atmospheric transmission will be found in Hudson's "Infrared Systems Engineering" (John Wiley & Sons, 1969). For infrared ranging and tracking, the phenomenon of scintillation (like twinkling) may affect accuracy.

For long atmospheric paths, such as 10 miles or more, the longer wavelengths, 5–8 μm and 12.5–14 μm, are largely absorbed.

*Compiled from several sources

Table 2-22.   TYPICAL CHARACTERISTICS
OF ALUMINIZED PHOSPHOR SCREENS*

Fluorescent screens are used in various electron devices, such as image tubes, cathode-ray tubes, and storage cathode-ray tubes, to convert electron energy into radiant energy. These viewing screens are comprised of many small-diameter (2 to 3 microns) phosphor crystals that emit light when bombarded by high-energy electrons. The spectral response of a phosphor screen is determined by its chemical and physical composition, deposition methods, and the tube processing procedures. Phosphor screens with given output characteristics have been categorized and assigned type numbers.

Phosphor screen P number	Chemical composition	Fluorescent color	Persistence classification	Typical peak wavelength, nanometers	Typical luminous equivalent, radiated lumens per radiated watt	Typical absolute efficiency, radiated watts per watt excitation	Typical quantum yield factor, photons per electron-volt
P1	$Zn_2SiO_4:Mn$	Yellow-green	Med.	525	520	0.06	0.026
P2	$ZnS:Cu$	Yellow-green	Med.	533	460	0.07	0.03
P3	$Zn_8BeSi_5O_{19}:Mn$	Yellow orange	Med.	603	380	0.041	0.02
P4	$ZnS:Ag+ZnCdS:Ag$ (all sulfide type)	White	Med. short	459	290	0.15	0.067
	Silicate-sulfide type	White	Med.	450	290	—	—
	Silicate type	White	Med.	410	240	—	—
P5	$CaWO_4:W$	Blue	Med. short	417	90	0.025	0.009
P6	$ZnS:Ag+ZnCdS:Ag$	White	Short	565	340	—	—
P7	$ZnS:Ag$ on $ZnCdS:Cu$	White (White decay)	Med. short Long	440 —	280 —	— —	— —
P10	KCl	(Dark trace)	Long	—	—	—	—
P11	$ZnS:Ag(Ni)$	Blue	Med. short	460	140	0.10	0.038
P12	$ZnMgF_2:Mn$	Orange	Long	590	410	—	—
P13	$MgSiO_3:Mn$	Red-orange	Med.	640	140	—	—
P14	$ZnS:Ag+ZnCdS:Cu$	Purple-blue (Yellow-orange decay)	Med. short Med.	440 —	250 —	— —	— —
P15	$ZnO:Zn$	Green	Short	390	250	0.051	0.02
P16	$CaMgSiO_3:Ce$	UV-blue	Very short	380	25	0.049	0.015
P17	$ZnO+ZnCdS:Cu$	Blue-white (Yellow decay)	Short Long	550 —	350 —	— —	— —
P18	$CaMgSiO_3:Ti+P3$	White	Med.	410	230	—	—
P19	$KMgF_2:Mn$	Orange	Long	590	390	0.0002	—
P20	$ZnCdS:Ag$	Yellow-green	Med.	560	480	0.14	0.063
P21	$MgF_2:Mn$	Red-orange	Med.	610	360	—	—
P22B	$ZnS:Ag$	Blue	Short	450	55	0.15	0.055
P22G	$Zn_2SiO_4:Mn$	Green	Med.	525	530	0.06	0.025
P22R	$Zn_3(PO_4)_2:Mn$	Red	Med.	645	150	0.05	0.022
P23	P4 type	White	Med. short	570	320	—	—
P24	$ZnO:Zn$ (Special)	Green	Short	510	360	0.026	0.011
P25	$CaSiO_3:Pb, Mn$	Orange	Med.	610	320	0.013	0.006
P26	ZnF	Orange	Very long	590	410	—	—
P27	$Zn_3(PO_4)_2:Mn$	Red-orange	Med.	635	60	—	—
P28	$ZnS:Ag, Cu$	Green-yellow	Long	550	500	—	—
P29	P2 & P25 type	—	—	—	—	—	—
P31	$ZnS:Cu$	Green	Med. short	522	230	0.22	—

Notes:

Since the response characteristics of phosphor screens depend on such variable parameters as chemical composition, particle size, deposition methods, and the tube-processing procedures, considerable departure from the data given on this chart is to be expected for individual screen samples.

With the exception of efficiency and quantum yield factor, the data are based primarily on JEDEC† Publication No. 16 entitled "Optical Characteristics of Cathode-Ray-Tube Screens."

Table 2-22.   TYPICAL CHARACTERISTICS
OF ALUMINIZED PHOSPHOR SCREENS *(Continued)*

Efficiency and quantum yield factor data on phosphor screen types shown in the table are derived directly from experimental measurements. For the remaining phosphors data from various published sources, primarily phosphor manufacturers, have been extrapolated to the stated units of measurement whenever possible.

Phosphor excitation is expressed in terms of the power dissipated by the electron beam in the phosphor layer proper, with corrections for power losses to the aluminum layer coating and to the glass substrate in the case where the electron beam completely penetrates the phosphor layer. Expressing the phosphor excitation in this manner minimizes the variations of efficiency ratings with accelerating potential. The ratings given are most accurate in the region from 8 kV to 12kV. Typical phosphor screen dead voltages, as a result of the aluminum coating, can be expected to be on the order of 1 kV to 3 kV.

Radiant output is expressed in terms of the total flux leaving the phosphor exit window. Luminance characteristics may be calculated from the data given, provided a radiating area is also specified and assuming a cosine-law radiance distribution (approximately valid only).

Input current levels are restricted to the linear response region for each phosphor with respect to both average and peak current densities.

Quantum yield factors are tabulated in terms of photons per electron-volt, making it possible to multiply the listed numerical factor by the selected effective excitation voltage to give the quantum yield in photons per electron.

†The Joint Electron Device Engineering Council is a cooperative effort of the Electronic Industries Association and the National Electrical Manufacturers Association.

*From: "Reference Data for Radio Engineers", 5th ed., Howard W. Sams & Co., Inc., Indianapolis, Indiana, 1968.

Table 2-23.   ELECTRICAL RESISTIVITY OF METALS AND ALLOYS
Table A.   PURE METALLIC ELEMENTS

Name	Resistivity		Temp coef of resist/ deg C[a]	Name	Resistivity		Temp coef of resist/ deg C[a]
	microhm-cm	ohm/ mil ft			microhm-cm	ohm/ mil ft	
Aluminum	2.7	16.2	0.004	Niobium	13.	78.	0.004
Antimony	40.	241.	0.003 6	Osmium	9.	54.	0.004
Beryllium	4.5	27.	0.025	Palladium	10.8	65.	0.003 5
Bismuth	115.	692.	0.004	Platinum	10.5	63.2	0.003 9
Cadmium	7.4	44.5	0.004	Plutonium	141.	848.	
Cerium	75.	451.	0.000 9	Potassium	7.	42.	
Cesium (liq)	20.	12.		Rhenium	19.	114.	0.004
Chromium	13.	78.	0.003	Rhodium	4.7	28.2	0.004 5
Cobalt	9.	54.	0.006	Rubidium	12.5	75.	
Copper	1.7	10.2	0.004	Selenium	12.	72.	
Gallium (liq)	17 4	105		Silver	1.6	9.65	0.003 8
Gold	2.4	14.4	0.003 5	Sodium	4.5	27.	
Hafnium	35.	210.	0.004	Strontium	23.	138.	
Indium	8.4	50.	0.005	Tantalum	13.	78.	0.003 5
Iridium	5.3	31.9	0.003 9	Thallium	18.	108.	
Iron	9.7	58.5	0.006	Thorium	18.	108.	0.003
Lead	21.	126.	0.004	Tin	11.4	68.6	0.004 5
Lithium	9.	54.	0.005	Titanium	43.	259.	0.003 5
Magnesium	4.5	27.	0.003	Tungsten	5.6	34.	0.004 5
Manganese	185.	1 113.		Uranium	30.	180.	
Mercury (liq)	96.	577.	0.000 9	Vanadium	25.	150.	0.003
Molybdenum	5.6	33.5	0.004	Zinc	5.9	35.5	0.004
Nickel	7.	42.	0.006	Zirconium	41.	247.	0.004 4

[a]Approximate value at room temperature. Usually much higher at high temperatures.

Table 2-23. ELECTRICAL RESISTIVITY OF METALS AND ALLOYS *(Continued)*

Table B. ALLOYS

| Composition | Name | Resistivity | | | | Temp coef of expansion/ deg C | Melting point, deg C |
		Microhm-cm	Ohm/mil-ft	Relative, pure copper = 1	Temp coef of resist/ deg C		
ALUMINUM ALLOYS							
97 Al, 2 Mg, .5 Cr	Alloy 5052	5	30	2.9	.004	.000 025	625
COPPER ALLOYS							
	Soft copper wire	1.72	10.3	1.01	.004	.000 017	1 080
Hard copper wire	6101-B317-64	1.8	10.8	1.06	.004		1 085
Copper-clad steel	30 HS, B227-65	35	210	20.6	.005		
98 Cu, 2 Ni	Alloy 30	5	30	2.9	.001 4	.000 017	1 090
94 Cu, 6 Ni	Cuprothal 60[a]	10	60	5.9	.001 4	.000 018	1 100
91 Cu, 8 Sn, .25 P	Phosphor bronze	13	65	7.65		.000 018	1 050
89 Cu, 11 Ni	Alloy 90	15	90	8.8	.000 5	.000 02	1 110
87 Cu, 13 Mn	Manganin	48	290	28.4	.000 01	.000 02	1 020
78 Cu, 22 Ni	Midohm[a] Cuprothal 180[a]	30	180	17.6	.000 2	.000 02	1 130
65 Cu, 35 Zn	Brass	7	42	4.1	.002	.000 018	940
64 Cu, 18 Zn, 18 Ni	Nickel, silver	28	168	16.5	.000 3	.000 018	1 110
57 Cu, 43 Ni	Constantan	49	294	28.8	.000 01	.000 015	1 220
IRON ALLOYS							
	Soft steel 1010	12	72	7.1	.006	.000 011	1 450
99 Fe, 1 C	Carbon steel	20	120	11.8	.005	.000 011	1 430
96 Fe, 4 Si	High silicon iron	59	354	34.7	.002	.000 013	1 410
81 Fe, 15 Cr, 4 Al	Alloy 750	125	750	73.5	.000 15	.000 015	1 520
74 Fe, 18 Cr, 8 Ni	18-8 Stainless steel	73	440	43.0	.000 94	.000 018	1 400
72 Fe, 22.5 Cr, 5.5 Al	Alloy 875	146	875	85.8	.000 02	.000 017	1 520
65 Fe, 35 Ni	Invar	81	485	47.6	.001 35	.000 001	1 425
62 Fe, 21 Ni, 12 Al, 5 Co	Alnico I[a]	75	450	44.1	.002	.000 015	
58 Fe, 42 Ni	Alloy 142	67	400	39.3	.001 2	.000 005	1 425
55 Fe, 37.5 Cr, 7.5 Al	High-resistance alloy	166	1 000	97.6	.001	.000 015	1 500
45 Fe, 35 Ni, 20 Cr	Chromax,[a] Chromel D[a]	100	600	58.8	.000 36	.000 016	1 380
NICKEL ALLOYS (>50%)							
80 Ni, 20 Cr	Nichrome V[a] Nikrothal 8[a] Chromel A[a]	108	650	63.6	.000 1	.000 016	1 375
72 Ni, 38 Fe	Hytemco[a]	20	120	11.8	.004 2	.000 015	1 425
68 Ni, 20 Cr, 8 Fe	Chromel AA[a]	117	700	68.7	.000 11	.000 014	1 390
60 Ni, 16 Cr, 24 Fe	Nichrome[a] Chromel C[a] Nicrothal 6[a]	112	675	66.0	.000 15	.000 015	1 350
95 Ni, 4 Mn, 1 Si	Alloy R63	22	135	13.0	.003	.000 015	1 400
67 Ni, 30 Cu, 1.4 Fe, 1 Mn	Monel	42	252	24.7	.002	.000 014	1 330
SILVER ALLOYS							
92.5 Ag, 7.5 Cu	Sterling silver	2	12	1.18	.004	.000 018	905
85 Ag, 15 Cd	Contact alloy	5	30	2.9	.004	.000 019	875

[a]Proprietary name.

Table 2-24. ELECTRICAL CONDUCTORS FOR GENERAL WIRING

Allowable Current-carrying Capacities in Amperes

The following table is extracted from various tables appearing in the National Electrical Code, 1971, prepared and copyrighted by the National Fire Protection Association. The NFPA standards apply primarily to wiring in buildings and distribution systems.

The capacities are based on room temperature of 30°C (86°F). For higher ambient temperatures the capacities are lower.

SYMBOLS:

W = moisture resistant	SIS = synthetic heat-resistant rubber
H = heat resistant	T = thermoplastic, a resin-base
A = asbestos	synthetic
FEP = fluorinated ethylene propylene	V = varnished cambric
N = nylon covering	X = polyethylene
R = rubber	Open = single conductor in free air
RU = higher-grade rubber, thinner insulation	Encl = not more than three conductors in raceway, cable, or direct burial

Size, AWG (or MCM)	INSULATIONS											
	Operating temperature, 60°C				Operating temperature, 75°C				Operating temperature, 90°C			
	RUW, T, TW				RH, RUH, XHHW, RHW, THW, THWN				TA, TBS, SA, SIS, AVB, FEP, FEPB, RHH, THHN, XHHW			
	Copper		Aluminum		Copper		Aluminum		Copper		Aluminum	
	Open	Encl	Open	Encl	Open	Encl	Open	Encl	Open	Encl	Open	Encl
14	20	15			20	15			30†	25†		
12	25	20	20	15	25	20	20	15	40†	30†	30†	25†
10	40	30	30	25	40	30	30	25	55†	40†	45†	30†
8	55	40	45	30	65	45	55	40	70	50	55	40
6	80	55	60	40	95	65	75	50	100	70	80	55
4	105	70	80	55	125	85	100	65	135	90	105	70
3	120	80	95	65	145	100	115	75	155	105	120	80
2	140	95	110	75	170	115	135	90	180	120	140	95
1	165	110	130	85	195	130	155	100	210	140	165	110
0	195	125	150	100	230	150	180	120	245	155	190	125
00	225	145	175	115	265	175	210	135	285	185	220	145
000	260	165	200	130	310	200	240	155	330	210	255	165
0000	300	195	230	155	360	230	280	180	385	235	300	185
(250)	340	215	265	170	405	255	315	205	425	270	330	215
(300)	375	240	290	190	445	285	350	230	480	300	375	240
(400)	455	280	355	225	545	335	425	270	575	360	450	290
(500)	515	320	405	260	620	380	485	310	660	405	515	330
(600)	575	355	455	285	690	420	545	340	740	455	585	370
(750)	655	400	515	320	785	475	620	385	845	500	670	405
(1,000)	780	455	625	375	935	545	750	445	1,000	585	800	480
(1,250)	890	495	710	405	1,065	590	855	485	1,130	645	905	530
(1,500)	980	520	795	435	1,175	625	950	520	1,260	700	1,020	580
(1,750)	1,070	545	875	455	1,280	650	1,050	545	1,370	735	1,125	615
(2,000)	1,155	560	960	470	1,385	665	1,150	560	1,470	775	1,220	650

Notes: The A types use asbestos insulation, and the maximum operating temperature is 90°C or above.

For general interior wiring, wire smaller than No. 14 should not be used, except for fixture wire, certain flexible cords, control circuits, etc.

Most types of flexible metal-clad cable and non-metallic sheathed cable are permitted for indoor wiring if properly installed close to the surface and not exposed to hazards; however, local codes may impose restrictions.

Overcurrent protection by fuses and circuit breakers is required on all circuits. Separate 20-ampere branch circuits are required for small domestic appliances; such circuits shall have no other outlets. Two such 20-ampere circuits are required in the kitchen, one in the laundry. No. 16 fixture wire or larger is required on 30-ampere circuits, and No. 12 fixture wire or taps for 50-ampere protection.

For automotive vehicle wiring see the "SAE Standards" in the latest annual SAE Handbook, Society of Automotive Engineers, New York, N.Y. 10017.

†For the smaller sizes of types FEP, FEPB, RHH, THHN, and XHHW, these capacities are not allowed. For these cases the allowable capacities are reduced to the following (in amperes):

Material	Open or enclosed	Size			Material	Open or enclosed	Size		
		10	12	14			10	12	14
Copper	Open	40	25	20	Aluminum	Open	30	20	—
Copper	Enclosed	30	20	15	Aluminum	Enclosed	25	15	—

Table 2-25. AMPERE CAPACITIES—HEAVY COPPER WIRE AND CABLE

Copper size, AWG (or MCM)	AIEE—IPCEA POWER CABLE AMPACITIES* 90°C conductor, 40°C ambient air, and 20°C ambient earth					NATIONAL ELECTRICAL CODE† for buildings 75°C conductor, 30°C ambient	
	Single conductor in air, one cable per support	Underground, direct burial, single-conductor cable	Underground, direct burial, 3-conductor cable	Underground duct; one 3-conductor cable	Underground duct bank; six 3-conductor cables	In free air	In conduit, one to three conductors
8	83	108	83	59	46	65	45
6	109	139	106	78	60	95	65
4	145	180	137	102	77	125	85
2	192	231	178	133	98	170	115
1	223	261	201	154	112	195	130
0	258	297	229	177	127	230	150
00	298	337	260	202	144	265	175
000	345	384	297	231	163	310	200
0000	400	434	335	264	185	360	230
(250)	445	472	367	292	202	405	255
(350)	552	569	442	354	242	505	310
(500)	695	690	531	429	289	620	380
(750)	898	847	648	529	348	785	475
(1,000)	1,076	980	729	599	390	935	545
(1,250)	1,228	1,083				1,065	590
(1,500)	1,367	1,176				1,175	625
(1,750)	1,493	1,257				1,280	650
(2,000)	1,606	1,325				1,385	665

†Maximum capacities depend on insulation.

*From: "Power Cable Ampacities", AIEE—IPCEA Publications S–135–1 and P–46–426. Ampere capacities in these five columns are for unshielded conductors for potentials to 3,000 volts, 100 percent load factor. At 75 percent load factor (max 1 hr/max 24 hr) the capacities underground are almost 15 percent higher. For higher conductor temperatures or lower ambient temperatures, the ampacities will be higher. For effects of temperature, shielding, voltage, or other cable arrangements, see the above source. At high frequencies the ampacities will be lower than above.

Table 2-26.   STANDARD FLAT CONDUCTOR CABLE*

Compact Cable for Aerospace, Communication, Instrument, and Automotive Applications

Flat conductor cable is made up of rolled flat conductors (equivalent round wire sizes 10 to 30 AWG), laminated between thin, flexible plastic-insulating films. Standard sizes cover the range from 2 to 57 conductors as shown in this table. Various insulating materials are used (see Table 1-84), and the standard thickness of the insulating layers is 2 mils, resulting in a very compact cable assembly. Connector plugs and receptacles are available. In addition to the standard cables shown in this table, there are high-temperature, high-density, and power cables of similar design.

All dimensions in the table are in inches; for mm multiply tabular dimension by 25.4.

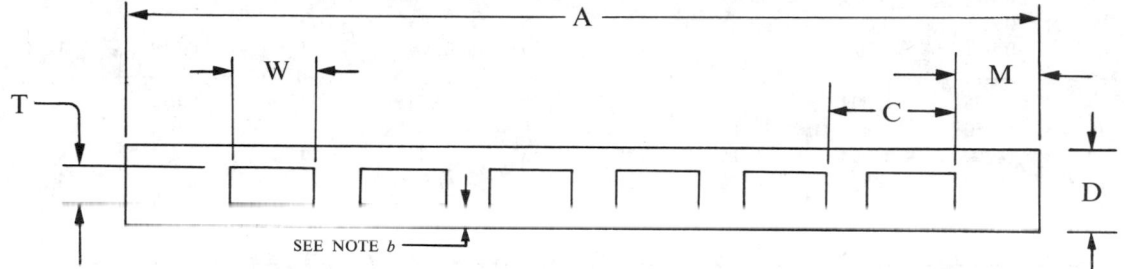

No.	Conductor Spacing "C" ±0.005	Size T ±0.0004	Size W ±0.0002	Nearest AWG wire size	Nominal conductor resistance, ohm/ 1000 ft[a]	Width "A" ±0.005	Margin "M" ±0.008	"D"	Weight, g/ft
4	0.050	0.003	0.025	30	111	0.275	0.050	0.010	1.2
7	0.050	0.003	0.025	30	111	0.425	0.050	0.010	1.9
17	0.050	0.003	0.025	30	111	0.925	0.050	0.010	4.2
27	0.050	0.003	0.025	30	111	1.425	0.050	0.010	6.5
37	0.050	0.003	0.025	30	111	1.925	0.050	0.010	8.8
47	0.050	0.003	0.025	30	111	2.425	0.050	0.010	11.2
57	0.050	0.003	0.025	30	111	2.925	0.050	0.010	13.5
3	0.075	0.004	0.025	29	83	0.290	0.057	0.011	1.3
6	0.075	0.004	0.025	29	83	0.515	0.057	0.011	2.4
12	0.075	0.004	0.025	29	83	0.965	0.057	0.011	4.5
18	0.075	0.004	0.025	29	83	1.415	0.057	0.011	6.6
25	0.075	0.004	0.025	29	83	1.940	0.057	0.011	9.1
31	0.075	0.004	0.025	29	83	2.465	0.057	0.011	11.5
38	0.075	0.004	0.025	29	83	2.915	0.057	0.011	13.6
3	0.075	0.003	0.040	28	69	0.290	0.050	0.010	1.3
6	0.075	0.003	0.040	28	69	0.515	0.050	0.010	2.4
12	0.075	0.003	0.040	28	69	0.965	0.050	0.010	4.5
18	0.075	0.003	0.040	28	69	1.415	0.050	0.010	6.7
25	0.075	0.003	0.040	28	69	1.940	0.050	0.010	9.2
32	0.075	0.003	0.040	28	69	2.465	0.050	0.010	11.7
38	0.075	0.003	0.040	28	69	2.915	0.050	0.010	13.8
2	0.100	0.004	0.065	25	32	0.265	0.050	0.011	1.5
4	0.100	0.004	0.065	25	32	0.465	0.050	0.011	2.7
9	0.100	0.004	0.065	25	32	0.965	0.050	0.011	5.9
14	0.100	0.004	0.065	25	32	1.465	0.050	0.011	9.1
19	0.100	0.004	0.065	25	32	1.965	0.050	0.011	12.3
24	0.100	0.004	0.065	25	32	2.465	0.050	0.011	15.5
29	0.100	0.004	0.065	25	32	2.965	0.050	0.011	18.6

[a]1000 ft = 304.8 m.
[b]The thickness of the insulation on each side of the conductors shall be uniform to within 0.001 in.

Table 2-26.   STANDARD FLAT CONDUCTOR CABLE *(Continued)*

	Conductor					Cable			
No.	Spacing "C" ±0.005	Size		Nearest AWG wire size	Nominal conductor resistance, ohm/ 1000 ft[a]	Width "A" ±0.005	Margin "M" ±0.008	"D"	Weight, g/ft
		T ±0.0004	W ±0.0002						
2	0.100	0.005	0.065	24	26	0.265	0.050	0.012	1.7
4	0.100	0.005	0.065	24	26	0.465	0.050	0.012	3.2
9	0.100	0.005	0.065	24	26	0.965	0.050	0.012	6.9
14	0.100	0.005	0.065	24	26	1.465	0.050	0.012	10.7
19	0.100	0.005	0.065	24	26	1.965	0.050	0.012	14.4
24	0.100	0.005	0.065	24	26	2.465	0.050	0.012	18.2
29	0.100	0.004	0.065	24	26	2.965	0.050	0.012	21.9
3	0.150	0.004	0.115	22	18	0.555	0.070	0.011	3.5
6	0.150	0.004	0.115	22	18	1.005	0.070	0.011	6.6
9	0.150	0.004	0.115	22	18	1.445	0.070	0.011	9.7
12	0.150	0.004	0.115	22	18	1.905	0.070	0.011	12.8
16	0.150	0.004	0.115	22	18	2.505	0.070	0.011	17.0
19	0.150	0.004	0.115	22	18	2.955	0.070	0.011	20.2
3	0.150	0.005	0.115	21	14	0.555	0.070	0.012	4.1
6	0.150	0.005	0.115	21	14	1.005	0.070	0.012	7.7
9	0.150	0.005	0.115	21	14	1.445	0.070	0.012	11.5
12	0.150	0.005	0.115	21	14	1.905	0.070	0.012	15.1
16	0.150	0.005	0.115	21	14	2.505	0.070	0.012	20.1
19	0.150	0.005	0.115	21	14	2.955	0.070	0.012	23.8

*From: "Flat Conductor Cable Technology", NASA Report SP–5043, 1968.

Table 2-27.   ELECTRICAL WIRE—FLEXIBLE CORDS AND FIXTURE WIRING

At Room Temperature, 30°C (86°F)

Description	National Electrical Code letter classification examples	Maximum amperes for AWG wire sizes				
		18	16	14	12	10
Flexible cord, various services, rubber insulation; also rubber fixture wire	C, K, PO, PD, PW, RF	5	7	15	20	25
Flexible fixture cord for high-temperature location; also heat-resistant fixture wire	CFC, AFC, CFPO, AFPO, CF, AF, SF, TF	6	8	17	23	28
Flexible cord, various services, silicone rubber, or thermoplastic	SO, SJ, SP, ST, SJT, SPT	7	10	15	20	25
Rubber and asbestos or neoprene heater cord	AFS, HC, HPD, HSJ, HS, HPN	10	15	20	30	35
Asbestos: flame- and moisture-resistant	AVPO, AVPD	17	22	28	36	47

Notes: Some of the above types are not furnished in the larger wire sizes.

Tinsel cord, size 27 AWG, is furnished with rubber (TP, TS) or thermoplastic (TPT, TST) insulation and is limited to 0.5 amperes.

Cords shall not be smaller than required for the rated current of the total connected equipment.

Tables 2-28, 2-29, 2-30, and 2-31.   SUPERCONDUCTIVITY*

B.W. ROBERTS

General Electric Research Laboratory, Schenectady, New York

The following tables on superconductivity include superconductive properties of chemical elements (Table 2-28), thin films (Table 2-29), a selected list of compounds and alloys (Table 2-30), and high-magnetic-field superconductors (Table 2-31).

The historically first observed and most distinctive property of a superconductive body is the near total loss of resistance at a critical temperature (T_c) that is characteristic of each material. Figure 1(a) below illustrates schematically two types of possible transitions. The sharp vertical discontinuity in resistance is indicative of that found for a single crystal of a very pure element or one of a few well annealed alloy compositions. The broad transition, illustrated by broken lines, suggests the transition shape seen for materials that are not homogeneous and contain unusual strain distributions. Careful testing of the resistivity limits for superconductors shows that it is less than 4×10^{-23} ohm-cm, while the lowest resistivity observed in metals is of the order of 10^{-13} ohm-cm. If one compares the resistivity of a superconductive body to that of copper at room temperature, the superconductive body is at least 10^{17} times less resistive.

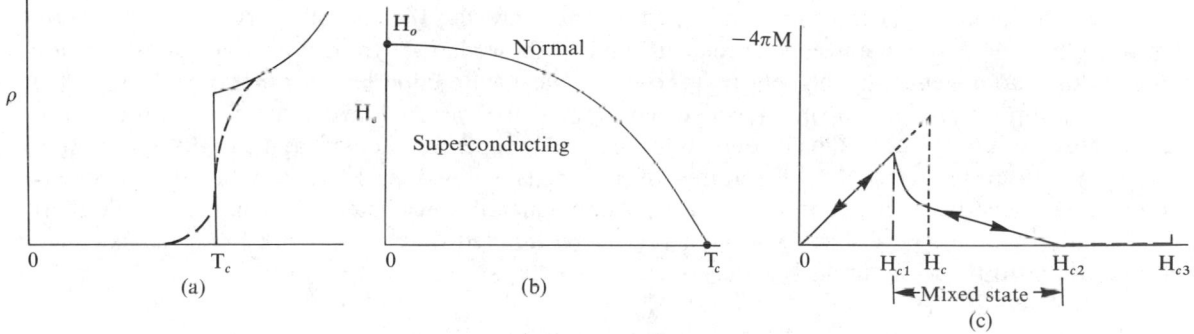

Figure 1.   PHYSICAL PROPERTIES OF SUPERCONDUCTORS

(a) Resistivity versus temperature for a pure and perfect lattice (solid line).
   Impure and/or imperfect lattice (broken line).
(b) Magnetic-field temperature dependence for Type-I or "soft" superconductors.
(c) Schematic magnetization curve for "hard" or Type-II superconductors.

The temperature interval ΔT_c, over which the transition between the normal and superconductive states takes place, may be of the order of as little as 2×10^{-5} °K *or* several °K in width, depending on the material state. The narrow transition width was attained in 99.9999 percent pure gallium single crystals.

A Type-I superconductor below T_c, as exemplified by a pure metal, exhibits perfect diamagnetism and excludes a magnetic field up to some critical field H_c, whereupon it reverts to the normal state as shown in the H-T diagram of Figure 1(b).

The difference in entropy near absolute zero between the superconductive and normal states relates directly to the electronic specific heat, γ: $(S_s - S_n)_{T \to 0} = -\gamma T$.

The magnetization of a typical high-field superconductor is shown in Figure 1(c). The discovery of the large current-carrying capability of Nb_3Sn and other similar alloys has led to an extensive study of the physical properties of these alloys. In brief, a high-field superconductor, or Type-II superconductor, passes from the perfect diamagnetic state at low magnetic fields to a mixed state and finally to a sheathed state before attaining the normal resistive state of the metal. The magnetic field values separating the four stages are given as H_{c1}, H_{c2}, and H_{c3}. The superconductive state below H_{c1} is perfectly diamagnetic, identical to the state of most pure metals of the "soft" or Type-I

*Prepared for Office of Standard Reference Data, National Bureau of Standards, by Standard Reference Data Center on Superconductive Materials, Schenectady, N.Y.

Tables 2-28, 2-29, 2-30, and 2-31.   SUPERCONDUCTIVITY *(Continued)*

superconductor. Between H_{c1} and H_{c2} a "mixed superconductive state" is found in which fluxons (a minimal unit of magnetic flux) create lines of normal superconductor in a superconductive matrix. The volume of the normal state is proportional to $-4\pi M$ in the "mixed state" region. Thus at H_{c2} the fluxon density has become so great as to drive the interior volume of the superconductive body completely normal. Between H_{c2} and H_{c3} the superconductor has a sheath of current-carrying superconductive material at the body surface, and above H_{c3} the normal state exists. With several types of careful measurement, it is possible to determine H_{c1}, H_{c2}, and H_{c3}. Table 2-31 contains some of the available data on high-field superconductive materials.

High-field superconductive phenomena are also related to specimen dimension and configuration. For example, the Type-I superconductor, Hg, has entirely different magnetization behavior in high magnetic fields when contained in the very fine sets of filamentary tunnels found in an un-processed Vycor glass. The great majority of superconductive materials are Type II. The elements in very pure form and a very few precisely stoichiometric and well annealed compounds are Type-I with the possible exceptions of vanadium and niobium.

Metallurgical Aspects. The sensitivity of superconductive properties to the material state is most pronounced and has been used in a reverse sense to study and specify the detailed state of alloys. The mechanical state, the homogeneity, and the presence of impurity atoms and other electron-scattering centers are all capable of controlling the critical temperature and the current-carrying capabilities in high-magnetic fields. Well annealed specimens tend to show sharper transitions than those that are strained or inhomogeneous. This sensitivity to mechanical state underlines a general problem in the tabulation of properties for superconductive materials. The occasional divergent values of the critical temperature and of the critical fields quoted for a Type-II superconductor may lie in the variation in sample preparation. Critical temperatures of materials studied early in the history of super-conductivity must be evaluated in light of the probable metallurgical state of the material, as well as the availability of less pure starting elements. It has been noted that recent work has given extended consideration to the metallurgical aspects of sample preparation.

REFERENCES

References to the data presented in this section, to additional entries of superconductive materials, and to those materials specifically tested and found non-superconductive to some low temperature may be found in the following publications:

"Superconductive Materials and Some of Their Properties", *Progress in Cryogenics*, B.W. Roberts, Vol. IV, Heywood and Co., 1964, pp. 160–231.

"Superconductive Materials and Some of Their Properties", B.W. Roberts, National Bureau of Standards Technical Notes 408 and 482, U.S. Government Printing Office, 1966 and 1969.

Table 2-28. PROPERTIES OF SUPERCONDUCTIVE ELEMENTS

Data have been selected generally from recent studies in which the sample purity and perfection appear to have been considered.

Element	T_c, °K		H_o, oersteds*		θ_D, °K†	γ, mJ mole^{-1} °K^{-2}‡
	Calorimetric	Magnetic	Calorimetric	Magnetic		
Al	1.183	1.196	104	99	420	1.36
Cd	0.518	0.56	29.6	30	209	0.688
Ga	1.087	1.091	59	51	317, 324.7	0.601, 0.596
Ga (β)		6.2				
Ga (γ)		7.62				
Hg (α)	4.16	4.154	380	410.9	87, 71.9	1.81
Hg (β)		3.949		339	93	1.37
In	3.407	3.4035	282.7	293	109	1.66
Ir		0.14		19	420	3.2
La (α)	4.80	4.9			142	10.0
La (β)	5.91	6.06		1,600	132	6.7
Mo	0.917	0.92	95	98	460	1.83
Nb	9.17	9.26	1,944	1,980	277	7.79
Os		0.655		65	500	2.35
Pa		1.4				
Pb	7.23	7.193		803	96.3	3.0
Re	1.699	1.698	188	198	415	2.35
Ru		0.49		66	550	3.0
Sb		2.6–2.7				
Sn	3.722	3.722	303	305.50	195	1.74
Ta	4.39	4.483	780	830	258	6.0
Tc		8.22, 7.92				
Th		1.368	131	162	168	4.65
Ti	0.42	0.39	56	100	425	3.32
Tl	2.38	2.39	176.5	171	78.5	1.47
U (α)		0.68, 0.23			206	12.2
U (pseudo-γ)		180 (extrapolated value)				
V	5.37	5.30	1,310	1,020	399	9.8
W		0.012		1.07	550	3.0
Zn	0.852	0.875	51.8	53	309	0.66
Zr		0.546		47	290	2.78
Zr (ω)		0.65				

*To convert oersteds to amperes/meter, multiply by 79.5775.
†θ_D indicates Debye temperature. For another data set see "Cryophysics", K. Mendelssohn, Interscience Book Co., 1960, p. 178.
‡"Specific Heats of Metals at Low Temperatures", D.H. Parkinson, Rep. Progr. Phys., 21:226–270, 1958.
 Also see "Low Temperature Specific Heat of Transition Metals and Alloys", F. Heiniger, E. Bucher, and J. Muller, Phys. Kondens. Materie, 5:243–284, 1966.

Table 2-29. THIN FILMS OF SUPERCONDUCTIVE ELEMENTS
Critical Temperatures

THIN FILMS FORMED AT VARIOUS TEMPERATURES		THIN FILMS FORMED UNDER HIGH PRESSURE		
Element	T_c, °K Magnetic	Element	T_c, °K Magnetic	Pressure
Al	1.3–3.7	Bi II	3.916	25,000 atm*
Be†	~6, ~8.4		3.90	25,200 atm
Bi	~6.0		3.86	26,800 atm
Ga	8.4, 7.2	Bi III	7.25	27,000–28,400 atm
In	3.94–4.25, 3.7	Ce	1.7	50 kbar
La	5.00–6.74	Ge	4.85–5.4	~120 kbar
Mo	~5	Se II	6.75, 6.95	~130 kbar
Nb	6.5–9.4	Si	7.1	120–130 kbar
Re	1.9–~7	Te	~3.3	~56,000 atm
Sn	4.6–4.7, 4.1	Tl (face-centered		
Ti	1.3 max.	cubic)	1.45	35 kbar
W	1.7–4.1	Tl (hexagonal,		
		close-packed)	1.95	35 kbar

*To convert atmospheres to newtons/meter², multiply by 101,325.
†$H_{c2} \gg 11,000$ (oersteds).

Table 2-30. SELECTED SUPERCONDUCTIVE COMPOUNDS AND ALLOYS

All compositions are denoted on an atomic basis, i.e., AB, AB_2, or AB_3 for compounds, unless noted. Solid solutions or odd compositions may be denoted as A_zB_{1-z} or A_zB. A series of three or more alloys is indicated as A_xB_{1-x} or by actual indication of the atomic fraction range, such as $A_{0-0.6}B_{1-0.4}$. The critical temperature of such a series of alloys is denoted by a range of values or possibly the maximum value.

The selection of the critical temperature from a transition in the effective permeability, or the change in resistance, or possibly the incremental changes in frequency observed by certain techniques is not often obvious from the literature. Most authors choose the mid-point of such curves as the probable critical temperature of the idealized material, while others will choose the highest temperature at which a deviation from the normal state property is observed. In view of the previous discussion concerning the variability of the superconductive properties as a function of purity and other metallurgical aspects, it is recommended that appropriate literature be checked to determine the most probable critical temperature or critical field of a given alloy.

A very limited amount of data on critical fields, H_o, is available for these compounds and alloys; these values are given at the end of the table.

SYMBOLS: n = number of normal carriers per cubic centimeter for semiconductor superconductors.

Substance	T_c, °K	Crystal structure type††	Substance	T_c, °K	Crystal structure type††
$Ag_xAl_yZn_{1-x-y}$	0.5–0.845		$Al_{\sim 0.8}Ge_{\sim 0.2}Nb_3$	20.7	A15
$Ag_7BF_4O_8$	0.15	Cubic	$AlLa_3$	5.57	DO_{19}
$AgBi_2$	3.0–2.78		Al_2La	3.23	C15
$Ag_7F_{0.25}N_{0.75}O_{10.25}$	0.85–0.90		Al_3Mg_2	0.84	Cubic, f.c.
Ag_7FO_8	0.3	Cubic	$AlMo_3$	0.58	A15
Ag_2F	0.066		$AlMo_6Pd$	2.1	
$Ag_{0.8-0.3}Ga_{0.2-0.7}$	6.5–8		AlN	1.55	B4
Ag_4Ge	0.85	Hex., c.p.	Al_2NNb_3	1.3	A13
$Ag_{0.438}Hg_{0.562}$	0.64	$D8_2$	$AlNb_3$	18.0	A15
$AgIn_2$	~ 2.4	C16	Al_xNb_{1-x}	<4.2–13.5	$D8_b$
$Ag_{0.1}In_{0.9}Te$			Al_xNb_{1-x}	12–17.5	A15
(n = 1.40×10^{22})	1.20–1.89	B1	$Al_{0.27}Nb_{0.73-0.48}V_{0-0.25}$	14.5–17.5	A15
$Ag_{0.2}In_{0.8}Te$			$AlNb_xV_{1-x}$	<4.2–13.5	
(n = 1.07×10^{22})	0.77–1.00	B1	AlOs	0.39	B2
AgLa (9.5 kbar)	1.2	B2	Al_3Os	5.90	
Ag_7NO_{11}	1.04	Cubic	AlPb (films)	1.2–7	
Ag_xPb_{1-x}	7.2 max.		Al_2Pt	0.48–0.55	C1
Ag_xSn_{1-x} (film)	2.0–3.8		Al_5Re_{24}	3.35	A12
Ag_xSn_{1-x}	1.5–3.7		Al_3Th	0.75	DO_{19}
$AgTe_3$	2.6	Cubic	$Al_xTi_yV_{1-x-y}$	2.05–3.62	Cubic
$AgTh_2$	2.26	C16	$Al_{0.108}V_{0.892}$	1.82	Cubic
$Ag_{0.03}Tl_{0.97}$	2.67		Al_xZn_{1-x}	0.5–0.845	
$Ag_{0.94}Tl_{0.06}$	2.32		$AlZr_3$	0.73	Ll_2
Ag_xZn_{1-x}	0.5–0.845		AsBiPb	9.0	
Al (film)	1.3–2.31		AsBiPbSb	9.0	
Al (1 to 21 katm)	1.170–0.687	Al	$As_{0.33}InTe_{0.67}$		
$AlAu_4$	0.4–0.7	Like Al3	(n = 1.24×10^{22})	0.85–1.15	B1
Al_2CMo_3	10.0	A13	$As_{0.5}InTe_{0.5}$		
Al_2CMo_3	9.8–10.2	A13 + trace 2nd phase	(n = 0.97×10^{22})	0.44–0.62	B1
			$As_{0.50}Ni_{0.06}Pd_{0.44}$	1.39	C2
Al_2CaSi	5.8		AsPb	8.4	
$Al_{0.131}Cr_{0.088}V_{0.781}$	1.46	Cubic	$AsPd_2$ (low-		
$AlGe_2$	1.75		temperature phase)	0.60	Hexagonal
$Al_{0.5}Ge_{0.5}Nb$	12.6	A15	$AsPd_2$ (high-temp. phase)	1.70	C22

††See key at end of table.

Table 2-30. SELECTED SUPERCONDUCTIVE COMPOUNDS AND ALLOYS (Continued)

Substance	T_c, °K	Crystal structure type††	Substance	T_c, °K	Crystal structure type††
$AsPd_5$	0.46	Complex	BW_2	3.1	Cl6
$AsRh$	0.58	B31	B_6Y	6.5–7.1	
$AsRh_{1.4-1.6}$	<0.03–0.56	Hexagonal	$B_{12}Y$	4.7	
$AsSn$	4.10		BZr	3.4	Cubic
$AsSn$			$B_{12}Zr$	5.82	
$(n = 2.14 \times 10^{22})$	3.41–3.65	B1	$BaBi_3$	5.69	Tetragonal
$As_{\sim 2}Sn_{\sim 3}$	3.5–3.6,		$Ba_xO_3Sr_{1-x}Ti$		
	1.21–1.17		$(n = 4.2-11 \times 10^{19})$	<0.1–0.55	
As_3Sn_4			$Ba_{0.13}O_3W$	1.9	Tetragonal
$(n = 0.56 \times 10^{22})$	1.16–1.19	Rhombohedral	$Ba_{0.14}O_3W$	<1.25–2.2	Hexagonal
Au_5Ba	0.4–0.7	$D2_d$	$BaRh_2$	6.0	Cl5
$AuBe$	2.64	B20	$Be_{22}Mo$	2.51	Cubic, like $Be_{22}Re$
Au_2Bi	1.80	Cl5			
Au_5Ca	0.34–0.38	$Cl5_b$	$Be_8Nb_5Zr_2$	5.2	
$AuGa$	1.2	B31	$Be_{0.98-0.92}Re_{0.02-0.08}$		
$Au_{0.40-0.92}Ge_{0.60-0.08}$	<0.32–1.63	Complex	(quenched)	9.5–9.75	Cubic
$AuIn$	0.4–0.6	Complex	$Be_{0.957}Re_{0.043}$	9.62	Cubic, like $Be_{22}Re$
$AuLu$	<0.35	B2			
$AuNb_3$	11.5	A15	$BeTc$	5.21	Cubic
$AuNb_d$	1.2	A2	$Be_{22}W$	4.12	Cubic, like $Be_{22}Re$
$Au_{0-0.3}Nb_{1-0.7}$	1.1–11.0				
$Au_{0.02-0.98}Nb_3Rh_{0.98-0.02}$	2.53–10.9	A15	$Be_{13}W$	4.1	Tetragonal
$AuNb_{3(1-x)}V_{3x}$	1.5–11.0	A15	Bi_3Ca	2.0	
$AuPb_2$	3.15		$Bi_{0.5}Cd_{0.13}Pb_{0.25}Sn_{0.12}$		
$AuPb_2$ (film)	4.3		(weight fractions)	8.2	
$AuPb_3$	4.40		$BiCo$	0.42–0.49	
$AuPb_3$ (film)	4.25		Bi_2Cs	4.75	Cl5
Au_2Pb	1.18, 6–7	Cl5	Bi_xCu_{1-x}		
$AuSb_2$	0.58	C2	(electrodeposited)	2.2	
$AuSn$	1.25	$D8_1$	$BiCu$	1.33 1.40	
Au_xSn_{1-x} (film)	2.0–3.8		$Bi_{0.019}In_{0.981}$	3.86	
Au_5Sn	0.7 1.1	A3	$Bi_{0.05}In_{0.95}$	4.65	α-phase
Au_3Te_5	1.62	Cubic	$Bi_{0.10}In_{0.90}$	5.05	α-phase
$AuTh_2$	3.08	Cl6	$Bi_{0.15-0.30}In_{0.85-0.70}$	5.3–5.4	α- and β-phases
$AuTl$	1.92		$Bi_{0.34-0.48}In_{0.66-0.52}$	4.0–4.1	
AuV_3	0.74	A15	Bi_3In_5	4.1	
Au_xZn_{1-x}	0.50 0.845		$BiIn_2$	5.65	β-phase
$AuZn_3$	1.21	Cubic	Bi_2Ir	1.7–2.3	
Au_xZr_y	1.7–2.8	A3	Bi_2Ir (quenched)	3.0–3.96	
$AuZr_3$	0.92	A15	BiK	3.6	
$BCMo_2$	5.4	Orthorhombic	Bi_2K	3.58	Cl5
$B_{0.03}C_{0.51}Mo_{0.47}$	12.5		$BiLi$	2.47	Ll_0, α-phase
$BCMo_2$	5.3–7.0	Orthorhombic	$Bi_{4-9}Mg$	0.7–~1.0	
BHf	3.1	Cubic	Bi_3Mo	3–3.7	
B_6La	5.7		$BiNa$	2.25	Ll_0
$B_{12}Lu$	0.48		$BiNb_3$ (high pressure		
BMo	0.5 (extrap-		and temperature)	3.05	A15
	olated)		$BiNi$	4.25	$B8_1$
BMo_2	4.74	Cl6	Bi_3Ni	4.06	Orthorhombic
BNb	8.25	B_f	$Bi_{1-0}Pb_{0-1}$	7.26–9.14	
BRe_2	2.80, 4.6		$Bi_{1-0}Pb_{0-1}$ (film)	7.25–8.67	
$B_{0.3}Ru_{0.7}$	2.58	$D10_2$	$Bi_{0.05-0.40}Pb_{0.95-0.60}$	7.35–8.4	Hexagonal, c.p., to ε-phase
$B_{12}Sc$	0.39				
BTa	4.0	B_f			
B_6Th	0.74		$BiPbSb$	8.9	

†† See key at end of table.

Table 2-30. SELECTED SUPERCONDUCTIVE COMPOUNDS AND ALLOYS *(Continued)*

Substance	T_c, °K	Crystal structure type††	Substance	T_c, °K	Crystal structure type††
$Bi_{0.5}Pb_{0.31}Sn_{0.19}$			$C_{0.44}Mo_{0.56}$	1.3	Bl
(weight fractions)	8.5		$C_{0.5}Mo_xNb_{1-x}$	10.8–12.5	Bl
$Bi_{0.5}Pb_{0.25}Sn_{0.25}$	8.5		$C_{0.6}Mo_{4.8}Si_3$	7.6	$D8_8$
$BiPd_2$	4.0		$CMo_{0.2}Ta_{0.8}$	7.5	Bl
$Bi_{0.4}Pd_{0.6}$	3.7–4	Hexagonal, ordered	$CMo_{0.5}Ta_{0.5}$	7.7	Bl
			$CMo_{0.75}Ta_{0.25}$	8.5	Bl
$BiPd$	3.7	Orthorhombic	$CMo_{0.8}Ta_{0.2}$	8.7	Bl
Bi_2Pd	1.70	Monoclinic, α-phase	$CMo_{0.85}Ta_{0.15}$	8.9	Bl
			CMo_xTi_{1-x}	10.2 max.	Bl
Bi_2Pd	4.25	Tetragonal, β-phase	$CMo_{0.83}Ti_{0.17}$	10.2	Bl
			CMo_xV_{1-x}	2.9–9.3	Bl
$BiPdSe$	1.0	C2	CMo_xZr_{1-x}	3.8–9.5	Bl
$BiPdTe$	1.2	C2	$C_{0.1-0.9}N_{0.9-0.1}Nb$	8.5–17.9	
$BiPt$	1.21	$B8_1$	$C_{0-0.38}N_{1-0.62}Ta$	10.0–11.3	
$BiPtSe$	1.45	C2	CNb (whiskers)	7.5–10.5	
$BiPtTe$	1.15	C2	$C_{0.984}Nb$	9.8	Bl
Bi_2Pt	0.155	Hexagonal	CNb (extrapolated)	~14	
Bi_2Rb	4.25	Cl5	$C_{0.7-1.0}Nb_{0.3-0}$	6–11	Bl
$BiRe_2$	1.9–2.2		CNb_2	9.1	
$BiRh$	2.06	$B8_1$	CNb_xTa_{1-x}	8.2–13.9	
Bi_3Rh	3.2	Orthorhombic, like NiB_3	CNb_xTi_{1-x}	<4.2–8.8	Bl
			$CNb_{0.6-0.9}W_{0.4-0.1}$	12.5–11.6	Bl
Bi_4Rh	2.7	Hexagonal	$CNb_{0.1-0.9}Zr_{0.9-0.1}$	4.2–8.4	Bl
Bi_3Sn	3.6–3.8		CRb_x (gold)	0.023–0.151	Hexagonal
$BiSn$	3.8		$CRe_{0.01-0.08}W$	1.3–5.0	
Bi_xSn_y	3.85–4.18		$CRe_{0.06}W$	5.0	
Bi_3Sr	5.62	Ll_2	CTa	~11 (extrapolated)	
Bi_3Te	0.75–1.0				
Bi_5Tl_3	6.4		$C_{0.987}Ta$	9.7	
$Bi_{0.26}Tl_{0.74}$	4.4	Cubic, disordered	$C_{0.848-0.987}Ta$	2.04–9.7	
			CTa (film)	5.09	Bl
$Bi_{0.26}Tl_{0.74}$	4.15	Ll_2, ordered?	CTa_2	3.26	L'_3
Bi_2Y_3	2.25		$CTa_{0.4}Ti_{0.6}$	4.8	Bl
Bi_3Zn	0.8–0.9		$CTa_{1-0.4}W_{0-0.6}$	8.5–10.5	Bl
$Bi_{0.3}Zr_{0.7}$	1.51		$CTa_{0.2-0.9}Zr_{0.8-0.1}$	4.6–8.3	Bl
$BiZr_3$	2.4–2.8		CTc (excess C)	3.85	Cubic
CCs_x	0.020–0.135	Hexagonal	$CTi_{0.5-0.7}W_{0.5-0.3}$	6.7–2.1	Bl
C_8K (gold)	0.55		CW	1.0	
$CGaMo_2$	3.7–4.1	Hexagonal, H-phase	CW_2	2.74	L'_3
			CW_2	5.2	Cubic, f.c.
$CHf_{0.5}Mo_{0.5}$	3.4	Bl	$CaIr_2$	6.15	Cl5
$CHf_{0.3}Mo_{0.7}$	5.5	Bl	$Ca_xO_3Sr_{1-x}Ti$		
$CHf_{0.25}Mo_{0.75}$	6.6	Bl	(n = 3.7–11.0 × 10¹⁹)	<0.1–0.55	
$CHf_{0.7}Nb_{0.3}$	6.1	Bl	$Ca_{0.1}O_3W$	1.4–3.4	Hexagonal
$CHf_{0.6}Nb_{0.4}$	4.5	Bl	$CaPb$	7.0	
$CHf_{0.5}Nb_{0.5}$	4.8	Bl	$CaRh_2$	6.40	Cl5
$CHf_{0.4}Nb_{0.6}$	5.6	Bl	$Cd_{0.3-0.5}Hg_{0.7-0.5}$	1.70–1.92	
$CHf_{0.25}Nb_{0.75}$	7.0	Bl	$CdHg$	1.77, 2.15	Tetragonal
$CHf_{0.2}Nb_{0.8}$	7.8	Bl	$Cd_{0.0075-0.05}In_{1-x}$	3.24–3.36	Tetragonal
$CHf_{0.9-0.1}Ta_{0.1-0.9}$	5.0–9.0	Bl	$Cd_{0.97}Pb_{0.03}$	4.2	
Ck (excess K)	0.55	Hexagonal	$CdSn$	3.65	
C_8K	0.39	Hexagonal	$Cd_{0.17}Tl_{0.83}$	2.3	
$C_{0.40-0.44}Mo_{0.60-0.56}$	9–13		$Cd_{0.18}Tl_{0.82}$	2.54	
CMo	6.5, 9.26		$CeCo_2$	0.84	Cl5
CMo_2	12.2	Orthorhombic	$CeCo_{1.67}Ni_{0.33}$	0.46	Cl5

††See key at end of table.

Table 2-30. SELECTED SUPERCONDUCTIVE COMPOUNDS AND ALLOYS (Continued)

Substance	T_c, °K	Crystal structure type††	Substance	T_c, °K	Crystal structure type††
$CeCo_{1.67}Rh_{0.33}$	0.47	C15	CuSSe	1.5–2.0	C18
$Ce_xGd_{1-x}Ru_2$	3.2–5.2	C15	$CuSe_2$	2.3–2.43	C18
$CeIr_3$,	3.34		CuSeTe	1.6–2.0	C18
$CeIr_5$	1.82		Cu_xSn_{1-x}	3.2–3.7	
$Ce_{0.005}La_{0.995}$	4.6		Cu_xSn_{1-x} (film) (made at 10°K)	3.6–7	
Ce_xLa_{1-x}	1.3–6.3				
$Ce_xPr_{1-x}Ru_2$	1.4–5.3	C15	Cu_xSn_{1-x} (film) (made at 300°K)	2.8–3.7	
Ce_xPt_{1-x}	0.7–1.55				
$CeRu_2$	6.0	C15	$CuTe_2$	<1.25–1.3	C18
$Co_xFe_{1-x}Si_2$	1.4 max.	C1	$CuTh_2$	3.49	C16
$CoHf_2$	0.56	$E9_3$	$Cu_{0-0.027}V$	3.9–5.3	A2
$CoLa_3$	4.28		Cu_xZn_{1-x}	0.5–0.845	
$CoLu_3$	~0.35		Er_xLa_{1-x}	1.4–6.3	
$Co_{0\ 0.01}Mo_{0.8}Re_{0.2}$	2–10		$Fe_{0-0.04}Mo_{0.8}Re_{0.2}$	1–10	
$Co_{0.02-0.10}Nb_3Rh_{0.98-0.90}$	2.28–1.90	A15	$Fe_{0.05}Ni_{0.05}Zr_{0.90}$	~3.9	
$Co_xNi_{1-x}Si_2$	1.4 max.	C1	Fe_3Th_7	1.86	D10
$Co_{0.5}Rh_{0.5}Si_2$	2.5		Fe_xTi_{1-x}	3.2 max.	Fe in α-Ti
$Co_xRh_{1-x}Si_2$	3.65 max.		Fe_xTi_{1-x}	3.7 max.	Fe in β-Ti
$Co_{~0.3}Sc_{~0.7}$	~0.35		$Fe_xTi_{0.6}V_{1-x}$	6.8 max.	
$CoSi_2$	1.40, 1.22	C1	FeU_6	3.86	$D2_c$
Co_3Th_7	1.83	$D10_2$	$Fe_{0.1}Zr_{0.9}$	1.0	A3
Co_xTi_{1-x}	2.8 max.	Co in α-Ti	$Ga_{0.5}Ge_{0.5}Nb_3$	7.3	A15
Co_xTi_{1-x}	3.8 max.	Co in β-Ti	$GaLa_3$	5.84	
$CoTi_2$	3.44	$E9_3$	Ga_2Mo	9.5	
CoTi	0.71	A2	$GaMo_3$	0.76	A15
CoU	1.7	B2, distorted	Ga_4Mo	9.8	
CoU_6	2.29	$D2_c$	GaN (black)	5.85	B4
$Co_{0.28}Y_{0.72}$	0.34		$GaNb_3$	14.5	A15
CoY_3	<0.34		$Ga_xNb_3Sn_{1-x}$	14–18.37	A15
$CoZr_2$	6.3	C16	$Ga_{0.7}Pt_{0.3}$	2.9	C1
$Co_{0.1}Zr_{0.9}$	3.9	A3	GaPt	1.74	B20
$Cr_{0.6}Ir_{0.4}$	0.4	Hexagonal, c.p.	GaSb (120 kbar, 77°K, annealed)	4.24	A5
$Cr_{0.65}Ir_{0.35}$	0.59	Hexagonal, c.p.			
$Cr_{0.7}Ir_{0.3}$	0.76	Hexagonal, c.p.	GaSb (unannealed)	~5.9	
$Cr_{0.72}Ir_{0.28}$	0.83		$Ga_{0.1}Sn_{1-0}$ (quenched)	3.47–4.18	
Cr_3Ir	0.45	A15	$Ga_{0-1}Sn_{1-0}$ (annealed)	2.6–3.85	
$Cr_{0-0.1}Nb_{1-0.9}$	4.6–9.2	A2	Ga_5V_2	3.55	Tetragonal, Mn_2Hg_5 type
$Cr_{0.80}Os_{0.20}$	2.5	Cubic			
Cr_xRe_{1-x}	1.2–5.2		GaV_3	16.8	A15
$Cr_{0.40}Re_{0.60}$	2.15	$D8_b$	$GaV_{2.1-3.5}$	6.3–14.45	A15
$Cr_{0.8-0.6}Rh_{0.2-0.4}$	0.5–1.10	A3	$GaV_{4.5}$	9.15	
Cr_3Ru (annealed)	3.3	A15	Ga_3Zr	1.38	
Cr_2Ru	2.02	$D8_b$	Gd_xLa_{1-x}	<1.0–5.5	
$Cr_{0.1-0.5}Ru_{0.9-0.5}$	0.34–1.65	A3	$Gd_xOs_2Y_{1-x}$	1.4–4.7	
Cr_xTi_{1-x}	3.6 max.	Cr in α-Ti	$Gd_xRu_2Th_{1-x}$	3.6 max.	C15
Cr_xTi_{1-x}	4.2 max.	Cr in β-Ti	GeIr	4.7	B31
$Cr_{0.1}Ti_{0.3}V_{0.6}$	5.6		Ge_2La	1.49, 2.2	Orthorhombic, distorted $ThSi_2$-type
$Cr_{0.0175}U_{0.9825}$	0.75	β-phase			
$Cs_{0.32}O_3W$	1.12	Hexagonal			
$Cu_{0.15}In_{0.85}$ (film)	3.75		$GeMo_3$	1.43	A15
$Cu_{0.04-0.08}In_{1-x}$	4.4		$GeNb_2$	1.9	
CuLa	5.85		$GeNb_3$ (quenched)	6–17	A15
Cu_xPb_{1-x}	5.7–7.7		$Ge_{0.29}Nb_{0.71}$	6	A15
CuS	1.62	B18	$Ge_xNb_3Sn_{1-x}$	17.6–18.0	A15
CuS_2	1.48–1.53	C18	$Ge_{0.5}Nb_3Sn_{0.5}$	11.3	

†† See key at end of table.

Table 2-30.   SELECTED SUPERCONDUCTIVE COMPOUNDS AND ALLOYS (Continued)

Substance	T_c, °K	Crystal structure type††	Substance	T_c, °K	Crystal structure type††
GePt	0.40	B31	InSb	2.1	
Ge_3Rh_5	2.12	Orthorhombic, related to $InNi_2$	$(InSb)_{0.95-0.10}Sn_{0.05-0.90}$ (various heat treatments)	3.8–5.1	
			$(InSb)_{0-0.07}Sn_{1-0.93}$	3.67–3.74	
Ge_2Sc	1.3		In_3Sn	~5.5	
Ge_3Te_4			In_xSn_{1-x}	3.4–7.3	
($n = 1.06 \times 10^{22}$)	1.55–1.80	Rhombohedral	$In_{0.82-1}Te$		
Ge_xTe_{1-x}			($n = 0.83-1.71 \times 10^{22}$)	1.02–3.45	B1
($n = 8.5-64 \times 10^{20}$)	0.07–0.41	B1	$In_{1.000}Te_{1.002}$	3.5–3.7	B1
GeV_3	6.01	A15	In_3Te_4		
Ge_2Y	3.80	C_c	($n = 0.47 \times 10^{22}$)	1.15–1.25	Rhombohedral
$Ge_{1.62}Y$	2.4		In_xTl_{1-x}	2.7–3.374	
$H_{0.33}Nb_{0.67}$	7.28	Cubic, b.c.	$In_{0.8}Tl_{0.2}$	3.223	
$H_{0.1}Nb_{0.9}$	7.38	Cubic, b.c.	$In_{0.62}Tl_{0.38}$	2.760	
$H_{0.05}Nb_{0.95}$	7.83	Cubic, b.c.	$In_{0.78-0.69}Tl_{0.22-0.31}$	3.18–3.32	Tetragonal
$H_{0.12}Ta_{0.88}$	2.81	Cubic, b.c.	$In_{0.69-0.62}Tl_{0.31-0.38}$	2.98–3.3	Cubic, f.c.
$H_{0.08}Ta_{0.92}$	3.26	Cubic, b.c.	Ir_2La	0.48	C15
$H_{0.04}Ta_{0.96}$	3.62	Cubic, b.c.	Ir_3La	2.32	$D10_2$
$HfN_{0.989}$	6.6	B1	Ir_3La_7	2.24	$D10_2$
$Hf_{0-0.5}Nb_{1-0.5}$	8.3–9.5	A2	Ir_5La	2.13	
$Hf_{0.75}Nb_{0.25}$	>4.2		Ir_2Lu	2.47	C15
$HfOs_2$	2.69	C14	Ir_3Lu	2.89	C15
$HfRe_2$	4.80	C14	IrMo	<1.0	A3
$Hf_{0.14}Re_{0.86}$	5.86	A12	$IrMo_3$	8.8	A15
$Hf_{0.99-0.96}Rh_{0.01-0.04}$	0.85–1.51		$IrMo_3$	6.8	$D8_b$
$Hf_{0-0.55}Ta_{1-0.45}$	4.4–6.5	A2	$IrNb_3$	1.9	A15
HfV_2	8.9–9.6	C15	$Ir_{0.4}Nb_{0.6}$	9.8	$D8_b$
Hg_xIn_{1-x}	3.14–4.55		$Ir_{0.37}Nb_{0.63}$	2.32	$D8_b$
HgIn	3.81		IrNb	7.9	$D8_b$
Hg_2K	1.20	Orthorhombic	$Ir_{0.02}Nb_3Rh_{0.98}$	2.43	A15
Hg_3K	3.18		$Ir_{0.05}Nb_3Rh_{0.95}$	2.38	A15
Hg_4K	3.27		$Ir_{0.287}O_{0.14}Ti_{0.573}$	5.5	$E9_3$
Hg_8K	3.42		$Ir_{0.265}O_{0.035}Ti_{0.65}$	2.30	$E9_3$
Hg_3Li	1.7	Hexagonal	Ir_xOs_{1-x}	0.3–0.98	
Hg_2Na	1.62	Hexagonal	(max.)–0.6		
Hg_4Na	3.05		IrOsY	2.6	C15
Hg_xPb_{1-x}	4.14–7.26		$Ir_{1.5}Os_{0.5}$	2.4	C14
HgSn	4.2		Ir_2Sc	2.07	C15
Hg_xTl_{1-x}	2.30–4.109		$Ir_{2.5}Sc$	2.46	C15
Hg_5Tl_2	3.86		$IrSn_2$	0.65–0.78	C1
Ho_xLa_{1-x}	1.3–6.3		Ir_2Sr	5.70	C15
$InLa_3$	9.83, 10.4	$L1_2$	$Ir_{0.5}Te_{0.5}$	~3	
$InLa_3$ (0–35, kbar)	9.75–10.55		$IrTe_3$	1.18	C2
$In_{1-0.86}Mg_{0-0.14}$	3.395–3.363		IrTh	<0.37	B_f
$InNb_3$			Ir_2Th	6.50	C15
(high pressure and temp.)	4–8, 9.2	A15	Ir_3Th	4.71	
$In_{0-0.3}Nb_3Sn_{1-0.7}$	18.0–18.19	A15	Ir_3Th_7	1.52	$D10_2$
$In_{0.5}Nb_3Zr_{0.5}$	6.4		Ir_5Th	3.93	$D2_d$
$In_{0.11}O_3W$	<1.25–2.8	Hexagonal	$IrTi_3$	5.40	A15
$In_{0.95-0.85}Pb_{0.05-0.15}$	3.6–5.05		IrV_2	1.39	A15
$In_{0.98-0.91}Pb_{0.02-0.09}$	3.45–4.2		IrW_3	3.82	
InPb	6.65		$Ir_{0.28}W_{0.72}$	4.49	
InPd	0.7	B2	Ir_2Y	2.18, 1.38	C15
InSb (quenched from			$Ir_{0.69}Y_{0.31}$	1.98, 1.44	C15
170 kbar into liquid N_2)	4.8	Like A5	$Ir_{0.70}Y_{0.30}$	2.16	C15

††See key at end of table.

Table 2-30.   SELECTED SUPERCONDUCTIVE COMPOUNDS AND ALLOYS *(Continued)*

Substance	T_c, °K	Crystal structure type††	Substance	T_c, °K	Crystal structure type††
Ir_2Y	1.09	C15	Mo_3Si	1.30	A15
Ir_2Y_3	1.61		$MoSi_{0.7}$	1.34	
Ir_xY_{1-x}	0.3–3.7		Mo_xSiV_{3-x}	4.54–16.0	A15
Ir_2Zr	4.10	C15	Mo_xTc_{1-x}	10.8–15.8	
$Ir_{0.1}Zr_{0.9}$	5.5	A3	$Mo_{0.16}Ti_{0.84}$	4.18, 4.25	
$K_{0.27-0.31}O_3W$	0.50	Hexagonal	$Mo_{0.913}Ti_{0.087}$	2.95	
$K_{0.40-0.57}O_3W$	1.5	Tetragonal	$Mo_{0.04}Ti_{0.96}$	2.0	Cubic
$La_{0.55}Lu_{0.45}$	2.2	Hexagonal, La type	$Mo_{0.025}Ti_{0.975}$	1.8	
			Mo_xU_{1-x}	0.7–2.1	
$La_{0.8}Lu_{0.2}$	3.4	Hexagonal, La Type	Mo_xV_{1-x}	0–~5.3	
			Mo_2Zr	4.27–4.75	C15
$LaMg_2$	1.05	C15	NNb (whiskers)	10–14.5	
LaN	1.35		NNb (diffusion wires)	16.10	
$LaOs_2$	6.5	C15	NNb (film)	6–9	B1
$LaPt_2$	0.46	C15	$N_{0.988}Nb$	14.9	B1
$La_{0.28}Pt_{0.72}$	0.54	C15	$N_{0.824-0.988}Nb$	14.4–15.3	B1
$LaRh_3$	2.60		$N_{0.70-0.795}Nb$	11.3–12.9	Cubic and tetragonal
$LaRh_5$	1.62				
La_7Rh_3	2.58	D10$_2$	NNb_xO_y	13.5–17.0	B1
$LaRu_2$	1.63	C15	NNb_xO_y	6.0–11	
La_3S_4	6.5	D7$_3$	$N_{100~42~w/o}Nb_{0~58~w/a}Ti$†	15–16.8	
La_3Se_4	8.6	D7$_3$	$N_{100~75~w/o}Nb_{0~25~w/o}Zr$†	12.5–16.35	
$LaSi_2$	2.3	C$_c$	NNb_xZr_{1-x}	9.8–13.8	B1
La_xY_{1-x}	1.7–5.4		$N_{0.93}Nb_{0.85}Zr_{0.15}$	13.8	B1
$LaZn$	1.04	B2	$N_xO_yTi_z$	2.9–5.6	Cubic
$LiPb$	7.2		$N_xO_yV_z$	5.8–8.2	Cubic
$LuOs_2$	3.49	C14	$N_{0.34}Re$	4–5	Cubic, f.c.
$Lu_{0.275}Rh_{0.725}$	1.27	C15	NTa	12–14	B1
$LuRh_5$	0.49			(extrapolated)	
$LuRu_2$	0.86	C14			
$Mg_{0.47}Tl_{0.53}$	2.75	B2	NTa (film)	4.84	B1
Mg_2Nb	5.6		$N_{0.6-0.987}Ti$	<1.17–5.8	B1
Mn_xTi_{1-x}	2.3 max.	Mn in α-Ti	$N_{0.82-0.99}V$	2.9–7.9	B1
Mn_xTi_{1-x}	1.1–3.0	Mn in β-Ti	NZr	9.8	B1
MnU_6	2.32	D2$_c$	$N_{0.906-0.984}Zr$	3.0–9.5	B1
MoN	12	Hexagonal	$Na_{0.28-0.35}O_3W$	0.56	Tetragonal
Mo_2N	5.0	Cubic, f.c.	$Na_{0.28}Pb_{0.72}$	7.2	
Mo_xNb_{1-x}	0.016–9.2		NbO	1.25	
Mo_3Os	7.2	A15	$NbOs_2$	2.52	A12
$Mo_{0.62}Os_{0.38}$	5.65	D8$_b$	Nb_3Os	1.05	A15
Mo_3P	5.31	DO$_e$	$Nb_{0.6}Os_{0.4}$	1.89, 1.78	D8$_b$
$Mo_{0.5}Pd_{0.5}$	3.52	A3	$Nb_3Os_{0.02-0.10}Rh_{0.98-0.90}$	2.42–2.30	A15
Mo_3Re	10.0		$Nb_{0.6}Pd_{0.4}$	1.60	D8$_f$ plus cubic
Mo_xRe_{1-x}	1.2–12.2		$Nb_3Pd_{0.02-0.10}Rh_{0.98-0.90}$	2.49–2.55	A15
$MoRe_3$	9.25, 9.89	A12	$Nb_{0.62}Pt_{0.38}$	4.21	D8$_b$
$Mo_{0.42}Re_{0.58}$	6.35	D8$_b$	Nb_3Pt	10.9	A15
$Mo_{0.52}Re_{0.48}$	11.1		Nb_5Pt_3	3.73	D8$_b$
$Mo_{0.57}Re_{0.43}$	14.0		$Nb_3Pt_{0.02-0.98}Rh_{0.98-0.02}$	2.52–9.6	A15
$Mo_{\sim0.60}Re_{0.395}$	10.6		$Nb_{0.38-0.18}Re_{0.62-0.82}$	2.43–9.70	A12
$MoRh$	1.97	A3	Nb_3Rh	2.64	A15
Mo_xRh_{1-x}	1.5–8.2	Cubic, b.c.	$Nb_{0.60}Rh_{0.40}$	4.21	D8$_b$ plus other
$MoRu$	9.5–10.5	A3	$Nb_3Rh_{0.98-0.90}Ru_{0.02-0.10}$	2.42–2.44	A15
$Mo_{0.61}Ru_{0.39}$	7.18	D8$_b$	Nb_xRu_{1-x}	1.2–4.8	
$Mo_{0.2}Ru_{0.8}$	1.66	A3	NbS_2	6.1–6.3	Hexagonal, NbSe$_2$ type
Mo_3Sb_4	2.1				

†w/o denotes weight percent.   ††See key at end of table.

Table 2-30.   SELECTED SUPERCONDUCTIVE COMPOUNDS AND ALLOYS (*Continued*)

Substance	T_c, °K	Crystal structure type††	Substance	T_c, °K	Crystal structure type††
NbS_2	5.0–5.5	Hexagonal, three-layer type	Os_2Zr	3.0	Cl4
			Os_xZr_{1-x}	1.50–5.6	
			PPb	7.8	
$Nb_3Sb_{0-0.7}Sn_{1-0.3}$	6.8–18	A15	$PPd_{3.0-3.2}$	<0.35–0.7	DO_{11}
$NbSe_2$	5.15–5.62	Hexagonal, NbS_2 type	P_3Pd_7 (high temperature)	1.0	Rhombohedral
$Nb_{1-1.05}Se_2$	2.2–7.0	Hexagonal, NbS_2 type	P_3Pd_7 (low temp.)	0.70	Complex
Nb_3Si	1.5	Ll_2	PRh	1.22	
Nb_3SiSnV_3	4.0		PRh_2	1.3	Cl
Nb_3Sn	18.05	A15	PW_3	2.26	DO
$Nb_{0.8}Sn_{0.2}$	18.18, 18.5	A15	Pb_2Pd	2.95	Cl6
Nb_xSn_{1-x} (film)	2.6–18.5		Pb_4Pt	2.80	Related to Cl6
$NbSn_2$	2.60	Orthorhombic	Pb_2Rh	2.66	Cl6
Nb_3Sn_2	16.6	Tetragonal	PbSb	6.6	
$NbSnTa_2$	10.8	A15	PbTe (plus 0.1 w/o Pb)†	5.19	
Nb_2SnTa	16.4	A15	PbTe (plus 0.1 w/o Tl)†	5.24–5.27	
$Nb_{2.5}SnTa_{0.5}$	17.6	A15	$PbTl_{0.27}$	6.43	
$Nb_{2.75}SnTa_{0.25}$	17.8	A15	$PbTl_{0.17}$	6.73	
$Nb_{3x}SnTa_{3(1-x)}$	6.0–18.0		$PbTl_{0.12}$	6.88	
NbSnTaV	6.2	A15	$PbTl_{0.075}$	6.98	
$Nb_2SnTa_{0.5}V_{0.5}$	12.2	A15	$PbTl_{0.04}$	7.06	
$NbSnV_2$	5.5	A15	$Pb_{1-0.26}Tl_{0-0.74}$	7.20–3.68	
Nb_2SnV	9.8	A15	$PbTl_2$	3.75–4.1	
$Nb_{2.5}SnV_{0.5}$	14.2	A15	Pb_3Zr_5	4.60	$D8_8$
Nb_xTa_{1-x}	4.4–9.2	A2	$PbZr_3$	0.76	A15
$NbTc_3$	10.5	A12	$Pd_{0.9}Pt_{0.1}Te_2$	1.65	C6
Nb_xTi_{1-x}	0.6–9.8		$Pd_{0.05}Ru_{0.05}Zr_{0.9}$	~9	
$Nb_{0.6}Ti_{0.4}$	9.8		$Pd_{2.2}S$ (quenched)	1.63	Cubic
Nb_xU_{1-x}	1.95 max.		$PdSb_2$	1.25	C2
$Nb_{0.88}V_{0.12}$	5.7	A2	PdSb	1.50	$B8_1$
$Nb_{0.75}Zr_{0.25}$	10.8		PdSbSe	1.0	C2
$Nb_{0.66}Zr_{0.33}$	10.8		PdSbTe	1.2	C2
$Ni_{0.3}Th_{0.7}$	1.98	$D10_2$	Pd_4Se	0.42	Tetragonal
$NiZr_2$	1.52		$Pd_{6-7}Se$	0.66	Like Pd_4Te
$Ni_{0.1}Zr_{0.9}$	1.5	A3	$Pd_{2.8}Se$	2.3	
$O_3Rb_{0.27-0.29}W$	1.98	Hexagonal	Pd_xSe_{1-x}	2.5 max.	
O_3SrTi ($n = 1.7-12.0 \times 10^{19}$)	0.12–0.37		PdSi	0.93	B31
O_3SrTi ($n = 10^{18}-10^{21}$)	0.05–0.47		PdSn	0.41	B31
O_3SrTi ($n = \sim 10^{20}$)	0.47		$PdSn_2$	3.34	
OTi	0.58		Pd_2Sn	0.41	C37
$O_3Sr_{0.08}W$	2–4	Hexagonal	Pd_3Sn_2	0.47–0.64	$B8_2$
$O_3Tl_{0.30}W$	2.0–2.14	Hexagonal	PdTe	2.3, 3.85	$B8_1$
OV_3Zr_3	7.5	$E9_3$	$PdTe_{1.02-1.08}$	2.56–1.88	$B8_1$
OW_3 (film)	3.35, 1.1	A15	$PdTe_2$	1.69	C6
OsReY	2.0	Cl4	$PdTe_{2.1}$	1.89	C6
Os_2Sc	4.6	Cl4	$PdTe_{2.3}$	1.85	C6
OsTa	1.95	A12	$Pd_{1.1}Te$	4.07	$B8_1$
Os_3Th_7	1.51	$Dl0_2$	$PdTh_2$	0.85	Cl6
Os_xW_{1-x}	0.9–4.1		$Pd_{0.1}Zr_{0.9}$	7.5	A3
OsW_3	~3		PtSb	2.1	$B8_1$
Os_2Y	4.7	Cl4	PtSi	0.88	B31
			PtSn	0.37	$B8_1$
			PtTe	0.59	Orthorhombic
			PtTh	0.44	B_f
			Pt_3Th_7	0.98	$Dl0_2$
			Pt_5Th	3.13	

†*w/o* denotes weight percent.   ††See key at end of table.

Table 2-30.   SELECTED SUPERCONDUCTIVE COMPOUNDS AND ALLOYS (Continued)

Substance	T_c, °K	Crystal structure type††	Substance	T_c, °K	Crystal structure type††
$PtTi_3$	0.58	A15	Ru_2Y	1.52	C14
$Pt_{0.02}U_{0.98}$	0.87	β-phase	Ru_2Zr	1.84	C14
$PtV_{2.5}$	1.36	A15	$Ru_{0.1}Zr_{0.9}$	5.7	A3
PtV_3	2.87–3.20	A15	SbSn	1.30–1.42, 1.42–2.37	B1 or distorted B1
$PtV_{3.5}$	1.26	A15			
$Pt_{0.5}W_{0.5}$	1.45	A1	$SbTi_3$	5.8	A15
Pt_xW_{1-x}	0.4–2.7		Sb_2Tl_7	5.2	
Pt_2Y_3	0.90		$Sb_{0.01-0.03}V_{0.99-0.97}$	3.76–2.63	A2
Pt_2Y	1.57, 1.70	C15	SbV_3	0.80	A15
Pt_3Y_7	0.82	$D10_2$	Si_2Th	3.2	C_c, α-phase
PtZr	3.0	A3	Si_2Th	2.4	C32, β-phase
$Re_{0.64}Ta_{0.36}$	1.46	A12	SiV_3	17.1	A15
$Re_{24}Ti_5$	6.60	A12	$Si_{0.9}V_3Al_{0.1}$	14.05	A15
Re_xTl_{1-x}	6.6 max.		$Si_{0.9}V_3B_{0.1}$	15.8	A15
$Re_{0.76}V_{0.24}$	4.52	$D8_b$	$Si_{0.9}V_3C_{0.1}$	16.4	A15
$Re_{0.92}V_{0.08}$	6.8	A3	$SiV_{2.7}Cr_{0.3}$	11.3	A15
$Re_{0.6}W_{0.4}$	6.0		$Si_{0.9}V_3Ge_{0.1}$	14.0	A15
$Re_{0.5}W_{0.5}$	5.12	$D8_b$	$SiV_{3.7}Mo_{0.3}$	11.7	A15
Re_2Y	1.83	C14	$SiV_{2.7}Nb_{0.3}$	12.8	A15
Re_2Zr	5.9	C14	$SiV_{2.7}Ru_{0.3}$	2.9	A15
Re_6Zr	7.40	A12	$SiV_{2.7}Ti_{0.3}$	10.9	A15
$Rh_{17}S_{15}$	5.8	Cubic	$SiV_{2.7}Zr_{0.3}$	13.2	A15
$Rh_{\sim0.24}Sc_{\sim0.76}$	0.88, 0.92		Si_2W_3	2.8, 2.84	
Rh_xSe_{1-x}	6.0 max.		$Sn_{0.174-0.104}Ta_{0.826-0.896}$	6.5–<4.2	A15
Rh_2Sr	6.2	C15	$SnTa_3$	8.35	A15, highly ordered
$Rh_{0.4}Ta_{0.6}$	2.35	$D8_b$			
$RhTc_2$	1.51	C2	$SnTa_3$	6.2	A15, partially ordered
$Rh_{0.67}Te_{0.33}$	0.49				
Rh_xTe_{1-x}	1.51 max.		$SnTaV_2$	2.8	A15
RhTh	0.36	B_f	$SnTa_2V$	3.7	A15
Rh_3Th_7	2.15	$D10_2$	Sn_xTe_{1-x}		
Rh_5Th	1.07		$(n = 10.5-20 \times 10^{20})$	0.07–0.22	B1
Rh_xTi_{1-x}	2.25–3.95		Sn_xTl_{1-x}	2.37–5.2	
$Rh_{0.02}U_{0.98}$	0.96		SnV_3	3.8	A15
RhV_3	0.38	A15	$Sn_{0.02-0.057}V_{0.98-0.943}$	2.87–~1.6	A2
RhW	~3.4	A3	$Ta_{0.025}Ti_{0.975}$	1.3	Hexagonal
RhY_3	0.65		$Ta_{0.05}Ti_{0.95}$	2.9	Hexagonal
Rh_2Y_3	1.48		$Ta_{0.05-0.75}V_{0.095-0.25}$	4.30–2.65	A2
Rh_3Y	1.07	C15	$Ta_{0.8-1}W_{0.2-0}$	1.2–4.4	A2
Rh_5Y	0.56		$Tc_{0.1-0.4}W_{0.9-0.6}$	1.25–7.18	Cubic
$RhZr_2$	10.8	C16	$Tc_{0.50}W_{0.50}$	7.52	α plus σ
$Rh_{0.005}Zr$ (annealed)	5.8		$Tc_{0.60}W_{0.40}$	7.88	σ plus α
$Rh_{0-0.45}Zr_{1-0.55}$	2.1–10.8		Tc_6Zr	9.7	A12
$Rh_{0.1}Zr_{0.9}$	9.0	Hexagonal, c.p.	$Th_{0-0.55}Y_{1-0.45}$	1.2–1.8	
Ru_2Sc	1.67	C14	$Ti_{0.70}V_{0.30}$	6.14	Cubic
Ru_2Th	3.56	C15	Ti_xV_{1-x}	0.2–7.5	
RuTi	1.07	B2	$Ti_{0.5}Zr_{0.5}$ (annealed)	1.23	
$Ru_{0.05}Ti_{0.95}$	2.5		$Ti_{0.5}Zr_{0.5}$ (quenched)	2.0	
$Ru_{0.1}Ti_{0.9}$	3.5		V_2Zr	8.80	C15
$Ru_xTi_{0.6}V_y$	6.6 max.		$V_{0.26}Zr_{0.74}$	≈5.9	
$Ru_{0.45}V_{0.55}$	4.0	B2	W_2Zr	2.16	C15
RuW	7.5	A3			

††See key at end of table.

Table 2-30.   SELECTED SUPERCONDUCTIVE COMPOUNDS AND ALLOYS (Continued)

CRITICAL FIELD DATA

Substance	H_o, oersteds	Substance	H_o, oersteds
Ag_2F	2.5	InSb	1,100
Ag_7NO_{11}	57	In_xTl_{1-x}	252–284
Al_2CMo_3	1,700	$In_{0.8}Tl_{0.2}$	252
$BaBi_3$	740	$Mg_{\sim 0.47}Tl_{\sim 0.53}$	220
Bi_2Pt	10	$Mo_{0.16}Ti_{0.84}$	<985
Bi_3Sr	530	$NbSn_2$	620
Bi_5Tl_3	>400	$PbTl_{0.27}$	756
CdSn	>266	$PbTl_{0.17}$	796
$CoSi_2$	105	$PbTl_{0.12}$	849
$Cr_{0.1}Ti_{0.3}V_{0.6}$	1,360	$PbTl_{0.075}$	880
$In_{1-0.86}Mg_{0-0.14}$	272.4–259.2	$PbTl_{0.04}$	864

KEY TO CRYSTAL STRUCTURE TYPES

"Struckturbericht" type*	Example	Class	"Struckturbericht" type*	Example	Class
A1	Cu	Cubic, f.c.	$C15_b$	$AuBe_5$	Cubic
A2	W	Cubic, b.c.	C16	$CuAl_2$	Tetragonal, b.c.
A3	Mg	Hexagonal, close packed	C18	FeS_2	Orthorhombic
A4	Diamond	Cubic, f.c.	C22	Fe_2P	Trigonal
A5	White Sn	Tetragonal, b.c.	C23	$PbCl_2$	Orthorhombic
A6	In	Tetragonal, b.c. (f.c. cell usually used)	C32	AlB_2	Hexagonal
			C36	$MgNi_2$	Hexagonal
A7	As	Rhombohedral	C37	Co_2Si	Orthorhombic
A8	Se	Trigonal	C49	$ZrSi_2$	Orthorhombic
A10	Hg	Rhombohedral	C54	$TiSi_2$	Orthorhombic
A12	α-Mn	Cubic, b.c.	C_c	Si_2Th	Tetragonal, b.c.
A13	β-Mn	Cubic	DO_3	BiF_3	Cubic, f.c.
A15	"β-W" (WO_3)	Cubic	DO_{11}	Fe_3C	Orthorhombic
B1	NaCl	Cubic, f.c.	DO_{18}	Na_3As	Hexagonal
B2	CsCl	Cubic	DO_{19}	Ni_3Sn	Hexagonal
B3	ZnS	Cubic	DO_{20}	$NiAl_3$	Orthorhombic
B4	ZnS	Hexagonal	DO_{22}	$TiAl_3$	Tetragonal
$B8_1$	NiAs	Hexagonal	DO_e	Ni_3P	Tetragonal, b.c.
$B8_2$	Ni_2In	Hexagonal	$D1_3$	Al_4Ba	Tetragonal, b.c.
B10	PbO	Tetragonal	$D1_c$	$PtSn_4$	Orthorhombic
B11	γ-CuTi	Tetragonal	$D2_1$	CaB_6	Cubic
B17	PtS	Tetragonal	$D2_c$	MnU_6	Tetragonal, b.c.
B18	CuS	Hexagonal	$D2_d$	$CaZn_5$	Hexagonal
B20	FeSi	Cubic	$D5_2$	La_2O_3	Trigonal
B27	FeB	Orthorhombic	$D5_8$	Sb_2S_3	Orthorhombic
B31	MnP	Orthorhombic	$D7_3$	Th_3P_4	Cubic, b.c.
B32	NaTl	Cubic, f.c.	$D7_b$	Ta_3B_4	Orthorhombic
B34	PdS	Tetragonal	$D8_1$	Fe_3Zn_{10}	Cubic, b.c.
B_f	δ-CrB	Orthorhombic	$D8_2$	Cu_5Zn_8	Cubic, b.c.
B_g	MoB	Tetragonal, b.c.	$D8_3$	Cu_9Al_4	Cubic
B_h	WC	Hexagonal	$D8_8$	Mn_5Si_3	Hexagonal
B_i	γ-MoC	Hexagonal	$D8_b$	CrFe	Tetragonal
C1	CaF_2	Cubic, f.c.	$D8_i$	Mo_2B_5	Rhombohedral
$C1_b$	MgAgAs	Cubic, f.c.	$D10_2$	Fe_3Th_7	Hexagonal
C2	FeS_2	Cubic	$E2_1$	$CaTiO_3$	Cubic
C6	CdI_2	Trigonal	$E9_3$	Fe_3W_3C	Cubic, f.c.
C11b	$MoSi_2$	Tetragonal, b.c.	$L1_0$	CuAu	Tetragonal
C12	$CaSi_2$	Rhombohedral	$L1_2$	Cu_3Au	Cubic
C14	$MgZn_2$	Hexagonal	L'_{2b}	ThH_2	Tetragonal, b.c.
C15	Cu_2Mg	Cubic, f.c.	L'_3	Fe_2N	Hexagonal

*See "Handbook of Lattice Spacing and Structures of Metals", W.B. Pearson, Vol. I, Pergamon Press, 1958, p. 79, and Vol. II, Pergamon Press, 1967, p. 3.

Table 2-31. HIGH CRITICAL MAGNETIC-FIELD SUPERCONDUCTIVE COMPOUNDS AND ALLOYS

With Critical Temperatures, H_{c1}, H_{c2}, H_{c3}, and the Temperature of Field Observations, T_{obs}

Substance	T_c, °K	H_{c1}, kg	H_{c2}, kg	H_{c3}, kg	T_{obs}, °K†
Al_2CMo_3	9.8–10.2	0.091	156		1.2
$AlNb_3$		0.375			
$Ba_xO_3Sr_{1-x}Ti$	<0.1–0.55	0.0039 max.			
$Bi_{0.5}Cd_{0.1}Pb_{0.27}Sn_{0.13}$			>24		3.06
Bi_xPb_{1-x}	7.35–8.4	0.122 max.	∼30 max.		4.2
$Bi_{0.56}Pb_{0.44}$	8.8		15		4.2
$Bi_{7.5\ w/o}Pb_{92.5\ w/o}‡$			2.32		
$Bi_{0.099}Pb_{0.901}$		0.29	2.8		
$Bi_{0.02}Pb_{0.98}$		0.46	0.73		
$Bi_{0.53}Pb_{0.32}Sn_{0.16}$			>25		3.06
$Bi_{1-0.93}Sn_{0-0.07}$			0–0.032		3.7
Bi_5Tl_3	6.4		>5.56		3.35
C_8K (excess K)	0.55		0.160 (H⊥c)		0.32
			0.730 (H‖c)		0.32
C_8K	0.39		0.025 (H⊥c)		0.32
			0.250 (H‖c)		0.32
$C_{0.44}Mo_{0.56}$	12.5–13.5	0.087	98.5		1.2
CNb	8–10	0.12	16.9		4.2
$CNb_{0.4}Ta_{0.6}$	10–13.6	0.19	14.1		1.2
CTa	9–11.4	0.22	4.6		1.2
$Ca_xO_3Sr_{1-x}Ti$	<0.1–0.55	0.002–0.004			
$Cd_{0.1}Hg_{0.9}$ (by weight)		0.23	0.34		2.04
$Cd_{0.05}Hg_{0.95}$		0.28	0.31		2.16
$Cr_{0.10}Ti_{0.30}V_{0.60}$	5.6	0.071	84.4		0
GaN	5.85	0.725			4.2
Ga_xNb_{1-x}			>28		4.2
GaSb (annealed)	4.24		2.64		3.5
$GaV_{1.95}$	5.3		73***		
$GaV_{2.1-3.5}$	6.3–14.45		230–300**		0
GaV_3		0.4	350***		0
			500**		
$GaV_{4.5}$	9.15		121*		0
Hf_xNb_y			>52–>102		1.2
Hf_xTa_y			>28–>86		1.2
$Hg_{0.05}Pb_{0.95}$		0.235	2.3		
$Hg_{0.101}Pb_{0.899}$		0.23	4.3		4.2
$Hg_{0.15}Pb_{0.85}$	∼6.75		>13		2.93
$In_{0.98}Pb_{0.02}$	3.45	0.1		0.12	2.76
$In_{0.96}Pb_{0.04}$	3.68	0.1	0.12	0.25	2.94
$In_{0.94}Pb_{0.06}$	3.90	0.095	0.18	0.35	3.12
$In_{0.913}Pb_{0.087}$	4.2	∼0.17	0.55	2.65	
$In_{0.316}Pb_{0.684}$		0.155	3.7		4.2
$In_{0.17}Pb_{0.83}$			2.8	5.5	4.2
$In_{1.000}Te_{1.002}$	3.5–3.7		1.2*		0
$In_{0.95}Tl_{0.05}$		0.263	0.263		3.3
$In_{0.90}Tl_{0.10}$		0.257	0.257		3.25
$In_{0.83}Tl_{0.17}$		0.242	0.39		3.21
$In_{0.75}Tl_{0.25}$		0.216	0.50		3.16
LaN	1.35	0.45			0.76
La_3S_4	6.5	≈0.15	>25		1.3
La_3Se_4	8.6	≈0.2	>25		1.25
$Mo_{0.52}Re_{0.48}$	11.1		14–21	22–33	4.2
			18–28	37–43	1.3
$Mo_{0.6\pm0.05}Re_{0.395}$	10.6		14–20	20–37	4.2
			19–26	26–37	1.3
$Mo_{\sim0.5}Tc_{\sim0.5}$			∼75*		0
$Mo_{0.16}Ti_{0.84}$	4.18	0.028	98.7*		0
			36–38		3.0
$Mo_{0.913}Ti_{0.087}$	2.95	0.060	∼15		4.2
$Mo_{0.1-0.3}U_{0.9-0.7}$	1.85–2.06		>25		
$Mo_{0.17}Zr_{0.83}$			∼30		
$N_{(12.8\ w/o)}Nb$	15.2		>9.5		13.2
NNb (wires)	16.1		153*		0
			132		4.2
			95		8
			53		12
NNb_xO_{1-x}	13.5–17.0		∼38		
NNb_xZr_{1-x}	9.8–13.8		4–>130		4.2
$N_{0.93}Nb_{0.85}Zr_{0.15}$	13.8		>130		4.2
$Na_{0.086}Pb_{0.914}$		0.19	6.0		
$Na_{0.016}Pb_{0.984}$		0.28	2.05		

‡w/o denotes weight percent.

Table 2-31. HIGH CRITICAL MAGNETIC-FIELD SUPERCONDUCTIVE COMPOUNDS AND ALLOYS (Continued)

Substance	T_c, °K	H_{c1}, kg	H_{c2}, kg	H_{c3}, kg	T_{obs}, °K†
Nb	9.15		2.020		1.4
			1.710		4.2
Nb		0.4–1.1	3–5.5		4.2
Nb (unstrained)		1.1–1.8	3.40	6–9.1	4.2
Nb (strained)		1.25–1.92	3.44	6.0–8.7	4.2
Nb (cold-drawn wire)		2.48	4.10	≈ 10	4.2
Nb (film)			> 25		4.2
NbSc			> 30		
Nb_3Sn		0.170	221		4.2
			70		14.15
			54		15
			34		16
			17		17
$Nb_{0.1}Ta_{0.9}$		0.084	0.154		4.195
$Nb_{0.2}Ta_{0.8}$			10		4.2
$Nb_{0.65-0.73}Ta_{0.02-0.10}Zr_{0.25}$			> 70–> 90		4.2
Nb_xTi_{1-x}			148 max.		1.2
			120 max.		4.2
$Nb_{0.222}U_{0.778}$		1.98	23		1.2
Nb_xZr_{1-x}			127 max.		1.2
			94 max.		4.2
O_3SrTi	0.43	.0049*	.504*		0
O_3SrTi	0.33	.00195*	.420*		0
$PbSb_{1\ w/o}$ (quenched)			> 1.5		4.2
$PbSb_{1\ w/o}$ (annealed)			> 0.7		4.2
$PbSb_{2.8\ w/o}$ (quenched)			> 2.3		4.2
$PbSb_{2.8\ w/o}$ (annealed)			> 0.7		4.2
$Pb_{0.871}Sn_{0.129}$		0.45	1.1		
$Pb_{0.965}Sn_{0.035}$		0.53	0.56		
$Pb_{1-0.26}Tl_{0-0.74}$	7.20–3.68		2–6.9*		0
$PbTl_{0.17}$	6.73		4.5*		0
$Re_{0.26}W_{0.74}$			> 30		
$Sb_{0.93}Sn_{0.07}$			0.12		3.7
SiV_3	17.0	0.55	156***		
Sn_xTe_{1-x}		0.00043–0.00236	0.005–0.0775		0.012–0.079
Ta (99.95%)		0.425	1.850		1.3
		0.325	1.425		2.27
		0.275	1.175		2.66
		0.090	0.375		3.72
$Ta_{0.5}Nb_{0.5}$			3.55		4.2
$Ta_{0.65-0}Ti_{0.35-1}$	4.4–7.8		> 14–138		1.2
$Ta_{0.5}Ti_{0.5}$			138		1.2
Te	~ 3.3	0.25*			0
Tc_xW_{1-x}	5.75–7.88		8–44		4.2
Ti				2.7	4.2
$Ti_{0.75}V_{0.25}$	5.3	0.029*	199*		0
$Ti_{0.775}V_{0.225}$	4.7	0.024*	172*		0
$Ti_{0.615}V_{0.385}$	7.07	0.050	~ 34		4.2
$Ti_{0.516}V_{0.484}$	7.20	0.062	~ 28		4.2
$Ti_{0.415}V_{0.585}$	7.49	0.078	~ 25		4.2
$Ti_{0.12}V_{0.88}$			17.3	28.1	4.2
$Ti_{0.09}V_{0.91}$			14.3	16.4	4.2
$Ti_{0.06}V_{0.94}$			8.2	12.7	4.2
$Ti_{0.03}V_{0.97}$			3.8	6.8	4.2
Ti_xV_{1-x}			108 max.		1.2
V	5.31	~ 0.8	~ 3.4		1.79
		~ 0.75	~ 3.15		2
		~ 0.45	~ 2.2		3
		~ 0.30	~ 1.2		4
$V_{0.26}Zr_{0.74}$	≈ 5.9	0.238			1.05
		0.227			1.78
		0.185			3.04
		0.165			3.5
W (film)	1.7–4.1		> 34		1

†Temperature of critical field measurement.
*Extrapolated.
**Linear extrapolation.
***Parabolic extrapolation.

Table 2-32. PROPERTIES OF DIELECTRICS*

Earle C. Gregg

For other data on dielectrics, see Index.

In most cases properties have been determined by ASTM (American Society for Testing and Materials) test methods at room temperature under standard conditions. Values will, in general, change considerably with temperature.

PLASTICS

Material	Dielectric constant, 10^6 cycles	Dielectric strength, volts/mil	Volume resistivity, ohms-cm, 23°C	Loss factor†
Allyl resin, cast	3.6–4.5	380	$>4 \times 10^{14}$	0.028–0.06
Aniline formadehyde resin, no filler	3.5–3.6	600–650	10^{16}–10^{17}	0.006–0.008
Casein	6.1–6.8	400–700		0.052
Cellulose acetate, molding	3.2–7.0	250–365	10^{10}–10^{13}	0.01–1.0
Cellulose acetate, sheet	4.0–5.5	250–300	10^{11}–10^{13}	0.04–0.06
Cellulose acetate butyrate	3.2–6.2	250–400	10^{10}–10^{12}	0.01–0.04
Cellulose nitrate (proxylin)	6.4	300–600	$(10–15) \times 10^{10}$	0.06–0.09
Cold-molded compound, inorganic, refractory		45		
Cold-molded compound, organic, non-refractory	6.0	85–115	1.3×10^{12}	0.07
Ethyl cellulose	2.8–3.9	350–500	10^{12}–10^{14}	0.01–0.06
Epoxy cast resin	3.62	400	10^{16}–10^{17}	0.019
Glycerol phthalate, cast, alkyd	3.7–4.0	300–350	$>10^{14}$	0.025–0.035
Melamine formaldehyde resins, molding				
Alpha cellulose filler	7.2–8.2	300–400	$10^{12} \times 10^{14}$	0.027–0.045
Asbestos filler	6.1–6.7	350–400	2.4×10^{11}	0.041–0.050
Cellulose filler	4.7–7.0	350–400		0.032–0.060
Flock filler		300–330		
Macerated fabric filler	6.5–6.9	250–350	10^9–10^{10}	0.036
Methyl methacrylate, cast	2.7–3.2	450–500	$>10^{15}$	0.02–0.03
Methyl methacrylate, molding	2.7–3.2	450–500	$>10^{14}$	0.02–0.03
Mica, glass-bonded, compression	7.4–7.85		10^{14}–10^{15}	0.0015–0.002
Mica, glass-bonded, injection	6.9–9.2		10^{14}–10^{17}	0.0015–0.012
Nylon (F.M. 3001)	3.5	470	4×10^{14}	0.03
Phenol formaldehyde resins, molding				
Asbestos filler	5.0–7.0	100–350	10^{10}–10^{12}	0.10–0.50
Glass fiber	6.6	140–370	7×10^{12}	0.02
Mica filler	4.2–5.2	300–460	10^{12}–10^{14}	0.005–0.04
No filler	4.5–5.0	300–400	10^{11}–10^{12}	0.015–0.03
Sisal felt	3–5	250–400	10^{11}–10^{12}	0.3–0.5
Wood flour filler	4.–7.	200–425	10^9–10^{13}	0.03–0.07
Phenol formaldehyde resins, cast				
Mineral filler	9–15	100–250	10^9–10^{12}	0.07–0.2
No filler	4.0–5.5	350–400	10^{14}–10^{15}	0.04–0.05
Polyacrylic ester (filled and vulcanized)		400–700	2×10^{11} at 70°C	
Polyester, cast resin, rigid	2.8–4.1	380–500	10^{14}	0.006–0.026
Polyester, cast resin, flexible	4.1–5.2	250–500		0.023–0.052
Polyester molding materials (glass fiber filler)	4.0–4.5	150–400	10^{12}–10^{14}	0.015–0.020
Polyethylene	2.3	460	1.6×10^{13}	<0.0005
Polymonochlorodifluoroethylene	2.5	400	1.2×10^{18}	0.010
Polystyrene molding	2.4–2.65	500–700	10^{17}–10^{19}	0.0001–0.0004
Polytetrafluoroethylene	2.0	480	$>10^{15}$	<0.0002–0.0005
Rubber, hard	2.8	470	2×10^{15}	0.06
Rubber, chlorinated	3 approx.		1.5×10^{13}	0.006
Rubber, modified, isomerized	2.4–2.7	620	5×10^{16}	0.0008–0.002
Shellac		200–600	1.8×10^9	
Silicone molding compound (glass-fiber-filled)	3.7	185	10^{11}–10^{13}	0.0017
Styrene, modified, molding (shock-resistant type)	2.4–3.8	300–600	10^{12}–10^{17}	0.0004–0.02
Urea formaldehyde resin, alpha cellulose filler	6.4–6.9	300–400	10^{12}–10^{13}	0.028–0.032
Vinyl butyral, flexible, unfilled	3.92	350	5×10^{10}	0.061
Vinyl butyral, rigid	3.33	400	$>10^{14}$	0.0065
Vinyl chloride, rigid	2.8–3.0	700–1300	$>10^{16}$	0.006–0.014
Vinyl chloride, flexible, filled	3.5–4.5	600–800	5×10^{14}	0.09–0.10
Vinyl chloride, flexible, unfilled	3.5–4.5	800–1000	5×10^{12}	0.09–0.10
Vinyl chloride acetate, rigid	3.0–3.1	425	10^{16}	0.018–0.019
Vinyl chloride acetate, flexible, unfilled	3.3–4.3	300–400	10^{11}–10^{13}	0.04–0.14
Vinyl chloride acetate, flexible, filled	3.3–4.3	250–350	10^{11}–10^{13}	0.04–0.14
Vinyl formal molding compound	3.0	490		0.023
Vinylidene chloride	3.0–4.0	350	10^{14}–10^{16}	0.05–0.08

†Power factor × dielectric constant equals loss factor.

Table 2-32.   PROPERTIES OF DIELECTRICS (Continued)

CERAMICS

Material	Dielectric constant, 10^6 cycles	Di-electric strength, volts/mil	Volume resistivity, ohms-cm, 23°C	Loss factor†
Alumina	4.5–8.4	40–160	10^{11}–10^{14}	0.0002–0.01
Corderite	4.5–5.4	40–250	10^{12}–10^{14}	0.004–0.012
Forsterite	6.2	240	10^{14}	0.0004
Porcelain (dry process)	6.0–8.0	40–240	10^{12}–10^{14}	0.003–0.02
Porcelain (wet process)	6.0–7.0	90–400	10^{12}–10^{14}	0.006–0.01
Porcelain, zircon	7.1–10.5	250–400	10^{13}–10^{15}	0.0002–0.008
Steatite	5.5–7.5	200–400	10^{13}–10^{15}	0.0002–0.004
Titanates (Ba, Sr, Ca, Mg, and Pb)	15–12,000	50–300	10^8–10^{15}	0.0001–0.02
Titanium dioxide	14–110	100–210	10^{12}–10^{18}	0.0002–0.005

†Power factor × dielectric constant equals loss factor.

GLASSES

Type	Dielectric constant, 100 mc, 20°C	Volume resistivity, megohm-cm, 350°C	Loss factor†
Corning 0010	6.32	10	0.015
Corning 0080	6.75	0.13	0.058
Corning 0120	6.65	100	0.012
Pyrex 1710	6.00	2,500	0.025
Pyrex 3320	4.71		0.019
Pyrex 7040	4.65	80	0.013
Pyrex 7050	4.77	16	0.017
Pyrex 7052	5.07	25	0.019
Pyrex 7060	4.70	13	0.018
Pyrex 7070	4.00	1,300	0.0048
Vycor 7230	3.83		0.0061
Pyrex 7720	4.50	16	0.014
Pyrex 7740	5.00	4	0.040
Pyrex 7750	4.28	50	0.011
Pyrex 7760	4.50	50	0.0081
Vycor 7900	3.9	130	0.0023
Vycor 7910	3.8	1,600	0.00091
Vycor 7911	3.8	4,000	0.00072
Corning 8870	9.5	5,000	0.0085
G. E. Clear (silica glass)	3.81	4,000–30,000	0.00038
Quartz (fused)	3.75–4.1 (1 mc)		0.0002 (1 mc)

†Power factor × dielectric constant equals loss factor.

*From: "CRC Handbook of Chemistry and Physics", 50th ed., R.C. Weast, Ed., The Chemical Rubber Co., 1969.

Table 2-33.   DIELECTRIC CONSTANTS FOR VARIOUS SOLIDS

Effect of High Frequencies

Room Temperature

Material	Frequency, cps			Material	Frequency, cps		
	10^3	10^6	10^9		10^3	10^6	10^9
Alkyd isocyanate foam	1.223	1.218	1.20	Nylon 66	3.75	3.33	3.16
Aluminum oxide	8.83	8.80	8.80	Nylon 610	3.50	3.14	3.0
Asbestos fiber	4.80	3.1	—	Plexiglas	3.12	2.76	—
Asphalt	2.66	2.58	2.55 (1×10^{10})	Polycarbonate	3.17	3.02	2.96
Balata	2.50	2.50	2.42	Polyether, chlorinated	3.1	3.0	2.9
Beeswax, yellow	2.66	2.53	2.45	Polyethylene	2.37	2.35	2.33
Buna S	2.66	2.56	2.52	Polyisobutylene	2.23	†	†
Butyl rubber compound	2.42	2.40	2.39	Polypropylene	2.25	†	†
Calcium titanate	167.7	†	†	Polyvinyliden fluoride	8.4	8.0	6.6
Cellulose nitrate	8.4	6.6	5.2	Porcelain (dry process)	5.36	5.08	5.04
Dichloronaphthalene	3.04	2.98	2.93	Rutile	100.0	†	†
Gutta percha	2.60	2.53	2.47	Selenium, amorphous	6.00	†	†
Hevea compound	36.	9.0	6.8	Shellac, natural	3.81	3.47	3.10
Lucite	2.84	2.63	2.58	Strontium titanate	233.0	232.0	232.0
Magnesium oxide	9.65	†	†	Teflon FEP	2.1	†	†
Magnesium silicate	5.98	5.97	5.96	Teflon PTFE	2.0	†	†
Magnesium titanate	13.9	†	†	Urea-formaldehyde	6.7	6.0	5.2
Methyl cellulose	6.8	5.7	4.3	Vinylite QYNA	3.10	2.88	2.85
Mica, ruby	5.4	†	†	Vinylite VYHH	3.12	2.91	2.83
Neoprene	6.60	6.26	4.5	Vinylite 5544	7.20	4.13	3.05

†No appreciable variation with frequency in the range 10^3–10^9 cps.

Table 2-34. ELECTRICAL CIRCUITS—COMMON TERMS AND EQUATIONS

Note: Most of these terms and relations are also applicable in analogous mechanical and fluid dynamic systems. Resistance is associated with friction, inductance, and capacitance with storage.

Variations of Ohm's law for d-c circuits:

$$E = IR = P/I = \sqrt{PR} = \text{potential, volts}$$
$$P = EI = I^2R = E^2/R = \text{power, watts}$$
$$I = E/R = P/E = \sqrt{P/R} = \text{current, amperes}$$
$$R = E/I = E^2/P = P/I^2 = \text{resistance, ohms}$$

In a-c circuits:

Power $= P = EI \cos \Phi$, watts ($\Phi = \arctan X/R$)

Inductive reactance $= X_L = 2\pi f L$ (L in henrys; f in hertz = cps)

Capacitative reactance $= X_C = \dfrac{1}{2\pi f C}$ (C in farads)

Total reactance $= X_T = X_L - X_C$ (X in ohms)

Impedance $= Z = \sqrt{R^2 + X_T^2} = \dfrac{1}{\text{Admittance}}$ (Z in ohms)

Series circuits:

Resistors in series: $R_T = R_1 + R_2 + R_3 + \ldots$

Capacitors in series: $C_T = \dfrac{1}{\dfrac{1}{C_1} + \dfrac{1}{C_2} + \dfrac{1}{C_3} + \ldots}$

Inductors in series: $L_T = L_1 + L_2 + L_3 + \ldots$

Resistance and reactance in series: $Z = \sqrt{R^2 + (X_L - X_C)^2}$

Series resonance: maximum current when $X_L = X_C$

Parallel circuits:

Resistors in parallel: $R_T = \dfrac{1}{\dfrac{1}{R_1} + \dfrac{1}{R_2} + \dfrac{1}{R_3} +}$

Capacitors in parallel: $C_T = C_1 + C_2 + C_3 + \ldots$

Inductors in parallel: $L_T = \dfrac{1}{\dfrac{1}{L_1} + \dfrac{1}{L_2} + \dfrac{1}{L_3} + \ldots}$

Resistance and reactance in parallel: $Z = \sqrt{(1/R)^2 + (1/X_L - 1/X_C)^2}$

Parallel resonance: maximum impedance when $X_L = X_C$

Q-factor of a capacitor:

$$Q = \frac{1}{2\pi f RC} \text{ (R and C in series)}$$

$$Q = 2\pi f RC \text{ (R and C in parallel)}$$

Q-factor of an inductor: $Q = \dfrac{2\pi f L}{R}$ (L and R in series)

Resonant frequency, L-C combination: $f = \dfrac{1}{2\pi\sqrt{LC}}$

Time constant, series R-C combination: $T = RC$ (T in seconds)

Time constant, series L-R combination: $T = \dfrac{L}{R}$

Table 2-35. PREFERRED NUMBERS FOR SIZES AND RATINGS

Used for Small Parts and Components

International preferred-number standards have been adopted by the ISO (International Organization for Standardization) and also by the USA Standards Institute.† These are based on fractional powers of 10, thus fixing the number of steps or values in the 10 to 100 scale, i.e., the $\frac{1}{5}$ power provides five steps, the $\frac{1}{6}$ power, six steps, etc.

These preferred-number steps are widely used in the electrical and electronics industries. The following table lists some of the most common series. Each series is named by the number of steps in the 10–100 scale.

	$10^{1/40}$	$10^{1/24}$	$10^{1/20}$	$10^{1/12}$	$10^{1/10}$	$10^{1/6}$	$10^{1/5}$
Exponent of 10							
Series name	40	24	20	12	10	6	5
Step multiplier	1.059	1.101	1.122	1.211	1.259	1.468	1.585
Approx. step size, %	6	10	12	20	25	45	60
Values, 10–100: (shift decimal point as required)	10	10	10	10	10	10	10
	11	11	11	—	—	—	—
	12	12	—	12	12	—	—
	13	13	13	—	—	—	—
	14	—	14	—	—	—	—
	15	15	—	15	—	15	—
	16	16	16	—	16	—	16
	17	—	—	—	—	—	—
	18	18	18	18	—	—	—
	19	—	—	—	—	—	—
	20	20	20	—	20	—	—
	21	—	—	—	—	—	—
	22	22	22	22	—	22	—
	23	—	—	—	—	—	—
	24	24	—	—	—	—	—
	25	—	25	—	25	—	25
	27	27	—	27	—	—	—
	28	—	28	—	—	—	—
	30	30	—	—	—	—	—
	32	—	32	—	32	—	—
	33	33	—	33	—	33	—
	35	35	35	—	—	—	—
	—	36	—	—	—	—	—
	37	—	—	—	—	—	—
	—	39	—	39	—	—	—
	40	—	40	—	40	—	40
	42	—	—	—	—	—	—
	—	43	—	—	—	—	—
	44	—	—	—	—	—	—
	—	—	45	—	—	—	—
	47	47	—	47	—	47	—
	50	—	50	—	50	—	—
	—	51	—	—	—	—	—
	52	—	—	—	—	—	—
	55	—	—	—	—	—	—
	—	56	56	56	—	—	—
	59	—	—	—	—	—	—
	62	62	—	—	—	—	—
	—	—	63	—	63	—	63
	66	—	—	—	—	—	—
	—	68	—	68	—	68	—
	69	—	—	—	—	—	—
	—	—	71	—	—	—	—
	73	—	—	—	—	—	—
	—	75	—	—	—	—	—
	77	—	—	—	—	—	—
	—	—	80	—	80	—	—
	82	82	—	82	—	—	—
	86	—	—	—	—	—	—
	—	—	89	—	—	—	—
	91	91	—	—	—	—	—
	96	—	—	—	—	—	—
	100	100	100	100	100	100	100

Example of tolerances: Components with 5% tolerance use Series 24 size-ratings; 10% components use Series 12; and 20% components use Series 6.

†Renamed American National Standards Institute in October 1969.

APPLICATION REFERENCES

"Components for Electronic Equipment, Preferred Values", IEC Standard 63 and USA Standard C 83.2.

"Guide to the Use of Preferred Numbers", ISO Standard R 17.

"Preferred Numbers", USA Standard Z 17.1.

"Series of Preferred Numbers", ISO Standard R 3.

Table 2-36.   RESISTOR COLOR CODE AND STANDARD VALUES

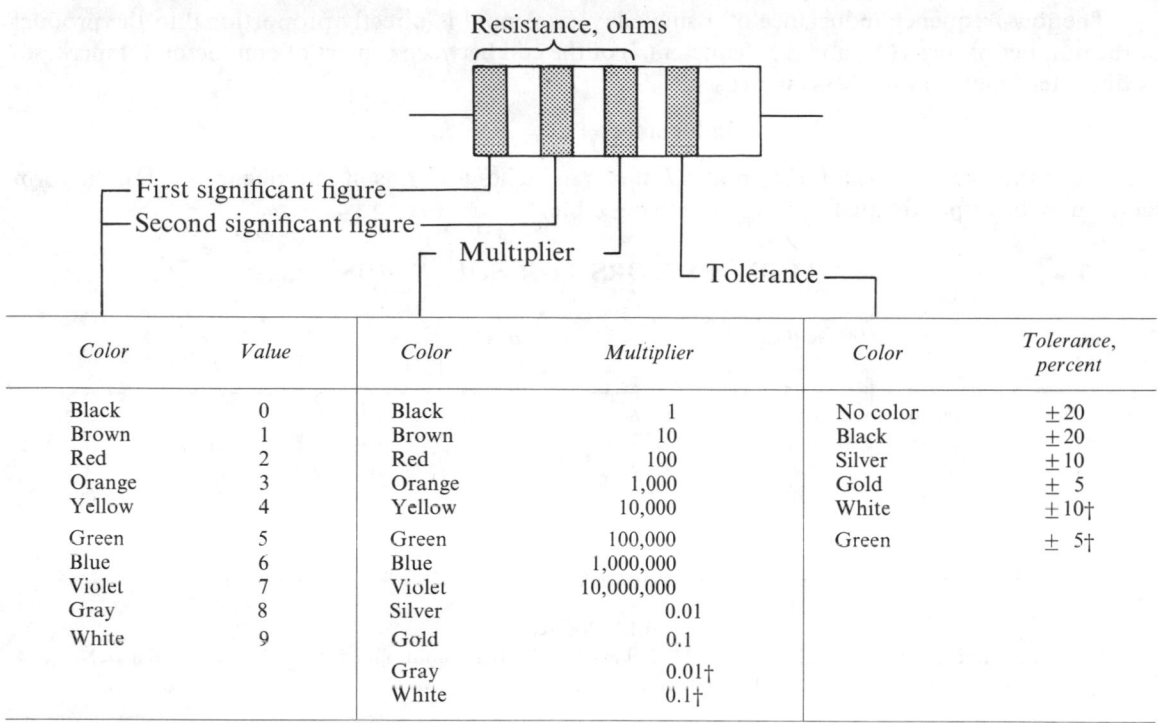

Color	Value	Color	Multiplier	Color	Tolerance, percent
Black	0	Black	1	No color	±20
Brown	1	Brown	10	Black	±20
Red	2	Red	100	Silver	±10
Orange	3	Orange	1,000	Gold	± 5
Yellow	4	Yellow	10,000	White	±10†
Green	5	Green	100,000	Green	± 5†
Blue	6	Blue	1,000,000		
Violet	7	Violet	10,000,000		
Gray	8	Silver	0.01		
White	9	Gold	0.1		
		Gray	0.01†		
		White	0.1†		

†Optional.

STANDARD RESISTOR VALUES, SIGNIFICANT FIGURES

±20%: 10, 15, 22, 33, 47, 68 (Series 6, Table 2-35).
±10%: 10, 12, 15, 18, 22, 27, 33, 39, 47, 56, 68, 82 (Series 12, Table 2-35).
 ±5%: 10, 11, 12, 13, 15, 16, 18, 20, 22, 24, 27, 30, 33, 36, 39, 43, 47, 51, 56, 62, 68, 75, 82, 91 (Series 24, Table 2-35).

Table 2-37.   CAPACITOR COLOR CODE

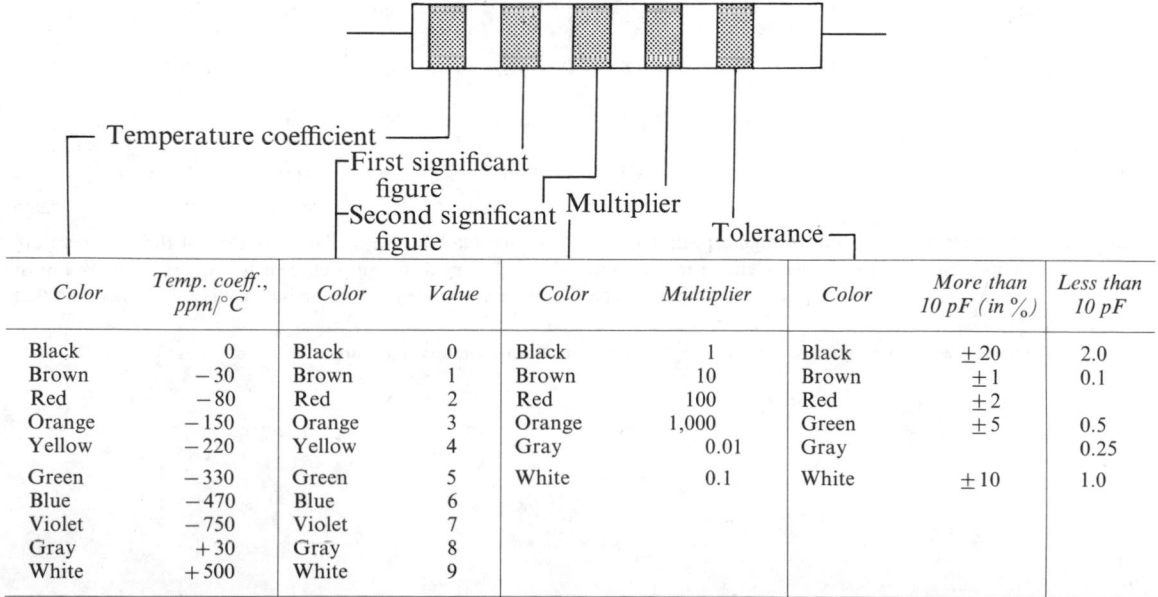

Color	Temp. coeff., ppm/°C	Color	Value	Color	Multiplier	Color	More than 10 pF (in %)	Less than 10 pF
Black	0	Black	0	Black	1	Black	±20	2.0
Brown	−30	Brown	1	Brown	10	Brown	±1	0.1
Red	−80	Red	2	Red	100	Red	±2	
Orange	−150	Orange	3	Orange	1,000	Green	±5	0.5
Yellow	−220	Yellow	4	Gray	0.01	Gray		0.25
Green	−330	Green	5	White	0.1	White	±10	1.0
Blue	−470	Blue	6					
Violet	−750	Violet	7					
Gray	+30	Gray	8					
White	+500	White	9					

Table 2-38.   INDUCTANCE OF A SINGLE-LAYER SOLENOID

The low-frequency inductance of a single-layer solenoid is directly proportional to the product of the number of turns (N) and the diameter (d) of the coil between centers of conductors; it increases as diameter/length ratio (d/l) is increased.

$$\text{Inductance} = L = FN^2d,$$

where F is the *form factor*, a function of d/l. If d is in inches and L is in microhenries (μH), the form factor may be approximated from the following table.

FORM FACTORS FOR SOLENOIDS

Ratio, d/l	F	Ratio, d/l	F	Ratio, d/l	F	Ratio, d/l	F
.02	.000 50	0.20	.004 6	2	.026	20	.062
.03	.000 75	0.30	.006 5	3	.032	30	.070
.04	.001 0	0.40	.008 4	4	.036	40	.075
.06	.001 5	0.60	.012	6	.044	60	.081
.10	.002 4	1.0	.017	10	.051	100	.088
.15	.003 6	1.5	.022	15	.058		

REFERENCE
"Radio Instruments and Measurements", D.H. Dellinger, L.E. Whittmore, and R.S. Ould, NBS Circular No. C74, National Bureau of Standards, 1924 (reprint 1937).

Table 2-39.   SILICON CARBIDE RECTIFIERS*

Electrical Properties of Typical Rectifiers at 30°C and 500°C

Unit type	Forward voltage (V, half wave av.)		Forward current (A, half wave av.)		Reverse voltage (V, peak)		Reverse current (ma, peak)	
	30°C	500°C	30°C	500°C	30°C	500°C	30°C	500°C
Power rectifier	4.2	2.8	1.5	1.5	200	150	0.01	0.40
Low-current instrument device	5.0	2.5	0.5	0.5	500	300	0.001	0.008
High voltage (stacked unit)	20.0	15.0	0.01	0.01	1000	750	0.001	0.005
Low-voltage blocking diode	5.1	4.3	15.0	15.0	120	100	0.8	1.6

Comparison with silicon rectifiers: Much higher peak current ratings are available with silicon rectifiers if their temperature is well below 200°C, but they cannot be operated much above 200°C. The peak forward current for SiC rectifiers is almost constant between 100°C and 500°C. The inherent resistance of SiC rectifiers to nuclear radiation is considerably higher than that of silicon rectifiers.

*From: Westinghouse *R & D Letter*, January, 1969, Westinghouse Astronuclear Laboratory, Pittsburgh, Pa. 15236.

Table 2-40. THERMAL EXPANSION COEFFICIENTS FOR MATERIALS USED IN INTEGRATED CIRCUITS*

Coefficient of Linear Thermal Expansion of Selected Materials per °K

For other values of this coefficient, see Tables 1-59, 1-60, and 1-61.

Note: Multiply all values by 10^{-6}.

Material	Temperature, °K						Material	Coef-ficient range	Tempera-ture range, °C
	300°	400°	500°	600°	700°	800°			
Aluminum	23.2	24.9	26.4	28.3	30.7	33.8	Aluminum oxide ceramic	6.0–7.0	25–300
Beryllium oxide	4.7	—	6.0	—	7.0	—	Brass	17.7–21.2	25–300
Copper	16.8	17.7	18.3	18.9	19.4	20.0	Kanthal A	13.9–15.1	20–900
Germanium	5.7	6.2	6.5	6.7	6.9	7.2	Kovar	5.0	25–300
Gold	14.1	14.5	15.	15.4	15.9	16.5	Pyrex glass	3.2	25–300
Indium	31.9	38.5	—	—	—	—	Pyroceram (#9608)	4–20	25–300
Lead	28.9	29.8	32.1	—	—	—	Pyroceram cement		
Molybdenum	5.0	5.2	5.3	5.4	5.5	5.7	Vitreous (#45)	4	0–300
Nickel	12.7	13.8	15.2	17.2	16.4	16.8	Devitrified	2.4	25–300
Platinum	8.9	9.2	9.5	9.7	10.0	10.2	Pyroceram cement		
Silicon	2.5	3.1	3.5	3.8	4.1	4.3	(#89, #95)	8–10	—
Silver	19.2	20.0	20.6	21.4	22.3	23.4	Silicon carbide	4.8	0–1,000
Tantalum	6.5	6.6	6.8	6.9	7.0	7.1	Silicon nitride		
Tin	21.2	24.2	27.5	—	—	—	α	2.9	25–1,000
Tungsten	4.5	4.6	4.6	4.7	4.7	4.8	β	2.25	25 1,000
Vitreous silica	.42	.56	.56	.55	.54	.54	Solder glass (Kimble CV-101)	809	0–300

*From: "Interconnections and Encapsulation", R.L. Beadles, Research Triangle Institute, *Integrated Silicon Device Technology*, Vol. XIV, Technical Report for January 1966–March 1967, Defense Documentation Center Report #AD654 630, May 1967, 182 pages.

Table 2-41. TYPICAL PROPERTIES OF BONDING WIRES USED IN INTEGRATED CIRCUITS

Type	Diameter, mils	Weight, mg/meter	Average minimum elongation, %		Breaking load, grams		Resistance, ohms/meter
			Stress-relieved	Annealed			
Gold	5	244	5	15	160		1.84
	3	88	5	15	105		5.32
	2	39	5	15	26		11.96
	1	9.78	5	6	6		47.6
	0.5	2.49	3	4	1.5		192
	0.2	0.39		2	0.24		796
			Hard	Annealed	Hard	Annealed	
Aluminum	5		1.5	15	400	200	2.23
	3		1.5	15	140	70	6.12
	2		1.3	12	65	32	13.9
	1		1.2	4	16	8	55.7
	0.5		1	3	4	2	233
Aluminum—1% Si	1		1	3	18	2.8	

Wedge bonding requires fully annealed wire; ball bonding requires stress-relieved wire.

*From: "Interconnections and Encapsulation", R.L. Beadles, Research Triangle Institute, *Integrated Silicon Device Technology*, Vol. XIV, Defense Documentation Center Report #AD654 630, May 1967.

Table 2-42. SUITABILITY OF PURE METALS AS OHMIC CONTACTS AND INTRACONNECTIONS IN INTEGRATED CIRCUITS*

For thermal properties of pure metals, see Table 1-60.

KEY: E—excellent   P—poor
      F—fair   TC—thermocompression
      G—good

Metal	Melting point, °C	Lowest eutectic temp, with Si, °C	Mechanical properties	Bulk resistivity, microhm-cm at 20°C	Resistance to HF	Heat of formation of oxide, kcal, 0°K	Adherence to SiO_2	Etchant availability	TC bonds with Au wire easily
Ag	960	830	E	1.6	E	−2.59	Low	Yes	Yes
Al	660	577	E	2.8	G	−376.7 (Al_2O_3)	Very high	Yes	Yes
Au	1 063	370	E	2.4	E	+39	Low	Yes	Yes
Be	1 280	1 090	P	4.0	P	−144	—	Yes	—
Ca	895	760	G	4.6	P	−151	—	Yes	—
Cd	320.9	—	F	6.83	P	−53.79	Low	Yes	—
Co	1 495	900	F	9.8	G	−51.0 (CoO)	High	Yes	No
Cr	1 880	1 320	—	12.9	E	−250 (Cr_2O_3)	Very high	Difficult	No
Cu	1 083	802	E	1.7	E	−34.98	Medium	Yes	Yes
Fe	1 530	1 200	G	10.	G	−177 (Fe_2O_3)	High	Yes	—
Ir	2 454	—	E	6.5	E	−38.4	Low	No	—
Mg	650	638	P	4.6	G	−136.13	High	Yes	Yes
Mo	2 625	1 410	F	5.7	E	−161.95 (MoO_3)	High	Yes	No
Ni	1 455	964	E	6.8	G	−51.7	High	Yes .	No
Os	2 700	—	P	60.2	E	−92	—	Yes	—
Pb	327.4	—	P	20.65	G	−45.25	Low	Yes	No
Pd	1 554	800	E	11.	E	−52.2	Low	Yes	Yes
Pt	1 773	830	E	10.	E	—	Low	Difficult	Yes
Rh	1 966	1 390	E	4.9	E	—	Low	No	Yes
Ru	2 500	1 490	P	7.6	G	—	—	No	—
Sn	232	232	P	11.0	G	−124.2	Low	Yes	—
Ta	2 996	1 385	G	12.45	P	−471 (Ta_2O_5)	High	Yes	No
Ti	1 725	860	G	42.0	G	−204 (TiO_2)	Very high	Difficult	No
V	1 710	1 380	G	24.8 to 26.0	P	−271 (V_2O_3)	High	Yes	No
W	3 410	1 400	P	5.6	E	−182.47 (WO_3)	High	Difficult	No
Zn	419.5	419.5	P	5.92	P	−76.0	Low	Yes	—
Zr	1 857	1 360	P	40.0	P	−244 (ZrO_2)	High	Yes	No

*From: "Intraconnections and Isolation", R.P. Donovan, Research Triangle Institute, *Integrated Silicon Device Technology*, Vol. XIII, Defense Documentation Center Report #AD655 081, May 1967.

Table 2-43. DIFFUSION COEFFICIENT D*

For Self-diffusion and Diffusion of Foreign Atoms

$$D = D_o e^{-\Delta E/kT}$$

Semiconductor and diffusing element	$D_o,$ cm^2/s	$\Delta E,$ eV	Temperature, deg C
Aluminum antimonide (AlSb)			
Al		≈ 1.8	
Cu	3.5×10^{-3}	0.36	150–500
Sb		≈ 1.5	
Zn	0.33 ± 0.15	1.93 ± 0.04	660–860
Cadmium selenide (CdSe)			
Se	2.6×10^{-3}	1.55	700–1800
Cadmium sulfide (CdS)			
Ag	$2.5 \times 10^{+1}$	1.2	250–500
Cd	3.4	2.0	750–1000
Cu	1.5×10^{-3}	0.76	450–750
Cadmium telluride (CdTe)			
Au	6.7×10^{11}	2.0	600–1000
In	4.1×10^{-1}	1.6	450–1000
Calcium ferrate (III) (CaFe$_2$O$_4$)			
Ca	30	3.7	
Fe	0.4	3.1	
α-Calcium metasilicate (CaSiO$_3$)			
Ca	7.4×10^4	4.8	
Gallium antimonide (GaSb)			
Ga	3.2×10^3	3.15	650–700
In	1.2×10^{-7}	0.53	400–650
Sb	3.4×10^4	3.44	650–700
	$8.7 \times 10^{+2}$	1.13	470–570
Sn	2.4×10^{-5}	0.80	320–570
Te	3.8×10^{-4}	1.2	400 650
Gallium arsenide (GaAs)			
Ag	2.5×10^{-3}	1.5	
	3.9×10^{-11}	0.33	
	4×10^{-4}	0.8 ± 0.05	500–1160
As	$4 \times 10^{?1}$	10.2 ± 1.2	1200–1250
Au	10^{-3}	1.0 ± 0.2	740–1024
Cd	0.05 ± 0.04	2.43 ± 0.06	868–1149
	$^a 5.0 \times 10^{-2}$	2.8^a	
Cu	0.03	0.52	100–600
Ga	1×10^7	5.60 ± 0.32	1125–1250
Li	0.53	1.0	250–400
Mg	1.4×10^{-4}	1.89	
	2.3×10^{-2}	2.6	740–1024
	$^a 2.6 \times 10^{-2}$	2.7^a	
Mn	$^a 6.5 \times 10^{-1}$	2.49^a	
	8.5×10^{-3}	1.7	740–1024
S	1.2×10^{-4}	1.8	
	$^a 1.6 \times 10^{-5}$	1.63^a	
	2.6×10^{-5}	1.86	
	4×10^3	4.04 ± 0.15	1000–1200
Se	3×10^3	4.16 ± 0.16	1000–1200
Sn	$^a 3.8 \times 10^{-2}$	2.7	
	6×10^{-4}	2.5	1069–1215
Zn	$^a 2.5 \times 10^{-1}$	3.0^a	
	3.0×10^{-7}	1.0	
	6.0×10^{-7}	0.6	
	15 ± 7	2.49 ± 0.05	800

^{a}Values obtained at the low concentration limit.

*Compiled from data in: "Diffusion in Semiconductors", B.I. Boltaks (trans. by J.I. Carasso), Infosearch Ltd. (distributed by Pion Ltd.; in the U.S.A. by Academic Press), 1963; "Physics of III–V Compounds", O. Madelung (trans. by D. Meyerhofer), John Wiley & Sons, Inc., 1964; and "Semiconductors and Semimetals", R.K. Willardson and A.C. Beer, Vol. 4, Academic Press, Inc., 1968.

Table 2-43.   DIFFUSION COEFFICIENT D (*Continued*)

Semiconductor and diffusing element	D_o, cm^2/s	ΔE, eV	Temperature, deg C
Gallium phosphide (GaP)			
Zn	1.0	2.1	700–1300
Germanium (Ge)			
Ag	4.4×10^{-2}	1.0	700–900
As	6.3	2.4	600–850
Au	2.2×10^{-2}	2.5	
B	1.6×10^{-9}	4.6	600–850
Cu	1.9×10^{-4}	0.18	600–850
Fe	1.3×10^{-1}	1.1	750–850
Ga	$4.0 \times 10^{+1}$	3.1	600–850
Ge	$8.7 \times 10^{+1}$	3.2	750–920
He	6.1×10^{-3}	0.69	750–850
In	3×10^{-2}	2.4	600–850
Li	1.3×10^{-4}	0.47	200–600
Ni	8×10^{-1}	0.9	700–875
P	2.5	2.5	600–850
Pb	—	3.6	600–850
Sb	4.0	2.4	600–850
Sn	1.7×10^{-2}	1.9	600–850
Zn	$1.0 \times 10^{+1}$	2.5	600–850
Indium antimonide (InSb)			
Ag	1.0×10^{-7}	0.25	
Au	$^a7 \times 10^{-4}$	0.32^a	140–510
Cd	$^a1.0 \times 10^{-5}$	1.1^a	250–500
	1.23×10^{-9}	0.52	442–519
	1.26	1.75	
	1.3×10^{-4}	1.2	360–500
Co	2.7×10^{-11}	0.39	
	10^{-7}	0.25	440–510
Cu	3.0×10^{-5}	0.37	
	$^a9.0 \times 10^{-4}$	1.08^a	
Fe	10^{-7}	0.25	440–510
Hg	$^a4.0 \times 10^{-6}$	1.17^a	
In	0.05	1.81	450–500
	1.8×10^{-9}	0.28	
Ni	10^{-7}	0.25	440–510
Sb	0.05	1.94	450–500
	1.4×10^{-6}	0.75	
Sn	5.5×10^{-8}	0.75	390–512
Te	1.7×10^{-7}	0.57	300–500
Zn	0.5	1.35	360–500
	1.6×10^{-6}	2.3 ± 0.3	360–500
	5.5	1.6	360–500
(Polycrystal)	1.7×10^{-7}	0.85	390–512
	$^a5.3 \times 10^7$	2.61	
(High concentration)	6.3×10^8	2.61	
	$^a8.7 \times 10^{-10}$	0.7^a	
(Conc. $= 2.2 \times 10^{20}$ cm^{-3})	9.0×10^{-10}	0	
(Single crystal)	1.4×10^{-7}	0.86	390–512
Indium arsenide (InAs)			
Cd	4.35×10^{-4}	1.17	600–900
Cu		0.52^a	
Ge	3.74×10^{-6}	1.17	600–900
Mg	1.98×10^{-6}	1.17	600–900
S	6.78	2.20	600–900
Se	12.55	2.20	600–900
Sn	1.49×10^{-6}	1.17	600–900
Te	3.43×10^{-5}	1.28	600–900
Zn	3.11×10^{-3}	1.17	600–900

^{a}Values obtained at the low concentration limit.

Table 2-43.   DIFFUSION COEFFICIENT D *(Continued)*

Semiconductor and diffusing element	D_o, cm^2/s	ΔE, eV	Temperature, deg C
Indium phosphide (InP)			
In	1×10^5	3.85 ± 0.05	850–1000
P	7×10^{10}	5.65 ± 0.06	850–1000
Iron oxide (Fe$_3$O$_4$)			
Fe	5.2	2.4	
Lead metasilicate (PbSiO$_3$)			
Pb	85	2.6	
Lead orthosilicate (PbSiO$_4$)			
Pb	8.2	2.0	
Mercury selenide (HgSe)			
Sb	6.3×10^{-5}	0.85	540–630
Nickel aluminate (NiAl$_2$O$_4$)			
Cr	1.17×10^{-3}	2.2	
Fe	1.33	3.5	
Nickel chromate (III) (NiCr$_2$O$_4$)			
Cr	0.74	3.1	
Cr	2.03×10^{-5}	1.9	
Fe	1.35×10^{-3}	2.6	
Ni	0.85	3.2	
Selenium (Se) (amorphous)			
Fe	1.1×10^{-5}	0.38	300–400
Ge	9.4×10^{-6}	0.39	300–400
In	5.2×10^{-6}	0.32	300–400
Sb	2.8×10^{-8}	0.29	300–400
Se	7.6×10^{-10}	0.14	300–400
Sn	4.8×10^{-8}	0.39	300–400
Te	5.4×10^{-6}	0.53	300–400
Tl	1.4×10^{-6}	0.35	300–400
Zn	3.8×10^{-7}	0.29	300–400
Silicon (Si)			
Al	8.0	3.5	1100–1400
Ag	2×10^{-3}	1.6	1100–1350
As	3.2×10^{-1}	3.5	1100–1350
Au	1.1×10^{-3}	1.1	800–1200
B	$1.0 \times 10^{+1}$	3.7	950–1200
Bi	$1.04 \times 10^{+3}$	4.6	1100–1350
Cu	4×10^{-2}	1.0	800–1100
Fe	6.2×10^{-3}	0.86	1000–1200
Ga	3.6	3.5	1150–1350
H$_2$	9.4×10^{-3}	0.47	1000–1200
He	1.1×10^{-1}	0.86	1000–1200
In	$1.65 \times 10^{+1}$	3.9	1100–1350
Li	9.4×10^{-3}	0.78	100–800
P	$1.0 \times 10^{+1}$	3.7	1100–1350
Sb	5.6	3.9	1100–1350
Tl	$1.65 \times 10^{+1}$	3.9	1100–1350
Silicon carbide (SiC)			
Al	2.0×10^{-1}	4.9	1800–2250
B	$1.6 \times 10^{+2}$	5.6	1850–2250
Cr	2.3×10^{-1}	4.8	1700–1900
Sulfur (S)			
S	2.8×10^{13}	2.0	>100
Tin zinc oxide (SnZn$_2$O$_4$)			
Sn	2×10^5	4.7	
Zn	37	3.3	
Zinc aluminate (ZnAl$_2$O$_4$)			
Zn	2.5×10^2	3.4	

Table 2-43.　DIFFUSION COEFFICIENT D　*(Continued)*

Semiconductor and diffusing element	D_o, cm^2/s	ΔE, eV	Temperature, deg C
Zinc chromate (III) ($ZnCr_2O_4$)			
Cr	8.5	3.5	
Zn	60	3.7	
Zinc ferrate (III) ($ZnFe_2O_4$)			
Fe	8.5×10^2	3.5	
Zn	8.8×10^2	3.7	
Zinc selenide (ZnSe)			
Cu	1.7×10^{-5}	0.56	200–570
Zinc sulfide (ZnS)			
Zn	$1.0 \times 10^{+16}$	6.50	>1030
	$1.5 \times 10^{+4}$	3.25	940–1030
	3.0×10^{-4}	1.52	<940

Table 2-44.　EFFECT OF IRRADIATION ON CERTAIN SEMICONDUCTORS

Material	Response to irradiation
n-type Ge	Converted to *p*-type
p-type Ge	Approaches a limiting hole concentration (7×10^{16} cm^{-3})
n & *p*-type Si	Carrier concentrations approach intrinsic value
n-type InSb	Approaches a limiting electron concentration (4×10^{16} cm^{-3})
p-type InSb	Converted to *n*-type
n-type GaSb	Converted to *p*-type
p-type GaSb	Approaches a limiting hole concentration ($\sim 10^{16}$ cm^{-3})
n-type AlSb	Approaches a limiting electron concentration
p-type AlSb	Converted to *n*-type
n-type InAs	Electron concentration appears to increase indefinitely
p-type InAs	Converted to *n*-type
n-type InP	Electron concentration decreases, no evidence of conversion
p-type CdTe	Converts to *n*-type

Table 2-45.　THIN-FILM RESISTORS*

Electrical and Physical Properties

For other properties of thin films, see Tables 2-57 and 2-58.

Film	Fabrication techniques	Sheet resistance, ohms/square	Temperature coefficient of resistance per °C	Linear coefficient of expansion per °C	Stability, % drift in 1,000 hrs at a given temperature
Carbon	Vacuum or vapor deposited	$10–10^7$	-500×10^{-6}	2.9×10^{-6}	
Nichrome	Vacuum deposited	$10–10^4$	$+50 \times 10^{-6}$	13×10^{-6}	2% at 200°C
Tantalum	Vacuum sputtered	$10–10^4$	$\pm 200 \times 10^{-6}$	6.6×10^{-6}	1% at 150°C
Electroless nickel	Chemical deposited	$20–10^5$	$\pm 250 \times 10^{-6}$	—	1% at 70°C
Tin oxide	Vapor decomposition	$10–5000$	$\pm 250 \times 10^{-6}$	—	4% at 500°C
Cermet (Cr-SiO)	Vacuum deposited (flash evaporation)	$200–10^5$	$\pm 250 \times 10^{-6}$	—	3% at 300°C

*From: "Resistance", Research Triangle Institute, *Integrated Silicon Device Technology*, Vol. I, Defense Documentation Center Report #AD408 961, June 1963.

Table 2-46. DISTRIBUTION COEFFICIENTS
During Crystal Growth from the Melt

Coefficient $k_c = C_s/C_l$,
where C_s = concentration in solid and C_l = concentration in liquid.

Semiconductor and doping element	Distribution coefficient, k_c	Semiconductor and doping element	Distribution coefficient, k_c	Semiconductor and doping element	Distribution coefficient, k_c
Aluminum antimonide (AlSb)*		**Gallium antimonide (GaSb)***		**Indium antimonide (InSb)***	
Mg	0.1	Zn	0.3	Zn	10
C	0.6		0.16		4.13
Si	0.045	Cd	0.02		3.38
Ge	0.026	In	≈1		3.0
Sn	$(2\text{–}8) \times 10^{-4}$	Si	≈1		2.3
Fe	0.02	Ge	0.32	Cd	0.26
Cu	0.01		0.2	Ga	2.4
Mn	0.01		0.08	Ge	0.045
Co	0.002	Sn	≈0.01	Sn	0.057
Zn	0.02	As	2–4	P	0.16
Cd	0.002	S	0.06	As	5.4
S	0.003	Se	0.18	S	0.1
Se	0.003		0.4	Se	0.5
Te	0.01		0.4		0.35
B	0.01–0.02	Te			0.17
Pb	0.01	Mn	0.05	Te	3.5
V	0.01	Fe	0.003		≈1
Ni	0.01	Ni	<0.02		0.54
Ti	0.01	Cu	0.002	Fe	0.04
In	≈1	**Germanium (Ge)†**		Cu	6.6×10^{-4}
Gallium arsenide (GaAs)*		Cu	1.5×10^{-5}	Ag	4.9×10^{-5}
Ag	0.1	Ag	10^{-5}	Au	1.9×10^{-6}
Mg	0.3	Au	3×10^{-5}	Te	5.2×10^{-4}
Ca	<0.02	Zn	10^{-2}	Ni	6.0×10^{-5}
Zn	0.1	B	~20	**Indium phosphide (InP)***	
	0.27–0.9	Al	10^{-1}	Ge	0.05
Cd	0.02	Ga	10^{-1}	Sn	0.03
Al	3	In	1.1×10^{-3}	S	≈0.8
In	0.1	Tl	4×10^{-5}	Se	≈0.6
C	0.8	P	1.2×10^{-1}	**Silicon (Si)‡**	
Si	0.1	As	4×10^{-2}	B	0.80
	0.14	Sb	3×10^{-3}	Al	0.0020
Ge	0.03	Fe	2×10^{-6}	Ga	0.0080
	0.018	Co	2×10^{-6}	In	4×10^{-4}
Sn	0.03	Ni	5×10^{-6}	Tl	—
Pb	<0.02	Pt	2×10^{-6}	P	0.35
P	2	**Indium arsenide (InAs)***		As	0.3
Sb	<0.02	Mg	0.7	Sb	0.023
S	0.3	Zn	0.77	Bi	7×10^{-4}
	0.5–1.0	Cd	0.13	Zn	~1×10^{-5}
Se	0.44–0.55	Si	0.40	Cu	4×10^{-4}
Te	0.3	Ge	0.07	Au	2.5×10^{-5}
	0.054–0.16	Sn	0.09	Pt	—
Cu	<0.002	S	1.0	Fe	8×10^{-6}
		Se	0.93	Ni	—
		Te	0.44	Co	8×10^{-6}
		Pb	<0.01	Li	0.01

*From: "Physics of III-V Compounds", O. Madelung (trans. by D. Meyerhofer), John Wiley & Sons, 1964.
†From: "An Introduction to Semiconductors", W.C. Dunlap, John Wiley & Sons, 1965.
‡From: "Silicon Technology", W. Runyon, McGraw-Hill Book Company, 1965.

Table 2-47.   COLOR DEPENDENCE OF SILICON DIOXIDE FILM THICKNESS*

This table gives a color chart for thermally grown silicon dioxide films observed perpendicularly under daylight fluorescent lighting.

Film thickness, microns	Order, 5460 Å	Color and comment	Film thickness, microns	Order, 5460 Å	Color and comment	Film thickness, microns	Order, 5460 Å	Color and comment
0.05_0		Tan	0.46_5		Red-violet	0.95		Dull yellow-green
0.07_5		Brown	0.47_6		Violet	0.97		Yellow to "yellowish"
0.10_0		Dark violet to red-violet	0.48_0		Blue-violet	0.99		Orange
0.12_5		Royal blue	0.49_3		Blue	1.00		Carnation pink
0.15_0		Light blue to metallic blue	0.50_2		Blue-green	1.02		Violet-red
			0.52_0		Green (broad)	1.05		Red-violet
0.17_5	I	Metallic to very light yellow-green	0.54_0		Yellow-green	1.06		Violet
			0.56_0	III	Green-yellow	1.07		Blue-violet
0.20_0		Light gold to yellow —slightly metallic	0.57_4		Yellow to "yellowish"†	1.10		Green
			0.58_5		Light-orange or yellow to pink borderline	1.11		Yellow-green
0.22_5		Gold with slight yellow-orange	0.60_0		Carnation pink	1.12	VI	Green
0.25_0		Orange to melon	0.63_0		Violet-red	1.18		Violet
0.27_5		Red-violet	0.68_0		"Bluish"‡	1.19		Red-violet
0.30_0		Blue to violet-blue	0.72	IV	Blue-green to green (quite broad)	1.21		Violet-red
0.31_0		Blue				1.24		Carnation pink to salmon
0.32_5		Blue to blue-green	0.77		"Yellowish"	1.25		Orange
0.34_5		Light green	0.80		Orange (rather broad for orange)	1.28		"Yellowish"
0.35_0		Green to yellow-green				1.32	VII	Sky blue to green-blue
0.36_5	II	Yellow-green	0.82		Salmon			
0.37_5		Green-yellow	0.85		Dull, light red-violet	1.40		Orange
0.39_0		Yellow	0.86		Violet	1.45		Violet
0.41_2		Light orange	0.87		Blue-violet	1.46		Blue-violet
0.42_6		Carnation pink	0.89		Blue	1.50	VIII	Blue
0.44_3		Violet-red	0.92	V	Blue-green	1.54		Dull yellow-green

†Not yellow but in the position where yellow is to be expected; at times it appears to be light, creamy grey or metallic.
‡Not blue but borderline between violet and blue-green; it appears more like a mixture between violet-red and blue-green and over-all looks greyish.

*From: W.A. Pliskin and E.E. Conrad, *IBM Journal of Research and Development* 8, 43, (1964).

Table 2-48.   TYPICAL FILM MATERIALS AND PROCESSES*

Application	Material	Process
Thin-film resistors	Nichrome (Ni–Cr)	Vacuum evaporation
	Tin oxide (SnO_2)	Vapor plating
	Tantalum–tantalum nitride (Ta–TaN)	Sputtering
	Cermet chromium-silicon monoxide (Cr–SiO)	Vacuum evaporation
Thick-film resistors	Palladium–silver	
	Palladium oxide–silver and glass	Silk screen
Thin-film dielectrics	Silicon monoxide (SiO)	Vacuum evaporation
	Silicon dioxide (SiO_2)	Vapor plating, high temperature, steam oxidation, reactive sputtering
	Aluminum oxide (Al_2O_3)	Vapor plating, anodization of aluminum films
	Tantalum oxide (Ta_2O_5)	Anodization of tantalum, reactive sputtering
Thick-film dielectrics	$BaTiO_3$ or TiO_2 and glass mixtures	Silk screen
Thin-film conductors	Chromium–gold	Vacuum evaporation
	Chromium–copper	Vacuum evaporation
	Aluminum	Vacuum evaporation
	Nickel on ceramic	Electroless plating
Thick-film conductors	Gold–platinum–glass	
	Gold–glass	Silk screen
	Silver–glass	

*From: "Reference Data for Radio Engineers", 5th ed., Howard W. Sams & Co., Inc., Indianapolis, Indiana, 1968.

REFERENCE

"Measurement Techniques for Thin Films", B. Schwartz and N. Schwartz, Eds., The Electrochemical Society, 30 E. 42nd St., New York, N.Y. 10017.

Table 2-49.   THIN-FILM DEPOSITION DATA*

ABBREVIATIONS:

DC—direct-current power
EB—electron beam
RF—radio-frequency power
Vit—vitreous

Name	Symbol	Melting point, deg C	Density, g/cm³	Temperature (deg C) at vapor press 10^{-8} torr	10^{-6} torr	10^{-4} torr	Evaporation techniques — Electron beam	Crucible	Coil	Boat	Basket	Sputter	Remarks
Aluminum	Al	660	2.7	677	812	1 010	Excellent	TiB_2-BN, ZrB_2-BN, Vit carbon	W	TiB_2-BN, Al_2O_3	W	RF	Alloys and wets. Tungsten-stranded superior
Aluminum fluoride	AlF, AlF_3	Subl	2.9	410 Sublimes	490	700	Poor	—	—	W, Ta	—	RF	Loses fluorine. Disproportionates
Alumina (α)	Al_2O_3	2 020	4.0	1 045	1 210	1 325	Excellent	W	W	W	W	RF-reactive	Sapphire excellent in EB. Chemically deposit. Index may change
Antimony	Sb	630	6.7	279	345	425	Poor	Al_2O_3	Mo, Ta	Mo, Ta	Mo, Ta	RF	—
Arsenic	As	814	5.7	107 Sublimes	150	210	Poor	Al_2O_3, BeO	—	Al_2O_3 Coated, C	—	—	Sublimes rapidly at low temperature. Toxic
Barium	Ba	725	3.5	545	627	735	Fair	Vit carbon	W	W, Ta, Mo	W	RF	Wets W/O alloying. Reacts with ceramics
Barium titanate	$BaTiO_3$	Dec	6.0	Decomposes			....	—	—	—	—	RF	Decomposes. Yields free Ba from single source—evaporate from two sources, flash from superheated W
Beryllium	Be	1 278	1.9	710	878	1 000	Excellent	BeO, C	W	W, Ta, Mo	W	—	Wets W/Mo/Ta. Oxides are toxic
Bismuth	Bi	271	9.8	330	410	520	Good	Al_2O_3, Vit carbon	W	W, Mo, Al_2O_3, Ta	W	DC, RF	Vapors are toxic. High resistivity. No shorting of baskets
Bismuth titanate	$Bi_2Ti_2O_7$	—	—	Decomposes			—	—	—	—	—	RF	Decomposes; evaporate from two sources
Boron	B	2 300	2.3	1 278	1 548	1 757	Excellent	C, Vit carbon	—	C	—	RF	Material explodes with rapid cooling
Boron carbide	B_4C	2 350	2.5	2 500 Sublimes	2 580	2 650	Excellent	—	—	—	—	RF	—
Cadmium	Cd	321	8.6	64	120	180	Poor	Al_2O_3, Quartz	—	W, Cb, Mo, Ta, Mo	W, Mo, Ta	DC, RF	Wets Cb. Use source heaters
Cadmium sulfide	CdS	1 405	4.8	Sublimes	357	~250	Fair	Quartz	—	—	—	RF	Sticking coefficient strongly affected by substrate temperature. Closed aperture should be used
Calcium	Ca	842	1.5	272	357	459	Poor	Al_2O_3, Quartz	W, Mo, Ta	W, Mo, Ta	W	—	Corrodes in air
Calcium fluoride	CaF_2	1 360	3.2	— Sublimes		~1 000	—	Quartz	W, Mo, Ta	W, Mo, Ta	W, Mo, Ta	RF	Rate control important

*Data furnished through courtesy of Sloan Technology Corporation.

Table 2-49.   THIN-FILM DEPOSITION DATA (Continued)

Name	Symbol	Melting point, deg C	Density, g/cm³	Temperature (deg C), at vapor press			Evaporation techniques					Sputter	Remarks
				10^{-8} torr	10^{-6} torr	10^{-4} torr	Electron beam	Crucible	Coil	Boat	Basket		
Calcium titanate	CaTiO₃	1 975	4.1	1 490	1 600	1 690	Poor	—	—	—	—	RF	Disproportionates except in sputtering
Carbon	C	Subl	1.8–2.3	1 657 Sublimes	1 867	2 137	Excellent	—	—	—	—	—	Arc evaporation
Cerium	Ce	795	6.8	970	1 150	1 380	Fair	Al₂O₃ BeO Vit carbon	W	W	W	DC RF	—
Cerium oxide	Ce₂O₃	1 692	6.9	—	—	—	Fair	Vit carbon	W	W	W	—	Alloys with source use .015–.020 W boat
Chromium	Cr	1 890	7.2	837	977	1 157	Good	Vit carbon	W	W	W	RF DC	Will sublime with controlled rate source
Cobalt	Co	1 495	8.9	850 Sublimes	990	1 200	Excellent	Al₂O₃ BeO	Plated	Plated	W	DC RF	Alloys with refractories
Copper	Cu	1 083	8.9	727	857	1 017	Excellent	Al₂O₃ TiB₂–BN	W Ta	W Ta Cb	W Ta	DC RF	Alloys with refractories
Gallium	Ga	30	5.9	619	742	907	Good	Al₂O₃ BeO Quartz	—	—	—	—	Alloys with refractories
Gallium arsenide	GaAs	1 238	5.3	—	—	—	Good	—	—	W Ta	W	RF	Compound decomposes. Chemically deposit
Gallium phosphide	GaP	1 540	4.1	—	770	920	Excellent	Quartz	—	W Ta	W	RF	Does not decompose. Alloy rate control important
Germanium	Ge	937	5.3	812	957	1 167	Excellent	Quartz Al₂O₃ Quartz	—	W C Ta Mo	W Mo	DC RF	—
Germanium oxide	GeO₂	1 086	6.2	—	—	—	Good	Al₂O₃ Quartz	—	Ta Mo	W Mo	RF-reactive	Similar to SiO
Gold	Au	1 062	19.3	807	947	1 132	Excellent	Al₂O₃ BN Vit carbon	W	W, Mo Coated Al₂O₃	W	DC RF	—
Indium	In	157	7.3	487	597	742	Excellent	Al₂O₃	—	W Mo	W	DC RF	Wets W and Cu; use Ta liner in guns
Indium antimonide	InSb	535	5.8	500	—	—	—	—	—	W Mo	—	RF	Dual source, flash
Indium arsenide	InAs	943	5.7	780	870	970	—	—	—	W	—	RF	Dual source, flash
Indium phosphide	InP	1 058	4.8	—	630	730	Excellent	—	—	W Ta W	W Ta	RF	Deposits P rich, flash evaporate
Iron	Fe	1 535	7.9	858	998	1 180	Excellent	Al₂O₃ BeO	W	W Ta W	W Ta	DC RF	Attacks W
Lead	Pb	328	11.3	342	427	497	Excellent	Al₂O₃ Quartz	W	W Ta Mo	W Ta	DC RF	Toxic, highly controlled rates required for superconductors

Table 2-49. THIN-FILM DEPOSITION DATA (Continued)

Name	Symbol	Melting point, deg C	Density, g/cm³	Temperature (deg C), at vapor press			Evaporation techniques					Sputter	Remarks
				10^{-8} torr	10^{-6} torr	10^{-4} torr	Electron beam	Crucible	Coil	Boat	Basket		
Lead stannate	PbSnO₃	1 115	8.1	670	780	905	Poor	Al₂O₃	—	Pt	Pt	—	Disproportionates
Lead telluride	PbTe	917	8.2	780	910	1 050	—	Al₂O₃	—	Pt	—	RF	Vapors toxic. Deposits Te rich. Use dual sources
Lithium	Li	179	0.53	227	307	407	Good	Al₂O₃ BeO	—	S.S.	—	—	Oxide destroys quartz
Lithium fluoride	LiF	870	2.6	875	1 020	1 180	Good	—	—	Ta Mo W	—	RF	Rate control important for optical films
Magnesium	Mg	651	1.7	185 Sublimes	247	327	Good	Al₂O₃ Vit carbon	W	W Mo Ta, Cb	W	DC RF	—
Magnesium fluoride	MgF₂	1 266	3.2	1 090	1 220	1 540	Excellent	Al₂O₃	—	W Ta Mo	W	RF	Rate control important for optical films
Manganese	Mn	1 244	7.2–7.4	507	572	647	Good	Al₂O₃ BeO	W	W Ta Mo	W	DC RF	Wets resistance sources
Molybdenum	Mo	2 610	10.2	1 592	1 822	2 117	Excellent	—	—	—	—	DC RF	Fine wire, evaporate with own resistance
Nichrome	Ni/Cr	1 350	8.2	847	987	1 217	Excellent	Al₂O₃ Vit carbon BeO	W	Al₂O₃ Coated	W Ta	DC RF	Alloys with refractories
Nickel	Ni	1 453	8.9	927	987	1 262	Excellent	Al₂O₃ BeO Vit carbon	W	W	W	DC RF	Alloys with refractories. May be electroplated
Niobium (Columbium)	Nb	2 468	8.6	1 728	1 977	2 287	Excellent	—	—	W	—	DC RF	Fine wire—will self evaporate. Attacks W source
Palladium	Pd	1 550	12.0	842	992	1 192	Excellent	Al₂O₃ BeO	W	W	W	DC RF	Alloys with refractories. Fine wire—rapid evaporation suggested
Permalloy	Ni/Fe	1 395	8.7	947	1 047	1 307	Good	Al₂O₃ Vit carbon	—	—	W	DC RF	Ni content low in film. Use 84% Ni source
Platinum	Pt	1 769	21.5	1 292	1 492	1 747	Excellent	C	W Pt	W	W	DC RF	Alloys with metals. Fine wire will self evaporate
Rhodium	Rh	1 966	12.4	1 277	1 472	1 707	Good	Vit carbon	W	W	W	DC RF	Use very low pressure for W source
Selenium	Se	217	4.8	89	125	170	Good	Al₂O₃ Vit carbon	W Mo	W Mo	W Mo	RF	Toxic
Silicon	Si	1 410	2.2	992	1 147	1 337	Excellent	Al₂O₃ Vit carbon BeO	W Mo	W Mo	—	DC RF	Heavy W boat alloys. SiO produced below 4×10^{-6}
Silicon dioxide	SiO₂	1 610–1 710	2.2–2.7	~850 Influenced by composition			Excellent	—	—	Ta	—	RF	Chemically deposit
Silicon monoxide	SiO	1 702	2.1	574 Sublimes	617	~600	Excellent	Ta	W	Ta	W	RF RF-reactive	Baffle box source best for resistance evaporation. Low rate suggested
Silver	Ag	961	10.5	574	617	684	Excellent	Al₂O₃	W	Ta	W	DC RF	Does not wet W

Table 2-49. THIN-FILM DEPOSITION DATA (Continued)

Name	Symbol	Melting point, deg C	Density, g/cm³	Temperature (deg C), at vapor press 10⁻⁸ torr	10⁻⁶ torr	10⁻⁴ torr	Evaporation techniques — Electron beam	Crucible	Coil	Boat	Basket	Sputter	Remarks
Sodium chloride	NaCl	801	2.2	675	835	1 014	Poor	Quartz	—	Ta	—	—	—
Sodium fluoride	NaF	988	2.6	945	1 080	1 200	Poor	BeO	—	W	—	RF	Furnace source deposit
Supermalloy	Ni/Fe/Mo	1 410	8.9	—	—	—	Good	—	Ta	Mo, Ta, W	—	RF, DC	Use dual sources. Permalloy and Mo
Tantalum	Ta	2 996	16.6	1 960	2 240	2 590	Excellent	—	Ta	—	—	DC	Fine wire—will self evaporate
Tantalum pentoxide	Ta₂O₅	1 800	8.2	1 550	1 780	1 920	Good	Vit carbon	W	Ta	W	RF, RF-reactive	—
Tellurium	Te	452	6.2	157	207	277	Poor	Al₂O₃, Quartz	W	W, Ta	W	RF, RF-reactive	Wets W/O alloying, toxic
Thallium	Tl	302	11.9	280	360	470	Poor	Al₂O₃, Quartz	—	W, Ta	W	—	Wets freely, very toxic
Tin	Sn	232	7.3	682	807	997	Excellent	Al₂O₃	W	Mo	W	DC, RF, RF-reactive	Wets Mo
Tin oxide	SnO₂	1 127	6.9	Sublimes		~600	Excellent	Al₂O₃	W	W	W	DC, RF, RF-reactive	Pyrolytic deposit
Titanium	Ti	1 675	4.5	1 067	1 235	1 453	Excellent	Vit carbon, TiC	—	W, Ta	W	DC, RF	—
Titanium dioxide	TiO₂	1 850	4.3	—	—	~1 000	Fair		W	W, Ta		RF-reactive	Sub-oxide, must be reoxidized to rutile
Tungsten	W	3 410	19.3	2 117	2 407	2 757	Good		—	C	—	DC, RF	Forms volatile oxides. Fine wire will self evaporate
Tungsten carbide	W₂C	2 860	15.7	1 480	1 720	2 120	Excellent		—		—	RF	
Tungsten trioxide	WO₃	1 473	7.2	1 120	1 290	1 460	Good		W, W	W, Pt	—	Reactive	Films oxidize
Uranium	U	1 132	19.0	1 132	1 327	1 582	Good		W	W	W	—	Alloys with W. Wets Mo
Vanadium	V	1 890	5.9	1 162	1 332	1 547	Excellent		W	Mo	W	—	
Zinc	Zn	419	7.1	127	177	250	Excellent	Al₂O₃, Quartz	W	W, Ta	W	DC, RF	Wets refractory metals, no reaction
Zinc fluoride	ZnF₂	872	4.9	790	905	1 035		Quartz	—	Pt, Ta	—	RF-reactive	Furnace deposit
Zinc oxide	ZnO	1 975	5.6	—	—	—	—	—	—	—	—	RF, RF-reactive	
Zinc sulfide	ZnS	1 830	4.1	—	—	~300	Good	Quartz	—	Ta, Mo	—	RF	Films partially decompose. Sticking coefficient varies with substrate temperature
Zirconium	Zr	1 852	6.5	1 477	1 702	1 987	Excellent	—	—	W	—	RF, DC	Alloys with W. Films oxidize readily
Zirconium oxide	ZrO₂	2 700	5.6	—	—	~2 200	Good	—	W	W	—	RF, RF-reactive	—

Table 2-50. PROPERTIES OF COMMONLY USED GLASS SUBSTRATE MATERIALS*

For conversion factors see Table 2-51.

	Soda lime	Alkali zinc borosilicate (microsheet)	Lime alumino-silicate (alkali-free)	Lime alumino-silicate (alkali-free)	Barium alumino-silicate (alkali-free)	Alkali borosilicate	96% silica	Fused silica	Fotoceram[a]	Synthetic sapphire[b]
Code number[c]	0080	0211	1715	1723	7055	7740	7900	7940	—	—
Annealing point, deg C	512	542	866	710	650	565	910	1 050	—	—
Softening point, deg C	696	720	1 060	910	872	820	1 500	1 580	700	2 040
Thermal exp coef, 10^{-6}/deg C 0–300 deg C	9.2	7.2	3.5	4.6	4.5	3.25	0.8	0.56	10.4	5.8
Thermal conductivity cal/cm·s·deg C at 25 deg C	0.002 3	—	0.002 3	0.003 2	0.003 6	0.002 7	0.003 8	0.003 4	0.005	0.098
cal/cm·s·deg C at 300 deg C	0.003 2	—	—	—	—	—	—	0.004 2	—	—
Density, g/cm^3	2.47	2.57	2.48	2.63	2.76	2.23	2.18	2.20	2.46	3.98
Dielectric constant, 100 hz 25 deg C	8.3	6.8	6.0	6.4	5.9	4.9	3.9	3.9	—	—
200 deg C	—	—	6.1	6.6	6.1	—	4.1	3.9	—	—
400 deg C	—	—	7.2	7.3	7.4	—	—	—	—	—
Dielectric constant, 1 Mhz 25 deg C	6.9	6.6	5.9	6.4	5.8	4.6	3.9	3.9	5.6	9.4–11.5
200 deg C	9.3	7.4	6.1	6.5	5.9	5.1	3.9	3.9	—	—
400 deg C	—	—	6.3	6.7	6.1	—	3.9	3.9	—	—
Loss tangent, 100 hz 25 deg C	0.078	0.01	0.001 8	0.000 8	0.001 1	0.027	0.000 6	0.000 06	—	—
200 deg C	—	—	0.004 7	0.002 4	0.006 2	—	0.07	0.000 3	—	—
400 deg C	—	—	0.2	0.09	0.19	—	—	—	—	—
Loss tangent, 1 Mhz 25 deg C	0.01	0.003 4 7	0.002 4	0.001 3	0.001 1	0.006 2	0.000 6	0.000 02	0.006	—
200 deg C	0.17	0.032	0.002 8	0.001 4	0.001 8	0.03	0.001	0.000 02	—	—
400 deg C	—	—	0.006 1	0.003 4	0.007 1	—	0.026	0.000 3	—	—
Log volume resistivity ohm-cm at 25 deg C	6.4	8.3	13.6	14.1	13.5	8.1	9.7	11.8	14.0	12.0
ohm-cm at 350 deg C	5.1	6.7	11.3	11.8	11.3	6.6	8.1	10.2	—	—
Dielectric strength, kV rms at 25 deg C	0.35	2.0	>10	>10	>10	2.0	7.0	>10	—	—
Weatherability, g/cm^2	>5	0.05–0.25	<0.01	<0.01	<0.01	0.05–0.25	<0.01	<0.01	—	Unaffected
Chemical durability, mg/cm^2 5% HCl, 24 hr	0.02	0.03	0.1	0.4	5.5	0.005	0.001	0.001	—	Unaffected
5% NaOH, 6 hr	0.5	2.0	1.2	0.3	3.7	1.1	1.1	0.7	—	Unaffected
0.02N Na$_2$CO$_3$, 6 hr	0.1	0.1	0.15	0.1	0.3	0.1	0.03	0.03	—	Unaffected

[a] Courtesy of Adolf Meller Company, Providence, R.I.

[b] Code numbers of Corning Glass Works, Corning, New York.

[c] Trade name, Corning Glass Works, Corning, New York.

*From: "Reference Data for Radio Engineers", 5th ed., Howard W. Sams & Co., Inc., Indianapolis, Indiana, 1968; and Corning Glass Works Bulletin, CEP-2/5M/9-62.

Table 2-51. PROPERTIES OF COMMONLY USED CERAMIC SUBSTRATE MATERIALS*

CONVERSION FACTORS:

For temperatures in K, add 273 to temperatures in °C.

For thermal conductivity in W/m·K, multiply cal/cm sec °C by 418.4.

For density in kg/m³, multiply g/cm³ by 1000.

Ceramic type	Alumina (96% Al_2O_3)	Alumina (99.7% Al_2O_3)**	Dense alumina (85% Al_2O_3)	Dense alumina (94% Al_2O_3 + CaO + SiO_2)	Dense alumina (96% Al_2O_3 + MgO + SiO_2)	Beryllia (BeO)	Dense beryllia (98% BeO)	Dense beryllia (99.5% BeO)	Titania (TiO_2)†	Barium titanate	Magnesium titanate**
Code number‡	—	—	576	719	614	—	735	754	192	—	—
Softening temp. (°C)	1650	>1650	1100	1500	1550	—	1600	1600	1600	1550	—
Melting point (°C)	2050	>2050	—	—	—	2550	—	—	1920	1700	—
Thermal exp. coef. (10^{-6}/°C)	6.4	5.0	6.5	6.2	6.4	7.5	6.1	6.0	8.3	8.1	7.5
Thermal conductivity cal/cm/sec/°C at 25°C	0.50	0.045	0.060	0.073	0.084	0.20	0.50	0.55	0.012	0.003	0.1
Density (gm/cm³)	3.7	3.9	3.4	3.58	3.7	3.08	2.9	2.88	4.0	5.5	3.6
Dielectric constant (1 MHz) at 25°C	10.0	9.5 (10 GHz)	8.3	8.9	9.3	7.1	6.3	6.4	85	10 to 10⁴	16 (10 GHz)
Loss tangent (1 MHz) at 25°C	0.0002	0.0001	0.0058	0.0018	0.0028	0.0006	0.0006	0.0006	0.0002	>0.02	0.0002
Log volume resistivity (ohm-cm) at 30°C	14.0	>14	10.7	12.8	10.0	>14	13.8	>14	14.0	12.0	>14
Dielectric strength at 25°C and 60 Hz (volts/mil)	—	—	230	230	230	—	255	260	100	—	—
Surface smoothness (microinches)§	10 to 30	10 to 30	10 to 30	10 to 30	10 to 30	10 to 30	10 to 30	10 to 30	0.1 to 30	10 to 30	10 to 30
Chemical resistance	excellent	excellent	excellent	excellent	excellent	excellent	excellent	excellent	good	good	good

†Also known as Rutile.

‡Code numbers of ALSIMAG brand ceramics, American Lava Corp., Chattanooga, Tennessee.

**Substrates used as dielectrics for microminiature microwave strip lines.

§1 microinch = 250 Å, or 0.025 μm.

*From: "Reference Data for Radio Engineers", 5th ed., Howard W. Sams & Co., Inc., Indianapolis, Indiana, 1968.

Figure 2-52.   CHANGE IN VOLUME RESISTIVITY FOR VARIOUS CERAMICS*

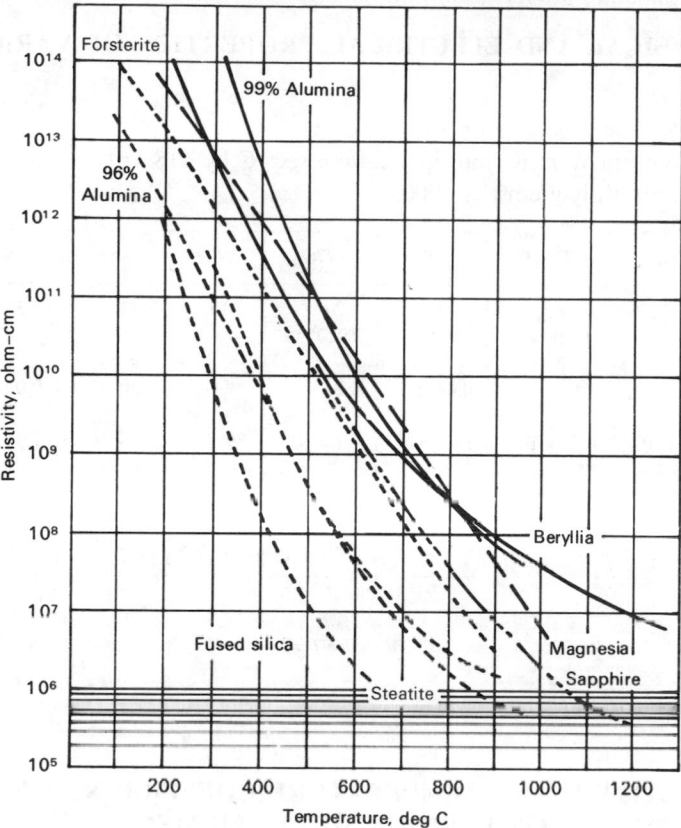

Figure 2-52.   Change in volume resistivity for various ceramics as a function of temperature.

*From: Properties of Ceramics for Electronic Applications, J.E. Comeforo, *Electron. Eng.*, April 1967.

Table 2-53.   DIELECTRIC STRENGTH OF VARIOUS INSULATING MATERIALS*

In rms V/mil at Frequencies from 60 hz to 100 Mhz

Material	Thick-ness, mils	Frequency						
		60 hz	1 khz	38 khz	180 khz	2 Mhz	18 Mhz	100 Mhz
Polystyrene, unpigmented	30	3 174	2 400	1 250	977	725	335	220
Polyethylene, unpigmented	30	1 091	965	500	460	343	180	132
Polytetrafluoroethylene (Teflon[a])	30	850	808	540	500	375	210	143
Monochlorotrifluoroethylene (Kel-F[c])	20	2 007	1 478	1 054	600	354	129	29[b]
Glass-bonded mica	32	712	643	—	360	207	121	76
Soda-lime glass	32	1 532	1 158	—	230	90	55	20[b]
Dry-process porcelain	32	232	226	—	90	83	71	60[b]
Steatite	32	523	427	—	300	80	58	56[b]
Forsterite (AlSiMag-243)	65	499	461	455	365	210	112	74
Alumina, 85% (AlSiMag-576[d])	55	298	298	253	253	178	112	69

[a]Trademark of E. I. du Pont de Nemours & Co., Inc., Wilmington, Del.
[b]Puncture with attendant volume heating effect.
[c]Trademark of Minnesota Mining and Manufacturing Co. (3M Co.), St. Paul, Minn.
[d]Trademark of American Lava Corp., Ridgefield, N.J.

*From: Frequency Dependence of Electric Field Strength, L.J. Frisco, *Electro-Tech.*, 68:110, August 1961.

Table 2-54.   THIN FILMS FOR INTEGRATED CIRCUITS

See Tables 2-45 and 2-49 for other properties of thin films.

Table A.   PHYSICAL AND ELECTRICAL PROPERTIES OF VARIOUS FILMS*

CONVERSION FACTORS:
   For temperatures in K, add 273 to temperatures in °C.
   For thermal conductivity in W/m·K, multiply cal/cm sec °C by 418.4.
   For density in kg/m^3, multiply g/cm^3 by 1000.

Characteristics	Ta	Ta$_2$O$_5$	Ti	TiO$_2$	SiO	Si	SiO$_2$	Al$_2$O$_3$	Al$_2$Si$_2$O$_7$
Density, g/cm^3	16.6	8.75	4.54	4.24	2.15	2.42	2.32	3.96	**
Melting point, °C	2996	1470	1668	1920	—	1420	1728	2040	**
Thermal conductivity, cal/°C sec cm	0.13	**	0.041	0.01	**	0.20	0.003	0.01	**
Thermal coefficient of expansion, per °C	6.46×10^{-6}	0.55×10^{-6}	8.4×10^{-6}	9.1×10^{-6}	**	7.6×10^{-6}	7.0×10^{-6}	8.0×10^{-6}	**
Resistivity, ohm-cm	12.5×10^{-6}	**	42×10^{-6}	10^{14} to 10^{25}	10^9	10^{-2} to 10^5	10^{14}	10^{11}	$>10^{14}$
Relative dielectric constant	—	25	—	100	1.8–5.8	12.0	3.78	10	8–9

**Unknown.

*From: "Capacitance", Research Triangle Institute, *Integrated Silicon Device Technology*, Vol. II, Defense Documentation Center Report #AD423 148, October 1963.

Table B.   RESISTIVITY AND TEMPERATURE COEFFICIENT OF RESISTANCE OF VARIOUS METAL FILMS*

SYMBOLS: ρ_f = electrical resistivity in microhm-cm
         TCR = temperature coefficient of resistance per °C

Metal	ρ_f before annealing, microhm-cm	ρ_f after annealing at 600°C, microhm-cm	Temperature coefficient of resistance	Bulk TCR, 0–100°C	Ratio of film TCR to bulk TCR
Au	22.2	4.95	$2,800 \times 10^{-6}$	3,400	.82
Pt	8.7	15.65	$2,500 \times 10^{-6}$	3,900	.64
Ir	12.8	42.5	$1,800 \times 10^{-6}$	4,000	.45
Rh	17.3	15.8	$2,000 \times 10^{-6}$	4,600	.43
Pd	20.3	20.8	$2,300 \times 10^{-6}$	3,700	.62
Ni	28.5	41.0	$5,000 \times 10^{-6}$	6,400	.78
Cr	172.5	62.0	600×10^{-6}	—	—
Ti	67.1	59.9	700×10^{-6}	5,400	.13
Zr	134.0	—	$<100 \times 10^{-6}$	4,400	.02
Mo	99.5	49.0	200×10^{-6}	3,300	.06
Ta	768.0	—	$<100 \times 10^{-6}$	3,100	.03
W	4,390.0	422.5	$<100 \times 10^{-6}$	4,800	.02
Al	.41	.36	$2,800 \times 10^{-6}$	4,300	.65

*From: "Resistance", Research Triangle Institute, *Integrated Silicon Device Technology*, Vol. I, Defense Documentation Center Report #AD408 961, June 1963.

Table 2-55.

MECHANICAL PROPERTIES OF THIN CONDENSED FILMS*

Intrinsic Stresses in Metal Films Approximately 1000-Å Thick

Metal	Substrate temp., °C[a]	Substrate material	Stress,[b] 10^9 dyne/cm^2	Sign[c]	Method[d]
Ag	90	Glass	0.75	T	C
	A	Copper	1.0	T	T
	A	Mica	0.2	T	C
	A	Copper	0.75	T	C
Al	A	Cellulose	1.2	T	ED
	A	Copper	0.1	T	C
Au	A	Cellulose	4.6	T	ED
	A	Quartz	2.9	T	P
	A	Copper	0.85	T	C
Cu	A	Copper	0.9	T	C
	25	Mica	0.2	T	B
	A	Cellulose	4.4	C	ED
	A	Copper	1.5	T	C
	75	Mica	0.1	T	C
	−150	Mica	3.6	T	C
Ni	A	Glass	5–8	T	P
	A	Copper	3.5	T	C
	A	Glass	7.7	T	C
	75	Mica	6.4	T	B
	175	Mica	2.6	T	B
	A	Mica	5–8	T	FR
Fe	75	Mica	10.5[e]	T	B
	175	Mica	5.9[e]	T	B
	75+A	Glass	9.6[e]	T	P
	A	Glass, silica	8.5[f]	T	P
	A	Copper	3.1	T	C
Permalloy	75	Glass	9	T	P
	75	Glass, mica	9	T	C
Sb	A	Copper	0.8	T	C
	A	Copper	0.25	T	C
Co	200	Glass	3.4	T	B
Pd	A	Copper	1.4	T	C
Mg	A	Copper	0		C
Bi	A	Copper	0		C
Zn	A	Copper	0		C
Pb	A	Nickel	0		C
Sn	A	Glass	0		C
In	A	Silica	0		C

[a] A means thermally floating at ambient temperature.
[b] For stress in N/m^2, multiply the value in dyne/cm by 10^{-10}.
[c] T refers to tension, C to compression.
[d] B, beam supported on both ends; C, cantilever beam; ED, electron diffraction; and FR, ferromagnetic resonance.
[e] Poisson-corrected.
[f] Poisson-corrected, 36.5° angle of incidence.

*From: "Physics of Thin Films", G. Hass and R.E. Thun, Eds., Vol. 3, Academic Press, 1966.

Table 2-56.   INTRINSIC STRESSES IN DIELECTRIC FILMS APPROXIMATELY 5000-Å THICK*

Material	Substrate temp., °C[a]	Substrate material	Stress,[b] 10^9 dyne/cm²	Sign[c]	Method[d]
ZnS	110	Glass	1.0	C	C
	A	Glass	(0.022)	C	C
	A	Mica		C	C
SiO	110	Glass	1.2	C	C
	A	Nickel	4	T	C
MgF$_2$	110	Glass	2.0	T	C
	75	Mica	2.2	T	B
	A	Glass	(0.11)	T	C
	A	Mica	(0.11)	T	C
	A	Glass	1	T	
LiF	110	Glass	0.4	T	C
	A	Cellulose	2.0	T	ED
	A	Glass	0.28	T	
	A	Mica	(0.023)	T	C
	A	Glass	(0.023)	T	C
CaF$_2$	110	Glass	0.2	T	C
	A	Mica	(<0.0003)	T	C
	A	Glass	(None)		C
Cryolite	A	Glass	(0.061)	T	C
	A	Glass	(0.06)	T	C
	A	Glass	0.5	T	
PbCl$_2$	50	Glass	0.18	T	C
	A	Glass	(0.014)	T	C
PbF$_2$	110	Glass	0.8	T	C
AgCl	A	Glass	(None)		C
AgF	A	Glass	(None)		C
AgI	A	Glass	(None)		C
BaF$_2$	A	Glass	(0.006)	T	C
BaO	50	Glass	0.15	C	C
Sb$_2$O$_3$	A	Glass	(0.004)	C	C
Sb$_2$S$_3$	A	Glass	(0.007)	T	C
Ce$_2$O$_3$	50	Glass	1.6	C	C
CeF$_3$	40	Glass	2.8	T	C
CdS	110	Glass	0.8	C	C
SnO$_2$	A	Glass	(0.008)	T	C
C	A	Glass	4.0	C	C
NaF	A	Glass	0.1	T	
B$_2$O$_3$	90	Glass	0.1	T	C
Chiolite	A	Glass	(0.029)	T	C
AlPh[e]	40	Glass	0.6	C	C
MgPh[e]	40	Glass	0.6	C	C
MoO$_3$	A	Glass	(0.013)	T	C
CuI	A	Glass	(None)		C
AlF$_3$	A	Glass	(None)		C
SrSO$_4$	A	Glass	(None)		C

[a] A, thermally floating at ambient temperature.
[b] For stress in N/m², multiply the value in dyne/cm by 10^{-10}. Values in parentheses are relative.
[c] C and T, compression or tension.
[d] B, end-supported beam; C, cantilever beam; and ED, electron-diffraction technique.
[e] Al and Mg phthalocyanine.
*From: "Physics of Thin Films", G. Hass and R.E. Thun, Eds., Vol. 3, Academic Press, 1966.

Table 2-57. SPUTTERING RATES FOR METALS USED IN THIN-FILM DEPOSITION*

Sputtering Yields in Atoms/Ion

DIRECT-CURRENT SPUTTERING YIELDS FOR MATERIALS BOMBARDED BY Ar⁺ IONS

Target	Bombarding energy, volts					
	200	600	1000	2000	5000	10 kv
Ag	1.6	3.4				8.8
Al	0.35	1.2				
Au	1.1	2.8				
Co	0.6	1.4				
Cr	0.7	1.3				
Cu	1.1	2.3	3.2	4.3	5.5	6.6
Fe	0.5	1.3	1.4	2.0‡	2.5‡	
Ge	0.5	1.2	1.5	2.0	3.0	
Mo	0.4	0.9	1.1			2.2
Nb	0.25	0.65				
Ni	0.7	1.5	2.1			
Os	0.4	0.95				
Pd	1.0	2.4				
Pt	0.6	1.6				
Re	0.4	0.9				
Rh	0.55	1.5				
Si	0.2	0.5	0.6	0.9	1.4	
Ta	0.3	0.6				
Th	0.3	0.7				
Ti	0.2	0.6				
U	0.35	1.0				
W	0.3	0.6				
Zr	0.3	0.75				
GaSb (111)	0.4	0.9	1.2			
SiC		1.8				

‡Type 304 stainless steel.

DIRECT-CURRENT SPUTTERING YIELDS FOR MATERIALS BOMBARDED BY Ne⁺ IONS

Target	Bombarding energy, volts		
	200	600	1000
Ag	1.0	2.0	
Al	0.2	0.8	
Au	0.6	1.2	
Co	0.4	1.0	
Cr	0.5	1.05	
Cu‡	0.8	2.0 (1.5)	1.8
Fe	0.4	1.0 (0.7)	0.8
Ge	0.3	0.8	
Mo	0.2	0.5 (0.3)	0.5
Nb	0.2	0.4	
Ni	0.5	1.3 (1.1)	1.2
Pd	0.6	1.3	
Pt	0.3	0.7	
Re	0.15	0.4	
Rh	0.4	0.8	
Si	0.0	0.5	
Ta	0.1	0.3	
Ti	0.2	0.45	
U	0.2	0.5	
V	0.2	0.55	
W	0.1	0.3	
Zr	0.2	0.4	
SiC		0.55	

‡2.6 at 5 kev and 3.0 at 10 kev.

DIRECT CURRENT SPUTTERING YIELDS FOR MATERIALS BOMBARDED BY Hg⁺ IONS

Target	Bombarding energy, volts			
	200	400	5 kv	10 kv
Ag	0.85		14	22.5
Al	0.15	0.45		
Au	1.0		13.5	21
C	0.05	0.15		
Co	0.2	0.6	4	6
Cr	0.2	0.8		
Cu	0.55	1.3	8.5	11.5
Fe	0.15	0.5	2.5	4
Ge	0.2	0.6	1.5 (1k)	
Mo	0.1	0.45	2.5	5
Nb	0.1	0.3		
Ni	0.25	0.7	5.5	7
Pd	0.5	1.3	7	9
Pt	0.5	1.4	7.5	10
Rh	0.35	1.0	5	7
Si	0.05	0.15		
Ta	0.2	0.5	2.5	3
Ti	0.1	0.4	2	2.5
V	0.1	0.3	2.5	3.5
W	0.15	0.55	2.5	2.5
Zr	0.15	0.35		

DIRECT-CURRENT SPUTTERING YIELDS FOR MATERIALS BOMBARDED BY Kr⁺ IONS

Target	Bombarding energy, volts		
	200	600	1000
Ag	1.4	3.9	6.4 (1075 volts)
Al	0.3	1.1	
Au	1.1	3.4	
C	0.05	0.2	
Co	0.3	1.3	
Cr	0.6	1.6	
Cu	0.9	2.8	3.4
Fe	0.4	1.2	1.4
Ge	0.4	1.4	
Mo	0.3	1.1	1.2
Nb	0.2	0.7	
Ni	0.5	1.5	1.7
Pd	1.0	2.6	
Pt	0.7	2.1	
Rh	0.5	1.7	
Si	0.1	0.6	
Ta	0.3	1.0	
Ti	0.2	0.5	
W	0.4	1.1	
Zr	0.2	0.7	
SiC		1.6	

Table 2-57.   SPUTTERING RATES FOR METALS
USED IN THIN-FILM DEPOSITION (Continued)

**DIRECT-CURRENT SPUTTERING YIELDS
FOR MATERIALS BOMBARDED BY Xe⁺ IONS**

Target	Bombarding energy, volts		
	200	600	1000
Ag	1.1	4.2	
Al	0.2	1.0	
Au	1.0	3.1	
C	0.04	0.2	
Co	0.4	1.3	
Cr	0.4	1.9	
Cu	0.8	2.4	3.6
Fe	0.3	1.2	1.8
Ge	0.3	1.2	1.6
Mo	0.3	1.1	1.6
Nb	0.2	0.6	
Ni	0.4	1.5	2.2
Pd	0.9	2.5	
Pt	0.7	2.2	
Rh	0.5	1.6	
Si	0.08	0.5	
Ta	0.3	1.0	
Ti	0.1	0.5	
W	0.4	1.2	
Zr	0.2	0.7	
SiC		1.6	

**DIRECT-CURRENT SPUTTERING YIELDS
FOR MATERIALS BOMBARDED BY He⁺ IONS**

Target	Bombarding energy, volts	
	200	600
Ag	0.08	0.2
Al	0.005	0.02
Au	0.02	0.08
C	0.02	0.09
Co	0.04	0.2
Cr	0.07	0.2
Cu	0.1	0.3
Fe	0.07	0.2
Ge	0.03	0.08
Mo	0.005	0.04
Nb	0.005	0.03
Ni	0.06	0.2
Pd	0.06	0.2
Pt	0.004	0.04
Rh	0.02	0.07
Si	0.05	0.2
Ta	0.002	0.01
Ti	0.04	0.08
W	0.001	0.008
SiC		0.16

**DIRECT-CURRENT SPUTTERING YIELDS
FOR MATERIALS BOMBARDED BY N₂⁺ IONS**

Target	Bombarding energy, kv		
	1	3	8
Cu	2.0	3.8	5.1
Fe	0.75	1.5	1.75
Mo	0.3	0.6	0.8
Ni	1.1	2.0	2.2
W	0.2	0.5	0.7

**SPUTTERING YIELDS
FOR MATERIALS BOMBARDED BY N⁺ IONS**

Target	Bombarding energy, kv		
	1	3	8
Cu	1.5	2.5	2.5
Fe	0.6	1.0	1.2
Mo	0.15	0.35	0.4
Ni	0.7	1.2	1.1
W	0.15	0.3	0.45

**SPUTTERING YIELDS IN MOLECULES PER ION
FOR INSULATORS BOMBARDED BY ARGON
USING RF**

Target	Mean ion energy, kv		
	1.1	2.0	2.9
SiO₂	0.16	0.39	0.50
Pyrex 7740	0.15	0.33	0.43
Al₂O₃	0.05	0.12	0.17

*From: "Deposition of Thin Films by Cathodic Sputtering", L.I. Maissel, *Physics of Thin Films*, G. Hass and R.E. Thun, Eds., Vol. 3, Academic Press, 1966.

Table 2-58.   ETCHANTS FOR SEMICONDUCTORS*

The compositions given in the table, with obvious exceptions, are reduced to integral proportions of liquid components in parts by volume. The percentage concentration of the reagents correspond to commercial ones, as follows:

HF 48	H₂O₂ 30	NH₄OH 30
HNO₃ 70	HCl 36	CH₃OH Absolute
Fuming HNO₃ 90	H₂SO₄ 97	CH₃COOH Glacial

$$HF\ 48 \qquad H_2O_2\ 30 \qquad NH_4OH\ 30$$
$$HNO_3\ 70 \qquad HCl\ 36 \qquad CH_3OH\ Absolute$$
$$Fuming\ HNO_3\ 90 \qquad H_2SO_4\ 97 \qquad CH_3COOH\ Glacial$$

No.	Material	Etchant	Conditions	Remarks
1	Ge	Iodine etch A 5 HF 10 HNO₃ 11 CH₃COOH with 30 mg I₂ dissolved	4 min	Polishing
2		CP4 15 HF 25 HNO₃ 15 CH₃COOH with 0.3 Br₂ dissolved	2 min	Chemical polishing
3		Dash Ge etch 2 HF 4 HNO₃ 15 CH₃COOH	20 sec to 1 min	Reveals dislocations after precipitation of lithium
4		No. 2 (Superoxol) 1 H₂O₂ 1 HF 4 H₂O	1–3 min	Develops etch figures
5		WAg 4 ml HF 2 ml HNO₃ 4 ml H₂O 0.2 g AgNO₃		(111) dislocation etch
6		Ferricyanide 100 ml deionized H₂O 9.7 g K₃Fe(CN)₆ 13.7 g KOH	1–10 min, 80°C	(111) dislocation etch (110)
7		(100) etch 75 ml HF 25 ml HNO₃ 50 ml H₂O 3 g Cu (NO₃)₂		(100) dislocation etch
8		Peroxide 1 ml HF 1 ml H₂O₂ (30%) 1 ml acetic acid	1 min	(100) dislocation etch
9	Si	Dash etch 1 ml HF 3 ml HNO₃ 10 ml CH₃COOH	1–16 hr	Yields deep dislocation pits (110) (100) (112)
10		CP4 3 HF 5 HNO₃ 3 CH₃COOH	2–3 min	Chemical polishing
11		NaOH or KOH, 1–30% solution	1–5 min, 50–100°C	Develops etch figures
12		Sirtl 1 g CrO₃ 2 ml H₂O Mix with 1 ml HF just before using	1–7 min	(111) dislocation etch
13		Copper etch 24 g copper nitrate 2.4 g bromine 600 ml HF 300 ml HNO₃ Mix 1 part to 10 parts deionized water to use	2 hr, ultrasonic	>0.02 ohm-cm (111) dislocation etch

*From: H.C. Gatos and M.C. Lavine, Chemical Behavior of Semiconductors: Etching Characteristics, "Progress in Semiconductors", A.F. Gibson and R.E. Burgess, Eds., Vol. 9, Temple Press Books Limited (CRC Press, U.S. distributor), 1965, pp. 1–46; and H.C. Gatos and M.C. Lavine, Chemical Behavior of Semiconductors: Etching Characteristics, *Lincoln Lab. Tech. Rpt.*, 293, January 1963.

Table 2-58. ETCHANTS FOR SEMICONDUCTORS (Continued)

No.	Material	Etchant	Conditions	Remarks
14	AlSb	1 HF 1 H_2O_2 followed by 1 HCl 1 H_2O followed by 1 HNO_3	1 min and 25 sec, respectively	Differentiates A and B surfaces
15	GaP	Cl_2 bubbled slowly through CH_3OH	CH_3OH is saturated with Cl_2, then sample immersed while gas is bubbling at least 20 min	Polishing
16		2 HNO_3 1 HCl	1–2 min	Polishing of some orientations
17	GaAs	5–20% Br_2 in CH_3OH		Polishing
18		1 HF 1 H_2O_2 1 H_2O	10 min	Differentiates A and B surfaces
19		1 HF 3 HNO_3 2 H_2O		Chemical polishing
20		2 HCl 1 HNO_3 2 H_2O		Etches A and B surfaces

21		HCl–HNO_3
		  2 ml HCl
		  1 ml HNO_3
		  2 ml deionized H_2O
			10 min	Mix fresh—shows pits on Ga side (111) dislocation etch

22		2 ml H_2O 8 mg $AgNO_3$ 1 g CrO_3 1 ml HF	10 min, 65°C	(100), Ga(111), (110), As(111), dislocation etch
23	GaSb	1 Br_2 10 CH_3OH	20 sec	Polishing; shallow pits on A surfaces

24		Modified CP4
		  2 HNO_3
		  1 HF
		  1 CH_3COOH
			15 sec	Chemical polishing; pits on A surfaces

25		1 H_2O_2 1 HCl 2 H_2O	1 min	Develops etch figures
26		0.2 M $Fe_2(SO_4)_3$ in conc. HCl	10 min, 60°C	Develops etch figures
27	InP	1 Br_2 10 CH_3OH	20 sec	Polishing
28		0.1 M $Fe_2(SO_4)_3$ in conc. HCl	1–5 min	Develops etch figures
29	InAs	15 HF 75 HNO_3 15 CH_3COOH 0.06 Br_2	5 sec	Etch pits on A surfaces
30		1 H_2SO_4 1 H_2O_2 1 H_2O	1 min at heat of mixing	Etch pits on A surfaces
31		HCl		Develops etch figures
32		Bromine–methanol 5–10% bromine in methanol by volume	13–30 sec	(111)

33		Silver nitrate–chromic acid
		  2 ml deionized H_2O
		  8 mg $AgNO_3$
		  1 g CrO_3
		  1 ml H
			10 min	(111)
				(110)
				(100)

Table 2-58.   ETCHANTS FOR SEMICONDUCTORS (Continued)

No.	Material	Etchant	Conditions	Remarks

34	InSb	Modified CP4
		  2 HF
		  1 HNO$_3$
		  1 CH$_3$COOH
			5–30 sec	Chemical polishing; pits on A surfaces

35		0.2 N ferric ion in 6 N HCl	30 min, 80°C	Develops etch figures
36		I$_2$ in CH$_3$OH		Polishing
37		1 HF 1 H$_2$O$_2$ 8 H$_2$O 0.4% n-butylthiobutane		Develops β-dislocation etch pits on A and B surfaces
38	ZnS	0.5 N K$_2$Cr$_2$O$_7$ in 16 M H$_2$SO$_4$	Natural cryst. 10 min, 95°C	Etch pits on A surfaces
39	ZnTe	3HF 2 H$_2$O$_2$ 1 H$_2$O	2 min	Preferential; differentiates A and B surfaces
40		3 HNO$_3$ 4 HF	8 sec; rinse in conc. HCl, then dist. H$_2$O	Chemical polishing
41	CdS	6 fuming HNO$_3$ 6 CH$_3$COOH 1 H$_2$O	2 min	Sharply defined hexagonal pits on B surfaces
42		Same as etchant No. 38	5–10 min, 95°C	Polishing with dislocation etch pits on B surfaces; shallow dishing on A surfaces
43		1 H$_2$SO$_4$ 100 H$_2$O 0.08 g Cr$_2$O$_3$	10 min, 80°C	Difference in pit formation on A and B surfaces
44	CdSe	30 HNO$_3$ 0.1 HCl 20 18 N H$_2$SO$_4$	8 sec, 40°C; rinse in conc. H$_2$SO$_4$ to dissolve Se film	A surfaces develop hexagonal pits
45		1 HNO$_3$ 3 HCl		Sharply bevelled pits on A surfaces
46	CdTe	Same as etchant No. 39	As in etchant No. 39	Etch pits on B surfaces; longer times will polish A surfaces

47		'E'
		  10 ml HNO$_3$
		  20 ml H$_2$O
		  4 g K$_2$Cr$_2$O$_7$
				Chemical polishing

48		EAg-1
		  10 ml 'E' Solution
		  0.5–10 mg AgNO$_3$
				Etch pit formation

49		7 sat. K$_2$Cr$_2$O$_7$ 3 H$_2$SO$_4$	2–3 min; rinse in H$_2$O, then in boiling 10% NaOH + 10%NaHSO$_3$, then in H$_2$O and in EDTA	
50	HgSe	50 HNO$_3$ 10 CH$_3$COOH 1 HCl 20 18 N H$_2$SO$_4$	10–15 min, 40°C	Polishing
51		6 HCl 3 HNO$_3$ 3 H$_2$O	Start with chemically polished surf. 5 min; film removed by immersing briefly in etchant No. 50, then brushing under H$_2$O; repeat process	B surfaces develop triangular figures
52	HgTe	6 HNO$_3$ 1 HCl 1 H$_2$O	10–15 min	Polishing
53		1 HCl 1 HNO$_3$	Start with chemically polished surf., three 1-min etchings with H$_2$O rinsing in between	Pits on A surfaces with background figures
54	CdTe and HgTe alloys	6 HNO$_3$ 1 HCl	Rinse in 1 HCl, 1 CH$_3$OH	Polishes CdTe (0.05)–HgTe (0.95); results vary with composition of alloy

Table 2-58. ETCHANTS FOR SEMICONDUCTORS *(Continued)*

No.	Material	Etchant	Conditions	Remarks
55	SiC (α and β)	Fused KOH (or NaOH)	2 min, 900°C	Dislocation etch
56		Fused borax	2 min, 1 000°C; remove excess borax by washing with NaOH	
57		Fused Na_2O_2	350–900°C	Develops etch figures
58	Ag_2Se	5 H_2SO_4 1 H_2O_2	5 min, 50°C; rinse in EDTA solution, then in dist. H_2O	Polishing of some orientations
59		2 sat. KOH 1 H_2O_2 2 ethylene glycol	Remove damage with etchant No. 58, then 2 min at 80°C	Polishing
60	Ag_2Te	3NH_4OH 2 H_2O_2	Film removed by brushing under water	Develops grain boundaries; polishing of some orientations
61	In_2Te_3	19 CH_3COOH sat. with citric acid	Rinse away etchant with excess of first component, followed by water wash	Chemical polishing
62	In–Sb–Te alloys	1 HF 3 HNO_3 4 CH_3COOH sat. with citric acid		Chemical polishing
63	Bi_2Se_3	1 H_2O_2 1 HCl		Chemical polishing; this etchant blackens Bi_2Te_3
64	Bi_2Te_3	1 HNO_3 1 HCl 2 sat. $K_2S_2O_8$		Reveals dislocations; also effective for $BiSbTe_3$
65	PbS	30 HCl 10 HNO_3 1 CH_3COOH	50°C for a few min, then rinse with 10% CH_3COOH	Chemical polishing
66		1 HCl 3–10% thiourea	1–10 min, 60°C	Reveals dislocations
67	PbSe	2–15% NaOH 1 sat. $Na_2S_2O_8$	10 min	Reveals dislocations
68		5 sat. KOH 1 H_2O_2 5 ethylene glycol	3 min, 40°C	Polishing
69	PbTe	2–15% NaOH 1 sat. $Na_2S_2O_8$	10 min	Reveals dislocations
70		Iodine etch 10 H_2O 5 g NaOH Add 0.2 I_2	5 min, 95°C	
71	SnTe	1–15% NaOH 1 sat. $Na_2S_2O_8$	10 min; rinse in dilute H_2O_2, then H_2O	Cleans; polishes some orientations
72	TlSe	1 HNO_3 1 HCl	1 min; rinse in H_2O, then in EDTA	Cleans
73	$MnTiO_3$	1 H_2SO_4 1 H_2O_2 1 H_2O	5 min, 60°C	Cleans

REFERENCE

For additional tabular data on etchants and etching, see "Handbook of Materials and Processes for Electronics", C.A. Harper, Ed., McGraw-Hill Book Company, 1970.

Table 2-59. PROPERTIES OF SEMICONDUCTORS*

COMPILED BY BRIAN RANDALL PAMPLIN

The term *semiconductor* is applied to a material in which electric current is carried by electrons or holes; its electrical conductivity when extremely pure rises exponentially with temperature and may be increased from this low "intrinsic" value by many orders of magnitude by "doping" with electrically active impurities.

Semiconductors are characterized by an energy gap in the allowed energies of electrons in the material that separates the normally filled energy levels of the *valence band* (where "missing" electrons behave as positively charged current carriers or "holes") and the *conduction band* (where electrons behave somewhat as a gas of free negatively charged carriers with an effective mass dependent on the material and the direction of the electrons' motion). This energy gap depends on the nature of the material and varies with direction in anisotropic crystals. It is slightly dependent on temperature and pressure, and this dependence is usually almost linear at normal temperatures and pressures.

Table A. GENERAL PROPERTIES OF SEMICONDUCTORS

Listed by Crystal Structure

Substance	Lattice parameters at room temperature, Å	Density, g/cc	Melting point, °K	Minimum room temperature energy gap, ev	Comparative thermal conductivity	Heat of formation, kcal/mole	Mobility (room temperature) Electrons cm²/volt-sec	Mobility (room temperature) Holes cm²/volt-sec	Remarks
PART A. TETRAHEDRAL SEMICONDUCTORS									
§ A1 Diamond Structure Elements (Strukturbericht symbol A4, Space Group Fd3m-O_h^7)									
C	3.5597	3.51	4300	5.4	2000	161	1800	1400	
Si	5.43072	2.3283	1685	1.107	1240	77.5	1900	500	
Ge	5.65754	5.3234	1231	0.67	640	69.5	3800	1820	
α-Sn	6.4912	5.765	503	0.08		64	2500	2400	
§ A2 Sphalerite (Zinc Blende) Structure Compounds (Strukturbericht symbol B3 Space Group F$\bar{4}$3m-T_d^2)									
I-VII Compounds									
CuF	4.255								
CuCl	5.4057	3.53	695						
CuBr	5.6905	4.72	770	2.94		115			
CuI	6.0427	5.63	878			105			
AgBr						102	4000		
AgI	6.473	5.67				93	30		
II-VI Compounds									
BeS	4.865	2.36							
BeSe	5.139	4.315							
BeTe	5.626	5.090							
BePo	8.38	7.3							
ZnO	4.63								See § A3
ZnS	5.4093	4.079	1920	3.54	140	114	180	5(400°C)	See also § A3
ZnSe	5.6676	5.42	1790	2.58	140	101	540	28	
ZnTe	6.101	5.72	1510	2.26	140	90	340	100	
ZnPo									
CdS	5.5818								See also § A3
CdSe	6.05								See § A3
CdTe	6.477	5.86	1370	1.44	55	81	1200	50	
CdPo									
HgS	5.8517	7.73	~2020						
HgSe	6.084	8.25	1070	0.30	10	59	20000		
HgTe	6.429	8.17	943	0.15	20	58	25000	350	
III-V Compounds									
BN	3.615	3.49	3000	~4	200	195			
BP(L.T.)	4.538	2.9		~6			500	70	
BAs	4.777								
AlP	5.451	2.85	1770	2.5					
AlAs	5.6622	3.81	1870	2.16		150	1200	420	
AlSb	6.1355	4.218	1330	1.60	600	140	200–400	550	
GaP	5.4505	4.13	1750	2.24	1100	152	300	100	
GaAs	5.65315	5.316	1510	1.35	370	128	8800	400	
GaSb	6.0954	5.619	980	0.67	270	118	4000	1400	
InP	5.86875	4.787	1330	1.27	800	134	4600	150	
InAs	6.05838	5.66	1215	0.36	290	114	33000	460	
InSb	6.47877	5.775	798	0.165	160	107	78000	750	

*Reprinted from: "Handbook of Chemistry and Physics", 53rd ed., R.C. Weast, Ed., The Chemical Rubber Co., 1972.

Table 2-59. PROPERTIES OF SEMICONDUCTORS *(Continued)*

Table A. GENERAL PROPERTIES OF SEMICONDUCTORS *(Continued)*

Substance	Lattice parameters at room temperature, Å		Density, g/cc	Melting point, °K	Minimum room temperature energy gap, ev	Comparative thermal conductivity	Heat of formation, kcal/mole	Mobility (room temperature)		Remarks
								Electrons	Holes	
								cm²/volt-sec		
Other Sphalerite Structure Compounds										
β-SiC	4.348		3.21	3070	2.3			4000		
Ga_2Te_3	5.899		5.75	1063	~1.0	~14	65			
In_2Te_3(H.T.)	6.150		5.8	940	~1.0	~8	47.4	~10		
$MgGeP_2$	5.652									
$ZnSnP_2$	5.65				2.1					
$ZnSnAs_2$(H.T.)	5.851		5.53	1050	~0.7	70				

§ A3 Wurtzite (Zincite) Structure Compounds (Strukturbericht symbol B4, Space Group P 6_3 mc-C_{6v}^4)

I-VII Compounds

Substance										
CuCl	3.91	6.42		T_c 680°K						
CuBr	4.06	6.66		T_c 658°K						
CuI	4.31	7.09								
AgI	4.580	7.494			2.63					
II-VI Compounds										
BeO	2.698	4.380		2800						
MgTe	4.54	7.39	3.85	~2800						
ZnO	3.24950	5.2069	5.66	2250	3.2	6	154	180		
ZnS	3.8140	6.2576	4.1	2100	3.67		110			
ZnSe	3.996	6.626								
ZnTe	4.27	6.99								
CdS	4.1348	6.7490	4.82	2020	2.42		96	400		
CdSe	4.299	7.010	5.66	1530	1.74		90	650		
CdTe	4.57	7.47			1.50					
III-V Compounds										
BP(H.T.)	3.562	5.900								
AlN	3.111	4.978	3.26	~2500	6.02		197			
GaN	3.180	5.166	6.10	1500	3.34		157			
InN	3.533	5.693	6.88	1200	2.0		133			
Other Wurtzite Structure Compounds										
SiC	3.076	5.048								
MnTe	4.078	6.701			~1.0					
Al_2S_3	3.579	5.829	2.55		4.1		426			
Al_2Se_3	3.890	6.30	3.91		3.1		367			

§ A4 Chalcopyrite Structure Compounds (Strukturbericht symbol $E1_1$, Space Group I $\bar{4}$ 2d $-D_{2d}^{12}$)

I-III-VI₂ Compounds

Substance										
$CuAlS_2$	5.323	10.44	3.47		2.5					
$CuAlSe_2$	5.617	10.92	4.70	1270	1.1					
$CuAlTe_2$	5.976	11.80	5.50	1160	0.88					
$CuGaS_2$	5.360	10.49	4.35							
$CuGaSe_2$	5.618	11.01	5.56	1310	0.96, 1.63					
$CuGaTe_2$	6.013	11.93	5.99	1150	0.82, 1.0					
$CuInS_2$	5.528	11.08	4.75		1.2					
$CuInSe_2$	5.785	11.57	5.77	1250	0.86, 0.92	37				
$CuInTe_2$	6.179	12.365	6.10	970	0.95	49				
$CuTlS_2$	5.580	11.17	6.32							
$CuTlSe_2$(L.T.)	5.844	11.65	7.11	680	1.07					
$CuFeS_2$	5.25	10.32		1150	0.53					
$CuFeSe_2$				850	0.16					
$CuLaS_2$	5.65	10.86								
$AgAlS_2$	5.707	10.28	3.94							
$AgAlSe_2$	5.968	10.77	5.07	1220	0.7					
$AgAlTe_2$	6.309	11.85	6.18	1000	0.56					
$AgGaS_2$	5.755	10.28	4.72							
$AgGaSe_2$	5.985	10.90	5.84	1120	1.66					
$AgGaTe_2$	6.301	11.96	6.05	990	1.1	10				
$AgInS_2$(L.T.)	5.828	11.19	5.00		1.9					
$AgInSe_2$	6.102	11.69	5.81	1053	1.18	30				
$AgInTe_2$	6.42	12.59	6.12	965	0.96, 0.52					
$AgFeS_2$	5.66	10.30	4.53							

Table 2-59. PROPERTIES OF SEMICONDUCTORS *(Continued)*

Table A. GENERAL PROPERTIES OF SEMICONDUCTORS *(Continued)*

Substance	Lattice parameters at room temperature, Å		Density, g/cc	Melting point, °K	Minimum room temperature energy gap, ev	Comparative thermal conductivity	Heat of formation, kcal/mole	Mobility (room temperature) Electrons Holes cm²/volt-sec		Remarks
II-IV-V₂ Compounds										
ZnSiP₂	5.400	10.441	3.39	1640	2.3			1000		
ZnGeP₂	5.465	10.771	4.17	1295	2.2					
CdSiP₂	5.678	10.431	4.00	~1470	2.2			1000		
CdGeP₂	5.741	10.775	4.48	~1060	1.8					
CdSnP₂	5.900	11.518			1.5					
ZnSiAs₂	5.61	10.88	4.70	~1350	1.7				50	
ZnGeAs₂	5.672	11.153	5.32	~1150	0.85	110				
ZnSnAs₂	5.8515	11.704	5.53	~ 910	0.65	150			300	Disorders at 910°K
CdSiAs₂	5.884	10.882								
CdGeAs₂	5.9427	11.2172	5.60	~ 903	0.53	40		70	25	Disorders at 903°K
CdSnAs₂	6.0944	11.9182	5.72	880	0.26	70		22000	250	

§ A5 "Defect Chalcopyrite" Structure Compounds (Strukturbericht symbol E3, Space Group I 4̄ − S₄²)

Substance	Lattice parameters		Density, g/cc	Melting point, °K	Minimum energy gap, ev	Comparative thermal conductivity	Heat of formation	Electrons	Holes	Remarks
ZnAl₂Se₄	5.503	10.90	4.37							
ZnAl₂Te₄(?)	5.104	12.05	4.95							
ZnGa₂S₄(?)	5.274	10.44	3.80							
ZnGa₂Se₄(?)	5.496	10.99	5.21							
ZnGa₂Te₄(?)	5.937	11.87	5.67		1.35					
ZnIn₂Se₄	5.711	11.42	5.44	1250	2.6					
ZnIn₂Te₄	6.122	12.24	5.83	1075	1.2					
CdAl₂S₄	5.564	10.32	3.06							
CdAl₂Se₄	5.747	10.68	4.54							
CdAl₂Te₄(?)	6.011	12.21	5.10							
CdGa₂S₄	5.577	10.08	4.03							
CdGa₂Se₄	5.743	10.73	5.32							
CdGa₂Te₄	6.093	11.81	5.77							
CdIn₂Te₄	6.205	12.41	5.9	1060	(1.26 or 0.9)	4000				
HgAl₂S₄	5.488	10.26	4.11							
HgAl₂Se₄	5.708	10.74	5.05							
HgAl₂Te₄(?)	6.004	12.11	5.81							
HgGa₂S₄	5.507	10.23	5.00							
HgGa₂Se₄	5.715	10.78	6.18							
HgIn₂Se₄	5.764	11.80	6.3	1100	0.6					
HgIn₂Te₄(?)	6.186	12.37	6.3	980	0.86		200			

§ A6 Other Tetrahedral Compounds

Substance	Lattice parameters		Density, g/cc	Melting point, °K	Minimum energy gap, ev	Comparative thermal conductivity	Heat of formation	Electrons	Holes	Remarks
α-SiC	3.0817 15.1183		3.21	3070	2.86			400		6H structure
Hg₅Ga₂Te₈	6.235									B3 with super lattice
Hg₅In₂Te₈	6.328				0.7			2000		B3 with super lattice

PART B. OCTAHEDRAL SEMICONDUCTORS

Halite Structure Semiconductors (Strukturbericht symbol B1, Space Group Fm3m − 0ₕ⁵)

Substance	Lattice parameters		Density, g/cc	Melting point, °K	Minimum energy gap, ev	Comparative thermal conductivity	Heat of formation	Electrons	Holes	Remarks
SnSe	6.020			1133						
SnTe	6.313			1080 (max)	0.5	91				
PbS	5.9362		7.61	1390	0.37	23	104	600	600	
PbSe	6.1243		8.15	1340	0.26	17	94	1000	900	
PbTe	6.454		8.16	1180	0.25	23	94	1600	600	
Selected Other Binary Chalcides										
BiSe	5.99		7.98	880	0.4					
BiTe	6.47									
EuSe	6.191			2300	1.8	2.4				
GdSe	5.771			2400						
NiD	4.1684		6.6	2260	2.0 or 3.7			4		
CdO	4.6953			1700	2.5	7	127	100		
SrS	6.0199		3.643	3000	4.1					

Table 2-59. PROPERTIES OF SEMICONDUCTORS *(Continued)*

Table A. GENERAL PROPERITES OF SEMICONDUCTORS *(Continued)*

Substance	Lattice parameters at room temperature, Å	Density, g/cc	Melting point, °K	Minimum room temperature energy gap, ev	Comparative thermal conductivity	Heat of formation, kcal/mole	Mobility (room temperature) Electrons	Holes	Remarks
Selected Ternary Compounds									
AgSbSe$_2$	5.786	6.60	910	0.58	10.5				
AgSbTe$_2$ (or	6.078	7.12	830	0.7, 0.27	8.6, 0.3				
Ag$_{19}$Sb$_{29}$Te$_{52}$)									
AgBiS$_2$(H.T.)	5.648								
AgBiSe$_2$(H.T.)	5.82								
AgBiTe$_2$(H.T.)	6.155								

Table B. SEMICONDUCTING PROPERTIES OF SELECTED MATERIALS

Substance	Minimum energy gap, ev — Room temperature	0°K	$\frac{dE_g}{dT}$ $\times 10^4$ ev/°C	$\frac{dE_g}{dP}$ $\times 10^6$ ev cm^2/kg	Density of states electron effective mass, m_{d_n} (m_o)	Electron mobility and temperature dependence — μ_n cm^2/volt-sec	$-x$	Density of states hole effective mass, m_{d_p} (m_o)	Hole mobility and temperature dependence — μ_p cm^2/volt-sec	$-x$	Dominant emission wavelength, 77°K, microns
Si	1.107	1.153	−2.3	−2.0	0.58	1,900	2.6	1.06	500	2.3	1.274
Ge	0.67	0.744	−3.7	+7.3	0.35	3,800	1.66	0.56	1,820	2.33	1.770
α-Sn	0.08	0.094	−0.5		0.02	2,500	1.65	0.3	2,400	2.0	—
Te	0.33				0.68	1,100		0.19	560		—
III-V Compounds											
AlAs	2.2	2.3				1,200			420		—
AlSb	1.6	1.7	−3.5	−1.6	0.09	200	1.5	0.4	500	1.8	—
GaP	2.24	2.40	−5.4	−1.7	0.35	300	1.5		150	1.5	0.59
GaAs	1.35	1.53	−5.0	+9.4	0.068	9,000	1.0	0.5	500	2.1	0.84
GaSb	0.67	0.78	−3.5	+12	0.050	5,000	2.0	0.23	1,400	0.9	1.6
InP	1.27	1.41	−4.6	+4.6	0.067	5,000	2.0		200	2.4	0.91
InAs	0.36	0.43	−2.8	+8	0.022	33,000	1.2	0.41	460	2.3	3.1
InSb	0.165	0.23	−2.8	+15	0.014	78,000	1.6	0.4	750	2.1	5.2
II-VI Compounds											
ZnO	3.2		−9.5	+0.6	0.38	180	1.5				0.37
ZnS	3.54		−5.3	+5.7		180			5 (400°C)		0.33
ZnSe	2.58	2.80	−7.2	+6		540			28		—
ZnTe	2.26			+6		340			100		—
CdO	2.5±.1		−6		0.1	120					—
CdS	2.42		−5	+3.3	0.165	400		0.8			0.49
CdSe	1.74	1.85	−4.6		0.13	650	1.0	0.6			0.68
CdTe	1.44	1.56	−4.1	+8	0.14	1,200		0.35	50		0.78
HgSe	0.30				0.030	20,000	2.0				—
HgTe	0.15		−1		0.017	25,000		0.5	350		—
Halite Structure Compounds											
PbS	0.37	0.28	+4		0.16	800		0.1	1,000	2.2	4.3
PbSe	0.26	0.16	+4		0.3	1,500		0.34	1,500	2.2	8.5
PbTe	0.25	0.19	+4	−7	0.21	1,600		0.14	750	2.2	6.5

Table 2-59. PROPERTIES OF SEMICONDUCTORS (Continued)
Table B. SEMICONDUCTING PROPERTIES OF SELECTED MATERIALS (Continued)

Substance	Minimum energy gap, ev		$\frac{dE_g}{dT}$ $\times 10^4$ ev/°C	$\frac{dE_g}{dP}$ $\times 10^6$ ev cm²/kg	Density of states electron effective mass, m_{d_n} (m_o)	Electron mobility and temperature dependence		Density of states hole effective mass, m_{d_p} (m_o)	Hole mobility and temperature dependence		Dominant emission wavelength, 77°K, microns
	Room temperature	0°K				μ_n cm²/volt-sec	$-x$		μ_p cm²/volt-sec	$-x$	
Others											
ZnSb	0.50	0.56			0.15	10					1.5
CdSb	0.45	0.57	−5.4		0.15	300			2,000	1.5	
Bi₂S₃	1.3					200			1,100		
Bi₂Se₃	0.27					600			675		
Bi₂Te₃	0.13		−0.95		0.58	1,200	1.68	1.07	510	1.95	
Mg₂Si		0.77	−6.4		0.46	400	2.5		70		
Mg₂Ge		0.74	−9			280	2		110		
Mg₂Sn	0.21	0.33	−3.5		0.37	320			260		
Mg₃Sb₂		0.32				20			82		
Zn₃As₂	0.93					10	1.1		10		
Cd₃As₂	0.13				0.046	15,000	0.88				
GaSe	2.05		3.8						20		
GaTe	1.66	1.80	−3.6			14	−5				
InSe	1.8					900					
TlSe	0.57		−3.9		0.3	30		0.6	20	1.5	
CdSnAs₂	0.23				0.05	25,000	1.7				
Ga₂Te₃	1.1	1.55	−4.8								
α-In₂Te₃	1.1	1.2			0.7				50	1.1	
β-In₂Te₃	1.0								5		
Hg₅In₂Te₀	0.5								11,000		
SnO₂									78		

Table C. VALENCE BANDS OF SEMICONDUCTORS
Room Temperature Data

Substance	Band curvature effective mass			Energy separation of "split-off" band, ev	Measured (light) hole mobility, cm²/volt-sec
	Heavy holes	Light holes	"Split-off" band holes		
	(Expressed as fraction of free electron mass)				

SEMICONDUCTORS WITH VALENCE BAND MAXIMUM AT CENTER OF BRILLOUIN ZONE ("Γ")

Substance	Heavy holes	Light holes	"Split-off" band holes	Energy separation of "split-off" band, ev	Measured (light) hole mobility, cm²/volt-sec
Si	0.52	0.16	0.25	0.044	500
Ge	0.34	0.043	0.08	0.3	1,820
Sn	0.3				2,400
AlAs					
AlSb	0.4			0.7	550
GaP				0.13	100
GaAs	0.8	0.12	0.20	0.34	400
GaSb	0.23	0.06		0.7	1,400
InP				0.21	150
InAs	0.41	0.025	0.083	0.43	460
InSb	0.4	0.015		0.85	750
CdTe	0.35				50
HgTe	0.5				350

SEMICONDUCTORS WITH MULTIPLE VALENCE BAND MAXIMA

Substance	Number of equivalent valleys and direction	Curvature effective masses		Anisotropy, $\frac{m_L}{K = m_T}$	Measured (light) hole mobility, cm²/volt-sec
		Longitudinal, m_L	Transverse, m_T		
PbSe	4 "L" [111]	0.095	0.047	2.0	1,500
PbTe	4 "L" [111]	0.27	0.02	10	750
Bi₂Te₃	6	0.207	~0.045	4.5	515

Table 2-59.   PROPERTIES OF SEMICONDUCTORS *(Continued)*
Table D. CONDUCTION BANDS OF SEMICONDUCTORS
Room Temperature Data

SINGLE VALLEY SEMICONDUCTORS

Substance	Energy gap, ev	Effective mass, m_o	Mobility, $cm^2/volt\text{-}sec$	Comments
GaAs	1.35	0.067	8,500	3(or 6?) equivalent [100] valleys 0.36 ev above this maximum with a mobility of ~50.
InP	1.27	0.067	5,000	3(or 6?) equivalent [100] valleys 0.4 ev above this minimum.
InAs	0.36	0.022	33,000	Equivalent valleys ~1.0 ev above this minimum.
InSb	0.165	0.014	78,000	
CdTe	1.44	0.11	1,000	4(or 8?) equivalent [111] valleys 0.51 ev above this minimum.

MULTIVALLEY SEMICONDUCTORS

Substance	Energy gap	Number of equivalent valleys and direction	Band curvature effective mass		Anistropy, $\dfrac{m_L}{K = m_T}$
			Longitudinal, m_L	Transverse, m_T	
Si	1.107	6 in [100] "Δ"	0.90	0.192	4.7
Ge	0.67	4 in [111] at "L"	1.588	0.0815	19.5
GaSb	0.67	as Ge (?)	~1.0	~0.2	~5
PbSe	0.26	4 in [111] at "L"	0.085	0.05	1.7
PbTe	0.25	4 in [111] at "L"	0.21	0.029	5.5
Bi_2Te_3	0.13	6			~0.05

Table 2-60.   ELASTICITY PARAMETERS FOR SOME CUBIC CRYSTALS*

SYMBOLS: γ_t = transverse frequencies γ_e = longitudinal frequencies c = velocity of light

Material	Stiffness constants in 10^{12} dynes/cm^2 at 300°K			Debye temperature, °K, (referred to 0°K)	Transverse and longitudinal lattice vibration frequencies		KNOOP hardness, kg/mm^2
	C_{11}	C_{12}	C_{44}		γ_t/c	γ_e/c	
Diamond	10.76	1.25	5.76	—	—	—	7000
Na	0.073	0.062	0.042	156	—	—	—
Li	0.135	0.114	0.088	335	—	—	—
Ge	1.285	0.483	0.680	374	—	—	700–880
Si	1.66	0.639	0.796	648	—	—	1100–1400
GaSb	0.885	0.404	0.433	266	—	—	450
InSb	0.672	0.367	0.302	203	185	197	220
MgO	2.86	0.87	1.48	—	—	—	370
NaCl	0.487	0.124	0.126	—	—	—	—
RbBr	0.317	0.042	0.039	—	—	—	—
RbI	0.256	0.031	0.029	—	—	—	—
CsBr	0.300	0.078	0.076	—	—	—	—
CsI	0.246	0.067	0.062	—	—	—	—
InP	—	—	—	322	307	351	540
InAs	0.833	0.453	0.396	249	219	243	380
AlSb	0.894	0.443	0.416	292	318	345	360
AlP	—	—	—	588	—	—	—
AlAs	—	—	—	417	—	—	—
GaAs	1.19	0.538	0.598	344	373	297	750
GaP	—	—	—	435	—	—	950
GaSb	8.849	4.037	4.325	266	231	240	—
LiF	1.112	0.420	0.628	—	—	—	—
KCl	0.403	0.066	0.063	—	—	—	—
BaF_2	0.891	0.400	0.254	—	—	—	—
K	0.037	0.031	0.019	91.1	—	—	—
Al	1.068	0.607	0.282	428	—	—	32
Au	1.923	1.631	0.420	162.4	—	—	60

*Compiled from several sources.

REFERENCES
"Physics of III-V Compounds", O. Madelung (trans. by D. Meyerhofer), John Wiley & Sons, 1964.
"Physics of III-V Compounds", *Semiconductors and Semimetals*, R. K. Willardson and A. C. Beer, Eds., Vol. 2, Academic Press, 1966.
"Semiconducting III-V Compounds", C. Hilsum and A. C. Rose-Innes, Pergamon Press, 1961.

Table 2-61.   ELECTRON PARAMAGNETIC RESONANCE*
Summary of Electron Paramagnetic Resonance Results in Compound Semiconductors

Host lattice	Paramagnetic atom	Spin S	g-value	Fine structure constant |a|, gauss	Hyperfine inter-action A, gauss	Remarks
GaAs	Manganese	$\frac{5}{2}$	2.003	15.5	-57	Fine structure not resolved
GaAs	Iron	$\frac{5}{2}$	2.046	374	—	Hyperfine structure not resolved
GaAs	Iron or defects	$\frac{5}{2}$	2.046	374	—	Spin density 10^2 times greater than iron density
GaAs	Nickel	$\frac{3}{2}$	2.106	—	—	One very broad line, about 130 G in half-width
GaAs	Zinc	$\frac{1}{2}$	8.1	—	—	Bound hole detected by uniaxial stress
GaAs	Cadmium	$\frac{1}{2}$	6.7	—	—	Bound hole detected by uniaxial stress
GaAs	Conduction elec.	$\frac{1}{2}$	0.52	—	—	About 100 G in half-width
InSb	Conduction elec.	$\frac{1}{2}$	50.7–48.8	—	—	g-value and line width change with concentration
GaP	Manganese	$\frac{5}{2}(?)$	2.002	—	60.5	Only one line observed
GaP	Iron	$\frac{5}{2}$	2.023	429	—	Hyperfine structure not resolved

*From: "Physics of III–V Compounds—Electron Paramagnetic Resonance", B. Goldstein, *Semiconductors and Semimetals*, R. K. Willardson and A. C. Beer, Vol. 2, Academic Press, 1966.

Figure 2-62.   RESISTIVITY VS. IMPURITY CONCENTRATIONS FOR VARIOUS SEMICONDUCTORS*

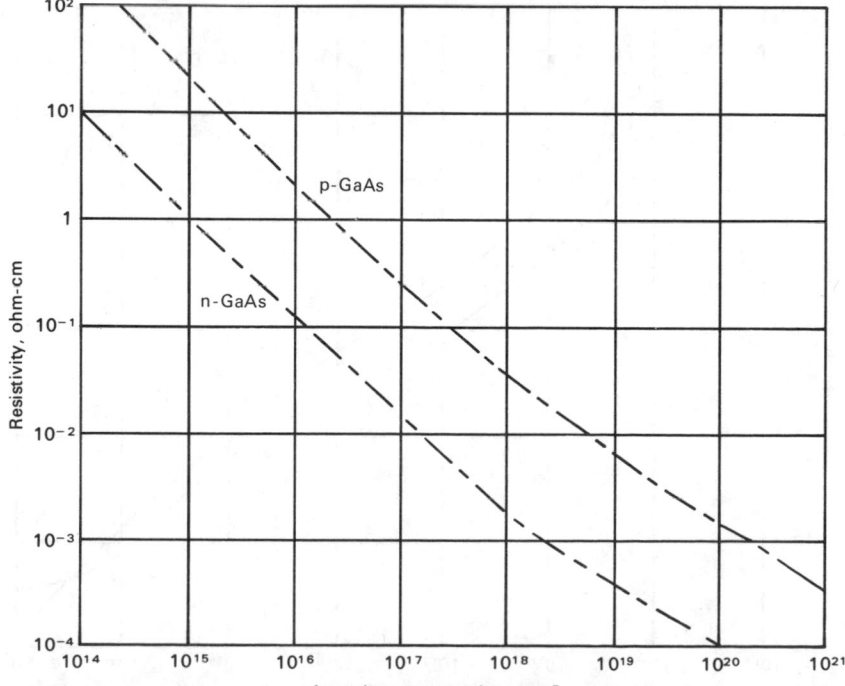

Figure 2-62A.   Resistivity vs. impurity concentration for GaAs at room temperature.

*From: S.M. Sze and J.C. Irvin, Resistivity, Mobility, and Impurity Levels in GaAs, Ge, and Si at 300 K, *Solid-State Electronics*, 11:599, 1968.

Figure 2-62. RESISTIVITY VS. IMPURITY CONCENTRATIONS FOR VARIOUS SEMICONDUCTORS *(Continued)*

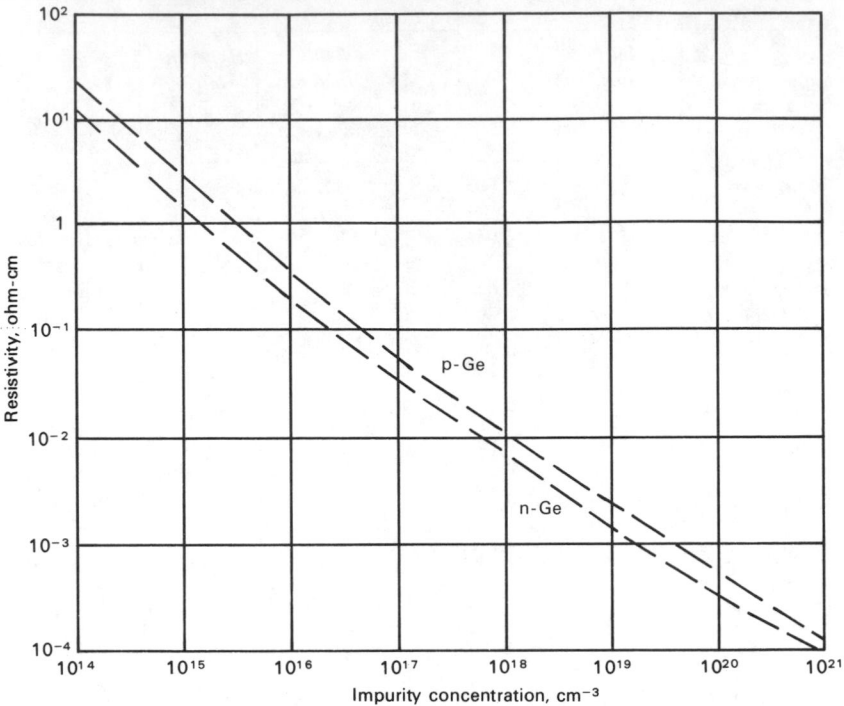

Figure 2-62B.   Resistivity vs. impurity concentration for Ge at room temperature.

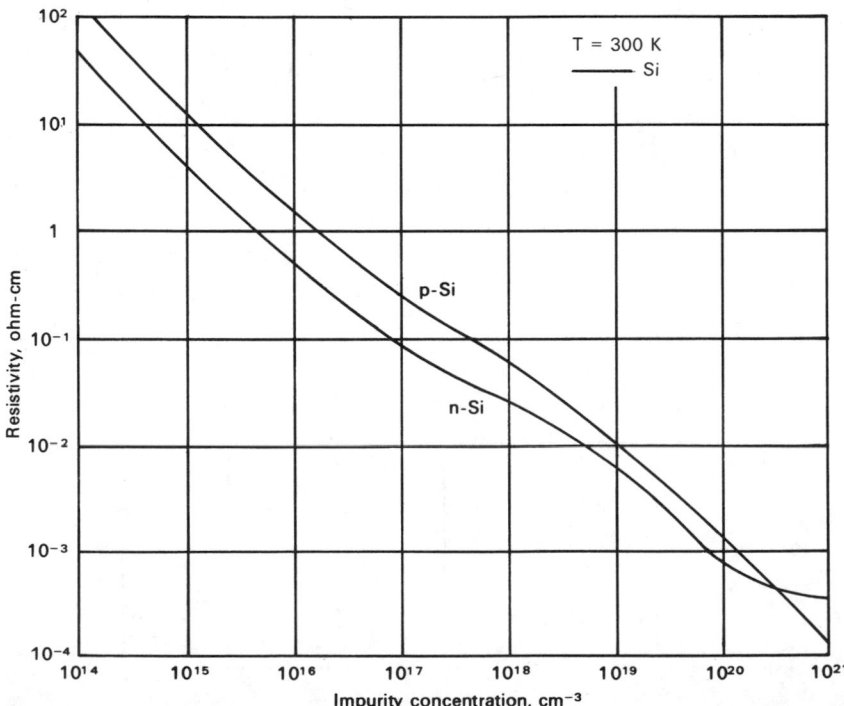

Figure 2-62C.   Resistivity vs. impurity concentration for Si at room temperature.

Table 2-63. LASER LINES STRONGLY ABSORBED BY THE ATMOSPHERE*

COURTESY OF WILLIAM EPPERS

Laser	λ, microns	Absorber
Atomic krypton	1.7843	H_2O
Atomic krypton	1.9211	H_2O
Tm^{+3}–$CaWO_4$	1.911	H_2O
Tm^{+3}–$CaWO_4$	1.916	H_2O
U^{+3}–SrF_2	2.472	H_2O
U^{+3}–CaF_2	2.511	H_2O
Atomic krypton	2.5234	H_2O
U^{+3}–BaF_2	2.556	H_2O
U^{+3}–CaF_2	2.613	H_2O
Atomic neon	3.391317	CH_4
Carbon monoxide	5.2 to 7	H_2O
Cesium	7.1021	H_2O
Atomic neon	18.3040	H_2O
Atomic neon	20.351	H_2O

*Reprinted from: Atmospheric Transmission, W.C. Eppers, in "Handbook of Lasers", R.J. Pressley, Ed., The Chemical Rubber Co., 1971.

Table 2-64. LASER LINES WITH WEAK TO MODERATE ABSORPTION BY THE ATMOSPHERE*

COURTESY OF WILLIAM EPPERS

Laser	λ	Comment
Ionized argon	4880 Å, 5145 Å	
Atomic neon (He–Ne)	6328 Å	
GaAs	8300 Å, 9200 Å	Close attention must be paid to temperature of operation; increased absorption occurs from approx. 8600 Å to 9250 Å
Ruby	6934 → 6945 Å	Strong H_2O absorptions can occur.
Nd^{+3}	$\approx 1.06 \mu$	Very low absorption
Atomic neon (He–Ne)	1.1523 μ 5 lines	Moderate H_2O absorption
CH_4 Raman shift of 1.06 μ	1.53 μ	
Er^{+3} (CaF_2) (glass)	1.55 → 1.65 μ	
Ho^{+3}–$CaWO_4$	2.04 μ	Mostly clear
Ho^{+3}–YAG	to	
Ho^{+3}–CaF_2	2.128 μ	
Atomic Xe	3.50704 μ	
DF	3.8 μ	
CO_2	10.6 μ	CO_2 Water absorption

*Reprinted from: Atmospheric Transmission, W.C. Eppers, in "Handbook of Lasers", R.J. Pressley, Ed., The Chemical Rubber Co., 1971.

Figure 2-65. ATMOSPHERIC ATTENUATION COEFFICIENTS*

$$T = \exp^{(-\sigma R)}$$

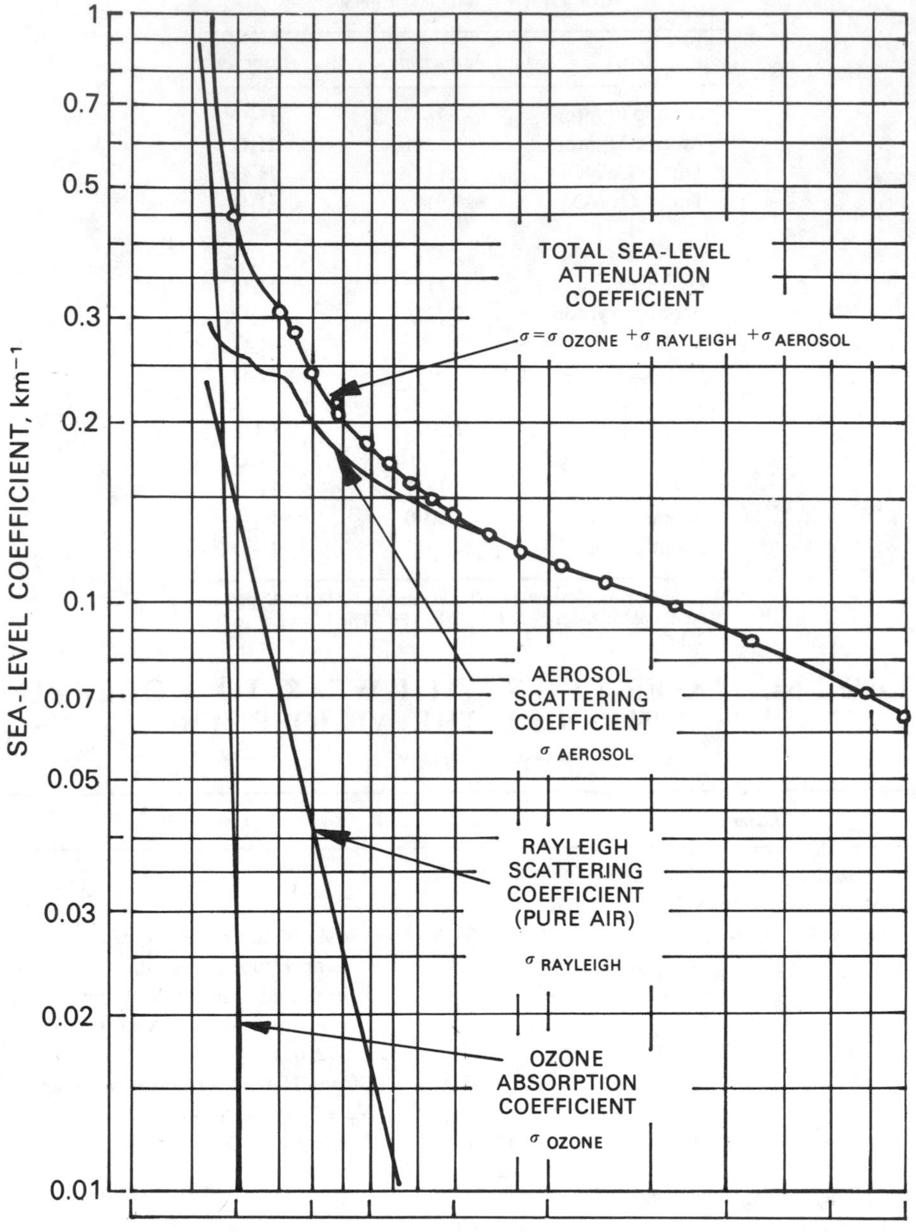

Figure 2-65. Calculated atmospheric attenuation coefficients for horizontal transmission at sea level in a model clear standard atmosphere.

*Adapted from data in: "Electro-Optics Handbook", RCA Corporation, 1968.

Figure 2-66. UNDERWATER TRANSMISSION OF LIGHT*

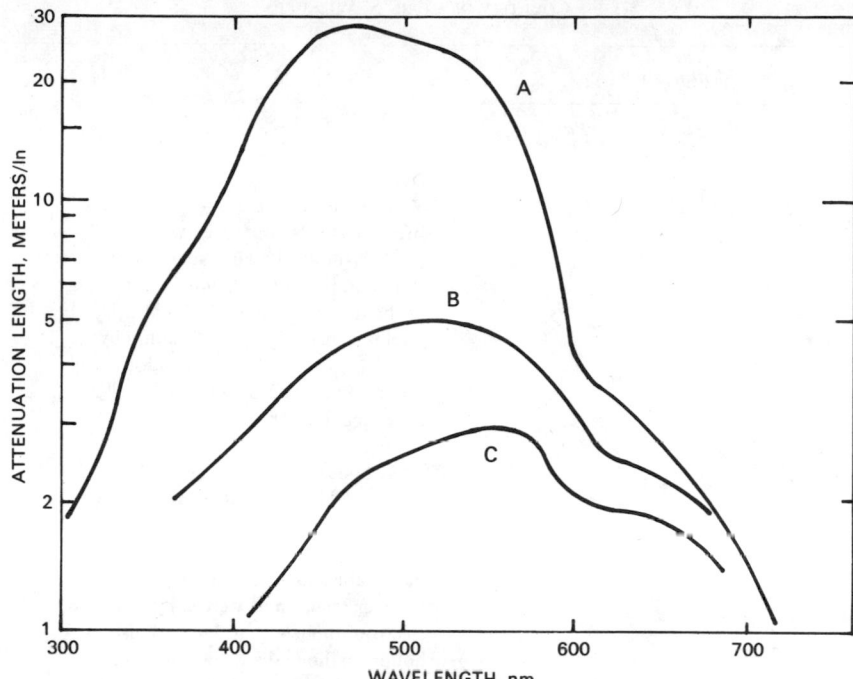

Figure 2-66. Attenuation length vs. wavelength for underwater transmission of light in (A) distilled water, (B) oceanic coastal water, and (C) typical lake water.

*From: E.O. Hulburt, *J. Opt. Soc. Amer.*, 35:698, 1945.

Table 2-67. LASER SOURCES FOR RAMAN SPECTROSCOPY*

Laser material	Laser wavelength, μ	Approximate power available
He-Ne (gas)	0.6328	10 to 100 mW
Ruby (crystal)	0.6943	100 mW to 2 W, cw, or continuously pulsed
Nd-YAG (crystal)	1.06	2 to 20 W
Nd-YAG (frequency doubled)	0.533	500 mW
CO_2 (gas)	10.6	2 to more than 200 W
Kr (gas)	0.5681	500 mW
Kr (gas)	0.6471	500 mW
A (gas)	0.4880	1 W
A (gas)	0.5145	1 W
Xe	0.8720	20 mW

*From: "Laser Raman Spectroscopy", S. Porto, *Ind. Res.*, 11:68, May 1969.

Table 2-68. NEUTRAL GAS LASER TRANSITIONS*

COURTESY OF COLIN S. WILLETT

Wave-length, μm	Identification		Notes	References
	Racah	Paschen		
NEON				
0.632 82	$5s'[1/2]_1^\circ - 3p'[3/2]_2$	$3s_2 - 2p_4$	CW. Strongest of the $3s_2 - 2p$ lines. In a 5:1 He–Ne mixture at 3.6 Torr-mm. The He–Ne mixture ratio and pressure depend on the bore of the discharge tube.[14] The Ne–$3s_2$ level is selectively excited mainly by He 2^1S_0 metastables in an endothermic excitation transfer reaction, as well as by direct electron impact.[6,9]	1–27
1.152 28	$4s'[1/2]_1^\circ - 3p'[3/2]_2$	$2s_2 - 2p_4$	CW. Strongest of the $2s - 2p$ lines. In a 10:1 He–Ne mixture at about 11 Torr-mm. The Ne–2s levels are selectively excited mainly by He 2^3S_1 metastables in an exothermic excitation transfer reaction as well as by direct electron impact.[9,11] Observed also in hollow cathode discharges in pure neon, and mixtures of neon and hydrogen, and neon and oxygen.	1, 5, 9, 11, 28, 29, 38, 54–65
3.390 3	$5s'[1/2]_1^\circ - 4p'[1/2]_1$	$3s_2 - 3p_2$	CW. In a 5:1 He–Ne mixture at a pD of 3.6 Torr-mm. The Ne–$3s_2$ level is selectively excited mainly by excitation transfer from He 2^1S_0 metastables in an endothermic reaction, as well as by direct electron impact from the ground state.[9,11]	30–33
3.391 3	$5s'[1/2]_1^\circ - 4p'[3/2]_2$	$3s_2 - 3p_4$	CW. Same as wavelength 3.390 3. This line exhibits very high gain.	3, 5, 7, 17, 21, 29, 30, 32, 34, 35, 36
XENON				
2.026 229	$5d[3/2]_1^\circ - 6p[3/2]_1$	$3d_2 - 2p_7$	CW. Superradiant line in even short discharge tubes. The gain in a 5-mm bore tube is more than 45 db/m. In a few mTorr of xenon, or in a 100:1 He–Xe mixture at a total pressure of about 10 Torr, $D = 5$ mm. To avoid cataphoretic effects in He–Xe mixture, rf-excitation is necessary. Clean-up of xenon is a problem.[49]	1, 29, 37–47, 53
3.508 0[a]	$5d[7/2]_3^\circ - 6p[5/2]_2$	$3d_4 - 2p_9$	CW. Superradiant line in even short discharge tubes, gain more than 70 db/m in a 2.6-mm tube.[49] In a few to tens of mTorr of xenon, and with helium to about 10 Torr with $D = 5$ mm. Clean-up of xenon is a problem.[49]	29, 31, 33, 39, 43–45, 47, 48–53

[a] Wavelengths of lines *in vacuo* are designated in italics.

*Condensed from: C.S. Willett, Neutral Gas Lasers, in "Handbook of Lasers", R.J. Pressley, Ed., The Chemical Rubber Co., 1971.

Table 2-68. NEUTRAL GAS LASER TRANSITIONS (Continued)

REFERENCES

1. W.R. Bennett, Jr., Inversion Mechanisms in Gas Lasers, in *Applied Optics, Supplement on Chemical Lasers*, pp. 3–33, 1965.
2. A.D. White and J.D. Rigden, The Effect of Super-radiance at 3.39 μ on the Visible Transitions in the He–Ne Maser, *Appl. Phys. Lett.*, 2:211–212, 1963.
3. A.L. Bloom, Observation of New Visible Gas Laser Transition by Removal of Dominance, *Appl. Phys. Lett.*, 2:101–102, 1963.
4. A.D. White and J.D. Rigden, Continuous Gas Maser Operation in the Visible, *Proc. IRE.*, 50:1697, 1962.
5. J.D. Rigden and A.D. White, The Interaction of Visible and Infra-red Maser Transitions in the Helium-neon System, *Proc. IEEE*, 51:943–945, 1963; also in "Quantum Electronics III", P. Grivet and N. Bloembergen, Eds., Columbia University Press, 1964, pp. 499–505.
6. A.D. White and E.I. Gordon, Excitation Mechanism and Current Dependence of Population Inversion in He–Ne Lasers, *Appl. Phys. Lett.*, 3:197–198, 1963.
7. E.I. Gordon and A.D. White, Similarity Laws for the Effect of Pressure and Discharge Diameter on Gain of He–Ne Lasers, *Appl. Phys. Lett.*, 3:199–201, 1963.
8. E. F. Labuda and E.I. Gordon, Microwave Determination of Average Electron Energy and Density in He–Ne Discharges, *J. Appl. Phys.*, 35:1647–1648, 1964.
9. R.T. Young, Jr., C.S. Willett, and R.T. Maupin, The Effect of Helium on Population Inversion in the He–Ne Laser, *J. Appl. Phys.*, 41:2936–2941, 1970.
10. R.T. Young, Jr., C.S. Willett, and R.T. Maupin, An Experimental Determination of the Relative Contributions of Resonance and Electron Impact Collision to the Excitation of Ne Atoms in He–Ne Laser Discharges, *Bull. Amer. Phys. Soc.*, 13:206, 1968.
11. C.R. Jones and W.W. Robertson, Temperature Dependence of Reaction of the Helium Metastable Atom, *Bull. Amer. Phys. Soc.*, 13:198, 1968. (At the 20th Gaseous Electronics Conference, San Francisco, Oct. 1967, material was presented on the potential barrier behavior of the reaction He* $2^1S_0 + Ne\ ^1S_0 \rightarrow He\ ^1S_0 + Ne\ 3s_2 - \Delta E_\infty$ (386 cm^{-1}), as well as the reaction involving He* 2^3S_1 metastables stated in the abstract in this reference.)
12. F.A. Korolev, A.I. Odintsov, and V.N. Mitsai, A Study of Certain Characteristics of a He–Ne Laser, *Opt. Spectry.*, 19:36–39, 1965.
13. N. Suzuki, Vacuum UV Measurements of Helium-neon Laser Discharge, *Japan. J. Appl. Phys.*, 4:285–291, 1965.
14. R.L. Field, Jr., Operating Characteristics of dc-Excited He–Ne Gas Lasers, *Rev. Sci. Instru.*, 38:1720–1722, 1967.
15. N. Suzuki, Spectroscopy of He–Ne Laser Discharges, *Japan. J. Appl. Phys., Supplement I* (Proceedings of the Conference on Photographic and Spectroscopic Optics, 1964), 4:642–647, 1965.
16. G.A. Gonchukov, et al., Temperature Effects in the He–Ne Laser, *Opt. Spectry.*, 20:601–602, 1966.
17. A.N. Alekseeva and D.V. Gordeev, The Effect of Longitudinal Magnetic Field on the Output of a Helium-neon Laser, *Opt. Spectry.*, 23:520–524, 1967.
18. P.W. Smith, The Output Power of a 6328-Å He–Ne Gas Laser, *IEEE J. Quant. Electronics*, QE-2:62–68, 1966.
19. P.W. Smith, On the Optimum Geometry of a 6328-Å Laser Oscillator, *IEEE J. Quant. Electronics*, QE-2:77–79, 1966.
20. G. Herziger, W. Holzapfel, and W. Seelig, Verstärkung einer He–Ne Gasentladung für die Laser Wellenlänge, $\lambda = 6328$ AE, *Z. für Physik.*, 189:385–400, 1966.
21. W.E. Bell and A.L. Bloom, Zeeman Effect at 3.39 microns in a Helium-neon Laser, *Appl. Optics*, 3:413–415, 1964.
22. I.M. Belusova, et al., Investigation of the Causes of Gas Temperature Effects on the Generation of Power of a He–Ne Laser at 6328 Å, *Opt. Spectry.*, 16:44–47, 1969.
23. L. Allen and D.G.C. Jones, "Principles of Gas Lasers", Plenum Press, 1967, pp. 73–103.
24. A.L. Bloom, "Gas Lasers", John Wiley & Sons, Inc., 1968, pp. 52–59.
25. A.D. White, Anomalous Behaviour of the 6402.84-Å Gas Laser, *Proc. IEEE*, 52:721, 1964.
26. A.L. Bloom and D.L. Hardwick, Operation of He–Ne Lasers in the Forbidden Resonator Region, *Phys. Lett.*, 20:373–375, 1966.
27. R.N. Zitter, 2s-2p and 3p-3s Neon Transitions in a Very Long Laser, *Bull. Amer. Phys. Soc.*, 9:500, 1964.
28. A. Javan, W.R. Bennett, Jr., and D.R. Herriott, Population Inversion and Continuous Optical Maser Oscillation in a Gas Discharge Containing a He–Ne Mixture, *Phys. Rev. Lett.*, 6:106–110, 1961.
29. W.R. Bennett, Jr., Gaseous Optical Masers, in *Appl. Optics, Supplement on Optical Masers*, O.S. Heavens, Ed., 1962, pp. 24–61.
30. W.L. Faust, et al., Noble Gas Optical Maser Lines at Wavelengths Between 2 and 35 μ, *Phys. Rev.*, 133:A1476–A1478, 1964.
31. R.A. McFarlane, et al., Gas Maser Operation at Wavelengths out to 28 microns in "Quantum Electronics III", P. Grivet and N. Bloembergen, Eds., Columbia University Press, 1964, pp. 573–586.
32. K. Bergman and W. Demtroder, A New Cascade Laser Transition in He–Ne Mixture, *Phys. Lett.*, 29A:94–95, 1969.
33. P.G. McMullin, Precise Wavelength Measurement of Infrared Optical Maser Lines, *Appl. Optics*, 3:641–642, 1964.
34. A.L. Bloom, W.E. Bell, and R.C. Rempel, Laser Action at 3.39 μ in a Helium-neon Mixture, *Appl. Optics*, 2:317–318, 1963.
35. D. Rosenberger, Superstrahlung in gepulsten Argon, Krypton und Xenon Entladungen, *Phys. Lett.*, 14:32, 1965.
36. F. Gires, H. Mayer, and M. Pailette, Sur quelques Transitions Présentant l'Effet Laser dans le Mélange Hélium-néon, *Compt. Rend.*, 256:3438–3439, 1963.
37. C.K.N. Patel, et al., Infrared Spectroscopy Using Stimulated Emission Techniques, *Phys. Rev. Lett.*, 9:102–104, 1962.
38. O. Andrade, M. Gallardo, and K. Bockasten, High-gain Laser Lines in Noble Gases, *Appl. Phys. Lett.*, 11:99–100, 1967.
39. W.T. Walter and J.M. Jarrett, Strong 3.27-μ Oscillation in Xenon, *Appl. Optics*, 3:780–790, 1964.
40. G.E. Courville, P.J. Walsh, and J.H. Wasko, Laser Action in Xe in Two Distinct Current Regions of ac and dc Discharges, *J. Appl. Phys.*, 35:2547–2548, 1964.
41. C.K.N. Patel, W.L. Faust, and R.A. McFarlane, High-gain Gaseous (Xe–He) Optical Maser, *Appl. Phys. Lett.*, 1:84–85, 1962.
42. C.K.N. Patel, R.A. McFarlane, and W.L. Faust, High-gain Medium for Gaseous Optical Masers, in "Quantum Electronics III", P. Grivet and N. Bloembergen, Eds., Columbia University Press, 1964, pp. 507–514.

Table 2-68. NEUTRAL GAS LASER TRANSITIONS *(Continued)*

43. W.L. Faust, et al., Gas Maser Spectroscopy in the Infrared, *Appl. Phys. Lett.*, 1:85–88, 1962.
44. A.A. Kuznetsov and D.I. Mash, Operating Conditions of an Optical Maser with a Helium-xenon Mixture in the Middle Infrared Region of the Spectrum, *Radio Eng. Electr. Phys.*, IC:319–320, 1965.
45. V.F. Moskalenko, E.P. Ostapchenko, and V.I. Pugnin, Mechanism of Xenon-level Population Inversion in the Positive Column of a Helium-xenon Mixture, *Opt. Spectry.*, 23:94–95, 1967.
46. V.N. Smiley, A.L. Lewis, and D.K. Forbes, Gain and Bandwidth Narrowing in a Regenerative He–Xe Laser Amplifier, *J. Opt. Soc. Amer.*, 55:1552–1553, 1965.
47. P.O. Clark, R.A. Hubach, and J.Y. Wada, Investigation of the dc-Excited Xenon Laser, Hughes Research Labs., Final Report JPL Contract NO950803, 1965.
48. L. Liberman, Sur la Structure Hyperfine de quelques Raies Laser Infrarouges de Xenon 129, *Compt. Rend.*, 266:236–239, 1968.
49. D.R. Armstrong, A Method for the Control of Gas Pressure in the Xenon Laser, *IEEE J. Quant. Electronics*, QE-4:968–969, 1968.
50. W.B. Bridges, High Optical Gain at 3.5 μ in Pure Xenon, *Appl. Phys. Lett.*, 3:45–47, 1963.
51. R.A. Paananen and D.L. Bobroff, Very High Gain Gaseous (Xe–He) Optical Maser at 3.5 μ, *Appl. Phys. Lett.*, 2:99–100, 1963.
52. P.O. Clark, Investigation of the Operating Characteristics of the 3.5 μ Xenon Laser, *IEEE J. Quant. Electronics*, QE-1:109–113, 1965.
53. A.A. Kuznetsov, et al., Operating Conditions of an Optical Quantum Generator (Laser) in Helium-neon and Xenon-helium Gas Mixtures, *Rad. Eng. Electr. Phys.*, 9:1576, 1964.
54. V.P. Chebotayev and V.V. Pokasov, Operation of a Laser on a Mixture of He–Ne with Discharge in a Hollow Cathode, *Radio Eng. Electr. Phys.*, 10:817–819, 1965.
55. C.G. Petrash and I.N. Knyazev, Study of Pulsed Laser Generation in Neon and Mixtures of Neon and Helium, *Sov. Phys. JETP*, 18:571–575, 1964.
56. R. der Agobian, et al., Cascades de Transitions Stimulées dans le Néon Pur, *Compt. Rend.*, 259:323–326, 1964.
57. V.P. Chebotayev, Effect of Hydrogen and Oxygen on the Operation of a Neon Maser, *Radio Eng. Electr. Phys.*, 10:316–318, 1965.
58. R. der Agobian, et al., Nouvelle Cascade de Transitions Stimulée du Néon, *Compt. Rend.*, 258:3661–3663, 1964.
59. W.R. Bennett, Jr., and J.W. Knutson, Jr., Simultaneous Laser Oscillation on the Neon Doublet at 1.1523 μ, *Proc. IEEE*, 52:861–862, 1964.
60. C.K.N. Patel, Optical Power Output in He–Ne and Pure Ne Maser, *J. Appl. Phys.*, 33:3194–3195, 1962.
61. V.P. Chebotayev, Operating Condition of an Optical Maser Containing a Helium-neon Mixture, *Radio Eng. Electr. Phys.*, 10:314–316, 1965.
62. H.A.H. Boot, D.M. Clunie, and R.S.A. Thorn, Pulsed Laser Operation in a High-pressure Helium-neon Mixture, *Nature*, 198:773–774, 1963.
63. J. Smith, Optical Maser Action in the Negative Glow Region of a Cold Cathode Glow Discharge, *J. Appl. Phys.*, 35:723–724, 1964.
64. K.S. Mustafin, V.A. Seleznev, and E.I. Shtyrkov, Stimulated Emission in the Negative Glow Region of a Glow Discharge, *Opt. Spectry.*, 21:429–430, 1966.
65. L.D. Schearer, Polarization Transfer Between Oriented Metastable Helium Atoms and Neon Atoms, *Phys. Lett.*, 27A:544–545, 1968.

Table 2-69. IONIZED GAS LASER TRANSITIONS*

<small>COURTESY OF WILLIAM B. BRIDGES AND ARTHUR N. CHESTER</small>

EXPLANATION:

CW	Continuous oscillation reported	S	Superradiant operation reported
E	Error in classification or wavelength	U	Unique or unusual excitation method
G	Gain measured	λ	Accurate wavelength measurement
P	Power output reported		

Wavelength,[a] μm	Measured value, μm	Ion	Identification Upper level — Lower level	References
			CADMIUM	
.325 031	.325 000 ± ?	Cd II	$5s^2\ ^2D_{3/2} - (^1S)5p^2P^\circ_{1/2}$	12-CW,P; 19-CW,G,P; 21-P
.441 563	.441 560 ± 70	Cd II	$5s^2\ ^2D_{5/2} - (^1S)5p^2P^\circ_{3/2}$	17; 11-CW,P; 18-CW,G,P; 20; 21-P

*Condensed from: W.B. Bridges and A.N. Chester, Ionized Gas Lasers, in "Handbook of Lasers", R.J. Pressley, Ed., The Chemical Rubber Co., 1971.

Table 2-69. IONIZED GAS LASER TRANSITIONS *(Continued)*

Wavelength,[a] μm	Measured value, μm	Ion	Identification Upper level—Lower level	References
ARGON				
.351 112	.351 113 ± 06	Ar III	$(^4S°)4p\,^3P_2 - (^4S°)4s\,^3S_1°$	
[.363 789]	.363 786 ± 04	Ar III	$(^2D°)4p\,^1F_3 - (^2D°)4s\,^1D_2°$	
.457 935	.457 936 ± 16	Ar II	$(^3P)4p\,^2S_{1/2}° - (^3P)4s\,^2P_{1/2}$	4; 3-G; 13-CW
.476 486	.476 488 ± 04	Ar II	$(^3P)4p\,^2P_{3/2}° - (^3P)4s\,^2P_{1/2}$	4; 3-G,P,S; 9-E; 13-CW; 14-U,P
.487 986	.487 986 ± 04	Ar II	$(^3P)4p\,^2D_{5/2}° - (^3P)4s\,^2P_{3/2}$	
.514 532	.514 533 ± 02	Ar II	$(^3P)4p\,^4D_{5/2}° - (^3P)4s\,^2P_{3/2}$	4-G,P; 3-G; 13-CW
KRYPTON				
.350 742	.350 742 ± 06	Kr III	$(^4S°)5p\,^3P_2 - (^4S°)5s\,^3S_1°$	
.406 737	.406 736 ± 06	Kr III	$(^2D°)5p\,^1F_3 - (^2D°)5s\,^1D_2°$	6; 8-CW; 10-CW,P; 15-CW,P
.461 915	.461 917 ± 10	Kr II	$(^3P)5p\,^2D_{5/2}° - (^3P)5s\,^2P_{3/2}$	5; 6-λ; 7-CW
.476 243	.476 244 ± 06	Kr II	$(^3P)5p\,^2D_{3/2}° - (^3P)5s\,^2P_{1/2}$	5; 2-U; 6-λ 1-CW
.568 188	.568 192 ± 04	Kr II	$(^3P)5p\,^4D_{5/2}° - (^3P)5s\,^2P_{3/2}$	5; 1-CW,P; 2-U; 6-λ; 7-CW
.647 088	.647 100 ± 50	Kr II	$(^3P)5p\,^4P_{5/2}° - (^3P)5s\,^2P_{3/2}$	5; 1-CW; 16-CW,P

[a] This column lists strong or characteristic laser lines in pure gas, giving the most accurate available wavelength for the given transition, in micrometers. In most cases this value of wavelength has been derived from the spontaneous emission spectroscopic literature, assuming the indicated classification for the line. The bracketed wavelength was taken from Reference 7.

REFERENCES

1. R. der Agobian, et al., Émission Stimulée en Régime Permanent dans le Spectra Visible du Krypton Ionisé, *Compt. Rend.*, 260:6327–6329, 1965.
2. W.E. Bell, Ring Discharge Excitation of Gas Ion Lasers, *Appl. Phys. Lett.*, 7:190–191, 1965.
3. W.R. Bennett, Jr., et al., Super-radiance, Excitation Mechanisms, and Quasi-cw Oscillation in the Visible Ar⁺ Laser, *Appl. Phys. Lett.*, 4:180–182, 1964.
4. W.B. Bridges, Laser Oscillation in Singly Ionized Argon in the Visible Spectrum, *Appl. Phys. Lett.*, 4:128–130, 1964; erratum: *Appl. Phys. Lett.*, 5:39, 1964.
5. W.B. Bridges, Laser Action in Singly Ionized Krypton and Xenon, *Proc. IEEE*, 52:843–844, 1964.
6. W.B. Bridges and A.N. Chester, Visible and UV Laser Oscillation at 118 Wavelengths in Ionized Neon, Argon, Krypton, Xenon, Oxygen, and Other Gases, *Appl. Opt.*, 4:573–580, 1965.
7. W.B. Bridges and A.N. Chester, Spectroscopy of Ion Lasers, *IEEE J. Quant. Electronics*, QE-1:66–84, 1965.
8. W.B. Bridges, R.J. Freiberg, and A.S. Halsted, New Continuous UV Ion Laser Transitions in Neon, Argon, and Krypton, *IEEE J. Quant. Electronics*, QE-3:339, 1967.
9. G. Convert, M. Armand, and P. Martinot-Lagarde, Effet Laser dans des Mélanges Mercure-Gaz Rares, *Compt. Rend.*, 258:3259–3260, 1964.
10. J.R. Fendley, Jr., Continuous UV Lasers, *IEEE J. Quant. Electronics*, QE-4:627 631, 1968.
11. G.R. Fowles and B.D. Hopkins, CW Laser Oscillation at 4416 Å in Cadmium, *IEEE J. Quant. Electronics*, QE-3:419, 1967.
12. J.P. Goldsborough, Continuous Laser Oscillation at 3250 Å in Cadmium Ion, *IEEE J. Quant. Electronics*, QE-5:133, 1969.
13. E.I. Gordon, E.F. Labuda, and W.B. Bridges, Continuous Visible Laser Action in Singly Ionized Argon, Krypton, and Xenon, *Appl. Phys. Lett.*, 4:178–180, 1964.
14. S.G. Kulagin, et al., States with Population Inversion in a Self-compressed Discharge, *JETP Lett.*, 3:6–8, 1966.
15. I.D. Latimer, High-power Quasi-CW Ultraviolet Ion Laser, *Appl. Phys. Lett.*, 13:333–335, 1968.
16. R. Paananen, Continuously-operated Ultraviolet Lasers, *Appl. Phys. Lett.*, 9:34–35, 1966.
17. W.T. Silfvast, G.R. Fowles, and B.D. Hopkins, Laser Action in Singly Ionized Ge, Sn, Pb, In, Cd, and Zn, *Appl. Phys. Lett.*, 8:318–319, 1966.
18. W.T. Silfvast, Efficient CW Laser Oscillation at 4416 Å in Cd (II), *Appl. Phys. Lett.*, 13:169–171, 1968.
19. W.T. Silfvast, New CW Metal-vapor Laser Transitions in Cd, Sn, and Zn, *Appl. Phys. Lett.*, 15:23–25, 1969.
20. Y. Sugawara and Y. Tokiwa, CW Laser Oscillations in Zn II and Cd II in Hollow Cathode Discharge, *Jap. J. Appl. Phys.*, 9:588–589, May 1970.
21. J.P. Goldsborough, Stable, Long-life CW Excitation of Helium-cadmium Lasers by dc Cataphoresis, *Appl. Phys. Lett.*, 15:159–161, 1969.

Table 2-70. MOLECULAR GAS LASER TRANSITIONS*

COURTESY OF MARTIN A. POLLACK

DIATOMIC MOLECULAR GAS LASERS

CO Laser. Transitions in the $X^1\Sigma^+$ electronic ground state include 219 lines with wavelengths from 5.086 91 to 6.663 2 μm.[a]

HCl Laser. Transitions include 60 lines with wavelengths from 3.707 1 to 27.508 μm.[a]

N_2 Laser. Transitions include 35 lines near 0.337 1 and 6 lines near 0.357 7 μm.[a]

HF Laser. Transitions include 65 wavelengths from 2.640 to 21.788 5 μm.[a]

NO Laser. Transitions in the $^2\pi_{\frac{1}{2}}$ and $^2\pi_{\frac{3}{2}}$ electronic ground states include 65 lines with wavelengths from 5.846 2 to 6.432 1 μm.[a]

[a]For a complete listing of these lines and references, see Molecular Gas Lasers, M.A. Pollack, in "Handbook of Lasers", R.J. Pressley, Ed., The Chemical Rubber Co., 1971.

TRIATOMIC MOLECULAR GAS LASER TRANSITIONS

CO_2 LASER

Wavelength,[b] λ_{vac}, μm	Frequency,[b] v, cm^{-1}	Transition	Wavelength,[b] λ_{vac}, μm	Frequency,[b] v, cm^{-1}	Transition
Transitions of (00^01)–(02^00) Band[c] (P Branch)			**Transitions of (00^01)–(10^00) Band[c] (P Branch)**		
9.552 428	1 046.854 2	P(20)	10.571 037[d]	945.981 0	P(18)
9.569 179	1 045.021 7	P(22)	10.591 035[d]	944.194 8	P(20)
9.586 227	1 043.163 3	P(24)	10.611 385	942.384 1	P(22)
9.603 573	1 041.279 1	P(26)	10.632 090	940.548 8	P(24)
9.621 219	1 039.369 3	P(28)	10.653 156	938.689 0	P(26)
9.639 166	1 037.434 1	P(30)			
Transitions of (00^01)–(10^00) Band[c] (R Branch)					
10.207 142	979.706 1	R(26)			
10.220 006	978.473 0	R(24)			
10.233 167	977.214 6	R(22)			
10.246 625	975.931 1	R(20)			

[b]Calculated from precise frequency measurements; see Ref. 2. The absolute accuracy is about ± 0.001 cm^{-1} or 0.000 01 μm. For relative accuracy all digits are significant.
[c]See Refs. 3, 20, and 21 and references cited therein.
[d]See Ref. 5.

HCN LASER

Wavelength, λ_{vac}, μm	Frequency, v, cm^{-1}	Vibrational transition	Rotational transition	Peak,[a] W	Remarks
128.629	77.743	$(12^{2d}0)$–$(05^{1d}0)$	R(25)	9	b,c,d,e
309.714 0	32.287 85	$(11^{1c}0)$–$(11^{1c}0)$	R(10)	0.4	b,c,d,e,f
310.887 0	32.166 03	$(11^{1c}0)$–(04^00)	R(10)	1.	b,c,d,e,f
335.183 1	29.834 4	(04^00)–(04^00)	R(9)		d,e,f
336.557 8	29.712 58	$(11^{1c}0)$–(04^00)	R(9)	0.6	b,c,d,e,f,g,h
372.528 3	26.843 60	(04^00)–(04^00)	R(8)	0.6	b,c,d,e,f

[a]See Refs. 18 and 22 and references cited therein for details. All transitions observed in a pulsed discharge. The unidentified transitions may not belong to HCN. Typical mixtures are $CH_4 + NH_3$, $CH_4 + N_2$, CH_3CN, $(CH_3)_2NH$, HCN or ICN (plus impurities).
[b]See Ref. 18. Observed in a pulsed (3 pps) discharge (200–900 amperes) in a 6.2 m long, 78 mm ID tube; pressure 0.2–0.3 Torr dimethylamine $[(CH_3)_2NH]$.
[c]See Ref. 15.
[d]See Refs. 17 and 12 for identification.
[e]Transitions also observed in a continuous discharge.
[f]See Ref. 9. Absolute frequency measurement.
[g]See Ref. 7.
[h]Splitting of this line has been observed by some authors; see Ref. 22.

*Condensed from: M.A. Pollack, Molecular Gas Lasers, in "Handbook of Lasers", R.J. Pressley, Ed., The Chemical Rubber Co., 1971.

Table 2-70.   MOLECULAR GAS LASER TRANSITIONS *(Continued)*

H₂O LASER[a]

Wavelength, λ_{vac}, μm	Frequency, ν, cm^{-1}	Vibrational transition	Rotational transition	Peak,[b] W	Remarks
27.970 7	357.516	(001)–(020)	6_{33}–5_{50}	3.0	b,d,e
33.029	302.76	(100)–(020)	5_{14}–4_{41}	7.0	b,c,d
47.468	210.67	(001)–(020)	6_{33}–6_{52}	0.06	b,c,d
78.443	127.481	(100)–(020)	8_{08}–8_{35}	0.007	b,d,e,f
79.087	126.44	(020)–(020)	8_{44}–8_{35}	0.006	b,c,d,f
118.591	84.323	(001)–(020)	6_{42}–6_{61}	0.001	b,c,d
220.230	45.407	(100)–(020)	5_{23}–5_{50}		c,d,g

[a]See Ref. 1 and references cited therein for details. All transitions observed in a pulsed discharge in H₂O.
[b]See Refs. 11 and 13.
[c]Wavelength from Ref. 1.
[d]Transitions also observed in a continuous discharge in H₂O.
[e]Wavelength from Ref. 4.
[f]Although lines at 117.04, 117.22, 117.40, 126.66, 126.87, 126.95, 127.07, 127.24, 127.73, 127.92, 128.08, and 128.31 cm^{-1} have been reported (Refs. 6 and 23), they are thought to be spurious responses associated with transitions at 116.87, 126.44, and 127.48 cm^{-1} (see Ref. 1).
[g]See Ref. 24.

N₂O LASER

Wavelength,[a] λ_{vac}, μm	Frequency,[a] ν, cm^{-1}	Transition	Remarks	Wavelength,[a] λ_{vac}, μm	Frequency,[a] ν, cm^{-1}	Transition	Remarks
Transitions of (00°1)–(10°0) Band (R Branch)				**Transitions of (00°1)–(10°0) Band (R Branch)**			
10.482 1	954.01	R(18)	b,c	10.862 9	920.56	P(21)	b,d,e
10.490 6	953.24	R(17)	b,c	10.873 6	919.65	P(22)	b,d,e
10.499 1	952.46	R(16)	b,c	10.884 4	918.75	P(23)	b,d,e
10.507 7	951.69	R(15)	b,c	10.895 2	917.83	P(24)	b,d,e
10.516 3	950.90	R(14)	b,c	10.906 1	916.92	P(25)	b,d,e
				10.917 0	916.00	P(26)	b,d,e
				10.928 0	915.08	P(27)	b,d,e

[a]Calculated from $\nu = 938.79 + 0.832\,82\,m - 0.001\,68\,m^2$, with $m = J + 1$. See Ref. 19.
[b]See Ref. 16. Observed in a continuous discharge N₂O–N₂ mixtures. Frequency selective resonator.
[c]See Ref. 8.   [d]See Ref. 19.   [e]See Ref. 14.

REFERENCES

1. W.S. Benedict, M.A. Pollack, and W.J. Tomlinson, The Water Vapor Laser, *IEEE J. Quant. Electronics*, QE5:108–124, 1969.
2. T.Y. Chang, Accurate Frequencies and Wavelengths of CO₂ Laser Lines, *Optics Communications*, 2(2):77–80, 1970.
3. P.K. Cheo, CO₂ Lasers, in "Lasers: A Series of Advances", A.K. Levine and A. DeMaria, Eds., Vol. III, Dekker, Marcel, Inc., 1971.
4. K.M. Evenson, et al., Absolute Frequency Measurements of the 28- and 78-μm CW Water Vapor Laser Lines, *Appl. Phys. Lett.*, 16:159–161, 1970.
5. K.M. Evenson, J.S. Wells, and L.M. Matarrese, Absolute Frequency Measurements of the CO₂ CW Laser, *Appl. Phys. Lett.*, 16:251–253, 1970.
6. W.L. Faust and R.A. McFarlane, private communication.
7. H.A. Gebbie, N.W.B. Stone, and F.D. Findlay, A Stimulated Emission Source at 0.34 Millimeter Wavelength, *Nature*, 202:685, 1964.
8. J.A. Howe, R-branch Laser Action in N₂O, *Phys. Lett.*, 17:252–253, 1965.
9. L.O. Hocker and A. Javan, Absolute Frequency Measurements on the New CW HCN Submillimeter Laser Lines, *Phys. Lett.*, 25A:489–490. 1967.
10. B. Hartmann, B. Kleman, and G. Spaongstedt, Water Vapor Laser Lines in the 7-μm Region, *IEEE J. Quant. Electronics*, QE-4:296, 1968.
11. W.Q. Jeffers and P.D. Coleman, The Far Infrared Stimulated Emission Spectrum of D₂O, *Proc. IEEE (Letters)*, 55:1222–1223, 1967.
12. D.R. Lide, Jr., and A.G. Maki, On the Explanation of the So-called CN Laser, *Appl. Phys. Lett.*, 11:62–64, 1967.
13. L.E.S. Mathias and A. Crocker, Stimulated Emission in the Far Infrared from Water Vapor and Deuterium Oxide Discharges, *Phys. Lett.*, 13:35–36, 1964.
14. L.E.S. Mathias, A. Crocker, and M.S. Wills, Laser Oscillations from Nitrous Oxide at Wavelengths around 10.9 μ, *Phys. Lett.*, 13:303–304, 1964.
15. L.E.S. Mathias, A. Crocker, and M.S. Wills, Laser Oscillations at Submillimeter Wavelengths from Pulsed Gas Discharges in Compounds of Hydrogen, Carbon, and Nitrogen, *Electronics Lett.*, 1:45–46, 1965.
16. G. Moeller and J.D. Rigden, Observation of Laser Action in the R-branch of CO₂ and N₂O Vibrational Spectra, *Appl. Phys. Lett.*, 8:69–70, 1966.
17. A.G. Maki, Assignment of Some DCN and HCN Laser Lines, *Appl. Phys. Lett.*, 12:122–123, 1968.
18. L.E.S. Mathias, A. Crocker, and M.S. Wills, Spectroscopic Measurements on the Laser Emission from Discharges in Compounds of Hydrogen, Carbon, and Nitrogen, *IEEE J. Quant. Electronics*, QE4:205–208, 1968.
19. C.K.N. Patel, CW Laser Action in N₂O (N₂-N₂O System), *Appl. Phys. Lett.*, 6:12–13, 1965.
20. C.K.N. Patel, Gas Lasers, "Lasers: A Series of Advances", A.K. Levine, Ed., Vol. II, Dekker, Marcel, Inc., 1968.
21. N.N. Sobolev and V.V. Sokovikov, CO₂ Lasers, *Sov. Phys. Uspekhi*, 10:153–170, 1967; also V.P. Tychinskii, Powerful Gas Lasers, *Sov. Phys. Uspekhi*, 10:131, 152, 1967.
22. H. Steffen and F.K. Kneubühl, Resonator Interferometry of Pulsed Submillimeter Wave Lasers, *IEEE J. Quant. Electronics*, QE4:992–1008, 1968.
23. W.Q. Jeffers and P.D. Coleman, Spiking and Time Behavior of a Pulsed Water-vapor Laser, *Appl. Phys. Lett.*, 10:7–9, 1967.
24. W.M. Muller and G.T. Flesher, Continuous Wave Submillimeter Oscillation in H₂O, D₂O, and CH₃CN, *Appl. Phys. Lett.*, 8:217–218, 1966.

Table 2-71. SUMMARY OF DYE LASER PROPERTIES*

COURTESY OF THOMAS F. DEUTSCH

Property		Typical values	Conditions	Comments
Wavelength		3400–11,750 Å	Flashlamp and/or laser pumped	A variety of dye-solvent combinations are available to span the entire wavelength range nearly continuously
Tuning		Up to 1760 Å	Prism, filter, grating, Q-switch in cavity	Length, concentration, and temperature of active medium also provide tuning. Solvent pH affects tuning range.
Spectral Width		15–150 Å ~ 0.5 Å ~ 0.01 Å	Broadband mirrors Grating in cavity Grating plus etalon	
Beam Divergence		2–5 mrad 0.5 mrad	Flashlamp and/or laser pumped Etalon in cavity	Dependent on uniformity of pumping
Efficiency		Up to 25 percent ~ 0.4 percent	Laser pumped Flashlamp pumped	Measured optical efficiency Electrical energy input to laser energy output
Output	Energy	2J (high)– 0.1J (typical)	Flashlamp pumped	
	Power	~ 2 MW 0.75–2.0 MW	20 MW pump Flashlamp pumped	Rhodamine 6G
	Repetition Rate	Up to 200 pps	Laser pumped	Pump laser rate limits
		20–50 pps	Linear flashlamp	System cooling and component failure are limits
		1 pps continuous	Annular flashlamp {Laser pumped small cavity}	Dye liquid flow transversely through cavity
Temporal	Pulse Duration	~ 20 nsec	Laser pumped	Follows pump pulse
		0.5μsec typical; up to 500 μsec achieved	Flashlamp pumped	Shorter than pump duration
	Mode-Locked		Mode-locked pump	Pump cavity length integral multiple of dye cavity
		Pulses < 10^{-9} sec	Flashlamp pumped with intracavity saturable absorber	
		Pulses < 10^{-11} sec	Mode-locked pump	Observed by two-photon fluorescence

*Reprinted from: T.F. Deutsch, Dye Lasers, in "Handbook of Lasers", R.J. Pressley, Ed., The Chemical Rubber Co., 1971.

Table 2-72. SPECTRAL RANGE COVERED BY SEMICONDUCTOR LASERS‡

COURTESY OF JACQUES I. PANKOVE

	λ (μ)	$h\nu$ (eV)	Injection	Electron beam	Optical	Avalanche
				Mode of excitation and reference		
ZnS	0.33	3.8		1	2	
†ZnO	0.37	3.4		3,4		
$Zn_{1-x}Cd_xS$	0.49–0.32	2.5–3.82			5	
ZnSe	0.46	2.7		6		
†CdS	0.49	2.5		4,7–11	12	
ZnTe	0.53	2.3		13		
GaSe	0.59	2.1		14		
$CdSe_{1-x}S_x$	0.49–0.68	2.5–1.8		15	16	
$CdSe_{0.95}S_{0.05}$	0.675	1.8			16	
CdSe	0.675	1.8		4,8,17	18,19	
†$Al_{1-x}Ga_xAs$	0.63–.90	2.0–1.4	20–29			
†$GaAs_{1-x}P_x$	0.61–0.90	2.0–1.4	30–34	35,36		
CdTe	0.785	1.6		37–39		
†GaAs	0.83–.91**	1.50–1.38	40–42	43–45	46,47	48
InP	0.91	1.36	49–50			51
$GaAs_{1-x}Sb_x$	0.95–1.5	1.4–0.83	*			
$CdSnP_2$	1.01	1.25		52		
$InAs_{1-x}P_x$	0.9–3.2	1.4–0.39	*			
$InAs_{0.94}P_{0.06}$	0.942	1.32	53			
$InAs_{0.51}P_{0.49}$	1.6	0.78	53			
GaSb	1.55	0.80	54–56	57		
$In_{1-x}Ga_xAs$	0.58–3.1	2.14–0.4	58			
$In_{0.65}Ga_{0.35}As$	1.77	0.70	58			
$In_{0.75}Ga_{0.25}As$	2.07	0.60	58			
Cd_3P_2	2.1	0.58			59	
InAs	3.1	0.39	60–63	64	65	
$InAs_{1-x}Sb_x$	3.1–5.4	0.39–0.23	66			
$InAs_{0.98}Sb_{0.02}$	3.19	0.39	66			
$Cd_{1-x}Hg_xTe$	3–15	0.41–0.08		67	67	
$Cd_{0.32}Hg_{0.68}Te$	3.8	0.33			67	
Te	3.72	0.334		68		
PbS	4.3	0.29	69	70		
InSb	5.2	0.236	71–73	74	75,76	77,78
PbTe	6.5	0.19	79	70		
$PbS_{1-x}Se_x$	3.9–8.5	0.32–0.146	80	81		
PbSe	8.5	0.146	82–85	70		
PbSnTe	28	0.045	86			
PbSnSe	8–31.2	0.155–0.040	87–89			

*Expected but not observed.
**Depending on temperature and doping.
†Have lased at room temperature.
‡Reprinted from: J.I. Pankove, Injection Lasers, in "Handbook of Lasers", R.J. Pressley, Ed., The Chemical Rubber Co., 1971.

REFERENCES

1. C. E. Hurwitz, *Appl. Phys. Lett.*, *9*, 116, 1966.
2. S. Wang and C. C. Chang, *Appl. Phys. Lett.*, *12*, 193, 1968.
3. F. H. Nicoll, *Appl. Phys. Lett.*, *9*, 13, 1966; *J. Appl. Phys.*, *39*, 4469, 1968.
4. J. R. Packard, D. A. Campbell, W. C. Tait, *J. Appl. Phys.*, *38*, 5255, 1967.
5. M. S. Brodin, P. I. Budnick, B. L. Vitrikhovskii, S. V. Zakrevskii, *Proc. Int. Conf. Phys. Semiconductors*, Moscow, 1968, p. 610.
6. O. V. Bogdankevich, M. M. Zverev, A. I. Krasilnikov, and A. N. Pechenov, *Phys. Stat. Sol.*, *19*, K5, 1967.
7. N. G. Basov, O. V. Bogdankevich, and A. G. Deviatkov, *Zh. Exp. i Teor. Fiz.*, *47*, 1588, 1964; *Sov. Phys. JETP*, *20*, 1067, 1965.
8. C. E. Hurwitz, *Appl. Phys. Lett.*, *8*, 121, 1966.

Table 2-72. SPECTRAL RANGE COVERED BY SEMICONDUCTOR LASERS *(Continued)*

9. F. H. Nicoll, *Appl. Phys. Lett.*, *10*, 69, 1967.
10. F. H. Nicoll, *RCA Review*, *29*, 379, 1968.
11. J. L. Brewster, *Appl. Phys. Lett.*, *13*, 385, 1968.
12. N. G. Basov, A. G. Grasyuk, I. G. Zubarev, and V. A. Katulin, *Sov. Phys. Solid State*, 7, 2932, 1966.
13. C. E. Hurwitz, *IEEE J. Quant. Electronics*, *3*, 333, 1967.
14. N. G. Basov, O. V. Bogdankevich, and A. N. Pechenov, *Sov. Phys. Dokl.*, *10*, 329, 1965.
15. C. E. Hurwitz, *Appl. Phys. Lett.*, *8*, 121, 1966.
16. D. L. Keune, J. A. Rossi, O. L. Gaddy, H. Merkelo, and N. Holonyak, Jr., *Appl. Phys. Lett.*, *14*, 99, 1969.
17. V. S. Vavilov and E. L. Nolle, *Proc. Int. Conf. Phys. Semiconductors*, Moscow, 1968, p. 600.
18. G. E. Stillman, M. D. Sirkis, J. A. Rossi, M. R. Johnson, and N. Holonyak, Jr., *Appl. Phys. Lett.*, *9*, 268, 1966.
19. N. Holonyak, Jr., M. R. Johnson, and D. L. Keune, *IEEE J. Quant. Electronics*, *QE-4*, 199, 1968.
20. H. Rupprecht, J. M. Woodall, G. D. Pettit, *IEEE J. Quant. Electronics*, *4*, 35, 1968.
21. W. Susaki, T. Sogo, T. Oku, *IEEE J. Quant. Electronics*, *4*, 423, 1968.
22. H. Nelson and H. Kressel, *Appl. Phys. Lett.*, *15*, 7, 1969.
23. H. Kressel and H. Nelson, *RCA Review*, *30*, 106, 1969.
24. I. Hayashi, M. B. Panish, and P. Foy, *IEEE J. Quantum. Electronics*, *5*, 211, 1969.
25. M. B. Panish, I. Hayashi, and S. Sumski, *IEEE J. Quant. Electronics*, *5*, 210, 1969.
26. M. B. Panish, I. Hayashi, and S. Sumski, *Appl. Phys. Lett.*, *16*, 326, 1970.
27. I. Hayashi and M. B. Panish, *J. Appl. Phys.*, *41*, 150, 1970.
28. Zh. I. Alferov, V. M. Andreev, V. I. Korolkov, E. L. Portnoi, and D. N. Tretyakov, *Sov. Phys. Semiconductors*, *2*, 1289, 1969.
29. Zh. I. Alferov, V. M. Andreev, E. L. Portnoi, and M. K. Trukan, *Sov. Phys. Semiconductors*, *3*, 1107, 1969.
30. N. Holonyak, Jr. and S. F. Bevacqua, *Appl. Phys. Lett.*, *1*, 82, 1962.
31. N. Holonyak, Jr., *Trans. TMS-AIME*, *230*, 276, 1964.
32. M. Pilkuhn and H. Rupprecht, *J. Appl. Phys.*, *36*, 684, 1965.
33. J. I. Pankove, H. Nelson, J. J. Tietjen, I. J. Hegyi, and H. P. Maruska, *RCA Review*, *28*, 560, 1967.
34. J. L. Pankove, I. J. Hegyi, *Proc. IEEE*, *56*, 324, 1968.
35. N. G. Basov, O. V. Bogdankevich, P. G. Eliseev, and B. M. Lavrushin, *Sov. Phys.—Solid State*, *8*, 1073, 1966.
36. L. N. Kurbatov, V. E. Mashchenko, N. N. Mochalkin, A. D. Britov, and A. I. Dirochka, *Proc. Int. Conf. Phys. Semiconductors*, Moscow, 1968, p. 587.
37. V. S. Vavilov and E. L. Nolle, *Dokl. Akad. Nauk SSSR*, *164*, 73, 1965.
38. V. S. Vavilov, E. L. Nolle, and V. D. Egorov, *Sov. Phys.—Solid State*, 7, 749 (1965); *ibid*, *9*, 657, 1967.
39. N. G. Basov, O. V. Bogdankevich, P. G. Eliseev, and B. M. Lavrushin, *Sov. Phys.—Solid State*, *8*, 1073, 1966.
40. M. I. Nathan, W. P. Dumke, G. Burns, F. H. Dill, Jr., and G. J. Lasher, *Appl. Phys. Lett.*, *1*, 62, 1962.
41. R. N. Hall, G. E. Fenner, J. D. Kingsley, T. J. Soltys, and R. O. Carlson, *Phys. Rev. Lett.*, *9*, 366, 1962.
42. T. M. Quist, R. H. Rediker, R. J. Keyes, W. E. Krag, B. Lax, A. L. McWhorter, and H. J. Zeiger, *Appl. Phys. Lett.*, *1*, 91, 1962.
43. C. E. Hurwitz and R. J. Keyes, *Appl. Phys. Lett.*, *5*, 139, 1964.
44. D. A. Cusano, *Appl. Phys. Lett.*, *7*, 151, 1965.
45. O. V. Bogdankevich, N. A. Borisov, I. V. Kryukova, and B. M. Lavrushin, *Sov. Phys. Semiconductors*, *2*, 845, 1969.
46. N. G. Basov, A. Z. Grasyuk, and V. A. Katulin, *Sov. Phys. Dokl.*, *10*, 343, 1965.
47. P. D. Dapkus, N. Holonyak, Jr., J. A. Rossi, F. V. Williams, and D. A. High, *J. Appl. Phys.*, *40*, 3300, 1969.
48. P. D. Southgate, *IEEE J. Quant. Electronics*, *QE-4*, 179, 1968.
49. K. Weiser, R. S. Levitt, *Appl. Phys. Lett.*, *2*, 176, 1963.
50. G. Burns, R. S. Levitt, M. I. Nathan, and K. Weiser, *Proc. IEEE*, *51*, 1148, 1963.
51. P. D. Southgate and R. T. Mazzochi, *Phys. Rev. Lett.*, *28A*, 216, 1968.
52. F. M. Berkovskii, N. A. Goryunova, V. M. Orlov, S. M. Ryvkin, V. I. Sokolova, E. V. Tsevkova, and G. P. Shpenkov, *Sov. Phys. Semiconductors*, *2*, 1027, 1969.
53. F. B. Alexander, V. R. Bird, D. R. Carpenter, G. W. Manley, P. S. McDermott, J. R. Peloke, H. F. Quinn, R. J. Riley, and L. R. Yetter, *Appl. Phys. Lett.*, *4*, 13, 1964.
54. C. Chipaux and E. Eymard, *Phys. Stat. Sol.*, *10*, 165, 1965.
55. I. V. Kryukova, *et al.*, *Sov. Phys. Solid State*, *8*, 822 and 1538, 1966.
56. B. Pistoulet and H. Mathieu, *Proc. Int. Conf. Phys. Semiconductors*, Moscow, 1968, p. 352.
57. C. Benoit-a-la Guillaume and J. M. Debever, *Compt. Rend. 258*, 2200, 1964.
58. I. Melngailis, A. J. Strauss and R. H. Rediker, *Proc. IEEE*, *51*, 1154, 1963.
59. S. G. Bishop, W. J. Moore and E. M. Swiggard, *Appl. Phys. Lett. 15*, 12, 1969.
60. I. Melngailis, *Appl. Phys. Lett. 2*, 176, 1963.
61. M. Rodot, P. Leroux Hugon, J. Besson and H. Lebloch, *L'Onde Electrique*, *45*, 1197, 1965.
62. I. Melngailis and R. H. Rediker, *J. Appl. Phys. 37*, 899, 1966.
63. M. Rodot, C. Vérié, Y. Marfaing, J. Besson and H. Lebloch, *IEEE J. Quant. Electronics*, *QE-2*, 586, 1966.
64. C. Benoit-a-la Guillaume and J. M. Debever, *Solid State Commun. 1*, 10, 1965.
65. I. Melngailis, *IEEE J. Quant. Electronics*, *QE-1*, 104, 1965.
66. N. G. Basov, A. V. Dudenkova, A. I. Krasil'nikov, V. V. Nikitin, and K. P. Fedoseev, *Sov. Phys. Solid State*, *8*, 847, 1966.
67. I. Melngailis and A. J. Strauss, *Appl. Phys. Lett. 8*, 179, 1966.
68. C. Benoit-a-la Guillaume and J. M. Debever, *Solid State Commun. 3*, 19, 1965.
69. J. F. Butler and A. R. Calawa, *J. Electrochem. Soc. 54*, 1056, 1965.
70. C. E. Hurwitz, A. R. Calawa and R. H. Rediker, *IEEE J. Quant. Electronics*, *QE-1*, 102, 1965.
71. C. Benoit-a-la Guillaume and P. Lavallard, *Solid State Commun. 1*, 148, 1963.
72. I. Melngailis, R. J. Phelan, and R. H. Rediker, *Appl. Phys. Lett. 5*, 99, 1964.
73. I. Melngailis, *Appl. Phys. Lett.*, *6*, 59, 1965.
74. C. Benoit-a-la Guillaume and J. M. Debever, *Solid State Commun.*, *2*, 145, 1964.
75. R. J. Phelan and R. H. Rediker, *Appl. Phys. Lett.*, *6*, 70, 1965.
76. R. J. Phelan, A. R. Calawa, R. H. Rediker, R. J. Keyes, and B. Lax, *Appl. Phys. Lett.*, *3*, 143, 1963.
77. A. P. Shotov, S. P. Grishechkina, B. D. Kopilovskii, and R. A. Muminov, *Sov. Phys. Solid State*, *8*, 865 and 1998, 1966.
78. A. P. Shotov, S. P. Grishechkina, and R. A. Muminov, *Proc. Int. Conf. Phys. Semiconductors*, Moscow, 1968, p. 539.
79. J. F. Butler, A. R. Calawa, R. J. Phelan, Jr., T. C. Harman, A. J. Strauss, and R. H. Rediker, *Appl. Phys. Lett. 5*, 75, 1964.
80. L. N. Krubatov, A. D. Britov, I. S. Aver'yanov, V. E. Mashchenko, N. N. Mochalkin, and A. I. Dirochka, *Sov. Phys. Semiconductors*, *2*, 1008, 1969.

Table 2-72. SPECTRAL RANGE COVERED BY SEMICONDUCTOR LASERS (Continued)

81. L. N. Kurbatov, V. E. Mashchenko, N. N. Mochalkin, A. D. Britov, and A. I. Dirochka, *Proc. Int. Conf. Phys. Semiconductors*, Moscow, 1968, p. 587.
82. J. F. Butler, A. R. Calawa, R. J. Phelan, Jr., A. J. Strauss, and R. H. Rediker, *Solid State Commun.*, 2, 303, 1964.
83. J. F. Butler, A. R. Calawa, and R. H. Rediker, *IEEE J. Quant. Electronics*, QE-1, 4, 1965.
84. J. M. Besson, J. F. Butler, A. R. Calawa, W. Paul, and R. H. Rediker, *Appl. Phys. Lett.*, 7, 206, 1965.
85. I. Chambouleyron, J. M. Besson, M. Balkanski, H. Rodot, and H. Abrales, *Proc. Int. Conf. Phys. Semiconductors*, Moscow, 1968, p. 546.
86. J. F. Butler and T. C. Harman, *Appl. Phys. Lett.*, 12, 347, 1968, and also *IEEE J. Quant. Electronics*, QE-5, 50, 1969.
87. J. F. Butler, A. R. Calawa, and T. C. Harman, *Appl. Phys. Lett.*, 9, 427, 1966.
88. T. C. Harman, A. R. Calawa, I. Melngailis, and J. O. Dimmock, *Appl. Phys. Lett.*, 14, 333, 1969.
89. A. R. Calawa, J. O. Dimmock, T. C. Harman, and I. Melngailis, *Phys. Rev. Lett.*, 23, 7, 1969.

Table 2-73. CONTINUOUS-WAVE INSULATING CRYSTAL LASERS*

COURTESY OF MARVIN J. WEBER

Laser ion	Host	Sensitizer ion(s)	Wavelength, μm	Temperature, K	Optical pump	Power, W	Efficiency,[a] percent	Reference
IRON GROUP IONS								
Cr^{3+}	Al_2O_3		0.694	300	Hg	2.4	~0.1	1,2
			0.694^b	4.2, 77	Ar laser			3
Ni^{2+}	MgF_2		1.67	85	W	1	0.2	4
Ni^{2+}	MnF_2		1.93	85	W			4
DIVALENT RARE EARTH IONS								
Dy^{2+}	CaF_2		2.36	77	W	1.2	0.06	5
			2.36	27	Sunlight			6
Tm^{2+}	CaF_2		1.12	27	Hg			7
TRIVALENT RARE EARTH IONS								
Nd^{3+}	$Ca(NbO_3)_2$		1.06	300	Xe	0.12	0.05	8
Nd^{3+}	$Ca_5(PO_4)_3F$		1.06	300	W	1.3	0.2	9
Nd^{3+}	$CaWO_4$		1.06	300	Xe	<0.1	~0.01	10
			1.06	300	Hg	~0.01	~0.01	11
			1.06	85	Hg	0.5	0.03	12
Nd^{3+}	LaF_3		1.04	300				13
Nd^{3+}	$YAlO_3$ (b axis)		1.06	300	Kr	35	0.8	14
			1.08	300	Kr	100	1.8	15
Nd^{3+}	$YAlO_3$ (c axis)		1.06	300	Kr	6.5	0.3	16
Nd^{3+}	$Y_3Al_5O_{12}$		1.06	300	W	~25	1.0	17
			1.06	300	Kr	250	2.1	18
			1.06	300	Kr	$1\,100^c$	2.0	19
			1.06	300	Plasma arc	200	0.2	20
			1.06	300	Na-doped Hg	0.5	0.2	21
			1.32	300	W	0.03	~0.01	22
		Cr^{3+}	1.06	300	Hg	10	0.4	23
Ho^{3+}	CaF_2	$Er^{3+}, Tm^{3+}, Yb^{3+}$	2.1	77	Xe			24
Ho^{3+}	Er_2O_3	Er^{3+}	2.12	77	W			25
Ho^{3+}	$Er_{1.5}Y_{1.5}Al_5O_{12}$	Er^{3+}	2.10	85	Hg, W			26
Ho^{3+}	$Y_3Al_5O_{12}$	Cr^{3+}	2.10	85	W			26
		Cr^{3+}	2.12	85	Hg, W			26
		Er^{3+}, Tm^{3+}	2.12	85	W	15	5	28
			2.12	77	W-I	20	3.5	27
Tm^{3+}	CaF_2	Er^{3+}	1.9	77	Xe			24
Tm^{3+}	Er_2O_3	Er^{3+}	1.93	77	W			29

[a] Overall efficiency: laser output/electrical energy input to pump lamp. Because the efficiency depends on a number of factors, such as rod quality, pump cavity efficiency, and output coupling efficiency, the original references should be consulted when comparing values.

[b] Nonspiking, single-mode operation.

[c] Multiple laser rods in series inside one resonant cavity.

*Reprinted from: M.J. Weber, Insulating Crystal Lasers, "Handbook of Lasers", R.J. Pressley, Ed., The Chemical Rubber Co., 1971.

Table 2-73.   CONTINUOUS-WAVE INSULATING CRYSTAL LASERS
(Continued)

Laser ion	Host	Sensitizer ion(s)	Wavelength, μm	Temperature, K	Optical pump	Power, W	Efficiency,[a] percent	Reference
TRIVALENT RARE EARTH IONS (Continued)								
Tm^{3+}	$Er_{1.5}Y_{1.5}Al_5O_{12}$	Er^{3+}	2.01	85	W			26
Tm^{3+}	$Y_3Al_5O_{12}$		2.01	85	W			26
		Cr^{3+}	2.01	85	Hg, W			26
ACTINIDE IONS								
U^{3+}	CaF_2		2.61	77	Hg	10^{-5}		30

REFERENCES

1. V. Evtuhov and J.K. Neeland, *J. Appl. Phys.*, 38:4051, 1967.
2. D. Roess, *IEEE J. Quant. Electronics*, QE-2:208, 1966.
3. M. Birnbaum, P.H. Wendzikowski, and C.L. Fincher, *Appl. Phys. Lett.*, 1b:436, 1970.
4. L.F. Johnson, H.J. Guggenheim, and R.A. Thomas, *Phys. Rev.*, 149:179, 1966.
5. R.J. Pressley and J.P. Wittke, *IEEE J. Quant. Electronics*, QE-3:1966, 1967.
6. Z.J. Kiss, H.R. Lewis, and R.C. Duncan, *Appl. Phys. Lett.*, 2:93, 1963.
7. R.C. Duncan and Z.J. Kiss, *Appl. Phys. Lett.*, 3:23, 1963.
8. Kh.S. Bagdasarov, M.M. Gritsenko, F.M. Zubkova, A.A. Kaminskii, A.M. Kevorkov, and L. Li, *Sov. Phys.-Cryst.*, 15:323, 1970.
9. R.H. Hopkins, et al., Technical Report AFAL-TR-69-239, Air Force Avionics Laboratory, 1969.
10. A.A. Kaminskii, L.S. Kornienko, G.V. Maksimova, V.V. Osiko, A.M. Prokhorov, and G.P. Shipulo, *Sov. Phys. JETP*, 22:22, 1966.
11. L.F. Johnson, G.D. Boyd, K. Nassau, and R.R. Soden, *Phys. Rev.*, 126:1406, 1962.
12. L.F. Johnson, in "Quantum Electronics Proceedings of the Third International Congress", P. Grivet and N. Bloembergen, Eds., Columbia University Press, 1964, p. 1021.
13. Yu. K. Voronko, M.V. Dmitruk, A.A. Kaminskii, V.V. Osiko, and V.N. Shpakov, *Sov. Phys. JETP*, 27:400, 1968.
14. G.A. Massey and J.M. Yarborough, *Appl. Phys. Lett.*, 18:576, 1971.
15. M. Bass and M.J. Weber, *Laser Focus*, 7:34, September 1971.
16. M.J. Weber, M. Bass, R.R. Monchamp, and E. Comperchio, Technical Report AFML-TR-70-258, Air Force Materials Laboratory, 1970.
17. J.E. Geusic, *NEREM Record*, 8:192, 1966.
18. W. Koechner, *Laser Focus*, 5:29, September 1969.
19. K. Erickson, *Laser Focus*, 6:16, 1970; and private communication.
20. J.E. Jackson and D.M. Yenni, Second Interim Technical Report Contract SRCR-66-4, May 1966.
21. R.G. Schlecht, C.H. Church, and D.A. Larson, *IEEE J. Quant. Electronics*, QE-2, 1966.
22. R.G. Smith, *IEEE J. Quant. Electronics*, QE-4:505, 1968.
23. R.J. Pressley, J.R. Collard, P.V. Goedertier, F. Sterzer, and W. Zernik, Technical Report AFAL-TR-66-129, June 1966.
24. Yu.K. Voronko, M.V. Dmitruk, T.M. Murina, and V.V. Osiko, *Izv. Akad. Nauk. SSSR, Neorgan. Materialy*, 5:506, 1969.
25. B.H. Soffer and R.H. Hoskins, *IEEE J. Quant. Electronics*, QE-2:253, 1966.
26. L.F. Johnson, J.E. Geusic, and L.G. Van Uitert, *Appl. Phys. Lett.*, 7:127, 1965.
27. D.P. Devor, B.H. Soffer, and G.E. Moss, Technical Report AFAL-TR-71-181, October 1971.
28. L.F. Johnson, J.E. Geusic, and L.G. Van Uitert, *Appl. Phys. Lett.*, 8:200, 1966.
29. B.H. Soffer and R.H. Hoskins, *Appl. Phys. Lett.*, 6:200, 1965.
30. G.D. Boyd, R.J. Collins, S.P.S. Porto, A. Yariv, and W.A. Hargreaves, *Phys. Rev. Lett.*, 8:269, 1962.

Table 2-74.   PROPERTIES OF SOLID-LASER MATERIALS*

The host crystal in which the best laser results were obtained is discussed in detail. Other host crystals are listed in the "Remarks" column.

Note: Numbers in parentheses designate temperatures in degrees Kelvin.

Laser material	Output wavelength, microns	Operating mode (and temperature, °K)	t_{spont}, milliseconds	Pulse threshold in joules (and temperature, °K)	Useful absorption regions, microns	Position of terminal level, E_1 (cm^{-1})	Laser transition	Remarks
$Cr^{3+}:Al_2O_3$ (ruby) 10^{19} atoms/cc	0.6934 (R_1, 77°) 0.6929 (R_2, 290°)	cw (77) pulsed (350) pulsed	3	~800 (77)	0.5–0.6 0.32–0.44 0.5–0.6 0.32–0.44	0 0	$^2E(\bar{E}) \rightarrow {}^4A_2$ $^2E(2\bar{A}) \rightarrow {}^4A_2$	"Spiking" observed in both pulsed and cw operation
$Cr^{3+}:Al_2O_3$ $n \sim 10^{20}$	0.701, 0.704	pulsed (77)				~100		Due to paired chromium ions
$U^{3+}:CaF_2$	2.613 2.438 2.511 2.223	pulsed (300) cw (~100) pulsed (77) pulsed (77) pulsed (77)	0.15 (77)	1 (77) 6 (77) 2000 (77)	~0.9	609 0	$^4I_{11/2} \rightarrow {}^4I_{9/2}$	"Spiking" present in pulsed but not in cw operation. Pulsed emission also observed in BaF_2 at 2.556 μ and in SrF_2 at 2.472 μ, 2.408 μ
$Nd^{3+}:CaWO_4$	1.065 1.063 1.066 1.058 1.064	cw (300) pulsed pulsed pulsed pulsed	~0.1 (77)	~1 (77) 14 (77) 6 (77) 80 (77) 7 (77)	0.57–0.6	~2000	$^4F_{3/2} \rightarrow {}^4I_{11/2}$	"Spiking" present in pulsed but not in cw operation. Laser emission from Nd^{3+} also observed in $SrMO_4$, $SrWO_4$, $CaMO_4$, $PbMO_4$, CaF_2, SrF_2, BaF_2, and LaF_3
$Nd^{3+}:Glass$	1.06	pulsed (300)		~50		~2000	$^4F_{3/2} \rightarrow {}^4I_{11/2}$	
$Pr^{3+}:CaWO_4$	1.047	pulsed (77)	0.05 (77)	15 (77)	0.45–0.5	37?	$^1G_4 \rightarrow {}^3H_4$	Pulsed laser emission of Pr^{3+} was also detected in $SrMO_4$

Table 2-74.   PROPERTIES OF SOLID-LASER MATERIALS (Continued)

Laser material	Output wavelength, microns	Operating mode (and temperature, °K)	t_{spont}, milliseconds	Pulse threshold in joules (and temperature, °K)	Useful absorption regions, microns	Position of terminal level, E_1 (cm^{-1})	Laser transition	Remarks
$Dy^{2+}:CaF_2$	2.36	cw (90)	~10 (77)	20 (77)	0.8–1.0	35	$^5I_7 \rightarrow {}^5I_8$	"Spiking" in pulsed operation but not in continuous
$Tm^{3+}:CaWO_4$	1.911 / 1.916	pulsed (77) / pulsed (77)		60 (77) / 73 (77)	{ 0.46–0.48 / 1.7–1.8	~325	$^3H_4 \rightarrow {}^3H_6$	1.918 μ emission also observed. Laser emission also observed in SrF_2
$Er^{3+}:CaWO_4$	1.612	pulsed (77)		800 (77)	0.38 / 0.52	375	$^4I_{13/2} \rightarrow {}^4I_{15/2}$	
$Ho^{3+}:CaWO_4$	2.046 / 2.059	pulsed (77) / pulsed (77)		80 (77) / 250 (77)	0.44–0.46	~230	$^5I_7 \rightarrow {}^5I_8$	
$Tm^{2+}:CaF_2$	1.116	pulsed (~4)	4	50 (4)	0.28–0.34 / 0.39–0.46 / 0.53–0.63	0	$^2F_{5/2} \rightarrow {}^2F_{7/2}$	
$Sm^{2+}:CaF_2$	0.708	pulsed (20)	0.002	0.01 (20)	0.425–0.5 / 0.59–0.65	263	$^5D_0 \rightarrow {}^7F_1$	No "spiking" in pulsed operation. Laser action at 0.6969 μ also observed in $SrF_2:Sm^{2+}$
$Yb^{3+}:Glass$	1.015	pulsed (77)	1.5	1300	~0.91 / ~0.95 / ~0.98		$^2F_{5/2} \rightarrow {}^2F_{7/2}$	
$Gd^{3+}:Glass$	0.3125	pulsed (77)	4 (300)		0.274 / 0.277		$^6P_{7/2} \rightarrow {}^8S_{7/2}$	
$Ho^{3+}:Glass$	$\lambda > 1.95\ \mu$	pulsed (77)	~0.7 (77)	3600 (77)	0.44–0.46		$^5I_7 \rightarrow {}^5I_8$	

*From: "The Laser", A. Yariv and J. P. Gordon, Proceedings of the IEEE, 51(1):13, The Institute of Electrical and Electronics Engineers, January 1963.

Table 2-75. PROPERTIES OF LINEAR ELECTROOPTICAL MATERIALS*

COURTESY OF IVAN P. KAMINOW AND EDWARD H. TURNER

The *refractive index* of a crystal is described by an ellipsoid (indicatrix)

$$B_{ij}X_iX_j = 1 \equiv B_{11}X_1^2 + B_{22}X_2^2 + B_{33}X_3^2 + 2B_{23}X_2X_3 + 2B_{13}X_1X_3 + 2B_{12}X_1X_2,$$

in which summation over repeated indices is understood and $B_{ij} = B_{ji}$. By definition

$$B_{ij} = \varepsilon_0 \partial E_i/\partial D_j = \left(\frac{1}{\varepsilon}\right)_{ij}$$

where ε_0 is the vacuum permittivity and ε is the relative dielectric constant.

The *electrooptic coefficient* $r_{ij,k} \equiv r_{lk}$ is defined by

$$\Delta B_{ij} = r_{ij,k}E_k$$

$$\Delta B_l = r_{lk}E_k,$$

in which the indices i, j, and k each cover the rectangular coordinate axes 1, 2, and 3, and $l = (ij)$ refers to the six reduced combinations $1 = (11)$, $2 = (22)$, $3 = (33)$, $4 = (23)$, $5 = (13)$, and $6 = (12)$.

If $r_{ij,h}$ is determined at constant strain—for example, by making a measurement at high frequencies well above acoustic resonances of the sample—the crystal is *clamped*, as indicated by the letter (S) or $r_{ij,k}^S$. If $r_{ij,k}$ is determined at constant stress—for example, at low frequencies well below the acoustic resonances of the sample—the crystal is *free*, as indicated by the letter T or $r_{ij,k}^T$.

In a principal axis system where $B_{ij} = 0$ for $i \neq j$, and $B_{ii} \equiv B_i \equiv 1/\varepsilon_i$

$$\Delta B_{ij} = -\Delta\varepsilon_{ij}/\varepsilon_i\varepsilon_j,$$

for $\Delta\varepsilon_{ij} \ll \varepsilon_i, \varepsilon_j$.

Typical accuracies for r_{lk} are $\pm15\%$. References containing more extensive wavelength and temperature dependence are indicated by λ and t, respectively. Electrooptic coefficients are stated in SI units (m/volt); to convert to CGS units (cm/statvolt), multiply by 3×10^4. Unless stated explicitly, the signs of r_{lk} have not been determined.

Table A. ABO₃-TYPE COMPOUNDS

Material and symmetry (critical temp, K)	Electrooptic coefficients			Refractive index		Dielectric constant
	$r_{13}, 10^{-12}$ m/V	$r_{lk}, 10^{-12}$ m/V	λ, µm	n_i	λ, µm	ε_l
LiNbO₃, 3m (1 470)	$(T)r_c = 19$	$(T)r_{22} = 7$	.633[23]	$n_1 = n_2 = 2.378\ 0$	45[32,λ]	$(T)\varepsilon_1 = \varepsilon_2 = 78[31,t]$
	$(T)r_c = 17.4$	$(T)r_{22} = 3.2$	.633[24]	2.271 6	.70[33,t,λ]	$(T)\varepsilon_3 = 32$
		$(T)r_{51} = 32$		2.237 0	1.00[35,t,λ]	$(S)\varepsilon_1 = \varepsilon_2 = 43[42]$
	$(T)r_{33} = +32.2$	$(T)r_{22} = 6.8$	.633[25,t]	2.197 4	2.00	$(S)\varepsilon_3 = 28$
	$(T)r_{13} = +10$			2.115 5	4.00	
	$(T)r_c = 18$	$(T)r_{22} = 6.7$	.633[28]	$n_3 = 2.277\ 2$	.45	
	$(T)r_c = 17$	$(T)r_{22} = 5.7$	1.15[28]	2.187 4	.70	
	$(T)r_c = 16$	$(T)r_{22} = 3.1$	3.39[28]	2.156 7	1.00	
	All$(T)r_{ij} > 0$		.633[26]	2.125 0	2.00	
	$(S)r_{33} = +30.8$	$(S)r_{22} = 3.4$	.633[29]	2.055 3	4.00	
	$(S)r_{13} = +8.6$	$(S)r_{51} = +28$				
	$(S)r_{33} = 28$	$(S)r_{22} = 3.1$	3.39[30]			
	$(S)r_{13} = 6.5$	$(S)r_{51} = 23$				
LiTaO₃, 3m (890)	$(T)r_c = 22$		.633[34]	$n_1 = n_2 = 2.183\ 4$	.60[36,λ]	$(T)\varepsilon_2 = \varepsilon_1 = 51[37]$
	$(S)r_{33} = 30.3$	$(S)r_{51} = 20$	.633[34]	2.130 5	1.20	$(T)\varepsilon_3 = 45$
	$(S)r_{13} = 7$	$(S)r_{22} \approx 1$		2.033 5	4.00	$(S)\varepsilon_2 = \varepsilon_1 = 41$
	$(S)r_{33} = 27$	$(S)r_{51} = 15$	3.39[30]	$n_3 = 2.187\ 8$	.60	$(S)\varepsilon_3 = 43$
	$(S)r_{13} = 4.5$	$(S)r_{22} \approx .3$		2.134 1	1.20	
	All$(T)r_{ij} > 0$		.633[27]	2.037 7	4.00	

*Condensed from: Ivan P. Kaminow and Edward H. Turner, Linear Electrooptic Materials, in "Handbook of Lasers", R.J. Pressley, Ed., The Chemical Rubber Co., 1971.

Table 2-75.   PROPERTIES OF LINEAR ELECTROOPTICAL MATERIALS
(Continued)

Material and symmetry (critical temp, K)	Electrooptic coefficients			Refractive index		Dielectric constant
	r_{13}, 10^{-12} m/V	r_{lk}, 10^{-12} m/V	λ, μm	n_i	λ, μm	ε_i
$Ba_2NaNb_5O_{15}$, $mm2$ (833)	$(T)r_c = 34$ $(T)r_{33} = 48$ $(T)r_{13} = 15$ $(T)r_{23} = 13$ $(S)r_{33} = +29$ $(S)r_{23} = 8$ $(S)r_{13} = 7$	$(T)r_{42} = 92$ $(T)r_{51} = 90$ $(S)r_{42} = 75$ $(S)r_{51} = 88$	.633[40] .633[39] .633[38]	$n_1 = 2.322$ $n_2 = 2.321$ $n_3 = 2.218$	.633[39,λ]	$(T)\varepsilon_1 = 235[41]$ $(T)\varepsilon_2 = 247$ $(T)\varepsilon_3 = 51$ $(S)\varepsilon_1 = 222$ $(S)\varepsilon_2 = 227$ $(S)\varepsilon_3 = 32$

Table B.   KDP- AND ADP-TYPE CRYSTALS

Material	T_c, K [1]	Electrooptic coefficients		Refractive index		Dielectric constant	
		r_{63}, 10^{-12} m/V	r_{41}, 10^{-12} m/V	n_3	n_1	ε_3	ε_1
KH_2PO_4 (KDP)	123	$(T)-10.5[2]$; $[3,\lambda;4,t]$ $(T)9.37[5]$ $(S)8.8[11]$; $[3,\lambda]$ $(S)8.15[12]$; $r_{63}<0[6]$	$+8.6[2]$ $r_{41}<0[7]$	1.47	$1.51[8,9-\lambda,t]$	$(T)21$ $(S)21$	$42[10]$ $44[13]$
KD_2PO_4 (DKDP)	222	$(T)26.4[14]$; $[3,\lambda;4,t]$ $(S)24.0[15]$; $.93r_{63}^T$ $[3,\lambda]$	8.8[16]	1.47	$1.51[9-\lambda,t]$	$(T)50[14]$ $(S)48$	$58[17]$
$NH_4H_2PO_4$ (ADP)	148[a]	$(T)-8.5[2]$; $[19,3,\lambda]$ $(S)5.5[2]$; $4.1[21]$; $[3,\lambda]$	$24.5[16]$, $23.1[20]$ $r_{41}<0[7]$	1.48	$1.53[8,9-\lambda,t]$	$(T)15$ $(S)14$	$56[10]$ $58[17]$
$ND_4D_2PO_4$ (DADP)	242[a]	$(T)11.9[18,\lambda]$; $[22,t]$			1.52[18]		

[a]Antiferroelectric transition.

Table C.   AB-TYPE COMPOUNDS

Material and symmetry	Electrooptic coefficients		Refractive index		Dielectric constant, ε_i
	r_{lk}, 10^{-12} m/V	λ, μm	n_i	λ, μm	
GaAs, $\bar{4}3m$	$(S)r_{41} = 1.2$ $(S)r_{41} = -1.5$ $(S+T)r_{41} = 1.2$ to 1.6 $(T)r_{41} = 1.0-1.2$ $(T)r_{41} = 1.6$ $(S)r_{41} = 1.5$	.9–1.08[43] 3.39[44] 1.0–3.0[45] 4.0–12.0[45] 10.6[47,48] Raman scat. [48,49]	$n_0 = 3.60$ $n_0 = 3.50$ $n_0 = 3.42$ $n_0 = 3.30$	.9[50] 1.02[50] 1.25[50] >5.0[51]	$(S)\varepsilon = 13.2[52]$ $(S)\varepsilon = 12.3[53]$ $(T)\varepsilon = 12.5[51]$
CdTe, $\bar{4}3m$	$(T)r_{41} = 6.8$ $(T)r_{41} = 6.8$ $(T)r_{41} = 5.5$ $(T)r_{41} = 5.0$	3.39[59] 10.6[59] 23.35[60] 27.95[60]	$n_0 = 2.82$ $n_0 = 2.60$ $n_0 = 2.58$ $n_0 = 2.53$	1.3[54] 10.6[61] 23.34[61] 27.95[61]	$(S)\varepsilon = 9.4[53]$
CdS, $6mm$	$(T)r_c = 4$ $(T)r_{51} = 3.7$ $(T)r_c = 5.5$ $(S)r_{33} = 2.4$ $(S)r_{13} = 1.1$	.589[62] .589[62] 10.6[47] .633[58] .633[58]	$n_3 = 2.48$ $n_1 = n_2 = 2.46$ $n_3 = 2.3$	.63[55] .63[55] 10.0[55]	$(T)\varepsilon_3 = 10.33[56]$ $(T)\varepsilon_1 = 9.35[56]$ $(S)\varepsilon_1 = 9.02[56]$; 8.7[57] $(S)\varepsilon_3 = 9.53[56]$; 9.25[57]

Table 2-75. PROPERTIES OF LINEAR ELECTROOPTICAL MATERIALS
(Continued)

REFERENCES

1. F. Jona and G. Shirane, "Ferroelectric Crystals", Macmillan Company, 1962.
2. R.O'B. Carpenter, The Electrooptic Effect in Crystals of the Dihydrogen Phosphate Type, Part III, Measurement of Coefficients, *J. Opt. Soc. Amer.*, 40:225–229, 1950; and Electrooptic Sound-on-film Modulator, *J. Opt. Soc. Amer.*, 25:1145–1148, 1953.
3. A.S. Vasilevskaya, The Electrooptic Properties of Crystals of KDP-type, *Sov. Phys.-Cryst.*, 11:644–647, 1967; J.F. Ward and P.A. Frankcen, Structure of Nonlinear Optical Phenomena in Potassium Dihydrogen Phosphate, *Phys. Rev.*, 133:A183–190, 1964.
4. A.S. Sonin, A.S. Vasilevskaya, and B.A. Strukov, Electrooptic Properties of Potassium Dihydrogen Phosphate and Deuterated Potassium Dihydrogen Phosphate in the Region of Their Phase Transitions, *Sov. Phys.-Solid State*, 8:2758–60, 1967.
5. O.G. Blokh, Dispersion of r_{63} for Crystals of ADP and KDP, *Sov. Phys.-Cryst.*, 7:509–511, 1962.
6. E.H. Turner, unpublished.
7. J.F. Ward and G.H.C. New, Optical Rectification in Ammonium Dihydrogen Phosphate, Potassium Dihydrogen Phosphate, and Quartz, *Proc. Roy. Soc.*, 4299:238–263, 1967.
8. F. Zernike, Jr., Refractive Indices of ADP and KDP between 0.2 and 1.5 μ, *J. Opt. Soc. Amer.*, 54:1215–1220, 1964; Correction, *ibid.*, 55:210E, 1965; V.N. Vishnevskii and I.V. Stefanski, Temperature Dependence on the Dispersion of the Refractivity of ADP and KDP Single Crystals, *Opt. and Spectr.*, 20:195–196, February 1966.
9. M. Yamazaki and T. Ogawa, Temperature Dependences of the Refractive Indices of $NH_4H_2PO_4$, KH_2PO_4, and Partially Deuterated KH_2PO_4, *J. Opt. Soc. Amer.*, 56:1407–1408, 1966; R.A. Phillips, Temperature Variation of the Index of Refraction of ADP, KDP, and Deuterated KDP, *J. Opt. Soc. Amer.*, 56:629–632, 1966.
10. D.A. Berlincourt, D.R. Curran, and H. Jaffe, "Physical Acoustics", W.P. Mason, Ed., Vol. 1, Pt. A, Academic Press, 1964, pp. 169–260.
11. R.D. Rosner, E.H. Turner, and I.P. Kaminow, Clamped Electrooptic Coefficients of KDP and Quartz, *Appl. Optics*, 6:778, 1967.
12. E.A. Ohm, A Linear Optical Modulator with High FM Sensitivity, *Appl. Optics*, 6:1233–1235, 1967.
13. I.P. Kaminow and G.O. Harding, Complex Dielectric Constant of KH_2PO_4 at 9.2 Gc/sec, *Phys. Rev.*, 129:1562–1566, 1963.
14. T.R. Sliker and S.R. Burlage, Some Dielectric and Optical Properties of KD_2PO_4, *J. Appl. Phys.*, 34:1837–1840, 1963.
15. T.M. Christmas and C.G. Wildey, Pulse Measurement of r_{63} in KD*P, *Electr. Lett.*, 6:152–153, March 1970.
16. J.H. Ott and T.R. Sliker, Linear Electrooptic Effects in KH_2PO_4 and Its Isomorphs, *J. Opt. Soc. Amer.*, 54:1442–1444, 1964.
17. I.P. Kaminow, Microwave Dielectric Properties of $NH_4H_2PO_4$, KH_2AsO_4, and Partially Deuterated KH_2PO_4, *Phys. Rev.*, 138:A1539–A1543, 1965.
18. R.S. Adhav, Linear Electro-optic Effects in Tetragonal Phosphates and Arsenates, *J. Opt. Soc. Amer.*, 59:414–418, 1969.
19. H. Koetser, Measurement of r_{63} for ADP up to Electric Breakdown, *Electronics Lett.*, 3:52–54, 1967.
20. J.M. Ley, Low-voltage Light-amplitude Modulation, *Electronics Lett.*, 2:12–13, 1966.
21. L. Silverstein and M. Sucher, Determination of the Pockels Electro-optic Coefficient in ADP at 5.5 Ghz, *Electronics Lett.*, 2:437–438, 1966.
22. A.S. Vasilevskaya, Electrooptic and Elastooptical Properties of Deuterated Ammonium Dihydrogen Phosphate Crystals, *Sov. Phys.-Solid State*, 8:2756–57, 1967.
23. P.V. Lenzo, E.G. Spencer, and K. Nassau, Electrooptic Coefficients in Lithium Niobate, *J. Opt. Soc. Amer.*, 56:633, 1966.
24. E. Bernal, G.D. Chen, and T.C. Lee, Low Frequency Electrooptic and Dielectric Constants of Lithium Niobate, *Phys. Lett.*, 21(3):259, 1966.
25. J.D. Zook, D. Chen, and G.N. Otto, Temperature Dependence and Model of the Electrooptic Effect in $LiNbO_3$, *Appl. Phys. Lett.*, 11(5):159, 1967.
26. K.F. Hulme, P.H. Davies, and V.M. Cound, The Signs of the Electrooptic Coefficients for Lithium Niobate, *J. Phys. C. (Solid-State Phys.)*, 2:855, 1969.
27. B. Luther-Davies, et al., The Signs of the Electrooptic Coefficients for Lithium Tantalate, *J. Phys. C. (Solid-State Phys.)*, 3:L106–L107, 1970.
28. P.H. Smakula and P.C. Claspy, The Electrooptic Effect in $LiNbO_3$ and KTN, *Trans. AIME*, 239:421, 1967.
29. E.H. Turner, High-frequency Electrooptic Coefficients of Lithium Niobate, *Appl. Phys. Lett.*, 8(11):303, 1966. Signs of coefficients were determined later; determination to be published.
30. E.H. Turner, Electrooptic Coefficients of Some Crystals at 3.39 microns, Paper Th A13, Optical Society of America, San Francisco, October 20, 1966.
31. K. Nassau, H.J. Levinstein, and G.M. Loiacono, Ferroelectric Lithium Niobate, 2. Preparation of Single Domain Crystals, *J. Phys. Chem. Solids*, 27:989, 1966.
32. G.D. Boyd, et al., $LiNbO_3$: an Efficient Phase Matchable Nonlinear Optical Material, *Appl. Phys. Lett.*, 15(11):234, 1964.
33. H. Iwasaki, et al., Piezoelectric and Optical Properties of $LiNbO_3$ Single Crystals, *Rev. Elec. Comm. Lab.*, 16(5–6):385, 1968.
34. P.V. Lenzo, et al., Electrooptic Coefficients and Elastic Wave Propagation in Single Domain Ferroelectric Lithium Tantalate, *Appl. Phys. Lett.*, 8(4):81, 1966.
35. G.D. Boyd, W.L. Bond, and H.L. Carter, Refractive Index as a Function of Temperature in $LiNbO_3$, *J. Appl. Phys.*, 38(4):1941, 1967.
36. W.L. Bond, Measurement of the Refractive Indices of Several Crystals, *J. Appl. Phys.*, 36:1674, 1965.
37. A.W. Warner, M. Onoe, and G.A. Coquin, Determination of Elastic and Piezoelectric Constants for Crystals in Class (3m), *J. Acoust. Soc. Amer.*, 42:1223, 1967.
38. E.H. Turner, unpublished.
39. S. Singh, D.A. Draegert, and J.E. Geusic, Optical and Ferroelectric Properties of Barium Sodium Niobate, *Phys. Rev.*, 2B:2709, October 1970.
40. R.L. Byer, et al., Nonlinear Optical Properties of $Ba_2NaNb_5O_{15}$ in the Tetragonal Phase, *J. Appl. Phys.*, 40(1):444, 1969.
41. A.W. Warner, G.A. Coquin, and J.L. Fink, Elastic and Piezoelectric Constants of $Ba_2NaNb_5O_{15}$, *J. Appl. Phys.*, 40:4353, 1969.
42. I.P. Kaminow and E.H. Turner, Electrooptic Light Modulators, *Proc. IEEE*, 54:1374, 1966.
43. S. Singh, private communication.
44. E.H. Turner and I.P. Kaminow, Electrooptical Effect in Gallium Arsenide, *J. Opt. Soc. Amer.*, 53:523, 1963.
45. E.H. Turner, to be published.
46. T.E. Walsh, Gallium Arsenide Electrooptic Modulators, *RCA Review*, XXVII:323, 1966.

Table 2-75. PROPERTIES OF LINEAR ELECTROOPTICAL MATERIALS
(Continued)

47. A. Yariv, C.A. Mead, and J.V. Parker, GaAs as an Electrooptic Modulator at 10.6 microns, *IEEE J. Quant. Electronics*, QE-2:243, 1966.

48. I.P. Kaminow, Measurements of the Electrooptic Effect in CdS, ZnTe, and GaAs at 10.6 microns, *IEEE J. Quant. Electronics*, QE-4:23, 1968.

49. A. Mooradian and A.L. McWhorter, Light Scattering from Plasmons and Phonons in GaAs, in "Light Scattering Spectra of Solids", G.B. Wright, Ed. Springer, 1969, p. 297.

50. W.D. Johnston, Jr., and I.P. Kaminow, Contributions to Optical Nonlinearity in GaAs as Determined from Raman Scattering Efficiencies, *Phys. Rev.*, 188:1209, 1969.

51. D.T.F. Marple, Refractive Index of GaAs, *J. Appl. Phys.*, 35:1241, 1964.

52. K.G. Hambleton, C. Hilsum, and B.R. Holeman, Determination of the Effective Ionic Charge of Gallium Arsenide from Direct Measurements of the Dielectric Constant, *Proc. Phys. Soc.*, 77:1147, 1961.

53. S. Jones and S. Mao, Further Investigation of the Dielectric Constant of Gallium Arsenide, *J. Appl. Phys.*, 39:4038, 1968.

54. C.J. Johnson, G.H. Sherman, and R. Weil, Far Infrared Measurement of the Dielectric Properties of GaAs and CdTe at 300 K and 8 K, *Appl. Optics*, 8:1667, 1969.

55. D.T.F. Marple, Refractive Index of ZnSe, ZnTe, and CdTe, *J. Appl. Phys.*, 35(3), 1964.

56. L.R. Shiozawa and J.M. Jost, Research on II–VI Compound Semiconductors, Clevite Corporation, Palo Alto, California, Report AD 620297, May 1965.

57. D. Berlincourt, H. Jaffe, and L.R. Shiozawa, Electroelastic Properties of the Sulfides, Selenides, and Tellurides of Zinc and Cadmium, *Phys. Rev.*, 129:1009, 1963.

58. A.S. Barker, Jr., and C.J. Summers, Infrared Dielectric Function of CdS, *J. Appl. Phys.*, 41:3552–3554, 1970.

59. E.H. Turner, to be published.

60. J.E. Kiefer and A. Yariv, Electrooptic Characteristics of CdTe at 3.39 and 10.6 μ, *Appl. Phys. Lett.*, 15(1):26, 1969.

61. C.J. Johnson, Electrooptic Effect in CdTe at 23.35 and 27.95 microns, *Proc. IEEE*, 56:1719, 1968.

62. O.G. Lorimer and W.G. Spitzer, Infrared Refractive Index and Absorption of InAs and Absorption of InAs and CdTe, *J. Appl. Phys.*, 36:1841, 1965.

63. D.J.A. Gainon, Linear Electrooptic Effect in CdS, *J. Opt. Soc. Amer.*, 54:270, 1964.

Table 2-76. SECOND HARMONIC COEFFICIENTS OF NON-LINEAR OPTICAL MATERIALS*

COURTESY OF S. SINGH

Material	Symmetry class	$10^{12} \times d_{il}^{2\omega}$, m/V	$10^2 \times \delta_{il}^{2\omega}$, m²/C	λ_1, μm	n^ω	$n^{2\omega}$	θ_m	References
SYSTEM—Orthorhombic								
Barium sodium niobate $Ba_2NaNb_3O_{15}$	$mm2—C_{2V}$	$d_{15} = 14.56 \pm 0.73$	$\delta_{15} = 2.35 \pm 0.11$	1.064 2	2.213 3	2.365 5		11,12,28
		$d_{24} = 13.83 \pm 0.73$	$\delta_{24} = 2.23 \pm 0.11$		2.214 0	2.367 2		
		$d_{31} = 14.56 \pm 0.73$	$\delta_{31} = 2.42 \pm 0.12$		2.256 7	2.250 2	73–75°	
		$d_{32} = 14.56 \pm 1.46$	$\delta_{32} = 2.41 \pm 0.16$		2.258 0	2.250 2	75–77°	
		$d_{33} = 20 \pm 1.46$	$\delta_{33} = 4.06 \pm 0.30$		2.170 0	2.250 2		
SYSTEM—Trigonal								
Lithium niobate $LiNbO_3$	$3m—C_{3V}$	$d_{22} = (3.07 \pm 0.28)$	$\delta_{22} = 0.497 \pm 0.045$	1.058 2	2.232 2	2.324 1		4,24
		$d_{31} = (5.82 \pm 0.85)$	$\delta_{31} = 1.04 \pm 0.15$	1.058 2	2.232 2	2.232 5		
		$d_{33} = (40.68 \pm 10.4)$	$\delta_{33} = 8.7 \pm 2.2$	1.058 2	2.154 4	2.232 5		
		$d_{31} = (5.01 \pm 0.47)$	$\delta_{31} = 0.92 \pm 0.09$	1.152	2.227 8	2.215 3	68 ± 1°	
		$d_{22} = (2.41 \pm 0.95)$	$\delta_{22} = 0.4 \pm 0.16$		2.227 8	2.303 7		
		$d_{31} = (5.16 \pm 0.8)$	$\delta_{31} = 0.95 \pm 0.15$	1.152 3	2.227 8	2.215 3		2
		$d_{15} = (6.28 \pm 0.63)$	$\delta_{15} = 1.18 \pm 0.12$	0.632 8 pump	2.185 4	2.285 4		7
		$d_{33} = (30.27 \pm 7.57)$	$\delta_{33} = 6.7 \pm 1.7$	1.152	2.150 6	2.215 3		16
		$d_{31} = (7.12 \pm 2.51)$	$\delta_{31} = 1.27 \pm 0.45$	1.058	2.232 2	2.232 5		13
Lithium tantalate $LiTaO_3$	$3m—C_{3V}$	$d_{22} = (2.08 \pm 0.24)$	$d_{22} = 0.48 \pm 0.06$	1.058 2	2.136 6	2.204 3		24
		$d_{31} = (1.28 + 0.24)$	$\delta_{31} = 0.29 + 0.06$		2.136 6	2.208 9		
		$d_{33} = (19.39 \pm 2.36)$	$\delta_{33} = 4.4 \pm 0.53$		2.140 6	2.208 9		
SYSTEM—Tetragonal								
Potassium lithium niobate $K_3Li_2Nb_5O_{15}$	$4mm—C_{4V}$	$d_{15} = 6.2 \pm 1.1$	$\delta_{15} = 1.19 \pm 0.2$	1.064 2	2.158 5	2.329 7		34,35
		$d_{31} = 7.0 \pm 1.5$	$\delta_{31} = 1.38 \pm 0.3$	1.064 2	2.205 7	2.198 0		
		$d_{33} = 12.7 \pm 1.8$	$\delta_{33} = 3.14 \pm 0.44$	1.064 2	2.111 3	2.198 0		
Potassium sodium barium niobate $K_{0.8}Na_{0.2}Ba_2Nb_5O_{15}$	$4mm—C_{4V}$	$d_{31} = 12.77$	$d_{31} = 2.22$	1.064 2	2.260 1	2.200 7		29
Ammonium dihydrogen phosphate (ADP) $NH_4H_2PO_4$	$\overline{4}2m—D_{2d}$	$d_{14} = 0.553 + 0.024$	$\delta_{14} = 3.19 + 0.14$	1.058 2	1.487 4	1.527 7		21–23
		$d_{36} = 0.558 \pm 0.028$	$\delta_{36} = 3.27 \pm 0.16$	1.058 2	1.506 7	1.481 6	41.9 ± 1°	
		$d_{14} = 0.482 \pm 0.024$	$\delta_{14} = 2.52 \pm 0.12$	0.694 3	1.497 3	1.549 8		21–23
		$d_{36} = 0.487 \pm 0.028$	$\delta_{36} = 2.57 \pm 0.15$	0.694 3	1.519 3	1.500 4	51.9 ± 1°	
		$d_{36} = 0.838 \pm 0.21$	$\delta_{36} = 4.3 \pm 1.08$	0.632 8	1.521 7	1.507 5		19
		$d_{36} = 0.544 \pm 0.14$	$\delta_{36} = 3.25 \pm 0.84$	1.15	1.503 64	1.479 4		
		$d_{36} = 0.29 \pm 0.17$	$\delta_{36} = 1.54 \pm 0.9$	0.694 3	1.519 3	1.500 4		17
		$d_{36} = 0.57 \pm 0.068$	$\delta_{36} = 2.93 \pm 0.35$	0.632 8	1.521 7	1.507 5		9
		$d_{36} = 0.58 \pm 0.09$	$\delta_{36} = 2.98 \pm 0.46$	0.632 8	1.521 7	1.507 5		3
		$d_{14} = 0.51 \pm 0.02$	$\delta_{14} = 2.68 \pm 0.11$	0.694 3	1.497 3	1.549 8		33
		$d_{36} = 0.59 \pm 0.02$	$\delta_{36} = 3.12 \pm 0.11$	0.694 3	1.519 3	1.500 4		
		$d_{36} = 0.56 \pm 0.08$	$\delta_{36} = 2.96 \pm 0.43$	0.694 3	1.519 3	1.500 4	48°	31
		$d_{36} = 0.49 \pm 0.07$	$\delta_{36} = 2.94 \pm 0.42$	1.15	1.503 6	1.479 4		2
		$d_{36} = 0.66 \pm 0.14$	$\delta_{36} = 3.65 \pm 0.77$	0.825 0	1.514 5	1.490 7	42°	20
Ammonium dideuterium phosphate (ADDP) $ND_4D_2PO_4$	$\overline{4}2m—D_{2d}$	$d_{36} = 0.52 \pm 0.08$	$\delta_{36} = 2.88 \pm 0.44$	0.694 3	1.513 8	1.492 6	47°	31
Potassium dihydrogen phosphate (KDP) KH_2PO_4	$\overline{4}2m—D_{2d}$	$d_{36} = 0.88 \pm 0.29$	$\delta_{36} = 5.75 \pm 1.89$	1.152 6	1.491 3	1.468 7		1,5
		$d_{14} = 0.49 \pm 0.02$	$\delta_{14} = 3.10 \pm 0.13$	1.058 2	1.475 1	1.512 3		21–23
		$d_{36} = 0.47$	$\delta_{36} = 3.02 \pm 0.2$	1.058 2	1.493 9	1.470 6	40.3 ± 1°	
		$d_{14} = 0.47 \pm 0.03$	$\delta_{14} = 2.71 \pm 0.17$	0.694 3	1.485 6	1.533 5		21,22
		$d_{36} = 0.47$	$\delta_{36} = 2.73$	0.694 3	1.505 8	1.487 4	50.4 ± 1°	

*Condensed from: S. Singh, Non-linear Optical Materials, in "Handbook of Lasers", R.J. Pressley, Ed., The Chemical Rubber Co., 1971.

Table 2-76. SECOND HARMONIC COEFFICIENTS OF NON-LINEAR OPTICAL MATERIALS *(Continued)*

Material	Symmetry class	$10^{12} \times d_{il}^{2\omega}$, m/V	$10^2 \times \delta_{il}^{2\omega}$, m²/C	λ_1, μm	n^ω	$n^{2\omega}$	θ_m	References
SYSTEM—Tetragonal								
Potassium dihydrogen	$\bar{4}2m—D_{2d}$	$d_{14} = 0.41 \pm 0.02$	$\delta_{14} = 2.36 \pm 0.12$	0.694 3	1.485 6	1.533 5		33
phosphate (KDP)		$d_{36} = 0.46 \pm 0.07$	$\delta_{36} = 3.01 \pm 0.46$	1.15	1.491 3	1.468 7		2
(Continued)		$d_{36} = 0.47 \pm .03$	$\delta_{36} = 3.01 \pm 0.2$	1.064 2	1.494 2	1.470 8		14
		$d_{36} = 0.46 \pm 0.04$	$\delta_{36} = 2.95 \pm 0.26$	1.058 2	1.493 9	1.470 6		13
Potassium dideuterium	$\bar{4}2m—D_{2d}$	$d_{14} = 0.5 \pm 0.02$	$\delta_{14} = 3.23 \pm 0.13$	1.058 2	1.478 9	1.508 5		22,23
phosphate (KDDP)		$d_{36} = 0.5 \pm 0.02$	$\delta_{36} = 3.24 \pm 0.13$	1.058 2	1.497 8	1.468 9		
		$d_{14} = 0.46 \pm 0.04$	$\delta_{14} = 2.71 \pm 0.23$	0.694 3	1.483 0	1.528 5		22
		$d_{36} = 0.5 \pm 0.02$	$\delta_{36} = 3.04 \pm 0.12$	0.694 3	1.502 2	1.485 5		
		$d_{14} = 0.4 \pm 0.05$	$\delta_{14} = 2.43 \pm 0.3$	0.694 3	1.483 0	1.528 5		33
		$d_{36} = 0.35$	$\delta_{36} = 2.08$	0.694 3	1.502 2	1.485 5	52°	32
SYSTEM—Cubic								
Cadmium telluride CdTe	$\bar{4}3m—Td$	$d_{14} = 16.7 \pm 6.3$	$\delta_{14} = 0.76 \pm 0.29$	10.6	2.69	2.71		25
Gallium arsenide GaAs	$\bar{4}3m—Td$	$d_{14} = 274 \pm 66$	$\delta_{14} = 1.4 \pm 0.34$	1.058 2	3.479	4.352		23
		$d_{14} = 387 \pm 126$	$\delta_{14} = 4.65 \pm 1.51$	10.6	3.27	3.30		27
		$d_{36} = 249 \pm 15$	$\delta_{36} = 1.27 \pm 0.08$	1.058	3.479	4.352		30
		$d_{14} = 188.5 \pm 19$	$\delta_{14} = 2.26 \pm 0.23$	10.6	3.27	3.30		36
		$d_{14} = 274.3 \pm 37.8$	$\delta_{14} = 1.4 \pm 0.19$	1.058	3.479	4.352		8
		$d_{14} = 134 \pm 42$	$\delta_{14} = 1.61 \pm 0.5$	10.6	3.27	3.30		18
		$d_{14} = 140 \pm 10$	$\delta_{14} = 0.72 \pm 0.05$	1.06	3.478	4.346		15
		$d_{14} = 100 \pm .521$	$\delta_{14} = 0.51 \pm 0.11$	1.06	3.478	4.346		26
		$d_{14} = 191 \pm 64$	$\delta_{14} = 1.17 \pm 0.39$	1.058	3.41	4.16		6
		$d_{14} = 137$	$\delta_{14} = 0.31$	0.843 5–.845 0	3.60	5.90		10
Indium antimonide InSb	$\bar{4}3m—Td$	$d_{14} = 520 \pm 47$	$\delta_{14} = 1.8$	1.058				8

REFERENCES

1. A. Ashkin, G.D. Boyd, and J.M. Dziedzic, Observation of Continuous Optical Harmonic Generation with Gas Masers, *Phys. Rev. Lett.*, 11:14–17, 1963.
2. J.E. Bjorkholm, Relative Measurements of the Optical Nonlinearities of KDP, ADP, LiNbO₃, and α-HIO₃, *IEEE J. Quant. Electronics*, QE-4:970–972, 1968; and J.E. Bjorkholm, Correction to Relative Measurements of the Optical Nonlinearities of KDP, ADP, LiNbO₃ and α-HIO₃, *IEEE J. Quant. Electronics*, QE-5:260, 1969.
3. J.E. Bjorkholm and A.E. Siegman, Accurate CW Measurements of Optical Second Harmonic Generation in Ammonium Dihydrogen Phosphate and Calcite, *Phys. Rev.*, 154:851–60, 1967.
4. G.D. Boyd, et al., LiNbO₃: an Efficient Phase Matchable Nonlinear Optical Material, *Appl. Phys. Lett.*, 5:234–236, 1964.
5. G.D. Boyd and D.A. Kleinman, Parametric Interaction of Focused Gaussian Light Beams, *J. Appl. Phys.*, 39:3597–3639, 1968.
6. R. Braunstein and N. Ockman, Interactions of Coherent Optical Radiation with Solids, Final Report Prepared for the Office of Naval Research, Department of the Navy, Washington, D.C. Contract #NONR–4128100, Arpa Order #306–62, August 1964.
7. R.L. Byer and S.E. Harris, Power and Bandwidth of Spontaneous Parametric Emission, *Phys. Rev.*, 168:1064–1068, 1968.
8. R.K. Chang, J. Ducuing, and N. Bloembergen, Dispersion of the Optical Nonlinearity in Semiconductors, *Phys. Rev. Lett.*, 15:415–418, 1965.
9. G.E. Francois, CW Measurement of the Optical Nonlinearity of Ammonium Dihydrogen Phosphate, *Phys. Rev.*, 143:597–600, 1966.
10. M. Garfinkel and W.F. Engeler, Sum Frequencies and Harmonic Generation in GaAs Lasers, *Appl. Phys. Lett.*, 3:178–180, 1963.
11. J.E. Geusic, et al., The Nonlinear Optical Properties of Ba₂NaNb₅O₁₅, *Appl. Phys. Lett.*, 11:269–271, 1967.
12. J.E. Geusic, et al., Erratum to The Nonlinear Optical Properties of Ba₂NaNb₅O₁₅, *Appl. Phys. Lett.*, 12:224, 1968.
13. W.F. Hagen and P.C. Magnante, Efficient Second Harmonic Generation with Diffraction Limited and High Spectral Radiance Nd-glass Lasers, *J. Appl. Phys.*, 40:219–224, 1969.

Table 2-76. SECOND HARMONIC COEFFICIENTS OF NON-LINEAR OPTICAL MATERIALS *(Continued)*

14. J. Jerphagnon and S.K. Kurtz, Optical Nonlinear Susceptibilities: Accurate Relative Values for Quartz, Ammonium Dihydrogen Phosphate, and Potassium Dihydrogen Phosphate, *Phys. Rev.*, 1B:1738–1744, 1970.

15. W.D. Johnston and I.P. Kaminow, Contributions to Optical Nonlinearity in GaAs as Determined from Raman Scattering Efficiencies, *Phys. Rev.*, 188:1209–1210, 1969.

16. D.A. Kleinman and R.C. Miller, Dependence of Second Harmonic Generation on the Position of the Focus, *Phys. Rev.*, 148:302–312, 1966.

17. P.D. Maker and R.W. Terhune, Study of Optical Effects due to an Induced Polarization Third Order in the Electric Field Strength, *Phys. Rev.*, 137A:801–818, 1965.

18. J.H. McFee, G.D. Boyd, and P.H. Schmidt, Redetermination of the Nonlinear Optical Coefficients of Te and GaAs, *Appl. Phys. Lett.*, 17:57–59, 1970.

19. D.H. McMahon and A.R. Franklin, Laser Focusing Effects on Second Harmonic Generation in ADP, *Appl. Phys. Lett.*, 6:14–16, 1965.

20. D.H. McMahon, Quantitative Nonlinear Optical Sum Frequency Experiments Using Incoherent Light, *J. Appl. Phys.*, 37:4832–39, 1966.

21. R.C. Miller and A. Savage, Harmonic Generation and Mixing of $CaWo_4$: Nd^{3+} and Ruby-pulsed Laser Beams in Piezoelectric Crystals, *Phys. Rev.*, 128:2175–79, 1962.

22. R.C. Miller, D.A. Kleinman, and A. Savage, Quantitative Studies of Optical Harmonic Generation in CdS, $BaTiO_3$, and KH_2PO_4-type Crystals, *Phys. Rev. Lett.*, 11:146–148, 1963, and private communication.

23. R.C. Miller, Optical Second Harmonic Generation in Piezoelectric Crystals, *Appl. Phys. Lett.*, 5:17–19, 1964, and private communication.

24. R.C. Miller and A. Savage, Temperature Dependence of the Optical Properties of Ferroelectric $LiNbO_3$ and $LiTaO_3$, *Appl. Phys. Lett.*, 9:169–171, 1966.

25. R.C. Miller and W.A. Nordland, Absolute Signs of Nonlinear Optical Coefficients of Polar Crystals, *Opt. Commun.*, 1:400–402, March 1970.

26. A. Mooradian and A.L. McWhorter, Light Scattering from Plasmons and Phonons in GaAs, *International Conference on Light Scattering in Solids* (New York University, 1968), G.B. Wright, Ed., Springer Publishing Company, 1969.

27. C.K.N. Patel, Optical Harmonic Generation in the Infra-red Using a CO_2 Laser, *Phys. Rev. Lett.*, 16:613–616, 1966.

28. S. Singh, D.A. Draegert, and J.E. Geusic, Optical and Ferroelectric Properties of Barium Sodium Niobate, *Phys. Rev.*, B2:2709, 2724, 1970.

29. A.W. Smith, G. Burns, and D.F. O'Kane, Optical and Ferroelectric Properties of $K_\alpha \cdot Na_{1-\alpha} Ba_2 \cdot Nb_5 O_{15}$, *J. Appl. Phys.*, 42:250–255, 1971.

30. R.A. Soref and H.W. Moos, Optical Second Harmonic Generation in ZnS-CdS and CdS-CdSe Alloys, *J. Appl. Phys.*, 35:2152–2158, 1964.

31. V.S. Suvorov, A.S. Sonin, and I.S. Rez, Some Nonlinear Optical Properties of Crystals of the KDP Group, *Sov. Physics JETP*, 26:33–37, January 1968.

32. V.S. Suvorov and A.S. Sonin, Nonlinear Optical Materials, *Sov. Phys.—Crystallography*, 11.711–723, 1967.

33. J.P. Vander Ziel and N. Bloembergen, Temperature Dependence of Optical Harmonic Generation in KH_2PO_4 Ferroelectrics, *Phys. Rev.*, 135.A1622–A1669, 1964.

34. L.G. Van Uitert, et al., A New and Stable Nonlinear Optical Material, *Appl. Phys. Lett.*, 11:161–163, 1967.

35. L.G. Van Uitert, et al., Erratum to A New and Stable Nonlinear Optical Material, *Appl. Phys. Lett.*, 12:224, 1968.

36. J.J. Wynne and N. Bloembergen, Measurement of the Lowest-order Nonlinear Susceptibility in III-V Semiconductors by Second Harmonic Generation with a CO_2 Laser, *Phys. Rev.*, 188:1211–1220, 1969.

Table 2-77.   MAGNETIC MATERIALS AND UNITS

Classifications of Materials and Their Magnetic Properties

Table A.   UNITS AND CONVERSIONS

Name	Symbol	cgs unit	Conversion, SI to cgs
Force	F	dyne	1 newton = 100,000 dynes
Magnetomotive force	$\mathscr{F}$	gilbert	1 ampere-turn = 0.4π gilbert
Power	P	ergs/sec	1 watt = 10^7 ergs/sec
Flux	Φ	maxwell	1 weber = 10^8 maxwell
Flux density[a]	B	gauss	1 tesla = 1 weber/meter^2 = 10,000 gauss
Magnetizing force[b]	H	oersted	1 ampere-turn/meter = 4π oersted/1,000

[a]Sometimes called intensity of field or magnetic induction. (Retentivity of magnetization is also expressed in gauss.)
[b]Sometimes called magnetic field strength.

Table B.   PROPERTIES OF MAGNETIC MATERIALS

Name	Symbols	Definition	Examples
Saturation flux density	B_s	Field intensity when material is saturated	Ferrite: 3,500 Cobalt-iron: 22,000
Remanence or retentivity	B_r	Magnetization remaining after mmf is removed	Ferrite: 1,000 Pure iron: 10,000 Co-Fe-V: 14,000
Coercivity	H_c	Coercive or magnetizing force to remove remanence	Mild steel: 50 Alnico 4: 700
Susceptibility	M/H	Ratio of magnetization to magnetizing force	
Permeability	μ, B/H	Ratio of flux density to field intensity, usually relative	Air: 1 Iron: 1,000 Nickel-iron: 100,000
Energy product	BH_{max}	Criterion of internal losses	Alnico 1: 1.4 Alnico 5: 5.0
Curie point	T_c	Temperature of high-permeability material above which thermal agitation overcomes molecular magnetic field	Ferrite: 150°C Silicon-iron: 700°C
Neel point	T_n	Temperature below which antiferromagnetics becomes spontaneously magnetized	Cesium: 13°K Chromium: 312°K

Table 2-77.   MAGNETIC MATERIALS AND UNITS *(Continued)*

Table C.   CLASSIFICATIONS AND USES OF MAGNETIC MATERIALS

Class or use	Properties	Typical examples
Ferromagnetic material	High permeability, 10 to 10^6; spontaneous magnetization below Curie point; magnetization in proportion to field strength	Iron (pure): $\mu = 8{,}000$ Silicon iron: $\mu = 7{,}000$ Mild steel: $\mu = 450$ Cast iron: $\mu = 90$ Cobalt: $\mu = 60$
Antiferromagnetic material; (or ferrimagnetic)	Antiparallel spin; transition at Curie point	Ferrous oxide; manganese oxide
Paramagnetic material	Magnetization in proportion to field strength. Permeability slightly above unity	Chromium; aluminum; beryllium; some salts
Diamagnetic material	Permeability slightly less than unity. Induced magnetism direction opposite to field	Copper; bismuth; zinc; silver; paraffin; wood
Ferrites (and garnets)	High permeability, 1 to 10; high resistivity; low power loss; high coercivity; low saturation flux density	High-frequency components; computer memories; magnetic ceramics
Electromagnet and transformer cores (soft)	High permeability; high resistivity; low energy product; low coercivity	Silicon iron
Permanent magnets (hard magnets)	High retentivity; high coercivity; low energy product if a-c	Iron, pure: $B_r = 20{,}000$ Co-Fe-V: $B_r = 14{,}000$ Cast iron: $B_r = 5{,}000$ MnZn ferrite: $B_r = 1{,}000$

Table 2-78.   PROPERTIES OF HIGH-PERMEABILITY MAGNETIC MATERIALS*

Properties are expressed in the cgs electromagnetic system (typical values).†

Name	Composition, %	Permeability		Coercivity, H_c, Oe	Retentivity, B_r, G	B, max, G	Resistivity, microhm-cm
		Initial	Maximum				
Iron, pure (laboratory conditions)	Annealed	25,000	350,000	0.05	12,000	14,000	9.7
Iron, Swedish		250	5,500	1.0	13,000	21,000	10
Iron, cast		100	600	4.5	5,300	20,000	30
Iron, silicon	4 Si, bal. Fe (hot-rolled)	500	7,000	0.3	7,000	20,000	50
Rhometal	36 Ni, bal. Fe	1,000	5,000	0.5	3,600	10,000	90
Permalloy 45	45 Ni, bal. Fe	2,500	25,000	0.3		16,000	45
Mumetal	71–78 Ni, 4.3–6 Cu, 0–2 Cr, bal. Fe	20,000	100,000	0.05	6,000	7,200	25–50
Supermalloy	79 Ni, 5 Mo, bal. Fe	100,000	1,000,000	0.002		8,000	60
HyMu80	80 Ni, bal. Fe	20,000	100,000	0.05		8,700	57
Alfenol	16 Al, bal. Fe	3,450	116,000	0.025	3,800	7,825	150
Permendure	50 Co, 1–2 V, bal. Fe	800	4,500	2.0	14,000	24,000	26
Sendust	10 Si, 5 Al, bal. Fe (cast)	30,000	120,000	0.05	5,000	10,000	60–80
Ferroxcube 3	Mn-Zn-Ferrite	1,000	1,500	0.1	1,000	3,000	$>10^6$
Ferroxcube 101	Ni-Zn-Ferrite	1,100		0.18	1,100	2,300	$>10^5$

†Values are approximate only and vary with heat treatment and mechanical working of the material.
*From: "Reference Data for Radio Engineers", 5th ed., Howard W. Sams & Co., Inc., Indianapolis, Indiana, 1968.

REFERENCES
"Ferromagnetism", R.N. Bozorth, D. Van Nostrand Company, Inc., 1951.
"Magnetic Materials in the Electrical Industry", P.R. Bardell, courtesy Philosophical Library, New York, N.Y., 1955.
Technical catalog data by Allegheny Ludlum Steel Corporation, Pittsburgh, Pa., and Ferroxcube Corporation of America, Saugerties, N.Y.

Table 2-79.   DATA ON METALLIC MAGNETIC-CORE MATERIALS*

Class or trade name	Composition, % (remainder is iron)	Characteristic property or application	Permeability		Direct-current saturation, kilogauss	Residual induction, kilogauss	Coercive force, oersteds	Resistivity, microhm-cm	Curie temperature, °C
			Initial	Maximum					
SILICON-IRON†									
Silicon-Iron	4 Si	Transformer	400	7,000	20	12	0.5	60	690
Hypersil							0.1		
Trancor 3X							to		
Silectron	3.5 Si	Grain oriented	1,500	35,000	20	13.7	0.3	50	750
Sendust	9.5 Si, 5.5 Al	High frequency, powder	30,000	120,000	10	5	0.05	80	—
COBALT-IRON†									
Hyperco	35 Co, 0.5 Cr	High saturation	650	10,000	24	>13	>1	28	970
Permendur 2V	49 Co, 2 V		800	4,500	24	14	2	25	980
NICKEL-IRON†									
Perminvar 45–25	45 Ni, 25 Co		400	2,000	15.5	3.3	1.2	20	715
Perminvar 7–70	70 Ni, 7 Co	"Constant"	850	4,000	12.5	2.4	0.6	15	650
Conpernik	50 Ni	permeability	1,500	2,000	16	—	—	45	—
Isoperm 36	36 Ni, 9 Cu		60	65	—	—	—	70	300
Isoperm 50	50 Ni	High frequency	90	100	16	—	—	40	500
Permalloy 45	45 Ni		2,700	23,000	16.5	8	0.3	45	440
Allegheny 4750	47 to 50 Ni		9,000	50,000	16	6.2†	0.08†	52	430
Armco 48	48 Ni	Combine good	—	—	16	—	—	—	—
Nicaloi	49 Ni	permeability	—	—	16	—	—	—	—
High Perm 49	49 Ni	and flux	5,000	50,000	16	6.5	0.03	43	475
Hipernik	50 Ni, Si, Mn	density	4,000	100,000	16	8†	0.03†	45	500
Monimax	47 Ni, 3 Mo		2,000	38,000	15	—	0.06	80	390
Sinimax	42 Ni, 3 Si	High resistivity	3,500	30,000	11	—	0.1	90	290
Permenorm 5000Z									
Permenite									
Deltamax									
Hypernik V			400	40,000	15.5	13	0.2	40	450
Orthonik			to	to	to	to	to	to	to
Orthonol	45 to 50 Ni		1,700	100,000	16	15	0.4	50	500
Permalloy 65	65 to 68 Ni	Rectangular hysteresis loop	1,500	250,000 to 600,000	13	13	0.03	20	600
Alloy 1040	72 Ni, 14 Cu, 3 Mo		40,000	100,000	6	2.5	0.02	55	290
Mumetal	77 Ni, 5 Cu, 2 Cr		20,000	100,000	8	6	0.05	60	400
Permalloy 78	78 Ni, 0.6 Mn		9,000	100,000	10.7	6	0.05	16	580
Mo-Permalloy 4–79	79 Ni, 4 Mo		20,000	75,000	8	5.5	0.05	55	—
Supermalloy	79 Ni, 5 Mo		55,000 to 150,000	500,000 to 1,000,000	6.8 to 7.8	—	0.002 to 0.05	65	400
		Highest permeability, low saturation							
Hymu 80	80 Ni		10,000	100,000	8	—	0.06	58	460
FERRITES‡									
3C3	MnZi	High-frequency transformers	2,200 ± 20%		4.6	3.5	0.1	60×10^6	150
3B7 and 3B9	MnZi	High-Q coils	2,300 ± 20%		4.6	3.0	0.2	100×10^6	170
3D3	NiZi	High-frequency	750 ± 20%		4.7	3.0	0.5	150×10^6	150
4C4	NiZi	High-Q coils	125 ± 20%		4.1	2.0	4.5	10×10^9	300

Note: The table shows characteristics as listed by the manufacturers. The parameters of different lots of material may vary considerably from the above values. In cases of residual induction and coercive force, the difference may amount to 50 percent.

†$B_m = 10,000$ gauss.

‡Data furnished by Ferroxcube Corporation of America, Saugerties, N.Y.

*From: "Evaluation of High-Performance Core Materials (Part 1)", S.R. Hoh, *Tele-Tech and Electronic Industries*, 12:86–89, 154–156, Copyright Chilton Co., October 1953.

Table 2-80. MAGNETIC PROPERTIES AND COMPOSITION OF PERMANENT MAGNETIC ALLOYS*

For conversion factors see Table 2-77.

Name	Composition,** weight percent					Remanence, B_r, gauss	Coercive force, H_c, oersteds	Maximum energy product, $(BH)_{max}$, gauss-oersteds $\times 10^{-6}$
	Al	Ni	Co	Cu	Other			
U.S.A.								
Alnico I	12	20–22	5			7,100	440	1.4
Alnico II	10	17	12.5	6		7,200	540	1.6
Alnico III	12	24–26		3		6,900	470	1.35
Alnico IV	12	27–28	5			5,500	700	1.3
Alnico V†	8	14	24	3		12,500	600	5.0
Alnico V DG†	8	14	24	3		13,100	640	6.0
Alnico VI†	8	15	24	3	1.25 Ti	10,500	750	3.75
Alnico VII†	8.5	18	24	3	5 Ti	7,200	1,050	2.75
Alnico XII	6	18	35		8 Ti	5,800	950	1.6
Carbon steel					1 Mn 0.9 C	10,000	50	0.2
Chromium steel					3.5 Cr 0.9 C 0.3 Mn	9,700	65	0.3
Cobalt steel			17		2.5 Cr 8 W 0.75 C	9,500	150	0.65
Cunico		21	29	50		3,400	660	0.80
Cunife		20		60		5,400	550	1.5
Ferroxdur 1			$BaFe_{12}O_{19}$			2,200	1,800	1.0
Ferroxdur 2			$BaF_{12}O_{19}$ (oriented)			3,840	2,000	3.5
Platinum-Cobalt			23		77 Pt	6,000	4,300	7.5
Remalloy			12		17 Mo	10,500	250	1.1
Silmanol	4.4				86.6 Ag 8.8 Mn	550	6,000	0.075
Tungsten steel					5 W 0.3 Mn 0.7 C	10,300	70	0.32
Vicalloy I			52		10 V	8,800	300	1.0
Vicalloy II (wire)			52		14 V	10,000	510	3.5
Germany								
Alni 90	12	21				8,000	350	1.2
Alni 120	13	27				6,000	570	1.2
Alnico 130	12	23	5			6,300	620	1.4
Alnico 160	11	24	12	4		6,200	700	1.6
Alnico 190	12	21	15	4		7,000	700	1.8
Alnico 250	8	19	23	4	6 Ti	6,500	1,000	2.2
Alnico 400†	9	15	23	4		12,000	650	4.8
Alnico 580† (semicolumnar)	9	15	23	4		13,000	700	6.0
Oerstit 800	9	18	19	4	4 Ti	6,600	750	1.95
Great Britain								
Alcomax I	7.5	11	25	3	1.5 Ti	12,000	475	3.5
Alcomax II	8	11.5	24	4.5		12,400	575	4.7
Alcomax IISC (semicolumnar)	8	11	22	4.5		12,800	600	5.15
Alcomax III	8	13.5	24	3	0.8 Nb	12,500	670	5.10
Alcomax IIISC (semicolumnar)	8	13.5	24	3	0.8 Nb	13,000	700	5.80
Alcomax IV	8	13.5	24	3	2.5 Nb	11,200	750	4.30
Alcomax IVSC (semicolumnar)	8	13.5	24	3	2.5 Nb	11,700	780	5.10
Alni, high B_r	13	24		3.5		6,200	480	1.25
Alni, normal						5,600	580	1.25
Alni, high H_c	12	32			0–0.5 Ti	5,000	680	1.25
Alnico, high B_r	10	17	12	6		8,000	500	1.70

*Reprinted from: "Handbook of Chemistry and Physics", 53rd ed., R.C. Weast, Ed., The Chemical Rubber Co., 1972.

Table 2-80. MAGNETIC PROPERTIES AND COMPOSITION OF PERMANENT MAGNETIC ALLOYS *(Continued)*

Name	Composition,** weight percent					Remanence, B_r, gauss	Coercive force, H_c, oersteds	Maximum energy product, $(BH)_{max}$, gauss-oersteds $\times 10^{-6}$
	Al	Ni	Co	Cu	Other			
Alnico, normal						7,250	560	1.70
Alnico, high H_c	10	20	13.5	6	0.25 Ti	6,600	620	1.70
Columax (columnar)	similar to Alcomax III or IV					13,000–14,000	700–800	7.0–8.5
Hycomax	9	21	20	1.6		9,500	830	3.3

†Cast anisotropic. Unmarked ones are cast isotropic.
**Remainder of unlisted composition is either iron or iron plus trace impurities.

Table 2-81. CAST PERMANENT MAGNETIC ALLOYS*

See Table 2-80 for magnetic properties of permanent magnetic alloys.

Name	Composition,** weight percent	Specific gravity, g/cc	Thermal expansion		Tensile strength		Remarks‡	Use
			$\frac{Cm \times 10^{-6}}{cm \times °C}$	Between °C	$\frac{Kg †}{mm^2}$	Form		
U.S.A.								
Alnico I	Al 12; Ni 20–22; Co 5	6.9	12.6	20–300	2.9	Cast	i.	Permanent magnets
Alnico II	Al 10; Ni 17; Cu 6; Co 12.5	7.1	12.4	20–300	2.1 45.7	Cast Sintered	i.	Temperature controls, magnetic toys, and novelties
Alnico III	Al 12; Ni 24–26; Cu 3	6.9	12	20–300	8.5	Cast	i.	Tractor magnetos
Alnico IV	Al 12; Ni 27–28; Co 5	7.0	13.1	20–300	6.3 42.1	Cast Sintered	i.	Application requiring high coercive force
Alnico V	Al 8; Ni 14; Co 24; Cu 3	7.3	11.3		3.8 35	Cast Sintered	a.	Application requiring high energy
Alnico V DG	Al 8; Ni 14; Co 24; Cu 3	7.3	11.3				a., c.	
Alnico VI	Al 8; Ni 15; Co 24; Cu 3; Ti 1.25	7.3	11.4		16.1	Cast	a.	Application requiring high energy
Alnico VII	Al 8.5; Ni 18; Cu 3; Co 24; Ti 5	7.17	11.4				a.	
Alnico XII	Al 6; Ni 18; Co 35; Ti 8	7.2	11	20–300				Permanent magnets
Comol	Co 12; Mo 17	8.16	9.3	20–300	88.6			Permanent magnets
Cunife	Cu 60; Ni 20	8.52			70.3			Permanent magnets
Cunico	Cu 50; Ni 21	8.31			70.3			Permanent magnets
Barium ferrite Feroxdur	Ba $Fe_{12}O_{19}$	4.7	10		70.3			Ceramics
Great Britain								
Alcomax I	Al 7.5; Ni 11; Co 25; Cu 3; Ti 1.5						a.	Permanent magnets
Alcomax II	Al 8; Ni 11.5; Co 24; Cu 4.5						a.	Permanent magnets

*Reprinted from: "Handbook of Chemistry and Physics", 53rd ed., R.C. Weast, Ed., The Chemical Rubber Co., 1972.

Table 2-81. CAST PERMANENT MAGNETIC ALLOYS *(Continued)*

Name	Composition,** weight percent	Specific gravity, g/cc	Thermal expansion		Tensile strength		Remarks‡	Use
			$\dfrac{Cm \times 10^{-6}}{cm \times °C}$	Between °C	$\dfrac{Kg †}{mm^2}$	Form		
Alcomax II SC	Al 8; Ni 11; Co 22; Cu 4.5	7.3					a., sc.	
Alcomax III	Al 8; Ni 13.5; Co 24; Nb 0.8	7.3					a.	Magnets for motors, loudspeakers
Alcomax IV	Al 8; Ni 13.5; Cu 3; Co 24; Nb 2.5							Magnets for cycle-dynamos
Columax	Similar to Alcomax III or IV						a., sc.	Permanent magnets, heat-treatable
Hycomax	Al 9; Ni 21; Co 20; Cu 1.6						a.	Permanent magnets
Alnico (high H_c)	Al 10; Ni 20; Co 13.5; Cu 6; Ti 0.25	7.3					i.	
Alnico (high B_r)	Al 10, Ni 17; Co 12; Cu 6	7.3					i.	
Alni (high H_c)	Al 12; Ni 32; Ti 0–0.5	6.9					i.	
Alni (high B_r)	Al 13; Ni 24; Cu 3.5						i.	
Germany								
Alnico 580	Al 9; Ni 15; Co 23; Cu 4						i.	
Alnico 400	Al 9; Ni 15; Co 23; Cu 4						a.	
Oerstit 800	Al 9; Ni 18; Co 19; Cu 4; Ti 4						i.	Permanent magnets
Alnico 250	Al 8; Ni 19; Co 23; Cu 4; Ti 6						i.	
Alnico 190	Al 12; Ni 21; Cu 4; Co 15						i.	
Alnico 130	Al 12; Ni 23; Co 5						i.	
Alni 120	Al 13; Ni 27						i.	
Alni 90	Al 12; Ni 21						i.	
Austria								
Alnico 160	Al 11; Ni 24; Co 12; Cu 4						i.	Permanent magnets, sintered

†kg/mm² × 1422.33 = lbs/in.² kg/mm² × 9.807 = MN/m².
‡i. = isotropic; a. = anisotropic; c. = columnar; sc. = semicolumnar.
**Iron is the additional alloying metal for each of the magnets listed.

Table 2-82.   FERROMAGNETIC MATERIALS—CURIE TEMPERATURES*
Upper Transition Temperature

Material	Curie temp,[a] K	Material	Curie temp,[a] K	Material	Curie temp,[a] K
AuFe	300	EuO	69.5	$GdNi_2$	85
Au_4Mn	263	EuS	16	$GdOs_2$	66
Au_4V	43	EuSe	4.58	$GdPd_2$	335
$BaFeO_3$	180	FeAl	923	GdZn	~280
$BiMnO_3$	103	Fe_3Al	773	$HgCr_2S_4$	36
$CdFe_2$	782	FeB	598	$HgCr_2Se_4$	120
$CeAl_2$	8	Fe_3C	483	Ho	20
$CeCo_3$	78	Fe_3Cr	1273	$HoAl_2$	27
$CeCo_5$	687	Fe-Ni		$HoCo_5$	1025
	737	4.5% Fe	683.0	$HoFe_2$	608
	464	19% Fe	834.0	$HoIr_2$	12
$CeFe_2$	235	23% Fe	876.0	HoNi	31
	878	50% Fe	786.1	$HoNi_2$	23
$CeFe_5$	228	Fe_2O_3	848	$HoNi_3$	66
Co	1390	ε-Fe_2O_3	483	$HoOs_2$	9
Co_2B	429	γ-Fe_2O_3	743	$LaCrO_3$	300
	433	Fe_3O_4	848–858	Mn (45% Al)	~653
Co_3B	747	FeP	215	Mn-Al-Co	Co rich: 370
$CoFeCoO_4$	450	Fe_2P (hexagonal)	278		Mn rich: 466
$CoFe_2O_4$	673–769	FePt	743	$MnAu_4$	360
CoPt	813	Fe_3Pt	453	MnB	578
CoS_2	110	FeRh	668	MnBi	633
Co_2VO_4	160	Fe-Si	1043.9–1012.6	Mn_3Ge	28
Cr	311	0.9–7.4 at. % Si		Mn_3Ge_2	300
CrO_2	391	Fe_3Si	808	Mn_3In	583
CuMnAl	433	Fe_3Sn	743	$MnNi_3$	750 (ordered)
Dy	85	Fe_2TiO_4	142		132 (dis-
$DyAl_2$	53	Fe-53.3% V	280		ordered)
	62	Fe_2Zr	628	Mn_3O_4	45
$DyCo_5$	1125	$Ga_{2-x}Fe_xO_3$		MnO-35.5%,	467.5
	966	$x = 1.08$	350	ZnO-15%,	
$DyCo_3$	450	$x = 1.20$	305	Fe_2O_3-49.5%	
$DyFe_2$	638	$x = 0.80$	205	MnO-26%,	368.5
	663	Gd	294	ZnO-24%,	
$DyIr_2$	23	$GdAl_2$	176	Fe_2O_3-50%	
DyN	22	$GdCl_3$	2.20	MnTe	260
DyNi	48	$GdCo_2$	404	Mn_2TiO_4	~77
$DyOs_2$	15	$GdCo_3$	612	Mn_2VO_4	62
Er	20	Gd-Dy		$NdAl_2$	65
$ErAl_2$	21	% Dy: 10	285	$NdCo_2$	116
$ErCo_2$	39, 36	% Dy: 50	226	NdGe	28
$ErCo_3$	401	% Dy: 61	193	$NdGe_2$	3.6
$ErFe_2$	473	% Dy: 87.5	120	NdNi	35
$ErIr_2$	3	$GdFe_2$	782	Ni	627
ErN	6		813		628.3
	5	$Gd_3Fe_5O_{12}$	564		633
ErNi	10		574.6	Ni-Cr	324
$ErNi_2$	14	$GdIr_2$	88	5.6 at. % Cr	
	21		90	$NiFe_2O_4$	858
$ErNi_3$	62	GdN	69–72	$NiMnO_3$	437
EuH_2	24	GdNi	73	Pd_3Fe	540

Table 2-82.   FERROMAGNETIC MATERIALS—CURIE TEMPERATURES
(Continued)

Material	Curie temp,[a] K	Material	Curie temp,[a] K	Material	Curie temp,[a] K
$PrAl_2$	34	SmIG	562	UFe_2	172
$PrCo_3$	349	SmNi	45	USe	160.5
PrGe	39	Tb	210	USe_2	13.1
$PrIr_2$	16	$TbAl_2$	121	UTe	103
PrNi	20	$TbCo_3$	506	YCo_3	301
$PrNi_2$	8	TbGa	155	YIG: Nd	548–568
$SmAl_2$	122	TbN	40	0–40% Nd_2O_3	
$SmCo_5$	1020	TbNi	50	$3Y_2O_3 \cdot Al_2O_3 \cdot 4Fe_2O_3$	415
	747	$TbNi_2$	46	$YbCo_5$	973
	1015			$ZrFe_2$	633

Note: Values in Tables 2-82 and 2-83 represent a small and somewhat random sample of those reported in the quoted source (see below). Users should consult the original source, which includes a large bibliography and keyed references.

[a]Curie temperature is defined as the point (temperature increasing) at which a material ceases to be ferromagnetic and becomes paramagnetic, i.e., the saturation decreases to zero. Pure iron is magnetic up to about 1455 deg F (790 deg C); above this temperature it is non-magnetic. Materials gain ferromagnetism on cooling but not always at the same temperature.

*From: Research Materials Information Center, Solid State Division, Oak Ridge National Laboratory, ORNL-RMIC-7 (Rev.), 1969. Oak Ridge National Laboratory is operated by Union Carbide Corporation for the U.S. Atomic Energy Commission.

REFERENCES

"Handbook of Chemistry and Physics", 53rd ed., R.C. Weast, Ed., The Chemical Rubber Co., 1972; gives magnetic susceptibilities of elements and compounds, 14 pages.

"American Institute of Physics Handbook", 2nd ed., D.E. Gray, Ed., McGraw-Hill Book Company, 1963; Sect. 5g, Magnetic Properties of Materials, contains 64 pages.

Table 2-83.   ANTIFERROMAGNETIC MATERIALS—NEEL TEMPERATURE*

Upper Transition Temperature

Material	Neel temp,[a] K	Material	Neel temp.[a] K	Material	Neel temp.[a] K	Material	Neel temp,[a] K
AuMn	493	CrGe	62	$FeWO_4$	66	β-MnS	110
Au_3Mn	140	$CrVO_4$	50	GdAs	25	(hexagonal)	
Ba_2CoWO_6	17	Cr_2WO_6	69	GdBi	28	$MnSO_4$	11.5
$BaFe_{12}O_{19}$	709.5	$CuCl_2$	70	GdP	15	MnTe	306.7
	(for H⊥c)	CuF_2	68.7	$GdPO_4$	225	$MnTe_2$	80
$CdCr_2O_4$	9	CuHo	27	GdSb	28	$MnUO_4$	12
Ce	13		(elec. res.)	Ho	133	$MnWO_4$	16
CeS	7	CuO	230	HoGe	18	Nd	12
CeSb	18	$CuSO_4$	34.5	$HoGe_2$	11	$NdSn_3$	4.7
CeSe	12	$CuWO_4$	90	HoSb	9	NdTe	13
CeTe	10	DyCu	64	$KCoF_3$	135	$NiCl_2$	50
$CoBr_2$	19	$DyCu_2$	24	$KCrF_3$	40	NiF_2	~73.2
$CoCO_3$	18	DyH_2	8	$KCuF_3$	243	$NiWO_4$	67
$CoCl_2$	25	Er	84	$KFeF_3$	113	$RbMnF_3$	54.5
CoF_2	37–45	Eu	103	$KMnF_3$	88		66
CoF_3	460	Fe_3Al	750	$LaVO_3$	137		82
$CoMoO_3$	391	$FeCl_2$	23.5	α-Mn	~100		83
$CoMoO_4$	5	$FeCl_3$	15	$MnCl_2$	1.96		~100
CoO	292	FeF_2	78.11	MnF_2	66.2	Sm	15
$CoSO_4$	12	FeO	186	MnF_3	47	Tb	230
$CoUO_4$	12	Fe_2O_3	259	MnO	118	Tm	51
$CoWO_4$	55	α-Fe_2O_3	963	MnO_2	92	UO_2	~30
Cr	312	FeS	593	Mn_2O_3	80	$ZnFe_2O_4$	15
CrF_2	53	$FeSO_4$	21	MnS	165		9
CrF_3	80	FeSn	373	β-MnS	160		
CrFe	308	$FeTiO_3$	68	(cubic)			

Note: Where there is more than one temperature reported, there has been no attempt to choose a best value.

[a]Neel temperature is defined as the point above which an antiferromagnetic or ferromagnetic material becomes paramagnetic. Above the Neel point the susceptibility decreases with increase in temperature.

*From: Research Materials Information Center, Solid State Division, Oak Ridge National Laboratory, ORNL-RMIC-7 (Rev.), 1969. Oak Ridge National Laboratory is operated by the Union Carbide Corporation for the U.S. Atomic Energy Commission.

Table 2-84.   PROPERTIES OF FERROFLUIDS*

Ferrofluids are liquid materials that exhibit a strong attraction to magnetic fields—yet retain their liquid property in the magnetized state. Ferrofluids are available in different base liquids including water, glycerides, esters, hydrocarbons, silicones, fluorocarbons, and others.

KEROSENE-BASE FERROFLUIDS

Property	Typical	Range available
Density	0.90 g/cc	up to 1.7 g/cc
Viscosity at 25°C	3 cp	up to 100 cp
Initial relative permeability	1.2	up to 3.0
Saturation magnetization	125 gauss	up to 1,000 gauss
Electrical resistivity at 60 hz	10^8 ohm-cm	—
Dielectric constant at 1 khz	20	2 to 100
Maximum operating temperature	212°F	up to 400°F
Surface tension	28 dynes/cm	—
Shelf life	No limit	No limit

*Courtesy of Ferrofluidics Corporation, 144 Middlesex Turnpike, Burlington, Mass. 01803.

Table 2-85.   NUMBER μ_B OF BOHR MAGNETONS PER MAGNETIC ATOM AND DATA ON SATURATION MAGNETIZATION AND CURIE POINTS*

Substance	Saturation magnetization M_s, gauss		$\mu_B(0°K)$, per formula unit	Ferromagnetic Curie temperature, $°K$
	Room temperature	$0°K$		
Fe	1707	1740	2.22	1043
Co	1400	1446	1.72	1400
Ni	485	510	0.606	631
Gd	—	2010	7.10	292
Dy	—	2920	10.0	85
Cu_2MnAl	500	(550)	(4.0)	710
MnAs	670	870	3.4	318
MnBi	620	680	3.52	630
Mn_4N	183	—	1.0	743
MnSb	710	—	3.5	587
MnB	152	163	1.92	578
CrTe	247	—	2.5	339
$CrBr_3$	—	—	—	37
CrO_2	515	—	2.03	392
$MnOFe_2O_3$	410	—	5.0	573
$FeOFe_2O_3$	480	—	4.1	858
$CoOFe_2O_3$	400	—	3.7	793
$NiOFe_2O_3$	270	—	2.4	858
$CuOFe_2O_3$	135	—	1.3	728
$MgOFe_2O_3$	110	—	1.1	713
UH_3	—	230	0.90	180
EuO	—	1920	6.8	69
$GdMn_2$	—	215	2.8	303
$Gd_3Fe_5O_{12}$	0	605	16.0	564
$Y_3Fe_5O_{12}$ (YIG)	130	200	5.0	560

*From: "Introduction to Solid-State Physics", C. Kittel, John Wiley & Sons, 1966, p. 461.

REFERENCES

"American Institute of Physics Handbook", 2nd ed., D.E. Gray, Ed., McGraw-Hill Book Company, 1963, Sect. 5g.

"Numerical Data and Functional Relationships in Physics, Chemistry, Astronomy, Geophysics, and Technology", 6th ed., K.H. Hellwege, Ed., Landolt-Börnstein, 1962.

Table 2-86. FERROMAGNETIC CURIE POINTS AND SATURATION MOMENTS OF AB$_5$ COMPOUNDS OF RARE-EARTH ELEMENTS (A), HAVING HEXAGONAL CaCu$_5$ STRUCTURE*

Per Formula Unit, at Temperatures Indicated

Rare-earth elements, A		Compounds, B		
Name	Symbol	Mn†	Co‡	Ni**
Yttrium	Y	490°K 2.21 μ_B (78°K) 1.38 μ_B (298°K)	995°K 8.2 μ_B (1.4°K)	~0 μ_B (1.4°K)
Cerium	Ce		687°K 7.4 μ_B (1.4°K)	
Praseodymium	Pr		9.9 μ_B (1.4°K)	
Neodymium	Nd		10.5 μ_B (1.4°K)	2.2 μ_B (1.4°K)
Samarium	Sm	440°K 1.72 μ_B (78°K) 1.40 μ_B (298°K)	1015°K 8.6 μ_B (1.4°K)	0.7 μ_B (1.4°K)
Gadolinium	Gd	465°K 6.23 μ_B (78°K) 2.89 μ_B (298°K)	1030°K 1.3 μ_B (1.4°K)	6.1 μ_B (1.4°K)
Terbium	Tb	445°K 6.18 μ_B (78°K) 2.66 μ_B (298°K)	0.7 μ_B (1.4°K)	7.0 μ_B (1.4°K)
Dysprosium	Dy	430°K 5.34 μ_B (78°K) 2.49 μ_B (298°K)	1125°K 1.5 μ_B (1.4°K)	7.7 μ_B (1.4°K)
Holmium	Ho	425°K 5.12 μ_B (78°K) 1.99 μ_B (298°K)	1025°K 1.9 μ_B (1.4°K)	7.2 μ_B (1.4°K)
Erbium	Er	415°K 3.74 μ_B (78°K) 1.63 μ_B (298°K)	1.3 μ_B (1.4°K)	7.7 μ_B (1.4°K)
Thulium	Tm		1.5 μ_B (1.4°K)	

*From: "Intermetallic Compounds—Magnetic Properties of Some Rare-Earth Compounds", T.H. Westbrook, John Wiley & Sons, 1967.
†"Magnetic Characteristics of Some Intermetallic Compounds between Manganese and the Lanthanide Metals", L.V. Cherry and W.E. Wallace, *J. Appl. Phys. Supp.*, 32:340, 1961.
‡Curie points from "Intermetallic Compounds between Lanthanons and Transmission Metals of the First Long Period II—Ferrimagnetism of AB$_5$ Cobalt Compounds", K. Nassau, L.V. Cherry, and W.E. Wallace, *J. Phys. Chem. Solids*, 16:131, 1960; moments from "Magnetic Moments of Intermatallic Compounds of Transition and Rare-Earth Elements", E.A. Nesbitt, H.J. Williams, J.H. Wernick, and R.C. Sherwood, *J. Appl. Phys.*, 33:1674, 1932.
**"Magnetic Moments of Intermetallic Compounds of Transition and Rare-Earth Elements", E.A. Nesbitt, H.J. Williams, J.H. Wernick, and R.C. Sherwood, *J. Appl. Phys.*, 33:1674, 1932.

Table 2-87.   PHYSICAL AND MAGNETIC CONSTANTS OF RARE-EARTH METALS*

For other properties of rare-earth metals, see Table 1-70.

Element	Atomic No., Z	Mol. weight	Density, g/cc	Cryst. form at room temp	Transition point, °C	M.P., °C	Curie point, θ_f, °K	Neel point, θ_N, °K	Asymptotic Curie point, θ_p, °K	Electronic state of 3+ ion — Number of 4f electrons	S	L	J	Magnetic moment in Bohr magnetons — Effective moment, 3+ ion	Effective moment, Metal	Saturation moment, gJ
Sc	21	44.96	2.992	h.c.p.	1335	1539										
Y	39	88.92	4.478	h.c.p.	1459	1509										
La	57	138.92	6.174	h.c.p.	310	920				0	0	0	0	0	0	
			6.186	f.c.c.	858											
Ce	58	140.13	6.771	f.c.c.	725	795		12.5	−45	1	$\frac{1}{2}$	3	$2\frac{1}{2}$	2.52	2.51	
Pr	59	140.92	6.782	hex.	798	935			−21	2	1	5	4	3.60	2.56	
Nd	60	144.27	7.004	hex.	862	1024		7.5	−15	3	$\frac{3}{2}$	6	$4\frac{1}{2}$	3.50	3.3–3.71	
Pm	61	(147)				1035				4	2	6	4			
Sm	62	150.35	7.536	rhomb.	917	1072		14.8	15	5	$\frac{5}{2}$	5	$2\frac{1}{2}$		1.74	
Eu	63	152.0	5.259	b.c.c.		826		(90)		6	3	3	0		8.3	
Gd	64	157.26	7.895	h.c.p.	1264	1512	289		310	7	$\frac{7}{2}$	0	$3\frac{1}{2}$	7.80	7.93	7.12
Tb	65	158.93	8.272	h.c.p.	1317	1356	218	230	235	8	3	3	6	9.74	9.62	9.25
Dy	66	162.51	8.536	h.c.p.		1407	90	179	151	9	$\frac{5}{2}$	5	$7\frac{1}{2}$	10.5	10.67	10.2
Ho	67	164.94	8.803	h.c.p.		1461	20	133	87	10	2	6	8	10.6	10.9	9.7†
Er	68	167.27	9.051	h.c.p.		1497	20	80 (53)	41.6	11	$\frac{3}{2}$	6	$7\frac{1}{2}$	9.6	10.0	8.3†
Tm	69	168.94	9.332	h.c.p.		1545	22	53	20	12	1	5	6	7.1	7.56	
Yb	70	173.04	6.977	f.c.c.	798	824				13	$\frac{1}{2}$	3	$3\frac{1}{2}$	4.4	0.0	
Lu	71	174.99	9.842	h.c.p.		1652				14	0	0	0	0	0	

†Determined by neutron diffraction.

*From: "Physics of Magnetism", S. Chickazumi, John Wiley & Sons, 1964.

Table 2-88. PRIMARY AND SECONDARY ELECTRON EMISSION

Primary emission. The Richardson-Dushman equation (given in Table 2-91) defines the increase in current density of electrons emitted from a metal surface as the temperature increases. It indicates that at a given temperature the flux is lower if the electron work function of the metal is high. Thus in electron tubes a low work function and a high operating temperature are desirable.

The work function is the energy that must be supplied to a body to make possible the primary emission of electrons across the surface barrier. Values are given in Table 2-89 for work functions of the elements as measured when the energy is supplied by heat, by light, or by contact with dissimilar metal. Some of these values are not yet well established (see Table 2-89 and References), but in many practical applications the surface is not a pure metallic element.

Secondary emission. This term is applied to the stream of electrons emitted from a solid when bombarded with primary electrons of considerable velocity. The ratio of secondary to primary electrons, called the "yield", may exceed unity, depending on the bombarded material and the velocity of the primary electrons, i.e., the voltage of the target with respect to the primary source. As the primary energy increases, the yield passes through unity and on to a maximum value, then decreases again. The following table gives typical values of the maximum yield and the corresponding voltage. One application of high-yield materials is for electron multipliers.

YIELD AND VOLTAGE FOR MAXIMUM SECONDARY EMISSION

Material	Maximum yield	Voltage at maximum yield	Material	Maximum yield	Voltage at maximum yield
Copper	1.3	600	Nickel	1.3	500
Diamond	2.8	750	Phosphor	3.	750
Glass (quartz)	3.	420	Platinum	1.8	700
Gold	1.4	800	SiO_2 (quartz)	2.5	400
Iron	1.3	400	Silver	1.5	800
Lead sulfide	1.2	500	Tantalum	1.3	600
MgO (crystal)	23.	1200	Tin	1.3	500
Molybdenum	1.3	370	Tungsten	1.4	650
NaBr (crystal)	24.	1800	Zinc sulfide	1.8	350
NaCl (crystal)	14.	1200			

REFERENCES

"Handbook of Chemistry and Physics", 53rd ed., R.C. Weast, Ed., The Chemical Rubber Co., 1972.

"American Institute of Physics Handbook", 2nd ed., D.E. Gray, Ed., McGraw-Hill Book Company, 1963.

"Handbook of Electronic Engineering", L.E.C. Hughes and F.W. Holland, Iliffe Books, London, 1967 (CRC Press, Cleveland).

"Reference Data for Radio Engineers", 5th ed., H.P. Westman, Ed., Howard W. Sams & Co., 1968.

"Advances in Electronics and Electron Physics", Vol. II, Academic Press, 1959.

Table 2-89. ELECTRON WORK FUNCTIONS OF THE ELEMENTS**

Herbert B. Michaelson

Preferred values for well outgassed polycrystalline materials are indicated by an asterisk (*). Because of the anisotropy and allotropy of work function, however, the designation "preferred value" has a rather limited meaning.

Values are expressed in electron-volts. Each value is followed by a reference to literature cited at the end of the table. The data is taken from literature published up to 1954.

Element	Thermionic work function	Photoelectric work function	Work function by contact potential method
Ag	3.09 (1) 3.56 (2) 4.08 (3) 4.31 (114)	3.67 (4) 4.1–4.75 (5) 4.50–4.52 (117) 4.56 (600°)* (6) 4.73 (20°)* (6) 4.75 (111) face (7) 4.81 (100) face (7)	4.21 (116) 4.33 (8) 4.35 (115) 4.44 (9) 4.47 (10)
Al	—	2.98 (4) 3.43 (11) 4.08* (12) 4.20 (113) 4.36 (13)	3.38 (9) 4.25 (115)
As	—	4.72 (118) 5.11 (14)	
Au	4.0–4.58 (2) 4.25 (114) 4.32 (3)	4.73 (740°C)* (15) 4.82 (20°C)* (15) 4.86–4.92 (5)	4.46 (9)
B		4.4–4.6 (16)	
Ba	2.11 (17)	2.48* (19) 2.49 (20) 2.52 (21)	1.73 (9) 2.39 (18)
Be	—	3.17 (14) 3.30 (22) 3.92* (23)	3.10 (9)
Bi	—	4.14 (11) 4.22–4.25* (24) 4.31 (25) 4.34 (118) 4.44 (22) 4.46 (47)	4.17 (9)
C	4.00 (26) 4.34 (27) 4.35 (119) 4.39 (28) 4.60* (120) 4.83 (121) 4.84 (122)	4.81 (29)	
Ca	2.24 (30)	2.42 (14) 2.706* (21) 2.76 (31) 3.20 (32) 3.21 (33)	3.33 (9)
Cd	—	3.68 (14) 3.73 (4) 3.94 (11) 4.07* (36) 4.099 (113)	4.00 (9) 4.49 (123)
Ce	2.6 (37)	2.84 (32)	
Co	4.40* (38) 4.41 (124)	3.90 (31) 4.12–4.25 (39)	4.21 (9)
Cr	4.58 (124) 4.60* (40) 4.7 (41)	4.37 (25)	4.38 (9)
Cs	1.81* (42)	1.9 (43) 1.96 (44)	
Cu	3.85 (1) 4.26 (2) 4.38 (3) 4.50 (114) 4.55† (45)	4.07 (4) 4.18 (31) 4.70 (125) 4.86 (111) face (46) 5.61 (100) face (46)	4.46* (9, 126) 4.61 (115) 4.86 (123)
Fe	4.04 (48) γ4.21 (38) γ4.31 (124) 4.47 (49) β4.48 (38) 4.76 (127)	3.91 (31) 3.92 (53) β4.62 (128) γ4.68 (128) α4.70 (128) 4.72 (50) 4.77 (51)	4.40 (9)
Ga	4.12 (14)	—	3.80 (9)
Ge	—	4.29 (31) 4.5 (52) 4.73 (14) 4.80 (16)	4.50 (9)
Hf	3.53 (54)		
Hg	—	4.30 (55) 4.52 (56) 4.53* (57, 58)	4.50 (9)
I	—	2.8 (rhombic) (129) 5.4 (monoclinic) (129) 6.8 (amorphous) (129)	
Ir	5.3 (130)	—	4.57 (9)
K	—	2.0 (59) 2.12 (22) 2.24* (44, 43) 2.26 (60)	1.60 (9)
La	3.3 (37)		
Li	—	2.28 (43) 2.42 (14)	1.40 (9) 2.49* (61)
Mg	—	2.74 (14) <3.0 (62) 3.59 (63) 3.62 (64) 3.68* (23) 3.79 (65)	3.58 (9) 3.78 (66)
Mn	3.83 (124)	3.76 (14)	4.14 (9)

† By field current method.
** From: "Handbook of Chemistry and Physics", 53rd ed., R.C. Weast, Ed., The Chemical Rubber Co., 1972.

Table 2-89. ELECTRON WORK FUNCTIONS
OF THE ELEMENTS (Continued)

Ele-ment	Thermionic work function		Photoelectric work function		Work function by contact potential method		Ele-ment	Thermionic work function		Photoelectric work function		Work function by contact potential method	
Mo	4.15	(67)	4.15	(67)	4.08	(75)	Sn	—		3.62	(4)	4.09	(9)
	4.17	(68)	4.34	(25)	4.48	(9)				3.87	(11)	4.64	(123)
	4.19	(69)								(liq.) 4.21	(92)		
	4.20*	(70)								γ4.38*	(92)		
	4.23	(121)								β4.50*	(92)		
	4.32	(71)					Sr	—		2.06	(93)		
	4.33	(72)								2.24	(14)		
	4.38	(73)								2.74*	(64)		
	4.44	(74)											
Na	—		2.06	(14)	1.60	(9)	Ta	4.03	(135)	4.05	(63)		
			2.25	(76)	2.26	(77)		4.07	(74)	4.12	(32)	3.96	(9)
			2.28*	(19)				4.10	(94)	4.16	(96)	4.25	(97)
			2.29	(23)				4.19*	(95)				
			2.47	(43)			Te	—		4.04	(117)	4.70	(9)
Nb	3.96	(34)								4.76	(16)		
	4.01*	(35)					Th	3.35*	(73)	3.38	(32)	3.46	(9)
Nd	3.3	(37)								3.47	(98)		
Ni	4.50	(124)	3.67	(4)	4.32	(9)				3.57	(11)		
	4.61	(38)	4.06	(31)	4.96	(79)	Ti	3.95	(124)	3.95	(14)	4.14	(9)
	4.63	(48)	4.87	(25)	(300°K)					4.17	(98)		
	5.03*	(78)	5.01*	(51)						4.45	(136)		
	5.1†	(45)	5.05 (623°K)	(131)			Tl	—		3.68	(107)	3.84	(9)
	5.24 (1150°K)	(131)	5.20 (1108°K)	(131)			U	3.27	(99)	3.63	(32)	4.32	(9)
Os					4.55	(9)	V	4.12	(124)	3.77	(14)	4.44	(9)
Pb			3.97	(4)	3.94	(9)	W	4.25	(100)	4.35 (310) face	(104)	4.38	(9)
			4.14	(11)				4.29 (116) face	(142)	4.49	(143)		
Pd	4.99*	(80)	4.97*	(80)	4.49	(9)		4.38 (111) face	(142)	4.50 (211) face	(104)		
Pr	2.7	(37)						4.39 (111) face	(140)	4.54	(105)		
Pt	4.72	(132)	4.09	(4)	4.52	(9)		4.39 (116) face	(140)	4.60	(32)		
	5.08	(26)	6.35	(84)	5.36*	(85)		4.45 (doped)	(138)				
	5.29	(81)						4.46	(101)				
	5.32*	(82)						4.52* (001) face					
	6.27	(83)							(26, 102, 142)				
Rb	—		2.09*	(44)				4.53	(103, 138)				
			2.16	(43)				4.56 (001) face	(140)				
Re	4.74	(133)						4.58 (110) face					
	5.1*	(86)	~5.0	(87)					(71, 73, 142)				
Rh	4.58	(88)	4.57	(88)	4.52	(9)		4.59 (001) face	(139)				
	4.80*	(89)						4.65 (112) face	(142)				
	4.9	(130)						4.68 (110) face	(140)				
								4.69 (112) face	(140)				
								5.01	(137)				
Ru	—		—		4.52	(9)	Zn	—		3.08	(4)	3.40	(111)
Sb	—		4.01	(90)	4.14	(9)				3.28	(106)	3.66	(9)
			4.60	(118)						3.32, 3.57	(108)	4.28	(112)
Se	—		4.62	(11)	4.42	(9)				3.60	(11)	4.65	(123)
			5.11	(14)						3.89	(31)		
Si	3.59	(28)			4.2	(91)				4.24	(109)		
	4.02	(134)	4.37–4.67	(117)						4.26 (0001) face	(110)		
										4.307	(113)		
Sm	3.2	(37)					Zr	4.12	(54)	3.73	(32)	3.60	(9)
								4.21*	(141)	4.33	(136)		

†By field current method.

Table 2-89. ELECTRON WORK FUNCTIONS
OF THE ELEMENTS *(Continued)*

REFERENCES

(1) A. Wehnelt and S. Seliger, Zeits. f. Physik **38**, 443 (1926).
(2) I. Ameiser, Zeits. f. Physik **69**, 111 (1931).
(3) A. Goetz, Zeits. f. Physik **43**, 531 (1927).
(4) P. Lukirsky and S. Prilesaev, Zeits. f. Physik **49**, 236 (1928).
(5) R. H. Fowler, Phys. Rev. **38**, 45 (1931).
(6) R. P. Winch, Phys. Rev. **37**, 1269 (1931).
(7) H. E. Farnsworth and R. P. Winch, Phys. Rev. **58**, 812 (1940).
(8) P. A. Anderson, Phys. Rev. **50**, 320 (1936).
(9) O. Klein and E. Lange, Zeits. f. Elektrochemie **44**, 542 (1938).
(10) P. A. Anderson, Phys. Rev. **59**, 1034 (1941).
(11) R. Hamer, Jour. Opt. Soc. Amer. **9**, 251 (1924).
(12) J. Brady and P. Jakobsmeyer, Phys. Rev. **49**, 670 (1936).
(13) E. Gaviola and J. Strong, Phys. Rev. **49**, 441 (1936).
(14) R. Schulze, Zeits, f Physik **92**, 212 (1934).
(15) L. W. Morris, Phys. Rev. **37**, 1263 (1931).
(16) L. Apker, E. Taft, and J. Dickey, Phys. Rev. **74**, 1462 (1948).
(17) A. L. Reimann, *Thermionic Emission* (Chapman and Hall, London, 1934), p. 37.
(18) P. A. Anderson, Phys. Rev. **47**, 958 (1935).
(19) R. J. Maurer, Phys. Rev. **57**, 653 (1940).
(20) R. J. Cashman and E. Bassoe, Phys. Rev. **55**, 63 (1939).
(21) N. C. Jamison and R. J. Cashman, Phys. Rev. **50**, 624 (1936).
(22) R. Suhrmann and A. Schallamach, Zeits. f. Physik **91**, 775 (1934).
(23) M. M. Mann, Jr. and L. A. DuBridge, Phys. Rev. **51**, 120 (1937).
(24) H. Jupnik, Phys. Rev. **60**, 884 (1941).
(25) H. C. Rentschler and D. E. Henry, Jour. Opt. Soc. Amer. **26**, 30 (1936).
(26) S. Dushman, "Thermal Emission of Electrons," International Critical Tables (1929) Vol. VI, pp. 53–4.
(27) A. L. Reiman, Proc. Phys. Soc. **50**, 496 (1938).
(28) A. Braun and G. Busch, Helv. Phys. Acta **20**, #1, 33 (1947).
(29) B. C. Roy, Proc. Roy. Soc. **A112**, 599 (1926).
(30) S. Dushman, Phys. Rev. **21**, 623 (1923).
(31) G. B. Welch, Phys. Rev. **32**, 657 (1928).
(32) C. Rentschler, D. E. Henry, and K. O. Smith, Rev. Sci. Instr. **3**, 794 (1932).
(33) I. Liben, Phys. Rev. **51**, 642 (1937).
(34) H. B. Wahlin and L. O. Sordahl, Phys. Rev. **45**, 886 (1934).
(35) A. L. Reimann and C. Kerr Grant, Phil. Mag. **22**, 34 (1936).
(36) H. Bomke, Ann. d. Physik **10**, 579 (1931).
(37) E. E. Schumaker, J. E. Harris, J. Amer. Chem. Soc. **48**, 3108 (1926).
(38) H. B. Wahlin, Phys. Rev. **61**, 509 (1942).
(39) A. B. Cardwell, Phys. Rev. **38**, 2033 (1931).
(40) H. B. Wahlin, Phys. Rev. **73**, 1458 (1948).
(41) H. Koesters, Zeits f. Physik **66**, 807 (1930).
(42) K. H. Kingdon, Rev. **25**, 892 (1925).
(43) A. R. Olpin, quoted by Hughes and DuBridge, *Photoelectric Phenomena* (1932).
(44) J. J. Brady, Phys. Rev. **41**, 613 (1932).
(45) W. P. Dyke, Thesis, University of Washington, (1946).
(46) N. Underwood, Phys. Rev. **47**, 502 (1935).
(47) A. H. Weber and C. J. Eisele, Phys. Rev. **59**, 473A (1941).
(48) W. Distler and G. Monch, Zeits. f. Physik **84**, 271 (1933).
(49) G. Siljeholm, Ann. d. Physik **10**, 178 (1931).
(50) A. B. Cardwell, Proc. Nat. Acad. Sci. **14**, 439 (1928).
(51) G. N. Glasoe, Phys. Rev. **38**, 1490 (1931).
(52) A. H. Smith, Phys. Rev. **75**, 953 (1949).
(53) G. B. Welch, Phys. Rev. **31**, 709A (1928).
(54) C. Zwikker, Phys. Zeits. **30**, 578 (1929).
(55) H. Cassel and A. Schneider, Naturwiss **22**, 464 (1934).
(56) D. Roller, W. H. Jordan, and C. S. Woodward, Phys. Rev. **38**, (1931).
(57) C. B. Kazda, Phys. Rev. **26**, 643 (1925).
(58) W. D. Hales, Phys. Rev. **32**, 950 (1928).
(59) H. E. Ives, J. Opt. Soc. Am. **8**, 551 (1924).
(60) H. Mayer, Ann. d. Phys. **29**, 129 (1937).
(61) P. A. Anderson, Phys. Rev. **75**, 1205 (1949).
(62) C. Kenty, Phys. Rev. **43**, 776(A), (1933).
(63) J. Cashman and S. Huxford, Phys. Rev. **48**, 734 (1935).
(64) R. J. Cashman and E. Bassoe, Phys. Rev. **53**, 919(A), (1938).
(65) R. J. Cashman, Phys. Rev. **54**, 971 (1938).
(66) P. A. Anderson, Phys. Rev. **54**, 753 (1938).
(67) L. A. DuBridge and W. W. Roehr, Phys. Rev. **42**, 52 (1932).
(68) H. B. Wahlin and J. A. Reynolds, Phys. Rev. **48**, 751 (1935).
(69) H. Grover, Phys. Rev. **52**, 982 (1937).
(70) R. W. Wright, Phys. Rev. **60**, 465 (1941).
(71) A. J. Ahearn, Phys. Rev. **44**, 277 (1933).

(72) H. Freitag and F. Kruger, Ann. d. Physik **21**, 697 (1934).
(73) C. Zwikker, Proc. Roy. Acad. Amsterdam **29**, 792 (1926).
(74) S. Dushman, H. N. Rowe, J. Ewald, C. A. Kidner, Phys. Rev. **25**, 338 (1925).
(75) C. W. Oatley, Proc. Roy. Soc. **A155**, 218 (1936).
(76) Z. Berkes, Math. Phys. Lapok **41**, 131 (1934).
(77) E. Patai, Zeits, f. Physik. **59**, 697 (1930).
(78) G. W. Fox and R. M. Bowie, Phys. Rev. **44**, 345 (1933).
(79) R. C. L. Bosworth, Trans. Far. Soc. **35**, 397 (1939).
(80) L. A. DuBridge and W. W. Roehr, Phys. Rev. **39**, 99 (1932).
(81) H. L. Van Velzer, Phys. Rev. **44**, 831 (1933).
(82) L. V. Whitney, Phys. Rev. **50**, 1154 (1936).
(83) L. A. DuBridge, Phys. Rev. **32**, 961 (1928).
(84) L. A. DuBridge, Phys. Rev. **31**, 236 (1928).
(85) C. W. Oatley, Proc. Phys. Soc. **51**, 318 (1939).
(86) C. Agte, H. Alterthum, K. Becker, G. Heyne, and K. Moers. Naturwiss **19**, 108 (1931).
(87) A. Englemann, Ann. d. Physik **17**, 185 (1933).
(88) E. H. Dixon, Phys. Rev. **37**, 60 (1931).
(89) H. B. Wahlin and L. V. Whitney, J. Chem. Phys. **6**, 594 (1938).
(90) V. Middel, Zeits. f. Physik **105**, 358 (1937).
(91) W. E. Meyerhof, Phys. Rev. **71**, 727 (1947).
(92) A. Goetz, Phys. Rev. **33**, 373 (1929).
(93) R. Doepel, Zeits. f. Physik **33**, 237 (1925).
(94) A. B. Cardwell, Phys. Rev. **47**, 628 (1935).
(95) M. D. Fiske, Phys. Rev. **61**, 513 (1942).
(96) A. B. Cardwell, Phys. Rev. **38**, 2041 (1931).
(97) W. Heinze, Zeits. f. Phys. **109**, 459 (1938).
(98) H. C. Rentschler and D. E. Henry, Trans. Electrochem. Soc. **87**, 289 (1945–6).
(99) W. L. Hole and R. W. Wright, Phys. Rev. **56**, 785 (1939).
(100) A. H. Warner, Proc. Nat. Acad. Sci. **13**, 56 (1927).
(101) G. M. Fleming and J. E. Henderson, Phys. Rev. **58**, 887 (1940).
(102) W. B. Nottingham, Phys. Rev. **47**, 806A (1935).
(103) H. Freitag and F. Krueger, Ann. d. Physik **21**, 697 (1934).
(104) C. E. Mendenhall and C. F. DeVoe, Phys. Rev. **51**, 346 (1937).
(105) A. H. Warner, Phys. Rev. **38**, 1871 (1931).
(106) A. Nitsche, Ann. d. Physik **14**, 463 (1932).
(107) R. Suhrmann and H. Csesch, Zeits. Chem. **B28**, 215 (1935).
(108) J. H. Dillon, Phys. Rev. **38**, 408 (1931).
(109) C. F. DeVoe, Phys. Rev. **50**, 481 (1936).
(110) W. Klug and H. Steyskal, Zeits. f. Physik **116**, 415 (1940).
(111) C. W. Oatley, Proc. Roy. Soc. **A155**, 218 (1936).
(112) P. A. Anderson, Phys. Rev. **57**, 122 ,(1940).
(113) R. Suhrmann and J. Pietrzyk, Zeits. f Physik **122**, 600 (1944).
(114) S. C. Jain and K. S. Krishnan, Proc. Roy. Soc. **A217**, 451 (1953).
(115) E. W. J. Mitchell and J. W. Mitchell, Proc. Roy. Soc. **A210**, 70 (1952).
(116) G. L. Weissler and T. N. Wilson, J. Appl. Phys. **24**, 472 (1953).
(117) S. M. Fainshtein, Zavodskaya Lab. **14**, 64 (1948).
(118) L. Apker, E. Taft and J. Dickey, Phys. Rev. **76**, 270 (1949).
(119) G. Glockler and J. W. Sausville, J. Electrochem. Soc. **95**, 292 (1949).
(120) H. F. Ivey, Phys. Rev. **76**, 567 (1949).
(121) S. B. L. Mathur, Natl. Inst. Sci. India **19**, 153 (1953).
(122) A. S. Bhatnager, Proc. Nat. Acad. Sci. India **A14**, 5 (1944).
(123) R. Hirschberg and E. Lange, Naturwiss. **39**, 131 (1952).
(124) K. S. Krishnan and S. C. Jain, Nature **170**, 759 (1952).
(125) R. Ito, J. Phys. Soc. Japan **6**, 188 (1951).
(126) P. A. Anderson, Phys. Rev. **76**, 388 (1949).
(127) S. B. L. Mathur, Proc. Natl. Inst. Sci. India **19**, 165 (1953).
(128) A. B. Cardwell, Phys. Rev. **92**, 554 (1953).
(129) D. C. West, Can. J. Phys. **31**, 691 (1953).
(130) O. A. Weinreich, Phys. Rev. **82**, 573 (1951).
(131) A. B. Cardwell, Phys. Rev. **76**, 125 (1949).
(132) K. Kondo, Rept. Inst. Sci. Tech., Tokyo Univ. **1**, 24 (1947).
(133) R. Levi and G. A. Espersen, Phys. Rev. **78**, 231 (1950).
(134) L. Esaki, J. Phys. Soc. Japan **8**, 347 (1953).
(135) R. J. Munick, W. B. LaBerge and E. A. Coomes, Phys. Rev. **80**, 887 (1950).
(136) H. Malamud and A. D. Krumbein, J. Appl. Phys. **25**, 591 (1954).
(137) G. M. Fleming and J. E. Henderson, Phys. Rev. **58**, 887 (1940).
(138) M. H. Nichols, Phys. Rev. **78**, 158 (1950).
(139) A. A. Brown, L. J. Neelands and H. E. Farnsworth, J. Appl. Phys. **21**, 1 (1950).
(140) C. Herring and M. H. Nichols, Rev. Mod. Phys. **21**, 185 (1949).
(141) A. Wahl, Phys. Rev. **82**, 574 (1951).
(142) G. F. Smith, Phys. Rev. **94**, 295 (1954).
(143) L. Apker, E. Taft and J. Dickey, Phys. Rev. **73**, 46 (1948).

Table 2-90.   SURFACE IONIZATION PROBABILITIES*

Surface ionization, resonance ionization, and contact ionization are synonymous terms used to describe ionization of atoms resulting from contact with a surface. This form of ionization is often observed in cesiated thermionic converters because of the low-ionization potential of cesium. The expression used to predict surface ionization probabilities was derived by Langmuir and Kingdon.[1] The derivation is based on the usual constraints of equilibrium but is found to apply quite accurately in highly non-equilibrium conditions. The equation is called the Saha-Langmuir equation and is given below in its simplest form for cesium ions leaving a surface.

$$\frac{f_+}{f_p} = \left\{1 + 2 \exp\frac{(3.8938 - V)(11{,}606)}{T}\right\}^{-1}$$

where f_+ = the flux of ions leaving the surface
f_p = the flux of incident atoms and ions
V = the effective work function in electron volts.
T = temperature, °K

The table below is a condensation of tabular information from Read.[2] Further discussions of the use and limitations of the Saha-Langmuir equation may be found in Zandberg and Ionov.[3]

SURFACE IONIZATION EFFICIENCY OF CESIUM, f_+/f_p

I = 3.89385 V

Temperature, °K

V, electron volts	1000°	1100°	1200°	1300°	1400°	1500°	1600°	1700°	1800°	1900°	2000°	2100°	2200°
2.00	.143E-9†	.105E-8	.556E-8	.227E-7	.761E-7	.217E-6	.541E-6	.121E-5	.249E-5	.474E-5	.844E-5	.142E-4	.229E-4
2.05	.255E-9	.178E-8	.901E-8	.355E-7	.115E-6	.319E-6	.778E-6	.171E-5	.344E-5	.643E-5	.113E-4	.188E-4	.298E-4
2.10	.455E-9	.302E-8	.146E-7	.555E-7	.174E-6	.470E-6	.112E-5	.240E-5	.475E-5	.872E-5	.151E-4	.248E-4	.389E-4
2.15	.813E-9	.512E-8	.237E-7	.868E-7	.264E-6	.691E-6	.161E-5	.338E-5	.655E-5	.118E-4	.202E-4	.326E-4	.506E-4
2.20	.145E-8	.867E-8	.384E-7	.136E-6	.399E-6	.102E-5	.231E-5	.476E-5	.904E-5	.161E-4	.269E-4	.430E-4	.658E-4
2.25	.259E-8	.147E-7	.624E-7	.212E-6	.604E-6	.150E-5	.332E-5	.669E-5	.125E-4	.218E-4	.360E-4	.567E-4	.857E-4
2.30	.464E-8	.249E-7	.101E-6	.331E-6	.914E-6	.221E-5	.477E-5	.941E-5	.172E-4	.296E-4	.481E-4	.748E-4	.112E-3
2.35	.828E-8	.422E-7	.164E-6	.517E-6	.138E-5	.325E-5	.685E-5	.132E-4	.238E-4	.402E-4	.643E-4	.986E-4	.145E-3
2.40	.148E-7	.715E-7	.266E-6	.808E-6	.209E-5	.478E-5	.985E-5	.186E-4	.328E-4	.545E-4	.860E-4	.130E-3	.189E-3
2.45	.264E-7	.121E-6	.431E-6	.126E-5	.317E-5	.704E-5	.142E-4	.262E-4	.453E-4	.740E-4	.115E-3	.171E-3	.246E-3
2.50	.472E-7	.205E-6	.700E-6	.197E-5	.480E-5	.104E-4	.203E-4	.369E-4	.625E-4	.100E-3	.154E-3	.226E-3	.320E-3
2.55	.843E-7	.348E-6	.113E-5	.308E-5	.726E-5	.153E-4	.292E-4	.519E-4	.863E-4	.136E-3	.205E-3	.298E-3	.417E-3
2.60	.151E-6	.590E-6	.184E-5	.482E-5	.110E-4	.225E-4	.420E-4	.730E-4	.119E-3	.185E-3	.274E-3	.392E-3	.543E-3
2.65	.269E-6	1.000E-6	.298E-5	.753E-5	.166E-4	.331E-4	.604E-4	.103E-3	.164E-3	.251E-3	.367E-3	.517E-3	.707E-3
2.70	.481E-6	.169E-5	.484E-5	.118E-4	.252E-4	.487E-4	.868E-4	.144E-3	.227E-3	.340E-3	.490E-3	.681E-3	.920E-3

†Read as 0.143×10^{-9}

Table 2-90.   SURFACE IONIZATION PROBABILITIES (Continued)

Temperature, °K

V, electron volts	1000°	1100°	1200°	1300°	1400°	1500°	1600°	1700°	1800°	1900°	2000°	2100°	2200°
2.75	.859E−6	.287E−5	.785E−5	.184E−4	.381E−4	.717E−4	.125E−3	.203E−3	.313E−3	.462E−3	.655E−3	.898E−3	.120E−2
2.80	.153E−5	.487E−5	.127E−4	.287E−4	.577E−4	.106E−3	.179E−3	.286E−3	.433E−3	.627E−3	.875E−3	.118E−2	.156E−2
2.85	.274E−5	.825E−5	.206E−4	.449E−4	.873E−4	.155E−3	.257E−3	.402E−3	.597E−3	.851E−3	.117E−2	.156E−2	.203E−2
2.90	.490E−5	.140E−4	.335E−4	.701E−4	.132E−3	.229E−3	.370E−3	.565E−3	.824E−3	.115E−2	.156E−2	.206E−2	.264E−2
2.95	.875E−5	.237E−4	.543E−4	.110E−3	.200E−3	.337E−3	.532E−3	.795E−3	.114E−2	.157E−2	.209E−2	.271E−2	.343E−2
3.00	.156E−4	.401E−4	.881E−4	.171E−3	.303E−3	.496E−3	.764E−3	.112E−2	.157E−2	.212E−2	.279E−2	.357E−2	.446E−2
3.05	.279E−4	.680E−4	.143E−3	.268E−3	.458E−3	.730E−3	.110E−2	.157E−2	.216E−2	.288E−2	.372E−2	.470E−2	.580E−2
3.10	.499E−4	.115E−3	.232E−3	.418E−3	.693E−3	.107E−2	.158E−2	.221E−2	.293E−2	.390E−2	.497E−2	.618E−2	.753E−2
3.15	.891E−4	.195E−3	.376E−3	.653E−3	.105E−2	.158E−2	.226E−2	.311E−2	.412E−2	.529E−2	.663E−2	.813E−2	.979E−2
3.20	.159E−3	.331E−3	.609E−3	.102E−2	.159E−2	.233E−2	.325E−2	.437E−2	.567E−2	.717E−2	.884E−2	.107E−1	.127E−1
3.25	.284E−3	.561E−3	.987E−3	.159E−2	.240E−2	.342E−2	.466E−2	.613E−2	.780E−2	.970E−2	.118E−1	.140E−1	.165E−1
3.30	.508E−3	.950E−3	.160E−2	.249E−2	.363E−2	.503E−2	.669E−2	.860E−2	.103E−1	.131E−1	.157E−1	.184E−1	.213E−1
3.35	.907E−3	.161E−2	.259E−2	.388E−2	.548E−2	.739E−2	.959E−2	.121E−1	.143E−1	.177E−1	.209E−1	.242E−1	.276E−1
3.40	.162E−2	.272E−2	.420E−2	.605E−2	.827E−2	.108E−1	.137E−1	.169E−1	.203E−1	.239E−1	.277E−1	.316E−1	.356E−1
3.45	.289E−2	.461E−2	.679E−2	.942E−2	.125E−1	.159E−1	.196E−1	.236E−1	.273E−1	.322E−1	.367E−1	.413E−1	.459E−1
3.50	.515E−2	.778E−2	.110E−1	.146E−1	.187E−1	.232E−1	.279E−1	.329E−1	.381E−1	.432E−1	.484E−1	.537E−1	.589E−1
3.55	.916E−2	.131E−1	.177E−1	.227E−1	.281E−1	.338E−1	.397E−1	.456E−1	.517E−1	.577E−1	.637E−1	.696E−1	.754E−1
3.60	.163E−1	.220E−1	.283E−1	.350E−1	.419E−1	.490E−1	.561E−1	.630E−1	.691E−1	.767E−1	.833E−1	.897E−1	.959E−1
3.65	.287E−1	.368E−1	.452E−1	.537E−1	.621E−1	.705E−1	.786E−1	.865E−1	.943E−1	.101E 0	.108E 0	.115E 0	.121E 0
3.70	.501E−1	.608E−1	.712E−1	.814E−1	.911E−1	.100E 0	.109E 0	.117E 0	.125E 0	.133E 0	.140E 0	.146E 0	.152E 0
3.75	.861E−1	.988E−1	.111E 0	.122E 0	.132E 0	.141E 0	.150E 0	.158E 0	.165E 0	.172E 0	.178E 0	.184E 0	.190E 0
3.80	.144E 0	.157E 0	.168E 0	.178E 0	.187E 0	.195E 0	.202E 0	.209E 0	.214E 0	.220E 0	.225E 0	.229E 0	.234E 0
3.85	.231E 0	.239E 0	.247E 0	.253E 0	.258E 0	.263E 0	.267E 0	.270E 0	.274E 0	.277E 0	.279E 0	.282E 0	.284E 0
3.90	.349E 0	.348E 0	.347E 0	.346E 0	.345E 0	.344E 0	.343E 0	.343E 0	.342E 0	.342E 0	.341E 0	.341E 0	.341E 0
3.95	.490E 0	.475E 0	.463E 0	.452E 0	.443E 0	.436E 0	.429E 0	.423E 0	.413E 0	.413E 0	.409E 0	.405E 0	.402E 0
4.00	.632E 0	.605E 0	.583E 0	.563E 0	.547E 0	.532E 0	.519E 0	.508E 0	.498E 0	.489E 0	.481E 0	.473E 0	.467E 0
4.05	.754E 0	.722E 0	.694E 0	.668E 0	.646E 0	.626E 0	.608E 0	.592E 0	.575E 0	.565E 0	.553E 0	.542E 0	.533E 0
4.10	.845E 0	.815E 0	.786E 0	.759E 0	.734E 0	.711E 0	.690E 0	.671E 0	.654E 0	.638E 0	.623E 0	.610E 0	.597E 0
4.15	.907E 0	.882E 0	.856E 0	.831E 0	.807E 0	.784E 0	.762E 0	.742E 0	.723E 0	.705E 0	.689E 0	.673E 0	.659E 0
4.20	.946E 0	.927E 0	.906E 0	.885E 0	.863E 0	.842E 0	.822E 0	.802E 0	.783E 0	.764E 0	.747E 0	.731E 0	.715E 0
4.25	.969E 0	.955E 0	.940E 0	.923E 0	.905E 0	.887E 0	.869E 0	.850E 0	.832E 0	.815E 0	.798E 0	.782E 0	.766E 0
4.30	.982E 0	.973E 0	.962E 0	.949E 0	.935E 0	.920E 0	.905E 0	.889E 0	.873E 0	.857E 0	.841E 0	.825E 0	.810E 0
4.35	.990E 0	.984E 0	.976E 0	.967E 0	.956E 0	.945E 0	.932E 0	.918E 0	.904E 0	.890E 0	.876E 0	.861E 0	.847E 0
4.40	.994E 0	.990E 0	.985E 0	.979E 0	.971E 0	.962E 0	.952E 0	.941E 0	.929E 0	.917E 0	.904E 0	.891E 0	.878E 0
4.45	.997E 0	.994E 0	.991E 0	.986E 0	.980E 0	.974E 0	.966E 0	.957E 0	.947E 0	.937E 0	.926E 0	.915E 0	.904E 0

Table 2-90.   SURFACE IONIZATION PROBABILITIES *(Continued)*

Temperature, °K

V, electron volts	1000°	1100°	1200°	1300°	1400°	1500°	1600°	1700°	1800°	1900°	2000°	2100°	2200°
4.50	.998E 0	.997E 0	.994E 0	.991E 0	.987E 0	.982E 0	.976E 0	.969E 0	.961E 0	.953E 0	.944E 0	.934E 0	.924E 0
4.55	.999E 0	.998E 0	.997E 0	.994E 0	.991E 0	.988E 0	.983E 0	.978E 0	.972E 0	.965E 0	.957E 0	.949E 0	.941E 0
4.60	.999E 0	.999E 0	.998E 0	.996E 0	.994E 0	.992E 0	.988E 0	.984E 0	.979E 0	.974E 0	.968E 0	.961E 0	.954E 0
4.65	1.000E 0	.999E 0	.999E 0	.998E 0	.996E 0	.994E 0	.992E 0	.989E 0	.985E 0	.981E 0	.976E 0	.970E 0	.964E 0
4.70	1.000E 0	1.000E 0	.999E 0	.999E 0	.998E 0	.996E 0	.994E 0	.992E 0	.989E 0	.986E 0	.982E 0	.977E 0	.972E 0
4.75	1.000E 0	1.000E 0	.999E 0	.999E 0	.998E 0	.997E 0	.996E 0	.994E 0	.992E 0	.989E 0	.986E 0	.983E 0	.979E 0
4.80	1.000E 0	1.000E 0	1.000E 0	.999E 0	.999E 0	.998E 0	.997E 0	.996E 0	.994E 0	.992E 0	.990E 0	.987E 0	.983E 0
4.85	1.000E 0	1.000E 0	1.000E 0	1.000E 0	.999E 0	.999E 0	.998E 0	.997E 0	.996E 0	.994E 0	.992E 0	.990E 0	.987E 0
4.90	1.000E 0	1.000E 0	1.000E 0	1.000E 0	1.000E 0	.999E 0	.999E 0	.998E 0	.997E 0	.996E 0	.994E 0	.992E 0	.990E 0
4.95	1.000E 0	1.000E 0	1.000E 0	1.000E 0	1.000E 0	.999E 0	.999E 0	.999E 0	.998E 0	.997E 0	.996E 0	.994E 0	.992E 0
5.00	1.000E 0	1.000E 0	1.000E 0	1.000E 0	1.000E 0	1.000E 0	.999E 0	.999E 0	.998E 0	.998E 0	.997E 0	.996E 0	.994E 0
5.05	1.000E 0	1.000E 0	1.000E 0	1.000E 0	1.000E 0	1.000E 0	1.000E 0	.999E 0	.999E 0	.998E 0	.998E 0	.997E 0	.996E 0
5.10	1.000E 0	1.000E 0	1.000E 0	1.000E 0	1.000E 0	1.000E 0	1.000E 0	.999E 0	.999E 0	.999E 0	.999E 0	.997E 0	.997E 0
5.15	1.000E 0	1.000E 0	1.000E 0	1.000E 0	1.000E 0	1.000E 0	1.000E 0	1.000E 0	.999E 0	.999E 0	.999E 0	.998E 0	.997E 0
5.20	1.000E 0	1.000E 0	1.000E 0	1.000E 0	1.000E 0	1.000E 0	1.000E 0	1.000E 0	1.000E 0	.999E 0	.999E 0	.999E 0	.998E 0
5.25	1.000E 0	1.000E 0	1.000E 0	1.000E 0	1.000E 0	1.000E 0	1.000E 0	1.000E 0	1.000E 0	.999E 0	.999E 0	.999E 0	.998E 0
5.30	1.000E 0	1.000E 0	1.000E 0	1.000E 0	1.000E 0	1.000E 0	1.000E 0	1.000E 0	1.000E 0	1.000E 0	.999E 0	.999E 0	.999E 0
5.35	1.000E 0	1.000E 0	1.000E 0	1.000E 0	1.000E 0	1.000E 0	1.000E 0	1.000E 0	1.000E 0	1.000E 0	1.000E 0	.999E 0	.999E 0
5.40	1.000E 0	1.000E 0	1.000E 0	1.000E 0	1.000E 0	1.000E 0	1.000E 0	1.000E 0	1.000E 0	1.000E 0	1.000E 0	1.000E 0	.999E 0
5.45	1.000E 0	1.000E 0	1.000E 0	1.000E 0	1.000E 0	1.000E 0	1.000E 0	1.000E 0	1.000E 0	1.000E 0	1.000E 0	1.000E 0	.999E 0
5.50	1.000E 0	1.000E 0	1.000E 0	1.000E 0	1.000E 0	1.000E 0	1.000E 0	1.000E 0	1.000E 0	1.000E 0	1.000E 0	1.000E 0	1.000E 0
5.55	1.000E 0	1.000E 0	1.000E 0	1.000E 0	1.000E 0	1.000E 0	1.000E 0	1.000E 0	1.000E 0	1.000E 0	1.000E 0	1.000E 0	1.000E 0
5.60	1.000E 0	1.000E 0	1.000E 0	1.000E 0	1.000E 0	1.000E 0	1.000E 0	1.000E 0	1.000E 0	1.000E 0	1.000E 0	1.000E 0	1.000E 0
5.65	1.000E 0	1.000E 0	1.000E 0	1.000E 0	1.000E 0	1.000E 0	1.000E 0	1.000E 0	1.000E 0	1.000E 0	1.000E 0	1.000E 0	1.000E 0
5.70	.100E 1	1.000E 0	1.000E 0	1.000E 0	1.000E 0	1.000E 0	1.000E 0	1.000E 0	1.000E 0	1.000E 0	1.000E 0	1.000E 0	1.000E 0
5.75	.100E 1	1.000E 0	1.000E 0	1.000E 0	1.000E 0	1.000E 0	1.000E 0	1.000E 0	1.000E 0	1.000E 0	1.000E 0	1.000E 0	1.000E 0
5.80	.100E 1	1.000E 0	1.000E 0	1.000E 0	1.000E 0	1.000E 0	1.000E 0	1.000E 0	1.000E 0	1.000E 0	1.000E 0	1.000E 0	1.000E 0
5.85	.100E 1	1.000E 0	1.000E 0	1.000E 0	1.000E 0	1.000E 0	1.000E 0	1.000E 0	1.000E 0	1.000E 0	1.000E 0	1.000E 0	1.000E 0
5.90	.100E 1	.100E 1	1.000E 0	1.000E 0	1.000E 0	1.000E 0	1.000E 0	1.000E 0	1.000E 0	1.000E 0	1.000E 0	1.000E 0	1.000E 0
5.95	.100E 1	.100E 1	1.000E 0	1.000E 0	1.000E 0	1.000E 0	1.000E 0	1.000E 0	1.000E 0	1.000E 0	1.000E 0	1.000E 0	1.000E 0
6.00	.100E 1	.100E 1	1.000E 0	1.000E 0	1.000E 0	1.000E 0	1.000E 0	1.000E 0	1.000E 0	1.000E 0	1.000E 0	1.000E 0	1.000E 0

*By permission of the General Electric Company © 1964.

REFERENCES

1. I. Langmuir and K.H. Kingdon, *Proc. Roy. Soc.* (London), A107:61, 1925.
2. P.L. Read, General Electric Research Lab. Report, No. 64-RL-3691-E, 1964.
3. E.Ya. Zandberg and N.I. Ionov, *Soviet Phys. Uspekhi*, 67(2):255, 1959.

Table 2-91. THEORETICAL FLOW OF ELECTRONS FROM A HEATED SURFACE*

The Richardson-Dushman[1,2] equation is the equation most often used to describe the flow of electrons emitted from a heated surface. The equation is

$$J = (1-R) A T^2 \exp{-\frac{\Phi}{kT}}$$

where $A = 4\pi mek^2/h^3 = 120.17$

J = current density in amp/cm^2

R = reflection coefficient usually set equal to 0

T = temperature in $°K$

e = electron charge in coulombs

h = Planck's constant

k = Boltzmann constant

m = mass of the electron

Φ = effective work function in electron volts

The equation is based on an ideal surface in equilibrium with its surroundings; thus the effective work function is not necessarily identical with the true work function encountered in practice. Despite these limitations it is generally accepted as a standard form of expressing the electron emission of metals. Many graphical and tabular solutions exist. Reprinted below is a typical tabulation from Read.[3] See this source for additional information.

Current Density in Amperes per Sq Cm

Electron volts	\(300°\)	\(400°\)	\(500°\)	\(600°\)	\(700°\)	\(800°\)	\(900°\)
0.5	.4307E −1†	.9638E 1	.2740E 3	.2730E 4	.1479E 5	.5445E 5	.1543E 6
0.6	.8998E −3	.5297E 0	.2690E 2	.3946E 3	.2819E 4	.1277E 5	.4249E 5
0.7	.1880E −4	.2911E −1	.2641E 1	.5704E 2	.5371E 3	.2992E 4	.1170E 5
0.8	.3928E −6	.1600E −2	.2593E 0	.8245E 1	.1023E 3	.7015E 3	.3224E 4
0.9	.8207E −8	.8791E −4	.2546E −1	.1192E 1	.1950E 2	.1645E 3	.8879E 3
1.0	.1715E −9	.4831E −5	.2499E −2	.1723E 0	.3716E 1	.3855E 2	.2445E 3
1.1	.3583E −11	.2655E −6	.2454E −3	.2490E −1	.7081E 0	.9038E 1	.6735E 2
1.2	.7486E −13	.1459E −7	.2409E −4	.3599E −2	.1349E 0	.2119E 1	.1855E 2
1.3	.1564E −14	.8019E −9	.2365E −5	.5203E −3	.2571E −1	.4967E 0	.5109E 1
1.4	.3268E −16	.4407E −10	.2322E −6	.7520E −4	.4899E −2	.1164E 0	.1407E 1
1.5	.6828E −18	.2422E −11	.2280E −7	.1087E −4	.9335E −3	.2730E −1	.3876E 0
1.6	.1427E −19	.1331E −12	.2238E −8	.1571E −5	.1779E −3	.6399E −2	.1068E 0
1.7	.2981E −21	.7314E −14	.2198E −9	.2271E −6	.3389E −4	.1500E −2	.2940E −1
1.8	.6228E −23	.4019E −15	.2157E −10	.3283E −7	.6459E −5	.3516E −3	.8098E −2
1.9	.1301E −24	.2209E −16	.2118E −11	.4745E −8	.1231E −5	.8244E −4	.2231E −2
2.0	.2719E −26	.1214E −17	.2080E −12	.6859E −9	.2345E −6	.1933E −4	.6143E −3
2.1	.5681E −28	.6671E −19	.2042E −13	.9915E −10	.4469E −7	.4530E −5	.1692E −3
2.2	.1187E −29	.3666E −20	.2004E −14	.1433E −10	.8515E −8	.1062E −5	.4660E −4
2.3		.2015E −21	.1968E −15	.2072E −11	.1622E −8	.2490E −6	.1284E −4
2.4		.1107E −22	.1932E −16	.2995E −12	.3092E −9	.5836E −7	.3535E −5
2.5		.6085E −24	.1897E −17	.4328E −13	.5891E −10	.1368E −7	.9737E −6
2.6		.3344E −25	.1862E −18	.6257E −14	.1123E −10	.3207E −8	.2682E −6
2.7		.1838E −26	.1828E −19	.9044E −15	.2139E −11	.7519E −9	.7387E −7
2.8		.1010E −27	.1795E −20	.1307E −15	.4076E −12	.1763E −9	.2035E −7
2.9		.5550E −29	.1762E −21	.1890E −16	.7767E −13	.4132E −10	.5604E −8
3.0		.3050E −30	.1730E −22	.2731E −17	.1480E −13	.9687E −11	.1543E −8
3.1			.1699E −23	.3948E −18	.2820E −14	.2271E −11	.4251E −9
3.2			.1668E −24	.5707E −19	.5373E −15	.5324E −12	.1171E −9
3.3			.1637E −25	.8249E −20	.1024E −15	.1248E −12	.3225E −10
3.4			.1607E −26	.1192E −20	.1951E −16	.2926E −13	.8882E −11
3.5			.1578E −27	.1724E −21	.3718E −17	.6858E −14	.2446E −11
3.6			.1549E −28	.2491E −22	.7084E −18	.1608E −14	.6738E −12
3.7			.1521E −29	.3601E −23	.1350E −18	.3769E −15	.1856E −12
3.8				.5205E −24	.2572E −19	.8836E −16	.5111E −13
3.9				.7524E −25	.4901E −20	.2071E −16	.1408E −13
4.0				.1088E −25	.9339E −21	.4856E −17	.3877E −14
4.1				.1572E −26	.1780E −21	.1138E −17	.1068E −14
4.2				.2272E −27	.3391E −22	.2669E −18	.2941E −15
4.3				.3285E −28	.6462E −23	.6256E −19	.8101E −16
4.4				.4748E −29	.1231E −23	.1467E −19	.2231E −16
4.5				.6863E −30	.2346E −24	.3438E −20	.6146E −17
4.6					.4471E −25	.8059E −21	.1693E −17
4.7					.8519E −26	.1889E −21	.4662E −18
4.8					.1623E −26	.4429E −22	.1284E −18
4.9					.3093E −27	.1038E −22	.3537E −19
5.0					.5894E −28	.2434E −23	.9741E −20
5.1					.1123E −28	.5706E −24	.2683E −20
5.2					.2140E −29	.1338E −24	.7389E −21
5.3					.4078E −30	.3136E −25	.2035E −21
5.4						.7351E −26	.5606E −22

†Read this as 0.4307×10^{-1}.

Table 2-91. THEORETICAL FLOW OF ELECTRONS FROM A HEATED SURFACE (Continued)

Electron volts	Temperature, °K						
	300°	400°	500°	600°	700°	800°	900°
5.5						.1723E−26	.1544E−22
5.6						.4040E−27	.4252E−23
5.7						.9471E−28	.1171E−23
5.8						.2220E−28	.3226E−24
5.9						.5205E−29	.8885E−25
6.0						.1220E−29	.2447E−25

Electron volts	Temperature, °K						
	1000°	1100°	1200°	1300°	1400°	1500°	1600°
0.5	.3629E 6	.7442E 6	.1375E 7	.2340E 7	.3733E 7	.5649E 7	.8186E 7
0.6	.1137E 6	.2591E 6	.5226E 6	.9584E 6	.1630E 7	.2606E 7	.3963E 7
0.7	.3563E 5	.9023E 5	.1987E 6	.3925E 6	.7114E 6	.1202E 7	.1919E 7
0.8	.1116E 5	.3142E 5	.7554E 5	.1608E 6	.3105E 6	.5546E 6	.9291E 6
0.9	.3498E 4	.1094E 5	.2872E 5	.6584E 5	.1355E 6	.2559E 6	.4499E 6
1.0	.1096E 4	.3809E 4	.1092E 5	.2697E 5	.5917E 5	.1180E 6	.2178E 6
1.1	.3434E 3	.1326E 4	.4152E 4	.1104E 5	.2583E 5	.5445E 5	.1055E 6
1.2	.1076E 3	.4618E 3	.1578E 4	.4523E 4	.1127E 5	.2512E 5	.5106E 5
1.3	.3372E 2	.1608E 3	.6001E 3	.1852E 4	.4922E 4	.1159E 5	.2472E 5
1.4	.1057E 2	.5599E 2	.2282E 3	.7587E 3	.2148E 4	.5346E 4	.1197E 5
1.5	.3310E 1	.1950E 2	.8674E 2	.3107E 3	.9378E 3	.2466E 4	.5796E 4
1.6	.1037E 1	.6788E 1	.3298E 2	.1273E 3	.4094E 3	.1138E 4	.2806E 4
1.7	.3250E 0	.2364E 1	.1254E 2	.5212E 2	.1787E 3	.5249E 3	.1359E 4
1.8	.1018E 0	.8230E 0	.4767E 1	.2135E 2	.7801E 2	.2421E 3	.6578E 3
1.9	.3191E −1	.2866E 0	.1812E 1	.8742E 1	.3405E 2	.1117E 3	.3185E 3
2.0	.9998E −2	.9978E −1	.6891E 0	.3580E 1	.1486E 2	.5153E 2	.1542E 3
2.1	.3133E −2	.3474E −1	.2620E 0	.1466E 1	.6488E 1	.2377E 2	.7466E 2
2.2	.9816E −3	.1210E −1	.9960E −1	.6006E 0	.2832E 1	.1097E 2	.3615E 2
2.3	.3076E −3	.4212E −2	.3787E −1	.2460E 0	.1236E 1	.5059E 1	.1750E 2
2.4	.9637E −4	.1467E −2	.1440E −1	.1007E 0	.5397E 0	.2334E 1	.8475E 1
2.5	.3020E −4	.5107E −3	.5474E −2	.4126E −1	.2356E 0	.1077E 1	.4103E 1
2.6	.9461E −5	.1778E −3	.2081E −2	.1690E −1	.1028E 0	.4967E 0	.1987E 1
2.7	.2964E −5	.6192E −4	.7912E −3	.6920E −2	.4489E −1	.2291E 0	.9619E 0
2.8	.9289E −6	.2156E −4	.3008E −3	.2834E −2	.1960E −1	.1057E 0	.4657E 0
2.9	.2910E −6	.7507E −5	.1144E −3	.1161E −2	.8554E −2	.4876E −1	.2255E 0
3.0	.9119E −7	.2614E −5	.4348E −4	.4754E −3	.3734E −2	.2250E −1	.1092E 0
3.1	.2857E −7	.9101E −6	.1653E −4	.1947E −3	.1630E −2	.1038E −1	.5286E −1
3.2	.8953E −8	.3169E −6	.6285E −5	.7974E −4	.7115E −3	.4787E −2	.2559E −1
3.3	.2805E −8	.1103E −6	.2390E −5	.3266E −4	.3106E −3	.2209E −2	.1239E −1
3.4	.8790E −9	.3842E −7	.9085E −6	.1338E −4	.1356E −3	.1019E −2	.6000E −2
3.5	.2754E −9	.1338E −7	.3454E −6	.5478E −5	.5918E −4	.4700E −3	.2905E −2
3.6	.8630E −10	.4658E −8	.1313E −6	.2243E −5	.2583E −4	.2168E −3	.1407E −2
3.7	.2704E −10	.1622E −8	.4993E −7	.9188E −6	.1128E −4	.1000E −3	.6810E −3
3.8	.8473E −11	.5648E −9	.1898E −7	.3763E −6	.4923E −5	.4614E −4	.3297E −3
3.9	.2655E −11	.1966E −9	.7217E −8	.1541E −6	.2149E −5	.2129E −4	.1597E −3
4.0	.8318E −12	.6847E −10	.2744E −8	.6312E −7	.9380E −6	.9820E −5	.7730E −4
4.1	.2606E −12	.2384E −10	.1043E −8	.2585E −7	.4095E −6	.4530E −5	.3743E −4
4.2	.8166E −13	.8302E −11	.3966E −9	.1059E −7	.1787E −6	.2090E −5	.1812E −4
4.3	.2559E −13	.2891E −11	.1508E −9	.4336E −8	.7802E −7	.9641E −6	.8774E −5
4.4	.8018E −14	.1006E −11	.5733E −10	.1776E −8	.3406E −7	.4448E −6	.4248E −5
4.5	.2512E −14	.3505E −12	.2180E −10	.7273E −9	.1487E −7	.2052E −6	.2057E −5
4.6	.7872E −15	.1220E −12	.8287E −11	.2979E −9	.6490E −8	.9466E −7	.9959E −6
4.7	.2466E −15	.4249E −13	.3151E −11	.1220E −9	.2833E −8	.4367E −7	.4822E −6
4.8	.7728E −16	.1479E −13	.1198E −11	.4996E −10	.1237E −8	.2014E −7	.2335E −6
4.9	.2421E −16	.5151E −14	.4554E −12	.2046E −10	.5398E −9	.9293E −8	.1130E −6
5.0	.7587E −17	.1794E −14	.1731E −12	.8381E −11	.2356E −9	.4287E −8	.5473E −7
5.1	.2377E −17	.6246E −15	.6583E −13	.3432E −11	.1029E −9	.1978E −8	.2650E −7
5.2	.7449E −18	.2175E −15	.2503E −13	.1406E −11	.4490E −10	.9124E −9	.1283E −7
5.3	.2334E −18	.7572E −16	.9515E −14	.5757E −12	.1960E −10	.4209E −9	.6212E −8
5.4	.7313E −19	.2637E −16	.3618E −14	.2358E −12	.8556E −11	.1942E −9	.3008E −8
5.5	.2291E −19	.9180E −17	.1375E −14	.9657E −13	.3735E −11	.8958E −10	.1456E −8
5.6	.7180E −20	.3197E −17	.5229E −15	.3955E −13	.1630E −11	.4132E −10	.7051E −9
5.7	.2250E −20	.1113E −17	.1988E −15	.1620E −13	.7117E −12	.1906E −10	.3414E −9
5.8	.7049E −21	.3876E −18	.7558E −16	.6634E −14	.3107E −12	.8794E −11	.1653E −9
5.9	.2209E −21	.1349E −18	.2874E −16	.2717E −14	.1356E −12	.4057E −11	.8003E −10
6.0	.6920E −22	.4699E −19	.1093E −16	.1113E −14	.5920E −13	.1872E −11	.3875E −10

Table 2-91.   THEORETICAL FLOW OF ELECTRONS FROM A HEATED SURFACE (*Continued*)

Electron volts	Temperature, °K					
	1700°	*1800°*	*1900°*	*2000°*	*2100°*	*2200°*
0.5	.1144E 8	.1550E 8	.2046E 8	.2642E 8	.3344E 8	.4161E 8
0.6	.5780E 7	.8135E 7	.1111E 8	.1479E 8	.1924E 8	.2455E 8
0.7	.2920E 7	.4269E 7	.6032E 7	.8277E 7	.1107E 8	.1449E 8
0.8	.1476E 7	.2241E 7	.3275E 7	.4633E 7	.6372E 7	.8549E 7
0.9	.7456E 6	.1176E 7	.1778E 7	.2593E 7	.3667E 7	.5045E 7
1.0	.3767E 6	.6171E 6	.9654E 6	.1452E 7	.2110E 7	.2977E 7
1.1	.1904E 6	.3239E 6	.5241E 6	.8126E 6	.1214E 7	.1757E 7
1.2	.9618E 5	.1700E 6	.2846E 6	.4549E 6	.6987E 6	.1037E 7
1.3	.4860E 5	.8920E 5	.1545E 6	.2546E 6	.4020E 6	.6116E 6
1.4	.2456E 5	.4682E 5	.8388E 5	.1425E 6	.2314E 6	.3609E 6
1.5	.1241E 5	.2457E 5	.4554E 5	.7978E 5	.1331E 6	.2130E 6
1.6	.6269E 4	.1289E 5	.2473E 5	.4466E 5	.7661E 5	.1257E 6
1.7	.3168E 4	.6767E 4	.1342E 5	.2500E 5	.4408E 5	.7416E 5
1.8	.1601E 4	.3551E 4	.7288E 4	.1399E 5	.2537E 5	.4376E 5
1.9	.8088E 3	.1864E 4	.3957E 4	.7833E 4	.1460E 5	.2582E 5
2.0	.4087E 3	.9782E 3	.2148E 4	.4384E 4	.8400E 4	.1524E 5
2.1	.2065E 3	.5133E 3	.1166E 4	.2454E 4	.4834E 4	.8991E 4
2.2	.1043E 3	.2694E 3	.6333E 3	.1374E 4	.2782E 4	.5305E 4
2.3	.5272E 2	.1414E 3	.3438E 3	.7690E 3	.1601E 4	.3130E 4
2.4	.2664E 2	.7420E 2	.1867E 3	.4305E 3	.9211E 3	.1847E 4
2.5	.1346E 2	.3894E 2	.1013E 3	.2410E 3	.5300E 3	.1090E 4
2.6	.6801E 1	.2044E 2	.5502E 2	.1349E 3	.3050E 3	.6432E 3
2.7	.3436E 1	.1073E 2	.2987E 2	.7550E 2	.1755E 3	.3795E 3
2.8	.1736E 1	.5629E 1	.1622E 2	.4226E 2	.1010E 3	.2240E 3
2.9	.8773E 0	.2954E 1	.8806E 1	.2366E 2	.5812E 2	.1322E 3
3.0	.4433E 0	.1550E 1	.4781E 1	.1324E 2	.3344E 2	.7798E 2
3.1	.2240E 0	.8137E 0	.2596E 1	.7412E 1	.1925E 2	.4602E 2
3.2	.1132E 0	.4270E 0	.1409E 1	.4149E 1	.1107E 2	.2715E 2
3.3	.5718E − 1	.2241E 0	.7651E 0	.2322E 1	.6373E 1	.1602E 2
3.4	.2889E − 1	.1176E 0	.4154E 0	.1300E 1	.3667E 1	.9455E 1
3.5	.1460E − 1	.6172E − 1	.2255E 0	.7277E 0	.2110E 1	.5579E 1
3.6	.7377E − 2	.3239E − 1	.1224E 0	.4073E 0	.1214E 1	.3292E 1
3.7	.3727E − 2	.1700E − 1	.6648E − 1	.2280E 0	.6988E 0	.1943E 1
3.8	.1833E − 2	.8922E − 2	.3609E − 1	.1276E 0	.4021E 0	.1146E 1
3.9	.9516E − 3	.4682E − 2	.1960E − 1	.7144E − 1	.2314E 0	.6764E 0
4.0	.4808E − 3	.2457E − 2	.1064E − 1	.3999E − 1	.1332E 0	.3991E 0
4.1	.2430E − 3	.1290E − 2	.5776E − 2	.2239E − 1	.7662E − 1	.2355E 0
4.2	.1228E − 3	.6768E − 3	.3136E − 2	.1253E − 1	.4409E − 1	.1390E 0
4.3	.6203E − 4	.3552E − 3	.1703E − 2	.7014E − 2	.2537E − 1	.8201E − 1
4.4	.3134E − 4	.1864E − 3	.9244E − 3	.3926E − 2	.1460E − 1	.4839E − 1
4.5	.1584E − 4	.9783E − 4	.5019E − 3	.2198E − 2	.8402E − 2	.2855E − 1
4.6	.8002E − 5	.5134E − 4	.2725E − 3	.1230E − 2	.4835E − 2	.1685E − 1
4.7	.4043E − 5	.2695E − 4	.1479E − 3	.6886E − 3	.2782E − 2	.9942E − 2
4.8	.2043E − 5	.1414E − 4	.8032E − 4	.3855E − 3	.1601E − 2	.5867E − 2
4.9	.1032E − 5	.7422E − 5	.4361E − 4	.2158E − 3	.9212E − 3	.3462E − 2
5.0	.5216E − 6	.3895E − 5	.2368E − 4	.1208E − 3	.5301E − 3	.2043E − 2
5.1	.2636E − 6	.2044E − 5	.1285E − 4	.6761E − 4	.3051E − 3	.1205E − 2
5.2	.1332E − 6	.1073E − 5	.6979E − 5	.3784E − 4	.1755E − 3	.7113E − 3
5.3	.6729E − 7	.5630E − 6	.3789E − 5	.2118E − 4	.1010E − 3	.4197E − 3
5.4	.3400E − 7	.2955E − 6	.2057E − 5	.1186E − 4	.5813E − 4	.2477E − 3
5.5	.1718E − 7	.1551E − 6	.1117E − 5	.6638E − 5	.3345E − 4	.1461E − 3
5.6	.8680E − 8	.8138E − 7	.6064E − 6	.3715E − 5	.1925E − 4	.8624E − 4
5.7	.4386E − 8	.4271E − 7	.3292E − 6	.2080E − 5	.1108E − 4	.5089E − 4
5.8	.2216E − 8	.2241E − 7	.1787E − 6	.1164E − 5	.6374E − 5	.3003E − 4
5.9	.1120E − 8	.1176E − 7	.9704E − 7	.6517E − 6	.3668E − 5	.1772E − 4
6.0	.5658E − 9	.6174E − 8	.5269E − 7	.3648E − 6	.2111E − 5	.1046E − 4

*By permission of the General Electric Company © 1964·

REFERENCES

1. O.W. Richardson, *Phil. Mag.*, 23:594, 1912.
2. S. Dushman, *Phys. Rev.*, 21:623, 1923.
3. P.L. Read, General Research Lab. Report, No. 64–RL–3605–E, 1964.

Table 2-92.   CESIUM VAPOR-PRESSURE DATA*

R. Breitwieser

Based on Best Fit of Vapor-pressure Data †

Temp, °K	Cesium vapor pressure		Mean free path, cm $P_c = 1000$	Vapor density, atoms/cm³	Evaporation rate, atoms/cm² sec	Atom arrival rate (equivalent current), amp/cm²
	N/m²	Torr				
250	.618 − 6‡	.464 − 8	.198 6	.179 9	.894 12	.143 − 6
260	.243 − 5	.182 − 7	.523 5	.677 9	.344 13	.552 − 6
270	.862 − 5	.646 − 7	.153 5	.231 10	.120 14	.192 − 5
280	.279 − 4	.209 − 6	.490 4	.722 10	.381 14	.611 − 5
290	.833 − 4	.625 − 6	.170 4	.208 11	.112 15	.179 − 4
300	.231 − 3	.173 − 5	.634 3	.558 11	.305 15	.488 − 4
310	.599 − 3	.450 − 5	.253 3	.140 12	.778 15	.125 − 3
320	.146 − 2	.110 − 4	.107 3	.332 12	.187 16	.300 − 3
330	.339 − 2	.254 − 4	.476 2	.744 12	.427 16	.683 − 3
340	.746 − 2	.560 − 4	.223 2	.159 13	.925 16	.148 − 2
350	.157 − 1	.118 − 3	.109 2	.325 13	.192 17	.307 − 2
360	.317 − 1	.237 − 3	.555 1	.637 13	.381 17	.611 − 2
370	.615 − 1	.461 − 3	.294 1	.120 14	.730 17	.117 − 1
380	.115 0	.864 − 3	.161 1	.220 14	.135 18	.216 − 1
390	.209 0	.157 − 2	.912 0	.388 14	.242 18	.388 − 1
400	.368 0	.276 − 2	.531 0	.666 14	.420 18	.674 − 1
410	.630 0	.472 − 2	.318 0	.111 15	.711 18	.114 0
420	.105 1	.788 − 2	.195 0	.181 15	.117 19	.188 0
430	.171 1	.128 − 1	.123 0	.288 15	.189 19	.302 0
440	.272 1	.204 − 1	.789 − 1	.448 15	.297 19	.476 0
450	.425 1	.319 − 1	.518 − 1	.684 15	.458 19	.733 0
460	.649 1	.487 − 1	.346 − 1	.102 16	.692 19	.111 1
470	.975 1	.731 − 1	.235 − 1	.150 16	.103 20	.165 1
480	.144 2	.108 0	.163 − 1	.217 16	.150 20	.241 1
490	.209 2	.157 0	.115 − 1	.309 16	.216 20	.346 1
500	.299 2	.224 0	.817 − 2	.433 16	.306 20	.490 1
510	.421 2	.316 0	.591 − 2	.599 16	.427 20	.684 1
520	.586 2	.440 0	.433 − 2	.817 16	.588 20	.942 1
530	.806 2	.604 0	.321 − 2	.110 17	.800 20	.128 2
540	.109 3	.820 0	.241 − 2	.147 17	.108 21	.172 2
550	.147 3	.111 1	.183 − 2	.193 17	.143 21	.229 2
560	.195 3	.146 1	.140 − 2	.252 17	.188 21	.302 2
570	.256 3	.192 1	.109 − 2	.326 17	.245 21	.393 2
580	.334 3	.250 1	.849 − 3	.417 17	.317 21	.508 2
590	.431 3	.323 1	.669 − 3	.529 17	.405 21	.649 2
600	.551 3	.413 1	.532 − 3	.665 17	.514 21	.824 2
610	.699 3	.524 1	.426 − 3	.830 17	.647 21	.104 3
620	.880 3	.660 1	.344 − 3	.103 18	.808 21	.129 3
630	.110 4	.825 1	.280 − 3	.126 18	.100 22	.160 3
640	.136 4	.102 2	.229 − 3	.154 18	.123 22	.198 3
650	.168 4	.126 2	.189 − 3	.187 18	.151 22	.242 3
660	.206 4	.154 2	.157 − 3	.226 18	.183 22	.294 3
670	.251 4	.188 2	.131 − 3	.271 18	.221 22	.355 3
680	.303 4	.227 2	.110 − 3	.323 18	.266 22	.426 3
690	.365 4	.273 2	.924 − 4	.383 18	.317 22	.508 3
700	.436 4	.327 2	.784 − 4	.451 18	.377 22	.604 3
710	.519 4	.390 2	.668 − 4	.530 18	.446 22	.714 3
720	.615 4	.461 2	.572 − 4	.619 18	.524 22	.840 3
730	.725 4	.544 2	.492 − 4	.720 18	.614 22	.983 3
740	.851 4	.638 2	.425 − 4	.833 18	.715 22	.115 4
750	.994 4	.746 2	.368 − 4	.960 18	.830 22	.133 4
760	.116 5	.868 2	.321 − 4	.110 19	.960 22	.154 4
770	.134 5	.101 3	.280 − 4	.126 19	.110 23	.177 4
780	.155 5	.116 3	.246 − 4	.144 19	.127 23	.203 4
790	.178 5	.134 3	.217 − 4	.163 19	.145 23	.232 4
800	.204 5	.153 3	.192 − 4	.185 19	.165 23	.264 4
810	.233 5	.175 3	.170 − 4	.208 19	.187 23	.300 4
820	.265 5	.199 3	.151 − 4	.234 19	.212 23	.339 4
830	.301 5	.226 3	.135 − 4	.263 19	.239 23	.383 4
840	.340 5	.255 3	.121 − 4	.294 19	.269 23	.430 4

*From: "Theoretical Background for Thermionic Conversion Including Space-charge Theory, Schottky Theory, and the Isothermal Diode Sheath Theory", W.B. Nottingham and R. Breitwieser, NASA Technical Note TN D-3324, 1966. See original source for full details and complete table.
†These data are given in "Thermodynamic Properties of Cesium up to 1500°K", NASA TN D-2906, 1965.
‡Read this as 0.618×10^{-6}.

Chemistry
and Applications

Table 3-1. PROPERTIES OF THE CHEMICAL ELEMENTS
Atmospheric Pressure at Room Temperature

Table 3-2 gives additional properties of the chemical elements.[a]

Name	Symbol	Atomic number	International at. wt.[b]	Specific gravity (or density)	Melting point, °C	Boiling point, °C	Specific heat at 25°C	Thermal conductivity, watt/cm °C
Actinium	Ac	89	(227)	(10.02)	1050.	3200.	—	—
Aluminum	Al	13	26.9815	2.70	660.	2441.	0.215	2.37
Americium	Am	95	(243)	11.7	994.	2607.	—	—
Antimony (Stibium)	Sb	51	121.75	6.69	630.	1750.	0.050	0.185
Argon	Ar	18	39.948	1.78 g/l	−189.	−186.	0.125	1.75×10^{-4}
Arsenic	As	33	74.9216	5.73 (gray)	815.[c]	613. (subl.)	0.079	—
Astatine	At	85	(210)	—	729.	2125.	—	—
Barium	Ba	56	137.34	3.5	725.	1630.	0.046	—
Berkelium	Bk	97	(247)	—	—	—	—	—
Beryllium	Be	4	9.0122	1.85	1285.	2475.	0.436	2.18
Bismuth	Bi	83	208.980	9.75	271.4	1560.	0.030	0.084
Boron	B	5	10.811	2.35	2300.	2550.	0.245	—
Bromine	Br	35	79.904	3.12 (liq.)	−7.2	58.8	0.11	0.43×10^{-1}
Cadmium	Cd	48	112.40	8.65	321.	767.	0.055	0.92
Calcium	Ca	20	40.08	1.55	840.	1485.	—	1.3
Californium	Cf	98	(251)	—	—	—	—	—
Carbon	C	6	12.01115					
Diamond				3.5	>3800.	4827.	0.124	1.5 (0°)
Graphite				2.1	>3500.	4200.	0.170	0.24
Cerium	Ce	58	140.12	6.77	798.	3257.	0.047	0.11
Cesium	Cs	55	132.905	1.87	28.6	678.	0.057	—
Chlorine	Cl	17	35.453	3.21 g/l	−101.	−34.6	0.114	0.86×10^{-4}
Chromium	Cr	24	51.996	7.2	1860.	2670.	0.110	0.91
Cobalt	Co	27	58.9332	8.9	1495.	2870.	0.10	0.69
Copper	Cu	29	63.546	8.96	1084.	2575.	0.092	3.98
Curium	Cm	96	(247)	—	—	—	—	—
Dysprosium	Dy	66	162.50	8.54	1409.	2335.	0.0414	0.10
Einsteinium	Es	99	(254)	—	—	—	—	—
Erbium	Er	68	167.26	9.05	1522.	2510.	0.04	0.096
Europium	Eu	63	151.96	5.25	822.	1597.	0.042	—
Fermium	Fm	100	(257)	—	—	—	—	—
Fluorine	F	9	18.9984	1.11 (liq.)	−219.6	−188.	0.197	2.63×10^{-4}
Francium	Fr	87	(223)	—	27.	677.	—	—
Gadolinium	Gd	64	157.25	7.90	1311.	3233.	0.055	0.088
Gallium	Ga	31	69.72	5.91	29.8	2300.	0.089	0.29 − 0.38
Germanium	Ge	32	72.59	5.32	937.	2830.	0.077	0.59
Gold (Aurum)	Au	79	196.967	19.32	1063.	2857.	0.031	3.15
Hafnium	Hf	72	178.49	13.29	2220.	4700.	0.035	0.220
Helium	He	2	4.0026	0.177 g/l	—	−269.	1.24	14.8×10^{-4}
Holmium	Ho	67	164.930	8.78	1470.	2720.	0.039	—
Hydrogen	H	1	1.00797	0.0899 g/l	−259.	−253.	3.41	18.4×10^{-4}
Indium	In	49	114.82	7.31	156.	2050.	0.056	0.24
Iodine	I	53	126.9044	4.93	113.5	184.4	0.102	43.5×10^{-4}
Iridium	Ir	77	192.2	22.42	2450.	4390.	0.031	1.47
Iron (Ferrum)	Fe	26	55.847	7.87	1536.	2870.	0.108	0.803
Krypton	Kr	36	83.80	3.73 g/l	−157.	−152.	0.059	0.94×10^{-4}
Lanthanum	La	57	138.91	6.17	920.	3454.	0.047	0.14
Lawrencium	Lr	103	(257)	—	—	—	—	—
Lead (Plumbum)	Pb	82	207.19	11.35	327.5	1750.	0.031	0.352
Lithium	Li	3	6.939	0.53	180.	1342.	0.84	0.71
Lutetium	Lu	71	174.97	9.84	1656.	3315.	0.037	—
Magnesium	Mg	12	24.312	1.74	650.	1090.	0.243	1.56

[a]See also "Handbook of Chemistry and Physics", 53rd ed., R.C. Weast, Ed., The Chemical Rubber Co., 1972.
[b]A value in parentheses is the mass number of the most stable isotope of the element.
[c]At 28 atm.

Table 3-1. PROPERTIES OF THE CHEMICAL ELEMENTS *(Continued)*

Name	Symbol	Atomic number	International at. wt.†	Specific gravity (or density)	Melting point, °C	Boiling point, °C	Specific heat at 25°C	Thermal conductivity, watt/cm °C
Manganese	Mn	25	54.9380	7.21–7.44	1244.	2060.	0.114	—
Mendelevium	Md	101	(256)	—	—	—	—	—
Mercury (Hydrargyrum)	Hg	80	200.59	13.546	−38.86	356.55	0.033	0.0839
Molybdenum	Mo	42	95.94	10.22	2620.	4651.	0.060	1.38
Neodymium	Nd	60	144.24	7.00	1010.	3127.	0.049	0.13
Neon	Ne	10	20.183	0.90 g/l	−249.	−246.	0.246	4.77×10^{-4}
Neptunium	Np	93	(237)	18.0–20.45	640.	3902.	0.296	—
Nickel	Ni	28	58.71	8.90	1453.	2914.	0.106	0.905
Niobium (Columbium)	Nb	41	92.906	8.57	2467.	4740.	0.064	0.53
Nitrogen	N	7	14.0067	1.251 g/l	−210.	−196.	0.249	2.55×10^{-4}
Nobelium	No	102	(254)	—	—	—	—	—
Osmium	Os	76	190.2	22.57	3025.	4225.	0.031	0.61
Oxygen	O	8	15.9994	1.43 g/l	−218.4	−183.	0.220	2.61×10^{-4}
Palladium	Pd	46	106.4	12.02	1550.	2927.	0.058	0.71
Phosphorus, white	P	15	30.9738	1.82	44.1	280.	0.18	—
Platinum	Pt	78	195.09	21.45	1770.	3825.	0.032	0.73
Plutonium	Pu	94	(244)	19.84	640.	3230.	0.032	0.08
Polonium	Po	84	(209)	9.32	254.	962.	0.030	—
Potassium (Kalium)	K	19	39.102	0.86	63.3	760.	0.180	0.99
Praseodymium	Pr	59	140.907	6.77	931.	3212.	0.046	0.12
Promethium	Pm	61	(145)	—	1080.	2460.	0.044	—
Protactinium	Pa	91	(231)	(15.37)	—	—	0.029	—
Radium	Ra	88	(226)	—	700.	1700.	0.029	—
Radon	Rn	86	(222)	9.73 g/l	−71.	−62.	0.0224	—
Rhenium	Re	75	186.2	21.0	3180.	5650.	0.033	0.71
Rhodium	Rh	45	102.905	12.41	1965.	3700.	0.058	1.50
Rubidium	Rb	37	85.47	1.532	39.	700.	0.086	—
Ruthenium	Ru	44	101.07	12.4	2400.	4100.	0.057	—
Samarium	Sm	62	150.35	7.54	1072.	1778.	0.047	—
Scandium	Sc	21	44.956	2.99	1539.	2832.	0.135	—
Selenium	Se	34	78.96	4.8	217.	700.	0.077	0.005
Silicon	Si	14	28.086	2.33	1411.	3280.	0.17	0.835
Silver (Argentum)	Ag	47	107.868	10.50	961.	2212.	0.057	4.27
Sodium (Natrium)	Na	11	22.9898	0.97	97.83	884.	0.293	1.34
Strontium	Sr	38	87.62	2.55	770.	1375.	0.072	—
Sulfur	S	16	32.064	1.96–2.07	113.	445.	0.175	26.4×10^{-4}
Tantalum	Ta	73	180.948	16.6	2980.	5365.	0.034	0.575
Technetium	Tc	43	(97)	(11.50)	2172.	4877.	0.058	—
Tellurium	Te	52	127.60	6.24	450.	990.	0.05	0.059
Terbium	Tb	65	158.924	8.23	1360.	3041.	0.0435	—
Thallium	Tl	81	204.37	11.85	304.	1480.	0.031	0.39
Thorium	Th	90	232.038	11.7	1750.	4800.	0.03	0.41
Thulium	Tm	69	168.934	9.31	1545.	1727.	0.0385	—
Tin (Stannum)	Sn	50	118.69	7.31	232.	2600.	0.054	0.67
Titanium	Ti	22	47.90	4.54	1670.	3290.	0.125	0.22
Tungsten (Wolfram)	W	74	183.85	19.3	3400.	5550.	0.032	1.78
Uranium	U	92	238.03	18.8	1132.	4140.	0.028	0.25
Vanadium	V	23	50.942	6.1	1900.	3400.	0.116	0.60
Xenon	Xe	54	131.30	5.89 g/l	−112.	−107.	0.038	5.2×10^{-4}
Ytterbium	Yb	70	173.04	6.97	824.	1193.	0.071	—
Yttrium	Y	39	88.905	4.46	1523.	3337.	0.0925	0.15
Zinc	Zn	30	65.37	7.	419.5	910.	0.093	1.21
Zirconium	Zr	40	91.22	6.53	1852.	4400.	0.067	0.227

†A value in parentheses is the mass number of the most stable isotope of the element.

Table 3-2. ADDITIONAL PROPERTIES OF THE CHEMICAL ELEMENTS*

Table 3-1 gives other properties of the chemical elements. For electron work functions see Table 2-89. For electrical resistivities see Table 2-23.

Name	Atomic number	Latent heat of fusion, cal/g	Coef of linear thermal expansion × 10^6, K^-1			Elasticity modulus, psi × 10^-6	First ionization potential, eV	Thermal neutron absorption cross section, barns^a
			100	300	500			
Actinium (227)	89	11	—	—	—	—	6.9	510.
Aluminum	13	95	12.5	24	27	10.0	5.984	0.24
Americium (243)	95	10	—	—	—	—	—	—
Antimony	51	38.5	9	9.5	10.5	11.3	8.639	5.7
Argon	18	6.7	—	—	—	—	15.755	0.66
Arsenic	33	88.5	—	4.7	—	—	9.81	4.3
Astatine	85						9.5	—
Barium	56	13.4	—	16	24	—	5.21	1.2
Berkelium	97	—	—	—	—	—	—	—
Beryllium	4	324	—	12	15	40–44	9.32	0.01
Bismuth	83	12.4	12	13	13.5	4.6	7.287	0.034
Boron	5	400	—	2	—	64	8.296	755
Bromine	35	16.2	—	—	—	—	11.84	6.7
Cadmium	48	13.2	26	30	38	8	8.991	2 450.
Calcium	20	52	17.5	23	26	3.2–3.8	6.111	0.44
Californium (251)	98	—	—	—	—	—	—	—
Carbon (Graphite)	6	—	—	—	—	0.7	11.256	0.004
Cerium	58	9	—	8	—	4.4	5.6	0.73
Cesium	55	3.8	—	97	—	—	3.893	30.0
Chlorine	17	2.16	—	—	—	—	13.01	34.
Chromium	24	79	3.5	6	9.5	36	6.764	3.1
Cobalt	27	66	—	12	13	30	7.86	38.
Columbium See Niobium								
Copper	29	49	10.5	16.5	18	17	7.724	3.8
Curium (247)	96	—	—	—	—	—	—	—
Dysprosium	66	26.4	—	9.0	—	9.2	6.8	950.
Einsteinium (254)	99	—	—	—	—	—	—	—
Erbium	68	24.6	—	9.0	—	10.6	6.08	170.
Europium	63	16.9	—	26	—	2.1	5.67	4 300.
Fermium	—	—	—	—	—	—	—	—
Fluorine	9	10.1	—	—	—	—	17.418	0.01
Francium	87	—	—	—	—	—	4	—
Gadolinium	64	16.4	—	4	—	8.1	6.16	46 000
Gallium	31	19.2	—	18	—	—	6	2.8
Germanium	32	114	2.5	5.6	6.5	—	7.88	2.45
Gold	79	15	11.5	14	15	10.8	9.22	98.8
Hafnium	72	34	—	6	—	20	7	105.
Helium	2	1.2	—	—	—	—	24.481	0.007
Holmium	67	—	—	—	—	9.7	—	65.
Hydrogen	1	15.0	—	—	—	—	13.595	0.33
Indium	49	6.8	25	33	—	—	5.785	191.
Iodine	53	15	—	93	—	—	10.454	7.0
Iridium	77	33	4	6.5	7.5	75	9	425.
Iron	26	65	6	12	14.5	28.5	7.87	2.6
Krypton	36	4.7	—	—	—	—	13.996	31.
Lanthanum	57	10	—	5	6.5	5.5	5.61	8.9
Lawrencium	103	—	—	—	—	—	—	—
Lead	82	5.5	25	29	32	2.0	7.415	0.18
Lithium	3	103	23	50	—	—	5.39	71.
Lutetium	71	26.4	—	—	—	12.2	—	112.

^a Values in parentheses apply only to that isotope for which the mass number is given following the name of the element. All other values of neutron cross section apply to the naturally occurring mixture of isotopes.

*Based largely on: "Handbook of Chemistry and Physics", 53rd ed., R.C. Weast, Ed., The Chemical Rubber Co., 1972; see this source for other properties of the chemical elements.

Table 3-2.   ADDITIONAL PROPERTIES OF THE CHEMICAL ELEMENTS (Continued)

Name	Atomic number	Latent heat of fusion, cal/g	Coef of linear thermal expansion × 10⁶, K⁻¹			Elasticity modulus, psi × 10⁻⁶	First ionization potential, eV	Thermal neutron absorption cross section, barns[a]
			100	300	500			
Magnesium	12	88.0	15	25	29	6.4	7.644	0.07
Manganese	25	64	11.5	23	28	23	7.432	13.3
Mendelevium	101	—	—	—	—	—	—	—
Mercury	80	2.7	—	—	—	—	10.43	375.
Molybdenum	42	69	3	5	5.5	40	7.10	2.7
Neodymium	60	13	—	7	7.5	5.5	5.51	46.
Neon	10	4.0	—	—	—	—	21.559	<2.8
Neptunium (237)	93	9.7	—					(170)
Nickel	28	71	6.5	13	15.5	31	7.633	4.6
Niobium	41	68	5	7	7.5	15	6.88	1.15
Nitrogen	7	6.2	—	—	—	—	14.53	1.9
Nobelium	102	—	—	—	—	—	—	—
Osmium	76	34	—	5	5.5	80	8.5	15.3
Oxygen	8	3.3	—	—	—	—	13.614	<0.000 2
Palladium	46	38	8.5	12	13	17	8.33	8.
Phosphorus	15	4.8	—	125	—	—	10.484	0.2
Platinum	78	24	6.8	8.9	9.5	21.3	9.0	8.8
Plutonium (244)	94	3	—	54	—	14	5.1	—
Polonium	84	11	—	—	—	—	8.43	—
Potassium	19	14.5	—	83	—	—	4.339	2.1
Praseodymium	59	17	—	5.	5.3	4.7	5.46	11.3
Promethium	61	—	—	—	—	6.1	—	—
Protactinium (231)	91	17	—	—	—	—	—	(200)
Radium (226)	88	10	—	—	—	—	5.277	(20)
Radon (222)	86	3.1	—	—	—	—	10.746	(0.7)
Rhenium	75	42	—	7	—	66.7	7.87	85.
Rhodium	45	50	5.0	8.3	9.3	42	7.46	150.
Rubidium	37	6.3	—	90	—	—	4.176	0.7
Ruthenium	44	60	—	9	—	60	7.364	2.6
Samarium	62	24.7	—	—	—	4.9	5.6	5 600.
Scandium	21	87	—	—	—	11.5	6.54	24.
Selenium	34	16	—	35	—	8.4	9.75	12.3
Silicon	14	430	—	2.5	3.5	16	8.149	0.160
Silver	47	26.5	14.3	19.0	20.6	10.5	7.574	63.
Sodium	11	27	45.7	70.0	—	—	5.138	.53
Strontium	38	25	—	—	—	—	5.692	1.21
Sulfur	16	9.2	42	63	—	—	10.357	0.52
Tantalum	73	41	5.2	6.6	6.9	27	7.88	21.
Technetium	43	56.7	—	—	—	—	7.28	22.
Tellurium	52	33	—	17	—	17	9.01	4.7
Terbium	65	23.6	—	7.0	—	8.3	5.98	46.
Thallium	81	5.0	24	29	32	—	6.106	3.4
Thorium	90	17	8.7	11.4	12.5	8.5	6.95	7.5
Thulium	69	26.0	—	—	—	11.0	5.81	127.
Tin	50	14.1	15.5	21	27.5	6	7.342	0.63
Titanium	22	100	4.4	8.6	9.8	16	6.82	5.8
Tungsten	74	46	2.7	4.4	4.6	50	7.98	19.
Uranium	92	12	10.6	13.5	17	24	6.08	7.7
Vanadium	23	98	4	8	—	19	6.74	5.
Xenon	54	4.2	—	—	—	—	12.127	35.
Ytterbium	70	12.7	—	25	26.3	2.6	6.2	37.
Yttrium	39	45	—	—	—	9.4	6.38	1.3
Zinc	30	27	23	30	32	12	9.391	1.10
Zirconium	40	54	3.9	5.5	6.2	13.7	6.84	0.18

[a]Values in parentheses apply only to that isotope for which the mass number is given following the name of the element. All other values of neutron cross section apply to the naturally occurring mixture of isotopes.

Table 3-3.   PERIODIC TABLE OF THE ELEMENTS†

KEY TO CHART

Atomic Number →	**50** +2 ← Oxidation States
Symbol →	Sn +4
Atomic Weight →	118.69
	-18 18 4 ← Electron Configuration

Main Table (cell format: Atomic Number, Oxidation States / Symbol / Atomic Weight / Electron Configuration)

1a	2a	3b	4b	5b	6b	7b	8	8	8	1b	2b	3a	4a	5a	6a	7a	0	Orbit
1 +1 H 1.008 1																	2 0 He 4.00260 2	K
3 +1 Li 6.94 2-1	4 +2 Be 9.01218 2-2											5 +3 B 10.81 2-3	6 C 12.011 2-4	7 N 14.0067 2-5	8 -2 O 15.9994 2-6	9 -1 F 18.9984 2-7	10 0 Ne 20.179 2-8	K-L
11 +1 Na 22.9898 2-8-1	12 +2 Mg 24.305 2-8-2											13 +3 Al 26.9815 2-8-3	14 +2+4-4 Si 28.086 2-8-4	15 +3+5-3 P 30.9738 2-8-5	16 +4+6-2 S 32.06 2-8-6	17 +1+5+7-1 Cl 35.453 2-8-7	18 0 Ar 39.948 2-8-8	K-L-M
19 +1 K 39.102 -8-8-1	20 +2 Ca 40.08 -8-8-2	21 +3 Sc 44.9559 -8-9-2	22 +2+3+4 Ti 47.90 -8-10-2	23 +2+3+4+5 V 50.9414 -8-11-2	24 +2+3+6 Cr 51.996 -8-13-1	25 +2+3+4+6+7 Mn 54.9380 -8-13-2	26 +2+3 Fe 55.847 -8-14-2	27 +2+3 Co 58.9332 -8-15-2	28 +2+3 Ni 58.71 -8-16-2	29 +1+2 Cu 63.546 -8-18-1	30 +2 Zn 65.37 -8-18-2	31 +3 Ga 69.72 -8-18-3	32 +2+4 Ge 72.59 -8-18-4	33 +3+5-3 As 74.9216 -8-18-5	34 +4+6-2 Se 78.96 -8-18-6	35 +1+5-1 Br 79.904 -8-18-7	36 0 Kr 83.80 -8-18-8	-L-M-N
37 +1 Rb 85.4678 -18-8-1	38 +2 Sr 87.62 -18-8-2	39 +3 Y 88.9059 -18-9-2	40 +4 Zr 91.22 -18-10-2	41 +3+5 Nb 92.9064 -18-12-1	42 +6 Mo 95.94 -18-13-1	43 +4+6+7 Tc 98.9062 -18-13-2	44 +3 Ru 101.07 -18-15-1	45 +3 Rh 102.9055 -18-16-1	46 +2+4 Pd 106.4 -18-18-0	47 +1 Ag 107.868 -18-18-1	48 +2 Cd 112.40 -18-18-2	49 +3 In 114.82 -18-18-3	50 +2+4 Sn 118.69 -18-18-4	51 +3+5-3 Sb 121.75 -18-18-5	52 +4+6-2 Te 127.60 -18-18-6	53 +1+5+7-1 I 126.9045 -18-18-7	54 0 Xe 131.30 -18-18-8	-M-N-O
55 +1 Cs 132.9055 18-8-1	56 +2 Ba 137.34 -18-8-2	57* +3 La 138.9055 18-9-2	72 +4 Hf 178.49 -32-10-2	73 +5 Ta 180.9479 -32-11-2	74 +6 W 183.85 -32-12-2	75 +4+6+7 Re 186.2 -32-13-2	76 +3+4 Os 190.2 -32-14-2	77 +3+4 Ir 192.22 -32-15-2	78 +2+4 Pt 195.09 -32-16-2	79 +1+3 Au 196.9665 -32-18-1	80 +1+2 Hg 200.59 -32-18-2	81 +1+3 Tl 204.37 -32-18-3	82 +2+4 Pb 207.2 -32-18-4	83 +3+5 Bi 208.9806 -32-18-5	84 +2+4 Po (209) -32-18-6	85 -1 At (210) -32-18-7	86 0 Rn (222) -32-18-8	-N-O-P
87 +1 Fr (223) -18-8-1	88 +2 Ra (226) -18-8-2	89** +3 Ac (227) -18-9-2	104 — 32-10-2	105														-O-P-Q

Transition Elements — Group 8

*Lanthanides

58 +3+4 Ce 140.12 -20-8-2	59 +3 Pr 140.9077 -21-8-2	60 +3 Nd 144.24 -22-8-2	61 +3 Pm (145) -23-8-2	62 +2+3 Sm 150.4 -24-8-2	63 +2+3 Eu 151.96 -25-8-2	64 +3 Gd 157.25 -25-9-2	65 +3 Tb 158.9254 -27-8-2	66 +3 Dy 162.50 -28-8-2	67 +3 Ho 164.9303 -29-8-2	68 +3 Er 167.26 -30-8-2	69 +3 Tm 168.9342 -31-8-2	70 +2+3 Yb 173.04 -32-8-2	71 +3 Lu 174.97 -32-9-2	-N-O-P

Orbit: -N-O-P

*Actinides

90 +4 Th 232.0381 -18-10-2	91 +4 Pa 231.0359 -20-9-2	92 +3+4+5+6 U 238.029 -21-9-2	93 +3+4+5+6 Np 237.0482 -22-9-2	94 +3+4+5+6 Pu (244) -24-8-2	95 +3+4+5+6 Am (243) -25-8-2	96 +3 Cm (247) -25-9-2	97 +3+4 Bk (247) -27-8-2	98 +3 Cf (251) -28-8-2	99 +3 Es (254) -29-8-2	100 Fm (257) -30-8-2	101 Md (256) -31-8-2	102 +2+3 No (254) -32-8-2	103 Lr (254) -32-9-2	-O-P-Q

Orbit: -O-P-Q

Numbers in parentheses are mass numbers of the most stable isotope of that element.

†From: "Handbook of Chemistry and Physics", 52nd ed., R.C. Weast, Ed., The Chemical Rubber Co., 1971.

Table 3-4.   AVAILABLE STABLE ISOTOPES OF THE ELEMENTS*

Element and mass No.	Natural abundance, percent	Element and mass No.	Natural abundance, percent	Element and mass No.	Natural abundance, percent
Hydrogen		Sulfur		Cobalt	
1	99.985	32	95.0	59	100.0
2	0.015	33	0.76		
		34	4.22	Nickel	
Helium		36	0.014	58	67.84
3	0.00013			60	26.23
4	~100.0	Chlorine		61	1.19
		35	75.53	62	3.66
Lithium		37	24.47	64	1.08
6	7.42				
7	92.58	Argon		Copper	
		36	0.34	63	69.09
Beryllium		38	0.06	65	30.91
9	100.0	40	99.60		
				Zinc	
Boron		Potassium		64	48.89
10	19.78	39	93.1	66	27.81
11	80.22	40[a]	0.01	67	4.11
		41	6.9	68	18.57
Carbon				70	0.62
12	98.89	Calcium			
13	1.11	40	96.97	Gallium	
		42	0.64	69	60.4
Nitrogen		43	0.14	71	39.6
14	99.63	44	2.06		
15	0.37	46	0.003	Germanium	
		48	0.18	70	20.52
Oxygen				72	27.43
16	99.76			73	7.76
17	0.04	Scandium		74	36.54
18	0.20	45	100.0	76	7.76
Fluorine		Titanium		Arsenic	
19	100.0	46	7.93	75	100.0
		47	7.28		
Neon		48	73.94	Selenium	
20	90.92	49	5.51	74	0.87
21	0.26	50	5.34	76	9.02
22	8.82			77	7.58
		Vanadium		78	23.52
Sodium		50[b]	0.24	80	49.82
23	100.0	51	99.76	82	9.19
Magnesium		Chromium		Bromine	
24	78.70	50	4.31	79	50.54
25	10.13	52	83.76	81	49.46
26	11.17	53	9.55		
		54	2.38	Krypton	
Aluminum				78	0.35
27	100.0	Manganese		80	2.27
		55	100.0	82	11.56
Silicon				83	11.55
28	92.21			84	56.90
29	4.70	Iron		86	17.37
30	3.09	54	5.82		
		56	91.66	Rubidium	
Phosphorus		57	2.19	85	72.15
31	100.0	58	0.33	87	27.85

*Abstracted from: "Handbook of Chemistry and Physics", 53rd ed., R.C. Weast, Ed., The Chemical Rubber Co., 1972.

Table 3-4.   AVAILABLE STABLE ISOTOPES OF THE ELEMENTS
(Continued)

Element and mass No.	Natural abundance, percent	Element and mass No.	Natural abundance, percent	Element and mass No.	Natural abundance, percent
Strontium		Cadmium (cont.)		Barium (cont.)	
84	0.56	113	12.26	136	7.81
86	9.86	114	28.86	137	11.30
87	7.02	116	7.58	138	71.66
88	82.56				
		Indium		Lanthanum	
Yttrium		113	4.28	138	0.09
89	100.0	115^c	95.72	139	99.91
Zirconium		Tin		Cerium	
90	51.46	112	0.96	136	0.193
91	11.23	114	0.66	138	0.250
92	17.11	115	0.35	140	88.48
94	17.40	116	14.30	142^d	11.07
96	2.80	117	7.61		
		118	24.03	Praseodymium	
Niobium		119	8.58	141	100.0
93	100.0	120	32.85		
		122	4.72	Neodymium	
Molybdenum		124	5.94	142	27.11
92	15.84			143	12.17
94	9.04	Antimony		144	23.85
95	15.72	121	57.25	145	8.30
96	16.53	123	42.75	146	17.22
97	9.46			148	5.73
98	23.78	Tellurium		150	5.62
100	9.63	120	0.09		
		122	2.46	Samarium	
Ruthenium		123	0.87	144	3.09
96	5.51	124	4.61	147^e	14.97
98	1.87	125	6.99	148^f	11.24
99	12.72	126	18.71	149^g	13.83
100	12.62	128	31.79	150	7.44
101	17.07	130	34.48	152	26.72
102	31.61			154	22.71
104	18.60	Iodine			
		127	100.0	Europium	
Rhodium				151	47.82
103	100.0	Xenon		153	52.18
		124	0.096		
Palladium		126	0.090	Gadolinium	
102	0.96	128	1.92	152^h	0.20
104	10.97	129	26.44	154	2.15
105	22.23	130	4.08	155	14.73
106	27.33	131	21.18	156	20.47
108	26.71	132	26.89	157	15.68
110	11.81	134	10.44	158	24.87
		136	8.87	160	21.90
Silver					
107	51.82			Terbium	
109	48.18	Cesium		159	100.0
		133	100.0		
Cadmium				Dysprosium	
106	1.22	Barium		156^i	0.052
108	0.88	130	0.101	158	0.090
110	12.39	132	0.097	160	2.29
111	12.75	134	2.42	161	18.88
112	24.07	135	6.59	162	25.53

Table 3-4.   AVAILABLE STABLE ISOTOPES OF THE ELEMENTS
(Continued)

Element and mass No.	Natural abundance, percent	Element and mass No.	Natural abundance, percent	Element and mass No.	Natural abundance, percent
Dysprosium (cont.)		Hafnium (cont.)		Platinum (cont.)	
163	24.97	179	13.75	195	33.8
164	28.18	180	35.24	196	25.3
				198	7.2
Holmium				Gold	
165	100.0	Tantalum		197	100.0
		180	0.012		
		181	99.988	Mercury	
Erbium				196	0.146
162	0.136	Tungsten		198	10.02
164	1.56	180	0.14	199	16.84
166	33.41	182	26.41	200	23.13
167	22.94	183	14.40	201	13.22
168	27.07	184	30.64	202	29.80
170	14.88	186	28.41	204	6.85
Thulium		Rhenium			
169	100.0	185	37.07	Thallium	
		187[l]	62.93	203	29.50
Ytterbium				205	70.50
168	0.135	Osmium			
170	3.03	184	0.018	Lead	
171	14.31	186	1.59	204	1.48
172	21.82	187	1.64	206	23.6
173	16.13	188	13.3	207	22.6
174	31.84	189	16.1	208	52.3
176	12.73	190	26.4		
		192	41.0	Bismuth	
Lutetium				209	100.0
175	97.40	Iridium			
176[j]	2.60	191	37.3	Thorium	
		193	62.7	232[n]†	100.0
Hafnium					
174[k]	0.18	Platinum		Uranium	
176	5.20	190[m]	0.013	234[o]†	0.0006
177	18.50	192	0.78	235[p]†	0.72
178	27.14	194	32.9	238[q]†	99.27

[a] Half-life $= 1.3 \times 10^9$ y.
[b] Half-life $> 10^{15}$ y.
[c] Half-life $= 5 \times 10^{14}$ y.
[d] Half-life $= 5 \times 10^{15}$ y.
[e] Half-life $= 1.06 \times 10^{11}$ y.
[f] Half-life $= 1.2 \times 10^{13}$ y.
[g] Half-life $= 4 \times 10^{14}$ y.
[h] Half-life $= 1.1 \times 10^{14}$ y.
[i] Half-life $= 2 \times 10^{14}$ y.

[j] Half-life $= 2.2 \times 10^{10}$ y.
[k] Half-life $= 4.3 \times 10^{15}$ y.
[l] Half-life $= 4 \times 10^{10}$ y.
[m] Half-life $= 6 \times 10^{11}$ y.
[n] Half-life $= 1.4 \times 10^{10}$ y.
[o] Half-life $= 2.5 \times 10^5$ y.
[p] Half-life $= 7.1 \times 10^8$ y.
[q] Half-life $= 4.5 \times 10^9$ y.
† Naturally occurring.

REFERENCE
"CRC Handbook of Radioactive Nuclides", Y. Wang, Ed., The Chemical Rubber Co., 1969, pp. 25–63.

Table 3-5.   VAPOR PRESSURE OF THE ELEMENTS*

RUDOLF LOEBEL

See Table 1-18 for pressure-temperature relations for saturated vapors and refrigerants.

This table lists the temperature in degrees Celsius (Centigrade) at which an element has a vapor pressure indicated by the headings of the columns. To convert pressures to SI units, 1 mm Hg (torr) = 133.3 N/m^2 and 1 atm = 0.1013 MN/m^2.

Element		mm Hg					atm				
		1	*10*	*100*	*400*	*760*	*2*	*5*	*10*	*20*	*40*
Aluminum	Al	1540	1780	2080	2320	2467	2610	2850	3050	3270	3530
Antimony	Sb		960	1280	1570	1750	1960	2490			
Arsenic	As	380	440	510	580	610					
Barium	Ba	860	1050	1300	1520	1640	1790	2030	2230		
Beryllium	Be	1520	1860	2300	2770	2970	3240	3730	4110	4720	5610
Bismuth	Bi		1060	1280	1450	1560	1660	1850	2000	2180	
Boron	B	2660	3030	3460	3810	4000					
Bromine	Br	−60	−30	+9	39	59	78	110			
Cadmium	Cd	393	486	610	710	765	830	930	1030	1120	1240
Calcium	Ca	800	970	1200	1390	1490	1630	1850	2020	2290	
Cesium	Cs		373	513	624	690					
Chlorine	Cl	−123	−101	−71	−46	−34	−17	+9	30	55	97
Chromium	Cr	1610	1840	2140	2360	2480	2630	2850	3010	3180	
Cobalt	Co	1910	2170	2500	2760	2870	3040	3270			
Copper	Cu		1870	2190	2440	2600	2760	3010	3500	3460	3740
Fluorine	F			−203	−193	−188	−180.7	−169.1	−159.6		
Gallium	Ga	1350	1570	1850	2060	2180	2320	2560	2730		
Germanium	Ge		2080	2440	2710	2830	2970	3200	3430		
Gold	Au	1880	2160	2520	2800	2940	3120	3490	3630	3890	
Indium	In				1960	2080	2230	2440	2600		
Iodine	I	40	72	115	160	185	216	265			
Iridium	Ir	2830	3170	3630	3960	4130	4310	4650			
Iron	Fe	1780	2040	2370	2620	2750	2900	3150	3360	3570	
Lanthanum	La				3230	3420	3620	3960	4270		
Lead	Pb	970	1160	1420	1630	1740	1880	2140	2320	2620	
Lithium	Li	750	890	1080	1240	1310	1420	1518			
Magnesium	Mg	620	740	900	1040	1110	1190	1330	1430	1560	
Manganese	Mn		1510	1810	2050	2100	2360	2580	2850		
Mercury	Hg			260	330	356.9	398	465	517	581	657
Molybdenum	Mo	3300	3770	4200	4580	4830	5050	5340	5680	5980	
Neodymium	Nd				2870	3100	3300	3680	3990		
Nickel	Ni	1800	2090	2370	2620	2730	2880	3120	3300	3310	
Palladium	Pd	1470	2290	2670	2950	3140	3270	3560	3840		
Phosphorus	P		127	199	253	283	319				
Platinum	Pt	2600	2940	3360	3650	3830	4000	4310	4570	4860	
Polonium	Po	472	587	752	890	960	1060	1200	1340		
Potassium	K			590	710	770	850	950	1110	1240	1420
Rhodium	Rh	2530	2850	3260	3590	3760	3930	4230	4440		
Rubidium	Rb		390	527	640	700					
Selenium	Se		429	547	640	685	750	850	920	1010	1120
Silver	Ag	1310	1540	1850	2060	2210	2360	2600	2850	3050	3300
Sodium	Na	440	546	700	830	890	980	1120	1230	1370	
Strontium	Sr	740	900	1100	1280	1380	1480	1670	1850	2030	
Sulfur	S		246	333	407	445	493	574	640	720	
Tellurium	Te	520	633	792	900	962	1030	1160	1250		
Thallium	Tl		1000	1210	1370	1470	1560	1750	1900	2050	2260
Tin	Sn	1610	1890	2270	2580	2750	2950	3270	3540	3890	
Titanium	Ti	2180	2480	2860	3100	3260	3400	3650	3800		
Tungsten	W	3980	4490	5160	5470	5940	6260	6670	7250	7670	
Uranium	U	2450	2800	3270	3620	3800	4040	4420			
Vanadium	V	2290	2570	2950	3220	3380	3540	3800			
Zinc	Zn		590	730	840	907	970	1090	1180	1290	

*From: "Handbook of Chemistry and Physics", 53rd ed., R.C. Weast, Ed., The Chemical Rubber Co., 1972.

Table 3-6. MELTING POINTS OF METALLIC COMPOUNDS*
Refractory and Ceramic Materials and Salts

Metal	Boride		Bromide		Carbide		Chloride		Fluoride		Iodide	
	K	deg F	K	deg F	K	deg F	K	deg F	K	deg F	K	deg F
Ag			AgBr 703	806			AgCl 728	851	AgF 708	815	AgI 831	1036
Al			AlBr$_3$ 371	207	Al$_4$C$_3$ 2000[a]	3600[a]	AlCl$_3$ 465	377	AlF$_3$ 1564[a]	2356[a]	AlI 464	376
B			BBr$_3$ 227	−51	B$_4$C 2720	4440	BCl$_3$ 166	−161	BF$_3$ 146	−196		
Ba	BaB$_4$ 2543	4118	BaBr$_2$ 1123	1562			BaCl$_2$ 1235	1764	BaF$_2$ 1627	2470	BaI$_2$ 1013	1364
Be	BeB$_2$ >2243	>3357	BeBr$_2$ 793	968	Be$_2$C >2375[a]	>3815[a]	BeCl$_2$ 713	824	BeF$_2$ 813	1004	BeI$_2$ 783	950
Bi			BiBr$_3$ 491	424			BiCl$_3$ 507	452	BiF$_3$ 1000	1341	BiI$_3$ 681	784
Ca			CaBr$_2$ 1003[a]	1346[a]			CaCl$_2$ 1055	1440	CaF$_2$ 1675	2555	CaI$_2$ 848	1067
Cd			CdBr$_2$ 841	1054			CdCl$_2$ 841	1054	CdF$_2$ 1373	2012	CdI$_2$ 423	302
Ce	CeB$_6$ 2463	3975					CeCl$_3$ 1095	1512	CeF$_3$ 1710	2618	CeI$_3$ 1025	1386
Cr	CrB$_2$ 2123	3362			Cr$_3$C$_2$ 2168	3440						
Cu			CuBr 777	939			CuCl 695	792	CuF$_2$ 1129	1573	CuI 878	1121
Fe			FeBr$_2$ 955	1754[a]	Fe$_3$C 2110	3339	FeCl$_2$ 945	1242	FeF$_3$ >1275	>1835		
In			InBr$_3$ 709	817			InCl 498	437	InF$_3$ 1443	2138	InI$_3$ 483	410
K			KBr 1008	1355			KCl 1043	1418	KF 1131	1576	KI 958	1265
Li			LiBr 823	1022			LiCl 883	1130	LiF 1119	1554	LiI 722	840
Mg			MgBr$_2$ 984	1312			MgCl$_2$ 987	1317	MgF$_2$ 1536	2305	MgI$_2$ <910	<1078
Mn							MnCl$_2$ 923	1202	MnF$_2$ 1129	1573		
Mo	MoB 2625	4250			Mo$_2$C 2963	4875			MoF$_6$ 290	63	MoI$_4$ 373[a]	212
Na			NaBr 1023	1382	NaC$_2$ 973	1292	NaCl 1073	1472	NaF 1267	1821	NaI 935	1224
Nb	NbB >2270	>3630			NbC 3770	6330						
Ni			NiBr$_2$ 1236	1765			NiCl$_3$ 1274	1834	NiF$_2$ 1273[a]	1832	NiI$_2$ 1070	1467

*Compiled from several sources.

Table 3-6. MELTING POINTS OF METALLIC COMPOUNDS (Continued)

Metal	Nitrate		Nitride		Oxide		Silicide		Sulfate		Sulfide	
	K	deg F	K	deg F	K	deg F	K	deg F	K	deg F	K	deg F
Ag	$AgNO_3$ 483	410			Ag_2O 573[a]	572			Ag_2SO_4 933	1220	Ag_2S 1098	1517
Al			AlN >2475	>4000	Al_2O_3 2322	3720			$Al_2(SO_4)_3$ 1043[a]	1418[a]	Al_2S_3 1373	2012
B			BN 3000	4945	B_2O_3 723	841					B_2S_4 663	734
Ba	$Ba(NO_3)_2$ 865	1098			BaO 2283	3649			$BaSO_4$ 1853	2876	BaS 1473	2192
Be			Be_3N_2 2513[a]	4064[a]	BeO 2725	4445			$BeSO_4$ 848[a]	1067[a]		
Bi					Bi_2O_3 1098	1516			$Bi(SO_4)_3$ 678[a]	761[a]	Bi_2S_3 1020	1377
Ca	$Ca(NO_3)_2$ 834	1042	Ca_3N_2 1468	2183	CaO 3183	5269			$CaSO_4$ 1723	3542		
Cd	$Cd(NO_3)_2$ 623	662			CdO 1773	2731			$CdSO_4$ 1273[b]	1832[b]	CdS 2023[b]	3182[b]
Ce					CeO_2 >2873	>4711			$Ce(SO_4)_2$ 468[a]	383[a]	CeS 2400	3860
Cr			CrN 1770[a]	2730[a]	Cr_2O_3 2603	4225	$CrSi_2$ 1843	2858				
Cu			Cu_3N 573[a]	572	Cu_2O 1508	2254	Cu_4Si 1123	1562	Cu_2S 1400	2060		
Fe					Fe_2O_3 1864	2895			$Fe_2(SO_4)_3$ 753[a]	896[a]	FeS 1468	2183
In					In_2O_3 2183	3469					In_2S_3 1323	1922
K	KNO_3 610	639			K_2O_3 703	806			K_2SO_4 1342	1956	K_2S 1113	1544
Li	$LiNO_3$ 527	489	Li_3N 1118[a]	1553	Li_2O >1975	>3095			Li_2SO_4 1132	1578	Li_2S 1198	1697
Mg					MgO 3098	5116	Mg_2Si 1375	2016	MgS >2275[a]	>3635[a]	$MgSO_4$ 1397[a]	2055[a]
Mn			—		MnO 1840	2852						
Mo					MoO_3 1068	1462	$MoSi_2$ 2553	3595			MoS_2 1458	2165
Na	$NaNO_3$ 583	590	Na_2N 573[a]	572[a]					Na_2SO_4 1157	1632	Na_2S 1453	2156
Nb			NbN 2323	3722	Nb_2O_5 1764	2715	$NbSi_2$ 2203	3505				
Ni					NiO 2257	3603			$NiSO_4$ 1121[a]	1558[a]	NiS 1070	1466

Table 3-6. MELTING POINTS OF METALLIC COMPOUNDS *(Continued)*

Metal	Compound											
	Boride		Bromide		Carbide		Chloride		Fluoride		Iodide	
	K	deg F	K	deg F	K	deg F	K	deg F	K	deg F	K	deg F
Pb			$PbBr_2$				$PbCl_2$		PbF_2		PbI_2	
			643	698			771	928	1095	1512	675	756
Pt			$PtBr_2$				$PtCl_2$				PtI_2	
			523^a	482^a			854^a	1078^a			633^a	680^a
Sb			$SbBr_3$				$SbCl_3$		SbF_3		SbI_3	
			370	207			346	164	565	558	443	338
Si					SiC				SiF_4			
					2970	4890			183	−130		
Sn			$SnBr_2$				$SnCl_2$		SnF_4		SnI_2	
			488	420			518	473	978^a	1300^a	593	608
Sr	SrB_6		$SrBr_2$		SrC_2		$SrCl_2$		SrF_2		SrI_2	
	2508	4055	916	1189	>1970	>3100	1148	1607	1736	2665	788	959
Ta	TaB		$TaBr_5$		TaC		$TaCl_5$		TaF_5			
	$>2270^a$	$>3630^a$	538	509	3813	6403	489	421	370	206		
Te			$TeBr_2$				$TeCl_2$					
			612	642			448	347				
Th	ThB_4		$ThBr_4$		ThC		$ThCl_4$		ThF_4			
	>2770	>4530	883	1130^a	2898	5250	1043	1418	1375	2015		
Ti	TiB_2		$TiBr_4$		TiC		$TiCl_4$		TiF_3		TiI_2	
	3253	5396	312	102	3433	5720	250	−9	1475	2195	873	1112
U	UB_2		UBr_4		UC		UCl_4		UF_4		UI_4	
	>1770	>2730	789	961	2863	4693	843	1058	1233	1760	779	943
V	VB_2				VC		VCl_4		VF_3		VI_2	
	2373	3812			3600	5120	245	−18	>1075	>1475	1048^a	1427
W	WB		WBr_5		WC		WCl_6					
	3133	5180	549	529	2900^a	4760^a	548	527				
Zn			$ZnBr_2$				$ZnCl_2$		ZnF_2		ZnI_2	
			667	741			548	527	1145	1602	719	835
Zr	ZrB_2		$ZrBr_2$		ZrC		$ZrCl_2$		ZrF_4		ZrI_4	
	3313	5505	$>625^a$	$>660^a$	3533	5900	623	662	873^a	1112^a	772	930

OTHER COMPOUNDS
Melting Point Temperatures in K (with deg F in Parentheses)

Ag_2CO_3—491 (424); $BaCO_3$—1653 (2516); $CaCO_3$—1613^b $(2444)^b$; $CdCO_3$—$<775^a$ $(<930^a)$; K_2CO_3—1170 (1647); Li_2CO_3—950 (1250); Na_2CO_3—1127 (1569); $PbCO_3$—588^a (600^a); $SrCO_3$—1770 (2726)

$AgNO_2$—413^a (284^a); $Ba(NO_2)_2$—540 (513); KNO_2—692 (786); $LiNO_2$—493 (428); $NaNO_2$—558 (545)

Ag_2Te—1228 (1750); Bi_2Te_3—861 (1091); $CdTe$—1314 (1906); $InTe$—965 (1277); $PtTe_2$—1523 (2282); Sb_2Te_3—891 (1145); $SnTe$—1053 (1436)

$BaTiO_3$—1891 (2944); $FeTiO_3$—1640 (2492)

$Ce_2(WO_4)_3$—1362 (1992); K_2WO_4—1203 (1706); $LiWO_4$—1015 (1368); $NaWO_4$—971 (1288)

Table 3-6.   MELTING POINTS OF METALLIC COMPOUNDS *(Continued)*

Metal	Nitrate		Nitride		Oxide		Silicide		Sulfate		Sulfide	
	K	*deg F*	*K*	*deg F*	*K*	*deg F*	*K*	*deg F*	*K*	*deg F*	*K*	*deg F*
Pb	$Pb(NO_3)_2$ 743[a]	878[a]			PbO 1159	1626			$PbSO_4$ 1443	2138	PbS 1387	2037
Pt											PtS_2 508[a]	455[a]
Sb					Sb_2O_3 928	1211					SbS_3 820	1016
Si			Si_3N_4 2175[b]	3450	SiO_2 1978	3100						
Sn					SnO 1353[a]	1973[a]			$SnSO_4$ >635	>680	SnS 1153	1616
Sr	$Sr(NO_3)_2$ 643	1058			SrO 2933	4819			$SrSO_4$ 1878	2921	SrS >2275	>3635
Ta			Ta_2N 3360	5595	Ta_2O_5 2100	3325	$TaSi_2$ 2670	4350			TaS_4 >1575	>2375
Tc					TeO_2 1006	1351						
Th			ThN 2903	4765	ThO_2 3493	5827					ThS_2 2198	3497
Ti			TiN 3200[a]	5790[a]	TiO_2 2113	3344	$TiSi_2$ 1813	2804				
U			UN 3123	5161	UO_2 3151	5212	USi_2 1970	3090			US_2 >1375	>2015
V			VN 2593	4208	V_2O_5 947	1245	VSi_2 2023	3182			V_2S_3 >875[a]	>1115[a]
W					WO_3 1744	2679	WSi_2 2320	3720			WS_2 1523[a]	2282[a]
Zn					ZnO 2248	3586			$ZnSO_4$ 873[a]	1112[a]		
Zr			ZrN 3250	5400	ZrO_2 3123	5161			$Zr(SO_4)_2$ 683	770	ZrS_2 1823	2822

[a] Decomposes or sublimes.
[b] Pressure above one atmosphere.

REFERENCES

"Handbook of Chemistry and Physics", 52nd ed., R.C. Weast, Ed., The Chemical Rubber Co., 1971.

"Melting Points of Metallic Elements and Selected Compounds", J.H. Charlesworth, Air Force Materials Laboratory Technical Report, AFML-TR-70-137, 1970.

NSRDS-NBS 15, "Molten Salts: Electrical Conductance, Density, and Viscosity Data", G.J. Janz, F.W. Dampier, G.R. Lakshminarayanan, P.K. Lorenz, and R.P.T. Tomkins, Vol. 1, National Bureau of Standards, October 1968.

Table 3-7.   COMMON CHEMICAL NAMES AND SYNONYMS

For names of alloys, see Table 1-58; minerals, Table 1-121; plastics, Table 1-83; pigments, Table 3-19; refractories, Table 1-222.

Common name	Alternate name	Common name	Alternate name
Acetone	2-Propanone or Dimethyl ketone, C_3H_6O	Litharge	Lead monoxide, PbO
Acrolein	Propenal, C_3H_4O	Lunar caustic	Silver nitrate, $AgNO_3$
Allene	Propadiene, C_3H_4	Magnesia	Magnesium oxide, MgO
Alum	Potassium aluminum sulfate, $K_2Al_2(SO_4)_4{\cdot}24H_2O$	Marble	Calcium carbonate, $CaCO_3$
		Marsh gas	Methane, CH_4
Alumina	Aluminum oxide, Al_2O_3	Milk of magnesia	Magnesium hydroxide, $Mg(OH)_2$
Alundum	Fused alumina, Al_2O_3	Moth balls	Naphthalene, $C_{10}H_{18}$
Aniline	Phenyl amine, $C_6H_5NH_2$	Muriate of potash	Potassium chloride, KCl
Aqua regia	Nitric acid and hydrochloric acid, $HNO_3 + 3HCl$	Muriatic acid	Hydrochloric acid, HCl
		Mustard gas	Yperite or 2,2'-dichlorodiethylsulfide, $C_4Cl_2H_8S$
Arsine	Arsenic hydride, AsH_3		
Aspirin	Acetyl-salicyclic acid, $C_6H_4(CO_2H)(OCOCH_3)$	Niter	Potassium nitrate, KNO_3
		Oil of vitriol	Concentrated sulfuric acid, H_2SO_4
Baking soda	Sodium bicarbonate, $NaHCO_3$	Oleum	Fuming sulfuric acid, $H_2SO_4 + SO_3$
Bentonite	Impure aluminum silicate	Paris green	Copper aceto-arsenite,
Blue vitriol	Copper sulfate, $CuSO_4{\cdot}5H_2O$		$Cu(C_2H_3O_2)_2{\cdot}3CuAs_2O_4$
Borax	Sodium tetraborate, $Na_2B_4O_7{\cdot}10H_2O$	Petrol	Gasoline
Brimstone	Sulfur, S	Phosgene	Carbonyl chloride, $COCl_2$
Butyric acid	Butanoic acid, $C_4H_8O_2$	Picric acid	Trinitrophenol, $C_6H_2(NO_2)_3OH$
Calamine	Zinc silicate, $2ZnO{\cdot}SiO_2{\cdot}H_2O$	Plaster of Paris	Calcium sulfate, $CaSO_4{\cdot}\frac{1}{2}H_2O$
Calomel	Mercurous chloride, HgCl	Plumbago	Graphite, C
Carbolic acid	Phenol, C_6H_5OH	Propyl alcohol	1-Propanol, C_3H_8O
Carbonic acid	Carbon dioxide, CO_2	Propylene	Propene, C_3H_6
Chalk	Calcium carbonate, $CaCO_3$	Propylene glycol	1,2-Propanediol, $C_3H_8O_2$
Chelen	Ethyl chloride, C_2H_5Cl	Prussiate of soda	Sodium ferrocyanide, $Na_3Fe(CN)_6{\cdot}H_2O$
Chili saltpeter	Sodium nitrate, $NaNO_3$	Prussic acid	Hydrocyanic acid, HCN
Chloride of lime	Calcium chloro-hypochlorite, $CaOCl_2$	Rochelle salt	Potassium sodium tartrate, $KNaC_4H_4O_6{\cdot}4H_2O$
Chromic acid	Chromium trioxide, CrO_3		
Cinnabar	Mercuric sulfide, HgS	Rouge	Ferric oxide, Fe_2O_3
Copperas	Ferrous sulfate, $FeSO_4{\cdot}7H_2O$	Saccharin	Benzoic sulfimide, $C_7H_5NO_3S$
Corrosive		Sal ammoniac	Ammonium chloride, NH_4Cl
sublimate	Mercuric chloride, $HgCl_2$	Saltpeter	Potassium nitrate, KNO_3
Corundum	Aluminum oxide, Al_2O_3	Sodium hyposulfite	Sodium thiosulfate, $Na_2S_2O_3{\cdot}5H_2O$
Epsom salts	Magnesium sulfate, $MgSO_4{\cdot}7H_2O$	Sugar, corn	Glucose, $C_6H_{12}O_6{\cdot}H_2O$
Ether	Diethyl (or ethyl) ether, $C_4H_{10}O$	Sugar, invert	Glucose and fructose, 50–50
Ethyne	Acetylene, C_2H_2	Sugar of lead	Lead acetate, $Pb(C_2H_3O_2)_2{\cdot}3H_2O$
Formalin	40% solution of formaldehyde in water, HCHO	Sulfuric ether	Diethyl ether, $(C_2H_5)_2O$
		Talc	Hydrated magnesium silicate, $Mg_3Si_4O_{11}{\cdot}H_2O$
Fuller's earth	Hydrated magnesium and aluminum silicates	Tannin	Tannic acid, $C_{76}H_{52}O_{46}$
Fusel oil	Mixed amyl alcohols, $C_5H_{11}OH$	Tartar emetic	Potassium antimonyl tartrate, $KSbC_4H_4O_7{\cdot}\frac{1}{2}H_2O$
Glauber salt	Sodium sulfate, $Na_2SO_4{\cdot}10H_2O$		
Guncotton	Nitrocellulose	Verdigris	Basic copper acetate,
Gypsum	Calcium sulfate, $CaSO_4{\cdot}2H_2O$		$2Cu(C_2H_3O_2)_2 + CuO(?)$
Hartshorn	Ammonium carbonate mixture	Vinegar acid	Acetic acid, $C_2H_4O_2$
Kaolin	Aluminum silicate, $Al_2O_3{\cdot}2SiO_2{\cdot}2H_2O$	Vitriol	Sulfuric acid, H_2SO_4
Kieselguhr	Siliceous earth, SiO_2	Water glass	Sodium silicates dissolved in water
Laughing gas	Nitrous oxide, N_2O	White arsenic	Arsenous oxide, As_2O_3
Lime	Calcium oxide, CaO	Whiting	Calcium carbonate, $CaCO_3$
Limonite	Ferric hydroxide, $Fe(OH)_3$	Wood alcohol	Methyl alcohol, CH_3OH

Table 3-8.   ELECTROCHEMICAL SERIES*

J.F. Hunsberger

Values Listed Are Standard Reduction Potentials

Table I.   ALPHABETICAL LISTING

Reaction	Potential, volts	Reaction	Potential, volts
$Ag^+ + e^- \rightleftharpoons Ag$	0.7996	$Ce^{+3} + 3e^- \rightleftharpoons Ce(Hg)$	−1.4373
$Ag^{+2} + e^- \rightleftharpoons Ag^{+1}(4f\,HClO_4)$	1.987	$Ce^{+4} + e^- \rightleftharpoons Ce^{+3}$	1.4430 (1.61)
$AgAc + e^- \rightleftharpoons Ag + Ac^-$	0.64	$Ce^{+4} + e^- \rightleftharpoons Ce^{+3}(0.5f\,H_2SO_4)$	1.4587
$AgBr + e^- \rightleftharpoons Ag + Br^-$	0.0713	$CeOH^{+3} + H^+ + e^- \rightleftharpoons Ce^{+3} + H_2O$	1.7134
$AgBrO_3 + e^- \rightleftharpoons Ag + BrO_3^-$	0.680	$Cl_2(g) + 2e^- \rightleftharpoons 2Cl^-$	1.3583
$AgC_2O_4 + 2e^- \rightleftharpoons Ag + C_2O_4^{-2}$	0.4776	$HClO + H^+ + e^- \rightleftharpoons 1/2Cl_2 + H_2O$	1.63
$AgCl + e^- \rightleftharpoons Ag + Cl^-$	0.2223	$HClO + H^+ + 2e^- \rightleftharpoons Cl^- + H_2O$	1.49
$AgCN + e^- \rightleftharpoons Ag + CN^-$	−0.02	$ClO^- + H_2O + 2e^- \rightleftharpoons Cl^- + 2OH^-$	0.90
$Ag_2CO_3 + 2e^- \rightleftharpoons 2Ag + CO_3^{-2}$	0.4769	$ClO_2 + e^- \rightleftharpoons ClO_2^-$	1.15
$Ag_2CrO_4 + 2e^- \rightleftharpoons 2Ag + CrO_4^{-2}$	0.4463	$ClO_2 + H^+ + e^- \rightleftharpoons HClO_2$	1.27
$Ag_4Fe(CN)_6 + 4e^- \rightleftharpoons 4Ag + Fe(CN)_6^{-4}$	0.1943	$HClO_2 + 2H^+ + 2e^- \rightleftharpoons HClO + H_2O$	1.64
$AgI + e^- \rightleftharpoons Ag + I^-$	−0.1519	$HClO_2 + 3H^+ + 3e^- \rightleftharpoons 1/2Cl_2 + 2H_2O$	1.63
$AgIO_3 + e^- \rightleftharpoons Ag + IO_3^-$	0.3551	$HClO_2 + 3H^+ + 4e^- \rightleftharpoons Cl^- + 2H_2O$	1.56
$Ag_2MoO_4 + 2e^- \rightleftharpoons 2Ag + MoO_4^{-2}$	0.49	$ClO_2^- + H_2O + 2e^- \rightleftharpoons ClO^- + 2OH^-$	0.59
$AgNO_2 + e \rightleftharpoons Ag + NO_2$	0.59	$ClO_2^- + 2H_2O + 4e^- \rightleftharpoons Cl^- + 4OH^-$	0.76
$Ag_2O + H_2O + 2e \rightleftharpoons 2Ag + 2OH^-$	0.342	$ClO_2(aq) + e^- \rightleftharpoons ClO_2^-$	0.954
$Ag_2O_3 + H_2O + 2e^- \rightleftharpoons 2AgO + 2OH^-$	0.74	$ClO_3^- + 2H^+ + e^- \rightleftharpoons ClO_2 + H_2O$	1.15
$2AgO + H_2O + 2e^- \rightleftharpoons Ag_2O + 2OH^-$	0.599	$ClO_3^- + 3H^+ + 2e^- \rightleftharpoons HClO_2 + H_2O$	1.21 (1.23)
$AgOCN + e^- \rightleftharpoons Ag + OCN^-$	0.41		
$Ag_2S + 2e \rightleftharpoons 2Ag + S$	−0.7051	$ClO_3^- + 6H^+ + 5e^- \rightleftharpoons 1/2Cl_2 + 3H_2O$	1.47
$Ag_2S + 2H^+ + 2e \rightleftharpoons 2Ag + H_2S$	−0.0366	$ClO_3^- + 6H^+ + 6e^- \rightleftharpoons Cl^- + 3H_2O$	1.45
$AgSCN + e^- \rightleftharpoons Ag + SCN^-$	0.0895	$ClO_3^- + H_2O + 2e^- \rightleftharpoons ClO_2^- + 2OH^-$	0.35
$Ag_2SeO_3 + 2e \rightleftharpoons 2Ag + SeO_3^{-2}$	0.3629	$ClO_3^- + 3H_2O + 6e^- \rightleftharpoons Cl^- + 6OH^-$	0.62
$Ag_2SO_4 + 2e^- \rightleftharpoons 2Ag + SO_4^{-2}$	0.653	$ClO_4^- + 2H^+ + 2e^- \rightleftharpoons ClO_3^- + H_2O$	1.19
$Ag_2(WO_4) + 2e^- \rightleftharpoons 2Ag + WO_4^{-2}$	0.466	$ClO_4^- + 8H^+ + 7e^- \rightleftharpoons 1/2Cl_2 + 4H_2O$	1.34
$Al^{+3} + 3e^- \rightleftharpoons Al\,(0.1f\,NaOH)$	−1.706	$ClO_4^- + 8H^+ + 8e^- \rightleftharpoons Cl^- + 4H_2O$	1.37
$H_2AlO_3^- + H_2O + 3e^- \rightleftharpoons Al + 4OH^-$	−2.35	$ClO_4^- + H_2O + 2e^- \rightleftharpoons ClO_3^- + 2OH^-$	0.17
$As + 3H^+ + 3e^- \rightleftharpoons AsH_3$	−0.54	$(CN)_2 + 2H^+ + 2e^- \rightleftharpoons 2HCN$	0.37
$As_2O_3 + 6H^+ + 6e^- \rightleftharpoons 2As + 3H_2O$	0.234	$2HCNO + 2H^+ + 2e^- \rightleftharpoons (CN)_2 + 2H_2O$	0.33
$HAsO_2 + 3H^+ + 3e^- \rightleftharpoons As + 2H_2O$	0.2475	$(CNS)_2 + 2e^- \rightleftharpoons 2CNS^-$	0.77
$AsO_2^- + 2H_2O + 3e \rightleftharpoons As + 4OH^-$	−0.68	$Co^{+2} + 2e^- \rightleftharpoons Co$	−0.28
$H_3AsO_4 + 2H^+ + 2e^- \rightleftharpoons HAsO_2 + 2H_2O(1f\,HCl)$	0.58	$Co^{+3} + e^- \rightleftharpoons Co^{+2}(3f\,HNO_3)$	1.842
$AsO_4^{-3} + 2H_2O + 2e^- \rightleftharpoons AsO_2^- + 4OH^-$	0.71	$CO_2 + 2H^+ + 2e^- \rightleftharpoons HCOOH$	−0.2
$AsO_4^{-3} + 2H_2O + 2e^- \rightleftharpoons AsO_2^- + 4OH^-\,(1f\,NaOH)$	−0.08	$2CO_2 + 2H^+ + 2e^- \rightleftharpoons H_2C_2O_4$	−0.49
$Au^+ + e^- \rightleftharpoons Au$	1.68	$Co(NH_3)_6^{+3} + e^- \rightleftharpoons Co(NH_3)_6^{+2}$	0.1
$Au^{+3} + 2e^- \rightleftharpoons Au^{+1}$	1.29	$Co(OH)_2 + 2e^- \rightleftharpoons Co + 2OH^-$	−0.73
$Au^{+3} + 3e \rightleftharpoons Au$	1.42	$Co(OH)_3 + e^- \rightleftharpoons Co(OH)_2 + OH^-$	0.2 (0.17)
$AuBr_2^- + e^- \rightleftharpoons Au + 2Br^-$	0.963	$Cr^{+2} + 2e \rightleftharpoons Cr$	−0.557
$AuBr_4^- + 3e^- \rightleftharpoons Au + 4Br^-$	0.858	$Cr^{+3} + e^- \rightleftharpoons Cr^{+2}$	−0.41
$AuCl_4^- + 3e^- \rightleftharpoons Au + 4Cl^-$	0.994	$Cr^{+3} + 3e^- \rightleftharpoons Cr$	−0.74
$Au(OH)_3 + 3H^+ + 3e^- \rightleftharpoons Au + 3H_2O$	1.45	$Cr^{+6} + 3e^- \rightleftharpoons Cr^{+3}(2f\,H_2SO_4)$	1.10
$H_4BO_3^- + 5H_2O + 8e^- \rightleftharpoons BH_4^- + 8OH^-$	−1.24	$Cr^{+6} + 3e^- \rightleftharpoons Cr^{+3}(1f\,NaOH)$	−0.12
$H_2BO_3^- + H_2O + 3e^- \rightleftharpoons B + 4OH^-$	−2.5	$Cr_2O_7^{-2} + 14H^+ + 6e^- \rightleftharpoons 2Cr^{+3} + 7H_2O$	1.33
$H_3BO_3 + 3H^+ + 3e^- \rightleftharpoons B + 3H_2O$	−0.73	$CrO_2^- + 2H_2O + 3e^- \rightleftharpoons Cr + 4OH^-$	−1.2
$Ba^{+2} + 2e^- \rightleftharpoons Ba$	−2.90	$HCrO_4^- + 7H^+ + 3e^- \rightleftharpoons Cr^{+3} + 4H_2O$	1.195
$Ba^{+2} + 2e^- \rightleftharpoons Ba(Hg)$	−1.570	$CrO_4^{-2} + 4H_2O + 3e^- \rightleftharpoons Cr(OH)_3 + 5OH^-$	−0.12
$Ba(OH)_2 \cdot 8H_2O + 2e^- \rightleftharpoons Ba + 2OH^- + 8H_2O$	−2.97	$Cr(OH)_3 + 3e^- \rightleftharpoons Cr + 3OH^-$	−1.3
$Be^{+2} + 2e^- \rightleftharpoons Be$	−1.70 (−1.85)	$Cs^+ + e^- \rightleftharpoons Cs$	−2.923
$Be_2O_3^{-2} + 3H_2O + 4e^- \rightleftharpoons 2Be + 6OH^-$	−2.28	$Cu^+ + e^- \rightleftharpoons Cu$	0.522
$Bi(Cl)_4^- + 3e^- \rightleftharpoons Bi + 4Cl^-$	0.168	$Cu^{+2} + 2CN^- + e^- \rightleftharpoons Cu(CN)_2^-$	1.12
$Bi_2O_3 + 3H_2O + 6e^- \rightleftharpoons 2Bi + 6OH^-$	−0.46	$Cu^{+2} + e \rightleftharpoons Cu^+$	0.158 (0.167)
$Bi_2O_4 + 4H^+ + 2e^- \rightleftharpoons 2BiO^+ + 2H_2O$	1.59		
$BiO^+ + 2H^+ + 3e^- \rightleftharpoons Bi + H_2O$	0.32	$Cu^{+2} + 2e^- \rightleftharpoons Cu$	0.3402
$BiOCl + 2H^+ + 3e^- \rightleftharpoons Bi + Cl^- + H_2O$	0.1583	$Cu^{+2} + 2e^- \rightleftharpoons Cu(Hg)$	0.345
$BiOOH + H_2O + 3e^- \rightleftharpoons Bi + 3OH^-$	−0.46	$CuI_2^- + e^- \rightleftharpoons Cu + 2I^-$	0.00
$Br_2(aq) + 2e^- \rightleftharpoons 2Br^-$	1.087	$Cu_2O + H_2O + 2e^- \rightleftharpoons 2Cu + 2OH^-$	−0.361
$Br_2(l) + 2e^- \rightleftharpoons 2Br^-$	1.065	$Cu(OH)_2 + 2e^- \rightleftharpoons Cu + 2OH^-$	−0.224
$HBrO + H^+ + e^- \rightleftharpoons 1/2Br_2 + H_2O$	1.59	$2Cu(OH)_2 + 2e^- \rightleftharpoons Cu_2O + 2OH^- + H_2O$	−0.09
$HBrO + H^+ + 2e^- \rightleftharpoons Br^- + H_2O$	1.33	$D^+ + e^- \rightleftharpoons 1/2D_2$	−0.0034
$2HBrO + 2H^+ + 2e^- \rightleftharpoons Br_2(l) + 2H_2O$	1.6	$2D^+ + 2e^- \rightleftharpoons D_2$	−0.044
$BrO^- + H_2O + 2e^- \rightleftharpoons Br^- + 2OH^-(1f\,NaOH)$	0.70	$Eu^{+3} + e^- \rightleftharpoons Eu^{+2}$	−0.43
$BrO_3^- + 6H^+ + 5e^- \rightleftharpoons 1/2Br_2 + 3H_2O$	1.52	$1/2F_2 + e^- \rightleftharpoons F^-$	2.85
$BrO_3^- + 6H^+ + 6e^- \rightleftharpoons Br^- + 3H_2O$	1.44	$1/2F_2 + H^+ + e^- \rightleftharpoons HF$	3.03
$BrO_3^- + 3H_2O + 6e^- \rightleftharpoons Br^- + 6OH^-$	0.61	$F_2 + 2e^- \rightleftharpoons 2F^-$	2.87
$C_6H_4O_2 + 2H^+ + 2e \rightleftharpoons C_6H_4(OH)_2$	0.6992	$F_2O + 2H^+ + 4e^- \rightleftharpoons H_2O + 2F^-$	2.1
$Ca^+ + e^- \rightleftharpoons Ca$	−3.02	$Fe^{+2} + 2e^- \rightleftharpoons Fe$	−0.409
$Ca^{+2} + 2e^- \rightleftharpoons Ca$	−2.76	$Fe^{+3} + 3e^- \rightleftharpoons Fe$	−0.036
Calomel Electrode, Molal KCl	0.2800	$Fe^{+3} + e^- \rightleftharpoons Fe^{+2}$	0.770
*Calomel Electrode, N KCl	0.2807	$Fe^{+3} + e^- \rightleftharpoons Fe^{+2}(1f\,HCl)$	0.770
*Calomel Electrode 0.1 N KCl	0.3337	$Fe^{+3} + e^- \rightleftharpoons Fe^{+2}(1f\,HClO_4)$	0.747
*Calomel Electrode, Sat'd. KCl	0.2415	$Fe^{+3} + e^- \rightleftharpoons Fe^{+2}(1f\,H_3PO_4)$	0.438
*Calomel Electrode, Sat'd NaCl	0.2360	$Fe^{+3} + e^- \rightleftharpoons Fe^{+2}(0.5f\,H_2SO_4)$	0.679
$Ca(OH)_2 + 2e^- \rightleftharpoons Ca + 2OH^-$	−3.02	$Fe(CN)_6^{-3} + e^- \rightleftharpoons Fe(CN)_6^{-4}(0.01f\,NaOH)$	0.46
$Cb_2O_5 + 10H^+ + 10e^- \rightleftharpoons 2Cb + 5H_2O$	−0.62	$Fe(CN)_6^{-3} + e^- \rightleftharpoons Fe(CN)_6^{-4}(1f\,H_2SO_4)$	0.69
$Cd^{+2} + 2e^- \rightleftharpoons Cd$	−0.4026	$FeO_4^{-2} + 8H^+ + 3e^- \rightleftharpoons Fe^{+3} + 4H_2O$	1.9
$Cd^{+2} + 2e^- \rightleftharpoons Cd(Hg)$	−0.3521	$Fe(OH)_3 + e^- \rightleftharpoons Fe(OH)_2 + OH^-$	−0.56
$Cd(OH)_2 + 2e^- \rightleftharpoons Cd(Hg) + 2OH^-$	−0.761 (−0.81)	$Fe\,(phenanthroline)_3^{+3} + e^- \rightleftharpoons Fe(ph)_3^{+2}$	1.14
$CdSO_4 \cdot 8/3H_2O + 2e^- \rightleftharpoons Cd(Hg) + CdSO_4(sat'd\,aq)$	−0.4346	$Fe\,(phenanthroline)_3^{+3} + e^- \rightleftharpoons Fe(ph)_3^{+2}(2f\,H_2SO_4)$	1.056
$Ce^{+3} + 3e^- \rightleftharpoons Ce$	−2.335	$Ga^{+3} + 3e^- \rightleftharpoons Ga$	−0.560
		$H_2GaO_3^- + H_2O + 3e^- \rightleftharpoons Ga + 4OH^-$	−1.22

Table 3-8. ELECTROCHEMICAL SERIES (Continued)

Reaction	Potential, volts	Reaction	Potential, volts
$GeO_2 + 2H^+ + 2e^- \rightleftharpoons GeO + H_2O$	−0.12	$NO_2^- + H_2O + e^- \rightleftharpoons NO + 2OH^-$	−0.46
$H_2GeO_3 + 4H^+ + 4e^- \rightleftharpoons Ge + 3H_2O$	−0.13	$2NO_2^- + 2H_2O + 4e^- \rightleftharpoons N_2O_2^{-2} + 4OH^-$	−0.18
$2H^+ + 2e^- \rightleftharpoons H_2$	0.0000	$2NO_2^- + 3H_2O + 4e^- \rightleftharpoons N_2O + 6OH^-$	0.15
$1/2H_2 + e^- \rightleftharpoons H^-$	−2.23	$NO_3^- + 3H^+ + 2e^- \rightleftharpoons HNO_2 + H_2O$	0.94
$2H_2O + 2e^- \rightleftharpoons H_2 + 2OH^-$	−0.8277	$NO_3^- + 4H^+ + 3e^- \rightleftharpoons NO + 2H_2O$	0.96
$H_2O_2 + 2H^+ + 2e^- \rightleftharpoons 2H_2O$	1.776	$2NO_3^- + 4H^+ + 2e^- \rightleftharpoons N_2O_4 + 2H_2O$	0.81
$HfO^{+2} + 2H^+ + 4e^- \rightleftharpoons Hf + H_2O$	−1.68	$NO_3^- + H_2O + 2e^- \rightleftharpoons NO_2^- + 2OH^-$	0.01
$HfO_2 + 4H^+ + 4e^- \rightleftharpoons Hf + 2H_2O$	−1.57	$2NO_3^- + 2H_2O + 2e^- \rightleftharpoons N_2O_4 + 4OH^-$	−0.85
$HfO(OH)_2 + H_2O + 4e^- \rightleftharpoons Hf + 4OH^-$	−2.60	$Np^{+3} + 3e^- \rightleftharpoons Np$	−1.9
$Hg^{+2} + 2e^- \rightleftharpoons Hg$	0.851	$Np^{+4} + e^- \rightleftharpoons Np^{+3}(1fHClO_4)$	0.155
$2Hg^{+2} + 2e^- \rightleftharpoons Hg_2^{+2}$	0.905	$Np^{+5} + e^- \rightleftharpoons Np^{+4}(1fHClO_4)$	0.739
$1/2Hg_2^{+2} + e^- \rightleftharpoons Hg$	0.7986	$Np^{+6} + e^- \rightleftharpoons Np^{+5}(1fHClO_4)$	1.137
$Hg_2^{+2} + 2e^- \rightleftharpoons 2Hg$	0.7961	$1/2O_2 + 2H^+(10^{-7}M) + 2e^- \rightleftharpoons H_2O$	0.815
$Hg_2(AcO)_2 + 2e^- \rightleftharpoons 2Hg + 2AcO^-$	0.5113	$O_2 + 2H^+ + 2e^- \rightleftharpoons H_2O_2$	0.682
$Hg_2Br_2 + 2e^- \rightleftharpoons 2Hg + 2Br^-$	0.1396	$O_2 + 4H^+ + 4e^- \rightleftharpoons 2H_2O$	1.229
$Hg_2Cl_2 + 2e^- \rightleftharpoons 2Hg + 2Cl^-$	0.2682	$O_2 + H_2O + 2e^- \rightleftharpoons HO_2^- + OH^-$	−0.076
$Hg_2Cl_2 + 2e \rightleftharpoons 2Hg + 2Cl^-(0.1f\ NaOH)$	0.3419 (0.268)	$O_2 + 2H_2O + 2e^- \rightleftharpoons H_2O_2 + 2OH^-$	−0.146
$Hg_2HPO_4 + H^+ + 2e^- \rightleftharpoons 2Hg + H_2PO_4^-$	0.639	$O_2 + 2H_2O + 4e^- \rightleftharpoons 4OH^-$	0.401
$Hg_2I_2 + 2e^- \rightleftharpoons 2Hg + 2I^-$	−0.0405	$O_3 + 2H^+ + 2e^- \rightleftharpoons O_2 + H_2O$	2.07
$Hg_2O + H_2O + 2e^- \rightleftharpoons 2Hg + 2OH^-$	0.123	$O_3 + H_2O + 2e^- \rightleftharpoons O_2 + 2OH^-$	1.24
$HgO + H_2O + 2e^- \rightleftharpoons Hg + 2OH^-$	0.0984	$O_{(g)} + 2H^+ + 2e^- \rightleftharpoons H_2O$	2.42
$Hg_2SO_4 + 2e^- \rightleftharpoons 2Hg + SO_4^{-2}$	0.6158	$OH + e^- \rightleftharpoons OH^-$	1.4
$HO_2 + H^+ + e^- \rightleftharpoons H_2O_2$	1.5	$HO_2^- + H_2O + 2e^- \rightleftharpoons 3OH^-$	0.87
$I_2 + 2e^- \rightleftharpoons 2I^-$	0.535	$OsO_4 + 8H^+ + 8e^- \rightleftharpoons Os + 4H_2O$	0.85
$I_3^- + 2e^- \rightleftharpoons 3I^-$	0.5338	$P + 3H^+ + 3e^- \rightleftharpoons PH_3(g)$	−0.04
$In^{+2} + e^- \rightleftharpoons In^{+1}$	−0.40	$P + 3H_2O + 3e^- \rightleftharpoons PH_3(g) + 3OH^-$	−0.87
$In^{+3} + e^- \rightleftharpoons In^{+2}$	−0.49	$Pb^{+2} + 2e^- \rightleftharpoons Pb$	−0.1263 (−0.126)
$In^{+3} + 2e^- \rightleftharpoons In^{+1}$	−0.40		
$In^{+3} + 3e^- \rightleftharpoons In$	−0.338	$Pb^{+2} + 2e^- \rightleftharpoons Pb(Hg)$	−0.1205
$H_3IO_6^{-2} + 2e^- \rightleftharpoons IO_3^- + 3OH^-$	~0.70	$PbBr_2 + 2e^- \rightleftharpoons Pb(Hg) + 2Br^-$	−0.275
$H_5IO_6 + H^+ + 2e^- \rightleftharpoons IO_3^- + 3H_2O$	~1.7	$PbCl_2 + 2e^- \rightleftharpoons Pb(Hg) + 2Cl^-$	−0.262
$HIO + H^+ + e^- \rightleftharpoons 1/2I_2 + H_2O$	1.45	$PbF_2 + 2e^- \rightleftharpoons Pb(Hg) + 2F^-$	−0.3444
$HIO + H^+ + 2e^- \rightleftharpoons I^- + H_2O$	0.99	$PbHPO_4 + H^+ + 2e^- \rightleftharpoons Pb(Hg) + HPO_4^-$	−0.2448
$IO^- + H_2O + 2e^- \rightleftharpoons I^- + 2OH^-$	0.49	$PbI_2 + 2e^- \rightleftharpoons Pb(Hg) + 2I^-$	−0.358
$IO_3^- + 6H^+ + 5e^- \rightleftharpoons 1/2I_2 + 3H_2O$	1.195	$PbO + H_2O + 2e^- \rightleftharpoons Pb + 2OH^-$	−0.576
$IO_3^- + 6H^+ + 6e^- \rightleftharpoons I^- + 3H_2O$	1.085	$PbO_2 + 4H^+ + 2e^- \rightleftharpoons Pb^{+2} + 2H_2O$	1.46
$2IO_3^- + 12H^+ + 10e^- \rightleftharpoons I_2 + 6H_2O$	1.19	$HPbO_2^- + H_2O + 2e^- \rightleftharpoons Pb + 3OH^-$	−0.54
$IO_3^- + 2H_2O + 4e^- \rightleftharpoons IO^- + 4OH^-$	0.56	$PbO_2 + H_2O + 2e^- \rightleftharpoons PbO + 2OH^-$	0.28
$IO_3^- + 3H_2O + 6e^- \rightleftharpoons I^- + 6OH^-$	0.26	$PbO_2 + SO_4^{-2} + 4H^+ + 2e^- \rightleftharpoons PbSO_4 + 2H_2O$	1.685
$IrCl_6^{-2} + e^- \rightleftharpoons IrCl_6^{-3}$	1.02	$PbSO_4 + 2e^- \rightleftharpoons Pb + SO_4^{-2}$	−0.356
$IrCl_6^{-3} + 3e^- \rightleftharpoons Ir + 6Cl^-$	0.77	$PbSO_4 + 2e^- \rightleftharpoons Pb(Hg) + SO_4^{-2}$	−0.3505
$Ir_2O_3 + 3H_2O + 6e^- \rightleftharpoons 2Ir + 6OH^-$	0.1	$Pd^{+2} + 2e^- \rightleftharpoons Pd$	0.83
$K^+ + e^- \rightleftharpoons K$	−2.924 (−2.923)	$Pd^{+2} + 2e^- \rightleftharpoons Pd(1fHCl)$	0.623
		$Pd^{+2} + 2e^- \rightleftharpoons Pd(4fHClO_4)$	0.987
$La^{+3} + 3e^- \rightleftharpoons La$	−2.37	$PdCl_4^{-2} + 2e^- \rightleftharpoons Pd + 4Cl^-$	0.623
$La(OH)_3 + 3e^- \rightleftharpoons La + 3OH^-$	−2.76	$PdCl_6^{-2} + 2e^- \rightleftharpoons PdCl_4^{-2} + 2Cl^-$	1.29
$Li^+ + e^- \rightleftharpoons Li$	−3.045 (−3.02)	$Pd(OH)_2 + 2e^- \rightleftharpoons Pd + 2OH^-$	0.1
		$H_2PO_2^- + e^- \rightleftharpoons P + 2OH^-$	−1.82
$Mg^{++} + 2e^- \rightleftharpoons Mg$	−2.375	$H_3PO_2 + H^+ + e^- \rightleftharpoons P + 2H_2O$	−0.51
$Mg(OH)_2 + 2e^- \rightleftharpoons Mg + 2OH^-$	−2.67	$H_3PO_3 + 2H^+ + 2e^- \rightleftharpoons H_3PO_2 + H_2O$	−0.50 (−0.59)
$Mn^{+2} + 2e^- \rightleftharpoons Mn$	−1.029		
$Mn^{+3} + e^- \rightleftharpoons Mn^{+2}$	1.51	$H_3PO_3 + 3H^+ + 3e^- \rightleftharpoons P + 3H_2O$	−0.49
$MnO_2 + 4H^+ + 2e^- \rightleftharpoons Mn^{+2} + 2H_2O$	1.208	$HPO_3^{-2} + 2H_2O + 2e^- \rightleftharpoons H_2PO_2^- + 3OH^-$	−1.65
$MnO_4^- + e^- \rightleftharpoons MnO_4^{-2}$	0.564	$HPO_3^{-2} + 2H_2O + 3e^- \rightleftharpoons P + 5OH^-$	−1.71
$MnO_4^- + 4H^+ + 3e^- \rightleftharpoons MnO_2 + 2H_2O$	+1.679	$H_3PO_4 + 2H^+ + 2e^- \rightleftharpoons H_3PO_3 + H_2O$	−0.276 (−0.2)
$MnO_4^- + 8H^+ + 5e^- \rightleftharpoons Mn^{+2} + 4H_2O$	1.491		
$MnO_4^- + 2H_2O + 3e^- \rightleftharpoons MnO_2 + 4OH^-$	0.588	$PO_4^{-3} + 2H_2O + 2e^- \rightleftharpoons HPO_3^{-2} + 3OH^-$	−1.05
$MnO_4^{-1} + 2H_2O + 3e^- \rightleftharpoons MnO_2 + 4OH^-$	0.58	$Pt^{+2} + 2e^- \rightleftharpoons Pt$	~1.2
$Mn(OH)_2 \rightleftharpoons Mn + 2OH^-$	−1.47	$PtCl_4^{-2} + 2e^- \rightleftharpoons Pt + 4Cl^-$	0.73
$Mn(OH)_3 + e^- \rightleftharpoons Mn(OH)_2 + OH^-$	−0.40	$PtCl_6^{-2} + 2e^- \rightleftharpoons PtCl_4^{-2} + 2Cl^-$	0.74
$H_2MoO_4 + 6H^+ + 6e^- \rightleftharpoons Mo + 4H_2O$	0.0	$Pt(OH)_2 + 2e^- \rightleftharpoons Pt + 2OH^-$	0.16
$N_2 + 2H_2O + 4H^+ + 2e^- \rightleftharpoons 2NH_3OH^+$	−1.87	$Pu^{+4} + e^- \rightleftharpoons Pu^{+3}(1fHClO_4)$	0.982
$3N_2 + 2H^+ + 2e^- \rightleftharpoons 2HN_3$	−3.1	$Pu^{+5} + e^- \rightleftharpoons Pu^{+4}(0.5fHCl)$	1.099
$N_2H_5^+ + 2e^- \rightleftharpoons 2NH_4^+$	1.27	$Pu^{+6} + e^- \rightleftharpoons Pu^{+5}(1fHClO_4)$	0.9184
$N_2O + 2H^+ + 2e^- \rightleftharpoons N_2 + H_2O$	1.77	$Pu^{+6} + e^- \rightleftharpoons Pu^{+4}(1fHCl)$	1.052
$H_2N_2O_2 + 2H^+ + 2e^- \rightleftharpoons N_2 + 2H_2O$	2.65	Quinhydrone Elec. $H^+, a = 1$	0.6995
$N_2O_4 + 2e^- \rightleftharpoons 2NO_2^-$	0.88	$Rb^+ + e^- \rightleftharpoons Rb$	−2.925 (−2.99)
$N_2O_4 + 2H^+ + 2e^- \rightleftharpoons 2HNO_2$	1.07		
$N_2O_4 + 4H^+ + 4e^- \rightleftharpoons 2NO + 2H_2O$	1.03	$Re^{+3} + 3e^- \rightleftharpoons Re$	0.3 ~
$Na^+ + e^- \rightleftharpoons Na$	−2.7109 (−2.712)	$ReO_4^- + 4H^+ + 3e^- \rightleftharpoons ReO_2 + 2H_2O$	0.51
		$ReO_2 + 4H^+ + 4e^- \rightleftharpoons Re + 2H_2O$	0.26
$Nb^{+5} + 2e \rightleftharpoons Nb^{+3}(2fHCl)$	0.344	$ReO_4^- + 2H^+ + e^- \rightleftharpoons ReO_{3(cc)} + 2H_2O$	0.768
$Nd^{+3} + 3e^- \rightleftharpoons Nd$	−2.246	$ReO_4^- + 4H_2O + 7e^- \rightleftharpoons Re + 8OH^-$	−0.81
$2NH_3OH^+ + H^+ + 2e^- \rightleftharpoons N_2H_5^+ + 2H_2O$	1.42	$ReO_4^- + 8H^+ + 7e^- \rightleftharpoons Re + 4H_2O$	0.367
$Ni^{+2} + 2e^- \rightleftharpoons Ni$	−0.23	$Rh^{+4} + e^- \rightleftharpoons Rh^{+3}$	1.43
$Ni(OH)_2 + 2e^- \rightleftharpoons Ni + 2OH^-$	−0.66	$Rh(Cl)_6^{-3} + 3e^- \rightleftharpoons Rh + 6Cl^-$	0.44
$NiO_2 + 4H^+ + 2e^- \rightleftharpoons Ni^{+2} + 2H_2O$	1.93	$Ru^{+3} + e^- \rightleftharpoons Ru^{+2}(0.1fHClO_4)$	−0.11
$NiO_2 + 2H_2O + 2e^- \rightleftharpoons Ni(OH)_2 + 2OH^-$	0.49	$Ru^{+3} + e^- \rightleftharpoons Ru^{+2}(1-6fHCl)$	−0.084
$2NO + 2e^- \rightleftharpoons N_2O_2^{-2}$	0.10	$Ru^{+4} + e^- \rightleftharpoons Ru^{+3}(0.1fHClO_4)$	0.49
$2NO + 2H^+ + 2e^- \rightleftharpoons N_2O + H_2O$	1.59	$Ru^{+4} + e^- \rightleftharpoons Ru^{+3}(2fHCl)$	0.858
$2NO + H_2O + 2e^- \rightleftharpoons N_2O + 2OH^-$	0.76	$RuO_2 + 4H^+ + 4e^- \rightleftharpoons Ru + 2H_2O$	−0.8
$HNO_2 + H^+ + e^- \rightleftharpoons NO + H_2O$	0.99	$RuO_4^- + e^- \rightleftharpoons RuO_4^{-2}$	0.59
$2HNO_2 + 4H^+ + 4e^- \rightleftharpoons H_2N_2O_2 + 2H_2O$	0.80	$RuO_{4(c)} + e^- \rightleftharpoons RuO_4^-$	1.00
$2HNO_2 + 4H^+ + 4e^- \rightleftharpoons N_2O + 3H_2O$	1.27 (1.29)	$S + 2e^- \rightleftharpoons S^{-2}$	−0.508
		$S + 2H^+ + 2e^- \rightleftharpoons H_2S_{(aq)}$	0.141
		$S + H_2O + 2e^- \rightleftharpoons HS^- + OH^-$	−0.478

Table 3-8. ELECTROCHEMICAL SERIES (Continued)

Reaction	Potential, volts	Reaction	Potential, volts
$S_2O_6^{-2} + 4H^+ + 2e^- \rightleftharpoons 2H_2SO_3$	0.6	$TeO_4^- + 8H^+ + 7e^- \rightleftharpoons Te + 4H_2O$	0.472
$S_2O_8^{-2} + 2e^- \rightleftharpoons 2SO_4^{-2}$	2.0	$H_6TeO_{6(s)} + 2H^+ + 2e^- \rightleftharpoons TeO_{2(s)} + 4H_2O$	1.02
	(2.05)	$Th^{+4} + 4e^- \rightleftharpoons Th$	-1.90
$S_4O_6^= + 2e^- \rightleftharpoons 2S_2O_3^{-2}$	0.09	$ThO_2 + 4H^+ + 4e^- \rightleftharpoons Th + 2H_2O$	-1.80
	(0.10)	$ThO_2 + 2H_2O + 4e^- \rightleftharpoons Th + 4OH^-$	-2.64
$Sb + 3H^+ + 3e^- \rightleftharpoons H_3Sb$	-0.51	$Ti^{+2} + 2e^- \rightleftharpoons Ti$	-1.63
$Sb^{+5} + 2e^- \rightleftharpoons Sb^{+3}(3.5f\,HCl)$	0.75	$Ti^{+3} + e^- \rightleftharpoons Ti^{+2}$	-2.0
$Sb_2O_3 + 6H^+ + 6e^- \rightleftharpoons 2Sb + 3H_2O$	0.1445	$TiO_2 + 4H^+ + 4e^- \rightleftharpoons Ti + 2H_2O$	-0.86
	(0.152)	$Ti(OH)^{+3} + H^+ + e^- \rightleftharpoons Ti^{+3} + H_2O$	0.06
$Sb_2O_5 + 4H^+ + 4e^- \rightleftharpoons Sb_2O_3 + 2H_2O$	0.69	$Tl^+ + e^- \rightleftharpoons Tl$	-0.3363
$Sb_2O_{5(s)} + 6H^+ + 4e^- \rightleftharpoons 2SbO^+ + 3H_2O$	0.64	$Tl^+ + e^- \rightleftharpoons Tl(Hg)$	-0.3338
$SbO^+ + 2H^+ + 3e^- \rightleftharpoons Sb + 2H_2O$	0.212	$Tl^{+3} + 2e^- \rightleftharpoons Tl^{+2}$	-0.37
$SbO_2^- + 2H_2O + 3e^- \rightleftharpoons Sb + 4OH^-$	-0.66	$Tl^{+3} + 2e^- \rightleftharpoons Tl^{+1}$	1.247
$SbO_3^- + H_2O + 2e^- \rightleftharpoons SbO_2^- + 2OH^-$	-0.59	$Tl^{+3} + 2e^- \rightleftharpoons Tl^{+1}(1f\,HCl)$	0.783
$Sc^{+3} + 3e^- \rightleftharpoons Sc$	-2.08	$TlBr + e^- \rightleftharpoons Tl(Hg) + Br^-$	-0.606
$Se + 2e^- \rightleftharpoons Se^{-2}$	-0.78	$TlCl + e^- \rightleftharpoons Tl(Hg) + Cl^-$	-0.555
$Se + 2H^+ + 2e^- \rightleftharpoons H_2Se(aq)$	-0.36	$TlI + e^- \rightleftharpoons Tl(Hg) + I^-$	-0.769
$H_2SeO_3 + 4H^+ + 4e^- \rightleftharpoons Se + 3H_2O$	0.74	$Tl_2O_3 + 3H_2O + 4e^- \rightleftharpoons 2Tl^+ + 6OH^-$	0.02
$SeO_3^{-2} + 3H_2O + 4e^- \rightleftharpoons Se + 6OH^-$	-0.35	$TlOH + e^- \rightleftharpoons Tl + OH^-$	-0.3445
$SeO_4^{-2} + 4H^+ + 2e^- \rightleftharpoons H_2SeO_3 + H_2O$	1.15	$Tl(OH)_3 + 2e^- \rightleftharpoons TlOH + 2OH^-$	-0.05
$SeO_4^{-2} + H_2O + 2e^- \rightleftharpoons SeO_3^{-2} + 2OH^-$	0.03	$Tl_2SO_4 + 2e^- \rightleftharpoons Tl(Hg) + SO_4^{-2}$	-0.4360
$SiF_6^{-2} + 4e^- \rightleftharpoons Si + 6F^-$	-1.2	$U^{+3} + 3e^- \rightleftharpoons U$	-1.8
$SiO_2 + 4H^+ + 4e^- \rightleftharpoons Si + 2H_2O$	-0.84	$U^{+4} + e^- \rightleftharpoons U^{+3}$	-0.61
$SiO_3^{-2} + 3H_2O + 4e^- \rightleftharpoons Si + 6OH^-$	-1.73	$U^{+4} + e^- \rightleftharpoons U^{+3}(1f\,HClO_4)$	-0.631
$Sn^{+2} + 2e^- \rightleftharpoons Sn$	-0.1364	$U^{+3} + e^- \rightleftharpoons U^{+4}(1f\,HCl)$	1.02
$Sn^{+4} + 2e^- \rightleftharpoons Sn^{+2}$	0.15	$U^{+6} + e^- \rightleftharpoons U^{+5}(1f\,HClO_4)$	0.063
$Sn^{+4} + 2e^- \rightleftharpoons Sn^{+2}(0.1f\,HCl)$	0.070	$UO_2^+ + 4H^+ + e^- \rightleftharpoons U^{+4} + 2H_2O$	0.62
$Sn^{+4} + 2e^- \rightleftharpoons Sn^{+2}(1f\,HCl)$	0.139	$UO_2^{+2} + e^- \rightleftharpoons UO_2^+$	0.062
$HSnO_2^- + H_2O + 2e^- \rightleftharpoons Sn + 3OH^-$	-0.79	$UO_2^{+2} + 4H^+ + 2e^- \rightleftharpoons U^{+4} + 2H_2O$	0.334
$Sn(OH)_6^{-2} + 2e^- \rightleftharpoons HSnO_2^- + 3OH^- + H_2O$	-0.96	$UO_2^{+2} + 4H^+ + 6e^- \rightleftharpoons U + 2H_2O$	-0.82
$2H_2SO_3 + H^+ + 2e^- \rightleftharpoons HS_2O_4^- + 2H_2O$	-0.08	$V^{+2} + 2e^- \rightleftharpoons V$	-1.2
$H_2SO_3 + 4H^+ + 4e^- \rightleftharpoons S + 3H_2O$	0.45	$V^{+3} + e^- \rightleftharpoons V^{+2}$	-0.255
$2SO_3^{-2} + 2H_2O + 2e^- \rightleftharpoons S_2O_4^{-2} + 4OH^-$	-1.12	$V^{+3} + e^- \rightleftharpoons V^{+4}(1f\,NaOH)$	-0.74
$2SO_3^{-2} + 3H_2O + 4e^- \rightleftharpoons S_2O_3^{-2} + 6OH^-$	-0.58	$VO^{+2} + 2H^+ + e^- \rightleftharpoons V^{+3} + H_2O$	0.337
$SO_4^{-2} + 4H^+ + 2e^- \rightleftharpoons H_2SO_3 + H_2O$	0.20	$VO_2^+ + 2H^+ + e^- \rightleftharpoons VO^{+2} + H_2O$	1.00
$2SO_4^{-2} + 4H^+ + 2e^- \rightleftharpoons S_2O_6^{-2} + H_2O$	-0.2	$V(OH)_4^+ + 2H^+ + e^- \rightleftharpoons VO^{+2} + 3H_2O$	1.00
$SO_4^{-2} + H_2O + 2e^- \rightleftharpoons SO_3^{-2} + 2OH^-$	-0.92	$V(OH)_4^+ + 4H^+ + 5e^- \rightleftharpoons V + 4H_2O$	-0.25
$Sr^{+2} + 2e^- \rightleftharpoons Sr$	-2.89	$W_2O_5 + 2H^+ + 2e^- \rightleftharpoons 2WO_2 + H_2O$	-0.04
$Sr^{+2} + 2e^- \rightleftharpoons Sr(Hg)$	-1.793	$WO_2 + 4H^+ + 4e^- \rightleftharpoons W + 2H_2O$	-0.12
$Sr(OH)_2 \cdot 8H_2O + 2e^- \rightleftharpoons Sr + 2OH^- + 8H_2O$	-2.99	$WO_3 + 6H^+ + 6e^- \rightleftharpoons W + 3H_2O$	-0.09
$Ta_2O_5 + 10H^+ + 10e^- \rightleftharpoons 2Ta + 5H_2O$	-0.71	$2WO_3 + 2H^+ + 2e^- \rightleftharpoons W_2O_5 + H_2O$	-0.03
$TcO_4^- + 4H^+ + 3e^- \rightleftharpoons TcO_{2(c)} + 2H_2O$	0.738	$Y^{+3} + 3e^- \rightleftharpoons Y$	-2.37
$Te + 2e^- \rightleftharpoons Te^{-2}$	-0.92	$Zn^{+2} + 2e^- \rightleftharpoons Zn$	-0.7628
$Te + 2H^+ + 2e^- \rightleftharpoons H_2Te(Ag)$	-0.69	$Zn^{+2} + 2e^- \rightleftharpoons Zn(Hg)$	-0.7628
	(-0.72)	$ZnO_2^{-2} + 2H_2O + 2e^- \rightleftharpoons Zn + 4OH^-$	-1.216
$Te^{+4} + 4e^- \rightleftharpoons Te(2.5f\,HCl)$	0.63	$ZnSO_4 \cdot 7H_2O + 2e^- \rightleftharpoons Zn(Hg) + SO_4^{-2}(Sat'd\,ZnSO_4)$	-0.7993
$TeO_2 + 4H^+ + 4e^- \rightleftharpoons Te + 2H_2O$	0.593	$ZrO_2 + 4H^+ + 4e^- \rightleftharpoons Zr + 2H_2O$	-1.43
$TeO_3^{-2} + 3H_2O + 4e^- \rightleftharpoons Te + 6OH^-$	-0.02	$ZrO(OH)_2 + H_2O + 4e^- \rightleftharpoons Zr + 4OH^-$	-2.32

Table II. NUMERICAL LISTING

This table is divided into two parts. The first part lists reduction reactions that are positive with respect to the potential of the Standard Hydrogen Electrode. In this first part of Table II the reduction reactions are written in increasing order of the positive potential, beginning with 0.00 and ending with +3.03 volts. The second part of this Table II lists reduction reactions having reduction potentials more negative than that of the Standard Hydrogen Electrode. This second part of the table lists reactions in their order of increasing negative potential, beginning with 0.00 and ending with −3.1 volts.

PART 1

Reaction	Potential, volts	Reaction	Potential, volts
$2H^+ + 2e^- \rightleftharpoons H_2$	0.0000	$Pd(OH)_2 + 2e^- \rightleftharpoons Pd + 2OH^-$	0.1
$CuI_2^- + e^- \rightleftharpoons Cu + 2I^-$	0.00	$Co(NH_3)_6^{+3} + e^- \rightleftharpoons Co(NH_3)_6^{+2}$	0.1
$H_2MoO_4 + 6H^+ + 6e^- \rightleftharpoons Mo + 4H_2O$	0.0	$Hg_2O + H_2O + 2e^- \rightleftharpoons 2Hg + 2OH^-$	0.123
$NO_3^- + H_2O + 2e^- \rightleftharpoons NO_2^- + 2OH^-$	0.01	$Sn^{+4} + 2e^- \rightleftharpoons Sn^{+2}(1f\,HCl)$	0.139
$Tl_2O_3 + 3H_2O + 4e^- \rightleftharpoons 2Tl^+ + 6OH^-$	0.02	$Hg_2Br_2 + 2e^- \rightleftharpoons 2Hg + 2Br^-$	0.1396
$SeO_4^{-2} + H_2O + 2e^- \rightleftharpoons SeO_3^{-2} + 2OH^-$	0.03	$S + 2H^+ + 2e^- \rightleftharpoons H_2S_{(aq)}$	0.141
$Ti(OH)^{+3} + H^+ + e^- \rightleftharpoons Ti^{+3} + H_2O$	0.06	$Sb_2O_3 + 6H^+ + 6e^- \rightleftharpoons 2Sb + 3H_2O$	0.1445
$UO_2^{+2} + e^- \rightleftharpoons UO_2^+$	0.062		(0.152)
$U^{+6} + e^- \rightleftharpoons U^{+5}(1f\,HClO_4)$	0.063	$2NO_2^- + 3H_2O + 4e^- \rightleftharpoons N_2O + 6OH^-$	0.15
$Sn^{+4} + 2e^- \rightleftharpoons Sn^{+2}(0.1f\,HCl)$	0.070	$ReO_4^- + 8H^+ + 7e^- \rightleftharpoons Re + 4H_2O$	0.15
$AgBr + e^- \rightleftharpoons Ag + Br^-$	0.0713	$Sn^{+4} + 2e^- \rightleftharpoons Sn^{+2}$	0.15
$AgSCN + e^- \rightleftharpoons Ag + SCN^-$	0.0895	$Np^{+4} + e^- \rightleftharpoons Np^{+3}(1f\,HClO_4)$	0.155
$S_4O_6^= + 2e^- \rightleftharpoons 2S_2O_3^{-2}$	0.09	$Cu^{+2} + e^- \rightleftharpoons Cu^+$	0.158
	(0.10)		(0.167)
$HgO + H_2O + 2e^- \rightleftharpoons Hg + 2OH^-$	0.0984	$BiOCl + 2H^+ + 3e^- \rightleftharpoons Bi + Cl^- + H_2O$	0.1583
$2NO + 2e^- \rightleftharpoons N_2O_2^{-2}$	0.10	$Pt(OH)_2 + 2e^- \rightleftharpoons Pt + 2OH^-$	0.16
$Ir_2O_3 + 3H_2O + 6e^- \rightleftharpoons 2Ir + 6OH^-$	0.1	$Bi(Cl)_4^- + 3e^- \rightleftharpoons Bi + 4Cl^-$	0.168

Table 3-8. ELECTROCHEMICAL SERIES (Continued)

Reaction	Potential, volts		Reaction	Potential, volts
$ClO_4^- + H_2O + 2e^- \rightleftharpoons ClO_3^- + 2OH^-$	0.17		$PtCl_4^{-2} + 2e^- \rightleftharpoons Pt + 4Cl^-$	0.73
$Ag_4Fe(CN)_6 + 4e^- \rightleftharpoons 4Ag + Fe(CN)_6^{-4}$	0.1943		$TcO_4^- + 4H^+ + 3e^- \rightleftharpoons TcO_{2(c)} + 2H_2O$	0.738
$SO_4^{-2} + 4H^+ + 2e^- \rightleftharpoons H_2SO_3 + H_2O$	0.20		$Np^{+5} + e^- \rightleftharpoons Np^{+4}(1f\,HClO_4)$	0.739
$Co(OH)_3 + e^- \rightleftharpoons Co(OH)_2 + OH^-$	0.2 (0.17)		$Ag_2O_3 + H_2O + 2e^- \rightleftharpoons 2AgO + 2OH^-$	0.74
			$H_2SeO_3 + 4H^+ + 4e^- \rightleftharpoons Se + 3H_2O$	0.74
$SbO^+ + 2H^+ + 3e^- \rightleftharpoons Sb + 2H_2O$	0.212		$PtCl_6^{-2} + 2e^- \rightleftharpoons PtCl_4^{-2} + 2Cl^-$	0.74
$AgCl + e^- \rightleftharpoons Ag + Cl^-$	0.2223		$Fe^{+3} + e^- \rightleftharpoons Fe^{+2}(1f\,HClO_4)$	0.747
*Calomel Electrode, Sat'd NaCl	0.2360		$Sb^{+5} + 2e^- \rightleftharpoons Sb^{+3}(3.5f\,HCl)$	0.75
$As_2O_3 + 6H^+ + 6e^- \rightleftharpoons 2As + 3H_2O$	0.234		$ClO_2^- + 2H_2O + 4e^- \rightleftharpoons Cl^- + 4OH^-$	0.76
*Calomel Electrode, Sat'd. KCl	0.2415		$NiO_2 + 2H_2O + 2e^- \rightleftharpoons Ni(OH)_2 + 2OH^-$	0.76
$HAsO_2 + 3H^+ + 3e^- \rightleftharpoons As + 2H_2O$	0.2475		$2NO + H_2O + 2e^- \rightleftharpoons N_2O + 2OH^-$	0.76
$IO_3^- + 3H_2O + 6e^- \rightleftharpoons I^- + 6OH^-$	0.26		$ReO_4^- + 2H^+ + e^- \rightleftharpoons ReO_{3(cc)} + 2H_2O$	0.768
$ReO_2 + 4H^+ + 4e^- \rightleftharpoons Re + 2H_2O$	0.26		$(CNS)_2 + 2e^- \rightleftharpoons 2CNS^-$	0.77
$Hg_2Cl_2 + 2e^- \rightleftharpoons 2Hg + 2Cl^-$	0.2682		$Fe^{+3} + e^- \rightleftharpoons Fe^{+2}$	0.770
Calomel Electrode, Molal KCl	0.2800		$Fe^{+3} + e^- \rightleftharpoons Fe^{+2}(1f\,HCl)$	0.770
$PbO_2 + H_2O + 2e^- \rightleftharpoons PbO + 2OH^-$	0.28		$IrCl_6^{-3} + 3e^- \rightleftharpoons Ir + 6Cl^-$	0.77
*Calomel Electrode, N KCl	0.2807		$Tl^{+3} + 2e^- \rightleftharpoons Tl^{+1}(1f\,HCl)$	0.783
$Re^{+3} + 3e^- \rightleftharpoons Re$	0.3 ~		$Hg_2^{+2} + 2e^- \rightleftharpoons 2Hg$	0.7961
$BiO^+ + 2H^+ + 3e^- \rightleftharpoons Bi + H_2O$	0.32		$1/2Hg_2^{+2} + e^- \rightleftharpoons Hg$	0.7986
$2HCNO + 2H^+ + 2e^- \rightleftharpoons (CN)_2 + 2H_2O$	0.33		$Ag^+ + e^- \rightleftharpoons Ag$	0.7996
*Calomel Electrode 0.1 N KCl	0.3337		$2HNO_2 + 4H^+ + 4e^- \rightleftharpoons H_2N_2O_2 + 2H_2O$	0.80
$UO_2^{+2} + 4H^+ + 2e^- \rightleftharpoons U^{+4} + 2H_2O$	0.334		$2NO_3^- + 4H^+ + 2e^- \rightleftharpoons N_2O_4 + 2H_2O$	0.81
$VO^{+2} + 2H^+ + e^- \rightleftharpoons V^{+3} + H_2O$	0.337		$1/2O_2 + 2H^+(10^{-7}M) + 2e^- \rightleftharpoons H_2O$	0.815
$Cu^{+2} + 2e^- \rightleftharpoons Cu$	0.3402		$Pd^{+2} + 2e^- \rightleftharpoons Pd$	0.83
$Hg_2Cl_2 + 2e^- \rightleftharpoons 2Hg + 2Cl^-(0.1fNaOH)$	0.3419 (0.268)		$OsO_4 + 8H^+ + 8e^- \rightleftharpoons Os + 4H_2O$	0.85
			$Hg^{+2} + 2e^- \rightleftharpoons Hg$	0.851
$Ag_2O + H_2O + 2e^- \rightleftharpoons 2Ag + 2OH^-$	0.342		$AuBr_4^- + 3e^- \rightleftharpoons Au + 4Br^-$	0.858
$Nb^{+5} + 3e^- \rightleftharpoons Nb^{+3}(2f\,HCl)$	0.344		$Ru^{+4} + e^- \rightleftharpoons Ru^{+3}(2f\,HCl)$	0.858
$Cu^{+2} + 2e^- \rightleftharpoons Cu(Hg)$	0.345		$TiO_2 + 4H^+ + 4e^- \rightleftharpoons Ti + 2H_2O$	0.86
$ClO_3^- + H_2O + 2e^- \rightleftharpoons ClO_2^- + 2OH^-$	0.35		$HO_2^- + H_2O + 2e^- \rightleftharpoons 3OH^-$	0.87
$AgIO_3 + e^- \rightleftharpoons Ag + IO_3$	0.3551		$N_2O_4 + 2e^- \rightleftharpoons 2NO_2^-$	0.88
$Ag_2SeO_3 + 2e^- \rightleftharpoons 2Ag + SeO_3^{-2}$	0.3629		$ClO^- + H_2O + 2e^- \rightleftharpoons Cl^- + 2OH^-$	0.90
$ReO_4^- + 8H^+ + 7e^- \rightleftharpoons Re + 4H_2O$	0.367		$2Hg^{+2} + 2e^- \rightleftharpoons Hg_2^{+2}$	0.905
$(CN)_2 + 2H^+ + 2e^- \rightleftharpoons 2HCN$	0.37		$Pu^{+6} + e^- \rightleftharpoons Pu^{+5}(1f\,HClO_4)$	0.9184
$O_2 + 2H_2O + 4e^- \rightleftharpoons 4OH^-$	0.401		$NO_3^- + 3H^+ + 2e^- \rightleftharpoons HNO_2 + H_2O$	0.94
$AgOCN + e^- \rightleftharpoons Ag + OCN^-$	0.41		$ClO_2(aq) + e^- \rightleftharpoons ClO_2^-$	0.954
$Fe^{+3} + e^- \rightleftharpoons Fe^{+2}(1f\,H_3PO_4)$	0.438		$NO_3^- + 4H^+ + 3e^- \rightleftharpoons NO + 2H_2O$	0.96
$Rh(Cl)_6^{-3} + 3e^- \rightleftharpoons Rh + 6Cl^-$	0.44		$AuBr_2^- + e^- \rightleftharpoons Au + 2Br^-$	0.963
$Ag_2CrO_4 + 2e^- \rightleftharpoons 2Ag + CrO_4^{-2}$	0.4463		$Pu^{+4} + e^- \rightleftharpoons Pu^{+3}(1f\,HClO_4)$	0.982
$H_2SO_3 + 4H^+ + 4e^- \rightleftharpoons S + 3H_2O$	0.45		$Pd^{+2} + 2e^- \rightleftharpoons Pd(4f\,HClO_4)$	0.987
$Fe(CN)_6^{-3} + e^- \rightleftharpoons Fe(CN)_6^{-4}(0.01fNaOH)$	0.46		$HIO + H^+ + 2e^- \rightleftharpoons I^- + H_2O$	0.99
$Ag_2(WO_4) + 2e^- \rightleftharpoons 2Ag + WO_4^{-2}$	0.466		$HNO_2 + H^+ + e^- \rightleftharpoons NO + H_2O$	0.99
$TeO_4^- + 8H^+ + 7e^- \rightleftharpoons Te + 4H_2O$	0.472		$AuCl_4^- + 3e^- \rightleftharpoons Au + 4Cl^-$	0.994
$Ag_2CO_3 + 2e^- \rightleftharpoons 2Ag + CO_3^{-2}$	0.4769		$RuO_{4(c)} + e^- \rightleftharpoons RuO_4^-$	1.00
$AgC_2O_4 + 2e^- \rightleftharpoons Ag + C_2O_4^{-2}$	0.4776		$VO_2^+ + 2H^+ + e^- \rightleftharpoons VO^{+2} + H_2O$	1.00
$Ag_2MoO_4 + 2e^- \rightleftharpoons 2Ag + MoO_4^{-2}$	0.49		$V(OH)_4^+ + 2H^+ + e^- \rightleftharpoons VO^{+2} + 3H_2O$	1.00
$IO^- + H_2O + 2e^- \rightleftharpoons I^- + 2OH^-$	0.49		$H_6TeO_{6(s)} + 2H^+ + 2e^- \rightleftharpoons TeO_{2(s)} + 4H_2O$	1.02
$NiO_2 + 2H_2O + 2e^- \rightleftharpoons Ni(OH)_2 + 2OH^-$	0.49		$IrCl_6^{-2} + e^- \rightleftharpoons IrCl_6^{-3}$	1.02
$Ru^{+4} + e^- \rightleftharpoons Ru^{+3}(0.1fHClO_4)$	0.49		$U^{+5} + e^- \rightleftharpoons U^{+4}(1f\,HCl)$	1.02
$ReO_4^- + 4H^+ + 3e^- \rightleftharpoons ReO_2 + 2H_2O$	0.51		$N_2O_4 + 4H^+ + 4e^- \rightleftharpoons 2NO + 2H_2O$	1.03
$Hg_2(AcO)_2 + 2e^- \rightleftharpoons 2Hg + 2AcO^-$	0.5113		$Pu^{+6} + 2e^- \rightleftharpoons Pu^{+4}(1f\,HCl)$	1.052
$Cu^+ + e^- \rightleftharpoons Cu$	0.522		$Fe(phenanthroline)_3^{+3} + e^- \rightleftharpoons Fe(ph)_3^{+2}(2f\,H_2SO_4)$	1.056
$I_3^- + 2e^- \rightleftharpoons 3I^-$	0.5338		$Br_{2(l)} + 2e^- \rightleftharpoons 2Br^-$	1.065
$I_2 + 2e^- \rightleftharpoons 2I^-$	0.535		$N_2O_4 + 2H^+ + 2e^- \rightleftharpoons 2HNO_2$	1.07
$IO_3^- + 2H_2O + 4e^- \rightleftharpoons IO^- + 4OH^-$	0.56		$IO_3^- + 6H^+ + 6e^- \rightleftharpoons I^- + 3H_2O$	1.085
$MnO_4^- + e^- \rightleftharpoons MnO_4^{-2}$	0.564		$Br_{2(aq)} + 2e^- \rightleftharpoons 2Br^-$	1.087
$MnO_4^{-1} + 2H_2O + 3e^- \rightleftharpoons MnO_2 + 4OH^-$	0.58		$Pu^{+5} + e^- \rightleftharpoons Pu^{+4}(0.5f\,HCl)$	1.099
$H_3AsO_4 + 2H^+ + 2e^- \rightleftharpoons HAsO_2 + 2H_2O(1f\,HCl)$	0.58		$Cr^{+6} + 3e^- \rightleftharpoons Cr^{+3}(2f\,H_2SO_4)$	1.10
$MnO_4^- + 2H_2O + 3e^- \rightleftharpoons MnO_2 + 4OH^-$	0.588		$Cu^{+2} + 2CN^- + e^- \rightleftharpoons Cu(CN)_2^-$	1.12
$AgNO_2 + e^- \rightleftharpoons Ag + NO_2^-$	0.59		$Np^{+6} + e^- \rightleftharpoons Np^{+5}(1f\,HClO_4)$	1.137
$ClO_3^- + H_2O + 2e^- \rightleftharpoons ClO^- + 2OH^-$	0.59		$Fe(phenanthroline)_3^{+3} + e^- \rightleftharpoons Fe(ph)_3^{+2}$	1.14
$RuO_4^- + e^- \rightleftharpoons RuO_4^{-2}$	0.59		$SeO_4^{-2} + 4H^+ + 2e^- \rightleftharpoons H_2SeO_3 + H_2O$	1.15
$TeO_2 + 4H_2 + 4e^- \rightleftharpoons Te + 2H_2O$	0.593		$ClO_2 + e^- \rightleftharpoons ClO_2^-$	1.15
$2AgO + H_2O + 2e^- \rightleftharpoons Ag_2O + 2OH^-$	0.599		$ClO_3^- + 2H^+ + e^- \rightleftharpoons ClO_2 + H_2O$	1.15
$S_2O_6^{-2} + 4H^+ + 2e^- \rightleftharpoons 2H_2SO_3$	0.6		$ClO_4^- + 2H^+ + 2e^- \rightleftharpoons ClO_3^- + H_2O$	1.19
$BrO_3^- + 3H_2O + 6e^- \rightleftharpoons Br^- + 6OH^-$	0.61		$2IO_3^- + 12H^+ + 10e^- \rightleftharpoons I_2 + 6H_2O$	1.19
$Hg_2SO_4 + 2e^- \rightleftharpoons 2Hg + SO_4^{-2}$	0.6158		$HCrO_4^- + 7H^+ + 3e^- \rightleftharpoons Cr^{+3} + 4H_2O$	1.195
$ClO_3^- + 3H_2O + 6e^- \rightleftharpoons Cl^- + 6OH^-$	0.62		$IO_3^- + 6H^+ + 5e^- \rightleftharpoons 1/2I_2 + 3H_2O$	1.195
$UO_2^+ + 4H^+ + e^- \rightleftharpoons U^{+4} + 2H_2O$	0.62		$Pt^{+2} + 2e^- \rightleftharpoons Pt$	~1.2
$Pd^{+2} + 2e^- \rightleftharpoons Pd(1f\,HCl)$	0.623		$MnO_2 + 4H^+ + 2e^- \rightleftharpoons Mn^{+2} + 2H_2O$	1.208
$PdCl_4^{-2} + 2e^- \rightleftharpoons Pd + 4Cl^-$	0.623		$ClO_3^- + 3H^+ + 2e \rightleftharpoons HClO_2 + H_2O$	1.21 (1.23)
$Te^{+4} + 4e^- \rightleftharpoons Te(2.5f\,HCl)$	0.63			
$Hg_2HPO_4 + H^+ + 2e^- \rightleftharpoons 2Hg + H_2PO_4^-$	0.639		$O_2 + 4H^+ + 4e^- \rightleftharpoons 2H_2O$	1.229
$AgAc + e^- \rightleftharpoons Ag + Ac^-$	0.64		$O_3 + H_2O + 2e^- \rightleftharpoons O_2 + 2OH^-$	1.24
$Sb_2O_{5(s)} + 6H^+ + 4e^- \rightleftharpoons 2SbO^+ + 3H_2O$	0.64		$Tl^{+3} + 2e^- \rightleftharpoons Tl^{+1}$	1.247
$Ag_2SO_4 + 2e^- \rightleftharpoons 2Ag + SO_4^{-2}$	0.653		$ClO_2 + H^+ + e^- \rightleftharpoons HClO_2$	1.27
$Fe^{+3} + e^- \rightleftharpoons Fe^{+2}(0.5f\,H_2SO_4)$	0.679		$2HNO_2 + 4H^+ + 4e^- \rightleftharpoons N_2O + 3H_2O$	1.27 (1.29)
$AgBrO_3 + e^- \rightleftharpoons Ag + BrO_3^-$	0.680			
$O_2 + 2H^+ + 2e^- \rightleftharpoons H_2O_2$	0.682		$N_2H_5^+ + 3H^+ + 2e^- \rightleftharpoons 2NH_4^+$	1.27
$Fe(CN)_6^{-3} + e^- \rightleftharpoons Fe(CN)_6^{-4}(1f\,H_2SO_4)$	0.69		$Au^{+3} + 2e^- \rightleftharpoons Au^{+1}$	~1.29
$Sb_2O_5 + 4H^+ + 4e^- \rightleftharpoons Sb_2O_3 + 2H_2O$	0.69		$PdCl_6^{-2} + 2e^- \rightleftharpoons PdCl_4^{-2} + 2Cl^-$	1.29
$C_6H_4O_2 + 2H^+ + 2e^- \rightleftharpoons C_6H_4(OH)_2$	0.6992		$HBrO + H^+ + 2e^- \rightleftharpoons Br^- + H_2O$	1.33
Quinhydrone Elec. H^+, a = 1	0.6995		$Cr_2O_7^{-2} + 14H^+ + 6e^- \rightleftharpoons 2Cr^{+3} + 7H_2O$	1.33
$BrO^- + H_2O + 2e^- \rightleftharpoons Br^- + 2OH^-(1f\,NaOH)$	0.70		$ClO_4^- + 8H^+ + 7e^- \rightleftharpoons 1/2Cl_2 + 4H_2O$	1.34
$H_3IO_6^{-2} + 2e^- \rightleftharpoons IO_3^- + 3OH^-$	~0.70		$Cl_{2(g)} + 2e^- \rightleftharpoons 2Cl^-$	1.3583
			$ClO_4^- + 8H^+ + 8e^- \rightleftharpoons Cl^- + 4H_2O$	1.37

Table 3-8. ELECTROCHEMICAL SERIES (Continued)

Reaction	Potential, volts	Reaction	Potential, volts
$OH + e^- \rightleftharpoons OH^-$	1.4	$HClO_2 + 3H^+ + 3e^- \rightleftharpoons 1/2Cl_2 + 2H_2O$	1.63
$Au^{+3} + 3e^- \rightleftharpoons Au$	1.42	$HClO + H^+ + e^- \rightleftharpoons 1/2Cl_2 + H_2O$	1.63
$2NH_3OH^+ + H^+ + 2e^- \rightleftharpoons N_2H_5^+ + 2H_2O$	1.42	$HClO_2 + 2H^+ + 2e^- \rightleftharpoons HClO + H_2O$	1.64
$Rh^{+4} + e^- \rightleftharpoons Rh^{+3}$	1.43	$MnO_4^- + 4H^+ + 3e^- \rightleftharpoons MnO_2 + 2H_2O$	1.679
$BrO_3^- + 6H^+ + 6e^- \rightleftharpoons Br^- + 3H_2O$	1.44	$Au^+ + e^- \rightleftharpoons Au$	1.68
$Ce^{+4} + e^- \rightleftharpoons Ce^{+3}$	1.4430 (1.61)	$PbO_2 + SO_4^{-2} + 4H^+ + 2e^- \rightleftharpoons PbSO_4 + 2H_2O$	1.685
$Au(OH)_3 + 3H^+ + 3e^- \rightleftharpoons Au + 3H_2O$	1.45	$H_5IO_6 + H^+ + 2e^- \rightleftharpoons IO_3^- + 3H_2O$	~1.7
$ClO_3^- + 6H^+ + 6e^- \rightleftharpoons Cl^- + 3H_2O$	1.45	$CeOH^{+3} + H^+ + e^- \rightleftharpoons Ce^{+3} + H_2O$	1.7134
$HIO + H^+ + e^- \rightleftharpoons 1/2I_2 + H_2O$	1.45	$N_2O + 2H^+ + 2e^- \rightleftharpoons N_2 + H_2O$	1.77
$Ce^{+4} + e^- \rightleftharpoons Ce^{+3}(0.5f\,H_2SO_4)$	1.4587	$H_2O_2 + 2H^+ + 2e^- \rightleftharpoons 2H_2O$	1.776
$PbO_2 + 4H^+ + 2e^- \rightleftharpoons Pb^{+2} + 2H_2O$	1.46	$Co^{+3} + e^- \rightleftharpoons Co^{+2}(3f\,HNO_3)$	1.842
$ClO_3^- + 6H^+ + 5e^- \rightleftharpoons 1/2Cl_2 + 3H_2O$	1.47	$FeO_4^{-2} + 8H^+ + 3e^- \rightleftharpoons Fe^{+3} + 4H_2O$	1.9
$HClO + H^+ + 2e^- \rightleftharpoons Cl^- + H_2O$	1.49	$NiO_2 + 4H^+ + 2e^- \rightleftharpoons Ni^{+2} + 2H_2O$	1.93
$MnO_4^- + 8H^+ + 5e^- \rightleftharpoons Mn^{+2} + 4H_2O$	1.491	$Ag^{+2} + e^- \rightleftharpoons Ag^{+1}(4f\,HClO_4)$	1.987
$HO_2 + H^+ + e^- \rightleftharpoons H_2O_2$	1.5	$S_2O_8^{-2} + 2e^- \rightleftharpoons 2SO_4^{-2}$	2.0 (2.05)
$Mn^{+3} + e^- \rightleftharpoons Mn^{+2}$	1.51	$O_3 + 2H^+ + 2e^- \rightleftharpoons O_2 + H_2O$	2.07
$BrO_3^- + 6H^+ + 5e^- \rightleftharpoons 1/2Br_2 + 3H_2O$	1.52	$F_2O + 2H^+ + 4e^- \rightleftharpoons H_2O + 2F^-$	2.1
$HClO_2 + 3H^+ + 4e^- \rightleftharpoons Cl^- + 2H_2O$	1.56	$O_{(g)} + 2H^+ + 2e^- \rightleftharpoons H_2O$	2.42
$Bi_2O_4 + 4H^+ + 2e^- \rightleftharpoons 2BiO^+ + 2H_2O$	1.59	$H_2N_2O_2 + 2H^+ + 2e^- \rightleftharpoons N_2 + 2H_2O$	2.65
$HBrO + H^+ + e^- \rightleftharpoons 1/2Br_2 + H_2O$	1.59	$1/2F_2 + e^- \rightleftharpoons F^-$	2.85
$2NO + 2H^+ + 2e^- \rightleftharpoons N_2O + H_2O$	1.59	$F_2 + 2e^- \rightleftharpoons 2F^-$	2.87
$2HBrO + 2H^+ + 2e^- \rightleftharpoons Br_{2(l)} + 2H_2O$	1.6	$1/2F_2 + H^+ + e^- \rightleftharpoons HF$	3.03

PART 2

Reaction	Potential, volts	Reaction	Potential, volts
$2H^+ + 2e^- \rightleftharpoons H_2$	0.0000	$In^{+2} + e^- \rightleftharpoons In^{+1}$	-0.40
$D^+ + e^- \rightleftharpoons 1/2D_2$	-0.0034	$In^{+3} + 2e^- \rightleftharpoons In^{+1}$	-0.40
$AgCN + e^- \rightleftharpoons Ag + CN^-$	-0.02	$Mn(OH)_3 + e^- \rightleftharpoons Mn(OH)_2 + OH^-$	-0.40
$TeO_3^{-2} + 3H_2O + 4e^- \rightleftharpoons Te + 6OH^-$	-0.02	$Cd^{+2} + 2e^- \rightleftharpoons Cd$	-0.4026
$2WO_3 + 2H^+ + 2e^- \rightleftharpoons W_2O_5 + H_2O$	-0.03	$Fe^{+2} + 2e^- \rightleftharpoons Fe$	-0.409
$Fe^{+3} + 3e^- \rightleftharpoons Fe$	-0.036	$Cr^{+3} + e^- \rightleftharpoons Cr^{+2}$	-0.41
$Ag_2S + 2H^+ + 2e^- \rightleftharpoons 2Ag + H_2S$	-0.0366	$Eu^{+3} + e^- \rightleftharpoons Eu^{+2}$	-0.43
$P + 3H^+ + 3e^- \rightleftharpoons PH_3(g)$	-0.04	$CdSO_4 \cdot 8/3H_2O + 2e^- \rightleftharpoons Cd(Hg) + CdSO_4 \text{ (sat'd aq)}$	-0.4346
$W_2O_5 + 2H^+ + 2e^- \rightleftharpoons 2WO_2 + H_2O$	-0.04	$Tl_2SO_4 + 2e^- \rightleftharpoons Tl(Hg) + SO_4^{-2}$	-0.4360
$Hg_2I_2 + 2e^- \rightleftharpoons 2Hg + 2I^-$	-0.0405	$Bi_2O_3 + 3H_2O + 6e^- \rightleftharpoons 2Bi + 6OH^-$	-0.46
$2D^+ + 2e^- \rightleftharpoons D_2$	-0.044	$BiOOH + H_2O + 3e^- \rightleftharpoons Bi + 3OH^-$	-0.46
$Tl(OH)_3 + 2e^- \rightleftharpoons TlOH + 2OH^-$	-0.05	$NO_2^- + H_2O + e^- \rightleftharpoons NO + 2OH^-$	-0.46
$O_2 + H_2O + 2e^- \rightleftharpoons HO_2^- + OH^-$	-0.076	$S + H_2O + 2e^- \rightleftharpoons HS^- + OH^-$	-0.478
$AsO_4^{-3} + 2H_2O + 2e^- \rightleftharpoons AsO_2^- + 4OH^-(1f\,NaOH)$	0.08	$H_4P_2O_6 + 3H^+ + 3e^- \rightleftharpoons P + 3H_2O$	-0.49
$2H_2SO_3 + H^+ + 2e^- \rightleftharpoons HS_2O_4^- + 2H_2O$	-0.08	$In^{+3} + e^- \rightleftharpoons In^{+2}$	-0.49
$Ru^{+3} + e^- \rightleftharpoons Ru^{+2}(1-6f\,HCL)$	-0.084	$2CO_2 + 2H^+ + 2e^- \rightleftharpoons H_2C_2O_4$	-0.49
$2Cu(OH)_2 + 2e^- \rightleftharpoons Cu_2O + 2OH^- + H_2O$	-0.09	$H_3PO_3 + 2H^+ + 2e^- \rightleftharpoons H_3PO_2 + H_2O$	-0.50 (-0.59)
$WO_3 + 6H^+ + 6e^- \rightleftharpoons W + 3H_2O$	-0.09		
$Ru^{+3} + e^- \rightleftharpoons Ru^{+2}(0.1f\,HClO_4)$	-0.11	$S + 2e^- \rightleftharpoons S^{-2}$	-0.508
$Cr^{+6} + 3e^- \rightleftharpoons Cr^{+3}(1f\,NaOH)$	-0.12	$H_3PO_3 + H^+ + e^- \rightleftharpoons P + 2H_2O$	-0.51
$CrO_4^{-2} + 4H_2O + 3e^- \rightleftharpoons Cr(OH)_3 + 5OH^-$	-0.12	$Sb + 3H^+ + 3e^- \rightleftharpoons H_3Sb$	-0.51
$GeO_2 + 2H^+ + 2e^- \rightleftharpoons GeO + H_2O$	-0.12	$As + 3H^+ + 3e^- \rightleftharpoons AsH_3$	-0.54
$WO_2 + 4H^+ + 4e^- \rightleftharpoons W + 2H_2O$	-0.12	$HPbO_2^- + H_2O + 2e^- \rightleftharpoons Pb + 3OH^-$	-0.54
$Pb^{+2} + 2e^- \rightleftharpoons Pb(Hg)$	-0.1205	$TlCl + e^- \rightleftharpoons Tl(Hg) + Cl^-$	-0.555
$Pb^{+2} + 2e^- \rightleftharpoons Pb$	-0.1263 (0.126)	$Cr^{+2} + 2e^- \rightleftharpoons Cr$	-0.557
$H_2GeO_3 + 4H^+ + 4e^- \rightleftharpoons Ge + 3H_2O$	-0.13	$Ga^{+3} + 3e^- \rightleftharpoons Ga$	-0.560
$Sn^{+2} + 2e^- \rightleftharpoons Sn$	-0.1364	$Fe(OH)_3 + e^- \rightleftharpoons Fe(OH)_2 + OH^-$	-0.56
$O_2 + 2H_2O + 2e^- \rightleftharpoons H_2O_2 + 2OH^-$	-0.146	$PbO + H_2O + 2e^- \rightleftharpoons Pb + 2OH^-$	-0.576
$AgI + e^- \rightleftharpoons Ag + I^-$	-0.1519	$2SO_3^{-2} + 3H_2O + 4e^- \rightleftharpoons S_2O_3^{-2} + 6OH^-$	-0.58
$2NO_2^- + 2H_2O + 4e^- \rightleftharpoons N_2O_2^{-2} + 4OH^-$	-0.18	$SbO_3^- + H_2O + 2e^- \rightleftharpoons SbO_2^- + 2OH^-$	-0.59
$CO_2 + 2H^+ + 2e^- \rightleftharpoons HCOOH$	-0.2	$TlBr + e^- \rightleftharpoons Tl(Hg) + Br^-$	-0.606
$2SO_4^{-2} + 4H^+ + 2e^- \rightleftharpoons S_2O_6^{-2} + 2H_2O$	-0.2	$U^{+4} + e^- \rightleftharpoons U^{+3}$	-0.61
$Cu(OH)_2 + 2e^- \rightleftharpoons Cu + 2OH^-$	-0.224	$Cb_2O_5 + 10H^+ + 10e^- \rightleftharpoons 2Cb + 5H_2O$	-0.62
$Ni^{+2} + 2e^- \rightleftharpoons Ni$	-0.23	$U^{+4} + e^- \rightleftharpoons U^{+3}(1\,fHClO_4)$	-0.631
$PbHPO_4 + H^+ + 2e^- \rightleftharpoons Pb(Hg) + HPO_4^-$	-0.2448	$Ni(OH)_2 + 2e^- \rightleftharpoons Ni + 2OH^-$	-0.66
$V(OH)_4^+ + 4H^+ + 5e^- \rightleftharpoons V + 4H_2O$	-0.25	$SbO_2^- + 2H_2O + 3e^- \rightleftharpoons Sb + 4OH^-$	-0.66
$V^{+3} + e^- \rightleftharpoons V^{+2}$	-0.255	$AsO_2^- + 2H_2O + 3e^- \rightleftharpoons As + 4OH^-$	-0.68
$PbCl_2 + 2e^- \rightleftharpoons Pb(Hg) + 2Cl^-$	-0.262	$*Te + 2H^+ + 2e^- \rightleftharpoons H_2Te(Ag)$	-0.69 (-0.72)
$PbBr_2 + 2e^- \rightleftharpoons Pb(Hg) + Br^-$	-0.275		
$H_3PO_4 + 2H^+ + 2e^- \rightleftharpoons H_3PO_3 + H_2O$	-0.276 (-0.2)	$Ag_2S + 2e^- \rightleftharpoons 2Ag + S^{-2}$	-0.7051
$Co^{+2} + 2e^- \rightleftharpoons Co$	-0.28	$Ta_2O_5 + 10H^+ + 10e^- \rightleftharpoons 2Ta + 5H_2O$	-0.71
$Tl^+ + e^- \rightleftharpoons Tl(Hg)$	-0.3338	$AsO_4^{-3} + 2H_2O + 2e^- \rightleftharpoons AsO_2^- + 4OH^-$	-0.71
$Tl^+ + e^- \rightleftharpoons Tl$	-0.3363	$Co(OH)_2 + 2e^- \rightleftharpoons Co + 2OH^-$	-0.73
$In^{+3} + 3e^- \rightleftharpoons In$	-0.338	$H_3BO_3 + 3H^+ + 3e^- \rightleftharpoons B + 3H_2O$	-0.73
$PbF_2 + 2e^- \rightleftharpoons Pb(Hg) + 2F^-$	-0.3444	$Cr^{+3} + 3e^- \rightleftharpoons Cr$	-0.74
$TlOH + e^- \rightleftharpoons Tl + OH^-$	-0.3445	$V^{+5} + e^- \rightleftharpoons V^{+4}(1f\,NaOH)$	-0.74
$SeO_3^{-2} + 3H_2O + 4e^- \rightleftharpoons Se + 6OH^-$	-0.35	$Cd(OH)_2 + 2e^- \rightleftharpoons Cd(Hg) + 2OH^-$	-0.761 (-0.81)
$PbSO_4 + 2e^- \rightleftharpoons Pb(Hg) + SO_4^{-2}$	-0.3505		
$Cd^{+2} + 2e^- \rightleftharpoons Cd(Hg)$	-0.3521	$Zn^{+2} + 2e^- \rightleftharpoons Zn$	-0.7628
$PbSO_4 + 2e^- \rightleftharpoons Pb + SO_4^{-2}$	-0.356	$Zn^{+2} + 2e^- \rightleftharpoons Zn(Hg)$	-0.7628
$PbI_2 + 2e^- \rightleftharpoons Pb(Hg) + 2I^-$	-0.358	$TlI + e^- \rightleftharpoons Tl(Hg) + I^-$	-0.769
$Se + 2H^+ + 2e^- \rightleftharpoons H_2Se(aq)$	-0.36	$Se + 2e^- \rightleftharpoons Se^{-2}$	-0.78
$Cu_2O + H_2O + 2e^- \rightleftharpoons 2Cu + 2OH^-$	-0.361	$HSnO_2^- + H_2O + 2e^- \rightleftharpoons Sn + 3OH^-$	-0.79
$Tl^{+3} + e^- \rightleftharpoons Tl^{+2}$	-0.37	$ZnSO_4 \cdot 7H_2O + 2e^- \rightleftharpoons Zn(Hg) + SO_4^{-2} \text{ (Sat'd ZnSO}_4)$	-0.7993
		$RuO_2 + 4H^+ + 4e^- \rightleftharpoons Ru + 2H_2O$	-0.8
		$ReO_4^- + 4H_2O + 7e^- \rightleftharpoons Re + 8OH^-$	-0.81

Table 3-8. ELECTROCHEMICAL SERIES (Continued)

Reaction	Potential, volts	Reaction	Potential, volts
$UO_2^{+2} + 4H^+ + 6e^- \rightarrow U + 2H_2O$	-0.82	$N_2 + 2H_2O + 4H^+ + 2e^- \rightarrow 2NH_3OH^+$	-1.87
$2H_2O + 2e^- \rightarrow H_2 + 2OH^-$	-0.8277	$Th^{+4} + 4e^- \rightarrow Th$	-1.90
$SiO_2 + 4H^+ + 4e^- \rightarrow Si + 2H_2O$	-0.84	$Np^{+3} + 3e^- \rightarrow Np$	-1.9
$2NO_3^- + 2H_2O + 2e^- \rightarrow N_2O_4 + 4OH^-$	-0.85	$Ti^{+3} + e^- \rightarrow Ti^{+2}$	-2.0
$TiO_2 + 4H^+ + 4e^- \rightarrow Ti + 2H_2O$	-0.86	$Sc^{+3} + 3e^- \rightarrow Sc$	-2.08
$P + 3H_2O + 3e^- \rightarrow PH_3(g) + 3OH^-$	-0.87	$1/2H_2 + e^- \rightarrow H^-$	-2.23
$SO_4^{-2} + H_2O + 2e^- \rightarrow SO_3^{-2} + 2OH^-$	-0.92	$Nd^{+3} + 3e^- \rightarrow Nd$	-2.246
$Te + 2e^- \rightarrow Te^{-2}$	-0.92	$Be_2O_3^{-2} + 3H_2O + 4e^- \rightarrow 2Be + 6OH^-$	-2.28
$Sn(OH)_6^{-2} + 2e^- \rightarrow HSnO_2^- + 3OH^- + H_2O$	-0.96	$ZrO(OH)_2 + H_2O + 4e^- \rightarrow Zr + 4OH^-$	-2.32
$Mn^{+2} + 2e^- \rightarrow Mn$	-1.029	$Ce^{+3} + 3e^- \rightarrow Ce$	-2.335
$PO_4^{-3} + 2H_2O + 2e^- \rightarrow HPO_3^{-2} + 3OH^-$	-1.05	$H_2AlO_3^- + H_2O + 3e^- \rightarrow Al + 4OH^-$	-2.35
$2SO_3^{-2} + 2H_2O + 2e^- \rightarrow S_2O_4^{-2} + 4OH^-$	-1.12	$Y^{+3} + 3e^- \rightarrow Y$	-2.37
$CrO_2^- + 2H_2O + 3e^- \rightarrow Cr + 4OH^-$	-1.2	$La^{+3} + 3e^- \rightarrow La$	-2.37
$SiF_6^{-2} + 4e^- \rightarrow Si + 6F^-$	-1.2	$Mg^{++} + 2e^- \rightarrow Mg$	-2.375
$V^{+2} + 2e^- \rightarrow V$	-1.2	$H_2BO_3^- + H_2O + 3e^- \rightarrow B + 4OH^-$	-2.5
$ZnO_2^- + 2H_2O + 2e^- \rightarrow Zn + 4OH^-$	-1.216	$HfO(OH)_2 + H_2O + 4e^- \rightarrow Hf + 4OH^-$	-2.60
$H_2GaO_3^- + H_2O + 3e^- \rightarrow Ga + 4OH^-$	-1.22	$ThO_2 + 2H_2O + 4e^- \rightarrow Th + 4OH^-$	-2.64
$H_2BO_3^- + 5H_2O + 8e^- \rightarrow BH_4^- + 8OH^-$	-1.24	$Mg(OH)_2 + 2e^- \rightarrow Mg + 2OH^-$	-2.67
$Cr(OH)_3 + 3e^- \rightarrow Cr + 3OH^-$	-1.3	$Na^+ + e^- \rightarrow Na$	-2.7109 (-2.712)
$ZrO_2 + 4H^+ + 4e^- \rightarrow Zr + 2H_2O$	-1.43		
$Ce^{+3} + 3e^- \rightarrow Ce(Hg)$	-1.4373	$Ca^{+2} + 2e^- \rightarrow Ca$	-2.76
$Mn(OH)_2 \rightarrow Mn + 2OH^-$	-1.47	$La(OH)_3 + 3e^- \rightarrow La + 3OH^-$	-2.76
$Ba^{+2} + 2e^- \rightarrow Ba(Hg)$	-1.570	$Sr^{+2} + 2e^- \rightarrow Sr$	-2.89
$HfO_2 + 4H^+ + 4e^- \rightarrow Hf + 2H_2O$	-1.57	$Ba^{+2} + 2e^- \rightarrow Ba$	-2.90
$Ti^{+2} + 2e^- \rightarrow Ti$	-1.63	$Cs^+ + e^- \rightarrow Cs$	-2.923
$HPO_3^{-2} + 2H_2O + 2e^- \rightarrow H_2PO_2^- + 3OH^-$	-1.65	$K^+ + e^- \rightarrow K$	-2.924 (-2.923)
$HfO^{+2} + 2H^+ + 4e^- \rightarrow Hf + H_2O$	-1.68	$Rb^+ + e^- \rightarrow Rb$	-2.925 (-2.99)
$Be^{+2} + 2e^- \rightarrow Be$	-1.70 (-1.85)		
$Al^{+3} + 3e^- \rightarrow Al(0.1f\ NaOH)$	-1.706	$Ba(OH)_2 \cdot 8H_2O + 2e^- \rightarrow Ba + 2OH^- + 8H_2O$	-2.97
$HPO_3^{-2} + 2H_2O + 3e^- \rightarrow P + 5OH^-$	-1.71	$Sr(OH)_2 \cdot 8H_2O + 2e^- \rightarrow Sr + 2OH^- + 8H_2O$	-2.99
$SiO_3^{-2} + 3H_2O + 4e^- \rightarrow Si + 6OH^-$	-1.73	$Ca(OH)_2 + 2e^- \rightarrow Ca + 2OH^-$	-3.02
$Sr^{+2} + 2e^- \rightarrow Sr(Hg)$	-1.793	$Ca^+ + e^- \rightarrow Ca$	-3.02
$ThO_2 + 4H^+ + 4e^- \rightarrow Th + 2H_2O$	-1.80	$Li^+ + e^- \rightarrow Li$	-3.045 (-3.02)
$U^{+3} + 3e^- \rightarrow U$	-1.8		
$H_2PO_2^- + e^- \rightarrow P + 2OH^-$	-1.82	$3N_2 + 2H^+ + 2e^- \rightarrow 2HN_3$	-3.1

*From: "Handbook of Chemistry and Physics", 53rd ed., R.C. Weast, Ed., The Chemical Rubber Co., 1972.

Table 3-9. ELECTROMOTIVE OR GALVANIC SERIES OF METALS AND ELECTROCHEMICAL POTENTIALS

Corrosion: In general, when dissimilar metals are exposed in a conducting solution, such as seawater, the more anodic metal will corrode, especially if the anodic area is relatively small. Anode areas nearer the cathode corrode more rapidly.

Metal	Potential, volts	Metal	Potential, volts
	Anodic or corroded end		**Anodic or corroded end**
Lithium	-3.04	Cadmium	-0.4
Rubidium	-2.93	Titanium	-0.33
Potassium	-2.92	Cobalt	-0.28
Barium	-2.90	Nickel	-0.23
Strontium	-2.89	Tin	-0.14
Calcium	-2.8	Lead	-0.126
Sodium	-2.71	HYDROGEN	0.00
Magnesium	-2.37	Copper	$+0.52$
Beryllium	-1.7	Silver	$+0.80$
Aluminum	-1.7	Mercury	$+0.85$
Manganese	-1.04	Palladium	$+1.0$
Zinc	-0.76	Platinum	$+1.2$
Chromium	-0.6	Gold	$+1.5$
Iron	-0.4		**Cathodic or noble-metal end**

Table 3-10.   GALVANIC EFFECTS OF METALS*

The first two columns below give the key to the metals represented by the numbers in the remaining columns. The corrosion of the metals in the second column may be accelerated to the degree indicated in columns A, B, C, and D when in contact with metals indicated in the latter columns. Such acceleration is likely to occur when both (a) the two metals are in electronic contact, and (b) there is a conducting electrolyte path between the two metals. The acceleration becomes more marked as the combined resistance of these two parts of the circuit decreases and may be entirely prevented by completely breaking either of those parts of the circuit: avoidance of the collection of water at the vital point prevents (b). Other methods of preventing or reducing the acceleration are as follows:

1. Dissimilarity may be destroyed by coating one metal with the other or one compatible with it.
2. Non-metallic coatings, such as paint films, oxide films and plastic coatings, may be applied to one or both metals. They are not usually completely impervious, and therefore they reduce rather than prevent the effect.
3. Inhibitors may be added to the electrolyte or may be incorporated in jointing compounds to reduce the effect.
4. Cathodic protection may reduce or prevent the effect.
5. The intensity of acceleration is reduced by decreasing the area in contact with the electrolyte of the metal in columns A, B, C, and D compared with that of the metal being affected.

Number	Metal	Acceleration of corrosion†			
		A—None	B—Slight	C—Marked	D—Severe
1	Au, Pt, Rh, Ag	1–19	—	—	—
2**	Ti	1–19	—	—	—
3**	Cr 2	1–19	—	—	—
4**	Fe 15	1–19	—	—	—
5	Ni 5	4–19	1, 2, 3	—	—
6**	Fe 14	3, 4, 9, 11–19 (5, 7, 8, 10)	—	1	—
7	Cu 10	7–19	5, 6, (3, 4)	1, 2	—
8	Cu 4, Cu 7	8–19	5, 6 (3, 4, 7)	1, 2	—
9	Ni 2	7–19	5 (3, 4, 6)	1, 2	—
10	Cu 2, Cu 8, Cu 12	11, 13–19	(3–9, 12)	1, 2	—
11**	Fe 13	11–19	(9)	1–8, 10	—
12	Sn 2, Pb 2, Pb 6	15–17, 19 (13, 14, 18)	9 (3–8, 10, 11)	1, 2	—
13	Fe 9, Fe 5	15 17, 19 (18)	14	1–12	—
14**	Al 4	17, 18 (19)	(3, 4, 6, 11)	2, 5, 9, 12, 13	1, 7, 8, 10
15**	Al 1, Al 2, Al 3	17, 18 (19)	(3, 4, 6, 11–13)	2, 5, 9	1, 7, 8, 10
16**	Al 11	17, 18 (19)	(4, 6, 11)	3, 5, 9, 12, 13	1, 2, 7, 8, 10
17	Cd 2	16–19	12, 15 (14)	1–11, 13	—
18	Zn 1, Zn 2	19 (16)	12, 17	1–11, 13–15	—
19	Mg 1, Mg 2	(15–18)	2–4, 6, 9–14	1, 5, 7, 8	—

†Numbers in parentheses denote that acceleration may be more marked in some circumstances than is indicated.
**The behavior of these metals is dependent on the maintenance of the natural oxide films on them.

*From: "Metals Reference Book", C.J. Smithells, Vol. 3, Butterworth & Co., Ltd. (London), 1967; Plenum Press (New York).

Table 3-11. CORROSION OF METALS

Metal	Subject to corrosion by	Resistant to
Aluminum and alloys	Acid solutions (except concentrated nitric and acetic); caustic and mild alkalies; sea water; saturated halogen vapors; mercury and its compounds; carbon tetrachloride; cobalt; copper and nickel compounds in solution	Air; water; ammonia; combustion products; halide refrigerants; dry steam; sulfur and its compounds; concentrated ammonium hydroxide; organic acids; most organic compounds
Cast iron	All water solutions, moist gases; dilute acids; acid-salt solutions	Dry gases except halogens; dry air; neutral water; dry soil; concentrated acids (nitric, sulfuric, phosphoric); weak or strong alkalies; organic acids
Chromium and high-chrome steels	Most strong acids (limited use with acetic, nitric, sulfuric, and phosphoric); most chlorides	Air; water; steam; weak acids; most inorganic salts; most alkalies; ammonia
Copper, red brass, and bronze	Mercury and its salts; aqueous ammonia; saturated halogen vapors; sulfur and sulfides; oxidizing acids (nitric, concentrated sulfuric, sulfurous); oxidizing salts (Hg, Ag, Cr, Fe, Cu); cyanides	Air; water; sea water; steam; sulfate and carbonate solutions; dry halogens; moist soils; alkaline solutions; refrigerants; petro-chemicals; non-oxidizing acids (acetic, hydrochloric, sulfuric)
Lead	Caustic solutions; halogens; acetic acid; calcium hydroxide; magnesium chloride; ferric chloride; sodium hypochlorite	Air; water; moist soil; ammonia; alcohols; sulfuric acid; ferrous chloride
Magnesium and alloys	Heavy metal; salts; all mineral acids (except hydrofluoric and chromic); sea water; fruit juices; milk	Most alkalies and organic compounds; air; water; soil; dry refrigerants; dry halogens
Monel metal and Ni-Cu alloys containing in excess of 50% nickel	Inorganic acids; sulfur; chlorine; acid solutions of ferric, stannic, or mercuric salts	Air; sea water; steam; food acids; neutral and alkaline salt solutions; dry gases; most alkalies; ammonia
Nickel and high-nickel steels	Inorganic acids; ammonia; mercury; oxidizing salts (Fe, Cu, Hg)	Air; water; steam; caustic and mild alkalies; organic acids; neutral and alkaline organic compounds; dry gases
Silver	Halogens and halogen acids; sulfur compounds; ammonia	Alkalies, including high-temperature caustic alkalies; hot concentrated organic acids; phosphoric and hydrofluoric acids
Steel, mild, low-alloy steels	Most acids; strong alkalies; salt water; sulfur and its compounds	Air; steam; ammonia; most alkalies; concentrated nitric acid; halide refrigerants
Tantalum	Hydrofluoric acid; concentrated sulfuric acid; strong alkalies	Nearly all salts; most acids; water; sea water; air; alcohols; hydrocarbons; sulfur
Tin	Inorganic acids; caustic solutions; halides	Most food acids; ammonia; neutral solutions
Titanium	Hydrochloric acid; sulfuric acid; hydrofluoric, oxalic, and formic acids; dangerously explosive in presence of nitric acid or liquid oxygen	Oxidizing media; air; water; sea water; aqueous chloride solutions; moist chlorine gas; sodium hypochlorite
Zinc	Acid or strong alkali solutions; sulfur dioxide; chlorides	Air; water; ammonia; dry common gases; refrigerants; gasoline
Zirconium	Concentrated sulfuric acid (hot); hydrofluoric acid; cupric and ferric chlorides	Solutions of alkalies and acids; aqua regia

Notes: Polished surfaces resist corrosion.
　　Non-uniformity within a metal tends to increase corrosion.
　　Stress (especially alternating stress) tends to increase corrosion.
　　Dissolved gases in water (especially oxygen) accelerate corrosion.

REFERENCES

"The Corrosion Handbook", H.H. Uhlig, sponsored by The Electrochemical Society, John Wiley & Sons, 1948.

"Metals Reference Book", C.J. Smithells, Vol. 2, Butterworth & Co., 1967.

"Underground Corrosion", M. Romanoff, National Bureau of Standards Circular 579, 1957.

"Corrosion Resistance of Metals and Alloys", F.L. LaQue and H.R. Copson, Eds., Reinhold Publishing Corporation, 1963.

"Handbook of Corrosion Resistant Piping", P.A. Schweitzer, Industrial Press, 1969.

Table 3-12. CORROSION-RATE RANGES EXPRESSED IN MILS PENETRATION PER YEAR*

1 Mil = 0.001 in.

| Metal | Acid solutions | | | Alkaline solutions | Neutral solutions | | Air |
| | Non-oxidizing | | Oxidizing | Sodium hydroxide, 5% | Fresh water | Sea water | Normal outdoor urban exposure |
	Sulfuric, 5%	Acetic, 5%	Nitric, 5%				
Aluminum	8–100	0.5–5	15–80	13,000	0.1	1–50	0–0.5
Zinc	High	600–800	High	15–200	0.5–10	0.5–10**	0–0.5
Tin	2–500**	2–500**	100–400	5–20	0–0.5	0.1	0–0.2
Lead	0–2	10–150**	100–6,000	5–500**	0.1–2	0.2–15	0–0.2
Iron	15–400**	10–400	1,000–10,000	0–0.2	0.1–10**	0.1–10**	1–8
Silicon iron	0–5	0–0.2	0–20	0–10	0–0.2	0–3	0–0.2
Stainless steel	0–100†	0–0.5	0–2	0–0.2	0–0.2	0–200†	0–0.2
Copper alloys	2–50**	2–15**	150–1500	2–5	0–1	0.2–15**	0–0.2
Nickel alloys	2–35**	2–10**	0.1–1500	0–0.2	0–0.2	0–1	0 0.2
Titanium	10–100	<0.1	0.1–1	<0.2	<0.1	<0.1	<0.1
Molybdenum	0–0.2	<0.1	High	<0.1	<0.1	<0.1	<0.1
Zirconium	<0.5	<0.1	<0.1	<0.1	<0.1	<0.1	<0.1
Tantalum	<0.1	<0.1	<0.1	<1	<0.1	<0.1	<0.1
Silver	0–1	<0.1	High	<0.1	<0.1	<0.1	<0.1
Platinum	<0.1	<0.1	<0.1	<0.1	<0.1	<0.1	<0.1
Gold	<0.1	<0.1	<0.1	<0.1	<0.1	<0.1	<0.1

Note: The corrosion-rate ranges for the solutions are based on temperatures up to 212°F.

†Aeration leads to passivity, scarcity of dissolved air to activity.

**Aeration leads to the higher rates in the range.

*From: "Corrosion Resistance of Metals and Alloys", F.L. LaQue and H.R. Copson, Eds., Reinhold Publishing Corporation, 1963.

Table 3-13. CORROSION RESISTANCE OF COPPER ALLOYS

Table A lists specific materials for which copper and its alloys are suitable; Table B lists materials for which these metals are not suitable in any use where direct contact is involved. These lists apply to pure copper and to the alloys of copper with zinc, tin, aluminum, and nickel. Included are such classes as the bronzes, red and yellow brass, Muntz metal, aluminum brass, copper nickel, and nickel silver.

Most common chemicals and fluids that are not included in the following lists should not be used with copper alloys without a careful check of suitability. A helpful guide is the list of 22 alloys and almost 200 materials in the bulletin "Copper and Copper Alloys for the Process Industries", published by the Copper Development Association, New York, N.Y.

Table A. MATERIALS SUITABLE FOR USE WITH COPPER ALLOYS

Acetone	Coffee	Methyl alcohol
Alcohols	Ethers	Methyl chloride, dry
Alumina	Ethyl alcohol	Nitrogen
Aluminum hydroxide	Fluorine refrigerants	Oxygen
Amyl alcohol	Gasoline	Paraffin
Asphalt	Gelatine	Potassium chromate
Barium carbonate	Glucose	Potassium sulfate
Barium sulfate	Glycerine	Propane
Benzine	Hydrocarbons, pure	Rosin
Benzol	Hydrogen	Sodium chromate
Borax	Hydrogen sulfide, dry	Sulfur chloride, dry
Butane	Kerosene	Sulfur dioxide, dry
Butyl alcohol	Lacquers	Sulfur trioxide, dry
Carbon dioxide, dry	Lacquer solvents	Toluene
Castor oil	Lime	Trichloroethylene, dry
Chloroform, dry	Magnesium hydroxide	Varnish

Table B. MATERIALS UNSUITABLE FOR USE WITH COPPER ALLOYS

Ammonia, moist†	Ferric sulfate	Potassium dichromate, acid
Ammonium chloride	Hydrocyanic acid	Silver salts
Ammonium hydroxide	Mercury	Sodium cyanide
Ammonium nitrate	Mercury salts	Sodium dichromate, acid
Chromic acid	Nitric acid	Sulfur, molten
Ferric chloride	Potassium cyanide	

Table 3-14.   CLASSIFICATION OF PAINTS AND COATINGS

For corrosion-resistant coatings see Table 3-20.

Materials and practices in the field of paints and coatings are diverse and changing rapidly with the shift from natural to synthetic materials. This table and those following quote data on well known practices and materials only, using industrial terminology. For more comprehensive data consult the references.

Classifications of coating materials are related to their uses, as indicated in the following table. Typical organic coatings are prepared with pigments and drying oils, with resins and solvents, or with latex or resin emulsions. The pigments are mixed for color, and dyes are sometimes added. Minor components may accelerate or retard drying. Many resin varnishes and lacquers are nearly transparent, but dyes, pigments, and metallic powders may be added for color effects. Plasticizers are used to improve the elastic qualities of the film.

The inexpensive coatings used in large quantities are simple formulations, but special uses may demand highly complex mixtures. Among the recognized and desirable properties are hiding power or opaqueness, light, heat, and chemical resistance, bleeding resistance, and ease of application. Other identified properties are penetration, chalking, gloss or texture, hardness, mold resistance, stain resistance, and package stability.

Class	Description or source	Forms and properties	Typical uses
Oil paint	Vegetable oil and mineral pigment	Linseed (or other) drying oil with opaque but colored (or white) pigment	Exterior and general painting; interior trim; industrial products
Latex paint	Partially polymerized liquid resins in water	Water paint for porous and absorbent materials; alkali resistant	Wood and masonry surfaces; wall board; sheathings and sidings; interiors
Varnish	Drying oil, resin, solvent, and drier	Usually transparent to amber in color; glossy and impervious	Furniture, woodwork, floors, boats, trim, sealers
Lacquer	Nitrocellulose, acrylic, or other resins in volatile solvent	Quick-drying by solvent evaporation	Automobiles; transport equipment; metal products; oven-dried finishes
Whitewash	Lime, water, and additives	Age-old water paint for decorative and reflective purposes	Inexpensive outdoor or barn paint
Cement-water	White cement, lime, and pigment	Water paint for smooth finish on coarse masonry	Concrete or tile finish or grout-coat
Glue or size	Gelatin, skin, bone, starch	Water solutions	Sealer for plaster, etc.
Bituminous	Petroleum; coal tar	Paints; hot mastic; emulsions	Underground; waterproofing; roofs; tanks
Fire retardant	Brominated resins; insulation; glass	Flame resistant and/or insulating or glazing	Prevention of flame spread over combustible surfaces or textiles
Chlorinated rubber	65% chlorine with poly-isoprene and plasticizer	Paints; chemical and water resistant	Traffic, swimming pool, and masonry paints; quick-drying industrial coatings
Strippable coating	Cellulose esters, or other resins, solvent and oil	Spray or hot-dip (oil migrates to interface)	Protection of machine parts and assemblies; "mothballing"
Fluorescent	Added fluorescent dyes	Absorbed violet reemitted as yellow, orange	High visibility for signs, hazard protection, advertising
Wax polish	Hard wax plus resin	Organic solvent, water emulsion, or both	Floors, furniture, automobiles, paneling

REFERENCES

"Organic Coatings—Properties, Selection, and Use", A.G. Roberts, National Bureau of Standards, Building Science Series 7, 1968.
"Outlines of Paint Technology", W.M. Morgans, Charles Griffin & Co., Ltd., 1969 (The Chemical Rubber Co.).
"Technology of Paints, Varnishes and Lacquers", C.R. Martens, Reinhold Publishing Corp., 1968.
"Treatise on Coatings" (12 vols.), R.R. Myers and J.S. Long, Eds., Vol. 1, Marcel Dekker, 1967.
"Principles of Surface Coating Technology", D.H. Parker, John Wiley & Sons, 1965.
"Organic Coating Technology", H.F. Payne, John Wiley & Sons, 1954, 2 vols.
"Organic Protective Coatings", W. von Fischer and E.G. Bobalek, Eds., Reinhold Publishing Corp., 1953.
"Asphalts and Allied Substances", H. Abraham, D. Van Nostrand Co., 1945, 2 vols.
"Raw Materials Index", National Paint, Varnish and Lacquer Association, Washington, D.C.
"Practical Emulsions", H. Bennett, J.L. Bishop, and M.F. Wulfinghoff, Vol. 2, Chemical Publishing Co., 1968; formulation of emulsion paints and coatings.

Table 3-15. ORGANIC COATINGS—DRYING OILS AND DRIERS

Drying oils contain glycerides or esters from a combination of glycerol with unsaturated fatty acids. The drying of a thin oil film is a complex chemical process. Oxygen plays a major part, and the process is retarded at low temperatures and in the absence of light. It is accelerated by the catalytic action compounds of certain metals (e.g., lead).

TYPICAL PROPERTIES OF DRYING OILS

Oil	Specific gravity	Viscosity, cp	Refractive index	Saponification value[a]	Acid value[b]	Iodine value[c]
Linseed (flax)						
Raw	0.93	40	1.48	192	2	177
Boiled	0.95	100	1.48		3	165
Blown	0.97	300	1.48		5	120
Tung	0.94		1.52	193	7	162
Fish	0.93		1.48	191	6	165
Tobacco seed	0.92		1.48	190	5	140
Soybean	0.93		1.47	190	2	135
Tall oil				169	160	

Note: Raw linseed oil dries slowly. Boiled oil, which has been heated with lead, cobalt, or manganese compounds to reduce drying time, is preferable for interior paints. Blown oil has been thickened by blowing air through hot oil.

Tung-nut oil is fast-drying and has high-water resistance.

Fish oil dries slowly and has a strong odor and darker color unless refined. It is heat-resistant and elastic and is used in high-temperature and roofing paints.

Soybean and other seed oils are semidrying but are used in oil blends to give elasticity and durability to the paint film.

Tall oil is a semidrying oil, a by-product of the sulfite papermaking process. It is expensive and is used with various synthetic resins.

[a] Saponification value is the quantity of potassium hydroxide required to saponify one gram of oil.
[b] Acid value is a quantitative index of the amount of free organic acid in an oil.
[c] Iodine value is a measure of the degree of unsaturation of the oil and, hence, of its drying properties.

Table 3-16. ORGANIC COATINGS—SOLVENTS AND THINNERS*

Typical Designations and Properties

Name	Alternative name	Specific gravity	Boiling range, deg C	Flash point, deg C (closed)	ASTM No.
PETROLEUM PRODUCTS					
Mineral spirits	Petroleum ether; ligroin	0.66	60–80	5	D484
Naphtha thinner	White spirits; turpentine substitute		150–190	40	D235
Aromatic naphtha		0.85			
COAL-TAR PRODUCTS					
Toluene	Toluol	0.87	110	5	D362
Xylene (3 isomers)		0.86	137–148	27	D364
Coal-tar naphtha	Heavy naphtha		<190	37	
CONIFEROUS TREE PRODUCTS					
Turpentine (gum, wood)		0.85	150–180	30–36	D13
Dipentene			175–195	54	
Pine oil			200–230		D802
ORGANIC COMPOUNDS					
Acetone		0.79	56	−18	D329
Amyl acetate	Banana oil	0.88	146	24	
Butanol	Butyl alcohol	0.81	118	25	D304
Denatured alcohol	Methylated ethanol	0.79	78	14	
Ethyl acetate	Acetic ester	0.90	77	5	
Isopropyl alcohol	2-Propanol	0.79	82	14	D770
Methanol	Wood alcohol	0.79	65	14	D1152
Propyl alcohol	n-Propanol	0.80	98	22	
Trichloroethylene		1.46	87	—	

*Compiled from several sources.

Table 3-17. ORGANIC COATINGS—NATURAL RESINS AND MATERIALS

Natural and Processed Materials Used in Paint, Varnish, and Protective Coatings and Finishes

See also Tables 3-18 and 3-19.

Common name	Alternate name	Source	Properties	Uses
Rosin	Colophony; gum and wood rosins; tall oil rosin; phenolic blend	U.S.A., France, Portugal, Spain; tapped from coniferous trees	Depends on processing; dispersion, adhesion, gloss	Wide variety of rosin products, e.g., ester gum, limed rosin, zinc resinate, maleic resin; for varnishes, enamels, lacquers
Copal resin	Class name for several resins	Africa, East Indies, Philippines	Hard, glossy film for varnish or enamel	Largely displaced by synthetic resins
Damar resin	Singapore; dipterocarpus	East Indies	Soft, non-yellowing, fume-resistant film	Specialty finishes; paper varnish (not widely used)
Lac	Shellac base; seed lac	India; insect excretion from tree sap	M.p., about 80 deg C; acid value, 70; iodine value, 20; soluble in alcohols; color, orange or darker; tough; fast-drying	Sealer and base coat; foundry patterns; white shellac on floors; largely displaced by synthetics
Rubber	Chlorinated rubber; cyclized rubber	Natural rubber latex	Tough; resistant to abrasion, moisture, chemicals, mildew, petroleum; quick-drying	In blended paints for chemical apparatus, marine uses, traffic areas, swimming pools, rust protection
Asphalt	Bitumen; gilsonite; manjak	Petroleum residues; some natural deposits, as in Trinidad	Softens at 46 deg C; good adhesion; waterproof and chemical-resistant	Roof and waterproof paints; corrosion protection; sound control
Pitch	Coal tar	Residue from coal-tar distillation	Impervious to water; high dielectric strength; heavy coating	Ship-bottom paint; moisture and corrosion protection; chemically resistant paint
Gum	Examples: gum arabic, gum tragacanth	Africa and Asia	Thickening agent and base for water colors	Seldom used except in artists' paints
Glue	Glue size	Skin, bone, animal, and fish scrap	Binder for water paint and ceiling white; adhesion; reduces "suction"	Older types of water paint and wall-size; sealers
Casein	Milk protein	Skimmed milk	Adhesion; thickening	Gel and thickener for water paints, including latex; wallboard base coat
Dextrin	Starch protein	Grain and seeds; potatoes	Adhesion; stiffener	Sizing and coating for fabric, paper, wallboard
Wax	Beeswax, paraffin wax, carnauba wax, etc.	Natural waxes; petroleum	Resistant to moisture, spray, marring, light abrasion; high gloss	Widely used in protective polishes; small amounts added to decorative enamels and varnishes

Table 3-18. ORGANIC COATINGS—SYNTHETIC RESINS

Classes, Modifications, Properties, and Applications of Synthetic Resins for Paints and Coatings

Class	Kinds and modifications	Properties or advantages	Usage
Acrylic	Emulsions and water-soluble polymers; thermoplastic or thermosetting, soluble in organic solvents	Thermoplastic; resistant to age, light, water, chemicals, oil; hardness; flexibility	Very widely used in both solution and emulsion; clear coatings; enamels; latex paints; automobile and metal finishes
Alkyd	Alkyd-oil; styrenated; modified; combined with other resins	Solubility; compatibility; durability; hardness; toughness; gloss retention; adhesion	Most widely used of all resins; air-dry or baked; paints, varnishes, enamels, primers
Amino	Urea and melamine; blends	Hardness; color retention; fast baking; durability	Durable finishes for automobiles, appliances, metal surfaces
Cellulose	Cellulose nitrate and others; also combinations with alkyd and amino resins	Flammable; compatibility with oils, plasticizers, and other resins	Quick-drying lacquers, enamels, dopes
Epoxy	Epoxy polyamide; esterified; amine catalyzed	Adhesion; toughness; chemical resistance; quick drying	Baking enamels; marine varnish; can coatings; chemical equipment
Fluorocarbon	Polytetrafluoroethylene; polychlorotrifluoroethylene	Resistant to moisture, heat, chemicals, fungi, and abrasion	Chemical-resistant linings for tanks and industrial equipment; electrical insulation; weatherproofing
Phenolic	Bakelite; modified oil-soluble	Resistant to chemicals and solvents; water- and weatherproof	Primer, sealer, varnish; structural and marine paints
Polyamide	Nylon; thermoplastic polyamides	Strength, toughness, abrasion resistance; oil and solvent resistance; good adhesion	Wear-resistant and chemical-resistant coatings for metal textiles, leather, paper, industrial equipment
	Glycol esters of unsaturated and saturated dibasic acids, such as with phthalic acid isomers or maleic or fumaric acids. Unsaturated acid types are usually used as copolymers with styrene or acrylic monomers	No volatile thinners need to be expelled in forming thick or thin films that are hard and adhesive	High-build coatings with or without fillers for wood, concrete, and other non-metallic substrates
Polyethylene	Low- and high-density; chlorinated unsaturated	Odorless, tasteless, non-toxic; waterproof; high strength; toughness; flexibility	Coatings for food cartons; wire insulation; flame-sprayed coatings on metals; textile coatings
Polystyrene	Used largely in copolymers and latex	Flexibility; adhesion; toughness; weather resistance	With butadiene or alkyd for latex paints; solution-type outdoor paints
Polyurethane	Urethane oil or alkyd; moisture-cured and two-package compositions	Flexibility; high gloss; abrasion and chemical resistance; toughness; dielectric strength	Varnishes for severe service; enamels; marine paints; textile coatings; wire coatings; concrete finish
Rubbers	Neoprene; butyl; nitrile; SBR, and copolymers of isoprene, butadiene, polypropylene	Variety of properties including chemical resistance, resilience, abrasion resistance, oil resistance	Coatings where resilience, elasticity, and abrasion resistance are needed
Silicone	Copolymer or modified with metal powder or frit	Stable in the 200–500 deg C range with heat-resistant pigments	Heat-stable and high temperature paints; corrosion-resistant paints
Vinyl	Acetate, chloride; copolymers; plastisols	Toughness; flexibility; chemical resistance; wear resistance; dielectric strength	Masonry finish; textured coatings; outdoor uses

Table 3-19. PIGMENTS FOR PAINTS AND COATINGS*

The term *pigments* was formerly applied to natural color materials, such as those used by artists. Surface coatings today are increasingly dependent on the synthetic resins. The following table covers most of the common classes of paint solids other than the synthetic resins (Table 3-17), the natural resins and gums (Table 3-18), and the water paint powders, such as lime and cement.

A paint mixture to be applied as a surface coating is often described in terms of two constituents—the pigment and the vehicle. The liquid vehicle wets the surface and dries or cures into a film, while the pigment, dispersed in the vehicle, functions largely as a radiation absorber and reflector, providing opacity or color. Common insoluble pigments are mixtures of very fine powders or flakes of metal oxides or salts and sometimes the metals themselves. Other constituents are added to the pigment or to the vehicle for special purposes, such as controlling the gloss, hardness, adhesion, abrasion resistance, and weather resistance of the film or providing corrosion resistance or increasing the dielectric strength.

For latex paints, and for solution paints such as shellac or asphalt paint, the dual-constituent classification in terms of a film-forming vehicle and a surface-hiding pigment is not applicable. The same is true for such water paints as whitewash, and for many of the complex formulations that often include both dispersions and solutions and even emulsions.

COMMON PAINT PIGMENTS

Name and composition	Typical particle size, μm	Specific gravity	Refractive index	Hiding power	Properties and uses	ASTM No.
WHITE PIGMENTS (OPAQUE)						
Titanium dioxide (anatase or rutile)	0.25	3.5	2.7	Excellent	Brilliant white; anatase form is chalking; non-toxic	D476
White lead (basic carbonate, about 68%), $2PbCO_3 \cdot Pb(OH)_2$; flake white		6.7	2.0	Good	Durable film; tends to darken; weather and water resistant; toxic; primer and undercoat	D81
Zinc oxide, ZnO; Chinese white	0.2–0.3	5.65	2.0	Good	Mildew resistant; highly durable; outdoor oil paints	D79
Zinc sulfide			2.37	Very good	Not widely used	D477
Antimony oxide	0.5–2.0	5.5	2.1	Good	Little used except in fire-retardant paint	
Lithopone (regular), approx 70% $BaSO_4$ and 30% ZnS	0.2–2.0	4.2	1.9	Fair	Interior oil and emulsion paints	D477
WHITE EXTENDER PIGMENTS (LOW REFRACTIVE INDEX)						
Calcium carbonate, $CaCO_3$; precipitated chalk; whiting	2–5	2.7	1.58	Poor	Ceiling white; undercoat	D1199
Calcium sulfate, $CaSO_4 \cdot 2H_2O$; gypsum		2.35	1.53	Poor	Filler; limited use in paints	
Magnesium silicate, $H_2Mg_3(SiO_3)_4$; talc		2.7	1.57	Poor	Flatting; anti-settling; chemically inert	D605
Barium sulfate, $BaSO_4$; barytes; barite		4.4	1.64	Poor	Undercoats; fillers; chemically stable	D602
China clay, $Al_2O_3 \cdot 2SiO_2 \cdot 2H_2O$; kaolin	1.0 (avg)	2.5	1.56	Poor	Undercoats; thickening agent	
Hydrated silica	1–10	2.2			Flatting; consistency control	D604

*This table has been compiled from several sources. While it includes many common pigments, it is in no sense a complete list. The synthetic organic pigments are entirely omitted, and many common mixtures and blends are not included. A more complete treatment should also give attention to the science of color, to phenomena of spectral absorption and reflectance, and to the various methods for specifying, matching, and measuring colors (see References, Table 3-14, and Tables 7-39 and 7-40).

Table 3-19. PIGMENTS FOR PAINTS AND COATINGS *(Continued)*

Name and composition	Typical particle size, μm	Specific gravity	Refractive index	Hiding power	Properties and uses	ASTM No.
BLACK, GRAY, BROWN PIGMENTS						
Carbon black; furnace black; impingement black	.02–.09		Opaque	Good	Very widely used; jet-black gloss	
Lampblack, C; oil black		1.8	Opaque	Good	Undercoats; primers; high adhesion; blue undertone; matte	D209
Graphite (amorphous)		2.3	Opaque	Good	Topcoat for steel structures; high coverage; durable	
Iron oxide black, Fe_3O_4; magnetite (synthetic)	0.5		Opaque	Good	Metal paints and fillers	D769
Bone black; ivory drop black	325 mesh (solution)	2.3	Opaque	Good	Black undercoat and filler	D210
Asphaltum; cut-back asphalt; petroleum paint (black to brown)			Opaque	Fair	High adhesion; automotive uses; roof paint; chemical and water resistance; also emulsions	
Blue lead (45% lead sulfate, 30% lead oxide); basic lead sulfate, sublimed			Opaque	Good	Structural steel primer (gray)	D405
Sienna and umber, largely Fe_2O_3; range of brown and gray (to red synthetics)		3.3	1.9	Fair	Widely used tinting colors; low cost	
COLORS						
Prussian blue (ferrocyanides); iron blue		1.8	1.55	Fair	Low cost; non-bleeding; high-temperature baking; light tints may fade	D261
Cobalt blue (cobalt and aluminum oxides)		3.8	1.7	Fair	Expensive; art finishes; highly stable and resistant	
Cobalt green (cobalt and zinc oxides)		4.0	1.95	Fair	Highly stable and resistant; not widely used	
Chrome green (yellow $PbCrO_4$ with $PbSO_4$ blended with iron blue)		5.1	2.5	Excellent	Widely used; wide range of greens; good color retention	D263; D212
Red lead (synthetic), Pb_3O_4 >85%	2.0	8.7	2.4	Excellent	Durable; good adhesion; corrosion protection; primer for steel; fades in sun	D83; D49
Iron oxide, Fe_2O_3; red hematite; red to maroon and brown		5.2	2.5	Excellent	Inexpensive; widely used	
Cadmium red, largely CdS		4.5	2.7	Excellent	Bright tones; non-bleeding high-temperature bake	
Manganese violet			1.7	Good		
Chrome yellow and orange, largely $PbCrO_4$ (with $PbSO_4$ light and PbO orange)		5.9	2.3	Excellent	Bright tints	D211
Zinc-yellow, ZnO and CrO_3; zinc chromate		3.5	1.9	Good	Exterior paints; resists darkening	D478
Yellow ochre, Fe_2O_3; ferrite yellow; natural or synthetic	0.5–1.5	3.5	2.0	Good	Inexpensive; low-temperature bake	D85
Cadmium yellow, largely CdS with CdSe; also lithopone with $BaSO_4$		4.3	2.4	Excellent	Interior oil and water paints	

Table 3-19. PIGMENTS FOR PAINTS AND COATINGS *(Continued)*

Name and composition	Typical particle size, μm	Specific gravity	Refractive index	Hiding power	Properties and uses	ASTM No.
ORGANIC TONERS	0.01–0.1	Variable	Varies with wavelength	Poor to excellent	Cover spectral range from red to violet; some with hiding power are used in mass tone or solid colors; most with low hiding power give vivid colors and are used in mixtures with titanium dioxide or aluminum flake	
METAL PIGMENTS						
Aluminum (leafing—flakes overlap)			Opaque	Excellent	Brilliant metallic finish; durable	D962
Aluminum (non-leafing)	100–200 mesh		Opaque	Good	Less brilliant; widely used; chemical and heat resistance; durable	D962
Aluminum (extra fine)	400 mesh		Opaque	Excellent	High hiding power; durable	D962
Zinc dust	325 mesh		Opaque	Good	Rust-inhibitive primer for steel and galvanized iron; weatherproof paints	D520
Copper bronze, 2% zinc			Opaque	Good	Decorative copper finish; fungicidal paint	
Gold bronze, Cu, Zn, and Al			Opaque	Good	Decorative color range, pale gold to red gold; exterior and interior	D267
FLUORESCENT PIGMENTS						
Organic-dyed resins (ultraviolet reflectors)			Variable	Fair	Maximum visibility in yellow-orange range; high-visibility coatings; safety paints	

Table 3-20. PAINTS AND COATINGS FOR CORROSION RESISTANCE

Corrosion is a very general term applied to a variety of processes by which metal surfaces are attacked and converted to oxides, sulfides, or other compounds. As much corrosion is electrochemical, methods of protection are based on isolation of the metal from the environment that contains oxygen, moisture, sulfur, etc., and on minimizing galvanic potentials. The methods by which these steps are accomplished are so diverse that no single summary can adequately cover them.

Galvanic corrosion is prevented by elimination of contact between metals, whether direct or through an electrolyte. A most common case is copper and iron. Surface treatment of steel with a phosphate wash or a primer, such as red lead or zinc chrome, may form a thin protective layer of stable compound, over which an impervious and durable coating will adhere.

The following table lists many of the common protective coatings, most of which are used on iron and steel, since they present the corrosion problem of greatest economic importance. Asphalt and red lead have long been widely used, but there is an increasing use of protective metals and of synthetic resins. For detailed discussions consult the references listed in Table 3-14.

Name or protective constituent	Description or composition	Advantages	Typical applications
Aluminum	Leafing or non-leafing paints, varnishes and primers (hot dipping)	High hiding; fume-resistant; low cost; high coverage; ultraviolet protection	One-coat protection for wide use; additive for asphalt and phenolic resin paints
Asphalt	Petroleum residue and solvents, with asbestos, vermiculite, and perlite	Low cost; moisture and chemical resistance; good adhesion	Thinned or hot; pipe lines, tanks, undercoat; buried structures
Barium potassium chromate	With barium and chromium oxides	High-strength film; elastic and durable	For steel or light metals; air-drying or baking types
Calcium plumbate	Calcium and lead oxides	Quick-drying; smooth; hard; salt-resistant	Primers for structural and galvanized steel
Carbon black	Furnace black or amorphous graphite	High hiding; low cost; high coverage; high chemical resistance	With black iron oxide for shop and foundry paint; topcoat varnish
Chlorinated rubber	10–25% in oil or alkyd paints or lacquers	Chemical resistance; low permeability; wide color range	Marine paint; machinery finish; high-build paints; chemical equipment
Etch primer	Phosphoric acid or phosphate; often with zinc chromate	Versatile water-solution wash as quick-dry primer	Clean-metal primers for good overcoat adhesion; various metals, including galvanized
Lacquer	Resin lacquers, as nitro-cellulose and acrylic	Transparent or colors; quick drying; cold or hot application	For highly finished metal surfaces, including brass
Lead cyanamide	Yellow $Pb(CN)_2$	Anodic protection	Primer for steel; with linseed oil and pigments
Pitch, coal-tar	Coal tar in aromatic solvent	Black, brown, green; glossy; hard; adhesive; chemical resistance	Emulsion topcoat; with epoxy or other resin for sea-water immersion
Phenolic resin	Air-dry varnish and baking lacquer	Very low permeability; good baking enamel; good adhesion	Chemically resistant paints; electric insulation varnish; tank linings
Red lead	Pb_3O_4 and PbO with linseed oil	Readily available; weather-resistant; good adhesion	Most widely used where linseed oil is common; outdoor ferrous structures
Red iron oxide	30–50% in oil or chlorinated rubber paints	Low cost; one or two coats without primer; high-hiding; durable	Ferrous metal primer or one-coat protection
Stainless steel flake	Barrier coat	Inherently corrosion resistant; good appearance	Metallic barrier for chemical resistance
Zinc chrome	10% or more in non-ferrous primers; with red iron oxide for ferrous metals	High-hiding, sealing, and adhesion; salt-resistant	On light metals; with red iron oxide on ferrous metals; alkyd paints

Table 3-20. PAINTS AND COATINGS FOR CORROSION RESISTANCE
(Continued)

Name or protec-tive constituent	Description or composition	Advantages	Typical applications
Zinc dust	75–95% Zn with binder; also with zinc oxide	One coat is common; resists high humidity; cathodic protection; flexible	Galvanized iron primer; synthetic resin or rubber vehicle
Zinc yellow	Basic potassium zinc chromate	Good penetration; neutralizes acid and releases chromate ions; colors available	Ferrous metal and light metal primers; often with iron oxide

General notes:
1. *Surface preparation* is very important prior to coating treatment. Solvent degreasing, alkali washing, sand or shot blasting, pickling, acid or phosphate dipping, flame treatment, wire brushing, and abrasive cleaning are among the methods that should receive consideration for surface preparation.
2. *Dual protective films* and even complex mixtures are often used for final-coat or one-coat protection. Examples are pigment with drying oil, pigment and alkyd or phenolic vehicle, powdered or leafing metal with polyvinyl chloride or other resin, pigment and chlorinated rubber, and pigment mixtures with both chlorinated rubber and drying oil. In each case there is at least a dual barrier coat, sometimes over a primer.
3. *High pigment content* (40–90%) is characteristic of most protective paints, since these solids act as barrier coats.
4. *Synthetic resins* are now being used in many corrosion-protective coatings. Alkyd, PVC, epoxy, vinyl, and certain copolymers are common in paints with various protective pigments.

REFERENCE
"Protective Coatings for Metals", 3rd ed., R.M. Burns and W.W. Bradley, ACS Monograph Series, Reinhold Publishing Corp., 1967.

Table 3-21. INORGANIC SURFACE COATINGS
For data on organic coatings, see Tables 3-14 to 3-20.

FUNCTIONS, MATERIALS, AND APPLICATION METHODS
Functions.   Inorganic coatings are applied to metals and to some other materials to protect and to modify the surface properties, as follows:
   Mechanical.   For resistance to abrasion, erosion, shock. For control of hardness, texture, smoothness, lubrication.
   Chemical.   For resistance to oxidation, corrosion, chemical reaction, diffusion.
   Thermal.   For heat insulation, ablation, control of heat transfer by conduction and by radiation.
   Electrical.   For control of electrical resistance, magnetic properties, thermionic performance.
   Radiation.   For control of light and radiative properties, including reflection, absorption, diffusion or transmission of radiation (total or spectral); for optical imaging by phosphor.
Materials.   Coating materials are used singly and in numerous combinations. Most coatings are chemical compounds of the metallic elements.
   Oxides of aluminum, silicon, calcium, magnesium, zirconium, chromium, beryllium, thorium, hafnium, nickel, etc.
   Other Oxygen Compounds.   The aluminates, silicates, chromates, zirconates.
   Hard Metal Compounds.   The carbides, nitrides, borides, silicides.
   Intermetallics.   The aluminides, beryllides, stannides.
   Combinations of Materials.   Classified under such common class names as enamels, glasses, ceramics, cermets, refractories, composites.
Application Methods.   These may include more than one of the following steps or processes:

Adhesion	Flame spraying	Spreading
Atomized spraying	Fusion or firing	Trowelling
Dipping (hot or cold)	Packed retorts	Vapor deposition (pyrolysis)
Electrophoresis	Painting; brushing	
Electroplating	Slip or slurry (fired)	

Examples.   These might include a very long list and a wide variety of uses, but those listed in the following table are illustrative.

Table 3-21. INORGANIC SURFACE COATINGS *(Continued)*

COMMON INORGANIC COATINGS

Coating name or function	Base or substrate	Coating description, processes, and typical compositions	Typical uses
Enamel (porcelain)	Steel	Surface coated with slip or slurry of powdered glasses and colors. Oxides of Si, Al, Na, B, and Ca predominate	Cookware, signs, tanks, tubs, pails, housewares
Enamel (vitreous)	Cast iron	Similar to above	Sanitary ware, structural decoration
Enamel	Aluminum	Low-melting frit (below 1050 deg F) containing PbO, Li_2O, SrO, TiO_2, etc. Aluminum alloy must be enameling grade	Architectural trim, siding, wall coverings; highway and advertising signs
Anodized	Aluminum	Anodic oxidation and coloring. Processes use phosphoric, sulfuric, or chromic acid	Architectural and vehicle trim; housewares; aircraft materials; handrails
Reflector	Glass, metal, or crystal	Mirror metals or paints for control of light or electromagnetic radiation	Solar reflection, optical instruments, thermal radiation control
Photo converter	Metal or glass	Amplifying, photoemissive, and photo-voltaic surfaces with compounds of Pb, Cd, Ge, and Cs	Solar cells; photomultipliers, amplifiers, detection, and measurement
Phosphor	Glass	Optical imaging, using zinc and cadmium sulfides, silicates	Cathode-ray and television screens; instrument displays, illumination tubes; particle detectors
Ablation coating	Metal	Wound or woven glass fibers with phenolic resin binder	Rockets, nosecones, re-entry vehicles
Lubricant	Metal pair	Lubricating powders and mixtures, usually oxides of Pb, Bi, Cd, or di-sulfides of Mo or W	High-temperature and space applications; journal and ball bearings; extrusion dies
Electrical insulation	Metal wire or surface	Enameled conductors baked from slurry-covered surface; thermionic layers	Large coils and magnets; structural separators
Diffusion coating	Metals	Reactive coating material diffused into substrate by heat treatment. Compounds of Al, Cr, Si, and Ti most common	Oxidation-resistant or hard-surface layers on metals and alloys
Ceramics and cermets	Metals and alloys	Sprayed ceramic mixtures and electro-deposited particles from stirred aqueous bath	High-temperature applications with short-term erosion
Fire retardants	Wood, fiber, or plastic	Compounds of P, Cl, Sb, Br, and B	Fireproofing of combustible surfaces
Thermal insulation	Metals	Thick, porous refractory and fiber coatings and bonded ceramic foams; refractory cements	High-temperature protection for structural materials

REFERENCES

"NASA Contributions to the Technology of Inorganic Coatings", J.D. Plunkett, NASA SP–5014, National Aeronautics and Space Administration, 1964.

"High-temperature Inorganic Coatings", J. Huminik, Reinhold Publishing Corporation (now Van Nostrand Reinhold Co.), 1963.

"High-temperature Materials and Technology", I.E. Campbell and E.M. Sherwood, Eds., sponsored by the Electrochemical Society, John Wiley & Sons, Inc., 1967.

"Aluminum, Fabrication and Finishing", K.R. Van Horn, Ed., Vol. 3, American Society for Metals, 1967.

"Porcelain Enamels", A.I. Andrews, Garrard Publishing Company, 1961.

Table 3-22. SUMMARY OF ELECTROPLATING PRACTICE*

Average Operating Conditions

Metal	Principal uses	Type of solution	Principal ingredients	Temp, °F	CD ASF	Volts	Cathode efficiency, %	Time to deposit, 0.001 in.
Cadmium	Protection	Cyanide	CdO, NaCN, brighteners	70–95	15–45	1–4	90	20 min
Chromium	Decorative Engineering (hard) Cylinder liners (porous)	Chromic acid	CrO_3, H_2SO_4	120	250	6–8	15	2 hr
Copper	Electroforming Undercoat for other metals Stop-off in case-hardening, etc.	Acid	$CuSO_4 \cdot 5H_2O$, H_2SO_4	75–120	15–40	1–2	100	35 min
		Cyanide	CuCN, NaCN, Na_2CO_3	75–100	5–15	1.5–3	50	90 min
		Rochelle	Above + rochelle salts	140–160	20–60	2–3	60	45 min
			Many other types, e.g., fluoborate, pyrophosphate, amine, all-potassium cyanide	—				
Gold	Decorative	Cyanide	$KAu(CN)_2$, K_2CO_3, KCN (Solutions vary considerably, depending on color wanted)	120–160	5–15	2–6	80	
Indium	Bearing surfaces	Cyanide	$InCl_3$, NaCN, addition agent	Room	10–150	—	40	—
		Sulfate	$In_2(SO_4)_3$, Na_2SO_4	Room	20	—	75	—
		Fluoborate	$In(BF_4)_3$, H_3BO_3, NH_4BF_4	70–90	50–100	—	50	—
Iron	Electroforming Repair	Chloride	$FeCl_2$, $CaCl_2$	190	60	—	95	20 min
		Sulfate	$FeSO_4(NH_4)_2SO_4$	Room	20	—	95	1 hr
Lead	Protection Bearing surfaces	Fluoborate	$Pb(BF_4)_2$, HBF_4, glue	Room	10–80	0.5	100	40 min
		Sulfamate	Pb sulfamate, sulfamic acid, addition agents	75–120	5–40	3–8	100	20 min
Nickel	Protection Decorative Electroforming Undercoat for Cr, etc.	Sulfate-chloride	$NiCl_2$, $NiSO_4$, NH_4 ion, H_3BO_3 (Formulations differ widely, depending on purpose)	75–100	Varies greatly	0.5–3	95	30 min
Rhodium	Decorative Optical	Sulfate Phosphate	Prepared salts	110–120	10–80	2.5–5	15	—
Silver	Decorative Protective Bearing surfaces	Cyanide	AgCN, KCN, K_2CO_3, CS_2 (Or Na in place of K)	80	5–15	1	100	
Tin	Protection Food and dairy Bearings Electrical To enable easy soldering	Sulfate	$SnSO_4$, H_2SO_4, addition agents	Room	40	1–3	90	15 min
		Fluoborate	$Sn(BF_4)_2$, HBF_4, addition agents	75–100	50	—	100	10 min
		Other acid electrolytes		—	—	—	—	—
		Stannate	Na_2- or $K_2Sn(OH)_6$, Na- or KOH	150–190	40	4–8	80	30 min
Zinc	Protection	Sulfate	$ZnSO_4$, NH_4Cl, addition agents	75–100	15–400	—	99	10 min
		Cyanide	$Zn(CN)_2$, NaCN, NaOH, brighteners	100	10–50	—	85	40 min

ALLOYS

Metal	Principal uses	Type of solution	Principal ingredients	Temp, °F	CD ASF	Volts	Cathode efficiency, %	Time to deposit, 0.001 in.
Brass	Rubber-bonding Decorative	Cyanide	CuCN, $Zn(CN)_2$, NaCN, Na_2CO_3	75–100	3–10	2–3	75	—
Bronze	Decorative Undercoat for chromium Stop-off for steel	Cyanide-stannate	CuCN, KCN, KOH, $K_2Sn(OH)_6$, rochelle salt	155	20–100	3–6	70	30 min
		Pyrophosphate-cyanide	$Sn_2P_2O_7$, KCN, CuCN, $K_4P_2O_7$, addition agents	140–180	20–70	2–5	70	30 min
Lead-tin	Bearings Solderability Electrotyping	Fluoborate	$Sn(BF_4)_2$, $Pb(BF_4)_2$, HBF_4, addition agents	Room	60	1–2	100	—
Tin-zinc	Solderability	Cyanide-stannate	$Zn(CN)_2$, KCN, KOH, $K_2Sn(OH)_6$	150	10–75	4–5	80–95	30 min

*From: "Standard Handbook for Electrical Engineers", 9th ed., A.E. Knowlton, Ed., McGraw-Hill Book Company, 1957.

REFERENCES

"Modern Electroplating", A. Gray, Ed., John Wiley & Sons, Inc., 1953.
"Plating and Finishing Guidebook-Directory", Finishing Publications, Inc. (published yearly).
H. Bandes, Trans. Electrochem. Soc., 88:263, 1945.
Technical Data Sheet No. 127, Metal & Thermit Corp., New York, 1954.
W.H. Safranek and C.L. Faust, Proc. Am. Electroplaters' Soc., 41:201, 1954.

Table 3-23. PROPERTIES AND USES OF ADSORBENTS

See Table 3-42 for ion-exchange resins.

For density in kg/m^3, multiply lb/ft^3 by 16.02.

Adsorbent	Shape of particles†	Bulk dry density, lb/cu ft	Surface area, sq m/g	Uses and method of regeneration
Active alumina	G	50	250	Drying gases and liquids; catalyst; catalyst support; defluoridation of alkylates; neutralization of lube oils. Can be regenerated.
	S	55	350	
Activated bauxite	C, G	~53	—	Decolorizing petroleum products and drying of gases. Regeneration by heating.
Aluminosilicates (Molecular sieves)	C, S, P	~44	770	Selective adsorption based on molecular size and shape; drying of gases and liquids; catalyst support. Regeneration by heating or elution.
Carbon or charcoal:				Water treatment; air and gas purification, gas masks and smoke filters; solvent recovery and purification; sugar refining; decolorizing of solutions; decolorizing natural products. Adsorbed gases can be evaporated.
Bone	G	20–30		
Coal	G	20–30	500–1200	
Petroleum	C	28–34	800–1100	
Shell	G	10–20		
Wood	G	10–35	625–1400	
Clay	P	30–45	225–300	Refining petroleum fractions; purifying vegetable oils, juices; catalyst base.
Fuller's earth	G	30–40	130–250	Uses same as for clay. Regeneration by washing and burning adsorbed organic matter.
Silica gel	G, P	~25	320	Drying of gases; adsorption, from solutions; hydrocarbons; catalyst base. Regeneration by evaporation of adsorbed liquid.
	S	50	650	

Notes:

Both surface areas and pore sizes are important in adsorption.

Distinction should be made between adsorption of gases by non-porous solids, such as smooth metals, and by porous solids, such as the adsorbents listed in this table. Non-porous inorganic block-solids have surface areas in the range below 10 m^2/g; their adsorption is correspondingly small but definitely measurable.

Gas and vapor adsorption increases as the temperature is reduced and the pressure of the gas is increased. The usual method for quantitative expression of this relationship is in terms of the adsorption at constant temperature. The adsorption isotherms are plotted with adsorption (e.g., g of adsorbate/g of adsorbent) on the ordinate and pressure (actual pressure/ saturation pressure) on the abscissas. These isotherms have several typical shapes, but in any case the adsorption is much higher if the gas is not highly superheated, i.e., it increases as saturation is approached at the given temperature. The BET* classification system for adsorption isotherms recognizes five types or shapes of such curves.

†Shape of particles indicated as follows: C, cylindrical pellets; G, granular; P, powder; S, spherical beads.
*Proposed by S. Brunauer, P.H. Emmet, and E. Teller in *J. Amer. Chem. Soc.*, 60:309, 1938, but still widely quoted.

REFERENCES

"Adsorption, Surface Area, and Porosity", S.J. Gregg and K.S.W. Sing, Academic Press, 1967.

"Activated Carbon", J.W. Hassler, Chemical Pub. Co., 1963.

"Symposium on Activated Carbon", Atlas Chemical Industries, 1968.

Table 3-24. WATER VAPOR EQUILIBRIUM CURVES OF SORBENTS OR DESICCANTS*

Sorbents used for the drying or dehumidification of air vary greatly in their performance. An *absorbent* changes either physically or chemically, or both, during the sorption process. As moisture from the atmosphere condenses into the fine pores of an *adsorbent*, the latent heat of evaporation is given off, but the adsorbent does not appear to change.

The quantity of water finally held by a given sorbent when equilibrium has been reached is dependent only on the relative humidity of the contacting air-vapor mixture. The relation of this mass ratio to the relative humidity is given in the following table. This ratio is largely independent of the temperature of the operation and of the vapor pressure at which the water exists in the air mixture.

Complete tables or charts on the performance of specific sorbents are available. Two additional facts about the sorption process are clearly shown by such charts but are not apparent from the table given here.

(1) For a given partial pressure of the water vapor in the air mixture, i.e., a given humidity ratio or equilibrium dewpoint, the equilibrium mass ratio decreases with increase in temperature.

(2) For a given relative humidity of contacting air, the mass ratio is only very slightly less at high temperature and high dewpoint (e.g., 125°F dry bulb) than it is at room temperature.

The tabular values here given are approximate for temperatures ranging from 0°F to 150°F. These are equilibrium data and should not be used for the design of dynamic systems.

Relative humidity, %	Mass ratio, water to sorbent						
	Absorbents			Adsorbents			
	LiCl	H_2SO_4	Triethylene glycol	Alumina gel	Silica gel	Molecular sieve	Activated alumina
10	.43	.49	—	.07	.05	.15	.04
20	1.35	.63	.05	.12	.10	.17	.06
30	—	.81	.09	.15	.16	.18	.08
40	—	1.00	.13	.17	.21	.18	.10
50	—	1.22	.21	.20	.26	.18	.12
60	—	—	.33	.25	.31	.19	.15
70	—	—	.51	.31	.35	.19	.17
80	—	—	.92	.36	.36	.19	.18
90	—	—	—	.41	.37	.20	.19
100	—	—	—	.42	.38	.20	.19

*Based on: "ASHRAE Handbook of Fundamentals", American Society of Heating, Refrigerating and Air-Conditioning Engineers, 1972.

Table 3-25.   ADHESIVES CLASSIFIED BY CHEMICAL COMPOSITION*

Group →	Natural	Thermoplastic	Thermosetting	Elastomeric	Alloys**
Types within group	Casein, blood albumin, hide, bone, fish, starch (plain and modified); rosin, shellac, asphalt; inorganic (sodium silicate, litharge-glycerin)	Polyvinyl acetate, polyvinyl alcohol, acrylic, cellulose nitrate, asphalt, oleo-resin	Phenolic, resorcinol, phenol-resorcinol, epoxy, epoxy-phenolic, urea, melamine, alkyd	Natural rubber, reclaim rubber, butadiene-styrene (GR-S), neoprene, acrylonitrile-butadiene (Buna-N), silicone	Phenolic-polyvinyl butyral, phenolic-polyvinyl formal, phenolic-neoprene rubber, phenolic-nitrile rubber modified epoxy
Most used form	Liquid, powder	Liquid, some dry film	Liquid, but all forms common	Liquid, some film	Liquid, paste, film
Common further classifications	By vehicle (water emulsion is most common but many types are solvent dispersions)	By vehicle (most are solvent dispersions or water emulsions)	By cure requirements (heat and/or pressure most common but some are catalyst types)	By cure requirements (all are common); also by vehicle (most are solvent dispersions or water emulsions)	By cure requirements (usually heat and pressure except some epoxy types); by vehicle (most are solvent dispersions or 100% solids); and by type of adherends or end-service conditions
Bond characteristics	Wide range, but generally low strength; good resistance to heat, chemicals; generally poor moisture resistance	Good to 200–500°F; poor creep strength; fair peel strength	Good to 200–500°F; good creep strength; fair peel strength	Good to 150–400°F; never melts completely; low strength; high flexibility	Balanced combination of properties of other chemical groups depending on formulation; generally higher strength over wider temperature range
Major type of use†	Household, general purpose, quick set, long shelf life	Unstressed joints; designs with caps, overlaps, stiffeners	Stressed joints at slightly elevated temperature	Unstressed joints on lightweight materials; joints in flexure	Where highest and strictest end-service conditions must be met; sometimes regardless of cost, as military uses
Materials most commonly bonded	Wood (furniture), paper, cork, liners, packaging (food), textiles, some metals and plastics. Industrial uses giving way to other groups	Formulation range covers all materials, but emphasis on non-metallics—especially wood, leather, cork, paper, etc.	Epoxy-phenolics for structural uses of most materials; others mainly for wood; alkyds for laminations; most epoxies are modified (alloys)	Few used "straight" for rubber, fabric, foil, paper, leather, plastics, films; also as tapes. Most modified with synthetic resins	Metals, ceramics, glass, thermosetting plastics; nature of adherends often not as vital as design or end-service conditions (i.e., high strength, temperature)

**Alloy, as used here, refers to formulations containing resins from two or more different chemical groups. There are also formulations that benefit from compounding two resin types from the same chemical group (e.g., epoxy-phenolic).

†Although some uses of the non-alloyed adhesives absorb a large percentage of the quantity of adhesives sold, the uses are narrow in scope; from the standpoint of diversified applications, the most important use of any group is the forming of adhesive alloys.

*From: "Encyclopedia of Engineering Materials and Processes", H.R. Clauser, Ed., Reinhold Publishing Corporation, 1963.

Table 3-26. TYPICAL APPLICATIONS OF ADHESIVES

SYMBOLS:

NATURAL ADHESIVES
A Casein
B Fish, hide, bone
C Asphalt

THERMOPLASTIC
G Polyvinyl acetate
H Cellulose nitrate (or acetate)
J Neoprene

THERMOSETTING
D Phenolic
E Epoxy

ELASTOMERIC
K Rubber (natural or reclaim)
L Buna N
M Blend or alloy[a]

Material to be bonded	To itself	To glass, metal, or ceramic	To plastic	To rubber	To wood, hardboard, or paper
Glass	E, G, H, M	E, H, M	K, M	K, M	E, G, K, M
Metal	D, E, M	E, H, M	E, H, K, M	E, K, M	D, E, G, K, M
Tile	C, H, K	E, H, M	E, H, M	K, M	C, G, K
Ceramic	E, H, M	E, H, M	K, M	K, M	E, G, K, M
Phenolic plastic	E, M	K, M	E, M	K, M	D, E, K
Vinyl plastic	L, M	L, M	M	K, M	K, M
Rubber	J, K, M	E, K, M	K, M	K, M	K, M
Wood	A, B, K	E, G, K, M	D, E, J, K	K, M	A, B, K
Paper	A, B, C, K	A, G, K	A, K	A, K	A, K
Fabrics	A, G, J, K	C, G, J, K	J, K, L	A, K	A, K
Felt	A, C, J, K	C, G, J, K	K, M	A, K	A, K
Leather	G, H, K	G, H, K	A, K	A, K	A, B, K

Notes:

Rubber cements or rubber and resin blends are among the most versatile adhesives.

For high strength and/or high temperature, a resin blend or alloy is preferred.

For porous material natural or thermoplastic adhesives are satisfactory and usually less expensive.

For any large-scale application special sources of information should be consulted, since there are many variations of each of the adhesives named and several not included in the above table.

[a]Typical resin blends and alloys include epoxy phenolic, vinyl phenolic, neoprene-rubber phenolic, nitrite-rubber phenolic, nylon phenolic, nylon epoxy, nitrile epoxy, and epoxy polyamide.

REFERENCES

"Adhesives for Metals", N.J. De Lollis, Industrial Press, 1970.
"Adhesives Handbook", J. Shields, Iliffe Books, 1970 (The Chemical Rubber Co., U.S. distributor).
"Aspects of Adhesion", D.J. Alner, Ed., University of London Press, 1965–68 (4 vols.).
"Handbook of Adhesives", I. Skeist, Ed., Reinhold Publishing Corp., 1962.
"Sealants", A. Damusis, Reinhold Publishing Corp., 1967.
"Structural Adhesives Bonding", M.J. Bodnar, Ed., Interscience Books, Inc., 1966.

Table 3-27.   PROPERTIES OF ADHESIVES*

Characteristics of Adhesive Materials and Joints

Property or quality	Typical examples	
	High values	Low values
TIME-DEPENDENT PROPERTIES		
Shelf-life	Animal glues; polyvinyl acetate; hot melts; water-based adhesives	One-component epoxy; nitrile and thiokol rubbers; poly-aromatics; polyurethane
Tack retention	Natural rubber; butyl rubber; epoxy	Ceramic; hot melts
Assembly time	Melamine formaldehyde; reclaim rubber	Epoxy; hot melts; phenolic-nitrile; polyamide
Curing or setting time	Epoxy; casein; acrylics	Solvent evaporation; hot melts
Creep resistance	Thermosetting resins; epoxy; ceramics	Cellulose acetate and nitrate
Resistance to biodeterioration	Ceramics; cellulose; rubbers; asphalt; acrylics; silicate	Casein; animal glue; starch
SOLVENT PROPERTIES		
Moisture resistance	Rubbers; epoxy; polystyrene; formaldehydes; ceramics; asphalt	Casein; cellulose acetate; animal glue; polyamide; acrylics
Alcohol resistance	Rubbers; phenolics; epoxy; vinyls	Animal glue; cellulose; acrylics; polyamide
Hydrocarbon resistance	Animal and casein glues; asphalt; epoxy	Rubbers; polyacrylate
Chemical resistance	Formaldehydes; polyurethane rubber; ceramics; asphalt	Phenolic; epoxy; cellulose; acrylics
MECHANICAL PROPERTIES		
Modulus of elasticity	Cellulose nitrate; glass ceramic	Cellulose acetate butyrate; silicone resins
Tensile strength	Cellulose acetate; melamine and urea formaldehydes; casein; nylon epoxy	Silicates; silicone resin or silicone rubber
Flexural strength	Casein; melamine formaldehyde	Polyvinylidene chloride
Compressive strength	Casein; melamine and phenol formaldehydes	PVC; polystyrene
Impact strength	Ethyl cellulose; silicone resins	Urea formaldehyde; casein
Peel strength	Phenolic; polyester; nylon epoxy; polyurethane rubber	Silicone rubber; cellulose; polyester; ceramic
Elongation	Rubbers; polyamide; vinyls	Epoxy; formaldehydes; cellulose acetate
THERMAL AND RADIATION PROPERTIES		
Heat (high temperature) resistance	Glass ceramic; silicates; phenolic epoxy; silicone resin; polyimide	PVC; rubbers; cellulose
Low-temperature stability	Nylon epoxy; silicone rubber; phenolic epoxy	Butyl rubber; polyacrylate
Thermal conductivity	Phenol formaldehyde	Rubbers; silicone resin
Thermal expansion coefficient	Cellulose acetate butyrate; waxes; reclaim rubber	Glass ceramic; silicate; melamine formaldehyde
ELECTRICAL PROPERTIES		
Volume resistivity	Polyvinyl chloride; natural rubber; glass ceramic	Casein; cellulose nitrate; water-based adhesives
Dielectric constant, 60 hz	Melamine formaldehyde	Butyl rubber
Dielectric constant, 1 Mhz	Polychloroprene rubber	Polystyrene
Power factor, 60 hz	Melamine formaldehyde	Polystyrene; natural rubber
Power factor, 1 Mhz	Polyvinyl chloride; cellulose nitrate	Mineral wax; silicone rubber
Dielectric strength	Vinyl chloride-vinylidene chloride copolymer; polyvinyl butyral	

*Compiled from several sources.

Table 3-28. THE pH SCALE

The pH value is numerically equal to the $\log_{10}$ of the hydrogen-ion concentration (with sign changed).

In acid solutions hydrogen-ion concentrations outbalance hydroxyl-ion concentrations. The reverse is true for basic, or alkaline, solutions. A solution becomes more acid when an electrolyte is added that contributes to the free hydrogen-ion concentration. It becomes more basic as the hydroxyl ions are increased.

pH value	Description	Hydrogen-ion concentration, moles per liter	pH value	Description	Hydrogen-ion concentration moles per liter
−1	Strong acid	10	7	Neutral (pure water)	10^{-7}
0		10^0	8		10^{-8}
1		10^{-1}	9		10^{-9}
2		10^{-2}	10		10^{-10}
3		10^{-3}	11		10^{-11}
4		10^{-4}	12		10^{-12}
5		10^{-5}	13		10^{-13}
6		10^{-6}	14	Strong base	10^{-14}

Table 3-29.   APPROXIMATE pH VALUES*

The following tables give approximate pH values for a number of substances such as acids, bases, foods, and biological fluids. All values are rounded off to the nearest tenth and are based on measurements made at 25°C. A few buffer systems with their pH values are also given.

Substance	pH value	Substance	pH value	Substance	pH value
ACIDS					
Hydrochloric, N	0.1	Oxalic, 0.1N	1.6	Acetic, 0.01N	3.4
Hydrochloric, 0.1N	1.1	Tartaric, 0.1N	2.2	Benzoic, 0.01N	3.1
Hydrochloric, 0.01N	2.0	Malic 0.1N	2.2	Alum, 0.1N	3.2
Sulfuric, N	0.3	Citric, 0.1N	2.2	Carbonic (saturated)	3.8
Sulfuric, 0.1N	1.2	Formic, 0.1N	2.3	Hydrogen sulfide, 0.1N	4.1
Sulfuric, 0.01N	2.1	Lactic, 0.1N	2.4	Arsenious (saturated)	5.0
Orthophosphoric, 0.1N	1.5	Acetic, N	2.4	Hydrocyanic, 0.1N	5.1
Sulfurous, 0.1N	1.5	Acetic, 0.1N	2.9	Boric, 0.1N	5.2
BASES					
Sodium hydroxide, N	14.0	Lime (saturated)	12.4	Magnesia (saturated)	10.5
Sodium hydroxide, 0.1N	13.0	Trisodium phosphate, 0.1 N	12.0	Sodium sesquicarbonate, 0.1M	10.1
Sodium hydroxide, 0.01N	12.0	Sodium carbonate, 0.1N	11.6	Ferrous hydroxide (saturated)	9.5
Potassium hydroxide, N	14.0	Ammonia, N	11.6	Calcium carbonate (saturated)	9.4
Potassium hydroxide, 0.1N	13.0	Ammonia, 0.1N	11.1	Borax, 0.1N	9.2
Potassium hydroxide, 0.01N	12.0	Ammonia, 0.01N	10.6	Sodium bicarbonate, 0.1N	8.4
Sodium metasilicate, 0.1N	12.6	Potassium cyanide, 0.1N	11.0		
BIOLOGIC MATERIALS					
Blood, plasma, human	7.3–7.5	Gastric contents, human	1.0–3.0	Milk, human	6.6–7.6
Spinal fluid, human	7.3–7.5	Duodenal contents, human	4.8–8.2	Bile, human	6.8–7.0
Blood, whole, dog	6.9–7.2	Feces, human	4.6–8.4		
Saliva, human	6.5–7.5	Urine, human	4.8–8.4		
FOODS					
Apples	2.9–3.3	Gooseberries	2.8–3.0	Potatoes	5.6–6.0
Apricots	3.6–4.0	Grapefruit	3.0–3.3	Pumpkin	4.8–5.2
Asparagus	5.4–5.8	Grapes	3.5–4.5	Raspberries	3.2–3.6
Bananas	4.5–4.7	Hominy (lye)	6.8–8.0	Rhubarb	3.1–3.2
Beans	5.0–6.0	Jams, fruit	3.5–4.0	Salmon	6.1–6.3
Beers	4.0–5.0	Jellies, fruit	2.8–3.4	Sauerkraut	3.4–3.6
Beets	4.9–5.5	Lemons	2.2–2.4	Shrimp	6.8–7.0
Blackberries	3.2–3.6	Limes	1.8–2.0	Soft drinks	2.0–4.0
Bread, white	5.0–6.0	Maple syrup	6.5–7.0	Spinach	5.1–5.7
Butter	6.1–6.4	Milk, cows	6.3–6.6	Squash	5.0–5.4
Cabbage	5.2–5.4	Olives	3.6–3.8	Strawberries	3.0–3.5
Carrots	4.9–5.3	Oranges	3.0–4.0	Sweet potatoes	5.3–5.6
Cheese	4.8–6.4	Oysters	6.1–6.6	Tomatoes	4.0–4.4
Cherries	3.2–4.0	Peaches	3.4–3.6	Tuna	5.9–6.1
Cider	2.9–3.3	Pears	3.6–4.0	Turnips	5.2–5.6
Corn	6.0–6.5	Peas	5.8–6.4	Vinegar	2.4–3.4
Crackers	6.5–8.5	Pickles, dill	3.2–3.6	Water, drinking	6.5–8.0
Dates	6.2–6.4	Pickles, sour	3.0–3.4	Wines	2.8–3.8
Eggs, fresh white	7.6–8.0	Pimento	4.6–5.2		
Flour, wheat	5.5–6.5	Plums	2.8–3.0		

*From: "Modern pH and Chlorine Control", W.A. Taylor & Co., by permission.

Table 3-30.   IONIZATION CONSTANTS FOR WATER (K_w)*

$-\log_{10} K_w$	Temperature, °C	$-\log_{10} K_w$	Temperature, °C
14.9435	0	13.8330	30
14.7338	5	13.6801	35
14.5346	10	13.5348	40
14.3463	15	13.3960	45
14.1669	20	13.2617	50
14.0000	24	13.1369	55
13.9965	25	13.0171	60

*From: "Handbook of Chemistry and Physics", 53rd ed., R.C. Weast, Ed., The Chemical Rubber Co., 1972.

Table 3-31.   IONIZATION CONSTANTS OF ACIDS IN WATER*

At Various Temperatures

Acids		Temperature, °C										
		0°	5°	10°	15°	20°	25°	30°	35°	40°	45°	50°
Formic	$K_A \cdot 10^4$	1.638	1.691	1.728	1.749	1.765	1.772	1.768	1.747	1.716	1.685	1.650
Acetic	$K_A \cdot 10^5$	1.657	1.700	1.729	1.745	1.753	1.754	1.750	1.728	1.703	1.670	1.633
Propionic	$K_A \cdot 10^5$	1.274	1.305	1.326	1.336	1.338	1.336	1.326	1.310	1.280	1.257	1.229
n-Butyric	$K_A \cdot 10^5$	1.563	1.574	1.576	1.569	1.542	1.515	1.484	1.439	1.395	1.347	1.302
Chloroacetic	$K_A \cdot 10^3$	1.528	—	1.488	—	—	1.379	—	—	1.230	—	—
Lactic	$K_A \cdot 10^4$	1.287	—	—	—	—	1.374	—	—	—	—	1.270
Glycollic	$K_A \cdot 10^4$	1.334	—	—	—	—	1.475	—	—	—	—	1.415
Oxalic	$K_{2A} \cdot 10^5$	5.91	5.82	5.70	5.55	5.40	5.18	4.92	4.67	4.41	4.09	3.83
Malonic	$K_{2A} \cdot 10^6$	2.140	2.165	2.152	2.124	2.076	2.014	1.948	1.863	1.768	1.670	1.575
Phosphoric	$K_A \cdot 10^3$	8.968	—	—	—	—	7.516	—	—	—	—	5.495
Phosphoric	$K_{2A} \cdot 10^8$	4.85	5.24	5.57	5.89	6.12	6.34	6.46	6.53	6.58	6.59	6.55
Boric	$K_A \cdot 10^{10}$	—	3.63	4.17	4.72	5.26	5.79	6.34	6.86	7.38	—	8.32
Carbonic	$K_{1A} \cdot 10^7$	2.64	3.04	3.44	3.81	4.16	4.45	4.71	4.90	5.04	5.13	5.19
Phenol-sulfonic	$K_{7A} \cdot 10^{10}$	4.45	5.20	6.03	6.92	7.85	8.85	9.89	10.94	12.00	13.09	14.16
Glycine	$K_{1A} \cdot 10^7$	—	3.82	3.99	4.17	4.32	4.46	4.57	4.66	4.73	4.77	4.79
Citric	$K_{1A} \cdot 10^4$	6.03	6.31	6.69	6.92	7.21	7.45	7.66	7.78	7.96	7.99	8.04
	$K_{2A} \cdot 10^5$	1.45	1.54	1.60	1.65	1.70	1.73	1.76	1.77	1.78	1.76	1.75
	$K_{3A} \cdot 10^7$	4.05	4.11	4.14	4.13	4.09	4.02	3.99	3.78	3.69	3.45	3.28

Reproducibility between various workers is about $\pm (0.01-0.02)10^5$.
All values are on the m-scale.

*From: "Handbook of Chemistry and Physics", 53rd ed., R.C. Weast, Ed., The Chemical Rubber Co., 1972.

Table 3-32.   IONIZATION CONSTANTS FOR DEUTERIUM OXIDE*

From 10 to 50°C

The subscript m indicates values on the molal scale, whereas the subscript c indicates values on the molar scale.

Temperature, °C	pK_m	pK_c
10	15.526	15.439
20	15.136	15.049
25	14.955	14.869
30	14.784	14.699
40	14.468	14.385
50	14.182	14.103

*From: "NBS Technical Note 400," National Bureau of Standards, U.S. Government Printing Office, 1966.

Table 3-33.   WATER SOLUTIONS*

Densities and Freezing Points of Water Solutions of Common Acids, Salts, and Other Compounds

For related data see Tables 1-48, 1-51, 3-36, 3-37, and 3-40.

SYMBOLS: $Wt\%$ = percentage of solute in the solution by weight

R_{wt} = weight ratio: solute/water

Sp gr = specific gravity 20/20

fp, deg C = freezing point, deg C

Wt%	R_{wt}	Acetic acid $(C_2H_4O_2)$		Acetone (C_3H_6O)		Ammonium chloride (NH_4Cl)		Ammonium hydroxide (NH_5O)	
		Sp gr	fp, deg C	Sp gr	fp, deg C	Sp gr	fp, deg C	Sp gr	fp, deg C
1	0.010 1	1.001 5	−.31	0.998 5	−.32	1.003 1	−.64	0.997 9	−.57
2	0.020 4	1.002 9	−.62	0.997 1	−.64	1.006 3	−1.27	0.995 7	−1.15
4	0.041 7	1.005 8	−1.25	0.994 3	−1.30	1.012 5	−2.57	0.991 5	−2.32
6	0.063 8	1.008 7	−1.88	0.991 7	−1.95	1.018 6	−3.94	0.987 4	−3.73
8	0.087 0	1.011 5	−2.54	0.989 1	−2.62	1.024 5	−5.40	0.983 3	−4.81
10	0.111 1	1.014 4	−3.21	0.986 7	−3.29	1.030 4	−6.94	0.979 3	−6.02
15	0.176 5	1.021 3	−4.96			1.044 7		0.969 6	−9.66
20	0.250 0	1.028 0	÷6.81			1.058 6		0.960 3	−14.41
25	0.333 3	1.034 3	−8.77					0.951 4	−20.70
30	0.428 6	1.040 3	−10.81					0.942 7	−28.91
35	0.538 5	1.045 7	−12.92					0.934 4	−39.96

Wt%	R_{wt}	Ammonium sulfate		Barium chloride		Calcium chloride		Citric acid $(C_6H_8O_7)$	
		Sp gr	fp, deg C	Sp gr	fp, deg C	Sp gr	fp, deg C	Sp gr	fp, deg C
1	0.010 1	1.005 9	−.33	1.008 8	−.23	1.008 3	−.44	1.004 0	−.11
2	0.020 4	1.011 9	−.63	1.017 7	−.46	1.016 6	−.88	1.008 1	−.23
4	0.041 7	1.023 8	−1.21	1.035 9	−.93	1.033 4	−1.82	1.016 5	−.47
6	0.063 8	1.035 6	−1.77	1.054 7	−1.43	1.050 5	−2.94	1.025 0	−.72
8	0.087 0	1.047 5	−2.32	1.074 0	−1.98	1.067 8	−4.28	1.033 6	−.97
10	0.111 1	1.059 3	−2.89	1.094 0	−2.57	1.085 4	−5.85	1.042 5	−1.23
15	0.176 5	1.088 5	−4.37	1.147 2	−4.30	1.131 1	−11.00	1.065 1	
20	0.250 0	1.117 4		1.205 2		1.179 6		1.088 3	
25	0.333 3			1.268 4		1.230 0		1.112 2	
30	0.428 6					1.283 4		1.136 8	
35	0.538 5					1.339 0			

Wt%	R_{wt}	Cupric sulfate		Ethanol		Ethylene glycol $(C_2H_6O_2)$		Ferric chloride	
		Sp gr	fp, deg C	Sp gr	fp, deg C	Sp gr	fp, deg C	Sp gr	fp, deg C
1	0.010 1	1.010 3	−.14	0.998 0	−.40	1.001 2	−.30	1.008 6	−.38
2	0.020 4	1.020 8	−.26	0.996 2	−.81	1.002 5	−.60	1.017 1	−.81
4	0.041 7	1.042 1	−.49	0.992 8	−1.65	1.004 9	−1.24	1.034 0	−1.49
6	0.063 8	1.063 9	−.70	0.989 6	−2.54	1.007 4	−1.92	1.051 1	−2.36
8	0.087 0	1.086 2	−.93	0.986 6	−3.47	1.010 0	−2.64	1.068 8	−3.45
10	0.111 1	1.109 0	−1.18	0.983 7	−4.48	1.012 5	−3.41	1.087 1	−4.77
15	0.176 5	1.169 0		0.976 9	−7.34			1.134 3	−9.33
20	0.250 0			0.970 2	−10.94			1.183 8	
25	0.333 3			0.963 2	−15.43			1.236 3	
30	0.428 6			0.955 6	−20.50			1.293 4	
35	0.538 5			0.946 8				1.355 3	

*Based largely on: "Handbook of Chemistry and Physics", 53rd ed., R.C. Weast, Ed., The Chemical Rubber Co., 1972; for viscosity, electrical conductivity, refractive index, and molar concentrations, see this source.

Table 3-33.   WATER SOLUTIONS *(Continued)*

Wt %	R_{wt}	Formic acid (CH_2O_2)		Glycerol ($C_3H_8O_3$)		Hydrochloric acid		Lead nitrate (PbN_2O_6)	
		Sp gr	fp, deg C	Sp gr	fp, deg C	Sp gr	fp, deg C	Sp gr	fp, deg C
1	0.010 1	1.002 7	−.39	1.002 4	−.20	1.005 0	−.98	1.008 8	−.13
2	0.020 4	1.005 4	−.81	1.004 7	−.41	1.009 9	−2.04	1.017 7	−.26
4	0.041 7	1.010 7	−1.67	1.009 5	−.85	1.019 8	−4.52	1.035 7	−.48
6	0.063 8	1.016 0	−2.53	1.014 3	−1.31	1.029 6	−7.54	1.054 3	−.69
8	0.087 0	1.021 1	−3.41	1.019 1	−1.81	1.039 5	−11.17	1.073 4	−.88
10	0.111 1	1.026 3	−4.30	1.024 0	−2.33	1.048 8	−15.47	1.093 1	−1.07
15	0.176 5	1.038 8	−6.60	1.036 3	−3.77	1.074 6		1.145 8	−1.55
20	0.250 0	1.051 0	−9.00	1.048 9	−5.46	1.100 3		1.204 3	−2.06
25	0.333 3	1.062 9	−11.51	1.061 6	−7.43	1.125 7		1.266 9	
30	0.428 6	1.074 8	−14.18	1.074 7	−9.72	1.151 0		1.330 2	
35	0.538 5	1.086 5	−17.04	1.087 8	−12.41	1.175 9			

Wt %	R_{wt}	Lithium chloride ($LiCl$)		Magnesium chloride		Magnesium sulfate		Manganese sulfate	
		Sp gr	fp, deg C	Sp gr	fp, deg C	Sp gr	fp, deg C	Sp gr	fp, deg C
1	0.010 1	1.005 9	−.83	1.008 2	−.55	1.010 2	−.18	1.009 8	−.15
2	0.020 4	1.011 7	−1.72	1.016 4	−1.05	1.020 4	−.35	1.019 6	−.29
4	0.041 7	1.023 3	−3.73	1.032 9	−2.26	1.041 1	−.70	1.039 6	−.57
6	0.063 8	1.034 8	−6.15	1.049 6	−3.78	1.062 1	−1.03	1.060 2	−.88
8	0.087 0	1.046 3	−9.09	1.066 5	−5.65	1.083 5	−1.39	1.081 3	−1.12
10	0.111 1	1.057 8	−12.59	1.083 5	−7.91	1.105 3	−1.82	1.103 1	−1.41
15	0.176 5	1.087 0	−23.39	1.127 2	−15.64	1.162 0	−1.82	1.160 6	−2.37
20	0.250 0	1.117 0		1.172 8		1.222 0		1.222 2	−3.77
25	0.333 3	1.148 3		1.220 6		1.285 4			
30	0.428 6	1.181 2		1.271 1					
35	0.538 5								

Wt %	R_{wt}	Methanol (CH_4O)		Nitric acid		Phosphoric acid		Potassium carbonate	
		Sp gr	fp, deg C	Sp gr	fp, deg C	Sp gr	fp, deg C	Sp gr	fp, deg C
1	0.010 1	0.998 2	−.56	1.005 4	−.54	1.005 5	−.24	1.009 0	−.34
2	0.020 4	0.996 5	−1.14	1.010 9	−1.11	1.011 0	−.46	1.018 0	−.66
4	0.041 7	0.993 1	−2.37	1.021 9	−2.33	1.021 8	−.94	1.036 2	−1.43
6	0.063 8	0.989 8	−3.71	1.032 9	−3.64	1.032 8	−1.44	1.054 7	−1.99
8	0.087 0	0.986 5	−5.13	1.044 6	−5.06	1.043 8	−2.08	1.073 4	−2.62
10	0.111 1	0.983 4	−6.57	1.056 3	−6.59	1.055 1	−2.77	1.092 4	−3.34
15	0.176 5	0.975 7	−10.51	1.086 1	−11.02	1.084 3	−4.70	1.141 1	−5.62
20	0.250 0	0.968 2	−15.05	1.116 9	−16.48	1.115 4	−6.99	1.191 9	−8.74
25	0.333 3	0.960 7	−20.18	1.149 0		1.148 2	−9.75	1.244 9	−13.00
30	0.428 6	0.953 1	−25.79	1.182 2		1.182 5	−13.08	1.300 1	−18.75
35	0.538 5	0.945 0	−31.81	1.215 8		1.218 4		1.357 5	

Table 3-33.   WATER SOLUTIONS *(Continued)*

Wt%	R_{wt}	Potassium chloride (KCl)		Potassium chromate (K_2CrO_4)		Potassium dichromate ($K_2Cr_2O_7$)		Potassium hydroxide (KOH)	
		Sp gr	fp, deg C	Sp gr	fp, deg C	Sp gr	fp, deg C	Sp gr	fp, deg C
1	0.010 1	1.006 4	−.46	1.008 0	−.16	1.007 0	−.19	1.010 1	
2	0.020 4	1.012 8	−.92	1.016 1	−.32	1.014 0	−.36	1.019 3	
4	0.041 7	1.025 7	−1.85	1.032 5	−.64	1.028 2		1.037 7	
6	0.063 8	1.038 7	−2.80	1.049 2	−.99	1.042 6		1.056 2	
8	0.087 0	1.051 9	−3.79	1.066 3	−1.32	1.057 3		1.074 8	
10	0.111 1	1.065 2	−4.81	1.083 6	−1.75	1.072 2		1.093 6	
15	0.176 5	1.099 3		1.128 7	−2.96			1.141 4	
20	0.250 0	1.134 8		1.176 5	−4.38			1.189 8	
25	0.333 3			1.227 3				1.240 5	
30	0.428 6			1.281 2				1.292 3	
35	0.538 5			1.338 5				1.345 8	

Wt%	R_{wt}	Potassium iodide (KI)		Potassium nitrate (KNO_3)		Potassium phosphate (KH_2PO_4)		Potassium sulfate	
		Sp gr	fp, deg C	Sp gr	fp, deg C	Sp gr	fp, deg C	Sp gr	fp, deg C
1	0.010 1	1.007 3	−.22	1.006 2	−.33	1.007 2	−.25	1.008 0	−.26
2	0.020 4	1.014 8	−.43	1.012 5	−.64	1.014 4	−.50	1.016 1	−.50
4	0.041 7	1.029 9	−.87	1.025 3	−1.22	1.028 8	−.97	1.032 4	−.95
6	0.063 8	1.045 5	−1.31	1.038 2	−1.76	1.043 2	−1.42	1.048 9	
8	0.087 0	1.061 5	−1.76	1.051 3	−2.27	1.057 7	−1.84	1.065 6	
10	0.111 1	1.078 0	−2.25	1.064 6	−2.75	1.072 2	−2.23	1.082 5	
15	0.176 5	1.121 3	−3.59	1.098 8					
20	0.250 0	1.168 1	−5.11	1.134 6					
25	0.333 3	1.218 6	−6.85						
30	0.428 6	1.273 4	−8.85						
35	0.538 5	1.333 3	−11.19						

Wt%	R_{wt}	Propylene glycol		Silver nitrate ($AgNO_3$)		Sodium acetate		Sodium bicarbonate ($NaHCO_3$)	
		Sp gr	fp, deg C	Sp gr	fp, deg C	Sp gr	fp, deg C	Sp gr	fp, deg C
1	0.010 1	1.000 8	−.24	1.008 7	−.20	1.005 1	−.43	1.007 2	−.43
2	0.020 4	1.001 6	−.49	1.017 3	−.39	1.010 2	−.88	1.014 5	−.83
4	0.041 7	1.003 1	−1.02	1.034 6	−.77	1.020 5	−1.82	1.028 9	−1.59
6	0.063 8	1.004 6	−1.58	1.052 4	−1.15	1.030 7	−2.85	1.043 2	−2.25
8	0.087 0	1.006 1		1.070 9	−1.52	1.041 0	−3.97		
10	0.111 1	1.007 5		1.090 1	−1.87	1.051 3	−5.17		
15	0.176 5			1.140 9	−2.70	1.077 4			
20	0.250 0			1.196 3		1.104 1			
25	0.333 3			1.256 8		1.131 4			
30	0.428 6			1.322 8		1.159 6			
35	0.538 5			1.395 8					

Table 3-33.   WATER SOLUTIONS *(Continued)*

Wt %	R_{wt}	Sodium bromide		Sodium carbonate		Sodium chloride (NaCl)		Sodium hydroxide	
		Sp gr	fp, deg C	Sp gr	fp, deg C	Sp gr	fp, deg C	Sp gr	fp, deg C
1	0.010 1	1.007 8	−.34	1.010 4	−.42	1.007 1	−.59	1.011 2	−.88
2	0.020 4	1.015 6	−.69	1.020 8	−.82	1.014 3	−1.19	1.022 3	−1.71
4	0.041 7	1.031 6	−1.39	1.041 6	−1.52	1.028 6	−2.41	1.044 5	−3.54
6	0.063 8	1.048 1	−2.14	1.062 5	−2.12	1.043 1	−3.69	1.066 7	−5.59
8	0.087 0	1.064 9	−2.93	1.083 5		1.057 8	−5.05	1.088 8	−7.91
10	0.111 1	1.082 3	−3.77	1.104 8		1.072 6	−6.54	1.111 0	−10.52
15	0.176 5	1.127 7	−6.16	1.159 4		1.110 5	−10.88	1.166 2	
20	0.250 0	1.176 6				1.147 8	−16.45	1.221 3	
25	0.333 3	1.229 4				1.190 9		1.276 0	
30	0.428 6	1.286 4						1.330 1	
35	0.538 5	1.348 6						1.382 5	

Wt %	R_{wt}	Sodium nitrate ($NaNO_3$)		Sodium sulfate		Sodium tartrate ($Na_2C_4H_0O_0$)		Sodium thiosulfate ($Na_2S_2O_3$)	
		Sp gr	fp, deg C	Sp gr	fp, deg C	Sp gr	fp, deg C	Sp gr	fp, deg C
1	0.010 1	1.006 7	−.40	1.009 1	−.32	1.007 0	−.23	1.008 3	
2	0.020 4	1.013 5	−.77	1.018 2	−.61	1.014 1	−.46	1.016 5	
4	0.041 7	1.027 1	−1.51	1.036 7	−1.13	1.028 4	−.89	1.033 3	
6	0.063 8	1.040 9	−2.28	1.055 3	−1.56	1.042 8		1.050 2	
8	0.087 0	1.055 0	−3.05	1.074 3		1.057 4		1.067 3	
10	0.111 1	1.069 3	−3.84	1.093 4		1.072 1		1.084 7	
15	0.176 5	1.106 3	−5.84	1.142 6		1.109 9		1.129 3	
20	0.250 0	1.144 9	−7.89	1.193 6		1.149 1		1.176 0	
25	0.333 3	1.185 4				1.190 0		1.225 0	
30	0.428 6	1.227 9						1.276 2	
35	0.538 5	1.272 6						1.329 7	

Wt %	R_{wt}	Sucrose ($C_{12}H_{22}O_{11}$)		Sulfuric acid		Zinc chloride ($ZnCl_2$)		Zinc sulfate	
		Sp gr	fp, deg C	Sp gr	fp, deg C	Sp gr	fp, deg C	Sp gr	fp, deg C
1	0.010 1	1.003 9	−.06	1.006 8	−.41			1.010 4	−.15
2	0.020 4	1.007 8	−.11	1.013 5	−.78	1.018 5		1.020 9	−.28
4	0.041 7	1.015 7	−.23	1.026 9	−1.52	1.036 8		1.042 1	−.53
6	0.063 8	1.023 7	−.35	1.040 4	−2.49	1.055 0		1.063 8	−.77
8	0.087 0	1.031 8	−.49	1.054 0	−3.55	1.073 3		1.086 1	−1.01
10	0.111 1	1.040 0	−.63	1.067 9	−4.70	1.083 7		1.109 1	−1.27
15	0.176 5	1.061 0	−1.01	1.103 9	−8.30	1.138 9		1.169 9	−2.07
20	0.250 0	1.082 9	−1.46	1.141 6	−13.58	1.188 4			
25	0.333 3	1.105 5	−2.00	1.180 4	−22.12	1.239 8			
30	0.428 6	1.129 0	−2.65	1.220 5	−36.29	1.294 6			
35	0.538 5	1.153 3	−3.44	1.262 0	−61.89	1.354 0			

Table 3-34. SOLUBILITY CHART**

SYMBOLS:

W = soluble in water
A = insoluble in water but soluble in acids
w = sparingly soluble in water but soluble in acids
a = insoluble in water and only sparingly soluble in acids
I = insoluble in both water and acids
d = decomposes in water

No.		Al	NH$_4$	Sb	Ba	Bi	Cd	Ca	Cr	Co	Cu	Au	Au'''	H	Fe''	Fe'''
1	Acetates —(C$_2$H$_3$O$_2$)	W Al(—)$_3$	W NH$_4$(—)		W Ba(—)$_2$	W Bi(—)$_3$	W Cd(—)$_2$	W Ca(—)$_2$	W Cr(—)$_3$	W Co(—)$_2$	W Cu(—)$_2$			W C$_2$H$_4$O$_2$	W Fe(—)$_2$	W Fe$_2$(—)$_6$
2	Arsenate —(AsO$_4$)	a Al(—)	W (NH$_4$)$_3$(—)	A Sb(—)	w Ba$_3$(—)$_2$	A Bi(—)	A Cd$_3$(—)$_2$	A Ca$_3$(—)$_2$		A Co$_3$(—)$_2$	A Cu$_3$(—)$_2$			W H$_3$AsO$_4$	A Fe$_3$(—)$_2$	A Fe(—)
3	Arsenite —(AsO$_3$)		W NH$_4$AsO$_2$	A Sb(—)				w Ca$_3$(—)$_2$		A Co$_3$H$_6$(—)$_4$	A CuH(—)					
4	Benzoate —(C$_7$H$_5$O$_2$)		W NH$_4$(—)		W Ba(—)$_2$	A Bi(—)$_3$	W Cd(—)$_2$	W Ca(—)$_2$		W Co(—)$_2$	w Cu(—)$_2$			W C$_7$H$_6$O$_2$	W Fe(—)$_2$	A Fe$_2$(—)$_6$
5	Bromide	W AlBr$_3$	W NH$_4$Br	d SbBr$_3$	W BaBr$_2$	W BiBr$_3$	W CdBr$_2$	W CaBr$_2$	W(I)* CrBr$_3$	W CoBr$_2$	W CuBr$_2$	W AuBr	W AuBr$_3$	W HBr	W FeBr$_2$	W FeBr$_3$
6	Carbonate		W (NH$_4$)$_2$CO$_3$		W BaCO$_3$		A CdCO$_3$	w CaCO$_3$	W CrCO$_3$	A CoCO$_3$					w FeCO$_3$	
7	Chlorate —(ClO$_3$)	W Al(—)$_3$	W NH$_4$(—)		W Ba(—)$_2$	W Bi(—)$_3$	W Cd(—)$_2$	W Ca(—)$_2$		W Co(—)$_2$	W Cu(—)$_2$			W HClO$_3$	W Fe(—)$_2$	W Fe(—)$_3$
8	Chloride	W AlCl$_3$	W NH$_4$Cl	W SbCl$_3$	W BaCl$_2$	d BiCl$_3$	W, CdCl$_2$	W CaCl$_2$	I CrCl$_3$	W CoCl$_2$	W CuCl$_2$	w AuCl	W AuCl$_3$	W HCl	W FeCl$_2$	W FeCl$_3$
9	Chromate —(CrO$_4$)		W (NH$_4$)$_2$(—)		A Ba(—)		A Cd(—)	W Ca(—)		A Co(—)						A Fe$_2$(—)
10	Citrate —(C$_6$H$_5$O$_7$)	W Al(—)	W (NH$_4$)$_3$(—)		w Ba$_3$(—)$_2$	A Bi(—)	W Cd$_3$(—)$_2$	w Ca$_3$(—)$_2$		w Co$_3$(—)$_2$				W C$_6$H$_8$O$_7$		W Fe(—)
11	Cyanide		W NH$_4$CN		W Ba(CN)$_2$	A Bi(CN)$_3$	A Cd(CN)$_2$	W Ca(CN)$_2$	W Cr(CN)$_2$	A Co(CN)$_2$	A Cu(CN)$_2$	w AuCN	W Au(CN)$_3$	W HCN	a Fe(CN)$_2$	
12	Ferricy'de —(Fe(CN)$_6$)		W (NH$_4$)$_3$(—)		w Ba$_3$(—)$_2$		A Cd$_3$(—)$_2$	A Ca$_3$(—)$_2$		I Co$_3$(—)$_2$	I Cu$_3$(—)$_2$			W H$_3$(—)	I Fe$_3$(—)$_2$	
13	Ferrocy'de —(Fe(CN)$_6$)	w Al$_4$(—)$_3$	W (NH$_4$)$_4$(—)		W Ba$_2$(—)		A Cd$_2$(—)	A Ca$_2$(—)		I Co$_2$(—)	I Cu$_2$(—)			W H$_4$(—)	W Fe$_2$(—)	a Fe$_4$(—)$_3$
14	Fluoride	W AlF$_3$	W NH$_4$F	W SbF$_3$	W BaF$_2$	W BiF$_3$	W CdF$_2$	A CaF$_2$	W(a)* CrF$_3$	W CoF$_2$	w CuF$_2$			W HF	W FeF$_2$	w FeF$_3$
15	Formate —(CHO$_2$)	W Al(—)$_3$	W NH$_4$(—)		W Ba(—)$_2$	W Bi(—)$_3$	W Cd(—)$_2$	W Ca(—)$_2$		W Co(—)$_2$	W Cu(—)$_2$			W CH$_2$O$_2$	W Fe(—)$_2$	W Fe(—)$_3$
16	Hydroxide	A Al(OH)$_3$	W NH$_4$OH		W Ba(OH)$_2$	A Bi(OH)$_3$	A Cd(OH)$_2$	W Ca(OH)$_2$	A Cr(OH)$_3$	A Co(OH)$_2$	A Cu(OH)$_2$	W AuOH	A Au(OH)$_3$		A Fe(OH)$_2$	A Fe(OH)$_3$
17	Iodide	W AlI$_3$	W NH$_4$I	d SbI$_3$	W BaI$_2$	A BiI$_3$	W CdI$_2$	W CaI$_2$	W CrI$_2$	W CoI$_2$	a CuI	a AuI	a AuI$_3$	W HI	W FeI$_2$	W FeI$_3$
18	Nitrate	W Al(NO$_3$)$_3$	W NH$_4$NO$_3$		W Ba(NO$_3$)$_2$	d Bi(NO$_3$)$_3$	W Cd(NO$_3$)$_2$	W Ca(NO$_3$)$_2$	W Cr(NO$_3$)$_3$	W Co(NO$_3$)$_2$	W Cu(NO$_3$)$_2$			W HNO$_3$	W Fe(NO$_3$)$_2$	W Fe(NO
19	Oxalate —(C$_2$O$_4$)	A Al$_2$(—)$_3$	W (NH$_4$)$_2$(—)		W Ba(—)	A Bi$_2$(—)$_3$	W Cd(—)	A Ca(—)	W Cr(—)	A Co(—)	A Cu(—)			W C$_2$H$_2$O$_4$	W Fe(—)	W Fe$_2$(—)
20	Oxide	a Al$_2$O$_3$		w Sb$_2$O$_3$	W BaO	A Bi$_2$O$_3$	A CdO	W CaO	a Cr$_2$O$_3$	A CoO	A CuO	W Au$_2$O	A Au$_2$O$_3$	W H$_2$O$_2$	A FeO	A Fe$_2$O$_3$
21	Phosphate	A AlPO$_4$	W NH$_4$H$_2$PO$_4$		A Ba$_3$(PO$_4$)$_2$	A BiPO$_4$	w Cd$_3$(PO$_4$)$_2$	w Ca$_3$(PO$_4$)$_2$	w Cr$_2$(PO$_4$)$_2$	A Co$_3$(PO$_4$)$_2$	A Cu$_3$(PO$_4$)$_2$			W H$_3$PO$_4$	A Fe$_3$(PO$_4$)$_2$	A FePO$_4$
22	Silicate, —(SiO$_3$)	I Al$_2$(—)$_3$			W Ba(—)		A Cd(—)	A Ca(—)		A Co$_2$SiO$_4$	w Cu(—)			W H$_2$SiO$_3$	I	
23	Sulfate	W Al$_2$(SO$_4$)$_3$	W (NH$_4$)$_2$SO$_4$	A Sb$_2$(SO$_4$)$_3$	A BaSO$_4$	d Bi$_2$(SO$_4$)$_3$	W CdSO$_4$	w CaSO$_4$	W(I)* Cr$_2$(SO$_4$)$_3$	W CoSO$_4$	W CuSO$_4$			W H$_2$SO$_4$	W FeSO$_4$	W Fe$_2$(SO$_4$)$_3$
24	Sulfide	d Al$_2$S$_3$	W (NH$_4$)$_2$S	A Sb$_2$S$_3$	d BaS	A Bi$_2$S$_3$	A CdS	w CaS	d Cr$_2$S$_3$	A CoS	A CuS	I Au$_2$S	I Au$_2$S$_3$	W H$_2$S	A FeS	d Fe$_2$S$_3$
25	Tartrate —(C$_4$H$_4$O$_6$)	w Al$_2$(—)$_3$	W (NH$_4$)$_2$(—)	W Sb$_2$(—)$_3$	w Ba(—)	A Bi$_2$(—)$_3$	A Cd(—)	w Ca(—)		w Co(—)	w Cu(—)			W C$_4$H$_6$O$_6$	w Fe(—)	W Fe$_2$(—)$_3$
26	Thiocy'te		W NH$_4$CNS		W Ba(CNS)$_2$			W Ca(CNS)		W Co(CNS)$_2$	d CuCNS			W CNSH	W Fe(CNS)$_2$	W Fe(CNS)$_3$

Table 3-34. SOLUBILITY CHART (Continued)

No.		Pb	Mg	Mn	Hg'	Hg''	Ni	K	Ag	Na	Sn''''	Sn''	Sr	Zn	Pt
1	Acetate —$(C_2H_3O_2)$	W $Pb(-)_2$	W $Mg(-)_2$	W $Mn(-)_2$	w $Hg(-)$	W $Hg(-)_2$	W $Ni(-)_2$	W $K(-)$	w $Ag(-)$	W $Na(-)$	W $Sn(-)_4$	d $Sn(-)_2$	W $Sr(-)_2$	W $Zn(-)_2$	
2	Arsenate —(AsO_4)	A $PbH(-)$	A $Mg_3(-)$	w $MnH(-)$	A $Hg_3(-)$	A $Hg_3(-)_2$	A $Ni_3(-)_2$	W $K_3(-)$	A $Ag_3(-)$	W $Na_3(-)$			w $SrH(-)$	A $Zn_3(-)_2$	
3	Arsenite —(AsO_3)		W $Mg_3(-)_2$	A $Mn_3H_6(-)$	A $Hg_3(-)$	A $Hg_3(-)$	A $Ni_3H_6(-)_4$	W K_3AsO_3	A $Ag_3(-)$	W $Na_2H(-)$		A $Sn_3(-)_2$	w $Sr_3(-)_2$		
4	Benzoate —$(C_7H_5O_2)$	w $Pb(-)_2$	W $Mg(-)_2$	W $Mn(-)_2$	w $Hg_2(-)_2$	W $Hg(-)_2$	W $Ni(-)_2$	W $K(-)$	w $Ag(-)$	W $Na(-)$				W $Zn(-)_2$	w
5	Bromide	W $PbBr_2$	W $MgBr_2$	W $MnBr_2$	A $HgBr$	W $HgBr_2$	W $NiBr_2$	W KBr	A $AgBr$	W $NaBr$	W $SnBr_4$	W $SnBr_2$	W $SrBr_2$	W $ZnBr_2$	W $PtBr_4$
6	Carbonate	A $PbCO_3$	A $MgCO_3$	A $MnCO_3$	A Hg_2CO_3		w $NiCO_3$	W K_2CO_3	A Ag_2CO_3	W Na_2CO_3			w $SrCO_3$	w $ZnCO_3$	
7	Chlorate —(ClO_3)	W $Pb(-)_2$	W $Mg(-)_2$	W $Mn(-)_2$	w $Hg(-)$	W $Hg(-)_2$	W $Ni(-)_2$	W $K(-)$	W $Ag(-)$	W $Na(-)$		W $Sn(-)_2$	W $Sr(-)_2$	W $Zn(-)_2$	
8	Chloride	W $PbCl_2$	W $MgCl_2$	W $MnCl_2$	a $HgCl$	W $HgCl_2$	W $NiCl_2$	W KCl	A $AgCl$	W $NaCl$	W $SnCl_4$	W $SnCl_2$	W $SrCl_2$	W $ZnCl_2$	W $PtCl_4$
9	Chromate —(CrO_4)	A $Pb(-)$	W $Mg(-)$		w $Hg_2(-)$	w $Hg(-)$	A $Ni(-)$	W $K_2(-)$	A $Ag_2(-)$	W $Na_2(-)$	w $Sn(-)_2$	W $Sn(-)$	w $Sr(-)$	w $Zn(-)$	
10	Citrate —$(C_6H_5O_7)$	W $Pb_3(-)_2$	W $Mg_3(-)_2$	w $MnH(-)$	w $Hg_3(-)$		W $Ni_3(-)_2$	W $K_3(-)$	W $Ag_3(-)$	W $Na_3(-)$			A $SrH(-)$	w $Zn_3(-)_2$	
11	Cyanide	w $Pb(CN)_2$	W $Mg(CN)_2$		$HgCN$	W $Hg(CN)_2$	W $Ni(CN)_2$	W KCN	A $AgCN$	W $NaCN$			W $Sr(CN)_2$	W $Zn(CN)_2$	I $Pt(CN)_2$
12	Ferricy'de —$Fe(CN)_6$	w $Pb_3(-)_2$	w $Mg_3(-)_2$			A $Hg_3(-)_2$	A $Ni_3(-)_2$	W $K_3(-)$	I $Ag_3(-)$	W $Na_3(-)$		A $Sn_3(-)_2$	W $Sr_3(-)_2$	W $Zn_3(-)_2$	
13	Ferrocy'de —$Fe(CN)_6$	w $Pb_2(-)$	W $Mg_2(-)$	A $Mn_2(-)$		I $Hg_2(-)_2$	I $Ni_2(-)$	W $K_4(-)$	I $Ag_4(-)$	W $Na_4(-)$		a $Sn_2(-)$	W $Sr_2(-)$	I $Zn_2(-)$	
14	Fluoride	w PbF_2	w MgF_2	A MnF_2	d HgF	d HgF_2	W NiF_2	W KF	W AgF	W NaF	W SnF_4	W SnF_2	w SrF_2	w ZnF_2	W PtF_4
15	Formate —(CHO_2)	W $Pb(-)_2$	W $Mg(-)_2$	W $Mn(-)_2$	w $Hg(-)$	W $Hg(-)_2$	W $Ni(-)_2$	W $K(-)$	w $Ag(-)$	W $Na(-)$			w $Sr(-)_2$	w $Zn(-)_2$	
16	Hydroxide	A $Pb(OH)_2$	A $Mg(OH)_2$	A $Mn(OH)_2$		A $Hg(OH)_2$	A $Ni(OH)_2$	W KOH		W $NaOH$	w $Sn(OH)_4$	A $Sn(OH)_2$	W $Sr(OH)_2$	A $Zn(OH)_2$	A $Pt(OH)_4$
17	Iodide	w PbI_2	W MgI_2	W MnI_2	A HgI	w HgI_2	W NiI_2	W KI	I AgI	W NaI	W SnI_4	d SnI_2	W SrI_2	W ZnI_2	I PtI_2
18	Nitrate	W $Pb(NO_3)_2$	W $Mg(NO_3)_2$	W $Mn(NO_3)_2$	W $HgNO_3$	W $Hg(NO_3)_2$	W $Ni(NO_3)_2$	W KNO_3	W $AgNO_3$	W $NaNO_3$		d $Sn(NO_3)_2$	W $Sr(NO_3)_2$	W $Zn(NO_3)_2$	W $Pt(NO_3)_4$
19	Oxalate —(C_2O_4)	A $Pb(-)$	w $Mg(-)$	w $Mn(-)$	a $Hg_2(-)$	w $Hg(-)$	A $Ni(-)$	W $K_2(-)$	A $Ag_2(-)$	W $Na_2(-)$		A $Sn(-)$	w $Sr(-)$	A $Zn(-)$	
20	Oxide	w PbO	A MgO	A MnO	A Hg_2O	A HgO	A NiO	W K_2O	A Ag_2O	d Na_2O	W SnO_2	A SnO	W SrO	W ZnO	A PtO
21	Phosphate	A $Pb_3(PO_4)_2$	A $Mg_3(PO_4)_2$	A $Mn_3(PO_4)_2$	A Hg_3PO_4	A $Hg_3(PO_4)_2$	A $Ni_3(PO_4)_2$	W K_3PO_4	A Ag_3PO_4	W Na_3PO_4		A $Sn_3(PO_4)_2$	A $Sr_3(PO_4)_2$	A $Zn_3(PO_4)_2$	
22	Silicate —(SiO_3)	A $Pb(-)$	A $Mg(-)$	I $Mn(-)$				W $K_2(-)$		W $Na_2(-)$			A $Sr(-)$	A $Zn(-)$	
23	Sulfate	W $PbSO_4$	W $MgSO_4$	W $MnSO_4$	W $HgSO_4$	W $HgSO_4$	W $NiSO_4$	W K_2SO_4	A Ag_2SO_4	W Na_2SO_4	W $Sn(SO_4)_2$	W $SnSO_4$	A $SrSO_4$	W $ZnSO_4$	W $Pt(SO_4)_2$
24	Sulfide	A PbS	d MgS	A MnS	I Hg_2S	I HgS	A NiS	W K_2S	A Ag_2S	W Na_2S	A SnS_2	A SnS	W SrS	A ZnS	I PtS
25	Tartrate —$(C_4H_4O_6)$	A $Pb(-)$	w $Mg(-)$	w $Mn(-)$	I $Hg_2(-)$		w $Ni(-)$	W $K_2(-)$	A $Ag_2(-)$	W $Na_2(-)$		w $Sn(-)$	w $Sr(-)$	w $Zn(-)$	
26	Thiocy'te —(CNS)	w $Pb(CNS)_2$	W $Mg(CNS)_2$	W $Mn(CNS)_2$	A $HgCNS$	W $Hg(CNS)_2$		W $KCNS$	I $AgCNS$	W $NaCNS$			W $Sr(CNS)_2$	W $Zn(CNS)_2$	

*Certain salts occur in two modifications.

**From: "Handbook of Chemistry and Physics", 53rd ed., R.C. Weast, Ed., The Chemical Rubber Co., 1972.

Table 3-35. SOLUBILITY PRODUCT*

The solubility product (or ion product constant) is the product of the concentrations of the ions in the saturated solution of a difficultly soluble salt. The concentrations are expressed as moles per liter of solution. The number of cations (or anions) resulting from the dissociation of one molecule of the salt appears in the formula for calculations of the solubility product as the exponent of the concentration of the cation (or anion).

If two solutions, each containing one of the ions of a difficultly soluble salt, are mixed, no precipitation takes place unless the product of the ion concentrations in the mixture is greater than the solubility product.

In a solution containing two salts that yield a common ion, the ratio of solubilities of the two salts is the ratio of the solubility products.

Substance	Solubility product at temperature noted, °C	Substance	Solubility product at temperature noted, °C
Aluminum hydroxide	4×10^{-13} (15°)	Lead iodide	7.47×10^{-9} (15°)
Aluminum hydroxide	1.1×10^{-15} (18°)	Lead iodide	1.39×10^{-8} (25°)
Aluminum hydroxide	3.7×10^{-15} (25°)	Lead oxalate	2.74×10^{-11} (18°)
Barium carbonate	7×10^{-9} (16°)	Lead sulfate	1.06×10^{-8} (18°)
Barium carbonate	8.1×10^{-9} (25°)	Lead sulfide	3.4×10^{-28} (18°)
Barium chromate	1.6×10^{-10} (18°)	Lithium carbonate	1.7×10^{-3} (25°)
Barium chromate	2.4×10^{-10} (28°)	Magnesium ammonium phosphate	2.5×10^{-13} (25°)
Barium fluoride	1.6×10^{-6} (9.5°)	Magnesium carbonate	2.6×10^{-5} (12°)
Barium fluoride	1.7×10^{-6} (18°)	Magnesium fluoride	7.1×10^{-9} (18°)
Barium fluoride	1.73×10^{-6} (25.8°)	Magnesium fluoride	6.4×10^{-9} (27°)
Barium iodate, $Ba(IO_3)_2 \cdot 2H_2O$	8.4×10^{-11} (10°)	Magnesium hydroxide	1.2×10^{-11} (18°)
Barium iodate, $Ba(IO_3)_2 \cdot 2H_2O$	6.5×10^{-10} (25°)	Magnesium oxalate	8.57×10^{-5} (18°)
Barium oxalate, $BaC_2O_4 \cdot 3\frac{1}{2}H_2O$	1.62×10^{-7} (18°)	Manganese hydroxide	4×10^{-14} (18°)
Barium oxalate, $BaC_2O_4 \cdot 2H_2O$	1.2×10^{-7} (18°)	Manganese sulfide	1.4×10^{-15} (18°)
Barium oxalate, $BaC_2O_4 \cdot \frac{1}{2}H_2O$	2.18×10^{-7} (18°)	Mercuric sulfide	4×10^{-53} to
Barium sulfate	0.87×10^{-10} (18°)		2×10^{-49} (18°)
Barium sulfate	1.08×10^{-10} (25°)	Mercurous bromide	1.3×10^{-21} (25°)
Barium sulfate	1.98×10^{-10} (50°)	Mercurous chloride	2×10^{-18} (25°)
Cadmium oxalate $CdC_2O_4 \cdot 3H_2O$	1.53×10^{-8} (18°)	Mercurous iodide	1.2×10^{-28} (25°)
Cadmium sulfide	3.6×10^{-29} (18°)	Nickel sulfide	1.4×10^{-24} (18°)
Calcium carbonate (calcite)	0.99×10^{-8} (15°)	Potassium acid tartrate $[K^+]$ $[HC_4H_4O_6^-]$	3.8×10^{-4} (18°)
Calcium carbonate (calcite)	0.87×10^{-8} (25°)		
Calcium fluoride	3.4×10^{-11} (18°)	Silver bromate	3.97×10^{-5} (20°)
Calcium fluoride	3.95×10^{-11} (26°)	Silver bromate	5.77×10^{-5} (25°)
Calcium iodate, $Ca(IO_3)_2 \cdot 6H_2O$	22.2×10^{-8} (10°)	Silver bromide	4.1×10^{-13} (18°)
		Silver bromide	7.7×10^{-13} (25°)
Calcium iodate, $Ca(IO_3)_2 \cdot 6H_2O$	64.4×10^{-8} (18°)	Silver carbonate	6.15×10^{-12} (25°)
Calcium oxalate, $CaC_2O_4 \cdot H_2O$	1.78×10^{-9} (18°)		
Calcium oxalate, $CaC_2O_4 \cdot H_2O$	2.57×10^{-9} (25°)	Silver chloride	0.21×10^{-10} (4.7°)
Calcium sulfate	1.95×10^{-4} (10°)	Silver chloride	0.37×10^{-10} (9.7°)
Calcium tartrate, $CaC_4H_4O_6 \cdot 2H_2O$	0.77×10^{-6} (18°)	Silver chloride	1.56×10^{-10} (25°)
		Silver chloride	13.2×10^{-10} (50°)
Cobalt sulfide	3×10^{-26} (18°)	Silver chloride	215×10^{-10} (100°)
Cupric iodate	1.4×10^{-7} (25°)		
Cupric oxalate	2.87×10^{-8} (25°)	Silver chromate	1.2×10^{-12} (14.8°)
Cupric sulfide	8.5×10^{-45} (18°)	Silver chromate	9×10^{-12} (25°)
Cuprous bromide	4.15×10^{-8} (18–20°)	Silver cyanide $[Ag^+][Ag(CN)_2^-]$	2.2×10^{-12} (20°)
Cuprous chloride	1.02×10^{-6} (18–20°)	Silver dichromate	2×10^{-7} (25°)
Cuprous iodide	5.06×10^{-12} (18–20°)	Silver hydroxide	1.52×10^{-8} (20°)
Cuprous sulfide	2×10^{-47} (16–18°)		
Cuprous thiocyanate	1.6×10^{-11} (18°)	Silver iodate	0.92×10^{-8} (9.4°)
Ferric hydroxide	1.1×10^{-36} (18°)	Silver iodide	0.32×10^{-16} (13°)
		Silver iodide	1.5×10^{-16} (25°)
Ferrous hydroxide	1.64×10^{-14} (18°)	Silver sulfide	1.6×10^{-49} (18°)
Ferrous oxalate	2.1×10^{-7} (25°)	Silver thiocyanate	0.49×10^{-12} (18°)
Ferrous sulfide	3.7×10^{-19} (18°)		
Lead carbonate	3.3×10^{-14} (18°)	Silver thiocyanate	1.16×10^{-12} (25°)
Lead chromate	1.77×10^{-14} (18°)	Strontium carbonate	1.6×10^{-9} (25°)
		Strontium fluoride	2.8×10^{-9} (18°)
Lead fluoride	2.7×10^{-8} (9°)	Strontium oxalate	5.61×10^{-8} (18°)
Lead fluoride	3.2×10^{-8} (18°)	Strontium sulfate	2.77×10^{-7} (2.9°)
Lead fluoride	3.7×10^{-8} (26.6°)		
Lead iodate	5.3×10^{-14} (9.2°)	Strontium sulfate	3.81×10^{-7} (17.4°)
Lead iodate	1.2×10^{-13} (18°)	Zinc hydroxide	1.8×10^{-14} (18–20°)
		Zinc oxalate, $ZnC_2O_4 \cdot 2H_2O$	1.35×10^{-9} (18°)
Lead iodate	2.6×10^{-13} (25.8°)	Zinc sulfide	1.2×10^{-23} (18°)

*From: "Handbook of Chemistry and Physics", 53rd ed., R.C. Weast, Ed., The Chemical Rubber Co., 1972.

Table 3-36.   CONDUCTANCES AND ACTIVITY COEFFICIENTS
EQUIVALENT CONDUCTANCES OF ELECTROLYTES IN AQUEOUS SOLUTIONS*

At 25°C

Compound	Infinite dilution	Concentration in gram equivalents per 1000 cm³							Compound	Infinite dilution	Concentration in gram equivalents per 1000 cm³						
		0.0005	0.001	0.005	0.01	0.02	0.05	0.1			0.0005	0.001	0.005	0.01	0.02	0.05	0.1
$AgNO_3$	133.36	131.36	130.51	127.20	124.76	121.41	115.24	109.14	$LaCl_3$	145.8	139.6	137.0	127.5	121.8	115.3	106.2	99.1
$BaCl_2$	139.98	135.96	134.34	128.02	123.94	119.09	111.48	105.19	$LiCl$	115.03	113.15	112.40	109.40	107.40	104.65	100.11	95.86
$CaCl_2$	135.84	131.93	130.36	124.25	120.36	115.65	108.47	102.4	$LiClO_4$	105.98	104.18	103.44	100.57	98.61	96.18	92.20	88.56
$Ca(OH)_2$	257.9			232.9	225.9	213.9			$MgCl_2$	129.40	125.61	124.11	118.31	114.55	110.04	103.08	97.10
$CuSO_4$	133.6	121.6	115.26	94.07	83.12	72.20	59.05	50.58	NH_4Cl	149.7		146.8	143.5	141.28	138.33	133.29	128.75
HCl	426.16	422.74	421.36	415.80	412.00	407.24	399.09	391.32	$NaCl$	126.45	124.50	123.74	120.65	118.51	115.51	111.06	106.74
KBr	151.9			146.09	143.43	140.48	135.68	131.39	$NaClO_4$	117.48	115.64	114.87	111.75	109.59	106.96	102.40	98.43
KCl	149.86	147.81	146.95	143.35	141.27	138.34	133.37	128.96	NaI	126.94	125.36	124.25	121.25	119.24	116.70	112.79	108.78
$KClO_4$	140.04	138.76	137.87	134.16	131.46	127.92	121.62	115.20	$NaOOCCH_3$	91.0	89.2	88.5	85.72	83.76	81.24	76.92	72.80
$K_3Fe(CN)_6$	174.5	166.4	163.1	150.7					$NaOOCC_2H_5$	85.9		83.5	80.9	79.1	76.6		
$K_4Fe(CN)_6$	184.5		167.24	146.09	134.83	122.82	107.70	97.87	$NaOOCC_3H_7$	82.70	81.04	80.31	77.58	75.76	73.39	69.32	65.27
$KHCO_3$	118.0	116.10	115.34	112.24	110.08	107.22			$NaOH$	247.8	245.6	244.7	240.8	238.0			
KI	150.38			144.37	142.18	139.45	134.97	131.11	Na_2SO_4	129.9	125.74	124.15	117.15	112.44	106.78	97.75	89.98
KIO_4	127.92	125.80	124.94	121.24	118.51	114.14	106.72	98.12	$SrCl_2$	135.80	131.90	130.33	124.24	120.24	115.54	108.25	102.19
KNO_3	144.96	142.77	141.84	138.48	135.82	132.41	126.31	120.40	$ZnSO_4$	132.8	121.4	114.53	95.49	84.91	74.24	61.20	52.64
$KReO_4$	128.20	126.03	125.12	121.31	118.49	114.49	106.40	97.40									

*From: "Handbook of Chemistry and Physics", 53rd ed., R.C. Weast, Ed., The Chemical Rubber Co., 1972.

EQUIVALENT CONDUCTANCE OF THE SEPARATE IONS*

Ion	Temperature, °C								Ion	Temperature, °C							
	0°	18°	25°	50°	75°	100°	128°	156°		0°	18°	25°	50°	75°	100°	128°	156°
K	40.4	64.6	74.5	115	159	206	263	317	$\frac{1}{2}SO_4$	41.	68.	79.	125	177	234	303	370
Na	26.	43.5	50.9	82	116	155	203	249	$\frac{1}{2}C_2O_4$	39.	63.	73.	115	163	213	273	336
NH_4	40.2	64.5	74.5	115	159	207	264	319	$\frac{1}{3}C_6H_5O_7$	36.	60.	70.	113	161	214		
Ag	32.9	54.3	63.5	101	143	188	245	299	$\frac{1}{4}Fe(CN)_6$	58.	95.	111.	173	244	321		
$\frac{1}{2}Ba$	33.	55.	65.	104	149	200	262	322									
$\frac{1}{2}Ca$	30.	51.	60.	98	142	191	252	312									
$\frac{1}{3}La$	35.	61.	72.	119	173	235	312	388									
Cl	41.1	65.5	75.5	116	160	207	264	318	H	240.	314.	350.	465	565	644	722	777
NO_3	40.4	61.7	70.6	104	140	178	222	263	OH	105.	172.	192.	284	360	439	525	592
$C_2H_3O_2$	20.3	34.0	40.8	67	96	130	171	211									

*From: "Smithsonian Physical Tables", 9th ed., W.E. Forsythe, Ed., The Smithsonian Institution, 1956.

ACTIVITY COEFFICIENTS OF ACIDS, BASES, AND SALTS*

The following coefficients are valid only at 25°C. The concentrations are expressed as molalities.

Name	0.1	0.2	0.3	0.4	0.5	0.6	0.7	0.8	0.9	1.0	Name	0.1	0.2	0.3	0.4	0.5	0.6	0.7	0.8	0.9	1.0
$AgNO_3$	0.734	0.657	0.606	0.567	0.536	0.509	0.485	0.464	0.446	0.429	KOH	0.798	0.760	0.742	0.734	0.732	0.733	0.736	0.742	0.749	0.756
$AlCl_3$	(0.337)	0.305	0.302	0.313	0.331	0.356	0.388	0.429	0.479	0.539	LiBr	0.796	0.766	0.756	0.752	0.753	0.758	0.767	0.777	0.789	0.803
$Al_2(SO_4)_3$	(0.0350)	0.0225	0.0176	0.0153	0.0143	0.0140	0.0142	0.0149	0.0159	0.0175	LiCl	0.790	0.757	0.744	0.740	0.739	0.743	0.748	0.755	0.764	0.774
$CdSO_4$	(0.150)	0.102	0.082	0.069	0.061	0.055	0.050	0.046	0.043	0.041	$LiClO_4$	0.812	0.794	0.792	0.798	0.808	0.820	0.834	0.852	0.869	0.887
$CrCl_3$	(0.331)	0.298	0.294	0.300	0.314	0.335	0.362	0.397	0.436	0.481	LiI	0.815	0.802	0.804	0.813	0.824	0.838	0.852	0.870	0.888	0.910
$Cr(NO_3)_3$	(0.319)	0.285	0.279	0.281	0.291	0.304	0.322	0.344	0.371	0.401	$LiNO_3$	0.788	0.752	0.736	0.728	0.726	0.727	0.729	0.733	0.737	0.743
$Cr_2(SO_4)_3$	(0.0458)	0.0300	0.0238	0.0207	0.0190	0.0182	0.0181	0.0185	0.0194	0.0208	LiOAc	0.784	0.742	0.721	0.709	0.700	0.691	0.689	0.688	0.688	0.689
CsBr	0.754	0.694	0.654	0.626	0.603	0.586	0.571	0.558	0.547	0.538	$MgSO_4$	(0.150)	0.108	0.088	0.076	0.068	0.062	0.057	0.054	0.051	0.049
CsCl	0.756	0.694	0.656	0.628	0.606	0.589	0.575	0.563	0.553	0.544	$MnSO_4$	(0.150)	0.106	0.085	0.073	0.064	0.058	0.053	0.049	0.046	0.044
CsI	0.754	0.692	0.651	0.621	0.599	0.581	0.567	0.554	0.543	0.533	NaBr	0.782	0.741	0.719	0.704	0.697	0.692	0.689	0.687	0.687	0.687
$CsNO_3$	0.733	0.655	0.602	0.561	0.528	0.501	0.478	0.458	0.439	0.422	NaCl	0.778	0.735	0.710	0.693	0.681	0.673	0.667	0.662	0.659	0.657
CsOAc	0.799	0.771	0.761	0.759	0.762	0.768	0.776	0.783	0.792	0.802	$NaClO_4$	0.775	0.729	0.701	0.683	0.668	0.656	0.648	0.641	0.635	0.629
$CuSO_4$	(0.150)	0.104	0.083	0.071	0.062	0.056	0.052	0.048	0.045	0.043	NaCNS	0.787	0.750	0.731	0.720	0.715	0.712	0.710	0.710	0.711	0.712
HBr	0.805	0.782	0.777	0.781	0.789	0.801	0.815	0.832	0.850	0.871	NaF	0.765	0.710	0.676	0.651	0.632	0.616	0.603	0.592	0.582	0.573
HCl	0.796	0.767	0.756	0.755	0.757	0.763	0.772	0.783	0.795	0.809	NaH_2PO_4	0.744	0.675	0.629	0.593	0.563	0.539	0.517	0.499	0.483	0.468
$HClO_4$	0.803	0.778	0.768	0.766	0.769	0.776	0.785	0.795	0.808	0.823	NaI	0.787	0.751	0.735	0.727	0.723	0.723	0.724	0.727	0.731	0.736
HI	0.818	0.807	0.811	0.823	0.839	0.860	0.883	0.908	0.935	0.963	$NaNO_3$	0.762	0.703	0.666	0.638	0.617	0.599	0.583	0.570	0.558	0.548
HNO_3	0.791	0.754	0.735	0.725	0.720	0.717	0.717	0.718	0.721	0.724	NaOAc	0.791	0.757	0.744	0.737	0.735	0.736	0.740	0.745	0.752	0.757
KBr	0.772	0.722	0.693	0.673	0.657	0.646	0.636	0.629	0.622	0.617	NaOH	0.766	0.727	0.708	0.697	0.690	0.685	0.681	0.679	0.678	0.678
KCl	0.770	0.718	0.688	0.666	0.649	0.637	0.626	0.618	0.610	0.604	$NiSO_4$	(0.150)	0.105	0.084	0.071	0.063	0.056	0.052	0.047	0.044	0.042
KCNS	0.769	0.716	0.685	0.663	0.646	0.633	0.623	0.614	0.606	0.599	RbBr	0.763	0.706	0.673	0.650	0.632	0.617	0.605	0.595	0.586	0.578
KF	0.775	0.727	0.700	0.682	0.670	0.661	0.654	0.650	0.646	0.645	RbCl	0.764	0.709	0.675	0.652	0.634	0.620	0.608	0.599	0.590	0.583
KI	0.778	0.733	0.707	0.689	0.676	0.667	0.660	0.654	0.649	0.645	RbI	0.762	0.705	0.671	0.647	0.629	0.614	0.602	0.591	0.583	0.575
KNO_3	0.739	0.663	0.614	0.576	0.545	0.519	0.496	0.476	0.459	0.443	$RbNO_3$	0.734	0.658	0.606	0.565	0.534	0.508	0.485	0.465	0.446	0.430
KOAc	0.796	0.766	0.754	0.750	0.751	0.754	0.759	0.766	0.774	0.783	RbOAc	0.796	0.767	0.756	0.753	0.755	0.759	0.766	0.773	0.782	0.792
											$TlNO_3$	0.702	0.606	0.545	0.500	....	....	....	....	....	....
											$ZnSO_4$	(0.150)	0.104	0.083	0.071	0.063	0.057	0.052	0.048	0.046	0.043

*From: "Handbook of Chemistry and Physics", 53rd ed., R.C. Weast, Ed., The Chemical Rubber Co., 1972.

Table 3-37.　HEAT OF DILUTION OF ACIDS*

Vivian B. Parker

ΔH_{diln}, the integral heat of dilution, is the change in enthalpy, per mole of solute, when a solution of concentration m_1 is diluted to a final finite concentration m_2. When the dilution is carried out by addition of an infinite amount of solvent, so the final solution is infinitely dilute, the enthalpy change is the integral heat of dilution to infinite dilution. Since Φ_L, the relative apparent molal enthalpy, is equal to and opposite in sign to this, only Φ_L is referred to here.

Φ_L, cal/mol, at 25 deg C (298.15 K)[a]

n	m	HF	HCl	HClO$_4$	HBr	HI	HNO$_3$	CH$_2$O$_2$	C$_2$H$_4$O$_2$
∞	0.00	0	0	0	0	0	0	0	0
500,000	.000111	300	5	5	5	5	5	9	40
100,000	.000555	900	10	10	9	9	11	13	50
50,000	.00111	1,300	16	14	13	12	15	20	53
20,000	.00278	1,800	25	22	22	20	23	23	55
10,000	.00555	2,130	34	30	31	29	31	25	58
7,000	.00793	2,250	40	35	37	34	36	26	59
5,000	.01110	2,360	47	40	44	41	42	26	61
4,000	.01388	2,450	54	43	49	46	46	27	62
3,000	.01850	2,550	60	47	56	52	51	28	62
2,000	.02775	2,700	74	54	68	63	59	28	63
1,500	.03700	2,812	85	58	77	71	65	29	64
1,110	.05000	2,927	97	62	89	81	73	29	65
1,000	.05551	2,969	102	62	92	84	76	29	65
900	.0617	2,989	107	63	97	88	78	30	66
800	.0694	3,015	113	64	102	92	81	31	67
700	.0793	3,037	120	65	108	96	84	32	68
600	.0925	3,057	129	65	115	102	88	32	68
555.1	.1000	3,060	133	65	119	105	89	32	69
500	.1110	3,077	140	65	124	108	92	32	70
400	.1388	3,097	156	64	135	116	97	33	72
300	.1850	3,126	176	61	150	125	103	34	76
277.5	.2000	3,129	182	59	155	128	105	35	79
200	.2775	3,142	212	50	176	140	117	36	82
150	.3700	3,148	242	36	197	154	118	39	88
111.0	.5000	3,156	280	18	225	170	119	42	97
100	.5551	3,160	295	+12	235	176	120	44	101
75	.7401	3,167	343	−14	270	194	121	49	113
55.51	1.0000	3,179	405	−48	314	223	121	54	130
50	1.1101	3,184	431	−61	331	234	121	56	147
40	1.3877	3,192	493	−91	379	260	121	60	155
37.00	1.5000	3,194	518	−103	398	269	121	62	162
30	1.8502	3,200	595	−138	455	301	124	65	183
27.75	2.0000	3,203	627	−149	477	315	126	66	192
25	2.2202	3,208	674	−162	510	336	130	67	204
22.20	2.5000	3,211	732	−173	550	365	139	68	218
20	2.7753	3,214	792	−182	590	396	149	69	233
18.50	3.0000	3,216	838	−187	624	427	159	69	245
15.86	3.500	3,221	946	−196	709	503	189	69	268
15	3.7004	3,227	988	−195	743	536	203	69	277
13.88	4.0000	3,234	1,052	−188	796	588	229	69	291
12.33	4.5000	3,246	1,171	−175	887	676	265	69	313
12	4.6255	3,249	1,190	−170	911	700	277	69	318
11.10	5.0000	3,256	1,271	−150	983	764	313	69	333
10	5.5506	3,265	1,396	−117	1,097	855	368	68	353
9.5	5.8427	3,269	1,462	−97	1,156	920	400	68	363
9.251	6.0000	3,272	1,498	−84	1,196	950	418	67	368
9.0	6.1674	3,274	1,535	−72	1,230	980	437	67	373
8.5	6.5301	3,278	1,618	−40	1,313	1,050	480	66	383
8.0	6.9383	3,282	1,710	+4	1,401	1,115	530	65	392
7.929	7.0000	3,283	1,725	11	1,416	1,130	538	65	394

[a] One calorie (thermochemical) equals 4.184 joules.

Table 3-37.   HEAT OF DILUTION OF ACIDS (Continued)

n	m	HF	HCl	HClO₄	HBr	HI	HNO₃	CH₂O₂	C₂H₄O₂
7.5	7.4008	3,286	1,820	61	1,497	1,210	595	63	402
7.0	7.9295	3,290	1,942	135	1,608	1,325	661	61	411
6.938	8.0000	3,291	1,960	146	1,622	1,340	667	61	412
6.5	8.5394	3,296	2,090	229	1,738	1,450	745	58	420
6.167	9.0000	3,302	2,202	306	1,845	1,570	805	55	426
6.0	9.2510	3,305	2,265	348	1,903	1,630	840	53	429
5.551	10.0000	3,316	2,447	481	2,078	1,820	940	49	436
5.5	10.0920	3,317	2,472	499	2,102	1,850	950	49	437
5.0	11.1012	3,335	2,721	730	2,344	2,100	1,098	43	445
4.5	12.3346	3,362	3,025	1,144	2,655	2,460	1,270	37	453
4.0	13.8765	3,400	3,404	1,574	3,089	2,960	1,495	29	462
3.700	15.0000	3,428	3,680	1,893	3,415	3,350	1,645	26	469
3.5	15.8589	3,450	3,882	2,150	3,668	3,660	1,770	21	473
3.25	17.0788	3,483	4,160	2,460	4,005	4,110	1,920	17	481
3.0	18.5020	3,520	4,460	2,880	4,370	4,630	2,101	13	488
2.775	20.0000	3,557	4,750	3,300	4,760	5,190	2,270	9	496
2.5	22.2024	3,607	5,180	4,000	5,300	6,000	2,520	+4	506
2.0	27.7530	3,712	6,200	5,500	6,050	.....	3,000	−5	520
1.5	37.0040	.....	8,240	.....	8,530	.....	3,770	−13	532
1.0	55.506	.....	10,900	.....	11,670	.....	4,715	+11	518
0.5	111.012	.....	.....	.....	.....	.....	.....	77	495
0.25	222.02	.....	.....	.....	.....	.....	.....	129	.....

*From: NSRDS—NBS 2, "Thermal Properties of Aqueous Uni-univalent Electrolytes", V.B. Parker, National Bureau of Standards, 1965.

Table 3-38.   HEATS OF SOLUTION*

Vivian B. Parker

$\Delta H^{\circ}_{\infty}$ 25 deg C for Uni-univalent Electrolytes in H_2O^a

Substance	State	$\Delta H^{\circ}_{\infty}$	Substance	State	$\Delta H^{\circ}_{\infty}$	Substance	State	$\Delta H^{\circ}_{\infty}$
		cal/mole			cal/mole			cal/mole
HF	g	−14,700	LiBr·2H₂O	c	−2,250	KCl	c	4,115
HCl	g	−17,888	LiBrO₃	c	340	KClO₃	c	9,890
HClO₄	l	−21,215	LiI	c	−15,130	KClO₄	c	12,200
HClO₄·H₂O	c	−7,875	LiI·H₂O	c	−7,090	KBr	c	4,750
HBr	g	−20,350	LiI·2H₂O	c	−3,530	KBrO₃	c	9,830
HI	g	−19,520	LiI·3H₂O	c	140	KI	c	4,860
HIO₃	c	2,100	LiNO₂	c	−2,630	KIO₃	c	6,630
HNO₃	l	−7,954	LiNO₂·H₂O	c	1,680	KNO₂	c	3,190
HCOOH	l	−205	LiNO₃	c	−600	KNO₃	c	8,340
CH₃COOH	l	−360				KC₂H₃O₂	c	−3,665
			NaOH	c	−10,637	KCN	c	2,800
NH₃	g	−7,290	NaOH·H₂O	c	−5,118	KCNO	c	4,840
NH₄Cl	c	3,533	NaF	c	218	KCNS	c	5,790
NH₄ClO₄	c	8,000	NaCl	c	928	KMnO₄	c	10,410
NH₄Br	c	4,010	NaClO₂	c	80			
NH₄I	c	3,280	NaClO₂·3H₂O	c	6,830	RbOH	c	−14,900
NH₄IO₃	c	7,600	NaClO₃	c	5,191	RbOH·H₂O	c	−4,310
NH₄NO₂	c	4,600	NaClO₄	c	3,317	RbOH·2H₂O	c	210
NH₄NO₃	c	6,140	NaClO₄·H₂O	c	5,380	RbF	c	−6,240
NH₄C₂H₃O₂	c	−570	NaBr	c	−144	RbF·H₂O	c	−100
NH₄CN	c	4,200	NaBr·2H₂O	c	4,454	RbF·1½H₂O	c	320
NH₄CNS	c	5,400	NaBrO₃	c	6,430	RbCl	c	4,130
CH₃NH₃Cl	c	1,378	NaI	c	−1,800	RbClO₃	c	11,410
(CH₃)₂NHCl	c	350	NaI·2H₂O	c	3,855	RbClO₄	c	13,560
N(CH₃)₄Cl	c	975	NaIO₃	c	4,850	RbBr	c	5,230
N(CH₃)₄Br	c	5,800	NaNO₂	c	3,320	RbBrO₃	c	11,700
N(CH₃)₄I	c	10,055	NaNO₃	c	4,900	RbI	c	6,000
			NaC₂H₃O₂	c	−4,140	RbNO₃	c	8,720
AgClO₄	c	1,760	NaC₂H₃O₂·3H₂O	c	4,700			
AgNO₂	c	8,830	NaCN	c	290	CsOH	c	−17,100
AgNO₃	c	5,400	NaCN·½H₂O	c	790	CsOH·H₂O	c	−4,900
			NaCN·2H₂O	c	4,440	CsF	c	−8,810
LiOH	c	−5,632	NaCNO	c	4,590	CsF·H₂O	c	−2,500
LiOH·H₂O	c	−1,600	NaCNS	c	1,632	CsF·1½H₂O	c	−1,300
LiF	c	1,130				CsCl	c	4,250
LiCl	c	−8,850	KOH	c	−13,769	CsClO₄	c	13,250
LiCl·H₂O	c	−4,560	KOH·H₂O	c	−3,500	CsBr	c	6,210
LiClO₄	c	−6,345	KOH·1½H₂O	c	−2,500	CsBrO₃	c	12,060
LiClO₄·3H₂O	c	7,795	KF	c	−4,238	CsI	c	7,970
LiBr	c	−11,670	KF·2H₂O	c	1,666	CsNO₃	c	9,560
LiBr·H₂O	c	−5,560						

^{a}25 deg C = 298.15 K. One calorie (thermochemical) = 4.184 joules.

*From: NSRDS—NBS 2, "Thermal Properties of Aqueous Uni-univalent Electrolytes", V.B. Parker, National Bureau of Standards, 1965.

Table 3-39.   CONCENTRATION OF ACIDS AND BASES*

Common Commercial Strengths

	Molecular weight	Moles per liter	Grams per liter	Percent by weight	Specific gravity[a]
ACIDS					
Acetic acid, glacial	60.05	17.4	1,045	99.5	1.05
Acetic acid	60.05	6.27	376	36	1.045
Butyric acid	88.1	10.3	912	95	0.96
Formic acid	46.02	23.4	1,080	90	1.20
	—	5.75	264	25	1.06
Hydriodic acid	127.9	7.57	969	57	1.70
	—	5.51	705	47	1.50
	—	0.86	110	10	1.1
Hydrobromic acid	80.92	8.89	720	48	1.50
	—	6.82	552	40	1.38
Hydrochloric acid	36.5	11.6	424	36	1.18
	—	2.9	105	10	1.05
Hydrocyanic acid	27.03	25	676	97	0.697
	—	0.74	19.9	2	0.996
Hydrofluoric acid	20.01	32.1	642	55	1.167
	—	28.8	578	50	1.155
Hydrofluosilicic acid	144.1	2.65	382	30	1.27
Hypophosphorous acid	66.0	9.47	625	50	1.25
	—	5.14	339	30	1.13
	—	1.57	104	10	1.04
Lactic acid	90.1	11.3	1,020	85	1.2
Nitric acid	63.02	15.99	1,008	71	1.42
	—	14.9	938	67	1.40
	—	13.3	837	61	1.37
Perchloric acid	100.5	11.65	1,172	70	1.67
	—	9.2	923	60	1.54
Phosphoric acid	98	14.7	1,445	85	1.70
Sulfuric acid	98.1	18.0	1,766	96	1.84
Sulfurous acid	82.1	0.74	61.2	6	1.02
BASES					
Ammonia water	17.0	14.8	252	28	0.898
Potassium hydroxide	56.1	13.5	757	50	1.52
	—	1.94	109	10	1.09
Sodium carbonate	106.0	1.04	110	10	1.10
Sodium hydroxide	40.0	19.1	763	50	1.53
	—	2.75	111	10	1.11

[a]For density in kg/m^3, multiply by 1000.

*Reprinted from: "The Merck Index" (1968), 8th ed., Merck and Co., Rahway, N.J., p. 1309, with permission of the copyright owners.

REFERENCE

For vapor-pressure curves for dilute solutions of hydrochloric, nitric, and sulfuric acids up to about 1000 psia, consult "Vapor-Pressure Data for Common Acids at High Temperatures", B.G. Staples, J.M. Procopio, Jr., and G.J. Su, *Chem. Eng.*, 77(25):113, 1970. The acid concentrations represented by these curves are for hydrochloric acid, 10–35% by weight; for nitric acid, 30–70% by weight; and for sulfuric acid, 0–70% by weight.

Table 3-40.   LOWERING OF VAPOR PRESSURE BY SALTS IN AQUEOUS SOLUTIONS*

The following table gives the reduction of the vapor pressure in millimeters of mercury due to the presence of the number of grammolecules of salt per liter of water given at the head of the columns. The temperature is 100°C, at which temperature the vapor pressure of pure water is 760 millimeters Hg.

Substance	0.5	1.0	2.0	3.0	4.0	5.0	6.0	8.0	10.0	Substance	0.5	1.0	2.0	3.0	4.0	5.0	6.0	8.0	10.0
$Al_2(SO_4)_3$	12.8	36.5								$LiOH$	15.9	37.4	78.1						
$AlCl_3$	22.5	61.0	179.0	318.0						Li_2CrO_4	16.4	32.6	74.0	120.0	171.0				
BaS_2O_6	6.6	15.4	34.4							$MgSO_4$	6.5	12.0	24.5	47.5					
$Ba(OH)_2$	12.3	22.5	39.0							$MgCl_2$	16.8	39.0	100.5	183.3	277.0	377.0			
$Ba(NO_3)_2$	13.5	27.0								$Mg(NO_3)_2$	17.6	42.0	101.0	174.8					
$Ba(ClO_3)_2$	15.8	33.3	70.5	108.2						$MgBr_2$	17.9	44.0	115.8	205.3	298.5				
$BaCl_2$	16.4	36.7	77.6							$MgH_2(SO_4)_2$	18.3	46.0	116.0						
$BaBr_2$	16.8	38.8	91.4	150.0	204.7					$MnSO_4$	6.0	10.5	21.0						
CaS_2O_8	9.9	23.0	56.0	106.0						$MnCl_2$	15.0	34.0	76.0	122.3	167.0	209.0			
$Ca(NO_3)_2$	16.4	34.8	74.6	139.3	161.7	205.4				NaH_2PO_4	10.5	20.0	36.5	51.7	66.8	82.0	96.5	126.7	157.1
$CaCl_2$	17.0	39.8	95.3	166.6	241.5	319.5				$NaHSO_4$	10.9	22.1	47.3	75.0	100.2	126.1	148.5	189.7	231.4
$CaBr_2$	17.7	44.2	105.8	191.0	283.3	308.3				$NaNO_3$	10.0	22.3	46.2	68.1	90.3	111.5	131.7	167.0	190.0
$CdSO_4$	4.1	8.9	18.1							$NaClO_3$	10.5	23.0	48.4	73.5	98.5	123.3	147.5	196.5	223.5
CdI_2	7.6	14.8	33.5	52.7						$(NaPO_3)_6$	11.6								
$CdBr_2$	8.6	17.8	36.7	55.7	80.0					$NaOH$	11.8	22.8	48.2	77.3	107.5	139.1	172.5	243.3	314.0
$CdCl_2$	9.6	18.8	36.7	57.0	77.3	99.0				$NaNO_2$	11.6	24.4	50.0	75.0	98.2	122.5	146.5	189.0	226.2
$Cd(NO_3)_2$	15.9	36.1	78.0	122.2						Na_2HPO_4	12.1	23.3	45.0	60.0	78.7	99.8	122.1		
$Cd(ClO_3)_2$	17.5									$NaHCO_3$	12.9	24.1	48.2	77.6	102.2	127.8	152.0	198.0	239.4
$CoSO_4$	5.5	10.7	22.9	45.5						Na_2SO_4	12.6	25.0	48.9	74.2					
$CoCl_2$	15.0	34.8	83.0	136.0	186.4					$NaCl$	12.3	25.2	52.1	80.0	111.0	143.0	176.5		
$Co(NO_3)_2$	17.3	39.2	89.0	152.0	218.7	282.0	332.0			$NaBrO_3$	12.1	25.0	54.1	81.3	108.8	136.0			
$FeSO_4$	5.8	10.7	24.0	42.4						$NaBr$	12.6	25.9	57.0	89.2	124.2	159.5	197.5	268.0	
H_3BO_3	6.0	12.3	25.1	38.0	51.0					NaI	12.1	25.6	60.2	99.5	136.7	177.5	221.0	301.5	370.0
H_3PO_4	6.6	14.0	28.6	45.2	62.0	81.5	103.0	146.9	189.5	$Na_4P_2O_7$	13.2	22.0							
H_3AsO_4	7.3	15.0	30.2	46.4	64.9					Na_2CO_3	14.3	27.3	53.5	80.2	111.0				
H_2SO_4	12.9	26.5	62.8	104.0	148.0	198.4	247.0	343.2		$Na_2C_2O_4$	14.5	30.0	65.8	105.8	146.0				
KH_2PO_4	10.2	19.5	33.3	47.8	60.5	73.1	85.2			Na_2WO_4	14.8	33.6	71.6	115.7	162.6				
KNO_3	10.3	21.1	40.1	57.6	74.5	88.2	102.1	126.3	148.0	Na_3PO_4	16.5	30.0	52.5						
$KClO_3$	10.6	21.6	42.8	62.1	80.0					$(NaPO_3)_3$	17.1	36.5							
$KBrO_3$	10.9	22.4	45.0							NH_4NO_3	12.8	22.0	42.1	62.7	82.9	103.8	121.0	152.2	180.0
$KHSO_4$	10.9	21.9	43.3	65.3	85.5	107.8	129.9	170.0		$(NH_4)_2SiFl_6$	11.5	25.0	44.5						
KNO_2	11.1	22.8	44.8	67.0	90.0	110.5	130.7	167.0	198.8	NH_4Cl	12.0	23.7	45.1	69.3	94.2	118.5	138.2	179.0	213.8
$KClO_4$	11.5	22.3								NH_4HSO_4	11.5	22.0	46.8	71.0	94.5	118.	139.0	181.2	218.0
KCl	12.2	24.4	48.8	74.1	100.9	128.5	152.2			$(NH_4)_2SO_4$	11.0	24.0	46.5	69.5	93.0	117.0	141.8		
$KHCO_3$	11.6	23.6	59.0	77.6	104.2	132.0	160.0	210.0	255.0	NH_4Br	11.9	23.9	48.8	74.1	99.4	121.5	145.5	190.2	228.5
KI	12.5	25.3	52.2	82.6	112.2	141.5	171.8	225.5	278.5	NH_4I	12.9	25.1	49.8	78.5	104.5	132.3	156.0	200.0	243.5
$K_2C_2O_4$	13.9	28.3	59.8	94.2	131.0					$NiSO_4$	5.0	10.2	21.5						
K_2WO_4	13.9	33.0	75.0	123.8	175.4	226.4				$NiCl_2$	16.1	37.0	86.7	147.0	212.8				
K_2CO_3	14.4	31.0	68.3	105.5	152.0	209.0	258.5	350.0		$Ni(NO_3)_2$	16.1	37.3	91.3	156.2	235.0				
KOH	15.0	29.5	64.0	99.2	140.0	181.8	223.0	309.5	387.8	$Pb(NO_3)_2$	12.3	23.5	45.0	63.0					
K_2CrO_4	16.2	29.5	60.0							$Sr(SO_3)_2$	7.2	20.3	47.0						
$LiNO_3$	12.2	25.9	55.7	88.9	122.2	155.1	188.0	253.4	309.2	$Sr(NO_3)_2$	15.8	31.0	64.0	97.4	131.4				
$LiCl$	12.1	25.5	57.1	95.0	132.5	175.5	219.5	311.5	393.5	$SrCl_2$	16.8	38.8	91.4	156.8	223.3	281.5			
$LiBr$	12.2	26.2	60.0	97.0	140.0	186.3	241.5	341.5	438.0	$SrBr_2$	17.8	42.0	101.1	179.0	267.0				
Li_2SO_4	13.3	28.1	56.8	89.0						$ZnSO_4$	4.9	10.4	21.5	42.1	66.2				
$LiHSO_4$	12.8	27.0	57.0	93.0	130.0	168.0				$ZnCl_2$	9.2	18.7	46.2	75.0	107.0	153.0	195.0		
LiI	13.6	28.6	64.7	105.2	154.5	206.0	264.0	357.0	445.0	$Zn(NO_3)_2$	16.6	39.0	93.5	157.5	223.8				
Li_2SiFl_6	15.4	34.0	70.0	106.0															

*From: "Smithsonian Physical Tables", 9th ed., W.E. Forsythe, Ed., The Smithsonian Institution, 1956.

Table 3-41. INDUSTRIAL SOLVENTS*
Properties at Atmospheric Pressure and Room Temperature

SYMBOLS:

Soluble in:
- A—alcohol
- B—benzene
- C—chloroform
- E—ether
- O—organic solvents
- W—water
- (s—slightly)

Applications:
- B—bitumens, asphalt, tar
- C—cellulose, esters, and ethers
- D—drugs and pharmaceuticals
- F—flavorings
- G—gum
- H—hydrocarbons
- I—insecticides
- K—inks, printing
- L—lacquer, varnish, shellac, paint
- O—oil, grease, fats
- P—plastics, resins
- R—rubber
- U—perfumes
- V—vegetable fibers
- W—waxes

Name	Formula	Mol wt	Sp gr	Sp heat[a]	Viscosity,[b] cp	Boiling point, deg C	Soluble in	Solvent applications
Acetic acid	CH_3CO_2H	60.05	1.05	.53	1.2	119.	A, B, E, W	G, O, R, V
Acetone	CH_3COCH_3	58.08	.79	.52	0.33	56.2	A, B, C, E, W	L, O, P, R
n-Amyl alcohol	$CH_3(CH_2)_4OH$	88.15	0.81	0.71	3.31	137.	A, B, E	C, L
Benzene	C_6H_6	78.11	.88	.41	.60	80.1	A, E, W-s	O, P, W
Benzyl acetate	$CH_3CO_2CH_2C_6H_5$	150.18	1.06	—	—	216.	A, W-s	C, K, L, O, P
Benzyl alcohol	$C_6H_5CH_2OH$	108.14	1.04	0.51	5.8	205	A, C, E, W	C, G, L, P, W
2-Butanone (Ethyl methyl ketone)	$CH_3CH_2COCH_3$	72.11	0.81	0.55	0.40	79.6	A, B, E, W	C, P
Butyl acetate	$CH_3CO_2(CH_2)_3CH_3$	116.16	0.88	—	6.71	127.	A, E, W-s	B, F
n-Butyl alcohol	$CH_3CH_2CH_2CH_2OH$	74.12	0.81	0.56	2.95	118.	A, E, W	C, L, O, P, W
Butyl ether	$CH_3(CH_2)_3O(CH_2)_3CH_3$	130.23	0.77	—	—	142	A, E	C, G, H, P
Carbon disulfide	CS_2	76.14	1.26	.24	.36	45	A, C, E, W-s	I, O, P, R
Carbon tetrachloride	CCl_4	153.82	1.59	.20	.91	76.8	A, B, C, E	L, O, P
Chlorobenzene	C_6H_5Cl	112.56	1.11	·0.30	.80	132	A, B, C, E	C, D, P, U
Chloroethane	C_5H_5Cl	64.52	0.90	0.37	—	13.1	A, E, W-s	F, O, P, W
Chloroform	$CHCl_3$	119.38	1.47	.25	.53	61.2	A, B, E, W-s	C, F, O, P, R, W
Cyclohexanol	$CH_2(CH_2)_4CHOH$	100.16	0.96	—	81	161.	A, B, E, W	C, L, O, R
Dibutyl phthalate	$C_{16}H_{22}O_4$	278.35	1.04	0.43	20.3	340	A, B, E	C, L
1,2-Dichlorobenzene	$C_6H_4Cl_2$	147.01	1.30	0.27	—	179	A, E	B, G, L, O, P
1,2-Dichloroethane	$ClCH_2CH_2Cl$	98.96	1.26	0.31	0.84	84	A, E, W-s	G, O, P, R, W
Dichloromethane	CH_2Cl_2	84.93	1.34	0.29	0.43	40	A, E, W-s	B, R, W
1,2-Dichloropropane	$CH_3CHClCH_2Cl$	112.99	1.16	0.22	0.01	96.2	W-s	G, O, P, W
Diethylene glycol	$HOCH_2CH_2OCH_2CH_2OH$	106.12	1.12	0.55	0.38	245	A, E, W	C, G, L, O, P
Dimethyl phthalate	$C_{10}H_{10}O_4$	194.18	1.19	—	17.2	282	A, B, E	C, P
p-Dioxane	$OCH_2CH_2OCH_2CH_2$	88.11	1.03	0.42	0.01	101	A, E, O, W	C, O, P
Ethanol (Ethyl alcohol)	C_2H_5OH	46.07	.79	.58	1.2	78.5	A, C, E, W	D, F, L, U
Ethyl acetate	$CH_3CO_2CH_2CH_3$	88.11	0.90	0.46	0.46	77.1	A, C, E, O, W	C, G, L, O, P
Ethyl ether	$CH_3CH_2OCH_2CH_3$	74.12	.72	.53	.23	34.6	A, B, C, W-s	C, G, O, P
Ethyl lactate (dl)	$CH_3CHOHCO_2C_2H_5$	118.13	1.03	—	—	155.	A, E, W	C, G, P
2-Furfuraldehyde (Furfural)	C_4H_3OCHO	96.08	1.16	0.42	1.49	162.	A, E, W	C, G, L, P
Glycerol (Glycerine)	$HOCH_2CHOHCH_2OH$	92.10	1.26	.62	1100.	290	A, E-s, W	D, F
Glycol	$HOCH_2CH_2OH$	62.07	1.11	0.58	0.21	198	A, E, W	C, D, P, W
Isopropyl alcohol	$CH_3CHOHCH_3$	60.09	0.79	0.60	0.02	82.4	A, E, W	G, L, O
Isopropyl ether	$(CH_3)_2CHOCH(CH_3)_2$	102.18	0.72	0.33	0.38	69.	A, E, W-s	C, O, W
Mesityl oxide	$(CH_3)_2C:CHCOCH_3$	98.14	0.86	0.52	0.60	130	A, E, W	C, L, P
Methanol (Methyl alcohol)	CH_3OH	32.04	.79	0.60	.56	65.0	A, E, W	C, G, O, P
Methyl acetate	$CH_3CO_2CH_3$	74.08	0.97	0.47	—	57	A, E, W	C, F, L
Methylal	$CH_2(OCH_3)_2$	76.10	0.89	—	—	45.5	A, E, O, W	C, P
Methylcyclohexane	C_7H_{14}	98.18	0.77	0.44	0.72	100.	A, E	O, R, W

[a]Specific heat is in Btu/lb·deg F or cal/g·deg C; for J/kg·K multiply by 4184.
[b]Viscosity is in centipoises; for N·s/m² divide by 1000; for lb/ft·s multiply by 0.67.
*Compiled from several sources.

Table 3-41. INDUSTRIAL SOLVENTS (Continued)

Name	Formula	Mol wt	Sp gr	Sp heat[a]	Vis-cosity,[b] cp	Boiling point, deg C	Soluble in	Solvent applications
4-Methyl-2-Pentanol	$(CH_3)_2CHCH_2CHOHCH_3$	102.18	0.80	—	—	130	A, E, W-s	G, O, P, W
4-Methyl-2-Pentanone	$(CH_3)_2CHCH_2COCH_3$	100.16	0.80	0.46	0.01	117.	A, B, E, W-s	C, G, L, O, P
Nitroethane	$CH_3CH_2NO_2$	75.07	1.04	—	—	115.	A, E, W	C, O, W
Pentachloroethane	Cl_3CCHCl_2	202.30	1.68	.027	2.5	162.	A, E	C, G, L, O, P
Pentasol	—	—	0.81	0.73	4.3	112–140	B, E, W	L, P
Propyl alcohol	$CH_3CH_2CH_2OH$	60.09	.80	.57	1.9	97.1	A, B, E, W	C, P, W
Propylene glycol (1,2-Propanediol)	$CH_3CHOHCH_2OH$	76.10	1.04	.59	230	189.	A, E, W	O, U, W
Propyl acetate	$CH_3CO_2CH_2CH_2CH_3$	102.13	0.89	—	—	102.	A, E, W-s	C, P
Pyridine	$N{:}CHCH{:}CHCH{:}CH$	79.10	0.98	0.43	974	116.	A, E, W	L, R
Styrene	$C_6H_5CH{:}CH_2$	104.14	0.91	0.32	0.77	145	A, E	P, R
1, 1, 2, 2-Tetrachloroethane	$Cl_2CHCHCl_2$	167.85	1.60	0.27	1.77	146	A, E, W-s	C, G, L, P
Tetrahydrofurfuryl alcohol	$C_4H_7OCH_2OH$	102.14	1.05	—	5.49	177	—	O, P, W
Tetralin	$C_6H_4CH_2(CH_2)_2CH_2$	132.20	0.97	—	—	207.	A, E	B, C, G, O
Toluene	$C_6H_5CH_3$	92.13	0.87	.41	590	111.	A, B, E	C, G, L, O, P
1, 1, 2-Trichloroethylene	$ClCH{:}CCl_2$	131.39	1.46	0.22	0.58	87	A, E, W-s	C, L, O, P, W
Triethylene glycol	$(CH_2OCH_2CH_2OH)_2$	150.18	1.13	—	—	276	A, E-s, W	C, G, P

[a]Specific heat is in Btu/lb·deg F or cal/g·deg C; for J/kg·K multiply by 4184.
[b]Viscosity is in centipoises; for N·s/m² divide by 1000; for lb/ft·s multiply by 0.67.

Table 3-42. ION-EXCHANGE RESINS*

SYMBOLS:

Manufacturers:

1. Dow Chemical Co., Midland, Mich. 48640
2. Diamond Shamrock Chemical Co., Redwood City, Calif. 94063
3. Ionac Chemical Corp., Birmingham, N. J. 08011
4. Nalco Chemical Co., Chicago, Ill. 60638
5. Rohm and Haas Co., Philadelphia, Pa. 19105

Characters:

S = strong
W = weak

Physical form:

b = beads
g = granules

Trade name	Manu-facturer	Char-acter	Active group	Matrix	Effective pH	Selectivity	Order of selectivity
ANION-EXCHANGE RESINS							
Dowex 1	1	S	Trimethyl benzyl ammonium	Polystyrene	0–14	Cl/H approx. 25	I, NO_3, Br, Cl, acetate, OH, F
Dowex 21 K	1	S	Trimethyl benzyl ammonium	Polystyrene	0–14	Cl/H approx. 1.5	I, NO_3, Br, Cl, acetate, OH, F
Duolite A-101 D	2	S	Trimethyl benzyl ammonium	Polystyrene	0–14	—	—
Ionac A-540	3	S	Trimethyl benzyl ammonium	Polystyrene	0–14	—	—
Amberlite IRA-400	5	S	Trimethyl benzyl ammonium	Polystyrene	0–14	—	—
Dowex 2	1	S	Dimethyl ethanol benzyl ammonium	Polystyrene	0–14	Cl/H approx. 1.5	I, NO_3, Br, Cl, acetate, OH, F
Duolite A-102 D	2	S	Dimethyl ethanol benzyl ammonium	Polystyrene	0–14	—	—
Ionac A-550	3	S	Dimethyl ethanol benzyl ammonium	Polystyrene	0–14	—	—
Amberlite IRA-410	5	S	Dimethyl ethanol benzyl ammonium	Polystyrene	0–14	—	—
Duolite A-30 B	2	W	Tertiary amine; quaternary ammonium	Epoxy poly-amine	0–9	—	—
Ionac A-300	3	W	Tertiary amine; quaternary ammonium	Epoxy poly-amine	0–12	—	—
Duolite A-6	2	W	Tertiary amine	Phenolic	0–5	—	—
Duolite A-7	2	W	Secondary amine	Phenolic	0–4	—	—
Amberlite IR-45	5	W	Primary, secondary, and tertiary amine	Polystyrene	0–9	—	—

*Condensed from: "Handbook of Chemistry and Physics", 53rd ed., R.C. Weast, Ed., The Chemical Rubber Co., 1972.

Table 3-42. ION-EXCHANGE RESINS *(Continued)*

Trade name	Manu-facturer	Total exchange capacity, meq/ml†	Total exchange capacity, meq/g†	Maximum thermal stability, °C	Physical form	Standard mesh range	Ionic form as shipped‡	Shipping density, lb/cu ft
				ANION-EXCHANGE RESINS				
Dowex 1	1	1.33	3.5	50 (OH⁻) 150 (Cl⁻)	b	20–50 (wet)	Cl⁻	44
Dowex 21 K	1	1.25	4.5	50 (OH⁻) 150 (Cl⁻)	b	20–50 (wet)	Cl⁻	43
Duolite A-101 D	2	1.4	4.2	60 (OH⁻) 100 (Cl⁻)	b	16–50	Cl⁻	—
Ionac A-540	3	1.0	3.6	60 (OH⁻) 100 (Salt)	b	16–50	Salt	43–66
Amberlite IRA-400	5	1.4	3.8	60 (OH⁻) 75 (Cl⁻)	b	—	Cl⁻	—
Dowex 2	1	1.33	3.5	30 (OH⁻) 150 (Cl⁻)	b	20–50 (wet)	Cl⁻	44
Duolite A-102 D	2	1.4	4.2	40 (OH⁻) 100 (Cl⁻)	b	16–50	Cl⁻	—
Ionac A-550	3	1.3	3.5	40 (OH⁻) 100 (Salt)	b	16–50	Salt	43–46
Amberlite IRA-410	5	1.4	3.3	40 (OH⁻) 75 (Cl⁻)	b	—	Cl⁻	—
Duolite A-30 B	2	2.6	8.7	80	b	16–50	Salt	—
Ionac A-300	3	1.8	5.5	40	g	16–50	Salt	19–21
Duolite A-6	2	2.4	7.6	60	g	16–50	Salt	—
Duolite A-7	2	2.4	9.1	40	g	16–50	Salt	—
Amberlite IR-45	5	1.9	5.2	100	b	—	Free base	—

Trade name	Manu-facturer	Char-acter	Active group	Matrix	Effec-tive pH	Selectivity	Order of selectivity
				CATION-EXCHANGE RESINS			
Dowex 50	1	S	Nuclear sulfonic acid	Polystyrene	0–14	Na/H approx. 1.2	Ag, Cs, Rb, K, NH₄, Na, H, Li, Ba, Sr, Ca, Mg, Be
Dowex MPC-1	4	S	Nuclear sulfonic acid	Polystyrene	0–14	—	—
Duolite C-20	2	S	Nuclear sulfonic acid	Polystyrene	0–14	—	—
Ionac C-240	3	S	Nuclear sulfonic acid	Polystyrene	0–14	—	—
Amberlite IR-120	5	S	Nuclear sulfonic acid	Polystyrene	0–14	—	—
Duolite C-3	2	S	Methylene sulfonic	Phenolic	0–9	—	—
Dowex CCR-1	4	W	Carboxylic	Phenolic	0–9	—	—
Duolite ES-63	2	W	Phosphonic	Polystyrene	4–14	—	—
Duolite ES-80	2	W	Carboxylic	Acrylic	6–14	—	—
Amberlite IRC-84	5	W	Carboxylic	Acrylic	4–14	—	—

Table 3-42.   ION-EXCHANGE RESINS *(Continued)*

Trade name	Manu-fac-turer	Total exchange capacity, meq/ml†	Total exchange capacity, meq/g†	Maximum thermal stability, °C	Physical form	Standard mesh range	Ionic form as shipped‡	Shipping density, lb/cu ft
			CATION-EXCHANGE RESINS					
Dowex 50	1	Na$^+$ 1.9 H$^+$ 1.7	Na$^+$ 4.8 H$^+$ 5.0	150	b	20–50 (wet)	H$^+$ or Na$^+$	H$^+$ 50 Na$^+$ 53
Dowex MPC-1	4	1.6–1.8 H$^+$ form	4.5–4.9 H$^+$ form	150	b	20–40 (wet)	Na$^+$	50
Duolite C-20	2	2.2	5.1	150	b	16–50	Na$^+$	—
Ionac C-240	3	1.9	4.6	140 (Na$^+$) 130 (H$^+$)	b	16–50	Na$^+$	50–55
Amberlite IR-120	5	1.9	4.4	120	b	—	Na$^+$ or H$^+$	—
Duolite C-3	2	1.1	2.9	60	g	16–50	H$^+$	—
Dowex CCR-1	4	—	—	38	g	20–50	H$^+$ (dry)	21
Duolite ES-63	2	3.3	6.6	100	b	16–50	H$^+$	—
Duolite ES-80	2	3.5	10.2	100	b	16–50	H$^+$	—
Amberlite IRC-84	5	3.5	10.3	120	b	—	H$^+$	—

†Meq = milligram equivalents. The equivalent weight of an element or ion is its atomic or formula weight divided by its valence. The exchange capacity of a resin is the number of milligram equivalents of ions in the solution that 1 g (dry) or 1 ml (wet) of the resin will exchange.
‡Changes in ionic form will cause resins to swell 5–25% or more.

APPLICATIONS OF ION-EXCHANGE RESINS

1. Replacement of deleterious by unobjectionable ions, as in water softening.
2. Concentration or recovery of a valuable substance, as the extraction of uranium and other metals from ores, and the concentration and purification of streptomycin.
3. Removal of harmful substances from waste effluent; for example, cyanide from plating baths, thiocyanate from gas-works liquors, and radioactive contaminants from wastes of nuclear power plants.
4. Removal of acids by OH-resins and alkalis by H-resins, as in the removal of formic acid from commercial formaldehyde solutions and the removal of acid catalyst from esterification mixtures.
5. Removal of both anions and cations, as in water deionization, sugar purification, and glycerol refining.
6. Other applications include analytical separations; treatment of wine, fruit juices, and milk; use in the manufacture of pulp and paper, pharmaceuticals, and petroleum products; in hydrometallurgy; in the isolation and purification of biochemical products; and as catalysts in chemical reactions.

REGENERATION OF ION-EXCHANGE RESINS

Ion exchange resins must be regenerated at intervals to restore their capacity to exchange ions. In water softening the resin is regenerated with a salt brine solution containing 10 to 20% NaCl. In formaldehyde purification the weak-base resins can be regenerated with dilute solutions of caustic soda, soda ash, or ammonia. In sugar purification sulfuric acid and caustic soda or ammonia are usually the reagents involved. Streptomycin is eluted from the resin with aqueous HCl. In the sulfuric acid leach process uranium is eluted with a solution of ammonium nitrate and nitric acid or sodium chloride and sulfuric acid.

REFERENCE
"Ion Exchange Polymers", I.M. Abrams and L. Benezra, *Encyclopedia of Polymer Science and Technology*, Vol. 7, John Wiley & Sons, 1967, pp. 692–742.

Table 3-43. PHYSICAL PROPERTIES OF HYDROCARBONS*

At Room Temperature and Pressure

For density in kg/m^3, multiply by 16.02. For specific heat in J/kg·K, multiply by 4184. For latent heat in J/kg, multiply by 2324.

Hydrocarbon	Formula	Mol. weight	Phase	Density, lb_m/ft^3	Specific heats, Btu/lb °F = cal/g °C			c_p/c_v, gas	Boiling	
					c_p, gas	c_p, liquid	c_v, gas		Temp, °F	Latent heat, Btu/lb_m
PARAFFINS OR ALKANES										
Methane	CH_4	16.043	gas	0.0415	0.526	—	0.400	1.315	−258.7	220.8
Ethane	C_2H_6	30.070	gas	0.0775	0.409	0.78	0.347	1.18	−127.5	210.5
Propane	C_3H_8	44.097	gas	0.114	0.388	0.58[a]	0.343	1.13	−43.8	183.3
n-Butane	C_4H_{10}	58.124	gas	36.14	0.397	0.55	0.361	1.10	31.1	165.5
iso-Butane	C_4H_{10}	58.124	gas	34.77	0.387[a]	0.56	0.348	1.11	10.9	157.5
n-Pentane	C_5H_{12}	72.151	liq	39.08	0.3974	0.557	0.3699	1.07	97.0	153.59
iso-Pentane	C_5H_{12}	72.151	liq	38.77	0.388	0.562	0.3605	1.076	82.2	145.66
Neopentane	C_5H_{12}	72.151	gas	38.27	0.391	0.543	0.3635	1.076	49.1	135.6
n-Hexane	C_6H_{14}	86.178	liq	41.14	0.3984	0.536	0.3753	1.062	155.7	144.7
Neohexane	C_6H_{14}	86.178	liq	40.51	0.3984	0.511	0.3753	1.062	121.5	132.6
n-Heptane	C_7H_{16}	100.206	liq	41.70	0.3992	0.525	0.3794	1.052	209.1	137.5
Triptane	C_7H_{16}	100.206	liq	43.07	0.3992	0.497	0.3794	1.052	177.6	124.6
n-Octane	C_8H_{18}	114.233	liq	44.14	0.3998	0.526	0.3824	1.046	258.3	131.65
iso-Octane	C_8H_{18}	114.233	liq	43.82	0.3998	0.489	0.3824	1.046	243.9	116.7
OLEFINS OR ALKENES										
Ethylene	C_2H_4	28.054	gas	0.0733	0.363	—	0.296	1.22	−154.7	207.6
Propylene	C_3H_6	42.081	gas	0.113	0.363	0.57[a]	0.316	1.15	−53.9	188.2
Butylene	C_4H_8	56.108	gas	37.12	0.371	0.53	0.334	1.11	21.2	168.0
iso-Butene	C_4H_8	56.108	gas	36.83	0.375	0.55[a]	0.335	1.12	19.6	168.7
n-Pentene	C_5H_{10}	70.135	liq	40.02	0.380		0.352	1.08	86.0	—
AROMATICS										
Benzene	C_6H_6	78.114	liq	55.18	0.342[b]	0.410	0.317	1.08	176.2	169.3
Toluene	C_7H_8	92.141	liq	55.31	0.347[b]	0.404	0.345	1.06	231.1	156.2
p-Xylene	C_8H_{10}	106.169	liq	53.75	—	0.407	—		281.1	146.1
OTHER HYDROCARBONS										
Acetylene	C_2H_2	26.038	gas	0.0675	0.689	—	0.547	1.26	−119.2	—
Naphthalene	$C_{10}H_8$	128.175	sol	71.48	0.325	—	—	—	424.4	64.1[c]

[a] Estimated value.

[b] At 250°F.

[c] Heat of fusion.

*Compiled from several sources.

Table 3-44.   COMBUSTION DATA FOR HYDROCARBONS*

For heating value in J/kg, multiply the value in Btu/lb$_m$ by 2324. For flame speed in m/s, multiply the value in ft/s by 0.3048.

Hydrocarbon	Formula	Higher heating value (vapor), Btu/lb$_m$	Theor. air/fuel ratio, by mass	Max flame speed, ft/sec	Adiabatic flame temp (in air), °F	Ignition temp (in air), °F	Flash point, °F	Flammability limits (in air), % by volume	
PARAFFINS OR ALKANES									
Methane	CH_4	23875	17.195	1.1	3484	1301	gas	5.0	15.0
Ethane	C_2H_6	22323	15.899	1.3	3540	968–1166	gas	3.0	12.5
Propane	C_3H_8	21669	15.246	1.3	3573	871	gas	2.1	10.1
n-Butane	C_4H_{10}	21321	14.984	1.2	3583	761	−76	1.86	8.41
iso-Butane	C_4H_{10}	21271	14.984	1.2	3583	864	−117	1.80	8.44
n-Pentane	C_5H_{12}	21095	15.323	1.3	4050	588	< −40	1.40	7.80
iso-Pentane	C_5H_{12}	21047	15.323	1.2	4055	788	< −60	1.32	9.16
Neopentane	C_5H_{12}	20978	15.323	1.1	4060	842	gas	1.38	7.22
n-Hexane	C_6H_{14}	20966	15.238	1.3	4030	478	−7	1.25	7.0
Neohexane	C_6H_{14}	20931	15.238	1.2	4055	797	−54	1.19	7.58
n-Heptane	C_7H_{16}	20854	15.141	1.3	3985	433	25	1.00	6.00
Triptane	C_7H_{16}	20824	15.141	1.2	4035	849	—	1.08	6.69
n-Octane	C_8H_{18}	20796	15.093	—	—	428	56	0.95	3.20
iso-Octane	C_8H_{18}	20770	15.093	1.1	—	837	10	0.79	5.94
OLEFINS OR ALKENES									
Ethylene	C_2H_4	21636	14.807	2.2	4250	914	gas	2.75	28.6
Propylene	C_3H_6	21048	14.807	1.4	4090	856	gas	2.00	11.1
Butylene	C_4H_8	20854	14.807	1.4	4030	829	gas	1.98	9.65
iso-Butene	C_4H_8	20737	14.807	1.2	—	869	gas	1.8	9.0
n-Pentene	C_5H_{10}	20720	14.807	1.4	4165	569	—	1.65	7.70
AROMATICS									
Benzene	C_6H_6	18184	13.297	1.3	4110	1044	12	1.35	6.65
Toluene	C_7H_8	18501	13.503	1.2	4050	997	40	1.27	6.75
p-Xylene	C_8H_{10}	18663	13.663	—	4010	867	63	1.00	6.00
OTHER HYDROCARBONS									
Acetylene	C_2H_2	21502	13.297	4.6	4770	763–824	gas	2.50	81
Naphthalene	$C_{10}H_8$	17303	12.932	—	4100	959	174	0.90	5.9

*Based largely on: "Gas Engineers' Handbook", American Gas Association, Inc., Industrial Press, 1967.

REFERENCES

"American Institute of Physics Handbook", 2nd ed., D.E. Gray, Ed., McGraw-Hill Book Company, 1963.

"Chemical Engineers' Handbook", 4th ed., R.H. Perry, C.H. Chilton, and S.D. Kirkpatrick, Eds., McGraw-Hill Book Company, 1963.

"Handbook of Chemistry and Physics", 53rd ed., R.C. Weast, Ed., The Chemical Rubber Company, 1972; gives the heat of combustion of 500 organic compounds.

"Handbook of Laboratory Safety", 2nd ed., N.V. Steere, Ed., The Chemical Rubber Company, 1971.

"Physical Measurements in Gas Dynamics and Combustion", Princeton University Press, 1954.

Table 3-45. PROPERTIES OF TYPICAL GASEOUS AND LIQUID COMMERCIAL FUELS*

For heating value in J/kg, multiply the value in Btu/lb$_m$ by 2324. For density in kg/m^3, multiply the value in lb/ft^3 by 16.02.

Gaseous fuels	Composition, percent by volume								Mol wt of fuel	Theor. air/fuel ratio by wt	Higher heating value, Btu/lb$_m$	Density, lb$_m$/ft^3
	H_2	N_2	O_2	CH_4	CO	CO_2	C_2H_4	C_6H_6				
Blast furnace gas	1.0	60.0	—	—	27.5	11.5	—	—	29.6	0.667	1,170	.075 5[a]
Blue water gas	47.3	8.3	0.7	1.3	37.0	5.4	—	—	16.4	3.759	6,550	.042 2[a]
Carb. water gas	40.5	2.9	0.5	10.2	34.0	3.0	6.1	2.8	18.3	7.299	11,350	.046 6[a]
Coal gas	54.5	4.4	0.2	24.2	10.9	3.0	1.5	1.3	12.1	10.87	16,500	.031 1[a]
Coke-oven gas	46.5	8.1	0.8	32.1	6.3	2.2	3.5	0.5	13.7	17.24	17,000	.032 6[a]
Natural gas (15.8% C_2H_6)	—	0.8	—	83.4	—	—	—	—	18.3	17.24	24,100	.045 1[a]
Producer gas	14.0	50.9	0.6	3.0	27.0	4.5	—	—	24.7	14.29	2,470	.063 6[a]

Liquid commercial fuels	Vapor		Gravity, API, 60°F	Distillation			Flash point, °F	Viscosity, centistokes, 100°F	Mol wt of fuel	Theor. air/fuel ratio by wt	Higher heating value, Btu/lb$_m$	Density, lb$_m$/ft^3
	c_p, 60°F	c_p/c_v, 60°F		10%, °F	90%, °F	End point, °F						
	(approximately)											
Gasoline	0.4	1.05	63	121	320	397	0	—	113	14.93	20,460	43.8[b]
Gasoline	0.4	1.05	63	118	330	410	0	—	126[c]	14.97	20,260	46.1[b]
Kerosene	0.4	1.05	41.9	370	510	546	130	—	154[c]	14.99	19,750	51.5[b]
Diesel oil (1-D)	0.4	1.05	42	—	550	—	100	1.4–2.5	170	15.02	19,240	54.6[b]
Diesel oil (2-D)	0.4	1.05	36	—	540–576	—	125	2.0–5.8	184	15.06	19,110	57.4[b]
Diesel oil (4-D)	0.4	1.05	—	—	550	—	130	5.8–26.4	198	14.93	18,830	59.9[b]

[a]Based on dry air at 25°C and 760 mm Hg.
[b]Based on H$_2$O at 60°F, 1 atm (ρ = 62.367 lb$_m$/ft^3).
[c]Estimated.

*Abridged from: "Engineering Experimentation", G.L. Tuve and L.C. Domholdt, McGraw-Hill Book Company, 1966; and "The Internal Combustion Engine", 2nd ed., C.F. Taylor and E.S. Taylor, International Textbook Co., 1961.

Table 3-46. TYPICAL COMMERCIAL LIQUID FUEL SPECIFICATIONS*

See Table 5-70 for liquid rocket-fuel specifications.

SYMBOLS: W = winter   F = fall   S = summer

Specification	Automotive gasolines[a]			Aviation gasolines[b]	Aviation turbine fuels[c]			
	Type A	Type B	Type C		Jet A	Jet B	JP-4	JP-5
Distillation:								
Fuel evaporated, °F, 10% min	W—140 F—149 S—158	W—140 F—149 S—158	167	167	400	—	250	400
Fuel evaporated, °F, 50% min	284	257	284	221	450	370	370	—
Fuel evaporated, °F, 90% min	392	356	392	225	—	470	470	—
Residue, % by vol., max	2	2	2	1.5	1.5	1.5	1.5	1.5
Distillation loss, % by vol., max	—	—	—	1.5	1.5	1.5	1.5	1.5
Gravity, deg API, min	—	—	—	—	39	45	45.0	36.0
Gravity, deg API, max	—	—	—	—	51	57	57.0	48.0
Gum, mg/100 ml, max	5	5	5	3.0	—	—	7	7
Sulfur, % by wt., max	—	—	—	0.05	0.3	0.3	0.4	0.4
Vapor pressure at 100°F, psi, max	W—15.0 F—11.5 S—10.0	W—15.0 F—11.5 S—10.0	W—15.0 F—11.5 S—10.0					
Freezing point, °F, max	—	—	—	−72	−36	−56	−76	−55
Lower heating value, Btu/lb$_m$, min	—	—	—	18,720[d]	18,400	18,400	18,400	18,300

[a]Octane number and rating method to be agreed upon between purchaser and seller.
[b]The various grades of aviation gasoline (80/87, 91/98, 100/130, 108/135, and 115/145) differ primarily in antiknock quality. The antiknock ratings, such as 115/145, are obtained from the aviation method (ASTM D614) and supercharge method (ASTM D909), respectively.
[c]Jet A-1 and JP-6 are cold-weather fuels corresponding to Jet A and JP-5, respectively.
[d]For grades 108/135 and 115/145 the minimum heat of combustion is 18,800.

*Abridged from: "Automotive Gasolines", ASTM 439–67T, American Society for Testing and Materials, 1967;
          "Aviation Gasolines", ASTM 910–67, American Society for Testing and Materials, 1967;
          "Aviation Turbine Fuels", ASTM D1655–67, American Society for Testing and Materials, 1967; and
          "Aviation Turbine Fuels", MIL–J–5624F, September 1962.

Table 3-47. TYPICAL COMMERCIAL LPG FUEL SPECIFICATIONS*

Specification	Commercial propane	Special-duty propane	Commercial butane	Commercial PB mixtures	Commercial hexanes	ASTM test methods
Vapor pressure at 100°F, psig, max	210	200	70	[a]	6.0[b]	D1267 or D2598
Volatile residue:						
Evap. temp. °F, 95% max, or	−37	−37	36	36	—	D1837
Butane and heavier, % max	2.5	2.5	—	—	—	D2163
Pentane and heavier, % max	—	—	2.0	2.0	—	D2163
Residue on evaporation of 100 ml, max	0.05 ml	0.05 ml	0.05 ml	0.05 ml	1.0 mg[c]	D2158
Sulfur content, g/100 ft³, max	15	10	15	15	30 ppm	D1266

[a]Must not exceed 200 psig; also must not exceed value calculated from the following: vapor pressure, max = 1167 − 1880 (Sp. gr. 60/60 °F).
[b]ASTM Method D323 only.
[c]ASTM Method D1353 only.

*Abridged from: "Liquified Petroleum (LP) Gases", ASTM D1835–67T, American Society for Testing and Materials, 1967;
          "Special Duty Propane", ASTM D2154–67, American Society for Testing and Materials, 1967; and
          "Commercial Hexanes", ASTM D1836–67T, American Society for Testing and Materials, 1967.

Table 3-48.   TYPICAL COMMERCIAL FUEL-OIL SPECIFICATIONS*

Specification	Fuel oils						Diesel fuel oils[e]		Gas turbine fuel oils[f]	
	No. 1	No. 2	No. 4	No. 5 light	No. 5 heavy	No. 6	No. 1-D	No. 2-D	No. 3-GT	No. 4-GT
Flash point, °F, min	100[d]	100[d]	130[d]	130[d]	130[d]	150	100[d]	125[d]	130[d]	150[d]
Pour point, °F, max	0	20	20	—	—	—	[b]	[b]	—	—
Water and sediment, % by vol., max	trace	0.10	0.50	1.00	1.00	2.00	trace	0.10	1.0	1.0
Carbon residue on 10% residuum, % max	0.15	0.35	—	—	—	—	0.15	0.35	—	—
Ash, % by wt., max	—	—	0.10	0.10	0.10	—	0.01	0.02	0.03	—
Distillation temp, °F										
10% max	420	440[a]	—	—	—	—	—	—	—	—
90% min		540					—	540	—	—
90% max	550	640					550	640	—	—
Saybolt viscosity, sec										
Universal at 100°F, min	—	32.6[c]	45	150	350	900[c]	—	32.6	45	45
Universal at 100°F, max	—	37.93[c]	125	300	750	9000[c]	34.4	40.1	—	—
Furol at 122°F, min	—	—	—	—	23[c]	45	—	—	—	—
Furol at 122°F, max	—	—	—	—	40[c]	300	—	—	300	300
Kinematic viscosity, centistokes										
At 100°F, min	1.4	2.0	5.8[c]	32[c]	75[c]	—	1.4	2.0	5.8[c]	5.8[c]
At 100°F, max	2.2	3.6	26.4[v]	65[v]	162[v]	—	2.5	4.3	—	—
At 122°F, min	—	—	—	—	42[c]	92[c]	—	—	—	—
At 122°F, max	—	—	—	—	81[c]	638[c]	—	—	638[c]	638[c]
Gravity, deg API, min	35	30	—	—	—	—	—	—	—	—

[a] For use in other than atomizing burners.

[b] For cold weather operation the pour point should be specified 10°F below ambient temperature at which engine is expected to operate (except where fuel-oil heating facilities are provided).

[c] For information only.

[d] Or legal.

[e] No. 4-D diesel fuel oil is very similar to No. 4 fuel oil.

[f] No. 1-GT and No. 2-GT gas-turbine fuels are very similar to No. 1 and No. 2 fuel oils.

*Abridged from: "Diesel Fuel Oils", ASTM D975–67, American Society for Testing and Materials, 1967; and "Fuel Oils", ASTM D396–67, American Society for Testing and Materials, 1967.

Table 3-49. CLASSIFICATION OF COALS*

Ash-free Basis

For heating value in J/kg, multiply the value in Btu/lb$_m$ by 2324.

Class and group	Fixed carbon,[a] dry, %		Volatile matter, dry, %		Heating value,[b] moist, Btu/lb$_m$		Agglomerating character
	Equal or greater than	Less than	Greater than	Equal or less than	Equal or greater than	Less than	
Anthracitic							
Meta-anthracite	98	—	—	2			
Anthracite	92	98	2	8			
Semianthracite	86	92	8	14	—	—	Non-agglomerating
Bituminous							
Low volatile	78	86	14	22	—	—	Commonly agglomerating
Medium volatile	69	78	22	31	—	—	Commonly agglomerating
High-volatile A	—	69	31	—	14,000	—	Commonly agglomerating
High-volatile B	—	—	—	—	13,000	14,000	Commonly agglomerating
High-volatile C	—	—	—	—	11,500	13,000	Commonly agglomerating
					10,500	11,500	Agglomerating
Subbituminous							
Subbituminous A	—	—	—	—	10,500	11,500	Non-agglomerating
Subbituminous B	—	—	—	—	9.500	10,500	
Subbituminous C	—	—	—	—	8,300	9,500	
Lignitic							
Lignite A	—	—	—	—	6,300	8,300	
Lignite B	—	—	—	—		6,300	

[a]Coals having 69% or more fixed carbon on the dry, ash-free basis shall be classified according to fixed carbon, regardless of heating value.

[b]Higher heating value, constant volume. The word *moist* refers to coal containing its natural inherent moisture but not including visible water on the surface of the coal.

*Compiled from several sources.

REFERENCES

"ASTM Standard D388", American Society for Testing and Materials, 1966.

A similar international classification under United Nations sponsorship is also used.

"Combustion Engineering", Rev. ed., G.R. Fryling, Ed., Combustion Engineering, Inc., 1966; includes data on wood and by-product fuels.

Table 3-50.   TYPICAL U.S. COMMERCIAL SOLID FUELS*

For tables on solid rocket fuels, see Index.

For heating value in J/kg, multiply the value in Btu/lb$_m$ by 2324.

Classification (See Table 3-49)	State, County	Proximate analysis,† %				Ultimate analysis,† %					Heating value,‡ Btu/lb$_m$
		Moisture	Volatile matter	Fixed carbon	Ash	Sulfur	Hydrogen	Carbon	Nitrogen	Oxygen	
Meta-anthracite	Rhode Island Newport	13.2	2.6	65.3	18.9	0.3	1.9	64.2	0.2	14.5	9,310
			3.8	96.2		0.4	0.6	94.7	0.3	4.0	13,720
Anthracite	Pennsylvania Lackawanna	4.3	5.1	81.0	9.6	0.8	2.9	79.7	0.9	6.1	12,880
			5.9	94.1		0.9	2.8	92.5	1.0	2.8	14,980
Semianthracite	Arkansas Johnson	2.6	10.6	79.3	7.5	1.7	3.8	81.4	1.6	4.0	13,880
			11.7	88.3		1.9	3.9	90.6	1.8	1.8	15,430
Low-volatile bituminous	West Va. Wyoming	2.9	17.7	74.0	5.4	0.8	4.6	83.2	1.3	4.7	14,400
			19.3	80.7		0.8	4.6	90.7	1.4	2.5	15,690
Medium-volatile bituminous	Pennsylvania Clearfield	2.1	24.4	67.4	6.1	1.0	5.0	81.6	1.4	4.9	14,310
			26.5	73.5		1.1	5.2	88.9	1.6	3.2	15,590
High-volatile A bituminous	West Va. Marion	2.3	36.5	56.0	5.2	0.8	5.5	78.4	1.6	8.5	14,040
			39.5	60.5		0.8	5.7	84.8	1.7	7.0	15,180
High-volatile B bituminous	Kentucky Muhlenburg	8.5	36.4	44.3	10.8	2.8	5.4	65.1	1.3	14.6	11,680
			45.0	55.0		3.4	5.5	80.6	1.7	8.8	14,460
High-volatile C bituminous	Illinois Sangamon	14.4	35.4	40.6	9.6	3.8	5.8	59.7	1.0	20.1	10,810
			46.6	53.4		5.0	5.6	78.6	1.3	9.5	14,230
Subbituminous A	Wyoming Sweetwater	16.9	34.8	44.7	3.6	1.4	6.0	60.4	1.2	27.4	10,650
			43.7	56.3		1.8	5.2	76.0	1.5	15.5	13,390
Subbituminous B	Wyoming Sheridan	22.2	33.2	40.3	4.3	0.5	6.9	53.9	1.0	33.4	9,610
			45.2	54.8		0.6	6.0	73.4	1.3	18.7	13,080
Subbituminous C	Colorado El Paso	25.1	30.4	37.7	6.8	0.3	6.2	50.5	0.7	35.5	8,560
			44.6	55.4		0.5	5.0	74.1	1.1	19.3	12,560
Lignite	North Dakota McLean	36.8	27.8	29.5	5.9	0.9	6.9	40.6	0.6	45.1	7,000
			48.4	51.6		1.6	5.0	70.9	1.1	21.4	12,230
High-temperature coke		5.0	1.3	83.7	10.0	0.8	0.5	82.0	1.0	0.7	12,200
Low-temperature coke		2.8	15.1	72.1	10.0	1.8	3.2	74.5	1.6	6.1	12,600
Beehive coke		0.5	1.8	86.0	11.7	1.0	0.7	84.4	1.2	0.5	12,527
By-product coke		0.8	1.4	87.1	10.7	1.0	0.7	85.0	1.3	0.5	12,690
High-temperature coke breeze		12.0	4.2	65.8	18.0	0.6	1.2	66.8	0.9	0.5	10,200
Petroleum coke		1.1	7.0	90.7	1.2	0.8	3.2	90.8	0.8	2.1	15,060
Pitch coke		0.3	1.1	97.6	1.0	0.5	0.6	96.6	0.7	0.3	14,097

†Analyses are "as received", except where moisture and ash percentages are omitted.
‡Higher heating values, constant volume.

*From: "ASTM Standard D388", American Society for Testing and Materials, 1966.

REFERENCE
"Combustion Engineering", rev. ed., G.R. Fryling, Ed., Combustion Engineering, Inc., 1966.

Table 3-51. HEAT OF COMBUSTION
OF VARIOUS ORGANIC COMPOUNDS*

At Atmospheric Pressure and 20°C

To convert heating values to J/kg, multiply by 2324.

Name	Formula	Mol wt	Physical state	$\dfrac{kcal}{g}$	$\dfrac{Btu}{lb}$
Acetaldehyde	CH_3CHO	44.1	liquid	6.33	11,394
Acetanilide	$CH_3CONHC_6H_5$	135.2	solid	7.47	13,446
Acetic acid	CH_3CO_2H	60.1	liquid	3.48	6,264
Acetic anhydride	$(CH_3CO)_2O$	102.1	liquid	4.23	7,614
Acetone	$(CH_3)_2CO$	58.1	liquid	7.34	13,212
Acetonitrile	CH_3CN	41.1	liquid	7.36	13,248
Allyl alcohol	$CH_2:CHCH_2OH$	58.1	liquid	7.61	13,698
p-Aminophenol	$HOC_6H_4NH_2$	109.1	solid	6.97	12,546
Amyl acetate	$C_4H_9CO_2C_2H_5$	132.2	liquid	7.89	14,202
Amyl alcohol (ferm.)	$(CH_3)_2CH_2CH_2CH_2OH$	88.2	liquid	9.00	16,200
Anthracene	$C_6H_4:(CH)_2:C_6H_4$	178.2	solid	9.54	17,172
Anthraquinone	$C_{14}H_8O_2$	208.2	solid	7.42	13,356
Azobenzene	$(C_6H_5N)_2$	182.2	solid	8.48	15,264
Benzaldehyde	C_6H_5CHO	106.1	liquid	7.93	14,274
Benzoic acid†	$C_6H_5CO_2H$	122.1	solid	6.32	11,376
Benzyl chloride	$C_6H_5CH_2Cl$	126.6	liquid	7.00	12,600
n-Butyl alcohol	C_4H_9OH	74.1	liquid	8.49	15,282
Camphor	$C_{10}H_{16}O$	152.2	solid	9.27	16,686
Carbon disulfide	CS_2	76.1	liquid	3.24	5,832
Chloroacetic acid	$ClCH_2CO_2H$	94.5	solid	1.81	3,258
Chloroform	$CHCl_3$	119.4	liquid	.747	1,345
Cyclohexanol	$CH_2(CH_2)_4CHOH$	100.2	liquid	8.89	16,002
Cycloheptane	$(CH_2)_7$	98.2	liquid	11.1	19,980
Cyclohexane	$(CH_2)_6$	84.2	liquid	11.1	19,980
Decane	$C_{10}H_{22}$	142.3	liquid	11.3	20,340
Diamyl ether	$(C_5H_{11})_2O$	158.3	liquid	10.2	18,360
Diamylene	$C_{10}H_{20}$	140.3	liquid	11.3	20,340
o-Dichlorobenzene	$C_6H_4Cl_2$	147.0	liquid	4.57	8,226
Diethyl ether	$(C_2H_5)_2O$	69.1	liquid	9.43	16,974
Dimethyl amine	$(CH_3)_2NH$	45.1	liquid	9.24	16,632
Dimethyl ether	$(CH_3)_2O$	46.1	gas	7.54	13,572
Dimethyl phthalate	$C_6H_4(CO_2CH_3)_2$	194.2	liquid	5.76	10,368
Dinitrotoluene	$C_6H_3(CH_3)(NO_2)_2-$ (1, 2, 4)	182.1	solid	4.68	8,424
Diphenyl	$(C_6H_5)_2$	154.2	solid	9.68	17,424
Diphenyl carbinol	$(C_6H_5)_2CHOH$	184.2	solid	8.76	15,768
Ethyl acetate	$CH_3CO_2C_2H_5$	88.1	liquid	6.09	10,962
Ethyl alcohol	C_2H_5OH	46.1	liquid	7.10	12,780
Ethylbenzene	$C_2H_5C_6H_5$	106.2	liquid	10.3	18,540
Ethyl benzoate	$C_6H_5CO_2C_2H_5$	150.2	liquid	7.31	13,158
Ethyl bromide	C_2H_5Br	109.0	vapor	3.12	5,616
Ethyl chloride	C_2H_5Cl	64.5	vapor	4.91	8,838
Ethylcycloheptane	$C_2H_5C_7H_{13}$	126.3	liquid	11.1	19,980
Fluorene	$(C_6H_4)_2:CH_2$	166.2	solid	9.53	17,154
Fluorobenzene	C_6H_5F	96.1	liquid	7.78	14,004
Formaldehyde	CH_2O	30.0	gas	4.47	8,046

Table 3-51. HEAT OF COMBUSTION
OF VARIOUS ORGANIC COMPOUNDS *(Continued)*

Name	Formula	Mol wt	Physical state	kcal g	Btu lb
Furfural	C_4H_3OCHO	96.1	liquid	5.82	10,476
Glycerol	$(CH_2OH)_2CHOH$	92.1	liquid	4.31	7,758
Hexachlorobenzene	C_6Cl_6	284.8	solid	1.79	3,222
Hexadecane	$C_{16}H_{34}$	226.4	solid	11.3	20,340
Indigo	$C_{16}H_{10}O_2N_2$	262.3	solid	6.92	12,456
Indole	C_8H_7N	117.1	solid	8.72	15,696
Isobutane	$(CH_3)_3CH$	58.1	gas	11.8	21,240
Isopropyl alcohol	$(CH_3)_2CHOH$	60.1	liquid	7.90	14,220
Lactose (anhydrous)	$C_{12}H_{22}O_{11}$	342.3	solid	3.94	7,092
Menthol	$C_{10}H_{20}O$	156.3	solid	9.65	17,370
Mesityl oxide	$(CH_3)_2C:CHCOCH_3$	98.1	liquid	8.63	15,534
Methyl acetate	$CH_3CO_2CH_3$	74.1	liquid	5.14	9,252
Methyl alcohol	CH_3OH	32.0	liquid	5.34	9,612
Methyl bromide	CH_3Br	95.0	vapor	1.94	3,492
Methyl chloride	CH_3Cl	50.5	gas	3.25	5,850
Methylcyclohexane	$CH_3C_6H_{11}$	98.2	liquid	11.1	19,980
Nicotine	$C_{10}H_{14}N_2$	186.3	liquid	7.66	13,788
Nitrobenzene	$C_6H_5NO_2$	123.1	liquid	6.00	10,800
Nitroethane	$C_2H_5NO_2$	75.1	liquid	4.29	7,722
Nitromethane	CH_3NO_2	61.0	liquid	2.78	5,004
Oleic acid	$C_{18}H_{34}O_2$	282.5	liquid	9.41	16,938
Palmitic acid	$C_{16}H_{32}O_2$	256.4	solid	9.35	16,830
Phenol	C_6H_5OH	94.1	solid	7.78	14,004
n-Propyl alcohol	C_3H_7OH	60.1	liquid	8.00	14,400
n-Propylbenzene	$C_3H_7C_6H_5$	120.2	liquid	10.4	18,720
Propylene glycol	$CH_3CHOHCH_2OH$	76.1	liquid	5.66	10,188
n-Propyltoluene	$C_6H_4(CH_3)(C_3H_7)—(1, 3)$	136.2	liquid	10.3	18,540
Pyridine	C_5H_5N	79.1	liquid	8.32	14,976
Quinone	$O:C_6H_4:O$	108.1	solid	6.07	10,926
Resorcinol	$C_6H_4(OH)_2$	110.1	solid	6.20	11,160
Starch	$(C_6H_{10}O_5)_x$	(162.1)	solid	25.8	46,440 (per kg)
Styrene	$C_6H_5CH:CH_2$	104.1	liquid	10.1	18,180
Sucrose	$C_{12}H_{22}O_{11}$	342.3	solid	3.94	7,092
Triaminotriphenyl carbinol	$(C_6H_4NH_2)_3COH$	305.4	solid	8.13	14,634
Tribenzyl amine	$(C_6H_5CH_2)_3N$	287.4	solid	9.61	17,298
Triisoamyl amine	$[(CH_3)_2CHCH_2CH_2]_3N$	227.4	liquid	10.8	19,440
2, 2, 4-Trimethylpentane	$(CH_3)_3C:CH_2CH(CH_3)_2$	114.2	liquid	11.4	20,520
Trinitroglycerol	$C_3H_5(NO_3)_3$	227.1	liquid	1.62	2,916
Trinitrotoluene	$C_6H_2(CH_3)(NO_2)_3—$ (1, 2, 4, 6)	227.1	solid	3.61	6,498
Triphenyl amine	$(C_6H_5)_3N$	245.3	solid	9.25	16,650
Triphenylbenzene	$C_6H_3(C_6H_5)_3—(1, 3, 5)$	306.4	solid	9.58	17,244
Triphenylmethane	$(C_6H_5)_3CH$	244.3	solid	9.78	17,604

†Accepted value by Int. Union of Pure and Applied Chem., Lyons, 1923.
*Adapted from: "Handbook of Chemistry and Physics", 53rd ed., R.C. Weast, Ed., The Chemical Rubber Co., 1972.

Table 3-52. PROPERTIES OF COMBUSTION GASES AT HIGH TEMPERATURES*

For Hydrocarbon Fuels of the Composition C_nH_{2n}

SYMBOLS AND UNITS:

P = total pressure, atmospheres. For psia multiply by 14.696. For kg/cm^2 multiply by 1.033. For MN/m^2 multiply by 0.1013.

T = temperature, degrees Rankine. For degrees Kelvin multiply by 5/9.

h = enthalpy, Btu/lb_m. For cal/g multiply by 0.5555. For J/kg multiply by 2324.

s = entropy, $Btu/lb_m °R$ = cal/g °K

k = specific heat ratio, c_p/c_v

a = sonic velocity, ft/sec. For m/sec multiply by 0.3048.

P, atm	T, °R	$h, \frac{Btu}{lb_m}$	$s, \frac{Btu}{lb_m °R}$	Mol wt	k, c_p/c_v	Sonic vel, ft/sec
\multicolumn{7}{} PURE AIR, N/O = 3.73, 0% EXCESS O₂						
0.1	2700.	634.8	2.272	28.90	1.268	2427.
	3600.	992.9	2.381	28.74	1.256	2797.
	4500.	1674.	2.543	27.56	1.264	3204.
	5400.	3084.	2.818	24.36	1.303	3790.
1.0	2700.	634.2	2.113	28.90	1.268	2426.
	3600.	956.8	2.215	28.83	1.255	2792.
	4500.	1453.	2.334	28.25	1.255	3153.
	5400.	2350.	2.506	26.52	1.273	3590.
20.0	2700.	633.8	1.907	28.90	1.268	2426.
	3600.	942.2	2.005	28.88	1.255	2789.
	4500.	1326.	2.099	28.65	1.250	3125.
	5400.	1894.	2.209	27.90	1.255	3476.
\multicolumn{7}{} PURE AIR, N/O = 3.73, 20% EXCESS O₂						
0.1	2700.	625.5	2.269	28.91	1.274	2438.
	3600.	947.1	2.370	28.84	1.261	2798.
	4500.	1575.	2.517	27.84	1.267	3190.
	5400.	2911.	2.778	24.77	1.304	3759.
1.0	2700.	625.3	2.111	28.91	1.274	2432.
	3600.	932.8	2.208	28.89	1.261	2795.
	4500.	1378.	2.314	28.47	1.258	3145.
	5400.	2225.	2.476	26.85	1.275	3570.
20.0	2700.	625.1	1.905	28.91	1.274	2432.
	3600.	926.2	2.001	28.91	1.260	2794.
	4500.	1278.	2.086	28.79	1.255	3123.
	5400.	1802.	2.187	28.17	1.258	3463.
\multicolumn{7}{} PURE AIR, N/O = 3.73, 100% EXCESS O₂						
0.1	2700.	668.3	2.232	31.43	1.222	2284.
	3600.	1031.	2.343	31.32	1.210	2630.
	4500.	1850.	2.525	29.69	1.217	3028.
	5400.	4463.	3.016	23.17	1.289	3865.
1.0	2700.	667.8	2.086	31.44	1.222	2284.
	3600.	1009.	2.193	31.39	1.210	2627.
	4500.	1541.	2.313	30.76	1.207	2963.
	5400.	2846.	2.551	27.65	1.231	3457.
20.0	2700.	667.4	1.897	31.44	1.222	2284.
	3600.	997.1	2.001	31.42	1.210	2625.
	4500.	1400.	2.096	31.23	1.203	3936.
	5400.	2446.	2.406	28.94	1.218	3362.

P, atm	T, °R	$h, \frac{Btu}{lb_m}$	$s, \frac{Btu}{lb_m °R}$	Mol wt	k, c_p/c_v	Sonic vel, ft/sec
\multicolumn{7}{} PURE OXYGEN, N/O = 0, 0% EXCESS O₂						
0.1	2700.	763.7	2.282	31.00	1.187	2268.
	3600.	1245.	2.432	30.61	1.178	2625.
	4500.	2516.	2.724	27.67	1.196	3109.
	5400.	5898.	3.365	20.53	1.278	4088.
1.0	2700.	762.3	2.134	31.01	1.187	2267.
	3600.	1187.	2.268	30.83	1.177	2614.
	4500.	1992.	2.456	29.38	1.181	2999.
	5400.	3832.	2.800	25.18	1.215	3600.
20.0	2700.	761.6	1.942	31.01	1.187	2267.
	3600.	1156.	2.067	30.95	1.176	2608.
	4500.	1702.	2.198	30.39	1.173	2939.
	5400.	2669.	2.382	28.52	1.183	3338.
\multicolumn{7}{} PURE OXYGEN, N/O = 0, 20% EXCESS O₂						
0.1	2700.	733.4	2.278	31.14	1.197	2271.
	3600.	1145.	2.405	30.98	1.186	2617.
	4500.	2225.	2.649	28.60	1.198	3062.
	5400.	5425.	3.253	21.37	1.279	4009.
1.0	2700.	732.9	2.131	31.14	1.197	2271.
	3600.	1115.	2.250	31.08	1.185	2612.
	4500.	1775.	2.402	30.12	1.185	2967.
	5400.	3464.	2.715	26.11	1.217	3538.
20.0	2700.	732.5	1.940	31.14	1.197	2271.
	3600.	1101.	2.056	31.12	1.185	2610.
	4500.	1565.	2.166	30.85	1.180	2925.
	5400.	2396.	2.321	29.37	1.186	3293.
\multicolumn{7}{} PURE OXYGEN, N/O = 0.0, 100% EXCESS O₂						
0.1	2700.	668.3	2.232	31.43	1.222	2284.
	3600.	1031.	2.343	31.32	1.210	2630.
	4500.	1850.	2.525	29.69	1.217	3028.
	5400.	4463.	3.016	23.17	1.289	3865.
1.0	2700.	667.8	2.086	31.44	1.222	2284.
	3600.	1009.	2.193	31.39	1.210	2627.
	4500.	1541.	2.313	30.76	1.207	2963.
	5400.	2846.	2.551	27.65	1.231	3457.
20.0	2700.	667.4	1.897	31.44	1.222	2284.
	3600.	997.1	2.001	31.42	1.210	2625.
	4500.	1400.	2.096	31.23	1.203	3936.
	5400.	2052.	2.215	30.25	1.206	3271.

Table 3-52. PROPERTIES OF COMBUSTION GASES AT HIGH TEMPERATURES *(Continued)*

P, atm	T, °R	$h, \dfrac{Btu}{lb_m}$	$s, \dfrac{Btu}{lb_m °R}$	Mol wt	k, c_p/c_v	Sonic vel, ft/sec
OXYGEN-ENRICHED AIR, N/O = 1.0, 0% EXCESS O$_2$						
0.1	2700.	690.7	2.299	29.78	1.227	2352.
	3600.	1101.	2.428	29.51	1.217	2717.
	4500.	2076.	2.654	27.51	1.231	3164.
	5400.	4407.	3.100	22.34	1.294	3944.
1.0	2700.	689.7	2.146	29.78	1.227	2352.
	3600.	1059.	2.262	29.65	1.215	2709.
	4500.	1707.	2.415	28.67	1.218	3083.
	5400.	3059.	2.670	25.77	1.246	3603.
20.0	2700.	689.2	1.946	29.78	1.227	2352.
	3600.	1036.	2.056	29.73	1.215	2704.
	4500.	1498.	2.167	29.35	1.211	3039.
	5400.	2267.	2.314	28.07	1.220	3415.
N/O = 1.0, 20% EXCESS O$_2$						
0.1	2700.	671.9	2.296	29.82	1.236	2359.
	3600.	1036.	2.410	29.71	1.224	2716.
	4500.	1888.	2.605	28.07	1.234	3136.
	5400.	4090.	3.027	22.98	1.296	3891.
1.0	2700.	671.5	2.143	29.82	1.236	2359.
	3600.	1013.	2.251	29.78	1.223	2712.
	4500.	1566.	2.379	29.11	1.223	3065.
	5400.	2822.	2.615	26.38	1.249	3565.
20.0	2700.	671.3	1.943	29.82	1.236	2359.
	3600.	1003.	2.049	29.81	1.223	2710.
	4500.	1410.	2.146	29.62	1.218	3033.
	5400.	2092.	2.275	28.60	1.223	3389.
N/O = 1.0, 100% EXCESS O$_2$						
0.1	2700.	632.9	2.270	29.91	1.257	2376.
	3600.	967.1	2.374	29.84	1.245	2733.
	4500.	1638.	2.526	28.74	1.249	3146.
	5400.	3452.	2.873	24.33	1.302	3791.
1.0	2700.	632.5	2.117	29.91	1.257	2375.
	3600.	951.6	2.217	29.88	1.245	2730.
	4500.	1416.	2.324	29.47	1.241	3070.
	5400.	2417.	2.511	27.39	1.260	3515.
20.0	2700.	632.3	1.918	29.91	1.257	2375.
	3600.	943.4	2.016	29.90	1.244	2729.
	4500.	1315.	2.105	29.78	1.238	3050.
	5400.	1871.	2.209	29.13	1.240	3380.

*This table is a sample of the data contained in "Thermodynamic Properties of Combustion Gases", by Jerry D. Pearson and Robert C. Fellinger, © 1966 by The Iowa State University Press (used by permission).

The original tables give h, s, M, k and a to five significant figures, covering ranges as follows:

Mixtures: N/O ratios 0, 0.5, 1.0, 2.0, 3.73.
Excess O_2: 0%, 20%, 50%, 100%.
Pressures: 0.01, 0.05, 0.10, 0.25, 0.50, 1.0, 1.5, 2, 3, 5, 7, 10, 15, 20, 25.
Temperatures: 2700°R to 6120°R, in 180° increments.

These tables apply to equilibrium products of combustion of any hydrocarbon fuel having an H/C atomic ratio of 2/1. (This ratio is a reasonable approximation for gasoline, kerosene and jet engine fuels.) Values for specific heat ratio k and sound velocity a are those for constant composition rather than for equilibrium. The component gases are assumed to follow the ideal gas law, $PV = RT$, but dissociation is taken into account. The volumetric analysis of dry air is assumed as nitrogen 78.11%, oxygen 20.95%, argon 0.94%.

Table 3-53. DETONATION VELOCITIES FOR VARIOUS GAS MIXTURES*

Mixture	T_1, °K	T_2, °K	P_1, atm	P_2/P_1	ρ_2/ρ_1	Detonation velocity Calc. m/sec	Detonation velocity Meas. m/sec
$(2H_2+O_2)$	291	3583	1	18.05		2806	2819
$(2H_2+O_2)+1O_2$	291	3390	1	17.4		2302	2314
$(2H_2+O_2)+3O_2$	291	2970	1	15.3		1925	1922
$(2H_2+O_2)+5O_2$	291	2620	1	14.13		1730	1700
$(2H_2+O_2)+1N_2$	291	3367	1	17.37		2378	2407
$(2H_2+O_2)+3N_2$	291	3003	1	15.63		2033	2055
$(2H_2+O_2)+5N_2$	291	2685	1	14.39		1850	1822
$(2H_2+O_2)+2H_2$	291	3314	1	17.25		3354	3273
$(2H_2+O_2)+4H_2$	291	2976	1	15.97		3627	3527
$(2H_2+O_2)+6H_2$	291	2650	1	14.18		3749	3532
H_2+O	283	3956		17.5	1.879	2629	{ 2810 2821
H_2+O	373	3981		12.9	1.864	2615	2790
H_2+O+5H	273	2596		14.4	1.79	3526	3530
H_2+O+5N	273	2596		14.4	1.79	1798	1822
H_2+O+5O	273	2596		14.4	1.79	1692	1707
$CO+O+$humidity	273	3852		17.2	1.887	1664	1676
$CO+O+$humidity	308	3748		15.6	1.88	1669	1738
$CO+H_2+O_2$	273	3900		17.3	1.881	1984	{ 2008 2143
$C_2H_2+3O_2$	273	4890		28.8	1.91	2120	2220
$C_2H_2+10O_2$	273	3560		22.0	1.84	1858	1850
$C_2H_2+O_2$	273	5570		54.5	1.84	3091	2961
$C_2N_2+O_2$	273	5960		58.2	1.837	2645	2728
$C_2N_2+O_2+2N_2$	273	4244		33.7	1.8	2214	2166
CH_4+O_2	273	3050		29.8	1.835	2477	2528
$C_2N_2+2O_2$	273	5150		34.8	1.914	2075	{ 2195 2321
CH_4+2O_2	273	4080		27.4	1.904	2220	{ 2287 2322
CH_4+4O_2	273	3570		23.4	1.86	2139	2166
$H+Cl$	273	3880		24.5	1.787	1851	1729
$H+Cl+H_2$	273	2400		14.7	1.73	2000	1855
N_2O+H_2	273	3933		25.9	1.865	2350	{ 2284 2305

*From: "Fundamentals of Gas Dynamics", H.W. Emmons, Ed., Vol. 3, High Speed Aerodynamics and Jet Propulsion Series, Copyright 1958 by Princeton University Press, pp. 654–655.

Table 3-54. PYROTECHNICS AND INCENDIARIES*

COURTESY OF F.J. HENDEL

Non-initiating or Non-priming

Compositions	Applications
Al powder 28% + KClO$_4$ 67% + polymer 5% B amorphous 24% + KNO$_3$ 71% + polymer 5% Mg powder 60% + fluorinated polyethylene 40%	First fire (between primer and main charge)
Fe$_2$O$_3$ + Al (Thermite) (in Mg shell) C$_6$H$_6$ + polymer (napalm) White phosphorus	Incendiary
PbCrO$_4$ 25–55% + Mn 30–45% + BaCrO$_4$ 2–5% W 30–60% + KClO$_4$ 5–10% + BaCrO$_4$ 30–60% + diatomaceous earth 5% Black powder	Time delay (time-fuse powder)
NaClO$_3$ or KClO$_3$ + polymer (1–5%)—generates O$_2$ NaNO$_2$ + NH$_4$Cl + stabilizer (MgO)—generates N$_2$ Composite solid propellants (well underoxidized) or black-powder mixture Pb (N$_3$)$_2$ 27.2% + B 8.8% + BaNO$_3$ 64%	Gas generators
Sr(NO$_3$)$_2$ + Mg + organic Cl compound + polymeric binder—red flare Ba(NO$_3$)$_2$ + Mg + Cu powder + organic Cl compound + binder—green flare Mg 52% + NaNO$_3$ 39% + polymer 9%—white light Al 36% + S 14% + KClO$_4$ 50%—white light	Light producers
Zn + hexachloroethane P (+ solid propellant) KClO$_3$ + NH$_4$Cl + lactose	Smoke producers (with or without dyes)

*From: "Explosive Ordnance for Space Vehicles and Missiles", F.J. Hendel, San Luis Obispo, California, 1969, p. 2-44. Published independently by F.J. Hendel; printed by Blake Printery.

Table 3-55. PROPERTIES OF EXPLOSIVES*

Courtesy of F.J. Hendel

Name	Formula	Color	Density (loading), g/cc	Shock sensitivity (drop), in.	Strength by Trauzl Block, cc	Propagation velocity, max., m/sec	Brisance by sand test, g	Melting point (cast), °C	Applications
DETONATING HIGH EXPLOSIVES									
Lead styphnate	$C_6H(NO_2)_3(O_2Pb)$	Deep yellow	3.1	2–3	120	4900–5200	10–21	Explodes	Primers and detonators
Lead azide	$Pb(N_3)_2$	White–buff	3.8	3–4	115	4000–5000	14–18	Explodes	Primers and detonators
Dextrinated PbN$_6$	$Pb(OH)_2$ + Dextrine 7%								
PETN	$C(CH_2ONO_2)_4$	White	1.6	6	500–560	8300	62	~140	Boosters; explosive cords
RDX-cyclonite	$(CH_2)_3N_3(NO_2)_3$	White	1.65	7	525	8400	61	200–204	Boosters; explosive cords
Tetryl	$C_6H_2(NO_2)_3NCH_3NO_2$	Lemon	1.57	8	375	7500	54	129–130	Boosters
Pentolite	PETN+TNT, 1:1	Light gray	1.6	9	345	7500	53	80–90	Bursting charge
Torpex	RDX+TNT+Al, 42:40:18	Gray	1.7	9	475	7300	58	90–95	Bursting charge
Tetrytol	Tetryl+TNT, 75:25	Light yellow	1.6	10	350	7300	50	70–90	Bursting charge
Tritonal	TNT+Al, 8:2	Gray	1.7	12	350	5500	42	80–90	Bursting charge
Composition C	RDX+Oil, 88:12	Brown	1.5	14	350	7400	47	200	Bursting charge
Amatol 50/50	NH_4NO_3+TNT, 1:1	Buff–brown	1.5	14	330	6500	38	80–85	Bursting charge
Amatol 80/20	NH_4NO_3+TNT, 4:1	Buff–brown	1.4	15	350	5400	32	76 (S.P.)	Bursting charge
Picratol	Expl. D+TNT, 1:1	Brown-yellow	1.6	18	330	7000	40	~150	Bursting charge
Explosive D	$C_6H_2(ONH_4)(NO_2)_3$	Lemon–brown	1.5	18	275	6500	35	265	Bursting charge
TNT	$C_6H_2CH_3(NO_2)_3$	Gray–light brown	1.55	14	235	6900	43	80	Bursting charge
LOW EXPLOSIVES—NON-DETONATING Fast-burning or Deflagrating									
Black powder A (or B)	KNO₃ (or NaNO₃) 74.0% / Charcoal 15.6% / S 10.4%	Black	1.74 at 25,000 psi / 1.89 at 75,000 psi	16	28.5	400	8	None, as it decomposes	Igniter; fuse powder

Table 3-55. PROPERTIES OF EXPLOSIVES (Continued)

Name	Formula	Color	Density (loading), g/cc	Shock sensitivity (drop), in.	Strength by Trauzl Block, cc	Propagation velocity, max, m/sec	Brisance by sand test, g	Melting point (cast), °C	Applications
Benite	Black powder gelatinized by nitrocellulose	Gray	← Similar to black powder →					None, as it decomposes	Igniter for smokeless powder
Smokeless powder, single base EC (Expl. Co.) Blank Fire	Nitrocellulose 13.25% N 80% Ba(NO₃)₂ 8% KNO₃ 8% Starch 3% Diphenylamine + aurine 1%	Gray	0.4	6	—	400	47	None, as it decomposes	Gun propellant; grenades
Smokeless powder, double base E.g., Bullseye	Nitrocellulose gelatinized with nitroglycerine + C (opacifier)	Black	← Similar to smokeless, single base →					None, as it decomposes	Gun propellant
Priming compositions for initiation	KClO₃ 53–45% Sb₂S₃ 17–22% Pb(SCN)₂ 25–33% Pb(N₃)₂ 0–30%		3.4	3–6		300		None, as it decomposes	Fuzes
	Zr + KClO₄ or Zr + BaCrO₄	Gray	2–2½	4–7		200		None, as it decomposes	Igniters for rockets

EXPLANATION OF TESTS:

Shock sensitivity: A sample of explosive (approx. 0.02 g) is subjected to the action of a falling weight of 4 lbs; the impact test value is the minimum height at which at least one of 10 trials results in explosion.

Strength: A sample of the explosive (approx. 10 g) is exploded in a cavity or a borehole in a lead block 200 mm in diameter and 200 mm in height; the Trauzl Block test shows the volume of additional cavity made by the explosive.

Brisance: A sample of the explosive (0.4 g) is exploded in a sand-test bomb containing 200 g of a special sand; the maximum net weight of sand that was crushed is termed the sand-test value.

*From: "Explosive Ordinance for Space Vehicles and Missiles", F.J. Hendel, San Luis Obispo, California, 1969, pp. 2-42–2-43. Published independently by F.J. Hendel; printed by Blake Printery.

Table 3-56. EQUILIBRIUM CONSTANTS FOR DISSOCIATION REACTIONS TO 5,000°K

For the reversible reaction a A + b B $\rightleftarrows$ c C + d D, the equilibrium constant K_p, in terms of partial pressures is defined as

$$K_p \equiv \frac{[C]^c [D]^d}{[A]^a [B]^b}, \text{ (atmospheres) }^{c+d-a-b}$$

where A, B, etc. are the constituents in the reaction

a, b, etc. are the stoichiometric coefficients of constituents A, B, etc., respectively

[A], [B], etc. are the partial pressures (in atmospheres) of constituents A, B, etc., respectively.

EXAMPLE: For $CO_2 \rightleftarrows CO + \frac{1}{2} O_2$ at 3,000°K:

$$K_p = 0.0339 = \frac{[CO][O_2]^{1/2}}{[CO_2]}, \text{ (atmospheres)}^{1/2}.$$

Temperature °K	°R	$\dfrac{[Br]^2}{[Br_2]}$	$\dfrac{[Cl]^2}{[Cl_2]}$	$\dfrac{[H]^2}{[H_2]}$	$\dfrac{[I]^2}{[I_2]}$	$\dfrac{[N]^2}{[N_2]}$	$\dfrac{[O]^2}{[O_2]}$	$\dfrac{[S]^2}{[S_2]}$	$\dfrac{[O_2][C]}{[CO_2]}$
300	540	1.00×10^{-28}	3.55×10^{-37}	5.89×10^{-71}	5.01×10^{-22}	7.94×10^{-119}	6.31×10^{-81}	1.20×10^{-70}	—
400	720	2.45×10^{-20}	1.20×10^{-26}	4.47×10^{-52}	2.51×10^{-15}	1.26×10^{-87}	2.51×10^{-59}	1.00×10^{-51}	—
600	1,080	6.17×10^{-12}	4.68×10^{-16}	3.89×10^{-33}	1.17×10^{-8}	1.58×10^{-56}	1.26×10^{-37}	9.76×10^{-33}	—
800	1,440	1.02×10^{-7}	1.01×10^{-10}	1.32×10^{-23}	2.57×10^{-5}	6.31×10^{-41}	7.94×10^{-27}	3.31×10^{-23}	—
1,000	1,800	3.55×10^{-5}	1.70×10^{-7}	7.41×10^{-18}	2.69×10^{-3}	1.26×10^{-31}	3.31×10^{-20}	1.82×10^{-17}	—
1,200	2,160	1.82×10^{-3}	2.45×10^{-5}	5.25×10^{-14}	6.17×10^{-2}	2.51×10^{-25}	7.94×10^{-16}	1.29×10^{-13}	—
1,400	2,520	3.02×10^{-2}	8.71×10^{-4}	3.09×10^{-11}	5.89×10^{-1}	7.94×10^{-21}	1.07×10^{-12}	7.25×10^{-11}	—
1,600	2,830	2.55×10^{-1}	1.29×10^{-2}	3.72×10^{-9}	3.02	1.58×10^{-17}	2.45×10^{-10}	8.32×10^{-9}	—
1,800	3,240	—	1.05×10^{-1}	1.57×10^{-7}	—	1.00×10^{-14}	1.69×10^{-8}	3.24×10^{-7}	7.08×10^{-25}
2,000	3,600	—	5.69×10^{-1}	3.19×10^{-6}	—	1.00×10^{-12}	5.04×10^{-7}	6.61×10^{-6}	1.02×10^{-21}
2,200	3,960	—	2.27	3.77×10^{-5}	—	5.01×10^{-11}	8.11×10^{-6}	7.76×10^{-5}	5.01×10^{-19}
2,400	4,320	—	7.21	2.96×10^{-4}	—	1.48×10^{-9}	8.36×10^{-5}	5.62×10^{-4}	8.90×10^{-17}
2,600	4,680	—	1.92×10	1.70×10^{-3}	—	2.51×10^{-8}	5.92×10^{-4}	3.31×10^{-3}	5.63×10^{-15}
2,800	5,040	—	4.47×10	7.67×10^{-3}	—	2.75×10^{-7}	3.20×10^{-3}	1.41×10^{-2}	2.46×10^{-13}
3,000	5,400	—	9.25×10	2.83×10^{-2}	—	2.24×10^{-6}	1.39×10^{-2}	5.13×10^{-2}	5.63×10^{-12}
3,200	5,760	—	—	8.89×10^{-2}	—	1.41×10^{-5}	5.13×10^{-2}	—	9.35×10^{-11}
3,500	6,300	—	—	3.90×10^{-1}	—	1.48×10^{-4}	2.65×10^{-1}	—	3.39×10^{-9}
4,000	7,200	—	—	2.81	—	3.47×10^{-3}	2.39	4.67	—
5,000	9,000	—	—	4.51×10	—	2.99×10^{-1}	5.19×10	7.15×10	—

Table 3-56. EQUILIBRIUM CONSTANTS FOR DISSOCIATION REACTIONS TO 5,000°K *(Continued)*

Temperature		$\dfrac{[CO][O_2]^{1/2}}{[CO_2]}$	$\dfrac{[H_2][O_2]^{1/2}}{[H_2O]}$	$\dfrac{[OH][H_2]^{1/2}}{[H_2O]}$	$\dfrac{[NO][O_2]^{1/2}}{[NO_2]}$	$\dfrac{[S_2]^{1/2}[O_2]}{[SO_2]}$	$\dfrac{[SO][O_2]^{1/2}}{[SO_2]}$	$\dfrac{[H_2]^{1/2}[Br]}{[HBr]}$
°K	°R							
300	540	1.91×10^{-45}	1.70×10^{-40}	5.01×10^{-44}	3.80×10^{-7}	4.68×10^{-60}	5.75×10^{-55}	4.27×10^{-10}
400	720	3.72×10^{-33}	5.50×10^{-30}	2.00×10^{-32}	1.15×10^{-4}	2.75×10^{-44}	2.09×10^{-40}	6.92×10^{-8}
600	1,080	8.51×10^{-21}	2.29×10^{-19}	1.00×10^{-20}	3.63×10^{-2}	1.66×10^{-28}	8.13×10^{-26}	1.17×10^{-5}
800	1,440	1.29×10^{-14}	5.25×10^{-14}	8.51×10^{-15}	6.61×10^{-1}	1.32×10^{-20}	1.58×10^{-18}	1.58×10^{-4}
1,000	1,800	6.31×10^{-11}	8.91×10^{-11}	2.95×10^{-11}	3.72	7.25×10^{-16}	3.80×10^{-14}	7.76×10^{-4}
1,200	2,160	1.76×10^{-8}	1.26×10^{-8}	6.76×10^{-9}	1.20×10	1.02×10^{-12}	3.16×10^{-11}	2.24×10^{-3}
1,400	2,520	1.00×10^{-6}	4.57×10^{-7}	3.39×10^{-7}	2.69×10	1.86×10^{-10}	3.89×10^{-9}	4.79×10^{-3}
1,600	2,830	1.93×10^{-5}	6.31×10^{-6}	6.31×10^{-6}	4.90×10	9.12×10^{-9}	1.41×10^{-7}	8.51×10^{-3}
1,800	3,240	2.04×10^{-4}	5.37×10^{-5}	6.46×10^{-5}	7.76×10	—	—	—
2,000	3,600	1.37×10^{-3}	3.02×10^{-4}	3.98×10^{-4}	1.18×10^{2}	2.09×10^{-6}	2.14×10^{-5}	—
2,200	3,960	6.41×10^{-3}	1.23×10^{-3}	1.82×10^{-3}	1.51×10^{2}	—	—	—
2,400	4,320	2.25×10^{-2}	3.89×10^{-3}	6.46×10^{-3}	1.95×10^{2}	7.94×10^{-5}	6.17×10^{-4}	—
2,600	4,680	6.22×10^{-2}	1.00×10^{-2}	1.82×10^{-2}	2.46×10^{2}	—	—	—
2,800	5,040	1.54×10^{-1}	2.34×10^{-2}	4.57×10^{-2}	3.02×10^{2}	1.05×10^{-3}	6.76×10^{-3}	—
3,000	5,400	3.39×10^{-1}	4.90×10^{-2}	1.00×10^{-1}	3.63×10^{2}	—	—	—
3,200	5,760	—	—	—	—	—	—	—
3,500	6,300	—	—	—	—	—	—	—
4,000	7,200	—	—	—	—	—	—	—
5,000	9,000	—	—	—	—	—	—	—

Temperature		$\dfrac{[H_2]^{1/2}[Cl_2]^{1/2}}{[HCl]}$	$\dfrac{[H_2]^{1/2}[I_2]^{1/2}}{[HI]}$	$\dfrac{[NO]}{[O_2]^{1/2}[N_2]^{1/2}}$	$\dfrac{[O_3]}{[O_2]^{3/2}}$	$\dfrac{[C][H_2]^2}{[CH_4]}$	$\dfrac{[N_2]^{1/2}[H_2]^{3/2}}{[NH_3]}$	$\dfrac{[HCN]}{[H_2]^{1/2}[N_2]^{1/2}[C]}$
°K	°R							
300	540	2.63×10^{-17}	3.80×10^{-2}	9.12×10^{-16}	5.13×10^{-29}	1.16×10^{-9}	—	1.74×10^{-21}
400	720	2.75×10^{-13}	5.89×10^{-2}	7.41×10^{-12}	7.08×10^{-23}	2.63×10^{-6}	—	7.59×10^{-16}
600	1,080	3.02×10^{-9}	1.15×10^{-1}	6.40×10^{-8}	1.05×10^{-16}	8.97×10^{-3}	—	3.24×10^{-10}
800	1,440	3.24×10^{-7}	1.55×10^{-1}	5.88×10^{-6}	1.29×10^{-13}	6.61×10^{-1}	—	2.09×10^{-7}
1,000	1,800	5.50×10^{-6}	1.91×10^{-1}	8.87×10^{-5}	9.33×10^{-12}	9.66	—	9.77×10^{-6}
1,200	2,160	3.66×10^{-5}	2.19×10^{-1}	5.41×10^{-4}	1.62×10^{-10}	6.01×10	—	1.26×10^{-4}
1,400	2,520	1.42×10^{-4}	2.40×10^{-1}	1.97×10^{-3}	1.29×10^{-9}	2.26×10^{2}	—	7.94×10^{-4}
1,600	2,830	3.95×10^{-4}	2.57×10^{-1}	5.19×10^{-3}	6.03×10^{-9}	6.17×10^{2}	—	3.09×10^{-3}
1,800	3,240	8.77×10^{-4}	—	1.10×10^{-2}	2.00×10^{-8}	1.35×10^{3}	3.55×10^{4}	8.71×10^{-3}
2,000	3,600	1.66×10^{-3}	—	2.02×10^{-2}	5.13×10^{-8}	2.50×10^{3}	5.25×10^{4}	2.04×10^{-2}
2,200	3,960	2.79×10^{-3}	—	3.32×10^{-2}	1.12×10^{-7}	4.12×10^{3}	7.25×10^{4}	—

Table 3-56. EQUILIBRIUM CONSTANTS FOR DISSOCIATION REACTIONS TO 5,000°K *(Continued)*

Temperature		$\dfrac{[H_2]^{1/2}[Cl_2]^{1/2}}{[HCl]}$	$\dfrac{[H_2]^{1/2}[I_2]^{1/2}}{[HI]}$	$\dfrac{[NO]}{[O_2]^{1/2}[N_2]^{1/2}}$	$\dfrac{[O_3]}{[O_2]^{3/2}}$	$\dfrac{[C][H_2]^2}{[CH_4]}$	$\dfrac{[N_2]^{1/2}[H_2]^{3/2}}{[NH_3]}$	$\dfrac{[HCN]}{[H_2]^{1/2}[N_2]^{1/2}[C]}$
°K	°R							
2,400	4,320	4.35×10^{-3}	—	5.01×10^{-2}	2.19×10^{-7}	6.32×10^{3}	9.34×10^{4}	—
2,600	4,680	6.30×10^{-3}	—	7.08×10^{-2}	3.80×10^{-7}	9.12×10^{3}	1.12×10^{5}	—
2,800	5,040	8.67×10^{-3}	—	9.57×10^{-2}	6.17×10^{-7}	1.25×10^{4}	1.35×10^{5}	—
3,000	5,400	1.15×10^{-2}	—	1.24×10^{-1}	9.33×10^{-7}	1.65×10^{4}	1.55×10^{5}	—
3,200	5,760	—	—	1.56×10^{-1}	1.35×10^{-6}	—	1.78×10^{5}	—
3,500	6,300	—	—	2.09×10^{-1}	2.09×10^{-6}	—	2.14×10^{5}	—
4,000	7,200	—	—	3.07×10^{-1}	3.98×10^{-6}	—	—	—
5,000	9,000	—	—	5.26×10^{-1}	9.55×10^{-6}	—	—	—

REFERENCE

For a general discussion and analysis of dissociation reactions, see "Combustion, Flames and Explosions of Gases", B. Lewis and G. Von Elbe, Academic Press, 1956.

Table 3-57. EQUILIBRIUM CONSTANTS FOR IONIZATION REACTIONS TO 50,000°K*

Temperature		$H \rightleftharpoons H^+ + e^-$	$C \rightleftharpoons C^+ + e^-$	$C^+ \rightleftharpoons C^{++} + e^-$	$C^{++} \rightleftharpoons C^{+++} + e^-$	$O \rightleftharpoons O^+ + e^-$	$O^+ \rightleftharpoons O^{++} + e^-$	$N \rightleftharpoons N^+ + e^-$	$N^+ \rightleftharpoons N^{++} + e^-$
°K	°R								
5,000	9,000	1.170×10^{-11}	3.394×10^{-9}	—	—	5.889×10^{-12}	4.677×10^{-33}	3.311×10^{-12}	3.467×10^{-28}
6,000	10,800	3.524×10^{-9}	4.088×10^{-7}	—	—	2.754×10^{-9}	7.943×10^{-27}	1.905×10^{-9}	8.130×10^{-23}
7,000	12,600	2.234×10^{-7}	1.324×10^{-5}	—	—	1.862×10^{-7}	2.512×10^{-22}	1.862×10^{-7}	6.309×10^{-19}
8,000	14,400	5.200×10^{-6}	1.870×10^{-4}	—	—	4.266×10^{-6}	5.754×10^{-19}	5.495×10^{-6}	5.371×10^{-16}
9,000	16,200	6.553×10^{-5}	1.345×10^{-3}	—	—	5.128×10^{-5}	2.239×10^{-16}	7.413×10^{-5}	1.000×10^{-13}
10,000	18,000	4.753×10^{-4}	8.292×10^{-3}	—	—	3.981×10^{-4}	2.630×10^{-14}	6.024×10^{-4}	6.456×10^{-12}
15,000	27,000	2.477×10^{-1}	1.657	—	—	2.512×10^{-1}	5.624×10^{-8}	3.802×10^{-1}	1.862×10^{-6}
20,000	36,000	8.561	3.406×10	6.266×10^{-3}	7.032×10^{-8}	6.918	1.203×10^{-4}	1.122×10	9.328×10^{-4}
30,000	54,000	1.880×10^{2}	5.494×10^{2}	2.147	1.396×10^{-3}	—	—	—	—
40,000	72,000	7.069×10^{2}	—	5.123×10	2.231×10^{-1}	—	—	—	—
50,000	90,000	1.517×10^{3}	—	3.836×10^{2}	9.174×10^{-1}	—	—	—	—

*Adapted from: "Combustion Engine Processes", 7th ed., L.C. Lichty, McGraw-Hill Book Company, 1967.

Table 3-58. HEAT LOST IN PRODUCTS OF COMBUSTION

This table gives approximate values for energy losses in the combustion products for common fuels burned with excess air (complete combustion), based on gas temperature and indexed by CO_2 content of the "dry" products. These total losses include both the sensible-heat loss and the hydrogen-moisture latent-heat loss, and they are expressed as percentages of the gross or high heating value of the fuel. It should be noted that in practical equipment it is almost impossible to avoid stratification and incomplete combustion at and near the stoichiometric fuel-air ratio; hence these tabular values will be low when the excess air is small (high CO_2).

EXIT GAS LOSSES FOR COMMON FUELS

% CO_2 in dry products	Losses in combustion products at the following temperature differences, intake to exit (°F)										
	300	400	500	600	700	800	900	1000	1200	1500	1800
Natural Gas (High Methane, 1,035 Btu/cu ft)											
4	25	30	35	40	46	52	57	63			
6	20	24	28	32	36	40	44	48	56		
8	18	21	24	27	30	33	36	39	45	55	65
10	17	19	21	24	27	29	32	34	40	48	56
11.9	16	18	20	22	24	27	29	31	36	43	51
Fuel Oil (19,300 Btu/lb)											
4	27	35	42	49	57	64					
6	20	26	31	36	41	46	51	56			
8	17	21	25	29	33	36	40	45			
10	15	19	22	25	28	31	35	39			
12	14	17	20	22	25	28	30	33			
14	13	15	18	20	22	25	27	29			
Propane (2,590 Btu/cu ft)											
4	26	32	37	43	49	56	63	69			
6	20	24	28	32	37	42	46	50	59		
8	17	21	24	27	30	34	37	41	47	59	
10	16	18	21	23	26	29	32	35	41	50	59
12	15	17	19	21	24	26	29	31	36	44	52
13.7	14	16	18	20	22	24	26	28	33	40	47
Butane (3,370 Btu/cu ft)											
4	25	31	38	44	51	57	64				
6	20	24	28	33	37	42	46	51	60		
8	17	20	24	27	30	34	37	41	49	59	
10	15	18	21	23	26	29	32	35	41	50	60
12	14	16	19	21	24	26	29	31	37	44	52
14.0	13	15	17	19	22	24	27	29	34	40	47
High-Btu Oil Gas (1,030 Btu/cu ft)											
4	21	26	32	37	43	48	54	60			
6	16	20	24	27	31	35	38	42	50	62	
8	14	17	20	22	25	28	31	34	40	50	59
10	12	15	17	19	22	24	27	29	34	42	50
12	11	13	15	17	19	21	23	25	30	37	44
Coke-Oven Gas (502 Btu/cu ft)											
4	23	27	31	35	39	44	48	52	60		
6	20	22	25	29	32	35	38	41	46	56	66
8	18	20	23	26	28	30	32	35	40	47	55

Table 3-58. HEAT LOST IN PRODUCTS OF COMBUSTION *(Continued)*

% CO_2 in dry products	Losses in combustion products at the following temperature differences, intake to exit (°F)										
	300	400	500	600	700	800	900	1000	1200	1500	1800
Carbureted Water Gas (550 Btu/cu ft)											
4	27	34	41	47	54	61					
6	21	26	31	35	40	44	49	54			
8	18	22	25	29	33	36	40	43			
10	16	19	22	25	28	31	34	37			
12	15	18	20	23	25	28	30	33			
14	14	16	18	21	23	26	28	30			
Water Gas (Coke) (287 Btu/cu ft)											
4	32	39	46	53	61						
6	25	30	35	40	45	51	56	61			
8	21	25	29	33	37	41	46	50			
10	19	23	26	29	33	36	39	43			
12	18	21	24	26	29	32	35	38			
14	17	20	22	24	27	30	32	35			
16	16	19	21	23	25	28	30	32			
Producer Gas (Buckwheat Anthracite) (143 Btu/cu ft)											
6	30	37	44	51	58	65					
8	24	30	35	41	47	52	57	63			
10	21	26	30	35	40	44	49	54			
12	19	23	28	31	35	39	43	47			
14	18	21	25	28	31	35	39	42			
16	17	20	23	26	29	32	36	38			
Coke (12,000 Btu/lb)											
4	31	40	50	60							
6	22	28	35	42	49	56	62				
8	17	22	27	32	38	43	48	53			
10	14	18	22	27	31	35	40	44			
12	12	15	19	23	26	29	33	36			
14	11	14	17	20	23	26	29	31			
16	10	12	15	18	20	23	25	28			
Anthracite Coal (13,000 Btu/lb)											
4	28	38	47	56	66						
6	20	26	32	39	45	51	57	63			
8	15	20	25	30	35	40	44	48			
10	13	16	20	24	28	32	36	40			
12	11	14	17	21	24	27	30	34			
14	9	12	15	18	21	24	26	29			
Bituminous Coal (14,000 Btu/lb)											
4	28	37	45	54	63						
6	20	26	32	38	44	49	55	62			
8	16	21	25	30	34	39	43	48			
10	13	17	21	24	28	32	36	40			
12	12	15	18	21	24	27	31	34			
14	11	13	16	19	22	24	27	30			

RELATION BETWEEN CO_2 AND O_2 IN PRODUCTS

Complete Combustion

	CO_2	O_2	CO_2	O_2	CO_2	O_2	CO_2	O_2	CO_2	O_2	CO_2	O_2
Natural gas	4	14.2	6	10.9	8	7.5	10	4.2				
Oil; gasoline	4	15.3	6	12.5	8	9.7	10	7.0	12	4.2		
Bituminous coal	4	16.4	6	14.1	8	11.8	10	9.6	12	7.3	14	5.1

Nuclides and
Nuclear Engineering

Table 4-1.   ESTIMATED AVAILABILITY
OF SEPARATED TRANSURANIUM ELEMENTS*

In the United States by 1975 and 1980

Element	Atomic number	Principal isotope(s)	Annual production rate, grams	
			1975	1980
Neptunium[a]	93	Np^{237}	2×10^5	5×10^5
Plutonium[a]	94	Pu^{239}	6×10^{6b}	2×10^{7b}
Americium[a]	95	$Am^{241,\,243}$	8×10^4	2×10^5
Curium[a]	96	Cm^{244}	1×10^{4c}	4×10^{4c}
Berkelium[d]	97	Bk^{249}	10^{-1}	
Californium[d]	98	Cf^{252}	1^e	
Einsteinium[d]	99	Es^{253}	10^{-2}	
		Es^{254}	10^{-5}	
		Es^{255}	10^{-6}	
Fermium[d]	100	Fm^{257}	10^{-11}	

[a] From U.S. civilian power reactors only.
[b] Includes higher-mass-number isotopes.
[c] Includes small quantity of higher-mass-number isotopes.
[d] From HFIR only.
[e] Excludes lower-mass-number isotopes that are present.

*Courtesy of Dr. Glenn T. Seaborg, Lawrence Radiation Laboratory, University of California.

Table 4-2.   TERMS USED IN NUCLEAR TECHNOLOGY

Following is a list of terms peculiar to the field and common terms assigned special meanings when applied to reactor technology (see also Table 2-1).

Absorbed dose. The energy imparted by directly or indirectly ionizing radiation per unit mass of irradiated material, expressed in rads.

Absorption. Transformation into other forms suffered by radiation passing through a material substance.

Accelerator. A device for accelerating charged particles to high velocities (e.g., betatron, cyclotron, Van de Graaff).

Alpha particle (or ray). Emitted in radioactive decay; two photons plus two neutrons, mass 4. Alpha particles are very easy to absorb.

Atomic number. The number of protons in the nucleus of an atom.

Barn. The unit for measuring cross section (q.v.).

Beta decay. The emission of an electron or positron from a radioactive atomic nucleus, entailing the change of a neutron into a proton.

BeV. 10^9 electronvolts.

Binding energy. Energy with which the constituent particles of an atom are bound together.

Buckling. A measure of the bending of the spatial distribution of the neutron flux. Critical buckling is the buckling in the critical system.

Capture. Absorption of a neutron with the excess energy emitted as gamma rays, resulting in an isotope of the original nucleus with mass number increased by one.

Critical mass. Least mass of a fissile material to maintain spontaneous chain reaction.

Cross section. Microscopic cross section is a measure of the ability of a nucleus to accept bombarding particles, or

$$\sigma = \frac{\text{neutron captures/cm}^2}{\text{target nuclei/cm}^2 \times \text{impinging neutrons/cm}^2} = \text{cm}^2/\text{nucleus (effective area/nucleus)}.$$

The unit is the barn. One barn equals 10^{-24} cm^2/nucleus.

Macroscopic cross section of a target material is $\Sigma = N\sigma$, cm^{-1}, where N is the number of nuclei per cm^3 of the target material. The number of neutrons per second interacting with one cm^3 of the material is the product of the neutron flux and the macroscopic cross section.

Curie. A unit of radioactivity, defined as 3.7×10^{10} disintegrations per second. It may be used to designate the strength of a radioisotope. The curie may be converted into watts for beta and gamma radiation: $W = 0.005\,92\ CiE$, where Ci is the curie-strength of the radioisotope and E is the average energy, in MeV of the radiation. (Note that the decay of a single atom is considered to be one disintegration, even if multiple emissions result.)

Table 4-2.   TERMS USED IN NUCLEAR TECHNOLOGY *(Continued)*

Decay. Diminution of a radioactive substance due to emissions.

Diffusion length. A measure of the effectiveness of a moderator, determined by measuring the flux in a diffusion material at various distances from a plane source. (See MIGRATION LENGTH.)

Elastic scattering. This "billiard ball" interaction predominates for neutron energies below 10 MeV and in the range of photon energies. The kinetic energies remain unchanged in an elastic scattering collision, but with inelastic scattering there is a loss of kinetic energy, which is expended in excitation of the target nucleus.

Electronvolt. A unit for describing the quality of high-energy radiation. It is the amount of energy an electron acquires in being accelerated through a field of one volt potential and is equal to 1.602×10^{-12} ergs, or 1.6×10^{-5} joules or watt-seconds.

Escape probability. The fraction of source neutrons that escape capture while being slowed down to a particular energy.

Fast neutrons. High-energy neutrons, such as > 10 keV, approx.

Fission. A nuclear reaction initiated by incoming neutrons and resulting in the splitting of the atomic nucleus into two "fission fragments", with liberation of neutrons and release of large amounts of energy. The masses of the two new atoms are less than the mass of the parent heavy atom.

Fluence. The total integrated flux of neutrons, or the time integral of flux density, expressed as neutrons per sq cm of the material subjected to neutron radiation.

Flux. The product of neutron density and velocity, equal to the total distance traveled in one second by all the neutrons present in a unit volume. It is thus a measure of neutron (or charged-particle) concentration.

Fusion. A nuclear reaction involving the combination of two atomic nuclei to produce a new atom.

Gamma ray. High-energy photons or quanta of electromagnetic-wave energy emitted in radioactive decay or in processes in which nuclei are produced. Gamma rays are similar but of much higher energy than X-rays.

Half-life. The time required for a radioactive mass to lose one-half of its radioactivity.

Ion. An electrically charged atom, one in which the total negative-electron charges do not balance the total positive charge of the nucleus. With more electrons the atom becomes a negative ion; with fewer it becomes a positive ion.

Isotope. An atom that differs from another atom of the same element only with respect to the number of neutrons; hence it differs in mass.

Isotope fuel. A radioactive nuclide encapsulated or contained in such a way that it may be used as a heat-power source for practical use.

Kerma. The total of the kinetic energy of directly ionizing particles ejected by the action of indirectly ionizing radiation per unit mass of the specified material. It is expressed in rads.

Labeling. The attaching of radioactive isotopes to a compound. The latter then becomes a "labeled compound".

Maximum credible accident. A reference design condition used for analysis of the safety of a given nuclear reactor plant. It normally refers to the joining of two heavy-hydrogen nuclei to form helium, with the release of a large amount of energy.

MeV. 10^6 electronvolts.

Migration length. A measure related to the total distance covered between the birth of a neutron by fission and its capture as a thermal neutron. This term and the terms *diffusion length* and *slowing-down length* are closely related.

Moderating ratio. The ratio of the slowing-down power to the macroscopic cross section. This is a measure of the moderating quality of a material.

Moderator. A material for slowing down high-energy neutrons with little tendency to capture them.

Neutron. An uncharged elementary particle of mass 1. It adds mass to an atom with negligible effect on chemical properties. It interacts with matter predominantly by collisions. Fast neutrons possess kinetic energy in the million ev range, while thermal neutrons have energies of one electron volt or less.

Nuclide. See RADIONUCLIDE.

Photoelectric effect. This takes place from the direct collision of a photon (light flux, zero mass) and an atom. The photon is annihilated, and an electron is ejected from one of the orbits, forming an ion. This effect is generally important at low energies.

Photon. An electromagnetic-wave quantity. Its energy is proportional to the frequency of the associated wave.

Poison. A material that captures neutrons readily (e.g., cadmium or boron). It may be used in a reactor for controlling neutron density (power).

Proton. A nuclear particle of mass 1 and charge $+1$.

Quality factor. A multiplying factor for reducing all ionizing radiation to a common scale of biological effect. If the flux density, or LET, is 3.5 or less (keV per μm in water) for X-rays, gamma rays, electrons, or positrons, the QF is unity. For neutrons of energy less than 10 keV, a QF of 3 may be assigned; for faster neutrons or protons, a QF of 10 would normally be safe.

Rad (rd). A unit of absorbed dose, equal to 0.01 J/kg or 100 ergs/gram. It is therefore specific energy, and also applies to kerma.

Table 4-2.   TERMS USED IN NUCLEAR TECHNOLOGY (Continued)

Radionuclide. A species of atom (isotope) that is radioactive, i.e., emits radiation because its neutron/proton ratio is too high or too low.

Rem. *Roentgen equivalent man*, the unit in which the dose equivalent is stated, in terms of biological effect; it is therefore the unit in which all dose-limiting rules and recommendations are expressed.

Rep. An almost-obsolete unit, slightly smaller than the rad.

Resonance. Peaks in the absorption spectrum (barns vs. eV) for a material. The absorption cross section at the resonance peaks may be very high, e.g., 20 000 barns for cadmium 113 at 0.17 eV.

Roentgen (R). An early unit applied to X-rays or gamma radiation in air. It is defined as the emission of ions carrying one electrostatic unit per 0.001 293 g of air, but it is equal to 88 erg/g of air (or 0.88 rad).

Scattering. Collisions of neutrons with nuclei that change the direction of motion of the neutron. If the collision adds internal energy, the process is called *inelastic scattering*; if it adds only kinetic energy to the nucleus, it is called *elastic scattering*. After sufficient elastic-scattering collisions, the kinetic energy of the neutron is reduced to the thermal state.

Stopping power. The average energy loss per unit path length when heavy charged particles impinge on a material and lose kinetic energy, largely by inelastic collisions.

Thermal neutrons. Neutrons in thermal equilibrium with their surroundings; in general thermal neutrons have energies less than one eV.

Tracer. A substance containing a radioisotope and used to determine location or path in a process (biological, industrial, etc.).

Transmutation. Changing of one element to another through nuclear rearrangement.

X-ray. Electromagnetic radiation of frequency between visible light and gamma rays; may be produced by high-energy electrons impinging on a metal target; accompanies electron shifts in the "shells" of the atoms.

Table 4-3.   NUCLEAR REACTIONS*

FISSION REACTIONS

Energy Released E (Mev), Prompt Neutrons ν, Ratio Values of Delayed to Prompt Neutrons β, per Thermal Fission

Isotope	Total energy of light fragments	Total energy of heavy fragments	Total energy of γ rays	Total energy of fission neutron	Total energy of beta rays	Total energy	ν	β
U^{233}	97	66	14	5	9	191	2.51	0.0026
U^{235}	98	67	15	4.9	9	194	2.47	0.0064
Pu^{239}	100	72	14	5.8	9	201	2.91	0.0021

BREEDING PROCESSES

$U^{238}+n = U^{239}+\gamma$; U^{239} 23 min. $Np^{239}+\beta-$; Np^{239} 2.3 days $Pu^{239}+\beta-$

$Th^{232}+n = Th^{233}+\gamma$; Th^{233} 23 min. $Pa^{233}+\beta-$; Pa^{233} 27.4 days $U^{233}+\beta-$

THERMONUCLEAR REACTIONS

$$D+D \; \substack{\rightarrow \\ \rightarrow} \quad \substack{T+P+4.0 \text{ Mev} \\ He^3+n+3.2 \text{ Mev}}$$

$$D+T \rightarrow He^4+n+17.6 \text{ Mev}$$

*From: S. Katcoff, "Nucleonics", Brookhaven National Laboratory, November 1960.

Table 4-4.   FISSILE AND FERTILE RADIOACTIVE MATERIALS

	Uranium 233	Uranium 235	Uranium 238	Plutonium 239	Thorium 232	Natural uranium ($>99\%$ U^{238})
Energy, Mev	4.8	4.4 and 4.6		5.1	4.0	4.2
Half-life, millions of years	0.162	713		0.024	13,900	4,510
Neutron binding energy, Mev	6.7	6.4		6.4	5.1	4.8
Critical energy for fission, Mev*	5.5	5.8		5.5	5.9	5.9
Fission cross section, barns†	527.	577.		742.		4.2
Capture cross section, barns†	54.	106.	2.71	287.		3.5
Total absorption, barns†	581.	683.	2.71	1,029.		7.7
Absorbed neutrons for fission, %	90.5	84.5	—	72.		

*Minimum energies of photons to cause fission.
†Data are for 2,200 m/sec thermal neutrons.

Table 4-5. NEUTRON SOURCES*

ALPHA SOURCES

Alpha source	Alpha energies, Mev	Half-life
Radium 226	7.683, 5.996, 5.305 . . . ., 4.59	1620 years
Actinium 227	7.36, 6.617, 6.273 . . . ., 4.942	22 years
Thorium 228	8.780, 6.775, 6.272 . . . ., 5.338	1.91 years
Uranium 232	8.780, 6.775, 6.272 . . . ., 5.261	74 years
Polonium 210	5.305	138.4 days
Radium D (Pb^{210})	5.305	19.4 years
Plutonium 238	5.495, 5.452	86.4 years
Plutonium 239	5.147, 5.134, 5.096	24360 years
Americium 241	5.534, 5.50, 5.477, 5.435	458 years
Americium 242	6.110, 6.066	100 years
Curium 242	6.110, 6.066	162.5 days

NEUTRON YIELD VS. ALPHA ENERGY

Alpha energy, Mev	Neutron yield from beryllium target, neutrons per million alphas
4.0	24
5.0	54
6.0	105
7.0	185

NEUTRON YIELDS OF VARIOUS TARGETS

The following table shows yields from different target materials when bombarded with the alphas from polonium^{210}, which has an alpha energy of 5.30 Mev.

Target	Yield, neutrons per million alphas
Lithium	2.7
Beryllium	77
Boron	22
Carbon	0.1
Fluorine	12

PRACTICAL YIELDS FROM ALPHA-NEUTRON SOURCES

Source	Yield, n/sec/curie
Ra–Be	$1.0–1.5 \times 10^7$
RaD–Be	2.5×10^6
Po–Be	2.5×10^6
Ac–Be	2.0×10^7
Pu^{239}–Be	2.2×10^6
RdTh–Be	2.0×10^7

GAMMA OUTPUTS OF ALPHA-NEUTRON SOURCES

Source	Gamma output, mrhm/curie	Gamma output, mrhm/10^6 n/sec
Ra–Be	974	54
RaD–Be	33	13.3
Po–Be	0.1	0.04
Ac–Be	145	5.5
Pu^{239}–Be	11	4.9
RdTh—Be	944	34

PROPERTIES OF TARGET MATERIALS

The following table gives the reaction energy and the approximate neutron yield for a number of target materials when the projectile particles are the 5.30 Mev alphas of polonium 210.

Also given is the Coulomb repulsion energy or electrostatic barrier: the alpha particle and the target nucleus are both positively charged; as a result, only the alpha particles that have sufficient energy to overcome the electrostatic barrier will enter the nucleus and initiate the reaction.

Target	Isotopic abundance	Reaction energy, Mev	Coulomb repulsion, Mev	Yield, n/sec/curie
Lithium 6	7.5%		1.64 ⎫	1.0×10^5
Lithium 7	92.5%	−2.79	⎬	
Beryllium 9	100 %	5.74	2.10	2.85×10^6
Boron 10	18.8%	1.37	2.57 ⎫	8.1×10^5
Boron 11	81.2%	0.27	2.52 ⎭	
Carbon 12	98.9%	−8.40	2.97 ⎫	3.7×10^3
Carbon 13	1.1%	2.36	2.92 ⎭	
Fluorine 19	100 %	−1.95	4.05	4.44×10^5

*From: "Neutron Sources and Their Characteristics", AECL Commercial Products, Technical Bulletin NS-2.

Table 4-6.   TOTAL CHAIN YIELD
FROM THERMAL NEUTRON FISSIONS IN U^{235}*

The total integrated chain yield is equivalent to the total fission yield of nuclei having a particular mass. The fission product given uniquely characterizes the total chain yield for the respective mass number.

Mass No.	Fission product	% yield	Mass No.	Fission product	% yield
72	Zn^{72} (49 hr)	0.000016	125	Sb^{125} (2.0 yr)	0.021
73	Ga^{73} (5.0 hr)	0.00011	126	Sb^{126} (9 hr)	0.05^a
77	As^{77} (38.7 hr)	0.0083	127	Sb^{127} (91 hr)	0.13^b
78	As^{78} (91 min)	0.021	128	Sn^{128} (57 min)	0.37
79	As^{79} (9.0 min)	0.056	129	I^{129} (1.7×10^7 yr)	0.9
81	Se^{81} (17.6 min)	0.14	130	Sb^{130} (10 min)	2.0
83	Kr^{83} (stable)	0.544	131	Xe^{131} (stable)	2.93, 2.88^a
84	Kr^{84} (stable)	1.00	132	Xe^{132} (stable)	4.38, 4.31^a
85	Rb^{85} (stable)	1.30	133	Cs^{133} (stable)	6.59, 6.49^a
86	Kr^{86} (stable)	2.02	134	Xe^{134} (stable)	8.06, 7.9^a
87	Rb^{87} (6×10^{10} yr)	2.49	135	Cs^{135} (2.6×10^6 yr)	6.41, 6.31^a
88	Sr^{88} (stable)	3.57^b	136	Xe^{136} (stable)	6.46, 6.36^a
89	Sr^{89} (51 days)	4.79	137	Cs^{137} (29 yr)	6.15, 6.05^a
90	Sr^{90} (28 yr)	5.77^b	138	Ba^{138} (stable)	5.74
91	Zr^{91} (stable)	5.84	139	Ba^{139} (84 min)	6.55^b
92	Zr^{92} (stable)	6.03	140	Ce^{140} (stable)	$6.44^{b,c}$
93	Zr^{93} (1.1×10^6 yr)	6.45	141	Ce^{141} (33 days)	~ 6.0
94	Zr^{94} (stable)	6.40	142	Ce^{142} (stable)	5.95
95	Mo^{95} (stable)	6.27	143	Nd^{143} (stable)	5.98^b
96	Zr^{96} (stable)	6.33	144	Nd^{144} (5×10^{15} yr)	5.67^b
97	Mo^{97} (stable)	6.09	145	Nd^{145} (stable)	3.95^b
98	Mo^{98} (stable)	5.78	146	Nd^{146} (stable)	3.07^b
99	Mo^{99} (66 hr)	6.06^b	147	Sm^{147} (1.3×10^{11} yr)	2.38
100	Mo^{100} (stable)	6.30	148	Nd^{148} (stable)	1.70^b
101	Ru^{101} (stable)	5.0	149	Sm^{149} (stable)	1.13
102	Ru^{102} (stable)	4.1	150	Nd^{150} (stable)	0.67^b
103	Ru^{103} (39.7 days)	3.0	151	Sm^{151} (80 yr)	0.45
104	Ru^{104} (stable)	1.8	152	Sm^{152} (stable)	0.285
105	Ru^{105} (4.45 hr)	0.90^b	153	Sm^{153} (47 hr)	0.15^b
106	Ru^{106} (1.01 yr)	0.38	154	Sm^{154} (stable)	0.077
107	Rh^{107} (22 min)	0.19	155	Sm^{155} (24 min)	0.033
109	Pd^{109} (13.4 hr)	0.030	156	Eu^{156} (15.4 days)	0.014^b
111	Ag^{111} (7.6 days)	0.019	157	Eu^{157} (15.4 hr)	0.0078
112	Pd^{112} (21 hr)	0.010^b	158	Eu^{158} (60 min)	0.002
115	Cd^{115} (53 hr) + Cd^{115}** (43 days)	0.011	159	Gd^{159} (18 hr)	0.00107^a
117	Cd^{117}** (3.0 hr)	0.011	161	Tb^{161} (6.9 days)	0.000076
121	Sn^{121} (27.5 hr)	0.015			

^{a}Average of values in references cited in original source.
^{b}E.P. Steinberg, ANL, Private Communication. Based on a yield of $(6.15 + 5.94)/2 = 6.05$ for Cs^{137}.
^{c}Measured absolute yield is 6.32%. The number 6.44% is used to normalize other yields.
**Metastable.
*From: "Reactor Physics Constants", 2nd ed., Argonne National Laboratory, ANL-5800, U.S. Atomic Energy Commission, July 1963. See this source for extensive references.

REFERENCES
"The Absolute Fission Yields of Twenty-eight Mass Chains in the Thermal Neutron Fission of U^{235}", J.A. Petruska, H.G. Thode, and R.H. Tomlinson, *Can. J. Phys.*, 33(11):693, November 1955.

"Survey of Radiochemical Studies of the Fission Process", E.P. Steinberg and L.E. Glendenin, *Proc., 1955 Geneva Conference*, 7: Paper No. 614, 3.

Figure 4-7.   YIELD VS. MASS NUMBER FOR U²³⁵*

Curves for Two Levels of Incident Neutron Energy

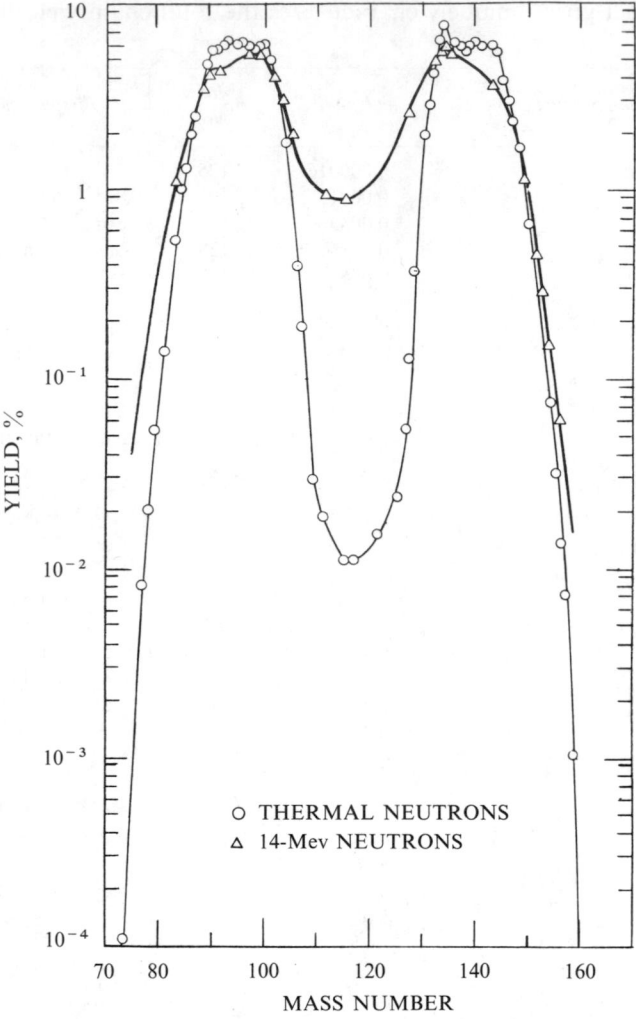

○ THERMAL NEUTRONS
△ 14-Mev NEUTRONS

*From: "Reactor Physics Constants", 2nd ed., Argonne National Labora-
tory, ANL-5800, U.S. Atomic Energy Commission, July 1963.

Table 4-8.   DELAYED NEUTRON FRACTION (β) FOR FAST AND THERMAL FISSION*

Following is a summary of the more useful values of the delayed neutron fraction, i.e., the relative number of delayed neutrons per fission, for the various isotopes.

Isotope	Fast fission	Thermal fission
U^{233}	0.0027 ± 0.0002	0.00264 ± 0.0002
U^{235}	0.0065 ± 0.0003	0.0065 ± 0.0003
U^{238}	0.0157 ± 0.0012	
Pu^{239}	0.0021 ± 0.0002	0.0021 ± 0.0002
Pu^{240}	0.0026 ± 0.0003	
Th^{232}	0.022	

*From: "Reactor Physics Constants", 2nd ed., Argonne
National Laboratory, ANL-5800, U.S. Atomic Energy Commission, July 1963.

Table 4-9.   CUMULATIVE YIELDS OF VARIOUS FISSION PRODUCTS FROM THERMAL NEUTRON FISSIONS IN U^{233}, U^{235}, AND Pu^{239}*

Data for U^{235} in Table 4-6 also apply but are not repeated here. Cumulative yields given below do not necessarily represent total chain yields, since a later member of the chain may also be formed directly from fission. In any disagreement between Tables 4-6 and 4-9, the former values are considered more reliable.

Mass No.	Fission product	U^{233} % yield	U^{235} % yield	Pu^{239} % yield[a,b,c]
72	Zn^{72} (49 hr)			0.00012
77	Ge^{77} (12 hr)	0.010	0.0031	
77	As^{77} (38.7 hr)	0.019		
78	Ge^{78} (86 min)		0.020	
81	Se^{81}** (57 min)		0.0084	
83	Se^{83} (25 min)		0.22	
83	Br^{83} (2.4 hr)	0.79	0.51	0.085
83	Kr^{83} (stable)	1.14		
84	Br^{84} (31.8 min)		0.90	
84	Kr^{84} (stable)	1.90		
85	Kr^{85} (10.3 yr)	0.56	0.293	
86	Kr^{86} (stable)	3.18		
87	Br^{87} (55 sec)		3.1	
89	Sr^{89} (51 days)	6.5	5.8	1.9
90	Sr^{90} (28 yr)			
91	Sr^{91} (9.7 hr)		5.8	2.4
91	Y^{91} (58 days)		5.3	3.0
91	Zr^{91} (stable)	6.53	5.35	
92	Zr^{92} (stable)	6.70		
93	Zr^{93} (1.1×10^6 yr)	7.10	5.4	
94	Y^{94} (16.5 min)			
94	Zr^{94} (stable)	6.82		
95	Zr^{95} (65 days)	5.9	6.2	5.9
95	Mo^{95} (stable)	6.10		
96	Zr^{96} (stable)	5.60		
97	Zr^{97} (17 hr)		5.9	
97	Mo^{97} (stable)	5.35		5.7
98	Mo^{98} (stable)	5.18		5.9
99	Mo^{99} (66 hr)	4.8		
100	Mo^{100} (stable)	4.40	~5.6	
101	Mo^{101} (14.6 min)	3.00		
101	Ru^{101} (stable)			
102	Mo^{102} (11.5 min)	2.37		
102	Ru^{102} (stable)		4.3	
103	Ru^{103} (39.7 days)	1.6		5.8
104	Ru^{104} (stable)			
105	Rh^{105} (35.3 hr)	0.96		3.9
106	Ru^{106} (1.0 yr)	0.28	0.030	5.0
109	Pd^{109} (~3.4 hr)	0.040		1.5
111	Ag^{111} (7.6 days)	0.025		0.27
112	Pd^{112} (21 hr)			0.10
115	Ag^{115} (21 min)	0.016	0.0077	
115	Cd^{115}** (43 days)	0.001	0.0007	0.003
115	Cd^{115} (53 hr)	0.019	0.0097	0.038
121	Sn^{121} (27.5 hr)	0.018		0.044
123	Sn^{123} (136 days)		0.0013	
125	Sn^{125} (9.6 days)	0.050	0.013	0.072
127	Sb^{127} (91 hr)		0.035	0.39
127	Te^{127}** (105 days)		0.35	
129	Te^{129}** (~7 days)		2.6	
131	Sb^{131} (23 min)		0.44	
131	Te^{131}** (30 hr)			
131	I^{131} (8.05 days)	2.7	3.1[a]	3.8
131	Xe^{131} (stable)	3.74		2.87
132	Te^{132} (77 hr)		4.7	5.2

Table 4-9. CUMULATIVE YIELDS OF VARIOUS FISSION PRODUCTS FROM THERMAL NEUTRON FISSIONS IN U^{233}, U^{235}, AND Pu^{239} *(Continued)*

Mass No.	Fission product	U^{233} % yield	U^{235} % yield	Pu^{239} % yield [a,b,c]
132	Xe^{132} (stable)	5.10		4.02
133	Sb^{133} (4.1 min)		4.0	
133	Te^{133}** (63 min)		4.9	5.3
133	I^{133} (21 hr)		6.9	5.27
133	Xe^{133} (5.27 days)		6.62	
133	Cs^{133} (stable)	6.18	6.9[a]	5.27
134	Te^{134} (44 min)			
134	Xe^{134} (stable)	6.54		5.69
135	I^{135} (6.7 hr)	5.1		5.8
135	Cs^{135} (2.6×10^6 yr)	>4.9	6.1[a]	5.53
136	I^{136} (86 sec)	1.7	3.1	2.1
136	Xe^{136} (stable)	<8.9		5.06
137	Cs^{137} (29 yr)	7.16	6.15	5.24
138	Cs^{138} (32 min)		5.74	
139	Ba^{139} (84 min)			5.7
140	Ba^{140} (12.8 days)	6.0	6.44[d]	5.68
140	Ce^{140} (stable)	5.6		5.68
141	Ce^{141} (33 days)			5.2
142	Ce^{142} (stable)	5.6		6.69
143	Ce^{143} (33 hr)		5.7	5.4
143	Nd^{143} (stable)	5.2		6.31
144	Ce^{144} (285 days)	4.1	6.0	5.28
144	Nd^{144} (stable)	4.0		5.29
145	Nd^{145} (stable)	3.0		4.24
146	Nd^{146} (stable)	2.3		3.53
147	Nd^{147} (11 days)	1.71	2.7	2.92
147	Sm^{147} (stable)	1.15		2.28
148	Nd^{148} (stable)			
149	Pm^{149} (5.6 hr)			
149	Sm^{149} (stable)	0.61	1.4	1.89
150	Nd^{150} (stable)	0.48		1.38
151	Sm^{151} (80 yr)	0.27		1.17
152	Sm^{152} (stable)	0.17		0.83
153	Sm^{153} (47 hr)	0.095		0.41
154	Sm^{154} (stable)	0.037		0.32
155	Sm^{155} (24 min)			
155	Eu^{155} (1.9 yr)		0.03	0.22
156	Sm^{156} (10 hr)		0.013	
156	Eu^{156} (15.4 hr)			0.12

[a] The Cs, Nd, and Sm yields were measured by Wiles, et al.[1] and corrected to 5.6% yield for Ba^{140}. For some of the Ce and Nd isotopes, the ratios of yields per Ref. 2 were used.

[b] Most values are from Ref. 3, corrected to a Ba^{140} yield of $1.06 \times 5.32\% = 5.68\%$.

[c] The xenon isotope yields were determined by Fleming, et al.[4] These were normalized to corrected Cs yields at mass 133.

[d] Measured value is 6.32%. The factor 1.06 is introduced to bring the total yield for light and heavy groups to 100% each.

** Metastable.

*From: "Reactor Physics Constants", 2nd ed., Argonne National Laboratory, ANL-5800, U.S. Atomic Energy Commission, July 1963. See this source for extensive references.

REFERENCES

1. "Some Cumulative Yields of Isotopes Formed in the Thermal Neutron Fission of Pu^{239}", D.M. Wiles, J.A. Petruska, and R.H. Tomlinson, *Can. J. Chem.*, 34(3):227, 1956.
2. "Rare-Earth Isotope Yields in the Fission of Pu^{239} by Pile Neutrons", L.M. Krizhansky, et al., *Soviet Journal of Atomic Energy* (English trans.), 2(3):334, 1957.
3. "Summary of Results of Fission-Yield Experiments", E.P. Steinberg and M.S. Freedman, National Nuclear Energy Series, Paper No. 204, Book 3, McGraw-Hill Book Company, 1951, pp. 1378–1390.
4. "The Relative Yields of the Isotopes of Xenon in Plutonium Fission", W.H. Fleming and H.G. Thode, *Can. J. Chem.*, 34(3):193, 1956.

Table 4-10. CUMULATIVE PERCENTAGE YIELDS FROM FISSION SPECTRUM NEUTRON-INDUCED FISSIONS IN Pu^{239}, U^{238}, AND Th^{232} [**]

Mass No.	Fission product	Pu^{239a}	U^{238b}	Th^{232c}	Mass No.	Fission product	Pu^{239a}	U^{238b}	Th^{232c}
72	Zn^{72} (49 hr)			0.00033	111	Ag^{111} (7.6 days)		0.073	0.052
73	Ga^{73} (5.0 hr)			0.00045	112	Pd^{112} (21 hr)	0.14	0.046	0.057
77	Ge^{77} (12 hr)			0.009	115	Cd^{115}* (43 days)		0.003	0.003
77	As^{77} (39 hr)		0.0038	0.020	115	Cd^{115} (53 hr)	0.069	0.037	0.072
83	Br^{83} (2.4 hr)			1.9	127	Sb^{127} (93 hr)		0.12	
83	Kr^{83} (stable)		0.40	1.99	131	I^{131} (8.05 days)			1.2
84	Kr^{84} (stable)		0.85	3.65	131	Xe^{131} (stable)		3.2	1.62
85	Kr^{85} (10.3 yr)		0.153	0.87	132	Te^{132} (77 hr)		4.7	2.4
86	Kr^{86} (stable)		1.38	6.0	132	Xe^{132} (stable)		4.7	2.87
89	Sr^{89} (51 days)		2.9	6.7	133	Cs^{133} (stable)		$5.5 (8.08)^d$	
90	Sr^{90} (28 yr)	2.2	3.2	6.8	134	Xe^{134} (stable)		6.6	5.38
91	Sr^{91} (9.7 hr)			7.2	135	Cs^{135} (2.6×10^6 yr)		6.0^d	
95	Zr^{95} (65 days)		5.7		136	Xe^{136} (stable)		5.9	5.65
97	Zr^{97} (17 hr)	5.2		5.2	137	Cs^{137} (29 yr)	6.6	$6.2 (7.11)^d$	6.3
99	Mo^{99} (67 hr)	5.9	6.3	2.7	140	Ba^{140} (12.8 days)	5.0	5.7	6.2
103	Ru^{103} (40 days)		6.6	0.16	141	Ce^{141} (33 days)			9.0
105	Rh^{105} (35 hr)			0.07	144	Ce^{144} (290 days)		4.9	7.1
106	Ru^{106} (1.0 yr)		2.7	0.042	153	Sm^{153} (47 hr)	0.48		
109	Pd^{109} (13.4 hr)	1.9	0.32	0.055	156	Eu^{156} (15.4 days)		0.066	

[a] Values from "Instrumentation for Energy Determination of High-Energy Particles", R.L. Shuey, UCRL-793, July 1950.

[b] Most values are averages of data in "Yields of Fission Products from U^{238} Irradiated with Fission Spectrum Neutrons", R.N. Keller, E.P. Steinberg, and L.E. Glendenin, *Phys. Rev.*, 94(4):969, 1954, and "Fission Yields in U^{238}", D.W. Engelkemier, et al., National Nuclear Energy Series, Paper No. 218, Book 3, McGraw–Hill Book Company, 1951, pp. 1375–77, normalized to absolute of Mo^{99} in Internal Memorandum, J. Terrell, et al., Los Alamos Scientific Laboratory, LADC–1463, 1953.

[c] Most values from "Radiochemical Studies on the Fission of Th^{232} with Pile Neutrons", A. Turkevich and J.B. Niday, *Phys. Rev.*, 84(1):52, 1951. Kr and Xe yields measured by T.J. Kennett and H.G. Thode, and reported by private communication to S. Katcoff.

[d] Cs^{133} and Cs^{135} yields derived from ratios to Cs^{137} measured by R.H. Tomlinson and reported by private communication to S. Katcoff.

[*] Metastable.

[**] From: "Reactor Physics Constants", 2nd ed., Argonne National Laboratory, ANL-5800, U.S. Atomic Energy Commission, July 1963. See this source for an extensive list of references.

Table 4-11. DELAYED NEUTRON YIELD FROM THERMAL NEUTRON-INDUCED FISSION IN U^{233}, U^{235}, AND Pu^{239}[*]

Isotope	Delayed neutrons/ fission	Group index, i	Half-life, T_i	Decay constant, λ_i	Relative abundance, a_i	Absolute group yield, %
U^{233}	0.0066 ± 0.0003	1	55.00 ± 0.54	0.0126 ± 0.0002	0.086 ± 0.003	0.057 ± 0.003
		2	20.57 ± 0.38	0.0337 ± 0.0006	0.299 ± 0.004	0.197 ± 0.009
		3	5.00 ± 0.21	0.139 ± 0.006	0.252 ± 0.040	0.166 ± 0.027
		4	2.13 ± 0.20	0.325 ± 0.030	0.278 ± 0.020	0.184 ± 0.016
		5	0.615 ± 0.242	1.13 ± 0.40	0.051 ± 0.024	0.034 ± 0.016
		6	0.277 ± 0.047	2.50 ± 0.42	0.034 ± 0.014	0.022 ± 0.009
U^{235}	0.0158 ± 0.0005	1	55.72 ± 1.28	0.0124 ± 0.0003	0.033 ± 0.003	0.052 ± 0.005
		2	22.72 ± 0.71	0.0305 ± 0.0010	0.219 ± 0.009	0.346 ± 0.018
		3	6.22 ± 0.23	0.111 ± 0.004	0.196 ± 0.022	0.310 ± 0.036
		4	2.30 ± 0.09	0.301 ± 0.012	0.395 ± 0.011	0.624 ± 0.026
		5	0.61 ± 0.083	1.13 ± 0.15	0.115 ± 0.009	0.182 ± 0.015
		6	0.23 ± 0.025	3.00 ± 0.33	0.042 ± 0.008	0.066 ± 0.008
Pu^{239}	0.0061 ± 0.0003	1	54.28 ± 2.34	0.0128 ± 0.0005	0.035 ± 0.009	0.021 ± 0.006
		2	23.04 ± 1.67	0.0301 ± 0.0022	0.298 ± 0.035	0.182 ± 0.023
		3	5.60 ± 0.40	0.124 ± 0.009	0.211 ± 0.048	0.129 ± 0.030
		4	2.13 ± 0.24	0.325 ± 0.036	0.326 ± 0.033	0.199 ± 0.022
		5	0.618 ± 0.213	1.12 ± 0.39	0.086 ± 0.029	0.052 ± 0.018
		6	0.257 ± 0.045	2.69 ± 0.47	0.044 ± 0.016	0.027 ± 0.010

[*] From: "Reactor Physics Constants", 2nd ed., Argonne National Laboratory, ANL-5800, U.S. Atomic Energy Commission, July 1963.

REFERENCE

"Delayed Neutrons from Fissionable Isotopes of Uranium, Plutonium, and Thorium", G.R. Keepin, et al., *Phys. Rev.*, 107(4):1044, 1957.

Table 4-12. MEAN ENERGIES OF DELAYED NEUTRON GROUPS FOR U^{235} FISSION*

Group index, i	Half-life, sec	Mean energy, kev		
		Hughes, et al. Ref. 1	Burgy, et al. Ref. 2	Batchelor Ref. 3
1	55.7	250 ± 60	300 ± 60	250 ± 20
2	22.7	560 ± 60	670 ± 60	460 ± 10
3	6.2	430 ± 60	650 ± 100	405 ± 20
4	2.3	620 ± 60	910 ± 90	450 ± 20
5	0.61	420 ± 60	400 ± 70	
6	0.23			

*From: "Reactor Physics Constants", 2nd ed., Argonne National Laboratory, ANL-5800, U.S. Atomic Energy Commission, July 1963. See this source for a complete list of references.

REFERENCES

1. "Energy of Delayed Neutrons from U^{235} Fissions", D.J. Hughes, et al., *Phys. Rev.*, 73(2):111, 1948.
2. "Energy of Delayed Neutrons from U^{235} Fissions", M. Burgy, et al., *Phys. Rev.*, 70(1 and 2):104, 1946.
3. "The Energy of Delayed Neutrons from Fission", R. Batchelor and H.R. McK. Hyder, *J. Nuclear Energy*, 3:7, 1956.

Table 4-13. PHOTONEUTRON HALF-LIVES AND YIELDS FROM U^{235} FISSION PRODUCTS IN D_2O*

The yield of photoneutrons for U^{235} fission products has been measured in D_2O.[a] Implicit in these measurements are the details of yield and half-life of delayed neutrons from U^{235} fissions. To each of the groups of photoneutrons there corresponds a half-life and a yield relative to the total yield of delayed neutrons, as shown in the following table.

Group	Half-life, T_j	Yield, %, relative to total delayed neutron yield[a]
1	307 hr	0.0065
2	53 hr	0.0163
3	4.4 hr	0.0515
4	1.65 hr	0.370
5	27 min	0.328
6	7.7 min	0.534
7	2.4 min	1.11
8	41 sec	3.24
9	2.5 sec	10.31

[a]"Yield of Photoneutrons from U^{235} Fission Products in Heavy Water", S. Bernstein, et al., *Phys. Rev.*, 71(9):573, 1947.
[b]Based on: "Energy of Delayed Neutrons from U^{235} Fissions", D.J. Hughes, et al., *Phys. Rev.*, 73(2):111, 1948.

*From: "Reactor Physics Constants", 2nd ed., Argonne National Laboratory, ANL-5800, U.S. Atomic Energy Commission, July 1963. See this source for an extensive list of references.

Table 4-14. DELAYED NEUTRON FRACTION IN Be*

Group No.	Mean life, τ_i	$\beta_i \times 10^5$
1	3814 min	0.0628
2	576 min	0.694
3	279 min	0.479
4	158 min	0.894
5	93 min	1.15
6	29 min	1.72
7	9.44 min	2.63
8	269 sec	1.51
9	80.7 sec	35.7
10	31.7 sec	166.0
11	6.50 sec	233.0
12	3.12 sec	14.3

*From: "Reactor Physics Constants", 2nd ed., Argonne National Laboratory, ANL-5800, U.S. Atomic Energy Commission, July 1963.

REFERENCE

"Physical Characteristics of Beryllium-Moderated Reactor", A.K. Krasin, et al., *Proc., 1958 Geneva Conference*, 12: Paper No. 2146, p. 571.

Table 4-15.   CROSS SECTIONS FOR NATURALLY OCCURRING ELEMENTS*

Thermal neutron cross sections for a most probable velocity of 2,200 m/sec (0.0253 ev or a wavelength of 1.80 Å) are given in terms of absorption and scattering (subscripts a and s). The total cross section (subscript t) determines the diminution of a neutron beam as it traverses a sample. The microscopic cross section applies to a single nucleus; the macroscopic cross section, $\Sigma = N\sigma$, is equivalent to the cross section per cm^3 for the target material, which contains N nuclei per cm^3. Because σ cm^2 is the effective area per single nucleus, i.e., the "cross section" per single nucleus, the dimensions of Σ are reciprocal length. One barn is a unit of 10^{-24} cm^2 per nucleus.

For density in kg/m^3, multiply the value in g/cm^3 by 1000. For density in lb/ft^3, multiply the value in g/cm^3 by 62.42.

SYMBOLS: $1-\bar{\mu}_0$ = transport scattering factor
ξ = logarithmic energy loss

Atomic No.	Element or compound	Atomic or mol. wt.	Density g/cm³	Nuclei per unit vol. ×10⁻²⁴	$1-\bar{\mu}_0$	ξ	Microscopic cross section, barn			Macroscopic cross section, cm⁻¹		
							σ_a	σ_s	σ_t	Σ_a	Σ_s	Σ_t
1	H	1.008	8.9ᵃ	5.3ᵃ	0.3386	1.000	0.33	38	38	1.7ᵃ	0.002	0.002
	H₂O	18.016	1	0.0335ᵇ	0.676	0.948	0.66	103	103	0.022	3.45	3.45
	D₂O	20.030	1.10	0.0331ᵇ	0.884	0.570	0.001	13.6	13.6	3.3ᵃ	0.449	0.449
2	He	4.003	17.8ᵃ	2.6ᵃ	0.8334	0.425	0.007	0.8	0.807	0.02ᵃ	2.1ᵃ	2.1ᵃ
3	Li	6.940	0.534	0.0463	0.9047	0.268	71	1.4	72.4	3.29	0.065	3.35
4	Be	9.013	1.85	0.1236	0.9259	0.209	0.010	7.0	7.01	124ᵃ	0.865	0.865
	BeO	25.02	3.025	0.0728ᵇ	0.939	0.173	0.010	6.8	6.8	73ᵃ	0.501	0.501
5	B	10.82	2.45	0.1364	0.9394	0.171	755	4	759	103	0.346	104
6	C	12.011	1.60	0.0803	0.9444	0.158	0.004	4.8	4.80	32ᵃ	0.385	0.385
7	N	14.008	0.0013	5.3ᵃ	0.9524	0.136	1.88	10	11.9	9.9ᵃ	50ᵃ	60ᵃ
8	O	16.000	0.0014	5.3ᵃ	0.9583	0.120	20ᵃ	4.2	4.2	0.000	21ᵃ	21ᵃ
9	F	19.00	0.0017	5.3ᵃ	0.9649	0.102	0.001	3.9	3.90	0.01ᵃ	20ᵃ	20ᵃ
10	Ne	20.183	0.0009	2.6ᵃ	0.9667	0.0968	<2.8	2.4	5.2	7.3ᵃ	6.2ᵃ	13.5ᵃ
11	Na	22.991	0.971	0.0254	0.9710	0.0845	0.525	4	4.53	0.013	0.102	0.115
12	Mg	24.32	1.74	0.0431	0.9722	0.0811	0.069	3.6	3.67	0.003	0.155	0.158
13	Al	26.98	2.699	0.0602	0.9754	0.0723	0.241	1.4	1.64	0.015	0.084	0.099
14	Si	28.09	2.42	0.0522	0.9762	0.0698	0.16	1.7	1.86	0.008	0.089	0.097
15	P	30.975	1.82	0.0354	0.9785	0.0632	0.20	5	5.20	0.007	0.177	0.184
16	S	32.066	2.07	0.0389	0.9792	0.0612	0.52	1.1	1.62	0.020	0.043	0.063
17	Cl	35.457	0.0032	5.3ᵃ	0.9810	0.0561	33.8	16	49.8	0.002	80ᵃ	0.003
18	Ar	39.944	0.0018	2.6ᵃ	0.9833	0.0492	0.66	1.5	2.16	1.7ᵃ	3.9	5.6ᵃ

Table 4-15. CROSS SECTIONS FOR NATURALLY OCCURRING ELEMENTS (Continued)

Atomic No.	Element or compound	Atomic or mol. wt.	Density, g/cm³	Nuclei per unit vol. ×10⁻²⁴	$1-\bar{\mu}_0$	ξ	Microscopic cross section, barn σ_a	σ_s	σ_t	Macroscopic cross section, cm⁻¹ Σ_a	Σ_s	Σ_t
19	K	39.100	0.87	0.0134	0.9829	0.0504	2.07	1.5	3.57	0.028	0.020	0.048
20	Ca	40.08	1.55	0.0233	0.9833	0.0492	0.44	3.0	3.44	0.010	0.070	0.080
21	Sc	44.96	2.5	0.0335	0.9852	0.0438	24	24	48	0.804	0.804	1.61
22	Ti	47.90	4.5	0.0566	0.9861	0.0411	5.8	4	9.8	0.328	0.226	0.555
23	V	50.95	5.96	0.0704	0.9869	0.0387	5	5	10.0	0.352	0.352	0.704
24	Cr	52.01	7.1	0.0822	0.9872	0.0385	3.1	3	6.1	0.255	0.247	0.501
25	Mn	54.94	7.2	0.0789	0.9878	0.0359	13.2	2.3	15.5	1.04	0.181	1.22
26	Fe	55.85	7.86	0.0848	0.9881	0.0353	2.62	11	13.6	0.222	0.933	1.15
27	Co	58.94	8.9	0.0910	0.9887	0.0335	38	7	45	3.46	0.637	4.10
28	Ni	58.71	8.90	0.0913	0.9887	0.0335	4.6	17.5	22.1	0.420	1.60	2.02
29	Cu	63.54	8.94	0.0848	0.9896	0.0309	3.85	7.2	11.05	0.326	0.611	0.937
30	Zn	65.38	7.14	0.0658	0.9897	0.0304	1.10	3.6	4.70	0.072	0.237	0.309
31	Ga	69.72	5.91	0.0511	0.9925	0.0283	2.80	4	6.80	0.143	0.204	0.347
32	Ge	72.60	5.36	0.0445	0.9909	0.0271	2.45	3	5.45	0.109	0.134	0.243
33	As	74.91	5.73	0.0461	0.9911	0.0264	4.3	6	10.3	0.198	0.277	0.475
34	Se	78.96	4.8	0.0366	0.9916	0.0251	12.3	11	23.3	0.450	0.403	0.853
35	Br	79.916	3.12	0.0235	0.9917	0.0247	6.7	6	12.7	0.157	0.141	0.298
36	Kr	83.80	0.0037	2.6ᵃ	0.9921	0.0236	31	7.2	38.2	81ᵃ	19ᵃ	99ᵃ
37	Rb	85.48	1.53	0.0108	0.9922	0.0233	0.73	12	12.7	0.008	0.130	0.138
38	Sr	87.63	2.54	0.0175	0.9925	0.0226	1.21	10	11.2	0.021	0.175	0.195
39	Yt	88.92	5.51	0.0373	0.9925	0.0223	1.31	3	4.3	0.049	0.112	1.160
40	Zr	91.22	6.4	0.0423	0.9927	0.0218	0.185	8	8.2	0.008	0.338	0.347
41	Nb	92.91	8.4	0.0545	0.9928	0.0214	1.16	5	6.16	0.063	0.273	0.336
42	Mo	95.95	10.2	0.0640	0.9931	0.0207	2.70	7	9.70	0.173	0.448	0.621
43	Tc	98	—	—	0.9932	0.0203	22	—	—	—	—	—
44	Ru	101.1	12.2	0.0727	0.9934	0.0197	2.56	6	8.56	0.186	0.436	0.622
45	Rh	102.91	12.5	0.0732	0.9935	0.0193	149	5	154	10.9	0.366	11.3
46	Pd	106.4	12.16	0.0689	0.9937	0.0187	8	3.6	11.6	0.551	0.248	0.799
47	Ag	107.88	10.5	0.0586	0.9938	0.0184	63	6	69	3.69	0.352	4.04
48	Cd	112.41	8.65	0.0464	0.9940	0.0178	2,450	7	2,457	114	0.325	114

Table 4-15. CROSS SECTIONS FOR NATURALLY OCCURRING ELEMENTS (Continued)

Atomic No.	Element or compound	Atomic or mol. wt.	Density, g/cm³	Nuclei per unit vol. × 10⁻²⁴	$1 - \bar{\mu}_0$	ξ	Microscopic cross section, barn σ_a	σ_s	σ_t	Macroscopic cross section, cm⁻¹ Σ_a	Σ_s	Σ_t
49	In	114.82	7.28	0.0382	0.9942	0.0173	191	2.2	193	7.30	0.084	7.37
50	Sn	118.70	6.5	0.0330	0.9944	0.0167	0.625	4	4.6	0.021	0.132	0.152
51	Sb	121.76	6.69	0.0331	0.9945	0.0163	5.7	4.3	10.0	0.189	0.142	0.331
52	Te	127.61	6.24	0.0295	0.9948	0.0155	4.7	5	9.7	0.139	0.148	0.286
53	I	126.91	4.93	0.0234	0.9948	0.0157	7.0	3.6	10.6	0.164	0.084	0.248
54	Xe	131.30	0.0059	2.7[a]	0.9949	0.0152	35	4.3	39.3	95[a]	12[a]	0.001
55	Cs	132.91	1.873	0.0085	0.9950	0.0150	28	20	48	0.238	0.170	0.408
56	Ba	137.36	3.5	0.0154	0.9951	0.0145	1.2	8	9.2	0.018	0.123	0.142
57	La	138.92	6.19	0.0268	0.9952	0.0143	8.9	15	24	0.239	0.403	0.642
58	Ce	140.13	6.78	0.0292	0.9952	0.0142	0.73	9	9.7	0.021	0.263	0.283
59	Pr	140.92	6.78	0.0290	0.9953	0.0141	11.3	4	15.3	0.328	0.116	0.444
60	Nd	144.27	6.95	0.0290	0.9954	0.0138	46	16	62	1.33	0.464	1.79
61	Pm	145	—	—	0.9954	0.0137	60	—	—	—	—	—
62	Sm	150.35	7.7	0.0309	0.9956	0.0133	5,600	5	5,605	173	0.155	173
	Sm₂O₃	348.70	7.43	0.0128[b]	0.974	0.076	16,500	22.6	16,500	211	0.289	211
63	Eu	152	5.22	0.0207	0.9956	0.0131	4,300	8	4,308	89.0	0.166	89.2
	Eu₂O₃	352.00	7.42	0.0127[b]	0.978	0.063	8,740	30.2	8,770	111	0.383	111
64	Gd	157.26	7.95	0.0305	0.9958	0.0127	46,000	—	—	1,403	—	—
65	Tb	158.93	8.33	0.0316	0.9958	0.0125	46	—	—	1.45	—	—
66	Dy	162.51	8.56	0.0317	0.9959	0.0122	950	100	1,050	30.1	3.17	33.3
	Dy₂O₃	372.92	7.81	0.0126[b]	0.993	0.019	2,200	2.4	2,414	27.7	2.7	30.4
67	Ho	164.94	8.76	0.0320	0.9960	0.0121	65	—	—	2.08	—	—
68	Er	167.27	9.16	0.0330	0.9960	0.0119	173	15	188	5.71	0.495	6.20
69	Tm	168.94	9.35	0.0333	0.9961	0.0118	127	7	134	4.23	0.233	4.46
70	Yb	173.04	7.01	0.0244	0.9961	0.0115	37	12	49	0.903	0.293	1.20
71	Lu	174.99	9.74	0.0335	0.9962	0.0114	112	—	113	3.75	—	—
72	Hf	178.5	13.3	0.0449	0.9963	0.0112	105	8	113	4.71	0.359	5.07
73	Ta	180.95	16.6	0.0553	0.9963	0.0110	21	5	26	1.16	0.277	1.44
74	W	183.86	19.3	0.0632	0.9964	0.0108	19.2	5	24.2	1.21	0.316	1.53
75	Re	186.22	20.53	0.0664	0.9964	0.0107	86	14	100	5.71	0.930	6.64
76	Os	190.2	22.48	0.0712	0.9965	0.0105	15.3	11	26.3	1.09	0.783	1.87

Table 4-15. CROSS SECTIONS FOR NATURALLY OCCURRING ELEMENTS (Continued)

Atomic No.	Element or compound	Atomic or mol. wt.	Density, g/cm^3	Nuclei per unit vol. $\times 10^{-24}$	$1 - \bar{\mu}_0$	ξ	Microscopic cross section, barn σ_a	σ_s	σ_t	Macroscopic cross section, cm^{-1} Σ_a	Σ_s	Σ_t
77	Ir	192.2	22.42	0.0703	0.9965	0.0104	440	—	—	30.9	—	—
78	Pt	195.09	21.37	0.0660	0.9966	0.0102	8.8	10	18.8	0.581	0.660	1.24
79	Au	197	19.32	0.0591	0.9966	0.0101	98.8	9.3	107.3	5.79	0.550	6.34
80	Hg	200.61	13.55	0.0407	0.9967	0.0099	380	20	400	15.5	0.814	16.3
81	Tl	204.39	11.85	0.0349	0.9967	0.0098	3.4	14	17.4	0.119	0.489	0.607
82	Pb	207.21	11.35	0.0330	0.9968	0.0096	0.170	11	11.2	0.006	0.363	0.369
83	Bi	209	9.747	0.0281	0.9968	0.0095	0.034	9	9	0.001	0.253	0.256
84	Po	210	9.24	0.0265	0.9968	0.0095	—	—	—	—	—	—
85	At	211	—	—	0.9968	0.0094	—	—	—	—	—	—
86	Rn	222	0.0097	2.6ᵃ	0.9970	0.0090	0.7	—	—	—	—	—
87	Fr	223	—	—	0.9980	0.0089	—	—	—	—	—	—
88	Ra	226.05	5	0.0133	0.9971	0.0088	20	—	—	0.266	—	—
89	Ac	227	—	—	0.9971	0.0088	510	—	—	—	—	—
90	Th	232.05	11.3	0.0293	0.9971	0.0086	7.56	12.6	20.2	0.222	0.369	0.592
91	Pa	231	15.4	0.0402	0.9971	0.0086	200	—	—	8.04	—	—
92	U	238.07	18.9	0.04783	0.9972	0.0084	7.68	8.3	16.0	0.367	0.397	0.765
	UO_2	270.07	10	0.0223ᵇ	0.9887	0.036	7.6	16.7	24.3	0.169	0.372	0.542
93	Np	237	—	—	0.9972	0.0084	170	—	—	—	—	—
94	Pu	239	19.74	0.0498	0.9972	0.0083	1,026	9.6	1,036	51.1	0.478	51.6
95	Am	242	—	—	0.9973	0.0082	8.000	—	—	—	—	—
96	Cm	245	—	—	0.9973	0.0081	500	—	—	—	—	—
97	Bk	249	—	—	0.9973	0.0081	900	—	—	—	—	—
98	Cf	249	—	—	0.9973	0.0079	160	—	—	—	—	—
99	E	253	—	—	0.9974	0.0079	—	—	—	—	—	—
100	Fm	256	—	—	0.9974	0.0078	—	—	—	—	—	—
101	Mv	260	—	—	0.9974	0.0077	—	—	—	—	—	—

ᵃValue has been multiplied by 10^5.

ᵇMolecules/cm³.

*From: "Reactor Physics Constants", 2nd ed., Argonne National Laboratory, ANL-5800, U.S. Atomic Energy Commission, July 1963. See this source for a complete list of references.

REFERENCES

"Neutron Cross Sections", 2nd ed., D.J. Hughes and R.B. Schwartz, Brookhaven National Laboratory, BNL–325, 1958. See also Supplement I to second edition, 1960.

For a table of scattering cross sections giving data for each isotope of the element, see "American Institute of Physics Handbook", 2nd ed., D.E. Gray, Ed., McGraw-Hill Book Company, 1963, pp. 8-148–8-167.

Table 4-16. DELAYED NEUTRON YIELD FROM FAST NEUTRON-INDUCED FISSION IN U^{233}, U^{235}, U^{238}, Pu^{239}, Pu^{240}, AND Th^{232}*

Isotope	Delayed neutrons/ fission	Group index, i	Half-life, T_i	Decay constant, λ_i	Relative abundance, a_i	Absolute group yield, %
U^{233}	0.0070 ± 0.0004	1	55.11 ± 1.86	0.0126 ± 0.0004	0.086 ± 0.003	0.06 ± 0.003
		2	20.74 ± 0.86	0.0334 ± 0.0014	0.274 ± 0.005	0.192 ± 0.009
		3	5.30 ± 0.19	0.131 ± 0.005	0.227 ± 0.035	0.159 ± 0.025
		4	2.29 ± 0.18	0.302 ± 0.024	0.317 ± 0.011	0.222 ± 0.012
		5	0.546 ± 0.108	1.27 ± 0.266	0.073 ± 0.014	0.051 ± 0.010
		6	0.221 ± 0.042	3.13 ± 0.675	0.023 ± 0.007	0.016 ± 0.005
U^{235}	0.0165 ± 0.0005	1	54.51 ± 0.94	0.0127 ± 0.0002	0.038 ± 0.003	0.063 ± 0.005
		2	21.84 ± 0.54	0.0317 ± 0.0008	0.213 ± 0.005	0.351 ± 0.011
		3	6.00 ± 0.17	0.115 ± 0.003	0.188 ± 0.016	0.310 ± 0.028
		4	2.23 ± 0.06	0.311 ± 0.008	0.407 ± 0.007	0.672 ± 0.023
		5	0.496 ± 0.029	1.40 ± 0.081	0.128 ± 0.008	0.211 ± 0.015
		6	0.179 ± 0.017	3.87 ± 0.369	0.026 ± 0.003	0.043 ± 0.005
U^{238}	0.0412 ± 0.0017	1	52.38 ± 1.29	0.0132 ± 0.0003	0.013 ± 0.001	0.054 ± 0.005
		2	21.58 ± 0.39	0.0321 ± 0.0006	0.137 ± 0.002	0.564 ± 0.025
		3	5.00 ± 0.19	0.139 ± 0.005	0.162 ± 0.020	0.667 ± 0.087
		4	1.93 ± 0.07	0.358 ± 0.014	0.388 ± 0.012	1.599 ± 0.081
		5	0.49 ± 0.023	1.41 ± 0.067	0.225 ± 0.013	0.927 ± 0.060
		6	0.172 ± 0.009	4.02 ± 0.214	0.075 ± 0.005	0.309 ± 0.024
Pu^{239}	0.0063 ± 0.0003	1	53.75 ± 0.95	0.0129 ± 0.0002	0.038 ± 0.003	0.024 ± 0.002
		2	22.29 ± 0.36	0.0311 ± 0.0005	$0.280 + 0.004$	0.176 ± 0.009
		3	5.19 ± 0.12	0.134 ± 0.003	0.216 ± 0.018	0.136 ± 0.013
		4	2.09 ± 0.08	0.331 ± 0.012	0.328 ± 0.010	0.207 ± 0.012
		5	0.549 ± 0.049	1.26 ± 0.115	0.103 ± 0.009	0.065 ± 0.007
		6	0.216 ± 0.017	3.21 ± 0.255	$0.035 + 0.005$	$0.022 + 0.003$
Pu^{240}	0.0088 ± 0.0006	1	53.56 ± 1.21	0.0129 ± 0.0004	0.028 ± 0.003	0.022 ± 0.003
		2	22.14 ± 0.38	0.0313 ± 0.0005	0.273 ± 0.004	0.238 ± 0.016
		3	5.14 ± 0.42	0.135 ± 0.011	0.192 ± 0.053	0.162 ± 0.044
		4	2.08 ± 0.19	0.333 ± 0.031	0.350 ± 0.020	0.315 ± 0.027
		5	0.511 ± 0.077	1.36 ± 0.205	0.128 ± 0.018	0.119 ± 0.018
		6	0.172 ± 0.033	4.04 ± 0.782	0.029 ± 0.006	0.024 ± 0.005
Th^{232}	0.0496 ± 0.0020	1	56.03 ± 0.95	0.0124 ± 0.0002	0.034 ± 0.002	0.169 ± 0.012
		2	20.75 ± 0.66	0.0334 ± 0.0011	0.150 ± 0.005	0.744 ± 0.037
		3	5.74 ± 0.24	0.121 ± 0.005	0.155 ± 0.021	0.769 ± 0.108
		4	2.16 ± 0.08	0.321 ± 0.011	0.446 ± 0.015	2.212 ± 0.110
		5	0.571 ± 0.042	1.21 ± 0.090	0.172 ± 0.013	0.853 ± 0.073
		6	0.211 ± 0.019	3.29 ± 0.297	0.043 ± 0.006	0.213 ± 0.031

*From: "Reactor Physics Constants", 2nd ed., Argonne National Laboratory, ANL-5800, U.S. Atomic Energy Commission, July 1963. See this source for a complete list of references.

Table 4-17. CHARACTERISTIC DECAY OF A RADIOISOTOPE*

At the end of each half-life interval, one-half of the starting material will remain; at the end of two half-lives, one-fourth; and at the end of four half-lives, one-sixteenth.

Half-lives	F†	Half-lives	F†	Half-lives	F†	Half-lives	F†
0.00	1.000	0.70	0.616	1.65	0.319	3.20	0.109
0.02	0.986	0.75	0.595	1.70	0.308	3.30	0.102
0.04	0.973	0.80	0.574	1.75	0.297	3.40	0.095
0.06	0.959	0.85	0.555	1.80	0.287	3.50	0.088
0.08	0.946	0.90	0.535	1.85	0.277	3.60	0.083
0.10	0.933	0.95	0.518	1.90	0.268	3.70	0.077
0.12	0.920	1.00	0.500	1.95	0.259	3.80	0.072
0.14	0.908	1.05	0.483	2.00	0.250	3.90	0.067
0.16	0.895	1.10	0.467	2.10	0.233	4.00	0.063
0.18	0.883	1.15	0.451	2.20	0.218	4.10	0.058
0.20	0.871	1.20	0.435	2.30	0.203	4.20	0.054
0.25	0.841	1.25	0.421	2.40	0.189	4.30	0.051
0.30	0.812	1.30	0.406	2.50	0.177	4.40	0.047
0.35	0.785	1.35	0.393	2.60	0.165	4.50	0.044
0.40	0.758	1.40	0.379	2.70	0.154	4.60	0.041
0.45	0.732	1.45	0.367	2.80	0.144	4.70	0.039
0.50	0.707	1.50	0.354	2.90	0.134	4.80	0.036
0.55	0.683	1.55	0.342	3.00	0.125	4.90	0.034
0.60	0.660	1.60	0.330	3.10	0.117	5.00	0.031
0.65	0.638						

†F = fraction remaining.
*From "CRC Handbook of Radioactive Nuclides", Y. Wang, Ed., The Chemical Rubber Co., 1969.

Table 4-18.   RADIOACTIVE ISOTOPES

The engineering uses of radioactive isotopes are limited; for more extensive data see the "CRC Handbook of Radioactive Nuclides", Y. Wang, Ed., The Chemical Rubber Co., 1969.

The atomic weight of an atom depends on the number of neutrons in the nucleus as well as the number of protons (and electrons) indicated by the atomic number. For each atomic number, which defines the element, there can be several atomic weights, depending on the number of neutrons. Atoms that differ only in the number of neutrons are known as isotopes or nuclides. While there are more than twenty anisotopic elements (one atomic weight), many others have two to ten isotopes. When there are too many or too few neutrons, the atom becomes unstable or radioactive, and a statistical "decay" occurs, involving the release of radiations of one kind or another. Isotopes may be produced artificially by fission or more likely by bombardment, as in a reactor or a cyclotron.

Isotope differences other than mass are very small and not easy to detect. In a few cases the physical or chemical properties of the isotopes are sufficiently different to permit separation by some common process. This is true with isotopes of boron, carbon, hydrogen, lithium, nitrogen, and oxygen. Separation methods include distillation, electrolysis, electromigration, and chemical exchange.

Table 4-18.   RADIOACTIVE ISOTOPES (Continued)
MASS NUMBER AND HALF-LIFE; COMMERCIALLY AVAILABLE ISOTOPES*

For stable isotopes, see Table 3-4.

ABBREVIATIONS: d = days
h = hours
y = years

Element and mass No.		Half-life	Element and mass No.		Half-life	Element and mass No.		Half-life
Hydrogen	3	12.3 y	Nickel	63	92 y	Ruthenium	106	1.0 y
Beryllium	7	53 d		65	2.56 h	Rhodium	102m	2.9 y
	10	2.7×10^6 y	Copper	61	3.3 h		105	36 h
Carbon	14	5.73×10^3 y		64	12.9 h	Palladium	103	17 d
Fluorine	18	1.8 h		67	61 h		109	13.5 h
Sodium	22	2.58 y	Zinc	65	245 d	Silver	105	40 d
	24	15.0 h		69m	14 h		110m	260 d
Magnesium	28	21.3 h	Gallium	66	9.5 h		111	7.5 d
Aluminum	26	7.4×10^5 y		67	78 h	Cadmium	109	1.3 y
Silicon	31	2.62 h		72	14.1 h		115m	43 d
Phosphorus	32	14.3 d	Germanium	68	282 d		115	2.3 d
	33	25 d		71	11 d	Indium	111	2.8 d
Sulfur	35	86.7 d		77	11 h		114m	50 d
Chlorine	36	3.0×10^5 y	Arsenic	74	18 d	Tin	113	118 d
	38	37.3 m		76	26.5 h		119m	250 d
Argon	37	35.1 d		77	39 h		121	27 h
Potassium	42	12.4 h	Selenium	75	120 d	Antimony	122	2.8 d
	43	22 h	Bromine	77	58 h		124	60 d
Calcium	45	165 d		82	35.3 h		125	2.7 y
	47	4.7 d	Krypton	79	34.9 h	Tellurium	125m	58 d
Scandium	43	3.9 h		83m	1.86 h		127m	105 d
	44m	2.4 d		85m	4.4 h		127	9.3 h
	44	4.0 h		85	10.76 y		129m	34 d
	46	84 d	Rubidium	83	83 d		129	1.1 h
	47	3.4 d		84	33 d		132	3.2 d
Titanium	44	48 y		86	18.7 d	Iodine	123	13 h
Vanadium	48	16.1 d	Strontium	85	64 d		124	4.2 d
	49	330 d		87m	2.8 h		125	60.2 d
Chromium	51	27.8 d		89	50.4 d		126	13.2 d
Manganese	52	5.7 d		90	28 y		129	1.6×10^7 y
	53	2.0×10^6 y	Yttrium	87	80 h		130	12.5 h
	54	303 d		88	108 d		131	8.05 d
	56	2.6 h		90	64.2 h		132	2.3 h
Iron	52	8.3 h		91	59 d		133	21 h
	55	2.7 y	Zirconium	95	65 d	Xenon	131m	12 d
	59	45 d		97	17 h		133	5.3 d
Cobalt	56	77.3 d	Niobium	95	35 d	Cesium	131	9.7 d
	57	267 d	Molybdenum	99	66 h		132	6.6 d
	58	71 d	Technetium	99m	6.0 h		134	2.1 y
	60	5.26 y		99	2.1×10^5 y		137	30 y
			Ruthenium	97	2.9 d	Barium	131	11.6 d
				103	40 d		133	7.2 y
							140	12.8 d

Table 4-18.   RADIOACTIVE ISOTOPES *(Continued)*

MASS NUMBER AND HALF-LIFE; COMMERCIALLY AVAILABLE ISOTOPES
(Continued)

Element and mass No.		Half-life	Element and mass No.		Half-life	Element and mass No.		Half-life
Lanthanum	140	40 h	Ytterbium	169	32 d	Thallium	204	3.8 y
Cerium	139	140 d		175	4.2 d	Lead	210	22 y
	141	32.5 d	Lutetium	177	6.8 d	Bismuth	206	6.24 d
	143	1.4 d	Hafnium	175	70 d		207	30 y
	144	285 d		181	45 d		210m	2.6×10^6 y
Praseodymium	142	19.2 h	Tantalum	182	115 d	Polonium	208	2.9 y
	143	13.7 d	Tungsten	181	130 d		210	138 d
Neodymium	147	11.1 d		185	74 d	Radon	222	3.8 d
	149	1.8 h		187	24 h	Radium	224	3.64 d
Promethium	147	2.7 y	Rhenium	183	70 d		226	1.62×10^3 y
	149	2.2 d		186	3.8 d		228	5.7 y
	151	1.2 d		188	17 h	Actinium	227	21.8 y
Samarium	153	2 d	Osmium	185	94 d	Thorium	228	1.91 y
Europium	152m	9.3 h		191m	14 h		230	7.6×10^4 y
	152	12.4 y		191	15 d	Protactinium	231	3.2×10^4 y
	154	16 y		193	32 h		233	27.4 d
	155	1.8 y	Iridium	192	74 d		234	6.7 h
Gadolinium	153	240 d		194	19 h	Uranium	232	72 y
	159	18 h	Platinum	193m	4.4 d		233	1.6×10^5 y
Terbium	160	72 d		197	20 h	Neptunium	237	2.1×10^6 y
	161	6.9 d	Gold	195	183 d	Plutonium	237	45.6 d
Dysprosium	165	2.3 h		198	2.7 d		239	2.4×10^4 y
Holmium	166m	1.2×10^3 y		199	3.15 d		240	6.7×10^3 y
Erbium	169	9.4 d	Mercury	197m	24 h	Americium	241	458 y
	171	7.5 h		197	65 h	Curium	242	163 d
Thulium	170	130 d		203	47 d		244	18 y
			Thallium	202	12 d			

*Data from 26-page table compiled by P.S. Baker, Oak Ridge National Laboratory, published in "CRC Handbook of Radioactive Nuclides", Y. Wang, Ed., The Chemical Rubber Co., 1969.

REFERENCES

A 300-page table of known stable and radioactive isotopes is given in the "CRC Handbook of Chemistry and Physics", 53rd ed., R.C. Weast, Ed., The Chemical Rubber Co., 1972.

For other data on isotopes, consult the "CRC Handbook of Radioactive Nuclides", Y. Wang, Ed., The Chemical Rubber Co., 1969.

Table 4-19. RADIOISOTOPES FOR INDUSTRIAL USE*

Typical Applications and Isotopes in Common Use

Isotopes	Applications	Isotopes	Applications
CHEMICAL INDUSTRY		**ELECTRICAL INDUSTRY**	
S^{35}, H^3	Efficiency of separation	Kr^{85}	Leak testing
Au^{198}, I^{131}, Na^{24}, Mn^{56}, Br^{82}, Cr^{51}	Thoroughness of mixing	Hg^{197}	Mercury-switch studies
I^{131}, Br^{82}, Na^{24}, H^3	Leak location	H^3, Kr^{85}, Pm^{147}	Luminous dials
Co^{60}, Cs^{137}	Gaging of liquid or solid levels	Co^{60}, Ni^{63}	Pre-ionization of gases in electronic tubes
Rb^{86}	Study of process-stream flow patterns	Sr^{90}	Power for navigational lights
Au^{198}	Location of pipe obstructions	**METALS INDUSTRY**	
Sb^{124}	Study of mass balances in refinery stream	Fe^{59}	Tracing blast furnace operations
C^{14}, H^3	Study of reaction mechanisms	H^3	Study of hydrogen embrittlement
Xe^{133}, Kr^{85}	Measurement of gas-flow velocities	Fe^{59}, Cu^{64}, Zn^{65}, Cr^{51}, Ni^{63}	Study of piston-ring and bearing wear
$ZrNb^{95}$, Co^{60}	Catalyst flow studies	Co^{60}	Control of discharge in coke ovens
C^{14}	Study of carbon deposits in fuel research	Sr^{90}, Kr^{85}, Tm^{170}, Eu^{155}, Ce^{144}, Cs^{137}	Thickness gaging
P^{32}, Co^{58}, Co^{60}, C^{14}	Drug-metabolism studies	Co^{60}	Measuring wear of firebrick linings
S^{35}	Study of vulcanizing process and tire wear	Co^{60}, Ir^{192}, Cs^{137}, Sm^{145}, Gd^{153}, Eu^{155}, Ce^{144}, Tm^{170}, Ta^{182}, Yb^{169}	Detecting thickness variation and defects in castings; weld inspection
C^{14}	Study of frictional forces in rubber		
Zn^{65}	Evaluation of plastic blood bags		
Ca^{45}, Na^{24}, Cl^{38}	Study of diffusion in glass	**TRANSPORTATION INDUSTRY**	
Kr^{85}, Xe^{133}	Determination of air pollution from refinery		
Sr^{90}, Kr^{85}	Control of rubber thickness on tire ply	Fe^{55}, Zn^{65}	Measurement of wear in pistons and bearings
Cs^{137}	Control of rock wool production	Xe^{133}	Studies of sediment and sand movement
Co^{60}, Cs^{137}	Initiation of chemical reactions; effecting of polymerization	Au^{198}	Evaluation of rail life
Sr^{90}, Kr^{85}	Elimination of static	Co^{60}, Ir^{192}, Cs^{137}	Gaging of automobile sheet steel
Co^{60}, Cs^{137}	Sterilization of medical supplies	H^3, Kr^{85}, Pm^{147}	Luminous locks and dials
Sr^{90}, Kr^{85}	Thickness gaging of paper and plastics		

*From: "CRC Handbook of Radioactive Nuclides", Y. Wang, Ed., The Chemical Rubber Co., 1969; see this source for discussion of methods.

Table 4-20.   STABILITY OF RADIOACTIVE COMPOUNDS*

Most users of compounds labeled with radioisotopes recognize that such compounds decompose on storage and that the decomposition is accelerated by self-irradiation. The degree of the decomposition in relation to the storage conditions of the compound and the measures that can be taken to control and minimize the rate of self-radiolysis are not always so well known. Information on this subject is largely empirical. Not only are a large number of labeled compounds—particularly organic compounds—extensively used as tracers, but many applications demand a very high purity. Fractions of a percent of radiochemical impurity can sometimes lead to incorrect deductions from a tracer investigation, and under these conditions the problem of decomposition by self-irradiation becomes a very serious one.

Compounds labeled with the pure beta-emitting radioisotopes—C^{14}, tritium, S^{35}, P^{32}, and Cl^{36}— are most commonly used in tracer investigations. Compounds labeled with the gamma-emitting radioisotopes, such as I^{125}, I^{131}, Co^{57}, Co^{58}, and Se^{75}, have special application in medicine. Some properties of these radionuclides are shown in Table A.

Table A. PHYSICAL PROPERTIES OF SOME RADIONUCLIDES

| Radionuclide | Half-life | Beta energy, Mev | | Specific activity, mCi/mA | | Daughter nuclide (stable) |
		Max.	Mean	Maximum	Common values for compounds	
H^3	12.26 years	0.018	0.0057	2.9×10^4	10^2–10^4	He^3
C^{14}	5700 years	0.159	0.050	64	1–10^2	N^{14}
S^{35}	87.2 days	0.167	0.049	1.5×10^6	1–10^2	Cl^{35}
Cl^{36}	3.03×10^5 years	0.714	0.3	1.2	10^{-3}–10^{-1}	Ar^{36}
P^{32}	14.3 days	1.71	0.69	9.3×10^6	10–10^2	S^{32}
I^{131}	8.04 days	0.81	0.19	1.7×10^7	10^2–10^4	Xe^{131}
I^{125}	60 days	Electron capture		2.2×10^6	10^2–10^4	Te^{125}
Co^{57}	270 days	Electron capture		4.9×10^5	10^3–10^5	Fe^{57}
Co^{58}	71 days	Electron capture $+\beta^+$		1.9×10^6	10^3–10^5	Fe^{58}
Se^{75}	121 days	Electron capture		1.1×10^6	10–10^3	As^{75}

Decomposition depends in part on the amount of energy absorbed by the compound during its useful life, so that, for a given amount of activity, the radiation energy emitted should be a guide to the seriousness of the problem. The problem of decomposition by self-irradiation might be expected to increase in magnitude as the series of pure beta emitters in Table A is descended, but in fact almost the reverse is true. This occurs largely for three reasons:

1. The fraction of energy absorbed is much less than unity for the more energetic beta emitters, such as P^{32}; on the other hand, almost complete total absorption of the beta energy occurs with tritium compounds. Gamma energy is, in general, little absorbed by the compound itself or by its immediate environs.
2. The decomposition also depends on the specific activity of the compound; as can be seen from Table A, the specific activities of tritiated compounds in current use are usually much higher than those for compounds labeled with other pure beta-emitting radionuclides.
3. The absorbed energy decreases exponentially with time; this is an important factor for compounds labeled with radionuclides having a short half-life, such as I^{131} or P^{32}.

The reason why labeled compounds decompose is not difficult to understand: the radiation energy will be commonly absorbed by the compound itself or by its environs. If the former occurs, then the excited molecules may break up in some manner; if the latter occurs, the radiation energy

Table 4-20.   STABILITY OF RADIOACTIVE COMPOUNDS *(Continued)*

can produce free radicals and other reactive species, which may then cause destruction of the molecules of the labeled compound.

The modes by which decomposition of labeled compounds can arise have been classified as shown in Table B.

Table B.   MODES OF DECOMPOSITION OF LABELED COMPOUNDS

Mode of decomposition	Cause	Method for control
Primary (internal)	Natural isotopic decay	None, for a given specific activity**
Primary (external)	Direct interaction of the radioactive emission (alpha, beta, or gamma) with molecules of the compound	Dispersal of the labeled molecules
Secondary	Interactions of excited products with molecules of the compound	Dispersal of active molecules; cooling to low temperatures; scavenging of free radicals
Chemical	Thermodynamic instability of compounds; poor choice of environment	Cooling to low temperatures; removal of harmful agents

** Note that dilution with the inactive form of the compound subsequent to preparation is not beneficial in this case.

Primary decomposition is the production of an impurity due to the disintegration of the unstable nucleus. Secondary decomposition is commonly the most damaging and the most difficult to control. Chemical decomposition (by oxidation, hydrolysis, etc.) is even more likely to occur with radioactive chemicals. It may also be necessary to guard against photochemical or micro-biological decomposition of the compound.

Tables reporting the various kinds of decomposition for a great variety of labeled compounds are available. Over 30 pages of such tables are given in the "CRC Handbook of Radioactive Nuclides" (see below); this source also cites a large number of references.

While it is true that no measurable decomposition has occurred in many labeled compounds that have been stored for months or even years, there are other compounds showing major decomposition in a matter of days. It is thus very important to take proper account of the condition and the stability of any compounds being used.

*Adapted from: "CRC Handbook of Radioactive Nuclides", Y. Wang, Ed., The Chemical Rubber Co., 1969.

Table 4-21. RADIOISOTOPE FUEL DATA

Robert T. Carpenter

This table represents radioisotopic fuel data only and does not include the capsule or containment materials that would reduce the overall heat source, specific power, and power density, depending on the application and the heat source design.

For data on the thermophysical, mechanical, chemical, biological, shielding, and electrical properties of these fuels, consult the listed references. Material compatibility data is also cited in these references.

Radioisotope	Weight of active isotope in compound	Half-life	Melting point, °C	Specific power, watts/g*	Power density, watts/cc
Co^{60}					
Metal	Pure Co^{60} (theoretical— not available)		1,495	17.7	156
Co-Ni[a]	17.5% Co^{60}	5.26 yr	1,487	3.1	27
Co-Ni[b]	35% Co^{60}		1,481	6.2	55
Sr^{90}					
Metal	55.0% Sr^{90}	28.0 yr	772	0.5	1.28
$SrTiO_3$	24.5% Sr^{90}		1,910	0.23	1.17
SrO	44.0% Sr^{90}		2,457	0.42	1.94
SrF_2	36.0% Sr^{90}		1,463	0.34	1.44
Sr_2TiO_4	31.5% Sr^{90}		1,860	0.30	1.48
Cs^{137}					
CsCl	28.9% Cs^{137}	30.0 yr	645	0.12	0.37
Cs_2SO_4	26.9% Cs^{137}		1,019	0.11	0.46
Ce^{144}					
Ce_2O_3	10.8% Ce^{144}		2,190	2.76	19
Ce_2O_2S	10.3% Ce^{144}	285 days	2,000	2.64	15.8
Ce_2S_3	9.4% Ce^{144}		1,890	2.41	12.5
CeF_3	9.0% Ce^{144}		1,437	2.31	14.3
Pm^{147}					
Metal	95% Pm^{147}	2.62 yr	865	0.31	2.30
Pm_2O_3	78% Pm^{147}		2,130	0.27	1.87
Tm^{170}					
Metal	16.3% Tm^{170} 2.9% Tm^{171}	128 days	1,545	2.24	20.5
Tm_2O_3	14.3% Tm^{170} 2.5% Tm^{171}	(Tm^{171}— 700 days)	2,375	2.24	19.7
Po^{210}					
Metal	95% Po^{210}	138 days	254	144	1,324
GdPo[c]			1,675 (GdPo)	1.6	16.6
GdPo[d]			1,675 (GdPo)	7.5	77.2
Pu^{238}					
Metal[e]	80% Pu^{238}	87 yr	575–615	0.45	6.8
PuO_2[f]	71% Pu^{238}		2,150	0.40	2.7
PuC	74.6% Pu^{238}		1,654	0.42	5.7
PuN	74.6% Pu^{238}		2,570	0.42	6.2
PuZr[g]	73% Pu^{238}		730	0.41	5.6
PuZr[h]	77% Pu^{238}		615	0.44	6.5
Cm^{242}					
Cm_2O_3–AmO_2	35.7% Cm^{242}	163 days	2,000	42.8	500
Cm^{244}					
Metal	95.5% Cm^{244}	18.1 yr	1,340	2.67	36
Cm_2O_3	86.9% Cm^{244}		1,950	2.42	26.1
Cm_2O_2S	84.4% Cm^{244}		2,000	2.35	23.3
CmF_3	77.5% Cm^{244}		1,406	2.15	21.1

Table 4-21.   RADIOISOTOPE FUEL DATA *(Continued)*

*For horsepower/lb fuel multiply these values of watts per gram by 0.61.

[a] 200 curies/g.	[e] Production grade.
[b] 400 curies/g.	[f] Microspheres; packing fraction 65%.
[c] 98% Ta matrix.	[g] 20% Zr alloy.
[d] 91% Ta matrix.	[h] 10% Zr alloy.

GENERAL REFERENCES

1. "Properties of Co^{60} and Cobalt Metal Fuel Forms", W.C. Winbley, Jr., DP–1051 (Rev 1), Savannah River Laboratory, E.I. du Pont de Nemours & Co., for U.S. Atomic Energy Commission, Oct. 1966.
2. "Strontium-90 Data Sheets", S.J. Rimshaw and E.E. Ketchen, ORNL–4188, Oak Ridge National Laboratory, Oak Ridge, Tenn., Dec. 1967.†
3. "Cesium-137 Data Sheets", S.J. Rimshaw and E.E. Ketchen, ORNL–4186, Oak Ridge National Laboratory, Oak Ridge, Tenn., Dec. 1967.†
4. "Promethium Isotopic Power Data Sheets", H.T. Fulham and H.H. Van Tuyl, BNWL–363, Battelle Northwest, Richland, Wash., Feb., 1967.†
5. "Properites of Thulium Metal and Oxide", P.K. Smith, J.R. Keski, and C.L. Angerman, DP–1114, Savannah River Laboratory, Aiken, S.C., June 1967.†
6. "Plutonium-238 and Polonium-210 Data Sheets", MLM–1441, Mound Laboratory, Monsanto Research Corp., Miamisburg, O., Sept. 29, 1967.†
7. "Curium Data Sheets", S.J. Rimshaw and E.E. Ketchen, ORNL–4187, Oak Ridge National Laboratory, Oak Ridge, Tenn., Dec. 1967.†
8. "Cerium-144 Data Sheets", S.J. Rimshaw and E.E. Ketchen, ORNL–4185, Oak Ridge National Laboratory, Oak Ridge, Tenn., Nov. 1967.†
9. Quarterly Report, BNWL–680, Pacific Northwest Laboratory, Division of Isotope Development Programs, Richland, Wash., Oct.–Dec. 1967.†

---

†Available from Clearinghouse for Federal Scientific and Technical Information, National Bureau of Standards, Springfield, Va. 22151.

Table 4-22. REACTOR MATERIALS

Each engineering material used in the construction of a nuclear reactor may serve one or more functions. In most cases the physical, mechanical, and thermal properties of the material are of some importance, as well as the nuclear properties. In the following list each function is briefly defined, and some of the important properties are noted.

Fuel. The fuel elements are uranium, plutonium, and thorium. The fissionable isotopes are U^{233}, U^{235}, and Pu^{239}. Although uranium has fourteen isotopes, all radioactive, natural uranium contains 99.28% of U^{238} and only 0.7% of U^{235}. The most common fuel is natural uranium, "enriched" with added U^{235}. Isotopes that can be converted into fissile fuels are called "fertile" materials (see Table 4-4). These include U^{238}, which forms U^{239} by capture of a neutron and undergoes transition to Pu^{239} by beta decompositions, and thorium 232, which is converted to Th^{233} by neutron fluxes and forms U^{233} by beta decay.

Reactor-fuel elements should be strong and corrosion-resistant and have high thermal conductivity. A minimum of fission products should enter the coolant. Alloying, cladding, or plating are required to attain these properties. Stainless steel, aluminum, titanium, molybdenum, beryllium, and zirconium are used, in addition to graphite and some ceramic materials.

General physical and mechanical properties of some of these materials are given in Table A.

Table A. REACTOR FUELS AND ASSOCIATED MATERIALS

Typical Properties

For specific heat in J/kg·K, multiply tabular values by 4187. For thermal conductivity in Btu/h·ft·deg F, multiply by 57.8. For N/m² multiply psi by 6895.

Material*	Specific gravity	Specific heat, 25°C	Thermal cond., $\frac{watts}{cm\,°C}$	Melting point, °C	Coef. lin. exp./°C $(\times 10^6)$	Tensile strength, psi	Modulus of elast., psi (millions)	Absorption cross section, σ_a, barns
Uranium	19.	0.03	0.25	1,132	46.	60,000.	23.	7.6
UO₂	10.96		0.08	2,600	10.	—	—	7.6
Plutonium	19.8	0.03	0.08	640.	54.		14.	—
Thorium	11.7	0.03	0.4	1,750.	11.5	32,000	9.	7.5
Beryllium	1.85	0.44	2.18	1,285	12.	30,000	42.	0.01
Zirconium	6.5	0.07	0.21	1,852	5.8	30,000	13.7	0.18
Cadmium	8.65	0.05	0.93	321	32.	10,300	8.	2,450.
Aluminum	2.7	0.21	2.37	660.	26.	13,000	10.	0.24
Stainless steel (347)	8.0	0.12	0.2	1,400.	16.	75,000	29.	3.0
Titanium	4.5	0.125	0.2	1,670.	8.5	9,000	16.	5.8
Graphite	2.2	0.17	0.24	>3,500	3.	100	0.7	0.004

*Some metals become radioactive and dangerous to handle after exposure to the neutron flux of a reactor, but these would not be used in the structure of a thermal reactor. Tungsten and cobalt are the worst in this respect, but tantalum, chromium, and manganese may also attain a high induced radioactivity, and copper a considerably lower one.

Moderator. For thermal reactors it is necessary to reduce the kinetic energy of the "fast" neutrons by successive scattering collisions. Materials effective for this "slowing down" of neutrons to the thermal range include graphite, water, heavy water, beryllium, and beryllium oxide and hydrocarbons (see Tables 4-26 and 4-28; also Tables 4-46 through 4-50). A moderator should be resistant to corrosion and stable at high temperatures and in high radiation flux. A moderator may serve as a fuel diluent or fuel container. In the latter case mechanical strength is required.

Reflector. Moderating materials that also scatter neutrons back into the core to reduce "leakage" are termed reflectors. A "bare reactor" is one without such reflectors. Savings in fuel and smaller critical size are attained with reflectors.

Control material. Control rods or other arrangements for varying the reactor output are of three classes: (1) *shim rods* for coarse control during startup; (2) *regulating rods* for small but quick control; and (3) *safety rods* or "scram" control for emergencies. The earlier methods used materials with large capture cross sections to absorb neutrons (cadmium or boron), but it is more economical to employ uranium, cobalt, or other material that produces a secondary product, such as a desired isotope. Other control methods involve the positioning of the fuel, moderator, or reflector elements. Typical properties of the more effective isotopes of the high-absorption materials are given in Table B.

Table 4-22.   REACTOR MATERIALS *(Continued)*

Table B.   PROPERTIES OF NEUTRON ABSORBERS (POISONS)

Isotope	Thermal microscopic cross sections, σ, barns	Major resonance	
		Energy, ev	σ_a, barns
Cadmium 113	20,000	0.18	7,200
Boron 10	3,850	None	
Samarium 149	41,000	0.096	16,000
Samarium 152	220	8.2	15,000
Gadolinium 155	56,000	2.6	1,400
Gadolinium 157	240,000	17.	1,000
Europium 151	7,800	0.46	11,000
Europium 153	450	2.46	3,000
Silver 107	31	16.6	630
Silver 109	87	5.1	12,500

Coolant. In some reactors it may be necessary to dissipate almost the entire heat output to a heat sink; in reactors for power generation, the primary coolant becomes the heat source for the power cycle, and two or more coolant circuits are involved. Practical coolants are water, liquid metals, and various gases. Heavy water is superior but costly. Organic liquids tend to leave surface deposits in the fuel section. High heat capacity and high thermal conductivity are desired, with low vapor pressure and stability under high temperature and radiation. For specific properties of materials in Table C, see the "Index of Properties", Section 1, and also Table 4-29.

Table C.   REACTOR COOLANTS

Liquids and liquid metals		Gases and vapors	
Diphenyl	NaK	Air	Mercury
Dowtherm®†	Potassium	Carbon dioxide	Neon
Gallium	Rubidium	Helium	Nitrogen
Heavy water	Sodium	Hydrogen	Steam
Lithium	Tin		
Mercury	Water		

†Dowtherm Heat Transfer Agent, Dow Chemical Co., Midland, Mich.

Shielding material. Design of shields for personnel protection depends on objectives, space and weight requirements, and types of radiation. A comparison of the common shielding materials is presented in Table 4-36, and Table 4-32 gives further data.

Table 4-23.   ENERGY LIBERATED BY FISSION

The following values are approximations for the energy quantities involved in fission of any one of the three common nuclear fuels—uranium 233 or 235 or plutonium 239—since there is a relatively small difference between these weights.

One pound of fissile material will liberate approximately 10 million kilowatt hours. The fission of one gram per day is roughly equal to one megawatt. The "burnup" of 1,000 kg (2,200 lb) of uranium would deliver 100 megawatts for about 25 years.

While almost all of the fission energy ultimately appears as heat within the matter surrounding the reaction, only about 80 percent of the fission energy is immediately converted to heat from the kinetic energy of the fission fragments. About 95 percent of the energy eventually appears as heat, but the 5 percent represented by the neutrinos essentially escapes from the reactor.

PERCENTAGE DISTRIBUTION OF FISSION ENERGY

Kinetic energy of fission fragments	82.5%
Instantaneous gamma-ray energy	3.5%
Gamma-rays from fission products	3.0%
Beta particles from fission products	3.5%
Kinetic energy of fission neutrons	2.5%
Neutrinos	5.0%

Table 4-24. DECAY OF REACTOR FISSION PRODUCTS*

Days from Shutdown, after Infinite Operation; Various Levels of Effective Energy (EE)

Time, days	Approximate decay, $MeV/W{\cdot}s$						
	Group 1 $EE = 0.4\ MeV$	Group 2 $EE = 0.8\ MeV$	Group 3 $EE = 1.3\ MeV$	Group 4 $EE = 1.7\ MeV$	Group 5 $EE = 2.2\ MeV$	Group 6 $EE = 2.5\ MeV$	Group 7 $EE = 2.8\ MeV$
0.1	1.5×10^9	1×10^{10}	1.1×10^9	4.3×10^9	1.2×10^9	3.4×10^8	1.7×10^9
0.2	1.5×10^9	9.5×10^9	6.8×10^8	3.5×10^9	7.4×10^8	3×10^8	9×10^8
0.4	1.45×10^9	8.8×10^9	3.7×10^8	3.1×10^9	3×10^8	2.9×10^8	2.8×10^8
0.7	1.35×10^9	8.0×10^9	2×10^8	3×10^9	1.15×10^8	2.8×10^8	4.7×10^7
1.0	1.3×10^9	7.5×10^9	1.2×10^8	3×10^9	8×10^7	2.7×10^8	
2.0	1.1×10^9	6.4×10^9	2.3×10^7	3×10^9	5.6×10^7	2.6×10^8	
4.0	8×10^8	5.4×10^9		2.9×10^9	4×10^7	2.4×10^8	
7.0	6×10^8	4.7×10^9		2.6×10^9	3×10^7	2.1×10^8	
10.	4.8×10^8	4×10^9		2.2×10^9	2.3×10^7	1.7×10^8	
20.	3×10^8	3.2×10^9		1.3×10^9	1.7×10^7	1.0×10^8	
40.	1.3×10^8	2.4×10^9		4×10^8	1.55×10^7	4×10^7	
70.	5×10^7	1.8×10^9		1×10^8	1.4×10^7	1.3×10^7	
100	2.4×10^7	1.5×10^9		3×10^7	1.3×10^7		

Note: These data for infinite-power operation may be used for estimating by calculation the sources resulting from finite-time operation before shutdown.

*Data from: "Reactor Shielding Design Manual", T. Rockwell III, USAEC Report No. TID–7004, U.S. Atomic Energy Commission, March 1956.

Table 4-25. HEAT PRODUCTION AFTER REACTOR SHUTDOWN*

Uranium Fuel Afterheat

Time after shutdown		Duration of full-power operation				
		1 hour	1 day	1 week	1 month	1 year
Seconds	Other units	Percentage of full-power after shutdown				
3×10^2	5 minutes	1.0				
6×10^2	10 minutes	0.8	1.4			
1.8×10^3	30 minutes	0.39	1.1	1.3		
3.6×10^3	1 hour	0.24	0.85	1.1		
7.2×10^3	2 hours	0.12	0.61	0.85	1.00	
2.88×10^4	8 hours	0.03	0.28	0.50	0.63	
8.64×10^4	1 day		0.09	0.27	0.41	0.58
2.6×10^5	3 days		0.03	0.13	0.28	0.42
6.1×10^5	1 week		0.010	0.08	0.18	0.31
1.21×10^6	2 weeks			0.03	0.10	0.22
2.6×10^6	30 days			0.015	0.06	0.16
7.8×10^6	90 days				0.02	0.08
3.1×10^7	1 year					0.015

*Based on: "Heat Generation in Irradiated Uranium", ANL–4790, Argonne National Laboratory.

Table 4-26.   MODERATOR MATERIALS*

Nuclear Characteristics of Potential Moderators

For density in lb/ft³, multiply the value in g/cm³ by 62.42.

Element or compound	A, atomic or molecular weight	Density,[a] g/cm³	N,[b] $\times 10^{-24}$	Scattering cross section σ_s epithermal, barns/atom	Absorption cross section σ_a, 0.025 ev, barns/atom or molecule
Hydrogen, H	1.008	0.0090	0.0054	20.3	0.33
Deuterium, D	2.02	0.0180	0.0054	3.3	0.00046
Helium, He	4.00	0.0180	0.0027	0.8	—
Beryllium, Be	9.01	1.85	0.124	6.1	0.009
Carbon (graphite), C	12.0	1.60	0.080	4.7	0.0045
Oxygen, O	16.0	0.014	0.0054	3.8	0.0002
Sodium, Na	23.0	0.97	0.0254	3.0	0.49
Magnesium, Mg	24.3	1.74	0.043	3.4	0.059
Aluminum, Al	27.0	2.70	0.060	1.35	0.215
Beryllia, BeO	25.0	3.025	0.073	9.9	0.009
Beryllium carbide, Be_2C	30.0	2.4	0.048	16.9	0.023
Beryllium fluoride, BeF_2	47.0	1.986	0.025	15.4	0.029
Light water, H_2O	18.0	1.00	0.033	44.4	0.66
Heavy water, D_2O	20.0	1.10	0.033	10.5	0.0011
Sodium hydroxide, NaOH	40.0	2.1	0.032	27.1	0.82
Zirconium hydride, ZrH_2	93.2	5.61	0.036[d]	48.6	0.84
Biphenyl, $C_{12}H_{10}$	154.2	0.87	0.0034	259.4	2.54
Polystyrene, $(CH)_n$	13.0	1.07	0.038[e]	25.0[g]	0.335[g]
Paraffin, $(CH_2)_n$	14.0	0.9	0.030[f]	45.3[h]	0.665[h]

Element or compound	Macroscopic absorption cross section, $\Sigma_a = N\sigma_a$ cm⁻¹	Macroscopic scattering cross section, $\Sigma_s = N\sigma_a$ cm⁻¹	Logarithmic mean energy loss/ collision,[c] ξ	Slowing-down power, $\xi\Sigma_s$	Moderating ratio, $\sigma_s\xi/\sigma_a$
Hydrogen, H	0.0018	0.11	1.000	0.11	61
Deuterium, D	2.5×10^{-6}	0.018	0.725	0.013	5,200
Helium, He	—	0.0022	0.425	0.0009	∞
Beryllium, Be	0.0011	0.76	0.206	0.16	145
Carbon (graphite), C	0.00036	0.38	0.158	0.060	165
Oxygen, O	1.1×10^{-6}	0.021	0.12	0.0025	230
Sodium, Na	0.012	0.076	0.083	0.0063	0.53
Magnesium, Mg	0.0025	0.15	0.073	0.011	4.4
Aluminum, Al	0.013	0.081	0.071	0.0058	0.45
Beryllia, BeO	0.00066	0.72	0.173	0.12	190
Beryllium carbide, Be_2C	0.0011	0.81	0.193	0.16	145
Beryllium fluoride, BeF_2	0.00074	0.39	0.151	0.058	84
Light water, H_2O	0.022	1.47	0.925	1.36	62
Heavy water, D_2O	36×10^{-6}	0.35	0.504	0.18	5,000
Sodium hydroxide, NaOH	0.026	0.87	0.77	0.67	26
Zirconium hydride, ZrH_2	0.030	1.75	0.84	1.47	49
Biphenyl, $C_{12}H_{10}$	0.00862	0.880	0.812	0.715	83
Polystyrene, $(CH)_n$	0.013	0.95	0.842	0.80	62
Paraffin, $(CH_2)_n$	0.020	1.36	0.913	1.24	62

[a] For the gases H_2, D_2, He, and O_2 an arbitrary density 100 times the density of NTP is assumed; i.e., a pressure on the order of 1,500 psi.
[b] Number of atoms or mole per cm³.
[c] $\xi = 1 - [(A-1)^2/2A] \ln [(A+1)/(A-1)]$.
[d] Below 800°C.   [e] Number of (CH) units/cm³.   [f] Number of (CH_2) units/cm³.   [g] Per (CH) unit.   [h] Per (CH_2) unit.

*From: "Reactor Handbook", 2nd ed., C.R. Tipton, Jr., Ed., Vol. 1: Materials, © 1960, Interscience Books, now a division of John Wiley & Sons, Inc. Reprinted by permission of John Wiley & Sons, Inc.

Table 4-27. GRAPHITE FOR REACTORS*

Commercial graphite for reactors is a mixture of crystalline graphite and cross-linking intercrystalline carbon. The physical properties that are measured are the result of contributions from both sources. Graphite for nuclear reactors is a material for neutron moderators, reflectors, thermal columns, and exponential piles. Its desirable properties for these uses include low neutron-capture cross section, high-temperature stability, and machinability.

The following table lists common properties of two typical nuclear graphite materials that differ primarily in purity. Many special grades are also available to meet such requirements as high density, low porosity, and high permeability.

Table A. TYPICAL PROPERTIES OF NUCLEAR GRAPHITE

Property	Grade A	Grade B
Slow neutron absorption, cross section per carbon atom $\times 10^{-27}$, mb[a]	3.95	4.5
Total ash content, percent	0.01	0.06
Boron content, ppm	0.2	0.3
Specific resistance, 10^{-5} ohm-cm		
Longitudinal	21	25
Transverse	48	42
Thermal conductivity, Btu·ft/hr·ft^2·deg F		
Longitudinal	141	116
Transverse	62	69
Coefficient of thermal expansion, 10^{-6}/deg F		
Longitudinal	0.3	1.0
Transverse	3.3	3.0
Bulk density, g/cc	1.73	1.71
Tensile strength, psi, longitudinal	1 400	1 300
Flexural strength, psi, longitudinal	3 400	2 700
Compressive strength, psi, longitudinal	5 000	5 000
Elastic modulus, 10^6 psi		
Longitudinal	2.3	1.8
Transverse	0.8	1.0

[a] mb = millibarns.

*Courtesy of Union Carbide Corporation, Carbon Products Division.

Graphite is available as a matrix material containing dispersions of uranium and/or thorium for high-temperature fuel elements or boron for control-rod and neutron-shielding purposes.

The comparative position of graphite as a neutron moderating and reflecting element is indicated in Table B. *Moderator figure-of-merit* is the ratio of the life of a fast neutron before absorption to the time required to slow it down to an energy of one electron volt. *Reflector figure-of-merit* is the ratio of the probability of thermal neutron scattering back into a reactor core to the probability of absorption.

Table B. COMPARATIVE MODERATING AND REFLECTOR CHARACTERISTICS*

	H_2O	D_2O	Be	BeO	Graphite
Moderator figure-of-merit	209	17 500	456	604	695
Reflector figure-of-merit	172	15 300	788	774	1 430

*Adapted from: "The Industrial Graphite Engineering Handbook", Carbon Products Division, Union Carbide Corporation (270 Park Avenue, New York, New York 10017), page 6.03.

Reactor graphite is made from petroleum coke, which is calcined, crushed, and screened, then mixed with a coal-tar pitch binder and fired at high temperature. Size of grains depends on the source of the raw material as well as on the processing. Physical properties and crystal orientation will be somewhat different for an extruded product than for a molded product. Impurities and trace elements will vary with the raw materials. The principal impurities are iron, silicon, calcium, and aluminum, each of these typically less than 500 ppm; other metals are typically less than 70 ppm.

Additional factors that affect the properties of the final product are size of the finished piece, the particle-size distribution in the original mix, the maximum processing temperature, and the temperature at which the properties are being measured. The tensile strength of graphite is about one-half its flexural strength, while the crushing strength is about twice the flexural strength. Strength increases with temperature, in the usual range, as do the specific heat and the coefficient of thermal expansion. Thermal conductivity decreases with increase in temperature.

Table 4-28.
NUCLEAR PROPERTIES OF WATER AND HEAVY WATER*

Quantity	Light water	Heavy water
Abundance, %	99.9849 to 99.9861	0.0139 to 0.0151
Molecular weight	18.016	20.028
Molecular density at 20°C, cm^{-3}	0.334×10^{24}	0.0332×10^{24}
Neutron macroscopic absorption cross section at 2,200 m/sec, cm^{-1}	0.0220	0.000040
Thermal neutron diffusion area, cm^2	8.12	—
For 0.16% H$_2$O content		13,500
Corrected for H$_2$O absorption		25,000
Thermal neutron macroscopic transport cross section, cm^{-1}	2.10	0.395
Average cosine of neutron scattering angle	0.34	0.15
Epithermal neutron macroscopic scattering cross section	1.49	0.349
Epithermal neutron macroscopic slowing-down cross section, cm^{-1}	1.38	0.178
Fission neutron age, cm^2		
To indium	30.4	100
To thermal	31.4	125

*From: "Reactor Handbook", 2nd ed., C.R. Tipton, Jr., Ed., Vol. 1: Materials, © 1960, Interscience Books, now a division of John Wiley & Sons, Inc. Reprinted by permission of John Wiley & Sons, Inc.

Table 4-29.   ACTIVATION DATA FOR SOME REACTOR COOLANTS*

Target isotope	Isotopic abundance, %	Activation cross section, barn	Radioactive product of reaction	Half-life	Energy of radiation, Mev (γ)	Gammas per disintegration of active atom
Na^{23}	100	0.53^a	Na^{24}	14.9 hr	2.76; 1.38	1; 1
K^{41}	6.8	1.15^a	K^{42}	12.4 hr	1.51	0.25
O^{18}	0.204	0.00021^a	O^{19}	29.4 sec	1.6	0.7
O^{16}	99.8	$\sim 0.019 \times 10^{-3\,b}$	N^{16}	7.4 sec	6.13; 7.10	0.76; 0.06
O^{17}	0.039	$\sim 0.0052 \times 10^{-3\,b}$	N^{17}	4.1 sec	1 (neutron)	1 (neutron)

^{a}Neutron activation cross section at 2200 meter/sec.
^{b}Fast (n, p) reactions; cross sections averaged over fission spectrum. Data are from "Production Cross Section of N^{16} and N^{17}", P.A. Roys and K. Shure, *Abstracts*, 1958 Annual Meeting of the American Nuclear Society, Los Angeles, Paper No. 3-8, June 1958. A previous determination in "The Production of N^{16} and N^{17} in the Cooling Water of the NRX Reactor", W.J. Henderson and P.R. Tunnicliffe, *Nuclear Sci. and Eng.*, 3(2):145, February 1958, gives cross sections of 0.0185×10^{-3} barns for N^{16}, and 0.0093×10^{-3} barns for N^{17}.

*From: "Reactor Physics Constants", 2nd ed., Argonne National Laboratory, ANL–5800, U.S. Atomic Energy Commission, July 1963.

REFERENCE
"Reactor Shielding Design Manual", T. Rockwell III, Ed., TID–7004, March 1956, Table 4.1.

Table 4-30. REPROCESSING URANIUM FUEL*

How Reprocessing Plants Fit Into the Uranium Fuel Cycle

Fuel reprocessing technology is being spurred by new AEC guidelines pertaining to the deposition of high-level radioactive wastes and the discharge of radioactive effluents from plants. These guidelines direct that all high-level wastes must be solidified within five years of production and buried within ten years; they propose that radiation-exposure levels for individuals remain under 5 mrem/yr incremental dosage from plant effluents.

When defined as a series of processing steps, the uranium fuel cycle consists of four vital recurring operations (mining is performed just once): fuel fabrication, burning, reprocessing, and enrichment. A fissionable U^{235} atom is likely to undergo these steps several times before being "burned".

In commercial power reactors the fuel consists almost entirely of uranium in metallic or oxide form, of which 3–4% is U^{235}, with the remainder being non-fissionable U^{238}. In the reactor only a fraction of the already dilute U^{235} is burned by the time thermal efficiency drops off, and the fuel assembly has to be replaced. Consequently, the "spent" fuel contains most of the valuable uranium, as well as plutonium and neptunium—fission products that are worth recovering.

The bulk of the remaining fission products pose an environmental hazard and an economic handicap. As worthless, radioactive materials, they must be sealed off from the biosphere for hundreds of years; as nuclear "poisons", they impede fission efficiency and cannot be allowed to build up in the fuel cycle.

The function of reprocessing plants is to conserve valuable uranium, plutonium, and neptunium, while converting worthless fission products into safe, disposable forms. Uranium is usually produced at reprocessing plants in the form of uranyl nitrate, which must be fluorinated to UF_6 prior to enrichment. A trend is developing for reprocessors to include fluorination as one of their services. Previously, this has been a separate, mini-step in the fuel cycle.

During enrichment, the concentration of U^{235} is brought back to the 3–4% level required for power generation; then the fuel goes back to the fabricator who prepares new fuel rods.

*From: *Chem. Eng.*, 78:19, August 23, 1971, Copyright © 1971, McGraw-Hill, Inc., New York, N.Y. 10036, with permission.

Table 4-31. RADIATION EXPOSURE AND SHIELDING**

Data on Hazards from Alpha, Beta, and Gamma Radiation

The following miscellaneous items of information on penetration and shielding are useful in evaluating certain radiation hazards and in planning personnel protection.

Gamma rays. As an approximation for the dose rate at a distance of one foot from a point source of gamma radiation, the value in rems per hr is six times the product of total gamma energy emitted per disintegration of the parent, in Mev, times number of curies of the parent nuclide. (Accuracy ± 20 percent from 0.07 to 4 Mev.)

The attenuation of dose rate with distance, from 100 curies of cobalt 60, is given in the following table. Account has been taken of air absorption and build-up factor as well as inverse-square attenuation.

Table A. DOSE RATE FROM 100 CURIES OF Co^{60}

Dose rate, rems/hr	Distance, feet	Dose rate, rems/hr	Distance, feet
1,500	1	0.0075	400
15	10	0.0012	800
0.6	50	0.0006	1,000
0.15	100		

Warning note: Hazardous beta rays may also be present. The above data provide the dose rates from gamma rays only.

Table 4-31.   RADIATION EXPOSURE AND SHIELDING *(Continued)*

Table B gives relative shield thickness for gamma radiation. The build-up factor to correct exponential to broad-beam radiation has been applied.

Table B.   SHIELD THICKNESS VS. GAMMA-DOSE TRANSMISSION†

Broad-beam transmission	Shield thickness, inches		
	Concrete,* 147 lb/cu ft	Iron	Lead

Radium (11 principal gammas, 0.24 to 2.20 Mev)

Broad-beam transmission	Concrete,* 147 lb/cu ft	Iron	Lead
0.1	10	3.1	1.6
0.01	19	6.2	3.5
0.001	28	9.1	5.5
0.0001	38	12.0	7.8
0.00001	47	15.3	10.2

Cobalt 60 (1.33 + 1.17 Mev per disintegration)

Broad-beam transmission	Concrete,* 147 lb/cu ft	Iron	Lead
0.1	11	3.2	1.7
0.01	19	6.0	3.3
0.001	27	8.8	4.8
0.0001	35	11.4	6.5
0.00001	43	14.6	8.1

Cesium 137 (0.66 Mev)

Broad-beam transmission	Concrete,* 147 lb/cu ft	Iron	Lead
0.1	8.5	2.6	0.85
0.01	15	4.7	1.7
0.001	22	6.8	2.5
0.0001	28	8.9	3.4
0.00001	34	11.0	4.2

Iridium 192 (gammas from 0.13 to 0.87 Mev, averaging 0.3 Mev)

Broad-beam transmission	Concrete,* 147 lb/cu ft	Iron	Lead
0.1	7		0.48
0.01	13		1.1
0.001	18.3		1.9
0.0001	24		2.6
0.00001	30		3.5

Gold 198 (0.41-, 0.68-, and 1.1- Mev gammas)

Broad-beam transmission	Concrete,* 147 lb/cu ft	Iron	Lead
0.1	6.6		0.35
0.01	12.0		0.83
0.001	17.4		1.7
0.0001	22.6		2.8
0.00001	28.0		4.3

Iodine 131 (0.08 to 0.723 Mev, predominantly 0.36 Mev)

Broad-beam transmission	Concrete,* 147 lb/cu ft	Iron	Lead
0.1	6		
0.01	12		
0.001	18		

Table 4-31. RADIATION EXPOSURE AND SHIELDING *(Continued)*

Table B. SHIELD THICKNESS VS. GAMMA-DOSE TRANSMISSION *(Continued)*

Barium 140 + Lanthanum 140 (0.030 to 2.5 Mev, averaging about 1.6 Mev)

Broad-beam transmission	Water	Aluminum	Iron	Lead	Uranium
0.1	25	9.8	3.4	1.6	0.87
0.01	44	18	6.4	3.4	2.0
0.001	64	27	9.2	5.2	3.0
0.0001	81	35	11.8	7.1	4.1
0.00001	104	44	14.8	9.0	5.2
0.000001		51			6.3

*After several mean free paths, every ten inches of concrete reduces the radiation by another factor of 10.
†From: "A Compendium of Information for Use in Controlling Radiation Emergencies", A. Brodsky and G.V. Beard, TID–8206 (rev). U.S. Atomic Energy Commission, 1960.

Alpha and beta particles. The energy required to just penetrate the 0.07 mm protective layer of skin is 7,500 kev for an alpha particle but only 70 kev for a beta particle. The alpha activity of several common sources is as follows:

Table C. ALPHA ACTIVITY IN PARTICLES PER MINUTE PER MICROGRAM

Plutonium	140,000.
Neptunium	1,519.
Natural uranium	1.5
Uranium 238	0.741
Thorium 232	0.247

The range of beta particles in air is about 12 ft per Mev. The air dose in rads per hour at one foot from a beta point source is about 200 times its value in curies (absorption neglected).

Beta-ray surface dose rates for several materials are as follows:

Table D. BETA-RAY SURFACE DOSE RATES†

Material	mrad/hr	Material	mrad/hr
Thorium, 4 to 5 years after separation	40	Uranium slug, natural	233
Tuballoy, D-38	200	UO_2, brown oxide	207
Oralloy		UF_4, green salt	179
40%	180	$UO_2(NO_3)_2 \cdot 6H_2O$	111
93%	140	UO_3, orange oxide	204
Plutonium 239		U_3O_8, black oxide	203
Nickel-coated	360	UO_2F_2	176
Uncoated	440	$Na_2U_2O_7$	167
Uranium 233			
1-month U^{232} build-up	7,000		
1-year U^{232} build-up	58,000		

†From: "A Compendium of Information for Use in Controlling Radiation Emergencies", A. Brodsky and G.V. Beard, TID–8206 (rev.), U.S. Atomic Energy Commission, 1960.

**Data from: "CRC Handbook of Radioactive Nuclides", Y. Wang, Ed., The Chemical Rubber Co., 1969.

Table 4-32. PROPERTIES OF SHIELDING MATERIALS*

Gamma-ray Mass-absorption Coefficients for Various Materials Used in Shielding

For density in lb/ft^3, multiply g/cm^3 by 62.42.

Material	Density, ρ, g/cm^3	Mass-absorption coefficient, μ, cm^{-1}			Material	Density, ρ, g/cm^3	Mass-absorption coefficient, μ, cm^{-1}		
		1 Mev	3 Mev	6 Mev			1 Mev	3 Mev	6 Mev
Air	0.001294	0.0000766	0.0000430	0.0000304	Flesh**	1	0.0699	0.0393	0.0274
Aluminum	2.7	0.166	0.0953	0.0718	Fuel oil				
Ammonia (liquid)	0.771	0.0612	0.0322	0.0221	(medium)	0.89	0.0716	0.0350	0.0239
Beryllium	1.85	0.104	0.0579	0.0392	Gasoline	0.739	0.0537	0.0299	0.0203
Beryllium carbide	1.9	0.112	0.0627	0.0429	Glass				
Beryllium oxide					Borosilicate	2.23	0.141	0.0805	0.0591
(hot-pressed blocks)	2.3	0.140	0.0789	0.0552	Lead (Hi-D)	6.4	0.439	0.257	0.257
Bismuth	9.80	0.700	0.409	0.440	Plate (avg)	2.4	0.152	0.0862	0.0629
Boral	2.53	0.153	0.0865	0.0678	Iron	7.86	0.470	0.282	0.240
Boron (amorphous)	2.45	0.144	0.0791	0.0679	Lead	11.34	0.797	0.468	0.505
Boron carbide					Lithium hydride				
(hot pressed)	2.5	0.150	0.0825	0.0675	(pressed powder)	0.70	0.0444	0.0239	0.0172
Bricks					Lucite (polymethyl				
Fire clay	2.05	0.129	0.0738	0.0543	methacrylate)	1.19	0.0816	0.0457	0.0317
Kaolin	2.1	0.132	0.0750	0.0552	Paraffin	0.89	0.0646	0.0360	0.0246
Silica	1.78	0.113	0.0646	0.0473	Rocks				
Carbon	2.25	0.143	0.0801	0.0554	Granite	2.45	0.155	0.0887	0.0654
Clay	2.2	0.130	0.0801	0.0590	Limestone	2.91	0.187	0.109	0.0824
Cements					Sandstone	2.40	0.152	0.0871	0.0641
Colemanite borated	1.95	0.128	0.0725	0.0528	Rubber				
Plain (1 Portland					Butadiene				
cement: 3 sand					copolymer	0.915	0.0662	0.0370	0.0254
mixture)	2.07	0.133	0.0760	0.0559	Natural	0.92	0.0652	0.0364	0.0248
Concretes					Neoprene	1.23	0.0813	0.0462	0.0333
Barytes	3.5	0.213	0.127	0.110	Sand	2.2	0.140	0.0825	0.0587
Barytes-boron frits	3.25	0.199	0.119	0.101	Type 347				
Barytes-limonite	3.25	0.200	0.119	0.0991	stainless steel	7.8	0.462	0.279	0.236
Barytes-lumnite-					Steel (1%C)	7.83	460	0.276	0.234
colemanite	3.1	0.189	0.112	0.0939	Uranium	18.7	1.46	0.813	0.881
Iron-Portland†	6.0	0.364	0.215	0.181	Uranium hydride	11.5	0.903	0.504	0.542
MO (ORNL mixture)	5.8	0.374	0.222	0.184	Water	1.0	0.0706	0.0396	0.0277
Portland (1 cement:					Wood				
2 sand: 4 gravel					Ash	0.51	0.0345	0.0193	0.0134
mixture)	2.2	0.141	0.0805	0.0592	Oak	0.77	0.0521	0.0293	0.0203
	2.4	0.154	0.0878	0.0646	White pine	0.67	0.0452	0.0253	0.0175

†Elemental composition, wt %: hydrogen, 1.0; oxygen, 52.9; silicon, 33.7; aluminum, 3.4; iron, 1.4; calcium, 4.4; magnesium, 0.2; carbon, 0.1; sodium, 1.6; potassium, 1.3.

**Composition, wt %: oxygen, 65.99; carbon, 18.27; hydrogen, 10.15; nitrogen, 3.05; calcium, 1.52; phosphorus, 1.02.

*From: "Reactor Handbook", 2nd ed., C.R. Tipton, Jr·, Ed., Vol. 1: Materials, Interscience Books, 1960.

Table 4-33. GAMMA-RAY SHIELD DESIGN

Effects of Ionizing Radiation

For data on radiation sources, dosimetry, and the protection of individuals, see Section 8.5. For data on the effects of radiation on engineering materials, see Tables 4-38 and 6-34.

The effects of radiation on human beings and the criteria for protection are usually discussed in terms of low-LET radiation, i.e., X-rays and gamma rays. Gamma rays of cobalt 60 and 200–250 kV X-rays have been used as the reference radiation. To obtain the dose equivalent in rems, in which all dose-limiting regulations are stated, the energy absorption in rads must be multiplied by QF, a quality factor that takes the relative biological effectiveness into account. The QF for all beta, gamma, and electron or positron radiation and X-rays is recommended as unity by the NCRP; for neutrons, protons, alpha particles, and fission fragments it is greater than unity and may be as high as 20.[a]

SHIELD DESIGN PROCEDURE

The design of gamma shields is straightforward. Given the amount of activity to be shielded, the source geometry (so that the amount of self-absorption can be estimated), and the energy of the gammas emitted, the shield thickness required to yield a given radiation dose rate at the shield surface can be calculated from the fundamental data for mass attenuation coefficients given in Tables 4-32 and 4-46. For thick shields a build-up factor (see Table 4-37) should be applied, i.e.,

$$D = \frac{SBe^{-\mu\rho t}}{K4\pi r^2}$$

where D = dose rate, rem/hr ρ = shield material density, g/cm^3
S = source strength, photon/s t = shield thickness, cm
B = build-up factor K = photons/s to give 1 rem/hr
μ = mass attenuation coefficient, cm^2/g r = shield radius for an equivalent point source, cm

Shields for seven typical radioactive isotopes can be designed, using Table 4-31, if the shielding material is concrete, iron, or lead. This table includes the build-up factor, i.e., the build-up of lower energy gammas as a consequence of degradation of the original, higher energy, gamma rays.

---

[a]Table 8-50 treats recommended radiation dose limits. See also "Basic Radiation Protection Criteria", NCRP Report No. 39, National Council on Radiation Protection and Measurements, January 1971, pages 80–86.

Table 4-34. NEUTRON ATTENUATION IN WATER*

This table gives fast neutron dose in water for a number of monoenergetic sources and a fission source. Calculations have been made by the moments method (from NDA 15C–42 and NDA 15C–60).

	Distance, r, in cm			
Source	30	60	90	120
	$4\pi r^2 D\ (r)$ Mrep cm^2/hr per N/s			
2 MeV	8×10^{-5}	1×10^{-7}		
4 MeV	8.4×10^{-4}	1×10^{-5}	7×10^{-8}	
6 MeV	2.5×10^{-3}	1.3×10^{-4}	5.5×10^{-6}	1.7×10^{-7}
8 MeV	3.6×10^{-3}	3×10^{-4}	2.1×10^{-5}	1.3×10^{-6}
10 MeV	3.9×10^{-3}	4.2×10^{-4}	4×10^{-5}	3.5×10^{-6}
14 MeV	4.4×10^{-3}	6.3×10^{-4}	7.5×10^{-5}	8×10^{-6}
Fission	2.7×10^{-4}	1×10^{-5}	4×10^{-7}	

*From: "The Attenuation of Gamma Rays and Neutrons in Reactor Shields", H. Goldstein, U.S. Atomic Energy Commission, May 1957.

REFERENCES

"Attenuation in Water of Radiation from the Bulk Shielding Reactor", ORNL–2518, Oak Ridge National Laboratory, July 1958.

"Calculation of Thermal Neutron Fluxes in Primary Shields", D.C. Anderson and K. Shure, WAPD–TM–193, November 1959

Table 4-35. FISSION REACTOR SHIELDING

Effects of Shield Thickness on Dose Rates

SHIELD DESIGN PROCEDURE

The design of even idealized reactor shields is rendered extremely complex by the generation of gamma rays by the inelastic scattering and capture of neutrons. The neutron dose at the outer surface of a shield of a given thickness is relatively insensitive to the composition of the shield if the materials are close to their theoretical density and the shield is free of voids; however, the gamma dose varies widely with the composition. Layers of heavy material for attenuating gamma rays should be placed far enough out in the shield so that they will not be a serious source of secondary gamma rays.

EXPERIMENTAL DATA ON TYPICAL SHIELDING

Extensive experiments at the Bulk Shielding Facility of the U.S. Atomic Energy Commission have yielded useful data, and reports from this source should be consulted. While many shields of complex make-up have been tested, it is possible to give useful estimates from the basic data on two common shielding materials—concrete and water. Roughly, 25.4 cm (10 in.) of concrete or 40.7 cm (16 in.) of water reduces the *gamma* radiation by a factor of ten. The attenuation per unit thickness for concrete will, of course, depend on the mix, but it remains fairly constant across a thick shield.

One set of test results at the Bulk Shielding Facility was reported by J.L. Meem and H.E. Hungerford in AEC Report No. ORNL 1147, April 1952. In these tests observations were made at distances of 20–120 cm from the surface of the reactor core. The core was equivalent to a nearly right circular cylinder of 33-cm radius, and the results were normalized to a uniform spherical source.

For a shield of *concrete*, an attenuation of 10/1 was produced for each increment of about 25 cm (10 in.) of concrete for *gamma* radiation and about 20 cm (7.9 in.) of concrete for *neutron* radiation, regardless of thickness.

For shielding by *water*, an attenuation from 10 rep/hr·W near the reactor wall to 0.001 rep/hr·W required 174 cm of water for gamma radiation but less than 60 cm of water for neutron radiation. In other terms, the mean thickness in this range for an attenuation of 10/1 by water shielding was slightly over 43 cm (17 in.) for gamma rays, but less than 15 cm (5.9 in.) for neutron radiation. In all of these tests, the outer layers of shield were less effective (per unit thickness) than those near the radiation source.

Other tests were made using a compound shield of water and 3 in. of iron near the core, plus nine 1-in. thicknesses of lead spaced from 5–75 cm from the reactor core surface. For this shield the gamma dose, in rep/hr·W, at 200 cm from the surface of the core was 3×10^{-7} rep/hr·W. The positions of the lead layers in several lead-water shields tested were close to that for minimum weight of an idealized, nearly spherical shield around a reactor having a core diameter of about 1 m. By adding 0.4% boron in the form of boric acid to the water in a lead-water shield, the generation of hydrogen-capture gammas in the water was suppressed sufficiently to reduce the gamma dose rate by a factor of over 4.

Similar data are given in other tables in this section. The general conclusion must be drawn that simple ratios for shield effectiveness, concrete vs. water or neutron vs. gamma radiation, should be used only for rough estimates.

Table 4-36.   COMPARATIVE VALUES OF SHIELDING MATERIALS

Equivalent Thickness for Various Shielding Materials

| Material | Specific gravity | Relaxation length in inches* | | Fast neutrons |
| | | Gamma rays | | |
		8 Mev	4 Mev	
Lead	11.3	0.75	0.95	3.5
Iron	7.8	1.73	1.45	2.4
Iron concrete	4.3	3.94	3.25	2.5
Aluminum	2.7	6.7	5.1	3.9
Concrete	2.3	7.1	5.5	4.7
Beryllium oxide	2.3	10.	7.1	3.5
Beryllium	1.85	12.	8.	3.6
Graphite	1.7	10.	7.5	3.6
Water	1.00	16.	12.	4.

Note the similarity of shielding materials for fast neutrons as compared with the wide differences when gamma rays are involved.

*Relaxation length is related to exponential decay and is defined as the thickness required to attenuate the radiation by a factor of $1/e = 1/2.72 = 0.368$.

Table 4-37.   GAMMA RAY BUILDUP FACTORS IN WATER AND LEAD*

| Plane isotropic source | E_0, MeV | | | | | | | |
	.5	1	2	3	4	6	8	10
B_E^{pl} **Water**								
$\mu_0 r = 1$	4.65	3.20	2.37	2.06	1.80	1.64	1.52	1.47
2	8.43	5.03	3.31	2.76	2.33	2.01	1.82	1.72
4	19.1	9.69	5.35	4.08	3.30	2.75	2.42	2.20
7	45.1	18.7	8.65	6.09	4.80	3.78	3.21	2.87
10	82.9	28.8	12.3	8.20	6.47	4.83	3.97	3.51
15	176.	50.4	18.8	11.8	8.91	6.73	5.15	4.49
B_r^{pl} **Water**								
$\mu_0 r = 1$	4.74	3.34	2.57	2.23	2.02	1.80	1.66	1.57
2	8.71	5.24	3.60	3.03	2.66	2.29	2.03	1.87
4	20.6	9.98	5.87	4.55	3.94	3.12	2.73	2.46
7	50.3	19.8	9.78	7.00	5.77	4.45	3.65	3.28
10	94.5	31.9	13.9	9.52	7.62	5.67	4.60	4.04
15	205.	56.8	21.3	13.9	10.7	7.60	6.05	5.21
B_E^{pl} **Lead**								
$\mu_0 r = 1$	1.37	1.58	1.59	1.53	1.44	1.30	1.23	1.18
2	1.54	1.88	1.94	1.86	1.78	1.54	1.41	1.31
4	1.79	2.46	2.70	2.55	2.42	2.13	1.86	1.67
7	2.12	3.22	3.82	3.89	3.74	3.44	3.00	2.60
10	2.37	3.85	4.93	5.44	5.46	5.63	5.00	4.35
15	2.61	4.59	6.83	8.19	9.43	13.0	12.5	11.6
B_r^{pl} **Lead**								
$\mu_0 r = 1$	1.38	1.61	1.67	1.61	1.49	1.37	1.28	1.22
2	1.55	1.92	2.11	2.03	1.86	1.66	1.47	1.38
4	1.80	2.52	2.91	2.86	2.64	2.38	2.05	1.84
7	2.14	3.30	4.11	4.27	4.21	4.09	3.53	3.06
10	2.43	4.07	5.37	5.97	6.26	6.95	6.20	5.48
15	2.73	5.15	7.40	8.86	10.9	16.7	17.8	16.1

*From: "The Attenuation of Gamma Rays and Neutrons in Reactor Shields", H. Goldstein, U.S. Atomic Energy Commission, May 1957.

Table 4-38.   RADIATION EFFECTS ON ENGINEERING MATERIALS

Dose Limits and Property Changes for Ionizing Radiation

For other units and terms see Table 4-2.

RADIATION TYPES AND MEASURING UNITS

Effects of radiation on materials depend on the type of radiation involved, its energy, and the duration of exposure. Changes in the material properties, such as strength of a solid or viscosity of a liquid, are dependent on the total radiation received and the type of radiation, i.e., gamma rays, neutrons, or other. Quantitative description of the radiation is actually related to the method for its measurement. By definition, dosimetry is concerned with the amount of radiation absorbed by the material in the test sample, but the measuring device uses a different material; hence, the measurement is indirect. For these reasons several units have been used for expressing the radiation dose, and direct conversions by numerical factors are not always applicable. The unit used in this table is the *rad* (rd), defined basically as 100 erg/g of the material under test.

CLASSES OF MATERIALS

Coverage in the following table is limited to several classes of common engineering materials, largely organic. A great many test results are available covering specific materials and various test conditions (see References). For biological effects and human safety requirements, see Section 8.5.

Metals and Alloys. Metals are only very slightly affected by gamma radiation, but they are damaged by energetic particles, which generate solid-state defects. When the fluence[a] is less than about 10^{17}, the radiation effects include (1) increases in strength, hardness, and electrical resistivity and (2) reductions in ductility and toughness. Metals in nuclear reactors are exposed to both high neutron and gamma dosage.

Ceramics. Gamma radiation can cause color changes in ceramics. Neutron radiation may cause the ceramic to expand, and the thermal conductivity is lowered.

[a] Refers to neutrons having more than one MeV of energy.

TYPICAL EFFECTS OF RADIATION ON MATERIALS

Class of material	Limiting dose, rd		Typical effect of radiation
	Negligible effect	Major (50%) effect	
Dielectrics			
Inorganic	10^8	10^{10}	Increased loss factor
Organic	10^7	10^9	Increased loss factor
Elastomers	10^6	10^8	Reduced strength, elasticity, durability; compression set
Fibers and textiles	10^6	10^8	Loss in tenacity and strength
Fuels (liquid)	10^8	10^9	Lowered volatility; increased gum
Glasses	10^5	10^7	Coloration; opaqueness
Lubricants (petroleum)	10^7	10^9	Increased viscosity; dark color; gas formation; lowered oxidation resistance
Paints and coatings	10^8	10^9	Cracking, blistering, peeling, porosity
Plastics[b]	10^7	10^9	Reduced strength; swelling; brittleness; gas evolution
Wood products	10^6	10^8	Loss of strength; "dry rot"

Note: Most organic materials are damaged by radiation dosage of 1 Mrd or more, regardless of the source or kind of radiation. Damage to semiconductors depends on nature and use of the semiconductor device.

Certain common materials and many assembled components are more sensitive to radiation than the above table would indicate, especially in prolonged exposure. Examples are foods, Teflon, Neoprene, explosives, capacitors (paper), rectifiers, and dry batteries.

Any application for which the exposure approaches the limits listed should be investigated with a goal of reducing the radiation or using less sensitive materials.

Effects of temperature are not considered in the above table.

[b] Some plastics are more radiation-resistant than others, e.g., PVC and polystyrene.

REFERENCES

"Effects of Radiation on Materials and Components", J.F. Kircher and R.E. Bowman, Reinhold Publishing Corp., 1964.
"Radiation Effects on Organic Materials", R.O. Bolt and J.G. Carroll, Eds., Academic Press, 1963.
"Reactor Shielding Design Manual", T. Rockwell, USAEC TID-7004, U.S. Atomic Energy Commission, March 1956.

Table 4-39.   LINEAR EXTRAPOLATION DISTANCE
FOR BLACK SPHERES: P_L APPROXIMATION OF λ_a*

A body is said to be black to neutrons if no neutron entering the body reenters the source medium (albedo zero). The blackbody is assumed to be placed in an isotropically and elastically scattering, non-absorbing, infinite, homogeneous medium. The linear extrapolation distance at the surface of a control rod is defined as the reciprocal of the logarithmic derivative of the flux evaluated at that surface.

SYMBOLS: a = radius of the sphere
P = various approximations for the linear extrapolation distance λ_a

a	P_L approximation							
	P_1	P_2	P_3	P_4	P_5	P_6	P_7	P_8
0.00	0.5774	∞	0.6395	∞	0.6537	∞	—	—
0.05	0.5774	16.7746	0.6894	8.7173	0.7499	6.0648	—	—
0.10	0.5774	8.7746	0.7306	4.7543	0.8211	3.4482	—	—
0.20	0.5774	4.7746	0.7922	2.7867	0.9116	2.1655	1.0140	1.3390
0.30	0.5774	3.4413	0.8332	2.1391	0.9585	1.7508	—	—
0.40	0.5774	2.7746	0.8598	1.8189	0.9806	1.5471	—	—
0.50	0.5774	2.3746	0.8764	1.6280	0.9886	1.4248	1.0291	1.1271
0.70	0.5774	1.9175	0.8909	1.4096	0.9842	1.2807	—	—
1.00	0.5774	1.5746	0.8891	1.2417	0.9604	1.1614	0.9874	—
1.50	0.5774	1.3079	0.8676	1.1008	0.9183	1.0507	—	—
2.00	0.5774	1.1746	0.8443	1.0222	0.8847	0.9843	0.9000	—
2.50	0.5774	1.0946	0.8246	0.9707	0.8589	0.9393	—	—
3.00	0.5774	1.0413	0.8087	0.9341	0.8389	0.9069	—	—
4.00	0.5774	0.9460	0.7853	0.8857	0.8104	0.8633	—	—
5.00	0.5774	0.9346	0.8552	0.7914	0.8356	0.8028	—	—
∞	0.5774	0.7746	0.6940	0.7297	0.7039	0.7197	0.7069	0.7159

*From: "Reactor Physics Constants", 2nd ed., Argonne National Laboratory, ANL–5800, U.S. Atomic Energy Commission, July 1963. See this source for an extensive list of references.

REFERENCE
"Linear Extrapolation Length for a Black Sphere and a Black Cylinder", B. Davison and S.A. Kushneriuk, MT-214, March 1946.

Table 4-40.   RESONANCE INTEGRALS FOR INFINITE DILUTION, BARNS*

When a resonance absorber is placed in a moderator at a concentration approaching zero (infinite dilution), its absorption per atom is unaffected by energy self-shielding between levels or by Doppler broadening. It absorbs neutrons in the slowing-down spectrum of the moderator.

SYMBOLS: (c) = capture integral   (f) = fission integral   (c−f) = capture plus fission integrals

Element	Natural abund., %	Measured		Calculated	No. of resonances	Lower energy, ev
		Activation	Absorption			
Li	—	—	28	—	—	0.4
B	—	—	280±40	—	—	0.49
N	—	—	4.8±2.4	—	—	0.49
F^{19}	100	—	2.3±0.5	—	—	0.49
Na^{23}	100	~0.24	0.27	—	—	0.4
Na^{23}	—	0.30±0.01	—	—	—	0.5
Mg	—	—	0.9	—	—	0.4
Mg	—	—	0.072^a	—	—	—
Al^{27}	100	—	$<0.18^a$	—	—	0.4
Al^{27}	100	~0.16	0.18	—	—	0.4
Si	—	—	0.5	—	—	0.4
P^{31}	100	—	<2	—	—	0.4
S	—	—	0.6	—	—	0.4
Cl	—	—	12	—	—	0.4
Cl	—	—	12.8±1.7	—	—	0.49
K	—	—	3.5±1.1	—	—	0.49
Ca	—	—	2	—	—	0.4
Sc^{45}	100	~10.7	—	—	—	0.4
Ti	—	—	3	—	—	0.4
Ti	—	—	3.8±0.9	—	—	0.49
V	—	—	4.1^a	—	—	—
V	—	—	3.3±0.8	—	—	0.49
V^{51}	99.76	~2.2	—	—	—	0.4
Cr	—	—	2.6±1.1	—	—	0.49
Mn^{55}	100	—	$15.4^{a,b}$	—	—	—
Mn^{55}	100	$15.6±0.6^c$	—	—	—	0.55
Mn^{55}	100	14.2±0.6	—	—	—	0.5
Mn^{55}	100	$14.0±0.3^d$	—	—	—	—
Mn^{55}	100	—	—	17^e	—	0.4
Fe	—	—	2.1	—	—	0.4
Fe	—	—	1.8^a	—	—	0.5
Fe	—	—	2.3±0.25	—	—	0.5
Fe	—	—	2.3±0.4	—	—	0.49
Co^{59}	100	$81±4^e$	—	—	—	0.5
Co^{59}	100	75±5	—	—	—	0.5
Co^{59}	100	72.3±5	—	—	—	0.4
Ni	—	3.7	4	—	—	0.64
Ni	—	3.13	$<3.1^a$	—	—	0.49
Ni	—	—	3.2±0.5	—	—	0.5
Cu	—	—	3.3±0.3	—	—	0.4
Cu	—	—	4	—	—	0.5
Cu	—	—	3.7±0.8	—	—	0.4
Cu^{53}	69.1	4.4	—	—	—	0.4
Cu^{53}	69.1	5.1±0.15	—	—	—	0.5
Cu^{55}	30.9	2.2	—	—	—	0.4
Cu^{55}	30.9	2.3±0.23	—	—	—	0.4
Zn	—	—	2	—	—	0.4
Zn	—	—	—	—	—	0.49
Ga	—	—	3.4±0.8	—	—	0.49
Ga^{69}	60.2	9.2	—	—	—	0.4
Ga^{71}	39.8	15	—	—	—	0.49
Ge	100	36.8	11.7±2.7	—	—	0.4
As^{75}	100	—	33	170^f	—	0.49
Se	—	—	3.5±2.9	—	—	0.4
Se^{77}	7.58	—	—	33^g	3	0.4
Se^{80}	49.82	—	—	0.9^g	3	0.49
Br	—	—	9.6±1.2	—	—	0.4
Br^{79}	50.52	147	118±14	—	—	0.4
Br^{81}	49.48	—	—	61.6^g	2	0.4

Table 4-40. RESONANCE INTEGRALS FOR INFINITE DILUTION, BARNS (Continued)

Element	Natural abund., %	Measured — Activation	Measured — Absorption	Calculated	No. of resonances	Lower energy, ev
Kr^{82}	11.56	—	—	200[g]	1	0.4
Kr^{83}	11.55	—	—	240[g]	2	0.4
Kr^{84}	56.90	—	—	3.4[g]	2	0.4
Rb	—	—	9.0±2.8	0.55[g]	4	0.49
Rb^{85}	72.15	—	—	—	—	0.4
Rb^{87}	27.85	—	—	0.17[g]	4	0.4
Sr	—	—	16	—	—	0.4
Sr	—	0.91	10.0±2.6	—	—	0.49
Y^{89}	100	—	—	0.73[g]	3	0.4
Y^{89}	100	—	—	—	—	0.4
Zr	—	—	3	—	—	0.4
Zr	—	—	0.69±0.09	0.63[a,h]	1	—
Zr	—	—	3.7±0.5	—	—	0.49
Zr^{91}	11.23	—	5.4±1.6[c]	—	—	0.55
Zr^{91}	11.23	—	—	5.6[g]	1	0.4
Nb^{93}	100	—	14[a]	16[a,i]	—	—
Mo	—	—	20[a]	22[a,i]	—	—
Mo^{95}	15.70	—	107[a,j]	107[a,i]	—	—
Mo^{96}	16.50	—	—	27[g]	—	0.4
Mo^{97}	9.45	—	—	15[g]	—	0.4
Mo^{98}	23.75	10.8±2.5	—	12.45	—	0.5
Mo^{98}	23.75	3.73±0.20	—	—	—	0.5
Mo^{100}	9.62	—	—	—	—	0.5
Tc^{99}	—	—	72[a]	—	—	—
Rh^{103}	100	656	575	1146[f]	—	0.4
Rh^{103}	100	—	—	1095[g]	5	0.4
Pd	—	—	23	—	—	0.4
Ag	—	—	>650	—	—	0.4
Ag	—	—	841[a]	761[a,i]	—	—
Ag^{107}	51.35	74	—	—	—	0.4
Ag^{109}	48.65	1160	—	—	—	0.4
Ag^{109}	48.65	—	1910[a,j]	1440[a,i]	—	0.4
Ag^{109}	—	—	3700[a]	3100[a,i]	—	—
In	—	—	—	—	—	—
In	—	—	—	3240[e]	—	0.4
In^{113}	4.23	—	—	271[e]	—	—
In^{115}	95.77	2640[k]	—	—	—	0.4
In^{115}	95.77	2630±133	—	—	—	—
In^{115}	95.77	—	—	3370[e]	—	—
Sn	—	—	8.8[a]	5.3[a,i]	—	0.49
Sn	—	—	—	—	—	0.4
Sn^{120}	32.97	—	5.7±0.7	0.82[g]	1	0.4
Sn^{124}	5.98	—	—	12[g]	1	0.49
Sb	—	—	106±13	—	—	0.5
Sb	—	—	115±12	—	—	0.4
Sb^{121}	57.25	162	—	203[g]	6	0.4
Sb^{121}	57.25	—	—	—	—	0.4
Sb^{123}	42.75	—	—	162[g]	4	0.4
Sb^{123}	42.75	—	—	—	—	0.49
Te	—	~138	~74[a,l]	>49[a,i]	1	0.4
Te	—	—	—	—	—	0.4
Te^{122}	2.46	—	50.0±6.0	56[g]	1	0.4
Te^{123}	0.87	—	—	6765[g]	3	0.4
Te^{124}	4.61	—	—	5.5[g]	1	0.4
Te^{125}	6.99	—	—	24[g]	5	0.4
Te^{126}	18.71	—	130±18	9.0[g]	1	0.4
Te^{128}	31.79	—	183[a]	10[g]	1	0.5
I^{127}	100	140	—	158[a,i]	—	—
I^{127}	100	—	—	—	—	0.4
I^{127}	100	—	—	—	—	0.6
I^{129}	—	36±4	—	173[m]	6	0.4
I^{129}	—	—	—	42[g]	—	0.4
I^{129}	—	—	—	—	—	0.4
Xe^{129}	26.44	—	—	260[g]	2	—
Xe^{131}	21.18	—	—	810[g]	3	0.6
Cs^{133}	100	—	—	394[a,i]	—	0.55
Cs^{133}	100	400±25[c]	504[a]	—	—	—
Cs^{135}	—	62±2	—	—	—	0.6
Ba	—	—	12.6±1.7	—	—	0.49

Table 4-40.   RESONANCE INTEGRALS FOR INFINITE DILUTION, BARNS (Continued)

Element	Natural abund., %	Measured		Calculated	No. of resonances	Lower energy, ev
		Activation	Absorption			
Ba^{136}	7.81	—	—	13[g]	2	0.4
La	—	—	11	13[g]	1	0.4
La^{139}	99.911	—	—	—	—	0.4
Ce	—	—	3.7±1.7	25[g]	7	0.49
Pr^{141}	100	—	—	—	—	0.4
Pr^{141}	100	15.5±3[c]	40±6	—	—	0.55
Nd	—	—	<218[a]	>181[a,h]	—	0.5
Nd^{143}	12.32	—	156[a,j]	>273[a,h]	—	—
Nd^{145}	8.29	—	—	2190[g]	9	0.4
Pm^{147}	—	—	—	—	—	—
Sm	—	—	1790±270	715[g]	13	0.49
Sm^{147}	15.07	—	—	—	—	0.4
Sm^{147}	15.07	2740±150[c]	<1385[a]	—	—	0.55
Sm^{152}	26.63	—	—	2210[g]	1	0.4
Sm^{152}	26.63	—	2960[a]	—	—	—
Eu^{151}	47.77	—	<6900[a]	7320[a,i]	—	0.49
Eu^{153}	52.23	—	1500[a]	1760[a,i]	—	0.13
Gd	—	—	67±8.0	393[g]	16	0.4
Tb^{159}	100	—	—	416[e]	—	0.5
Dy^{164}	28.18	420±50	—	—	—	0.49
Dy^{164}	28.18	482±33[d]	720±70	—	—	0.4
Lu	—	—	—	—	—	0.5
Lu^{175}	97.40	463±15[d]	2800±600	2050[a,i]	—	0.5
Lu^{176}	2.60	887±65[d]	2900[a]	—	—	—
Hf	—	—	—	750[a,i]	—	0.4
Hf	—	—	—	—	—	—
Hf^{180}	35.44	21.8	1110[a]	—	—	0.49
Ta	—	—	474±62	555[n]	—	0.4
Ta	—	—	—	353[e]	—	—
Ta^{181}	99.988	590	—	—	—	0.49
W	—	—	290±35	—	—	—
W	—	—	340[a]	—	—	—
W^{182}	26.4	—	—	584[e]	—	—
W^{183}	14.4	—	—	414[e]	—	—
W^{184}	30.6	355	—	8.6[e]	—	0.4
W^{186}	28.4	—	—	481[e]	—	—
W^{186}	28.4	—	—	—	—	0.4
W^{186}	28.4	396±59[d]	—	1630[f]	—	0.4
Re^{185}	37.07	1160	180±20	310[n]	—	0.49
Re^{187}	62.93	305	2000±490	—	—	0.49
Os	—	—	—	—	—	0.4
Ir	—	—	—	7680[f]	—	0.4
Ir^{191}	38.5	3500[k]	—	976[f]	—	0.4
Ir^{193}	61.5	1370[k]	—	—	—	—
Pt	—	—	69	114[a,l]	—	0.5
Pt	—	—	~189[a,l]	1591[a,i]	—	0.4
Au^{197}	100	—	1558[a]	—	—	0.64
Au^{197}	100	1553±40	—	—	—	0.4
Au^{197}	100	1558	—	1584[m]	—	—
Au^{197}	100	1553	—	—	—	0.5
Hg	—	—	73±5	—	—	0.49
Hg	—	—	72.4±8	—	—	0.4
Tl^{203}	29.50	125	—	—	—	0.4
Tl^{205}	70.50	0.5	—	—	—	0.5
Pb	—	—	0.1	—	—	0.4
Pb	—	—	0.13[a]	—	—	0.4
Bi^{209}	100	—	0.5	—	—	0.5
Bi^{209}	100	—	0.07[a]	—	—	—
Th^{233}	100	—	110[a,j]	108[a,i]	—	0.4
Th^{232}	100	8.5	—	—	—	—
Th^{232}	100	500±150[o]	—	96[m]	—	0.5
Th^{233}	100	400±100[o]	—	—	—	0.4
Th^{233}	—	700±200[o]	—	—	—	0.5
Pa^{233}	—	770±90	—	—	—	0.4
Pa^{233}	—	—	—	—	—	0.6

Table 4-40. RESONANCE INTEGRALS FOR INFINITE DILUTION, BARNS (Continued)

Element	Natural abund., %	Measured Activation	Measured Absorption	Calculated	No. of resonances	Lower energy, ev
U	—	—	224±40	—	—	0.49
U233	—	—	100±4 (c)	—	—	0.4
U233	—	—	1000±200 (c+f)	—	—	0.4
U233	—	—	—	813 (f)[p]	—	—
U233	—	—	—	812 (f)[p]	—	0.4
U233	—	870 (f)	—	—	—	—
U234	0.0057	—	—	700±70[o]	—	0.4
U235	0.714	271 (f)	—	—	—	0.4
U235	0.714	274±11(f)	—	—	—	0.5
U235	0.714	—	—	340 (f)[p]	—	—
U236	—	415±40[c]	—	—	—	0.55
U236	—	—	400±50[o]	—	—	0.4
U236	—	350±90	—	340±40[q]	—	0.4
U238	99.3	280±10[o]	—	—	—	0.4
U238	99.3	—	287[q]	—	—	—
U238	99.3	281±20	—	—	—	0.4
U238	99.3	—	—	279±7[q]	—	0.4
Np237	—	—	955[a,b]	—	—	—
Pu239	—	327±22 (f)	—	3600±1000 (c+f)[o]	—	0.15
Pu239	—	—	—	—	—	0.5
Pu239	—	—	—	1500±300 (c)[o]	—	0.15
Pu239	—	—	—	230 (f)[p]	—	—
Pu240	—	—	—	—	—	0.5
Pu240	—	8700±800	11,450[a,b]	8450[a,i]	—	0.4
Pu240	—	—	—	—	—	0.4
Pu240	—	9000±3000	—	8000±400	—	0.2
Pu240	—	—	10,000±2800	—	—	—
Pu240	—	—	—	8565[e]	—	0.5
Pu241	—	557±33 (f)	—	—	—	0.5
Pu242	—	1275±30	—	—	—	0.5
Am243	—	2290±50	—	—	—	0.5

[a] The resonance integral given is not significantly dependent on what cutoff is used since the material has a cross section dependence that is close to 1/v in the cutoff region.

[b] Values were deduced from measurements in the DIMPLE Maxwellian spectrum and with the GLEEP oscillator.

[c] Gold resonance integral (including 1/v) of 1534b used as standard (private communication from F. Feiner, KAPL, 1961).

[d] Private communication from J. L. Crandall, Savannah River, 1960. Work done by G. M. Jacks.

[e] Calculated by P. J. Persiani of Argonne from level parameters given in Supplement 1, Second Edition, BNL–325, (1960); includes unresolved-level corrections. The Dy^{164} value includes the bound level. The parameters for this level were obtained from R. Sher, BNL (private communication).

[f] Estimated from parameters of the first large resonance and the thermal cross section; does not include correction for unresolved levels.

[g] Calculated from level parameters in the second edition of BNL–325, or its supplement (1960). The number of levels listed is the number of resonance levels for which separate calculations were carried out. If the number of resolved levels is three or more, the resonance integral listed includes a contribution from the unresolved resonances, calculated using average resonance parameters.

[h] Calculated from parameters of the first resonance only; includes unresolved-resonance contributions.

[i] Calculated from parameters given in BNL–325; includes unresolved-level contributions.

[j] Measurements on single solutions only, and corrected for screening.

[k] A resonance near thermal leads to considerable dependence on the details of the cadmium absorber that was used.

[l] Only one sample of these materials was available and the estimated screening was large. Values listed must be treated with caution.

[m] Calculated by P. J. Persiani of Argonne from level parameters given in "Reports to the AEC Nuclear Cross Sections Advisory Group, Argonne National Laboratory, September 19–21, 1960", J. A. Harvey, WASH–1029, October 1960.

[n] Estimated from level parameters; does not include corrections for unresolved levels.

[o] Values preferred from analysis of available data.

[p] Numerical integration of the fission cross section curves.

[q] Calculated from level parameters and the thermal cross section; includes unresolved-level corrections.

REFERENCES

"Resonance Capture Integrals", R.L. Macklin and H.S. Pomerance, *Proc. 1955 Geneva Conference*, 5(Paper No. 833):96.

"Yields and Effective Cross Sections of Fission Products and Pseudo Fission Products", W.H. Walker, CRRP–913, March 1960.

*From: "Reactor Physics Constants", 2nd ed., Argonne National Laboratory, ANL–5800, U.S. Atomic Energy Commission, July 1963. See this source for an extensive list of references.

Table 4-41. RESONANCE ABSORPTION PROBABILITIES*

For Single Breit-Wigner Resonances: Homogeneous Mixtures of H and U^{238}

E_0, ev	H/U atom ratio = 1				H/U atom ratio = 10		
	NN	Mod. NN	NW	Mod. NW	NN	Mod. NN	NW

INTEGRATIONS OVER ENTIRE ENERGY RANGE

E_0, ev	NN	Mod. NN	NW	Mod. NW	NN	Mod. NN	NW
6.7	0.2104	0.07708	0.1806		0.06071	0.06215	0.06102
21.0	0.08780	0.07539	0.08401		0.02406	0.02955	0.02728
36.9	0.05625	0.11065	0.06946		0.01522	0.02654	0.02240
66.3	0.02236	0.02613	0.02595	0.0253	0.005961	0.006967	0.008183
81.3	0.006521	0.006300	0.005562	0.00627	0.001672	0.001675	0.001697
90	0.001137	0.001129	0.000968	0.00113	0.0002019	0.0002020	0.0002004
103.5	0.01390	0.02135	0.02174	0.0202	0.003691	0.004840	0.006807
117.5	0.008259	0.008361	0.008563	0.00840	0.002177	0.002248	0.002670
146	0.001782	0.001763	0.001498	0.00177	0.000419	0.0004189	0.000417
166	0.002976	0.002938	0.002632	0.00294	0.000761	0.000762	0.000796
192	0.005957	0.009141	0.01245	0.01043	0.001573	0.002006	0.003768
212	0.004684	0.005283	0.007058		0.001257	0.001362	0.002163
242	0.003842	0.004239	0.005788		0.001010	0.001077	0.001765
258	0.000900	0.000895	0.000763		0.000206	0.0002064	0.000207
278	0.002913	0.003041	0.003842		0.000763	0.000787	0.001170
297	0.002638	0.002743	0.003479		0.000690	0.000710	0.001037
368	0.002234	0.002574	0.004520		0.000585	0.000635	0.001315
418	0.001717	0.001827	0.002867		0.000446	0.000463	0.000841

INTEGRATIONS OVER ONE PRACTICAL WIDTH AROUND E_0

E_0, ev	NN	Mod. NN	NW	Mod. NW	NN	Mod. NN	NW
6.7	0.1775	0.07416	0.1468		0.05676	0.05799	0.05706
21.0	0.07281	0.06417	0.06892		0.02242	0.02715	0.02565
36.9	0.04651	0.08037	0.05974		0.01418	0.02371	0.02137
66.3	0.01840	0.02090	0.02196		0.005548	0.006412	0.007770
81.3	0.005333	0.005185	0.004399		0.001551	0.001553	0.001577
90	0.000901	0.000896	0.000734		0.000178	0.000178	0.000176
103.5	0.01142	0.01620	0.01924		0.003435	0.004419	0.006551
117.5	0.006783	0.006851	0.007079		0.002024	0.002086	0.002518
146	0.001454	0.001442	0.001172		0.000385	0.000385	0.000383
166	0.002439	0.002414	0.002097		0.000705	0.000706	0.000741
192	0.004893	0.006928	0.01137		0.001464	0.001834	0.003658
212	0.003846	0.004244	0.006206		0.001169	0.001259	0.002077
242	0.003154	0.003419	0.005088		0.000939	0.000997	0.001694
258	0.000733	0.000730	0.000597		0.000189	0.000189	0.000190
278	0.002391	0.002477	0.003315		0.000709	0.000730	0.001116
297	0.002165	0.002235	0.002999		0.000671	0.000690	0.001008
368	0.001833	0.002058	0.004110		0.000544	0.000587	0.001274
418	0.001409	0.001482	0.002552		0.000426	0.000442	0.000832

Note: $\Gamma\gamma = 0.025$ ev was assumed for all resonances. No corrections were made for Doppler broadening or for interference between resonance scattering and potential scattering.

*From: "Reactor Physics Constants", 2nd ed., Argonne National Laboratory, ANL–5800, U.S. Atomic Energy Commission, July 1963. See this source for a complete list of references.

REFERENCES

"Resonance Absorption in Homogeneous Mixtures", K.T. Spinney, *J. Nuclear Energy*, 6(1 and 2): 53, 1957.

"Some Refinements in the Calculation of Resonance Integrals", J. Chernick and R. Vernon, *Nuclear Sci. and Eng.*, 4(5): 649–672, November 1958.

Table 4-42. EQUILIBRIUM FISSION PRODUCT GAMMA-RAY SPECTRA*

The equilibrium condition assumes that the reactor has been operating for a sufficiently long time, so that fission product activities are essentially at their saturation values.

Energy group	Spectrum A				Spectrum B		
	Energy range, Mev	Effective energy, Mev	M(E)		Effective energy, Mev	M(E)	
			Mev[a] per fission	Mev/sec watt[b] of reactor power		Mev[a] per fission	Mev/sec watt[b] of reactor power
1	0.1–0.4	0.4	0.645	2.0×10^{10}	1	5.16	1.6×10^{11}
2	0.4–0.9	0.8	3.87	1.2×10^{11}			
3	0.9–1.35	1.3	0.645	2.0×10^{10}	2	1.737	5.38×10^{10}
4	1.35–1.8	1.7	1.06	3.3×10^{10}			
5	1.8–2.2	2.18	0.677	2.1×10^{10}	3	0.322	1.0×10^{10}
6	2.2–2.6	2.5	0.290	9.0×10^{9}			
7	2.6	2.8	0.032	1.0×10^{9}			
Total			7.219	2.24×10^{11}			

[a]3.1×10^{10} fissions/watt sec assumed.
[b]Data from: "Reactor Shielding Design Manual", T. Rockwell III, Ed., TID–7004, March 1956, Table 3-5.

*From: "Reactor Physics Constants", 2nd ed., Argonne National Laboratory, ANL–5800, U.S. Atomic Energy Commission, July 1963; gives an extensive list of references.

Table 4-43. PROMPT FISSION GAMMA-RAY SPECTRA*

Spectrum A[a]			Spectrum B		Spectrum C	
E, Mev	N(E), γ/fission	M(E), Mev/fission	E, Mev	M(E), Mev/fission	E, Mev	M(E), Mev/fission
0.5	3.1	1.55	—	—	—	—
1.0	1.9	1.90	1.0	3.450	1.0	3.451
1.5	0.84	1.26	—	—	—	—
2.0	0.55	1.10	2.0	2.360	2.0	3.085
2.5	0.29	0.725	—	—	—	—
3.0	0.15	0.450	3.0	1.175	—	—
3.5	0.062	0.217	—	—	—	—
4.0	0.065	0.260	4.0	0.477	4.0	1.035
4.5	0.024	0.108	—	—	—	—
5.0	0.019	0.095	5.0	0.203	—	—
5.5	0.017	0.094	—	—	—	—
6.0	0.007	0.042	6.0	0.136	6.0	0.256
6.5	0.004	0.026	7.0	0.026	—	—
—	7.028	7.827	—	7.827	—	7.827

Note: In addition, Bertini, et al. ("Basic Gamma-Ray Data for ART Heat Deposition Calculations", H.W. Bertini, et al., ORNL–2113, October 3, 1956) estimate the effective prompt capture gamma-ray source from U^{235} in a thermal reactor to be 1.18 Mev per fission, based on a value of 0.184 for α of U^{235}.

[a]Reproduced from: "Reactor Shielding Design Manual", T. Rockwell III, Ed., TID–7004, March 1956, Table 3.2 (from data in "Prompt Fission Product Gamma Rays from Uranium-235", R.L. Gamble, Reactor Shielding Information Meeting, May 12–13, 1955, Fort Belvoir, Virginia, WASH–292, Pt. 3, 28, September 1955).

*From: "Reactor Physics Constants", 2nd ed., Argonne National Laboratory, ANL–5800, U.S. Atomic Energy Commission, July 1963; gives an extensive list of references.

Table 4-44.   TABULATION OF GAMMA-RAY SPECTRA FROM THERMAL-NEUTRON CAPTURE*

Target nucleus	Photons/100 captures							Highest energy gamma ray, Mev	Average[a] number of photons per capture
	0–1 Mev	1–2 Mev	2–3 Mev	3–5 Mev	5–7 Mev	7–9 Mev	>9 Mev		
1 H	0	0	100	0	0	0	0	2.230	—
1 D	0	0	0	0	100	0	0	6.244	—
2 He	—	—	—	—	—	—	—	—	—
3 Li	—	—	—	—	~40	~60	0	7.26	—
4 Be	0	0	0	54	73	0	0	6.80	—
5 B^{10}	0	0	0	>110	>28	≥6	0.8	11.43	—
6 C	0	0	0	100	0	0	0	4.95	1.3
7 N^{14}	—	—	—	>54	≥11	15	12	10.833	—
8 O	—	—	—	—	—	—	—	—	—
9 F	—	>100[b]	—		≥111	0	0	6.600	—
10 Ne	—	—	—	—	—	—	—	—	—
11 Na	>96	127	187	70	31	0	0	6.41	<2
12 Mg			>28	>72	10	3.3	0.57	11.089	—
13 Al	>236	195[c]	69	62	19	19	0	7.724	~2
14 Si	>100	63	30	89	11	4.1	0.1	10.59	—
15 P	>290	97	55	98	27	7.2	0	7.94	—
16 S	>70	32	72	70	44	6.5	0	8.64	—
17 Cl	>49	85	41	47	55	24	0	8.55	3.1
18 A	—	—	—	—	—	—	—	—	—
19 K	>100	81	57	106	37	4.7	0	9.36	—
20 Ca	>14	191	77	83	64	1.8	0	7.83	—
21 Sc	>52	59	38	65	29	12	0	8.85	—
22 Ti	>54	160	16	24	78	1.3	0.2	10.47	—
23 V	>83	132[d]	11.4	21	67	16	0	7.98	2.5
24 Cr	>85	41	21	12	23	39	6.4	9.716	>2
25 Mn	>125[e]	91[f]	60[g]	50	34	17	0	7.261	2.6
26 Fe	>75	60	27	23	25	38	2.1	10.16	1.7
27 Co	>61	26	17	42	52	8.5	0	7.486	—
28 Ni	>84	40	23	23	34	62	0.8	8.997	—
29 Cu	>68	47	26	30	27	43	0	7.914	2.6
30 Zn	>156	93	67	48	29	16	1	9.51	—
31 Ga	—	—	—	57	34	0.2	0	7.73	—
32 Ge	—	—	—	—	—	—	—	—	—
33 As	—	—	—	47	22	1	0	7.30	2.7
34 Se	—	—	—	65	27	9	1.7	10.483	—
35 Br	—	—	—	79	41	6.4	0	7.879	—
36 Kr	—	—	—	—	—	—	—	—	—
37 Rb	—	—	—	—	—	—	—	—	—
38 Sr	—	—	—	69	51	14	0.1	9.22	—
39 Y	>71	23	6	50	59	0	0	6.850	—
40 Zr	—	—		113	35	4	0	8.66	—
41 Nb	—	—		54	14	0.4	0	7.19	2.6
42 Mo	>137	>18	—	84	26	3	0.03	9.15	—
43 Tc	—	—	—	—	—	—	—	—	—
44 Ru	—	—	—	—	—	—	—	—	—
45 Rh	>91	99	61	38	10	0	0	6.792	—
46 Pd	—	—	—	—	—	—	—	—	—
47 Ag	>92	87	64	70	17	0.5	0	7.27	2.9
48 Cd	>135	92	96	73	17	1	0.1	9.046	4.1
49 In	>102[h]	197[i]	78[j]	36	4	0	0	5.86	3.3
50 Sn	>216	153	67	139	33	4	0.4	9.35	—
51 Sb	150	99	58	36	12	0	0	6.80	—
52 Te	>58	—	—	—	—	—	—	—	—
53 I	>30	—	—	97	22	0	0	6.71	—
54 Xe	—	—	—	—	—	—	—	—	—

Table 4-44. TABULATION OF GAMMA-RAY SPECTRA FROM THERMAL-NEUTRON CAPTURE *(Continued)*

Target nucleus	Photons/100 captures							Highest energy gamma ray, Mev	Average[a] number of photons per capture
	0–1 Mev	1–2 Mev	2–3 Mev	3–5 Mev	5–7 Mev	7–9 Mev	>9 Mev		
55 Cs	>46	—	—	61	25	0	0	6.702	—
56 Ba	—	—	—	75	14	1.4	0.1	9.23	—
57 La	>21.7	8.2	—	>35	>12.5	0	0	5.045	—
58 Ce	—	—	—	—	—	—	—	—	—
59 Pr	—	—	—	34	8	0	0	5.83	—
60 Nd	>105	—	—	—	—	—	—	—	—
61 Pm	—	—	—	—	—	—	—	—	—
62 Sm	>167	150	109	45	5	1	0	7.89	5.6
63 Eu	106	153	109	56	6.5	0	0	6.05	—
64 Gd	194	117	100	23	34	0.3	0	7.33	3.9
65 Tb	>7	—	—	—	—	—	—	—	—
66 Dy	90	102	106	43	10	0	0	5.87	—
67 Ho	—	98	77	49	8	0	0	6.1	—
68 Er	225	145	133	103	14	0	0	6.680	—
69 Tm	—	91	73	55	10	0	0	6.5	—
70 Yb	—	—	—	—	—	—	—	—	—
71 Lu	—	—	—	—	—	—	—	—	—
72 Hf	>137	137	85	52	12	0.5	0	7.62	—
73 Ta	>137	99	66	55	5	0	0	6.04	—
74 W	>68	82	59	53	15	0.5	0	7.42	—
75 Re	124	88	62	51	10.5	0	0	6.14	—
76 Os	—	—	—	—	—	—	—	—	—
77 Ir	98	85	58	51	19.6	0	0	6.088	—
78 Pt	>109	92	64	45	15	1	0	7.920	—
79 Au	>100	69	33	68	38	0.1	0	6.494	3.5
80 Hg	>94	122	55	86	41	0	0	6.446	3.3
81 Tl	—	—	—	76	62	0	0	6.54	—
82 Pb	0	0	0	0	7	93	0	7.38	—
83 Bi	0	0	0	100	0	0	0	4.17	—
84 Po	—	—	—	—	—	—	—	—	—
85 At	—	—	—	—	—	—	—	—	—
86 Rn	—	—	—	—	—	—	—	—	—
87 Fr	—	—	—	—	—	—	—	—	—
88 Ra	—	—	—	—	—	—	—	—	—
89 Ac	—	—	—	—	—	—	—	—	—
90 Th	>118	140	64	17	0	0		4.92	—
91 Pa	—	—	—	—	—	—	—	—	—
92 U[238]	254	178	91	34	0	0		4.062	—
93 Np	—	—	—	—	—	—	—	—	—
94 Pu	—	—	—	—	—	—	—	—	—

[a] All values in this column are taken from C. O. Muehlhause, *Phys. Rev.* 79:277, 1950. They may therefore differ from the multiplicities deduced from the spectra directly.

[b] Includes a 100%, 1.63-Mev gamma ray in Ne^{20} following 11s decay of F^{20}.

[c] Includes a 100%, 1.78-Mev gamma ray in Si^{28} following 2.3m decay of Al^{28}.

[d] Includes a 99.75%, 1.43-Mev gamma ray in Cr^{52} following 3.8m decay of V^{52}.

[e] Includes a 99%, 0.85-Mev gamma ray in Fe^{56} following 2.6h decay of Mn^{56}.

[f] Includes a 23%, 1.80-Mev gamma ray in Fe^{56} following 2.6h decay of Mn^{56}.

[g] Includes a 15%, 2.13-Mev gamma ray in Fe^{56} following 2.6h decay of Mn^{56}.

[h] Includes a 5%, 0.41-Mev and an 8%, 0.82-Mev gamma ray in Sn^{116} following 54m decay of In^{116}.

[i] Includes a 36%, 1.09-Mev and a 63%, 1.29-Mev gamma ray in Sn^{116} following 54m decay of In^{116}.

[j] Includes an 11%, 2.12-Mev gamma ray in Sn^{116} following 54m decay of In^{116}.

*From: "Reactor Physics Constants", 2nd ed., Argonne National Laboratory, ANL–5800, U.S. Atomic Energy Commission, July 1963; see Table 8–3, p. 631, for an extensive list of references.

REFERENCES

"Compilation of Thermal Neutron Capture γ-Rays", G.A. Bartholomew and L.A. Miggs, CRGP–784, AECL–669, 1958.

"Atlas of γ-Ray Spectra from Radiative Capture of Thermal Neutrons", L.V. Groshev, et al., Atomizdat, Moscow, 1958 (translation by J.B. Sykes, Pergamon Press, New York, 1959).

Table 4-45. GAMMA-RAY ACTIVITY DUE TO THERMAL NEUTRON CAPTURE**

Following are activation data for isotopes produced by capture of thermal neutrons. Gamma-ray activity data are given for both primary and possible secondary products of neutron capture.

Radioactive isotope	Half-life	E_γ, Mev	Yield-% per disintegration	Parent isotope	Isotopic-% or half-life	Activation cross section, barn
B^{12}	0.027 sec	~4.5	~4	B^{11}	81.2	<50 mb
N^{16}	7.4 sec	6.1 / 7.1	55 / 20	N^{15}	0.37	24±8 µb
O^{19}	29.4 sec	1.4	70	O^{18}	0.20	0.21±0.04 mb
F^{20}	10.7 sec	1.63	100	F^{19}	100	9±2 mb
Ne^{23}	40.2 sec	0.44 / 1.65	29 / 1	Ne^{22}	8.8	36±15 mb
Na^{24}	15.1 hr	2.76 / 1.38 / 4.14	99.96 / 99.96 / 0.04	Na^{23}	100	0.53±0.02
Mg^{27}	9.45 min	1.015 / 0.181 / 0.834	29 / 1 / 70	Mg^{26}	11.3	26±2 mb
Al^{28}	2.3 min	1.78	100	Al^{27}	100	0.21±0.02
Si^{31}	2.65 hr	1.26	0.07	Si^{30}	3.12	0.11±0.01
S^{37}	5.04 min	3.12	90	S^{36}	0.016	0.14±0.04
Cl^{38*}	1.0 sec	0.66 / 2.15	100 / 47	Cl^{37}	24.5	~5 mb
Cl^{38}	37.5 min	1.65	31	Cl^{37} / Cl^{38*}	24.5 / 1 sec	0.56±0.12
A^{41}	1.83 hr	1.37	99.3	A^{40}	99.6	0.53±0.02
K^{42}	12.46 hr	1.53	18	K^{41}	6.91	1.5±0.11
Ca^{47}	4.9 days	1.31	71	Ca^{46}	0.0033	0.25±0.10
Ca^{49}	8.8 min	0.82 / 0.49 / 3.10 / 4.05	5 / 5 / 89.78 / 9.88	Ca^{48}	0.185	1.1±0.1
Sc^{46*}	19.5 sec	0.14	100	Sc^{45}	100	10±4
Sc^{46}	85 days	1.12 / 0.89	99.9 / 100	Sc^{45} / Sc^{46*}	100 / 19.5 sec	12±4
Sc^{47}	3.43 days	0.16	74	Ca^{47}	4.9 days	<50 mb
Ti^{51}	5.79 min	0.605 / 0.928 / 0.323	1.4 / 4.2 / 95.8	Ti^{50}	5.25	0.14±0.03
V^{52}	3.76 min	1.44	100	V^{51}	99.8	4.5±0.9
Cr^{51}	27.8 days	0.32	10	Cr^{50}	4.31	13.5±1.4
Mn^{56}	2.58 hr	2.98 / 2.13 / 2.65 / 1.81 / 0.845	0.4 / 15 / 1.8 / 24 / 99	Mn^{55}	100	13.4±0.3
Fe^{59}	45.1 days	0.191 / 1.29 / 1.10	3 / 43 / 57	Fe^{58}	0.31	0.9±0.2
Co^{60*}	10.5 min	0.058	99.7	Co^{59}	100	16±3
Co^{60}	5.3 yr	1.17 / 1.33	99.9 / 100	Co^{59} / Co^{60*}	100 / 10.5 min (99.7%)	20±3
Co^{61}	1.65 hr	0.07	100	Co^{60}	5.28 yr	6±2
Ni^{65}	2.56 hr	0.37 / 1.49 / 1.12	4.1 / 24.9 / 18.1	Ni^{64}	1.16	1.6±0.2
Cu^{64}	12.87 hr	1.34	0.4	Cu^{63}	69	3.9±0.8
Cu^{66}	5.15 min	1.05	9	Cu^{65}	31	1.8±0.4
Zn^{65}	245 days	1.11	45.5	Zn^{64}	48.9	0.5±0.1
Zn^{69*}	13.8 hr	0.439	100	Zn^{68}	18.6	97±10 mb
As^{77}	38.8 hr	0.524	0.52	Ge^{77*}	52 sec (85.8%)	

**From: "Reactor Physics Constants", 2nd ed., Argonne National Laboratory, ANL-5800, U.S. Atomic Energy Commission, July 1963.

Table 4-45. GAMMA-RAY ACTIVITY DUE TO THERMAL NEUTRON CAPTURE (*Continued*)

Radio-active isotope	Half-life	E_γ, Mev	Yield-% per disintegration	Parent isotope	Isotopic-% or half-life	Activation cross section, barn
As^{77}	38.8 hr	0.246	1.51	Ge^{77}	12 hr	
Se^{75}	119.9 days	0.121^a 0.265 0.280 0.136 0.402	24.7 58.7 17.8 59.9 17.8	Se^{74}	0.87	26±6
Se^{77*}	17.5 sec	0.162	100	Se^{76}	9	7±3
Se^{79*}	3.9 min	0.096	100	Se^{78}	23.5	0.4
Zn^{71*}	3 hr	0.38 0.49 0.61	100 100 100	Zn^{70}	0.62	?
Zn^{71}	2.2 min	0.51	100	Zn^{70}	0.62	85±20 mb
Ga^{70}	21 min	1.04	0.8	Ga^{69}	60.2	1.4±0.3
Ga^{72}	14.2 hr	0.69^a 0.89 1.86 2.51	28.2 105 30.3 53.1	Ga^{71}	39.8	5.0±1.0
Ge^{75*}	48 sec	0.139	100	Ge^{74}	36.7	?
Ge^{75}	82 min	0.63^a 0.264	0.80 11.15	Ge^{74} Ge^{75*}	36.7 48 sec	0.45±0.08
Ge^{77*}	52 sec	0.159	14.2	Ge^{76}	7.7	30±20 mb
Ge^{77}	12 hr	0.56^a 0.79 1.36 2.0	21.6 4.3 4 1	Ge^{76} Ge^{77*}	7.7 52 sec (14.2%)	0.2±0.1
As^{76}	26.8 hr	2.05 1.40 1.20 0.64 0.55	1 2 9.22 8.78 39.78	As^{75}	100	4.2±0.8

Radio-active isotope	Half-life	E_γ, Mev	Yield-% per disintegration	Parent isotope	Isotopic-% or half-life	Activation cross section, barn
Se^{81*}	56.8 min	0.103	100	Se^{80}	49.8	30±10 mb
Se^{83}	25 min	0.176 0.950	100 100	Se^{82}	9.2	4±2 mb
Br^{83}	2.33 hr	0.046	20	Se^{83*} Se^{83}	67 sec 25 min	
Kr^{83*}	1.88 hr	0.032 0.009	100 100	Br^{83}	2.33 hr	
Br^{80*}	4.5 hr	0.048	100	Br^{79}	50.54	2.9±0.5
Br^{80}	18.5 min	0.036 0.62	100 14	Br^{79} Br^{80*}	50.54 4.5 hr	8.5±1.4
Br^{82}	35.9 hr	0.688^a 0.817 1.469	152.4 108 69.8	Br^{81}	49.5	3.5±0.5
Kr^{79}	34.5 hr	0.217^a 0.398 0.833	34.4 12.7 5.84	Kr^{78}	0.35	2.0±0.5
Kr^{85*}	4.4 hr	0.305 0.150	22 78	Kr^{84}	57	0.10±0.03
Kr^{87}	78 min	2.57 2.05 0.85 0.403	21.9 3.57 13.57 86.9	Kr^{86}	17.4	60±20 mb
Rb^{86*}	1.02 min	0.56	100	Rb^{85}	72.15	?
Rb^{86}	18.6 days	1.08	10	Rb^{85} Rb^{86*}	72.15 1.02 min	0.72±0.15
Rb^{88}	17.8 min	2.68 0.91 1.85 4.89	6.1 13.6 24 Small	Rb^{87}	27.85	0.12±0.03

Table 4-45.   GAMMA-RAY ACTIVITY DUE TO THERMAL NEUTRON CAPTURE (*Continued*)

Radioactive isotope / Half-life	E_γ, Mev	Yield-% per disintegration	Parent isotope	Isotopic-% or half-life	Activation cross section, barn	Radioactive isotope / Half-life	E_γ, Mev	Yield-% per disintegration	Parent isotope	Isotopic-% or half-life	Activation cross section, barn
Sr^{85*}, 70 min	0.233 0.007 0.225	1.3 84.7 84.7	Sr^{84}	0.55	?	Mo^{101}, 14.6 min	0.083 0.896 2.08	93.5 3.5 8.5			
Sr^{85}, 64 days	0.150 0.513	14 100	Sr^{84} Sr^{85*}	0.55 70 min (86%)	10±0.3	Tc^{99*}, 6.04 hr	0.142 0.002 0.140	1 99 99	Mo^{99}	67 hr (94%)	
Sr^{87*}, 2.8 hr	0.388	100	Sr^{86}	9.87	1.3±0.4	Tc^{101}, 14 min	0.307[a] 0.939 0.545 0.130	87 3 10.4 3.5	Mo^{101}	14.6 min	
Zr^{95}, 63.3 days	0.754 0.722	54 43	Zr^{94}	17.4	0.09±0.05	Ru^{97}, 2.8 days	0.57 0.325 0.216	1 3.5 95.5	Ru^{96}	5.5	10±4 mb
Zr^{97}, 17 hr	1.15 1.72 0.58	80 20 80	Zr^{96}	2.8	0.10±0.05	Tc^{97*}, 92 days	0.099 0.090	100 100	Ru^{97}	2.8 days	
Nb^{94*}, 6.6 min	0.0414 0.90	99.9 ~0.1	Nb^{93}	100	1.0±0.5	Ru^{103}, 39.7 days	0.610 0.495 0.055	6.45 90.5 1.09	Ru^{102}	31.5	1.2±0.3
Nb^{95*}, 84 hr	0.235	100	Zr^{95}	63.3 days (3%)		Ru^{105}, 4.5 hr	0.730	100	Ru^{104}	18.7	0.7±0.2
Nb^{95}, 35 days	0.770	99	Nb^{95*} Zr^{95}	84 hr 63.3 days (97%)		Rh^{103*}, 54 min	0.040	100	Ru^{103} Pd^{103}	41 days 17 days (94%)	
Nb^{97*}, 60 sec	0.747	100	Zr^{97}	17 hr		Rh^{105*}, 40 sec	0.130	100	Ru^{105}	4.5 hr	
Nb^{97}, 74 min	0.665	100	Nb^{97*}	60 sec		Rh^{105}, 36.5 hr	0.310	30	Rh^{105*}	40 sec	
Mo^{93*}, 6.8 hr	0.26 0.68 1.48	100 100 100	Mo^{92}	15.86	<6 mb	Rh^{104*}, 4.4 min	0.051	100	Rh^{103}	100	12±2
Mo^{99}, 67 hr	0.780[a] 0.740 0.377 0.041 0.181	8 6 1 5 6	Mo^{98}	23.75	0.45±0.10	Rh^{104}, 42 sec	1.53 1.24 0.556	0.0065 0.11 1.98	Rh^{104*} Rh^{103}	4.4 min 100	140±30
Mo^{101}, 14.6 min	0.515[a]	96	Mo^{100}	9.62	0.50±0.05	Pd^{103}, 17 days	0.498 0.362 0.298	4 2.84 2.05	Pd^{102}	0.96	4.8±1.5

Table 4-45. GAMMA-RAY ACTIVITY DUE TO THERMAL NEUTRON CAPTURE *(Continued)*

Radioactive isotope	Half-life	E_γ, Mev	Yield-% per disintegration	Parent isotope	Isotopic-% or half-life	Activation cross section, barn
Pd^{109*}	4.8 min	0.17	100	Pd^{108}	26.71	?
Ag^{109*}	39.2 sec	0.0875	100	Pd^{109} Cd^{109}	13.6 hr 1.30 yr	
Pd^{111}	22 min	0.73 0.65 0.56	<1 <1 <1	Pd^{110}	11.81	0.3±0.1
Ag^{108}	2.3 min	0.60 0.62 0.43	0.22 0.8 0.28	Ag^{107}	51.35	44±9
Ag^{110*}	270 days	0.883[a] 0.945 1.382 1.519	203 48 40 8	Ag^{109}	48.65	2.8±0.5
Ag^{111}	7.5 days	0.340 0.243	8 1	Pd^{111}	22 min	
Cd^{107}	6.7 hr	0.846	<1	Cd^{106}	1.21	1±0.5
Ag^{107*}	44.3 sec	0.094	100	Cd^{107}	6.7 hr	
Cd^{111*}	48.6 min	0.148 0.247	100 100	Cd^{110}	12.4	0.2±0.1
Cd^{113*}	5.1 yr	0.265	0.1	Cd^{112}	24.1	30±15 mb
Cd^{115*}	43 days	1.30 0.935	1 2	Cd^{114}	28.86	0.14±0.03
Cd^{115}	53 hr	0.033 0.523 0.49 0.26	1 23.8 12.4 2.32	Cd^{114}	28.86	1.1±0.3
In^{115*}	4.5 hr	0.335	95	Cd^{115}	53.0 hr	1.5±0.3
Cd^{117*}	2.9 hr	0.84 1.27 1.55 0.425	61 30 9 61	Cd^{116}	7.58	

Radioactive isotope	Half-life	E_γ, Mev	Yield-% per disintegration	Parent isotope	Isotopic-% or half-life	Activation cross section, barn
Cd^{117*}	2.9 hr	0.281	91	Cd^{117*}	2.9 hr	
In^{117*}	1.9 hr	0.312 0.160	22 23	Cd^{117}	50 min	
In^{117}	66 min	0.562 0.712 0.160	91 9 91	In^{117*}	1.9 hr (22%)	?
Cd^{117}	50 min	0.425 0.281	100 100	Cd^{116}	7.58	
Sn^{117*}	14 days	0.562 0.160	100 100	In^{117}	66 min	56±12
In^{114*}	50 days	0.192 0.722 0.556	96.5 3.5 3.5	In^{113}	4.2	
In^{114}	72 sec	1.300	0.09	In^{113}	4.2	2±0.6
In^{116*}	54.2 min	0.137[a] 1.49 0.40 1.27 2.09	3 21 25 129 25	In^{115}	95.8	145±15
Sn^{113}	112 days	0.400	100	Sn^{112}	0.95	1.3±0.3
In^{113*}	1.75 hr	0.258 0.393 0.135	11 89 11	Sn^{113}	112 days	
Sn^{117*}	14 days	0.162 0.320 0.159	99 1 99	Sn^{116}	14.24	6±2 mb
Sn^{119*}	275 days	0.065 0.024	100 100	Sn^{118}	24.01	10±6 mb
Sn^{123}	40 min	0.153	100	Sn^{122}	4.71	0.16±0.04
Sn^{125}	9.5 min	1.39 0.326	1.9 99.7	Sn^{124}	5.98	0.2±0.1

Table 4-45. GAMMA-RAY ACTIVITY DUE TO THERMAL NEUTRON CAPTURE *(Continued)*

Radioactive isotope	Half-life	E_γ, Mev	Yield-% per disintegration	Parent isotope	Isotopic-% or half-life	Activation cross section, barn
Sb^{125}	2.4 yr	0.637[a] 0.175 0.465 0.0354	22 11 60 11	Sn^{125} Sn^{125}	9.5 min 9.5 days	
Te^{125}	58 days	0.110 0.0354	100 100	Sb^{125}	2.4 yr	
Sn^{125}	9.5 days	1.97 0.81 1.07	1.2 1 3.75	Sn^{124}	5.98	4 ± 2 mb
Sb^{122*}	3.5 min	0.0753 0.0607	100 100	Sb^{121}	57.25	?
Sb^{122}	2.8 days	1.137 0.686 1.258 0.566	0.73 3.4 0.66 66.3	Sb^{121} Sb^{122*}	57.25 3.5 min	6.8 ± 1.5
Sb^{124*}	21 min	0.0185	100	Sb^{123}	42.75	30 ± 15 mb
Sb^{124*}	1.3 min	0.012	100	Sb^{123}	42.75	30 ± 15 mb
Sb^{124}	60.9 days	1.37[a] 2.09 0.967 0.644 0.725	3.75 49.2 2.16 108.6 16	Sb^{123} Sb^{124*} Sb^{124*}	42.75 21 min 1.3 min	2.5 ± 0.5
Te^{121*}	154 days	0.0818 0.214	100 100	Te^{120}	0.089	?
Te^{121}	17 days	0.575 0.506	87 13	Te^{121*}	154 days	
Te^{123*}	104 days	0.0887 0.159	100 100	Te^{122}	2.46	1.1 ± 0.5
Te^{125*}	58 days	0.110 0.0354	100 100	Te^{124}	4.61	5 ± 3
Te^{127*}	105 days	0.089 0.0585	98 1.5	Te^{126}	18.71	90 ± 20 mb
Te^{127}	9.35 hr	0.418 0.370 0.170	Small Small Small	Te^{126} Te^{127*}	18.71 105 days (98%)	0.8 ± 0.2
Te^{129*}	33 days	0.106	100	Te^{128}	31.8	15 ± 5
Te^{129}	74 min	1.12 0.21 0.72 0.475 0.027	10.4 1.7 2 17.1 98	Te^{128} Te^{129*}	31.8 33 days	0.13 ± 0.03
Te^{131*}	30 hr	0.180[a] 0.239 1.12 0.099 0.446	21.7 70 90 84 50	Te^{130}	34.5	<8 mb
Te^{131}	24.8 min	0.773 0.446 0.147 0.099 0.051	5 45 60 40 40	Te^{130} Te^{131*}	34.5 30 hr (21.7%)	0.22 ± 0.05
I^{128}	24.98 min	0.540 0.455 0.990	1.8 17.3 0.2	I^{127}	100	5.5 ± 0.5
Xe^{125}	18 hr	0.243 0.056 0.187	95 5 5	Xe^{124}	0.096	?
I^{125}	60 days	0.035	100	Xe^{125}	18 hr	
Xe^{127}	36.4 days	0.368 0.170	10 85	Xe^{126}	0.09	?
Xe^{129*}	8 days	0.200 0.145 0.056 0.0400	85 2 4 100	Xe^{128}	1.919	?

Table 4-45. GAMMA-RAY ACTIVITY DUE TO THERMAL NEUTRON CAPTURE (Continued)

Radioactive isotope	Half-life	E_γ, Mev	Yield-% per disintegration	Parent isotope	Isotopic-% or half-life	Activation cross section, barn
Xe129*	8 days	0.196	100			
Xe133*	2.3 days	0.232	100	Xe132	26.89	?
Xe133	5.27 days	0.081	100	Xe132	26.89	0.2±0.1
				Xe133*	2.3 days	
Xe135*	15.6 min	0.52	100	Xe134	10.44	?
Xe135	9.2 hr	0.370	2	Xe134	10.44	0.2±0.1
		0.620	3	Xe135*	15.6 min	
		0.250	97			
Cs134*	3.2 hr	0.137	0.8	Cs133	100	17±4 mb
		0.0105	98.2			
		0.127	98.2			
Cs134	2.3 yr	0.200^a	13	Cs133	100	26±5
		0.801	210	Cs134*	3.2 hr	
		1.17	4.02			
		1.37	6.70			
Ba131	11.6 days	1.03^a	2.9	Ba130	0.101	10.1±1.0
		0.82	5.6			
		0.495	79.7			
		0.245	17.2			
		0.122	61			
Ba133*	38.8 hr	0.276	100	Ba132	0.097	?
		0.012	100			
Ba133	7.2 yr	0.360	96.5	Ba132	0.097	7±2
		0.070	3.5	Ba133*	38.8 hr	
		0.292	3.5			
Ba135*	28.7 hr	0.081	100	Ba134	2.42	?
Ba137*	2.60 min	0.268	100	Ba136	7.81	?
		0.661	100			
Ba139	85 min	1.43	19	Ba138	71.66	0.5±0.1
		0.163	85			
La140	40.2 hr	2.57^a	20	La139	99.9	8.4±1.7
La140	40.2 hr	0.920	40			
		0.486	73			
		1.60	76			
		0.130	26			
Ce137*	35 hr	0.255	100	Ce136	0.193	0.6±0.2
Ce137	8.7 hr	0.445	3	Ce136	0.193	6.3±1.5
				Ce137*	35 hr	
Ce139*	55 sec	0.740	100	Ce138	0.250	7±5 mb^b
Ce139	140 days	0.1665	100	Ce138	0.25	0.6±0.3
				Ce139*	55 sec	
Ce141	32 days	0.1449	75	Ce140	88.48	0.31±0.10
Ce143	32 hr	1.10^a	6	Ce142	11.07	0.95±0.05
		0.861	17			
		0.351	24			
		0.294	21			
		0.057	36			
Pr142	19.3 hr	1.61	2.8	Pr141	100	10±3
		0.69	3			
Nd147	11.06 days	0.318	20	Nd146	17.26	1.8±0.6
		0.165	3			
		0.532	25			
		0.092	60			
Nd149	2.0 hr	0.65^a	15	Nd148	5.74	3.7±1.2
		0.124	81			
		0.285	68			
Pm149	52 hr	0.285	100	Nd149	2 hr	
		1	100			
Nd151	12 min	0.73^a	40	Nd150	5.63	?
		0.085	60			
		0.117	120			
		0.421	60			
		1.14	100			

Table 4-45. GAMMA-RAY ACTIVITY DUE TO THERMAL NEUTRON CAPTURE (Continued)

Radioactive isotope	Half-life	E_γ, Mev	Yield-% per disintegration	Parent isotope	Isotopic-% or half-life	Activation cross section, barn
Pm^{151}	27.5 hr	0.715[a] 0.340 0.177 0.275 0.069	10 42 115 70 105	Nd^{151}	12 min	
Sm^{145}	400 days	0.0613	100	Sm^{144}	3.16	<2
Sm^{153}	47 hr	0.17 0.07 0.1	25 15 53	Sm^{152}	26.63	140±40
Sm^{155}	24 min	1.05 0.246	100 100	Sm^{154}	22.53	5.5±1.1
Eu^{155}	1.7 yr	0.102 0.084 0.018	50 27 27	Sm^{155}	24 min	
Eu^{152}	9.2 hr	1.39[a] 0.983 0.344 0.122	1.54 21.4 2.8 15.2	Eu^{151}	47.8	1400±300[b]
Gd^{153}	236 days	0.1	100	Gd^{152}	0.2	<125
Gd^{159}	18 hr	0.364 0.23 0.136 0.079 0.056	10 6 4 2 12	Gd^{158}	24.87	4±2
Gd^{161}	3.63 min	0.360 0.102 0.316 0.060	10 90 90 10	Gd^{160}	21.90	0.8±0.3
Tb^{161}	7 days	0.106[a] 0.0573 0.0277 0.0783	2 18 27 5	Gd^{161}	3.63 min	
Tb^{160}	73 days	0.093[a] 0.976 1.45 0.466 0.297	80 63 26 39 63	Tb^{159}	100	>22
Dy^{165*}	1.25 min	0.108 0.515 0.361	90 6 4	Dy^{164}	28.18	510±20
Dy^{165}	139 min	0.279[a] 0.361 1.02 0.71 0.094	1 40 8 2 10	Dy^{164} Dy^{165*}	28.18 1.25 min (90%)	2100±300[b]
Ho^{166}	27.2 hr	.378 0.080	1 48	Ho^{165}	100	60±12
Er^{169}	9.4 days	0.0084	15	Er^{168}	27.1	2±0.4
Er^{171}	7.5 hr	0.308[a] 0.013	95 3	Er^{170}	14.9	9±2
Tm^{170}	127 days	0.126 0.118 0.005	40 55 65	Tm^{169}	100	118±30
Yb^{169}	32 days	0.084 0.0084[a] 0.094	24 60 60	Yb^{168}	0.14	11000±3000[b]
Yb^{175}	4.1 days	0.118 0.20 0.31 0.396[a] 0.144 0.251 0.138 0.114	60 50 40 10 5 2 3 13	Yb^{174}	31.84	60±40

Table 4-45. GAMMA-RAY ACTIVITY DUE TO THERMAL NEUTRON CAPTURE (Continued)

Radioactive isotope	Half-life	E_γ, Mev	Yield-% per disintegration	Parent isotope	Isotopic-% or half-life	Activation cross section, barn
Yb^{177}	1.8 hr	1.24^a	4	Yb^{176}	12.73	5.5±1
		0.140	1			
		0.147	9			
		0.118	3			
Lu^{176*}	3.7 hr	0.089	90	Lu^{175}	97.4	35±15
Lu^{177}	6.7 days	0.321	2	Lu^{176}	2.6	4000±800
		0.208	2	Yb^{177}	1.8 hr	
		0.072	3			
		0.250	3			
		0.113	5			
Hf^{175}	70 days	0.430	1.5	Hf^{174}	0.19	?
		0.089	13.7			
		0.342	97.5			
Hf^{180*}	5.5 hr	0.444	80	Hf^{179}	13.75	?
		0.333	20			
		0.216	20			
		0.093	80			
		0.0576	100			
Hf^{181}	44.6 days	0.482^a	89.3	Hf^{180}	35.25	10±3
		0.137	110			
		0.0039	1			
		0.346	13.2			
Ta^{182*}	16.5 min	0.180	100	Ta^{181}	100	30±10 mb
Ta^{182}	115 days	0.229^a	45	Ta^{181}	100	19±7
		0.100	65	Ta^{182*}	16.5 min	
		0.084	55			
		1.289	94			
W^{181}	140 days	0.152	0.176	W^{180}	0.135	10±10
		0.136	0.114			
W^{185*}	1.7 min	1.30	100	W^{184}	30.6	2.1±0.6
		1.65	100			
W^{187}	24 hr	0.686	12	W^{186}	28.4	34±7
W^{187}	24 hr	0.480	8			
		0.206	5			
		0.072	3			
		0.132	3			
Re^{186}	91 hr	0.768	0.05	Re^{185}	37.07	100±20
		0.123	2			
		0.137	22			
Re^{188*}	18.7 min	0.060	100	Re^{187}	62.93	?
		0.105	100			
Re^{188}	17 hr	1.96^a	0.20	Re^{187}	62.93	75±15
		0.633	1	Re^{188*}	18.17 min	
		0.155	9			
Os^{185}	95 days	0.879^a	15	Os^{184}	0.018	<200
		0.233	2			
		0.160	2			
		0.646	85			
Os^{191}	14 hr	0.074	100	Os^{190}	26.4	?
Ir^{191*}	4.9 sec	0.042	100	Os^{191}	16 days	
Os^{193}	31.5 hr	0.129	100	Os^{192}	41	1.6±0.4
		0.387^a	5			
		0.460	5			
		0.281	12			
Ir^{192*}	1.45 min	0.139	7	Ir^{191}	38.5	260±100
		0.073	36			
Ir^{192}	74.5 days	0.056	100	Ir^{191}	38.5	700±200
		0.485^a	208	Ir^{192*}	1.45 min	
		0.613	23			
		1.06	0.04			
Ir^{194}	19 hr	2.05^a	0.14	Ir^{193}	61.5	130±30
		1.66	8.9			
		1.18	9.4			
		0.643	18.5			

Table 4-45.   GAMMA-RAY ACTIVITY DUE TO THERMAL NEUTRON CAPTURE *(Continued)*

Radioactive isotope	Half-life	E_γ, Mev	Yield-% per disintegration	Parent isotope	Isotopic-% or half-life	Activation cross section, barn
Ir^{194}	19 hr	0.328	19.3			
Pt^{193*}	3.5 days	0.130	100	Pt^{192}	0.78	90±40
Pt^{195*}	~6 days	0.130 0.031 0.099	60 40 40	Pt^{194}	32.8	?
Pt^{197}	18 hr	0.279 0.191 0.077	0.9 10.6 99.1	Pt^{196}	25.4	0.30±0.10
Pt^{199}	31 min	0.96a 0.54 0.197 0.246 0.316	3 41 81 42 41	Pt^{198}	7.2	3.9±0.8
Au^{199}	3.15 days	0.209 0.050 0.159	3.7 20.6 89.9	Pt^{199}	31 min	
Au^{198}	2.7 days	1.089 0.4118	0.16 99.018	Au^{197}	100	96±10
Hg^{197*}	24 hr	0.164 0.133	96.6 96.6	Hg^{196}	0.146	?
Hg^{197}	65 hr	0.191 0.077	1.2 100	Hg^{197*}	24 hr (96.6%)	

Radioactive isotope	Half-life	E_γ, Mev	Yield-% per disintegration	Parent isotope	Isotopic-% or half-life	Activation cross section, barn
Au^{197*}	7.4 sec	0.407 0.130 0.277	2 1.4 1.4	Hg^{197*}	24 hr (3.4%)	
Hg^{203}	45.4 days	0.279	100	Hg^{202}	29.8	3.8±0.8
Hg^{205}	5.5 min	0.203	Small	Hg^{204}	6.85	0.43±0.10
Tl^{204}	3 yr	0.38	0.008	Tl^{203}	29.5	8±3
Th^{233}	23.5 min	0.662 0.448 0.350 0.172 0.098	0.05 0.10 0.004 0.03 0.25	Th^{232}	100	7.34±0.15
Pa^{233}	26.95 days	0.417a 0.104 0.313 0.058 0.028	20 35 80 111 39	Th^{233}	23.5 min	
U^{239}	23.5 min	0.075	100	U^{238}	99.274	2.76±0.09
Np^{239}	2.33 days	0.067a 0.285 0.105 0.228	62 87 30 4	U^{239}	23.5 min	

a Effective value.
b Reactor neutrons.
* Metastable.

REFERENCES

"Neutron Cross Sections", D. J. Hughes and R. B. Schwartz, Brookhaven National Laboratory, BNL-325, July 1958.

"Trilinear Chart of Nuclides", 2nd Rev., W. H. Sullivan, USAEC, January 1957.

"The Reactor Handbook", Vol. I: Physics, AECD-3645, 1955, pp. 158ff.

Table 4-46.   TOTAL MASS ATTENUATION COEFFICIENTS*

In cm^2/g

The following table gives the total gamma-ray attenuation coefficient, μ/ρ, for some common materials The product of these numbers and the density of the material gives the familiar cross sections, μ, Table 4-47.

Material	Gamma-ray energy, Mev						
	0.1	0.2	0.5	1.0	2	5	10.0
H	.295	.243	.173	.126	.0876	.0502	.0321
Be	.132	.109	.0773	.0565	.0394	.0234	.0161
C	.149	.122	.0870	.0636	.0444	.0270	.0194
N	.150	.123	.0869	.0636	.0445	.0273	.0200
O	.151	.123	.0870	.0636	.0445	.0276	.0206
Na	.151	.118	.0833	.0608	.0427	.0274	.0215
Mg	.160	.122	.0860	.0627	.0442	.0286	.0228
Al	.161	.120	.0840	.0614	.0432	.0282	.0229
Si	.172	.125	.0869	.0635	.0447	.0296	.0243
P·	.174	.122	.0846	.0617	.0436	.0290	.0242
S	.188	.127	.0874	.0635	.0448	.0302	.0255
A	.188	.117	.0790	.0573	.0407	.0279	.0241
K	.215	.127	.0852	.0618	.0438	.0305	.0267
Ca	.238	.132	.0876	.0634	.0451	.0316	.0280
Fe	.344	.138	.0828	.0595	.0424	.0313	.0294
Cu	.427	.147	.0820	.0585	.0418	.0316	.0305
Mo	1.03	.225	.0851	.0575	.0414	.0344	.0359
Sn	1.58	.303	.0886	.0568	.0408	.0355	.0383
I	1.83	.339	.0913	.0571	.0409	.0361	.0394
W	4.21	.708	.125	.0640	.0437	.0409	.0465
Pt	4.75	.795	.135	.0659	.0445	.0418	.0477
Tl	5.16	.866	.143	.0675	.0452	.0423	.0484
Pb	5.29	.896	.145	.0684	.0457	.0426	.0489
U	1.06	1.17	.176	.0757	.0484	.0446	.0511
Air	.151	.123	.0868	.0655	.0445	.0274	.0202
NaI	1.57	.305	.0901	.0577	.0412	.0347	.0366
H$_2$O	.167	.136	.0966	.0706	.0493	.0301	.0219
Concrete^a	.169	.124	.0870	.0635	.0445	.0287	.0229
Tissue	.163	.132	.0936	.1683	.0478	.0292	.0212

^{a}Type 04.

*From: "Reactor Physics Constants", 2nd ed., Argonne National Laboratory, ANL–5800, U.S. Atomic Energy Commission, July 1963.

Table 4-47.   TOTAL GAMMA-RAY ATTENUATION CROSS SECTIONS*

In cm⁻¹

Material	Density, g/cm³	Gamma-ray energy, Mev						
		0.1	0.2	0.5	1.0	2	5	10.0
Be	1.85	.244	.202	.1430	.1045	.0729	.0433	.0298
C	2.25	.335	.275	.1958	.1431	.0999	.0608	.0437
Na	.9712	.147	.115	.0809	.0590	.0415	.0266	.0209
Mg	1.741	.279	.212	.1497	.1092	.0770	.0498	.0397
Al	2.70	.435	.324	.2268	.1658	.1166	.0761	.0618
Si	2.42	.416	.303	.2103	.1537	.1082	.0716	.0588
P	1.83	.318	.223	.1548	.1129	.0798	.0531	.0443
S	2.07	.389	.263	.1809	.1314	.0927	.0625	.0328
K	0.87	.187	.110	.0741	.0538	.0381	.0265	.0232
Ca	1.55	.369	.205	.1358	.0983	.0699	.0490	.0434
Fe	7.86	2.704	1.085	.6508	.4677	.3333	.2460	.2311
Cu	8.933	3.814	1.313	.7325	.5226	.3734	.2823	.2725
Mo	9.01	9.280	2.027	.7668	.5181	.3730	.3099	.3190
Sn	7.298	11.53	2.211	.6466	.4145	.2978	.2591	.2795
I	4.94	9.040	1.675	.4510	.2821	.2020	.1783	.1946
W	19.3	81.25	13.66	2.413	1.235	.8434	.7894	.8975
Pt	21.37	101.51	16.99	2.885	1.408	.9510	.8933	1.019
Tl	11.86	61.20	10.27	1.696	.8005	.5361	.5017	.5740
Pb	11.34	59.99	10.16	1.644	.7757	.5182	.4831	.5545
U	18.7	19.82	21.88	3.291	1.416	.9051	.8340	.9556
NaI	3.667	5.757	1.118	.3304	.2116	.1511	.1272	.1342
H₂O	1.00	.167	.136	.0966	.0706	.0493	.0301	.0219
Concreteᵃ	2.35	.397	.291	.2045	.1492	.1046	.0674	.0538

ᵃType 04.

*From: "Reactor Physics Constants", 2nd ed., Argonne National Laboratory, ANL–5800, U.S. Atomic Energy Commission, July 1963.

Table 4-48.   ENERGY ABSORPTION MASS ATTENUATION COEFFICIENT*

In cm^2/g

Only a fraction of the events represented by the total attenuation cross section actually removes the gamma-ray. In particular, Compton scattering can cause a change in the direction and the energy of a photon without absorbing it. The energy absorption mass attenuation coefficient (μ_e/ρ) is a measure of the fraction of the gamma-ray energy that is converted from radiant energy into heat. The product of these coefficients and of the density of the material gives the energy absorption cross section, Table 4-49.

Material	Gamma-ray energy, Mev						
	0.1	0.2	0.5	1.0	2	5	10.0
H	.0411	.0531	.0591	.0557	.0467	.0318	.0255
Be	.0183	.0237	.0264	.0248	.0210	.0151	.0118
C	.0215	.0267	.0297	.0280	.0237	.0177	.0145
N	.0224	.0267	.0297	.0280	.0236	.0180	.0151
O	.0233	.0271	.0297	.0280	.0238	.0183	.0157
Na	.0289	.0266	.0284	.0268	.0229	.0185	.0168
Mg	.0335	.0278	.0293	.0276	.0237	.0194	.0180
Al	.0373	.0275	.0286	.0270	.0232	.0192	.0182
Si	.0435	.0286	.0290	.0274	.0236	.0198	.0189
P	.0501	.0292	.0290	.0271	.0234	.0200	.0195
S	.0601	.0310	.0300	.0279	.0242	.0209	.0206
A	.0729	.0302	.0272	.0252	.0220	.0195	.0197
K	.0909	.0340	.0295	.0272	.0237	.0214	.0219
Ca	.111	.0367	.0304	.0279	.0244	.0222	.0231
Fe	.225	.0489	.0294	.0261	.0231	.0227	.0250
Cu	.310	.0594	.0296	.0260	.0229	.0231	.0261
Mo	.922	.141	.0348	.0263	.0233	.0262	.0316
Sn	1.469	.222	.0403	.0268	.0233	.0276	.0339
I	1.726	.260	.0433	.0274	.0236	.0283	.0353
W	4.112	.631	.0786	.0353	.0271	.0335	.0426
Pt	4.645	.719	.0892	.0375	.0280	.0343	.0438
Tl	5.057	.791	.0972	.0393	.0288	.0349	.0446
Pb	5.193	.821	.0994	.0402	.0293	.0352	.0450
U	9.63	1.096	.132	.0482	.0324	.0374	.0474
Air	.0233	.0268	.0297	.0280	.0238	.0181	.0153
NaI	1.466	.224	.0410	.0273	.0235	.0268	.0325
H$_2$O	.0253	.0300	.0330	.0311	.0264	.0198	.0165
Concrete	.0416	.0289	.0296	.0278	.0239	.0194	.0177
Tissue	.0271	.0293	.0320	.0300	.0256	.0192	.0160

*From: "Reactor Physics Constants", 2nd ed., Argonne National Laboratory, ANL–5800, U.S. Atomic Energy Commission, July 1963.

Table 4-49. GAMMA-RAY ABSORPTION CROSS SECTION*

In cm^{-1}

Material	Density, g/cm³	0.1	0.2	0.5	1.0	2	5	10.0
Be	1.85	.0339	.0438	.0488	.0459	.0389	.0279	.0218
C	2.25	.0484	.0601	.0668	.0630	.0533	.0398	.0326
Na	.9712	.0281	.0258	.0276	.0260	.0222	.0180	.0163
Mg	1.741	.0583	.0484	.0510	.0481	.0413	.0338	.0313
Al	2.70	.1007	.0743	.0772	.0729	.0626	.0518	.0491
Si	2.42	.1053	.0692	.0702	.0663	.0571	.0479	.0457
P	1.83	.0917	.0534	.0531	.0496	.0428	.0366	.0357
S	2.07	.1244	.0642	.0621	.0578	.0501	.0433	.0426
K	0.87	.0791	.0296	.0257	.0237	.0206	.0186	.0191
Ca	1.55	.172	.0569	.0471	.0432	.0378	.0344	.0358
Fe	7.86	1.769	.3844	.2311	.2051	.1816	.1784	.1965
Cu	8.933	2.769	.5306	.2644	.2323	.2046	.2064	.2332
Mo	9.01	8.307	1.270	.3135	.2370	.2099	.2361	.2847
Sn	7.298	10.721	1.620	.2941	.1956	.1700	.2014	.2474
I	4.94	8.704	1.284	.2139	.1354	.1166	.1398	.1744
W	19.3	79.362	12.178	1.517	.6813	.5320	.6466	.8222
Pt	21.37	99.264	15.365	1.906	.8014	.5984	.7330	.9360
Tl	11.86	59.976	9.381	1.153	.4661	.3416	.4139	.5290
Pb	11.34	58.889	9.310	1.127	.4559	.3323	.3992	.5103
U	18.7	180.08	20.495	2.468	.9013	.6059	.6994	.8864
NaI	3.667	5.376	.8214	.1503	.1001	.0862	.0983	.1192
H_2O	1.00	.0253	.0300	.0330	.0311	.0264	.0198	.0165
Concrete	2.35	.0978	.0679	.0697	.0653	.0562	.0456	.0416

*From: "Reactor Physics Constants", 2nd ed., Argonne National Laboratory, ANL–5800, U.S. Atomic Energy Commission, July 1963.

Table 4-50. REMOVAL CROSS SECTIONS FOR VARIOUS MATERIALS*

The removal cross section is a measure of the ability of a material to remove fast neutrons for shielding attenuation. It is most often applied to a wall of solid material between the fission source and a layer of water or hydrogenous material. The solid wall reduces the neutron energy to such an extent that it will be thermalized and captured in the water.

SYMBOLS: σ_R = microscopic removal cross section, barns/atom
Σ_R = macroscopic removal cross section per cm

Material	σ_R, barn	N_0 at 20°C, atom/cm³	Σ_R, cm^{-1}	Material	σ_R, barn	N_0 at 20°C, atom/cm³	Σ_R, cm^{-1}
Hydrogen	1.00 ± 0.05	—	—	Lead	3.53 ± 0.30	0.0330	0.116
Deuterium	0.92 ± 0.10[a]	—	—	Bismuth	3.49 ± 0.35	0.0282	0.098
Lithium	1.01 ± 0.04	0.0460 × 10²⁴	0.046	Uranium	3.6 ± 0.4	0.0473	0.17
Beryllium	1.07 ± 0.06	0.120	0.128	Boric oxide (B_2O_3)	4.30 ± 0.41	—	—
Boron	0.97 ± 0.10	0.139	0.135	Boron carbide (B_4C)	5.1 ± 0.4	—	—
Carbon (graphite)	0.72 ± 0.05	0.113	0.081	Fluorothene (C_2F_3Cl)	6.66 ± 0.8	—	—
Oxygen	0.92 ± 0.05	—	—	Heavy water (D_2O)	2.76 ± 0.11	—	—
Fluorine	1.29 ± 0.06	—	—	Lithium fluoride (LiF)	2.43 ± 0.34	—	—
Aluminum	1.31 ± 0.05	0.0603	0.079	Oil (CH_2)	2.84 ± 0.11	—	—
Chlorine	1.2 ± 0.8	—	—	Paraffin ($C_{30}H_{62}$)	80.5 ± 5.2	—	—
Iron	1.98 ± 0.08	0.0848	0.168	Perfluoroheptane (C_7F_{16})	26.3 ± 0.8	—	—
Nickel	1.89 ± 0.10	0.0913	0.173				
Copper	2.04 ± 0.11	0.0846	0.173				
Zirconium	2.36 ± 0.12	0.0423	0.10				
Tungsten	3.13 ± 0.25	0.0631	0.198				

[a]Calculated: $\sigma_R(D_2O) = 2.76$ b.

*From: "Reactor Physics Constants", 2nd ed., Argonne National Laboratory, ANL–5800, U.S. Atomic Energy Commission, July 1963.

REFERENCE

"Effective Neutron Removal Cross Sections for Shielding", G.T. Chapman and C.L. Storrs, AECD–3978 (ORNL–1843), September 19, 1955.

Table 4-51.　CRITICAL MASS AND ITS PARAMETERS

Critical mass, the minimum quantity of fissile material capable of sustaining a fission chain, varies greatly with reactor design (as in the 1 to 100 kg range); thus no simple formula involving the parameters is possible. A number of theoretical models have been set up, and the calculation procedure described, but the ultimate answers are experimental·†

Among the factors or conditions determining criticality are:

1. The fuel and fuel enrichment;
2. Absorption and leakage of neutrons;
3. Size and shape of the system.

More specifically, the parameters are set up in terms of more narrowly defined factors, such as the following:

p = *resonance escape probability*, the fraction of source neutrons that escape capture while being slowed down to a particular energy level. (The term "resonance" refers to the resonance region of the absorber.)

f = *thermal utilization*, the ratio of thermal neutrons absorbed in the fuel to the total thermal neutrons absorbed

n = *liberation ratio*, the number of neutrons liberated per neutron absorbed in the fuel

VALUES OF n

	Natural uranium	U^{233}	U^{235}	Pu^{239}
Fast neutrons	1.09	2.60	2.18	2.74
Thermal neutrons	1.33	2.27	2.06	2.10

e = *fast-fission factor*, the ratio of fast neutrons slowing down, to those produced by thermal-neutron fissions

$k\infty$ = *infinite multiplication factor*, ratio of neutrons from fission to neutrons absorbed in the preceding generation, in a system of infinite size

The following table illustrates the wide variation in critical mass for cylindrical and slab cores, unreflected or water-reflected, contained in either stainless steel or aluminum. It will be noted that the critical mass varies in the range of 1 kg to 46 kg.

---

†For extensive tables of critical mass data, see "Reactor Physics Constants", 2nd ed., Argonne National Laboratory, ANL–5800, U.S. Atomic Energy Commission, July 1963, Sections 3 and 4.

Table 4-51.   CRITICAL MASS AND ITS PARAMETERS *(Continued)*

Cylindrical and Slab Cores Containing UO_2F_2–H_2O Solutions, Various Concentrations; Uranium about 93% Enriched in U^{235} *

ALUMINUM-WALLED CYLINDERS AND SLABS

Water-reflected

Core diameter, cm	H/U^{235} atom ratio	U^{235} concentration, g/cm³ sol.	Critical core height, cm	Critical mass, kg U^{235}
15.2	27.1	0.8288	89.3	13.47
	43.2	0.537	70.1	6.87
	58.8	0.415	71.8	5.44
16.5	26.2	0.827	44.5	7.91
	44.3	0.5376	38.7	4.45
	78.7	0.315	42.6	2.87
	119.0	0.212	52.6	2.39
20.3	29.9	0.759	20.7	5.09
	49.5	0.488	18.8	2.97
	78.7	0.315	19.4	1.98
	192.0	0.134	28.1	1.22
	290.0	0.0881	40.1	1.15
25.4	27.1	0.8288	12.4	5.2
	43.2	0.537	12.5	3.40
	51.5	0.470	11.4	2.72
	127	0.199	14.4	1.45
	328.7	0.0787	22.4	0.893
	499	0.0522	35.2	0.0930
38.1	27.1	0.8288	7.7	7.3
	52.9	0.459	7.90	4.14
	221	0.116	11.30	1.49
	499	0.0522	16.90	1.01
	755	0.0343	27.10	1.02
	999	0.0260	44.30	1.31
76.2	27.1	0.8288	5.0±1	18.9±3.8
	44.3	0.5376	4.8±1	11.8±2.5
	72.4	0.3423	5.5±1	8.6±1.6
50.8 × 50.8 sq.	27.1	0.8288	6.3±1	13.5±2.2
	44.3	0.5376	4.3±1	6.0±1.4
	72.4	0.3423	6.2±1	5.5±0.9

Bare (Continued)

Core diameter, cm	H/U^{235} atom ratio	U^{235} concentration, g/cm³ sol.	Critical core height, cm	Critical mass, kg U^{235}
30.5	44.3	0.5376	23.2	9.1
	50.1	0.480	22.6	7.92
	55.4	0.437	22.7	7.25
	60.8	0.402	22.7	6.67
	331	0.0779	32.8	1.86
38.1	27.1	0.8288	18.5	17.5
	44.3	0.5376	17.9	11.0
	50.1	0.480	17.9	9.79
	74.6	0.3314	16.8	6.4
	169.0	0.151	18.5	3.18
	328.7	0.0787	21.7	1.95
	331	0.0779	22.9	2.03
	499	0.0522	27.4	1.63
	755	0.0343	43.6	1.70
50.8	27.1	0.8288	15.8	26.5
	44.3	0.5376	15.0	16.3
	50.1	0.480	15.4	14.9
	60.8	0.402	15.3	12.5
	73.4	0.3370	15.2	10.4
	325	0.0791	18.7	2.97
76.2	44.3	0.5376	13.7	33.6
	50.1	0.480	13.8	30.2
	72.4	0.3423	13.9±0.5	21.6±0.7
	331	0.0779	16.3	5.79
50.8 × 50.8 sq.	27.1	0.8288	15±1	32±2
	72.4	0.3423	14.3	12.6
	331	0.0779	17.9	3.60

Bare

Core diameter, cm	H/U^{235} atom ratio	U^{235} concentration, g/cm³ sol.	Critical core height, cm	Critical mass, kg U^{235}
22.3	44.3	0.5376	219	45.8
	50.1	0.480	202.2	37.8
	55.4	0.437	171.6	29.2
	60.8	0.402	162.5	25.4
	66.1	0.373	159.8	23.2
	71.5	0.350	163.2	22.2
25.4	27.1	0.8288	38.9	16.4
	44.3	0.5376	35.1	9.6
	50.1	0.480	34.8	8.40
	52.9	0.459	34.0	7.90
	55.4	0.437	34.3	7.60
	60.8	0.402	34.1	6.96
	66.1	0.373	34.1	6.45
	73.4	0.3370	33.7	5.8
	83.1	0.300	34.4	5.22
	169	0.151	41.2	3.15
	328	0.0785	147.8	5.83
	331	0.0779	170.1	6.72

Partially Water-Reflected[a]

Core diameter, cm	H/U^{235} atom ratio	U^{235} concentration, g/cm³ sol.	Critical core height, cm	Critical mass, kg U^{235}
15.2	44.3	0.5376	75.0	7.34
19.1	44.3	0.5376	25.7	3.93
20.3	44.3	0.5376	23.6	4.12
	51.5	0.470	23.8	3.64
	72.4	0.3423	23.3	2.59
25.4	43.2	0.537	17.3	4.71
	72.4	0.3423	16.7	2.90
38.1	74.6	0.3314	12.0	4.5
50.8	72.4	0.3423	10.6	7.3
76.2	72.4	0.3423	9.2	14.4
76.2 × 152.4[b]	57.0	0.4240	8.4	41.5

Table 4-51. CRITICAL MASS AND ITS PARAMETERS *(Continued)*

Core diameter, cm	H/U^{235} atom ratio	U^{235} concentration, g/cm³ sol.	Critical core height, cm	Critical mass, kg U^{235}	Core diameter, cm	H/U^{235} atom ratio	U^{235} concentration, g/cm³ sol.	Critical core height, cm	Critical mass, kg U^{235}

STAINLESS STEEL CYLINDERS

Core diameter, cm	H/U^{235} atom ratio	U^{235} concentration, g/cm³ sol.	Critical core height, cm	Critical mass, kg U^{235}	Core diameter, cm	H/U^{235} atom ratio	U^{235} concentration, g/cm³ sol.	Critical core height, cm	Critical mass, kg U^{235}
			Bare				Water-Reflected *(Continued)*		
22.9	74.6	0.3314	59.0	8.1	20.3 *(Cont.)*	226	0.114	36.3	1.34
						320	0.0805	60.1	1.57
25.4	43.9	0.538	32.3	8.80					
	62.7	0.396	31.7	6.36	25.4	31.6	0.724	15.3	5.61
	86.4	0.288	31.9	4.65		43.9	0.538	14.9	4.06
	123.2	0.205	34.3	3.56		62.7	0.396	15.2	3.05
	174	0.148	38.7	2.90		86.4	0.288	15.4	2.25
	226	0.114	46.6	2.69		123.2	0.205	16.8	1.74
						174	0.148	18.1	1.36
30.5	174	0.148	24.7	2.67		226	0.114	20.0	1.15
	226	0.114	26.2	2.18		320	0.0805	25.0	1.02
	320	0.0805	30.3	1.78					
	499	0.0522	48.9	1.86	30.5	62.6	0.394	12.3	3.53
						174	0.148	14.9	1.61
38.1	56.7	0.424	17.1	8.26		226	0.114	16.5	1.37
	221	0.116	19.5	2.58		320	0.0805	18.5	1.09
	499	0.0522	27.0	1.61		499	0.0522	26.3	1.00
	755	0.0343	41.7	1.63		755	0.0343	48.7	1.22
50.8	119	0.212	14.3	6.14	38.1	56.7	0.424	10.1	4.88
	221	0.116	15.7	3.69		221	0.116	13.0	1.72
	329	0.0787	17.4	2.77		499	0.0522	20.0	1.15
	499	0.0522	20.5	2.17		755	0.0343	28.8	1.13
	755	0.0343	26.7	1.86					
			Water-Reflected				Cadmium-Lined (0.44 g/cm²) Water-Reflected		
15.2	44.3	0.5376	118.4	11.59	20.3	43.9	0.538	51.5	8.98
						62.7	0.396	48.4	6.21
16.5	31.6	0.724	49.0	7.59		86.4	0.288	56.5	5.28
	43.9	0.538	47.1	5.42					
	86.4	0.288	53.8	3.31	22.9	31.6	0.724	29.0	8.62
						43.9	0.538	28.8	6.36
17.8	31.6	0.724	34.0	6.11		62.7	0.396	28.3	4.61
	43.9	0.538	32.7	4.37		86.4	0.288	29.2	3.45
	86.4	0.288	36.0	2.57		123.2	0.205	32.0	2.69
	174	0.148	57.3	2.10		174	0.148	37.3	2.26
20.3	31.6	0.724	22.6	5.31	25.4	31.6	0.724	21.1	7.74
	43.9	0.538	21.9	3.82		43.9	0.538	21.4	5.83
	56.7	0.424	22.2	3.05		62.7	0.396	22.0	4.42
	86.4	0.288	23.8	2.22		86.4	0.288	22.0	3.21
	123.2	0.205	26.0	1.73		221	0.116	28.7	1.69
	174	0.148	30.1	1.44					
					30.5	56.7	0.424	15.8	4.89
						174	0.148	19.5	2.10
						226	0.114	20.9	1.74
						499	0.052	32.8	1.25

[a]Assemblies have no top reflector.
[b]This vessel was coated with Unichrome; others were coated with Heresite.

*From: "Reactor Physics Constants", 2nd ed., Argonne National Laboratory, ANL–5800, U.S. Atomic Energy Commission, July 1963.

Energy Engineering
and Transport

Table 5-1. THERMODYNAMIC NONFLOW PROCESS EQUATIONS

For a System Containing a Perfect Gas with Constant Specific Heats

$$_1Q_2 = {}_1W_2 + (U_2 - U_1)$$

For flow processes see Table 5-7.

Process	Constant pressure	Constant volume	Isothermal	Isentropic $S = $ constant	Polytropic $pV^n = $ constant
p, V, T $pV = mRT\ddagger$ $pv = RT\ddagger$	$p = $ constant $p = p_1 = p_2$ $\dfrac{V}{T} = $ constant $\dfrac{V_1}{T_1} = \dfrac{V_2}{T_2}$	$V = $ constant $V = V_1 = V_2$ $\dfrac{p}{T} = $ constant $\dfrac{p_1}{T_1} = \dfrac{p_2}{T_2}$	$T = $ constant $T = T_1 = T_2$ $pV = $ constant $p_1V_1 = p_2V_2$	$p_1V_1^k = p_2V_2^k$ $= $ constant $\dfrac{T_2}{T_1} = \left(\dfrac{p_2}{p_1}\right)^{\frac{k-1}{k}}$ $\dfrac{T_2}{T_1} = \left(\dfrac{V_1}{V_2}\right)^{k-1}$	$p_1V_1^n = p_2V_2^n$ $= $ constant $\dfrac{T_2}{T_1} = \left(\dfrac{p_2}{p_1}\right)^{\frac{n-1}{n}}$ $\dfrac{T_2}{T_1} = \left(\dfrac{V_1}{V_2}\right)^{n-1}$
Specific heat c $= {}_1Q_2/(T_2 - T_1)\ddagger$	$c_p = \left(\dfrac{kR\ddagger}{(k-1)}\right)$	$c_v = \left(\dfrac{R\ddagger}{(k-1)}\right)$	∞	0	$c_n = c_v\dfrac{(k-n)}{(1-n)}$
Exponent n for polytropic process	0	∞	1	$k = \left(\dfrac{c_p\dagger}{c_v}\right)$	$n = $ any value
Quantity of heat $_1Q_2 = \int T\,dS$ positive for heat into system from surroundings	$mc_p(T_2 - T_1)$ $c_p(p/R)(V_2 - V_1)$ $\dfrac{k}{k-1}\,{}_1W_2$ $H_2 - H_1$	$mc_v(T_2 - T_1)$ $\dfrac{V(p_2 - p_1)}{k-1}$ $U_2 - U_1$	$\begin{cases} p_1V_1\ln(V_2/V_1) \\ p_1V_1\ln(p_1/p_2) \\ mRT\ln(V_2/V_1) \\ mRT\ln(p_1/p_2) \\ {}_1Q_2 = {}_1W_2 \end{cases}$	Adiabatic 0	$mc_n(T_2 - T_1)$
Quantity of work $_1W_2 = \int p\,dV$ positive for work done by system on surroundings	$p(V_2 - V_1)$ $mR(T_2 - T_1)$ $\dfrac{k-1}{k}\,{}_1Q_2$	0		$\dfrac{p_2V_2 - p_1V_1}{1-k}$ $mc_v(T_1 - T_2)$ $U_1 - U_2$ $\dfrac{p_1V_1}{k-1}\left[1 - \left(\dfrac{p_2}{p_1}\right)^{\frac{k-1}{k}}\right]$	$\dfrac{p_2V_2 - p_1V_1}{1-n}$ $\dfrac{mR(T_2 - T_1)}{1-n}$ $\dfrac{p_1V_1}{n-1}\left[1 - \left(\dfrac{p_2}{p_1}\right)^{\frac{n-1}{n}}\right]$
Internal energy $U_2 - U_1$ $dU = mc_v\,dT\dagger$	$mc_v(T_2 - T_1)\dagger$ $\dfrac{p(V_2 - V_1)\ddagger}{(k-1)}$	$mc_v(T_2 - T_1)\dagger$ $\dfrac{V(p_2 - p_1)\ddagger}{(k-1)}$	0	$mc_v(T_2 - T_1)\dagger$ $\dfrac{p_2V_2 - p_1V_1\ddagger}{(k-1)}$	$mc_v(T_2 - T_1)\dagger$ $\dfrac{p_2V_2 - p_1V_1\ddagger}{(k-1)}$
Enthalpy $H_2 - H_1$ $dH = mc_p\,dT\dagger$	$mc_p(T_2 - T_1)\dagger$ $\dfrac{kp(V_2 - V_1)\ddagger}{(k-1)}$	$mc_p(T_2 - T_1)\dagger$ $\dfrac{kV(p_2 - p_1)\ddagger}{(k-1)}$	0	$mc_p(T_2 - T_1)\dagger$ $\dfrac{k(p_2V_2 - p_1V_1)\ddagger}{(k-1)}$	$mc_p(T_2 - T_1)\dagger$ $\dfrac{k(p_2V_2 - p_1V_1)\ddagger}{(k-1)}$
Entropy $S_2 - S_1$ $dS = \dfrac{dH}{T} - \dfrac{V\,dp\dagger}{T}$ $= \dfrac{dU}{T} + \dfrac{p\,dV\dagger}{T}$	$mc_p\ln(T_2/T_1)$ $mc_p\ln(V_2/V_1)$	$mc_v\ln(T_2/T_1)$ $mc_v\ln(p_2/p_1)$	$mR\ln(V_2/V_1)$ $mR\ln(p_1/p_2)$	0	$mc_n\ln(T_2/T_1)$

†Valid in general, not only for process or processes listed, and not only for perfect gases.
‡Valid in general for perfect gases, not only for process listed.

Table 5-2. SPECIFIC-HEAT RATIO FUNCTIONS

SYMBOLS: $k = c_p/c_v$, the ratio of the specific heats of a gas or vapor

k	$\sqrt{k}$	$\dfrac{1}{k}$	$\dfrac{1}{\sqrt{k}}$	$\dfrac{1}{k-1}$	$\dfrac{k}{k-1}$	$\sqrt{\dfrac{k}{k-1}}$	$\dfrac{k-1}{k}$	$\sqrt{\dfrac{k-1}{k}}$	$\dfrac{k}{k+1}$	$\sqrt{\dfrac{k}{k+1}}$	$\dfrac{k+1}{k}$	$\sqrt{\dfrac{k+1}{k}}$
1.10	1.049	.9091	.9535	10.00	11.00	3.317	.0909	.3015	.5238	.7238	1.909	1.382
1.15	1.072	.8696	.9325	6.67	7.67	2.769	.1304	.3612	.5349	.7314	1.870	1.367
1.20	1.095	.8333	.9129	5.00	6.00	2.450	.1667	.4083	.5455	.7386	1.833	1.354
1.25	1.118	.8000	.8944	4.00	5.00	2.236	.2000	.4472	.5556	.7454	1.800	1.342
1.26	1.123	.7937	.8909	3.85	4.85	2.201	.2064	.4543	.5575	.7467	1.794	1.339
1.27	1.127	.7874	.8874	3.70	4.70	2.169	.2126	.4611	.5595	.7480	1.787	1.337
1.28	1.131	.7813	.8839	3.57	4.57	2.138	.2188	.4677	.5614	.7493	1.781	1.335
1.29	1.136	.7752	.8805	3.45	4.45	2.109	.2248	.4741	.5633	.7506	1.775	1.332
1.30	1.140	.7692	.8771	3.33	4.33	2.082	.2308	.4804	.5652	.7518	1.769	1.330
1.31	1.145	.7634	.8737	3.23	4.23	2.056	.2366	.4865	.5671	.7531	1.763	1.328
1.32	1.149	.7576	.8704	3.13	4.13	2.031	.2424	.4924	.5690	.7543	1.758	1.326
1.33	1.153	.7519	.8671	3.03	4.03	2.008	.2481	.4981	.5708	.7555	1.752	1.324
1.34	1.158	.7463	.8639	2.94	3.94	1.985	.2537	.5037	.5727	.7567	1.746	1.321
1.35	1.162	.7407	.8607	2.86	3.86	1.964	.2593	.5092	.5745	.7579	1.741	1.319
1.36	1.167	.7353	.8575	2.78	3.78	1.944	.2647	.5145	.5763	.7591	1.735	1.317
1.37	1.170	.7299	.8544	2.70	3.70	1.924	.2701	.5197	.5781	.7603	1.730	1.315
1.38	1.175	.7246	.8513	2.63	3.63	1.906	.2754	.5248	.5798	.7615	1.725	1.313
1.39	1.179	.7194	.8482	2.56	3.56	1.888	.2806	.5297	.5816	.7626	1.719	1.311
1.40	1.183	.7143	.8452	2.50	3.50	1.871	.2857	.5345	.5833	.7638	1.714	1.309
1.67	1.291	.6000	.7746	1.50	2.50	1.581	.4000	.6325	.6250	.7906	1.600	1.265

k	$\dfrac{k+1}{k-1}$	$\sqrt{\dfrac{k+1}{k-1}}$	$\dfrac{k-1}{k+1}$	$\sqrt{\dfrac{k-1}{k+1}}$	$\dfrac{2}{k-1}$	$\sqrt{\dfrac{2}{k-1}}$	$\dfrac{k-1}{2}$	$\sqrt{\dfrac{k-1}{2}}$	$\dfrac{k+1}{2}$	$\sqrt{\dfrac{k+1}{2}}$	$\left(\dfrac{k+1}{2}\right)^{\frac{1}{k-1}}$	$\left(\dfrac{k+1}{2}\right)^{\frac{k}{k-1}}$
1.10	21.00	4.583	.0476	.2182	20.00	4.472	.0500	.2236	1.050	1.025	1.629	1.710
1.15	14.33	3.786	.0698	.2641	13.25	3.639	.0755	.2748	1.075	1.037	1.619	1.742
1.20	11.00	3.317	.0909	.3015	10.00	3.162	.1000	.3162	1.100	1.049	1.611	1.772
1.25	9.00	3.000	.1111	.3333	8.00	2.828	.1250	.3536	1.125	1.061	1.602	1.802
1.26	8.69	2.948	.1150	.3392	7.69	2.774	.1300	.3606	1.130	1.063	1.600	1.808
1.27	8.41	2.900	.1189	.3449	7.41	2.722	.1350	.3674	1.135	1.065	1.598	1.814
1.28	8.14	2.854	.1228	.3504	7.14	2.673	.1400	.3742	1.140	1.068	1.597	1.820
1.29	7.90	2.810	.1266	.3559	6.90	2.626	.1450	.3808	1.145	1.070	1.595	1.826
1.30	7.67	2.769	.1304	.3612	6.67	2.582	.1500	.3873	1.150	1.072	1.593	1.832
1.31	7.45	2.730	.1342	.3663	6.45	2.540	.1550	.3937	1.155	1.075	1.592	1.838

Table 5-2.   SPECIFIC-HEAT RATIO FUNCTIONS (Continued)

k	$\frac{k+1}{k-1}$	$\sqrt{\frac{k+1}{k-1}}$	$\frac{k-1}{k+1}$	$\sqrt{\frac{k-1}{k+1}}$	$\frac{2}{k-1}$	$\sqrt{\frac{2}{k-1}}$	$\frac{k-1}{2}$	$\sqrt{\frac{k-1}{2}}$	$\frac{k+1}{2}$	$\sqrt{\frac{k+1}{2}}$	$\left(\frac{k+1}{2}\right)^{\frac{1}{k-1}}$	$\left(\frac{k+1}{2}\right)^{\frac{k}{k-1}}$
1.32	7.25	2.693	.1379	.3714	6.25	2.500	.1600	.4000	1.160	1.077	1.590	1.845
1.33	7.06	2.657	.1416	.3763	6.06	2.462	.1650	.4062	1.165	1.079	1.589	1.851
1.34	6.88	2.623	.1453	.3812	5.88	2.425	.1700	.4123	1.170	1.082	1.587	1.857
1.35	6.71	2.591	.1489	.3859	5.71	2.390	.1750	.4183	1.175	1.084	1.585	1.863
1.36	6.56	2.560	.1525	.3906	5.56	2.357	.1800	.4243	1.180	1.086	1.584	1.869
1.37	6.41	2.531	.1561	.3951	5.41	2.325	.1850	.4301	1.185	1.089	1.582	1.875
1.38	6.26	2.503	.1597	.3996	5.26	2.294	.1900	.4359	1.190	1.091	1.581	1.881
1.39	6.13	2.476	.1632	.4040	5.13	2.265	.1950	.4416	1.195	1.093	1.579	1.887
1.40	6.00	2.450	.1667	.4083	5.00	2.236	.2000	.4472	1.200	1.095	1.577	1.893
1.67	4.00	2.000	.2500	.5000	3.00	1.732	.3333	.5774	1.333	1.155	1.540	2.053

k	$\left(\frac{k+1}{2}\right)^{\frac{k+1}{2(k-1)}}$	$\frac{2}{k+1}$	$\sqrt{\frac{2}{k+1}}$	$\left(\frac{2}{k+1}\right)^{\frac{1}{k-1}}$	$\left(\frac{2}{k+1}\right)^{\frac{k}{k-1}}$	$\left(\frac{2}{k+1}\right)^{\frac{k+1}{k-1}}$	$\left(\frac{2}{k+1}\right)^{\frac{k+1}{2(k-1)}}$	$k\left[\left(\frac{2}{k+1}\right)^{\frac{k+1}{2(k-1)}}\right]$	$\frac{2k^2}{k-1}$	$\frac{2k^2}{k-1}\left[\left(\frac{2}{k+1}\right)^{\frac{k+1}{k-1}}\right]$
1.10	1.669	.9524	.9759	.6139	.5847	.3589	.5991	.6590	24.20	8.686
1.15	1.679	.9302	.9645	.6175	.5742	.3547	.5955	.6854	17.63	6.254
1.20	1.689	.9091	.9535	.6209	.5645	.3505	.5920	.7104	14.40	5.047
1.25	1.699	.8889	.9428	.6243	.5549	.3464	.5886	.7357	12.50	4.331
1.26	1.701	.8850	.9407	.6250	.5531	.3456	.5879	.7408	12.21	4.221
1.27	1.703	.8811	.9387	.6256	.5512	.3449	.5872	.7458	11.95	4.120
1.28	1.705	.8772	.9366	.6263	.5494	.3441	.5866	.7508	11.71	4.026
1.29	1.707	.8734	.9345	.6269	.5475	.3433	.5859	.7558	11.48	3.940
1.30	1.709	.8696	.9325	.6276	.5457	.3425	.5852	.7608	11.27	3.859
1.31	1.711	.8658	.9305	.6282	.5439	.3417	.5846	.7658	11.07	3.783
1.32	1.713	.8621	.9285	.6289	.5421	.3409	.5839	.7708	10.89	3.713
1.33	1.715	.8584	.9265	.6295	.5404	.3402	.5832	.7757	10.72	3.647
1.34	1.716	.8547	.9245	.6302	.5386	.3394	.5826	.7807	10.56	3.585
1.35	1.718	.8511	.9225	.6308	.5369	.3387	.5819	.7856	10.41	3.527
1.36	1.720	.8475	.9206	.6314	.5351	.3379	.5813	.7905	10.28	3.472
1.37	1.722	.8439	.9186	.6321	.5334	.3371	.5806	.7955	10.15	3.420
1.38	1.724	.8403	.9167	.6327	.5317	.3364	.5800	.8004	10.02	3.372
1.39	1.726	.8368	.9148	.6333	.5300	.3356	.5793	.8053	9.91	3.326
1.40	1.728	.8333	.9129	.6339	.5283	.3349	.5787	.8102	9.80	3.282
1.67	1.778	.7500	.8660	.6495	.4871	.3164	.5625	.9375	8.33	2.637

Table 5-3. DIFFERENTIAL THERMODYNAMIC RELATIONS

Function	Differential	Maxwell relation	Relations resulting by holding one variable constant	Equivalent expressions
$\Delta u = q - W$	$du = T\,ds - p\,dv$	$\left(\dfrac{\partial T}{\partial v}\right)_s = -\left(\dfrac{\partial p}{\partial s}\right)_v$	$\left(\dfrac{\partial u}{\partial v}\right)_s = -p$ $\left(\dfrac{\partial u}{\partial s}\right)_v = T$	$\left(\dfrac{\partial u}{\partial v}\right)_s = \left(\dfrac{\partial F}{\partial v}\right)_T$ $\left(\dfrac{\partial u}{\partial s}\right)_v = \left(\dfrac{\partial h}{\partial s}\right)_p$
$h = u + pv$ $h = enthalpy$	$dh = T\,ds + v\,dp$	$\left(\dfrac{\partial T}{\partial p}\right)_s = \left(\dfrac{\partial v}{\partial s}\right)_p$	$\left(\dfrac{\partial h}{\partial p}\right)_s = v$ $\left(\dfrac{\partial h}{\partial s}\right)_p = T$	$\left(\dfrac{\partial h}{\partial p}\right)_s = \left(\dfrac{\partial G}{\partial p}\right)_T$ $\left(\dfrac{\partial h}{\partial s}\right)_p = \left(\dfrac{\partial u}{\partial s}\right)_v$
$F = u - Ts$ $F = Helmholtz\ function$	$dF = -s\,dT - p\,dv$	$\left(\dfrac{\partial s}{\partial v}\right)_T = \left(\dfrac{\partial p}{\partial T}\right)_v$	$\left(\dfrac{\partial F}{\partial v}\right)_T = -p$ $\left(\dfrac{\partial F}{\partial T}\right)_v = -s$	$\left(\dfrac{\partial F}{\partial v}\right)_T = \left(\dfrac{\partial u}{\partial v}\right)_s$ $\left(\dfrac{\partial F}{\partial T}\right)_v = \left(\dfrac{\partial G}{\partial T}\right)_p$
$G = h - Ts$ $G = Gibbs\ function$	$dG = -s\,dT + v\,dp$	$\left(\dfrac{\partial s}{\partial p}\right)_T = -\left(\dfrac{\partial v}{\partial T}\right)_p$	$\left(\dfrac{\partial G}{\partial p}\right)_T = v$ $\left(\dfrac{\partial G}{\partial T}\right)_p = -s$	$\left(\dfrac{\partial G}{\partial p}\right)_T = \left(\dfrac{\partial h}{\partial p}\right)_s$ $\left(\dfrac{\partial G}{\partial T}\right)_p = \left(\dfrac{\partial F}{\partial T}\right)_v$

$$du = c_v\,dT + \left[\,T\left(\frac{\partial p}{\partial T}\right)_v - p\,\right]dv$$

$$dh = c_p\,dT - \left[\,T\left(\frac{\partial v}{\partial T}\right)_p - v\,\right]dp$$

$$ds = c_v\,\frac{dT}{T} + \left(\frac{\partial p}{\partial T}\right)_v dv = c_p\,\frac{dT}{T} - \left(\frac{\partial v}{\partial T}\right)_p dp$$

$$dq = c_v\,dT + T\left(\frac{\partial p}{\partial T}\right)_v dv = c_p\,dT - T\left(\frac{\partial v}{\partial T}\right)_p dp$$

$$c_v = \left(\frac{\partial q}{\partial T}\right)_v = T\left(\frac{\partial s}{\partial T}\right)_v = \left(\frac{\partial u}{\partial T}\right)_v$$

$$c_p = \left(\frac{\partial q}{\partial T}\right)_p = T\left(\frac{\partial s}{\partial T}\right)_p = \left(\frac{\partial h}{\partial T}\right)_p$$

$$c_p - c_v = T\left(\frac{\partial v}{\partial T}\right)_p\left(\frac{\partial p}{\partial T}\right)_v$$

REFERENCE

For additional relations see "American Institute of Physics Handbook", 2nd ed., D.E. Gray, Ed., McGraw-Hill Book Company, 1963, pp. 4-22 to 4-28.

Table 5-4. THERMODYNAMIC CYCLE EFFICIENCIES

SYMBOLS: $\eta_c = (T_H - T_L)/T_H$ = Carnot cycle efficiency, percent

$\eta_o = 1 - r^{1-k}$ = Otto cycle efficiency, percent

$\eta_D = 1 - r^{1-k} \dfrac{(S^k - 1)}{k(S-1)}$ = Diesel cycle efficiency, percent

$\eta_\beta = 1 - r_p^{(1-k)/k}$ = Brayton (or Joule) cycle efficiency, percent

where T_H = absolute temperature of energy reservoir from which energy is drawn

T_L = absolute temperature of energy reservoir to which energy is rejected

r = compression ratio for Otto and Diesel cycles, maximum volume/minimum volume

r_p = pressure ratio for Brayton (or Joule) cycle, maximum pressure/minimum pressure

S = cut-off ratio for Diesel cycle, volume at end of constant pressure heat addition process/minimum volume

$k = c_p/c_v$ = specific heat ratio

OTTO CYCLE EFFICIENCY, η_o, PERCENT

$k = \dfrac{c_p}{c_v}$	Compression ratio, r												
	5	6	7	8	9	10	11	12	13	14	15	20	50
1.30	38.3	41.6	44.2	46.4	48.3	49.9	51.3	52.5	53.7	54.7	55.6	59.3	69.1
1.35	43.1	46.6	49.4	51.7	53.7	55.3	56.8	58.1	59.3	60.3	61.2	65.0	74.6
1.40	47.5	51.2	54.1	56.5	58.5	60.2	61.7	63.0	64.2	65.2	66.1	69.8	79.1
5/3	65.8	69.7	72.7	75.0	76.9	78.5	79.8	80.9	81.9	82.7	83.6	86.4	92.6

DIESEL CYCLE EFFICIENCY, η_D, PERCENT

$k = \dfrac{c_p}{c_v}$	Cut-off ratio, S	Compression ratio, r											
		5	10	14	15	16	17	18	19	• 20	25	30	50
1.30	2	30.6	43.6	49.0	50.1	51.0	51.9	52.7	53.5	54.2	57.2	59.5	65.2
1.30	3	24.7	38.9	44.7	45.9	46.9	47.9	48.8	49.6	50.3	53.6	56.0	62.3
1.30	4	19.9	34.9	41.2	42.4	43.5	44.5	45.5	46.3	47.2	50.6	53.2	59.9
1.30	5	15.7	31.5	38.1	39.4	40.5	41.6	42.6	43.5	44.4	48.0	50.8	57.8
1.35	2	34.7	48.7	54.4	55.5	56.5	57.4	58.3	59.1	59.8	62.8	65.1	70.8
1.35	3	28.2	43.6	49.9	51.1	52.2	53.2	54.1	55.0	55.8	59.1	61.6	67.9
1.35	4	22.7	39.4	46.1	47.4	48.6	49.6	50.6	51.6	52.4	56.0	58.7	65.5
1.35	5	18.0	35.6	42.8	44.1	45.4	46.5	47.6	48.6	49.5	53.3	56.2	63.3
1.40	2	38.5	53.4	59.3	60.4	61.4	62.3	63.2	63.9	64.7	67.7	70.0	75.5
1.40	3	31.4	48.0	54.6	55.8	56.9	58.0	58.9	59.8	60.6	64.0	66.5	72.7
1.40	4	25.4	43.5	50.6	51.9	53.2	54.3	55.3	56.3	57.2	60.8	63.6	70.3
1.40	5	20.1	39.4	47.1	48.5	49.8	51.0	52.1	53.2	54.1	58.0	61.0	68.2

BRAYTON (OR JOULE) CYCLE EFFICIENCY, η_β, PERCENT

$k = \dfrac{c_p}{c_v}$	Pressure ratio, r_p												
	3	4	5	6	7	8	9	10	12	14	15	20	50
1.30	22.4	27.4	31.0	33.9	36.2	38.1	39.8	41.2	43.6	45.5	46.5	49.9	59.5
1.35	24.8	30.2	34.1	37.2	39.6	41.7	43.4	45.0	47.5	49.4	50.4	54.0	63.7
1.40	27.0	32.7	36.9	40.1	42.7	44.8	46.6	48.2	50.8	52.9	53.9	57.5	67.3
5/3	35.6	42.6	47.5	51.2	54.1	56.5	58.5	60.2	63.0	65.1	66.1	69.8	79.1

Table 5-4.   THERMODYNAMIC CYCLE EFFICIENCIES *(Continued)*

CARNOT CYCLE EFFICIENCY, η_c, PERCENT[a]

T_L		T_H, K (R)										
K	R	200 (360)	300 (540)	400 (720)	500 (900)	1 000 (1 800)	1 500 (2 700)	2 000 (3 600)	2 500 (4 500)	3 000 (5 400)	4 000 (7 200)	5 000 (9 000)
100	180	50.0	66.7	75.0	80.0	90.0	93.3	95.0	96.0	96.7	97.5	98.0
200	360	0	33.3	50.0	60.0	80.0	86.7	90.0	92.0	93.3	95.0	96.0
300	540	—	0	25.0	40.0	70.0	80.0	85.0	88.0	90.0	92.5	94.0
400	720	—	—	0	20.0	60.0	73.3	80.0	84.0	86.7	90.0	92.0
500	900	—	—	—	0	50.0	66.7	75.0	80.0	83.3	87.5	90.0
1 000	1 800	—	—	—	—	0	33.3	50.0	40.0	66.7	75.0	80.0

[a]These values are valid for any reversible cycle with heat addition at T_H and heat rejection at T_L. Stirling and Ericcson cycles with ideal regeneration meet this requirement, for example.

Otto Cycle.   The Otto cycle consists of isentropic compression, constant-volume heat addition, isentropic expansion, and constant-volume heat rejection.

Diesel Cycle.   The Diesel cycle consists of isentropic compression, constant-pressure heat addition, isentropic expansion, and constant-volume heat rejection.

Brayton Cycle.   The Brayton, or Joule, cycle consists of isentropic compression, constant-pressure heat addition, isentropic expansion, and constant-pressure heat rejection.

Carnot Cycle.   The Carnot cycle consists of isothermal compression (with heat rejection), isentropic compression, isothermal expansion (with heat addition), and isentropic expansion.

Stirling Cycle.   The Stirling cycle consists of isothermal compression (with heat rejection), constant-volume heat addition, isothermal expansion (with heat addition), and constant-volume heat rejection.

Ericcson Cycle.   The Ericcson cycle consists of isothermal compression (with heat rejection), constant-pressure heat addition, isothermal expansion (with heat addition), and constant-pressure heat rejection.

Table 5-5. HEAT OF FUSION OF SOME INORGANIC COMPOUNDS*

Rudolf Loebel

For heat of fusion in J/kg, multiply values in cal/g by 4184. For heat of fusion in J/mol, multiply values in cal/g·mol (= cal/mol) by 4.184. For melting point in K, add 273.15 to values in °C. Values in parentheses are of uncertain reliability.

Compound	Formula	Melting point, °C	Heat of fusion Btu/lb	Heat of fusion cal/g	Heat of fusion cal/g mole
Actinium[227]	Ac	1050±50	(20.)	(11.0)	(3400)
Aluminum	Al	658.5	170.	94.5	2550
Aluminum bromide	Al$_2$Br$_6$	87.4	18.2	10.1	5420
Aluminum chloride	Al$_2$Cl$_6$	192.4	114.	53.6	19600
Aluminum iodide	Al$_2$I$_6$	190.9	17.6	9.8	7960
Aluminum oxide	Al$_2$O$_3$	2045.0	(461.)	(256.0)	(26000)
Antimony	Sb	630	70.4	39.1	4770
Antimony pentachloride	SbCl$_5$	4.0	14.4	8.0	2400
Antimony tribromide	SbBr$_3$	96.8	17.5	9.7	3510
Antimony trichloride	SbCl$_3$	73.3	23.9	13.3	3030
Antimony trioxide	Sb$_4$O$_6$	655.0	(83.3)	(46.3)	(26990)
Antimony trisulfide	Sb$_4$S$_6$	546.0	59.4	33.0	11200
Argon	Ar	−190.2	13.1	7.25	290
Arsenic	As	816.8	(39.6)	(22.0)	(6620)
Arsenic pentafluoride	AsF$_5$	−80.8	29.7	16.5	2800
Arsenic tribromide	AsBr$_3$	30.0	16.0	8.9	2810
Arsenic trichloride	AsCl$_3$	−16.0	23.9	13.3	2420
Arsenic trifluoride	AsF$_3$	−6.0	34.0	18.9	2486
Arsenic trioxide	As$_4$O$_6$	312.8	40.0	22.2	8000
Barium	Ba	725	23.9	13.3	1830
Barium bromide	BaBr$_2$	846.8	39.4	21.9	6000
Barium chloride	BaCl$_2$	959.8	46.6	25.9	5370
Barium fluoride	BaF$_2$	1286.8	30.8	17.1	3000
Barium iodide	BaI$_2$	710.8	(31.1)	(17.3)	(6800)
Barium nitrate	Ba(NO$_3$)$_2$	594.8	(40.7)	(22.6)	(5900)
Barium oxide	BaO	1922.8	168.	93.2	13800
Barium phosphate	Ba$_3$(PO$_4$)$_2$	1727	55.6	30.9	18600
Barium sulfate	BaSO$_4$	1350	74.9	41.6	9700
Beryllium	Be	1278	468.	250.0	—
Beryllium bromide	BeBr$_2$	487.8	(47.9)	(26.6)	(4500)
Beryllium chloride	BeCl$_2$	404.8	(54)	(30)	(3000)
Beryllium oxide	BeO	2550.0	1223.	679.7	17000
Bismuth	Bi	271	21.6	12.0	2505
Bismuth trichloride	BiCl$_3$	223.8	14.8	8.2	2600
Bismuth trifluoride	BiF$_3$	726.8	(41.9)	(23.3)	(6200)
Bismuth trioxide	Bi$_2$O$_3$	815.8	26.3	14.6	6800
Boron	B	2300	(882)	(490)	(5300)
Boron tribromide	BBr$_3$	−48.8	(5.2)	(2.9)	(700)
Boron trichloride	BCl$_3$	−107.8	(7.7)	(4.3)	(500)
Boron trifluoride	BF$_3$	−128.0	12.6	7.0	480
Boron trioxide	B$_2$O$_3$	448.8	142.	78.9	5500
Bromine	Br$_2$	−7.2	29.0	16.1	2580
Bromine pentafluoride	BrF$_5$	−61.4	12.7	7.07	1355
Cadmium	Cd	320.8	23.2	12.9	1460
Cadmium bromide	CdBr$_2$	567.8			
Cadmium chloride	CdCl$_2$	567.8	51.8	28.8	5300
Cadmium fluoride	CdF$_2$	1110	(64.6)	(35.9)	(5400)
Cadmium iodide	CdI$_2$	386.8	18.0	10.0	3660
Cadmium sulfate	CdSO$_4$	1000	41.2	22.9	4790
Calcium	Ca	851	100.	55.7	2230
Calcium bromide	CaBr$_2$	729.8	37.6	20.9	4180
Calcium carbonate	CaCO$_3$	1282	(227)	(126)	(12700)
Calcium chloride	CaCl$_2$	782	99	55	6100
Calcium fluoride	CaF$_2$	1382	94.5	52.5	4100
Calcium metasilicate	CaSiO$_3$	1512	208.	115.4	13400
Calcium nitrate	Ca(NO$_3$)$_2$	560.8	56.2	31.2	5120
Calcium oxide	CaO	2707	(393.)	(218.1)	(12240)
Calcium sulfate	CaSO$_4$	1297	88.6	49.2	6700
Carbon dioxide	CO$_2$	−57.6	77.8	43.2	1900
Carbon monoxide	CO	−205	12.8	7.13	199.7
Cyanogen	C$_2$N$_2$	−27.2	71.3	39.6	2060
Cyanogen chloride	CNCl	−5.2	65.5	36.4	2240
Cerium	Ce	775	27.2	15.1	2120
Cesium	Cs	28.3	6.7	3.7	500
Cesium chloride	CsCl	641.8	38.5	21.4	3600
Cesium nitrate	CsNO$_3$	406.8	29.9	16.6	3250

Table 5-5. HEAT OF FUSION OF SOME INORGANIC COMPOUNDS (Continued)

Compound	Formula	Melting point, °C	Heat of fusion Btu/lb	cal/g	cal/g mole
Chlorine	Cl$_2$	−103±5	41.0	22.8	1531
Chromium	Cr	1890	112.	62.1	3660
Chromium (II) chloride	CrCl$_2$	814	119.	65.9	7700
Chromium (III) sequioxide	Cr$_2$O$_3$	2279	49.7	27.6	4200
Chromium trioxide	CrO$_3$	197	67.9	37.7	3770
Cobalt	Co	1490	112.	62.1	3640
Cobalt (II) chloride	CoCl$_2$	727	102.	56.9	7390
Copper	Cu	1083	88.2	49.0	3110
Copper (II) chloride	CuCl$_2$	430	44.5	24.7	4890
Copper (I) chloride	CuCl	429	47.5	26.4	2620
Copper (I) cyanide	Cu$_2$(CN)$_2$	473	(54.2)	(30.1)	(5400)
Copper (I) iodide	CuI	587	(24.5)	(13.6)	(2600)
Copper (II) oxide	CuO	1446	63.7	35.4	2820
Copper (I) oxide	Cu$_2$O	1230	(168.)	(93.6)	(13400)
Copper (I) sulfide	Cu$_2$S	1129	62.3	34.6	5500
Dysprosium	Dy	1407	45.4	25.2	4100
Erbium	Er	1496	44.1	24.5	4100
Europium	Eu	826	29.5	16.4	2500
Europium trichloride	EuCl$_3$	622	(37.6)	(20.9)	(8000)
Fluorine	F$_2$	−219.6	11.5	6.4	244.0
Gadolinium	Gd	1312	42.8	23.8	3700
Gallium	Ga	29	(34.4)	19.1	1336
Germanium	Ge	959	(206.)	(114.3)	(8300)
Gold	Au	1063	(27.5)	15.3	3030
Hafnium	Hf	2214	(61.4)	(34.1)	(6000)
Holmium	Ho	1461	44.6	24.8	4100
Hydrogen	H$_2$	−259.25	24.8	13.8	28
Hydrogen bromide	HBr	−86.96	12.8	7.1	575.1
Hydrogen chloride	HCl	−114.3	23.4	13.0	476.0
Hydrogen fluoride	HF	−83.11	98.5	54.7	1094
Hydrogen iodide	HI	−50.91	9.7	5.4	686.3
Hydrogen nitrate	HNO$_3$	−47.2	17.1	9.5	601
Hydrogen oxide (water)	H$_2$O	0	138.	79.72	1436
Deuterium oxide	D$_2$O	3.78	136.	75.8	1516
Hydrogen peroxide	H$_2$O$_2$	−0.7	15.4	8.58	2920
Hydrogen selenate	H$_2$SeO$_4$	57.8.	42.8	23.8	3450
Hydrogen sulfate	H$_2$SO$_4$	10.4	43.2	24.0	2360
Hydrogen sulfide	H$_2$S	−85.6	30.2	16.8	5683
Hydrogen sulfide, di-	H$_2$S$_2$	−89.7	49.1	27.3	1805
Hydrogen telluride	H$_2$Te	−49.0	23.2	12.9	1670
Indium	In	156.3	12.2	6.8	781
Iodine	I$_2$	112.9	25.7	14.3	3650
Iodine chloride (α)	ICl	17.1	29.5	16.4	2660
Iodine chloride (β)	ICl	13.8	23.9	13.3	2270
Iron	Fe	1530.0	115.	63.7	3560
Iron carbide	Fe$_3$C	1226.8	123.	68.6	12330
Iron (III) chloride	Fe$_2$Cl$_6$	303.8	114.	63.2	20500
Iron (II) chloride	FeCl$_2$	677	111.	61.5	7800
Iron (II) oxide	FeO	1380	(193.)	(107.2)	(7700)
Iron oxide	Fe$_3$O$_4$	1596	257.	142.5	33000
Iron pentacarbonyl	Fe(CO)$_5$	−21.2	29.7	16.5	3250
Iron (II) sulfide	FeS	1195	102.	56.9	5000
Lanthanum	La	920	31.3	17.4	2400
Lead	Pb	327.3	10.6	5.9	1224
Lead bromide	PbBr$_2$	487.8	21.1	11.7	4290
Lead chloride	PbCl$_2$	497.8	36.5	20.3	5650
Lead fluoride	PbF$_2$	823	13.7	7.6	1860
Lead iodide	PbI$_2$	412	32.2	17.9	5970
Lead molybdate	PbMoO$_4$	1065	(127.)	70.8	(25800)
Lead oxide	PbO	890	22.7	12.6	2820
Lead sulfate	PbSO$_4$	1087	56.9	31.6	9600
Lead sulfide	PbS	1114	31.1	17.3	4150
Lithium	Li	178.8	285.	158.5	1100
Lithium bromide	LiBr	552	60.1	33.4	2900
Lithium chloride	LiCl	614	136.	75.5	3200
Lithium fluoride	LiF	896	(164.)	(91.1)	(2360)
Lithium hydroxide	LiOH	462	186.	103.3	2480
Lithium iodide	LiI	440	(19.1)	(10.6)	(1420)
Lithium metasilicate	Li$_2$SiO$_3$	1177	144.	80.2	7210
Lithium molybdate	Li$_2$MoO$_4$	705	43.4	24.1	4200
Lithium nitrate	LiNO$_3$	250	158.	87.8	6060
Lithium orthosilicate	Li$_4$SiO$_4$	1249	109.	60.5	7430

Table 5-5. HEAT OF FUSION OF SOME INORGANIC COMPOUNDS (Continued)

Compound	Formula	Melting point, °C	Heat of fusion		
			Btu/lb	cal/g	cal/g mole
Lithium sulfate	Li₂SO₄	857	49.7	27.6	3040
Lithium tungstate	Li₂WO₄	742	(46.1)	(25.6)	(6700)
Lutetium	Lu	1651	47.3	26.3	4600
Magnesium	Mg	650	150.	88.9	2160
Magnesium bromide	MgBr₂	711	81.0	45.0	8300
Magnesium chloride	MgCl₂	712	149.	82.9	8100
Magnesium fluoride	MgF₂	1221	170.	94.7	5900
Magnesium oxide	MgO	2642	826.	459.0	18500
Magnesium silicate	MgSiO₃	1524	254.	146.4	14700
Magnesium sulfate	MgSO₄	1327	52.0	28.9	3560
Manganese	Mn	1220	113.	62.7	3450
Manganese dichloride	MnCl₂	650	105.	58.4	7340
Manganese metasilicate	MnSiO₃	1274	(113.)	(62.6)	(8200)
Manganese (II) oxide	MnO	1784	330.	183.3	13000
Manganese oxide	Mn₃O₄	1590	(307.)	(170.4)	(39000)
Mercury	Hg	−39	4.9	2.7	557.2
Mercury bromide	HgBr₂	241	19.6	10.9	3960
Mercury chloride	HgCl₂	276.8	27.5	15.3	4150
Mercury iodide	HgI₂	250	17.8	9.9	4500
Mercury sulfate	HgSO₄	850	8.6	(4.8)	(1440)
Molybdenum	Mo	2622	(123.)	(68.4)	(6600)
Molybdenum dichloride	MoCl₂	726.8	54.4	30.2	6000
Molybdenum hexafluoride	MoF₆	17	21.4	11.9	2500
Molybdenum trioxide	MoO₃	795	(31.1)	(17.3)	(2500)
Neodymium	Nd	1020	21.2	11.8	1700
Neon	Ne	−248.6	6.89	3.8	77.4
Nickel	Ni	1452	129.	71.5	4200
Nickel chloride	NiCl₂	1030	257.	142.5	18470
Nickel subsulfide	Ni₃S₂	790	46.4	25.8	5800
Niobium	Nb	2496	(124.)	(68.9)	(6500)
Niobium pentachloride	NbCl₅	211	55.4	30.8	8400
Niobium pentoxide	Nb₂O₅	1511	164.	91.0	24200
Nitric oxide	NO	−163.7	32.9	18.3	549.5
Nitrogen	N₂	−210	11.1	6.15	172.3
Nitrogen tetroxide	N₂O₄	−13.2	108.	60.2	5540
Nitrous oxide	N₂O	−90.9	53.9	35.5	1563
Osmium	Os	2700	(66.1)	(36.7)	(7000)
Osmium tetroxide (white)	OsO₄	41.8	16.6	9.2	2340
Osmium tetroxide (yellow)	OsO₄	55.8	27.9	15.5	4060
Oxygen	O₂	−218.8	5.9	3.3	106.3
Palladium	Pd	1555	69.5	38.6	4120
Phosphoric acid	H₃PO₄	42.3	46.4	25.8	2520
Phosphoric acid, hypo-	H₄P₂O₆	54.8	92.2	51.2	8300
Phosphorus acid, hypo-	H₃PO₂	17.3	63.0	35.0	2310
Phosphorus acid, ortho-	H₃PO₃	73.8	67.3	37.4	3070
Phosphorus oxychloride	POCl₃	1.0	36.5	20.3	3110
Phosphorus pentoxide	P₄O₁₀	569.0	108.	60.1	17080
Phosphorus trioxide	P₄O₆	23.7	27.5	15.3	3360
Phosphorus, yellow	P₄	44.1	8.6	4.8	600
Platinum	Pt	1770	43.4	24.1	4700
Potassium	K	63.4	26.3	14.6	574
Potassium borate, meta-	KBO₂	947	(124.)	(69.1)	(5660)
Potassium bromide	KBr	742	75.6	42.0	5000
Potassium carbonate	K₂CO₃	897	102.	56.4	7800
Potassium chloride	KCl	770	155.	85.9	6410
Potassium chromate	K₂CrO₄	984	64.1	35.6	6920
Potassium cyanide	KCN	623	(96.7)	(53.7)	(3500)
Potassium dichromate	K₂Cr₂O₇	398	53.6	29.8	8770
Potassium fluoride	KF	875	201.	111.9	6500
Potassium hydroxide	KOH	360	(63.5)	(35.3)	(1980)
Potassium iodide	KI	682	44.5	24.7	4100
Potassium nitrate	KNO₃	338	50.6	28.1	2840
Potassium peroxide	K₂O₂	490	99.5	55.3	6100
Potassium phosphate	K₃PO₄	1340	75.4	41.9	8900
Potassium pyrophosphate	K₄P₂O₇	1092	76.3	42.4	14000
Potassium sulfate	K₂SO₄	1074	83.5	46.4	8100
Potassium thiocyanate	KSCN	179	41.6	23.1	2250
Praseodymium	Pr	931	34.2	19.0	2700
Rhenium	Re	3167±60	(76.3)	(42.4)	(7900)
Rhenium heptoxide	Re₂O₇	296	54.2	30.1	15340
Rhenium hexafluoride	ReF₆	19.0	29.9	16.6	5000

Table 5-5. HEAT OF FUSION OF SOME INORGANIC COMPOUNDS (Continued)

Compound	Formula	Melting point, °C	Heat of fusion Btu/lb	Heat of fusion cal/g	Heat of fusion cal/g mole
Rubidium	Rb	38.9	11.0	6.1	525
Rubidium bromide	RbBr	677	40.3	22.4	3700
Rubidium chloride	RbCl	717	65.5	36.4	4400
Rubidium fluoride	RbF	833	71.1	39.5	4130
Rubidium iodide	RbI	638	25.2	14.0	2990
Rubidium nitrate	$RbNO_3$	305	16.4	9.1	1340
Samarium	Sm	1072	31.1	17.3	2600
Scandium	Sc	1538	152.	84.4	3800
Selenium	Se	217	27.7	15.4	1220
Selenium oxychloride	$SeOCl_3$	9.8	11.0	6.1	1010
Silane, hexafluoro-	Si_2F_6	−28.6	41.2	22.9	3900
Silicon	Si	1427	607.	337.0	9470
Silicon dioxide (Cristobalite)	SiO_2	2100	63.0	35.0	2100
Silicon dioxide (Quartz)	SiO_2	1470	102.	56.7	3400
Silicon tetrachloride	$SiCl_4$	−67.7	19.4	10.8	1845
Silver	Ag	961	45.0	25.0	2700
Silver bromide	AgBr	430	20.9	11.6	2180
Silver chloride	AgCl	455	39.6	22.0	3155
Silver cyanide	AgCN	350	36.9	20.5	2750
Silver iodide	AgI	557	17.1	9.5	2250
Silver nitrate	$AgNO_3$	209	29.2	16.2	2755
Silver sulfate	Ag_2SO_4	657	(24.7)	(13.7)	(4280)
Silver sulfide	Ag_2S	841	24.3	13.5	3360
Sodium	Na	97.8	49.3	27.4	630
Sodium borate, meta-	$NaBO_2$	966	242.	134.6	8660
Sodium bromide	NaBr	747	107.	59.7	6140
Sodium carbonate	Na_2CO_3	854	119.	66.0	7000
Sodium chlorate	$NaClO_3$	255	89.5	49.7	5290
Sodium chloride	NaCl	800	222.	123.5	7220
Sodium cyanide	NaCN	562	(160.)	(88.9)	(4360)
Sodium fluoride	NaF	992	300.	166.7	7000
Sodium hydroxide	NaOH	322	90.0	50.0	2000
Sodium iodide	NaI	662	63.2	35.1	5340
Sodium molybdate	Na_2MoO_4	687	31.5	17.5	3600
Sodium nitrate	$NaNO_3$	310	79.6	44.2	3760

Compound	Formula	Melting point, °C	Heat of fusion Btu/lb	Heat of fusion cal/g	Heat of fusion cal/g mole
Sodium peroxide	Na_2O_2	460	135.	75.1	5860
Sodium phosphate, meta-	$NaPO_3$	988	(87.5)	(48.6)	(4960)
Sodium pyrophosphate	$Na_2P_2O_7$	970	(92.7)	(51.5)	(13700)
Sodium silicate, aluminum-	$NaAlSi_3O_8$	1107	90.2	50.1	13150
Sodium silicate, di-	$Na_2Si_2O_5$	884	83.5	46.4	8460
Sodium silicate, meta-	Na_2SiO_3	1087	152.	84.4	10300
Sodium sulfate	Na_2SO_4	884	73.8	41.0	5830
Sodium sulfide	Na_2S	920	(27.7)	15.4	(1200)
Sodium thiocyanate	NaSCN	323	98.6	54.8	4450
Sodium tungstate	Na_2WO_4	702	35.3	19.6	5800
Strontium	Sr	757	45.0	25.0	2190
Strontium bromide	$SrBr_2$	643	34.7	19.3	4780
Strontium chloride	$SrCl_2$	872	47.7	26.5	4100
Strontium fluoride	SrF_2	1400	61.2	34.0	4260
Strontium oxide	SrO	2430	290.	161.2	16700
Sulfur (monatomic)	S	119	16.6	9.2	295
Sulfur dioxide	SO_2	−73.2	58.0	32.2	2060
Sulfur trioxide (α)	SO_3	16.8	46.4	25.8	2060
Sulfur trioxide (β)	SO_3	32.3	65.0	36.1	2890
Sulfur trioxide (γ)	SO_3	62.1	142.	79.0	6310
Tantalum	Ta	2996±50	(62.3)	34.6 to 41.5	(7500)
Tantalum pentachloride	$TaCl_5$	206.8	45.2	25.1	9000
Tantalum pentoxide	Ta_2O_5	1877	195.	108.6	48000
Tellurium	Te	453	45.5	25.3	3230
Terbium	Tb	1356	44.3	24.6	3900
Thallium	Tl	302.4	9.0	5.0	1030
Thallium bromide, mono-	TlBr	460	37.8	21.0	5990
Thallium carbonate	Tl_2CO_3	273	17.1	9.5	4400
Thallium chloride, mono-	TlCl	427	31.9	17.7	4260
Thallium iodide, mono-	TlI	440	16.9	9.4	3125
Thallium nitrate	$TlNO_3$	207	15.5	8.6	2290
Thallium sulfate	Tl_2SO_4	632	19.6	10.9	5500
Thallium sulfide	Tl_2S	449	12.2	6.8	3000

Table 5-5. HEAT OF FUSION OF SOME INORGANIC COMPOUNDS (Continued)

Compound	Formula	Melting point, °C	Heat of fusion Btu/lb	cal/g	cal/g mole
Thorium	Th	1845	(<35.6)	(<19.8)	(<4600)
Thorium chloride	ThCl$_4$	765	111.	51.6	22500
Thorium dioxide	ThO$_2$	2952	1984.	1102.0	291100
Thulium	Tm	1545	46.8	26.0	4400
Tin	Sn	231.7	25.9	14.4	1720
Tin bromide, di-	SnBr$_2$	231.8	(11.0)	(6.1)	(1720)
Tin bromide, tetra-	SnBr$_4$	29.8	12.2	6.8	3000
Tin chloride, di-	SnCl$_2$	247	28.8	16.0	3050
Tin chloride, tetra-	SnCl$_4$	-33.3	15.1	8.4	2190
Tin iodide, tetra-	SnI$_4$	143.4	(12.4)	(6.9)	(4330)
Tin oxide	SnO	1042	(84.2)	(46.8)	(6400)
Titanium	Ti	1800	(188.)	(104.4)	(5000)
Titanium bromide, tetra-	TiBr$_4$	38	(10.1)	(5.6)	(2060)
Titanium chloride, tetra-	TiCl$_4$	-23.2	21.4	11.9	2240
Titanium dioxide	TiO$_2$	1825	(257.)	(142.7)	(11400)
Titanium oxide	TiO	991	394	219	14000
Tungsten	W	3387	(82.4)	(45.8)	(8420)
Tungsten dioxide	WO$_2$	1270	108.	60.1	13940
Tungsten hexafluoride	WF$_6$	-0.5	10.8	6.0	1800
Tungsten tetrachloride	WCl$_4$	327	33.1	18.4	6000
Tungsten trioxide	WO$_3$	1470	108.	60.1	13940
Uranium[235]	U	~1133	36	20	3700
Uranium tetrachloride	UCl$_4$	590	48.8	27.1	10300
Vanadium	V	1917	(126)	(70)	(4200)
Vanadium dichloride	VCl$_2$	1027	118.	65.6	8000
Vanadium oxide	VO	2077	403.	224.0	15000
Vanadium pentoxide	V$_2$O$_5$	670	154.	85.5	15560
Xenon	Xe	-111.6	10.1	5.6	740
Ytterbium	Yb	823	22.9	12.7	2200
Yttrium	Y	1504	83.0	46.1	4100
Yttrium oxide	Y$_2$O$_3$	2227	199.	110.7	25000
Zinc	Zn	419.4	43.9	24.4	1595
Zinc chloride	ZnCl$_2$	283	(73.1)	(40.6)	(5540)
Zinc oxide	ZnO	1975	98.8	54.9	4470
Zinc sulfide	ZnS	1745	168.	(93.3)	(9100)
Zirconium	Zr	1857	(108)	(60)	(5500)
Zirconium dichloride	ZrCl$_2$	727	81.0	45.0	7300
Zirconium oxide	ZrO$_2$	2715	304.	168.8	20800

*From: "Handbook of Chemistry and Physics", 53rd ed., R.C. Weast, Ed., The Chemical Rubber Co., 1972.

Table 5-6. THERMAL ENERGY STORAGE BY HEAT OF FUSION

For design of energy-storage units at specific desired temperatures, as in air-conditioning installations, eutectic mixtures of salts offer many possibilities.

For heat of fusion in J/kg, multiply values in cal/g by 4184. For melting point in K, add 273.15 to values in °C.

Material	Formula	Melting point		Heat of fusion	
		°C	°F	Cal/g	Btu/lb
Magnesium oxide	MgO	2800	5072	460	828
Beryllium oxide	BeO	2547	4617	680	1224
50% CaO–50% MgO		2300	4172	340	612
Boron	B	2170	3938	490	882
Aluminum oxide	Al_2O_3	2045	3713	260	468
Magnesium silicate	$MgSiO_3$	1525	2777	146	263
Nickel	Ni	1455	2651	71.6	129
Silicon	Si	1400	2552	431	776
$CoSi_3$		1306	2383	238	428
Beryllium	Be	1280	2336	250	450
Cuprous oxide	Cu_2O	1229	2244	93.6	168
Calcium borate	$CaOB_2O_3$	1162	2124	141	254
Mg_2Si		1106	2023	270	486
Sodium silicate (meta)	Na_2SiO_3	1089	1992	102	184
Aluminum sodium fluoride	Na_3AlF_6	1000	1832	99.1	178
Sodium fluoride	NaF	995	1823	180	324
Potassium borate (meta)	KBO_2	947	1737	69.6	125
Potassium carbonate	K_2CO_3	896	1645	56.4	102
Sodium sulfate	Na_2SO_4	890	1634	40.8	73
Potassium fluoride	KF	860	1580	115	207
Sodium carbonate	Na_2CO_3	854	1567	70	136
Lithium fluoride	LiF	850	1560	250	450
Calcium	Ca	845	1553	55.0	99
Sodium chloride	NaCl	800	1472	116	209
Calcium chloride	$CaCl_2$	782	1440	60	108
Sodium bromide	NaBr	750	1382	59.3	107
Vanadium oxide	V_2O_5	670	1238	85.5	154
Aluminum	Al	660	1220	96	173
Magnesium	Mg	650	1202	88.9	160
Lithium chloride	LiCl	610	1130	75.5	136
Calcium nitrate	$Ca(NO_3)_2$	561	1042	31.1	56.0
Magnesium potassium chloride	$KCl \cdot MgCl_2$	487	909	74.7	134
50% LiF–50% BeF_2		360	680	153	275
Lithium	Li	180	356	150	286
Sodium	Na	98	208	27.4	49.3
Calcium nitrate	$Ca(NO_3)_2 \cdot 4H_2O$	41	106	50	90
Sodium phosphate	$Na_2HPO_4 \cdot 12H_2O$	34.6	94	66.8	120
Sodium carbonate	$Na_2CO_3 \cdot 10H_2O$	32–36	90–97	63.8	115
Sodium sulfate	$Na_2SO_4 \cdot 10H_2O$	32	90	57.7	104
Lithium nitrate	$LiNO_3 \cdot 3H_2O$	29.9	86	70.7	127
Calcium chloride	$CaCl_2 \cdot 6H_2O$	29–39	84–102	41.6	75
Ice	H_2O	0	32	80	144
Dry ice	CO_2	−109.5	−78	137	246

Table 5-7. FLOW EQUATIONS FOR A PERFECT GAS †

SYMBOLS:

A = cross-sectional area of flow passage
a = speed of sound
D = diameter of flow passage
D_h = hydraulic diameter of the constant-area passage
F = impulse function = $pA + \rho AV^2$
f = friction factor
g_c = conversion factor in Newton's law of motion (unity for any consistent set of units)
H = enthalpy
k = specific heat ratio (constant)
L = length of flow passage
L_{max} = length of passage (duct) where the Mach number reaches unity
M = Mach number

p = pressure
R = gas constant
S = entropy
T = temperature, absolute
V = velocity
ρ = mass density

SUPERSCRIPTS:
0 = stagnation state
$*$ = critical state, $M = 1$

SUBSCRIPTS:
o = standard conditions
x = upstream supersonic side of a shock
y = downstream side of a shock

A. FORMULAS USED TO CALCULATE THE ONE-DIMENSIONAL COMPRESSIBLE FLOW FUNCTIONS OF A PERFECT GAS FOR A FRICTIONLESS CONSTANT-AREA FLOW PROCESS WITH HEAT TRANSFER, Table 5-8

1.
$$\frac{p^{0*}}{p^0} = \frac{1+kM^2}{k+1}\left(\frac{k+1}{2}\right)^{k/(k-1)}\left(1+\frac{k-1}{2}M^2\right)^{-k/(k-1)}$$

2.
$$\frac{p}{p^*} = \frac{k+1}{1+kM^2}$$

3.
$$\frac{p}{p^{0*}} = \left(\frac{2}{k+1}\right)^{k/(k-1)}\frac{k+1}{1+kM^2}$$

4.
$$\frac{T^0}{T^{0*}} = \frac{2(k+1)M^2}{(1+kM^2)^2}\left(1+\frac{k-1}{2}M^2\right)$$

5.
$$\frac{T}{T^*} = M^2\left(\frac{k+1}{1+kM^2}\right)^2$$

6.
$$\frac{T}{T^{0*}} = \frac{2(k+1)M^2}{(1+kM^2)^2}$$

7.
$$\frac{\rho^*}{\rho} = \frac{V}{a^*} = \frac{(k+1)M^2}{1+kM^2}$$

8.
$$\frac{\rho V^2}{2g_c p^{0*}} = \frac{kM^2}{2}\left(\frac{2}{k+1}\right)^{k/(k-1)}\frac{k+1}{1+kM^2}$$

9.
$$\frac{S^*-S}{R} = -\frac{k}{k-1}\ln M^2 - \frac{k+1}{k-1}\ln\left(\frac{k+1}{1+kM^2}\right)$$

†From: "Air Tables", D. Jordan and M.D. Mintz, McGraw-Hill Book Company, 1965. This book contains complete tables computed from these equations for six values of the specific heat ratio k = 1.20 to k = 1.67, for perfect gases and for real air.

Table 5-7.   FLOW EQUATIONS FOR A PERFECT GAS *(Continued)*

B. FORMULAS USED TO CALCULATE THE ONE-DIMENSIONAL COMPRESSIBLE FLOW FUNCTIONS OF A PERFECT GAS FOR AN ADIABATIC CONSTANT-AREA FLOW PROCESS WITH FRICTION, Table 5-9

1. $$\frac{p^{0*}}{p^0} = M\left[\frac{2}{k+1}\left(1+\frac{k-1}{2}M^2\right)\right]^{-(k+1)/[2(k-1)]}$$

2. $$\frac{p^*}{p} = M\left[\frac{2}{k+1}\left(1+\frac{k-1}{2}M^2\right)\right]^{1/2}$$

3. $$\frac{p^{0*}}{p} = M\left(\frac{k+1}{2}\right)^{(k+1)/[2(k-1)]}\left(1+\frac{k-1}{2}M^2\right)^{1/2}$$

4. $$\frac{T}{T^*} = \frac{k+1}{2}\left(1+\frac{k-1}{2}M^2\right)^{-1}$$

5. $$\frac{\rho^*}{\rho} = \frac{V}{a^*} = M\left[\frac{k+1}{2}\left(1+\frac{k-1}{2}M^2\right)^{-1}\right]^{1/2}$$

6. $$\frac{\rho V^2}{2g_c p^{0*}} = \frac{kM}{2}\left(\frac{2}{k+1}\right)^{(k+1)/[2(k-1)]}\left(1+\frac{k-1}{2}M^2\right)^{-1/2}$$

7. $$\frac{F^*}{F} = \frac{M}{1+kM^2}\left[2(k+1)\left(1+\frac{k-1}{2}M^2\right)\right]^{1/2}$$

8. $$4f\frac{L_{max}}{D_h} = \frac{1-M^2}{kM^2} + \frac{k+1}{2k}\ln\left[\frac{k+1}{2}M^2\left(1+\frac{k-1}{2}M^2\right)^{-1}\right]$$

9. $$\frac{S^*-S}{R} = -\frac{1}{2}\ln M^2 - \frac{1}{2}\frac{k+1}{k-1}\ln\left[\frac{k+1}{2}\left(1+\frac{k-1}{2}M^2\right)^{-1}\right]$$

C. FORMULAS USED TO CALCULATE THE ONE-DIMENSIONAL COMPRESSIBLE FLOW FUNCTIONS OF A PERFECT GAS FOR AN ISOTHERMAL CONSTANT-AREA FLOW PROCESS, Table 5-10

1. $$\frac{p^{0*}}{p^0} = k^{1/2}M\left[\frac{2k}{3k-1}\left(1+\frac{k-1}{2}M^2\right)\right]^{-k/(k-1)}$$

2. $$\frac{p^*}{p} = \frac{\rho^*}{\rho} = \frac{V}{a^*} = k^{1/2}M$$

3. $$\frac{p^{0*}}{p} = k^{1/2}M\left(1+\frac{k-1}{2k}\right)^{k/(k-1)}$$

4. $$\frac{T^0}{T^{0*}} = \frac{2k}{3k-1}\left(1+\frac{k-1}{2}M^2\right)$$

5. $$\frac{\rho V^2}{2g_c p^{0*}} = \frac{1}{2}k^{1/2}M\left(1+\frac{k-1}{2k}\right)^{-k/(k-1)}$$

6. $$\frac{F^*}{F} = \frac{2k^{1/2}M}{1+kM^2}$$

7. $$4f\frac{L_{max}}{D_h} = \frac{1-kM^2}{kM^2} + \ln(kM^2)$$

8. $$\frac{S^*-S}{R} = -\ln(k^{1/2}M)$$

Table 5-7.   FLOW EQUATIONS FOR A PERFECT GAS *(Continued)*

D.   FORMULAS USED TO CALCULATE ONE-DIMENSIONAL, COMPRESSIBLE FLOW FUNCTIONS OF A PERFECT GAS FOR AN ISENTROPIC FLOW PROCESS, Table 5-11

1.
$$\frac{p}{p^0} = \left(1 + \frac{k-1}{2}M^2\right)^{-k/(k-1)}$$

2.
$$\frac{T}{T^0} = \left(1 + \frac{k-1}{2}M^2\right)^{-1}$$

3.
$$\frac{\rho}{\rho^0} = \left(1 + \frac{k-1}{2}M^2\right)^{-1/(k-1)}$$

4.
$$\frac{\rho V^2}{2g_c p^0} = \frac{kM^2}{2}\left(1 + \frac{k-1}{2}M^2\right)^{-k/(k-1)}$$

5.
$$\frac{A^*}{A} = M\left[\frac{2}{k+1}\left(1 + \frac{k-1}{2}M^2\right)\right]^{-(k+1)/[2(k-1)]}$$

6.
$$\frac{F^*}{F} = \frac{M}{1 + kM^2}\left[2(k+1)\left(1 + \frac{k-1}{2}M^2\right)\right]^{1/2}$$

7.
$$\frac{V}{a^*} = M\left(\frac{k+1}{2}\right)^{1/2}\left(1 + \frac{k-1}{2}M^2\right)^{-1/2}$$

8.
$$\frac{\rho V}{p}\left(\frac{RT^0}{g_c}\right)^{1/2} = M\left[k\left(1 + \frac{k-1}{2}M^2\right)\right]^{1/2}$$

9.
$$\frac{\rho V}{p^0}\left(\frac{RT^0}{g_c}\right)^{1/2} = Mk^{1/2}\left(1 + \frac{k-1}{2}M^2\right)^{-(k+1)/[2(k-1)]}$$

E.   FORMULAS USED TO CALCULATE THE ONE-DIMENSIONAL COMPRESSIBLE FLOW FUNCTIONS OF A PERFECT GAS FOR A NORMAL-SHOCK FLOW PROCESS, Table 5-11B

1.
$$M_y = \left[\left(M_x^2 + \frac{2}{k-1}\right)\left(\frac{2k}{k-1}M_x^2 - 1\right)^{-1}\right]^{1/2}$$

2.
$$\frac{p_y^0}{p_x^0} = \left[\frac{k+1}{2}M_x^2\left(1 + \frac{k-1}{2}M_x^2\right)^{-1}\right]^{k/(k-1)}\left(\frac{2k}{k+1}M_x^2 - \frac{k-1}{k+1}\right)^{-1/(k-1)}$$

3.
$$\frac{p_x}{p_y} = \left(\frac{2k}{k+1}M_x^2 - \frac{k-1}{k+1}\right)^{-1}$$

4.
$$\frac{p_x}{p_x^0} = \left(1 + \frac{k-1}{2}M_x^2\right)^{-k/(k-1)}$$

5.
$$\frac{T_x}{T_y} = \frac{(k+1)^2 M_x^2}{2(k-1)}\left[\left(1 + \frac{k-1}{2}M_x^2\right)\left(\frac{2k}{k-1}M_x^2 - 1\right)\right]^{-1}$$

6.
$$\frac{T_x}{T_x^0} = \left(1 + \frac{k-1}{2}M_x^2\right)^{-1}$$

7.
$$\frac{\rho_x}{\rho_y} = \frac{V_y}{V_x} = \left(\frac{2kM_x^2}{k-1} - 1\right)\left(1 + \frac{k-1}{2}M_x^2\right)\left\{\left(\frac{2kM_x^2}{k+1} - \frac{k-1}{k+1}\right)\left[\frac{(k+1)^2}{2(k-1)}M_x^2\right]\right\}^{-1}$$

8.
$$\frac{S_y - S_x}{R} = \frac{k}{k-1}\ln\left[\frac{2}{(k+1)M_x^2} + \frac{k-1}{k+1}\right] + \frac{1}{k-1}\ln\left(\frac{2k}{k+1}M_x^2 - \frac{k-1}{k+1}\right)$$

Table 5-8. FLOW FUNCTIONS OF A PERFECT GAS— RAYLEIGH LINE†

For explanation of symbols, see Table 5-7. For equations see Table 5-7A.

The following table gives one-dimensional compressible flow for the frictionless constant-area flow process with heat transfer, $k = 1.4$.

M	p^{0*}/p^0	p/p^*	p/p^{0*}	T^0/T^{0*}	T/T^*	T/T^{0*}	$\frac{\rho^*}{\rho} = \frac{V}{a^*}$	$\frac{\rho V^2}{2g_c p^{0*}}$	$\frac{S^*-S}{R}$	M
0.00	0.7887	2.400	1.268	0.0000000	0.0000000	0.0000000	0.0000000	0.0000000	—	0.00
0.05	0.7901	2.392	1.263	0.01192	0.01430	0.01192	0.005979	0.002211	15.74	0.05
0.10	0.7942	2.367	1.250	0.04678	0.05602	0.04668	0.02367	0.008753	10.95	0.10
0.15	0.8009	2.327	1.229	0.1020	0.1218	0.1015	0.05235	0.01936	8.213	0.15
0.20	0.8100	2.273	1.201	0.1736	0.2066	0.1722	0.09091	0.03362	6.340	0.20
0.25	0.8212	2.207	1.166	0.2568	0.3044	0.2537	0.1379	0.05101	4.955	0.25
0.30	0.8343	2.131	1.126	0.3469	0.4089	0.3407	0.1918	0.07094	3.887	0.30
0.35	0.8489	2.049	1.082	0.4389	0.5141	0.4284	0.2510	0.09280	3.046	0.35
0.40	0.8646	1.961	1.036	0.5290	0.6151	0.5126	0.3137	0.1160	2.374	0.40
0.45	0.8810	1.870	0.9878	0.6139	0.7080	0.5900	0.3787	0.1400	1.834	0.45
0.50	0.8976	1.778	0.9392	0.6914	0.7901	0.6584	0.4444	0.1644	1.400	0.50
0.55	0.9141	1.686	0.8907	0.7599	0.8599	0.7166	0.5100	0.1886	1.051	0.55
0.60	0.9300	1.596	0.8430	0.8189	0.9167	0.7639	0.5745	0.2124	0.7717	0.60
0.65	0.9450	1.508	0.7967	0.8683	0.9608	0.8007	0.6371	0.2356	0.5507	0.65
0.70	0.9587	1.423	0.7520	0.9085	0.9929	0.8274	0.6975	0.2579	0.3781	0.70
0.75	0.9708	1.343	0.7093	0.9401	1.014	0.8450	0.7552	0.2793	0.2459	0.75
0.80	0.9810	1.266	0.6687	0.9639	1.025	0.8546	0.8101	0.2996	0.1477	0.80
0.85	0.9892	1.193	0.6303	0.9810	1.029	0.8571	0.8620	0.3188	0.07810	0.85
0.90	0.9952	1.125	0.5941	0.9921	1.025	0.8538	0.9110	0.3369	0.03270	0.90
0.95	0.9988	1.060	0.5601	0.9981	1.015	0.8455	0.9569	0.3539	0.007714	0.95
1.00	1.0000	1.0000	0.5283	1.0000	1.0000	0.8333	1.000	0.3698	0.000000	1.00
1.05	0.9988	0.9436	0.4985	0.9984	0.9816	0.8180	1.040	0.3847	0.006903	1.05
1.10	0.9952	0.8909	0.4706	0.9939	0.9603	0.8003	1.078	0.3986	0.02618	1.10
1.15	0.9892	0.8417	0.4446	0.9872	0.9369	0.7807	1.113	0.4116	0.05593	1.15
1.20	0.9810	0.7958	0.4204	0.9787	0.9118	0.7599	1.146	0.4237	0.09453	1.20
1.25	0.9706	0.7529	0.3978	0.9689	0.8858	0.7382	1.176	0.4351	0.1406	1.25
1.30	0.9582	0.7130	0.3767	0.9580	0.8592	0.7160	1.205	0.4456	0.1930	1.30
1.35	0.9439	0.6758	0.3570	0.9464	0.8323	0.6936	1.232	0.4554	0.2507	1.35
1.40	0.9279	0.6410	0.3386	0.9343	0.8054	0.6712	1.256	0.4646	0.3128	1.40
1.45	0.9105	0.6086	0.3215	0.9218	0.7787	0.6490	1.280	0.4732	0.3787	1.45
1.50	0.8916	0.5783	0.3055	0.9093	0.7525	0.6271	1.301	0.4812	0.4476	1.50
1.55	0.8716	0.5500	0.2906	0.8967	0.7268	0.6057	1.321	0.4887	0.5190	1.55
1.60	0.8506	0.5236	0.2766	0.8842	0.7017	0.5848	1.340	0.4956	0.5926	1.60
1.70	0.8063	0.4756	0.2513	0.8597	0.6538	0.5448	1.375	0.5083	0.7444	1.70
1.80	0.7599	0.4335	0.2290	0.8363	0.6089	0.5075	1.405	0.5194	0.9003	1.80
1.90	0.7126	0.3964	0.2094	0.8141	0.5673	0.4728	1.431	0.5292	1.059	1.90
2.00	0.6653	0.3636	0.1921	0.7934	0.5289	0.4408	1.455	0.5379	1.218	2.00
2.50	0.4501	0.2462	0.1300	0.7101	0.3787	0.3156	1.538	0.5689	1.997	2.50
3.00	0.2920	0.1765	0.09323	0.6540	0.2803	0.2336	1.588	0.5873	2.717	3.00
3.50	0.1877	0.1322	0.06986	0.6158	0.2142	0.1785	1.620	0.5990	3.370	3.50
4.00	0.1216	0.1026	0.05418	0.5891	0.1683	0.1403	1.641	0.6068	3.960	4.00
4.50	0.07999	0.08177	0.04320	0.5698	0.1354	0.1128	1.656	0.6123	4.494	4.50
5.00	0.05367	0.06667	0.03522	0.5556	0.1111	0.09259	1.667	0.6163	4.982	5.00
5.50	0.03675	0.05536	0.02925	0.5447	0.09272	0.07727	1.675	0.6193	5.430	5.50
6.00	0.02568	0.04669	0.02467	0.5363	0.07849	0.06541	1.681	0.6216	5.843	6.00
7.00	0.01326	0.03448	0.01822	0.5244	0.05826	0.04855	1.690	0.6248	6.582	7.00
8.00	0.007319	0.02649	0.01399	0.5165	0.04491	0.03743	1.695	0.6269	7.230	8.00
9.00	0.004276	0.02098	0.01108	0.5110	0.03565	0.02971	1.699	0.6284	7.805	9.00

†Condensed from: "Air Tables", D. Jordan and M.D. Mintz, McGraw–Hill Book Company, 1965. This book contains 850 pages of tables and examples of their use. The tables are computed from the equations given in Table 5-7 herewith, for six values of the specific heat ratio k = 1.20 to k = 1.67, for perfect gases and for real air.

Table 5-9. FLOW FUNCTIONS OF A PERFECT GAS—FANNO LINE†

For explanation of symbols, see Table 5-7. For equations see Table 5-7B.

The following table gives one-dimensional compressible flow for an adiabatic constant-area flow process with friction, $k = 1.4$.

M	p^{0*}/p^0	p^*/p	p^{0*}/p	T/T^*	$\dfrac{\rho^*}{\rho} = \dfrac{V}{a^*}$	$\dfrac{\rho V^2}{2g_c p^0}$	F^*/F	$\dfrac{4fL_{max}}{D_h}$	$\dfrac{S^*-S}{R}$	M
0.00	0.00000	0.000000	0.00000	1.200	0.00000	0.000000	0.00000	—	—	0.00
0.05	0.08627	0.04565	0.08642	1.199	0.05476	0.02025	0.1092	280.0	2.450	0.05
0.10	0.1718	0.09138	0.1730	1.198	0.1094	0.04047	0.2163	66.92	1.762	0.10
0.15	0.2557	0.1372	0.2598	1.195	0.1639	0.06063	0.3193	27.93	1.364	0.15
0.20	0.3374	0.1833	0.3470	1.190	0.2182	0.08070	0.4166	14.53	1.086	0.20
0.25	0.4162	0.2296	0.4347	1.185	0.2722	0.1006	0.5068	8.483	0.8766	0.25
0.30	0.4914	0.2763	0.5230	1.179	0.3257	0.1204	0.5889	5.299	0.7105	0.30
0.35	0.5624	0.3234	0.6122	1.171	0.3788	0.1401	0.6625	3.452	0.5755	0.35
0.40	0.6289	0.3709	0.7022	1.163	0.4313	0.1595	0.7273	2.308	0.4638	0.40
0.45	0.6903	0.4190	0.7932	1.153	0.4833	0.1787	0.7835	1.566	0.3706	0.45
0.50	0.7464	0.4677	0.8853	1.143	0.5345	0.1977	0.8315	1.069	0.2926	0.50
0.55	0.7968	0.5170	0.9787	1.132	0.5851	0.2164	0.8717	0.7281	0.2271	0.55
0.60	0.8416	0.5671	1.073	1.119	0.6348	0.2348	0.9049	0.4908	0.1724	0.60
0.65	0.8806	0.6179	1.170	1.107	0.6837	0.2528	0.9318	0.3246	0.1272	0.65
0.70	0.9138	0.6696	1.267	1.093	0.7318	0.2706	0.9532	0.2081	0.09018	0.70
0.75	0.9412	0.7221	1.367	1.079	0.7789	0.2880	0.9696	0.1273	0.06055	0.75
0.80	0.9632	0.7756	1.468	1.064	0.8251	0.3051	0.9818	0.07229	0.03752	0.80
0.85	0.9798	0.8301	1.571	1.048	0.8704	0.3219	0.9904	0.03633	0.02046	0.85
0.90	0.9912	0.8856	1.676	1.033	0.9146	0.3382	0.9960	0.01451	0.008824	0.90
0.95	0.9979	0.9422	1.784	1.017	0.9578	0.3542	0.9991	0.003278	0.002143	0.95
1.00	1.0000	1.000	1.893	1.0000	1.000	0.3698	1.0000	0.0000000	0.0000000	1.00
1.05	0.9980	1.059	2.004	0.9832	1.041	0.3850	0.9992	0.002714	0.002027	1.05
1.10	0.9921	1.119	2.118	0.9662	1.081	0.3998	0.9970	0.009935	0.007894	1.10
1.15	0.9828	1.181	2.235	0.9490	1.120	0.4143	0.9936	0.02053	0.01730	1.15
1.20	0.9705	1.243	2.353	0.9317	1.158	0.4283	0.9893	0.03364	0.02999	1.20
1.25	0.9553	1.307	2.475	0.9143	1.195	0.4420	0.9843	0.04858	0.04569	1.25
1.30	0.9378	1.373	2.598	0.8969	1.231	0.4553	0.9788	0.06483	0.06420	1.30
1.35	0.9182	1.440	2.725	0.8794	1.266	0.4682	0.9728	0.08199	0.08529	1.35
1.40	0.8969	1.508	2.854	0.8621	1.300	0.4807	0.9666	0.09974	0.1088	1.40
1.45	0.8742	1.578	2.986	0.8448	1.333	0.4928	0.9601	0.1178	0.1345	1.45
1.50	0.8502	1.649	3.121	0.8276	1.365	0.5046	0.9536	0.1361	0.1623	1.50
1.55	0.8254	1.722	3.259	0.8105	1.395	0.5160	0.9469	0.1543	0.1919	1.55
1.60	0.7998	1.796	3.400	0.7937	1.425	0.5271	0.9403	0.1724	0.2233	1.60
1.70	0.7476	1.949	3.690	0.7605	1.482	0.5482	0.9272	0.2078	0.2909	1.70
1.80	0.6949	2.109	3.993	0.7282	1.536	0.5680	0.9145	0.2419	0.3639	1.80
1.90	0.6430	2.276	4.308	0.6969	1.586	0.5865	0.9023	0.2743	0.4416	1.90
2.00	0.5926	2.449	4.637	0.6667	1.633	0.6039	0.8907	0.3050	0.5232	2.00
2.50	0.3793	3.423	6.480	0.5333	1.826	0.6752	0.8427	0.4320	0.9695	2.50
3.00	0.2362	4.583	8.674	0.4286	1.964	0.7263	0.8087	0.5222	1.443	3.00
3.50	0.1473	5.935	11.23	0.3478	2.064	0.7633	0.7847	0.5864	1.915	3.50
4.00	0.09329	7.483	14.17	0.2857	2.138	0.7907	0.7675	0.6331	2.372	4.00
4.50	0.06038	9.231	17.47	0.2376	2.194	0.8112	0.7549	0.6676	2.807	4.50
5.00	0.04000	11.18	21.16	0.2000	2.236	0.8269	0.7454	0.6938	3.219	5.00
5.50	0.02712	13.33	25.23	0.1702	2.269	0.8391	0.7381	0.7140	3.607	5.50
6.00	0.01880	15.68	29.69	0.1463	2.295	0.8488	0.7323	0.7299	3.974	6.00
7.00	0.009602	21.00	39.75	0.1111	2.333	0.8629	0.7241	0.7528	4.646	7.00
8.00	0.005260	27.13	51.35	0.08696	2.359	0.8724	0.7187	0.7682	5.248	8.00
9.00	0.003056	34.07	64.50	0.06977	2.377	0.8791	0.7148	0.7790	5.791	9.00

†Condensed from: "Air Tables", D. Jordan and M.D. Mintz, McGraw–Hill Book Company, 1965. This book contains 850 pages of tables and examples of their use. The tables are computed from the equations given in Table 5-7 herewith, for six values of the specific heat ratio k = 1.20 to k = 1.67, for perfect gases and for real air.

Table 5-10. FLOW FUNCTIONS OF A PERFECT GAS— ISOTHERMAL[†]

For explanation of symbols, see Table 5-7. For equations see Table 5-7C.

The following table gives one-dimensional compressible flow for an isothermal constant-area flow process, $k = 1.4$.

SYMBOLS:

D_h = hydraulic diameter of the constant-area passage

F = impulse function = $pA + \rho A V^2$

f = friction factor

g_c = conversion factor in Newton's law of motion (unity for any consistent set of units)

L_{max} = length of passage (duct) where the Mach number reaches unity

M = Mach number

p = pressure

R = gas constant

S = entropy

T = temperature, absolute

V = velocity

ρ = mass density

SUPERSCRIPTS:

0 = stagnation state

* = critical state

M	p^{0*}/p^0	p^*/p	p^{0*}/p	T^0/T^{0*}	$\dfrac{\rho V^2}{2g_c p^{0*}}$	F^*/F	$\dfrac{4fL_{max}}{D_h}$	$\dfrac{S^*-S}{R}$	M
0.0000	0.00000	0.00000	0.00000	0.8750	0.000000	0.00000	—	—	0.0000
0.0100	0.01888	0.01183	0.01888	0.8750	0.003707	0.02366	7133.	4.437	0.0100
0.0500	0.09424	0.05916	0.09441	0.8754	0.01854	0.1179	279.1	2.827	0.0500
0.1000	0.1875	0.1183	0.1888	0.8767	0.03707	0.2334	66.16	2.134	0.1000
0.1500	0.2788	0.1775	0.2832	0.8789	0.05561	0.3441	27.29	1.729	0.1500
0.2000	0.3672	0.2366	0.3776	0.8820	0.07415	0.4482	13.97	1.441	0.2000
0.2500	0.4520	0.2958	0.4720	0.8859	0.09268	0.5440	7.992	1.218	0.2500
0.3000	0.5322	0.3550	0.5664	0.8907	0.1112	0.6305	4.865	1.036	0.3000
0.3500	0.6072	0.4141	0.6609	0.8964	0.1298	0.7070	3.068	0.8816	0.3500
0.4000	0.6764	0.4733	0.7553	0.9030	0.1483	0.7733	1.968	0.7481	0.4000
0.4500	0.7394	0.5324	0.8497	0.9104	0.1668	0.8297	1.267	0.6303	0.4500
0.5000	0.7959	0.5916	0.9441	0.9187	0.1854	0.8765	0.8073	0.5249	0.5000
0.5500	0.8455	0.6508	1.038	0.9279	0.2039	0.9143	0.5021	0.4296	0.5500
0.6000	0.8882	0.7099	1.133	0.9380	0.2224	0.9441	0.2989	0.3426	0.6000
0.6500	0.9239	0.7691	1.227	0.9489	0.2410	0.9665	0.1655	0.2625	0.6500
0.7000	0.9529	0.8283	1.322	0.9607	0.2595	0.9825	0.08085	0.1884	0.7000
0.7500	0.9751	0.8874	1.416	0.9734	0.2781	0.9929	0.03095	0.1194	0.7500
0.8000	0.9909	0.9466	1.511	0.9870	0.2966	0.9985	0.006257	0.05491	0.8000
0.8452	1.0000	1.0000	1.596	1.0000	0.3133	1.0000	0.0000000	0.000000	0.8452

[†]Condensed from: "Air Tables", D. Jordan and M.D. Mintz, McGraw–Hill Book Company, 1965. This book contains 850 pages of tables and examples of their use. The tables are computed from the equations given in Table 5-7 herewith, for six values of the specific heat ratio $k = 1.20$ to $k = 1.67$, for perfect gases and for real air.

Table 5-11. COMPRESSIBLE FLOW AERODYNAMIC TABLES*

Ratios of Properties for $c_p/c_v = 1.4$; No Heat Transfer; $c_p = 0.24$

For equations see Tables 5-7D and 5-7E.

Table A. ISENTROPIC SUBSONIC FLOW, $c_p/c_v \doteq 1.4$

SYMBOLS *(see also Table 5-7)*:

v = Prandtl–Meyer angle (angle through which a supersonic stream is turned to expand from $M = 1$ to $M > 1$), deg

μ = Mach angle, $\sin^{-1}\dfrac{1}{M}$, deg

A group of digits followed by a superscript n means that the decimal point should be shifted n places to the right. If n is negative, the decimal point should be shifted to the left. For example, 0.7000^{-4} means 0.00007000.

M	$\dfrac{p}{p^\circ}$	$\dfrac{\rho}{\rho^\circ}$	$\dfrac{T}{T^\circ}$	$\sqrt{1-M^2}$	$\dfrac{\rho V^2}{2p^\circ}$	$\dfrac{A}{A^*}$	$\dfrac{V}{a^*}$
0	1.0000	1.0000	1.0000	1.0000	0	∞	0
.01	.9999	1.0000	1.0000	1.0000	$.7000^{-4}$	57.8738	.01095
.02	.9997	.9998	.9999	.9998	$.2799^{-3}$	28.9421	.02191
.03	.9994	.9996	.9998	.9995	$.6296^{-3}$	19.3005	.03286
.04	.9989	.9992	.9997	.9992	$.1119^{-2}$	14.4815	.04381
.05	.9983	.9988	.9995	.9987	$.1747^{-2}$	11.5914	.05476
.06	.9975	.9982	.9993	.9982	$.2514^{-2}$	9.6659	.06570
.07	.9966	.9976	.9990	.9975	$.3418^{0}$	8.2915	.07664
.08	.9955	.9968	.9987	.9968	$.4460^{-2}$	7.2616	.08758
.09	.9944	.9960	.9984	.9959	$.5638^{-2}$	6.4613	.09851
.10	.9930	.9950	.9980	.9950	$.6951^{-2}$	5.8218	.10944
.11	.9916	.9940	.9976	.9939	$.8399^{-2}$	5.2992	.12035
.12	.9900	.9928	.9971	.9928	$.9979^{-2}$	4.8643	.13126
.13	.9883	.9916	.9966	.9915	$.1169^{-1}$	4.4969	.14217
.14	.9864	.9903	.9961	.9902	$.1353^{-1}$	4.1824	.15306
.15	.9844	.9888	.9955	.9887	$.1550^{-1}$	3.9103	.16395
.16	.9823	.9873	.9949	.9871	$.1760^{-1}$	3.6727	.17482
.17	.9800	.9857	.9943	.9854	$.1983^{-1}$	3.4635	.18569
.18	.9776	.9840	.9936	.9837	$.2217^{-1}$	3.2779	.19654
.19	.9751	.9822	.9928	.9818	$.2464^{-1}$	3.1123	.20739
.20	.9725	.9803	.9921	.9798	$.2723^{-1}$	2.9635	.21822
.21	.9697	.9783	.9913	.9777	$.2994^{-1}$	2.8293	.22904
.22	.9668	.9762	.9904	.9755	$.3276^{-1}$	2.7076	.23984
.23	.9638	.9740	.9895	.9732	$.3569^{-1}$	2.5968	.25063
.24	.9607	.9718	.9886	.9708	$.3874^{-1}$	2.4956	.26141
.25	.9575	.9694	.9877	.9682	$.4189^{-1}$	2.4027	.27217
.26	.9541	.9670	.9867	.9656	$.4515^{1}$	2.3173	.28291
.27	.9506	.9645	.9856	.9629	$.4851^{-1}$	2.2385	.29364
.28	.9470	.9619	.9846	.9600	$.5197^{-1}$	2.1656	.30435
.29	.9433	.9592	.9835	.9570	$.5553^{-1}$	2.0979	.31504
.30	.9395	.9564	.9823	.9539	$.5919^{-1}$	2.0351	.32572
.31	.9355	.9535	.9811	.9507	$.6293^{-1}$	1.9765	.33637
.32	.9315	.9506	.9799	.9474	$.6677^{-1}$	1.9219	.34701
.33	.9274	.9475	.9787	.9440	$.7069^{-1}$	1.8707	.35762
.34	.9231	.9445	.9774	.9404	$.7470^{-1}$	1.8229	.36822
.35	.9188	.9413	.9761	.9367	$.7879^{-1}$	1.7780	.37879
.36	.9143	.9380	.9747	.9330	$.8295^{-1}$	1.7358	.38935
.37	.9098	.9347	.9733	.9290	$.8719^{-1}$	1.6961	.39988
.38	.9052	.9313	.9719	.9250	$.9149^{-1}$	1.6587	.41039
.39	.9004	.9278	.9705	.9208	$.9587^{-1}$	1.6234	.42087

Table 5-11. COMPRESSIBLE FLOW AERODYNAMIC TABLES (Continued)

Table A. SUBSONIC FLOW (Continued)

M	$\dfrac{p}{p^{\circ}}$	$\dfrac{\rho}{\rho^{\circ}}$	$\dfrac{T}{T^{\circ}}$	$\sqrt{1-M^{2}}$	$\dfrac{\rho V^{2}}{2p^{\circ}}$	$\dfrac{A}{A^{*}}$	$\dfrac{V}{a^{*}}$
.40	.8956	.9243	.9690	.9165	.1003	1.5901	.43133
.41	.8907	.9207	.9675	.9121	.1048	1.5587	.44177
.42	.8857	.9170	.9659	.9075	.1094	1.5289	.45218
.43	.8807	.9132	.9643	.9028	.1140	1.5007	.46257
.44	.8755	.9094	.9627	.8980	.1187	1.4740	.47293
.45	.8703	.9055	.9611	.8930	.1234	1.4487	.48326
.46	.8650	.9016	.9594	.8879	.1281	1.4246	.49357
.47	.8596	.8976	.9577	.8827	.1329	1.4018	.50385
.48	.8541	.8935	.9560	.8773	.1378	1.3801	.51410
.49	.8486	.8894	.9542	.8717	.1426	1.3595	.52433
.50	.8430	.8852	.9524	.8660	.1475	1.3398	.53452
.51	.8374	.8809	.9506	.8602	.1525	1.3212	.54469
.52	.8317	.8766	.9487	.8542	.1574	1.3034	.55483
.53	.8259	.8723	.9468	.8480	.1624	1.2865	.56493
.54	.8201	.8679	.9449	.8417	.1674	1.2703	.57501
.55	.8142	.8634	.9430	.8352	.1724	1.2550	.58506
.56	.8082	.8589	.9410	.8285	.1774	1.2403	.59507
.57	.8022	.8544	.9390	.8216	.1825	1.2263	.60505
.58	.7962	.8498	.9370	.8146	.1875	1.2130	.61501
.59	.7901	.8451	.9349	.8074	.1925	1.2003	.62492
.60	.7840	.8405	.9328	.8000	.1976	1.1882	.63481
.61	.7778	.8357	.9307	.7924	.2026	1.1767	.64466
.62	.7716	.8310	.9286	.7846	.2076	1.1657	.65448
.63	.7654	.8262	.9265	.7766	.2127	1.1552	.66427
.64	.7591	.8213	.9243	.7684	.2177	1.1452	.67402
.65	.7528	.8164	.9221	.7599	.2227	1.1356	.68374
.66	.7465	.8115	.9199	.7513	.2276	1.1265	.69342
.67	.7401	.8066	.9176	.7424	.2326	1.1179	.70307
.68	.7338	.8016	.9153	.7332	.2375	1.1097	.71268
.69	.7274	.7966	.9131	.7238	.2424	1.1018	.72225
.70	.7209	.7916	.9107	.7141	.2473	1.0944	.73179
.71	.7145	.7865	.9084	.7042	.2521	1.0873	.74129
.72	.7080	.7814	.9061	.6940	.2569	1.0806	.75076
.73	.7016	.7763	.9037	.6834	.2617	1.0742	.76019
.74	.6951	.7712	.9013	.6726	.2664	1.0681	.76958
.75	.6886	.7660	.8989	.6614	.2711	1.0624	.77894
.76	.6821	.7609	.8964	.6499	.2758	1.0570	.78825
.77	.6756	.7557	.8940	.6380	.2804	1.0519	.79753
.78	.6691	.7505	.8915	.6258	.2849	1.0471	.80677
.79	.6625	.7452	.8890	.6131	.2894	1.0425	.81597
.80	.6560	.7400	.8865	.6000	.2939	1.0382	.82514
.81	.6495	.7347	.8840	.5864	.2983	1.0342	.83426
.82	.6430	.7295	.8815	.5724	.3027	1.0305	.84335
.83	.6365	.7242	.8789	.5578	.3069	1.0270	.85239
.84	.6300	.7189	.8763	.5426	.3112	1.0237	.86140
.85	.6235	.7136	.8737	.5268	.3153	1.0207	.87037
.86	.6170	.7083	.8711	.5103	.3195	1.0179	.87929
.87	.6106	.7030	.8685	.4931	.3235	1.0153	.88818
.88	.6041	.6977	.8659	.4750	.3275	1.0129	.89703
.89	.5977	.6924	.8632	.4560	.3314	1.0108	.90583
.90	.5913	.6870	.8606	.4359	.3352	1.0089	.91460
.91	.5849	.6817	.8579	.4146	.3390	1.0071	.92332
.92	.5785	.6764	.8552	.3919	.3427	1.0056	.93201
.93	.5721	.6711	.8525	.3676	.3464	1.0043	.94065
.94	.5658	.6658	.8498	.3412	.3500	1.0031	.94925
.95	.5595	.6604	.8471	.3122	.3534	1.0022	.95781
.96	.5532	.6551	.8444	.2800	.3569	1.0014	.96633
.97	.5469	.6498	.8416	.2431	.3602	1.0008	.97481
.98	.5407	.6445	.8389	.1990	.3635	1.0003	.98325
.99	.5345	.6392	.8361	.1411	.3667	1.0001	.99165
1.00	.5283	.6339	.8333	.0000	.3698	1.0000	1.00000

Table 5-11. COMPRESSIBLE FLOW AERODYNAMIC TABLES (Continued)

Table B. SUPERSONIC FLOW—ISENTROPIC FLOW AND NORMAL SHOCKS, $c_p/c_v \cong 1.4$

Although the values with the subscript y are for normal shock waves, the values $\frac{p_y}{p_x}$, $\frac{\rho_y}{\rho_x}$, $\frac{T_y}{T_x}$, and $\frac{t_y^\circ}{t_x^\circ}$ may also be used for oblique shock waves, provided that $M \sin \theta$ is used instead of M_x in the first column, where M is the upstream Mach number and θ is the angle between the shock wave and the upstream flow direction.

$k = 1.4$

M or M_x	$\frac{p}{p^\circ}$	$\frac{\rho}{\rho^\circ}$	$\frac{T}{T^\circ}$	$\sqrt{M^2-1}$	$\frac{\rho V^2}{2p^\circ}$	$\frac{A}{A^*}$	$\frac{V}{a^*}$	ν	μ	M_y	$\frac{p_y}{p_x}$	$\frac{\rho_y}{\rho_x}$	$\frac{T_y}{T_x}$	$\frac{p_y^\circ}{p_x^\circ}$	$\frac{p_x}{p_y^\circ}$
1.00	0.5283	0.6339	0.8333	0	0.3698	1.000	1.00000	0	90.00	1.000	1.000	1.000	1.000	1.000	0.5283
1.01	.5221	.6287	.8306	.1418	.3728	1.000	1.00831	.04473	81.93	.9901	1.023	1.017	1.007	1.000	.5221
1.02	.5160	.6234	.8278	.2000	.3758	1.000	1.01658	.1257	78.64	.9805	1.047	1.033	1.013	1.000	.5160
1.03	.5099	.6181	.8250	.2468	.3787	1.001	1.02481	.2294	76.14	.9720	1.071	1.050	1.020	1.000	.5100
1.04	.5039	.6129	.8222	.2857	.3815	1.001	1.03300	.3510	74.06	.9620	1.095	1.067	1.026	.9999	.5039
1.05	.4979	.6077	.8193	.3202	.3842	1.002	1.04114	.4874	72.25	.9531	1.120	1.084	1.033	.9999	.4980
1.06	.4919	.6024	.8165	.3516	.3869	1.003	1.04925	.6367	70.63	.9444	1.144	1.101	1.039	.9997	.4920
1.07	.4860	.5972	.8137	.3807	.3895	1.004	1.05731	.7973	69.16	.9360	1.169	1.118	1.046	.9996	.4861
1.08	.4800	.5920	.8108	.4079	.3919	1.005	1.06533	.9680	67.81	.9277	1.194	1.135	1.052	.9994	.4803
1.09	.4742	.5869	.8080	.4337	.3944	1.006	1.07331	1.148	66.55	.9196	1.219	1.152	1.059	.9992	.4746
1.10	.4684	.5817	.8052	.4583	.3967	1.008	1.08124	1.336	65.38	.9118	1.245	1.169	1.065	.9989	.4689
1.11	.4626	.5766	.8023	.4818	.3990	1.010	1.08913	1.532	64.28	.9041	1.271	1.186	1.071	.9986	.4632
1.12	.4568	.5714	.7994	.5044	.4011	1.011	1.09699	1.735	63.23	.8966	1.297	1.203	1.078	.9982	.4576
1.13	.4511	.5663	.7966	.5262	.4032	1.013	1.10479	1.944	62.25	.8892	1.323	1.221	1.084	.9978	.4521
1.14	.4455	.5612	.7937	.5474	.4052	1.015	1.11256	2.160	61.31	.8820	1.350	1.238	1.090	.9973	.4467
1.15	.4398	.5562	.7908	.5679	.4072	1.017	1.12029	2.381	60.41	.8750	1.376	1.255	1.097	.9967	.4413
1.16	.4343	.5511	.7879	.5879	.4090	1.020	1.12797	2.607	59.55	.8632	1.403	1.272	1.103	.9961	.4360
1.17	.4287	.5461	.7851	.6074	.4108	1.022	1.13561	2.839	58.73	.8565	1.430	1.290	1.109	.9953	.4307
1.18	.4232	.5411	.7822	.6264	.4125	1.025	1.14321	3.074	57.94	.8549	1.458	1.307	1.115	.9946	.4255
1.19	.4178	.5361	.7793	.6451	.4141	1.026	1.15077	3.314	57.18	.8435	1.485	1.324	1.122	.9937	.4204
1.20	.4124	.5311	.7764	.6633	.4157	1.030	1.15828	3.558	56.44	.8422	1.513	1.342	1.128	.9928	.4154
1.21	.4070	.5262	.7735	.6812	.4171	1.033	1.16575	3.806	55.74	.8350	1.541	1.359	1.134	.9918	.4104
1.22	.4017	.5213	.7706	.6989	.4185	1.037	1.17319	4.057	55.05	.8300	1.570	1.376	1.141	.9907	.4055
1.23	.3964	.5164	.7677	.7162	.4198	1.040	1.18057	4.312	54.39	.8241	1.598	1.394	1.147	.9896	.4006
1.24	.3912	.5115	.7648	.7332	.4211	1.043	1.18792	4.569	53.75	.8183	1.627	1.411	1.153	.9884	.3958

Table 5-11. COMPRESSIBLE FLOW AERODYNAMIC TABLES (Continued)

Table B. SUPERSONIC FLOW – ISENTROPIC FLOW AND NORMAL SHOCKS (Continued)

M or M_x	$\dfrac{p}{p^\circ}$	$\dfrac{\rho}{\rho^\circ}$	$\dfrac{T}{T^\circ}$	$\sqrt{M^2-1}$	$\dfrac{\rho V^2}{2p^\circ}$	$\dfrac{A}{A^*}$	$\dfrac{V}{a^*}$	ν	μ	M_y	$\dfrac{p_y}{p_x}$	$\dfrac{\rho_y}{\rho_x}$	$\dfrac{T_y}{T_x}$	$\dfrac{p^\circ_y}{p^\circ_x}$	$\dfrac{p_x}{p^\circ_y}$
1.25	.3861	.5067	.7619	.7500	.4223	1.047	1.19523	4.830	53.13	.8126	1.656	1.429	1.159	.9871	.3911
1.30	.3609	.4829	.7474	.8307	.4270	1.066	1.23114	6.170	50.28	.7860	1.805	1.516	1.191	.9794	.3685
1.35	.3370	.4598	.7329	.9069	.4299	1.089	1.26601	7.561	47.79	.7618	1.960	1.603	1.223	.9697	.3475
1.40	.3142	.4374	.7184	.9798	.4311	1.115	1.29987	8.987	45.58	.7397	2.120	1.690	1.255	.9582	.3280
1.45	.2927	.4158	.7040	1.050	.4308	1.144	1.33272	10.438	43.60	.7196	2.286	1.776	1.287	.9448	.3098
1.50	.2724	.3950	.6897	1.118	.4290	1.176	1.36458	11.905	41.81	.7011	2.458	1.862	1.320	.9298	.2930
1.55	.2533	.3750	.6754	1.184	.4259	1.212	1.39546	13.381	40.18	.6841	2.636	1.947	1.354	.9132	.2773
1.60	.2353	.3557	.6614	1.249	.4216	1.250	1.42539	14.861	38.68	.6684	2.820	2.032	1.388	.8952	.2628
1.65	.2184	.3373	.6475	1.312	.4162	1.292	1.45439	16.338	37.31	.6540	3.010	2.115	1.423	.8760	.2493
1.70	.2026	.3197	.6337	1.375	.4098	1.338	1.48247	17.810	36.03	.6405	3.205	2.198	1.458	.8557	.2368
1.75	.1878	.3029	.6202	1.436	.4026	1.386	1.50966	19.273	34.85	.6281	3.406	2.279	1.495	.8346	.2251
1.80	.1740	.2868	.6068	1.497	.3947	1.439	1.53598	20.725	33.75	.6165	3.613	2.359	1.532	.8127	.2142
1.85	.1612	.2715	.5936	1.556	.3862	1.495	1.56145	22.163	32.72	.6057	3.826	2.438	1.569	.7902	.2040
1.90	.1492	.2570	.5807	1.616	.3771	1.555	1.58609	23.586	31.76	.5956	4.045	2.516	1.608	.7674	.1945
1.95	.1381	.2432	.5680	1.674	.3677	1.619	1.60993	24.992	30.85	.5862	4.270	2.592	1.647	.7442	.1856
2.00	.1278	.2300	.5556	1.732	.3579	1.688	1.63299	26.380	30.00	.5774	4.500	2.667	1.688	.7209	.1773
2.10	.1094	.2058	.5313	1.847	.3376	1.837	1.67687	29.097	28.44	.5613	4.978	2.812	1.770	.6742	.1622
2.20	$.9352^{-1}$	.1841	.5081	1.960	.3169	2.005	1.71791	31.732	27.04	.5471	5.480	2.951	1.857	.6281	.1489
2.30	$.7997^{-1}$	.1646	.4859	2.071	.2961	2.193	1.75629	34.283	25.77	.5344	6.005	3.085	1.947	.5833	.1371
2.40	$.6840^{-1}$	.1472	.4647	2.182	.2758	2.403	1.79218	36.746	24.62	.5231	6.553	3.212	2.040	.5401	.1266
2.50	$.5853^{-1}$	.1317	.4444	2.291	.2561	2.637	1.82574	39.124	23.58	.5130	7.125	3.333	2.138	.4990	.1173
2.60	$.5012^{-1}$	.1179	.4252	2.400	.2371	2.896	1.85714	41.415	22.62	.5039	7.720	3.449	2.238	.4601	.1089
2.70	$.4295^{-1}$	.1056	.4068	2.508	.2192	3.183	1.88653	43.621	21.74	.4956	8.338	3.559	2.343	.4236	.1014
2.80	$.3685^{-1}$	$.9463^{-1}$	.3894	2.615	.2022	3.500	1.91404	45.746	20.92	.4882	8.980	3.664	2.451	.3895	$.9461^{-1}$
2.90	$.3165^{-1}$	$.8489^{-1}$	.3729	2.722	.1863	3.850	1.93981	47.790	20.17	.4814	9.645	3.763	2.563	.3577	$.8848^{-1}$
3.00	$.2722^{-1}$	$.7623^{-1}$	.3571	2.828	.1715	4.235	1.96396	49.757	19.47	.4752	10.33	3.857	2.679	.3283	$.8291^{-1}$
3.50	$.1311^{-1}$	$.4523^{-1}$	.2899	3.354	.1124	6.790	2.06419	58.530	16.60	.4512	14.13	4.261	3.315	.2129	$.6157^{-1}$
4.00	$.6586^{-2}$	$.2766^{-1}$	.2381	3.873	$.7376^{-1}$	10.72	2.13809	65.785	14.48	.4350	18.50	4.571	4.047	$.1388^{-1}$	$.4747^{-1}$
4.50	$.3455^{-2}$	$.1745^{-1}$	.1980	4.387	$.4898^{-1}$	16.56	2.193600	71.832	12.84	.4236	23.46	4.812	4.875	$.9170^{-1}$	$.3768^{-1}$
5.00	$.1890^{-2}$	$.1134^{-1}$	.1667	4.899	$.3308^{-1}$	25.00	2.236068	76.920	11.54	.4152	29.00	5.000	5.800	$.6172^{-1}$	$.3062^{-1}$

Table 5-11. COMPRESSIBLE FLOW AERODYNAMIC TABLES (Continued)

Table B. SUPERSONIC FLOW—ISENTROPIC FLOW AND NORMAL SHOCKS (Continued)

M or M_x	$\dfrac{p}{p^\circ}$	$\dfrac{\rho}{\rho^\circ}$	$\dfrac{T}{T^\circ}$	$\sqrt{M^2-1}$	$\dfrac{\rho V^2}{2p^\circ}$	$\dfrac{A}{A^*}$	$\dfrac{V}{a^*}$	ν	μ	M_y	$\dfrac{p_y}{p_x}$	$\dfrac{\rho_y}{\rho_x}$	$\dfrac{T_y}{T_x}$	$\dfrac{p_y^\circ}{p_x^\circ}$	$\dfrac{p_x}{p_y^\circ}$
5.50	.1075 $^{-2}$	.7578 $^{-2}$	.1418	5.408	.2276 $^{-1}$	36.87	2.269127	81.245	10.48	.4990	35.13	5.149	6.822	.4236 $^{-1}$	.2537 $^{-1}$
6.00	.6334 $^{-3}$	.5194 $^{-2}$	.1220	5.916	.1596 $^{-1}$	53.18	2.295276	84.955	9.594	.4042	41.83	5.268	7.941	.2965 $^{-1}$	.2136 $^{-1}$
6.50	.3855 $^{-3}$	.3643 $^{-2}$	.1058	6.423	.1140 $^{-1}$	75.13	2.316264	88.169	8.850	.4004	49.13	5.365	9.156	.2115 $^{-1}$	.1823 $^{-1}$
7.00	.2416 $^{-3}$	.2609 $^{-2}$	.9259 $^{-1}$	6.928	.8286 $^{-2}$	104.1	2.333333	90.973	8.213	.3974	57.00	5.444	10.47	.1535 $^{-1}$	.1574 $^{-1}$
7.50	.1554 $^{-3}$	.1904 $^{-2}$	.8163 $^{-1}$	7.433	.6120 $^{-2}$	141.8	2.347382	93.440	7.662	.3949	65.46	5.510	11.88	.1133 $^{-1}$	.1372 $^{-1}$
8.00	.1024 $^{-3}$	.1414 $^{-2}$	.7246 $^{-1}$	7.937	.4589 $^{-2}$	190.1	2.359071	95.625	7.181	.3929	74.50	5.565	13.39	.8488 $^{-2}$	.1207 $^{-1}$
8.50	.6898 $^{-4}$	.1066 $^{-2}$	.6472 $^{-1}$	8.441	.3489 $^{-2}$	251.1	2.368892	97.573	6.756	.3912	84.13	5.612	14.99	.6449 $^{-2}$	.1070 $^{-1}$
9.00	.4739 $^{-4}$	.8150 $^{-3}$	.5814 $^{-1}$	8.944	.2687 $^{-2}$	327.2	2.377217	99.319	6.379	.3898	94.33	5.651	16.69	.4964 $^{-2}$	.9546 $^{-2}$
9.50	.3314 $^{-4}$	.6313 $^{-3}$	.5249 $^{-1}$	9.447	.2094 $^{-2}$	421.1	2.384332	100.892	6.042	.3886	105.1	5.685	18.49	.3866 $^{-2}$	.8572 $^{-2}$
10.00	.2356 $^{-4}$	.4948 $^{-3}$	.4762 $^{-1}$	9.950	.1649 $^{-2}$	535.9	2.390457	102.32	5.739	.3876	116.5	5.714	20.39	.3045 $^{-2}$	.7739 $^{-2}$
11.00	.1245 $^{-4}$	.3137 $^{-3}$	.3968 $^{-1}$	10.95	.1054 $^{-2}$	841.9	2.400397	104.80	5.216	.3859	141.0	5.762	24.47	.1945 $^{-2}$	.6400 $^{-2}$
12.00	.6922 $^{-5}$	.2063 $^{-3}$	.3356 $^{-1}$	11.96	.6978 $^{-3}$	1276	2.408040	106.88	4.780	.3847	167.8	5.799	28.94	.1287 $^{-2}$	.5380 $^{-2}$
13.00	.4023 $^{-5}$	.1400 $^{-3}$	.2874 $^{-1}$	12.96	.4759 $^{-3}$	1876	2.414039	108.65	4.412	.3837	197.0	5.828	33.81	.8771 $^{-3}$	.4586 $^{-2}$
14.00	.2428 $^{-5}$	.9760 $^{-4}$	.2488 $^{-1}$	13.96	.3331 $^{-3}$	2685	2.4188316	110.18	4.096	.3829	228.5	5.851	39.06	.6138 $^{-3}$	.3956 $^{-2}$
15.00	.1515 $^{-5}$	.6968 $^{-4}$	.2174 $^{-1}$	14.97	.2386 $^{-3}$	3755	2.4227186	111.51	3.823	.3823	262.3	5.870	44.69	.4395 $^{-3}$	.3446 $^{-2}$
16.00	.9731 $^{-6}$	.5079 $^{-4}$	.1916 $^{-1}$	15.97	.1744 $^{-3}$	5145	2.4259137	112.68	3.583	.3817	298.5	5.885	50.72	.3212 $^{-3}$	.3030 $^{-2}$
18.00	.4328 $^{-6}$	.2848 $^{-4}$	.1520 $^{-1}$	17.97	.9815 $^{-4}$	9159	2.4308053	114.63	3.185	.3810	377.8	5.909	63.94	.1807 $^{-3}$	.2395 $^{-2}$
20.00	.2091 $^{-6}$	.1694 $^{-4}$	.1235 $^{-1}$	19.98	.5885 $^{-4}$	1538 $^{+1}$	2.4343225	116.20	2.866	.3804	466.5	5.926	78.72	.1078 $^{-3}$	.1940 $^{-2}$
25.00	.4454 $^{-7}$	.5611 $^{-5}$	.7937 $^{-2}$	24.98	.1948 $^{-4}$	4631 $^{+1}$	2.4397502	119.03	2.292	.3795	729.0	5.952	122.5	.3586 $^{-4}$	.1242 $^{-2}$
30.00	.1254 $^{-7}$	.2269 $^{-5}$	.5525 $^{-2}$	29.98	.7897 $^{-5}$	1144 $^{+2}$	2.4427138	120.93	1.910	.3790	1050	5.967	175.9	.1453 $^{-4}$	.8626 $^{-3}$
40.00	.1688 $^{-8}$	.5417 $^{-6}$	.3115 $^{-2}$	39.99	.1890 $^{-5}$	4785 $^{+2}$	2.4456714	123.30	1.433	.3786	1867	5.981	312.1	.3477 $^{-5}$	.4853 $^{-3}$
50.00	.3553 $^{-9}$	.1780 $^{-6}$	.1996 $^{-2}$	49.99	.6217 $^{-6}$	1456 $^{+3}$	2.4470439	124.73	1.146	.3783	2917	5.988	487.1	.1144 $^{-5}$	.3106 $^{-3}$
60.00	.9937 $^{-10}$	.7165 $^{-7}$	.1387 $^{-2}$	59.99	.2504 $^{-6}$	3615 $^{+3}$	2.4477905	125.68	.9550	.3782	4200	5.992	700.9	.4606 $^{-6}$	.2157 $^{-3}$
70.00	.3382 $^{-10}$	.3318 $^{-7}$	.1019 $^{-2}$	69.99	.1160 $^{-6}$	7804 $^{+3}$	2.4482410	126.36	.8185	.3782	5717	5.994	953.7	.2134 $^{-6}$	.1585 $^{-3}$
80.00	.1329 $^{-10}$	.1703 $^{-7}$	.7806 $^{-3}$	79.99	.5954 $^{-7}$	1521 $^{+4}$	2.4485335	126.88	.7162	.3781	7467	5.995	1245	.1095 $^{-6}$	.1214 $^{-3}$
90.00	.5831 $^{-11}$	.9452 $^{-8}$	.6169 $^{-3}$	89.99	.3306 $^{-7}$	2739 $^{+4}$	2.4487341	127.27	.6366	.3781	9450	5.996	1576	.6082 $^{-7}$	.9588 $^{-4}$
100.00	.2790 $^{-11}$	.5583 $^{-8}$	.4998 $^{-3}$	100.00	.1953 $^{-7}$	4637 $^{+4}$	2.4488776	127.59	.5730	.3781	1167 $^{+1}$	5.997	1945	.3593 $^{-7}$	.7765 $^{-4}$

*From: "Equations, Tables, and Charts for Compressible Flow", Report 1135, National Advisory Committee for Aeronautics, Ames Aeronautical Laboratory, 1953.

Table 5-12. CONSERVATION EQUATIONS OF A VISCOUS, HEAT-CONDUCTING FLUID*

In Curvilinear Orthogonal Coordinates

NOMENCLATURE:

e_{ij} = components of rate of strain tensor

E = internal energy per unit mass

f = scalar

$\mathbf{F}$ = body force per unit volume

$h_1 h_2 h_3$ = scale factors

H = static enthalpy per unit mass

H_t = total enthalpy, $H_t = H + V^2/2$

k = thermal conductivity

p = static pressure

$\mathbf{q}$ = heat-flux vector

q_β = heat-flux component normal to the surface

t = time

T = temperature

u = velocity component in α direction

v = velocity component in β direction

$\mathbf{V}$ = velocity vector

w = velocity component in γ direction

W = heat generation per unit volume

α, β, γ = orthogonal coordinates

κ = bulk viscosity

λ = second viscosity coefficient

μ = shear viscosity

ρ = density

τ = viscous stress tensor

I. Introduction

Although formulation of the conservation equations of a viscous, heat-conducting fluid in curvilinear orthogonal coordinates is well known through vector and tensor analysis (Refs. 1 and 2), a complete, written-out set of equations, including the energy equation, is not readily available in any given source. The momentum equation was given by Goldstein (Ref. 3) in curvilinear orthogonal coordinates for an incompressible, constant-property fluid, and by Tsien (Ref. 4) for a compressible, variable-property fluid. Only the commonly used special cases of the set of equations in rectangular, cylindrical, and spherical coordinates appear in the literature (e.g., Ref. 5). The purpose of this paper is to briefly present the complete set of equations in stationary, curvilinear orthogonal coordinates. For convenience in expressing the equations in various coordinates, scale factors for eleven coordinate systems are tabulated.

II. Conservation Equations

Three forms of the energy equation are considered, one form of which may be best suited for a particular application. These relations involve the total enthalpy H_t,

$$\rho \frac{\partial H_t}{\partial t} + \rho (\mathbf{V} \cdot \nabla) H_t = \frac{\partial p}{\partial t} - \nabla \cdot \mathbf{q} + \nabla \cdot (\tau \cdot \mathbf{V}) + \mathbf{F} \cdot \mathbf{V} + W$$

the internal energy E,

$$\rho \frac{\partial E}{\partial t} + \rho (\mathbf{V} \cdot \nabla) E + p \nabla \cdot \mathbf{V} = - \nabla \cdot \mathbf{q} + \tau : (\nabla \mathbf{V}) + W$$

and the enthalpy H,

$$\rho \frac{\partial H}{\partial t} + \rho (\mathbf{V} \cdot \nabla) H - \left[\frac{\partial p}{\partial t} + (\mathbf{V} \cdot \nabla) p \right] = - \nabla \cdot \mathbf{q} + \tau : (\nabla \mathbf{V}) + W$$

To complete the set of conservation equations, the continuity and momentum equations are, respectively,

$$\frac{\partial \rho}{\partial t} + \nabla \cdot (\rho \mathbf{V}) = 0$$

$$\frac{\partial \mathbf{V}}{\partial t} + (\mathbf{V} \cdot \nabla) \mathbf{V} = - \frac{1}{\rho} \nabla p + \frac{\mathbf{F}}{\rho} + \frac{1}{\rho} \nabla \cdot \tau$$

The quantities that appear in these equations are identified in the nomenclature.

---

*From: "JPL Technical Report 32–1332", L.H. Back, Jet Propulsion Laboratory, California Institute of Technology, 1968.

Table 5-12. CONSERVATION EQUATIONS OF A VISCOUS, HEAT-CONDUCTING FLUID *(Continued)*

By use of the same notation as used by Goldstein (Ref. 3), the orthogonal coordinates are taken as α, β, and γ such that the elements of length at α, β, and γ in the directions of increasing α, β, and γ are $h_1 d\alpha$, $h_2 d\beta$, and $h_3 d\gamma$, respectively. The differential arc length ds is, then,

$$(ds)^2 = h_1^2 (d\alpha)^2 + h_2^2 (d\beta)^2 + h_3^2 (d\gamma)^2$$

If u, v, and w are components of the velocity vector $\mathbf{V}$ in direction of increasing α, β, and γ, the continuity equation is

$$\frac{\partial \rho}{\partial t} + \frac{1}{h_1 h_2 h_3}$$
$$\times \left[\frac{\partial}{\partial \alpha} (h_2 h_3 \rho u) + \frac{\partial}{\partial \beta} (h_1 h_3 \rho v) + \frac{\partial}{\partial \gamma} (h_1 h_2 \rho w) \right] = 0$$

The momentum equation written in the α, β, and γ directions is

$$\alpha: \frac{\partial u}{\partial t} + \frac{u}{h_1} \frac{\partial u}{\partial \alpha} + \frac{v}{h_2} \frac{\partial u}{\partial \beta} + \frac{w}{h_3} \frac{\partial u}{\partial \gamma} + \frac{uv}{h_1 h_2} \frac{\partial h_1}{\partial \beta} + \frac{uw}{h_1 h_3} \frac{\partial h_1}{\partial \gamma}$$
$$- \frac{v^2}{h_1 h_2} \frac{\partial h_2}{\partial \alpha} - \frac{w^2}{h_1 h_3} \frac{\partial h_3}{\partial \alpha} = -\frac{1}{\rho} \frac{1}{h_1} \frac{\partial p}{\partial \alpha} + \frac{F_\alpha}{\rho} + \frac{1}{\rho} (\nabla \cdot \tau)_\alpha$$

$$\beta: \frac{\partial v}{\partial t} + \frac{u}{h_1} \frac{\partial v}{\partial \alpha} + \frac{v}{h_2} \frac{\partial v}{\partial \beta} + \frac{w}{h_3} \frac{\partial v}{\partial \gamma} + \frac{vu}{h_1 h_2} \frac{\partial h_2}{\partial \alpha} + \frac{vw}{h_2 h_3} \frac{\partial h_2}{\partial \gamma}$$
$$- \frac{u^2}{h_1 h_2} \frac{\partial h_1}{\partial \beta} - \frac{w^2}{h_2 h_3} \frac{\partial h_3}{\partial \beta} = -\frac{1}{\rho} \frac{1}{h_2} \frac{\partial p}{\partial \beta} + \frac{F_\beta}{\rho} + \frac{1}{\rho} (\nabla \cdot \tau)_\beta$$

$$\gamma: \frac{\partial w}{\partial t} + \frac{u}{h_1} \frac{\partial w}{\partial \alpha} + \frac{v}{h_2} \frac{\partial w}{\partial \beta} + \frac{w}{h_3} \frac{\partial w}{\partial \gamma} + \frac{wu}{h_1 h_3} \frac{\partial h_3}{\partial \alpha} + \frac{wv}{h_2 h_3} \frac{\partial h_3}{\partial \beta}$$
$$- \frac{u^2}{h_1 h_3} \frac{\partial h_1}{\partial \gamma} - \frac{v^2}{h_2 h_3} \frac{\partial h_2}{\partial \gamma} = -\frac{1}{\rho} \frac{1}{h_3} \frac{\partial p}{\partial \gamma} + \frac{F_\gamma}{\rho} + \frac{1}{\rho} (\nabla \cdot \tau)_\gamma$$

The components of the divergence of the symmetric viscous stress tensor τ in the α, β, and γ direction (Ref. 6)[1]

are:

$$(\nabla \cdot \tau)_\alpha =$$
$$\frac{1}{h_1 h_2 h_3} \left[\frac{\partial}{\partial \alpha} (h_2 h_3 \tau_{\alpha\alpha}) + \frac{\partial}{\partial \beta} (h_1 h_3 \tau_{\alpha\beta}) + \frac{\partial}{\partial \gamma} (h_1 h_2 \tau_{\gamma\alpha}) \right]$$
$$+ \tau_{\alpha\beta} \frac{1}{h_1 h_2} \frac{\partial h_1}{\partial \beta} + \tau_{\gamma\epsilon} \frac{1}{h_1 h_3} \frac{\partial h_1}{\partial \gamma}$$
$$- \tau_{\beta\beta} \frac{1}{h_1 h_2} \frac{\partial h_2}{\partial \alpha} - \tau_{\gamma\gamma} \frac{1}{h_1 h_3} \frac{\partial h_3}{\partial \alpha}$$

$$(\nabla \cdot \tau)_\beta =$$
$$\frac{1}{h_1 h_2 h_3} \left[\frac{\partial}{\partial \alpha} (h_2 h_3 \tau_{\alpha\beta}) + \frac{\partial}{\partial \beta} (h_1 h_3 \tau_{\beta\beta}) + \frac{\partial}{\partial \gamma} (h_1 h_2 \tau_{\beta\gamma}) \right]$$
$$+ \tau_{\alpha\beta} \frac{1}{h_1 h_2} \frac{\partial h_2}{\partial \alpha} + \tau_{\beta\gamma} \frac{1}{h_2 h_3} \frac{\partial h_2}{\partial \gamma}$$
$$- \tau_{\alpha\alpha} \frac{1}{h_1 h_2} \frac{\partial h_1}{\partial \beta} - \tau_{\gamma\gamma} \frac{1}{h_2 h_3} \frac{\partial h_3}{\partial \beta}$$

$$(\nabla \cdot \tau)_\gamma =$$
$$\frac{1}{h_1 h_2 h_3} \left[\frac{\partial}{\partial \alpha} (h_2 h_3 \tau_{\gamma\alpha}) + \frac{\partial}{\partial \beta} (h_1 h_3 \tau_{\beta\gamma}) + \frac{\partial}{\partial \gamma} (h_1 h_2 \tau_{\gamma\gamma}) \right]$$
$$+ \tau_{\gamma\alpha} \frac{1}{h_1 h_3} \frac{\partial h_3}{\partial \alpha} + \tau_{\beta\gamma} \frac{1}{h_2 h_3} \frac{\partial h_3}{\partial \beta}$$
$$- \tau_{\alpha\alpha} \frac{1}{h_1 h_3} \frac{\partial h_1}{\partial \gamma} - \tau_{\beta\beta} \frac{1}{h_2 h_3} \frac{\partial h_2}{\partial \gamma}$$

The components of the viscous stress tensor for a Stokes' fluid are related to the components of the rate of strain tensor by

$$\tau_{\alpha\alpha} = \lambda \nabla \cdot \mathbf{V} + \mu e_{\alpha\alpha}$$
$$\tau_{\beta\beta} = \lambda \nabla \cdot \mathbf{V} + \mu e_{\beta\beta}$$
$$\tau_{\gamma\gamma} = \lambda \nabla \cdot \mathbf{V} + \mu e_{\gamma\gamma}$$
$$\tau_{\alpha\beta} = \tau_{\beta\alpha} = \mu e_{\alpha\beta}$$
$$\tau_{\alpha\gamma} = \tau_{\gamma\alpha} = \mu e_{\alpha\gamma}$$
$$\tau_{\beta\gamma} = \tau_{\gamma\beta} = \mu e_{\beta\gamma}$$

where the divergence of the velocity vector is

$$\nabla \cdot \mathbf{V} = \frac{1}{h_1 h_2 h_3} \left[\frac{\partial}{\partial \alpha} (h_2 h_3 u) + \frac{\partial}{\partial \beta} (h_1 h_3 v) + \frac{\partial}{\partial \gamma} (h_1 h_2 w) \right]$$

[1] The h_1, h_2, and h_3 used by Love are the reciprocals of those used herein.

Table 5-12.   CONSERVATION EQUATIONS
OF A VISCOUS, HEAT-CONDUCTING FLUID *(Continued)*

and the components of the rate of strain tensor are (Ref. 3):

$$\frac{1}{2}\,e_{\alpha\alpha} = \frac{1}{h_1}\frac{\partial u}{\partial \alpha} + \frac{v}{h_1 h_2}\frac{\partial h_1}{\partial \beta} + \frac{w}{h_3 h_1}\frac{\partial h_1}{\partial \gamma}$$

$$\frac{1}{2}\,e_{\beta\beta} = \frac{1}{h_2}\frac{\partial v}{\partial \beta} + \frac{w}{h_2 h_3}\frac{\partial h_2}{\partial \gamma} + \frac{u}{h_1 h_2}\frac{\partial h_2}{\partial \alpha}$$

$$\frac{1}{2}\,e_{\gamma\gamma} = \frac{1}{h_3}\frac{\partial w}{\partial \gamma} + \frac{u}{h_1 h_3}\frac{\partial h_3}{\partial \alpha} + \frac{v}{h_2 h_3}\frac{\partial h_3}{\partial \beta}$$

$$e_{\alpha\beta} = \frac{h_2}{h_1}\frac{\partial}{\partial \alpha}\left(\frac{v}{h_2}\right) + \frac{h_1}{h_2}\frac{\partial}{\partial \beta}\left(\frac{u}{h_1}\right)$$

$$e_{\alpha\gamma} = \frac{h_1}{h_3}\frac{\partial}{\partial \gamma}\left(\frac{u}{h_1}\right) + \frac{h_3}{h_1}\frac{\partial}{\partial \alpha}\left(\frac{w}{h_3}\right)$$

$$e_{\beta\gamma} = \frac{h_3}{h_2}\frac{\partial}{\partial \beta}\left(\frac{w}{h_3}\right) + \frac{h_2}{h_3}\frac{\partial}{\partial \gamma}\left(\frac{v}{h_2}\right)$$

The second viscosity coefficient λ is related to the shear viscosity μ (first viscosity coefficient) by $\lambda = -2/3\,\mu$ if the bulk viscosity coefficient defined by $\kappa = \lambda + 2/3\,\mu$ is *zero*. Otherwise, λ is given by

$$\lambda = \kappa - \frac{2}{3}\,\mu$$

In the various forms of the energy equations, the operator $(\mathbf{V}\cdot\nabla)$ applied to a scalar f, such as H_t, E, p, or H, gives the convection of that quantity by the flow,

$$(\mathbf{V}\cdot\nabla)f = u\frac{1}{h_1}\frac{\partial f}{\partial \alpha} + v\frac{1}{h_2}\frac{\partial f}{\partial \beta} + w\frac{1}{h_3}\frac{\partial f}{\partial \gamma}$$

The divergence of the heat flux vector $\mathbf{q}$ is

$$\nabla\cdot\mathbf{q} = \frac{1}{h_1 h_2 h_3}\left[\frac{\partial}{\partial \alpha}\,(h_2 h_3 q_\alpha) + \frac{\partial}{\partial \beta}\,(h_1 h_3 q_\beta) + \frac{\partial}{\partial \gamma}\,(h_1 h_2 q_\gamma)\right]$$

In particular, if the heat flux vector is given by Fourier's heat-conduction law, $\mathbf{q} = -k\,\nabla T$, then the components are

$$q_\alpha = -k\frac{1}{h_1}\frac{\partial T}{\partial \alpha}, \qquad q_\beta = -k\frac{1}{h_2}\frac{\partial T}{\partial \beta},$$

$$q_\gamma = -k\frac{1}{h_3}\frac{\partial T}{\partial \gamma}$$

The rate at which work is done by body forces is, simply,

$$\mathbf{F}\cdot\mathbf{V} = F_\alpha u + F_\beta v + F_\gamma w$$

The rate at which work is done by the viscous stresses is given by

$$\nabla\cdot(\tau\cdot\mathbf{V}) = \frac{1}{h_1 h_2 h_3}\left\{\frac{\partial}{\partial \alpha}\,[h_2 h_3\,(\tau_{\alpha\alpha}u + \tau_{\beta\alpha}v + \tau_{\gamma\alpha}w)]\right.$$

$$+ \frac{\partial}{\partial \beta}\,[h_1 h_3\,(\tau_{\alpha\beta}u + \tau_{\beta\beta}v + \tau_{\gamma\beta}w)]$$

$$\left.+ \frac{\partial}{\partial \gamma}\,[h_1 h_2\,(\tau_{\alpha\gamma}u + \tau_{\beta\gamma}v + \tau_{\gamma\gamma}w)]\right\}$$

Lastly, the rate of dissipation of energy takes the form

$$\tau:(\nabla\mathbf{V}) = \tau_{\alpha\alpha}\left(\frac{1}{h_1}\frac{\partial u}{\partial \alpha} + \frac{v}{h_1 h_2}\frac{\partial h_1}{\partial \beta} + \frac{w}{h_1 h_3}\frac{\partial h_1}{\partial \gamma}\right)$$

$$+ \tau_{\beta\beta}\left(\frac{1}{h_2}\frac{\partial v}{\partial \beta} + \frac{u}{h_1 h_2}\frac{\partial h_2}{\partial \alpha} + \frac{w}{h_2 h_3}\frac{\partial h_2}{\partial \gamma}\right)$$

$$+ \tau_{\gamma\gamma}\left(\frac{1}{h_3}\frac{\partial w}{\partial \gamma} + \frac{u}{h_1 h_3}\frac{\partial h_3}{\partial \alpha} + \frac{v}{h_2 h_3}\frac{\partial h_3}{\partial \beta}\right)$$

$$+ \tau_{\alpha\beta}\left(\frac{1}{h_2}\frac{\partial u}{\partial \beta} + \frac{1}{h_1}\frac{\partial v}{\partial \alpha} - \frac{v}{h_1 h_2}\frac{\partial h_2}{\partial \alpha} - \frac{u}{h_1 h_2}\frac{\partial h_1}{\partial \beta}\right)$$

$$+ \tau_{\alpha\gamma}\left(\frac{1}{h_3}\frac{\partial u}{\partial \gamma} + \frac{1}{h_1}\frac{\partial w}{\partial \alpha} - \frac{w}{h_1 h_3}\frac{\partial h_3}{\partial \alpha} - \frac{u}{h_1 h_3}\frac{\partial h_1}{\partial \gamma}\right)$$

$$+ \tau_{\beta\gamma}\left(\frac{1}{h_3}\frac{\partial v}{\partial \gamma} + \frac{1}{h_2}\frac{\partial w}{\partial \beta} - \frac{w}{h_2 h_3}\frac{\partial h_3}{\partial \beta} - \frac{v}{h_2 h_3}\frac{\partial h_2}{\partial \gamma}\right)$$

This rate of dissipation of energy term usually appears in the literature as Φ.

Table 1 presents descriptive information on a number of orthogonal coordinate systems for which the conservation equations can be readily written by use of the foregoing relations. The last two entries in Table 1, in which the coordinates are taken along and normal to the surface, are useful in analyzing internal and external boundary-layer flows. For many flow problems in these coordinates, the dominant viscous stress is the shear stress that lies in the plane of $\beta = $ const ($\tau_{\alpha\beta}$ for a two-dimensional flow and $\tau_{\alpha\beta}$, $\tau_{\gamma\beta}$ for a three-dimensional flow), and the important heat-flux component is normal to the surface, q_β.

Table 5-12.   CONSERVATION EQUATIONS OF A VISCOUS, HEAT-CONDUCTING FLUID (*Continued*)

Table 1.   COORDINATE SYSTEMS AND SCALE FACTORS

1. Orthogonal coordinate system, and 2. orthogonal coordinates α, β, γ	Rectangular coordinates			Scale factors $h_1\ h_2\ h_3$			Coordinate configuration
	x	y	z	h_1	h_2	h_3	
Cylindrical r, θ, z	$r \cos \theta$	$r \sin \theta$	z	1	r	1	
Spherical r, ϕ, θ	$r \cos \theta \sin \phi$	$r \sin \theta \sin \phi$	$r \cos \phi$	1	r	$r \sin \phi$	
Parabolic cylindrical ξ, η, z	$\frac{1}{2}(\xi^2 - \eta^2)$	$\xi \eta$	z	$\sqrt{\xi^2 + \eta^2}$	$\sqrt{\xi^2 + \eta^2}$	1	$\xi = \text{Const}$, $\eta = \text{Const}$ Confocal Parabolas
Paraboloidal ξ, η, ϕ	$\xi \eta \cos \phi$	$\xi \eta \sin \phi$	$\frac{1}{2}(\xi^2 - \eta^2)$	$\sqrt{\xi^2 + \eta^2}$	$\sqrt{\xi^2 + \eta^2}$	$\xi \eta$	$\xi = \text{Const}$, $\eta = \text{Const}$ Confocal Parabolas
Elliptic cylindrical ξ, η, z	$a \cosh \xi \cos \eta$ $a = \text{const}$	$a \sinh \xi \sin \eta$	z	$a\sqrt{\sinh^2 \xi + \sin^2 \eta}$	$a\sqrt{\sinh^2 \xi + \sin^2 \eta}$	1	$\xi = \text{Const}$ — Ellipse, $\eta = \text{Const}$ — Hyperbola

Table 5-12. CONSERVATION EQUATIONS OF A VISCOUS, HEAT-CONDUCTING FLUID (Continued)

Table 1. COORDINATE SYSTEMS AND SCALE FACTORS (Continued)

Coordinate system						
Prolate spheroidal ξ, η, ϕ	$a \sinh \xi \sin \eta \cos \phi$ $a = \text{const}$	$a \sinh \xi \sin \eta \sin \phi$	$a \cosh \xi \cos \eta$	$a \sqrt{\sinh^2 \xi + \sin^2 \eta}$	$a \sqrt{\sinh^2 \xi + \sin^2 \eta}$	$a \sinh \xi \sin \eta$
Oblate spheroidal ξ, η, ϕ	$a \cosh \xi \cos \eta \cos \phi$ $a = \text{const}$	$a \cosh \xi \cos \eta \sin \phi$	$a \sinh \xi \sin \eta$	$a \sqrt{\sinh^2 \xi + \sin^2 \eta}$	$a \sqrt{\sinh^2 \xi + \sin^2 \eta}$	$a \cosh \xi \cos \eta$
Bipolar ξ, η, z	$\dfrac{a \sinh \eta}{\cosh \eta - \cos \xi}$ $a = \text{const}$	$\dfrac{a \sin \xi}{\cosh \eta - \cos \xi}$	z	$\dfrac{a}{\cosh \eta - \cos \xi}$	$\dfrac{a}{\cosh \eta - \cos \xi}$	1
Toroidal ξ, η, ϕ	$\dfrac{a \sinh \eta \cos \phi}{\cosh \eta - \cos \xi}$ $a = \text{const}$	$\dfrac{a \sinh \eta \sin \phi}{\cosh \eta - \cos \xi}$	$\dfrac{a \sin \xi}{\cosh \eta - \cos \xi}$	$\dfrac{a}{\cosh \eta - \cos \xi}$	$\dfrac{a}{\cosh \eta - \cos \xi}$	$\dfrac{a \sinh \eta}{\cosh \eta - \cos \xi}$
Local coordinates along surface (Ref. 3) x, y, z	—	—	—	$1 + \kappa y$	1	1
Local coordinates along surface (Ref. 3) Symmetric about axis x, y, ϕ	—	—	—	$1 + \kappa y$	1	r

Note, $r(x,y)$

Table 5-12.   CONSERVATION EQUATIONS
OF A VISCOUS, HEAT-CONDUCTING FLUID *(Continued)*

REFERENCES

1. "Laminar Flow Theory", P.A. Lagerstrom, *Theory of Laminar Flows, Vol. IV, High-Speed Aerodynamics and Jet Propulsion*, F.K. Moore, Ed., Princeton University Press, 1964.
2. "Methods of Theoretical Physics", P.M. Morse and H. Feshbach, Part I, McGraw-Hill Book Company, 1953.
3. "Modern Developments in Fluid Dynamics", S. Goldstein, Vol. I, Oxford University Press, 1938.
4. "The Equations of Gas Dynamics", H.S. Tsien, *Fundamentals of Gas Dynamics, Vol. III, High-Speed Aerodynamics and Jet Propulsion*, H.W. Emmons, Ed., Princeton University Press, 1958.
5. "Transport Phenomena", R.B. Bird, W.E. Stewart, and E.N. Lightfoot, John Wiley & Sons, 1960.
6. "Treatise on the Mathematical Theory of Elasticity", A.E.H. Love, Dover Publications, 1944, p. 90.

Figure 5-13.   SHOCK-WAVE AND FLOW-DEFLECTION ANGLES FOR VARIOUS UPSTREAM MACH NUMBERS*

Perfect Gas, k = 1.4

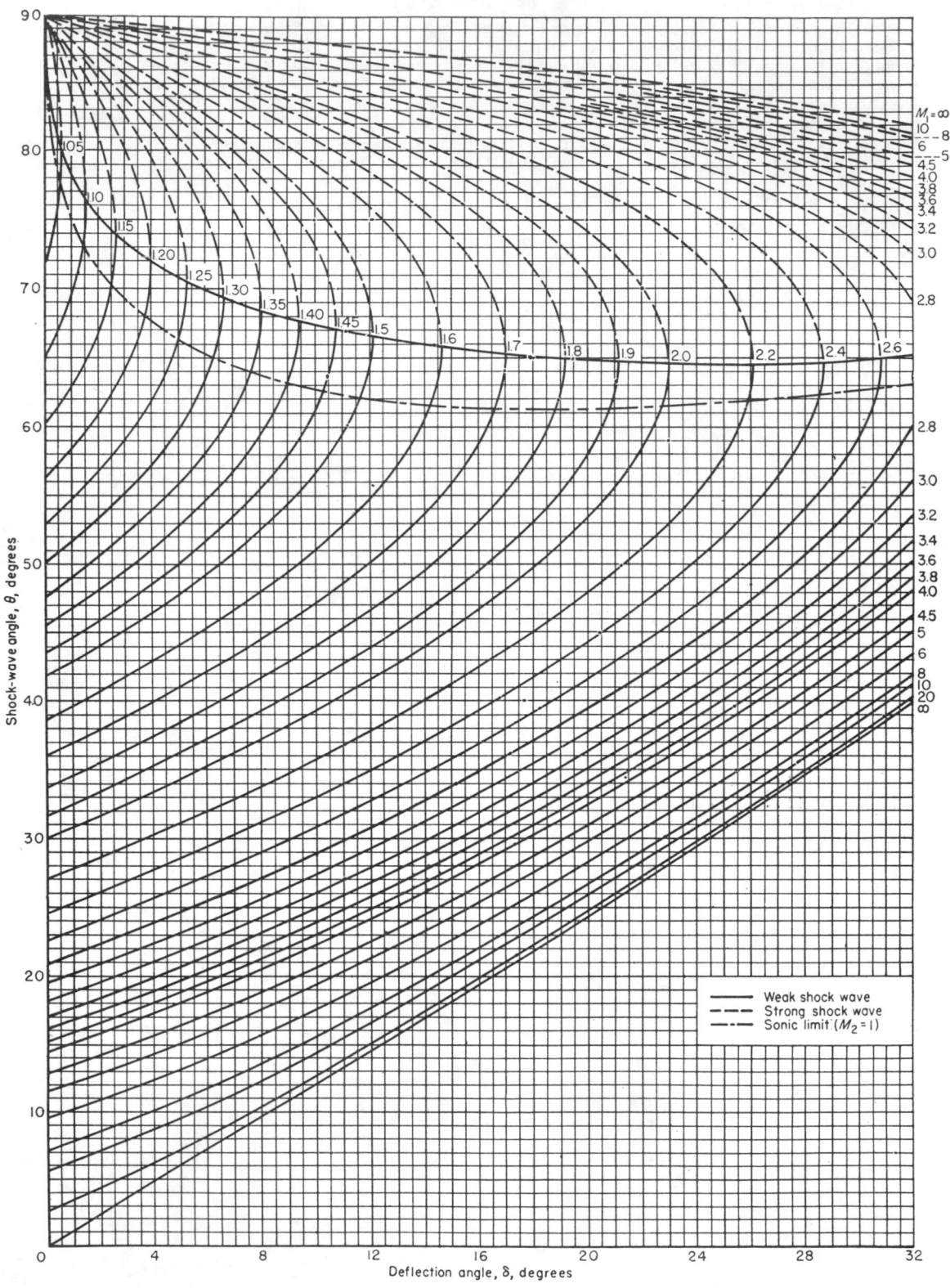

Figure 5-13. SHOCK-WAVE AND FLOW-DEFLECTION ANGLES FOR VARIOUS UPSTREAM MACH NUMBERS *(Continued)*

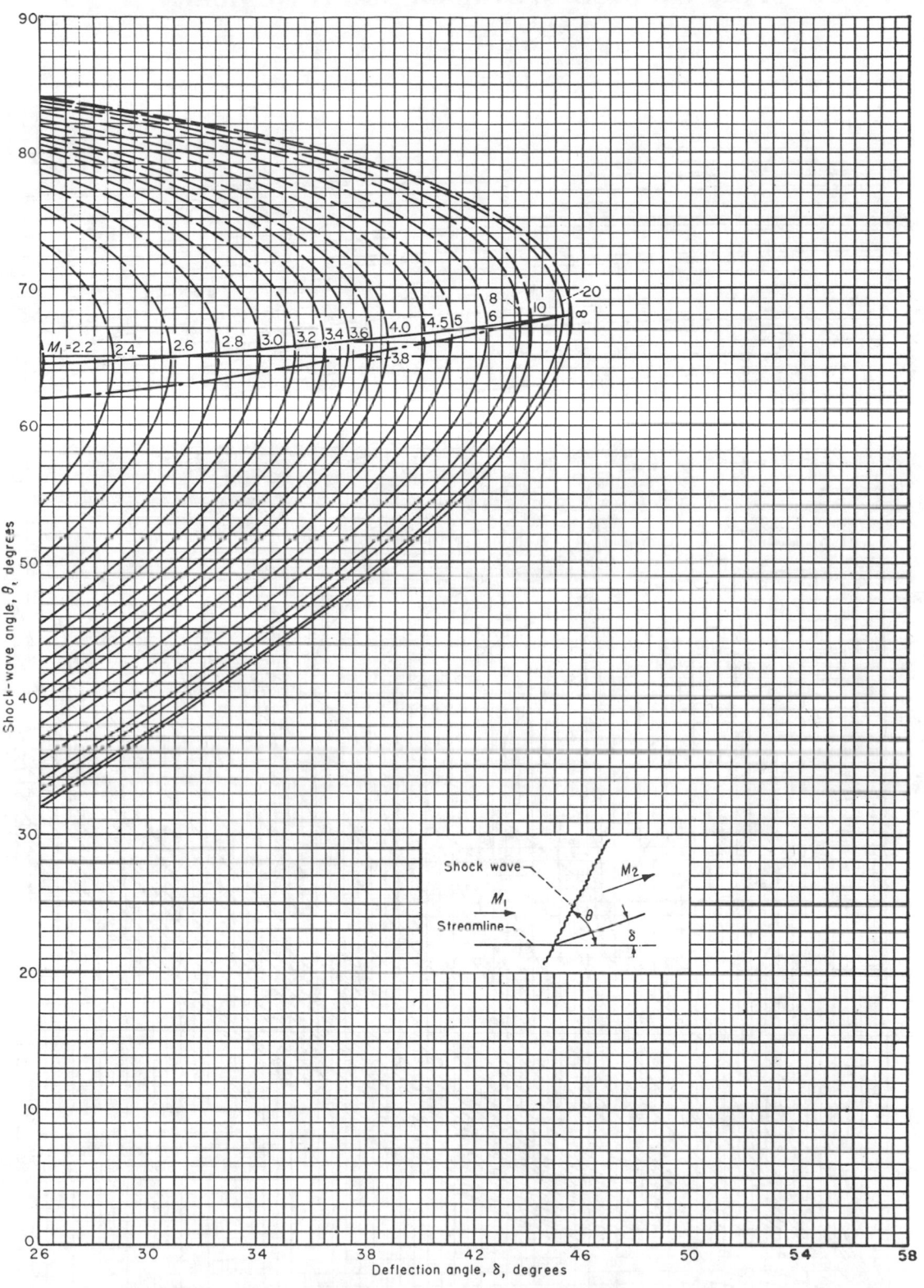

*From: "Equations, Tables, and Charts for Compressible Flow", NACA Report 1135, Ames Aeronautical Laboratory, 1953.

Figure 5-14.   PRESSURE COEFFICIENT ACROSS SHOCK WAVES AND FLOW-DEFLECTION ANGLE FOR VARIOUS UPSTREAM MACH NUMBERS*

Perfect Gas, k = 1.4

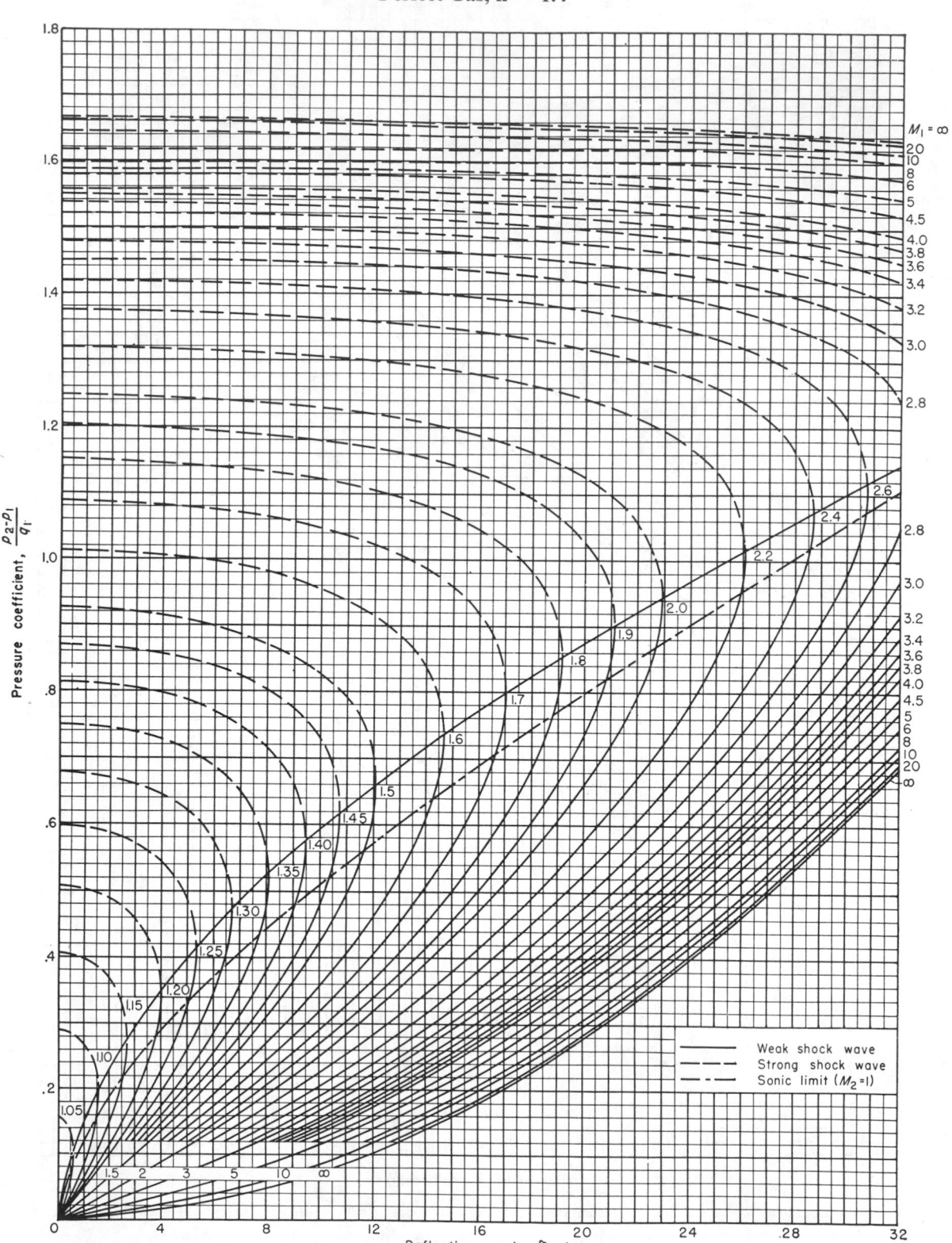

Figure 5-14.   PRESSURE COEFFICIENT ACROSS SHOCK WAVES
AND FLOW-DEFLECTION ANGLE
FOR VARIOUS UPSTREAM MACH NUMBERS *(Continued)*

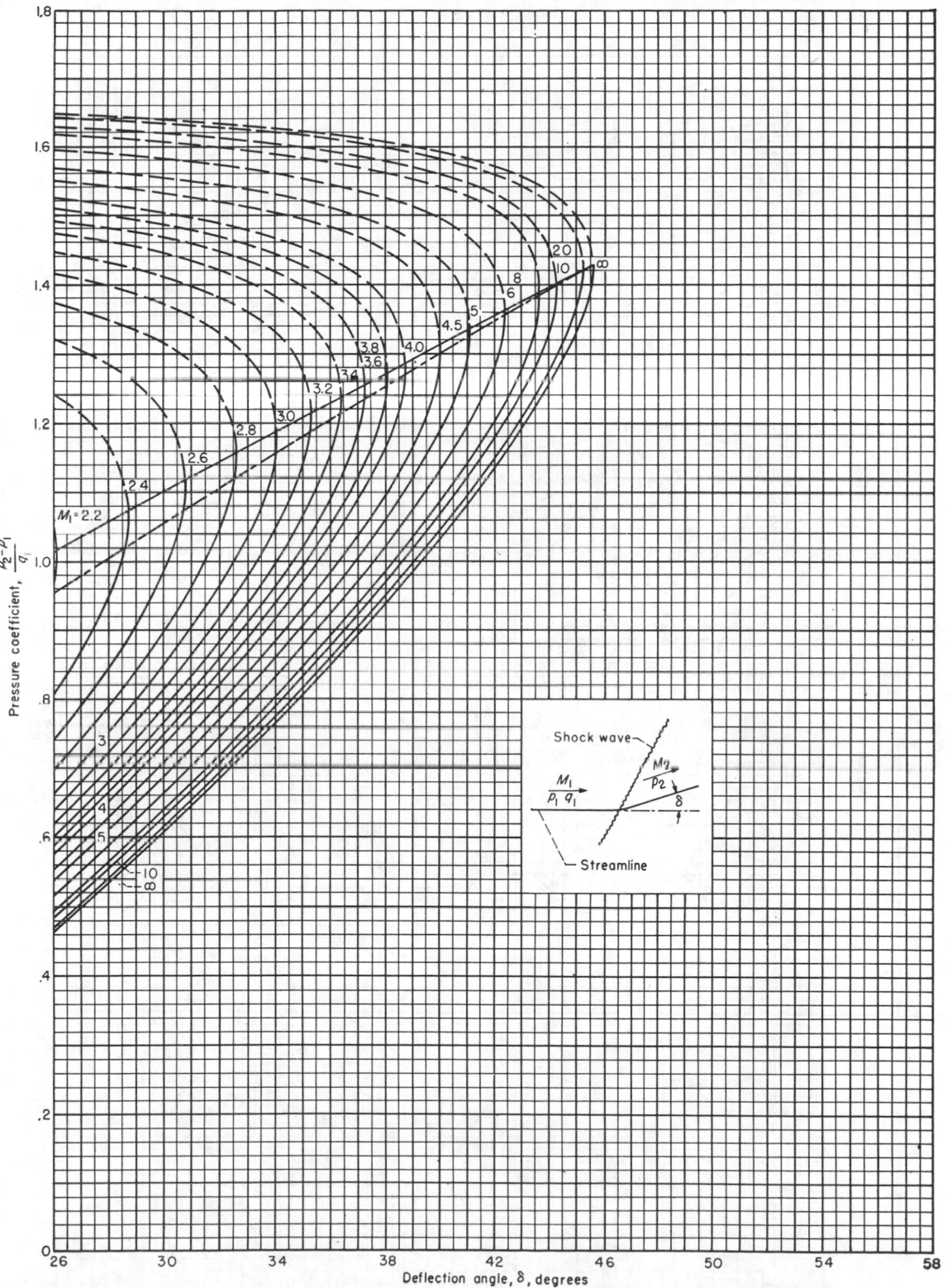

*From: "Equations, Tables, and Charts for Compressible Flow", NACA Report 1135, Ames Aeronautical Laboratory, 1953.

Figure 5-15.   **MACH NUMBER DOWNSTREAM OF A SHOCK WAVE
AND FLOW-DEFLECTION ANGLE
FOR VARIOUS UPSTREAM MACH NUMBERS***

Perfect Gas, k = 1.4

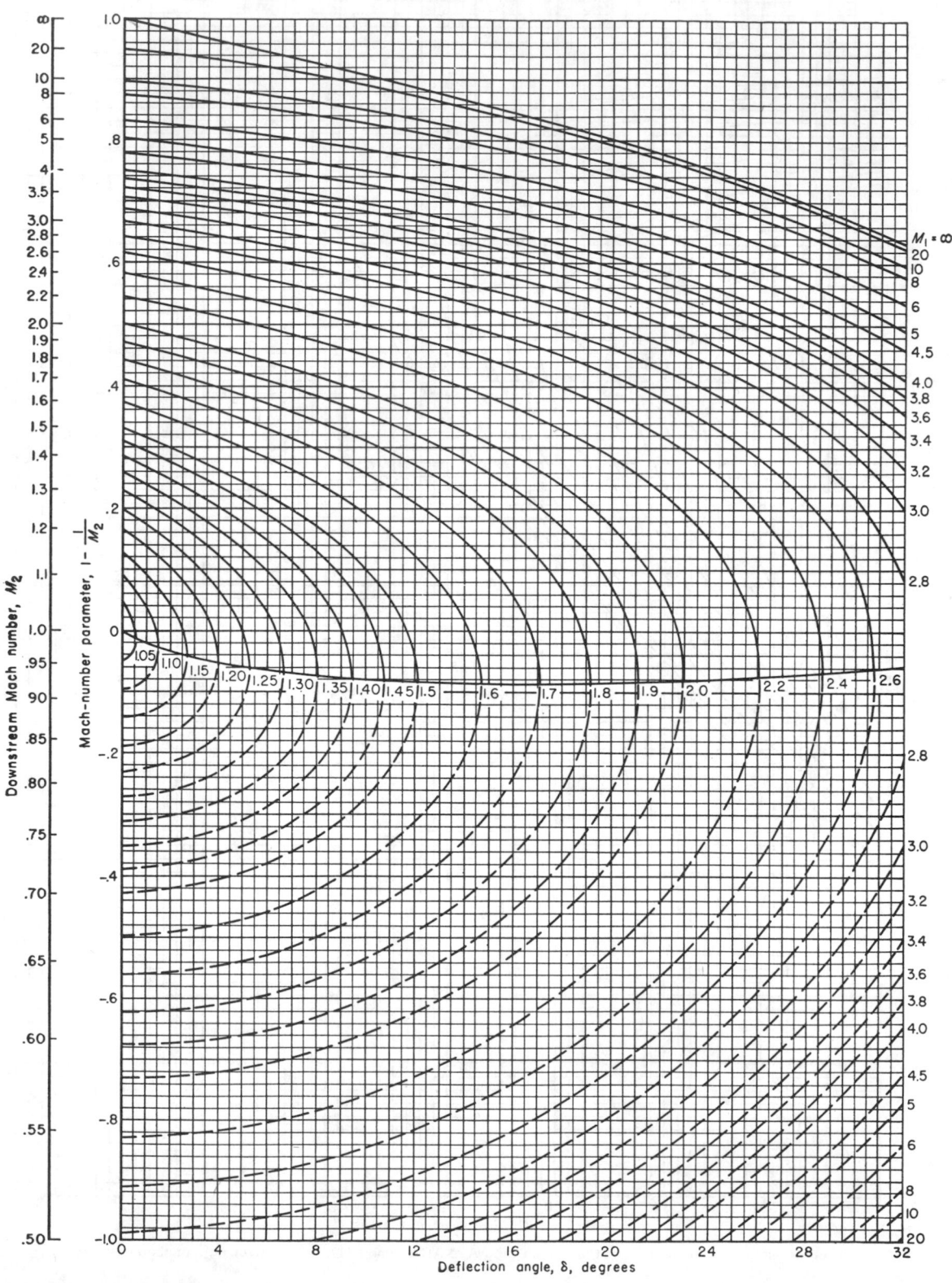

Figure 5-15.   MACH NUMBER DOWNSTREAM OF A SHOCK WAVE AND FLOW-DEFLECTION ANGLE FOR VARIOUS UPSTREAM MACH NUMBERS *(Continued)*

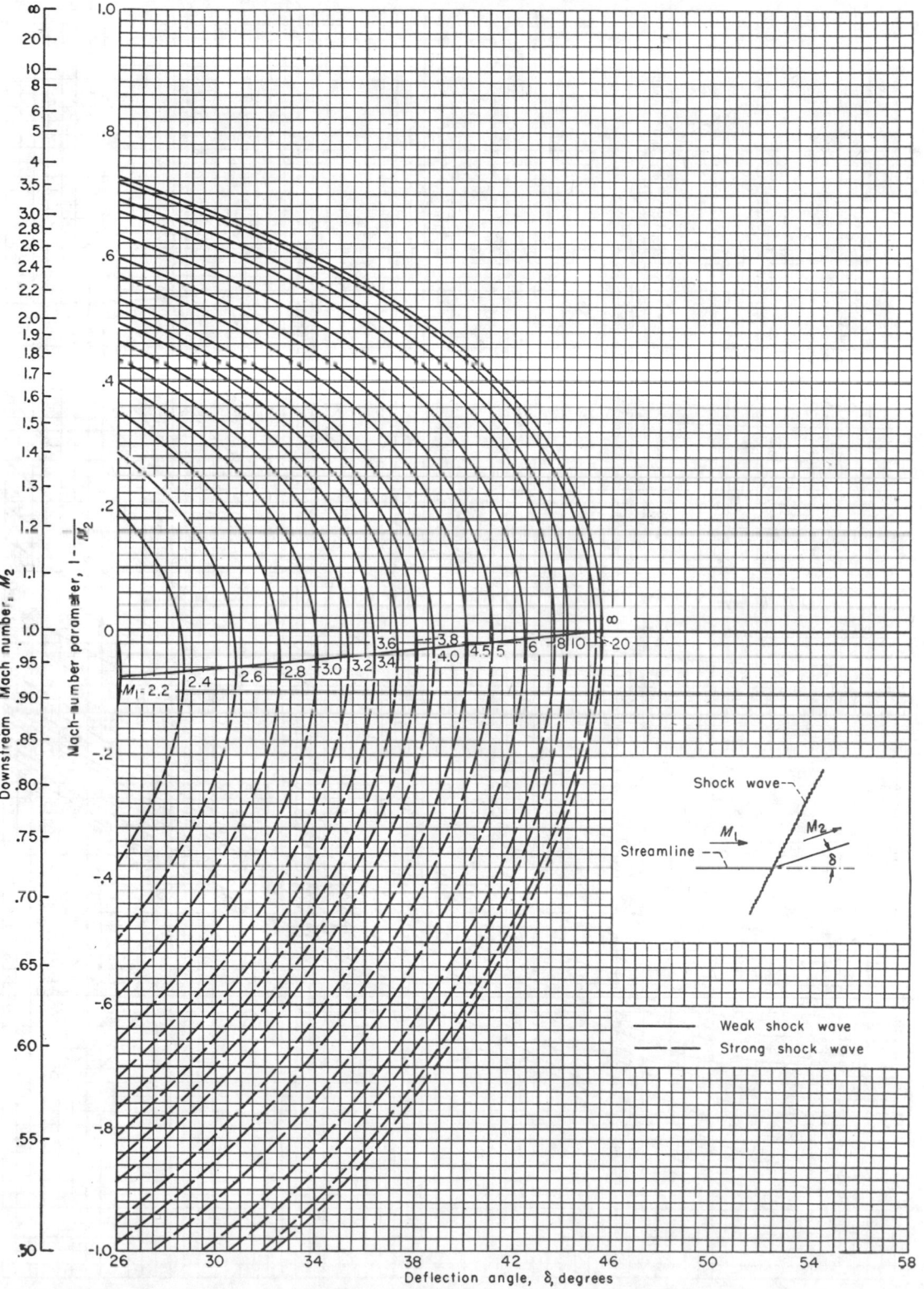

*From: "Equations, Tables, and Charts for Compressible Flow", NACA Report 1135, Ames Aeronautical Laboratory, 1953.

Figure 5-16.   **SHOCK-WAVE AND CONE SEMIVERTEX ANGLES FOR VARIOUS UPSTREAM MACH NUMBERS***

Perfect Gas, k = 1.405

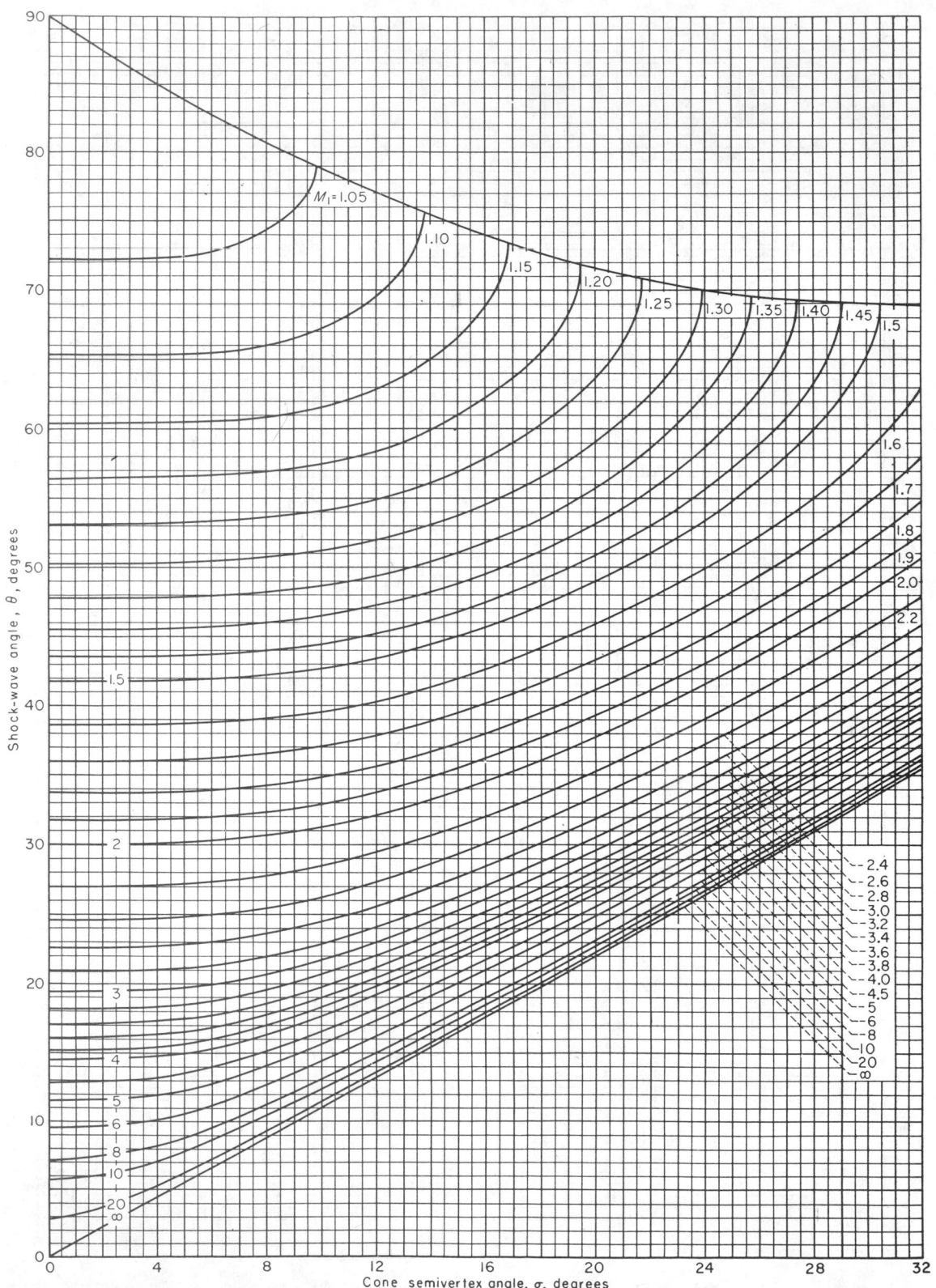

Figure 5-16.   SHOCK-WAVE AND CONE SEMIVERTEX ANGLES FOR VARIOUS UPSTREAM MACH NUMBERS *(Continued)*

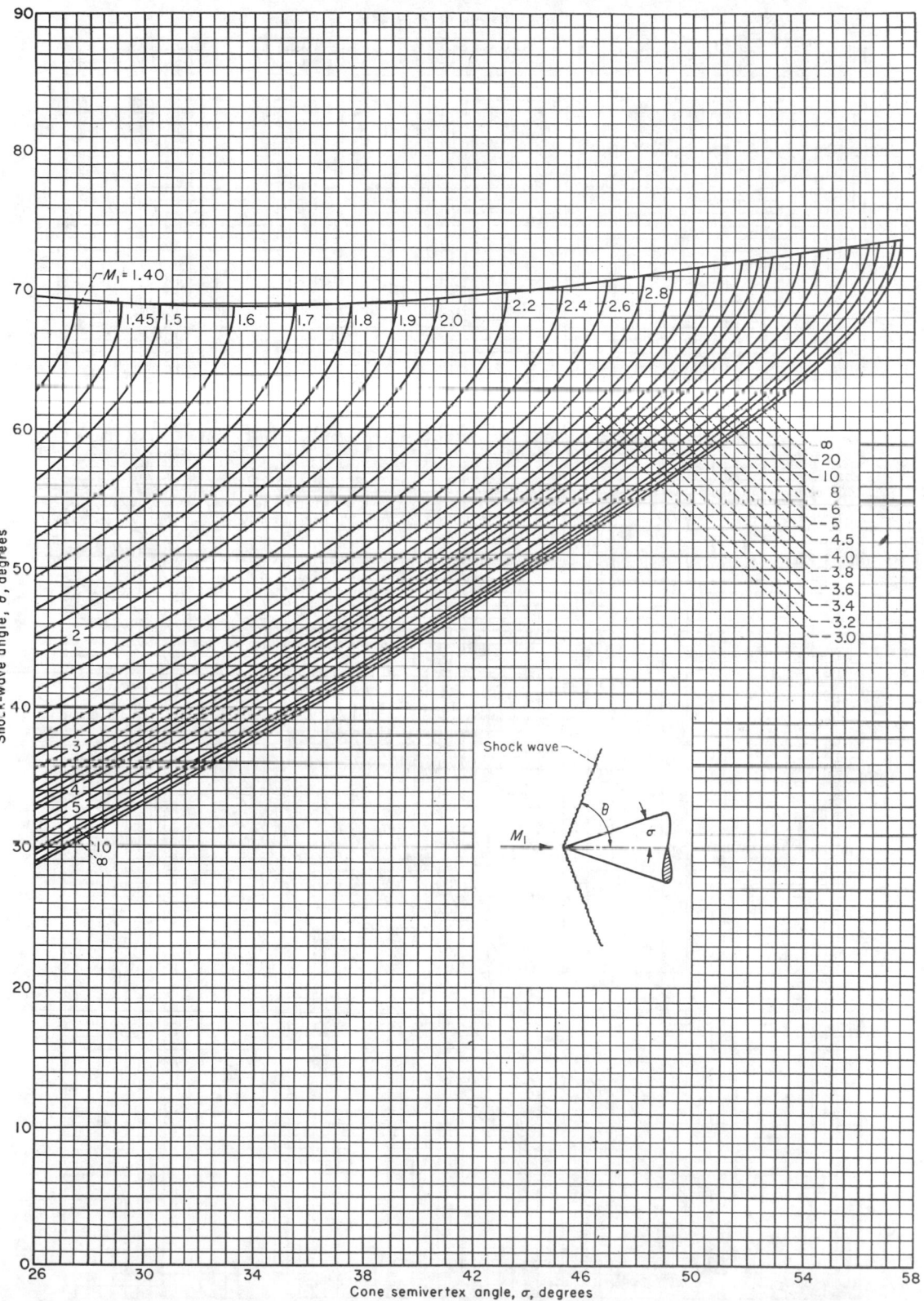

Figure 5-17.   **SURFACE PRESSURE COEFFICIENT
AND CONE SEMIVERTEX ANGLE
FOR VARIOUS UPSTREAM MACH NUMBERS***

Perfect Gas, k = 1.405

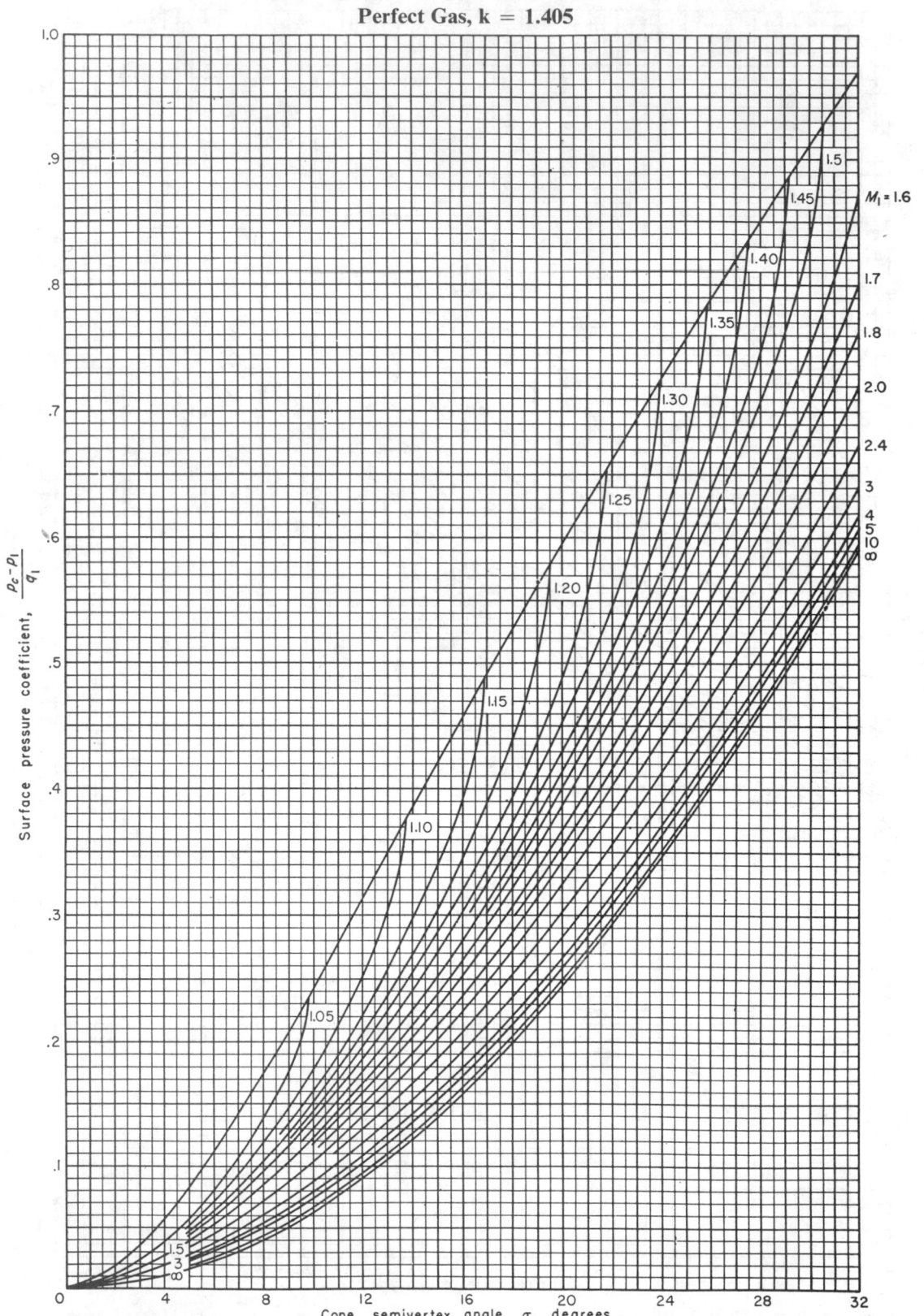

Figure 5-17.   SURFACE PRESSURE COEFFICIENT
AND CONE SEMIVERTEX ANGLE
FOR VARIOUS UPSTREAM MACH NUMBERS *(Continued)*

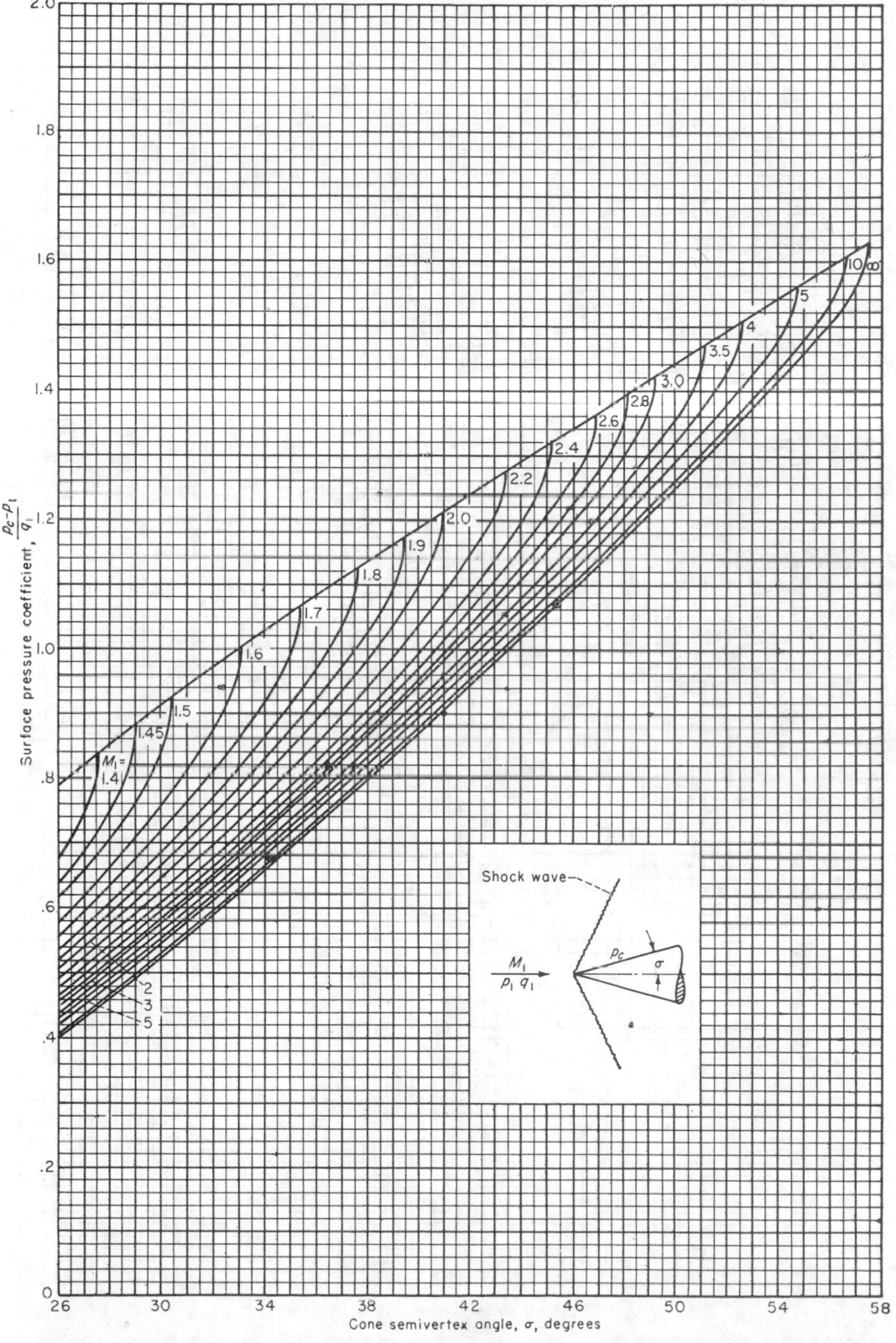

Figure 5-18.   **MACH NUMBER AT THE SURFACE OF A CONE
AND CONE SEMIVERTEX ANGLE
FOR VARIOUS UPSTREAM MACH NUMBERS***

Perfect Gas, k = 1.405

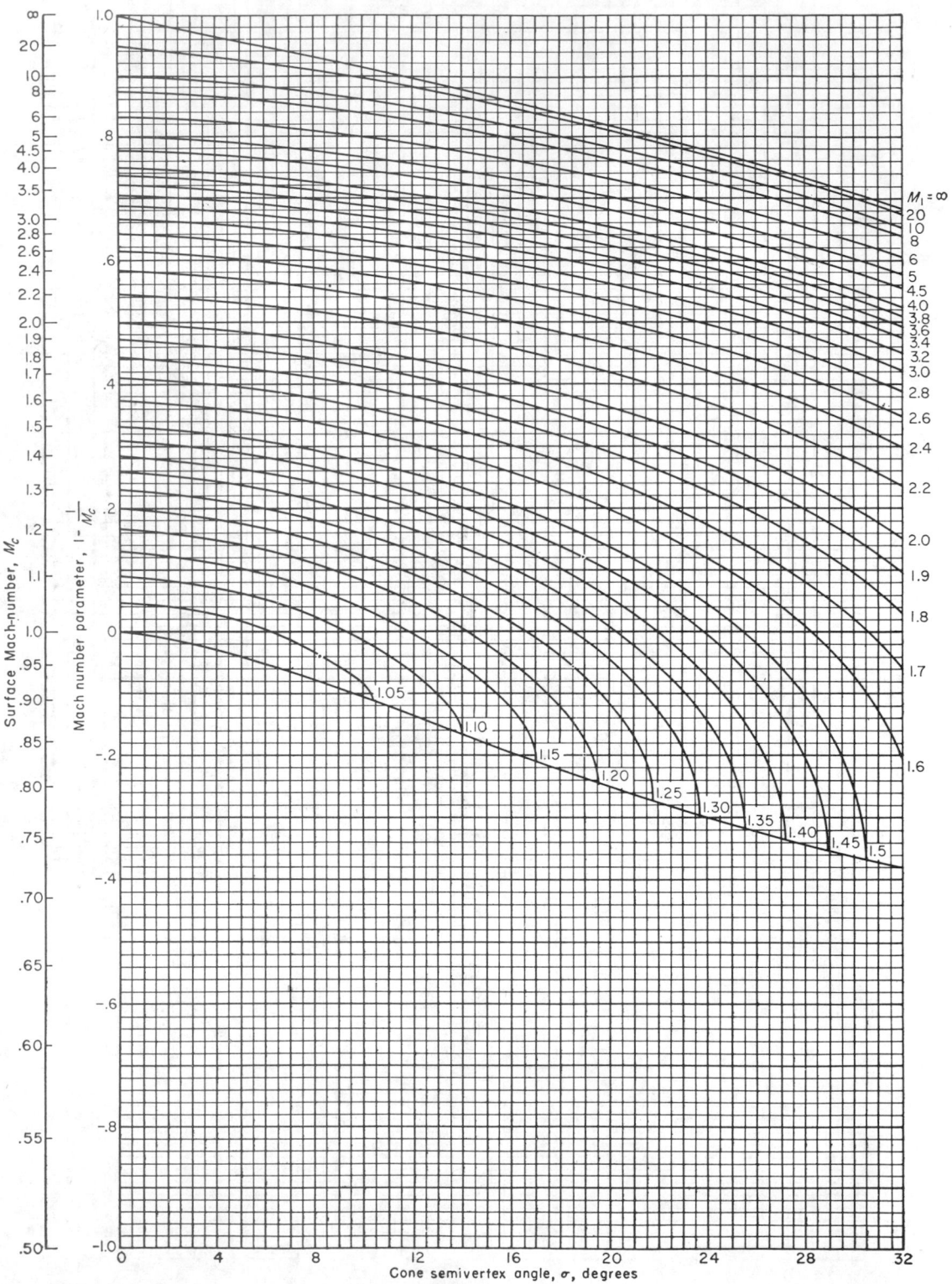

Cone semivertex angle, σ, degrees

Figure 5-18.   **MACH NUMBER AT THE SURFACE OF A CONE
AND CONE SEMIVERTEX ANGLE
FOR VARIOUS UPSTREAM MACH NUMBERS** *(Continued)*

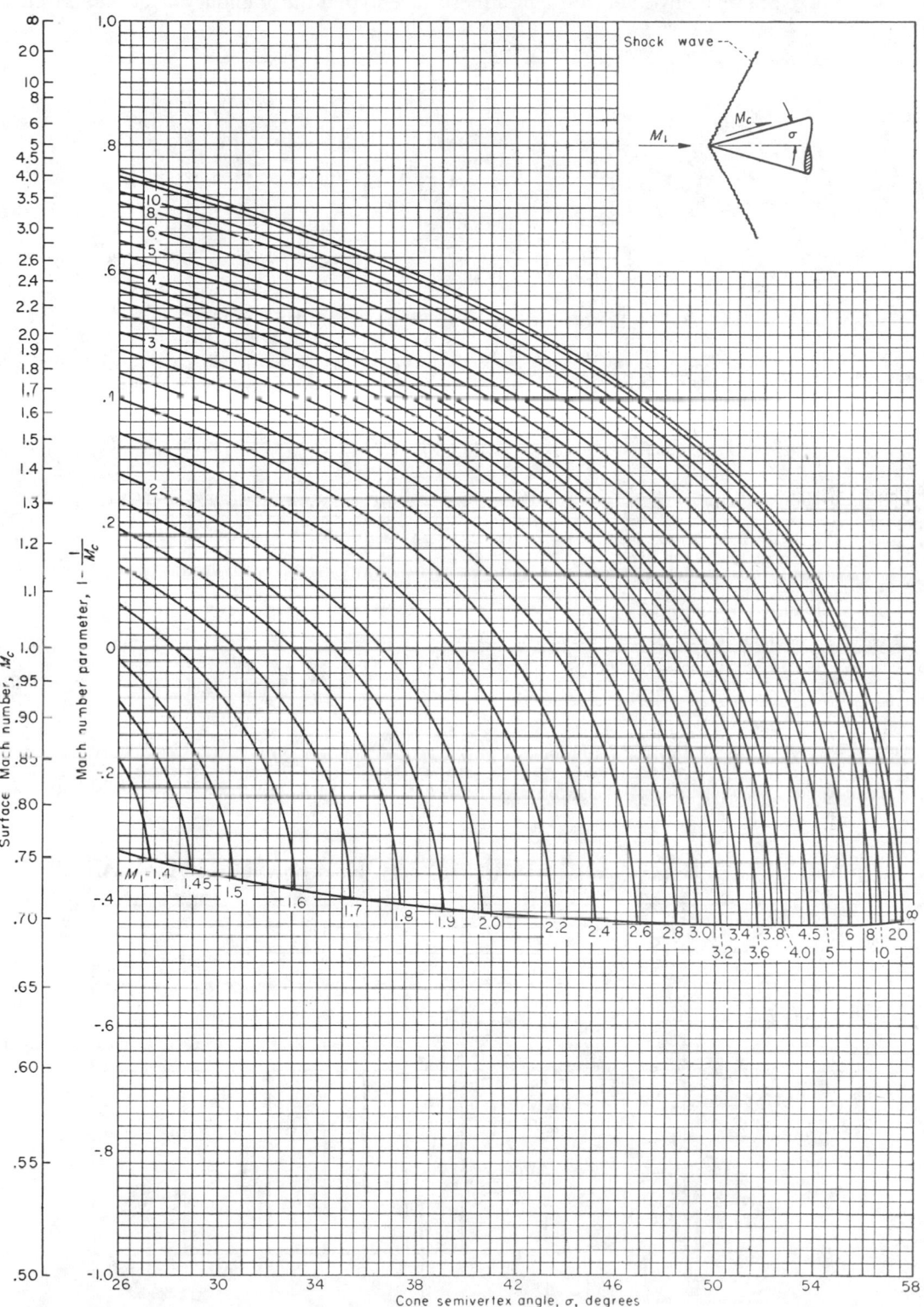

*From: "Equations, Tables, and Charts for Compressible Flow", NACA Report 1135, Ames Aeronautical Laboratory, 1953.

Table 5-19.   ADMISSIBLE ROUGHNESS OF FLUID-FLOW SURFACES*

The following table furnishes a guide for the necessary surface finish on aircraft, ships, and turbine-compressor blading to obtain a "hydraulically smooth" surface so that the boundary layer is unaffected by the roughness.

Item	Description	Length, l m (ft)	Velocity, w			Kinematic viscosity, $10^6 \times v$		$R = \dfrac{wl}{v}$	Admissible roughness, k_{adm}	
			km/h	m/sec	ft/sec	m²/sec	ft²/sec		mm	in.
Ship's hull	large fast	250 (820)	56 30 knots	15	49	1.0	10	4×10^9	0.007	0.00028
	small slow	50 (165)	18 10 knots	5	165	1.0	10	3×10^8	0.02	0.0008
Airship	—	250 (820)	120	33	1100	15	150	5×10^8	0.05	0.002
Aeroplane (wing)	large fast	4† (13)	600	166	545	15	150	5×10^7	0.01	0.0004
	small slow	2 (6.5)	200	55	180	15	150	8×10^6	0.025	0.001
Compressor blades	slow	0.1 (0.33)	—	150	490	15	150	1×10^6	0.01	0.0004
Model wings	small	0.2 (0.65)	144	40	130	15	150	5×10^5	0.05	0.002
Steam turbine blades	high pressure $t = 300\ C\ (\sim 550°F)$	10 mm (0.4 in.)	—	200	650	0.4	4	5×10^6	0.0002	0.000008
	high pressure $t = 500\ C(\sim 950°F)$	10 mm (0.4 in.)	—	200	650	0.8	8	2.5×10^6	0.0005	0.00002
	low pressure	100 mm	—	400	1300	8	80	5×10^6	0.002	0.00008

†Chord.

*From: "Boundary Layer Theory", 6th ed., H. Schlichting (trans. by J. Kestin), McGraw–Hill Book Company, 1967; see this source for an extended discussion.

Figure 5-20.   DRAG COEFFICIENTS FOR A SMOOTH PLATE*

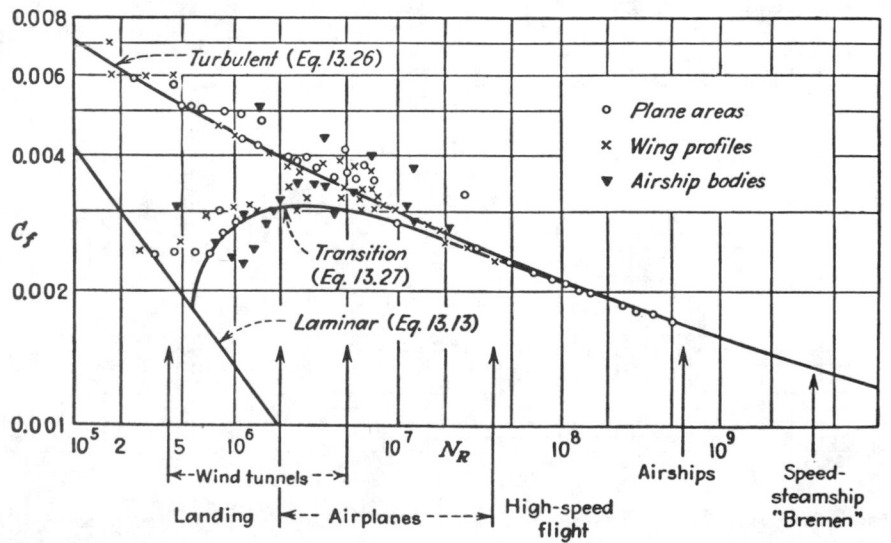

*Adapted from: "NACA Tech. Mem. 1218", National Advisory Committee for Aeronautics, 1949, p. 117.

Table 5-21. AERODYNAMIC DRAG COEFFICIENTS

See also Figure 5-14 and Table 5-22.

The drag coefficient is defined as the ratio of the drag force per unit area to the velocity pressure:

$$F_D = C_D A \rho V^2 / 2.$$

For geometric or irregular shapes A is the projected area normal to the velocity, and for parallel flat surfaces it is the total area. (Coefficients are for smooth surfaces.)

Table A. SIMPLE SHAPES—VARIOUS REYNOLDS NUMBERS

| Reynolds No. | Drag coefficients, C_D | | |
	Sphere	*Circular cylinder*	*Flat plate (parallel)*
$<0.75‡$	24/R		
1	25	10	
10	5	2	
100	1	1.5	
1,000	.5	1	
10^4	.45	1	
10^5	.5	1.4	
CRITICAL	(.15)	(.3)	
10^6	.3	.4	.004†
10^7	.4		.003
10^8	—		.0022
10^9	—		.0017

† Unstable region from Reynolds No. 10^5 to 10^7. Coefficient may vary from .002 to .0045, depending on character of stream.
‡ Stokes flow.

Table B. COMPLEX SHAPES—WIND-TUNNEL TESTS OF MODELS†
High Reynolds No., Above Critical

Model	*Drag coefficients, C_D*
Square or circular plate	1.2
Circular cylinder L/D = 6 (endwise)	0.8
Round-nose cylinder, L/D = 6	0.2
Ellipsoid (endwise)	0.06
Hemispherical cup (round upstream)	0.4
Hemispherical cup (flat upstream)	1.3
Structural angle (common edge upstream)	1.5
Airfoil section (A = chord × span)	$\sim .02^{2, 3}$
Airship body (L/D = 5, A = total)	0.05
Nacelle or float	0.07
Parachutist or ski jumper (A = 8 ft^2)	1.0
Motorcycle and rider (A = 6 ft^2)	0.9
Automobiles (approx. only)	
Open convertible	0.9
Sedan or coupe	0.5
Fastback, streamlined	0.25
Bus or railcar	0.5
Faired locomotive or car	0.4
Streamlined train (short)	0.15
Car within train	0.15

† Based on projected area unless otherwise stated.

REFERENCES

1. "Aerodynamic Drag", S.F. Hoerner, The Otterbein Press, 1950; gives a very complete summary of drag data (turbulent, laminar, and supersonic ranges).
2. "Summary of Air Foil Data", I.H. Abbott and A.E. Von Doenhoff, U.S. National Committee on Aeronautics Report No. 824, U.S. Government Printing Office, 1945 [i.e., 1949]; gives extensive airfoil data.
3. "Theory of Wing Sections", I.H. Abbott and A.E. Von Doenhoff, McGraw-Hill Book Company, 1949.
4. For tabular data on air-flow resistance of stored grain and seed, see "Agricultural Engineers Yearbook", American Society of Agricultural Engineers, and "Resistance of Grains and Seeds to Air Flow", *Agr. Eng.*, 34:616–619, September 1953.

Table 5-22.   AUTOMOBILE DRAG COEFFICIENT ESTIMATES*

Drag rating values in parentheses are for use in the equation
drag coefficient, $C_d = 0.16 + 0.0095 \Sigma$ (summation) drag ratings.
Σ drag ratings must include one rating from each of the nine categories listed (A through I).

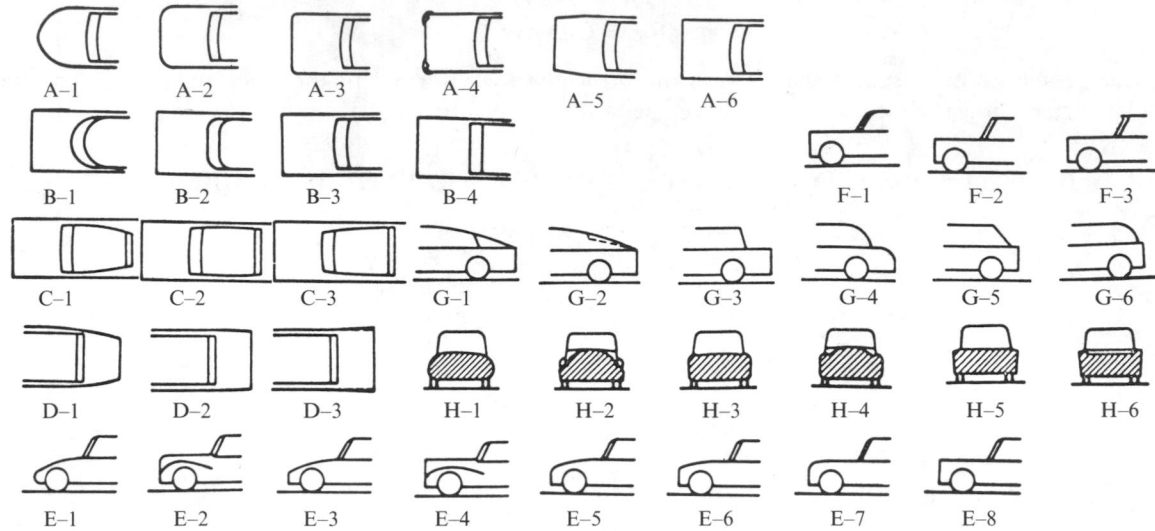

A. Plan view, front end
A–1 Approximately semi-circular (1)
A–2 Well-rounded outer quarters (2)
A–3 Rounded corners without protuberances (3)
A–4 Rounded corners with protuberances (4)
A–5 Squared tapering-in corners (5)
A–6 Squared constant-width front (6)

B. Plan view, windshield[c]
B–1 Full wrap-round (approximately semi-circular) (1)
B–2 Wrapped-round ends (2)
B–3 Bowed (3)
B–4 Flat (4)

C. Plan view, roof
C–1 Well- or medium-tapered to rear (1)
C–2 Tapering to front and rear (max. width at BC post) or approximately constant width (2)
C–3 Tapering to front (max. width at rear) (3)

D. Plan view, lower rear end
D–1 Well- or medium-tapered to rear (1)
D–2 Small taper to rear or constant width (2)
D–3 Outward taper (or flared-out fins) (3)

E. Side elevation, front end[b]
E–1 Low, rounded front, sloping up (1)
E–2 High, tapered, rounded hood (1)
E–3 Low, squared front, sloping up (2)
E–4 High, tapered, squared hood (2)
E–5 Medium-height, rounded front, sloping up (3)
E–6 Medium-height, squared front, sloping up (4)
E–7 High, rounded front, with horizontal hood (4)
E–8 High, squared front, with horizontal hood (5)

F. Side elevation, windshield peak
F–1 Rounded (1)
F–2 Squared (including flanges or gutters) (2)
F–3 Forward-projecting peak (3)

G. Side elevation, rear roof/trunk[d]
G–1 Fastback (roof line continuous to tail) (1)
G–2 Semi-fastback (with discontinuity in line to tail) (2)
G–3 Squared roof with trunk rear-edge squared (3)
G–4 Rounded roof with rounded trunk (4)
G–5 Squared roof with short or no trunk (4)
G–6 Rounded roof with short or no trunk (5)

H. Front elevation, cowl and fender cross-section at windshield[a]
H–1 Flush hood and fenders, well-rounded body sides (1)
H–2 High cowl, low fenders (2)
H–3 Hood flush with rounded-top fenders (3)
H–4 High cowl, with rounded-top fenders (3)
H–5 Hood flush with square-edged fenders (4)
H–6 Depressed hood, with high square-edged fenders (5)

Table 5-22. AUTOMOBILE DRAG COEFFICIENT ESTIMATES *(Continued)*

I. Underbody

Underbody drag rating values must be estimated from vehicle examination. The following can be used as a guide:

Drag rating (1): For integral, flush floor, little projecting mechanism. Examples: Porsche coupe, Citroen DS19, Saab 96.

Drag rating (2): Intermediate between descriptions for (1) and (3). Examples: Volkswagen 1300.

Drag rating (3): Integral, projecting structure and mechanism. Examples: Mercedes 300 SE, Ford Falcon and Mustang.

Drag rating (4): Intermediate between descriptions for (3) and (5). Examples: Oldsmobile Toronado, Jenson 541, Ferrari 300 GTB.

Drag rating (5): Deep chassis. Examples: Typical U.S. full size automobiles with separate body and frame construction.

Note: Throughout table, the words *taper* or *tapered* refer to the plan view.

[a] Fender mirrors; include in protuberances if at the fender leading end; otherwise add 1.

[b] Add: 3 for separate fenders; 4 for open front to fenders (above bumper level): 2 for raised built-in headlamps; 4 for small separate headlamps; 7 for large separate headlamps.

[c] Add: 1 for upright windshield; 1 for prominent flanges or rain gutters.

[d] Add: 3 for high fins or sharp, longitudinal edges to trunk; 2 for separate fenders. *Note:* In all the ratings in this column, the trunk is assumed to be rounded laterally.

*Adapted from: "A Method of Estimating Automobile Drag Coefficients", R.G.S. White, SAE Paper No. 690189, Society of Automotive Engineers, 1969.

Table 5-23. FORCES PRODUCED BY WIND

For wind velocity in m/s, multiply velocity in mph by 0.44704.

| | Wind velocity, mph | | Wind pressure based on projected area | | | |
| | | | On flat surface | | On cylindrical surface | |
Actual	Measured, 3-cup anemometer	Measured, 4-cup anemometer	lbf/ft^2	N/m^2	lbf/ft^2	N/m^2
10	9	10	0.42	20.1	0.25	12.0
20	20	23	1.7	81.4	1.0	48.0
30	31	36	3.8	182	2.3	110
40	42	50	6.7	321	4.0	192
50	54	64	10.5	503	6.3	302
60	65	77	15.1	723	9.0	431
70	76	91	20.6	986	12.3	589
80	88	105	26.8	1 280	16.0	766
90	99	119	34.0	1 630	20.3	972
100	110	132	42.0	2 010	25.0	1 200
110	121	146	50.8	2 430	30.3	1 450
120	133	160	60.5	2 900	36.0	1 720
130	144	173	71.0	3 400	42.3	2 020
140	155	187	82.3	3 940	49.0	2 350
150	167	201	94.5	4 520	56.3	2 700

Since 1932 the U.S. Weather Bureau has published all wind velocity data on the basis of actual wind velocity. For pressures on structures, it is recommended that the "fastest single-mile velocities" reported by the U.S. Weather Bureau be multiplied by 1.3 to obtain a gust velocity.

Actual pressures depend on size and other factors, but it is usual to use a direct relation with the square of the air velocity: $P = KA^2$. The pressure ratio for cylindrical/flat surfaces is shown here as approximately $\frac{3}{5}$. The Electronics Industries Association specifies $\frac{2}{3}$ for this ratio. In any case the wind-pressure forces on structures are larger than the impact pressure of the wind, due to suction forces on the leeward side. The pressures are also higher on tall structures. The net force is usually 40% to 100% higher than that due only to the stagnation pressure (impact) of the wind. The factor used for the flat plate in the above table is 1.68.

Table 5-24.   REYNOLDS NUMBERS FOR FLOW OF WATER AND AIR IN PIPE AND TUBING—SI UNITS

SYMBOLS AND UNITS:

$Re = \rho VD/\mu = VD/v$ = Reynolds number

For perfect gases $(P = \rho RT)$, $Re = PVD/\mu RT$

ρ = density of fluid, lb_m/ft^3

V = mean velocity of fluid, ft/sec

D = inside diameter of pipe or tubing, ft

μ = absolute viscosity, lb_m/sec ft

$v = \mu/\rho$ = kinematic viscosity, ft^2/sec^a

P = absolute pressure, lb_f/ft^2

R = gas constant, ft lb_f/lb_m °R

T = absolute temperature, °R

Temperature		$v,^a$ (ft^2/sec) $\times 10^5$	Product of velocity and diameter, VD, ft^2/sec									
°C	°F		1	2	3	4	5	6	7	8	9	10
REYNOLDS NUMBERS FOR WATER $(Re \times 10^{-5})^b$												
0	32	1.88	0.53	1.06	1.60	2.13	2.66	3.19	3.72	4.26	4.79	5.32
21	70	1.08	0.93	1.85	2.78	3.70	4.63	5.56	6.48	7.41	8.33	9.26
38	100	0.74	1.35	2.70	4.05	5.41	6.76	8.11	9.46	10.8	12.2	13.5
66	150	0.47	2.13	4.26	6.38	8.51	10.6	12.8	14.9	17.0	19.1	21.3
93	200	0.33	3.03	6.06	9.09	12.1	15.2	18.2	21.2	24.2	27.3	30.3
149	300	0.21	4.76	9.52	14.3	19.0	23.8	28.6	33.3	38.1	42.9	47.6
REYNOLDS NUMBERS FOR AIR AT ATMOSPHERIC PRESSURE $(Re \times 10^{-4})^c$												
−32	−25	11.1	0.90	1.81	2.70	3.60	4.50	5.40	6.30	7.21	8.11	9.01
−18	0	12.5	0.80	1.60	1.84	3.20	4.00	4.80	5.60	6.40	7.20	8.00
0	32	14.1	0.71	1.42	2.13	2.84	3.55	4.26	4.96	5.67	6.38	7.09
21	70	16.2	0.62	1.23	1.85	2.47	3.09	3.70	4.32	4.94	5.56	6.17
38	100	18.1	0.55	1.10	1.66	2.21	2.76	3.31	3.87	4.42	4.97	5.52
93	200	23.4	0.43	0.86	1.28	1.71	2.14	2.56	2.99	3.42	3.85	4.27
204	400	37.9	0.26	0.53	0.79	1.06	1.32	1.60	1.85	2.11	2.37	2.64
316	600	54.0	0.19	0.37	0.56	0.74	0.93	1.11	1.30	1.48	1.67	1.85
427	800	70.5	0.14	0.28	0.43	0.57	0.71	0.85	0.99	1.13	1.28	1.43
538	1000	91.6	0.11	0.22	0.33	0.44	0.55	0.66	0.76	0.87	0.98	1.09

[a]To convert ft^2/s to m^2/s, multiply by 0.0929.

[b]Viscosity and thus Reynolds numbers for water are only slightly affected by pressure.

[c]Absolute viscosity for air is only slightly affected by pressure, but density and thus Reynolds numbers are directly proportional to pressure. For common pressures multiply the Reynolds numbers in the table by the air pressure in atmospheres.

FLUID FRICTION—PIPES, VALVES, AND FITTINGS
Symbols and Equations

The loss of pressure P_L in an incompressible fluid-flow system is ordinarily expressed in terms of the dynamic pressure $\rho V^2/2$, the length-to-diameter ratio L/D, and the friction factor f:

$$P_L = f\frac{L}{D}\frac{\rho V^2}{2},$$

where ρ is the mass density and V is the average fluid velocity. Expressed as a head loss h_L:

$$h_L = \frac{P_L}{w} = f\frac{L}{D}\frac{\rho V^2}{2w} = f\frac{L}{D}\frac{V^2}{2g},$$

w is the weight-per-unit volume, g is the local acceleration due to gravity, and $V^2/2g$ is the velocity head.

For English units P_L is in lbf/ft^2 and h_L is in feet of fluid flowing at the average weight-per-unit volume. For SI units P_L is in N/m^2.

Friction losses in a length L of pipe or conduit are computed with the aid of Figure 5-25 for the determination of f. Losses due to fittings and valves are added either by adding equivalent L/D values or by a separate calculation in terms of the loss of velocity head or velocity pressure in each fitting or change of section.

For estimating friction in non-circular conduits, the hydraulic diameter, area/perimeter, should be used in place of D.

Typical values of equivalent L/D and fL/D are given in Table 5-26. These are based essentially on the common steel-pipe systems for water, steam, gas, and compressed air in the turbulent flow range.

Figure 5-25.   FRICTION FACTOR BASED ON RELATIVE ROUGHNESS FOR VARIOUS KINDS AND SIZES OF PIPE*

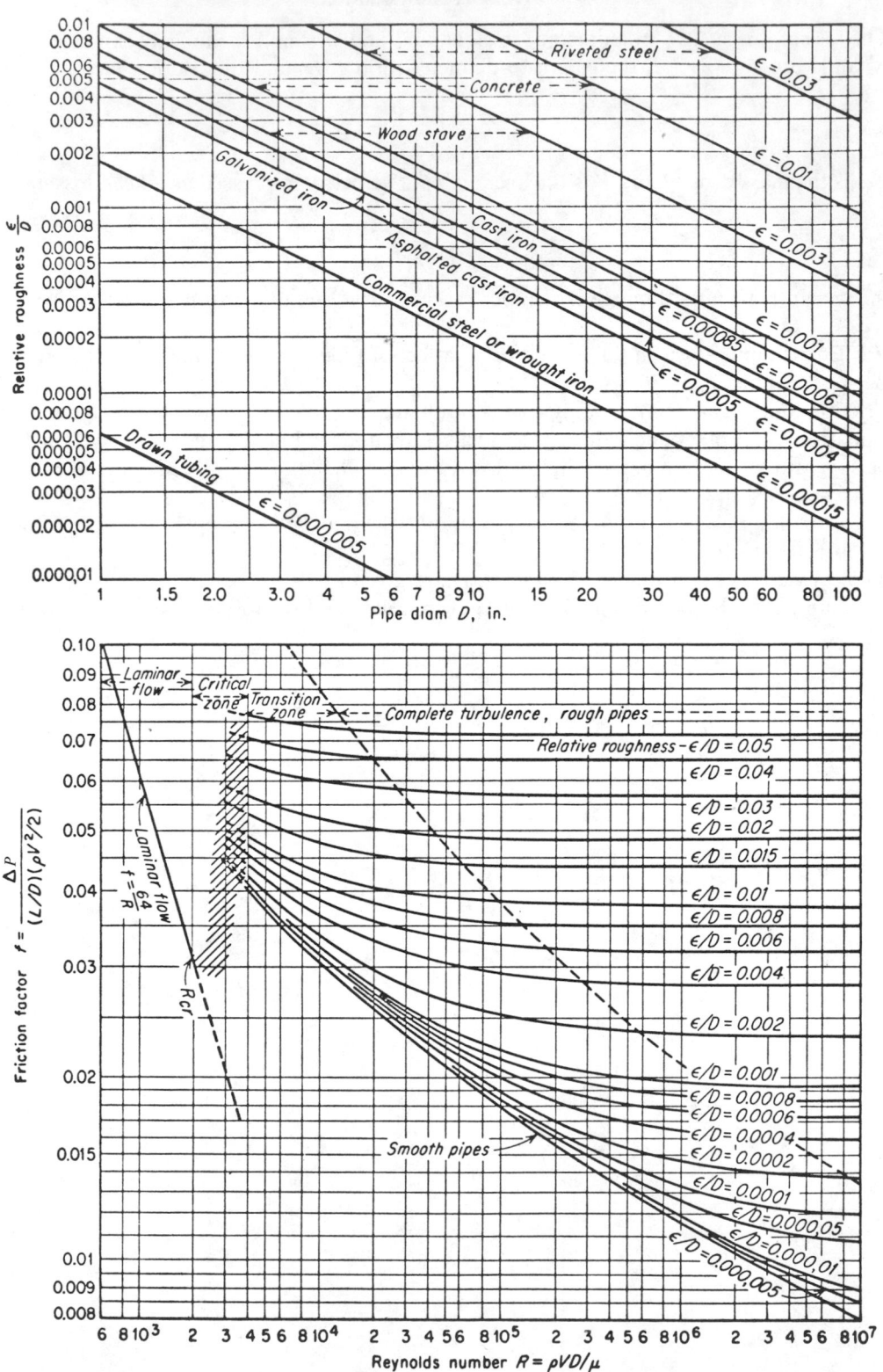

*From: ENGINEERING EXPERIMENTATION by G.L. Tuve and L.C. Domholdt. Copyright © 1966 by McGraw-Hill, Inc. Used with permission of McGraw-Hill Book Company. Adapted from: L.F. Moody, *Trans. ASME*, 66:671, 1944.

Table 5-26. FRICTION OF VALVES AND FITTINGS*

In Terms of L/D or f L/D

		Description		Equivalent length in pipe diameters, L/D
Globe valves	Conventional	With no obstruction in flat, bevel, or plug-type seat	Fully open	340
Angle valves	Conventional	With no obstruction in flat, bevel, or plug-type seat	Fully open	145
		With wing- or pin-guided disc	Fully open	200
Gate valves	Conventional wedge disc, double disc, or plug disc		Fully open	13
			Three-quarters open	35
			One-half open	160
			One-quarter open	900
Check valves	Conventional swing	0.5† . . . Fully open		135
	Clearway swing	0.5† . . . Fully open		50
Butterfly valves (8-inch and larger)			Fully open	40
Cocks	Straight-through	Rectangular plug port area equal to 100% of pipe area	Fully open	18
	Three-way	Rectangular plug port area equal to 80% of pipe area	Flow straight through, fully open	44
			Flow through branch	140
Fittings	90-degree standard elbow			30
	45-degree standard elbow			16
	90-degree long radius elbow			20
	90-degree street elbow			50
	45-degree street elbow			26
	Square-corner elbow			57
	Standard tee	With flow through run		20
		With flow through branch		60
	Close pattern return bend			50
Pipe	90-degree pipe bends		Radius = 1 pipe diameter	20
			Radius = 2D	12
			Radius = 4D	16
			Radius = 8D	24

	Description	Loss in terms of $f L/D$‡
	Sudden enlargement	
	$d_1/d_2 = 0.5$	0.56
	$= 0.7$	0.26
	$= 0.9$	0.04
	Sudden contraction	
	$d_2/d_1 = 0.5$	0.34
	$= 0.7$	0.21
	$= 0.9$	0.04
	Square-edge entrance, negligible velocity of approach	0.50
	Slightly rounded entrance	0.25
	Rounded flow-nozzle entrance	0.04

†Minimum calculated pressure drop (psi) across valve to provide sufficient flow to lift disc fully.
‡D is the smaller diameter.

*Extracted from: "Technical Paper No. 410—Flow of Fluids", 11th printing. Courtesy of Crane Co.

Table 5-27.   VACUUM TECHNOLOGY APPLICATIONS

Vacuum Pumps and Vacuum Measurement

Typical absolute pressure		Typical applications or engineering uses	Vacuum pump or source of vacuum	Pressure gage
in. Hg	torr (mm Hg)			
20–30	640–960	Industrial air-exhaust; pneumatic conveying systems; vacuum cleaning	Centrifugal fan; rotary-vane pump	Spring gage
6–20	150–640	Automotive brakes and devices; refrigeration compressors; evaporators	Engine intake; piston, rotary, centrifugal or jet pump	Bourdon tube gage
3–6	75–150	Evaporators; desalination; evaporative vacuum cooling (foods)	Centrifugal compressor; rotary or jet pump	Mercury manometer
1–3	25–75	Steam (power-plant) condensers—summer; degassing molten metals	Steam ejectors	Electromechanical transducer (strain gage, capacitor, inductor, etc.)
0.5	12.7	Steam condensers—winter; vacuum furnaces and reactors	Multi-stage steam ejectors	
	1	Freeze drying (food and drugs); lamp bulbs	Single-stage mechanical pump	Compression (McLeod) gage
	10^{-2}	High-altitude environmental testing; vacuum furnaces and melting; sputtering	Multi-stage mechanical pump	Thermal conductivity gage
	10^{-4}	Evaporation of metals for film deposition; mirrors; electron-beam furnace; cryogenic insulation	Oil diffusion pump	Radiometer gage
	10^{-6}	Electron tubes; space simulation	Diffusion pump with liquid nitrogen trap	Ionization gage
	10^{-8}	Space simulation chambers; traveling wave tubes; computer elements	Diffusion pumps with traps and metal seals	Ionization gage
	10^{-10}	Out-gassing of materials	Diffusion pump and baked system	Ionization gage
	10^{-12}	Space vacuum	Ion pump; sorption and cryogenic pumping	Mass spectrometer

REFERENCES

"Vacuum Technology and Space Simulation", NASA SP105, National Aeronautics and Space Administration, 1966.
"Vacuum Science and Engineering", C.M. Van Atta, McGraw-Hill Book Company, 1965.
"Materials of High Vacuum Technology", W. Espe, Pergamon Press, Inc., 1966–1968 (3 volumes).

Table 5-28.   FLOW IN VACUUM SYSTEMS

Relationships of Flow Conductance, Throughput, and Pressure Drop

In design calculations involving vacuum pump capacity, piping or conduit sizes, and pumpdown time, it is necessary to evaluate the pressure loss for a given section of the flow system. The two extreme cases are the long tube of uniform cross-section and the thin-plate orifice. Intermediate are short tubes of various L/D ratios. It is often possible to estimate the performance of a rather complex system, including many fittings, in terms of these basic elements by taking series and parallel arrangements into account.

For any component of a gas-flow system, there is a definite relationship between the rate of flow and the pressure differential from entrance to exit. Four consistent patterns of flow are recognized, and the relation of gas-flow rate to differential pressure is as follows:

1. *Supersonic or choked flow*, in which the flow rate is proportional to the upstream pressure only and unaffected by downstream pressure.
2. *Turbulent flow*, in which the flow rate is proportional to the square root of the pressure differential.
3. *Laminar or viscous flow*, in which the flow rate is directly proportional to the pressure differential.
4. *Molecular flow*, in which the flow rate is directly proportional to the pressure differential but unaffected by the absolute pressure.

Table 5-28.   FLOW IN VACUUM SYSTEMS *(Continued)*

A given pattern of flow will exist only under certain conditions. Extensive "critical" regions of mixed or unstable flow pattern may be encountered as the type of flow changes from one to another. In all cases the rates of flow are dependent on the nature and properties of the gas and on the dimensional properties of the flow path and its components.

In most vacuum-flow calculations the state of the gas is assumed to follow the simple perfect-gas law, $PV = RT$, and the same assumption is made here. The common flow equations are based on average properties across the stream; they take no account of transverse gradients of pressure, temperature, or velocity. Sidewall pressure-tap measurements are the basis for the comparisons between computed and actual performance (see Table 5-27).

In high-vacuum engineering it is accepted practice to compute the flow rates by means of the conductance analogy, as used for the flow of electricity or heat, i.e., Ohm's law. Mixed metric units are almost invariably used, with flow expressed in liters per second (l/s) and pressure in torr (mm Hg). Flow rate is called *throughput* Q, a mass-flow quantity in units of torr·l/s. Volume flow rate per unit of pressure is treated as a conductance C, l/s. The Ohm's-law analogy $Q = CP$ is used for three different cases:

1. for any pipe or fitting $Q = CP$, where C is the volume-rate conductance at the average pressure $P_{avg} = \dfrac{P_1 + P_2}{2}$ and $P = \Delta P = P_1 - P_2$.

2. for a vacuum pump intake $Q = CP$, where C is the volume-rate conductance or *pumping speed* at the inlet pressure P_i, and $P = P_i$.

3. for *any* designated cross-section $Q = CP$, where C is the volume-rate conductance at the pressure P_a measured at that cross-section (called the *speed* at the section) and $P = P_s$.

The computation of the conductance or "speed" C, l/s per unit of pressure depends on the pattern of flow, whether molecular, laminar, or transitional. Turbulent flow, or even supersonic flow, may exist for a short time at the beginning of pumpdown in a high-vacuum system, but these are not decisive factors in the piping design. Turbulent flow may be important in low-vacuum systems.

Molecular Flow

Within the various phases of vacuum engineering, all four of the usual flow regimes will be encountered. In the high-vacuum conduits, where the mean free path of the gas molecules is much larger than the diameter of the conduit, molecular flow exists. True molecular flow depends (statistically) on the paths of the gas molecules only, and therefore on the geometry of the system and the absolute temperature. For a given gas in molecular flow, the volume flow rate through a "hole" (thin-plate orifice between large containers) reduces to:

$$C = 3.64 \, A \sqrt{\frac{T}{M}}, \text{ l/s,}$$

where A is area in cm^2, T is temperature in K, and M is the molecular weight of the gas. For air at room temperature and a circular thin-plate *orifice*, this becomes $C = 9.15 \, D^2$, l/s (Table A). For a *long, round conduit* with end effects neglected, the corresponding equations are

$$C = 3.8 \sqrt{\frac{T}{M}} \, D^3/L \text{ and } C = 12.2 \, D^3/L, \text{ l/s.}$$

These volume flow rates per unit of pressure are treated as conductances.

Table A.   MOLECULAR FLOW THROUGH AN ORIFICE[a]

Conductance, or Volume Rate of Flow, l/s, through a Thin-plate Orifice of Diameter D, cm

D	C	D	C
0.2	0.366	4	146.
0.5	2.29	6	329.
1.0	9.15	8	586.
1.5	20.6	10	915.
2.0	36.6	20	3 660.
3.0	82.4		

[a]Computed from: $C = 9.15 \, D^2$, for air at room temperature. For other temperatures or other gases, use $2.86 \, D^2 \sqrt{\dfrac{T}{M}}$, where T is in K and M is the molecular weight.

If orifice diameter is given in inches, multiply tabular values by 13.7 to obtain conductance in cfm.

Table 5-28.   FLOW IN VACUUM SYSTEMS (Continued)

For parallel conduits, the conductances are additive—$C_1 + C_2 + C_3 \cdots = C$; in series,

$$\frac{1}{C_1} + \frac{1}{C_2} + \frac{1}{C_3} \cdots = \frac{1}{C}.$$

For sections of conduit less than about 100 diameters long, the end effects are taken into account by modifying the long-pipe equation:

$$C = 12.2\ D^3 \Big/ \Big(L + \frac{4}{3}\ D \Big).$$

Flow rates or conductances of conduits or pipes to 0.2–20 cm diameter and 1–500 cm long are given in Table B. These values include the end correction.

Table B.   MOLECULAR FLOW—ROUND PIPES[a]

Conductance or Volume Rate of Flow, l/s, for Air at Room Temperature

Length, cm	Internal diameter, cm										
	0.2	0.5	1:0	1.5	2	3	4	6	8	10	20
1	.077	.915	5.24	13.7	26.6	66.0	123.	293.			
2	.043	.571	3.66	10.3	21.0	55.0	115.	264.	494.	796.	
3	.030	.415	2.82	8.22	17.3	47.0	93.7	240.	457.	748.	
4	.024	.321	2.29	6.85	14.7	41.2	83.6	220.	426.	705.	3180.
5	.019	.269	1.93	5.88	12.7	36.6	75.5	203.	400.	667.	3080.
6	.016	.228	1.67	5.14	11.2	33.0	68.9	188.	376.	632.	2980.
8	.012	.173	1.31	4.11	9.12	27.5	58.6	165.	336.	573.	2810.
10	.010	.141	1.08	3.42	7.69	23.5	50.9	146.	303.	524.	2660.
15	.006	.096	.746	2.42	5.51	17.4	38.4	115.	244.	431.	2340.
20	.005	.072	.572	1.87	4.30	13.7	30.8	94.1	204.	366.	2090.
30	.003	.050	.389	1.29	2.99	9.69	22.1	69.3	154.	282.	1720.
40	.002	.038	.297	.980	2.29	7.49	17.2	54.9	123.	229.	1460.
50	.002	.030	.238	.784	1.85	6.10	14.1	45.4	103.	193.	1270.
60		.025	.199	.663	1.56	5.15	12.0	38.8	88.5	166.	1130.
80		.019	.150	.502	1.18	3.92	9.15	29.9	68.9	131.	900.
100		.015	.120	.403	.950	3.17	7.41	24.4	56.5	108.	770.
150		.010	.081	.270	.639	2.14	5.03	16.7	38.9	74.7	552.
200		.076	.061	.204	.481	1.62	3.80	12.7	29.7	57.2	430.
300			.041	.137	.326	1.08	2.56	8.56	20.1	38.9	299.
400			.031	.103	.244	.823	1.93	6.46	15.2	29.5	229.
500			.024	.082	.195	.660	1.56	5.19	12.2	23.8	185.

[a] Computed from: $C_M = 12.2\ D^3/(L + 1.33\ D)$. This assumes that the overall conductance is the sum of the tube conductance plus the conductance of an orifice of the same diameter. For other temperatures or other gases, use $C = 3.8\ \sqrt{\dfrac{T}{M}}\ D^3/(L + 1.33\ D)$, where T is in K and M is the molecular weight.

   If pipe dimensions are given in inches, multiply tabular values by 13.7 to obtain conductance in cfm (P in torr and μ in poises).

Laminar Flow

   When the absolute pressure is above 1 torr, the flow is usually laminar rather than molecular. Turbulent flow seldom occurs in a high-vacuum system. Turbulent and laminar flows are covered in Tables 5-24 to 5-26. In order to facilitate calculations in terms of conductance C and throughput Q (where $Q = CP$), the values of C are given in Table C in l/s for laminar flow in "long" conduits up to $D = 20$ cm.

   This table gives the flow of air at room temperature by the equation:

$$C = 180\ D^4\ P/L,$$

where $P_{avg} = 1.0$ torr. This equation is based on the usual Poiseuille formula for pressure drop ΔP, in long conduits with viscous flow:

$$\Delta P = 128\ \mu Lq/\pi D^4,$$

in which q is the volume rate of flow. In consistent metric units with μ in poises (dyne·s/cm^2 or g/cm·s), the pressure is in dyne/cm^2, and the flow rate is in cm^3/s. For throughput in torr l/s the conductance becomes $C = 0.0327\ D^4 P/\mu L$.

Table 5-28.   FLOW IN VACUUM SYSTEMS *(Continued)*

Throughput is related to mass flow rate by a simple constant in which the perfect gas equation is used in the form:

$$\text{density} = \frac{P_{avg}}{RT} = \frac{P_1 + P_2}{2RT}.$$

Table C.   LAMINAR FLOW IN LONG PIPES[a]

Conductance or Volume Rate of Flow for Round Conduits, l/s, at Average Pressure of 1.0 torr

Length-diameter ratio	Internal diameter, cm										
	0.2	0.5	1.0	1.5	2	3	4	6	8	10	20
4	.360	5.62	45.0	152	360	1 215	2 880	9.740			
5	.288	4.50	36.0	122	288	972	2 300	7 780			
6	.240	3.74	30.0	99.6	240	810	1 920	6 500	15 400[b]		
8	.180	2.81	22.5	76.0	180	608	1 440	4 860	11 500		
10	.144	2.25	18.0	60.8	143	486	1 152	3 890	9 200	18 000[b]	
15	.096	1.50	12.0	40.5	96.0	324	768	2 600	6 140	12 000	
20	.072	1.12	9.00	30.4	72.0	243	576	1 950	4 600	9 000	
30	.048	.750	6.00	20.2	48.0	162	384	1 300	3 070	6 000	
40	.036	.562	4.50	15.2	36.0	122	288	974	2 300	4 500	36 000[b]
50	.029	.450	3.60	12.2	28.8	97.2	230	778	1 840	3 600	28 800
60	.024	.374	3.00	9.96	24.0	81.0	192	650	1 540	3 000	24 000
80	.018	.281	2.25	7.60	18.0	60.8	144	486	1 150	2 250	18 000
100	.014	.225	1.80	6.08	14.3	48.6	115	260	920	1 800	14 300
150	.010	.150	1.20	4.05	9.60	32.4	76.8	195	614	1 200	9 600
200	.007	.112	.900	3.04	7.20	24.3	57.6	130	460	900	7 200
300	.005	.075	.600	2.02	4.80	16.2	38.4	97.0	307	600	4 800
400	.004	.056	.450	1.52	3.60	12.2	28.8	77.8	230	450	3 600
500	.003	.045	.360	1.22	2.88	9.7	23.0	65.0	184	360	2 880

[a]Computed from: $C_L = \dfrac{180\,D^4}{L}\dfrac{(P_1 + P_2)}{2}$ for air at room temperature. For other temperatures or other gases, use $C = 0.0328\,D^4 P_{avg}/\mu L$, where μ is in poises

   If pipe dimensions are given in inches, multiply tabular values by 34.7 to obtain conductance in cfm (P in torr and μ in poises).

[b]Flow pattern probably becomes turbulent above about $C = 2\,000\,D/P$.

Transition Flow

   In a long conduit the transition from laminar flow to molecular flow usually occurs between 0.001 and 1.0 torr. By equating the formulas for laminar flow and molecular-flow conductance, it is found that the two curves intersect at about $DP_{avg} = 0.068$. This might be considered the midpoint of the transition range. Within this range both the values from Table B and Table C are lower than the actual conductance. Two methods for obtaining an additive correction are indicated in Table D. For approximate estimates, it may be assumed that the transition range covers two decades, i.e., from $DP_{avg} = 0.0068$ to $DP_{avg} = 0.68$.

Table D.   TRANSITION FLOW EXAMPLES

Method 1: Add values from Tables B and C.

Method 2: Compute $P_{avg}\,D$ and select correction factor from second column, Table D. Multiply value from Table B by this correction factor, and add the result to the value from Table C.

   Example: Find the conductances of a pipe 1.0 cm in diameter and 100 cm long, assuming transition flow. Explore the pressure range from 100 torr to 1×10^{-4} torr.

   Solution: Molecular flow conductance is unaffected by P_{avg} and $C = 0.120$ torr, from Table B. Laminar flow conductance varies directly as P_{avg}, and $C = 1.80$ at 1.0 torr, from Table C. Values at other pressures are given in column 3.

   Conclusion: In most cases conductance in the transition range between molecular and laminar flow is closely approximated by using the sum of the laminar-flow conductance and the molecular-flow conductance. If $DP_{avg} < 0.001$, use Table B only, molecular flow. If $DP_{avg} > 10$, use Table C only, laminar flow.

Table 5-28.   FLOW IN VACUUM SYSTEMS *(Continued)*

Table D.   TRANSITION FLOW EXAMPLES *(Continued)*

Product DP_{avg}, torr cm	Correction factor[a]	C from Table C	C_{corr}, method 1	C_{corr}, method 2	Product DP_{avg}, torr cm	Correction factor[a]	C from Table C	C_{corr}, method 1	C_{corr}, method 2
1×10^{-4}	.994	.0002	.120	.119	0.1	.816	.18	.30	.28
2×10^{-4}	.989	.0004	.120	.119	0.2	.813	.36	.48	.46
3×10^{-4}	.984	.0005	.120	.118	0.3	.812	.54	.66	.64
6×10^{-4}	.970	.0011	.121	.118	0.6	.811	1.08	1.20	1.18
1×10^{-3}	.954	.0018	.122	.118	1.0	.811	1.80	1.92	1.90
2×10^{-3}	.927	.0036	.124	.118	2.0	.810	3.60	3.72	3.70
3×10^{-3}	.908	.0054	.125	.114	3.0	.810	5.40	5.52	5.50
6×10^{-3}	.876	.0108	.131	.116	6.0	.810	10.8	10.9	10.9
1×10^{-2}	.857	.0180	.138	.120	10.	.810	18.0	18.1	18.1
2×10^{-2}	.836	.0360	.156	.136	20.	.810	36.0	36.1	36.1
3×10^{-2}	.828	.0540	.174	.153	30.	.810	54.0	54.1	54.1
6×10^{-2}	.820	.1080	.228	.206	60.	.810	108.	108.	108.
					100.	.810	180.	180.	180.

[a]Computed from $[(1 + 256\ DP_{avg})/(1 + 316\ DP_{avg})]$, for air at room temperature.

EXAMPLES:

Example 1.   The piping in a high-vacuum system includes a 100-cm length of 6-cm ID tubing. Estimate the flow conditions for room temperature air, as follows:
1.   Within what approximate range of P_{avg} pressures does "transition flow" probably exist?
2.   What is the conductance at each extreme of this P_{avg} range and at the midpoint of this range?

Solution.   Assume the range extends from $DP_{avg} = .0068$ to $DP_{avg} = .68$, mean $DP_{avg} = .068$.
1.   Minimum $P_{avg} = .0068/6 = .00113 = 1.13 \times 10^{-3}$ torr
     Maximum $P_{avg} = .68/6 = .113 = 1.13 \times 10^{-1}$ torr
     Mean $P_{avg} = .068/6 = .0113 = 1.13 \times 10^{-2}$ torr
2.   Minimum P_{avg}, molecular flow, from Table B, conductance $= 24.4$ l/s
     Maximum P_{avg}, laminar flow, from Table C, conductance $= 264.$ l/s
     Mean P_{avg}, transition flow, from Table D, conductance $= 26.4 + .85 \times 24.4 = 47.1$ l/s

Example 2.   The 6 cm $\times$ 100 cm pipe section of Example 1 is connected directly to a vacuum pump that operates at a "pumping speed" of 15 l/s at 1×10^{-4} torr. The piping system includes other components, and at the load end (farthest from the pump), a pressure of 2×10^{-4} torr is desired. Assume that there is no leakage or outgassing in the piping system, and molecular flow throughout. Determine the following:
1.   Pressure at the inlet end of the 6×100 cm pipe. (Conductance, C_P).
2.   Speed and throughput at the load end (entrance to piping system).
3.   Minimum conductance of the "other components", C_O.

Solution.   Designate the pump inlet as section 1, the inlet of the 6×100 cm pipe as section 2, and the load connection as section 3. $S =$ speed. $C =$ conductance. $P =$ pressure. $Q =$ throughput. As Q is a mass-flow quantity, it is constant.
1. $Q = P_1 S_1 = P_3 S_3 = C_P(P_2 - P_1) = 15 \times 1 \times 10^{-4} = 1.5 \times 10^{-3}$ torr·l/s. From Table B: $C_P = 24.4$ l/s

$$P_2 - P_1 = \frac{Q}{C} = 1.5 \times 10^{-3}/24.4 = 6.15 \times 10^{-5} \text{ torr}$$

$$P_2 = 10^{-4} + 6.15 \times 10^{-5} = 1.615 \times 10^{-4} \text{ torr}$$

2. $P_3 = 2 \times 10^{-4}$ torr; $S_3 = \dfrac{Q}{P_3} = 1.5 \times 10^{-3}/2 \times 10^{-4} = 7.5$ l/s;

     throughput $Q = 1.5 \times 10^{-3}$ torr l/s
3. $C_O(P_3 - P_2) = Q$; $C_O = 1.5 \times 10^{-3}/(2 \times 10^{-4} - 1.615 \times 10^{-4}) = 39.0$ l/s

REFERENCES

"Introduction to the Theory and Practice of High-Vacuum Technology", L. Ward and J.P. Bunn, Butterworths, 1967.
"Vacuum Science and Engineering", C.M. Van Atta, McGraw-Hill Book Company, 1965.
"Vacuum Engineering", Holkeboer, Jones, Pagano and Santeler, Boston Technical Publishers, 1967.
"High Vacuum Engineering", A.E. Barrington, Prentice-Hall, 1963.

Table 5-29. THERMAL CONDUCTIVITY OF METALS*

100°K to 3000°K

Values in this table are in watts/cm °K. To convert to Btu/hr ft °R, multiply the tabular values by 57.818. These data apply only to metals of purity of at least 99.9%. In the table the third significant figure is for smoothness and is not indicative of the degree of accuracy.

Note: Values in parentheses are for liquid state.

Metal	Temperature, °K									
	100°	200°	273°	300°	400°	500°	600°	700°	800°	900°
Aluminum	3.0	2.37	2.36	2.37	2.40	2.37	2.32	2.26	2.20	2.13
Cadmium	1.03	0.993	0.975	0.968	0.947	0.920	(0.420)	(0.490)	(0.559)	
Chromium	1.58	1.11	0.948	0.903	0.873	0.848	0.805	0.757	0.713	0.678
Copper	4.83	4.13	4.01	3.98	3.92	3.88	3.83	3.77	3.71	3.64
Gold	3.45	3.27	3.18	3.15	3.12	3.09	3.04	2.98	2.92	2.85
Iron	1.32	0.94	0.835	0.803	0.694	0.613	0.547	0.487	0.433	0.380
Lead	0.396	0.366	0.355	0.352	0.338	0.325	0.312	(0.174)	(0.190)	(0.203)
Magnesium	1.69	1.59	1.57	1.56	1.53	1.51	1.49	1.47	1.46	1.45
Mercury			(0.078)	(0.084)	(0.098)	(0.109)	(0.120)	(0.127)	(0.130)	
Molybdenum	1.79	1.43	1.39	1.38	1.34	1.30	1.26	1.22	1.18	1.15
Nickel	1.58	1.06	0.94	0.905	0.801	0.721	0.655	0.653	0.674	0.696
Niobium (Columbium)	0.552	0.526	0.533	0.537	0.552	0.567	0.582	0.598	0.613	0.629
Platinum	0.79	0.748	0.734	0.730	0.722	0.719	0.720	0.723	0.729	0.737
Silver	4.50	4.30	4.28	4.27	4.20	4.13	4.05	3.97	3.89	3.82
Tantalum	0.592	0.575	0.574	0.575	0.578	0.582	0.586	0.590	0.594	0.598
Tin	0.85	0.733	0.682	0.666	0.622	0.596	(0.323)	(0.343)	(0.364)	(0.384)
Titanium	0.312	0.245	0.224	0.219	0.204	0.197	0.194	0.194	0.197	0.202
Tungsten	2.35	1.97	1.82	1.78	1.62	1.49	1.39	1.33	1.28	1.24
Zinc	1.32	1.26	1.22	1.21	1.16	1.11	1.05	(0.499)	(0.557)	(0.615)
Zirconium	0.332	0.252	0.232	0.227	0.216	0.210	0.207	0.209	0.216	0.226

Metal	Temperature, °K									
	1000°	1100°	1200°	1400°	1600°	1800°	2000°	2200°	2600°	3000°
Aluminum	(0.93)	(0.96)	(0.99)							
Chromium	0.653	0.636	0.624	0.611						
Copper	3.57	3.50	3.42							
Gold	2.78	2.71	2.62							
Iron	0.326	0.297	0.282	0.309	0.327					
Lead	(0.215)									
Magnesium	(0.84)	(0.91)	(0.98)							
Molybdenum	1.12	1.08	1.05	0.996	0.946	0.907	0.880	0.858	0.825	
Nickel	0.718	0.739	0.761	0.804						
Niobium (Columbium)	0.644	0.659	0.675	0.705	0.735	0.764	0.791	0.815		
Platinum	0.748	0.760	0.775	0.807	0.842	0.877	0.913			
Silver	3.74	3.66	3.58							
Tantalum	0.602	0.606	0.610	0.618	0.626	0.634	0.640	0.647	0.658	0.665
Tin	(0.405)	(0.425)	(0.446)	(0.487)						
Titanium	0.207	0.213	0.220	0.236	0.253	0.271				
Tungsten	1.21	1.18	1.15	1.11	1.07	1.03	1.00	0.98	0.94	0.915
Zinc	(0.673)	(0.730)								
Zirconium	0.237	0.248	0.257	0.275	0.290	0.302	0.313			

*From: NSRDS–NBS–16, "Thermal Conductivity of Selected Materials", Part 2, C.Y. Ho, R.W. Powell, P.E. Liley, National Standard Reference Data System-National Bureau of Standards, February 1968.

Table 5-30.   THERMAL CONDUCTIVITY OF METALS AT CRYOGENIC TEMPERATURES*

As Recommended by National Standard Reference Data System-N.B.S.

For other thermal conductivities and specific heats, see Tables 1-60 to 1-63.

Values in this table are in watts/cm °K. To convert to Btu/hr ft °R, multiply the tabular values by 57.818. These data apply only to metals of purity of at least 99.9%. In the table the third significant figure is for smoothness and is not indicative of the degree of accuracy.

Temperature °K	Temperature °R	Aluminum	Cadmium	Chromium	Copper	Gold	Iron	Lead	Magnesium	Molybdenum
1	1.8	7.8	48.7	0.401	28.7	4.4	0.75	27.7	1.30	0.146
2	3.6	15.5	89.3	0.802	57.3	8.9	1.49	42.4	2.59	0.292
3	5.4	23.2	104	1.20	85.5	13.1	2.24	34.0	3.88	0.438
4	7.2	30.8	92.0	1.60	113	17.1	2.97	22.4	5.15	0.584
5	9	38.1	69.0	1.99	138	20.7	3.71	13.8	6.39	0.730
6	10.8	45.1	44.2	2.38	159	23.7	4.42	8.2	7.60	0.876
7	12.6	51.5	28.0	2.77	177	26.0	5.13	4.9	8.75	1.02
8	14.4	57.3	18.0	3.14	189	27.5	5.80	3.2	9.83	1.17
9	16.2	62.2	12.2	3.50	195	28.2	6.45	2.3	10.8	1.31
10	18	66.1	8.87	3.85	196	28.2	7.05	1.78	11.7	1.45
11	19.8	69.0	6.91	4.18	193	27.7	7.62	1.46	12.5	1.60
12	21.6	70.8	5.56	4.49	185	26.7	8.13	1.23	13.1	1.74
13	23.4	71.5	4.67	4.78	176	25.5	8.58	1.07	13.6	1.88
14	25.2	71.3	4.01	5.04	166	24.1	8.97	0.94	14.0	2.01
15	27	70.2	3.55	5.27	156	22.6	9.30	0.84	14.3	2.15
16	28.8	68.4	3.16	5.48	145	20.9	9.56	0.77	14.4	2.28
18	32.4	63.5	2.62	5.81	124	17.7	9.88	0.66	14.3	2.53
20	36	56.5	2.26	6.01	105	15.0	9.97	0.59	13.9	2.77
25	45	40.0	1.79	6.07	68	10.2	9.36	0.507	12.0	3.25
30	54	28.5	1.56	5.58	43	7.6	8.14	0.477	9.5	3.55
35	63	21.0	1.41	5.03	29	6.1	6.81	0.462	7.4	3.62
40	72	16.0	1.32	4.30	20.5	5.2	5.55	0.451	5.7	3.51
45	81	12.5	1.25	3.67	15.3	4.6	4.50	0.442	4.57	3.26
50	90	10.0	1.20	3.17	12.2	4.2	3.72	0.435	3.75	3.00
60	108	6.7	1.13	2.48	8.5	3.8	2.65	0.424	2.74	2.60
70	126	5.0	1.08	2.08	6.7	3.58	2.04	0.415	2.23	2.30
80	144	4.0	1.06	1.82	5.7	3.52	1.68	0.407	1.95	2.09
90	162	3.4	1.04	1.68	5.14	3.48	1.46	0.401	1.78	1.92
100	180	3.0	1.03	1.58	4.83	3.45	1.32	0.396	1.69	1.79

Temperature °K	Temperature °R	Nickel	Niobium	Platinum	Silver	Tantalum	Tin	Titanium	Tungsten	Zinc	Zirconium
1	1.8	0.64	0.251	2.31	39.4	0.115		0.0144	14.4	19.0	0.111
2	3.6	1.27	0.501	4.60	78.3	0.230		0.0288	28.7	37.9	0.223
3	5.4	1.91	0.749	6.79	115	0.345	297	0.0432	42.6	55.5	0.333
4	7.2	2.54	0.993	8.8	147	0.459	181	0.0576	55.6	69.7	0.442
5	9	3.16	1.23	10.5	172	0.571	117	0.0719	67.1	77.8	0.549
6	10.8	3.77	1.46	11.8	187	0.681	76	0.0863	76.2	78.0	0.652
7	12.6	4.36	1.67	12.6	193	0.788	52	0.101	82.4	71.7	0.748
8	14.4	4.94	1.86	12.9	190	0.891	36	0.115	85.3	61.8	0.837
9	16.2	5.49	2.04	12.8	181	0.989	26	0.129	85.1	51.9	0.916
10	18	6.00	2.18	12.3	168	1.08	19.3	0.144	82.4	43.2	0.984
11	19.8	6.48	2.30	11.7	154	1.16	14.8	0.158	77.9	36.4	1.04
12	21.6	6.91	2.39	10.9	139	1.24	11.6	0.172	72.4	30.8	1.08
13	23.4	7.30	2.46	10.1	124	1.30	9.3	0.186	66.4	26.1	1.11
14	25.2	7.64	2.49	9.3	109	1.36	7.6	0.200	60.4	22.4	1.13
15	27	7.92	2.50	8.4	96	1.40	6.3	0.214	54.8	19.4	1.13
16	28.8	8.15	2.49	7.6	85	1.44	5.3	0.227	49.3	16.9	1.12
18	32.4	8.45	2.42	6.1	66	1.47	4.0	0.254	40.0	13.3	1.08
20	36	8.56	2.29	4.9	51	1.47	3.2	0.279	32.6	10.7	1.01
25	45	8.15	1.87	3.15	29.5	1.36	2.22	0.337	20.4	6.9	0.85
30	54	6.95	1.45	2.28	19.3	1.16	1.76	0.382	13.1	4.9	0.74
35	63	5.62	1.16	1.80	13.7	0.99	1.50	0.411	8.9	3.72	0.65
40	72	4.63	0.97	1.51	10.5	0.87	1.35	0.422	6.5	2.97	0.58
45	81	3.91	0.84	1.32	8.4	0.78	1.23	0.416	5.07	2.48	0.535
50	90	3.36	0.76	1.18	7.0	0.72	1.15	0.401	4.17	2.13	0.497
60	108	2.63	0.66	1.01	5.5	0.651	1.04	0.377	3.18	1.71	0.442
70	126	2.21	0.61	0.90	4.97	0.616	0.96	0.356	2.76	1.48	0.403
80	144	1.93	0.58	0.84	4.71	0.603	0.91	0.339	2.56	1.38	0.373
90	162	1.72	0.563	0.81	4.60	0.596	0.88	0.324	2.44	1.34	0.350
100	180	1.58	0.552	0.79	4.50	0.592	0.85	0.312	2.35	1.32	0.332

*From: NSRDS-NBS-8 and NSRDS-NBS-16, "Thermal Conductivity of Selected Materials", C.Y. Ho, R.W. Powell, P.E. Liley, National Standard Reference Data System-National Bureau of Standards, Part 1, 1966, Part 2, 1968.

Table 5-31. CRYOGENIC THERMAL INSULATION*

For other thermal conductivities, see Tables 1-110, 5-32, and the Index.

DESCRIPTION AND ADVANTAGES

CLASS

1 **Liquid and vapor shields.** Very low-temperature, valuable, or dangerous liquids, such as helium or fluorine, are often shielded by an intermediate cryogenic liquid or vapor container that must in turn be insulated by one of the methods described below.

2 **Multilayer reflecting shields.** Foil or aluminized plastic alternated with paper-thin glass- or plastic-fiber sheets; lowest conductivity, low density, and heat storage; good stability; minimum support structure.

3 **Opacified evacuated powders.** Contain metallic flakes to reduce radiation; conform to irregular shapes.

4 **Evacuated dielectric powders.** Very fine powders of low-conductivity adsorbent; moderate vacuum requirement; minimum fire hazard in oxygen.

5 **Vacuum flasks (Dewar).** Tight shield-space with highly reflecting walls and high vacuum; minimum heat capacity; rugged; small thickness.

6 **Gas-filled powders.** Same powders as Class 4 but with air or inert gas; low cost; easy application; no vacuum requirement.

7 **Expanded foams.** Very light foamed plastic; inexpensive; minimum weight but bulky; self-supporting.

8 **Porous fiber blankets.** Blanket material of fine fibers, usually glass; minimum cost and easy installation but not an adequate insulation for most cryogenic applications.

INSULATION PROPERTIES

Class	Descriptive name	Approximate density		Approximate specific heat		Range of mean conductivities		Interspace pressure, mm Hg[a]
		$\frac{lbm}{ft^3}$	$\frac{kg}{m^3}$	$\frac{Btu}{lbm \cdot deg\ F}$	$\frac{kJ}{kg \cdot K}$	$\frac{Btu}{hr \cdot ft \cdot deg\ F}$	$\frac{mW}{m \cdot K}$	
2	Multilayer	5	80	.22	0.92	.000023–.00012	0.04–0.2	10^{-4}
3	Opacified powder	7	110	.23	0.96	.00015–.0004	0.26–0.7	10^{-4}
4	Evacuated powder	6	100	.25	1.05	.00057–.00115	1.0–2.0	10^{-4}
5	Vacuum flask	—	—	—	—	.0029	5.0	10^{-6}
6	Gas-filled powder	6	100	.25	1.05	.001–.004	1.7–7.0	760
7	Expanded foam	2	30	0.4	1.67	.0029–.020	5.0–35	760
8	Fiber blanket	8	130	0.5	2.09	.02–.026	35–45	760

[a] For N/m² multiply by 133.32.

STRUCTURAL SUPPORT

For those insulating materials and constructions requiring structural support, the relative strengths, weights, heat capacities, and conductivities of the supporting materials are important.

Material	Tensile yield strength S, 1000's psi[b]	Density, ρ		Specific heat, c_p		Mean thermal conductivity k,[c] 20–300 °K		Relative		
		$\frac{lbm}{ft^3}$	$\frac{kg}{m^3}$	$\frac{Btu}{lbm \cdot deg\ F}$	$\frac{kJ}{kg \cdot K}$	$\frac{Btu}{hr \cdot ft \cdot deg\ F}$	$\frac{W}{m \cdot K}$	$\frac{S}{k}$	$\frac{\rho}{k}$	$\frac{c_p \rho}{S}$
Aluminum alloy	50	170	2720	.22	0.92	50	86	1	3	.75
"K" Monel	100	520	8330	.13	0.54	10	17	10	52	.68
Stainless steel	100	500	8010	.12	0.50	5.4	9.3	18	93	.60
Titanium alloy	100	625	10010	.06	0.25	3.5	6.1	29	180	.37
Nylon	15	70	1120	.4	1.67	.17	0.29	88	41	1.9
Teflon	2	120	1920	.25	1.05	.14	0.24	14	86	15.0

[b] For MN/m² multiply tabulated values in 1000's psi by 6.8948.
[c] For solid members. Perforation and lamination used to reduce conduction.

*Compiled from several sources.

REFERENCE

"Thermal Insulation Systems", NASA SP–5027, National Aeronautics and Space Administration, 1967.

Table 5-32. HEAT TRANSMISSION
THROUGH BUILDING STRUCTURES

For other data on thermal conductivities, see Tables 1-110, 5-31, and the Index.

Temperature control within structures requires the use of accurate and consistent data on heat transfer coefficients. Extensive tables have been compiled by the American Society of Heating, Refrigerating and Air-Conditioning Engineers ("ASHRAE Handbook of Fundamentals", Chap. 26), and these are accepted as standard. The following tables give some of the basic coefficients (Table A) and examples of overall coefficients for common wall and roof constructions (Table B) from this source.

Table A. BASIC COEFFICIENTS

Emissivities: Aluminum foil, 1 surface, 0.05; 2 surfaces, 0.03. Aluminum paint, 0.50. Non-metallic surface, 0.90. (Blackbody = 1.0).

Coefficients in Btu/hr·ft^2·deg F (W/m^2·K): Indoor non-metallic surfaces, still air; vertical, 1.46 (8.29); horizontal ceiling, winter, 1.63 (9.26); summer, 1.08 (6.13). Vertical air spaces: non-metallic, 1.04 (5.91); foil on both surfaces, 0.34 (1.93). Horizontal air space: non-metallic surfaces, winter, 1.18 (6.70); summer, 1.01 (5.74). Outdoor surface coefficients, winter, 6.0 (34) for 15 m/h (6.7 m/s) wind; summer, 4.0 (23) for 7.5 m/h (3.35 m/s) wind.

For density in kg/m^3, multiply values in lbm/ft^3 by 16.018.

Conductivity, k, Btu in./ft^2·hr·deg F (at 75 deg F)

Material	Density, $\frac{lbm}{ft^3}$	Thermal conductivity		Material	Density, $\frac{lbm}{ft^3}$	Thermal conductivity	
		$\frac{Btu \cdot in.}{hr \cdot ft^2 \cdot deg\,F}$	$\frac{W}{m \cdot K}$			$\frac{Btu \cdot in.}{hr \cdot ft^2 \cdot deg\,F}$	$\frac{W}{m \cdot K}$
Asbestos cement board	120	4.0	0.58	Plaster, cement-sand	116	5.0	0.72
Blanket of batt insulation	3	0.27	0.039	Plaster, gypsum-perlite	45	1.5	0.22
Brick, common	120	5.0	0.72	Plaster, vermiculite	45	1.7	0.25
Brick, face	130	9.0	1.30	Plywood	34	0.80	0.115
Cement mortar	116	5.0	0.72	Redwood bark fill	4	0.27	0.039
Concrete, gypsum-fiber	51	1.66	0.24	Sheathing, wood fiber	22	0.41	0.059
Concrete, lightweight	40	1.15	0.17	Shredded wood, cemented	22	0.60	0.086
Corkboard	7	0.27	0.039	Vermiculite	8	0.47	0.068
Hardboard, wood fiber	65	1.40	0.20	Wood, oak, maple	45	1.10	0.159
Mineral fiberboard	18	0.35	0.05	Wood or cane fiberboard	15	0.35	0.050
Mineral wool fill	3	0.28	0.04	Wood, pine, fir	32	0.80	0.115

Table B. OVERALL COEFFICIENTS (TRANSMITTANCE) AIR TO AIR

Structure	U, $W/m^2 \cdot K$	U, $Btu/hr \cdot ft^2 \cdot deg\,F$
Single-glass window	6.2	1.1
Double-insulating glass with $\frac{1}{2}$-in. air space, or storm window	3.2	0.56
Frame wall: wood sheathing, lath, and sand plaster	1.5	0.26
Frame wall: insulating sheathing, lath, and lightweight plaster	1.1	0.19
Frame wall: full fibrous insulating between studs	0.40	0.07
Brick veneer, insulating sheathing, lightweight plaster	1.2	0.21
Brick wall, 8-in. solid, sand plaster	2.6	0.45
Brick wall, 8-in., gypsum lath, furred, lightweight plaster	1.5	0.27
Clay tile wall, 8-in. hollow, gypsum lath, furred, lightweight plaster	1.3	0.23
Masonry cavity wall: 4-in. face brick, air space, 4-in. cinder block, gypsum lath, furred, lightweight plaster, two aluminum foil surfaces or $\frac{3}{4}$-in. insulation	0.74	0.13
Roof, pitched, shingle, unventilated rafter space, lath and plaster ceiling	1.7	0.30
Roof, pitched, asbestos cement, slate or tile shingles on wood sheathing, plastered ceiling, 3-in. insulation between joists	0.40	0.07
Roof, flat, metal deck, plaster ceiling on gypsum board	1.9	0.33
Roof, flat, preformed 2-in. insulating slab deck, acoustical ceiling on gypsum board	0.74	0.13

Table 5-33.   LOGARITHMIC MEAN

Logarithmic Mean $\bar{N} = (N_1 - N_2)/\ln(N_1/N_2)$

The logarithmic mean is often used, as in heat-transfer calculations for temperature differences and for areas. The notation $LMTD$ for logarithmic mean temperature difference is commonly used:

$$LMTD = (\Delta T_1 - \Delta T_2)/\ln(\Delta T_1/\Delta T_2)$$

The table can be extended to greater values of N_1, N_2, and $\bar{N}$ by multiplying all values by a common multiplier. For example, for $N_1 = 6\,000$ and $N_2 = 1\,000$

$$\bar{N} = 100(\bar{N})_{N_1 = 60, N_2 = 10} = 100(27.9) = 2\,790$$

For $N_1 = 0.06$ and $N_2 = 0.01$,

$$\bar{N} = 0.01(\bar{N})_{N_1 = 6, N_2 = 1} = 0.01(2.79) = 0.027\,9$$

N_1	N_2									
	1	2	3	4	5	6	7	8	9	10
1	1.000	1.443	1.820	2.164	2.485	2.791	3.083	3.366	3.641	3.909
2	1.443	2.000	2.466	2.885	3.274	3.641	3.991	4.328	4.654	4.971
3	1.820	2.466	3.000	3.476	3.915	4.328	4.721	5.098	5.461	5.814
4	2.164	2.885	3.476	4.000	4.481	4.933	5.361	5.771	6.166	6.410
5	2.485	3.274	3.915	4.481	5.000	5.484	5.944	6.383	6.805	7.213
6	2.791	3.641	4.328	4.933	5.484	6.000	6.487	6.952	7.399	7.830
7	3.083	3.991	4.721	5.361	5.944	6.487	7.000	7.489	7.958	8.411
8	3.366	4.328	5.098	5.771	6.383	6.952	7.489	8.000	8.490	8.963
9	3.641	4.654	5.461	6.166	6.805	7.399	7.958	8.490	9.000	9.491
10	3.909	4.971	5.814	6.410	7.213	7.830	8.411	8.963	9.491	10.000
20	6.342	7.817	8.961	9.941	10.820	11.628	12.383	13.096	13.776	14.427
30	8.526	10.340	11.726	12.904	13.953	14.912	15.804	16.644	17.442	18.205
40	10.572	12.685	14.284	15.634	16.831	17.922	18.933	19.883	20.782	21.640
50	12.525	14.912	16.706	18.213	19.543	20.752	21.871	22.919	23.909	24.853
60	14.410	17.053	19.027	20.679	22.134	23.452	24.669	25.808	26.883	27.905
70	16.241	19.126	21.271	23.059	24.630	26.051	27.360	28.584	29.738	30.834
80	18.028	21.145	23.451	25.369	27.051	28.568	29.966	31.269	32.497	33.663
90	19.779	23.117	25.579	27.621	29.408	31.019	32.499	33.879	35.178	36.410
100	21.498	25.051	27.662	29.824	31.712	33.411	34.972	36.425	37.791	39.086
200	37.559	42.995	46.908	50.102	52.862	55.325	57.571	59.648	61.591	63.424
300	52.421	59.473	64.492	68.558	72.051	75.153	77.970	80.566	82.987	85.264
400	66.595	75.118	81.139	85.990	90.141	93.816	97.144	100.204	103.051	105.723
500	80.295	90.306	97.146	102.727	107.488	111.693	115.492	118.979	122.219	125.255
600	93.639	104.843	112.677	118.947	124.282	128.985	133.228	137.117	140.724	144.101
700	106.700	119.155	127.832	134.759	140.642	145.819	150.483	154.753	158.710	162.410
800	119.528	133.190	142.678	150.236	156.645	162.278	167.345	171.981	176.272	180.282
900	132.159	146.990	157.264	165.433	172.349	178.420	183.878	188.865	193.478	197.786
1 000	144.620	160.589	171.626	180.387	187.795	194.292	200.127	205.455	210.380	214.976

Table 5-34.   HEAT TRANSFER PROPERTIES OF FLUIDS*

To obtain this table, data on the fluid properties were selected from References 1–15, then plotted and averaged to yield the values given here for c_p, k, μ, and ρ. These values may differ slightly from those given elsewhere in this handbook.

UNITS AND CONVERSION FACTORS:

c_p = specific heat at constant pressure; it is the same for Btu/lb·deg F and cal/g·deg C. To convert to J/k·K, multiply by 4 184.

k = thermal conductivity in Btu/hr·ft·deg F. For W/m·K multiply by 1.73.

μ = dynamic viscosity in lb/ft·hr. For N·s/m² multiply by 0.000 413.

ρ = density in lb/ft³. For kg/m³ multiply by 16.02.

crit G = approximate mass velocity in lb/sec·ft² in a one-inch diameter conduit that results in a critical Reynolds number, taken as 2 000.

$\dfrac{c_p \mu}{k}$ = Prandtl number, which is dimensionless.

Temperature, deg F	c_p	k	μ	ρ	crit G	$\dfrac{c_p\mu}{k}$	$\left(\dfrac{c_p\mu}{k}\right)^{0.4}$	$\left(\dfrac{c_p\mu}{k}\right)^{0.67}$
Water								
32	1.029 3	0.337	4.32	62.54	28.8	13.2	2.80	5.59
100	0.999	0.365	1.62	62.2	10.8	4.43	1.81	2.71
200	1.003 9	0.393	0.738	60.2	4.92	1.88	1.287	1.524
400	1.075	0.382	0.32	53.62	0.976	0.91	0.963	0.939
600	1.525	0.293	0.215	42.37	0.934	1.08	1.03	1.053
Aqueous solution 30 % ethylene glycol								
60	0.882	0.276	6.04	64.9	40.2	19.6	3.295	7.28
100	0.900	0.285	3.27	64.3	21.8	10.3	2.54	4.74
200	0.934	0.292	1.23	62.1	8.2	3.93	1.73	2.49
300	0.970	0.285	0.692	59.2	4.61	2.355	1.408	1.77
Ethylene glycol								
60	0.556	0.169	62.1	69.4	414	204	8.4	34.8
100	0.581	0.159 5	25.1	68.7	167.3	91.4	6.09	20.3
200	0.644	0.135	5.67	66.2	37.8	27.05	3.74	9.04
300	0.706	0.111	2.295	63.3	15.3	14.6	2.92	5.97
H₂ (liquid)								
−430	1.91	0.063 6	0.044 7	4.67	0.298	1.345	1.126	1.218
−410	4.44	0.079 6	0.020 4	3.69	0.136	1.135	1.052	1.088
N₂ (liquid)								
−210	0.500	0.041	0.162	34.5	1.08	1.975	1.313	1.575
−110	0.474	0.095	0.756	54.0	5.04	3.75	1.696	2.41
NH₃ (liquid)								
0	1.08	0.29	0.567	42.0	3.71	2.075	1.499	1.96
100	1.17	0.29	0.172	35.6	1.14	0.694	0.864	0.784
Dowtherm A								
200	0.432	0.086 3	2.71	62.6	18.05	13.56	2.84	5.67
400	0.600	0.105	1.14	56.8	7.60	6.51	2.115	3.49
600	0.700	0.103 7	0.727	50.5	4.85	4.90	1.89	2.885
Methyl alcohol								
0	0.57	0.124	2.80	51.3	18.70	12.87	2.78	5.50
100	0.615	0.120 5	1.15	48.1	7.67	5.87	2.03	3.258
200	0.65	0.117	0.666	43.1	4.45	3.70	1.687	2.39

*From: A.P. Fraas and M.N. Ozisik, "Heat Exchanger Design". Copyright © 1965 by John Wiley & Sons, Inc.

Table 5-34. HEAT TRANSFER PROPERTIES OF FLUIDS (Continued)

Temperature, deg F	c_p	k	μ	ρ	crit G	$\dfrac{c_p\mu}{k}$	$\left(\dfrac{c_p\mu}{k}\right)^{0.4}$	$\left(\dfrac{c_p\mu}{k}\right)^{0.67}$
Refrigerant 11								
0	0.198	0.06	1.639	98.27	10.92	5.40	1.96	3.08
100	0.212	0.053	0.920	90.19	6.14	3.68	1.685	2.38
200	0.225	0.046	0.637	80.94	4.25	3.12	1.576	2.135
Gasoline								
0	0.447	0.110	2.60	49.7	17.35	10.58	2.57	4.81
200	0.565	0.103	0.745	42.7	4.97	4.08	1.748	2.55
400	0.683	0.096 7	0.336	36.8	2.24	2.37	1.413	1.78
Kerosene								
0	0.430	0.101	17.1	52.5	114	72.8	5.55	17.5
200	0.545	0.095	1.59	47.4	10.6	9.12	2.42	4.37
400	0.655	0.089 2	0.625	42.4	4.17	4.58	1.839	2.76
600	0.745	0.082 9	0.31	38.1	2.07	2.78	1.506	1.978
SAE 10 petroleum lubricating oil								
0	0.411	0.093 75	4 730	55.6	31 500	20 750	54.0	760
200	0.52	0.088 4	11.88	52.25	79.25	69.7	5.45	16.9
300	0.575	0.085 2	4.503	48.75	30.0	22.5	3.48	7.98
HTS (NaNO₃, KNO₃, KNO₂)								
400	1.93	0.34	18.15	120.6	121	103	6.4	22.0
500	1.89	0.34	12.0	118.2	80.0	66.7	5.37	16.5
600	1.85	0.35	7.02	115.8	46.8	37.1	4.25	10.15
ORNL molten salt number 14 (10.9 % NaF, 43.5 % KF, 44.5 % LiF, 1.1 % UF₄)								
1 100	0.488	2.3	12.6	132.1	84.0	2.67	1.482	1.925
1 200	0.488	2.44	8.8	129.0	58.7	1.76	1.254	1.458
ORNL molten salt number 30 (50 % NaF, 46 % ZrF₄, 4 % UF₄)								
1 200	0.26	1.38	17.0	207.8	113.5	3.2	1.594	2.17
1 300	0.255 5	1.405	12.8	204.6	85.4	2.33	1.403	1.757
1 400	0.250 8	1.5	10.6	201.3	70.7	1.77	1.257	1.464
Sodium								
200	0.330 5	49.1	1.725	57.9	11.5	0.011 6	0.168	0.051
400	0.319 9	46.7	1.095	56.4	7.3	0.007 5	0.141 5	0.033 5
600	0.311 5	43.8	0.797	54.6	5.31	0.005 67	0.128	0.034
800	0.304 9	40.1	0.61	53.0	4.07	0.004 64	0.117	0.028
1 000	0.302	37.2	0.56	51.2	3.73	0.004 55	0.115 5	0.027 5
1 200	0.301 1	35.0	0.475	49.1	3.16	0.004 08	0.111	0.025 7
1 400	0.303 3	32.7	0.415	47.7	2.765	0.003 85	0.108	0.024 5
NaK (56 % Na, 44 % K)								
200	0.270	14.9	1.36	55.3	9.100	0.024 6	0.227	0.084
400	0.260	15.3	0.92	53.8	5.92	0.015 5	0.188	0.061
600	0.255	15.7	0.71	52.1	4.48	0.011 5	0.168	0.050
800	0.251	16.0	0.5	50.6	3.57	0.008 6	0.149	0.041
1 000	0.250	16.0	0.49	49.0	3.04	0.008 3	0.147	0.040
1 200	0.251	16.0	0.41	47.2	2.54	0.008 0	0.145	0.039
Potassium								
800	0.183	22.8	0.51	46.1	3.16	0.004 1	0.111	0.025
1 000	0.182	21.1	0.414	44.4	2.55	0.003 6	0.106	0.023
1 200	0.183	19.5	0.354	42.9	2.20	0.003 3	0.102	0.022
1 400	0.187	18.0	0.322	41.5	2.00	0.003 3	0.102	0.022

Table 5-34.   HEAT TRANSFER PROPERTIES OF FLUIDS *(Continued)*

Temperature, deg F	c_p	k	μ	ρ	crit G	$\dfrac{c_p\mu}{k}$	$\left(\dfrac{c_p\mu}{k}\right)^{0.4}$	$\left(\dfrac{c_p\mu}{k}\right)^{0.67}$
Lithium								
400	1.042 5	26.8	1.31	31.65	8.73	0.051	0.304	0.136
600	1.02	24.9	1.08	31.0	7.2	0.044 3	0.287 5	0.124
800	1.005 7	22.1	0.95	30.4	6.33	0.043 2	0.284 5	0.122
1 000	0.996 2	17.6	0.84	29.6	5.6	0.047 6	0.296	0.130
Mercury								
0	0.033 8	5.64	4.435	851.9	29.6	0.026 6	0.234	0.089
200	0.032 6	6.00	2.957	833.4	19.7	0.016 1	0.192	0.072
400	0.032 4	7.3	2.43	818.4	16.2	0.010 8	0.163 5	0.049
600	0.034 2	7.9	2.27	802.6	15.15	0.009 83	0.157	0.046

GASES (AT ATMOSPHERIC PRESSURE)

Temperature, deg F	c_p	k	μ	ρ	crit G	$\dfrac{c_p\mu}{k}$	$\left(\dfrac{c_p\mu}{k}\right)^{0.4}$	$\left(\dfrac{c_p\mu}{k}\right)^{0.67}$
Air								
− 200	0.239 2	0.007 9	0.025 2	0.153	0.168	0.780	0.904	0.847 3
0	0.240 0	0.014	0.041 5	0.086 4	0.276	0.711	0.872 5	0.796 5
200	0.241 4	0.018 4	0.051 9	0.060 2	0.346	0.685	0.859 5	0.777
400	0.245 1	0.022 4	0.062 4	0.046 2	0.416	0.683	0.858 5	0.776
600	0.250 5	0.026 3	0.072 0	0.037 5	0.480	0.686	0.860	0.777 8
800	0.256 7	0.030 0	0.080 5	0.031 6	0.537	0.688	0.861	0.779 5
1 000	0.263	0.033 2	0.088 4	0.027 2	0.589	0.700	0.867	0.788 2
1 200	0.269 2	0.036 3	0.096 0	0.023 9	0.640	0.712	0.873	0.797 5
1 400	0.275 5	0.039 1	0.103 5	0.021 4	0.691	0.728	0.881	0.809 2
H₂								
− 400	2.46	0.014	0.004 3	0.045	0.028 65	0.756	0.894	0.83
− 200	2.975	0.055	0.013 1	0.010 5	0.087 3	0.709	0.869	0.792
0	3.385	0.092	0.020 4	0.005 9	0.136	0.738	0.891	0.825 5
200	3.45	0.122	0.024 8	0.004 15	0.165 5	0.700	0.867	0.788
400	3.46	0.152	0.029 7	0.003 2	0.198	0.676	0.855	0.77
600	3.47	0.18	0.034 2	0.002 6	0.228	0.659	0.846	0.757
800	3.48	0.207	0.039 4	0.002 1	0.262 5	0.662	0.848	0.76
1 000	3.48	0.223	0.042 1	0.001 86	0.280 5	0.656	0.845	0.755
1 200	3.49	0.241	0.046 1	0.001 65	0.307	0.667	0.85	0.763
1 400	3.50	0.257	0.049 7	0.001 47	0.331	0.676	0.855	0.77
He								
− 200	1.25	0.052	0.039 5	0.021	0.263 5	0.949	0.979 3	0.965 7
0	1.25	0.08	0.043 4	0.012	0.289 2	0.678	0.856	0.772
200	1.25	0.098 5	0.054 5	0.083	0.363 5	0.691	0.862 5	0.782
400	1.25	0.118	0.066	0.064	0.44	0.699	0.866 5	0.788
600	1.25	0.137	0.077	0.005 1	0.513	0.702	0.868	0.790
800	1.25	0.156	0.088	0.004 4	0.587	0.705	0.869 5	0.792
1 000	1.25	0.176	0.099	0.003 7	0.66	0.703	0.868 5	0.791
1 200	1.25	0.194	0.109	0.003 3	0.727	0.702	0.868	0.790
1 400	1.25	0.212	0.119	0.002 9	0.794	0.701	0.867 5	0.789
Argon								
0	0.124	0.009	0.049	0.113 5	0.327	0.674	0.854	0.769
200	0.124	0.012	0.064	0.079 2	0.426	0.661	0.847 5	0.759
400	0.124	0.014 7	0.078	0.060 7	0.520	0.658	0.846	0.756
600	0.124	0.017 2	0.090 5	0.049 2	0.603	0.652	0.843	0.752
800	0.124	0.019 4	0.102	0.041 5	0.681	0.652	0.843	0.752
1 000	0.124	0.021 8	0.112 5	0.035 8	0.750	0.640	0.837	0.742
1 200	0.124	0.023 4	0.122 5	0.031 4	0.817	0.649	0.841	0.750
1 400	0.124	0.025 2	0.131 5	0.028 1	0.876	0.648	0.841	0.749

Table 5-34. HEAT TRANSFER PROPERTIES OF FLUIDS *(Continued)*

Temperature, deg F	c_p	k	μ	ρ	crit G	$\dfrac{c_p\mu}{k}$	$\left(\dfrac{c_p\mu}{k}\right)^{0.4}$	$\left(\dfrac{c_p\mu}{k}\right)^{0.67}$
Neon								
0	0.246			0.060				
200	0.246	0.032 4	0.088 4	0.042	0.589	0.670	0.841	0.766
400	0.246	0.038 4	0.104 0	0.032	0.685	0.668	0.852	0.764
600	0.246	0.043 8	0.119 0	0.026	0.795	0.668	0.852	0.764
800	0.246	0.048 8	0.132 5	0.022	0.885	0.668	0.852	0.764
1 000	0.246	0.053 5	0.145 0	0.019	0.968	0.666	0.850	0.762
1 200	0.246	0.058 5	0.158 5	0.016 6	1.060	0.666	0.850	0.762
1 400	0.246	0.062 5	0.170 0	0.014 9	1.135	0.666	0.850	0.752
CO_2								
0	0.19	0.007 7	0.031	0.131 5	0.206 7	0.765	0.898	0.837
200	0.218	0.012 7	0.043 3	0.091 5	0.288 5	0.743	0.888	0.820 5
400	0.238	0.017 7	0.054 8	0.070 2	0.365	0.737	0.885	0.816
600	0.255 4	0.022 6	0.065 2	0.057 0	0.435	0.736	0.884 5	0.815
800	0.268 4	0.027 3	0.074	0.048 0	0.493	0.727	0.880	0.808 5
1 000	0.279 3	0.031 7	0.082 7	0.041 5	0.551	0.728	0.881	0.809
1 200	0.289 8	0.035 8	0.091	0.036 4	0.606	0.735	0.884	0.814 5
1 400	0.297 5	0.039 6	0.098 8	0.032 5	0.658	0.742	0.887 5	0.82
NH_3								
0	0.522	0.011 7	0.021 3	0.044 1	0.142	0.633	0.833	0.737
200	0.532	0.019 2	0.030 3	0.030 7	0.202	0.84	0.932 7	0.890 5
400	0.574	0.028 0	0.039 4	0.023 6	0.263	0.807	0.917 7	0.867
600	0.625	0.039 7	0.047 9	0.019 2	0.319	0.755	0.894	0.829
800	0.675	0.053 7	0.055 7	0.016 1	0.371	0.700	0.867	0.788
CH_4								
0	0.507	0.015 7	0.023 7	0.045 5	0.158	0.765	0.898 5	0.836 5
200	0.579	0.025 5	0.031 7	0.031 7	0.211	0.720	0.877	0.803 5
400	0.674	0.035 8	0.038	0.024 3	0.253	0.715	0.874 5	0.800
600	0.772	0.050 5	0.044	0.019 7	0.293	0.672	0.853	0.767
Freon 11								
0	0.124	0.004 12	0.023 2	0.039 8	0.154 5	0.698	0.866	0.787
100	0.134	0.005 19	0.027 4	0.032 2	0.182 5	0.707	0.870 5	0.793 5
200	0.145	0.006 27	0.031 2	0.027 8	0.208	0.722	0.878	0.805

Note: Data from References 1–15 were selected, plotted, and averaged to yield the values given here for c_p, k, μ, and ρ.

GENERAL REFERENCES

"Standards of the Tubular Heat Exchanger Manufacturers Association", 1959.

"Heat Exchangers", The Patterson-Kelley Co., 1960.

"Flow of Fluids through Valves, Fittings, and Pipe", Technical Paper No. 410, Crane Co., 1957.

R.B. Scott, "Cryogenic Engineering", D. Van Nostrand Co., 1959.

J.A. Lane, et al., "Fluid Fuel Reactors", Addison-Wesley Publishing Company, 1958.

J. Hilsenrath and Y.S. Touloukian, "The Viscosity, Thermal Conductivity, and Prandtl Number for Air, O_2, N_2, NO, H_2, CO, CO_2, H_2O, He, and A", *Trans. ASME*, 76:967, 1954.

R.C. Reid and T.K. Sherwood, "The Properties of Gases and Liquids", McGraw-Hill Book Company, 1958.

R.G. Vines, "Measurement of the Thermal Conductivities of Gases at High Temperatures", *J. Heat Transfer, Trans. ASME*, 82(2):48, 1960.

F.G. Keyes and D.J. Sandell, Jr., "New Measurements of the Heat Conductivity of Steam and Nitrogen", *Trans. ASME*, 72:767, 1950.

J.H. Keenan and J. Kaye, "Gas Tables", John Wiley & Sons, Inc., 1948.

W.D. Weatherford, Jr., et al., "Properties of Inorganic Energy-Conversion and Heat-Transfer Fluids for Space Applications", U.S. Air Force, WADD Technical Report, 61–96, 1961.

"Handbook of Chemistry and Physics", 44th ed., The Chemical Rubber Publishing Co., 1962.

N.A. Lange, "Handbook of Chemistry", 10th ed., McGraw-Hill Book Company, 1961.

J.H. Perry, et al., "Chemical Engineers' Handbook", 4th ed., McGraw-Hill Book Company, 1963.

U.S. Atomic Energy Commission, "Reactor Handbook, Materials", Vol. 1, Interscience Publishers, Inc., 1960.

Table 5-35. CONVECTION HEAT TRANSFER—
AIR IN ROUND TUBES*

Approximate forced-convection coefficients are for air at atmospheric pressure flowing in a long round tube, pipe, or duct, turbulent range.

Inside diameter, round pipe, in.	Convection coefficient, h, Btu/ft² hr °F											
	COOLING: Mean air temperature 70°F						HEATING: Mean air temperature 200°F					
	Mean air velocity, fps						Mean air velocity, fps					
	3	5	10	20	50	100	3	5	10	20	50	100
0.5				6.7	13.9	24.2				5.9	12.2	21.2
1.0			3.3	5.9	12.0	21.3			2.9	5.2	10.6	18.8
2.0		1.7	2.9	5.1	10.5	18.3		1.5	2.6	4.5	9.2	16.1
3.0	1.0	1.6	2.7	4.7	9.7	16.9		1.4	2.3	4.1	8.5	14.8
6.0	0.9	1.3	2.3	4.0	8.5	14.7	0.8	1.2	2.0	3.5	7.4	12.9
12.0	0.8	1.2	2.0	3.5	7.3	12.8	0.7	1.0	1.8	3.1	6.5	11.2

Notes:

These coefficients were computed from: $Nu = 0.023\ Re^{.8}\ Pr^{.4}$, or $h = .023\ \dfrac{k}{D}\left(\dfrac{DV\rho}{\mu}\right)^{.8}\left(\dfrac{c\mu}{k}\right)^{.4}$. Air properties and velocities are for 70°F and 200°F mean air temperatures. In a specific problem the air properties (c, k, μ, and ρ) should be evaluated at the film temperature, usually assumed as the average between the mean wall temperature and the mean mixed fluid temperature. Hence the tabular values above are useful for estimating, but more precise values should be computed from the formula, if required.

*Computed; data from several sources.

Table 5-36. CONVECTION HEAT TRANSFER—
WATER IN ROUND TUBES*

Approximate forced-convection coefficients are for water in turbulent flow through long round pipes or tubes.

Inside diameter, round pipe, in.	Convection coefficient, h, Btu/ft² hr °F											
	Av. temp. 40°F				Av. temp. 100°F				Av. temp. 150°F			
	Mean water velocity, fps				Mean water velocity, fps				Mean water velocity, fps			
	2	5	10	20	2	5	10	20	2	5	10	20
0.25		1040	1810	3160	725	1510	2640	4610	900	1872	3274	5716
0.50	437	905	1580	2770	638	1320	2310	4040	791	1637	2864	5010
1.00	363	782	1400	2400	530	1142	2040	3500	659	1416	2530	4340
1.50	350	727	1270	2200	511	1060	1850	3210	634	1314	2294	3980
2.00	331	689	1200	2100	482	1005	1750	3060	598	1246	2170	3794
3.00	306	634	1100	1910	446	925	1610	2780	553	1147	1996	3447

Notes:

These coefficients were computed from $Nu = .023\ Re^{.8}Pr^{.4}$. Water properties were measured at the arithmetical average of the mean wall temperature and the mean mixed-fluid temperature.

As viscous flow is approached (Re < 6000), the coefficients may be erratic. Turbulence promoters and spirals within the tube will increase the heat transfer; at the lower Reynolds numbers this increase may even reach 100 percent. Coefficients for coiled tubes are slightly higher than those in this table.

*Computed; data from several sources.

Table 5-37.   REYNOLDS NUMBERS FOR FLOW OF WATER AND AIR IN PIPE AND TUBING—SI UNITS

$$Re = \rho VD/\mu = VD/\nu = \text{Reynolds number, dimensionless}$$

SYMBOLS AND UNITS:

For perfect gases $(P = \rho RT)$, $Re = PVD/\mu RT$

ρ = density of fluid, kg/m^3
V = mean velocity of fluid, m/s
D = inside diameter of pipe or tubing, m
μ = absolute viscosity, N·s/m^2 (=kg/m·s)
ν = μ/ρ = kinematic viscosity, m^2/s
P = absolute pressure, N/m^2
R = gas constant, J/kg·K
T = absolute temperature, K

Temperature			$\nu \times 10^6$, m^2/s	Product of velocity and diameter, VD, m^2/s								
deg C	K	deg F		1	2	3	4	5	6	7	8	9
REYNOLDS NUMBERS FOR WATER $(Re \times 10^{-6})$[a]												
0	273.15	32	1.787	0.560	1.12	1.68	2.24	2.80	3.36	3.92	4.48	5.04
20	293.15	68	1.004	0.996	1.99	2.99	3.98	4.98	5.98	6.97	7.97	8.96
40	313.15	104	0.658	1.52	3.04	4.56	6.08	7.60	9.12	10.6	12.2	13.7
60	333.15	140	0.474	2.11	4.22	6.33	8.44	10.5	12.7	14.8	16.9	19.0
80	353.15	176	0.364	2.75	5.49	8.24	11.0	13.7	16.5	19.2	22.0	24.7
100	373.15	212	0.294	3.40	6.80	10.2	13.6	17.0	20.4	23.8	27.2	30.6
150	423.15	302	0.198	5.05	10.1	15.2	20.2	25.3	30.3	35.4	40.4	45.5
200	473.15	392	0.327	3.06	6.12	9.17	12.2	15.3	18.3	21.4	24.5	27.5
REYNOLDS NUMBERS FOR AIR AT ATMOSPHERIC PRESSURE $(Re \times 10^{-3})$[b]												
−73.15	200	−99.67	7.51	133.	266.	399.	533.	666.	799.	932.	1 065.	1 198.
−48.15	225	−54.67	9.33	107.	214.	322.	429.	536.	643.	750.	857.	965.
−23.15	250	−9.67	11.3	88.5	177.	265.	354.	442.	531.	619.	708.	796.
+1.85	275	+35.33	13.4	74.6	149.	224.	299.	373.	448.	522.	597.	672.
26.85	300	80.33	15.7	63.7	127.	191.	255.	318.	382.	446.	510.	573.
51.85	325	125.33	18.1	55.2	110.	166.	221.	276.	331.	387.	442.	497.
76.85	350	170.33	20.6	48.5	97.1	146.	194.	243.	291.	340.	388.	437.
101.85	375	215.33	23.2	43.1	86.2	129.	172.	216.	259.	302.	345.	388.
126.85	400	260.33	25.9	38.6	77.2	116.	154.	193.	232.	270.	309.	347.
226.85	500	440.33	37.8	26.5	52.9	79.4	106.	132.	159.	185.	212.	238.
326.85	600	620.33	51.3	19.5	39.0	58.5	78.0	97.5	117.	136.	156.	175.
526.85	800	980.33	82.2	12.2	24.3	36.5	48.7	60.8	73.0	85.2	97.3	109.
726.85	1 000	1 340.33	118.	8.47	16.9	25.4	33.9	42.4	50.8	59.3	67.8	76.3
926.85	1 200	1 700.33	157.	6.37	12.7	19.1	25.4	31.8	38.2	44.6	51.0	57.3
1 126.85	1 400	2 060.33	201.	4.98	9.95	14.9	19.9	24.9	29.9	34.8	39.8	44.8
1 326.85	1 600	2 420.33	247.	4.05	8.10	12.1	16.2	20.2	24.3	28.3	32.4	36.4
1 526.85	1 800	2 780.33	297.	3.37	6.73	10.1	13.5	16.8	20.2	23.6	26.9	30.3

[a] Viscosity and density, and thus Reynolds numbers for water, are only slightly affected by pressure.
[b] Absolute viscosity for air is only slightly affected by pressure, but density, and thus Reynolds numbers, are directly proportional to pressure for perfect gases. For air at pressures other than atmospheric, multiply the Reynolds numbers in the air table by the pressure in atmospheres.

Table 5-38. CONVECTION HEAT TRANSFER— AIR FLOW OVER SURFACES*

These approximate forced-convection coefficients are for air at atmospheric pressure flowing over external surfaces. The coefficients were computed from $h = \text{constant} \times \dfrac{k}{D} \times \left(\dfrac{DV\rho}{\mu}\right)^{.6}\left(\dfrac{c\mu}{k}\right)^{.3}$. Coefficients are for 70°F mean air temperature and are somewhat high for mean temperatures above 100°F. Air properties and air velocity were measured at the mean mixed-air temperature.

	Convection coefficient, h,[a] $Btu/ft^2\ hr\ °F$			
Shapes, sizes, and applications	Air velocity through free area, fps			
	5	10	20	50
Over a flat surface[b]	1.8	3.0	5.0	10.0
Over one row of $\frac{1}{4}$-inch tubes[c]	9.1	13.9	21.0	36.5
Over one row of $\frac{5}{8}$-inch tubes[c]	6.3	9.6	14.6	25.4
Over one row of 1-inch tubes[c]	5.3	8.0	12.1	21.0
Over many rows of $\frac{5}{8}$-inch tubes, staggered (with or without fins)[d]	10.1	15.4	23.4	40.4
Over many rows of 1-inch tubes, staggered (with or without fins)[d]	8.4	12.8	19.4	33.5
Over many rows of 1-inch tubes, in line[d]	5.9	9.0	13.6	23.6

Note: Surface area to include fins, if any. Capacities will be reduced if fin-temperature gradients are appreciable. Local turbulence produced by pin-fins, corrugated surfaces, or otherwise will increase the heat-transfer coefficient. All coefficients apply to sensible heat transfer only, no condensation on the surfaces.

[a] For $W/m^2{\cdot}K$ multiply by 5.674.
[b] Angle between stream and surface less than 45°.
[c] Coefficients apply to normal air-stream turbulence and uniform velocity over the face area. High local turbulence increases the coefficient (even as much as 50% for a special turbulence grid mounted within 3 tube diameters upstream).
[d] Coefficients are based on logarithmic mean temperature difference air-to-surface. It is assumed that the surface temperature does not vary greatly in the direction of air flow.

*Computed; data from several sources.

Table 5-39. HEAT TRANSFER FROM SURFACES— CONVECTION PLUS RADIATION

Approximate values of the combined surface coefficient of heat transfer for natural convection and radiation in room-temperature air. Convection computed from $h_c = K\left(\dfrac{\Delta t}{D}\right)^{.25}$ Radiation computed from Stefan-Boltzmann law assuming total hemispherical radiation.

	Combined surface coefficient, $h_c + h_r$, $Btu/ft^2\ hr\ °F$[a]											
Orientation and size	A: Non-metallic surface Mean temp diff: Ts–Ta[b]				B: Dull metal Mean temp diff: Ts–Ta				C: Polished metal Mean temp diff: Ts–Ta			
	50	100	200	300	50	100	200	300	50	100	200	300
Horizontal cylinders												
$\frac{1}{4}$-inch diam.	3.0	3.4	4.1	4.8	2.4	2.8	3.3	3.7	2.0	2.2	2.5	2.8
$\frac{1}{2}$-in. diam.	2.7	3.1	3.7	4.3	2.1	2.5	2.9	3.3	1.7	1.9	2.1	2.3
1-in. diam.	2.4	2.8	3.4	4.0	1.9	2.2	2.6	2.9	1.4	1.6	1.8	2.0
2-in. diam.	2.2	2.6	3.1	3.7	1.7	1.9	2.3	2.6	1.2	1.4	1.6	1.7
3-in. diam.	2.1	2.4	2.9	3.5	1.6	1.8	2.1	2.5	1.1	1.2	1.4	1.5
6-in. diam.	2.0	2.3	2.8	3.3	1.4	1.6	1.9	2.3	0.9	1.0	1.2	1.3
12-in. diam.	1.8	2.1	2.6	3.1	1.3	1.5	1.8	2.1	0.8	0.9	1.0	1.1
Vertical walls	1.6	1.9	2.5	3.0	1.2	1.4	1.7	2.0	0.7	0.8	0.9	1.0

Surface A: Non-metallic, any color; paint, paper, glass, masonry, oxides. Emissivity 0.90.
Surface B: Bright metallic paint; dull but clean copper, zinc, aluminum. Emissivity 0.45.
Surface C: Clean and polished brass, copper, aluminum, tin, stainless steel, nickel or chrome plate. Emissivity 0.05.
Notes: For *vertical* cylinders the tabular values are slightly high.
For *large*, vertical heated or cooled walls the values in the last line of this table are slightly high.
For small, horizontal heated plates facing upward or cooled plates facing downward, the values in the last line are approximate, but lower values should be used for hot surfaces facing downward or cold surfaces facing upward.
For a single row of horizontal or vertical tubes with wide spacing, the tabular values are only slightly high.
Approximation for 2- and 3-row natural convection coils: use values in table section "C", and add radiation to or from the rectangular envelope enclosing the coil.
Approximation for outdoor conditions, dry weather: add 50 percent to the tabular value (section "A") for each 5 mph estimated average wind velocity.

[a] For $W/m^2{\cdot}K$ multiply by 5.674.
[b] Temperature differences, degrees Fahrenheit.

Table 5-40.   RADIATION AREA FACTORS

The following table gives values of F_a in the Stefan-Boltzmann equation, to account for the geometric shape and orientation of two radiating and absorbing surfaces in air[†]:

$$q = \sigma F_\varepsilon \, F_a \, A \, \left[\left(\frac{T_1}{100} \right)^4 - \left(\frac{T_2}{100} \right)^4 \right]$$

For infinite parallel planes, infinite concentric cylinders, and completely enclosed bodies, F_a is always unity.

UNITS: F_ε and F_a are dimensionless. See Table 2-8 for values of F_ε.
  For q in Btu/ft^2·hr use
    $\sigma \doteq 0.1713$, A in ft^2, T in deg R.
  For q in W/cm^2, use
    $\sigma = 0.000\,567$, A in cm^2, T in K.

Description	Area factors, F_a					
	$D/L = 16$	$D/L = 8$	$D/L = 4$	$D/L = 2$	$D/L = 1$	$D/L = 0.5$
Parallel and equal disks of diameter D or squares of side D, when distance apart is L	0.88	0.78	0.61	0.38	0.17	0.056
Parallel and equal *narrow* rectangles; length of smaller side D, distance apart L	0.94	0.88	0.78	0.62	0.41	0.236
Perpendicular equal rectangles with a common side; length of common side L, other side D	0.04	0.06	0.10	0.15	0.20	0.241

[†] If additional surfaces are involved (reflecting, absorbing, or re-radiating), or if the surrounding gas is absorbent, a more complex treatment is required.

REFERENCE

"Thermal Radiation Heat Transfer", NASA SP-164, National Aeronautics and Space Administration, Vol. 1 1968, Vol. 2 1969.

Table 5-41.   HEAT RADIATION—LOW TEMPERATURES

Net Radiant Heat Exchange and Radiation Coefficient in the Non-luminous Range

Values are computed from the Stefan–Boltzmann law. For q_r in W/m², multiply values in Btu/ft²·hr by 3.1525. For h_r in W/m²·K, multiply values in Btu/ft²·hr·deg F by 5.6745. For temperatures in K, multiply values in deg R by 5/9. For emissivity and area factors, see Tables 2-8 and 5-40.

Net heat exchange q_r ($Btu/ft^2\ hr$) and radiation coefficient h_r ($Btu/ft^2\ hr\ °F$) at receiver temperatures of

Temperature of heat source			0 °R −460 °F −273 °C		300 °R −160 °F −107 °C		400 °R −60 °F −51 °C		450 °R −10 °F −23 °C		500 °R 40 °F +4 °C		550 °R 90 °F 32 °C		600 °R 140 °F 60 °C	
°R	°F	°C	q_r	h_r	q_r	h_r	q_r	h_r	q_r	h_r	q_r	h_r	q_r	h_r	q_r	h_r
300	−160	−107.	13.89	.046	—	—	—	—	—	—	—	—	—	—	—	—
350	−110	−79.	25.70	.073	11.83	.237	—	—	—	—	—	—	—	—	—	—
400	−60	−51.	43.84	.110	29.97	.300	—	—	—	—	—	—	—	—	—	—
450	−10	−23.	70.23	.156	56.36	.376	26.39	.528	—	—	—	—	—	—	—	—
500	+40	+4.	107.0	.214	93.17	.466	63.20	.632	36.81	.736	—	—	—	—	—	—
520	60	15.	125.2	.241	111.3	.506	81.38	.678	54.99	.786	18.18	.909	—	—	—	—
540	80	27.	145.6	.270	131.7	.549	101.7	.727	75.40	.838	38.59	.965	—	—	—	—
560	100	38.	168.4	.301	154.5	.594	124.5	.779	98.20	.893	61.39	1.02	11.68	1.17	—	—
580	120	49.	193.8	.334	179.9	.643	149.9	.833	123.5	.951	86.77	1.08	37.07	1.24	—	—
600	140	60.	221.9	.370	208.0	.694	178.1	.891	151.7	1.01	114.9	1.15	65.21	1.30	—	—
620	160	71.	253.0	.408	239.2	.748	209.2	.951	182.8	1.07	146.0	1.22	96.32	1.38	31.11	1.56
640	180	82.	287.3	.449	273.5	.804	243.5	1.01	217.1	1.14	180.3	1.29	130.6	1.45	65.38	1.63
660	200	93.	325.0	.492	311.1	.864	281.1	1.08	254.7	1.21	217.9	1.36	168.2	1.53	103.0	1.72
680	220	104.	366.2	.538	352.3	.927	322.3	1.15	296.0	1.28	259.2	1.44	209.5	1.61	144.2	1.80
700	240	116.	411.2	.587	397.3	.993	367.3	1.22	341.0	1.36	304.1	1.52	254.5	1.70	189.3	1.89
750	290	143.	541.9	.722	528.0	1.17	498.0	1.42	471.7	1.57	434.9	1.74	385.2	1.92	319.9	2.13
800	340	171.	701.5	.877	687.6	1.38	657.7	1.64	631.2	1.80	594.5	1.98	544.8	2.18	479.6	2.40
850	390	199.	894.0	1.05	880.2	1.60	850.2	1.89	823.8	2.06	787.0	2.25	737.3	2.46	672.1	2.69
900	440	227.	1123.	1.25	1110.	1.85	1080.	2.16	1053.	2.34	1016.	2.54	967.0	2.76	901.7	3.01
950	490	254.	1395.	1.47	1381.	2.12	1351.	2.46	1324.	2.65	1288.	2.86	1238.	3.10	1173.	3.35
1000	540	282.	1713.	1.71	1699.	2.43	1669.	2.78	1642.	2.99	1606.	3.21	1556.	3.46	1491.	3.73
1100	640	338.	2508.	2.28	2494.	3.12	2464.	3.52	2437.	3.75	2401.	4.00	2351.	4.27	2286.	4.57
1200	740	393.	3551.	2.96	3538.	3.93	3508.	4.38	3481.	4.64	3444.	4.92	3395.	5.22	3329.	5.55
1300	840	449.	4892.	3.76	4878.	4.88	4848.	5.39	4821.	5.67	4785.	5.98	4735.	6.31	4670.	6.67
1400	940	504.	6580.	4.70	6566.	5.97	6536.	6.54	6509.	6.85	6472.	7.19	6423.	7.56	6358.	7.95
1500	1040	560.	8671.	5.78	8657.	7.21	8627.	7.84	8600.	8.19	8564.	8.56	8514.	8.96	8448.	9.39

Table 5-42. HEAT RADIATED
FROM HIGH-TEMPERATURE SURFACES

For emissivity and area factors, see Tables 2-8 and 5-40.

Net radiant heat q_r, from high-temperature surfaces to room-temperature environment (540°R, 80°F, 300 K, 27°C). Values are computed from the Stefan–Boltzmann law. For temperature in K, multiply values in °R by 5/9. For radiation in kW/m², multiply values in W/cm² by 10.

Temperature of heat source			Net radiation		Temperature of heat source			Net radiation	
°R	°F	°C	Btu ft² hr	watts cm²	°R	°F	°C	Btu ft² hr	watts cm²
1400	940	504	6,435	2.029	2500	2040	1116	66,770	21.05
1420	960	516	6,819	2.150	2550	2090	1143	72,280	22.78
1440	980	527	7,220	2.276	2600	2140	1171	78,130	24.63
1460	1000	538	7,638	2.408	2650	2190	1199	84,330	26.58
1480	1020	549	8,073	2.545	2700	2240	1227	90,890	28.65
1500	1040	560	8,526	2.688	2750	2290	1254	97,820	30.84
1520	1060	571	8,998	2.837	2800	2340	1282	105,140	33.15
1540	1080	582	9,489	2.991	2850	2390	1310	112,870	35.58
1560	1100	593	9,999	3.152	2900	2440	1338	121,010	38.15
1580	1120	604	10,530	3.319	2950	2490	1366	129,590	40.85
1600	1140	616	11,080	3.493	3000	2540	1393	138,610	43.70
1620	1160	627	11,650	3.673	3100	2640	1449	158,050	49.83
1640	1180	638	12,247	3.861	3200	2740	1504	179,475	56.58
1660	1200	649	12,860	4.054	3300	2840	1560	203,000	64.00
1680	1220	660	13,500	4.256	3400	2940	1616	228,770	72.12
1700	1240	671	14,160	4.464	3500	3040	1671	256,910	81.00
1720	1260	682	14,850	4.680	3600	3140	1727	287,570	90.66
1740	1280	693	15,560	4.904	3700	3240	1782	320,890	101.2
1760	1300	704	16,290	5.135	3800	3340	1838	357,040	112.6
1780	1320	716	17,050	5.375	3900	3440	1893	396,150	124.9
1800	1340	727	17,838	5.623	4000	3540	1949	438,380	138.2
1820	1360	738	18,650	5.879	4100	3640	2004	483,900	152.6
1840	1380	749	19,490	6.144	4200	3740	2060	532,880	168.0
1860	1400	760	20,360	6.418	4300	3840	2116	585,490	184.6
1880	1420	771	21,250	6.700	4400	3940	2171	641,900	202.4
1900	1440	782	22,180	6.992	4500	4040	2227	702,290	221.4
1920	1460	793	23,135	7.293	4600	4140	2282	766,840	241.7
1940	1480	804	24,120	7.603	4700	4240	2338	835,740	263.5
1960	1500	816	25,140	7.924	4800	4340	2393	909,180	286.6
1980	1520	827	26,180	8.254	4900	4440	2449	987,360	311.3
2000	1540	838	27,260	8.594	5000	4540	2504	1,070,500	337.5
2050	1590	866	30,110	9.491	5200	4740	2615	1,252,400	394.8
2100	1640	893	33,170	10.45	5400	4940	2726	1,456,500	459.2
2150	1690	921	36,460	11.49	5600	5140	2837	1,684,500	531.0
2200	1740	949	39,980	12.60	5800	5340	2949	1,938,400	611.1
2250	1790	977	43,760	13.79	6000	5540	3060	2,219,900	699.8
2300	1840	1004	47,790	15.06	6200	5740	3171	2,531,100	797.9
2350	1890	1032	52,100	16.42	6400	5940	3282	2,873,800	906.0
2400	1940	1060	56,690	17.87	6600	6140	3393	3,250,300	1025.
2450	1990	1088	61,570	19.41	6800	6340	3504	3,662,500	1155.

Table 5-43. APPROXIMATE BOILING AND CONDENSING COEFFICIENTS*

Fluids and conditions	Saturation pressure		Saturation temperature		Temperature difference, saturation to surface		Approximate coefficient h†	
	$\frac{lbf}{in.^2}$	$\frac{MN}{m^2}$	deg F	K	deg F	K	$\frac{Btu}{hr \cdot ft^2 \cdot deg F}$	$\frac{kW}{m^2 \cdot K}$
EVAPORATION INSIDE TUBES (NUCLEATE BOILING)								
Water								
Pure water, free convection	14.7	0.101	212	373	30	16.7	8,000	45.4
Pure water, free convection	14.7	0.101	212	373	20	11.1	3,000	17.0
Pure water, free convection	14.7	0.101	212	373	10	5.56	1,000	5.67
Pure water, free convection	250	1.72	400	478	20	11.1	7,000	39.7
Pure water, forced convection	14.7	0.101	212	373	20	11.1	12,000	68.1
Boiler water (normal contamination)‡	14.7	0.101	212	373	30	16.7	2,000	11.4
Boiler water (normal contamination)‡	14.7	0.101	212	373	20	11.1	1,400	7.94
Boiler water (normal contamination)‡	14.7	0.101	212	373	10	5.56	500	2.84
Boiler water (normal contamination)‡	250	1.72	400	478	20	11.1	2,500	14.2
Ammonia refrigerant								
Pure ammonia	35	0.241	5	258	20	11.1	1,100	6.24
Refrigerant evaporator (normal contamination)‡	35	0.241	5	258	—		400–700	2.27–3.97
R–12 refrigerant								
Pure R–12	47	0.324	35	275	20	11.1	800	4.54
Refrigerant evaporator (normal contamination)‡	47	0.324	35	275	—		200–400	1.14–2.27
CONDENSATION OUTSIDE TUBES								
Water								
Pure steam, film condensation	14.7	0.101	212	373	30	16.7	2,700	15.3
Pure steam, film condensation	14.7	0.101	212	373	20	11.1	3,500	19.9
Pure steam, film condensation	2	0.014	125	325	20	11.1	2,500	14.2
Pure steam, dropwise condensation	14.7	0.101	212	373	20	11.1	10,000†	56.7
Condenser (normal contamination)‡	14.7	0.101	212	373	30	16.7	1,700	9.65
Condenser (normal contamination)‡	2	0.014	125	325	20	11.1	1,500	8.51
Ammonia refrigerant								
Pure ammonia	165	1.14	85	303	20	11.1	2,000	11.4
Condenser (normal contamination)‡	165	1.14	85	303	—		500–1,00	2.84–5.67
R–12 refrigerant								
Pure R–12	105	0.724	85	303	20	11.1	800	4.54
Condenser (normal contamination)‡	105	0.724	85	303	—		200–350	1.14–1.99

Note: All coefficients are sharply affected by the condition of the surface, contamination, the degree of local turbulence in the vapor or liquid, and the extent of flooding or submergence in the liquid. The given values should be considered only as approximations for estimating or checking purposes.
†Multiply coefficient values by 0.48824 to obtain cal/hr sq cm °C or by 4.8824 to obtain kcal/hr sq m °C.
‡Normal contamination in steam is usually air and noncondensable gases; in refrigerant, air, gases, and oil are likely to be present.

*Based on: "Engineering Experimentation", by G.L. Tuve and L.C. Domholdt. Copyright © 1966 by McGraw-Hill, Inc. Used with permission of McGraw-Hill Book Company.

Table 5-44. MASS TRANSFER BY DIFFUSION

Eddy Diffusion in Air and Water in Turbulent Flow Streams

Tables 5-45, 5-46, and 5-47 furnish data for the two most common cases of diffusion, i.e., diffusion of gases and vapors into air and diffusion of solutes into water. The mass transfer coefficient k_c is proportional to the fluid velocity and is a function of the Schmidt No., $\mu/\rho D$, where D is the diffusion constant. Values of D for various materials and assigned temperatures are given in Tables 5-45, 5-46, and 5-47. The mass transfer coefficient is analogous to the heat transfer coefficient in forced convection, in that it is the coefficient by which the driving force is multiplied to obtain the flux per unit area:

$$q_h/A_1 = h \, \Delta t \text{ and } q_m/A_1 = k_c \, \Delta C$$

In mass transfer the driving force is a difference in concentration, but it may also be expressed as a mole fraction (dimensionless) or, for gases, as a partial pressure.

CALCULATION PROCEDURE (TURBULENT FLOW)

(1) **To calculate the mass transfer**

mass flux = mass transfer coefficient × driving force

$$q_m/A_1 = k_c (c_{a1} - \bar{c}_a)$$

where q_m/A_1 = mass flux: lb moles/hr ft^2

k_c = mass transfer coefficient, $\dfrac{\text{lb moles of } (a) \text{ transferred}}{\text{hr ft}^2 \text{ (lb moles/ft}^3)}$ = ft/hr

c_{a1} = concentration of component a at the boundary, lb moles/ft^3

$\bar{c}_a$ = average value of the concentration of component a in the flowing stream, lb moles/ft^3

For diffusion into gases (alternate variables, partial pressure or mole fraction):
    Driving force variable, p_a, the partial pressure of component a, given in atm:

$$\frac{q_m}{A_1} = \frac{k_c}{RT}(p_{a1} - \bar{p}_a)$$

where R = 0.73008 ft^3 atm/°R (lb mole)

T = temperature, degrees Rankine

Driving force variable, y_a, the mole fraction of component a, dimensionless:

$$\frac{q_m}{A_1} = \frac{k_c P}{RT}(y_{a1} - \bar{y}_a)$$

P = total pressure, atm

For diffusion into liquids (alternate variable, mole fraction):
    Driving force variable, x_a, the mole fraction of component a, dimensionless:

$$\frac{q_m}{A_1} = k_c \bar{\rho} (x_{a1} - \bar{x}_a)/M$$

where $\bar{\rho}$ = average density of a mixture of (a) and (b), lb/ft^3

$\overline{M}$ = average molecular weight of a mixture of (a) and (b).

(2) **To obtain the mass transfer coefficient k_c**

$$k_c = j_D \, V/Sc^{1.5}$$

where V = the velocity of flow, ft/hr

Sc = Schmidt No. from Tables 5-45, 5-46, or 5-47.

j_D = Colburn "j-factor" (dimensionless) for mass transfer from the following table.

Table 5-44. MASS TRANSFER BY DIFFUSION *(Continued)*

FACTORS j_D FOR MASS TRANSFER BY EDDY DIFFUSION*

Turbulent Flow at Various Reynolds Numbers j_D, ft/hr

Reynolds No., $LV\rho/\mu$	Flow geometry			Reynolds No., $LV\rho/\mu$	Flow geometry		
	Along a flat surface	Over a single cylinder	Over a single sphere		Along a flat surface	Over a single cylinder	Over a single sphere
2,000		0.014	0.016	20,000	0.0051	0.0054	0.0059
4,000		0.010	0.012	40,000	0.0044	0.0040	0.0044
6,000		0.0089	0.010	60,000	0.0041		
8,000		0.0079	0.0088	80,000	0.0038		
10,000	0.0059	0.0072	0.0080	100,000	0.0037		
				200,000	0.0032		

j_D factors were calculated from the following equations:

For flat plates: $j_D = 0.039\,(\mathrm{Re}_L)^{-0.206}$

For single cylinders: $j_D = 0.335\,\mathrm{Re}^{-0.417}$

For single spheres: $j_D = 0.422\,\mathrm{Re}^{-0.431}$

Note that the Schmidt No. is the ratio of kinematic viscosity to molecular diffusivity, a relation of momentum transfer to mass transfer. The Prandtl No., $c\,\mu/k$, is a relation of momentum transfer to heat transfer. The Reynolds No., which relates viscous forces and inertia forces, is an index of turbulence. The Sherwood No., $k_c\,L/D$, is analogous to the Nusselt No., $h\,L/k$. One alternative for determining the mass transfer coefficient is to use the relation

$$Sh = \mathrm{const}\ Re^a\ Sc^b,$$

which is analogous to $Nu = \mathrm{const}\ Re^c\ Pr^d$ (see Tables 5-35 and 5-36). In the above table the Colburn "*j*-factor" analog is used instead.

*Adapted from: "ASHRAE Handbook of Fundamentals", American Society of Heating, Refrigerating and Air-Conditioning Engineers, 1972.

REFERENCES

"Momentum, Heat and Mass Transfer", C.O. Bennett and J.E. Myers, McGraw-Hill Book Company, 1962.

"Perry's Chemical Engineers' Handbook", 4th ed., R.H. Perry, C.H. Chilton, and S.D. Kirkpatrick, Eds., McGraw-Hill Book Company, 1963, Section 17.

Table 5-45.   DIFFUSION OF SOLUTES INTO WATER*

Dilute Solutions at 20°C

Diffusion constant, D, English—for ft^2/hr multiply by 10^{-5}. Diffusion constant, D, metric—for m^2/s multiply by 10^{-9}.

Substance	Diffusion constant, D†		Schmidt number, $\left(\dfrac{\mu}{\rho D}\right)$‡	Substance	Diffusion constant, D†		Schmidt number, $\left(\dfrac{\mu}{\rho D}\right)$‡
	English	Metric			English	Metric	
H_2	19.8	5.13	196	H_2SO_4	6.70	1.73	581
O_2	6.97	1.80	558	NaOH	5.84	1.51	666
CO_2	6.85	1.77	568	NaCl	5.22	1.35	744
NH_3	6.81	1.76	571	Ethyl alcohol	3.87	1.00	1005
N_2	6.35	1.64	613	Acetic acid	3.41	0.88	1140
Acetylene	6.04	1.56	644	Phenol	3.25	0.84	1200
Cl_2	4.72	1.22	824	Glycerol	2.79	0.72	1400
HCl	10.2	2.64	381	Sucrose	1.74	0.45	2230
HNO_3	10.1	2.6	390				

†The following relationship may be used to estimate the effect of temperature on the diffusion constant

$$\frac{D_1}{D_2} = \frac{T_1}{T_2}\frac{\mu_2}{\mu_1}$$

where T = temperature, °K

μ = solution viscosity, centipoises.

The diffusion constant varies with concentration because of the changes in viscosity and the degree of ideality of the solution.

‡Based on $\mu/\rho = 0.01005$ sq cm/sec for water at 20°C. Applies only for dilute solutions.

*Compiled and computed from several sources.

Table 5-46.   DIFFUSION OF WATER VAPOR INTO AIR*

Values of Diffusion Constant and Schmidt Number

Temp, °C	Diffusion constant, D		$\left(\dfrac{\mu}{\rho D}\right)$†
	sq ft/hr	sq cm/sec	
0	0.844	0.218	0.608
10	0.898	.232	.610
20	0.952	.246	.612
30	1.01	.260	.614
40	1.06	.275	.615
50	1.12	.290	.616
60	1.18	.305	.618
70	1.24	.321	.619
80	1.30	.337	.619

†The values of $\left(\dfrac{\mu}{\rho D}\right)$ were calculated using the viscosity and density of dry air.

Thus the values apply only when the diffusing water vapor is very dilute.

*Compiled and computed from several sources.

REFERENCES

"The Chemical Engineers' Handbook", 4th ed., R.H. Perry, C.H. Chilton, and S.D. Kirkpatrick, Eds., McGraw-Hill Book Company, 1963.

"International Critical Tables of Numerical Data", National Research Council, Vol. V, McGraw-Hill Book Company, 1929.

"Landolt-Börnstein Physikalisch-Chemische Tabellen", Julius Springer, Berlin, 1923–36.

"Principals of Unit Operations", A.S. Foust, L.A. Wenzel, C.W. Clump, L. Maus, and L.B. Anderson, John Wiley & Sons, Inc., 1960.

Tables 5-47. DIFFUSION OF GASES AND VAPORS INTO AIR

Values of Diffusion Constant and Schmidt Number

At 1 atm Pressure

Substance	Diffusion constant, D, sq ft/hr		Diffusion constant, D, sq cm/sec		$\left(\dfrac{\mu}{\rho D}\right)$†	
	0°C	25°C	0°C	25°C	0°C	25°C
H_2	2.37	2.76	0.611	0.712	0.217	0.216
NH_3	0.766	0.886	0.198	0.229	0.669	0.673
N_2	0.691		0.178		0.744	
O_2	0.689	0.80	0.178	0.206	0.744	0.748
CO_2	0.550	0.635	0.142	0.164	0.933	0.940
CS_2	0.36	0.414	0.094	0.107	1.41	1.44
Methyl alcohol	0.513	0.615	0.132	0.159	1.00	0.969
Formic acid	0.509	0.615	0.131	0.159	1.01	0.969
Acetic acid	0.411	0.515	0.106	0.133	1.25	1.16
Ethyl alcohol	0.394	0.461	0.102	0.119	1.30	1.29
Chloroform	0.352		0.091		1.46	
Diethylamine	0.342	0.406	0.0884	0.105	1.50	1.47
n-Propyl alcohol	0.329	0.387	0.085	0.100	1.56	1.54
Propionic acid	0.328	0.383	0.0846	0.099	1.57	1.56
Methyl acetate	0.325	0.387	0.0840	0.100	1.58	1.54
Butylamine	0.318	0.391	0.0821	0.101	1.61	1.53
Ethyl ether	0.304	0.360	0.0786	0.093	1.69	1.66
Benzene	0.291	0.341	0.0751	0.088	1.76	1.75
Ethyl acetate	0.277	0.330	0.0715	0.085	1.85	1.81
Toluene	0.274	0.325	0.0709	0.084	1.87	1.83
n-Butyl alcohol	0.272	0.348	0.0703	0.090	1.88	1.71
i-Butyric acid	0.263	0.313	0.0679	0.081	1.95	1.90
Chlorobenzene		0.283		0.073		2.11
Aniline	0.236	0.279	0.0610	0.072	2.17	2.14
Xylene	0.228	0.275	0.059	0.071	2.25	2.17
Amyl alcohol	0.228	0.271	0.0589	0.070	2.25	2.20
n-Octane	0.195	0.232	0.0505	0.060	2.62	2.57
Naphthalene	0.199	0.20	0.0513	0.052	2.58	2.96

†Based on $\dfrac{\mu}{\rho}$ = 0.1325 sq cm/sec for air at 0°C and 0.1541 sq cm/sec for air at 25°C; applies only when the diffusing gas or vapor is very dilute.

Table 5-48.   PERMEABILITY OF MATERIALS TO WATER VAPOR*

For Comparative Estimates of Water-Vapor Transmission through Walls

Values in this table give the permeance of each material for typical thickness. The equation $W = nA\Delta P$ is applicable, where W = rate of vapor transmission, n = permeance, A = area of flow path, ΔP = vapor-pressure difference, from one side to the other, i.e., corresponding to the dewpoints on the two sides.

Values in the table are results of laboratory tests either by the dry-cup method of ASTM E96 or some other test method. The permeance is in grain/hr·ft²·in. Hg (=perms); corresponding units should be used for area and vapor-pressure difference to obtain W in grain/hr. For permeance in SI units, kg/s·N, multiply the tabulated values by 57.213×10^{-12}.

| Material | Permeance | | Material | Permeance | |
	Dry-cup method	Other methods		Dry-cup method	Other methods
STRUCTURAL MATERIALS			**FELTS AND BUILDING AND ROOFING PAPERS**		
Acrylic, glass-fiber reinforced sheet, 56 mil	0.12		Blanket thermal insulation back-up paper, asphalt-coated (31)	0.4	0.6 4.2
Asbestos-cement board, 0.2 in. thick	0.54	—	Duplex sheet, asphalt-laminated, aluminum foil one side (43)	0.002	0.176
Brick masonry, 4 in. thick	—	0.8	Felt, 15 lb asphalt (70)	1.0	5.6
Concrete, 1:2:4 mix, per inch	—	3.2	Felt, 15 lb tar (70)	4.0	18.2
Concrete block, 8 in. cored, limestone aggregate	—	2.4	Kraft paper and asphalt-laminated, reinforced 30–120–30 (34)	0.3	1.8
Gypsum sheathing, $\frac{1}{2}$ in. asphalt impregnated	20	—	Roll roofing, saturated and coated (326)	0.05	0.24
Hardboard, $\frac{1}{8}$ in. standard	—	11	Sheathing paper, asphalt-saturated, uncoated (22)	3.3	20.2
Hardboard, $\frac{1}{8}$ in. tempered	—	5	Single-kraft, double-infused (16)	31	42
Plaster on plain gypsum lath, with studs	—	20	Vapor-barrier paper, asphalt-saturated, coated (43)	0.2–0.3	0.6
Plaster on metal lath, $\frac{3}{4}$ in.	—	15	**LIQUID-APPLIED COATING MATERIALS**		
Plywood, Douglas-fir, exterior, $\frac{1}{4}$ in. thick	—	0.7	*Paint—2 coats*		
Plywood, Douglas-fir, interior $\frac{1}{4}$ in. thick		1.9	Aluminum varnish on wood	0.3–0.5	—
Polyester, glass-fiber reinforced sheet, 48 mil	0.05	—	Asphalt paint on plywood	—	0.4
Tile masonry, glazed, 4 in. thick	—	0.12	Enamels on smooth plaster	—	0.5–1.5
			Sealers or flat paint on interior-insulation board	—	0.9–2.1
THERMAL INSULATIONS, PER INCH THICKNESS			Various primers plus 1 coat flat oil paint on plaster		1.6–3.0
Corkboard	2.1–2.6	—	*Paint—3 coats*		
Expanded polystyrene, bead	2.0–5.8	—	Asphalt cut-back mastic, $\frac{1}{16}$ in. dry	0.14	—
Expanded polystyrene, extruded	1.2	—	Asphalt cut-back mastic, $\frac{3}{16}$ in. dry	0.0	—
Expanded polyurethane (R–11 blown)	0.4–1.6	—	Chloro-sulfonated polyethylene mastic, 3.5 oz/sq ft	1.7	—
PLASTIC AND METAL FOILS AND FILMS			Chloro-sulfonated polyethylene mastic, 7.0 oz/sq ft	0.06	—
Aluminum foil, 0.35 mil	0.05	—	Exterior paint, white lead-zinc oxide and oil on wood	0.9	—
Aluminum foil, 1 mil	0.0	—	Hot melt asphalt, 2 oz/sq ft	0.5	—
Cellulose acetate, 10 mil	0.46	—	Hot melt asphalt, 3.5 oz/sq ft	0.1	—
Polyester, 1 mil	0.7	—	Polyvinyl-acetate latex coating, 4 oz/sq ft	5.5	—
Polyethylene, 2 mil	0.16	—	Styrene-butadiene latex coating, 2 oz/sq ft	11	—
Polyethylene, 6 mil	0.06	—			
Polyethylene, 10 mil	0.03	—			
Polyvinylchloride, plasticized, 4 mil	0.8–1.4	—			
Polyvinylchloride, unplasticized, 2 mil	0.68	—			

*Based on: "ASHRAE Handbook of Fundamentals", American Society of Heating, Refrigerating and Air-Conditioning Engineers, 1972.

Table 5-49. RATE OF EVAPORATION FOR DRYING OF MATERIALS IN TRAYS*

In Pounds of Water/hr/ft² ; at Air Velocity of 300 ft/min

For kg/s·m² multiply tabulated values by 0.001 356 2. For velocity in m/s, multiply values in ft/min by 0.005 08.

Relative humidity	Temperature, deg F (K)				Velocity correction†	
	100 (311)	150 (339)	200 (366)	300 (422)	Velocity, ft/min	Rate factor
10%	0.11	0.18	0.24	0.38	200	0.7
20%	0.09	0.14	0.19	0.30	300	1.0
30%	0.07	0.11	0.15	0.23	400	1.3
40%	0.06	0.09	0.12	0.18	500	1.5
50%	0.05	0.07	0.09	0.14	600	1.7
60%	0.04	0.05	0.07	0.11	700	2.0
70%	0.03	0.04	0.05	0.08	800	2.2
80%	0.02	0.02	0.03	0.05	900	2.4
90%	0.01	0.01	0.02	0.02		

Note: Table applies to tray drying of a material in the constant-rate drying period.
†Velocity correction varies with the 0.8 power of air velocity.

*Based largely on: "ASHRAE Guide and Data Book: Applications", American Society of Heating, Refrigerating and Air-Conditioning Engineers, 1968.

Table 5-50. MAINTENANCE OF CONSTANT HUMIDITY*

Saturated solutions (with excess of solid phase) can be used for maintaining a constant relative humidity if adequate contact between the solution and the air is provided. The following salts will maintain the specified relative humidity at the temperature indicated.

Solid phase	°C	°F	% Relative humidity	Solid phase	°C	°F	% Relative humidity
$H_3PO_4 \cdot \frac{1}{2}H_2O$	25	77	9	$NaNO_2$	20	68	66
$ZnCl_2 \cdot 1\frac{1}{2}H_2O$	20	68	10	$NaC_2H_3O_2 \cdot 3H_2O$	20	68	76
$LiCl \cdot H_2O$	20	68	15	$Na_2S_2O_3 \cdot 5H_2O$	20	68	78
$KC_2H_3O_2$	20	68	20	NH_4Cl	20	68	79
KF	100	212	23	$(NH_4)_2SO_4$	20	68	81
$NaBr$	100	212	23	KBr	20	68	84
$CaCl_2 \cdot 6H_2O$	20	68	33	Tl_2SO_4	105	221	85
CrO_3	20	68	35	$KHSO_4$	20	68	86
$K_2CO_3 \cdot 2H_2O$	25	77	43	$Na_2CO_3 \cdot 10H_2O$	25	77	87
$Ca(NO_3)_2 \cdot 4H_2O$	25	77	51	K_2CrO_4	20	68	88
$NaHSO_4 \cdot H_2O$	20	68	52	$NaBrO_3$	20	68	92
$Mg(NO_3)_2 \cdot 6H_2O$	25	77	52	$Na_2SO_4 \cdot 10H_2O$	20	68	93
$NaClO_3$	100	212	54	$Na_2HPO_4 \cdot 12H_2O$	20	68	95
$NaBr \cdot 2H_2O$	20	68	58	NaF	100	212	97
$Mg(C_2H_3O_2)_2 \cdot 4H_2O$	20	68	65	$Pb(NO_3)_2$	20	68	98

*From: "International Critical Tables", Vol. 1, McGraw-Hill Book Company, 1926, p. 67. Reprinted by permission of McGraw-Hill Book Company and National Academy of Sciences.

Table 5-51. REGAIN OF HYGROSCOPIC MATERIALS*

Moisture Content Expressed in Percent of Dry Weight
of the Substance at Various Relative Humidities

| Classification | Material | Description | Relative humidity—percent (75°F) | | | | | | | | | Authority |
			10	20	30	40	50	60	70	80	90	
Natural textile fibers	Cotton	Sea island—roving	2.5	3.7	4.6	5.5	6.6	7.9	9.5	11.5	14.1	Hart-shorne
	Cotton	American—cloth	2.6	3.7	4.4	5.2	5.9	6.8	8.1	10.0	14.3	Schloesing
	Cotton	Absorbent	4.8	9.0	12.5	15.7	18.5	20.8	22.8	24.3	25.8	Fuwa
	Wool	Australian merino—skein	4.7	7.0	8.9	10.8	12.8	14.9	17.2	19.9	23.4	Hart-shorne
	Silk	Raw chevennes—skein	3.2	5.5	6.9	8.0	8.9	10.2	11.9	14.3	18.3	Schloesing
	Linen	Table cloth	1.9	2.9	3.6	4.3	5.1	6.1	7.0	8.4	10.2	Atkinson
	Linen	Dry spun—yarn	3.6	5.4	6.5	7.3	8.1	8.9	9.8	11.2	13.8	Sommer
	Jute	Average of several grades	3.1	5.2	6.9	8.5	10.2	12.2	14.4	17.1	20.2	Storch
	Hemp	Manila and sisal—rope	2.7	4.7	6.0	7.2	8.5	9.9	11.6	13.6	15.7	Fuwa
Rayons	Viscose nitro-cellulose	Average skein	4.0	5.7	6.8	7.9	9.2	10.8	12.4	14.2	16.0	Robertson
	Cuprammonium Cellulose acetate		0.8	1.1	1.4	1.9	2.4	3.0	3.6	4.3	5.3	Robertson
Paper	M.F. Newsprint	Wood pulp—24% ash	2.1	3.2	4.0	4.7	5.3	6.1	7.2	8.7	10.6	NBS
	H.M.F. Writing	Wood pulp—3% ash	3.0	4.2	5.2	6.2	7.2	8.3	9.9	11.9	14.2	NBS
	White bond	Rag—1% ash	2.4	3.7	4.7	5.5	6.5	7.5	8.8	10.8	13.2	NBS
	Com. ledger	75% rag—1% ash	3.2	4.2	5.0	5.6	6.2	6.9	8.1	10.3	13.9	NBS
	Kraft wrapping	Coniferous	3.2	4.6	5.7	6.6	7.6	8.9	10.5	12.6	14.9	NBS
Misc. organic materials	Leather	Sole-oak—tanned	5.0	8.5	11.2	13.6	16.0	18.3	20.6	24.0	29.2	Phelps
	Catgut	Racquet strings	4.6	7.2	8.6	10.2	12.0	14.3	17.3	19.8	21.7	Fuwa
	Glue	Hide	3.4	4.8	5.8	6.6	7.6	9.0	10.7	11.8	12.5	Fuwa
	Rubber	Solid tires	0.11	0.21	0.32	0.44	0.54	0.66	0.76	0.88	0.99	Fuwa
	Wood	Timber (average)	3.0	4.4	5.9	7.6	9.3	11.3	14.0	17.5	22.0	Forest P. Lab.
	Soap	White	1.9	3.8	5.7	7.6	10.0	12.9	16.1	19.8	23.8	Fuwa
	Tobacco	Cigarette	5.4	8.6	11.0	13.3	16.0	19.5	25.0	33.5	50.0	Ford
Foodstuffs	White bread		0.5	1.7	3.1	4.5	6.2	8.5	11.1	14.5	19.0	Atkinson
	Crackers		2.1	2.8	3.3	3.9	5.0	6.5	8.3	10.9	14.9	Atkinson
	Macaroni		5.1	7.4	8.8	10.2	11.7	13.7	16.2	19.0	22.1	Atkinson
	Flour		2.6	4.1	5.3	6.5	8.0	9.9	12.4	15.4	19.1	Bailey
	Starch		2.2	3.8	5.2	6.4	7.4	8.3	9.2	10.6	12.7	Atkinson
	Gelatin		0.7	1.6	2.8	3.8	4.9	6.1	7.6	9.3	11.4	Atkinson
Misc. inorganic materials	Asbestos fiber	Finely divided	0.16	0.24	0.26	0.32	0.41	0.51	0.62	0.73	0.84	Fuwa
	Silica gel		5.7	9.8	12.7	15.2	17.2	18.8	20.2	21.5	22.6	Fuwa
	Domestic coke		0.20	0.40	0.61	0.81	1.03	1.24	1.46	1.67	1.89	Selvig
	Activated charcoal	Steam activated	7.1	14.3	22.8	26.2	28.3	29.2	30.0	31.1	32.7	Fuwa
	Sulfuric acid	H_2SO_4	33.0	41.0	47.5	52.5	57.0	61.5	67.0	73.5	82.5	Mason

*From: "ASHRAE Guide and Data Book: Applications", American Society of Heating, Refrigerating and Air-Conditioning Engineers, 1971.

Figure 5-52.   ENERGY CONVERSIONS*

Directions and Methods for Energy Conversion

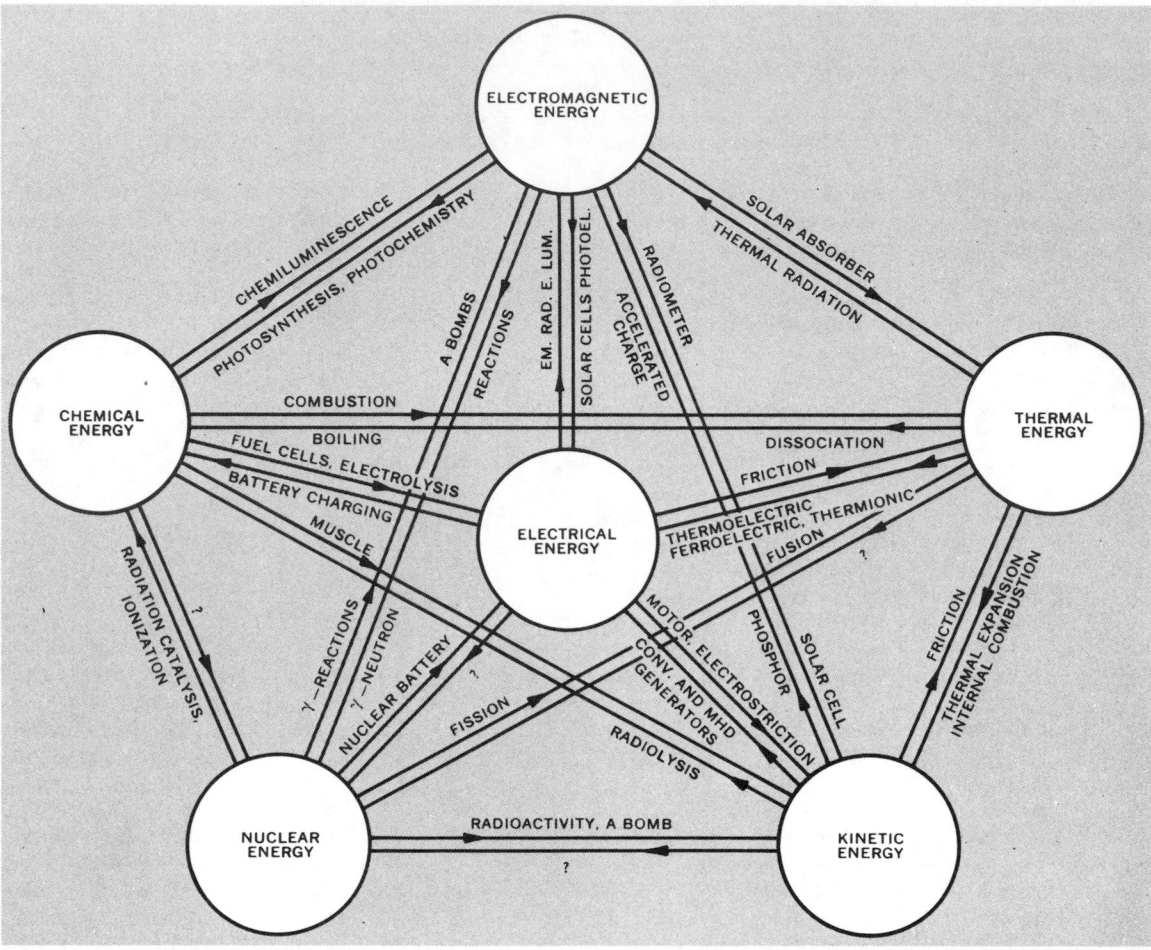

Figure 5-52. Energy conversion chart. The circles represent the different forms of energy and the arrows the ways of converting energy from one form to another.

*From: M.A. Kettani, "Direct Energy Conversion," Addison-Wesley Publishing Company, Inc., 1970, p. 6.

Table 5-53.   CONSUMPTION OF ENERGY RESOURCES BY MAJOR SOURCES AND CONSUMING SECTORS*

10^{12} Btu

	1966	*1963*	*1959[a]*	*1953*	*1947*
Household and commercial					
Anthracite	143	103.0	192.2	457.2	812.8
Bituminous and lignite	573	671.0	814.9	1 614.8	2 585.5
Natural gas	5 945	5 026.8	4 023.8	2 293.7	1 125.0
Petroleum[b]	5 769	5 257.8	4 718.6	3 391.2	2 250.9
Hydropower	0	0.0	0.0	0.0	0.0
Nuclear	0	0.0	0.0	0.0	0.0
Total	12 430	11 058.6	9 749.5	7 756.9	6 774.2
Industrial					
Anthracite	88	56.7	54.9	48.3	284.7
Bituminous and lignite	5 806	5 014.6	4 691.8	6 056.9	7 013.6
Natural gas	8 105	6 775.9	5 689.8	4 303.5	2 874.7
Petroleum[b]	4 334	3 994.0	3 458.1	3 092.0	2 489.7
Hydropower	0	0.0	0.0	0.0	0.0
Nuclear	0	0.0	0.0	0.0	0.0
Total	18 333	15 841.2	13 894.6	13 500.7	12 662.7
Transportation					
Anthracite	0	0.0	7.4	12.7	23.9
Bituminous and lignite	18	19.1	99.8	796.3	3 006.2
Natural gas	553	438.6	361.6	238.4	0.0
Petroleum[b]	12 785	11 506.1	9 923.1	8 157.8	5 760.5
Hydropower	0	0.0	0.0	0.0	0.0
Nuclear	0	0.0	0.0	0.0	0.0
Total	13 356	11 963.8	10 391.9	9 205.2	8 790.6
Electricity generation, utilities					
Anthracite	56	54.7	67.0	91.4	89.5
Bituminous and lignite	6 341	5 016.9	3 989.4	2 714.1	1 994.4
Natural gas	2 692	2 217.9	1 684.0	1 070.4	386.1
Petroleum[b]	905	599.8	545.9	577.1	468.0
Hydropower	2 060	1 910.0	1 695.0	1 550.0	1 459.0
Nuclear	58	31.1	11.7	0.0	0.0
Total	12 112	9 830.4	7 993.0	6 003.0	4 397.0
Miscellaneous					
Anthracite	3	146.2	156.0	101.6	13.3
Bituminous and lignite	0	0.0	0.0	0.0	0.0
Natural gas	0	383.8	231.1	250.0	132.6
Petroleum[b]	599	592.3	1 101.3	879.9	397.9
Hydropower	0	0.0	0.0	0.0	0.0
Nuclear	0	0.0	0.0	0.0	0.0
Total	602	1 122.3	1 488.4	1 231.5	543.8
Total gross energy					
Anthracite	290	360.6	477.5	711.2	1 224.2
Bituminous and lignite	12 740	10 721.6	9 595.9	11 182.1	14 599.7
Natural gas	17 295	14 843.0	11 990.3	8 156.0	4 518.4
Petroleum[b]	24 392	21 950.0	19 747.0	16 098.0	11 367.0
Hydropower	2 060	1 910.0	1 695.0	1 550.0	1 459.0
Nuclear	58	31.1	11.7	0.0	0.0
Total	56 835	49 816.3	43 517.4	37 697.3	33 168.3

[a]Total gross energy includes Alaska and Hawaii beginning in 1959.
[b]Natural gas liquids included with petroleum.

*Data from: "Minerals Yearbook: 1966. Volumes I–II. Metals, Minerals, and Fuels", U.S. Bureau of Mines, 1967; and "Petroleum Facts and Figures", American Petroleum Institute, 1967.

Table 5-54.   ENERGY AND POWER IN THE UNITED STATES*

Energy Sources, Production, and Consumption

Item	1970[a]	1965	1960	1950[b]
Total horsepower of all prime movers, 10^6 hp[c]	20 419[d]	15 096	11 008	4 868
Automotive[e]	19 325[d]	14 306	10 367	4 404
Electric central stations[f]	435[d]	307	217	88
Aircraft[g]	183[d]	55	37	22
Factories and mines	99[d]	88	77	55
Railroads and ships	76[d]	68	71	134
Farms[h]	301[d]	272	239	165
Production of all fuels, 10^{12} Btu	59 568[d]	47 944	40 121	32 937
Bituminous and lignitic coals[j]	15 848[d]	13 417	10 886	13 527
Crude petroleum	19 579[d]	16 521	14 935	11 449
Natural gas[k]	24 295[d]	17 628	13 822	6 841
Installed electric generating capacity, 10^6 kW[m]	360 000	254 519	185 815	83 000
Private utilities	263 000	178 000	128 000	55 000
Public and co-op	78 000	59 000	40 000	14 000
Industrial plants	19 000	18 000	18 000	14 000
Electric energy production, 10^9 kWhr	1 638	1 158	842	389
Private utilities	1 183	810	579	267
Public and co-op	347	246	175	62
Industrial plants	108	102	88	60
Electric utility customers, thousands	72 485	65 558	58 870	44 986
Residential or domestic	64 018	57 596	51 446	37 553
Water power capacity, 10^3 kW[d]	50 248	44 490	33 180	18 675
Undeveloped water power, 10^3 kW[d]	128 900	124 087	114 200	87 604
Gas utility industry, sales, 10^6 therms	164 682	189 803	92 876	42 090
Residential	49 044	39 990	31 881	13 839
Gas utility customers, thousands	41 775	37 338	33 054	24 001
Residential customers	38 361	34 341	30 418	22 146
Mileage of gas utility, thousands	891 624[d]	767 520	630 950	387 470
Long-distance transmission	248 071[d]	211 240	183 660	113 050

[a] 1970 data are preliminary, unless otherwise stated.
[b] Excludes Alaska and Hawaii.
[c] Includes water, wind, and animal power, but not electric motors.
[d] These data are for 1969. Data for 1970 not available.
[e] Passenger cars, buses, trucks, and motorcycles.
[f] As of July 1.
[g] Includes private planes. Inadequate data, not strictly comparable.
[h] Includes 1.2 million hp in work animals and 20 000 hp in windmills.
[j] Includes lignite.
[k] Marketed production from wells and dry and liquid gas, including transmission and storage losses.
[m] Utilities and industrial plants. Does not include farms, isolated buildings, or very small plants.

*Compiled from: "Statistical Abstract of the United States: 1971", 92nd ed., U.S. Bureau of the Census, 1971; see this source for references.
  (Available from Superintendent of Documents, price—$5.50.)

Table 5-55.   ELECTRIC-POWER GENERATING CAPACITY*

ESTIMATES OF INSTALLED CAPACITY:
  1968, approximately 300,000 megawatts; 1990, approximately 1,000,000 megawatts
GROWTH-RATE EXPERIENCE:
  Last 30 years, 7 percent per year; each decade, capacity doubled

INSTALLATIONS PLANNED FOR SERVICE AFTER JAN. 1, 1968

Type	100–499 MW size	Larger than 500 MW
Conventional hydro	70	6
Pumped storage	29	0
Fossil fueled	103	84
Nuclear fueled	8	66
Total	210	156

ESTIMATES OF FUTURE ELECTRIC-POWER LOADS

Preliminary Projections by Advisory Committees of the Federal Power Commission

Region	Ratio of loads			
	1980/1970		1990/1970	
	Peak demand	Total	Peak demand	Total
Northeast	1.8	1.8	3.2	3.2
East central	1.9	1.8	3.4	3.4
Southeast	2.1	2.2	4.1	4.1
West central	2.0	2.0	3.8	3.8
South central	2.3	2.3	4.5	4.7
West	2.0	2.0	4.0	4.0

*From: "Steam Power Plant Site Selection", Office of Science and Technology, U.S. Government Printing Office, 1969 (Superintendent of Documents, $1.25).

Table 5-56.   FULL-LOAD CURRENT OF ELECTRIC MOTORS

Typical Approximate Values in Amperes at Full Voltage

Motor rating, hp[a]	Direct-current motors			Single-phase a-c motors		Three-phase a-c motors						
						Induction				Synchronous		
	115v	230v	550v	115v	230v	220v	440v	550v	2300v	220v	440v	2300v
$\frac{1}{4}$	4.6	2.3		7.4	3.7	2.0	1.0	0.8				
$\frac{1}{2}$	6.6	3.3	1.4	10.2	5.1	2.8	1.4	1.1				
$\frac{3}{4}$	8.8	4.3	1.8	13.0	6.3	3.3	1.8	1.4				
1	12.6	6.3	2.6	18.4	9.2	5.0	2.5	2.0				
2	16.4	8.3	3.4	34	17	6.5	3.3	2.6				
3	24	12	5.0	34	17	9	4.5	4				
5	40	20	8.3	56	28	15	7.5	6				
$7\frac{1}{2}$	58	29	12	80	40	22	11	9				
10	76	38	16	100	50	27	14	11				
15	112	56	23			40	20	16				
20	148	74	31			52	26	21				
25	184	92	38			64	32	26	7.0	54	27	5.4
30	220	110	46			78	39	31	8.5	65	33	6.5
40	292	146	61			104	52	41	10.5	86	43	8
50	360	180	75			125	63	50	13	108	54	10
60	430	215	90			150	75	60	16	128	64	12
75	536	268	111			185	93	74	19	161	81	15
100		355	148			246	123	98	25	211	106	20
125		443	184			310	155	124	31	264	132	25
150		534	220			360	180	144	37		158	30
200		712	295			480	240	195	48		210	40
250						600	300	240	57		266	51
300						716	358	286	69		318	61
350						832	416	333	79		370	71
400						948	474	379	91		422	80
450						1,068	534	427	102		473	91
500							590	472	108		524	100
600							704	563	135		627	120
700							803	656	157		729	140
800							934	747	179		830	159
900							1,046	836	200		932	178
1,000							1,160	927	222		1,034	198

Note: Line currents are approximately the same for squirrel-cage or wound-rotor motors. In either case the values will be higher than above if speed is less than 1,200 rpm; for example, add 5% for 720 rpm and add 25% for 277 rpm.

[a] For watts multiply by 746.

Table 5-57. COMPARISON OF BATTERY TYPES

For temperature in K, see Table 9-22 for conversion. For output in kJ/kg, multiply the values in watt-hr/lb by 7.9367.

Name	Type	Anode	Cathode	Electrolyte	Nominal cell voltage	Temp range, °F	Typical output, watt-hr/lb	Cycle life if recharged (50% discharge)
Leclanché or carbon-zinc	Primary	Zinc	$C + MnO_2$	$NH_4Cl-ZnCl_2$	1.5	40–130	2–30	—
Mercury	Primary	Zinc	HgO	$KOH-K_2Zn_2O_3$	1.35 and 1.4	40–130	50	—
Silver oxide	Primary	Zinc	Ag_2O or AgO	KOH or NaOH	1.5	0–130	30–80	100–300
Alkaline or manganese zinc	Primary and rechargeable	Zinc or zinc–Hg	MnO_2	KOH or NaOH	1.5	0–130		50–100
Lalande	Primary	Zinc	CuO	NaOH	0.65		20	—
Nickel-cadmium	Rechargeable (secondary)	Cadmium	$Ni(OH)_2$	KOH	1.25	0–115	6–15	100–2,000
Silver-cadmium	Rechargeable	Cadmium	Ag_2O_2 or AgO	KOH	1.1	−40–100	10–40	300–1,000
Lead-acid or Planté	Rechargeable	Lead	PbO_2	H_2SO_4	2.0	−40–120	7–26	100–400
Nickel-iron or Edison	Rechargeable	Iron	NiO_2	KOH	1.2		10–15	100–3,000
Cuprous chloride	Activated	Magnesium	Cu_2Cl_2	Sea water	1.2	−80–150	20–40	—
Silver chloride	Activated	Magnesium	AgCl	Sea water	1.4	−80–150	20–80	—

Notes:

Mercury oxide and zinc batteries are important commercially (called RM batteries for Signal Corps walkie-talkies). They have the following characteristics: very flat voltage curve; good heavy-drain characteristics; high capacity per unit volume and weight; long dry-storage life (90% at 4 years); suited for continuous service; available in miniature (button-type); usable, at light loads, down to −40 deg F and up to 160 deg F; withstand pressure, vibration, acceleration, impact; and often used as voltage-reference sources, 1.35 V/cell.

Silver oxide and zinc batteries may be dry stored and activated immediately prior to use. They have a high capacity per unit weight and are non-magnetic. Special batteries may be charged, but their high-current performance is inferior to the primary type.

Manganese oxide and zinc batteries sustain voltage at high current drain. They are usable to −5 deg F, are inexpensive, and have a long shelf-life.

Lalande copper oxide and zinc batteries are commercially used for crossing and semaphore signals, approach lighting, etc., in sizes of 75–1000 Ahr (ampere hour). They have a flat voltage-time curve and easy field replacement. KOH electrolyte is substituted when service is much below atmospheric freezing (but above 0 deg F).

Nickel-cadmium batteries may be recharged many times and tolerate overcharge.

Silver-cadmium batteries have a low cell voltage but high output per unit volume and weight. Their chief characteristics are long charge-cycle life, rapid charge, non-magnetic, and no residual field.

Table 5-58. PROBABLE EFFICIENCIES OF FUEL CELLS AT VARIOUS LOADS*

Percentage of maximum power	Efficiencies in percent			
	Present dissolved methanol type cells	Future dissolved methanol type cells	Present hydrazine type cells	Diesel engine
10	43	72	68	22
20	36	67	65	27
40	28	60	58	29
60	23	56	50	31
80	19	49	43	28
100	12	35	30	25

*From: Carl Berger, Ed., HANDBOOK OF FUEL CELL TECHNOLOGY, © 1968, Prentice-Hall, Inc. By permission.

Table 5-59.   PROTOTYPE BATTERIES*

For energy density in kJ/kg, multiply the values in Whr/lb by 7.9367. For energy density in MJ/m³, multiply the values in Whr/in.³ by 219.69. For power density in W/kg, multiply the values in W/lb by 2.2046.

Type	Composition (charged)			Average energy density			Maximum power density, W/lb	Cell potential, v		Cell life	Remarks	Manufacturer or developer
	+ Cathode	− Anode	Electrolyte	Whr/lb		Whr/in.³		Open	Discharging			
				Theoretical	Actual							
Zinc-nickel-S†	Ni	Zn	KOH	220	40–50							Yardney Electric
Zinc-air	Air	Zn	KOH	464	50–60 to 100	2–3	30 (40 est.)	1.65	0.9–1.2 to 1.4	Several hundred cycles	Several approaches to avoid dendritic zinc	GE, General Atomics (Gulf), Yardney, ESB
Zinc-air-P‡	Air	Zn	KOH	464	150		20 (40 est.)	1.65	0.9–1.2			Leesona-Moos
Magnesium-air-P	Air	Mg	Aqueous solutions of NaCl, CaCl, LiCl, MgCl	345	500 with local H_2O refills				1.3			GE
Sodium-sulfur-S	S	Na	Ceramic	345	84–100 est. to 150	4 est. 8	100 est. to 200	2	1.75	Indefinite	Operates at high temperature (250–300°C), self-maintaining once brought to operating temperature	Ford
Sodium-air-S	Air	Na	NaOH	930	160–215 (4-hr discharge)	5	40–55 (4-hr discharge)	2.60	2.3–2.4	Indefinite	2-step process involving complex sodium amalgam—operates at 130°C	Atomic International (Northern American Aviation)

Table 5-59. PROTOTYPE BATTERIES (Continued)

Type	Composition (charged)			Average energy density Whr/lb			Maximum power density, W/lb	Cell potential, v		Cell life	Remarks	Manufacturer or developer
	+ Cathode	− Anode	Electrolyte	Theoretical	Actual	Whr/in.³		Open	Discharging			
Sodium-air-P	Air	Na	NaOH	930			~40	1.9–2.0	1.4		Sodium-mercury amalgam, ambient temperature	Western Reserve
Calcium-air-P	Air	Ca	KOH									Yardney
Lithium-moist air-S	Air + H_2O	Li	Non-aqueous	2566								Globe-Union
Aluminum-air-P	Air	Al	KOH		240–400		30 av. 48–75 peak	2.7	1.1		Power controlled by adjustment of liquid level	Zaromb Research Corp.
Iron-air-S	Air	Fe		650								Westinghouse
Dry tape-P	Halogen systems	Li	Organic		200 (ex hardware)				3.0–3.1		Proprietary development	Monsanto
		Mg	Aqueous		100 (ex hardware)				2.0			
Dry tape-P	Ag_2O_2	Zn	KOH		25–30							Monsanto
Lithium-chlorine-S	Cl_2	Li	Lithium-chloride	1200	100 est. to 300		75 est. to 150	3.5	3.2	Indefinite	Operates at 650°C	GM
Lithium-tellurium-S	Li_2Te	Li	LiCl-LiF eutectic	270	90		140	1.94	1.67–1.79	Indefinite	Operates at 471°C	Argonne National Laboratory
Lithium-nickel chloride-S	$NiCl_2$	Li	Propylene carbonate or K-phosphofluoride	435	est. 100		Low <20	2.50			Room temperature	Gulton

Table 5-59.   PROTOTYPE BATTERIES (Continued)

Type	Composition (charged)			Average energy density			Maximum power density, W/lb	Cell potential v		Cell life	Remarks	Manufacturer or developer
	+ Cathode	− Anode	Electrolyte	Whr/lb		Whr/in.³		Open	Discharging			
				Theoretical	Actual							
Lithium-nickel fluoride-S	NiF_2	Li	Same as above	620	90–100		Low <70	2.5			Room temperature	Gulton
Lithium-silver difluoride-S	AgF_2	Li	Butyrolactone, KPF_6	678			Low			Has undergone 50 cycles at 90% depth of discharge	Room temperature	Whittaker
Lithium-silver chloride-S	AgCl	Li	Li salts in PC, other organics	231	30 est. 90		Low		2.3–2.8	50–200 cycles	Room temperature	Lockheed, Electrochimica
Lithium-copper fluoride-P	CuF_2	Li	Li salts in PC, other organics	746	57–80 est. 110, 200		Low	3.6	2.3–2.8		Room temperature	Lockheed, ESB, Electrochimica
Lithium-copper chloride-S	$CuCl_2$	Li	Li salts in PC, other organics	503	25 est. 175		Low	3.1	2.3–2.8		Room temperature	Mallory, Electrochimica

†S—secondary.
‡P—primary.

*From: "Power Sources", *SAE Journal*, 76(12):64, December 1968.

Table 5-60.   TYPICAL COMPLETE FUEL-CELL SYSTEMS*

For mass in kg, multiply values in pounds by 0.4536. For volume in m³, multiply values in ft³ by 0.028 317.

Output, kW	Mass (excluding fuel), lb	Volume, cu ft	Fuel and oxidant	Electrode and electrolyte	Conditions	Operating application
UNION CARBIDE						
0.3[a]	33 (including fuel)	0.87	Hydrazine, air	Carbon plated on nickel circulating KOH	Ambient	Commercial
1.0–2.5[b]	42 + auxiliaries	3.36	Hydrogen, oxygen		120°F–150°F	Commercial
3.74–9.4[c]		28	Hydrogen, oxygen		120°F–150°F	
94	3650		Hydrogen, oxygen		120°F–150°F	GM Electrovan
ALLIS-CHALMERS						
0.5			Methanol, oxygen	Porous nickel electrodes, platinized anode, silver cathode, asbestos matrix held KOH	Ambient	Demonstration
3					Ambient	Demonstration
4.5	1190	34	Hydrazine, air JP–150 (reformer)		Reformer 1450°F Cell 160–180°F	U.S. Army
2.5 (overload)	169 (complete)	< 5.3	Hydrogen, oxygen		190°F	Space
15		20 + auxiliaries	Propane, oxygen		Ambient	Mounted on tractor
MONSANTO RESEARCH						
0.06	14.5	0.35	Hydrazine, air	Circulating KOH	200°F	
5	200		Hydrazine, air		200°F	
60	1080		Hydrazine, air		200°F	U.S. Army truck
PRATT & WHITNEY (UNITED AIRCRAFT)						
0.5	82.75	2.63	JP–4 (reformer)	Asbestos matrix, KOH		Army (battery charger)
1			Hydrogen, oxygen		~500°F, 55 psi	NASA (LM)
2			Hydrogen, oxygen		~500°F, 55 psi	NASA (Apollo)
GENERAL ELECTRIC						
0.03	7	0.3	Lithium hydride, air	Ion-exchange membrane, sulfonic acid	35–110°F	Military field use
0.06	10	0.7	Active metal/ hydrogen, air		35–110°F	Military field use
0.2	60	—	Hydrogen, air		35–110°F	
1.0	70 + auxiliaries		Hydrogen, oxygen		35–110°F	NASA (Gemini type)
1.5	140	8	JP–4 reformer, air		35–110°F	Battery charger
ENERGY CONVERSION LTD. (U.K.)						
6	1300	2813	Methanol, air	—	—	Electric truck
CHLORIDE GROUP (U.K.)						
—	—	—	Hydrazine, air	—	—	Demonstration
SHELL (U.K.)						
—	—	—	Methanol, air	Acid electrolyte	—	Demonstration
ASEA (SWEDEN)						
50	—	—	Ammonia (reformer)		—	Submarine
VARTA A.G. (GERMANY)						
2	—	—	Hydrogen, oxygen		—	Fork-lift truck

[a]H2R–1–278.
[b]E2F–1–468.
[c]E2F–4–864.

*From: "Power Sources", *SAE Journal*, 76(12):73, December 1968.

Table 5-61.   FUEL-CELL CHARACTERISTICS

For current density in A/m², multiply values in amp/sq ft by 10.764. For output in m³/kW, multiply values in ft³/kW by 0.0283. For output in g/W, multiply values in lb/kW by 0.4536.

Type	Fuel	Oxidant	Electrode	Electrolyte	Operating temperature	Operating pressure	Open-circuit voltage — Theoretical	Open-circuit voltage — Measured	Current density, amp/sq ft	Output, Whr/lb	Output ft³/kW	Output lb/kW	% efficiency thermal to electric
Ion-exchange membrane	Hydrogen	Oxygen or air	Activated metal	(Solid) ion-exchange membrane	50°F above ambient −65 to +165°F	Atmospheric	1.1	1 at 0.8 volt and 10 ma	22 at 0.8 volt	100 (measured)	3.5–5	250–500	60 at 15–20 amp/sq ft
Redox	Liquefied fuel	Oxygen or air	Porous metal	Liquids	70–85°C	Approx atmospheric	~1		200	1,200 with air 1,600 with pure oxygen	5	50–75	
Carbox	HCO-petroleum hydrocarbons (kerosene)	Oxygen or air	Porous metal	Fused carbonate	500–800°C	Atmospheric	~1	0.7	60				64
Hydrox	Hydrogen	Oxygen	Porous metal		200–250°C, 400–500°F	10–55 atm, 400–600 psi	1.1	1.0	Up to 1,000		0.25–1.2	40–90	
Thermal regenerative	Hydrogen	Group I metals	Fuel metal and nickel	Fused group I chlorides	608 or 1004°F	200–500 mm, Hg abs	0.75	0.72	245 at 0.72 volt with lithium	~5			Carnot: 40
Solar regenerative	Nitric oxide	Chlorine	Carbon-disk	Liquid nitrosyl chloride	70°F	15 psig	0.21	0.21	2 at 0.1 volt				
Low temp-pressure	Hydrogen	Oxygen	Specially processed carbon	12 molar solution of KOH	70–150°F	1–5 atm	1.2	1.12	100 at 0.95 volt and 104°F and oxygen at 5 atm	1,620	3.5–5	250–500	75

Table 5-62. ELECTROMOTIVE FORCE AND COMPOSITION OF VOLTAIC CELLS*

STANDARD CELLS

Name of cell	Negative pole	Solution	Positive pole	Depolarizer	Electromotive force, volts
Weston normal	Cadmium amalgam	Saturated solution of $CdSO_4$	Mercury	Paste of Hg_2SO_4 and $CdSO_4$	1.0183 at 20°C
Clark standard	Zinc amalgam	Saturated solution of $ZnSO_4$	Mercury	Paste of Hg_2SO_4 and $ZnSO_4$	1.4328 at 15°C

Temperature equations (temperature in °C):

Clark cell: $E_t = 1.4328[1 - 0.00119(t-15) - 0.000007(t-15)^2]$ volt

Weston cell: $E_t = 1.0183[1 - 0.0000406(t-20) - 0.00000095(t-20)^2 + 0.00000001(t-20)^3]$ volt

DOUBLE FLUID CELLS

Name of cell	Negative pole	Solution	Positive pole	Solution	Electromotive force, volts
Bunsen	Amal. zinc	1 part H_2SO_4 to 12 parts H_2O	Carbon	Fuming nitric acid	1.94
Bunsen	Amal. zinc	1 part H_2SO_4 to 12 parts H_2O	Carbon	HNO_3, density 1.38	1.86
Bichromate	Amal. zinc	12 parts $K_2Cr_2O_7$ to 25 parts H_2SO_4 and 100 parts H_2O	Carbon	1 part H_2SO_4 to 12 parts H_2O	2.00
Bichromate	Amal. zinc	1 part H_2SO_4 to 12 parts H_2O	Carbon	12 parts $K_2Cr_2O_7$ to 100 parts H_2O	2.03
Daniell	Amal. zinc	1 part H_2SO_4 to 4 parts H_2O	Copper	Saturated solution of $CuSO_4 + 5H_2O$	1.06
Daniell	Amal. zinc	5% solution of $ZnSO_4 + 6H_2O$	Copper	Saturated solution of $CuSO_4 + 5H_2O$	1.08
Daniell	Amal. zinc	1 part NaCl to 4 parts H_2O	Copper	Saturated solution of $CuSO_4 + 5H_2O$	1.05
Grove	Amal. zinc	1 part H_2SO_4 to 12 parts H_2O	Platinum	Fuming nitric acid	1.93
Grove	Amal. zinc	Solution of $ZnSO_4$	Platinum	HNO_3, density 1.33	1.66

*From: "Smithsonian Physical Tables", 9th ed., W.E. Forsythe, Ed., The Smithsonian Institution, 1956.

Table 5-63. CHARACTERISTICS OF NUCLEAR BATTERIES*

	Constant-current charging			Contact-potential difference	Junction	Photo-junction	Thermo-junction
Radioactive material	Sr^{90}	H^3	Kr^{85}	H^3	Sr^{90}	Pm^{147}	Po^{210}
Half-life	25 yr	12 yr	10 yr	12 yr	25 yr	2.6 yr	138 days
Quantity	10 Mc	1 curie	1 curie	1.5 Mc/cell	50 Mc	4.5 curies	3,000 curies
Size	1 cu in.	1 cu in.	5 cu in.	1 cu in.		0.2×0.7 in. diam	5.5×4.75 in. diam
Weight	6 oz	1 oz	14 oz	1.5 oz		0.6 oz less shielding	5 lb
Current amp	10^{-12}	6×10^{-10}	10^{-9}	10^{-10}	5×10^{-6}	0.25–1 volt, 20μw	5 watts
Voltage	14 kv	1 kv	1 kv	100 volts (66 cells)	0.2 volt		
Development status	Sr batteries in production; prototypes of H, Kr batteries under test			Development complete, but not in production	Development complete, but not in production	Development complete	Prototypes completed; larger units being investigated
Manufacturers	Radiation Research Corp.; Patterson Moos Div.; Universal Winding Co.			Tracerlab, Inc.	RCA	Elgin National Watch Co.	Mound Laboratory; Martin Company

*Reprinted from *Electronics*, March 20, 1959; copyright McGraw-Hill, Inc., 1959.

Table 5-64. ESTIMATED FUTURE TRENDS IN AIR-JET PROPULSION*

Characteristic	Reheat turbojet			Reheat turbofan			Turboramjet			Ramjet		
Mach number capability	0	2	4	0	2	5	0	4	8	2	5	10
Specific weight, static, lb/lb thrust	0.08	0.10	0.12	0.11	0.13	0.16	—	0.13	0.20	—	—	—
Fuel specific impulse, at maximum thrust, lb_f sec/lb_m {JP-6	1800	1700	1350	2400	2200	1300	1800	1600	—	1530	1440	—
{H_2	5000	4700	3760	6600	6100	3600	5000	4400	1025	4240	4000	750
Air specific impulse at maximum thrust, I_a, lb_f sec/lb_m	120	100	65	55	60	55	120	73	30	60	80	17
Thrust coefficient (based on free-stream inlet captive area)	—	3.92	1.07	—	2.0	0.74	—	1.2	0.25	2.35	1.0	0.11
Applications	Fighters			Supersonic transport			Hypersonic cruise			Missiles		
	Bombers			Air-breathing boost			Air-breathing boost for hypersonic and space systems			Hypersonic staged transports		
	Booster recovery			Booster recovery						Recoverable two-stage boosters		
	Reentry vehicles			Reentry vehicles			Reentry vehicles			Reentry vehicles		

*From: "Handbook of Astronautical Engineering", H.H. Koelle, McGraw-Hill Book Company, 1961.

Table 5-65.   BASIC ROCKET PROPERTIES AND THEIR RELATIONSHIPS*

COURTESY OF F.J.HENDEL

SYMBOLS AND UNITS:

A_t = throat area of rocket (thrust) chamber, in.^{2}
C = effective exhaust velocity of propellant gases, ft/sec
C^* = characteristic velocity of propellant gases, ft/sec
C_F = thrust coefficient
F = thrust of the rocket, lb$_f$
g_c = gravitational conversion factor, $32.174 \dfrac{\text{lb}_m \text{ ft}}{\text{lb}_f \text{ sec}^2}$

I_s = specific impulse (lb$_f$ thrust per lb$_m$ propellant per sec)
L^* = characteristic length of rocket chamber, in.
P_c = pressure of combustion in rocket chamber, psia
V_c = volume of rocket chamber (up to the throat), in.^{3}
w_s = specific flow (consumption) of propellant, lb$_m$/lb$_f$ sec
$w_t = \dot{w}$ = mass flow rate of propellant (oxidizer+fuel), lb$_m$/sec

Property	C	C^*	C_f	F	I_s	L^*	w_s
Effective exhaust velocity, C (ft/sec)	—	C^*C_F	C^*C_F	Fg_c/w_t	$I_s g_c$	$P_c V_c C_F g_c/L^*w_t$	g_c/w_s
Characteristic velocity, C^* (ft/sec)	C/C_F	$P_c A_t g_c$	$Fg_c/C_F w_t$	$P_c A_t g_c I_s/F$	$I_s g_c/C_F$	$P_c V_c g_c/L^*w_t$	$P_c A_t g_c/Fw_s$
Thrust coefficient, C_F	C/C^*	C/C^*	—	$F/P_c A_t$	$I_s g_c/C^*$	$L^*w_t I_s/P_c V_c$	$w_t/P_c A_t w_s$
Thrust (theoretical), F (lb$_f$)	$w_t C/g_c$	$P_c A_t g_c/C^*$	$P_c A_t C_F$	—	$w_t I_s$	$P_c V_c C_F/L^*$	w_t/w_s
Specific impulse, I_s (lb$_f$ sec/lb$_m$)	C/g_c	C^*C_F/g_c	$P_c A_t C_F/w_t$	F/w_t	—	$P_c V_c C_F/L^*w_t$	$1/w_s$
Characteristic chamber length, L^* (in.)	$P_c V_c C/FC^*$	$P_c V_c g_c/w_t C^*$	$P_c V_c C_F/F$	$P_c V_c C_F/F$	$P_c V_c C_F/w_t I_s$	—	$P_c V_c C_F w_s/w_t$
Specific propellant flow (consumption), w_s (lb$_m$/lb$_f$ sec)	g_c/C	g_c/C^*C_F	$w_t/P_c A_t C_F$	w_t/F	$1/I_s$	$L^*w_t/P_c V_c C_F$	—

*Based largely on: "How Basic Rocket Properties Are Related", F.J. Hendel, *Chem. Eng.* (N.Y.), 68:101, March 6, 1961.

Table 5-66. TOTAL PRESSURE RECOVERY PR IN TERMS OF KINETIC RAM EFFICIENCY η_R AND MACH NUMBER M

$$PR = \frac{P_2/P_o}{\{[(k-1)/2]\, M^2 + 1\}^{k/(k-1)}}$$

$$\eta_R = \frac{1 - (P_o/P_2)^{(k-1)/k}}{1 - \{[(k-1)/2]\, M^2 + 1\}^{-1}}$$

$$PR = \left[(1-\eta_R)\frac{k-1}{2} M^2 + 1\right]^{\frac{k}{1-k}},$$

where $k = c_p/c_v$ = specific heat ratio = 1.4
M = Mach number of diffuser in ambient air
P_o = static pressure of ambient air
P_2 = static pressure at the end of subsonic diffusion

Mach No., M	Kinetic ram efficiency, η_R										
	0.50	0.60	0.70	0.80	0.85	0.90	0.95	0.96	0.97	0.98	0.99
1.0	.7164	.7639	.8155	.8717	.9017	.9330	.9658	.9725	.9792	.9861	.9930
1.5	.4915	.5603	.6420	.7396	.7956	.8572	.9251	.9395	.9542	.9691	.9844
2.0	.3080	.3784	.4710	.5948	.6726	.7639	.8717	.8956	.9203	.9457	.9725
2.5	.1828	.2419	.3281	.4579	.5480	.6622	.8088	.8430	.8791	.9172	.9574
3.0	.1058	.1498	.2206	.3409	.4332	.5603	.7396	.7840	.8260	.8836	.9395
3.5	.06086	.09155	.1454	.2477	.3344	.4644	.6673	.7209	.7802	.8458	.9188
4.0	.03529	.05588	.09486	.1770	.2536	.3784	.5948	.6560	.7256	.8048	.8956
4.5	.02077	.03435	.06183	.1254	.1899	.3042	.5244	.5913	.6694	.7614	.8703
5.0	.01247	.02138	.04048	.08839	.1410	.2419	.4579	.5283	.6131	.7163	.8430
6.0	.00479	.00869	.01783	.04407	.07705	.1498	.3409	.4124	.5044	.6245	.7840
7.0	.00201	.00379	.00824	.02241	.04222	.09155	.2477	.3142	.4057	.5345	.7209
8.0	.00091	.00176	.00401	.01175	.02350	.05588	.1770	.2353	.3206	.4503	.6560
9.0	.00044	.00087	.00205	.00637	.01338	.03435	.1253	.1740	.2500	.3744	.5913
10.0	.00023	.00046	.00110	.00358	.00781	.02138	.08839	.1278	.1930	.3080	.5283

Table 5-67. TYPICAL SOLID PROPELLANT FORMULATIONS*

Material	Vinyl polyester		Polyurethane		Rubber		Polybutadiene copolymer	Polysulfide
NH_4NO_3	—	72.50	—	—	—	86.50	—	—
NH_4ClO_4	64.60	—	62.00	70.00	90.00	—	68.00	71.00
$KClO_4$	11.40	—	—	12.00	—	—	—	—
Binder	23.60	25.50	21.40	17.50	7.00	11.30	16.00	26.00
Plasticizer	—	—	—	—	2.90	—	—	—
Aluminum	—	—	15.50	—	—	—	16.00	—
Catalyst	0.25	—	—	—	0.10	2.20	—	2.00
Additives	0.15	2.00	1.10	0.50	—	—	—	1.00

Values are mass percentages.
*Adapted from: "Handbook of Astronautical Engineering", H.H. Koelle, McGraw-Hill Book Company, 1961.

Table 5-68. TYPICAL SOLID PROPELLANT EXHAUST-GAS COMPOSITION*

Gas product†	Types of propellants						
	Double base	Asphalt, $KClO_4$	Rubber, NH_4NO_3	Vinyl polyester, NH_4ClO_4	Polysulfide, NH_4ClO_4	Polyurethane	
						Without Al, NH_4ClO_4	With Al, NH_4ClO_4
CO_2	27.7	3.3	15.63	20.46	16.66	27.9	1.31
CO	23.4	44.0	2.48	0.06	3.90	9.6	28.69
H_2	7.6	27.4	28.62	0.06	4.04	0.8	38.39
H_2O	26.2	9.9	31.51	48.24	41.54	24.9	6.60
N_2	15.1	0.1	21.76	10.37	9.20	8.6	6.39
HCl	—	—	—	20.75	18.42	—	12.00
$Cu\ (s)$	—	—	—	—	—	—	0.02
$Cu\ (l)$	—	—	—	—	—	—	0.07
$Al_2O_3\ (s)$	—	—	—	—	—	—	6.53
Cr_2O_3	—	—	—	0.03	—	0.05	—
CrO_2	—	—	—	0.03	—	—	—
CuO	—	—	—	—	—	0.02	—
HS	—	—	—	—	0.03	—	—
H_2S	—	—	—	—	2.59	—	—
S_2	—	—	—	—	2.02	—	—
SO_2	—	0.2	—	—	1.57	—	—
S	—	—	—	—	0.03	—	—
KCl	—	15.1	—	—	—	6.6	—

†Expressed in percent exhausted to 14.7 psia.
*From: "Handbook of Astronautical Engineering", H.H. Koelle, McGraw-Hill Book Company, 1961.

Table 5-69.   OXIDIZERS FOR SOLID PROPELLANTS*

COURTESY OF F.J. HENDEL

Ranked in Order of Available Oxygen

Available oxygen, % by wt	Name	Formula	Specific gravity	Development status and (remarks)
100	Frozen oxygen	Solid O_2		(Cryogenic)
73	Lithium ozonide	LiO_3		Distant future
66	Nitronium perchlorate or nitryl perchlorate	NO_2ClO_4	2.22	(Very hygroscopic)
61	Lithium superoxide	LiO_2		Future
60	Lithium perchlorate	$LiClO_4$	2.43	(Very hygroscopic)
58	Lithium nitrate	$LiNO_3$	2.38	(Hygroscopic)
57	Magnesium perchlorate	$Mg(ClO_4)_2$	2.60	(Very hygroscopic)
56	Sodium ozonide	NaO_3		Future
54	Oxonium perchlorate or perchloric acid, 84.8%	H_3OClO_4	1.88	(Hygroscopic; melts at 122°F)
54	Calcium perchlorate	$Ca(ClO_4)_2$		(Not attractive)
52	Sodium perchlorate	$NaClO_4$		(Hygroscopic)
47	Hydrogen peroxide	Solid H_2O_2	1.71	Future
46	Potassium perchlorate	$KClO_4$	2.52	Was used
46	Potassium ozonide	KO_3		(Not attractive)
46	Calcium superoxide	$Ca(O_2)_2$		(Not attractive)
45	Sodium chlorate	$NaClO_3$	2.49	(Not attractive)
44	Sodium superoxide	NaO_2		(Not attractive)
42	Hydroxyl amine perchlorate	$NH_2OH \cdot HClO_4$		Was used
41	Hydrazine (hydrazinium) diperchlorate	$N_2H_6(ClO_4)_2$	2.2	(Hygroscopic)
40	Potassium nitrate	KNO_3	2.1	(Not attractive)
34	Ammonium perchlorate	NH_4ClO_4	1.95	Widely used
24	Hydrazine perchlorate	$N_2H_5ClO_4$		(Detonable)
20	Ammonium nitrate	NH_4NO_3	1.73	In gas generators
8.4	Hydrazine nitrate	$N_2H_5NO_3$		(Melts at 159°F; sublimes at 284°F)
3.5	Nitroglycerine	$C_3H_5(NO_3)_3$	1.59	(Melts at 56°F)

*From: "Review of Solid Propellants for Space Exploration", F.J. Hendel, NASA Technical Memorandum No. 33–254, Jet Propulsion Laboratory, 1965; adapted from "Gaseous Environment During Space Missions", F.J. Hendel, *J. Spacecraft and Rockets*, 1:353, 1964.

Table 5-70.   PROPERTIES OF ROCKET PROPELLANTS

COURTESY OF WILLIAM TOMAZIC

For data on solid propellants, see Tables 5-67, 5-68, and 5-69.

Propellant	Formula	Boiling point, °F	Freezing point, °F	Critical temperature, °F	Critical pressure, psia	Specific gravity†	Stability	Acceptable materials	Toxicity
FUELS Hydrogen	H_2	−423	−434	−400	188	.071 at N.B.P.	Stable	Non-corrosive; however, only materials with good low-temp properties should be used	Non-toxic
Ammonia	NH_3	−28.1	−107.9	270.3	1652	.611	Stable	Teflon, 18–8 SS, aluminum, Kel-F, polyethylene	50 ppm permissible for 8-hr exposure; over 1,720 ppm may be fatal for short exposure —less than ½ hr
Hydrazine	N_2H_4	236.3	34.7	716	2135	1.008	Explosive decomposition catalyzed by metal oxides; thermal decomposition above 320°F	Teflon, 18–8 SS, Kel-F, polyethylene, aluminum	Toxic; mean lethal dose approx 570 ppm; 1 ppm allowable for 40-hr week exposure
Monomethyl hydrazine (MMH)	N_2H_6C	192.5	−62.5	594	1195	.874 at 77°F	Fairly stable under 500°F	18–8 SS, aluminum, Teflon, Kel-F	Toxic; mean lethal dose approx 74 ppm
Unsymmetrical dimethyl-hydrazine (UDMH)	$N_2H_8C_2$	147	−71.0	480	880	.792	Stable at ordinary temp; decomposition may become violent above 500°F	Mild steel, 18–8 SS, aluminum, Kel-F, Teflon, polyethylene	Less toxic than hydrazine; more toxic than ammonia
RP-1	$C_{11.74}H_{21.83}$	10% distillation: 365 min 410 max	−40 max	—	—	.815 max .801 min	Stable at ordinary temp; cracking begins at about 700°F	Essentially all common metals; neoprene, asbestos, fluorocarbons, epoxies	Comparable to kerosene
Methane	CH_4	−258.7	−299.2	−116.3	673	.415 at −263°F	Stable	Essentially all common metals; neoprene, asbestos, fluorocarbons, epoxies	Essentially non-toxic
Propane	C_3H_8	−43.9	−304.8	206.3	619	.585 at −48°F	Stable	Essentially all common metals; neoprene, asbestos, fluorocarbons, epoxies	Essentially non-toxic

Table 5-70.   PROPERTIES OF ROCKET PROPELLANTS (*Continued*)

Propellant	Formula	Boiling point, °F	Freezing point, °F	Critical temperature, °F	Critical pressure, psia	Specific gravity†	Stability	Acceptable materials	Toxicity
FUELS (Cont.) Diborane	B_2H_6	−134.6	−264.7	62.1	581	.435 at N.B.P.	Decomposes slowly at room temp: decomposes rapidly to elements above 570°F	Most metals; organic materials that have no functional groups and are not saturated	Toxicity comparable to phosgene; mean lethal dose approx 150 ppm; maximum for continuous exposure approx. 0.1 ppm
OXIDIZERS Oxygen	O_2	−297.4	−361.8	−181.0	730.4	1.143 at N.B.P.	Stable	Non-corrosive; however, metals with good low-temperature properties should be used; many non-metals usable; new materials must be screened for compatability	Non-toxic
Fluorine	F_2	−306.6	−363.3	−200.5	808.5	1.505 at N.B.P.	Stable	18–8 SS, nickel, copper, monel preferred; thorough cleanliness very important; no non-metallic materials entirely resistant to attack	Highly toxic and corrosive to body tissues; approx 25 ppm produces toxic symptoms in less than 1 minute
Nitrogen tetroxide	N_2O_4	70.1	11.8	316.4	1450	1.45	Stable	Mild steel (when dry), stainless steels, aluminum, Teflon, Kel-F, polyethylene	Toxic; comparable to chlorine; 5 ppm maximum allowable for 40-hr week exposure
Chlorine trifluoride	ClF_3	53.2	−105.4	345	838	1.825	Stable	18–8 SS, nickel, monel preferred; most common metals possible; must be clean; Teflon possible	Highly toxic; comparable to F_2; approx 20 ppm produces toxic symptom within 15 min
Inhibited red fuming nitric acid	.835 HNO_3 .140 NO_2 .020 H_2O .005 HF	Approx 140	Approx −80	—	—	1.56	Subject to thermal decomposition; rate dependent on temp	18–8 SS, aluminum, polyethylene, Teflon, Kel-F	Toxic and corrosive to body tissues; approx 5 ppm allowable for 40-hr week exposure

Table 5-70.   PROPERTIES OF ROCKET PROPELLANTS (*Continued*)

Propellant	Formula	Boiling point, °F	Freezing point, °F	Critical temperature, °F	Critical pressure, psia	Specific gravity†	Stability	Acceptable materials	Toxicity
OXIDIZERS (Cont.) Oxygen difluoride	OF_2	−229	−371	−72	719	1.521 at N.B.P.	Stable at ambient temp; appreciable decomposition rate above 480°F	18–8 SS, copper, aluminum, monel, nickel; thorough cleanliness important; nonmetallic materials generally unsuitable	Highly toxic and corrosive to body tissues; mean lethal dose approx 8 ppm for 5-min exposure
MONOPROPELLANTS Hydrogen peroxide	.9 H_2O_2 .1 H_2O	286	11.3	—	—	1.39	May decompose rapidly in presence of impurities or catalytic materials, such as carbon, steel, or copper	Aluminum (pure and 25 alloy), tin, glass, polyethylene, Teflon, Kel-F	Not toxic in usual sense, but a strong primary irritant to skin and respiratory passages
Ethylene oxide	C_2H_4O	51	−171	384	1043	.87	Generally stable, but polymerization rate is increased in presence of some materials	18–8 SS, aluminum, mild steel, copper, nylon, Teflon	Relatively toxic; 100 ppm permissible for 8-hr exposure
Nitromethane	CH_3NO_2	214.2	−20.2	599	915	1.14	May detonate under conditions of confinement, heating, and mechanical impact	18–8 SS, aluminum, Kel-F, polyethylene	Relatively toxic; 50 to 200 ppm permissible for 8-hr exposure
n-Propyl nitrate	$C_3H_7NO_3$	230.9	< −150	585	588	1.057	Relatively stable; insensitive to mechanical or thermal shock	18–8 SS, aluminum, polyethylene, Teflon, nylon, Orlon, Dacron, Kel-F, Mylar	No serious toxicity problem

†At 68°F unless noted.

REFERENCES

"Liquid Propellants Handbook", Battelle Memorial Institute Staff, Aug. 1957.
"Liquid Propellant Manual", Liquid Propellant Information Agency, Johns Hopkins University.

Table 5-71. LIQUID ROCKET PROPELLANT THEORETICAL PERFORMANCE **

Chamber Pressure = 1,000 psia

SYMBOLS: S.E. = shifting equilibrium   *F.C. = frozen composition*

Fuel and oxidizer	S.E. or F.C.	Ratio: O/F	Combustion temp, °R	Bulk specific gravity	C^*, ft/sec	Specific impulse, $\frac{lb_f}{lb_m \, sec}$, sea level	ε,† area ratio, sea level	Vacuum specific impulse			
								ε-20†	ε-40	ε-80	ε-160
F: Hydrogen	S.E.	3.50	4910	.262	7972	388.4	7.86	439.1	450.8	459.6	466.3
O: Oxygen	F.C.	3.50	4910	.262	7934	386.3	7.80	436.4	447.9	456.6	463.1
	S.E.	5.00	5935	.325	7797	387.6	8.76	441.0	455.2	466.4	475.1
	F.C.	5.00	5935	.325	7685	378.0	8.32	429.0	441.8	451.8	459.4
	S.E.	6.00	6295	.362	7577	382.1	9.44	436.4	452.2	464.8	475.0
	F.C.	6.00	6295	.362	7436	367.0	8.50	417.1	430.1	440.3	448.3
F: RP-1	S.E.	2.40	6539	1.003	5936	299.6	9.48	342.3	354.9	365.1	373.6
O: Oxygen	F.C.	2.40	6539	1.003	5800	285.0	8.31	323.5	333.4	341.1	347.2
	S.E.	2.60	6631	1.010	5896	300.3	10.03	344.1	357.8	369.1	378.4
	F.C.	2.60	6631	1.010	5754	283.1	8.39	321.6	331.5	339.4	345.6
	S.E.	2.80	6671	1.017	5841	299.0	10.44	343.5	358.4	370.7	381.0
	F.C.	2.80	6671	1.017	5697	280.6	8.44	318.9	328.8	336.7	343.0
F: Ammonia	S.E.	1.30	5459	.881	5869	291.5	8.74	331.7	342.4	350.9	357.5
O: Oxygen	F.C.	1.30	5459	.881	5781	284.8	8.41	323.5	333.4	341.1	347.1
	S.E.	1.40	5533	.890	5846	293.6	9.21	334.9	346.5	355.7	363.0
	F.C.	1.40	5533	.890	5743	283.2	8.45	321.7	331.6	339.4	345.5
	S.E.	1.50	5524	.898	5775	289.4	9.12	330.0	341.3	350.3	357.4
	F.C.	1.50	5524	.898	5676	279.9	8.46	318.0	327.8	335.5	341.5
F: Hydrazine	S.E.	.75	5913	1.061	6253	309.6	8.56	351.8	362.7	371.2	377.8
O: Oxygen	F.C.	.75	5913	1.061	6143	301.0	8.13	341.1	350.9	358.4	364.2
	S.E.	.925	6133	1.069	6202	312.8	9.39	357.1	369.7	379.8	387.8
	F.C.	.925	6133	1.069	6062	297.7	8.26	337.7	347.7	355.5	361.5
	S.E.	1.00	6149	1.071	6143	311.3	9.77	356.3	369.9	380.8	389.6
	F.C.	1.00	6149	1.071	6003	294.9	8.28	334.6	334.6	352.4	358.4
F: UDMH	S.E.	1.44	6291	.964	6159	307.5	8.95	350.4	362.2	371.8	379.7
O: Oxygen	F.C.	1.44	6291	.964	6028	295.5	8.18	335.1	344.9	352.5	358.4
	S.E.	1.70	6496	.979	6092	309.6	9.87	354.5	368.2	379.3	388.3
	F.C.	1.70	6496	.979	5942	292.1	8.32	331.5	341.6	349.5	355.6
	S.E.	2.10	6523	.998	5903	302.6	10.58	348.0	363.9	377.6	389.5
	F.C.	2.10	6523	.998	5754	283.3	8.41	321.8	331.8	339.7	345.9
F: Hydrogen	S.E.	4.50	5504	.322	8434	403.7	7.01	452.8	462.4	469.4	474.6
O: Fluorine	F.C.	4.50	5504	.322	8327	397.7	6.87	445.4	454.6	461.3	466.2
	S.E.	8.00	7133	.463	8365	410.9	8.01	464.4	476.4	485.2	491.8
	F.C.	8.00	7133	.463	8125	388.4	6.92	435.3	444.3	450.8	455.5
	S.E.	16.00	8698	.687	8035	403.7	9.32	460.8	476.8	489.1	498.3
	F.C.	16.00	8698	.687	7675	365.4	6.72	408.6	416.6	422.3	426.4
F: Ammonia	S.E.	3.10	8192	1.158	7197	358.3	8.64	406.7	418.7	427.6	434.3
O: Fluorine	F.C.	3.10	8192	1.158	6913	330.5	6.95	370.6	378.4	384.0	388.1
	S.E.	3.30	8264	1.171	7184	359.4	8.92	408.7	421.2	430.7	437.7
	F.C.	3.30	8264	1.171	6889	329.2	6.92	369.0	376.7	382.3	386.3
	S.E.	3.50	8298	1.183	7152	356.3	8.49	403.8	415.2	423.7	430.2
	F.C.	3.50	8298	1.183	6856	327.4	6.89	366.9	374.4	379.9	383.9
F: Hydrazine	S.E.	2.00	8321	1.290	7288	362.5	8.68	411.8	424.1	433.4	440.4
O: Fluorine	F.C.	2.00	8321	1.290	7005	334.9	6.94	375.5	383.4	389.1	393.2
	S.E.	2.33	8496	1.309	7263	364.6	9.19	415.5	428.9	439.1	446.8
	F.C.	2.33	8496	1.309	6957	332.2	6.88	372.2	379.9	385.5	389.5
	S.E.	2.50	8527	1.317	7226	361.6	8.79	410.6	422.7	431.8	438.7
	F.C.	2.50	8527	1.317	6917	330.0	6.85	369.7	377.3	382.7	386.7
F: UDMH	S.E.	1.90	7069	1.142	6583	335.7	9.92	384.2	398.8	410.3	419.4
O: Fluorine	F.C.	1.90	7069	1.142	6278	304.7	7.69	344.1	353.1	359.9	365.1
	S.E.	2.45	7447	1.188	6586	340.6	10.65	391.0	407.3	420.3	430.5
	F.C.	2.45	7447	1.188	6284	303.8	7.49	342.5	351.0	357.4	362.2
	S.E.	2.70	7542	1.205	6531	338.3	10.53	387.9	404.0	417.4	428.6
	F.C.	2.70	7542	1.205	6265	302.4	7.42	340.7	349.0	355.2	359.8

Table 5-71. LIQUID ROCKET PROPELLANT THEORETICAL PERFORMANCE (Continued)

Fuel and oxidizer	S.E. or F.C.	Ratio: O/F	Combustion temp, °R	Bulk specific gravity	C*, ft/sec	Specific impulse, $\frac{lb_f}{lb_m\ sec}$, sea level	ε,† area ratio, sea level	Vacuum specific impulse			
								ε–20†	ε–40	ε–80	ε–160
F: Ammonia	S.E.	1.80	5027	1.023	5420	266.1	8.26	301.9	310.8	317.6	322.9
O: Nitrogen tetroxide	F.C.	1.80	5027	1.023	5379	263.8	8.18	299.0	307.7	314.3	319.4
	S.E.	2.00	5204	1.043	5448	270.2	8.63	307.1	316.8	324.3	330.2
	F.C.	2.00	5204	1.043	5366	263.6	8.26	299.0	307.9	314.7	320.0
	S.E.	2.20	5150	1.061	5339	264.2	8.58	300.3	309.7	317.0	322.6
	F.C.	2.20	5150	1.061	5266	258.6	8.25	293.3	302.0	308.6	313.8
F: Hydrazine	S.E.	1.10	5672	1.190	5887	289.6	8.30	328.5	338.2	345.8	351.5
O: Nitrogen tetroxide	F.C.	1.10	5672	1.190	5801	283.7	8.05	321.3	330.4	337.3	342.6
	S.E.	1.33	5870	1.210	5852	292.1	8.85	332.4	343.2	351.7	358.3
	F.C.	1.33	5870	1.210	5728	280.8	8.16	318.3	327.5	334.6	340.1
	S.E.	1.40	5879	1.215	5817	291.8	9.11	332.6	343.8	352.6	359.6
	F.C.	1.40	5879	1.215	5691	279.0	8.17	316.3	325.5	332.6	338.1
	S.E.	1.50	5861	1.223	5753	288.6	9.14	329.0	340.2	349.0	356.0
	F.C.	1.50	5861	1.223	5629	276.0	8.18	313.0	322.1	329.1	334.6
F: Monomethyl	S.E.	1.60	5675	1.149	5752	282.3	8.26	320.4	330.2	338.0	344.4
hydrazine	F.C.	1.60	5675	1.149	5678	277.2	7.97	313.8	322.5	329.2	334.2
O: Nitrogen tetroxide	S.E.	2.00	6061	1.180	5764	287.9	8.93	327.9	338.9	347.7	354.8
	F.C.	2.00	6061	1.180	5636	276.3	8.19	313.4	322.6	329.7	335.2
	S.E.	2.20	6113	1.193	5714	288.4	9.40	329.2	341.0	350.4	358.1
	F.C.	2.20	6113	1.193	5575	273.6	8.24	310.4	319.6	326.8	332.4
	S.E.	2.40	6109	1.205	5644	287.0	9.95	328.6	341.4	351.6	360.0
	F.C.	2.40	6109	1.205	5504	270.3	8.27	306.7	315.8	323.0	328.6
F: UDMH	S.E.	2.05	5879	1.126	5715	281.8	8.45	320.2	330.3	338.4	345.1
O: Nitrogen tetroxide	F.C.	2.05	5879	1.126	5622	274.9	8.05	311.4	320.2	327.1	332.3
	S.E.	2.70	6185	1.169	5642	286.1	9.67	327.0	339.2	349.0	357.0
	F.C.	2.70	6185	1.169	5501	270.1	8.27	306.5	315.7	322.9	328.5
	S.E.	2.85	6183	1.178	5598	285.1	10.02	326.6	339.4	349.7	358.2
	F.C.	2.85	6183	1.178	5456	268.0	8.29	304.2	313.3	320.5	326.1
F: .50 Hydrazine	S.E.	1.50	5688	1.144	5774	283.7	8.30	322.0	331.8	339.6	345.9
.50 UDMH	F.C.	1.50	5688	1.144	5696	278.3	8.00	315.1	323.9	330.6	335.8
O: Nitrogen tetroxide	S.E.	2.00	6061	1.184	5734	288.9	9.29	329.6	341.1	350.4	357.9
	F.C.	2.00	6061	1.184	5599	274.7	8.23	311.6	320.8	328.0	333.6
	S.E.	2.20	6057	1.197	5658	287.3	9.88	328.9	341.6	351.7	359.9
	F.C.	2.20	6057	1.197	5520	271.0	8.26	307.5	316.6	323.8	329.4
F: Ammonia	S.E.	3.60	6345	1.325	5690	276.0	7.51	310.8	318.2	323.7	327.7
O: Chlorine	F.C.	3.60	6345	1.325	5562	266.3	7.01	298.8	305.2	309.9	313.3
trifluoride	S.E.	3.90	6503	1.347	5709	278.3	7.70	313.9	321.7	327.4	331.6
	F.C.	3.90	6503	1.347	5566	266.4	6.99	298.9	305.2	309.9	313.3
	S.E.	4.20	6595	1.367	5699	277.6	7.75	313.5	321.6	327.6	332.0
	F.C.	4.20	6595	1.367	5551	265.5	6.96	297.7	304.0	308.6	311.9
F: Hydrazine	S.E.	2.33	6766	1.457	5982	291.9	7.75	329.4	337.7	343.8	348.3
O: Chlorine	F.C.	2.33	6766	1.457	5829	279.2	7.03	313.3	320.1	325.0	328.6
trifluoride	S.E.	2.85	7054	1.496	5969	293.7	8.20	332.6	341.6	348.4	353.4
	F.C.	2.85	7054	1.496	5793	277.1	6.97	310.8	317.4	322.2	325.7
	S.E.	3.00	7092	1.506	5951	291.9	7.98	330.2	339.3	346.2	351.3
	F.C.	3.00	7092	1.506	5771	276.0	6.95	309.4	316.0	320.7	324.2
F: Monomethyl	S.E.	2.50	6314	1.385	5669	281.6	8.60	319.7	329.4	336.7	342.4
hydrazine	F.C.	2.50	6314	1.385	5465	264.3	7.52	298.1	305.5	311.1	315.3
O: Chlorine	S.E.	2.75	6464	1.407	5684	283.1	8.74	321.7	331.6	339.2	345.1
trifluoride	F.C.	2.75	6464	1.407	5465	264.1	7.47	297.7	305.0	310.5	314.6
	S.E.	3.00	6559	1.426	5684	283.7	8.90	322.8	333.1	341.0	347.1
	F.C.	3.00	6559	1.426	5452	263.1	7.41	296.4	303.6	309.0	313.1
	S.E.	3.25	6599	1.444	5662	280.3	8.50	318.6	329.0	337.3	344.0
	F.C.	3.25	6599	1.444	5422	261.4	7.35	294.3	301.3	306.6	310.5

Table 5-71.   LIQUID ROCKET PROPELLANT THEORETICAL PERFORMANCE (Continued)

Fuel and oxidizer	S.E. or F.C.	Ratio: O/F	Combustion temp, °R	Bulk specific gravity	C*, ft/sec	Specific impulse, lb_f / lb_m sec, sea level	ε,† area ratio, sea level	Vacuum specific impulse			
								ε–20†	ε–40	ε–80	ε–160
F: UDMH	S.E.	2.70	6363	1.335	5566	277.5	8.83	315.7	325.8	333.6	339.7
O: Chlorine	F.C.	2.70	6363	1.335	5382	261.6	7.77	295.7	303.6	309.6	314.2
trifluoride	S.E.	2.85	6444	1.349	5576	278.4	8.91	316.9	327.1	335.1	341.3
	F.C.	2.85	6444	1.349	5384	261.5	7.73	295.5	303.3	309.2	313.8
	S.E.	3.00	6507	1.362	5580	279.0	9.01	317.8	328.2	336.3	342.6
	F.C.	3.00	6507	1.362	5380	261.1	7.69	294.9	302.7	308.5	313.0
F: RP–1	S.E.	4.50	5636	1.273	5239	261.8	9.05	298.5	308.9	317.4	324.5
O: IRFNA	F.C.	4.50	5636	1.273	5144	254.0	8.55	288.8	298.1	305.4	311.2
	S.E.	4.90	5679	1.286	5212	263.3	9.47	300.8	311.9	320.9	328.3
	F.C.	4.90	5679	1.286	5103	252.3	8.62	287.0	296.3	303.7	309.6
	S.E.	5.30	5654	1.297	5157	262.8	10.03	301.1	313.0	322.7	330.7
	F.C.	5.30	5654	1.297	5046	249.7	8.65	284.1	293.4	300.7	306.6
F: N_2H_4	S.E.	1.30	5310	1.231	5642	277.3	8.29	314.6	323.9	331.1	336.7
O: IRFNA	F.C.	1.30	5310	1.231	5583	273.6	8.15	310.1	319.1	325.9	331.2
	S.E.	1.50	5438	1.248	5625	278.9	8.63	317.1	327.0	334.9	340.9
	F.C.	1.50	5438	1.248	5531	271.6	8.25	308.1	317.2	324.3	329.7
	S.E.	1.70	5373	1.263	5505	273.7	8.76	311.4	321.4	329.3	335.4
	F.C.	1.70	5373	1.263	5415	266.0	8.27	301.8	310.7	317.7	323.0
F: Monomethyl	S.E.	2.25	5546	1.224	5513	273.1	8.68	310.8	320.9	329.0	335.5
hydrazine	F.C.	2.25	5546	1.224	5428	266.9	8.33	303.0	312.2	319.3	324.9
O: IRFNA	S.E.	2.50	5620	1.240	5486	274.5	9.03	312.9	323.6	332.1	339.1
	F.C.	2.50	5620	1.240	5377	264.9	8.41	300.9	310.1	317.4	323.1
	S.E.	2.75	5596	1.254	5413	273.7	9.59	312.8	324.4	333.6	341.1
	F.C.	2.75	5596	1.254	5300	261.3	8.45	296.8	306.1	313.3	319.0
F: UDMH	S.E.	2.70	5580	1.198	5453	270.4	8.74	307.8	318.0	326.3	333.0
O: IRFNA	F.C.	2.70	5580	1.198	5367	264.2	8.37	300.0	309.1	316.3	321.9
	S.E.	3.00	5665	1.216	5430	272.1	9.11	310.3	321.1	329.8	336.9
	F.C.	3.00	5665	1.216	5321	262.3	8.46	298.1	307.4	314.7	320.5
	S.E.	3.30	5650	1.231	5364	271.8	9.66	310.8	322.3	331.7	339.3
	F.C.	3.30	5650	1.231	5249	259.0	8.51	294.4	303.7	311.0	316.8
F: .50 Hydrazine	S.E.	2.00	5476	1.210	5530	273.2	8.57	310.7	320.6	328.5	334.8
.50 UDMH	F.C.	2.00	5476	1.210	5454	268.0	8.28	304.1	313.2	320.2	325.7
O: IRFNA	S.E.	2.25	5576	1.228	5510	275.0	8.92	313.2	323.7	332.1	338.8
	F.C.	2.25	5576	1.228	5406	266.1	8.38	302.2	311.4	318.6	324.3
	S.E.	2.50	5553	1.244	5431	274.0	9.47	313.0	324.4	333.4	340.7
	F.C.	2.50	5553	1.244	5321	262.1	8.42	297.8	306.9	314.1	319.8
F: Methane	S.E.	5.50	8334	1.047	7171	359.1	9.06	408.8	421.8	431.6	439.0
O: .826 Fluorine	F.C.	5.50	8334	1.047	6881	329.0	6.95	368.9	376.7	382.3	386.5
.174 Oxygen	S.E.	5.75	8411	1.057	7183	360.2	9.12	410.2	423.4	433.4	440.9
	F.C.	5.75	8411	1.057	6883	328.8	6.91	368.6	376.3	381.8	385.9
	S.E.	6.00	8404	1.067	7139	357.2	8.84	406.3	420.0	431.1	439.7
	F.C.	6.00	8404	1.067	6842	326.7	6.89	366.2	373.8	379.3	383.2
F: Propane	S.E.	4.20	8186	1.112	6984	349.1	8.96	397.2	409.6	419.0	426.2
O: .76 Fluorine	F.C.	4.20	8186	1.112	6718	321.8	7.03	361.1	369.0	374.7	378.9
.24 Oxygen	S.E.	4.55	8377	1.127	7043	352.4	8.98	401.0	413.6	423.1	430.3
	F.C.	4.55	8377	1.127	6758	323.2	6.96	362.4	370.1	375.7	379.8
	S.E.	4.80	8352	1.136	6987	348.9	8.71	396.7	410.2	421.2	429.7
	F.C.	4.80	8352	1.136	6705	320.5	6.94	359.3	366.9	372.4	376.4
F: Diborane	S.E.	3.60	8280	.991	7333	371.6	9.98	426.3	444.3	459.7	473.1
O: Oxygen difluoride	F.C.	3.60	8280	.991	7093	343.5	7.61	387.8	397.9	405.7	411.6
	S.E.	4.00	8512	1.020	7336	372.1	10.01	426.9	444.9	460.3	473.5
	F.C.	4.00	8512	1.020	7083	342.8	7.58	386.9	396.9	404.5	410.4
	S.E.	4.40	8674	1.046	7316	371.3	10.03	426.0	444.0	459.5	472.7
	F.C.	4.40	8674	1.046	7054	341.1	7.54	384.9	394.8	402.3	408.1

Table 5-71. LIQUID ROCKET PROPELLANT THEORETICAL PERFORMANCE (Continued)

MONOPROPELLANT

Monopropellant	S.E. or F.C.	Combustion temp, °R	Bulk specific gravity	C^*, ft/sec	Specific impulse, $\frac{lb_f}{lb_m\ sec}$, sea level	ε,† area ratio, sea level	Vacuum specific impulse			
							ε–20†	ε–40	ε–80	ε–160
.9 Hydrogen peroxide	S.E.	1854	1.352	3084	148.2	7.22	167.3	172.4	176.9	180.9
.1 Water	F.C.	1854	1.352	3084	148.2	7.22	166.7	170.6	173.5	175.8
Hydrazine	S.E.	1631	1.004	4054	198.5	8.39	226.1	233.9	240.6	246.5
	F.C.	1631	1.004	4009	189.2	6.38	211.0	214.9	condensation in nozzle	
Ethylene oxide	S.E.	2300	.883	3893	199.0	10.45	228.8	239.1	248.1	256.0
	F.C.	2300	.883	3825	188.4	8.37	213.8	220.3	225.2	229.1
Nitromethane	S.E.	4439	1.115	5025	245.4	8.10	278.5	287.5	295.4	302.4
	F.C.	4439	1.115	5013	243.5	7.71	275.0	282.2	287.6	291.6
n-Propyl nitrate	S.E.	2432	1.051	4121	209.6	10.23	240.8	251.5	260.8	269.1
	F.C.	2432	1.051	3989	191.2	7.06	214.6	219.3	222.8	condensation in nozzle

†$\varepsilon = A_e/A_t$, the supersonic area ratio, nozzle exit area to throat area.
*The symbol C^* is used for characteristic velocity of propellant gases, ft/sec.

**Contributed by William A. Tomazic, Bonnie J. McBride, and Sanford Gordon, National Aeronautics and Space Administration, Lewis Research Center, Cleveland, Ohio.

REFERENCES

"A General IBM 704 or 7090 Computer Program for Computation of Chemical Equilibrium Compositions, Rocket Performance, and Chapman-Jouguet Detonations", F.J. Zeleznik and S. Gordon, NASA TN D-1454, Oct. 1962.

"A General IBM 704 or 7090 Computer Program for Computation of Chemical Equilibrium Compositions, Rocket Performance, and Chapman-Jouguet Detonations, Supplement 1, Assigned Area Ratio Performance", S. Gordon and F.J. Zeleznik, NASA TN D-1737, Oct. 1963.

Table 5-72. MAJOR U.S. SPACE PROJECTS*

Project	Description
AEROS	A 300-pound probe for studies of upper atmosphere and the solar ultraviolet.
APOLLO	Three-man spacecraft for lunar landing and exploration; lunar staytime up to 66 hours.
ATS	Geostationary satellite to study the technology of the stationary orbit.
BIOSATELLITE	Unmanned scientific biological satellite to test effects of spaceflight on various plant and animal forms, including primates, for periods up to 30 days.
COMSAT	Satellite to provide basis for operational communications satellite system.
DISCOVERER	Satellite test bed for testing space effects on bioastronautic specimens, materials, and equipment.
EARLY BIRD	First version of COMSAT; will provide 240 two-way telephone channels between United States and Europe.
ECHO	Inflated balloon; passive communications satellite.
EXPLORER	Unmanned scientific satellite to explore the atmosphere, ionosphere, magnetosphere, and near-Earth region of space.
GEMINI	Two-man spacecraft for Earth-orbital flight up to two weeks; to test orbital rendezvous.
GEOS	A 500-pound satellite for calibrating satellite altimeters, tracking units, and radars.
HEAO (A,B)	Unmanned 21 000-pound spacecraft for surveys of X-, gamma, and cosmic rays.
HELIOS	Automated spacecraft for studies in interplanetary space close to the sun.
HEOS	Small satellite for studying high-latitude magnetosphere around northern neutral point.
IMP	Unmanned interplanetary satellite to measure magnetic fields, cosmic rays, and solar wind.
LEM	Part of Apollo mission; lands two men on surface of moon, followed by rendezvous with orbiting command module.
LUNAR ORBITER	Unmanned scientific satellite to orbit moon, taking high-resolution pictures.
MACS	Medium-altitude military communications satellite.
MARINER	Unmanned 900-pound craft for explorations of the environments of Venus and Mercury.
MERCURY	First U.S. manned spacecraft; consisted of two suborbital and four orbital flights.
MTS	Meteoroid technology satellite for measuring meteoroid velocities and impacts.
NIMBUS	Weather satellite for studying the atmosphere from earth orbit.
PIONEER	A 500-pound probe for interplanetary studies beyond Mars and near Jupiter.
RAE	Radio astronomy explorer for studies of celestial-source radio signals.
RANGER	Lunar hard-landing spacecraft; TV pictures of lunar surface before impact.
RELAY	Low-altitude active communications satellite.
SAS	Small astronomy satellite for sky surveys of high-energy gamma radiation.
SATAR	Low-cost Atlas pod-mounted scientific satellite.
SECOR	Geodetic satellite to provide reference for ground triangulation.
SKYLAB	Orbital workshop with docking facilities for manned spacecraft, allowing long-duration studies.
SURVEYOR	Lunar soft-landing spacecraft; scientific survey of lunar surface after landing, in preparation for Apollo.
SYNCOM	High-altitude active communications satellite.
TELSTAR	Low-altitude active communications satellite to test broadband microwave space communications.
TIROS	Meteorological satellite.
TRANSIT	Navigational satellite that provides highly accurate position data to ships and aircraft.
VELA SATELLITES	X- and gamma-ray detection satellite system for nuclear explosions in space.
VIKING	Unmanned interplanetary spacecraft to orbit and land a capsule on Mars.

*Data from various NASA sources.

Table 5-73.   COMPARATIVE SPEEDS FOR SPACE TRAVEL*

Condition	Speed		g sec
	Metric	English	
Fall 4.9 m (16.1 ft)	9.8 m/sec	32.2 ft/sec (22 mph)	1
Automobile (nominal speed)	100 km/hr	62.1 mph	2.8
Commercial jet airplane (nominal speed)	1,000 km/hr	621 mph	28
X-15 rocket airplane (nominal speed)	1.61 km/sec	1.00 mps (3,609 mph)	163
Earth orbit (at 200 km)—tangential velocity	7.78 km/sec	4.84 mps	790
Escape the earth—radial velocity	11.2 km/sec	6.95 mps	1,140
Sun orbit (at 150×10^6 km)—tangential velocity	29.8 km/sec	18.5 mps	3,020
Escape the sun (from 150×10^6 km)—radial velocity	42.4 km/sec	26.4 mps	4,320

Note: To attain altitude near the earth requires 143 g sec/100 km or 180 g sec/100 miles. To attain 10^6 km/hr requires 7.85 g hr. To attain 10^6 mph requires 12.6 g hr. To attain one-half the speed of light requires 5.9 g months.

*From: "Human Factors in Technology", E. Bennett, J. Degan, and J. Spiegel, Eds., McGraw-Hill Book Company, 1963.

Table 5-74.   PLANETARY-MISSION VELOCITY REQUIREMENTS*

In Kilometers per Second

For planetary-satellite missions circular orbits in 1,000 km altitude and rocket braking have been assumed. If aerodynamic braking is used instead, 1.5 km/sec for Mars and 2.6 km/sec for Venus can be saved.

Maneuver	No recovery				Plus return				
	Mars satellite	Venus satellite	Mars soft landing	Venus soft landing	Mars satellite	Venus satellite	Mars soft landing	Venus soft landing	Planetoid in Mars orbit, soft landing
Earth launch:									
Ideal minimum launch	11.589	11.441	11.589	11.441	11.589	11.441	11.589	11.441	11.589
Rotational gain	0.3	0.3	0.3	0.3	0.3	0.3	0.3	0.3	0.3
g loss	1.42	1.42	1.42	1.42	1.42	1.42	1.42	1.42	1.42
Drag loss	0.16	0.16	0.16	0.16	0.16	0.16	0.16	0.16	0.16
Maneuvering	0.05	0.05	0.05	0.05	0.05	0.05	0.05	0.05	0.05
Maneuvering transfer	0.4	0.5	0.4	0.5	0.4	0.5	0.4	0.05	0.4
At target arrival:									
Ideal minimum	2.01	3.104			2.01	3.104			2.55
g loss	0.1	0.15			0.1	0.15			
Maneuvering	0.05	0.05	0.01	0.01	0.05	0.05	0.01	0.01	0.05
Target launch:									
Ideal minimum launch					2.01	3.104	5.64	10.5	2.55
Rotational gain							0.2		
g loss					0.05	0.1	0.3	1.5	
Drag loss							0.15	0.2	
Maneuvering					0.05	0.05	0.05	0.05	0.05
Maneuvering transfer					0.04	0.05	0.04	0.05	0.04
Earth-landing maneuver					0.05	0.05	0.05	0.05	0.05
Total	15.50	16.60	13.40	13.35	18.05	20.40	19.80	26.15	18.95

*From: "Handbook of Astronautical Engineering", H.H. Koelle, McGraw-Hill Book Company, 1961.

Table 5-75. DATA ON NASA SPACECRAFT LAUNCH VEHICLES

Vehicle	Stages and engines	Propellant	Thrust, 1,000 lb	Maximum diameter, feet	Height, feet	Payload, lbs, ETR* 135 naut mi	300 naut mi	Escape	Mars/Venus	First NASA launch
Scout	1. Algol II–B 2. Castor II 3. Antares X–259 4. Altair X–258 (4. FW–4S)††	Solid Solid Solid Solid Solid	100.9** 60.7 20.9 5.7 (5.9)	3.3 2.6 2.5 1.7	29.8 20.3 9.5 4.8 } 64.4		300 (320)	—	—	1 July 1960
Delta	1. Thor (DM–21) 2. AJ–10–118A 3. Altair X–258	LOX/RP–1 IRFNA/UDMH Solid	172** 7.5 5.8	8 4.56 1.50	59.50 16.45 5.00 } 31.0		880	150	120	13 May 1960 (6 Nov. 1965; first improved Delta)
TAD (Thrust Augmented Delta)	1. Thor (DM–21) plus three TX–354 "strap-ons" 2. AJ–10–118A 3.-FW–4D	LOX/RP Solid IRFNA/UDMH Solid	170** 162** 7.8 5.9	8 33" each	109		1,300	250	220	13 May 1960
Thor—Agena	1. Thor (DM–21) 2. Agena	LOX/RP IRFNA/UDMH	170** 16	8	76		1,600	—	—	29 Sept. 1962
TAT (Thrust Augmented Thor–Agena)	1. Thor (DSV–3L), three TX–354 "strap-ons" 2. Agena	LOX/RP–1 Solid IRFNA/UDMH	170** 160** 16	8 33" each 5	110		2,200	—	—	14 Oct. 1965
Atlas—Agena	1. Atlas (boost and sust.) 1a. Atlas (sust. only) 2. Agena	LOX/RP–1 LOX/RP–1 IRFNA/UDMH	388** 80 16	10	104	—	6,000	850	550	23 Aug. 1961
Atlas—Centaur	1. Atlas (boost and sust.) 1a. Atlas (sust. only) 2. Centaur (2 RL–10)	LOX/RP–1 LOX/RP–1 LOX/LH$_2$	400** 60 30	10	117	—	9,900	2,700	1,700	8 May 1962
Saturn IB	1. S–IB (Eight H–1) 2. S–IVB (One J–2)	LOX/RP–1 LOX/LH$_2$	1,640 230	21.6	142	40,000	§	—	§	26 Feb. 1966
Saturn V	1. S–IC (Five F–1) 2. S–II (Five J–2) 3. S–IVB (One J–2)	LOX/RP–1 LOX/LH$_2$ LOX/LH$_2$	7,500 1,125 230	33	281†	285,200‡	§	100,000	§	9 Nov. 1967

*ETR—Eastern Test Range.
**At sea level.
§Payload capability dependent on mission and payload configuration.
†365 feet with Apollo spacecraft and launch escape system.
††Alternate vehicle configuration with performance indicated.
‡For Apollo mission only: 285,000 lb payload includes S-IVB and IU weight in orbit.

Table 5-76. TYPICAL ROCKET ENGINE APPLICATIONS*

	Equivalent vacuum vehicle velocity increment, fps	Typical thrust, lb	Thrust application	Typical propellant	Range of delivered specific impulse, sec	Burn time	Significant special features
Space booster stage	4,000–10,000	100,000– several million	Single constant thrust burn	Composite solid-perchlorate; LOX/kerosene; N_2O_4/amine fuel	250–265	1–8 min	Thrust-vector control; tested in cluster
Space-maneuverable upper stage	8,000–20,000	5,000– 200,000	Several starts, 3 to 1 throttling	LOX/H_2; N_2O_4/amine fuel	300–450	1–10 min	Very high nozzle area ratio; thrust-vector control; vacuum environment
Spacecraft attitude control	100–3,000	0.001– 20,000	Thousands of pulses of varying duration	Cold gas; hydrazine; subliming propellant; N_2O_4/N_2H_4	70 (N_2) to 290 (N_2O_4–MMH) sec	100–10,000 sec cumulative	Very short pulses (0.02 sec); high reliability; vacuum environment; no maintenance in orbit
ICBM booster	4,000–10,000	100,000– 500,000	Constant propellant flow; single burn	Composite solid	240–250 (sea level)	$\frac{1}{2}$–2 min	Thrust-vector control; long storage in ready-to-operate condition
ICBM upper stage	8,000–15,000	15,000– 100,000	Constant propellant flow; single burn with precise thrust termination	Composite solid	260–300 (altitude)	$\frac{1}{2}$–2 min	Thrust-vector control; special thrust termination devices; long storage in ready-to-operate condition
ABM upper stage	6,000–14,000	40,000– 150,000	Constant thrust; 1 or 2 burns	Composite solid or storable liquid	260–310 (altitude)	1 min	High side and axial accelerations; operate at zero g
Air-launched guided missile	1,000–6,000	2,000– 20,000	Constant or stepped thrust	Composite solid (NH_4CO_4); prepackaged storable liquid (N_2H_4/amine)	235–250 (sea level)	2–25 sec	Variable ambient temperatures; severe vibration environment; high average density impulse; long storage life
Unguided antitank weapon	500–2,000	1,000– 25,000		Smokeless solid (usually double-base type)	200–240 (sea level)	0.1–0.5 sec	Simplicity, reliability, low cost; wide range of environmental conditions; smokeless exhaust; tube launching
Aircraft super-performance powerplant	1,000	1,500– 10,000	10 to 1 throttling; multiple restart	H_2O_2–JP4	220–225	2–8 min	Simple engine replacement and servicing; long time between overhauls; operation during aircraft maneuvers
Aircraft takeoff assist	500	250–5,000	Constant thrust	Nitrate or perchlorate-type solid	200–245 (sea level)	5–15 sec	Low cost and simplicity; long storage life

*From: "Choosing a Rocket Engine", G.P. Sutton, *Space/Aeronautics*, December 1968. Copyright © 1968 by Ziff-Davis Publishing Co. Reprinted by permission of *Space/Aeronautics* Magazine and Ziff-Davis Publishing Co.

Table 5-77. THE SUN*

The sun, a typical star, is nature's nuclear reactor, converting its mass into energy through the fusion of hydrogen nucleii into helium. The expansive force of fusion is counter-balanced by the massive pressure of hundreds of thousands of miles of super-condensed gas pulled by gravitational force toward the sun's center. While the radiation released by the solar furnace is the ultimate source of much of the earth's available energy, it is also a hazard to space travel. Concentrations of radiation emitted by solar eruptions pose an obstacle to be overcome by man in space. These solar eruptions, which occur near sunspots, also produce ionospheric disturbances, magnetic storms, and auroral displays. Sunspots are regions of violent expansion, or solar storms.

Diameter	864,000 mi	Semi-diameter at mean distance	16' 00"	
Volume	1.4122×10^{33} cm^3	Apparent stellar magnitude	-26.78	
(Earth = 1)	1,304,000	Absolute stellar magnitude	$+4.79$	
Mass	1.989×10^{33} g	Spectral type	G2V	
(Earth = 1)	332,488	Internal constitution;		
Mean density	1.409 g/cm^3	central values		
Surface area (Earth = 1)	11,920	Temperature	24,500,000 °F	
Surface gravity	27,398 cm/sec^2	Density	98 g/cm^2	
(Earth = 1)	27.94	Pressure	2.0×10^{17} dyn/cm^2	
Escape velocity	385 mi/sec	Solar constant (radiation)	2.00 cal/cm^2/min	
Mean equatorial horizontal		Inclination of solar equator to		
parallax	8".79415	ecliptic	7° 15'	
Mean distance from Earth	92,956,000 mi	Mean sunspot period	11.04 yr	
Perihelion distance	91,600,000 mi	Temperatures	°K	°F
Aphelion distance	94,750,000 mi	Effective	5800	9980
Radiation emitted	3.90×10^{33} erg/sec	Photosphere	6110	10500
Radiation emitted at surface	6.41×10^{10}	Chromospheric kinetic	8000	13900
	erg/cm^2/sec	Outer corona		
Moment of inertia	6.0×10^{53} g	(millions of degrees)	0.9–2.4	1.62–4.32
Angular momentum	1.7×10^{48}	Large central sunspots,		
	g/cm^2/sec	umbra	4400	7430
Rotational energy	2.5×10^{42} erg	Angular velocity (16° latitude)	2.865×10^{-6} rad/sec	
Polar magnetic field		Mean synodic period	27.2753 days	
(at spot minimum)	1 or 2 gauss	Axial rotation period	25.38 days	
Polar magnetic flux				
(at spot minimum)	8×10^{21} maxwell			

*Reprinted by permission of the General Electric Company © 1971.

Table 5-78. THE MOON*

Mean diameter	2,160 mi	Period of moon's nodes	18.6 yr
Volume	2.20×10^{25} cm^3	Earth to lunar mass ratio	81.303?
Mass	7.35×10^{25} g	Mean photographic magnitude of	
Mean density	3.34 g/cm^3	full moon	-11^m91
Surface area	3.80×10^{17} cm^2	Mean visual magnitude of full	
Surface gravity	162.2 cm/sec^2	moon, 5800 A	-12^m74
Escape velocity (full moon)	1.37 mi/sec	Mean brightness of full moon	0.247 candles/cm^2
Escape velocity (new moon)	1.52 mi/sec	Visible spherical albedo	0.073
Mean earth equatorial		Libration in longitude	$\pm 7° 53' 51"$
horizontal parallax	57' 1".97	Libration in latitude	$\pm 6° 50' 45"$
Mean lunar equatorial		Diurnal libration (parallactic)	$\pm 57'$
horizontal parallax	15' 32".63	Physical libration	$\pm 0.04°$ lat.,
Inclination of lunar orbital			$\pm 0.02°$ long.
plane to ecliptic	5.1453964 degrees	Angular velocity earth-moon	
Orbital eccentricity (1900.0)	0.05900489	system	2.661699×10^{-6}
Inclination of lunar equator to			rad/sec
the ecliptic	1° 32'	Mean orbital speed (1900.0)	3370.3 ft/sec
Orbit semi-major axis	239,578 mi	Probable dust size	5–60 microns
Mean distance from Earth	238,857 mi	Temperature variation	100 to 389°K
Synodic month (1900.0), period			or -279 to
between full moons	29.530588 days		$+243$°F
Sidereal month (1900.0), true		Atmosphere	$<10^{-13}$ earth
period of revolution	27.3216610 days		atmospheres
Nodal or draconic month		Magnetic field	$<10^{-2}$ gauss
(1900.0), period from one		Craters: total number visible	>300,000
node back to the same	27.21220 days	Highest mountains:	
		Leibnitz Mountains	29,000 ft
		Doerfel Mountains	18,325 ft

*Reprinted by permission of the General Electric Company © 1971.

Table 5-79.   THE PLANETS: PHYSICAL DATA†

Our solar system includes an assemblage of 9 planets, 31 moons, thousands of asteroids, scores of comets, and millions of meteors. The planets are classified into two groups: inner planets (Mercury, Venus, Earth, Mars) and outer planets (Jupiter, Saturn, Uranus, Neptune, Pluto).

In the inner group Venus most closely approximates the size of Earth. The apparent deposits of frost that appear at the poles during the Martian winter indicate that Mars' atmosphere contains water vapor. Speculation concerning possible life on Mars has been stimulated by the seasonal color changes in the equatorial regions and the linear markings, called canals, on the surface of the planet.

The four largest planets (Jupiter, Saturn, Uranus, and Neptune) have thick atmospheres, which probably consist mostly of methane and ammonia, lying on top of thick layers of hydrogen. Pluto, which is approximately 40 times farther from the sun than Earth, yet less than half Earth's size, may not be a true planet, but an escaped satellite of Neptune. Pluto does not mark the outer boundary of our solar system. Some comets spend most of their lifetimes at great distances beyond Pluto. There is also some evidence indicating the existence of one or more planets outside the bounds of the nine discovered to date.

Property	Mercury	Venus	Earth	Mars	Jupiter	Saturn	Uranus	Neptune	Pluto
Mean diameter, mi	3010	7650	7926	4240	86,900	71,600	29,400	26,800	3560
Mean diameter, km	4842	12,322	12,742	6664	139,785	115,064	47,402	43,070	5734
Mean diameter, Earth = 1	0.38	0.967	1.000	0.532	10.97	9.03	3.72	3.38	0.45
Volume, cm³	5.95×10^{25}	9.62×10^{26}	1.083×10^{27}	1.624×10^{26}	1.425×10^{30}	8.33×10^{29}	5.41×10^{28}	4.55×10^{28}	1.08×10^{27}
Surface area, cm²	7.38×10^{17}	4.76×10^{18}	5.101×10^{18}	1.46×10^{18}	6.15×10^{20}	4.18×10^{20}	7.04×10^{19}	5.84×10^{19}	1.03×10^{18}
Mass, g (excluding satellites)	3.225×10^{26}	4.87×10^{27}	5.977×10^{27}	6.45×10^{26}	1.898×10^{30}	5.69×10^{29}	8.67×10^{28}	1.029×10^{29}	4.78×10^{27}?
Mass, Earth = 1	0.0543	0.8136	1.0000	0.1069	318.35	95.3	14.58	17.26	<0.1?
Mean density, g/cm³(H₂O = 1)	5.46	5.1	5.517	3.97	1.334	0.684	1.60	2.25	<5.5?
Oblateness, $(r_e - r_p)/r_e$	0.0	0.0	0.0034	0.0052	0.062	0.096	0.06	0.02	?
Escape velocity, ft/sec	13,800	33,800	36,700	16,400	200,000	121,000	72,200	82,000	<17,300
Escape velocity, km/sec	4.3	10.4	11.6	5.1	61.0	36.7	22.4	25.6	<5.3
Surface gravity, ft/sec²	11.8	28.5	32.2284	12.3	85.3	36.8	30.8	49.2	<0.5?
Surface gravity, Earth = 1	0.38	0.87	1.00	0.39	2.65	1.17	1.05	1.23	<0.5?
Mean dist. from sun, million mi	36.0	67.2	93.0	141.5	484	887	1785	2797	3670
Mean dist. from sun, A.U.	0.387099	0.723332	1.000000	1.52369	5.2028	9.540	19.18	30.07	39.44
Closest dist. to sun, million mi	28.6	66.8	91.4	128.4	460	837	1700	2775	2761
Closest dist. to sun, A.U.	0.3078	0.718	0.984	1.38	4.95	9.00	18.26	29.8	29.7
Max. dist. to sun, million mi	43.4	67.7	94.6	154.9	507	936.5	1860	2820	4610
Max. dist. to sun, A.U.	0.467	0.728	1.018	1.665	5.46	10.06	20.00	30.3	49.5
Mean orbital velocity, ft/sec	157,100	115,000	97,600	79,200	42,800	31,650	22,300	17,810	15,550
Mean orbital velocity, mi/hr	107,000	78,400	66,600	53,900	29,200	21,580	15,200	12,140	10,600
Sidereal period, days	87.9686	224.700	365.257	686.980	4332.587	10759.20	30,685	60,188	90,700

Table 5-79. THE PLANETS: PHYSICAL DATA *Continued*

Property	Mercury	Venus	Earth	Mars	Jupiter	Saturn	Uranus	Neptune	Pluto
Sidereal period, tropical years	0.24085	0.61521	1.00004	1.88089	11.86223	29.45772	84.013	164.79	248.4
Synodic period, days	115.88	583.92	—	779.94	398.88	378.09	369.66	367.49	366.74
Sidereal mean daily motion, sec	14732.420	5767.67	3548.193	1886.519	299.128	120.456	42.234	21.532	14.283
Orbital eccentricity	0.205615	0.006820	0.016750	0.093312	0.048332	0.055890	0.0471	0.0085	0.2494
Orbital inclination to ecliptic, deg/min/sec	7/00/10.6 + 6.3T*	3/23/37.1 + 3.6T*	—	1/51/01.1 − 2.3T*	1/18/31.4 − 20.5T*	2/29/33.1 − 14T*	0/46/21 + 3T*	1/46/45 − 34T*	17/10/
Period of axial rotation, days/hrs/min/sec	58/15/30/?	242.982	0/23/56/4.1	0/24/37/22.7	0/9/50/30	0/10/14/24	0/10/49/?	0/15/48/?	6/9/?/?
Inclination of equator to orbit plane, deg/min	7?/?	174	23/27	23/59	3/05	26/44	97/55	28/48	?
Maximum stellar magnitude	−1.9	−4.4	−3.8**	−2.8	−2.5	−0.4	+5.7	+7.85	+14.87
Visual albedo	0.059	0.85	0.40	0.15	0.73	0.76	0.93	0.84	0.15
Color index	+0.93	+0.79	+0.2	+1.41	+0.6	+0.9	+0.55	+0.43	+0.80
Lowest visible surface	Solid	Cloud	Solid	Solid	Cloud	Cloud	Cloud	Cloud	?
Max. visible surface temp., °K	611	307 (600 surface)	295	260	145	25	88	80	80–62
Number of known satellites	0	0	1	2	12	10	5	2	0

*T = number of Julian centuries of 36,525 days, at the midnight beginning the day, that have elapsed since mean noon on 1900 January 0 at the Greenwich meridian.

**As seen from the sun.

†Reprinted by permission of the General Electric Company © 1971.

REFERENCE

The NASA series Handbooks of the Physical Properties of the Planets, including "Mars", SP-3030, 1967: "Jupiter", SP-3031, 1967; and "Venus", SP-3039, 1967.

Table 5-80. THE TWENTY BRIGHTEST STARS*

More than half of the stars are known multiple systems. Stars known to have companions are indicated by the symbol †. For distance in meters, multiply values in light years by 9.4606×10^{15}.

Star	Visual magnitude (appar.)	Distance, light years	Diameter (Sun = 1)	Luminosity (Sun = 1)
Sirius†	−1.43	8.7	2	23
Canopus	−0.73	100	30	1,400
Alpha Centauri†	−0.27 (comb.)	4.3	1	2 (comb.)
Arcturus	−0.06	38	24	110
Vega	+0.04	26.5	3.2	55
Capella†	+0.09 (comb.)	50	16	145 (comb.)
Rigel†	+0.15	900	40	57,000
Procyon†	+0.37	11.3	2	7
Achernar	+0.53	120	8	700
Beta Centauri†	+0.66	490	10	10,000
Betelgeuse	+0.4 (var.)	520	550–920	14,000 (var.)
Altair	+0.80	16.5	1.5	11
Aldebaran†	+0.85	55	36	160
Alpha Crucis†	+0.87 (comb.)	370	8	4,900 (comb.)
Antares†	+0.98	520	220	9,000
Spica†	+1.00 (comb.)	220	7	1,700 (comb.)
Fomalhaut	+1.16	23	2	11
Pollux	+1.16	33	16	33
Deneb	+1.26	1,500	22	57,000
Beta Crucis	+1.31	490	8	5,800

*Reprinted by permission of the General Electric Company © 1971.

Table 5-81.   COMPOSITION OF APOLLO 12 LUNAR SAMPLES*

Average of analyses of nine samples of crystalline rocks and analyses of five individual samples of breccias and fines.

Component	Average, crystalline rocks	Sample 12070[a]	Sample 12073[b]	Sample 12010[c]	Sample 12033[d]	Sample 12013[e]
ELEMENTS						
Silicon (Si), percent	18.5	19.6	19.1	20	19.2	28.5
Iron (Fe), percent	16.6	13.2	13.0	15.2	12.4	7.8
Calcium (Ca), percent	7.6	7.1	8.2	7.0	8.2	4.5
Aluminum (Al), percent	5.9	7.4	7.9	6.1	8.5	6.3
Magnesium (Mg), percent	7.2	7.2	6.6	6.6	6.5	3.6
Sodium (Na), percent	.33	.30	.29	.39	.40	.51
Titanium (Ti), percent	2.3	1.9	1.9	2.2	1.6	.72
Barium (Ba), ppm	72	420	510	180	720	2 150
Chromium (Cr), ppm	3 750	2 800	2 800	3 050	2 100	1 050
Cobalt (Co), ppm	40	42	30	39	34	13
Lithium (Li), ppm	5.5	11	25	7	15	100
Manganese (Mn), ppm	2 050	1 900	1 500	1 400	1 800	950
Nickel (Ni), ppm	54	200	350	80	140	105
Potassium (K), ppm	540	1 500	2 100	1 300	3 240	1.66[f]

*From: "Apollo 12 Preliminary Science Report", NASA SP-235, National Aeronautics and Space Administration.

Table 5-81. COMPOSITION OF APOLLO 12 LUNAR SAMPLES
(Continued)

Component	Average, crystal-line rocks	Sample 12070[a]	Sample 12073[b]	Sample 12010[c]	Sample 12033[d]	Sample 12013[e]
ELEMENTS (Continued)						
Rubidium (Rb), ppm	0.64	3.2	4.9	2.0	7.5	33
Scandium (Sc), ppm	50	47	42	50	33	21
Strontium (Sr), ppm	145	170	230	145	260	150
Vanadium (V), ppm	88	64	50	92	37	13
Ytterbium (Yb), ppm	—	—	—	—	12	20
Yttrium (Y), ppm	51	130	180	87	260	240
Zirconium (Zr), ppm	170	670	1 200	380	950	2 200
OXIDES						
Silicon dioxide (SiO_2), wt %	40	42	41	43	41	61
Iron oxide (FeO), wt %	21.3	17	17	19.5	16	10
Aluminum oxide (Al_2O_3), wt %	11.2	14	15	11.5	16	12
Calcium oxide (CaO), wt %	10.7	10	11.5	10	11.5	6.3
Magnesium oxide (MgO), wt %	11.7	12	11	11	11	6.0
Titanium dioxide (TiO_2), wt %	3.7	3.1	3.1	3.7	2.6	1.2
Chromium oxide (Cr_2O_3), wt %	.55	.41	.41	.43	.31	.13
Sodium oxide (Na_2O), wt %	.45	.40	.30	.33	.34	.69
Manganese oxide (MnO), wt %	.26	.25	.19	.18	.23	.12
Potassium oxide (K_2O), wt %	.065	.18	.25	.16	.39	2.0
Zirconium dioxide (ZrO_2), wt %	.023	.09	.16	.05	.13	.30
Nickel monoxide (NiO), wt %	—	.025	.044	—	.018	.013

Note: The Apollo 12 samples may be contrasted with those from Apollo 11 as follows:

1. While still old by terrestrial standards, the Apollo 12 rocks are approximately 1 billion years younger than those from Apollo 11.

2. Whereas the Apollo 11 collection contained approximately half microbreccias, there are only two breccias out of 45 rocks in the Apollo 12 return.

3. The amount of solar-wind material in the Apollo 12 fines is considerably lower than that in the Apollo 11 fines.

4. Chemically, the "nonearthly" character of the Apollo 11 samples (high refractory element concentration and low volatile element concentration) is also noted in the Apollo 12 samples, but to a lesser degree.

5. The chemical composition of the Apollo 12 fine material equals that of the breccias, but does not equal that of the crystalline rocks; this was not observed in the Apollo 11 collection.

[a] Fine material
[b] Breccia.
[c] Breccia (?).
[d] Light-colored fines; sample 12033 also contains 44 ppm Nb.
[e] Sample 12013 also contains 30 ppm Pb, 15 ppm B, and 170 ppm Nb.
[f] In percent.

REFERENCE

For extensive data on properties of lunar materials, see "Proceedings of the Second Lunar Science Conference", (January 1971), 3 volumes, MIT Press, 1972.

Table 5-82.   PERFORMANCE OF REFRIGERATION SYSTEMS
IDEAL PERFORMANCE

The following table gives the approximate ideal performance of typical systems for pumping heat from a lower to a higher temperature level. This performance is expressed in terms of the ratio of refrigerating effect or refrigeration load to the energy input (in identical units) and is called the *coefficient of performance*. Similar ratios in mixed units are often used, i.e., tons per kilowatt or Btu/hr of refrigeration per horsepower. Although no actual temperatures are indicated in the table, it is assumed that the heat-rejection (condenser) temperature is near that of the atmosphere.

The *figure of merit* for a thermoelectric material depends on its electrical and thermal properties. A high Seebeck coefficient, high thermal conductivity, and low electrical resistivity are desired. The figure of merit for most thermoelectric materials is well below $Z = 0.05$.

IDEAL OR THEORETICAL PERFORMANCE

Temperature difference, evaporator to condenser	Ideal coefficient of performance		
	Ideal Carnot cycle	Ideal compression machine, refrigerant 12 [†]	Thermoelectric system, figure of merit $Z = 0.10$
10	53		6.1
20	26		2.8
40	12.5	11.1 (0.425)‡	1.0
60	8.0	6.8 (0.694)	
80	5.7	4.7 (1.00)	
100	4.4	3.5 (1.34)	
120	3.5	2.7 (1.74)	

ACTUAL PERFORMANCE

Although the performances shown in the above table cannot be attained, the table serves to emphasize the great reduction in economy as the temperature difference between the heat-absorption level and the heat-rejection level is increased. The properties of the working fluids, or the thermoelectric materials, also affect the economy. Tables 5-83 and 5-84 illustrate that the theoretical horsepower necessary to drive a compression machine may vary as much as two to one by a change in the refrigerant.

In estimating the power required to operate a refrigeration compressor, the value of one horsepower per ton is often used. This represents a coefficient of performance of about 4.7. Aside from the operating temperatures, the main factors tending to lower the coefficient of performance of a compression system are the temperature differences required for heat transfer, pressure drops in the piping, heat losses, and mechanical friction. None of these are taken into account in the above table.

The actual power to drive a refrigeration compressor is obtained by dividing the horsepower/ton by the product of the compression efficiency and the mechanical efficiency of the compressor. To obtain the electrical input for a motor-driven compressor, the motor and drive efficiencies must be taken into account. For rough estimates the values in parentheses in the above table (hp/ton) may be multiplied by 1.5 to give numerical values of either horsepower to drive the compressor or kilowatts input to the driving motor. For large high-efficiency units this factor (1.5) is high, while for small, inexpensive units it is low.

---

†Isentropic compression; see Tables 5-83, 5-84, and 5-85.
‡Numbers in parentheses are horsepower per ton. For kilowatts per ton, multiply by 0.746.

Table 5-83. COMPARATIVE REFRIGERANT PERFORMANCE PER TON*

Based on 5°F Evaporation and 86°F Condensation; Ideal Isentropic Cycle

Saturation suction vapor was assumed except for Refrigerants 113, 114, and 115. In these cases enough suction superheat was assumed to give saturated discharge vapor.

No.	Name	Evaporator pressure, psig	Condensing pressure, psig	Compression ratio	Net refrigerating effect, Btu/lb	Refrigerant circulated, lb/min	Liquid circulated, cu in./min	Specific volume of suction gas, cu ft/lb	Compressor displacement, cfm	Horsepower, hp	Coefficient of performance	Comp. discharge temp., °F
170	Ethane	221.3	661.1	2.86	58.6	3.41	342.9	0.53	1.82	1.953	2.41	122
744A	Nitrous oxide	294.3	922.3	3.03	85.2	2.35	71.2	0.28	0.66	1.310	3.60	
744	Carbon dioxide	317.5	1031.0	3.15	55.5	3.62	167.1	0.27	0.96	1.840	2.56	151
13B1	Bromotrifluoromethane	63.2	247.1	3.36	29.3	6.86	123.8	0.38	2.63	1.030	4.25	124
1270	Propylene	37.0	167.0	3.51	173.0	1.1	61.5	2.61	3.03	1.046	4.51	108
290	Propane	27.2	140.5	3.70	121.0	1.65	94.0	2.48	4.09	1.030	4.58	97
502	22/115 azeotrope	36.0	175.1	3.75	45.7	4.38	99.4	0.82	3.61	1.079	4.37	99
22	Chlorodifluoromethane	28.2	158.2	4.03	70.0	2.86	67.4	1.24	3.55	1.011	4.66	128
115	Chloropentafluoroethane	24.0	135.8	3.89	29.1	6.88	151	0.77	5.30	1.17	4.02	86
717	Ammonia	19.6	154.5	4.94	474.4	0.422	19.6	8.15	3.44	0.989	4.76	210
500	12/152a azeotrope	16.4	112.9	4.12	60.6	3.30	80.3	1.50	4.95	1.01	4.65	105
12	Dichlorodifluoromethane	11.8	93.3	4.08	50.0	4.00	85.6	1.46	5.83	1.002	4.70	101
40	Methyl chloride	6.5	80.0	4.48	150.2	1.33	40.9	4.47	5.95	0.962	4.90	172
600a	Isobutane	3.3†	44.8	4.54	111.5	1.79	91.0	6.41	11.50	1.083	4.36	80
764	Sulfur dioxide	5.9†	51.8	5.63	141.4	1.41	26.6	6.42	9.09	0.968	4.87	191
630	Methylamine	9.9†	46.8	6.13	304.0	0.66	28.2	15.54	10.23	0.978	4.81	
600	Butane	13.2†	26.9	5.07	128.6	1.56	75.9	9.98	15.52	0.953	4.95	88
114	Dichlorotetrafluoroethane	16.1†	22.0	5.42	43.1	4.64	89.2	4.34	20.14	1.049	4.49	86
21	Dichlorofluoromethane	19.2†	16.5	5.96	89.4	2.24	45.7	9.13	20.43	0.941	5.01	142
160	Ethyl chloride	20.5†	12.4	5.83	142.3	1.45	45.8	17.06	24.82	0.906	5.21	106
631	Ethylamine	23.1†	10.0	7.40	225.5	0.89	349.0	32.32	38.67	0.855	5.52	
11	Trichlorofluoromethane	23.9†	3.5	6.19	66.8	2.99	56.6	12.21	36.54	0.938	5.03	111
611	Methyl formate	26.3†	1.6†	7.74	189.2	1.06	29.9	48.25	51.00			
610	Ethyl ether	26.9†	4.9†	8.20	126.3	1.58	62.9	35.00	55.40	0.822	5.74	
30	Methylene chloride	27.6†	9.5†	8.60	134.6	1.49	30.9	49.90	74.30	0.963	4.90	205
113	Trichlorotrifluoroethane	27.9†	13.9†	8.02	53.7	3.73	66.5	27.38	102.03	0.973	4.84	
1130	Dichloroethylene	28.3†	15.8†	8.42	114.3	1.75	38.3	63.60	111.20	0.973	4.83	86
1120	Trichloroethylene	29.6†	26.2†	11.65	91.7	2.18	41.6	229.40	502.00	0.980	4.82	

†Inches of mercury vacuum.

*From: "ASHRAE Handbook of Fundamentals", American Society of Heating, Refrigerating and Air-Conditioning Engineers. 1972.

Table 5-84.   EFFECT OF TEMPERATURE ON REFRIGERATION*

The following tables give refrigerating capacity and theoretical horsepower per ton with constant compressor displacement.

Table A.   CAPACITY IN Btu PER HOUR

Refrigerants 12 and 22

Evaporating temperature, °F	Refrigerating capacity for condensing temperatures			
	60°F	80°F	100°F	120°F
REFRIGERANT 12				
−20	5,240	4,790	4,320	3,841
0	8,280	7,590	6,890	6,160
20	12,590	11,580	10,550	9,470
40	18,510	17,080	15,620	14,090
REFRIGERANT 22				
−20	8,672	7,987	7,279	6,536
0	13,518	12,479	11,402	10,277
20	20,303	18,782	17,203	15,553
40	29,553	27,386	25,139	22,789

Table B.   ISENTROPIC HORSEPOWER PER TON OF REFRIGERATION

Evaporating temperature, °F	Theoretical horsepower for condensing temperatures			
	60°F	80°F	100°F	120°F
REFRIGERANT 12				
−20	1.03	1.36	1.76	2.24
0	0.70	0.99	1.33	1.73
20	0.43	0.68	0.97	1.31
40	0.20	0.42	0.67	0.95
REFRIGERANT 22				
−20	1.03	1.37	1.77	2.24
0	0.71	1.01	1.34	1.74
20	0.44	0.69	0.98	1.32
40	0.20	0.42	0.67	0.97

*From: "ASHRAE Handbook of Fundamentals", American Society of Heating, Refrigerating and Air-Conditioning Engineers, 1972.

Table 5-85.   REFRIGERANT PERFORMANCE AT VARIOUS TEMPERATURES*

Following is the calculated theoretical performance per ton of refrigeration based on isentropic compression and saturated vapor at suction.

No.	Name	Evaporating pressure, psig	Condensing pressure, psig	Compression ratio	Net refrigerating effect, Btu/lb	Refrigerant circulated, lb/min	Specific volume of vapor cu.ft/lb	Compressor displacement, cfm	Horsepower
		−130°F EVAPORATING, −40°F CONDENSING							
13	Monochlorotrifluoromethane	11.43†	72.7	9.6	45.8	4.366	3.605	15.74	1.68
170	Ethane	2.24†	93.2	8.3	155.1	1.29	8.16	10.52	1.6
1150	Ethylene	16.16	195.7	6.8	141.4	1.42	3.72	5.27	1.13
		−100°F EVAPORATING, −30°F CONDENSING							
13	Monochlorotrifluoromethane	7.6	93.9	4.8	46.4	4.31	1.56	6.74	1.13
22	Monochlorodifluoromethane	25.04†	4.9	8.2	90.8	2.20	18.43	40.59	1.07
170	Ethane	16.6	123.2	4.3	153.8	1.30	3.87	5.03	1.13
		−76°F EVAPORATING, 5°F CONDENSING							
12	Dichlorodifluoromethane	23.24†	11.8	8.1	59.3	3.37	10.22	34.46	1.17
13	Chlorotrifluoromethane	26.16	177.1	4.7	39.6	5.05	0.89	4.50	1.34
22	Chlorodifluoromethane	18.85†	23.2	7.9	84.4	2.37	8.60	20.38	1.18
170	Ethane	40.40	221.3	4.3	131.1	1.53	2.30	3.51	1.3
290	Propane	17.40†	27.2	6.8	146.4	1.37	14.80	20.22	1.11
717	Ammonia	23.40†	19.6	10.7	533.5	0.37	75.30	28.24	1.24
		−40°F EVAPORATING, 68°F CONDENSING							
12	Dichlorodifluoromethane	11.0†	67.6	8.8	49.3	4.06	3.88	15.71	1.56
22	Chlorodifluoromethane	0.5	117.3	8.7	70.7	2.83	3.30	9.32	1.57
290	Propane	1.5	109.0	7.6	119.7	1.67	6.13	10.24	
717	Ammonia	8.7†	109.6	11.9	279.3	0.42	24.86	10.38	1.60
744	Carbon dioxide	131.2	815.9	5.7	75.9	2.64	0.61	1.61	
		−10°F EVAPORATING, 100°F CONDENSING							
11	Trichlorofluoromethane	26.0†	8.8	12.2	62.1	3.22	18.08	58.26	1.40
12	Dichlorodifluoromethane	4.5	117.2	6.9	45.1	4.43	1.97	8.74	1.53
21	Dichlorofluoromethane	22.9†	25.3	11.6	84.0	2.38	13.43	31.96	1.40
22	Chlorodifluoromethane	16.5	195.9	6.8	64.2	3.12	1.68	5.24	1.54
717	Ammonia	9.0	197.2	8.9	553.3	0.44	11.50	5.07	1.49
		20°F EVAPORATING, 80°F CONDENSING							
12	Dichlorodifluoromethane	21.0	84.2	2.8	53.0	3.77	1.10	4.15	0.68
22	Chlorodifluoromethane	43.0	143.6	2.7	73.3	2.73	0.94	2.56	0.69
40	Methyl chloride	14.5	71.6	3.0	154.5	1.29	3.31	4.29	0.66

Table 5-85. REFRIGERANT PERFORMANCE AT VARIOUS TEMPERATURES (Continued)

No.	Name	Evaporating pressure, psig	Condensing pressure, psig	Compression ratio	Net refrigerating effect, Btu/lb	Refrigerant circulated, lb/min	Specific volume of vapor, cu ft/lb	Compressor displacement, cfm	Horsepower
		20°F EVAPORATING, 80°F CONDENSING (Cont.)							
290	Propane	40.8	128.1	2.6	129.1	1.55	1.90	2.94	
600	Butane	6.3†	22.9	3.2	136.6	1.46	7.23	10.59	
717	Ammonia	33.5	138.3	3.2	485.8	0.41	5.91	2.43	0.67
764	Sulfur dioxide	2.5	45.0	3.5	144.5	1.38	4.49	6.21	
		40°F EVAPORATING, 100°F CONDENSING							
11	Trichlorofluoromethane	15.6†	8.8	3.3	68.1	2.94	5.43	15.95	0.62
12	Dichlorodifluoromethane	37.0	117.2	2.6	50.3	3.97	0.77	3.07	0.66
21	Dichlorofluoromethane	4.8†	25.3	3.2	90.0	2.22	4.13	9.17	0.63
22	Chlorodifluoromethane	68.5	195.9	2.5	68.9	2.90	0.66	1.91	0.68
40	Methyl chloride	28.6	102.0	2.7	149.1	1.34	2.29	3.07	0.63
290	Propane	63.3	172.0	2.4	121.5	1.65	1.37	2.25	
600	Butane	3.0	37.5	2.9	131.8	1.52	4.77	7.24	
718	Water	29.7†	28.0†	7.8	1010.7	0.20	2445.0	483.80	
764	Sulfur dioxide	12.4	69.8	3.1	135.5	1.48	2.89	4.26	

†Inches of mercury vacuum.

*From: "ASHRAE Handbook of Fundamentals", American Society of Heating, Refrigerating and Air-Conditioning Engineers, 1972.

Table 5-86. REFRIGERATED STORAGE

The following table gives temperatures commonly used for storage spaces for various perishable commodities. Other factors besides temperature must be taken into account in refrigerated storage, such as method of packing and arrangement, humidity, ventilation, moisture and frost protection, turnover, and method of chilling. For discussions of these and other factors, including refrigerated transport, see the latest edition of "ASHRAE Guide and Data Book: Applications".

Many perishables, such as prepared foods, keep best at freezer temperatures. Zero °F is a common temperature, but most foods are not damaged by lower temperatures.

STORAGE TEMPERATURES, °F

RETAIL DISPLAY CASES (Refrigerated air temperature)

Produce	35–45
Dairy products	35–42
Meat, unwrapped	34–38
Meat, packaged	28–36
Frozen foods	−20 to +10

COLD STORAGE OF FOODS

Melons, cucumbers, squash, pumpkin, eggplant, peppers, okra, olives, green beans, potatoes, grapefruit, lemons, limes, dried or whole milk	45–50
Butter, cheese, margarine, bacon, sausage, cranberries, beer	35–40
Root vegetables, greens, lettuce, celery, cauliflower, corn, apples, cherries, peaches, apricots, pears, plums, grapes, dried fruits, berries, candy, chocolate	32

GREENHOUSE AND NURSERY STOCK

Cut flowers:	
Lilies and orchids	40–50
Others	31–35
Decorative greens	32
Bulbs:	
Lily, tulip	31–33
Most others (except iris)	40–45
Iris	80–85
Nursery stock	32–35

CLOTHING AND FURS

To inactivate insects	34–40
To kill moths	0–5

Table 5-87. LOW-TEMPERATURE COOLING BATHS

Cryogenic and Refrigerating Fluids—Atmospheric Pressure

Liquid	Boiling point			Liquid	Boiling point		
	K	deg C	deg F		K	deg C	deg F
Helium	4.2	−268.95	−452.1	Refrigerant 14 (CF_4)	145.	−128.	−198.
Hydrogen	20.4	−252.7	−422.8	Refrigerant 13 ($CClF_3$)	192.	−81.	−114.
Neon	27.1	−246.	−410.8	Carbon dioxide (CO)[a]	195.	−78.	−108.5
Nitrogen	77.4	−195.8	−320.4	Propylene (C_3H_6)	225.	−48	−54.
Air	79.	−194.	−317.	Refrigerant 502 (Azeotrope)	227.	−46.	−50.
Argon	87.3	−185.9	−302.6	Propane (C_3H_8)	225.	−48.	−54.
Oxygen	90.2	−183.0	−297.5	Refrigerant 22 ($CHClF_2$)	232.	−41	−41.
Methane (CH_4)	111.	−162.	−259.	Refrigerant 12 (CCl_2F_2)	243.	−30	−22.

Notes:

Low-temperature baths are conveniently prepared by adding dry ice or liquid nitrogen to acetone, ether, chloroform, or one of the alcohols, stirring the mixture until a slush is formed. Bath temperatures will then be fixed by the freezing point and will range from about −82 deg F for chloroform to −197 deg F for propyl alcohol (see Table 1-45). The lowest bath temperature with dry ice is −108 deg F.

Minimum temperatures attainable (with difficulty) using water-ice and salt mixtures are as follows: NaCl (23.3%), −6 deg F; and $CaCl_2$ (30%), −67 deg F. See Table 3-33 for other mixtures and proportions.

Warning:

Adequate safety precautions are necessary when using low-temperature baths. In addition to skin "burns", toxic and explosion hazards may be present. In the initial cooling of a bath from room temperature, using dry ice or liquid nitrogen, the evolution of gas may be so rapid as to approach an explosion. Adequate venting and room ventilation are necessary. Liquid nitrogen absorbs and condenses oxygen (shown by bluish color) and may also produce "liquid air" in a heat-transfer device, such as a vacuum cold trap. Later re-evaporation of the air calls for adequate venting, or explosive forces are produced. Bath liquids and vapors may be toxic or irritant to skin or eyes (e.g., acetone or ketone) and contact should be avoided.

[a]Solid; sublimes.

Table 5-88. APPLICATIONS OF CRYOGENICS

Cryogenic applications are based largely on the use of liquefied "permanent" gases. Following the long period of development of gas-liquefaction systems (1875–1935) and the improvement of cryogenic insulation, the uses of the liquefied gases have expanded rapidly. Included in the partial list of applications given below are several uses of liquefied gases that are as yet in the early stages of development. The future rate or extent of each of these developments is not predictable, but the prospects are sufficient to warrant their inclusion.

Table A. APPLICATIONS AT CRYOGENIC TEMPERATURES

Type of use	Brief description	Gases	Minimum temperature K	Minimum temperature deg F
Space simulation	Pressures of 10^{-10} to 10^{-14} torr by cryopumping. Plasma engines	He	5	−451
High-capacity electromagnets	High-intensity electromagnets with low-power consumption at cryogenic temperatures; compact, high-strength electromagnets with superconducting coils of special alloys (e.g., niobium–tin)	He	5	−451
Superconducting power cables	Long-distance and high-capacity underground power transmission with negligible I^2R losses (or superconductors)	He	5	−451
Cryotrons	Thin-film (or wire-wound) superconducting switching elements used as memory, logic element, etc. in computers	He	5	−451
Frictionless bearings	Superconductors at cryogenic temperatures support bearing loads by magnetic repulsion	He	5	−451
Electronic "noise" suppression	Thermal noise in amplifiers suppressed at cryogenic temperatures; cryogenically cooled masers	He H_2	5 20	−451 −423
High-capacity electric motors and generators	Increased capacity of conventional designs at low temperatures; compact motors for liquid-cryogen pumps by special design using superconducting motor and ac or pulsed dc in stator windings	He C_3H_8	5 225	−451 −54
Cryogenic infrared detectors	Photoconductor sensors highly sensitive to wavelengths above 10 μm at cryogenic temperatures	H_2 He	20 5	−423 −451
Gas liquefaction separation, rectification, refining	Large-scale production of liquid natural gas; natural-gas separations	N_2 He	77 5	−320 −451
Modification of "radiation damage" to materials	Alleviation of damage to materials resulting from neutron bombardment	He H_2	5 20	−451 −423
Bubble chambers	Experiments involving elementary particles and nucleation in slightly superheated cryogenic liquid	H_2	20	−423
Rocket propulsion	Liquid hydrogen fuel. Liquid oxygen oxidizer for liquid or solid fuels	H_2 O_2	20 90	−423 −297
Neutron moderator	D_2O ice, cooled by liquid hydrogen, used for slowing down of fast neutrons in a thermal reactor	D_2O-H_2	20	$(-423)^a$
Food freezing	Food frozen by liquid nitrogen spray or food cartons conveyed through liquid nitrogen. N_2 also used in frozen food transport	N_2	77	−320
Cryobiology	Cryogenic long-time preservation of blood, sperm, and other bio-materials	N_2	77	−320
Cryosurgery	Used for cryogenic brain surgery for treatment of Parkinson's disease; eye surgery	CO_2	200	−100
High vacuum and "cold traps"	Outgassing of high-vacuum systems, removal of oil vapors and residuals. Deposition of thin metal films under high vacuum; semiconductor manufacture	CO_2 N_2 H_2	195 77 20	−108 −320 −423

^{a}Heavy-water ice cooled to 20 K (36 deg R).

Table 5-88.   APPLICATIONS OF CRYOGENICS *(Continued)*

Type of use	Brief description	Gases	Minimum temperature	
			K	deg F
Reduction of chemical reaction rate	Excessive rates of chemical reaction at usual temperature may be slowed as desired by very low temperatures	CO_2 (bath) N_2	195 77	−108 −320
Refrigerated transport	Dry ice in trucks, railway cars, ships, planes, and in insulated cartons (liquid nitrogen for some transport)	CO_2 N_2	195[b] 77	−108[b] −320
Cutting-tool coolant	Low-temperature heat removal at cutting-tool face when machining superalloys	CO_2 (bath) N_2	195 77	−108 −320
Metal working and fabrication	Stretching, forming, and rolling of metals at cryogenic temperatures. Precipitation hardening and work hardening of austenitic steels	CO_2 (bath) N_2	195 77	−108 −320

[b] Solid CO_2 sublimes at 195 K, −108.5 deg F.

Table B.   RE-EVAPORATED GASES

Type of use	Typical applications	Gases
Oxygen steel processes	Oxygen lances in open-hearth, converter, or electric-furnace operation; oxygen-enriched blast in blast furnace	O_2
Chemical industries	Nitrogen for raw material in chemical manufacture. Hydrogenation processes	N_2 H_2
Combustion processes	Fuel gases with air, or oxygen-enriched combustion processes	H_2 CH_4 C_3H_8 C_4H_{10}
Flame cutting and welding	Gas welding and cutting processes, metals, and alloys	O_2 H_2 C_2H_2

REFERENCE

Advances in Cryogenic Engineering Series, Plenum Publishing Corporation, 15 volumes.

Table 5-89. CRYOGENIC AND REFRIGERATING LIQUIDS—SI UNITS*

Fluid	Boiling point K	deg R	deg C	deg F	Density, kg/m³	Volume ratio (to room temp), gas/liq	Latent heat of vaporization, kJ/kg	Specific heat, c_p liquid, kJ/kg·K	gas, kJ/kg·K	Viscosity liquid, mN·s/m²	gas, mN·s/m²	Thermal conductivity liquid, mW/m·K	gas, mW/m·K	Dielectric constant
Air	79	142	−194	−318	875	740:1	205	1.97	1.02	0.080 6	0.006 61		7.44	
Argon	87	157	−186	−303	1 400	840:1	162	1.14	0.531	0.252	0.008 27	123	6.05	1.52
Carbon dioxide[a]	195	350	−79	−110	1 560	730:1	572	1.33	0.795				14.7[c]	1.59[c]
Ethane	185	334	−88	−126	548	420:1	488	2.51	1.05	0.245	0.007 36	135	25.1[d]	1.43
Fluorine	85	154	−188	−306	1 500	880:1	166	1.55	0.812				7.20	
Helium 3	3.20	5.76	−270	−454	58.9	600:1	8.48	4.60		0.001 62		17.1		
Helium 4	4.215	7.59	−269	−452	125	600:1	20.7	4.56	6.82	0.003 57	0.001 28	27.0	9.69	1.049 2
Hydrogen	20	36.7	−253	−423	71.0	800:1	446	9.79	11.9	0.013 1	0.001 05	118	15.6	1.226
Methane	111	201	−162	−259	424	550:1	509	3.45	1.66	0.119		111		1.68
Neon	27	48.8	−246	−411	1 200	1 400:1	86.7	1.84	1.17	0.124	δ.004 55	130	9.86	
Nitrogen	77.4	139.2	−196	−320	810	700:1	198	2.04	1.08	0.158	0.005 54	139	7.18	1.434
Oxygen[e]	90.2	162.	−183	−297	1 140	860:1	213	1.90	1.40					1.51
Propane	231	416	−42	−44	581	310:1	425	2.20	1.21					1.27
Propylene	225	406	−48	−54	614	330:1	437	2.59	1.52					
Refrigerant 12	243	438	−30	−22	1 490	280:1	165	0.891	0.456	0.371	0.010 9			2.13
Refrigerant 13	192	346	−81	−114	1 520	350:1	149	0.895	0.413			60.5		
Refrigerant 13B1	215	388	−58	−72	1 990	310:1	119	1.69	0.849					
Refrigerant 22	232	419	−41	−41	1 410	380:1	234	1.05	0.464	0.351	0.010 5	110[b]	7.78	2.44

[a] Sublimes.
[b] At 253.15 K.
[c] At 273.15 K.
[d] At 324.82 K.
[e] Solidifies at −218 deg C, 55 K.
*Compiled from several sources.

Table 5-90.   STRENGTH OF METALS AT CRYOGENIC TEMPERATURES*
Test Data on Twenty-five Alloys

For other data on low-temperature properties of materials, see Tables 1-60, 1-63, and 5-30.

KEY:
UTS = ultimate tensile strength, kpsi
YS = yield strength, kpsi
Elong = percentage elongation in 2 inches
Notch ratio = strength ratio: notched/unnotched
Joint eff = percentage strength ratio welded/clear specimen

Specimens were mostly $\frac{1}{16}$-in. sheet, 2-in. gage length, cut longitudinal to rolling direction. Tests were made in triplicate. The test temperatures were as follows: 297–300 K = room temperature; 200 K = −100 deg F; 144 K = −200 deg F; 77 K = −320 deg F; 20 K = −423 deg F; 5 K = −450 deg F.
For MN/m², multiply kpsi by 6.894 8.

Temp, K	UTS, kpsi	YS, kpsi	Elong, %, 2 in.	Notch ratio	Joint eff, %	Temp, K	UTS, kpsi	YS, kpsi	Elong, %, 2 in.	Notch ratio	Joint eff, %
ALUMINUM ALLOYS						ALUMINUM ALLOYS (Continued)					
2014: Al+Cu 4.5%, Mn 1%, Si 1%, Mg 0.5%						5456: Al+Mg 5.0%, Mn 0.75%, Zr <0.25%					
300	70	63	9.7	0.99	66	300	57	45	8.7	0.92	84
200	73	68	9.5	1.00	60	200	57	45	9.3	0.90	84
144	76	71	9.3	0.99	59	144	62	47	11.7	0.91	86
77	84	76	11.7	0.93	63	77	74	53	13.0	0.79	85
20	96	79	13.6	0.88	82	20	87	57	8.7	0.75	74
5	97	82	10.4	0.83	74	5	86	57	9.5	0.73	74
2020: Al+Cu 4.5%, Li 1.1%, Mn 0.5%						7002: Al+Zn 3.35%, Mg 2.07%, Cu 0.88%					
300	79	75	8.0	0.67	—	300	67	57	16.7	1.05	74
200	82	76	6.3	0.65	—	200	71	61	18.0	1.07	73
144	86	83	3.0	0.60	—	144	74	64	18.8	1.08	73
77	95	88	4.0	0.52	—	77	83	70	19.8	1.03	68
20	101	93	2.3	0.50	—	20	104	77	18.9	0.86	56
5	104	95	3.6	0.50							
2119: Al+Cu 5.9%, Fe 0.15%, Ti 0.15%						7075: Al+Zn 5.6%, Mg 2.5%, Cu 1.6%, Cr 0.3%					
300	60	43	9.0	0.93	—	300	79	74	9.2	0.90	—
200	63	44	9.5	0.92	—	200	83	77	8.7	0.82	—
144	66	47	10.2	0.90	—	144	85	80	6.7	0.78	—
77	76	53	12.2	0.86	—	77	94	88	5.2	0.68	—
20	88	43	16.5	0.62	—	20	101	95	3.2	0.56	—
2219: Al+Cu 6.0%, Mn 0.33%, Fe 0.16%						7079: Al+Zn 4.5%, Mg 3.3%, Cu 0.6%					
300	65	52	9.8	0.92	66	300	76	67	9.0	1.00	—
200	69	55	9.3	0.92	67	200	80	68	9.0	0.84	—
144	73	58	10.0	0.92	70	144	86	75	7.0	0.78	—
77	83	64	12.1	0.90	75	77	93	84	4.0	0.68	—
20	96	79	15.3	0.81	77	20	101	94	3.0	0.56	—
5	94	69	12.0	0.68	80	5	102	93	2.5	0.53	—
5052: Al+Mg 2.5%, Fe 0.45%, Cr 0.25%						7178: Al+Zn 7%, Mg 3%, Cu 2%, Cr 0.3%					
300	34	25	10.6	0.97	95	300	94	88	7.5	0.67	—
200	36	26	15.1	1.01	93	200	96	93	4.0	0.65	—
144	39	27	20.1	0.96	98	144	100	96	4.0	0.60	—
77	52	29	30.0	0.93	98	77	109	104	1.2	0.41	—
20	73	37	26.5	0.88	92	20	117	113	1.0	0.32	—
5	72	33	27.0	0.83	95						
5086: Al+Mg 4.0%, Fe 0.5%, Mn 0.45%						MAGNESIUM ALLOYS					
						LA-91: Mg+Li 9.0%, Al 1.0%					
300	47	37	10.4	1.00	90	297	23	20	36.8	0.95	100
200	48	38	12.0	0.99	86	200	32	23	11.2	0.82	—
144	52	39	15.4	0.98	89	144	34	23	14.3	0.77	—
77	64	44	25.0	0.89	93	77	36	25	15.5	0.77	85
20	85	47	20.2	0.76	80	20	46	36	20.5	0.75	88
5	80	49	23.4	0.74	86	5	48	41	11.5	—	

*From: "Effects of Low Temperatures on Structural Metals", NASA SP-5012, National Aeronautics and Space Administration, 1964 (revised edition, 1968).

Table 5.90.   STRENGTH OF METALS AT CRYOGENIC TEMPERATURES
(Continued)

Temp, K	UTS, kpsi	YS, kpsi	Elong, %, 2 in.	Notch ratio	Joint eff, %	Temp, K	UTS, kpsi	YS, kpsi	Elong, %, 2 in.	Notch ratio	Joint eff, %
MAGNESIUM ALLOYS *(Continued)*						**NICKEL ALLOYS** *(Continued)*					
LA-141: Mg + Li 14.5%, Al 1.5%						**Waspaloy-H:[a] Ni + Cr 19%, Co 14%, Mo 4.3%, Ti 3.0%,**					
297	20	18	23.7	1.06	95	**Al 1.3%**					
200	28	21	10.8	0.94	88	300	178	116	26.3	0.81	79
144	30	23	13.7	0.94	87	200	193	128	20.5	0.81	86
77	33	28	13.8	0.95	87	144	203	139	18.8	0.79	83
20	43	39	14.3	0.86	98	77	205	142	15.0	0.80	92
						20	197	154	10.2	0.85	97
ALLOY STEELS											
20-Cb, Carpenter: Fe + Ni 25%, Cr 20%, Cu 3.5%, Mo 2.5%						**K Monel-H:[a] Ni + Cu 29%, Al 3%, Fe 1.5%, Mn 1.0%**					
300	95	55	33.3	0.92	101	300	148	106	22.7	0.92	67
200	109	63	36.2	0.93	102	200	156	111	24.0	0.96	70
144	120	71	35.7	0.92	105	144	165	119	24.7	0.93	72
77	154	87	64.0	0.82	101	77	177	128	30.7	0.92	76
20	163	104	30.1	0.89	117	20	192	137	28.3	0.90	83
A286-N:[a] Fe + Ni 26%, Cr 16%, Ti 2%						**René 41-H: Ni + Cr 20%, Co 10%, Mo 10%, Ti 3.0%, Fe 3.0%**					
300	93	42	37.3	0.86	100	300	199	147	16.0	—	77
200	104	48	38.8	0.88	101	200	201	152	14.0	—	86
144	115	57	43.0	0.87	101	144	210	167	12.0	—	81
77	144	68	71.0	0.80	99	77	229	179	12.0	—	85
20	161	81	47.3	0.82	96	20	239	199	9.0	—	85
A286-H:[a] (Same as above)						**TITANIUM ALLOYS**					
300	140	94	22.0	0.94	71	**TiAlV: Ti + Al 5.9%, V 4%, Fe 0.12%**					
200	153	101	25.7	0.92	74	297	140	133	11.0	1.02	100
144	162	110	28.2	0.90	79	200	161	158	9.3	1.00	99
77	191	122	40.7	0.82	72	144	178	177	6.5	0.99	100
20	218	137	28.5	0.83	71	77	218	214	13.0	0.82	102
						20	240	240	1.7	0.61	96
AISI 202-N:[a] Fe + Cr 18%, Mn 8%, Ni 5%						5	242	—	0.2	0.62	100
300	101	47	56.8	—	106						
200	156	70	40.7	—	101	**TiAlSn: Ti + Al 5.2%, Sn 2.4%, Fe 0.32%**					
144	176	78	43.7	—	100	297	134	128	12.8	1.20	102
77	231	88	51.7	—	100	200	157	152	11.8	1.10	100
20	—	—	—	—	—	144	172	169	9.3	1.10	100
5	206	111	25.0	—	91	77	213	207	14.0	0.81	101
						20	234	234	5.0	0.66	104
Maraging H:[a] Fe + Ni 18%, Co 8%, Mo 5%						5	235	—	1.3	0.62	100
300	254	245	2.8	1.09	94						
200	275	266	2.8	1.08	100	**TiVCr: Ti + V 13.4%, Cr 11.3%, Al 2.8%, Fe 0.18%**					
144	288	274	3.0	1.06	102	297	137	137	13.3	1.20	106
77	321	309	2.5	0.90	99	200	182	182	6.0	1.10	106
20	365	355	3.2	0.41	86	144	218	215	4.5	0.96	106
						77	285	282	2.5	0.54	60
NICKEL ALLOYS						20	289	—	0.7	0.40	34
Inconel X-H:[a] Ni + Cr 15%, Fe 7%, Ti 2.5%, Mn 1.0%,						5	301	—	0.0	—	—
Al 0.7%											
300	180	125	25.3	0.90	67						
200	194	132	26.7	0.87	69						
144	203	136	27.2	0.84	71						
77	220	139	32.0	0.79	72						
20	224	140	28.0	0.82	79						

Notes:

Only the major alloying elements are listed herewith; for minor alloying elements, see original data.

Data on specimens cut transverse to rolling direction are given in original source, as are data on aged and cold-worked alloys.

Note that 5, 20, and 77 K are approximately the atmospheric boiling points of liquid helium, hydrogen, and nitrogen, respectively; 200 K is just above the evaporation temperature of dry ice.

Notch ratios of less than one indicate that the material is weakened by the presence of stress-raising defects.

[a] N—annealed; H—thermally age hardened, welded before aging.

Mechanics, Structures, and Machines

Figure 6-1. DYNAMIC RESPONSE OF RC ELEMENT
TO A STEP-CHANGE INPUT

For a simple first-order system with resistance and capacitance in series, the response to a sudden change in input is not immediate but involves a time lag. The "time constant" or index of this lag is $\tau = RC$, in seconds, which is the time required to attain 63.2 percent of the step change. In terms of "percent incomplete", or the ratio of the remaining response to the total step, the equation for the condition at any time t is

$$\frac{(\text{final condition}) - (\text{condition at time } t)}{(\text{final condition}) - (\text{initial condition})} = e^{-t/\tau}$$

Figure 6-1 shows this relationship, plotted on linear and semi-log coordinates.

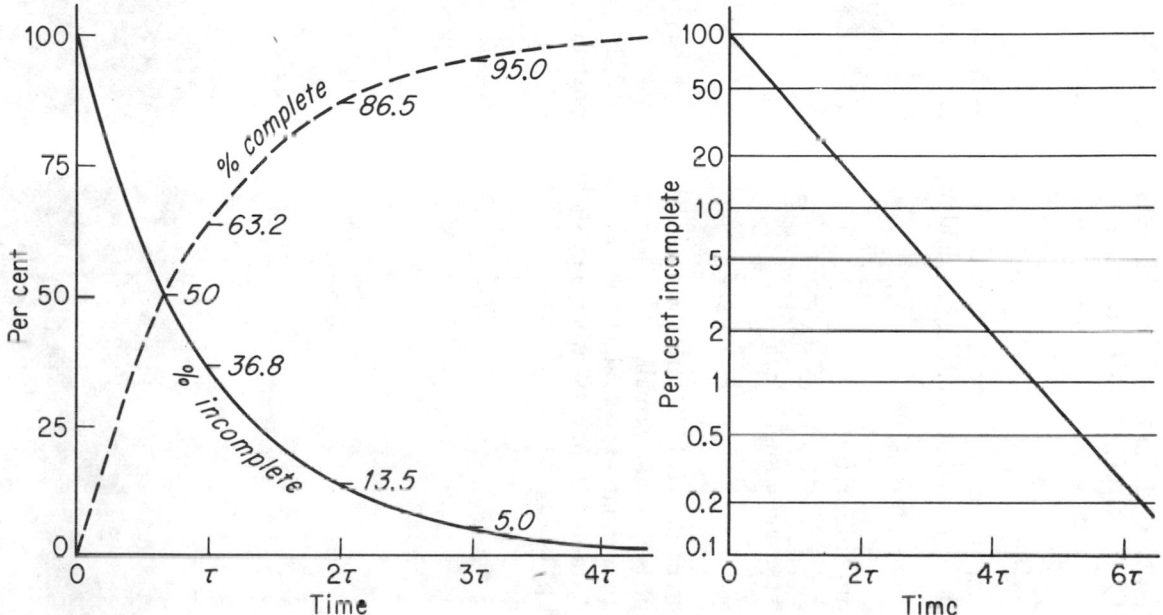

Figure 6-1. Response of a time-constant (linear) element to a step-change input. (Dimensionless representation.)*

Figure 6-2. DYNAMIC RESPONSE OF RC ELEMENT TO A SINE-WAVE INPUT

For frequency response of second-order systems, see Section 11.

If a linear system is subjected to an oscillatory sine-wave input, the frequency of the response remains the same as the frequency of the input; there is a fixed time lag; and the response is a sine wave of lesser amplitude. Both the amplitude ratio and the phase lag can be expressed in terms of τ, the time constant (RC, seconds), as follows:

$$\text{Amplitude ratio} = \frac{1}{\sqrt{1+(2\pi f \tau)^2}}$$

$$\text{Phase lag} = \tan^{-1} 2\pi f \tau$$

where f is the frequency, cps.

If the frequency is numerically equal to $1/2\pi\tau$, the amplitude ratio is $1/\sqrt{2}$, and the phase lag is 45°. If the product $f\tau$ is small, the amplitude of the output wave is nearly the same as that of the input, and the phase lag is also small.

On logarithmic coordinates the amplitude-ratio curve approximates two straight lines intersecting at $1/2\pi\tau$; this value is known as the "corner frequency". This provides a method for sketching the amplitude-ratio or gain curve when only the time constant τ is known.

Figure 6-2 shows the shape of the amplitude-ratio and phase-lag curves.

Figure 6-2. DYNAMIC RESPONSE OF RC ELEMENT TO A SINE-WAVE INPUT (*Continuea*)

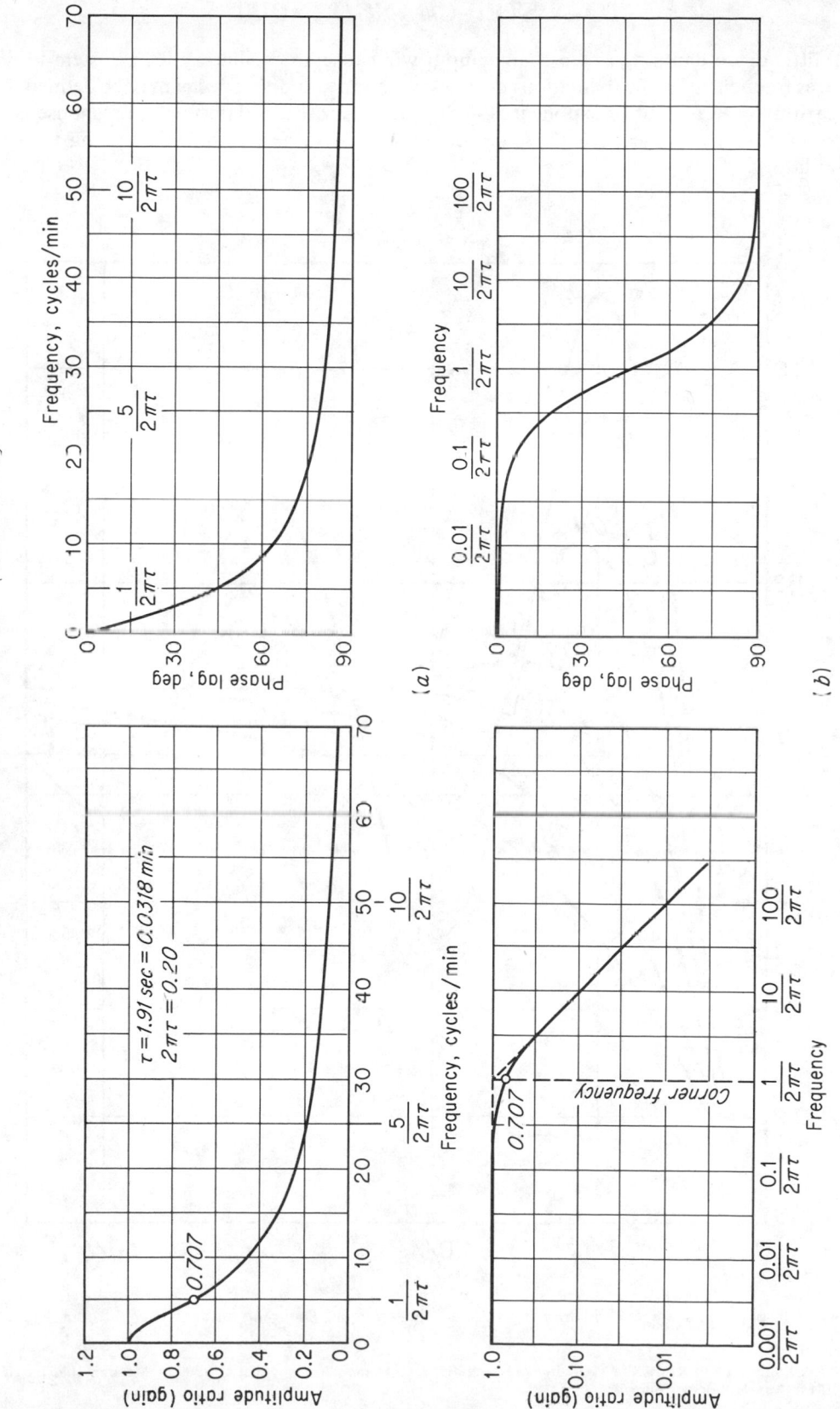

Figure 6-2. Response of a time-constant (linear) element to a sine-wave input. (*a*) Rectangular coordinates. (*b*) Bode logarithmic diagrams.*

*From: ENGINEERING EXPERIMENTATION by G.L. Tuve and L.C. Domholdt. Copyright © 1966 by McGraw-Hill Book Company.

Figure 6-3. DYNAMIC RESPONSE OF RCL SYSTEM TO A STEP-CHANGE INPUT

With little or no damping a step-change input will cause an oscillatory RCL system to respond at its natural frequency f_n. The oscillations decrease with time, and this decay may be defined in terms of the logarithmic decrement or exponential decay ratio. At critical damping the response is similar to that of a linear system subjected to the same step input. With large amounts of damping, the response is non-oscillatory.

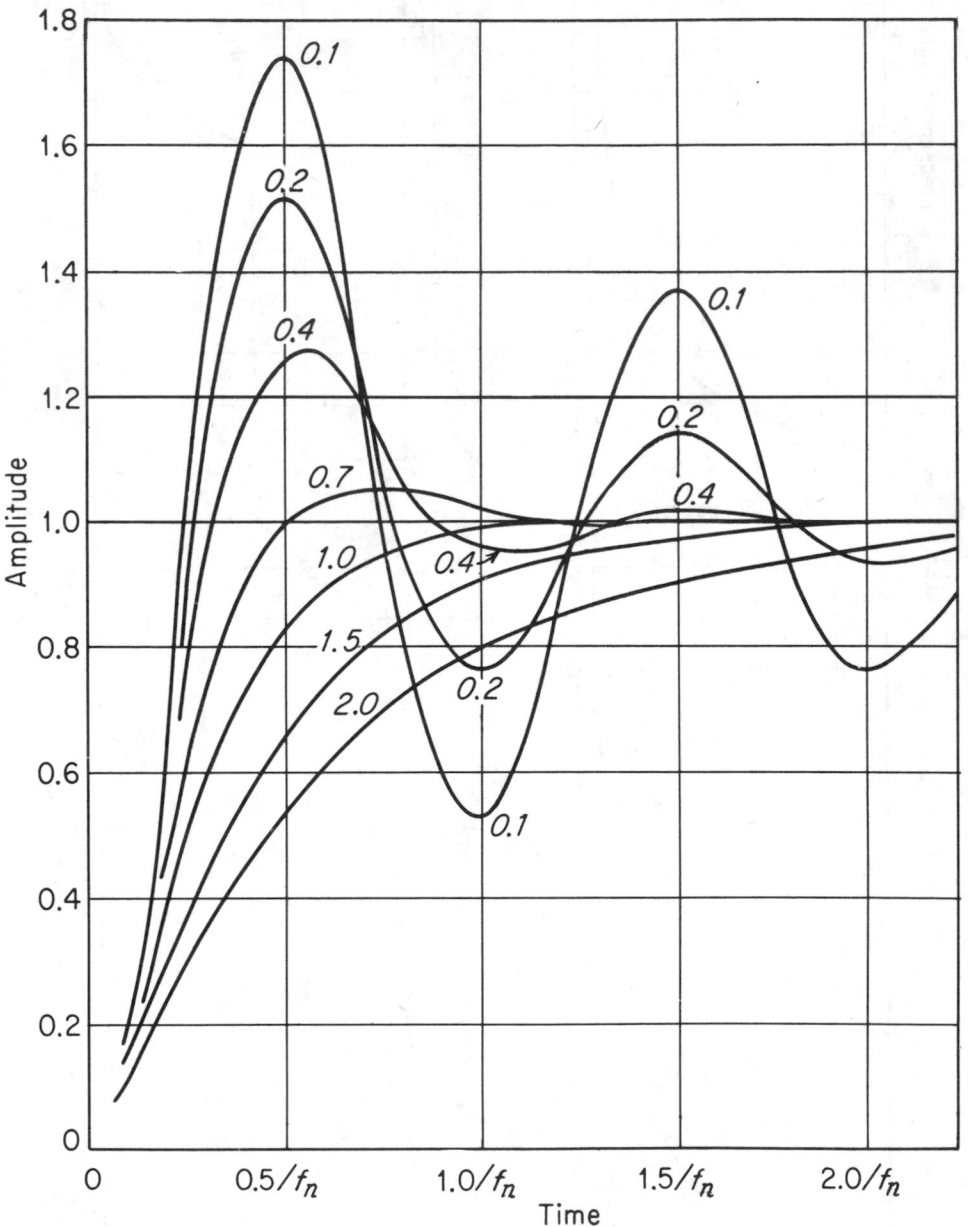

Figure 6-3. Response of a simple oscillatory system to a unit step input. Damping ratios, $z = c/c_c$, from 0.1 to 2.0.*

*From: ENGINEERING EXPERIMENTATION by G.L. Tuve and L.C. Domholdt. Copyright © 1966 by McGraw-Hill, Inc. Used with permission of McGraw-Hill Book Company.

Figure 6-4. AMPLITUDE RESPONSE—SECOND-ORDER SYSTEM

If the input frequency is low, the response of an oscillatory system will almost duplicate the input. At the higher frequencies the response will depend on the ratio of actual damping c to critical damping c_c. For the electrical system critical damping is $2\sqrt{L/C}$; for the mechanical mass-spring-damper system the critical damping is $2\sqrt{km}$, where k is the spring constant and m is the mass.

Figures 6-4 and 6-5 show the response of a simple oscillatory (RCL) system to a sine-wave input, with damping ratios c/c_c from 0.1 to 20.0.

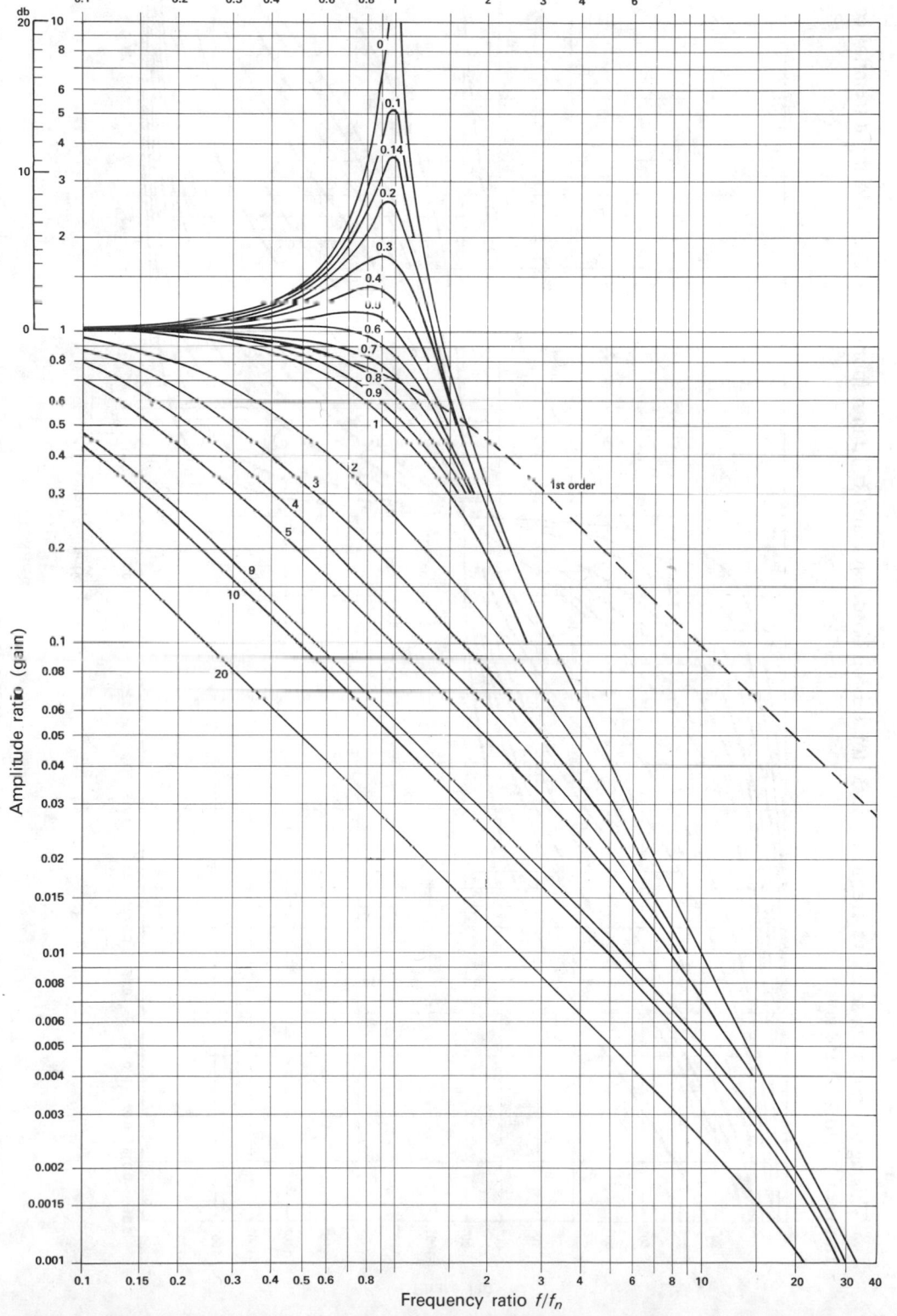

Figure 6-5.   PHASE RESPONSE—SECOND-ORDER SYSTEM

This figure shows phase lag vs. frequency ratio for a second-order system (RCL) in response to a sine-wave input (**semilog coordinates**). Damping ratios of 0.1 to 20.0 are given.

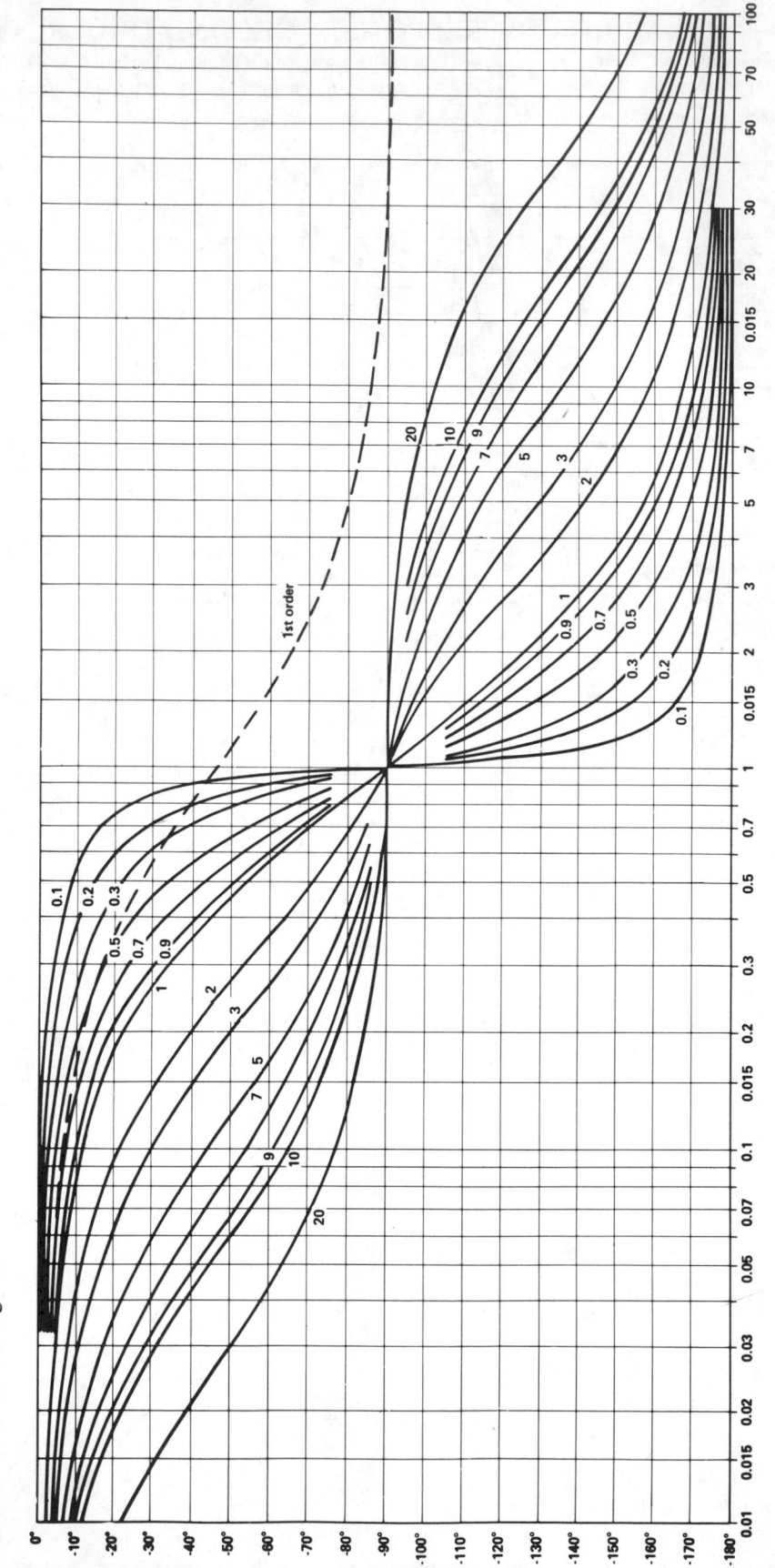

Frequency ratio f/f_n

Phase lag

REFERENCE

For large-scale curves giving values for damping ratios to 20, frequency ratios to 40, and gains as low as 0.001, see "Handbook of the Engineering Sciences", J.H. Potter, Ed., Vol. 2, D. Van Nostrand Co., 1967, pp. 786–787.

Table 6-6.
FREQUENCY-RESPONSE APPROXIMATIONS AND CORRECTIONS

When the magnitude of the output-input ratio and the phase-angle response are each plotted against frequency on logarithmic coordinates, the work of obtaining the transfer function becomes largely a matter of graphical addition and subtraction. (Semilog plots may also be used.)

A simplification is attained by treating separately each of the four basic types of factors in the transfer function and by starting with straight-line approximations of the actual curves.*

Corrections from the straight-line approximations, to obtain the actual curves, are given in the following table and also in Figure 6-7.

VALUES OF LOG MAGNITUDE AND ANGLES OF $(1+j\omega T)^{-1}$

The corner frequency ω_{cf} is used as the index, i.e., $1/(1+j\omega T) = 1/(1+j\omega/\omega_{cf})$. Range is one decade above and below ω_{cf}.

ω_{cf}	Exact magnitude, db	Value of the asymptote, db	Error, db	Angle, degrees
0.10	−0.04	0	−0.04	−5.7
0.50	−0.97	0	−0.97	−26.6
0.76	−2.00	0	−2.00	−45.0
1.00	−3.01	0	−3.01	
1.31	−4.35	−2.35	−2.00	
2.00	−6.99	−6.02	−0.97	−63.4
4.00	−12.30	−12.04	−0.26	
10.00	−20.04	−20.00	−0.04	84.3

*For a full discussion of the method, with examples, see. "Feedback Control System Analysis and Synthesis", 2nd ed., J.J. D'Azzo and C.H. Houpis, McGraw Hill Book Company, 1966, pp. 278–303.

Figure 6-7.
CORRECTIONS TO THE LOG MAGNITUDE AND PHASE DIAGRAM*

For $[1 + j2\zeta\omega/\omega_n + (j\omega/\omega_n)^2]^{-1}$

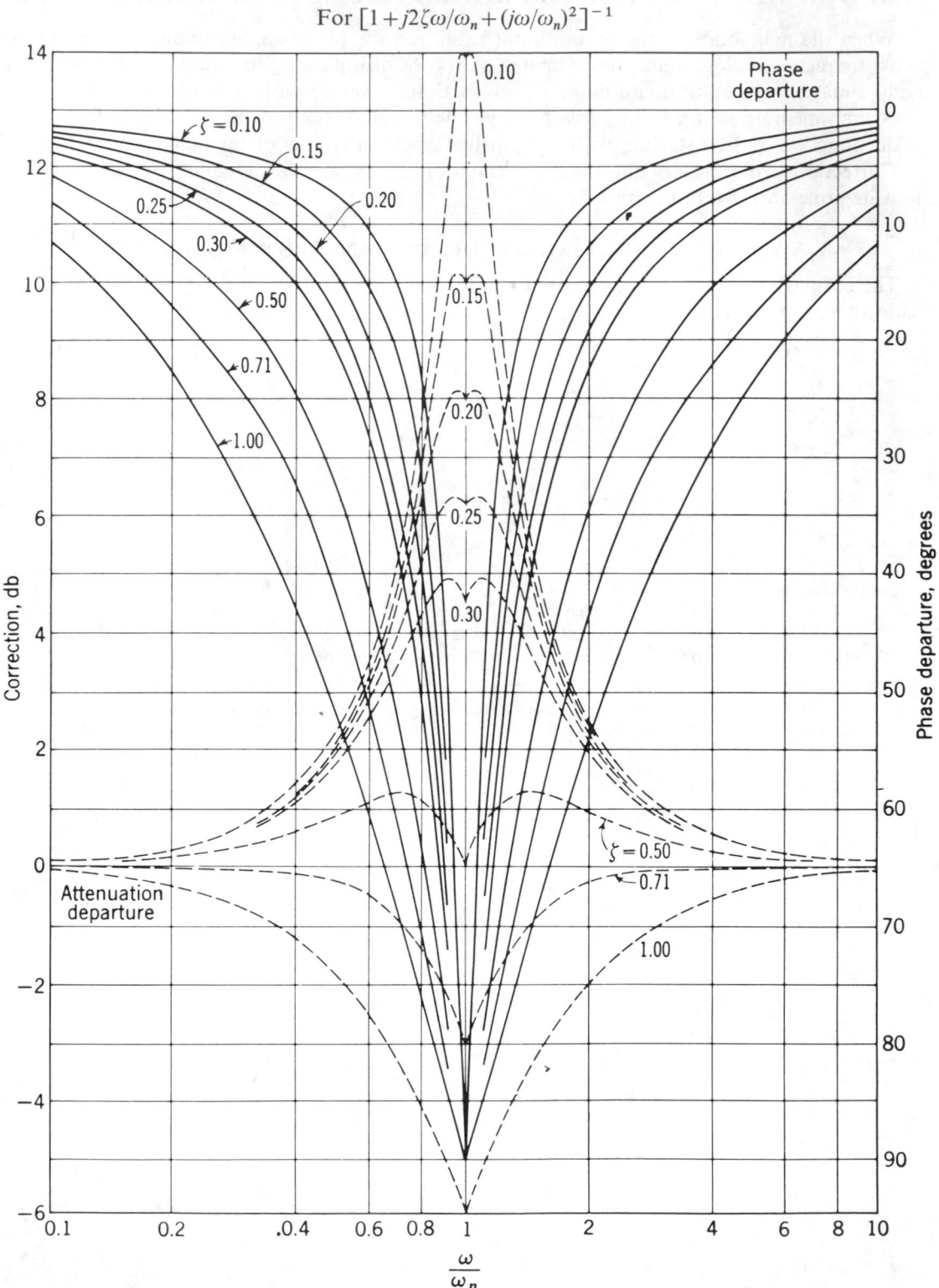

*Reproduced from: "Theory of Servomechanisms", H.M. James, N.B. Nichols, and R.S. Phillips, McGraw-Hill Book Company, 1947.

Table 6-8. MACHINERY VIBRATIONS*

See Table 6-9 for vibration in transport vehicles.

The vibrations of rotating machines of bearings, bases, etc. may be measured and reported in terms of displacement, velocity, or acceleration. Amplitude is measured by using a seismic pickup with a soft spring. The coil-and-magnet pickup senses velocity, while the accelerometer usually employs either a piezoelectric or a strain-gage transducer.

Many authorities prefer to rate machine vibration in terms of rms velocities, approximately as follows:

> Very smooth: 0.03 in./sec or less
> Good or fair: 0.1 to 0.2 in./sec
> Very rough: 0.5 to 0.8 in./sec

In this type of evaluation, the quality of smoothness or roughness is almost independent of frequency. This is very convenient when a velocity-type pickup is used exclusively. It is sometimes claimed, however, that higher vibration velocities can be tolerated in the intermediate frequency range (for example, 30 to 150 cps) than at lower or higher frequencies.

For steady periodic vibration in a single direction, the conversions from vibration velocity to acceleration or to displacement (or vice versa) are based on the assumption of simple harmonic motion (sine wave), such as is observed with the mass-spring assembly. A convenient form of the equations for these conversions is

$$V_{max} = 2\pi f x$$
$$a_{max} = 4\pi^2 f^2 x,$$

where x is the half-amplitude of the wave, i.e., one-half the peak-to-peak displacement, f is the vibration frequency, V_{max} is the maximum velocity, and a_{max} is the acceleration. For rms values multiply the maximum values by 0.707; for "average" values multiply the maximum values by 0.637.

Since vibration magnitude, or peak-to-peak displacement, is more easily visualized than either velocity or acceleration, it is valuable to have tabular data in these terms also. The following table provides a rough guide for all three measured properties, assuming simple harmonic vibration as obtained with the mass-and-spring or with the usual "shake table". Any tabular values for vibration tolerance should be used with care, since the allowable vibration depends so much on the type and use of the machine. For example, the tabular values herewith would be too high to be tolerated in a machine tool, especially in the case of precision grinders, boring machines, and others.

MACHINERY VIBRATION RATINGS
Vibration Ranges in Terms of Displacement, Velocity, and Acceleration

SYMBOLS: x = displacement, mil, peak-to-peak (1 mil = 0.001 in. = 0.0254 mm)
v = velocity, in./s, rms; for m/s multiply by 0.0254
a = acceleration, in./s^2, rms; for m/s^2 multiply by 0.0254

Quality		100	500	1,200	1,800	3,600	5,000	10,000
	rpm				Frequency or speed			
	cps	1.67	8.3	20	30	60	83.	167.
Very smooth	x	4.0	1.4	.57	.38	.19	.14	.05
	v	.015	.025	.025	.025	.025	.025	.02
	a	.16	1.3	3.1	4.7	9.4	13.	21.
Smooth	x	8.0	2.2	1.13	.76	.38	.27	.11
	v	.03	.04	.05	.05	.05	.05	.04
	a	.32	2.1	6.3	9.4	19.	26.	42.
Fair to average	x	27	7.1	3.4	2.2	1.1	.8	.37
	v	.10	.13	.15	.15	.15	.15	.14
	a	1.0	6.8	19.	29.	56.	78.	150.
Somewhat rough; should be corrected	x	68	17.	8.0	5.3	2.6	1.9	.86
	v	.25	.32	.35	.35	.35	.35	.32
	a	2.6	17.	44.	68.	130.	180.	330.
Very rough; intolerable	x	120	33	15	10.	4.9	3.5	1.6
	v	.45	.60	.65	.65	.65	.64	.59
	a	4.7	43.	82.	120.	240.	330.	620.

*Compiled from several sources.

REFERENCES

"Shock and Vibration Handbook", C.M. Harris and C.E. Crede, Vol. 3, McGraw-Hill Book Company, 1961.
"Vibration: An Indicating Tool", R. Baxter and D. Bernhard, *Mech. Eng.*, 90(3):36–41, March 1968.

Table 6-9. VIBRATION IN TYPICAL TRANSPORT VEHICLES*

For peak amplitude in mm, multiply by 25.4.

Vehicle	Range of frequencies, hertz	Approximate peak amplitude, inches	Nature of excitation	Usual choice of isolator resonant frequency
Ships	0–15	0.02	Engine vibration in diesel or reciprocating steam drive	6 hertz for vibration isolation in commercial vessels; 27 to 30 hertz for shock isolation on naval vessels. These latter mounts amplify most vibrations to some extent.
	0–33	0.01	Propeller-blade frequency = (propeller rpm) × (number of blades)/60	
Piston-engine aircraft	0–60	0.01	Engine vibrations	Above 20 hertz. Amplitude of vibrations varies with location in aircraft. Landing shock can be neglected.
	0–100	0.01	Propeller vibrations. Aerodynamic vibrations due to buffeting	
Turboprop aircraft	0–60	0.01	Engine vibrations = (engine rpm)/60	9 hertz
	0–100	0.01	Propeller vibrations. Also aerodynamic vibrations due to buffeting and turbulence	
Jet aircraft	< 500	0.001	Audible noise frequencies due to jet wake and combustion turbulence; very little engine vibration	9 hertz
Passenger automobiles	1	6	Suspension resonance	25 hertz will usually avoid resonance with wheel hop and suspension resonant frequencies.
	8–12	0.02	Unsprung weight resonance (wheel hop)	
	> 20	0.002	Irregular transient vibrations due to resonances of structural members with road roughnesses	
Automobile trucks	4	5	Suspension resonance	Above 20 hertz and should not correspond with any structural resonance. It is not advisable to attempt to isolate suspension and unsprung weight resonances.
	20	0.05	Unsprung weight resonance	
	> 80	0.005	Structural resonances	
Military tanks	1–3	2	Suspension resonance	Similar to automobile truck
	Depends on speed	—	Track-laying frequency ≈ 17.6 (speed in mph)/ (tread spacing in inches)	
	> 100	0.001	Structural resonances	
Railroad trains	Broad and erratic		Similar to automobiles with additional excitations from rail joints and from side slop in rail trucks and draft gear	20 hertz has been successful in railroad applications. Shock with velocity changes up to 100 inches per second in direction of train occurs when coupling cars or starting freight trains.

*From: "Reference Data for Radio Engineers", 5th ed., Howard W. Sams & Co., Inc., Indianapolis, Indiana, 1968.

Table 6-10.   HELICAL STEEL SPRINGS*

Compression or Tension

The upper figure is the load in pounds at 100 000 psi (689 MN/m^2) stress by the "corrected" stress equation. The lower figure gives spring stiffness in lb/in. per single coil, based on a shear modulus of 11.5×10^6 psi (79.3 GN/m^2). The stiffness is independent of load. Both figures may be adjusted in direct proportion to selected stress or modulus. For multicoil springs divide the stiffness per coil by the number of active coils. For load in N, multiply the values in lbf by 4.4482. For stiffness in N/m, multiply the values in lbf/in. by 175.13.

HELICAL SPRINGS—LOAD AND STIFFNESS

Wire diam, in.	Outside diameter of coil, in.												
	$\frac{1}{8}$	$\frac{3}{16}$	$\frac{1}{4}$	$\frac{5}{16}$	$\frac{3}{8}$	$\frac{7}{16}$	$\frac{1}{2}$	$\frac{9}{16}$	$\frac{5}{8}$	$\frac{11}{16}$	$\frac{3}{4}$	$\frac{7}{8}$	1
.010	.305 9.47												
.012	.522 20.7	.350 5.43											
.014	.823 40.3	560 10.5	422 4.18										
.016	1.21 72.9	.824 18.4	.626 7.37										
.018	1.71 124	1.17 30.7	889 12.1										
.020	2.32 200	1.60 48.6	1.22 18.9	.972 9.06									
.022	3.03 306	2.12 73.9	1.60 28.2	1.30 13.7									
.024	3.90 464	2.72 108	2.09 41.4	1.68 19.9	1.41 11.1	1.21 6.67							
.028	5.98 965	4.25 216	3.29 80.6	2.65 38.0	2.24 21.1	1.92 12.9	1.69 8.37	1.51 5.79					
.032		6.25 398	4.84 146	3.95 68.0	3.33 37.5	2.85 22.4	2.51 14.8	2.24 10.0	2.03 7.27				
.037		9.42 785	7.41 280	6.01 128	5.07 69.6	4.41 42.2	3.88 27.3	3.44 18.5	3.10 13.3	2.83 9.77	2.61 7.43		
.043		14.4 1,618	11.4 559	9.33 249	7.91 134	6.82 80.0	6.07 51.6	5.36 34.9	4.89 25.0	4.44 18.3	4.10 14.0	3.52 8.60	
.049			16.4 1,019	13.7 452	11.6 240	10.1 141	8.84 90.8	7.95 60.7	7.14 43.2	6.54 32.0	6.00 24.1	5.17 14.6	4.55 9.66
.055			22.8 1,767	18.9 768	16.2 401	14.1 234	12.4 149	11.2 101	10.0 70.4	9.23 51.9	8.50 39.5	7.32 23.9	6.40 15.6
.063				27.9 1,461	23.8 721	20.9 429	18.6 272	16.6 181	15.1 128	13.8 92.6	12.7 70.0	10.9 42.3	9.60 27.5
.071				38.7 2,563	33.4 1,295	29.4 741	26.1 463	23.6 307	21.3 215	19.5 156	18.0 117	15.5 70.6	13.6 45.3
.080					47.1 2,300	41.3 1,285	36.9 795	33.4 525	30.2 364	27.8 262	25.7 196	22.2 117	19.5 75.6
.090						57.9 2,236	51.7 1,368	46.9 890	42.9 617	39.0 442	36.2 329	31.2 195	27.6 125
.100						77.5 3,726	69.7 2,248	63.3 1,449	58.0 993	53.0 707	49.2 525	42.6 309	37.6 197
.112							95.6 3,886	87.4 2,462	79.7 1,678	73.8 1,186	68.1 871	59.3 509	52.2 323
.125							130 6,701	119 4,205	109 2,817	101 1,977	93.8 1,436	81.9 832	72.5 525
.148									175 6,350	163 4,380	151 2,166	133 1,802	119 1,119
.177											250 7,485	222 4,150	198 2,526
.207												341 8,857	307 5,266
.244													484 11,805

Table 6-10.　HELICAL STEEL SPRINGS　(Continued)

Wire diam, in.	\multicolumn{13}{c}{Outside diameter of coil, in.}												
	$1\frac{1}{8}$	$1\frac{1}{4}$	$1\frac{3}{8}$	$1\frac{1}{2}$	$1\frac{3}{4}$	2	$2\frac{1}{4}$	$2\frac{1}{2}$	$2\frac{3}{4}$	3	$3\frac{1}{4}$	$3\frac{1}{2}$	4
.055	5.70												
	10.7												
.063	8.55	7.75											
	18.8	13.6											
.071	12.2	11.1	9.97	9.21									
	31.5	22.4	16.4	12.6									
.080	17.3	15.7	14.2	13.1									
	51.4	37.0	27.0	20.5									
.090	24.7	22.2	20.2	18.6									
	85.1	60.3	44.3	33.5									
.100	33.5	30.5	27.7	25.5	22.1	19.3							
	132	94.5	69.0	52.4	32.1	20.9							
.112	47.0	42.4	38.6	35.8	30.7	27.0	24.1						
	218	152	112	84.5	51.7	33.5	23.1						
.125	65.0	58.8	53.9	49.4	42.6	37.6	33.4	30.2					
	350	246	180	135	82.0	53.3	36.6	26.2					
.148	106	96.3	88.0	81.2	70.5	62.0	55.1	49.7	45.3	41.7			
	741	514	375	279	168	109	74.3	53.1	39.1	29.7			
.177	178	163	149	137	119	105	94.0	84.7	77.2	71.0	65.5	61.2	
	1,664	1,145	819	609	361	234	159	113	83.3	63.2	48.4	38.4	
.207	279	255	235	217	188	166	150	134	122	113	104	97.2	85.2
	3,419	2,320	1,658	1,218	716	457	310	218	160	121	93.6	74.1	48.5
.244	444	406	376	306	349	269	241	218	199	183	170	158	139
	7,462	4,988	3,498	2,562	1,488	941	631	443	322	243	188	148	95.9
.283		614	574	534	467	415	374	338	309	285	263	247	216
		10,199	7,086	5,154	2,919	1,820	1,210	849	616	460	352	276	180
.331			882	824	733	652	589	534	491	453	421	391	343
			15,207	10,785	6,048	3,713	2,449	1,693	1,217	911	694	543	348
.375				1,166	1,038	932	844	767	704	652	605	567	497
				19,966	10,903	6,652	4,302	2,960	2,124	1,575	1,195	933	597
.437					1,587	1,433	1,307	1,201	1,102	1,019	951	888	783
					23,202	13,859	8,861	6,005	4,243	3,132	2,367	1,835	1,166
.500						2,073	1,897	1,742	1,617	1,500	1,396	1,310	1,160
						26,645	16,817	11,268	7,880	5,736	4,325	3,330	2,097
.562							2,623	2,426	2,251	2,096	1,956	1,832	1,628
							29,977	19,853	13,751	9,957	7,412	5,686	3,546
.625								3,239	3,010	2,825	2,649	2,491	2,221
								33,289	22,872	16,386	12,135	9,240	5,710

*From: H.F. Ross, *Trans. ASME*, 69:727, 1947. For a more complete table, see the original source.

Table 6-11.   ULTRASONIC ENERGY AND APPLICATIONS*

Table A.   APPLICATIONS OF ULTRASONICS

Processes	Typical frequencies, kc/s
HIGH POWER RANGE[a]	
Surface cleaning; grease and film removal	15–60
Emulsifying; homogenizing; production of dispersions	10–40
Degassing of liquids and molten metals (grain refinement)	10–40
Stimulating mechanical processes: mixing, diffusion, defoaming, atomizing, drying, plastic sealing, particle agglomeration, flow of powders, adhesion in soldering and welding, grain refinement in casting	10–500
Cutting and forming; impact grinding of brittle materials; abrasive cutting; die forming with reduced friction	15–100
Stimulating chemical processes; combustion and other reactions	10–200
LOW POWER RANGE (ABSORPTION AND ECHO)	
Inspection and flow detection	500–5 000
Pulse-echo counting and inspection	700–10 000
Medical examination, diagnosis, and therapy	500–2 500
Measurement and control (flow, thickness, density, liquid level, viscosity)	500–20 000
Sonic detection and ranging; command signaling; delay lines	10 000–20 000

[a]The desired effects in this range are largely accomplished by cavitation (in a liquid) or by vibrations and high accelerations that affect materials in contact with each other. *Cavitation* is bubble formation at a nucleus (such as a particle), followed by bubble growth and collapse. High pressures and temperatures occur at the instant of collapse, and the number of bubbles collapsing can be millions per second. Cavitation is suppressed by high static pressure and varies with liquid temperature. As the ultrasonic frequency is increased, up to a practical limit of about 10^7 cps (hz), the sound intensity must be increased to pass the threshold at which cavitation begins. Intensities very much above the threshold are not advantageous. Cavitation is increased in a liquid of low viscosity, low vapor pressure, and high surface tension.

Transmission and matching of acoustics power to the load is often not simple; in fact this step is an art in itself (see references).

Table B.   GENERATORS OR TRANSDUCERS

Type of generator	Typical limits		
	Mechanical power,[a] W	Frequency, kc/s	Efficiency,[b] %
Air whistles	75	40	15
Jet-edge vibrators (gas or liquid)	50	20	15
Cavity resonators	500	12	15
Sirens (jet interruption)	1 000	25	70
Piezoelectric-quartz		5 000	90
Piezoelectric-ceramic (e.g., barium titanate)	4 000	5 000	75
Magnetostrictive (Ni, Fe, Co, ferrites)	5 000	90	50
Electron tube	1 000	30	50
Rotating alternator	25 000	25	50

[a]Power intensity per unit area at point of use depends on methods for transmission and focusing.
[b]The efficiencies of available equipment for industrial use are often much below these values.

*Compiled from several sources.

REFERENCES

"High-Intensity Ultrasonics: Industrial Applications", B. Brown and J.E. Goodman, D. Van Nostrand Co., 1965.

"Sources of High-Intensity Ultrasound", L.D. Rozenberg, Ed., 2 volumes, Plenum Press, 1968.

"Ultrasonic Engineering", J.R. Frederick, John Wiley & Sons, Inc., 1965.

"Ultrasonic Machining of Intractable Materials", A.I. Markov, Ed., Butterworth & Co., 1967.

"Ultrasonics: Theory and Application", G.L. Gooberman, English Universities Press, 1968.

Table 6-12. VIBRATION AND SHOCK BY GROUND MOTION

Recorded ground accelerations from earthquakes have seldom exceeded 0.3 g; from trains, machines, and vehicles they are seldom over 0.1 g. Very large detonations that may fracture rock occur around 2 g. For commercial blasting the "safe" distances for various weights of charge are roughly as follows:

Weight of explosive charge, lb	Safe distance, ft (no structural damage)	Weight of explosive charge, lb	Safe distance, ft (no structural damage)
1	10	10	50
2	18	50	150
5	30	500	600

The response of a structure to ground motion can only be approximated by careful analysis. The natural period of vibration and the inherent damping are important. If the damping is low, the structure will be subjected to relatively large forces in the event of strong ground motion. Ordinary buildings have 5 to 15 percent of critical damping. Bare steel structures may exhibit as low as 1 to 3 percent of critical damping, and concrete structures, 8 to 10 percent. Periods of vibration will vary from 0.1 second for a solid one-story building up to 8 seconds for the Empire State Building. (A rough rule is 0.1 second per story, but measured values are usually lower.)

REFERENCE
For a brief discussion and extensive references see "Shock and Vibration Handbook", C.M. Harris and C.E. Crede, Vol. 3, McGraw-Hill Book Company, 1961, Sects. 49–50.

Table 6-13. APPROXIMATE SEISMIC SHOCK LIMITS FOR BUILDING STRUCTURES*

Limits of Displacement for Vibration Frequencies of 5 to 40 cps

Frequency, f, cps	5	6	8	10	20	30	40
Probable limiting displacement (L, in.) for safety from damage	.04	.028	.015	.01	.0025	.001	—
Definite damage above displacement of	.40	.28	.15	.10	.025	.011	.006

Note: The above values define a "caution zone" above g = 0.01 with a "damage zone" above g = 1.0, where g = $4\pi^2 f^2 (L/12)/32.2$.

*From: "Seismic Effects of Quarry Blasting", U.S. Bureau of Mines Bulletin 442.

Table 6-14. EARTHQUAKE SEVERITY AND FREQUENCY

Modified Mercalli intensity scale	Human reactions and observed damage	Number of earthquakes in 50 years within each category	
		World	California
8	Considerable damage to ordinary substantial buildings	53	1
7	Damage negligible in buildings built to resist; some damage to other structures	670	4
6	Felt by all; slight damage to poorly constructed buildings	—	45
5	Felt indoors by nearly all, outdoors by many	—	—
4	Felt indoors by many, outdoors by few	—	—

Table 6-15. CORRECTION OF ACCELERATION DUE TO GRAVITY*

The following table gives the acceleration of gravity at sea level at various latitudes. The length of the simple pendulum with a period of two seconds, that is, which beats seconds, has been added to the table.

To correct the acceleration of gravity for altitude, subtract 0.0003086 cm/sec^2 for each meter of altitude; subtract 0.000003086 ft/sec^2 for each foot of altitude.

Latitude, degrees	Acceleration due to gravity		Length of seconds pendulum	
	cm/sec^2	ft/sec^2	cm	in.
0	978.039	32.0878	99.0961	39.0141
5	978.078	32.0891	99.1000	39.0157
10	978.195	32.0929	99.1119	39.0204
15	978.384	32.0991	99.1310	39.0279
20	978.641	32.1076	99.1571	39.0382
25	978.960	32.1180	99.1894	39.0509
30	979.329	32.1302	99.2268	39.0656
31	979.407	32.1327		
32	979.487	32.1353		
33	979.569	32.1380		
34	979.652	32.1407		
35	979.737	32.1435	99.2681	39.0819
36	979.822	32.1463		
37	979.908	32.1491		
38	979.995	32.1520		
39	980.083	32.1549		
40	980.171	32.1578	99.3121	39.0992
41	980.261	32.1607		
42	980.350	32.1636		
43	980.440	32.1666		
44	980.531	32.1696		
45[a]	980.621	32.1725	99.3577	39.1171
46	980.711	32.1755		
47	980.802	32.1785		
48	980.892	32.1814		
49	980.981	32.1844		
50	981.071	32.1873	99.4033	39.1351
51	981.159	32.1902		
52	981.247	32.1931		
53	981.336	32.1960		
54	981.422	32.1988		
55	981.507	32.2016	99.4475	39.1525
56	981.592	32.2044		
57	981.675	32.2071		
58	981.757	32.2098		
59	981.839	32.2125		
60	981.918	32.2151	99.4891	39.1689
65	982.288	32.2272	99.5266	39.1836
70	982.608	32.2377	99.5590	39.1964
75	982.868	32.2463	99.5854	39.2068
80	983.059	32.2525	99.6047	39.2144
85	983.178	32.2564	99.6168	39.2191
90	983.217	32.2577	99.6207	39.2207

[a] "Standard gravity" is 980.665 cm/sec or 32.174 ft/sec.

*Based on the formula of the U.S. Coast and Geodetic Survey.

Table 6-16.   COEFFICIENT OF FRICTION— IDENTICAL METALS*
COURTESY OF EDMOND E. BISSON AND DONALD H. BUCKLEY

The following table gives coefficients of kinetic sliding friction for polycrystalline pure metals in contact with themselves.

Metal	Coefficient of friction			
	Lubricated		Unlubricated	
	Oil or grease†	Solid film MoS₂	Dry-sliding in air	Vacuum, with surfaces cleaned
BODY-CENTERED CUBIC				
Iron on iron	0.15	.04–.08	1.0	Seizure (2)
Tantalum on tantalum	0.1	.04–.08	1.0	Seizure
Molybdenum on molybdenum	0.1	.04–.08	1.2	Seizure
Tungsten on tungsten	0.1	.04–.08	0.3	3.0 (3)
Chromium on chromium	0.34	.04–.08	0.4 (4)	—
FACE-CENTERED CUBIC				
Copper on copper	0.08	.04–.08	1.2–1.5	Seizure (6)
Nickel on nickel	0.28	.04–.08	0.8	Seizure
Silver on silver	0.55	.04–.08	1.5	Seizure
Gold on gold	0.2	.04–.08	2.0 (8)	Seizure
Aluminum on aluminum	0.12	.04–.08	1.0	Seizure
Platinum on platinum	0.25	.04–.08	1.2	Seizure
Rhodium on rhodium	0.1	.04–.08	0.4	3.0–5.0 (6)
Iridium on iridium	0.1	—	0.4	4.0
Lead on lead	0.1	—	2.0	—
HEXAGONAL				
Beryllium on beryllium	0.1	.04–.08	0.4	0.5 (6)
Magnesium on magnesium	0.08	.04–.08	0.4	0.6 (6)
Lanthanum on lanthanum	0.1	—	0.4 (6)	0.3 (6)
Titanium on titanium	0.1	.04–.08	0.6 (7)	1.2 (6)
Zirconium on zirconium	0.1	.04–.08	0.6	0.5 (6)
Rhenium on rhenium	0.1	.04–.08	0.4 (7)	0.3 (6)
Osmium on osmium	0.1	—	0.3	0.6 (6)
Ruthenium on ruthenium	0.1	—	0.3	0.5 (6)
Thallium on thallium	0.1	—	0.3 (6)	0.4 (6)
Cobalt on cobalt	0.1	.04–.08	0.5	0.4 (6)
Cadmium on cadmium	0.05	.04–.08	0.8 (4)	—
Zinc on zinc	0.04	—	0.9 (4)	—
RHOMBOHEDRAL				
Bismuth	—		0.9 (4)	—

†Paraffinic oil plus 1% lauric acid.

Metal	Surface film	Coefficient of friction
Copper on copper	Oxide	0.8 (4)
	Sulfide	0.7 (4)
Steel on steel	Oxide	
	Fe₂O₃	0.6 (9)
	Fe₃O₄	0.4 (9)
	Sulfide (FeS)	0.5 (9)
	Chloride (FeCl₂)	0.1 (9)
	Oleic Acid	0.1 (9)
	Graphite	0.1 (9)
	Teflon (PTFE)	0.04 (4)

*Values are principally from Reference 1 and NASA data, except where indicated by other reference numbers in parentheses.

Table 6-16. COEFFICIENT OF FRICTION— IDENTICAL METALS *(Continued)*

REFERENCES

1. "The Friction and Lubrication of Solids", F.P. Bowden and D. Tabor, Parts I and II, Oxford Press, Part I: 1958, Part II: 1964.
2. "Influence of Chemisorbed Films on Adhesion and Friction of Clean Iron", D.H. Buckley, NASA TN D–4775, 1968.
3. "Influence of Chemisorbed Films of Various Gases on Adhesion and Friction of Tungsten", D.H. Buckley, *J. Appl. Phys.*, 39(9):4224–4233, 1968.
4. "Friction and Wear", I.V. Kragelskii, Butterworths, 1965; available as an English translation.
5. "The Influence of Crystal Structure, Orientation and Solubility on the Adhesion and Sliding of Various Metal Single Crystals in Vacuum (10^{-11} Torr)", D.H. Buckley, ASTM SP–431, 1968, pp. 248–271.
6. "The Influence of Crystal Structure and Some Properties of Hexagonal Metals on Friction and Adhesion", D.H. Buckley and R.L. Johnson, *Wear*, 11:405–419, 1968.
7. "Friction and Wear of Materials", E. Rabinowicz, John Wiley & Sons, 1965.
8. "The Lubrication of Gold", M. Antler, *Wear*, 6:44–65, 1963.
9. "Advanced Bearing Technology", E.E. Bisson and W.J. Anderson, NASA SP–38, 1964.

Table 6-17. COEFFICIENT OF FRICTION— IDENTICAL ALLOY PAIRS

COURTESY OF EDMOND E. BISSON AND DONALD H. BUCKLEY

Coefficients of Kinetic Sliding Friction for Pairs of Identical Metal Alloys

Alloy	Coefficient of friction*			
	Lubricated		Unlubricated	
	Oil or grease†	Solid film MoS_2	Dry-sliding in air	Vacuum, with surface cleaned
1020 steel on 1020 steel	0.1	.04–.08	0.5 (1)	Seizure
52100 steel on 52100 steel	0.1	.04–.08	0.5 (1)	5.0 (1)
440 C S.S. on 440 C S.S.	0.1	.04–.08	0.4 (1)	2.5
304 S.S. on 304 S.S.	0.1	.04–.08	0.9 (1)	Seizure
Cast iron on cast iron	0.1	.04–.08	0.3 (4)	—
M–1 tool steel on M–1 tool steel	0.1	.04–.08	0.5 (1)	—
Brass on brass	0.1 (2)	.04–.08 (2)	0.4 (4)	—
Rene 41 on Rene 41	0.1	.04–.08	0.4 (1)	4.0
Inconel on Inconel	0.1	.04–.08	0.8 (1)	Seizure
Hastelloy D on Hastelloy D	0.1	—	0.7 (1)	Seizure
Cermet K 162 B on Cermet K 162 B	0.1	.04–.08	0.2 (1)	1.0
Stellite Star J on Stellite Star J	0.1	.04–.08	0.3 (1)	0.5 (1)
Co–25 Mo on Co–25 Mo	0.08	0.04	0.5 (3)	0.3
Ti–12 Sn on Ti–12 Sn	—	—	0.8 (3)	0.6 (3)
Ti–16 Al on Ti–16 Al	—	—	0.5 (3)	0.3 (3)

†Lubricated with a mineral oil containing oxidation and corrosion inhibitors.

*Data from NASA—Lewis Research Center, except where indicated by reference numbers in parentheses.

REFERENCES

1. "Advanced Bearing Technology", E.E. Bisson and W.J. Anderson, NASA SP–38, 1964.
2. "Friction and Wear", I.V. Kragelskii, Butterworths, 1965; available as an English translation.
3. "Friction and Wear of Hexagonal Metals and Alloys as Related to Crystal Structure and Lattice Parameters in Vacuum", D.H. Buckley and R.L. Johnson, *ASLE Trans.*, 9:121–135, 1966.
4. "Friction and Wear of Materials", E. Rabinowicz, John Wiley & Sons, 1965.

Table 6-18.   COEFFICIENT OF FRICTION—DISSIMILAR METALS

COURTESY OF EDMOND E. BISSON AND DONALD H. BUCKLEY

Following are coefficients of kinetic sliding friction for dissimilar pure metals in contact with each other.

Metal couple	Coefficient of friction*			
	Lubricated		Unlubricated	
	Oil or grease†	Solid film MoS$_2$	Dry-sliding in air	Vacuum, with surface cleaned
Aluminum on iron	0.1	.04–.08	1.1 (2)	Seizure
Aluminum on zinc	0.1 (1)	.04–.08	0.8 (2)	—
Cadmium on aluminum	0.1 (1)	.04–.08	0.6 (2)	—
Cadmium on bismuth	0.1 (1)	.04–.08	0.8 (2)	—
Cadmium on iron	0.1	.04–.08	0.6 (2)	—
Cadmium on zinc	0.1 (1)	.04–.08	0.6 (2)	—
Cobalt on iron	0.1	.04–.08	0.5 (2)	0.7 (3)
Cobalt on copper	0.1	.04–.08	0.9 (2)	—
Cobalt on aluminum	0.1 (1)	.04–.08	1.0 (2)	—
Copper on cadmium	0.1 (1)	.04–.08	0.9 (2)	—
Copper on zinc	0.1 (1)	.04–.08	0.9 (2)	—
Copper on iron	0.1	.04–.08	1.0 (2)	5.0 (1)
Copper on nickel	0.1	.04–.08	1.2 (4)	2.0 (4)
Copper on tungsten	0.1	.04–.08	0.4 (4)	0.5 (4)
Nickel on tungsten	0.1	.04–.08	0.3 (4)	4.0 (4)
Zinc on iron	0.1 (1)	.04–.08	0.9 (2)	—
Zinc on antimony	0.1 (1)	.04–.08	0.9 (2)	—
Zinc on bismuth	0.1 (1)	.04–.08	0.7 (2)	—

†Lubricated with mineral oil containing oxidation and corrosion inhibitors.

Material combination	Coefficient of friction*	
	Dry-sliding	Boundary lubrication‡
Hard steel on Babbitt (ASTM 1)	0.33 (6)	0.16 (6)
Hard steel on Babbitt (ASTM 8)	0.35 (6)	0.14 (6)
Hard steel on Babbitt (ASTM 10)	—	0.13 (6)
Monel on SAE 52100 bearing steel	0.4 (5)	0.33 (5)
Beryllium copper on SAE 52100 bearing steel	0.8 (5)	0.10 (5)
Brass on SAE 52100 bearing steel	0.5 (5)	0.12 (5)
Bronze on SAE 52100 bearing steel	0.3 (5)	0.17 (5)
Gray cast iron on SAE 52100 bearing steel	0.6 (5)	0.29 (5)
Nodular iron on SAE 52100 bearing steel	0.5 (5)	0.17 (5)
Nichrome V on SAE 52100 bearing steel	0.3 (5)	0.13 (5)
24ST-aluminum on SAE 52100 bearing steel	0.3 (5)	0.17 (5)

‡Paraffinic oil with oxidation and corrosion inhibitor.

*Values from NASA data, except where indicated by reference numbers in parentheses.

REFERENCES

1. "Friction and Wear", I.V. Kragelskii, Butterworths, 1965; available as an English translation.
2. "Surface Friction of Clean Metals—A Basic Factor in the Metal Cutting Process", H. Ernst and M.E. Merchant, *Proc. Conf. Friction and Surface Finish*, MIT, June 1940, pp. 76–101.
3. "Marked Influence of Crystal Structure on the Friction and Wear Characteristics of Cobalt and Cobalt Base Alloys in Vacuum to 10^{-9} Millimeter of Mercury", D.H. Buckley and R.L. Johnson, NASA TN D–2523, 1964.
4. "The Influence of Crystal Structure, Orientation and Solubility on the Adhesion and Sliding of Various Metal Single Crystals in Vacuum (10^{-11} Torr)", D.H. Buckley, ASTM SP–431, pp. 248–271, 1968.
5. "Investigation of Wear and Friction Properties Under Sliding Conditions of Some Materials, Suitable for Cages of Rolling-Contact Bearings", R.L. Johnson, M.A. Swikert, and E.E. Bisson, NACA Report 1062, 1952.
6. "Studies in Boundary Lubrication", W.E. Campbell, *Trans. ASME*, 61(7):633–641, 1939.

Table 6-19. COEFFICIENT OF FRICTION—SINGLE CRYSTALS*

COURTESY OF EDMOND E. BISSON AND DONALD H. BUCKLEY

The following tables give coefficients of kinetic sliding friction for single crystals—metals and non-metals.

METALS

Metal	Atomic plane	Coefficient of friction	
		In air, 20°C	In vacuum, clean
Single crystal copper on single crystal copper	(100)	0.60	>40
	(110)	0.40	>40
	(111)	0.21	21.0
Single crystal cobalt on polycrystalline cobalt	(0001)	0.40	0.35
	(10$\bar{1}$0)	—	0.80
Single crystal magnesium on polycrystalline magnesium	(0001)	0.30	0.40
	(10$\bar{1}$0)	—	0.90
Single crystal rhenium on polycrystalline rhenium	(0001)	0.20	0.29
	(1010)	0.25	0.38
Single crystal beryllium on polycrystalline beryllium	(0001)	0.45	0.48
	(10$\bar{1}$0)	0.46	0.51
Single crystal titanium on polycrystalline titanium	(0001)	0.48	0.56
	(10$\bar{1}$0)	0.25	0.36
Single crystal tungsten on single crystal tungsten	(100)	0.60	3.0 (2)
	(110)	0.41	1.9 (2)
	(210)	0.40	1.3 (2)

NON-METALS

Material	Atomic plane and direction		Coefficient of friction	
			In air, 20°C	In vacuum, clean
Diamond on diamond	(100)	<100>	0.15 (3, 4)	—
	(100)	<110>	0.05 (3, 4)	—
	(111)	—	0.05 (3, 4)	0.9 (4)
Sapphire on sapphire	(0001)	<11$\bar{2}$0>	0.15	0.50
	(0001)	<10$\bar{1}$0>	—	0.96
	(10$\bar{1}$0)	<11$\bar{2}$0>	0.20	0.93
	(10$\bar{1}$0)	<0001>	—	1.00
Diamond on magnesium oxide	(100)	<100>	0.07	—
Diamond on lithium fluoride	(100)	<110>	0.24	0.80
Diamond on potassium fluoride	(100)	<110>	0.71	—
Diamond on sodium chloride	(100)	<110>	0.47–0.70	—
Diamond on potassium bromide	(100)	<110>	0.85	—

*Data from Reference 1 unless otherwise indicated by reference numbers in parentheses.

REFERENCES

1. "The Influence of the Atomic Nature of Crystalline Materials on Friction", D.H. Buckley, *ASLE Trans.*, 2:89–100, 1968.
2. "Influence of Film of Various Gases on Adhesion and Friction of Tungsten", D.H. Buckley, *J. Appl. Phys.*, 39(9):4224–4233, 1968.
3. "The Abrasion of Diamond", M. Seal, *Roy. Soc. Proc.*, Series A, 248:379–393, 1958.
4. "The Friction and Lubrication of Solids", F.P. Bowden and D. Tabor, Part I, Oxford University Press, 1958.

Table 6-20.   COEFFICIENT OF FRICTION—NON-METALS*
COURTESY OF EDMOND E. BISSON AND DONALD H. BUCKLEY

Listed below are coefficients of kinetic sliding friction for plastics and other non-metals in identical pairs and on steel.

Material	Coefficient of friction			
	Dry-sliding in air		Vacuum, with clean surface	
	On itself	On steel	On itself	On steel
Teflon (PTFE)	0.1	0.04	—	.2–.3 (7)
Nylon	0.15–0.25	0.2	—	—
Perspex	0.8	0.5	—	—
Polystyrene	0.5	0.3	—	—
PCFE	0.2	0.08	—	0.3 (7)
Polyimide	—	0.25 (7)	0.5 (7)	0.2 (7)
Bakelite	0.3	0.30 (8)	—	—
Titanium carbide	0.2	0.5	0.9	—
Glass	1.0	0.6	—	—
Diamond	0.1 (2)	0.1	0.9	—
Sapphire	0.2 (3)	0.15 (4)	0.8 (5)	0.2 (5)
Mica	1.0	—	—	—
Carbon	0.2	0.15	—	0.4 (6)
Graphite	0.1	0.1	0.8	0.3 (6)

*Data from Reference 1 unless otherwise indicated by reference numbers in parentheses.

REFERENCES

1. "The Friction and Lubrication of Solids", F.P. Bowden and D. Tabor, Part I, Oxford University Press, 1954.
2. "The Abrasion of Diamond", M. Seal, *Roy. Soc. Proc.*, Series A, 248:379–393, 1958.
3. "Friction and Wear of Single Crystals", R.P. Steijn, *Wear-Usure-Verschleiss*, 7(1):48–66, 1964.
4. "Friction and Wear of Single-Crystal Sapphire Sliding on Steel", S.J. Duwell, *J. Appl. Phys.*, 33:2691–2698, 1962.
5. "Friction Characteristics in Vacuum of Single and Polycrystalline Aluminum Oxide in Contact with Themselves and with Various Metals", D.H. Buckley, *ASLE Trans.*, 10:134–145, 1965.
6. "Mechanism of Lubrication for Solid Carbon Materials in Vacuum to 10^{-9} Millimeters of Mercury", D.H. Buckley and R.L. Johnson, *ASLE Trans.*, 7(1):91–100, 1964.
7. "Degradation of Polymeric Composition in Vacuum to 10^{-9} mm Hg in Evaporation and Sliding Friction Experiments", D.H. Buckley and R.L. Johnson, *SPE Trans.*, 4(4):1–9, 1964.
8. M.B. Peterson and J.F. Murray, Requirements of Materials for Sliding and Rolling Contacts, "Boundary Lubrication", American Society of Mechanical Engineers, 1969, Chapter 9.

Table 6-21.   COEFFICIENT OF FRICTION—LUBRICATING POWDERS
COURTESY OF EDMOND E. BISSON AND DONALD H. BUCKLEY

Listed below are coefficients of kinetic sliding friction for steel on steel (SAE 4620 on SAE 1020) with various powders between the surfaces.

Powder	Coefficient of friction†	Powder	Coefficient of friction†
Cadmium iodide, CdI_2	0.06	Zinc stearate, $Zn (C_{18}H_{35}O_2)_2$	0.11
Cadmium chloride, $CdCl_2$	0.07	Cobalt chloride, $CoCl_2$	0.10
Tungsten disulfide, WS_2	0.08	Mercury iodide, HgI_2	0.18
Silver sulfate, Ag_2SO_4	0.14	Copper bromide, $CuBr_2$	0.06
Lead iodide, PbI_2	0.28	Silver iodide, AgI	0.25

†Data compiled from: "Advanced Bearing Technology", E.E. Bisson and W.J. Anderson, NASA SP–38, 1964.

Table 6-22.   HARDNESS AND ABRASIVE WEAR OF METALS

COURTESY OF EDMOND E. BISSON AND DONALD H. BUCKLEY

PURE METALS

Metal	Hardness,[†] kg/mm²	Relative abrasive wear resistance[‡]	Metal	Hardness,[†] kg/mm²	Relative abrasive wear resistance[‡]
Lead	4		Platinum	110	
Tin	10		Nickel	140	
Cadmium	20		Cobalt	150	
Aluminum	30	Increasing resistance to abrasive wear	Chromium	220	Increasing resistance to abrasive wear
Zinc	40		Titanium	250	
Gold	50		Rhodium	260	
Silver	60		Molybdenum	280	
Copper	80		Beryllium	330	
Palladium	100		Tungsten	440	
Zirconium	110				

†Values are approximate, from Reference 2. Some values may represent maximum attainable work hardnesses.
‡Data obtained from References 1 and 2.

VARIOUS ALLOYS

Alloy	Hardness, kg/mm²	Relative abrasive wear resistance**	Alloy	Hardness, kg/mm²	Relative abrasive wear resistance**
Babbitt	20		Tool steel	700–1,000	
Aluminum alloy	50	Increasing resistance to abrasive wear	Carburized steel	900	Increasing resistance to abrasive wear
Bronze	80		Nitrided steel	900	
Bearing steel	700–950		Tungsten carbide		
			Cermet	1,600	

**Data obtained from Reference 3. Resistance to abrasion by silica.

REFERENCES

1. "The Effect of Heat Treatment and Work Hardening on the Resistance to Abrasive Wear of Some Alloy Steels", M.M. Kruschov and M.A. Babichev. A translation from the Russian of: "Friction and Wear in Machinery", Vol. 19, published by ASME in 1965.
2. "Investigations in the Wear of Metals" (in Russian), M.M. Kruschov and M.A. Babichev, USSR Academy of Sciences, Moscow, 1960.
3. "Friction and Wear of Materials", E. Rabinowicz, John Wiley & Sons, 1965.

Table 6-23.   FRICTIONAL BEHAVIOR OF LOOSE POWDERS*

Coefficients of Friction with Specimens of Nickel-base Alloy; Load, 17 lb; Sliding Velocity, 16.7 ft/min

	Oxide	Molybdate	Tungstate	Sulfate	Silicate		Oxide	Molybdate	Tungstate	Sulfate	Silicate
Cu	0.42	—	0.41	—	—	W	0.54	—	—	—	—
Fe	0.70	0.42	0.49	—	—	Pb	0.11	0.29	0.35	—	0.19
Ni	0.70	0.29	0.51	—	—	Ag	—	0.28	—	0.14	—
Co	0.29	—	—	—	—	K	—	0.20	—	0.17	—
Zn	0.35	—	—	—	—	Na	—	0.17	0.19	0.70	0.30
Mo	0.21	—	—	—	—	B	0.11	—	—	—	—

*From: "Consideration of Lubricants for Temperatures Above 1 000 deg F", M.B. Peterson, J.J. Florek, and S.F. Murray, *ASLE Trans.*, 2(2): 225–234, 1960.

Table 6-24. WEAR OF METAL SURFACES
Causes, Classifications, and Avoidance of Mechanical Wear

Metal surface failures are mechanical or chemical in nature and frequently both. Several types may be classified and causes recognized. Common terminology associates the wear of metals with relative motion by sliding, rolling, abrasion, or impact. It is advantageous to identify the factors that accelerate the wear and the means by which failures may be prevented.

In general, a metal surface that is to resist wear must be hard and smooth and must operate under clean, non-corrosive conditions. Lubricating films (liquid, gas, powder) greatly reduce friction and prevent wear, but the conditions are complex with thousands of variations. Wear is minimized by low surface pressure and the maintenance of moderate operating temperature. Vibration, impact, distortion, and abrasion by hard particles are among the unfavorable conditions.

METAL SURFACE FAILURE BY WEAR
Descriptions and Prevention

Common names	Brief description	Typical examples	Causes	Superior materials	Prevented or reduced by
Scoring, scuffing, adhesive wear	Abrasion of surface asperities	Bearings, cylinder liners, shafts, gear teeth	Lubrication failure; high temperature	Hard, smooth-surfaced, incompatible pairs	Non-alloying, non-soluble pairs; low pressure; cooling; oil additives
Abrasion; grinding wear	Scratch tearing by hard particles	Grinders; crushers; earth and well tools; slurry handling; chute liners; conveyors	Impact and metal removal; abrasive foreign particles	Hard facing alloys; martensitic steels; chill-cast white irons; non-machinable alloys; high chromium cast iron	Remove or reduce source of abrasive contaminants
Spalling, galling	Intermittent local pressure welding	High-load and high-temperature bearings	Subsurface fatigue; high contact pressure and temperature	Alloy steels; high temperature alloys	High hot-hardness; EP lubricants; additive MoS_2, graphite
Fretting; frettage corrosion	Minute reciprocating motion with corrosive action	Cams, rockers, ball and roller bearings, king pins; joints subjected to vibration	Loose fits and wear	High-carbon, martensitic steel; surface-hardened alloy steel	Protection from oxygen; surface hardening; lubrication; improved fits
Impact wear; gauging wear; vibration wear	Momentary stress beyond elastic limit; fatigue	Power hammers; impact mills; power shovels; sandblast nozzles; frogs and switches; crushers	Cumulative deformation; fatigue cracking	Hard facing alloys; manganese steel; austenitic steels	Selection of alternate, hard materials of high compressive strength
Fluid erosion	Surface damage by fluid or fluid mixture	Propellers, pumps, engine cylinders, turbines; valve seats; jet devices	Fluid impact; cavitation	CoWCr alloys; hard face overlays containing WC	Reduction in fluid velocity, and of changes in direction of flow (fewer fittings in piping)
Corrosive pitting	Chemical action under stress concentration	Gear teeth; impellers; chemical equipment	Repetitive stresses and surface fatigue; corrosive substances; thermal stresses	High Ni, Cr, Si; high fatigue strength steels	High surface hardness; corrosion protection; reduced stress concentration

REFERENCES
"Handbook of Mechanical Wear", C. Lipson and L.V. Colwell, Eds., University of Michigan Press, 1961.
"Metals Handbook", Vol. 1: Properties and Selection, American Society for Metals, 1961.
"Composite Engineering Laminates", A.G.H. Dietz, Ed., MIT Press, 1969.
"Interdisciplinary Approach to Friction and Wear", P.M. Ku, Ed. (Symposium), NASA SP181, 1968.

Table 6-25.   COEFFICIENT OF FRICTION—DRY SURFACES*

Zero Wear Conditions; Low Pressure

SYMBOLS: H_m = microhardness, kg/mm^2 (Vickers)
μ = coefficient of friction

Material and No.	H_m	μ	Material and No.	H_m	μ
COMMON TEST SURFACE, HIGH HARDNESS					
Sliding Material 52100 Steel; H_m = 746 kg/mm^2; Paired with					
Aluminum 43	60.7	1.42	Aluminum 220	124.5	0.79
Aluminum 356	62.1	1.40	Steel 1060	397	0.73
Free-cut Invar 36, annealed	184	1.28	Monel C nickel alloy	184	0.73
Aluminum 355	90.5	1.21	Steel 8214	220	0.71
Stainless steel 321	224	1.16	Steel 8620	216	0.70
Stainless steel 347	252	1.15	Steel 1045	468	0.67
Aluminum 112	117	1.08	Steel 4150	276	0.67
Aluminum 195	96.8	1.07	Stainless steel 440C	296	0.66
Stainless steel 302	270	1.00	Steel 5130LL	260	0.62
Hy-Mu "80" nickel alloy	270	1.00	Steel 4140LL	384	0.57
Stainless steel 410	270	0.85	Sint. Fe-Cu 1, 15% Cu 7.1	220	0.47
Steel 1085	359	0.81	Sint Fe Cu 2, 20% Cu 5.8 6.2	190	0.43
Steel 1018	199	0.80	Sint steel 1, AISI 316L	220	0.34
Stainless steel 303EZ	296	0.79	Sint bronze 2, ASTM		
			B255, Type 11, 7.0	150	0.31
Steel 4620	242	0.79	Sint bronze 1, ASTM B202,		
			Type 11, Cl A, 6.4–6.8	135	0.26
COMMON TEST SURFACE, MEDIUM HARDNESS					
Sliding Material 302 Stainless Steel; H_m = 270 kg/mm^2; Paired with					
Aluminum 356	62.1	1.78	Steel 1060	397	0.88
Aluminum 43	60.7	1.67	Stainless steel 303EZ	296	0.86
Stainless steel 321	224	1.47	Steel 8214	220	0.86
Stainless steel 347	252	1.33	Steel 5130LL	260	0.84
Free-cut Invar 36, annealed	184	1.33	Steel 8620	216	0.83
Hy-Mu "80" nickel alloy	270	1.18	Steel 1018	199	0.80
Aluminum 195	96.8	1.17	Steel 4140LL	384	0.78
Aluminum 112	117	1.16	Steel 1045	468	0.71
Aluminum 355	90.5	1.11	Nylatron GS plastic	88	0.60
Steel 1085	359	1.03	Polystyrene plastic	45	0.60
Stainless steel 302	270	1.02	Zytel 101 nylon plastic	76	0.60
Stainless steel 410	270	1.00	Nylatron G plastic	88	0.57
Monel C nickel alloy	184	0.99	HD polyethylene 3 plastic		
Steel 4620	242	0.96	(Sp gr = .941)	8.2	0.40
Aluminum 220	124.5	0.92	Delrin acetal plastic	88	0.36
Stainless steel 440C	296	0.90	Teflon fluorocarbon plastic	16	0.09
Steel 4150	276	0.90			
COMMON TEST SURFACE, SOFT					
Sliding Material Brass; H_m = 115 kg/mm^2; Paired with					
Aluminum 43	60.7	1.12	Steel 4620	242	0.74
Aluminum 356	62.1	1.12	Steel 4140LL	384	0.73
Aluminum 355	90.5	1.11	Stainless steel 440C	296	0.72
Aluminum 220	124.5	0.95	Steel 1060	397	0.72
Aluminum 195	96.8	0.94	Stainless steel 302	270	0.70
Aluminum 112	117	0.92	Steel 8214	220	0.69
Monel C nickel alloy	184	0.85	Steel 8620	216	0.69
Stainless steel 347	252	0.82	Steel 1085	359	0.68
Hy-Mu "80" nickel alloy	270	0.81	Steel 4150	276	0.67
Steel 52100	220	0.80	Stainless steel 410	270	0.67
Stainless steel 321	224	0.78	Steel 1045	468	0.66
Stainless steel 300EZ	296	0.77	Steel 5130LL	260	0.62
Free-cut Invar 36,			Steel 1018	199	0.60
annealed, nickel alloy	184	0.76			

*Adapted from: "Designing for Zero Wear", R.G. Bayer, A.T. Shalkey, and A.R. Wayson *Machine Design*, 41(1):142; January 1969.

Table 6-26. COEFFICIENTS OF FRICTION FOR BOUNDARY LUBRICATION*

Zero Wear Conditions; Low Pressure

SYMBOLS:

H_m = microhardness, kg/mm^2 (Vickers)

μ_1 = coefficient of friction of two surfaces with a boundary lubrication of Socony Vacuum Gargoyle PE–797

μ_2 = coefficient of friction of two surfaces with a boundary lubrication of Esso Standard Millcot K–50

μ_{avg} = average of μ_1 and μ_2

Material and No.	H_m	μ_1	μ_2	μ_{avg}	Material and No.	H_m	μ_1	μ_2	μ_{avg}
COMMON TEST SURFACE, HIGH HARDNESS									
Sliding Material 52100 Steel; H_m = 746 kg/mm^2; Paired with									
Aluminum 220	124.5	.25	.23	.24	Steel 1085	359	.20	.14	.17
Sint Fe-Cu 2, 20% Cu					Steel 4620	242	.19	.15	.17
5.8–6.2	190	.21	.26	.235	Sint bronze 1, ASTM				
Free-cut Invar 36,					B202, Type 11, Cl A				
annealed	184	.24	.18	.21	6.4–6.8	135	.23	.11	.17
Stainless steel 303EZ	296	.21	.19	.20	Aluminum 43	60.7	.12	.22	.17
Aluminum 112	117	.25	.15	.20	Aluminum 356	62.1	.13	.20	.165
Sint bronze 2, ASTM					Steel 1045	468	.15	.17	.16
B255, Type 11, 7.0	150	.20	.20	.20	Steel 5130LL	260	.16	.16	.16
Steel 1018	199	.18	.21	.195	Stainless steel 347	252	.16	.15	.155
Steel 8620	216	.22	.17	.195	Stainless steel 440C	296	.18	.13	.155
Aluminum 356	62.1	.22	.17	.195	Stainless steel 321	224	.17	.13	.15
Sint Fe-Cu 1,					Stainless steel 410	270	.15	.15	.15
15% Cu 7.1	220	.20	.19	.195	Steel 4150	276	.15	.15	.15
Steel 4140LL	384	.21	.17	.19	Aluminum 195	96.8	.17	.13	.15
Steel 52100	220	.21	.16	.185	Sint steel 1, AISI 316L	220	.15	.15	.15
Steel 8214	220	.17	.19	.18	Hy-Mu "80" nickel alloy,				
Stainless steel 302	270	.19	.16	.175	annealed	270	.16	.13	.145
Steel 1060	397	.14	.21	.175	Monel C nickel alloy	184	.12	.14	.13
COMMON TEST SURFACE, MEDIUM HARDNESS									
Sliding Material 302 Stainless Steel; H_m = 270 kg/mm^2; Paired with									
Polystyrene plastic	45	.30	.31	.305	Aluminum 112	117	.20	.14	.18
HD Polyethylene 3					Free-cut Invar 36,				
plastic	8.2	.24	.30	.27	annealed	184	.16	.19	.175
Zytel 101 nylon plastic	76	.27	.27	.27	Steel 1085	359	.18	.15	.165
Aluminum 43	60.7	.24	.24	.24	Steel 1045	468	.16	.14	.15
Nylatron GS plastic	80	.24	.24	.24	Steel 5130LL	260	.16	.14	.15
Nylatron G plastic	88	.22	.24	.23	Monel C nickel alloy	184	.15	.15	.15
Steel 4620	242	.18	.15	.165	Steel 321	224	.15	.14	.145
Hy-Mu "80" nickel alloy	270	.16	.17	.165	Steel 8214	220	.16	.15	.155
Delrin acetal plastic	88	.15	.18	.165	Aluminum 220	124.5	.17	.14	.155
Steel 4150	276	.17	.15	.165	Stainless steel 347	252	.15	.15	.15
Stainless steel 302	270	.16	.15	.155	Stainless steel 410	270	.16	.14	.15
Stainless steel 303EZ	296	.16	.15	.155	Steel 1018	199	.14	.16	.15
Steel 1060	397	.16	.15	.155	Aluminum 195	96.8	.15	.14	.145
Aluminum 356	62.1	.18	.21	.195	Stainless steel 440C	296	.13	.15	.14
Aluminum 355	90.5	.17	.20	.185	Steel 4140LL	384	.14	.14	.14
Steel 8620	216	.21	.15	.18	Teflon fluorocarbon	16	.15	.11	.13

Table 6-26.   COEFFICIENTS OF FRICTION
FOR BOUNDARY LUBRICATION *(Continued)*

Material and No.	H_m	μ_1	μ_2	μ_{avg}	Material and No.	H_m	μ_1	μ_2	μ_{avg}
COMMON TEST SURFACE, SOFT									
Sliding Material Brass; H_m = 115 kg/mm^2; Paired with									
Steel 4620	242	.26	.27	.265	Steel 4150	276	.25	.19	.22
Aluminum 220	124.5	.26	.27	.265	Monel C nickel alloy	184	.22	.22	.22
Aluminum 356	62.1	.24	.25	.245	Stainless steel 347	252	.19	.23	.21
Aluminum 43	60.7	.23	.25	.24	Steel 8214	220	.22	.20	.21
Aluminum 112	117	.23	.25	.24	Aluminum 195	96.8	.17	.25	.21
Steel 4140LL	384	.22	.24	.23	Stainless steel 302	270	.22	.19	.205
Steel 5130LL	260	.26	.20	.23	Stainless steel				
Steel 52100	220	.26	.20	.23	300EZ	296	.22	.19	.205
Free-cut Invar 36,					Stainless steel 410	270	.23	.17	.20
annealed	184	.26	.20	.23	Steel 1085	359	.20	.20	.20
Hy-Mu "80" nickel alloy	270	.25	.21	.23	Stainless steel 321	224	.23	.13	.18
Aluminum 355	90.5	.22	.24	.23	Steel 8620	216	.21	.17	.18
Steel 1060	397	.22	.23	.225	Stainless steel 440C	296	.18	.16	.17
Steel 1018	199	.24	.20	.22	Steel 1045	468	.20	.12	.16

*Adapted from: "Designing for Zero Wear", R.G. Bayer, A.T.Shalkey, and A.R. Wayson, *Machine Design*, 41(1):142, January 9, 1969.

Table 6-27.   ROLLING FRICTION

A coefficient of friction is defined as $f = P/W$, the ratio of the frictional force (or resistance) to the normal load; for static or sliding friction, f is dimensionless. In the case of rolling friction, it is more practical to substitute $P = kW/r$, where r is the distance between the point of contact and the perpendicular resisting or propelling force, P. For practical purposes k may be considered as the coefficient of rolling friction when r is unity. Thus, k is not dimensionless.

These definitions imply that f and k do not vary with the load or the velocity. Since r may be considered as the length of a lever arm, it is evident that the resistance P will vary inversely as the radius of the wheel, ball, or roller.

Consider three practical cases: (1) a heavy disk, ball, or cylinder rolling on a flat surface; (2) ball or roller bearings; (3) rollers under skids for moving heavy loads. In the first two examples an average value of k would be used; in example 3 the two values of k might be widely different, and it would be necessary to write $P = (k_1 + k_2)W/r$.

Since rolling friction is caused by slight deflection of the materials and roughness of the surfaces, values of k will increase as the surfaces become softer and more irregular. Deflections, such as those of a tire, a resilient roller or ball, or a soft or rough road surface, will increase the friction. While friction tends to decrease as velocity increases, these changes are difficult to separate from the effects of other variables, such as the drag of grease or water, the effects of dust and sand on the surface, and the windage of rotating elements.

Some typical values of f, reported in the literature, are given in the following table, but in several cases the descriptions are inadequate.

Values of k in $P = kW/r$. If two rolling surfaces are involved, the values should be added as $P = (k_1 + k_2)W/r$.

D = 2r,		Wheel or roller	Rolling on	k
in.	cm			
>0.5	>1.3	Steel, hard, polished	Steel, hard, polished	.000 2–.000 4
>0.5	>1.3	Iron or steel	Iron or steel	.002
2–10	5–25	Hardwood	Hardwood	.02
>0.5	>1.3	Steel, smooth, clean	Steel, smooth, clean	.000 5–.002
>0.5	>1.3	Steel, smooth, oiled	Steel, smooth, oiled	.001–.002
>1	>2.5	Steel, rusty	Steel, rusty	.005–.01
>10	>25	Tire, pneumatic[a]	Concrete, smooth	.01–.03

Note: Loads are assumed to be less than those causing permanent set of surface material.

Extensive experiments on clean, smooth surfaces indicate that if the simple equation $P = kW/r$ is used, the value of k increases as the load, W, is increased; k decreases as the radius, r, is increased. There is more or less deformation, especially on the first pass. Its magnitude depends on the load and on the elastic and plastic properties of the surfaces. After many passes (as in a rolling bearing), equilibrium is established, but friction work remains, as elastic hysteresis and edge slip.

[a]Normal recommended inflation and load.

REFERENCE

"The Friction and Lubrication of Solids", F.P. Bowden and D. Tabor, Vol. 2, Oxford University Press, 1964, Chapters 15 and 24.

Table 6-28. COEFFICIENTS OF STATIC AND SLIDING FRICTION*

Reference letters indicate the lubricant used; numbers in parentheses give sources (see References).

KEY TO LUBRICANTS USED:

a = oleic acid
b = Atlantic spindle oil (light mineral)
c = castor oil
d = lard oil
e = Atlantic spindle oil plus 2% oleic acid
f = medium mineral oil
g = medium mineral oil plus $\frac{1}{2}$% oleic acid
h = stearic acid
i = grease (zinc oxide base)
j = graphite
k = turbine oil plus 1% graphite
l = turbine oil plus 1% stearic acid

m = turbine oil (medium mineral)
n = olive oil
p = palmitic acid
q = ricinoleic acid
r = dry soap
s = lard
t = water
u = rape oil
v = 3-in-1 oil
w = octyl alcohol
x = triolein
y = 1% lauric acid in paraffin oil

Materials	Static		Sliding	
	Dry	Greasy	Dry	Greasy
Hard steel on hard steel	0.78 (1)	0.11 (1, a)	0.42 (2)	0.029 (5, h)
		0.23 (1, b)		0.081 (5, c)
		0.15 (1, c)		0.080 (5, i)
		0.11 (1, d)		0.058 (5, j)
		0.0075 (18, p)		0.084 (5, d)
		0.0052 (18, h)		0.105 (5, k)
				0.096 (5, l)
				0.108 (5, m)
				0.12 (5, a)
Mild steel on mild steel	0.74 (19)		0.57 (3)	0.09 (3, a)
				0.19 (3, u)
Hard steel on graphite	0.21 (1)	0.09 (1, a)		
Hard steel on Babbitt (ASTM 1)	0.70 (11)	0.23 (1, b)	0.33 (6)	0.16 (1, b)
		0.15 (1, c)		0.06 (1, c)
		0.08 (1, d)		0.11 (1, d)
		0.085 (1, e)		
Hard steel on Babbitt (ASTM 8)	0.42 (11)	0.17 (1, b)	0.35 (11)	0.14 (1, b)
		0.11 (1, c)		0.065 (1, c)
		0.09 (1, d)		0.07 (1, d)
		0.08 (1, e)		0.08 (11, h)
Hard steel on Babbitt (ASTM 10)		0.25 (1, b)		0.13 (1, b)
		0.12 (1, c)		0.06 (1, c)
		0.10 (1, d)		0.055 (1, d)
		0.11 (1, e)		
Mild steel on cadmium silver				0.097 (2, f)
Mild steel on phosphor bronze			0.34 (3)	0.173 (2, f)
Mild steel on copper lead				0.145 (2, f)
Mild steel on cast iron		0.183 (15, c)	0.23 (6)	0.133 (2, f)
Mild steel on lead	0.95 (11)	0.5 (1, f)	0.95 (11)	0.3 (11, f)
Nickel on mild steel			0.64 (3)	0.178 (3, x)
Aluminum on mild steel	0.61 (8)		0.47 (3)	
Magnesium on mild steel			0.42 (3)	
Magnesium on magnesium	0.6 (22)	0.08 (22, y)		
Teflon on Teflon	0.04 (22)			0.04 (22, f)
Teflon on steel	0.04 (22)			0.04 (22, f)
Tungsten carbide on tungsten carbide	0.2 (22)	0.12 (22, a)		
Tungsten carbide on steel	0.5 (22)	0.08 (22, a)		
Tungsten carbide on copper	0.35 (23)			
Tungsten carbide on iron	0.8 (23)			
Bonded carbide on copper	0.35 (23)			
Bonded carbide on iron	0.8 (23)			
Cadmium on mild steel			0.46 (3)	
Copper on mild steel	0.53 (8)		0.36 (3)	0.18 (17, a)
Nickel on nickel	1.10 (16)		0.53 (3)	0.12 (3, w)
Brass on mild steel	0.51 (8)		0.44 (6)	
Brass on cast iron			0.30 (6)	
Zinc on cast iron	0.85 (16)		0.21 (7)	
Magnesium on cast iron			0.25 (7)	
Copper on cast iron	1.05 (16)		0.29 (7)	
Tin on cast iron			0.32 (7)	
Lead on cast iron			0.43 (7)	

Table 6-28. COEFFICIENTS OF STATIC AND SLIDING FRICTION
(Continued)

Materials	Static		Sliding	
	Dry	Greasy	Dry	Greasy
Aluminum on aluminum	1.05 (16)		1.4 (3)	
Glass on glass	0.94 (8)	0.01 (10, p)	0.40 (3)	0.09 (3, a)
		0.005 (10, q)		0.116 (3, v)
Carbon on glass			0.18 (3)	
Garnet on mild steel			0.39 (3)	
Glass on nickel	0.78 (8)		0.56 (3)	
Copper on glass	0.68 (8)		0.53 (3)	
Cast iron on cast iron	1.10 (16)		0.15 (9)	0.070 (9, d)
				0.064 (9, n)
Bronze on cast iron			0.22 (9)	0.077 (9, n)
Oak on oak (parallel to grain)	0.62 (9)		0.48 (9)	0.164 (9, r)
				0.067 (9, s)
Oak on oak (perpendicular)	0.54 (9)		0.32 (9)	0.072 (9, s)
Leather on oak (parallel)	0.61 (9)		0.52 (9)	
Cast iron on oak			0.49 (9)	0.075 (9, n)
Leather on cast iron			0.56 (9)	0.36 (9, t)
				0.13 (9, n)
Laminated plastic on steel			0.35 (12)	0.05 (12, t)
Fluted rubber bearing on steel				0.05 (13, t)

*From: "Standard Handbook for Mechanical Engineers", 7th ed., T. Baumeister, Ed., McGraw-Hill Book Company, 1967.

REFERENCES

(1) Campbell, *Trans. ASME*, 1939; (2) Clarke, Lincoln, and Sterrett, *Proc. API*, 1935; (3) Beare and Bowden, *Phil. Trans. Roy. Soc.*, 1935; (4) Dokos, *Trans. ASME*, 1946; (5) Boyd and Robertson, *Trans. ASME*, 1945; (6) Sachs, *zeit. f. angew. Math. und Mech.*, 1924; (7) Honda and Yama la, *Jour. I of M*, 1925; (8) Tomlinson, *Phil. Mag.*, 1929; (9) Morin, *Acad. Roy. des Sciences*, 1838; (10) Claypoole, *Trans. ASME*, 1943; (11) Tabor, *Jour. Applied Phys.*, 1945; (12) Eyssen, General Discussion on Lubrication, *ASME*, 1937; (13) Brazier and Holland-Bowyer, General Discussion on Lubrication, *ASME*, 1937; (14) Burwell, *Jour. SAE*, 1942; (15) Stanton, "Friction", Longmans; (16) Ernst and Merchant, Conference on Friction and Surface Finish, M.I.T., 1940; (17) Gongwer, Conference on Friction and Surface Finish, M.I.T., 1940; (18) Hardy and Bircumshaw, *Proc. Roy. Soc.*, 1925; (19) Hardy and Hardy, *Phil. Mag.*, 1919; (20) Bowden and Young, *Proc. Roy. Soc.*, 1951; (21) Hardy and Doubleday, *Proc. Roy. Soc.*, 1923; (22) Bowden and Tabor, "The Friction and Lubrication of Solids", Oxford; (23) Shooter, *Research*, 4, 1951.

Table 6-29. TIRE FRICTION AND STOPPING DISTANCE

From assumed values for tire friction coefficients and car design, it is possible to compute a braking distance for full stop from a given steady car speed. To this must be added the perception-reaction distance. Extensive tests on brake-reaction time show that most drivers can react in one-half second when fully attentive (test conditions). Great variations in road surfaces and conditions, in tire design and conditions, and in vehicle design and brake conditions also make it dangerous to specify either average or minimum stopping distance.

The following values are based on a perception-reaction time of 2.5 seconds, friction coefficients as shown, and an "average" passenger car of the 1950's. They should be considered only as rough guides to define the problem.

Distances in Feet (Meters)

Conditions and results	Original car speed, mi/hr (km/hr)									
	30	(48)	40	(64)	50	(80)	60	(97)	70	(113)
DRY PAVEMENT										
Coefficient of friction	.62		.60		.58		.56		.55	
Perception-reaction distance	110	(33.5)	150	(45.7)	185	(56.4)	220	(67.0)	260	(79.2)
Braking distance	50	(15.2)	90	(27.4)	145	(44.2)	215	(65.5)	300	(91.4)
Total distance to stop	160	(48.8)	240	(73.2)	330	(100)	435	(132)	560	(171)
WET PAVEMENT										
Coefficient of friction	.36		.33		.31		.30		.29	
Perception-reaction distance	110	(33.6)	150	(45.7)	185	(56.4)	220	(67.0)	260	(79.2)
Braking distance	86	(26.2)	170	(51.8)	280	(85.3)	415	(126)	585	(178)
Total distance to stop	196	(59.7)	320	(97.5)	465	(142)	635	(194)	845	(258)

Table 6-30. SAE VISCOSITY NUMBERS FOR CRANKCASE OILS*

| SAE No. | Official values† | | | | For information only | | | | | |
| | At 0°F, centipoises | | At 210°F, centistokes | | At 0°F, centistokes | | At 0°F, Saybolt Universal Seconds | | At 210°F, Saybolt Universal Seconds | |
	Min‡	Max	Min	Max	Min	Max	Min‡	Max	Min	Max
5W		1,200	3.9			1,300		6,000	39	
10W	1,200	2,400	3.9		1,300	2,600	6,000	12,000	39	
20W	2,400	9,600	3.9		2,600	10,500	12,000	48,000	39	
20			5.7	9.6					45	58
30			9.6	12.9					58	70
40			12.9	16.8					70	85
50			16.8	22.7					85	110

†Official values at 0°F from ASTM D2602, "Method of Test for Apparent Viscosity of Motor Oils at Low Temperatures Using Cold Cranking Simulator", 1967, and at 210°F from ASTM D445, "Method of Test for Viscosity of Transparent and Opaque Liquids".

‡Minimum values at 0°F may be waived provided viscosity at 210°F is not below 4.2 centistokes (40 Saybolt universal seconds) or 5.7 centistokes (45 Saybolt Universal Seconds) for 10W and 20W, respectively.

*Adapted from: "Crankcase Oil Viscosity Classification", SAE Recommended Practice J300a, Society of Automotive Engineers, rev. April 1967.

Table 6-31. LUBRICANT CLASSIFICATIONS BY VISCOSITY*

ABBREVIATIONS AND SYMBOLS:

SUS = Saybolt Universal Seconds   S = spindle applications
cSt = centistokes   W = slideways applications
H = hydraulic-system applications

Classification number	Viscosity	Temperature, °F	Viscosity	Temperature, °F
American Gear Manufacturers Association (AGMA)				
1	180–240 SUS	100		
2	280–360 SUS	100		
3	490–700 SUS	100		
4	700–1,000 SUS	100		
5			80–105 SUS	210
6			105–125 SUS	210
7 comp[a]			125–150 SUS	210
8 comp[a]			150–190 SUS	210
8A comp[a]			190–250 SUS	210
American Society of Lubrication Engineers (ASLE)				
H–150	135–165 SUS	100	28.7–35.4 cSt	100
H–215	194–236 SUS	100	41.9–51.0 cSt	100
H–315	284–346 SUS	100	61.4–74.9 cSt	100
H–700	630–770 SUS	100	136–167 cSt	− 100
S–32	29–35 SUS	100	1.1–2.7 cSt	100
S–60	54–66 SUS	100	8.6–12.0 cSt	100
S–105	95–115 SUS	100	19.4–24.1 cSt	100
W–150	135–165 SUS	100	28.7–35.4 cSt	100
W–315	284–346 SUS	100	61.4–74.9 cSt	100
W–1000	900–1,100 SUS	100	195–238 cSt	100
Society of Automotive Engineers (SAE): transmission and axle lubricants[d]				
75	15,000 SUS max	0	40 SUS min	210
80[b]	15,000–100,000 SUS	0	40 SUS min	210
90[c]			75–120 SUS	210
140			120–200 SUS	210
250			200 SUS min	210

[a] Oils marked *comp* are those compounded with 3 to 10% of acidless tallow or other suitable animal fats.

[b] May waive minimum at 0°F if viscosity at 210°F is greater than 48 SUS.

[c] May waive maximum at 210°F if viscosity at 0°F is 750,000 SUS or less (extrapolated).

[d] From: "1970 SAE Handbook"; viscosity values at 0°F are under revision.

*From: "Lubrication and Lubricants", *Encyclopedia of Chemical Technology*, R.E. Kirk and D.F. Othmer, Eds., Vol. 12, John Wiley & Sons, 1967.

Table 6-32.   REPRESENTATIVE PETROLEUM LUBRICATING OILS*

For viscosity in m²/s, multiply values in centistokes by 10^{-6}. For density in kg/m³, multiply values in g/ml by 1000. For temperature in K, see Table 9-22 for conversion.

SYMBOLS:

R = rust inhibitor	P = pour-point depressant
O = oxidation inhibitor	W = antiwear
D = detergent-dispersant	EP = extreme pressure
VI = viscosity-index improver	F = antifoam

Type	Viscosity				Flash point, °F	Pour point, °F	Density, g/ml at 60°F	Viscosity index	Common additives
	Centistokes		Saybolt Universal Seconds						
	100°F	210°F	100°F	210°F					
Automotive: automobile, truck, and marine reciprocating engines									
SAE									
10W	41	6.0	190	46	410	−15	0.870	102	R, O, D, VI,
20W	71	8.5	330	54	440	−10	0.885	96	P, W, F
30	114	11.3	530	64	460	−5	0.891	92	
40	173	14.8	800	77	475	0	0.890	90	
50	270	19.7	1250	97	490	10	0.902	90	
10W–30	76	12.7	354	69	410	−35	0.876	141	
20W–40	93	13.7	430	73	410	−35	0.892	135	
Gear: automotive and industrial gear units									
SAE									
75	47	7.3	220	50	380	−10	0.900	121	EP, O, R, P,
80	69	7.9	320	52	365	−25	0.934	78	F
90	287	20.4	1330	100	450	−10	0.930	91	
140	725	34	3350	160	500	0	0.937	82	
250	1220	47	5660	220	490	5	0.940	83	
Automatic transmission: automotive hydraulic systems									
	40	7.3	185	50	360	−55	0.887	141	R, O, W, F, VI, P
Turbine: steam turbines, electric motors, steel mills, and industrial circulating systems									
Light	32	5.4	150	44	410	0	0.872	109	R, O, W
Medium	65	8.2	300	53	455	10	0.877	105	
Heavy	99	10.8	460	62	470	10	0.885	100	
Hydraulic: machine-tool hydraulic systems, aircraft hydraulic systems (extra-low temperature)									
Light	32	4.8	150	42	370	−45	0.887	64	R, O, W
Medium	67	7.3	310	50	405	−15	0.895	66	
Heavy	196	14.0	910	74	495	10	0.901	70	
Extra-low temperature	14	5.2	74	43	230	−80	0.844	226	R, O, W, VI, P
Machine oil: spindles, machine tools, and general once-through lubrication									
	22	3.9	105	39	350	25	0.881	80	None
	44	6.0	205	46	390	25	0.898	80	
	66	7.0	305	49	365	10	0.915	83	
	110	9.9	510	59	390	5	0.915	25	
	200	15.5	930	80	455	15	0.890	80	
Aviation: reciprocating aircraft engines									
Grade									
80	140	15.0	650	78	465	−15	0.887	114	D, VI, F
100	238	21.0	1100	103	515	−10	0.892	110	
Refrigeration: refrigerator compressors and low temperatures									
	33	4.9	154	42	375	−45	0.888	63	None
	68	7.4	314	50	415	−40	0.894	70	

*Abstracted from: "Lubrication and Lubricants", R.E. Lee, Jr., and E.R. Booser, *Encyclopedia of Chemical Technology*, R.E. Kirk and D.F. Othmer, Eds., Vol. 12, John Wiley & Sons, 1968, pp. 566–567.

Table 6-33. PROPERTIES OF REPRESENTATIVE SYNTHETIC OILS*

For viscosity in m^2/s, multiply values in centistokes by 10^{-6}. For temperature in K, see Table 9-22 for conversion.

Type	Viscosity, centistokes			Pour point, °F	Flash point, °F	Typical uses
	210°F	100°F	−65°F			
Silicones						
SF–96(50)[a]	16	38	460	−65	600+	Wide-temperature hydraulic
SF–96(1000)[a]	270	680	7,000	−55	600+	and damping fluids
SF–1038(100)[a]	30	75	1,400	−100	560	High-temperature greases
F–50[a]	16	52	2,500	−100	550	Aircraft, missile, and vacuum lubrication and hydraulic fluid
Organic esters						
MIL-O-6085	3.5	13.5	10,000	−90	450	Aircraft hydraulic and instrument fluid
MIL-L-7808	3.8	15.2	11,600	−80	450	Jet engines
MIL-L-9236	3.6	15.8	20,450	−80	460	High-temperature jet engines
Polyglycol						
LB-300-X[b]	11	65		−40	490	Textile and nonsludging
50-HB-2000[b]	72	433		25	440	high-temperature industrial use
Celluguard 200[c]				−20	None	Water-base fire-resistant hydraulic fluid
Phosphate						
Tricresyl phosphate	4.4	35.1		−15	465	Fire-resistant fluids for die
Pydraul F-9[d]	5.8	54		−5	430	casting, air compressors,
Cellulube 220[c]	5.1	47		0	500	and hydraulic systems
Skydrol 500A[d]	3.9	11.5	2,500	−85	355	Aircraft hydraulic fluid
Polybutene						
L-10[d]	4.4	23.6		−50	250	Electrical and refrigerator
H-35[e]	80	1,710		0	295	oils, kilns and ovens
Polyphenyl ether						
VRT-E[f]	6.0	61		10	465	Radiation-resistant and high-temperature aircraft use
Silicate						
OS-45[d]	4.0	12.2	2,230	−85	370	Electronic coolant, military hydraulic fluid
Chlorinated aromatic						
Aroclor 1248[d]	3.1	48		20	380	Fire-resistant, usually blended with phosphate
Fluorochemical						
Fluorolube S[g]	4.6	24.1			None	Oxygen compressors, liquid-oxygen systems

[a] Supplied by General Electric Co.
[b] Supplied by Union Carbide Chemicals Co.
[c] Supplied by Stauffer Chemical Co.
[d] Supplied by Monsanto Chemical Co.
[e] Supplied by Amoco Chemicals Co.
[f] Supplied by Shell Oil Co.
[g] Supplied by Hooker Chemical Co.

*Abstracted from: "Lubrication and Lubricants", R.E. Lee, Jr., and E.R. Booser, *Encyclopedia of Chemical Technology*, R.E. Kirk and D.F. Othmer, Eds., Vol. 12, John Wiley & Sons, 1968, p. 578.

Table 6-34. EFFECT OF NUCLEAR RADIATION ON LUBRICATING OILS

Table A. EFFECT OF RADIATION ON CONVENTIONAL PETROLEUM TURBINE OILS*

For radiation dose absorbed in MJ/kg, divide values in millions of rads by 100. For viscosity in m^2/S, multiply values in SUS at 210°F by 4.635×10^{-6}; multiply values in SUS at 100°F by 4.667×10^{-6}. For density in kg/m^3, see Table 9-27 for conversion from gravity in degrees API. For temperature in K, see Table 9-22 for conversion.

	Radiation dosage, rads, in millions				
	0	9	45	90	270
Viscosity, Saybolt Universal Seconds					
At 100°F	152.4	155.1	162.8	174.5	224
At 210°F	43.8	44.1	44.6	45.5	49.2
Viscosity index	108	110	110	109	112
Gravity, °API	31.5	31.5	31.3	31.4	30.7
Color, ASTM	1.0	1.0	1.25	1.25	2.5
Flash point, open-cup, °F	430	410	405	415	295
Fire point, open-cup, °F	485	470	475	485	475
Pour point, °F	0	0	0	10	10
Neutralization No., mg of KOH/g	0.09	0.04	0.03	0.01	0.01

*From: "The Effects of High Energy Ionizing Radiation on Turbine Oil Performance Characteristics", C.F. Kottcamp, R.P. Nejak, and R.T. Kern, *ASLE Trans.*, 2(1):7–12, 1959.

Table B. RADIATION TOLERANCE LIMIT FOR SEVERAL OIL TYPES†

For radiation dose absorbed in J/kg, divide values in rads by 100.

Oil	Tolerance limit, for 25% increase in 100 deg F viscosity, rads	Reference	Oil	Tolerance limit, for 25% increase in 100 deg F viscosity, rads	Reference
Petroleum			**Synthetic**		
SAE 10 automotive	1.3×10^8	1	Diester	1.1×10^8	4
SAE 20 automotive	$0.4–2.4 \times 10^8$	1	Polypropylene oxide	1.0×10^8	1
Light turbine oil	1.5×10^8	3	Alkylbenzene	5×10^8	1
Machine oil	2.5×10^8	3	Methyl silicone	$< 1.0 \times 10^8$	1
Marine engine oil	1.0×10^8	3	Methyl phenyl silicone	1×10^8	2
Steam cylinder oil	1.3×10^8	3	Tetraaryl silicate	0.6×10^8	1
			Dichlorobiphenyl	3×10^8	1

†Abstracted from: "Lubrication and Lubricants", R.E. Lee, Jr., and E.R. Booser, *Encyclopedia of Chemical Technology*, R.E. Kirk and D.F. Othmer, Eds., Vol. 12, John Wiley & Sons, 1968, p. 612.

REFERENCES

1. "Radiation Damage to Liquids and Organic Materials", V.P. Calkins, *Nuclear Engineering Handbook*, McGraw-Hill Book Company, 1958, pp. 126–148.
2. "Effects of Nuclear Radiation on Hydrocarbon Oils, Greases, and Some Synthetic Fluids", V.W. David and R. Irving, Conference on Lubrication and Wear, Institution of Mechanical Engineers, London, October, 1957.
3. J.G. Carroll and S.R. Calish, *Lubrication Eng.*, 13(7):388–392, 1957.
4. W.L.R. Rice, *Lubrication Eng.*, 16(4):157–160, 1960.

Table 6-35. USEFUL RANGE OF SOLIDS FOR DRY LUBRICATION*

For load in MN/m², multiply values in psi by 6.8948×10^{-3}.

Lubricant	Load,[b] psi	Temperature, °F
Molybdenum disulfide	2,000 to yield point of metal	−300 to +750[a]
Graphite powder	2,000 to 100,000	−300 to +1,200
Tungsten disulfide powder	2,000 to yield point of metal	low not investigated, max 850[a]
Polytetrafluoroethylene (PTFE) powder and sintered shapes	150 to 3,000	−300 to +500
Polytetrafluoroethylene and other plastics containing fillers	50 to 4,000	max 500
Organic binder coatings, graphite—MoS₂ type	2,000 to yield point of metal	−100 to 500
Inorganic binder coatings, graphite—MoS₂ type	2,000 to yield point of metal	−300 to 1,000

[a] In inert and reducing atmospheres these solids are used at temperatures above 2000°F.

[b] Speed affects the load capacity to a marked degree, but solid lubricants should only be considered at low speeds. As speed increases, load-carrying capacity decreases.

*From: "Solid and Bonded-Film Lubricants", A. DiSapio and H.S. Gerstung, *Machine Design*: The Bearings Reference Issue, 38(6):8–11, March 10, 1966.

Table 6-36. VISCOSITY-PRESSURE COEFFICIENTS FOR SEVERAL PETROLEUM OILS*

The following viscosity-pressure coefficients are used in the approximate equation

$$\mu_p = \mu_o \, e^{Bp}$$

where μ_p = absolute viscosity at pressure p. For N·s/m² (= kg/m·s) divide values in centipoises by 1000.

μ_o = absolute viscosity at atmospheric pressure (essentially zero pressure)

e = base of natural logarithms, 2.71828 . . .

B = viscosity-pressure coefficient in in.²/lbf. For m²/N multiply tabulated values by 1.4503×10^{-4}.

p = pressure in psi for B values as tabulated

Example: For a paraffinic oil having an absolute viscosity of 12.62 centipoise at 210°F at atmospheric pressure, the viscosity at 100,000 psi will be approximately

$$\mu_p = 12.62e\,(0.000095)\,(100,000) \approx 1.7 \times 10^5 \text{ centipoise.}$$

Property	Oil type					
	Paraffinic			Naphthenic		
Absolute viscosity, centipoise at 1 atm						
100 deg F (311 K)	48.0	144.3	523	49.9	131.4	485
210 deg F (372 K)	5.94	12.62	27.6	5.07	8.32	15.49
Viscosity index	96	99	91	23	8	−42
Molecular weight	450	576	700	349	383	400

Temperature, deg F (K)	Average viscosity-pressure coefficient B					
32 (273)	.000199	.000197	.000265	.000292	.00034	.00048
77 (298)	.000145	.000147	.000184	.000214	.000258	.000323
100 (311)	.000129	.000136	.000161	.000187	.000218	.000283
210 (372)	.000087	.000095	.000109	.000111	.000131	.000154
425 (491)	.000057	.000064	.000067	.000059	.000071	.000081

*For additional information see "Viscosity and Density of Over 40 Lubricating Fluids of Known Composition at Pressures to 150,000 psi and Temperatures to 425°F", Vols. I and II, American Society of Mechanical Engineers (ASME), 1953.

Figure 6-37.   PROPERTIES OF SECTIONS*

SYMBOLS: I = moment of inertia

$\dfrac{I}{c}$ = section modulus

r = radius of gyration

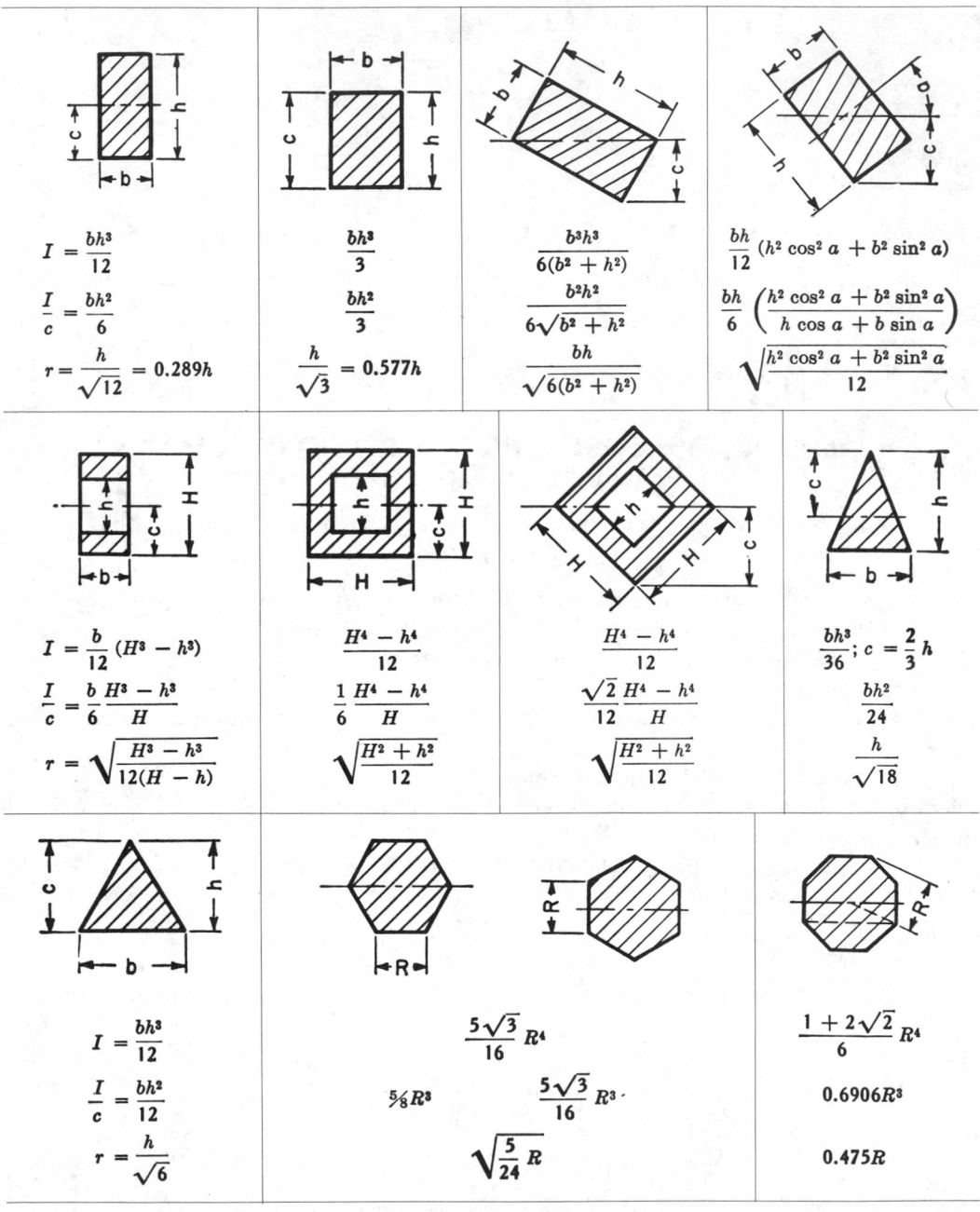

$$I = \frac{bh^3}{12}$$

$$\frac{I}{c} = \frac{bh^2}{6}$$

$$r = \frac{h}{\sqrt{12}} = 0.289h$$

$$\frac{bh^3}{3}$$

$$\frac{bh^2}{3}$$

$$\frac{h}{\sqrt{3}} = 0.577h$$

$$\frac{b^3h^3}{6(b^2 + h^2)}$$

$$\frac{b^2h^2}{6\sqrt{b^2 + h^2}}$$

$$\frac{bh}{\sqrt{6(b^2 + h^2)}}$$

$$\frac{bh}{12}(h^2 \cos^2 a + b^2 \sin^2 a)$$

$$\frac{bh}{6}\left(\frac{h^2 \cos^2 a + b^2 \sin^2 a}{h \cos a + b \sin a}\right)$$

$$\sqrt{\frac{h^2 \cos^2 a + b^2 \sin^2 a}{12}}$$

$$I = \frac{b}{12}(H^3 - h^3)$$

$$\frac{I}{c} = \frac{b}{6}\frac{H^3 - h^3}{H}$$

$$r = \sqrt{\frac{H^3 - h^3}{12(H - h)}}$$

$$\frac{H^4 - h^4}{12}$$

$$\frac{1}{6}\frac{H^4 - h^4}{H}$$

$$\sqrt{\frac{H^2 + h^2}{12}}$$

$$\frac{H^4 - h^4}{12}$$

$$\frac{\sqrt{2}}{12}\frac{H^4 - h^4}{H}$$

$$\sqrt{\frac{H^2 + h^2}{12}}$$

$$\frac{bh^3}{36}; c = \frac{2}{3}h$$

$$\frac{bh^2}{24}$$

$$\frac{h}{\sqrt{18}}$$

$$I = \frac{bh^3}{12}$$

$$\frac{I}{c} = \frac{bh^2}{12}$$

$$r = \frac{h}{\sqrt{6}}$$

$$\frac{5\sqrt{3}}{16}R^4$$

$$\tfrac{5}{8}R^3$$

$$\frac{5\sqrt{3}}{16}R^3$$

$$\sqrt{\frac{5}{24}}R$$

$$\frac{1 + 2\sqrt{2}}{6}R^4$$

$$0.6906R^3$$

$$0.475R$$

Square, axis same as first rectangle, side $= h$; $I = h^4/12$; $I/c = h^3/6$; $r = 0.289h$.
Square, diagonal taken as axis: $I = h^4/12$; $I/c = 0.1179h^3$; $r = 0.289h$.

Figure 6-37. PROPERTIES OF SECTIONS *(Continued)*

Section	Moment of inertia	Section modulus	Radius of gyration
Equilateral Polygon A = area R = rad circumscribed circle r = rad inscribed circle n = no. sides a = length of side Axis as in preceding section of octagon	$I = \dfrac{A}{24}(6R^2 - a^2)$ $= \dfrac{A}{48}(12r^2 + a^2)$ $= \dfrac{AR^2}{4}$ (approx)	$\dfrac{I}{c} = \dfrac{I}{r}$ $= \dfrac{I}{R \cos \dfrac{180°}{n}}$ $= \dfrac{AR}{4}$ (approx)	$\sqrt{\dfrac{6R^2 - a^2}{24}} \approx \dfrac{R}{2}$ $\sqrt{\dfrac{12r^2 + a^2}{48}}$

| | $I = \dfrac{6b^2 + 6bb_1 + b_1^2}{36(2b + b_1)}h^3$

 $c = \dfrac{1}{3}\dfrac{3b + 2b_1}{2b + b_1}h$ | $\dfrac{I}{c} = \dfrac{6b^2 + 6bb_1 + b_1^2}{12(3b + 2b_1)}h^2$ | $\dfrac{h\sqrt{12b^2 + 12bb_1 + 2b_1^2}}{6(2b + b_1)}$ |

| | $I = \dfrac{BH^3 + bh^3}{12}$

 $\dfrac{I}{c} = \dfrac{BH^3 + bh^3}{6H}$ | | $\sqrt{\dfrac{BH^3 + bh^3}{12(BH + bh)}}$ |

| | $I = \dfrac{BH^3 - bh^3}{12}$

 $\dfrac{I}{c} = \dfrac{BH^3 - bh^3}{6H}$ | | $\sqrt{\dfrac{BH^3 - bh^3}{12(BH - bh)}}$ |

| | $I = \tfrac{1}{3}(Bc_1^3 - B_1h^3 + bc_2^3 - b_1h_1^3)$

 $c_1 = \dfrac{1}{2}\dfrac{aH^2 + B_1d^2 + b_1d_1(2H - d_1)}{aH + B_1d + b_1d_1}$ | | $\sqrt{\dfrac{I}{(Bd + bd_1) + a(h + h_1)}}$ |

| | | | $I = \tfrac{1}{3}(Bc_1^3 - bh^3 + ac_2^3)$

 $c_1 = \dfrac{1}{2}\dfrac{aH^2 + bd^2}{aH + bd}$

 $c_2 = H - c_1$

 $r = \sqrt{\dfrac{I}{[Bd + a(H - d)]}}$ |

| | $I = \dfrac{\pi d^4}{64} = \dfrac{\pi r^4}{4} = \dfrac{A}{4}r^2$

 $= 0.05d^4$ (approx) | $\dfrac{I}{c} = \dfrac{\pi d^3}{32} = \dfrac{\pi r^3}{4} = \dfrac{A}{4}r$

 $= 0.1d^3$ (approx) | $\dfrac{r}{2} = \dfrac{d}{4}$ |

Figure 6-37.   PROPERTIES OF SECTIONS *(Continued)*

Section	*Moment of inertia*	*Section modulus*	*Radius of gyration*
$d_m = \frac{1}{2}(D + d)$ $s = \frac{1}{2}(D - d)$	$I = \frac{\pi}{64}(D^4 - d^4)$ $= \frac{\pi}{4}(R^4 - r^4)$ $= \frac{1}{4}A(R^2 + r^2)$ $= 0.05(D^4 - d^4)$ (approx)	$\frac{I}{c} = \frac{\pi}{32}\frac{D^4 - d^4}{D}$ $= \frac{\pi}{4}\frac{R^4 - r^4}{R}$ $= 0.8d_m^2 s$ (approx) when $\frac{s}{d_m}$ is very small	$\frac{\sqrt{R^2 + r^2}}{2} = \frac{\sqrt{D^2 + d^2}}{4}$
	$I = r^4\left(\frac{\pi}{8} - \frac{8}{9\pi}\right)$ $= 0.1098r^4$	$\frac{I}{c_2} = 0.1908r^3$ $\frac{I}{c_1} = 0.2587r^3$ $c_1 = 0.4244r$	$\frac{\sqrt{9\pi^2 - 64}}{6\pi}r = 0.264r$
	$I = 0.1098(R^4 - r^4)$ $- \frac{0.283R^2r^2(R - r)}{R + r}$ $= 0.3tr_1^3$ (approx) when $\frac{t}{r_1}$ is very small	$c_1 = \frac{4}{3\pi}\frac{R^2 + Rr + r^2}{R + r}$ $c_2 = R - c_1$	$\sqrt{\frac{2I}{\pi(R^2 - r^2)}}$ $= 0.31r_1$ (approx)
	$I = \frac{\pi a^3 b}{4} = 0.7854a^3b$	$\frac{I}{c} = \frac{\pi a^2 b}{4} = 0.7854a^2b$	$\frac{a}{2}$
	$I = \frac{\pi}{4}(a^3b - a_1^3b_1)$ $= \frac{\pi}{4}a^2(a + 3b)t$ (approx)	$\frac{I}{c} = \frac{\pi}{4}a(a + 3b)t$ (approx)	$\sqrt{\frac{I}{(\pi ab - a_1b_1)}} =$ $\frac{a}{2}\sqrt{\frac{a + 3b}{a + b}}$ (approx)
	$I = \frac{1}{12}\left[\frac{3\pi}{16}d^4 + b(h^3 - d^3) + b^3(h - d)\right]$ $\frac{I}{c} = \frac{1}{6h}\left[\frac{3\pi}{16}d^4 + b(h^3 + d^3) + b^3(h - d)\right]$		$\sqrt{\frac{I}{\pi\frac{d^2}{4} + 2b(h - d)}}$ (approx)
	$I = \frac{t}{4}\left(\frac{\pi B^3}{16} + B^2h + \frac{\pi Bh^2}{2} + \frac{2}{3}h^3\right)$ $h = H - \frac{1}{2}B$ $\frac{I}{c} = \frac{2I}{H + t}$		$\sqrt{\frac{I}{2\left(\tau\frac{B}{4} + h\right)t}}$

Figure 6-37. PROPERTIES OF SECTIONS *(Continued)*

Section	Moment of inertia and section modulus	Radius of gyration
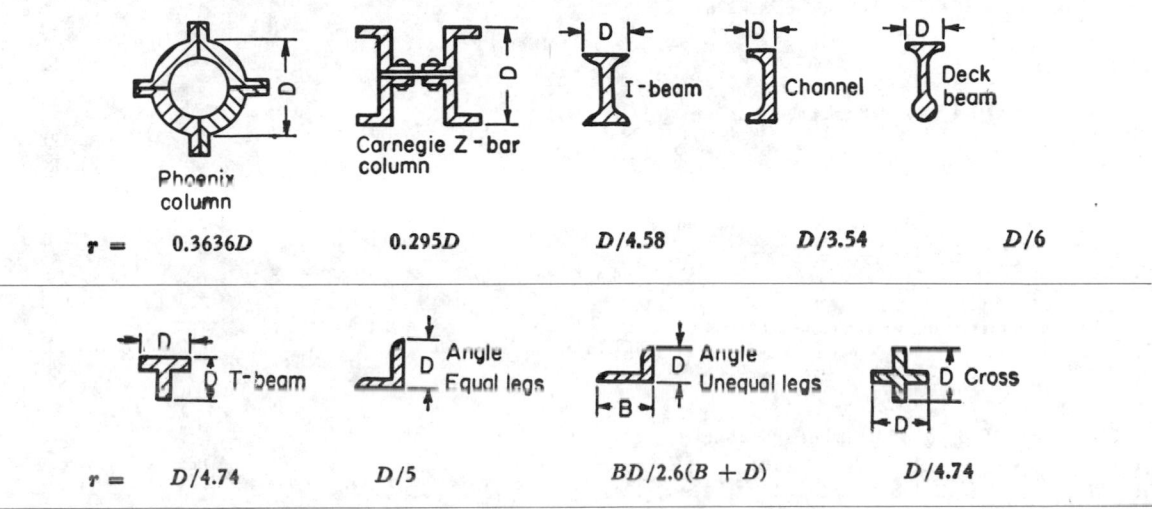 Corrugated sheet iron, parabolically curved	$I = \dfrac{64}{105}(b_1 h_1^3 - b_2 h_2^3)$, where $h_1 = \frac{1}{2}(H + t)$ ⎮ $b_1 = \frac{1}{4}(B + 2.6t)$ $h_2 = \frac{1}{2}(H - t)$ ⎮ $b_2 = \frac{1}{4}(B - 2.6t)$ $\dfrac{I}{c} = \dfrac{2I}{H + t}$	$r = \sqrt{\dfrac{3I}{t(2B + 5.2H)}}$

Approximate values of *least* radius of gyration r

Phoenix column	Carnegie Z-bar column	I-beam	Channel	Deck beam
$r = \quad 0.3636D$	$0.295D$	$D/4.58$	$D/3.54$	$D/6$

T-beam	Angle Equal legs	Angle Unequal legs	Cross
$r = \quad D/4.74$	$D/5$	$BD/2.6(B + D)$	$D/4.74$

TERMINOLOGY FOR BEAMS:

Moment of inertia $= I = \int y^2 dm$

Section modulus $= I/c$, where c is the distance from the outer fiber to the neutral axis

Bending moment $= M = SI/c$, where S is the elastic unit stress at the outer fiber

Radius of gyration $= r = \sqrt{I/A}$, where A is the area of the section. (See dimensions indicated on each sketch.)

---

*From: "Standard Handbook for Mechanical Engineers", 7th ed., T. Baumeister, Ed., McGraw-Hill Book Company, 1967.

Table 6-38.   SAFE LOADS FOR PIPE COLUMNS

Following are approximate maximum axial loads for "Standard Pipe" (Schedule 40) used as columns, plain or filled with concrete. For load in kN, multiply tabulated values in thousands of pounds by 4.4482.

Description	Maximum load in thousands of pounds for column lengths, ft						
	6	7	8	9	10	11	12
3-in. pipe, OD 3.5 in., wall thickness 0.216 in.							
Plain	25	24	23	21	20	18	16
3½-in. pipe, OD 4.0 in., wall thickness 0.226 in.							
Plain	31	30	29	28	26	24	23
Filled	38	37	36	35	33	31	30
4-in. pipe, OD 4.5 in., wall thickness 0.237 in.							
Plain	38	37	36	34	33	31	30
Filled	46	45	44	43	41	40	38
5-in. pipe, OD 5.563 in., wall thickness 0.258 in.							
Plain	52	51	50	49	48	47	46
Filled	66	65	64	63	62	60	59
6-in. pipe, OD 6⅝ in., wall thickness 0.280 in.							
Plain	69	68	67	66	65	64	63
Filled	88	87	86	85	84	83	82
8-in. pipe, OD 8⅝ in., wall thickness 0.322 in.							
Plain	105	104	103	102	101	100	99
Filled	126	125	124	123	122	121	120
10-in. pipe, OD 10¾ in., wall thickness 0.365 in.							
Plain	150	149	148	147	146	145	144
Filled	203	202	201	200	199	198	197

Table 6-39.   LOADS FOR NAILS AND SCREWS IN WOOD*

Table A.   GROUPING OF WOOD SPECIES

The holding power of wood fastenings depends largely on the density of the wood, hard to soft. Loads increase with the specific gravity of wood.

Group No.	Typical species	Specific gravity
1	Hickory, hard maple, oak, beech, birch	.62–.75
2	Douglas fir (larch), Southern pine, gum	.51–.55
3	Hemlock, spruce, Northern pine, cypress, redwood	.42–.48
4	Balsam fir, cedar, Engelmann spruce, soft white pine	.31–.41

*These examples indicate the ranges of the allowable loads specified in the extensive tables in "National Design Specification for Stress-Grade Lumber and Its Fastenings", National Forest Products Association, 1971.

Table 6-39. LOADS FOR NAILS AND SCREWS IN WOOD *(Continued)*
Table B. SIZES AND ALLOWABLE DESIGN LOADS

The following table gives the normal range of loading in pounds for joints secured by nails, spikes, wood screws, lag screws, or bolts. For load in kg, multiply tabular values by 0.453 6.

SYMBOLS: L = length; D = diameter

Fastener description (seasoned wood only)	Nominal gage	Length and diameter, in.	Wood species, group			
			1	*2*	*3*	*4*
Nails,[a] allowable withdrawal load per individual nail, per in. penetration into side grain	6 d	2 × .113	47–76	29–34	18–25	9–17
	8 d	2.5 × .131	55–87	34–39	21–29	10–20
	10 d	3 × .148	62–78	38–44	23–33	12–22
	12 d	3.25 × .148	62–78	38–44	23–33	12–22
	16 d	3.5 × .162	68–85	42–49	25–36	13–24
Spikes, allowable withdrawal load per individual spike per in. penetration into side grain	12	3.25 × .192	80–127	49–57	30–42	15–29
	16	3.5 × .207	86–138	53–61	33–46	16–31
	20	4 × .225	94–150	58–67	35–50	18–33
	40	5 × .263	110–174	68–79	41–58	21–39
Nails and spikes, allowable lateral load in total lb shear for penetration from 10 diam for Group 1 to 14 diam for Group 4, into final member	6 d	2 × .113	78	63	51	41
	8 d	2.5 × .131	97	78	64	51
	10 d	3 × .148	116	94	77	62
	12 d	3.25 × .148	116	94	77	62
	16	3.5 × .207	191	155	126	101
	20	4 × .225	218	176	144	116
	40	5 × .263	276	223	182	146
Wood screws,[b] allowable withdrawal load per in. penetration to full length of threaded section	6	.67 L × .138	151–222	102–118	69–91	38–66
	8	.67 L × .164	180–263	121–141	82–108	45–79
	10	.67 L × .190	208–306	141–164	95–125	53–91
	12	.67 L × .216	237–347	160–186	109–142	59–103
	16	.67 L × .268	294–430	199–231	135–176	73–128
Wood screws,[c] allowable lateral load for penetration 7 times the shank diameter into final member	8	7 D × .164	129	106	87	68
	10	7 D × .190	173	143	117	91
	12	7 D × .216	224	185	151	118
	16	7 D × .268	345	284	233	181
Lag screws,[d] allowable withdrawal load lb per in. penetration into side grain	$\frac{3}{8}$	10 D × .375	421–561	313–356	235–287	145–226
	$\frac{1}{2}$	10 D × .500	523–697	389–443	291–357	180–281
	$\frac{5}{8}$	10 D × .625	618–824	460–524	344–421	213–332
	$\frac{3}{4}$	10 D × .75	708–944	528–601	395–483	244–380
Lag bolts, allowable lateral load using $\frac{1}{2}$-in. metal side pieces; penetration into side grain, single shear; load parallel to grain[e]	$\frac{5}{16}$	4 × .312	410	355	290	235
	$\frac{3}{8}$	6 × .375	630	545	490	430
	$\frac{1}{2}$	8 × .5	1 140	985	880	775
	$\frac{3}{4}$	10 × .75	2 540	2 190	1 970	1 625
	1	12 × 1	4 520	3 900	3 290	2 630
Bolted joints, double shear (joint consisting of 3 members; 2 side members, each $\frac{1}{2}$ the thickness of main member. Bolt length given for main-member thickness only.)	$\frac{1}{2}$	1.5 × .5	820–1 120			
	$\frac{5}{8}$	3 × .625	1 700–2 340			
	$\frac{3}{4}$	4.5 × .75	2 500–3 440			
	$\frac{7}{8}$	7.5 × .875	3 390–4 670			
	1	11.5 × 1	4 460–6 140			

[a] Loads for threaded, hardened steel nails, in 6 d–20 d sizes, are the same as for common nails.

[b] Wood screws shall not be driven with a hammer. Soap or other lubricant may be used, and lead holes are permitted, usually 70 percent of root diameter of thread. Spacing of screws shall be such as to prevent splitting.

[c] Tabular values are for screws inserted into side grain. With metal side plates loads may be increased 25 percent. If screw is inserted in end grain, allowable loads are to be reduced 33 percent.

[d] Penetration of threaded portion about 10 D, but withdrawal resistance approximates tensile (root) strength for penetrations as follows: species group 1 = 7 D; group 2 = 8 D; group 3 = 10 D; group 4 = 11 D. For penetration into end grain, allowable loads are reduced 25 percent. Lead holes for threaded section about 75 percent of shank diameter. Lag screws shall not be driven with a hammer.

[e] For load perpendicular to grain, the allowable loads are much less.

Table 6-40.   UNIT STRESSES FOR LAMINATED TIMBERS

The following table gives the range of allowable unit stresses for "structural glued-laminated timber", in which the grain of all laminations is approximately parallel, net finished widths $2\frac{1}{4}$ in. to $14\frac{1}{2}$ in. These values are for dry conditions of use (moisture content <16%) and normal loading duration. Lower unit stresses are specified for wet conditions of use, for curved members, and for deep and slender beams (see References). Ranges indicate various grades and also numbers of laminations.

For stress and modulus of elasticity in MN/m², multiply values in psi by 0.0068948.

Laminations of	Modulus of elasticity, millions of psi	Allowable unit stresses, psi					
		Bending, load parallel†	Bending, load perpendicular	Tension, parallel to grain	Compression, parallel‡ to grain	Compression, perpendicular to grain	Horizontal shear
Douglas fir or larch	1.82	900–2,300	1,500–2,600	1,800–2,400	1,200–1,800	385–450	195
Douglas fir, Coast region	1.80	1,200–2,600	2,200–2,600	1,600	1,500	450	165
Southern pine	1.80	1,000–2,200	1,800–2,600	2,200–2,600	1,800–2,000	385–450	200
California redwood	1.30	1,100–2,300	1,400–2,200	1,800–2,200	1,800–2,200	325	125
Ash, commercial white	1.60	1,200–2,450	—	1,400–2,300	1,600–2,200	610	230
Birch, sweet or yellow	1.80	600–1,250	—	1,500–2,450	1,800–2,400	610	230
Cottonwood, Eastern	1.10	1,500–3,100	—	750–1,250	900–1,200	180	110
Hickory	2.00	1,200–2,450	—	1,900–3,100	2,200–3,000	730	260
Maple, hard	1.80	1,100–2,300	—	1,500–2,450	1,800–2,400	610	230
Oak, red or white	1.60		—	1,350–2,300	1,500–2,000	610	230

†Parallel or perpendicular with respect to the wide face of the laminations.

REFERENCES

"National Design Specification for Stress-Grade Lumber and Its Fastenings", National Forest Products Association, Washington, D.C., 1971.
"Standards for Structural Glued-Laminated Members Assembled with WWPA Grades of Douglas Fir and Larch Lumber", Western Wood Products Association, 1966.
"Standard Specifications for Structural Glued-Laminated Douglas Fir Timber", West Coast Lumber Inspection Bureau, 1963.
"Standard Specifications, Structural Glued-Laminated California Redwood Timber", California Redwood Association, 1965.
"Standard Specification for the Design and Fabrication of Hardwood Glued-Laminated Lumber for Structural, Marine, and Vehicular Uses", Southern Hardwood Producers, Inc., Appalachian Hardwood Manufacturers, Inc., and Northern Hardwood and Pine Manufacturers Association.
"Standard Specifications for Structural Glued Laminated Timber", American Institute of Timber Construction, 1970.

Table 6-41. SOIL MECHANICS—CLASSES OF SOILS

Comparison of Classification Systems

Classification System									
American Society for Testing and Materials	Colloids†	Clay	Silt	Fine sand	Coarse sand	Fine gravel	Medium gravel	Coarse gravel	Boulders
American Association of State Highway Officials Soil Classification	Colloids†	Clay	Silt	Fine sand	Coarse sand	Fine gravel	Coarse gravel	Cobbles	
U.S. Department of Agriculture Soil Classification	Clay	Silt	Very fine sand / Fine sand / Medium sand / Coarse sand / Very coarse sand	Fine gravel	Coarse gravel				
Federal Aviation Agency Soil Classification	Clay	Silt	Fine sand	Coarse sand	Gravel				
Unified Soil Classification (Corps of Engineers, Department of the Army, and Bureau of Reclamation)	Fines (silt or clay)	Fine sand	Medium sand	Coarse sand	Fine gravel	Coarse gravel	Cobbles		
Massachusetts Institute of Technology	Clay	Silt (Fine / Medium / Coarse)	Sand (Fine / Medium / Coarse)	Gravel					

†Colloids are included in clay fraction in test reports.

Table 6-42.
SOIL MECHANICS—VOLUME AND WEIGHT RELATIONSHIPS
Density, Porosity, and Moisture

REPRESENTATIVE SPECIFIC GRAVITIES OF SOME DENSE SOILS

Soil	Specific gravity
Normal inorganic clay	2.70
Silt	2.70
Loess	2.70
Sand	2.65
Diatomaceous earth	2.65
Bentonite clay	2.34

RELATIVE DENSITY OF SANDS

Very dense	85–100%
Dense	65–85%
Medium	35–65%
Loose	15–35%

POROSITY OF SOILS IN NATURAL STATE†

Porosity is the percentage ratio, volume of voids to total volume. *Void ratio* is the ratio of volume of voids to volume of moist solids. For unit weights in kg/m^3, multiply values in g/cm^3 by 1000.

Description	Poros-ity, %	Void ratio	Water content, %	Unit weights			
				g/cm^3		lb/ft^3	
				dry	sat	dry	sat
Uniform sand, loose	46	0.85	32	1.43	1.89	90	118
Uniform sand, dense	34	0.51	19	1.75	2.09	109	130
Mixed-grained sand, loose	40	0.67	25	1.59	1.99	99	124
Mixed-grained sand, dense	30	0.43	16	1.86	2.16	116	135
Glacial till, very mixed-grained	20	0.25	9	2.12	2.32	132	145
Soft glacial clay	55	1.2	45	—	1.77	—	110
Stiff glacial clay	37	0.6	22	—	2.07	—	129
Soft slightly organic clay	66	1.9	70	—	1.58	—	98
Soft very organic clay	75	3.0	110	—	1.43	—	89
Soft bentonite	84	5.2	194	—	1.27	—	80

†From: "Soil Mechanics in Engineering Practice", 2nd ed., K. Terzaghi and R.B. Peck, John Wiley & Sons, 1967, p. 28, Table 6.3.

DEGREE OF SATURATION, DECIMAL

Description	Degree of saturation
Dry soil	0.0
Humid soil	<0.25
Damp soil	0.26–0.50
Moist soil	0.51–0.75
Wet soil	0.76–0.99
Saturated	1.00

Table 6-43.   SOIL MECHANICS—PLASTICITY AND WATER CONTENT OF SOILS

Moisture, Consistency, and Atterberg Limits

DEFINITIONS:

Plastic limit. The percentage of water content by weight at which soil crumbles when rolled into thin threads.

Liquid limit. The percentage of water content by weight at which soil has a very small but measurable shearing strength.

Plasticity index. The arithmetical difference between the liquid limit and the plastic limit.

TERMINOLOGY FOR DEGREE OF PLASTICITY

Description	Range of plasticity index
Non-plastic	0–5
Moderately plastic	5–15
Plastic	15–40
Highly plastic	>40

ATTERBERG LIMITS FOR CLAYS

Clay	Adsorbed ion	Plastic limit	Liquid limit
Kaolinite	Sodium	26	52
	Calcium	36	73
Illite	Sodium	34	61
	Calcium	40	90
Montmorillonite	Sodium	97	900
	Calcium	63	177

Table 6-44.
SOIL MECHANICS—COMPARATIVE PROPERTIES OF CLAY SOILS

Consistency, Sensitivity, Brittleness, and Activity

DEFINITIONS:

Consistency. The relative compressive strength of the unconfined clay.

Sensitivity. The ratio of unconfined compressive strengths (undisturbed material to remolded material).

Brittleness. The percentage of strain at failure in an unconfined compression test.

Activity. The ratio of the plasticity index to the clay fraction, i.e. (liquid limit − plastic limit)/(percentage by weight of particles smaller than 2 microns).

Consistency		Sensitivity		Brittleness		Activity	
Description	Range	Description	Range	Description	Range	Description	Range
Very soft	<.25	Insensitive	<2	Brittle	3–8	Inactive	<.75
Soft	.25–.50	Moderately		Semi-brittle	8–14	Normal	.75–1.25
Medium	.50–1.0	sensitive	2–4	Plastic	14–20	Active	>1.25
Stiff	1.0–2.0	Sensitive	4–8				
Very stiff	2.0–4.0	Very sensitive	8–16				
Hard	>4.0	Slightly quick	16–32				
		Medium quick	32–64				
		Quick	>64				

Table 6-45.   SOIL MECHANICS—DENSITY AND PENETRATION

Density and consistency of soil may be determined during boring and sampling operations by the standard penetration test. In this test the number of blows required per foot of penetration of a standard sampling spoon is observed. A 140-lb hammer is used, with a 30-in. drop; the sampling spoon is 2.0 in. OD and 1.375 in. ID. The density and consistency are characterized in the following table.

Relative density of sand		Relative consistency of clay	
No. of blows per ft	Description	No. of blows per ft	Description
0–4	Very loose	0–2	Very soft
4–10	Loose	2–4	Soft
10–30	Medium	4–8	Medium
30–50	Dense	8–15	Stiff
>50	Very dense	15–30	Very stiff
		>30	Hard

Note: For sands and silts a correction for depth (and overburden pressure) or a correction for location of the water table may be required.

Table 6-46.   SOIL MECHANICS—PERMEABILITY OF SOILS

The *coefficient of permeability*, k, is a simple index of the water conductivity or drainage characteristics of a soil. It is determined by using a *permeameter*, which is an arrangement for measuring the flow through a soil sample under a given hydraulic head.

It is assumed that the rate of flow is directly proportional to the head h (laminar-type flow). The coefficient is based on the gross area of the flow conduit. The coefficient of permeability is a measure of the velocity of flow per unit of head, or the velocity, based on gross area, at which the head loss is unity. In metric units the dimensions are

$$q = VA. \qquad \frac{cm^3}{sec} = \frac{cm}{sec} \times cm^2$$

$$q = khA. \qquad \frac{cm^3}{sec} = \left(\frac{cm}{sec \times cm}\right)(cm)\,(cm^2).$$

In the field the permeability is determined by a pumping test in which the hydraulic gradients at fixed distances from the pumping point are measured while a constant and known rate of flow is being maintained.

Natural soil deposits often consist of layers, each of which has a different permeability. The flow resistances are additive (Ohm's law), and the individual coefficients of permeability are analogous to individual conductivities. An average coefficient of permeability can thus be computed.

The following table gives a classification of soils in terms of the numerical values of the coefficient of permeability. The units of k are cm/sec for a head of one centimeter.

DESCRIPTIVE RANGES OF PERMEABILITY

Degree of permeability	Values of k, cm/sec	Degree of permeability	Values of k, cm/sec
High	>.1	Very low	10^{-5}–10^{-7}
Medium	.1–.001	Practically	
Low	.001–.00001	impervious	$<10^{-7}$

The value of $k = 0.1$ is typical for a medium to coarse, clean sand or a sand and gravel mixture. Homogeneous clays are often nearly impervious. The three intermediate ranges characterize very fine sands, silts, and mixtures, including those with organic materials.

For soils with a coefficient greater than 0.001, the values of k can be readily determined, using either the constant-head permeameter or a pumping test. The falling-head permeameter can be used both for these and for finer soils, but as the fineness increases so does the importance of a large amount of experience in making the tests.

Table 6-47. BUILDING FOUNDATIONS AND PILES

Typical Building Code Requirements for Soil Bearing Pressures and Design of Pile Foundations

Table A. ALLOWABLE BEARING PRESSURES

Soil descriptions[a]	Maximum bearing pressure[b]	
	tons/ft^2	MN/m^2
Rock, bedrock, solid, massive, sound	100	1 380
Rock, sound foliated (limestone, slate)	40	550
Rock, hard sedimentary (hard shale, sandstone, conglomerate)	25	245
Rock, broken bedrock	10	138
Hardpan	10	138
Gravel, very compact	10	138
Gravel, compact	6	83
Gravel, loose	4	55
Sand or silty gravel	1–3	14–40

[a]Descriptive terminology used in building codes is far from uniform, but the terms here given occur in several codes.

[b]Each of these values is to be found in several of the building codes of large U.S. cities or government authorities. Modifications or corrections are recognized for conditions such as depth of stratum, depth of embedment, and extent of overburden. For gravel and sand the allowable bearing pressures are sometimes stated in terms of on-site soil test results.

Table B. ALLOWABLE DESIGN STRESSES FOR LOADED PILES

Widely Used Specifications in Building Codes

KEY: BS = basic stress for clear timber as per ASTM D25 and D2555
CS = compressive strength of concrete
YS = yield strength of steel

Type of pile	No. of components	Allowable unit stress, psi[a]		
		Core	Reinforcing steel	Steel shell
Steel H-section	1	35% YS	—	—
Timber	1	60% BS	—	—
Precast concrete	2	22.5% CS	35% YS	—
Cast-in-place concrete	3	22.5% CS	35% YS	35% YS
Steel pipe (filled)	3	22.5% CS	35% YS	35% YS

Notes:

Most piles are designed as "short" columns, in compression only; for long piles in soft soil, a "long-column" treatment may be necessary.

Estimation of axial-load distribution along the pile may be included in a building code, but there is little agreement on the basis for estimating. The percentage of the total load being carried at the tip of the pile is usually assumed to be less than one-third; for shorter piles in coarse material on bedrock, it may even be 100 percent.

Possible pile damage, due to overdriving, obstructions, or defects, should be considered in the design decisions.

Pile spacing is often specified in the building codes, with 24- or 30-in. (0.61–0.76 m) as minimum for cylindrical piles.

Success of a pile foundation depends on the ability of the soil to support the pile, which, in turn, depends on the adhesion and shear properties of the soil. The use of tapered piles and rigid-body caps is common. In any case the limitations on eventual settling should be examined, especially in locations where settling has been a problem, as in Mexico City.

For soil conditions producing very heavy corrosion, it may be necessary to avoid steel piles.

For extreme loads the caisson-type pile, terminating in a recess in bedrock, is recommended.

[a]Many city codes specifying numerical values for allowable stresses and variations among cities are occasionally 2 to 1 or even more. A few codes use percentages other than those here quoted. A limit of 1 000 psi (6 895 MN/m^2) is often specified for timber piling.

REFERENCES

"Building Code Requirements for Reinforced Concrete", ACI 318, American Concrete Institute.

"Design Manual", DM7, Bureau of Yards and Docks, U.S. Navy.

"Foundation Piling: A Survey of Practice", Building Research Advisory Board, Report No. 4, National Academy of Sciences–National Research Council.

"Southern Standard Building Code" and "Uniform Building Code".

"Specification for the Design Fabrication and Erection of Structural Steel for Buildings", American Institute of Steel Construction.

Consult also (1) *Journal of the Soil Mechanics and Foundations Division, Proceedings of the ASCE*, (2) "Proceedings of the International Conferences on Soil Mechanics and Foundation Engineering", and (3) city building codes.

Table 6-48. SOIL CHARACTERISTICS PERTINENT TO ROADS AND AIRFIELDS*

The following table is based on the "Unified Soil Classification System" used by the military services and the Bureau of Reclamation. Other classification systems include those of the Highway Research Board and the Civil Aeronautics Administration. See also Table 6-41 for comparisons.

KEY:

A—excellent	*F*—poor	*SR*—steel-wheeled rollers
B—good to excellent	*G*—poor to very poor	*SF*—sheepsfoot roller
C—good	*NS*—not suitable	*RT*—rubber-tired equipment
D—fair to good	*CT*—crawler-type tractor	*M*—close control of moisture required
E—fair to poor		

Soil class and symbol	Value as foundation[a]		Compressibility and expansion	Compaction equipment[b]	Dry weight,[c] lb/ft^3	Field CBR[d]	Subgrade modulus,[e] k, psi/in.
	X	Y					
GW—well graded gravels or gravel-sand mixtures, little or no fines	A	C	Almost none	CT, RT, SR	125–140	60–80	>300
GP—poorly graded gravels or gravel-sand mixtures, little or no fines	B	E	Almost none	CT, RT, SR	110–130	25–60	>300
GM—silty gravels, gravel-sand-silt mixtures	B	D	Very slight	RT, SF, M	130–145	40–80	>300
	C	F	Slight	RT, SF	120–140	20–40	200–300
GC—clayey gravels, gravel-sand-clay mixtures	C	F	Slight	RT, SF	120–140	20–40	200–300
SW—well graded sands or gravelly sands, little or no fines	C	F	Almost none	CT, RT	110–130	20–40	200–300
SP—poorly graded sands or gravelly sands, little or no fines	D	F–NS	Almost none	CT, RT	100–120	10–25	200–300
SM—silty sands, sand-silt mixtures	C	F	Very slight	RT, SF, M	120–135	20–40	200–300
	D	NS	Slight to medium	RT, SF	105–130	10–20	200–300
SC—clay-like sands, sand-clay mixtures	D	NS	Slight to medium	RT, SF	105–130	10–20	200–300
ML—inorganic silts and very fine sands, rock flour, silty, or clay-like fine sands, or clay-like silts with slight plasticity	E	NS	Slight to medium	RT, SF, M	100–125	5–15	100–200
CL—inorganic clays of low to medium plasticity, gravelly clays, sandy clays, silty clays, lean clays	E	NS	Medium	RT, SF	100–125	5–15	100–200
OL—organic silts and organic silt-clays of low plasticity	F	NS	Medium to high	RT, SF	90–105	4–8	100–200
MH—inorganic silts, micaceous or diatomaceous, fine, sandy or silty soils, elastic silts	F	NS	High	SF	80–100	4–8	100–200
CH—inorganic clays of high plasticity, fat clays	G	NS	High	SF	90–110	3–5	50–100
OH—organic clays of medium to high plasticity, organic silts	G	NS	High	SF	80–105	3–5	50–100
PT—peat and other highly organic soils	NS	NS	Very high	None			

*Condensed from: "Unified Soil Classification System", Appendix B: Characteristics of Soil Groups Pertaining to Roads and Airfields, U.S. Corps of Engineers, Technical Memorandum No. 3-357, Vol. III, USAE Waterways Experiment Station, March 1953.

Table 6-48.   SOIL CHARACTERISTICS PERTINENT TO ROADS AND AIRFIELDS (Continued)

[a]Values in column X are for subgrades and base courses not subject to frost action, while those in column Y apply to base directly under bituminous pavement. The A rating in column Y has been reserved for base materials consisting of high-quality processed crushed stone.
[b]The equipment listed will usually produce the required densities with a reasonable number of passes when moisture conditions and thickness of lift are properly controlled. In some instances several types of equipment are listed, because variable soil characteristics within a given soil group may require different equipment. A combination of two types may be necessary.

Processed Base Materials and Other Angular Materials.   Steel-wheeled rollers are recommended for hard angular materials with limited fines or screenings. Rubber-tired equipment is recommended for softer materials subject to degradation.
Finishing.   Rubber-tired equipment is recommended for rolling during final shaping operations for most soils and processed materials.
Equipment Size.   The following sizes of equipment are necessary to assure the high densities required for airfield construction

Crawler-type tractor—total weight in excess of 30 000 lb.
Rubber-tired equipment—wheel load in excess of 15 000 lb. Wheel loads as high as 40 000 lb may be necessary to obtain the required densities for some materials (based on contact pressure of approximately 65–150 psi).
Sheepsfoot roller—unit pressure (on 6–12 sq in. foot) to be in excess of 250 psi and unit pressures as high as 650 psi may be necessary to obtain the required densities for some materials. The area of the feet should be at least 5 percent of the total peripheral area of the drum, using the diameter measured to the faces of the feet.

[c]Unit dry weights are for compacted soil at optimum moisture content for modified AASHO compactive effort.
[d]The percentage ratio of the resistance to penetration developed by a subgrade soil to that developed by a specimen of standard crushed rock base material is called the California bearing ratio (CBR). It was developed by the California Division of Highways and is applied in connection with the design of flexible pavements.
[e]The subgrade modulus is a test ratio representing the slope of the stress-strain (or pressure-settlement) diagram. It is actually the ratio of the pressure on the 30-in. test plate in psi to the settlement in inches when the latter reaches a specified value (either 0.05 or 0.10 in.).

Table 6-49.   ALLOWABLE SOIL PRESSURES FOR COLUMN FOOTINGS

Allowable bearing capacities or foundation soil loadings are often specified in local building codes. These are applicable to the design of ordinary extended footings, and the loads and capacities are those at footing grade. There are certain assumptions—that no weaker layer exists within the stress zone and that the stress zones do not overlap. Mats or continuous footings cannot be loaded as heavily. Even for preliminary estimates a relation between allowable bearing capacity and standard penetration resistance by test (ASTM D1586) is a much better basis than mere soil-texture classification. The following table gives typical ranges for the bearing loads as related to the penetration resistance determined in a soil test. The required blows per foot for a 2-in. OD sampling spoon represent the test data.

ESTIMATING BEARING LOADS FROM RESULTS OF PENETRATION TESTS

Numbers in this table represent the minimum number of hammer blows per linear foot by standard penetration test ASTM D1586.

Description of soil below footing grade	Maximum soil load, thousands of psi								
	2	3	4	5	6	8	10	12	14
Gravel, sand, clay—well graded mixture		3	4	6	8	14	22	30	40
Gravel, sand, clay—poorly graded mixture		6	8	12	15	24	36	50	65
Gravel, sand—well graded		3	5	8	10	18	30	55	
Sandy clay	4	7	9	12	16	25	38	50	
Inorganic clay, no silt	8	12	15	20	25	40	60		
Fine sand, uniform	6	10	15	22	30	50			
Organic clay	16	30	65						
Fine silt	15	25	45						
Rough Guides									
Sand, gravel, clay—stiff mixture									50
Compact gravel mixture								55	
Sandy clay							45		
Firm gravel						20			
Hard clay					35				
Firm, coarse sand				20					
Loose gravel				10					
Medium sandy clay			10						
Medium stiff clay		15							
Firm silt	15								

Table 6-50.   TYPICAL PROPERTIES OF FOUNDATION ROCKS*

For strength in MN/m², multiply values in psi by 0.0068948.

Name of rock	Compressive strength, psi†	Shear strength, psi	Tangent of angle of internal friction
Andesite	18,900	4,060	5.7
Basalt	24,600	4,500	7.4
Diorite	12,200	2,000	9.2
Gneiss	16,100	2,500	10.0
Granite	22,300	3,250	11.8
Graywacke	8,700	1,700	6.5
Limestone	15,300	2,150	12.6
Sandstone	13,000	2,450	7.0
Schist	10,000	1,260	15.7
Shale	10,300	1,160	19.7
Siltstone	4,000	720	7.7
Tuff	430	100	4.9

†Uniaxial compressive strength for a specimen having a length-to-diameter ratio of 2. The product of the principal stresses is assumed to be linear.

*From: U.S. Bureau of Reclamation Report SP–39, August 1953.

Table 6-51.   STATIC MECHANICAL PROPERTIES OF ROCK*

For strength in MN/m², multiply tabulated values in thousands of psi by 6.8948. For modulus of elasticity in GN/m², multiply tabulated values in millions of psi by 6.8948.

Rock type and number of test groups†	Compressive strength of test group, thousands of psi			Modulus of rupture of test group, thousands of psi			Modulus of elasticity of test group, millions of psi				
							Static‡		Dynamic**		
	Max	Min	>50% of data within	Max	Min	>50% of data within	Max	Min	Max	Min	>50% of data within
Amphibolite (13)	74.9	30.4	—	7.4	4.0	—	—	—	15.1	6.7	—
Basalt (9)	52.0	11.8	—	6.6	2.1	—	—	—	12.4	5.9	—
Dibase (10)	51.8	23.2	40–50	8.0	4.5	5.0–5.5	—	—	13.9	10.2	—
Diorite (11)	48.3	22.5	25–35	7.3	2.0	—	—	—	6.12	3.6	4.0–5.0
Dolomite (10)	52.0	9.0	—	3.8	2.5	—	—	—	12.3	3.2	—
Gneiss (15)	36.4	22.2	25–35	3.1	1.2	2.0–3.0	—	—	15.1	3.5	7.0–10
Granite (17)	42.6	23.0	25–35	3.9	1.2	2.0–3.0	11.0	2.5	11.9	1.5	—
Greenstone (11)	45.5	16.6	—	6.7	1.7	—	9.0	6.9	15.2	3.4	10–13
Limestone§ (46)	37.6	5.3	20–30	5.2	0.4	—	12.0	4.2	14.1	1.2	—
Marble (8)	34.5	6.7	30–35	3.3	1.7	—	—	—	—	—	—
Marlstone (15)	28.2	10.4	10–20	4.8	0.4	—	4.8	0.6	7.0	1.5	—
Quartzite (11)	91.2	21.1	—	6.4	1.2	—	—	—	—	—	—
Sandstone (48)	34.1	4.8	—	3.6	0.6	0.6–1.0	7.3	1.4	8.0	0.8	—
Shale (18)	33.5	10.9	10–20	4.2	0.3	—	7.6	1.7	9.9	1.5	1.5–2.5
Siltstone (8)	45.8	5.0	—	5.0	1.1	—	—	—	9.3	1.0	—

†The number following rock type indicates the number of petrographically distinct groups tested in uniaxial compression. A smaller number of groups were tested for the other listed mechanical properties.
‡Tangent modulus of elasticity at the midstrength point.
**Determined by the resonant bar velocity method.
§Excluding chalk and coral rock.

*From: U.S. Bureau of Mines Reports of Investigations 3891, 4459, 5130, and 5244.

Table 6-52.   PROPERTIES OF PLASTER OF PARIS*

The following table gives variations in mechanical properties of pure gypsum (hydrous calcium sulfate) plaster with proportions of water in mix.†

For density in kg/m^3, multiply values in lbm/ft^3 by 16.018. For Young's modulus and strength in MN/m^2, multiply values in psi by 0.0068948. The unit *g/liter* is identical to kg/m^3.

Density, lb/ft^3	Proportion by mass, %			Retarder,‡ g/liter of water	Young's modulus, psi	Compressive strength, psi	Tensile strength, psi	Set time, minutes
	Water	Plaster	Diatomaceous earth					
75	60	100	0	0.50	1,000,000	2000	—	—
71	65	100	0	0.25	870,000	1700	274	7
67	70	100	0	0.25	730,000	1500	230	7
62	80	100	0	0.20	600,000	1100	—	11
58	90	100	0	0	500,000	800	145	6.5
53	100	100	0	0	440,000	700	—	10
40	110	100	0	0	368,000	470	103	7
46	120	100	0	0	316,000	400	82	9
43	130	100	0	0	273,000	330	72	9.5
40	140	100	0	0	225,000	280	69	9.5
38	150	100	0	0	190,000	240	—	10
37	160	100	8	0	164,000	200	51	12
36	170	100	8	0	131,000	170	—	12

†Poisson's ratio 0.24 and independent of mix.
‡Sodium citrate.

*From: "The Development of a Model for a Mine Structure", K. Barron and G.E. Larocque, *Proc. Rock. Mech. Symp.*, Mines Branch of the Department of Energy, Mines and Resources, Ontario, Canada.

Table 6-53.   CONCRETE FOR VARIOUS TYPES OF CONSTRUCTION*

For cement factor in kg/m^3, multiply the tabulated value in sacks (94 lb) per cubic yard by 55.767. For aggregate size in mm, multiply values in inches by 25.4.

Type of construction	Typical structures	Consistency	Cement factor, sacks per cu yd	Maximum size of aggregate, in.
Massive	Dams, heavy piers, large open foundations	Stiff	$2\frac{1}{2}$–5	3–6
Semimassive	Piers, heavy walls, foundations, heavy arches, girders	Stiff; medium	4–6	2–3
Heavy building	Large structural members, small piers, medium footings; wide to moderately wide spacing of reinforcement	Medium wet	5–7	1–2
Light	Small structural members, thin slabs, small columns, heavily reinforced sections, closely spaced reinforcement	Wet	$5\frac{1}{2}$–7	$\frac{1}{2}$–1

*From: "Composition and Properties of Concrete", G.E. Troxell and H.E. Davis, McGraw-Hill Book Company, 1968.

Table 6-54. TYPES OF PORTLAND CEMENT AND INCREASE IN STRENGTH OF CONCRETE WITH AGE

Composition and Properties

Portland cement is a mixture consisting mainly of calcium and silicon oxides (as calcium silicates) in powder form with particle sizes largely in the range 10–50 microns. While the composition is approximately 65% CaO, 21% SiO_2, and 7% Al_2O_3 (balance from other oxides), the actual compounds are $3CaO \cdot SiO_2$, $2CaO \cdot SiO_2$, $3CaO \cdot Al_2O_3$, and $4CaO \cdot Al_2O_3 \cdot Fe_2O_3$. The proportions of these actual compounds will vary with the type of cement. When water is added and the mixture hardens, the heat of hydration is approximately 100 cal/g cement (180 Btu/lb). The heat is released gradually as hydration proceeds, with a shrinkage of about 8% compared with the total volume of cement-plus-water in the original mix. Shrinkage is minimum if original water content is low. To accelerate the setting of cement, especially in cold weather, calcium chloride (2% or less of weight of cement) often is added. In very hot weather retarders are sometimes used to delay hardening.

Standard ASTM classi-fication	Character and use	Compressive strength,[a] psi/1000, at age of				
		7 days	28 days	3 months	1 year	5 years
I	Common; general use	3.0	4.3	5.1	5.5	5.7
II	General use; moderate resistance to sulfate attack; lower heat of hydration	2.6	4.2	5.2	5.9	6.4
III	High early strength; shorter curing time	3.8	4.7	5.1	5.4	5.5
IV	Develops less heat; slow-curing; resists cracking; high resistance to sulfate attack	1.5	3.5	5.2	6.0	6.5
V	Highest resistance to sulfate attack	2.5	4.1	5.3	6.1	6.7

Note: Although strengths will vary greatly with materials, proportions, and curing conditions, these are typical.

[a]For strength in MN/m^2, multiply tabulated values in thousands of psi by 6.8948.

Table 6-55. COMPRESSIVE STRENGTH OF CONCRETE*

The following are typical 28-day compressive strengths with variation of water-cement ratios. Strengths are based on tests of 6- by 12-in. cylinders, moist-cured. For strength in MN/m^2, multiply tabulated values in psi by 0.0068948.

Water-cement ratio		Probable average compressive strength at 28 days, psi	
Gal per sack of cement	By wt	Non-air-entrained concrete	Air-entrained concrete[†]
4	0.35	6,000	4,800
5	0.44	5,000	4,000
6	0.53	4,000	3,200
7	0.62	3,200	2,600
8	0.71	2,500	2,000
9	0.80	2,000	1,600

[†]Air-entrained concrete is over 3% air by volume. Strength is reduced by increase in air content.

*From: "Recommended Practice for Selecting Proportions for Concrete", ACI 613–54, American Concrete Institute, *Proc. ACI*, 26(1):49–64, September 1954. For sizes of aggregate and amount of water per cu yd of concrete, see original source.

Table 6-56.   PROPERTIES OF SPECIAL CONCRETES

A great many varieties of aggregates have been used for concrete, dependent largely on the materials available. In general high density results in high strength and high thermal conductivity, and vice versa, although such variables as water/cement ratio, percentage of fines, and curing conditions may result in wide differences in properties with the same materials. The following table gives typical examples.

Description; type of aggregate	*Approximate density, lb/ft³*		*Compressive strength, lb/in.²*	*Thermal conductivity, Btu/hr·ft·deg F*
	aggregate	*concrete*		
Frost-resisting; 1% CaCl₂; normal aggregates	110	140	4 500	1.0
Frost-resisting porous; 6% air entrainment	100	130	3 500	0.85
Lightweight, with expanded shale or clay	50	75	1 500	0.25
Lightweight, with foamed slag	40	75	1 000	0.20
Cinder concrete, fine and coarse	50	80	1 000	0.25
Pulverized-fuel ash	60	85	1 200	0.25
Lightweight refractory concrete with aluminous cement	35	65	3 500	0.20
Lightweight, insulating, with perlite	10	35	250	0.15
Lightweight, insulating, with expanded vermiculite	8	30	150	0.10

Table 6-57.   EFFECTS OF VARIOUS SUBSTANCES ON UNPROTECTED CONCRETE*

In addition to attack by corrosion (e.g., by acids) and other chemical combination (with sulfates and chlorides), there are many types of erosion of concrete. Concrete is highly susceptible to erosion by cavitation occurring in water-flow structures.

Substance	*Effect on unprotected concrete*
Petroleum oils: heavy, light, and volatile	None
Coal-tar distillates	None or very slight
Inorganic acids	Disintegration
Organic materials	
Acetic acid	Slow disintegration
Oxalic and dry carbonic acids	None
Carbonic acid in water	Slow attack
Lactic and tannic acids	Slow attack
Vegetable oils	Slight or very slight attack
Inorganic salts	
Sulfates of calcium, sodium, magnesium, potassium, aluminum, iron	Active attack
Chlorides of sodium, potassium	None
Chlorides of magnesium, calcium	Slight attack
Miscellaneous	
Milk	Slow attack
Silage juices	Slow attack
Molasses, corn syrup, and glucose	Slight attack

*From: "Concrete Manual", 7th ed., U.S. Bureau of Reclamation, 1966.

Section 7

Environmental and Bioengineering

Table 7-1. COMPONENTS OF THE ATMOSPHERE*

For properties of dry air at sea level, see Table 1-1. For atmospheric conditions at high altitudes, see Tables 7-4 and 7-5.

AVERAGE COMPOSITION OF DRY AIR

For most engineering applications the following accepted values for the "average" composition of the atmosphere are adequate. These values are for sea level or any land elevation. Proportions remain essentially constant to 50 000 ft (15 240 m) altitude.

Gas	Molecular weight	Percentage by volume, mol fraction	Percentage by weight
Nitrogen	$N_2 = 28.016$	78.09	75.55
Oxygen	$O_2 = 32.000$	20.95	23.13
Argon	$Ar = 39.944$	0.93	1.27
Carbon dioxide	$CO_2 = 44.010$	0.03	0.05
		100.00	100.00

For many engineering purposes the percentages 79% N_2–21% O_2 by volume and 77% N_2–23% O_2 by weight are sufficiently accurate, the argon being considered as nitrogen with an adjustment of molecular weight to 28.16.

Other gases in the atmosphere constitute less than 0.003% (actually 27.99 parts per million by volume), as given in the following table.

MINOR CONSTITUENTS OF DRY AIR

Gas	Molecular weight	Parts per million	
		By volume	By weight
Neon	$Ne = 20.183$	18.	12.9
Helium	$He = 4.003$	5.2	0.74
Methane	$CH_4 = 16.04$	2.2	1.3
Krypton	$Kr = 83.8$	1.	3.0
Nitrous oxide	$N_2O = 44.01$	1.	1.6
Hydrogen	$H_2 = 2.0160$	0.5	0.03
Xenon	$Xe = 131.3$	0.08	0.37
Ozone	$O_3 = 48.000$	0.01	0.02
Radon	$Rn = 222.$	(0.06×10^{-12})	

Minor constituents may also include dust, pollen, bacteria, spores, smoke particles, SO_2, H_2S, hydrocarbons, and larger amounts of CO_2 and ozone, depending on weather, volcanic activity, local industrial activity, and concentration of human, animal, and vehicle population. In certain enclosed spaces the minor constituents will vary considerably with industrial operations and with occupancy by humans, plants, or animals.

The above data do not include water vapor, which is an important constituent in all normal atmospheres. See Table 1-3 for properties of moist air; see Table 7-17 for weather conditions at various latitudes.

---

*Compiled from several sources.

Table 7-2.   U.S. STANDARD ATMOSPHERE, TO 300,000 ft*
45° North Latitude, July

For variations of high altitude atmosphere with latitude and season, see Table 7-4.

SYMBOLS:

Z, ft = geometric altitude, feet
Z, m = geometric altitude, meters
H, ft = geopotential altitude, feet
t, °F = temperature, degrees Fahrenheit
t, °C = temperature, degrees Celsius
P, in. Hg = pressure, inches of mercury. For atmospheres multiply by 0.0334210. For psia multiply by 0.491154. For kN/m^2 multiply by 3.3864.

ρ, English = density. For lb$_m$/ft^3 multiply by 10^{-3}.
ρ, metric = density. For kg/m^3 multiply by 10^{-3}.
V_s, fps = speed of sound, ft/sec. For m/sec multiply by 0.3048.
μ = viscosity. For lb$_m$/ft sec multiply by 10^{-5}. For N·s/m^2 (= kg/m·s) multiply by 10^{-6} and by 14.882.
k = thermal conductivity. For Btu/sec ft °R multiply by 10^{-5}. For W/m·K multiply by 622.65.

Z, ft	H, ft	Z, m	t, °F	t, °C	P, in. Hg	ρ, English	ρ, metric	V_s, fps	μ	k
0	0	0	73.5	23.1	29.93	74.4	1,192.	1,132	1.228	.417
1,000	1,000	305	70.7	21.4	28.89	72.2	1,157	1,129	1.223	.415
2,000	2,000	610	68.0	20.0	27.89	70.1	1,123	1,126	1.218	.413
3,000	3,000	915	65.2	18.4	26.91	68.0	1,090	1,123	1.213	.411
4,000	3,999	1,220	62.4	16.9	25.96	65.9	1,057	1,120	1.209	.409
5,000	4,999	1,525	59.7	15.4	25.05	63.9	1,025	1,117	1.204	.407
6,000	5,998	1,830	57.0	13.9	24.16	62.0	994	1,114	1.199	.405
7,000	6,998	2,135	53.9	12.2	23.30	60.1	964	1,111	1.193	.403
8,000	7,997	2,440	50.4	10.2	22.46	58.4	936	1,107	1.187	.401
9,000	8,996	2,745	46.9	8.3	21.65	56.6	908	1,103	1.180	.398
10,000	9,995	3,050	43.4	6.3	20.86	55.0	881	1,100	1.174	.396
11,000	10,994	3,355	40.0	4.4	20.09	53.3	855	1,096	1.168	.393
12,000	11,993	3,660	36.6	2.6	19.35	51.7	829	1,092	1.162	.391
13,000	12,992	3,965	33.2	.7	18.63	50.1	803	1,088	1.155	.389
14,000	13,991	4,270	29.8	−1.2	17.94	48.6	779	1,085	1.149	.386
15,000	14,989	4,575	26.5	−3.1	17.26	47.1	755	1,081	1.143	.384
16,000	15,988	4,880	23.1	−4.9	16.61	45.6	731	1,077	1.137	.381
17,000	16,986	5,185	19.7	−6.8	15.97	44.2	708	1,073	1.130	.379
18,000	17,984	5,490	16.4	−8.7	15.36	42.8	686	1,070	1.124	.377
19,000	18,983	5,795	13.0	−10.6	14.77	41.4	664	1,066	1.118	.374
20,000	19,981	6,100	9.61	−12.5	14.19	40.1	643	1,062	1.111	.372
21,000	20,979	6,405	6.02	−14.4	13.63	38.8	622	1,058	1.105	.369
22,000	21,977	6,710	2.44	−16.4	13.10	37.6	602	1,054	1.098	.367
23,000	22,975	7,015	−1.14	−18.4	12.57	36.4	583	1,050	1.091	.364
24,000	23,972	7,320	−4.72	−20.4	12.07	35.2	564	1,046	1.084	.361
25,000	24,970	7,625	−8.31	−22.4	11.58	34.0	545	1,042	1.077	.359
26,000	25,968	7,930	−11.9	−24.4	11.11	32.9	527	1,037	1.070	.356
27,000	26,965	8,235	−15.5	−26.4	10.65	31.8	510	1,033	1.063	.353
28,000	27,962	8,540	−19.0	−28.3	10.21	30.7	492	1,029	1.056	.351
29,000	28,960	8,845	−22.6	−30.4	9.79	29.7	476	1,025	1.049	.348
30,000	29,957	9,150	−26.2	−32.3	9.38	28.7	460	1,021	1.043	.346
32,000	31,951	9,760	−33.3	−36.3	8.60	26.7	428	1,012	1.029	.340
34,000	33,945	10,370	−40.4	−40.2	7.87	24.9	399	1,004	1.014	.335
36,000	35,938	10,980	−47.6	−44.2	7.19	23.1	371	995	1.000	.330
38,000	37,931	11,590	−54.7	−48.2	6.56	21.5	344	987	.986	.325
40,000	39,923	12,200	−61.8	−52.1	5.98	19.9	319.	978	.972	.319
45,000	44,903	13,725	−71.5	−57.5	4.71	16.1	258	966	.952	.312
50,000	49,880	15,250	−71.5	−57.5	3.70	12.7	203	966	.952	.312
55,000	54,855	16,775	−71.5	−57.5	2.91	9.95	159	966	.952	.312
60,000	59,828	18,300	−68.8	−56.0	2.29	7.78	125	969	.957	.314
65,000	64,798	19,825	−65.6	−54.2	1.81	6.08	97.5	973	.964	.316
70,000	69,766	21,350	−62.3	−54.4	1.43	4.77	76.4	977	.970	.319
75,000	74,731	22,875	−59.0	−50.6	1.13	3.75	60.3	981	.977	.321
80,000	79,694	24,400	−55.8	−48.8	.898	2.95	47.2	985	.984	.324
85,000	84,655	25,925	−52.5	−46.9	.714	2.33	37.3	989	.990	.326
90,000	89,613	27,450	−48.7	−44.8	.569	1.84	29.4	994	.998	.329
100,000	99,523	30,500	−37.3	−38.5	.364	1.14	18.3	1,008	1.021	.337
125,000	124,255	38,125	−4.6	−20.3	.126	.368	5.90	1,046	1.084	.361
150,000	148,929	45,750	29.3	−1.50	.047	.129	2.06	1,084	1.148	.386
175,000	173,544	53,375	32.5	.28	.019	.0503	.807	1,088	1.154	.388
200,000	198,100	61,000	−1.2	−18.4	.0071	.0205	.329	1,050	1.091	.364
250,000	247,039	76,250	−116.2	−82.3	.0007	.0028	.0444	909	.857	.278
300,000	295,746	91,500	−156.6	−104.8	.00004	.0002	.0025	855	.770	.247

*Condensed from: "U.S. Standard Atmosphere Supplements", U.S. Government Printing Office, 1966.

Table 7-3.   U.S. STANDARD ATMOSPHERE, 1962*

Middle-Latitude, Year-Round Mean Conditions

For variations with geography and season, see Tables 7-2, 7-4, and 7-5.

The "U.S. Standard Atmosphere, 1962" was developed to serve the aerospace community as a mean basis for design and operation of vehicles and for general scientific considerations. For all practical purposes it is in agreement with the ICAO Standard Atmosphere over the altitude range that they have in common.[a]

Several earlier standards existed (see below); for realistic tables accounting for departures from the mean conditions due to geography, season, time of day, and solar activity, the "U.S. Standard Atmosphere, 1966" should be consulted.

The 1962 standard and the 1966 supplement (almost 300 pages each) were both cosponsored by the National Aeronautics and Space Administration, the U.S. Air Force, the former U.S. Weather Bureau, and COESA.[b]

The "U.S. Standard Atmosphere, 1962" agrees in general but differs in detail from the COSPAR International Reference Atmosphere, 1961.[c] Among the earlier standards were the COESA standard, "U.S. Extension to the ICAO Standard Atmosphere, 1958"; "ICAO Standard, 1952," adopted by NACA and published as NACA Report 1235, 1955; "ARDC Model Atmosphere, 1959"; "U.S.S.R. Standard Atmosphere, 1960". A later standard is "COSPAR International Reference Atmosphere, 1965".

BASIS OF THE 1962 TABLES[d]

These tables depict idealized year-round mean conditions at 45°N latitude, for the range of solar activity that occurs between sunspot minimum and sunspot maximum. In this model the atmosphere id defined in terms of temperature. The low-altitude temperature patterns are shown with abrupt changes at the following geometric altitudes (in km): 11, 20, 32, 47, 52, 61, and 79. Variations with season, weather, latitude, and solar activity are not included in this mean table (see other tables and their references).

The "U.S. Standard Atmosphere, 1962" treats air as a clean, dry, perfect-gas mixture $(c_p/c_v - 1.40)$, having a molecular weight to 90 km of 28.964 4 (C-12 scale). The principal sea-level constituents are assumed to be N_2—78.084%, O_2—20.947 6%, Ar—0.934%, CO_2—0.031 4%, Ne—0.001 818%, He—0.000 524%, methane—0.000 2%.

Assigned mean conditions at sea level are as follows:

$P = 0.101\ 325\ 0\ MN/m^2 = 2\ 116.22\ psf = 14.696\ psi$

$T = 288.15\ K = 15\ deg\ C = 59\ deg\ F$

$\rho = 1.225\ 0\ kg/m^3 = 0.076\ 474\ lb/ft^3$

$g = 980.655\ m/s^2 = 32.174\ 1\ ft/s^2$

$R = 8.314\ 32\ J/mol\cdot K = 1\ 545.31\ ft\ lb/lb\cdot mol\cdot deg\ R.$

The viscosities as tabulated are based on the equation:

$$\mu = 1.458 \times 10^{-6} T^{1.5}/T + 110.4,$$

where temperature is in Kelvin.

The thermal conductivities as tabulated are based on the equation:

$$k = 6.325 \times 10^{-7} T^{1.5}/T + 245.4 \times 10^{-(12/T)},$$

where temperature is in Kelvin.

The velocity of sound is based on the equation:

$$V_s = 1.4[c_p/c_v(RT/M_o)].$$

---

[a] ICAO—International Civil Aviation Organization.

[b] COESA—Committee on Extension to the Standard Atmosphere, a group of 30 U.S. scientific and engineering organizations, including APL-Johns Hopkins, Battelle, Boeing, FAA, Smithsonian Institution, JPL-Cal. Tech., Lockheed, NBS, and NRL.

[c] COSPAR—Committee on Space Research.

[d] For a 30-page discussion, see the original source.

*Condensed and computed from: "U.S. Standard Atmosphere, 1962", U.S. Government Printing Office, 1962.

Table 7-3.   U.S. STANDARD ATMOSPHERE, 1962 *(Continued)*

Table A.   U.S. STANDARD ATMOSPHERE—ENGLISH UNITS

Altitude		Temperature, t		Pressure, P		Density, ρ, lb/ft^3 $\times 10^5$	Gravity, g, ft/s^2	Molecular weight, M	Velocity of sound, V_s, fps	Viscosity, μ, $lb/ft \cdot s$ $\times 10^5$	Conductivity, k, $kcal \times 10^5$ $m \cdot s \cdot K$
Geometric, Z, ft	Geopotential, H, ft	deg F	deg R	in. Hg	psia						
0	0	59.00	518.7	29.92	14.7	7 647.	32.174	28.96	1 116.	1.202	.406 7
500	500	57.22	516.9	29.38	14.4	7 536.	32.173	28.96	1 115.	1.199	.405 5
1 000	1 000	55.43	515.1	28.86	14.2	7 426.	32.171	28.96	1 113.	1.196	.404 2
1 500	1 500	53.65	513.3	28.33	13.9	7 317.	32.169	28.96	1 111.	1.193	.403 0
2 000	2 000	51.87	511.5	27.82	13.7	7 210.	32.168	28.96	1 109.	1.190	.401 7
2 500	2 500	50.09	509.8	27.32	13.4	7 104.	32.166	28.96	1 107.	1.186	.400 2
3 000	3 000	48.30	508.0	26.82	13.2	6 998.	32.165	28.96	1 105.	1.183	.399 2
3 500	3 499	46.52	506.2	26.33	12.9	6 895.	32.163	28.96	1 103.	1.180	.398 0
4 000	3 999	44.74	504.4	25.84	12.7	6 792.	32.162	28.96	1 101.	1.177	.396 7
4 500	4 499	42.96	502.6	25.37	12.5	6 690.	32.160	28.96	1 099.	1.173	.395 4
5 000	4 999	41.17	500.8	24.90	12.2	6 590.	32.159	28.96	1 097.	1.170	.394 2
5 500	5 499	39.39	499.1	24.43	12.0	6 491.	32.157	28.96	1 095.	1.167	.392 9
6 000	5 998	37.61	497.3	23.98	11.8	6 393.	32.156	28.96	1 093.	1.164	.391 7
6 500	6 498	35.83	495.5	23.53	11.6	6 296.	32.154	28.96	1 091.	1.160	.390 4
7 000	6 998	34.05	493.7	23.09	11.3	6 200.	32.152	28.96	1 089.	1 157.	.389 1
7 500	7 497	32.26	491.9	22.66	11.1	6 105.	32.151	28.96	1 087.	1.154	.387 9
8 000	7 997	30.48	490.2	22.23	10.9	6 012.	32.149	28.96	1 085.	1.150	.386 6
8 500	8 497	28.70	488.4	21.81	10.7	5 919.	32.148	28.96	1 083.	1.147	.385 3
9 000	8 996	26.92	486.6	21.39	10.5	5 828.	32.146	28.96	1 081.	1.144	.384 0
9 500	9 496	25.14	484.8	20.98	10.3	5 738.	32.145	28.96	1 079.	1.140	.382 8
10 000	9 995	23.36	483.0	20.58	10.1	5 648.	32.145	28.96	1 077.	1.137	.381 5
11 000	10 994	19.79	479.5	19.80	9.72	5 473.	32.140	28.96	1 073.	1.130	.379 0
12 000	11 993	16.23	475.9	19.03	9.34	5 302.	32.137	28.96	1 069.	1.124	.376 4
13 000	12 992	12.67	472.3	18.30	8.99	5 135.	32.134	28.96	1 065.	1.117	.373 8
14 000	13 991	9.12	468.8	17.58	8.63	4 973.	32.131	28.96	1 061.	1.110	.371 3
15 000	14 989	5.55	465.2	16.89	8.29	4 814.	32.128	28.96	1 057.	1.104	.368 7
16 000	15 988	+1.99	461.6	16.23	7.97	4 659.	32.125	28.96	1 053.	1.097	.366 1
17 000	16 986	−1.58	458.1	15.58	7.65	4 508.	32.122	28.96	1 049.	1.090	.363 6
18 000	17 984	−5.14	454.5	14.95	7.34	4 361.	32.119	28.96	1 045.	1.083	.361 0
19 000	18 983	−8.70	451.0	14.35	7.05	4 217.	32.115	28.96	1 041.	1.076	.358 4
20 000	19 981	−12.2	447.4	13.76	6.76	4 077.	32.112	28.96	1 037.	1.070	.355 8
21 000	20 979	−15.8	443.9	13.20	6.48	3 941.	32.109	28.96	1 033.	1.063	.353 2
22 000	21 977	−19.4	440.3	12.65	6.21	3 808.	32.106	28.96	1 029.	1.056	.350 6
23 000	22 975	−22.9	436.7	12.12	5.95	3 679.	32.103	28.96	1 024.	1.049	.348 0
24 000	23 972	−26.5	433.2	11.61	5.70	3 553.	32.100	28.96	1 020.	1.042	.345 4
25 000	24 970	−30.0	429.6	11.09	5.44	3 431.	32.097	28.96	1 016.	1.035	.342 8
26 000	25 968	−33.6	426.1	10.64	5.22	3 311.	32.094	28.96	1 012.	1.028	.340 1
27 000	26 965	−37.2	422.5	10.18	5.00	3 195.	32.091	28.96	1 008.	1.021	.337 5
28 000	27 962	−40.7	418.6	9.74	4.78	3 082.	32.088	28.96	1 003.	1.014	.334 9
29 000	28 960	−44.3	415.4	9.31	4.57	2 973.	32.085	28.96	999.	1.007	.332 2
30 000	29 957	−47.8	411.8	8.90	4.37	2 866.	32.082	28.96	995.	1.000	.329 6
32 000	31 951	−54.9	404.7	8.12	3.99	2 661.	32.08	28.96	987.	0.986	.324 6
34 000	33 945	−62.0	397.6	7.40	3.63	2 468.	32.07	28.96	978.	0.971	.319 0
36 000	35 938	−69.2	390.5	6.73	3.30	2 285.	32.06	28.96	969.	0.956	.313 7
38 000	37 931	−69.7	390.0	6.12	3.05	2 079.	32.06	28.96	968.	0.955	.313 3
40 000	29 923	−69.7	390.0	5.56	2.73	1 890.	32.05	28.96	968.	0.955	.313 3
45 000	44 903	−69.7	390.0	4.375	2.148	1 487.	32.04	28.96	968.	0.955	.313 3
50 000	49 880	−69.7	390.0	3.444	1.691	1 171.	32.02	28.96	968.	0.955	.313 3
55 000	54 855	−69.7	390.0	2.712	1.332	922.	32.00	28.96	968.	0.955	.313 3
60 000	59 828	−69.7	390.0	2.135	1.048	726.	31.99	28.96	968.	0.955	.313 3

Table 7-3.   U.S. STANDARD ATMOSPHERE, 1962 (Continued)

Altitude		Temperature, t		Pressure, P		Density, ρ, lb/ft^3 $\times 10^5$	Gravity, g, ft/s^2	Molec- ular weight, M	Velocity of sound, V_s, fps	Vis- cosity, μ, $lb/ft \cdot s$ $\times 10^5$	Conduc- tivity, k, $kcal \times 10^5$ $m \cdot s \cdot K$
Geometric, Z, ft	Geopotential, H, ft	deg F	deg R	in. Hg	psia						
65 000	64 798	-69.7	390.0	1.681	0.825	572.	31.97	28.96	968.	0.955	.313 3
70 000	69 766	-67.4	392.2	1.325	0.651	447.9	31.96	28.96	971.	0.960	.315 0
75 000	74 731	-64.7	394.8	1.046	0.514	351.1	31.94	28.96	974.	0.966	.317 0
80 000	79 694	-62.0	397.7	0.827	0.406	273.1	31.93	28.96	978.	0.971	.319 1
85 000	84 655	-59.2	400.4	0.655	0.322	217.0	31.91	28.96	981.	0.977	.321 1
90 000	89 613	-56.5	403.1	0.520	0.255	171.0	31.90	28.96	984.	0.982	.323 1
100 000	99 523	-51.1	408.6	0.329	0.162	106.8	31.87	28.96	991.	0.993	.327 2
200 000	198 100	-2.6	457.0	0.006	0.003	1.696	31.57	28.96	1 048.	1.088	.362 8
300 000	295 745	-126.8	332.9	3.7×10^{-5}	1.8×10^{-5}	1.5×10^{-2}	31.27	28.96	895.	0.834	.269 8
400 000	392 471	$+233.9$	693.6	6.3×10^{-7}	3.1×10^{-7}	1.2×10^{-4}	30.97	27.97	—	—	—
500 000	488 291	1 203.	1 663.	1.4×10^{-7}	0.7×10^{-7}	1.0×10^{-5}	30.68	26.86	—	—	—
1 000 000	954 232	2 125.	2 584.	5.1×10^{-9}	2.5×10^{-9}	2.0×10^{-7}	29.30	22.52	—	—	—
1 500 000	1 399 317	2 221.	2 681.	5.5×10^{-10}	2.7×10^{-10}	1.8×10^{-8}	28.00	18.68	—	—	—
2 000 000	1 824 911	2 251.	2 710.	9.2×10^{-11}	4.5×10^{-11}	2.6×10^{-9}	26.79	16.77	—	—	—

Table B.   U.S. STANDARD ATMOSPHERE—METRIC UNITS

Altitude		Temperature, t		Pressure, P		Density, ρ, kg/m^3	Gravity, g, m/s^2	Molec- ular weight, M	Velocity of sound, V_s, m/s	Vis- cosity, μ, $kg/m \cdot s$ $\times 10^5$	Conduc- tivity, k, $kcal \times 10^5$ $m \cdot s \cdot K$
Geometric, Z, m	Geopotential, H, m	deg C	K	mbar	mm Hg						
0	0	15.00	288.15	1 013.3	760.	1.225	980.7	28.96	340.3	1.789	.605 3
200	200	13.70	286.85	989.5	742.	1.202	980.6	28.96	339.5	1.783	.602 9
400	400	12.40	285.55	966.1	724.	1.179	980.5	28.96	338.8	1.777	.600 4
600	600	11.10	284.25	943.2	707.	1.156	980.5	28.96	338.0	1.771	.598 0
800	800	9.80	282.95	920.8	691.	1.134	980.4	28.96	337.2	1.764	.595 5
1 000	1 000	8.50	281.65	898.8	674.	1.112	980.4	28.96	336.4	1.758	.593 1
1 200	1 200	7.20	280.35	877.2	658.	1.090	980.3	28.96	335.7	1.752	.590 6
1 400	1 400	5.90	279.05	856.0	642.	1.069	980.2	28.96	334.9	1.745	.588 1
1 600	1 600	4.60	277.75	835.3	627.	1.048	980.2	28.96	334.1	1.739	.585 7
1 800	1 799	3.30	276.45	814.9	611.	1.027	980.1	28.96	333.3	1.732	.583 2
2 000	1 999	2.00	275.15	795.0	596.	1.007	980.0	28.96	332.5	1.726	.580 7
2 200	2 199	0.70	273.86	775.5	582.	0.987	980.0	28.96	331.7	1.720	.578 4
2 400	2 399	-0.59	272.56	756.3	567.	0.967	979.9	28.96	331.0	1.713	.575 9
2 600	2 599	-1.89	271.26	737.6	553.	0.947	979.9	28.96	330.2	1.707	.573 3
2 800	2 799	-3.19	269.96	719.2	539.	0.928	979.8	28.96	329.4	1.700	.570 8
3 000	2 999	-4.49	268.66	701.2	526.	0.909	979.7	28.96	328.6	1.694	.568 3
3 200	3 198	-5.79	267.36	683.6	513.	0.891	979.7	28.96	327.8	1.687	.565 8
3 400	3 398	-7.09	266.06	666.3	500.	0.872	979.6	28.96	327.0	1.681	.563 4
3 600	3 598	-8.39	264.76	649.4	487.	0.854	979.6	28.96	326.2	1.674	.560 9
3 800	3 798	-9.69	263.47	632.8	475.	0.837	979.5	28.96	325.4	1.668	.558 4
4 000	3 997	-10.98	262.17	616.6	462.	0.819	979.4	28.96	324.6	1.661	.555 9
4 200	4 197	-12.3	260.87	600.7	451.	0.802	979.4	28.96	323.8	1.655	.553 4
4 400	4 397	-13.6	259.57	585.2	439.	0.785	979.3	28.96	323.0	1.648	.550 8
4 600	4 597	-14.9	258.27	570.0	428.	0.769	979.3	28.96	322.2	1.642	.548 3
4 800	4 796	-16.2	256.97	555.1	416.	0.752	979.2	28.96	321.4	1.635	.545 8
5 000	4 996	-17.5	255.68	540.5	405.	0.736	979.1	28.96	320.5	1.628	.543 3
5 200	5 196	-18.8	254.38	526.2	395.	0.721	979.1	28.96	319.7	1.622	.540 8
5 400	5 395	-20.1	253.08	512.3	384.	0.705	979.0	28.96	318.9	1.615	.538 3
5 600	5 595	-21.4	251.78	498.6	374.	0.690	978.9	28.96	318.1	1.608	.535 7
5 800	5 795	-22.7	250.48	485.2	364.	0.675	978.9	28.96	317.3	1.602	.533 2

Table 7-3.   U.S. STANDARD ATMOSPHERE, 1962 *(Continued)*

Altitude Geometric, Z, m	Altitude Geopotential, H, m	Temperature, t deg C	Temperature, t K	Pressure, P mbar	Pressure, P mm Hg	Density, ρ, kg/m³	Gravity, g, m/s²	Molecular weight, M	Velocity of sound, V_s', m/s	Viscosity, μ, kg/m·s × 10⁵	Conductivity, k, kcal × 10⁵ m·s·K
6 000	5 994	−24.0	249.19	472.2	354.	0.660	978.8	28.96	316.5	1.595	.530 7
6 200	6 194	−25.3	247.89	459.4	345.	0.646	978.8	28.96	315.6	1.588	.528 2
6 400	6 394	−26.6	246.59	446.9	335.	0.631	978.7	28.96	314.8	1.582	.525 6
6 600	6 593	−27.9	245.29	434.7	326.	0.617	978.6	28.96	314.0	1.575	.523 1
6 800	6 793	−29.2	244.00	422.7	317.	0.604	978.6	28.96	313.1	1.568	.520 5
7 000	6 992	−30.5	242.70	411.1	308.	0.590	978.5	28.96	312.3	1.561	.518 0
7 500	7 491	−33.7	239.46	383.0	287.	0.572	978.4	28.96	310.2	1.544	.511 6
8 000	7 990	−36.9	236.22	356.5	267.	0.526	978.2	28.96	308.1	1.527	.505 2
8 500	8 489	−40.2	232.97	331.6	249.	0.496	978.1	28.96	306.0	1.510	.498 8
9 000	8 987	−43.4	229.73	308.0	231.	0.467	977.9	28.96	303.8	1.493	.492 4
9 500	9 486	−46.7	226.49	285.8	214.	0.440	977.7	28.96	301.7	1.475	.485 9
10 000	9 984	−49.9	223.25	265.0	199.	0.414	977.6	28.96	299.5	1.458	.479 4
11 000	10 981	−56.4	216.77	227.0	170.	0.365	977.3	28.96	295.2	1.422	.466 4
12 000	11 977	−56.5	216.65	194.0	146.	0.312	977.0	28.96	295.1	1.422	.466 2
13 000	12 973	−56.5	216.65	165.8	124.	0.267	976.7	28.96	295.1	1.422	.466 2
14 000	13 969	−56.5	216.65	141.7	106.	0.228	976.4	28.96	295.1	1.422	.466 2
15 000	14 965	−56.5	216.65	121.1	90.8	0.195	976.1	28.96	295.1	1.422	.466 2
16 000	15 960	−56.5	216.65	103.5	77.7	0.166	975.8	28.96	295.1	1.422	.466 2
17 000	16 955	−56.5	216.65	88.5	66.4	0.142	975.4	28.96	295.1	1.422	.466 2
18 000	17 949	−56.5	216.65	75.7	56.7	0.122	975.1	28.96	295.1	1.422	.466 2
19 000	18 943	−56.5	216.65	64.7	48.5	0.104	974.8	28.96	295.1	1.422	.466 2
20 000	19 937	−56.5	216.65	55.3	41.4	0.088 9	974.5	28.96	295.1	1.422	.466 2
22 000	21 924	−54.6	218.57	40.5	30.4	0.064 5	973.9	28.96	296.4	1.433	.470 2
24 000	23 910	−52.6	220.56	29.7	22.3	0.046 9	973.3	28.96	297.8	1.444	.474 2
26 000	25 894	−50.6	222.54	21.9	16.4	0.034 3	972.7	28.96	299.1	1.454	.478 2
28 000	27 877	−48.6	224.53	16.2	12.1	0.025 1	972.1	28.96	300.4	1.465	.482 0
30 000	29 859	−46.6	226.51	12.0	9.0	0.018 4	971.5	28.96	301.7	1.475	.485 9
35 000	34 808	−36.6	236.51	5.75	4.31	0.008 5	970.0	28.96	308.3	1.529	.505 8
40 000	39 750	−22.8	250.35	2.87	2.15	0.004 0	968.4	28.96	317.2	1.601	.533 0
50 000	49,610	−2.5	270.65	0.798	0.598	0.001 0	965.4	28.96	329.8	1.704	.572 1
100 000	98 451	−63.1	210.02	3×10^{-4}	2.6×10^{-4}	5×10^{-7}	950.1	28.88	—	—	—
200 000	193 898	+962.8	1 236.	1.3×10^{-6}	1×10^{-6}	3.3×10^{-10}	921.7	25.56	—	—	—
400 000	376 312	1 214.	1 487.	4×10^{-8}	3×10^{-8}	6.5×10^{-12}	867.9	19.94	—	—	—
700 000	630 530	1 234.	1 508.	1.2×10^{-9}	8.9×10^{-10}	1.5×10^{-13}	795.6	16.17	—	—	—

Table 7-4. VARIATIONS OF HIGH-ALTITUDE ATMOSPHERE*

Values for 20,000 ft Geometric Altitude (6,100 Meters)

SYMBOLS:

H, ft = geopotential altitude, feet. For meters multiply by 0.3048.

t, °F = temperature, degrees Fahrenheit

t, °C = temperature, degrees Celsius

P, psia = atmospheric pressure, psia. For atmospheres multiply by 0.068046. For kN/m² multiply by 6.8948.

ρ = density. For lb_m/ft^3 multiply by 10^{-3}. For kg/m³ multiply by 10^{-3} and by 16.018.

V_s, fps = speed of sound, ft/sec. For m/sec multiply by 0.3048.

μ = viscosity. For lb_m/ft sec multiply by 10^{-5}. For N·s/m² (= kg/m·s) multiply by 10^{-5} and by 1.4882.

k = thermal conductivity. For Btu/sec ft °R multiply by 10^{-5}. For W/m·K multiply by 0.62265.

Latitude, °N	Season	H, ft	t, °F	t, °C	P, psia	ρ	V_s, fps	μ	k
30	Winter	19,953	−1.20	−18.4	6.92	40.8	1050	1.091	.364
30	Summer	19,953	18.89	−7.9	7.06	39.8	1072	1.129	.378
45	Winter	19,981	−21.99	−30.0	6.62	40.8	1026	1.051	.349
45	Summer	19,981	9.61	−12.4	6.97	40.1	1062	1.111	.372
60	Winter (cold)	20,006	−39.36	−39.7	6.39	41.0	1005	1.016	.336
60	Winter (warm)	20,006	−39.36	−39.7	6.39	41.0	1005	1.016	.336
60	Summer	20,006	−5.01	−20.6	6.79	40.3	1045	1.083	.361
75	Winter (cold)	20,026	−48.66	−44.8	6.24	41.0	994	.998	.329
75	Winter (warm)	20,026	−48.66	−44.8	6.24	41.0	994	.998	.329
75	Summer	20,026	−12.73	−24.9	6.71	40.5	1036	1.069	.355

*Condensed from: "U.S. Standard Atmosphere Supplements", U.S. Government Printing Office, 1966.

Table 7-5. EXTREMELY HIGH-ALTITUDE CONDITIONS

The following table gives temperature and molecular weight of atmosphere above 390,000 ft (120 km) geometric altitude (summer model).† For pressure in N/m², multiply values in millibars by 100.

Altitudes, geometric		Exospheric temperature,‡ 600°K			Exospheric temperature,‡ 2100°K		
km	ft	Temp, °K	Molecular weight	Pressure, millibars	Temp, °K	Molecular weight	Pressure, millibars
120	394,000	380.0	26.77	2.30×10^{-5}	379.6	26.76	2.27×10^{-5}
140	459,000	483.9	25.40	5.92×10^{-6}	819.9	25.78	8.16×10^{-6}
160	525,000	535.0	23.90	2.00×10^{-6}	1142.6	25.07	4.55×10^{-6}
180	590,000	561.1	22.32	7.80×10^{-7}	1375.3	24.47	2.93×10^{-6}
200	656,000	574.1	20.79	3.36×10^{-7}	1544.8	23.92	2.03×10^{-6}
250	820,000	593.7	17.85	5.53×10^{-8}	1796.5	22.65	9.49×10^{-7}
300	984,000	598.4	16.06	1.19×10^{-8}	1942.0	21.50	5.01×10^{-7}
400	1,312,000	599.9	11.30	9.91×10^{-10}	2058.4	19.55	1.70×10^{-7}
1000	3,280,000	600.0	1.28	5.44×10^{-11}	2100.0	13.29	1.94×10^{-9}

†Atmospheric density varies from 9.05×10^{-14} to 1.93×10^{-8} kg/m³.
‡Exospheric temperature depends on solar activity and season.

REFERENCES

"U.S. Standard Atmosphere Supplements", U.S. Government Printing Office, 1966.
"Extension of Tables for the U.S. Standard Atmosphere", R.A. Minzner, CRL-67-0335, USAF, 1967.

Table 7-6. WEATHER DATA FOR SELECTED U.S. CITIES*

For temperature data see Table 7-17. For solar data see Table 7-19.

	Location	Sun total,[a] %	Wind,[b] avg, mi/hr	Rain total,[c] in.	Relative[d] humidity, % 7 a.m.	Relative[d] humidity, % 1 p.m.	Precipitation,[e] days	Snow per season,[f] in.	32 deg F or less,[g] days	Particulate pollution,[h] µg/m³
AL	Mobile	61	9.5	68.13	85	57	123	0.4	21	
AK	Juneau	31	8.6	54.62	87	81	220	102.7	149	
AZ	Phoenix	86	5.8	7.20	55	33	35		16	123
AR	Little Rock	62	8.2	48.66	84	56	102	5.5	68	
CA	Los Angeles	73	7.3	12.63	77	61	36			104
	Sacramento	78	8.5	16.29	83	65	57		20	
	San Francisco	67	10.5	18.69	84	69	62		4	59
CO	Denver	70	9.1	14.81	68	40	87	58.3	164	135
CT	Hartford	57	9.1	42.92	78	53	125	56.3	138	69
DE	Wilmington	58	9.0	44.56	79	55	116	22.6	104	127
DC	Washington	58	9.4	40.78	73	51	111	18.2	82	77
FL	Jacksonville	60	8.9	53.36	87	55	116	0.1	11	
	Miami	66	9.0	59.76	84	61	128			69
GA	Atlanta	61	9.2	47.14	83	57	115	1.8	62	85
HI	Honolulu	69	11.5	21.89	75	71	101			42
ID	Boise	67	9.0	11.43	70	53	90	21.7	127	105
IL	Chicago	57	10.3	33.18	75	58	120	38.4	119	145
	Peoria	58	10.4	34.84	83	62	109	22.4	133	
IN	Indianapolis	59	9.8	39.25	83	61	121	20.6	122	118
IA	Des Moines	60	11.3	30.37	82	63	105	31.9	140	103
KS	Wichita	65	12.8	28.41	80	55	83	14.3	112	67
KY	Louisville	58	8.3	41.32	82	59	122	18.0	98	127
LA	New Orleans	61	8.4	53.90	88	63	112	0.2	12	75
ME	Portland	59	8.7	42.85	80	60	126	73.7	162	
MD	Baltimore	58	9.8	43.05	77	53	111	25.1	103	118
MA	Boston	60	13.2	42.77	72	55	128	42.2	100	
MI	Detroit	54	10.1	30.95	78	58	131	31.6	125	129
	Sault Ste. Marie	47	9.8	31.22	86	67	164	104.0	179	
MN	Duluth	55	11.8	28.97	79	62	134	76.8	186	76
	Minneapolis– St. Paul	58	10.7	24.78	81	61	112	44.4	160	74
MS	Jackson	59	7.9	50.82	90	58	106	1.8	54	80
MO	Kansas City	64	10.0	34.07	77	57	99	20.1	100	116
	St. Louis	59	9.5	35.31	81	57	105	17.4	107	213
MT	Great Falls	64	13.2	14.07	66	50	99	55.2	150	
NE	Omaha	62	11.0	27.56	79	59	98	32.0	137	115
NV	Reno	80	6.4	7.15	69	44	48	25.3	188	
NH	Concord	54	6.7	38.80	80	53	125	62.7	169	35
NJ	Atlantic City	56	10.9	42.36	80	54	112	17.8	121	
NM	Albuquerque	77	8.8	8.13	57	37	58	9.8	118	
NY	Albany	54	8.8	35.08	79	56	133	63.9	146	
	Buffalo	53	12.5	35.65	79	63	165	87.2	138	
	New York	59	9.5	42.37	72	56	121	29.9	82	131
NC	Charlotte	66	7.5	43.38	84	53	110	5.4	76	108
	Raleigh	61	7.9	43.58	84	52	113	7.4	88	
ND	Bismarck	62	10.8	15.15	77	55	96	38.1	188	82
OH	Cincinnati	58	7.1	39.51	80	57	132	19.0	99	117
	Cleveland	52	10.9	35.35	79	62	154	50.3	128	128
	Columbus	55	8.5	36.67	81	59	134	28.0	125	106
OK	Oklahoma City	67	13.2	30.82	82	56	82	9.5	85	70
OR	Portland	46	7.7	37.18	87	73	153	8.4	44	84
PA	Philadelphia	58	9.6	42.48	76	54	116	21.9	108	129
	Pittsburgh	53	9.4	36.14	79	57	149	46.0	128	165

*Federal Government data, published 1971. For monthly data see "Statistical Abstract of the United States: 1971", 92nd ed., U.S. Bureau of the Census, 1971.

Table 7-6. WEATHER DATA FOR SELECTED U.S. CITIES (*Continued*)

	Location	Sun total,[a] %	Wind,[b] avg, mi/hr	Rain total,[c] in.	Relative[d] humidity, % 7 a.m.	Relative[d] humidity, % 1 p.m.	Precipi- tation,[e] days	Snow per season,[f] in.	32 deg F or less,[g] days	Particulate pollution,[h] µg/m³
PR	San Juan	64	8.5	64.21	82	66	205			
RI	Providence	57	11.0	42.13	73	54	124	40.9	126	80
SC	Columbia	64	7.0	46.82	86	50	109	1.2	55	
SD	Sioux Falls	63	11.1	25.16	81	59	93	41.6	173	69
TN	Memphis	65	9.2	49.73	82	57	103	6.1	60	76
	Nashville	58	7.8	45.15	85	57	118	11.8	75	98
TX	Dallas	65	10.9	34.55	79	55	80	2.1	37	84
	El Paso	83	9.9	7.89	52	35	44	4.4	63	
	Houston	62	10.8	45.95	87	60	103	0.4	13	88
UT	Salt Lake City	70	8.6	13.90	67	47	87	55.1	141	90
VT	Burlington	52	8.9	33.21	77	59	149	73.8	167	46
VA	Norfolk	63	10.6	44.94	79	58	115	7.5	56	94
	Richmond	61	7.7	44.21	82	53	113	15.0	86	
WA	Seattle	45	9.5	38.94	84	75	161	15.8	32	61
	Spokane	57	8.4	17.19	76	62	115	54.6	139	
WV	Charleston	48	6.6	44.43	82	55	147	28.4	101	213
WI	Milwaukee	56	11.8	29.51	81	64	120	43.4	149	124
WY	Cheyenne	64	12.7	15.06	63	40	99	51.6	171	32

Note: Data are largely from airport stations, through 1969.

[a] Average percentage of possible sunshine.
[b] Average annual windspeed.
[c] Average annual precipitation from 1931–60.
[d] Generally the highest and lowest humidity, during the day.
[e] Average number of days with precipitation of 0.01 inches or more.
[f] Snow and sleet, total inches for season, average for period of record.
[g] Mean annual number of days during which the temperature falls to freezing.
[h] Average particulate concentration, biweekly samples, 1969 only, National Air Pollution Control Administration.

Table 7-7. AVERAGE AMOUNTS OF THE ELEMENTS IN THE EARTH'S CRUST*

In Grams Per Metric Ton or Parts Per Million

Element	Quantity	Element	Quantity	Element	Quantity
O	466,000	N	46	Br	1.6
Si	277,200	Ce	46	Ho	1.2
Al	81,300	Sn	40	Eu	1.1
Fe	50,000	Y	28	Sb	1?
Ca	36,300	Nd	24	Tb	0.9
Na	28,300	Nb	24	Lu	0.8
K	25,900	Co	23	Tl	0.6
Mg	20,900	La	18	Hg	0.5
Ti	4,400	Pb	16	I	0.3
H	1,400	Ga	15	Bi	0.2
P	1,180	Mo	15	Tm	0.2
Mn	1,000	Th	12	Cd	0.15
S	520	Cs	7	Ag	0.1
C	320	Ge	7	In	0.1
Cl	314	Sm	6.5	Se	0.09
Rb	310	Gd	6.4	Ar	0.04
F	300	Be	6	Pd	0.01
Sr	300	Pr	5.5	Pt	0.005
Ba	250	Sc	5	Au	0.005
Zr	220	As	5	He	0.003
Cr	200	Hf	4.5	Te	0.002?
V	150	Dy	4.5	Rh	0.001
Zn	132	U	4	Re	0.001
Ni	80	B	3	Ir	0.001
Cu	70	Yb	2.7	Os	0.001?
W	69	Er	2.5	Ru	0.001?
Li	65	Ta	2.1		

*From: "Principles of Geochemistry", B. Mason, John Wiley & Sons, 1952.

Table 7-8. COMPOSITION OF SEA WATER

Table A. ELEMENTS IN SOLUTION

Excluding Dissolved Gases

Element	Concentration, parts per million (approximate)	Percent by weight	Element	Concentration, parts per million (approximate)	Percent by weight
Oxygen	857,000.	85.7000	Potassium	380.	0.0380
Hydrogen	108,000.	10.8000	Bromine	65.	0.0065
Chlorine	19,000.	1.9000	Carbon	30.	0.0030
Sodium	10,500.	1.0500	Strontium	13.	0.0013
Magnesium	1,275.	0.1275	Boron	4.6	0.00046
Sulfur	885.	0.0885	Silicon	2.	0.0002
Calcium	400.	0.0400	Fluorine	1.3	0.00013
			Aluminum	1.	0.0001

Table B. IONIC CONSTITUENTS

Anions in Sea Water of 34.4 Salinity per Mil, or 3.44 Percent, by Weight

Tabular values are given in percent by weight.

Chloride	1.897	Bromide	0.0065
Sulfate	0.265	Borate	0.0027
Bicarbonate	0.014		

Table C. ARTIFICIAL SEA WATER

To simulate the physical properties, a 3.4 percent solution of sodium chloride or natural sea salt may be used.

For a more exact chemical reproduction the following is an average.

Salt	Grams
NaCl	25.
MgCl$_2$	3.
MgSO$_4$ (or NaSO$_4$)	4.
CaCl$_2$	1.
KCl	0.7
NaHCO$_3$	0.2
NaBr (or KBr)	0.1
TOTAL	34.0

Note: Add water to make 1 kilogram.

USN Specification 44T27b–1940 is as follows:

NaCl	23 g
Na$_2$SO$_4$·10H$_2$O	8 g
Stock solution	20 ml

Sterile distilled water to 1 liter.

The stock solution is as follows:

Magnesium chloride	550 g
Calcium chloride	110 g
Potassium bromide	45 g
Potassium chloride	10 g

Sterile distilled water to 1 liter.

Table 7-9.   MEAN ANNUAL OCEAN TEMPERATURES AND VARIATIONS

Approximate mean surface temperatures for the Atlantic, Pacific, and Indian oceans are given. Seasonal variations at mid-latitudes may be 10°F or more. Between 40°N and 40°S and at a depth of 800 ft (244 m), the mean temperatures are about 15°F lower than at the surface; at 2000 ft (610 m) the mean temperatures are about 25°F lower than at the surface. For temperature in K, add 273.15 to the values in deg C.

Latitude	Temperature		Latitude	Temperature	
	°C	°F		°C	°F
75° North	−1.	30.	15° South	25.	77.
60° North	4.	39.	30° South	20.	68.
45° North	11.	52.	45° South	10.	50.
30° North	22.	72.	60° South	3.	37.
15° North	26.	79.	75° South	−1.7	29.
Equator	27.	80.			

REFERENCE

For ocean temperature, density, and salinity data at various locations, see "Handbook of Oceanographic Tables", SP–68, U.S. Naval Oceanographic Office, 1966.

Table 7-10.   FREEZING POINT AND DENSITY OF SEA WATER

For temperature in K, add 273.15 to the values in deg C.

Salinity, parts per thousand	Freezing point		Temperature of maximum density	
	°C	°F	°C	°F
0	0.0	32.0	3.98	39.2
10	−0.53	31.05	1.72	35.1
20	−1.08	30.05	−0.6	31.3
30	−1.63	29.07	−2.4	27.6
40	−2.20	28.08	−4.5	24.0

Table 7-11.   COMPRESSIBILITY OF SEA WATER*

For a Salinity of 35% at 0°C

For pressure in MN/m^2, divide the values in bars by 10. For specific volume in m^3/kg, divide values in cc/g by 1000. For density in kg/m^3, multiply values in lb/cu ft by 16.018.

Pressure			Specific volume, cc/g	Density, lb/cu ft
bars	standard atmospheres	psia		
1.013	1.00	14.7	.9726	64.18
100	98.7	1,450	.9682	64.47
200	197.3	2,900	.9639	64.77
300	296.1	4,350	.9597	65.06
400	394.8	5,800	.9557	65.32
500	493.5	7,250	.9517	65.57
600	592.1	8,700	.9479	65.88
700	690.8	10,150	.9442	66.09
800	789.5	11,600	.9406	66.34
900	888.2	13,050	.9371	66.62
1,000	986.9	14,500	.9337	66.84

*Based on data of Bjerkness and Sandstrom, Carnegie Institution, Washington, 1910.

Table 7-12.   SEA-CONDITION TERMINOLOGY AND WIND VELOCITY*

For wave height in meters, multiply values in feet by 0.3048.

Beaufort number	Wind speed				Seaman's term	U.S. Weather Bureau term	Effects observed at sea	Effects observed on land	World Meteorological Organization Code	
	knots	mph	meters per second	km per hour					Term and height of waves, feet	Code
0	<1	<1	0.0–0.2	<1	Calm		Sea like mirror	Calm; smoke rises vertically	Calm, glassy, 0	0
1	1–3	1–3	0.3–1.5	1–5	Light air	Light	Ripples with appearance of scales; no foam crests	Smoke drift indicates wind direction; vanes do not move	Rippled, 0–1	1
2	4–6	4–7	1.6–3.3	6–11	Light breeze	Light	Small wavelets; crests of glassy appearance, not breaking	Wind felt on face; leaves rustle; vanes begin to move	Smooth, 1–2	2
3	7–10	8–12	3.4–5.4	12–19	Gentle breeze	Gentle	Large wavelets; crests begin to break; scattered whitecaps	Leaves, small twigs in constant motion; light flags extended	Slight, 2–4	3
4	11–16	13–18	5.5–7.9	20–28	Moderate breeze	Moderate	Small waves, becoming longer; numerous whitecaps	Dust, leaves, and loose paper raised up; small branches move	Moderate, 4–8	4
5	17–21	19–24	8.0–10.7	29–38	Fresh breeze	Fresh	Moderate waves, taking longer form; many whitecaps; some spray	Small trees in leaf begin to sway	Rough, 8–13	5
6	22–27	25–31	10.8–13.8	39–49	Strong breeze	Strong	Larger waves forming; whitecaps everywhere; more spray	Larger branches of trees in motion; whistling heard in wires	Very rough, 13–20	6
7	28–33	32–38	13.9–17.1	50–61	Moderate gale	Strong	Sea heaps up; white foam from breaking waves begins to be blown in streaks	Whole trees in motion; resistance felt in walking against wind	Very rough, 13–20	6
8	34–40	39–46	17.2–20.7	62–74	Fresh gale	Gale	Moderately high waves of greater length; edges of crests begin to break into spindrift; foam is blown in well-marked streaks	Twigs and small branches broken off trees; progress generally impeded	High, 20–30	7
9	41–47	47–54	20.8–24.4	75–88	Strong gale	Gale	High waves; sea begins to roll; dense streaks of foam; spray may reduce visibility	Slight structural damage occurs; slate blown from roofs	High, 20–30	7
10	48–55	55–63	24.5–28.4	89–102	Whole gale	Whole gale	Very high waves with overhanging crests; sea takes white appearance as foam is blown in very dense streaks; rolling is heavy and visibility reduced	Seldom experienced on land; trees broken or uprooted; considerable structural damage occurs	Very high, 30–45	8
11	56–63	64–72	28.5–32.6	103–117	Storm	Whole gale	Exceptionally high waves; sea covered with white-foam patches; visibility still more reduced		Very high, 30–45	8
12	64–71	73–82	32.7–36.9	118–133	Hurricane	Hurricane	Air filled with foam; sea completely white with driving spray; visibility greatly reduced	Very rarely experienced on land; usually accompanied by widespread damage	Phenomenal, over 45	9
13	72–80	83–92	37.0–41.4	134–149						
14	81–89	93–103	41.5–46.1	150–166						
15	90–99	104–114	46.2–50.9	167–183						
16	100–108	115–125	51.0–56.0	184–201						
17	109–118	126–136	56.1–61.2	202–220						

Note: Since January 1, 1955, weather map symbols have been based on wind speed in knots at five-knot intervals, rather than on Beaufort number.

*From: "Handbook of Oceanographic Tables", SP-68, U.S. Naval Oceanographic Office, 1966.

Table 7-13.   OCEAN WAVE PREDICTIONS*

The following table gives the minimum time that the wind must blow to form waves of significant height and period. The fetch, or distance, swept by the wind is in nautical miles.

SYMBOLS AND UNITS:

T = time, hours

H = height, feet. For meters multiply by 0.3048.

P = period, seconds

1 nautical mile = 1.151 statute miles = 1852 m

1 knot = 1 nautical mile per hour = 0.5144 m/s

Fetch, nautical miles	Beaufort number†											
	4 (5 knots)			6 (25 knots)			8 (37 knots)			10 (52 knots)		
	T	H	P	T	H	P	T	H	P	T	H	P
20	6.2	3.2	2.9	4.7	7.0	3.8	3.9	10.0	4.4	3.2	14.0	5.2
40	10.3	3.9	3.6	7.8	9.0	4.6	6.5	14.0	5.4	5.4	21.0	6.3
60	14.0	4.0	4.0	10.2	10.3	5.1	8.7	17.0	6.0	7.4	25.0	7.0
80	17.0	4.0	4.2	13.0	11.0	5.6	11.0	18.9	6.6	9.3	28.0	7.7
100	20.0	4.0	4.4	15.1	11.4	6.0	12.8	20.5	6.9	11.0	32.0	8.1
150	27.1	4.2	5.0	20.1	12.0	6.5	17.0	22.5	7.8	14.5	36.3	9.0
200	33.5	4.3	5.6	25.4	12.2	7.1	21.5	23.5	8.5	18.1	40.0	9.8
300	47.0	4.4	6.3	34.1	13.1	8.0	29.0	25.0	9.5	24.3	45.0	11.1
400	—	—	—	42.2	13.5	8.6	35.6	26.0	10.2	30.2	47.5	12.0
500	—	—	—	49.2	13.8	9.1	42.1	27.5	10.9	35.5	49.0	12.7
600	—	—	—	56.3	13.8	9.5	47.7	27.5	11.3	40.3	50.0	13.3
700	—	—	—	—	—	—	53.2	27.5	11.8	45.4	50.5	14.0
800	—	—	—	—	—	—	59.2	27.5	12.3	50.6	51.5	14.5

†See Table 7-12.

*From: U.S. Navy Hydrographic Office, Publications 603 and 604.

Table 7-14.   PERIODS OF OCEAN WAVES

Period range, seconds	Generating forces	Common names	Remarks
<0.1	Wind	Capillary waves	Ripples; restoring force is surface tension
0.1–1.	Wind	Ultra-gravity waves	Restoring forces surface tension and gravity
1–30	Wind; storm	Gravity waves	Short-crested; non-uniform; more peaked than sine waves; skewed in direction of propagation. Deep-water sine-wave swell in absence of wind
30–10,000	Wind; storm; earthquake	Infra-gravity waves; long-period waves	Include some "seiches" or occasional oscillations at natural frequency of body of water in deep basin. Tidal waves common in the Pacific, have periods 10–30 min; wavelengths several miles
12 or 24 hr	Moon and sun	Tides	Predictable; occur one hour later each day; components with longer periods; influenced by configuration of basin
>24 hr	Storm, sun, moon	Trans-tidal waves	Storm-tides not periodic; oscillations persist after wind-stress is relaxed

Table 7-15.   THIRTY-TWO COMPASS POINTS

Names and Angular Equivalents, Clockwise, Starting North to East

Point No.	Name	Deg	Min	Sec	Deg	Point No.	Name	Deg	Min	Sec	Deg
0	North	0	00	00	0	17	S by W	191	15	00	191.25
1	N by E	11	15	00	11.25	18	SSW	202	30	00	202.50
2	NNE	22	30	00	22.50	19	SW by S	213	45	00	213.75
3	NE by N	33	45	00	33.75	20	SW	225	00	00	225.00
4	NE	45	00	00	45.00	21	SW by W	236	15	00	236.25
5	NE by E	56	15	00	56.25	22	WSW	247	30	00	247.50
6	ENE	67	30	00	67.50	23	W by S	258	45	00	258.75
7	E by N	78	45	00	78.75	24	West	270	00	00	270.00
8	East	90	00	00	90.00	25	W by N	281	15	00	281.25
9	E by S	101	15	00	101.25	26	WNW	292	30	00	292.50
10	ESE	112	30	00	112.50	27	NW by W	303	45	00	303.75
11	SE by E	123	45	00	123.75	28	NW	315	00	00	315.00
12	SE	135	00	00	135.00	29	NW by N	326	15	00	326.25
13	SE by S	146	15	00	146.25	30	NNW	337	30	00	337.50
14	SSE	157	30	00	157.50	31	N by W	348	45	00	348.75
15	S by E	168	45	00	168.75	32	North	360	00	00	360.00
16	South	180	00	00	180.00						

Table 7-16.   LATITUDE AND LONGITUDE LENGTH OF A DEGREE*

The following table lists lengths of one degree latitude and longitude at various distances (°lat) from the equator.

Length of 1° latitude				From equator location, °lat	Length of 1° longitude			
Nautical mile	Statute mile	Feet	Kilo-meter		Nautical mile	Statute mile	Feet	Kilo-meter
59.702	68.703	362,752	110.567	0	60.109	69.172	365,226	111.321
59.707	68.709	362,781	110.576	5	59.882	68.910	363,844	110.900
59.720	68.724	362,863	110.601	10	59.202	68.128	359,714	109.641
59.743	68.750	363,001	110.643	15	58.074	66.830	352,863	107.553
59.773	68.785	363,185	110.699	20	56.506	65.026	343,336	104.649
59.810	68.828	363,411	110.768	25	54.510	62.729	331,207	100.952
59.853	68.878	363,674	110.848	30	52.100	59.955	316,561	96.488
59.902	68.934	363,969	110.938	35	49.293	56.725	299,507	91.290
59.953	68.993	364,281	111.033	40	46.110	53.063	280,170	85.396
60.006	69.053	364,602	111.131	45	42.575	48.994	258,690	78.849
60.059	69.114	364,924	111.229	50	38.714	44.551	235,229	71.698
60.111	69.174	365,239	111.325	55	34.555	39.765	209,960	63.996
60.159	69.229	365,531	111.414	60	30.131	34.674	183,077	55.802
60.203	69.280	365,800	111.496	65	25.474	29.314	154,780	47.177
60.241	69.324	366,029	111.566	70	20.620	23.729	125,288	38.188
60.272	69.359	366,216	111.623	75	15.606	17.959	94,826	28.903
60.295	69.385	366,354	111.665	80	10.472	12.051	63,628	19.394
60.309	69.401	366,440	111.691	85	5.257	6.049	31,939	9.735
60.313	69.406	366,466	111.699	90	0.000	0.000	0	0

*From: "Handbook of Oceanographic Tables", SP–68, U.S. Naval Oceanographic Office, 1966.

Table 7-17.   WEATHER CONDITIONS FOR ENVIRONMENTAL DESIGN*

For additional data on atmospheric conditions, see Section 7.1.

Latitudes are given to the nearest degree.

Degree-days are the yearly totals, 65 deg F base, i.e., for all days when the mean temperature was below 65 deg F. For monthly and yearly degree-days see the "ASHRAE Guide and Data Book".

Winter temperatures are the temperatures equaled or exceeded 99% (and 97.5%) of the hours during the coldest months.

Summer temperatures represent the highest 1% (or 2.5 or 5%) hourly dry-bulb (DB) or wet-bulb (WB) temperatures during the warmest months. In a normal season there would be about 30 hours above the 1% design temperature and 150 hours above the 5% dry-bulb design temperature. In most locations the 1% design wet-bulb temperature is about 2 degrees above the 5% dry-bulb value.

For elevation in meters, multiply tabulated values in feet by 0.3048. For temperature in K, see Table 9-22 for conversion.

Location	Lati-tude	Eleva-tion, ft	Degree-days heating	Temperatures, deg F					
				Winter		Summer			
				99%	97.5%	1% DB	1% WB	5% DB	5% WB
USA									
Alabama, Birmingham	33	610	2551	19	22	97	94	93	77
Alaska, Anchorage	61	90	10864	−25	−20	73	70	67	59
Alaska, Juneau	58	17	9075	−7	−4	75	71	68	62
Arizona, Phoenix	34	1117	1765	31	34	108	106	104	75
Arkansas, Little Rock	35	257	3219	19	23	99	96	94	78
California, Los Angeles	34	312	2061	42	44	94	90	87	69
California, Sacramento	39	17	2502	30	32	100	97	94	69
California, San Francisco	38	52	3015	42	44	80	77	73	61
Colorado, Denver	40	5283	6283	−2	3	92	90	89	63
Connecticut, Hartford	42	15	6235	1	5	90	88	85	74
Delaware, Wilmington	40	78	4930	12	15	93	90	87	76
District of Columbia	39	14	4224	16	19	94	92	90	76
Florida, Jacksonville	31	24	1239	29	32	96	94	92	79
Florida, Miami	26	7	214	44	47	92	90	89	79
Florida, Tampa	28	19	683	36	39	92	91	90	79
Georgia, Atlanta	34	1005	2961	18	23	95	92	90	76
Hawaii, Honolulu	21	7	0	60	62	87	85	84	73
Idaho, Boise	44	2842	5809	4	10	96	93	91	65
Illinois, Chicago	42	594	6639	−3	1	94	91	88	75
Indiana, Indianapolis	40	793	5699	0	4	93	91	88	76
Iowa, Des Moines	42	948	6588	−7	−3	95	92	89	76
Kansas, Topeka	39	877	5182	3	6	99	96	94	77
Kentucky, Louisville	38	474	4660	8	12	96	93	91	77
Louisiana, New Orleans	30	3	1385	32	35	93	91	90	79
Maine, Portland	44	61	7511	−5	0	88	85	81	71
Maryland, Baltimore	39	146	4654	12	15	94	91	89	77
Massachusetts, Boston	42	15	5634	6	10	91	88	85	73
Michigan, Detroit, (Met.)	42	633	6232	4	8	92	88	85	74
Michigan, Sault Ste. Marie	47	721	9048	−12	−8	83	81	78	69
Minnesota, Duluth	47	1426	10000	−19	−15	85	82	79	69
Minnesota, Minneapolis/St. Paul	45	822	8382	−14	−10	92	89	86	74
Mississippi, Jackson	32	330	2239	21	24	98	96	94	78
Missouri, Kansas City	39	742	4711	4	8	100	97	94	76
Missouri, St. Louis	39	465	4900	7	11	96	94	92	77
Montana, Billings	46	3567	7049	−10	−6	94	91	88	65
Montana, Butte	46	5526	—	−24	−16	86	83	80	57
Nebraska, Omaha	41	978	6612	−5	−1	97	94	91	76
Nevada, Las Vegas	36	2162	2709	23	26	108	106	104	70
New Hampshire, Manchester	43	253	—	−5	1	92	89	86	73
New Jersey, Atlantic City	40	11	4812	14	18	91	88	85	76
New Mexico, Albuquerque	35	5310	4348	14	17	96	94	92	64
New York, Albany	43	277	6875	−5	0	91	88	85	73

*Condensed with permission from: "ASHRAE Handbook of Fundamentals" and "ASHRAE Guide and Data Book", American Society of Heating, Refrigerating and Air-Conditioning Engineers, 1972 and 1968, respectively. Extended data for over 1000 locations are given in these sources.

Table 7-17.   WEATHER CONDITIONS
FOR ENVIRONMENTAL DESIGN (Continued)

Location	Lati-tude	Eleva-tion, ft	Degree-days heating	Temperatures, deg F						
				Winter		Summer				
				99%	97.5%	1% DB	1% WB	5% DB	5% WB	
USA (Cont.)										
New York, Buffalo	43	705	7062	3	6	88	86	83	72	
New York, NYC-LaGuardia	41	19	4811	12	16	93	90	87	75	
North Carolina, Durham	36	406	—	15	19	94	92	89	76	
North Dakota, Bismarck	47	1647	8851	−24	−19	95	91	88	70	
Ohio, Cincinnati	39	761	4410	8	12	94	92	90	76	
Ohio, Cleveland	41	777	6351	2	7	91	89	86	74	
Oklahoma, Tulsa	36	650	3860	12	16	102	99	96	77	
Oregon, Portland	46	57	4635	26	29	91	88	84	67	
Pennsylvania, Philadelphia	40	7	5144	11	15	93	90	87	76	
Pennsylvania, Pittsburgh	41	749	5987	7	11	90	88	85	73	
Rhode Island, Providence	42	55	5954	6	10	89	86	83	74	
South Carolina, Columbia	34	217	2484	20	23	98	96	94	78	
South Dakota, Sioux Falls	44	1420	7839	−14	−10	95	92	89	74	
Tennessee, Memphis	35	263	3232	17	21	98	96	94	78	
Tennessee, Nashville	36	577	3578	12	16	97	95	92	77	
Texas, Dallas	33	481	2363	19	24	101	99	97	78	
Texas, Houston	30	158	1396	29	33	96	94	92	79	
Texas, San Antonio	30	792	1546	25	30	99	97	96	76	
Utah, Salt Lake City	41	4220	6052	5	9	97	94	92	65	
Vermont, Burlington	45	331	8269	−12	−7	88	85	83	71	
Virginia, Richmond	38	162	3865	14	18	96	93	91	77	
Washington, Seattle	48	14	4424	28	32	81	79	76	64	
Washington, Spokane	48	2357	6655	−2	4	93	90	87	63	
West Virginia, Charleston	38	939	4476	9	14	92	90	88	74	
Wisconsin, LaCrosse	44	652	7589	−12	−8	90	88	85	75	
Wisconsin, Milwaukee	43	672	7635	−6	−2	90	87	84	73	
Wyoming, Cheyenne	41	6126	7381	−6	−2	89	86	83	61	
CANADA										
Alberta, Edmonton	54	2219	10268	−29	−26	86	83	80	65	
Br. Columbia, Vancouver	49	16	5515	15	19	80	78	76	65	
Manitoba, Winnipeg	50	786	10679	−28	−25	90	87	84	72	
Ontario, Ottawa	45	339	8735	−17	−13	90	87	84	73	
Ontario, Toronto	44	578	6827	−3	1	90	87	85	73	
Quebec, Montreal	46	98	8203	−16	−10	88	86	84	73	
Quebec, Quebec	47	245	9372	−19	−13	86	82	79	71	
OTHER COUNTRIES										
Argentina, Buenos Aires	35	89	—	32	34	91	89	86	75	
Australia, Sydney	34	138	—	40	42	89	84	80	72	
Brazil, Sao Paulo	24	2608	—	42	46	86	84	82	74	
France, Paris	49	164	—	22	25	89	86	83	67	
Germany, Berlin	52	187	—	7	12	84	81	78	66	
Hong Kong, Hong Kong	22	109	—	48	50	92	91	90	80	
India, Calcutta	23	21	—	52	54	98	97	96	82	
Italy, Rome	42	377	—	30	33	94	92	89	72	
Japan, Tokyo	36	19	—	26	28	91	89	87	79	
Mexico, Mexico City	19	7575	—	37	39	83	81	79	59	
Netherlands, Amsterdam	52	5	—	20	23	79	76	73	63	
Soviet Union, Moscow	56	505	—	−11	−6	84	81	78	65	
Spain, Madrid	40	2188	—	25	28	93	91	89	67	
Sweden, Stockholm	59	146	—	5	8	78	74	72	60	
United Kingdom, London	51	149	—	24	26	82	79	76	65	

Table 7-18.   AVERAGE EARTH TEMPERATURES*

The following table gives average earth temperatures for use in computing heating requirements for basements, tunnels, underground storage, underground shelters, or any structures not over 10 ft (3.05 m) below grade. These temperatures are the integrated average from surface to 10-ft (3.05 m) depth, calculated from observed earth temperatures. Earth thermal diffusivity has been assumed at 0.025 ft^2/hr (0.645 × 10^{-6} m^2/s). Maximum summer temperatures, obtained by the same method, are included to show the yearly range in each location and to provide data for summer temperature control.

For temperature in K, see Table 9-22 for conversion.

Earth temperature station	Integrated minimum, °F	Integrated maximum, °F	Earth temperature station	Integrated minimum, °F	Integrated maximum, °F
Bozeman, Mont.	32	56	Urbana, Ill.	42	68
Madison, S.D.	33	61	Faucett, Mo.	43	65
St. Paul, Minn.	34	62	Sikeston, Mo	43	71
Burlington, Vt.	35	63	Seattle, Wash.	45	61
Ft. Collins, Colo.	36	64	Corvallis, Oreg.	46	66
Huntley, Mont.	36	64	Lexington, Ky.	46	70
Ottawa, Ont., Canada	36	59	Decatur, Ala.	48	71
Pullman, Wash.	36	60	Union, S.C.	48	70
East Lansing, Mich.	37	63	Jackson, Tenn.	49	71
Moscow, Idaho	37	57	Pawhuska, Okla.	50	74
Burlington, Iowa	38	71	Lake Hefner, Okla.	51	77
West Lafayette, Ind.	38	66	Calhoun, S.C.	52	76
Ithaca, N.Y.	39	59	Raleigh, N.C.	52	73
Lemont, Ill.	39	65	State Univ., Miss.	55	79
Lincoln, Nebr.	39	69	Auburn, Ala.	56	74
Pendleton, Oreg.	39	67	Davis, Calif.	56	76
Coshocton, Ohio	40	64	Athens, Ga.	57	77
Norfolk, Nebr.	40	66	Tempe, Ariz.	59	81
Salt Lake City, Utah	40	63	Temple, Texas	59	83
Columbus, Ohio	41	65	Tifton, Ga.	62	80
Manhattan, Kansas	41	69	Tucson, Ariz.	65	85
Kansas City, Mo.	42	66	Brawley, Calif.	68	90
New Brunswick, N.J.	42	65	Gainesville, Fla.	69	80
Upper Marlboro, Md.	42	70			

*From: "ASHRAE Guide and Data Book: Applications", American Society of Heating, Refrigerating and Air-Conditioning Engineers, 1968.

Table 7-19. MAXIMUM SOLAR ALTITUDE, RADIATION INTENSITY, AND HEAT GAIN*

For additional data on solar radiation, see Section 2.1.

24° to 56° North Latitude

	N latitude, deg	Jan. 21	Feb. 21	March 21	April 21	May 21	June 21	July 21	Aug. 21	Sept. 21	Oct. 21	Nov. 21	Dec. 21
Solar altitude at noon, deg[1]	24	46.0	55.2	66.0	77.6	86.0	89.5	86.6	78.3	66.0	55.5	46.2	42.6
Max normal irradiation[2]	24	320	323	317	300	287	280	279	285	301	311	314	317
Max heat gain,[3] horiz skylight	24	215	249	275	284	282	279	278	276	267	244	213	199
Max heat gain,[3] window	24	253	243	234	229	218	212	214	220	222	235	249	252
Max heat gain,[3] half-day[4]	24	1064	1095	981	953	950	936	933	916	939	1050	1046	1019
Solar altitude at noon, deg	32	38.0	47.2	58.0	69.6	78.0	81.5	78.6	70.3	58.0	47.5	38.2	34.6
Max normal irradiation	32	309	316	312	298	286	280	278	283	296	304	303	304
Max heat gain, horiz skylight	32	176	217	252	272	277	276	273	265	244	213	175	158
Max heat gain, window	32	249	248	227	228	220	214	216	219	218	239	244	246
Max heat gain, half-day	32	974	1091	1054	965	983	972	964	929	1004	1044	955	947
Solar altitude at noon, deg	40	30.0	39.2	50.0	51.6	70.0	73.5	70.6	62.3	50.0	39.5	30.2	26.6
Max normal irradiation	40	293	306	306	294	284	278	276	279	290	293	287	284
Max heat gain, horiz skylight	40	133	180	223	253	265	267	262	247	215	177	132	113
Max heat gain, window	40	254	247	236	225	220	215	216	216	226	238	250	253
Max heat gain, half-day	40	903	1035	1100	1003	1002	1019	986	961	1045	989	884	831
Solar altitude at noon, deg	48	22.0	31.2	42.0	53.6	62.0	65.4	62.6	54.3	42.0	31.5	22.2	18.6
Max normal irradiation	48	267	291	297	289	280	275	272	273	280	278	260	250
Max heat gain, horiz skylight	48	85	138	188	227	247	252	245	222	182	136	85	64
Max heat gain, window	48	245	251	239	219	218	215	214	211	228	233	240	233
Max heat gain, half-day	48	735	936	1111	1086	1058	1098	1044	1040	1050	895	719	623
Solar altitude at noon, deg	56	14.0	23.2	34.0	45.6	54.0	57.4	54.6	46.3	34.0	23.5	14.2	10.6
Max normal irradiation	56	216	267	284	280	275	270	267	265	266	252	210	180
Max heat gain, horiz skylight	56	39	91	149	196	222	230	221	193	144	90	40	23
Max heat gain, window	56	205	244	241	224	215	213	211	215	231	234	200	171
Max heat gain, half-day	56	489	819	1080	1144	1120	1158	1101	1095	1013	776	476	330

*Adapted from: "ASHRAE Handbook of Fundamentals", American Society of Heating, Refrigerating, and Air-Conditioning Engineers, 1972.

Table 7-19.   MAXIMUM SOLAR ALTITUDE, RADIATION INTENSITY, AND HEAT GAIN (*Continued*)

[1]Solar altitude. Maximum height of the sun, degrees above the horizon at noon (zero azimuth).

[2]Maximum normal irradiation. Maximum radiant heat in Btu/hr·ft² for surfaces perpendicular to the sun's rays at noon (average moisture). For W/m² multiply by 3.152 5.

[3]Maximum heat gain. Solar heat transmitted through sunlit, single-glass window or skylight (overall heat transfer coefficient $U = 1.06$ Btu/hr·ft²·deg F = 6.01 W/m²·K). The window is vertical and faces the direction for which the rate of heat gain is maximum; the skylight is horizontal. Tabulated values are in Btu/hr·ft²; for W/m² multiply by 3.152 5.

[4]Maximum heat gain, half-day. Total solar heat transmitted in one-half day, forenoon or afternoon, vertical window (as in item 3) that faces the direction for which the half-day heat gain is maximum. Tabulated values are Btu/ft² per half day; for MJ/m² per half day, multiply by 0.011 35.

Solar constant. The intensity or flux density of solar radiation just beyond the earth's atmosphere, at the mean earth-sun distance, as adopted for the International Pyrheliometric Scale, 1956. This constant is 121 cal/hr·cm² = 445 Btu/hr·ft² = 1 400 W/m². (See also Table 2-4.)

Spectral energy distribution (outer space):

9 percent of total energy in ultraviolet range, between 0.29 and 0.40 μm

40 percent of total energy in visible range, between 0.4 and 0.7 μm

51 percent of total energy in infrared range, between 0.7 and 3.5 μm

Peak intensity of 700 Btu/hr·ft²·μm(2 210 MW/m³) at 0.48 μm (green); resembles blackbody radiation at approximately 11 250·R (6 250 K)

Solar intensity in Btu/hr·ft² (W/m²) outside the earth's atmosphere (approximate):

Jan. 21	457	(1 440)	July 21	430	(1 360)
Feb. 21	451	(1 420)	Aug. 21	434	(1 370)
March 21	447	(1 410)	Sept. 21	441	(1 390)
April 21	439	(1 380)	Oct. 21	448	(1 410)
May 21	433	(1 370)	Nov. 21	451	(1 420)
June 21	428	(1 350)	Dec. 21	457	(1 440)

Shading coefficients. The ratio of the actual heat gain of the window to the heat gain of a sunlit single-glass window (approximate):

$\frac{3}{32}$" or $\frac{1}{8}$" regular sheet glass	1.00
$\frac{1}{4}$" heat-absorbing plate glass	0.67
$\frac{1}{4}$" gray plate glass	0.75
Indoor Venetian blinds, closed	0.60
Roller shades, light color, drawn	0.35
Heavy drapes, drawn	0.60
30° sun screen (high sun)	0.35

Table 7-20. AIR CONDITIONING LOADS—HEAT FROM OCCUPANTS*

Heat per Occupant, Btu/hr

For watts multiply tabulated values by 0.292 88.

Degree of activity	Typical application	Total heat, adults (male)	Total heat,[a] adjusted	Sensible heat, adjusted	Latent heat, adjusted
Seated at rest	Theater—matinee	390	330	225	105
	Theater—evening	390	350	245	105
Seated, very light work	Offices, hotels, apartments	450	400	245	155
Moderately active office work	Offices, hotels, apartments	475	450	250	200
Standing, light work; or walking slowly	Department store, retail store	550	450	250	200
Walking; seated; standing	Drug store, bank	550	500	250	250
Sedentary activity	Restaurant[b]	490	550	275	275
Light-bench work	Factory	800	750	275	475
Moderate dancing	Dance hall	900	850	305	545
Walking 3 mph; moderately heavy work	Factory	1,000	1,000	375	625
Bowling[c]	Bowling alley				
Heavy work	Factory	1,500	1,450	580	870

[a] Adjusted heat gain is based on normal percentage of men, women, and children for the application listed, with the postulate that the gain from an adult female is 85 percent of that for an adult male, and that the gain from a child is 75 percent of that for an adult male.
[b] Adjusted total heat value for sedentary activity, restaurant, includes 60 Btu per hour for food per individual (30 Btu sensible and 30 Btu latent).
[c] For bowling, figure one person per alley actually bowling, and all others as sitting (400 Btu per hour) or standing (550 Btu per hour).

*From: "ASHRAE Handbook of Fundamentals", American Society of Heating, Refrigerating and Air-Conditioning Engineers, 1972.

Table 7-21. AIR REQUIRED TO REMOVE ODORS*

Following are the required dilutions to remove objectionable body odors, i.e., required fresh air per person (laboratory conditions). For air space in m^3, multiply values in ft^3 by 0.028 317. For air supply in m^3/s, multiply values in cfm by $4.719\,5 \times 10^{-4}$.

Type of occupant	Air space per person, cu ft	Odor-free air supply, cfm per person	Type of occupant	Air space per person, cu ft	Odor-free air supply, cfm per person
Heating season with or without recirculation. Air not conditioned.			**Heating season with or without recirculation. Air not conditioned.** (Cont.)		
Sedentary adults of average socio-economic status	100	25	Children attending private grade schools	100	22
	200	16			
	300	12	**Heating season. Air humidified by means of centrifugal humidifier. Water atomization rate 8 to 10 gph. Total air circulation 30 cfm per person.**		
	500	7			
Laborers	200	23			
Grade-school children of average socio-economic status	100	29	Sedentary adults	200	12
	200	21			
	300	17	**Summer season. Air cooled and dehumidified by means of a spray dehumidifier. Spray water changed daily. Total air circulation 30 cfm per person.**		
	500	11			
Grade-school children of lower socio-economic status	200	38	Sedentary adults	200	<4

*From: "ASHRAE Handbook of Fundamentals", American Society of Heating, Refrigerating and Air-Conditioning Engineers, 1972.

Table 7-22.   AIR REQUIRED FOR VENTILATION*

Many local codes list minimum requirements for ventilation of occupied spaces, but there is a difference between minimum tolerable ventilation and that desired or recommended. The following table gives values for the amount of odor-free outdoor air, minimum and recommended. These figures represent common practice. They may be compared with values given in Table 7-21 for the air required to remove odors under laboratory test conditions.

For purposes of good distribution of conditioned air, the *total* air supply required will usually be higher than the fresh air requirement and may not even be reducible to the terms of air quantity per occupant.

For m^3/s per person, multiply values in cfm per person by $4.719\ 5 \times 10^{-4}$. For m^3/s per m^2 of floor area, multiply values in cfm per sq ft of floor by 5.08×10^{-3}.

OUTDOOR AIR REQUIREMENTS[a]

Application	Smoking	Cfm per person[b] Recommended	Cfm per person[b] Minimum[c]	Cfm per sq ft of floor[b] Minimum[c]
Apartment:				
Average	Some	20	10	
Deluxe	Some	20	10	
Banking space	Occasional	10	$7\frac{1}{2}$	
Barber shops	Considerable	15	10	
Beauty parlors	Occasional	10	$7\frac{1}{2}$	
Brokers' board rooms	Very heavy	50	20	
Cocktail bars		40	25	
Corridors (supply or exhaust)				0.25
Department stores	None	$7\frac{1}{2}$	5	0.05
Directors' rooms	Extreme	50	30	
Drug stores[e]	Considerable	10	$7\frac{1}{2}$	
Factories[d, f]	None	10	$7\frac{1}{2}$	0.10
Five-and-ten-cent stores	None	$7\frac{1}{2}$	5	
Funeral parlors	None	10	$7\frac{1}{2}$	
Garages[d]				1.0
Hospitals:				
Operating rooms[f, g]	None			2.0
Private rooms	None	30	25	0.33
Wards	None	20	10	
Hotel rooms	Heavy	30	25	0.33
Kitchens:				
Residence				2.0
Restaurant				4.0
Laboratories[e]	Some	20	15	
Meeting rooms	Very heavy	50	30	1.25
Offices:				
General	Some	15	10	0.25
Private	None	25	15	0.25
Private	Considerable	30	25	0.25
Restaurants:				
Cafeteria[c]	Considerable	12	10	
Dining room[e]	Considerable	15	12	
Schoolrooms[d]	None			
Shop, retail	None	10	$7\frac{1}{2}$	
Theater[d]	None	$7\frac{1}{2}$	5	
Theater	Some	15	10	
Toilets[d] (exhaust)				2.0

[a] Taken from present-day practice.
[b] This is contaminant-free air.
[c] *When minimum is used, take the larger of the two.*
[d] See local codes which may govern.
[e] May be governed by exhaust.
[f] May be governed by special sources of contamination or local codes.
[g] All outside air recommended to overcome explosion hazard of anesthetics.

*From: "ASHRAE Handbook of Fundamentals", American Society of Heating, Refrigerating and Air-Conditioning Engineers, 1972.

Table 7-23. AIR DILUTION TO REDUCE HEALTH AND FIRE HAZARDS
Industrial Ventilation to Control Hazards from Gases or Volatile Liquids

For any hazardous gas or vapor released into a confined space, there is a minimum rate of dilution by ventilation air to maintain the environment within the contamination limits specified for protection against health hazard or fire hazard. Specifications are usually stated in parts per million by volume for toxic substances. Flammable limits are listed in parts per hundred (volume %, Tables 3-44 and 8-77).

The dilution air shown in the 100% column in the following table assumes perfect distribution and mixing. These values represent dilution to the specified limits, i.e., the TLV and lower flammable limit. They are actually inadequate for safety, however, because there is always some stratification, stagnation, and unequal mixing. Larger amounts are usually required by specifications. Fire safety codes often call for 400% or 500%.

VENTILATION FOR HAZARD CONTROL

The following table lists fresh ventilation air required *per* ft^3 of contaminant vapor released (standard conditions, 70 deg F and 14.7 psia, or 21 deg C and 760 mm Hg).

Pollutant	Chemical formula	Density,[a] if liquid, kg/m^3	Specific volume of vapor[b] m^3/kg	Specific volume of vapor[b] ft^3/lb	Specified limits TLV,[c] ppm	Specified limits Lower flammable limit, vol. %	Ventilation air[d] Health hazard, ft^3 (100%)	Ventilation air[d] Fire hazard, ft^3 (100%)
Acetic acid	$C_2H_4O_2$	1 049.	0.402	6.44	10		100 000	
Acetone	C_3H_6O		0.415	6.65	1 000	2.6	1 000	38.5
Acetylene	C_2H_2		0.930	14.9		2.5		40.0
Ammonia	NH_3		1.423	22.0	50	15.5	20 000	6.4
Benzene	C_6H_6	878.	0.309	4.95	25	1.3	40 000	76.9
Butyl acetate	$C_2H_{12}O_2$	1 058.	0.208	3.33	150	1.4	6 670	71.4
Butyl alcohol	$C_4H_{10}O$	1 038.	0.326	5.22	100	1.7	10 000	58.8
Carbon disulfide	CS_2	1 260.	0.317	5.08	20	1.2	50 000	83.3
Carbon monoxide	CO		0.861	13.8	50	12.5	20 000	8.0
Carbon tetrachloride	CCl_4	1 584.	0.157	2.52	10		100 000	
Chloroform	$CHCl_3$	1 464.	0.202	3.24	50		20 000	
Ethanol	C_2H_6O	788.	0.524	8.40	1 000	3.3	1 000	30.3
Ethyl acetate	$C_4H_8O_2$	900.	0.275	4.40	400	2.2	2 500	45.5
Ethylene	C_2H_4		0.867	13.9		3.0		33.3
Ethyl ether	$C_4H_{10}O$	720.	0.326	5.22	400	1.8	2 500	55.6
Formaldehyde	CH_2O		0.805	12.9	5	7.0	200 000	14.3
Gasoline			0.212	3.4		1.4		71.4
Hydrogen	H_2		12.043	195.		4.0		25.0
Hydrogen sulfide	H_2S		0.711	11.4	10	4.3	100 000	23.3
Mercury	Hg	13 560.	0.120	1.93	0.1		10 000 000	
Methane	CH_4		1.510	24.3		5.0		20.0
Methanol	CH_4O	788.	0.755	12.1	200	6.0	5 000	16.7
Naphtha (painter's)		708.	0.218	3.5	500	1.	2 000	100.
Nitric acid	HNO_3	1 404.	0.383	6.14	2		500 000	
Nitric oxide	NO		0.799	12.8	25		40 000	
Ozone	O_3		0.503	8.06	0.1		10 000 000	
Phenol	C_6H_6O	1 548.	0.256	4.11	5		200 000	
Propane	C_3H_8		0.549	8.8	1 000	2.1	1 000	47.6
Propyl alcohol	C_3H_8O	799.	0.402	6.44	200	2.3	5 000	43.5
Sulfur dioxide	SO_2		0.378	6.05	5		200 000	
Toluene	C_7H_8	866.	0.262	4.20	200	1.27	5 000	78.7
Turpentine		870.			100	0.8	10 000	125.
Xylene	C_8H_{10}	862.	0.228	3.65	100	1.0	10 000	100.

Table 7-23.   AIR DILUTION TO REDUCE HEALTH AND FIRE HAZARDS
(Continued)

[a]For approximate specific gravity, divide the value in this column by 1 000. For lb/gal, multiply the value in this column by 0.008 34.

[b]All specific volumes are for the gaseous state at atmospheric pressure and 25 deg C, as computed from the perfect gas law.

[c]TLV is the *threshold limit value*, parts of pollutant vapor or gas per million parts of air, by volume. This is the time-weighted average concentration for a normal workday to which it is believed that workers may be exposed without adverse effect.

[d]These volumes are based on perfect distribution and mixing; more is actually required for safety.

EXAMPLES:

Example 1.   A lacquering process liberates 90 ft^3 (13.55 lb) of acetone vapor per hour into an industrial workroom. At what minimum rate in cfm must the room be ventilated, assuming perfect distribution and mixing, to keep the room air (at a distance from the process) within the TLV of 1 000 ppm and to eliminate explosion hazards? The ventilating air supply already contains 50 ppm of acetone.

Solution: For health, minimum ft^3/min $= \dfrac{\text{contaminant, ft}^3/\text{min}}{\text{TLV ratio} - \text{supply ratio}} = \dfrac{1.5}{(1\ 000 - 50)/10^6} = 1\ 580.$

For fire protection, minimum ft^3/min $= \dfrac{1.5}{(2.6 - 0.002)/10^2} = 57.8.$

A safe specification might call for at least 3 000 ft^3/min, to allow for inadequate air distribution.

Example 2.   One pound of propane is liberated into air. Assuming perfect mixing, how large a space would be completely filled with an explosive mixture?

Solution: $\dfrac{8.8}{2.1/100} = 420$ ft^3.

Table 7-24.   EQUIVALENT COLD WEATHER CONDITIONS

Body-chilling effect depends on wind velocity. The four values in each vertical column are approximately equivalent.

Wind velocity, mph	Cool	Cold	Bitter cold	Flesh freezes quickly
2	55°F	20°F	−10°F	−50°F
5	60°F	37°F	+10°F	−25°F
10	65°F	40°F	20°F	−10°F
30	70°F	50°F	32°F	+10°F

Notes:
   The hands feel painfully cold when the skin temperature reaches 50°F.
   The feet feel painfully cold when the skin temperature reaches 55°F.
   The following conditions are of approximately equal comfort for a person dressed in warm winter clothing:

Sleeping	60°F
Resting, seated	50°F
Light work	20°F
Moderate work	−10°F

Table 7-25. ROOM AIR VELOCITIES FROM SUPPLY OUTLETS

Now that almost all occupied spaces are supplied with conditioned air for temperature control and ventilation, it is often necessary to specify air-velocity limits. A convenient equation for estimating the residual velocity at some distance from a supply outlet is

$$V_x = K V_o \sqrt{A/X},$$

which states that the maximum (centerline) residual velocity V_x at the distance X from the face of an outlet where the original velocity was V_o, varies directly as the size of the outlet and inversely as the normal distance from it. In Table A K is in ft$^{-\frac{1}{2}}$; for K in m$^{-\frac{1}{2}}$, multiply the tabulated values by 1.811 29.

Table A. CENTERLINE VELOCITY CONSTANT, K

Based on Net Flow Area of Outlet, A

Type of outlet	K
Free opening, round or square	6.
Grilles and grids, free area 40% or more	5
Long slots (length/width > 40)	4.
Ceiling circular outlets, narrow annulus	3.
Perforated panels (free area < 5%)	2.5

Note: Wide angle deflection vanes may reduce the throw up to 50%.

The above equation and applicable constants provide a method for determining the maximum air velocity in a plane parallel to the outlet face, at any selected normal distance X, from the outlet. For a symmetrical outlet this maximum velocity occurs along the centerline axis. Where it may also be important to estimate the residual velocity at other points in this parallel plane, the following data may be applied.

The total conical angle of a free jet is about 22°. In any plane parallel to the outlet face, a velocity equal to one-half the maximum velocity, i.e., $V_x/2$, exists at approximately one-half the distance from the centerline to the periphery of the 22° cone. Velocities at other points in the parallel plane can be estimated from Table B.

Table B. AIR VELOCITY PATTERN ACROSS THE STREAM

In the following table the velocity at any point in a plane parallel to the outlet face (grille face) is given as a percentage of the maximum centerline velocity V_x in that plane. Radial distances are expressed in terms of the radius of the 22° jet-cone circle in the same plane.

Radius, % of max	Velocity, % of V_x	Radius, % of max	Velocity, % of V_x
10	97	60	37
20	89	70	25
30	77	80	16
40	64	90	9
50	50		

REFERENCE

Effects of outlet location, interference of walls or adjacent jets, radial velocity gradients, room-air entrainment and other factors are covered in "ASHRAE Guide and Data Book: Systems and Equipment", American Society of Heating, Refrigerating and Air-Conditioning Engineers, 1969; this source also gives extensive references.

Table 7-26. ENERGY COST ESTIMATES FOR SPACE HEATING AND COOLING

Fuel and power costs for space heating or cooling can be estimated from weather data (Table 7-17) and computed loads (Tables 2-78, 7-18, and 7-20) by methods suggested in the following material.

Fuel Consumption—Degree-day Method. The efficiency of fuel utilization will be in the range 60 to 80 percent, depending on the fuel, the combustion and heat transfer equipment, and the size, condition, and loading of the plant. After selecting the probable efficiency and determining the design heating load, the following values of fuel consumption, applied with local fuel prices, will provide an estimate of seasonal costs.

FUEL CONSUMPTION PER DEGREE-DAY
For Each 1,000 Btu/hr "Design" Heat Loss

Fuel quantity	Average efficiency of heating plant		
	60	70	80
Hundreds of cu ft of 1,000 Btu gas	.00572	.00490	.00429
Gallons of 141,000 Btu oil	.00405	.00347	.00304
Pounds of 12,000 Btu coal	.0476	.0408	.0357

These values apply to 70° design temperature difference (70°F indoors, 0°F outdoors). For other temperature differences apply the following factors.

Temp diff	Corr factor	Temp diff	Corr factor
90	0.778	60	1.167
80	0.875	50	1.400

Example: Estimate the cost of heating a commercial building in a location where the design temperature is +10°F if the calculated design heat loss at this temperature is 250,000 Btu/hr and natural gas fuel is to be used, costing $0.75 per 1,000 cu ft (1,000 Btu/cu ft); estimated plant efficiency is 70 percent. The degree-days per season for this location are 5,000.

Solution: Cost per season, dollars

= (unit fuel cost) (fuel consumption factor) (design heat loss, thousands) (degree-days)

= (0.075) (.0049 × 1.167) (250) (5,000)

= $535 per season.

Electrical House Heating. Experience has indicated that electrical heating installations for residences use *less* energy than would be estimated from the direct load-degree-day calculation. The "NEMA Manual for Electric House Heating" and REA Bulletin 142-1 "Electric House Heating" suggest details of calculation. An approximation is obtained by the following:

$$\text{kwhr per season} = \frac{(\text{Btu/hr design heat loss}) (\text{degree-days/season})}{12,000}$$

This value is for 70°F indoors, 0°F outdoors design conditions. For higher or lower design temperature differences, the result may be corrected in direct proportion.

Residential Cooling. There are so many variables in summer residential cooling, including the weather, the type and location of residential structure, and the habits of the occupants, that only a rough estimate of the probable cost of operating a residential cooling system can be attained. Assuming that a careful study has been made to properly size the equipment, such an estimate will be obtained from the following:

kwhr per season = size of unit in tons × kw per ton × operating hr per season.

Table 7-26. ENERGY COST ESTIMATES FOR SPACE HEATING AND COOLING *(Continued)*

The total load per ton, including auxiliaries, is about 1.10 kw for a water-cooled unit and 1.40 for an air-cooled unit. If a water-cooled unit is used, an additional cost is the 40 to 70 gallons of cooling water per hour of operation for each ton of machine capacity (depending on water temperature).

Example: A typical power consumption for a properly-sized air-cooled unit of 5-ton capacity serving a residence in Dallas, Texas, would be

$$\text{kwhr per season} = 5 \times 1.40 \times 1,400 = 9,800.$$

RESIDENTIAL COOLING

Hours of Operation Per Season

City	Hours of operation	City	Hours of operation
Jasksonville, Fla.	1,600	Atlanta, Ga.	750
New Orleans, La.	1,500	Cleveland, Ohio	450
Dallas, Texas	1,400	Chicago, Ill.	400
St. Louis, Mo.	1,000	Minneapolis, Minn.	350
Fresno, Calif.	900	New York, N.Y.	350
Washington, D.C.	800	Boston, Mass.	200

REFERENCE

"ASHRAE Guide and Data Book: Applications", American Society of Heating, Refrigerating and Air-Conditioning Engineers, 1968, Chap. 54: Estimating Fuel or Energy Consumption.

Table 7-27. CONDENSATION ON WINDOWS AND WALLS*

Relative Humidity in Percent at Which Visible Condensation Appears on Inside Surface

Outdoor temp, °F	Transmittance = U†				
	.20	.40	.60	.80	1.00
−30	65	42	27	15	8
−20	68	45	30	18	11
−10	71	50	35	23	15
0	74	55	40	29	20
10	77	60	46	35	26
20	81	66	53	42	34
30	85	73	61	51	43

Caution: The maximum relative-humidity limit for a non-homogeneous wall will be lower than might be inferred from the average transmittance, because of spots where the actual transmittance is above the average. In any case the relative humidity must be that adjacent to the cooler wall areas.

†Btu/hr ft² °F temperature difference between indoor and outdoor air.

*Adapted from: "ASHRAE Handbook of Fundamentals", American Society of Heating, Refrigerating and Air-Conditioning Engineers, 1972.

Table 7-28. SURVIVAL IN COLD WATER

Based on Voluntary Tolerance Tests

Water temperature, °F	Survival time
40	15 min
50	30 min
60	60 min

Table 7-29.   MINIMUM ILLUMINATION LEVELS

For data on indoor lighting, see Section 2.1.

This table is intended as a guide for practical minimum levels of illumination at the surfaces being observed. Values are in footcandles, including both general and supplementary illumination, with minimum glare. These values are generally below the ideal or optimum ranges. Accepted levels of illumination are increasing, and there is a divergence in the recommendations from various sources.

Illuminance† in footcandles = flux density in lumens/ft²	Location or task
10	Corridors; warehouses; bulk shipping areas; parking garages
20	Stairways; washrooms; storage rooms; trains and buses; elevators; locker rooms
30	General office and filing; assembly rooms, active storage; library stacks; lobbies; waiting rooms; gymnasiums; shipping rooms
40	Home, school, and general office desks; large laboratories; laundries; recreation rooms; assembly shops; general reading areas; restaurants; art galleries; department stores
50	Secretarial and library desks; workbench and machine locations; kitchens; classrooms; banks; chalkboards; retail specialty shops; conference rooms
75	Sewing; inspection; small machining; woodworking; sketching and drawing; special laboratory observation; close office work; small parts storage; floodlighted displays
100	Art and general drafting; machine and appliance assembly; close inspection; composing and printing; accounting offices; layout and scribing; instrument repairs; bank tellers; mail sorting
125	Fine work on dark materials, e.g., cutting, sewing, inspecting; wood finishing; showcases; clothes pressing; supermarkets
150	Special display lighting; detail drawing; color printing; fine assembly
200	Dental and minor surgical; fine inspection or appraisal; intricate machining or repair
1,000 and above	Operating tables; unusual displays; welding

†For units of lumens/sq meter of lux, multiply by 10.76. One phot = 10,000 lux = lumens/sq cm.

Table 7-30.   ILLUMINATION RANGE FOR NORMAL HUMAN VISION

Object	Luminance, milli-lamberts	Illumination, foot candles
New snow in sunlight	9,000	2,700
Blue sky on clear day	1,500	450
White book page, good light	15	5
White book page, 1 candle at 1 ft	1	0.3
Book page, barely readable	.09	0.025
Treeless outdoors, full moon	.0015	0.0005

Table 7-31. LUMINOSITY FOR THE AVERAGE NORMAL EYE

Approximate Relative Sensitivity in Percent of Maximum; Photopic, 2-degree Field

	Violet		Blue		Green			Yellow				Red			
Wavelength, microns	.42	.44	.46	.48	.50	.52	.54	.56	.58	.60	.62	.64	.66	.68	.70
Relative luminosity, %	.004	.023	.06	.14	.32	.71	.95	.99	.87	.63	.38	.17	.06	.017	.004

Note: One micron equals 10,000 Angstrom units. One centimeter equals 10,000 microns.

Table 7-32. CHARACTERISTICS OF LIGHT, LIGHT SOURCES, AND LIGHTING MATERIALS*

Characteristic	Dimensional unit	Equipment	Technique
LIGHT			
Wavelength	Nanometer	Interference grating	Laboratory
Color	None	Spectrophotometer and colorimeter	Laboratory
Flux density (illumination)	Lumens per square foot (foot-candles)	Photometer	Laboratory or field
Orientation of polarization	Degree (angle)	Analyzing Nicol prism	Laboratory
Degree of polarization†	Percent (dimensionless ratio)	Polarization photometer	Laboratory
LIGHT SOURCES			
Energy radiated	Joule per square meter	Calibrated radiometer	Laboratory
Color temperature	Kelvin (K)	Colorimeter or filtered photometer	Laboratory or field
Luminous intensity	Candela	Photometer	Laboratory or field
Luminance (photometric brightness)	Candelas per unit area	Photometer or luminance meter	Laboratory or field
Spectral energy distribution	Joule per nanometer	Spectrometer	Laboratory
Power consumption	Watt	Wattmeter, or voltmeter and ammeter for dc, and unity power factor ac circuits	Laboratory or field
Light output (total flux)	Lumen	Integrating sphere photometer	Laboratory
Zonal distribution	Lumens or candelas	Distribution or goniophotometer	Laboratory
LIGHTING MATERIALS			
Reflectance	Percent (dimensionless ratio)	Reflectometer	Laboratory or field
Transmittance	Percent (dimensionless ratio)	Photometer	Laboratory or field
Spectral reflectance and transmittance	Percent (at specific wavelengths)	Spectrophotometer	Laboratory
Optical density	Dimensionless number	Densitometer	Laboratory

†Committee on Testing Procedures for Illumination Characteristics of the IES: "Resolution on Reporting Polarization," *Illum. Eng.*, 58:386, 1963.

*From: "IES Lighting Handbook", 5th ed., Illuminating Engineering Society, 1972.

Table 7-33. TYPICAL ENERGY DISTRIBUTION FOR ELECTRIC LAMPS

Operated at Rated Wattage

This table applies to single lamps in room-temperature environment. Fluorescent lamps are sensitive to temperature changes and air motion.

Types and distribution of energy	Incandescent, 100 w	Fluorescent cool white, 40 w	Mercury, 400 w	Infrared lamp	High-pressure sodium, 400 w
Visible light, %	11	18	14	8	26
Ultraviolet, %†	—	1	2	—	—
Infrared, %	71	28	46	67	37
Conduction-convection, %	18	30	27	25	22
Ballast (heat), %	—	23	11	—	15
Luminous efficacy, source only	18	79	54	—	110
Overall luminous efficacy, lumens/watt	18	61	48	—	94

†Sunlamps, bactericidal, and "black" ultraviolet lamps are omitted. The ultraviolet output of these lamps ranges from 2 to 10 percent, depending on size and type.

Table 7-34. APPROXIMATE LUMINANCE (PHOTOMETRIC BRIGHTNESS) OF VARIOUS LIGHT SOURCES*

Light source		Approximate average luminance†
Natural light sources		
Sun ⎫ As observed from	At meridian	160,000
Sun ⎬ the earth's surface	Near horizon	600
Moon ⎭	Bright spot	0.25
Clear sky	Average brightness	0.8
Lightning flash		8,000,000
Combustion sources		
Candle flame (sperm)	Bright spot	1.0
Kerosene flame (flat wick)	Bright spot	1.2
Illuminating gas flame	Fish-tail burner	0.4
Welsbach mantle	Bright spot	6.2
Acetylene flame	Mees burner	10.5
Photoflash lamps		16,000–40,000 peak
Nuclear sources		
Atomic fission bomb	0.1 millisecond after firing—90-ft diameter ball	200,000,000
Self-luminous paints		0.00003–0.000017
Incandescent electric lamps		
Carbon filament	3.15 lumens per watt	52
Tungsten filament	Gas-filled lamp, 20 lumens per watt	1,200
Tungsten filament	1,200-watt projection lamp, 31.5 lumens per watt	3,300
Blackbody at 6500°K		300,000
Blackbody at 4000°K		25,000
Blackbody at 2042°K		60
60-watt inside-frosted bulb		12
25-watt inside-frosted bulb		5
60-watt "white" bulb		3
Fluorescent lamps		
T-17 bulb, cool white	420 ma, low-loading	0.43
T-12 bulb, cool white	430 ma, medium-loading	0.82
T-12 bulb, cool white	800 ma, high-loading	1.13
T-12 bulb, cool white	1,500 ma, extra high-loading	1.70
Electroluminescent lamps		
Green color at 120 volts, 60 cycles		0.0027
Electric arcs		
Plain carbon arc	Positive crater	13,000–16,000
High-intensity carbon arc	13.6-mm rotating positive carbon	70,000–150,000
Electric arc lamps		
High-pressure mercury arc	Type H38, 10 atmospheres	180
High-intensity mercury short arc	Type SAH1000A, 30 atmospheres	24,000 (425,000 peak)
Water-cooled high-pressure mercury arc	Type H6, 75 atmospheres	13,000 (30,000 peak)
Clear-glass neon tube	15 mm 60 ma	0.16
Clear-glass fluorescent		
Daylight and white	15 mm 60 ma	0.50
Green	15 mm 60 ma	0.95
Blue and gold	15 mm 60 ma	0.30
Pink and coral	15 mm 60 ma	0.20
Sodium-arc lamp	10,000 lumen size	5.50
Electronic flash tubes		100,000–300,000

†These values are in candela/cm²; for candela/m² multiply by 10,000; for footlamberts multiply by 2919; for candela/ft² multiply by 930.

*From: "IES Lighting Handbook", 5th ed., Illuminating Engineering Society, 1972.

Table 7-35.   LUMINOUS CHARACTERISTICS OF RADIANT SOURCES*

Luminance, Chromaticity Coordinates, and Color Temperature

Source	Luminance, cd/cm^2	Chromaticity coordinates		Correlated color temperature, K
		x	y	
Sun's disk[1]				
Measured above atmosphere ($m = 0$)	200,000	0.318	0.330	6200
Measured near sea level for air masses[2]				
$m = 1$	150,000	0.331	0.344	5600
$m = 2$	125,000	0.343	0.357	5100
$m = 3$	100,000	0.356	0.369	4700
$m = 4$	80,000	0.368	0.379	4400
$m = 5$	65,000	0.380	0.388	4100
Sky[3]				
Clear-blue sky	0.06 to 0.4	0.262	0.270	15,000
		0.247	0.251	30,000
Partly-cloudy sky	0.1 to 0.4	0.294	0.309	8000
		0.279	0.291	10,000
Overcast sky	0.2 to 0.5	0.313	0.329	6500
Blackbody[4]				
Gold point	0.1073	0.607	0.381	1336
Platinum point	60.0	0.522	0.414	2042
Rhodium point	170.1	0.502	0.415	2233
Iridium point	1253.0	0.458	0.410	2716
Tungsten filament[5]				
Vacuum lamp (10 lm/W)	200	0.477	0.414	2400–2600
Gas-filled lamp (20 lm/W)	1200	0.444	0.406	2700–3100
750-W projection lamp (25 lm/W)	2400	0.430	0.402	3000–3200
1200-W projection lamp (30 lm/W)	3300	0.417	0.396	3200–3400
60-W inside-frosted bulb (14 lm/W)	12	0.452	0.409	2800
60-W "white" bulb (14 lm/W)	3	0.452	0.409	2800
CIE source A (500-W)	800	0.448	0.407	2854
CIE source B (500-W)	200	0.349	0.352	4870
CIE source C (500-W)	125	0.310	0.316	6740
Carbon arc	15,000–100,000			2900–6500
Mercury arc (H6) (1000 W; 65 lm/W)	30,000			
Fluorescent lamps (T-12; 40-W; 0.430 amp)				
Standard warm white	0.7	0.437	0.405	3000
White	0.7	0.409	0.398	3500
Standard cool white	0.7	0.372	0.374	4200
Daylight	0.6	0.311	0.338	6500
Xenon (high pressure, XBO 1600-W DC)	65,000			6000
Zirconium (300-W)	4500	0.423	0.399	3200

[1] Chromaticity coordinates from Moon (1940).

[2] The air masses $m = 1, 2, \ldots, 5$ are equivalent to the secants of the following angles between the zenith and the sun: 0° ($m = 1$), 60° ($m = 2$), 70.5° ($m = 3$); 75.5° ($m = 4$), 78.5° ($m = 5$).

[3] For clear blue and partly cloudy skies the range of minimum luminance is given (Paix 1963). Chromaticity coordinates and correlated color temperature are estimated from results reported by Judd, MacAdam, and Wyszecki (1964).

[4] From Pivovonski and Nagel (1961).

[5] From *IES Lighting Handbook*, Illum. Eng. Soc., 1959.

Table 7-36. SOURCES OF ULTRAVIOLET RADIATION*

$\mu W/cm^2$; Bandwidth, 3 nm

Safety note: When ultraviolet sources are used in an occupied room, the ozone level must be kept very low by ventilation. Ultraviolet radiation at wavelengths below 310 nm is especially dangerous to the eyes.

Wave-length, nm	Sunlight		Sunlamp	Fluorescent sunlamp	Black-light	Fluorescent blacklight	Mercury arc	Xenon arc	Hydrogen arc	Fluorescent daylight
	July 17	Dec. 27								
266	—	—	3.8	—	—		6.3	0.9	0.61	—
272	—	—	2.9	—	—		12.1	1.6	0.58	—
278	—	—	4.0	—	—		17.1	3.3	0.54	—
284	—	—	4.3	0.9	0.08		20.7	7.0	0.52	—
290	—	—	16.3	7.5	0.09	0.5	27.5	15.6	0.49	—
296	0.31	0.06	63.9	15.3	0.11	0.8	44.2	25.6	0.45	—
302	2.8	0.09	163	22.9	0.17	1.0	60.4	35.4	0.42	—
308	13.0	0.64	11.2	26.9	0.24	1.3	35.5	52.2	0.38	—
314	32.3	3.1	343	30.3	0.55	7.1	109.5	63.3	0.34	4.5
320	50.1	7.2	13.8	23.7	1.3	6.9	49.9	72.9	0.31	—
326	82.8	15.1	12.8	18.4	2.4	13.1	33.0	81.3	0.28	—
332	102.0	21.4	13.6	14.0	4.4	22.7	31.1	92.9	0.26	—
338	98.5	23.7	13.0	10.0	5.5	31.6	25.7	99.7	0.24	—
344	113	30.0	9.9	6.8	7.5	39.2	21.8	107	0.22	—
350	129	36.2	8.1	4.7	10.3	43.1	19.9	118	0.20	0.14
356	146	44.3	9.2	3.3	12.7	41.4	20.3	130	0.19	0.42
362	151	48.8	11.6	2.4	15.8	34.7	35.1	139	0.18	0.69
368	198	69.1	333	2.2	18.2	27.8	221	146	0.17	1.9
374	180	65.2	7.8	1.3	20.6	19.1	95.1	155	0.16	0.96
380	230	84.0	7.9	1.1	22.5	12.7	70.0	169	0.16	1.24
386	196	78.5	5.1	0.6	22.0	7.2	68.2	177	0.15	1.65
392	224	90.1	16.1	0.5	19.2	4.9	68.9	187	0.13	2.00
398	265	113	5.4	0.5	13.6	2.8	69.2	198	0.13	2.60
404	378	166	202	6.4	8.7	9.8	303	188	0.13	14.1
410	386	175	49.3	0.5	4.5	1.6	213	196	0.13	4.17

IDENTIFICATION OF SOURCES:

Sunlight measured at Stamford, Conn. (approx. 41°N—73½°W), 1957.
Sunlamp: G.E. bulb-RS, 275 w, 8¼ in. distance.
Fluorescent sunlamp: Westinghouse F 20 T 12/sun, 20 w, 3 in. distance.
Blacklight, purple X: G.E., 250 A21/60 250 w, 3 in. distance.
Fluorescent blacklight: G.E., F 15 T8/BL 15 w, 3 in. distance.
Mercury arc: G.E., AH6, 1000 w, 10 in. distance.
Xenon arc: Hanovia, 507 C 1, 10 in. distance.
Hydrogen arc: Beckman Instruments, air-cooled, 3 in. distance.
Fluorescent daylight: G.E., F 15 T 8/D, 15 w, 3 in. distance.

*From: "Ultraviolet Spectral Energy Distributions of Natural Sunlight and Accelerated Test Light Sources", R.C. Hirt, R.G. Schmitt, N.D. Searle, and A.P. Sullivan, *J. Opt. Soc. Amer.*, 50:706, July 1960.

Table 7-37.   ILLUMINATION ON A HORIZONTAL OR VERTICAL PLANE (POINT SOURCE)*

Figures in parentheses represent the angle in degrees between the light ray and the normal axis. To convert from feet to meters, multiply by 0.3048. For illumination in lux (lm/m²), multiply tabular values by 10.764.

Normal distance, source to plane, ft	Illumination flux density in foot-candles, lumens/ft², at a distance (in the plane) from the normal axis of											
	0 (axis)	1 ft	2 ft	3 ft	4 ft	5 ft	6 ft	8 ft	10 ft	15 ft	20 ft	32 ft
FOR EACH 100 CANDLEPOWER												
4	6.25	5.71 (14)	4.47 (27)	3.20 (37)	2.21 (45)	1.52 (51)	1.07 (56)	.559 (63)	.320 (68)	.107 (75)	.047 (79)	.012 (83)
6	2.78	2.67 (9)	2.37 (18)	1.99 (27)	1.60 (34)	1.26 (40)	.982 (45)	.600 (53)	.378 (59)	.142 (68)	.066 (73)	.017 (79)
8	1.56	1.53 (7)	1.43 (14)	1.28 (21)	1.12 (27)	.953 (32)	.800 (37)	.552 (45)	.381 (51)	.163 (62)	.080 (68)	.022 (76)
10	1.00	.985 (5.7)	.943 (11)	.879 (17)	.801 (22)	.716 (27)	.631 (31)	.470 (39)	.334 (45)	.171 (56)	.089 (63)	.027 (73)
12	.694	.687 (4.8)	.668 (9)	.634 (14)	.593 (18)	.546 (23)	.497 (27)	.400 (34)	.315 (40)	.169 (51)	.094 (59)	.030 (69)
14	.510	.506 (4.1)	.495 (8)	.477 (12)	.454 (16)	.426 (20)	.396 (23)	.334 (30)	.275 (36)	.162 (47)	.096 (55)	.033 (66)
16	.391	.388 (3.6)	.382 (7)	.371 (11)	.357 (14)	.339 (17)	.321 (21)	.280 (27)	.238 (32)	.152 (43)	.095 (51)	.035 (63)
18	.309	.307 (3.2)	.303 (6)	.297 (9)	.287 (13)	.276 (16)	.264 (18)	.236 (24)	.206 (29)	.140 (40)	.092 (48)	.036 (61)
20	.250	.249 (2.9)	.246 (5.7)	.242 (9)	.236 (11)	.228 (14)	.219 (17)	.200 (22)	.179 (27)	.128 (37)	.088 (45)	.037 (58)
25	.160	.160 (2.3)	.158 (4.6)	.157 (7)	.154 (9)	.151 (11)	.147 (14)	.138 (18)	.128 (22)	.101 (31)	.076 (39)	.037 (52)
30	.111	.111 (1.9)	.111 (3.8)	.109 (5.7)	.108 (8)	.107 (9)	.105 (11)	.100 (15)	.095 (18)	.080 (27)	.064 (34)	.036 (47)
40	.063	.062 (1.4)	.062 (2.9)	.062 (4.3)	.062 (5.7)	.061 (7)	.060 (9)	.059 (11)	.057 (14)	.051 (21)	.045 (27)	.030 (39)
50	.040	.040 (1.2)	.040 (2.3)	.040 (3.4)	.040 (4.6)	.039 (5.7)	.039 (7)	.039 (9)	.038 (11)	.035 (16)	.032 (22)	.024 (33)
60	.028	.028 (1.0)	.028 (1.9)	.028 (2.9)	.028 (3.8)	.027 (4.8)	.027 (5.7)	.027 (8)	.027 (9)	.025 (14)	.024 (18)	.019 (28)
FOR EACH 100,000 CANDLEPOWER												
80	15.6	15.6 (0.7)	15.6 (1.4)	15.6 (2.2)	15.6 (2.9)	15.5 (3.6)	15.5 (4.3)	15.4 (5.7)	15.3 (7)	14.8 (11)	14.3 (14)	12.5 (22)
100	10.0	10.0 (0.6)	9.99 (1.2)	9.99 (1.7)	9.98 (2.3)	9.96 (2.9)	9.95 (3.4)	9.90 (4.6)	9.85 (5.7)	9.66 (9)	9.44 (11)	8.63 (18)
125	6.40	6.40 (0.5)	6.40 (0.9)	6.40 (1.4)	6.39 (1.8)	6.39 (2.3)	6.38 (2.8)	6.36 (3.7)	6.34 (4.6)	6.26 (7)	6.16 (9)	5.83 (14)
150	4.44	4.44 (0.4)	4.44 (0.8)	4.44 (1.2)	4.44 (1.5)	4.44 (1.9)	4.43 (2.3)	4.42 (3.0)	4.41 (3.8)	4.38 (5.7)	4.32 (8)	4.16 (12)
175	3.27	3.27 (0.3)	3.27 (0.7)	3.26 (1.0)	3.26 (1.3)	3.26 (1.6)	3.26 (2.0)	3.26 (2.6)	3.25 (3.3)	3.23 (4.9)	3.20 (7)	3.11 (10)
200	2.50	2.50 (0.3)	2.50 (0.6)	2.50 (0.9)	2.50 (1.2)	2.50 (1.4)	2.50 (1.7)	2.49 (2.3)	2.49 (2.9)	2.48 (4.3)	2.46 (5.7)	2.41 (9)

*Condensed from: "IES Lighting Handbook", 5th ed., Illuminating Engineering Society, 1972.

Table 7-38. VISUAL SPECTRAL EFFICIENCY VALUES*

Unity at Wavelength of Maximum Luminous Efficacy

Wavelength, λ, nanometers†	Values		Wavelength, λ, nanometers†	Values	
	Photopic‡	Scotopic§		Photopic‡	Scotopic§
380	0.00004	0.000589	590	.757	.0655
390	.00012	.002209	600	.631	.03315
400	.0004	.00929	610	.503	.01593
410	.0012	.03484	620	.381	.000737
420	.0040	.0966	630	.265	.003335
430	.0116	.1998	640	.175	.001497
440	.023	.3281	650	.107	.000677
450	.038	.455	660	.061	.0003129
460	.060	.567	670	.032	.0001480
470	.091	.676	680	.017	.0000715
480	.139	.793	690	.0082	.00003533
490	.208	.904	700	.0041	.00001780
500	.323	.982	710	.0021	.00000914
510	.503	.997	720	.00105	.00000478
520	.710	.935	730	.00052	.000002546
530	.862	.811	740	.000025	.000001379
540	.954	.650	750	.00012	.000000760
550	.995	.481	760	.00006	.000000425
560	.995	.3288	770	.00003	.0000002413
570	.952	.2076	780	.000015	.0000001390
580	.870	.1212			

†Nanometers are the same as millimicrons.
‡Photopic vision occurs at the higher luminance levels and invloves mainly the cones in the fovea.
§Scotopic vision occurs at low levels of field luminance and is peripheral vision by the rod-type photoreceptors.

*From: "IES Lighting Handbook", 5th ed., Illuminating Engineering Society, 1972.

Table 7-39. REFLECTANCE AND APPEARANCE OF COLORS

The following table gives the percentage reflectance in daylight of typical glossy- or smooth-surface finishes on wood, paper, metal, and other materials. So-called "fluorescent" colors are not included.

Descriptive name	Light reflected, percent	Federal color No.†	Descriptive name	Light reflected, percent	Federal color No.†
White	85	—	DARK COLORS		
LIGHT COLORS			Light navy gray	28	16251
Light ivory	75	13711	Medium green	25	14277
Soft yellow	75	13695	Vivid orange	23	12246
Cream; off-white	75	—	Clear blue	19	15177
Light buff; light gray	70	—	Radiation purple	15	17142
Peach	64	12648	Wood finish, oak or walnut	15	—
Suntan	60	13613	Medium navy gray	14	16187
Light green	55	14516	Light red	13	—
Light blue	50	15526	Medium brown	10	—
MEDIUM COLORS			Wood finish, red mahogany	9	—
Brilliant yellow	58	13538	Fire red	7	11105
Highlight buff	55	13578	Passive green	7	14077
Clear green	50	—	Maroon; dark green	7	—
Pearl gray	46	16492	Deep navy gray	7	16081
Wood finish, maple	40	—	Marine Corps green	4	14052
			Dark brown	4	—

†Federal Color Standards No. 595. Color numbers beginning with digit 1 are gloss finish.

APPEARANCE OF COLORS

Intense or saturated hues protrude.

Incandescent lighting tends to dull the green, blue, lavender, and purple shades. Olive green appears brown.

"Cool white" fluorescent lamps make reds appear less bright, and they somewhat dull yellow and blue shades. Illuminated by "warm white" fluorescent lighting, the shades of gray and olive green are changed, and the blues and purples appear dull.

Colored light sources change the hue of most pigment colors, except those that are nearly the same color as the light.

Table 7-40. STANDARD ILLUMINATION SOURCES
FOR COLORIMETRY

The International Commission on Illumination (CIE) has adopted a trichromatic system for mathematical specification of the color of light sources. The three standard sources for colorimetry are as follows:

Source A is a tungsten lamp operated at 2854°K. For purposes of computation, the data for a blackbody at 2854°K are used (Planckian radiator).

Source B, the British daylight standard, has a correlated color temperature of approximately 4870°K. It consists of source A and a specified filter. This approximates noon sunlight.

Source C approximates average daylight and has a correlated color temperature of approximately 6500°K. It consists of source A and a specified filter.

RELATIVE ENERGY DISTRIBUTION OF SOURCES A, B, AND C

	Relative energy†				*Relative energy†*				*Relative energy†*		
Wave-length, nano-meters	*A*	*B*	*C*	*Wave-length, nano-meters*	*A*	*B*	*C*	*Wave-length, nano-meters*	*A*	*B*	*C*
320	1.93	0.02	0.01	490	53.91	96.50	120.70	630	150.83	101.00	88.00
340	3.59	2.40	2.70	500	59.86	94.20	112.10	640	157.98	102.20	87.80
360	6.15	9.60	12.90	510	66.06	90.70	102.30	650	165.03	103.93	88.20
380	9.79	22.40	33.00	520	72.50	89.50	96.90	660	171.96	105.00	87.90
390	12.09	31.30	47.40	530	79.13	92.20	98.00	670	178.77	104.90	86.30
400	14.71	41.30	63.30	540	85.95	96.90	102.10	680	185.43	103.90	84.00
410	17.68	52.10	80.60	550	92.91	101.00	105.20	690	191.93	101.60	80.20
420	21.00	63.20	98.10	560	100.00	102.80	105.30	700	198.26	99.10	76.30
430	24.67	73.10	112.40	570	107.18	102.60	102.30	710	204.41	96.20	72.40
440	28.70	80.80	121.50	580	114.44	101.00	97.80	720	210.36	92.90	68.30
450	33.09	85.40	124.00	590	121.73	99.20	93.20	730	216.12	89.40	64.40
460	37.82	88.30	123.10	600	129.04	98.00	89.70	740	221.66	86.90	61.50
470	42.87	92.00	123.80	610	136.34	98.50	88.40	750	227.00	85.20	59.20
480	48.25	95.20	123.90	620	143.62	99.70	88.10	760	232.11	84.70	58.10

†Arbitrary units. Source A at 560 nanometers (millimicrons) = 100.

REFERENCES

"Color Science", G. Wyszecki and W.S. Stiles, John Wiley & Sons, 1967, p. 32.
"IES Lighting Handbook", 5th ed., Illuminating Engineering Society, 1972.

Table 7-41.   SPECTRAL TRANSMITTANCE OF FLOAT GLASS*

UV, Visible, and IR Transmittances for Panes ¼-in. Thick, Soda-lime Glass

Wave-length, nm[a]	Transmittance, percent			
	Clear, polished, ¼-in. plate	Gray tint, ¼-in. Solargray®	Bronze tint, ¼-in. Solarbronze®	Blue green, ¼-in. Solex®
ULTRAVIOLET REGION				
300	1.1	0.0	0.0	0.0
320	0.3	0.0	0.0	0.0
340	29.4	2.1	0.3	0.7
360	74.8	33.3	22.8	33.3
380	79.3	44.2	34.1	44.3
Total solar	68.	34.	26.	35.
VISIBLE REGION				
400	87.2	53.8	48.5	70.6
440	87.3	43.1	44.3	71.6
480	89.3	40.6	43.3	79.0
520	89.5	38.1	45.2	79.8
560	88.7	46.1	54.2	76.5
600	87.0	38.3	51.7	69.4
640	84.4	38.6	51.5	60.3
680	81.3	49.4	54.4	50.1
720	77.9	59.6	55.7	40.5
Total solar	88.	41.	51.	75.
INFRARED REGION				
800	71.4	51.3	46.7	27.1
900	65.8	43.6	39.0	18.8
1 000	63.4	40.5	36.3	15.8
1 100	62.9	38.9	35.2	15.3
1 200	63.8	38.2	35.8	16.3
1 400	69.9	44.4	43.7	21.1
1 600	73.8	54.3	55.3	32.6
1 800	75.9	56.3	57.8	40.0
2 000	77.1	58.5	58.9	40.3
Total solar	67.	46.	42.	22.

[a]One nanometer $= 1 \times 10^{-9}$ meters $= 1 \times 10^{-3}$ micrometers.

*Courtesy of Pittsburgh Plate Glass Industries.

Table 7-42.   SOUND FREQUENCIES AND SCALES

For data on loudness and noise, see Tables 7-45 to 7-53. For data on speech communication, see Tables 9-2 and 9-3.

A pure tone or musical note is produced if the vibrations consist of a single frequency within the audible range. A "harmonic" is a partial tone, the frequency of which is an integral multiple of the fundamental or lowest frequency. A harmonic series consists of integral multiples.

If the frequency is doubled, the tone rises one octave on the musical scale. The common musical scale in Western countries has twelve half-tones, with an interval or frequency ratio between successive tones of $\sqrt[12]{2}$. In this equally tempered scale the intervals are given the following names.

Name	Frequency ratio	Name	Frequency ratio
Semitone	1.05946	Perfect fifth	1.49831
Whole tone or major second	1.12246	Minor sixth	1.58740
Minor third	1.18921	Major sixth	1.68179
Major third	1.25992	Minor seventh	1.78180
Perfect fourth	1.33484	Major seventh	1.88775
Augmented fourth†	1.41421	Octave	2.00000

†Also called diminished fifth.

In a complex tone the ear judges the pitch by the lowest frequency, or fundamental, and interprets the tone quality in terms of the accompanying higher frequencies, or overtones. The audible range, which is about 16–20,000 hz, varies among individuals and among animals.

Very complex sounds may be given special names, such as white noise. A "noise" is defined as an unwanted sound, and it may or may not have a prevailing frequency. Electronic generators are available for generating sound combinations of almost any desired pattern.

Engineering specifications and measurements are most likely to deal with sound intensity or energy, but the direct measurement is that of sound pressure in decibels expressed on a logarithmic scale.[a] Loudness is the human subjective interpretation of sound intensity and is expressed in phons (or sones) (see Tables 7-47 to 7-49). Sounds are analyzed with respect to frequency through the use of electronic band pass filters, e.g., octave bands or $\frac{1}{3}$-octave bands, with special attention to five octaves in the auditory range, 125–4000 hz.

The following diagram provides a comparison of ranges in the acoustic spectrum, as represented by average human voices and by the usual string and wind instruments of the orchestra. Frequencies for the C-scale on the piano are also given, based on the tuning of this instrument to American Standard Pitch, A = 440 hz (formerly C at 256 hz).

[a]Sound measurement techniques are covered by standard codes; see ANSI–S1.2 and entire ANSI S-series of over 30 standards.

Table 7-42. SOUND FREQUENCIES AND SCALES (Continued)

THE ACOUSTIC SPECTRUM IN HERTZ*

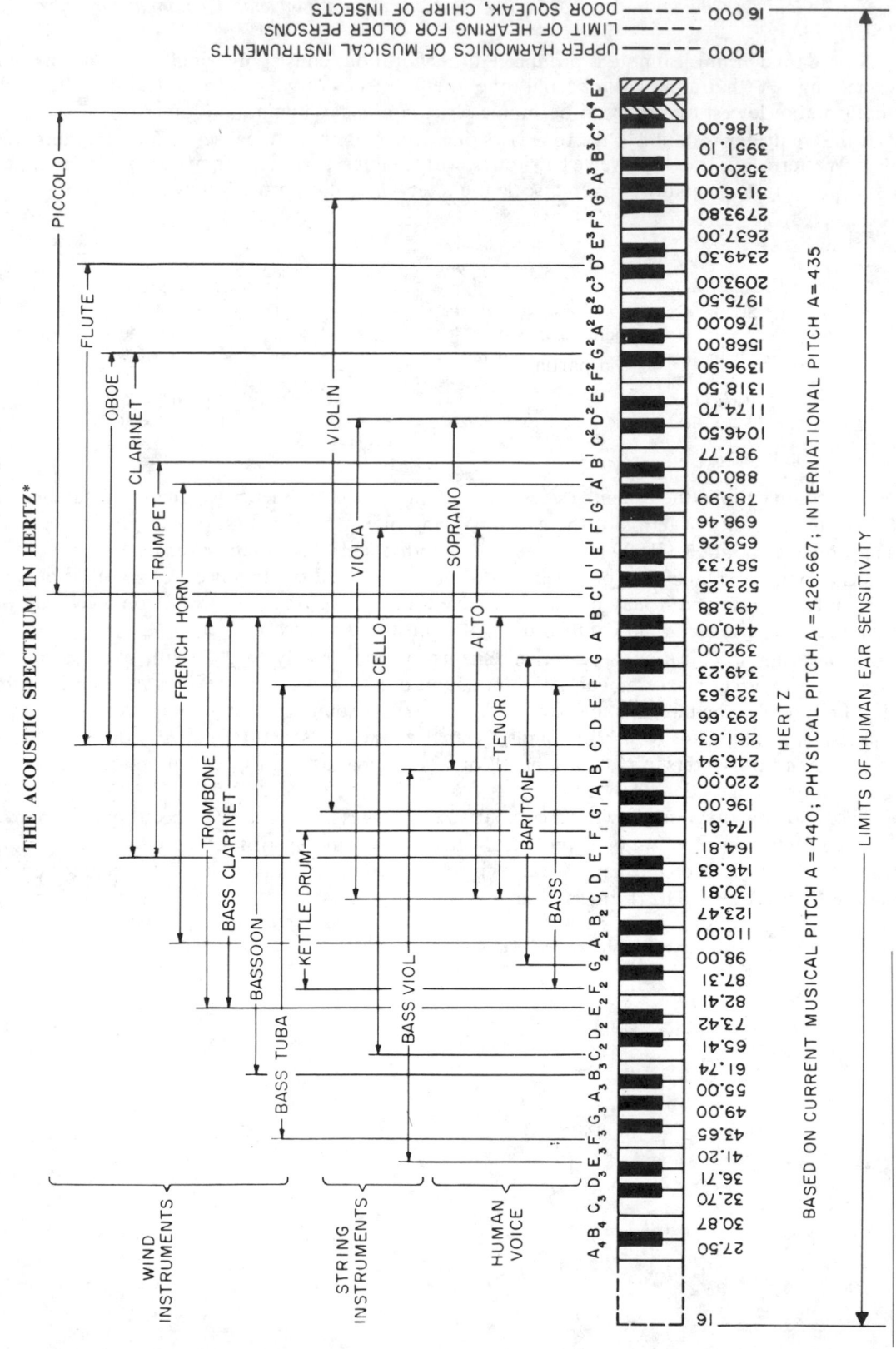

Note: The term *hertz* is identical to cycles per second (cps).

*From: "Reference Data for Radio Engineers", 5th ed., Howard W. Sams & Co., Inc., Indianapolis, Indiana, 1968, p. 35-16.

Table 7-43. ELASTIC AND SOUND PROPERTIES OF SOLIDS*

Conversion factors: For N/m^2, multiply values in psi by 6 895. For ft/s, multiply values in m/s by 3.281.

Material	Modulus of elasticity in tension, millions of psi	Poisson's ratio	Accoustic velocity, m/s			Acoustic impedance, $V_p \times$ density, millions of $kg/m^2 \cdot s$
			Bulk, V_p (plane)	Longi-tudinal, V_b (bar)	Shear, V_s	
METALS						
Aluminum	10.	.35	6 400	5 100	3 050	17.3
Beryllium	44.5	.05	12 890		8 880	24.
Bismuth			2 200	1 800	1 100	21.5
Cadmium			2 800	2 400	1 500	24.
Copper	18.	.37	4 800	3 700	2 325	44.5
Gold	10.8	.42	3 240	2 030	1 200	62.5
Iron	30.	.29	5 950	5 160	3 230	46.5
Lead	2.3	.43	2 160	1 250	750	23.
Magnesium	6.1	.31	5 760	4 920	3 050	10.
Molybdenum			6 250	5 400	3 350	64.
Nickel	31.	.34	5 500	4 800	2 990	51.
Platinum	24.	.30	3 500	2 800	1 700	85.
Silver	11.	.38	3 650	2 650	1 640	38.
Tin	30.	.34	3 320	2 730	1 670	24.5
Titanium	8.	.32	6 000	5 080	3 120	27.
Tungsten	52.	.35	5 250	4 400	2 890	103.
Zinc	15.	.25	4 200	3 830	2 450	30.
ALLOYS						
Brass (70–30)	15.		4 600	3 450	2 110	36.
Cast iron	13		4 500	4 480	2 800	33.
Constantan			5 200	4 270	2 630	46.
Duralumin (2014)	10.5	.33	6 320	5 150	3 130	
Monel	26.		5 350	4 400	2 720	45.
Nickel silver	18.		4 760	3 575	2 160	40.
PLASTICS						
Acrylic	0.4		2 700			3.2
Polyamide	0.3	.4	2 600			2.9
Polyethylene	0.1	.46	1 950			
Polystyrene	0.5		2 500			2.8
MISCELLANEOUS						
Barium titanate			5 500	4 300		3.1
Brick	3.		4 300			7.4
Cork			500			1.0
Ice			3 980			3.5
Glass, crown			5 200			13.
Glass, flint	9.5	.22	4 500			16.
Glass, Pyrex	9.	.24	5 650			
Glass, quartz	10.4	.16	5 500			12.
Glass, silica	10.	.17	5 900			13.
Granite			4 000			11.
Marble	8.		3 800			10.5
Paraffin wax			1 500			2.3
Porcelain			5 900			13.5
Rubber			1 700			3.
Slate			4 500			13.5
Tissue, human[a]			1 530			1.6
Tungsten carbide			6 700		4 000	80.
Wood, oak	1.6		4 000			2.

[a]For data on sound velocity and absorption in human tissues, see "Tabular Data of the Velocity and Absorption of High-Frequency Sound in Mammalian Tissues", D.E. Goldman and T.F. Hueter, *J. Acous. Soc. Amer.*, 28(1): 35, January 1956.

*Compiled from several sources.

Table 7-44.   VELOCITY OF SOUND IN BAR-SHAPED SOLIDS*
Longitudinal Direction

Material	Velocity[a] m/s	Velocity[a] f/s	Material	Velocity[a] m/s	Velocity[a] f/s
Metals:			Crystals:		
Aluminum	5,240	17,200	Quartz X-cut	5,440	17,850
Antimony	3,400	11,200	Ammonium dihydrogen phosphate		
Bismuth	1,790	5,880	($NH_4H_2PO_4$) 45° Z-cut	3,280	10,770
Brass	3,420	11,200	Rochelle salt (sodium potassium		
Cadmium	2,400	7,880	tartrate, $KNaC_4H_4O_6 \cdot 4H_2O$)		
Constantan	4,300	14,100	45° Y-cut	2,470	8,100
Copper	3,580	11,750	Calcium fluoride (CaF_2, fluorite) X-cut	6,740	22,100
German silver	3,580	11,750	Sodium chloride (NaCl, rock salt) X-cut	4,510	14,800
Gold	2,030	6,650	Sodium bromide (NaBr) X-cut	2,790	9,150
Iridium	4,790	15,700	Potassium chloride (KCl, sylvite) X-cut	4,140	13,600
Iron	5,170	16,950	Potassium bromide (KBr) X-cut	3,380	11,100
Lead	1,250	4,100	Glasses:		
Magnesium	4,900	16,100	Heavy flint	3,490	11,440
Manganese	3,830	12,570	Extra-light flint	4,550	14,930
Nickel	4,760	15,620	Crown	5,300	17,400
Platinum	2,800	9,200	Heaviest crown	4,710	15,440
Silver	2,640	8,550	Quartz	5,370	17,600
Steel	5,050	16,600	Granite	3,950	12,970
Tantalum	3,350	11,000	Ivory	3,010	9,880
Tin	2,730	8,950	Marble	3,810	12,500
Tungsten	4,310	14,150	Slate	4,510	14,800
Zinc	3,810	12,500	Woods: Elm	1,010	3,320
Cork	500	1,640	Oak	4,100	13,480

[a]Sound velocities in bulk material are somewhat higher (see Table 7-43).

*Data from: "Wavelengths of Sound", B.W. Henvis, *Electronics*, 20:134–136, March 1947; copyright McGraw-Hill, Inc., 1947.

Table 7-45.   TYPICAL SOUND PRESSURE LEVELS
Measurements by Standard Noise Meter

Sound level, db	Relative sound power	Typical location or source
10	0.001	Soundproof vault; threshold of audibility
20	0.01	Whisper or rustle; sound picture studio
25		Broadcast studio; church; very quiet office or study
30	0.10	Country residence; empty concert hall
35		Drama theater; sleeping area; large table conference room; voice range 10–30 ft
40	1.00	Private office; library; movie theater; hospital
45		Classroom; auditorium; conference room
50	10	Average office; hotel lobby; bank
55		Department store; laboratory
60	100	Busy dining room, kitchen; very noisy office; telephone use difficult
65		Typing and accounting office
70	1,000	City street; automobile
75		Busy machine shop; raised voice necessary for communication at 2 ft; telephone use unsatisfactory
80	10,000	Motor bus; noisy factory
85		Vehicular tunnel; voice communication nearly impossible
90	100,000	Superhighway; New York subway
95		Large motor trucks
100	1,000,000	Busy woodworking stop
110	10,000,000	Riveting shop
120	100,000,000	Propeller plane take off; thunder
130	1,000,000,000	Jet plane at 100 ft; space rocket, 1 mile; 1-minute hearing damage; pain—use earplugs

Table 7-46.   COMPARING OR ADDING NOISE LEVELS*

For data on decibel conversion, see Table 8-24.

While it is possible to indicate an absolute level of sound energy in watts, or sound pressure in microbars, all practical measurements are comparative. Sound levels and ear response cover such a great range that it is convenient to use a logarithmic scale, using a dimensionless unit, the *bel.*[a] In terms of power, the bel is defined as the logarithm to the base 10 of the ratio of W (the sound power in question) to W_o (a reference level of sound power), i.e.,

$$\text{power level in bels} = \log_{10}(W/W_o).$$

The *decibel* (one-tenth of a bel) is the preferred unit:

$$\text{power level in decibels} = 10\log_{10}(W/W_o).$$

As sound pressure is usually proportional to the square root of the sound power, the sound pressure level is commonly expressed as:[b]

$$\text{pressure level in decibels} = 20\log_{10}(P/P_o).$$

The reference levels most widely used correspond roughly to the threshold of audibility for the average human ear:[c]

$$W_o = 10^{-12}\ \text{watts};\ P_o = 0.0002\ \text{microbars.}^d$$

Unless otherwise stated, all values are rms (effective) quantities.[e]

The human ear will not ordinarily detect sound-level differences of less than one decibel; thus, engineers usually prefer to express noise levels to the nearest decibel or half decibel.

Comparing two sound measurements[f] as a positive decibel difference, the pressure ratio and the power ratio may be computed from the previously discussed relationships. It is convenient to show these ratios in tabular form.

ADDITION OF TWO SOUND-METER MEASUREMENTS

When two noise sources operate simultaneously but have been separately evaluated, the cumulative noise level may be obtained readily from the power levels. Usually it is desired to find how much a noise source raises the decibel level above the level existing before it became operative. A direct answer, without the need for finding the energy ratios, may be obtained from the following table. Using the difference between the two previous sound-meter measurements, find the increment to be added to the higher reading.

Difference, db	Increment, db	Difference, db	Increment, db	Difference, db	Increment, db
0.0	3.0	5.5	1.1	10.5	0.4
0.5	2.8	6.0	1.0	11.0	0.3
1.0	2.6	6.5	0.9	11.5	0.3
1.5	2.4	7.0	0.8	12.0	0.3
2.0	2.2	7.5	0.7	12.5	0.2
2.5	1.9	8.0	0.6	13.0	0.2
3.0	1.7	8.5	0.6	13.5	0.2
3.5	1.6	9.0	0.5	14.0	0.2
4.0	1.4	9.5	0.5	14.5	0.2
4.5	1.3	10.0	0.4	15.0	0.1
5.0	1.2				

[a] A scale based on Naperian logarithms is used in certain rare cases; the unit is then designated as the *neper.*
[b] In some sound fields the sound pressure is *not* proportional to the square root of the sound power.
[c] Other reference levels, used in certain cases, are $W_o = 10^{-13}$ watts and $P_o = 1$ microbar.
[d] A microbar is 0.1 newton per square meter (N/m^2) or 1.0 dyne per square cm (dyn/cm^2). One atmosphere is 1 013 250 microbars, or 101 325 N/m^2.
[e] The tabular values apply to any two sound meter readings taken with the same response network, i.e., A(40), B(70), or C(flat), or to any single reading (giving ratios above the reference levels).
[f] Each measurement may be the result of many observations, as the measurement of noise levels as specified in the applicable test codes almost always involves multiple readings. Accurate results depend also on the character of the room. Completely reflecting (reverberant) or completely absorbing (anechoic) rooms are unusual in field testing.

*Compiled and computed.

Table 7-47.   LOUDNESS LEVELS

Equal-loudness Contours at Usual Test Frequencies

The following table presumes "free-field" sound measurement by noise meter. In the decibel scale, which is $10 \log_{10}$ of the ratio of two quantities, the loudness and the sound pressure are numerically equal at 1,000 cps. These curves are not fully standardized; see equal-loudness contours in ISO Recommendation R226–1961.

Sound pressure level, db	Loudness in phons†							
	125 cps	250 cps	500 cps	1,000 cps	2,000 cps	4,000 cps	8,000 cps	10,000 cps
10	—	—	—	10.0	18.0	18.0		
20	—	6.3	16.0	20.0	28.0	28.0	11.0	
30	4.0	18.0	26.5	30.0	37.0	36.5	20.5	17.0
40	17.0	31.0	38.5	40.0	45.5	45.0	29.5	26.0
50	34.0	45.5	52.0	50.0	55.0	54.0	38.0	35.0
60	52.0	59.5	64.5	60.0	64.0	63.5	47.0	43.5
70	70.0	72.5	76.0	70.0	73.5	72.5	56.0	53.5
80	86.0	84.5	86.0	80.0	84.5	83.0	66.0	63.5
90	98.0	95.5	96.0	90.0	95.0	94.5	77.0	73.5
100	108.0	105.5	105.0	100.0	106.0	106.0	88.0	85.5
110	118.0	115.5	113.0	110.0	117.0	117.5	101.5	98.0

†Loudness is a physiological response, not a physical property. Sensitivity of the ear to sound pressure is lower at the extreme frequencies; hence, the loudness in phons is usually less than the sound meter reading. The *sone*, a larger unit of loudness, is sometimes used. A 40-db sound at 1,000 cps has a loudness of 40 phons or one sone.

Table 7-48.   CONVERSION OF LOUDNESS LEVEL IN PHONS TO LOUDNESS IN SONES*

A simplified relation between the loudness in sones and the loudness level in phons has been standardized internationally (ISO/R131–1959). This relation is a good approximation to the psycho-acoustical data and is useful for engineering purposes, but it should not be expected to be accurate enough for research on the subjective aspects of hearing.

The relation is

$$S = 2^{(P-40)/10}$$

where S is the loudness in sones and P is the loudness level in phons.

Phons	0	+1	+2	+3	+4	+5	+6	+7	+8	+9
	LOUDNESS IN SONES									
20	.25	.27	.29	.31	.33	.35	.38	.41	.44	.47
30	.50	.54	.57	.62	.66	.71	.76	.81	.87	.93
40	1	1.07	1.15	1.23	1.32	1.41	1.52	1.62	1.74	1.87
50	2	2.14	2.30	2.46	2.64	2.83	3.03	3.25	3.48	3.73
60	4	4.29	4.59	4.92	5.28	5.66	6.06	6.50	6.96	7.46
70	8	8.6	9.2	9.8	10.6	11.3	12.1	13.0	13.9	14.9
80	16	17.1	18.4	19.7	21.1	22.6	24.3	26.0	27.9	29.9
90	32	34.3	36.8	39.4	42.2	45.3	48.5	52.0	55.7	59.7
100	64	68.6	73.5	78.8	84.4	90.5	97	104	111	119
110	128	137	147	158	169	181	194	208	223	239
120	256	274	294	315	338	362	388	416	446	478

*From: "Handbook of Noise Measurement", 6th ed., A.P.G. Peterson and E.E. Gross, Jr., General Radio Company, 1967, pp. 49–52.

Table 7-49. PROCEDURE FOR CALCULATING LOUDNESS*

This table is used to calculate the loudness from octave-band levels of the preferred series. The procedure is as follows:

1. From the table find the proper loudness index for each band level.
2. Add all the loudness indexes (ΣS).
3. Multiply this sum by 0.3.
4. Add this product to 0.7 of the index for the band that has the largest index ($0.3 \, \Sigma S + 0.7 \, S_{max}$). This value is the total loudness in sones.
5. This total loudness is then converted to loudness level in phons by the relation shown in the two right-hand columns of the table.

The calculated loudness is labeled sones (OD), and the loudness level is labeled phons (OD) to designate that they have been calculated from octave-band levels (O) and for a diffuse field (D).

A similar calculation can be made for third-octave bands, which are labeled TD.

For steady noises having a broad frequency spectrum, the loudness calculated by means of the tables, which are based on Stevens's[1] method, agrees reasonably well with direct assessments made by loudness balances against a 1000-hz tone.

Band level, db	Band loudness index									Loudness, sones	Loudness level, phons
	31.5	63	125	250	500	1000	2000	4000	8000		
20						.18	.30	.45	.61	.25	20
25					.21	.40	.55	.73	.94	.35	25
30				.16	.49	.67	.87	1.10	1.35	.50	30
31				.21	.55	.73	.94	1.18	1.44	.54	31
32				.26	.61	.80	1.02	1.27	1.54	.57	32
33				.31	.67	.87	1.10	1.35	1.64	.62	33
34			.07	.37	.73	.94	1.18	1.44	1.75	.66	34
35			.12	.43	.80	1.02	1.27	1.54	1.87	.71	35
36			.16	.49	.87	1.10	1.35	1.64	1.99	.76	36
37			.21	.55	.94	1.18	1.44	1.75	2.11	.81	37
38			.26	.62	1.02	1.27	1.54	1.87	2.24	.87	38
39			.31	.69	1.10	1.35	1.64	1.99	2.38	.93	39
40		.07	.37	.77	1.18	1.44	1.75	2.11	2.53	1.00	40
41		.12	.43	.85	1.27	1.54	1.87	2.24	2.68	1.07	41
42		.16	.49	.94	1.35	1.64	1.99	2.38	2.84	1.15	42
43		.21	.55	1.04	1.44	1.75	2.11	2.53	3.0	1.23	43
44		.26	.62	1.13	1.54	1.87	2.24	2.68	3.2	1.32	44
45		.31	.69	1.23	1.64	1.99	2.38	2.84	3.4	1.41	45
46	.07	.37	.77	1.33	1.75	2.11	2.53	3.0	3.6	1.52	46
47	.12	.43	.85	1.44	1.87	2.24	2.68	3.2	3.8	1.62	47
48	.16	.49	.94	1.56	1.99	2.38	2.84	3.4	4.1	1.74	48
49	.21	.55	1.04	1.69	2.11	2.53	3.0	3.6	4.3	1.87	49
50	.26	.62	1.13	1.82	2.24	2.68	3.2	3.8	4.6	2.00	50
51	.31	.69	1.23	1.96	2.38	2.84	3.4	4.1	4.9	2.14	51
52	.37	.77	1.33	2.11	2.53	3.0	3.6	4.3	5.2	2.30	52
53	.43	.85	1.44	2.24	2.68	3.2	3.8	4.6	5.5	2.46	53
54	.49	.94	1.56	2.38	2.84	3.4	4.1	4.9	5.8	2.64	54
55	.55	1.04	1.69	2.53	3.0	3.6	4.3	5.2	6.2	2.83	55
56	.62	1.13	1.82	2.68	3.2	3.8	4.6	5.5	6.6	3.03	56
57	.69	1.23	1.96	2.84	3.4	4.1	4.9	5.8	7.0	3.25	57
58	.77	1.33	2.11	3.0	3.6	4.3	5.2	6.2	7.4	3.48	58
59	.85	1.44	2.27	3.2	3.8	4.6	5.5	6.6	7.8	3.73	59
60	.94	1.56	2.44	3.4	4.1	4.9	5.8	7.0	8.3	4.00	60

Table 7-49. PROCEDURE FOR CALCULATING LOUDNESS *(Continued)*

Band level, db	Band loudness index									Loud-ness, sones	Loudness level, phons
	31.5	*63*	*125*	*250*	*500*	*1000*	*2000*	*4000*	*8000*		
61	1.04	1.69	2.62	3.6	4.3	5.2	6.2	7.4	8.8	4.29	61
62	1.13	1.82	2.81	3.8	4.6	5.5	6.6	7.8	9.3	4.59	62
63	1.23	1.96	3.0	4.1	4.9	5.8	7.0	8.3	9.9	4.92	63
64	1.33	2.11	3.2	4.3	5.2	6.2	7.4	8.8	10.5	5.28	64
65	1.44	2.27	3.5	4.6	5.5	6.6	7.8	9.3	11.1	5.66	65
66	1.56	2.44	3.7	4.9	5.8	7.0	8.3	9.9	11.8	6.06	66
67	1.69	2.62	4.0	5.2	6.2	7.4	8.8	10.5	12.6	6.50	67
68	1.82	2.81	4.3	5.5	6.6	7.8	9.3	11.1	13.5	6.96	68
69	1.96	3.0	4.7	5.8	7.0	8.3	9.9	11.8	14.4	7.46	69
70	2.11	3.2	5.0	6.2	7.4	8.8	10.5	12.6	15.3	8.00	70
71	2.27	3.5	5.4	6.6	7.8	9.3	11.1	13.5	16.4	8.6	71
72	2.44	3.7	5.8	7.0	8.3	9.9	11.8	14.4	17.5	9.2	72
73	2.62	4.0	6.2	7.4	8.8	10.5	12.6	15.3	18.7	9.8	73
74	2.81	4.3	6.6	7.8	9.3	11.1	13.5	16.4	20.0	10.6	74
75	3.0	4.7	7.0	8.3	9.9	11.8	14.4	17.5	21.4	11.3	75
76	3.2	5.0	7.4	8.8	10.5	12.6	15.3	18.7	23.0	12.1	76
77	3.5	5.4	7.8	9.3	11.1	13.5	16.4	20.0	24.7	13.0	77
78	3.7	5.8	8.3	9.9	11.8	14.4	17.5	21.4	26.5	13.9	78
79	4.0	6.2	8.8	10.5	12.6	15.3	18.7	23.0	28.5	14.9	79
80	4.3	6.7	9.3	11.1	13.5	16.4	20.0	24.7	30.5	16.0	80
81	4.7	7.2	9.9	11.8	14.4	17.5	21.4	26.5	32.9	17.1	81
82	5.0	7.7	10.5	12.6	15.3	18.7	23.0	28.5	35.3	18.4	82
83	5.4	8.2	11.1	13.5	16.4	20.0	24.7	30.5	38	19.7	83
84	5.8	8.8	11.8	14.4	17.5	21.4	26.5	32.9	41	21.1	84
85	6.2	9.4	12.6	15.3	18.7	23.0	28.5	35.3	44	22.6	85
86	6.7	10.1	13.5	16.4	20.0	24.7	30.5	38	48	24.3	86
87	7.2	10.9	14.4	17.5	21.4	26.5	32.9	41	52	26.0	87
88	7.7	11.7	15.3	18.7	23.0	28.5	35.3	44	56	27.9	88
89	8.2	12.6	16.4	20.0	24.7	30.5	38	48	61	29.9	89
90	8.8	13.6	17.5	21.4	26.5	32.9	41	52	66	32.0	90
91	9.4	14.8	18.7	23.0	28.5	35.3	44	56	71	34.3	91
92	10.1	16.0	20.0	24.7	30.5	38	48	61	77	36.8	92
93	10.9	17.3	21.4	26.5	32.9	41	52	66	83	39.4	93
94	11.7	18.7	23.0	28.5	35.3	44	56	71	90	42.2	94
95	12.6	20.0	24.7	30.5	38	48	61	77	97	45.3	95
96	13.6	21.4	26.5	32.9	41	52	66	83	105	48.5	96
97	14.8	23.0	28.5	35.3	44	56	71	90	113	52.0	97
98	16.0	24.7	30.5	38	48	61	77	97	121	55.7	98
99	17.3	26.5	32.9	41	52	66	83	105	130	59.7	99
100	18.7	28.5	35.3	44	56	71	90	113	139	64.0	100
101	20.3	30.5	38	48	61	77	97	121	149	68.6	101
102	22.1	32.9	41	52	66	83	105	130	160	73.5	102
103	24.0	35.3	44	56	71	90	113	139	171	78.8	103
104	26.1	38	48	61	77	97	121	149	184	84.4	104
105	28.5	41	52	66	83	105	130	160	197	90.5	105
106	31.0	44	56	71	90	113	139	171	211	97	106
107	33.9	48	61	77	97	121	149	184	226	104	107
108	36.9	52	66	83	105	130	160	197	242	111	108
109	40.3	56	71	90	113	139	171	211	260	119	109

Table 7-49. PROCEDURE FOR CALCULATING LOUDNESS *(Continued)*

Band level, db	Band loudness index									Loud-ness, sones	Loudness level, phons
	31.5	*63*	*125*	*250*	*500*	*1000*	*2000*	*4000*	*8000*		
110	44	61	77	97	121	149	184	226	278	128	110
111	49	66	83	105	130	160	197	242	298	137	111
112	54	71	90	113	139	171	211	260	320	147	112
113	59	77	97	121	149	184	226	278	343	158	113
114	65	83	105	130	160	197	242	298	367	169	114
115	71	90	113	139	171	211	260	320		181	115
120	105	130	160	197	242	298	367			256	120
125	149	184	226	278	343					362	125

To illustrate this procedure, consider the following calculations based on octave-band measurements of noise in a factory:

Octave band No.	Octave band, hz	Band level, db	Band loudness index
15	31.5	78	4
18	63	76	5
21	125	78	8
24	250	82	13
27	500	81	14
30	1000	80	16
33	2000	80	20
36	4000	73	15
39	8000	65	11

$$\Sigma S = \text{ sum of band loudness indexes} = 106$$
$$S_m = \text{ maximum band loudness index} = 20$$
$$0.3\,\Sigma S = 31.8$$
$$0.7\,S_m = 14$$

$$0.3\,\Sigma S + 0.7\,S_m = 46 \quad \text{sones } (OD)^2$$
$$\text{or computed loudness level} = 95 \quad \text{phons } (OD)^2$$

For a quick check to find which band contributes most to the loudness, add 3 db to the band level in the second octave, 6 db to the third, 9 db to the fourth, and so on. Then the highest shifted level is usually the dominant band. This check will often be all that is needed to tell where to start in a noise reduction program, if one does not have the loudness calculation charts at hand. This check is not reliable if the levels are low and the low-frequency bands dominate.

Another and more elaborate loudness calculation procedure has been developed by Zwicker[3] for third-octave analysis. It is not at all clear, however, that this more difficult calculation results in a calculated loudness that is in better agreement with subjective data.

---

[1]The method used here is that given in "Procedure for Calculating Loudness: Mark VI", S.S. Stevens, *J. Acous. Soc. Amer.*, 33(11):1577–1585, November 1961. Chart paper No. 31460-A (Codex Book Company, Norwood, Massachusetts 02062) is available for this calculation when the older series of octave bands is used.

[2]OD = octave diffuse (an octave-band analysis for a diffuse field).

[3]"Ein Verfahren zur Berechnung der Lautstärke", E. Zwicker, *Acustica* 10(1):304–308, 1960.

*From: "Handbook of Noise Measurement", 6th ed., A.P.G. Peterson and E.E. Gross, Jr., General Radio Company, 1967, pp. 49–52.

REFERENCE

"Procedure for the Computation of Loudness of Noise", USAS Standard S3.4–1968, American National Standards Institute.

Table 7-50.   AIRBORNE SOUND TRANSMISSION THROUGH WALLS, WINDOWS, AND FLOORS*

Table A.   TRANSMISSION LOSS THROUGH SINGLE, SOLID WALLS (MASS LAW)

The following are approximate, average measured values for airborne sound attenuation. Edge effects are neglected.

Weight of wall, lb/ft^2	2	5	10	20	30	50	75	100	200
Weight of wall, kg/m^2	9.8	24.4	48.8	97.6	146	244	366	488	976
Average sound attenuation,[a] db	26	32	36	39	42	45	48	50	54

Note: Attenuation of high-frequency sounds is more, and that of low-frequency sounds is much less, than the "average" given in the above table.

The critical frequency of the wall or of its component panels may affect the attenuation, especially if that critical frequency is within the range 50–4000 hz.

An uninterrupted air space or a highly porous layer will increase the sound attenuation 5–20 db, depending on thickness and construction. Resilient separating layers are also effective.

[a] The usual "average" is the direct arithmetical average of the db readings at the test frequencies or bands, 125–4000 hz.

Table B.   TRANSMISSION LOSS THROUGH COMPOSITE WALLS

The sound-insulating value of a composite wall with air spaces is somewhat better than that of a solid wall of equal weight (Table A). The improvement is greater for the sound frequencies above 1000 hz than for those below 250 hz.

The following are examples representing test results on composite construction.

Description	Weight		Average sound attenuation, db[a]
	lb/ft^2	kg/m^2	
Plate glass, two panes, $\frac{1}{4}$-in., 6-in. air space, sealed	7.1	34.6	43
Wood studs, 2 × 4-in., 16-in. centers, $\frac{1}{2}$-in. gypsum plaster on $\frac{3}{4}$-in. gypsum lath, both sides	15.2	74.1	44
Fiberboard lath, 2 × 8-in. joists, and gypsum plaster ceiling, sub-floor, and finished floor of pine	15.5	75.5	47
Steel studs, 3-in., 16-in. centers, $\frac{7}{8}$-in. gypsum plaster on metal lath, both sides	19.6	95.5	38
Wood studs, 2 × 4-in., *staggered*, $\frac{7}{8}$-in. gypsum plaster on metal lath	19.8	96.6	50
Two *isolated* walls 2-in. solid gypsum tile, 4-in. air space	20.4	327.	59
Cinder block, 4-in. (102 mm), plastered both sides	36.	577.	46
Concrete slab floor, 4-in., suspended ceiling, rigid	47.	754.	53
Same 4-in. concrete slab floor and ceiling, resilient suspension	47.	754.	58
Two walls of 3-in. cinder block with 3-in. air space between	54.	865.	65

TRANSMISSION LOSS THROUGH FLOORS AND CEILINGS

Attenuation of *airborne sound* through floors may be estimated from the test data for walls. In general, for wood-joist construction with single wood floor and plastered ceiling the attenuation is about 35 db. With carpet on heavy padding, or with 2-in. (51 mm) mineral wool between joists, the attenuation is increased by 5–10 db; if both are used, the attenuation may exceed 45 db. If the mass is increased by pouring a 2-in. layer of dry sand between the joists, the attenuation, with carpeted floor, will be about 48 db.

Impact-noise transmission is often more important than airborne noise transmission. Common impact noises are footsteps, kitchen noises, and office-machine noises. While these noises vary too much for tabular classification, there is a widely accepted test standard for impact-noise transmission. Tests are made with a standard tapping machine set on the floor; the results are expressed in terms of the noise spectrum curve as measured in the room below. Although there is no simple interpretation of these results, noise meter readings not in excess of 65–70 db at lower frequencies and 45–55 db at higher frequencies represent adequate attenuation.

*Compiled from various sources.

Table 7-50.   AIRBORNE SOUND TRANSMISSION THROUGH WALLS, WINDOWS, AND FLOORS (Continued)

Table C.   SOUND ATTENUATION BY DOORS AND WINDOWS

Material	Sound attenu-ation, db
Single ordinary door	25
Heavy single door, packed clearances	30
Heavy double doors, minimum clearances	35
Single glazed window, ordinary	15–20
Double glazing, fixed, sealed	30

Edge-sealing and caulking of fixed joints is very important, as small holes and cracks quickly reduce the acoustical insulating value of a wall.

The approximate effect of a window or a door in a highly insulating wall is shown in Table D.

Table D.   LOSS OF SOUND INSULATION BY WALL OPENINGS

Resulting Loss in db

Window or door area, % of total	Difference in insulating values[a]				
	10	20	30	40	50
1	—	3	11	20	30
2	1	5	14	23	33
5	2	7	17	27	37
10	3	10	20	30	40
20	4	13	23	33	43

[a] For example, this is the difference between the sound attenuation for walls (Table A or B) and the sound attenuation for doors or windows (Table C), in db.

BUILDING CODES AND TEST STANDARDS

Several European countries have adopted minimum values for the sound-insulating qualities of building walls, especially for apartment buildings. Although such standards exist or have been proposed for the United States and Canada, they remain largely for voluntary adoption. In the several adopted European codes the average sound attenuation required for a wall between two residential apartments, for example, ranges from 48–52 db; this is the equivalent of an 8-in. solid brick wall.

There is an increasing tendency to abandon the "average" attenuation requirement in favor of a series of "grading curves," each covering the entire test-frequency range. For example, the United Kingdom "Grade I" for airborne sound insulation by walls between dwellings calls for 36 db attenuation at 100 hz, increasing (straight line) to 56 db at 1600 hz and above. This curve is slightly below the "house standard," which requires 40 db attenuation at 100 hz.

Several European codes specify maximum allowable sound levels in impact noise tests with the standard tapping machine. This machine taps fast enough (about 10 per second) so that an ordinary noise meter can be used, but there are various interpretations of the results, especially at the higher frequencies. The range of "allowable" noise readings at 3000 hz and above will range from 40–55 db.

The International Standards Organization's recommendation ISO R140 specifies "Field and Laboratory Measurements of Airborne and Impact Sound Transmission" (see also British Standards, No. 2750). The ASTM Recommended Practice specifications E90 and E336 cover the measurement of airborne sound transmission and define a sound transmission class (STC) curve or contour.

Some building codes specify a maximum tolerable level of *extraneous* noise. A typical specification for auditoriums and classrooms requires that, when empty, the maximum airborne level of extraneous noise transmitted from outside the room shall not exceed 30 db.

REFERENCES

"Acoustics in Architectural Design", L.L. Doelle, NRC 8358, National Research Council of Canada, Division of Building Research, 1965.
"Building Physics: Acoustics", H.J. Purkis, Pergamon Press, 1966.
"Noise Reduction", L.L. Beranek, Ed., McGraw-Hill Book Company, 1960.
"Sound Insulation of Wall and Floor Constructions", Report BMS–144, National Bureau of Standards, 1955 (Supplement, 1956).

Table 7-51. SOUND TRANSMISSION THROUGH PARTITION WALLS*
Dry-wall Construction

The following table presents typical results selected from approximately 100 tests in the National Research Council of Canada Building Research series. The tests were conducted in accordance with ASTM E90–66T, "Recommended Practice for Laboratory Measurement of Airborne Sound Transmission Loss of Building Partitions." The gypsum board samples represented extremes of density and thickness (as allowed under CSA Standards); they included the fire-resistant and the vinyl-covered types, with no significant differences in performance.

Number	Description of wall	Trans-mission loss, nominal average	Transmission loss, db, at octave frequencies of[a]					
			125	250	500	1 000	2 000	4 000
	SINGLE-LEAF OR BOARD ON ONE SIDE OF STUDS ONLY							
1	$\frac{3}{8}$-in. plasterboard[b]	26	12	17	23	28	33	23
2	$\frac{1}{2}$-in. plasterboard[b]	28	15	20	25	30	33	27
3	$\frac{5}{8}$-in. plasterboard[b]	29	16	21	27	31	29	30
4	22-gage galvanized iron	27	13	17	22	28	34	39
5	$\frac{1}{2}$-in. plasterboard $+\frac{3}{16}$-in. plywood[c]	28	16	20	25	29	32	31
6	$\frac{1}{2}$-in. plasterboard $+\frac{1}{2}$-in. fiberboard[d]	30	16	22	28	33	32	30
	TWO-LEAF WALLS—WALLBOARD BOTH SIDES							
7	$\frac{1}{2}$-in. plasterboard, $3\frac{5}{8}$-in. space	35	15	28	32	41	46	38
8	$\frac{1}{2}$-in. plasterboard, 2-in. fill[e]	40	22	33	35	43	47	40
9	Staggered studs $+$ 2-in. fill[f]	46	26	37	45	50	50	47
10	Plasterboard $+$ plywood $+$ fiberboard[g]	46	22	36	46	54	56	56
11	$\frac{5}{8}$-in. plasterboard, $3\frac{5}{8}$-in. space	36	21	25	34	41	36	42
12	$\frac{5}{8}$-in. plasterboard, 4-in. fill[h]	45	28	40	45	48	40	45
13	Staggered studs $+$ no fill[j]	39	27	25	37	46	38	49
14	Staggered studs, 4-in. fill[k]	46	32	36	47	50	42	52
15	Multilayer, steel studs, glass fiber[m]	53	36	45	49	55	55	56

[a] The quoted report included test data on the following additional frequencies: 160, 200, 315, 400, 630, 800, 1 250, 1 600, 2 500, 3 150, and 5 000 hz.
[b] Joints taped for plasterboard walls.
[c] Sheets joined by contact cement on faces.
[d] Plasterboard and wood fiberboard laminated with gypsum joint compound.
[e] Glass fiber batts, 2-in. thick, between studs; $\frac{1}{2}$-in. plasterboard.
[f] Staggered wood studs, 2-in. $\times$ 4-in., $3\frac{5}{8}$-in. space, 2-in. thick glass fiber batts, $\frac{1}{2}$-in. plasterboard, both faces.
[g] $\frac{1}{2}$-in. plasterboard and $\frac{3}{16}$-in. plywood on both faces; 2-in. $\times$ 4-in. wood studs on 16-in. centers; $\frac{1}{2}$-in. wood fiberboard between studs.
[h] 4-in. low-density glasswool batts, compressed into stud space between $3\frac{5}{8}$-in. steel-channel studs.
[j] 2-in. $\times$ 4-in. wood studs, staggered, $3\frac{5}{8}$-in. space, $\frac{5}{8}$-in. plasterboard, no fill.
[k] Same as No. 13 but 4-in. low-density glasswool batts in space between 2-in. $\times$ 4-in. wood studs.
[m] Three $\frac{1}{2}$-in. layers, one $\frac{5}{8}$-in. layer plasterboard; $3\frac{5}{8}$-in. steel channel studs, 24-in. centers; $2\frac{1}{2}$-in. glass fiber.

GENERAL NOTES

With studs of low torsional rigidity, such as steel channels, sound transmission via the studs appears to be negligible.

The simpler constructions have been tested by several laboratories, and results have been found to be reasonably reproducible.

The test specimens were mounted and caulked into an opening 10-ft wide by 8-ft high that separated two reverberation rooms. The test signal consisted of third-octave bands of "pink" noise.

The test results were not critically sensitive to normal variations in thickness, density, or techniques of erection.

Several of the additional tests included multi-layer constructions and the additions of resilient bars, horizontal or vertical. Highest sound attenuation was 53 db (test No. 15 in table).

*Data from: "Transmission Loss of Plasterboard Walls", T.D. Northwood, Building Research Note 66, Division of Building Research, National Research Council, revised July 1970.

Table 7-52.   SOUND-ABSORPTION COEFFICIENTS

Sound-absorption data on various materials are presented here in terms of the acoustic properties of rooms; these data are also valuable for other applications, such as the acoustic treatments for ducts and tunnels. Architectural applications include the control of noise level in offices, stores, restaurants, and other occupied spaces, and the control of reverberation time and elimination of echoes in concert and lecture halls.

Non-absorbent surfaces. Solid walls, floors, or ceilings finished with glazed tile, marble, terrazzo, very smooth concrete, or with linoleum, rubber, cork, or plastic tile cemented directly to concrete, are not sound absorbent. The absorption coefficient (in percentage of incident energy absorbed) is seldom more than 1 or 2 percent at all wavelengths from 125 hz to 4,000 hz.

Poor sound absorbers. Brick surfaces, painted concrete blocks, hardwood floors, gypsum board, or smooth nonporous plaster (lime or gypsum) are all poor absorbers. With solid structural backing these finishes seldom afford as much as 10 percent absorption at any wavelength from 125 hz to 4,000 hz. Ordinary window areas may absorb up to 25 percent of the low-frequency sounds (125 to 250 hz) but much less at the higher frequencies.

Sound-absorbing materials. Carpets, drapes, and upholstered seats are fair-to-good sound absorbers.

| Materials | Absorption coefficients, % (percentage of incident energy absorbed) | | | | | |
| | Sound frequency, hz | | | | | |
	125	250	500	1,000	2,000	4,000
Heavy carpet on concrete or solid floor	2	6	14	37	60	65
Heavy carpet on heavy hairfelt or elastic pad	8	25	50	60	65	65
Heavy drapes (1 lb/sq yd) draped 2 to 1 area	10	30	50	75	70	65
Light hung fabric	3	4	11	17	24	35
Unoccupied wood or metal chairs[a]	15	20	25	40	40	30
Unoccupied upholstered seating[a]	45	55	65	70	70	60
Full audience, occupying upholstered seats[a]	70	75	85	95	90	85

[a] Equivalent values based on floor area.

ACOUSTIC TREATMENTS

The following data apply mainly to acoustic ceiling treatments, but other surfaces may be similarly treated. These data do not apply to through-transmission of sound. Low sound transmission accompanies high density or weight of material.

| Class description[b] | Absorption coefficients, % (percentage of incident energy absorbed) | | | | | |
| | Sound frequency, hz | | | | | |
	125	250	500	1,000	2,000	4,000
Porous, lightweight fiber board or tile $\frac{3}{4}$ in. thick; perforated or fissured; painted; on $\frac{3}{4}$ in. furring strips	20	58	61	80	80	68
Porous fireproof mineral tile, $\frac{5}{8}$ in. thick; perforated or fissured; painted; direct solid application to structural surface	10	26	70	89	75	60
Porous fireproof mineral tile, $\frac{3}{4}$ in. thick, drop ceiling or large air space; metal supports	70	66	72	92	88	75
Bonded wood or mineral fibers; thickness about $1\frac{1}{2}$ in.	41	59	88	85	76	65
Perforated metal pans or hardboard backed with $1\frac{1}{2}$ in. loose pad	36	56	87	94	74	56

[b] Data on each class based on five or more commercial products.

Table 7-53. ROOM PROPORTIONS FOR GOOD ACOUSTICS

Room dimensions should be separated by one-third or two-thirds of an octave ($2^{.33}$ or $2^{.67}$) in order to guard against resonance and standing waves. The cubical room is the least desirable. Following are typical dimensions for well proportioned rooms. In some cases it may still be necessary to introduce oblique surfaces for best acoustics.

For linear dimensions in meters, multiply values in feet by 0.3048. For volumes in m^3, multiply values in ft³ by 0.028317.

Ceiling height, ft	Average room			Long room			Low-ceilinged room		
	Width, ft	Length, ft	Volume, cu ft	Width, ft	Length, ft	Volume, cu ft	Width, ft	Length, ft	Volume, cu ft
10	12.5†	16	2,000†	12.5	32	4,000†	25	32	8,000†
12	15.†	19	3,400†	15.	38	6,850†	30	38	13,700
14	17.5†	22	5,400†	17.5	44	10,800	35	44	21,600
16	20†	25.5	8,150†	20	50	16,000	40	50	32,000
18	29	45	23,500	23	57	23,600	45	57	46,100
20	32	50	32,000	25	63	31,500	50	63	63,000
22	35	55	42,000	28	69	42,500	55	69	83,300
24	38	60	55,000	30	76	54,600	60	76	109,200
26	42	65	71,000	33	82	70,300	65	82	138,500
28	45	70	87,300	35	88	86,500	70	88	173,000
30	48	75	108,000	38	95	108,500	75	95	214,000
32	51	80	131,000	40	101	129,000	80	101	258,000
34	55	85	159,000	43	107	156,000	85	107	310,000
36	58	90	188,000	45	114	185,000	90	114	370,000
38	61	95	220,000	48	120	219,000	95	120	433,000
40	64	100	256,000	50	126	252,000	100	126	504,000

†These might be classed as "small" music rooms, conference rooms, classrooms, lounges, etc.

Table 7-54. TYPICAL SOUND-REVERBERATION TIMES FOR AUDITORIUMS

For optimum speech clarity a short reverberation time is desired; for music a much longer reverberation time is preferred. The requirements for a talk studio are much different from those of a hall intended for organ concerts; hence, the design and treatment become a compromise, except for single-usage auditoriums.

GENERAL USAGE CLASSES:

    Class I: conference rooms, sound studios, and classrooms
    Class II: music and lecture halls; churches
    Class III: speech and music auditoriums; theaters
    Class IV: larger churches; municipal auditoriums

REVERBERATION TIME VS. ROOM VOLUME

Class of use							
I		*II*		*III*		*IV*	
Volume, ft³	*Time, s*	*Volume, ft³*	*Time, s*	*Volume, ft³*	*Time, s*	*Volume, ft³*	*Time, s*
2 000	0.4	10 000	0.9	40 000	1.2	60 000	1.4
4 000	0.5	20 000	1.0	60 000	1.3	100 000	1.5
6 000	0.6	40 000	1.1	100 000	1.4	200 000	1.6
10 000	0.7	60 000	1.2	200 000	1.5	400 000	1.7
20 000	0.8	100 000	1.3	400 000	1.6	600 000	1.8

Notes:

Reverberation time is the elapsed time for the sound to decay to inaudibility after the source of the sound is suddenly stopped.

Reverberation time increases with room size. It is reduced by increasing the sound absorption of the surfaces and contents of the room. In existing auditoriums little can be done to increase reverberation time except through the use of electronic sound systems.

It is assumed that reflecting surfaces have been properly designed, that room proportions conform to acoustic requirements, and that no unfavorable echoes exist.

In most large auditoriums a total room volume of about 250 ft³ per audience seat is desirable; in rooms designed for speech, a volume as small as 100 ft³ per seat may be preferred.

The reverberation time for extremely large churches is correspondingly longer than any given in the above table. For example, recent extensive measurements in St. Peter's in Rome (the largest church, 20 million ft³) gave reverberation times in the range of 6–7 seconds for single tones up to about 3000 hz (Ref. 3), and over 4 seconds for the highest note on the piano or the piccolo.

GENERAL REFERENCES

1. "Acoustical Designing in Architecture", V.O. Knudsen and C.M. Harris, John Wiley & Sons, 1950.
2. "Acoustics, Noise, and Buildings", P.H. Parkin and H.R. Humphreys, Frederick A. Praeger, 1958.
3. "Acoustics of St. Peter's and Patriarchal Basilicas in Rome", R.S. and H.K. Shankland, *J. Acoust. Soc. Amer.*, 50:2, 1971.
4. "Music, Acoustics, and Architecture", L.L. Beranek, John Wiley & Sons, 1962.

Table 7-55.   SOME U.S. TRENDS AFFECTING ENGINEERS

Table A.   ENGINEERING INDUSTRIES

Description	1950	1960	1965	1967	1968	1969	1970	Source
Manufacturing								
Value of shipments, billions of dollars		369.6	492.0	548.5	604.6	656.7	666.6	1
Total employees, millions	15.2	16.8	18.1	19.4	19.8	20.0	19.4	4,5
Index of manufacturing capacity,								
1957–59 = 100	84	135	164	186	197	208		2
Output/capacity, percent	90	81	89	85	85	84	77[a]	2
Durable goods, percent of product value		51	54	55	55	55	53	1
National income from manufacturing, percent			31	30	30	29		3
Production index, 1957–59 = 100	75	109	143	158	166	173	168[a]	1
Construction								
Value of new construction, billions of dollars	33.8	53.9	72.3	76.2	84.7	90.9	91.3	1
Total employees, millions	2.33	2.89	3.19	3.21	3.44	3.35	3.29	5
Composite cost index, 1957–59 = 100[b]	78.3	94.1	105.7		121.6	132.3	148.6	1
Public construction, percent of total	20.4	29.4	30.5	33.6	32.7	30.9	30.9	1
Non-farm residential, percent of total	53.9	40.0	36.4	31.1	34.0	32.7	32.1	1
Highways and streets, percent of total	6.3	10.1	10.4	11.2	11.0	10.2	10.9	1
Transportation								
Total employees, millions		2.55	2.53			2.71	2.69	5
Freight, billions of ton-miles								
Railroads	628	595	721	742	757	780[a]		7
Inland waterways	163	220	262	281	291	300[a]		7
Motor vehicles	173	285	359	389	396	404[a]		7
Oil pipelines	129	229	306	361	391	411[a]		7
Airways	0.3	0.8	1.9	2.6	2.9	3.2[a]		7
Intercity passengers, billions of passenger-miles								
Railroads	32.5	21.6	17.6	15.3	13.3	12.3[a]		7
Buses (excludes school buses)	26.4	19.3	23.8	24.9	24.5	24.9[a]		7
Airways, public and private	10.1	34.0	58.1	87.2	101.	111.[a]		7
Private automobiles	438.	706.	818.	890.	936.	977.[a]		7
Communication								
Postal revenues, billions of dollars	1.7	3.3	4.5	5.1	5.7	6.3	6.5	8
Total postal employees, thousands	501.	563.	596.		731.	739.	741.	8
Postal service, pieces handled, billions	45.1	63.7	71.9	78.4	79.5	82.0	84.9	8
Telephones per hundred population	28	41	48	52	54	56	58	8
Telephone employees, thousands	565	627	655		734	794		9
Commercial broadcast stations, number	2 336.	4 218.	4 867.	5 100.	5 236.	5 309.		9
Mining and Minerals (including coal, oil, gas)								
Value of mineral production, billions of dollars		18.0	21.5	23.7	25.0	26.9	29.5[a]	6
Employees working daily, average, millions		1.05	0.91	0.90	0.90	0.89[a]		6

Table B.   PEOPLE AND EMPLOYMENT

Description	1950	1960	1965	1967	1968	1969	1970	Source
U.S. Population								
Total population, millions	152.3	180.7	194.2	198.6	200.6	202.6	204.8	1
Civilian resident population, millions	150.8	178.1	191.5	195.2	197.0	199.1	201.6	1
Urban areas, percent of total	64.0	69.9					73.5	1
Number of households, millions	42.9	52.8	57.3		60.4	61.8	62.9	1
Employment and Production								
Labor force, 16 and over, millions	63.9	72.1	77.2	80.8	82.3	84.2	85.9	5
Employed civilians, millions	58.9	65.8	71.1	74.4	75.9	77.9	78.6	5
Unemployed, millions	3.3	3.9	3.4	3.0	2.8	2.8	4.1	5
Employed in goods-related industries, millions	22.5	24.4	25.9	27.5	28.0	28.7	27.9	5
Gross national product (GNP), billions of								
dollars	284.8	503.7	684.9	793.9	865.0	931.4	976.5[a]	3
GNP in 1958 constant dollars, billions	355.3	487.7	617.8	675.2	707.2	727.1	724.1[a]	3

[a]Preliminary data.
[b]Department of Commerce data.

Table 7-55.   SOME U.S. TRENDS AFFECTING ENGINEERS *(Continued)*

Table B.   PEOPLE AND EMPLOYMENT *(Continued)*

Description	1950	1960	1965	1967	1968	1969	1970	Source
Employment and Production *(Continued)*								
Engineers in private industry, thousands			759.	836.		864.		5
Engineers in federal government, thousands		61.		81.	83.	84.		10
Technicians in private industry, thousands			646.	735.		772.		5
Physical scientists in private industry, thousands			109.	118.		127.		5
Education								
High school graduates, thousands	1 200.	1 864.	2 665.	2 680.	2 702.	2 839.	2 906.	12
College graduates, first degree, thousands	432.	392.	534.	595.	672.	764.	784.	12
Engineering graduates, first degree, thousands	52.7	37.8	36.7	36.2	38.0	40.0	43.0	11,12
Engineering degrees, M.S., thousands	4.9	7.2	12.1	13.9	15.2	14.9	15.5	11,12
Engineering degrees, Ph.D., thousands	0.5	0.8	2.1	2.6	2.9	3.4	3.7	11,12
Engineering technician certificates, 2-yr, thousands		7.6	5.6	6.1	6.3	6.5	7.7	11

SOURCES OF DATA

1. Department of Commerce, Bureau of the Census.
2. Federal Reserve Bulletin.
3. Department of Commerce, Office of Business Economics.
4. Office of Management and the Budget.
5. Department of Labor, Bureau of Labor Statistics.
6. Department of Interior, Bureau of Mines.
7. Interstate Commerce Commission.
8. U.S. Postal Service.
9. Federal Communications Commission.
10. National Science Foundation.
11. Engineering Manpower Commission.
12. Office of Education, Department of Health, Education, and Welfare.

Table 7-56.   MUSCULAR STRENGTH AND MAN POWER

Manual lifting or operation of levers, knobs, cranks, and handwheels should require minimum strength so that the weakest user can accomplish the task; however, much depends on body position and convenience of access and grip. Following are approximate forces that can be applied by a young adult male.

Task	Force
Lever operation at convenient height; operator standing or seated and braced; steady push or pull	100 lb
Foot-pedal pressure; operator seated and braced; extended leg position	170 lb
Lifting with leg muscles; knee angle 150°; horizontal bar grip	600 lb
Lifting tote box; back muscles	200 lb
Turning large handwheel through large angle	100 in.-lb
Turning small handwheel; full grip of one hand	80 in.-lb
Turning large knob; full-finger grip only	20 in.-lb

Mechanical power available in heavy labor over extended periods is about 0.1 hp; for a well trained man the physiological limit may reach 0.5 hp for steady work, 0.6 hp or more for a few minutes, and as high as 2.0 hp for a few seconds. A draft horse can actually work for extended periods at the rate of one horsepower.

Table 7-57. BODY SIZE OF YOUNG ADULTS
Dimensions in Inches; Nude

Because a set of body dimensions based on "averages" seldom applies to any one individual, such data must be used with caution. Published data are largely based on records applying to military personnel; hence they apply to young adults. Clothing makes a considerable difference, especially for males in cold environments. Many engineering designs must be based more nearly on a maximum or mimimum dimension than on the average. If work place or equipment dimensions are to be adjustable, a table such as this provides a general guide.

	Males			Females		
	5th %	Median	95th %	5th %	Median	95th %
Approximate weight, lb	132	155	200	100	122	150
Height	63	68	73	60	64	68.5
Vertical reach†	76	82	89	69	76	81.5
Horizontal spread reach†	58	64	68	54	59	65
Forward reach†	29	33	35	25	30	32
Crotch to floor	30	32.5	36	24	29	31
Shoulder width	16	18	20	14	17	19
Hip width	13	14	15.5	12.5	14	15.5
Foot length	9.8	10.5	11.3	8.7	9.5	10.2
Head circumference	21.5	22.5	23.5	20.4	21.5	22.7
Chest circumference	35	39	43	30	33.5	37
Waist circumference	28	33	37	23.5	25.5	28.5
Thigh circumference	20	23	25	19	22	24
Calf circumference	13	14.5	16	11.7	13.5	15
Wrist circumference	6.3	7	7.5	5.5	6.5	6.9
Knee clearance to floor	20	21.5	23	17	21	22
Body surface area, sq in.	2,400	2,800	3,200	2,150	2,300	2,800

†Functional reach.

Table 7-58.   PROPERTIES OF THE SKIN*
"Average" Man

Surface area	20 sq ft	Sensation *(Cont.)*	
Specific gravity	1.1	Shivering cold	86°F
Water content	70–75%	Uncomfortably cold	<88°F
Thickness	.02–0.2 in.	Comfortably warm	93°F
Thermal conductivity, k,		Uncomfortably warm	>95°F
room temperature	$\dfrac{.0015 \text{ cal}}{\text{cm sec }°C}$	Pain threshold	113°F (45°C)
		Infrared emissivity	0.99
Sensation vs "mean" skin		Solar reflectivity:	
temperature (entire body):		Very white	42%
Extremely cold	84°F	Medium-colored	20%
		Very black	10%

*Condensed from: "Bioastronautics Data Book", P. Webb, Ed., NASA SP–3006, National Aeronautics and Space Administration, 1964.

Table 7-59. ORGANS OF STANDARD MAN*

Mass and Effective Radius of Organs of the Adult Human Body

In order to calculate the concentration of the radionuclide in an organ, using external counting techniques, the mass of an organ must be known. The mass of an organ may be estimated on the basis of scanning, radiography, physical examination, or other data. Values of organ mass for a 70-kg "standard" man are given in the following table.

Organ	Mass (m), g	Percent of total body	Effective radius (X), cm
Total body	70 000	100	30
Muscle	30 000	43	30
Skin and subcutaneous tissue[a]	6 100	8.7	0.1
Fat	10 000	14	20
Skeleton			
without bone marrow	7 000	10	5
red marrow	1 500	2.1	
yellow marrow	1 500	2.1	
Blood	5 400	7.7	
Gastrointestinal tract[b]	2 000	2.9	30
Contents of gastrointestinal tract			
lower large intestine	150		5
stomach	250		10
small intestine	1 100		30
upper large intestine	135		5
Liver	1 700	2.4	10
Brain	1 500	2.1	15
Lungs (2)	1 000	1.4	10
Lymphoid tissue	700	1.0	
Kidneys (2)	300	0.43	7
Heart	300	0.43	7
Spleen	150	0.21	7
Urinary bladder	150	0.21	
Pancreas	70	0.10	5
Salivary glands (6)	50	0.071	
Testes (2)	40	0.057	3
Spinal cord	30	0.043	1
Eyes (2)	30	0.043	0.25
Thyroid gland	20	0.029	3
Teeth	20	0.029	
Prostate gland	20	0.029	3
Adrenal glands or suprarenal glands (2)	20	0.029	3
Thymus	10	0.014	
Ovaries (2)	8	0.011	3
Hypophysis (pituitary)	0.6	8.6×10^{-6}	0.5
Pineal gland	0.2	2.9×10^{-6}	0.04
Parathyroids (4)	0.15	2.1×10^{-6}	0.06
Miscellaneous (blood vessels, cartilage, nerves, etc.)	390	0.56	

[a]The mass of the skin alone is taken to be 2000 g.
[b]Does not include contents of the gastrointestinal tract.

*From: International Commission on Radiation Protection, Committee II, 1959, Permissible Dose for Internal Radiation, *Health Phys.*, 3:1, 1960. (Courtesy of *Health Physics*.)

REFERENCE

"Handbook of Radioactive Nuclides", Y. Wang, Ed., The Chemical Rubber Co., 1969.

Table 7-60.   TYPICAL OXYGEN INPUT AND HEAT OUTPUT OF HUMAN ADULTS

Activity	Oxygen used, lb/hr	Heat output	
		kcal/hr	Btu/hr
SEDENTARY			
Asleep	.05	72	280
Lying, relaxed	.06	80	320
Sitting, at rest	.07	100	400
LIGHT ACTIVITY			
Writing, seated	.07	108	430
Drafting, standing	.08	114	460
Rapid typing, seated	.09	140	550
LIGHT WORK			
At machine or bench, standing	.12	175	700
Slow walking	.15	225	900
Ironing	.17	260	1,050
Walking, 50-lb load, 3 mph	.20	300	1,200
MEDIUM-TO-HEAVY WORK			
Swimming, leisurely	.24	375	1,500
Cycling, 10–12 mph	.30	450	1,800
Swimming, crawl, 1 mph	.35	530	2,200
Digging, shoveling, sawing	.38	570	2,300
Climbing stairs, 2 steps per second	.40	600	2,400
VERY HEAVY WORK			
Slow climbing, 33% grade, with load	.5	750	3,000
Wrestling	.5	800	3,200
Running, 8 mph	.6	900	3,600
Maximum work for 10 minutes	.8	1,200	4,800

Note: The above tabular values are approximate for young, healthy males of 150-lb weight. Values for lighter-weight persons, women, or children are lower. Trained athletes or men accustomed to heavy work can produce higher values.

REFERENCE

"Bioastronautics Data Book", P. Webb, Ed., NASA SP–3006, National Aeronautics and Space Administration, 1964, Section 10.

Table 7-61.   TYPICAL WATER BALANCE

Adult Male; Light Activity; Comfortable Room

Processes and materials	ml/day	lb/day
INTAKE		
Liquids	1,500	3.3
Water in foods	700	1.54
Metabolic	300	.66
Total	2,500	5.50
OUTPUT		
Urine	1,500	3.3
Feces	100	.22
Respiratory and evaporation	900	1.98
Total	2,500	5.50

Note: Total water quantities may be increased 100 to 200% in hot, dry climate, depending on activity.

Table 7-62.   COMMISSIONERS 1958 STANDARD ORDINARY MORTALITY TABLE*

Age, x	Number living, l_x	Number dying, d_x	Deaths per 1,000	Age, x	Number living, l_x	Number dying, d_x	Deaths per 1,000
0	10,000,000	70,800	7.08	50	8,762,306	72,902	8.32
1	9,929,200	17,475	1.76	51	8,689,404	79,160	9.11
2	9,911,725	15,066	1.52	52	8,610,244	85,758	9.96
3	9,896,659	14,449	1.46	53	8,524,486	92,832	10.89
4	9,882,210	13,835	1.40	54	8,431,654	100,337	11.90
5	9,868,375	13,322	1.35	55	8,331,317	108,307	13.00
6	9,855,053	12,812	1.30	56	8,223,010	116,849	14.21
7	9,842,241	12,401	1.26	57	8,106,161	125,970	15.54
8	9,829,840	12,091	1.23	58	7,980,191	135,663	17.00
9	9,817,749	11,879	1.21	59	7,844,528	145,830	18.59
10	9,805,870	11,865	1.21	60	7,698,698	156,592	20.34
11	9,794,005	12,047	1.23	61	7,542,106	167,736	22.24
12	9,781,958	12,325	1.26	62	7,374,370	179,271	24.31
13	9,769,633	12,896	1.32	63	7,195,099	191,174	26.57
14	9,756,737	13,562	1.39	64	7,003,925	203,394	29.04
15	9,743,175	14,225	1.46	65	6,800,531	215,917	31.75
16	9,728,950	14,983	1.54	66	6,584,614	228,749	34.74
17	9,713,967	15,737	1.62	67	6,355,865	241,777	38.04
18	9,698,230	16,390	1.69	68	6,114,088	254,835	41.68
19	9,681,840	16,846	1.74	69	5,859,253	267,241	45.61
20	9,664,994	17,300	1.79	70	5,592,012	278,426	49.79
21	9,647,694	17,655	1.83	71	5,313,586	287,731	54.15
22	9,630,039	17,912	1.86	72	5,025,855	294,766	58.65
23	9,612,127	18,167	1.89	73	4,731,089	299,289	63.26
24	9,593,960	18,324	1.91	74	4,431,800	301,894	68.12
25	9,575,636	18,481	1.93	75	4,129,906	303,011	73.37
26	9,557,155	18,732	1.96	76	3,826,895	303,014	79.18
27	9,538,423	18,981	1.99	77	3,523,881	301,997	85.70
28	9,519,442	19,324	2.03	78	3,221,884	299,829	93.06
29	9,500,118	19,760	2.08	79	2,922,055	295,683	101.19
30	9,480,358	20,193	2.13	80	2,626,372	288,848	109.98
31	9,460,165	20,718	2.19	81	2,337,524	278,983	119.35
32	9,439,447	21,239	2.25	82	2,058,541	265,902	129.17
33	9,418,208	21,850	2.32	83	1,792,639	249,858	139.38
34	9,396,358	22,551	2.40	84	1,542,781	231,433	150.01
35	9,373,807	23,528	2.51	85	1,311,348	211,311	161.14
36	9,350,279	24,685	2.64	86	1,100,037	190,108	172.82
37	9,325,594	26,112	2.80	87	909,929	168,455	185.13
38	9,299,482	27,991	3.01	88	741,474	146,997	198.25
39	9,271,491	30,132	3.25	89	594,477	126,303	212.46
40	9,241,359	32,622	3.53	90	468,174	106,809	228.14
41	9,208,737	35,362	3.84	91	361,365	88,813	245.77
42	9,173,375	38,253	4.17	92	272,552	72,480	265.93
43	9,135,122	41,382	4.53	93	200,072	57,881	289.30
44	9,093,740	44,741	4.92	94	142,191	45,026	316.66
45	9,048,999	48,412	5.35	95	97,165	34,128	351.24
46	9,000,587	52,473	5.83	96	63,037	25,250	400.56
47	8,948,114	56,910	6.36	97	37,787	18,456	488.42
48	8,891,204	61,794	6.95	98	19,331	12,916	668.15
49	8,829,410	67,104	7.60	99	6,415	6,415	1,000.00

*Reproduced by permission of The Society of Actuaries.

Environmental Protection and Human Safety

Table 8-1.   ENVIRONMENTAL QUALITY CONTROL

Early U.S. environmentalists, such as Theodore Roosevelt and Gifford Pinchot, succeeded in obtaining legal support for the conservation movement. By the 1950's federal legislation and agencies had included protection of water resources and some attention to air pollution. Gradually, the problems of solid-waste disposal, radiation protection, noise abatement, and chemical pollution hazards were recognized as federal concerns, and legal safeguards began to be established.

By the end of 1970 most of the federal control activity was coordinated under the Environmental Protection Agency. The National Environmental Policy Act of 1969 also had created the Council on Environmental Quality. This organization, in the Executive Office of the President, is charged with assisting the President in preparing an annual environmental quality report and with making recommendations for national policies to improve the environment. The Council also is empowered to analyze conditions and trends in the quality of the environment and to conduct investigations relating to the environment. The annual report "Environmental Quality" is transmitted to Congress in August.[a]

A partial list of agencies, laws, and acts of federal origin is offered to assist in the identification of environmental data and information issued by the federal government. For listings of current publications consult the "Monthly Catalog of United States Government Publications".

FEDERAL AGENCIES, LAWS, AND ORDERS; PARTIAL LIST, 1960–1972

THE TOTAL ENVIRONMENT

Council on Environmental Quality. Established by the President in May 1969, succeeding the Environmental Quality Council.

Environmental Control Administration (Department of Health, Education, and Welfare). Deals with environmental health problems.

Environmental Data Service (National Oceanic and Atmospheric Administration, Department of Commerce). Issues a bimonthly bulletin.

Environmental Health Service (Department of Health, Education, and Welfare). Includes Community Environmental Management Bureau.

Environmental Protection Agency. Administers all phases of pollution control. Annual budget, $1 billion plus. In addition to its offices of Air, Water, Solid Waste, Pesticides, and Radiation, it has ten regional offices and an Office of International Affairs.

Environmental Quality Council. Established by the President in July 1969 as the first independent, cabinet-level organization charged with a broad overview of environmental problems; abolished by Reorganization Plan No. 2, 1970.

Environmental Quality Improvement Act, 1970, PL 91-224. Created the Office of Environmental Quality and added the following responsibilities to the Council on Environmental Quality: (1) review environmental monitoring, (2) evaluate effects of technology, and (3) assist federal agencies in development of environmental standards.

Environmental Research Laboratories, Boulder, Colorado. Investigates and reports on environmental subjects for NOAA.

Environmental Science Services Administration. Monitors and inspects pollution load and distribution. Formerly a component of the Department of Commerce; became major part of NOAA.

Executive Order No. 11514, March 5, 1970. Instructed all federal agencies in conformity with the National Environmental Policy Act. Listed the responsibilities of the Council on Environmental Quality.

National Environmental Policy Act, 1969, PL 91-190. Set up a national policy on environment with directions for implementation. Established Council on Environmental Quality and outlined its functions.

National Oceanic and Atmospheric Administration (Department of Commerce). Selected marine, atmospheric, and weather projects.

Office of Environmental Quality. Consists of the supporting staff office serving the Council on Environmental Quality and its Chairman and Director. Annual budget, $2 million plus.

AIR POLLUTION CONTROL

Air Pollution Control Act, 1962. Authorized the Surgeon General to study health aspects of air pollution.

Air Pollution Control Office. Now called the Air Programs Office (*q.v.*); formerly a division of the EPA. See Annual Reports, e.g., Senate Document 92-11, May 3, 1971.

---

[a]These annual reports (about 300 pages each), perhaps the best available summaries of the interests and progress of the federal government in all fields of pollution control, are available from the Superintendent of Documents.

Table 8-1.   ENVIRONMENTAL QUALITY CONTROL *(Continued)*

FEDERAL AGENCIES, LAWS, AND ORDERS; PARTIAL LIST, 1960–1972 *(Continued)*

Air Programs Office (Environmental Protection Agency). Publishes material treating all aspects of air-pollution control. Formerly the Air Pollution Control Office (*q.v.*).

Air Quality Act of 1967, PL 90-148. Elaborated upon and amended the 1963–65 act.

Clean Air Act, 1963. Encouraged state and local pollution agencies and provided grants for control programs. Amended October 1965 to permit national regulation of air pollution from new motor vehicles.

Clean Air Amendments, 1970, PL 91-604. Contained major amendments and additions to the 1963, 1965, and 1967 acts.

Executive Order No. 11507, February 4, 1970. Stipulated that projects or installations owned or leased by the federal government must be designed and operated to meet air and water-quality standards.

National Air Pollution Control Administration. Moved from HEW to EPA December 2, 1970. Later renamed Air Programs Office (*q.v.*).

Task Force on Air Pollution. Created by the President on November 18, 1969.

WATER POLLUTION CONTROL

Clean Water Restoration Act, 1966. Provided federal money for building water-treatment facilities.

Executive Order No. 11574, December 23, 1970. Implemented a permit program to regulate discharge of pollutants into navigable waters; charged several federal agencies with making this permit system effective for pollution control.

Federal Water Pollution Control Act, 1956. Amended 1961, 1965, 1971. Authorized planning technical assistance and grants for state and municipal programs and construction.

Federal Water Pollution Control Administration (Department of the Interior). Set up in 1966 as successor to programs in HEW. Changed to Federal Water Quality Administration in 1970 (*q.v.*).

Federal Water Quality Administration (Department of the Interior). Sets water-quality standards for all interstate and coastal waters. Publishes "Inventory of Municipal Waste Facilities" (1940, 1945, 1957, and 1962 inventories by PHS). Created in 1970.

Water Quality Improvement Act, 1970. Set up the Federal Water Quality Administration and provided tighter controls over oil pollution and pollution from vessel dumping.

Water Quality Office (Environmental Protection Agency). Succeeded a similar office in FWQA.

OTHER POLLUTION CONTROLS

Atomic Energy Commission. Implements and enforces radiation standards by licensing authority. Regulates radioactive waste disposal.

Bureau of Solid Waste Management (Department of Health, Education, and Welfare). Treats environmental health aspects of waste disposal.

Federal Aviation Administration (Department of Transportation). Administers aircraft-noise-abatement laws and rules.

Federal Radiation Council. Established by Executive Order of the President. Published about 20 reports and memoranda during the 1960's dealing with radiation protection. Established "protective action guides".

Occupational Safety and Health Administration (Department of Labor). Administers safety and health standards, with special reference to the Williams-Steiger Act of January 28, 1971, which prescribes environmental conditions in industry.

Pesticides Research Program (Food and Drug Administration, Department of Health, Education, and Welfare). Sets limits for pesticides in foods.

Solid Waste Management Programs Office (Environmental Protection Agency). Reports and makes recommendations concerning solid-waste problems.

Walsh-Healey Act. Comprises federal code of occupational noise regulations listing permissible noise exposures; amended January 24, 1970.

Table 8-2.   FEDERAL ENVIRONMENTAL AGENCIES*
Federal Agencies That, by Law or Special Expertise, Comment on Various Types of Environmental Impacts

AIR

Air Quality and Air Pollution Control

Department of Agriculture: Forest Service (effects on vegetation)

Department of Health, Education, and Welfare (health aspects)

Environmental Protection Agency: Air Pollution Control Office

Department of the Interior:
  Bureau of Mines (fossil and gaseous fuel combustion)
  Bureau of Sport Fisheries and Wildlife (wildlife)

Department of Transportation:
  Assistant Secretary for Systems Development and Technology (auto emissions)
  Coast Guard (vessel emissions)
  Federal Aviation Administration (aircraft emissions)

Weather Modification

Department of Commerce: National Oceanic and Atmospheric Administration

Department of Defense: Department of the Air Force

Department of the Interior: Bureau of Reclamation

ENERGY

Environmental Aspects of Electric Energy Generation and Transmission

Atomic Energy Commission (nuclear power)

Environmental Protection Agency:
  Water Quality Office
  Air Pollution Control Office

Department of Agriculture: Rural Electrification Administration (rural areas)

Department of Defense: Army Corps of Engineers (hydro-facilities)

Federal Power Commission (hydro-facilities and transmission lines)

Department of Housing and Urban Development (urban areas)

Department of the Interior (facilities on federal lands)

Natural Gas Energy Development, Transmission, and Generation

Federal Power Commission (natural gas production, transmission, and supply)

Department of the Interior:
  Geological Survey
  Bureau of Mines

HAZARDOUS SUBSTANCES

Toxic Materials

Department of Commerce: National Oceanic and Atmospheric Administration

Department of Health, Education, and Welfare (health aspects)

Environmental Protection Agency

Department of Agriculture:
  Agricultural Research Service
  Consumer and Marketing Service

Department of Defense

Department of the Interior: Bureau of Sport Fisheries and Wildlife

Pesticides

Department of Agriculture:
  Agricultural Research Service (biological controls, food and fiber production)
  Consumer and Marketing Service
  Forest Service

Department of Commerce:
  National Marine Fisheries Service
  National Oceanic and Atmospheric Administration

Environmental Protection Agency: Office of Pesticides

Department of the Interior:
  Bureau of Sport Fisheries and Wildlife (effects on fish and wildlife)
  Bureau of Land Management

Department of Health, Education, and Welfare (health aspects)

Herbicides

Department of Agriculture:
  Agricultural Research Service
  Forest Service

Environmental Protection Agency: Office of Pesticides

Department of Health, Education, and Welfare (health aspects)

Department of the Interior:
  Bureau of Sport Fisheries and Wildlife
  Bureau of Land Management
  Bureau of Reclamation

Transportation and Handling of Hazardous Materials

Department of Commerce:
  Maritime Administration
  National Marine Fisheries Service
  National Oceanic and Atmospheric Administration (impact on marine life)

Department of Defense:
  Armed Services Explosive Safety Board
  Army Corps of Engineers (navigable waterways)

Department of Health, Education, and Welfare: Office of the Surgeon General (health aspects)

Department of Transportation:
  Federal Highway Administration Bureau of Motor Carrier Safety
  Coast Guard
  Federal Railroad Administration
  Federal Aviation Administration
  Assistant Secretary for Systems Development and Technology
  Office of Hazardous Materials
  Office of Pipeline Safety

Environmental Protection Agency (hazardous substances)

Atomic Energy Commission (radioactive substances)

*From: R.E. Train, Federal Actions Affecting the Environment—Guidelines (Appendix II), *Federal Register*, April 27, 1971, pp. 7724–29.

Table 8-2. FEDERAL ENVIRONMENTAL AGENCIES *(Continued)*

LAND USE AND MANAGEMENT

Coastal Areas: wetlands, estuaries, waterfowl refuges, and beaches

Department of Agriculture: Forest Service
Department of Commerce:
   National Marine Fisheries Service (impact on marine life)
   National Oceanic and Atmospheric Administration (impact on marine life)
Department of Transportation: Coast Guard (bridges, navigation)
Department of Defense: Army Corps of Engineers (beaches, dredge and fill permits, Refuse Act permits)
Department of the Interior:
   Bureau of Sport Fisheries and Wildlife
   National Park Service
   U.S. Geological Survey (coastal geology)
   Bureau of Outdoor Recreation (beaches)
Department of Agriculture: Soil Conservation Service (soil stability, hydrology)
Environmental Protection Agency: Water Quality Office

Historic and Archeological Sites

Department of the Interior: National Park Service
Advisory Council on Historic Preservation
Department of Housing and Urban Development (urban areas)

Flood Plains and Watersheds

Department of Agriculture:
   Agricultural Stabilization and Research Service
   Soil Conservation Service
   Forest Service
Department of the Interior:
   Bureau of Outdoor Recreation
   Bureau of Reclamation
   Bureau of Sport Fisheries and Wildlife
   Bureau of Land Measurement
   U.S. Geological Survey
Department of Housing and Urban Development (urban areas)
Department of Defense: Army Corps of Engineers

Mineral Land Reclamation

Appalachian Regional Commission
Department of Agriculture: Forest Service
Department of the Interior:
   Bureau of Mines
   Bureau of Outdoor Recreation
   Bureau of Sport Fisheries and Wildlife
   Bureau of Land Management
   U.S. Geological Survey
Tennessee Valley Authority

Parks, Forests, and Outdoor Recreation

Department of Agriculture:
   Forest Service
   Soil Conservation Service
Department of the Interior:
   Bureau of Land Management
   National Park Service
   Bureau of Outdoor Recreation
   Bureau of Sport Fisheries and Wildlife

Department of Defense: Army Corps of Engineers
Department of Housing and Urban Development (urban areas)

Soil and Plant Life, Sedimentation, Erosion, and Hydrologic Conditions

Department of Agriculture:
   Soil Conservation Service
   Agricultural Research Service
   Forest Service
Department of Defense: Army Corps of Engineers (dredging, aquatic plants)
Department of Commerce: National Oceanic and Atmospheric Administration
Department of the Interior:
   Bureau of Land Management
   Bureau of Sport Fisheries and Wildlife
   Geological Survey
   Bureau of Reclamation

NOISE

Noise Control and Abatement

Department of Health, Education, and Welfare (health aspects)
Department of Commerce: National Bureau of Standards
Department of Transportation:
   Assistant Secretary for Systems Development and Technology
   Federal Aviation Administration (Office of Noise Abatement)
Environmental Protection Agency (Office of Noise)
Department of Housing and Urban Development (urban land use aspects, building materials standards)

PHYSIOLOGICAL HEALTH AND HUMAN WELL-BEING

Chemical Contamination of Food Products

Department of Agriculture: Consumer and Marketing Service
Department of Health, Education, and Welfare (health aspects)
Environmental Protection Agency: Office of Pesticides (economic poisons)

Food Additives and Food Sanitation

Department of Health, Education, and Welfare (health aspects)
Environmental Protection Agency: Office of Pesticides (economic poisons, e.g., pesticide residues)
Department of Agriculture: Consumer Marketing Service (meat and poultry products)

Microbiological Contamination

Department of Health, Education, and Welfare (health aspects)

Radiation and Radiological Health

Department of Commerce: National Bureau of Standards

Table 8-2.   FEDERAL ENVIRONMENTAL AGENCIES *(Continued)*

Atomic Energy Commission
Environmental Protection Agency: Office of Radiation
Department of the Interior: Bureau of Mines (uranium
    mines)

Sanitation and Waste Systems
Department of Health, Education, and Welfare (health
    aspects)
Department of Defense: Army Corps of Engineers
Environmental Protection Agency:
    Solid Waste Office
    Water Quality Office
Department of Transportation: U.S. Coast Guard (ship
    sanitation)
Department of the Interior:
    Bureau of Mines (mineral waste and recycling, mine acid
        wastes, urban solid wastes)
    Bureau of Land Management (solid wastes on public lands)
    Office of Saline Water (demineralization of liquid wastes)

Shellfish Sanitation
Department of Commerce:
    National Marine Fisheries Service
    National Oceanic and Atmospheric Administration
Department of Health, Education, and Welfare (health
    aspects)
Environmental Protection Agency: Office of Water Quality

TRANSPORTATION

Air Quality
Environmental Protection Agency: Air Pollution Control
    Office
Department of Transportation: Federal Aviation
    Administration
Department of the Interior:
    Bureau of Outdoor Recreation
    Bureau of Sport Fisheries and Wildlife
Department of Commerce: National Oceanic and
    Atmospheric Administration (meteorological
    conditions)

Water Quality
Environmental Protection Agency: Office of Water Quality
Department of the Interior: Bureau of Sport Fisheries and
    Wildlife
Department of Commerce: National Oceanic and
    Atmospheric Administration (impact on marine life
    and ocean monitoring)
Department of Defense: Army Corps of Engineers
Department of Transportation: Coast Guard

URBAN

**Congestion in Urban Areas, Housing, and Building
    Displacement**
Department of Transportation: Federal Highway
    Administration
Office of Economic Opportunity
Department of Housing and Urban Development
Department of the Interior: Bureau of Outdoor Recreation

**Environmental Effects with Special Impact in Low-income
    Neighborhoods**
Department of the Interior: National Park Service
Office of Economic Opportunity
Department of Housing and Urban Development (urban
    areas)
Department of Commerce (economic development area):
    Economic Development Administration
Department of Transportation: Urban Mass Transportation
    Administration

Rodent Control
Department of Health, Education, and Welfare (health
    aspects)
Department of Housing and Urban Development (urban
    areas)

Urban Planning
Department of Transportation: Federal Highway
    Administration
Department of Housing and Urban Development
Environmental Protection Agency
Department of the Interior:
    Geological Survey
    Bureau of Outdoor Recreation
Department of Commerce: Economic Development
    Administration

WATER
Water Quality and Water Pollution Control
Department of Agriculture:
    Soil Conservation Service
    Forest Service
Department of the Interior:
    Bureau of Reclamation
    Bureau of Land Management
    Bureau of Sports Fisheries and Wildlife
    Bureau of Outdoor Recreation
    Geological Survey
    Office of Saline Water
Environmental Protection Agency: Water Quality Office
Department of Health, Education, and Welfare (health
    aspects)
Department of Defense:
    Army Corps of Engineers
    Department of the Navy (ship pollution control)
Department of Transportation: Coast Guard (oil spills, ship
    sanitation)
Department of Commerce: National Oceanic and
    Atmospheric Administration

Marine Pollution
Department of Commerce: National Oceanic and
    Atmospheric Administration
Department of Transportation: Coast Guard
Department of Defense:
    Army Corps of Engineers
    Office of Oceanographer of the Navy

Table 8-2.   FEDERAL ENVIRONMENTAL AGENCIES *(Continued)*

River and Canal Regulation and Stream Channelization
Department of Agriculture: Soil Conservation Service
Department of Defense: Army Corps of Engineers
Department of the Interior:
  Bureau of Reclamation
  Geological Survey
  Bureau of Sport Fisheries and Wildlife
Department of Transportation: Coast Guard

WILDLIFE
Environmental Protection Agency
Department of Agriculture:
  Forest Service
  Soil Conservation Service
Department of the Interior:
  Bureau of Sport Fisheries and Wildlife
  Bureau of Land Management
  Bureau of Outdoor Recreation

Table 8-3. MOTOR VEHICLE POLLUTION LIMITATIONS

Actual federal regulation of automobile emissions began with the 1968 model year and applied only to manufacturers' new cars. In 1970 the Department of Health, Education, and Welfare indicated the standards given in the following table.

STANDARDS AND PROJECTIONS BY HEW
Limitations on Vehicle Exhaust

Exhaust	Limitations, in g/mi[a]			
	Prior to controls	1970 standards	Proposed 1975	Goal for 1980
Hydrocarbons	11	2.2	0.5	0.25
Carbon monoxide	80	23	11.0	4.7
Oxides of nitrogen	4		0.9	0.4
Particulates	0.3		0.1	0.03

[a]For data concerning emissions from light-duty vehicles, see Table 8-9.

Confusion and contradiction as to proposed standards have arisen from several causes, as follows:

1. Inadequate testing and sampling procedures
2. Lack of production-line monitoring
3. Conflicting state and local regulations
4. Disregard of feasibility
5. Responsibilities shifted by reorganizations
6. Frequent changes in proposals and laws

The Clean Air Act Amendments of 1970 (PL 91-604, December 31, 1970) gave the Administrator of the National Air Pollution Control Administration rather broad powers and responsibilities, including that of prescribing and revising regulations "in accordance with" its section on motor-vehicle standards. That section specifically required the following:

A. Model year 1975: "a reduction of at least 90 percentum from emissions of *carbon monoxide* and *hydrocarbons* allowable under the standards under this section applicable to light-duty vehicles and engines manufactured in model year 1970."

B. Model year 1976: "a reduction of at least 90 percentum from the average of emissions of *oxides of nitrogen* actually measured from light-duty vehicles manufactured during model year 1971 which are not subject to any federal or state emission standard for oxides of nitrogen. Such average of emissions shall be determined by the Administrator on the basis of measurements made by him."

There has been much argument over these provisions (A and B) of the Clean Air Act, based on methods and programs for sampling and measurement and on technical and economic feasibility. One major problem in future control of motor-vehicle emissions is that lead alkyls in gasoline, and the chlorine and bromine compounds that must accompany them, will influence very seriously the capability to control emissions to meet the proposed 1975–76 standards.

The standards listed in the following table have been proposed by the National Air Pollution Control Administration.

NAPCA RECOMMENDED LIMITS
Based on Proposed 1972 Test Procedure

Exhaust	Standards, g/mi		
	1972	1973	1975
Hydrocarbons	2.9		0.5
Carbon monoxide	37		11.0
Oxides of nitrogen		3	0.9
Particulates			0.1

Table 8-4. SOURCES OF POLLUTION AND METHODS FOR CONTROL*

Table A.   POLLUTION SOURCES

General class	Comments
AIR POLLUTION	
Combustion-engine exhaust	Internal-combustion-powered transportation is the major source of carbon monoxide, oxides of nitrogen, and hydrocarbons.
Chimney gases	Electric central stations are principally responsible for the oxides of sulfur.
Discharged particulates	Industry is the principal source of particulate matter and miscellaneous pollutants.
Natural and agricultural	Wind-eroded soil, pollen spores, bacteria, sea-salt nuclei, volcanic dust, gases.
WATER POLLUTION	
Sewage and oxygen-demanding wastes	Ordinarily these wastes are reduced to stable compounds through the action of aerobic bacteria that require and obtain oxygen from the water. At excessive residue levels the resultant oxygen reduction can have a serious impact on the life in the water.
Infectious agents	Although modern disinfection techniques have greatly reduced the dangers from disease-causing organisms in water, incomplete elimination of these agents from sewage and domestic water supplies poses a continuing health hazard.
Plant nutrients	Problems from algae blooms and excessive plant growth are increasingly troublesome in many of the major lakes, and effective remedial measures are largely lacking.
Organic chemicals	Insecticides, pesticides, and detergents have caused vast kills of fish and wild life.
Other minerals and chemicals	Includes chemical residues, petrochemicals, salts, acids, silts, and sludges.
Land-erosion sediments	These fill channels and reservoirs, necessitate additional treatment of water supplies, and blanket fish nests.
Radioactive substances	Anticipated large increase in nuclear-power reactors poses a serious challenge.
Thermal pollution	Heat from power and industrial plants increases water temperature with serious consequences for fish and other aquatic life.
LAND POLLUTION	
Solid wastes	Annual cost of refuse collection and disposal is already over $3 billion, increasing some 4% per year, and public health is also involved. Crop and livestock residues are another problem.
Soil contamination	Includes radioactive materials and pesticides.

Table B.   BASIC METHODS FOR POLLUTION CONTROL

Recovery and reuse—both of used air, water, and land and of the pollutants.

Waste treatment—including both modification of contaminants in waste and their removal and disposition.

Product modification—by introducing pollution-reducing properties into potentially contaminating materials.

Process change—by modifying contaminant-using processes to prevent or reduce release.

Elimination—by preventing the contaminant from entering the environment. (It should be noted that, if not reused, the contaminant must be disposed of in some form.)

Dispersion—by distributing waste discharge over larger areas or through greater volume.

Dilution—by artificially augmenting the volume of the environment.

Detention—by holding polluted material for gradual or later release.

Diversion—by transporting the waste to another location for discharge.

Environmental treatment—to remove pollutants or diminish their effects.

Desensitization—of pollution receptors. (Screening junkyards is an example in the land environment. Vaccination against waterborne diseases exemplifies desensitization in the water regime.)

*Based on: "Waste Management and Control", Publication 1400, Committee on Pollution, National Academy of Sciences–National Research Council, Washington, D.C., 1966.

Table 8-5. SUMMARY OF AIR POLLUTION SOURCES*

Category	Examples	Pollutants
Chemical plants	Petroleum refineries, pulp mills, superphosphate fertilizer plants, cement mills	Hydrogen sulfide, oxides of sulfur, fluorides, organic vapors, particles, odors
Crop spraying and dusting	Pest and weed control	Organic phosphates, chlorinated hydrocarbons, arsenic, lead
Crushing, grinding, screening	Road-mix plants	Mineral and organic particulates
Demolition	Urban renewal	
Field burning	Stubble and slash burning	Smoke, fly ash, and soot
Frost-damage control	Smudge pots	
Fuel burning	Home heating and power plants	Oxides of sulfur and of nitrogen, carbon monoxide, smoke, and odors
Fuel fabrication	Gaseous diffusion	Fluoride
Inks	Photogravure and printing	
Metallurgical plants	Smelters, steel mills, aluminum refineries	Metal fumes (lead, arsenic, and zinc), fluorides, and oxides of sulfur
Milling	Grain elevators	
Motor vehicles	Autos, buses, and trucks	Oxides of sulfur and of nitrogen, carbon monoxide, smoke, and odors
Nuclear-device testing	Atmospheric explosions	Radioactive fallout (strontium–90, cesium–137, carbon–14)
Nuclear fission	Nuclear reactors	Argon–41
Ore preparation	Crushing, grinding, and screening	Uranium and beryllium dust
Refuse burning	Community and apartment-house incinerators, open-burning dumps	Oxides of sulfur, smoke, fly ash, organic vapors, and odors
Solvent cleaning	Dry cleaning, degreasing	
Spent-fuel processing	Chemical separation	Iodine–131
Spray painting	Automobile assembly, furniture and appliances finishing	Hydrocarbons and other organic vapors
Waste recovery	Metal scrap yards, auto-body burning, rendering plants	Smoke, soot, organic vapors, and odors

*Excerpted by special permission from: CHEMICAL ENGINEERING, October 14, 1968, Copyright © 1968, by McGraw-Hill, Inc., New York, N.Y. 10036.

Table 8-6. AIR POLLUTANT EMISSIONS*
Estimates of Nationwide Emissions During 1969, Millions of Tons

Source	CO	Particulates	SO_x	HC	NO_x
Transportation	111.0	0.7	1.0	19.7	14.1
Fuel combustion in stationary sources	1.8	7.2	24.4	0.9	9.6
Industrial processes	11.5	14.4	7.5	3.5	0.2
Solid waste disposal	7.9	1.4	0.2	2.0	0.4
Miscellaneous	18.2	11.4	0.2	9.2	2.0

*Estimates from Office of Air Programs, Environmental Protection Agency.

Table 8-7. POLLUTANT EMISSIONS FROM METROPOLITAN AREAS*

The following table gives 1967–68 data from the larger metropolitan statistical areas. Almost 60 percent of the urban population of the United States is included.

Standard metropolitan statistical area	Population	Area, mi²	Emissions, 10^3 tons/yr				
			SO_x	Particulate	HC	NO_x	CO
New York–New Jersey[a]	15 420 000	6 930	1 590	231	NA	NA	5 297
Chicago[b]	7 500 000	4 660	1 780	586	NA	NA	2 726
Los Angeles[c]	7 070 000	41 000	168	103	1 268	471	4 997
Philadelphia[d]	5 550 000	4 590	1 168	241	468	406	2 691
San Francisco[e]	4 500 000	7 000	157	77	788	188	2 520
Detroit[f]	4 090 000	2 680	786	241	481	300	1 896
Cleveland[g]	3 030 000	3 500	819	304	NA	NA	1 384
Washington, D.C.	2 720 000	2 270	247	35	310	135	1 259
Boston	2 700 000	1 280	424	82	87	168	921
Pittsburgh	2 520 000	3 050	934	387	95	267	915
St. Louis[h]	2 410 000	4 500	662	176	326	181	1 643
Hartford and New Haven[i]	2 290 000	2 650	337	56	123	134	846
Seattle and Tacoma[j]	2 010 000	15 000	255	33	165	74	907
Houston and Galveston[k]	2 000 000	7 800	144	156	292	213	1 100
Milwaukee[l]	1 730 000	2 630	243	100	83	111	619
Minneapolis–St. Paul[m]	1 660 000	2 830	215	46	NA	NA	960
Cincinnati[n]	1 660 000	2 620	428	123	55	130	537
Buffalo	1 320 000	1 470	410	140	93	130	470
Denver[o]	1 230 000	10 300	31	33	NA	NA	616
Kansas City[p]	1 230 000	3 200	125	60	233	NA	744
Providence[q]	1 200 000	1 000	118	23	54	64	435
Indianapolis	1 050 000	3 080	164	78	74	69	757
Dayton[r]	880 000	2 310	108	94	64	62	367
Louisville[s]	840 000	1 390	303	128	46	44	305
Birmingham[t]	750 000	1 120	34	205	64	26	253
Steubenville[u]	370 000	1 530	638	155	NA	NA	152

NA—not available. SMSA—standard metropolitan statistical area.

[a]Includes New York SMSA less Suffolk County, Jersey City SMSA, Patterson-Clifton-Passaic SMSA, and Somerset, Middlesex, and Monmouth Counties, N.J., and Fairfield County, Conn.
[b]Includes Chicago and Gary-Hammond-East Chicago SMSA's.
[c]Includes Los Angeles-Long Beach, Oxnard-Ventura, Anaheim-Santa Ana-Garden Grove, San Bernardino-Riverside-Ontario, and San Diego SMSA's.
[d]Includes Philadelphia SMSA, Trenton SMSA, and Wilmington SMSA less Cecil County.
[e]Includes San Francisco-Oakland SMSA, San Jose SMSA, Vallejo-Napa SMSA, and Sonoma County.
[f]Plus St. Clair County.
[g]Includes Cleveland, Lorain-Elyria, Akron, and Canton SMSA's.
[h]Plus Monroe County, Illinois.
[i]Plus Hampden County, less Tolland County.
[j]Plus Skagit, Whatcom, Thurston, Mason, Kitsap, Jefferson, and Clallam Counties.
[k]Plus Chambers and Waller Counties.
[l]Includes Milwaukee SMSA, Racine SMSA, Kenosha SMSA, and Walworth County.
[m]Plus Carver and Scott Counties.
[n]Plus Butler County.
[o]Plus Larimer and Weld Counties.
[p]Plus Leavenworth County, Kentucky.
[q]Includes Providence-Pawtucket-Warwick SMSA, Falls River SMSA, New Bedford SMSA, and Newport County.
[r]Plus Darke and Sheby Counties.
[s]Plus Oldham and Bullitt Counties.
[t]Less Shelby and Walker Counties.
[u]Includes Steubenville-Weirton and Wheeling SMSA's.

*Reprinted from: "Air Quality Criteria for Carbon Monoxide", AP–62, National Air Pollution Control Administration, March 1970.

Table 8-8. AIR POLLUTION FROM POWER AND HEATING PLANTS*

Following are approximate average rates of emission from chimneys and stacks when fossil fuel is burned under typical conditions.

Table A. PARTICULATE EMISSION—COAL FUEL

	Pulverized coal	Cyclone burner	Spreader stoker	Other stoker	Hand fired
Total particulates per ton of coal burned, lb	16A†	2A†	13A†	5A†	20
20 microns or larger, %	48	17	79	86	0
Collection efficiencies, %					
(a) Electrostatic precipitator	80–99.9	65–99	—	—	—
(b) Cyclone separator‡	65–75	30–40	85–90	90–95	—
(c) Low-resistance cyclone	40–60	20–30	70–80	75–85	—
(d) Settling chamber**	—	—	20–30	25–50	—

†Multiply the percent ash (A) in coal by this number to obtain the pound particulate per ton. For kg/kg divide by 2 000.
‡High-efficiency type.
**Expanded chimney bases.

Table B. GASEOUS POLLUTANTS

Fuel and use	Aldehydes	Carbon monoxide	Hydro-carbons (CH_4)	Oxides of nitrogen[y]	SO_2
Coal fuel[x]					
Power plant	0.005	0.5	0.2	20.	38S[c]
Industrial	0.005	3.	1.	20.	38S[c]
Commercial and					
domestic	0.005	50.	10.	8.	38S[c]
Oil fuel[x]					
Power plant	0.6	0.04	3.2	104.	157S[c]
Industrial[a]	2.	2.	2.	72.	157S[c]
Industrial[b]	2.	2.	3.	12.	157S[c]
Natural gas fuel[x]					
Power plant	1.	—	—	390.	0.4
Industrial	2.	0.4	—	214.	0.4
Commercial and					
domestic	—	0.4	—	116.	0.4

[a]Residual oil fuel.
[b]Distillate oil fuel.
[c]Multiply the percent sulfur (S) in fuel by this number.
[x]Pounds per ton of coal burned. For kg/kg divide by 2 000.
 Pounds per 1 000 gal oil burned. For kg/m³ multiply by 0.119 83.
 Pounds per million ft³ of gas burned. For kg/m³ multiply by 16.018×10^{-6}.
[y]Nitrogen oxide per kwhr (approximate): coal fuel, 8.6 lb; oil, 7.6 lb; gas, 4.1 lb.

*From: "Steam Power Plant Site Selection", Office of Science and Technology, 1969; Superintendent of Documents, $1.25.

Table 8-9.   EMISSIONS FROM LIGHT-DUTY VEHICLES*

Values apply to vehicles with a gross vehicle weight of 6 000 pounds or less. Vehicles were "on the road," assuming 50% urban and 50% rural driving, and were tested according to federal LDV procedures. Cars used were not over one year old.

For approximate emissions in pounds per car per year, on the road, multiply any tabular value by 30.

Pollutant	Uncontrolled	Federal standards		
	1963 model	1968 model	1970 model	1971 model
	Emissions in grams per vehicle mile			
Exhaust				
Hydrocarbons	8.87	2.98	1.91	1.91
Carbon monoxide	44.47	20.30	13.30	13.30
Oxides of nitrogen (as NO_2)	5.26	6.60	6.60	6.60
Crankcase blowby				
Hydrocarbons	3.15[a]	0.0	0.00	0.00
Evaporation				
Hydrocarbons	2.77	2.77	2.77	0.49
Total				
Hydrocarbons	14.79[a]	5.75	4.67	2.40
Carbon monoxide	44.47	20.30	13.30	13.30
Oxides of nitrogen	5.26	6.60	—	—
Oxides of nitrogen (as NO_2)	—	—	6.60	6.60

[a] Applies only to pre-1963 cars.

*From: Federal Motor Vehicle Certification Data, Public Health Service, Consumer Protection and Environmental Health Service, 1968–69.

Table 8-10.   AIR POLLUTANTS FROM AUTOMOTIVE VEHICLES*

Average On-the-road Emission Rates—All Cars and Trucks

The following table gives the probable average emissions per vehicle-mile, based on pre-1972 systems and expected progress in emissions controls to 1990. It is expected that hydrocarbon and carbon monoxide emission rates will be greatly reduced, but nitrogen oxides will increase somewhat. These values are averages for on-the-road driving.

For rough estimates of total emissions, it may be assumed that the average vehicle travels 9 500 miles per year (1970) and that the total number of vehicle-miles within a given region increases about 19 percent per five-year interval.

ESTIMATES PROJECTED TO 1990

Pollutant	Emissions in grams per vehicle-mile					
	1965	1970	1975	1980	1985	1990
Carbon monoxide						
Urban	87.6	67.9	40.9	27.6	25.7	25.6
Rural	35.6	28.1	17.5	11.9	10.7	10.8
Hydrocarbons						
Urban	20.4	14.6	7.39	4.20	3.71	3.69
Rural	13.2	9.49	4.78	2.65	2.18	2.18
Nitrogen oxides						
Urban	5.76	6.27	6.84	7.08	7.08	7.06
Rural	6.53	7.10	7.81	8.18	8.27	8.31

Note: Actual emission rates from a given vehicle will depend on the mode of operation at that time. Variations of emissions from the exhaust are roughly as follows:

Cruising at freeway speeds: exhaust of hydrocarbons and CO is low, but there is a moderate emission of nitrogen oxides.

Heavy acceleration: exhaust of CO is high, but the emission of hydrocarbons and nitrogen oxides is only moderate.

Idling engine: exhaust of hydrocarbons and CO is high, but the emission of nitrogen oxides is very low.

*From: "Determination of Air Pollutant Emissions from Gasoline-Powered Motor Vehicles", AP–66, National Air Pollution Control Administration, 1970.

Table 8-11. SOURCES OF HYDROCARBON EMISSIONS

1967–68 Data from 22 Metropolitan Statistical Areas

Table A. TOTAL HYDROCARBON EMISSIONS IN EACH AREA*

Location[a]	Population	Area, mi^2	Emissions, 10^3 tons/yr
Los Angeles	7 100 000	41 000	1 270
Philadelphia	5 500 000	4 590	470
San Francisco	4 500 000	7 000	790
Detroit	4 090 000	2 680	480
Washington, D.C.	2 700 000	2 270	310
Boston	2 700 000	1 280	87
Pittsburgh	2 520 000	3 050	95
St. Louis	2 410 000	4 500	330
Hartford	2 290 000	2 650	120
Dallas	2 187 000	8 000	143
Seattle	2 010 000	15 000	170
Houston	2 000 000	7 800	292
Milwaukee	1 730 000	2 630	83
Cincinnati	1 660 000	2 620	55
Buffalo	1 300 000	1 470	130
Kansas City	1 230 000	3 200	230
Providence	1 200 000	1 000	54
Indianapolis	1 050 000	3 080	74
San Antonio	982 000	7 320	71
Dayton	880 000	2 310	64
Louisville	840 000	1 390	46
Birmingham	750 000	1 120	64

[a]Defined on the basis of standard metropolitan statistical areas (SMSA); these may include substantial areas that are rural in nature and thus of low population density.

*From: "Reference Book of Nationwide Emissions", National Air Pollution Control Administration.

Table B. EMISSIONS BY SOURCE CATEGORY*

Source category	Percentage of total area hydrocarbon emissions	
	Average	Range
Transportation	70.2	37–90+
Process losses	19.9	1–63
Refuse disposal	7.1	0.4–26
Fuel combustion	2.8	0–18
Industrial	(2.2)	
Domestic	(0.5)	
Power plants	(0.1)	

*From: "Sources and Air Pollutant Emission Patterns in Major Metropolitan Areas", D.V. Mason et al., paper presented at the 62nd Annual Meeting of the Air Pollution Control Association, New York, June 22–26, 1969.

Table 8-12.   NATIONWIDE HYDROCARBON EMISSIONS, 1968*

Emissions are given in millions of tons/year; for millions of kg/year multiply by 907.

Source	Emissions, 10^6 tons/yr		Percentage of total	
Transportation	16.6		51.9	
Motor vehicles		15.6		48.8
Gasoline		15.2		47.5
Diesel		0.4		1.3
Aircraft	0.3		0.9	
Railroads	0.3		0.9	
Vessels	0.1		0.3	
Non-highway use of motor fuels	0.3		1.0	
Fuel combustion in stationary sources	0.7		2.2	
Coal	0.2		0.6	
Fuel oil	0.1		0.3	
Natural gas	N^a		N^a	
Wood	0.4		1.3	
Industrial processes	4.6		14.4	
Solid waste disposal	1.6		5.0	
Miscellaneous	8.5		26.5	
Forest fires	2.2		6.9	
Structural fires	0.1		0.3	
Coal refuse burning	0.2		0.6	
Agricultural burning	1.7		5.3	
Organic solvent evaporation	3.1		9.7	
Gasoline marketing	1.2		3.7	
Total	32.0		100.0	

[a]N—negligible.

*From: "Nationwide Inventory of Air Pollutant Emission, 1968", U.S. Department of Health, Education, and Welfare (Public Health Service, Environmental Health Service, National Air Pollution Control Administration), 1970.

Table 8-13.   TYPICAL RATES OF EMISSION FROM AIRCRAFT AND DIESELS*

Source	Carbon monoxide	Hydrocarbons (as C)	Oxides of nitrogen (as NO_2)
Aircraft, emissions in pounds per flight[a]			
4-engine conventional jet[b]	35	10	23
4-engine fan jet[b]	20.6	29.0	9.2
2-engine turboprop	2.0	2.0	1.1
4-engine turboprop	9.0	1.2	5.0
2-engine, piston-engine	268.0	50.0	12.6
4-engine, piston-engine	652.0	120.0	30.8
Diesel engines[c]	60	136	222

[a]A flight is defined as a combination of a takeoff and a landing.
[b]No water injection on takeoff. For 3-engine aircraft multiply these data by 0.75; and for 2-engine aircraft multiply these data by 0.5.
[c]Emissions in pounds per 1 000 diesel fuel.

*From: "Compilation of Air Pollutant Emission Factors", R.L. Duprey, AP–42, National Air Pollution Control Administration, 1968.

Table 8-14.   AIR POLLUTION BY SULFUR OXIDES*

For data on pollutant emissions from metropolitan areas, see Table 8-7.

Sulfur-oxide pollution originates mainly from the burning of coals and heavy fuel oils that contain sulfur; the industrial processes of sulfuric acid manufacture, paper making, and smelting of sulfur-rich ores are also sources of sulfur-oxide pollution. High atmospheric concentrations, therefore, are found in the communities where these operations exist. Air-quality regulations usually demand a short-period observed maximum of less than 1 ppm in the atmosphere and a 24-hour average of 0.10 ppm or less. Small amounts of sulfur trioxide, sulfuric acid vapor, and sulfates accompany the sulfur dioxide, and analyses by different methods may not fully agree on the "sulfur oxides as SO_2."

SULFUR DIOXIDE EMISSIONS TO THE ATMOSPHERE
Estimated Total Emissions by Source, 1963 and 1966

Process	1963		1966	
	Tons[a]	Percent of total emissions	Tons[a]	Percent of total emissions
Burning of coal				
Power generation (211 189 000 tons, 1963 data)	9 580 000	41.0	11 925 000	41.6
Other combustion (112 630 000 tons, 1963 data)	4 449 000	19.0	4 700 000	16.6
Subtotal	14 029 000	60.0	16 625 000	58.2
Combustion of petroleum products				
Residual oil	3 703 000	15.9	4 386 000	15.3
Other products	1 114 000	4.8	1 218 000	4.3
Subtotal	4 817 000	20.7	5 604 000	19.6
Refinery operations	1 583 000	6.8	1 583 000	5.5
Smelting of ores	1 735 000	7.4	3 500 000	12.2
Coke processing	462 000	2.0	500 000	1.8
Sulfuric acid manufacture	451 000	1.9	550 000	1.9
Coal refuse banks	183 000	0.8	100 000	0.4
Refuse incineration	100 000	0.4	100 000	0.4
Total emissions	23 360 000	100.0	28 562 000	100.0

Note: Some emission rates of sulfur compounds from various processes are approximately as follows:

Burning of coal: lb SO_2/ton = 38 × percent sulfur by weight.
Burning of fuel oil: lb SO_2/1 000 gal = 159 × percent sulfur by weight.
Diesel engine exhaust: SO_2/1 000 gal = 40, based on 0.3% S in oil.
Sulfuric acid manufacture: 20–70 lb SO_2/ton of 100% acid.
Copper smelting:[b] 1 250 lb SO_2/ton of concentrated ore.
Lead smelting:[b] 660 lb SO_2/ton of concentrated ore.
Zinc smelting:[b] 530 lb SO_2/ton of concentrated ore.
Sulfite paper making—recovery furnace: 40 lb SO_2/ton of air-dried pulp.[c]
Coke drying: 0.25 lb SO_2 plus SO_3/ton of product.

[a]A small amount of this tonnage is converted to sulfuric-acid mist before discharge to the atmosphere. The rest eventually is oxidized and/or washed out. Only under unusual meteorologic conditions does accumulation occur.
[b]These are for primary smelting processes.
[c]Assumes 90% recovery of SO_2.

*Based on data from: "Air Quality Criteria for Sulfur Oxides", AP–50, National Air Pollution Control Administration, 1969; and "Control Techniques for Sulfur Oxide Air Pollutants", AP–52, National Air Pollution Control Administration, 1969.

Table 8-15. AIR POLLUTION BY NITROGEN OXIDES

See also Tables 8-7, 8-9, and 8-14.

The nitrogen oxides become a pollution problem in congested areas, especially where there is heavy motor traffic or large electric power plants. Although the natural nitrogen cycle of organic growth releases perhaps 25 times as much as the man-made emissions, the level of nitrogen oxides in urban areas will reach 50 parts per billion (ppb) as compared with a natural background level of 1 ppb. Nitric oxide (NO) combines with oxygen to form nitrogen dioxide (NO_2) and then progressively forms nitric acid and nitrates in the atmosphere. (See Table 8-22 for concentrations of photochemical oxidants.) Residence time in the atmosphere depends on weather, but it is only a few days.

Man-made nitrogen oxides arise mainly from high-temperature combustion, largely in automotive engines and large power-boiler furnaces. Almost four times as much originates from combustion of petroleum and natural gas as from coal (see Table 8-6).

NITROGEN OXIDE EMISSIONS FROM STATIONARY SOURCES*
Estimates of Total Emissions, 1968

Source	Total emissions, tons/yr
Fuel combustion	
Coal	4 000 000
Fuel oil	1 110 000
Natural gas	4 640 000
Wood	230 000
Solid waste burning	
Open burning	450 000
Incinerators	106 000
Coal waste banks	190 000
Forest burning	1 200 000
Agricultural burning	280 000
Structural fires	23 000
Industrial processes	200 000
Total	12 429 000 tons or 11 300 000 metric tons

Note: In utility plants it is estimated that the rates of emission of nitrogen oxides are approximately as follows:
Coal fuel, 20 lb/ton
Fuel oil, 104 lb/1 000 gal
Natural gas, 390 lb/million ft^3

*From: "Reference Book of Nationwide Emissions", National Air Pollution Control Administration.

REFERENCES

"Control Techniques for Nitrogen Oxide Emissions from Stationary Sources", AP–67, National Air Pollution Control Administration, March 1970.

"Determination of Air Pollutant Emissions from Gasoline-powered Motor Vehicles", AP–66, National Air Pollution Control Administration, 1970.

Table 8-16.   CARBON MONOXIDE EMISSIONS, 1968*

Source	Ton/yr, thousands	Percentage of total	Source	Ton/yr, thousands	Percentage of total
Mobile fuel combustion		63.8	Forest fires		7.2
Motor vehicles			Wild fires	4 740	
Gasoline	59 000		Prescribed burning	2 480	
Diesel	160		Agricultural burning	8 250	8.2
Aircraft[a]	2 400		Industrial processes		9.7
Railroad	120		Blast-furnace sinter plants	2 400	
Vessels	310		Gray-iron cupolas	3 600	
Non-highway users	1 800		Basic oxygen furnaces	100	
Stationary fuel combustion		1.9	Beehive coke ovens	20	
Coal	770		Kraft recovery furnaces, lime kilns	830	
Fuel oil	50		Carbon black	350	
Natural gas	3		Petroleum catalytic cracking units	2 200	
Wood	1 010		Fluid coking burners	160	
Solid-waste disposal		7.8	Methanol	4	
Incineration	800		Formaldehyde	34	
Open burning	3 400				
Conical burners	3 600				
Coal-refuse fires	1 200	1.2	**Total**	100 041	
Structural fires	250	0.2			

Note: Although several other industrial processes are emitters of CO, adequate data are not available.

[a]Includes emissions during cruising

*From: "Reference Book of Nationwide Emissions", National Air Pollution Control Administration.

Table 8-17.   PARTICULATE POLLUTION—SOURCES AND CONTROL*

Industry or process	Source of emissions	Particulate matter	Method of control
Iron and steel mills	Blast furnaces, steel-making furnaces, sintering machines	Iron oxide, dust, smoke	Cyclones, baghouses, electrostatic precipitators, wet collectors
Gray iron foundries	Cupolas, shake-out systems, core making	Iron oxide, dust, smoke, oil, grease, metal fumes	Scrubbers, dry centrifugal collectors
Metallurgical (non-ferrous)	Smelters and furnaces	Smoke, metal fumes, oil, grease	Electrostatic precipitators, fabric filters
Petroleum refineries	Catalyst regenerators, sludge incinerators	Catalyst dust, ash from sludge	High-efficiency cyclones, electrostatic precipitators, scrubbing towers, baghouses
Portland cement	Kilns, dryers, materials-handling systems	Alkali and process dusts	Fabric filters, electrostatic precipitator, mechanical collectors
Kraft paper mills	Chemical-recovery furnaces, smelt tanks, lime kilns	Chemical dusts	Electrostatic precipitators, venturi scrubbers
Acid manufacture—phosphoric, sulfuric	Thermal processes, phosphate rock acidulating, grinding and handling systems	Acid mist, dust	Electrostatic precipitators, mesh mist eliminators
Coke manufacturing	Charging and discharging oven cells, quenching, materials handling	Coal and coke dusts, coal tars	Meticulous design, operation, and maintenance
Glass and glass fiber	Raw-materials handling, glass furnaces, fiberglass forming and curing	Sulfuric acid mist, raw materials dusts, alkaline oxides, resin aerosols	Glass fabric filters, afterburners
Coffee processing	Roasters, spray dryers, waste heat boilers, coolers, conveying equipment	Chaff, oil aerosols, ash from chaff burning, dehydrated coffee dusts	Cyclones, afterburners, fabric filters

*From: "Control Techniques for Particulate Air Pollutants", AP–51, National Air Pollution Control Administration, January 1969.

Table 8-18.   TYPICAL RATES OF PARTICULATE EMISSION*

Emission Factors for Selected Categories of Uncontrolled Sources

Emission source	Emission factor
Natural gas combustion	
Power plants	15 lb/million ft^3 of gas burned
Industrial boilers	18 lb/million ft^3 of gas burned
Domestic and commercial furnaces	19 lb/million ft^3 of gas burned
Distillate oil combustion	
Industrial and commercial furnaces	15 lb/thousand gallons of oil burned
Domestic furnaces	8 lb/thousand gallons of oil burned
Residual oil combustion	
Power plants	10 lb/thousand gallons of oil burned
Industrial and commercial furnaces	23 lb/thousand gallons of oil burned
Coal combustion	
Cyclone furnaces	2X (ash percent) lb/ton of coal burned
Other pulverized-coal-fired furnaces	13–17X (ash percent) lb/ton of coal burned
Spreader stokers	13X (ash percent) lb/ton of coal burned
Other stokers	2–5X (ash percent) lb/ton of coal burned
Incineration	
Municipal incinerator (multiple chamber)	17 lb/ton of refuse burned
Commercial incinerator (multiple chamber)	3 lb/ton of refuse burned
Commercial incinerator (single chamber)	10 lb/ton of refuse burned
Flue-fed incinerator	28 lb/ton of refuse burned
Domestic incinerator (gas-fired)	15 lb/ton of refuse burned
Open burning of municipal refuse	16 lb/ton of refuse burned
Motor vehicles	
Gasoline-powered engines	12 lb/thousand gallons of gasoline burned
Diesel-powered engines	110 lb/thousand gallons of diesel fuel burned
Gray-iron cupola furnaces	17.4 lb/ton of metal charged
Cement manufacturing	38 lb/barrel of cement produced
Kraft pulp mills	
Smelt tank	20 lb/ton of dried pulp produced
Lime kiln	94 lb/ton of dried pulp produced
Recovery furnaces[a]	150 lb/ton of dried pulp produced
Sulfuric acid manufacturing	0.3–7.5 lb acid mist/ton of acid produced
Steel manufacturing	
Open-hearth furnaces	1.5–20 lb/ton of steel produced
Electric-arc furnaces	15 lb/ton of metal charged
Aircraft, 4-engine jet	7.4 lb/flight
Food and agricultural	
Coffee roasting, direct-fired	7.6 lb/ton of green coffee beans
Cotton ginning and incineration of trash	11.7 lb/bale of cotton
Feed and grain mills	6 lb/ton of product
Secondary metal industry	
Aluminum smelting, chlorination-lancing	1 000 lb/ton of chlorine
Brass and bronze smelting, reverberatory furnace	26.3 lb/ton of metal charged

[a]With primary stack gas scrubber.

*From: "Air Quality Criteria for Particulate Matter", AP–49, National Air Pollution Control Administration, January 1969; and "Control Techniques for Particulate Air Pollutants", AP–51, National Air Pollution Control Administration, January 1969.

Table 8-19. AIR POLLUTION IN VARIOUS U.S. CITIES*

Total Suspended Particulates in One Location in Each City

State and city		Micrograms per cu meter			Milligrams 1,000 ft³, average	State and city		Micrograms per cu meter			Milligrams 1,000 ft³, average
		min	max	avg				min	max	avg	
Ala.,	Birmingham[a]	49	505	149	4.2	N.J.,	Atlantic City	31	142	79	2.2
	Huntsville	31	166	80	2.3		Newark	41	246	110	3.1
Ariz.,	Phoenix	93	377	201	5.7	N.Mex.,	Albuquerque	39	421	160	4.5
	Tucson	49	219	118	3.3	N.Y.,	Albany	28	216	86	2.4
Calif.,	Los Angeles	40	251	127	3.6		Buffalo	37	254	115	3.3
	San Bernardino	52	316	173	4.9		New York	75	431	215	6.1
	San Diego	35	282	87	2.5		Rochester	47	219	94	2.7
	San Francisco	29	158	72	2.0		Schenectady	37	275	99	2.8
Colo.,	Denver	86	673	191	5.4		Syracuse	48	269	123	3.5
Conn.,	Hartford	45	185	103	2.9	N.C.,	Durham	31	172	106	3.0
	New Haven	38	167	85	2.4		Raleigh	16	121	58	1.6
Del.,	Wilmington	68	621	168	4.8	Ohio,	Akron	37	288	132	3.7
D.C.,	Washington	63	231	116	3.3		Cincinnati	69	204	121	3.4
Fla.,	Miami	24	154	62	1.8		Cleveland	72	344	156	4.4
	Tampa	45	184	102	2.9		Columbus	37	227	112	3.2
Ga.,	Atlanta	33	186	104	2.9		Dayton	48	198	114	3.2
	Savannah	16	145	80	2.3		Toledo	45	412	114	3.2
Hawaii,	Honolulu	24	88	42	1.2	Okla.,	Tulsa	27	108	65	1.8
Ill.,	Chicago	67	257	144	4.1	Oreg.,	Portland	26	166	87	2.5
	Joliet	35	480	136	3.9	Pa.,	Erie	59	214	100	2.8
Ind.,	Hammond	30	581	186	5.3		Philadelphia	73	308	156	4.4
	Indianapolis	93	367	174	4.9		Pittsburgh	53	402	185	5.2
	South Bend	52	236	126	3.6	R.I.,	Providence	23	422	129	3.7
Iowa,	Cedar Rapids	39	319	137	3.9	S.C.,	Charleston	22	131	69	2.0
	Des Moines	40	351	127	3.9	Tenn.,	Knoxville	52	314	124	3.5
Kans.,	Topeka	32	379	93	2.6		Memphis	59	229	133	3.8
La.,	New Orleans	49	187	96	2.7		Nashville	58	285	133	3.8
Maine,	Portland	34	118	69	2.0	Tex.,	Amarillo	16	710	86	2.4
Md.,	Baltimore	43	292	137	3.9		Dallas	30	324	88	2.5
	Hagerstown	51	135	97	2.8		El Paso	51	294	136	3.8
Mass.,	Boston	42	180	114	3.2		Ft. Worth	41	184	93	2.6
	Worcester	32	161	74	2.1		Houston	37	199	100	2.8
Mich.,	Detroit	52	404	131	3.7		Wichita Falls	50	342	140	4.0
	Flint	26	504	105	3.0	Utah,	Salt Lake City	51	293	123	3.5
	Grand Rapids	44	588	177	5.0	Vt.,	Burlington	18	87	55	1.6
Minn.,	Duluth	16	128	64	1.8	Va.,	Norfolk	47	301	119	3.4
	Minneapolis	24	166	77	2.2		Richmond	53	138	83	2.4
	St. Paul	32	277	117	3.3	Wash.,	Seattle	29	129	63	1.8
Mo.,	Kansas City	64	465	160	4.5		Spokane	34	191	93	2.6
	St. Louis	42	180	118	3.3	W.Va.,	Charleston	42	459	192	5.5
Nebr.,	Omaha	26	326	109	3.1	Wisc.,	Madison	20	230	85	2.4
							Milwaukee	41	297	129	3.7

[a]A pollutant level of over 770 mg/m³ was reached in Birmingham on November 16, 1971.

*From: "Air Pollution Measurements of the National Air Sampling Network: Analysis of Suspended Particulates in 1963", Public Health Service, Robert A. Taft Sanitary Engineering Center, Cincinnati, Ohio.

Figure 8-20. CHARACTERISTICS OF PARTICLES AND PARTICLE DISPERSOIDS**

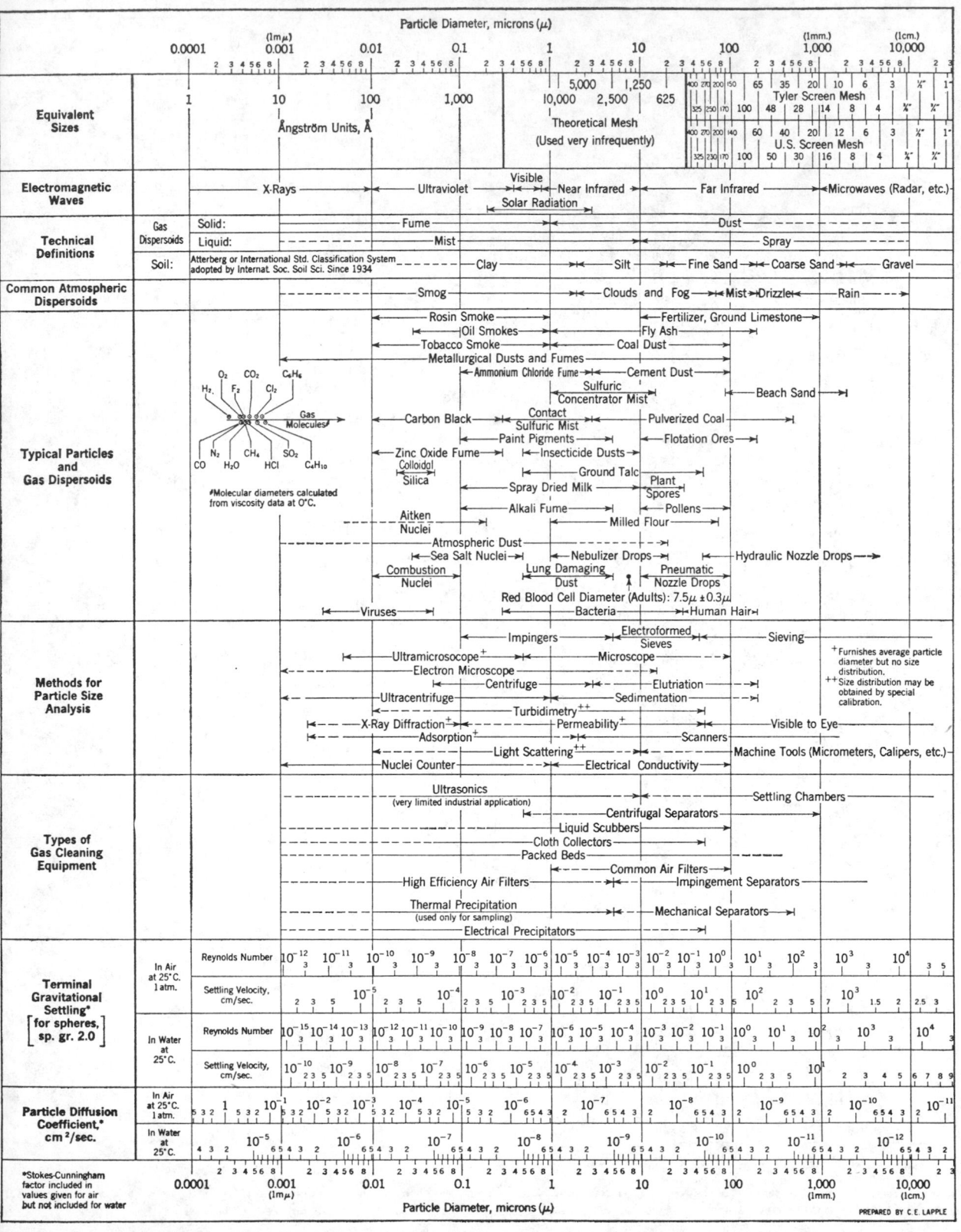

**From: C.E. Lapple, *Stanford Res. Inst. J.*, 5:95, 1961. Reprinted by permission of the author and the publisher.

Table 8-21.   VISUAL RANGE IN A POLLUTED ATMOSPHERE*

Relation Between Equivalent Visual Range and Particle Concentration

Mass concentration, $\mu g/m^3$	Scattering coefficient due to aerosol, b_{scat}/m	Equivalent visual range, km	Equivalent visual range, miles
10	0.3×10^{-4}	120.0	75.00
30	1.0×10^{-4}	40.0	25.00
100	3.3×10^{-4}	12.0	7.50
300	10.0×10^{-4}	4.0	2.50
1 000	33.0×10^{-4}	1.2	0.75

Note: These experimental data are in substantial agreement with other reported measurements; see "Air Quality Criteria for Particulate Matter", AP–49, National Air Pollution Control Administration, January 1969.

*From: R.J. Charlson, Atmospheric Aerosol Research at the University of Washington, *J. Air Pollut. Contr. Assoc.*, 18:652–654, 1968.

Table 8-22.   CONCENTRATIONS OF PHOTOCHEMICAL OXIDANTS*

SOURCE AND MEASUREMENT OF OXIDANTS

Solar radiation initiates a series of reactions between the common constituents of polluted air. Among the products are oxidants that are detrimental to biological systems. The products include ozone, nitric acid, nitrates, aldehydes, and ketones. These oxidants are more active than atmospheric oxygen, and their net effect is measured by exposing some sensitive material to the atmosphere in question. A widely used sensitive material is potassium iodide, which gives a colorimetric indication.

A large number of continuous monitoring stations have been established since 1960. The following table summarizes some of the concentrations recorded in metropolitan areas.

OXIDANT CONCENTRATIONS IN SELECTED CITIES, 1964–67

Station	Days of observation	Percentage of total days with maximum hourly average equal to or greater than 0.15 ppm	0.10 ppm	0.05 ppm	Maximum hourly average, ppm	Peak concentration, ppm
Pasadena	728	41.1	55.1	75.0	0.46	0.67
Los Angeles	730	30.1	48.5	74.0	0.58	0.65
San Diego	623	5.6	20.9	70.6	0.38	0.46
Denver[a]	285	4.9	17.9	79.3	0.25	0.31
St. Louis	582	2.4	10.1	62.2	0.35	0.85
Philadelphia	556	2.3	10.9	41.9	0.21	0.25
Sacramento	711	2.3	14.6	62.3	0.26	0.45
Cincinnati	613	1.6	9.0	52.0	0.26	0.32
Santa Barbara	723	1.5	10.5	70.5	0.25	0.28
Washington, D.C.	577	1.2	11.3	54.2	0.21	0.24
San Francisco	647	0.9	4.5	28.6	0.18	0.22
Chicago	530	0	4.5	50.8	0.13	0.19

Note: A maximum hourly average concentration of ozone in the range 0.15–0.25 ppm has been reported for several U.S. cities.

[a]Eleven months of data beginning February 1965.

*From: "Air Quality Criteria for Photochemical Oxidants", AP–63, National Air Pollution Control Administration, March 1970 (200 pages).

Table 8-23.   CONTROL OF ODOR POLLUTION

In odor control there are two alternative procedures. The much preferred procedure is the removal of odorous constituents at their source or from effluents near the source. Once the odors reach a zone of human occupancy, the methods of control are usually limited to dilution, air washing, or adsorption in a large air-handling system, or masking by release of another odorant within the room.

As it may be very difficult to remove all perceptible traces of a given odorous pollutant, the question of maximum "allowable" concentration arises. For toxic vapors there are established limits, but the presence of almost any distinct odor will give rise to complaints. As far as health hazards are concerned, the threshold limit values (TLV's) listed by the American Conference of Governmental Industrial Hygienists (ACGIH) furnish the most often quoted guide.[a]

A special problem occurs when the dangerous or irritant level is below the odor-perception threshold. A number of such substances are listed in the following table.

ODOR THRESHOLDS AND HAZARD THRESHOLDS
Volume of Pollutant Vapor per Million Volumes of Polluted Air

Chemical compound	Approximate olfactory threshold, ppm	Probable health-hazard threshold, ppm	Chemical compound	Approximate olfactory threshold, ppm	Probable health-hazard threshold, ppm
Acrolein	15.	0.1	Methanol	6 000.	200.
Benzene	60.	25.	Methyl acetate	200.	200.
Camphor	16.	2.	Methyl formate	2 000.	100.
Carbon tetrachloride	200.	10.	Ozone	0.1	0.1
1,2-Dichloroethane	110.	50.	Sulfur dioxide	30.	5.
Dioxane	170.	100.	Trichloroethylene	200.	100.

DILUTION

For occupied rooms the amount of fresh-air ventilation is determined by the odor-control requirement. This may demand as high as 40 cfm per person if smoking is heavy. Actual design figures for most occupied spaces are in the range of 10–30 cfm per person. Control of the carbon-dioxide level requires not more than 5 cfm of fresh air makeup per occupant. In very severe weather the higher rates are costly and may overtax the heating or cooling system. It is not uncommon to supplement the dilution makeup by deodorizing the recirculated air with activated carbon, ultraviolet radiation, or ozone treatment. For an air-dilution system the true index of fresh-air ventilation required for odor control is obtained from a laboratory test for the "number of thresholds" for dilution to the barely perceptible odor level.

Dilution of a smoke or gas plume by high-chimney discharge has been widely studied, and tables and formulas are available for calculations. Since odor intensity at ground level is highly dependent on meteorological conditions, conservative design is necessary to avoid accidents of high contamination from inversions or high winds.

ADSORPTION

Activated carbon is by far the most often used adsorbent for removal of odors from air. At face velocities of about 50 fpm, a $\frac{1}{2}$-inch bed of activated charcoal will remove about 95 percent of odorous pollutants (including cigarette smoke), but deeper filters are often preferred. Effectiveness is not uniform for all contaminants, but it is satisfactory for most pollutants of the air in occupied spaces. Retentivity is low for ammonia, ethylene, hydrogen sulfide, and formaldehyde. For non-industrial spaces the yearly usage is about 1–3 pounds of activated carbon per occupant.

[a]The TLV's of approximately 1 000 chemical compounds are listed in the "Handbook of Laboratory Safety", 2nd ed., N.V. Steere, Ed., The Chemical Rubber Co., 1971. Both allowable concentrations and olfactory thresholds are listed in "Odour Pollution of Air", W. Summer, International Textbook Co., 1971.

Table 8-23.   CONTROL OF ODOR POLLUTION *(Continued)*

CHEMICAL REACTION

Destruction by oxidation is the most common chemical treatment. Potassium permanganate is used in both liquid-spray and solid-bed systems. Although ozone will destroy odors, ozone generators must be used with caution; ozone is toxic above about 0.1 ppm in air. For industrial odor destruction chlorine is used, and various combustion methods may be employed with furnaces or surface combustion. Catalytic combustion, which is especially useful when the stream is rich in combustible pollutants, has been widely used for treating engine-exhaust gases.

Specific chemical reactions are available for destroying certain odorous pollutants. Examples are the use of ammonia for treating streams containing sulfur dioxide and processes using ferric hydroxide or ferric oxide for removing hydrogen sulfide. Several acid-base reactions are common. Catalysts are used in many of the chemical-reaction processes.

Chlorophyll has been credited with deodorizing properties. It assists in oxidizing, and it has an affinity for hydrogen sulfide. Chlorophyll is often combined with a masking odor.

Ultraviolet irradiation destroys odors by generating nascent oxygen through dissociation, which both promotes immediate oxidation of certain pollutants and sterilizes the air. The effective radiation is of shorter wavelength than 0.25 micrometers; to maximize this emission it is necessary to use lamp envelopes of very pure quartz and polished reflectors of very pure aluminum. Other effects of ultraviolet irradiation that contribute to the deodorizing include ozone generation. Ozone concentration in the treated air must be kept well below the toxicity level (0.1 ppm). Hence, an ultraviolet irradiation system must be very carefully designed and its operation fully monitored.

MASKING AND COUNTERACTION

Certain pairs of odors, when smelled together, are counteractive, and the olfactory effects are diminished almost to the point of odor disappearance. Experimenters have reported that typical pairs are butyric acid and oil of juniper, pyridine and oil of wintergreen, iodoform and coffee, camphor and oil of juniper, and certain combinations with benzene, toluene, and xylene.

Masking is a quite different type of odor pairing, in which the offensive odor is hidden by a more pleasant odor, but the intensity is probably undiminished. Esters are used to mask odors of rubbers and plastics; aromatic oils and perfumes are used to "freshen" an occupied space. In some odorous industrial processes the addition of masking agents, such as ethyl acetate, vanillin, or oil of lemon, becomes a part of the process, reducing the offensiveness of the atmosphere. Certain masking agents and other additives tend to reduce olfactory sensitivity; this paralysis is accepted as a reduction in odor intensity.

WASHING, ABSORPTION, AND CONDENSATION

Soluble or emulsifiable vapors are effectively removed by liquid sprays or bubble towers, usually with water or a water solution. As the odor-laden liquid must be renewed or regenerated, its final disposal may be a problem. Fine sprays with turbulent or counterflow air streams, porous, wetted media, or multiple trays will increase the effectiveness of washers and scrubbers. It is important that the water spray be cold, probably 15 deg C or lower, as the solubility of gases in liquids is much less at the higher temperatures. Among the odorous vapors that are effectively removed because of their high solubility in water are acetone, sulfur dioxide, hydrogen sulfide, phenol, and butyric acid. Insoluble or only slightly soluble, the following compounds are not removed by washing: the mercaptans, skatole, indole, furfural, and the ketones.

Air mixed with odorous vapors can be purified by condensing the vapors below their dew point, whether they are soluble or not. Either spray or surface condensers may be used, depending somewhat on the dew-point temperature. The odor intensity from the cold condensate may not be high, but its disposal, nevertheless, poses an odor problem.

Table 8-24. ODOR TECHNOLOGY

ODOR INTENSITY AND MEASUREMENT

Odor Intensity

The following arbitrary, descriptive scale of odor intensity is usually used:

1. Imperceptible	3. Slight; acceptable	5. Strong; intolerable
2. Threshold	4. Moderate; objectionable	6. Dangerous

It is possible to recognize about 25 barely perceptible steps of odor intensity between "threshold" and "strong."

The *number of thresholds* is the ratio of mixture volume to sample volume when a sample of odorous air is diluted with odorless air until the odor of the mixture is barely perceptible.

All odor intensities are presumed to be expressed in terms of the odor perception of the "average" observer. Actually, there are very large, individual differences in odor perception and evaluation. High odor intensities of various substances are associated with high vapor pressure, volatility, and solubility. A few of the odorous substances are only slightly soluble in water, but they are highly soluble in alcohol or ether; among these are benzene, toluene, xylene, camphor, iodoform, skatole, indole, and naphthalene.

Odor Panels

A panel of experts or trained odor observers is most often used in odor testing. Panels of six or more members are necessary to be representative, and it is advantageous to include various viewpoints, such as that of a chemist, physicist, statistician, physiologist, etc.; persons of ages 20 to 50 are preferred, either sex. Olfactory fatigue from occupation or residence in a contaminated atmosphere must be avoided. Colds, medication, heavy smoking, or allergies will affect the competence of the panel members. Training should include special attention to the reporting terminology, and test conditions must be fully outlined. Long breaks in the fresh air are necessary, and alcohol, smoking, and eating of strong foods are not allowable on test days. Test surroundings should be as non-adsorbent as possible, with hard, smooth, clean surfaces; no rugs or draperies are allowed.

Odor Instruments

At least a dozen instruments for determining the "number of thresholds" have been described and used. These differ mainly in the manner of introducing the sample into the nose of the observer.

The Moncrieff thermal instrument depends on the fact that odor adsorption is an exothermal process. Two identical thermistors are used in the arms of a sensitive bridge. One thermistor is chemically clean, while the other is coated with an adsorbing material, preferably a protein. When odorous air is passed over the assembly, the bridge is unbalanced by the heating of the coated thermistor. This measured unbalance increases with increase in the odor intensity of the sample (see reference 2).

For measurement of air pollutants, odorous or otherwise, several instrumental methods have been devised (see reference 3). Tracer techniques, infrared analysis, absorption or adsorption, surface tension, and colorimetry are some of the methods used. Many tests apply to single, specific pollutants, such as sulfur dioxide, ozone, nitrogen oxides, or hydrogen sulfide. Semiconductor, thin-film adsorbers with electronic instrumentation are promising, but there is as yet no real substitute for the human observer when odorous levels are involved.

ODOR-CLASS TERMINOLOGY

Descriptive names	Examples	Descriptive names	Examples
Fragrant; sweet; flowery; fruity	Violet; geranium; rose; orange; mint; pineapple	Spicy	Cloves; anise; cinnamon; nutmeg
Appetizing	Cooking meat; baking bread	Sour; acid	Sour milk; acetic acid
Ethereal	Ethanol; ethyl ether	Burnt	Tar; wood smoke; burnt sugar
Pungent	Ammonia; hydrogen chloride	Close; suffocating	Crowd smell; high-carbon-dioxide-content air
Resinous	Turpentine; eucalyptus	Unpleasant; stinky	Skunk; animal odors
Aromatic	Camphor; benzene; mothballs	Foul; nauseous; disgusting; caprylic; putrid	Hydrogen sulfide; feces; carrion

Chemical Relationships[a]

The only odorous chemical elements are the halogens, phosphorus, arsenic, and ozone. Electronegative and non-metallic elements are odor-producing. Polymerization reduces odor. Strong odor is often accompanied by volatility and chemical reactivity. Some substances change odor on dilution.

Compounds in which an element functions at a valency lower than its maximum usually have offensive odors, such as H_2S. In ring compounds the number of ring members often determines the odor. Esters have fragrant, fruity odors. Ketones generally have pleasant odors. Nitrogen compounds frequently have an "animal" odor. Many sulfur compounds have offensive odors. Compounds of phosphorus, bismuth, and arsenic often have garlicky odors.

---

[a]As reported by Moncrieff, reference 2.

REFERENCES

"ASHRAE Handbook of Fundamentals", American Society of Heating, Refrigerating and Air-Conditioning Engineers, 1967, Chapter 12; and "ASHRAE Guide and Data Book: Systems", American Society of Heating, Refrigerating and Air-Conditioning Engineers, 1970, Chapter 31.

"The Chemical Senses", R.W. Moncrieff, Leonard Hill Books, London, 1967 (The Chemical Rubber Co., U.S. distributor).

"Odour Pollution of Air", W. Summer, International Textbook Co., 1971.

"Handbook of Laboratory Safety", 2nd ed., N.V. Steere, Ed., The Chemical Rubber Co., 1971.

Table 8-25. COMMUNITY STANDARDS FOR AIR QUALITY

Standards are being developed by several federal and state organizations for use by states and communities. Air quality standards relate to health, damage to materials and vegetation, and also to visibility and esthetic values. Both the concentration of contaminants and the duration of exposure must be considered in setting standards.

Contaminants of major concern at this time include carbon monoxide, oxides of sulfur and of nitrogen, ozone and oxidants, acids, fluoride, toxic metals such as lead, and total particulates.

The following tables quote some existing standards and indicate the severe effects at higher concentrations.

Table A. TYPICAL AIR QUALITY STANDARDS

The quantities of pollutant in parts per million given in this table represent legal restrictions. They do not necessarily define the levels above which there will be sensory irritation, damage to vegetation, reduction in visibility, or other similar effects.

CARBON MONOXIDE (CO)
  120 ppm: California—1 hr
  <60 ppm: New York—1 hr
  30 ppm: California—8 hr
  <15 ppm: New York—8 hr
  5.2 ppm: USSR—one time
  0.9 ppm: USSR—24 hr

SULFUR DIOXIDE (SO$_2$)
  1.0 ppm: California; Dade Co., Fla.—1 hr
  0.5 ppm: Colorado—1 hr
  0.3 ppm: California—8 hr
  0.25 ppm: New York, Montana—1 hr
  0.20 ppm: West Germany—$\frac{1}{2}$ hr continuous
  0.19 ppm: USSR—one time
  0.10 ppm: New York, Colorado, Montana—24 hr
  0.10 ppm: Dade Co., Florida—8 hr
  0.06 ppm: USSR—24 hr

NITROGEN DIOXIDE (NO$_2$)
  0.5 ppm: West Germany—$\frac{1}{2}$ hr continuous,
      "nitrous gases"
  0.25 ppm: California—1 hr
  0.19 ppm: USSR—one time

NITROGEN DIOXIDE (NO$_2$) (Cont.)
  0.10 ppm: Colorado—1%, 3 months,
      "nitrogen oxides"
  0.06 ppm: USSR—24 hr

HYDROGEN SULFIDE (H$_2$S)
  0.10 ppm: California—1 hr; West Germany—
      $\frac{1}{2}$ hr continuous
  <0.10 ppm: New York (all sub-regions)
  0.03 ppm: Montana—$\frac{1}{2}$ hr
  0.006 ppm: USSR—one time

TOTAL OXIDANTS
  <0.152 ppm: New York—1 hr
  0.10 ppm: California—1 hr
  0.11 ppm: Colorado—3 months time
  0.05 ppm: New York—24 hr

SUSPENDED PARTICLES
  500 ppm: USSR—one time
  250 ppm: Oregon—heavy industry
  200 ppm: Montana—more than 1% days/yr
  150 ppm: Oregon residential; USSR 24 hr
  120 ppm: Colorado—3 months average

Table B. SEVERE EFFECTS OF INDIVIDUAL POLLUTANTS

CARBON MONOXIDE (CO)
  Discomfort: 900 ppm, 1 hr; 100 ppm, 9 hr
  Severe distress: 100 ppm, 15 hr
  Lethal: 4,000 ppm, less than 1 hr

SULFUR DIOXIDE (SO$_2$)
  Odor and taste: 0.4 ppm, in seconds
  Affects pulmonary function: 1.6 ppm, 10 min
  Severe distress: 5–10 ppm, 10 min
  Plant leaf symptoms: 0.28 ppm, 24 hr

NITROGEN DIOXIDE (NO$_2$)
  Odor: 5 ppm, 5 sec
  Severe distress, small animals: 20–100 ppm
  Lethal to man: 500 ppm
  Plant leaf symptoms: 3 ppm

HYDROGEN SULFIDE (H$_2$S)
  Odor: 0.025 ppm
  Affects central nervous system: 0.1 ppm, 1 hr
  Pathological: 100 ppm
  Lethal: 400–700 ppm

OZONE (O$_3$), THE MAJOR OXIDANT
  Odor: 0.02 ppm, instantaneous
  Discomfort: 0.05 ppm
  Affects pulmonary function: 0.6–0.8 ppm
  Severe distress: 1.5–2.0 ppm

TOTAL OXIDANTS
  Ozone and photochemical smog at concentrations
      of 0.05 to 0.15 ppm will produce eye irritation,
      damage to vegetation, and reduced visibility

Table 8-26. TESTS FOR TOXIC CONTAMINANTS IN AIR

Colorimetric indicators are very widely used for testing possible toxic contamination in industrial atmospheres. Glass indicating tubes containing solid chemicals are largely replacing the liquid reagent methods and those using chemically treated papers.

The following table lists the measuring ranges of indicator tubes from four different manufacturers for detection and measurement of some common contaminants. Although the operational procedure is simple (breaking the two sealed ends of the glass tube and pumping air through it), many precautions are required. Detailed instructions for both testing and interpretation should be ascertained in each case and followed carefully. The storage life of some of the colorimetric tubes is six months or less, but in most cases it is a year or more at room temperature. Some contaminants interfere with the measurement of others in the same atmosphere. Results by either color matching or calibrated length of stain are subject to some variation with the test conditions, the observer, and the sampling.

MEASURING RANGES OF GLASS-TUBE COLORIMETRIC INDICATORS

Material	Threshold limit value[a]	Available colorimeters[b]
Acetone	1,000	0.05–5.0% K; 100–12,000 S
Acetylene	—	50–1,000 K; 3–600 M; 500–3,000 S
Ammonia	50	20–700 K; 10–1,500 M; 25–700 S
Benzene	25–C	10–310 K; 5–200 M; 15–420 S
Carbon monoxide	50	5–100,000 B; 25–6,000 K; 10–3,000 M; 10–300 S
Carbon tetrachloride	10	5–300 K; 10–200 M; 10–100 S
Chlorine	1–C	1–40 K; 0.5–20 M; 0.02–30 S
Dichlorodifluoromethane (Refrigerant 12)	1,000	25–3,000 M
Ethanol	1,000	.04–5.0% K; 200–10,000 M; 100–3,000 S
Ethyl acetate	400	.01–5.0% K; 200–3,000 S
Ethylene	—	.05–100 K; .5–100 M; 0.5–2.0% S
Hydrogen chloride	5–C	2–500 M; 2–30 S
Hydrogen fluoride	3	0.5–5 M; 0.5–7.5 S
Hydrogen sulfide	10	1–650 B; 5–160 K; 1–800 M; 1–600 S
Lead dust and fume	0.2 mg/m^3	0.05–6.3 mg/m^3 M
Mercury	0.1 mg/m^3	0.1–2 mg/m^3 K; 0.5–2 mg/m^3 M; 0.1–2 mg/m^3 S
Methanol	200	0.01–6% K; 100–10,000 M; 100–3,000 S
Methyl chloroform (1,1,1–trichloroethane)	350	50–800 K; 100–700 M; 50–300 S
Nitrogen dioxide	5–C	1–1,000 K; 0.1–50 M; 0.5–10 S
Ozone	0.1	0.05–5 M; 0.05–1.4 S
Phenol	5	0.5–10 S
Phosgene	0.1	0.05–50 K; 0.1–10 M; 0.25–75 S
Sulfur dioxide	5	1–27,000 B; 5–300 K; 1–400 M; 1–20 S
Toluene	200	1–1,000 K; 10–800 M; 5–400 S
Xylene	100	50–5,000 K; 10–800 M; 25–1,860 S

[a]Threshold limit value is the accepted limit of concentration for repeated human exposure. When marked –C, it is the ceiling limit that should never be exceeded. Numerical values are parts per million unless otherwise shown.

[b]Initials indicate the following suppliers:
 B—Bacharach Industrial Instrument Co., Pittsburgh, Pa. 15208
 K—Union Industrial Equipment Corp., Fall River, Mass. 02720 (Kitagawa)
 M—Mine Safety Appliances Co., Pittsburgh, Pa. 15208
 S—Acme Protection Equipment Corp., South Haven, Mich. 49090 (Scott-Draeger)

REFERENCES

"Chemical Detectors", E.E. Campbell and H.E. Miller, Los Alamos Scientific Laboratory, LAMS–2378, Vol I, 1961, Vol. II, 1964.

"CRC Handbook of Analytical Toxicology", I. Sunshine, Ed., The Chemical Rubber Co., 1969.

Table 8-27.   ESTIMATED PRESENT AND FUTURE U.S. WATER REQUIREMENTS*

TOTAL WATER USAGE

The tabulated values are billions of gallons per day, based on 1,100 billion gallons per day, average fresh water supply. For m^3/day multiply tabulated values by 3.7854×10^6.

Use	Year 1954			Year 2000		
	Gross withdrawal	Consumptive use	Return	Gross withdrawal	Consumptive use	Return
Irrigation	176.1	103.9	72.2	184.5	126.3	58.2
Municipal	16.7	2.1	14.6	42.2	5.5	36.7
Manufacturing	31.9	2.8	29.1	229.2	20.8	208.4
Mining	1.5	0.3	1.2	3.4	0.7	2.7
Power-plant cooling	74.1	0.4	73.7	429.4	2.9	426.5
Total	300.3	109.5	190.8	888.7	156.2	732.5

REGIONAL WATER USAGE FOR IRRIGATION

The following water requirements are based on estimations by Resources for the Future, Inc. Tabulated values are in thousands of acre-feet per year. For m^3/year multiply tabulated values by 1,233.5. One acre-ft/yr = 893 gal/day.

Water resource region	Total groundwater and surface-water requirements					Surface-water requirement at estimated future efficiency			Additional surface-water requirement at estimated future efficiency	
	At current efficiency[a]			At estimated future efficiency						
	1957[b]	1980	2000	1980	2000	1957	1980	2000	1980	2000
Eastern										
New England	54	96	297	80	247	49	73	225	24	176
Delaware-Hudson	220	312	583	292	437	110	146	219	36	109
Chesapeake Bay	84	156	1,675	132	1,256	77	121	1,156	44	1,079
Southeast	1,743	3,022	12,196	2,374	8,711	906	1,234	4,530	328	3,624
Eastern Great Lakes	44	66	713	54	546	32	39	399	7	367
Western Great Lakes	67	142	1,330	120	997	58	103	857	45	799
Ohio	84	186	2,178	158	1,633	59	111	1,143	52	1,084
Cumberland	8	11	113	9	85	7	8	78	1	71
Tennessee	44	60	354	57	275	40	52	253	12	213
Upper Mississippi	100	243	6,081	212	4,561	56	119	2,554	63	2,488
Lower Mississippi	1,592	2,295	4,409	2,020	3,351	605	768	1,273	163	668
Lower Missouri	21	45	1,650	39	1,350	18	34	1,161	16	1,143
Lower Arkansas, White, Red	1,560	2,177	3,847	1,880	3,148	437	526	881	89	444
Subtotal	5,621	8,811	35,426	7,427	26,597	2,454	3,334	14,729	880	12,275
Western										
Upper Missouri	17,717	20,747	32,040	16,346	22,331	15,237	14,058	19,205	−1,179[c]	3,968
Upper Arkansas, White, Red	4,748	4,876	6,488	4,035	4,922	1,329	1,130	1,378	−199	49
Western Gulf	11,840	11,789	15,689	9,226	10,914	5,328	4,613	6,003	−715	675
Upper Rio Grande-Pecos	6,576	7,079	8,053	5,522	5,315	2,959	2,761	2,923	−198	−36[c]
Colorado	20,262	20,335	20,708	16,071	14,696	14,386	12,214	11,904	−2,172	−2,482
Great Basin	9,444	9,481	9,742	7,449	6,727	8,216	6,481	5,852	−1,735	−2,364
Pacific Northwest	16,182	16,837	21,053	13,680	15,132	14,887	12,586	13,921	−2,301	−966
Central Pacific	35,924	35,031	35,623	30,223	27,241	17,962	16,623	16,345	−1,339	−1,617
South Pacific	5,044	4,486	4,635	3,671	3,371	2,522	2,019	2,023	−503	−499
Subtotal	127,737	130,661	154,031	106,223	110,649	82,826	72,485	79,554	−10,341[c]	−3,272[c]
United States	133,358	139,472	189,457	113,650	137,246	85,280	75,819	94,283	−9,461[c]	9,003

[a] Assuming no increase in efficiency of application and transmission of irrigation water.
[b] Based on adequate irrigation of all land under irrigation.
[c] Negative values indicate a net decrease in water requirements resulting from estimated increased efficiency in use and transmission of irrigation water.

*From: "Waste Management and Control", Publication 1400, Committee on Pollution, National Academy of Sciences–National Research Council, Washington, D.C., 1966.

Table 8-28. SURFACE WATER CRITERIA FOR PUBLIC WATER SUPPLIES*

Units in Parts Per Million Unless Otherwise Stated

SYMBOLS AND UNITS:

N = narrative discussion of the criteria in the original report

pc/l = picocuries per liter. For disintegrations per $s \cdot m^3$, multiply values in pc/l by 3.7.

1 mg/l = 1 g/m^3 = 0.001 kg/m^3

100 ml = 10^{-4} m^3

Constituent or characteristic	Permissible criteria**	Desirable criteria**	Constituent or characteristic	Permissible criteria**	Desirable criteria**
PHYSICAL			**INORGANIC CHEMICALS** (*Cont.*)		
Color (platinum-cobalt standard)	75 color units	<10 color units	Uranyl ion†	5 mg/l	Absent
Odor	N	Virtually absent	Zinc†	5 mg/l	Virtually absent
Temperature†	N	N	**ORGANIC CHEMICALS**		
Turbidity	N	Virtually absent	Carbon chloroform		
MICROBIOLOGICAL			extract† (CCE)	0.15 mg/l	<0.04 mg/l
Coliform organisms	10,000/100 ml‡	<100/100 ml‡	Cyanide†	0.20 mg/l	Absent
Fecal coliforms	2,000/100 ml‡	<20/100 ml‡	Methylene blue active		
INORGANIC CHEMICALS			substances†	0.5 mg/l	Virtually absent
Alkalinity	N	N	Oil and grease†	Virtually absent	Absent
Ammonia	0.5 mg/l	<0.01 mg/l			
Arsenic†	0.05 mg/l	Absent	*Pesticides:*		
Barium†	1.0 mg/l	Absent	Aldrin†	0.017 mg/l	Absent
Boron†	1.0 mg/l	Absent	Chlordane†	0.003 mg/l	Absent
Cadmium†	0.01 mg/l	Absent	DDT†	0.042 mg/l	Absent
Chloride†	250 mg/l	<25 mg/l	Dieldrin†	0.017 mg/l	Absent
Chromium, hexavalent†	0.05 mg/l	Absent	Endrin†	0.001 mg/l	Absent
Copper†	1.0 mg/l	Virtually absent	Heptachlor†	0.018 mg/l	Absent
Dissolved oxygen	≥4 mg/l	Near saturation	Heptachlor		
	≥3 mg/l		epoxide†	0.018 mg/l	Absent
Fluoride†	N	N	Lindane†	0.056 mg/l	Absent
Hardness†	N	N	Methoxylchlor†	0.035 mg/l	Absent
Iron (filterable)	0.3 mg/l	Virtually absent	Organic phosphates		
Lead†	0.05 mg/l	Absent	plus carbamates†	0.1 mg/l§	Absent
Manganese (filterable)†	0.05 mg/l	Absent	Toxaphene†	0.005 mg/l	Absent
Nitrates plus nitrites†	10 mg/l as N	Virtually absent	*Herbicides:*		
pH (range)	6.0–8.5	N	2, 4-D+2, 4, 5-T		
Phosphorus†	N	N	+2, 4, 5-TP†	0.1 mg/l	Absent
Selenium†	0.01 mg/l	Absent	Phenols†	0.001 mg/l	Absent
Silver†	0.05 mg/l	Absent	**RADIOACTIVITY**		
Sulfate†	250 mg/l	<50 mg/l	Gross beta†	1,000 pc/l	<100 pc/l
Total dissolved solids (filtered residue)†	500 mg/l	<200 mg/l	Radium-226†	3 pc/l	<1 pc/l
			Strontium-90†	10 pc/l	<2 pc/l

†The defined treatment process has little effect on this constituent.

‡Microbiological limits are monthly arithmetic averages based on an adequate number of samples. Total coliform limit may be relaxed if fecal coliform concentration does not exceed the specified limit.

**"Permissible criteria" define raw surface waters that "will allow the production of" a public supply meeting drinking-water standards (see PHS Publication 956). "Desirable criteria" define raw surface waters that "represent high-quality water in all respects for use as public-water supplies".

§As parathion in cholinesterase inhibition. It may be necessary to resort to even lower concentrations for some compounds or mixtures.

*From: "Report of the Committee on Water Quality Criteria", Federal Water-Pollution Control Administration, 1968. This table is accompanied by seven pages of discussion (Sect. II), in which each of the listed constituents is discussed.

REFERENCE

For a listing of water resources and published reports relating to specific local areas, consult "Water Data for Metropolitan Areas", Geological Survey Water Supply Paper 1871, 1968; Superintendent of Documents, $1.50, 400 pages. This summary covers 222 statistical areas in the United States, including all larger cities in each state. It gives the federally supported activities in each area, sources of water supplies, and references to publications containing further data.

Table 8-29. PROCESSES UTILIZED IN POLLUTION CONTROL*
Application Examples in Natural or Engineered Systems

Process or phenomenon	Application examples
Sedimentation	Suspended particles removed from flowing or ponded water; particles and droplets removed from atmosphere; gases or liquids having different temperature, salinity, or sediment load separated gravimetrically; used in clarification of water for municipal and industrial supply; used to separate solid and liquid fractions of domestic sewage for treatment; used to remove fly ash, etc., from industrial gases
Filtration	Particulate matter in atmosphere reduced by trees and other vegetation; surface strata of soil acts as biofilter to remove dissolved and suspended organic solids; standard method of treating domestic and industrial water supplies (sand filter) and of treating organic fraction of domestic and industrial wastes (biological filter)
Coagulation and chemical precipitation	Standard method of clarifying domestic and industrial water supplies and of treating industrial-waste waters; common method of softening domestic and industrial water supplies
Flocculation and coalescence	Particles in flowing and stored water and in air, agglomerated, often leading to precipitation; commonly involved in treatment of domestic and industrial water supplies and in waste-water treatment
Mixing, diffusion, and dilution	Aerosols and gases mixed in atmosphere; dilution by mixing waste streams used in industrial waste-water disposal
Flotation	Floatable solids separated from flowing and ponded water; used in separating solids from liquids in domestic and industrial waste waters
Ion-exchange	Ions interchanged between percolating water and soil or groundwater; widely used in softening and deionizing water for domestic and industrial supplies
Oxidation	Oxidation of gases and particulates in atmosphere; burning of vegetation; stabilization of organic matter; widely used in domestic and industrial refuse incineration
Biodegradation	Normal process of reduction of unstable organic matter in soil, water, and air to inorganic salts, gases, and water; used in treatment of organic solids in domestic and industrial wastes
Adsorption	Widespread use in water treatment for domestic and industrial use; used to remove volatile and nonreactive radionuclides; widespread use in removal of odors from air; widespread use in control of humidity
Evaporation and prevention of evaporation	Salts concentrated in soil or water; limited use in reducing volume of liquid wastes
Sterilization and disinfection	Ultraviolet radiation reduces microorganisms. Ozone reduces microorganisms. Widely used in treating domestic and industrial water supplies and in treating waste waters containing organic matter
Microstraining	Bivalves remove bacteria and other food water. Plant roots reject salts in water. Commonly used in separating fine solids from liquid wastes
Dissolution	Leaching and transport of mineral salts and organics; leaching of solid-waste accumulations; commonly used in treating industrial wastes
Capillarity	Soil-moisture and plant-liquid transfer; limited use for storing liquid wastes in unsaturated soils

*Condensed from: "Waste Management and Control", Publication 1400, Committee on Pollution, National Academy of Sciences–National Research Council, Washington, D.C., 1966.

REFERENCE

For data on removal of particulates from air and gaseous mixtures, see "Adhesion of Dust and Powder", A.D. Zimon, Plenum Press, 1969.

Table 8-30. WASTEWATER TREATMENT PROCESSES*

Classification—Processes	Used for removing	Removal efficiency, %
BIOLOGICAL		
Conventional secondary treatment	⌠ Suspended solids	90
(trickling filter and activated sludge)	⎨ Organic matter	90–BOD†
	⌡ Bacteria, etc.	—
Aerobic process modifications	Conventional treatment plus nutrients	—
Anaerobic dentrification	Nitrate nitrogen	80–95
Algae harvesting	Nitrate nitrogen and phosphorus	50–90
CHEMICAL		
Ammonia stripping	Ammonia nitrogen	80–95
Ion exchange	Nitrates and phosphates	80–92
Electrodialysis	Dissolved salts	10–40
Chemical precipitation	Suspended solids and phosphates	88–95
PHYSICAL		
Activated carbon adsorption	Organic compounds and suspended solids	90–98
Sorption	Phosphates	90–98
Sedimentation	Suspended solids	
Filtration	Suspended solids	50–90
Reverse osmosis	Dissolved salts	65–95
Distillation	Dissolved salts	90–98
Foam separation	Detergents	—
Land application	Suspended and organic and inorganic matter	—

†Biochemical oxygen demand.

*Excerpted by special permission from: CHEMICAL ENGINEERING, October 14, 1968, Copyright © 1968, by McGraw-Hill, Inc., New York, N.Y. 10036.

Table 8-31. EFFECTS OF WATER-TREATMENT PROCESSES*

The relative degree of effectiveness of conventional water-treatment methods is indicated on an arbitrary scale. Increasing relative effectiveness is designated by positive numbers. Adverse effects are shown by negative numbers. Parentheses indicate that the effects are indirect.

Property	Aeration	Coagulation and sedimen-tation (C & S)	Lime-soda softening and sedi-mentation	Slow sand filtration without C & S	Rapid sand filtration preceded by C & S	Disinfection (chlorination)
Bacteria	0	+2	(+3)[a,b]	+4	+4	+4
Color	0	+3	0	+2	+4	0
Turbidity	0	+3	(+2)[b]	+4[c]	+4	0
Odor and taste	+2[d]	(+1)	(+2)[b]	+2	(+2)	+4[e]
						−2[f]
Hardness	+1	(−2)[g]	+4[k]	0	(−2)[g]	0
Corrosiveness	+3[h]	(−2)[j]	0	0	(−2)[j]	0
	−3[i]					
Iron and manganese	+3	+1[m]	(+2)	+4[m]	+4[m]	0

[a] When very high pH values are produced by excess lime treatment.
[b] By inclusion in precipitates.
[c] But filters clog too rapidly at high turbidities.
[d] Not including chlorophenol tastes.
[e] When break-point chlorination is employed or superchlorination is followed by dechlorination.
[f] When (e) is not employed in the presence of intense odors and tastes.
[g] Some coagulants convert carbonates into sulfates.
[h] By removal of carbon dioxide.
[i] By addition of oxygen when it is low.
[j] Some coagulants release carbon dioxide.
[k] Variable; some metals are attacked at high pH values.
[m] After aeration.

*Based on: "Water and Wastewater Engineering", G.M. Fair, J.C. Geyer, and D.A. Okun, Vol. 2, John Wiley & Sons, Inc., 1968.

Table 8-32.   RELATIVE EFFICIENCIES OF SEWAGE TREATMENTS*

Treatment operation or process	5-day, 20 C, BOD[a]	Removal of suspended solids, %	Bacteria	COD[b]
Fine screening	5–10	2–20	10–20	5–10
Chlorination of raw or settled sewage	15–30	—	90–95	—
Plain sedimentation	25–40	40–70	25–75	20–35
Chemical precipitation	50–85	70–90	40–80	40–70
Trickling filtration preceded and followed by plain sedimentation	50–95	50–92	90–95	50–80
Activated-sludge treatment preceded and followed by plain sedimentation	55–95	55–95	90–98	50–80
Stabilization ponds	90–95	85–95	95–98	70–80
Chlorination of biologically treated sewage	—	—	98–99	—

[a] Biological oxygen demand.
[b] Chemical oxygen demand.

*From: "Water and Wastewater Engineering", G.M. Fair, J.C. Geyer, and D.A. Okun, Vol. 2, John Wiley & Sons, Inc. 1968.

Table 8-33.   UNIT QUANTITIES OF SEWAGE FLOW*

Tabular values are given in gallons per day per person. For ft^3 multiply by 0.133; for m^3 multiply by 0.003 8.

Location or establishment	gal/day/person
Small dwellings and cottages with seasonal occupancy	50
Single-family dwellings	75
Multiple-family dwellings (apartments)	60
Rooming houses	40
Boarding houses	50
Additional kitchen wastes for nonresident boarders	10
Hotels without private baths	50
Hotels with private baths (2 persons per room)	60
Restaurants (toilet and kitchen wastes per patron)	7–10
Restaurants (kitchen wastes per meal served)	2½–3
Additional for bars and cocktail lounges	2
Tourist camps or trailer parks with central bathhouse	35
Tourist courts or mobile-home parks with individual bath units	50
Resort camps (night and day) with limited plumbing	50
Luxury camps	100–150
Work or construction camps (semipermanent)	50
Day camps (no meals served)	15
Day schools without cafeterias, gymnasiums, or showers	15
Day schools with cafeterias but without gymnasiums or showers	20
Day schools with cafeterias, gymnasiums, and showers	25
Boarding schools	75–100
Day workers at schools and offices (per shift)	15
Hospitals	150–250+
Institutions other than hospitals	75–125
Factories (gal per person per shift, exclusive of industrial wastes)	15–35
Picnic parks (toilet wastes only, gal per picnicker)	5
Picnic parks with bathhouses, showers, and flush toilets	10
Swimming pools and bathhouses	10
Luxury residences and estates	100–150
Country clubs (per resident member)	100
Country clubs (per nonresident member present)	25
Motels (per bed space)	40
Motels with bath, toilet, and kitchen wastes	50
Drive-in theaters (per car space)	5
Movie theaters (per auditorium seat)	5
Airports (per passenger)	3–5
Self-service laundries (gal per wash, i.e., per customer)	50
Stores (per toilet room)	400
Service stations (per vehicle served)	10

*From: "Manual of Septic Tank Practice", Public Health Service Publication 526, revised 1967.

Table 8-34.
TRACE ELEMENT TOLERANCES FOR IRRIGATION WATERS*
$1\ mg/l = 1\ g/m^3 = 0.001\ kg/m^3$

Element	Water used continuously on all soils, mg/l	Short-term use on fine textured soils only, mg/l	Element	Water used continuously on all soils, mg/l	Short-term use on fine textured soils only, mg/l
Aluminum	1.0	20.0	Lead	5.0	20.0
Arsenic	1.0	10.0	Lithium	5.0	5.0
Beryllium	0.5	1.0	Manganese	2.0	20.0
Boron	0.75	2.0	Molybdenum	0.005	0.05
Cadmium	0.005	0.05	Nickel	0.5	2.0
Chromium	5.0	20.0	Selenium	0.05	0.05
Cobalt	0.2	10.0	Vanadium	10.0	10.0
Copper	0.2	5.0	Zinc	5.0	10.0

*Abstracted from and discussed in: "Report of the Committee on Water Quality Criteria", Federal Water Pollution Control Administration, 1968.

Table 8-35. TOLERANCE OF CROPS TO SALT*

Name of crop	Salinity causing a reduction in yield[a] 10%	50%	Name of crop	Salinity causing a reduction in yield[a] 10%	50%
VEGETABLE CROPS			**FIELD CROPS** *(Cont.)*		
Beets	8.	11.5	Soybean	5.25	9.
Spinach	5.5	8.	Sesbania	3.75	9.
Tomato	4.	8.	Rice (paddy)	5.	8.
Broccoli	4.	8.	Corn	5.	7.
Cabbage	2.5	7.	Broad bean	3.5	6.5
Potato	2.5	6.	Flax	2.75	6.5
Corn	2.5	6.	Beans	1.25	3.
Sweet potato	2.5	6.	**FORAGE CROPS**		
Lettuce	2.	5.	Bermuda grass	13.	18.
Bell pepper	2.	5.	Wheatgrass, tall	11.	18.
Onion	2.	4.	Wheatgrass, crested	6.	18.
Carrot	1.25	4.	Fescue, tall	7.	14.5
Beans	1.25	2.5	Barley hay	8.25	13.5
FIELD CROPS			Perennial rye	8.	13.
Barley	12.	17.5	Harding grass	8.	13.
Sugar beets	10.	16.	Bird's foot trefoil	6.	10.
Cotton	10.	16.	Beardless wild rye	4.	11.
Safflower	6.75	14.	Alfalfa	3.	8.
Wheat	7.	14.	Orchard grass	2.5	8.
Sorghum	5.75	12.	Meadow foxtail	2.	6.5
			Clover, alsike and red	2.	4.

[a]The salinity is expressed as the electrical conductivity value of the saturation extract of the soil in millimho/cm at 25 deg C. For conductivity in 1/ohm·m, divide values in millimho/cm by 10.

*Abstracted from and discussed in: "Report of the Committee on Water Quality Criteria", Federal Water Pollution Control Administration, 1968.

Table 8-36.   POLLUTION BY SOLID WASTES*
TYPICAL U.S. URBAN WASTE QUANTITIES

Location	Description	Year	Waste, lb/day/capita
U.S. average	Solid wastes collected	1968	4.9
U.S. average	Total urban wastes	1968	8.2
Atlanta	All refuse collected	1967	3.2
Cincinnati	All refuse collected	1967	3.8
Hartford	All refuse collected	1967	5.4
Los Angeles	All refuse collected	1965	6.5
Los Angeles	All refuse collected	1967	6.9
New Orleans	Total refuse	1967	5.7
New York	All refuse collected	1965	4.1
Seattle	All refuse collected	1965	4.1
Washington	All refuse collected	1965	4.2
Washington	All refuse collected	1967	4.8

Note: In general, these data apply to waste-collection agencies only; they do not include private disposal or incineration of wastes, sewage-treatment solids, street and catch-basin cleanings, and demolition refuse.

Origins and Quantities of Municipal Wastes

Total discarded waste materials in U.S. cities, as measured and reported in the late 1960's, varied between 5 and 10 pounds per capita per day, depending largely on the size of the city and the types and activities of its businesses and industry. For European cities the reported averages were usually less than one-half of this amount. The total of solid wastes has increased markedly since 1950, and numerous studies have been made of the composition of the wastes and of the methods used for disposal. Activities toward reclamation of valuable materials from solid wastes or toward wider use of heat from combustible wastes have been very slow. Aside from paper and rags, recycling has been practiced in only a few scattered experiments.

In most of the tables in this section, the quantities are given by weight; for transportation and storage, volume also is important, and few comparative data are available. Methods for compacting waste have been confined largely to burning or to segregation and separate handling of bulky materials, such as packaging materials and tree trimmings.

Tabular values of waste per capita are lower than actual total waste, as construction and demolition wastes, landscaping refuse, sewage sludge, scrapped cars, and several kinds of industrial wastes often are omitted. In large cities the combined household and commercial wastes (trash and garbage) are well over one-half the total for the urban area; the actual figures, however, depend somewhat on how the industrial, construction, and demolition wastes are counted and what disposal methods are used.

Table 8-37.   SOLID WASTES FROM PACKAGING MATERIALS*
Classifications and U.S. Consumption—1966

| Material | Millions of lb/yr | | lb/capita/day |
	Approximate division	Total	
Paper products		50 300	0.70
Newsprint	18 000		
Paperboard	16 000		
Other paper	16 300		
Metals[a]		14 300	0.20
Wood		8 100	0.11
Nailed boxes	6 600		
Wirebound and veneer	1 300		
Cooperage	200		
Glass		16 500	0.23
Beverage	7 000		
Food	6 000		
Cosmetic and other	3 500		
Plastic		2 200	0.03
Polyethylene	1 100		
Cellophane	700		
Other	400		
Miscellaneous		12 000	0.17
Total		103 400	1.44

[a] Including aerosol cans.

*Table summarized largely from U.S. Public Health Service Publication No. 1855, 1969.

Table 8-38.   COMPOSITION OF SOLID WASTES

Most reports of the composition of solid waste have been obtained through urban waste-collection agencies, which list the kinds or sources of waste (paper, glass, metals, etc.) with some data on combustible content, reclaimability, and disposal. In most cities at least one-half of the waste is combustible, with an average heating value of about 4 000 Btu/lb (9 300 000 J/kg). Overall moisture content of waste as collected is often about one-third of the total weight.

The composition of collected waste depends somewhat on the policies of the collection agencies concerning the segregation of the several kinds of waste. One general survey is reported in Table A. The "1968 National Survey of Solid Waste Practices", reported for the American Public Works Association, arrived at a higher national average figure of 7.92 lb/capita/day.

Table A.   AVERAGE SOLID WASTE COLLECTED*

Kind of waste	lb/capita/day		
	Urban	Rural	National, average
Household, collected separately	1.26	0.72	1.14
Commercial, collected separately	0.46	0.11	0.38
Combined collections, H and C	2.63	2.60	2.63
Industrial	0.65	0.37	0.59
Demolition, construction	0.23	0.02	0.18
Street and alley	0.11	0.03	0.09
Miscellaneous	0.38	0.08	0.31
Total	5.72	3.93	5.32

*From: "The National Solid Waste Survey", HEW–PHS, Environmental Control Administration, 1968.

Analysis of waste from each of the categories in Table A will vary widely. One composite study published in 1971, but based on 1963–68 data, arrived at a composition of "mixed municipal refuse," which consisted of 55% paper, 14% food wastes, 9% each of metallics and glass, and only 1% plastics. A similar nationwide estimate made in a report to HEW–PHS gave 60% paper, 8.5% food, 8% each of metals and glass (and ceramics), and 3.5% of plastics. Another study of 21 U.S. cities listed only 44% paper, 18% food, 9% each of metals and glass, and 3% of "plastics, rubber and leather." A 1965 study of California waste management listed 70% combustibles (4 000 Btu/lb), including 54% paper and only 1% synthetics. The non-combustibles included 15% garbage, 10% metal, glass, and ceramics, and 10% "ashes and dirt." In some reports the types of paper are listed separately, with "newspapers and magazines" and "corrugated and packing" materials contributing about equally. An interesting item in a 1965 report was "junk mail, 3.03%." Garden refuse has been reported separately in several surveys, at about 10%.

In Europe the collected wastes contain much less paper and less waste metal. In summer the kitchen and other organic wastes are 25–50 percent; some of the northern countries use coal in winter, with a resulting high percentage of ashes.

The ultimate chemical composition of urban refuse has been of minor concern, except when considering problems of effective combustion. Sample data are given in Table B.

Table B.   CHEMICAL COMPOSITION OF MUNICIPAL REFUSE*
U.S. East Coast

Constituent	Percentage[a]	Constituent	Percentage[a]
Moisture	28.0	Nitrogen	0.5
Carbon	25.0	Sulfur	0.1
Oxygen	21.1	Metal, glass, ceramics,	
Hydrogen	3.3	and ash	22.0

[a] As received basis. Note that this gives a moisture-and-ash-free analysis of twice the above percentages for C, O, H, N, and S.

*From: "Refuse Reduction Processes", Surgeon General's Conference on Solid Waste Management for Metropolitan Washington, U.S. Public Health Service Publication No. 1729, 1967.

Two solid wastes not usually included in municipal collections are fly ash from power plants and sewage sludge from treatment plants. Although analysis of fly ash depends on the source of the sample, collected or emitted, it is composed almost entirely of oxides (largely silica, alumina, and iron oxide); as much as 10% of unburned fuel may be entrained. Analysis of raw sewage, converted to the dry basis, is roughly two-thirds organic and one-third mineral. After the sludge is well-digested, these percentages are approximately reversed; even the filter cake is about one-fourth water.

Table 8-39. DISPOSAL OF SOLID WASTES

Over two-thirds of U.S. urban waste disposal is by direct dumping on land; however, only about 8% of total waste goes to sanitary landfills, as compared to almost 60% to open dumps. Most solid waste adds to the air pollution problem by odors or smoke. Typical disposal percentages are shown in the following table.

SOLID WASTE DISPOSAL, 1968[a]

Disposal method	Percent
Open dumps, burned or unburned	59.4
On-site incineration	15.0
Sanitary landfills	7.9
Wigwam burners	7.3
Municipal incineration	5.2
Composting, hog feeding, treatment plants	5.2

[a]Based on: "Nationwide Inventory of Air Pollution Emission", HEW–PHS, National Air Pollution Control Administration, 1970.

The transported wastes usually are collected weekly, and less than one-half of industrial and commercial wastes is collected publicly. About one-half of all waste is incinerated (open or closed fires), and roughly 50 ft^3 of air is required for complete combustion per pound of waste. Since more than 100% excess air will be supplied in all cases, the burning of each ton of refuse will discharge at least 200 000 ft^3 of gaseous mixture to pollute the surrounding atmosphere. From small incinerators in particular, the pollutants include oxides of nitrogen and sulfur, plus acids, aldehydes, and ammonia.

Although there has been some use of heat from the burning of urban waste in furnaces and incinerators, most of these operating plants are in Europe, since fuel is comparatively cheap in the United States. Reported figures show less than 2 lb of steam generated per lb of refuse, and less than 200 kWhr per ton.

A 1966 survey of about 400 landfill sites for waste disposal showed a median expected future life of about seven years. In 190 cities the present sites would be usable for less than six more years.

The number of operating municipal incinerators is comparatively small (estimated about 250 in 1966). One of the problems in operation is the fact that over one-half of the incinerator residue is glass and tin cans; the balance contains iron, ceramics, other metals, stone, and brick. Ash and inorganic fines are usually less than 15 percent of the residue unless the original waste is well segregated.

Table 8-40. OCEAN DUMPING AND FISH KILLS IN U.S. COASTAL WATERS*

Coastal area	Waste quantities, thousands of tons							
	Barge-delivered from cities, annual averages[a]			1968 quantities				
	1964–8	1959–63	1954–8	Dredge spoils	Indus-trial wastes	Sewage sludge	Demo-lition, other	Total
Atlantic	6 220	5 454	3 200	15 808	3 013	4 477	589[b]	23 887
Gulf	520	172	56	15 300	696	—		15 996
Pacific	682	188	170	7 320	981	—	26	8 327
Total	7 422	5 814	3 426	38 428	4 690	4 477	615	48 210

[a]Excludes dredge spoils, radioactive wastes, and military explosives.
[b]Includes 15 000 tons of explosives dumped into the Atlantic.

REPORTED FISH KILLS

The largest single cause of reported fish kills was sewerage systems. Oil spillage, which depends on accidents, killed 4.3 million fish in 1968, but only 40 000 in 1970. Other major sources were insecticides and agricultural poisons, food-industry wastes, manure-silage drainage, mining wastes, and chemical industry wastes.

Fish reported killed, millions per year:

1970—21.9	1968—14.8
1969—40.6	1965—11.4

*Abstracted from: "Ocean Dumping, A National Policy", Executive Office of the President, Council on Environmental Quality, October 1970; and "Pollution-caused Fish Kills", Environmental Protection Agency, Annual Report, 1969.

Table 8-41.   U.S. AGRICULTURAL WASTES FROM LIVESTOCK

A government estimate placed the total annual production of solid wastes by livestock in the United States in 1965 at about 1.14 billion tons.[1] Another statement in the same publication claimed that "the waste production of domestic animals in the United States is equivalent to that of a human population of 1 900 million." It appears that about one-half of the animals are kept in concentrated groups, as in large barns or feedlots.[a] Concurrently, the annual human consumption of meat in this country was roughly 200 pounds per capita (about 50% beef).

The population of animals for food sources (meat, dairy products, and eggs) is largely concentrated in the North Central region, with numbers of animals distributed approximately as shown by the following table.

[a]In 1964 there were over 800 large feedlots with a capacity of 2 000 or more cattle each.

U.S. FARM ANIMAL POPULATION IN MILLIONS*

Animals	Region					Total[a]
	North Atlantic	South Atlantic	North Central	South Central	Western	
Cattle	4.8	7.1	47.7	26.7	19.9	106.2
Hogs	0.8	4.2	40.1	5.0	1.1	51.2
Chickens	188.	1 058.	202.	1 090.	165.	2 703.

[a]A total of over 3 million horses and mules, 26 million sheep, and about 100 million turkeys should be added.

*Data from: "Pollution Implications of Animal Wastes", R.C. Loehr, Federal Water Pollution Control Administration, July 1968.

Within each region the pollution implications with respect to the actual human population have been tabulated. The following table shows the equivalent populations in terms of animal wastes for the total United States for 1960.

ANIMAL WASTES VS. HUMAN WASTES*
Equivalent U.S. Population in Millions

Animal population	Animal population in terms of equivalent people[a]		
	Total dry solids basis	Total nitrogen basis	Biological oxygen demand (BOD) basis
Cattle	1 600	1 040	655
Hogs	74	89	106
Chickens	135	196	196
Total	1 809.	1 325.	957.

[a]In terms of equivalent waste contribution from people, units are in terms of millions of people or equivalent people.

*Data from: "Pollution Implications of Animal Wastes", R.C. Loehr, Federal Water Pollution Control Administration, July 1968.

REFERENCE

"Wastes in Relation to Agriculture and Forestry", C.H. Wadleigh, USDA Miscellaneous Publication No. 1065, U.S. Department of Agriculture, 1968.

Table 8-42. OCCUPATIONAL NOISE REGULATIONS

For information concerning maximum noise levels and hearing damage risk levels, see Tables 8-45 and 8-46, respectively.

Duration of exposure, hours/day	Sound level, db, A	Duration of exposure, hours/day	Sound level, db, A
8	90	1.5	102
6	92	1	105
4	95	0.5	110
3	97	0.25 or less	115
2	100	Impulse or impact	140 (peak)

(WALSH-HEALEY) FEDERAL REGULATIONS AS AMENDED JANUARY 24, 1970
PERMISSIBLE NOISE EXPOSURE[a]

1. Protection against the effects of noise exposure shall be provided when the sound levels exceed those shown in the table, as measured on the A scale of a standard sound-level meter at slow response.[b]

2. When employees are subjected to sound levels exceeding those listed in the table, feasible administrative or engineering controls shall be utilized. If such controls fail to reduce sound levels within the levels of the table, personal protective equipment shall be provided and used to reduce sound levels within the levels of the table.

3. If the variations in noise level involve maxima at intervals of one second or less, it is to be considered continuous.

4. In all cases where the sound levels exceed the values shown herein, a continuing, effective hearing conservation program shall be administered.

[a] When the daily noise exposure is composed of two or more periods of noise exposure of different levels, their combined effect should be considered, rather than the individual effect of each. If the sum of the fractions $C1/T1 + C2/T2 \ldots Cn/Tn$ exceeds unity, then the mixed exposure should be considered to exceed the limit value. Cn indicates the total time of exposure at a specified noise level, and Tn indicates the total time of exposure permitted at that level.

[b] When noise levels are determined by octave-band analysis the equivalent A-weighted sound level may be determined by plotting the individual pressure-level readings on a standard graph of equivalent sound-level contours (see code and standard graph herewith). The A-weighted sound level corresponding to the point of highest penetration into the sound-level contours is used to determine the exposure limits specified in the preceding table.

Table 8-42.   OCCUPATIONAL NOISE REGULATIONS *(Continued)*

GUIDELINES

In applying the provisions of this act, some explanatory assistance is available in Bulletin 334 of the Workplace Standards Administration of the Bureau of Labor Standards, entitled "Guidelines to the Department of Labor's Occupational Noise Standards for Federal Supply Contracts". The following material has been abstracted from Bulletin 334.

The permissible basic intensity of 90 dbA (db—decibel; A—A scale) is considered to be the upper limit of an 8-hour daily dose that will not produce disabling loss of hearing in more than 20 percent of the exposed population.

The following chart permits the readings from an octave-band analyzer to be converted to corresponding dbA values in the above table. The octave-band readings are to be plotted on this standard graph; the A-weighted sound level corresponding to the highest penetration into the sound-level contours is used to determine the exposure limits as specified in the table. Usually, this result is very close to that by direct dbA measurement.

STANDARD GRAPH

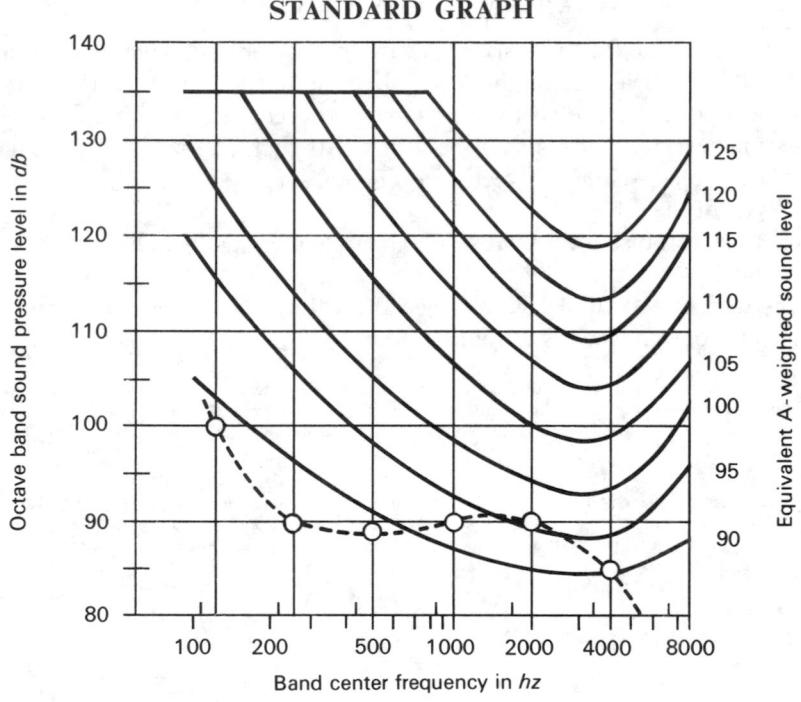

Example: The dotted-line curve shown with test points has been inserted to demonstrate how a rather unusual series of octave-band measurements at or below the 90 dbA curve may contain one point (2 000 hz) where the reading falls on the 95 dbA curve. The sound level for use in the table must then be taken as 95 dbA.

When a sound-level meter set on the A scale, with slow response, fluctuates upward from a generally steady reading at intervals of one second or less, the high reading shall be used for application in the table (see No. 3 in previous discussion).

The Walsh-Healey regulations speak of the use of "feasible engineering or administrative controls" (see No. 2). Bulletin 334 lists such engineering controls as maintenance, substitution of less noisy machines or processes, vibration dampening, reduction of sound transmission, and isolation of the noise source or of the operator. Administrative controls include changes in work schedules, division of high-noise work between two or more operators, and consideration of the noise factor in production planning.

Bulletin 334 discusses ear protection and a hearing conservation program (see No. 4), with audiometric tests at regular intervals and the keeping of records on each employee. Annual noise surveys are recommended, and an 8-step compliance plan is outlined.

Table 8-43. TYPICAL FREEWAY NOISE LEVELS*

The following noise levels are based on vehicle speed and traffic density. The noise levels are given in decibels on the A scale, dbA. They apply to passenger cars only for level road at grade.

Vehicle speed, mph	Vehicle spacing, ft/vehicle	Traffic density		Approximate "average" noise level	
		vehicle/mi	vehicle/hr	100 ft	1 000 ft
70	25	211	14 800	80	70
70	50	106	7 400	77	67
70	100	53	3 700	74	64
70	200	26	1 850	71	61
65	25	211	13 700	78	68
65	50	106	6 850	75	65
65	100	53	3 425	72	62
65	200	26	1 710	69	59
60	25	211	12 700	77	67
60	50	106	6 350	74	64
60	100	53	3 175	71	61
60	200	26	1 580	68	58
50	25	211	10 550	74	64
50	50	106	5 275	71	61
50	100	53	2 640	68	58
50	200	26	1 320	65	55
40	25	211	8 450	71	61
40	50	106	4 225	68	58
40	100	53	2 110	65	55
40	200	26	1 055	62	52

Note: This table is based on the empirical equation:

$$\text{Noise, dbA} = 10 \log_{10} \frac{\text{vehicle/hr} \times 100}{\text{observer distance, ft}} + 20 \log_{10} \text{mph}$$

For 10% trucks, add 4 db; for 20% trucks, add 8 db.

Distances are from the middle of a single-lane roadway. At a 1 000-ft distance the correction for a 2- or 3-lane road is negligible.

For a divided highway with wide median strip, the two noise values should be determined separately and added logarithmically. (The sum never exceeds the higher noise level by more than 3 db.)

Higher noise levels are produced by large trucks, motorcycles, acceleration to pass, poor mufflers or tire treads, and rough pavement. At a 100-ft distance an elevated or a depressed roadway is about 8 db less noisy.

The major vehicle categories are passenger cars and large trucks, with the latter 10–15 db more noisy (at 50-ft distance). Deviations within each group are usually less than 10 db except for extreme deviant violators.

The two major noise sources are tires and exhaust. For new passenger cars at uniform speed on a smooth road, these two are roughly equal. Research efforts to reduce noise from each should be continued, but the greatest engineering possibilities for noise-nuisance control are in design, location, and proximity of the highways.

Noise spectrum measurements indicate the highest levels of vehicle noise in the frequency range 63–250 cps. For new passenger cars in good condition, this maximum level is in the 70–80 db range, measured at 25-ft distance, depending on operating conditions and road surface. The corresponding levels for trucks and sports cars are in the 80–90 db range and for motorcycles up to 100 db (this 100-db value could easily be reduced).

Noise reduction by closed residential buildings is in the range 19–26 db, depending on frequency (lesser reduction at lower frequency).

Noise level as measured on the A-scale of a noise meter correlates well with human reactions to highway noise and is simpler than the loudness methods.

Human reactions to highway noise are highly variable with individuals and classes, type of activity in the location, the time of day, and general attitudes toward freeways and trucks. In general, average highway noise of less than 70 db is rated as acceptable by residents living near a freeway. Convenience for access to recreation was listed by both men and women as the most important advantage of the freeway.

Objective noise-test limits for discriminating the flagrant violators from the normal population are recommended as 77 db for passenger cars and 88 db for trucks, large buses, and motorcycles. The standard test procedure involves full-throttle operation from 30 mph, on level road, and 50-ft measurement distance.

*Data in this table are based on "Highway Noise—Measurement, Simulation, and Mixed Reactions", Report 78, National Cooperative Highway Research Program, 1969 (also see References). The research was sponsored by the American Association of State Highway Officials in cooperation with the U.S. Bureau of Public Roads.

REFERENCES

Publications of National Cooperative Highway Research Program, available from Highway Research Board, National Academy of Sciences, 2101 Constitution Ave., Washington, D.C. 20418.

Report 78: Highway Noise—Measurement, Simulation, and Mixed Reactions.
Report 49: National Survey of Transportation Attitudes and Behavior.
Report 24: Urban Travel Patterns for Airports, Shopping Centers, and Industrial Plants.
"Traffic Noise", R.J. Stephenson and G.H. Vulkan, *J. Sound Vib.*, 7(2):247, 1968.
"Preliminary Investigation of Vehicular Noise Associated with Super Highways", W.E. Clark, V.A. Clark, and W.J. Galloway, *Abstr. J. Acoust. Soc. Amer.*, 29, 1957.
"Prediction of Noise from Motor Vehicles in Freely Flowing Traffic", W.J. Galloway and W.E. Clark, Paper L28, in *Proc. Fourth Int. Congr. Acoust.*, Copenhagen, August 1962.
"Experimental Study of the Airborne Noise Generated by Passenger Automobile Tires", F.M. Wiener, *Noise Control*, 6(4):13, 1960.
"Measurement and Evaluation of Exhaust Noise of Over-the-Road Trucks", D.B. Callaway, *SAE Trans.*, 62:151, 1954.

Table 8-44. MAXIMUM SOUND LEVELS—OUTDOOR MACHINERY*

The following table gives maximum sound levels for engine-powered equipment. The sound levels are expressed as maximum db at 50 ft, "A" weighting network, fast response. Sound level readings are taken as follows:

1. The ambient noise level must be at least 10 db below the minimum test reading.
2. The test area must be open, with no reflecting surfaces within 50 ft of the equipment.
3. The microphone is mounted 4 ft above ground level and 50 ft from the center of the equipment.
4. The specified sound levels are the highest obtained on the loudest side of the equipment.

Max sound level, db	Type of equipment
90	Mobile construction and industrial equipment, powered by internal combustion engines: bulldozers, loaders, power shovels, and cranes
90	Engine-powered equipment of 20 hp or less, intended for *infrequent commercial* use in residential areas
72	Engine-powered equipment of 20 hp or less, intended for frequent use in residential areas: lawn mowers, garden tools, snow removal equipment

Notes:
SAE Recommended Practice J919 specifies methods for measuring maximum sound levels at the operator station, but no maximum limits are specified. Octave-band analysis is included in these tests.

Sound level tests for passenger cars and light trucks are specified in SAE Standard J986a; for heavy trucks and buses, in J672a; and for truck cab interiors, in J336.

*From: "Maximum Sound Levels for Engine-Powered Equipment", SAE Recommended Practice J952, Society of Automotive Engineers.

Table 8-45. MAXIMUM NOISE LIMITS

Three separate criteria are commonly recognized in evaluating noise limits in occupied spaces, viz., speech interference, discomfort, and hearing damage. In normal room conversation speech intelligibility is reduced when the background noise is above 55 db. Comfort depends to a considerable extent on whether or not the occupant expects quiet; hence the maximum background noise for comfort might be 40 db in a private office or a residence and as high as 70 db in a factory workroom. Prolonged exposure to noises above 85 to 90 db may cause deafness.

LIMITS FOR DEAFNESS-AVOIDANCE AND COMFORT*

Maximum Permissible Sound Pressure Level, db

Octave band	Occasional exposure, 1 hr or less	Repeated exposure, months	Noisy (noise expected)	Quiet (quiet expected)
38–75	125	115	100	80
75–150	120	110	95	70
150–300	120	110	90	60
300–600	120	105	85	55
600–1,200	115	100	75	50
1,200–2,400	110	95	65	50
2,400–4,800	105	90	60	50
4,800–9,600	110	95	55	45

*From: U.S. Navy Bureau of Ships Report No. 371–N–12.

Table 8-46.   HEARING DAMAGE RISK LEVELS*

The noise level ratings given here apply only to exposure during a regular working day for a number of years and to steady noises. They do not apply to impact or impulsive sounds, such as gunfire, for which the information regarding hearing damage is not yet adequate to provide risk criteria.

These risk levels have been set so that, on the average, persons will not suffer serious hearing losses if the levels for the exposure durations are less than those indicated. Because some susceptible persons may suffer significant losses, however, it is wise to reduce the noise level or provide ear protection if the levels approach those shown in the table.

SOUND PRESSURE LEVELS IN db (re 20 μN/m^2)

Duration per day	Octave band center, hz						
	125	250	500	1000	2000	4000	7000
8 hr	96	92	88	86	85	85	86
4 hr	103	96	91	88	86	85	87
2 hr	110	101	94	91	88	87	90
1 hr	118	107	99	95	91	90	95
30 min	126	114	105	100	95	93	99
15 min	135	122	112	106	99	98	104
7 min	135	135	122	114	105	104	111
3 min	135	135	134	124	113	111	120
<1.5 min	135	135	135	134	124	121	130

Notes:
If the level in any one octave band exceeds that in the two adjacent bands by more than 5 db, the indicated damage-risk level for that band should be reduced by 5 db. This rule is used to account for noises that have sharp peaks in the spectra.

Those concerned with the problem of noise-induced hearing loss should request the latest information on this subject from the Research Center, Subcommittee on Noise of the Committee on Conservation of Hearing of the American Academy of Ophthalmology and Otolaryngology, 3819 Maple Ave., Dallas, Texas 75219.

*From: "Handbook of Noise Measurement", 6th ed., A.P.G. Peterson and E.E. Gross, Jr., General Radio Company, 1967, p. 68.

REFERENCE
"Hazardous Exposure to Intermittent and Steady-State Noise", K.D. Kryter, W.D. Ward, J.D. Miller, and D.H. Eldredge, *J. Acous. Soc. Amer.*, 39(3):451–464, March 1966: a report for the Committee on Hearing, Bioacoustics, and Biomechanics of the National Academy of Sciences and the National Research Council. This report also has risk criteria for multiple exposures per day and for pure tones and should be consulted for these when necessary.

Table 8-47.   PROGRESSIVE LOSS OF HEARING WITH AGE (PRESBYCUSIS)*

Hearing loss in four frequency bands is expressed in decibels below the normal acuity of young men and women.

Age group, years	Decibels below normal acuity			
	440–512 cps	880–1024 cps	1760–2048 cps	3520–4096 cps
30–40	3.65	4.05	5.5	13.85
40–50	7.55	8.65	11.65	26.55
50–60	12.4	14.8	21.7	38.45

Note: Averaged data from three other sources give somewhat lesser values for hearing loss.

*From: National Health Survey; monaural listening, both ears tested, scores averaged; one thousand or more subjects in each age group.

Figure 8-48. TYPICAL NOISE CONTROL METHODS AND RESULTS*

Frequencies in hertz or cps

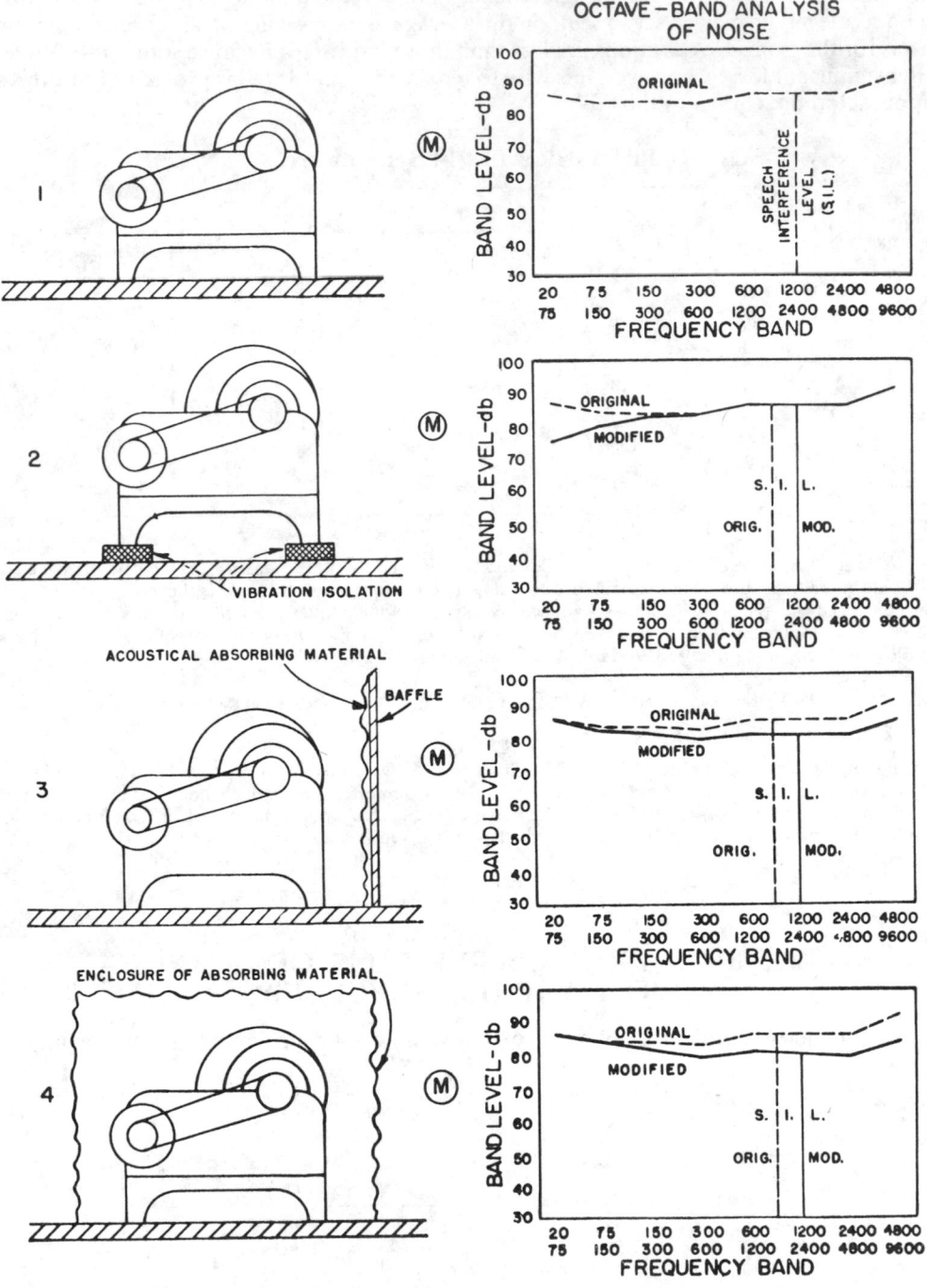

Figures 1 through 4. Examples to illustrate the possible noise reduction effects of some noise control measures.

Figure 8-48. TYPICAL NOISE CONTROL METHODS
AND RESULTS (Continued)

Frequencies in hertz or cps

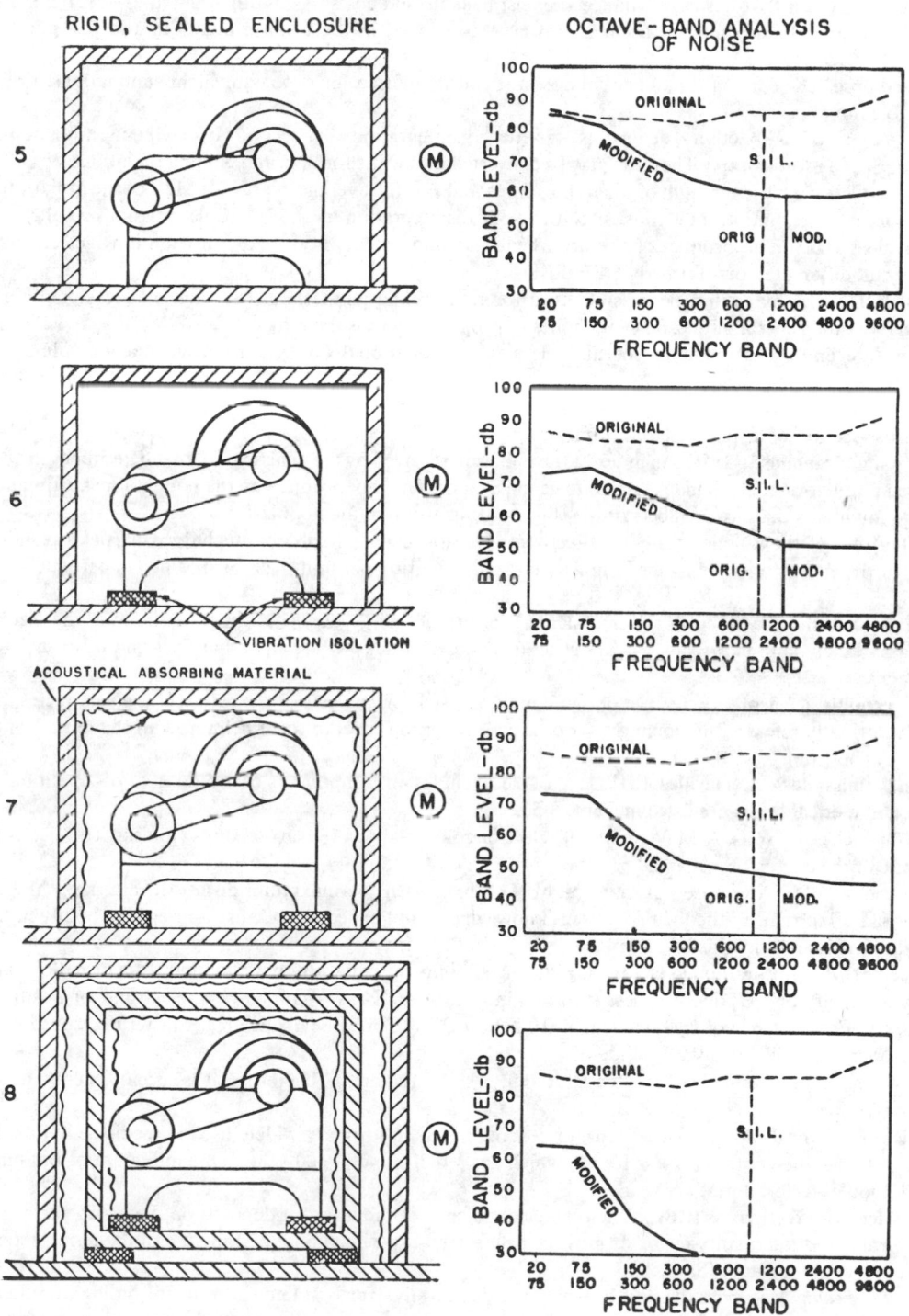

Figures 5 through 8. Examples to illustrate the noise reduction possible by the use of enclosures.

*From: "Handbook of Noise Measurement", 6th ed., A.P.G. Peterson and E.E. Gross, Jr., General Radio Company, 1967, pp. 167–168.

Table 8-49. RADIATION PROTECTION TERMINOLOGY

For additional terms used in nuclear technology, see Table 4-2.

Absorbed dose (AD). The energy imparted by directly or indirectly ionizing radiation per unit mass of irradiated material, expressed in rads.

Acceptable risk. A risk that has been balanced against benefits and costs. As in other activities, it is determined by personal decision. It also may be applied, however, to groups in the total population, and this goes beyond individual evaluation.

Accumulated dose. The summation of annual doses. It is usually used for dose equivalent summations, to define a permissible limit.

Curie (Ci). A unit of radioactivity, defined as 3.7×10^{10} disintegrations per second. It may be used for designating the strength of a radioisotope. The curie may be converted into watts for beta and gamma radiation: $W = 0.005\,92\ CiE$, where Ci is the curie-strength of the radioisotope and E is the average energy, in MeV of the radiation. (Note that the decay of a single atom is considered to be one disintegration, even if multiple emissions result.)

Dose equivalent (DE). The product of the absorbed dose and the quality factor, distribution factor, and other necessary modifying factors (formerly RBE dose).

Dose limit (DL). The maximum permissible dose for non-occupational exposure, such as occasionally exposed individuals or the general public. Since it applies to people, its unit is the rem.

Emergency dose limit. The maximum permissible single occupational dose, either accidental or voluntary (life-saving, etc.).

Fluence. The time integral of flux density or fluence rate. The latter is defined as the number of particles that enter a unit cross-section of a sphere per unit time.

Genetically significant dose (GSD). An index of the radiation received by the genetic pool that determines the progeny of a given population, i.e., an index related to genetic injury. The genetic injury to the population is estimated from data on individuals that are members of the child-bearing group in the population.

Kerma. The total of the kinetic energy of directly ionizing particles released by uncharged particles reacting with matter, or the energy released by the indirectly ionizing radiation per unit mass of specified material, e.g., "tissue kerma."

Linear energy transfer (or stopping power) (LET). The rate of energy deposition measured along the track of an ionizing particle. Gamma rays and X-rays generate low LET tracks, while alpha particles and neutrons give high LET tracks.

Maximum credible accident. A reference design condition used for analysis of the safety of a given nuclear reactor plant. It normally refers to the joining of two heavy-hydrogen nuclei to form helium, with the release of a large amount of energy.

Maximum permissible dose equivalent (MPD; MPDE). Includes all components of radiation of occupational origin. Specified numerical limits are listed in Table 8-50.

MeV—Million electron volts. A MeV equals 10^6 electronvolts or about 16 watt-seconds (joules). The electronvolt is a measure of the energy of a charged particle.

Permissible dose (PD). Any dose less than the MPD. Most often the maximum dose is meant, i.e., "the dose of ionizing radiation that, in the light of present knowledge, is not expected to cause appreciable bodily injury to a person at any time during his lifetime."

Quality factor (QF). A multiplying factor for reducing all ionizing radiation to a common scale of biological effect. If the flux density or LET is 3.5 or less (keV/μm in water) for X-rays, gamma rays, electrons, or positrons, the QF is unity. For neutrons of energy less than 10 keV, a QF of 3 may be assigned; for faster neutrons or protons a QF of 10 usually will be on the safe side.

Rad (r). This is a unit of absorbed dose, equal to 0.01 J/kg or 100 erg/g. Therefore, it is specific energy, and it also applies to kerma.

Radium (Ra). A natural radioactive element (discovered in 1898), formerly widely used in medicine. Of its thirteen known isotopes, the common one is Ra^{226}, which has a half-life of 1 600 years. One milligram of radium yields about 4.2 joules (one calorie) per year.

Relative biological effectiveness (RBE). The biological potency of radiation; it depends on the tissue, the total dose, the dose rate, and the quality factor. It must be determined experimentally, and the use of the term preferably is restricted to experimental radiobiology.

Rem. *Roentgen equivalent man*, the unit in which the dose equivalent is stated, in terms of biological effect; it is, therefore, the unit in which all dose-limiting rules and recommendations are expressed. (In many cases rems = rads.)

Rep. An almost obsolete unit, slightly smaller than the rad.

Roentgen (R). An early unit applied to X-rays or gamma radiation in air. It is defined as the emission of ions carrying one electrostatic unit per 0.001 293 g of air, but it is equal to 88 ergs/g of air (or 0.88 rad).

Tolerance dose (TD). The use of this term is discouraged, because a dose that can be "tolerated" is nevertheless injurious.

Table 8-50.   RECOMMENDED RADIATION DOSE LIMITS*

1971 Specifications of the National Council on Radiation Protection and Measurements (NCRP)

Condition	Rate
Maximum permissible dose equivalent for occupational exposure	
Combined whole-body occupational exposure	
Prospective annual limit	5 rem in any one year[a]
Retrospective annual limit	10–15 rem in any one year[b]
Long-term accumulation to age N years	$(N-18) \times 5$ rem
Skin	15 rem in any one year
Hands	75 rem in any one year (25/qtr)
Forearms	30 rem in any one year (10/qtr)
Other organs, tissues, and organ systems	15 rem in any one year (5/qtr)
Fertile women (with respect to fetus)	0.5 rem in gestation period[c]
Dose limits for the public, or occasionally exposed individuals	
Individual or occasional	0.5 rem in any one year[d]
Students	0.1 rem in any one year
Population dose limits	
Genetic	0.17 rem average per year
Somatic	0.17 rem average per year[e]
Emergency dose limits—life-saving	
Individual (older than 45 years if possible)	100 rem
Hands and forearms	200 rem, additional (300 rem, total)
Emergency dose limits—less urgent	
Individual	25 rem[f]
Hands and forearms	100 rem, total
Family of radioactive patients	
Individual (under age 45)	0.5 rem in any one year
Individual (over age 45)	5 rem in any one year

[a]The critical organs are considered to be the gonads, the lens of the eye, and red bone marrow.

[b]Measured or estimated retrospective doses, well distributed over time.

[c]Fertile women should be employed only in situations where the annual dose limit is unlikely to exceed 2 or 3 rem. During the entire gestation period the maximum permissible dose equivalent to the fetus from occupational exposure of the expectant mother should not exceed 0.5 rem.

[d]The limit of 0.5 rem/year applies to the integration of contributions from all sources, excluding natural and medical radiation.

[e]The idealized objective is to have public exposure, in addition to that from natural radiation, as close to zero as is reasonably possible.

[f]This applies under less stressful circumstances where it is still desirable to enter a hazardous area to protect facilities, to eliminate further escape of effluents, or to control fires.

*Data from: "Basic Radiation Protection Criteria", NCRP Report No. 39, National Council on Radiation Protection and Measurements, January 1971.

Table 8-51. RADIATION PROTECTION STANDARDS

IMPORTANT DATES IN THE DEVELOPMENT OF RADIATION PROTECTION STANDARDS*

1915	British Roentgen Society proposals for radiation protection
1921	British adopt radiation protection recommendations
1922	American Roentgen Ray Society adopts radiation protection rules
1928	Unit of X-ray intensity proposed by Second International Congress of Radiology
1928	International Committee on X-ray and Radium Protection established
1928	First international recommendations on radiation protection adopted by Second International Congress of Radiology
1929	Advisory Committee on X-ray and Radium Protection established (United States of America)
1931	The roentgen adopted as a unit of X-radiation
1934	Tolerance dose of 0.1 roentgen/day recommended by Advisory Committee on X-ray and Radium Protection (March)
1934	Tolerance dose on 0.2 roentgen/day recommended by International Committee on X-ray and Radium Protection (July)
1941	Advisory Committee on X-ray and Radium Protection recommends 0.1 μCi permissible body burden for radium
1946	Advisory Committee on X-ray and Radium Protection reorganized as the National Committee on Radiation Protection
1949	National Committee on Radiation Protection lowers basic maximum permissible dose (MPD) for radiation workers to 0.3 rem/week. Risk-benefit philosophy introduced
1950	International Commission on Radiological Protection and International Commission on Radiological Units reorganized from pre-war committees
1950	International Commission on Radiological Protection adopts basic MPD of 0.3 rem/week for radiation workers
1953	International Commission on Radiological Units introduces concept of *absorbed dose*
1956	National Academy of Sciences and International Commission on Radiological Protection recommend lower basic permissible dose for radiation workers of 5 rem/year
1957	National Committee on Radiation Protection and Measurements introduces age proration concept for occupational exposure and 0.5 rem/year for individuals in population
1959	International Commission on Radiological Protection recommends limitation of genetically significant dose to population of 5 rem in 30 years
1964	Federal Radiation Council introduces concept of protective action guides
1971	National Council on Radiation Protection and Measurements recommends same value of 15 rem/year for all non-critical organs

*From: "Radiation Protection Standards", L.S. Taylor, The Chemical Rubber Co., 1971; see this source for an extensive bibliography.

RADIATION PROTECTION PUBLICATIONS

Since so many agencies have been involved in the development of U.S. radiation protection standards, the following list is offered to indicate some of the contributions. The role of the NCRP is discussed first, and the participating agencies are then listed alphabetically.

NCRP **National Council on Radiation Protection and Measurements.** Formed as a committee in 1946, succeeding the ACRP (*q.v.*), and was chartered by Congress in 1964 as a Council and a cooperating organization with ICRP, FRC, and others. By 1971 there were 28 organizations collaborating with the Council, and it had issued over 40 reports and major publications on radiation protection. The Council functions through 34 scientific committees; its published results and recommendations are widely accepted and form the basis for most of the rules and standards on radiation protection now current in the United States. The groups collaborating with NCRP include most of the concerned divisions of the U.S. government, as well as the scientific societies and trade associations that have some direct concern with radiation protection.

ABCC **Atomic Bomb Casualty Commission.** Initiated by AEC, through NAS, the ABCC combined the efforts of U.S. and Japanese scientists in a continuing study of atomic-bomb survivors.

ACRP **Advisory Committee on X-ray and Radium Protection.** Formed in 1929, it was a small group with representation from AMA, NBS, and manufacturers. It issued a major report in 1931 and issued several more reports throughout the 1930's. In 1946 ACRP became the National Committee on Radiation Protection.

AEC **Atomic Energy Commission.** A post-war development that took over the protection activities of the Manhattan District, the AEC has been very active in the field and has published reports and also regulations governing the operations under its jurisdiction.

Table 8-51.   RADIATION PROTECTION STANDARDS *(Continued)*

ARRS   American Roentgen Ray Society. Published a set of recommendations for radiation protection in 1920. Active committee work and issuance of reports on the subject have continued since.

BMRC   British Medical Research Council. An active group that published an independent report in 1956, "The Hazards to Man of Nuclear and Allied Radiations" (HMS–Cmd 9780, 1956).

BRS   British Roentgen Society. Adopted a pioneering resolution in 1915 that steps be taken toward the adoption of stringent rules for the personal safety of operators conducting roentgen-ray examinations.

FRC   Federal Radiation Council. Established by executive order of the President in 1959 to coordinate the work of federal agencies and advise the President. It has worked in close consultation with the NAS and the HCRP. In 1960 the FRC began publishing standards and reports (available from Superintendent of Documents).

IAEA   International Atomic Energy Agency. A large organization within the family of the United Nations. It has issued hundreds of papers and reports, largely on nuclear science and its peaceful uses, but also on safety and on legal aspects.

ICRP   International Committee on X-ray and Radium Protection. Organized in 1928 and did major work in the ten years following. Its recommendations covered working conditions, shielding, and permissible dose levels (0.2 roentgen/day) for radiation workers. Beginning in 1950, there have been at least fifteen reports on later ICRP activities (published largely by Pergamon Press, London).

ICRU   International Commission on Radiation Units and Measurements. A large international committee, formed in the 1920's.

JCAE   Joint Committee on Atomic Energy. A U.S. Congress committee. It has held several hearings dealing with fallout, waste disposal, and radiation safety and standards; at least 25 documents on these activities are available (from the Superintendent of Documents).

NACOR   National Advisory Committee on Radiation. Under the U.S. Public Health Service, NACOR was organized in 1958, largely as an outgrowth of the ICAE hearings. It functioned effectively in the public-health field and issued reports up to the late 1960's.

NAS–NRC   National Academy of Sciences–National Research Council. Government-chartered organizations. On invitation they have taken an active part in radiation-protection research, especially since 1956, and have published a dozen or more reports.

NBS   National Bureau of Standards. Participated in much of the radiation-protection activity; over the years of 1931–64 it has published more than 30 NBS handbooks in the field. These publications matched closely the recommendations promulgated by the NCRP.

TPC   Tri-partite Conferences. A series of three meetings (1949, 1950, 1953) with representation from the United States, Canada, and England. The proceedings were held classified.

UNSCEAR   United Nations Scientific Committee on the Effects of Atomic Radiation. Especially active through 1958–1966, when several reports were published by the United Nations. These dealt largely with fallout and its possible biological effects and with the effects of nuclear explosions.

Miscellaneous Organizations. Other organizations have functioned effectively in radiation protection, including several in the medical, biological, and standardization fields. Many similar activities are carried on in other nations and within states and provinces. Of particular interest are the codes and standards of the ASA and its successors (the USASI and the ANSI), the ISO (International Standards Organization), and the IEC (International Electrotechnical Commission), and the ASTM. (See latest ANSI and ASTM lists of U.S. standards.)

Table 8-52.   EXISTING SOURCES OF RADIATION HAZARDS

Exposures and Dose Rates

For other data concerning radiation dosages, see Tables 8-50, 8-55, and 8-59.

HIGH LET RADIATION

All dose limits are expressed in terms of low LET (linear energy transfer) radiation, such as X-rays and gamma rays. The relative biological effectiveness of high-LET radiation, such as alpha rays and fast neutrons, may be much higher, and appropriate quality factors (QF) must be applied. The QF is as high as 11 for 0.5–1.0 MeV neutrons and may reach 20 for alpha particles or fission fragments.

NATURAL RADIATION SOURCES (COSMIC-RAY AND GROUND GAMMA)[a]

Sea level, cosmic rays, latitude of Florida: 30 mrem/yr

Sea level, mid-latitude: 33 mrem/yr

Sea level, latitude of Alaska: 35 mrem/yr

One-mile elevation, mid-latitude: 70 mrem/yr

Average, for U.S. citizen, cosmic rays: 40 mrem/yr

Gamma rays from ground, outdoors:

At Dallas, Texas, approx 30 mrem/yr

At Denver, Colorado, approx 130 mrem/yr

Net total natural gamma radiation for U.S. citizen, cosmic plus ground: 60–200 mrem/yr. In the future, if many aircraft fly at 60 000 ft or more, the increased dose from natural radiation might become significant, especially for the crews.

MAN-MADE RADIATION SOURCES

Medical. Total medical irradiation (as a genetically significant dose) is probably between 50 and 70 mrem per year.

Occupational. Since large exposures occasionally are encountered by some radiation workers, military personnel, and radiology technicians, special protection and isolation to limited-access areas are required. The time pattern of all intermittent exposures is important, because the body has a great capacity for gradual recovery from widely spaced exposures. In any massive-dose exposure special protection for the eyes and the gonads is important, and the whole body should be shielded as much as possible (see data on shielding).

Fallout. Where fallout is high, the integrated dose over a 70-year life span has been estimated (in 1962) at 215 mrem per capita; the average increment of dose from fallout is estimated at 4% of natural radiation.

Atomic power plants. This dosage is estimated as only a small fraction of that due to fallout.

Miscellaneous sources. As excessive dosage can readily occur to a specific individual, as in occupational or military assignments, protection and controls are highly important. Some television receivers, self-luminous signs, markers and dials, certain industrial equipment, and fluoroscopes should be subjected to meaningful controls. In some cases areas of restricted access must be designated, with a complete clothing change at the boundary.

---

[a]In addition to the above, the presence of radionuclides in the air, in drinking water, and in foods, plus the contributions of uranium and thorium and their decay products and of radiopotassium and radio carbon, will add to the total natural dose. It is unlikely that for any individual in the United States the annual dose is less than 100 or more than 400 mrem. The maximum (400) is not regarded currently as sufficient to require separate consideration in the control of total dosage from all possible natural sources, as no validated deleterious effect has been shown from a 400-mrem annual dose.

Table 8-53.   LIMITS FOR HUMAN EXPOSURE
TO IONIZING RADIATION

Regulatory Limits and Other Guides for Protecting Against Hazards of Radioactivity

For other data concerning radiation dose limits, see Tables 8-50 and 8-54.

The recommended permissible genetic dose to the general population to age 30 (one generation) is given in the following table.

Table A.   MAXIMUM DOSAGE LIMITS—GENERAL

Medical sources	4.5 rem/generation
Background sources	3.0 rem/generation
All other sources	5.0 rem/generation
Total, all sources	12.5 rem/generation

It has been suggested that all consumer items ever distributed should not expose the public to an average genetic dose of more than 0.3 rem/generation.

Since air and water intake may represent continuous dosage, the permissible concentrations of radionuclides in these media become important. Extensive tables for these limits are given in "Code of Federal Regulations: Standards for Protection Against Radiation", USAEC (also see References below).

Regulatory limits. Certain legal boundaries have been established, applying particularly to employers who operate with hazardous materials. International recommendations are designated in the publications of the International Commission on Radiological Protection. In applying legal regulations, records of previous exposure become necessary. Uncertainties as to cumulative exposure call for a reduction in the usual limits allowed, and the "Code of Federal Regulations" (USAEC) specifies limits as given in the following table.

Table B.   CFR LIMITS, REM PER CALENDAR QUARTER

Whole body; head and trunk; gonads; lenses of eyes; active blood-forming organs	1.25
Skin of whole body	7.5
Feet and ankles; hands and forearms	18.75

Regulations are usually related to a license to use radioactive materials, from the AEC or individual states. Information is obtainable from: Director of Regulations, U.S. Atomic Energy Commission, Washington, D.C. 20545.

Emergencies. Unexpected situations or "planned-special-exposure" operations call for modifications of usual recommended practices. When such exposures exceed 12 rem, a major emergency is indicated.

REFERENCES

"CRC Handbook of Radioactive Nuclides", Y. Wang, Ed., The Chemical Rubber Co., 1969.

"Principles of Radiation Protection—A Textbook of Health Physics", K.Z. Morgan and J.E. Turner, Eds., John Wiley & Sons, 1967.

Table 8-54. RADIATION LIMITS—VARIOUS SOURCES

Table A. MAXIMUM PERMISSIBLE NEUTRON FLUX

The following table gives the maximum permissible neutron flux to deliver 5 rems in a 2,000-hour work year. While biological effectiveness and properties of the tissues have been taken into account, the allowable dosage is seen to involve the neutron radiation spectrum of the actual total flux to which each individual is exposed. Hence the permissible total flux must be a composite estimate.

Neutron energy, Mev	Flux,* $\dfrac{neutrons}{sq\ cm\ sec}$	Neutron energy, Mev	Flux,* $\dfrac{neutrons}{sq\ cm\ sec}$
Thermal	670.	1.0	18.
0.0001	500.	2.5	20.
0.005	570.	5.0	18.
0.02	280.	7.5	17.
0.1	80.	10.	17.
0.5	30.	>10.	10.

*Data from: "NBS Handbook 75", National Bureau of Standards, U.S. Government Printing Office, 1961.

Table B. AIR FORCE LIMITS FOR PLUTONIUM

"The Handbook of Peacetime Nuclear Accidents" (Strategic Air Command) gives the following limits for land contamination from plutonium nuclear-weapons accidents or criticality bursts. Some authorities state that these limits are much too high.

Condition	Limit in $\mu g/m^2$
Releasable for controlled area for continuous residence	0–100
Safe for continuous occupancy	1,000
Safe for approximately 8 days	4,000
Safe for 3 to 4 days	5,000

Table C. EMERGENCY LEVELS FOR FOOD AND WATER†

Under emergency conditions (and not recommended for peacetime use) are the following upper-limit levels for food and water that could be ingested.

Recommended as	Duration of ingestion	μCi per cm^3	Disintegrations per min per cu m
Preferable	10 days	0.0035	7,700
Acceptable	10 days	0.09	200,000
Preferable	30 days	0.0011	2,600
Acceptable	30 days	0.03	70,000

†From: United States Atomic Energy Commission TID–8206, rev. 1960.

Table 8-55.   RADIATION DOSIMETRY AND BIOLOGICAL RESPONSE*

SYMBOLS:

T = total body   r = absorbed dose in rad (or rep)

L = local   One rad = 0.10 J/kg

Dose, r	Dose rate	Exposure	Biological response
0.3	Weekly	T	Probably none (permissible dose)
1	Daily (for years)	T	Leukopenia
1.5	Weekly	L	Probably none (permissible dose for hands and fingers)
25	Single dose	L	Chromosome break in tumor cells (tissue culture)
50–100	In accumulated small dose	L	Gene mutations to double spontaneous rate per generation
60	Single dose	L	Depression of phosphate activity
200	Single dose	T	Nausea
300	Single dose	L	Erythema dose for 100 kV (small field)
300–500	Single dose	T	LD_{50} for man
300–600	Single dose	L (ovaries)	Sterilization in female
400	Single dose	L	Reversible epilation
400–500	10–50 r/day	T	Clinical recovery
500	Single dose	L	Erythema dose for 200 kV (small field)
600–800	Single dose	L (testes)	Sterilization in male
600 900	300 r/day or small doses	L	Radiation cataract
1 000	Single dose	L	Erythema dose for radium
1 000–1 500	200–300 r/day	L	Epiphyseal retardation
1 000–2 500	200–300 r/day	L	Response of markedly radiosensitive cancer
1 500–2 000	200–300 r/day	L	Cessation of salivary glandular functions
1 800–2 000	200–300 r/day	L (stomach)	Achlorhydria
2 000	Single dose	L	Erythema dose for 2 MeV
2 000–3 000	200–300 r/day	L (kidney)	Radiation nephritis
2 500–6 000	200–300 r/day	L	Response of moderately radiosensitive cancer
2 700–3 000	Single dose	L	Moist desquamation, but healing of skin; 100-kV radiation (small field)
3 600–5 000	200–300 r/day	L	Limits of skin (single portal, 200 kV, 5 × 5-cm field)
4 000–5 000	200–300 r/day	L	Limits of nervous tissue
5 000–6 000	200–300 r/day	L	Limits of gastrointestinal tract
5 000–7 000	200–300 r/day	L	Moist desquamation, but healing of skin (single portal, 2 000 to 3 000-kV radiation, 10 × 10-cm field)
~ 50 000	10–100 r/day	L	Carcinogenic

*Based on: "Radiation Dosimetry", G.J. Hine and G.L. Brownell, Eds., Academic Press, 1956, Table II, Chapt. 3, Section 3-V-D.

Table 8-56. RADIATION FROM COMMON SOURCES*

Source	Rate
Cosmic rays (at sea level)	30 mr/year
K^{40} in blood	30 mr/year
Gammas from natural radioactivity of earth	30–100 mr/year
Gammas from a radium watch dial (locally, at wrist)	100 mr/day
Chest X-ray (with film)	50–200 mr/exposure
Dental X-ray (with film)	4–5 r/film
X-ray examination (fluoroscopic)	15–25 r/min
Radiation therapy (local, to destroy tissue)	500–10 000 r
Foot X-ray machines, shoe stores	1 r/min
Normal lifetime dose, without special exposures, for a person in the "prereactor era"	~ 10–40 r/lifetime
Dose expected to kill about one-half of the persons exposed to it, if received in a short time[a]	400–500 rem
Natural inland water radium content	0.4–4 $\mu\mu$c/liter
Maximum permissible concentration of radium in drinking water	40 $\mu\mu$c/liter
Spring water at Shimane, Japan, contains radium to the extent of	~ 1 μc/liter
Deposits near natural springs are as high as	1–3 mc/g
Luminous watch dial contains	~ 5 μc radium
The average human body contains natural C^{14} totaling	~ 0.01 μc
The average human body contains natural K^{40} totaling	~ 0.5 μc

[a] Known as "semilethal dose", SLD, or LD_{50}.

*Data from: "Reactor Shielding Design Manual", T. Rockwell III, USAEC Report No. TID-7004, U.S. Atomic Energy Commission, March 1956.

Table 8-57. EXPECTED EFFECTS OF ACUTE WHOLE-BODY RADIATION DOSES*

Acute dose, rems[†]	Probable effect
0–50	No obvious effect, except possibly minor blood changes
80–120	Vomiting and nausea for about 1 day in 5 to 10 percent of exposed personnel. Fatigue but no serious disability
130–170	Vomiting and nausea for about 1 day, followed by other symptoms of radiation sickness in about 25 percent of personnel. No deaths anticipated
180–220	Vomiting and nausea for about 1 day, followed by other symptoms of radiation sickness in about 50 percent of personnel. No deaths anticipated
270–330	Vomiting and nausea in nearly all personnel on first day, followed by other symptoms of radiation sickness. About 20 percent deaths within 2 to 6 weeks after exposure; survivors convalescent for about 3 months
400–500	Vomiting and nausea in all personnel on first day, followed by other symptoms of radiation sickness. About 50 percent deaths within 1 month; survivors convalescent for about 6 months
550–750	Vomiting and nausea in all personnel within 4 hr from exposure, followed by other symptoms of radiation sickness. Up to 100 percent deaths; few survivors convalescent for about 6 months
1,000	Vomiting and nausea in all personnel within 1 to 2 hr. Probably no survivors from radiation sickness
5,000	Incapacitation almost immediately. All personnel will be fatalities within a week

Note: The U.S. Atomic Energy Commission TID-7652 reports the following predicted life expectancy for man exposed to daily doses of radiation, starting at age 20: 0.8 rad/day, 20-year expectancy; 0.4 rad/day, 30-year expectancy.

[†] Received in less than 1 week.
*From: "The Effects of Nuclear Weapons", S. Glasstone, Ed., U.S. Department of Defense, June 1957.

Table 8-58. RADIONUCLIDES IN AIR AND WATER*

Maximum Permissible Concentrations

The concentrations of various radionuclides given below under Section A are generally used as the limiting average concentrations in air or drinking water to which an employee may be exposed 40 hours per week during a lifetime of employment. Values in Section B are reduced, representing maximum average concentrations in air and water that may occur in unrestricted areas.

Additional restrictions and special conditions, as well as other Title 10 regulations, are also covered in 10 CFR 20 (see footnote below).

Isotope	In water, $\mu Ci/ml$	
	A	B
If it is known that Sr^{90}, I^{125}, I^{126}, I^{129}, I^{131}, (I^{133}, Section B only), Pb^{210}, Po^{210}, At^{211}, Ra^{223}, Ra^{224}, Ra^{226}, Ac^{227}, Ra^{228}, Th^{230}, Pa^{231}, Th^{232}, $Th^{natural}$, Cm^{248}, Cf^{254}, and Fm^{256} are not present	9×10^{-5}	3×10^{-6}
If it is known that Sr^{90}, I^{125}, I^{126}, I^{129}, (I^{131}, I^{133}, Section B only), Pb^{210}, Po^{210}, Ra^{223}, Ra^{226}, Ra^{228}, Pa^{231}, $Th^{natural}$, Cm^{248}, Cf^{254}, and Fm^{256} are not present	6×10^{-5}	2×10^{-6}
If it is known that Sr^{90}, I^{129}, (I^{125}, I^{126}, I^{131}, Section B only), Pb^{210}, Ra^{226}, Ra^{228}, Cm^{248}, and Cf^{254} are not present	2×10^{-5}	6×10^{-7}
If it is known that (I^{129}, Section B only), Ra^{226}, and Ra^{228} are not present	3×10^{-6}	1×10^{-7}

	In air, $\mu Ci/ml$	
	A	B
If it is known that alpha emitters and Sr^{90}, I^{129}, Pb^{210}, Ac^{227}, Ra^{228}, Pa^{230}, Pu^{241}, and Bk^{249} are not present	3×10^{-9}	1×10^{-10}
If it is known that alpha emitters and Pb^{210}, Ac^{227}, Ra^{228}, and Pu^{241} are not present	3×10^{-10}	1×10^{-11}
If it is known that alpha emitters and Ac^{227} are not present	3×10^{-11}	1×10^{-12}
If it is known that Ac^{227}, Th^{230}, Pa^{231}, Pu^{238}, Pu^{239}, Pu^{240}, Pu^{242}, Pu^{244}, Cm^{248}, Cf^{249}, and Cf^{251} are not present	3×10^{-12}	1×10^{-13}

*From: "Code of Federal Regulations", Title 10, Part 20, Standards for Protection Against Radiation, U.S. Atomic Energy Commission, revised August 1966.

Table 8-59. NEUTRON AND GAMMA FLUENCES*

Flux of Neutrons or Gammas to Produce 1 mrem/hr

Values are given as a function of neutron energy.

Neutron energy, MeV	Neutrons/ $cm^2.s$	Gammas/ $cm^2.s$	Neutron energy, MeV	Neutrons/ $cm^2.s$	Gammas/ $cm^2.s$
Thermal (0.02 eV)	4 800	—	1.0	87	5.2×10^5
0.000 1	2 400	—	2.5	87	2.6×10^5
0.005	2 400	8×10^4	5.0	67	1.5×10^5
0.02	1 600	1.5×10^6	7.5	53	1.1×10^5
0.10	530	6.4×10^6	10.0	48	0.92×10^5
0.5	147	9.4×10^5			

*Based on data from: "Reactor Shielding Design Manual", T. Rockwell III, USAEC Report No. TID-7004, U.S. Atomic Energy Commission, March 1956.

Table 8-60. PERSONNEL MONITORING FOR RADIATION EXPOSURE*

ROBERT RADTKE

Personnel monitoring is the determination of the amount of ionizing radiation to which an individual has been exposed. The determination is commonly made by the use of photographic films, pocket ionization chambers, thermoluminescent materials, or radiophotoluminescent materials.

When the use of radioactive material is under AEC or state regulations, personnel monitoring is required for the following:

1. Each individual who receives, or is likely to receive, a dose in any calendar quarter in excess of 25 percent of the applicable value in the table below.
2. Each individual under 18 years of age who receives, or is likely to receive, a dose in in any calendar quarter in excess of 5 percent of the applicable value in the table below.
3. Each individual who enters an area where a major portion of his body could receive in any one hour a dose in excess of 100 millirem.**

REMS PER CALENDAR QUARTER †

1. Whole body, head and trunk, active blood-forming organs, lenses of eyes, or gonads	$1\frac{1}{4}$
2. Hands and forearms, feet and ankles	$18\frac{3}{4}$
3. Skin of whole body	$7\frac{1}{2}$

The most widely used method of monitoring personnel for radiation exposure is the photographic film, or "film badge". Basically the method is a comparison of the film exposure due to known and unknown amounts of radiation. The physical principles of the photographic-film response and the practical problems encountered in the use of photographic film for personnel monitoring are extensively covered in the literature.

The pocket ionization chamber is a device used for specific applications in personnel monitoring. It is a small pencil-sized instrument with an ionization chamber that discharges a capacitor in the presence of ionizing radiation. The rate of discharge depends primarily on the intensity of the incident radiation. The ionization chamber may be either a direct- or indirect-reading instrument, but either type can be read at short intervals.

Thermoluminescent materials, when heated, will release an amount of light that is proportional to the ionizing-radiation exposure. This property forms the basis for their use in dosimetry and personnel monitoring. Thermoluminescent dosimetry (TLD) systems are available to individuals providing their own personnel-monitoring programs, or the TLD monitoring may be purchased as a service.

Radiophotoluminescent (RPL) materials in personnel monitoring are commonly called glasses. Measurement of either the intensity of luminescence or changes in optical absorption of the RPL materials are proportional to the ionizing-radiation exposure. Table 8-61 gives a comparison of the various personnel-monitoring techniques.

**Title 10, Code of Federal Regulations, Part 20.202, April 1967, and The Council of State Governments, Suggested State Regulations for Control of Radiation, Sec. C. 202, Chicago, 1964.
†Title 10, Code of Federal Regulations, Part 20.101(a) April 1967.

*From: "CRC Handbook of Radioactive Nuclides", Y. Wang, Ed., The Chemical Rubber Co., 1969.

Table 8-61.   PERSONNEL-MONITORING DETECTORS*

ROBERT RADTKE

Detector	Radiation detected	Range	Minimum energy detected	Advantages	Possible disadvantages
Film	Gamma Beta Thermal neutron Fast neutron	0.01 to 10,000 rem	20 kev for gamma rays, 200 kev for beta rays	Inexpensive Gives estimate of integrated dose Provides permanent record	Moderate directional dependence Strong energy dependence for low-energy X-rays False readings produced by heat, pressure, and certain vapors
Pocket ionization chambers	Gamma Beta minus gamma Thermal neutron Fast neutron minus gamma	0.001 to 2,000 roentgens/hr	30 kev for gamma rays, 20 kev for fast neutron	Yield fairly accurate information quickly Small size, low directional dependence Reasonably uniform in response to radiation in the energy range of 50 kev to 2 Mev Economical for long-term use Require little maintenance Reusable	No permanent record Frequent reading, tabulation, and recharging may be required Subject to accidental discharge (through shock and, sometimes, electrical leakage) Range of measurement limited; full scale ranges from 0.2 to 2,000 R available
Thermo-luminescent dosimetry (TLD)	Gamma Beta Thermal neutron Fast neutron	10^5 rad	20 kev	Indefinite shelf life within the useful range Small size and low directional dependence Small energy dependence Reusable Inexpensive Gives estimate of integrated dose over long periods	Limited TLD systems supplied as commercial service Cancellation of dose upon reading Dose range depends upon sensitivity of reader Radiations detected depend on type of thermoluminescent material
Radiophoto-luminescent (RPL)	Gamma Beta Thermal neutron Fast neutron	0.01 to 10^6 rad	40 kev	Indefinite shelf life within the useful range Small size, low directional dependence Reusable Gives estimate of integrated dose over long periods, with unlimited number of interval measurements Provides a permanent record	Primary use today in civil defense and military dosimetry Commercial service limited Number of RPL systems available is limited Luminescent contaminations on glass surface are possible Build-up of RPL immediately after short-time exposure

*From: "CRC Handbook of Radioactive Nuclides", Y. Wang, Ed., The Chemical Rubber Co., 1969.

Table 8-62. PERMISSIBLE LEVELS FOR EXPOSURE TO LASER RADIATION*

COURTESY OF A.M. CLARKE

Safe values for chronic and acute exposure to the various optical sources have been proposed by the AMD, American Conference of Government Industrial Hygienists (ACGIH),[1] the Air Force,[2] and the Army and Navy;[3] the American National Standards Institute Z-136 Standards Committee on the "Safe Use of Lasers and Masers" presently is coordinating an effort of all laboratories and industrial groups to set useful, safe, and unrestrictive guidelines for "maximum permissible exposure."[4]

The Departments of the Army and Navy in TB MED 279/NAVMED P-5052-35[3] define the following "maximum exposure level(s) not expected to cause detectable bodily injury," measured at the cornea:

Q-switched laser	1×10^{-7} J/cm^2
Non-Q-switched laser	1×10^{-6} J/cm^2
(1 ms duration)	
CW-laser (He-Ne/Ar)	1×10^{-6} W/cm^2

for all wavelengths between 0.4 μm and 1.5 μm. A "safety factor of 2 is recommended for use in field evaluation and training exercises," and "for long-term work with lasers, such as in the laboratory, a safety factor of 10 should be used."

CO$_2$(10.6 μm) CW laser   100 mW/cm^2

"Use of safety factors is not deemed necessary for the CO$_2$ CW laser."

The Department of the Air Force, in AFM 161-8,[2] indicates considerably higher "permissible exposure levels" entering the eye:

Ruby laser (Q-switched)	7.5×10^{-6} J
(non-Q-switched)	1×10^{-4} J
Nd laser (Q-switched)	45×10^{-6} J
(non-Q-switched)	5×10^{-4} J
Helium–neon and argon laser	
(488, 515, 633 nm)	

<1 ms	20 mW
1–10 ms	10 mW
10 ms to 1 s	5 mW

CO$_2$ laser

<1 ms	8 W/cm^2
10–50 ms	3 W/cm^2
50–250 ms	1 W/cm^2

To compare the Army/Navy "permissible levels" with those of the Air Force, using the most extreme condition of complete dark adaptation, a pupil diameter of 8 mm (area of 0.5 cm^2) must be used.

Although the Air Force values are intended for battlefield conditions, and the Army/Navy values for peacetime operations, the differences are quite marked. The ratio of the levels are for Air Force to Army/Navy:

Q-switched laser	75
Non-Q-switched laser	200
Argon/He–Ne CW laser	10^4
CO$_2$ laser	10

The envisioned environmental conditions during exposure should be considered when comparing the values, and the Air Force values are "permissible" under the conditions for which they are established; however, they should not be considered as "safe" for casual exposure of any type or for levels of exposure for the general public except in times of emergency.[5]

The American Conference of Government Industrial Hygienists (ACGIH) endorses the following "safe" irradiance at the cornea for the worst case, or nighttime pupil diameter of 7 mm:

Q-switched ruby laser	1×10^{-8} J/cm^2
Non-Q-switched ruby laser	1×10^{-7} J/cm^2
Continuous wave exposure	1×10^{-5} W/cm^2

Table 8-62. PERMISSIBLE LEVELS FOR EXPOSURE
TO LASER RADIATION *(Continued)*

Again using the Army/Navy values for comparison, the ACGIH values are, by a factor of 10, more lenient in the CW case. The British Ministry of Aviation,[6] the Australian Department of Supply,[7] and other foreign and domestic government agencies have also established "safe" limits, but they are not included here.

Sliney,[8] of the U.S. Army Environmental Hygiene Agency, which has been instrumental in conversion of certain damage values to "safe" values for the Army and the ACGIH, has indicated that, for practical application, the single Army values at the cornea are sufficient.

To clear up the confusion immediately apparent to an uninitiated "safety" officer attempting to apply the several values of laser "permissible exposure" levels, the ANSI Z-136 Standards Committee[4] is working to provide a uniform, standard manner for specifying "safe and permissible" levels of exposure in both the near and far field and in the extended image and limiting-image cases.

For UV sources the American Medical Association Council on Physical Therapy[9] recommends that for wavelengths of less than 0.4 μm the 8-hour working day irradiation level be held below 0.5 μm W/cm^2.

For more detailed information, the reader should consult a recent review[10] and the references.

---

*Reprinted from: A.M. Clarke, Ocular Hazards, in "Handbook of Lasers", R.J. Pressley, Ed., The Chemical Rubber Co., 1971.

REFERENCES

1. "Threshold Limit Values for Physical Agents", Supplement 7, American Council of Government Industrial Hygienists, 1968 (published in *Laser Focus*, 4:50, 1968).
2. "Laser Health Hazards Control", AFM 161-8, Department of the Air Force, 1 April, 1969.
3. "Control of Hazards to Health from Laser Radiation", TB MED 279/NAVMED P-5052-33, Departments of the Army and the Navy, 24 February, 1969.
4. "Safe Use of Lasers and Masers", Z-136 Standards Committee, American National Standards Institute (in preparation).
5. Bell, H.E., and Townsend, A.R., The Philosophy of a Regulation, *Arch. Environ. Health,* 18:416, 1969.
6. "Laser Systems—Code of Practice", British Ministry of Aviation, 1965.
7. "Defense Standards", Admin. Memo No. 29, Department of Supply, Commonwealth of Australia, 1966.
8. Sliney, D.H., Evaluation Hazards—and Controlling Them, *Laser Focus*, 5:39, 1969.
9. "Permissible Limit for Continuous Ultraviolet Exposure", Council on Physical Therapy, American Medical Association, 1948.
10. Clarke, A.M., Ocular Hazards from Lasers and Other Optical Sources, *Crit. Rev. Environ. Control,* 1(3):307, 1970.

Table 8-63.   CHEMICAL HAZARD INFORMATION*

SYMBOLS:
C = ceiling limit
S = skin absorption possible
T = tentative value
W = reacts violently with water
x = estimate

EXTINGUISHING AGENTS:
1 = water
2 = foam
2a = alcohol foam
3 = CO_2
4 = see Note at end of table

Chemical name	Concentration in air, 1966 TLV (Threshold Limit Values)		Principal effects of inhalation exposures above TLV	Relative hazard to health from concentrated short-term exposure					NFPA hazard identification signals		Extin-guishing agents
				Eye contact	Inhalation	Skin penetration	Skin irritation	Ingestion	Health	Fire	
	ppm	mg/m³							(4 is high; 0 is low)		
				(5 is high; 1 is low)							
Acetaldehyde	200	360	Irritant	3	5	2	2	2	2	4	1,3,4
Acetamide	—	—	—	1x	1	1x	3	1	—	—	—
Acetic acid	10	25	Irritant	5	3	3	3	2	2	2	1,2a,3
Acetic anhydride (Diacetyl monoxide)	5	20	Irritant	5	4	2	2	2	2	2,W	3
Acetone	1 000	2 400	Narcosis	2	3	1	1	1	1	3	1,2a,3
Acetonitrile	40	70	Toxic	2	4	2	1	2	2	3	2a,3
Acrolein	0.1	0.25	Irritant	5	4	4	5	4	3	3	2a,3
Acrylic acid	—	—	—	4	1	3	4	2	—	—	1,2a
Acrylonitrile	20S	45S	Toxic	2	4	3	1	4	4	3	2a,3
Allyl alcohol	2S	5S	—	2	4	4	1	4	3	3	1,2a,3
Aluminum chloride, anhydrous	—	—	—	5			3	2	—	—	4
Ammonia, anhydrous	50	35	Irritant	4	5		3	—	3	1	4
Ammonium hydroxide	—	—	—	4	5	3	3	3	—	—	—
Aniline	5S	19S	Toxic	3	4	4	1	3	3	2	1,2,2a,3
Antimony compounds	—	0.5	Toxic	—	—	—	—	—	—	—	—
Arsenic compounds	—	0.5	Toxic	—	—	—	—	—	—	—	—
Arsine (Arsenic hydride)	0.05	0.2	Toxic	Highly toxic					—	—	—
Barium chloride	—	—	—	Extremely toxic					—	—	—
Barium hydroxide	—	0.5	Toxic	4x	—	1x	4x	3x	—	—	—
Benzene	C25S	C80S	Toxic	2	4	2	2	2	2	3	2,3
Benzoyl peroxide	—	5T	Toxic	5x	4x	2x	4x	3x	1	4	1,2,3
Benzyl alcohol	—	—	—	4	2	2	3	2	2	1	2a
Beryllium	—	0.002	Toxic	1x	5	—	1	—	4	1	—
Bromine	0.1	0.7	Irritant	5x	2x	1x	5x	—	4	0	—
Bromomethane (Methyl bromide)	C20S	C80S	Toxic	1x	5x	—	4x	—	3	1	—
Bromotrifluoromethane (Freon-13B1®)	1 000	6 100	Narcosis	—	—	—	—	—	—	—	—
1,3-Butadiene	1 000	2 000	Narcosis	2x	2	1x	1x	—	2	4	4
Butane	—	—	—	—	—	—	3x	—	1	4	4
2-Butanone (Methyl ethyl ketone)	200	590	Irritant	2	4	2	1	2	1	3	2a
2-Butoxyethanol (Butyl Cellosolve®; Dowanol® EB; Ethylene glycol monobutyl ether)	50S	240S	Toxic	5	1	3	3	2	—	—	1,2a,3
Butyl acetate	150T	710T	Irritant, narcosis	2	3	1	1	1	1	3	2,2a,3
Butyl alcohol (1-Butanol)	100	300	Narcosis	3	1	2	1	2	1	3	1,2a,3
Cadmium soluble salts and metal dust	—	.02T	Toxic	—	—	—	—	—	—	—	—

*Condensed from: "Handbook of Laboratory Safety", 2nd ed., N.V. Steere, Ed., The Chemical Rubber Co., 1971. For an extended listing of 1 100 compounds, see this source.

Table 8-63. CHEMICAL HAZARD INFORMATION *(Continued)*

Chemical name	Concentration in air, 1966 TLV (Threshold Limit Values) ppm	mg/m³	Principal effects of inhalation exposures above TLV	Eye contact	Inhalation	Skin penetration	Skin irritation	Ingestion	NFPA Health	Fire	Extinguishing agents
				(5 is high; 1 is low)					*(4 is high; 0 is low)*		
Calcium carbide	—	—	—	4x	2x	1x	3x	2x	1	4,W	—
Calcium chloride	—	—	—	2x	—	1x	2x	2x	—	—	—
Calcium hydroxide	—	5	Irritant	4	3x	1x	4	2	—	—	—
Camphor	2	—	Toxic	4x	2x	2x	3x	2x	2	2	1,2,3
Carbon black	—	3.5T	Toxic	—	—	—	—	—	—	—	—
Carbon dioxide	5 000	9 000	Toxic	—	3x	—	—	—	—	—	—
Carbon disulfide	20S	60S	Toxic	3x	5	3x	2x	3x	2	3	1,3
Carbon monoxide	50T	55T	Toxic	—	5	—	—	—	2	4	4
Carbon tetrachloride	10S	65S	Toxic	3x	5	2	2x	2	—	—	—
Cellulose nitrate	—	—		Combustion produces toxic oxides of nitrogen					solution 1 / 3; solid 2 / 3		—
Chlordane	—	0.5S	Toxic	1	3x	3x	2	3	—	—	—
Chlorine	C1T	C3T	Irritant	5	5	—	3	—	3	0	—
Chloroethane	1 000	2 600	Narcosis	Narcotic properties; overexposure may be toxic					2	4	3,4
Chloroethylene (Vinyl chloride)	C500	C1,300	Narcosis, toxic	2x	3x	2x	2x	2x	2	4	3
Chloroform	C50	C240	Toxic	2x	3	2x	1	2	—	—	—
Chloromethane (Methyl chloride)	C100	C210	Toxic	1x	3x	—	1x	—	2	4	—
Chloroprene (2-Chloro-1,3-butadiene)	25S	90S	Toxic	2x	4	2x	2x	2x	—	—	2a,4
3-Chloropropene (Allyl chloride)	1	3	Irritant	3	3	3	3	3	3	3	2a
α-Chlorotoluene (Benzyl chloride)	1	5	Toxic	—	—	—	—	—	2	2	—
Chromic salts, soluble	—	0.5	Toxic	—	—	—	—	—	—	—	—
Chromium trioxide (Chromic acid)	—	0.1	Irritant, toxic	5x	4	2x	4	4x	1	0	—
Citric acid	—	—		5	2x	2x	2x	2x	—	—	—
Coal-tar pitch volatiles	—	0.2T	Carcinogenic, toxic	3x	2x	1x	2x	1x	—	—	—
m-Cresol	5S	22S	Toxic	4	1	3	4	3	2	1	1,3
Cyanamide	—	—	—	Very irritating and caustic					—	—	1,3
Cyanides	—	5S	Toxic	—	—	—	—	—	—	—	—
Cyanogen	—	—	—	Highly toxic; exposure limit of 10 ppm is recommended							
Diborane	0.1	0.1	Toxic	Highly toxic					3	4,W	4
1,2-Dibromoethane (Ethylene bromide; Ethylene dibromide)	C25S	C190S	Toxic	2x	3x	—	4x	—	—	—	—
Dibutylamine (Di-*n*-butylamine)	—	—	—	4	5	3	3	3	3	2	1,2,2a,3
o-Dichlorobenzene	C50	C300	Toxic	3x	3x	2x	2x	2x	2	2	1,2,3
1,2-Dichloroethane (Ethylene dichloride; Ethylene chloride)	50	200	Toxic	Respiratory and conjunctival irritant; causes narcosis					2	3	—
Diethylamine	25	75	Irritant	5	4	3	2	3	3	3	2a,3
Dimethylamine	10	18	Toxic	4x	5x	—	4x	—	3	4	3,4

Table 8-63. CHEMICAL HAZARD INFORMATION *(Continued)*

Chemical name	Concentration in air, 1966 TLV (Threshold Limit Values) ppm	mg/m³	Principal effects of inhalation exposures above TLV	Relative hazard to health from concentrated short-term exposure (5 is high; 1 is low) Eye contact	Inhalation	Skin penetration	Skin irritation	Ingestion	NFPA hazard identification signals (4 is high; 0 is low) Health	Fire	Extinguishing agents
N,N-Dimethylformamide	10S	30S	Toxic	2	2	2	2	2	1	2	1,2,2a,3
p-Dioxane (1,4-Dioxane)	100S	360S	Toxic	2	3	2	1	2	2	3	1,2a
2-Ethoxyethanol	200S	740S	Toxic	1	2	—	2	—	—	—	2a
Ethylene oxide (Oxirane)	50	90	Irritant, toxic	5	3	—	5	—	2	4	1,2a,3,4
Ethyl ether (Diethyl ether)	400	1 200	Narcosis	1	4	1	1	2	2	4	2a,3,4
Fluorine	0.1	0.2	Irritant	5	5	—	5	—	4	0,W	4
Formaldehyde	C5	C6	Irritant	4	3	4	4	3	—	4	1,4
Formic acid	5T	9	Irritant	4	4x	2x	4	2	3	2	1,2a,3
n-Heptane	500	2 000	Narcosis	1	2x	1x	1	1	1	3	2,3
Hexane	500	1 800	Narcosis	Impurities must be considered in evaluating health hazard					1	3	2,3
Hydrazine	1S	1.3S	Toxic	4x	4x	3x	3x	3x	3	3	1,3
Hydrochloric acid	—	—	—	4	5	3x	5	3x	3	0	—
Hydrocyanic acid 96%	—	—	—	4x	5x	4x	2x	4x	4	4	1
Hydrofluoric acid	—	—	—	5	5	4x	5	4x	—	—	—
Hydrogen cyanide	10S	11S	Toxic	4x	5	—	2x	—	4	4	3
Hydrogen fluoride	3S	2S	Toxic	5	5	—	5	—	4	0	—
Hydrogen peroxide 90%	1	1.4	Irritant	5x	—	4x	5x	4x	2	0	—
Hydrogen sulfide	10	15	Toxic	—	5x	—	—	—	3	4	4
Iodine	C0.1	C1	Irritant	5x	4x	3x	4x	4x	—	—	—
Isobutyl alcohol	100	300	Irritant	3	2	2	1	2	1	3	1,2a,3
Isoprene	—	—	—	3x	3x	—	2x	—	2	4	3,4
Lead	—	0.2	Toxic	1x	5x	1x	1x	2x	—	—	—
Lead acetate	—	—	—	2x	5x	1x	1x	2x	—	—	—
Mercuric chloride	—	—	—	4	4	2	4	4	—	—	—
Mercury	—	0.1S	Toxic	Vapor toxic; salts are general cellular poisons					—	—	—
Methanol (Methyl alcohol)	200	260	Narcosis, toxic	2	2	2	1	1	1	3	2a,3
Methylamine	10T	12T	Irritant	5x	5x	—	4x	—	3	4	3,4
Naphtha (coal tar)	100	400									
Naphtha, varnish makers' and painters' regular	500	2 000	Narcosis	—	—	—	—	—	1	3	2,3
Naphthalene	10	50	Toxic	2	1	2x	2	2	2	2	1,3
Nickel—metal and solid compounds	—	1	Toxic, carcinogenic	May cause dermatitis					—	—	—
Nickel carbonyl	0.001	0.007	Carcinogenic	Exposure may be fatal or carcinogenic					—	—	—
Nicotine	—	0.5S	Toxic	—	—	—	—	—	4	1	2a,3
Nitric acid	2	5	Irritant, toxic	4x	5x	3x	4x	4x	2	0	—
Nitrobenzene	1S	5S	Toxic	—	—	—	—	—	3	2	1,2,3
Nitrogen dioxide (Nitrogen peroxide)	C5	C9	Irritant, toxic	3x	5	—	3x	—	—	—	—
n-Octane	500	2,350	Narcosis	3x	2x	1x	1x	1x	0	3	2,3
Oxalic acid	—	1T	Toxic	3x	—	2x	3x	3			

Table 8-63. CHEMICAL HAZARD INFORMATION (Continued)

Chemical name	Concentration in air, 1966 TLV (Threshold Limit Values)		Principal effects of inhalation exposures above TLV	Relative hazard to health from concentrated short-term exposure					NFPA hazard identification signals		Extinguishing agents
	ppm	mg/m³		Eye contact	Inhalation	Skin penetration	Skin irritation	Ingestion	Health	Fire	
				(5 is high; 1 is low)					(4 is high; 0 is low)		
Ozone	0.1	0.2	Toxic	Very irritating; lethal in a few minutes in concentrations over 1 700 ppm					—	—	—
Pentane	1 000	2 950	Narcosis	May be narcotic in high concentrations					1	4	2,3,4
Pentyl acetate (n-Amyl acetate)	100	525	Irritant	1	3	1	2	2	1	3	—
Pentyl alcohol (Amyl alcohol; 1-Pentanol)	100	360	Narcosis	4	1	2	1	2	1	3	2a
Perchloric acid	—	—	—	5x	4x	2x	4x	4x	3	0	—
Phenol (Carbolic acid)	5S	19S	Toxic	4	1	3	4	2	3	2	1,2a,3
Phosgene (Carbonyl chloride)	0.1	0.4	Toxic	Highly toxic					—	—	—
Phosphoric acid	—	1	Irritant	4	—	2x	4	3x	—	—	—
Phosphorus, white or yellow	—	0.1	Toxic	5x	5	4x	5x	4x	3	3	1
Picric acid (2,4,6-Trinitrophenol)	—	0.1S	Toxic	—	—	—	—	—	2	4	1
Potassium dichromate	—	—	—	Corrosive poison					—	—	—
Potassium hydroxide	—	—	—	5	—	1x	5	2	3	0	—
Potassium permanganate	—	—	—	3	—	1x	3	2	0	0	—
Propane	1 000	1 800	Narcosis		3x			—	1	4	4
Propyl alcohol (n-Propyl alcohol; propanol)	200	500	Narcosis	2	2	2	1	2	1	3	2a
Propylene oxide	100	240	Toxic	2	5	3	1	2	2	4	2a,3,4
Pyridine	5	15	Toxic	3	3	3x	2	2	2	3	1,3
Selenium compounds	—	0.2	Toxic	—	—	—	—	—	—	—	—
Silica, quartz	—	—	—	2	4	1x	1	1	—	—	—
Silver—metal and soluble compounds	—	0.01	Toxic	—	—	—	—	—	—	—	—
Silver nitrate	—	—	—	5x	—	2x	4x	4x	1	0	—
Sodium	—	—	—	5x	—	—	4x	—	3	1,W	—
Sodium carbonate	—	—	—	3x	—	1x	2x	2x	—	—	—
Sodium hydroxide (Caustic soda)	—	2	Irritant	5	—	1x	5	3x	3	0	—
Sodium nitrite				2x	—	1x	1x	3	—	—	—
Sodium silicate	—	—	—	4x	—	1x	3x	2x	—	—	—
Stannic chloride	—	2	Toxic	4x	—	1x	4x	4x	—	—	—
Stibine	0.1	0.5	Toxic	Very toxic					—	—	—
Styrene	C100	C420	Irritant, narcosis	2	2	2x	2	2	2	3	2,3
Sulfur dioxide	5	13	Irritant	4x	4	—	4x	—	3	0	—
Sulfuric acid	—	1	Irritant	4	—	2x	4	4	3	0,W	—
Tetrachloroethylene (Perchloroethylene)	100	670	Narcosis	2x	—	3	1	2x	—	—	—
Tetraethyllead	—	0.075S	Toxic	—	—	—	—	—	3	2	—
Titanium tetrachloride	—	—	—	5x	5x	2x	4x	3x	—	W	—
Toluene	200	750	Irritant	4	3	2	2	2	2	3	2,3
Trichloroethylene	100	535	Narcosis, toxic	2	3	1	3	2	—	—	—
Turpentine	100	560	Irritant, narcosis	1	2x	2x	2	2	1	3	2,3

Table 8-63.   CHEMICAL HAZARD INFORMATION *(Continued)*

Chemical name	Concentration in air, 1966 TLV (Threshold Limit Values)		Principal effects of inhalation exposures above TLV	Relative hazard to health from concentrated short-term exposure					NFPA hazard identification signals		Extin-guishing agents
				Eye contact	Inhalation	Skin penetration	Skin irritation	Ingestion	Health	Fire	
	ppm	mg/m³		*(5 is high; 1 is low)*					*(4 is high; 0 is low)*		
Valeric acid	—	—	—	4	1	3	4	2	—	—	—
Warfarin	—	0.1	Toxic	—	—	—	—	—	—	—	—
Zinc chloride	—	1T	Irritant	5x	—	2x	4x	3x	2	0	—
Zinc oxide fume	—	5	Fume fever	—	—	—	—	—	—	—	—

Note: While carbon dioxide, dry chemical, and in some instances water spray may be used to extinguish *small* gas fires, these agents generally are not recommended in larger gas fires. The discharge of gas or volatile liquid will continue unless shut off promptly and may create a more serious explosion hazard. Generally, water is used to keep the surroundings cool until the leak can be shut off or until the volatile liquid has completely burned.

REFERENCES

"Handbook of Analytical Toxicology", I. Sunshine, Ed., The Chemical Rubber Co., 1969.
"Practical Toxicology of Plastics", R. Lefaux, Iliffe Books Ltd., 1968.
Williams-Steiger Occupational Safety and Health Act of 1970, Chapter XVII, Title 29, *Code of Federal Regulations*, April 13, 1971 (amended August 13, 1971), Part 1910 (Occupational Safety and Health Standards); published in the *Federal Register*, Vol. 36, No. 105, May 29, 1971.

Table 8-64.   TOXICITY OF CARBON DIOXIDE IN AIR*

One Atmosphere Total Pressure

Concentrations below 0.5 percent produce no definite effects with exposures even as long as 40 days. For concentrations above 1.5 percent, the toxic effects begin to appear within an hour or less, as shown by the following table.

Effects	Minimum percent concentration for effects described		
	10-minute exposure	40-minute exposure	80-minute exposure
Mild physiological strain; perceptible changes; slight hearing loss; added respiration	2.0	1.6	1.4
Air hunger; mental depression; nausea; headache; decreased visual discrimination	3.5	2.8	2.4
Dizziness and stupor leading to unconsciousness	7.9	6.8	6.1

Note: Typical volumetric analysis of *exhaled* air is 4.1% carbon dioxide, 16.4% oxygen, and 79.5% nitrogen.

*From: "Bioastronautics Data Book", P. Webb, Ed., NASA SP–3006, U.S. Government Printing Office 1964.

Table 8-65.   EFFECTS OF OXYGEN IN LOW-PRESSURE ATMOSPHERES

Physiological Reactions Based on Continuous Exposure (One Week or More)

Total pressure of atmosphere, psia	Normal or sea-level equivalent	Unimpaired performance		Limits of human toleration	
		Minimum oxygen	Maximum oxygen	Minimum oxygen	Maximum oxygen
		Oxygen percentages by volume			
7	48	36	80	16	—
8	40	30	68	14	—
9	36	27	60	12	90
10	32	24	54	10	80
11	28	22	49	9	72
12	26	20	44	8	65
13	24	18	40	8	59
14	22	16	37	7	56
14.7	21	15	36	7	54

Notes:
A deficiency of oxygen in the tissues is known as *hypoxia.*
The low figures shown as minimum oxygen toleration cannot be attained without acclimatization.

Table 8-66.   TOXICITY OF HIGH-OXYGEN ATMOSPHERES*

In the partial pressure range of 400–760 mm Hg, the respiratory and nervous-system symptoms predominate. Onset of toxic signs depends on partial pressure of the oxygen, approximately as given in the following table.

Partial pressure of O_2		Approximate time to onset of symptoms, hr	Partial pressure of O_2		Approximate time to onset of symptoms, hr
mm Hg	psia		mm Hg	psia	
160	3.1	00 (sea level atmosphere)	500	9.7	28
			600	11.6	19
200	3.9	170	700	13.5	14
300	5.8	60	800	15.5	10
400	7.7	40			

Note: At pressures above 2,000 mm Hg, severe symptoms such as dizziness, fainting, and convulsions may appear in one hour or less.

*From: "Bioastronautics Data Book", P. Webb, Ed., NASA SP-3006, U.S. Government Printing Office, 1964.

Table 8-67.   TOXICITY OF CARBON MONOXIDE*
Percentage Concentration[a]

	Exposure time	
	10 min	60 min
Danger of collapse and death	0.60	0.15
Headache, dizziness, nausea	0.20	0.045
Aircraft maximum exposure (USAF)	0.06	0.01
Aircraft maximum exposure (USN)	0.015	0.01

Note: The usual TLV (threshold limit value) for 8-hr exposure in industry is 0.01 percent or 100 ppm. Some authorities consider this too high. Studies have been made of carbon-monoxide exposure of commuters in city traffic. Brief exposures (2–3 min) in the range of 100–150 ppm (0.01–0.015%) have been reported for several U.S. cities. Exposures for 20 minutes or more are rarely higher than 50 ppm; average concentrations on the busiest streets seldom exceed 30 ppm (0.003%).

[a]For parts per million move decimal point 4 places to the right.
*Data from: "Bioastronautics Data Book", P. Webb, Ed., NASA SP-3006, U.S. Government Printing Office, 1964.

Table 8-68. MEASUREMENTS OF ALCOHOL CONCENTRATION IN BLOOD*

The following test methods are used to determine the concentration of alcohol in blood:

1. Aeration
2. Diffusion
3. Distillation
4. Gas chromatography
5. Reaction with alcohol dehydrogenase
6. Analysis of expired gas

Under prescribed conditions all these techniques will give reliable ethanol assays. Aeration and diffusion are best suited for rapid semiquantitative analyses, analysis of expired gas for rapid quantitative analyses, and the remainder for the more specific quantitative determinations required for forensic purposes.

No significant endogenous ethanol is present in biological materials. Table A gives the estimated blood ethanol concentration one might expect in a healthy person following the absorption of all the ingested ethanol. Table B shows the expected pharmacological responses with given levels of blood alcohol.

Table A. APPROXIMATE PERCENT OF ALCOHOL CONCENTRATION IN BLOOD†

Body weight, lb	Total number of drinks‡									
	1	2	3	4	5	6	7	8	9	10
100	0.038	0.075	0.113	0.150	0.188	0.225	0.263	0.300	0.338	0.375
120	0.031	0.063	0.094	0.125	0.156	0.188	0.219	0.250	0.281	0.313
140	0.027	0.054	0.080	0.107	0.134	0.161	0.188	0.214	0.241	0.268
150	0.025	0.051	0.075	0.101	0.126	0.151	0.176	0.201	0.226	0.251
160	0.023	0.047	0.070	0.094	0.117	0.141	0.164	0.188	0.211	0.234
180	0.021	0.042	0.063	0.083	0.104	0.125	0.146	0.167	0.188	0.208
200	0.019	0.038	0.056	0.075	0.094	0.113	0.131	0.150	0.169	0.188
220	0.017	0.034	0.051	0.068	0.085	0.102	0.119	0.136	0.153	0.170
240	0.016	0.031	0.047	0.063	0.078	0.094	0.109	0.125	0.141	0.156

†If these drinks were not taken within one hour, deduct one drink from the total number of drinks for each hour that elapsed between the first and last drink.

‡A drink is defined as one ounce of 100-proof "hard liquor" (whiskey, vodka, gin, etc.), or twelve ounces of 4 percent beer, or three ounces of fortified wine.

Table B. PHARMACOLOGICAL EFFECT OF ALCOHOL ON THE BRAIN

General range of blood alcohol concentration‡	Area of brain affected	Usual symptoms and signs produced
0.01–0.10%	Frontal lobe	Reaction colored by individual's personality Removal of inhibitions, loss of self-control, weakening of will power Development of euphoria, feeling of well-being, exaltation, increased confidence, expansiveness, generosity, altered judgment, increased good fellowship, loquaciousness, dulling of attention
0.10–0.20%	Psychomotor area	Apraxis, agraphia, ataxia, tremors, slurred speech, loss of skill
0.10–0.30%	Somestheto-psychic area	Dulled or distorted sensibilities
0.15–0.35%	Cerebellum	Disturbance of equilibrium
0.20–0.30%	Visuo-psychic areas	Disturbance of color perception, dimensions, form, motion, distance; diplopia
0.25–0.40%	Diencephalon	Apathy, inertia, tremors, cessation of automatic movements, sweating, dilation of surface capillaries, stupor, coma
0.40–0.50%	Medulla	Depression of respiration, peripheral collapse, subnormal temperature, death

‡*Caution:* The described symptoms and signs generally occur within the stated ranges, but no range sharply delineates the described symptoms.

Table 8-68.   MEASUREMENTS OF ALCOHOL CONCENTRATION IN BLOOD *(Continued)*

Ethanol is rapidly absorbed. On an empty stomach this is complete in 60 to 90 minutes. When food is present in the stomach, a longer period is required, depending on the amount and type of food. Once absorbed, ethanol is distributed throughout the body in proportion to the water content of each tissue or fluid. After absorption 90 percent of the ethanol is metabolized, primarily by the liver. For each individual the metabolic rate is relatively constant and independent of the amount of alcohol in the body (100 mg ethanol/kg body weight on the average). Excretion, primarily in the urine and the breath, disposes of the remainder of the absorbed ethanol.

Incorrect clinical diagnoses have resulted from the erroneous assumption that the odor of an alcoholic beverage on the breath or clothes of a person, coupled with staggering gait and/or slurred speech, is presumptive evidence of ethanol intoxication. Natural disease or trauma may cause these same symptoms, and one drink of an alcoholic beverage is enough to give its characteristic odor to the breath. For these reasons chemical analysis of a blood specimen is desirable. The result of this analysis will indicate whether or not an alcoholic beverage has been imbibed and, if so, the amount that has been absorbed.

Thirty-eight states have enacted legislation defining what is considered "under the influence of an alcoholic beverage." These laws vary and are based on the following recommendation of the Committee on Alcohol and Drugs of The National Safety Council.

1. "Below 0.05 percent alcohol in the blood: no influence by alcohol within the meaning of the law;
2. "Between 0.05 and 0.10 percent: a liberal, wide zone; alcoholic influence usually is present, but courts of law are advised to consider the behavior of the individual and circumstances leading to the arrest in making their decision;
3. "0.10 percent: definite evidence of 'under the influence', since every individual with this concentration would have lost to a measurable extent some of that clearness of intellect and control of himself that he would normally possess."

---

*From: "CRC Handbook of Clinical Laboratory Data", 2nd ed., W.R. Faulkner, J.W. King, and H.C. Damm, Eds., The Chemical Rubber Co., 1968, pp. 363–365.

Table 8-69. PHYSICAL PROPERTIES OF ANESTHETICS*

Name and molecular formula	Molecular weight	Class	Density	Vapor density (air = 1), 25°C	Boiling point, °C	Melting point, °C	Refractive index	Ignition temp, °C		Inflammable range in %	
								Air	O_2	Air	O_2
Bromoform $CHBr_3$	252.8	Brominated hydrocarbon	2.902^{15}_{15}		149.5	8.3	$n_D^{15}1.6005$				
Chloroform $CHCl_3$	119.4	Chlorinated hydrocarbon	1.4916^{18}_{4}	4.12	61.3	−63.5	$n_D^{25}1.4433$	Oxidized by flame to phosgene and HCl			
Cyclopropane $(CH_2)_3$	42.08	Cyclic hydrocarbon	1.478		−32.8	−126.6		498	454	2.40–10.3	2.48–60.0
Diethyl ether $(C_2H_5)_2O$	74.12	Saturated ether	0.7142^{20}_{20}	2.56	34.6	Freezes	$n_D^{20}1.3526$	304	182	1.85–36.5	2.10–82.0
Divinyl ether $(C_2H_3)_2O$	70.09	Unsaturated ether	0.774^{20}_{20}	2.42	28.4	−116.2	$n_D^{20}1.3989$	360	327	1.70–27.0	1.85–85.5
Ethyl chloride C_2H_5Cl	64.5	Halogenated hydrocarbon	0.9028^{15}_{4}	2.23	12.6	−138.7	$n_D^{10}1.3742$	517	468	4.0–14.8	4.05–67.2
Ethylene C_2H_4	28.06	Unsaturated hydrocarbon	0.001260^{0}_{4} 760 mm	0.97	−103.9	−169.4	$n_D^{100}1.363$	490	485	3.05–28.6	2.90–79.9
Ethyl vinyl ether $C_2H_5OC_2H_3$	72.1	Unsaturated ether		2.49	35.8						2.1 (lower limit)
Halothane $C_2HBrClF_3$	197.4	Saturated halogenated hydrocarbon	1.862^{20}_{4}		50.2						
Methoxyflurane $C_3H_4Cl_2F_2O$	165	Saturated halogenated hydrocarbon	1.4224^{25}		104.7						
Nitrous oxide N_2O	44.02	Oxide of nitrogen		1.53	−88.5	−90.81		Will support combustion only			50+
Trichloroethylene C_2HCl_3	131.4	Unsaturated chlorinated hydrocarbon	1.462^{15}	4.53	86.7	−73	$n_D^{17}1.47914$	419		Below 32°C nonflammable; above 32°C 15%	10.3– 64.5

*Adapted from: "CRC Handbook of Analytical Toxicology", I. Sunshine, Ed., The Chemical Rubber Co., 1969; see this source for complete table.

Table 8-70. TOXICITY OF PESTICIDES*

Estimated Relative Acute Toxic Hazard of Pesticides to Pesticide Appliers

The acute toxicity of a compound as measured by the oral or cutaneous LD_{50} to experimental animals is a good guide to its probable toxicity in man. It is not necessarily the final indicator of relative toxicity or of relative hazard; human experience is the final arbiter of the relative hazard of a compound.

All compounds in this table are hazardous to man in some degree. Toxicity to experimental animals was the principal basis for categorizing the compounds, but some modifications have been made based on human experience.

Most dangerous	Dangerous	Less dangerous	
Antu	Acrolein	Arsenic trioxide	Lethane 384
Chloropicrin	Aldrin	Baygon	Lindane
Cycloheximide	Azodrin	BHC	Mesurol
Demeton	Bidrin	Binapacryl	Methyl trithion
Dimefox	Bomyl	Bromoxynil	Nabam
Disulfoton	Carbophenothion	C-56	Naled
Isobenzan	Coumaphos	Cacodylic acid	Nemacide
Norbormide	DDVP	Carbaryl	Pyrolan
Para-Oxon	Dexon	Ceresan M	Ruelene
Phorate	Dieldrin	Chlordane	Strobane
Phosdrin	Dimetilan	Ciodrin	Tetrachlorothiophene
Sodium fluoroacetate	Dinitrobutylphenol	DEF	Toxaphene
Temik	Dinitrocresol	2,4-DEP	Zytron
TEPP	Dinitrocyclohexyl-	Di-allate	
Thionazin	phenol	Diazinon	
Warfarin	Dioxathion	Dimethoate	
	Endrin	Dinitrocyclohexyl phenol,	
	EPN	dicyclohexyl amine salt	
	Ethion	Dursban	
	GC-6,506	Endosulfan	
	Isolan	Fenthion	
	Matacil	Guthion	
	Methyl demeton	Heptachlor	
	Methyl parathion	Imidan	
	Nicotine	Ioxynil	
	Phosphamidon	Kepon	
	Zectran	Lanstan	

Note: Over 100 other pesticides, including DDT, are considered somewhat toxic.

*From: "CRC Handbook of Analytical Toxicology", I. Sunshine, Ed., The Chemical Rubber Co., 1969; includes complete data on safety of pesticides.

Table 8-71.   FIRE LOSSES AND SOURCES OF FIRE-PROTECTION DATA

The usual basic library for fire-protection information is the ten-volume NFPA "National Fire Codes" and its condensed counterpart, the NFPA "Fire Protection Handbook" (see References). The NFPA handbook has twenty sections, titled as follows:

1. Loss of Life and Property
2. Fire Protection and Allied Organizations
3. Fire Protection Standards, Laws, and Regulations
4. Characteristics and Behavior of Fire
5. Fire Hazard Properties of Materials
6. Storage and Handling of Materials
7. Process Hazards
8. Buildings and Other Structures
9. Building Equipment and Facilities
10. Public Fire Services
11. Organization for Private Protection
12. Hydraulic Calculations and Tests
13. Water Supplies and Services
14. Detection, Alarm, and Fire Guard Systems
15. Extinguishing Agents
16. Fire Protection Systems—Sprinklers
17. Fixed Fire and Hazard Protection Systems
18. Portable Fire Extinguishers
19. Transportation Equipment and Stationary Engines
20. Miscellaneous Data

General information on the extent and distribution of fire losses is given in the following table. This information emphasizes the need for more effective engineering design and planning for fire protection for all structures, vehicles, planes, and ships.

U.S. FIRE LOSSES*

	1960	1965	1969
Estimated U.S. Fire Losses			
Total fire losses, billions of dollars	1.11	1.46	1.95
Total fire losses, per capita, dollars	6.20	7.60	9.67
Urban Fires Reported			
Number of cities reporting	926	852	886
Fires reported, total, thousands	923	841	937
Total losses, millions of dollars	279	348	466
Building fires, total, thousands	437	332	337
Origins of Building Fires (Estimated)			
Heating and cooking, percent of total	23.3	16.7	14.0
Smoking and matches, percent of total	15.8	17.8	11.5
Electrical, percent of total	14.6	16.1	14.0
Open flames and sparks, percent of total	6.5	6.9	6.2
Children and matches, percent of total	4.4	6.3	8.1
Probably incendiary, percent of total	2.8	3.7	5.8
Forest Fires, Federal Areas			
Total number of fires	12 090	9 073	10 112
Area burned, thousands of acres	622	146	4 112
Forest Fires, State and Private Lands			
Total number of fires	91 297	104 611	103 239
Area burned, thousands of acres	3 856	2 506	2 577

*Based on: "Statistical Abstract of the United States, 1971", U.S. Department of Commerce, Bureau of the Census. Data from Insurance Information Yearbook, National Board of Fire Underwriters, National Fire Protection Association, Forest Service, and the U.S. Department of Agriculture.

REFERENCES

Fire Technology, a quarterly dealing with fire-protection engineering and research; sponsored by the Society of Fire Protection Engineers and published by the National Fire Protection Association.

"Fire Protection Handbook", 13th ed., National Fire Protection Association, 1969.

"Handbook of Industrial Loss Prevention", 2nd ed., McGraw-Hill Book Company, 1967.

"National Fire Codes", National Fire Protection Association, 10 volumes.

Publications of the Underwriters' Laboratories, Inc., and Factory Mutual Engineering Corporation.

Table 8-72. COLOR CODES AND SIGNALS

In general, the widely adopted color codes follow the pattern used in traffic lights: green for go, safe, or normal; red for stop or danger; and amber or yellow for caution. Instrument scales, pilot lights, control panels, and service boxes are often coded by these colors. Railroad signal lights add a blue caution signal and a purple stop signal in certain cases. Hazardous materials are coded by blue for a health hazard, red for a fire hazard, and yellow for an explosive or reactive hazard.

Code colors should contrast with the background in both hue and brightness. An observer with some color blindness will probably distinguish between red, blue, and green, but these might be confused with grays or yellows. Against a dark background fluorescent pigment colors are recommended. "International orange" can be seen at great distances.

The following table summarizes three common color codes.

Color	National Safety Council	Interstate Commerce Commission (for shipments)	Code for Identification of Piping
Red	Fire protection equipment	Red on white: poisons, explosives, tear gas, poisonous gases Black on red: flammable liquids and fireworks	Fire protection equipment
Green	Safe materials	Black on green: compressed gases	Safe materials: e.g., cold water, compressed air, vacuum
Yellow	—	Black on yellow: inflammable solids; oxidizing materials	Dangerous materials: e.g., high-pressure steam, combustible liquids and gases
Blue	Protective materials	—	Protective materials: e.g., coolants
Purple	Valuable materials	—	Highly valuable materials

Table 8-73. REQUIRED WATER FLOWS AND FIRE HYDRANTS*

Table A. REQUIRED FLOWS FOR FIRE PROTECTION

Population	Required fire flow for average city, gpm	Duration of fire flow, hr	Population	Required fire flow for average city, gpm	Duration of fire flow, hr
1 000	1 000	4	22 000	4 500	10
1 500	1 250	5	27 000	5 000	10
2 000	1 500	6	33 000	5 500	10
3 000	1 750	7	40 000	6 000	10
4 000	2 000	8	55 000	7 000	10
5 000	2 250	9	75 000	8 000	10
6 000	2 500	10	95 000	9 000	10
10 000	3 000	10	120 000	10 000	10
13 000	3 500	10	150 000	11 000	10
17 000	4 000	10	200 000	12 000	10

Note: With over 200 000 population a flow of 12 000 gallons with a reserve of 2 000–3 000 (gpm) additional for a second fire is necessary. In residential areas the required fire flows are dependent on the character and congestion of the buildings. In districts where about one-third of the lots in a block are of small area and low height, at least 500 gpm is required; if the buildings are of larger area or of greater height, up to 1 000 gpm is required. Where districts are more closely built, or where the buildings consist of high-value residences, apartments, tenements, dormitories, or similar structures, 1 500–3 000 gpm is required. In densely built districts with three-story and higher buildings, up to 6 000 gpm is required.

Table B. HYDRANT DISTRIBUTION FOR FIRE FLOWS

Fire flow required, gpm	Average area per hydrant, ft²	Fire flow required, gpm	Average area per hydrant, ft²
1 000	120 000	7 000	70 000
2 000	110 000	8 000	60 000
3 000	100 000	9 000	55 000
4 000	90 000	10 000	48 000
5 000	85 000	11 000	43 000
6 000	80 000	12 000	40 000

*Courtesy of Insurance Services Office.

Table 8-74.   SURFACE FLAMMABILITY OF WOODS

Tests on Bare Wood, Plywood, Wallboards, and Painted Wood Surfaces
BASIS OF TEST DATA

A report titled "Surface Flammability of Various Wood-Base Building Materials" was issued by the Forest Products Laboratory in 1959.[a] A similar report on painted wood surfaces appeared in 1963 in the *Official Digest* of the FSPT,[b] and this material was reprinted in the *NFPA Quarterly* in 1964.[c] All tests were made in a tunnel furnace by a uniform method.[c] From these tests three index numbers were computed and tabulated. In each case an index number of 100 was assigned to the performance of a standard red-oak specimen; the comparative index numbers were determined as follows:

1. **Flame-spread index.** If the flame spread was faster than for the standard red-oak specimen, the index was determined by the length of time for the flames to reach the end of the test specimen, as compared with the standard time (18.4 min) for the flame to reach the end of the red-oak specimen. For a flame spread slower than that for red oak, the index was based on the ratio of the distances reached by the flames on the two specimens in the standard 18.4 min period (87 in. for red oak).
2. **Index of heat contributed.** This index was based on the readings of thermocouples in the furnace stack.
3. **Smoke density index.** This index was obtained from the readings of a photoelectric smoke meter in the furnace stack.

A zero reference for both the smoke-density and heat-contributed indices was established by using a test specimen of asbestos millboard.

It is apparent that this is an arbitrary test procedure that gives approximate comparative results only. Tests of identical specimens often varied ten points or more on the index scales (sometimes with one above 100 and the other below 100). It must be concluded that the various woods and finishes cannot be ranked in absolute order and that the tests serve rather to establish classes or categories of surface-flammability performance. This also is to be expected from the fact that the specimens within a class will vary, i.e., no two trees are exactly alike and no two manufacturers of wallboard of the same description will produce identical specimens.

Although specific index numbers are quoted in the following table, they must be considered as approximate test results rather than absolute ratings.

---

[a] "Surface Flammability of Various Wood-Base Building Materials", H.D. Bruce and L.E. Downs, Forest Products Laboratory Report No. 2140, U.S. Department of Agriculture, Forest Service, 1959.
[b] *Official Digest*, Federation of Societies for Paint Technology, August 1963.
[c] W.C. Eickner and C.C. Peters, Surface Flammability of Wood Coatings, *National Fire Protection Association Quarterly*, April 1964.

Description of test specimens	lb/ft³	Moisture, %	Index numbers			Evaluation of flammability
			Flame spread	Heat contributed	Smoke density	
Tests on 1-in. lumber						
Alder	29.7	6.5	121	121	112	8
Aspen	27.3	6.1	121	124	72	7
Bald cypress	29.6	6.9	112	109	389	14
Basswood	28.0	5.4	128	128	79	3
Beech	46.8	5.8	101	140	132	24
Birch	39.4	5.6	96	94	86	26
Cedar, red	33.8	8.9	109	94	224	16
Chestnut	29.0	6.1	120	92	17	10
Cottonwood	27.3	5.0	134	135	125	1
Elm, slippery	38.5	6.1	89	108	151	29
Fir, Douglas	27.1	5.9	116	77	81	11
Fir, white	29.9	6.7	115	96	109	12
Hemlock	29.0	7.4	108	95	78	17

Table 8-74.   SURFACE FLAMMABILITY OF WOODS *(Continued)*

Description of test specimens	lb/ft³	Moisture, %	Index numbers			Evaluation of flammability
			Flame spread	Heat contributed	Smoke density	
Tests on 1-in. lumber *(Continued)*						
Larch	34.9	6.8	106	98	56	19
Mahogany	29.7	6.6	104	70	34	22
Mahogany, Philippine	35.3	5.4	106	81	55	20
Maple, sugar	41.7	6.2	95	83	76	28
Oak, red	39.0	5.0	100	100	100	25
Oak, white	40.8	6.4	95	91	40	27
Pine, Northern white	23.9	5.7	132	104	193	2
Pine, ponderosa	27.7	6.5	114	102	230	13
Pine, Southern yellow	29.2	6.8	102	115	158	23
Pine, sugar	24.0	6.2	125	98	262	4
Pine, western white	25.9	6.0	123	113	274	6
Poplar, yellow	31.0	5.7	124	125	155	5
Redwood	25.6	6.4	121	68	188	9
Spruce, sitka	26.8	6.8	112	80	57	15
Sweet gum	32.3	6.8	105	105	68	21
Walnut, black	37.4	5.1	107	114	85	18
Tests on plywood						
¼-in., 3 ply, fir, interior, protein glue	30.7	6.3	123	119	136	
¼-in., 3 ply, fir, interior, resin glue	34.6	5.0	121	122	81	
⅜-in., 3 ply, fir, exterior, resin glue	33.6	4.8	114	112	96	
⅜-in., 3 ply, fir, exterior, paint A, 1 coat	34.8	4.8	83	58	477	
⅜-in., 3 ply, fir, exterior, paint A, 2 coats	33.9	5.1	53	19	968	
⅜-in., 3 ply, fir, exterior, paint B, 1 coat	34.1	5.0	65	21	747	
⅜-in., 3 ply, fir, exterior, paint B, 2 coats	34.1	5.0	34	10	1 143	
⅜-in., 3 ply, fir, exterior, paint C, 2 coats	34.1	5.0	81	38	1 000	
⅜-in., 3 ply, fir, exterior, paper, plastic overlay	36.	5.0	105	112	141	
Tests on fiberboard						
Four different ½-in. insulating fiberboards	19.	5.0	125	91	105	
Three heavier fiberboards	32.	4.0	122	161	—	

Table 8-74. SURFACE FLAMMABILITY OF WOODS *(Continued)*

Description of test specimens	lb/ft³	Moisture, %	Index numbers			Evaluation of flammability
			Flame spread	*Heat contributed*	*Smoke density*	
Tests on hardboard						
Willow, oak, wax, resin, 0.22 in.	51.4	3.2	119	177	131	
Fir, pine, wax, resin (4 makes)	61.4	4.3	96	169	407	
Fir, redwood, dense, oil-tempered	70.8	3.1	90	165	657	
Tests on particle boards						
Nine wood-chip boards	36–67	3.8–6.6	85–104	80–150	55–650	

CONCLUSIONS FROM TEST RESULTS

Index numbers obtained by flame-spread tests of 29 species of structural lumber by the tunnel-furnace method gave a graduated scale from the 130 range for flammable softwoods, such as white pine, cottonwood, basswood, and poplar, to the 90 range for the dense hardwoods, such as white oak, birch, elm, and sugar maple. Although the "heat-contributed" index followed the flame-spread index roughly, the smoke density was dependent on other factors. The flame-spread index of $\frac{1}{4}$-in. and $\frac{3}{8}$-in. fir plywood was somewhat higher than for a 1-in. fir board. The flame-spread index for either structural or insulating fiberboard differed little from those for the softwoods.

The flammability of pressed hardboards varied from the softwood range for the lighter boards (50 lb/ft³) to the hardwood range for oil-tempered board of high density (65 lb/ft³).

The particle boards tested were no more flammable than hardwood; however, the range was rather wide, as there is a great variety of such boards.

Many tests were made on painted specimens, most of them with $\frac{3}{8}$-in. Douglas-fir plywood as a base. For some coatings, such as interior oil-base paints and enamels, varnish, shellac, lacquer, and asphalt paint, the flammability was changed little from that of bare plywood, but the smoke and heat were increased. The effectiveness of fire-retardant paints was clearly demonstrated, especially if two coats were used or an intumescent (foaming) composition was employed. Most of the fire-retardant paints reduced the flammability of the fir-plywood specimen to that of the least-flammable hardwood. The best of the fire-retardant paints reduced the flammability index to the 30 range, on a scale of red oak = 100 and asbestos millboard = 0.

Table 8-75.   COMBUSTION RETARDANTS FOR SOLID MATERIALS

Fire and Flame Retardants for Combustible, Structural Materials, Sheets, Fabrics, and Plastics

For thermal failure, flammability, and explosive-mixture data, see Tables 1-80, 3-44, 8-77, and 8-78.

Variables. Conditions for the combustion of solids in air are so diverse that most data and tests provide only relative values. Among the variables, in addition to the chemical composition, are size, shape, and orientation of the materials and the heat source, initial and ambient temperatures, humidity and moisture content, intensity of radiation, air motion, and duration of the exposure. In addition, the material itself may agglomerate, melt, vaporize, decompose, distort, or intumesce (swell), thus greatly altering the access of oxygen to support combustion.

Fire retardants. In most cases a fire-retardant treatment also will raise the maximum service temperature, and both objectives deserve attention. Oxygen supply is largely determined by surface/volume ratio; three classes are quickly recognized: (1) bulk solids and shapes; (2) sheets and fabrics; and (3) particulates and dusts. The common combustible solids—wood, paper, fabrics, and plastics—each present their special problems. The methods of application also are distinctive, e.g., compounding (filling), coating, and impregnating.

The chemistry of fire retardants deals largely with six elements—phosphorus, chlorine, bromine, boron, antimony, and nitrogen. As many treatments involve two or more retardants, a great variety of compounds of these six elements have been used. Inert and refractory coatings of oxides or mineral powders comprise another class of fire retardants, merging into the insulating protective coverings as the thickness and porosity are increased. Many kinds of fillers are used in plastics and rubbers; these usually reduce the flammability, but residues of plasticizers and solvents have opposite effects.

Standard tests. Arbitrary methods of testing for flammability, fire resistance, or fire endurance have been prescribed by the ASTM,[a] the Underwriters' Laboratories, and similar agencies in other countries.

Tabular summary. The following table presents a partial list of the fire-retardant treatments for wood, paper, textiles, and plastics. Many of these same materials are used in fire-retardant treatment or compounding for rubbers, asphalts, and bitumens. Not included in this table are several of the incombustible, or reflective, coatings that may be applied to the surfaces of combustible materials.

A single table hardly can suggest the many complexities of the methods for fire retardation. Flammability is reduced by stabilizing a material to reduce the decomposition into volatile, combustible products. Some treatments increase the residual char, which then acts as a barrier against propagation of the flame. Certain additives melt and form a hard skin; others decompose and give a protective blanket of inert gas. These are complex processes that are not easily analyzed, but there is a vast literature reporting the efforts and the specific results. It should be mentioned that when certain fire-retarding compounds are used together, their effectiveness increases. Examples are combinations of phosphorus compounds with chlorine compounds and antimony in combination with halogens.

---

[a] See the ASTM Index for flammability tests, such as Nos. D568, D635, D777, D1230, D2859, E162, and E286. See also fire tests, such as D1360–61, E84, etc.

REFERENCES

"The Chemistry and Uses of Fire Retardants", J.W. Lyons, John Wiley & Sons, Inc., 1970.
"Fireproofing", J.H. Goundry, American Elsevier Publishing Company, Inc., 1970.

Table 8-75. COMBUSTION RETARDANTS FOR SOLID MATERIALS
(Continued)

CHEMICAL FIRE RETARDANTS FOR SOLIDS

KEY FOR METHODS OF TREATMENT: I = immersion or impregnation; M = mixture-compounded; S = surface application

Class of compound	Examples	Fire retardant for			
		Wood[b]	Paper[c]	Cellulose fabrics	Plastics
Phosphorus compounds					
Phosphoric acid compounds	TCP,DCP,TPP,THPC	I,S		I	M
Ammonium phosphate	$(NH_4)_2PO_4$;$NH_4H_2PO_4$	M	I,M	I	M
p-halogen compounds	PCl_3	I			M
Bromine compounds					
Organic	CHBr aromatics		M,S	I	M
Bromides	$MgBr_2$;$ZnBr_2$;NH_4Br	I			
Chlorine compounds					
Organic	Chlorinated paraffin	M	M,S	I	M
Zinc chloride	$ZnCl_2$	I			
Boron compounds					
Borax; boric acid	$Na_2B_2O_7$;H_3BO_3	S	M,S	I	
Nitrogen compounds					
Ammonium compounds	$(NH_4)_2SO_4$;$(NH_4)_2HPO_4$	S	M		M
Miscellaneous materials					
Silicates	Sodium silicate	I,S	S,M		
Fire-retardant paints		S			
Titanium salts	$(TiCl_4$;$Ti_2(SO_4)_3$				M
Antimony compounds (often with halogens)					
Oxide	$Sb_4O_6(Sb_2O_3)$	M	M,S	I	M
Chloride	$SbCl_3$	I			
Organic		M	M	I	

[b] Including fiber wallboards.
[c] Including cover stock, cardboards, and boxboards.

Table 8-76. TYPICAL FIRE-RESISTANCE RATINGS FOR REINFORCED CONCRETE CONSTRUCTIONS

Ratings are dependent on protective cover of concrete over steel. For length in millimeters, multiply values in inches by 25.4.

Fire-resistance rating, hours[†]	Structural members	Cover over steel, inches
1	Beams, medium size	$\frac{3}{4}-\frac{7}{8}$
	Slabs, unrestrained	1
2	Beams, medium size	$1-1\frac{1}{4}$
	Slabs, unrestrained	$1\frac{1}{2}$
3	Beams, medium size	$1\frac{1}{2}$
	Columns, 12–14 in.	$1\frac{1}{2}$
	Slabs, unrestrained	2
	(Solid concrete walls, 6-in.)	—
4	Columns, 12–14 in.	2
	Columns, >16 in.	$1\frac{1}{2}$
	Slabs, unrestrained	2

Note:

Concrete with aggregates containing more than 30% quartz, chert, flint, or granite has an inferior fire rating unless mesh is used.

High-cement content or light aggregates improve the fire rating.

Cover is defined as the minimum distance from steel to exposed surface.

Prestressed concrete members require more cover for a given rating.

[†] The standard test for fire-resistance rating for a large wall specifies measurement of the time required for a temperature rise of 250°F on the unexposed face when the other face is subjected to a "standard fire".

Table 8-77. FLAMMABILITY LIMITS FOR GASES AND VAPORS IN AIR*

At Atmospheric Pressure and Approximately Room Temperature[a]

For combustion data on common hydrocarbons, see Table 3-44.

Compound	Formula	Flammability limits in air, % of total volume		Compound	Formula	Flammability limits in air, % of total volume	
		Lower (lean)	Upper (rich)			Lower (lean)	Upper (rich)
Acetaldehyde	C_2H_4O	4.	57.	Ethyl bromide	C_2H_5Br	6.7	11.3
Acetone	C_3H_6O	2.6	12.8	(Bromoethane)			
(2-Propanone)				Ethyl chloride	C_2H_5Cl	3.6	15.4
Acetonitrile	C_2H_3N	4.5	16.	(Chloroethane)			
Acrolein	C_3H_4O	2.7	40.	Ethyl ether	$C_4H_{10}O$	1.7	48.
Acrylonitrile	C_3H_3N	3.	17.	(Diethyl ether)			
Allyl alcohol	C_3H_6O	2.5	18.	Ethyl formate	$C_3H_6O_2$	2.7	16.5
Allyl bromide	C_3H_5Br	1.35	7.3	Ethyl nitrite	$C_2H_5NO_2$	3.	50.
(3-Bromopropene)				Ethylene dichloride	$C_2H_4Cl_2$	6.2	16.
Allyl chloride	C_3H_5Cl	3.3	11.2	(1,2-Dichloroethane)			
(3-Chloropropene)				Ethylene oxide	C_2H_4O	3.	90.
Ammonia	NH_3	15.5	27.	(Oxirane)			
Amyl chloride	$C_5H_{11}Cl$	1.5	8.6	Hydrocyanic acid	HCN	5.6	41.
(1-Chloropentane)				Hydrogen	H_2	4.	75.
1,3-Butadiene	C_4H_6	2.	12.	Hydrogen sulfide	H_2S	4.3	45.5
2-Butoxyethanol	$C_6H_{14}O_2$	1.1	12.7	Isopropyl acetate	$C_5H_{10}O_2$	1.8	7.8
Butyl acetate	$C_6H_{12}O_2$	1.7	10.	Methyl acetate	$C_3H_6O_2$	3.1	16.
Butyl alcohol	$C_4H_{10}O$	1.4	18.	Methyl alcohol	CH_4O	7.3	40.
(1-Butanol)				Methylamine	CH_5N	4.9	20.8
Butylamine	$C_4H_{11}N$	1.7	9.8	Methyl bromide	CH_3Br	10.	16.
Butyl chloride	C_4H_9Cl	1.8	10.1	(Bromomethane)			
(1-Chlorobutane)				Methyl butyl ketone	$C_6H_{12}O$	1.25	8.
Butyl ether	$C_8H_{18}O$	1.5	7.6	Methyl chloride	CH_3Cl	8.1	19.5
Carbon disulfide	CS_2	1.	50.	Methyl ether	C_2H_6O	2.	20.
Carbon monoxide	CO	12.	74.	Methyl ethyl ketone	C_4H_8O	1.8	11.
Chloroprene	C_4H_5Cl	4.	20.	Methyl formate	$C_2H_4O_2$	5.	22.7
Crotonaldehyde	C_4H_6O	2.1	15.5	n-Propyl acetate	$C_5H_{10}O_2$	2.0	8.
Cyanogen	C_2N_2	6.	42.	n-Propyl alcohol	C_3H_8O	2.1	13.5
Cyclohexane	C_6H_{12}	1.3	8.3	Propylamine	C_3H_9N	2.	10.4
Cyclopropane	C_3H_6	2.4	10.4	Propyl chloride	C_3H_7Cl	2.6	11.1
n-Decane	$C_{10}H_{22}$	.7	5.4	(1-Chloropropane)			
Deuterium	D_2	5.	75.	Propylene dichloride	$C_3H_6Cl_2$	3.4	14.5
Diborane	B_2H_6	.9	98.	(1,2-Dichloropropane)			
Dichlorobenzene	$C_6H_4Cl_2$	2.2	9.2	Propylene oxide	C_3H_6O	2.	22.
Diethylamine	$C_4H_{11}N$	1.8	10.1	Pyridine	C_5H_5N	1.8	12.4
Dimethylamine	C_2H_7N	2.8	14.4	Triethylamine	$C_6H_{15}N$	1.2	8.
Dioxane	$C_4H_8O_2$	2.	22.2	Trimethylamine	C_3H_9N	2.	11.6
2-Ethoxyethanol	$C_4H_{10}O_2$	1.7	15.6	Vinyl acetate	$C_6H_6O_2$	2.6	21.7
Ethyl acetate	$C_4H_8O_2$	2.5	11.5	Vinyl chloride	C_2H_3Cl	4.	22.
Ethyl alcohol	C_2H_6O	4.3	19.	(Chloroethylene)			
Ethylamine	C_2H_7N	3.5	14.	Xylene	C_8H_{10}	1.	7.

[a]Flammable or explosive limits in pure air will differ greatly at other temperatures or pressures. In general, the effect of increasing temperature or pressure is to widen the flammable range.

*Compiled from several sources.

REFERENCE

Williams-Steiger Occupational Safety and Health Act of 1970, Chapter XVII, Title 29, *Code of Federal Regulations*, April 13, 1971 (amended August 13, 1971), Part 1910 (Occupational Safety and Health Standards); published in the *Federal Register*, Vol. 36, No. 105, May 29, 1971.

Table 8-78. DUST EXPLOSION CHARACTERISTICS

The following table is based on laboratory-test results by the U.S. Bureau of Mines on dried samples of fine dusts (passing 200-mesh sieve).* The values below probably represent "the most hazardous conditions" for these materials.

Type of dust	Ignition temperature of dust cloud, °C	Minimum igniting energy, joules	Minimum explosive concentration, oz/cu ft	Maximum explosion pressure, psig	Maximum rate of pressure rise, psi/sec	Terminal oxygen concentration, %†	Relative explosion hazard
AGRICULTURAL							
Alfalfa	530	0.320	0.105	92	2,200	—	Moderate
Cereal grass	550	0.800	0.250	52	500	—	Weak
Coffee	720	0.160	0.085	53	300	—	Weak
Corn	400	0.040	0.055	95	6,000	—	Strong
Corncob	480	0.080	0.040	110	3,100	—	Strong
Cornstarch	390	0.030	0.040	115	9,000	—	Severe
Cotton linters	520	1,920	0.500	48	150	—	Moderate
Cottonseed	530	0.120	0.055	96	3,000	—	Moderate
Grain, mixed	430	0.030	0.055	115	5,500	—	Strong
Grass seed	490	0.260	0.290	34	400	—	Weak
Malt, brewers	400	0.035	0.055	92	4,400	—	Strong
Milk, skim	490	0.050	0.050	83	2,100	—	Strong
Peanut hull	460	0.050	0.045	82	4,700	—	Strong
Peat, sphagnum	460	0.050	0.045	84	2,200	—	Strong
Pecan nutshell	440	0.050	0.030	106	4,400	—	Strong
Potato starch	440	0.025	0.045	97	8,000	—	Severe
Rice	440	0.050	0.050	93	2,600	—	Strong
Soy flour	550	0.100	0.060	111	1,600	15	Moderate
Sugar, powdered	370	0.030	0.045	91	1,700	—	Strong
Wheat flour	440	0.060	0.050	104	4,400	—	Strong
Wheat straw	470	0.050	0.055	99	6,000	—	Strong
CARBONACEOUS							
Charcoal, hardwood mix, volatile content 27.1%	530	0.020	0.140	100	1,800	18	Strong
Coal, Ill., No. 7, volatile content 48.6%	600	0.050	0.040	84	1,800	15	Strong
Coal, Pa. (Pittsburgh), volatile content 37.0%	610	0.060	0.055	83	2,300	17	Strong
Gilsonite, Utah, volatile content 86.5%	580	0.025	0.020	78	3,700	—	Severe
Lignite, Calif., volatile content 60.4%	450	0.030	0.030	90	8,000	—	Severe
Pitch, coal tar, volatile content 58.1%	710	0.020	0.035	88	6,000	—	Severe
METALS							
Aluminum	650	0.015	0.045	100	10,000†	2	Severe
Copper	900	—	—	—	—	—	Fire
Iron	420	0.020	0.100	46	6,000	10	Strong
Magnesium	520	0.020	0.020	94	10,000†	0	Severe
Tin	630	0.080	0.190	37	1,300	15	Moderate
Titanium	460	0.010	0.045	80	10,000†	0	Severe
Uranium	20	0.045	0.060	53	3,400	0	Severe
Zinc	600	0.640	0.480	48	1,800	9	Weak

Table 8-78.   DUST EXPLOSION CHARACTERISTICS *(Continued)*

Type of dust	Ignition temperature of dust cloud, °C	Minimum igniting energy, joules	Minimum explosive concentration, oz/cu ft	Maximum explosion pressure, psig	Maximum rate of pressure rise, psi/sec	Terminal oxygen concentration, %†	Relative explosion hazard
PLASTICS							
Acetal resin (polyformaldehyde)	440	0.020	0.035	89	4,100	11	Severe
Acrylic polymer resin							
Methyl methacrylate-ethyl acrylate	480	0.010	0.030	85	6,000	11	Severe
Alkyd resin							
Alkyd molding compound	500	0.120	0.155	15	150	15	Weak
Amino resin							
Urea-formaldehyde molding compound	450	0.080	0.075	89	3,600	17	Strong
Cellulose fillers							
Wood flour	430	0.020	0.035	110	5,500	17	Severe
Cellulose resin							
Ethyl cellulose molding compound	320	0.010	0.025	102	6,000	11	Severe
Epoxy resin	530	0.020	0.020	86	6,000	12	Severe
Phenolic resin							
Phenol-formaldehyde molding compound	500	0.020	0.030	92	10,000†	14	Severe
Rayon (viscose) flock	520	0.240	0.055	88	1,700	—	Moderate
Rubber, synthetic	320	0.030	0.030	93	3,100	15	Severe

†The terminal oxygen concentration is the limiting oxygen concentration in air-CO_2 atmosphere required to prevent ignition of dust clouds by electric spark.

*See U.S. Bureau of Mines Reports of Investigations, Nos. 5624, 5753, 5971, 6516.

Table 8-79.   U.S. MOTOR VEHICLE STATISTICS

Table A.   AUTOMOTIVE SALES, REGISTRATION, AND SCRAPPING

Numbers in Millions

Year	Registration[a]	Factory sales[b]	Scrapped[c]	Year	Registration[a]	Factory sales[b]	Scrapped[c]
1971	112.01			1958	68.30	5.14	4.18
1970	108.98	8.24		1957	67.13	7.22	4.27
1969	105.10	10.15		1956	65.15	6.92	4.95
1968	100.88	10.72	7.02	1955	62.69	9.17	4.39
1967	96.93	8.98	7.49	1954	58.51	6.60	4.35
1966	94.19	10.33	6.93	1953	56.22	7.32	4.05
1965	90.36	11.05	7.07	1952	53.27	5.54	3.77
1964	86.31	9.29	6.83	1951	51.91	6.77	3.77
1963	82.75	9.10	5.95	1950	49.16	8.00	3.23
1962	79.17	8.17	5.42	1949	44.14[c]	6.25	3.15
1961	75.95	6.77	4.98	1948	40.57[c]		
1960	73.87	7.87	4.78	1947	37.36[c]	4.98	1.51
1959	71.36	6.73	5.11				

[a]U.S. Department of Transportation.
[b]Automobile Manufacturers Association (includes foreign sales).
[c]Does not include publicly owned vehicles.

Table B.   SURVIVAL RATE OF MOTOR VEHICLES

The following data on the survival of motor vehicles were published by the Environmental Control Administration, 1968.

Years after original registration	Surviving vehicles, %	Years after original registration	Surviving vehicles, %
14	18	8	81
12	33	6	95
10	56		

Table C.   ANNUAL ADDITIONS OF PASSENGER CARS*

In 1970 it was estimated (at the University of Michigan) that 82% of all families in the United States owned one or more automobiles.

Year	Thousands of cars		Year	Thousands of cars	
	Domestic production	Imports		Domestic production	Imports
1970	6 550	2 013	1966	8 605	913
1969	8 224	1 847	1965	9 335	559
1968	8 849	1 620	1964	7 745	516
1967	7 413	1 021	1963	7 644	409

*Data from the U.S. Department of Commerce.

MOTOR VEHICLE ACCIDENTS*

There are approximately one-seventh as many reported vehicular accidents annually in the United States as there are registered motor vehicles. The fatalities average about one for each 275 accidents; hence, a fatality occurs about once in 20 million vehicle miles (1960–70 data). Nevertheless, about one death in forty results directly from an automobile accident, and the ratio of non-fatal injuries to traffic deaths is more than eight to one. On the basis of total population, the annual traffic

Table 8-79.   U.S. MOTOR VEHICLE STATISTICS *(Continued)*

death rate is roughly one in 4 000. The annual death rate from all causes is slightly above one in one hundred. Motor-vehicle fatalities approach one-half of all accidental deaths. For women automobile injuries occur less often than injuries at home. Contrary to popular opinion, the densely populated states do not have the highest overall rate of accidents per capita. For vehicle accidents of all kinds the deaths in Alaska, Wyoming, New Mexico, and Nevada are twice as great per thousand population as they are in New York, New Jersey, Pennsylvania, and Illinois.

The economic losses by serious automobile accidents are very high. Based on a sample of police-reported serious or fatal automobile accidents (1967), the average families involved suffered an average total loss of actually more than twice the total reparations from insurance and other settlements.[a] As to the overall costs of insurance, in the late 1960's, the insurance companies were paying out for losses close to 60% of premiums; in some earlier years the percentage was as low as 40.

[a]As reported in "Economic Consequences of Automobile Accident Injuries", Vol. 1, Department of Transportation, 1970.

	1950	1960	1965	1967	1968	1969	1970
Number of accidents, millions	8.3	10.4	13.2	13.7	14.6	15.5	16.4[a]
Number of deaths, thousands	34.76	38.1	49.2	52.9	54.9	56.0	55.0
Fatalities per 100 000 population[b]	23.1	23.4	21.3	25.4	26.7	27.5	26.8
Motor vehicle deaths as % of all U.S. accidental deaths from all causes	38.1	40.7	45.5		47.3		
Number of personal injuries, thousands		3 078.	4 100.		4 400.		5 100
Collisions, all kinds, % of accidents		75	75		76	73	72
Fatal accidents at night, 6 p.m.–6 a.m., % of deaths		57	57		56	57	56
Insurance paid, bodily injury, millions of dollars	396.	1 700.	2 460.	2 580.	2 800.	3 100.	

Note: The Department of Transportation has estimated that the overall direct cost of each fatal accident in 1967 was about $12 000 (plus loss of future earnings of at least $40 000).

[a]Preliminary data.

[b]It should be noted that from 1950–1970 the ownership of automobiles increased from 59% of all families to over 80% of all families.

*Based largely on: "Vital Statistics of the United States", HEW–PHS annual reports; data from National Safety Council, Department of Transportation, and annual insurance reports.

Table 8-80.   ACCELERATION TOLERANCE OF THE HUMAN BODY

Response to acceleration depends on magnitude, direction, and rate of change of the acceleration, and the posture and support of the body. Physiological acceleration is defined as total reactive force per unit of body mass, $a = f/m$. The common accelerations are up, down, forward, and backward. Approximate typical or average acceleration tolerance, i.e., the duration tolerated without grayout or blackout, is given in the following table.

MAXIMUM VALUES OF ACCELERATION TOLERATED, g†

Direction of body motion	Duration, seconds		
	6	60	300
Down	5	3.1	2.4
Up	10.1	7	5
Forward	20	11	7.5
Backward	17	9	6.5

†Moderate rate of onset; not an impact.

BODILY EFFECTS OF VIBRATION INTENSITY
For a Seated Subject, Upright

Intensity	Direction of acceleration	Result
2 g		Visual acuity reduced
4 g		Body movements difficult
4 g	Downward	Eye hemorrhage probable
5 g	Upward	Loss of muscular control
12 g	Forward	Breathing difficult

Table 8-81.   LIMIT OF POSSIBLE SURVIVAL IN DECELERATION FROM A GIVEN VELOCITY*

The survival limit shown by the following table is based essentially on free falls that have been survived. This survival limit represents a force of 175 to 200 g. Any data on impact experience of humans should be used with extreme care because of the great variations in biophysical factors.

Velocity		Equivalent free-fall distance, ft†	Minimum deceleration distance for survival, in.
fps	*mph*		
15	10.5	(4)	0.25
30	20.5	15	1.0
50	34.	42	2.5
75	51.	120	6.
100	68.	210	10.

†Approximate, for entire human body in sea-level air.

Notes:

A severe (fatal) auto accident from 60 mph might represent a deceleration in one to three inches, or 500 to 1,500 g.

Parachute-opening shock might be 10 g, from 80 mph and a deceleration distance of 15 ft with landing at about 12 mph, 2.5 g.

Head impacts equivalent to a fall of 3 ft and deceleration in 1.5 in. (about 20 g) might be tolerated with helmet protection.

Abrupt deceleration greater than 10 g may produce some injury, especially if it is a free fall. Transverse deceleration, seated and supported, can reach 30 to 40 g without severe injury.

A fall of 70 ft into a fireman's net, stopped in about 3 ft, represents an impact shock of about 15 g.

*Condensed from: "Bioastronautics Data Book", P. Webb, Ed., NASA SP-3006, U.S. Government Printing Office, 1964, Chart No. 5-2.

Table 8-82.   EFFECTS OF SINUSOIDAL MECHANICAL VIBRATION

Effects will vary with onset, duration, and frequency. Frequencies of 4 to 8 cps are most objectionable.

EFFECTS OF VIBRATION INTENSITY

For a Seated Subject, with Seat Belt; Frequencies Less Than 100 cps

Intensity of vibration	Effect on human subject
.005–0.10 g	Perceptible
.05–0.5 g	Unpleasant
0.6–1.0 g	Very annoying
1–2 g	Tolerable but alarming
3 or 4 g	Injurious if prolonged

EFFECTS OF VIBRATION FREQUENCY

The greatest intensities of discomfort (or pain) occur in various frequency ranges for different parts of the body. Resonant responses are often the cause of these intense sensations. The following are typical reactions.

Nature of discomfort	Frequency, cps
Respiration difficulty; air hunger	3–8
Abdominal pain or contractions	4–12
Muscular tightness and contractions	4–20
Chest pains	5–9
Lumbo-sacral pain	5–12
Speech difficulty	>5
Jaw and throat discomfort	6–16
Urge to defecate or urinate	10–18

Table 8-83. EFFECTS OF ELECTRIC CURRENT ON MAN*

GENERAL EFFECTS OF VARIOUS CURRENTS

Effect	Milliamperes					
	Direct current		Alternating current			
			60-hz		10 000-hz	
	Men	Women	Men	Women	Men	Women
Slight sensation on hand	1	0.6	0.4	0.3	7	5
Perception threshold, median	5.2	3.5	1.1	0.7	12	8
Shock—not painful and muscular control not lost	9	6	1.8	1.2	17	11
Painful shock—muscular control lost by $\frac{1}{2}\%$	62	41	9	6	55	37
Painful shock—let-go threshold, median	76	51	16	10.5	75	50
Painful and severe shock—breathing difficult, muscular control lost by 99.5%	90	60	23	15	94	63
Possible ventricular fibrillation						
Three-second shocks	500	500	675	675		
Short shocks (T in seconds)			$165/\sqrt{T}$	$165/\sqrt{T}$		
Capacitor discharges	50†	50†				

†Energy in watt-seconds.

SUMMARY OF LETHAL EFFECTS OF ELECTRIC CURRENT ON MAN

If long continued, currents in excess of one's let-go current may produce collapse, unconsciousness, and death.

Currents flowing through the chest, the head, or nerve centers controlling respiration may produce respiratory inhibition. Respiratory inhibition is dangerous, because paralysis of the respiratory organs may last for a considerable period even after interruption of the current, and the approved method of artificial resuscitation *must* be applied promptly to prevent suffocation.

Ventricular fibrillation is caused by moderately small currents that produce over-stimulation of the heart rather than physical damage to that vital organ. When fibrillation occurs, the rhythmic pumping action of the heart ceases, and death usually follows in a few minutes.

Heart standstill may be caused by relatively high currents.

Relatively high currents may produce fatal damage to the central nervous system.

Relatively high currents may produce deep burns, and currents sufficient to materially raise body temperature produce immediate death.

Victims who have been revived sometimes die suddenly without apparent cause. This is thought to be due to: (*a*) aggravation of pre-existing conditions, (*b*) the result of hemorrhages affecting vital centers, or (*c*) the effects of shock to the nervous system. Delayed death also may be due to burns or other complications.

*Abstracted from: "Deleterious Effects of Electric Shock", C.F. Dalziel, *Record of Proceedings: Meeting of Experts on Electrical Accidents and Related Matters*, International Labor Office, 1961.

Section **9**

Communication
and Computation

Table 9-1. THE GREEK AND RUSSIAN ALPHABETS

Greek letter	Greek name	English equivalent	Russian letter	English equivalent	Greek letter	Greek name	English equivalent	Russian letter	English equivalent
A α	Alpha	(ä)	А а	(ä)	N ν	Nu	(n)	Р р	(r)
B β	Beta	(b)	Б б	(b)	Ξ ξ	Xi	(ks)	С с	(s)
Γ γ	Gamma	(g)	В в / Г г	(v) / (g)	O o	Omicron	(o)	Т т	(t)
Δ δ	Delta	(d)	Д д	(d)	Π π	Pi	(p)	У у	(ōo)
E ε	Epsilon	(e)	Е е / Ж ж	(ye) / (zh)	P ρ	Rho	(r)	Ф ф	(f)
Z ζ	Zeta	(z)	З з	(z)	Σ σ ς	Sigma	(s)	Х х	(kh)
H η	Eta	(ā)	И и	(i, ē)				Ц ц	(ts)
Θ θ	Theta	(th)	Й й	(ē) *	T τ	Tau	(t)	Ч ч	(ch)
I ι	Iota	(ē)	К к	(k)	Y υ	Upsilon	(ü, ōo)	Ш ш	(sh)
K κ	Kappa	(k)	Л л	(l)	Φ φ	Phi	(f)	Щ щ	(shch)
			М м	(m)	X χ	Chi	(H)	Ъ ъ	†
Λ λ	Lambda	(l)	Н н	(n)				Ы ы	(ē)
			О о	(o, o)	Ψ ψ	Psi	(ps)	Ь ь	‡
								Э э	(e)
M μ	Mu	(m)	П п	(p)	Ω ω	Omega	(ō)	Ю ю	(ū)
								Я я	(yä)

*Used only as a second vowel in a diphthong.
†Indicates nonpalatalization of a preceding consonant.
‡Indicates palatalization of a preceding consonant.

Table 9-2. NOISE CRITERIA FOR COMMUNICATION

From studies of speech interference and noise tolerance, a series of noise-criteria (NC) curves was developed. These have come into wide use for specifying or evaluating maximum noise levels in buildings in terms of octave-band measurements. These criteria apply to steady background noises that do not contain noticeable pure tones. The NC = 10 curve is the approximate threshold of hearing for continuous noise.

OCTAVE-BAND NOISE CRITERIA

Mid frequency of octave band →	62.5	125	250	500	1,000	2,000	4,000	8,000
Noise criterion level	Octave-band sound pressure level, db							
NC = 10	49	36	26	17	12	—	—	—
NC = 20	51	41	33	27	22	19	16	15
NC = 25	54	45	38	31	27	24	21	20
NC = 30	58	49	42	36	32	29	27	26
NC = 35	61	53	46	40	37	34	32	31
NC = 40	64	57	51	44	42	39	37	36
NC = 45	68	60	55	48	47	44	43	41
NC = 50	71	64	58	54	51	49	47	46
NC = 55	74	68	63	58	56	54	53	51
NC = 60	78	72	67	63	61	59	57	56
NC = 65	—	75	71	68	66	64	63	61
NC = 70	—	79	75	73	71	69	68	67

Note: If a one-third octave-band meter is used, reduce all values in the body of this table by 5 db.
Example: A performance specification might require that the steady "equipment noise" in a given location (unoccupied) shall not exceed a certain NC curve. This designated NC number would normally be at least 5 db below the expected noise-meter reading when the room is occupied (see Table 7-45). It is usually understood that no single octave-band reading may exceed the specified NC value. The speech-interference level is the arithmetic average of the octave-band readings centered at 500, 1,000, and 2,000 hz.

REFERENCES

"Criteria for Noise in Buildings", L.L. Beranek, *Noise Control*, 3(1):19–27, January 1957.
"ASHRAE Guide and Data Book: Systems and Equipment", American Society of Heating, Refrigerating and Air-Conditioning Engineers, 1970, Chapter 33—Sound and Vibration Control.

Table 9-3. SPEECH COMMUNICATION INTERFERENCE

The Speech-Interference Level (SIL) is a quantity defined as the arithmetic mean of the sound-pressure levels within the three octave bands centered at 500, 1,000, and 2,000 cps for typical broadband background noises.

Table A. MAXIMUM SPEECH-INTERFERENCE LEVELS*

Distance to talker, ft	Acoustic absorption of area, sabines	Shout, db	Very loud, db	Raised voice, db	Normal, db
$\frac{1}{2}$	Any value	90	84	78	72
1	Any value	84	78	72	66
2	Any value	78	72	66	60
4	Below 100	78	72	66	60
	Above 100	72	66	60	54
8 or more	Below 100	78	72	66	60
	100–400	72	66	60	54
	400–1,600	66	60	54	48

Note: A sabine is a unit of absorption equal to the absorption of one square foot of surface that is totally sound absorbent. As more absorption is present, less interference can be tolerated, because the speech energy is partially absorbed. The above table is for the male voice—decrease 5 db for the female voice. Speech energy below 200 cps or above 7,000 cps contributes very little to speech intelligibility.

*From: "Human Engineering Guide for Equipment Designers", 2nd ed., W.E. Woodson, University of California Press, 1954. Reprinted by permission of The Regents of the University of California.

Table B. SPEECH-COMMUNICATION CRITERIA†

Since reactions to background noise are dependent on location and type of activity, the following interpretation of the SIL scale is helpful.

SIL, db	Voice level and distance	Nature of possible communication	Type of working area
45	Normal voice at 10 ft	Relaxed conversation	Private offices; conference rooms
55	Normal voice at 3 ft; raised voice at 6 ft; very loud voice at 12 ft	Continuous communication in work areas	Business, secretarial, control rooms of test cells, etc.
65	Raised voice at 2 ft; very loud voice at 4 ft; shouting at 8 ft	Intermittent communication	
75	Very loud voice at 1 ft; shouting at 2–3 ft	Minimal communication (danger signals; restricted prearranged vocabulary desirable)	

†From: "Handbook of Acoustic Noise Control", W.A. Rosenblith and K.N. Stevens, Vol. II: "Noise and Man", WADC TR 52–204, Wright Air Development Center, Wright Patterson Air Force Base, June 1953,

REFERENCES

"Bioastronautics Data Book", P. Webb, Ed., NASA SP-3006, National Aeronautics and Space Administration, 1964.
"Human Engineering Guide to Equipment Design", C.T. Morgan, J.S. Cook, III, D. Chapanis, and M.W. Lund, Eds., McGraw-Hill Book Company, 1963.

Table 9-4. VISUAL INFORMATION DISPLAY CONSOLES*

Typical uses for computer-serviced visual-display systems. Requirements for cathode-ray tube, console displays with computer connection. Alphanumeric and graphical presentation.

DISPLAY-CONSOLE USERS AND FEATURES

SYMBOLS:
  A—more than one font
  B—special inputs, e.g., light pen, video
  C—special outputs (hard copy, etc.)
  D—plotting capability desired

User	Number of characters	Screen size, in.	Color	Remarks
Airline-reservations clerks	500	10	No	C
Auto-rental dispatcher	500	10	No	C
Banking and finance specialists	1,000	14	No	
Bank managers	1,000	14	No	C
Bank tellers	500	10	No	
Communications				
Message-center operator (alphanumeric)	5,000	14	No	A, B, C
Design engineers	$\approx 5,000$	≈ 20	Yes?	B, C, D
Fleet-vehicle location and contents				
controllers	1,000	14	No	C
Hospital administrators—nurses	≈ 500	10–14	No	C
Industrial administrators (personnel records,				
group-insurance files, financial admini-				
stration)	1,000	14	No	C
Industrial government management personnel				
Budgeting	2,000	14	No	C, D
Process control	500	14	?	C, D
Production control	500	14	?	C, D
Inventory management	1,000	14	No	C
Librarian (indexing and abstracting)	5,000	14–16	No	A, C
Librarian (library search)	1,000	14	No	C
Mathematicians	$\approx 5,000$	≈ 20	Yes?	B, C, D
Military administrators	1,000	14	No	C
Military commanders				
Photographic and map interpretation	5,000	20	Yes	A, B, C, D
Cryptography and translation	2,000	14	?	A, C
Planning and war games	Max possible	20	Yes	B, C, D
Railroad car and shipment-disposition				
controllers	1,000	14	No	C
Research engineers	1,000	14	No	A, C
State and local government personnel				
(vehicle license identification, traffic and				
criminal office records)	500	10	No	C
Stock brokers				
Portfolio analysis	1,000	14	No	C, D
Stock quotations	500	10–14	No	
Systems operators, monitors, paging clerks,				
security personnel	500	14–20 or more	Yes?	B, C
Text editors	500	14	No	
Utilities, telephone-service representatives	500	10	No	C

*Adapted from: "Visual Information Display Systems: A Survey", NASA SP–5049, National Aeronautics and Space Administration, 1968.

Table 9-5. SUGGESTIONS FOR VISUAL DISPLAY

Alphabetical or Digital

Following are letter heights and digit sizes for various viewing distances. Good light and good contrast are desired.

Viewing distance, ft	2	3	5	10	20	30	50	100	200	300	500	1000
Character height, in.	0.20	0.22	0.28	0.40	0.66	1.0	1.6	3.1	6.0	9.0	15.	30.

Notes:
    Character height/width ratio should usually be 1.5/1, never taller than 2/1 or wider than 1/1.
    Avoid crowding; observe symmetry. Use simple, bold character styles. Do not mix lettering styles or capital letters with lower case. Avoid glossy surface finishes.

Table 9-6. VOICE-TRANSMISSION FACILITIES*

Number of Voice Circuits, Present and Future

Transmission method	Maximum frequency used	Voice circuits per system	Maximum voice circuits per route
Carrier on paired cable	260 Khz	12–24	1,000–2,000
Digital on paired cable	1.5 Mhz	24	4,800
Carrier on coaxial cable	2.7 Mhz	600	1,800
	8.3 Mhz	1,860	5,580
	18.0 Mhz	3,600	32,900
Carrier on microwave radio	4.2 Ghz	600	3,000
	6.4 Ghz	1,800	14,000
	11.7 Ghz	100	400
Satellite	30 Ghz	1,500	20,000
Millimeter wave guide	100 Ghz	5,000	250,000
Optical-guide laser	1,000,000 Ghz	?	10,000,000

*From: "The Changing Criteria for Optimizing Engineering Designs", P.S. Myers, SAE–690303, Society of Automotive Engineers, 1969, p. 4.

Table 9-7. MODULATION SYSTEMS

Terminology and Characteristics

Modulation is the process of modifying the characteristics of a carrier wave in accordance with a message signal.

CONTINUOUS MODULATION SYSTEMS

These are analog systems in which some parameter in the information channel bears a direct relation to the information to be transmitted.

AM: In *amplitude modulation* the amptitude of a high-frequency carrier is varied by the modulating signal.

DSB: In *double-sideband* amplitude modulation the radio-frequency envelope follows the waveform of the modulating signal, and the upper and lower bands are symmetrical. In one modification of this, the output is passed through a filter that attenuates one sideband differently from the other; the result is known as *vestigial-sideband modulation*.

SSB: *Single-sideband* transmission utilizes a filter that eliminates the signal on one side of the carrier frequency.

Detection: In *coherent* or *synchronous detection* of either DSB or SSB signals, the detector must be supplied with a carrier wave that is synchronized with the transmitter wave. In *envelope detection* the modulated carrier enters a half-wave rectifier from which the output is filtered to produce the message signal. If the signal/noise ratio is low, envelope detection cannot be used.

PM: *Phase modulation* is a system in which the instantaneous phase angle of the carrier signal is varied in accordance with the amplitude of the modulating signal.

FM: In *frequency modulation* the instantaneous carrier frequency is varied in accordance with the amplitude of the modulating signal. It is therefore a form of angular modulation. It has many modifications or varieties, including those in which the control of carrier frequency by the modulator is indirect. The FM system lends itself to several methods for signal/noise ratio improvement. Wide-band FM is much superior to AM with respect to noise interference.

PULSE MODULATION SYSTEMS

There are several pulse-modulation systems, all of which are based on the sampling principle in which frequent samples of the message function are used to modulate some parameter of the carrier. The unmodulated carrier is a uniform series of identical pulses. The sampling rate must be higher than twice the highest signal frequency.

PAM: In *pulse-amplitude modulation* the amplitude of the recurring pulses is changed by the instantaneous samples of the message.

PTM: *Pulse-time modulation* is of several kinds. The *duration* (length or width) of the pulse may be changed by the message sample, or the *position* of the uniform short pulse on the time scale may be changed.

PFM: *Pulse-frequency modulation* is similar to continuous FM, but the carrier wave is now a series of pulses, and the modulating or message signal is a series of samples from the continuous message function.

PCM: The term *pulse-code modulation* is commonly applied to a digital-code system in which the presence or absence of a pulse is used to convey information. Groups of pulses (or spaces) constitute codes with various numbers of discrete levels, e.g., a 4-digit code with 16 levels or a 10-digit code with 1,024 levels (2^n). Number systems other than binary or decimal may be used, thus greatly increasing the amount of information transmitted (see Section 9-3). The great advantages of pulse-code systems have helped to make possible the vast developments in such fields as space communication and computers as well as the great increases in channel capacity for long-distance communication, and also some of the outstanding advances in automatic control.

One great advantage of pulse-type communication is the comparative freedom from noise problems. A signal/noise ratio of the order of 20 db represents a threshold above which interference is negligible, and the presence of a pulse can be detected even if the ratio approaches unity.

Analysis of problems and development of equipment in the fields of coding, decoding, error detection, and error correction have been major activities in communications and computer engineering in the past decade (see References).

REFERENCES

"Frequency Modulation Engineering", C.E. Tibbs and G.G. Johnstone, John Wiley & Sons, 1956.

IEEE Trans. on Information Theory, all volumes (since 1963).

IRE Trans. on Information Theory, all volumes (1955–1962).

"Modulation, Noise and Spectral Analysis", P.F. Panter, McGraw–Hill Book Company, 1965.

Table 9-8. DETERMINATION OF NECESSARY BANDWIDTH*

FCC Rules and Regulations, Part 2, January 1964

The necessary bandwidth is the minimum value of bandwidth sufficient to insure the transmission of information at the rate and with the quality required for the system employed. Emissions needed for satisfactory functioning of the receiving equipment such as the carrier in reduced-carrier systems, or a vestigial sideband, are included in the necessary bandwidth.

SYMBOLS:

B_n = necessary bandwidth in hertz

B = telegraph speed in bauds

N = maximum possible number of black-plus-white elements to be transmitted per second, in facsimile and television

M = maximum modulation frequency in hertz

C = subcarrier frequency in hertz

D = half the difference between the maximum and minimum values of the instantaneous frequency; instantaneous frequency is the rate of change of phase

t = pulse duration in seconds

K = an overall numerical factor that varies according to the emission and depends on the allowable signal distortion

Class of emission	Description	Necessary bandwidth, hertz	Examples and details	Designation of emission
		AMPLITUDE MODULATION		
A1	Continuous-wave telegraphy	$B_n = BK$ where $K = 5$ for fading circuits $K = 3$ for nonfading circuits	Morse code at 25 words per minute, $B = 20$, $K = 5$ Bandwidth: 100 hertz	0.1A1
			Four-channel time-division multiplex, 7-unit code, 42.5 bauds per channel, $B = 170$, $K = 5$ Bandwidth: 850 hertz	0.85A1
A2	Telegraphy, modulated by an audio frequency	$B_n = BK + 2M$ where $K = 5$ for fading circuits $K = 3$ for nonfading circuits	Morse code at 25 words per minute, $B = 20$, $M = 1,000$, $K = 5$ Bandwidth: 2,100 hertz	2.1A2
A3	Telephony	$B_n = M$ for single sideband $B_n = 2M$ for double sideband	Double-sideband telephony, $M = 3,000$ Bandwidth: 6,000 hertz	6A3
			Single-sideband telephony, reduced carrier, $M = 3,000$ Bandwidth: 3,000 hertz	3A3A
			Telephony, two independent sidebands, $M = 3,000$ Bandwidth: 6,000 hertz	6A3B
A3	Sound broadcasting	$B_n = 2M$ M may vary between 4,000 and 10,000, depending on quality desired	Speech and music, $M = 4,000$ Bandwidth: 8,000 hertz	8A3
A4	Facsimile, carrier modulated by tone and by keying	$B_n = KN + 2M$ where $K = 1.5$	The total number of picture elements (black plus white) transmitted per second is equal to the circumference of the cylinder multiplied by the number of lines per unit length and by the speed of rotation of the cylinder in rps. Diameter of cylinder = 70 millimeters Number of lines per millimeter = 5 Speed of rotation = 1 rotation per second $N = 1,100$ $M = 1,900$ Bandwidth: 5,450 hertz	5.45A4
A5 and F3	Television (visual and aural)	Refer to relevant CCIR documents for bandwidths of commonly used television systems	Number of lines = 525 Number of lines per second = 15,750 Video bandwidth: 4.2 megahertz Total visual bandwidth: 5.75 megahertz FM aural bandwidth including guard bands: 250,000 hertz Total bandwidth: 6 megahertz	5,750A5C 250F3
A9	Composite transmission	$B_n = 2M$ (double sideband)	Television relay, video limited to 4 megahertz, audio on 6.5 megahertz FM subcarrier, subcarrier deviation = 50 kilohertz M = subcarrier frequency plus its maximum deviation = 6.55×10^6 Bandwidth: 13.1×10^6 hertz	13,100A9

Table 9-8. DETERMINATION OF NECESSARY BANDWIDTH *(Continued)*

Class of emission	Description	Necessary bandwidth, hertz	Examples and details	Designation of emission
		AMPLITUDE MODULATION *(Continued)*		
A9	Composite transmission	$B_n = 2M$ (double-sideband)	Microwave relay system providing 10 telephone channels occupying baseband between 4 and 164 kilohertz $M = 164,000$ Bandwidth: 328,000 hertz	328A9
		FREQUENCY MODULATION		
F1	Frequency-shift telegraphy	$B_n = 2.6D + 0.55B$ for $1.5 \leq 2D/B \leq 5.5$ $B_n = 2.1D + 1.9B$ for $5.5 \leq 2D/B \leq 20$	Four-channel time-division multiplex with 7-unit code, 42.5 bauds per channel, $B = 170$, $D = 200$; $2D/B = 2.35$; therefore the first equation in column 2 applies Bandwidth: 613 hertz	0.6F1
F3	Commercial telephony	$B_n = 2M + 2DK$ K is normally 1, but under certain conditions a higher value may be necessary	For an average case of commercial telephony, $D = 15,000$, $M = 3,000$ Bandwidth: 36,000 hertz	36F3
F3	Sound broad-casting	$B_n = 2M + 2DK$	$D = 75,000$, $M = 15,000$, and assuming $K = 1$ Bandwidth: 180,000 hertz	180F3
F4	Facsimile	$B_n = KN + 2M + 2D$ where $K = 1.5$	(See Facsimile, Amplitude Modulation) Diameter of cylinder = 70 millimeters Number of lines per millimeter = 5 Speed of rotation = 1 rotation per second $N = 1,100$ $M = 1,900$ $D = 10,000$ Bandwidth: 25,450 hertz	25.5F4
F6	Four-frequency diplex telegraphy	If channels are not synchronized, $B_n = 2.6D + 2.75B$, where B is the speed of the higher-speed channel If channels are synchronized, the bandwidth is as for F1, B being the speed of either channel	Four-frequency diplex system with 400-hertz spacing between frequencies, channels not synchronized, 170 bauds keying in each channel, $D = 600$, $B = 170$ Bandwidth: 2,027 hertz	2.05F6
F9	Composite transmission	$B_n = 2M + 2DK$	Microwave relay system providing 240 telephone channels occupying baseband between 60 and 1,050 kilohertz $M = 1.05 \times 10^6$ $D = 2.35 \times 10^6$ Bandwidth: 6.8×10^6 hertz	6,800F9
F9	Composite transmission	$B_n = 2M + 2DK$	TV microwave relay, aural program on 7.5 megahertz subcarrier; subcarrier deviation plus or minus 150 kilohertz $M = $ subcarrier frequency plus maximum deviation $= (7.5$ plus $0.15) \times 10^6$ $D = 1 \times 10^6$ (visual) plus 0.3×10^6 (aural) Bandwidth: 17.9×10^6 hertz	17,900F9
F9	Composite transmission	$B_n = 2M + 2DK$ where $K = 1$	Stereophonic FM broadcasting (United States system) with multiplexed sub-sidiary communications subcarrier, $M = 75,000$, $D = 75,000$ Bandwidth: 300,000 hertz	300F9
		PULSE MODULATION		
P0	Unmodulated pulse	$B_n = 2K/t$ K depends on the ratio of pulse duration to pulse rise time. Its value usually falls between 1 and 10, and in many cases it does not need to exceed 6	$t = 3 \times 10^{-6}$, $K = 6$ Bandwidth: 4×10^6 hertz	4,000P0
P2 or P3	Modulated pulse	Bandwidth depends on the particular types of modulation used, many of these being still in the development stage		
P9	Composite transmission	$B_n = 2K/t$ where $K = 1.6$	Microwave relay, pulse-position modulated by 36-channel baseband; pulse width at half amplitude = 0.4 microsecond Bandwidth: 8×10^6 hertz	8,000P9

*From: "Reference Data for Radio Engineers", 5th ed., Howard W. Sams & Co., Inc., Indianapolis, Indiana, 1968.

Table 9-9. INTERNATIONAL TELEVISION STANDARDS*

For a summary of frequencies allocated to broadcast services, see Table 2-13.

Standards are designated by letter and used (or proposed for use) in the various countries as follows:

M—United States, Canada, Mexico, Panama, Japan, Korea, Iran, Saudi Arabia

A—Great Britain, Ireland

B—Netherlands, Switzerland, Italy, Spain, Portugal, Sweden, Norway, Denmark, Finland, Austria, West Germany, Morocco, Nigeria, Rhodesia, India, Pakistan, Australia, New Zealand

D—U.S.S.R., Czechoslovakia, Poland, Hungary, Romania, Bulgaria

E—France, Monaco

Other systems are little used, but are distributed as follows: *C* and *F* in Belgium; *F* in Luxembourg; *G* in Italy and West Germany; *K* in U.S.S.R.; *L* in France; *N* in Argentina.

	A	M	N	B	C	G	H	I	D, K	L	F	E
Lines/frame	405	525	625	625	625	625	625	625	625	625	819	819
Fields/sec	50	60	50	50	50	50	50	50	50	50	50	50
Interlace	2/1	2/1	2/1	2/1	2/1	2/1	2/1	2/1	2/1	2/1	2/1	2/1
Frames/sec	25	30	—	25	25	25	25	25	25	25	25	25
Lines/sec	10,125	15,750	—	15,625	15,625	15,625	15,625	15,625	15,625	15,625	20,475	20,475
Aspect ratio[1]	4/3	4/3	—	4/3	4/3	4/3	4/3	4/3	4/3	4/3	4/3	4/3
Video band (Mhz)	3	4.2	4.2	5	5	5	5	5.5	6	6	5	10
RF band (Mhz)	5	6	6	7	7	8	8	8	8	8	7	14
Visual polarity[2]	+	—	—	—	+	—	—	—	—	+	+	+
Sound modulation	A3	F3	—	F3	A3	F3	F3	F3	F3	F3	A3	A3
Pre-emphasis in microseconds	—	75	—	50	50	50	50	50	50	—	50	—
Deviation (khz)	—	25	—	50	—	50	50	50	50	—	—	—
Gamma of picture signal	0.45	0.45	—	0.5	0.5	0.5	0.5	0.5	0.5	0.5	0.5	0.6

[1]In all systems the scanning sequence is from left to right and top to bottom.

[2]All visual carriers are amplitude modulated. Positive polarity indicates that an increase in light intensity causes an increase in radiated power. Negative polarity (as used in the US—Standard *M*) means that a decrease in light intensity causes an increase in radiated power.

*From: International Radio Consultative Committee (CCIR), Report 308, Geneva, 1963.

Table 9-10. ATTENUATION OF VIDEO-LINE FACILITIES*

At 4.5 Megacycles

Type	Shielding	Loss per 1,000 ft, db
Shielded video pairs:		**At 75°F**
16 PSV–S	Spiral inner tape	3.65
16 PSV–L	Longitudinal inner tape	3.52
16 PEV–L	Longitudinal inner tape, expanded polyethylene	3.52
Office cabling:		
Paired:		**At 70°F**
720	Single braid	10.0
754A	Single braid	5.3
754B	Single braid	5.3
754D	Double braid	5.2
754E	Double braid	5.2
KT51	Double braid	6.6
KT52	Double braid	5.3
Coaxial:		
KS–8086	Double braid	11.0
724	Double braid	5.2

*From: "NAB Engineering Handbook", 5th ed., McGraw-Hill Book Company, 1960. Reprinted by permission of the National Association of Broadcasters.

Table 9-11. COMPARISON OF OPEN-WIRE SOLID-DIELECTRIC LINES AND WAVE GUIDES*

Characteristics	Parallel open wires	Bead-supported	Stub-supported	Solid dielectric (polyethylene)	Coaxial rubber	Wave guide	Pulse cables
Insulation resistance	O.K. in dry air, drops in rain	O.K. in dry air or nitrogen	O.K. dry or slightly wet	O.K. dry or wet	O.K. dry, drops when wet. Varies with temperature	O.K.	Same as rubber
Corona voltage and dielectric strength	Depends on dry air and spacing, low	Depends on dry air, depends on size, low	Depends on size, low	Depends on size, high	Depends on dry air, size, medium high	Depends on size, dry air, pressure, low	Depends on size, dry air, high
Dielectric constant	Low, air (1)	Low, air plus beads (1+)	Low, air (1)	Medium (2.3)	Medium high (3–4)	—	High (5–6)
Power factor	Low in dry air, goes up in wet	Low in dry air, goes up in wet	O.K. dry or slightly wet	O.K. dry or wet	High dry, higher wet. No good for HF	—	Same as rubber
D-c conductor resistance	Low, goes down with size	Same	Same	Higher than concentrics, goes down with size	High, goes down with size	Low, goes down with size	High, goes down with size
Ease of installation	Easy	More difficult	Difficult	Easier than concentric	Same as solid dielectric	Same as bead-supported	Same as rubber
Ease of maintenance	Easy	Very difficult	Not very	Easy	No	Easy	Same as rubber
Flexible	Somewhat, depends on size	Some, depends on size	No	Yes, flexibility goes down with size	Yes, more than solid dielectric	No	Same as rubber
Mechanically strong	Yes, but separators are weak points	Yes, beads are weak points	Yes, but stubs are fragile	Yes, but can be mistreated	Yes	Yes, but can be crushed	Yes
Can be sealed	Yes	Yes, but you have to do the right job	Yes	Yes	Yes	Yes	Yes
Self-shielding (electrical)	No	Yes	Yes	Yes, depends on the braid construction	Same as solid dielectric	Yes	Same as solid dielectric
Fire resistive	Yes	Yes	Yes	Yes, except largest size	Not completely	Yes	Same as rubber
High-temperature resistant	Yes, depends on material of separators	Yes, depends on rubber gaskets or solder	Yes, same as bead-supported	180–200°F for short time	220–250°F for short time	Yes, depends on rubber gasket or solder	Same as rubber
Transmission at high frequency	Limited by spacing	Good at low, medium frequency, losses high at microwave	Good at high frequency not at low, need spec for each	Good, at low and medium not so good at microwave	Poor, loss too high	Best, but each frequency needs spec guide	Poor, loss too high
Stand shock vibration	Good	Fair, beads are weak points	Good, stubs are weak points	Very good	Very good	Very good	Very good
Low temperatures	Yes	Yes, but allow for expansion	Same as beaded	Yes	Yes	Yes, but allow for expansion	Yes

*From: "NAB Engineering Handbook", 5th ed., McGraw-Hill Book Company, 1960. Reprinted by permission of the National Association of Broadcasters.

Table 9-12. PATTERNS, GAINS, AND AREAS OF TYPICAL ANTENNAS*

Type	Configuration	Pattern	Power gain over isotropic	Effective area
Electric doublet		$\cos \theta$	1.5	$1.5\, \lambda^2/4\pi$
Magnetic doublet or loop		$\sin \theta$	1.5	$1.5\, \lambda^2/4\pi$
Half-wave dipole		$\dfrac{\cos (\pi/2 \sin \theta)}{\cos \theta}$	1.64	$1.64\, \lambda^2/4\pi$
Half-wave dipole and screen		$2 \sin (S° \cos \beta)$	6.5	$1.64\, \lambda^2/\pi$
Turnstile array		$\dfrac{\sin (n\, S°/2 \sin \beta)}{n \sin (S°/2 \sin \beta)}$	n or $2L/\lambda$	$n\, \lambda^2/4\pi$ or $L\lambda/2\pi$
Loop array		$\dfrac{\cos \beta \sin (n\, S°/2 \sin \beta)}{n \sin (S°/2 \sin \beta)}$	n or $2L/\lambda$	$n\, \lambda^2/4\pi$ or $L\, \lambda/2\pi$
Optimum horn $L \geq a^2/\lambda$		Half-power width $70\, \lambda/a$ degrees (H plane) $51\, \lambda/b$ degrees (E plane)	$10\, ab/\lambda^2$	$0.81\, ab$
Parabola		Half-power width $70\, \lambda/d$ degrees	$2\, \pi\, d^2/\lambda^2$	$d^2/2$

*From: "NAB Engineering Handbook", 5th ed., McGraw-Hill Book Company, 1960. Reprinted by permission of the National Association of Broadcasters.

Table 9-13. MICROWAVE-ANTENNA POWER GAINS AND BEAMWIDTHS*

Parabola diameter, ft	2,000 Mc		7,000 Mc		13,000 Mc	
	Antenna gain, db	Half-power beamwidth, deg	Antenna gain, db	Half-power beamwidth, deg	Antenna gain, db	Half-power beamwidth, deg
2	19	18	30	5	36	2.5
4	25	9	36	2.5	42	1.5
6	29	6	40	1.75	46	0.875
8	31.5	4	43	1.25	48	0.65
10	33.5	3	44.5	1.00	50	0.50

*From: "NAB Engineering Handbook", 5th ed., McGraw-Hill Book Company, 1960. Reprinted by permission of the National Association of Broadcasters.

Table 9-14. UNITS AND THEIR CONVERSION

Policy of This Edition

In each table in this handbook, the numerical values are preferably expressed in those units most commonly used by U.S. engineers working in the specific field, but SI metric units have also been added. In some cases two tables are given, one in English units, one in metric. In other tables parallel columns showing figures in both units are used, or the conversion factors are listed.

In a general engineering handbook complete consistency in units, abbreviations, and symbols is hardly possible, or even desirable. Such consistency would quickly defeat the objective of providing quick access to numbers of maximum immediate usefulness. Within each special field of engineering, the technical societies and industry associations have developed certain uniform practices and standards; if tables and data are given only in units that are foreign to these prevailing standards, convenience is sacrificed. In any case the practical demands of compilation and new typesetting costs, and the usual requirement of a copyright owner that reprinted material should not be changed, may well govern the units used in any given table.

The present edition of this handbook reflects the changes in abbreviations, symbols, and forms that are resulting from the efforts to reduce the diversity of practices from one specialty to another and from one nation to another. Recommendations of the International Organization for Standardization (ISO-R 1000) and of the "Metric Practice Guide", adopted by ASTM, NBS, APL, and others, have focused attention on the diversity of so-called standards.

Since the United States is the only major industrial nation that has not yet converted to metric units, some legal requirements in that direction are to be expected. It is now a contradiction to speak of the "English" system of units, and for some time to come U.S. engineers must accommodate to a wide use of conversions from one set of units to another. The extensive conversion tables that follow are offered with this expectation.

In spite of major efforts to unify engineering practices, there are many good reasons for retaining several means of expressing a physical quantity. For ease of learning and communication a descriptive name is better than one arbitrarily assigned, such as Hz for cps, celsius for centigrade, and torr for mm Hg; an opposite trend is prevalent at this time. Numerical scales directly related to the physical phenomena and to the method of their measurement have an advantage in the laboratory or field and will not soon be abandoned. Examples are barometric pressure in mm or in. of mercury, viscosity in seconds Saybolt, the calorie or the Btu, and even the "coefficients" of expansion, friction, diffusion, attenuation, and reflection. Symbols, abbreviations, and even the units themselves are not infrequently subject to change; note, for example, the now preferred dB in place of the well established db; elimination of widely used abbreviations, such as kwh, cps, gpm, cc, and psi; and revised values for the second, the calorie, or the atomic weights. Users of this handbook are invited to call attention to places where consistency could be improved without sacrificing the objectives.

Of the many named units that might have more than one value, this book uses (unless otherwise stated) the thermochemical gram-caloric (4.184 J), the thermochemical Btu (1 054.35 J), the avoirdupois pound and ounce, the statute mile (5 280 ft), the short ton (2 000 lb), the U.S. liquid gallon (231 in.^{3}), and the electrical horsepower (746 W).

Rather than present a special and condensed table of engineering conversion factors, the editors have chosen to reprint the large table that has been developed over the years for the "Handbook of Chemistry and Physics" (Table 9-18 in this handbook). Certain specialized conversion factors and tables have been included.

THE METRIC INTERNATIONAL SYSTEM (SI)

Moves toward an international system of metric units are now following each other in quick succession, so a table of conversion factors for the most common units is given herewith (see also Table 9-15). Perhaps the most definite are the moves toward the SI standards already initiated by the National Bureau of Standards, the various military services, the National Aeronautics and Space Administration, and other U.S. Government research groups. The American Society for Testing and

Table 9-14.   UNITS AND THEIR CONVERSION *(Continued)*

Materials has declared in favor of SI units and will give other units only a secondary place in all newly issued ASTM Standards.[a] Other major engineering societies have committees to explore the adoption of SI units and are holding many meetings for discussion among members.

Whatever the decisions about converting to the metric system, the actual process will require many years, as can readily be seen from the experiences of other countries; in Great Britain, for example, even the single conversion to decimal monetary units and coinage moves very slowly. The practices and standards among the metric-system countries are far from uniform; no real international system exists among them.

Mere conversion of present U.S. specifications, drawings, tools, machines, and stock sizes, to equivalent metric units (so-called "soft" conversion) will not in any sense result in an "international" system. Instead, a "hard" conversion representing the abandonment of the $\frac{1}{2}$-fractional system in favor of a $\frac{1}{10}$-fractional system is necessary to attain the real advantages of the metric system. This means re-sizing of all round and sheet stock, lumber, bolts, screws, nails, wires, gears, containers, modules, and sub-assemblies, plus all the tools and machines related thereto. A long period of double-stocking must follow. The entire change is made the more difficult by the great penetration of U.S. products and materials into the markets of the world, e.g., airplanes and military equipment, production, and construction machinery. This is not to mention the problem of the individual engineer, technician, and user, who visualizes all his size relationships in inches and feet and his weights in pounds. Realistically, more than one generation will be required for the educational conversion alone.

In presenting data in international standard metric units throughout this edition, the practices and forms used in the "Metric Practice Guide" have been carefully followed.[a] Certain conventions used in the "Metric Practice Guide" are not consistent with those originally adopted for this handbook, nor with ANSI standards. Special attention is directed to the following conventions:

1.   For degrees Kelvin the degree symbol is omitted; for example, 50 K, not 50°K.
2.   For multiplication a center point is used; for example, the unit of dynamic viscosity is abbreviated as N·s/m², not N s/m² or N × s/m².
3.   Symbols for SI units are not capitalized unless the unit is derived from a proper name, as N for Sir Isaac Newton; however, *unabbreviated* units are not capitalized, such as newton, kelvin, hertz.

CONVERSION FACTORS TO SI STANDARD UNITS

For more complete conversions to SI units, see Tables 9-15 and 9-16.

To convert	To	Multiply by	To convert	To	Multiply by
ACCELERATION			POWER		
feet/second²	meters/second²	0.3048	Btu/second	watt	1054.350
AREA			foot-pounds/		
square feet	square meters	0.09290304	second	watt	1.355818
ENERGY			horsepower	watt	746.
Btu (mean)	joule	1055.87	PRESSURE		
calorie (mean)	joule	4.19002	atmosphere	Newtons/meter²	101325.0
electron volt	joule	1.60210×10^{-19}	bar	Newtons/meter²	100000.
foot-pound	joule	1.355818	kilograms/cm²	Newtons/meter²	98066.50
watthour	joule	3600.	pounds/in.²	Newtons/meter²	6894.757
FORCE			torr (mm Hg,		
dyne	Newton	0.00001	0°C)	Newtons/meter²	133.322
kilogram	Newton	9.80665	VISCOSITY		
pound	Newton	4.448222	centipoise	Newton-second/	
LENGTH				meter²	0.001
foot	meter	0.3048000	pounds/foot	Newton-second/	
mil	meter	0.0000254	second	meter²	1.488164
mile (U.S.			VOLUME		
statute)	meter	1609.344	cubic foot	cubic meter	0.02831685
MASS			gallon (U.S.		
pound	kilogram	0.4535924	liquid)	cubic meter	0.003785412
slug	kilogram	14.59390			
ton (2000 lb)	kilogram	907.1847			

[a]See "Metric Practice Guide", ASTM Standard E 380—70, American Society for Testing and Materials, 1970.

Table 9-15.   INTERNATIONAL SYSTEM (SI) METRIC UNITS

BASIC UNITS—MKS

Length	meter	m	Electric current	ampere	A
Mass	kilogram	kg	Thermodynamic temperature	kelvin	K
Time	second	s	Luminous intensity	candela	cd

DERIVED UNITS

Property	Units†	Abbreviations and dimensions	
Acceleration	meter per second squared	m/s^2	
Activity (of radioactive source)	1 per second	s^{-1}	
Angular acceleration	radian per second squared	rad/s^{-1}	
Angular velocity	radian per second	rad/s	
Area	square meter	m^2	
Density	kilogram per cubic meter	kg/m^3	
Dynamic viscosity	newton-second per sq meter	$N \cdot s/m^2$	
Electric capacitance	farad	F	$(A \cdot s/V)$
Electric charge	coulomb	C	$(A \cdot s)$
Electric field strength	volt per meter	V/m	
Electric resistance	ohm		(V/A)
Entropy	joule per kelvin	J/K	
Force	newton	N	$(kg \cdot m/s^2)$
Frequency	hertz	hz	(s^{-1})
Illumination	lux	lx	(lm/m^2)
Inductance	henry	H	$(V \cdot s/A)$
Kinematic viscosity	sq meter per second	m^2/s	
Luminance	candela per sq meter	cd/m^2	
Luminous flux	lumen	lm	$(cd \cdot sr)$
Magnetomotive force	ampere	A	
Magnetic field strength	ampere per meter	A/m	
Magnetic flux	weber	Wb	$(V \cdot s)$
Magnetic flux density	tesla	T	(Wb/m^2)
Power	watt	W	(J/s)
Pressure	newton per square meter	N/m^2	
Radiant intensity	watt per steradian	W/sr	
Specific heat	joule per kilogram kelvin	$J/kg\ K$	
Thermal conductivity	watt per meter kelvin	$W/m\ K$	
Velocity	meter per second	m/s	
Volume	cubic meter	m^3	
Voltage, potential difference, electromotive force	volt	V	(W/A)
Wave number	1 per meter	m^{-1}	
Work, energy, quantity of heat	joule	J	$(N \cdot m)$

PREFIX NAMES OF MULTIPLES AND SUBMULTIPLES OF UNITS

Decimal equivalent	Prefix	Pronun- ciation	Symbol	Exponential expression
1,000,000,000,000	tera	tĕr′á	T	10^{+12}
1,000,000,000	giga	jĭ′gá	G	10^{+9}
1,000,000	mega	mĕg′á	M	10^{+6}
1,000	kilo	kĭl′ō	k	10^{+3}
100	hecto	hĕk′tō	h	10^{+2}
10	deka	dĕk′á	da	10
0.1	deci	dĕs′ĭ	d	10^{-1}
0.01	centi	sĕn′tĭ	c	10^{-2}
0.001	milli	mĭl′ĭ	m	10^{-3}
0.000 001	micro	mī′krō	μ	10^{-6}
0.000 000 001	nano	năn′ō	n	10^{-9}
0.000 000 000 001	pico	pē′kō	p	10^{-12}
0.000 000 000 000 001	femto	fĕm′tō	f	10^{-15}
0.000 000 000 000 000 001	atto	ăt′tō	a	10^{-18}

Table 9-15. INTERNATIONAL SYSTEM (SI) METRIC UNITS *(Continued)*

DEFINITIONS OF THE MOST IMPORTANT INTERNATIONAL SYSTEM (SI) UNITS

The *ampere* (unit of electric current) is the constant current that, if maintained in two straight parallel conductors of infinite length, of negligible circular sections, and placed 1 meter apart in a vacuum, will produce between these conductors a force equal to 2×10^{-7} newton per meter of length.

The *candela* is the luminous intensity, in the direction of the normal, of a black body surface 1/600,000 square meter in area, at the temperature of solidification of platinum under a pressure of 101,325 newtons per square meter.

The *coulomb* (unit of quantity of electricity) is the quantity of electricity transported in 1 second by a current of 1 ampere.

The *ephemeris second* (unit of time) is exactly 1/31 556 925.974 7 of the tropical year of 1900, January, 0 days, and 12 hours ephemeris time.

The *farad* (unit of electric capacitance) is the capacitance of a capacitor between the plates of which there appears a difference of potential of 1 volt when it is charged by a quantity of electricity equal to 1 coulomb.

The *henry* (unit of electric inductance) is the inductance of a closed circuit in which an electromotive force of 1 volt is produced when the electric current in the circuit varies uniformly at a rate of 1 ampere per second.

The *International Practical Kelvin Temperature Scale* of 1960 and the *International Practical Celsius Temperature Scale* of 1960 are defined by a set of interpolation equations based on the following reference temperatures:

	K	deg C
Oxygen, liquid-gas equilibrium	90.18	−182.97
Water, solid-liquid equilibrium	273.15	0.00
Water, solid-liquid-gas equilibrium	273.16	0.01
Water, liquid-gas equilibrium	373.15	100.00
Zinc, solid-liquid equilibrium	692.655	419.505
Sulfur, liquid-gas equilibrium	717.75	444.6
Silver, solid-liquid equilibrium	1233.95	960.8
Gold, solid-liquid equilibrium	1336.15	1063.0

The *joule* (unit of energy) is the work done when the point of application of 1 newton is displaced a distance of 1 meter in the direction of the force.

The *kelvin* (unit of thermodynamic temperature) is the fraction 1/273.16 of the thermodynamic temperature of the triple point of water. The decision was made at the 13th General Conference on Weights and Measures on October 13, 1967, that the name of the unit of thermodynamic temperature would be changed from *degree Kelvin* (symbol: $°K$) to *kelvin* (symbol: K). The name (*kelvin*) and symbol (K) are to be used for expressing temperature intervals. The former convention that expressed a temperature interval in *degrees Kelvin* or, abbreviated, *deg K* is dropped. However, the old designations are acceptable temporarily as alternatives to the new ones. One may also express temperature intervals in *degrees Celsius*.

The *kilogram* (unit of mass) is the mass of a particular cylinder of platinum iridium alloy, called the International Prototype Kilogram, which is preserved in a vault at Sèvres, France, by the International Bureau of Weights and Measures.

Length: The name *micron*, for a unit of length equal to 10^{-6} meter, and the symbol μ that has been used for it were dropped by action of the 13th General Conference on Weights and Measures on October 13, 1967. The symbol μ is to be used solely as an abbreviation for the prefix *micro-*, standing for the multiplication by 10^{-6}. Thus the length previously designated as 1 micron should be designated 1 μm.

The *lumin* (unit of luminous flux) is the luminous flux emitted in a solid angle of 1 steradian by a uniform point source having an intensity of 1 candela.

Table 9-15. INTERNATIONAL SYSTEM (SI) METRIC UNITS *(Continued)*

The *newton* (unit of force) is that force that gives to a mass of 1 kilogram an acceleration of 1 meter per second.

The *ohm* (unit of electric resistance) is the electric resistance between two points of a conductor when a constant difference of potential of 1 volt, applied between these two points, produces in this conductor a current of 1 ampere, this conductor not being the source of any electromotive force.

The *meter* (unit of length) is the length of exactly 1 650 763.73 wavelengths of the radiation in vacuum corresponding to the unperturbed transition between the levels $2p_{10}$ and $5d_5$ of the atom of krypton 86, the orange-red line.

The *second* is the unit of time of the International System of Units. The definition adopted at the October 13, 1967, meeting of the 13th General Conference on Weights and Measures is: "The second is the duration of 9 192 631 770 periods of the radiation corresponding to the transition between the two hyperfine levels of the fundamental state of the atom of cesium 133." The frequency (9 192 631 770 hz), which the definition assigns to the cesium radiation, was carefully chosen to make it impossible, by any existing experimental evidence, to distinguish the new second from the *ephemeris second* based on the earth's motion. Therefore no changes need to be made in data stated in terms of the old standard in order to convert them to the new one. The atomic definition has two important advantages over the previous definition: (1) it can be realized (i.e., generated by a suitable clock) with sufficient precision, ± 1 part per hundred billion (10^{11}) or better, to meet the most exacting demands of modern metrology; and (2) it is available to anyone who has access to or who can build an atomic clock controlled by the specified cesium radiation.‡ In addition one can compare other high-precision clocks directly with such a standard in a relatively short time—an hour or so compared against years with the astronomical standard. Laboratory-type atomic clocks are complex and expensive, so that most clocks and frequency generators will continue to be calibrated against a standard such as the NBS Frequency Standard, controlled by a cesium atomic beam, at the Radio Standards Laboratory in Boulder, Colorado. In most cases the comparison will be by way of the standard-frequency and time-interval signals broadcast by NBS radio stations WWV, WWVH, WWVB, and WWVL.

The *volt* (unit of electric potential difference and electromotive force) is the difference of electric potential between two points of a conducting wire carrying a constant current of 1 ampere, when the power dissipated between these points is equal to 1 watt.

The *watt* (unit of power) is the power that gives rise to the production of energy at the rate of 1 joule per second.

The *weber* (unit of magnetic flux) is the magnetic flux that, linking a circuit of one turn, produces in it an electromotive force of 1 volt as it is reduced to zero at a uniform rate in 1 second.

---

†According to SI terminology, the following should be treated as obsolete:

angstrom (now 100 picometers or 0.1 nanometer)	kiloton (now gigagram)
	liter (now cubic decimeter)
bar (now 100 kilonewtons/meter²)	metric ton (now megagram)
kiloliter (now cubic meter)	micron (now micrometer)

‡A description of such clocks is given in "Atomic Frequency Standards", *NBS Tech. News Bull.*, 45:8–11, January 1961.

For more recent developments and technical details, see R.E. Beehler, R.C. Mockler, and J.M. Richardson, "Cesium Beam Atomic Time and Frequency Standards", *Metrologia*, 1:114–131, July 1965.

Table 9-16. CONVERSIONS TO SI UNITS

International Metric System

This table can be used for conversion of any quantity in English units to corresponding SI units to five significant figures (without the use of a calculator). Exact values are shown in boldface. Unless otherwise stated, values are in thermochemical calorie, thermochemical Btu, and avoirdupois mass units.[a]

Instruction: Shift decimal as required for each digit in the original quantity and add the converted results.

Example: Convert an acceleration of 15.30 ft/s² to m/s².

Solution: From first line of table, 3.048 0 + 1.524 0 + 0.091 44 = 4.663 4 m/s².

	1	2	3	4	5	6	7	8	9
ACCELERATION									
foot/second² to meter/second², m/s²	**0.304 8**	0.609 6	0.914 4	1.219 2	1.524 0	1.828 8	2.133 6	2.438 4	2.743 2
g's (free fall, standard) to meter/second², m/s²	**9.806 65**	19.613	29.420	39.227	49.033	58.840	68.647	78.453	88.260
inch/second² to meter/second², m/s²	**0.025 4**	0.050 8	0.076 2	0.101 6	0.127 0	0.152 4	0.177 8	0.203 2	0.228 6
AREA									
acre to meter², m²	4 046.856	8 093.7	12 141	16 187	20 234	24 281	28 328	32 375	36 422
circular mil to meter², m²	$5.067\,075 \times 10^{-10}$	10.134×10^{-10}	15.201×10^{-10}	20.268×10^{-10}	25.335×10^{-10}	30.402×10^{-10}	35.470×10^{-10}	40.537×10^{-10}	45.604×10^{-10}
foot² to meter², m²	**0.092 903 04**	0.185 81	0.278 71	0.371 61	0.464 52	0.557 42	0.650 32	0.743 22	0.836 13
inch² to meter², m²	**0.000 645 16**	**0.001 290 32**	**0.001 935 48**	**0.002 580 64**	**0.003 225 80**	**0.003 870 96**	**0.004 516 12**	**0.005 161 28**	**0.005 806 44**
mile² (U.S. statute) to meter², m²	2 589 988	5 180 000	7 770 000	10 360 000	12 950 000	15 540 000	18 130 000	20 720 000	23 310 000
yard² to meter², m²	**0.836 127 36**	1.672 3	2.508 4	3.344 5	4.180 6	5.016 8	5.852 9	6.689 0	7.525 1
BENDING MOMENT OR TORQUE									
ounce-force-inch to newton-meter, N·m	0.007 061 552	0.014 123	0.021 185	0.028 246	0.035 308	0.042 369	0.049 431	0.056 492	0.063 554
pound-force-inch to newton-meter, N·m	0.112 984 8	0.225 97	0.338 95	0.451 94	0.564 92	0.677 91	0.790 89	0.903 88	1.016 9
pound-force-foot to newton-meter, N·m	1.355 818	2.711 6	4.067 5	5.423 3	6.779 1	8.134 9	9.490 7	10.847	12.202

[a]The thermochemical calorie is exactly 4.184 joules by definition. The international steam table (IT) calorie is exactly 4.186 8 joules by definition. The thermochemical Btu is 1 054.350 joules. Each Btu is defined in terms of the corresponding calorie by 1 Btu/lbm·R ≡ 1 cal/g·K.

Table 9-16. CONVERSIONS TO SI UNITS (Continued)

	1	2	3	4	5	6	7	8	9
DENSITY (MASS/VOLUME)									
grain/gallon to kilogram/meter³, kg/m³	0.017 118 06	0.034 236	0.051 354	0.068 472	0.085 590	0.102 71	0.119 83	0.136 94	0.154 06
ounce/gallon to kilogram/meter³, kg/m³	7.489 152	14.978	22.467	29.957	37.446	44.935	52.424	59.913	67.402
ounce/inch³ to kilogram/meter³, kg/m³	1 729.994	3 460.0	5 190 0	6 920 0	8 650.0	10 380	12 110	13 840	15 570
pound-mass/foot³ to kilogram/meter³, kg/m³	16.018 46	32.037	48.055	64.074	80.092	96.111	112.13	128.15	144.17
pound-mass/inch³ to kilogram/meter³, kg/m³	27 679.90	55 360	83 040	110 720	138 400	166 030	193 760	221 440	249 120
pound-mass/gallon to kilogram/meter³, kg/m³	119.826 4	239.65	359.43	479.31	599.13	718.95	838.78	958.61	1 078.4
slug/foot³ to kilogram/meter³, kg/m³	515.378 8	1 030.8	1 546.1	2 061.5	2 576.9	3 092 3	3 607.7	4 123.0	4 638.4
ELECTRICITY AND MAGNETISM									
ampere-hour to coulomb, C	3 600	7 200	10 800	14 400	18 000	21 600	25 200	28 800	32 400
faraday (based on C-12) to coulomb, C	96 487.00	192 970	289 460	385 950	482 440	578 920	675 410	771 900	868 380
gauss to tesla, T	0.000 1	0.000 2	0.000 3	0.000 4	0.000 5	0.000 6	0.000 7	0.000 8	0.000 9
gilbert to ampere-turn	0.795 774 7	1.591 5	2.387 3	3.183 1	3.978 9	4.774 6	5.570 4	6.366 2	7.162 0
oersted to ampere-meter, A/m	79.577 47	159.15	238.73	318.31	397.89	477.45	557.04	636.62	716.20
unit pole to weber, Wb	$1.256\,637 \times 10^{-7}$	$2.513\,3 \times 10^{-7}$	$3.769\,9 \times 10^{-7}$	$5.026\,5 \times 10^{-7}$	$6.283\,2 \times 10^{-7}$	$7.539\,8 \times 10^{-7}$	$8.796\,5 \times 10^{-7}$	10.053×10^{-7}	11.310×10^{-7}
ENERGY AND WORK									
British thermal unit to joule, J	1 054.350	2 108.7	3 163.1	4 217.4	5 271.8	6 326 1	7 380.5	8 434.8	9 489.2
British thermal unit (IT) to joule, Jᵃ	1 055.056	2 110.1	3 165.2	4 220.2	5 275.3	6 330 3	7 385.4	8 440.4	9 495.5
calorie to joule, J	4.184	8.368	12.552	16.736	20.920	25.104	29.288	33.472	37.656
calorie (IT) to joule, Jᵃ	4.186 8	8.373 6	12.560 4	16.747 2	20.934 0	25.120 8	29.307 6	33.494 4	37.681 2
electron volt to joule, J	$1.602\,10 \times 10^{-19}$	$3.204\,2 \times 10^{-19}$	$4.806\,3 \times 10^{-19}$	$6.408\,4 \times 10^{-19}$	$8.010\,5 \times 10^{-19}$	$9.612\,6 \times 10^{-19}$	11.215×10^{-19}	12.817×10^{-19}	14.419×10^{-19}
foot-pound-force to joule, J	1.355 818	2.711 6	4.067 5	5.423 3	6.779 1	8.134 9	9.490 7	10.847	12.202
kilowatt-hour to joule, J	3 600 000	7 200 000	10 800 000	14 400 000	18 000 000	21 600 000	25 200 000	28 800 000	32 400 000
horsepower-hour to joule, J	2 684 520	5 369 039	8 053 559	10 738 078	13 422 593	16 107 117	18 791 637	21 476 156	24 160 676

ᵃThe thermochemical calorie is exactly 4.184 joules by definition. The international steam table (IT) calorie is exactly 4.186 8 joules by definition. The thermochemical Btu is 1 054.350 joules by definition. Each Btu is defined in terms of the corresponding calorie by 1 Btu/lbm·R ≡ 1 cal/g·K.

Table 9-16. CONVERSIONS TO SI UNITS (Continued)

	1	2	3	4	5	6	7	8	9
FLOW RATE									
foot³/minute to meter³/second, m³/s	0.000 471 947 4	0.000 943 89	0.001 415 8	0.001 887 8	0.002 359 7	0.002 831 7	0.003 303 6	0.003 775 6	0.004 247 5
foot³/second to meter³/second, m³/s	0.028 316 85	0.056 634	0.084 951	0.113 27	0.141 58	0.169 90	0.198 22	0.226 53	0.254 85
gallon (U.S. liquid)/day to meter³/second, m³/s	$4.381\,264 \times 10^{-8}$	$8.762\,5 \times 10^{-8}$	13.144×10^{-8}	17.525×10^{-8}	21.906×10^{-8}	26.288×10^{-8}	30.669×10^{-8}	35.050×10^{-8}	39.431×10^{-8}
gallon (U.S. liquid)/minute to meter³/second, m³/s	0.000 063 090 20	0.000 126 18	0.000 189 27	0.000 252 36	0.000 315 45	0.000 378 54	0.000 441 63	0.000 504 72	0.000 567 81
pound-mass/hour to kilogram/second, kg/s	0.000 125 997 9	0.000 252 00	0.000 377 99	0.000 503 99	0.000 629 99	0.000 755 99	0.000 881 99	0.001 007 98	0.001 133 98
pound-mass/minute to kilogram/second, kg/s	0.007 559 873	0.015 120	0.022 680	0.030 239	0.037 799	0.045 359	0.052 919	0.060 479	0.068 039
FORCE									
kilogram-force to newton, N	9.806 65	19.613	29.420	39.227	49.033	58.840	68.647	78.453	88.260
ounce-force to newton, N	0.278 014 0	0.556 03	0.834 04	1.112 1	1.390 1	1.668 1	1.946 1	2.224 1	2.502 1
pound-force to newton, N	4.448 222	8.896 4	13.345	17.793	22.241	26.689	31.138	35.586	40.034
HEAT									
SPECIFIC HEAT CAPACITY									
British thermal unit/pound-mass-deg F to joule/kilogram-kelvin, J/kg·K	4 184	8 368	12 552	16 736	20 920	25 104	29 288	33 472	37 656
British thermal unit (IT)/pound-mass-deg F to joule/kilogram-kelvin, J/kg·K[a]	4 186.8	8 373.6	12 560.4	16 747.2	20 934.0	25 120.8	29 307.6	33 494.4	37 681.2
calorie/gram-deg C to joule/kilogram-kelvin, J/kg·K	4 184	8 368	12 552	16 736	20 920	25 104	29 288	33 472	37 656
ENERGY/MASS (ENTHALPY, ETC.)									
British thermal unit/pound-mass to joule/kilogram, J/kg	2 324.444	4 648.9	6 973.3	9 297.8	11 622	13 947	16 271	18 596	20 920
British thermal unit (IT)/pound-mass to joule/kilogram, J/kg[a]	2 326	4 652	6 978	9 304	11 630	13 956	16 282	18 608	20 934
calorie/gram to joule/kilogram, J/kg	4 184	8 368	12 552	16 736	20 920	25 104	29 288	33 472	37 656

[a]The thermochemical calorie is exactly 4.184 joules by definition. The international steam table (IT) calorie is exactly 4.186 8 joules by definition. The thermochemical Btu is 1 054.350 joules. Each Btu is defined in terms of the corresponding calorie by 1 Btu/lbm·R ≡ 1 cal/g·K.

Table 9-16.   CONVERSIONS TO SI UNITS (*Continued*)

	1	2	3	4	5	6	7	8	9
THERMAL CONDUCTIVITY									
British thermal unit/hour-foot-deg F to watt/meter-kelvin, W/m·K	1.729 577	3.459 2	5.188 7	6.918 3	8.647 9	10.377	12.107	13.837	15.566
British thermal unit (IT)/hour-foot-deg F to watt/meter-kelvin, W/m·K[a]	1.730 735	3.461 5	5.192 2	6.922 9	8.653 7	10.384	12.115	13.846	15.577
British thermal unit-inch/hour-foot²-deg F to watt/meter-kelvin, W/m·K	0.144 131 4	0.288 26	0.432 39	0.576 53	0.720 66	0.854 79	1.008 9	1.153 1	1.297 2
British thermal unit (IT)-inch/hour-foot²-deg F to watt/meter-kelvin, W/m·K[a]	0.144 227 9	0.288 46	0.432 68	0.576 91	0.721 14	0.855 37	1.009 60	1.153 82	1.298 05
calorie/second-centimeter-deg C to watt/meter-kelvin, W/m·K	418.4	836.8	1 255.2	1 673.6	2 092.0	2 510.4	2 928.8	3 347.2	3 765.6
ENERGY PER UNIT AREA									
British thermal unit/foot² to joule/meter², J/m²	11 348.93	22 698	34 047	45 396	56 745	68 094	79 443	90 791	102 140
calorie/centimeter² to joule/meter², J/m²	41 840	83 680	125 520	167 360	209 200	251 040	292 880	334 720	376 560
THERMAL DIFFUSIVITY foot²/hour to meter²/second, m²/s	0.000 025 806 4	0.000 051 612 8	0.000 077 419 2	0.000 103 256	0.000 129 032	0.000 154 838 4	0.000 180 644 8	0.000 206 451 2	0.000 232 257 6
THERMAL RESISTANCE deg F-hour-foot²/British thermal unit to kelvin-meter²/watt, K·m²/W	0.176 228 0	0.352 46	0.528 68	0.704 91	0.881 14	1.057 4	1.233 6	1.409 8	1.586 1
THERMAL CONDUCTANCE British thermal unit/hour-foot²-deg F to watt/meter²-kelvin, W/m²·K	5.674 466	11.349	17.023	22.698	28.372	34.047	39.721	45.396	51.070

[a]The thermochemical calorie is exactly 4.184 joules by definition. The international steam table (IT) calorie is exactly 4.186 8 joules by definition. The thermochemical Btu is 1 054.350 joules. Each Btu is defined in terms of the corresponding calorie by 1 Btu/lbm·R ≡ 1 cal/g·K.

Table 9-16. CONVERSIONS TO SI UNITS (Continued)

	1	2	3	4	5	6	7	8	9
THERMAL CONDUCTANCE (continued)									
British thermal unit/second-foot²-deg F to watt/meter²-kelvin, W/m²·K	20 428.08	40 856	61 284	81 712	102 140	122 570	143 000	163 420	183 850
calorie/second-centimeter²-deg C to watt/meter²-kelvin, W/m²·K	41 840	83 680	125 520	167 360	209 200	251 040	292 880	334 720	376 560
LENGTH									
caliber to meter, m	**0.000 254**	**0.000 508**	**0.000 762**	**0.001 016**	**0.001 270**	**0.001 524**	**0.001 778**	**0.002 032**	**0.002 286**
fathom to meter, m	**1.828 8**	**3.657 6**	**5.486 4**	**7.315 2**	**9.144 0**	**10.972 8**	**12.801 6**	**14.630 4**	**16.459 2**
foot to meter, m	**0.304 8**	**0.609 6**	**0.914 4**	**1.219 2**	**1.524 0**	**1.828 8**	**2.133 6**	**2.438 4**	**2.743 2**
inch to meter, m	**0.025 4**	**0.050 8**	**0.076 2**	**0.101 6**	**0.127 0**	**0.152 4**	**0.177 8**	**0.203 2**	**0.228 6**
light year to meter, m	$9.460\,550 \times 10^{15}$	18.921×10^{15}	28.382×10^{15}	37.842×10^{15}	47.303×10^{15}	56.763×10^{15}	66.224×10^{15}	75.684×10^{15}	85.145×10^{15}
mil to meter, m	**0.000 025 4**	**0.000 050 8**	**0.000 076 2**	**0.000 101 6**	**0.000 127 0**	**0.000 152 4**	**0.000 177 8**	**0.000 203 2**	**0.000 228 6**
mile (U.S. nautical) to meter, m	1 852	3 704	5 556	7 408	9 260	11 112	12 964	14 816	16 668
mile (U.S. statute) to meter, m	1 609.344	3 218.7	4 828.0	6 437.4	8 046.7	9 656.1	11 265	12 875	14 484
rod to meter, m	**5.029 2**	**10.058 4**	**15.087 6**	**20.116 8**	**25.146 0**	**30.175 2**	**35.204 4**	**40.233 6**	**45.262 8**
yard to meter, m	**0.914 4**	**1.828 8**	**2.743 2**	**3.657 6**	**4.572 0**	**5.486 4**	**6.400 8**	**7.315 2**	**8.229 6**
MASS									
grain to kilogram, kg	**0.000 064 798 91**	**0.000 129 60**	**0.000 194 40**	**0.000 259 20**	**0.000 324 00**	**0.000 388 80**	**0.000 453 60**	**0.000 518 40**	**0.000 583 20**
ounce-mass to kilogram, kg	0.028 349 52	0.056 699	0.085 049	0.113 40	0.141 75	0.170 10	0.198 45	0.226 80	0.255 15
ounce-mass (troy or apothecary) to kilogram, kg	0.031 103 48	0.062 207	0.093 310	0.124 41	0.155 52	0.186 62	0.217 72	0.248 83	0.279 93
pound-mass to kilogram, kg	**0.453 592 37**	0.907 18	1.360 8	1.814 4	2.268 0	2.721 6	3.175 1	3.628 7	4.082 3
pound-mass (troy or apothecary) to kilogram, kg	0.373 241 7	0.746 48	1.119 7	1.493 0	1.866 2	2.239 5	2.612 7	2.985 9	3.359 2
slug to kilogram, kg	14.593 90	29.188	43.782	58.376	72.970	87.563	102.16	116.75	131.35
ton (long, 2 240 lb$_m$) to kilogram, kg	1 016.047	2 032.1	3 048.1	4 064.2	5 080.2	6 096.3	7 112.3	8 128.4	9 144.4
ton (short, 2 000 lb$_m$) to kilogram, kg	907.184 7	1 814.4	2 721.6	3 628.7	4 535.9	5 443.1	6 350.3	7 257.5	8 164.7

Table 9-16. CONVERSIONS TO SI UNITS (Continued)

	1	2	3	4	5	6	7	8	9
POWER									
British thermal unit/second to watt, W	1 054.350	2 108.7	3 163.	4 217.4	5 271.8	6 326.1	7 380.5	8 434.8	9 489.2
British thermal unit/minute to watt, W	17.572 50	35.145	52.713	70.290	87.863	105.44	123.01	140.58	158.15
British thermal unit/hour to watt, W	0.292 875 1	0.585 75	0.878 63	1.171 5	1.464 4	1.757 3	2.050 1	2.343 0	2.635 9
British thermal unit (IT)/hour to watt, W[a]	0.293 071 1	0.586 14	0.879 21	1.172 3	1.465 4	1.758 4	2.051 5	2.344 6	2.637 6
calorie/second to watt, W	4.184	8.368	12.552	16.736	20.920	25.104	29.288	33.472	37.656
calorie/minute to watt, W	0.069 733 33	0.139 47	0.209 20	0.278 93	0.348 67	0.418 40	0.488 13	0.557 87	0.627 60
foot-pound-force/second to watt, W	1.355 818	2.711 6	4.067 5	5.423 3	6.779 1	8.134 9	9.490 7	10.847	12.202
foot-pound-force/minute to watt, W	0.022 596 97	0.045 194	0.067 791	0.090 388	0.112 98	0.135 58	0.158 18	0.180 78	0.203 37
foot-pound-force/hour to watt, W	0.000 376 616 1	0.000 753 23	0.001 129 8	0.001 505 5	0.001 883 1	0.002 259 7	0.002 636 3	0.003 012 9	0.003 389 5
horsepower (550 ft·lb$_f$/s) to watt, W	745.699 9	1 491.4	2 237.1	2 982.8	3 728.5	4 474.2	5 219.9	5 965.6	6 711.3
horsepower (electric) to watt, W	746.	1 492.	2 238.	2 984.	3 730.	4 476.	5 222.	5 968.	6 714.
tons of refrigeration to watt, W	3 516.853	7 033.7	10 551	14 067	17 584	21 101	24 618	28 135	31 652
POWER/AREA									
British thermal unit/foot²-second to watt/meter², W/m²	11 348.93	22 698	34 047	45 396	56 745	68 094	79 443	90 791	102 140
British thermal unit/foot²-minute to watt/meter², W/m²	189.148 9	378.30	567.45	756.60	945.74	1 134.9	1 324.0	1 513.2	1 702.3
British thermal unit/foot²-hour to watt/meter², W/m²	3.152 481	6.305 0	9.457 4	12.610	15.762	18.915	22.067	25.220	28.372
British thermal unit/inch²-second to watt/meter², W/m²	1 634 246	3 268 500	4 902 700	6 537 000	8 171 200	9 805 500	11 440 000	13 074 000	14 708 000
calorie/centimeter²-minute to watt/meter², W/m²	697.333 3	1 394.7	2 092.0	2 789.3	3 486.7	4 184.0	4 881.3	5 578.7	6 276.0

[a]The thermochemical calorie is exactly 4.184 joules by definition. The international steam table (IT) calorie is exactly 4.186 8 joules by definition. The thermochemical Btu is 1 054.350 joules. Each Btu is defined in terms of the corresponding calorie by 1 Btu/lbm·R ≡ 1 cal/g·K.

Table 9-16. CONVERSIONS TO SI UNITS (*Continued*)

	1	2	3	4	5	6	7	8	9
PRESSURE OR STRESS (FORCE/AREA)									
atmosphere (normal = 760 torr) to newton/meter², N/m²	101 325	202 650	303 975	405 300	506 625	607 950	709 275	810 600	911 925
bar to newton/meter², N/m²	100 000	200 000	300 000	400 000	500 000	600 000	700 000	800 000	900 000
foot of water (39.2 F) to newton/meter², N/m²	2 988.980	5 978.0	8 966.9	11 956	14 945	17 934	20 923	23 912	26 901
inch of mercury (32 F) to newton/meter², N/m²	3 386.389	6 772.8	10 159	13 546	16 932	20 318	23 705	27 091	30 478
inch of water (39.2 F) to newton/meter², N/m²	249.082 0	498.16	747.25	996.33	1 245.4	1 494.5	1 743.6	1 992.7	2 241.7
inch of water (60 F) to newton/meter², N/m²	248.840 0	497.68	746.52	995.36	1 244.2	1 493.0	1 741.9	1 990.7	2 239.6
kilogram-force/centimeter² to newton/meter², N/m²	98 066.5	196 133	294 199.5	392 266	490 332.5	588 399	686 465.5	784 532	882 598.5
millimeter of mercury (0 C), torr, to newton/meter², N/m²	133.322 4	266.64	399.97	533.29	666.61	799.93	933.26	1 066.6	1 199.9
pound-force/foot² to newton/meter², N/m²	47.880 26	95.761	143.64	191.52	239.40	287.28	335.16	383.04	430.92
pound-force/inch² (psi) to newton/meter², N/m²	6 894.757	13 790	20 684	27 579	34 474	41 369	48 263	55 158	62 053
TEMPERATURE (*see* Tables 9-21–9-24)									
VELOCITY									
foot/hour to meter/second, m/s	0.000 084 666 67	0.000 169 33	0.000 254 00	0.000 338 67	0.000 423 33	0.000 508 00	0.000 592 67	0.000 677 33	0.000 762 00
foot/minute to meter/second, m/s	0.005 08	0.010 16	0.015 24	0.020 32	0.025 40	0.030 48	0.035 56	0.040 64	0.045 72
foot/second to meter/second, m/s	0.304 8	0.609 6	0.914 4	1.219 2	1.524 0	1.828 8	2.133 6	2.438 4	2.743 2
inch/second to meter/second, m/s	0.025 4	0.050 8	0.076 2	0.101 6	0.127 0	0.152 4	0.177 8	0.203 2	0.228 6
kilometer/hour to meter/second, m/s	0.277 777 8	0.555 56	0.833 33	1.111 1	1.388 9	1.666 7	1.944 4	2.222 2	2.500 0
knot (international) to meter/second, m/s	0.514 444 4	1.028 9	1.543 3	2.057 8	2.572 2	3.086 7	3.601 1	4.115 6	4.630 0

Table 9-16. CONVERSIONS TO SI UNITS (Continued)

	1	2	3	4	5	6	7	8	9
VELOCITY (continued)									
mile/hour (U.S. statute) to meter/second, m/s	0.447 04	0.894 08	1.341 12	1.788 16	2.235 20	2.682 24	3.129 28	3.576 32	4.023 36
mile/minute (U.S. statute) to meter/second, m/s	26.822 4	53.644 8	80.467 2	107.289 6	134.112 0	160.934 4	187.756 8	214.587 2	241.401 6
mile/second (U.S. statute) to meter/second, m/s	1 609.344	3 218.7	4 828.0	6 437.4	8 046.7	9 656.1	11 265	12 875	14 484
VISCOSITY									
DYNAMIC OR ABSOLUTE, μ									
centipoise to newton-second/meter², N·s/m²	0.001	0.002	0.003	0.004	0.005	0.006	0.007	0.008	0.009
pound-mass/foot-second to newton-second/meter², N·s/m²	1.488 164	2.976 3	4.464 5	5.952 7	7.440 8	8.929 0	10.417	11.905	13.393
pound-force-second/foot² to newton-second/meter², N·s/m²	47.880 26	95.761	143.64	191.52	239.40	287.28	335.16	383.04	430.92
slug/foot-second to newton-second/meter², N·s/m²	47.880 26	95.761	143.64	191.52	239.40	287.28	335.16	383.04	430.92
KINEMATIC, ν									
centistoke to meter²/second, m²/s	1×10^{-6}	2×10^{-6}	3×10^{-6}	4×10^{-6}	5×10^{-6}	6×10^{-6}	7×10^{-6}	8×10^{-6}	9×10^{-6}
foot²/second to meter²/second, m²/s	0.092 903 04	0.185 81	0.278 71	0.371 61	0.464 52	0.557 42	0.650 32	0.743 22	0.836 12
VOLUME									
acre-foot to meter³, m³	1 233.482	2 457.0	3 700.4	4 933.9	6 167.4	7 400.9	8 634.4	9 867.9	11 101
barrel (oil, 42 gal) to meter³, m³	0.158 987 3	0.317 97	0.476 96	0.635 95	0.794 94	0.953 92	1.112 9	1.271 9	1.430 9
board foot to meter³, m³	0.002 359 737	0.004 719 5	0.007 079 2	0.009 438 9	0.011 799	0.014 158	0.016 518	0.018 878	0.021 238
bushel (U.S.) to meter³, m³	0.035 239 07	0.070 478	0.105 72	0.140 96	0.176 20	0.211 43	0.246 67	0.281 91	0.317 15
foot³ to meter³, m³	0.028 316 85	0.056 634	0.084 951	0.113 27	0.141 58	0.169 90	0.198 22	0.226 53	0.254 85
gallon (U.S. liquid) to meter³, m³	0.003 785 412	0.007 570 8	0.011 356	0.015 142	0.018 927	0.022 712	0.026 498	0.030 283	0.034 069

Table 9-16. CONVERSIONS TO SI UNITS (*Continued*)

	1	2	3	4	5	6	7	8	9
VOLUME (*Continued*)									
inch³ to meter³, m³	0.000 016 387 06	0.000 032 774	0.000 049 161	0.000 065 548	0.000 081 935	0.000 098 322	0.000 114 71	0.000 131 10	0.000 147 48
ounce (U.S. fluid) to meter³, m³	0.000 029 573 53	0.000 059 147	0.000 088 721	0.000 118 29	0.000 147 87	0.000 177 44	0.000 207 01	0.000 236 59	0.000 266 16
peck (U.S.) to meter³, m³	0.008 809 768	0.017 620	0.026 429	0.035 239	0.044 049	0.052 859	0.061 668	0.070 478	0.079 288
quart (U.S. liquid) to meter³, m³	0.000 946 352 9	0.001 892 7	0.002 839 1	0.003 785 4	0.004 731 8	0.005 678 1	0.006 624 5	0.007 570 8	0.008 517 2
yard³ to meter³, m³	0.764 554 9	1.529 1	2.293 7	3.058 2	3.822 8	4.587 3	5.351 9	6.116 4	6.881 0
VOLUME/MASS (SPECIFIC VOLUME)									
foot³/pound to meter³/kilogram, m³/kg	0.062 427 96	0.124 86	0.187 28	0.249 71	0.312 14	0.374 57	0.437 00	0.499 42	0.561 85

Table 9-17. FUNDAMENTAL PHYSICAL CONSTANTS*

B. N. Taylor, W. H. Parker, and D. N. Langenberg

The numbers in parentheses are the standard deviation uncertainties in the last digits of the quoted value, computed on the basis of internal consistency.

Quantity	Symbol	Value	Error, ppm	Units SI	Units cgs
Velocity of light	c	2.9979250(10)	0.33	10^8 m sec^{-1}	10^{10} cm sec^{-1}
Fine-structure constant, $[\mu_0 c^2/4\pi](e^2/hc)$	α α^{-1}	7.297351(11) 137.03602(21)	1.5 1.5	10^{-3}	10^{-3}
Electron charge	e	1.6021917(70) 4.803250(21)	4.4 4.4	10^{-19} C	10^{-20} emu 10^{-10} esu
Planck's constant	h $\hbar = h/2\pi$	6.626196(50) 1.0545919(80)	7.6 7.6	10^{-34} J·sec 10^{-34} J·sec	10^{-27} erg·sec 10^{-27} erg·sec
Avogadro's number	N	6.022169(40)	6.6	10^{26} kmole^{-1}	10^{23} mole^{-1}
Atomic mass unit	amu	1.660531(11)	6.6	10^{-27} kg	10^{-24} g
Electron rest mass	m_e $m_e{}^*$	9.109558(54) 5.485930(34)	6.0 6.2	10^{-31} kg 10^{-4} amu	10^{-28} g 10^{-4} amu
Proton rest mass	M_p $M_p{}^*$	1.672614(11) 1.00727661(8)	6.6 0.08	10^{-27} kg amu	10^{-24} g amu
Neutron rest mass	M_n $M_n{}^*$	1.674920(11) 1.00866520(10)	6.6 0.10	10^{-27} kg amu	10^{-24} g amu
Ratio of proton mass to electron mass	M_p/m_e	1836.109(11)	6.2		
Electron charge to mass ratio	e/m_e	1.7588028(54) 5.272759(16)	3.1 3.1	10^{11} C kg^{-1}	10^7 emu g^{-1} 10^{17} csu g^{-1}
Magnetic flux quantum, $[c]^{-1}(hc/2e)$	Φ_0 h/e	2.0678538(69) 4.135708(14) 1.3795234(46)	3.3 3.3 3.3	10^{-15} T·m^2 10^{-15} J·sec C^{-1}	10^{-7} G·cm^2 10^{-7} erg·sec emu^{-1} 10^{-17} erg·sec esu^{-1}
Quantum of circulation	$h/2m_e$ h/m_e	3.636947(11) 7.273894(22)	3.1 3.1	10^{-4} J·sec kg^{-1} 10^{-4} J·sec kg^{-1}	erg sec g^{-1} erg·sec g^{-1}
Faraday constant, Ne	F	9.648670(54) 2.892599(16)	5.5 5.5	10^7 C kmole^{-1}	10^3 emu mole^{-1} 10^{14} esu mole^{-1}
Rydberg constant, $[\mu_0 c^2/4\pi]^2(m_e e^4/4\pi\hbar^3 c)$	R_∞	1.09737312(11)	0.10	10^7 m^{-1}	10^5 cm^{-1}
Bohr radius, $[\mu_0 c^2/4\pi]^{-1}(\hbar^2/m_e e^2) = \alpha/4\pi R_\infty$	a_0	5.2917715(81)	1.5	10^{-11} m	10^{-9} cm
Classical electron radius, $[\mu_0 c^2/4\pi](e^2/m_e c^2) = \alpha^3/4\pi R_\infty$	r_0	2.817939(13)	4.6	10^{-15} m	10^{-13} cm
Electron magnetic moment in Bohr magnetons	μ_e/μ_B	1.0011596389(31)	0.0031		
Bohr magneton, $[c](e\hbar/2m_e c)$	μ_B	9.274096(65)	7.0	10^{-24} J T^{-1}	10^{-21} erg G^{-1}
Electron magnetic moment	μ_e	9.284851(65)	7.0	10^{-24} J T^{-1}	10^{-21} erg G^{-1}
Gyromagnetic ratio of protons in H_2O	$\gamma_p{}'$ $\gamma_p{}'/2\pi$	2.6751270(82) 4.257597(13)	3.1 3.1	10^8 rad sec^{-1}·T^{-1} 10^7 Hz T^{-1}	10^4 rad sec^{-1}·G^{-1} 10^3 Hz G^{-1}
γ_p corrected for diamagnetism of H_2O	γ_p $\gamma_p/2\pi$	2.6751965(82) 4.257707(13)	3.1 3.1	10^8 rad sec^{-1}·T^{-1} 10^7 Hz T^{-1}	10^4 rad sec^{-1}·G^{-1} 10^3 Hz G^{-1}
Magnetic moment of protons in H_2O in Bohr magnetons	$\mu_p{}'/\mu_B$	1.52099312(10)	0.066	10^{-3}	10^{-3}
Proton magnetic moment in Bohr magnetons	μ_p/μ_B	1.52103264(46)	0.30	10^{-3}	10^{-3}
Proton magnetic moment	μ_p	1.4106203(99)	7.0	10^{-26} J T^{-1}	10^{-23} erg G^{-1}
Magnetic moment of protons in H_2O in nuclear magnetons	$\mu_p{}'/\mu_n$	2.792709(17)	6.2		
$\mu_p{}'/\mu_n$ corrected for diamagnetism of H_2O	μ_p/μ_n	2.792782(17)	6.2		
Nuclear magneton, $[c](e\hbar/2M_p c)$	μ_n	5.050951(50)	10	10^{-27} J T^{-1}	10^{-24} erg G^{-1}
Compton wavelength of the electron, $h/m_e c$	λ_C $\lambda_C/2\pi$	2.4263096(74) 3.861592(12)	3.1 3.1	10^{-12} m 10^{-12} m	10^{-10} cm 10^{-10} cm

*Reprinted from: *Rev. Mod. Phys.*, 41:375, 1969. Reprinted by permission of the publisher, American Institute of Physics.

Table 9-17. FUNDAMENTAL PHYSICAL CONSTANTS *(Continued)*

Quantity	Symbol	Value	Error, ppm	Units SI	Units cgs
Compton wavelength of the proton, $h/M_p c$	$\lambda_{C,p}$	1.3214409(90)	6.8	10^{-15} m	10^{-13} cm
	$\lambda_{C,p}/2\pi$	2.103139(14)	6.8	10^{-16} m	10^{-14} cm
Compton wavelength of the neutron, $h/M_n c$	$\lambda_{C,n}$	1.3196217(90)	6.8	10^{-15} m	10^{-13} cm
	$\lambda_{C,n}/2\pi$	2.100243(14)	6.8	10^{-16} m	10^{-14} cm
Gas constant	R_0	8.31434(35)	42	10^3 J kmole$^{-1} \cdot$K^{-1}	10^7 erg mole$^{-1} \cdot$K^{-1}
Boltzmann's constant, R_0/N	k	1.380622(59)	43	10^{-23} J K^{-1}	10^{-16} erg K^{-1}
Stefan–Boltzmann constant, $\pi^2 k^4/60h^3 c^2$	σ	5.66961(96)	170	10^{-8} W m^{-2} K^4	10^{-5} erg sec$^{-1} \cdot$cm$^{-2} \cdot$K^{-4}
First radiation constant, $8\pi hc$	c_1	4.992579(38)	7.6	10^{-24} J$\cdot$m	10^{-15} erg$\cdot$cm
Second radiation constant, hc/k	c_2	1.438833(61)	43	10^{-2} m$\cdot$K	cm$\cdot$K
Gravitational constant	G	6.6732(31)	460	10^{-11} N$\cdot$m^2 kg^{-2}	10^{-8} dyn$\cdot$cm^{2}g^{-2}
kx-unit-to-angstrom conversion factor, $\Lambda = \lambda(\text{Å})/\lambda(\text{kxu}); \lambda(\text{Cu}K\alpha_1) \equiv$ 1.537400 kxu	Λ	1.0020764(53)	5.3		
Å*-to-angstrom conversion factor, $\Lambda = \lambda(\text{Å})/\lambda(\text{Å*}); \lambda(\text{W}K\alpha_1) \equiv$ 0.2090100 Å*	Λ^*	1.0000197(56)	5.6		

*Note that the unified atomic mass scale ^{12}C $\equiv$ 12 has been used throughout, that amu = atomic mass unit, C = coulomb, G = gauss, Hz = hertz = cycles/sec, J = joule, K = kelvin (degrees kelvin), T = tesla (10^4 G), V = volt, and W = watt. In cases where formulas for constants are given (e.g., R_∞), the relations are written as the product of two factors. The second factor, in parentheses, is the expression to be used when all quantities are expressed in cgs units, with the electron charge in electrostatic units. The first factor, in brackets, is to be included only if all quantities are expressed in SI units. We remind the reader that with the exception of the auxiliary constants which have been taken to be exact, the uncertainties of these constants are correlated, and therefore the general law of error propagation must be used in calculating additional quantities requiring two or more of these constants.

ENERGY CONVERSION FACTORS

Quantity	Value	Unit	Error, ppm
1 kg	5.609538(24)	10^{29} MeV	4.4
1 amu	931.4812(52)	MeV	5.5
Electron mass	0.5110041(16)	MeV	3.1
Proton mass	938.2592(52)	MeV	5.5
Neutron mass	939.5527(52)	MeV	5.5
1 electron volt	1.6021917(70)	10^{-19} J 10^{-12} erg	4.4
	2.4179659(81)	10^{14} Hz	3.3
	8.065465(27)	10^5 m^{-1} 10^3 cm^{-1}	3.3
	1.160485(49)	10^4 K	42
Energy–wavelength conversion	1.2398541(41)	10^{-6} eV$\cdot$m 10^{-4} eV$\cdot$cm	3.3
Rydberg constant, R_∞	2.179914(17)	10^{-18} J 10^{-11} erg	7.6
	13.605826(45)	eV	3.3
	3.2898423(11)	10^{15} Hz	0.35
	1.578936(67)	10^5 K	43
Bohr magneton, μ_B	5.788381(18)	10^{-5} eV T^{-1}	3.1
	1.3996108(43)	10^{10} Hz T^{-1}	3.1
	46.68598(14)	m$^{-1} \cdot$T^{-1} 10^{-2} cm$^{-1} \cdot$T^{-1}	3.1
	0.671733(29)	K T^{-1}	43
Nuclear magneton, μ_n	3.152526(21)	10^{-8} eV T^{-1}	6.8
	7.622700(42)	10^6 Hz T^{-1}	5.5
	2.542659(14)	10^{-2} m$^{-1} \cdot$T^{-1} 10^{-4} cm$^{-1} \cdot$T^{-1}	5.5
	3.65846(16)	10^{-4} K T^{-1}	44
Gas constant, R_0	8.20562(35)	10^{-2} m$^3 \cdot$atm kmole$^{-1} \cdot$K^{-1}	42
Standard volume of ideal gas, V_0	22.4136	m^3 kmole^{-1}	

Table 9-18.   CONVERSION FACTORS*
From Various U.S. Government and IUPAC Publications and from Calculations Based on Values Given in These Publications

To convert from	To	Multiply by
Abamperes	Amperes	10
"	E.M. cgs. units of current	1
"	E.S. cgs. units	2.997930×10^{10}
"	Faradays (chem.)/sec	1.036377×10^{-4}
"	Faradays (phys.)/sec	1.036086×10^{-4}
"	Statamperes	2.997930×10^{10}
Abamperes/cm	E.M. cgs. units of surface charge density	1
"	E.S. cgs. units	2.997930×10^{10}
Abamperes/sq. cm.	Amperes/circ. mil	5.0670748×10^{-5}
"	Amperes/sq. cm	10
"	Amperes/sq. inch	64.516
Abampere-turns	Ampere-turns	10
Abampere-turns/cm	Ampere-turns/cm	10
Abcoulombs	Ampere-hours	0.0027777
"	Coulombs	10
"	Electronic charges	6.24196×10^{19}
"	E.M. cgs. units of charge	1
"	E.S. cgs. units	2.997930×10^{10}
"	Faradays (chem.)	1.036377×10^{-4}
"	Faradays (phys.)	1.036086×10^{-4}
"	Statcoulombs	2.997930×10^{10}
Abfarads	E.M. cgs. units of capacitance	1
"	E.S. cgs. units	8.987584×10^{20}
"	Farads	1×10^{9}
"	Microfarads	1×10^{15}
"	Statfarads	8.987584×10^{20}
Abhenries	E.M. cgs. units of induction	1
"	E.S. cgs. units	1.112646×10^{-21}
"	Henries	1×10^{-9}
Abmhos	E.M. cgs. units of conductance	1
"	E.S. cgs. units	8.987584×10^{20}
"	Megamhos	1000
"	Mhos	1×10^{9}
"	Statmhos	8.987584×10^{20}
Abohms	E.M. cgs. units of resistance	1
"	Megohms	1×10^{-15}
"	Microhms	0.001
"	Ohms	1×10^{-9}
"	Statohms	1.112646×10^{-21}
Abohm-cm	Circ. mil-ohms/ft	0.0060153049
"	E.M. cgs. units of resistivity	1
"	Microhm-inches	0.00039370079
"	Ohm-cm	1×10^{-9}
Abvolts	Microvolts	0.01
"	Millivolts	1×10^{-5}
"	Volts	1×10^{-8}
"	Volts (Int.)	9.99670×10^{-9}
Abvolts/cm	E.M. cgs. units of electric field intensity	1
"	E.S. cgs. units	3.335635×10^{-11}
"	Volts/cm	1×10^{-8}
"	Volts/inch	2.54×10^{-8}
"	Volts/meter	1×10^{-6}
Acres	Sq. cm	40468564
"	Sq. ft	43560
"	Sq. ft. (U.S. Survey)	43559.826
"	Sq. inches	6272640
"	Sq. kilometers	0.0040468564
"	Sq. links (Gunter's)	1×10^{5}
"	Sq. meters	4046.8564
"	Sq. miles (statute)	0.0015625
"	Sq. perches	160
"	Sq. rods	160
"	Sq. yards	4840
Acre-feet	Cu. feet	43560
"	Cu. meters	1233.4818
"	Cu. yards	1613.333
Acre-inches	Cu. feet	3630
"	Cu. meters	102.79033
Acre-inches	Gallons (U.S.)	27154.286
Amperes	Abamperes	0.1
"	Amperes (Int.)	1.000165
"	Cgs. units of current	1
"	Mks. units of current	1
"	Coulombs/sec	1
"	Coulombs (Int.)/sec	1.000165
"	Faradays (chem.)/sec	1.036377×10^{-5}
"	Faradays (phys.)/sec	1.036086×10^{-5}
"	Statamperes	2.997930×10^{9}
Amperes (Int.)	Amperes	0.999835
"	Coulombs/sec	0.999835
"	Coulombs (Int.)/sec	1
"	Faradays (chem.)/sec	1.03623×10^{-5}
"	Faradays (phys.)/sec	1.03592×10^{-5}
Amperes/meter	Cgs. units of surface current density	0.01
"	E.M. cgs. units	0.001
"	E.S. cgs. units	2.997930×10^{7}
"	Mks. units	1
Amperes/sq. meter	Cgs. units of volume current density	0.0001
"	E.M. cgs. units	1×10^{-5}
"	E.S. cgs. units	299793.0
"	Mks. units	1
Amperes/sq. mil	Abamperes/sq. cm	15500.031
"	Amperes/sq. cm	1.5500031×10^{6}
Ampere-hours	Abcoulombs	360
"	Coulombs	3600
"	Faradays (chem.)	0.0373096
"	Faradays (phys.)	0.0372991
Ampere-turns	Cgs. units of magnetomotive force	1.2566371
"	E.M. cgs. units	1.2566371
"	E.S. cgs. units	3.767310×10^{10}
"	Gilberts	1.2566371
Ampere-turns/weber	Cgs. units of reluctance	1.256637×10^{-8}
"	E.M. cgs. units	1.256637×10^{-8}
"	E.S. cgs. units	1.129413×10^{19}
"	Gilberts/maxwell	1.256637×10^{-8}
Ångstrom units	Centimeters	1×10^{-8}
"	Inches	3.9370079×10^{-9}
"	Microns	0.0001
"	Millimicrons	0.1
"	Wave length of orange-red line of krypton 86	0.000165076373
"	Wave length of red line of cadmium	0.000155316413
Ares	Acres	0.024710538
"	Sq. dekameters	1
"	Sq. feet	1076.3910
"	Sq. ft. (U.S. Survey)	1076.3867
"	Sq. meters	100
"	Sq. miles	3.8610216×10^{-5}
Atmospheres	Bars	1.01325
"	Cm. of Hg (0°C.)	76
"	Cm. of H_2O (4°C.)	1033.26
"	Dynes/sq. cm	1.01325×10^{6}
"	Ft. of H_2O (39.2°F.)	33.8995
"	Grams/sq. cm	1033.23
"	In. of Hg (32°F.)	29.9213
"	Kg./sq. cm	1.03323
"	Mm. of Hg (0°C.)	760
"	Pounds/sq. inch	14.6960
"	Tons (short)/sq. ft	1.05811
"	Torrs	760
Atomic mass units (chem.)	Electron volts	9.31395×10^{8}
"	Grams	1.66024×10^{-24}
Atomic mass units (phys.)	Electron volts	9.31141×10^{8}
"	Grams	1.65979×10^{-24}

*From: "Handbook of Chemistry and Physics", 53rd ed., R.C. Weast, Ed., The Chemical Rubber Co., 1972.

Table 9-18. CONVERSION FACTORS (*Continued*)

To convert from	To	Multiply by	To convert from	To	Multiply by
Bags (Brit.)	Bushels (Brit.)	3	B.t.u.	Kw.-hours (Int.)	0.000292827
Barns	Sq. cm	1×10^{-24}	"	Liter-atm	10.4053
Barrels (Brit.)*	Bags (Brit.)	1.5	"	Tons of refrig. (U.S. std.)	3.46995×10^{-6}
"	Barrels (U.S., dry)	1.415404	"	Watt-seconds	1054.35
"	Barrels (U.S., liq.)	1.372513	"	Watt-seconds (Int.)	1054.18
"	Bushels (Brit.)	4.5	B.t.u. (IST.)	B.t.u.	1.00065
"	Bushels (U.S.)	4.644253	B.t.u. (mean)	B.t.u.	1.00144
"	Cu. feet	5.779568	"	B.t.u. (IST.)	1.00078
"	Cu. meters	0.1636591	"	B.t.u. (39°F.)	0.996415
"	Gallons (Brit.)	36	"	B.t.u. (60°F.)	1.00113
"	Liters	163.6546	"	Hp.-hours	0.000393317
Barrels (petroleum, U.S.)	Cu. feet	5.614583	"	Joules	1055.87
"	Gallons (U.S.)	42	"	Kg.-meters	107.669
"	Liters	158.98284	"	Kw.-hours	0.000293297
Barrels (U.S., dry)	Barrels (U.S. liq.)	0.969696	"	Kw.-hours (Int.)	0.000293248
"	Bushels (U.S.)	3.2812195	"	Liter-atm	10.4203
"	Cu. feet	4.083333	"	Watt-hours	0.293297
"	Cu. inches	7056	"	Watt-hours (Int.)	0.293248
"	Cu. meters	0.11562712	B.t.u. (39°F.)	B.t.u.	1.00504
"	Quarts (U.S., dry)	105	"	B.t.u. (IST.)	1.00439
Barrels (U.S., liq.)	Barrels (U.S., dry)	1.03125	"	B.t.u. (mean)	1.00360
"	Barrels (wine)	1	"	B.t.u. (60°F)	1.00473
"	Cu. feet	4.2109375	"	Joules	1059.67
"	Cu. inches	7276.5	B.t.u. (60°F.)	B.t.u.	1.00031
"	Cu. meters	0.11924047	"	B.t.u. (IST.)	0.999657
"	Gallons (Brit.)	26.22925	"	B.t.u. (mean)	0.998873
"	Gallons (U.S., liq.)	31.5	"	B.t.u. (39°F.)	0.995291
"	Liters	119.23713	B.t.u./hr.	Cal., *kg*./hr.	0.251996
Bars	Atmospheres	0.986923	"	Ergs/sec.	2.928751×10^{6}
"	Baryes	1×10^{6}	"	Foot-pounds/hr.	777.649
"	Cm. of Hg (0°C.)	75.0062	"	Horsepower	0.000392752
"	Dynes/sq. cm.	1×10^{6}	"	Horsepower (boiler)	2.98563×10^{-5}
"	Ft. of H₂O (60°F.)	33.4883	"	Horsepower (electric)	0.000392594
"	Grams/sq. cm.	1019.716	"	Horsepower (metric)	0.000398199
"	In. of Hg (32°F.)	29.5300	"	Kilowatts	0.000292875
"	Kg./sq. cm.	1.019716	"	Lb. ice melted/hr.	0.0069714
"	Millibars	1000	"	Tons of refrig. (U.S. comm.)	8.32789×10^{-5}
"	Pounds/sq. inch	14.5038	"	Watts	0.292875
Baryes	Atmospheres	9.86923×10^{-7}	B.t.u./min.	Cal., *kg*./min.	0.251996
"	Bars	1×10^{-6}	"	Ergs/sec.	1.75725×10^{8}
"	Dynes/sq. cm.	1	"	Foot-pounds/min.	777.649
"	Grams/sq. cm.	0.001019716	"	Horsepower	0.0235651
"	Millibars	0.001	"	Horsepower (boiler)	0.00179138
Bels	Decibels	10	"	Horsepower (electric)	0.0235556
Board feet	Cu. cm.	2359.7372	"	Horsepower (metric)	0.0238920
"	Cu. feet	0.083333	"	Joules/sec.	17.5725
"	Cu. inches	144	"	Kg.-meters/min.	107.514
Bolts of cloth	Linear feet	120	"	Kilowatts	0.0175725
"	Meters	36.576	"	Lb. ice melted/hr.	0.41828
Bougie decimales	Candles (Int.)	1.00	"	Tons of refrig. (U.S. comm.)	0.00499673
B.t.u.	B.t.u. (IST.)**	0.999346	"	Watts	17.5725
"	B.t.u. (mean)	0.998563	B.t.u. (mean)/min.	B.t.u. (mean)/hr.	60
"	B.t.u. (39°F.)	0.994982	"	Cal., *kg*. (mean)/hr.	15.1197
"	B.t.u. (60°F.)	0.999689	"	Cal., *kg*. (mean)/min.	0.251996
"	Cal. *gm*.	251.99576	"	Ergs/sec.	1.75978×10^{8}
"	Cal., *gm*. (IST.)	251.831	"	Foot-pounds/min.	778.768
"	Cal., *gm*. (mean)	251.634	"	Horsepower	0.0235990
"	Cal., *gm*. (20°C.)	252.122	"	Horsepower (boiler)	0.00179396
"	Cu. cm.-atm.	10405.6	"	Horsepower (electric)	0.0235895
"	Ergs	1.05435×10^{10}	"	Horsepower (metric)	0.0239264
"	Foot-poundals	25020.1	"	Joules/sec.	17.5978
"	Foot-pounds	777.649	"	Kg.-meters/min.	107.669
"	Gram-cm.	1.07514×10^{7}	"	Kilowatts	0.0175978
"	Hp.-hours	0.000392752	"	Lb. ice-melted/hr.	0.41888
"	Hp.-years	4.48347×10^{-8}	B.t.u./lb.	Cal., *gm*./gram	0.555555
"	Joules	1054.35	"	Cu. cm.-atm./gram	22.9405
"	Joules (Int.)	1054.18	"	Cu. ft.-atm./lb.	0.367471
"	Kg.-meters	107.514	"	Cu. ft.-(lb./sq. in.)/lb.	5.40034
"	Kw.-hours	0.000292875	"	Foot-pounds/lb.	777.649
			"	Hp.-hr./lb.	0.000392752

*Barrel (Brit., liq.) = Barrel (Brit., dry).
**International Steam Table.

Table 9-18. CONVERSION FACTORS (Continued)

To convert from	To	Multiply by
B.t.u./lb	Joules/gram	2.32444
B.t.u. (mean)/lb	Cal., gm. (mean)/gram	0.555555
"	Cu. cm.-atm./gram	22.9735
"	Foot-pounds/lb	778.768
"	Hp.-hr./lb	0.000393317
"	Joules/gram	2.32779
B.t.u./sec	B.t.u./hr	3600
"	B.t.u./min	60
"	Cal., kg./hr	907.185
"	Cal., kg./min	15.1197
"	Cheval-vapeur	1.43352
"	Ergs/sec	1.05435×10^{10}
"	Foot-pounds/sec	777.649
"	Horsepower	1.41391
"	Horsepower (boiler)	0.107483
"	Horsepower (electric)	1.41334
"	Horsepower (metric)	1.43352
"	Kg.-meters/sec	107.514
"	Kilowatts	1.05435
"	Kilowatts (Int.)	1.05418
"	Watts	1054.35
"	Watts (Int.)	1054.18
B.t.u. (mean)/sec	Ergs/sec	1.05587×10^{10}
"	Foot-pounds/sec	778.768
"	Horsepower	1.41594
"	Horsepower (boiler)	0.107637
"	Horsepower (electric)	1.41537
"	Horsepower (metric)	1.43558
"	Watts	1055.87
B.t.u./sq. ft	Cal., gm./sq. cm	0.271246
B.t.u./sq.ft. × min.)	Hp./sq. ft	0.0235651
"	Kw./sq. ft	0.0175725
"	Watts/sq. in	0.122031
Buckets (Brit.)	Cu. cm	18184.35
"	Gallons (Brit.)	4
Bushels (Brit.)	Bags (Brit.)	0.333333
"	Bushels (U.S.)	1.032056
"	Cu. cm	36368.70
"	Cu. feet	1.284348
"	Cu. inches	2219.354
"	Dekaliters	3.636768
"	Gallons (Brit.)	8
"	Hectoliters	0.3636768
"	Liters	36.36768
Bushels (U.S.)*	Barrels (U.S.), dry	0.3047647
"	Bushels (Brit.)	0.9689395
"	Cu. cm	35239.07
"	Cu. feet	1.244456
"	Cu. inches	2150.42
"	Cu. meters	0.03523907
"	Cu. yards	0.04609096
"	Gallons (U.S., dry)	8
"	Gallons (U.S., liq.)	9.309177
"	Liters	35.23808
"	Ounces (U.S., fluid)	1191.575
"	Pecks (U.S.)	4
"	Pints (U.S., dry)	64
"	Quarts (U.S., dry)	32
"	Quarts (U.S., liq.)	37.23671
Butts (Brit.)	Bushels (U.S.)	13.53503
"	Cu. feet	16.84375
"	Cu. meters	0.4769619
"	Gallons (U.S.)	126
Cable lengths	Fathoms	120
"	Feet	720
"	Meters	219.456

To convert from	To	Multiply by
Calories, gm.**	B.t.u.	0.0039683207
"	B.t.u. (IST.)	0.00396573
"	B.t.u. (mean)	0.00396262
"	B.t.u. (39°F.)	0.00394841
"	B.t.u. (60°F.)	0.00396709
"	Cal., gm. (IST.)	0.999346
"	Cal., gm. (mean)	0.998563
"	Cal., gm. (15°C.)	0.999570
"	Cal., gm. (20°C.)	1.00050
"	Cal., kg.	0.001
"	Cal., kg. (IST.)	0.000999346
"	Cal., kg. (mean)	0.000998563
"	Cal., kg. (15°C.)	0.000999570
"	Cal., kg. (20°C.)	0.00100050
"	Cu. cm.-atm	41.2929
"	Cu. ft.-atm	0.00145821
"	Ergs	4.184×10^{7}
"	Foot-poundals	99.2878
"	Foot-pounds	3.08596
"	Gram-cm	42664.9
"	Hp.-hours	1.55857×10^{-6}
"	Joules	4.184
"	Joules (Int.)	4.18331
"	Kg.-meters	0.426649
"	Kw.-hours	1.162222×10^{-6}
"	Liter-atm	0.0412017
"	Watt-hours	0.001162222
"	Watt-hours (Int.)	0.00116209
"	Watt-seconds	4.184
Calories, gm. (mean)	B.t.u.	0.00397103
"	Cal., gm	1.00144
"	Cal., gm. (IST.)	1.00078
"	Cal., gm. (20°C.)	1.00194
"	Cal., kg. (mean)	0.001
"	Cu. cm.-atm	41.3523
"	Cu. ft.-atm	0.00146034
"	Ergs	4.19002×10^{7}
"	Foot-poundals	99.4308
"	Foot-pounds	3.09040
"	Hp.-hours	1.56081×10^{-6}
"	Joules	4.19002
"	Joules (Int.)	4.18933
"	Kg.-meters	0.427263
"	Kw.-hours	1.16390×10^{-6}
"	Liter-atm	0.0413511
"	Watt-seconds	4.19002
Calories, gm. (15°C.)	B.t.u.	0.00397003
"	Cal., gm	1.00043
"	Cal., gm. (IST.)	0.999776
"	Cal., gm. (mean)	0.998992
"	Cal., gm. (20°C.)	1.00093
"	Joules	4.18580
"	Joules (Int.)	4.18511
Calories, gm. (20°C.)	B.t.u.	0.00396633
"	Cal., gm	0.999498
"	Cal., gm. (IST.)	0.998845
"	Cal., gm. (mean)	0.998061
"	Cal., gm. (15°C.)	0.999068
"	Joules	4.18190
"	Joules (Int.)	4.18121
Calories, kg	B.t.u.	3.9683207
"	B.t.u. (IST.)	3.96573
"	B.t.u. (mean)	3.96262
"	B.t.u. (60°F.)	3.96709
"	Cal., gm	1000
"	Cal., kg. (mean)	0.998563
"	Cal., kg. (15°C.)	0.999570
"	Cal., kg. (20°C.)	1.00050
"	Cu. cm.-atm	41292.86

*Stricken or struck bushel. A heaped bushel for apples of 2747.715 cu. inches was established by the U.S. Court of Customs Appeals of Feb. 15, 1912. A heaped bushel equal to 1¼ stricken bushels is also known.

**This is the calorie as defined by the U.S. National Bureau of Standards and is equal to 4.18400 joules.

Table 9-18. CONVERSION FACTORS (Continued)

To convert from	To	Multiply by
Calories, *kg*	Ergs	4.184×10^{10}
"	Foot-poundals	99287.8
"	Foot-pounds	3085.96
"	Gram-cm	4.26649×10^7
"	Hp.-hours	0.00155857
"	Joules	4184
"	Kw.-hours	0.001162222
"	Liter-atm	41.2917
"	Watt-hours	1.162222
Calories, *kg.* (mean)	B.t.u.	3.97403
"	B.t.u. (IST.)	3.97144
"	B.t.u. (mean)	3.9683207
"	B.t.u. (60°F.)	3.97280
"	Cal., *gm.*	1001.44
"	Cal., *gm.* (IST.)	1000.78
"	Cal., *gm.* (mean)	1000
"	Cal., *gm.* (15°C.)	1000.10
"	Cal., *gm.* (20°C.)	1001.94
"	Ergs	4.19002×10^{10}
"	Foot-poundals	99430.8
"	Foot-pounds	3090.40
"	Gram-cm	4.27263×10^7
"	Hp.-hours	0.00156081
"	Joules	4190.02
"	Kg.-meters	427.263
"	Kw.-hours (Int.)	0.00116370
"	Liter-atm	41.3511
"	Watt-hours	1.16390
Cal., *gm.*/°C	B.t.u./°F	0.00220462
"	Joules/°F	2.324444
"	Joules (Int.)/°F	2.32406
Cal., *gm.*/gram	B.t.u./lb	1.8
"	Foot-pounds/lb	1399.77
"	Joules/gram	4.184
"	Watt-hours/gram	0.001162222
Cal., *gm.*/(gram × °C)	B.t.u./(lb. × °C.)	1.8
"	B.t.u./(lb. × °F.)	1
"	Cal., *kg.*/(kg. × °C.)	1
"	Joules/(gram × °C.)	4.184
"	Joules/(lb. × °F.)	1054.35
Cal., *gm.*/hr	B.t.u./hr	0.0039683207
"	Ergs/sec	11622.222
"	Watts	0.001162222
Cal., *gm.* (mean)/hr	B.t.u. (mean)/hr	0.0039683207
"	Ergs/sec	11639.0
"	Watts	0.00116390
Cal., *kg.*/hr	Watts	1.162222
Cal., *gm.*/min	B.t.u./min	0.0039683207
"	Ergs/sec	697333.3
"	Watts	0.069733
Cal., *gm.* (mean)/min	B.t.u. (mean)/min	0.0039683207
"	Ergs/sec	698337
"	Joules/sec	0.0698337
"	Watts	0.0698337
Cal., *kg.*/min	Kg. ice melted/min	0.012548
"	Lb. ice melted/min	0.027665
"	Watts	69.7333
Cal., *gm.*/sec	B.t.u./sec	0.0039683207
"	Ergs/sec	4.184×10^7
"	Foot-pounds/sec	3.08596
"	Horsepower	0.00561084
"	Watts	4.184
Cal., *gm.* (mean)/sec	Ergs/sec	4.19002×10^7
"	Watts	4.19002
Cal., *gm.*/(sec. × sq. cm.)	B.t.u./(hr. × sq. ft.)	13272.1
"	Cal., *gm.*/(hr. × sq. cm.)	3600
"	Watts/sq. cm	4.184
Cal., *gm.*/(sec. × sq. cm. × °C.)	B.t.u./(hr. × sq. ft. × °F.)	7373.38
Cal., *gm.*/sq. cm	B.t.u./sq. ft	3.68669

To convert from	To	Multiply by
Cal., *gm.*-cm. / (hr. × sq. cm. × °C.)	B.t.u.-ft. / (hr. × sq. ft. × °F.)	0.0671969
"	B.t.u.-inch / (hr. × sq. ft. × °F.)	0.806363
Cal., *gm.*-cm./sq. cm.	B.t.u.-inch/sq. ft.	1.4514530
Cal., *gm.*-sec.	Planck's constant	6.31531×10^{33}
Cal., *gm.*-sec./Avog. No. (chem.)	Planck's constant	1.04849×10^{10}
Cal., *gm.*-sec./Avog. No. (phys.)	Planck's constant	1.04821×10^{10}
Candles (English)	Candles (Int.)	1.04
"	Hefner units	1.16
Candles (German)	Candles (English)	1.01
"	Candles (Int.)	1.05
"	Hefner units	1.17
Candles (Int.)	Candles (English)	0.96
"	Candles (German)	0.95
"	Candles (pentane)	1.00
"	Hefner units	1.11
"	Lumens (Int.)/steradian	1
Candles (pentane)	Candles (Int.)	1.00
Candles/sq. cm	Candles/sq. inch	6.4516
"	Candles/sq. meter	10000
"	Foot-lamberts	2918.6351
"	Lamberts	3.1415927
Candles/sq. ft	Candles/sq. inch	0.0069444
"	Candles/sq. meter	10.763910
"	Foot-lamberts	3.1415927
"	Lamberts	0.0033815822
Candles/sq. inch	Candles/sq. cm.	0.15500031
"	Candles/sq. foot	144
"	Foot-lamberts	452.38934
"	Lamberts	0.48694784
Candle power (spher.)	Lumens	12.566370
Carats (parts of gold per 24 of mixture)	Milligrams/gram	41.6666
Carats (1877)	Grains	3.168
"	Milligrams	205.3
Carats (metric)	Grains	3.08647
"	Grams	0.2
"	Milligrams	200
Carcel units	Candles (Int.)	9.61
Centals	Kilograms	45.359237
"	Pounds	100
Centares	Ares	0.01
"	Sq. feet	10.763910
"	Sq. inches	1550.0031
"	Sq. meters	1
"	Sq. yards	1.1959900
Centigrams	Grains	0.15432358
"	Grams	0.01
Centiliters	Cu. cm.	10.00028
"	Cu. inches	0.6102545
"	Liters	0.01
"	Ounces (U.S., fluid)	0.3381497
Centimeters	Ångström units	1×10^8
"	Feet	0.032808399
"	Feet (U.S. Survey)	0.032808333
"	Hands	0.098425197
"	Inches	0.39370079
"	Links (Gunter's)	0.049709695
"	Links (Ramden's)	0.032808399
"	Meters	0.01
"	Microns	10000
"	Miles (naut., Int.)	5.3995680×10^{-6}
"	Miles (statute)	6.2137119×10^{-6}
"	Millimeters	10
"	Millimicrons	1×10^7
"	Mils	393.70079
"	Picas (printer's)	2.3710630
"	Points (printer's)	28.452756

Table 9-18. CONVERSION FACTORS (Continued)

To convert from	To	Multiply by	To convert from	To	Multiply by
Centimeters	Rods	0.0019883878	Circumferences	Minutes	21600
"	Wave length of orange-red line of krypton 86	16507.6373	"	Radians	6.2831853
"	Wave length of red line of cadmium	15531.6413	"	Seconds	1296000
"			Cords	Cord-feet	8
"	Yards	0.010936133	"	Cu. feet	128
Cm. of Hg (0°C.)	Atmospheres	0.013157895	"	Cu. meters	3.6245734
"	Bars	0.0133322	Cord-feet	Cords	0.125
"	Dynes/sq. cm	13332.2	"	Cu. feet	16
"	Ft. of H_2O (4°C.)	0.446050	Coulombs	Abcoulombs	0.1
"	Ft. of H_2O (60°F.)	0.446474	"	Ampere-hours	0.0002777
"	In. of Hg (0°C.)	0.39370079	"	Ampere-seconds	1
"	Kg./sq. meter	135.951	"	Coulombs (Int.)	1.000165
"	Pounds/sq. ft	27.8450	"	Electronic charge	6.24196×10^{18}
"	Pounds/sq. inch	0.193368	"	E.M. cgs. units of electric charge	0.1
"	Torrs	10	"	E.S. cgs. units of electric charge	2.997930×10^9
Cm. of H_2O (4°C.)	Atmospheres	0.000967814	"	Faradays (chem.)	1.036377×10^{-5}
"	Dynes/sq. cm	980.638	"	Faradays (phys.)	1.036086×10^{-5}
"	Pounds/sq. inch	0.0142229	"	Mks. units of electric charge	1
Centimeters/sec.	Feet/min	1.9685039	"	Statcoulombs	2.997930×10^9
"	Feet/sec	0.032808399	Coulombs/cu. meter	E.M. cgs. units of volume charge density	1×10^{-7}
"	Kilometers/hr	0.036			
"	Kilometers/min	0.0006	"	E.S. cgs. units	2997.930
"	Knots (Int.)	0.019438445	Coulombs/sq. cm	Abcoulombs/sq. cm	0.1
"	Meters/min	0.6	"	Cgs. units of polarization, and surface charge density	1
"	Miles/hr	0.022369363			
"	Miles/min	0.00037282272	Cubic centimeters	Board feet	0.00042377600
Cm./(sec.)(sec.)	Kilometers/(hr. × sec.)	0.036	"	Bushels (Brit.)	2.749617×10^{-5}
"	Miles/(hr. × sec.)	0.022369363	"	Bushels (U.S.)	2.837759×10^{-5}
Centimeters/year	Inches/year	0.30370079	"	Cu. feet	3.5314667×10^{-5}
Centipoises*	Grams/(cm. × sec.)	0.01	"	Cu. inches	0.061023744
"	Poises	0.01	"	Cu. meters	1×10^{-6}
"	Pound/(ft. × hr.)	2.4190883	"	Cu. yards	1.3079506×10^{-6}
"	Pounds/(ft. × sec.)	0.00067196898	"	Drachms (Brit., fluid)	0.28156080
Centistokes*	Stokes	0.01	"	Drams (U.S., fluid)	0.27051218
Chains (Gunter's)	Centimeters	2011.68	"	Gallons (Brit.)	0.0002199604
"	Chains (Ramden's)	0.66	"	Gallons (U.S., dry)	0.00022702075
"	Feet	66	"	Gallons (U.S., liq.)	0.00026417205
"	Feet (U.S. Survey)	65.999868	"	Gills (Brit.)	0.007039020
"	Furlongs	0.1	"	Gills (U.S.)	0.0084535058
"	Inches	792	"	Liters	0.000000079
"	Links (Gunter's)	100	"	Ounces (Brit., fluid)	0.03519510
"	Links (Ramden's)	66	"	Ounces (U.S., fluid)	0.033814023
"	Meters	20.1168	"	Pints (U.S., dry)	0.0018161660
"	Miles (statute)	0.0125	"	Pints (U.S., liq.)	0.0021133764
"	Rods	4	"	Quarts (Brit.)	0.0008798775
"	Yards	22	"	Quarts (U.S., dry)	0.00090808298
Chains (Ramden's)	Centimeters	3048	"	Quarts (U.S., liq.)	0.0010566882
"	Chains (Gunter's)	1.515151	Cu. cm./gram	Cu. ft./lb	0.016018463
"	Feet	100	Cu. cm./sec	Cu. ft./min	0.0021188800
"	Feet (U.S. Survey)	99.999800	"	Gal. (U.S.)/min	0.015850323
Cheval-vapeur	Horsepower (metric)	1	"	Gal. (U.S.)/sec	0.00026417205
Cheval-vapeur-heures	Joules	2647795	Cu. cm.-atm	B.t.u.	9.61019×10^{-5}
Circles	Degrees	360	"	B.t.u. (mean)	9.59637×10^{-5}
"	Grades	400	"	Cal., gm	0.0242173
"	Minutes	21600	"	Cal., gm. (mean)	0.0241824
"	Radians	6.2831853	"	Cu. ft.-atm	3.5314667×10^{-5}
"	Signs	12	"	Joules	0.101325
Circular inches	Circular mm	645.16	"	Watt-hours	2.81458×10^{-5}
"	Sq. cm	5.0670748	Cu. cm.-atm./gram	B.t.u./lb	0.0435911
"	Sq. inches	0.78539816	"	Cal., gm./gram	0.0242173
Circular mm	Sq. cm	0.0078539816	"	Cu. ft.-(lb./sq. in.)/lb	0.235406
"	Sq. inches	0.0012173696	"	Ft.-lb./lb	33.8985
"	Sq. mm	0.78539816	"	Joules/gram	0.101325
Circular mils	Circular inches	1×10^{-6}	"	Kg.-meters/gram	0.0103323
"	Sq. cm	5.0670748×10^{-6}	"	Kw.-hr./gram	2.81458×10^{-8}
"	Sq. inches	7.8539816×10^{-7}	Cubic decimeters	Cu. cm	1000
"	Sq. mm	0.00050670748	"	Cu. feet	0.035316667
"	Sq. mils	0.78539816	"	Cu. inches	61.023744
Circumferences	Degrees	360	"	Cu. meters	0.001
"	Grades	400			

*See also Table 9-26.

Table 9-18. CONVERSION FACTORS (*Continued*)

To convert from	To	Multiply by	To convert from	To	Multiply by
Cubic decimeters.....	Cu. yards................	0.0013079506	Cubic inches........	Ounces (U.S., fluid)......	0.55411255
"	Liters................	0.999972	"	Pecks (U.S.)............	0.0018601017
Cubic dekameters....	Cu. decimeters...........	1×10^6	"	Pints (U.S., dry)........	0.029761628
"	Cu. feet................	35314.667	"	Pints (U.S., liq.)........	0.034632035
" ...	Cu. inches..............	6.1023744×10^7	"	Quarts (U.S., dry).......	0.014880814
" ...	Cu. meters..............	1000	"	Quarts (U.S., liq.).......	0.017316017
"	Liters................	999972	Cu. in. of H_2O (4°C)..	Pounds of H_2O........	0.0361263
Cubic feet..........	Acre-feet..............	2.2956841×10^{-5}	Cu. in. of H_2O (60°F.).	Pounds of H_2O...........	0.0360916
"	Board feet.............	12	Cubic meters........	Acre-feet...............	0.00081071319
"	Bushels (Brit.).........	0.7786049	"	Barrels (Brit.)..........	6.110261
"	Bushels (U.S.)..........	0.80356395	"	Barrels (U.S., dry)......	8.648490
"	Cords (wood)...........	0.0078125	"	Barrels (U.S., liq.)......	8.3864145
"	Cord-feet..............	0.0625	"	Bushels (Brit.).........	27.49617
"	Cu. centimeters...........	28316.817	"	Bushels (U.S.)...........	28.377593
"	Cu. meters.............	0.028316847	"	Cu. cm................	1×10^6
"	Gallons (U.S., dry)......	6.4285116	"	Cu. feet...............	35.314667
"	Gallons (U.S., liq.)......	7.4805195	"	Cu. inches..............	61023.74
"	Liters................	28.31605	"	Cu. yards..............	1.3079506
"	Ounces (Brit., fluid).....	996.6143	"	Gallons (Brit.).........	219.9694
"	Ounces (U.S., fluid).....	957.50649	"	Gallons (U.S., liq.)......	264.17205
"	Pints (U.S., liq.).......	59.844156	"	Hogshead..............	4.1932072
"	Quarts (U.S., dry).....	25.714047	"	Liters................	999.972
"	Quarts (U.S., liq.).....	29.922078	"	Pints (U.S., liq.).......	2113.3764
Cu. ft. of H_2O (39.2°F.)..........	Pounds of H_2O......	62.4262	"	Quarts (U.S., liq.).......	1056.6882
Cu. ft. of H_2O (60°F.)	Pounds of H_2O......	62.3663	"	Steres................	1
Cu. ft./hr..........	Acre-feet/hr...........	2.2956841×10^{-5}	Cu. meters/min.....	Gal. (Brit.)/min........	219.9694
"	Cu. cm./sec...........	7.8657907	"	Gal. (U.S.)/min........	264.17205
"	Cu. ft./day...........	24	"	Liters/min.............	999.972
"	Gal. (U.S.)/hr........	7.4805195	Cu. millimeters	Cu. cm................	0.001
"	Liters/hr.............	28.31605	"	Cu. inches..............	6.1023744×10^{-5}
Cu. ft./min..........	Acre-feet/hr...........	0.0013774105	"	Cu. meters.............	1×10^{-9}
"	Acre-feet/min.........	2.2956841×10^{-5}	"	Minims (Brit.).........	0.01689365
"	Cu. cm./sec...........	471.94744	"	Minims (U.S.)..........	0.016230731
"	Cu. ft./hr............	60	Cu. yards..........	Bushels (Brit.).........	21.02233
"	Gal. (U.S.)/min........	7.4805195	"	Bushels (U.S.)..........	21.696227
"	Liters/sec............	0.4719342	"	Cu. cm................	764554.86
Cu. ft./lb..........	Cu. cm./gram.........	62.427961	"	Cu. feet...............	27
"	Millimeters/gram.......	62.42621	"	Cu. inches..............	46.656
Cu. ft./sec..........	Acre-inches/hr........	0.99173553	"	Cu. meters.............	0.76455486
"	Cu. cm./sec...........	28316.847	"	Gallons (Brit.).........	168.1787
"	Cu. yards/min.........	2.222222	"	Gallons (U.S., dry)......	173.56981
"	Gal. (U.S.)/min........	448.83117	"	Gallons (U.S., liq.)......	201.97403
"	Liters/min............	1698.963	"	Liters................	764.5335
"	Liters/sec............	28.31605	"	Quarts (Brit.).........	672.7146
Cu. ft. of H_2O (60°F.)/sec......	Lb. of H_2O/min......	3741.98	"	Quarts (U.S., dry)......	694.27926
Cu. ft.-atm..........	B.t.u................	2.72130	"	Quarts (U.S., liq.)......	807.89610
"	Cal., *gm*.............	685.756	Cu. yd./min..........	Cu. ft./sec...........	0.45
"	Cu. cm.-atm...........	28316.847	"	Gal. (U.S.)/sec........	3.3662338
"	Cu. ft.-(lb/sq. in.).......	14.6960	"	Liters/sec............	12.74222
"	Foot-pounds............	2116.22	Cubits..........	Centimeters............	45.72
"	Hp.-hours.............	0.00106880	"	Feet................	1.5
"	Joules................	2869.20	"	Inches...............	18
"	Kg.-meters.............	292.577	Daltons (chem.)......	Grams...............	1.66024×10^{-24}
"	Kw.-hours.............	0.000797001	Daltons (phys.)......	Grams...............	1.65979×10^{-24}
Cubic inches........	Barrels (Brit.).........	0.0001001292	Days (mean solar)....	Days (sidereal)..........	1.00273791
"	Barrels (U.S., dry).....	0.00014172336	"	Hours (mean solar)......	24
"	Board feet.............	0.0069444	"	Hours (sidereal)........	24.065710
"	Bushels (Brit.).........	0.0004505815	"	Years (calendar)........	0.0027397260
"	Bushels (U.S.)..........	0.00046502544	"	Years (sidereal)........	0.0027378031
"	Cu. cm................	16.387064	"	Years (tropical)........	0.0027379093
"	Cu. feet...............	0.00057870370	Days (sidereal)......	Days (mean solar).......	0.99726957
"	Cu. meters.............	1.6387064×10^{-5}	"	Hours (mean solar)......	23.934470
"	Cu. yards..............	2.1433470×10^{-5}	"	Hours (sidereal)........	24
"	Drams (U.S., fluid).....	4.4329004	"	Minutes (mean solar).....	1436.0682
"	Gallons (Brit.).........	0.003604652	"	Minute (sidereal).......	1440
"	Gallons (U.S., dry)......	0.0037202035	"	Second (sidereal).......	86400
"	Gallons (U.S., liq.)......	0.0043290043	"	Years (calendar)........	0.0027322454
"	Liters................	0.01638661	"	Years (sidereal)........	0.0027303277
"	Milliliters.............	16.38661	"	Years (tropical)........	0.0027304336
"	Ounces (Brit., fluid)......	0.5767444	Decibels..........	Bels................	0.1
			Decimeters..........	Centimeters............	10

Table 9-18. CONVERSION FACTORS (Continued)

To convert from	To	Multiply by	To convert from	To	Multiply by
Decimeters	Feet	0.32808399	Dynes/sq. cm	Cm. of H_2O (4°C.)	0.001019745
"	Feet (U.S. Survey)	0.328083333	"	Grams/sq. cm	0.001019716
"	Inches	3.9370079	"	In. of Hg (32°F.)	2.95300×10^{-5}
"	Meters	0.1	"	In. of H_2O (4°C.)	0.000401474
Decisteres	Cu. meters	0.1	"	Kg./sq. meter	0.01019716
Degrees	Circles	0.0027777	"	Poundals/sq. in	0.00046664510
"	Minutes	60	"	Pounds/sq. in	1.450377×10^{-5}
"	Quadrants	0.0111111	Dyne-centimeters	Ergs	1
"	Radians	0.017453293	"	Foot-poundals	2.3730360×10^{-6}
"	Seconds	3600	"	Foot-pounds	7.37562×10^{-8}
Degrees/cm	Radians/cm	0.017453293	"	Gram-cm	0.001019716
Degrees/foot	Radians/cm	0.00057261458	"	Inch-pounds	8.85075×10^{-7}
Degrees/inch	Radian/cm	0.0068713750	"	Kg.-meters	1.019716×10^{-8}
Degrees/min	Degrees/sec	0.0166666	"	Newton-meters	1×10^{-7}
"	Radians/sec	0.00029088821			
"	Revolutions/sec	4.629629×10^{-5}	Electron volts	Ergs	1.60209×10^{-12}
Degrees/sec	Radians/sec	0.017453293	"	Grams	1.78253×10^{-33}
"	Revolutions/min	0.166666	Electronic charges	Abcoulombs	1.60209×10^{-20}
"	Revolutions/sec	0.0027777	"	Coulombs	1.60209×10^{-19}
Dekaliters	Pecks (U.S.)	1.135136	"	Statcoulombs	4.80296×10^{-10}
"	Pints (U.S., dry)	18.16217	Electronic charges/kg.	Statcoulombs/dyne	4.89766×10^{-16}
Dekameters	Centimeters	1000	E.S. cgs. units of induction flux	E.M. cgs. units	2.997930×10^{10}
"	Feet	32.808399	E.S. cgs. units of magnetic charge	E.M. cgs. units	2.997930×10^{10}
"	Feet (U.S. Survey)	32.808333	E.S. cgs. units of magnetic field intensity	E.M. cgs. units	3.335635×10^{-11}
"	Inches	393.70079			
"	Kilometers	0.01	Ells	Centimeters	114.3
"	Meters	10	"	Inches	45
"	Yards	10.93613	Ergs	B.t.u.	9.48451×10^{-11}
Demals	Gram-equiv./cu. decimeter	1	"	Cal., gm	2.39006×10^{-8}
Drachms (Brit., fluid)	Cu. cm	3.551631	"	Cal., ky	2.39006×10^{-11}
"	Cu. inches	0.2167338	"	Cal., kg. (20°C.)	2.39126×10^{-11}
"	Drams (U.S., fluid)	0.9607594	"	Cu. cm.-atm	9.86923×10^{-7}
"	Milliliters	3.551531	"	Cu. ft.-atm	3.48529×10^{-11}
Drams (apoth. *or* troy)	Drams (avdp.)	2.1942857	"	Cu. ft.-(lb./sq. in.)	5.12196×10^{-10}
"	Grains	60	"	Dyne-cm	1
"	Grams	3887.9346	"	Electron volts	6.24196×10^{11}
"	Ounces (apoth. *or* troy)	0.125	"	Foot-poundals	2.3730360×10^{-6}
"	Ounces (avdp.)	0.13714286	"	Foot-pounds	7.37562×10^{-8}
"	Scruples (apoth.)	3	"	Gram-cm	0.001019716
Drams (avdp.)	Drams (apoth. *or* troy)	0.455729166	"	Joules	1×10^{-7}
"	Grains	27.34375	"	Joules (Int.)	9.99835×10^{-8}
"	Grams	1.7718452	"	Kw.-hours	2.777777×10^{-14}
"	Ounces (apoth. *or* troy)	0.056966146	"	Kg.-meters	1.019716×10^{-8}
"	Ounces (avdp.)	0.0625	"	Liter-atm	9.86895×10^{-10}
"	Pennyweights	1.1393229	"	Watt-sec	1×10^{-7}
"	Pounds (apoth. *or* troy)	0.0047471788	Ergs/(gram-mol. × °C.)	Foot-pounds/(lb.-mol. × °F.)	1.85863×10^{-5}
"	Pounds (avdp.)	0.00390625	Ergs/sec	B.t.u./min	5.69071×10^{-9}
"	Scruples (apoth.)	1.3671875	"	Cal., gm./min	1.43403×10^{-6}
Drams (U.S., fluid)	Cu. cm	3.6067162	"	Dyne-cm./sec	1
"	Cu. inches	0.22558594	"	Foot-pounds/min	4.42537×10^{-6}
"	Drachms (Brit., fluid)	1.040843	"	Gram-cm./sec	0.001019716
"	Gills (U.S.)	0.03125	"	Horsepower	1.34102×10^{-10}
"	Milliliters	3.696588	"	Joules/sec	1×10^{-7}
"	Minims (U.S.)	60	"	Kilowatts	1×10^{-10}
"	Ounces (U.S., fluid)	0.125	"	Watts	1×10^{-7}
"	Pints (U.S., liq.)	0.0078125	Ergs/sq. cm	Dynes/cm	1
Dynes	Grains	0.01573663	"	Ergs/sq. mm	0.01
"	Grams	0.001019716	Ergs/sq. mm	Dynes/cm	100
"	Newtons	0.00001	"	Ergs/sq. cm	100
"	Poundals	7.2330138×10^{-5}	Erg-sec	Planck's constant	1.50932×10^{26}
"	Pounds	2.248089×10^{-6}			
Dynes/cm	Ergs/sq. cm	1	Farads	Abfarads	1×10^{-9}
"	Ergs/sq. mm	0.01	"	E.M. cgs. units	1×10^{-9}
"	Grams/cm	0.001019716	"	E.S. cgs. units	8.987584×10^{11}
"	Poundals/inch	0.00018371855	"	Farads (Int.)	1.000495
Dynes/cu. cm	Grams/cu. cm	0.001019716	"	Microfarads	1×10^{6}
"	Poundals/cu. inch	0.0011852786	"	Statfarads	8.98758×10^{11}
Dynes/sq. cm	Atmospheres	9.86923×10^{-7}	Farads (Int.)	Farads	0.999505
"	Bars	1×10^{-6}	Fathoms	Centimeters	182.88
"	Baryes	1			
"	Cm. of Hg (0°C.)	7.50062×10^{-5}			

Table 9-18. CONVERSION FACTORS (Continued)

To convert from	To	Multiply by	To convert from	To	Multiply by
Fathoms	Feet	6	Feet/(sec. × sec.)	Meters/(sec. × sec.)	0.3048
"	Inches	72	"	Miles/(hr. × sec.)	0.68181818
"	Meters	1.8288	Firkins (Brit.)	Bushels (Brit.)	1.125
"	Miles (naut., Int.)	0.00098747300	"	Cu. cm	40914.79
"	Miles (statute)	0.001136363	"	Cu. feet	1.444892
"	Yards	2	"	Firkins (U.S.)	1.200949
Feet	Centimeters	30.48	"	Gallons (Brit.)	9
"	Chains (Gunter's)	0.01515151	"	Liters	40.91364
"	Fathoms	0.166666	"	Pints (Brit.)	72
"	Feet (U.S. Survey)	0.99999800	Firkins (U.S.)	Barrels (U.S., dry)	0.29464286
"	Furlongs	0.00151515	"	Barrels (U.S., liq.)	0.28571429
"	Inches	12	"	Bushels (U.S.)	0.96678788
"	Meters	0.3048	"	Cu. feet	1.203125
"	Microns	304800	"	Firkins (Brit.)	0.8326747
"	Miles (naut., Int.)	0.00016457883	"	Liters	34.06775
"	Miles (statute)	0.000189393	"	Pints (U.S., liq.)	72
"	Rods	0.060606	Foot-candles	Lumens/sq. ft	1
"	Ropes (Brit.)	0.05	"	Lumens/sq. meter	10.763910
"	Yards	0.333333	"	Lux	10.763910
Feet (U.S. Survey)	Centimeters	30.480061	"	Milliphots	1.0763910
"	Chains (Gunter's)	0.015151545	Foot-lamberts	Candles/sq. cm	0.00034262591
"	Chains (Ramden's)	0.010000020	"	Candles/sq. ft	0.31830989
"	Feet	1.0000020	"	Millilamberts	1.0763910
"	Inches	12.000024	"	Lamberts	0.0010763910
"	Links (Gunter's)	1.5151545	"	Lumens/sq. ft	1
"	Links (Ramden's)	1.0000020	Foot-poundals	B.t.u.	3.99678 × 10⁻⁵
"	Meters	0.30480061	"	B.t.u. (IST.)	3.99417 × 10⁻⁵
"	Miles (statute)	0.00018939432	"	B.t.u. (mean)	3.99104 × 10⁻⁵
"	Rods	0.060606182	"	Cal., gm	0.0100717
"	Yards	0.33333400	"	Cal., gm. (IST.)	0.0100651
Feet of air (1 atm., 60°F.)	Atmospheres	3.6083 × 10⁻⁵	"	Cal., gm. (mean)	0.0100573
"	Ft. of Hg (32°F.)	0.00089970	"	Cu. cm.-atm	0.415890
"	Ft. of H₂O (60°F.)	0.0012244	"	Cu. ft.-atm	1.46870 × 10⁻⁵
"	In. of Hg (32°F.)	0.0010796	"	Dyne-cm	4.2140110 × 10⁵
"	Pounds/sq. inch	0.00053027	"	Ergs	4.2140110 × 10⁵
Feet of Hg (32°F.)	Cm. of Hg (0°C.)	30.48	"	Foot-pounds	0.0310810
"	Ft. of H₂O (60°F.)	13.6085	"	Hp.-hours	1.56974 × 10⁻⁸
"	In. of Hg (60°F.)	163.302	"	Joules	0.042140110
"	Ounces/sq. inch	94.3016	"	Joules (Int.)	0.0421332
"	Pounds/sq. inch	5.89385	"	Kg.-meters	0.00429710
Feet of H₂O (4°C.)	Atmospheres	0.0294990	"	Kw.-hours	1.17056 × 10⁻⁸
"	Cm. of Hg (0°C.)	2.24192	"	Liter-atm	0.000415879
"	Dynes/sq. cm	29889.8	"	B.t.u.	0.00128593
"	Grams/sq. cm	30.4791	Foot-pounds	B.t.u. (IST.)	0.00128509
"	In. of Hg (32°F.)	0.882646	"	B.t.u. (mean)	0.00128408
"	Kg./sq. meter	304.791	"	Cal., gm	0.324048
"	Pounds/sq. inch	0.433515	"	Cal., gm. (IST.)	0.323836
Feet/hour	Cm./hr	30.48	"	Cal., gm. (mean)	0.323582
"	Cm./min	0.508	"	Cal., gm. (20°C.)	0.324211
"	Cm./sec	0.0084666	"	Cal., kg	0.000324048
"	Feet/min	0.0166666	"	Cal., kg. (IST.)	0.000323836
"	Inches/hr	12	"	Cal., kg. (mean)	0.000323582
"	Kilometers/hr	0.0003048	"	Cu. ft.-atm	0.000472541
"	Kilometers/min	5.08 × 10⁻⁶	"	Dyne-cm	1.35582 × 10⁷
"	Knots (Int.)	0.0001645788	"	Ergs	1.35582 × 10⁷
"	Miles/hr	0.000189393	"	Foot-poundals	32.1740
"	Miles/min	3.156565 × 10⁻⁶	"	Gram-cm	13825.5
"	Miles/sec	5.2609428 × 10⁻⁸	"	Hp.-hours	5.05050 × 10⁻⁷
Feet/minute	Cm./sec	0.508	"	Joules	1.35582
"	Feet/sec	0.0166666	"	Kg.-meters	0.138255
"	Kilometers/hr	0.018288	"	Kw.-hours	3.76616 × 10⁻⁷
"	Meters/min	0.3048	"	Kw.-hours (Int.)	3.76554 × 10⁻⁷
"	Meters/sec	0.00508	"	Liter-atm	0.0133805
"	Miles/hr	0.01136363	"	Newton-meters	1.3558180
Feet/second	Cm./sec	30.48	"	Lb. H₂O evap. from and at 212°F.	1.3245 × 10⁻⁶
"	Kilometers/hr	1.09728	"	Watt-hours	0.000376616
"	Kilometers/min	0.018288	Foot-pounds/hr	B.t.u./min	2.14321 × 10⁻⁵
"	Meters/min	18.288	"	B.t.u. (mean)/min	2.14013 × 10⁻⁵
"	Miles/hr	0.68181818	"	Cal., gm./min	0.00540080
"	Miles/min	0.01136363	"	Cal., gm. (mean)/min	0.00539304
Feet/(sec. × sec.)	Kilometers/(hr. × sec.)	1.09728	"	Ergs/min	2.25970 × 10⁵

Table 9-18. CONVERSION FACTORS (Continued)

To convert from	To	Multiply by	To convert from	To	Multiply by
Foot-pounds/hr	Foot-pounds/min	0.0166666	Gallons (U.S., dry)	Cu. inches	268.8025
"	Horsepower	5.050505×10^{-7}	"	Gallons (U.S., liq.)	1.16364719
"	Horsepower (metric)	5.12055×10^{-7}	"	Liters	4.404760
"	Kilowatts	3.76616×10^{-7}	Gallons (U.S., liq.)	Acre-feet	3.0688833×10^{-6}
"	Watts	0.000376616	"	Barrels (U.S., liq.)	0.031746032
"	Watts (Int.)	0.000376554	"	Barrels (petroleum, U.S.)	0.023809524
Foot-pounds/min	B.t.u./sec	2.14321×10^{-5}	"	Bushels (U.S.)	0.10742088
"	B.t.u. (mean)/sec	2.14013×10^{-5}	"	Cu. centimeters	3785.4118
"	Cal., gm./sec	0.00540080	"	Cu. feet	0.133680555
"	Cal., gm. (mean)/sec	0.00539304	"	Cu. inches	231
"	Ergs/sec	2.25970×10^{5}	"	Cu. meters	0.0037854118
"	Foot-pounds/sec	0.0166666	"	Cu. yards	0.0049511317
"	Horsepower	3.030303×10^{-5}	"	Gallons (Brit.)	0.8326747
"	Horsepower (metric)	3.07233×10^{-5}	"	Gallons (U.S., dry)	0.85936701
"	Joules/sec	0.0225970	"	Gallons (wine)	1
"	Joules (Int.)/sec	0.0225932	"	Gills (U.S.)	32
"	Kilowatts	2.25970×10^{-5}	"	Liters	3.785306
"	Watts	0.0225970	"	Minims (U.S.)	61440
Foot-pounds/lb	B.t.u./lb	0.00128593	"	Ounces (U.S., fluid)	128
"	B.t.u. (IST.)/lb	0.00128509	"	Pints (U.S., liq.)	8
"	B.t.u. (mean)/lb	0.00128408	"	Quarts (U.S., liq.)	4
"	Cal., gm./gm.	0.000714404	Gallons (U.S.) of H₂O (4°C.)	Lb. of H₂O in air	8.33585
"	Cal., gm. (IST.)/gram	0.000713937	Gallons (U.S.) of H₂O (60°F.)	Lb. of H₂O in air	8.32823
"	Cal., gm. (mean)/gram	0.000713377	Gallons (U.S.)/day	Cu. ft./hr	0.0055700231
"	Hp.-hr./lb	5.05050×10^{-7}	Gallons (Brit.)/hr	Cu. meters/min	7.576812×10^{-5}
"	Joules/gram	0.00298907	Gallons (U.S.)/hr	Acre-feet/hr	3.0688833×10^{-6}
"	Kg.-meters/gram	0.000304800	"	Cu. ft./hr	0.1336805
"	Kw.-hr./gram	8.30296×10^{-10}	"	Cu. meters/min	6.3090197×10^{-5}
Foot-pounds/sec	B.t.u./min	0.0771556	"	Cu. yd./min	8.2518861×10^{-5}
"	B.t.u. (mean)/min	0.0770447	"	Liters/hr	3.785306
"	B.t.u./sec	0.00128593	Gal. (Brit.)/sec	Cu. cm./sec	4546.087
"	B.t.u. (mean)/sec	0.00128408	Gal. (U.S.)/sec	Cu. cm./sec	3785.4118
"	Cal., gm./sec	0.324048	"	Cu. ft./min	8.020833
"	Cal., gm. (mean)/sec	0.323582	"	Cu. yd./min	0.29706790
"	Ergs/sec	1.35582×10^{7}	"	Liters/min	227.1183
"	Gram-cm./sec	13825.5	Gammas	Grams	1×10^{-6}
"	Horsepower	0.00181818	"	Micrograms	1
"	Joules/sec	1.35582	Gausses	E.M. cgs. units of magnetic flux density	1
"	Kilowatts	0.00135582	"	E.S. cgs. units	3.335635×10^{-11}
"	Watts	1.35582	"	Gausses (Int.)	0.999670
"	Watts (Int.)	1.35559	"	Maxwells/sq. cm.	1
Furlongs	Centimeters	20116.8	"	Lines/sq. cm	1
"	Chains (Gunter's)	10	"	Lines/sq. inch	6.4516
"	Chains (Ramden's)	6.6	Gausses (Int.)	Gausses	1.000330
"	Feet	660	Gausses/oersted	E.M. cgs. units of permeability	1
"	Inches	7920	"	E.S. cgs. units	1.112646×10^{-21}
"	Meters	201.168	Geepounds	Slugs	1
"	Miles (naut., Int.)	0.10862203	"	Kilograms	14.5939
"	Miles (statute)	0.125	Gigameters	Meters	1×10^{9}
"	Rods	40	Gilberts	Abampere-turns	0.079577472
"	Yards	220	"	Ampere-turns	0.79577472
Gallons (Brit.)	Barrels (Brit.)	0.027777	"	E.M. cgs. units of mmf., or magnetic potential	1
"	Bushels (Brit.)	0.125	"	E.S. cgs. units	2.997930×10^{10}
"	Cu. centimeters	4546.087	"	Gilberts (Int.)	1.000165
"	Cu. feet	0.1605436	Gilberts (Int.)	Gilberts	0.999835
"	Cu. inches	277.4193	Gilberts/cm	Ampere-turns/cm	0.79577472
"	Drachms (Brit. fluid)	1280	"	Ampere-turns/in	2.0212678
"	Firkins (Brit.)	0.111111	"	Oersteds	1
"	Gallons (U.S., liq.)	1.200949	Gilberts/maxwell	Ampere-turns/weber	7.957747×10^{7}
"	Gills (Brit.)	32	"	E.M. cgs. units of reluctance	1
"	Liters	4.545960	"	E.S. cgs. units	8.987584×10^{20}
"	Minims (Brit.)	76800	Gills (Brit.)	Cu. cm	142.0652
"	Ounces (Brit., fluid)	160	"	Gallons (Brit.)	0.03125
"	Ounces (U.S., fluid)	153.7215	"	Gills (U.S.)	1.200949
"	Pecks (Brit.)	0.5	"	Liters	0.1420613
"	Lb. of H₂O (62°F.)	10	"	Ounces (Brit., fluid)	5
Gallons (U.S., dry)	Barrels (U.S., dry)	0.038095592			
"	Barrels (U.S., liq.)	0.036941181			
"	Bushels (U.S.)	0.125			
"	Cu. centimeters	4404.8828			
"	Cu. feet	0.15555700			

Table 9-18. CONVERSION FACTORS (Continued)

To convert from	To	Multiply by	To convert from	To	Multiply by
Gills (Brit.)	Ounces (U.S., fluid)	4.803764	Grams/cu. cm	Pounds/gal. (U.S., dry)	9.7111064
"	Pints (Brit.)	0.25	"	Pounds/gal. (U.S., liq.)	8.3454044
Gills (U.S.)	Cu. cm	118.29412	Grams/cu. meter	Grains/cu. ft.	0.43699572
"	Cu. inches	7.21875	Grams/liter	Parts/million*	1000
"	Drams (U.S., fluid)	32	"	Lb./cu. ft.	0.06242621
"	Gallons (U.S., liq.)	0.03125	"	Lb./gal. (U.S.)	8.345171×10^{-3}
"	Gills (Brit.)	0.8326747	Grams/milliliter	Grams/cu. cm	0.999972
"	Liters	0.1182908	"	Pounds/cu. ft.	62.42621
"	Minims (U.S.)	1920	"	Pounds/gallon (U.S.)	8.345171
"	Ounces (U.S., fluid)	4	Grams/sq. cm	Atmospheres	0.000967841
"	Pints (U.S., liq.)	0.25	"	Bars	0.000980665
"	Quarts (U.S., liq.)	0.125	"	Cm. of Hg. (0°C.)	0.0735559
Grades	Circles	0.0025	"	Dynes/sq. cm	980.665
"	Circumferences	0.0025	"	In. of Hg (32°F.)	0.0289590
"	Degrees	0.9	"	Kg./sq. meter	10
"	Minutes	54	"	Mm. of Hg (0°C.)	0.735559
"	Radians	0.015707963	"	Poundals/sq. inch	0.457623
"	Revolutions	0.0025	"	Pounds/sq. inch	0.014223343
"	Seconds	3240	Grams/ton (long)	Milligrams/kg	0.98420653
Grains	Carats (metric)	0.32399455	Grams/ton (short)	Milligrams/kg	1.1023113
"	Drams (apoth. or troy)	0.016666	Grams-cm	B.t.u.	9.30113×10^{-8}
"	Drams (avdp.)	0.036571429	"	B.t.u. (IST.)	9.29505×10^{-8}
"	Dynes	63.5460	"	B.t.u. (mean)	9.28776×10^{-8}
"	Grams	0.06479891	"	Cal., gm	2.34385×10^{-5}
"	Milligrams	64.79891	"	Cal., gm. (IST.)	2.34231×10^{-5}
"	Ounces (apoth. or troy)	0.0020833	"	Cal., gm. (mean)	2.34048×10^{-5}
"	Ounces (avdp.)	0.0022857143	"	Cal., gm (15°C.)	2.34284×10^{-5}
"	Pennyweights	0.041666	"	Cal., gm, (20°C.)	2.34502×10^{-5}
"	Pounds (apoth. or troy)	0.000173611	"	Cal., kg	2.34385×10^{-8}
"	Pounds (avdp.)	0.00014285714	"	Cal., kg. (IST.)	2.34231×10^{-8}
"	Scruples (apoth.)	0.05	"	Cal., kg. (mean)	2.34048×10^{-8}
"	Tons (metric)	6.479891×10^{-8}	"	Dyne-cm	980.665
Grains/cu. ft.	Grams/cu. meter	2.2883519	"	Ergs	980.665
Grains/gal. (U.S.)	Parts/million*	17.11854	"	Foot-poundals	0.00232715
"	Pounds/million gal.	142.8571	"	Foot-pounds	7.2330138×10^{-5}
Grams	Carats (metric)	5	"	Hp.-hours	3.65303×10^{-11}
"	Decigrams	10	"	Joules	9.80665×10^{-5}
"	Dekagrams	0.1	"	Kw.-hours	2.72407×10^{-11}
"	Drams (apoth. or troy)	0.25720597	"	Kw.-hours (Int.)	2.72362×10^{-11}
"	Drams (avdp.)	0.56438339	"	Newton-meters	9.80665×10^{-5}
"	Dynes	980.665	"	Watt-hours	2.72407×10^{-8}
"	Grains	15.432358	Gram-cm./sec	B.t.u./sec	9.30113×10^{-8}
"	Kilograms	0.001	"	Cal., gm./sec	2.34385×10^{-5}
"	Micrograms	1×10^{6}	"	Ergs-sec	980.665
"	Myriagrams	0.0001	"	Foot-pounds/sec	7.2330138×10^{-5}
"	Ounces (apoth. or troy)	0.032150737	"	Horsepower	1.31509×10^{-7}
"	Ounces (avdp.)	0.035273962	"	Joules/sec	9.80665×10^{-5}
"	Pennyweights	0.64301493	"	Kilowatts	9.80665×10^{-8}
"	Poundals	0.0709316	"	Kilowatts (Int.)	9.80503×10^{-8}
"	Pounds (apoth. or troy)	0.0026792289	"	Watts	9.80665×10^{-5}
"	Pounds (avdp.)	0.0022046226	Gram/sq. cm	Pounds/sq. inch	0.000341717
"	Scruples (apoth.)	0.77161792	Gram wt.-sec,/sq. cm.	Poises	980.665
"	Tons (metric)	1×10^{-6}	Gravitational con-		
Grams/cm	Dynes/cm	980.665	stants	Cm./(sec. × sec.)	980.621
"	Grams/inch	2.54	"	Ft./(sec. × sec.)	32.1725
"	Kg./km	100			
"	Kg./meter	0.1	Hands	Centimeters	10.16
"	Poundals/inch	0.180166	"	Inches	4
"	Pounds/ft.	0.067196898	Hectares	Acres	2.4710538
"	Pounds/inch	0.0055997415	"	Ares	100
"	Tons (metric)/km	0.1	"	Sq. cm	1×10^{8}
Grams/(cm. × sec.)	Poises	1	"	Sq. feet	107639.10
"	Lb./(ft. × sec.)	0.06719690	"	Sq. meters	10000
Grams/cu. cm	Dynes/cu. cm	980.665	"	Sq. miles	0.0038610216
"	Grains/milliliter	15.43279	"	Sq. rods	395.36861
"	Grams/milliliter	1.000028	Hectograms	Grams	100
"	Poundals/cu. inch	1.16236	"	Poundals	7.09316
"	Pounds/circ. mil-ft.	3.4049170×10^{-7}	"	Pounds (apoth or troy)	0.26792289
"	Pounds/cu. ft.	62.427961	"	Pounds (avdp.)	0.22046226
"	Pounds/cu. inch	0.036127292	Hectoliters	Bushels (Brit.)	2.749694
"	Pounds/gal. (Brit.)	10.02241	"	Bushels (U.S.)	2.837839

*Based on density of 1 gram/ml.

Table 9-18. CONVERSION FACTORS (Continued)

To convert from	To	Multiply by
Hectoliters	Cu. cm	$1{,}00028 \times 10^5$
"	Cu. feet	3.531566
"	Gallons (U.S., liq.)	26.41794
"	Liters	100
"	Ounces (U.S.) fluid	3381.497
"	Pecks (U.S.)	11.35136
Hectometers	Centimeters	10000
"	Decimeters	1000
"	Dekameters	10
"	Feet	328.08399
"	Meters	100
"	Rods	19.883878
"	Yards	109.3613
Hectowatts	Watts	100
Hefner units	Candles (English)	0.86
"	Candles (German)	0.85
"	Candles (Int.)	0.90
"	10-cp. pentane candles	0.090
Henries	Abhenries	1×10^9
"	E.M. cgs. units	1×10^9
"	E.S. cgs. units	1.112646×10^{-18}
"	Henries (Int.)	0.999505
"	Millihenries	1000
"	Mks. (r or nr) units	1
"	Stathenries	1.112646×10^{-12}
Henries (Int.)	Henries	1.000495
Henries/meter	Cgs. units of permeability	795774.72
"	E.M. cgs. units	795774.72
"	E.S. cgs. units	8.854156×10^{-16}
"	Gausses/oersted	795774.72
"	Mks. (nr) units	0.079577472
"	Mks. (r) units	1
Hogsheads	Butts (Brit.)	0.5
"	Cu. feet	8.421875
"	Cu. inches	14553
"	Cu. meters	0.23848094
"	Gallons (Brit.)	52.458505
"	Gallons (U.S.)	63
"	Gallons (wine)	63
"	Liters	238.47427
Horsepower*	B.t.u. (mean)/hr	2542.48
"	B.t.u./min	42.4356
"	B.t.u. (mean)/sec	0.706243
"	Cal., gm./hr	6.41616×10^5
"	Cal., gm. (IST.)/hr	6.41196×10^5
"	Cal., gm. (mean)/hr	6.40693×10^5
"	Cal., gm./min	10693.6
"	Cal., gm. (IST.)/min	10686.6
"	Cal., gm. (mean)/min	10678.2
"	Ergs/sec	7.45700×10^9
"	Foot-pounds/hr	1980000
"	Foot-pounds/min	33000
"	Foot-pounds/sec	550
"	Horsepower (boiler)	0.0760181
"	Horsepower (electric)	0.999598
"	Horsepower (metric)	1.01387
"	Joules/sec	745.700
"	Kilowatts	0.745700
"	Kilowatts (Int.)	0.745577
"	Tons of refrig. (U.S., comm.)	0.21204
"	Watts	745.700
Horsepower (boiler)	B.t.u. (mean)/hr	33445.7
"	Cal., gm./min	140671.6
"	Cal., gm. (mean)/min	140469.4
"	Cal., gm. (15°C.)/min	140611.1
"	Cal., gm., (20°C.)/min	140742.2
"	Ergs/sec	9.80950×10^{10}
"	Foot-pounds/min	434107
"	Horsepower	13.1548
"	Horsepower (electric)	13.1495

To convert from	To	Multiply by
Horsepower (boiler)	Horsepower (metric)	13.3372
"	Horsepower (water)	13.1487
"	Joules/sec	9809.50
"	Kilowatts	9.80950
"	Lb. H_2O evap. per hr. from and at 212°F	34.5
Horsepower (electric)	B.t.u./hr	2547.16
"	B.t.u. (IST.)/hr	2545.50
"	B.t.u. (mean)/hr	2543.50
"	Cal., gm./sec	178.298
"	Cal., kg./hr	641.874
"	Ergs/sec	7.46×10^9
"	Foot-pounds/min	33013.3
"	Foot-pounds/sec	550.221
"	Horsepower	1.00040
"	Horsepower (boiler)	0.0760487
"	Horsepower (metric)	1.0142777
"	Horsepower (water)	0.999942
"	Joules/sec	746
"	Kilowatts	0.746
"	Watts	746
Horsepower (metric)	B.t.u./hr	2511.31
"	B.t.u. (IST.)/hr	2509.66
"	B.t.u. (mean)/hr	2507.70
"	Cal., gm./hr	6.32838×10^5
"	Cal., gm. (IST.)/hr	6.32425×10^5
"	Cal., gm. (mean)/hr	6.31929×10^5
"	Ergs/sec	7.35499×10^9
"	Foot-pounds/min	32548.6
"	Foot-pounds/sec	542.476
"	Horsepower	0.986320
"	Horsepower (boiler)	0.0749782
"	Horsepower (electric)	0.985923
"	Horsepower (water)	0.985866
"	Kg.-meters/sec	75
"	Kilowatts	0.735499
"	Watts	735.499
Horsepower (water)	Foot-pounds/min	33015.2
"	Horsepower	1.00016
"	Horsepower (boiler)	0.0760531
"	Horsepower (electric)	1.00006
"	Horsepower (metric)	1.01434
"	Kilowatts	0.746043
Horsepower-hours	B.t.u.	2546.14
"	B.t.u. (IST.)	2544.47
"	B.t.u. (mean)	2542.48
"	Cal., gm.	641616
"	Cal., gm. (IST.)	641196
"	Cal., gm. (mean)	640693
"	Foot-pounds	1.98×10^6
"	Joules	2.68452×10^6
"	Kg.-meters	273745
"	Kw.-hours	0.745700
"	Watt-hours	745.700
Hp.-hr./lb	B.t.u./lb	2546.14
"	Cal., gm./gram	1414.52
"	Cu. ft.-(lb./sq. in.)/lb	13750
"	Foot-pounds/lb	1980000
"	Joules/gram	5918.35
Hours (mean solar)	Days (mean solar)	0.0416666
"	Days (sidereal)	0.041780746
"	Hours (sidereal)	1.00273791
"	Minutes (mean solar)	60
"	Minutes (sidereal)	60.164275
"	Seconds (mean solar)	3600
"	Seconds (sidereal)	3609.8565
"	Weeks (mean calendar)	0.0059523809
Hours (sidereal)	Days (mean solar)	0.41552899
"	Days (sidereal)	0.0416666
"	Hours (mean solar)	0.99726957
"	Minutes (mean solar)	59.836174

*Mechanical horsepower, equal to 550 ft-lb/sec.

Table 9-18.   CONVERSION FACTORS *(Continued)*

To convert from	To	Multiply by	To convert from	To	Multiply by
Hours (sidereal)......	Minutes (sidereal)........	60	Joules (abs.).........	Cal., *kg.* (mean)..........	0.000238662
Hundredweights (long)	Kilograms................	50.802345	" "	Cu. ft.-atm.............	0.000348529
"	Pounds....................	112	" "	Ergs...................	1×10^7
"	Quarters (Brit., long)......	4	" "	Foot-poundals...........	23.730360
"	Quarters (U.S., long).....	0.2	" "	Foot-pounds.............	0.737562
"	Tons (long)...............	0.05	" "	Gram-cm...............	10197.16
Hundredweights (short)...........	Kilograms................	45.359237	" "	Hp.-hours..............	3.72506×10^{-7}
"	Pounds (advp.)...........	100	" "	Joules (Int.)............	0.999835
"	Quarters (Brit., short).....	4	" "	Kg.-meters.............	0.1019716
"	Quarters (U.S., short).....	0.2	" "	Kw.-hours.............	2.7777×10^{-7}
"	Tons (long)...............	0.044642857	" "	Liter-atm.............	0.00986895
"	Tons (metric).............	0.045359237	" "	Volt-coulombs (Int.).......	0.999835
"	Tons (short).............	0.05	" "	Watt-hours (abs.)........	0.0002777777
			" "	Watt-hours (Int.)........	0.000277732
Inches.............	Ångström units..........	2.54×10^8	" "	Watt-sec..............	1
"	Centimeters.............	2.54	" "	Watt-sec. (Int.).........	0.999835
"	Chains (Gunter's)........	0.00126262	Joules (Int.).........	B.t.u.................	0.000948608
"	Cubits.................	0.055555	"	B.t.u. (IST.)..........	0.000947988
"	Fathoms...............	0.013888	"	B.t.u. (mean)..........	0.000947244
"	Feet..................	0.083333	"	Cal. *gm.*.............	0.239045
"	Feet (U.S. Survey)........	0.083333167	"	Cal., *gm.* (IST.)........	0.238888
"	Links (Gunter's).........	0.126262	"	Cal., *gm.* (mean).......	0.238702
"	Links (Ramden's)........	0.083333	"	C.h.u................	0.000527004
"	Meters................	0.0254	"	C.h.u. (IST.)..........	0.000526660
"	Mils..................	1000	"	C.h.u. (mean).........	0.000526247
"	Picas (printer's).........	6.0225	"	Cu. cm.-atm..........	9.87086
"	Points (printer's)........	72.27000	"	Cu. ft.-atm..........	0.000348586
"	Wave length of orange-red line of krypton 86......	41929.399	"	Dyne-cm.............	1.000165×10^7
"	Wave length of the red line of cadmium............	39450.369	"	Ergs................	1.000165×10^7
"	Yards................	0.027777	"	Foot-poundals........	23.73428
Inches of Hg (32°F.)..	Atmospheres...........	0.0334211	"	Foot-pounds..........	0.737684
"	Bars................	0.0338639	"	Gram-cm............	10198.8
"	Dynes/sq. cm.........	33863.9	"	Joules (abs.).........	1.000165
"	Ft. of air (1 atm., 60°F.)...	926.24	"	Kw.-hours..........	2.77824×10^{-7}
"	Ft. of H$_2$O (39.2°F.).....	1.132957	"	Liter-atm...........	0.00987058
"	Grams/sq. cm........	34.5316	"	Volt-coulombs........	1.000165
"	Kg./sq. meter.........	345.316	"	Volt-coulombs (Int.)......	1
"	Mm. of Hg (60°F.)......	25.4	"	Watt-sec............	1.000165
"	Ounces/sq. inch.........	7.85847	"	Watt-sec. (Int.).......	1
Inches of Hg (32°F.)..	Pounds/sq. ft.........	70.7262	Joules/(abcoulomb × °F.)...	Joules/(coulomb × °C.)...	0.18
Inches of Hg (60°F.)..	Atmospheres...........	0.0333269	Joules/amp.-hr......	Joules/abcoulomb.......	0.002777
"	Dynes/sq. cm.........	39768.5	"	Joules/statcoulomb........	9.265653×10^{-14}
"	Grams/sq. cm........	34.4343	Joules/coulomb.....	Joules/abcoulomb........	10
"	Mm. of Hg (60°F.)......	25.4	"	Volts...............	1
"	Ounces/sq. inch.........	7.83633	Joules/(coulomb × °F.)...	Joules/(coulomb × °C.)...	1.8
"	Pounds/sq. ft.........	70.5269	Joules/°C.........	B.t.u./°F...........	0.000526917
Inches of H$_2$O(4°C.)..	Atmospheres...........	0.0024582	"	Cal., *gm.*/°C...........	0.239006
"	Dynes/sq. cm.........	2490.82	"	Cal., *gm.* (mean)/°C......	0.238662
"	In. of Hg (32°F.).......	0.0735539	Joules/electronic charge...........	Joules/abcoulomb........	6.24196×10^{19}
"	Kg./sq. meter.........	25.3993	Joules/(electronic charge × °C.)...	Joules/(coulomb × °C.)...	6.24196×10^{18}
"	Ounces/sq. ft.........	83.2350	Joules/(gram × °C.)..	B.t.u./(lb. × °F.)......	0.239006
"	Ounces/sq. inch.........	0.578020	"	Cal., *gm.*/(gram × °C.)....	0.239006
"	Pounds/sq. ft.........	5.20218	Joules (Int.)/(gram °C.)...............	B.t.u./(lb. × °F.)......	0.239045
"	Pounds/sq. inch.........	0.03612628	"	Cal., *gm.* (mean)/(gram × °C.)..................	0.238702
Inches/hr..........	Cm./hr.............	2.54	Joules/sec. (abs.).....	B.t.u./min............	0.0569071
"	Feet/hr.............	0.0833333	"	Cal., *gm.*/min........	14.3403
"	Miles/hr.............	1.578282×10^{-5}	"	Cal., *kg.*/min........	0.0143403
Inches/min.........	Cm./hr.............	152.4	"	Cal., *kg.* (mean)/min......	0.0143197
"	Feet/hr.............	5	"	Dyne-cm./sec.........	1×10^7
"	Miles/hr.............	0.000946969	"	Ergs/sec............	1×10^7
			"	Foot-pounds/sec........	0.737562
Joules (abs.)........	B.t.u................	0.000948451	"	Gram-cm./sec.........	10197.16
" "	B.t.u. (IST.)..........	0.000947831	"	Horsepower..........	0.00134102
" "	B.t.u. (mean).........	0.000947088	"	Watts...............	1
" "	Cal., *gm.*...........	0.239006	"	Watts (Int.)..........	0.999835
" "	Cal., *gm.* (IST.)........	0.238849	Joules (Int.)/sec......	B.t.u./min............	0.0569165
" "	Cal., *gm.* (mean)........	0.238662			
" "	Cal., *gm.* (15°C.)........	0.238903			
" "	Cal., *gm.* (20°C.)........	0.239126			

Table 9-18. CONVERSION FACTORS (Continued)

To convert from	To	Multiply by
Joules (Int.)/sec......	B.t.u. (mean)/min........	0.0568347
"	Cal., gm./min..........	14.3427
"	Cal., kg./min..........	0.0143427
"	Dyne-cm./sec......	1.000165 × 10⁷
"	Ergs/sec......	1.000165 × 10⁷
"	Foot-pounds/min......	44.2610
"	Foot-pounds/sec......	0.737684
"	Gram-cm./sec......	10198.8
"	Horsepower......	0.00134124
"	Watts......	1.000165
"	Watts (Int.)......	1
Kilderkins (Brit.)......	Cu. cm....	81829.57
"	Cu. feet....	2.889784
"	Cu. inches....	4993.55
"	Cu. meters....	0.08182957
"	Gallons (Brit.)....	18
Kilograms.........	Drams (apoth. or troy)....	257.20597
"	Drams (avdp.)....	564.38339
"	Dynes....	980665
"	Grains....	15432.358
"	Hundredweights (long)....	0.019684131
"	Hundredweights (short)....	0.022046226
"	Ounces (apoth. or troy)....	32.150737
"	Ounces (avdp.)....	35.273962
"	Pennyweights....	643.01493
"	Poundals....	70.931635
"	Pounds (apoth. or troy)....	2.6792289
"	Pounds (avdp.)....	2.2046226
"	Quarters (Brit., long)....	0.078736522
"	Quarters (U.S. long)....	0.0039368261
"	Scruples (apoth.)....	771.61792
"	Slugs....	0.06852177
"	Tons (long)....	0.00098420653
"	Tons (metric)....	0.001
"	Tons (short)....	0.0011023113
Kilograms/cu. meter..	Grams/cu. cm....	0.001
"	Lb./cu. ft....	0.062427961
"	Lb./cu. inch....	3.6127202 × 10⁻⁵
Kg. of ice melted/hr.	Tons of refrig. (U.S., comm.)....	0.026336
Kilograms/sq. cm....	Atmospheres....	0.967841
"	Bars....	0.980665
"	Cm. of Hg (0°C.)....	73.5559
"	Dynes/sq. cm....	980665
"	Ft. of H₂O (39.2°F.)....	32.8093
"	In. of Hg (32°F.)....	28.9590
"	Pounds/sq. inch....	14.223343
Kilograms/sq. meter..	Atmospheres....	9.67841 × 10⁻⁵
"	Bars....	9.80665 × 10⁻⁵
"	Dynes/sq. cm....	98.0665
"	Ft. of H₂O (39.2°F.)....	0.00328093
"	Grams/sq. cm....	0.1
"	In. of Hg (32°F.)....	0.00289590
"	Mm. of Hg (0°C.)....	0.0735559
"	Pounds/sq. ft....	0.20481614
"	Pounds/sq. in....	0.0014223343
Kilograms/sq. mm...	Pounds/sq. ft....	204816.14
"	Pounds/sq. in....	1422.3343
"	Tons (short)/sq. in....	0.71116716
Kilogram sq. cm....	Pounds sq. ft....	0.0023730360
"	Pounds sq. in....	0.34171719
Kilogram-meters....	B.t.u. (mean)....	0.00928776
"	Cal., gm. (mean)....	2.34048
"	Cal., kg. (mean)....	0.00234048
"	Cu. ft.-atm....	0.00341790
"	Dynes-cm....	9.80665 × 10⁷
"	Ergs....	9.80665 × 10⁷
"	Foot-poundals....	232.715
"	Foot-pounds....	7.23301
"	Gram-cm....	100000

To convert from	To	Multiply by
Kilogram-meters.....	Hp.-hours....	3.65304 × 10⁻⁶
"	Joules....	9.80665
"	Joules (Int.)....	9.80503
"	Kw.-hours....	2.72407 × 10⁻⁶
"	Liter-atm....	0.0967814
"	Newton-meters....	9.80665
"	Watt-hours....	0.00272407
"	Watt-hours (Int.)....	0.00272362
Kilogram-meters/sec..	Watts....	9.80665
Kilolines....	Maxwells....	1000
"	Webers....	1 × 10⁻⁵
Kiloliters....	Cu. centimeters....	1.000028 × 10⁶
"	Cu. feet....	35.31566
"	Cu. inches....	61025.45
"	Cu. meters....	1.000028
"	Cu. yards....	1.307987
"	Gallons (Brit.)....	219.9755
"	Gallons (U.S., dry)....	227.0271
"	Gallons (U.S., liq.)....	264.1794
"	Liters....	1000
Kilometers	Astronomical units....	6.68878 × 10⁻⁹
"	Centimeters....	100000
"	Feet....	3280.8399
"	Feet (U.S. Survey)....	3280.833
"	Light years....	1.05702 × 10⁻¹³
"	Meters....	1000
"	Miles (naut., Int.)....	0.53995680
"	Miles (statute)....	0.62137119
"	Myriameters....	0.1
"	Rods....	198.83878
"	Yards....	1093.6133
Kilometers/hr....	Cm./sec....	27.7777
"	Feet/hr....	3280.8399
"	Feet/min....	54.680665
"	Knots (Int.)....	0.53995680
"	Meters/sec....	0.277777
"	Miles (statute)/hr....	0.62137119
Kilometers/(hr. × sec.)....	Cm./(sec. × sec.)....	27.7777
"	Ft./(sec. × sec.)....	0.91134449
"	Meters/(sec. × sec.)....	0.277777
Kilometers/min....	Cm./sec....	1666.666
"	Feet/min....	3280.8399
"	Kilometers/hr....	60
"	Knots (Int.)....	32.397408
"	Miles/hr....	37.282272
"	Miles/min....	0.62137119
Kilovolts/cm....	Abvolts/cm....	1 × 10¹¹
"	Microvolts/meter....	1 × 10¹¹
"	Millivolts/meter....	1 × 10⁸
"	Statvolts/cm....	3.335635
"	Volts/inch....	2540
Kilowatts....	B.t.u....	3414.43
"	B.t.u. (IST.)/hr....	3412.19
"	B.t.u. (mean)/hr....	3409.52
"	B.t.u. (mean)/min....	56.8253
"	B.t.u. (mean)/sec....	0.947088
"	Cal., gm. (mean)/hr....	859184
"	Cal., gm. (mean)/min....	14319.7
"	Cal., gm. (mean)/sec....	238.662
"	Cal., kg. (mean)/hr....	859.184
"	Cal., kg. (mean)/min....	14.3197
"	Cal., kg. (mean)/sec....	0.238662
"	Cu. ft.-atm./hr....	1254.70
"	Ergs/sec....	1 × 10¹⁰
"	Foot-poundals/min....	1.42382 × 10⁶
"	Foot-pounds/hr....	2.65522 × 10⁶
"	Foot-pounds/min....	44253.7
"	Foot-pounds/sec....	737.562
"	Gram-cm./sec....	1.019716 × 10⁷
"	Horsepower....	1.34102

Table 9-18. CONVERSION FACTORS *(Continued)*

To convert from	To	Multiply by	To convert from	To	Multiply by
Kilowatts	Horsepower (boiler)	0.101942	Lamberts	Candles/sq. cm	0.31830989
"	Horsepower (electric)	1.34048	"	Candles/sq. ft	295.71956
"	Horsepower (metric)	1.35962	"	Candles/sq. inch	2.0536081
"	Joules/hr	3.6×10^6	"	Foot-lamberts	929.0304
"	Joules (IST.)/hr	3.59941×10^6	"	Lumens/sq. cm	1
"	Joules/sec	1000	Lasts (Brit.)	Liters	2909.414
"	Kg.-meters/hr	3.67098×10^5	Leagues (naut., Brit.)	Feet	18240
"	Kilowatts (Int.)	0.999835	"	Kilometers	5.559552
"	Watts (Int.)	999.835	"	Leagues (naut., Int.)	1.0006393
Kilowatts (Int.)	B.t.u./hr	3414.99	"	Leagues (statute)	1.151515
"	B.t.u. (IST.)/hr	3412.76	"	Miles (statute)	3.454545
"	B.t.u. (mean)/hr	3410.08	Leagues (naut., Int.)	Fathoms	3038.0577
"	B.t.u. (mean)/min	56.8347	"	Feet	18228.346
"	B.t.u. (mean)/sec	0.947244	"	Kilometers	5.556
"	Cal., *gm.* (mean)/hr	859326	"	Leagues (statute)	1.1507794
"	Cal., *gm.* (mean)/min	14322.1	"	Miles (statute)	3.4523383
"	Cal., *kg.*/hr	860.563	Leagues (statute)	Fathoms	2640
"	Cal., *kg.* (IST.)/hr	860	"	Feet	15840
"	Cal., *kg.* (mean)/hr	859.326	"	Kilometers	4.828032
"	Cu. cm.-atm./hr	3.55351×10^7	"	Leagues (naut., Int.)	0.86897625
"	Cu. ft.-atm./hr	1254.91	"	Miles (naut., Int.)	2.6069287
"	Ergs/sec	1.000165×10^{10}	"	Miles (statute)	3
"	Foot-poundals/min	1.42406×10^6	Light years	Astronomical units	63279.5
"	Foot-pounds/min	44261.0	"	Kilometers	9.46055×10^{12}
"	Foot-pounds/sec	737.684	"	Miles (statute)	5.87851×10^{12}
"	Gram-cm./sec	1.01988×10^7	Lines	Maxwells	1
"	Horsepower	1.34124	Lines (Brit.)	Centimeters	0.211666
"	Horsepower (boiler)	0.101959	"	Inches	0.083333
"	Horsepower (electric)	1.34070	Lines/sq. cm	Gausses	1
"	Horsepower (metric)	1.35985	Lines/sq. inch	Gausses	0.15500031
"	Joules/hr	3.60059×10^6	"	Webers/sq. inch	1×10^{-8}
"	Joules (Int.)/hr	3.6×10^6	Links (Gunter's)	Chains (Gunter's)	0.01
"	Kg.-meters/hr	367158	"	Feet	0.66
"	Kilowatts	1.000165	"	Feet (U.S. Survey)	0.65999868
Kilowatt-hours	B.t.u. (mean)	3409.52	"	Inches	7.92
"	Cal., *gm.* (mean)	859184	"	Meters	0.201168
"	Foot-pounds	2.65522×10^6	"	Miles (statute)	0.000125
"	Hp.-hours	1.34102	"	Rods	0.04
"	Joules	3.6×10^6	Links (Ramden's)	Centimeters	30.48
"	Kg.-meters	367098	"	Chains (Ramdens)	0.01
"	Lb. H$_2$O evap. from and at 212°F.	3.5168	"	Feet	1
"			"	Inches	12
"	Watt-hours	1000	Liters	Bushels (Brit.)	0.02749694
"	Watt-hours (Int.)	999.835	"	Bushels (U.S.)	0.02837839
Kilowatt-hours (Int.)	B.t.u. (mean)	3410.08	"	Cu. centimeters	1000.028
"	Cal., *gm.* (IST.)	860000	"	Cu. feet	0.03531566
"	Cal., *gm.* (mean)	859326	"	Cu. inches	61.02545
"	Cu. cm.-atm	3.55351×10^7	"	Cu. meters	0.001000028
"	Cu. ft.-atm	1254.91	"	Cu. yards	0.001307987
"	Foot-pounds	2.65566×10^6	"	Drams (U.S., fluid)	270.5198
"	Hp.-hours	1.34124	"	Gallons (Brit.)	0.2199755
"	Joules	3.60059×10^6	"	Gallons (U.S., dry)	0.2270271
"	Joules (Int.)	3.6×10^6	"	Gallons (U.S., liq.)	0.2641794
"	Kg.-meters	367158	"	Gills (Brit.)	7.039217
Kw.-hr./gram	B.t.u./lb	1.54876×10^6	"	Gills (U.S.)	8.453742
"	B.t.u. (IST.)/lb	1.54774×10^6	"	Hogsheads	0.004193325
"	B.t.u. (mean)/lb	1.54653×10^6	"	Minims (U.S.)	16231.19
"	Cal., *gm.*/gram	860421	"	Ounces (Brit., fluid)	35.19609
"	Cal., *gm.* (mean)/gram	859184	"	Ounces (U.S., fluid)	33.81497
"	Cu. cm.-atm./gram	3.55292×10^7	"	Pecks (Brit.)	0.1099878
"	Cu. ft.-atm./lb	569124	"	Pecks (U.S.)	0.1135136
"	Hp.-hr./lb	608.277	"	Pints (Brit.)	1.759804
"	Joules/gram	3.6×10^6	"	Pints (U.S., dry)	1.816217
Knots (Int.)	Cm./sec	51.4444	"	Pints (U.S., liq.)	2.113436
"	Feet/hr	6076.1155	"	Quarts. (Brit.)	0.8799021
"	Feet/min	101.26859	"	Quarts (U.S., dry)	0.9081084
"	Feet/sec	1.6878099	"	Quarts (U.S., liq.)	1.056718
"	Kilometers/hr	1.852	Liters/min	Cu. ft./min	0.03531566
"	Meters/min	30.8666	"	Cu. ft./sec	0.0005885943
"	Meters/sec	0.514444	"	Gal. (U.S., liq.)/min	0.2641794
"	Miles (naut., Int.)/hr	1	Liters/sec	Cu. ft./min	2.118939
"	Miles (statute)/hr	1.1507794	"	Cu. ft./sec	0.03531566

Table 9-18.   CONVERSION FACTORS (Continued)

To convert from	To	Multiply by
Liters/sec............	Cu. yards/min..........	0.07847923
"	Gal. (U.S., liq.)/min.....	15.85077
"	Gal. (U.S., liq.)/sec.....	0.2641794
Liter-atm..........	B.t.u................	0.0961045
"	B.t.u. (IST.)............	0.0960417
"	B.t.u. (mean)...........	0.0959664
"	Cal., gm............	24.2179
"	Cal., gm. (IST.)........	24.2021
"	Cal., gm. (mean)........	24.1831
"	Cu. ft.-atm............	0.0353157
"	Foot-poundals..........	2404.55
"	Foot-pounds...........	74.7356
"	Hp.-hours.............	3.77452×10^{-5}
"	Joules...............	101.328
"	Joules (Int.)...........	101.311
"	Kg.-meters............	10.3326
"	Kw.-hours.............	2.81466×10^{-5}
Liter-atm. (lat. 45°)...	Joules...............	101.323
Lumens............	Candle power (spher.).....	0.079577472
Lumens (at 5550 Å)..	Watts...............	0.0014705882
Lumens/sq. cm.....	Lamberts............	1
"	Phots...............	1
Lumens/(sq. cm. × steradian)........	Lamberts............	3.1415927
Lumens/sq. ft.....	Foot-candles..........	1
"	Foot-lamberts.........	1
"	Lumens/sq. meter......	10.763910
Lumens/(sq. ft. × steradian)........	Millilamberts.........	3.3815822
Lumens/sq. meter....	Foot-candles..........	0.09290304
"	Lumens/sq. ft........	0.09290304
"	Phots...............	0.0001
Lux................	Foot-candles..........	0.09290304
"	Lumens/sq. meter......	1
"	Phots...............	0.0001
Maxwells...........	E.M. cgs. units of induction flux	1
"	E.S. cgs. units.........	3.335635×10^{-11}
"	Gauss-sq. cm..........	1
"	Lines...............	1
"	Maxwells (Int.)........	0.999670
"	Volt-seconds..........	1×10^{-8}
"	Webers..............	1×10^{-8}
Maxwells (Int.).....	Maxwells............	1.000330
Maxwells/sq. cm....	Maxwells/sq. in.......	6.4516
"	Maxwells (Int.)/sq. cm.	0.999670
Maxwells (Int.)/sq. cm..............	Maxwells/sq. cm........	1.000330
Maxwells/sq. inch...	Maxwells/sq. cm........	0.15500031
Megalines..........	Maxwells............	1×10^6
Megmhos/cm........	Abmhos/cm...........	0.001
"	Megmho/inch cube.......	2.54
"	(Microhm-cm.)$^{-1}$.......	1
Megmhos/inch......	Megmhos/cm...........	0.39370079
"	(Microhm-inches)$^{-1}$....	1
Megohms..........	Microhms............	1×10^{12}
"	Ohms...............	1×10^6
"	Statohms............	1.112646×10^{-6}
Megohms^{-1}........	Micromhos..........	1
Meters............	Ångström units........	1×10^{10}
"	Centimeters..........	100
"	Chains (Gunter's)......	0.049709695
"	Chains (Ramden's).....	0.032808399
"	Fathoms.............	0.54680665
"	Feet...............	3.2808399
"	Feet (U.S. Survey)......	3.280833
"	Furlongs............	0.0049709695
"	Inches..............	39.370079
"	Kilometers...........	0.001
"	Links (Gunter's)......	4.9709695
Meters............	Links (Ramden's)........	3.2808399
"	Megameters..........	1×10^{-6}
"	Miles (naut., Brit.).....	0.00053961182
"	Miles (naut., Int.).....	0.00053995680
"	Miles (statute).......	0.00062137119
"	Millimeters..........	1000
"	Millimicrons.........	1×10^9
"	Mils...............	39370.079
"	Rods...............	0.19883878
"	Yards..............	1.0936133
Meters of Hg (0°C.)..	Atmospheres..........	1.3157895
"	Ft. of H_2O (60°F.).....	44.6474
"	In. of Hg (32°F.).....	39.370079
"	Kg./sq. cm..........	1.35951
"	Pounds/sq. inch.......	19.3368
Meters/hr..........	Feet/hr.............	3.2808399
"	Feet/min............	0.054680665
"	Knots (Int.)...........	0.00053995680
"	Miles (statute)/hr......	0.00062137119
Meters/min.........	Cm./sec.............	1.000000
"	Feet/min............	3.2808399
"	Feet/sec............	0.054680665
"	Kilometers/hr........	0.06
"	Knots (Int.)...........	0.032397408
"	Miles (statute)/hr......	0.037282272
Meters/sec.........	Feet/min............	196.85039
"	Feet/sec............	3.2808399
"	Kilometers/hr........	3.6
"	Kilometers/min.......	0.06
"	Miles (statute)/hr......	2.2369363
Meters/(sec. × sec.).	Kilometers/(hr. × sec.)....	3.6
"	Miles/(hr. × sec.)......	2.2369363
Meter-candles.......	Lumens/sq. meter.....	1
Mhos..............	Abmhos..............	1×10^{-9}
"	Cgs. units of conductance.	1
"	E.M. cgs. units........	1×10^{-9}
"	E.S. cgs. units........	8.987584×10^{11}
"	Mhos (Int.)..........	1.000495
"	Mks. (r or nr) units....	1
"	Ohms^{-1}............	1
"	Siemen's units........	1
"	Statmhos............	8.987584×10^{11}
Mhos (Int.).......	Abmhos..............	9.99505×10^{-10}
"	Mhos...............	0.999505
Mhos/meter........	Abmhos/cm...........	1×10^{-11}
"	Mhos (Int.)/meter....	1.000495
Mho-ft./circ. mil.....	Mhos/cm............	6.0153049×10^6
Microfarads........	Abfarads............	1×10^{-15}
"	Farads..............	1×10^{-6}
"	Statfarads...........	8.987584×10^5
Micrograms........	Grams..............	1×10^{-6}
"	Milligrams..........	0.001
Microhenries.......	Henries.............	1×10^{-6}
"	Stathenries..........	1.112646×10^{-18}
Microhms..........	Abohms.............	1000
"	Megohms............	1×10^{-12}
"	Ohms..............	1×10^{-6}
"	Statohms............	1.112646×10^{-18}
Microhm-cm........	Abohm-cm...........	1000
"	Circ. mil-ohms/ft......	6.0153049
"	Microhm-inches.......	0.39370079
"	Ohm-cm.............	1×10^{-6}
Microhm-inches.....	Circ. mil-ohms/ft......	15.278875
"	Michrom-cm..........	2.54
Micromicrofarads....	Farads..............	1×10^{-12}
Micromicrons.......	Ångström units........	0.01
"	Centimeters..........	1×10^{-10}
"	Inches..............	$3.9370079 \times 10^{-11}$
"	Meters..............	1×10^{-12}
"	Microns.............	1×10^{-6}
Microns...........	Ångström units........	10000

Table 9-18. CONVERSION FACTORS (Continued)

To convert from	To	Multiply by	To convert from	To	Multiply by
Microns............	Centimeters............	0.0001	Milligrams..........	Grains............	0.015432358
"	Feet............	3.2808399×10^{-6}	"	Grams............	0.001
"	Inches............	3.9370079×10^{-5}	"	Ounces (apoth. or troy).....	3.2150737×10^{-5}
"	Meters............	1×10^{-6}	"	Ounces (avdp.)......	3.5273962×10^{-5}
"	Millimeters............	0.001	"	Pennyweights............	0.00064301493
"	Millimicrons............	1000	"	Pounds (apoth. or troy)....	2.6792289×10^{-6}
Miles (naut., Brit.)...	Cable lengths (Brit.)......	8.4444	"	Pounds (avdp.)...........	2.2046226×10^{-6}
" ...	Fathoms............	1013.333	"	Scruples (apoth.)..........	0.00077161792
" ...	Feet............	6080	Milligrams/assay ton.	Milligrams/kg......	34.285714
" ...	Meters............	1853.184	"	Ounces (troy)/ton (avdp.).	1
" ...	Miles (Adm., Brit.)......	1	Milligrams/gm......	Dynes/cm......	0.980665
" ...	Miles (naut., Int.)......	1.0006393	"	Pounds/inch........	5.5997415×10^{-6}
" ...	Miles (statute)......	1.151515	Milligrams/gram......	Carats (parts gold per 24 of mixture)	0.024
Miles (naut., Int.)....	Cable lengths............	8.4390493	"	Grams/ton (short).......	907.18474
"	Fathoms............	1012.6859	"	Milligrams/assay ton......	29.166666
"	Feet............	6076.1155	"	Ounces (avdp.)/ton (long)..	35.84
"	Feet (U.S. Survey)......	6076.1033	"	Ounces (avdp.)/ton (short).	32
"	Kilometers............	1.852	"	Ounces (troy)/ton (long)...	32.6666
"	Leagues (naut., Int.)......	0.333333	"	Ounces (troy)/ton (short)..	29.1666
"	Meters............	1852	Milligrams/inch......	Dynes/cm......	0.386089
"	Miles (geographical).......	1	"	Dynes/inch......	0.980665
"	Miles (naut. Brit.)......	0.99936110	"	Grams/cm......	0.00039370079
"	Miles (statute)...........	1.1507794	"	Grams/inch......	0.0001
Miles (statute)......	Centimeters............	160934.4	Milligrams/kg......	Pounds (avdp.)/ton (short)	0.002
"	Chains (Gunter's).........	80	Milligrams/liter....	Grains/gal. (U.S.)......	0.05841620
"	Chains (Ramden's).......	52.8	"	Grams/liter......	0.001
"	Feet............	5280	"	Parts/million*......	1
"	Feet (U.S. Survey)......	5279.9894	"	Lb./cu. ft............	6.242621×10^{-5}
"	Furlongs............	8	Milligrams/mm......	Dynes/cm......	9.80665
"	Inches............	63360	Millihenries..........	Abhenries............	1×10^{6}
"	Kilometers............	1.609344	"	Henries............	0.001
"	Light years............	1.70111×10^{-13}	"	Stathenries............	1.112646×10^{-15}
"	Links (Gunter's).........	8000	Millilamberts........	Candles/sq. cm......	0.00031830989
"	Meters............	1609.344	"	Candles/sq. inch......	0.0020536081
"	Miles (naut., Brit.).......	0.86842105	"	Foot-lamberts............	0.9290304
"	Miles (naut., Int.)......	0.86897624	"	Lamberts............	0.001
"	Myriameters............	0.1609344	"	Lumens/sq. cm......	0.001
"	Rods............	320	"	Lumens/sq. ft......	0.9290304
"	Yards............	1760	Milliliters............	Cu. cm............	1.000028
Miles/hr........	Cm./sec............	44.704	"	Cu. inches............	0.06102545
"	Feet/hr............	5280	"	Drams (U.S., fluid)......	0.2705198
"	Feet/min............	88	"	Gills (U.S.)............	0.008453742
"	Feet/sec............	1.466666	"	Liters............	0.001
"	Kilometers/hr............	1.609344	"	Minims (U.S.)............	16.23119
"	Knots (Int.)............	0.86897624	"	Ounces (Brit., fluid)......	0.03519609
"	Meters/min............	26.8224	"	Ounces (U.S., fluid)......	0.03381497
"	Miles/min............	0.0166666	"	Pints (Brit.)............	0.001759804
Miles/(hr. $\times$ min.)..	Cm./(sec. $\times$ sec.)......	0.7450666	"	Pints (U.S., liq.)......	0.002113436
Miles/(hr. $\times$ sec.)....	Cm./(sec. $\times$ sec.)......	44.704	Millimeters............	Ångström units............	1×10^{7}
"	Ft./(sec. $\times$ sec.)......	1.466666	"	Centimeters............	0.1
"	Kilometers/(hr. $\times$ sec.)....	1.609344	"	Decimeters............	0.01
"	Meters/(sec. $\times$ sec.)......	0.44704	"	Dekameters............	0.0001
Miles/min........	Cm./sec............	2682.24	"	Feet............	0.0032808399
"	Feet/hr............	316800	"	Inches............	0.039370079
"	Feet/sec............	88	"	Meters............	0.001
"	Kilometers/min............	1.609344	"	Microns............	1000
"	Knots (Int.)............	52.138574	"	Mils............	39.370079
"	Meters/min............	1609.344	"	Wave length of orange-red line of krypton 86	1650.76373
"	Miles/hr............	60	"	Wave length of red line of cadmium	1553.16413
Millibars........	Atmospheres............	0.000986923	Millimeters of Hg (0°C.)............	Atmospheres............	0.0013157895
"	Bars............	0.001	"	Bars............	0.00133322
"	Baryes............	1000	"	Dynes/sq. cm......	1333.224
"	Dynes/sq. cm......	1000	"	Grams/sq. cm......	1.35951
"	Grams/sq. cm......	1.019716	"	Kg./sq. meter......	13,5951
"	In. of Hg (32°F.)........	0.0295300	"	Pounds/sq. ft......	2.78450
"	Pounds/sq. ft......	2.088543	"	Pounds/sq. inch........	0.0193368
"	Pounds/sq. inch........	0.0145038	"	Torrs............	1
Milligrams........	Carats (1877)............	0.004871			
"	Carats (metric)...........	0.005			
"	Drams (apoth. or troy)....	0.00025720597			
"	Drams (advp.)............	0.00056438339			

*Density of 1 gram per milliliter of solvent.

Table 9-18. CONVERSION FACTORS (Continued)

To convert from	To	Multiply by
Millimicrons........	Ångström units.........	10
"	Centimeters.........	1×10^{-7}
"	Inches.........	3.9370079×10^{-8}
"	Microns.........	0.001
"	Millimeters.........	1×10^{-6}
Milliphots.........	Foot-candles.........	0.9290304
"	Lumens/sq. ft.........	0.9290304
"	Lumens/sq. meter.........	10
"	Lux.........	10
"	Phots.........	0.001
Millivolts.........	Statvolts.........	3.335635×10^{-6}
"	Volts.........	0.001
Minims (Brit.).........	Cu. cm.........	0.05919385
"	Cu. inches.........	0.003612230
"	Milliliters.........	0.05919219
"	Ounces (Brit., fluid).........	0.0020833333
"	Scruples (Brit., fluid).........	0.05
Minims (U.S.).........	Cu. cm.........	0.061611520
"	Cu. inches.........	0.0037597656
"	Drams (U.S., fluid).........	0.0166666
"	Gallons (U.S., liq.).........	1.6276042×10^{-5}
"	Gills (U.S.).........	0.0005208333
"	Liters.........	6.160979×10^{-5}
"	Milliliters.........	0.06160979
"	Ounces (U.S., fluid).........	0.002083333
"	Pints (U.S., liq.).........	0.0001302083
Minutes (angular).........	Degrees.........	0.0166666
"	Quadrants.........	0.000185185
"	Radians.........	0.00029088821
"	Seconds (angular).........	60
Minutes (mean solar).........	Days (mean solar).........	0.0006944444
"	Days (sidereal).........	0.00069634577
"	Hours (mean solar).........	0.0166666
"	Hours (sidereal).........	0.016712298
"	Minutes (sidereal).........	1.00273791
Minutes (sidereal).........	Days (mean solar).........	0.00069254831
"	Minutes (mean solar).........	0.99726957
"	Months (mean calendar).........	2.2768712×10^{-5}
"	Seconds (sidereal).........	60
Minutes/cm.........	Radians/cm.........	0.00029099921
Months (lunar).........	Days (mean solar).........	29.530588
"	Hours (mean solar).........	708.73411
"	Minutes (mean solar).........	42524.047
"	Second (mean solar).........	2.5514428×10^{6}
"	Weeks (mean calendar).........	4.2186554
Months (mean calendar).........	Days (mean solar).........	30.416666
"	Hours (mean solar).........	730
"	Months (lunar).........	1.0300055
"	Weeks (mean calendar).........	4.3452381
"	Years (calendar).........	0.08333333
"	Years (sidereal).........	0.083274845
"	Years (tropical).........	0.083278075
Myriagrams.........	Grams.........	10000
"	Kilograms.........	10
"	Pounds (avdp.).........	22.046226
Newtons.........	Dynes.........	100000
"	Pounds.........	0.22480894
Newton-meters.........	Dyne-cm.........	1×10^{7}
"	Gram-cm.........	10197.162
"	Kg.-meters.........	0.10197162
"	Pound-feet.........	0.73756215
Noggins (Brit.).........	Cu. cm.........	142.0652
"	Gallons (Brit.).........	0.03125
"	Gills (Brit.).........	1
Oersteds.........	Ampere-turns/inch.........	2.0212678
"	Ampere-turns/meter.........	79.577472
"	E.M. cgs. units of magnetic field intensity	1
"	E.S. cgs. units.........	2.997930×10^{10}

To convert from	To	Multiply by
Oersteds.........	Gilberts/cm.........	1
"	Oersteds (Int.).........	1.000165
Oersteds (Int.).........	Oersteds.........	0.999835
Ohms.........	Abohms.........	1×10^{9}
"	Cgs. units of resistance....	1
"	Megohms.........	1×10^{-6}
"	Microhms.........	1×10^{6}
"	Ohms (Int.).........	0.999505
"	Statohms.........	1.112646×10^{-12}
Ohms (Int.).........	Ohms.........	1.000495
Ohms (mil, foot).....	Circ. mil-ohms/ft.........	1
"	Ohm-cm.........	1.6624261×10^{-7}
Ohm-cm.........	Circ. mil-ohms/ft.........	6.0153049×10^{6}
"	Microhm-cm.........	1×10^{6}.
"	Ohm-inches.........	0.39370079
Ohm-inches.........	Ohm-cm.........	2.54
Ohm-meters.........	Abohm-cm.........	1×10^{11}
"	E.M. cgs. units.........	1×10^{11}
"	E.S. cgs. units.........	1.112646×10^{-10}
"	Mks. units.........	1
"	Statohm-cm.........	1.112646×10^{-10}
Ounces (apoth. or troy).........	Dekagrams.........	1.7554286
"	Drams (apoth. or troy)....	8
"	Drams (avdp.).........	17.554286
"	Grains.........	480
"	Grams.........	31.103486
"	Milligrams.........	31103.486
"	Ounces (avdp.).........	1.0971429
"	Pennyweights.........	20
"	Pounds (apoth. or troy)....	0.0833333
"	Pounds (avdp.).........	0.068571429
"	Scruples (apoth.).........	24
"	Tons (short).........	3.4285714×10^{-5}
Ounces (avdp.).........	Drams (apoth. or troy)....	7.291666
"	Drams (avdp.).........	16
"	Grains.........	437.5
"	Grams.........	28.349523
"	Hundredweights (long)....	0.00055803571
"	Hundredweights (short)....	0.000625
"	Ounces (apoth. or troy)....	0.9114583
"	Pennyweights.........	18.229100
"	Pounds (apoth. or troy)....	0.075954861
"	Pounds (avdp.).........	0.0625
"	Scruples (apoth.).........	21.875
"	Tons (long).........	2.7901786×10^{-5}
"	Tons (metric).........	2.8349527×10^{-5}
"	Tons (short).........	3.125×10^{-5}
Ounces (Brit., fluid).....	Cu. cm.........	28.41305
"	Cu. inches.........	1.733870
"	Drachms (Brit., fluid).....	8
"	Drams (U.S., fluid).........	7.686075
"	Gallons (Brit.).........	0.00625
"	Milliliters.........	28.41225
"	Minims (Brit.).........	480
"	Ounces (U.S., fluid).........	0.9607594
Ounces (U.S., fluid).....	Cu. cm.........	29.573730
"	Cu. inches.........	1.8046875
"	Cu. meters.........	2.9573730×10^{-5}
"	Drams (U.S., fluid).........	8
"	Gallons (U.S., dry).........	0.0067138047
"	Gallons (U.S., liq.).........	0.0078125
"	Gills (U.S.).........	0.25
"	Liters.........	0.029572702
"	Minims (U.S.).........	480
"	Ounces (Brit., fluid).........	1.040843
"	Pints (U.S., liq.).........	0.0625
"	Quarts (U.S., liq.).........	0.03125
Ounces/sq. inch......	Dynes/sq. cm.........	4309.22
"	Grams/sq. cm.........	4.3941849
"	In. of H_2O (39.2°F.).........	1.73004
"	In. of H_2O (60°F.).........	1.73166

Table 9-18. CONVERSION FACTORS (Continued)

To convert from	To	Multiply by	To convert from	To	Multiply by
Ounces/sq. inch	Pounds/sq. ft	9	Pints (Brit.)	Minims (Brit.)	9600
"	Pounds/sq. inch	0.0625	"	Ounces (Brit., fluid)	20
Ounces (avdp.)/ton (long)	Milligrams/kg	27.901786	"	Pints (U.S., dry)	1.032056
Ounces (avdp.)/ton (short)	Milligrams/kg	31.25	"	Pints (U.S., liq.)	1.200949
Paces	Centimeters	76.2	"	Quarts (Brit.)	0.5
"	Chains (Gunter's)	0.0378788	"	Scruples (Brit., fluid)	480
"	Chains (Ramden's)	0.025	Pints (U.S., dry)	Bushels (U.S.)	0.015625
"	Feet	2.5	"	Cu. cm	550.61047
"	Hands	7.5	"	Cu. inches	33.6003125
"	Inches	30	"	Gallons, (U.S., dry)	0.125
"	Ropes (Brit.)	0.125	"	Gallons (U.S., liq.)	0.14545590
Palms	Centimeters	7.62	"	Liters	0.5505951
"	Chains (Ramden's)	0.0025	"	Pecks (U.S.)	0.0625
"	Cubits	0.1666666	"	Quarts (U.S., dry)	0.5
"	Feet	0.25	Pints (U.S., liq.)	Cu. cm	473.17647
"	Hands	0.75	"	Cu. feet	0.016710069
"	Inches	3	"	Cu. inches	28.875
Parsecs	Kilometers	3.08374×10^{13}	"	Cu. yards	0.00061889146
"	Miles (statute)	1.91615×10^{13}	"	Drams (U.S., fluid)	128
Parts/million*	Grains/gal. (Brit.)	0.07015488	"	Gallons (U.S., liq.)	0.125
"	Grains/gal. (U.S.)	0.05841620	"	Gills (U.S.)	4
"	Grams/liter	0.001	"	Liters	0.4731632
"	Milligrams/liter	1	"	Milliliters	473.1632
Pecks (Brit.)	Bushels (Brit.)	0.25	"	Minims (U.S.)	7680
"	Coombs (Brit.)	0.0625	"	Ounces (U.S., fluid)	16
"	Cu. cm	9092.175	"	Pints (Brit.)	0.8326747
"	Cu. inches	554.8385	"	Quarts (U.S., liq.)	0.5
"	Gallons (Brit.)	2	Planck's constant	Erg-seconds	6.6255×10^{-27}
"	Gills (Brit.)	64	"	Joule-seconds	6.6255×10^{-34}
"	Hogsheads	0.03812537	"	Joule-sec./Avog. No. (chem.)	3.9905×10^{-10}
"	Kilderkins (Brit.)	0.111111	Points (printer's)	Centimeters	0.03514598
"	Liters	9.091920	"	Inches	0.013837
"	Pints (Brit.)	16	"	Picas	0.0833333
"	Quarterns (Brit., dry)	4	Poises**	Cgs. units of absolute viscosity	1
"	Quarters (Brit., dry)	0.03125	"	Grams/(cm. × sec.)	1
"	Quarts (Brit.)	8	Poise-cu. cm./gram	Sq. cm./sec	1
"	Quarts (U.S., dry)	8.256449	Poise-cu. ft./lb	Sq. cm./sec	62.427960
Pecks (U.S.)	Barrels (U.S., dry)	0.076191185	Poise-cu. in./gram	Sq. cm./sec	16.387064
"	Bushels (U.S.)	0.25	Poles/sq. cm	E.M. cgs. units of magnetization	1
"	Cu. cm	8809.7675	Pottles (Brit.)	Gallons (Brit.)	0.5
"	Cu. feet	0.311114005	"	Liters	2.272980
"	Cu. inches	537.605	Poundals	Dynes	13825.50
"	Gallons (U.S., dry)	2	"	Grams	14.09808
"	Gallons (U.S., liq.)	2.3272944	"	Pounds (avdp.)	0.0310810
"	Liters	8.809521	Pounds (apoth. or troy)	Drams (apoth. or troy)	96
"	Pints (U.S., dry)	16	"	Drams (avdp.)	210.65143
"	Quarts (U.S., dry)	8	"	Grains	5760
Pennyweights	Drams (apoth. or troy)	0.4	"	Grams	373.24172
"	Drams (avdp.)	0.87771429	"	Kilograms	0.37324172
"	Grains	24	"	Ounces (apoth. or troy)	12
"	Grams	1.55517384	"	Ounces (avdp.)	13.165714
"	Ounces (apoth. or troy)	0.05	"	Pennyweights	240
"	Ounces (avdp.)	0.054857143	"	Pounds (avdp.)	0.8228571
"	Pounds (apoth. or troy)	0.0041666	"	Scruples (apoth.)	288
"	Pounds (avdp.)	0.0034285714	"	Tons (long)	0.00036734694
Perches (masonry)	Cu. feet	24.75	"	Tons (metric)	0.00037324172
Phots	Foot-candles	929.0304	"	Tons (short)	0.00041142857
"	Lumens/sq. cm	1	Pounds (avdp.)	Drams (apoth. or troy)	116.6666
"	Lumens/sq. meter	10000	"	Drams (avdp.)	256
"	Lux	10000	"	Grains	7000
Picas (printer's)	Centimeters	0.42175176	"	Grams	453.59237
"	Inches	0.166044	"	Hundredweights (long)	0.00892857
Pints (Brit.)	Cu. cm	568.26092	"	Hundredweights (short)	0.01
"	Gallons (Brit.)	0.125	"	Kilograms	0.45359237
"	Gills (Brit.)	4	"	Ounces (apoth. or troy)	14.583333
"	Gills (U.S.)	4.803797	"	Ounces (avdp.)	16
"	Liters	0.5682450			

*Based on density of 1 gram/ml for the solvent.
**For viscosity conversions see Tables 9-26 and 9-30.

Table 9-18. CONVERSION FACTORS (Continued)

To convert from	To	Multiply by	To convert from	To	Multiply by
Pounds (avdp.)	Pennyweights	291.6666	Quarterns (Brit., liq.)	Gallons (Brit.)	0.03125
"	Poundals	32.1740	"	Liters	0.1420613
"	Pounds (apoth. or troy)	1.215277	Quarters (U.S., long)	Kilograms	254.0117272
"	Scruples (apoth.)	350	"	Pounds (avdp.)	560
"	Slugs	0.0310810	Quarters (U.S., short)	Kilograms	226.796185
"	Tons (long)	0.00044642857	"	Pounds	500
"	Tons (metric)	0.00045359237	Quarts (Brit.)	Cu. cm	1136.522
"	Tons (short)	0.0005	"	Cu. inches	69.35482
Pounds of H_2O evap. from and at 212°F	B.t.u.	970.9	"	Gallons (Brit.)	0.25
"	B.t.u. (IST.)	970.2	"	Gallons (U.S., liq.)	0.3002373
"	B.t.u. (mean)	969.4	"	Liters	1.136490
"	Joules	1.0237×10^6	"	Quarts (U.S., dry)	1.032056
"	Joules (Int.)	1.0234×10^6	"	Quarts (U.S., liq.)	1.200949
Pounds/cu. ft	Grams/cu. cm	0.016018463	Quarts (U.S., dry)	Bushels (U.S.)	0.03125
"	Kg./cu. meter	16.018463	"	Cu. cm	1101.2209
Pounds/cu. inch	Grams/cu. cm	27.679905	"	Cu. feet	0.038889251
"	Grams/liter	27.68068	"	Cu. inches	67.200625
"	Kg./cu. meter	27679.905	"	Gallons (U.S., dry)	0.25
Pounds/gal. (Brit.)	Pounds/cu. ft	6.228839	"	Gallons (U.S., liq.)	0.29091180
Pounds/gal. (U.S., liq.)	Grams/cu. cm	0.11982643	"	Liters	1.1011801
"	Pounds/cu. ft	7.4805195	"	Pecks (U.S.)	0.125
Pounds/inch	Grams/cm	178.57967	"	Pints (U.S., dry)	2
"	Grams/ft	5443.1084	Quarts (U.S., liq.)	Cu. cm	946.35295
"	Grams/inch	453.59237	"	Cu. feet	0.033420136
"	Ounces/cm	6.2992	"	Cu. inches	57.75
"	Ounces/inch	16	"	Drams (U.S., fluid)	256
"	Pounds/meter	39.370079	"	Gallons (U.S., dry)	0.21484175
Pounds/minute	Kilograms/hr	27.2155422	"	Gallons (U.S., liq.)	0.25
"	Kilograms/min	0.45359237	"	Gills (U.S.)	8
Pounds of H_2O (39.2°F.)/min	Cu. ft./min	0.01601891	"	Liters	0.9463264
"	Gal. (U.S.)/min	0.1198298	"	Ounces (U.S., fluid)	32
"	Liters/min	0.45350237	"	Pints (U.S., liq.)	2
Pounds/sq. ft	Atmospheres	0.000472541	"	Quarts (Brit.)	0.8326747
"	Bars	0.000478803	"	Quarts (U.S., dry)	0.8593670
"	Cm. of Hg (0°C.)	0.0359131	Quintals (metric)	Grams	100000
"	Dynes/sq. cm	478.803	"	Hundredweights (long)	1.9684131
"	Ft. of air (1 atm., 60°F.)	13.006	"	Kilograms	100
"	Grams/sq. cm	0.48824276	"	Pounds (avdp.)	220.46226
"	In. of Hg (32°F.)	0.0141390			
"	In. of H_2O (39.2°F.)	0.192227	Radians	Circumferences	0.15915494
"	Kg./sq. meter	4.8824276	"	Degrees	57.295779
"	Mm. of Hg (0°C.)	0.359131	"	Minutes	3437.7468
Pounds/sq. inch	Atmospheres	0.0680460	"	Quadrants	0.63661977
"	Bars	0.0680476	"	Revolutions	0.15915494
"	Cm. of Hg (0°C.)	5.17149	"	Seconds	206264.81
"	Cm. of H_2O (4°C.)	70.3089	Radians/cm	Degrees/cm	57.295779
"	Dynes/sq. cm	68947.6	"	Degrees/ft	1746.3754
"	Grams/sq. cm	70.306958	"	Degrees/inch	145.53128
"	In. of Hg (32°F.)	2.03602	"	Minutes/cm	3437.7468
"	In. of H_2O (39.2°F.)	27.6807	Radians/sec	Degrees/sec	57.295779
"	Kg./sq. cm	0.070306958	"	Revolutions/min	9.5492966
"	Mm. of Hg (0°C.)	51.7149	"	Revolutions/sec	0.15915494
Pound wt.-sec./sq. ft	Poises	478.803	Radians/(sec. × sec.)	Revolutions/(min. × min.)	572.95779
Pound wt.-sec./sq. in	Poises	68947.6	"	Revolutions/(min. × sec.)	9.5492966
Puncheons (Brit.)	Cu. meters	0.31797510	"	Revolutions/(sec. × sec.)	0.15915494
"	Gallons (Brit.)	69.94467	Register tons	Cu. feet	100
"	Gallons (U.S.)	84	"	Cu. meters	2.8316847
			Revolutions	Degrees	360
Quadrants	Minutes	5400	"	Grades	400
"	Radians	1.5707963	"	Quadrants	4
Quarterns (Brit., dry)	Buckets (Brit.)	0.125	"	Radians	6.2831853
"	Bushels (Brit.)	0.0625	Reyns*	Centipoises	6.89476×10^6
"	Cu. cm	2273.044	Rhes	Poises^{-1}	1
"	Gallons (Brit.)	0.5	Rods	Centimeters	502.92
"	Liters	2.272980	"	Chains (Gunter's)	0.25
"	Pecks (Brit.)	0.25	"	Chains (Ramden's)	0.165
Quarterns (Brit., liq.)	Cu. cm	142.0652	"	Feet	16.5
			"	Feet (U.S. Survey)	16.499967
			"	Furlongs	0.025
			"	Inches	198
			"	Links (Gunter's)	25

*For viscosity conversions see Tables 9-26 and 9-30.

Table 9-18.   CONVERSION FACTORS *(Continued)*

To convert from	To	Multiply by	To convert from	To	Multiply by
Rods	Links (Ramden's)	16.5	Sq. centimeters	Sq. mm	100
"	Meters	5.0292	"	Sq. mils	155000.31
"	Miles (statute)	0.003125	"	Sq. rods	3.9536861×10^{-6}
"	Perches	1	"	Sq. yards	0.00011959900
"	Yards	5.5	Sq. chains (Gunter's)	Acres	0.1
Rods (Brit., volume)	Cu. feet	1000	"	Sq. feet	4356
"	Cu. meters	28.316847	"	Sq. ft. (U.S. Survey)	4355.9826
Roods (Brit.)	Acres	0.25	"	Sq. inches	627264
"	Ares	10.117141	"	Sq. links (Gunter's)	10000
"	Sq. perches	40	"	Sq. meters	404.68564
"	Sq. yards	1210	"	Sq. miles	0.00015625
Ropes (Brit.)	Feet	20	"	Sq. rods	16
"	Meters	6.096	"	Sq. yards	484
"	Yards	6.6666666	Sq. chains (Ramden's)	Acres	0.22956841
			"	Sq. feet	10000
Scruples (apoth.)	Drams (apoth. *or* troy)	0.333333	"	Sq. ft. (U.S. Survey)	9999.9600
"	Drams (avdp.)	0.73142857	"	Sq. inches	1.44×10^6
"	Grains	20	"	Sq. links (Ramden's)	10000
"	Grams	1.2959782	"	Sq. meters	929.0304
"	Ounces (apoth. *or* troy)	0.041666	"	Sq. miles	0.00035870064
"	Ounces (avdp.)	0.045714286	"	Sq. rods	36.730946
"	Pennyweights	0.833333	"	Sq. yards	1111.111
"	Pounds (apoth. *or* troy)	0.003472222	Sq. decimeters	Sq. cm	100
"	Pounds (avdp.)	0.0028571429	"	Sq. inches	15.500031
Scruples (Brit., fluid)	Minims (Brit.)	20	Square degrees	Steradians	0.00030461742
Seams (Brit.)	Bushels (Brit.)	8	Sq. dekameters	Acres	0.024710538
"	Cu. feet	10.27479	"	Ares	1
"	Liters	290.9414	"	Sq. meters	100
Seconds (angular)	Degrees	0.000277777	"	Sq. yards	119.59900
"	Minutes	0.0166666	Sq. feet	Acres	2.295684×10^{-5}
"	Radians	4.8481368×10^{-6}	"	Ares	0.0009290304
Seconds (mean solar)	Days (mean solar)	1.1574074×10^{-5}	"	Sq. cm	929.0304
"	Days (sidereal)	1.1605763×10^{-5}	"	Sq. chains (Gunter's)	0.00022956841
"	Hours (mean solar)	0.0002777777	"	Sq. ft. (U.S. Survey)	0.99999600
"	Hours (sidereal)	0.00027853831	"	Sq. inches	144
"	Minutes (mean solar)	0.0166666	"	Sq. links (Gunter's)	2.2956841
"	Minutes (sidereal)	0.016712298	"	Sq. meters	0.09290304
"	Seconds (sidereal)	1.00273791	"	Sq. miles	3.5870064×10^{-8}
Seconds (sidereal)	Days (mean solar)	1.1542472×10^{-5}	"	Sq. rods	0.0036730946
"	Days (sidereal)	1.1574074×10^{-5}	"	Sq. yards	0.111111
"	Hours (mean solar)	0.00027701932	Sq. feet (U.S. Survey)	Acres	$2.29569330 \times 10^{-5}$
"	Hours (sidereal)	0.000277777	"	Sq. centimeters	929.03412
"	Minutes (mean solar)	0.016621159	"	Sq. chains (Ramden's)	0.00010000040
"	Minutes (sidereal)	0.0166666	"	Sq. feet	1.0000040
"	Seconds (mean solar)	0.99726957	Sq. hectometers	Sq. meters	10000
Siemen's units	*Same as Mhos*		Sq. inches	Circ. mils	1273239.5
Skeins	Feet	360	"	Sq. cm	6.4516
"	Meters	109.728	"	Sq. chains (Gunter's)	1.5942251×10^{-6}
Slugs	Geepounds	1	"	Sq. decimeters	0.064516
"	Kilograms	14.5939	"	Sq. feet	0.0069444
"	Pounds (avdp.)	32.1740	"	Sq. ft. (U.S. Survey)	0.0069444167
Slugs/cu. ft.	Grams/cu. cm	0.515379	"	Sq. links (Gunter's)	0.01594225
Space (entire)	Hemispheres	2	"	Sq. meters	0.00064516
"	Steradians	12.566371	"	Sq. miles	$2.4909767 \times 10^{-10}$
Spans	Centimeters	22.86	"	Sq. mm	645.16
"	Fathoms	0.125	"	Sq. mils	1×10^6
"	Feet	0.75	Sq. inches/sec	Sq. cm./hr.	23225.76
"	Inches	9	"	Sq. cm./sec.	6.4516
"	Quarters (Brit. linear)	1	"	Sq. ft./min	0.416666
Spherical right angles	Hemispheres	0.25	Sq. kilometers	Acres	247.10538
"	Spheres	0.125	"	Sq. feet	1.0763910×10^7
"	Steradians	1.5707963	"	Sq. ft. (U.S. Survey)	1.0763867×10^7
Sq. centimeters	Ares	1×10^{-6}	"	Sq. inches	1.5500031×10^9
"	Circ. mm	127.32395	"	Sq. meters	1×10^6
"	Circ. mils	197352.52	"	Sq. miles	0.38610216
"	Sq. chains (Gunter's)	2.4710538×10^{-7}	"	Sq. yards	1.1959900×10^6
"	Sq. chains (Ramden's)	1.0763910×10^{-7}	Sq. links (Gunter's)	Acres	1×10^{-5}
"	Sq. decimeters	0.01	"	Sq. cm	404.68564
"	Sq. feet	0.0010763910	"	Sq. chains (Gunter's)	0.0001
"	Sq. ft. (U.S. Survey)	0.0010763867	"	Sq. feet	0.4356
"	Sq. inches	0.15500031	"	Sq. ft. (U.S. Survey)	0.43559826
"	Sq. meters	0.0001	"	Sq. Inches	62.7264

Table 9-18. CONVERSION FACTORS *(Continued)*

To convert from	To	Multiply by
Sq. links (Ramden's)..	Acres	2.2956841×10^{-5}
"	Sq. feet	1
Sq. meters	Acres	0.00024710538
"	Ares	0.01
"	Hectares	0.0001
"	Sq. cm	10000
"	Sq. feet	10.763910
"	Sq. inches	1550.0031
"	Sq. kilometers	1×10^{-6}
"	Sq. links (Gunter's)	24.710538
"	Sq. links (Ramden's)	10.763910
"	Sq. miles	3.8610216×10^{-7}
"	Sq. mm	1×10^{6}
"	Sq. rods	0.039536861
"	Sq. yards	1.1959900
Sq. miles	Acres	640
"	Hectares	258.99881
"	Sq. chains (Gunter's)	6400
"	Sq. feet	2.7878288×10^{7}
"	Sq. ft. (U.S. Survey)	2.78288×10^{7}
"	Sq. kilometers	2.5899881
"	Sq. meters	2589988.1
"	Sq. rods	102400
"	Sq. yards	3.0976×10^{6}
Sq. millimeters	Circ. mm	1.2732395
"	Circ. mils	1973.5252
"	Sq. cm	0.01
"	Sq. inches	0.0015500031
"	Sq. meters	1×10^{-6}
Sq. mils	Circ. mils	1.2732395
"	Sq. cm	6.4516×10^{-6}
"	Sq. inches	1×10^{-6}
"	Sq. mm	0.00064516
Sq. rods	Acres	0.00625
"	Ares	0.2529285264
"	Hectares	0.002529285264
"	Sq. cm	252928.5204
"	Sq. feet	272.25
"	Sq. ft. (U.S. Survey)	272.24891
"	Sq. inches	39204
"	Sq. links (Gunter's)	625
"	Sq. links (Ramden's)	272.25
"	Sq. meters	25.29285264
"	Sq. miles	9.765625×10^{-6}
"	Sq. yards	30.25
Sq. yards	Acres	0.00020661157
"	Ares	0.0083612736
"	Hectares	8.3612736×10^{-5}
"	Sq. cm	8361.2736
"	Sq. chains (Gunter's)	0.0020661157
"	Sq. chains (Ramden's)	0.0009
"	Sq. feet	9
"	Sq. ft. (U.S. Survey)	8.9999640
"	Sq. inches	1296
"	Sq. links (Gunter's)	20.661157
"	Sq. links (Ramden's)	9
"	Sq. meters	0.83612736
"	Sq. miles	$3.228305785 \times 10^{-7}$
"	Sq. perches (Brit.)	0.033057851
"	Sq. rods	0.033057851
Statamperes	Abamperes	3.335635×10^{-11}
"	Amperes	3.335635×10^{-10}
"	E.M. cgs. units of current	3.335635×10^{-11}
"	E.S. cgs. units	1
Statcoulombs	Ampere-hours	9.265653×10^{-14}
"	Coulombs	3.335635×10^{-10}
"	Electronic charges	2.082093×10^{9}
"	E.M. cgs. units of electric charge	3.335635×10^{-11}
Statfarads	E.M. cgs. units of capacitance	1.112646×10^{-21}
"	E.S. cgs. units	1
Statfarads	Farads	1.112646×10^{-12}
"	Microfarads	1.112646×10^{-6}
Stathenries	Abhenries	8.987584×10^{20}
"	E.M. cgs. units of inductance	8.987584×10^{20}
"	E.S. cgs. units	1
"	Henries	8.987584×10^{11}
"	Millihenries	8.987584×10^{14}
Statohms	Abohms	8.987584×10^{20}
"	E.S. cgs. units	1
"	Ohms	8.987584×10^{11}
Statvolts	Abvolts	2.997930×10^{10}
"	Volts	299.7930
Statvolts/cm	Volts/cm	299.7930
"	Volts/inch	761.4742
Statvolts/inch	Volts/cm	118.0287
Steradians	Hemispheres	0.15915494
"	Solid angles	0.079577472
"	Spheres	0.079577472
"	Spher. right angles	0.63661977
"	Square degrees	3282.8063
Steres	Cubic meters	1
"	Decisteres	10
"	Dekasteres	0.1
"	Liters	999.972
Stilbs	Candles/sq. cm	1
"	Candles/sq. inch	6.4516
"	Lamberts	3.1415927
Stokes*	Cgs. units of kinematic viscosity	1
"	Sq. cm./sec.	1
"	Sq. inches/sec	0.15500031
"	Poise cu. cm./gram	1
Stones (Brit., legal)	Centals (Brit.)	0.14
Tons (long)	Dynes	9.96402×10^{8}
"	Hundredweights (long)	20
"	Hundredweights (short)	22.4
"	Kilograms	1016.0469
"	Ounces (avdp.)	35840
"	Pounds (apoth. or troy)	2722.22
"	Pounds (avdp.)	2240
"	Tons (metric)	1.0160469
"	Tons (short)	1.12
Tons (metric)	Dynes	9.80665×10^{8}
"	Grams	1×10^{6}
"	Hundredweights (short)	22.046226
"	Kilograms	1000
"	Ounces (avdp.)	35273.962
"	Pounds (apoth. or troy)	2679.2289
"	Pounds (avdp.)	2204.6226
"	Tons (long)	0.98420653
"	Tons (short)	1.1023113
Tons (short)	Dynes	8.89644×10^{8}
"	Hundredweights (short)	20
"	Kilograms	907.18474
"	Ounces (avdp.)	32000
"	Pounds (apoth. or troy)	2430.555
"	Pounds (avdp.)	2000
"	Tons (long)	0.89285714
"	Tons (metric)	0.90718474
Tons of refrig. (U.S., comm.)	B.t.u. (IST.)/hr	12000
"	B.t.u. (IST.)/min	200
"	Cal., *kg.* (IST.)/hr	3023.949
"	Horsepower	4.71611
"	Kg. of ice melted/hr	37.971
"	Lb. of ice melted/hr	83.711
Tons of refrig. (U.S., std.)	B.t.u. (IST.)	288000
"	B.t.u. (mean)	287774
"	Cal., *kg.* (IST.)	72574.8

*For viscosity conversions see Tables 9-26 and 9-30.

Table 9-18. CONVERSION FACTORS (Continued)

To convert from	To	Multiply by	To convert from	To	Multiply by
Tons of refrig. (U.S., std.)	Cal., *kg.* (mean)	72517.9	Watts (Int.)	Ergs/sec	1.000165×10^7
"	Lb. of ice melted	2009.1	"	Joules (Int.)/sec	1
Tons (long)/sq. ft.	Atmospheres	1.05849	"	Watts	1.000165
"	Dynes/sq. cm.	1.07252×10^6	Watts/sq. cm.	B.t.u./(hr. × sq. ft.)	3172.10
"	Grams/sq. cm.	1093.6638	"	Cal., *gm.*/(hr. × sq. cm.)	860.421
"	Pounds/sq. ft.	2240	"	Ft.-lb./(min. × sq. ft.)	41113.1
Tons (short)/sq. ft.	Atmospheres	0.945082	Watts/sq. in.	B.t.u./(hr. × sq. ft.)	491.677
"	Dynes/sq. cm.	957.605	"	Cal., *gm.*/(hr. × sq. cm.)	133.365
"	Grams/sq. cm.	976.486	"	Ft.-lb./(min. × sq. ft.)	6372.54
"	Pounds/sq. inch	13.8888	Watt-hours	B.t.u.	3.41443
Tons (long)/sq. in.	Atmospheres	152.423	"	B.t.u. (mean)	3.40952
"	Dynes/sq. cm.	1.54443×10^8	"	Cal., *gm.*	860.421
"	Grams/sq. cm.	157487.59	Watt-hours	Cal., *kg.* (mean)	0.859184
Tons (short)/sq. in.	Dynes/sq. cm.	1.37895×10^8	"	Cal., *gm.* (mean)	859.184
"	Kg./sq. mm.	1406.139	"	Foot-pounds	2655.22
"	Pounds/sq. inch	2000	"	Hp.-hours	0.00134102
Torrs (*or* Tors)	Millimeters of Hg (0°C.)	1	"	Joules	3600
Townships (U.S.)	Acres	23040	"	Joules (Int.)	3599.41
"	Sections	36	"	Kg.-meters	367.098
"	Sq. miles	36	"	Kw.-hours	0.001
Tuns	Gallons (U.S.)	252	"	Watt-hours (Int.)	0.999835
"	Hogsheads	4	Watt-sec.	Foot-pounds	0.737562
			"	Gram-cm.	10197.16
Volts	Abvolts	1×10^8	"	Joules	1
"	Mks. (r *or* nr) units	1	"	Liter-atm.	0.00986895
"	Statvolts	0.003335635	"	Volt-coulombs	1
"	Volts (Int.)	0.999670	Wave length of orange-red line of krypton 86	Ångström units	6057.80211
Volts (Int.)	Volts	1.000330	"	Millimeters	0.000605780211
Volts/°C.	Joules/(coulomb × °C.)	1	Wave length of red line of cadmium	Ångström units	6438.4696
Volt-coulombs	Joules (Int.)	0.999835	"	Millimeters	0.00064384696
Volt-coulombs (Int.)	Joules	1.000165	Webers	Cgs. units of induction flux	1×10^8
Volt-electronic charge-seconds	Planck's constant	2.41814×10^{14}	"	E.M. cgs. units of induction flux	1×10^8
Volt-faraday (chem.)-seconds	Planck's constant	1.45650×10^{38}	"	Lines	1×10^8
Volt-faraday (phys.)-seconds	Planck's constant	1.45690×10^{38}	"	Maxwells	1×10^8
Volt-seconds	Maxwells	1×10^8	"	Mks. units of induction flux	1
			"	Mks(nr). units of magnetic charge	0.079577472
Watts	B.t.u./hr.	3.41443	"	Mks(r) units of magnetic charge	1
"	B.t.u. (mean)/hr.	3.40952	"	Volt-seconds	1
"	B.t.u. (mean)/min.	0.0568253	Webers/sq. cm.	Gausses	1×10^8
"	B.t.u./sec.	0.000948451	"	Lines/sq. cm.	1×10^8
"	B.t.u. (mean)/sec.	0.000947088	"	Lines/sq. inch	6.4516×10^8
"	Cal., *gm.*/hr.	860.421	Webers/sq. in.	Gausses	1.5500031×10^7
"	Cal., *gm.* (mean)/hr.	859.184	Weeks (mean calendar)	Days (mean solar)	7
"	Cal., *gm.* (20°C.)/hr.	860.853	"	Days (sidereal)	7.0191654
"	Cal., *gm.*/min.	14.3403	"	Hours (mean solar)	168
"	Cal., *gm.* (IST.)/min.	14.3310	"	Hours (sidereal)	168.45997
"	Cal., *gm.* (mean)/min.	14.3197	"	Minutes (mean solar)	10080
"	Cal., *kh.*/min.	0.0143403	"	Minutes (sidereal)	10107.598
"	Cal., *kg.* (IST.)/min.	0.0143310	"	Months (lunar)	0.23704235
"	Cal., *kg.* (mean)/min.	0.0143197	"	Months (mean calendar)	0.23013699
"	Ergs/sec.	1×10^7	"	Years (calendar)	0.019178082
"	Foot-pounds/min.	44.2537	"	Years (sidereal)	0.019164622
"	Horsepower	0.00134102	"	Years (tropical)	0.019165365
"	Horsepower (boiler)	0.000101942	Weys (Brit., mass.)	Pounds (avdp.)	252
"	Horsepower (elec.)	0.00134048			
"	Horsepower (metric)	0.00135962	Yards	Centimeters	91.44
"	Joules/sec.	1	"	Chains (Gunter's)	0.4545454
"	Kilowatts	0.001	"	Chains (Ramden's)	0.03
"	Liter-atm./hr.	35.5282	"	Cubits	2
Watts (Int.)	B.t.u./hr.	3.41499	"	Fathoms	0.5
"	B.t.u. (mean)/hr.	3.41008	"	Feet	3
"	B.t.u./min.	0.569165	"	Feet (U.S. Survey)	2.9999940
"	B.t.u. (mean)/min.	0.0568347	"	Furlongs	0.00454545
"	Cal., *gm.*/hr.	860.563	"	Inches	36
"	Cal., *gm.* (mean)/hr.	859.326	"	Meters	0.9144
"	Cal., *kg.*/min.	0.0143427	"	Poles (Brit.)	0.181818
"	Cal., *kg.* (IST.)/min.	0.0143333			
"	Cal., *kg.* (mean)/min.	0.0143221			

Table 9-18. CONVERSION FACTORS *(Continued)*

To convert from	To	Multiply by	To convert from	To	Multiply by
Yards...............	Quarters (Brit., linear)....	4	Years (sidereal)......	Days (sidereal)..........	366.25640
"	Rods....................	0.181818	"	Years (calendar)..........	1.0007024
"	Spans...................	4	"	Years (tropical)..........	1.0000388
Years (calendar).....	Days (mean solar)........	365	Years (tropical)......	Days (mean solar)........	365.24219
"	Hours (mean solar).......	8760	"	Days (sidereal)..........	366.24219
"	Minutes (mean solar).....	525600	"	Hours (mean solar).......	8765.8126
"	Months (lunar)..........	12.360065	"	Hours (sidereal)..........	8789.8126
"	Months (mean calendar)...	12	"	Months (mean calendar)...	12.007963
"	Seconds (mean solar).....	3.1536×10^7	"	Seconds (mean solar).....	3.1556926×10^7
"	Weeks (mean calendar)....	52.142857	"	Seconds (sidereal).........	3.1643326×10^7
"	Years (sidereal)..........	0.99929814	"	Weeks (mean calendar)....	52.177456
"	Years (tropical)..........	0.99933690	"	Years (calendar)..........	1.0006635
Years (leap)........	Days (mean solar).......	366	"	Years (sidereal)..........	0.99996121
Years (sidereal)......	Days (mean solar).......	365.25636			

DEFINED VALUES AND EQUIVALENTS

Meter...	**(m)**	1 650 763.73 wave lengths in vacuo of the unperturbed transition $2p_{10}$ — $5d_5$ in ^{86}Kr
Kilogram...	**(kg)**	mass of the international kilogram at Sèvres, France
Second...	**(s)**	1/31 556 925.974 7 of the tropical year at 12^h ET, 0 January 1900
Degree Kelvin..	**(°K)**	defined in the thermodynamic scale by assigning 273.16 °K to the triple point of water (freezing point, 273.15 °K = 0 °C)
Unified atomic mass unit................................	**(u)**	1/12 the mass of an atom of the ^{12}C nuclide
Mole..	**(mol)**	amount of substance containing the same number of atoms as 12 g of pure ^{12}C
Standard acceleration of free fall.....................	**(g$_n$)**	9.806 65 m s^{-2}, 980.665 cm s^{-2}
Normal atmospheric pressure............................	**(atm)**	101 325 N m^{-2}, 1 013 250 dyn cm^{-2}
Thermochemical calorie.................................	**(cal$_{th}$)**	4.1840 J, 4.1840 × 10^7 erg
International Steam Table calorie......................	**(cal$_{IT}$)**	4.1868 J, 4.1868 × 10^7 erg
Liter...	**(l)**	0.001 m^3, 1000 cm^3 (recommended by GCWM, 1964)
Inch..	**(in)**	0.0254 m, 2.54 cm
Pound (avdp)..	**(lb)**	0.453 592 37 kg, 453.592 37 g

Table 9-19. INCH–MILLIMETER CONVERSIONS—EXACT

$\frac{1}{64}$ in. to 10 in.

EVEN INCHES

Inches	1	2	3	4	5	6	7	8	9	10
Millimeters	25.4	50.8	76.2	101.6	127.0	152.4	177.8	203.2	228.6	254.0

All of the above values are exact, based on the accepted conversion 1 in. = 25.4000 mm; hence the exact millimeter equivalent of any decimal-multiple or decimal-fractional value may be obtained by shifting the decimal point.

FRACTIONAL INCHES

16ths	32nds	64ths	Decimal equivalents, in. (exact)	Millimeters (exact)	16ths	32nds	64ths	Decimal equivalents, in. (exact)	Millimeters (exact)
		1	0.015 625	0.396 875			33	0.515 625	13.096 875
	1	2	0.031 25	0.793 750		17	34	0.531 25	13.493 750
		3	0.046 875	1.190 625			35	0.546 875	13.890 625
1	2	4	0.062 5	1.587 500	9	18	36	0.562 5	14.287 500
		5	0.078 125	1.984 375			37	0.578 125	14.684 375
	3	6	0.093 75	2.381 250		19	38	0.593 75	15.081 250
		7	0.109 375	2.778 125			39	0.609 375	15.478 125
2	4	8	0.125 0	3.175 000	10	20	40	0.625 0	15.875 000
		9	0.140 625	3.571 875			41	0.640 625	16.271 875
	5	10	0.156 25	3.968 750		21	42	0.656 25	16.668 750
		11	0.171 875	4.365 625			43	0.671 875	17.065 625
3	6	12	0.187 5	4.762 500	11	22	44	0.687 5	17.462 500
		13	0.203 125	5.159 375			45	0.703 125	17.859 375
	7	14	0.218 75	5.556 250		23	46	0.718 75	18.256 250
		15	0.234 375	5.953 125			47	0.734 375	18.653 125
4	8	16	0.250 0	6.350 000	12	24	48	0.750 0	19.050 000
		17	0.265 625	6.746 875			49	0.765 625	19.446 875
	9	18	0.281 25	7.143 750		25	50	0.781 25	19.843 750
		19	0.296 875	7.540 625			51	0.796 875	20.240 625
5	10	20	0.312 5	7.937 500	13	26	52	0.812 5	20.637 500
		21	0.328 125	8.334 375			53	0.828 125	21.034 375
	11	22	0.343 75	8.731 250		27	54	0.843 75	21.431 250
		23	0.359 375	9.128 125			55	0.859 375	21.828 125
6	12	24	0.375 0	9.525 000	14	28	56	0.875 0	22.225 000
		25	0.390 625	9.921 875			57	0.890 625	22.621 875
	13	26	0.406 25	10.318 750		29	58	0.906 25	23.018 750
		27	0.421 875	10.715 625			59	0.921 875	23.415 625
7	14	28	0.437 5	11.112 500	15	30	60	0.937 5	23.812 500
		29	0.453 125	11.509 375			61	0.953 125	24.209 375
	15	30	0.468 75	11.906 250		31	62	0.968 75	24.606 250
		31	0.484 375	12.303 125			63	0.984 375	25.003 125
8	16	32	0.500 0	12.700 000	16	32	64	1.000 0	25.400 000

Table 9-20. DEGREES—RADIANS CONVERSIONS*

DEGREES—RADIANS

The table gives in radians the angle that is expressed in degrees and minutes at the side and top. Angles expressed to the nearest minute and second can readily be converted to radians by adding to the equivalent of the whole number of degrees the equivalents of the minutes and seconds found on the third and fourth pages of this table.

°	00′	10	20	30	40	50
0	0.00000	0.00291	0.00582	0.00873	0.01164	0.01454
1	0.01745	0.02036	0.02327	0.02618	0.02909	0.03200
2	0.03491	0.03782	0.04072	0.04363	0.04654	0.04945
3	0.05236	0.05527	0.05818	0.06109	0.06400	0.06690
4	0.06981	0.07272	0.07563	0.07854	0.08145	0.08436
5	0.08727	0.09018	0.09308	0.09599	0.09890	0.10181
6	0.10472	0.10763	0.11054	0.11345	0.11636	0.11926
7	0.12217	0.12508	0.12799	0.13090	0.13381	0.13672
8	0.13963	0.14254	0.14544	0.14835	0.15126	0.15417
9	0.15708	0.15999	0.16290	0.16581	0.16872	0.17162
10	0.17453	0.17744	0.18035	0.18326	0.18617	0.18908
11	0.19199	0.19490	0.19780	0.20071	0.20362	0.20653
12	0.20944	0.21235	0.21526	0.21817	0.22108	0.22398
13	0.22689	0.22980	0.23271	0.23562	0.23853	0.24144
14	0.24435	0.24725	0.25016	0.25307	0.25598	0.25889
15	0.26180	0.26471	0.26762	0.27053	0.27343	0.27634
16	0.27925	0.28216	0.28507	0.28798	0.29089	0.29380
17	0.29671	0.29961	0.30252	0.30543	0.30834	0.31125
18	0.31416	0.31707	0.31998	0.32289	0.32579	0.32870
19	0.33161	0.33452	0.33743	0.34034	0.34325	0.34616
20	0.34907	0.35197	0.35488	0.35779	0.36070	0.36361
21	0.36652	0.36943	0.37234	0.37525	0.37815	0.38106
22	0.38397	0.38688	0.38979	0.39270	0.39561	0.39852
23	0.40143	0.40433	0.40724	0.41015	0.41306	0.41597
24	0.41888	0.42170	0.42470	0.42761	0.43051	0.43342
25	0.43633	0.43924	0.44215	0.44506	0.44797	0.45088
26	0.45379	0.45669	0.45960	0.46251	0.46542	0.46833
27	0.47124	0.47415	0.47706	0.47997	0.48287	0.48578
28	0.48869	0.49160	0.49451	0.49742	0.50033	0.50324
29	0.50615	0.50905	0.51196	0.51487	0.51778	0.52069
30	0.52360	0.52651	0.52942	0.53233	0.53523	0.53814
31	0.54105	0.54396	0.54687	0.54978	0.55269	0.55560
32	0.55851	0.56141	0.56432	0.56723	0.57014	0.57305
33	0.57596	0.57887	0.58178	0.58469	0.58759	0.59050
34	0.59341	0.59632	0.59923	0.60214	0.60505	0.60796
35	0.61087	0.61377	0.61668	0.61959	0.62250	0.62541
36	0.62832	0.63123	0.63414	0.63705	0.63995	0.64286
37	0.64577	0.64868	0.65159	0.65450	0.65741	0.66032
38	0.66323	0.66613	0.66904	0.67195	0.67486	0.67777
39	0.68068	0.68359	0.68650	0.68941	0.69231	0.69522
40	0.69813	0.70104	0.70395	0.70686	0.70977	0.71268
41	0.71558	0.71849	0.72140	0.72431	0.72722	0.73013
42	0.73304	0.73595	0.73886	0.74176	0.74467	0.74758
43	0.75049	0.75340	0.75631	0.75922	0.76213	0.76504
44	0.76794	0.77085	0.77376	0.77667	0.77958	0.78249
45	0.78540	0.78831	0.79122	0.79412	0.79703	0.79994
46	0.80285	0.80576	0.80867	0.81158	0.81449	0.81740
47	0.82030	0.82321	0.82612	0.82903	0.83194	0.83485
48	0.83776	0.84067	0.84358	0.84648	0.84939	0.85230
49	0.85521	0.85812	0.86103	0.86394	0.86685	0.86976

*Reprinted from: "Handbook of Tables for Mathematics", 4th ed., S.M. Selby, Ed., The Chemical Rubber Co., 1970.

Table 9-20. DEGREES—RADIANS CONVERSIONS *(Continued)*

DEGREES—RADIANS *(Continued)*

°	00'	10	20	30	40	50
50	0.87266	0.87557	0.87848	0.88139	0.88430	0.88721
51	0.89012	0.89303	0.89594	0.89884	0.90175	0.90466
52	0.90757	0.91048	0.91339	0.91630	0.91921	0.92212
53	0.92502	0.92793	0.93084	0.93375	0.93666	0.93957
54	0.94248	0.94539	0.94830	0.95120	0.95411	0.95702
55	0.95993	0.96284	0.96575	0.96866	0.97157	0.97448
56	0.97738	0.98029	0.98320	0.98611	0.98902	0.99193
57	0.99484	0.99775	1.00066	1.00356	1.00647	1.00938
58	1.01229	1.01520	1.01811	1.02102	1.02393	1.02684
59	1.02974	1.03265	1.03556	1.03847	1.04138	1.04429
60	1.04720	1.05011	1.05302	1.05592	1.05883	1.06174
61	1.06465	1.06756	1.07047	1.07338	1.07629	1.07920
62	1.08210	1.08501	1.08792	1.09083	1.09374	1.09665
63	1.09956	1.10247	1.10538	1.10828	1.11119	1.11410
64	1.11701	1.11992	1.12283	1.12574	1.12865	1.13156
65	1.13446	1.13737	1.14028	1.14319	1.14610	1.14901
66	1.15192	1.15483	1.15774	1.16064	1.16355	1.16646
67	1.16937	1.17228	1.17519	1.17810	1.18101	1.18392
68	1.18682	1.18973	1.19264	1.19555	1.19846	1.20137
69	1.20428	1.20719	1.21009	1.21300	1.21591	1.21882
70	1.22173	1.22464	1.22755	1.23046	1.23337	1.23627
71	1.23918	1.24209	1.24500	1.24791	1.25082	1.25373
72	1.25664	1.25955	1.26245	1.26536	1.26827	1.27118
73	1.27409	1.27700	1.27991	1.28282	1.28573	1.28863
74	1.29154	1.29445	1.29736	1.30027	1.30318	1.30609
75	1.30900	1.31191	1.31481	1.31772	1.32063	1.32354
76	1.32645	1.32936	1.33227	1.33518	1.33809	1.34099
77	1.34390	1.34681	1.34972	1.35263	1.35554	1.35845
78	1.36136	1.36427	1.36717	1.37008	1.37299	1.37590
79	1.37881	1.38172	1.38463	1.38754	1.39045	1.39335
80	1.39626	1.39917	1.40208	1.40499	1.40790	1.41081
81	1.41372	1.41663	1.41953	1.42244	1.42535	1.42826
82	1.43117	1.43408	1.43699	1.43990	1.44281	1.44571
83	1.44862	1.45153	1.45444	1.45735	1.46026	1.46317
84	1.46608	1.46899	1.47189	1.47480	1.47771	1.48062
85	1.48353	1.48644	1.48935	1.49226	1.49517	1.49807
86	1.50098	1.50389	1.50680	1.50971	1.51262	1.51553
87	1.51844	1.52135	1.52425	1.52716	1.53007	1.53298
88	1.53589	1.53880	1.54171	1.54462	1.54753	1.55043
89	1.55334	1.55625	1.55916	1.56207	1.56498	1.56789
90	1.57080	1.57371	1.57661	1.57952	1.58243	1.58534
91	1.58825	1.59116	1.59407	1.59698	1.59989	1.60279
92	1.60570	1.60861	1.61152	1.61443	1.61734	1.62025
93	1.62316	1.62607	1.62897	1.63188	1.63479	1.63770
94	1.64061	1.64352	1.64643	1.64934	1.65225	1.65515
95	1.65806	1.66097	1.66388	1.66679	1.66970	1.67261
96	1.67552	1.67842	1.68133	1.68424	1.68715	1.69006
97	1.69297	1.69588	1.69879	1.70170	1.70460	1.70751
98	1.71042	1.71333	1.71624	1.71915	1.72206	1.72497
99	1.72788	1.73078	1.73369	1.73660	1.73951	1.74242
100	1.74533	1.74824	1.75115	1.75406	1.75696	1.75987
101	1.76278	1.76569	1.76860	1.77151	1.77442	1.77733
102	1.78024	1.78314	1.78605	1.78896	1.79187	1.79478
103	1.79769	1.80060	1.80351	1.80642	1.80932	1.81223
104	1.81514	1.81805	1.82096	1.82387	1.82678	1.82969
105	1.83260	1.83550	1.83841	1.84132	1.84423	1.84714
106	1.85004	1.85296	1.85587	1.85878	1.86168	1.86459
107	1.86750	1.87041	1.87332	1.87623	1.87914	1.88205
108	1.88496	1.88786	1.89077	1.89368	1.89659	1.89950
109	1.90241	1.90532	1.90823	1.91114	1.91404	1.91695
110	1.91986	1.92277	1.92568	1.92859	1.93150	1.93441

Table 9-20. DEGREES—RADIANS CONVERSIONS *(Continued)*

DEGREES—RADIANS *(Continued)*

Deg	Radians	Deg	Radians	Min	Radians	Sec	Radians
90	1.57080	**150**	2.61799	**0**	0.00000	**0**	0.00000
91	1.58825	151	2.63545	1	0.00029	1	0.00000
92	1.60570	152	2.65290	2	0.00058	2	0.00001
93	1.62316	153	2.67035	3	0.00087	3	0.00001
94	1.64061	154	2.68781	4	0.00116	4	0.00002
95	1.65806	**155**	2.70526	**5**	0.00145	**5**	0.00002
96	1.67552	156	2.72271	6	0.00175	6	0.00003
97	1.69297	157	2.74017	7	0.00204	7	0.00003
98	1.71042	158	2.75762	8	0.00233	8	0.00004
99	1.72788	159	2.77507	9	0.00262	9	0.00004
100	1.74533	**160**	2.79253	**10**	0.00291	**10**	0.00005
101	1.76278	161	2.80998	11	0.00320	11	0.00005
102	1.78024	162	2.82743	12	0.00349	12	0.00006
103	1.79769	163	2.84489	13	0.00378	13	0.00006
104	1.81514	164	2.86234	14	0.00407	14	0.00007
105	1.83260	**165**	2.87979	**15**	0.00436	**15**	0.00007
106	1.85005	166	2.89725	16	0.00465	16	0.00008
107	1.86750	167	2.91470	17	0.00495	17	0.00008
108	1.88496	168	2.93215	18	0.00524	18	0.00009
109	1.90241	169	2.94961	19	0.00553	19	0.00009
110	1.91986	**170**	2.96706	**20**	0.00582	**20**	0.00010
111	1.93732	171	2.98451	21	0.00611	21	0.00010
112	1.95477	172	3.00197	22	0.00640	22	0.00011
113	1.97222	173	3.01942	23	0.00669	23	0.00011
114	1.98968	174	3.03687	24	0.00698	24	0.00012
115	2.00713	**175**	3.05433	**25**	0.00727	**25**	0.00012
116	2.02458	176	3.07178	26	0.00756	26	0.00013
117	2.04204	177	3.08923	27	0.00785	27	0.00013
118	2.05949	178	3.10669	28	0.00814	28	0.00014
119	2.07694	179	3.12414	29	0.00844	29	0.00014
120	2.09440	**180**	3.14159	**30**	0.00873	**30**	0.00015
121	2.11185	190	3.31613	31	0.00902	31	0.00015
122	2.12930	200	3.49066	32	0.00931	32	0.00016
123	2.14676	210	3.66519	33	0.00960	33	0.00016
124	2.16421	220	3.83972	34	0.00989	34	0.00016
125	2.18166	**230**	4.01426	**35**	0.01018	**35**	0.00017
126	2.19911	240	4.18879	36	0.01047	36	0.00017
127	2.21657	250	4.36332	37	0.01076	37	0.00018
128	2.23402	260	4.53786	38	0.01105	38	0.00018
129	2.25147	270	4.71239	39	0.01134	39	0.00019
130	2.26893	**280**	4.88692	**40**	0.01164	**40**	0.00019
131	2.28638	290	5.06145	41	0.01193	41	0.00020
132	2.30383	300	5.23599	42	0.01222	42	0.00020
133	2.32129	310	5.41052	43	0.01251	43	0.00021
134	2.33874	320	5.58505	44	0.01280	44	0.00021
135	2.35619	**330**	5.75959	**45**	0.01309	**45**	0.00022
136	2.37365	340	5.93412	46	0.01338	46	0.00022
137	2.39110	350	6.10865	47	0.01367	47	0.00023
138	2.40855	360	6.28319	48	0.01396	48	0.00023
139	2.42601	370	6.45772	49	0.01425	49	0.00024
140	2.44346	**380**	6.63225	**50**	0.01454	**50**	0.00024
141	2.46091	390	6.80678	51	0.01484	51	0.00025
142	2.47837	400	6.98132	52	0.01513	52	0.00025
143	2.49582	410	7.15585	53	0.01542	53	0.00026
144	2.51327	420	7.33038	54	0.01571	54	0.00026
145	2.53073	**430**	7.50492	**55**	0.01600	**55**	0.00027
146	2.54818	440	7.67945	56	0.01629	56	0.00027
147	2.56563	450	7.85398	57	0.01658	57	0.00028
148	2.58309	460	8.02851	58	0.01687	58	0.00028
149	2.60054	470	8.20305	59	0.01716	59	0.00029
150	2.61799	**480**	8.37758	**60**	0.01745	**60**	0.00029

Table 9-20. DEGREES—RADIANS CONVERSIONS (*Continued*)

DEGREES, MINUTES, AND SECONDS TO RADIANS

Units in degrees, minutes or seconds	Degrees to radians	Minutes to radians	Seconds to radians
10	0.174 5329	0.002 9089	0.000 0485
20	0.349 0659	0.005 8178	0.000 0970
30	0.523 5988	0.008 7266	0.000 1454
40	0.698 1317	0.011 6355	0.000 1939
50	0.872 6646	0.014 5444	0.000 2424
60	1.047 1976	0.017 4533	0.000 2909
70	1.221 7305	(0.020 3622)	(0.000 3394)
80	1.396 2634	(0.023 2711)	(0.000 3879)
90	1.570 7963	(0.026 1800)	(0.000 4364)
100	1.745 3293	....	....
200	3.490 6585	....	....
300	5.235 9878	....	....

where n = 1, 2, 3, 4, etc. n (100°) = n (1.745 3293)

RADIANS TO DEGREES AND DECIMALS

Radians	Degrees	Radians	Degrees	Radians	Degrees	Radians	Degrees
1	57.2958	0.1	5.7296	0.01	0.5730	0.001	0.0573
2	114.5916	.2	11.4592	.02	1.1459	.002	.1146
3	171.8873	.3	17.1887	.03	1.7189	.003	.1719
4	229.1831	.4	22.9183	.04	2.2918	.004	.2292
5	286.4789	.5	28.6479	.05	2.8648	.005	.2865
6	343.7747	.6	34.3775	.06	3.4377	.006	.3438
7	401.0705	.7	40.1070	.07	4.0107	.007	.4011
8	458.3662	.8	45.8366	.08	4.5837	.008	.4584
9	515.6620	.9	51.5662	.09	5.1566	.009	.5157
10	572.9578	1.0	57.2958	.10	5.7296	.010	.5730

RADIANS—DEGREES

Multiples and Fractions of π Radians in Degrees

Radians	Radians	Degrees	Radians	Radians	Degrees	Radians	Radians	Degrees
π	3.1416	180	$\pi/2$	1.5708	90	$2\pi/3$	2.0944	120
2π	6.2832	360	$\pi/3$	1.0472	60	$3\pi/4$	2.3562	135
3π	9.4248	540	$\pi/4$	0.7854	45	$5\pi/6$	2.6180	150
4π	12.5664	720	$\pi/5$	0.6283	36	$7\pi/6$	3.6652	210
5π	15.7080	900	$\pi/6$	0.5236	30	$5\pi/4$	3.9270	225
6π	18.8496	1080	$\pi/7$	0.4488	25.714	$4\pi/3$	4.1888	240
7π	21.9911	1260	$\pi/8$	0.3927	22.5	$3\pi/2$	4.7124	270
8π	25.1327	1440	$\pi/9$	0.3491	20	$5\pi/3$	5.2360	300
9π	28.2743	1620	$\pi/10$	0.3142	18	$7\pi/4$	5.4978	315
10π	31.4159	1800	$\pi/12$	0.2618	15	$11\pi/6$	5.7596	330

Table 9-21. TEMPERATURE CONVERSION*

This table permits one to convert from degrees Celsius to degrees Fahrenheit or from degrees Fahrenheit to degrees Celsius. The conversion is accomplished by first locating in a column printed in boldface type the number that is to be converted. If the number to be converted is in degrees Fahrenheit, one may find its equivalent in degrees Celsius by reading to the left. If the number to be converted is in degrees Celsius, one may find its equivalent in degrees Fahrenheit by reading to the right. Degrees Celsius are identical to degrees Centigrade; however, the word Celsius is preferred for international use.

The approved international symbolic abbreviation for degrees Celsius is °C; for degrees Fahrenheit it is °F. Absolute zero on the Celsius scale is $-273.15°C$; on the Fahrenheit scale it is $-459.67°F$. The relation between degrees Fahrenheit and degrees Celsius may be expressed by

$$°C = 5/9(°F - 32) \text{ or}$$
$$°F = 9/5(°C) + 32.$$

To °C	←°F or °C→	To °F	To °C	←°F or °C→	To °F	To °C	←°F or °C→	To °F
−273.15	−459.67	—	−92.78	−135	−211	−28.33	−19	−2.2
−267.78	−450	—	−90	−130	−202	−27.78	−18	−0.4
−262.22	−440	—	−87.22	−125	−193	−27.22	−17	1.4
−256.67	−430	—	−84.44	−120	−184	−26.67	−16	3.2
−251.11	−420	—	−81.67	−115	−175	−26.11	−15	5
−245.56	−410	—	−78.89	−110	−166	−25.56	−14	6.8
−240	−400	—	−76.11	−105	−157	−25	−13	8.6
−234.44	−390	—	−73.33	−100	−148	−24.44	−12	10.4
−228.89	−380	—	−70.56	−95	−139	−23.89	−11	12.2
−223.33	−370	—	−67.78	−90	−130	−23.33	−10	14
−217.78	−360	—	−65	−85	−121	−22.78	−9	15.8
−212.22	−350	—	−62.22	−80	−112	−22.22	−8	17.6
−206.67	−340	—	−59.44	−75	−103	−21.67	−7	19.4
−201.11	−330	—	−56.67	−70	−94	−21.11	−6	21.2
−195.56	−320	—	−53.89	−65	−85	−20.56	−5	23
190	310	—	−51.11	−60	−76	−20	−4	24.8
−184.44	−300	—	−48.33	−55	−67	−19.44	−3	26.6
−178.89	−290	—	−45.56	−50	−58	18.89	−2	28.4
−173.33	−280	—	−42.78	−45	−49	−18.33	−1	30.2
−167.78	−270	−454	−40	−40	−40	−17.78	0	32
−162.22	260	436	−39.44	−39	38.2	−17.22	1	33.8
−156.67	−250	−418	−38.89	−38	−36.4	−16.67	2	35.6
−151.11	−240	−400	−38.33	−37	−34.6	−16.11	3	37.4
−145.56	−230	−382	−37.78	−36	−32.8	−15.56	4	39.2
−140	−220	−364	−37.22	−35	−31	−15	5	41
−134.44	−210	−346	−36.67	−34	−29.2	−14.44	6	42.8
−131.67	−205	−337	−36.11	−33	−27.4	−13.89	7	44.6
−128.89	−200	−328	−35.56	−32	−25.6	−13.33	8	46.4
−126.11	−195	−319	−35	−31	−23.8	−12.78	9	48.2
−123.33	−190	−310	−34.44	−30	−22	−12.22	10	50
−120.56	−185	−301	−33.89	−29	−20.2	−11.67	11	51.8
−117.78	−180	−292	−33.33	−28	−18.4	−11.11	12	53.6
−115	−175	−283	−32.78	−27	−16.6	−10.56	13	55.4
−112.22	−170	−274	−32.22	−26	−14.8	−10	14	57.2
−109.44	−165	−265	−31.67	−25	−13	−9.44	15	59
−106.67	−160	−256	−31.11	−24	−11.2	−8.89	16	60.8
−103.89	−155	−247	−30.56	−23	−9.4	−8.33	17	62.6
−101.11	−150	−238	−30	−22	−7.6	−7.78	18	64.4
−98.33	−145	−229	−29.44	−21	−5.8	−7.22	19	66.2
−95.56	−140	−220	−28.89	−20	−4	−6.67	20	68

*Condensed from: "Handbook of Chemistry and Physics", 53rd ed., R.C. Weast, Ed., The Chemical Rubber Co., 1972.

Table 9-21.   TEMPERATURE CONVERSION *(Continued)*

To °C	←°F or °C→	To °F	To °C	←°F or °C→	To °F	To °C	←°F or °C→	To °F
	To convert			*To convert*			*To convert*	
−6.11	21	69.8	23.89	75	167	54.44	130	266
−5.56	22	71.6	24.44	76	168.8	55	131	267.8
−5	23	73.4	25	77	170.6	55.56	132	269.6
−4.44	24	75.2	25.56	78	172.4	56.11	133	271.4
−3.89	25	77	26.11	79	174.2	56.67	134	273.2
−3.33	26	78.8	26.67	80	176	57.22	135	275
−2.78	27	80.6	27.22	81	177.8	57.78	136	276.8
−2.22	28	82.4	27.78	82	179.6	58.33	137	278.6
−1.67	29	84.2	28.33	83	181.4	58.89	138	280.4
−1.11	30	86	28.89	84	183.2	59.44	139	282.2
−0.56	31	87.8	29.44	85	185	60	140	284
0	32	89.6	30	86	186.8	60.56	141	285.8
.56	33	91.4	30.56	87	188.6	61.11	142	287.6
1.11	34	93.2	31.11	88	190.4	61.67	143	289.4
1.67	35	95	31.67	89	192.2	62.22	144	291.2
2.22	36	96.8	32.22	90	194	62.78	145	293
2.78	37	98.6	32.78	91	195.8	63.33	146	294.8
3.33	38	100.4	33.33	92	197.6	63.89	147	296.6
3.89	39	102.2	33.89	93	199.4	64.44	148	298.4
4.44	40	104	34.44	94	201.2	65	149	300.2
5	41	105.8	35	95	203	65.56	150	302
5.56	42	107.6	35.56	96	204.8	66.11	151	303.8
6.11	43	109.4	36.11	97	206.6	66.67	152	305.6
6.67	44	111.2	36.67	98	208.4	67.22	153	307.4
7.22	45	113	37.22	99	210.2	67.78	154	309.2
7.78	46	114.8	37.78	100	212	68.33	155	311
8.33	47	116.6	38.33	101	213.8	68.89	156	312.8
8.89	48	118.4	38.89	102	215.6	69.44	157	314.6
9.44	49	120.2	39.44	103	217.4	70	158	316.4
			40	104	219.2	70.56	159	318.2
10	50	122	40.56	105	221	71.11	160	320
10.56	51	123.8	41.11	106	222.8	71.67	161	321.8
11.11	52	125.6	41.67	107	224.6	72.22	162	323.6
11.67	53	127.4	42.22	108	226.4	72.78	163	325.4
12.22	54	129.2	42.78	109	228.2	73.33	164	327.2
12.78	55	131	43.33	110	230	73.89	165	329
13.33	56	132.8	43.89	111	231.8	74.44	166	330.8
13.89	57	134.6	44.44	112	233.6	75	167	332.6
14.44	58	136.4	45	113	235.4	75.56	168	334.4
15	59	138.2	45.56	114	237.2	76.11	169	336.2
15.56	60	140	46.11	115	239	76.67	170	338
16.11	61	141.8	46.67	116	240.8	77.22	171	339.8
16.67	62	143.6	47.22	117	242.6	77.78	172	341.6
17.22	63	145.4	47.78	118	244.4	78.33	173	343.4
17.78	64	147.2	48.33	119	246.2	78.89	174	345.2
18.33	65	149	48.89	120	248	79.44	175	347
18.89	66	150.8	49.44	121	249.8	80	176	348.8
19.44	67	152.6	50	122	251.6	80.56	177	350.6
20	68	154.4	50.56	123	253.4	81.11	178	352.4
20.56	69	156.2	51.11	124	255.2	81.67	179	354.2
21.11	70	158	51.67	125	257	82.22	180	356
21.67	71	159.8	52.22	126	258.8	82.78	181	357.8
22.22	72	161.6	52.78	127	260.6	83.33	182	359.6
22.78	73	163.4	53.33	128	262.4	83.89	183	361.4
23.33	74	165.2	53.89	129	264.2	84.44	184	363.2

Table 9-21. TEMPERATURE CONVERSION (Continued)

To convert			To convert			To convert		
To °C	←°F or °C→	To °F	To °C	←°F or °C→	To °F	To °C	←°F or °C→	To °F
85	185	365	115.56	240	464	146.11	295	563
85.56	186	366.8	116.11	241	465.8	146.67	296	564.8
86.11	187	368.6	116.67	242	467.6	147.22	297	566.6
86.67	188	370.4	117.22	243	469.4	147.78	298	568.4
87.22	189	372.2	117.78	244	471.2	148.33	299	570.2
87.78	190	374	118.33	245	473	148.89	300	572
88.33	191	375.8	118.89	246	474.8	149.44	301	573.8
88.89	192	377.6	119.44	247	476.6	150	302	575.6
89.44	193	379.4	120	248	478.4	150.56	303	577.4
90	194	381.2	120.56	249	480.2	151.11	304	579.2
90.56	195	383	121.11	250	482	151.67	305	581
91.11	196	384.8	121.67	251	483.8	152.22	306	582.8
91.67	197	386.6	122.22	252	485.6	152.78	307	584.6
92.22	198	388.4	122.78	253	487.4	153.33	308	586.4
92.78	199	390.2	123.33	254	489.2	153.89	309	588.2
93.33	200	392	123.89	255	491	154.44	310	590
93.89	201	393.8	124.44	256	492.8	155	311	591.8
94.44	202	395.6	125	257	494.6	155.56	312	593.6
95	203	397.4	125.56	258	496.4	156.11	313	595.4
95.56	204	399.2	126.11	259	498.2	156.67	314	597.2
96.11	205	401	126.67	260	500	157.22	315	599
96.67	206	402.8	127.22	261	501.8	157.78	316	600.8
97.22	207	404.6	127.78	262	503.6	158.33	317	602.6
97.78	208	406.4	128.33	263	505.4	158.89	318	604.4
98.33	209	408.2	128.89	264	507.2	159.44	319	606.2
98.89	210	410	129.44	265	509	160	320	608
99.44	211	411.8	130	266	510.8	160.56	321	609.8
100	212	413.6	130.56	267	512.6	161.11	322	611.6
100.56	213	415.4	131.11	268	514.4	161.67	323	613.4
101.11	214	417.2	131.67	269	516.2	162.22	324	615.2
101.67	215	419	132.22	270	518	162.78	325	617
102.22	216	420.8	132.78	271	519.8	163.33	326	618.8
102.78	217	422.6	133.33	272	521.6	163.89	327	620.6
103.33	218	424.4	133.89	273	523.4	164.44	328	622.4
103.89	219	426.2	134.44	274	525.2	165	329	624.2
104.44	220	428	135	275	527	165.56	330	626
105	221	429.8	135.56	276	528.8	166.11	331	627.8
105.56	222	431.6	136.11	277	530.6	166.67	332	629.6
106.11	223	433.4	136.67	278	532.4	167.22	333	631.4
106.67	224	435.2	137.22	279	534.2	167.78	334	633.2
107.22	225	437	137.78	280	536	168.33	335	635
107.78	226	438.8	138.33	281	537.8	168.89	336	636.8
108.33	227	440.6	138.89	282	539.6	169.44	337	638.6
108.89	228	442.4	139.44	283	541.4	170	338	640.4
109.44	229	444.2	140	284	543.2	170.56	339	642.2
110	230	446	140.56	285	545	171.11	340	644
110.56	231	447.8	141.11	286	546.8	171.67	341	645.8
111.11	232	449.6	141.67	287	548.6	172.22	342	647.6
111.67	233	451.4	142.22	288	550.4	172.78	343	649.4
112.22	234	453.2	142.78	289	552.2	173.33	344	651.2
112.78	235	455	143.33	290	554	173.89	345	653
113.33	236	456.8	143.89	291	555.8	174.44	346	654.8
113.89	237	458.6	144.44	292	557.6	175	347	656.6
114.44	238	460.4	145	293	559.4	175.56	348	658.4
115	239	462.2	145.56	294	561.2	176.11	349	660.2

Table 9-21.   TEMPERATURE CONVERSION *(Continued)*

To °C	←°F or °C→	To °F	To °C	←°F or °C→	To °F	To °C	←°F or °C→	To °F
176.67	**350**	662	210	**410**	770	271.11	**520**	968
177.22	**351**	663.8	211.11	**412**	773.6	272.22	**522**	971.6
177.78	**352**	665.6	212.22	**414**	777.2	273.33	**524**	975.2
178.33	**353**	667.4	213.33	**416**	780.8	274.44	**526**	978.8
178.89	**354**	669.2	214.44	**418**	784.4	275.56	**528**	982.4
179.44	**355**	671	215.56	**420**	788	276.67	**530**	986
180	**356**	672.8	216.67	**422**	791.6	277.78	**532**	989.6
180.56	**357**	674.6	217.78	**424**	795.2	278.89	**534**	993.2
181.11	**358**	676.4	218.89	**426**	798.8	280	**536**	996.8
181.67	**359**	678.2	220	**428**	802.4	281.11	**538**	1000.4
182.22	**360**	680	221.11	**430**	806	282.22	**540**	1004
182.78	**361**	681.8	222.22	**432**	809.6	283.33	**542**	1007.6
183.33	**362**	683.6	223.33	**434**	813.2	284.44	**544**	1011.2
183.89	**363**	685.4	224.44	**436**	816.8	285.56	**546**	1014.8
184.44	**364**	687.2	225.56	**438**	820.4	286.67	**548**	1018.4
185	**365**	689	226.67	**440**	824	287.78	**550**	1022
185.56	**366**	690.8	227.78	**442**	827.6	288.89	**552**	1025.6
186.11	**367**	692.6	228.89	**444**	831.2	290	**554**	1029.2
186.67	**368**	694.4	230	**446**	834.8	291.11	**556**	1032.8
187.22	**369**	696.2	231.11	**448**	838.4	292.22	**558**	1036.4
187.78	**370**	698	232.22	**450**	842	293.33	**560**	1040
188.33	**371**	699.8	233.33	**452**	845.6	294.44	**562**	1043.6
188.89	**372**	701.6	234.44	**454**	849.2	295.56	**564**	1047.2
189.44	**373**	703.4	235.56	**456**	852.8	296.67	**566**	1050.8
190	**374**	705.2	236.67	**458**	856.4	297.78	**568**	1054.4
190.56	**375**	707	237.78	**460**	860	298.89	**570**	1058
191.11	**376**	708.8	238.89	**462**	863.6	300	**572**	1061.6
191.67	**377**	710.6	240	**464**	867.2	301.11	**574**	1065.2
192.22	**378**	712.4	241.11	**466**	870.8	302.22	**576**	1068.8
192.78	**379**	714.2	242.22	**468**	874.4	303.33	**578**	1072.4
193.33	**380**	716	243.33	**470**	878	304.44	**580**	1076
193.89	**381**	717.8	244.44	**472**	881.6	305.56	**582**	1079.6
194.44	**382**	719.6	245.56	**474**	885.2	306.67	**584**	1083.2
195	**383**	721.4	246.67	**476**	888.8	307.78	**586**	1086.8
195.56	**384**	723.2	247.78	**478**	892.4	308.89	**588**	1090.4
196.11	**385**	725	248.89	**480**	896	310	**590**	1094
196.67	**386**	726.8	250	**482**	899.6	311.11	**592**	1097.6
197.22	**387**	728.6	251.11	**484**	903.2	312.22	**594**	1101.2
197.78	**388**	730.4	252.22	**486**	906.8	313.33	**596**	1104.8
198.33	**389**	732.2	253.33	**488**	910.4	314.44	**598**	1108.4
198.89	**390**	734	254.44	**490**	914	315.56	**600**	1112
199.44	**391**	735.8	255.56	**492**	917.6	316.67	**602**	1115.6
200	**392**	737.6	256.67	**494**	921.2	317.78	**604**	1119.2
200.56	**393**	739.4	257.78	**496**	924.8	318.89	**606**	1122.8
201.11	**394**	741.2	258.89	**498**	928.4	320	**608**	1126.4
201.67	**395**	743	260	**500**	932	321.11	**610**	1130
202.22	**396**	744.8	261.11	**502**	935.6	322.22	**612**	1133.6
202.78	**397**	746.6	262.22	**504**	939.2	323.33	**614**	1137.2
203.33	**398**	748.4	263.33	**506**	942.8	324.44	**616**	1140.8
203.89	**399**	750.2	264.44	**508**	946.4	325.56	**618**	1144.4
204.44	**400**	752	265.56	**510**	950	326.67	**620**	1148
205.56	**402**	755.6	266.67	**512**	953.6	327.78	**622**	1151.6
206.67	**404**	759.2	267.78	**514**	957.2	328.89	**624**	1155.2
207.78	**406**	762.8	268.89	**516**	960.8	330	**626**	1158.8
208.89	**408**	766.4	270	**518**	964.4	331.11	**628**	1162.4

Table 9-21. TEMPERATURE CONVERSION *(Continued)*

To convert			To convert			To convert		
To °C	←°F or °C→	To °F	To °C	←°F or °C→	To °F	To °C	←°F or °C→	To °F
332.22	**630**	1166	393.33	**740**	1364	454.44	**850**	1562
333.33	**632**	1169.6	394.44	**742**	1367.6	455.56	**852**	1565.6
334.44	**634**	1173.2	395.56	**744**	1371.2	456.67	**854**	1569.2
335.56	**636**	1176.8	396.67	**746**	1374.8	457.78	**856**	1572.8
336.67	**638**	1180.4	397.78	**748**	1378.4	458.89	**858**	1576.4
337.78	**640**	1184	398.89	**750**	1382	460	**860**	1580
338.89	**642**	1187.6	400	**752**	1385.6	461.11	**862**	1583.6
340	**644**	1191.2	401.11	**754**	1389.2	462.22	**864**	1587.2
341.11	**646**	1194.8	402.22	**756**	1392.8	463.33	**866**	1590.8
342.22	**648**	1198.4	403.33	**758**	1396.4	464.44	**868**	1594.4
343.33	**650**	1202	404.44	**760**	1400	465.56	**870**	1598
344.44	**652**	1205.6	405.56	**762**	1403.6	466.67	**872**	1601.6
345.56	**654**	1209.2	406.67	**764**	1407.2	467.78	**874**	1605.2
346.67	**656**	1212.8	407.78	**766**	1410.8	468.89	**876**	1608.8
347.78	**658**	1216.4	408.89	**768**	1414.4	470	**878**	1612.4
348.89	**660**	1220	410	**770**	1418	471.11	**880**	1616
350	**662**	1223.6	411.11	**772**	1421.6	472.22	**882**	1619.6
351.11	**664**	1227.2	412.22	**774**	1425.2	473.33	**884**	1623.2
352.22	**666**	1230.8	413.33	**776**	1428.8	474.44	**886**	1626.8
353.33	**668**	1234.4	414.44	**778**	1432.4	475.56	**888**	1630.4
354.44	**670**	1238	415.56	**780**	1436	476.67	**890**	1634
355.56	**672**	1241.6	416.67	**782**	1439.6	477.78	**892**	1637.6
356.67	**674**	1245.2	417.78	**784**	1443.2	478.89	**894**	1641.2
357.78	**676**	1248.8	418.89	**786**	1446.8	480	**896**	1644.8
358.89	**678**	1252.4	420	**788**	1450.4	481.11	**898**	1648.4
360	**680**	1256	421.11	**790**	1454	482.22	**900**	1652
361.11	**682**	1259.6	422.22	**792**	1457.6	483.33	**902**	1655.6
362.22	**684**	1263.2	423.33	**794**	1461.2	484.44	**904**	1659.2
363.33	**686**	1266.8	424.44	**796**	1464.8	485.56	**906**	1662.8
364.44	**688**	1270.4	425.56	**798**	1468.4	486.67	**908**	1666.4
365.56	**690**	1274	426.67	**800**	1472	487.78	**910**	1670
366.67	**692**	1277.6	427.78	**802**	1475.6	488.89	**912**	1673.6
367.78	**694**	1281.2	428.89	**804**	1479.2	490	**914**	1677.2
368.89	**696**	1284.8	430	**806**	1482.8	491.11	**916**	1680.8
370	**698**	1288.4	431.11	**808**	1486.4	492.22	**918**	1684.4
371.11	**700**	1292	432.22	**810**	1490	493.33	**920**	1688
372.22	**702**	1295.6	433.33	**812**	1493.6	494.44	**922**	1691.6
373.33	**704**	1299.2	434.44	**814**	1497.2	495.56	**924**	1695.2
374.44	**706**	1302.8	435.56	**816**	1500.8	496.67	**926**	1698.8
375.56	**708**	1306.4	436.67	**818**	1504.4	497.78	**928**	1702.4
376.67	**710**	1310	437.78	**820**	1508	498.89	**930**	1706
377.78	**712**	1313.6	438.89	**822**	1511.6	500	**932**	1709.6
378.89	**714**	1317.2	440	**824**	1515.2	501.11	**934**	1713.2
380	**716**	1320.8	441.11	**826**	1518.8	502.22	**936**	1716.8
381.11	**718**	1324.4	442.22	**828**	1522.4	503.33	**938**	1720.4
382.22	**720**	1328	443.33	**830**	1526	504.44	**940**	1724
383.33	**722**	1331.6	444.44	**832**	1529.6	505.56	**942**	1727.6
384.44	**724**	1335.2	445.56	**834**	1533.2	506.67	**944**	1731.2
385.56	**726**	1338.8	446.67	**836**	1536.8	507.78	**946**	1734.8
386.67	**728**	1342.4	447.78	**838**	1540.4	508.89	**948**	1738.4
387.78	**730**	1346	448.89	**840**	1544	510	**950**	1742
388.89	**732**	1349.6	450	**842**	1547.6	511.11	**952**	1745.6
390	**734**	1353.2	451.11	**844**	1551.2	512.22	**954**	1749.2
391.11	**736**	1356.8	452.22	**846**	1554.8	513.33	**956**	1752.8
392.22	**738**	1360.4	453.33	**848**	1558.4	514.44	**958**	1756.4

Table 9-21. TEMPERATURE CONVERSION (Continued)

	To convert			To convert			To convert	
To °C	←°F or °C→	To °F	To °C	←°F or °C→	To °F	To °C	←°F or °C→	To °F
515.56	960	1760	576.67	1070	1958	637.78	1180	2156
516.67	962	1763.6	577.78	1072	1961.6	638.89	1182	2159.6
517.78	964	1767.2	578.89	1074	1965.2	640	1184	2163.2
518.89	966	1770.8	580	1076	1968.8	641.11	1186	2166.8
520	968	1774.4	581.11	1078	1972.4	642.22	1188	2170.4
521.11	970	1778	582.22	1080	1976	643.33	1190	2174
522.22	972	1781.6	583.33	1082	1979.6	644.44	1192	2177.6
523.33	974	1785.2	584.44	1084	1983.2	645.56	1194	2181.2
524.44	976	1788.8	585.56	1086	1986.8	646.67	1196	2184.8
525.56	978	1792.4	586.67	1088	1990.4	647.78	1198	2188.4
526.67	980	1796	587.78	1090	1994	648.89	1200	2192
527.78	982	1799.6	588.89	1092	1997.6	650	1202	2195.6
528.89	984	1803.2	590	1094	2001.2	651.11	1204	2199.2
530	986	1806.8	591.11	1096	2004.8	652.22	1206	2202.8
531.11	988	1810.4	592.22	1098	2008.4	653.33	1208	2206.4
532.22	990	1814	593.33	1100	2012	654.44	1210	2210
533.33	992	1817.6	594.44	1102	2015.6	655.56	1212	2213.6
534.44	994	1821.2	595.56	1104	2019.2	656.67	1214	2217.2
535.56	996	1824.8	596.67	1106	2022.8	657.78	1216	2220.8
536.67	998	1828.4	597.78	1108	2026.4	658.89	1218	2224.4
537.78	1000	1832	598.89	1110	2030	660	1220	2228
538.89	1002	1835.6	600	1112	2033.6	661.11	1222	2231.6
540	1004	1839.2	601.11	1114	2037.2	662.22	1224	2235.2
541.11	1006	1842.8	602.22	1116	2040.8	663.33	1226	2238.8
542.22	1008	1846.4	603.33	1118	2044.4	664.44	1228	2242.4
543.33	1010	1850	604.44	1120	2048	665.56	1230	2246
544.44	1012	1853.6	605.56	1122	2051.6	666.67	1232	2249.6
545.56	1014	1857.2	606.67	1124	2055.2	667.78	1234	2253.2
546.67	1016	1860.8	607.78	1126	2058.8	668.89	1236	2256.8
547.78	1018	1864.4	608.89	1128	2062.4	670	1238	2260.4
548.89	1020	1868	610	1130	2066	671.11	1240	2264
550	1022	1871.6	611.11	1132	2069.6	672.22	1242	2267.6
551.11	1024	1875.2	612.22	1134	2073.2	673.33	1244	2271.2
552.22	1026	1878.8	613.33	1136	2076.8	674.44	1246	2274.8
553.33	1028	1882.4	614.44	1138	2080.4	675.56	1248	2278.4
554.44	1030	1886	615.56	1140	2084	676.67	1250	2282
555.56	1032	1889.6	616.67	1142	2087.6	677.78	1252	2285.6
556.67	1034	1893.2	617.78	1144	2091.2	678.89	1254	2289.2
557.78	1036	1896.8	618.89	1146	2094.8	680	1256	2292.8
558.89	1038	1900.4	620	1148	2098.4	681.11	1258	2296.4
560	1040	1904	621.11	1150	2102	682.22	1260	2300
561.11	1042	1907.6	622.22	1152	2105.6	683.33	1262	2303.6
562.22	1044	1911.2	623.33	1154	2109.2	684.44	1264	2307.2
563.33	1046	1914.8	624.44	1156	2112.8	685.56	1266	2310.8
564.44	1048	1918.4	625.56	1158	2116.4	686.67	1268	2314.4
565.56	1050	1922	626.67	1160	2120	687.78	1270	2318
566.67	1052	1925.6	627.78	1162	2123.6	688.89	1272	2321.6
567.78	1054	1929.2	628.89	1164	2127.2	690	1274	2325.2
568.89	1056	1932.8	630	1166	2130.8	691.11	1276	2328.8
570	1058	1936.4	631.11	1168	2134.4	692.22	1278	2332.4
571.11	1060	1940	632.22	1170	2138	693.33	1280	2336
572.22	1062	1943.6	633.33	1172	2141.6	694.44	1282	2339.6
573.33	1064	1947.2	634.44	1174	2145.2	695.56	1284	2343.2
574.44	1066	1950.8	635.56	1176	2148.8	696.67	1286	2346.8
575.56	1068	1954.4	636.67	1178	2152.4	697.78	1288	2350.4

Table 9-21. TEMPERATURE CONVERSION *(Continued)*

To convert			To convert			To convert		
To °C	←°F or °C→	To °F	To °C	←°F or °C→	To °F	To °C	←°F or °C→	To °F
698.89	**1290**	2354	760	**1400**	2552	843.33	**1550**	2922
700	**1292**	2357.6	761.11	**1402**	2555.6	848.89	**1560**	2840
701.11	**1294**	2361.2	762.22	**1404**	2559.2	854.44	**1570**	2858
702.22	**1296**	2364.8	763.33	**1406**	2562.8	860	**1580**	2876
703.33	**1298**	2368.4	764.44	**1408**	2566.4	865.56	**1590**	2894
704.44	**1300**	2372	765.56	**1410**	2570	871.11	**1600**	2912
705.56	**1302**	2375.6	766.67	**1412**	2573.6	876.67	**1610**	2930
706.67	**1304**	2379.2	767.78	**1414**	2577.2	882.22	**1620**	2948
707.78	**1306**	2382.8	768.89	**1416**	2580.8	887.78	**1630**	2966
708.89	**1308**	2386.4	770	**1418**	2584.4	893.33	**1640**	2984
710	**1310**	2390	771.11	**1420**	2588	898.89	**1650**	3002
711.11	**1312**	2393.6	772.22	**1422**	2591.6	904.44	**1660**	3020
712.22	**1314**	2397.2	773.33	**1424**	2595.2	910	**1670**	3038
713.33	**1316**	2400.8	774.44	**1426**	2598.8	915.56	**1680**	3056
714.44	**1318**	2404.4	775.56	**1428**	2602.4	921.11	**1690**	3074
715.56	**1320**	2408	776.67	**1430**	2606	926.67	**1700**	3092
716.67	**1322**	2411.6	777.78	**1432**	2609.6	932.22	**1710**	3110
717.78	**1324**	2415.2	778.89	**1434**	2613.2	937.78	**1720**	3128
718.89	**1326**	2418.8	780	**1436**	2616.8	943.33	**1730**	3146
720	**1328**	2422.4	781.11	**1438**	2620.4	948.89	**1740**	3164
721.11	**1330**	2426	782.22	**1440**	2624	954.44	**1750**	3182
722.22	**1332**	2429.6	783.33	**1442**	2627.6	960	**1760**	3200
723.33	**1334**	2433.2	784.44	**1444**	2631.2	965.56	**1770**	3218
724.44	**1336**	2436.8	785.56	**1446**	2634.8	971.11	**1780**	3236
725.56	**1338**	2440.4	786.67	**1448**	2638.4	976.67	**1790**	3254
726.67	**1340**	2444	787.78	**1450**	2642	982.22	**1800**	3272
727.78	**1342**	2447.6	788.89	**1452**	2645.6	987.78	**1810**	3290
728.89	**1344**	2451.2	790	**1454**	2649.2	993.33	**1820**	3308
730	**1346**	2454.8	791.11	**1456**	2652.8	998.89	**1830**	3326
731.11	**1348**	2458.4	792.22	**1458**	2656.4	1004.44	**1840**	3344
732.22	**1350**	2462	793.33	**1460**	2660	1010	**1850**	3362
733.33	**1352**	2465.6	794.44	**1462**	2663.6	1015.56	**1860**	3380
734.44	**1354**	2469.2	795.56	**1464**	2667.2	1021.11	**1870**	3398
735.56	**1356**	2472.8	796.67	**1466**	2670.8	1026.67	**1880**	3416
736.67	**1358**	2476.4	797.78	**1468**	2674.4	1032.22	**1890**	3434
737.78	**1360**	2480	798.89	**1470**	2678	1037.78	**1900**	3452
738.89	**1362**	2483.6	800	**1472**	2681.6	1043.33	**1910**	3470
740	**1364**	2487.2	801.11	**1474**	2685.2	1048.89	**1920**	3488
741.11	**1366**	2490.8	802.22	**1476**	2688.8	1054.44	**1930**	3506
742.22	**1368**	2494.4	803.33	**1478**	2692.4	1060	**1940**	3524
743.33	**1370**	2498	804.44	**1480**	2696	1065.56	**1950**	3542
744.44	**1372**	2501.6	805.56	**1482**	2699.6	1071.11	**1960**	3560
745.56	**1374**	2505.2	806.67	**1484**	2703.2	1076.67	**1970**	3578
746.67	**1376**	2508.8	807.78	**1486**	2706.8	1082.22	**1980**	3596
747.78	**1378**	2512.4	808.89	**1488**	2710.4	1087.78	**1990**	3614
748.89	**1380**	2516	810	**1490**	2714	1093.33	**2000**	3632
750	**1382**	2519.6	811.11	**1492**	2717.6	1098.89	**2010**	3650
751.11	**1384**	2523.2	812.22	**1494**	2721.2	1104.44	**2020**	3668
752.22	**1386**	2526.8	813.33	**1496**	2724.8	1110	**2030**	3686
753.33	**1388**	2530.4	814.44	**1498**	2728.4	1115.56	**2040**	3704
754.44	**1390**	2534	815.56	**1500**	2732	1121.11	**2050**	3722
755.56	**1392**	2537.6	821.11	**1510**	2750	1126.67	**2060**	3740
756.67	**1394**	2541.2	826.67	**1520**	2768	1132.22	**2070**	3758
757.78	**1396**	2544.8	832.22	**1530**	2786	1137.78	**2080**	3776
758.89	**1398**	2548.4	837.78	**1540**	2804	1143.33	**2090**	3794

Table 9-21. TEMPERATURE CONVERSION *(Continued)*

To convert			To convert			To convert		
To °C	←°F or °C→	To °F	To °C	←°F or °C→	To °F	To °C	←°F or °C→	To °F
1148.89	2100	3812	1579.44	2875	5207	2343.33	4250	7682
1154.44	2110	3830	1593.33	2900	5252	2357.22	4275	7727
1160	2120	3848	1607.22	2925	5297	2371.11	4300	7772
1165.56	2130	3866	1621.11	2950	5342	2385.00	4325	7817
1171.11	2140	3884	1635	2975	5387	2398.89	4350	7862
1176.67	2150	3902	1648.89	3000	5432	2412.78	4375	7907
1182.22	2160	3920	1662.78	3025	5477	2426.67	4400	7952
1187.78	2170	3938	1676.67	3050	5522	2440.56	4425	7997
1193.33	2180	3956	1690.56	3075	5567	2454.44	4450	8042
1198.89	2190	3974	1704.44	3100	5612	2468.33	4475	8087
1204.44	2200	3992	1718.33	3125	5657	2482.22	4500	8132
1210	2210	4010	1732.22	3150	5702	2496.11	4525	8177
1215.56	2220	4028	1746.11	3175	5747	2510.00	4550	8222
1221.11	2230	4046	1760	3200	5792	2523.89	4575	8267
1226.67	2240	4064	1773.89	3225	5837	2537.78	4600	8312
1232.22	2250	4082	1787.78	3250	5882	2551.67	4625	8357
1237.78	2260	4100	1801.67	3275	5927	2565.56	4650	8402
1243.33	2270	4118	1815.56	3300	5972	2579.44	4675	8447
1248.89	2280	4136	1829.44	3325	6017	2593.33	4700	8492
1254.44	2290	4154	1843.33	3350	6062	2607.22	4725	8537
1260	2300	4172	1857.22	3375	6107	2621.11	4750	8582
1265.56	2310	4190	1871.11	3400	6152	2635.00	4775	8627
1271.11	2320	4208	1885.00	3425	6197	2648.89	4800	8672
1276.67	2330	4226	1898.89	3450	6242	2662.78	4825	8717
1282.22	2340	4244	1912.78	3475	6287	2676.67	4850	8762
1287.78	2350	4262	1926.67	3500	6332	2690.55	4875	8807
1293.33	2360	4280	1940.56	3525	6377	2704.44	4900	8852
1298.89	2370	4298	1954.44	3550	6422	2718.33	4925	8897
1304.44	2380	4316	1968.33	3575	6467	2732.22	4950	8942
1310	2390	4334	1982.22	3600	6512	2746.11	4975	8987
1315.56	2400	4352	1996.11	3625	6557	2760.00	5000	9032
1221.11	2410	4370	2010.00	3650	6602	2787.78	5050	9122
1326.67	2420	4388	2023.89	3675	6647	2815.56	5100	9212
1332.22	2430	4406	2037.78	3700	6692	2843.33	5150	9302
1337.78	2440	4424	2051.67	3725	6737	2871.11	5200	9392
1343.33	2450	4442	2065.56	3750	6782	2898.89	5250	9482
1348.89	2460	4460	2079.44	3775	6827	2926.67	5300	9572
1354.44	2470	4478	2093.33	3800	6872	2954.44	5350	9662
1360	2480	4496	2107.22	3825	6917	2982.22	5400	9752
1365.56	2490	4514	2121.11	3850	6962	3010.00	5450	9842
1371.11	2500	4532	2135.00	3875	7007	3037.78	5500	9932
1385	2525	4577	2148.89	3900	7052	3065.56	5550	10022
1398.89	2550	4622	2162.78	3925	7097	3093.33	5600	10112
1412.78	2575	4667	2176.67	3950	7142	3121.11	5650	10202
1426.67	2600	4712	2190.56	3975	7187	3148.89	5700	10292
1440.56	2625	4757	2204.44	4000	7232	3176.67	5750	10382
1454.44	2650	4802	2218.33	4025	7277	3204.44	5800	10472
1468.33	2675	4847	2232.22	4050	7322	3232.22	5850	10562
1482.22	2700	4892	2246.11	4075	7367	3260.00	5900	10652
1496.11	2725	4937	2260.00	4100	7412	3287.78	5950	10742
1510	2750	4982	2273.89	4125	7457	3315.56	6000	10832
1523.89	2775	5027	2287.78	4150	7502	3593.33	6500	11732
1537.78	2800	5072	2301.67	4175	7547	3871.11	7000	12632
1551.67	2825	5117	2315.56	4200	7592	4148.89	7500	13532
1565.56	2850	5162	2329.44	4225	7537	4426.67	8000	14432

Table 9-22.   TEMPERATURE CONVERSION—DEG F TO K

$$K = (5/9)(deg\ F + 459.67)$$

deg F	K	deg F	K	deg F	K	deg F	K	deg F	K	deg F	K
−459.67	0	−412	26.48	−220	133.14	−28	239.82	20	266.48	68	293.15
−459	.37	−411	27.04	−215	135.91	−27	240.37	21	267.04	69	293.71
−458	.93	−410	27.59	−210	138.69	−26	240.93	22	267.59	70	294.26
−457	1.48	−409	28.15	−205	141.47	−25	241.48	23	268.15	71	294.82
−456	2.04	−408	28.70	−200	144.25	−24	242.04	24	268.71	72	295.37
−455	2.59	−407	29.26	−195	147.02	−23	242.59	25	269.26	73	295.93
−454	3.15	−406	29.81	−190	149.80	−22	243.15	26	269.82	74	296.48
−453	3.70	−405	30.37	−185	152.58	−21	243.71	27	270.37	75	297.04
−452	4.26	−404	30.92	−180	155.36	−20	244.26	28	270.93	76	297.59
−451	4.82	−403	31.48	−175	158.13	−19	244.82	29	271.48	77	298.15
−450	5.37	−402	32.04	−170	160.91	−18	245.37	30	272.04	78	298.71
−449	5.93	−401	32.59	−165	163.69	−17	245.93	31	272.59	79	299.26
−448	6.48	−400	33.15	−160	166.47	−16	246.48	32	273.15	80	299.82
−447	7.04	−395	35.92	−155	169.24	−15	247.04	33	273.71	81	300.37
−446	7.59	−390	38.70	−150	172.02	−14	247.59	34	274.26	82	300.93
−445	8.15	−385	41.48	−145	174.80	−13	248.15	35	274.82	83	301.48
−444	8.70	−380	44.26	−140	177.58	−12	248.71	36	275.37	84	302.04
−443	9.26	−375	47.03	−135	180.35	−11	249.26	37	275.93	85	302.59
−442	9.82	−370	49.81	−130	183.13	−10	249.82	38	276.48	86	303.15
−441	10.37	−365	52.59	−125	185.91	−9	250.37	39	277.04	87	303.71
−440	10.93	−360	55.37	−120	188.69	−8	250.93	40	277.59	88	304.26
−439	11.48	−355	58.14	−115	191.46	−7	251.48	41	278.15	89	304.82
−438	12.04	−350	60.92	−110	194.24	−6	252.04	42	278.71	90	305.37
−437	12.59	−345	63.70	−105	197.02	−5	252.59	43	279.26	91	305.93
−436	13.15	−340	66.48	−100	199.80	−4	253.15	44	279.82	92	306.48
−435	13.70	−335	69.25	−95	202.57	−3	253.71	45	280.37	93	307.04
−434	14.26	−330	72.03	−90	205.35	−2	254.26	46	280.93	94	307.59
−433	14.82	−325	74.81	−85	208.13	−1	254.82	47	281.48	95	308.15
−432	15.37	−320	77.59	−80	210.91	0	255.37	48	282.04	96	308.71
−431	15.93	−315	80.36	−75	213.68	1	255.93	49	282.59	97	309.26
−430	16.48	−310	83.14	−70	216.46	2	256.48	50	283.15	98	309.82
−429	17.04	−305	85.92	−65	219.24	3	257.04	51	283.71	99	310.37
−428	17.59	−300	88.70	−60	222.02	4	257.59	52	284.26	100	310.93
−427	18.15	−295	91.47	−55	224.79	5	258.15	53	284.82	101	311.48
−426	18.70	−290	94.25	−50	227.57	6	258.71	54	285.37	102	312.04
−425	19.26	−285	97.03	−45	230.35	7	259.26	55	285.93	103	312.59
−424	19.81	−280	99.81	−40	233.13	8	259.82	56	286.48	104	313.15
−423	20.37	−275	102.58	−39	233.71	9	260.37	57	287.04	105	313.71
−422	20.92	−270	105.36	−38	234.26	10	260.93	58	287.59	106	314.26
−421	21.48	−265	108.14	−37	234.82	11	261.48	59	288.15	107	314.82
−420	22.04	−260	110.92	−36	235.37	12	262.04	60	288.71	108	315.37
−419	22.59	−255	113.69	−35	235.93	13	262.59	61	289.26	109	315.93
−418	23.15	−250	116.47	−34	236.48	14	263.15	62	289.82	110	316.48
−417	23.70	−245	119.25	−33	237.04	15	263.71	63	290.37	111	317.04
−416	24.26	−240	122.03	−32	237.59	16	264.26	64	290.93	112	317.59
−415	24.81	−235	124.80	−31	238.15	17	264.82	65	291.48	113	318.15
−414	25.37	−230	127.58	−30	238.71	18	265.37	66	292.04	114	318.71
−413	25.92	−225	130.36	−29	239.26	19	265.93	67	292.59	115	319.26

Table 9-22.   TEMPERATURE CONVERSION—DEG F TO K *(Continued)*

deg F	K	deg F	K	deg F	K	deg F	K	deg F	K	deg F	K
116	319.82	166	347.59	216	375.37	266	403.15	460	510.93	960	788.71
117	320.37	167	348.15	217	375.93	267	403.71	470	516.48	970	794.26
118	320.93	168	348.71	218	376.48	268	404.26	480	522.04	980	799.82
119	321.48	169	349.26	219	377.04	269	404.82	490	527.59	990	805.37
120	322.04	170	349.82	220	377.59	270	405.37	500	533.15	1 000	810.93
121	322.59	171	350.37	221	378.15	271	405.93	510	538.71	1 010	816.48
122	323.15	172	350.93	222	378.71	272	406.48	520	544.26	1 020	822.04
123	323.71	173	351.48	223	379.26	273	407.04	530	549.82	1 030	827.59
124	324.26	174	352.04	224	379.82	274	407.59	540	555.37	1 040	833.15
125	324.82	175	352.59	225	380.37	275	408.15	550	560.93	1 050	838.71
126	325.37	176	353.15	226	380.93	276	408.71	560	566.48	1 060	844.26
127	325.93	177	353.71	227	381.48	277	409.26	570	572.04	1 070	849.82
128	326.48	178	354.26	228	382.04	278	409.82	580	577.59	1 080	855.37
129	327.04	179	354.82	229	382.59	279	410.37	590	583.15	1 090	860.93
130	327.59	180	355.37	230	383.15	280	410.93	600	588.71	1 100	866.48
131	328.15	181	355.93	231	383.71	281	411.48	610	594.26	1 110	872.04
132	328.71	182	356.48	232	384.26	282	412.04	620	599.82	1 120	877.59
133	329.26	183	357.04	233	384.82	283	412.59	630	605.37	1 130	883.15
134	329.82	184	357.59	234	385.37	284	413.15	640	610.93	1 140	888.71
135	330.37	185	358.15	235	385.93	285	413.71	650	616.48	1 150	894.26
136	330.93	186	358.71	236	386.48	286	414.26	660	622.04	1 160	899.82
137	331.48	187	359.26	237	387.04	287	414.82	670	627.59	1 170	905.37
138	332.04	188	359.82	238	387.59	288	415.37	680	633.15	1 180	910.93
139	332.59	189	360.37	239	388.15	289	415.93	690	638.71	1 190	916.48
140	333.15	190	360.93	240	388.71	290	416.48	700	644.26	1 200	922.04
141	333.71	191	361.48	241	389.26	291	417.04	710	649.82	1 210	927.59
142	334.26	192	362.04	242	389.82	292	417.59	720	655.37	1 220	933.15
143	334.82	193	362.59	243	390.37	293	418.15	730	660.93	1 230	938.71
144	335.37	194	363.15	244	390.93	294	418.71	740	666.48	1 240	944.26
145	335.93	195	363.71	245	391.48	295	419.26	750	672.04	1 250	949.82
146	336.48	196	364.26	246	392.04	296	419.82	760	677.59	1 260	955.37
147	337.04	197	364.82	247	392.59	297	420.37	770	683.15	1 270	960.93
148	337.59	198	365.37	248	393.15	298	420.93	780	688.71	1 280	966.48
149	338.15	199	365.93	249	393.71	299	421.48	790	694.26	1 290	972.04
150	338.71	200	366.48	250	394.26	300	422.04	800	699.82	1 300	977.59
151	339.26	201	367.04	251	394.82	310	427.59	810	705.37	1 310	983.15
152	339.82	202	367.59	252	395.37	320	433.15	820	710.93	1 320	988.71
153	340.37	203	368.15	253	395.93	330	438.71	830	716.48	1 330	994.26
154	340.93	204	368.71	254	396.48	340	444.26	840	722.04	1 340	999.82
155	341.48	205	369.26	255	397.04	350	449.82	850	727.59	1 350	1 005.37
156	342.04	206	369.82	256	397.59	360	455.37	860	733.15	1 360	1 010.93
157	342.59	207	370.37	257	398.15	370	460.93	870	738.71	1 370	1 016.48
158	343.15	208	370.93	258	398.71	380	466.48	880	744.26	1 380	1 022.04
159	343.71	209	371.48	259	399.26	390	472.04	890	749.82	1 390	1 027.59
160	344.26	210	372.04	260	399.82	400	477.59	900	755.37	1 400	1 033.15
161	344.82	211	372.59	261	400.37	410	483.15	910	760.93	1 410	1 038.71
162	345.37	212	373.15	262	400.93	420	488.71	920	766.48	1 420	1 044.26
163	345.93	213	373.71	263	401.48	430	494.26	930	772.04	1 430	1 049.82
164	346.48	214	374.26	264	402.04	440	499.82	940	777.59	1 440	1 055.37
165	347.04	215	374.82	265	402.59	450	505.37	950	783.15	1 450	1 060.93

Table 9-22.   TEMPERATURE CONVERSION—DEG F TO K *(Continued)*

deg F	K	deg F	K	deg F	K	deg F	K	deg F	K	deg F	K
1 460	1 066.48	1 960	1 344.26	3 150	2 005.37	4 400	2 699.82	6 300	3 755.37	8 800	5 144.26
1 470	1 072.04	1 970	1 349.82	3 175	2 019.26	4 425	2 713.70	6 350	3 783.15	8 850	5 172.04
1 480	1 077.59	1 980	1 355.37	3 200	2 033.15	4 450	2 727.59	6 400	3 810.93	8 900	5 199.82
1 490	1 083.15	1 990	1 360.93	3 225	2 047.04	4 475	2 741.48	6 450	3 838.71	8 950	5 227.59
1 500	1 088.71	2 000	1 366.48	3 250	2 060.93	4 500	2 755.37	6 500	3 866.48	9 000	5 255.37
1 510	1 094.26	2 025	1 380.37	3 275	2 074.82	4 525	2 769.26	6 550	3 894.26	9 050	5 283.15
1 520	1 099.82	2 050	1 394.26	3 300	2 088.70	4 550	2 783.15	6 600	3 922.04	9 100	5 310.93
1 530	1 105.37	2 075	1 408.15	3 325	2 102.59	4 575	2 797.04	6 650	3 949.82	9 150	5 338.71
1 540	1 110.93	2 100	1 422.04	3 350	2 116.48	4 600	2 810.93	6 700	3 977.59	9 200	5 366.48
1 550	1 116.48	2 125	1 435.93	3 375	2 130.37	4 625	2 824.82	6 750	4 005.37	9 250	5 394.26
1 560	1 122.04	2 150	1 449.82	3 400	2 144.26	4 650	2 838.70	6 800	4 033.15	9 300	5 422.04
1 570	1 127.59	2 175	1 463.70	3 425	2 158.15	4 675	2 852.59	6 850	4 060.93	9 350	5 449.82
1 580	1 133.15	2 200	1 477.59	3 450	2 172.04	4 700	2 866.48	6 900	4 088.71	9 400	5 477.59
1 590	1 138 71	2 225	1 491.48	3 475	2 185.93	4 725	2 880.37	6 950	4 116.48	9 450	5 505.37
1 600	1 144.26	2 250	1 505.37	3 500	2 199.82	4 750	2 894.26	7 000	4 144.26	9 500	5 533.15
1 610	1 149.82	2 275	1 519.26	3 525	2 213.70	4 775	2 908.15	7 050	4 172.04	9 550	5 560.93
1 620	1 155.37	2 300	1 533.15	3 550	2 227.59	4 800	2 922.04	7 100	4 199.82	9 600	5 588.71
1 630	1 160.93	2 325	1 547.04	3 575	2 241.48	4 825	2 935.93	7 150	4 227.59	9 650	5 616.48
1 640	1 166.48	2 350	1 560.93	3 600	2 255.37	4 850	2 949.82	7 200	4 255.37	9 700	5 644.26
1 650	1 172.04	2 375	1 574.82	3 625	2 269.26	4 875	2 963.70	7 250	4 283.15	9 750	5 672.04
1 660	1 177.59	2 400	1 588.70	3 650	2 283.15	4 900	2 977.59	7 300	4 310.93	9 800	5 699.82
1 670	1 183.15	2 425	1 602.59	3 675	2 297.04	4 925	2 991.48	7 350	4 338.71	9 850	5 727.59
1 680	1 188.71	2 450	1 616.48	3 700	2 310.93	4 950	3 005.37	7 400	4 366.48	9 900	5 755.37
1 690	1 194.26	2 475	1 630.37	3 725	2 324.82	4 975	3 019.26	7 450	4 394.26	9 950	5 783.15
1 700	1 199.82	2 500	1 644.26	3 750	2 338.70	5 000	3 033.15	7 500	4 422.04	10 000	5 810.93
1 710	1 205.37	2 525	1 658.15	3 775	2 352.59	5 050	3 060.93	7 550	4 449.82		
1 720	1 210.93	2 550	1 672.04	3 800	2 366.48	5 100	3 088.71	7 600	4 477.59		
1 730	1 216.48	2 575	1 685.93	3 825	2 380.37	5 150	3 116 48	7 650	4 505.37		
1 740	1 222.04	2 600	1 699.82	3 850	2 394.26	5 200	3 144.26	7 700	4 533.15		
1 750	1 227.59	2 625	1 713.70	3 875	2 408.15	5 250	3 172.04	7 750	4 560.93		
1 760	1 233.15	2 650	1 727.59	3 900	2 422.04	5 300	3 199.82	7 800	4 588.71		
1 770	1 238.71	2 675	1 741.48	3 925	2 435.93	5 350	3 227.59	7 850	4 616.48		
1 780	1 244.26	2 700	1 755.37	3 950	2 449.82	5 400	3 255.37	7 900	4 644.26		
1 790	1 249.82	2 725	1 769.26	3 975	2 463.70	5 450	3 283.15	7 950	4 672.04		
1 800	1 255.37	2 750	1 783.15	4 000	2 477.59	5 500	3 310.93	8 000	4 699.82		
1 810	1 260.93	2 775	1 797.04	4 025	2 491.48	5 550	3 338.71	8 050	4 727.59		
1 820	1 266.48	2 800	1 810.93	4 050	2 505.37	5 600	3 366.48	8 100	4 755.37		
1 830	1 272.04	2 825	1 824.82	4 075	2 519.26	5 650	3 394.26	8 150	4 783.15		
1 840	1 277.59	2 850	1 838.70	4 100	2 533.15	5 700	3 422.04	8 200	4 810.93		
1 850	1 283.15	2 875	1 852.59	4 125	2 547.04	5 750	3 449.82	8 250	4 838.71		
1 860	1 288.71	2 900	1 866.48	4 150	2 560.93	5 800	3 477.59	8 300	4 866.48		
1 870	1 294.26	2 925	1 880.37	4 175	2 574.82	5 850	3 505.37	8 350	4 894.26		
1 880	1 299.82	2 950	1 894.26	4 200	2 588.70	5 900	3 533.15	8 400	4 922.04		
1 890	1 305.37	2 975	1 908.15	4 225	2 602.59	5 950	3 560.93	8 450	4 949.82		
1 900	1 310.93	3 000	1 922.04	4 250	2 616.48	6 000	3 588.71	8 500	4 977.59		
1 910	1 316.48	3 025	1 935.93	4 275	2 630.37	6 050	3 616.48	8 550	5 005.37		
1 920	1 322.04	3 050	1 949.82	4 300	2 644.26	6 100	3 644.26	8 600	5 033.15		
1 930	1 327.59	3 075	1 963.70	4 325	2 658.15	6 150	3 672.04	8 650	5 060.93		
1 940	1 333.15	3 100	1 977.59	4 350	2 672.04	6 200	3 699.82	8 700	5 088.71		
1 950	1 338.71	3 125	1 991.48	4 375	2 685.93	6 250	3 727.59	8 750	5 116.48		

Table 9-23.   WIDE-RANGE TEMPERATURE CONVERSION FROM FAHRENHEIT TO KELVIN OR CELSIUS*

Table A (for Kelvin) and Table B (for Celsius) enable the user to estimate quickly the Kelvin or Celsius equivalent of any Fahrenheit temperature to 90 000 deg F. Conversion of any temperature to 100 000 deg F to any desired accuracy can be obtained by simple addition, using Table A or B for the first digit and Table C for additional digits. See examples.

Examples:

106 deg F = (100 + 6) deg F = ? K	**3.96 deg F = (3 + 0.9 + 0.06) deg F = ? deg C**	**− 103 deg F = (− 100 − 3) deg F = ? K**
100, Table A: 310.927 7 . . .	3.00, Table B: − 16.111 1 . . .	− 100, Table A: + 199.816 6 . . .
6, Table C: 3.333 3 . . .	0.90, Table C: 0.500 0 . . .	− 3, Table C: − 1.666 6 . . .
106 deg F = 314.261 1 . . . K	0.06, Table C: + 0.033 3 . . .	− 103 deg F = 198.150 0 . . . K
	3.96 deg F = − 15.577 7 . . . deg C	

Table A.   TEMPERATURE CONVERSION FROM FAHRENHEIT TO KELVIN[a]

$$T, K = (T, \deg F + 459.67)(5/9) \qquad 0 \deg F = 255.372\ 222\dots K$$

Range, deg F	Fahrenheit temperature								
	1	2	3	4	5	6	7	8	9
− 100 to − 400	199.816	144.261	88.705	33.150					
− 10 to − 90	249.816	244.261	238.705	233.150	227.594	222.038	216.483	210.927	205.372
− 1 to − 9	254.816	254.261	253.705	253.150	252.594	252.038	251.483	250.927	250.372
1 to 9	255.927	256.483	257.038	257.594	258.150	258.705	259.261	259.816	260.372
10 to 90	260.927	266.483	272.038	277.594	283.150	288.705	294.261	299.816	305.372
100 to 900	310.927	366.483	422.038	477.594	533.150	588.705	644.261	699.816	755.372
1 000 to 9 000	810.927	1 366.483	1 922.038	2 477.594	3 033.150	3 588.705	4 144.261	4 699.816	5 255.372
10 000 to 90 000	5 810.927	11 366.483	16 922.038	22 477.594	28 033.150	33 588.705	39 144.261	44 699.816	50 255.372

Table B.   TEMPERATURE CONVERSION FROM FAHRENHEIT TO CELSIUS[a]

$$T, \deg C = (T, \deg F − 32)(5/9) \qquad 0 \deg F = − 17.777\dots \deg C$$

Range, deg F	Fahrenheit temperature								
	1	2	3	4	5	6	7	8	9
− 100 to − 400	− 73.333	− 128.888	− 184.444	− 240.000					
− 10 to − 90	− 23.333	− 28.888	− 34.444	− 40.000	− 45.555	− 51.111	− 56.666	− 62.222	− 67.777
− 1 to − 9	− 18.333	− 18.888	− 19.444	− 20.000	− 20.555	− 21.111	− 21.666	− 22.222	− 22.777
1 to 9	− 17.222	− 16.666	− 16.111	− 15.555	− 15.000	− 14.444	− 13.888	− 13.333	− 12.777
10 to 90	− 12.222	− 6.666	− 1.111	+ 4.444	+ 10.000	+ 15.555	+ 21.111	+ 26.666	+ 32.222
100 to 900	+ 37.777	+ 93.333	+ 148.888	204.444	260.000	315.555	371.111	426.666	482.222
1 000 to 9 000	537.777	1 093.333	1 648.888	2 204.444	2 760.000	3 315.555	3 871.111	4 426.666	4 982.222
10 000 to 90 000	5 537.777	11 093.333	16 648.888	22 204.444	27 760.000	33 315.555	38 871.111	44 426.666	49 982.222

Table C.   TEMPERATURE INCREMENT CONVERSION FROM FAHRENHEIT TO KELVIN OR CELSIUS[a]

$$\Delta T, K = \Delta T, \deg C = (\Delta T, \deg F)(5/9)$$

Range, deg F	Fahrenheit temperature increment								
	1	2	3	4	5	6	7	8	9
0.01 to 0.09	0.005	0.011	0.016	0.022	0.027	0.033	0.038	0.044	0.050
0.1 to 0.9	0.055	0.111	0.166	0.222	0.277	0.333	0.388	0.444	0.500
1 to 9	0.555	1.111	1.666	2.222	2.777	3.333	3.888	4.444	5.000
10 to 90	5.555	11.111	16.666	22.222	27.777	33.333	38.888	44.444	50.000
100 to 900	55.555	111.111	166.666	222.222	277.777	333.333	388.888	444.444	500.000
1 000 to 9 000	555.555	1 111.111	1 666.666	2 222.222	2 777.777	3 333.333	3 888.888	4 444.444	5 000.000

[a]The final digit in every Kelvin value, Celsius value, and Kelvin/Celsius increment repeats ad infinitum. The user should carry them out and/or round them off to suit his needs.

Table 9-24. WIDE-RANGE TEMPERATURE CONVERSION FROM CELSIUS TO RANKINE OR FAHRENHEIT*

Table A (for Rankine) and Table B (for Fahrenheit) enable the user to estimate quickly the Rankine or Fahrenheit equivalent of any Celsius temperature to 90 000 deg C. Conversion of any temperature to 100 000 deg C to any desired accuracy can be obtained by simple addition, using Table A or B for the first digit and Table C for additional digits. See examples.

Examples:

106 C = (100 + 6) C = ? R	3.96 C = (3 + 0.9 + 0.06) C = ? F	−103 C = (−100 − 3) C = ? R = ? F
100, Table A: 671.67	3.00, Table B: 37.4	−100, Table A: 311.67
6, Table C: 10.8	0.90, Table C: 1.62	−3, Table C: −5.4
	0.06, Table C: 0.108	
106 C = 682.47 R		−103 C = +306.27 R
	3.96 C = 39.128 F	−100, Table B: −148.0
		3, Table C: −5.4
		−103 C = −153.4 F

Table A. TEMPERATURE CONVERSION FROM CELSIUS TO RANKINE[a]
T, deg R = (T, deg C + 273.15)(9/5)   0 deg C = 491.67 deg R (exactly)

Range, deg C	Celsius temperature								
	1	2	3	4	5	6	7	8	9
−100 to −200	311.67	131.67							
−10 to −90	473.67	455.67	437.67	419.67	401.67	383.67	365.67	347.67	329.67
−1 to −9	489.87	488.07	486.27	484.47	482.67	480.87	479.07	477.27	475.47
1 to 9	493.47	495.27	497.07	498.87	500.67	502.47	504.27	506.07	507.87
10 to 90	509.67	527.67	545.67	563.67	581.67	599.67	617.67	635.67	653.67
100 to 900	671.67	851.67	1 031.67	1 211.67	1 391.67	1 571.67	1 751.67	1 931.67	2 111.67
1 000 to 9 000	2 291.67	4 091.67	5 891.67	7 691.67	9 491.67	11 291.67	13 091.67	14 891.67	16 691.67
10 000 to 90 000	18 491.67	36 491.67	54 491.67	72 491.67	90 491.67	108 491.67	126 491.67	144 491.67	162 491.67

Table B. TEMPERATURE CONVERSION FROM CELSIUS TO FAHRENHEIT[a]
T, deg F = (T, deg C)(9/5) + 32   0 deg C = 32 deg F

Range, deg C	Celsius temperature								
	1	2	3	4	5	6	7	8	9
−100 to −200	−148.0	−328.0							
−10 to −90	+14.0	−4.0	−22.0	−40.0	−58.0	−76.0	−94.0	−112.0	−130.0
−1 to −9	+30.2	+28.4	+26.6	+24.8	+23.0	+21.2	+19.4	+17.6	+15.8
1 to 9	33.8	35.6	37.4	39.2	41.0	42.8	44.6	46.4	48.2
10 to 90	50.0	68.0	86.0	104.0	122.0	140.0	158.0	176.0	194.0
100 to 900	212.0	392.0	572.0	752.0	932.0	1 112.0	1 292.0	1 472.0	1 652.0
1 000 to 9 000	1 832.0	3 632.0	5 432.0	7 232.0	9 032.0	10 832.0	12 632.0	14 432.0	16 232.0
10 000 to 90 000	18 032.0	36 032.0	54 032.0	72 032.0	90 032.0	108 032.0	126 032.0	144 032.0	162 032.0

Table C. TEMPERATURE INCREMENT CONVERSION FROM CELSIUS TO RANKINE OR FAHRENHEIT[a]
ΔT, deg R = ΔT, deg F = (ΔT, deg C)(9/5)

Range, deg C	Celsius temperature increment								
	1	2	3	4	5	6	7	8	9
0.01 to 0.09	0.018	0.036	0.054	0.072	0.090	0.108	0.126	0.144	0.162
0.1 to 0.9	0.18	0.36	0.54	0.72	0.90	1.08	1.26	1.44	1.62
1 to 9	1.8	3.6	5.4	7.2	9.0	10.8	12.6	14.4	16.2
10 to 90	18	36	54	72	90	108	126	144	162
100 to 900	180	360	540	720	900	1 080	1 260	1 440	1 620
1 000 to 9 000	1 800	3 600	5 400	7 200	9 000	10 800	12 600	14 400	16 200

[a]All values are exact.

Table 9-25. MERCURY MANOMETER CONVERSION FACTORS

Equivalent Pressures in psi and Inches of Water per Inch of Mercury

For values in N/m^2, multiply values in psi by 6894.8.

Observed temperature of mercury column		Equivalent values				
				Inches of water		
		psi	Mercury column at 32°F	60°F; 15.56°C	68°F; 20°C	77°F; 25°C
°F	°C					
0	−17.78	0.49275	1.0032	13.652	13.663	13.680
10	−12.22	0.49225	1.0022	13.638	13.649	13.665
20	−6.67	0.49175	1.0012	13.625	13.636	13.652
30	−1.11	0.49126	1.0002	13.611	13.622	13.638
32	0.00	0.49116	1.0000	13.609	13.620	13.635
35	1.67	0.49101	0.9997	13.604	13.615	13.631
40	4.44	0.49076	0.9992	13.598	13.609	13.624
45	7.22	0.49051	0.9987	13.591	13.602	13.618
50	10.00	0.49026	0.9982	13.584	13.595	13.611
55	12.78	0.49002	0.9977	13.577	13.588	13.604
60	15.56	0.48977	0.9972	13.570	13.581	13.597
65	18.33	0.48952	0.9967	13.564	13.575	13.590
68	20.00	0.48938	0.9964	13.560	13.570	13.586
70	21.11	0.48928	0.9962	13.557	13.568	13.584
75	23.89	0.48904	0.9957	13.550	13.561	13.577
77	25.00	0.48894	0.9955	13.547	13.558	13.574
80	26.67	0.48879	0.9952	13.543	13.554	13.570
85	29.44	0.48854	0.9947	13.536	13.547	13.563
90	32.22	0.48830	0.9942	13.530	13.541	13.556
95	35.00	0.48805	0.9937	13.523	13.534	13.549
100	37.78	0.48780	0.9932	13.516	13.527	13.543
110	43.33	0.48732	0.9922	13.502	13.513	13.529
120	48.89	0.48683	0.9912	13.489	13.500	13.515
130	54.44	0.48634	0.9902	13.475	13.486	13.502
140	60.00	0.48585	0.9892	13.462	13.472	13.488

Table 9-26. VISCOSITY CONVERSIONS—LIQUIDS

Absolute viscosity				Kinematic viscosity						
Centi-poises	$\dfrac{lb_m}{ft\ sec}$	$\dfrac{N\text{-}sec}{m^2}$	Mac-Michael	Centi-stokes	$\dfrac{ft^2}{sec}$	$\dfrac{m^2}{sec}$	Saybolt Universal Seconds	Redwood standard seconds	Engler degrees	Ford 3
25	.0168	.025	—	25	.000269	.000025	115	112	3.4	
50	.0336	.050	185	50	.000538	.000050	230	225	6.4	28
75	.0504	.075	240	75	.000807	.000075	346	335	9.4	38
100	.0672	.100	295	100	.001076	.000100	462	448	13	49
125	.0840	.125	350	125	.001346	.000125	577	558	16	61
150	.1008	.150	405	150	.001615	.000150	692	673	19	73
175	.1176	.175	465	175	.001884	.000175	807	786	22	84
200	.1344	.200	520	200	.002153	.000200	923	898	26	95
225	.1512	.225	575	225	.002422	.000225	1038	1010	29	106
250	.1680	.250	625	250	.002691	.000250	1153	1120	32	116
275	.1848	.275	675	275	.002960	.000275	1269	1233	35	125
300	.2016	.300	725	300	.003229	.000300	1384	1345	39	135
325	.2184	.325	780	325	.003498	.000325	1500	1457	43	144
350	.2352	.350	840	350	.003767	.000350	1616	1570	45	152
375	.2520	.375	900	375	.004037	.000375	1731	1683	49	161
400	.2688	.400	960	400	.004306	.000400	1846	1797	53	170
425	.2856	.425	1020	425	.004575	.000425	1960	1910	56	178
450	.3024	.450	1080	450	.004844	.000450	2075	2023	60	188
475	.3192	.475	1160	475	.005113	.000475	2190	2136	63	198
500	.3360	.500	1210	500	.005382	.000500	2307	2240	67	208
525	.3528	.525	1270	525	.005651	.000525	2422	2355	71	217
550	.3696	.550	1330	550	.005920	.000550	2537	2460	75	226
575	.3864	.575	1385	575	.006189	.000575	2652	2575	79	234
600	.4032	.600	1440	600	.006458	.000600	2766	2690	82	242
625	.4200	.625	1500	625	.006728	.000625	2882	2800	85	250
650	.4368	.650	1560	650	.006997	.000650	2999	2915	88	258
675	.4536	.675	1620	675	.007266	.000675	3116	3030	91	266
700	.4704	.700	1680	700	.007535	.000700	3230	3143	94	274
725	.4872	.725	1740	725	.007804	.000725	3345	3256	97	282
750	.5040	.750	1800	750	.008073	.000750	3460	3370	101	290

Notes:
   In all equations relating viscosity to other physical properties or processes (e.g., fluid flow or heat transfer), the consistent units—fps, MKS, or cgs—should be used. (See Table 9-30 for conversion factors.) In consistent units kinematic viscosity is equal to absolute viscosity divided by density.
   A rotating-cup instrument (MacMichael) gives a reading directly convertible to absolute viscosity. The short-tube viscometers give readings directly convertible to kinematic viscosity.
   Values in this table are approximate. The conversion factors vary slightly with temperature; if highly accurate conversions are required, the "ASTM Viscosity Tables" should be used.
   The U.S. standard for light liquids is the Saybolt Universal short-tube viscometer. For heavy road oils and fuel oils the Saybolt Furol instrument is used. For the same oil the Saybolt Universal Seconds are approximately ten times the Saybolt Furol values.
   The Redwood No. 1 is the English viscometer; the Engler, the German viscometer. Engler degrees represent the ratio of the outflow time of the liquid under test to that for a like volume of water at the same temperature. Engler seconds are sometimes given; this value is about 51 times the Engler degrees.

REFERENCE

   For a conversion chart including several other viscometer scales, see "Handbook of Chemistry", 10th ed., N.A. Lange, Ed., McGraw-Hill Book Company, 1961, p.1844.

Table 9-27. GRAVITY CONVERSIONS—LIQUIDS

	Liquids lighter than water					Liquids heavier than water					
Sp gr	$\frac{lb}{ft^3}$	$\frac{kg}{m^3}$	$\frac{lb}{gal}$	°Bé	°API	Sp gr	$\frac{lb}{ft^3}$	$\frac{kg}{m^3}$	$\frac{lb}{gal}$	°Bé	°Twad	°Brix
1.00	62.38	999.2	8.34	10.0	10.0	1.00	62.38	999.2	8.34	0	0	0
.99	61.76	989.2	8.25	11.4	11.4	1.01	63.00	100.9	8.41	1.44	2	2.5
.98	61.14	979.2	8.17	12.9	12.9	1.02	63.63	101.9	8.50	2.84	4	5.1
.97	60.51	969.2	8.09	14.3	14.4	1.03	64.26	102.9	8.58	4.22	6	7.6
.96	59.89	959.2	8.00	15.8	15.9	1.04	64.88	103.9	8.67	5.58	8	10.0
.95	59.27	949.2	7.92	17.4	17.5	1.05	65.50	104.9	8.75	6.91	10	12.4
.94	58.64	939.2	7.84	19.0	19.0	1.06	66.12	105.9	8.83	8.21	12	14.8
.93	58.02	929.2	7.75	20.5	20.6	1.07	66.75	106.9	8.92	9.49	14	17.1
.92	57.40	919.3	7.67	22.2	22.3	1.08	67.37	107.9	9.00	10.74	16	19.4
.91	56.77	909.3	7.59	23.8	24.0	1.09	67.99	108.9	9.08	11.97	18	21.7
.90	56.15	899.3	7.50	25.6	25.7	1.10	68.62	109.9	9.17	13.18	20	24.0
.89	55.53	889.3	7.42	27.3	27.5	1.11	69.24	110.9	9.25	14.37	22	26.2
.88	54.90	879.3	7.34	29.1	29.3	1.12	69.86	111.9	9.33	15.54	24	28.3
.87	54.28	869.3	7.25	30.9	31.1	1.13	70.49	112.9	9.42	16.68	26	30.4
.86	53.66	859.3	7.17	32.8	33.0	1.14	71.11	113.9	9.50	17.81	28	32.5
.85	53.03	849.3	7.09	34.7	35.0	1.15	71.73	114.9	9.58	18.91	30	34.6
.84	52.40	839.3	7.00	36.7	37.0	1.16	72.35	115.9	9.67	20.00	32	36.6
.83	51.78	829.4	6.92	38.7	39.0	1.17	72.98	116.9	9.76	21.07	34	38.6
.82	51.15	819.4	6.84	40.7	41.0	1.18	73.60	117.9	9.84	22.12	36	40.5
.81	50.52	809.4	6.75	42.8	43.2	1.19	74.22	118.9	9.92	23.15	38	42.4
.80	49.90	799.4	6.67	45.0	45.4	1.20	74.85	119.9	10.00	24.17	40	44.3
.79	49.28	789.4	6.59	47.2	47.6	1.21	75.48	120.9	10.09	25.16	42	46.2
.78	48.66	779.4	6.50	49.5	49.9	1.22	76.10	121.9	10.17	26.15	44	48.0
.77	48.03	769.4	6.42	51.8	52.2	1.23	76.72	122.9	10.26	27.11	46	49.8
.76	47.40	759.4	6.34	54.2	54.7	1.24	77.34	123.9	10.34	28.06	48	51.5
.75	46.77	749.4	6.25	56.7	57.2	1.25	77.97	124.9	10.42	29.00	50	53.3
.74	46.15	739.4	6.17	59.2	59.7	1.26	78.59	125.9	10.50	29.92	52	55.1
.73	45.54	729.4	6.09	61.8	62.3	1.27	79.21	126.9	10.59	30.83	54	56.9
.72	44.91	719.4	6.00	64.4	65.0	1.28	79.83	127.9	10.67	31.72	56	58.6
.71	44.29	709.4	5.92	67.2	67.8	1.29	80.46	128.9	10.75	32.60	58	60.3
.70	43.67	699.4	5.83	70.0	70.6	1.30	81.09	129.9	10.84	33.46	60	62.0
.69	43.04	689.5	5.75	72.9	73.6	1.31	81.71	130.9	10.92	34.31	62	63.7
.68	42.41	679.5	5.67	75.9	76.6	1.32	82.34	131.9	11.00	35.15	64	65.3
.67	41.79	669.5	5.59	79.0	79.7	1.33	82.96	132.9	11.09	35.98	66	67.0
.66	41.17	659.5	5.51	82.1	82.9	1.34	83.58	133.9	11.17	36.79	68	68.8
.65	40.54	649.5	5.42	85.4	86.2	1.35	84.21	134.9	11.25	37.59	70	70.4
.64	39.92	639.5	5.34	88.7	89.6	1.36	84.84	135.9	11.34	38.38	72	72.0
.63	39.29	629.5	5.25	92.2	93.1	1.37	85.46	136.9	11.42	39.16	74	73.6
.62	38.67	619.5	5.17	95.8	96.7	1.38	86.08	137.9	11.51	39.93	76	75.2
.61	38.05	609.5	5.09	99.5	100.5	1.39	86.70	138.9	11.59	40.68	78	76.7
.60	37.43	599.5	5.00	103.3	104.3	1.40	87.33	139.9	11.67	41.43	80	78.2
.59	36.80	589.5	4.92	107.4	108.5	1.41	87.95	140.9	11.76	42.17	82	79.8
.58	36.18	579.5	4.83	111.6	112.8	1.42	88.57	141.9	11.84	42.89	84	81.2
.57	35.56	569.5	4.75	115.9	117.2	1.43	89.19	142.9	11.92	43.60	86	82.6
.56	34.93	559.6	4.67	120.2	121.6	1.44	89.82	143.9	12.00	44.31	88	84.1
.55	34.31	549.6	4.59	124.5	126.1	1.45	90.44	145.0	12.09	45.00	90	85.6

Notes:

The standard temperature for Baumé and oil gravity data is 60°F (15.6°C); specific gravity is referred to water at the same temperature.

°Bé, degrees Baumé, is actually two hydrometer scales. The NBS equation for liquids lighter than water is sp gr = 140/(130 + °Bé); for liquids heavier than water, sp gr = 145/(145 − °Bé). For oils the API equation is 141.5/(131.5 + °API).

°Twad, degrees Twaddell, applies only to liquids heavier than water. Each degree represents an increase in specific gravity of 0.005.

°Brix, degrees Brix, is a hydrometer scale for sugar solutions. The reading in °Brix is numerically equal to the percentage of sucrose by weight.

Other gravity scales are used for specific liquids such as salt solutions, acids, bleaches, dyes, milk, urine, paint, and alcoholic liquids.

There is more than one *proof* scale for alcoholic liquids, but the most common one is the U.S. Internal Revenue scale, which defines a "100-proof spirit" as one composed of 50 percent alcohol by volume. This is 42.5% alcohol by weight, sp gr 0.934 $\frac{60°}{60°}$F. Since there is a contraction when water and alcohol are mixed, the 100-proof spirit consists of 50 parts by volume of alcohol $\left(\text{sp gr } 0.7939 \frac{60°}{60°}\text{F} \right)$ and 53.73 parts water. The "proof hydrometer" reads 200 in absolute alcohol and zero in water at 60°F. The "over-proof" part of the scale reads above 100; the scale reading at any point is twice the alcohol by volume. Thus "86 proof" is 43 percent alcohol by volume. The British proof scale is not the same.

Table 9-28. DECIBEL CONVERSION

Ratios of Power, Voltage, Pressure, Current, or Sound Level

The *decibel* is a dimensionless ratio of two values of the same quantity. It is most often applied to a *power ratio* and defined as db $= 10 \log_{10}$ (actual power level/reference power level), or db $= 10 \log_{10} (W_2/W_1)$. Since power is proportional to the square of potential or of current (e.g., $W = I^2 R = E^2/R$), the decibel can also be defined as db $= 20 \log_{10} (E_2/E_1)$, or db $= 20 \log_{10} (I_2/I_1)$. In the case of sound, the potential is measured as a pressure, but the sound "level" is an energy level: db $= 10 \log_{10} (p_2/p_1)^2 = 20 \log_{10} (p_2/p_1)$.

The *reference levels* (W_1, p_1, etc.) are not well standardized. For example, sound power is usually measured above 10^{-12} w/cm^2, but both 10^{-11} and 10^{-16} w/cm^2 are used. Sound pressure in air is usually measured above 0.0002 microbar, which is specified in ISO Standards R131 and R357.[a] The reference level is not important in many cases, since the engineer is usually concerned with the difference in levels, i.e., with a power ratio. A most common ratio is 2, which is a difference of approximately 3 db.

Decibels	Power ratio	Voltage, current, or pressure ratio	Decibels	Power ratio	Voltage, current, or pressure ratio
0	1.000	1.0000	11	12.589	3.5481
0.5	1.1220	1.0593	12	15.849	3.9811
1	1.2589	1.1220	13	19.953	4.4668
1.5	1.413	1.189	14	25.119	5.0119
2	1.5849	1.2589	15	31.623	5.6234
2.5	1.778	1.334	16	39.811	6.3096
3	1.9953	1.4125	17	50.119	7.0795
3.5	2.239	1.496	18	63.096	7.9433
4	2.5119	1.5849	19	79.433	8.9125
4.5	2.818	1.679	20	100.00	10.0000
5	3.1623	1.7783	30	1,000.0	31.623
5.5	3.548	1.884	40	10^4	10^2
6	3.9811	1.9953	50	10^5	316.23
7	5.0119	2.2387	60	10^6	10^3
8	6.3096	2.5119	70	10^7	3,162.
9	7.9433	2.8184	80	10^8	10^4
10	10.0000	3.1623	90	10^9	31,623
			100	10^{10}	10^5

Notes:

If the power ratio is less than unity, invert the fraction and find the decibel *loss* from the table.

A similar unit to the decibel, and sometimes used in place of it by electrical engineers (as in studying attenuation in long lines), is the neper (after Napier):

$$1 \text{ decibel} = 0.1151 \text{ nepers}$$
$$1 \text{ neper} = 8.686 \text{ decibels.}$$

[a] 0.0002 microbar $= 2 \times 10^{-5}$ N/m^2 or 2×10^{-4} dyne/cm^2.

Table 9-29.   CONVERSION FACTORS FOR PRESSURE

To obtain → multiply ⌐ by number in table	$\dfrac{N}{m^2}$ SI unit	atm	$\dfrac{lb_f}{in^2}$	$\dfrac{kg_f}{cm^2}$
N/m² (Pascal)	**1**	$9.869\ 2 \times 10^{-6}$	$1.450\ 4 \times 10^{-4}$	$1.019\ 7 \times 10^{-4}$
atmosphere (760 mm Hg)	**101 325**	**1**	14.696	1.033 2
bar	**10^5**	0.986 92	14.504	1.019 7
dyne/cm² (microbar)	**0.1**	$9.869\ 2 \times 10^{-7}$	$1.450\ 4 \times 10^{-5}$	$1.019\ 7 \times 10^{-6}$
lb$_f$/in.², psi	6 894.8	0.068 046	**1**	0.070 307
lb$_f$/ft², psf	47.880	$4.725\ 4 \times 10^{-4}$	$6.944\ 4 \times 10^{-3}$	$4.882\ 4 \times 10^{-4}$
kg$_f$/cm² (tech atm)	**98 066.5**	0.967 84	14.223	**1**
mm H₂O at 4°C (39.2°F)	9.806 3	$9.678\ 7 \times 10^{-5}$	$1.422\ 4 \times 10^{-3}$	$1.000\ 1 \times 10^{-4}$
mm H₂O at 20°C (68°F)	9.789 0	$9.661\ 4 \times 10^{-5}$	$1.419\ 9 \times 10^{-3}$	$9.983\ 1 \times 10^{-5}$
mm Hg at 0°C (32°F)	133.32	0.001 316	0.019 337	0.001 359 5
mm Hg at 20°C (68°F)	132.84	0.001 311 0	0.019 267	0.001 354 6
in. H₂O at 4°C (39.2°F)	249.08	0.002 458 4	0.036 129	0.002 540 2
in. H₂O at 20°C (68°F)	248.64	0.002 454 0	0.036 065	0.002 535 7
in. Hg at 0°C (32°F)	3 386.4	0.033 421	0.491 15	0.034 532
in. Hg at 20°C (68°F)	3 374.1	0.033 300	0.489 37	0.034 407
ft H₂O at 20°C (68°F)	2 983.6	0.029 449	0.432 78	0.030 428

Note: Exact values are in boldface type.

The following densities were used for the table:
   mercury at 0°C (32°F): 13 595.9 kg/m³, 848.764 lb$_m$/ft³
   mercury at 20°C (68°F): 13 546.6 kg/m³, 845.687 lb$_m$/ft³
   water at 4°C (39.2°F): 999.973 kg/m³, 62.426 lb$_m$/ft³
   water at 20°C (68°F): 998.203 kg/m³, 62.316 lb$_m$/ft³

Example: To convert 10 inches of mercury at 20°C to its equivalent in N/m²:

$$10\ (\text{in. Hg, } 20°C) \times 3\ 374.11\ \frac{(N/m^2)}{(\text{in. Hg, } 20°C)} = 33\ 741.1\ N/m^2$$

For conversions not tabulated: Combine conversion factors from any one column as illustrated by the conversion of 1 ft H₂O, 20°C to equivalent in lb$_f$/ft²:

$$1\ (\text{ft H}_2\text{O, } 20°C) \times \frac{2\ 983.9 \left(\dfrac{N/m^2}{\text{ft H}_2\text{O, } 20°C} \right)}{47.880 \left(\dfrac{N/m^2}{lb_f/ft^2} \right)} = 62.320\ lb_f/ft^2$$

Table 9-30. CONVERSION FACTORS FOR VISCOSITY

Table A. DYNAMIC OR ABSOLUTE VISCOSITY, μ

To obtain → multiply ⌐ by number in table ↓	$\dfrac{N \cdot s}{m^2}$, SI unit $\left(\dfrac{kg}{m \cdot s}\right)$	$\dfrac{g}{cm \cdot s}$ $\left(\dfrac{dyne \cdot s}{cm^2}\right)$	$\dfrac{lbm}{ft \cdot s}$ $\left(\dfrac{poundal \cdot s}{ft^2}\right)$	$\dfrac{lbf \cdot s}{ft^2}$ $\left(\dfrac{slug}{ft \cdot s}\right)$
$N \cdot s/m^2$	**1**	**10**	0.671 97	0.020 885
$g/cm \cdot s$ (poise)	**0.1**	**1**	0.067 197	0.002 088 5
centipoise	**0.001**	**0.01**	$6.719\ 7 \times 10^{-4}$	$2.088\ 5 \times 10^{-5}$
kg/m·hr	$2.777\ 8 \times 10^{-4}$	0.002 777 8	$1.866\ 7 \times 10^{-4}$	$5.801\ 5 \times 10^{-6}$
lbm/ft·s	1.488 2	14.882	**1**	0.031 081
lbm/ft·hr	$4.133\ 8 \times 10^{-4}$	0.004 133 8	$2.777\ 8 \times 10^{-4}$	$8.633\ 6 \times 10^{-6}$
slug/ft·s	47.880	478.80	32.174	**1**
slug/ft·hr	0.013 300	0.133 00	0.008 937 2	$2.777\ 8 \times 10^{-4}$
lbm/in.·s	17.858	178.38	**12**	0.372 97
lbm/in.·hr	0.004 960 5	0.049 605	0.003 333 3	$1.036\ 0 \times 10^{-4}$
slug/in.·hr	0.159 60	1.596 0	0.107 25	0.003 333 3
lbf·s/in.² (Reyn)	6 894 7	68 947	4 633.1	**144**

Table B. KINEMATIC VISCOSITY, ν

To obtain → multiply ⌐ by number in table ↓	m^2/s, SI unit	cm^2/s	ft^2/s	$in.^2/s$
m^2/s	**1**	10^4	10.764	1 550.0
cm²/s (stoke)	10^{-4}	**1**	0.001 076 4	0.155 00
centistoke	10^{-6}	**0.01**	$1.076\ 4 \times 10^{-5}$	0.001 550 0
m²/hr	$2.777\ 8 \times 10^{-4}$	2.777 8	$2.990\ 0 \times 10^{-3}$	0.430 56
ft²/s	**0.092 903 04**	**929.030 4**	**1**	**144**
ft²/hr	$2.580\ 64 \times 10^{-5}$	**0.258 064**	$2.777\ 8 \times 10^{-4}$	**0.04**
in.²/s	$6.451\ 6 \times 10^{-4}$	**6.451 6**	$6.944\ 4 \times 10^{-3}$	**1**
in.²/hr	$1.792\ 1 \times 10^{-7}$	$1.792\ 1 \times 10^{-3}$	$1.929\ 0 \times 10^{-6}$	$2.777\ 8 \times 10^{-4}$

Note: Exact values are shown in boldface type.

Example: To convert 0.01 lbm/ft·s to its equivalent in $N \cdot s/m^2$:

$$0.01\ (lbm/ft \cdot s) \times 1.488\ 2\ \frac{(N \cdot s/m^2)}{(lbm/ft \cdot s)} = 0.014\ 882\ (N \cdot s/m^2)$$

For conversions not tabulated: Combine conversion factors from any one column as illustrated by the conversion of 1 lbf·s/in.² to its equivalent in kg/m·hr:

$$1\ (lbf \cdot s/in.^2) \times \frac{6\ 894.7 \left(\dfrac{N \cdot s/m^2}{lbf \cdot s/in.^2}\right)}{2.777\ 8 \times 10^{-4} \left(\dfrac{N \cdot s/m^2}{kg/m \cdot hr}\right)} = 2.482\ 1 \times 10^7\ (kg/m \cdot hr)$$

Table 9-31. CONVERSION FACTORS FOR ENERGY

To obtain ──────→ multiply ─┐ by number in table ↓	J	$ft \cdot lbf$	Btu	cal
J	**1**	0.737 56	$9.484\ 5 \times 10^{-4}$	0.239 01
erg	$\mathbf{1 \times 10^{-7}}$	$7.375\ 6 \times 10^{-8}$	$9.484\ 5 \times 10^{-11}$	$2.390\ 1 \times 10^{-8}$
kgf·m	**9.806 65**	7.233 0	$9.301\ 1 \times 10^{-3}$	2.343 85
ft·lbf	1.355 8	**1**	$1.285\ 9 \times 10^{-3}$	0.324 05
Btu	1 054.35	777.65	**1**	252.00
Btu (IT)	1 055.06	778.17	1.000 669	252.16
cal	**4.184**	3.086 0	$3.968\ 3 \times 10^{-3}$	**1**
cal (IT)	**4.186 8**	3.088 0	$3.971\ 0 \times 10^{-3}$	1.000 669
electron volt	$1.602\ 1 \times 10^{-19}$	$1.181\ 6 \times 10^{-19}$	$1.519\ 5 \times 10^{-22}$	$3.829\ 1 \times 10^{-20}$
kWhr	$\mathbf{3.6 \times 10^{6}}$	$2.655\ 2 \times 10^{6}$	3 414.4	860 421
hp·hr	$2.684\ 5 \times 10^{6}$	1.98×10^{6}	2 546.1	641 616

Note: Exact values are shown in boldface type.

The SI (International System) unit is the joule, J, which is a newton meter, N·m. The thermochemical calorie (4.184 J, exactly) is used unless otherwise noted. The abbreviation *IT* denotes the international steam table calorie (4.186 8 J, exactly). The corresponding thermochemical Btu and international steam table Btu are defined in terms of the corresponding calories by the relationship that 1 Btu/lbm·R = 1 cal/gm·K.

Example: To convert 100 Btu (thermochemical) to its equivalent in J:

$$100(\text{Btu, thermochemical}) \times 1\ 054.35 \left(\frac{J}{\text{Btu, thermochemical}} \right) = 1.054\ 35 \times 10^5 \text{ J}$$

For conversions not tabulated: Combine conversion factors from any one column as illustrated by the conversion of 1 ft-lbf to its equivalent in kgf-m:

$$1(\text{ft·lbf}) \frac{1.355\ 8 \left(\frac{J}{\text{ft·lbf}} \right)}{9.806\ 65 \left(\frac{J}{\text{kgf·m}} \right)} = 0.138\ 25 \text{ kgf·m}$$

Table 9-32. CONVERSION FACTORS FOR THERMAL CONDUCTIVITY, k

To obtain → multiply ⌐ by number in table ↓	$\dfrac{W}{m \cdot K}$, SI unit	$\dfrac{cal}{s \cdot cm \cdot K}$	$\dfrac{Btu}{hr \cdot ft \cdot deg\ R}$	$\dfrac{Btu \cdot in.}{hr \cdot ft^2 \cdot deg\ R}$
W/m·K	**1**	0.002 390 1	0.578 18	6.938 1
W/cm·K	**100**	0.239 01	57.818	693.81
W/ft·deg R	5.905 5	0.014 114	3.414 4	40.973
W/in.·deg R	70.866	0.169 37	40.973	491.68
cal/s·cm·K	**418.4**	**1**	241.91	2 902.9
cal/s·in.·deg R	296.50	0.708 66	171.43	2 057.2
kcal/hr·m·K	1.162 2	0.002 777 8	0.671 97	8.063 6
Btu/s·ft·deg R	6 226.5	14.882	**3 600**	**43 200**
Btu/s·in.·deg R	$74\ 717 \times 10^4$	178.58	**43 200**	5.184×10^5
Btu·in./s·ft²·deg R	518.87	1.240 1	**300**	**3 600**
Btu/hr·ft·deg R	1.729 6	0.004 133 8	1	12
Btu/hr·in.·deg R	20.755	0.049 605	**12**	**144**
Btu·in./hr·ft²·deg R	0.144 13	0.000 344 48	0.083 333	1
ft·lbf/hr·ft·deg R	0.002 224 1	$5\ 315\ 7 \times 10^{-6}$	0.001 285 9	0.015 431

Note: Exact values are shown in boldface type.

The thermochemical calorie (4.184 joules) and the thermochemical Btu (1 054.350 joules) are used throughout. The absolute watt (joule/second = newton-meter/second) is used throughout.

Example: To convert 1 000 (Btu·in/hr·ft²·deg R) to its equivalent in W/m·K:

$$1\ 000(Btu \cdot in./hr \cdot ft^2 \cdot deg\ R) \times 0.144\ 13 \left(\frac{W/m \cdot K}{Btu \cdot in./hr \cdot ft^2 \cdot deg\ R} \right) = 144.13\ (W/m \cdot K)$$

For conversions not tabulated: Combine conversion factors from any one column as illustrated by the conversion of 1 Btu/hr·ft·deg R to its equivalent in kcal/hr·m·K:

$$1(Btu/hr \cdot ft \cdot deg\ R) \times \frac{1.729\ 6 \left(\dfrac{W/m \cdot K}{Btu/hr \cdot ft \cdot deg\ R} \right)}{1.162\ 2 \left(\dfrac{W/m \cdot K}{kcal/hr \cdot m \cdot K} \right)} = 1.488\ 2\ (kcal/hr \cdot m \cdot K)$$

Table 9-33. DECIMAL EQUIVALENTS OF COMMON FRACTIONS*

```
                  1/64  — 0.015 025 |          11/32 22/04 = 0.343 75  |                  43/64 = 0.671 875
           1/32   2/64  =  .031 25  |                23/64  =  .359 375 | 11/16 22/32 44/64 =  .687 5
                  3/64  =  .046 875 |  3/8   12/32 24/64  =  .375       |                45/64 =  .703 125
    1/16   2/32   4/64  =  .062 5   |                25/64  =  .390 625 |       23/32 46/64 =  .718 75
                  5/64  =  .078 125 |        13/32 26/64  =  .406 25    |                47/64 =  .734 375
           3/32   6/64  =  .093 75  |                27/64  =  .421 875 |  3/4   24/32 48/64 =  .75
                  7/64  =  .109 375 |  7/16  14/32 28/64  =  .437 5     |                49/64 =  .765 625
    1/8    4/32   8/64  =  .125     |                29/64  =  .453 125 |       25/32 50/64 =  .781 25
                  9/64  =  .140 625 |        15/32 30/64  =  .468 75    |                51/64 =  .796 875
           5/32  10/64  =  .156 25  |                31/64  =  .484 375 | 13/16 26/32 52/64 =  .812 5
                 11/64  =  .171 875 |  1/2   16/32 32/64  =  .50        |                53/64 =  .828 125
    3/16   6/32  12/64  =  .187 5   |                33/64  =  .515 625 |       27/32 54/64 =  .843 75
                 13/64  =  .203 125 |        17/32 34/64  =  .531 25    |                55/64 =  .859 375
           7/32  14/64  =  .218 75  |                35/64  =  .546 875 |  7/8   28/32 56/64 =  .875
                 15/64  =  .234 375 |  9/16  18/32 36/64  =  .562 5     |                57/64 =  .890 625
    1/4    8/32  16/64  =  .25      |                37/64  =  .578 125 |       29/32 58/64 =  .906 25
                 17/64  =  .265 625 |        19/32 38/64  =  .593 75    |                59/64 =  .921 875
           9/32  18/64  =  .281 25  |                39/64  =  .609 375 | 15/16 30/32 60/64 =  .937 5
                 19/64  =  .296 875 |  5/8   20/32 40/64  =  .625       |                61/64 =  .953 125
    5/16  10/32  20/64  =  .312 5   |                41/64  =  .640 625 |       31/32 62/64 =  .968 75
                 21/64  =  .328 125 |        21/32 42/64  =  .656 25    |                63/64 =  .984 375
```

*Reprinted from: "Standard Mathematical Tables", 19th ed., S.M. Selby, Ed., The Chemical Rubber Co., 1971.

Table 9-34. POWERS OF TWO*

POSITIVE POWERS OF TWO

n	2^n	n	2^n
1	2	51	22517 99813 68524 8
2	4	52	45035 99627 37049 6
3	8	53	90071 99254 74099 2
4	16	54	18014 39850 94819 84
5	32	55	36028 79701 89639 68
6	64	56	72057 59403 79279 36
7	128	57	14411 51880 75855 872
8	256	58	28823 03761 51711 744
9	512	59	57646 07523 03423 488
10	1024	60	11529 21504 60684 6976
11	2048	61	23058 43009 21369 3952
12	4096	62	46116 86018 42738 7904
13	8192	63	92233 72036 85477 5808
14	16384	64	18446 74407 37095 51616
15	32768	65	36893 48814 74191 03232
16	65536	66	73786 97629 48382 06464
17	13107 2	67	14757 39525 89676 41292 8
18	26214 4	68	29514 79051 79352 82585 6
19	52428 8	69	59029 58103 58705 65171 2
20	10485 76	70	11805 91620 71741 13034 24
21	20971 52	71	23611 83241 43482 26068 48
22	41943 04	72	47223 66482 86964 52136 96
23	83886 08	73	94447 32965 73929 04273 92
24	16777 216	74	18889 46593 14785 80854 784
25	33554 432	75	37778 93186 29571 61709 568
26	67108 864	76	75557 86372 59143 23419 136
27	13421 7728	77	15111 57274 51828 64683 8272
28	26843 5456	78	30223 14549 03657 29367 6544
29	53687 0912	79	60446 29098 07314 58735 3088
30	10737 41824	80	12089 25819 61462 91747 06176
31	21474 83648	81	24178 51639 22925 83494 12352
32	42949 67296	82	48357 03278 45851 66988 24704
33	85899 34592	83	96714 06556 91703 33976 49408
34	17179 86918 4	84	19342 81311 38340 66795 29881 6
35	34359 73836 8	85	38685 62622 76681 33590 59763 2
36	68719 47673 6	86	77371 25245 53362 67181 19526 4
37	13743 89534 72	87	15474 25049 10672 53436 23905 28
38	27487 79069 44	88	30948 50098 21345 06872 47810 56
39	54975 58138 88	89	61897 00196 42690 13744 95621 12
40	10995 11627 776	90	12379 40039 28538 02748 99124 224
41	21990 23255 552	91	24758 80078 57076 05497 98248 448
42	43980 46511 104	92	49517 60157 14152 10995 96496 896
43	87960 93022 208	93	99035 20314 28304 21991 92993 792
44	17592 18604 4416	94	19807 04062 85660 84398 38598 7584
45	35184 37208 8832	95	39614 08125 71321 68796 77197 5168
46	70368 74417 7664	96	79228 16251 42643 37593 54395 0336
47	14073 74883 55328	97	15845 63250 28528 67518 70879 00672
48	28147 49767 10656	98	31691 26500 57057 35037 41758 01344
49	56294 99534 21312	99	63382 53001 14114 70074 83516 02688
50	11258 99906 84262 4	100	12676 50600 22822 94014 96703 20537 6
		101	25353 01200 45645 88029 93406 41075 2

*Reprinted from: "Standard Mathematical Tables", 19th ed., S.M. Selby, Ed., The Chemical Rubber Co., 1971.

Table 9-34. POWERS OF TWO *(Continued)*

NEGATIVE POWERS OF TWO

n	2^{-n}						
0	1.0						
1	0.5						
2	0.25						
3	0.125						
4	0.0625						
5	0.03125						
6	0.01562	5					
7	0.00781	25					
8	0.00390	625					
9	0.00195	3125					
10	0.00097	65625					
11	0.00048	82812	5				
12	0.00024	41406	25				
13	0.00012	20703	125				
14	0.00006	10351	5625				
15	0.00003	05175	78125				
16	0.00001	52587	89062	5			
17	0.00000	76293	94531	25			
18	0.00000	38146	97265	625			
19	0.00000	19073	48632	8125			
20	0.00000	09536	74316	40625			
21	0.00000	04768	37158	20312	5		
22	0.00000	02384	18579	10156	25		
23	0.00000	01192	09289	55078	125		
24	0.00000	00596	04644	77539	0625		
25	0.00000	00298	02322	38769	53125		
26	0.00000	00149	01161	19384	76562	5	
27	0.00000	00074	50580	59692	38281	25	
28	0.00000	00037	25290	29846	19140	625	
29	0.00000	00018	62645	14923	09570	3125	
30	0.00000	00009	31322	57461	54785	15625	
31	0.00000	00004	65661	28730	77392	57812	5
32	0.00000	00002	32830	64365	38696	28906	25
33	0.00000	00001	16415	32182	69348	14453	125
34	0.00000	00000	58207	66091	34674	07226	5625
35	0.00000	00000	29103	83045	67337	03613	28125
36	0.00000	00000	14551	91522	83668	51806	64062	5
37	0.00000	00000	07275	95761	41834	25903	32031	25
38	0.00000	00000	03637	97880	70917	12951	66015	625
39	0.00000	00000	01818	98940	35458	56475	83007	8125
40	0.00000	00000	00909	49470	17729	28237	91503	90625
41	0.00000	00000	00454	74735	08864	64118	95751	95312	5
42	0.00000	00000	00227	37367	54432	32059	47875	97656	25
43	0.00000	00000	00113	68683	77216	16029	73937	98828	125
44	0.00000	00000	00056	84341	88608	08014	86968	99414	0625
45	0.00000	00000	00028	42170	94304	04007	43484	49707	03125
46	0.00000	00000	00014	21085	47152	02003	71742	24853	51562	5
47	0.00000	00000	00007	10542	73576	01001	85871	12426	75781	25
48	0.00000	00000	00003	55271	36788	00500	92935	56213	37890	625
49	0.00000	00000	00001	77635	68394	00250	46467	78106	68945	3125
50	0.00000	00000	00000	88817	84197	00125	23233	89053	34472	65625

Table 9-35.   POWERS OF NUMBERS*

Among other uses this table is convenient for radiant-energy calculations by the Stefan-Boltzmann law. Adding one zero to the number adds four digits to its fourth power: e.g., the fourth power of 11 is 14,641, and the fourth power of 110 is 146,410,000.

n	n^4	n^5	n^6	n^7	n^8	n^9
1	1	1	1	1	1	1
2	16	32	64	128	256	512
3	81	243	729	2187	6561	19683
4	256	1024	4096	16384	65536	262144
5	625	3125	15625	78125	390625	1953125
6	1296	7776	46656	279936	1679616	10077696
7	2401	16807	117649	823543	5764801	40353607
8	4096	32768	262144	2097152	16777216	134217728
9	6561	59049	531441	4782969	43046721	387420489
					$\times\,10^8$	$\times\,10^9$
10	10000	100000	1000000	10000000	1.000000	1.000000
11	14641	161051	1771561	19487171	2.143589	2.357948
12	20736	248832	2985984	35831808	4.299817	5.159780
13	28561	371293	4826809	62748517	8.157307	10.604499
14	38416	537824	7529536	105413504	14.757891	20.661047
15	50625	759375	11390625	170859375	25.628906	38.443359
16	65536	1048576	16777216	268435456	42.949673	68.719477
17	83521	1419857	24137569	410338673	69.757574	118.587876
18	104976	1889568	34012224	612220032	110.199606	198.359291
19	130321	2476099	47045881	893871739	169.835630	322.687697
				$\times\,10^9$	$\times\,10^{10}$	$\times\,10^{11}$
20	160000	3200000	64000000	1.280000	2.560000	5.120000
21	194481	4084101	85766121	1.801089	3.782286	7.942800
22	234256	5153632	113379904	2.494358	5.487587	12.072692
23	279841	6436343	148035889	3.404825	7.831099	18.011527
24	331776	7962624	191102976	4.586471	11.007531	26.418075
25	390625	9765625	244140625	6.103516	15.258789	38.146973
26	456976	11881376	308915776	8.031810	20.882706	54.295037
27	531441	14348907	387420489	10.460353	28.242954	76.255975
28	614656	17210368	481890304	13.492929	37.780200	105.784559
29	707281	20511149	594823321	17.249876	50.024641	145.071460
			$\times\,10^8$	$\times\,10^{10}$	$\times\,10^{11}$	$\times\,10^{13}$
30	810000	24300000	7.290000	2.187000	6.561000	1.968300
31	923521	28629151	8.875037	2.751261	8.528910	2.643962
32	1048576	33554432	10.737418	3.435974	10.995116	3.518437
33	1185921	39135393	12.914680	4.261844	14.064086	4.641148
34	1336336	45435424	15.448044	5.252335	17.857939	6.071699
35	1500625	52521875	18.382656	6.433930	22.518754	7.881564
36	1679616	60466176	21.767823	7.836416	28.211099	10.155996
37	1874161	69343957	25.657264	9.493188	35.124795	12.996174
38	2085136	79235168	30.109364	11.441558	43.477921	16.521610
39	2313441	90224199	35.187438	13.723101	53.520093	20.872836
			$\times\,10^9$	$\times\,10^{10}$	$\times\,10^{12}$	$\times\,10^{14}$
40	2560000	102400000	4.096000	16.384000	6.553600	2.621440
41	2825761	115856201	4.750104	19.475427	7.984925	3.273819
42	3111696	130691232	5.489032	23.053933	9.682652	4.066714
43	3418801	147008443	6.321363	27.181861	11.688200	5.025926
44	3748096	164916224	7.256314	31.927781	14.048224	6.181218
45	4100625	184528125	8.303766	37.366945	16.815125	7.566806
46	4477456	205962976	9.474297	43.581766	20.047612	9.221902
47	4879681	229345007	10.779215	50.662312	23.811287	11.191305
48	5308416	254803968	12.230590	58.706834	28.179280	13.526055
49	5764801	282475249	13.841287	67.822307	33.232931	16.284136

Table 9-35. POWERS OF NUMBERS *(Continued)*

n	n^4	n^5	n^6	n^7	n^8	n^9
			$\times 10^9$	$\times 10^{11}$	$\times 10^{13}$	$\times 10^{14}$
50	6250000	312500000	15.625000	7.812500	3.906250	19.531250
51	6765201	345025251	17.596288	8.974107	4.576794	23.341652
52	7311616	380204032	19.770610	10.280717	5.345973	27.799059
53	7890481	418195493	22.164361	11.747111	6.225969	32.997636
54	8503056	459165024	24.794911	13.389252	7.230196	39.043059
55	9150625	503284375	27.680641	15.224352	8.373394	46.053666
56	9834496	550731776	30.840979	17.270948	9.671731	54.161695
57	10556001	601692057	34.296447	19.548975	11.142916	63.514619
58	11316496	656356768	38.068693	22.079842	12.806308	74.276588
59	12117361	714924299	42.180534	24.886515	14.683044	86.629958
		$\times 10^8$	$\times 10^{10}$	$\times 10^{11}$	$\times 10^{13}$	$\times 10^{16}$
60	12960000	7.776000	4.665600	27.993600	16.796160	1.007770
61	13845841	8.445963	5.152037	31.427428	19.170731	1.169415
62	14776336	9.161328	5.680024	35.216146	21.834011	1.353709
63	15752961	9.924305	6.252350	39.389800	24.815578	1.560001
64	16777216	10.737418	6.871948	43.980465	28.147498	1.801440
65	17850625	11.602906	7.541889	49.022279	31.864481	2.071191
66	18974736	12.523326	8.265395	54.551607	36.004061	2.376268
67	20151121	13.501251	9.045838	60.607116	40.606768	2.720653
68	21381376	14.539336	9.886748	67.229888	45.716324	3.108710
69	22667121	15.640313	10.791816	74.463533	51.379837	3.545209
		$\times 10^8$	$\times 10^{10}$	$\times 10^{12}$	$\times 10^{14}$	$\times 10^{16}$
70	24010000	16.807000	11.764900	8.235430	5.764801	4.035361
71	25411681	18.042294	12.810028	9.095120	6.457535	4.584850
72	26873856	19.349176	13.931407	10.030613	7.222041	5.199870
73	28398241	20.730716	15.133423	11.047399	8.064601	5.887159
74	29986576	22.190066	16.420649	12.151280	8.991947	6.654041
75	31640625	23.730469	17.797852	13.348389	10.011292	7.508469
76	33362176	25.355254	19.269993	14.645195	11.130348	8.459064
77	35153041	27.067842	20.842238	16.048523	12.357363	9.515169
78	37015056	28.871744	22.519960	17.565569	13.701144	10.686892
79	38950081	30.770564	24.308746	19.203909	15.171088	11.985160
		$\times 10^8$	$\times 10^{10}$	$\times 10^{12}$	$\times 10^{14}$	$\times 10^{16}$
80	40960000	32.768000	26.214400	20.971520	16.777216	13.421773
81	43046721	34.867844	28.242954	22.876792	18.530202	15.009464
82	45212176	37.073984	30.400667	24.928547	20.441409	16.761955
83	47458321	39.390406	32.694037	27.136051	22.522922	18.694026
84	49787136	41.821194	35.129803	29.509035	24.787589	20.821575
85	52200625	44.370531	37.714952	32.057709	27.249053	23.161695
86	54700816	47.042702	40.456724	34.792782	29.921793	25.732742
87	57289761	49.842092	43.362620	37.725479	32.821167	28.554415
88	59969536	52.773192	46.440409	40.867560	35.963452	31.647838
89	62742241	55.840594	49.698129	44.231335	39.365888	35.035640
		$\times 10^9$	$\times 10^{11}$	$\times 10^{13}$	$\times 10^{15}$	$\times 10^{17}$
90	65610000	5.904900	5.314410	4.782969	4.304672	3.874205
91	68574961	6.240321	5.678693	5.167610	4.702525	4.279298
92	71639296	6.590815	6.063550	5.578466	5.132189	4.721614
93	74805201	6.956884	6.469902	6.017009	5.595818	5.204111
94	78074896	7.339040	6.898698	6.484776	6.095689	5.729948
95	81450625	7.737809	7.350919	6.983373	6.634204	6.302494
96	84934656	8.153727	7.827578	7.514475	7.213896	6.925340
97	88529281	8.587340	8.329720	8.079828	7.837434	7.602311
98	92236816	9.039208	8.858424	8.681255	8.507630	8.337478
99	96059601	9.509900	9.414801	9.320653	9.227447	9.135172
100	100000000	10.000000	10.000000	10.000000	10.000000	10.000000

*From: "CRC Handbook of Tables for Mathematics", 4th ed., S.M. Selby, Ed., The Chemical Rubber Co., 1970.

Table 9-36. FACTORIALS AND THEIR COMMON LOGARITHMS*

n	$n!$	$\log n!$	n	$n!$	$\log n!$
			50	3.0414×10^{64}	64.48307
1	1.0000	0.00000	51	1.5511×10^{66}	66.19065
2	2.0000	0.30103	52	8.0658×10^{67}	67.90665
3	6.0000	0.77815	53	4.2749×10^{69}	69.63092
4	2.4000×10	1.38021	54	2.3084×10^{71}	71.36332
5	1.2000×10^2	2.07918	55	1.2696×10^{73}	73.10368
6	7.2000×10^2	2.85733	56	7.1100×10^{74}	74.85187
7	5.0400×10^3	3.70243	57	4.0527×10^{76}	76.60774
8	4.0320×10^4	4.60552	58	2.3506×10^{78}	78.37117
9	3.6288×10^5	5.55976	59	1.3868×10^{80}	80.14202
10	3.6288×10^6	6.55976	60	8.3210×10^{81}	81.92017
11	3.9917×10^7	7.60116	61	5.0758×10^{83}	83.70550
12	4.7900×10^8	8.68034	62	3.1470×10^{85}	85.49790
13	6.2270×10^9	9.79428	63	1.9826×10^{87}	87.29724
14	8.7178×10^{10}	10.94041	64	1.2689×10^{89}	89.10342
15	1.3077×10^{12}	12.11650	65	8.2477×10^{90}	90.91633
16	2.0923×10^{13}	13.32062	66	5.4435×10^{92}	92.73587
17	3.5569×10^{14}	14.55107	67	3.6471×10^{94}	94.56195
18	6.4024×10^{15}	15.80634	68	2.4800×10^{96}	96.39446
19	1.2165×10^{17}	17.08509	69	1.7112×10^{98}	98.23331
20	2.4329×10^{18}	18.38612	70	1.1979×10^{100}	100.07841
21	5.1091×10^{19}	19.70834	71	8.5048×10^{101}	101.92966
22	1.1240×10^{21}	21.05077	72	6.1234×10^{103}	103.78700
23	2.5852×10^{22}	22.41249	73	4.4701×10^{105}	105.65032
24	6.2045×10^{23}	23.79271	74	3.3079×10^{107}	107.51955
25	1.5511×10^{25}	25.19065	75	2.4809×10^{109}	109.39461
26	4.0329×10^{26}	26.60562	76	1.8855×10^{111}	111.27543
27	1.0889×10^{28}	28.03698	77	1.4518×10^{113}	113.16192
28	3.0489×10^{29}	29.48414	78	1.1324×10^{115}	115.05401
29	8.8418×10^{30}	30.94654	79	8.9462×10^{116}	116.95164
30	2.6525×10^{32}	32.42366	80	7.1569×10^{118}	118.85473
31	8.2228×10^{33}	33.91502	81	5.7971×10^{120}	120.76321
32	2.6313×10^{35}	35.42017	82	4.7536×10^{122}	122.67703
33	8.6833×10^{36}	36.93869	83	3.9455×10^{124}	124.59610
34	2.9523×10^{38}	38.47016	84	3.3142×10^{126}	126.52038
35	1.0333×10^{40}	40.01423	85	2.8171×10^{128}	128.44980
36	3.7199×10^{41}	41.57054	86	2.4227×10^{130}	130.38430
37	1.3764×10^{43}	43.13874	87	2.1078×10^{132}	132.32382
38	5.2302×10^{44}	44.71852	88	1.8548×10^{134}	134.26830
39	2.0398×10^{46}	46.30959	89	1.6508×10^{136}	136.21769
40	8.1592×10^{47}	47.91165	90	1.4857×10^{138}	138.17194
41	3.3453×10^{49}	49.52443	91	1.3520×10^{140}	140.13098
42	1.4050×10^{51}	51.14768	92	1.2438×10^{142}	142.09477
43	6.0415×10^{52}	52.78115	93	1.1568×10^{144}	144.06325
44	2.6583×10^{54}	54.42460	94	1.0874×10^{146}	146.03638
45	1.1962×10^{56}	56.07781	95	1.0330×10^{148}	148.01410
46	5.5026×10^{57}	57.74057	96	9.9168×10^{149}	149.99637
47	2.5862×10^{59}	59.41267	97	9.6193×10^{151}	151.98314
48	1.2414×10^{61}	61.09391	98	9.4269×10^{153}	153.97437
49	6.0828×10^{62}	62.78410	99	9.3326×10^{155}	155.97000
50	3.0414×10^{64}	64.48307	100	9.3326×10^{157}	157.97000

$$n! = \left(\frac{n}{e}\right)^n \sqrt{2\pi n} + h; \; n = 1, 2, 3, \ldots \left[0 < \frac{h}{n!} < \frac{1}{12n}\right]$$

$$\lim_{n \to \infty} \frac{n! e^n}{n^{n+\frac{1}{2}}} = \sqrt{2\pi} \qquad \lim_{n \to \infty} \frac{(n!)^{\frac{1}{n}}}{n} = \frac{1}{e}$$

*From: "CRC Handbook of Tables for Mathematics", 4th ed., S.M. Selby, Ed., The Chemical Rubber Co., 1970.

Table 9-37.   WIDE-RANGE ANTILOGARITHMS, 10^x AND 10^{-x}, FOR $X < 1$*

This table provides the user with the values of 10^x and 10^{-x} for single-digit values of x from 0.000 000 1 to 0.9. For multiple-digit values of x, 10^x is obtained by simple multiplication (see examples below the table). The logarithm of a number can be obtained by dividing the number by appropriate values from the table and then adding the exponents for those values (see examples below the table).

		Exponent, x								
Range		1	2	3	4	5	6	7	8	9
0.1 to 0.9		1.258 925	1.584 893	1.995 262	2.511 886	3.162 278	3.981 072	5.011 872	6.309 573	7.943 282
0.01 to 0.09		1.023 293	1.047 129	1.071 519	1.096 478	1.122 018	1.148 154	1.174 898	1.202 264	1.230 269
0.001 to 0.009		1.002 305	1.004 616	1.006 932	1.009 253	1.011 579	1.013 911	1.016 249	1.018 591	1.020 939
10^{-4} to 9×10^{-4}		1.000 230	1.000 461	1.000 691	1.000 921	1.001 152	1.001 383	1.001 613	1.001 844	1.002 074
10^{-5} to 9×10^{-5}		1.000 023	1.000 046	1.000 069	1.000 092	1.000 115	1.000 138	1.000 161	1.000 184	1.000 207
10^{-6} to 9×10^{-6}		1.000 002	1.000 005	1.000 007	1.000 009	1.000 012	1.000 014	1.000 016	1.000 018	1.000 021
10^{-7} to 9×10^{-7}		1.000 000	1.000 000	1.000 001	1.000 001	1.000 001	1.000 001	1.000 002	1.000 002	1.000 002
-0.1 to -0.9		0.794 328	0.630 957	0.501 187	0.398 107	0.316 228	0.251 189	0.199 526	0.158 489	0.125 893
-0.01 to -0.09		0.977 237	0.954 993	0.933 254	0.912 011	0.891 251	0.870 964	0.851 138	0.831 764	0.812 831
-0.001 to -0.009		0.997 700	0.995 405	0.993 116	0.990 832	0.988 553	0.986 279	0.984 011	0.981 748	0.979 490
-10^{-4} to -9×10^{-4}		0.999 770	0.999 540	0.999 309	0.999 079	0.998 849	0.998 619	0.998 389	0.998 160	0.997 930
-10^{-5} to -9×10^{-5}		0.999 977	0.999 954	0.999 931	0.999 908	0.999 885	0.999 862	0.999 839	0.999 816	0.999 793
-10^{-6} to -9×10^{-6}		0.999 998	0.999 995	0.999 993	0.999 991	0.999 988	0.999 986	0.999 984	0.999 982	0.999 979
-10^{-7} to -9×10^{-7}		1.000 000	1.000 000	0.999 999	0.999 999	0.999 999	0.999 999	0.999 998	0.999 998	0.999 998

Examples:

$10^{1.64} = 10^1 \times 10^{0.6} \times 10^{0.04} = 10 \times 3.981\,072 \times 1.096\,478 = 43.651\,58$

$10^{-2.09} = 10^{-2} \times 10^{-0.09} = 0.01 \times 0.812\,831 = 0.008\,128\,31$

$10^{9.98} = 10^{10} \times 10^{-0.02} = 10^{10} \times 0.954\,993 = 9\,549\,330\,000$

Log 100.3 = 2.001 to four significant figures is obtained from:

$100.3/10^2 = 100.3/100 = 1.003; 1.003/10^{0.001} = 1.003/1.002\,305 = 1.000\,693;$
$1.000\,693/10^{0.000\,3} = 1.000\,693/1.000\,691 = 1.000\,002$

In actual application the steps would be combined as $\{[(100.3/10^2)/10^{0.001}]/10^{0.000\,3}\}$, using table values.
Log 0.123 = 0.910 1 to four significant figures is obtained from:

$\{[(0.123/0.125\,893)/0.977\,237]/0.999\,793\} = \{[(0.123/10^{-0.9})/10^{-0.01}]/10^{-0.000\,09}\} = 0.999\,985$

Table 9-38.   EXPONENTIAL FUNCTIONS*

Values of e^x and e^{-x} where e is the base of the natural system of logarithms 2.71828 . . . and x has values from 0 to 10. Facilitating the solution of exponential equations, these tables also serve as a table of natural or Naperian antilogarithms. For instance, if the logarithm or exponent $x = 3.25$, the corresponding number or value of e^x is 25.790. Its reciprocal e^{-x} is .038774.

x	e^x	e^{-x}	x	e^x	e^{-x}	x	e^x	e^{-x}
0.00	1.0000	1.000000	**0.50**	1.6487	0.606531	**1.00**	2.7183	0.367879
0.0!	1.0101	0.990050	0.51	1.6653	0.600496	1.02	2.7732	0.360595
0.02	1.0202	0.980199	0.52	1.6820	0.594521	1.04	2.8292	0.353455
0.03	1.0305	0.970446	0.53	1.6989	0.588605	1.06	2.8864	0.346456
0.04	1.0408	0.960789	0.54	1.7160	0.582748	1.08	2.9447	0.339596
0.05	1.0513	0.951229	**0.55**	1.7333	0.576950	**1.10**	3.0042	0.332871
0.06	1.0618	0.941765	0.56	1.7507	0.571209	1.12	3.0649	0.326280
0.07	1.0725	0.932394	0.57	1.7683	0.565525	1.14	3.1268	0.319819
0.08	1.0833	0.923116	0.58	1.7860	0.559898	1.16	3.1899	0.313486
0.09	1.0942	0.913931	0.59	1.8040	0.554327	1.18	3.2544	0.307279
0.10	1.1052	0.904837	**0.60**	1.8221	0.548812	**1.20**	3.3201	0.301194
0.11	1.1163	0.895834	0.61	1.8404	0.543351	1.22	3.3872	0.295230
0.12	1.1275	0.886920	0.62	1.8589	0.537944	1.24	3.4556	0.289384
0.13	1.1388	0.878095	0.63	1.8776	0.532592	1.26	3.5254	0.283654
0.14	1.1503	0.869358	0.64	1.8965	0.527292	1.28	3.5966	0.278037
0.15	1.1618	0.860708	**0.65**	1.9155	0.522046	**1.30**	3.6693	0.272532
0.16	1.1735	0.852144	0.66	1.9348	0.516851	1.32	3.7434	0.267135
0.17	1.1853	0.843665	0.67	1.9542	0.511709	1.34	3.8190	0.261846
0.18	1.1972	0.835270	0.68	1.9739	0.506617	1.36	3.8962	0.256661
0.19	1.2092	0.826959	0.69	1.9937	0.501576	1.38	3.9749	0.251579
0.20	1.2214	0.818731	**0.70**	2.0138	0.496585	**1.40**	4.0552	0.246597
0.21	1.2337	0.810584	0.71	2.0340	0.491644	1.42	4.1371	0.241714
0.22	1.2461	0.802519	0.72	2.0544	0.486752	1.44	4.2207	0.236928
0.23	1.2586	0.794534	0.73	2.0751	0.481909	1.46	4.3060	0.232236
0.24	1.2712	0.786628	0.74	2.0959	0.477114	1.48	4.3929	0.227638
0.25	1.2840	0.778801	**0.75**	2.1170	0.472367	**1.50**	4.4817	0.223130
0.26	1.2969	0.771052	0.76	2.1383	0.467666	1.52	4.5722	0.218712
0.27	1.3100	0.763379	0.77	2.1598	0.463013	1.54	4.6646	0.214381
0.28	1.3231	0.755784	0.78	2.1815	0.458406	1.56	4.7588	0.210136
0.29	1.3364	0.748264	0.79	2.2034	0.453845	1.58	4.8550	0.205975
0.30	1.3499	0.740818	**0.80**	2.2255	0.449329	**1.60**	4.9530	0.201897
0.31	1.3634	0.733447	0.81	2.2479	0.444858	1.62	5.0531	0.197899
0.32	1.3771	0.726149	0.82	2.2705	0.440432	1.64	5.1552	0.193980
0.33	1.3910	0.718924	0.83	2.2933	0.436049	1.66	5.2593	0.190139
0.34	1.4049	0.711770	0.84	2.3164	0.431711	1.68	5.3656	0.186374
0.35	1.4191	0.704688	**0.85**	2.3396	0.427415	**1.70**	5.4739	0.182684
0.36	1.4333	0.697676	0.86	2.3632	0.423162	1.72	5.5845	0.179066
0.37	1.4477	0.690734	0.87	2.3869	0.418952	1.74	5.6973	0.175520
0.38	1.4623	0.683861	0.88	2.4109	0.414783	1.76	5.8124	0.172045
0.39	1.4770	0.650509	0.89	2.4351	0.410656	1.78	5.9299	0.168638
0.40	1.4918	0.670320	**0.90**	2.4596	0.406570	**1.80**	6.0496	0.165299
0.41	1.5068	0.663650	0.91	2.4843	0.402524	1.82	6.1719	0.162026
0.42	1.5220	0.657047	0.92	2.5093	0.398519	1.84	6.2965	0.158817
0.43	1.5373	0.650509	0.93	2.5345	0.394554	1.86	6.4237	0.155673
0.44	1.5527	0.644036	0.94	2.5600	0.390628	1.88	6.5535	0.152590
0.45	1.5683	0.637628	**0.95**	2.5857	0.386741	**1.90**	6.6859	0.149569
0.46	1.5841	0.631284	0.96	2.6117	0.382893	1.92	6.8210	0.146607
0.47	1.6000	0.625002	0.97	2.6379	0.379083	1.94	6.9588	0.143704
0.48	1.6161	0.618783	0.98	2.6645	0.375311	1.96	7.0993	0.140858
0.49	1.6323	0.612626	0.99	2.6912	0.371577	1.98	7.2427	0.138069

Table 9-38. EXPONENTIAL FUNCTIONS (Continued)

x	e^x	e^{-x}	x	e^x	e^{-x}	x	e^x	e^{-x}
2.00	7.3891	0.135335	**4.75**	115.58	0.008652	**7.50**	1808.0	0.000553
2.05	7.7679	0.128735	4.80	121.51	0.008230	7.55	1900.7	0.000526
2.10	8.1662	0.122456	4.85	127.74	0.007828	7.60	1998.2	0.000501
2.15	8.5849	0.116484	4.90	134.29	0.007447	7.65	2100.6	0.000476
2.20	9.0250	0.110803	4.95	141.17	0.007083	7.70	2208.3	0.000453
2.25	9.4877	0.105399	**5.00**	148.41	0.006738	**7.75**	2321.6	0.000431
2.30	9.9742	0.100259	5.05	156.02	0.006409	7.80	2440.6	0.000410
2.35	10.486	0.095369	5.10	164.02	0.006097	7.85	2565.7	0.000390
2.40	11.023	0.090718	5.15	172.43	0.005799	7.90	2697.3	0.000371
2.45	11.588	0.086294	5.20	181.27	0.005517	7.95	2835.6	0.000353
2.50	12.182	0.082085	**5.25**	190.57	0.005248	**8.00**	2981.0	0.000336
2.55	12.807	0.078082	5.30	200.34	0.004992	8.05	3133.8	0.000319
2.60	13.464	0.074274	5.35	210.61	0.004748	8.10	3294.5	0.000304
2.65	14.154	0.070651	5.40	221.41	0.004517	8.15	3463.4	0.000289
2.70	14.880	0.067206	5.45	232.76	0.004296	8.20	3641.0	0.000275
2.75	15.643	0.063928	**5.50**	244.69	0.004087	**8.25**	3827.6	0.000261
2.80	16.445	0.060810	5.55	257.24	0.003888	8.30	4023.9	0.000249
2.85	17.288	0.057844	5.60	270.43	0.003698	8.35	4230.2	0.000238
2.90	18.174	0.055023	5.65	284.29	0.003518	8.40	4447.1	0.000225
2.95	19.106	0.052340	5.70	298.87	0.003346	8.45	4675.1	0.000214
3.00	20.086	0.049787	**5.75**	314.19	0.003183	**8.50**	4914.8	0.000204
3.05	21.115	0.047359	5.80	330.30	0.003028	8.55	5166.8	0.000194
3.10	22.198	0.045049	5.85	347.23	0.002880	8.60	5431.7	0.000184
3.15	23.336	0.042852	5.90	365.04	0.002739	8.65	5710.1	0.000175
3.20	24.533	0.040762	5.95	383.75	0.002606	8.70	6002.9	0.000167
3.25	25.790	0.038774	**6.00**	403.43	0.002479	**8.75**	6310.7	0.000159
3.30	27.113	0.036883	6.05	424.11	0.002379	8.80	6634.2	0.000151
3.35	28.503	0.035084	6.10	445.86	0.002243	8.85	6974.4	0.000143
3.40	29.964	0.033373	6.15	468.72	0.002134	8.90	7332.0	0.000136
3.45	31.500	0.031746	6.20	492.75	0.002030	8.95	7707.9	0.000130
3.50	33.115	0.030197	**6.25**	518.01	0.001931	**9.00**	8103.1	0.000123
3.55	34.813	0.028725	6.30	544.57	0.001836	9.05	8518.5	0.000117
3.60	36.598	0.027324	6.35	572.49	0.001747	9.10	8955.3	0.000112
3.65	38.475	0.025991	6.40	601.85	0.001662	9.15	9414.4	0.000106
3.70	40.447	0.024724	6.45	632.70	0.001581	9.20	9897.1	0.000101
3.75	42.521	0.023518	**6.50**	665.14	0.001503	**9.25**	10405	0.000096
3.80	44.701	0.022371	6.55	699.24	0.001430	9.30	10938	0.000091
3.85	46.993	0.021280	6.60	735.10	0.001360	9.35	11499	0.000087
3.90	49.402	0.020242	6.65	772.78	0.001294	9.40	12088	0.000083
3.95	51.935	0.019255	6.70	812.41	0.001231	9.45	12708	0.000079
4.00	54.598	0.018316	**6.75**	854.06	0.001171	**9.50**	13360	0.000075
4.05	57.397	0.017422	6.80	897.85	0.001114	9.55	14045	0.000071
4.10	60.340	0.016573	6.85	943.88	0.001060	9.60	14765	0.000068
4.15	63.434	0.015764	6.90	992.27	0.001008	9.65	15522	0.000064
4.20	66.686	0.014996	6.95	1043.1	0.000959	9.70	16318	0.000061
4.25	70.105	0.014264	**7.00**	1096.6	0.000912	**9.75**	17154	0.000058
4.30	73.700	0.013569	7.05	1152.9	0.000868	9.80	18034	0.000056
4.35	77.478	0.012907	7.10	1212.0	0.000825	9.85	18958	0.000053
4.40	81.451	0.012277	7.15	1274.1	0.000785	9.90	19930	0.000050
4.45	85.627	0.011679	7.20	1339.4	0.000747	9.95	20952	0.000048
4.50	90.017	0.011109	**7.25**	1408.1	0.000710	**10.00**	22026	0.000045
4.55	94.632	0.010567	7.30	1480.3	0.000676			
4.60	99.484	0.010052	7.35	1556.2	0.000643			
4.65	104.58	0.009562	7.40	1636.0	0.000611			
4.70	109.95	0.009095	7.45	1719.9	0.000581			

*Condensed from: "CRC Handbook of Tables for Mathematics", 4th ed., S.M. Selby, Ed., The Chemical Rubber Co., 1970.

Table 9-39. WIDE-RANGE EXPONENTIAL FUNCTIONS, e^x AND e^{-x} *

This table provides the user with the values of e^x and e^{-x} for single-digit values of x from 0.000 001 to 90. For multiple-digit values of x, e^x is obtained by simple multiplication (see examples below the table). The natural logarithm of a number can be obtained by dividing the number by appropriate values from the table and then adding the exponents for those values (see examples below the table).

Exponent, x

Range	1	2	3	4	5	6	7	8	9
10 to 90	$2.202\,647\times10^4$	$4.851\,652\times10^8$	$1.068\,647\times10^{13}$	$2.353\,853\times10^{17}$	$5.184\,706\times10^{21}$	$1.142\,007\times10^{26}$	$2.515\,439\times10^{30}$	$5.540\,622\times10^{34}$	$1.220\,403\times10^{39}$
1 to 9	2.718 282	7.389 056	$2.008\,554\times10$	$5.459\,815\times10$	$1.484\,132\times10^2$	$4.034\,288\times10^2$	$1.096\,633\times10^3$	$2.980\,958\times10^3$	$8.103\,084\times10^3$
0.1 to 0.9	1.105 171	1.221 403	1.349 859	1.491 825	1.648 721	1.822 119	2.013 753	2.225 541	2.459 603
0.01 to 0.09	1.010 050	1.020 201	1.030 455	1.040 811	1.051 271	1.061 837	1.072 508	1.083 287	1.094 174
10^{-3} to 9×10^{-3}	1.001 001	1.002 002	1.003 005	1.004 008	1.005 013	1.006 018	1.007 025	1.008 032	1.009 041
10^{-4} to 9×10^{-4}	1.000 100	1.000 200	1.000 300	1.000 400	1.000 500	1.000 600	1.000 700	1.000 800	1.000 900
10^{-5} to 9×10^{-5}	1.000 010	1.000 020	1.000 030	1.000 040	1.000 050	1.000 060	1.000 070	1.000 080	1.000 090
10^{-6} to 9×10^{-6}	1.000 001	1.000 002	1.000 003	1.000 004	1.000 005	1.000 006	1.000 007	1.000 008	1.000 009
-10 to -90	$4.539\,993\times10^{-5}$	$2.061\,154\times10^{-9}$	$9.357\,623\times10^{-14}$	$4.248\,354\times10^{-18}$	$1.928\,750\times10^{-22}$	$8.756\,511\times10^{-27}$	$3.975\,450\times10^{-31}$	$1.804\,851\times10^{-35}$	$8.194\,013\times10^{-40}$
-1 to -9	$3.678\,794\times10^{-1}$	$1.353\,353\times10^{-1}$	$4.978\,707\times10^{-2}$	$1.831\,564\times10^{-2}$	$6.737\,947\times10^{-3}$	$2.478\,752\times10^{-3}$	$9.118\,820\times10^{-4}$	$3.354\,626\times10^{-4}$	$1.234\,098\times10^{-4}$
-0.1 to -0.9	0.904 837	0.818 731	0.740 818	0.670 320	0.606 531	0.548 812	0.496 585	0.449 329	0.406 570
-0.01 to -0.09	0.990 050	0.980 199	0.970 446	0.960 789	0.951 229	0.941 765	0.932 394	0.923 116	0.913 931
-10^{-3} to -9×10^{-3}	0.999 000	0.998 002	0.997 004	0.996 008	0.995 012	0.994 018	0.993 024	0.992 032	0.991 040
-10^{-4} to -9×10^{-4}	0.999 900	0.999 800	0.999 700	0.999 600	0.999 500	0.999 400	0.999 300	0.999 200	0.999 100
-10^{-5} to -9×10^{-5}	0.999 990	0.999 980	0.999 970	0.999 960	0.999 950	0.999 940	0.999 930	0.999 920	0.999 910
-10^{-6} to -9×10^{-6}	0.999 999	0.999 998	0.999 997	0.999 996	0.999 995	0.999 994	0.999 993	0.999 992	0.999 991

Examples:

$e^{1.64} = e^1 \times e^{0.6} \times e^{0.04} = 2.718\,282 \times 1.822\,119 \times 1.040\,811 = 5.155\,172$

$e^{-2.09} = e^{-2} \times e^{-0.09} = 1.353\,353\times10^{-1} \times 0.913\,931 = 0.123\,687$

$e^{9.98} = e^{10} \times e^{-0.02} = 2.202\,647\times10^4 \times 0.980\,199 = 2.159\,032\times10^4 = 21\,590.32$

Ln 1.35 = 0.300 1 to four significant figures is obtained from: $[(1.35/1.349\,859)/1.000\,100] = [(1.35/e^{0.3})/e^{0.000\,1}] = 1.000\,004$

Table 9-40. NUMERICAL CONSTANTS

π Constants

$$\pi = 3.14159\ 26535\ 89793\ 23846\ 26433\ 83279\ 50288\ 41971\ 69399\ 37510$$
$$1/\pi = 0.31830\ 98861\ 83790\ 67153\ 77675\ 26745\ 02872\ 40689\ 19291\ 48091$$
$$\pi^2 = 9.86960\ 44010\ 89358\ 61883\ 44909\ 99876\ 15113\ 53136\ 99407\ 24079$$
$$\log_e \pi = 1.14472\ 98858\ 49400\ 17414\ 34273\ 51353\ 05871\ 16472\ 94812\ 91531$$
$$\log_{10} \pi = 0.49714\ 98726\ 94133\ 85435\ 12682\ 88290\ 89887\ 36516\ 78324\ 38044$$
$$\log_{10} \sqrt{2\pi} = 0.39908\ 99341\ 79057\ 52478\ 25035\ 91507\ 69595\ 02099\ 34102\ 92127$$

Logarithmic Constants

$$e = 2.71828\ 18284\ 59045\ 23536\ 02874\ 71352\ 66249\ 77572\ 47093\ 69995$$
$$1/e = 0.36787\ 94411\ 71442\ 32159\ 55237\ 70161\ 46086\ 74458\ 11131\ 03176$$
$$e^2 = 7.38905\ 60989\ 30650\ 22723\ 04274\ 60575\ 00781\ 31803\ 15570\ 55184$$
$$M = \log_{10} e = 0.43429\ 44819\ 03251\ 82765\ 11289\ 18916\ 60508\ 22943\ 97005\ 80366$$
$$1/M = \log_e 10 = 2.30258\ 50929\ 94045\ 68401\ 79914\ 54684\ 36420\ 76011\ 01488\ 62877$$
$$\log_{10} M = 9.63778\ 43113\ 00536\ 78912\ 29674\ 98565\ - 10$$

Miscellaneous π and e Constants

$$\pi^e = 22.45915\ 77183\ 61045\ 47342\ 71522$$
$$e^\pi = 23.14069\ 26327\ 79269\ 00572\ 90864$$
$$e^{-\pi} = 0.04321\ 39182\ 63772\ 24977\ 44177$$
$$e^{\frac{1}{2}\pi} = 4.81047\ 73809\ 65351\ 65547\ 30357$$
$$i^i = e^{-\frac{1}{2}\pi} = 0.20787\ 95763\ 50761\ 90854\ 69556$$

Numerical Constants

$$\sqrt{2} = 1.41421\ 35623\ 73095\ 04880\ 16887\ 24209\ 69807\ 85696\ 71875\ 37694$$
$$\sqrt[3]{2} = 1.25992\ 10498\ 94873\ 16476\ 72106\ 07278\ 22835\ 05702\ 51464\ 70150$$
$$\log_e 2 = 0.69314\ 71805\ 59945\ 30941\ 72321\ 21458\ 17656\ 80755\ 00134\ 36025$$
$$\log_{10} 2 = 0.30102\ 99956\ 63981\ 19521\ 37388\ 94724\ 49302\ 07001\ 09001\ 16010$$
$$\sqrt{3} = 1.73205\ 08075\ 68877\ 29352\ 74463\ 41505\ 87236\ 69428\ 05253\ 81038$$
$$\sqrt[3]{3} = 1.44224\ 95703\ 07408\ 38232\ 16383\ 10780\ 10958\ 83918\ 69253\ 49935$$
$$\log_e 3 = 1.09861\ 22886\ 68109\ 69139\ 52452\ 36922\ 52570\ 46471\ 90557\ 82274$$
$$\log_{10} 3 = 0.47712\ 12547\ 19662\ 43729\ 50279\ 03255\ 11530\ 92001\ 28864\ 19069$$

Miscellaneous

$$\text{Euler's Constant } \gamma = 0.57721\ 56649\ 01532\ 86061$$
$$\log_e \gamma = -0.54953\ 93129\ 81644\ 82234$$
$$\text{Golden Ratio } \phi = 1.61803\ 30887\ 49894\ 84820\ 45868\ 34365\ 63811\ 77203\ 09180$$

Numbers Containing π

$$\pi = 3.14159\ 26536 \qquad \log_{10}\pi = 0.49714\ 98727 \qquad \log_e \pi = 1.11179\ 08858$$

	Number	Logarithm		Number	Logarithm
π	3.1415 927	0.4971 499	π^2	9.8696 044	0.9942 997
2π	6.2831 853	0.7981 799	$2\pi^2$	19.7392 088	1.2953 297
3π	9.4247 780	0.9742 711	$4\pi^2$	39.4784 176	1.5963 597
4π	12.5663 706	1.0992 099	$1/\pi^2$	0.1013 212	9.0057 003 − 10
8π	25.1327 412	1.4002 399	$1/(2\pi^2)$	0.0506 606	8.7046 703 − 10
$\pi/2$	1.5707 963	0.1961 199	$1/(4\pi^2)$	0.0253 303	8.4036 403 − 10
$\pi/3$	1.0471 976	0.0200 286	$\sqrt{\pi}$	1.7724 539	0.2485 749
$\pi/4$	0.7853 982	9.8950 899 − 10	$\sqrt{\pi/4}$ or	0.8862 269	9.9475 449 − 10
$\pi/6$	0.5235 988	9.7189 986 − 10	$\sqrt{\pi/2}$		
$\pi/8$	0.3926 991	9.5940 599 − 10	$\sqrt{\pi/4}$	0.4431 135	9.6465 149 − 10
$2\pi/3$	2.0943 951	0.3210 586	$\sqrt{\pi/2}$	1.2533 141	0.0980 599
$4\pi/3$	4.1887 902	0.6220 886	$\sqrt{2/\pi}$	0.7978 846	9.9019 401 − 10
$1/\pi$	0.3183 099	9.5028 501 − 10	π^3	31.0062 767	1.4914 496
$2/\pi$	0.6366 198	9.8038 801 − 10	$\sqrt[3]{\pi}$	1.4645 919	0.1657 166
$4/\pi$	1.2732 395	0.1049 101	$1/\sqrt[3]{\pi}$	0.6827 841	9.8342 834 − 10
$1/(2\pi)$	0.1591 549	9.2018 201 − 10	$\sqrt[3]{\pi^2}$	2.1450 294	0.3314 332
$1/(4\pi)$	0.0795 775	8.9007 901 − 10	$1/\sqrt{\pi}$	0.5641 896	9.7514 251 − 10
$1/(6\pi)$	0.0530 516	8.7246 989 − 10	$1/\sqrt{2\pi}$	0.3989 423	9.6009 101 − 10
$1/(8\pi)$	0.0397 887	8.5997 601 − 10	$2/\sqrt{\pi}$	1.1283 792	0.0524 551
$\pi/180$	0.0174 533	8.2418 774 − 10			
$180/\pi$	57.2957 795	1.7581 226			

Change of Base

$$\log_a x = \log_b x / \log_b a$$
$$\log_{10} x = \log_e x / \log_e 10 \qquad\qquad \log_e x = \log_{10} x / \log_{10} e$$
$$\log_e x = 1/M \log_{10} x = 2.30258\ 50930 \log_{10} x$$
$$\log_{10} x = M \log_e x = 0.43429\ 44819 \log_e x$$

Table 9-41. MATHEMATICAL CONSTANTS

For Use on a Digital Computer

Constant	Decimal (Base 10)				Octal (Base 8)					
π CONSTANTS										
π	3.14159	26535	89793	23846	3.1103	7552	4210	2643	0215	1423
π^{-1}	0.31830	98861	83790	67153	0.2427	6301	5562	3442	0251	2376
$\sqrt{\pi}$	1.77245	38509	05516	02729	1.6133	7611	0664	7366	5247	4703
π^2	9.86960	44010	89358	61883	11.6751	7144	6762	1357	1322	2556
$\sqrt{2\pi}$	2.5062	82746	31000	50241	2.4033	1143	7754	2340	5454	5371
$(\pi/2)^{\frac{1}{2}}$	1.25331	41373	15500	25120	1.2015	4461	7766	1160	2626	2574
$\pi^{-\frac{1}{2}}$	0.56418	95835	47756	28694	0.4406	7272	4041	2333	3210	6561
$(2\pi)^{-\frac{1}{2}}$	0.39894	22804	01432	67793	0.3142	0424	6365	0331	2043	2077
$\pi^{\frac{1}{3}}$	1.46459	18875	61523	26302	1.3556	7576	3461	0113	3612	7621
$\log_{10} \pi$	0.49714	98726	94133	85435	0.3764	2466	6306	7216	7300	1457
$\ln \pi$	1.14472	98858	49400	17414	1.1120	6404	4347	5033	6413	6537
πe	8.53973	42226	73567		10.4242	6005	5056	5072		
π/e	1.15572	73497	90921		1.1175	6677	3047	0733		
π^e	22.45915	77183	61045	47342	26.3530	5534	1601	0421	1613	1026
e CONSTANTS										
e	2.71828	18284	59045	23536	2.5576	0521	3050	5355	1246	5277
e^{-1}	0.36787	94411	71442	32159	0.2742	6530	6613	1674	6761	5272
e^π	23.14069	26327	79269	00572	27.1100	2156	5411	1471	4754	6647
$e^{-\pi}$	0.4321	39182	63772	24977	0.0261	0021	1732	6307	3706	4257
$e^{\pi/2}$	4.81047	73809	65351	65547	4.6367	5562	0526	2327	6476	2132
$\log_{10} e$	0.43429	44819	03251	82765	0.3362	6754	2511	5624	1614	5232
NUMERICAL CONSTANTS										
$\sqrt{2}$	1.41421	35623	73095	04880	1.3240	4746	3177	1674	6220	4262
$\sqrt[3]{2}$	1.25992	10498	94873	16477						
$\sqrt{3}$	1.73205	08075	68877	29641	1.5666	3656	4130	2312	5167	0145
$\sqrt[3]{3}$	1.44224	95703	07408	38232						
$\log_{10} 2$	0.30102	99956	63981	19251						
$\ln 2$	0.69314	71805	59945	30941	0.5427	1027	7575	0717	3632	5711
$\log_{10} 3$	0.47712	12547	19662	43729						
$\ln 3$	1.09861	22886	68109	69139						
$\ln 10$	2.30258	50929	94045	68401	2.2327	3067	3552	5242	5405	5651
$\log_2 10$					3.24464	741136				
EULER'S CONSTANT: γ										
γ	0.57721	56649	01532	86060	0.4474	2147	7067	6660	6172	2321
e^γ	1.78107	24179	90197	98522	1.6177	2134	5261	1526	5761	
$e^{-\gamma}$	0.56145	94835	66885	16903	0.4373	5717	0177	1345	7454	
$\log \gamma$	−0.23866	18912	16832	38945	−0.1721	4362	0631	1753	0063	
$\ln \gamma$	−0.54953	93129	81644	82234	−0.4312	7233	6021	7532	2777	

Table 9-42. RANDOM DIGITS*

*Reprinted by permission from: "The Compleat Strategyst", J.D. Williams. Copyright 1954. McGraw-Hill Book Company, Inc.

Table 9-43. RANDOM UNITS*

Line \ Column	(1)	(2)	(3)	(4)	(5)	(6)	(7)	(8)	(9)	(10)	(11)	(12)	(13)	(14)
1	10480	15011	01536	02011	81647	91646	69179	14194	62590	36207	20969	99570	91291	90700
2	22368	46573	25595	85393	30995	89198	27982	53402	93965	34095	52666	19174	39615	99505
3	24130	48360	22527	97265	76393	64809	15179	24830	49340	32081	30680	19655	63348	58629
4	42167	93093	06243	61680	07856	16376	39440	53537	71341	57004	00849	74917	97758	16379
5	37570	39975	81837	16656	06121	91782	60468	81305	49684	60672	14110	06927	01263	54613
6	77921	06907	11008	42751	27756	53498	18602	70659	90655	15053	21916	81825	44394	42880
7	99562	72905	56420	69994	98872	31016	71194	18738	44013	48840	63213	21069	10634	12952
8	96301	91977	05463	07972	18876	20922	94595	56869	69014	60045	18425	84903	42508	32307
9	89579	14342	63661	10281	17453	18103	57740	84378	25331	12566	58678	44947	05585	56941
10	85475	36857	43342	53988	53060	59533	38867	62300	08158	17983	16439	11458	18593	64952
11	28918	69578	88231	33276	70997	79936	56865	05859	90106	31595	01547	85590	91610	78188
12	63553	40961	48235	03427	49626	69445	18663	72695	52180	20847	12234	90511	33703	90322
13	09429	93969	52636	92737	88974	33488	36320	17617	30015	08272	84115	27156	30613	74952
14	10365	61129	87529	85689	48237	52267	67689	93394	01511	26358	85104	20285	29975	89868
15	07119	97336	71048	08178	77233	13916	47564	81056	97735	85977	29372	74461	28551	90707
16	51085	12765	51821	51259	77452	16308	60756	92144	49442	53900	70960	63990	75601	40719
17	02368	21382	52404	60268	89368	19885	55322	44819	01188	65255	64835	44919	05944	55157
18	01011	54092	33362	94904	31273	04146	18594	29852	71585	85030	51132	01915	92747	64951
19	52162	53916	46369	58586	23216	14513	83149	98736	23495	64350	94738	17752	35156	35749
20	07056	97628	33787	09998	42698	06691	76988	13602	51851	46104	88916	19509	25625	58104
21	48663	91245	85828	14346	09172	30168	90229	04734	59193	22178	30421	61666	99904	32812
22	54164	58492	22421	74103	47070	25306	76468	26384	58151	06646	21524	15227	96909	44592
23	32639	32363	05597	24200	13363	38005	94342	28728	35806	06912	17012	64161	18296	22851
24	29334	27001	87637	87308	58731	00256	45834	15398	46557	41135	10367	07684	36188	18510
25	02488	33062	28834	07351	19731	92420	60952	61280	50001	67658	32586	86679	50720	94953
26	81525	72295	04839	96423	24878	82651	66566	14778	76797	14780	13300	87074	79666	95725
27	29676	20591	68086	26432	46901	20849	89768	81536	86645	12659	92259	57102	80428	25280
28	00742	57392	39064	66432	84673	40027	32832	61362	98947	96067	64760	64584	96096	98253
29	05366	04213	25669	26422	44407	44048	37937	63904	45766	66134	75470	66520	34693	90449
30	91921	26418	64117	94305	26766	25940	39972	22209	71500	64568	91402	42416	07844	69618
31	00582	04711	87917	77341	42206	35126	74087	99547	81817	42607	43808	76655	62028	76630
32	00725	69884	62797	56170	86324	88072	76222	36086	84637	93161	76038	65855	77919	88006
33	69011	65797	95876	55293	18988	27354	26575	08625	40801	59920	29841	80150	12777	48501
34	25976	57948	29888	88604	67917	48708	18912	82271	65424	69774	33611	54262	85963	03547
35	09763	83473	73577	12908	30883	18317	28290	35797	05998	41688	34952	37888	38917	88050
36	91567	42595	27958	30134	04024	86385	29880	99730	55536	84855	29080	09250	79656	73211
37	17955	56349	90999	49127	20044	59931	06115	20542	18059	02008	73708	83517	36103	42791
38	46503	18584	18845	49618	02304	51038	20655	58727	28168	15475	56942	53389	20562	87338
39	92157	89634	94824	78171	84610	82834	09922	25417	44137	48413	25555	21246	35509	20468
40	14577	62765	35605	81263	39667	47358	56873	56307	61607	49518	89656	20103	77490	18062
41	98427	07523	33362	64270	01638	92477	66969	98420	04880	45585	46565	04102	46880	45709
42	34914	63976	88720	82765	34476	17032	87589	40836	32427	70002	70663	88863	77775	69348
43	70060	28277	39475	46473	23219	53416	94970	25832	69975	94884	19661	72828	00102	66794
44	53976	54914	06990	67245	68350	82948	11398	42878	80287	88267	47363	46634	06541	97809
45	76072	29515	40980	07391	58745	25774	22987	80059	39911	96189	41151	14222	60697	59583
46	90725	52210	83974	29992	65831	38857	50490	83765	55657	14361	31720	57375	56228	41546
47	64364	67412	33339	31926	14883	24413	59744	92351	97473	89286	35931	04110	23726	51900
48	08962	00358	31662	25388	61642	34072	81249	35648	56891	69352	48373	45578	78547	81788
49	95012	68379	93526	70765	10593	04542	76463	54328	02349	17247	28865	14777	62730	92277
50	15664	10493	20492	38391	91132	21999	59516	81652	27195	48223	46751	22923	32261	85653

Table 9-43. RANDOM UNITS (Continued)

Line	(1)	(2)	(3)	(4)	(5)	(6)	(7)	(8)	(9)	(10)	(11)	(12)	(13)	(14)
51	16408	81899	04153	53381	79401	21438	83035	92350	36693	31238	59649	91754	72772	02338
52	18629	81953	05520	91962	04739	13092	97662	24822	94730	06496	35090	04822	86772	98289
53	73115	35101	47498	87637	99016	71060	88824	71013	18735	20286	23153	72924	35165	43040
54	57491	16703	23167	49323	45021	33132	12544	41035	80780	45393	44812	12515	98931	91202
55	30405	83946	23792	14422	15059	45799	22716	19792	09983	74353	68668	30429	70735	25499
56	16631	35006	85900	98275	32388	52390	16815	69298	82732	38480	73817	32523	41961	44437
57	96773	20206	42559	78985	05300	22164	24369	54224	35083	19687	11052	91491	60383	19746
58	38935	64202	14349	82674	66523	44133	00697	35552	35970	19124	63318	29686	03387	59846
59	31624	76384	17403	53363	44167	64486	64758	75366	76554	31601	12614	33072	60332	92325
60	78919	19474	23632	27889	47914	02584	37680	20801	72152	39339	34806	08930	85001	87820
61	03931	33309	57047	74211	63445	17361	62825	39908	05607	91284	68833	25570	38818	46920
62	74426	33278	43972	10119	89917	15665	52872	73823	73144	88662	88970	74492	51805	99378
63	09066	00903	20795	95452	92648	45454	09552	88815	16553	51125	79375	97596	16296	66092
64	42238	12426	87025	14267	20979	04508	64535	31355	86064	29472	47689	05974	52468	16834
65	16153	08002	26504	41744	81959	65642	74240	56302	00033	67107	77510	70625	28725	34191
66	21457	40742	29820	96783	29400	21840	15035	34537	33310	06116	95240	15957	16572	06004
67	21581	57802	02050	89728	17937	37621	47075	42080	97403	48626	68995	43805	33386	21597
68	55612	78095	83197	33732	05810	24813	86902	60397	16489	03204	88525	42786	05260	92532
69	44657	66999	99324	51281	84463	60563	79312	93454	68876	25471	93911	25650	12682	73572
70	91340	84979	46949	81973	37949	61023	43997	15263	80644	43942	89203	71795	99533	50501
71	91227	21199	31935	27022	84067	05462	35216	14486	29891	68607	41867	14951	91696	85065
72	50001	38140	66321	19924	72163	09538	12151	06878	91903	18749	34405	56087	82790	70925
73	65390	05224	72958	28609	81406	39147	25549	48542	42627	45233	57202	94617	23772	07896
74	27504	96131	83944	41575	10573	08619	64482	73923	36152	05184	94142	25299	84387	34925
75	37169	94851	39117	89632	00959	16487	65536	49071	39782	17095	02330	74301	00275	48280
76	11508	70225	51111	38351	19444	66499	71945	05422	13442	78675	84081	66938	93654	59894
77	37449	30362	06604	54690	04052	53115	62757	95348	78662	11163	81651	50245	34971	52924
78	46515	70331	85922	38329	57015	15765	97161	17869	45349	61796	66345	81073	49106	79860
79	30986	81223	42416	58353	21532	30502	32305	80482	05174	07901	54339	58861	74818	46942
80	63798	64995	46583	09765	44160	78128	83991	42865	92520	83531	80377	35909	81250	54238
81	82486	84846	99254	67632	43218	50076	21361	64816	51202	88124	41870	52689	51275	83556
82	21885	32906	92431	09060	64297	51674	64126	62570	26123	05155	59194	52799	28225	85762
83	60336	98782	07408	53458	13564	59089	26445	29789	85205	41001	12535	12133	14645	23541
84	43937	46891	24010	25560	86355	33941	25786	54990	71899	15475	95434	98227	21824	19585
85	97656	63175	89303	16275	07100	92063	21942	18611	47348	20203	18534	03862	78095	50136
86	03299	01221	05418	38982	55758	92237	26759	86367	21216	98442	08303	56613	91511	75928
87	79626	06486	03574	17668	07785	76020	79924	25651	83325	88428	85076	72811	22717	50585
88	85636	68335	47539	03129	65651	11977	02510	26113	99447	68645	34327	15152	55230	93448
89	18039	14367	61337	06177	12143	46609	32989	74014	64708	00533	35398	58408	13261	47908
90	08362	15656	60627	36478	65648	16764	53412	09013	07832	41574	17639	82163	60859	75567
91	79556	29068	04142	16268	15387	12856	66227	38358	22478	73373	88732	09443	82558	05250
92	92608	82674	27072	32534	17075	27698	98204	63863	11951	34648	88022	56148	34925	57031
93	23982	25835	40055	67006	12293	02753	14827	22235	35071	99704	37543	11601	35503	85171
94	09915	96306	05908	97901	28395	14186	00821	80703	70426	75647	76310	88717	37890	40129
95	50937	33300	26695	62247	69927	76123	50842	43834	86654	70959	79725	93872	28117	19233
96	42488	78077	69882	61657	34136	79180	97526	43092	04098	73571	80799	76536	71255	64239
97	46764	86273	63003	93017	31204	36692	40202	35275	57306	55543	53203	18098	47625	88684
98	03237	45430	55417	63282	90816	17349	88298	90183	36600	78406	06216	95787	42579	90730
99	86591	81482	52667	61583	14972	90053	89534	76036	49199	43716	97548	04379	46370	28672
100	38534	01715	94964	87288	65680	43772	39560	12918	86537	62738	19636	51132	25739	56947

Table 9-43.   RANDOM UNITS (Continued)

Column Line	(1)	(2)	(3)	(4)	(5)	(6)	(7)	(8)	(9)	(10)	(11)	(12)	(13)	(14)
101	13284	16834	74151	92027	24670	36665	00770	22878	02179	51602	07270	76517	97275	45960
102	21224	00370	30420	03883	96648	89428	41583	17564	27395	63904	41548	49197	82277	24120
103	99052	47887	81085	64933	66279	80432	65793	83287	34142	13241	30590	97760	35848	91983
104	00199	50993	98603	38452	87890	94624	69721	57484	67501	77638	44331	11257	71131	11059
105	60578	06483	28733	37867	07936	98710	98539	27186	31237	80612	44488	97819	70401	95419
106	91240	18312	17441	01929	18163	69201	31211	54288	39296	37318	65724	90401	79017	62077
107	97458	14229	12063	59611	32249	90466	33216	19358	02591	54263	88449	01912	07436	50813
108	35249	38646	34475	72417	60514	69257	12489	51924	86871	92446	36607	11458	30440	52639
109	38980	46600	11759	11900	46743	27860	77940	39298	97838	95145	32378	68038	89351	37005
110	10750	52745	38749	87365	58959	53731	89295	59062	39404	13198	59960	70408	29812	83126
111	36247	27850	73958	20673	37800	63835	71051	84724	52492	22342	78071	17456	96104	18327
112	70994	66986	99744	72438	01174	42159	11392	20724	54322	36923	70009	23233	65438	59685
113	99638	94702	11463	18148	81386	80431	90628	52506	02016	85151	88598	47821	00265	82525
114	72055	15774	43857	99805	10419	76939	25993	03544	21560	83471	43989	90770	22965	44247
115	24038	65541	85788	55835	38835	59399	13790	35112	01324	39520	76210	22467	83275	32286
116	74976	14631	35908	28221	39470	91548	12854	30166	09073	75887	36782	00268	97121	57676
117	35553	71628	70189	26436	63407	91178	90348	55359	80392	41012	36270	77786	89578	21059
118	35676	12797	51434	82976	42010	26344	92920	92155	58807	54644	58581	95331	78629	73344
119	74815	67523	72985	23183	02446	63594	98924	20633	58842	85961	07648	70164	34994	67662
120	45246	88048	65173	50989	91060	89894	36063	32819	68559	99221	49475	50558	34698	71800
121	76509	47069	86378	41797	11910	49672	88575	97966	32466	10083	54728	81972	58975	30761
122	19689	90332	04315	21358	97248	11188	39062	63312	52496	07349	79178	33692	57352	72862
123	42751	35318	97513	61537	54955	08159	00337	80778	27507	95478	21252	12746	37554	97775
124	11946	22681	45045	13964	57517	59419	58045	44067	58716	58840	45557	96345	33271	53464
125	96518	48688	20996	11090	48396	57177	83867	86464	14342	21545	46717	72364	86954	55580
126	35726	58643	76869	84622	39098	36083	72505	92265	23107	60278	05822	46760	44294	07672
127	39737	42750	48968	70536	84864	64952	38404	94317	65402	13589	01055	79044	19308	83623
128	97025	66492	56177	04049	80312	48028	26408	43591	75528	65341	49044	95495	81256	53214
129	62814	08075	09788	56350	76787	51591	54509	49295	85830	59860	30883	89660	96142	18354
130	25578	22950	15227	83291	41737	79599	96191	71845	86899	70694	24290	01551	80092	82118
131	68763	69576	88991	49662	46704	63362	56625	00481	73323	91427	15264	06969	57048	54149
132	17900	00813	64361	60725	88974	61005	99709	30666	26451	11528	44323	34778	60342	60388
133	71944	60227	63551	71109	05624	43836	58254	26160	32116	63403	35404	57146	10909	07346
134	54684	93691	85132	64399	29182	44324	14491	55226	78793	34107	30374	48429	51376	09559
135	25946	27623	11258	65204	52832	50880	22273	05554	99521	73791	85744	29276	70326	60251
136	01353	39318	44961	44972	91766	90262	56073	06606	51826	18893	83448	31915	97764	75091
137	99083	88191	27662	99113	57174	35571	99884	13951	71057	53961	61448	74909	07322	80960
138	52021	45406	37945	75234	24327	86978	22644	87779	23753	99926	63898	54886	18051	96314
139	78755	47744	43776	83098	03225	14281	83637	55984	13300	52212	58781	14905	46502	04472
140	25282	69106	59180	16257	22810	43609	12224	25643	89884	31149	85423	32581	34374	70873
141	11959	94202	02743	86847	79725	51811	12998	76844	05320	54236	53891	70226	38632	84776
142	11644	13792	98190	01424	30078	28197	55583	05197	47714	68440	22016	79204	06862	94451
143	06307	97912	68110	59812	95448	43244	31262	88880	13040	16458	43813	89416	42482	33939
144	76285	75714	89585	99296	52640	46518	55486	90754	88932	19937	57119	23251	55619	23679
145	55322	07589	39600	60866	63007	20007	66819	84164	61131	81429	60676	42807	78286	29015
146	78017	90928	90220	92503	83375	26986	74399	30885	88567	29169	72816	53357	15428	86932
147	44768	43342	20696	26331	43140	69744	82928	24988	94237	46138	77426	39039	55596	12655
148	25100	19336	14605	86603	51680	97678	24261	02464	86563	74812	60069	71674	15478	47642
149	83612	46623	62876	85197	07824	91392	58317	37726	84628	42221	10268	20692	15699	29167
150	41347	81666	82961	60413	71020	83658	02415	33322	66036	98712	46795	16308	28413	05417

Table 9-43. RANDOM UNITS *(Continued)*

Column / Line	(1)	(2)	(3)	(4)	(5)	(6)	(7)	(8)	(9)	(10)	(11)	(12)	(13)	(14)
151	38128	51178	75096	13609	16110	73533	42564	59870	29399	67834	91055	89917	51096	89011
152	60950	00455	73254	96067	50717	13878	03216	78274	65863	37011	91283	33914	91303	49326
153	90524	17320	29832	96118	75792	25326	22940	24904	80523	38928	91374	55597	97567	38914
154	49897	18278	67160	39408	97056	43517	84426	59650	20247	19293	02019	14790	02852	05819
155	18494	99209	81060	19488	65596	59787	47939	91225	98768	43688	00438	05548	09443	82897
156	65373	72984	30171	37741	70203	94094	87261	30056	58124	70133	18936	02138	59372	09075
157	40653	12843	04213	70925	95360	55774	76439	61768	52817	81151	52188	31940	54273	49032
158	51638	22238	56344	44587	83231	50317	74541	07719	25472	41602	77318	15145	57515	07633
159	69742	99303	62578	83575	30337	07488	51941	84316	42067	49692	28616	29101	03013	73449
160	58012	74072	67488	74580	47992	69482	58624	17106	47538	13452	22620	24260	40155	74716
161	18348	19855	42887	08279	43206	47077	42637	45606	00011	20662	14642	49984	94509	56380
162	59614	09193	58064	29086	44385	45740	70752	05663	49081	26960	57454	99264	24142	74648
163	75688	28630	39210	52897	62748	72658	98059	67202	72789	01869	13496	14663	87645	89713
164	13941	77802	69101	70061	35460	34576	15412	81304	58757	35498	94830	75521	00603	97701
165	96656	86420	96475	86458	54463	96419	55417	41375	76886	19008	66877	35934	59801	00497
166	03363	82042	15942	14549	38324	87094	19069	67590	11087	68570	22591	65232	85915	91499
167	70366	08390	69155	25496	13240	57407	91407	49160	07379	34444	94567	66035	38918	65708
168	47870	36605	12927	16043	53257	93796	52721	73120	48025	76074	95605	67422	41646	14557
169	79504	77606	22761	30518	28373	73898	30550	76684	77366	32276	04690	61667	64798	66276
170	46967	74841	50923	15339	37755	98995	40162	89561	69199	42257	11647	47603	48779	97907
171	14558	50769	35444	59030	87516	48193	02945	00922	48189	04724	21263	20892	92955	90251
172	12440	25057	01132	38611	28135	68089	10954	10097	54243	06460	50856	65435	79377	53890
173	32293	29938	68653	10497	98919	46587	77701	99119	93165	67788	17638	23097	21468	36992
174	10640	21875	72462	77981	56550	55999	87310	69643	45124	00349	25748	00844	96831	30651
175	47615	23169	39571	56972	20628	21788	51736	33133	72696	32605	41569	76148	91544	21121
176	16948	11128	71624	72754	49084	96303	27830	45817	67867	18062	87453	17226	72904	71474
177	21258	61002	66634	70335	02448	17354	83432	49608	66520	06442	59664	20420	39201	69549
178	15072	48853	15178	30730	47481	48490	41436	25015	49932	20474	53821	51015	79841	32405
179	99154	57412	09858	05671	70055	71479	63520	31357	56068	06739	34165	70685	04184	25250
180	08759	61089	23706	32994	35426	36666	63988	98844	37533	08269	27021	45886	22835	78451
181	67323	57839	61114	62192	47547	58023	64630	34886	98777	75442	95592	06141	45096	73117
182	09255	13986	84834	20764	72206	89393	34548	93438	88730	61805	78955	18952	46436	58740
183	36304	74712	00374	10107	85061	69228	81969	92216	03568	39630	81869	52824	50937	27954
184	15884	67429	86612	47367	10242	44880	12060	44309	46629	55105	66793	93173	00480	13311
185	18745	32031	35303	08134	33925	03044	59929	95418	04917	57596	24878	61733	92834	64454
186	72934	40086	88292	65728	38300	42323	64068	98373	48971	09049	59943	36538	05976	82118
187	17626	02944	20910	57662	80181	38579	24580	90529	52303	50436	29401	57824	86039	81062
188	27117	61399	50967	41399	81636	16663	15634	79717	94696	59240	25543	97989	63306	90946
189	93995	18678	90012	63645	85701	85269	62263	68331	00389	72571	15210	20769	44686	96176
190	67392	89421	09623	80725	62620	84162	87368	29560	00519	84545	08004	24526	41252	14521
191	04910	12261	37566	80016	21245	69377	50420	85658	55263	68667	78770	04533	14513	18099
192	81453	20283	79929	59839	23875	13245	46808	74124	74703	35769	95588	21014	37078	39170
193	19480	75790	48539	23703	15537	48885	02861	86587	74539	65227	90799	58789	96257	02708
194	21456	13162	74608	81011	55512	07481	93551	72189	76261	91206	89941	15132	37738	59284
195	89406	20912	46189	76376	25538	87212	20748	12831	57166	35026	16817	79121	18929	40628
196	09866	07414	55977	16419	01101	69343	13305	94302	80703	57910	36933	57771	42546	03003
197	86541	24681	23421	13521	28000	94917	07423	57523	97234	63951	42876	46829	09781	58160
198	10414	96941	06205	72222	57167	83902	07460	69507	10600	08858	07685	44472	64220	27040
199	49942	06683	41479	58982	56288	42853	92196	20632	62045	78812	35895	51851	83534	10689
200	23995	68882	42291	23374	24299	27024	67460	94783	40937	16961	26053	78749	46704	21983

*From: "CRC Handbook of Tables for Mathematics", 4th ed., S.M. Selby, Ed., The Chemical Rubber Co., 1970.

Table 9-44.   DERIVATIVES*

In the following formulas u, v, w represent functions of x, while a, c, n represent fixed real numbers. All arguments in the trigonometric functions are measured in radians, and all inverse trigonometric and hyperbolic functions represent principal values.†

1. $\dfrac{d}{dx}(a) = 0$

2. $\dfrac{d}{dx}(x) = 1$

3. $\dfrac{d}{dx}(au) = a\dfrac{du}{dx}$

4. $\dfrac{d}{dx}(u + v - w) = \dfrac{du}{dx} + \dfrac{dv}{dx} - \dfrac{dw}{dx}$

5. $\dfrac{d}{dx}(uv) = u\dfrac{dv}{dx} + v\dfrac{du}{dx}$

6. $\dfrac{d}{dx}(uvw) = uv\dfrac{dw}{dx} + vw\dfrac{du}{dx} + uw\dfrac{dv}{dx}$

7. $\dfrac{d}{dx}\left(\dfrac{u}{v}\right) = \dfrac{v\dfrac{du}{dx} - u\dfrac{dv}{dx}}{v^2} = \dfrac{1}{v}\dfrac{du}{dx} - \dfrac{u}{v^2}\dfrac{dv}{dx}$

8. $\dfrac{d}{dx}(u^n) = nu^{n-1}\dfrac{du}{dx}$

9. $\dfrac{d}{dx}(\sqrt{u}) = \dfrac{1}{2\sqrt{u}}\dfrac{du}{dx}$

10. $\dfrac{d}{dx}\left(\dfrac{1}{u}\right) = -\dfrac{1}{u^2}\dfrac{du}{dx}$

11. $\dfrac{d}{dx}\left(\dfrac{1}{u^n}\right) = -\dfrac{n}{u^{n+1}}\dfrac{du}{dx}$

12. $\dfrac{d}{dx}\left(\dfrac{u^n}{v^m}\right) = \dfrac{u^{n-1}}{v^{m+1}}\left(nv\dfrac{du}{dx} - mu\dfrac{dv}{dx}\right)$

13. $\dfrac{d}{dx}(u^n v^m) = u^{n-1}v^{m-1}\left(nv\dfrac{du}{dx} + mu\dfrac{dv}{dx}\right)$

14. $\dfrac{d}{dx}[f(u)] = \dfrac{d}{du}[f(u)] \cdot \dfrac{du}{dx}$

† Let $y = f(x)$ and $\dfrac{dy}{dx} = \dfrac{d[f(x)]}{dx} = f'(x)$ define respectively a function and its derivative for any value x in their common domain. The differential for the function at such a value x is accordingly defined as

$$dy = d[f(x)] = \dfrac{dy}{dx}dx = \dfrac{d[f(x)]}{dx}dx = f'(x)\,dx$$

Each derivative formula has an associated differential formula. For example, formula 6 above has the differential formula

$$d(uvw) = uv\,dw + vw\,du + uw\,dv$$

---

*From: "Handbook of Tables for Mathematics", 4th ed., S.M. Selby, Ed., The Chemical Rubber Co., 1970.

Table 9-44. DERIVATIVES *(Continued)*

15. $\dfrac{d^2}{dx^2}[f(u)] = \dfrac{df(u)}{du} \cdot \dfrac{d^2u}{dx^2} + \dfrac{d^2f(u)}{du^2} \cdot \left(\dfrac{du}{dx}\right)^2$

16. $\dfrac{d^n}{dx^n}[uv] = \binom{n}{0}v\dfrac{d^nu}{dx^n} + \binom{n}{1}\dfrac{dv}{dx}\dfrac{d^{n-1}u}{dx^{n-1}} + \binom{n}{2}\dfrac{d^2v}{dx^2}\dfrac{d^{n-2}u}{dx^{n-2}}$

$$+ \cdots + \binom{n}{k}\dfrac{d^kv}{dx^k}\dfrac{d^{n-k}u}{dx^{n-k}} + \cdots + \binom{n}{n}u\dfrac{d^nv}{dx^n}$$

where $\binom{n}{r} = \dfrac{n!}{r!(n-r)!}$ the binomial coefficient, n non-negative integer and $\binom{n}{0} = 1$.

17. $\dfrac{du}{dx} = \dfrac{1}{\dfrac{dx}{du}}$ if $\dfrac{dx}{du} \neq 0$

18. $\dfrac{d}{dx}(\log_a u) = (\log_a e)\dfrac{1}{u}\dfrac{du}{dx}$

19. $\dfrac{d}{dx}(\log_e u) = \dfrac{1}{u}\dfrac{du}{dx}$

20. $\dfrac{d}{dx}(a^u) = a^u(\log_e a)\dfrac{du}{dx}$

21. $\dfrac{d}{dx}(e^u) = e^u\dfrac{du}{dx}$

22. $\dfrac{d}{dx}(u^v) = vu^{v-1}\dfrac{du}{dx} + (\log_e u)u^v\dfrac{dv}{dx}$

23. $\dfrac{d}{dx}(\sin u) = \dfrac{du}{dx}(\cos u)$

24. $\dfrac{d}{dx}(\cos u) = -\dfrac{du}{dx}(\sin u)$

25. $\dfrac{d}{dx}(\tan u) = \dfrac{du}{dx}(\sec^2 u)$

26. $\dfrac{d}{dx}(\cot u) = -\dfrac{du}{dx}(\csc^2 u)$

27. $\dfrac{d}{dx}(\sec u) = \dfrac{du}{dx}\sec u \cdot \tan u$

28. $\dfrac{d}{dx}(\csc u) = -\dfrac{du}{dx}\csc u \cdot \cot u$

29. $\dfrac{d}{dx}(\text{vers } u) = \dfrac{du}{dx}\sin u$

30. $\dfrac{d}{dx}(\arcsin u) = \dfrac{1}{\sqrt{1-u^2}}\dfrac{du}{dx},$ $\left(-\dfrac{\pi}{2} \leq \arcsin u \leq \dfrac{\pi}{2}\right)$

Table 9-44. DERIVATIVES *(Continued)*

31. $\dfrac{d}{dx}(\text{arc cos } u) = -\dfrac{1}{\sqrt{1-u^2}}\dfrac{du}{dx}$, $\quad (0 \le \text{arc cos } u \le \pi)$

32. $\dfrac{d}{dx}(\text{arc tan } u) = \dfrac{1}{1+u^2}\dfrac{du}{dx}$, $\quad \left(-\dfrac{\pi}{2} < \text{arc tan } u < \dfrac{\pi}{2}\right)$

33. $\dfrac{d}{dx}(\text{arc cot } u) = -\dfrac{1}{1+u^2}\dfrac{du}{dx}$, $\quad (0 \le \text{arc cot } u \le \pi)$

34. $\dfrac{d}{dx}(\text{arc sec } u) = \dfrac{1}{u\sqrt{u^2-1}}\dfrac{du}{dx}$, $\quad \left(0 \le \text{arc sec } u < \dfrac{\pi}{2}, -\pi \le \text{arc sec } u < -\dfrac{\pi}{2}\right)$

35. $\dfrac{d}{dx}(\text{arc csc } u) = -\dfrac{1}{u\sqrt{u^2-1}}\dfrac{du}{dx}$, $\quad \left(0 < \text{arc csc } u \le \dfrac{\pi}{2}, -\pi < \text{arc csc } u \le -\dfrac{\pi}{2}\right)$

36. $\dfrac{d}{dx}(\text{arc vers } u) = \dfrac{1}{\sqrt{2u-u^2}}\dfrac{du}{dx}$, $\quad (0 \le \text{arc vers } u \le \pi)$

37. $\dfrac{d}{dx}(\sinh u) = \dfrac{du}{dx}(\cosh u)$

38. $\dfrac{d}{dx}(\cosh u) = \dfrac{du}{dx}(\sinh u)$

39. $\dfrac{d}{dx}(\tanh u) = \dfrac{du}{dx}(\text{sech}^2 u)$

40. $\dfrac{d}{dx}(\coth u) = -\dfrac{du}{dx}(\text{csch}^2 u)$

41. $\dfrac{d}{dx}(\text{sech } u) = -\dfrac{du}{dx}(\text{sech } u \cdot \tanh u)$

42. $\dfrac{d}{dx}(\text{csch } u) = -\dfrac{du}{dx}(\text{csch } u \cdot \coth u)$

43. $\dfrac{d}{dx}(\sinh^{-1} u) = \dfrac{d}{dx}[\log(u + \sqrt{u^2+1})] = \dfrac{1}{\sqrt{u^2+1}}\dfrac{du}{dx}$

44. $\dfrac{d}{dx}(\cosh^{-1} u) = \dfrac{d}{dx}[\log(u + \sqrt{u^2-1})] = \dfrac{1}{\sqrt{u^2-1}}\dfrac{du}{dx}$, $\quad (u > 1, \cosh^{-1} u > 0)$

45. $\dfrac{d}{dx}(\tanh^{-1} u) = \dfrac{d}{dx}\left[\dfrac{1}{2}\log\dfrac{1+u}{1-u}\right] = \dfrac{1}{1-u^2}\dfrac{du}{dx}$, $\quad (u^2 < 1)$

46. $\dfrac{d}{dx}(\coth^{-1} u) = \dfrac{d}{dx}\left[\dfrac{1}{2}\log\dfrac{u+1}{u-1}\right] = \dfrac{1}{1-u^2}\dfrac{du}{dx}$, $\quad (u^2 > 1)$

47. $\dfrac{d}{dx}(\text{sech}^{-1} u) = \dfrac{d}{dx}\left[\log\dfrac{1+\sqrt{1-u^2}}{u}\right] = -\dfrac{1}{u\sqrt{1-u^2}}\dfrac{du}{dx}$, $\quad (0 < u < 1, \text{sech}^{-1} u > 0)$

48. $\dfrac{d}{dx}(\text{csch}^{-1} u) = \dfrac{d}{dx}\left[\log\dfrac{1+\sqrt{1+u^2}}{u}\right] = -\dfrac{1}{|u|\sqrt{1+u^2}}\dfrac{du}{dx}$

Table 9-44. DERIVATIVES *(Continued)*

49. $\dfrac{d}{dq}\displaystyle\int_p^q f(x)\,dx = f(q), \qquad [p \text{ constant}]$

50. $\dfrac{d}{dp}\displaystyle\int_p^q f(x)\,dx = -f(p), \qquad [q \text{ constant}]$

51. $\dfrac{d}{da}\displaystyle\int_p^q f(x,a)\,dx = \int_p^q \dfrac{\partial}{\partial a}[f(x,a)]\,dx + f(q,a)\dfrac{dq}{da} - f(p,a)\dfrac{dp}{da}$

Table 9-45. FACTS FROM ALGEBRA*

FACTORS AND EXPANSIONS

$(a \pm b)^2 = a^2 \pm 2ab + b^2.$

$(a \pm b)^3 = a^3 \pm 3a^2b + 3ab^2 \pm b^3.$

$(a \pm b)^4 = a^4 \pm 4a^3b + 6a^2b^2 \pm 4ab^3 + b^4.$

$a^2 - b^2 = (a - b)(a + b).$

$a^2 + b^2 = (a + b\sqrt{-1})(a - b\sqrt{-1}).$

$a^3 - b^3 = (a - b)(a^2 + ab + b^2).$

$a^3 + b^3 = (a + b)(a^2 - ab + b^2).$

$a^4 + b^4 = (a^2 + ab\sqrt{2} + b^2)(a^2 - ab\sqrt{2} + b^2).$

$a^n - b^n = (a - b)(a^{n-1} + a^{n-2}b + \ldots + b^{n-1}).$

$a^n - b^n = (a + b)(a^{n-1} - a^{n-2}b + \ldots - b^{n-1}),$
$\qquad\qquad\qquad\qquad$ for even values of *n.*

$a^n + b^n = (a + b)(a^{n-1} - a^{n-2}b + \ldots + b^{n-1}),$
$\qquad\qquad\qquad\qquad$ for odd values of *n.*

$a^4 + a^2b^2 + b^4 = (a^2 + ab + b^2)(a^2 - ab + b^2).$

$(a + b + c)^2 = a^2 + b^2 + c^2 + 2ab + 2ac + 2bc.$

$(a + b + c)^3 = a^3 + b^3 + c^3 + 3a^2(b + c) + 3b^2(a + c) +$
$\qquad\qquad\qquad\qquad\qquad 3c^2(a + b) + 6abc.$

$(a + b + c + d + \ldots)^2 = a^2 + b^2 + c^2 + d^2 + \ldots +$
$2a(b + c + d + \ldots) + 2b(c + d + \ldots) + 2c(d + \ldots) + \ldots$

POWERS AND ROOTS

$a^x \times a^y = a^{(x+y)}.$
$\qquad\qquad$ $a^0 = 1\,[\text{if } a \neq 0]$
$\qquad$ $(ab)^x = a^x b^x.$

$\dfrac{a^x}{a^y} = a^{(x-y)}.$
$\qquad\qquad$ $a^{-x} = \dfrac{1}{a^x}.$
$\qquad\qquad$ $\left(\dfrac{a}{b}\right)^x = \dfrac{a^x}{b^x}.$

$(a^x)^y = a^{xy}.$
$\qquad\qquad$ $a^{\frac{1}{x}} = \sqrt[x]{a}.$
$\qquad\qquad$ $\sqrt[x]{ab} = \sqrt[x]{a}\,\sqrt[x]{b}.$

$\sqrt[x]{\sqrt[y]{a}} = \sqrt[xy]{a}.$
$\qquad\qquad$ $a^{\frac{x}{y}} = \sqrt[y]{a^x}.$
$\qquad\qquad$ $\sqrt[x]{\dfrac{a}{b}} = \dfrac{\sqrt[x]{a}}{\sqrt[x]{b}}.$

PROPORTION

If $\quad \dfrac{a}{b} = \dfrac{c}{d},\quad$ then $\quad \dfrac{a+b}{b} = \dfrac{c+d}{d},$

$\dfrac{a-b}{b} = \dfrac{c-d}{d}, \qquad \dfrac{a-b}{a+b} = \dfrac{c-d}{c+d}.$

*From: "Standard Mathematical Tables", 19th ed., S.M. Selby, Ed., The Chemical Rubber Co., 1971.

Table 9-46.   INTEGRALS—ELEMENTARY FORMS*

1. $\displaystyle\int a\,dx = ax$

2. $\displaystyle\int a \cdot f(x)\,dx = a \int f(x)\,dx$

3. $\displaystyle\int \phi(y)\,dx = \int \frac{\phi(y)}{y'}\,dy,$ where $y' = \dfrac{dy}{dx}$

4. $\displaystyle\int (u + v)\,dx = \int u\,dx + \int v\,dx,$ where u and v are any functions of x

5. $\displaystyle\int u\,dv = u \int dv - \int v\,du = uv - \int v\,du$

6. $\displaystyle\int u\frac{dv}{dx}dx = uv - \int v\frac{du}{dx}dx$

7. $\displaystyle\int x^n\,dx = \frac{x^{n+1}}{n+1},$ except $n = -1$

8. $\displaystyle\int \frac{f'(x)\,dx}{f(x)} = \log f(x),$ $(df(x) = f'(x)\,dx)$

9. $\displaystyle\int \frac{dx}{x} = \log x$

10. $\displaystyle\int \frac{f'(x)\,dx}{2\sqrt{f(x)}} = \sqrt{f(x)},$ $(df(x) = f'(x)\,dx)$

11. $\displaystyle\int e^x\,dx = e^x$

12. $\displaystyle\int e^{ax}\,dx = e^{ax}/a$

13. $\displaystyle\int b^{ax}\,dx = \frac{b^{ax}}{a\log b},$ $(b > 0)$

14. $\displaystyle\int \log x\,dx = x\log x - x$

15. $\displaystyle\int a^x \log a\,dx = a^x,$ $(a > 0)$

16. $\displaystyle\int \frac{dx}{a^2 + x^2} = \frac{1}{a}\tan^{-1}\frac{x}{a}$

---

*Reprinted from: "Standard Mathematical Tables", 20th ed., S.M. Selby, Ed., The Chemical Rubber Co., 1972.

Table 9-46. INTEGRALS—ELEMENTARY FORMS *(Continued)*

17. $\displaystyle\int \frac{dx}{a^2 - x^2} = \begin{cases} \dfrac{1}{a}\tanh^{-1}\dfrac{x}{a} \\[2mm] \text{or} \\[2mm] \dfrac{1}{2a}\log\dfrac{a+x}{a-x}, & (a^2 > x^2) \end{cases}$

18. $\displaystyle\int \frac{dx}{x^2 - a^2} = \begin{cases} -\dfrac{1}{a}\coth^{-1}\dfrac{x}{a} \\[2mm] \text{or} \\[2mm] \dfrac{1}{2a}\log\dfrac{x-a}{x+a}, & (x^2 > a^2) \end{cases}$

19. $\displaystyle\int \frac{dx}{\sqrt{a^2 - x^2}} = \begin{cases} \sin^{-1}\dfrac{x}{|a|} \\[2mm] \text{or} \\[2mm] -\cos^{-1}\dfrac{x}{|a|}, & (a^2 > x^2) \end{cases}$

20. $\displaystyle\int \frac{dx}{\sqrt{x^2 \pm a^2}} = \log\left(x + \sqrt{x^2 \pm a^2}\right)$

21. $\displaystyle\int \frac{dx}{x\sqrt{x^2 - a^2}} = \frac{1}{|a|}\sec^{-1}\frac{x}{a}$

22. $\displaystyle\int \frac{dx}{x\sqrt{a^2 \pm x^2}} = -\frac{1}{a}\log\left(\frac{a + \sqrt{a^2 \pm x^2}}{x}\right)$

Table 9-47.   SERIES*

The expression in parentheses following certain of the series indicates the region of convergence. If not otherwise indicated it is to be understood that the series converges for all finite values of x.

BINOMIAL

$$(x + y)^n = x^n + nx^{n-1}y + \frac{n(n-1)}{2!} x^{n-2}y^2$$

$$+ \frac{n(n-1)(n-2)}{3!} x^{n-3}y^3 + \cdots \quad (y^2 < x^2)$$

$$(1 \pm x)^n = 1 \pm nx + \frac{n(n-1)x^2}{2!} \pm \frac{n(n-1)(n-2)x^3}{3!} + \cdots \text{ etc.} \quad (x^2 < 1)$$

$$(1 \pm x)^{-n} = 1 \mp nx + \frac{n(n+1)x^2}{2!} \mp \frac{n(n+1)(n+2)x^3}{3!} + \cdots \text{ etc.} \quad (x^2 < 1)$$

$$(1 \pm x)^{-1} = 1 \mp x + x^2 \mp x^3 + x^4 \mp x^5 + \cdots \qquad (x^2 < 1)$$
$$(1 \pm x)^{-2} = 1 \mp 2x + 3x^2 \mp 4x^3 + 5x^4 \mp 6x^5 + \cdots \qquad (x^2 < 1)$$

REVERSION OF SERIES

Let a series be represented by

$$y = a_1 x + a_2 x^2 + a_3 x^3 + a_4 x^4 + a_5 x^5 + a_6 x^6 + \cdots \quad (a_1 \neq 0)$$

to find the coefficients of the series

$$x = A_1 y + A_2 y^2 + A_3 y^3 + A_4 y^4 + \cdots$$

$$A_1 = \frac{1}{a_1} \qquad A_2 = -\frac{a_2}{a_1^3} \qquad A_3 = \frac{1}{a_1^5}(2a_2^2 - a_1 a_3)$$

$$A_4 = \frac{1}{a_1^7}(5a_1 a_2 a_3 - a_1^2 a_4 - 5a_2^3)$$

$$A_5 = \frac{1}{a_1^9}(6a_1^2 a_2 a_4 + 3a_1^2 a_3^2 + 14a_2^4 - a_1^3 a_5 - 21a_1 a_2^2 a_3)$$

$$A_6 = \frac{1}{a_1^{11}}(7a_1^3 a_2 a_5 + 7a_1^3 a_3 a_4 + 84a_1 a_2^3 a_3 - a_1^4 a_6 - 28a_1^2 a_2^2 a_4 - 28a_1^2 a_2 a_3^2 - 42a_2^5)$$

$$A_7 = \frac{1}{a_1^{13}}(8a_1^4 a_2 a_6 + 8a_1^4 a_3 a_5 + 4a_1^4 a_4^2 + 120a_1^2 a_2^3 a_4$$

$$+ 180a_1^2 a_2^2 a_3^2 + 132a_2^6 - a_1^5 a_7$$

$$- 36a_1^3 a_2^2 a_5 - 72a_1^3 a_2 a_3 a_4 - 12a_1^3 a_3^3 - 330a_1 a_2^4 a_3)$$

TAYLOR

1. $f(x) = f(a) + (x - a)f'(a) + \dfrac{(x - a)^2}{2!} f''(a) + \dfrac{(x - a)^3}{3!} f'''(a)$

$$+ \cdots + \frac{(x - a)^n}{n!} f^{(n)}(a) + \cdots \text{ (Taylor's Series)}$$

---

*From: "CRC Handbook of Tables for Mathematics", 4th ed., S.M. Selby, Ed., The Chemical Rubber Co., 1970.

Table 9-47. SERIES *(Continued)*

(Increment form)

2. $f(x + h) = f(x) + hf'(x) + \dfrac{h^2}{2!} f''(x) + \dfrac{h^3}{3!} f'''(x) + \cdots$

$$= f(h) + xf'(h) + \frac{x^3}{2!} f''(h) + \frac{x^3}{3!} f'''(h) + \cdots$$

3. If $f(x)$ is a function possessing derivatives of all orders throughout the interval $a \leq x \leq b$, then there is a value X, with $a < X < b$, such that

$$f(b) = f(a) + (b - a)f'(a) + \frac{(b - a)^2}{2!} f''(a) + \cdots$$

$$+ \frac{(b - a)^{n-1}}{(n - 1)!} f^{(n-1)}(a) + \frac{(b - a)^n}{n!} f^{(n)}(X)$$

$$f(a + h) = f(a) + hf'(a) + \frac{h^2}{2!} f''(a) + \cdots + \frac{h^{n-1}}{(n - 1)!} f^{(n-1)}(a)$$

$$+ \frac{h^n}{n!} f^{(n)}(a + \theta h), \quad b = a + h, \; 0 < \theta < 1.$$

or

$$f(x) = f(a) + (x - a)f'(a) + \frac{(x - a)^2}{2!} f''(a) + \cdots + (x - a)^{n-1} \frac{f^{(n-1)}(a)}{(n - 1)!} + R_n,$$

where

$$R_n = \frac{f^{(n)}[a + \theta \cdot (x - a)]}{n!} (x - a)^n, \quad 0 < \theta < 1.$$

The above forms are known as Taylor's series with the remainder term.

4. *Taylor's series for a function of two variables*

If $\left(h \dfrac{\partial}{\partial x} + k \dfrac{\partial}{\partial y} \right) f(x, y) = h \dfrac{\partial f(x, y)}{\partial x} + k \dfrac{\partial f(x, y)}{\partial y}$;

$$\left(h \frac{\partial}{\partial x} + k \frac{\partial}{\partial y} \right)^2 f(x, y) = h^2 \frac{\partial^2 f(x, y)}{\partial x^2} + 2hk \frac{\partial^2 f(x, y)}{\partial x \, \partial y} + k^2 \frac{\partial^2 f(x, y)}{\partial y^2}$$

etc., and if $h \left(\dfrac{\partial}{\partial x} + k \dfrac{\partial}{\partial y} \right)^n f(x, y)\Big|_{\substack{x=a \\ y=b}}$ with the bar and subscripts means that after differentiation we are to replace x by a and y by b,

$$f(a + h, b + k) = f(a, b) + \left(h \frac{\partial}{\partial x} + k \frac{\partial}{\partial y} \right) f(x, y)\Big|_{\substack{x=a \\ y=b}} + \cdots$$

$$+ \frac{1}{n!} \left(h \frac{\partial}{\partial x} + k \frac{\partial}{\partial y} \right)^n f(x, y)\Big|_{\substack{x=a \\ y=b}} + \cdots$$

MACLAURIN

$$f(x) = f(0) + xf'(0) + \frac{x^2}{2!} f''(0) + \frac{x^3}{3!} f'''(0) + \cdots + x^{n-1} \frac{f^{(n-1)}(0)}{(n - 1)!} + R_n,$$

where

$$R_n = \frac{x^n f^{(n)}(\theta x)}{n!}, \quad 0 < \theta < 1.$$

Table 9-47. SERIES *(Continued)*

EXPONENTIAL

$$e = 1 + \frac{1}{1!} + \frac{1}{2!} + \frac{1}{3!} + \frac{1}{4!} + \cdots$$

$$e^x = 1 + x + \frac{x^2}{2!} + \frac{x^3}{3!} + \frac{x^4}{4!} + \cdots \qquad \text{(all real values of } x)$$

$$a^x = 1 + x \log_e a + \frac{(x \log_e a)^2}{2!} + \frac{(x \log_e a)^3}{3!} + \cdots$$

$$e^x = e^a \left[1 + (x - a) + \frac{(x - a)^2}{2!} + \frac{(x - a)^3}{3!} + \cdots \right]$$

LOGARITHMIC

$$\log_e x = \frac{x - 1}{x} + \frac{1}{2}\left(\frac{x - 1}{x}\right)^2 + \frac{1}{3}\left(\frac{x - 1}{x}\right)^3 + \cdots \qquad (x > \tfrac{1}{2})$$

$$\log_e x = (x - 1) - \tfrac{1}{2}(x - 1)^2 + \tfrac{1}{3}(x - 1)^3 - \cdots \qquad (2 \geq x > 0)$$

$$\log_e x = 2\left[\frac{x - 1}{x + 1} + \frac{1}{3}\left(\frac{x - 1}{x + 1}\right)^3 + \frac{1}{5}\left(\frac{x - 1}{x + 1}\right)^5 + \cdots\right] \qquad (x > 0)$$

$$\log_e (1 + x) = x - \tfrac{1}{2}x^2 + \tfrac{1}{3}x^3 - \tfrac{1}{4}x^4 + \cdots \qquad (-1 < x < 1)$$

$$\log_e (n + 1) - \log_e (n - 1) = 2\left[\frac{1}{n} + \frac{1}{3n^3} + \frac{1}{5n^5} + \cdots\right]$$

$$\log_e (a + x) = \log_e a + 2\left[\frac{x}{2a + x} + \frac{1}{3}\left(\frac{x}{2a + x}\right)^3 + \frac{1}{5}\left(\frac{x}{2a + x}\right)^5 + \cdots\right]$$

$$(a > 0, -a < x < +\infty)$$

$$\log_e \frac{1 + x}{1 - x} = 2\left[x + \frac{x^3}{3} + \frac{x^5}{5} + \cdots + \frac{x^{2n-1}}{2n - 1} + \cdots\right], \qquad -1 < x < 1$$

$$\log_e x = \log_e a + \frac{(x - a)}{a} - \frac{(x - a)^2}{2a^2} + \frac{(x - a)^3}{3a^3} - + \cdots, \qquad 0 < x \leq 2a$$

TRIGONOMETRIC

$$\sin x = x - \frac{x^3}{3!} + \frac{x^5}{5!} - \frac{x^7}{7!} + \cdots \qquad \text{(all real values of } x)$$

$$\cos x = 1 - \frac{x^2}{2!} + \frac{x^4}{4!} - \frac{x^6}{6!} + \cdots \qquad \text{(all real values of } x)$$

$$\tan x = x + \frac{x^3}{3} + \frac{2x^5}{15} + \frac{17x^7}{315} + \frac{62x^9}{2835} + \cdots + \frac{2^{2n}(2^{2n} - 1)B_{2n}}{(2n)!} x^{2n-1} + \cdots,$$

$$\left[x^2 < \frac{\pi^2}{4}, \text{ and } B_n \text{ represents the } n\text{'th Bernoulli number.}\right]$$

$$\cot x = \frac{1}{x} - \frac{x}{3} - \frac{x^3}{45} - \frac{2x^5}{945} - \frac{x^7}{4725} - \cdots - \frac{2^{2n}B_{2n}}{(2n)!} x^{2n-1} - \cdots,$$

$$[x^2 < \pi^2, \text{ and } B_n \text{ represents the } n\text{'th Bernoulli number.}]$$

Table 9-47.   SERIES *(Continued)*

$$\sec x = 1 + \frac{x^2}{2} + \frac{5}{24}x^4 + \frac{61}{720}x^6 + \frac{277}{8064}x^8 + \cdots + \frac{E_{2n}x^{2n}}{(2n)!} + \cdots,$$

$$\left[x^2 < \frac{\pi^2}{4}, \text{ and } E_n \text{ represents the } n\text{'th Euler number.}\right]$$

$$\csc x = \frac{1}{x} + \frac{x}{6} + \frac{7}{360}x^3 + \frac{31}{15,120}x^5 + \frac{127}{604,800}x^7 + \cdots$$

$$+ \frac{2(2^{2n-1} - 1)}{(2n)!}B_{2n}x^{2n-1} + \cdots,$$

$$[x^2 < \pi^2, \text{ and } B_n \text{ represents } n\text{'th Bernoulli number.}]$$

$$\sin x = x\left(1 - \frac{x^2}{\pi^2}\right)\left(1 - \frac{x^2}{2^2\pi^2}\right)\left(1 - \frac{x^2}{3^2\pi^2}\right)\cdots \qquad (x^2 < \infty)$$

$$\cos x = \left(1 - \frac{4x^2}{\pi^2}\right)\left(1 - \frac{4x^2}{3^2\pi^2}\right)\left(1 - \frac{4x^2}{5^2\pi^2}\right)\cdots \qquad (x^2 < \infty)$$

$$\sin^{-1} x = x + \frac{x^3}{2\cdot3} + \frac{1\cdot3}{2\cdot4\cdot5}x^5 + \frac{1\cdot3\cdot5}{2\cdot4\cdot6\cdot7}x^7 + \cdots \qquad \left(x^2 < 1, -\frac{\pi}{2} < \sin^{-1} x < \frac{\pi}{2}\right)$$

$$\cos^{-1} x = \frac{\pi}{2} - \left(x + \frac{x^3}{2\cdot3} + \frac{1\cdot3}{2\cdot4\cdot5}x^5 + \frac{1\cdot3\cdot5x^7}{2\cdot4\cdot6\cdot7} + \cdots\right) \qquad (x^2 < 1, 0 < \cos^{-1} x < \pi)$$

$$\tan^{-1} x = x - \frac{x^3}{3} + \frac{x^5}{5} - \frac{x^7}{7} + \cdots \qquad (x^2 < 1)$$

$$\tan^{-1} x = \frac{\pi}{2} - \frac{1}{x} + \frac{1}{3x^2} - \frac{1}{5x^5} + \frac{1}{7x^7} - \cdots \qquad (x > 1)$$

$$\tan^{-1} x = -\frac{\pi}{2} - \frac{1}{x} + \frac{1}{3x^2} - \frac{1}{5x^5} + \frac{1}{7x^7} - \cdots \qquad (x < -1)$$

$$\cot^{-1} x = \frac{\pi}{2} - x + \frac{x^3}{3} - \frac{x^5}{5} + \frac{x^7}{7} - \cdots \qquad (x^2 < 1)$$

$$\log_e \sin x = \log_e x - \frac{x^2}{6} - \frac{x^4}{180} - \frac{x^6}{2835} - \cdots \qquad (x^2 < \pi^2)$$

$$\log_e \cos x = -\frac{x^2}{2} - \frac{x^4}{12} - \frac{x^6}{45} - \frac{17x^8}{2520} - \cdots \qquad \left(x^2 < \frac{\pi^2}{4}\right)$$

$$\log_e \tan x = \log_e x + \frac{x^2}{3} + \frac{7x^4}{90} + \frac{62x^6}{2835} + \cdots \qquad \left(x^2 < \frac{\pi^2}{4}\right)$$

$$e^{\sin x} = 1 + x + \frac{x^2}{2!} - \frac{3x^4}{4!} - \frac{8x^5}{5!} - \frac{3x^6}{6!} + \frac{56x^7}{7!} + \cdots$$

$$e^{\cos x} = e\left(1 - \frac{x^2}{2!} + \frac{4x^4}{4!} - \frac{31x^6}{6!} + \cdots\right)$$

$$e^{\tan x} = 1 + x + \frac{x^2}{2!} + \frac{3x^3}{3!} + \frac{9x^4}{4!} + \frac{37x^5}{5!} + \cdots \qquad \left(x^2 < \frac{\pi^2}{4}\right)$$

$$\sin x = \sin a + (x - a)\cos a - \frac{(x - a)^2}{2!}\sin a$$

$$- \frac{(x - a)^3}{3!}\cos a + \frac{(x - a)^4}{4!}\sin a + \cdots$$

Table 9-47.   SERIES *(Continued)*

HYPERBOLIC AND INVERSE HYPERBOLIC

Table of expansion of certain functions into power series

$$\sinh x = x + \frac{x^3}{3!} + \frac{x^5}{5!} + \frac{x^7}{7!} + \cdots + \frac{x^{2n+1}}{(2n+1)!} + \cdots \qquad |x| < \infty$$

$$\cosh x = 1 + \frac{x^2}{2!} + \frac{x^4}{4!} + \frac{x^6}{6!} + \cdots + \frac{x^{2n}}{(2n)!} + \cdots \qquad |x| < \infty$$

$$\tanh x = x - \frac{1}{3} x^3 + \frac{2}{15} x^5 - \frac{17}{315} x^7 + \frac{62}{2835} x^9 - \cdots$$

$$+ \frac{(-1)^{n+1} 2^{2n}(2^{2n} - 1)}{(2n)!} B_{2n} x^{2n-1} \pm \cdots \text{ (1)} \quad |x| < \frac{\pi}{2}$$

$$\coth x = \frac{1}{x} + \frac{x}{3} - \frac{x^3}{45} + \frac{2x^5}{945} - \frac{x^7}{4725} + \cdots + \frac{(-1)^{n+1} 2^{2n}}{(2n)!} B_{2n} x^{2n-1} \pm \cdots \text{ (1)}$$

$$0 < |x| < \pi$$

$$\operatorname{sech} x = 1 - \frac{1}{2!} x^2 + \frac{5}{4!} x^4 - \frac{61}{6!} x^6 + \frac{1385}{8!} x^8 - \cdots + \frac{(-1)^n}{(2n)!} E_{2n} x^{2n} \pm \cdots \text{ (2)}$$

$$|x| < \frac{\pi}{2}$$

$$\operatorname{cosech} x = \frac{1}{x} - \frac{x}{6} + \frac{7x^3}{360} - \frac{31x^5}{15,120} + \cdots + \frac{2(-1)^n(2^{2n-1} - 1)}{(2n)!} B_{2n} x^{2n-1} + \cdots \text{ (1)}$$

$$0 < |x| < \pi$$

$$\arg\sinh x = x - \frac{1}{2\cdot 3} x^3 + \frac{1\cdot 3}{2\cdot 4\cdot 5} x^5 - \frac{1\cdot 3\cdot 5}{2\cdot 4\cdot 6\cdot 7} x^7 + \cdots$$

$$+ (-1)^n \cdot \frac{1\cdot 3\cdot 5(2n-1)}{2\cdot 4\cdot 6\ldots 2n(2n+1)} x^{2n+1} \pm \cdots \quad |x| < 1$$

$$\arg\cosh x = \pm \left[\ln(2x) - \frac{1}{2\cdot 2x^2} - \frac{1\cdot 3}{2\cdot 4\cdot 4x^4} - \frac{1\cdot 3\cdot 5}{2\cdot 4\cdot 6\cdot 6x^6} - \cdots \right] \qquad x > 1$$

$$\arg\tanh x = x + \frac{x^3}{3} + \frac{x^5}{5} + \frac{x^7}{7} + \cdots + \frac{x^{2n+1}}{2n+1} + \cdots \qquad |x| < 1$$

$$\arg\coth x = \frac{1}{x} + \frac{1}{3x^3} + \frac{1}{5x^5} + \frac{1}{7x^7} + \cdots + \frac{1}{(2n+1)x^{2n+1}} + \cdots \qquad |x| > 1$$

---

(1) B_n denotes Bernoulli's numbers.
(2) E_n denotes Euler's numbers.

FOURIER

(Also see Index for Cosine and Sine Transforms)

1. If $f(x)$ is a bounded periodic function of period $2L$ (i.e. $f(x + 2L) = f(x)$), and satisfies the *Dirichlet conditions*:

   a) In any period $f(x)$ is continuous, except possibly for a finite number of jump discontinuities.

Table 9-47. SERIES (Continued)

b) In any period $f(x)$ has only a finite number of maxima and minima. Then $f(x)$ may be represented by the *Fourier series*

$$\frac{a_0}{2} + \sum_{n=1}^{\infty} \left(a_n \cos \frac{n\pi x}{L} + b_n \sin \frac{n\pi x}{L} \right),$$

where a_n and b_n are as determined below. This series will converge to $f(x)$ at every point where $f(x)$ is continuous, and to

$$\frac{f(x^+) + f(x^-)}{2}$$

(i.e. the average of the left-hand and right-hand limits) at every point where $f(x)$ has a jump discontinuity.

$$a_n = \frac{1}{L} \int_{-L}^{L} f(x) \cos \frac{n\pi x}{L} \, dx, \quad n = 0, 1, 2, 3, \ldots;$$

$$b_n = \frac{1}{L} \int_{-L}^{L} f(x) \sin \frac{n\pi x}{L} \, dx, \quad n = 1, 2, 3, \ldots$$

We may also write

$$a_n = \frac{1}{L} \int_{\alpha}^{\alpha + 2L} f(x) \cos \frac{n\pi x}{L} \, dx \text{ and } b_n = \frac{1}{L} \int_{\alpha}^{\alpha + 2L} f(x) \sin \frac{n\pi x}{L} \, dx,$$

where α is any real number. Thus if $\alpha = 0$,

$$a_n = \frac{1}{L} \int_{0}^{2L} f(x) \cos \frac{n\pi x}{L} \, dx, \quad n = 0, 1, 2, 3, \ldots;$$

$$b_n = \frac{1}{L} \int_{0}^{2L} f(x) \sin \frac{n\pi x}{L} \, dx, \quad n = 1, 2, 3, \ldots$$

2. If in addition to the above restrictions, $f(x)$ is even (i.e. $f(-x) = f(x)$), the Fourier series reduces to

$$\frac{a_0}{2} + \sum_{n=1}^{\infty} a_n \cos \frac{n\pi x}{L}.$$

That is, $b_n = 0$. In this case, a simpler formula for a_n is

$$a_n = \frac{2}{L} \int_{0}^{L} f(x) \cos \frac{n\pi x}{L} \, dx, \quad n = 0, 1, 2, 3, \ldots$$

3. If in addition to the restrictions in (1), $f(x)$ is an odd function (i.e. $f(-x) = -f(x)$), then the Fourier series reduces to

$$\sum_{n=1}^{\infty} b_n \sin \frac{n\pi x}{L}.$$

That is, $a_n = 0$. In this case, a simpler formula for the b_n is

$$b_n = \frac{2}{L} \int_{0}^{L} f(x) \sin \frac{n\pi x}{L} \, dx, \quad n = 1, 2, 3, \ldots$$

Table 9-47.   SERIES *(Continued)*

4. If in addition to the restrictions in (2) above, $f(x) = -f(L - x)$, then a_n will be 0 for all even values of n, including $n = 0$. Thus in this case, the expansion reduces to

$$\sum_{m=1}^{\infty} a_{2m-1} \cos \frac{(2m - 1)\pi x}{L}.$$

5. If in addition to the restrictions in (3) above, $f(x) = f(L - x)$, then b_n will be 0 for all even values of n. Thus in this case, the expansion reduces to

$$\sum_{m=1}^{\infty} b_{2m-1} \sin \frac{(2m - 1)\pi x}{L}.$$

(The series in (4) and (5) are known as *odd-harmonic series*, since only the odd harmonics appear. Similar rules may be stated for even-harmonic series, but when a series appears in the even-harmonic form, it means that $2L$ has not been taken as the smallest period of $f(x)$. Since any integral multiple of a period is also a period, series obtained in this way will also work, but in general computation is simplified if $2L$ is taken to be the smallest period.)

6. If we write the Euler definitions for $\cos \theta$ and $\sin \theta$, we obtain the complex form of the Fourier Series known either as the "Complex Fourier Series" or the "Exponential Fourier Series" of $f(x)$. It is represented as

$$f(x) = \tfrac{1}{2} \sum_{n=-\infty}^{n=+\infty} c_n e^{i\omega_n x},$$

where

$$c_n = \frac{1}{L} \int_{-L}^{L} f(x) e^{-i\omega_n x} dx, \quad n = 0, \pm 1, \pm 2, \pm 3, \dots$$

with $\omega_n = \dfrac{n\pi}{L}, \quad n = 0, \pm 1, \pm 2, \dots$

The set of coefficients $\{c_n\}$ is often referred to as the Fourier spectrum.

7. If both sine and cosine terms are present and if $f(x)$ is of period $2L$ and expandable by a Fourier series, it can be represented as

$$f(x) = \frac{a_0}{2} + \sum_{n=1}^{\infty} c_n \sin\left(\frac{n\pi x}{L} + \phi_n\right), \text{ where } a_n = c_n \sin \phi_n,$$

$$b_n = c_n \cos \phi_n, \quad c_n = \sqrt{a_n^2 + b_n^2}, \quad \phi_n = \arctan\left(\frac{a_n}{b_n}\right)$$

It can also be represented as

$$f(x) = \frac{a_0}{2} + \sum_{n=1}^{\infty} c_n \cos\left(\frac{n\pi x}{L} + \phi_n\right), \text{ where } a_n = c_n \cos \phi_n,$$

$$b_n = -c_n \sin \phi_n, \quad c_n = \sqrt{a_n^2 + b_n^2}, \quad \phi_n = \arctan\left(-\frac{b_n}{a_n}\right)$$

where ϕ_n is chosen so as to make a_n, b_n, and c_n hold.

Table 9-47.   SERIES *(Continued)*

8. The following table of trigonometric identities should be helpful for developing Fourier Series.

	n	n even	n odd	$n/2$ odd	$n/2$ even
$\sin n\pi$	0	0	0	0	0
$\cos n\pi$	$(-1)^n$	$+1$	-1	$+1$	$+1$
$*\sin \dfrac{n\pi}{2}$		0	$(-1)^{(n-1)/2}$	0	0
$*\cos \dfrac{n\pi}{2}$		$(-1)^{n/2}$	0	-1	$+1$
$\sin \dfrac{n\pi}{4}$			$\dfrac{\sqrt{2}}{2}(-1)^{(n^2+4n+11)/8}$	$(-1)^{(n-2)/4}$	0

*A useful formula for $\sin \dfrac{n\pi}{2}$ and $\cos \dfrac{n\pi}{2}$ is given by

$$\sin \frac{n\pi}{2} = \frac{(i)^{n+1}}{2}[(-1)^n - 1] \text{ and } \cos \frac{n\pi}{2} = \frac{(i)^n}{2}[(-1)^n + 1], \text{ where } i^2 = -1.$$

AUXILIARY FORMULAS FOR FOURIER SERIES

$$1 = \frac{4}{\pi}\left[\sin \frac{\pi x}{k} + \frac{1}{3}\sin \frac{3\pi x}{k} + \frac{1}{5}\sin \frac{5\pi x}{k} + \cdots\right] \qquad [0 < x < k]$$

$$x = \frac{2k}{\pi}\left[\sin \frac{\pi x}{k} - \frac{1}{2}\sin \frac{2\pi x}{k} + \frac{1}{3}\sin \frac{3\pi x}{k} - \cdots\right] \qquad [-k < x < k]$$

$$x = \frac{k}{2} - \frac{4k}{\pi^2}\left[\cos \frac{\pi x}{k} + \frac{1}{3^2}\cos \frac{3\pi x}{k} + \frac{1}{5^2}\cos \frac{5\pi x}{k} + \cdots\right] \qquad [0 < x < k]$$

$$x^2 = \frac{2k^2}{\pi^3}\left[\left(\frac{\pi^2}{1} - \frac{4}{1}\right)\sin \frac{\pi x}{k} - \frac{\pi^2}{2}\sin \frac{2\pi x}{k} + \left(\frac{\pi^2}{3} - \frac{4}{3^3}\right)\sin \frac{3\pi x}{k}\right.$$
$$\left. - \frac{\pi^2}{4}\sin \frac{4\pi x}{k} + \left(\frac{\pi^2}{5} - \frac{4}{5^3}\right)\sin \frac{5\pi x}{k} + \cdots\right] \quad [0 < x < k]$$

$$x^2 = \frac{k^2}{3} - \frac{4k^2}{\pi^2}\left[\cos \frac{\pi x}{k} - \frac{1}{2^2}\cos \frac{2\pi x}{k} + \frac{1}{3^2}\cos \frac{3\pi x}{k} - \frac{1}{4^2}\cos \frac{4\pi x}{k} + \cdots\right]$$
$$[-k < x < k]$$

$$1 - \frac{1}{3} + \frac{1}{5} - \frac{1}{7} + \cdots = \frac{\pi}{4}$$

$$1 + \frac{1}{2^2} + \frac{1}{3^2} + \frac{1}{4^2} + \cdots = \frac{\pi^2}{6}$$

$$1 - \frac{1}{2^2} + \frac{1}{3^2} - \frac{1}{4^2} + \cdots = \frac{\pi^2}{12}$$

$$1 + \frac{1}{3^2} + \frac{1}{5^2} + \frac{1}{7^2} + \cdots = \frac{\pi^2}{8}$$

$$\frac{1}{2^2} + \frac{1}{4^2} + \frac{1}{6^2} + \frac{1}{8^2} + \cdots = \frac{\pi^2}{24}$$

Table 9-48. FOURIER EXPANSIONS
FOR BASIC PERIODIC FUNCTIONS*

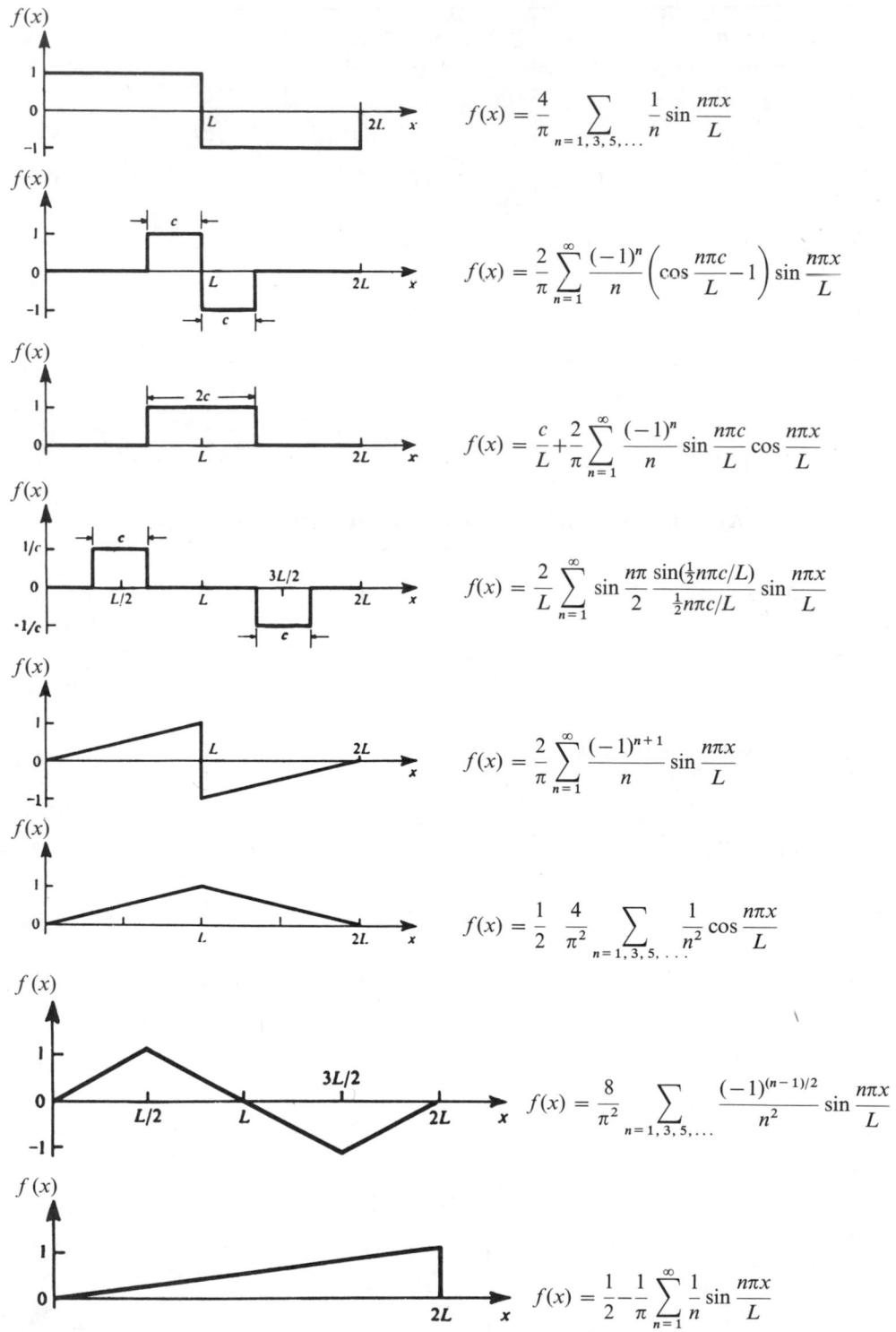

$$f(x) = \frac{4}{\pi} \sum_{n=1,3,5,\ldots} \frac{1}{n} \sin \frac{n\pi x}{L}$$

$$f(x) = \frac{2}{\pi} \sum_{n=1}^{\infty} \frac{(-1)^n}{n} \left(\cos \frac{n\pi c}{L} - 1 \right) \sin \frac{n\pi x}{L}$$

$$f(x) = \frac{c}{L} + \frac{2}{\pi} \sum_{n=1}^{\infty} \frac{(-1)^n}{n} \sin \frac{n\pi c}{L} \cos \frac{n\pi x}{L}$$

$$f(x) = \frac{2}{L} \sum_{n=1}^{\infty} \sin \frac{n\pi}{2} \frac{\sin(\frac{1}{2}n\pi c/L)}{\frac{1}{2}n\pi c/L} \sin \frac{n\pi x}{L}$$

$$f(x) = \frac{2}{\pi} \sum_{n=1}^{\infty} \frac{(-1)^{n+1}}{n} \sin \frac{n\pi x}{L}$$

$$f(x) = \frac{1}{2} - \frac{4}{\pi^2} \sum_{n=1,3,5,\ldots} \frac{1}{n^2} \cos \frac{n\pi x}{L}$$

$$f(x) = \frac{8}{\pi^2} \sum_{n=1,3,5,\ldots} \frac{(-1)^{(n-1/2)}}{n^2} \sin \frac{n\pi x}{L}$$

$$f(x) = \frac{1}{2} - \frac{1}{\pi} \sum_{n=1}^{\infty} \frac{1}{n} \sin \frac{n\pi x}{L}$$

---

*Extracted from: "Differential Equations in Engineering Problems", M.G. Salvadori and R.J. Schwarz, by permission of Prentice-Hall, Inc., 1954, pp. 372–373.

Table 9-48.   FOURIER EXPANSIONS
FOR BASIC PERIODIC FUNCTIONS *(Continued)*

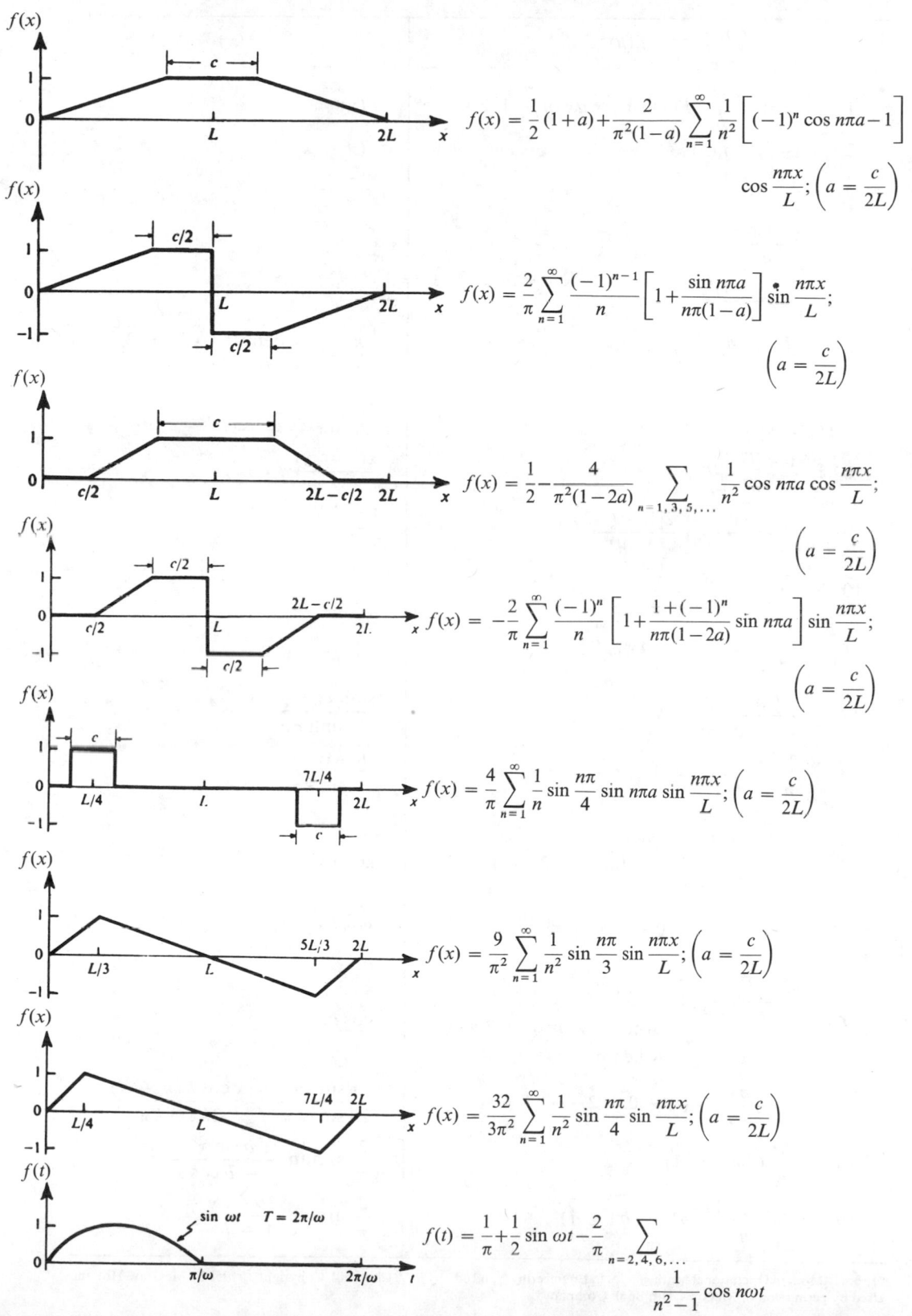

$$f(x) = \frac{1}{2}(1+a) + \frac{2}{\pi^2(1-a)} \sum_{n=1}^{\infty} \frac{1}{n^2} \left[(-1)^n \cos n\pi a - 1 \right] \cos \frac{n\pi x}{L}; \left(a = \frac{c}{2L} \right)$$

$$f(x) = \frac{2}{\pi} \sum_{n=1}^{\infty} \frac{(-1)^{n-1}}{n} \left[1 + \frac{\sin n\pi a}{n\pi(1-a)} \right] \sin \frac{n\pi x}{L}; \left(a = \frac{c}{2L} \right)$$

$$f(x) = \frac{1}{2} - \frac{4}{\pi^2(1-2a)} \sum_{n=1, 3, 5, \ldots} \frac{1}{n^2} \cos n\pi a \cos \frac{n\pi x}{L}; \left(a = \frac{c}{2L} \right)$$

$$f(x) = -\frac{2}{\pi} \sum_{n=1}^{\infty} \frac{(-1)^n}{n} \left[1 + \frac{1+(-1)^n}{n\pi(1-2a)} \sin n\pi a \right] \sin \frac{n\pi x}{L}; \left(a = \frac{c}{2L} \right)$$

$$f(x) = \frac{4}{\pi} \sum_{n=1}^{\infty} \frac{1}{n} \sin \frac{n\pi}{4} \sin n\pi a \sin \frac{n\pi x}{L}; \left(a = \frac{c}{2L} \right)$$

$$f(x) = \frac{9}{\pi^2} \sum_{n=1}^{\infty} \frac{1}{n^2} \sin \frac{n\pi}{3} \sin \frac{n\pi x}{L}; \left(a = \frac{c}{2L} \right)$$

$$f(x) = \frac{32}{3\pi^2} \sum_{n=1}^{\infty} \frac{1}{n^2} \sin \frac{n\pi}{4} \sin \frac{n\pi x}{L}; \left(a = \frac{c}{2L} \right)$$

$$f(t) = \frac{1}{\pi} + \frac{1}{2} \sin \omega t - \frac{2}{\pi} \sum_{n=2, 4, 6, \ldots} \frac{1}{n^2-1} \cos n\omega t$$

Table 9-49.   FOURIER TRANSFORMS

FINITE SINE TRANSFORMS*

	$f_s(n)$	$F(x)$
1	$f_s(n) = \displaystyle\int_0^\pi F(x) \sin nx\, dx\ (n = 1, 2, \cdots)$	$F(x)$
2	$(-1)^{n+1} f_s(n)$	$F(\pi - x)$
3	$\dfrac{1}{n}$	$\dfrac{\pi - x}{\pi}$
4	$\dfrac{(-1)^{n+1}}{n}$	$\dfrac{x}{\pi}$
5	$\dfrac{1 - (-1)^n}{n}$	1
6	$\dfrac{2}{n^2} \sin \dfrac{n\pi}{2}$	$\begin{cases} x & \text{when } 0 < x < \pi/2 \\ \pi - x & \text{when } \pi/2 < x < \pi \end{cases}$
7	$\dfrac{(-1)^{n+1}}{n^3}$	$\dfrac{x(\pi^2 - x^2)}{6\pi}$
8	$\dfrac{1 - (-1)^n}{n^3}$	$\dfrac{x(\pi - x)}{2}$
9	$\dfrac{\pi^2(-1)^{n-1}}{n} - \dfrac{2[1 - (-1)^n]}{n^3}$	x^2
10	$\pi(-1)^n \left(\dfrac{6}{n^3} - \dfrac{\pi^2}{n} \right)$	x^3
11	$\dfrac{n}{n^2 + c^2} [1 - (-1)^n e^{c\pi}]$	e^{cx}
12	$\dfrac{n}{n^2 + c^2}$	$\dfrac{\sinh c(\pi - x)}{\sinh c\pi}$
13	$\dfrac{n}{n^2 - k^2}\ (k \neq 0, 1, 2, \cdots)$	$\dfrac{\sin k(\pi - x)}{\sin k\pi}$
14	$\begin{cases} \dfrac{\pi}{2} & \text{when } n = m \\ 0 & \text{when } n \neq m \end{cases} (m = 1, 2, \cdots)$	$\sin mx$
15	$\dfrac{n}{n^2 - k^2} [1 - (-1)^n \cos k\pi]$ $(k \neq 1, 2, \cdots)$	$\cos kx$
16	$\begin{cases} \dfrac{n}{n^2 - m^2} [1 - (-1)^{n+m}] \\ \qquad \text{when } n \neq m = 1, 2, \cdots \\ 0 \qquad \text{when } n = m \end{cases}$	$\cos mx$
17	$\dfrac{n}{(n^2 - k^2)^2}\ (k \neq 0, 1, 2, \cdots)$	$\dfrac{\pi \sin kx}{2k \sin^2 k\pi} - \dfrac{x \cos k(\pi - x)}{2k \sin k\pi}$
18	$\dfrac{b^n}{n}\ (\lvert b \rvert \leq 1)$	$\dfrac{2}{\pi} \arctan \dfrac{b \sin x}{1 - b \cos x}$
19	$\dfrac{1 - (-1)^n}{n} b^n\ (\lvert b \rvert \leq 1)$	$\dfrac{2}{\pi} \arctan \dfrac{2b \sin x}{1 - b^2}$

Table 9-49. FOURIER TRANSFORMS *(Continued)*

FINITE COSINE TRANSFORMS*

	$f_c(n)$	$F(x)$
1	$f_c(n) = \displaystyle\int_0^\pi F(x) \cos nx\, dx \quad (n = 0, 1, 2, \cdots)$	$F(x)$
2	$(-1)^n f_c(n)$	$F(\pi - x)$
3	0 when $n = 1, 2, \cdots;\ f_c(0) = \pi$	1
4	$\dfrac{2}{n} \sin \dfrac{n\pi}{2};\ f_c(0) = 0$	$\begin{cases} 1 \text{ when } 0 < x < \pi/2 \\ -1 \text{ when } \pi/2 < x < \pi \end{cases}$
5	$-\dfrac{1 - (-1)^n}{n^2};\ f_c(0) = \dfrac{\pi^2}{2}$	x
6	$\dfrac{(-1)^n}{n^2};\ f_c(0) = \dfrac{\pi^2}{6}$	$\dfrac{x^2}{2\pi}$
7	$\dfrac{1}{n^2};\ f_c(0) = 0$	$\dfrac{(\pi - x)^2}{2\pi} - \dfrac{\pi}{6}$
8	$3\pi^2 \dfrac{(-1)^n}{n^2} - 6 \dfrac{1 - (-1)^n}{n^4};\ f_c(0) = \dfrac{\pi^4}{4}$	x^3
9	$\dfrac{(-1)^n e^c \pi - 1}{n^2 + c^2}$	$\dfrac{1}{c} e^{cx}$
10	$\dfrac{1}{n^2 + c^2}$	$\dfrac{\cosh c(\pi - x)}{c \sinh c\pi}$
11	$\dfrac{k}{n^2 - k^2} [(-1)^n \cos \pi k - 1]$ $(k \neq 0, 1, 2, \cdots)$	$\sin kx$
12	$\dfrac{(-1)^{n+m} - 1}{n^2 - m^2};\ f_c(m) = 0 \quad (m = 1, 2, \cdots)$	$\dfrac{1}{m} \sin mx$
13	$\dfrac{1}{n^2 - k^2} \quad (k \neq 0, 1, 2, \cdots)$	$-\dfrac{\cos k(\pi - x)}{k \sin k\pi}$
14	0 when $n = 1, 2, \cdots;$ $f_c(m) = \dfrac{\pi}{2} \quad (m = 1, 2, \cdots)$	$\cos mx$

*From: "Modern Operational Mathematics in Engineering", 2nd ed., by R.V. Churchill. Copyright © 1958 by McGraw-Hill, Inc. Used by permission of McGraw-Hill Book Company.

Table 9-49.   FOURIER TRANSFORMS *(Continued)*

FOURIER SINE TRANSFORMS[1]

	$F(x)$	$f_s(\alpha)$
1	$\begin{cases} 1 & (0 < x < a) \\ 0 & (x > a) \end{cases}$	$\sqrt{\dfrac{2}{\pi}}\left[\dfrac{1 - \cos \alpha}{\alpha}\right]$
2	$x^{p-1}\ (0 < p < 1)$	$\sqrt{\dfrac{2}{\pi}}\ \dfrac{\Gamma(p)}{\alpha^p}\ \sin \dfrac{p\pi}{2}$
3	$\begin{cases} \sin x & (0 < x < a) \\ 0 & (x > a) \end{cases}$	$\dfrac{1}{\sqrt{2\pi}}\left[\dfrac{\sin[a(1 - \alpha)]}{1 - \alpha} - \dfrac{\sin[a(1 + \alpha)]}{1 + \alpha}\right]$
4	e^{-x}	$\sqrt{\dfrac{2}{\pi}}\left[\dfrac{\alpha}{1 + \alpha^2}\right]$
5	$xe^{-x^2/2}$	$\alpha e^{-\alpha^2/2}$
6	$\cos \dfrac{x^2}{2}$	$\sqrt{2}\left[\sin \dfrac{\alpha^2}{2}\ C\!\left(\dfrac{\alpha^2}{2}\right) - \cos \dfrac{\alpha^2}{2}\ S\!\left(\dfrac{\alpha^2}{2}\right)\right]^{*}$
7	$\sin \dfrac{x^2}{2}$	$\sqrt{2}\left[\cos \dfrac{\alpha^2}{2}\ C\!\left(\dfrac{\alpha^2}{2}\right) + \sin \dfrac{\alpha^2}{2}\ S\!\left(\dfrac{\alpha^2}{2}\right)\right]^{*}$

*$C(y)$ and $S(y)$ are the Fresnel integrals

$$C(y) = \frac{1}{\sqrt{2\pi}} \int_0^y \frac{1}{\sqrt{t}} \cos t\, dt,$$

$$S(y) = \frac{1}{\sqrt{2\pi}} \int_0^y \frac{1}{\sqrt{t}} \sin t\, dt.$$

[1] More extensive tables of the Fourier sine and cosine transforms can be found in Fritz Oberhettinger, "Tabellen zur-Fourier Transformation," Springer (1957).

FOURIER COSINE TRANSFORMS

	$F(x)$	$f_c(\alpha)$
1	$\begin{cases} 1 & (0 < x < a) \\ 0 & (x > a) \end{cases}$	$\sqrt{\dfrac{2}{\pi}}\ \dfrac{\sin a\alpha}{\alpha}$
2	$x^{p-1}\ \ (0 < p < 1)$	$\sqrt{\dfrac{2}{\pi}}\ \dfrac{\Gamma(p)}{\alpha^p}\ \cos \dfrac{p\pi}{2}$
3	$\begin{cases} \cos x & (0 < x < a) \\ 0 & (x > a) \end{cases}$	$\dfrac{1}{\sqrt{2\pi}}\left[\dfrac{\sin[a(1 - \alpha)]}{1 - \alpha} + \dfrac{\sin[a(1 + \alpha)]}{1 + \alpha}\right]$
4	e^{-x}	$\sqrt{\dfrac{2}{\pi}}\left(\dfrac{1}{1 + \alpha^2}\right)$
5	$e^{-x^2/2}$	$e^{-\alpha^2/2}$
6	$\cos \dfrac{x^2}{2}$	$\cos\left(\dfrac{\alpha^2}{2} - \dfrac{\pi}{4}\right)$
7	$\sin \dfrac{x^2}{2}$	$\cos\left(\dfrac{\alpha^2}{2} + \dfrac{\pi}{4}\right)$

Table 9-49.   FOURIER TRANSFORMS *(Continued)*

MISCELLANEOUS TRANSFORMS[1]

	$F(x)$	$f(\alpha)$
1	$\dfrac{\sin ax}{x}$	$\begin{cases} \sqrt{\dfrac{\pi}{2}} & \lvert\alpha\rvert < a \\ 0 & \lvert\alpha\rvert > a \end{cases}$
2	$\begin{cases} e^{iwx} & (p < x < q) \\ 0 & (x < p,\; x > q) \end{cases}$	$\dfrac{i}{\sqrt{2\pi}}\dfrac{e^{ip(w+\alpha)} - e^{iq(w+\alpha)}}{(w+\alpha)}$
3	$\begin{cases} e^{-cx+iwx} & (x > 0) \\ 0 & (x < 0) \end{cases}$ $(c > 0)$	$\dfrac{i}{\sqrt{2\pi}(w + \alpha + ic)}$
4	e^{-px^2} $R(p) > 0$	$\dfrac{1}{\sqrt{2p}}\,e^{-\alpha^2/4p}$
5	$\cos px^2$	$\dfrac{1}{\sqrt{2p}}\cos\left[\dfrac{\alpha^2}{4p} - \dfrac{\pi}{4}\right]$
6	$\sin px^2$	$\dfrac{1}{\sqrt{2p}}\cos\left[\dfrac{\alpha^2}{4p} + \dfrac{\pi}{4}\right]$
7	$\lvert x\rvert^{-p}$ $(0 < p < 1)$	$\sqrt{\dfrac{2}{\pi}}\,\dfrac{\Gamma(1-p)\sin\dfrac{p\pi}{2}}{\lvert\alpha\rvert^{(1-p)}}$
8	$\dfrac{e^{-a\lvert x\rvert}}{\sqrt{\lvert x\rvert}}$	$\dfrac{\sqrt{\sqrt{(a^2+\alpha^2)} + a}}{\sqrt{a^2+\alpha^2}}$
9	$\dfrac{\cosh ax}{\cosh \pi x}$ $(-\pi < a < \pi)$	$\sqrt{\dfrac{2}{\pi}}\,\dfrac{\cos\dfrac{a}{2}\cosh\dfrac{\alpha}{2}}{\cosh\alpha + \cos a}$
10	$\dfrac{\sinh ax}{\sinh \pi x}$ $(-\pi < a < \pi)$	$\dfrac{1}{\sqrt{2\pi}}\dfrac{\sin a}{\cosh\alpha + \cos a}$
11	$\begin{cases} \dfrac{1}{\sqrt{a^2 - x^2}} & (\lvert x\rvert < a) \\ 0 & (\lvert x\rvert > a) \end{cases}$	$\sqrt{\dfrac{\pi}{2}}\,J_0(a\alpha)$
12	$\dfrac{\sin[b\sqrt{a^2+x^2}]}{\sqrt{a^2+x^2}}$	$\begin{cases} 0 & (\lvert\alpha\rvert > b) \\ \sqrt{\dfrac{\pi}{2}}\,J_0(a\sqrt{b^2-\alpha^2}) & (\lvert\alpha\rvert < b) \end{cases}$
13	$\begin{cases} P_n(x) & (\lvert x\rvert < 1) \\ 0 & (\lvert x\rvert > 1) \end{cases}$	$\dfrac{i^n}{\sqrt{\alpha}}\,J_{n+\frac{1}{2}}(\alpha)$
14	$\begin{cases} \dfrac{\cos[b\sqrt{a^2-x^2}]}{\sqrt{a^2-x^2}} & (\lvert x\rvert < a) \\ 0 & (\lvert x\rvert > a) \end{cases}$	$\sqrt{\dfrac{\pi}{2}}\,J_0(a\sqrt{a^2+b^2})$
15	$\begin{cases} \dfrac{\cosh[b\sqrt{a^2-x^2}]}{\sqrt{a^2-x^2}} & (\lvert x\rvert < a) \\ 0 & (\lvert x\rvert > a) \end{cases}$	$\sqrt{\dfrac{\pi}{2}}\,J_0(a\sqrt{\alpha^2-b^2})$

[1]More extensive tables of Fourier transforms can be found in W. Magnus and F. Oberhettinger, "Formulas and Theorems of the Special Functions of Mathematical Physics," pp. 116–120. Chelsea (1949).

Table 9-50.   LAPLACE OPERATIONS*

	$F(t)$	$f(s)$
1	$F(t)$	$\int_0^\infty e^{-st} F(t)\,dt$
2	$AF(t) + BG(t)$	$Af(s) + Bg(s)$
3	$F'(t)$	$sf(s) - F(+0)$
4	$F^{(n)}(t)$	$s^n f(s) - s^{n-1} F(+0) - s^{n-2} F'(+0) - \cdots \\ \qquad\qquad - F^{(n-1)}(+0)$
5	$\int_0^t F(\tau)\,d\tau$	$\dfrac{1}{s} f(s)$
6	$\int_0^t \int_0^r F(\lambda)\,d\lambda\,d\tau$	$\dfrac{1}{s^2} f(s)$
7	$\int_0^t F_1(t - \tau) F_2(\tau)\,d\tau = F_1 * F_2$	$f_1(s)\, f_2(s)$
8	$tF(t)$	$-f'(s)$
9	$t^n F(t)$	$(-1)^n f^{(n)}(s)$
10	$\dfrac{1}{t} F(t)$	$\int_s^\infty f(x)\,dx$
11	$e^{at} F(t)$	$f(s - a)$
12	$F(t - b)$, where $F(t) = 0$ when $t < 0$	$e^{-bs} f(s)$
13	$\dfrac{1}{c} F\left(\dfrac{t}{c}\right)$	$f(cs)$
14	$\dfrac{1}{c} e^{(bt)/c}\, F\left(\dfrac{t}{c}\right)$	$f(cs - b)$
15	$F(t + a) = F(t)$	$\dfrac{\int_0^a e^{-st} F(t)\,dt}{1 - e^{-as}}$
16	$F(t + a) = -F(t)$	$\dfrac{\int_0^a e^{-st} F(t)\,dt}{1 + e^{-as}}$
17	$F_1(t)$, the half-wave rectification of $F(t)$ in No. 16	$\dfrac{f(s)}{1 - e^{-as}}$
18	$F_2(t)$, the full-wave rectification of $F(t)$ in No. 16	$f(s) \coth \dfrac{as}{2}$
19	$\displaystyle\sum_1^m \dfrac{p(a_n)}{q'(a_n)}\, e^{a_n t}$	$\dfrac{p(s)}{q(s)},\ q(s) = (s - a_1)(s - a_2)\cdots(s - a_m)$
20	$e^{at} \displaystyle\sum_{n=1}^r \dfrac{\phi^{(r-n)}(a)}{(r - n)!}\dfrac{t^{n-1}}{(n - 1)!} + \cdots$	$\dfrac{p(s)}{q(s)} = \dfrac{\phi(s)}{(s - a)^r}$

Table 9-51.   LAPLACE TRANSFORMS**

	$f(s)$	$F(t)$
1	$\dfrac{1}{s}$	$n(t)$, unit step function
2	$\dfrac{1}{s^2}$	t
3	$\dfrac{1}{s^n}$ $(n = 1, 2, \ldots)$	$\dfrac{t^{n-1}}{(n-1)!}$
4	$\dfrac{1}{\sqrt{s}}$	$\dfrac{1}{\sqrt{\pi t}}$
5	$s^{-3/2}$	$2\sqrt{\dfrac{t}{\pi}}$
6	$s^{-[n+(1/2)]}$ $(n = 1, 2, \ldots)$	$\dfrac{2^n t^{n-(1/2)}}{1\cdot 3\cdot 5\cdots(2n-1)\sqrt{\pi}}$
7	$\dfrac{\Gamma(k)}{s^k}$ $(k \geq 0)$	t^{k-1}
8	$\dfrac{1}{s-a}$	e^{at}
9	$\dfrac{1}{(s-a)^2}$	te^{at}
10	$\dfrac{1}{(s-a)^n}$ $(n = 1, 2, \ldots)$	$\dfrac{1}{(n-1)!}\,t^{n-1}e^{at}$
11	$\dfrac{\Gamma(k)}{(s-a)^k}$ $(k \geq 0)$	$t^{k-1}e^{at}$
12*	$\dfrac{1}{(s-a)(s-b)}$	$\dfrac{1}{a-b}(e^{at} - e^{bt})$
13*	$\dfrac{s}{(s-a)(s-b)}$	$\dfrac{1}{a-b}(ae^{at} - be^{bt})$
14*	$\dfrac{1}{(s-a)(s-b)(s-c)}$	$-\dfrac{(b-c)e^{at} + (c-a)e^{bt} + (a-b)e^{ct}}{(a-b)(b-c)(c-a)}$
15	$\dfrac{1}{s^2+a^2}$	$\dfrac{1}{a}\sin at$
16	$\dfrac{s}{s^2+a^2}$	$\cos at$
17	$\dfrac{1}{s^2-a^2}$	$\dfrac{1}{a}\sinh at$
18	$\dfrac{s}{s^2-a^2}$	$\cosh at$

*Here a, b, and (in 14) c represent distinct constants.

**From: "Modern Operational Mathematics in Engineering", 2nd ed., by R.V. Churchill. Copyright © 1958 by McGraw-Hill, Inc. Used by permission of McGraw-Hill Book Company.

Table 9-51.   LAPLACE TRANSFORMS *(Continued)*

	$f(s)$	$F(t)$
19	$\dfrac{1}{s(s^2 + a^2)}$	$\dfrac{1}{a^2}(1 - \cos at)$
20	$\dfrac{1}{s^2(s^2 + a^2)}$	$\dfrac{1}{a^3}(at - \sin at)$
21	$\dfrac{1}{(s^2 + a^2)^2}$	$\dfrac{1}{2a^3}(\sin at - at \cos at)$
22	$\dfrac{s}{(s^2 + a^2)^2}$	$\dfrac{t}{2a}\sin at$
23	$\dfrac{s^2}{(s^2 + a^2)^2}$	$\dfrac{1}{2a}(\sin at + at \cos at)$
24	$\dfrac{s^2 - a^2}{(s^2 + a^2)^2}$	$t \cos at$
25	$\dfrac{s}{(s^2 + a^2)(s^2 + b^2)}\ (a^2 \neq b^2)$	$\dfrac{\cos at - \cos bt}{b^2 - a^2}$
26	$\dfrac{1}{(s - a)^2 + b^2}$	$\dfrac{1}{b}e^{at}\sin bt$
27	$\dfrac{s - a}{(s - a)^2 + b^2}$	$e^{at}\cos bt$
27.1	$\dfrac{1}{[(s + a)^2 + b^2]^n}$	$\dfrac{-e^{-at}}{4^{n-1}b^{2n}}\displaystyle\sum_{r=1}^{n}\binom{2n - r - 1}{n - 1}(-2t)^{r-1}\dfrac{d^r}{dt^r}[\cos(bt)]$
27.2	$\dfrac{s}{[(s + a)^2 + b^2]^n}$	$\dfrac{e^{-at}}{4^{n-1}b^{2n}}\left\{\displaystyle\sum_{r=1}^{n}\binom{2n - r - 1}{n - 1}\right.$ $(-2t)^{r-1}\dfrac{d^r}{dt^r}[a\cos(bt) + b\sin(bt)]$ $-2b\displaystyle\sum_{r=1}^{n-1}r\binom{2n - r - 2}{n - 1}$ $\left.(-2t)^{r-1}\dfrac{d^r}{dt^r}[\sin bt]\right\}$
28	$\dfrac{3a^2}{s^3 + a^3}$	$e^{-at} - e^{(at)/2}\left(\cos\dfrac{at\sqrt{3}}{2} - \sqrt{3}\sin\dfrac{at\sqrt{3}}{2}\right)$
29	$\dfrac{4a^3}{s^4 + 4a^4}$	$\sin at \cosh at - \cos at \sinh at$
30	$\dfrac{s}{s^4 + 4a^4}$	$\dfrac{1}{2a^2}\sin at \sinh at$

Table 9-51.   LAPLACE TRANSFORMS *(Continued)*

	$f(s)$	$F(t)$
31	$\dfrac{1}{s^4 - a^4}$	$\dfrac{1}{2a^3}(\sinh at - \sin at)$
32	$\dfrac{s}{s^4 - a^4}$	$\dfrac{1}{2a^2}(\cosh at - \cos at)$
33	$\dfrac{8a^3 s^2}{(s^2 + a^2)^3}$	$(1 + a^2 t^2)\sin at - \cos at$
34*	$\dfrac{1}{s}\left(\dfrac{s-1}{s}\right)^n$	$L_n(t) = \dfrac{e^t}{n!}\dfrac{d^n}{dt^n}(t^n e^{-t})$
35	$\dfrac{s}{(s-a)^{3/2}}$	$\dfrac{1}{\sqrt{\pi t}}e^{at}(1 + 2at)$
36	$\sqrt{s-a} - \sqrt{s-b}$	$\dfrac{1}{2\sqrt{\pi t^3}}(e^{bt} - e^{at})$
37	$\dfrac{1}{\sqrt{s}+a}$	$\dfrac{1}{\sqrt{\pi t}} - ae^{a^2 t}\operatorname{erfc}(a\sqrt{t})$
38	$\dfrac{\sqrt{s}}{s-a^2}$	$\dfrac{1}{\sqrt{\pi t}} + ae^{a^2 t}\operatorname{erf}(a\sqrt{t})$
39	$\dfrac{\sqrt{s}}{s+a^2}$	$\dfrac{1}{\sqrt{\pi t}} - \dfrac{2a}{\sqrt{\pi}}e^{-a^2 t}\displaystyle\int_0^{a\sqrt{t}}e^{\lambda^2}d\lambda$
40	$\dfrac{1}{\sqrt{s}(s-a^2)}$	$\dfrac{1}{a}e^{a^2 t}\operatorname{erf}(a\sqrt{t})$
41	$\dfrac{1}{\sqrt{s}(s+a^2)}$	$\dfrac{2}{a\sqrt{\pi}}e^{-a^2 t}\displaystyle\int_0^{a\sqrt{t}}e^{\lambda^2}d\lambda$
42	$\dfrac{b^2 - a^2}{(s-a^2)(b+\sqrt{s})}$	$e^{a^2 t}[b - a\operatorname{erf}(a\sqrt{t})] - be^{b^2 t}\operatorname{erfc}(b\sqrt{t})$
43	$\dfrac{1}{\sqrt{s}(\sqrt{s}+a)}$	$e^{a^2 t}\operatorname{erfc}(a\sqrt{t})$
44	$\dfrac{1}{(s+a)\sqrt{s+b}}$	$\dfrac{1}{\sqrt{b-a}}e^{-at}\operatorname{erf}(\sqrt{b-a}\sqrt{t})$
45	$\dfrac{b^2 - a^2}{\sqrt{s}(s-a^2)(\sqrt{s}+b)}$	$e^{a^2 t}\left[\dfrac{b}{a}\operatorname{erf}(a\sqrt{t}) - 1\right] + e^{b^2 t}\operatorname{erfc}(b\sqrt{t})$
46†	$\dfrac{(1-s)^n}{s^{n+(1/2)}}$	$\dfrac{n!}{(2n)!\sqrt{\pi t}}H_{2n}(\sqrt{t})$
47	$\dfrac{(1-s)^n}{s^{n+(3/2)}}$	$-\dfrac{n!}{\sqrt{\pi}(2n+1)!}H_{2n+1}(\sqrt{t})$

*$L_n(t)$ is the Laguerre polynomial of degree n.

†$H_n(x)$ is the Hermite polynomial, $H_n(x) = e^{x^2}\dfrac{d^n}{dx^n}(e^{-x^2})$.

Table 9-51.　LAPLACE TRANSFORMS *(Continued)*

	$f(s)$	$F(t)$
48†	$\dfrac{\sqrt{s+2a}}{\sqrt{s}} - 1$	$ae^{-at}[I_1(at) + I_0(at)]$
49	$\dfrac{1}{\sqrt{s+a}\,\sqrt{s+b}}$	$e^{-(1/2)(a+b)t}I_0\left(\dfrac{a-b}{2}\,t\right)$
50	$\dfrac{\Gamma(k)}{(s+a)^k(s+b)^k}\ (k \geq 0)$	$\sqrt{\pi}\left(\dfrac{t}{a-b}\right)^{k-(1/2)}e^{-(1/2)(a+b)t}I_{k-(1/2)}\left(\dfrac{a-b}{2}\,t\right)$
51	$\dfrac{1}{(s+a)^{1/2}(s+b)^{3/2}}$	$te^{-(1/2)(a+b)t}\left[I_0\left(\dfrac{a-b}{2}\,t\right) + I_1\left(\dfrac{a-b}{2}\,t\right)\right]$
52	$\dfrac{\sqrt{s+2a} - \sqrt{s}}{\sqrt{s+2a} + \sqrt{s}}$	$\dfrac{1}{t}\,e^{-at}I_1(at)$
53	$\dfrac{(a-b)^k}{(\sqrt{s+a} + \sqrt{s+b})^{2k}}\ (k > 0)$	$\dfrac{k}{t}\,e^{-(1/2)(a+b)t}I_k\left(\dfrac{a-b}{2}\,t\right)$
54	$\dfrac{(\sqrt{s+a} + \sqrt{s})^{-2\nu}}{\sqrt{s}\,\sqrt{s+a}}\ (\nu > -1)$	$\dfrac{1}{a^\nu}\,e^{-(1/2)(at)}I_\nu\left(\dfrac{1}{2}\,at\right)$
55	$\dfrac{1}{\sqrt{s^2+a^2}}$	$J_0(at)$
56	$\dfrac{(\sqrt{s^2+a^2} - s)^\nu}{\sqrt{s^2+a^2}}\ (\nu > -1)$	$a^\nu J_\nu(at)$
57	$\dfrac{1}{(s^2+a^2)^k}\ (k > 0)$	$\dfrac{\sqrt{\pi}}{\Gamma(k)}\left(\dfrac{t}{2a}\right)^{k-(1/2)}J_{k-(1/2)}(at)$
58	$(\sqrt{s^2+a^2} - s)^k\ (k > 0)$	$\dfrac{ka^k}{t}\,J_k(at)$
59	$\dfrac{(s - \sqrt{s^2-a^2})^\nu}{\sqrt{s^2-a^2}}\ (\nu > -1)$	$a^\nu I_\nu(at)$
60	$\dfrac{1}{(s^2-a^2)^k}\ (k > 0)$	$\dfrac{\sqrt{\pi}}{\Gamma(k)}\left(\dfrac{t}{2a}\right)^{k-(1/2)}I_{k-(1/2)}(at)$
61	$\dfrac{e^{-ks}}{s}$	$S_k(t) = \begin{cases}0 \text{ when } 0 < t < k \\ 1 \text{ when } t > k\end{cases}$
62	$\dfrac{e^{-ks}}{s^2}$	$\begin{cases}0 \text{ when } 0 < t < k \\ t - k \text{ when } t > k\end{cases}$
63	$\dfrac{e^{-ks}}{s^\mu}\ (\mu > 0)$	$\begin{cases}0 \qquad\qquad \text{ when } 0 < t < k \\ \dfrac{(t-k)^{\mu-1}}{\Gamma(\mu)} \text{ when } t > k\end{cases}$
64	$\dfrac{1 - e^{-ks}}{s}$	$\begin{cases}1 \text{ when } 0 < t < k \\ 0 \text{ when } t > k\end{cases}$

†$I_n(x) = i^{-n}J_n(ix)$, where J_n is Bessel's function of the first kind.

Table 9-51. LAPLACE TRANSFORMS *(Continued)*

	$f(s)$	$F(t)$
65	$\dfrac{1}{s(1 - e^{-ks})} = \dfrac{1 + \coth \frac{1}{2}ks}{2s}$	$S(k, t) = n$ when $\quad (n - 1)k < t < nk\,(n = 1, 2, \ldots)$
66	$\dfrac{1}{s(e^{ks} - a)}$	$\begin{cases} 0 \text{ when } 0 < t < k \\ 1 + a + a^2 + \cdots + a^{n-1} \\ \quad \text{when } nk < t < (n + 1)k\,(n = 1, 2, \ldots) \end{cases}$
67	$\dfrac{1}{s} \tanh ks$	$M(2k, t) = (-1)^{n-1}$ $\quad$ when $2k(n - 1) < t < 2kn$ $\qquad\qquad\qquad (n = 1, 2, \ldots)$
68	$\dfrac{1}{s(1 + e^{-ks})}$	$\dfrac{1}{2} M(k, t) + \dfrac{1}{2} = \dfrac{1 - (-1)^n}{2}$ $\quad$ when $(n - 1)k < t < nk$
69*	$\dfrac{1}{s^2} \tanh ks$	$H(2k, t)$
70	$\dfrac{1}{s \sinh ks}$	$2S(2k, t + k) - 2 - 2(n - 1)$ $\quad$ when $(2n - 3)k < t < (2n - 1)k \ \ (t > 0)$
71	$\dfrac{1}{s \cosh ks}$	$M(2k, t + 3k) + 1 - 1 + (-1)^n$ $\quad$ when $(2n - 3)k < t < (2n - 1)k \ \ (t > 0)$
72	$\dfrac{1}{s} \coth ks$	$2S(2k, t) - 1 = 2n - 1$ $\qquad$ when $2k(n - 1) < t < 2kn$
73	$\dfrac{k}{s^2 + k^2} \coth \dfrac{\pi s}{2k}$	$\lvert \sin kt \rvert$
74	$\dfrac{1}{(s^2 + 1)(1 - e^{-\pi s})}$	$\begin{cases} \sin t \text{ when } (2n - 2)\pi < t < (2n - 1)\pi \\ 0 \quad \text{when } (2n - 1)\pi < t < 2n\pi \end{cases}$
75	$\dfrac{1}{s} e^{-k/s}$	$J_0(2\sqrt{kt})$
76	$\dfrac{1}{\sqrt{s}} e^{-k/s}$	$\dfrac{1}{\sqrt{\pi t}} \cos 2\sqrt{kt}$
77	$\dfrac{1}{\sqrt{s}} e^{k/s}$	$\dfrac{1}{\sqrt{\pi t}} \cosh 2\sqrt{kt}$
78	$\dfrac{1}{s^{3/2}} e^{-k/s}$	$\dfrac{1}{\sqrt{\pi k}} \sin 2\sqrt{kt}$
79	$\dfrac{1}{s^{3/2}} e^{k/s}$	$\dfrac{1}{\sqrt{\pi k}} \sinh 2\sqrt{kt}$
80	$\dfrac{1}{s^{\mu}} e^{-k/s}(\mu > 0)$	$\left(\dfrac{t}{k}\right)^{(\mu-1)/2} J_{\mu-1}(2\sqrt{kt})$

*$H(2k, t) = k + (r - k)(-1)^n$ where $t = 2kn + r; 0 \le r < 2k; n = 0, 1, 2, \ldots$.

Table 9-51. LAPLACE TRANSFORMS *(Continued)*

	$f(s)$	$F(t)$
81	$\dfrac{1}{s^{\mu}}\,e^{k/s}\,(\mu > 0)$	$\left(\dfrac{t}{k}\right)^{(\mu-1)/2} I_{\mu-1}(2\sqrt{kt})$
82	$e^{-k\sqrt{s}}\,(k > 0)$	$\dfrac{k}{2\sqrt{\pi t^3}}\exp\left(-\dfrac{k^2}{4t}\right)$
83	$\dfrac{1}{s}\,e^{-k\sqrt{s}}\,(k \geq 0)$	$\operatorname{erfc}\left(\dfrac{k}{2\sqrt{t}}\right)$
84	$\dfrac{1}{\sqrt{s}}\,e^{-k\sqrt{s}}\,(k \geq 0)$	$\dfrac{1}{\sqrt{\pi t}}\exp\left(-\dfrac{k^2}{4t}\right)$
85	$s^{-3/2}e^{-k\sqrt{s}}\,(k \geq 0)$	$2\sqrt{\dfrac{t}{\pi}}\exp\left(-\dfrac{k^2}{4t}\right) - k\operatorname{erfc}\left(\dfrac{k}{2\sqrt{t}}\right)$
86	$\dfrac{ae^{-k\sqrt{s}}}{s(a + \sqrt{s})}\,(k \geq 0)$	$-e^{ak}e^{a^2t}\operatorname{erfc}\left(a\sqrt{t} + \dfrac{k}{2\sqrt{t}}\right) + \operatorname{erfc}\left(\dfrac{k}{2\sqrt{t}}\right)$
87	$\dfrac{e^{-k\sqrt{s}}}{\sqrt{s}(a + \sqrt{s})}\,(k \geq 0)$	$e^{ak}e^{a^2t}\operatorname{erfc}\left(a\sqrt{t} + \dfrac{k}{2\sqrt{t}}\right)$
88	$\dfrac{e^{-k\sqrt{s(s+a)}}}{\sqrt{s(s + a)}}$	$\begin{cases} 0 & \text{when } 0 < t < k \\ e^{-(1/2)(at)}I_0(\tfrac{1}{2}a\sqrt{t^2 - k^2}) & \text{when } t > k \end{cases}$
89	$\dfrac{e^{-k\sqrt{s^2+a^2}}}{\sqrt{s^2 + a^2}}$	$\begin{cases} 0 & \text{when } 0 < t < k \\ J_0(a\sqrt{t^2 - k^2}) & \text{when } t > k \end{cases}$
90	$\dfrac{e^{-k\sqrt{s^2-a^2}}}{\sqrt{s^2 - a^2}}$	$\begin{cases} 0 & \text{when } 0 < t < k \\ I_0(a\sqrt{t^2 - k^2}) & \text{when } t > k \end{cases}$
91	$\dfrac{e^{-k(\sqrt{s^2+a^2}-s)}}{\sqrt{s^2 + a^2}}\,(k \geq 0)$	$J_0(a\sqrt{t^2 + 2kt})$
92	$e^{-ks} - e^{-k\sqrt{s^2+a^2}}$	$\begin{cases} 0 & \text{when } 0 < t < k \\ \dfrac{ak}{\sqrt{t^2 - k^2}}J_1(a\sqrt{t^2 - k^2}) & \text{when } t > k \end{cases}$
93	$e^{-k\sqrt{s^2+a^2}} - e^{-ks}$	$\begin{cases} 0 & \text{when } 0 < t < k \\ \dfrac{ak}{\sqrt{t^2 - k^2}}I_1(a\sqrt{t^2 - k^2}) & \text{when } t > k \end{cases}$
94	$\dfrac{a^{\nu}e^{-k\sqrt{s^2-a^2}}}{\sqrt{s^2 + a^2}(\sqrt{s^2 + a^2} + s)^{\nu}}$ $(\nu > -1)$	$\begin{cases} 0 & \text{when } 0 < t < k \\ \left(\dfrac{t - k}{t + k}\right)^{(1/2)\nu}J_{\nu}(a\sqrt{t^2 - k^2}) & \text{when } t > k \end{cases}$
95	$\dfrac{1}{s}\log s$	$\Gamma'(1) - \log t\;[\Gamma'(1) = -0.5772]$
96	$\dfrac{1}{s^k}\log s\,(k > 0)$	$t^{k-1}\left\{\dfrac{\Gamma'(k)}{[\Gamma(k)]^2} - \dfrac{\log t}{\Gamma(k)}\right\}$
97	$\dfrac{\log s}{s - a}\,(a > 0)$	$e^{at}[\log a - \operatorname{Ei}(-at)]$

Table 9-51. LAPLACE TRANSFORMS (*Continued*)

	$f(s)$	$F(t)$
98	$\dfrac{\log s}{s^2 + 1}$	$\cos t \, \mathrm{Si}(t) - \sin t \, \mathrm{Ci}(t)$
99	$\dfrac{s \log s}{s^2 + 1}$	$-\sin t \, \mathrm{Si}(t) - \cos t \, \mathrm{Ci}(t)$
100	$\dfrac{1}{s} \log (1 + ks) \, (k > 0)$	$-\mathrm{Ei}\left(-\dfrac{t}{k}\right)$
101	$\log \dfrac{s - a}{s - b}$	$\dfrac{1}{t} (e^{bt} - e^{at})$
102	$\dfrac{1}{s} \log (1 + k^2 s^2)$	$-2\mathrm{Ci}\left(\dfrac{t}{k}\right)$
103	$\dfrac{1}{s} \log (s^2 + a^2) \quad (a > 0)$	$2 \log a - 2\mathrm{Ci}(at)$
104	$\dfrac{1}{s^2} \log (s^2 + a^2) \quad (a > 0)$	$\dfrac{2}{a} [at \log a + \sin at - at \, \mathrm{Ci}(at)]$
105	$\log \dfrac{s^2 + a^2}{s^2}$	$\dfrac{2}{t} (1 - \cos at)$
106	$\log \dfrac{s^2 - a^2}{s^2}$	$\dfrac{2}{t} (1 - \cosh at)$
107	$\arctan \dfrac{k}{s}$	$\dfrac{1}{t} \sin kt$
108	$\dfrac{1}{s} \arctan \dfrac{k}{s}$	$\mathrm{Si}(kt)$
109	$e^{k^2 s^2} \mathrm{erfc}\,(ks) \quad (k > 0)$	$\dfrac{1}{k\sqrt{\pi}} \exp\left(-\dfrac{t^2}{4k^2}\right)$
110	$\dfrac{1}{s} e^{k^2 s^2} \mathrm{erfc}\,(ks) \quad (k > 0)$	$\mathrm{erf}\left(\dfrac{t}{2k}\right)$
111	$e^{ks} \mathrm{erfc}\,(\sqrt{ks}) \quad (k > 0)$	$\dfrac{\sqrt{k}}{\pi \sqrt{t}(t + k)}$
112	$\dfrac{1}{\sqrt{s}} \mathrm{erfc}\,(\sqrt{ks})$	$\begin{cases} 0 & \text{when } 0 < t < k \\ (\pi t)^{-1/2} & \text{when } t > k \end{cases}$
113	$\dfrac{1}{\sqrt{s}} e^{ks} \mathrm{erfc}\,(\sqrt{ks}) \, (k > 0)$	$\dfrac{1}{\sqrt{\pi(t + k)}}$
114	$\mathrm{erf}\left(\dfrac{k}{\sqrt{s}}\right)$	$\dfrac{1}{\pi t} \sin (2k\sqrt{t})$
115	$\dfrac{1}{\sqrt{s}} e^{k^2/s} \mathrm{erfc}\left(\dfrac{k}{\sqrt{s}}\right)$	$\dfrac{1}{\sqrt{\pi t}} e^{-2k\sqrt{t}}$

Table 9-51.   LAPLACE TRANSFORMS *(Continued)*

	$f(s)$	$F(t)$
115.1	$-e^{as}\,\mathrm{Ei}(-as)$	$\dfrac{1}{t+a}\,;\,(a>0)$
115.2	$\dfrac{1}{a}+se^{as}\,\mathrm{Ei}(-as)$	$\dfrac{1}{(t+a)^2}\,;\,(a>0)$
115.3	$\left[\dfrac{\pi}{2}-\mathrm{Si}(s)\right]\cos s+\mathrm{Ci}(s)\sin s$	$\dfrac{1}{t^2+1}$
116*	$K_0(ks)$	$\begin{cases}0 & \text{when } 0<t<k\\ (t^2-k^2)^{-1/2} & \text{when } t>k\end{cases}$
117	$K_0(k\sqrt{s})$	$\dfrac{1}{2t}\exp\left(-\dfrac{k^2}{4t}\right)$
118	$\dfrac{1}{s}\,e^{ks}K_1(ks)$	$\dfrac{1}{k}\sqrt{t(t+2k)}$
119	$\dfrac{1}{\sqrt{s}}\,K_1(k\sqrt{s})$	$\dfrac{1}{k}\exp\left(-\dfrac{k^2}{4t}\right)$
120	$\dfrac{1}{\sqrt{s}}\,e^{k/s}K_0\left(\dfrac{k}{s}\right)$	$\dfrac{2}{\sqrt{\pi t}}\,K_0(2\sqrt{2kt})$
121	$\pi e^{-ks}I_0(ks)$	$\begin{cases}[t(2k-t)]^{-1/2} & \text{when } 0<t<2k\\ 0 & \text{when } t>2k\end{cases}$
122	$e^{-ks}I_1(ks)$	$\begin{cases}\dfrac{k-t}{\pi k\sqrt{t(2k-t)}} & \text{when } 0<t<2k\\ 0 & \text{when } t>2k\end{cases}$

* $K_n(x)$ is Bessel's function of the second kind for the imaginary argument.

REFERENCES

Several additional transforms, especially those involving other Bessel functions, can be found in the following sources:

"Fourier Integrals for Practical Applications", G.A. Campbell and R.M. Foster, D. Van Nostrand Company, 1948. In these tables, only those entries containing the condition $0<g$ or $k<g$, where g is our t, are Laplace transforms.

"Formulaire pour le calcul symbolique", N.W. McLachlan and P. Humbert, Gauthier—Villars, Paris, 1947.

"Tables of Integral Transforms", Bateman Manuscript Project, California Institute of Technology, A. Erdélyi and W. Magnus, Eds., McGraw-Hill Book Company, 1954; based on notes left by Harry Bateman.

Table 9-52.   SOLID ANGLES*

COMPILED BY DR. A. VICTOR MASKET

I. SOLID ANGLE SUBTENDED BY THE CIRCULAR DISC

Let the point in space at which the solid angle is subtended be 0, an origin of coordinate systems. The normal to the plane of the disc from 0 defines the z-axis, and the normal to the symmetry axis of the disc defines the x-axis. *All lengths are considered normalized relative to the radius of the disc as a unit of length.* The solid angle is represented by the symbol $\Omega(1;\rho,z)$, where ρ is the distance (along the x-axis) from 0 to the symmetry axis, and z is the distance (along the z-axis) from 0 to the plane of the disc. From the general expression for solid angles given by $\Omega = \int\int \sin\theta\, d\theta\, d\phi$ in spherical polar coordinates, three contour integrals have been derived; they correspond to the cases with ρ less than, equal to, or greater than the radius of the disc as a unit of length.

Case 1. $\rho < 1$; the z-axis pierces the interior of the disc.

The contour integral employed for digital computation is

$$(1) \qquad \Omega(1;\rho,z) = 2\cdot\int_0^\pi f_1^2[1+f_1^2+(1+f_1^2)^{1/2}]^{-1}\, d\phi \qquad (\rho<1),$$

where $f_1 = S_1/z$, $S_1 = \rho\cos\phi+(1-\rho^2\sin^2\phi)^{1/2}$. When $\rho = 0$, $\Omega(1;0,z) = 2\pi[1-z/(1+z^2)^{1/2}]$, exactly.

Case 2. $\rho = 1$; the z-axis touches the periphery of the disc.

$$(2) \qquad \Omega(1;1,z) = 2\cdot\int_0^{\pi/2} f_2^2[1+f_2^2+(1+f_2^2)^{1/2}]^{-1}\, d\phi,$$

with $f_2 = (2\cos\phi)/z$. In elliptic integrals the formula is given exactly by $\Omega(1;1,z) = \pi - kzK(k)$, where $k^2 = 4/(z^2+4)$ and $K(k) \equiv F(\pi/2,k)$ is the complete elliptic integral of the first kind.

Case 3. $\rho > 1$; the z-axis is external to the disc.

The computational form, where $S_2 = \rho\cos\phi-(1-\rho^2\sin^2\phi)^{1/2}$, is

$$(3) \qquad \Omega(1;\rho,z) = 2z\int_{(\rho>1)}^{\arcsin(1/\rho)} \frac{(S_1^2-S_2^2)\, d\phi}{[(z^2+S_1^2)(z^2+S_2^2)]^{1/2}[(z^2+S_1^2)^{1/2}+(z^2+S_2^2)^{1/2}]}$$

The solid angle subtended by the circular disc is expressed most generally in elliptic integrals. In conformity with the definitions adopted for ρ and z,

$$(4) \qquad \Omega(1;\rho,z) = 2\pi - 2[k'K(k)\sin\alpha+K(k)\cdot E(k',\alpha) - F(k',\alpha)\{K(k)-E(k)\}],$$

where $k^2 = 4\rho/[(1+\rho)^2+z^2]$, $(k')^2 = 1-k^2$, $\alpha = (\pi-\phi)/2$, and $K(k)$ and $E(k)$ are the complete elliptic integrals, and $F(k',\alpha)$ and $E(k',\alpha)$ are the corresponding (incomplete) elliptic integrals of the first and second kind. The formula given by (4) is valid for all values of ρ and z; it is not convenient for computation. When $z > 100$, it is practical to use $\Omega \sim (\pi/d^2)\cdot\cos\gamma$, where d is the normalized distance from 0 to the center of the disc and γ is the projection angle.

II. SOLID ANGLE SUBTENDED BY THE RIGHT CIRCULAR CYLINDER

The tabulated values consist of the solid angles subtended only by the lateral surface at points in the plane of a circular flat (base) external to a cylinder of given radius and height. It will be shown that these values alone or taken in conjunction with the values of solid angles subtended by the circular disc are sufficient to cover the general case of any point in space external to the right circular cylinder. *All lengths are normalized with respect to the radius of the cylinder as a unit of length.* The variables of the table are h, the normalized height of the cylinder, and ρ, the normalized distance from the point (at which the solid angle is subtended) to the axis of the cylinder. With these definitions the solid angle is represented by $\Omega(1,h;\rho)$.

Let the point in the plane of a flat be 0, the origin of coordinate systems. The z-axis is the normal to the plane, and the x-axis is the line in the plane from 0 to the axis of the cylinder. The solid angle subtended at 0 by the lateral surface of the cylinder is given by the contour integral.

Table 9-52. SOLID ANGLES *(Continued)*

(5) $$\Omega(1,h;\rho) = 2h \int_0^{\arcsin(1/\rho)} [h^2 + S_2^2]^{-1/2} \, d\phi \qquad\qquad [\rho \geq 1],$$

where ϕ is the azimuthal angle of the associated spherical polar system.

If a point is not located in a plane determined by a flat of the cylinder, then it is either interior or exterior to the space between the planes of the flats. In the first case a plane through the point normal to the cylindrical axis determines two smaller cylinders and two solid angles in accordance with (5) that are added. For the second case, where a flat is viewed from 0 in addition to the lateral surface, let the normalized height of the cylinder be h and the normalized distance to the plane of the viewed flat be H. Then the total solid angle is

(6) $$\Omega = \Omega(1,h+H;\rho) - \Omega(1,H;\rho) + \Omega(1;\rho,H),$$

where ρ is the normalized distance from 0 to the axis of the cylinder.

III. USE OF TABLES

A. Reading the Entry Numbers (steradians)

Each entry is given to six significant figures followed by an exponent for the power of 10 as a multiplier. For example,

$$3.73950\text{--}3 \equiv 3.73950 \times 10^{-3} \equiv 0.00373950;$$
$$5.65798\text{--}0 \equiv 5.65798 \times 10^0 \equiv 5.65798.$$

B. Solid Angle Subtended by a Disc

To find the solid angle subtended by an end-window counter 3 cm in radius at a point 15 cm above its plane and 18 cm from its axis, we have

$$\Omega(1;\rho,z) = \Omega(1;18/3,15/3) = \Omega(1;6,5) = 0.0331611.$$

C. Solid Angle Subtended by the Lateral Surface of a Right Circular Cylinder

To find the solid angle subtended by a Geiger counter 3 cm in radius and 30 cm long at a point in the plane of a flat at 25.5 cm from the axis, we have

$$\Omega(1,h;\rho) = \Omega(1,30/3;25.5/3) = \Omega(1,10;8.5) = 0.186909.$$

D. Fraction of Radiation Entering a Circular Aperture from a Distributed Plane Source

Consider the problem of estimating the fraction of radiation entering a circular end-window counter from a thin-disc source whose plane is parallel to the window. Let b be the normalized radius of the source relative to the counter; let s be the normalized distance between the (parallel) axes of the source and window; and let z be the normalized distance between the planes of the window and source.

Case 1. $s = 0$: coaxial.

Divide the source into N concentric annuli of width $\Delta\rho_i$, $i = 1,2,3 \ldots, N$. Each point in a given annulus has subtended approximately the same solid angle $\Omega(1;\rho_i,z)$ by the window. The area of each annulus is $2\pi\rho_i\Delta\rho_i$. Hence the fraction is given by the summation

(6) $$G_1 = \frac{1}{2\pi b^2} \sum_{i=1}^{N} \Omega(1;\rho_i,z)\rho_i\Delta\rho_i.$$

Case 2. $s \geq b$: non-coaxial and no vertical overlap.

Divide the source into N concentric strips using the point common to the axis of the window and the plane of the source as a center. The area of each strip is $2\alpha_i\rho_i\Delta\rho_i$, where $\alpha_i = \arccos[(\rho_i^2 + S^2 - b^2)/2S\rho_i]$. The fraction is given by

(7) $$G_2 = \frac{1}{2\pi^2 b^2} \sum_{i=1}^{N} \Omega(1;\rho_i,z)\alpha_i\rho_i\Delta\rho_i.$$

Case 3. $s < b$: non-coaxial with vertical overlap. Apply Cases 1 and 2.

In the above calculations self-absorption, scattering, non-uniformity of source strength, and non-isotropic emission have been omitted. Corrections for these effects cannot be treated here.

Table 9-52. SOLID ANGLES (Continued)

I. SOLID ANGLE SUBTENDED BY THE CIRCULAR DISC OF UNIT RADIUS

z / ρ	1.0	2.0	3.0	4.0	5.0	6.0	7.0	8.0	9.0	10.0
0.0	1.84030 0	6.63334–1	3.22432–1	1.87600–1	1.22015–1	8.54895–2	6.31492–2	4.85195–2	3.84296–2	3.11823–2
0.1	1.83198 0	6.61650–1	3.21986–1	1.87442–1	1.21947–1	8.54556–2	6.31305–2	4.85085–2	3.84227–2	3.11777–2
0.2	1.80709 0	6.56633–1	3.20651–1	1.86969–1	1.21742–1	8.53539–2	6.30746–2	4.84753–2	3.84018–2	3.11639–2
0.3	1.76590 0	6.48380–1	3.18446–1	1.86185–1	1.21403–1	8.51848–2	6.29816–2	4.84201–2	3.83670–2	3.11409–2
0.4	1.70895 0	6.37049–1	3.15396–1	1.85096–1	1.20929–1	8.49490–2	6.28518–2	4.83430–2	3.83185–2	3.11089–2
0.5	1.63710 0	6.22856–1	3.11538–1	1.83710–1	1.20325–1	8.46475–2	6.26855–2	4.82442–2	3.82562–2	3.10677–2
0.6	1.55174 0	6.06068–1	3.06918–1	1.82038–1	1.19594–1	8.42812–2	6.24833–2	4.81238–2	3.81802–2	3.10175–2
0.7	1.45479 0	5.86992 1	3.01590–1	1.80093–1	1.18738–1	8.38517–2	6.22456–2	4.79822–2	3.80908–2	3.09584–2
0.8	1.34883 0	5.65969–1	2.95612–1	1.77888–1	1.17763–1	8.33605–2	6.19732–2	4.78197–2	3.79881–2	3.08904–2
0.9	1.23698 0	5.43360–1	2.89051–1	1.75441–1	1.16673–1	8.28094–2	6.16668–2	4.76365–2	3.78722–2	3.08136–2
1.00	1.12269 0	5.19535–1	2.81976–1	1.72768–1	1.15474–1	8.22003–2	6.13272–2	4.74331–2	3.77433–2	3.07281–2
1.25	8.48210–1	4.57034–1	2.62506–1	1.65219–1	1.12039–1	8.04395–2	6.03399–2	4.68396–2	3.73662–2	3.04775–2
1.50	6.19102–1	3.94380–1	2.41380–1	1.56679–1	1.08058–1	7.83691–2	5.91679–2	4.61304–2	3.69135–2	3.01757–2
1.75	4.47467–1	3.35486–1	2.19628–1	1.47457–1	1.03639–1	7.60318–2	5.78301–2	4.53147–2	3.63901–2	2.98253–2
2.00	3.25802–1	2.82710–1	1.98115–1	1.37847–1	9.88928–2	7.34729–2	5.63469–2	4.44027–2	3.58011–2	2.94290–2
2.25	2.41045–1	2.37015–1	1.77498–1	1.28115–1	9.39249–2	7.07383–2	5.47399–2	4.34049–2	3.51524–2	2.89906–2
2.50	1.81769–1	1.98380–1	1.58222–1	1.18484–1	8.88347–2	6.78732–2	5.30310 2	4.23328 2	3.44501–2	2.85133–2
2.75	1.39726–1	1.66206–1	1.40544–1	1.09128–1	8.37106–2	6.49209–2	5.12417–2	4.11975–2	3.37003–2	2.80006–2
3.00	1.09364–1	1.39646–1	1.24571–1	1.00181–1	7.86288–2	6.19210–2	4.93929–2	4.00103–2	3.29093–2	2.74563–2
3.25	8.70214–2	1.17811–1	1.10300–1	9.17341 2	7.36525 2	5.89096 2	4.75045 2	3.87827–2	3.20839–2	2.68842–2
3.50	7.02786–2	9.98737–2	9.76532–2	8.38425–2	6.88315–2	5.59182 2	4.55950–2	3.75251–2	3.12302–2	2.62881–2
3.75	5.75168–2	8.51134–2	8.65120–2	7.65315–2	6.42035–2	5.29731–2	4.36807–2	3.62475–2	3.03542–2	2.56719–2
4.00	4.76364–2	7.29297 2	7.67350–2	6.98043–2	5.97951–2	5.00967–2	4.17765 2	3.49592–2	2.94618–2	2.50392–2
4.25	3.98770–2	6.28312–2	6.81754–2	6.36476–2	5.56236–2	4.73067–2	3.98953–2	3.36689 2	2.85585 2	2.43935 2
4.50	3.37040–2	5.44214–2	6.06901–2	5.80367–2	5.16981–2	4.46165–2	3.80479–2	3.23842–2	2.76496–2	2.37383–2
4.75	2.87348–2	4.73822–2	5.41457–2	5.29395–2	4.80209 2	4.20364 2	3.62437–2	3.11117–2	2.67394–2	2.30767–2
5.00	2.46916–2	4.14597–2	4.84209–2	4.83198–2	4.45896–2	3.95730–2	3.44900–2	2.98574–2	2.58325–2	2.24118–2
5.25	2.13695–2	3.64505–2	4.34075–2	4.41397–2	4.13980–2	3.72301–2	3.27926–2	2.86266–2	2.49327–2	2.17463–2
5.50	1.86154–2	3.21920–2	3.90107–2	4.03613–2	3.84369–2	3.50094–2	3.11558–2	2.74234–2	2.40434–2	2.10828–2
5.75	1.63135–2	2.85533–2	3.51474 2	3.69480 2	3.56953 2	3.29105–2	2.95827–2	2.62513–2	2.31677–2	2.04236–2
6	1.43749–2	2.54291–2	3.17460–2	3.38650–2	3.31611–2	3.09318–2	2.80753–2	2.51131–2	2.23081–2	1.97707–2
7	9.08291 3	1.65909 2	2.16464 2	2.42250–2	2.48393–2	2.41460–2	2.27104–2	2.09340–2	1.90688–2	1.72572–2
8	6.09755–3	1.13744–2	1.52974–2	1.77274–2	1.88341–2	1.89341–2	1.83577–2	1.73801–2	1.62019–2	1.49563–2
9	4.28845–3	8.11658–3	1.11544–2	1.32651–2	1.44862–2	1.49641–2	1.48841–2	1.44236–2	1.37283–2	1.29076–2
10	3.12931 3	5.98473–3	8.35619–3	1.01338–2	1.13073–2	1.19398–2	1.21324–2	1.19953–2	1.16294–2	1.11176–2
11	2.35275–3	4.53449–3	6.40752–3	7.88872–3	8.95332–3	9.62443–3	9.95559–3	1.00137–2	9.86657–3	9.57422–3
12	1.81318–3	3.51520–3	5.01328–3	6.24568–3	7.18572–3	7.83797–3	8.22949–3	8.40019–3	8.39434–3	8.25465–3
13	1.42670–3	2.77865–3	3.99173–3	5.02016–3	5.83979–3	6.44701–3	6.85434–3	7.08506–3	7.16773–3	7.13201–3
14	1.14266–3	2.23358–3	3.22741–3	4.09003–4	4.80096–3	5.35343–3	5.75213–3	6.01019–3	6.14578–3	6.17921–3
15	9.29263–4	1.82180–3	2.64488–3	3.37280–3	3.98878–3	4.48521–3	4.86267–3	5.12814–3	5.29301–3	5.37102–3
16	7.65850–4	1.50506–3	2.19353–3	2.81177–3	3.34614–3	3.78929–3	4.13974–3	4.40082–3	4.57949–3	4.68495–3
18	5.38052–4	1.06128–3	1.55599–3	2.01056–3	2.41611–3	2.76681–3	3.05991–3	3.29539–3	3.47548–3	3.60413–3
20	3.92328–4	7.75889–4	1.14245–3	1.48481–3	1.79711–3	2.07510–3	2.31616–3	2.51927–3	2.68476–3	2.81414–3
22	2.94811–4	5.84171–4	8.62906–4	1.12637–3	1.37065–3	1.59267–3	1.79034–3	1.96240–3	2.10853–3	2.22910–3
24	2.27108–4	4.50685–4	6.67349–4	8.74012–4	1.06799–3	1.24707–4	1.40964–3	1.55456–3	1.68127–3	1.78968–3
26	1.78644–4	3.54918–4	5.26543–4	6.91403–4	8.47617–4	9.93601–3	1.12808–3	1.25014–3	1.35919–3	1.45496–3
28	1.43043–4	2.84450–4	4.22636–4	5.56117–4	6.83550–4	8.03776–4	9.15822–4	1.01895–3	1.11262–3	1.19651–3
30	1.16307–4	2.31453–4	3.44313–4	4.53822–4	5.59000–4	6.58989–4	7.53044–4	8.40575–4	9.21127–4	9.94393–4
32	9.58388–5	1.90836–4	2.84176–4	3.75075–4	4.62812–4	5.46736–4	6.26282–4	7.00975–4	7.70439–4	8.34397–4
34	7.99048–5	1.59187–4	2.37245–4	3.13491–4	3.87386–4	4.58434–4	5.26194–4	5.90291–4	6.50415–4	7.06326–4

Table 9-52. SOLID ANGLES (Continued)

I. SOLID ANGLE SUBTENDED BY THE CIRCULAR DISC OF UNIT RADIUS (Continued)

ρ \ z	11	12	13	14	15	16	17	18	19	20
0.0	2.58037–2	2.17036–2	1.85072–2	1.59675–2	1.39163–2	1.22360–2	1.08424–2	9.67389–3	8.68443–3	7.83929–3
0.1	2.58006–2	2.17014–2	1.85056–2	1.59662–2	1.39153–2	1.22353–2	1.08419–2	9.67344–3	8.68408–3	7.83899–3
0.2	2.57911–2	2.16947–2	1.85007–2	1.59626–2	1.39126–2	1.22332–2	1.08402–2	9.67211–3	8.68300–3	7.83812–3
0.3	2.57754–2	2.16836–2	1.84926–2	1.59566–2	1.39080–2	1.22296–2	1.08374–2	9.66988–3	8.68121–3	7.83665–3
0.4	2.57534–2	2.16680–2	1.84812–2	1.59481–2	1.39016–2	1.22246–2	1.08335–2	9.66677–3	8.67869–3	7.83461–3
0.5	2.57251–2	2.16479–2	1.84667–2	1.59372–2	1.38933–2	1.22182–2	1.08285–2	9.66277–3	8.67547–3	7.83198–3
0.6	2.56907–2	2.16235–2	1.84489–2	1.59240–2	1.38832–2	1.22104–2	1.08223–2	9.65788–3	8.67153–3	7.82877–3
0.7	2.56500–2	2.15947–2	1.84279–2	1.59083–2	1.38713–2	1.22012–2	1.08151–2	9.65211–3	8.66688–3	7.82497–3
0.8	2.56033–2	2.15615–2	1.84037–2	1.58903–2	1.38576–2	1.21906–2	1.08067–2	9.64546–3	8.66151–3	7.82060–3
0.9	2.55505–2	2.15240–2	1.83763–2	1.58699–2	1.38421–2	1.21786–2	1.07973–2	9.63793–3	8.65544–3	7.81565–3
1.00	2.54916–2	2.14822–2	1.83458–2	1.58471–2	1.38247–2	1.21651–2	1.07867–2	9.62951–3	8.64865–3	7.81011–3
1.25	2.53188–2	2.13593–2	1.82561–2	1.57801–2	1.37737–2	1.21256–2	1.07556–2	9.60471–3	8.62864–3	7.79379–3
1.50	2.51101–2	2.12106–2	1.81474–2	1.56988–2	1.37117–2	1.20775–2	1.07178–2	9.57453–3	8.60427–3	7.77390–3
1.75	2.48672–2	2.10371–2	1.80202–2	1.56036–2	1.36390–2	1.20211–2	1.06734–2	9.53907–3	8.57562–3	7.75051–3
2.00	2.45914–2	2.08395–2	1.78751–2	1.54946–2	1.35556–2	1.19563–2	1.06221–2	9.49837–3	8.54272–3	7.72368–3
2.25	2.42852–2	2.06195–2	1.77132–2	1.53730–2	1.34625–2	1.18838–2	1.05649–2	9.45265–3	8.50573–3	7.69344–3
2.50	2.39506–2	2.03784–2	1.75354–2	1.52391–2	1.33599–2	1.18038–2	1.05017–2	9.40198–3	8.46471–3	7.65986–3
2.75	2.35893–2	2.01171–2	1.73420–2	1.50931–2	1.32478–2	1.17163–2	1.04324–2	9.34649–3	8.41974–3	7.62304–3
3.00	2.32037–2	1.98372–2	1.71343–2	1.49359–2	1.31266–2	1.16216–2	1.03575–2	9.28636–3	8.37096–3	7.58305–3
3.25	2.27964–2	1.95402–2	1.69131–2	1.47681–2	1.29971–2	1.15202–2	1.02770–2	9.22171–3	8.31846–3	7.53998–3
3.50	2.23696–2	1.92275–2	1.66794–2	1.45902–2	1.28596–2	1.14124–2	1.01912–2	9.15272–3	8.26236–3	7.49393–3
3.75	2.19257–2	1.89009–2	1.64344–2	1.44032–2	1.27146–2	1.12984–2	1.01004–2	9.07957–3	8.20280–3	7.44498–3
4.00	2.14672–2	1.85618–2	1.61790–2	1.42076–2	1.25627–2	1.11786–2	1.00048–2	9.00242–3	8.13992–3	7.39323–3
4.25	2.09963–2	1.82118–2	1.59143–2	1.40043–2	1.24041–2	1.10533–2	9.90500–3	8.92150–3	8.07387–3	7.33880–3
4.50	2.05153–2	1.78524–2	1.56414–2	1.37938–2	1.22396–2	1.09230–2	9.80063–3	8.83700–3	8.00477–3	7.28179–3
4.75	2.00262–2	1.74851–2	1.53611–2	1.35770–2	1.20696–2	1.07881–2	9.69227–3	8.74910–3	7.93279–3	7.22232–3
5.00	1.95314–2	1.71113–2	1.50747–2	1.33545–2	1.18945–2	1.06487–2	9.58018–3	8.65801–3	7.85808–3	7.16052–3
5.25	1.90326–2	1.67323–2	1.47829–2	1.31270–2	1.17150–2	1.05055–2	9.46466–3	8.56394–3	7.78079–3	7.09649–3
5.50	1.85318–2	1.63496–2	1.44869–2	1.28953–2	1.15315–2	1.03587–2	9.34596–3	8.46710–3	7.70107–3	7.03033–3
5.75	1.80306–2	1.59645–2	1.41875–2	1.26599–2	1.13445–2	1.02086–2	9.22437–3	8.36768–3	7.61910–3	6.96220–3
6	1.75307–2	1.55780–2	1.38856–2	1.24216–2	1.11546–2	1.00557–2	9.10016–3	8.26590–3	7.53501–3	6.89221–3
7	1.55713–2	1.40401–2	1.26691–2	1.14513–2	1.03741–2	9.42281–3	8.58240–3	7.83921–3	7.18077–3	6.59605–3
8	1.37257–2	1.25564–2	1.14713–2	1.04791–2	9.58060–3	8.77072–3	8.04325–3	7.39062–3	6.80522–3	6.27977–3
9	1.20370–2	1.11663–2	1.03261–2	9.53327–3	8.79628–3	8.11783–3	7.49719–3	6.93168–3	6.41759–3	5.95072–3
10	1.05231–2	9.89186–3	9.25505–3	8.63315–3	8.03868–3	7.47864–3	6.95626–3	6.47231–3	6.02601–3	5.61563–3
11	9.18660–3	8.74249–3	8.27073–3	7.79187–3	7.31996–3	6.86414–3	6.43004–3	6.02070–3	5.63744–3	5.28034–3
12	8.01888–3	7.71835–3	7.37790–3	7.01642–3	6.64783–3	6.28199–3	5.92569–3	5.58331–3	5.25753–3	4.94971–3
13	7.00581–3	6.81362–3	6.57595–3	6.30927–3	6.02638–3	5.73695–3	5.44809–3	5.16484–3	4.89063–3	4.62770–3
14	6.13076–3	6.01915–3	5.86083–3	5.66965–3	5.45688–3	5.23144–3	5.00025–3	4.76848–3	4.53994–3	4.31732–3
15	5.37672–3	5.32420–3	5.22639–3	5.09465–3	4.93859–3	4.76613–3	4.58366–3	4.39620–3	4.20760–3	4.02076–3
16	4.72752–3	4.71769–3	4.66549–3	4.58004–3	4.46938–3	4.34033–3	4.19862–3	4.04890–3	3.89490–3	3.73950–3
18	3.68640–3	3.72795–3	3.73463–3	3.71212–3	3.66576–3	3.60038–3	3.52028–3	3.42913–3	3.33008–3	3.22572–3
20	2.90978–3	2.97464–3	3.01202–3	3.02537–3	3.01809–3	2.99345–3	2.95448–3	2.90392–3	2.84421–3	2.77746–3
22	2.32514–3	2.39814–3	2.44991–3	2.48250–3	2.49803–3	2.49863–3	2.48638–3	2.46323–3	2.43101–3	2.39133–3
24	1.88015–3	1.95338–3	2.01033–3	2.05219–3	2.08026–3	2.09590–3	2.10051–3	2.09544–3	2.08201–3	2.06141–3
26	1.53749–3	1.60705–3	1.66415–3	1.70943–3	1.74369–3	1.76778–3	1.78263–3	1.78915–3	1.78827–3	1.78088–3
28	1.27054–3	1.33474–3	1.38935–3	1.43473–3	1.47132–3	1.49967–3	1.52037–3	1.53404–3	1.54134–3	1.54288–3
30	1.06019–3	1.11851–3	1.16939–3	1.21303–3	1.24968–3	1.27967–3	1.30337–3	1.32121–3	1.33362–3	1.34106–3
32	8.92664–4	9.45152–4	9.91852–4	1.03284–3	1.06825–3	1.09827–3	1.12315–3	1.14316–3	1.15859–3	1.16976–3
34	7.57846–4	8.04868–4	8.47346–4	8.85295–4	9.18776–4	9.47896–4	9.72800–4	9.93667–4	1.01070–3	1.02411–3

Table 9-52. SOLID ANGLES (Continued)

I. SOLID ANGLE SUBTENDED BY THE CIRCULAR DISC OF UNIT RADIUS (Continued)

ρ \ z	22	26	30	40	50	60	70	80	90	100
0.0	6.48085-3	4.64218-3	3.48775-3	1.96256-3	1.25626-3	8.72483-4	6.41043-4	4.90816-4	3.87815-4	3.14136-4
0.1	6.48065-3	4.64207-3	3.48769-3	1.96256-3	1.25625-3	8.72479-4	6.41041-4	4.90815-4	3.87814-4	3.14135-4
0.2	6.48005-3	4.64177-3	3.48752-3	1.96250-3	1.25623-3	8.72468-4	6.41035-4	4.90812-4	3.87812-4	3.14134-4
0.3	6.47905-3	4.64125-3	3.48723-3	1.96241-3	1.25619-3	8.72450-4	6.41026-4	4.90806-4	3.87809-4	3.14132-4
0.4	6.47765-3	4.64053-3	3.48683-3	1.96228-3	1.25614-3	8.72425-4	6.41012-4	4.90798-4	3.87804-4	3.14128-4
0.5	6.47586-3	4.63961-3	3.48630-3	1.96212-3	1.25607-3	8.72393-4	6.40995-4	4.90788-4	3.87797-4	3.14124-4
0.6	6.47366-3	4.63848-3	3.48567-3	1.96192-3	1.25599-3	8.72353-4	6.40973-4	4.90775-4	3.87790-4	3.14119-4
0.7	6.47107-3	4.63715-3	3.48492-3	1.96168-3	1.25589-3	8.72306-4	6.40948-4	4.90761-4	3.87780-4	3.14113-4
0.8	6.46808-3	4.63562-3	3.48405-3	1.96140-3	1.25578-3	8.72252-4	6.40919-4	4.90744-4	3.87770-4	3.14106-4
0.9	6.46469-3	4.63388-3	3.48306-3	1.96109-3	1.25565-3	8.72190-4	6.40886-4	4.90724-4	3.87758-4	3.14098-4
1.00	6.46089-3	4.63192-3	3.48196-3	1.96074-3	1.25551-3	8.72120-4	6.40847-4	4.90701-4	3.87743-4	3.14089-4
1.25	6.44972-3	4.62618-3	3.47871-3	1.95971-3	1.25509-3	8.71817-4	6.40738-4	4.90638-4	3.87704-4	3.14063-4
1.50	6.43610-3	4.61916-3	3.47475-3	1.95845-3	1.25457-3	8.71668-4	6.40603-4	4.90559-4	3.87654-4	3.14030-4
1.75	6.42006-3	4.61090-3	3.47007-3	1.95696-3	1.25396-3	8.71373-4	6.40444-4	4.90466-4	3.87596-4	3.13992-4
2.00	6.40164-3	4.60142-3	3.46470-3	1.95525-3	1.25327-3	8.71034-4	6.40261-4	4.90358-4	3.87529-4	3.13945-4
2.25	6.38086-3	4.59069-3	3.45861-3	1.95331-3	1.25247-3	8.70649-4	6.40053-4	4.90236-4	3.87453-4	3.13898-4
2.50	6.35778-3	4.57871-3	3.45181-3	1.95115-3	1.25154-3	8.70218-4	6.39822-4	4.90100-4	3.87368-4	3.13842-4
2.75	6.33242-3	4.56556-3	3.44434-3	1.94876-3	1.25056-3	8.69743-4	6.39565-4	4.89950-4	3.87274-4	3.13780-4
3.00	6.30482-3	4.55123-3	3.43619-3	1.94616-3	1.24949-3	8.69225-4	6.39282-4	4.89784-4	3.87171-4	3.13713-4
3.25	6.27507-3	4.53574-3	3.42730-3	1.94333-3	1.24833-3	8.68661-4	6.38977-4	4.89605-4	3.87059-4	3.13639-4
3.50	6.24321-3	4.51911-3	3.41788-3	1.94028-3	1.24706-3	8.68052-4	6.38649-4	4.89411-4	3.86938-4	3.13560-4
3.75	6.20927-3	4.50136-3	3.40773-3	1.93702-3	1.24572-3	8.67399-4	6.38296-4	4.89204-4	3.86808-4	3.13475-4
4.00	6.17334-3	4.48252-3	3.39695-3	1.93353-3	1.24428-3	8.66703-4	6.37919-4	4.88983-4	3.86670-4	3.13384-4
4.25	6.13546-3	4.46260-3	3.38553-3	1.92984-3	1.24275-3	8.65962-4	6.37517-4	4.88747-4	3.86523-4	3.13287-4
4.50	6.09570-3	4.44163-3	3.37349-3	1.92593-3	1.24114-3	8.65178-4	6.37093-4	4.88497-4	3.86366-4	3.13184-4
4.75	6.05414-3	4.41964-3	3.36083-3	1.92182-3	1.23943-3	8.64350-4	6.36644-4	4.88233-4	3.86201-4	3.13075-4
5.00	6.01086-3	4.39668-3	3.34759-3	1.91751-3	1.23764-3	8.63479-4	6.36171-4	4.87956-4	3.86027-4	3.12961-4
5.25	5.96589-3	4.37273-3	3.33375-3	1.91299-3	1.23576-3	8.62564-4	6.35675-4	4.87664-4	3.85844-4	3.12841-4
5.50	5.91932-3	4.34786-3	3.31933-3	1.90820-3	1.23379-3	8.61607-4	6.35155-4	4.87359-4	3.85654-4	3.12716-4
5.75	5.87124-3	4.32207-3	3.30436-3	1.90333-3	1.23174-3	8.60607-4	6.34612-4	4.87039-4	3.85454-4	3.12585-4
6	5.82171-3	4.29540-3	3.28885-3	1.89822-3	1.22960-3	8.59565-4	6.34047-4	4.86706-4	3.85245-4	3.12448-4
7	5.61063-3	4.18064-3	3.22162-3	1.87584-3	1.22021-3	8.54978-4	6.31551-4	4.85235-4	3.84324-4	3.11842-4
8	5.38247-3	4.05444-3	3.14684-3	1.85055-3	1.20953-3	8.49735-4	6.28692-4	4.83548-4	3.83200-4	3.11145-4
9	5.14198-3	3.91890-3	3.06552-3	1.82258-3	1.19764-3	8.43859-4	6.25477-4	4.81648-4	3.82072-4	3.10359-4
10	4.89367-3	3.77612-3	2.97879-3	1.79214-3	1.18454-3	8.37371-4	6.21917-4	4.79538-4	3.80745-4	3.09484-4
11	4.64164-3	3.62813-3	2.88741-3	1.75948-3	1.17034-3	8.30295-4	6.18020-4	4.77224-4	3.79287-4	3.08521-4
12	4.38948-1	3.47681-3	2.79265-3	1.72482-3	1.15511-3	8.22659-4	6.13800-4	4.74711-4	3.77702-4	3.07473-4
13	4.14026-3	3.32389-3	2.69539-3	1.68841-3	1.13893-3	8.14491-4	6.09266-4	4.72004-4	3.75990-4	3.06340-4
14	3.89651-3	3.17092-3	2.59653-3	1.65051-3	1.12185-3	8.05819-4	6.04431-4	4.69109-4	3.74157-4	3.05123-4
15	3.66024-3	3.01926-3	2.49689-3	1.61136-3	1.10399-3	7.96676-4	5.99310-4	4.66033-4	3.72204-4	3.03826-4
16	3.43295-3	2.87006-3	2.39723-3	1.57119-3	1.08543-3	7.87092-4	5.93915-4	4.62782-4	3.70134-4	3.02449-4
18	3.00925-3	2.58263-3	2.20044-3	1.48869-3	1.04648-3	7.66730-4	5.82360-4	4.55791-4	3.65662-4	2.99466-4
20	2.62993-3	2.31411-3	2.01048-3	1.40470-3	1.00563-3	7.44990-4	5.69882-4	4.48163-4	3.60770-4	2.96190-4
22	2.29533-3	2.06752-3	1.83037-3	1.32068-3	9.63516-4	7.22132-4	5.56601-4	4.39987-4	3.55489-4	2.92639-4
24	2.00320-3	1.84398-3	1.66202-3	1.23786-3	9.20632-4	6.98409-4	5.42637-4	4.31314-4	3.49851-4	2.88830-4
26	1.74991-3	1.64331-3	1.50644-3	1.15722-3	8.77511-4	6.74062-4	5.28109-4	4.22205-4	3.43889-4	2.84782-4
28	1.53121-3	1.46444-3	1.36392-3	1.07951-3	8.34596-4	6.49315-4	5.13133-4	4.12722-4	3.37638-4	2.80515-4
30	1.34281-3	1.30580-3	1.23426-3	1.00528-3	7.92265-4	6.24378-4	4.97820-4	4.02926-4	3.31133-4	2.76049-4
32	1.18065-3	1.16560-3	1.11692-3	9.34894-4	7.50829-4	5.99437-4	4.82275-4	3.92875-4	3.24407-4	2.71406-4
34	1.04102-3	1.04193-3	1.01112-3	8.68556-4	7.10546-4	5.74662-4	4.66596-4	3.82627-4	3.17493-4	2.66604-4

Table 9-52. SOLID ANGLES *(Continued)*

II. SOLID ANGLE SUBTENDED BY THE LATERAL SURFACE OF A RIGHT CIRCULAR CYLINDER OF UNIT RADIUS

ρ \ h	1.0	2.0	3.0	4.0	5.0	6.0	7.0	8.0	9.0	10.0
1.2	1.89875 0	1.95124 0	1.96168 0	1.96540 0	1.96713 0	1.96807 0	1.96864 0	1.96901 0	1.96926 0	1.96945 0
1.4	1.42410 0	1.54292 0	1.56907 0	1.57861 0	1.58310 0	1.58556 0	1.58705 0	1.58802 0	1.58869 0	1.58917 0
1.6	1.10050 0	1.27070 0	1.31287 0	1.32879 0	1.33638 0	1.34057 0	1.34312 0	1.34478 0	1.34593 0	1.34675 0
1.8	8.67005–1	1.06851 0	1.12506 0	1.14727 0	1.15805 0	1.16404 0	1.16771 0	1.17011 0	1.17176 0	1.17295 0
2.0	6.94977–1	9.10596–1	9.78962–1	1.00702 0	1.02091 0	1.02872 0	1.03352 0	1.03667 0	1.03885 0	1.04042 0
2.2	5.66248–1	7.83743–1	8.61087–1	8.94347–1	9.11171–1	9.20739–1	9.26663–1	9.30574–1	9.33286–1	9.35241–1
2.4	4.68412–1	6.80031–1	7.63616–1	8.01317–1	8.20839–1	8.32085–1	8.39103–1	8.43758–1	8.46997–1	8.49338–1
2.6	3.92871–1	5.94203–1	6.81600–1	7.22945–1	7.44887–1	7.57704–1	7.65771–1	7.71151–1	7.74908–1	7.77631–1
2.8	3.33637–1	5.22511–1	6.11697–1	6.55902–1	6.79965–1	6.94229–1	7.03287–1	7.09365–1	7.13628–1	7.16725–1
3.0	2.86501–1	4.62166–1	5.51530–1	5.97861–1	6.23737–1	6.39312–1	6.49299–1	6.56043–1	6.60793–1	6.64256–1
3.2	2.48476–1	4.11026–1	4.99338–1	5.47130–1	5.74517–1	5.91261–1	6.02107–1	6.09480–1	6.14698–1	6.18516–1
3.4	2.17413–1	3.67416–1	4.53777–1	5.02444–1	5.31049–1	5.48821–1	5.60453–1	5.68417–1	5.74081–1	5.78240–1
3.6	1.91744–1	3.30009–1	4.13789–1	4.62830–1	4.92381–1	5.11038–1	5.23382–1	5.31895–1	5.37982–1	5.42468–1
3.8	1.70310–1	2.97746–1	3.78529–1	4.27523–1	4.57768–1	4.77174–1	4.90154–1	4.99174–1	5.05659–1	5.10457–1
4.0	1.52241–1	2.69771–1	3.47309–1	3.95909–1	4.26622–1	4.46646–1	4.60188–1	4.69672–1	4.76528–1	4.81622–1
4.2	1.36877–1	2.45391–1	3.19561–1	3.67491–1	3.98469–1	4.18987–1	4.33018–1	4.42922–1	4.50123–1	4.55497–1
4.4	1.23708–1	2.24044–1	2.94812–1	3.41854–1	3.72920–1	3.93818–1	4.08267–1	4.18549–1	4.26069–1	4.31706–1
4.6	1.12339–1	2.05265–1	2.72666–1	3.18654–1	3.49654–1	3.70826–1	3.85627–1	3.96245–1	4.04056–1	4.09938–1
4.8	1.02459–1	1.88673–1	2.52789–1	2.97600–1	3.28403–1	3.49752–1	3.64841–1	3.75753–1	3.83831–1	3.89941–1
5.0	9.38206–2	1.73954–1	2.34894–1	2.78444–1	3.08938–1	3.30378–1	3.45695–1	3.56863–1	3.65180–1	3.71501–1
5.2	8.62255–2	1.60844–1	2.18740–1	2.60971–1	2.91065–1	3.12519–1	3.28008–1	3.39393–1	3.47924–1	3.54439–1
5.4	7.95131–2	1.49123–1	2.04118–1	2.44998–1	2.74616–1	2.96015–1	3.11626–1	3.23193–1	3.31913–1	3.38605–1
5.6	7.35522–2	1.38608–1	1.90849–1	2.30366–1	2.59448–1	2.80731–1	2.96417–1	3.08130–1	3.17017–1	3.23869–1
5.8	6.82353–2	1.29142–1	1.78778–1	2.16934–1	2.45432–1	2.66548–1	2.82265–1	2.94094–1	3.03124–1	3.10120–1
6.0	6.34731–2	1.20594–1	1.67771–1	2.04581–1	2.32460–1	2.53362–1	2.69072–1	2.80987–1	2.90137–1	2.97262–1
6.5	5.35358–2	1.02539–1	1.44178–1	1.77743–1	2.03974–1	2.24190–1	2.39735–1	2.51747–1	2.61112–1	2.68493–1
7.5	3.95575–2	7.66348–2	1.09462–1	1.37269–1	1.60119–1	1.78567–1	1.93338–1	2.05150–1	2.14626–1	2.22272–1
8.5	3.04119–2	5.93586–2	8.57040–2	1.08813–1	1.28530–1	1.45042–1	1.58716–1	1.69982–1	1.79257–1	1.86909–1
9.5	2.41050–2	4.72910–2	6.88047–2	8.81595–2	1.05149–1	1.19793–1	1.32262–1	1.42802–1	1.51679–1	1.59152–1
10.5	1.95736–2	3.85425–2	5.63913–2	7.27542–2	8.74333–2	1.00380–1	1.11658–2	1.21398–1	1.29769–1	1.36944–1
11.5	1.62091–2	3.20044–2	4.70231–2	6.09902–2	7.37342–2	8.51825–2	9.53431–2	1.04281–1	1.12096–1	1.18904–1
12.5	1.36428–2	2.69931–2	3.97889–2	5.18218–2	6.29492–2	7.30946–2	8.22391–2	9.04085–2	9.76595–2	1.04066–1
13.5	1.16408–2	2.30692–2	3.40916–2	4.45484–2	5.43227–2	6.33428–2	7.15781–2	7.90322–2	8.57343–2	9.17306–2
14.5	1.00492–2	1.99405–2	2.95278–2	3.86876–2	4.73248–2	5.53752–2	6.28045–2	6.96043–2	7.57866–2	8.13787–2
15.5	8.76301–3	1.74061–2	2.58174–2	3.38999–2	4.15762–2	4.87904–2	5.55086–2	6.17160–2	6.74144–2	7.26186–2
16.5	7.70879–3	1.53249–2	2.27613–2	2.99409–2	3.68004–2	4.32920–2	4.93837–2	5.50582–2	6.03113–2	6.51492–2
17.5	6.83395–3	1.35952–2	2.02149–2	2.66314–2	3.27925–2	3.86575–2	4.41973–2	4.93940–2	5.42399–2	5.87361–2
18.5	6.09999–3	1.21422–2	1.80714–2	2.38377–2	2.93981–2	3.47178–2	3.97707–2	4.45396–2	4.90150–2	5.31947–2
19.5	5.47822–3	1.09099–2	1.62504–2	2.14588–2	2.64994–2	3.13425–2	3.59652–2	4.03511–2	4.44902–2	4.83781–2
20.5	4.94688–3	9.85584–3	1.46904–2	1.94168–2	2.40052–2	2.84302–2	3.26717–2	3.67145–2	4.05486–2	4.41685–2
21.5	4.48925–3	8.94732–3	1.33441–2	1.76515–2	2.18443–2	2.59010–2	2.98037–2	3.35387–2	3.70964–2	4.04706–2
22.5	4.09231–3	8.15873–3	1.21742–2	1.61152–2	1.99603–2	2.36911–2	2.72920–2	3.07505–2	3.40574–2	3.72067–2
23.5	3.74577–3	7.46988–3	1.11513–2	1.47701–2	1.83083–2	2.17497–2	2.50807–2	2.82902–2	3.13696–2	3.43129–2
24.5	3.44146–3	6.86465–3	1.02518–2	1.35860–2	1.68518–2	2.00352–2	2.31243–2	2.61092–2	2.89819–2	3.17365–2
25.5	3.17278–3	6.33004–3	9.45665–3	1.25382–2	1.55613–2	1.85140–2	2.13857–2	2.41674–2	2.68519–2	2.94337–2
26	3.05009–3	6.08584–3	9.09327–3	1.20590–2	1.49707–2	1.78170–2	2.05881–2	2.32755–2	2.58723–2	2.83732–2
28	2.62417–3	5.23774–3	7.83055–3	1.03920–2	1.29134–2	1.53859–2	1.78017–2	2.01541–2	2.24375–2	2.46470–2
30	2.28161–3	4.55522–3	6.81299–3	9.04732–3	1.12510–2	1.34174–2	1.55405–2	1.76147–2	1.96353–2	2.15984–2
32	2.00200–3	3.99785–3	5.98150–3	7.94708–3	9.88899–3	1.18020–2	1.36812–2	1.55222–2	1.73212–2	1.90748–2
34	1.77080–3	3.53680–3	5.29326–3	7.03556–3	8.75929–3	1.04603–2	1.21346–2	1.37786–2	1.53892–2	1.69635–2

Table 9-52. SOLID ANGLES (Continued)

II. SOLID ANGLE SUBTENDED BY THE LATERAL SURFACE OF A RIGHT CIRCULAR CYLINDER OF UNIT RADIUS (Continued)

ρ \ h	11	12	13	14	15	16	17	18	19	20
1.2	1.96958 0	1.96968 0	1.96976 0	1.96983 0	1.96988 0	1.96992 0	1.96995 0	1.96998 0	1.97001 0	1.97003 0
1.4	1.58952 0	1.58979 0	1.59000 0	1.59016 0	1.59030 0	1.59041 0	1.59050 0	1.59058 0	1.59064 0	1.59069 0
1.6	1.34735 0	1.34782 0	1.34818 0	1.34846 0	1.34870 0	1.34889 0	1.34904 0	1.34917 0	1.34929 0	1.34938 0
1.8	1.17383 0	1.17450 0	1.17503 0	1.17544 0	1.17578 0	1.17606 0	1.17628 0	1.17648 0	1.17664 0	1.17678 0
2.0	1.04159 0	1.04248 0	1.04317 0	1.04372 0	1.04417 0	1.04453 0	1.04484 0	1.04509 0	1.04531 0	1.04549 0
2.2	9.36695-1	9.37807-1	9.38675-1	9.39365-1	9.39923-1	9.40381-1	9.40761-1	9.41079-1	9.41349-1	9.41580-1
2.4	8.51083-1	8.52418-1	8.53462-1	8.54292-1	8.54964-1	8.55516-1	8.55974-1	8.56358-1	8.56683-1	8.56962-1
2.6	7.79665-1	7.81223-1	7.82442-1	7.83414-1	7.84200-1	7.84846-1	7.85382-1	7.85833-1	7.86214-1	7.86541-1
2.8	7.19044-1	7.20824-1	7.22218-1	7.23330-1	7.24231-1	7.24971-1	7.25587-1	7.26103-1	7.26542-1	7.26916-1
3.0	6.66855-1	6.68852-1	6.70420-1	6.71671-1	6.72687-1	6.73521-1	6.74215-1	6.74798-1	6.75293-1	6.75716-1
3.2	6.21388-1	6.23600-1	6.25338-1	6.26728-1	6.27856-1	6.28784-1	6.29556-1	6.30205-1	6.30756-1	6.31228-1
3.4	5.81377-1	5.83798-1	5.85704-1	5.87230-1	5.88470-1	5.89490-1	5.90340-1	5.91055-1	5.91663-1	5.92183-1
3.6	5.45862-1	5.48487-1	5.50556-1	5.52216-1	5.53565-1	5.54678-1	5.55605-1	5.56385-1	5.57048-1	5.57616-1
3.8	5.14097-1	5.16920-1	5.19150-1	5.20940-1	5.22398-1	5.23600-1	5.24603-1	5.25448-1	5.26166-1	5.26782-1
4.0	4.85500 1	4.88517-1	4.90898 1	4.92816-1	4.94380-1	4.95671-1	4.96749-1	4.97658 2	4.98431-1	4.99094-1
4.2	4.59601-1	4.62799-1	4.65334 1	4.67376 1	4.69011-1	4.70422-1	4.71574-1	4.72545-1	4.73373-1	4.74082-1
4.4	4.36025-1	4.39399-1	4.42080-1	4.44243-1	4.46011-1	4.47475-1	4.48699-1	4.49732-1	4.50613-1	4.51369-1
4.6	4.14461-1	4.18004-1	4.20825-1	4.23106-1	4.24973-1	4.26520-1	4.27815-1	4.28909 1	4.29842-1	4.30643-1
4.8	3.94656-1	3.98360-1	4.01316-1	4.03710-1	4.05672-1	4.07300-1	4.08665-1	4.09819-1	4.10803-1	4.11649-1
5.0	3.76396-1	3.80253-1	3.83339-1	3.85841-1	3.87897-1	3.89604-1	3.91036-1	3.92249-1	3.93284-1	3.94174-1
5.2	3.59504-1	3.63505-1	3.66714 1	3.69322-1	3.71467-1	3.73252-1	3.74750-1	3.76020-1	3.77104-1	3.78038-1
5.4	3.43827-1	3.47965-1	3.51291-1	3.54000-1	3.56232-1	3.58091-1	3.59654-1	3.60979-1	3.62113-1	3.63089-1
5.6	3.29236-1	3.33503-1	3.36941-1	3.39746-1	3.42062-1	3.43993-1	3.45618-1	3.46998-1	3.48179-1	3.49198-1
5.8	3.15621-1	3.20008-1	3.23552-1	3.26450-1	3.28845-1	3.30846-1	3.32532-1	3.33966-1	3.35193-1	3.36252-1
6.0	3.02886-1	3.07384-1	3.11028-1	3.14014-1	3.16487-1	3.18555-1	3.20300-1	3.21785-1	3.23058-1	3.24157-1
6.5	2.74376-1	2.79120-1	2.82988-1	2.86175-1	2.88826-1	2.91051-1	2.92935-1	2.94543-1	2.95924-1	2.97120-1
7.5	2.28488-1	2.33581-1	2.37791-1	2.41298-1	2.44243-1	2.46736-1	2.48862-1	2.50686-1	2.52261-1	2.53630-1
8.5	1.93246-1	1.98524-1	2.02945-1	2.06671-1	2.09832-1	2.12530-1	2.14847-1	2.16848-1	2.18586-1	2.20103 1
9.5	1.65452-1	1.70779-1	1.75302-1	1.79157-1	1.82461-1	1.85305-1	1.87766-1	1.89906-1	1.91775-1	1.93415-1
10.5	1.43092-1	1.48367-1	1.52901 1	1.56811-1	1.60193-1	1.63131-1	1.65693-1	1.67935-1	1.69905-1	1.71643-1
11.5	1.24825-1	1.29971-1	1.34449-1	1.38352-1	1.41761-1	1.44746-1	1.47369-1	1.49681-1	1.51724-1	1.53537-1
12.5	1.09712-1	1.14680-1	1.19051-1	1.22899-1	1.26291-1	1.29286-1	1.31937-1	1.34289-1	1.36381-1	1.38246-1
13.5	0.70772-2	1.01834-1	1.06063-1	1.09821-1	1.13162-1	1.16130-1	1.18787-1	1.21154 1	1.23272 1	1.25171-1
14.5	8.64177-2	9.09463-2	9.50095-2	9.86522-2	1.01917-1	1.04846-1	1.07474-1	1.09835-1	1.11960-1	1.13875-1
15.5	7.73521-2	8.16447-2	8.55292-2	8.90399-2	9.22107-2	9.50741-2	9.76608-2	9.99992-2	1.02115-1	1.04032-1
16.5	6.95864-2	7.36430-2	7.73426-2	8.07111-2	8.37747-2	8.65597-2	8.90911-2	9.13926-2	9.34864-2	9.53927-2
17.5	6.28905 2	6.67161-2	7.02299-2	7.34510-2	7.63996-2	7.90965-2	8.15620-2	8.38159-2	8.58768-2	8.77622-2
18.5	5.70820-2	6.06853-2	6.40160-2	6.70883-2	6.99176-2	7.25201-2	7.49123-2	7.71105-2	7.91301-2	8.09861-2
19.5	5.20154-2	5.54067-2	5.85598-2	6.14847-2	6.41931-2	6.66977-2	6.90116-2	7.11481-2	7.31200-2	7.49401-2
20.5	4.75729-2	5.07639-2	5.37463-2	5.65273-2	5.91155-2	6.15207-2	6.37533-2	6.58240-2	6.77436-2	6.95226-2
21.5	4.36589-2	4.66616-2	4.94814-2	5.21232-2	5.45934-2	5.68993-2	5.90492-2	6.10517-2	6.29156-2	6.46497-2
22.5	4.01949-2	4.30213-2	4.56871-2	4.81955-2	5.05509-2	5.27589-2	5.48259-2	5.67589-2	5.85649 2	6.02515-2
23.5	3.71163-2	3.97781-2	4.22987-2	4.46797-2	4.69243-2	4.90367-2	5.10217-2	5.28849-2	5.46320-2	5.62691-2
24.5	3.43692-2	3.68778-2	3.92617-2	4.15219-2	4.36603-2	4.56799-2	4.75844-2	4.93782-2	5.10660-2	5.26528-2
25.5	3.19088-2	3.42748-2	3.65306-2	3.86764-2	4.07133-2	4.26435-2	4.44696-2	4.61952-2	4.78239-2	4.93598-2
26	3.07741-2	3.30726-2	3.52674-2	3.73584-2	3.93464-2	4.12332-2	4.30211-2	4.47131-2	4.63125-2	4.78230-2
28	2.67791-2	2.88310-2	3.08012-2	3.26887-2	3.44934-2	3.62159-2	3.78574-2	3.94197-2	4.09046-2	4.23147-2
30	2.35008-2	2.53397-2	2.71135-2	2.88210-2	3.04614-2	3.20348-2	3.35416-2	3.49826-2	3.63590-2	3.76724-2
32	2.07801-2	2.24347-2	2.40370-2	2.55855-2	2.70794-2	2.85184-2	2.99023-2	3.12314-2	3.25064-2	3.37281-2
34	1.84991-2	1.99938-2	2.14460-2	2.28543-2	2.42179-2	2.55360-2	2.68084-2	2.80350-2	2.92161-2	3.03521-2

Table 9-52. SOLID ANGLES *(Continued)*

II. SOLID ANGLE SUBTENDED BY THE LATERAL SURFACE OF A RIGHT CIRCULAR CYLINDER OF UNIT RADIUS *(Continued)*

ρ \ h	22	26	30	40	50	60	70	80	90	100
1.2	1.97006 0	1.97011 0	1.97014 0	1.97017 0	1.97019 0	1.97020 0	1.97021 0	1.97021 0	1.97021 0	1.97021 0
1.4	1.59078 0	1.59090 0	1.59098 0	1.59108 0	1.59112 0	1.59115 0	1.59116 0	1.59117 0	1.59118 0	1.59119 0
1.6	1.34953 0	1.34974 0	1.34987 0	1.35004 0	1.35012 0	1.35016 0	1.35019 0	1.35021 0	1.35022 0	1.35023 0
1.8	1.17700 0	1.17730 0	1.17749 0	1.17774 0	1.17786 0	1.17792 0	1.17796 0	1.17798 0	1.17800 0	1.17801 0
2.0	1.04579 0	1.04619 0	1.04644 0	1.04677 0	1.04692 0	1.04701 0	1.04706 0	1.04709 0	1.04711 0	1.04713 0
2.2	9.41951-1	9.42453-1	9.42769-1	9.43186-1	9.43380-1	9.43485-1	9.43548-1	9.43589-1	9.43617-1	9.43638-1
2.4	8.57409-1	8.58016-1	8.58397-1	8.58901-1	8.59135-1	8.59262-1	8.59339-1	8.59388-1	8.59422-1	8.59447-1
2.6	7.87066-1	7.87778-1	7.88226-1	7.88818-1	7.89093-1	7.89242-1	7.89333-1	7.89391-1	7.89431-1	7.89460-1
2.8	7.27520-1	7.28338-1	7.28853-1	7.29535-1	7.29851-1	7.30023 1	7.30127-1	7.30194-1	7.30240-1	7.30273-1
3.0	6.76398-1	6.77323-1	6.77906-1	6.78678-1	6.79036-1	6.79231-1	6.79348-1	6.79424-1	6.79477-1	6.79514-1
3.2	6.31988-1	6.33021-1	6.33672-1	6.34534-1	6.34934-1	6.35152-1	6.35283-1	6.35368-1	6.35427-1	6.35469-1
3.4	5.93021-1	5.94161-1	5.94880-1	5.95832-1	5.96275-1	5.96516-1	5.96661-1	5.96756-1	5.96820-1	5.96867-1
3.6	5.58533-1	5.59779-1	5.60566-1	5.61609-1	5.62095-1	5.62359-1	5.62518-1	5.62622-1	5.62693-1	5.62743-1
3.8	5.27776-1	5.29129-1	5.29984-1	5.31118-1	5.31646-1	5.31933-1	5.32107-1	5.32219-1	5.32297-1	5.32352-1
4.0	5.00164-1	5.01624-1	5.02546-1	5.03771-1	5.04342-1	5.04652-1	5.04840-1	5.04962-1	5.05045-1	5.05105-1
4.2	4.75229-1	4.76794-1	4.77783-1	4.79100-1	4.79713-1	4.80047-1	4.80248-1	4.80380-1	4.80469-1	4.80534-1
4.4	4.52591-1	4.54260-1	4.55317-1	4.56724-1	4.57380-1	4.57737-1	4.57953-1	4.58093-1	4.58189-1	4.58258-1
4.6	4.31940-1	4.33713-1	4.34837-1	4.36334-1	4.37032-1	4.37413-1	4.37643-1	4.37793-1	4.37895-1	4.37969-1
4.8	4.13020-1	4.14896-1	4.16086-1	4.17673-1	4.18414-1	4.18818-1	4.19063-1	4.19221-1	4.19330-1	4.19408-1
5.0	3.95617-1	3.97595-1	3.98851-1	4.00528-1	4.01312-1	4.01739-1	4.01997-1	4.02166-1	4.02281-1	4.02363-1
5.2	3.79553-1	3.81631-1	3.82953-1	3.84719-1	3.85545-1	3.85996-1	3.86268-1	3.86446-1	3.86567-1	3.86654-1
5.4	3.64674-1	3.66852-1	3.68238-1	3.70093-1	3.70962-1	3.71436-2	3.71723-1	3.71909-1	3.72037-1	3.72129-1
5.6	3.50852-1	3.53128-1	3.54579-1	3.56522-1	3.57433-1	3.57930-1	3.58231-1	3.58427-1	3.58561-1	3.58658-1
5.8	3.37975-1	3.40348-1	3.41862-1	3.43893-1	3.44846-1	3.45367-1	3.45682-1	3.45887-1	3.46028-1	3.46129-1
6.0	3.25947-1	3.28415-1	3.29993-1	3.32111-1	3.33106-1	3.33650-1	3.33979-1	3.34193-1	3.34340-1	3.34446-1
6.5	2.99070-1	3.01771-1	3.03503-1	3.05837-1	3.06936 1	3.07538-1	3.07903-1	3.08140-1	3.08303-1	3.08420-1
7.5	2.55877-1	2.59013-1	2.61041-1	2.63793-1	2.65097-1	2.65813-1	2.66248-1	2.66531-1	2.66726-1	2.66866-1
8.5	2.22609-1	2.26138-1	2.28440-1	2.31591-1	2.33094-1	2.33923-1	2.34427-1	2.34756-1	2.34982-1	2.35145-1
9.5	1.96141-1	2.00019-1	2.02572-1	2.06099-1	2.07795-1	2.08734-1	2.09307-1	2.09681-1	2.09939-1	2.10123-1
10.5	1.74552-1	1.78731-1	1.81512-1	1.85391-1	1.87273-1	1.88320-1	1.88959-1	1.89378-1	1.89667-1	1.89874-1
11.5	1.56591-1	1.61027-1	1.64010-1	1.68216-1	1.70276-1	1.71427-1	1.72133-1	1.72595-1	1.72915-1	1.73144-1
12.5	1.41411-1	1.46059-1	1.49220-1	1.53727-1	1.55956-1	1.57209-1	1.57978-1	1.58484-1	1.58834-1	1.59085-1
13.5	1.28415-1	1.33232-1	1.36547-1	1.41328-1	1.43718-1	1.45068-1	1.45901-1	1.46449-1	1.46828-1	1.47102-1
14.5	1.17169-1	1.22117-1	1.25561-1	1.30589-1	1.33130-1	1.34575-1	1.35469-1	1.36059-1	1.36467-1	1.36762-1
15.5	1.07352-1	1.12393-1	1.15943-1	1.21192-1	1.23876-1	1.25410-1	1.26364-1	1.26994-1	1.27432-1	1.27748-1
16.5	9.87155-2	1.03817-1	1.07453-1	1.12896-1	1.15712-1	1.17333-1	1.18344-1	1.19015-1	1.19481-1	1.19818-1
17.5	9.10694-2	9.62022-2	9.99031-2	1.05515-1	1.08454-1	1.10158-1	1.11224-1	1.11933-1	1.12427-1	1.12785-1
18.5	8.42621-2	8.93999-2	9.31475-2	9.89041-2	1.01956-1	1.03738-1	1.04858-1	1.05605-1	1.06126-1	1.06504-1
19.5	7.81718-2	8.32920-2	8.70693-2	9.29470-2	9.61031-2	9.79583-2	9.91302-2	9.99138-2	1.00462-1	1.00859-1
20.5	7.26993-2	7.77824-2	8.15742-2	8.75508-2	9.08012-3	9.27260-2	9.39474-2	9.47666-2	9.53407-2	9.57578-2
21.5	6.77631-2	7.27924-2	7.65849-2	8.26398-2	8.59752-2	8.79652-2	8.92341-2	9.00877-2	9.06873-2	9.11235-2
22.5	6.32951-2	6.82567-2	7.20379-2	7.81515-2	8.15629-2	8.36139-2	8.49281-2	8.58149-2	8.64392-2	8.68943-2
23.5	5.92381-2	6.41207-2	6.78797-2	7.40341-2	7.75126-2	7.96204-2	8.09777-2	8.18966-2	8.25451-2	8.30184-2
24.5	5.55437-2	6.03378-2	6.40655-2	7.02440-2	7.37812-2	7.59416-2	7.73398-2	7.82896-2	7.89615-2	7.94528-2
25.5	5.21704-2	5.68689-2	6.05571-2	6.67446-2	7.03324-2	7.25412-2	7.39780-2	7.49576-2	7.56521-1	7.61610-2
26	5.05929-2	5.52413-2	5.89072-2	6.50939-2	6.87040-2	7.09355-2	7.23909-2	7.33849-2	7.40905-2	7.46079-2
28	4.49208-2	4.93585-2	5.29206-2	5.90727-2	6.27538-2	6.50659-2	6.65901-2	6.76386-2	6.83867-2	6.89374-2
30	4.01167-2	4.43342-2	4.77750-2	5.38500-2	5.75745-2	5.99521-2	6.15364-2	6.26346-2	6.34223-2	6.40044-2
32	3.60162-2	4.00114-2	4.33199-2	4.92838-2	5.30278-2	5.54564-2	5.70926-2	5.82354-2	5.90598-2	5.96715-2
34	2.24916-2	3.62675-2	3.94376-2	4.52648-2	4.90070-2	5.14733-2	5.31534-2	5.43361-2	5.51942-2	5.58335-2

*Abridged from: "Tables of Solid Angles", A.V.H. Masket and W.C. Rodgers, U.S. Atomic Energy Commission Report TID–14975, July 1962.

TABLES OF STATISTICAL PROBABILITY

Mathematical probability deals with the random or chance variation of numerical data. When the probability or statistical chance is expressed numerically (percentage or decimal), it is a specific likelihood representing the ratio of chances in favor to total chances available. In the usual probability-distribution graphs probability is represented by an area under the frequency curve.

Measures of central tendency are the *mean* (μ) or arithmetical average, the *median* or middle value, and the *mode* or most frequent value.

Measures of dispersion are the individual *deviation* (x), which is the difference between the mean and the specific value under consideration; the *standard deviation* (σ), which is the square root of the mean of the squares of the deviations (rms†); and the *variance*, which is the square of the standard deviation (σ^2). The *range* is the spread between the smallest and the largest items of data.

Frequency (y) is a measure of the importance of a given value in terms of the frequency of its occurrence. It is commonly expressed as the frequency of occurrence of stated values of the deviation from y_o, the mean, but it also refers to the frequency of occurrence of a given magnitude in original data.

Frequency distribution describes the frequency of occurrence of the various numerical values, or the frequency of occurrence of stated deviations from the mean. A frequency distribution is represented mathematically, by equations, curves, or tables. The distributions covered by the following tables include the normal, the binomial, the Poisson, the t, the F, and the chi-square distributions.

Statistical sample is a random sample representative of all the original data. The data being sampled are referred to collectively as the *population* or the *universe*.

Statistical significance is a general term for the assumed importance of the probability. The range of probabilities used to describe a very likely occurrence is often expressed as a percentage between 90 and 99.9. For very unlikely occurrences the probabilities between 0.1 percent and 10 percent are examined. The arbitrarily selected percentages are often called *confidence limits*, implying that there is a 100 percent probability, representing full or complete certainty. The borderline between a "significant" probability and one that is not significant must be arbitrarily selected. (See Student's t-Distribution.)

Degrees of freedom (N) refer to the number of independent properties of a sample. It is usually $n-1$, where n is the number of data items.

Table 9-53. Normal or Gaussian Probability Distribution.‡ This "continuous" distribution is applicable to a population or universe for which the number of items of data is infinitely large and in which the deviations from the mean are random and unrelated. It applies also to a large representative sample of such data, such as 50 or 100 items; the larger the sample, the closer the approximation. For this symmetrical distribution the mean, the median, and the mode all coincide and are represented by the maximum ordinate y_o. The table gives normalized values, in which the maximum probability area under the curve is unity; the deviations are measured in units of σ, the mean deviation, and the maximum ordinate is $\dfrac{1}{\sqrt{2\pi}} = 0.39894$.

Table 9-54. Student's t-Distribution. The t-test is widely used to evaluate the significance of differences, such as the difference between the means of two samples and the difference between a sample mean and the population mean. At the top of each column in the table is that probability that the difference would exist by chance alone. The probability of a match or a fit decreases as t increases. Two common borderline values are $p = 0.01$ and $p = 0.05$. For example, if the computed value of t is larger than the one given in the column headed *0.05*, the interpretation might be as follows: the probability that this difference is due to chance alone is less than one in twenty; hence the difference is significant and is due to factors other than pure chance. The ratio t must be correctly computed.** For comparing a sample with a known parent population, it is the ratio of the difference

---

†For linear correlation by a line of regression, using least squares, the standard deviation of the points from the line is called the "standard error of estimate".
‡Also called the normal error function and the normal frequency curve.
**Consult a textbook on statistics; see References.

TABLES OF STATISTICAL PROBABILITY *(Continued)*

between the sample mean and the population mean to the standard deviation of the mean of the parent population (corrected for sample size):

$$t = \frac{mean_1 - mean_2}{\sigma/\sqrt{N}}.$$

The t-distribution approaches the normal distribution as the number of degrees of freedom approaches infinity, but in any case the means themselves are assumed to be normally distributed.

Table 9-55. Chi-Square Distribution. This is another test for the significance of differences by evaluation of the spread of the data. There are several ways to apply the chi-square test. One involves the ratio of the squares of the two standard deviations:

$$\text{chi-square} = N(\sigma_1/\sigma_2)^2.$$

For example, if it is desired to test the variability or dispersion for a sample when that of a parent population is known, a value of chi-square ($= N$ times the ratio of variances) larger than the one in the *0.10* column would mean that there are fewer than ten chances in one hundred that the sample represented the parent population, and that its larger variability occurred purely by chance. As the chi-square increases, the probability of matching or agreement decreases.

Another application of the chi-square table uses the summation of the squares of frequency differences for goodness of fit with the parent distribution:

$$\text{chi-square} = \frac{\Sigma(y - y_c)^2}{y_c},$$

where the values of y_c are those of the comparison standard.

A very useful application of the chi-square test is in the evaluation of attribute data where there are a number of classes and the expectations in the different classes are unequal.

Table 9-56. *F*-Distribution. This distribution is used for testing dispersion in terms of variance. One use for the F-test is to determine whether two samples, possibly of different sizes, drawn independently from two normal populations, actually represent populations with identical standard deviations. Here F is the ratio of the variance of the *samples*:

$$F = \sigma_1^2/\sigma_2^2.$$

In the tables, since the two sample sizes and degrees of freedom may be unequal, the additional variable is accommodated by setting up a separate table for each probability value, the p values used here being in the range 0.001 to 0.10. The borderline between a significant difference and one that is not significant must be selected, and the table with that probability value is used. If the value of F is larger than the corresponding one in that table, the probability that the two samples came from like populations is even less than that selected as a borderline.

Table 9-57. Binomial Distribution. This is a "discrete" distribution representing the probabilities of "success" in N trials for a population or sample in which only two outcomes are possible, but for which the eventual outcome is fixed and known if an infinite number of trials are made.† This eventual outcome is fixed by the conditions, such as 0.5 for one face of a coin or 0.1667 for one face of a six-sided die. Values in the body of the table represent the cumulative probability of X or more successes in N trials. In applications to acceptance or attribute sampling, the table gives the probability of X or more acceptances (or rejections) in a single sample of N items. In either case the known or fixed probability of the result (success or failure) for the entire population is represented by p.

Table 9-58. Poisson Distribution. This is a discrete distribution approximating the binomial when the total number of items of data (the populations) is very large, but the probability (p) is very small and the sample is small compared with the population.

---

†Outcomes might be expressed as success or failure, yes or no, hit or miss, accept or reject, heads or tails, plus or minus, one or zero.

TABLES OF STATISTICAL PROBABILITY *(Continued)*

The Poisson cumulative probability, i.e., the probability that X is greater than or equal to X', is expressed as

$$P = \sum_{X = X'}^{\infty} \frac{e^{-Np}(Np)^x}{X'},$$

for specified values of X' and Np.

The table is arranged in terms of the product, Np, where N is the sample size and p is the fixed probability for the entire population. For this distribution $Np = \mu = \sigma^2$, i.e., both the mean and the variance are equal to Np. The standard deviation is $\sigma = \sqrt{Np}$. Values in the body of the table represent the cumulative probability of X or more successes in N trials (the same as for the binomial table); or in sampling, the values represent the probability of X or more acceptances in sample of N items. In either case the fixed probability for the whole population is p.

REFERENCES

"Applied General Statistics", 3rd ed., F.E. Croxton, D.J. Cowden and S. Klein, Eds., Prentice-Hall, Inc., 1967.
"Biometrika Tables for Statisticians", E.S. Pearson and H.O. Hartley, Eds., Vol. 1, Cambridge University Press, 1962.
"CRC Handbook of Chemistry and Physics", 50th ed., R.C. Weast, Ed., The Chemical Rubber Co., 1969.
"CRC Handbook of Probability and Statistics", 2nd ed., W.H. Beyer, Ed., The Chemical Rubber Co., 1968.
"CRC Handbook of Tables for Mathematics", 4th ed., S.M. Selby, Ed., The Chemical Rubber Co., 1970.
"CRC Standard Mathematical Tables", 17th ed., S.M. Selby, Ed., The Chemical Rubber Co., 1969.
"Geigy Scientific Tables", 6th ed., Geigy Chemical Corporation, 1962.
"Navord Report", No. 3369, Ordinance Test Station, 1955.
"Quality Control Handbook", J.M. Juran, Ed., McGraw-Hill Book Company, 1962.
"Statistics", A.D. Rickmers and H.N. Todd, Eds., McGraw-Hill Book Company, 1967.

Table 9-53.   ORDINATES AND AREAS FOR NORMAL OR GAUSSIAN PROBABILITY DISTRIBUTION

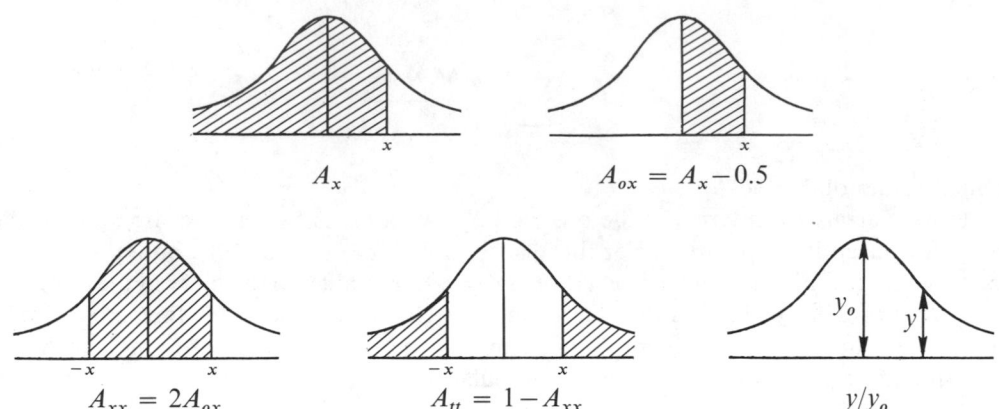

$$A_x \qquad\qquad A_{ox} = A_x - 0.5$$

$$A_{xx} = 2A_{ox} \qquad A_{tt} = 1 - A_{xx} \qquad y/y_o$$

SYMBOLS:

x = deviation from the mean (or from zero error). One unit of x equals one standard deviation.

y = frequency of occurrence of the deviation ("probability density")

$$= f(x) = \frac{1}{\sqrt{2\pi}} e^{-x^2/2}$$

σ = standard deviation or error (rms) = $\sqrt{\dfrac{\Sigma x^2}{n}}$

y_o = frequency of occurrence of mean value

y/y_o = relative frequency in terms of mean frequency

A_x = area under curve from $-\alpha$ to x = $\displaystyle\int_{-\infty}^{+x} \frac{1}{\sqrt{2\pi}} e^{-x^2/2}\, dx$

A_{ox} = area under curve from zero to x

A_{xx} = area under curve from $-x$ to $+x$ = probability of occurrence of values deviating from mean value in range from $-x$ to $+x$

A_{tt} = residual area, in the two "tails" = $1 - A_{xx}$

Note: All areas in the table are based on a transformation of the variable such that $A_{-\infty}^{+\infty} = 1$, with decimal values shown in the table.

Table 9-53. ORDINATES AND AREAS FOR NORMAL
OR GAUSSIAN PROBABILITY DISTRIBUTION *(Continued)*

INDEPENDENT VARIABLE = DEVIATION = x/σ

x/σ	A_{xx}	A_{ox}	A_x	A_{tt}	y	y/y_o
.00	.0000	.0000	.5000	1.0000	.3989	1.0000
.05	.0398	.0199	.5199	.9602	.3984	.9986
.10	.0797	.0398	.5398	.9203	.3971	.9950
.15	.1192	.0596	.5596	.8808	.3945	.9888
.20	.1585	.0793	.5793	.8415	.3910	.9802
.25	.1774	.0987	.5987	.8026	.3867	.9692
.30	.2358	.1179	.6179	.7642	.3814	.9560
.35	.2736	.1368	.6368	.7263	.3752	.9405
.40	.3108	.1554	.6554	.6892	.3683	.9231
.45	.3472	.1736	.6736	.6527	.3605	.9037
.50	.3829	.1915	.6915	.6171	.3521	.8825
.55	.4176	.2088	.7088	.5823	.3429	.8596
.60	.4515	.2257	.7257	.5485	.3332	.8353
.65	.4844	.2422	.7422	.5157	.3230	.8096
.70	.5161	.2580	.7580	.4839	.3123	.7827
.75	.5468	.2734	.7734	.4533	.3011	.7548
.80	.5763	.2881	.7881	.4237	.2897	.7262
.85	.6046	.3023	.8023	.3953	.2780	.6968
.90	.6319	.3159	.8159	.3681	.2661	.6670
.95	.6578	.3289	.8289	.3421	.2541	.6368
1.00	.6827	.3413	.8413	.3173	.2420	.6065
1.05	.7062	.3531	.8531	.2938	.2299	.5762
1.10	.7286	.3643	.8643	.2714	.2179	.5461
1.15	.7498	.3749	.8749	.2502	.2059	.5162
1.20	.7698	.3849	.8849	.2302	.1942	.4868
1.25	.7887	.3944	.8944	.2113	.1826	.4578
1.30	.8064	.4032	.9032	.1936	.1714	.4296
1.35	.8229	.4115	.9115	.1771	.1604	.4020
1.40	.8384	.4192	.9192	.1616	.1497	.3753
1.45	.8530	.4265	.9265	.1470	.1394	.3495
1.50	.8664	.4332	.9332	.1336	.1295	.3247
1.55	.8788	.4394	.9394	.1212	.1200	.3008
1.60	.8904	.4452	.9452	.1096	.1109	.2780
1.65	.9010	.4505	.9505	.0990	.1023	.2563
1.70	.9108	.4554	.9554	.0892	.0940	.2376
1.75	.9198	.4599	.9599	.0802	.0863	.2163
1.80	.9281	.4641	.9641	.0720	.0790	.1979
1.85	.9356	.4678	.9678	.0644	.0721	.1806
1.90	.9426	.4713	.9713	.0574	.0656	.1645
1.95	.9488	.4744	.9744	.0512	.0596	.1494
2.00	.9545	.4772	.9772	.0455	.0540	.1353
2.05	.9596	.4798	.9798	.0404	.0488	.1223
2.10	.9642	.4821	.9821	.0358	.0440	.1040
2.15	.9684	.4842	.9842	.0316	.0396	.0992
2.20	.9722	.4861	.9861	.0278	.0355	.0890

Table 9-53. ORDINATES AND AREAS FOR NORMAL OR GAUSSIAN PROBABILITY DISTRIBUTION (Continued)

INDEPENDENT VARIABLE = DEVIATION = x/σ (Continued)

x/σ	A_{xx}	A_{ox}	A_x	A_{tt}	y	y/y_o
2.25	.9756	.4878	.9878	.0244	.0317	.0796
2.30	.9786	.4893	.9893	.0214	.0283	.0709
2.35	.9812	.4906	.9906	.0188	.0252	.0632
2.40	.9836	.4918	.9918	.0164	.0224	.0561
2.45	.9858	.4929	.9929	.0143	.0198	.0497
2.50	.9876	.4938	.9938	.0124	.0175	.0439
2.55	.9892	.4946	.9946	.0108	.0155	.0387
2.60	.9907	.4953	.9953	.0093	.0136	.0341
2.65	.9920	.4960	.9960	.0080	.0119	.0299
2.70	.9930	.4965	.9965	.0070	.0104	.0261
2.75	.9940	.4970	.9970	.0060	.0091	.0228
2.80	.9948	.4974	.9974	.0051	.0079	.0198
2.85	.9956	.4978	.9978	.0044	.0069	.0172
2.90	.9962	.4981	.9981	.0037	.0060	.0150
2.95	.9968	.4984	.9984	.0032	.0051	.0129
3.00	.9973	.4987	.9987	.0027	.0044	.0111

INDEPENDENT VARIABLE = PROBABILITY = A_{xx}

x/σ	A_{xx}	A_{ox}	A_x	A_{tt}	y	y/y_o
.005	.005	.002	.502	.995	.399	.999
.013	.010	.005	.505	.990	.399	.999
.063	.050	.025	.525	.950	.398	.998
.126	.100	.050	.550	.900	.396	.990
.189	.150	.075	.575	.850	.392	.982
.253	.200	.100	.600	.800	.386	.965
.319	.250	.125	.625	.750	.379	.950
.385	.300	.150	.650	.700	.370	.925
.454	.350	.175	.675	.650	.360	.900
.524	.400	.200	.700	.600	.348	.870
.598	.450	.225	.725	.550	.334	.835
.674	.500	.250	.750	.500	.318	.795
.755	.550	.275	.775	.450	.300	.749
.842	.600	.300	.800	.400	.280	.702
.935	.650	.325	.825	.350	.258	.643
1.036	.700	.350	.850	.300	.233	.583
1.150	.750	.375	.875	.250	.206	.516
1.282	.800	.400	.900	.200	.176	.440
1.440	.850	.425	.925	.150	.142	.355
1.645	.900	.450	.950	.100	.103	.257
1.960	.950	.475	.975	.050	.058	.146
2.054	.960	.480	.980	.040	.048	.121
2.170	.970	.485	.985	.030	.038	.095
2.326	.980	.490	.990	.020	.027	.066
2.576	.990	.495	.995	.010	.014	.036
2.748	.995	.497	.997	.005	.009	.022
3.090	.999	.499	.999	.001	.003	.008

Table 9-54. STUDENT'S t-DISTRIBUTION*

Values of t at Specified Levels of Significance; Residual Area A_{tt}, Two Tails

SYMBOLS:　　N = degrees of freedom　　　　P = probability of agreement

P / N	.50	.40	.30	.20	.10	.05	.02	.01	.005	.001
1	1.000	1.376	1.963	3.078	6.314	12.706	31.821	63.657	127.32	636.619
2	.816	1.061	1.386	1.886	2.920	4.303	6.965	9.925	14.089	31.598
3	.765	.978	1.250	1.638	2.353	3.182	4.541	5.841	7.453	12.924
4	.741	.941	1.190	1.533	2.132	2.776	3.747	4.604	5.598	8.610
5	.727	.920	1.156	1.476	2.015	2.571	3.365	4.032	4.773	6.869
6	.718	.906	1.134	1.440	1.943	2.447	3.143	3.707	4.317	5.959
7	.711	.896	1.119	1.415	1.895	2.365	2.998	3.499	4.029	5.408
8	.706	.889	1.108	1.397	1.860	2.306	2.896	3.355	3.832	5.041
9	.703	.883	1.100	1.383	1.833	2.262	2.821	3.250	3.690	4.781
10	.700	.879	1.093	1.372	1.812	2.228	2.764	3.169	3.581	4.587
11	.697	.876	1.088	1.363	1.796	2.201	2.718	3.106	3.497	4.437
12	.695	.873	1.083	1.356	1.782	2.179	2.681	3.055	3.428	4.318
13	.694	.870	1.079	1.350	1.771	2.160	2.650	3.012	3.372	4.221
14	.692	.868	1.076	1.345	1.761	2.145	2.624	2.977	3.326	4.140
15	.691	.866	1.074	1.341	1.753	2.131	2.602	2.947	3.286	4.073
16	.690	.865	1.071	1.337	1.746	2.120	2.583	2.921	3.252	4.015
17	.689	.863	1.069	1.333	1.740	2.110	2.567	2.898	3.222	3.965
18	.688	.862	1.067	1.330	1.734	2.101	2.552	2.878	3.197	3.922
19	.688	.861	1.066	1.328	1.729	2.093	2.539	2.861	3.174	3.883
20	.687	.860	1.064	1.325	1.725	2.086	2.528	2.845	3.153	3.850
21	.686	.859	1.063	1.323	1.721	2.080	2.518	2.831	3.135	3.819
22	.686	.858	1.061	1.321	1.717	2.074	2.508	2.819	3.119	3.792
23	.685	.858	1.060	1.319	1.714	2.069	2.500	2.807	3.104	3.767
24	.685	.857	1.059	1.318	1.711	2.064	2.492	2.797	3.090	3.745
25	.684	.856	1.058	1.316	1.708	2.060	2.485	2.787	3.078	3.725
26	.684	.856	1.058	1.315	1.706	2.056	2.479	2.779	3.067	3.707
27	.684	.855	1.057	1.314	1.703	2.052	2.473	2.771	3.056	3.690
28	.683	.855	1.056	1.313	1.701	2.048	2.467	2.763	3.047	3.674
29	.683	.854	1.055	1.311	1.699	2.045	2.462	2.756	3.038	3.659
30	.683	.854	1.055	1.310	1.697	2.042	2.457	2.750	3.030	3.646
40	.681	.851	1.050	1.303	1.684	2.021	2.423	2.704	2.971	3.551
60	.679	.848	1.046	1.296	1.671	2.000	2.390	2.660	2.915	3.460
120	.677	.845	1.041	1.289	1.658	1.980	2.358	2.617	2.860	3.373
∞	.674	.842	1.036	1.282	1.645	1.960	2.326	2.576	2.807	3.291

*Abridged from: "Statistical Tables for Biological, Agricultural, and Medical Research", 6th ed., R.A. Fisher and F. Yates, published by Oliver and Boyd, by permission of the authors and publishers; and "Biometrika Tables for Statisticians", E.S. Pearson and H.O. Hartley, Eds., Vol. 1, Cambridge University Press, 1962.

Table 9-55.   CHI-SQUARE DISTRIBUTION*

Values of Chi Square at Specified Levels of Significance; Single Tail

P = probability of agreement

SYMBOLS: N = degrees of freedom

N	.995	.990	.975	.950	.900	.750	.500	.250	.100	.050	.025	.010	.005	.001
1	—	.0002	.001	.0039	.0158	.102	.455	1.32	2.71	3.84	5.02	6.63	7.88	10.83
2	.0100	.0201	.0506	.103	.211	.575	1.39	2.77	4.61	5.99	7.38	9.21	10.6	13.82
3	.0717	.115	.216	.352	.584	1.21	2.37	4.11	6.25	7.81	9.35	11.3	12.8	16.27
4	.207	.297	.484	.711	1.06	1.92	3.36	5.39	7.78	9.49	11.1	13.3	14.9	18.47
5	.412	.554	.831	1.15	1.61	2.67	4.35	6.63	9.24	11.1	12.8	15.1	16.7	20.52
6	.676	.872	1.24	1.64	2.20	3.45	5.35	7.84	10.6	12.6	14.4	16.8	18.5	22.46
7	.989	1.24	1.69	2.17	2.83	4.25	6.35	9.04	12.0	14.1	16.0	18.5	20.3	24.32
8	1.34	1.65	2.18	2.73	3.49	5.07	7.34	10.2	13.4	15.5	17.5	20.1	22.0	26.13
9	1.73	2.09	2.70	3.33	4.17	5.90	8.34	11.4	14.7	16.9	19.0	21.7	23.6	27.88
10	2.16	2.56	3.25	3.94	4.87	6.74	9.34	12.5	16.0	18.3	20.5	23.2	25.2	29.59
11	2.60	3.05	3.82	4.57	5.58	7.58	10.3	13.7	17.3	19.7	21.9	24.7	26.8	31.26
12	3.07	3.57	4.40	5.23	6.30	8.44	11.3	14.8	18.5	21.0	23.3	26.2	28.3	32.91
13	3.57	4.11	5.01	5.89	7.04	9.30	12.3	16.0	19.8	22.4	24.7	27.7	29.8	34.53
14	4.07	4.66	5.63	6.57	7.79	10.2	13.3	17.1	21.1	23.7	26.1	29.1	31.3	36.12
15	4.60	5.23	6.26	7.26	8.55	11.0	14.3	18.2	22.3	25.0	27.5	30.6	32.8	37.70
16	5.14	5.81	6.91	7.96	9.31	11.9	15.3	19.4	23.5	26.3	28.8	32.0	34.3	39.25
17	5.70	6.41	7.56	8.67	10.1	12.8	16.3	20.5	24.8	27.6	30.2	33.4	35.7	40.79
18	6.26	7.01	8.23	9.39	10.9	13.7	17.3	21.6	26.0	28.9	31.5	34.8	37.2	42.31
19	6.84	7.63	8.91	10.1	11.7	14.6	18.3	22.7	27.2	30.1	32.9	36.2	38.6	43.82
20	7.43	8.26	9.59	10.9	12.4	15.5	19.3	23.8	28.4	31.4	34.2	37.6	40.0	45.32
21	8.03	8.90	10.3	11.6	13.2	16.3	20.3	24.9	29.6	32.7	35.5	38.9	41.4	46.80
22	8.64	9.54	11.0	12.3	14.0	17.2	21.3	26.0	30.8	33.9	36.8	40.3	42.8	48.27
23	9.26	10.2	11.7	13.1	14.8	18.1	22.3	27.1	32.0	35.2	38.1	41.6	44.2	49.73
24	9.89	10.9	12.4	13.8	15.7	19.0	23.3	28.2	33.2	36.4	39.4	43.0	45.6	51.18
25	10.5	11.5	13.1	14.6	16.5	19.9	24.3	29.3	34.4	37.7	40.6	44.3	46.9	52.62
26	11.2	12.2	13.8	15.4	17.3	20.8	25.3	30.4	35.6	38.9	41.9	45.6	48.3	54.05
27	11.8	12.9	14.6	16.2	18.1	21.7	26.3	31.5	36.7	40.1	43.2	47.0	49.6	55.48
28	12.5	13.6	15.3	16.9	18.9	22.7	27.3	32.6	37.9	41.3	44.5	48.3	51.0	56.89
29	13.1	14.3	16.0	17.7	19.8	23.6	28.3	33.7	39.1	42.6	45.7	49.6	52.3	58.30
30	13.8	15.0	16.8	18.5	20.6	24.5	29.3	34.8	40.3	43.8	47.0	50.9	53.7	59.70

*From: "Biometrika Tables for Statisticians", E.S. Pearson and H.O. Hartley, Eds., Vol. 1, Cambridge University Press, 1962.

REFERENCE

For more complete tables see "CRC Handbook of Probability and Statistics", 2nd ed., W.H. Beyer, Ed., The Chemical Rubber Co., 1968.

Table 9-56. F-DISTRIBUTION*

For m and n Degrees of Freedom; $p = .001$ to $.100$; Single Tail

Table A. $p = .001$

n\m	1	2	3	4	5	6	7	8	9	10	15	30	60	∞
1	4053†	5000†	5404†	5625†	5764†	5859†	5929†	5981†	6023†	6056†	6158†	6261†	6313†	6366†
2	998.5	999.0	999.2	999.2	999.3	999.3	999.4	999.4	999.4	999.4	999.4	999.5	999.5	999.5
3	167.0	148.5	141.1	137.1	134.6	132.8	131.6	130.6	129.9	129.2	127.4	125.4	124.5	123.5
4	74.14	61.25	56.18	53.44	51.71	50.53	49.66	49.00	48.47	48.05	46.76	45.43	44.75	44.05
5	47.18	37.12	33.20	31.09	29.75	28.84	28.16	27.64	27.24	26.92	25.91	24.87	24.33	23.79
6	35.51	27.00	23.70	21.92	20.81	20.03	19.46	19.03	18.69	18.41	17.56	16.67	16.21	15.75
7	29.25	21.69	18.77	17.19	16.21	15.52	15.02	14.63	14.33	14.08	13.32	12.53	12.12	11.70
8	25.42	18.49	15.83	14.39	13.49	12.86	12.40	12.04	11.77	11.54	10.84	10.11	9.73	9.33
9	22.86	16.39	13.90	12.56	11.71	11.13	10.70	10.37	10.11	9.89	9.24	8.55	8.19	7.81
10	21.04	14.91	12.55	11.28	10.48	9.92	9.52	9.20	8.96	8.75	8.13	7.47	7.12	6.76
12	18.64	12.97	10.80	9.63	8.89	8.38	8.00	7.71	7.48	7.29	6.71	6.09	5.76	5.42
15	16.59	11.34	9.34	8.25	7.57	7.09	6.74	6.47	6.26	6.08	5.54	4.95	4.64	4.31
30	13.29	8.77	7.05	6.12	5.53	5.12	4.82	4.58	4.39	4.24	3.75	3.22	2.92	2.59
60	11.97	7.76	6.17	5.31	4.76	4.37	4.09	3.87	3.69	3.54	3.08	2.55	2.25	1.89
∞	10.83	6.91	5.42	4.62	4.10	3.74	3.47	3.27	3.10	2.96	2.51	1.99	1.66	1.00

†Multiply these entries by 100.

Table B. $p = .005$

n\m	1	2	3	4	5	6	7	8	9	10	15	30	60	∞
1	16211	20000	21615	22500	23056	23437	23715	23925	24091	24224	24630	25044	25253	25465
2	198.5	199.0	199.2	199.2	199.3	199.3	199.4	199.4	199.4	199.4	199.4	199.5	199.5	199.5
3	55.55	49.80	47.47	46.19	45.39	44.84	44.43	44.13	43.88	43.69	43.08	42.47	42.15	41.83
4	31.33	26.28	24.26	23.15	22.46	21.97	21.62	21.35	21.14	20.97	20.44	19.89	19.61	19.32
5	22.78	18.31	16.53	15.56	14.94	14.51	14.20	13.96	13.77	13.62	13.15	12.66	12.40	12.14
6	18.63	14.54	12.92	12.03	11.46	11.07	10.79	10.57	10.39	10.25	9.81	9.36	9.12	8.88
7	16.24	12.40	10.88	10.05	9.52	9.16	8.89	8.68	8.51	8.38	7.97	7.53	7.31	7.08
8	14.69	11.04	9.60	8.81	8.30	7.95	7.69	7.50	7.34	7.21	6.81	6.40	6.18	5.95
9	13.61	10.11	8.72	7.96	7.47	7.13	6.88	6.69	6.54	6.42	6.03	5.62	5.41	5.19
10	12.83	9.43	8.08	7.34	6.87	6.54	6.30	6.12	5.97	5.85	5.47	5.07	4.86	4.64
12	11.75	8.51	7.23	6.52	6.07	5.76	5.52	5.35	5.20	5.09	4.72	4.33	4.12	3.90
15	10.80	7.70	6.48	5.80	5.37	5.07	4.85	4.67	4.54	4.42	4.07	3.69	3.48	3.26
30	9.18	6.35	5.24	4.62	4.23	3.95	3.74	3.58	3.45	3.34	3.01	2.63	2.42	2.18
60	8.49	5.79	4.73	4.14	3.76	3.49	3.29	3.13	3.01	2.90	2.57	2.19	1.96	1.69
∞	7.88	5.30	4.28	3.72	3.35	3.09	2.90	2.74	2.62	2.52	2.19	1.79	1.53	1.00

Table C. $p = .010$

n\m	1	2	3	4	5	6	7	8	9	10	15	30	60	∞
1	4052	4999.5	5403	5625	5764	5859	5928	5982	6022	6056	6157	6261	6313	6366
2	98.50	99.00	99.17	99.25	99.30	99.33	99.36	99.37	99.39	99.40	99.43	99.47	99.48	99.50
3	34.12	30.82	29.46	28.71	28.24	27.91	27.67	27.49	27.35	27.23	26.87	26.50	26.32	26.13
4	21.20	18.00	16.69	15.98	15.52	15.21	14.98	14.80	14.66	14.55	14.20	13.84	13.65	13.46
5	16.26	13.27	12.06	11.39	10.97	10.67	10.46	10.29	10.16	10.05	9.72	9.38	9.20	9.02
6	13.75	10.92	9.78	9.15	8.75	8.47	8.26	8.10	7.98	7.87	7.56	7.23	7.06	6.88
7	12.25	9.55	8.45	7.85	7.46	7.19	6.99	6.84	6.72	6.62	6.31	5.99	5.82	5.65
8	11.26	8.65	7.59	7.01	6.63	6.37	6.18	6.03	5.91	5.81	5.52	5.20	5.03	4.86
9	10.56	8.02	6.99	6.42	6.06	5.80	5.61	5.47	5.35	5.26	4.96	4.65	4.48	4.31
10	10.04	7.56	6.55	5.99	5.64	5.39	5.20	5.06	4.94	4.85	4.56	4.25	4.08	3.91
12	9.33	6.93	5.95	5.41	5.06	4.82	4.64	4.50	4.39	4.30	4.01	3.70	3.54	3.36
15	8.68	6.36	5.42	4.89	4.56	4.32	4.14	4.00	3.89	3.80	3.52	3.21	3.05	2.87
30	7.56	5.39	4.51	4.02	3.70	3.47	3.30	3.17	3.07	2.98	2.70	2.39	2.21	2.01
60	7.08	4.98	4.13	3.65	3.34	3.12	2.95	2.82	2.72	2.63	2.35	2.03	1.84	1.60
∞	6.63	4.61	3.78	3.32	3.02	2.80	2.64	2.51	2.41	2.32	2.04	1.70	1.47	1.00

Table 9-56. F-DISTRIBUTION (Continued)

Table D. $p = .025$

n \ m	1	2	3	4	5	6	7	8	9	10	15	30	60	∞
1	647.8	799.5	864.2	899.6	921.8	937.1	948.2	956.7	963.3	968.6	984.9	1001	1010	1018
2	38.51	39.00	39.17	39.25	39.30	39.33	39.36	39.37	39.39	39.40	39.43	39.46	39.48	39.50
3	17.44	16.04	15.44	15.10	14.88	14.73	14.62	14.54	14.47	14.42	14.25	14.08	13.99	13.90
4	12.22	10.65	9.98	9.60	9.36	9.20	9.07	8.98	8.90	8.84	8.66	8.46	8.36	8.26
5	10.01	8.43	7.76	7.39	7.15	6.98	6.85	6.76	6.68	6.62	6.43	6.23	6.12	6.02
6	8.81	7.26	6.60	6.23	5.99	5.82	5.70	5.60	5.52	5.46	5.27	5.07	4.96	4.85
7	8.07	6.54	5.89	5.52	5.29	5.12	4.99	4.90	4.82	4.76	4.57	4.36	4.25	4.14
8	7.57	6.06	5.42	5.05	4.82	4.65	4.53	4.43	4.36	4.30	4.10	3.89	3.78	3.67
9	7.21	5.71	5.08	4.72	4.48	4.32	4.20	4.10	4.03	3.96	3.77	3.56	3.45	3.33
10	6.94	5.46	4.83	4.47	4.24	4.07	3.95	3.85	3.78	3.72	3.52	3.31	3.20	3.08
12	6.55	5.10	4.47	4.12	3.89	3.73	3.61	3.51	3.44	3.37	3.18	2.96	2.85	2.72
15	6.20	4.77	4.15	3.80	3.58	3.41	3.29	3.20	3.12	3.06	2.86	2.64	2.52	2.40
30	5.57	4.18	3.59	3.25	3.03	2.87	2.75	2.65	2.57	2.51	2.31	2.07	1.94	1.79
60	5.29	3.93	3.34	3.01	2.79	2.63	2.51	2.41	2.33	2.27	2.06	1.82	1.67	1.48
∞	5.02	3.69	3.12	2.79	2.57	2.41	2.29	2.19	2.11	2.05	1.83	1.57	1.39	1.00

Table E. $p = .050$

n \ m	1	2	3	4	5	6	7	8	9	10	15	30	60	∞
1	161.4	199.5	215.7	224.6	230.2	234.0	236.8	238.9	240.5	241.9	245.9	250.1	252.2	254.3
2	18.51	19.00	19.16	19.25	19.30	19.33	19.35	19.37	19.38	19.40	19.43	19.46	19.48	19.50
3	10.13	9.55	9.28	9.12	9.01	8.94	8.89	8.85	8.81	8.79	8.70	8.62	8.57	8.53
4	7.71	6.94	6.59	6.39	6.26	6.16	6.09	6.04	6.00	5.96	5.86	5.75	5.69	5.63
5	6.61	5.79	5.41	5.19	5.05	4.95	4.88	4.82	4.77	4.74	4.62	4.50	4.43	4.36
6	5.99	5.14	4.76	4.53	4.39	4.28	4.21	4.15	4.10	4.06	3.94	3.81	3.74	3.67
7	5.59	4.74	4.35	4.12	3.97	3.87	3.79	3.73	3.68	3.64	3.51	3.38	3.30	3.23
8	5.32	4.46	4.07	3.84	3.69	3.58	3.50	3.44	3.39	3.35	3.22	3.08	3.01	2.93
9	5.12	4.26	3.86	3.63	3.48	3.37	3.29	3.23	3.18	3.14	3.01	2.86	2.79	2.71
10	4.96	4.10	3.71	3.48	3.33	3.22	3.14	3.07	3.02	2.98	2.85	2.70	2.62	2.54
12	4.75	3.89	3.49	3.26	3.11	3.00	2.91	2.85	2.80	2.75	2.62	2.47	2.38	2.30
15	4.54	3.68	3.29	3.06	2.90	2.79	2.71	2.64	2.59	2.54	2.40	2.25	2.16	2.07
30	4.17	3.32	2.92	2.69	2.53	2.42	2.33	2.27	2.21	2.16	2.01	1.84	1.74	1.62
60	4.00	3.15	2.76	2.53	2.37	2.25	2.17	2.10	2.04	1.99	1.84	1.65	1.53	1.39
∞	3.84	3.00	2.60	2.37	2.21	2.10	2.01	1.94	1.88	1.83	1.67	1.46	1.32	1.00

Table F. $p = .100$

n \ m	1	2	3	4	5	6	7	8	9	10	15	30	60	∞
1	39.86	49.50	53.59	55.83	57.24	58.20	58.91	59.44	59.86	60.19	61.22	62.26	62.79	63.33
2	8.53	9.00	9.16	9.24	9.29	9.33	9.35	9.37	9.38	9.39	9.42	9.46	9.47	9.49
3	5.54	5.46	5.39	5.34	5.31	5.28	5.27	5.25	5.24	5.23	5.20	5.17	5.15	5.13
4	4.54	4.32	4.19	4.11	4.05	4.01	3.98	3.95	3.94	3.92	3.87	3.82	3.79	3.76
5	4.06	3.78	3.62	3.52	3.45	3.40	3.37	3.34	3.32	3.30	3.24	3.17	3.14	3.10
6	3.78	3.46	3.29	3.18	3.11	3.05	3.01	2.98	2.96	2.94	2.87	2.80	2.76	2.72
7	3.59	3.26	3.07	2.96	2.88	2.83	2.78	2.75	2.72	2.70	2.63	2.56	2.51	2.47
8	3.46	3.11	2.92	2.81	2.73	2.67	2.62	2.59	2.56	2.54	2.46	2.38	2.34	2.29
9	3.36	3.01	2.81	2.69	2.61	2.55	2.51	2.47	2.44	2.42	2.34	2.25	2.21	2.16
10	3.29	2.92	2.73	2.61	2.52	2.46	2.41	2.38	2.35	2.32	2.24	2.16	2.11	2.06
12	3.18	2.81	2.61	2.48	2.39	2.33	2.28	2.24	2.21	2.19	2.10	2.01	1.96	1.90
15	3.07	2.70	2.49	2.36	2.27	2.21	2.16	2.12	2.09	2.06	1.97	1.87	1.82	1.76
30	2.88	2.49	2.28	2.14	2.05	1.98	1.93	1.88	1.85	1.82	1.72	1.61	1.54	1.46
60	2.79	2.39	2.18	2.04	1.95	1.87	1.82	1.77	1.74	1.71	1.60	1.48	1.40	1.29
∞	2.71	2.30	2.08	1.94	1.85	1.77	1.72	1.67	1.63	1.60	1.49	1.34	1.24	1.00

*From: "Biometrika Tables for Statisticians", E.S. Pearson and H.O. Hartley, Eds., Vol. 1, Cambridge University Press, 1962.

REFERENCE

For more complete tables see "CRC Handbook of Probability and Statistics", 2nd ed., W.H. Beyer, Ed., The Chemical Rubber Co., 1968, pp. 304–310.

Table 9-57.   BINOMIAL DISTRIBUTION—
CUMULATIVE PROBABILITIES: P*

SYMBOLS:

N = number of trials or size of a sample

p = fixed probability of the outcome for the entire population (success or failure, whichever is less)

P = cumulative probability of observing X or more successes within N

N	X	.05	.10	.15	.20	.25	.30	.35	.40	.45	.50
2	1	.0975	.1900	.2775	.3600	.4375	.5100	.5775	.6400	.6975	.7500
	2	.0025	.0100	.0225	.0400	.0625	.0900	.1225	.1600	.2025	.2500
3	1	.1426	.2710	.3859	.4880	.5781	.6570	.7254	.7840	.8336	.8750
	2	.0072	.0280	.0608	.1040	.1562	.2160	.2818	.3520	.4252	.5000
	3	.0001	.0010	.0034	.0080	.0156	.0270	.0429	.0640	.0911	.1250
4	1	.1855	.3439	.4780	.5904	.6836	.7399	.8215	.8704	.9005	.9375
	2	.0140	.0523	.1095	.1808	.2617	.3483	.4370	.5248	.6090	.6875
	3	.0005	.0037	.0120	.0272	.0508	.0837	.1265	.1792	.2415	.3125
	4	.0000	.0001	.0005	.0016	.0039	.0081	.0150	.0256	.0410	.0625
5	1	.2262	.4095	.5563	.6723	.7627	.8319	.8840	.9222	.9497	.9688
	2	.0226	.0815	.1648	.2627	.3672	.4718	.5716	.6630	.7438	.8125
	3	.0012	.0086	.0266	.0579	.1035	.1631	.2352	.3174	.4069	.5000
	4	.0000	.0005	.0022	.0067	.0156	.0308	.0540	.0870	.1312	.1875
	5	.0000	.0000	.0001	.0003	.0010	.0024	.0053	.0102	.0185	.0312
6	1	.2649	.4686	.6229	.7379	.8220	.8824	.9246	.9533	.9723	.9844
	2	.0328	.1143	.2235	.3447	.4661	.5798	.6809	.7667	.8364	.8906
	3	.0022	.0158	.0473	.0989	.1694	.2557	.3529	.4557	.5585	.6562
	4	.0001	.0013	.0059	.0170	.0376	.0705	.1174	.1792	.2553	.3438
	5	.0000	.0001	.0004	.0016	.0046	.0109	.0223	.0410	.0692	.1094
	6	.0000	.0000	.0000	.0001	.0002	.0007	.0018	.0041	.0083	.0156
7	1	.3017	.5217	.6794	.7903	.8665	.9176	.9510	.9720	.9848	.9922
	2	.0444	.1497	.2834	.4233	.5551	.6706	.7662	.8414	.8976	.9375
	3	.0038	.0257	.0738	.1480	.2436	.3529	.4677	.5801	.6836	.7734
	4	.0002	.0027	.0121	.0333	.0706	.1260	.1998	.2898	.3917	.5000
	5	.0000	.0002	.0012	.0047	.0129	.0288	.0556	.0963	.1529	.2266
	6	.0000	.0000	.0001	.0004	.0013	.0038	.0090	.0188	.0357	.0625
8	1	.3366	.5695	.7275	.8322	.8999	.9424	.9681	.9832	.9916	.9961
	2	.0572	.1869	.3428	.4967	.6329	.7447	.8309	.8936	.9368	.9648
	3	.0058	.0381	.1052	.2031	.3215	.4482	.5722	.6846	.7799	.8555
	4	.0004	.0050	.0214	.0563	.1138	.1941	.2936	.4059	.5230	.6367
	5	.0000	.0004	.0029	.0104	.0273	.0580	.1061	.1737	.2604	.3633
	6	.0000	.0000	.0002	.0012	.0042	.0113	.0253	.0498	.0885	.1445
	7	.0000	.0000	.0000	.0001	.0004	.0013	.0036	.0085	.0181	.0352
9	1	.3698	.6126	.7684	.8658	.9249	.9596	.9793	.9899	.9954	.9980
	2	.0712	.2252	.4005	.5638	.6997	.8040	.8789	.9295	.9615	.9805
	3	.0084	.0530	.1409	.2618	.3993	.5372	.6627	.7682	.8505	.9102
	4	.0006	.0083	.0339	.0856	.1657	.2703	.3911	.5174	.6386	.7461
	5	.0000	.0009	.0056	.0196	.0489	.0988	.1717	.2666	3.786	.5000
	6	.0000	.0001	.0006	.0031	.0100	.0253	.0536	.0994	.1658	.2539
	7	.0000	.0000	.0000	.0003	.0013	.0043	.0112	.0250	.0498	.0898
	8	.0000	.0000	.0000	.0000	.0001	.0004	.0014	.0038	.0091	.0195

Table 9-57. BINOMIAL DISTRIBUTION—
CUMULATIVE PROBABILITIES: *P (Continued)*

N	X	.05	.10	.15	.20	.25	.30	.35	.40	.45	.50
10	1	.4013	.6513	.8031	.8926	.9437	.9718	.9865	.9940	.9975	.9990
	2	.0861	.2639	.4557	.6242	.7560	.8507	.9140	.9536	.9767	.9893
	3	.0115	.0702	.1798	.3222	.4744	.6172	.7384	.8327	.9004	.9453
	4	.0010	.0128	.0500	.1209	.2241	.3504	.4862	.6177	.7340	.8281
	5	.0001	.0016	.0099	.0328	.0781	.1503	.2485	.3669	.4956	.6230
	6	.0000	.0001	.0014	.0064	.0197	.0473	.0949	.1662	.2616	.3770
	7	.0000	.0000	.0001	.0009	.0035	.0106	.0260	.0548	.1020	.1719
	8	.0000	.0000	.0000	.0001	.0004	.0016	.0048	.0123	.0274	.0547
	9	.0000	.0000	.0000	.0000	.0000	.0001	.0005	.0017	.0045	.0107
12	1	.4596	.7176	.8578	.9313	.9683	.9862	.9943	.9978	.9992	.9998
	2	.1184	.3410	.5565	.7251	.8416	.9150	.9576	.9804	.9917	.9968
	3	.0196	.1109	.2642	.4417	.6093	.7472	.8487	.9166	.9579	.9807
	4	.0022	.0256	.0922	.2054	.3512	.5075	.6533	.7747	.8655	.9270
	5	.0002	.0043	.0239	.0726	.1576	.2763	.4167	.5618	.6956	.8062
	6	.0000	.0005	.0046	.0194	.0544	.1178	.2127	.3348	.4731	.6128
	7	.0000	.0001	.0007	.0039	.0143	.0386	.0846	.1582	.2607	.3872
	8	.0000	.0000	.0001	.0006	.0028	.0095	.0255	.0573	.1117	.1938
	9	.0000	.0000	.0000	.0001	.0004	.0017	.0056	.0153	.0356	.0730
	10	.0000	.0000	.0000	.0000	.0000	.0002	.0008	.0028	.0079	.0193
15	1	.5367	.7941	.9126	.9648	.9866	.9953	.9984	.9995	.9999	1.0000
	2	.1710	.4510	.6814	.8329	.9198	.9647	.9858	.9948	.9983	.9995
	3	.0362	.1841	.3958	.6020	.7639	.8732	.9383	.9729	.9893	.9963
	4	.0055	.0556	.1773	.3518	.5387	.7031	.8273	.9095	.9576	.9824
	5	.0006	.0127	.0617	.1642	.3135	.4845	.6481	.7827	.8796	.9408
	6	.0001	.0022	.0168	.0611	.1484	.2784	.4357	.5968	.7392	.8491
	7	.0000	.0003	.0036	.0181	.0566	.1311	.2452	.3902	.5478	.6964
	8	.0000	.0000	.0006	.0042	.0173	.0500	.1132	.2131	.3465	.5000
	9	.0000	.0000	.0001	.0008	.0042	.0152	.0422	.0950	.1818	.3036
	10	.0000	.0000	.0000	.0001	.0008	.0037	.0124	.0338	.0769	.1509
	11	.0000	.0000	.0000	.0000	.0001	.0007	.0028	.0093	.0255	.0592
	12	.0000	.0000	.0000	.0000	.0000	.0001	.0005	.0019	.0063	.0176
20	1	.6415	.8784	.9612	.9885	.9968	.9992	.9998	1.0000	1.0000	1.0000
	2	.2642	.6083	.8244	.9308	.9757	.9924	.9979	.9995	.9999	1.0000
	3	.0755	.3231	.5951	.7939	.9087	.9645	.9879	.9964	.9991	.9998
	4	.0159	.1330	.3523	.5886	.7748	.8929	.9556	.9840	.9951	.9987
	5	.0026	.0432	.1702	.3704	.5852	.7625	.8818	.9490	.9811	.9941
	6	.0003	.0113	.0673	.1958	.3828	.5836	.7546	.8744	.9447	.9793
	7	.0000	.0024	.0219	.0867	.2142	.3920	.5834	.7500	.8701	.9423
	8	.0000	.0004	.0059	.0321	.1018	.2277	.3990	.5841	.7480	.8684
	9	.0000	.0001	.0013	.0100	.0409	.1133	.2376	.4044	.5857	.7483
	10	.0000	.0000	.0002	.0026	.0139	.0480	.1218	.2447	.4086	.5881
	11	.0000	.0000	.0000	.0006	.0039	.0171	.0532	.1275	.2493	.4119
	12	.0000	.0000	.0000	.0001	.0009	.0051	.0196	.0565	.1308	.2517
	13	.0000	.0000	.0000	.0000	.0002	.0013	.0060	.0210	.0580	.1316
	14	.0000	.0000	.0000	.0000	.0000	.0003	.0015	.0065	.0214	.0577
	15	.0000	.0000	.0000	.0000	.0000	.0000	.0003	.0016	.0064	.0207

Note: Individual binomial probability terms can be obtained by subtraction, i.e.,

$$P_X = \sum_{X}^{N} (P_X) - \sum_{X+1}^{N} (P_X).$$

*Condensed from: "CRC Handbook of Tables for Mathematics", 4th ed., S.M. Selby, Ed., The Chemical Rubber Co., 1970.

Table 9-58. POISSON DISTRIBUTION—
CUMULATIVE PROBABILITIES: P^*

SYMBOLS:

N = number of trials or size of the sample

p = fixed probability of the outcome for entire population (success or failure, whichever is less)

P = cumulative probability of X or more successes (or failures) within N

X' \ Np	0.1	0.2	0.3	0.4	0.5	0.6	0.7	0.8	0.9	1.0	1.1	1.2
1	.095	.181	.259	.330	.394	.451	.503	.551	.593	.632	.667	.699
2	.005	.018	.037	.062	.090	.122	.156	.191	.228	.264	.301	.337
3	.000	.001	.004	.008	.014	.023	.034	.047	.063	.080	.100	.121
4	.000	.000	.000	.001	.002	.003	.006	.009	.014	.019	.026	.034

X' \ Np	1.3	1.4	1.5	1.6	1.7	1.8	1.9	2.0	2.1	2.2	2.3	2.4
1	.728	.753	.777	.798	.817	.835	.850	.865	.878	.889	.900	.909
2	.373	.408	.442	.475	.507	.537	.566	.594	.620	.645	.669	.692
3	.143	.167	.191	.217	.243	.269	.296	.323	.350	.377	.404	.430
4	.043	.054	.066	.079	.093	.109	.125	.143	.161	.181	.201	.221
5	.011	.014	.019	.024	.030	.036	.044	.053	.062	.073	.084	.096
6	.002	.003	.005	.006	.008	.010	.013	.017	.020	.025	.030	.036

X' \ Np	2.5	2.6	2.7	2.8	2.9	3.0	3.1	3.2	3.3	3.4	3.5	3.6
1	.918	.926	.933	.939	.945	.950	.955	.959	.963	.967	.970	.973
2	.713	.733	.751	.769	.785	.801	.815	.829	.841	.853	.864	.874
3	.456	.482	.506	.531	.554	.577	.599	.620	.641	.660	.679	.697
4	.242	.264	.286	.308	.330	.353	.375	.398	.420	.442	.463	.485
5	.109	.123	.137	.152	.168	.185	.202	.219	.237	.256	.275	.294
6	.042	.049	.057	.065	.074	.084	.094	.105	.117	.130	.142	.156
7	.014	.017	.021	.024	.029	.034	.039	.045	.051	.058	.065	.073
8	.004	.005	.007	.008	.010	.012	.014	.017	.020	.023	.027	.031

X' \ Np	3.7	3.8	3.9	4.0	4.1	4.2	4.3	4.4	4.5	4.6	4.7	4.8
1	.975	.978	.980	.982	.983	.985	.986	.988	.989	.990	.991	.992
2	.884	.893	.901	.908	.916	.922	.928	.934	.939	.944	.948	.952
3	.715	.731	.747	.762	.776	.790	.803	.815	.826	.837	.848	.858
4	.506	.527	.547	.567	.586	.605	.623	.641	.658	.674	.690	.706
5	.313	.332	.352	.371	.391	.410	.430	.449	.468	.487	.505	.524
6	.170	.184	.199	.215	.231	.247	.263	.280	.297	.314	.332	.349
7	.082	.091	.101	.111	.121	.133	.144	.156	.169	.182	.195	.209
8	.035	.040	.045	.051	.057	.064	.071	.079	.087	.095	.104	.113
9	.014	.016	.019	.021	.025	.028	.032	.036	.040	.045	.050	.056
10	.005	.006	.007	.008	.010	.011	.013	.015	.017	.020	.022	.025

X' \ Np	4.9	5.0	5.1	5.2	5.3	5.4	5.5	5.6	5.7	5.8	5.9	6.0
1	.993	.993	.994	.995	.995	.996	.996	.996	.997	.997	.997	.998
2	.956	.960	.963	.966	.969	.971	.973	.976	.978	.979	.981	.983
3	.867	.875	.884	.891	.898	.905	.912	.918	.923	.929	.933	.938
4	.721	.735	.749	.762	.775	.787	.798	.809	.820	.830	.840	.849
5	.542	.560	.577	.594	.611	.627	.643	.658	.673	.687	.701	.715
6	.367	.384	.402	.419	.437	.454	.471	.488	.505	.522	.538	.554
7	.223	.238	.253	.268	.283	.298	.314	.330	.346	.362	.378	.394
8	.123	.133	.144	.155	.167	.178	.191	.203	.216	.229	.242	.256
9	.062	.068	.075	.082	.089	.097	.106	.114	.123	.133	.143	.153
10	.028	.032	.036	.040	.044	.049	.054	.059	.065	.071	.077	.084
11	.012	.014	.016	.018	.020	.023	.025	.028	.031	.035	.039	.043
12	.005	.006	.006	.007	.008	.010	.011	.013	.014	.016	.018	.020

Table 9-58. POISSON DISTRIBUTION—
CUMULATIVE PROBABILITIES: P (Continued)

Np / X'	6.1	6.2	6.3	6.4	6.5	6.6	6.7	6.8	6.9	7.0	7.1	7.2
1	.998	.998	.998	.998	.999	.999	.999	.999	.999	.999	.999	.999
2	.984	.985	.987	.988	.989	.990	.991	.991	.992	.993	.993	.994
3	.942	.946	.950	.954	.957	.960	.963	.966	.968	.970	.973	.975
4	.858	.866	.874	.881	.888	.895	.901	.907	.913	.918	.923	.928
5	.728	.741	.753	.765	.776	.787	.798	.808	.818	.827	.836	.845
6	.570	.586	.601	.616	.631	.645	.659	.673	.686	.699	.712	.724
7	.410	.426	.442	.458	.474	.489	.505	.520	.535	.550	.565	.580
8	.270	.284	.298	.313	.327	.342	.357	.372	.386	.401	.416	.431
9	.163	.174	.185	.197	.208	.220	.233	.245	.258	.271	.284	.297
10	.091	.098	.106	.114	.123	.131	.140	.150	.151	1.70	.180	.190
11	.047	.051	.056	.061	.067	.073	.079	.085	.092	.099	.106	.113
12	.022	.025	.028	.031	.034	.037	.041	.045	.049	.053	.058	.063

Np / X'	7.3	7.4	7.5	7.6	7.7	7.8	7.9	8.0	8.2	8.4	8.6	8.8
1	.999	.999	.999	1.000	1.000	1.000	1.000	1.000	1.000	1.000	1.000	1.000
2	.994	.995	.995	.996	.996	.996	.997	.997	.998	.998	.998	.999
3	.976	.978	.980	.981	.983	.984	.985	.986	.988	.990	.991	.993
4	.933	.937	.941	.945	.948	.952	.955	.958	.963	.968	.972	.976
5	.853	.861	.868	.875	.882	.888	.895	.900	.911	.921	.930	.938
6	.736	.747	.759	.769	.780	.790	.799	.809	.826	.843	.858	.872
7	.594	.608	.622	.635	.649	.662	.674	.687	.710	.733	.754	.774
8	.446	.461	.475	.490	.504	.519	.533	.547	.575	.601	.627	.652
9	.311	.324	.338	.352	.366	.380	.394	.408	.435	.463	.491	.518
10	.201	.212	.224	.235	.247	.259	.271	.283	.309	.334	.360	.386
11	.121	.129	.138	.147	.156	.165	.174	.184	.205	.226	.248	.271
12	.068	.074	.079	.085	.092	.098	.105	.112	.127	.143	.160	.178
13	.036	.039	.043	.046	.050	.055	.059	.064	.074	.085	.097	.110
14	.018	.020	.022	.024	.026	.029	.031	.034	.041	.048	.056	.064

Np / X'	9.0	10	11	12	13	14	15	16	17	18	19	20
2	.999	1.000	1.000	1.000	1.000	1.000	1.000	1.000	1.000	1.000	1.000	1.000
3	.994	.997	.999	1.000	1.000	1.000	1.000	1.000	1.000	1.000	1.000	1.000
4	.979	.990	.995	.998	.999	1.000	1.000	1.000	1.000	1.000	1.000	1.000
5	.945	.971	.985	.992	.996	.998	.999	1.000	1.000	1.000	1.000	1.000
6	.884	.933	.963	.980	.989	.995	.997	.999	.999	1.000	1.000	1.000
7	.793	.870	.921	.954	.974	.986	.992	.996	.998	.999	1.000	1.000
8	.676	.780	.857	.911	.946	.968	.982	.990	.995	.997	.999	.999
9	.544	.667	.768	.845	.900	.938	.963	.978	.987	.993	.996	.998
10	.413	.542	.660	.758	.834	.891	.930	.957	.974	.985	.991	.995
11	.294	.417	.540	.653	.748	.824	.882	.923	.951	.970	.982	.989
12	.197	.303	.421	.538	.647	.740	.815	.873	.915	.945	.965	.979
13	.124	.208	.311	.424	.537	.642	.732	.807	.865	.908	.939	.961
14	.074	.136	.219	.319	.427	.536	.637	.726	.799	.857	.902	.934
15	.042	.084	.146	.228	.325	.430	.534	.633	.719	.792	.850	.895
16	.022	.049	.093	.156	.236	.331	.432	.533	.629	.713	.785	.844
17	.011	.027	.056	.101	.165	.244	.336	.434	.532	.625	.708	.779
18	.005	.014	.032	.063	.110	.173	.251	.341	.436	.531	.622	.703
19	.002	.007	.018	.037	.070	.117	.181	.258	.345	.438	.531	.619
20	.001	.004	.009	.021	.043	.077	.125	.188	.264	.349	.439	.530
21	.000	.002	.005	.012	.025	.048	.083	.132	.195	.269	.353	.441
22	.000	.001	.002	.006	.014	.029	.053	.089	.139	.201	.275	.356
23	.000	.000	.001	.003	.008	.017	.033	.058	.095	.145	.207	.279
24	.000	.000	.001	.002	.004	.009	.020	.037	.063	.101	.151	.213
25	.000	.000	.000	.001	.002	.005	.011	.022	.041	.068	.107	.157
26	.000	.000	.000	.000	.001	.003	.006	.013	.025	.045	.073	.112

Note: Individual Poisson-probability terms can be obtained by subtraction, i.e.,

$$P_x = \sum_{x}^{N} (P_x) - \sum_{x+1}^{N} (P_x).$$

*Condensed from: "CRC Handbook of Tables for Mathematics", 4th ed., S.M. Selby, Ed., The Chemical Rubber Co., 1970.

Table 9-59.   CRITICAL VALUES FOR THE SIGN TEST*

Two-tail Percentage Points for the Binomial for $p = .5$

The observations in a random sample of size n from X and those of the same size from Y are paired according to the order of observation: (X_i, Y_i), $i = 1, 2, \ldots, n$. The differences $d_i = X_i - Y_i$ are calculated for each of the n pairs. The null hypothesis is that the difference d_i has a distribution with median zero, i.e., the true proportion of positive (negative) signs is equal to $p = \frac{1}{2}$. Thus the test is whether X and Y have the same median. The probability of x positive (negative) signs is given by the binomial probability function

$$f(x) = f(x;n, p = \tfrac{1}{2}) = \binom{n}{x}\left(\frac{1}{2}\right)^{n}.$$

This table gives the critical value k such that

$$P(x \leq k) = \sum_{x=0}^{k} \binom{n}{x}\left(\frac{1}{2}\right)^{n} < \frac{\alpha}{2}.$$

n	1%	5%	10%	25%	n	1%	5%	10%	25%
1					46	13	15	16	18
2					47	14	16	17	19
3				0	48	14	16	17	19
4				0	49	15	17	18	19
5			0	0	50	15	17	18	20
6		0	0	1	51	15	18	19	20
7		0	0	1	52	16	18	19	21
8	0	0	1	1	53	16	18	20	21
9	0	1	1	2	54	17	19	20	22
10	0	1	1	2	55	17	19	20	22
11	0	1	2	3	56	17	20	21	23
12	1	2	2	3	57	18	20	21	23
13	1	2	3	3	58	18	21	22	24
14	1	2	3	4	59	19	21	22	24
15	2	3	3	4	60	19	21	23	25
16	2	3	4	5	61	20	22	23	25
17	2	4	4	5	62	20	22	24	25
18	3	4	5	6	63	20	23	24	26
19	3	4	5	6	64	21	23	24	26
20	3	5	5	6	65	21	24	25	27
21	4	5	6	7	66	22	24	25	27
22	4	5	6	7	67	22	25	26	28
23	4	6	7	8	68	22	25	26	28
24	5	6	7	8	69	23	25	27	29
25	5	7	7	9	70	23	26	27	29
26	6	7	8	9	71	24	26	28	30
27	6	7	8	10	72	24	27	28	30
28	6	8	9	10	73	25	27	28	31
29	7	8	9	10	74	25	28	29	31
30	7	9	10	11	75	25	28	29	32
31	7	9	10	11	76	26	28	30	32
32	8	9	10	12	77	26	29	30	32
33	8	10	11	12	78	27	29	31	33
34	9	10	11	13	79	27	30	31	33
35	9	11	12	13	80	28	30	32	34
36	9	11	12	14	81	28	31	32	34
37	10	12	13	14	82	28	31	33	35
38	10	12	13	14	83	29	32	33	35
39	11	12	13	15	84	29	32	33	36
40	11	13	14	15	85	30	32	34	36
41	11	13	14	16	86	30	33	34	37
42	12	14	15	16	87	31	33	35	37
43	12	14	15	17	88	31	34	35	38
44	13	15	16	17	89	31	34	36	38
45	13	15	16	18	90	32	35	36	39

For values of n larger than 90, approximate values of r may be found by taking the nearest integer less than $(n-1)/2 - k\sqrt{n+1}$, where k is 1.2879, 0.9800, 0.8224, 0.5752 for the 1, 5, 10, 25% values, respectively.

*From: "CRC Handbook of Tables for Mathematics", 4th ed., S.M. Selby, Ed., The Chemical Rubber Co., 1970.

Table 9-60.
FACTORS FOR COMPUTING CONTROL LIMITS*

A.   CONTROL CHARTS FOR MEASUREMENT

If the process mean and standard deviation, μ and σ, are known, and it is assumed that the underlying distribution is normal, it is possible to assert with probability $1-\alpha$ that the mean of a random sample of size n will fall between $\bar{x} - z_{\alpha/2}\dfrac{\sigma}{\sqrt{n}}$ and $\bar{x} + z_{\alpha/2}\dfrac{\sigma}{\sqrt{n}}$. These two limits on $\bar{x}$ provide upper and lower control limits. In actual practice μ and σ are usually unknown, and it is necessary to estimate their values from a large sample taken while the process is "in control". The central line of an $\bar{x}$ chart is given by μ, and the lower and upper three-sigma control limits are given by $\mu - A\sigma$ and $\mu + A\sigma$, respectively, where $A = \dfrac{3}{\sqrt{n}}$ and n is the sample size. Where the population parameters are unknown, it is necessary to estimate these parameters on the basis of preliminary samples. If k samples are used, each of size n, denote the mean of the ith sample by $\bar{x}_i$ and the grand mean of the k sample means by $\bar{\bar{x}}$, i.e.,

$$\bar{\bar{x}} = \frac{1}{k}\sum_{i=1}^{k}\bar{x}_i.$$

Denote the range of the ith sample by R_i and by $\bar{R}$ the mean of the k sample ranges, i.e.,

$$\bar{R} = \frac{1}{k}\sum_{i=1}^{k}R_i.$$

Since $\bar{\bar{x}}$ is an unbiased estimate of the population mean μ, the central line for the $\bar{x}$ chart is given by $\bar{\bar{x}}$. The statistic R does not provide an unbiased estimate of σ, but $A_2\bar{R}$ is an unbiased estimate of $\dfrac{3\sigma}{\sqrt{n}}$. The constant multiplier A_2 depends on the assumption of normality. Thus, the central line and the lower and upper three-sigma limits, LCL and UCL, for an x chart (with μ and σ estimated from past data) are given by

$$\text{central line} = \bar{\bar{x}}$$
$$\text{LCL} = \bar{\bar{x}} - A_2\bar{R}$$
$$\text{UCL} = \bar{\bar{x}} + A_2\bar{R}.$$

The central line and control limits of an R chart are based on the distribution of the range of samples of size n from a normal population. The mean and standard deviation of the sampling distribution of R are given by $d_2\sigma$ and $d_3\sigma$, respectively, when σ is known. Here d_2 and d_3 are constants that depend on the size of the sample. The set of control-chart values for an R chart (with σ known) is given by

$$\text{central line} = d_2\sigma$$
$$\text{LCL} = D_1\sigma$$
$$\text{UCL} = D_2\sigma,$$

where $D_1 = d_2 - 3d_3$ and $D_2 = d_2 + 3d_3$.

If σ is unknown, the control chart values for an R chart are given by

$$\text{central line} = \bar{R}$$
$$\text{LCL} = D_3\bar{R}$$
$$\text{UCL} = D_4\bar{R},$$

where $D_3 = \dfrac{D_1}{d_2}$ and $D_4 = \dfrac{D_2}{d_2}$.

Table 9-60.
FACTORS FOR COMPUTING CONTROL LIMITS *(Continued)*

The central line and control limits of an s chart are based on estimates obtained from the samples. A pooled estimate of the population variance is obtained from the k samples, i.e.,

$$s_p^2 = \frac{\sum_i (n_i - 1)s_i^2}{\sum_i (n_i - 1)}, \, i = 1, 2, \ldots, k.$$

If the sample sizes are all equal, the pooled estimate is

$$s_p^2 = \frac{1}{k} \sum_i s_i^2.$$

The control chart values for an s chart are given by

$$\begin{aligned} \text{central line} &= C_2' s_p \\ \text{LCL} &= B_2' s_p \\ \text{UCL} &= B_4' s_p. \end{aligned}$$

If one uses the biased estimator of the variance s_p', as is often done in quality control work, the control-chart values are given by

$$\begin{aligned} \text{central line} &= c_2 s_p' \\ \text{LCL} &= B_2 s_p' \\ \text{UCL} &= B_4 s_p'. \end{aligned}$$

B. CONTROL CHARTS FOR ATTRIBUTES

Control limits for a fraction-defective chart are based on the sampling theory for proportions, using the normal curve approximation to the binomial. If k samples are taken, the estimator of p is given by

$$\bar{p} = \frac{\sum_i x_i}{\sum_i n_i}, \, i = 1, 2, \ldots, k,$$

where x_i is the number of defectives in the ith sample of size n_i. The central line and control limits of a fraction-defective chart based on analysis of past data are given by

$$\text{central line} = \bar{p}$$

$$\text{LCL} = \bar{p} - 3 \sqrt{\frac{\bar{p}(1 - \bar{p})}{n_i}}$$

$$\text{UCL} = \bar{p} + 3 \sqrt{\frac{\bar{p}(1 - \bar{p})}{n_i}}.$$

When the sample sizes are approximately equal, n_i is replaced by $\bar{n} = \frac{1}{k} \sum_i n_i$.

Equivalent to the p chart for the fraction defective is the control chart for the number of defective. Here, if p is estimated by $\bar{p}$, the control-chart values for a number-of-defectives chart are given by

$$\begin{aligned} \text{central line} &= \bar{n}\bar{p} \\ \text{LCL} &= \bar{n}\bar{p} - 3 \sqrt{\bar{n}\bar{p}(1 - \bar{p})} \\ \text{UCL} &= \bar{n}\bar{p} + 3 \sqrt{\bar{n}\bar{p}(1 - \bar{p})}. \end{aligned}$$

Table 9-60.
FACTORS FOR COMPUTING CONTROL LIMITS *(Continued)*

In many cases it is necessary to control the number of defects per unit C, where C is taken to be a value of a random variable having a Poisson distribution. If k is the number of units available for estimating λ, the parameter of the Poisson distribution, and if C_i is the number of defects in the ith unit, then λ is estimated by

$$\bar{C} = \frac{1}{k} \sum_{i=1}^{k} C_i,$$

and the control-chart values for the C chart are

$$\text{central line} = \bar{C}$$
$$\text{LCL} = \bar{C} - 3\sqrt{\bar{C}}$$
$$\text{UCL} = \bar{C} + 3\sqrt{\bar{C}}.$$

This table presents values of the factors for computing control limits for various sample sizes n.

Number of observations in sample, n	$\bar{X}$ chart		R chart			s chart			$\hat{\sigma}$ chart (biased)		
	Factors for control limits		Factor for central line	Factors for control limits		Factor for central line	Factors for control limits		Factor for central line	Factors for control limits	
	A	A_2	d_2	D_3	D_4	c'_2	B'_2	B'_4	c_2	B_2	B_4
2	2.121	1.880	1.128	0	3.267	0.798	0	2.298	0.5642	0	3.267
3	1.732	1.023	1.693	0	2.575	0.886	0	2.111	0.7236	0	2.568
4	1.500	0.729	2.059	0	2.282	0.921	0	1.982	0.7979	0	2.266
5	1.342	0.577	2.326	0	2.115	0.940	0	1.889	0.8407	0	2.089
6	1.225	0.483	2.534	0	2.004	0.951	0.085	1.817	0.8686	0.030	1.970
7	1.134	0.419	2.704	0.076	1.924	0.960	0.158	1.762	0.8882	0.118	1.882
8	1.061	0.373	2.847	0.136	1.864	0.965	0.215	1.715	0.9027	0.185	1.815
9	1.000	0.337	2.970	0.184	1.816	0.969	0.262	1.676	0.9139	0.239	1.761
10	0.949	0.308	3.078	0.223	1.777	0.973	0.302	1.644	0.9227	0.284	1.716
11	0.905	0.285	3.173	0.256	1.744	0.976	0.336	1.616	0.9300	0.321	1.679
12	0.866	0.266	3.258	0.284	1.716	0.977	0.365	1.589	0.9359	0.354	1.646
13	0.832	0.249	3.336	0.308	1.692	0.980	0.392	1.568	0.9410	0.382	1.618
14	0.802	0.235	3.407	0.329	1.671	0.981	0.414	1.548	0.9453	0.406	1.594
15	0.775	0.223	3.472	0.348	1.652	0.982	0.434	1.530	0.9490	0.428	1.572
16	0.750	0.212	3.532	0.364	1.636	0.984	0.454	1.514	0.9523	0.448	1.552
17	0.728	0.203	3.588	0.379	1.621	0.984	0.469	1.499	0.9551	0.466	1.534
18	0.707	0.194	3.640	0.392	1.608	0.986	0.486	1.486	0.9576	0.482	1.518
19	0.688	0.187	3.689	0.404	1.596	0.986	0.500	1.472	0.9599	0.497	1.503
20	0.671	0.180	3.735	0.414	1.586	0.987	0.513	1.461	0.9619	0.510	1.490
21	0.655	0.173	3.778	0.425	1.575	0.988	0.525	1.451	0.9638	0.523	1.477
22	0.640	0.167	3.819	0.434	1.566	0.988	0.536	1.440	0.9655	0.534	1.466
23	0.626	0.162	3.858	0.443	1.557	0.989	0.546	1.432	0.9670	0.545	1.455
24	0.612	0.157	3.895	0.452	1.548	0.989	0.556	1.422	0.9684	0.555	1.445
25	0.600	0.153	3.931	0.459	1.541	0.990	0.566	1.414	0.9696	0.565	1.435

*From: "CRC Handbook of Tables for Mathematics", 4th ed., S.M. Selby, Ed., The Chemical Rubber Co., 1970.

Table 9-61. COMPOUND INTEREST—SINGLE PAYMENT OR LUMP SUM, VALUES OF F_1*

SYMBOLS:

$F_1 = (1+i)^n$

= compound amount factor

= total accumulated amount at the end of n years if a principal of one dollar is invested today at compound interest at the given percentage rate

$1/F_1$ = present worth factor

= the fraction of one dollar that must be invested as the principal today to accumulate one dollar at the end of n years, including compound interest at the given percentage rate

Periods, n	Rate, i				
	.005 ($\frac{1}{2}$%)	.01 (1%)	.02 (2%)	.025 ($2\frac{1}{2}$%)	.03 (3%)
1	1.0050 0000	1.0100 0000	1.0200 0000	1.0250 0000	1.0300 0000
2	1.0100 2500	1.0201 0000	1.0404 0000	1.0506 2500	1.0609 0000
3	1.0150 7513	1.0303 0100	1.0612 0800	1.0768 9063	1.0927 2700
4	1.0201 5050	1.0406 0401	1.0824 3216	1.1038 1289	1.1255 0881
5	1.0252 5125	1.0510 1005	1.1040 8080	1.1314 0821	1.1592 7407
6	1.0303 7751	1.0615 2015	1.1261 6242	1.1596 9342	1.1940 5230
7	1.0355 2940	1.0721 3535	1.1486 8567	1.1886 8575	1.2298 7387
8	1.0407 0704	1.0828 5671	1.1716 5938	1.2184 0290	1.2667 7008
9	1.0459 1058	1.0936 8527	1.1950 9257	1.2488 6297	1.3047 7318
10	1.0511 4013	1.1046 2213	1.2189 9442	1.2800 8454	1.3439 1638
11	1.0563 9583	1.1156 6835	1.2433 7431	1.3120 8666	1.3842 3387
12	1.0616 7701	1.1268 2503	1.2682 4179	1.3448 8882	1.4257 6089
13	1.0669 8620	1.1380 9328	1.2936 0663	1.3785 1104	1.4685 3371
14	1.0723 2113	1.1494 7421	1.3194 7876	1.4129 7382	1.5125 8972
15	1.0776 8274	1.1609 6896	1.3458 6834	1.4482 9817	1.5579 6742
16	1.0830 7115	1.1725 7864	1.3727 8571	1.4845 0562	1.6047 0644
17	1.0884 8651	1.1843 0443	1.4002 4142	1.5216 1826	1.6528 4763
18	1.0939 2894	1.1961 4748	1.4282 4625	1.5596 5872	1.7024 3306
19	1.0993 9858	1.2081 0895	1.4568 1117	1.5986 5019	1.7535 0605
20	1.1048 9558	1.2201 9004	1.4859 4740	1.6386 1644	1.8061 1123
21	1.1104 2006	1.2323 9194	1.5156 6634	1.6795 8185	1.8602 9457
22	1.1159 7216	1.2447 1586	1.5459 7967	1.7215 7140	1.9161 0341
23	1.1215 5202	1.2571 6302	1.5768 9926	1.7646 1068	1.9735 8651
24	1.1271 5978	1.2697 3465	1.6084 3725	1.8087 2595	2.0327 9411
25	1.1327 9558	1.2824 3200	1.6406 0599	1.8539 4410	2.0937 7793
26	1.1384 5955	1.2952 5631	1.6734 1811	1.9002 9270	2.1565 9127
27	1.1441 5185	1.3082 0888	1.7068 8648	1.9478 0002	2.2212 8901
28	1.1498 7261	1.3212 9097	1.7410 2421	1.9964 9502	2.2879 2768
29	1.1556 2197	1.3345 0388	1.7758 4469	2.0464 0739	2.3565 6551
30	1.1614 0008	1.3478 4892	1.8113 6158	2.0975 6758	2.4272 6247
31	1.1672 0708	1.3613 2740	1.8475 8882	2.1500 0677	2.5000 8035
32	1.1730 4312	1.3749 4068	1.8845 4059	2.2037 5694	2.5750 8276
33	1.1789 0833	1.3886 9009	1.9222 3140	2.2588 5086	2.6523 3524
34	1.1848 0288	1.4025 7699	1.9606 7603	2.3153 2213	2.7319 0530
35	1.1907 2689	1.4166 0276	1.9998 8955	2.3732 0519	2.8138 6245
36	1.1966 8052	1.4307 6878	2.0398 8734	2.4325 3532	2.8982 7833
37	1.2026 6393	1.4450 7647	2.0806 8509	2.4933 4870	2.9852 2668
38	1.2086 7725	1.4595 2724	2.1222 9879	2.5556 8242	3.0747 8348
39	1.2147 2063	1.4741 2251	2.1647 4477	2.6195 7448	3.1670 2698
40	1.2207 9424	1.4888 6373	2.2080 3966	2.6850 6384	3.2620 3779
41	1.2268 9821	1.5037 5237	2.2522 0046	2.7521 9043	3.3598 9893
42	1.2330 3270	1.5187 8989	2.2972 4447	2.8209 9520	3.4606 9589
43	1.2391 9786	1.5339 7779	2.3431 8936	2.8915 2008	3.5645 1677
44	1.2453 9385	1.5493 1757	2.3900 5314	2.9638 0808	3.6714 5227
45	1.2516 2082	1.5648 1075	2.4378 5421	3.0379 0328	3.7815 9584
46	1.2578 7892	1.5804 5885	2.4866 1129	3.1138 5086	3.8950 4372
47	1.2641 6832	1.5962 6344	2.5363 4352	3.1916 9713	4.0118 9503
48	1.2704 8916	1.6122 2608	2.5870 7039	3.2714 8956	4.1322 5188
49	1.2768 4161	1.6283 4834	2.6388 1179	3.3532 7680	4.2562 1944
50	1.2832 2581	1.6446 3182	2.6915 8803	3.4371 0872	4.3839 0602

*Condensed from: "Handbook of Tables for Mathematics", 4th ed., S.M. Selby, Ed., The Chemical Rubber Co., 1970.

Table 9-61. COMPOUND INTEREST—SINGLE PAYMENT OR LUMP SUM, VALUES OF F_1 *(Continued)*

Periods, n	Rate, i				
	.035 (3½%)	.04 (4%)	.045 (4½%)	.05 (5%)	.055 (5½%)
1	1.0350 0000	1.0400 0000	1.0450 0000	1.0500 0000	1.0550 0000
2	1.0712 2500	1.0816 0000	1.0920 2500	1.1025 0000	1.1130 2500
3	1.1087 1788	1.1248 6400	1.1411 6613	1.1576 2500	1.1742 4138
4	1.1475 2300	1.1698 5856	1.1925 1860	1.2155 0625	1 2388 2465
5	1.1876 8631	1.2166 5290	1.2461 8194	1.2762 8156	1.3069 6001
6	1.2292 5533	1.2653 1902	1.3022 6012	1.3400 9564	1.3788 4281
7	1.2722 7926	1.3159 3178	1.3608 6183	1.4071 0042	1.4546 7916
8	1.3168 0904	1.3685 6905	1.4221 0061	1.4774 5544	1.5346 8651
9	1.3628 9735	1.4233 1181	1.4860 9514	1.5513 2822	1.6190 9427
10	1.4105 9876	1.4802 4428	1.5529 6942	1.6288 9463	1.7081 4446
11	1.4599 6972	1.5394 5406	1.6228 5305	1.7103 3936	1.8020 9240
12	1.5110 6866	1.6010 3222	1.6958 8143	1.7958 5633	1.9012 0749
13	1.5639 5606	1 6650 7351	1.7721 9610	1.8856 4914	2.0057 7390
14	1.6186 9452	1.7316 7645	1.8519 4492	1.9799 3160	2.1160 9146
15	1.6753 4883	1.8009 4351	1.9352 8244	2.0789 2818	2.2324 7649
16	1.7339 8604	1.8729 8125	2.0223 7015	2.1828 7459	2.3552 6270
17	1.7946 7555	1.9479 0050	2.1133 7681	2.2920 1832	2.4848 0215
18	1.8574 8920	2.0258 1652	2.2084 7877	2.4066 1923	2.6214 6627
19	1.9225 0132	2.1068 4918	2.3078 6031	2.5269 5020	2.7656 4691
20	1.9897 8886	2.1911 2314	2.4117 1402	2.6532 9771	2.9177 5749
21	2.0594 3147	2.2787 6807	2.5202 4116	2.7859 6259	3.0782 3415
22	2.1315 1158	2.3699 1879	2.6336 5201	2.9252 6072	3.2475 3703
23	2.2061 1448	2.4647 1554	2.7521 6635	3.0715 2376	3.4261 5157
24	2.2833 2849	2.5633 0416	2.8760 1383	3.2250 9994	3.6145 8990
25	2.3632 4498	2.6658 3633	3.0054 3446	3.3863 5494	3.8133 9235
26	2.4459 5856	2.7724 6978	3.1406 7901	3.5556 7269	4.0231 2893
27	2.5315 6711	2.8833 6858	3.2820 0956	3.7334 5632	4.2444 0102
28	2.6201 7196	2.9987 0332	3.4296 9999	3.9201 2914	4.4778 4307
29	2.7118 7798	3.1186 5145	3.5840 3649	4.1161 3560	4.7241 2444
30	2.8067 9370	3.2433 9751	3.7453 1813	4.3219 4238	4.9839 5129
31	2.9050 3148	3.3731 3341	3.9138 5745	4.5380 3949	5.2580 8461
32	3.0067 0759	3.5080 5875	4.0899 8104	4.7649 4147	5.5472 6238
33	3.1119 4235	3.6483 8110	4.2740 3018	5.0031 8854	5.8523 6181
34	3.2208 6033	3.7943 1634	4.4663 6154	5.2533 4797	6.1742 4171
35	3.3335 9045	3.9460 8899	4.6673 4781	5.5160 1537	6.5138 2501
36	3.4502 6611	4.1039 3255	4.8773 7846	5.7918 1614	6.8720 8538
37	3.5710 2543	4.2680 8986	5.0968 6049	6.0814 0694	7.2500 5008
38	3.6960 1132	4.4388 1345	5.3262 1921	6.3854 7729	7.6488 0283
39	3.8253 7171	4.6163 6599	5.5658 9908	6.7047 5115	8.0694 8699
40	3.9592 5972	4.8010 2063	5.8163 6454	7.0399 8871	8.5133 0877
41	4.0978 3381	4.9930 6145	6.0781 0094	7.3919 8815	8.9815 4076
42	4.2412 5799	5.1927 8391	6.3516 1548	7.7615 8756	9.4755 2550
43	4.3897 0202	5.4004 9527	6 6374 3818	8.1496 6693	9.9966 7940
44	4.5433 4160	5.6165 1508	6.9361 2290	8.5571 5028	10.5464 9677
45	4.7023 5855	5.8411 7568	7.2482 4843	8.9850 0779	11.1265 5409
46	4.8669 4110	6.0748 2271	7.5744 1961	9 4342 5818	11.7385 1456
47	5.0372 8404	6.3178 1562	7.9152 6849	9.9059 7109	12.3841 3287
48	5.2135 8898	6.5705 2824	8.2714 5557	10.4012 6965	13.0652 6017
49	5.3960 6459	6.8333 4937	8.6436 7107	10.9213 3313	13.7838 4948
50	5.5849 2686	7.1066 8335	9.0326 3627	11.4673 9979	14.5419 6120

Table 9-61. COMPOUND INTEREST—SINGLE PAYMENT OR LUMP SUM, VALUES OF F_1 *(Continued)*

Periods, n	Rate, i				
	.06 (6%)	.065 ($6\frac{1}{2}$%)	.07 (7%)	.075 ($7\frac{1}{2}$%)	.08 (8%)
1	1.0600 0000	1.0650 0000	1.0700 0000	1.0750 0000	1.0800 0000
2	1.1236 0000	1.1342 2500	1.1449 0000	1.1556 2500	1.1664 0000
3	1.1910 1600	1.2079 4963	1.2250 4300	1.2422 9688	1.2597 1200
4	1.2624 7696	1.2864 6635	1.3107 9601	1.3354 6914	1 3604 8896
5	1.3382 2558	1.3700 8666	1.4025 5173	1.4356 2933	1.4693 2808
6	1.4185 1911	1.4591 4230	1.5007 3035	1.5433 0153	1.5868 7432
7	1.5036 3026	1.5539 8655	1.6057 8148	1.6590 4914	1.7138 2427
8	1.5938 4807	1.6549 9567	1.7181 8618	1.7834 7783	1.8509 3021
9	1.6894 7896	1.7625 7039	1.8384 5921	1.9172 3866	1.9990 0463
10	1.7908 4770	1.8771 3747	1.9671 5136	2.0610 3156	2.1589 2500
11	1.8982 9856	1.9991 5140	2.1048 5195	2.2156 0893	2.3316 3900
12	2.0121 9647	2.1290 9624	2.2521 9159	2.3817 7960	2.5181 7012
13	2.1329 2826	2.2674 8750	2.4098 4500	2.5604 1307	2.7196 2373
14	2.2609 0396	2.4148 7418	2.5785 3415	2.7524 4405	2.9371 9362
15	2.3965 5819	2.5718 4101	2.7590 3154	2.9588 7735	3.1721 6911
16	2.5403 5168	2.7390 1067	2.9521 6375	3.1807 9315	3.4259 4264
17	2.6927 7279	2.9170 4637	3.1588 1521	3.4193 5264	3.7000 1805
18	2.8543 3915	3.1066 5438	3.3799 3228	3.6758 0409	3.9960 1950
19	3.0255 9950	3.3085 8691	3.6165 2754	3.9514 8940	4.3157 0106
20	3.2071 3547	3.5236 4506	3.8696 8446	4.2478 5110	4.6609 5714
21	3.3995 6360	3.7526 8199	4.1405 6237	4.5664 3993	5.0338 3372
22	3.6035 3742	3.9966 0632	4.4304 0174	4.9089 2293	5.4365 4041
23	3.8197 4966	4.2563 8573	4.7405 2986	5.2770 9215	5.8714 6365
24	4.0489 3464	4.5330 5081	5.0723 6695	5.6728 7406	6.3411 8074
25	4.2918 7072	4.8276 9911	5.4274 3264	6.0983 3961	6.8484 7520
26	4.5493 8296	5.1414 9955	5.8073 5292	6.5557 1508	7.3963 5321
27	4.8223 4594	5.4756 9702	6.2138 6763	7.0473 9371	7.9880 6147
28	5.1116 8670	5.8316 1733	6.6488 3836	7.5759 4824	8.6271 0639
29	5.4183 8790	6.2106 7245	7.1142 5705	8.1441 4436	9.3172 7490
30	5.7434 9117	6.6143 6616	7.6122 5504	8.7549 5519	10.0626 5689
31	6.0881 0064	7.0442 9996	8.1451 1290	9.4115 7683	10.8676 6944
32	6.4533 8668	7.5021 7946	8.7152 7080	10.1174 4509	11.7370 8300
33	6.8405 8988	7.9898 2113	9.3253 3075	10.8762 5347	12.6760 4964
34	7.2510 2528	8.5091 5950	9.9781 1354	11.6919 7248	13.6901 3361
35	7.6860 8679	9.0022 5487	10.6765 8148	12.5688 7042	14.7853 4420
36	8.1472 5200	9.6513 0143	11.4239 4219	13.5115 3570	15.9681 7134
37	8.6360 8712	10.2786 3603	12.2236 1814	14.5249 0088	17.2456 2558
38	9.1542 5235	10.9467 4737	13.0792 7141	15.6142 6844	18.6252 7563
39	9.7035 0749	11.6582 8595	13.9948 2041	16.7853 3858	20.1152 9768
40	10.2857 1794	12.4160 7453	14.9744 5784	18.0442 3897	21.7245 2150
41	10.9028 6101	13.2231 1938	16.0226 6989	19.3975 5689	23.4624 8322
42	11.5570 3267	14.0826 2214	17.1442 5678	20.8523 7366	25.3394 8187
43	12.2504 5463	14.9979 9258	18.3443 5475	22.4163 0168	27.3666 4042
44	12.9854 8191	15.9728 6209	19.6284 5959	24.0975 2431	29.5559 7166
45	13.7646 1083	17.0110 9813	21.0024 5176	25.9048 3863	31.9204 4939
46	14.5904 8748	18.1168 1951	22.4726 2338	27.8477 0153	34.4740 8534
47	15.4659 1673	19.2944 1278	24.0457 0702	29.9362 7915	37.2320 1217
48	16.3938 7173	20.5485 4961	25.7289 0651	32.1815 0008	40.2105 7314
49	17.3775 0403	21.8842 0533	27.5299 2997	34.5951 1259	43.4274 1899
50	18.4201 5427	23.3066 7868	29.4570 2506	37.1897 4603	46.9016 1251

Table 9-61. COMPOUND INTEREST—SINGLE PAYMENT OR LUMP SUM, VALUES OF F_1 *(Continued)*

Periods, n	Rate, i				
	.085 (8.5%)	.09 (9%)	.10 (10%)	.12 (12%)	.14 (14%)
1	1.085 000	1.090 000	1.100 000	1.120 000	1.140 000
2	1.177 225	1.188 100	1.210 000	1.254 400	1.299 600
3	1.277 289	1.295 029	1.331 000	1.404 928	1.481 544
4	1.385 859	1.411 582	1.464 100	1.573 352	1.688 960
5	1.503 657	1.538 624	1.610 510	1.762 342	1.925 415
6	1.631 468	1.677 100	1.771 561	1.973 823	2.194 973
7	1.770 142	1.828 039	1.948 717	2.210 681	2.502 269
8	1.920 604	1.992 563	2.143 589	2.475 963	2.852 586
9	2.083 856	2.171 893	2.357 948	2.773 079	3.251 948
10	2.260 983	2.367 364	2.593 742	3.105 848	3.707 221
11	2.453 167	2.580 426	2.853 117	3.478 550	4.226 232
12	2.661 686	2.812 665	3.138 428	3.895 976	4.817 905
13	2.887 930	3.065 805	3.452 271	4.363 493	5.492 411
14	3.133 404	3.341 727	3.797 498	4.887 112	6.261 349
15	3.399 743	3.642 482	4.177 248	5.473 566	7.137 938
16	3.688 721	3.970 306	4.594 973	6.130 393	8.137 249
17	4.002 262	4.327 633	5.054 470	6.866 041	9.276 464
18	4.342 455	4.717 120	5.559 917	7.689 966	10.575 169
19	4.711 563	5.141 661	6.115 909	8.612 761	12.055 692
20	5.112 046	5.604 411	6.727 500	9.646 293	13.743 489
21	5.546 570	6.108 808	7.400 250	10.803 848	15.667 578
22	6.018 028	6.658 600	8.140 275	12.100 309	17.861 038
23	6.529 561	7.257 874	8.954 302	13.552 347	20.361 584
24	7.084 573	7.911 083	9.849 732	15.178 628	23.212 205
25	7.686 762	8.623 080	10.834 705	17.000 063	26.461 914

Periods, n	Rate, i			
	.15 (15%)	.16 (16%)	.18 (18%)	.20 (20%)
1	1.150 000	1.160 000	1.180 000	1.200 000
2	1.322 500	1.345 600	1.392 400	1.440 000
3	1.520 875	1.560 896	1.643 032	1.728 000
4	1.749 006	1.810 639	1.938 778	2.073 600
5	2.011 357	2.100 342	2.287 758	2.488 320
6	2.313 061	2.436 396	2.699 554	2.985 984
7	2.660 020	2.826 220	3.185 474	3.583 181
8	3.059 023	3.278 415	3.758 859	4.299 817
9	3.517 876	3.802 961	4.435 454	5.159 780
10	4.045 558	4.411 435	5.233 835	6.191 736
11	4.652 391	5.117 265	6.175 926	7.430 083
12	5.350 250	5.936 027	7.287 592	8.916 100
13	6.152 787	6.885 791	8.599 359	10.699 320
14	7.075 706	7.987 518	10.147 243	12.839 184
15	8.137 061	9.265 520	11.973 747	15.407 021
16	9.357 620	10.748 004	14.129 022	18.488 425
17	10.761 263	12.467 684	16.672 246	22.186 110
18	12.375 453	14.462 514	19.673 250	26.623 331
19	14.231 771	16.776 516	23.214 435	31.947 998
20	16.366 536	19.460 758	27.393 033	38.337 597
21	18.821 517	22.574 480	32.323 779	46.005 116
22	21.644 744	26.186 396	38.142 059	55.206 139
23	24.891 456	30.376 219	45.007 629	66.247 367
24	28.625 174	35.236 414	53.109 002	79.496 840
25	32.918 950	40.874 241	62.668 622	95.396 208

Table 9-62. COMPOUND INTEREST—UNIFORM SERIES OR ANNUITY, VALUES OF F_P*

Present Worth and Capital Recovery

SYMBOLS:

$F_P = [1-(1+i)^{-n}]/i$

= present worth factor

= present capital value of equal payments of one dollar annually for n years, when these payments accumulate at compound interest at the given percentage rate

$1/F_P$ = capital recovery factor

= fraction of one dollar that must be invested annually for n years to pay off or recover one dollar in present value

= annuity payable for n years for each dollar of capital invested today, with compound interest at the given annual percentage rate

= annual mortgage loan payment required for n years for each dollar of principal of the loan today

Periods, n	.005 ($\frac{1}{2}$%)	.01 (1%)	.02 (2%)	.025 (2$\frac{1}{2}$%)	.03 (3%)
1	0.9950 2488	0.9900 9901	0.9803 9216	0.9756 0976	0.9708 7379
2	1.9850 9938	1.9703 9506	1.9415 6094	1.9274 2415	1.9134 6970
3	2.9702 4814	2.9409 8521	2.8838 8327	2.8560 2356	2.8286 1135
4	3.9504 9566	3.9019 6555	3.8077 2870	3.7619 7421	3.7170 9840
5	4.9258 6633	4.8534 3124	4.7134 5951	4.6458 2850	4.5797 0719
6	5.8963 8441	5.7954 7647	5.6014 3089	5.5081 2536	5.4171 9144
7	6.8620 7404	6.7281 9453	6.4719 9107	6.3493 9060	6.2302 8296
8	7.8229 5924	7.6516 7775	7.3254 8144	7.1701 3717	7.0196 9219
9	8.7790 6392	8.5660 1758	8.1622 3671	7.9708 6553	7.7861 0892
10	9.7304 1186	9.4713 0453	8.9825 8501	8.7520 6393	8.5302 0284
11	10.6770 2673	10.3676 2825	9.7868 4805	9.5142 0871	9.2526 2411
12	11.6189 3207	11.2550 7747	10.5753 4122	10.2577 6460	9.9540 0399
13	12.5561 5131	12.1337 4007	11.3483 7375	10.9831 8497	10.6349 5533
14	13.4887 0777	13.0037 0304	12.1062 4877	11.6909 1217	11.2960 7314
15	14.4166 2465	13.8650 5252	12.8492 6350	12.3813 7773	11.9379 3509
16	15.3399 2502	14.7178 7378	13.5777 0931	13.0550 0266	12.5611 0203
17	16.2586 3186	15.5622 5127	14.2918 7188	13.7121 9772	13.1661 1847
18	17.1727 6802	16.3982 6858	14.9920 3125	14.3533 6363	13.7535 1308
19	18.0823 5624	17.2260 0850	15.6784 6201	14.9788 9134	14.3237 9911
20	18.9874 1915	18.0455 5297	16.3514 3334	15.5891 6229	14.8774 7486
21	19.8879 7925	18.8569 8313	17.0112 0916	16.1845 4857	15.4150 2414
22	20.7840 5896	19.6603 7934	17.6580 4820	16.7654 1324	15.9369 1664
23	21.6756 8055	20.4558 2113	18.2922 0412	17.3321 1048	16.4436 0839
24	22.5628 6622	21.2433 8726	18.9139 2560	17.8849 8583	16.9355 4212
25	23.4456 3803	22.0231 5570	19.5234 5647	18.4243 7642	17.4131 4769
26	24.3240 1794	22.7952 0366	20.1210 3576	18.9506 1114	17.8768 4242
27	25.1980 2780	23.5596 0759	20.7068 9780	19.4640 1087	18.3270 3147
28	26.0676 8936	24.3164 4316	21.2812 7236	19.9648 8866	18.7641 0823
29	26.9330 2424	25.0657 8530	21.8443 8466	20.4535 4991	19.1884 5459
30	27.7940 5397	25.8077 0822	22.3964 5555	20.9302 9259	19.6004 4135
31	28.6507 9997	26.5422 8537	22.9377 0152	21.3954 0741	20.0004 2849
32	29.5032 8355	27.2695 8947	23.4683 3482	21.8491 7796	20.3887 6553
33	30.3515 2592	27.9896 9255	23.9885 6355	22.2918 8094	20.7657 9178
34	31.1955 4818	28.7026 6589	24.4985 9172	22.7237 8628	21.1318 3668
35	32.0353 7132	29.4085 8009	24.9986 1933	23.1451 5734	21.4872 2007
36	32.8710 1624	30.1075 0504	25.4888 4248	23.5562 5107	21.8322 5250
37	33.7025 0372	30.7995 0994	25.9694 5341	23.9573 1812	22.1672 3544
38	34.5298 5445	31.4846 6330	26.4406 4060	24.3486 0304	22.4924 6159
39	35.3530 8900	32.1630 3298	26.9025 8883	24.7303 4443	22.8082 1513
40	36.1722 2786	32.8346 8611	27.3554 7924	25.1027 7505	23.1147 7197
41	36.9872 9141	33.4996 8922	27.7994 8945	25.4661 2200	23.4123 9997
42	37.7982 9991	34.1581 0814	28.2347 9358	25.8206 0683	23.7013 5920
43	38.6052 7354	34.8100 0806	28.6615 6233	26.1664 4569	23.9819 0213
44	39.4082 3238	35.4554 5352	29.0799 6307	26.5038 4945	24.2542 7392
45	40.2071 9640	36.0945 0844	29.4901 5987	26.8330 2386	24.5187 1254
46	41.0021 8547	36.7272 3608	29.8923 1360	27.1541 6962	24.7754 4907
47	41.7932 1937	37.3536 9909	30.2865 8196	27.4674 8255	25.0247 0783
48	42.5803 1778	37.9739 5949	30.6731 1957	27.7731 5371	25.2667 0664
49	43.3635 0028	38.5880 7871	31.0520 7801	28.0713 6947	25.5016 5693
50	44.1427 8635	39.1961 1753	31.4236 0589	28.3623 1168	25.7297 6401

Table 9-62. COMPOUND INTEREST—UNIFORM SERIES OR ANNUITY, VALUES OF F_P *(Continued)*

Periods, n	Rate, i				
	.035 (3½%)	.04 (4%)	.045 (4½%)	.05 (5%)	.055 (5½%)
1	0.9661 8357	0.9615 3846	0.9569 3780	0.9523 8095	0.9478 6730
2	1.8996 9428	1.8860 9467	1.8726 6775	1.8594 1043	1.8463 1971
3	2.8016 3698	2.7750 9103	2.7489 6435	2.7232 4803	2.6979 3338
4	3.6730 7921	3.6298 9522	3.5875 2570	3.5459 5050	3.5051 5012
5	4.5150 5238	4.4518 2233	4.3899 7674	4.3294 7667	4.2702 8448
6	5.3285 5302	5.2421 3686	5.1578 7248	5.0756 9207	4.9955 3031
7	6.1145 4398	6.0020 5467	5.8927 0094	5.7863 7340	5.6829 6712
8	6.8739 5554	6.7327 4487	6.5958 8607	6.4632 1276	6.3345 6599
9	7.6076 8651	7.4353 3161	7.2687 9050	7.1078 2168	6.9521 9525
10	8.3166 0532	8.1108 9578	7.9127 1818	7.7217 3493	7.5376 2583
11	9.0015 5104	8.7604 7671	8.5289 1692	8.3064 1422	8.0925 3633
12	9.6633 3433	9.3850 7376	9.1185 8078	8.8632 5164	8.6185 1785
13	10.3027 3849	9.9856 4785	9.6828 5242	9.3935 7299	9.1170 7853
14	10.9205 2028	10.5631 2293	10.2228 2528	9.8986 4094	9.5896 4790
15	11.5174 1090	11.1183 8743	10.7395 4573	10.3796 5804	10.0375 8094
16	12.0941 1681	11.6522 9561	11.2340 1505	10.8377 6956	10.4621 6203
17	12.6513 2059	12.1656 6885	11.7071 9143	11.2740 6625	10.8646 0856
18	13.1896 8173	12.6592 9697	12.1599 9180	11.6895 8690	11.2460 7447
19	13.7098 3742	13.1339 3940	12.5932 9359	12.0853 2086	11.6076 5352
20	14.2124 0330	13.5903 2634	13.0079 3645	12.4622 1034	11.9503 8248
21	14.6979 7420	14.0291 5995	13.4047 2388	12.8211 5271	12.2752 4406
22	15.1671 2484	14.4511 1533	13.7844 2476	13.1630 0258	12.5831 6973
23	15.6204 1047	14.8568 4167	14.1477 7489	13.4885 7388	12.8750 4239
24	16.0583 6760	15.2469 6314	14.4954 7837	13.7986 4179	13.1516 9895
25	16.4815 1459	15.6220 7994	14.8282 0896	14.0939 4457	13.4139 3266
26	16.8903 5226	15.9827 6918	15.1466 1145	14.3751 8530	13.6624 9541
27	17.2853 6451	16.3295 8575	15.4513 0282	14.6430 3362	13.8980 9991
28	17.6670 1885	16.6630 6322	15.7428 7351	14.8981 2726	14.1214 2172
29	18.0357 6700	16.9837 1463	16.0218 8853	15.1410 7358	14.3331 0116
30	18.3920 4541	17.2920 3330	16.2888 8854	15.3724 5103	14.5337 4517
31	18.7362 7576	17.5884 9356	16.5443 9095	15.5928 1050	14.7239 2907
32	19.0688 6547	17.8735 5150	16.7888 9086	15.8026 7667	14.9041 9817
33	19.3902 0818	18.1476 4567	17.0228 6207	16.0025 4921	15.0750 6936
34	19.7006 8423	18.4111 9776	17.2467 5796	16.1929 0401	15.2370 3257
35	20.0006 6110	18.6646 1323	17.4610 1240	16.3741 9429	15.3905 5220
36	20.2904 9381	18.9082 8195	17.6660 4058	16.5468 5171	15.5360 6843
37	20.5705 2542	19.1425 7880	17.8622 3979	16.7112 8734	15.6739 9851
38	20.8410 8736	19.3678 6423	18.0499 9023	16.8678 9271	15.8047 3793
39	21.1024 9987	19.5844 8484	18.2296 5572	17.0170 4067	15.9286 6154
40	21.3550 7234	19.7927 7388	18.4015 8442	17.1590 8635	16.0461 2469
41	21.5991 0371	19.9930 5181	18.5661 0949	17.2943 6796	16.1574 6416
42	21.8348 8281	20.1856 2674	18.7235 4975	17.4232 0758	16.2629 9920
43	22.0626 8870	20.3707 9494	18.8742 1029	17.5459 1198	16.3630 3242
44	22.2827 9102	20.5488 4129	19.0183 8305	17.6627 7331	16.4578 5063
45	22.4954 5026	20.7200 3970	19.1563 4742	17.7740 6982	16.5477 2572
46	22.7009 1813	20.8846 5356	19.2883 7074	17.8800 6650	16.6329 1537
47	22.8994 3780	21.0429 3612	19.4147 0884	17.9810 1571	16.7136 6386
48	23.0912 4425	21.1951 3088	19.5356 0654	18.0771 5782	16.7902 0271
49	23.2765 6450	21.3414 7200	19.6512 9813	18.1687 2173	16.8627 5139
50	23.4556 1787	21.4821 8462	19.7620 0778	18.2559 2546	16.9315 1790

Table 9-62. COMPOUND INTEREST—UNIFORM SERIES OR ANNUITY, VALUES OF F_P (Continued)

Periods, n	Rate, i				
	.06 (6%)	.065 (6½%)	.07 (7%)	.075 (7½%)	.08 (8%)
1	0.9433 9623	0.9389 6714	0.9345 7944	0.9302 3256	0.9259 2593
2	1.8333 9267	1.8206 2642	1.8080 1817	1.7955 6517	1.7832 6475
3	2.6730 1195	2.6484 7551	2.6243 1604	2.6005 2574	2.5770 9699
4	3.4651 0561	3.4257 9860	3.3872 1126	3.3493 2627	3.3121 2684
5	4.2123 6379	4.1556 7944	4.1001 9744	4.0458 8490	3.9927 1004
6	4.9173 2433	4.8410 1356	4.7665 3966	4.6938 4642	4.6228 7966
7	5.5823 8144	5.4845 1977	5.3892 8940	5.2966 0132	5.2063 7006
8	6.2097 9381	6.0887 5096	5.9712 9851	5.8573 0355	5.7466 3894
9	6.8016 9227	6.6561 0419	6.5152 3225	6.3788 8703	6.2468 8791
10	7.3600 8705	7.1888 3022	7.0235 8154	6.8640 8096	6.7100 8140
11	7.8868 7458	7.6890 4246	7.4986 7434	7.3154 2415	7.1389 6426
12	8.3838 4394	8.1587 2532	7.9426 8630	7.7352 7827	7.5360 7802
13	8.8526 8296	8.5997 4208	8.3576 5074	8.1258 4026	7.9037 7594
14	9.2949 8393	9.0138 4233	8.7454 6799	8.4891 5373	8.2442 3698
15	9.7122 4899	9.4026 6885	9.1079 1401	8.8271 1975	8.5594 7869
16	10.1058 9527	9.7677 6418	9.4466 4860	9.1415 0674	8.8513 6916
17	10.4772 5969	10.1105 7670	9.7632 2299	9.4339 5976	9.1216 3811
18	10.8276 0348	10.4324 6638	10.0590 8691	9.7060 0908	9.3718 8714
19	11.1581 1649	10.7347 1022	10.3355 9524	9.9590 7821	9.6035 9920
20	11.4699 2122	11.0185 0725	10.5940 1425	10.1944 9136	9.8181 4741
21	11.7640 7662	11.2849 8333	10.8355 2733	10.4134 8033	10.0168 0316
22	12.0415 8172	11.5351 9562	11.0612 4050	10.6171 9101	10.2007 4366
23	12.3033 7898	11.7701 3673	11.2721 8738	10.8066 8931	10.3710 5895
24	12.5503 5753	11.9907 3871	11.4693 3400	10.9829 6680	10.5287 5828
25	12.7833 5616	12.1978 7673	11.6535 8318	11.1469 4586	10.6747 7619
26	13.0031 6610	12.3923 7251	11.8257 7867	11.2994 8452	10.8099 7795
27	13.2105 3414	12.5749 9766	11.9867 0904	11.4413 8095	10.9351 6477
28	13.4061 6428	12.7464 7668	12.1371 1125	11.5733 7763	11.0510 7849
29	13.5907 2102	12.9074 8984	12.2776 7407	11.6961 6524	11.1584 0601
30	13.7648 3115	13.0586 7591	12.4090 4118	11.8103 8027	11.2577 8334
31	13.9290 8599	13.2006 3465	12.5318 1419	11.9166 3839	11.3497 9939
32	14.0840 4339	13.3339 2925	12.6465 5532	12.0154 7757	11.4349 9944
33	14.2302 2901	13.4590 8850	12.7537 9002	12.1074 2099	11.5138 8837
34	14.3681 4114	13.5766 0892	12.8540 0936	12.1929 4976	11.5869 3367
35	14.4982 4636	13.6869 5673	12.9476 7230	12.2725 1141	11.6545 6822
36	14.6209 8713	13.7905 6970	13.0352 0776	12.3465 2224	11.7171 9279
37	14.7367 8031	13.8878 5887	13.1170 1660	12.4153 6952	11.7751 7851
38	14.8460 1916	13.9792 1021	13.1934 7345	12.4794 1351	11.8288 6899
39	14.9490 7468	14.0649 8611	13.2649 2846	12.5389 8931	11.8785 8240
40	15.0462 9687	14.1455 2687	13.3317 0884	12.5944 0866	11.9246 1333
41	15.1380 1592	14.2211 5199	13.3941 2041	12.6459 6155	11.9672 3457
42	15.2245 4332	14.2921 6140	13.4524 4898	12.6939 1772	12.0066 0867
43	15.3061 7294	14.3588 3708	13.5069 6167	12.7385 2811	12.0432 3951
44	15.3831 8202	14.4214 4327	13.5579 0810	12.7800 2615	12.0770 7362
45	15.4558 3209	14.4802 2842	13.6055 2159	12.8186 2898	12.1084 0150
46	15.5243 6990	14.5354 2575	13.6500 2018	12.8545 3858	12.1374 0880
47	15.5890 2821	14.5872 5422	13.6916 0764	12.8879 4287	12.1642 6741
48	15.6500 2661	14.6359 1946	13.7304 7443	12.9190 1662	12.1891 3649
49	15.7075 7227	14.6816 1451	13.7667 9853	12.9479 2244	12.2121 6341
50	15.7618 6064	14.7245 2067	13.8007 4629	12.9748 1157	12.2334 8464

*Condensed from: "CRC Handbook of Tables for Mathematics", 4th ed., S.M. Selby, Ed., The Chemical Rubber Co., 1970.

Table 9-63. COMPOUND INTEREST—UNIFORM SERIES OR ANNUITY, VALUES OF F_A*

Compound Amount and Sinking Fund

SYMBOLS:

$$F_A = [(1+i)^n - 1]/i$$

= total amount accumulated by equal payments of one dollar annually for n years, when these payments draw compound interest at the given percentage rate

$1/F_A$ = sinking fund factor

= fraction of one dollar that must be set aside each year to accumulate to one dollar at the end of n years, including interest compounded annually at the given percentage rate

Periods, n	Rate, i				
	.005 ($\frac{1}{2}$%)	.01 (1%)	.02 (2%)	.025 (2$\frac{1}{2}$%)	.03 (3%)
1	1.0000 0000	1.0000 0000	1.0000 0000	1.0000 0000	1.0000 0000
2	2.0050 0000	2.0100 0000	2.0200 0000	2.0250 0000	2.0300 0000
3	3.0150 2500	3.0301 0000	3.0604 0000	3.0756 2500	3.0909 0000
4	4.0301 0013	4.0604 0100	4.1216 0800	4.1525 1563	4.1836 2700
5	5.0502 5063	5.1010 0501	5.2040 4016	5.2563 2852	5.3091 3581
6	6.0755 0188	6.1520 1506	6.3081 2096	6.3877 3673	6.4684 0988
7	7.1058 7939	7.2135 3521	7.4342 8338	7.5474 3015	7.6624 6218
8	8.1414 0879	8.2856 7056	8.5829 6905	8.7361 1590	8.8923 3605
9	9.1821 1583	9.3685 2727	9.7546 2843	9.9545 1880	10.1591 0613
10	10.2280 2641	10.4622 1254	10.9497 2100	11.2033 8177	11.4638 7931
11	11.2791 6654	11.5668 3467	12.1687 1542	12.4834 6631	12.8077 9569
12	12.3355 6237	12.6825 0301	13.4120 8973	13.7955 5297	14.1920 2956
13	13.3972 4018	13.8093 2804	14.6803 3152	15.1404 4179	15.6177 9045
14	14.4642 2639	14.9474 2132	15.9739 3815	16.5189 5284	17.0863 2416
15	15.5365 4752	16.0968 9554	17.2934 1692	17.9319 2666	18.5989 1389
16	16.6142 3026	17.2578 6449	18.6392 8525	19.3802 2483	20.1568 8130
17	17.6973 0141	18.4304 4314	20.0120 7096	20.8647 3045	21.7615 8774
18	18.7857 8791	19.6147 4757	21.4123 1238	22.3863 4871	23.4144 3537
19	19.8797 1685	20.8108 9504	22.8405 5863	23.9460 0743	25.1168 6844
20	20.9791 1544	22.0190 0399	24.2973 6980	25.5446 5761	26.8703 7449
21	22.0840 1101	23.2391 9403	25.7833 1719	27.1832 7405	28.6764 8572
22	23.1944 3107	24.4715 8598	27.2989 8354	28.8628 5590	30.5367 8030
23	24.3104 0323	25.7163 0183	28.8449 6321	30.5844 2730	32.4528 8370
24	25.4319 5524	26.9734 6485	30.4218 6247	32.3490 3798	34.4264 7022
25	26.5591 1502	28.2431 9950	32.0302 9972	34.1577 6393	36.4592 6432
26	27.6919 1059	29.5256 3150	33.6709 0572	36.0117 0803	38.5530 4225
27	28.8303 7015	30.8208 8781	35.3443 2383	37.9120 0073	40.7096 3352
28	29.9745 2200	32.1290 9669	37.0512 1031	39.8598 0075	42.9309 2252
29	31.1243 9461	33.4503 8766	38.7922 3451	41.8562 9577	45.2188 5020
30	32.2800 1658	34.7848 9153	40.5680 7921	43.9027 0316	47.5754 1571
31	33.4414 1666	36.1327 4045	42.3794 4079	46.0002 7074	50.0026 7818
32	34.6086 2375	37.4940 6785	44.2270 2961	48.1502 7751	52.5027 5852
33	35.7816 6686	38.8690 0853	46.1115 7020	50.3540 3445	55.0778 4128
34	36.9605 7520	40.2576 9862	48.0338 0160	52.6128 8531	57.7301 7652
35	38.1453 7807	41.6602 7560	49.9944 7763	54.9282 0744	60.4620 8181
36	39.3361 0497	43.0768 7836	51.9943 6719	57.3014 1263	63.2759 4427
37	40.5327 8549	44.5076 4714	54.0342 5453	59.7339 4794	66.1742 2259
38	41.7354 4942	45.9527 2361	56.1149 3962	62.2272 9664	69.1594 4927
39	42.9441 2666	47.4122 5085	58.2372 3841	64.7829 7906	72.2342 3275
40	44.1588 4730	48.8863 7336	60.4019 8318	67.4025 5354	75.4012 5973
41	45.3796 4153	50.3752 3709	62.6100 2284	70.0876 1737	78.6632 9753
42	46.6065 3974	51.8789 8946	64.8622 2330	72.8398 0781	82.0231 9645
43	47.8395 7244	53.3977 7936	67.1594 6777	75.6608 0300	85.4838 9234
44	49.0787 7030	54.9317 5715	69.5026 5712	78.5523 2308	89.0484 0911
45	50.3241 6415	56.4810 7472	71.8927 1027	81.5161 3116	92.7198 6139
46	51.5757 8498	58.0458 8547	74.3305 6447	84.5540 3443	96.5014 5723
47	52.8336 6390	59.6263 4432	76.8171 7576	87.6678 8530	100.3965 0095
48	54.0978 3222	61.2226 0777	79.3535 1928	90.8595 8243	104.4083 9598
49	55.3683 2138	62.8348 3385	81.9405 8966	94.1310 7199	108.5406 4785
50	56.6451 6299	64.4631 8218	84.5794 0145	97.4843 4879	112.7968 6729

Table 9-63. COMPOUND INTEREST—UNIFORM SERIES OR ANNUITY, VALUES OF F_A *(Continued)*

Periods, n	Rate, i				
	.035 (3½%)	.04 (4%)	.045 (4½%)	.05 (5%)	.055 (5½%)
1	1.0000 000	1.0000 000	1.0000 000	1.0000 000	1.0000 000
2	2.0350 000	2.0400 000	2.0450 000	2.0500 000	2.0550 000
3	3.1062 250	3.1216 000	3.1370 250	3.1525 000	3.1680 250
4	4.2149 429	4.2464 640	4.2781 911	4.3101 250	4.3422 664
5	5.3624 659	5.4163 226	5.4707 097	5.5256 313	5.5810 910
6	6.5501 522	6.6329 755	6.7168 917	6.8019 128	6.8880 510
7	7.7794 075	7.8982 945	8.0191 518	8.1420 085	8.2668 938
8	9.0516 868	9.2142 263	9.3800 136	9.5491 089	9.7215 730
9	10.3684 958	10.5827 953	10.8021 142	11.0265 643	11.2562 595
10	11.7313 932	12.0061 071	12.2882 094	12.5778 925	12.8753 538
11	13.1419 919	13.4863 514	13.8411 788	14.2067 872	14.5834 982
12	14.6019 616	15.0258 055	15.4640 318	15.9171 265	16.3855 907
13	16.1130 303	16.6268 377	17.1599 133	17.7129 828	18.2867 981
14	17.6769 864	18.2919 112	18.9321 094	19.5986 320	20.2925 720
15	19.2956 809	20.0235 876	20.7840 543	21.5785 636	22.4086 635
16	20.9710 297	21.8245 311	22.7193 367	23.6574 918	24.6411 400
17	22.7050 157	23.6975 124	24.7417 069	25.8403 004	26.9904 027
18	24.4996 913	25.6454 129	26.8550 837	28.1323 847	29.4812 048
19	26.3571 805	27.6712 294	29.0635 625	30.5390 039	32.1026 711
20	28.2796 818	29.7780 786	31.3714 228	33.0659 541	34.8683 180
21	30.2694 707	31.9692 017	33.7831 368	35.7192 518	37.7860 755
22	32.3280 022	34.2479 608	36.3033 780	38.5052 144	40.8643 097
23	34.4604 137	36.6178 886	38.9370 300	41.4304 751	44.1118 467
24	36.6665 282	39.0826 041	41.6891 963	44.5019 989	47.5379 983
25	38.9498 567	41.6459 083	44.5652 101	47.7270 988	51.1525 882
26	41.3131 017	44.3117 446	47.5706 446	51.1134 538	54.9059 805
27	43.7590 602	47.0842 144	50.7113 236	54.6691 264	58.9891 094
28	46.2906 273	49.9675 830	53.9933 332	58.4025 828	63.2335 105
29	48.9107 993	52.9662 863	57.4230 332	62.3227 119	67.7113 535
30	51.6226 773	56.0849 378	61.0070 697	66.4388 475	72.4354 780
31	54.4294 710	59.3283 353	64.7523 878	70.7607 899	77.4194 293
32	57.3345 025	62.7014 687	68.6662 452	75.2988 294	82.6774 979
33	60.3412 101	66.2095 274	72.7562 263	80.0637 708	88.2247 603
34	63.4531 524	69.8579 085	77.0302 565	85.0669 594	94.0771 221
35	66.6740 127	73.6522 240	81.4966 180	90.3203 074	100.2513 638
36	70.0076 032	77.5983 138	86.1639 658	95.8363 227	106.7651 888
37	73.4578 693	81.7022 464	91.0413 443	101.6281 389	113.6372 742
38	77.0288 947	85.9703 363	96.1382 048	107.7095 458	120.8873 242
39	80.7249 060	90.4091 497	101.4644 240	114.0950 231	128.5361 271
40	84.5502 777	95.0255 157	107.0303 231	120.7997 742	136.6056 141
41	88.5095 375	99.8265 363	112.8466 876	127.8397 630	145.1189 228
42	92.6073 713	104.8195 978	118.9247 885	135.2317 511	154.1004 636
43	96.8486 293	110.0123 817	125.2764 040	142.9933 387	163.5759 891
44	101.2383 313	115.4128 770	131.9138 422	151.1430 056	173.5726 685
45	105.7816 729	121.0293 920	138.8499 651	159.7001 559	184.1191 653
46	110.4840 314	126.8705 677	146.0982 135	168.6851 637	195.2457 194
47	115.3509 725	132.9453 904	153.6726 331	178.1194 218	206.9842 339
48	120.3882 566	139.2632 060	161.5879 016	188.0253 929	219.3683 668
49	125.6018 456	145.8337 343	169.8593 572	198.4266 626	232.4336 270
50	130.9979 102	152.6670 837	178.5030 283	209.3479 957	246.2174 764

Table 9-63. COMPOUND INTEREST—UNIFORM SERIES OR ANNUITY, VALUES OF F_A (Continued)

Periods, n	Rate, i				
	.06 (6%)	.065 ($6\frac{1}{2}$%)	.07 (7%)	.075 ($7\frac{1}{2}$%)	.08 (8%)
1	1.0000 000	1.0000 000	1.0000 000	1.0000 000	1.0000 000
2	2.0600 000	2.0650 000	2.0700 000	2.0750 000	2.0800 000
3	3.1836 000	3.1992 250	3.2149 000	3.2306 250	3.2464 000
4	4.3746 160	4.4071 746	4.4399 430	4.4729 219	4.5061 120
5	5.6370 930	5.6936 410	5.7507 390	5.8083 910	5.8666 010
6	6.9753 185	7.0637 276	7.1532 907	7.2440 203	7.3359 290
7	8.3938 376	8.5228 699	8.6540 211	8.7873 219	8.9228 034
8	9.8974 679	10.0768 565	10.2598 026	10.4463 710	10.6366 276
9	11.4913 160	11.7318 522	11.9779 887	12.2298 488	12.4875 578
10	13.1807 949	13.4944 225	13.8164 480	14.1470 875	14.4865 625
11	14.9716 426	15.3715 600	15.7835 993	16.2081 191	16.6454 875
12	16.8699 412	17.3707 114	17.8884 513	18.4237 280	18.9771 265
13	18.8821 377	19.4998 076	20.1406 429	20.8055 076	21.4952 966
14	21.0150 659	21.7672 951	22.5504 879	23.3659 207	24.2149 203
15	23.2759 699	24.1821 693	25.1290 220	26.1183 647	27.1521 139
16	25.6725 281	26.7540 103	27.8880 536	29.0772 421	30.3242 830
17	28.2128 798	29.4930 210	30.8402 173	32.2580 352	33.7502 257
18	30.9056 525	32.4100 674	33.9990 325	35.6773 879	37.4502 437
19	33.7599 917	35.5167 218	37.3789 648	39.3531 919	41.4462 632
20	36.7855 912	38.8253 087	40.9954 923	43.3046 813	45.7619 643
21	39.9927 267	42.3489 537	44.8651 768	47.5525 324	50.4229 214
22	43.3922 903	46.1016 357	49.0057 392	52.1189 724	55.4567 552
23	46.9958 277	50.0982 420	53.4361 409	57.0278 953	60.8932 956
24	50.8155 774	54.3546 278	58.1766 708	62.3049 874	66.7647 592
25	54.8645 120	58.8876 786	63.2490 377	67.9778 615	73.1059 400
26	59.1563 827	63.7153 777	68.6764 704	74.0762 011	79.9544 151
27	63.7057 657	68.8568 772	74.4838 233	80.6319 162	87.3507 684
28	68.5281 116	74.3325 743	80.6976 909	87.6793 099	95.3388 298
29	73.6397 983	80.1641 916	87.3465 293	95.2552 582	103.9659 362
30	79.0581 862	86.3748 640	94.4607 863	103.3994 025	113.2832 111
31	84.8016 774	92.9892 302	102.0730 414	112.1543 577	123.3458 680
32	90.8897 780	100.0335 302	110.2181 543	121.5659 345	134.2135 374
33	97.3431 647	107.5357 096	118.9334 251	131.6833 796	145.9506 204
34	104.1837 546	115.5255 308	128.2587 648	142.5596 331	158.6266 701
35	111.4347 799	124.0346 903	138.2368 784	154.2516 056	172.3168 037
36	119.1208 667	133.0969 451	148.9134 598	166.8204 760	187.1021 480
37	127.2681 187	142.7482 466	160.3374 020	180.3320 117	203.0703 198
38	135.9042 058	153.0268 826	172.5610 202	194.8569 126	220.3159 454
39	145.0584 581	163.9736 300	185.6402 916	210.4711 810	238.9412 210
40	154.7619 656	175.6319 159	199.6351 120	227.2565 196	259.0565 187
41	165.0476 836	188.0479 904	214.6095 698	245.3007 586	280.7810 402
42	175.9505 446	201.2711 098	230.6322 397	264.6983 155	304.2435 234
43	187.5075 772	215.3537 320	247.7764 965	285.5506 891	329.5830 053
44	199.7580 319	230.3517 245	266.1208 513	307.9669 908	356.9496 457
45	212.7435 138	246.3245 866	285.7493 108	332.0645 151	386.5056 174
46	226.5081 246	263.3356 848	306.7517 626	357.9693 537	418.4260 668
47	241.0986 121	281.4525 043	329.2243 860	385.8170 553	452.9001 521
48	256.5645 288	300.7469 170	353.2700 930	415.7533 344	490.1321 643
49	272.9584 006	321.2954 666	378.9989 995	447.9348 345	530.3427 374
50	290.3359 046	343.1796 720	406.5289 295	482.5299 471	573.7701 564

*Condensed from: "CRC Handbook of Tables for Mathematics", 4th ed., S.M. Selby, Ed., The Chemical Rubber Co., 1970.

Table 9-64.   NUMBER SYSTEMS
AND CHANGE OF BASE*

POSITIONAL NOTATION

In our ordinary system of writing numbers, the value of any digit depends on its position in the number. The value of a digit in any position is ten times the value of the same digit one position to the right, or one-tenth the value of the same digit one position to the left. For example,

$$173.246 = 1 \times 10^2 + 7 \times 10^1 + 3 + 2 \times \frac{1}{10} + 4 \times \frac{1}{10^2} + 6 \times \frac{1}{10^3}.$$

There is no reason that a number other than 10 cannot be used as the *base*, or *radix*, of the number system. In fact, bases of 2, 8, and 16 are commonly used in working with digital computers. When the base used is not clear from the context, it is usually indicated as a parenthesized subscript or merely as a subscript. Thus

$$743_{(8)} = 7 \times 8^2 + 4 \times 8 + 3 = 7 \times 64 + 4 \times 8 + 3 = 448 + 32 + 3 = 483_{(10)}$$
$$1011.101_{(2)} = 1 \times 2^3 + 0 \times 2^2 + 1 \times 2 + 1 + 1 \times \tfrac{1}{2} + 0 \times \tfrac{1}{4} + 1 \times \tfrac{1}{8} = 11.625_{(10)}.$$

CHANGE OF BASE

In this section it is assumed that all calculations will be performed in base 10, since this is the only base in which most people can easily compute. However, there is no logical reason that some other base could not be used for the computations.

To convert a number from another base into base 10:

Simply write down the digits of the number, with each one multiplied by its appropriate positional value. Then perform the indicated computations in base 10, and write down the answer.

To convert a number from base 10 into another base:

The part of the number to the left of the point and the part to the right must be operated on separately.

For the integer part (the part to the left of the point):

a. Divide the number by the new base, getting an integer quotient and remainder.
b. Write the remainder as the last digit of the number in the new base.
c. Using the quotient from the last division in place of the original number, repeat the above two steps until the quotient becomes zero.

For the fractional part (the part to the right of the point):

a. Multiply the number by the new base.
b. Write down the integral part of the product as the first digit of the fractional part in the new base.
c. Using the fractional part of the last product in place of the original number, repeat the above two steps until the product becomes an integer, or until the desired number of places have been computed.

Table 9-64. NUMBER SYSTEMS
AND CHANGE OF BASE (*Continued*)

Examples:

These examples show a convenient method of arranging the computations.

1. *Convert $103.118_{(10)}$ to base 8.*

$$
\begin{array}{r|r|l}
8 & 103 & 7 \\
8 & 12 & 4 \\
& 1 &
\end{array}
\qquad 147.074324\ldots
$$

$$
\begin{array}{r}
.118 \\
\underline{8} \\
.944 \\
\underline{8} \\
7.552 \\
\underline{8} \\
4.416 \\
\underline{8} \\
3.328 \\
\underline{8} \\
2.624 \\
\underline{8} \\
4.992
\end{array}
$$

The calculation of the fractional part could be carried out as far as desired. It is a non-terminating fraction that will eventually repeat itself.

$$103.118_{(10)} = 147.074324\ldots_{(8)}$$

The calculations may be further shortened by not writing the multiplier and divisor at each step of the algorithm, as shown in the next example.

2. *Convert $275.824_{(10)}$ to base 5.*

$$
\begin{array}{r|r|l}
5 & 275 & 0 \\
& 55 & 0 \\
& 11 & 1 \\
& 2 &
\end{array}
\qquad
\begin{array}{r}
.824 \\
4.120 \\
0.600 \\
3.000
\end{array}
$$

$$275.824_{(10)} = 2100.403_{(5)}$$

To convert from one base to another (neither of which is 10):

The easiest procedure is usually to convert first to base 10, and then to the desired base. However, there are two exceptions to this:

1. If computational facility is possessed in either of the bases, it may be used instead of base 10, and the appropriate one of the above methods applied.
2. If the two bases are different powers of the same number, the conversion may be done digit-by-digit to the base that is the common root of both bases, and then digit-by-digit back to the other base.

Example: Convert $127.653_{(8)}$ to base 16. (For base 16, the letters A–F are used for the digits $10_{(10)}-15_{(10)}$.)

The first step is to convert the number to base 2, simply by converting each digit to its binary equivalent:

$$127.653_{(8)} = 001\ 010\ 111 \cdot 110\ 101\ 011_{(2)}$$

Now by simply regrouping the binary number into groups of four binary digits, starting at the point, we convert to base 16:

$$127.653_{(16)} = 101\ 0111 \cdot 1101\ 0101\ 1_{(2)} = 57.D58_{(16)}$$

Table 9-65. BINARY, OCTAL, AND DECIMAL NUMBERS*

$10^{\pm n}$ IN OCTAL SCALE

10^n	n	10^{-n}	10^n	n	10^{-n}
1	0	1.000 000 000 000 000	112 402 762 000	10	0.000 000 000 006 676
12	1	0.063 146 314 631 463	1 351 035 564 000	11	0.000 000 000 000 537
144	2	0.005 075 341 217 270	16 432 451 210 000	12	0.000 000 000 000 043
1 750	3	0.000 406 111 564 570	221 441 634 520 000	13	0.000 000 000 000 003
23 420	4	0.000 032 155 613 530	2 657 142 036 440 000	14	0.000 000 000 000 000
303 240	5	0.000 002 476 132 610	34 327 724 461 500 000	15	0.000 000 000 000 000
3 641 100	6	0.000 000 206 157 364	434 157 115 760 200 000	16	0.000 000 000 000 000
46 113 200	7	0.000 000 015 327 745	5 432 127 413 542 400 000	17	0.000 000 000 000 000
575 360 400	8	0.000 000 001 257 143	67 405 553 164 731 000 000	18	0.000 000 000 000 000
7 346 545 000	9	0.000 000 000 104 560			

2^n IN DECIMAL SCALE

n	2^n	n	2^n	n	2^n
0.001	1.00069 33874 62581	0.01	1.00695 55500 56719	0.1	1.07177 34625 36293
0.002	1.00138 72557 11335	0.02	1.01395 94797 90029	0.2	1.14869 83549 97035
0.003	1.00208 16050 70677	0.03	1.02101 21257 07193	0.3	1.23114 44133 44916
0.004	1.00277 64359 01078	0.04	1.02811 38266 56067	0.4	1.31950 79107 72894
0.005	1.00347 17485 09503	0.05	1.03526 49238 41377	0.5	1.41421 35623 73095
0.006	1.00416 75432 38973	0.06	1.04246 57608 41121	0.6	1.51571 65665 10398
0.007	1.00486 38204 23785	0.07	1.04971 66836 23067	0.7	1.62450 47927 12471
0.008	1.00556 05803 98468	0.08	1.05701 80405 61380	0.8	1.74110 11265 92248
0.009	1.00625 78234 97782	0.09	1.06437 01824 53360	0.9	1.86606 59830 73615

$n \log_{10} 2$, $n \log_2 10$ IN DECIMAL SCALE

n	$n \log_{10} 2$	$n \log_2 10$	n	$n \log_{10} 2$	$n \log_2 10$
1	0.30102 99957	3.32192 80949	6	1.80617 99740	19.93156 85693
2	0.60205 99913	6.64385 61898	7	2.10720 99696	23.25349 66642
3	0.90308 99870	9.96578 42847	8	2.40823 99653	26.57542 47591
4	1.20411 99827	13.28771 23795	9	2.70926 99610	29.89735 28540
5	1.50514 99783	16.60964 04744	10	3.01029 99566	33.21928 09489

ADDITION AND MULTIPLICATION TABLES

Binary Scale

Addition	Multiplication
$0 + 0 = 0$	$0 \times 0 = 0$
$0 + 1 = 1 + 0 = 1$	$0 \times 1 = 1 \times 0 = 0$
$1 + 1 = 10$	$1 \times 1 = 1$

Octal Scale

Addition

0	01	02	03	04	05	06	07
1	02	03	04	05	06	07	10
2	03	04	05	06	07	10	11
3	04	05	06	07	10	11	12
4	05	06	07	10	11	12	13
5	06	07	10	11	12	13	14
6	07	10	11	12	13	14	15
7	10	11	12	13	14	15	16

Multiplication

1	02	03	04	05	06	07
2	04	06	10	12	14	16
3	06	11	14	17	22	25
4	10	14	20	24	30	34
5	12	17	24	31	36	43
6	14	22	30	36	44	52
7	16	25	34	43	52	61

MATHEMATICAL CONSTANTS IN OCTAL SCALE

$\pi = (3.11037\ 552421)_{(8)}$ $e = (2.55760\ 521305)_{(8)}$ $\gamma = (0.44742\ 147707)_{(8)}$

$\pi^{-1} = (0.24276\ 301556)_{(8)}$ $e^{-1} = (0.27426\ 530661)_{(8)}$ $\log_e \gamma = -(0.43127\ 233602)_{(8)}$

$\sqrt{\pi} = (1.61337\ 611067)_{(8)}$ $\sqrt{e} = (1.51411\ 230704)_{(8)}$ $\log_2 \gamma = -(0.62573\ 030645)_{(8)}$

$\log_e \pi = (1.11206\ 404435)_{(8)}$ $\log_{10} e = (0.33626\ 754251)_{(8)}$ $\sqrt{2} = (1.32404\ 746320)_{(8)}$

$\log_2 \pi = (1.51544\ 163223)_{(8)}$ $\log_2 e = (1.34252\ 166245)_{(8)}$ $\log_e 2 = (0.54271\ 027760)_{(8)}$

$\sqrt{10} = (3.12305\ 407267)_{(8)}$ $\log_2 10 = (3.24464\ 741136)_{(8)}$ $\log_e 10 = (2.23273\ 067355)_{(8)}$

---

*From: "CRC Handbook of Tables for Mathematics", 4th ed., S.M. Selby, Ed., The Chemical Rubber Co., 1970.

Table 9-66. OCTAL-DECIMAL INTEGER CONVERSION*

	0	1	2	3	4	5	6	7
0000	0000	0001	0002	0003	0004	0005	0006	0007
0010	0008	0009	0010	0011	0012	0013	0014	0015
0020	0016	0017	0018	0019	0020	0021	0022	0023
0030	0024	0025	0026	0027	0028	0029	0030	0031
0040	0032	0033	0034	0035	0036	0037	0038	0039
0050	0040	0041	0042	0043	0044	0045	0046	0047
0060	0048	0049	0050	0051	0052	0053	0054	0055
0070	0056	0057	0058	0059	0060	0061	0062	0063
0100	0064	0065	0066	0067	0068	0069	0070	0071
0110	0072	0073	0074	0075	0076	0077	0078	0079
0120	0080	0081	0082	0083	0084	0085	0086	0087
0130	0088	0089	0090	0091	0092	0093	0094	0095
0140	0096	0097	0098	0099	0100	0101	0102	0103
0150	0104	0105	0106	0107	0108	0109	0110	0111
0160	0112	0113	0114	0115	0116	0117	0118	0119
0170	0120	0121	0122	0123	0124	0125	0126	0127
0200	0128	0129	0130	0131	0132	0133	0134	0135
0210	0136	0137	0138	0139	0140	0141	0142	0143
0220	0144	0145	0146	0147	0148	0149	0150	0151
0230	0152	0153	0154	0155	0156	0157	0158	0159
0240	0160	0161	0162	0163	0164	0165	0166	0167
0250	0168	0169	0170	0171	0172	0173	0174	0175
0260	0176	0177	0178	0179	0180	0181	0182	0183
0270	0184	0185	0186	0187	0188	0189	0190	0191
0300	0192	0193	0194	0195	0196	0197	0198	0199
0310	0200	0201	0202	0203	0204	0205	0206	0207
0320	0208	0209	0210	0211	0212	0213	0214	0215
0330	0216	0217	0218	0219	0220	0221	0222	0223
0340	0224	0225	0226	0227	0228	0229	0230	0231
0350	0232	0233	0234	0235	0236	0237	0238	0239
0360	0240	0241	0242	0243	0244	0245	0246	0247
0370	0248	0249	0250	0251	0252	0253	0254	0255

	0	1	2	3	4	5	6	7
0400	0256	0257	0258	0259	0260	0261	0262	0263
0410	0264	0265	0266	0267	0268	0269	0270	0271
0420	0272	0273	0274	0275	0276	0277	0278	0279
0430	0280	0281	0282	0283	0284	0285	0286	0287
0440	0288	0289	0290	0291	0292	0293	0294	0295
0450	0296	0297	0298	0299	0300	0301	0302	0303
0460	0304	0305	0306	0307	0308	0309	0310	0311
0470	0312	0313	0314	0315	0316	0317	0318	0319
0500	0320	0321	0322	0323	0324	0325	0326	0327
0510	0328	0329	0330	0331	0332	0333	0334	0335
0520	0336	0337	0338	0339	0340	0341	0342	0343
0530	0344	0345	0346	0347	0348	0349	0350	0351
0540	0352	0353	0354	0355	0356	0357	0358	0359
0550	0360	0361	0362	0363	0364	0365	0366	0367
0560	0368	0369	0370	0371	0372	0373	0374	0375
0570	0376	0377	0378	0379	0380	0381	0382	0383
0600	0384	0385	0386	0387	0388	0389	0390	0391
0610	0392	0393	0394	0395	0396	0397	0398	0399
0620	0400	0401	0402	0403	0404	0405	0406	0407
0630	0408	0409	0410	0411	0412	0413	0414	0415
0640	0416	0417	0418	0419	0420	0421	0422	0423
0650	0424	0425	0426	0427	0428	0429	0430	0431
0660	0432	0433	0434	0435	0436	0437	0438	0439
0670	0440	0441	0442	0443	0444	0445	0446	0447
0700	0448	0449	0450	0451	0452	0453	0454	0455
0710	0456	0457	0458	0459	0460	0461	0462	0463
0720	0464	0465	0466	0467	0468	0469	0470	0471
0730	0472	0473	0474	0475	0476	0477	0478	0479
0740	0480	0481	0482	0483	0484	0485	0486	0487
0750	0488	0489	0490	0491	0492	0493	0494	0495
0760	0496	0497	0498	0499	0500	0501	0502	0503
0770	0504	0505	0506	0507	0508	0509	0510	0511

0000   0000
to   to
0777   0511
(Octal) (Decimal)

Octal Decimal
10000- 4096
20000- 8192
30000-12288
40000-16384
50000-20480
60000-24576
70000-28672

	0	1	2	3	4	5	6	7
1000	0512	0513	0514	0515	0516	0517	0518	0519
1010	0520	0521	0522	0523	0524	0525	0526	0527
1020	0528	0529	0530	0531	0532	0533	0534	0535
1030	0536	0537	0538	0539	0540	0541	0542	0543
1040	0544	0545	0546	0547	0548	0549	0550	0551
1050	0552	0553	0554	0555	0556	0557	0558	0559
1060	0560	0561	0562	0563	0564	0565	0566	0567
1070	0568	0569	0570	0571	0572	0573	0574	0575
1100	0576	0577	0578	0579	0580	0581	0582	0583
1110	0584	0585	0586	0587	0588	0589	0590	0591
1120	0592	0593	0594	0595	0596	0597	0598	0599
1130	0600	0601	0602	0603	0604	0605	0606	0607
1140	0608	0609	0610	0611	0612	0613	0614	0615
1150	0616	0617	0618	0619	0620	0621	0622	0623
1160	0624	0625	0626	0627	0628	0629	0630	0631
1170	0632	0633	0634	0635	0636	0637	0638	0639
1200	0640	0641	0642	0643	0644	0645	0646	0647
1210	0648	0649	0650	0651	0652	0653	0654	0655
1220	0656	0657	0658	0659	0660	0661	0662	0663
1230	0664	0665	0666	0667	0668	0669	0670	0671
1240	0672	0673	0674	0675	0676	0677	0678	0679
1250	0680	0681	0682	0683	0684	0685	0686	0687
1260	0688	0689	0690	0691	0692	0693	0694	0695
1270	0696	0697	0698	0699	0700	0701	0702	0703
1300	0704	0705	0706	0707	0708	0709	0710	0711
1310	0712	0713	0714	0715	0716	0717	0718	0719
1320	0720	0721	0722	0723	0724	0725	0726	0727
1330	0728	0729	0730	0731	0732	0733	0734	0735
1340	0736	0737	0738	0739	0740	0741	0742	0743
1350	0744	0745	0746	0747	0748	0749	0750	0751
1360	0752	0753	0754	0755	0756	0757	0758	0759
1370	0760	0761	0762	0763	0764	0765	0766	0767

	0	1	2	3	4	5	6	7
1400	0768	0769	0770	0771	0772	0773	0774	0775
1410	0776	0777	0778	0779	0780	0781	0782	0783
1420	0784	0785	0786	0787	0788	0789	0790	0791
1430	0792	0793	0794	0795	0796	0797	0798	0799
1440	0800	0801	0802	0803	0804	0805	0806	0807
1450	0808	0809	0810	0811	0812	0813	0814	0815
1460	0816	0817	0818	0819	0820	0821	0822	0823
1470	0824	0825	0826	0827	0828	0829	0830	0831
1500	0832	0833	0834	0835	0836	0837	0838	0839
1510	0840	0841	0842	0843	0844	0845	0846	0847
1520	0848	0849	0850	0851	0852	0853	0854	0855
1530	0856	0857	0858	0859	0860	0861	0862	0863
1540	0864	0865	0866	0867	0868	0869	0870	0871
1550	0872	0873	0874	0875	0876	0877	0878	0879
1560	0880	0881	0882	0883	0884	0885	0886	0887
1570	0888	0889	0890	0891	0892	0893	0894	0895
1600	0896	0897	0898	0899	0900	0901	0902	0903
1610	0904	0905	0906	0907	0908	0909	0910	0911
1620	0912	0913	0914	0915	0916	0917	0918	0919
1630	0920	0921	0922	0923	0924	0925	0926	0927
1640	0928	0929	0930	0931	0932	0933	0934	0935
1650	0936	0937	0938	0939	0940	0941	0942	0943
1660	0944	0945	0946	0947	0948	0949	0950	0951
1670	0952	0953	0954	0955	0956	0957	0958	0959
1700	0960	0961	0962	0963	0964	0965	0966	0967
1710	0968	0969	0970	0971	0972	0973	0974	0975
1720	0976	0977	0978	0979	0980	0981	0982	0983
1730	0984	0985	0986	0987	0988	0989	0990	0991
1740	0992	0993	0994	0995	0996	0997	0998	0999
1750	1000	1001	1002	1003	1004	1005	1006	1007
1760	1008	1009	1010	1011	1012	1013	1014	1015
1770	1016	1017	1018	1019	1020	1021	1022	1023

1000   0512
to   to
1777   1023
(Octal) (Decimal)

Table 9-66.   OCTAL-DECIMAL INTEGER CONVERSION *(Continued)*

	0	1	2	3	4	5	6	7
2000	1024	1025	1026	1027	1028	1029	1030	1031
2010	1032	1033	1034	1035	1036	1037	1038	1039
2020	1040	1041	1042	1043	1044	1045	1046	1047
2030	1048	1049	1050	1051	1052	1053	1054	1055
2040	1056	1057	1058	1059	1060	1061	1062	1063
2050	1064	1065	1066	1067	1068	1069	1070	1071
2060	1072	1073	1074	1075	1076	1077	1078	1079
2070	1080	1081	1082	1083	1084	1085	1086	1087
2100	1088	1089	1090	1091	1092	1093	1094	1095
2110	1096	1097	1098	1099	1100	1101	1102	1103
2120	1104	1105	1106	1107	1108	1109	1110	1111
2130	1112	1113	1114	1115	1116	1117	1118	1119
2140	1120	1121	1122	1123	1124	1125	1126	1127
2150	1128	1129	1130	1131	1132	1133	1134	1135
2160	1136	1137	1138	1139	1140	1141	1142	1143
2170	1144	1145	1146	1147	1148	1149	1150	1151
2200	1152	1153	1154	1155	1156	1157	1158	1159
2210	1160	1161	1162	1163	1164	1165	1166	1167
2220	1168	1169	1170	1171	1172	1173	1174	1175
2230	1176	1177	1178	1179	1180	1181	1182	1183
2240	1184	1185	1186	1187	1188	1189	1190	1191
2250	1192	1193	1194	1195	1196	1197	1198	1199
2260	1200	1201	1202	1203	1204	1205	1206	1207
2270	1208	1209	1210	1211	1212	1213	1214	1215
2300	1216	1217	1218	1219	1220	1221	1222	1223
2310	1224	1225	1226	1227	1228	1229	1230	1231
2320	1232	1233	1234	1235	1236	1237	1238	1239
2330	1240	1241	1242	1243	1244	1245	1246	1247
2340	1248	1249	1250	1251	1252	1253	1254	1255
2350	1256	1257	1258	1259	1260	1261	1262	1263
2360	1264	1265	1266	1267	1268	1269	1270	1271
2370	1272	1273	1274	1275	1276	1277	1278	1279

	0	1	2	3	4	5	6	7
2400	1280	1281	1282	1283	1284	1285	1286	1287
2410	1288	1289	1290	1291	1292	1293	1294	1295
2420	1296	1297	1298	1299	1300	1301	1302	1303
2430	1304	1305	1306	1307	1308	1309	1310	1311
2440	1312	1313	1314	1315	1316	1317	1318	1319
2450	1320	1321	1322	1323	1324	1325	1326	1327
2460	1328	1329	1330	1331	1332	1333	1334	1335
2470	1336	1337	1338	1339	1340	1341	1342	1343
2500	1344	1345	1346	1347	1348	1349	1350	1351
2510	1352	1353	1354	1355	1356	1357	1358	1359
2520	1360	1361	1362	1363	1364	1365	1366	1367
2530	1368	1369	1370	1371	1372	1373	1374	1375
2540	1376	1377	1378	1379	1380	1381	1382	1383
2550	1384	1385	1386	1387	1388	1389	1390	1391
2560	1392	1393	1394	1395	1396	1397	1398	1399
2570	1400	1401	1402	1403	1404	1405	1406	1407
2600	1408	1409	1410	1411	1412	1413	1414	1415
2610	1416	1417	1418	1419	1420	1421	1422	1423
2620	1424	1425	1426	1427	1428	1429	1430	1431
2630	1432	1433	1434	1435	1436	1437	1438	1439
2640	1440	1441	1442	1443	1444	1445	1446	1447
2650	1448	1449	1450	1451	1452	1453	1454	1455
2660	1456	1457	1458	1459	1460	1461	1462	1463
2670	1464	1465	1466	1467	1468	1469	1470	1471
2700	1472	1473	1474	1475	1476	1477	1478	1479
2710	1480	1481	1482	1483	1484	1485	1486	1487
2720	1488	1489	1490	1491	1492	1493	1494	1495
2730	1496	1497	1498	1499	1500	1501	1502	1503
2740	1504	1505	1506	1507	1508	1509	1510	1511
2750	1512	1513	1514	1515	1516	1517	1518	1519
2760	1520	1521	1522	1523	1524	1525	1526	1527
2770	1528	1529	1530	1531	1532	1533	1534	1535

2000 → 1024
to
2777 → 1535
(Octal) (Decimal)

Octal	Decimal
10000-	4096
20000-	8192
30000-	12288
40000-	16384
50000-	20480
60000-	24576
70000-	28672

	0	1	2	3	4	5	6	7
3000	1536	1537	1538	1539	1540	1541	1542	1543
3010	1544	1545	1546	1547	1548	1549	1550	1551
3020	1552	1553	1554	1555	1556	1557	1558	1559
3030	1560	1561	1562	1563	1564	1565	1566	1567
3040	1568	1569	1570	1571	1572	1573	1574	1575
3050	1576	1577	1578	1579	1580	1581	1582	1583
3060	1584	1585	1586	1587	1588	1589	1590	1591
3070	1592	1593	1594	1595	1596	1597	1598	1599
3100	1600	1601	1602	1603	1604	1605	1606	1607
3110	1608	1609	1610	1611	1612	1613	1614	1615
3120	1616	1617	1618	1619	1620	1621	1622	1623
3130	1624	1625	1626	1627	1628	1629	1630	1631
3140	1632	1633	1634	1635	1636	1637	1638	1639
3150	1640	1641	1642	1643	1644	1645	1646	1647
3160	1648	1649	1650	1651	1652	1653	1654	1655
3170	1656	1657	1658	1659	1660	1661	1662	1663
3200	1664	1665	1666	1667	1668	1669	1670	1671
3210	1672	1673	1674	1675	1676	1677	1678	1679
3220	1680	1681	1682	1683	1684	1685	1686	1687
3230	1688	1689	1690	1691	1692	1693	1694	1695
3240	1696	1697	1698	1699	1700	1701	1702	1703
3250	1704	1705	1706	1707	1708	1709	1710	1711
3260	1712	1713	1714	1715	1716	1717	1718	1719
3270	1720	1721	1722	1723	1724	1725	1726	1727
3300	1728	1729	1730	1731	1732	1733	1734	1735
3310	1736	1737	1738	1739	1740	1741	1742	1743
3320	1744	1745	1746	1747	1748	1749	1750	1751
3330	1752	1753	1754	1755	1756	1757	1758	1759
3340	1760	1761	1762	1763	1764	1765	1766	1767
3350	1768	1769	1770	1771	1772	1773	1774	1775
3360	1776	1777	1778	1779	1780	1781	1782	1783
3370	1784	1785	1786	1787	1788	1789	1790	1791

	0	1	2	3	4	5	6	7
3400	1792	1793	1794	1795	1796	1797	1798	1799
3410	1800	1801	1802	1803	1804	1805	1806	1807
3420	1808	1809	1810	1811	1812	1813	1814	1815
3430	1816	1817	1818	1819	1820	1821	1822	1823
3440	1824	1825	1826	1827	1828	1829	1830	1831
3450	1832	1833	1834	1835	1836	1837	1838	1839
3460	1840	1841	1842	1843	1844	1845	1846	1847
3470	1848	1849	1850	1851	1852	1853	1854	1855
3500	1856	1857	1858	1859	1860	1861	1862	1863
3510	1864	1865	1866	1867	1868	1869	1870	1871
3520	1872	1873	1874	1875	1876	1877	1878	1879
3530	1880	1881	1882	1883	1884	1885	1886	1887
3540	1888	1889	1890	1891	1892	1893	1894	1895
3550	1896	1897	1898	1899	1900	1901	1902	1903
3560	1904	1905	1906	1907	1908	1909	1910	1911
3570	1912	1913	1914	1915	1916	1917	1918	1919
3600	1920	1921	1922	1923	1924	1925	1926	1927
3610	1928	1929	1930	1931	1932	1933	1934	1935
3620	1936	1937	1938	1939	1940	1941	1942	1943
3630	1944	1945	1946	1947	1948	1949	1950	1951
3640	1952	1953	1954	1955	1956	1957	1958	1959
3650	1960	1961	1962	1963	1964	1965	1966	1967
3660	1968	1969	1970	1971	1972	1973	1974	1975
3670	1976	1977	1978	1979	1980	1981	1982	1983
3700	1984	1985	1986	1987	1988	1989	1990	1991
3710	1992	1993	1994	1995	1996	1997	1998	1999
3720	2000	2001	2002	2003	2004	2005	2006	2007
3730	2008	2009	2010	2011	2012	2013	2014	2015
3740	2016	2017	2018	2019	2020	2021	2022	2023
3750	2024	2025	2026	2027	2028	2029	2030	2031
3760	2032	2033	2034	2035	2036	2037	2038	2039
3770	2040	2041	2042	2043	2044	2045	2046	2047

3000 → 1536
to
3777 → 2047
(Octal) (Decimal)

Table 9-66.   OCTAL-DECIMAL INTEGER CONVERSION *(Continued)*

	0	1	2	3	4	5	6	7
4000	2048	2049	2050	2051	2052	2053	2054	2055
4010	2056	2057	2058	2059	2060	2061	2062	2063
4020	2064	2065	2066	2067	2068	2069	2070	2071
4030	2072	2073	2074	2075	2076	2077	2078	2079
4040	2080	2081	2082	2083	2084	2085	2086	2087
4050	2088	2089	2090	2091	2092	2093	2094	2095
4060	2096	2097	2098	2099	2100	2101	2102	2103
4070	2104	2105	2106	2107	2108	2109	2110	2111
4100	2112	2113	2114	2115	2116	2117	2118	2119
4110	2120	2121	2122	2123	2124	2125	2126	2127
4120	2128	2129	2130	2131	2132	2133	2134	2135
4130	2136	2137	2138	2139	2140	2141	2142	2143
4140	2144	2145	2146	2147	2148	2149	2150	2151
4150	2152	2153	2154	2155	2156	2157	2158	2159
4160	2160	2161	2162	2163	2164	2165	2166	2167
4170	2168	2169	2170	2171	2172	2173	2174	2175
4200	2176	2177	2178	2179	2180	2181	2182	2183
4210	2184	2185	2186	2187	2188	2189	2190	2191
4220	2192	2193	2194	2195	2196	2197	2198	2199
4230	2200	2201	2202	2203	2204	2205	2206	2207
4240	2208	2209	2210	2211	2212	2213	2214	2215
4250	2216	2217	2218	2219	2220	2221	2222	2223
4260	2224	2225	2226	2227	2228	2229	2230	2231
4270	2232	2233	2234	2235	2236	2237	2238	2239
4300	2240	2241	2242	2243	2244	2245	2246	2247
4310	2248	2249	2250	2251	2252	2253	2254	2255
4320	2256	2257	2258	2259	2260	2261	2262	2263
4330	2264	2265	2266	2267	2268	2269	2270	2271
4340	2272	2273	2274	2275	2276	2277	2278	2279
4350	2280	2281	2282	2283	2284	2285	2286	2287
4370	2288	2289	2290	2291	2292	2293	2294	2295
4370	2296	2297	2298	2299	2300	2301	2302	2303

	0	1	2	3	4	5	6	7
4400	2304	2305	2306	2307	2308	2309	2310	2311
4410	2312	2313	2314	2315	2316	2317	2318	2319
4420	2320	2321	2322	2323	2324	2325	2326	2327
4430	2328	2329	2330	2331	2332	2333	2334	2335
4440	2336	2337	2338	2339	2340	2341	2342	2343
4450	2344	2345	2346	2347	2348	2349	2350	2351
4460	2352	2353	2354	2355	2356	2357	2358	2359
4470	2360	2361	2362	2363	2364	2365	2366	2367
4500	2368	2369	2370	2371	2372	2373	2374	2375
4510	2376	2377	2378	2379	2380	2381	2382	2383
4520	2384	2385	2386	2387	2388	2389	2390	2391
4530	2392	2393	2394	2395	2396	2397	2398	2399
4540	2400	2401	2402	2403	2404	2405	2406	2407
4550	2408	2409	2410	2411	2412	2413	2114	2415
4560	2416	2417	2418	2419	2420	2421	2422	2423
4570	2424	2425	2426	2427	2428	2429	2430	2431
4600	2432	2433	2434	2435	2436	2437	2438	2439
4610	2440	2441	2442	2443	2444	2445	2446	2447
4620	2448	2449	2450	2451	2452	2453	2454	2455
4630	2456	2457	2458	2459	2460	2461	2462	2463
4640	2464	2465	2466	2467	2468	2469	2470	2471
4650	2472	2473	2474	2475	2476	2477	2478	2479
4660	2480	2481	2482	2483	2484	2485	2486	2487
4670	2488	2489	2490	2491	2492	2493	2494	2495
4700	2496	2497	2498	2499	2500	2501	2502	2503
4710	2504	2505	2506	2507	2508	2509	2510	2511
4720	2512	2513	2514	2515	2516	2517	2518	2519
4730	2520	2521	2522	2523	2524	2525	2526	2527
4740	2528	2529	2530	2531	2532	2533	2534	2535
4750	2536	2537	2538	2539	2540	2541	2542	2543
4760	2544	2545	2546	2547	2548	2549	2550	2551
4770	2552	2553	2554	2555	2556	2557	2558	2259

4000 to 4777 (Octal) = 2048 to 2559 (Decimal)

Octal	Decimal
10000-	4096
20000-	8192
30000-	12288
40000-	16384
50000-	20480
60000-	24576
70000-	28672

	0	1	2	3	4	5	6	7
5000	2560	2561	2562	2563	2564	2565	2566	2567
5010	2568	2569	2570	2571	2572	2573	2574	2575
5020	2576	2577	2578	2579	2580	2581	2582	2583
5030	2584	2585	2586	2587	2588	2589	2590	2591
5040	2592	2593	2594	2595	2596	2597	2598	2599
5050	2600	2601	2602	2603	2604	2605	2606	2607
5060	2608	2609	2610	2611	2612	2613	2614	2615
5070	2616	2617	2618	2619	2620	2621	2622	2623
5100	2624	2625	2626	2627	2628	2629	2630	2631
5110	2632	2633	2634	2635	2636	2637	2638	2639
5120	2640	2641	2642	2643	2644	2645	2646	2647
5130	2648	2649	2650	2651	2652	2653	2654	2655
5140	2656	2657	2658	2659	2660	2661	2662	2663
5150	2664	2665	2666	2667	2668	2669	2670	2671
5160	2672	2673	2674	2675	2676	2677	2678	2679
5170	2680	2681	2682	2683	2684	2685	2686	2687
5200	2688	2689	2690	2691	2692	2693	2694	2695
5210	2696	2697	2698	2699	2700	2701	2702	2703
5220	2704	2705	2706	2707	2708	2709	2710	2711
5230	2712	2713	2714	2715	2716	2717	2718	2719
5240	2720	2721	2722	2723	2724	2725	2726	2727
5250	2728	2729	2730	2731	2732	2733	2734	2735
5260	2736	2737	2738	2739	2740	2741	2742	2743
5270	2744	2745	2746	2747	2748	2749	2750	2751
5300	2752	2753	2754	2755	2756	2757	2758	2759
5310	2760	2761	2762	2763	2764	2765	2766	2767
5320	2768	2769	2770	2771	2772	2773	2774	2775
5330	2776	2777	2778	2779	2780	2781	2782	2783
5340	2784	2785	2786	2787	2788	2789	2790	2791
5350	2792	2793	2794	2795	2796	2797	2798	2799
5360	2800	2801	2802	2803	2804	2805	2806	2807
5370	2808	2809	2810	2811	2812	2813	2814	2815

	0	1	2	3	4	5	6	7
5400	2816	2817	2818	2819	2820	2821	2822	2823
5410	2824	2825	2826	2827	2828	2829	2830	2831
5420	2832	2833	2834	2835	2836	2837	2838	2839
5430	2840	2841	2842	2843	2844	2845	2846	2847
5440	2848	2849	2850	2851	2852	2853	2854	2855
5450	2856	2857	2858	2859	2860	2861	2862	2863
5460	2864	2865	2866	2867	2868	2869	2870	2871
5470	2872	2873	2874	2875	2876	2877	2878	2879
5500	2880	2881	2882	2883	2884	2885	2886	2887
5510	2888	2889	2890	2891	2892	2893	2894	2895
5520	2896	2897	2898	2899	2900	2901	2902	2903
5530	2904	2905	2906	2907	2908	2909	2910	2911
5540	2912	2913	2914	2915	2916	2917	2918	2919
5550	2920	2921	2922	2923	2924	2925	2926	2927
5560	2928	2929	2930	2931	2932	2933	2934	2935
5570	2936	2937	2938	2939	2940	2941	2942	2943
5600	2944	2945	2946	2947	2948	2949	2950	2951
5610	2952	2953	2954	2955	2956	2957	2958	2959
5620	2960	2961	2962	2963	2964	2965	2966	2967
5630	2968	2969	2970	2971	2972	2973	2974	2975
5640	2976	2977	2978	2979	2980	2981	2982	2983
5650	2984	2985	2986	2987	2988	2989	2990	2991
5660	2992	2993	2994	2995	2996	2997	2998	2999
5670	3000	3001	3002	3003	3004	3005	3006	3007
5700	3008	3009	3010	3011	3012	3013	3014	3015
5710	3016	3017	3018	3019	3020	3021	3022	3023
5720	3024	3025	3026	3027	3028	3029	3030	3031
5730	3032	3033	3034	3035	3036	3037	3038	3039
5740	3040	3041	3042	3043	3044	3045	3046	3047
5750	3048	3049	3050	3051	3052	3053	3054	3055
5760	3056	3057	3058	3059	3060	3061	3062	3063
5770	3064	3065	3066	3067	3068	3069	3070	3071

5000 to 5777 (Octal) = 2560 to 3071 (Decimal)

Table 9-66.   OCTAL-DECIMAL INTEGER CONVERSION (Continued)

	0	1	2	3	4	5	6	7
6000	3072	3073	3074	3075	3076	3077	3078	3079
6010	3080	3081	3082	3083	3084	3085	3086	3087
6020	3088	3089	3090	3091	3092	3093	3094	3095
6030	3096	3097	3098	3099	3100	3101	3102	3103
6040	3104	3105	3106	3107	3108	3109	3110	3111
6050	3112	3113	3114	3115	3116	3117	3118	3119
6060	3120	3121	3122	3123	3124	3125	3126	3127
6070	3128	3129	3130	3131	3132	3133	3134	3135
6100	3136	3137	3138	3139	3140	3141	3142	3143
6110	3144	3145	3146	3147	3148	3149	3150	3151
6120	3152	3153	3154	3155	3156	3157	3158	3159
6130	3160	3161	3162	3163	3164	3165	3166	3167
6140	3168	3169	3170	3171	3172	3173	3174	3175
6150	3176	3177	3178	3179	3180	3181	3182	3183
6160	3184	3185	3186	3187	3188	3189	3190	3191
6170	3192	3193	3194	3195	3196	3197	3198	3199
6200	3200	3201	3202	3203	3204	3205	3206	3207
6210	3208	3209	3210	3211	3212	3213	3214	3215
6220	3216	3217	3218	3219	3220	3221	3222	3223
6230	3224	3225	3226	3227	3228	3229	3230	3231
6240	3232	3233	3234	3235	3236	3237	3238	3239
6250	3240	3241	3442	3243	3244	3245	3246	3247
6260	3248	3249	3250	3251	3252	3253	3254	3255
6270	3256	3257	3258	3259	3260	3261	3262	3263
6300	3264	3265	3266	3267	3268	3269	3270	3871
6310	3272	3273	3274	3275	3276	3277	3278	3279
6320	3280	3281	3282	3283	3284	3285	3286	3287
6330	3288	3289	3290	3291	3292	3293	3294	3295
6340	3296	3297	3298	3299	3300	3301	3302	3003
6350	3304	3305	3306	3307	3308	3309	3310	3311
6360	3312	3313	3314	3315	3316	3317	3318	3319
6370	3320	3321	3322	3323	3324	3325	3326	3327

	0	1	2	3	4	5	6	7
6400	3328	3329	3330	3331	3332	3333	3334	3335
6410	3336	3337	3338	3339	3340	3341	3342	3343
6420	3344	3345	3346	3347	3348	3349	3350	3351
6430	3352	3353	3354	3355	3356	3357	3358	3359
6440	3360	3361	3362	3363	3364	3365	3366	3367
6450	3368	3369	3370	3371	3372	3373	3374	3375
6460	3376	3377	3378	3379	3380	3381	3382	3383
6470	3384	3385	3386	3387	3388	3389	3390	3391
6500	3392	3393	3394	3395	3396	3397	3398	3399
6510	3400	3401	3402	3403	3404	3405	3406	3407
6520	3408	3409	3410	3411	3412	3413	3414	3415
6530	3416	3417	3418	3419	3420	3421	3422	3423
6540	3424	3425	3426	3427	3428	3429	3430	3431
6550	3432	3433	3434	3435	3436	3437	3438	3439
6560	3440	3441	3442	3443	3444	3445	3446	3447
6570	3448	3449	3450	3451	3452	3453	3454	3455
6600	3456	3457	3458	3459	3460	3461	3462	3463
6610	3464	3465	3466	3467	3468	3469	3470	3471
6620	3472	3473	3474	3475	3476	3477	3478	3479
6630	3480	3481	3482	3483	3484	3485	3486	3487
6640	3488	3489	3490	3491	3492	3493	3494	3495
6650	3496	3497	3498	3499	3500	3501	3502	3503
6660	3504	3505	3506	3507	3508	3509	3510	3511
6670	3512	3513	3514	3515	3516	3517	3518	3519
6700	3520	3521	3522	3523	3524	3525	3526	3527
6710	3528	3529	3530	3531	3532	3533	3534	3535
6720	3536	3537	3538	3539	3540	3541	3542	3543
6730	3544	3545	3546	3547	3548	3549	3550	3551
6740	3552	3553	3554	3555	3556	3557	3558	3559
6750	3560	3561	3562	3563	3564	3655	3566	3567
6760	3568	3569	3570	3571	3572	3573	3574	3575
6770	3576	3577	3578	3579	3580	3581	3582	3583

	0	1	2	3	4	5	6	7
7000	3584	3585	3586	3587	3588	3589	3590	3591
7010	3592	3593	3594	3595	3596	3597	3598	3599
7020	3600	3601	3602	3603	3604	3605	3606	3607
7030	3608	3609	3610	3611	3612	3613	3614	3615
7040	3616	3617	3618	3619	3620	3621	3622	3623
7050	3624	2625	3626	3627	3628	3629	3630	3631
7060	3632	3633	3634	3635	3636	3637	3638	3639
7070	3640	3641	3642	3643	3644	3645	3646	3647
7100	3648	3649	3650	3651	3652	3653	3654	3655
7110	3656	2657	2658	2659	3660	3661	3662	3663
7120	3664	3665	3666	3667	3668	3669	3670	3671
7130	3672	3673	3674	3675	3676	3677	3678	3679
7140	3680	3681	3682	3683	3684	3685	3686	3687
7150	3688	2689	2690	3691	3692	3693	3694	3695
7160	3696	3697	3698	3699	3700	3701	3702	3703
7170	3704	3705	3706	3707	3708	3709	3710	3711
7200	3712	3713	3714	3715	3716	3717	3718	3719
7210	3720	3721	3722	3723	3724	3725	3726	3727
7220	3728	3729	3730	3731	3732	3733	3734	3735
7230	3736	3737	3738	3739	3740	3741	3742	3743
7240	3744	3745	3746	3747	3748	3749	3750	3751
7250	3752	3753	3754	3755	3756	3757	3758	3759
7260	3760	3761	3762	3763	3764	3765	3766	3767
7270	3768	3769	3770	3771	3772	3773	3774	3775
7300	3776	3777	3778	3779	3780	3781	3782	3783
7310	3784	3785	3786	3787	3788	3789	3790	3791
7320	3792	3893	3794	3795	3796	3797	3798	3799
7330	3800	3801	3802	3803	3804	3805	3806	3807
7340	3808	3809	3810	3811	3812	3813	3814	3815
7350	3816	3817	3818	3819	3820	3821	3822	3823
7360	3824	3825	3826	3827	3828	3829	3830	3831
7370	3832	3833	3834	3835	3836	3837	3838	3839

	0	1	2	3	4	5	6	7
7400	3840	3841	3482	3843	3844	3845	3846	3847
7410	3848	3849	3850	3851	3852	3853	3854	3855
7420	2856	3857	3858	3859	3860	3861	3862	3863
7430	3864	3865	3866	3867	3868	3869	3870	3871
7440	3872	3873	3874	3875	3876	3877	3878	3879
7450	3880	3881	3882	3883	3884	3885	3886	3887
7460	3888	3889	3890	3891	3892	3893	3894	3895
7470	3896	3897	3898	3899	3900	3901	3902	3903
7500	3904	3905	3906	3907	3908	3909	3910	3911
7510	3912	3913	3914	3915	3016	3917	3918	3919
7520	3920	3921	3922	3923	3924	3925	3926	3927
7530	3928	3929	3930	3931	3932	3933	3934	3935
7540	3036	3037	3038	3039	3940	3941	3942	3943
7550	3944	2945	3946	3947	3948	3949	3950	3951
7560	3952	3953	3954	3955	3956	3957	3958	3959
7570	3960	3961	3962	3963	3964	3965	3966	3967
7600	3968	3969	4970	3971	3972	3973	3974	3975
7610	3976	3977	3978	3979	3980	3981	3982	3983
7620	3984	3985	3986	3987	3988	3989	3990	3991
7630	3992	3993	3994	3995	3996	3997	3998	3999
7640	4000	4001	4002	4003	4004	4005	4006	4007
7650	4008	4009	4010	4011	4012	4013	4014	4015
7660	4016	4017	4018	4019	4020	4021	4022	4023
7670	4024	4025	4026	4027	4028	4029	4030	4031
7700	4032	4033	4034	4035	4036	4037	4038	4039
7710	4040	4041	4042	4043	4044	4045	4046	4047
7720	4048	4049	4050	4051	4052	4053	4054	4055
7730	4056	4057	4058	4059	4060	4061	4062	4063
7740	4064	4065	4066	4067	4068	4069	4070	4071
7750	4072	4073	4074	4075	4076	4077	4078	4079
7760	4080	4081	4082	4083	4084	4085	4086	4087
7770	4088	4089	4090	4091	4092	4093	4094	4095

Side reference panels:

6000	3072
to	to
6777	3583
(Octal)	(Decimal)

Octal	Decimal
10000-	4096
20000-	8192
30000-	12288
40000-	16384
50000-	20480
60000-	24576
70000-	28672

7000	3584
to	to
7777	4095
(Octal)	(Decimal)

Table 9-67. OCTAL-DECIMAL FRACTION CONVERSION*

This table covers the entries from $(.000)_8$ to $(.377)_8$. For entries from $(.400)_8$ to $(.777)_8$, it should be recognized that $(.400)_8$ is $(.500)_{10}$. Hence if $(\cdot637)_8$ is desired, find $(.237)_8$ in the table, namely, $(.310456)_{10}$ and add $(.50000)_{10}$ for $(.400)_8$. Thus $(.637)_8 = (.237)_8 + (.400)_8 = (.310456)_{10} + (.50000)_{10} = (.810456)_{10}$.

Octal	Dec	Octal	Dec	Octal	Dec	Octal	Dec
.000	.000000	.100	.125000	.200	.250000	.300	.375000
.001	.001953	.101	.126953	.201	.251953	.301	.376953
.002	.003906	.102	.128906	.202	.253906	.302	.378906
.003	.005859	.103	.130859	.203	.255859	.303	.380859
.004	.007812	.104	.132812	.204	.257812	.304	.382812
.005	.009765	.105	.134765	.205	.259765	.305	.384765
.006	.011718	.106	.136718	.206	.261718	.306	.386718
.007	.013671	.107	.138671	.207	.263671	.307	.388671
.010	.015625	.110	.140625	.210	.265625	.310	.390625
.011	.017578	.111	.142578	.211	.267578	.311	.392578
.012	.019531	.112	.144531	.212	.269531	.312	.394531
.013	.021484	.113	.146484	.213	.271484	.313	.396484
.014	.023437	.114	.148437	.214	.273437	.314	.398437
.015	.025390	.115	.150390	.215	.275390	.315	.400490
.016	.027343	.116	.152343	.216	.277343	.316	.402343
.017	.029296	.117	.154296	.217	.279296	.317	.404296
.020	.031250	.120	.156250	.220	.281250	.320	.406250
.021	.033203	.121	.158203	.221	.283203	.321	.408203
.022	.035156	.122	.160156	.222	.285156	.322	.410156
.023	.037109	.123	.162109	.223	.287109	.323	.412109
.024	.039062	.124	.164062	.224	.289062	.324	.414062
.025	.041015	.125	.166015	.225	.291015	.325	.416015
.026	.042968	.126	.167968	.226	.292968	.326	.417968
.027	.044921	.127	.169921	.227	.294921	.327	.419921
.030	.046875	.130	.171875	.230	.294875	.330	.421875
.031	.048828	.131	.173828	.231	.298828	.331	.423828
.032	.050781	.132	.175781	.232	.300781	.332	.425781
.033	.052734	.133	.177734	.233	.302734	.333	.427734
.034	.054687	.134	.179687	.234	.304687	.334	.429687
.035	.056640	.135	.181640	.235	.306640	.335	.431640
.036	.058593	.136	.183593	.236	.308593	.336	.433593
.037	.060546	.137	.185546	.237	.310546	.337	.435546
.040	.062500	.140	.187500	.240	.312500	.340	.437500
.041	.064453	.141	.189453	.241	.314453	.341	.439453
.042	.066406	.142	.191406	.242	.316406	.342	.441406
.043	.068359	.143	.193359	.243	.318359	.343	.443359
.044	.070312	.144	.195312	.244	.320312	.344	.445312
.045	.072265	.145	.197265	.245	.322265	.345	.447265
.046	.074218	.146	.199218	.246	.324218	.346	.449218
.047	.076171	.147	.201171	.247	.326171	.347	.451171
.050	.078125	.150	.203125	.250	.328125	.350	.453125
.051	.080078	.151	.205078	.251	.330078	.351	.455078
.052	.082031	.152	.207031	.252	.332031	.352	.457031
.053	.083984	.153	.208984	.253	.333984	.353	.458984
.054	.085937	.154	.210937	.254	.335937	.354	.460937
.055	.087890	.155	.212890	.255	.337890	.355	.462890
.056	.089843	.156	.214843	.256	.339843	.356	.464843
.057	.091796	.157	.216796	.257	.341796	.357	.466796
.060	.093750	.160	.218750	.260	.343750	.360	.468750
.061	.095703	.161	.220703	.261	.345703	.361	.470703
.062	.097656	.162	.222656	.262	.347656	.362	.472656
.063	.099609	.163	.224609	.263	.349609	.363	.474609
.064	.101562	.164	.226562	.264	.351562	.364	.476562
.065	.103515	.165	.228515	.265	.353515	.365	.478515
.066	.105468	.166	.230468	.266	.355468	.366	.480468
.067	.107421	.167	.232421	.267	.357421	.367	.482421
.070	.109375	.170	.234375	.270	.359375	.370	.484375
.071	.111328	.171	.236328	.271	.361328	.371	.486328
.072	.113281	.172	.238281	.272	.363281	.372	.488281
.073	.115234	.173	.240234	.273	.365234	.373	.490234
.074	.117187	.174	.242187	.274	.367187	.374	.492187
.075	.119140	.175	.244140	.275	.369140	.375	.494140
.076	.121093	.176	.246093	.276	.371093	.376	.496093
.077	.123046	.177	.248046	.277	.373046	.377	.498046

Table 9-67.   OCTAL-DECIMAL FRACTION CONVERSION *(Continued)*

Octal	Dec	Octal	Dec	Octal	Dec	Octal	Dec
.000000	.000000	.000100	.000244	.000200	.000488	.000300	.000732
.000001	.000004	.000101	.000247	.000201	.000492	.000301	.000736
.000002	.000007	.000102	.000251	.000202	.000495	.000302	.000740
.000003	.000011	.000103	.000255	.000203	.000499	.000303	.000743
.000004	.000015	.000104	.000259	.000204	.000503	.000304	.000747
.000005	.000019	.000105	.000263	.000205	.000507	.000305	.000751
.000006	.000022	.000106	.000267	.000206	.000511	.000306	.000755
.000007	.000026	.000107	.000270	.000207	.000514	.000307	.000759
.000010	.000030	.000110	.000274	.000210	.000518	.000310	.000762
.000011	.000034	.000111	.000278	.000211	.000522	.000311	.000766
.000012	.000038	.000112	.000282	.000212	.000526	.000312	.000770
.000013	.000041	.000113	.000286	.000213	.000530	.000313	.000774
.000014	.000045	.000114	.000289	.000214	.000534	.000314	.000778
.000015	.000049	.000115	.000293	.000215	.000537	.000315	.000782
.000016	.000053	.000116	.000297	.000216	.000541	.000316	.000785
.000017	.000057	.000117	.000301	.000217	.000545	.000317	.000789
.000020	.000061	.000120	.000305	.000220	.000549	.000320	.000793
.000021	.000064	.000121	.000308	.000221	.000553	.000321	.000797
.000022	.000068	.000122	.000312	.000222	.000556	.000322	.000801
.000023	.000072	.000123	.000316	.000223	.000560	.000323	.000805
.000024	.000076	.000124	.000320	.000224	.000564	.000324	.000808
.000025	.000080	.000125	.000324	.000225	.000568	.000325	.000812
.000026	.000083	.000126	.000328	.000226	.000572	.000326	.000816
.000027	.000087	.000127	.000331	.000227	.000576	.000327	.000820
.000030	.000091	.000130	.000335	.000230	.000579	.000330	.000823
.000031	.000095	.000131	.000339	.000231	.000583	.000331	.000827
.000032	.000099	.000132	.000343	.000232	.000587	.000332	.000831
.000033	.000102	.000133	.000347	.000233	.000591	.000333	.000835
.000034	.000106	.000134	.000350	.000234	.000595	.000334	.000839
.000035	.000110	.000135	.000354	.000235	.000598	.000335	.000843
.000036	.000114	.000136	.000358	.000236	.000602	.000336	.000846
.000037	.000118	.000137	.000362	.000237	.000606	.000337	.000850
.000040	.000122	.000140	.000366	.000240	.000610	.000340	.000854
.000041	.000125	.000141	.000370	.000241	.000614	.000341	.000858
.000042	.000129	.000142	.000373	.000242	.000617	.000342	.000862
.000043	.000133	.000143	.000377	.000243	.000621	.000343	.000865
.000044	.000137	.000144	.000381	.000244	.000625	.000344	.000869
.000045	.000141	.000145	.000385	.000245	.000629	.000345	.000873
.000046	.000144	.000146	.000389	.000246	.000633	.000346	.000877
.000047	.000148	.000147	.000392	.000247	.000637	.000347	.000881
.000050	.000152	.000150	.000396	.000250	.000640	.000350	.000885
.000051	.000156	.000151	.000400	.000251	.000644	.000351	.000888
.000052	.000160	.000152	.000404	.000252	.000648	.000352	.000892
.000053	.000164	.000153	.000408	.000253	.000652	.000353	.000896
.000054	.000167	.000154	.000411	.000254	.000656	.000354	.000900
.000055	.000171	.000155	.000415	.000255	.000659	.000355	.000904
.000056	.000175	.000156	.000419	.000256	.000663	.000356	.000907
.000057	.000179	.000157	.000423	.000257	.000667	.000357	.000911
.000060	.000183	.000160	.000427	.000260	.000671	.000360	.000915
.000061	.000186	.000161	.000431	.000261	.000675	.000361	.000919
.000062	.000190	.000162	.000434	.000262	.000679	.000362	.000923
.000063	.000194	.000163	.000438	.000263	.000682	.000363	.000926
.000064	.000198	.000164	.000442	.000264	.000686	.000364	.000930
.000065	.000202	.000165	.000446	.000265	.000690	.000365	.000934
.000066	.000205	.000166	.000450	.000266	.000694	.000366	.000938
.000067	.000209	.000167	.000453	.000267	.000698	.000367	.000942
.000070	.000213	.000170	.000457	.000270	.000701	.000370	.000946
.000071	.000217	.000171	.000461	.000271	.000705	.000371	.000949
.000072	.000221	.000172	.000465	.000272	.000709	.000372	.000953
.000073	.000225	.000173	.000469	.000273	.000713	.000373	.000957
.000074	.000228	.000174	.000473	.000274	.000717	.000374	.000961
.000075	.000232	.000175	.000476	.000275	.000720	.000375	.000965
.000076	.000236	.000176	.000480	.000276	.000724	.000376	.000968
.000077	.000240	.000177	.000484	.000277	.000728	.000377	.000972

Table 9-67.   OCTAL-DECIMAL FRACTION CONVERSION *(Continued)*

Octal	Dec	Octal	Dec	Octal	Dec	Octal	Dec
.000400	.000976	.000500	.001220	.000600	.001464	.000700	.001708
.000401	.000980	.000501	.001224	.000601	.001468	.000701	.001712
.000402	.000984	.000502	.001228	.000602	.001472	.000702	.001716
.000403	.000988	.000503	.001232	.000603	.001476	.000703	.001720
.000404	.000991	.000504	.001235	.000604	.001480	.000704	.001724
.000405	.000995	.000505	.001239	.000605	.001483	.000705	.001728
.000406	.000999	.000506	.001243	.000606	.001487	.000706	.001731
.000407	.001003	.000507	.001247	.000607	.001491	.000707	.001735
.000410	.001007	.000510	.001251	.000610	.001495	.000710	.001739
.000411	.001010	.000511	.001255	.000611	.001499	.000711	.001743
.000412	.001014	.000512	.001258	.000612	.001502	.000712	.001747
.000413	.001018	.000513	.001262	.000613	.001506	.000713	.001750
.000414	.001022	.000514	.001266	.000614	.001510	.000714	.001754
.000415	.001026	.000515	.001270	.000615	.001514	.000715	.001758
.000416	.001029	.000516	.001274	.000616	.001518	.000716	.001762
.000417	.001033	.000517	.001277	.000617	.001522	.000717	.001766
.000420	.001037	.000520	.001281	.000620	.001525	.000720	.001770
.000421	.001041	.000521	.001285	.000621	.001529	.000721	.001733
.000422	.001045	.000522	.001289	.000622	.001533	.000722	.001777
.000423	.001049	.000523	.001293	.000623	.001537	.000723	.001781
.000424	.001052	.000524	.001296	.000624	.001541	.000724	.001785
.000425	.001056	.000525	.001300	.000625	.001544	.000725	.001789
.000426	.001060	.000526	.001304	.000626	.001548	.000726	.001792
.000427	.001064	.000527	.001308	.000627	.001552	.000727	.001796
.000430	.001068	.000530	.001312	.000630	.001556	.000730	.001800
.000431	.001071	.000531	.001316	.000631	.001560	.000731	.001804
.000432	.001075	.000532	.001319	.000632	.001564	.000732	.001808
.000433	.001079	.000533	.001323	.000633	.001567	.000733	.001811
.000434	.001083	.000534	.001327	.000634	.001571	.000734	.001815
.000435	.001087	.000535	.001331	.000635	.001575	.000735	.001819
.000436	.001091	.000536	.001335	.000636	.001579	.000736	.001823
.000437	.001094	.000537	.001338	.000637	.001583	.000737	.001827
.000440	.001098	.000540	.001342	.000640	.001586	.000740	.001831
.000441	.001102	.000541	.001346	.000641	.001590	.000741	.001834
.000442	.001106	.000542	.001350	.000642	.001594	.000742	.001838
.000443	.001110	.000543	.001354	.000643	.001598	.000743	.001842
.000444	.001113	.000544	.001358	.000644	.001602	.000744	.001846
.000445	.001117	.000545	.001361	.000645	.001605	.000745	.001850
.000446	.001121	.000546	.001365	.000646	.001609	.000746	.001853
.000447	.001125	.000547	.001369	.000647	.001613	.000747	.001857
.000450	.001129	.000550	.001373	.000650	.001617	.000750	.001861
.000451	.001132	.000551	.001377	.000651	.001621	.000751	.001865
.000452	.001136	.000552	.001380	.000652	.001625	.000752	.001869
.000453	.001140	.000553	.001384	.000653	.001628	.000753	.001873
.000454	.001144	.000554	.001388	.000654	.001632	.000754	.001876
.000455	.001148	.000555	.001392	.000655	.001636	.000755	.001880
.000456	.001152	.000556	.001396	.000656	.001640	.000756	.001884
.000457	.001155	.000557	.001399	.000657	.001644	.000757	.001888
.000460	.001159	.000560	.001403	.000660	.001647	.000760	.001892
.000461	.001163	.000561	.001407	.000661	.001651	.000761	.001895
.000462	.001167	.000562	.001411	.000662	.001655	.000762	.001899
.000463	.001171	.000563	.001415	.000663	.001659	.000763	.001903
.000464	.001174	.000564	.001419	.000664	.001663	.000764	.001907
.000465	.001178	.000565	.001422	.000665	.001667	.000765	.001911
.000466	.001182	.000566	.001426	.000666	.001670	.000766	.001914
.000467	.001186	.000567	.001430	.000667	.001674	.000767	.001918
.000470	.001190	.000570	.001434	.000670	.001678	.000770	.001922
.000471	.001194	.000571	.001438	.000671	.001682	.000771	.001926
.000472	.001197	.000572	.001441	.000672	.001686	.000772	.001930
.000473	.001201	.000573	.001445	.000673	.001689	.000773	.001934
.000474	.001205	.000574	.001449	.000674	.001693	.000774	.001937
.000475	.001209	.000575	.001453	.000675	.001697	.000775	.001941
.000476	.001213	.000576	.001457	.000676	.001701	.000776	.001945
.000477	.001216	.000577	.001461	.000677	.001705	.000777	.001949

*From: "CRC Handbook of Tables for Mathematics", 4th ed., S.M. Selby, Ed., The Chemical Rubber Co., 1970.

Table 9-68. HEXADECIMAL AND DECIMAL CONVERSION*

The following tables aid in converting hexadecimal (base 16) numbers to decimal, and the reverse. Note that the base 16 digits for the decimal values 10–15 are represented by the letters A–F, respectively.

This table provides direct conversion of decimal and hexadecimal numbers in these ranges:

HEXADECIMAL	DECIMAL
000 to FFF	0000 to 4095

For numbers outside the range of the table, add the values in the right-hand columns to the table figures.

HEXADECIMAL	DECIMAL
1000	4096
2000	8192
3000	12288
4000	16384
5000	20480
6000	24576
7000	28672
8000	32768
9000	36864
A000	40960
B000	45056
C000	49152
D000	53248
E000	57344
F000	61440

	0	1	2	3	4	5	6	7	8	9	A	B	C	D	E	F
00_	0000	0001	0002	0003	0004	0005	0006	0007	0008	0009	0010	0011	0012	0013	0014	0015
01_	0016	0017	0018	0019	0020	0021	0022	0023	0024	0025	0026	0027	0028	0029	0030	0031
02_	0032	0033	0034	0035	0036	0037	0038	0039	0040	0041	0042	0043	0044	0045	0046	0047
03_	0048	0049	0050	0051	0052	0053	0054	0055	0056	0057	0058	0059	0060	0061	0062	0063
04_	0064	0065	0066	0067	0068	0069	0070	0071	0072	0073	0074	0075	0076	0077	0078	0079
05_	0080	0081	0082	0083	0084	0085	0086	0087	0088	0089	0090	0091	0092	0093	0094	0095
06_	0096	0097	0098	0099	0100	0101	0102	0103	0104	0105	0106	0107	0108	0109	0110	0111
07_	0112	0113	0114	0115	0116	0117	0118	0119	0120	0121	0122	0123	0124	0125	0126	0127
08_	0128	0129	0130	0131	0132	0133	0134	0135	0136	0137	0138	0139	0140	0141	0142	0143
09_	0144	0145	0146	0147	0148	0149	0150	0151	0152	0153	0154	0155	0156	0157	0158	0159
0A_	0160	0161	0162	0163	0164	0165	0166	0167	0168	0169	0170	0171	0172	0173	0174	0175
0B_	0176	0177	0178	0179	0180	0181	0182	0183	0184	0185	0186	0187	0188	0189	0190	0191
0C_	0192	0193	0194	0195	0196	0197	0198	0199	0200	0201	0202	0203	0204	0205	0206	0207
0D_	0208	0209	0210	0211	0212	0213	0214	0215	0216	0217	0218	0219	0220	0221	0222	0223
0E_	0224	0225	0226	0227	0228	0229	0230	0231	0232	0233	0234	0235	0236	0237	0238	0239
0F_	0240	0241	0242	0243	0244	0245	0246	0247	0248	0249	0250	0251	0252	0253	0254	0255
10_	0256	0257	0258	0259	0260	0261	0262	0263	0264	0265	0266	0267	0268	0269	0270	0271
11_	0272	0273	0274	0275	0276	0277	0278	0279	0280	0281	0282	0283	0284	0285	0286	0287
12_	0288	0289	0290	0291	0292	0293	0294	0295	0296	0297	0298	0299	0300	0301	0302	0303
13_	0304	0305	0306	0307	0308	0309	0310	0311	0312	0313	0314	0315	0316	0317	0318	0319
14_	0320	0321	0322	0323	0324	0325	0326	0327	0328	0329	0330	0331	0332	0333	0334	0335
15_	0336	0337	0338	0339	0340	0341	0342	0343	0344	0345	0346	0347	0348	0349	0350	0351
16_	0352	0353	0354	0355	0356	0357	0358	0359	0360	0361	0362	0363	0364	0365	0366	0367
17_	0368	0369	0370	0371	0372	0373	0374	0375	0376	0377	0378	0379	0380	0381	0382	0383
18_	0384	0385	0386	0387	0388	0389	0390	0391	0392	0393	0394	0395	0396	0397	0398	0399
19_	0400	0401	0402	0403	0404	0405	0406	0407	0408	0409	0410	0411	0412	0413	0414	0415
1A_	0416	0417	0418	0419	0420	0421	0422	0423	0424	0425	0426	0427	0428	0429	0430	0431
1B_	0432	0433	0434	0435	0436	0437	0438	0439	0440	0441	0442	0443	0444	0445	0446	0447
1C_	0448	0449	0450	0451	0452	0453	0454	0455	0456	0457	0458	0459	0460	0461	0462	0463
1D_	0464	0465	0466	0467	0468	0469	0470	0471	0472	0473	0474	0475	0476	0477	0478	0479
1E_	0480	0481	0482	0483	0484	0485	0486	0487	0488	0489	0490	0491	0492	0493	0494	0495
1F_	0496	0497	0498	0499	0500	0501	0502	0503	0504	0505	0506	0507	0508	0509	0510	0511

Table 9-68. HEXADECIMAL AND DECIMAL CONVERSION *(Continued)*

	0	1	2	3	4	5	6	7	8	9	A	B	C	D	E	F
20__	0512	0513	0514	0515	0516	0517	0518	0519	0520	0521	0522	0523	0524	0525	0526	0527
21__	0528	0529	0530	0531	0532	0533	0534	0535	0536	0537	0538	0539	0540	0541	0542	0543
22__	0544	0545	0546	0547	0548	0549	0550	0551	0552	0553	0554	0555	0556	0557	0558	0559
23__	0560	0561	0562	0563	0564	0565	0566	0567	0568	0569	0570	0571	0572	0573	0574	0575
24__	0576	0577	0578	0579	0580	0581	0582	0583	0584	0585	0586	0587	0588	0589	0590	0591
25__	0592	0593	0594	0595	0596	0597	0598	0599	0600	0601	0602	0603	0604	0605	0606	0607
26__	0608	0609	0610	0611	0612	0613	0614	0615	0616	0617	0618	0619	0620	0621	0622	0623
27__	0624	0625	0626	0627	0628	0629	0630	0631	0632	0633	0634	0635	0636	0637	0638	0639
28__	0640	0641	0642	0643	0644	0645	0646	0647	0648	0649	0650	0651	0652	0653	0654	0655
29__	0656	0657	0658	0659	0660	0661	0662	0663	0664	0665	0666	0667	0668	0669	0670	0671
2A__	0672	0673	0674	0675	0676	0677	0678	0679	0680	0681	0682	0683	0684	0685	0686	0687
2B__	0688	0689	0690	0691	0692	0693	0694	0695	0696	0697	0698	0699	0700	0701	0702	0703
2C__	0704	0705	0706	0707	0708	0709	0710	0711	0712	0713	0714	0715	0716	0717	0718	0719
2D__	0720	0721	0722	0723	0724	0725	0726	0727	0728	0729	0730	0731	0732	0733	0734	0735
2E__	0736	0737	0738	0739	0740	0741	0742	0743	0744	0745	0746	0747	0748	0749	0750	0751
2F__	0752	0753	0754	0755	0756	0757	0758	0759	0760	0761	0762	0763	0764	0765	0766	0767
30__	0768	0769	0770	0771	0772	0773	0774	0775	0776	0777	0778	0779	0780	0781	0782	0783
31__	0784	0785	0786	0787	0788	0789	0790	0791	0792	0793	0794	0795	0796	0797	0798	0799
32__	0800	0801	0802	0803	0804	0805	0806	0807	0808	0809	0810	0811	0812	0813	0814	0815
33__	0816	0817	0818	0819	0820	0821	0822	0823	0824	0825	0826	0827	0828	0829	0830	0831
34__	0832	0833	0834	0835	0836	0837	0838	0839	0840	0841	0842	0843	0844	0845	0846	0847
35__	0848	0849	0850	0851	0852	0853	0854	0855	0856	0857	0858	0859	0860	0861	0862	0863
36__	0864	0865	0866	0867	0868	0869	0870	0871	0872	0873	0874	0875	0876	0877	0878	0879
37__	0880	0881	0882	0883	0884	0885	0886	0887	0888	0889	0890	0891	0892	0893	0894	0895
38__	0896	0897	0898	0899	0900	0901	0902	0903	0904	0905	0906	0907	0908	0909	0910	0911
39__	0912	0913	0914	0915	0916	0917	0918	0919	0920	0921	0922	0923	0924	0925	0926	0927
3A__	0928	0929	0930	0931	0932	0933	0934	0935	0936	0937	0938	0939	0940	0941	0942	0943
3B__	0944	0945	0946	0947	0948	0949	0950	0951	0952	0953	0954	0955	0956	0957	0958	0959
3C__	0960	0961	0962	0963	0964	0965	0966	0967	0968	0969	0970	0971	0972	0973	0974	0975
3D__	0976	0977	0978	0979	0980	0981	0982	0983	0984	0985	0986	0987	0988	0989	0990	0991
3E__	0992	0993	0994	0995	0996	0997	0998	0999	1000	1001	1002	1003	1004	1005	1006	1007
3F__	1008	1009	1010	1011	1012	1013	1014	1015	1016	1017	1018	1019	1020	1021	1022	1023
40__	1024	1025	1026	1027	1028	1029	1030	1031	1032	1033	1034	1035	1036	1037	1038	1039
41__	1040	1041	1042	1043	1044	1045	1046	1047	1048	1049	1050	1051	1052	1053	1054	1055
42__	1056	1057	1058	1059	1060	1061	1062	1063	1064	1065	1066	1067	1068	1069	1070	1071
43__	1072	1073	1074	1075	1076	1077	1078	1079	1080	1081	1082	1083	1084	1085	1086	1087
44__	1088	1089	1090	1091	1092	1093	1094	1095	1096	1097	1098	1099	1100	1101	1102	1103
45__	1104	1105	1106	1107	1108	1109	1110	1111	1112	1113	1114	1115	1116	1117	1118	1119
46__	1120	1121	1122	1123	1124	1125	1126	1127	1128	1129	1130	1131	1132	1133	1134	1135
47__	1136	1137	1138	1139	1140	1141	1142	1143	1144	1145	1146	1147	1148	1149	1150	1151
48__	1152	1153	1154	1155	1156	1157	1158	1159	1160	1161	1162	1163	1164	1165	1166	1167
49__	1168	1169	1170	1171	1172	1173	1174	1175	1176	1177	1178	1179	1180	1181	1182	1183
4A__	1184	1185	1186	1187	1188	1189	1190	1191	1192	1193	1194	1195	1196	1197	1198	1199
4B__	1200	1201	1202	1203	1204	1205	1206	1207	1208	1209	1210	1211	1212	1213	1214	1215
4C__	1216	1217	1218	1219	1220	1221	1222	1223	1224	1225	1226	1227	1228	1229	1230	1231
4D__	1232	1233	1234	1235	1236	1237	1238	1239	1240	1241	1242	1243	1244	1245	1246	1247
4E__	1248	1249	1250	1251	1252	1253	1254	1255	1256	1257	1258	1259	1260	1261	1262	1263
4F__	1264	1265	1266	1267	1268	1269	1270	1271	1272	1273	1274	1275	1276	1277	1278	1279
50__	1280	1281	1282	1283	1284	1285	1286	1287	1288	1289	1290	1291	1292	1293	1294	1295
51__	1296	1297	1298	1299	1300	1301	1302	1303	1304	1305	1306	1307	1308	1309	1310	1311
52__	1312	1313	1314	1315	1316	1317	1318	1319	1320	1321	1322	1323	1324	1325	1326	1327
53__	1328	1329	1330	1331	1332	1333	1334	1335	1336	1337	1338	1339	1340	1341	1342	1343
54__	1344	1345	1346	1347	1348	1349	1350	1351	1352	1353	1354	1355	1356	1357	1358	1359
55__	1360	1361	1362	1363	1364	1365	1366	1367	1368	1369	1370	1371	1372	1373	1374	1375
56__	1376	1377	1378	1379	1380	1381	1382	1383	1384	1385	1386	1387	1388	1389	1390	1391
57__	1392	1393	1394	1395	1396	1397	1398	1399	1400	1401	1402	1403	1404	1405	1406	1407
58__	1408	1409	1410	1411	1412	1413	1414	1415	1416	1417	1418	1419	1420	1421	1422	1423
59__	1424	1425	1426	1427	1428	1429	1430	1431	1432	1433	1434	1435	1436	1437	1438	1439
5A__	1440	1441	1442	1443	1444	1445	1446	1447	1448	1449	1450	1451	1452	1453	1454	1455
5B__	1456	1457	1458	1459	1460	1461	1462	1463	1464	1465	1466	1467	1468	1469	1470	1471
5C__	1472	1473	1474	1475	1476	1477	1478	1479	1480	1481	1482	1483	1484	1485	1486	1487
5D__	1488	1489	1490	1491	1492	1493	1494	1495	1496	1497	1498	1499	1500	1501	1502	1503
5E__	1504	1505	1506	1507	1508	1509	1510	1511	1512	1513	1514	1515	1516	1517	1518	1519
5F__	1520	1521	1522	1523	1524	1525	1526	1527	1528	1529	1530	1531	1532	1533	1534	1535

Table 9-68. HEXADECIMAL AND DECIMAL CONVERSION *(Continued)*

	0	1	2	3	4	5	6	7	8	9	A	B	C	D	E	F
60__	1536	1537	1538	1539	1540	1541	1542	1543	1544	1545	1546	1547	1548	1549	1550	1551
61__	1552	1553	1554	1555	1556	1557	1558	1559	1560	1561	1562	1563	1564	1565	1566	1567
62__	1568	1569	1570	1571	1572	1573	1574	1575	1576	1577	1578	1579	1580	1581	1582	1583
63__	1584	1585	1586	1587	1588	1589	1590	1591	1592	1593	1594	1595	1596	1597	1598	1599
64__	1600	1601	1602	1603	1604	1605	1606	1607	1608	1609	1610	1611	1612	1613	1614	1615
65__	1616	1617	1618	1619	1620	1621	1622	1623	1624	1625	1626	1627	1628	1629	1630	1631
66__	1632	1633	1634	1635	1636	1637	1638	1639	1640	1641	1642	1643	1644	1645	1646	1647
67__	1648	1649	1650	1651	1652	1653	1654	1655	1656	1657	1658	1659	1660	1661	1662	1663
68__	1664	1665	1666	1667	1668	1669	1670	1671	1672	1673	1674	1675	1676	1677	1678	1679
69__	1680	1681	1682	1683	1684	1685	1686	1687	1688	1689	1690	1691	1692	1693	1694	1695
6A__	1696	1697	1698	1699	1700	1701	1702	1703	1704	1705	1706	1707	1708	1709	1710	1711
6B__	1712	1713	1714	1715	1716	1717	1718	1719	1720	1721	1722	1723	1724	1725	1726	1727
6C__	1728	1729	1730	1731	1732	1733	1734	1735	1736	1737	1738	1739	1740	1741	1742	1743
6D__	1744	1745	1746	1747	1748	1749	1750	1751	1752	1753	1754	1755	1756	1757	1758	1759
6E__	1760	1761	1762	1763	1764	1765	1766	1767	1768	1769	1770	1771	1772	1773	1774	1775
6F__	1776	1777	1778	1779	1780	1781	1782	1783	1784	1785	1786	1787	1788	1789	1790	1791
70__	1792	1793	1794	1795	1796	1797	1798	1799	1800	1801	1802	1803	1804	1805	1806	1807
71	1808	1809	1810	1811	1812	1813	1814	1815	1816	1817	1818	1819	1820	1821	1822	1823
72__	1824	1825	1826	1827	1828	1829	1830	1831	1832	1833	1834	1835	1836	1837	1838	1839
73__	1840	1841	1842	1843	1844	1845	1846	1847	1848	1849	1850	1851	1852	1853	1854	1855
74__	1856	1857	1858	1859	1860	1861	1862	1863	1864	1865	1866	1867	1868	1869	1870	1871
75__	1872	1873	1874	1875	1876	1877	1878	1879	1880	1881	1882	1883	1884	1885	1886	1887
76__	1888	1889	1890	1891	1892	1893	1894	1895	1896	1897	1898	1899	1900	1901	1902	1903
77__	1904	1905	1906	1907	1908	1909	1910	1911	1912	1913	1914	1915	1916	1917	1918	1919
78__	1920	1921	1922	1923	1924	1925	1926	1927	1928	1929	1930	1931	1932	1933	1934	1935
79__	1936	1937	1938	1939	1940	1941	1942	1943	1944	1945	1946	1947	1948	1949	1950	1951
7A__	1952	1953	1954	1955	1956	1957	1958	1959	1960	1961	1962	1963	1964	1965	1966	1967
7B__	1968	1969	1970	1971	1972	1973	1974	1975	1976	1977	1978	1979	1980	1981	1982	1983
7C__	1984	1985	1986	1987	1988	1989	1990	1991	1992	1993	1994	1995	1996	1997	1998	1999
7D__	2000	2001	2002	2003	2004	2005	2006	2007	2008	2009	2010	2011	2012	2013	2014	2015
7E__	2016	2017	2018	2019	2020	2021	2022	2023	2024	2025	2026	2027	2028	2029	2030	2031
7F__	2032	2033	2034	2035	2036	2037	2038	2039	2040	2041	2042	2043	2044	2045	2046	2047
80__	2048	2049	2050	2051	2052	2053	2054	2055	2056	2057	2058	2059	2060	2061	2062	2063
81__	2064	2065	2066	2067	2068	2069	2070	2071	2072	2073	2074	2075	2076	2077	2078	2079
82__	2080	2081	2082	2083	2084	2085	2086	2087	2088	2089	2090	2091	2092	2093	2094	2095
83__	2096	2097	2098	2099	2100	2101	2102	2103	2104	2105	2106	2107	2108	2109	2110	2111
84	2112	2113	2114	2115	2116	2117	2118	2119	2120	2121	2122	2123	2124	2125	2126	2127
85__	2128	2129	2130	2131	2132	2133	2134	2135	2136	2137	2138	2139	2140	2141	2142	2143
86__	2144	2145	2146	2147	2148	2149	2150	2151	2152	2153	2154	2155	2156	2157	2158	2159
87__	2160	2161	2162	2163	2164	2165	2166	2167	2168	2169	2170	2171	2172	2173	2174	2175
88__	2176	2177	2178	2179	2180	2181	2182	2183	2184	2185	2186	2187	2188	2189	2190	2191
89__	2192	2193	2194	2195	2196	2197	2198	2199	2200	2201	2202	2203	2204	2205	2206	2207
8A__	2208	2209	2210	2211	2212	2213	2214	2215	2216	2217	2218	2219	2220	2221	2222	2223
8B__	2224	2225	2226	2227	2228	2229	2230	2231	2232	2233	2234	2235	2236	2237	2238	2239
8C__	2240	2241	2242	2243	2244	2245	2246	2247	2248	2249	2250	2251	2252	2253	2254	2255
8D__	2256	2257	2258	2259	2260	2261	2262	2263	2264	2265	2266	2267	2268	2269	2270	2271
8E__	2272	2273	2274	2275	2276	2277	2278	2279	2280	2281	2282	2283	2284	2285	2286	2287
8F__	2288	2289	2290	2291	2292	2293	2294	2295	2296	2297	2298	2299	2300	2301	2302	2303
90__	2304	2305	2306	2307	2308	2309	2310	2311	2312	2313	2314	2315	2316	2317	2318	2319
91__	2320	2321	2322	2323	2324	2325	2326	2327	2328	2329	2330	2331	2332	2333	2334	2335
92__	2336	2337	2338	2339	2340	2341	2342	2343	2344	2345	2346	2347	2348	2349	2350	2351
93__	2352	2353	2354	2355	2356	2357	2358	2359	2360	2361	2362	2363	2364	2365	2366	2367
94__	2368	2369	2370	2371	2372	2373	2374	2375	2376	2377	2378	2379	2380	2381	2382	2383
95__	2384	2385	2386	2387	2388	2389	2390	2391	2392	2393	2394	2395	2396	2397	2398	2399
96__	2400	2401	2402	2403	2404	2405	2406	2407	2408	2409	2410	2411	2412	2413	2414	2415
97__	2416	2417	2418	2419	2420	2421	2422	2423	2424	2425	2426	2427	2428	2429	2430	2431
98__	2432	2433	2434	2435	2436	2437	2438	2439	2440	2441	2442	2443	2444	2445	2446	2447
99__	2448	2449	2450	2451	2452	2453	2454	2455	2456	2457	2458	2459	2460	2461	2462	2463
9A__	2464	2465	2466	2467	2468	2469	2470	2471	2472	2473	2474	2475	2476	2477	2478	2479
9B__	2480	2481	2482	2483	2484	2485	2486	2487	2488	2489	2490	2491	2492	2493	2494	2495
9C__	2496	2497	2498	2499	2500	2501	2502	2503	2504	2505	2506	2507	2508	2509	2510	2511
9D__	2512	2513	2514	2515	2516	2517	2518	2519	2520	2521	2522	2523	2524	2525	2526	2527
9E__	2528	2529	2530	2531	2532	2533	2534	2535	2536	2537	2538	2539	2540	2541	2542	2543
9F__	2544	2545	2546	2547	2548	2549	2550	2551	2552	2553	2554	2555	2556	2557	2558	2559

Table 9-68. HEXADECIMAL AND DECIMAL CONVERSION *(Continued)*

	0	1	2	3	4	5	6	7	8	9	A	B	C	D	E	F
A0__	2560	2561	2562	2563	2564	2565	2566	2567	2568	2569	2570	2571	2572	2573	2574	2575
A1__	2576	2577	2578	2579	2580	2581	2582	2583	2584	2585	2586	2587	2588	2589	2590	2591
A2__	2592	2593	2594	2595	2596	2597	2598	2599	2600	2601	2602	2603	2604	2605	2606	2607
A3__	2608	2609	2610	2611	2612	2613	2614	2615	2616	2617	2618	2619	2620	2621	2622	2623
A4__	2624	2625	2626	2627	2628	2629	2630	2631	2632	2633	2634	2635	2636	2637	2638	2639
A5__	2640	2641	2642	2643	2644	2645	2646	2647	2648	2649	2650	2651	2652	2653	2654	2655
A6__	2656	2657	2658	2659	2660	2661	2662	2663	2664	2665	2666	2667	2668	2669	2670	2671
A7__	2672	2673	2674	2675	2676	2677	2678	2679	2680	2681	2682	2683	2684	2685	2686	2687
A8__	2688	2689	2690	2691	2692	2693	2694	2695	2696	2697	2698	2699	2700	2701	2702	2703
A9__	2704	2705	2706	2707	2708	2709	2710	2711	2712	2713	2714	2715	2716	2717	2718	2719
AA__	2720	2721	2722	2723	2724	2725	2726	2727	2728	2729	2730	2731	2732	3733	2734	2735
AB__	2736	2737	2738	2739	2740	2741	2742	2743	2744	2745	2746	2747	2748	2749	2750	2751
AC__	2752	2753	2754	2755	2756	2757	2758	2759	2760	2761	2762	2763	2764	2765	2766	2767
AD__	2768	2769	2770	2771	2772	2773	2774	2775	2776	2777	2778	2779	2780	2781	2782	2783
AE__	2784	2785	2786	2787	2788	2789	2790	2791	2792	2793	2794	2795	2796	2797	2798	2799
AF__	2800	2801	2802	2803	2804	2805	2806	2807	2808	2809	2810	2811	2812	2813	2814	2815
B0__	2816	2817	2818	2819	2820	2821	2822	2823	2824	2825	2826	2827	2828	2829	2830	2831
B1__	2832	2833	2834	2835	2836	2837	2838	2839	2840	2841	2842	2843	2844	2845	2846	2847
B2__	2848	2849	2850	2851	2852	2853	2854	2855	2856	2857	2858	2859	2860	2861	2862	2863
B3__	2864	2865	2866	2867	2868	2869	2870	2871	2872	2873	2874	2875	2876	2877	2878	2879
B4__	2880	2881	2882	2883	2884	2885	2886	2887	2888	2889	2890	2891	2892	2893	2894	2895
B5__	2896	2897	2898	2899	2900	2901	2902	2903	2904	2905	2906	2907	2908	2909	2910	2911
B6__	2912	2913	2914	2915	2916	2917	2918	2919	2920	2921	2922	2923	2924	2925	2926	2927
B7__	2928	2929	2930	2931	2932	2933	2934	2935	2936	2937	2938	2939	2940	2941	2942	2943
B8__	2944	2945	2946	2947	2948	2949	2950	2951	2952	2953	2954	2955	2956	2957	2958	2959
B9__	2960	2961	2962	2963	2964	2965	2966	2967	2968	2969	2970	2971	2972	2973	2974	2975
BA__	2976	2977	2978	2979	2980	2981	2982	2983	2984	2985	2986	2987	2988	2989	2990	2991
BB__	2992	2993	2994	2995	2996	2997	2998	2999	3000	3001	3002	3003	3004	3005	3006	3007
BC__	3008	3009	3010	3011	3012	3013	3014	3015	3016	3017	3018	3019	3020	3021	3022	3023
BD__	3024	3025	3026	3027	3028	3029	3030	3031	3032	3033	3034	3035	3036	3037	3038	3039
BE__	3040	3041	3042	3043	3044	3045	3046	3047	3048	3049	3050	3051	3052	3053	3054	3055
BF__	3056	3057	3058	3059	3060	3061	3062	3063	3064	3065	3066	3067	3068	3069	3070	3071
C0__	3072	3073	3074	3075	3076	3077	3078	3079	3080	3081	3082	3083	3084	3085	3086	3087
C1__	3088	3089	3090	3091	3092	3093	3094	3095	3096	3097	3098	3099	3100	3101	3102	3103
C2__	3104	3105	3106	3107	3108	3109	3110	3111	3112	3113	3114	3115	3116	3117	3118	3119
C3__	3120	3121	3122	3123	3124	3125	3126	3127	3128	3129	3130	3131	3132	3133	3134	3135
C4__	3136	3137	3138	3139	3140	3141	3142	3143	3144	3145	3146	3147	3148	3149	3150	3151
C5__	3152	3153	3154	3155	3156	3157	3158	3159	3160	3161	3162	3613	3164	3165	3166	3167
C6__	3168	3169	3170	3171	3172	3173	3174	3175	3176	3177	3178	3179	3180	3181	3182	3183
C7__	3184	3185	3186	3187	3188	3189	3190	3191	3192	3193	3194	3195	3196	3197	3198	3199
C8__	3200	3201	3202	3203	3204	3205	3206	3207	3208	3209	3210	3211	3212	3213	3214	3215
C9__	3216	3217	3218	3219	3220	3221	3222	3223	3224	3225	3226	3227	3228	3229	3230	3231
CA__	3232	3233	3234	3235	3236	3237	3238	3239	3240	3241	3242	3243	3244	3245	3246	3247
CB__	3248	3249	3250	3251	3252	3253	3254	3255	3256	3257	3258	3259	3260	3261	3262	3263
CC__	3264	3265	3266	3267	3268	3269	3270	3271	3272	3273	3274	3275	3276	3277	3278	3279
CD__	3280	3281	3282	3283	3284	3285	3286	3287	3288	3289	3290	3291	3292	3293	3294	3295
CE__	3296	3297	3298	3299	3300	3301	3302	3303	3304	3305	3306	3307	3308	3309	3310	3311
CF__	3312	3313	3314	3315	3316	3317	3318	3319	3320	3321	3322	3323	3324	3325	3326	3327
D0__	3328	3329	3330	3331	3332	3333	3334	3335	3336	3337	3338	3339	3340	3341	3342	3343
D1__	3344	3345	3346	3347	3348	3349	3350	3351	3352	3353	3354	3355	3356	3357	3358	3359
D2__	3360	3361	3362	3363	3364	3365	3366	3367	3368	3369	3370	3371	3372	3373	3374	3375
D3__	3376	3377	3378	3379	3380	3381	3382	3383	3384	3385	3386	3387	3388	3389	3390	3391
D4__	3392	3393	3394	3395	3396	3397	3398	3399	3400	3401	3402	3403	3404	3405	3406	3407
D5__	3408	3409	3410	3411	3412	3413	3414	3415	3416	3417	3418	3419	3420	3421	3422	3423
D6__	3424	3425	3426	3427	3428	3429	3430	3431	3432	3433	3434	3435	3436	3437	3438	3439
D7__	3440	3441	3442	3443	3444	3445	3446	3447	3448	3449	3450	3451	3452	3453	3454	3455
D8__	3456	3457	3458	3459	3460	3461	3462	3463	3464	3465	3466	3467	3468	3469	3470	3471
D9__	3472	3473	3474	3475	3476	3477	3478	3479	3480	3481	3482	3483	3484	3485	3486	3487
DA__	3488	3489	3490	3491	3492	3493	3494	3495	3496	3497	3498	3499	3500	3501	3502	3503
DB__	3504	3505	3506	3507	3508	3509	3510	3511	3512	3513	3514	3515	3516	3517	3518	3519
DC__	3520	3521	3522	3523	3524	3525	3526	3527	3528	3529	3530	3531	3532	3533	3534	3535
DD__	3536	3537	3538	3539	3540	3541	3542	3543	3544	3545	3546	3547	3548	3549	3550	3551
DE__	3552	3553	3554	3555	3556	3557	3558	3559	3560	3561	3562	3563	3564	3565	3566	3567
DF__	3568	3569	3570	3571	3572	3573	3574	3575	3576	3577	3578	3579	3580	3581	3582	3583

Table 9-68.   HEXADECIMAL AND DECIMAL CONVERSION *(Continued)*

	0	1	2	3	4	5	6	7	8	9	A	B	C	D	E	F
E0__	3584	3585	3586	3587	3588	3589	3590	3591	3592	3593	3594	3595	3596	3597	3598	3599
E1__	3600	3601	3602	3603	3604	3605	3606	3607	3608	3609	3610	3611	3612	3613	3614	3615
E2__	3616	3617	3618	3619	3620	3621	3622	3623	3624	3625	3626	3627	3628	3629	3630	3631
E3__	3632	3633	3634	3635	3636	3637	3638	3639	3640	3641	3642	3643	3644	3645	3646	3647
E4__	3648	3649	3650	3651	3652	3653	3654	3655	3656	3657	3658	3659	3660	3661	3662	3663
E5__	3664	3665	3666	3667	3668	3669	3670	3671	3672	3673	3674	3675	3676	3677	3678	3679
E6__	3680	3681	3682	3683	3684	3685	3686	3687	3688	3689	3690	3691	3692	3693	3694	3695
E7__	3696	3697	3698	3699	3700	3701	3702	3703	3704	3705	3706	3707	3708	3709	3710	3711
E8__	3712	3713	3714	3715	3716	3717	3718	3719	3720	3721	3722	3723	3724	3725	3726	3727
E9__	3728	3729	3730	3731	3732	3733	3734	3735	3736	3737	3738	3739	3740	3741	3742	3743
EA__	3744	3745	3746	3747	3748	3749	3750	3751	3752	3753	3754	3755	3756	3757	3758	3759
EB__	3760	3761	3762	3763	3764	3765	3766	3767	3768	3769	3770	3771	3772	3773	3774	3775
EC__	3776	3777	3778	3779	3780	3781	3782	3783	3784	3785	3786	3787	3788	3789	3790	3791
ED__	3792	3793	3794	3795	3796	3797	3798	3799	3800	3801	3802	3803	3804	3805	3806	3807
EE__	3808	3809	3810	3811	3812	3813	3814	3815	3816	3817	3818	3819	3820	3821	3822	3823
EF__	3824	3825	3826	3827	3828	3829	3830	3831	3832	3833	3834	3835	3836	3837	3838	3839
F0__	3840	3841	3842	3843	3844	3845	3846	3847	3848	3849	3850	3851	3852	3853	3854	3855
F1__	3856	3857	3858	3859	3860	3861	3862	3863	3864	3865	3866	3867	3868	3869	3870	3871
F2__	3872	3873	3874	3875	3876	3877	3878	3879	3880	3881	3882	3883	3884	3885	3886	3887
F3__	3888	3889	3890	3891	3892	3893	3894	3895	3896	3897	3898	3899	3900	3901	3902	3903
F4__	3904	3905	3906	3907	3908	3909	3910	3911	3912	3913	3914	3915	3916	3917	3918	3919
F5__	3920	3921	3922	3923	3924	3925	3926	3927	3928	3929	3930	3931	3932	3933	3934	3935
F6__	3936	3937	3938	3939	3940	3941	3942	3943	3944	3945	3946	3947	3948	3949	3950	3951
F7__	3952	3953	3954	3955	3956	3957	3958	3959	3960	3961	3962	3963	3964	3965	3966	3967
F8__	3968	3969	3970	3971	3972	3973	3974	3975	3976	3977	3978	3979	3980	3981	3982	3983
F9__	3984	3985	3986	3987	3988	3989	3990	3991	3992	3993	3994	3995	3996	3997	3998	3999
FA__	4000	4001	4002	4003	4004	4005	4006	4007	4008	4009	4010	4011	4012	4013	4014	4015
FB__	4016	4017	4018	4019	4020	4021	4022	4023	4024	4025	4026	4027	4028	4029	4030	4031
FC__	4032	4033	4034	4035	4036	4037	4038	4039	4040	4041	4042	4043	4044	4045	4046	4047
FD__	4048	4049	4050	4051	4052	4053	4054	4055	4056	4057	4058	4059	4060	4061	4062	4063
FE__	4064	4065	4066	4067	4068	4069	4070	4071	4072	4073	4074	4075	4076	4077	4078	4079
FF__	4080	4081	4082	4083	4084	4085	4086	4087	4088	4089	4090	4091	4092	4093	4094	4095

*From: "CRC Handbook of Tables for Mathematics", 4th ed., S.M. Selby, Ed., The Chemical Rubber Co., 1970.

Table 9-69. HEXADECIMAL AND DECIMAL INTEGER CONVERSION*

	8		7		6		5		4		3		2		1
Hex	Decimal	Hex	Decimal	Hex	Decimal	Hex	Decimal	Hex	Decimal	Hex	Decimal	Hex	Decimal	Hex	Decimal
0	0	0	0	0	0	0	0	0	0	0	0	0	0	0	0
1	268,435,456	1	16,777,216	1	1,048,576	1	65,536	1	4,096	1	256	1	16	1	1
2	536,870,912	2	33,554,432	2	2,097,152	2	131,072	2	8,192	2	512	2	32	2	2
3	805,306,368	3	50,331,648	3	3,145,728	3	196,608	3	12,288	3	768	3	48	3	3
4	1,073,741,824	4	67,108,864	4	4,194,304	4	262,144	4	16,384	4	1,024	4	64	4	4
5	1,342,177,280	5	83,886,080	5	5,242,880	5	327,680	5	20,480	5	1,280	5	80	5	5
6	1,610,612,736	6	100,663,296	6	6,291,456	6	393,216	6	24,576	6	1,536	6	96	6	6
7	1,879,048,192	7	117,440,512	7	7,340,032	7	458,752	7	28,672	7	1,792	7	112	7	7
8	2,147,483,648	8	134,217,728	8	8,388,608	8	524,288	8	32,768	8	2,048	8	128	8	8
9	2,415,919,104	9	150,994,944	9	9,437,184	9	589,824	9	36,864	9	2,304	9	144	9	9
A	2,684,354,560	A	167,772,160	A	10,485,760	A	655,360	A	40,960	A	2,560	A	160	A	10
B	2,952,790,016	B	184,549,376	B	11,534,336	B	720,896	B	45,056	B	2,816	B	176	B	11
C	3,221,225,472	C	201,326,592	C	12,582,912	C	786,432	C	49,152	C	3,072	C	192	C	12
D	3,489,660,928	D	218,103,808	D	13,631,488	D	851,968	D	53,248	D	3,328	D	208	D	13
E	3,758,096,384	E	234,881,024	E	14,680,064	E	917,504	E	57,344	E	3,584	E	224	E	14
F	4,026,531,840	F	251,658,240	F	15,728,640	F	983,040	F	61,440	F	3,840	F	240	F	15
	8		7		6		5		4		3		2		1

TO CONVERT HEXADECIMAL TO DECIMAL

1. Locate the column of decimal numbers corresponding to the left-most digit or letter of the hexadecimal; select from this column and record the number that corresponds to the position of the hexadecimal digit or letter.
2. Repeat step 1 for the next (second from the left) position.
3. Repeat step 1 for the units (third from the left) position.
4. Add the numbers selected from the table to form the decimal number.

To convert integer numbers greater than the capacity of the table, use the techniques below.
    Successive cumulative multiplication from left to right, adding units position.

Example: $D34_{16} = 3380_{10}$

$$
\begin{array}{rr}
D = & 13 \\
 & \times 16 \\
\hline
 & 208 \\
3 = & +3 \\
\hline
 & 211 \\
 & \times 16 \\
\hline
 & 3376 \\
4 = & +4 \\
\hline
 & 3380
\end{array}
$$

EXAMPLE
Conversion of Hexadecimal
Value   D34
1. D   3328
2. 3   48
3. 4   4
4. Decimal   3380

TO CONVERT DECIMAL TO HEXADECIMAL

1. (a) Select from the table the highest decimal number that is equal to or less than the number to be converted.
    (b) Record the hexadecimal of the column containing the selected number.
    (c) Subtract the selected decimal from the number to be converted.
2. Using the remainder from step 1(c), repeat all of step 1 to develop the second position of the hexadecimal (and a remainder).
3. Using the remainder from step 2, repeat all of step 1 to develop the units position of the hexadecimal.
4. Combine terms to form the hexadecimal number.
Divide and collect the remainder in reverse order.

Example: $3380_{10} = X_{16}$

$$
\begin{array}{r}
16 \,|\, 3380 \quad \text{remainder} \\
16 \,|\, 211 \quad\searrow 4 \\
16 \,|\, 13 \quad\searrow 3 \\
\quad\searrow D
\end{array}
$$

$3380_{10} = D34_{16}$

EXAMPLE
Conversion of Decimal
Value   3380
1. D   −3328
52
2. 3   −48
4
3. 4   −4
4. Hexa- decimal   D34

*From: "CRC Handbook of Tables for Mathematics", 4th ed., S.M. Selby, Ed., The Chemical Rubber Co., 1970.

Table 9-70.   HEXADECIMAL AND DECIMAL FRACTION CONVERSION*

	1		2		3				4			
Hex	Decimal	Hex	Decimal		Hex	Decimal			Hex	Decimal Equivalent		
.0	.0000	.00	.0000	0000	.000	.0000	0000	0000	.0000	.0000	0000	0000	0000
.1	.0625	.01	.0039	0625	.001	.0002	4414	0625	.0001	.0000	1525	8789	0625
.2	.1250	.02	.0078	1250	.002	.0004	8828	1250	.0002	.0000	3051	7578	1250
.3	.1875	.03	.0117	1875	.003	.0007	3242	1875	.0003	.0000	4577	6367	1875
.4	.2500	.04	.0156	2500	.004	.0009	7656	2500	.0004	.0000	6103	5156	2500
.5	.3125	.05	.0195	3125	.005	.0012	2070	3125	.0005	.0000	7629	3945	3125
.6	.3750	.06	.0234	3750	.006	.0014	6484	3750	.0006	.0000	9155	2734	3750
.7	.4375	.07	.0273	4375	.007	.0017	0898	4375	.0007	.0001	0681	1523	4375
.8	.5000	.08	.0312	5000	.008	.0019	5312	5000	.0008	.0001	2207	0312	5000
.9	.5625	.09	.0351	5625	.009	.0021	9726	5625	.0009	.0001	3732	9101	5625
.A	.6250	.0A	.0390	6250	.00A	.0024	4140	6250	.000A	.0001	5258	7890	6250
.B	.6875	.0B	.0429	6875	.00B	.0026	8554	6875	.000B	.0001	6784	6679	6875
.C	.7500	.0C	.0468	7500	.00C	.0029	2968	7500	.000C	.0001	8310	5468	7500
.D	.8125	.0D	.0507	8125	.00D	.0031	7382	8125	.000D	.0001	9836	4257	8125
.E	.8750	.0E	.0546	8750	.00E	.0034	1796	8750	.000E	.0002	1362	3046	8750
.F	.9375	.0F	.0585	9375	.00F	.0036	6210	9375	.000F	.0002	2888	1835	9375
	1		2			3				4		

TO CONVERT .ABC HEXADECIMAL TO DECIMAL

Find .A in position 1   .6250
Find .0B in position 2   .0429 6875
Find .00C in position 3   .0029 2968 7500
.ABC Hex is equal to   .6708 9843 7500

TO CONVERT .13 DECIMAL TO HEXADECIMAL

1. Find .1250 next lowest to   .1300
   Subtract   −.1250   = .2 Hex
2. Find .0039 0625 next lowest to   .0050 0000
                                     −.0039 0625   = .01
3. Find .0009 7656 2500   .0010 9375 0000
                                     −.0009 7656 2500   = .004
4. Find .0001 0681 1523 4375   .0001 1718 7500 0000
                                     −.0001 0681 1523 4375 = .0007
                                      .0000 1037 5976 5625 = .2147 Hex

5. .13 Decimal is approximately equal to ⟶↑

To convert fractions beyond the capacity of table, use the techniques below.

HEXADECIMAL FRACTION TO DECIMAL

Convert the hexadecimal fraction to its decimal equivalent using the same technique as for integer numbers. Divide the results by 16^n (n is the number of fraction positions).

Example: $.8A7_{16} = .540771_{10}$

$8A7_{16} = 2215_{10}$

$16^3 = 4096$ $4096 \overline{)2215.000000}$ .540771

DECIMAL FRACTION TO HEXADECIMAL

Collect integer parts of product in the order of calculation.

Example: $.5408_{10} = .8A7_{16}$

        .5408
      × 16
8 ← 　[8] .6528
      × 16
A ← [10] .4448
      × 16
7 ← 　[7] .1168

*From: "CRC Handbook of Tables for Mathematics", 4th ed., S.M. Selby, Ed., The Chemical Rubber Co., 1970.

Table 9-71.   HEXADECIMAL ADDITION AND SUBTRACTION*

Example: 6 + 2 = 8,   8 − 2 = 6,   and   8 − 6 = 2

	1	2	3	4	5	6	7	8	9	A	B	C	D	E	F
1	02	03	04	05	06	07	08	09	0A	0B	0C	0D	0E	0F	10
2	03	04	05	06	07	08	09	0A	0B	0C	0D	0E	0F	10	11
3	04	05	06	07	08	09	0A	0B	0C	0D	0E	0F	10	11	12
4	05	06	07	08	09	0A	0B	0C	0D	0E	0F	10	11	12	13
5	06	07	08	09	0A	0B	0C	0D	0E	0F	10	11	12	13	14
6	07	08	09	0A	0B	0C	0D	0E	0F	10	11	12	13	14	15
7	08	09	0A	0B	0C	0D	0E	0F	10	11	12	13	14	15	16
8	09	0A	0B	0C	0D	0E	0F	10	11	12	13	14	15	16	17
9	0A	0B	0C	0D	0E	0F	10	11	12	13	14	15	16	17	18
A	0B	0C	0D	0E	0F	10	11	12	13	14	15	16	17	18	19
B	0C	0D	0E	0F	10	11	12	13	14	15	16	17	18	19	1A
C	0D	0E	0F	10	11	12	13	14	15	16	17	18	19	1A	1B
D	0E	0F	10	11	12	13	14	15	16	17	18	19	1A	1B	1C
E	0F	10	11	12	13	14	15	16	17	18	19	1A	1B	1C	1D
F	10	11	12	13	14	15	16	17	18	19	1A	1B	1C	1D	1E

*From: "CRC Handbook of Tables for Mathematics", 4th ed., S.M. Selby, Ed., The Chemical Rubber Co., 1970.

Table 9-72.   HEXADECIMAL MULTIPLICATION*

Example: 2 × 4 = 08,   F × 2 = 1E

	1	2	3	4	5	6	7	8	9	A	B	C	D	E	F
1	01	02	03	04	05	06	07	08	09	0A	0B	0C	0D	0E	0F
2	02	04	06	08	0A	0C	0E	10	12	14	16	18	1A	1C	1E
3	03	06	09	0C	0F	12	15	18	1B	1E	21	24	27	2A	2D
4	04	08	0C	10	14	18	1C	20	24	28	2C	30	34	38	3C
5	05	0A	0F	14	19	1E	23	28	2D	32	37	3C	41	46	4B
6	06	0C	12	18	1E	24	2A	30	36	3C	42	48	4E	54	5A
7	07	0E	15	1C	23	2A	31	38	3F	46	4D	54	5B	62	69
08	08	10	18	20	28	30	38	40	48	50	58	60	68	70	78
9	09	12	1B	24	2D	36	3F	48	51	5A	63	6C	75	7E	87
A	0A	14	1E	28	32	3C	46	50	5A	64	6E	78	82	8C	96
B	0B	16	21	2C	37	42	4D	58	63	6E	79	84	8F	9A	A5
C	0C	18	24	30	3C	48	54	60	6C	78	84	90	9C	A8	B4
D	0D	1A	27	34	41	4E	5B	68	75	82	8F	9C	A9	B6	C3
E	0E	1C	2A	38	46	54	62	70	7E	8C	9A	A8	B6	C4	D2
F	0F	1E	2D	3C	4B	5A	69	78	87	96	A5	B4	C3	D2	E1

*From: "CRC Handbook of Tables for Mathematics", 4th ed., S.M. Selby, Ed., The Chemical Rubber Co., 1970.

Table 9-73.   BOOLEAN THEOREMS

$$A + 0 = A$$
$$A \cdot 1 = A$$
$$A + A = A$$
$$A \cdot A = A$$
$$A + 1 = 1$$
$$A \cdot 0 = 0$$
$$A + AB = A$$
$$\bar{\bar{A}} = A$$
$$\overline{A + B} = \bar{A}\bar{B}$$

$$\overline{AB} = \bar{A} + \bar{B}$$
$$(A + B) + C = A + (B + C)$$
$$(AB)C = A(BC)$$
$$A + \bar{A}B = A + B$$
$$A(\bar{A} + B) = AB$$
$$(A + B)(\bar{A} + C) = AC + \bar{A}B$$
$$(\overline{AC + B\bar{C}}) = \bar{A}C + \bar{B}\bar{C}$$
$$(A + C)(B + \bar{C}) = (\bar{A} + C)(\bar{B} + \bar{C})$$

EXPLANATION:

These Boolean theorems (sometimes called switching theorems) are used in problems involving binary states. The two states may be considered as functional propositions, true or false (hence the alternate name "propositional calculus"). But in physical devices, such as switches, controls, or computers, the two states may be on or off, short circuit or open circuit, high voltage or low voltage, or presence or absence of a hole in a card or tape, and the digits 1 and 0 are arbitrarily used.

In these theorems each of the variables can represent an arbitrary function. One method for manipulating forms in switching algebra is to use a map (see Table 9-76).

Since the use of symbols in Boolean algebra has not yet been fully standardized, the following is a detailed explanation of the symbols used in the above table.

SYMBOLS:

A, B, and C are variables.

The bar above the variable indicates the negation of the variable, e.g., $\bar{A}$ means "not A".

The plus sign, $+$, is used for the *or* function. This function does *not* obey the conventional arithmetical rules for sums.

The multiplication sign, $\cdot$, is used for the *and* function, sometimes called conjunction. This function obeys the conventional arithmetical rules for products. Thus if the binary values are taken arithmetically as one and zero, $1 \cdot 1 = 1$, and $1 \cdot 0 = 0$. But, in Boolean notation, $1 + 1 = 1$, which is not correct by arithmetical notation.

If a variable (e.g., a switch) can have only two states, designated as 1 or 0, it follows that $\bar{1}$ is equivalent to 0, and $\bar{0}$ is equivalent to 1.

Table 9-74.   TYPICAL LOGIC DEVICES*

Major class	Subclass	Gates	Flip-flops	Inverters	Amplifiers	Speed
Electronic	Diode logic	×				Medium
	Transistor logic†	×	×	×	×	High
	Integrated circuits	×	×	×	×	Medium-high
	Tunnel diode	×	×			Very high
	Cryotron	×	×			Medium
	Neuristor	×	×		×	Medium
	Vacuum-tube circuits	×	×	×	×	Medium-high
Relay	—	×	×	×	×	Slow
Magnetic	Magnetic-core logic	×	×	×		Medium
Mechanical		×	×	×	×	Slow
Pneumatic or fluidic	—	×	×	×	×	Slow

Note: The essential characteristic that a digital logic element must possess is that it operate in two or more well defined and distinct states. Binary devices predominate.

†Resistor-transistor logic, diode-transistor logic, transistor-transistor logic, current switching, etc.
*From: "Reference Data for Radio Engineers", 5th ed., Howard W. Sams & Co., Inc., Indianapolis, Indiana, 1968.

Figure 9-75.   APPLICATIONS AND FUNCTIONS OF TWO VARIABLES*

Table of Combinations

AND	OR	A	B	X
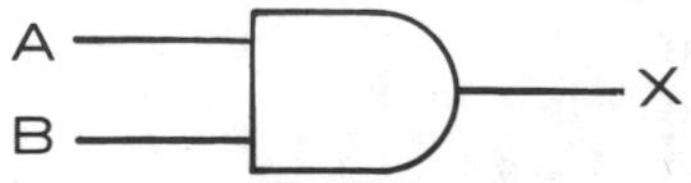 A, B → X	 A, B → X	H H L L	H L H L	H L L L
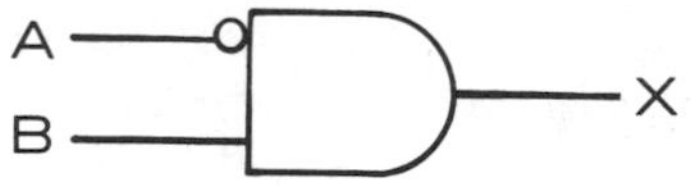 A, B → X	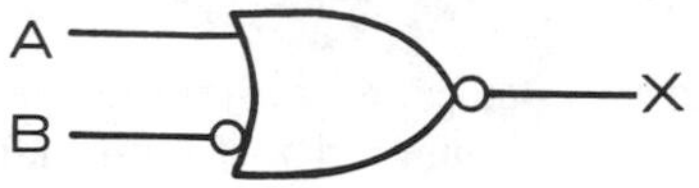 A, B → X	H H L L	H L H L	L L H L
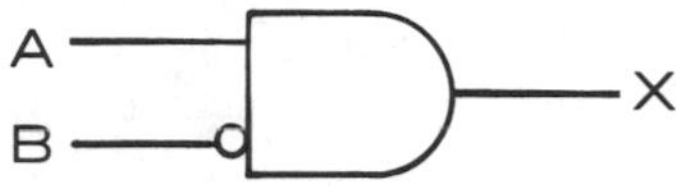 A, B → X	 A, B → X	H H L L	H L H L	L H L L
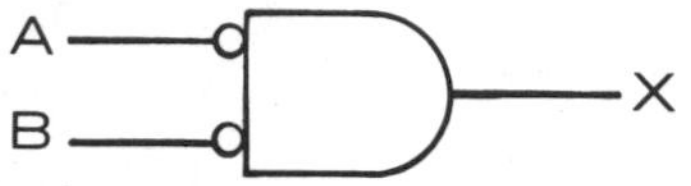 A, B → X	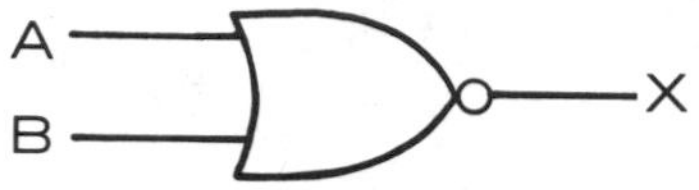 A, B → X	H H L L	H L H L	L L L H
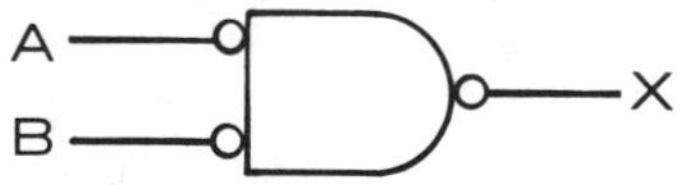 A, B → X	 A, B → X	H H L L	H L H L	H H H L
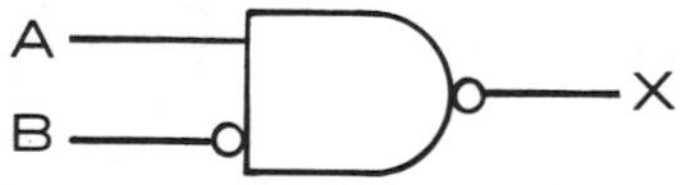 A, B → X	 A, B → X	H H L L	H L H L	H H H H
 A, B → X	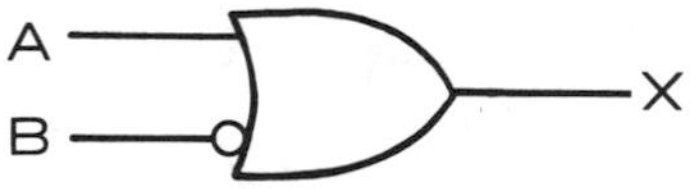 A, B → X	H H L L	H L H L	H H L H
 A, B → X	 A, B → X	H H L L	H L H L	L L H H

*From: MIL–STD 806B, February 1962.

Table 9-76. KARNAUGH MAPS*

For 2, 3, 4, 5, 6, 7, and 8 Variables

These maps are widely used for minimizing the functions of two to eight variables. The map is a pictorial description of a table of combinations, so that a map of n variables contains 2^n squares, each square representing one of the fundamental products. Digit zero in each of the maps represents the case of all variables that equal zero.

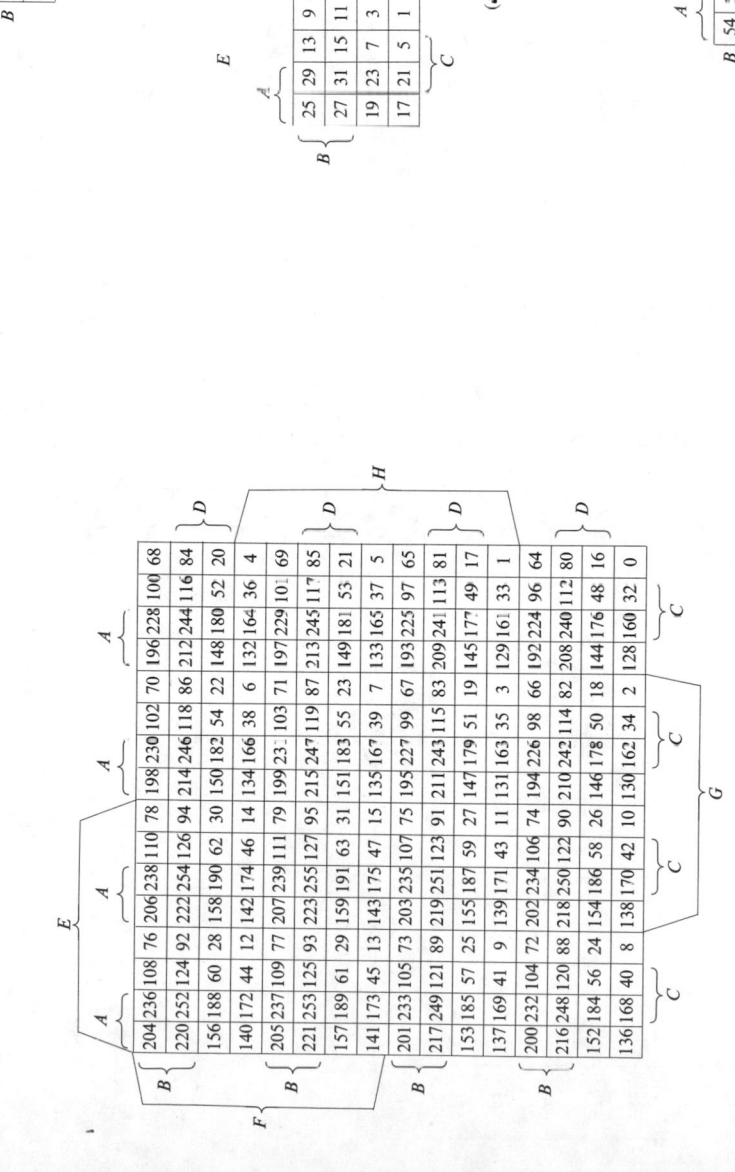

*From: "Logical Design of Digital Computers", M. Phister, Jr., John Wiley & Sons, 1963, pp. 84–85.

Table 9-77. HARVARD CHART

FOUR VARIABLES

	A	B	C	D	AB	AC	AD	BC	BD	CD	ABC	ABD	ACD	BCD	ABCD
M_0	0	0	0	0	0	0	0	0	0	0	0	0	0	0	0
M_1	0	0	0	1	0	0	1	0	1	1	0	1	1	1	1
M_2	0	0	1	0	0	1	0	1	0	2	1	0	2	2	2
M_3	0	0	1	1	0	1	1	1	1	3	1	1	3	3	3
M_4	0	1	0	0	1	0	0	2	2	0	2	2	0	4	4
M_5	0	1	0	1	1	0	1	2	3	1	2	3	1	5	5
M_6	0	1	1	0	1	1	0	3	2	2	3	2	2	6	6
M_7	0	1	1	1	1	1	1	3	3	3	3	3	3	7	7
M_8	1	0	0	0	2	2	2	0	0	0	4	4	4	0	8
M_9	1	0	0	1	2	2	3	0	1	1	4	5	5	1	9
M_{10}	1	0	1	0	2	3	2	1	0	2	5	4	6	2	10
M_{11}	1	0	1	1	2	3	3	1	1	3	5	5	7	3	11
M_{12}	1	1	0	0	3	2	2	2	2	0	6	6	4	4	12
M_{13}	1	1	0	1	3	2	3	2	3	1	6	7	5	5	13
M_{14}	1	1	1	0	3	3	2	3	2	2	7	6	6	6	14
M_{15}	1	1	1	1	3	3	3	3	3	3	7	7	7	7	15

THREE VARIABLES

	A	B	C	AB	AC	BC	ABC
M_0	0	0	0	0	0	0	0
M_1	0	0	1	0	1	1	1
M_2	0	1	0	1	0	2	2
M_3	0	1	1	1	1	3	3
M_4	1	0	0	2	2	0	4
M_5	1	0	1	2	3	1	5
M_6	1	1	0	3	2	2	6
M_7	1	1	1	3	3	3	7

Table 9-77.   HARVARD CHART *(Continued)*

FIVE VARIABLES

	A	B	C	D	E	AB	AC	AD	AE	BC	BD	BE	CD	CE	DE	ABC	ABD	ABE	ACD	ACE	ADE	BCD	BCE	BDE	CDE	ABCD	ABCE	ABDE	ACDE	BCDE	ABCDE
M_0	0	0	0	0	0	0	0	0	0	0	0	0	0	0	0	0	0	0	0	0	0	0	0	0	0	0	0	0	0	0	0
M_1	0	0	0	0	1	0	0	0	1	0	0	1	0	1	1	0	0	1	0	1	1	0	1	1	1	0	1	1	1	1	1
M_2	0	0	0	1	0	0	0	1	0	0	1	0	1	0	2	0	1	0	1	0	2	1	0	2	2	1	0	2	2	2	2
M_3	0	0	0	1	1	0	0	1	1	0	1	1	1	1	3	0	1	1	1	1	3	1	1	3	3	1	1	3	3	3	3
M_4	0	0	1	0	0	0	1	0	0	1	0	0	2	2	0	1	0	0	2	2	0	2	2	0	4	2	2	0	4	4	4
M_5	0	0	1	0	1	0	1	0	1	1	0	1	2	3	1	1	0	1	2	3	1	2	3	1	5	2	3	1	5	5	5
M_6	0	0	1	1	0	0	1	1	0	1	1	0	3	2	2	1	1	0	3	2	2	3	2	2	6	3	2	2	6	6	6
M_7	0	0	1	1	1	0	1	1	1	1	1	1	3	3	3	1	1	1	3	3	3	3	3	3	7	3	3	3	7	7	7
M_8	0	1	0	0	0	1	0	0	0	2	2	2	0	0	0	2	2	2	0	0	0	4	4	4	0	4	4	4	0	8	8
M_9	0	1	0	0	1	1	0	0	1	2	2	3	0	1	1	2	2	3	0	1	1	4	5	5	1	4	5	5	1	9	9
M_{10}	0	1	0	1	0	1	0	1	0	2	3	2	1	0	2	2	3	2	1	0	2	5	4	6	2	5	4	6	2	10	10
M_{11}	0	1	0	1	1	1	0	1	1	2	3	3	1	1	3	2	3	3	1	1	3	5	5	7	3	5	5	7	3	11	11
M_{12}	0	1	1	0	0	1	1	0	0	3	2	2	2	2	0	3	2	2	2	2	0	6	6	4	4	6	6	4	4	12	12
M_{13}	0	1	1	0	1	1	1	0	1	3	2	3	2	3	1	3	2	3	2	3	1	6	7	5	5	6	7	5	5	13	13
M_{14}	0	1	1	1	0	1	1	1	0	3	3	2	3	2	2	3	3	2	3	2	2	7	6	6	6	7	6	6	6	14	14
M_{15}	0	1	1	1	1	1	1	1	1	3	3	3	3	3	3	3	3	3	3	3	3	7	7	7	7	7	7	7	7	15	15
M_{16}	1	0	0	0	0	2	2	2	2	0	0	0	0	0	0	4	4	4	4	4	4	0	0	0	0	8	8	8	8	0	16
M_{17}	1	0	0	0	1	2	2	2	3	0	0	1	0	1	1	4	4	5	4	5	5	0	1	1	1	8	9	9	9	1	17
M_{18}	1	0	0	1	0	2	2	3	2	0	1	0	1	0	2	4	5	4	5	4	6	1	0	2	2	9	8	10	10	2	18
M_{19}	1	0	0	1	1	2	2	3	3	0	1	1	1	1	3	4	5	5	5	5	7	1	1	3	3	9	9	11	11	3	19
M_{20}	1	0	1	0	0	2	3	2	2	1	0	0	2	2	0	5	4	4	6	6	4	2	2	0	4	10	10	8	12	4	20
M_{21}	1	0	1	0	1	2	3	2	3	1	0	1	2	3	1	5	4	5	6	7	5	2	3	1	5	10	11	9	13	5	21
M_{22}	1	0	1	1	0	2	3	3	2	1	1	0	3	2	2	5	5	4	7	6	6	3	2	2	6	11	10	10	14	6	22
M_{23}	1	0	1	1	1	2	3	3	3	1	1	1	3	3	3	5	5	5	7	7	7	3	3	3	7	11	11	11	15	7	23
M_{24}	1	1	0	0	0	3	2	2	2	2	2	2	0	0	0	6	6	6	4	4	4	4	4	4	0	12	12	12	8	8	24
M_{25}	1	1	0	0	1	3	2	2	3	2	2	3	0	1	1	6	6	7	4	5	5	4	5	5	1	12	13	13	9	9	25
M_{26}	1	1	0	1	0	3	2	3	2	2	3	2	1	0	2	6	7	6	5	4	6	5	4	6	2	13	12	14	10	10	26
M_{27}	1	1	0	1	1	3	2	3	3	2	3	3	1	1	3	6	7	7	5	5	7	5	5	7	3	13	13	15	11	11	27
M_{28}	1	1	1	0	0	3	3	2	2	3	2	2	2	2	0	7	6	6	6	6	4	6	6	4	4	14	14	12	12	12	28
M_{29}	1	1	1	0	1	3	3	2	3	3	2	3	2	3	1	7	6	7	6	7	5	6	7	5	5	14	15	13	13	13	29
M_{30}	1	1	1	1	0	3	3	3	2	3	3	2	3	2	2	7	7	6	7	6	6	7	6	6	6	15	14	14	14	14	30
M_{31}	1	1	1	1	1	3	3	3	3	3	3	3	3	3	3	7	7	7	7	7	7	7	7	7	7	15	15	15	15	15	31

Measurement and Instrumentation

Table 10–1. INSTRUMENT TRANSDUCERS

Table A classifies instrument transducers, giving the common or descriptive name for each, a brief statement of the principle and nature of the device, and a statement of the basic quantity measured. Table B lists almost one hundred properties or characteristics to be measured, with typical transducers used for each measurement.

Almost all transducers (except counters) respond to one of four basic inputs, viz., displacement, force, temperature, or radiation (and derivatives). Each of these four inputs may, in turn, be utilized to change the resistance, inductance, or capacitance of a passive electric circuit or the output of a voltage generator. Hence a number of choices are possible for measuring any one property or quantity, especially with the inclusion of special circuits for differentiating, integrating, and damping. Any particular requirements of range, sensitivity, accuracy, and dynamic response can almost always be met by the instrument designer.

A single table such as this can only list common applications and examples for guidance in studying the extensive technical literature and for interpreting suppliers' literature. This table does not distinguish between equilibrium measurements and the great variety of dynamic measurements. It should also be emphasized that most transducers permit a choice of readout, i.e., indicating, digital, recording, integrating, or combinations of these.

These reasons of versatility, coupled with that of the ready availability of transducer instruments in a variety of grades, have resulted in the almost complete displacement of other instrumentation by the electrical transducer.

Of the commercial instrument transducers, about one-half of the market is represented either by measurements of displacement (or dimension) or measurements of force (or pressure). The two other large uses are those of temperature measurement and of fluid-flow measurement.

TRANSDUCER DEVELOPMENT

New and improved transducers are in process of rapid development, with emphasis on compactness, ease of calibration, and improvement in the ranges available in a given type or design. Environmental specifications are becoming more demanding, especially as regards vibration. There are many new developments in radiation measurement (see Section 10.2).

New circuitry is available for dealing with zero shift and predictable error. Printed circuits, rigid mounting and encapsulation, and contributions from solid state science are greatly improving the transducer, while there are parallel improvements in readout instrumentation.

Table 10-1. INSTRUMENT TRANSDUCERS (Continued)

Table A. CLASSIFICATION OF TRANSDUCERS

Name or class	Nature and principle	Basic measurement
EXTERNALLY POWERED TRANSDUCERS (PASSIVE)		
VARIABLE RESISTANCE		
1. Variable resistor	Slider or contact varies resistance in potentiometer or bridge circuit	Displacement, linear or angular
2. Resistance thermometer	Wire or thermistor, with large temperature-coefficient of resistivity	Temperature
3. Resistance strain gage	Resistance of a wire grid; foil or semiconductor changed by stress	Displacement, strain
4. Hot-wire meter	Heated wire or film (constant temperature or constant current) in fluid stream	Temperature (fluid velocity inferred)
5. Radiation bolometer	Radiation focused on resistance-thermometer sensor	Temperature (total radiation inferred)
6. Thermistor radiometer	Radiation focused on thermistor bolometer	Temperature (total radiation inferred)
7. Thickness gage	Resistance between contacts depends on thickness and resistivity of separating material	Dimension
8. Photoconductive cell	Radiation on photoresistive element	Radiation
9. Photoemissive or photomultiplier tube	Radiation causes electron emission and current (amplification available)	Radiation (illumination)
10. Ionization gage	Glow-discharge tube in high-frequency field: asymmetry generates voltage	Displacement
11. Resistance hygrometer	Resistivity of conductive strip changed by moisture	Partial pressure (humidity)
VARIABLE CAPACITANCE		
12. Adjustable capacitor	Capacitance varied by changing distance between plates or area of plates	Displacement
13. Capacitance bridge pickup	Modification of No. 12 using a-c bridge: high sensitivity	Displacement
14. Dielectric gage	Capacitance varied by changing position or thickness of dielectric	Displacement, dimension
15. Dielectric thermometer	Variation of capacitance with temperature of dielectric	Temperature
16. Condenser microphone	Capacitance between diaphragm and fixed electrode varied by sound pressure	Displacement
VARIABLE INDUCTANCE		
17. Air-gap gage	Self-inductance or mutal inductance changed by varying the magnetic path	Displacement, thickness
18. Differential transformer	Transformer with differential secondaries and movable magnetic core	Displacement
19. Reluctance pickup	Reluctance of magnetic circuit varied by positioning or core material	Displacement
20. Eddy-current gage	Inductance of a-c coil varied by position of eddy-current plate	Displacement
21. Magnetostriction gage	Magnetic properties varied by pressure and stress	Force
22. Hall-effect transducer	Magnetic field interacts with current through semiconductor to produce voltage at right angle	Field strength
23. Inductance bridge pick up	Modification of No. 17 using inductance bridge	Displacement
SELF-GENERATING TRANSDUCERS		
24. Moving magnet-and-coil generator	Relative movement of coil and magnet varies output voltage	Displacement velocity. linear or angular
25. Thermocouple and thermopile	Pairs of dissimilar metals or semiconductors generate voltage if terminals not at same temperature	Temperature
26. Piezoelectric pickup	Quartz or other crystal mounted in compression, bending, or twisting	Force
27. Photovoltaic cell	Layer-built semiconductor cell or transistor generates voltage from radiation	Radiation. light
28. Radiation counter (special class)	Gas counters collect charge released by ionizing radiation	Radiation. radioactive or nuclear

Table 10-1. INSTRUMENT TRANSDUCERS (Continued)

Table B. EXAMPLES OF TRANSDUCER APPLICATIONS

Table B lists some typical applications of instrument transducers and common types of transducers used for each measurement. Since the properties of materials play a large part in all measurement techniques, the tables that give such properties will suggest other applications of transducers (consult the Index of Properties, Section 1). Another major field of transducer use not listed here is in energy measurements.

Quantity to be measured	Transducers — Common examples	See Table A numbers
Acceleration, angular	Unbonded strain gage; force balance	3, 12, 19, 26
Acceleration, linear	Seismic potentiometer; piezoelectric accelerometer	3, 12, 19, 26
Altitude	Capsule or bellows altimeter	1, 18
Angle	Variable reluctance pickup	18, 19
Blast pressure	Piezoelectric pickup	3, 12, 26
Count, events	Stroboscope; electronic-pulse counter	8, 27
Count, particles	Photoconductive cell; photovoltaic element	
Current, stream	Rotating-current meter; impact-tube meter	1, 3, 7
Density, gas	Hot-wire meter	24
Dewpoint	Photovoltaic cell	4
Dielectric constant	Dielectric gage	11, 27
Dimension, linear	Differential transformer; slide-wire resistor	14
Dimension, micrometer	Capacitance gage; unbonded strain gage	1, 3, 7
Displacement, angular	Slide-wire potentiometer; inductance gage; eddy-current gage; electrolytic gage	3, 12, 13, 14
Displacement, linear	Differential transformer; slide-wire resistor; reluctance pickup	1, 3, 17, 19, 20; 1, 3, 12, 17, 18, 19, 20
Distance	Slide-wire resistor	1, 3, 12, 17, 20
Duration, time	Tuning fork	
Emissivity	Thermopile radiometer	5, 6, 25
Field strength, magnetic	Hall-effect pickup	22

Quantity to be measured	Transducers — Common examples	See Table A numbers
Film thickness	Dielectric gage	7, 12, 14, 17
Flow, gas or vapor	Differential head meter (orifice, nozzle, pitot); hot-wire or thermocouple anemometer	3, 4, 18, 19, 25
Flow, liquid	Differential head meter (orifice, venturi); turbine flowmeter	1, 3, 18
Flow, open-channel	Rotating-current meter	24
Force	Reluctance pickup; carbon pile; strain gage; magnetostrictive gage	3, 12, 17, 18, 19, 21, 23, 26
Frequency	Moving-coil generator; stroboscope	17, 19, 24
Gamma rays	Geiger counter	28
Hardness	Indenter (displacement)	3, 12, 17, 18
Head		(See pressure)
Heat flow	Thermopile sandwich	25
Humidity, air	Resistance hygrometer; thermocouple physchrometer	2, 11, 27
Infrared	Thermistor bolometer; thermopile photoconductive cell	5, 6, 8, 25
Ionization		10
Jerk		3, 12, 26
Level, liquid	Dielectric gage; capacitance gage	12, 14
Light intensity	Photovoltaic cell	5, 8, 25, 27
Load, force or weight	Bonded strain gage (strut); inductor (elastic element); proving ring (displacement)	1, 3, 12, 13, 17, 18, 19, 21, 26

Table 10-1. INSTRUMENT TRANSDUCERS (Continued)

Table B. EXAMPLES OF TRANSDUCER APPLICATIONS (Continued)

Quantity to be measured	Transducers — Common examples	See Table A numbers
Moisture, in solids	Resistance gage; dielectric gage; nuclear magnetic resonance	1, 14
Noise	Condenser microphone; piezoelectric crystal	16, 26
Nuclear radiation	Geiger counter	28
Particle counting	Ionization gage	10, 28
Position, angular	Resistance gage; differential transformer	1, 12, 18, 19
Position, absolute	Contact potentiometer	1, 12, 18, 19
Position, linear	Bourdon-tube potentiometer	1, 3, 18, 19
Pressure, differential	Bellows gage	1, 12, 18, 19
Pressure, dynamic	Piezoelectric pickup; strain gage	1, 12, 18, 19
Pressure, gage	Reluctance gage; strain-gage pickup; capacitor (elastic element)	3, 12, 16, 19, 26
Pressure, impact	Pitot and bellows with differential transformer	1, 3, 12, 18
Radiation, light (optical)†	Optical-target thermopile; photo-resistive gage	12, 18, 19
Radiation, nuclear†	Geiger counter	8, 9, 27
Radiation, total†	Photomultiplier tube; bolometer	10, 28
Reflectivity	Radiometer	5, 6, 8, 25
Rotational speed	Reluctance pickup	5, 6, 25
Rugosity	Moving-coil tracer	19, 24
Shock	Ceramic crystal	19, 24
Sound pressure	Piezoelectric; condenser microphone	16, 26
Speed, rotational	Moving-coil tachometer; pulse counter; stroboscopic counter	19, 24
Strain	Wire or foil strain gage	3
Temperature	Thermocouple; wire resistance; thermistor	2, 5, 6, 8, 15, 25
Thickness, metal	Eddy-current gage; capacitance pick-up; contact gage (resistance); ultra-sonic probe; isotope gage	1, 3, 12, 13, 14, 17
Time	Synchronous motor; tuning fork	3, 12, 18, 26
Torque	Strain gage	4
Turbulence, fluid	Hot-wire pickup	10
Vacuum, high	Ionization gage	
Vacuum, low	Corrugated diaphragm	
Velocity, linear	Moving-coil generator	
Vibration acceleration	Piezoelectric crystal; strain gage (force)	3, 12, 18, 26
Vibration displacement	Reluctance gage; seismic vibrometer	18, 19
Vibration frequency	Calibrated oscilloscope; stroboscopic counter	
Vibration velocity	Magnet and coil	24
Viscosity	Drag-cup torque meter; falling-ball displacement gage	3, 18
Voltage	Moving-coil meter or galvanometer	19, 20, 24
Weight	Strain gage; force balance	3, 12, 17, 18, 26

†For additional data on radiation transducers, see Section 10.2.

REFERENCE

For specific data on 1,250 models of transducers, see "ISA Transducer Compendium", E.J. Minnar, Ed., Instrument Society of America, 1963 (distributed by Plenum Press).

Tables 10-2 to 10-7. THERMOCOUPLE CALIBRATION

The following thermocouple temperature-emf tables are condensed from USA Standard for Temperature Measurement Thermocouples, C96. 1–1969, published by the Instrument Society of America, 530 William Penn Place, Pittsburgh, Pa. This code was approved by the American National Standards Institute and prepared with the assistance and approval of representatives of the major technical societies and federal departments in the United States. The code includes specifications for thermocouples and extension wires and their fabrication, protection, and installation. Suppliers will furnish calibrated thermocouples and color-coded wire conforming with these specifications.

The following table gives the standard and the special limits of error for protected thermocouples; these apply only within the limits of temperatures recommended for the various wire sizes, as shown in the table.

Table 10–2. TEMPERATURE AND ERROR LIMITS FOR THERMOCOUPLES*

Thermo-couple type	Name (positive wire first)	Thermocouple limits of error, °F			Upper temp limits, °F, for AWG wire size				
		Range, °F	Standard	Special	8	14	20	24	28
E	Chromel-constantan	32 to 600 600 to 1600	±3 ±.50%	— —	1600	1200	1000	800	800
J	Iron-constantan	32 to 530 530 to 1400	±1 ±.75%	±2 ±.38%	1400	1100	900	700	700
K	Chromel-Alumel® †	32 to 530 530 to 2300	±4 ±.75%	±2 ±.38%	2300	2000	1800	1600	1600
R, S	Platinum-rhodium	32 to 1000 1000 to 2700	±5 ±.50%	±2.5 ±.25%	—	—	—	2700	—
T	Copper-constantan	−150 to −75 −75 to +200 200 to 700	±2.0% ±1.5 ±.75%	±1% ±.75 ±.38%	—	700	500	400	400

†Chromel–Alumel Thermocouple Alloys, Hoskins Manufacturing Co., Detroit, Mich.

*From: USA Standard for Temperature Measurement Thermocouples, C96.1–1969, Instrument Society of America.

Table 10–3. COPPER-CONSTANTAN THERMOCOUPLE CALIBRATION*

Type-T Thermocouples; Electromotive Force in Absolute Millivolts; Reference Junction at 32°F

Temp, °F	0°	10°	20°	30°	40°	50°	60°	70°	80°	90°
					Millivolts					
−200	−4.111	−4.246	−4.377	−4.504	−4.627	−4.747	−4.863	−4.974	−5.081	−5.185
−100	−2.559	−2.730	−2.897	−3.062	−3.223	3.380	−3.533	−3.684	−3.829	−3.972
(−)0	−0.670	−0.872	−1.072	−1.270	−1.463	−1.654	−1.842	−2.026	−2.207	−2.385
(+)0	−0.670	0.463	0.254	−0.042	+0.171	0.389	0.609	0.832	1.057	1.286
100	1.517	1.751	1.987	2.226	2.467	2.711	2.958	3.207	3.458	3.712
200	3.967	4.225	4.486	4.749	5.014	5.280	5.550	5.821	6.094	6.370
300	6.647	6.926	7.208	7.491	7.776	8.064	8.352	8.642	8.935	9.229
400	9.525	9.823	10.123	10.423	10.726	11.030	11.336	11.643	11.953	12.263
500	12.575	12.888	13.203	13.520	13.838	14.157	14.477	14.799	15.122	15.447
600	15.773	16.101	16.429	16.758	17.089	17.421	17.754	18.089	18.425	18.761
700	19.100	19.439	19.779	20.120	20.463	20.805	—		—	—

CORRECTION TABLE FOR REFERENCE JUNCTION OTHER THAN 32°F

Note: Correction should be added to observed emf before entering the above table.

Reference junction, °F	35	40	45	50	55	60	65	70	75	80	85	90	95
Correction, millivolts	.064	.171	.280	.389	.499	.609	.720	.832	.944	1.057	1.171	1.286	1.401

*From: USA Standard for Temperature Measurement Thermocouples, C96.1–1969, Instrument Society. of America.

REFERENCES

"Handbook of Chemistry and Physics", 53rd ed., R.C. Weast, Ed., The Chemical Rubber Co., 1972; gives tables in one-degree intervals and in degrees Centigrade and Fahrenheit.

"NBS Circular 561", National Bureau of Standards, 1955.

Table 10–4. IRON-CONSTANTAN THERMOCOUPLE CALIBRATION*

Type-J Thermocouples; Electromotive Force in Absolute Millivolts; Reference Junction at 32°F

Temp, °F	0°	10°	20°	30°	40°	50°	60°	70°	80°	90°
					Millivolts					
−300	−7.52	−7.66	−7.79	—	—	—	—	—	—	—
−200	−5.76	−5.96	−6.16	−6.35	−6.53	−6.71	−6.89	−7.06	−7.22	−7.38
−100	−3.49	−3.73	−3.97	−4.21	−4.44	−4.68	−4.90	−5.12	−5.34	−5.55
(−)0	−0.89	−1.16	−1.43	−1.70	−1.96	−2.22	−2.48	−2.74	−2.99	−3.24
(+)0	−0.89	−0.61	−0.34	−0.06	+0.22	0.50	0.79	1.07	1.36	1.65
100	1.94	2.23	2.52	2.82	3.11	3.41	3.71	4.01	4.31	4.61
200	4.91	5.21	5.51	5.81	6.11	6.42	6.72	7.03	7.33	7.64
300	7.94	8.25	8.56	8.87	9.17	9.48	9.79	10.10	10.41	10.72
400	11.03	11.34	11.65	11.96	12.26	12.57	12.88	13.19	13.50	13.81
500	14.12	14.42	14.73	15.04	15.34	15.65	15.96	16.26	16.57	16.88
600	17.18	17.49	17.80	18.11	18.41	18.72	19.03	19.34	19.64	19.95
700	20.26	20.56	20.87	21.18	21.48	21.79	22.10	22.40	22.71	23.01
800	23.32	23.63	23.93	24.24	24.55	24.85	25.16	25.47	25.78	26.09
900	26.40	26.70	27.02	27.33	27.64	27.95	28.26	28.58	28.89	29.21
1000	29.52	29.84	30.16	30.48	30.80	31.12	31.44	31.76	32.08	32.40
1100	32.72	33.05	33.37	33.70	34.03	34.36	34.68	35.01	35.35	35.68
1200	36.01	36.35	36.69	37.02	37.36	37.71	38.05	38.39	38.74	39.08
1300	39.43	39.78	40.13	40.48	40.83	41.19	41.54	41.90	42.25	42.61
1400	42.96	43.32	43.68	44.03	44.39	44.75	45.10	45.46	45.82	46.18

CORRECTION TABLE FOR REFERENCE JUNCTION OTHER THAN 32°F

Note: Correction should be added to observed emf before entering the above table.

Reference junction, °F	35	40	45	50	55	60	65	70	75	80	85	90	95
Correction, millivolts	0.08	0.22	0.36	0.50	0.65	0.79	0.93	1.07	1.22	1.36	1.51	1.65	1.80

*From: USA Standard for Temperature Measurement Thermocouples, C96.1−1969, Instrument Society of America. For References see Table 10−3.

Table 10–5.
CHROMEL-CONSTANTAN THERMOCOUPLE CALIBRATION*

Type-E Thermocouples; Electromotive Force in Absolute Millivolts; Reference Junction at 32°F

Temp, °F	0°	10°	20°	30°	40°	50°	60°	70°	80°	90°
					Millivolts					
−200	−6.40	−6.62	−6.83	−7.04	−7.24	−7.44	−7.62	−7.80	−7.97	−8.14
−100	−3.94	−4.21	−4.47	−4.73	−4.98	−5.23	−5.48	−5.72	−5.95	−6.18
(−)0	−1.02	−1.33	−1.64	−1.94	−2.24	−2.54	−2.83	−3.11	−3.39	−3.67
(+)0	−1.02	−0.71	−0.39	−0.07	0.26	0.59	0.92	1.26	1.59	1.93
100	2.27	2.62	2.97	3.32	3.68	4.04	4.40	4.77	5.13	5.50
200	5.87	6.25	6.62	7.00	7.38	7.76	8.15	8.54	8.93	9.32
300	9.71	10.11	10.51	10.91	11.31	11.71	12.11	12.52	12.93	13.34
400	13.75	14.17	14.59	15.00	15.42	15.84	16.26	16.68	17.10	17.52
500	17.95	18.38	18.81	19.23	19.66	20.09	20.52	20.95	21.39	21.82
600	22.25	22.69	23.13	23.57	24.00	24.44	24.88	25.32	25.76	26.20
700	26.65	27.09	27.53	27.97	28.42	28.86	29.31	29.75	30.19	30.64
800	31.09	31.54	31.98	32.43	32.87	33.32	33.77	34.22	34.67	35.12
900	35.57	36.02	36.47	36.92	37.37	37.82	38.26	38.71	39.16	39.61
1000	40.06	40.51	40.96	41.41	41.86	42.31	42.76	43.21	43.66	44.11
1100	44.56	45.01	45.46	45.91	46.36	46.81	47.26	47.71	48.15	48.60
1200	49.04	49.49	49.93	50.37	59.82	51.27	51.72	52.16	52.61	53.05
1300	53.50	53.94	54.38	54.83	55.27	55.71	56.15	56.59	57.03	57.48
1400	57.92	58.36	58.80	59.24	59.68	60.11	60.55	60.99	61.43	61.86
1500	62.30	62.74	63.17	63.60	64.04	64.47	64.90	65.34	65.77	66.20
1600	66.63	67.03	67.48	67.91	68.34	68.76	69.19	69.62	70.05	70.47

CORRECTION TABLE FOR REFERENCE JUNCTION OTHER THAN 32°F

Note: Correction should be added to observed emf before entering the above table.

Reference junction, °F	35	40	45	50	55	60	65	70	75	80	85	90	95
Correction, millivolts	0.10	0.26	0.42	0.59	0.76	0.92	1.09	1.26	1.42	1.59	1.76	1.93	2.10

*From: USA Standard for Temperature Measurement Thermocouples, C96.1−1969, Instrument Society of America. For References see Table 10−3.

Table 10–6. CHROMEL-ALUMEL THERMOCOUPLE CALIBRATION*

Type-K Thermocouples; Electromotive Force in Absolute Millivolts; Reference Junction at 32°F

Temp. °F	0°	10°	20°	30°	40°	50°	60°	70°	80°	90°
					Millivolts					
−300	−5.51	−5.60	—	—	—	—	—	—	—	−5.41
−200	−4.29	−4.44	−4.58	−4.71	−4.84	−4.96	−5.08	−5.20	−5.30	−5.41
−100	−2.65	−2.84	−3.01	−3.19	−3.36	−3.52	−3.69	−3.84	−4.00	−4.15
(−)0	−0.68	−0.89	−1.10	−1.30	−1.50	−1.70	−1.90	−2.09	−2.28	−2.47
(+)0	−0.68	−0.47	−0.26	−0.04	+0.18	0.40	0.62	0.84	1.06	1.29
100	1.52	1.74	1.97	2.20	2.43	2.66	2.89	3.12	3.36	3.59
200	3.82	4.05	4.28	4.51	4.74	4.97	5.20	5.42	5.65	5.87
300	6.09	6.31	6.53	6.76	6.98	7.20	7.42	7.64	7.87	8.09
400	8.31	8.54	8.76	8.98	9.21	9.43	9.66	9.88	10.11	10.34
500	10.57	10.79	11.02	11.25	11.48	11.71	11.94	12.17	12.40	12.63
600	12.86	13.09	13.32	13.55	13.78	14.02	14.25	14.48	14.71	14.95
700	15.18	15.41	15.65	15.88	16.12	16.35	16.59	16.82	17.06	17.29
800	17.53	17.76	18.00	18.23	18.47	18.70	18.94	19.18	19.41	19.65
900	19.89	20.13	20.36	20.60	20.84	21.07	21.31	21.54	21.78	22.02
1000	22.26	22.49	22.73	22.97	23.20	23.44	23.68	23.91	24.15	24.39
1100	24.63	24.86	25.10	25.34	25.57	25.81	26.05	26.28	26.52	26.75
1200	26.98	27.22	27.45	27.69	27.92	28.15	28.39	28.62	28.86	29.09
1300	29.32	29.56	29.79	30.02	30.25	30.49	30.72	30.95	31.18	31.42
1400	31.65	31.88	32.11	32.34	32.57	32.80	33.02	33.25	33.48	33.71
1500	33.93	34.16	34.39	34.62	34.84	35.07	33.29	35.52	35.75	35.97
1600	36.19	36.42	36.64	36.87	37.09	37.31	37.54	37.76	37.98	38.20
1700	38.43	38.65	38.87	39.09	39.31	39.53	39.75	39.96	40.18	40.40
1800	40.62	40.84	41.05	41.27	41.49	41.70	41.92	42.14	42.35	42.57
1900	42.78	42.99	43.21	43.42	43.63	43.85	44.06	44.27	44.49	44.70
2000	44.91	45.12	45.33	45.54	45.75	45.96	46.17	46.38	46.58	46.79
2100	47.00	47.21	47.41	47.62	47.82	48.03	48.23	48.44	48.64	48.85
2200	49.05	49.25	49.45	49.65	49.86	50.06	50.26	50.46	50.65	50.85
2300	51.05	51.25	51.45	51.64	51.84	52.03	52.23	52.42	52.62	52.81

CORRECTION TABLE FOR REFERENCE JUNCTION OTHER THAN 32°F

Note: Correction should be added to observed emf before entering the above table.

Reference junction, °F	35	40	45	50	55	60	65	70	75	80	85	90	95
Correction, millivolts	0.07	0.18	0.29	0.40	0.51	0.62	0.73	0.84	0.95	1.06	1.18	1.29	1.40

*From: USA Standard for Temperature Measurement Thermocouples, C96.1–1969, Instrument Society of America. For References see Table 10–3.

Table 10–7. PLATINUM-RHODIUM THERMOCOUPLE CALIBRATION

Electromotive Force in Absolute Millivolts; Reference Junction at 32°F

Table A.   PLATINUM-13% RHODIUM THERMOCOUPLES*

Type-R Thermocouples

Temp. °F	0°	10°	20°	30°	40°	50°	60°	70°	80°	90°
					Millivolts					
0	—	—	—	—	0.024	0.055	0.086	0.119	0.152	0.186
100	0.220	0.255	0.291	0.327	0.363	0.400	0.438	0.476	0.516	0.556
200	0.596	0.637	0.678	0.721	0.763	0.807	0.850	0.894	0.939	0.984
300	1.030	1.075	1.121	1.167	1.214	1.261	1.309	1.357	1.406	1.455
400	1.504	1.553	1.603	1.653	1.703	1.754	1.805	1.856	1.908	1.960
500	2.012	2.065	2.117	2.170	2.223	2.277	2.330	2.384	2.438	2.493
600	2.547	2.602	2.657	2.712	2.768	2.823	2.879	2.935	2.991	3.047
700	3.103	3.160	3.217	3.273	3.330	3.387	3.445	3.502	3.560	3.618
800	3.677	3.735	3.794	3.852	3.911	3.970	4.029	4.087	4.146	4.205
900	4.264	4.324	4.384	4.443	4.503	4.563	4.624	4.685	4.746	4.807
1000	4.868	4.930	4.991	5.053	5.115	5.176	5.238	5.301	5.363	5.426
1100	5.488	5.551	5.614	5.677	5.741	5.805	5.869	5.933	5.996	6.060
1200	6.125	6.188	6.252	6.317	6.381	6.446	6.511	6.577	6.642	6.706
1300	6.773	6.838	6.904	6.970	7.037	7.103	7.169	7.235	7.302	7.369
1400	7.436	7.503	7.571	7.639	7.706	7.774	7.842	7.911	7.979	8.047
1500	8.116	8.184	8.253	8.322	8.391	8.460	8.530	8.599	8.669	8.739
1600	8.809	8.879	8.949	9.019	9.090	9.161	9.232	9.303	9.374	9.445
1700	9.516	9.587	9.659	9.730	9.802	9.874	9.946	10.019	10.092	10.164

Table 10–7.
PLATINUM-RHODIUM THERMOCOUPLE CALIBRATION *(Continued)*

Temp, °F	0°	10°	20°	30°	40°	50°	60°	70°	80°	90°
					Millivolts					
1800	10.237	10.310	10.383	10.456	10.529	10.603	10.676	10.749	10.823	10.898
1900	10.973	11.048	11.122	11.197	11.273	11.348	11.424	11.499	11.575	11.651
2000	11.726	11.802	11.878	11.954	12.029	12.105	12.182	12.258	12.335	12.411
2100	12.488	12.564	12.641	12.718	12.795	12.871	12.948	13.025	13.102	13.178
2200	13.255	13.332	13.409	13.486	13.564	13.641	13.718	13.795	13.872	13.949
2300	14.027	14.104	14.181	14.258	14.335	14.412	14.490	14.567	14.644	14.721
2400	14.798	14.875	14.952	15.029	15.107	15.184	15.261	15.338	15.415	15.492
2500	15.568	15.645	15.722	15.800	15.877	15.954	16.031	16.108	16.185	16.263
2600	16.340	16.417	16.494	16.571	16.648	16.725	16.802	16.880	16.957	17.033
2700	17.110	17.186	17.263	17.340	17.416	17.493	17.569	17.646	17.723	17.799

*From: USA Standard for Temperature Measurement Thermocouples, C96.1–1969, Instrument Society of America.

Table B.   PLATINUM-10% RHODIUM THERMOCOUPLES*

Type-S Thermocouples

Temp, °F	0°	10°	20°	30°	40°	50°	60°	70°	80°	90°
					Millivolts					
0	—	—	—	—	0.024	0.056	0.087	0.120	0.153	0.187
100	0.221	0.256	0.291	0.327	0.364	0.401	0.439	0.477	0.516	0.555
200	0.595	0.635	0.676	0.717	0.758	0.800	0.843	0.886	0.929	0.973
300	1.017	1.061	1.106	1.151	1.196	1.242	1.287	1.334	1.380	1.427
400	1.474	1.521	1.569	1.616	1.664	1.712	1.761	1.809	1.858	1.907
500	1.956	2.005	2.055	2.105	2.155	2.205	2.255	2.306	2.357	2.407
600	2.458	2.510	2.561	2.613	2.664	2.716	2.768	2.820	2.872	2.924
700	2.977	3.029	3.082	3.135	3.188	3.240	3.293	3.347	3.400	3.453
800	3.506	3.560	3.614	3.667	3.721	3.775	3.829	3.883	3.937	3.991
900	4.046	4.100	4.155	4.210	4.264	4.319	4.374	4.430	4.485	4.540
1000	4.596	4.651	4.707	4.763	4.818	4.874	4.930	4.987	5.043	5.099
1100	5.156	5.212	5.269	5.326	5.383	5.440	5.497	5.555	5.612	5.669
1200	5.726	5.784	5.842	5.899	5.957	6.015	6.073	6.131	6.190	6.248
1300	6.307	6.365	6.424	6.483	6.542	6.601	6.660	6.719	6.778	6.838
1400	6.897	6.957	7.017	7.076	7.136	7.196	7.257	7.317	7.377	7.438
1500	7.498	7.559	7.620	7.681	7.742	7.803	7.864	7.925	7.987	8.048
1600	8.110	8.172	8.234	8.296	8.358	8.420	8.482	8.545	8.607	8.670
1700	8.732	8.795	8.858	8.921	8.984	9.048	9.111	9.174	9.238	9.302
1800	9.365	9.429	9.493	9.557	9.621	9.686	9.750	9.815	9.879	9.944
1900	10.009	10.074	10.139	10.204	10.269	10.334	10.400	10.465	10.531	10.597
2000	10.662	10.728	10.794	10.860	10.926	10.992	11.058	11.124	11.190	11.257
2100	11.323	11.389	11.456	11.522	11.589	11.655	11.722	11.789	11.855	11.922
2200	11.989	12.055	12.122	12.189	12.256	12.322	12.389	12.456	12.523	12.590
2300	12.657	12.724	12.790	12.857	12.924	12.991	13.058	13.124	13.191	13.258
2400	13.325	13.391	13.458	13.525	13.591	13.658	13.725	13.791	13.858	13.924
2500	13.991	14.058	14.124	14.191	14.257	14.324	14.390	14.457	14.523	14.589
2600	14.656	14.722	14.789	14.855	14.921	14.988	15.054	15.120	15.186	15.253
2700	15.319	15.385	15.451	15.517	15.583	15.649	15.715	15.781	15.847	15.913

CORRECTION TABLE FOR REFERENCE JUNCTION OTHER THAN 32°F FOR Pt-Rh COUPLES

Note: Correction should be added to observed emf before entering the above tables.

Reference junction, °F	35	40	45	50	55	60	65	70	75	80	85	90	95
Correction, millivolts	.009	.024	.038	.055	.070	.086	.102	.119	.134	.152	.168	.186	.202

*From: USA Standard for Temperature Measurement Thermocouples, C96.1–1969, Instrument Society of America.

REFERENCES

"Thermocouples for Temperatures above 1500°C", E.D. Zysk and A.R. Robertson, *Instrumentation Tech.,* 8(11):30, 1961.

"Handbook of Chemistry and Physics", 53rd ed., R.C. Weast, Ed., The Chemical Rubber Co., 1972; gives tables in one-degree intervals and in degrees Centigrade and Fahrenheit.

"NBS Circular 561", National Bureau of Standards, 1955.

Table 10–8. CONSTANTS OF SOME PIEZOELECTRIC MATERIALS*

Physical property	Quartz, 0° X-cut	Lithium sulfate, 0° Y-cut	Barium titanate, type B	Lead zirconate-titanate		Lead meta-niobate	Units
				PZT-4	PZT-5		
Density, ρ	2.65	2.06	5.6	7.6	7.7	5.8	10^3 kg/m^3
Acoustic impedance, ρc	15.2	11.2	24	30.0	28.0	16	10^6 kg/m^{2}s
Frequency thickness constant, ft	2870	2730	2740	2000	1800	1400	khz/mm
Maximum operating temperature	550	75	70–90	250	290	500	°C
Dielectric constant	4.5	10.3	1700	1300	1700	225	—
Electromechanical coupling factor for thickness mode, k_{33}	0.1	0.35	0.48	0.64	0.675	0.42	—
Electromechanical coupling factor for radial mode, k_p	0.1	—	0.33	0.58	0.60	0.07	—
Elastic quality factor, Q	10^6	—	400	500	75	11	—
Piezoelectric modulus for thickness mode, d_{33}	2.3	16	149	285	374	85	10^{-12} m/V
Piezoelectric pressure constant, g_{33}	58	175	14.0	26.1	24.8	42.5	10^{-3}(V/m)/(N/m^2)
Volume resistivity at 25°C	$>10^{12}$	—	$>10^{11}$	$>10^{12}$	$>10^{13}$	10^9	
Curie temperature	575	—	115	320	365	550	°C
Young's modulus, E	8.0	—	11.8	8.15	6.75	2.9	10^{10} N/m^2
Rated dynamic tensile strength	—	—	—	3500	4000	—	psi

Note: The properties of the ceramic materials can vary with slight changes in composition and processing; hence the values that are shown should not be taken as exact.

*From: "Ultrasonic Engineering", J.R. Frederick, John Wiley & Sons, 1965, p. 66.

REFERENCE

For technical data on specific commercial piezoelectric materials consult "Handbook of Materials and Processes for Electronics", C.A. Harper, Ed., McGraw-Hill, 1970.

Table 10–9. CHARACTERISTICS OF STRAIN GAUGE MATERIALS
Room Temperature

Material	Gauge factor	Temp coef per °C	Material	Gauge factor	Temp coef per °C
Constantan	2.0	0.6061	Nickel	−12.1	0.006
Copper	2.6	0.004	Phosphor bronze	1.9	0.003
Isoelastic	3.5	0.00047	Platinum	6.0	0.0035
Karma	2.0	—	Platinum-tungsten	4.0	
Manganin	0.47	0.0000	Silicon	120	0.005–0.007
Monel	1.9	0.002	Silver	2.9	0.004
Nichrome	2.1–2.5	0.0003			

Table 10–10. SUMMARY OF ULTRASONIC IMAGING-DETECTION METHODS*

Methods for visualizing ultrasonic energy over an extended area closely parallel the practices of visual and X-ray image-forming systems. Sensitivities of various methods are listed in this table. A discussion of the electronic methods (with bibliography) is given in the original source.*

Technique	Approximate threshold sensitivity, w/cm^2	Technique	Approximate threshold sensitivity, w/cm^2
PHOTOGRAPHIC AND CHEMICAL METHODS		OPTICAL AND MECHANICAL TECHNIQUES	
Direct action on film	1–5	Optical detection of density variations	3×10^{-4}
Use of photographic paper in developer	1	Optical detection of acoustic birefringence	10^{-1}
Film in iodine solution	1	Optical detection of liquid surface	
Starch plate in iodine solution	1	deformation	10^{-6}
Color changes caused by chemical action	0.5–1	Mechanical alignment of flakes in liquid	2.8×10^{-7}
THERMAL EFFECTS		ELECTRONIC METHODS	
Thermal-sensitive color changes	1	Mechanical movement of transducer	
Phosphor-persistence changes	0.05–0.2	or object to form an image	—
Extinction of luminescence	>1	Probe detection of potential on back	
Stimulation of luminescence	—	of piezoelectric receiver	—
Change in photoemission	0.1	Electronic scanning of piezoelectric	
Change in electrical conductivity	0.1–0.2	receiver	$10^{-7}–10^{-9}$
Thermocouple and thermistor detectors	0.1	Electronic scanning of piezoresistive	
		receiver	10^{-7}

*From: "Ultrasound Image-Converter Systems Utilizing Electron-Scanning Techniques", J.E. Jacobs, *IEEE Transactions on Sonics and Ultrasonics*, SU–15:3, July 1968, pp. 146–152.

Table 10–11. CHECKLIST OF EXPERIMENTAL ERRORS

The following checklist is offered to assist in the prevention of errors when planning and setting up an experimental project.

CHECKLIST FOR MEASUREMENT TECHNIQUES*

Category of error or inadequacy	Items to be examined and checked
Instrument selection	Range, sensitivity and scale length, scale graduation, precision or accuracy (quality), resolution or least count, damping, response or time constant
Instrument condition	Leveling, zero setting, friction, pen or pointer drag, hysteresis, wear and backlash, yielding of supports, capillarity
Instrument calibration	Calibration before and after use, accuracy at calibration points, amplification and linearity, repeatability, zero drift, damping
Environment	Temperature, heat transfer and radiation, humidity, air motion, ambient or barometric pressure, vibration, local gravity, leakage, grounding or short-circuiting, "noise", accessibility and convenience of setup
Observational	Parallax, accidental scale-reading errors, inaccurate estimates of average reading, poor timing or non-simultaneous readings, inaccurate interpolation, inaccurate conversion of units, pure mistakes
Test planning	Nonequilibrium conditions, inadequate control of variables, nonrepresentative samples, poor scheduling, nontypical conditions

The environment in which an instrument is used may affect its accuracy. Temperature, barometric pressure, and humidity affect many instruments; in some cases local gravity, vibration and noise, local radiation or heat exchange, or fluid leakage may also cause instrument errors. Examples of environmental and application errors are stem emergence of thermometers, changes in orifice area with temperature, errors in potential readings caused by resistance of leads or piping, and circuit disturbances caused by low-impedance (bypass) meters.

Because recorded test data are usually small samples, every precaution must be taken to insure that "representative" samples are used. Statistical analysis of the data will often assist in the final evaluation.

*From "Engineering Experimentation", G.L. Tuve and L.C. Domholdt. Copyright © 1966 by McGraw-Hill, Inc. Used with permission of McGraw-Hill Book Company.

Table 10–12. PARTICLE-SIZE MEASUREMENTS

The accuracy of any method for particle-size measurement and classification is dependent on (1) representative samples and (2) separation of agglomerated particles. Experimental techniques must ensure that both of these requirements are met before numerical data on particle sizes can be accepted.

Sieve Size Range

There is no general agreement on a standard sieve series. The International Standard Series (ISO R565) is based on a factor of two with intermediate sizes at two. ASTM sieve number designations give the approximate number of openings per linear inch of screen. In industry there is a natural tendency to use the round numbers, such as 100 and 200. For specifying or describing the size range of a powder, it is usually advantageous to indicate not more than three or four sieve sizes. In any case the numerical values obtained by sieve analysis will depend on the quality and condition of the sieves. Quality standards and tolerances are specified by ASTM E11. (See also ANSI Standard Z23.1 and ASTM B214, D185, D480, and E 161.) Some difference in sieve analysis of the same material by different laboratories must be expected.

Table A. SIZE OPENINGS OF U.S. STANDARD TEST SIEVES

1 000 microns = 1 mm; 1 000 mils = 1 inch

Sieve size No.	Sieve openings		Sieve size No.	Sieve openings	
	microns	mils		microns	mils
10	2 000	78.7	60	250[a]	9.85
12	1 680	66.1	70	210	8.28
14	1 410	55.5	80	177[a]	6.97
16	1 190	46.9	100	149	5.87
18	1 000	39.4	120	125[a]	4.92
20	841	33.1	140	105	4.13
25	707[a]	27.8	170	88[a]	3.55
30	595	23.4	200	74	2.92
35	500[a]	19.7	230	63[a]	2.48
40	420	16.5	270	53	2.09
45	354[a]	13.9	325	44[a]	1.73
50	297	11.7	400[b]	37[c]	1.46

[a] ISO Standard.

[b] The 400-mesh standard sieve is seldom used.

[c] Micromesh sieves are available for finer particles, down to about 5 microns.

Table 10–12. PARTICLE-SIZE MEASUREMENTS (*Continued*)

Table B. METHODS FOR MEASURING PARTICLES

Name or identification	Description or character of operation	Typical particle size range, microns	Typical uses	Advantages
Micromesh sieves		5-100	Loose powders too fine for standard sieve analysis	Simple and familiar equipment
Air sedimentation column or classifier		1-150	Loose or de-agglomerated powders in low micron range	Accurate for spherical particles; full classification of particle sizes; suited to all powders
Liquid sedimentation column	Measurement of sediment accumulation vs. settling time	1-200	Heavy, larger particles	Particles must be well dispersed and mixed
Liquid sedimentation column	Density readings in suspension (sampling or hydrometer)	<2 000	Soil samples (ASTM D422)	Simple and rapid
Turbidimetric sedimentation	Settling time of particles in water measured by photoelectric turbidity meter	0-20	Very fine metal and carbide powders (ASTM B430)	Rapid and simple test; data shown by recorder
Centrifugal sedimentation	Time vs. quantity of sediment in centrifuge tube	0.01-10	Very fine suspensions	Test conditions can be varied to suit sample
Electronic particle counter (Coulter)	Pulse counter gives number and size of particles in liquid flowing through orifice	0.5-800	Liquid suspensions	Rapid; large counts obtainable for statistical accuracy
Membrane filtering	Particulates from given volume of gas deposited on filter	0.1-100	Analysis of particulates in air or gas (ASTM D2009)	Quantity and size of particles measurable
Microscopic examination	Count and size measurements of spread particles	Depends on magnification	Subsieve size distribution (ASTM E20)	Very small sample; slow and tedious

Many methods are in use for measuring the sizes of particles not suitable for sieve analysis. Certain methods are applicable only to powders; others are used for particles in air, water, or other fluids. Table B provides only a rough guide to the available methods. For detailed instructions and discussions of results, consult the references given.

REFERENCES

"Symposium on Particle Size Measurement", ASTM Technical Publication 234, 1959; see also latest edition of Index to ASTM Standards.

"Fine Particle Measurement", C. Orr, Jr., and J. M. Dalla Valle, Macmillan Company, 1960.

"Particle Size Determination", R. D. Calle, Interscience Books, Inc., 1955.

"Particle Size: Measurement, Interpretation, and Application", R. R. Irani and C. F. Calles, John Wiley & Sons, 1963.

"Effect of Non-Uniformity and Particle Shape on 'Average Particle Size' ", H. Green, *J. Franklin Inst.*, 204:713, 1927.

Table 10-13. STANDARD TEST SIEVES (WIRE CLOTH)*

Sieve Designation		Nominal Sieve Opening in	Permissible Variation of Average Opening from Standard Sieve Designation	Max. Opening Size for Not More than 5 Percent of Openings	Maximum Individual Opening	Nominal Wire Diameter, mm[a]
Standard	Alternative					
(1)	(2)	(3)	(4)	(5)	(6)	(7)
125 mm	5 in.	5	±3.7 mm	130.0 mm	130.9 mm	8.0
106 mm	4.24 in.	4.24	±3.2 mm	110.2 mm	111.1 mm	6.40
100 mm	4 in.	4	±3.0 mm	104.0 mm	104.8 mm	6.30
90 mm	$3\frac{1}{2}$ in.	3.5	±2.7 mm	93.6 mm	94.4 mm	6.08
75 mm	3 in.	3	±2.2 mm	78.1 mm	78.7 mm	5.80
63 mm	$2\frac{1}{2}$ in.	2.5	±1.9 mm	65.6 mm	66.2 mm	5.50
53 mm	2.12 in.	2.12	±1.6 mm	55.2 mm	55.7 mm	5.15
50 mm	2 in.	2	±1.5 mm	52.1 mm	52.6 mm	5.05
45 mm	$1\frac{3}{4}$ in.	1.75	±1.4 mm	46.9 mm	47.4 mm	4.85
37.5 mm	$1\frac{1}{2}$ in.	1.5	±1.1 mm	39.1 mm	39.5 mm	4.59
31.5 mm	$1\frac{1}{4}$ in.	1.25	±1.0 mm	32.9 mm	33.2 mm	4.23
26.5 mm	1.06 in.	1.06	±0.8 mm	27.7 mm	28.0 mm	3.90
25.0 mm	1 in.	1	±0.8 mm	26.1 mm	26.4 mm	3.80
22.4 mm	$\frac{7}{8}$ in.	0.875	±0.7 mm	23.4 mm	23.7 mm	3.50
19.0 mm	$\frac{4}{4}$ in.	0.750	±0.6 mm	19.9 mm	20.1 mm	3.30
16.0 mm	$\frac{5}{8}$ in.	0.625	±0.5 mm	16.7 mm	17.0 mm	3.00
13.2 mm	0.530 in.	0.530	±0.41 mm	13.83 mm	14.05 mm	2.75
12.5 mm	$\frac{1}{2}$ in.	0.500	±0.39 mm	13.10 mm	13.31 mm	2.67
11.2 mm	$\frac{7}{16}$ in.	0.438	±0.35 mm	11.75 mm	11.94 mm	2.45
9.5 mm	$\frac{3}{8}$ in.	0.375	±0.30 mm	9.97 mm	10.16 mm	2.27
8.0 mm	$\frac{5}{16}$ in.	0.312	±0.25 mm	8.41 mm	8.58 mm	2.07
6.7 mm	0.265 in.	0.265	±0.21 mm	7.05 mm	7.20 mm	1.87
6.3 mm	$\frac{1}{4}$ in.	0.250	±0.20 mm	6.64 mm	6.78 mm	1.82
5.6 mm	No. $3\frac{1}{2}$	0.223	±0.18 mm	5.90 mm	6.04 mm	1.68
4.75 mm	No. 4	0.187	±0.15 mm	5.02 mm	5.14 mm	1.54
4.00 mm	No. 5	0.157	±0.13 mm	4.23 mm	4.35 mm	1.37
3.35 mm	No. 6	0.132	±0.11 mm	3.55 mm	3.66 mm	1.23
2.80 mm	No. 7	0.111	±0.095 mm	2.975 mm	3.070 mm	1.10
2.36 mm	No. 8	0.0937	±0.080 mm	2.515 mm	2.600 mm	1.00
2.00 mm	No. 10	0.0787	±0.070 mm	2.135 mm	2.215 mm	0.900
1.70 mm	No. 12	0.0661	±0.060 mm	1.820 mm	1.890 mm	0.810
1.40 mm	No. 14	0.0555	±0.050 mm	1.505 mm	1.565 mm	0.725
1.18 mm	No. 16	0.0469	±0.045 mm	1.270 mm	1.330 mm	0.650
1.00 mm	No. 18	0.0394	±0.040 mm	1.080 mm	1.135 mm	0.580
850 μm	No. 20	0.0331	±35 μm	925 μm	970 μm	0.510
710 μm	No. 25	0.0278	±30 μm	775 μm	815 μm	0.450
600 μm	No. 30	0.0234	±25 μm	660 μm	695 μm	0.390
500 μm	No. 35	0.0197	±20 μm	550 μm	585 μm	0.340
425 μm	No. 40	0.0165	±19 μm	471 μm	502 μm	0.290
355 μm	No. 45	0.0139	±16 μm	396 μm	425 μm	0.247
300 μm	No. 50	0.0117	±14 μm	337 μm	363 μm	0.215
250 μm	No. 60	0.0098	±12 μm	283 μm	306 μm	0.180
212 μm	No. 70	0.0083	±10 μm	242 μm	263 μm	0.152
180 μm	No. 80	0.0070	±9 μm	207 μm	227 μm	0.131
150 μm	No. 100	0.0059	±8 μm	174 μm	192 μm	0.110
125 μm	No. 120	0.0049	±7 μm	147 μm	163 μm	0.091
106 μm	No. 140	0.0041	±6 μm	126 μm	141 μm	0.076
90 μm	No. 170	0.0035	±5 μm	108 μm	122 μm	0.064
75 μm	No. 200	0.0029	±5 μm	91 μm	103 μm	0.053
63 μm	No. 230	0.0025	±4 μm	77 μm	89 μm	0.044
53 μm	No. 270	0.0021	±4 μm	66 μm	76 μm	0.037
45 μm	No. 325	0.0017	±3 μm	57 μm	66 μm	0.030
38 μm	No. 400	0.0015	±3 μm	48 μm	57 μm	0.025

* The average diameter of the warp and of the shoot wires, taken separately, of the cloth of any sieve shall not deviate from the nominal values by more than the following:

Sieves coarser than 600 μm	5 percent
Sieves 600 to 125 μm	$7\frac{1}{2}$ percent
Sieves finer than 125 μm	10 percent

*From: "CRC Handbook of Chemistry and Physics", 53rd ed., R.C. Weast, Ed., The Chemical Rubber Co., 1972.

Tables 10–14 to 10–20. FLUID METERING

Fluid-metering data in Tables 10–14 to 10–20 are for differential-head meters, for which the following flow equation applies:

$$q = YFCA \sqrt{2gh}$$

where

q = volume rate of flow, upstream

Y = compressibility factor, Table 10–14

F = velocity-of-approach factor, Table 10–15

C = coefficient of discharge, Tables 10–16, 10–17, and 10–18

K = flow coefficient CF

A = throat area, corrected, Table 10–19

g = acceleration of gravity

h = differential head measured across the two pressure taps, in terms of the metered fluid at upstream density

D = upstream pipe diameter, inside

d = throat diameter for area A

β = diameter ratio d/D

R_D = pipe Reynolds number, upstream

Consistent units must be used, usually ft-lb-sec.

For construction details of meters, see "Instruments and Apparatus", ASME Power Test Codes 19.5, The American Society of Mechanical Engineers, and references therein.

Table 10–14. FLOWMETER CORRECTION FACTOR FOR COMPRESSIBILITY, Y OR Y_a

For Use in the Equation q = YFCA $\sqrt{2gh}$
Based on Density of Gas at Upstream Pressure Tap

SYMBOLS:

$$Y = 1 - \left[0.41 + 0.35\left(\frac{d}{D}\right)^4\right]\left(1 - \frac{p_2}{p_1}\right)\frac{1}{k}, \text{ for square-edged orifices}$$

$$Y_a = \left[\frac{k}{k-1}\frac{1-(p_2/p_1)^{(k-1)/k}}{1-(p_2/p_1)}\frac{1-(d/D)^4}{1-(d/D)^4(p_2/p_1)^{2/k}}\right]^{\frac{1}{2}}\left(\frac{p_2}{p_1}\right)^k, \text{ for venturi meters and flow nozzles}$$

d/D = metering element diameter/pipe diameter

$\dfrac{p_1 - p_2}{p_1}$ = pressure difference/upstream pressure

Pressure ratio, $\dfrac{p_1 - p_2}{p_1}$	Y, for square-edged orifices						Y_a, for venturi meters and flow nozzles					
	Diameter ratio, d/D						Diameter ratio, d/D					
	0.25	0.50	0.60	0.70	0.75	0.80	0.25	0.50	0.60	0.70	0.75	0.80
.02	0.994	0.994	0.994	0.993	0.993	0.992	0.989	0.988	0.987	0.984	0.981	0.978
.04	0.986	0.986	0.986	0.985	0.985	0.984	0.978	0.976	0.974	0.969	0.964	0.958
.06	0.980	0.980	0.979	0.979	0.977	0.976	0.967	0.965	0.961	0.955	0.948	0.938
.08	0.974	0.974	0.973	0.972	0.970	0.969	0.956	0.953	0.948	0.940	0.932	0.919
.10	0.970	0.969	0.968	0.964	0.962	0.961	0.945	0.941	0.935	0.925	0.915	0.900
.12	0.964	0.963	0.962	0.959	0.956	0.053	0.933	0.928	0.921	0.909	0.898	0.881
.14	0.958	0.956	0.954	0.952	0.948	0.946	0.922	0.916	0.907	0.895	0.881	0.863
.16	0.952	0.950	0.948	0.944	0.940	0.937	0.909	0.903	0.894	0.880	0.865	0.845
.18	0.947	0.944	0.942	0.937	0.933	0.929	0.897	0.890	0.882	0.865	0.849	0.828
.20	0.941	0.938	0.935	0.930	0.925	0.920	0.885	0.878	0.867	0.851	0.834	0.810

Note: Values apply for gases having a specific heat ratio $k = 1.4$, but they may be used for gases for which $k = 1.3$ with a maximum error of less than 1.0 percent.

*From: "Fluid Meters, Their Theory and Application", The American Society of Mechanical Engineers (ASME). Used by permission.

Table 10–15. FLOWMETER CORRECTION FACTOR FOR VELOCITY OF APPROACH, F

For Use in the Equation q = YFCA $\sqrt{2gh}$ (see p. 988)

SYMBOLS: d/D = metering element diameter/pipe diameter

$$F = 1/\sqrt{1-(d/D)^4}$$

d/D	F	d/D	F	d/D	F	d/D	F	d/D	F	d/D	F
0.20	1.0008	0.30	1.0040	0.40	1.0130	0.50	1.0328	0.60	1.0719	0.70	1.1472
0.21	1.0010	0.31	1.0047	0.41	1.0144	0.51	1.0356	0.61	1.0774	0.71	1.1579
0.22	1.0012	0.32	1.0053	0.42	1.0159	0.52	1.0387	0.62	1.0832	0.72	1.1694
0.23	1.0014	0.33	1.0060	0.43	1.0175	0.53	1.0420	0.63	1.0895	0.73	1.1818
0.24	1.0017	0.34	1.0068	0.44	1.0193	0.54	1.0454	0.64	1.0962	0.74	1.1951
0.25	1.0020	0.35	1.0076	0.45	1.0212	0.55	1.0492	0.65	1.1033	0.75	1.2095
0.26	1.0023	0.36	1.0085	0.46	1.0232	0.56	1.0531	0.66	1.1109	0.76	1.2250
0.27	1.0027	0.37	1.0095	0.47	1.0253	0.57	1.0574	0.67	1.1191	0.77	1.2418
0.28	1.0031	0.38	1.0106	0.48	1.0276	0.58	1.0619	0.68	1.1278	0.78	1.2600
0.29	1.0036	0.39	1.0118	0.49	1.0310	0.59	1.0667	0.69	1.1372	0.79	1.2799

Table 10–16. COEFFICIENT OF DISCHARGE, C, FOR ASME LONG-RADIUS FLOW NOZZLES*

For Use in the Equation q = YFCA $\sqrt{2gh}$ (see p. 988)

SYMBOLS: β = ratio of orifice diameter to pipe diameter

R_D = Reynolds number in approach pipe

R_D \ β	.2	.3	.4	.5	.6	.7	.8
500	.885	.861	.854	.848	.831	—	—
1,000	.913	.893	.888	.885	.882	.879	.873
2,000	.936	.918	.913	.911	.909	.907	.899
5,000	.958	.944	.939	.937	.935	.932	.924
10,000	.970	.958	.955	.952	.951	.947	.941
20,000	.978	.971	.968	.965	.963	.960	.955
50,000	.986	.982	.979	.977	.976	.974	.970
100,000	.991	.987	.984	.984	.984	.982	.978
200,000	—	.990	.988	.987	.987	.986	.984
500,000	—	—	.992	.991	.990	.989	.987
1,000,000	—	—	.993	.993	.992	.990	.989
10,000,000	—	—	—	—	.993	.992	.991

Note: Pressure taps at 1.0 D and 0.5 D. Velocity-of-approach factor not included.

Pressure Taps. The above data are for "pipe taps" located 1-pipe diameter upstream and ½-pipe diameter downstream. While "throat taps" in the nozzle itself are more difficult to install, they do, under certain conditions, give more accurate flow-rate measurements. For additional data on nozzle coefficients with throat taps, see K. C. Cotton, J. A. Carcich, and P. Schofield, "Experience with Throat-tap Nozzles for Accurate Flow Measurement", ASME Paper No. 1971-WA/PTC-1. See also "Fluid Meters: Their Theory and Application", American Society of Mechanical Engineers, 1971, 273 pages.

*From: "Flow Measurement", ASME Power Test Code 19.5.4–1959, The American Society of Mechanical Engineers. Used by permission.

Table 10–17. FLOW COEFFICIENTS, K, FOR SQUARE-EDGED ORIFICES*

FLANGE TAPS†

R_D / β	1,000	2,000	5,000	10,000	50,000	1,000,000	1,000	2,000	5,000	10,000	50,000	10,000,000
		2-in. Pipe, D = 2.067 in.						4-in. Pipe, D = 4.026 in.				
.100	.6134	.6085	.6056	.6046	.6037	.6035	.6118	.6035	.5988	.5971	.5959	.5956
.200		.6104	.6026	.6000	.5979	.5974		.6151	.6030	.5990	.5958	.5950
.300		.6223	.6091	.6047	.6012	.6004		.6317	.6128	.6065	.6015	.6002
.400			.6241	.6166	.6107	.6093			.6291	.6188	.6105	.6085
.500			.6521	.6384	.6275	.6249			.6631	.6431	.6271	.6233
.600			.7031	.6773	.6567	.6518			.7292	.6891	.6569	.6494
.700				.7429	.7053	.6963			.8454	.7685	.7071	.6925
.750										.8295	.7461	.7262

TAPS AT 1 D AND ½ D‡

R_D / β	1,000	2,000	5,000	10,000	50,000	1,000,000	1,000	2,000	5,000	10,000	50,000	10,000,000
		2-in. Pipe, D = 2.067 in.						4-in. Pipe, D = 4.026 in.				
.100	.6169	.6097	.6034	.6002	.5959	.5932	.6087	.6040	.5998	.5976	.5948	.5930
.200	.6196	.6114	.6041	.6004	.5954	.5923	.6126	.6068	.6016	.5990	.5955	.5933
.300	.6285	.6183	.6094	.6049	.5988	.5950	.6215	.6138	.6070	.6036	.5991	.5962
.400	.6448	.6321	.6208	.6152	.6076	.6028	.6375	.6274	.6184	.6138	.6077	.6039
.500		.6563	.6420	.6348	.6253	.6193		.6510	.6391	.6331	.6251	.6200
.600			.6781	.6690	.6568	.6492			.6745	.6666	.6560	.6493
.700			.7402	.7276	.7108	.7002			.7370	.7252	.7094	.6994
.750			.7922	.7754	.7530	.7389			.7914	.7745	.7518	.7376

VENA-CONTRACTA TAPS**

R_D / β	1,000	2,000	5,000	10,000	50,000	1,000,000	1,000	2,000	5,000	10,000	50,000	10,000,000
		2-in. Pipe, D = 2.067 in.						4-in. Pipe, D = 4.026 in.				
.100	.6115	.6071	.6031	.6011	.5984	.5967	.6088	.6043	.6003	.5983	.5957	.5940
.200	.6170	.6103	.6044	.6014	.5974	.5949	.6161	.6094	.6035	.6005	.5965	.5940
.300	.6273	.6182	.6101	.6061	.6007	.5973	.6268	.6177	.6097	.6056	.6002	.5968
.400	.6441	.6322	.6216	.6163	.6092	.6047	.6438	.6319	.6213	.6160	.6089	.6045
.500		.6564	.6427	.6357	.6265	.6207		.6562	.6425	.6356	.6263	.6205
.600			.6791	.6699	.6576	.6498			.6790	.6698	.6575	.6497
.700			.7415	.7283	.7105	.6994			.7419	.7285	.7106	.6993
.750			.7903	.7733	.7506	.7363			.7916	.7741	.7509	.7363

10-in. PIPE, D = 10.136 in.

R_D / β	5,000	10,000	50,000	10,000,000	5,000	10,000	50,000	1,000,000	5,000	10,000	50,000	10,000,000
	Flange Taps				Taps at 1 D and ½ D				Vena-contracta Taps			
.100	.5998	.5953	.5926	.5919	.5989	.5974	.5955	.5943	.5993	.5973	.5946	.5930
.200	.6112	.6035	.5972	.5958	.6010	.5991	.5966	.5950	.6031	.6002	.5962	.5937
.300	.6215	.6110	.6025	.6006	.6065	.6037	.6001	.5979	.6095	.6055	.6001	.5967
.400	.6411	.6246	.6114	.6083	.6175	.6137	.6086	.6054	.6213	.6159	.6088	.6044
.500	.6917	.6570	.6291	.6226	.6379	.6326	.6256	.6212	.6424	.6355	.6263	.6205
.600		.7247	.6626	.6478	.6729	.6656	.6559	.6498	.6791	.6698	.6575	.6497
.700		.8474	.7201	.6899	.7390	.7264	.7096	.6990	.7432	.7294	.7110	.6994
.750		.9370	.7408	.7202	.8034	.7825	.7545	.7370	.7956	.7770	.7521	.7366

†Flange taps have the center of each tap one inch from the nearer face of the orifice.
‡Taps at 1 D and ½ D are located 1-pipe diameter upstream and ½-pipe diameter downstream.
**Vena-contracta taps are located 1-pipe diameter upstream and at the vena contracta downstream.

*Abridged from: "Flow Measurement", ASME Power Test Code 19.5.4–1959, The American Society of Mechanical Engineers. Used by permission.

Table 10–18. DISCHARGE COEFFICIENTS FOR HERSCHEL-TYPE VENTURI TUBES*

For Use in the Equation q = YFCA $\sqrt{2\,gh}$†; Expressed as a Function of the Pipe Reynolds Number

SYMBOLS: R_D = Reynolds number in approach pipe
C = discharge coefficient
%T = percent tolerance

R_D	C	%T	R_D	C	%T
40,000	.957	2.62	100,000	.976	1.54
60,000	.966	2.18	200,000	.984	0.71
80,000	.972	1.85	250,000	.984	0.71
			and over		

Note: These coefficients apply when the diameter ratio is 0.25 to 0.75, for pipes 2 inches and larger.
*From: "Flow Measurement", ASME Power Test Code 19.5.4–1959, The American Society of Mechanical Engineers. Used by permission.
†For explanation of metering equation, see p. 988.

Table 10–19. ORIFICE- OR NOZZLE-AREA CORRECTIONS*

For Use in the Equation q = YFCA $\sqrt{2gh}$†; Percentage Correction Due to Thermal Expansion

Material	Temperature, °F					
	200°	400°	600°	800°	1000°	1200°
Bronze	0.25	0.67	—	—	—	—
304 Stainless steel	0.23	0.63	1.04	1.45	1.89	2.33
Monel	0.20	0.54	0.90	1.28	1.69	2.09
Carbon steel	0.17	0.47	0.78	1.10	1.48	1.85
5% Chromium- molybdenum	0.15	0.42	0.70	1.00	1.33	1.66
430 Stainless steel	0.14	0.38	0.64	0.93	1.22	1.52

Note: The use of bronze in piping is restricted to temperatures below 406°F.
*From: "Flow Measurement", ASME Power Test Code 19.5.4–1959, The American Society of Mechanical Engineers. Used by permission.
†For explanation of metering equation, see p. 988.

Table 10–20. OVERALL PRESSURE LOSS DUE TO FLOWMETERS*

For Use in the Equation q = YFCA $\sqrt{2gh}$†; Pressure Loss in Percent of Measured Differential Pressure

Diameter ratio	Orifice	Flow nozzle	Venturi tube with 15° recovery cone	Herschel-type venturi tube	Diameter ratio	Orifice	Flow nozzle	Venturi tube with 15° recovery cone	Herschel-type venturi tube
β	O	F	V	H	β	O	F	V	H
.2	93	93	29	15	.5	73	63	15	10
.3	88	85	23	13	.6	63	52	13	10
.4	81	74	19	11	.7	52	41	12	10

*From: "Flow Measurement", ASME Power Test Code 19.5.4–1959, The American Society of Mechanical Engineers. Used by permission.
†For explanation of metering equation, see p. 988.

Table 10–21. OSCILLOSCOPE SINE-WAVE PATTERNS

Lissajous Figures for Various Phase Angles and Frequency Ratios

Frequency ratio, V/H	Phase difference								
	Zero	$\frac{\pi}{4} = 45°$	$\frac{\pi}{2} = 90°$	$\frac{3\pi}{4} = 135°$	$\pi = 180°$	$\frac{5\pi}{4} = 225°$	$\frac{3\pi}{2} = 270°$	$\frac{7\pi}{4} = 315°$	$2\pi = 360°$
$\frac{1}{1}$									
$\frac{2}{1}$									
$\frac{3}{1}$									
$\frac{4}{1}$									
$\frac{5}{1}$									
$\frac{3}{2}$ (1.5)									
$\frac{5}{2}$ (2.5)									
$\frac{4}{3}$ (1.33)									
$\frac{5}{3}$ (1.67)									
$\frac{5}{4}$ (1.25)									

Other angles	
$\pi/3 = 60°$	$2\pi/3 = 120°$
$\pi/3 = 60°$	$2\pi/3 = 120°$
$\pi/8 = 22.5°$	$3\pi/8 = 67.5°$

Table 10–22. RELIABILITY ENGINEERING AND SPECIFICATIONS

Reliability measurement and prediction has become increasingly important with the increase in the number of situations where replacement is impossible, very difficult, or economically prohibitive. This is true not only for space and aircraft equipment and military situations, but it is also emphasized by the trend toward miniaturized sub-assemblies and, in consumer goods, by the prohibitive cost of repair services.

Following are a partial list of U.S. Government publications and specifications and a list of recent books that contain data and extensive references.

GENERAL INFORMATION GOVERNMENT PUBLICATIONS

"Definitions for Reliability Engineering", MIL–STD–721.
"Requirements for Reliability Program", MIL–STD–785.
"Reliability Design Handbook", Bureau of Ships, NAVSHIPS–94501.
"Reliability Analysis Data for Systems and Components Design Engineers", PB–181080.
"Reliability Program for Material and Equipment", AR–705–25.
"Sampling Procedure and Table for Life and Reliability Testing", DOD H–108.
"Techniques for Reliability Measurement and Prediction Based on Field Failure Data", TR–80.

SPECIFIC APPLICATIONS

"Reliability of Military Electronic Equipment", AGREE, Office of the Assistant Secretary of Defense, June 4, 1957 (Superintendant of Documents).

"Reliability Requirements for Ground Electronic Equipment", RADC–2623; also USAF BLTN 2629.

"Prediction and Measurement of Air Force Ground Electronic Reliability", (ASTIA), AD–148977.

"General Specification for Reliability and Longevity Requirements, Electronic Equipment", MIL–R–26667.

"Reliability Requirements for Development of Electronic Subsystems for Equipment", MIL–R–26484.

"Reliability Requirements for Shipboard and Ground Electronic Equipment", MIL–R–22732.

"General Specification for Reliability Assurance for Production Acceptance of Avionic Equipment", MIL–R–23094.

"Reliability Engineering Program Provisions for Space System Contractors", M–REL–M–131–62.

"Reliability Program Provisions for Space System Contractors", NASA NPC 250–1.

"Procedures for Prediction and Reporting Prediction of Reliability of Weapon Systems", MIL–STD–756.

"Naval Weapons Requirements, Reliability Evaluation", WR–41 (BUWEPS).

REFERENCES

"Reliability Handbook", W.G. Ireson, Ed., McGraw-Hill Book Company, 1966.
"Statistical Theory of Reliability", M. Zelen, Ed., University of Wisconsin Press, 1963.
"Reliability Engineering", W.H. Von Alven, Ed., Prentice-Hall, 1964.

Table 10–23. GRAPHICAL CORRELATION OF EXPERIMENTAL DATA

Functions That Can Be Plotted as Straight Lines for Least-squares Curve-fitting Analysis

Function	Plotting variables		Slope	Intercepts	
	x, abscissa	y, ordinate	m, Δ ord/ Δ abs	abscissa $(y = 0)$	b, ordinate $(x = 0)$
$y = a + bx$	x	y	b	$-a/b$	a
$y = a + b/x$	$1/x$	y	b	$-a/b$	a
$y = a + bx^m$	x^m	y	b	$-a/b$	a
$y = a + b/x^m$	$1/x^m$	y	b	$-a/b$	a
$y = a + b[f(x)]$	$f(x)$	y	b	$-a/b$	a
$y = a + b/[f(x)]$	$1/f(x)$	y	b	$-a/b$	a
$y = bx^m$	x^m	y	b	0	0
$\log y = \log b + m \log x$	$\log x$	$\log y$	m	$-(\log b)/m$	$\log b$
$y = ae^{bx}$	e^{bx}	y	a	0	0
$\ln y = \ln a + bx$	x	$\ln y$	b	$-(\ln a)/b$	$\ln a$
$\log y = \log a + b(\log e)x$	x	$\log y$	$b \log e$	$-(\log a)/b \log e$	$\log a$
$y = aA^{bx}$	A^{bx}	y	a	0	0
$\log y = \log a + b(\log A)x$	x	$\log y$	$b \log A$	$-\log a/b \log A$	$\log a$
$y = x/(a + bx)$ $x/y = a + bx$	x	x/y	b	a	$-a/b$
$y = a + bx + cx^2$ $(y - a)/x = b + cx$	x	$(y - a)/x$	c	$-b/c$	b
$y = x/a + bx + cx^2$ $[(x/y) - a]/x = b + cx$	x	$[(x/y) - a]/x$	c	$-b/c$	b
$y = ae^{b + cx} = ae^b e^{cx}$	e^{cx}	y	ae^b	0	0
$\ln y - \ln a = b + cx$	x	$\ln y - \ln a$	c	$-b/c$	b
$y = ae^{x^m(b + cx)}$ $(\ln y - \ln a)/x^m = b + cx$	x	$(\ln y - \ln a)/x^m$	c	$-b/c$	b

Least-squares curve fitting amounts to determining the straight line passing through the average of x and the average of y, i.e., the center of gravity of the data, with the slope m, which minimizes the sum of the squares of the deviations of the data from the line. The equation of the line is $y = mx + b$,

where $m = \dfrac{\Sigma xy - n\bar{x}\bar{y}}{\Sigma x^2 - n\bar{x}^2}$

$b = \bar{y} - m\bar{x}$

$n =$ the number of data points.

Table 10–24. CLASSIFICATION OF RADIATION INSTRUMENTS

The term "radiation detector" is widely used to refer both to the sensitive elements, or transducers or pickups that quantitatively "detect" radiation, and to the complete instruments employing such sensitive elements.

Broadly speaking, any element or device that is sensitive to radiation in any range of the electromagnetic spectrum is a radiation detector (see Table 2–3). Engineering applications in the various wavelength or frequency bands are so diverse, however, that a variety of detection and measurement methods have evolved, each to serve within its own field of application. The data given in this section deal only with measurement of short-wavelength, high-frequency radiation, i.e., no wavelengths longer than that of the far infrared are considered here. Certain data on detection of long-wavelength radiation, in the ranges of electronic communication (such as 1 kc to 10,000 Mc), are included in Table 9–7.

Because the human senses measure radiation intensity as heat or light and spectral intensity as color, we might expect radiation detectors to be of two types: (1) instruments that measure total radiation over a wide band, and (2) detectors with peak sensitivity over a narrow band of wavelengths. Actually there is a third class of detectors for radiation in both the luminous and the ionizing ranges; these are usually called *quantum detectors.*

While the human body is not equipped to "measure" radiation even as short as the ultraviolet, profound biological effects are produced by ultraviolet, X-rays, gamma rays, and all of the other particulate radiations (see Tables 2–1 and 2–3 and Sections 8.5 and 10.2).

The following table is a list of common names for several of the important types of radiation detectors in the spectral ranges above indicated.

RADIATION TRANSDUCERS AND INSTRUMENTS

Common name	Essential principles or materials	Common name	Essential principles or materials
NARROW-SPECTRUM INSTRUMENTS		TOTAL RADIATION FLUX METERS	
Photoconductive detector	Photons change electrical resistance of sulfide or other crystal	Thermoelectric radiometer	Radiation focused on thermo-couple or thermopile
Photographic film	Silver salts and special dyes changed by photons	Deflecting radiometer	Absorbing and reflecting surfaces or bimetal
Geiger-Mueller counter	Ionization pulse counter with photoemission	Radiation pyrometer	Matching fields or disappearing filament
Ionization chamber	Usually a d-c flow-rate particle or radiation meter	Conductivity bolometer	Total radiation inferred from resistance thermometer
Proportional counter	Ionization chamber in proportional region, with gas multiplication	Golay pneumatic cell	Gas thermometer with photocell readout
Scintillation counter	Pulse counter actuated by excited phosphor with photomultiplier	Radiation calorimeter	Measurement of total heat storage
Photovoltaic detector	Photoemissive target with potential measurement		

Table 10–25. APPLICATIONS OF RADIATION MEASUREMENT

For other data on radiation, see Section 2.1.

Among the quantities to be measured, the rate of energy flux and its spectral distribution are usually of the first importance. Properties of the materials are involved, including temperature, emissivity, absorptivity, reflectivity, transmissivity, and refractive index. Changes of state and chemical changes must be taken into account.

Radiation measurement is a typical example of a technique that has been developed in different ways by several independent groups. Thus the following list of applications is designed to suggest sources of information useful to anyone studying radiation measurement.

Table A. FIELDS OF APPLICATION

Typical Fields and Equipment in Which Radiation Measurements Are Involved

IMAGING AND MAPPING

Field scanning	Photoelastic analysis
Fire detection	Photography
Flow detection	Prospecting
Infrared mapping	Space tracking
Infrared photography	Telephotography
Microscopy	Television

CHEMICAL AND BIOLOGICAL

Chemical processing
Diagnostic studies
Health and comfort studies
Personnel monitoring (safety)
Radiation therapy
Spectroscopy
Tracer studies

TOTAL RADIATION (Control of Temperature and Heat Transfer)

Boilers, evaporators, dryers
Building heating and lighting
Direct-energy conversion
Electrical power equipment
Engines and gas turbines
Furnaces, ovens, reactors
Heat exchangers
Meteorological studies
Refrigeration and cryogenics
Solar-heat utilization
Structures and outdoor storage

Table B. TYPICAL NARROW-SPECTRAL RANGES

Wavelength Ranges for Some Special Applications of Radiation Measurement

Uses	Wavelength range, microns	Uses	Wavelength range, microns
Germanium-photoconductor applications	5–20	Ultraviolet "black light"	0.32–0.37
Infrared spectroscopy	0.85–15	Ultraviolet photography	0.25–0.37
Infrared-beam "telephones"	0.8–3.5	Ultraviolet therapy	0.2–0.32
Phosphorography (stimulation peaks)	1–1.3	Ultraviolet germicidal lamps	0.15–0.3
Infrared searchlights and image converters	0.75–1.25	Ultraviolet spectrographic-calibration	
Infrared photography, including diagnostic and microscopic	0.8–1.2	standards	0.09–1.9

Table 10-26. INFRARED DETECTORS

See also Tables 10-24 and 10-26 and Section 2.1.

Infrared measuring devices are of two general types: (1) instruments with thermal detectors or transducers, and (2) instruments with quantum detectors based on photoeffects, on semiconductors, or on photographic film.

It should be noted that many detectors that are sensitive in the visual range may also be sensitive to infrared radiation. The effective spectral range can be controlled by various filters and windows. Most materials and devices used primarily for detection and measurement in the optical range have been omitted in the following discussion. Among these are the instruments or methods dependent on photochemical reactions, luminescence, and stimulation of phosphorescence. The Golay-pneumatic cell*, which is essentially a gas thermometer or radiometer with a photocell readout, is another available instrument.

Thermal instruments measure the temperature changes of a "black" radiation target, employing thermal expansion, change of electrical resistivity with temperature, or thermoelectric-voltage generation. Total radiation is inferred from the temperature change, while the problems of instrument design are concerned with the "blackness" of the target, its protection and longevity, heat leakage, thermal inertia (time constant), and calibration.

Most *quantum detectors* employ photoconductors that show a change in electrical resistance with incident radiation. Photovoltaic generators are also used, and the photoelectro-magnetic generator (PEM) is very similar to the Hall-effect transducer. In the PEM detector the semiconductor is located in a magnetic field, and current flow is produced by the absorption of photons at the front surface of the semiconductor.

Photoemissive detectors comprise a large family of instruments for infrared, visible, and other radiations.† Many commercial devices are available, including phototubes with a very narrow spectral response, electron multipliers, and scintillation counters. One large group of commercial phototubes has a peak sensitivity in the visible range from 0.35 to 0.55 microns. Peak sensitivities in the infrared range, 0.8 microns or above, are also available.

Infrared photographic film is now produced in all major industrial countries. Most common is the film having an upper spectral limit of about 0.8 microns, but several films are available with sensitivities to 1.1 microns or above.

The following table lists some of the characteristics of the common photoconductors. The sensitive element is usually a narrow strip of approximately 0.5 cm in length and .01 sq cm in area. Peak-wavelength response is in the range of 2 to 5 microns, except for the doped germanium and silicon detectors. Time response and peak-operating temperature will depend on design.

---

*Supplied by Eppley Laboratory, Inc., Sheffield Ave., Newport, R.I. 02840. For further information on the Golay-pneumatic cell, see "A Pneumatic Infrared Detector", M.J.E. Golay, *Rev. Sci. Inst.*, 18:357, 1947.

†For further information on photoemissive detectors, see "Photoemissive Materials", A.H. Sommer, John Wiley & Sons, 1968.

CHARACTERISTICS OF TYPICAL PHOTOCONDUCTOR CELLS

Material	Temper-ature	Time response, microsec	Peak response, microns	Cutoff, microns	Resistance, megohms		Sensitivity
					Dark	Light	
Lead sulfide	25°C		2.4	3.5	1–4		0.3 mv/μW
Indium antimonide	77°K		5.	5.4		.01–.06	2.3 mv/μW
Germanium-gold	77°K	.05	5.3	8.3	0.5		
Germanium-copper	5°K	<1.	24.	29.	1.0		
Germanium-silicon-gold	50°K	<1.	5.5	10.3	1,000		
Lead selenide	25°C	2	4.	4.8	0.2		
Cadmium sulfide and selenide	25°C	Long	0.6	0.8	Low		High

REFERENCES

"Infrared Radiation", A. Vaško, The Chemical Rubber Co., 1968.

"Fundamentals of Infrared Technology", M.R. Holter, Ed., Macmillan Company, 1962.

"Infrared Physics and Engineering", J.A. Jamieson, et al., McGraw-Hill Book Company, 1963.

Table 10–27. INFRARED DETECTORS—QUANTUM TYPE

TYPICAL CHARACTERISTICS AND USE OF VARIOUS SEMICONDUCTOR PHOTON DETECTORS

Modes of Operation

Photovoltaic (PV) or self-generating detectors. These may be operated into a low impedance and characterized by the observed short-circuit current, or into a high impedance and characterized by the open-circuit voltage. Amplification may be obtained by using secondary emitters in stages with the result designated a photomultiplier.

Photoconductive (PC) detectors. These are the most widely used types for infrared. Acting as a low conductivity variable resistor, the detector is connected in series with a load resistor and bias battery, and the signal is measured across the load resistor. Cryogenic cooling is necessary for high response to long wavelengths.

Performance and Merit

Most photon detectors exhibit a curve of increasing spectral response up to a certain maximum, followed by a rather abrupt decrease and cutoff. Performance must be defined in terms of the temperature of the source (blackbody), or it may be defined for monochromatic or narrow-band radiation. The usefulness of the detector signal depends on the noise, and the limiting performance, a signal-to-noise ratio of unity, is called the noise equivalent power (NEP), which is defined as the sinusoidal radiant power that will give an rms signal voltage equal to the rms noise voltage. The statement that the NEP shown by test of a detector is 5×10^{-11} watts (500K, 800, 5) means that the test was made with a blackbody source at $500°K$, 800 cps center frequency and 5 cps bandwidth. The reference area and temperature must also be stated, probably 1 cm^2 and $295°K$ ($22°C$, $71.6°F$). The reciprocal $1/NEP$ increases with the quality of the detector, and often is called the detectivity, D.

The most definitive figure of merit is the reciprocal NEP corrected for area and called specific responsivity or detectivity, "dee star": $D^* = \sqrt{A}/NEP$ (reference bandwidth 1 cps).

Other performance data include the resistance and the time constant. The most common photon detectors (photoconductive) are high resistance, and a time constant of less than one microsecond can be attained.

Operating Temperature

Thermal noise is greatly reduced at low temperatures. The performance of a photon detector is improved by operating it at a very low temperature. The common minimum temperatures are the phase-transition temperatures of carbon dioxide ($195°K$, $-78°C$, $-108°F$), of nitrogen ($77°K$, $-196°C$, $-321°F$), and sometimes even of helium ($4.2°K$).

Amplification and Output

Signal processing, display and recording of IR detector signals are similar to those for other electrical signals, and a great variety of methods and instruments are available. Although transistors, printed circuits, and even microelectronics may be used, simpler and less costly instrumentation is often adequate. Very elaborate systems have been developed for military and space applications.

Infrared Spectral Analysis

Engineering uses of IR spectral data include chemical analysis and the monitoring of streams of gases and liquids representing input or output of processes. Objectives may be inspection and quality control, proportioning of inputs, maintenance of safety standards, detection of leakage or of contamination, etc. The equipment is an industrial modification of laboratory assemblies representing a radiation source, monochromator, probably a modulator or chopper, followed by sample cells, detectors, amplifiers and output recorder. The detector will probably be either a photoconductive device, a thermopile radiometer or a conductivity bolometer.

Optical Systems

Since lenses have spectral limitations, mirrors are widely used in infrared systems.

Conversion from the infrared to a visible image can be accomplished in many different ways, including direct "image converters" using phosphors, but those methods based on electronic scanning and television techniques are the most versatile. Military applications are obvious, and have been highly developed. Medical diagnosis, nondestructive testing, and large-scale earth surveys are other uses.

For target identification, tracking, and similar applications, background suppression is necessary. Sometimes optical filters are adequate. "Space filtering" by the use of a rotating reticle allows the substitution of electrical filtering for direct spectral filtering and gives great opportunities for special designs to fit the particular case. Optical modulation can also be arranged to provide directional and distance information about a target.

Table 10-27. INFRARED DETECTORS—QUANTUM TYPE (*Continued*)

PROPERTIES OF QUANTUM-TYPE DETECTORS

Basic detector material	Film or single crystal	Dark resistance, ohms	Approx. RC, time constant, μ sec	Operating temperature	Wavelengths[a] Peak response, microns	Wavelengths[a] High cutoff, microns	D*, specific responsivity, 500K-f-1, (f = cps), cm $\sqrt{}$ cps/watts
PV = photovoltaic							
Indium antimonide, InSb	Crystal	1-50K	< 2	77°K	5.2	5.6	3-20 x 10^9 (900)
Indium arsenide, InAs	Crystal	30-60K	< 2	Room	3.3	3.7	1-3 x 10^8 (900)
Gallium arsenide, GaAs	Crystal	4.5M	1000	Room	0.8	0.9	4.5 x 10^11 (400)
PC = photoconductive							
Lead sulfide, PbS	Film	0.3-5M	1-2000	Room	2-3	2.5-3.5	4.5 x 10^8 (90)
				195°K	2.5-3.5	3-5	4 x 10^9 (90)
				77°K	2.7-4	4-6	4 x 10^9 (90)
Lead selenide, PbSe	Film	1-10M	2-40	Room	3.5-4	4.3-5.7	10^8 (90)
				77°K	4.5-6	5.7-7.5	2 x 10^9 (90)
Lead telluride, PbTe	Film	30-100M	5-25	77°K	4-5	5-5.7	10^9 (90)
Indium antimonide, InSb		20-2000	0.2-10	Room	6-6.5	6.5-7.7	
				77°K	5-5.5	5.3-6	5 x 10^9 (900)
Tellurium, Te	Crystal	2K	60	77°K	3.5	3.8	6 x 10^10 (900)
Germanium-gold, Ge:Au	Crystal	1-10M	< 1	77°K	5-5.5	7-9	3-7 x 10^9 (900)
Germanium-zinc, Ge:Zn	Crystal	0.3M	< 0.1	5°K	20-35	30-40	3-4 x 10^9 (800)
Germanium-copper, Ge:Cu	Crystal		< 0.1	< 20°K	20	27	
Germanium-cadmium, Ge:Cd	Crystal			< 25°K	16	21.5	

[a] Peak response and other properties vary with impurities or doping.

REFERENCES

1. "Infrared System Engineering," R.D. Hudson, Wiley-Interscience, 1969.
2. "Elements of Infrared Technology," P.W. Kruse, L.D. McGlauchlin, R.B. McQuistan, John Wiley & Sons, 1962.
3. "Infrared Methods," G.K.T. Conn, D.G. Avery, Academic Press, 1960.

Table 10–28. TYPICAL PHOTODETECTOR PARAMETERS†

Courtesy of Donald E. Bode

Detector material	Mode[a]	Operating temperature, K	Recommended spectral range, μm	Spectral peak, μm	Detectivity range $D^*(\lambda m, fm)$, $cmHz^{1/2}W^{-1}$	Time constant range, seconds	Remarks
Si	A	295	0.4 to 1.1	0.9	—	1×10^{-8}	NEP** = 5 $\times 10^{-14}$ W/Hz$^{1/2}$ with preamp
Ge	A	253	0.9 to 1.5	1.5	—	3×10^{-9}	NEP = 0.5 $- 1.0 \times 10^{-12}$ W/Hz$^{1/2}$ with preamp
PbS	PC	295	1.0 to 3.0	2.4	0.7 to 1.5×10^{11}	1 to 5×10^{-4}	
PbS	PC	193	1.0 to 3.5	2.7	2.0 to 7.0×10^{11}	5×10^{-3}	
PbS	PC	77	1.0 to 4.0	3.2	0.8 to 2.0×10^{11}	3×10^{-3}	
PbSe	PC	295	1.0 to 4.5	3.7	0.3 to 1.2×10^{10}	2×10^{-6}	
PbSe	PC	193	1.0 to 5.1	4.4	1.5 to 4.0×10^{10}	3×10^{-5}	
PbSe	PC	77	1.0 to 6.5	5.0	1.0 to 3.0×10^{10}	4×10^{-5}	
InAs	PV	295	1.0 to 3.6	3.5	0.25 to 1.3×10^{10}	RC	$C \approx 100$ to 300 pf/mm²
InAs	PV	193	1.0 to 3.4	3.3	0.4 to 3.0×10^{11}	RC	
InAs	PV	77	1.0 to 3.1	3.0	3.0 to 8.0×10^{11}	RC	
InSb	PV	77	2.0 to 5.4	5.3	0.5 to 1.1×10^{11}	RC	
InSb	PC	77	2.0 to 5.4	5.3	2.5 to 5.0×10^{10}	5×10^{-6}	
Ge:Au	PC	77	2.0 to 7.0	5.0	3.0 to 6.0×10^9	0.2 to 5×10^{-8}	$D^*(10.6\mu)$ $\sim 10^7$ cm Hz$^{1/2}$ watt^{-1}
Ge:Hg	PC	<28	2.0 to 13.8	11.0	0.7 to 1.5×10^{10}	1 to 7×10^{-8}	Time constants below 1 nano- second are possible
Ge:Cd	PC	<21	2.0 to 23	22.0	0.7 to 1.5×10^{10}	1 to 7×10^{-8}	
Ge:Cu	PC	<15	2.0 to 28	24.0	0.7 to 1.5×10^{10}	1 to 7×10^{-8}	
Ge:Zn	PC	<12	2.0 to 38	35.0	0.7 to 1.5×10^{10}	1 to 5×10^{-8}	
Hg$_{0.8}$Cd$_{0.2}$Te	PC	77	8.0 to 13	12 ± 1	2 to 6×10^9	1 to 3×10^{-7}	Time constants of 10^{-8} sec- onds possible with lower D^*
Hg$_{0.8}$Cd$_{0.2}$Te	PV	77	8.0 to 13	12 ± 1	3 to 9×10^9	RC	Can be less than 5 nano- seconds

Note: D^* values reported when looking at a 295 K background with a 2π steradian field of view, and the detector used under optimum electrical load. A mismatch of detector to load can result in a lower D^*. A reduced effective background condition can improve D^*.
**Noise equivalent power.

[a] A–Avalanche Diode
PC–Photoconductive
PV–Photovoltaic

†Reprinted from: D.E. Bode, Optical Detectors, in "Handbook of Lasers", R.J. Pressley, Ed., The Chemical Rubber Co., 1971.

Table 10-29. CHARACTERISTICS OF STANDARD PHOTOSURFACES*

S number[1]	Principal photocathode components[2]	Entrance window material	Photocathode supporting substrate[3]	Typical luminous sensitivity[4], μa/lumen	Typical photocathode dark current[5] at 25°C, amp/cm²
S1	Ag-O-Cs	Visible-light-transmitting glass[6]	Entrance window or opaque material[7]	25	10^{-11}–10^{-13}
S3	Ag-O-Rb	Visible-light-transmitting glass[6]	Opaque material[7]	6.5	10^{-12}
S4	Cs-Sb	Visible-light-transmitting glass[6]	Opaque material[7]	40	10^{-14}
S5	Cs-Sb	Ultraviolet-transmitting glass	Opaque material[7]	40	10^{-14}
S8	Cs-Bi	Visible-light-transmitting glass[6]	Opaque material[7]	3	10^{-14}–10^{-15}
S9	Cs-Sb	Visible-light-transmitting glass[6]	Entrance window	30	10^{-14}
S10	Ag-Bi-O-Cs	Visible-light-transmitting glass[6]	Entrance window	40	10^{-13}–10^{-14}
S11	Cs-Sb	Visible-light-transmitting glass[6]	Entrance window	60	10^{-14}–10^{-15}
S13	Cs-Sb	Fused silica	Entrance window	60	10^{-14}–10^{-15}
S17	Cs-Sb	Visible-light-transmitting glass[6]	Opaque-reflecting material[7]	125	10^{-14}–10^{-15}
S19	Cs-Sb	Fused silica	Opaque material[7]	40	10^{-14}
S20	Sb-K-Na-Cs	Visible-light-transmitting glass[6]	Entrance window	150	10^{-15}–10^{-16}
S21	Cs-Sb	Ultraviolet-transmitting glass	Entrance window	30	10^{-14}
UV[8]	Cs-Te	Sapphire	Opaque material[7]	0	—

[1] The S number is the designation of the spectral response characteristic of the device and includes the transmission of the device window material.

[2] Principal components of the photocathode are listed without regard to order of processing or relative proportions.

[3] When the supporting substrate is the entrance window, an intermediate semitransparent electrically conductive layer may be used.

[4] Corresponding to the specific absolute-response curves using a 2354°K color-temperature tungsten-lamp test source.

[5] Specific dark current excludes direct-current leakage.

[6] Lime glass and Kovar-sealing borosilicate glass are commonly used for visible-light-transmitting glass

[7] The opaque material used as the supporting substrate for photocathodes in which the input radiation is incident on the same side as the emitted photoelectrons is usually metallic in nature.

[8] An S-number designation has not yet been assigned to this experimental "solar-blind" photoemissive surface.

*From: "Reference Data for Radio Engineers", 5th ed., Howard W. Sams & Co., Inc., Indianapolis, Indiana, 1968.

Figure 10–30. SENSITIVITY OF PHOTOEMISSIVE SURFACES*

The *S* number designates the spectral response characteristic of the device and includes the transmission of the device window material.

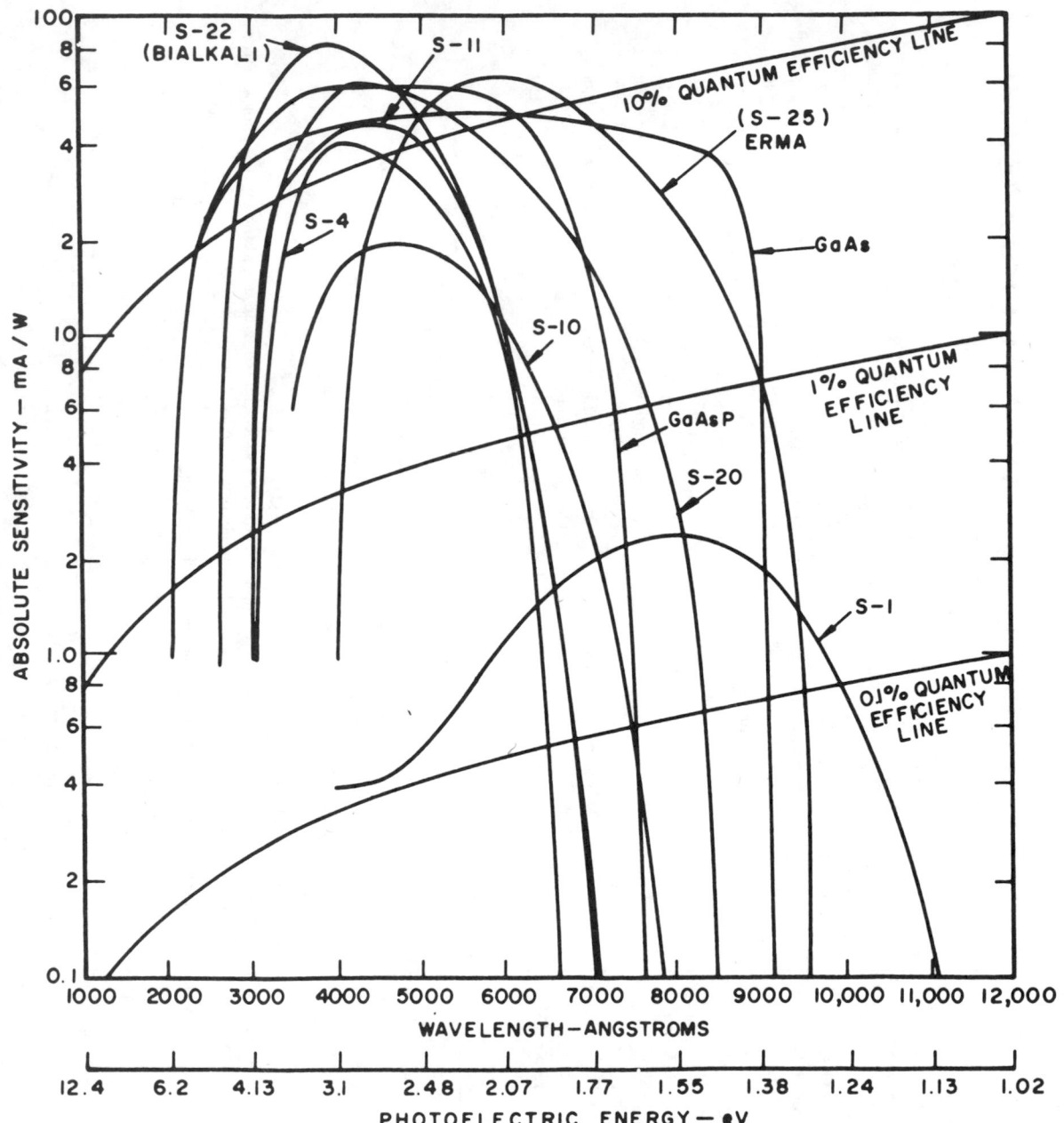

Figure 10–30 Absolute sensitivity *vs.* wavelength for photoemissive surfaces available in commercial tubes.

*Courtesy of RCA.

Table 10–31. EMISSIVITY CORRECTIONS FOR OPTICAL PYROMETERS*

The following corrections are to be added to the reading of an instrument that is calibrated for blackbody radiators. The corrections are based on the energy balance equation solved for T_t, in the following form:

$$T_t = \frac{T_r}{1 + T_r \left[\lambda \ln \left(e_\lambda / e_{\lambda c}\right)/C_2\right]}$$

where

T_t = true temperature, absolute
T_r = pyrometer reading, absolute
λ = wavelength for which pyrometer is calibrated
e_λ = emissivity of radiator at wavelength λ
$e_{\lambda c}$ = emissivity for which the pyrometer is calibrated at wavelength λ (1.0 for blackbody)
C_2 = constant in the Wien equation, 14 388 $\mu m \cdot K$ = 25 898 $\mu m \cdot deg\ R$

Emissivity of the radiator (at 0.65 μm, red)

Instrument reading		0.05	0.10	0.20	0.30	0.40	0.50	0.60	0.70	0.80	0.90
deg R	deg F	Correction to be added, T_t-T_r, deg R = deg F									
1460	1000	180	134	91	67	51	38	28	19	12	6
1510	1050	193	144	98	72	54	41	30	21	13	6
1560	1100	207	155	105	77	58	43	32	22	14	6
1610	1150	222	165	112	82	62	46	34	23	15	7
1660	1200	237	176	119	88	66	49	36	25	16	7
1710	1250	252	187	127	93	70	52	38	26	17	8
1760	1300	268	199	135	99	74	55	41	28	18	8
1810	1350	285	211	143	105	78	59	43	30	19	9
1860	1400	302	224	151	111	83	62	45	31	20	9
1910	1450	320	237	160	117	88	66	48	33	21	10
1960	1500	339	250	168	123	92	69	50	35	22	10
2010	1550	358	264	177	130	97	73	53	37	23	11
2060	1600	377	278	187	137	102	76	56	39	24	11
2110	1650	398	293	196	144	107	80	59	41	25	12
2160	1700	419	308	206	151	113	84	61	43	26	12
2210	1750	440	323	216	158	118	88	64	45	28	13
2260	1800	462	339	227	166	124	92	67	47	29	14
2310	1850	485	356	238	173	129	97	70	49	30	14
2360	1900	509	373	249	181	135	101	74	51	32	15
2410	1950	533	390	260	189	141	105	77	53	33	15
2460	2000	558	408	271	197	147	110	80	55	34	16
K	deg C	Correction to be added, T_t - T_r, K = deg C									
800	527	97	73	49	36	27	21	15	10	7	3
850	577	110	82	56	41	31	23	17	12	7	3
900	627	125	93	63	46	35	26	19	13	8	4
950	677	140	104	70	52	39	29	21	15	9	4
1000	727	157	116	78	58	43	32	24	16	10	5
1050	777	174	129	87	64	48	36	26	18	11	5
1100	827	192	142	96	70	52	39	29	20	12	6
1150	877	212	156	105	77	57	43	31	22	13	6
1200	927	233	171	115	84	63	47	34	24	15	7
1250	977	255	187	125	91	68	51	37	26	16	7
1300	1027	278	203	136	99	74	55	40	28	17	8
1350	1077	302	221	147	107	80	60	43	30	19	9
1400	1127	327	239	159	115	86	64	47	32	20	9
1450	1177	354	258	171	124	93	69	50	35	22	10
1500	1227	382	277	184	133	99	74	54	37	23	11
1600	1327	442	319	211	153	113	84	61	42	26	12
1700	1427	508	365	240	173	129	96	69	48	30	14
1800	1527	580	415	271	195	145	108	78	54	33	16
1900	1627	658	468	305	219	162	120	87	60	37	17
2000	1727	742	525	340	244	181	134	97	67	41	19

*Computed.

Table 10–32. NUCLEAR-RADIATION DETECTORS

Instruments for the measurement of nuclear radiation have a wide variety of uses, and their character varies accordingly. For many uses, such as personnel monitoring or radioactive surveys, portability, compactness, and simplicity are important (see Tables 8–61 and 10–34). Other instrument characteristics are more important in such laboratory applications as radiochemistry or the medical uses of radionuclides. Uses for nuclear detectors include those in nuclear-power plants, in fallout surveys and contamination monitoring, in dosimetry studies, and in radioactive dating (see Section 4).

Nuclear-radiation detectors produce a detectable signal resulting from the ionization of the medium receiving radiation or from the capacity of the radiation to excite the atoms of certain materials (scintillation). The signal is amplified by electronic or photoelectric means, one common output being a counter served by a pulse amplifier.

The following table lists some of the characteristics of common detection methods.

COMMON NUCLEAR-DETECTION INSTRUMENTS*

Detector	Sensitive medium	Output signal, volts	Resolving time, seconds	Energy resolution, percent	Advantages	Disadvantages
Ion chamber	Gas	10^{-6}	10^{-4}	—	Low energy dependence Simple to operate	Slow response Low sensitivity
Proportional counter	Gas	10^{-2}	10^{-6}	15	Rapid response	Requires stable high-voltage supply
G-M counter	Gas	1	10^{-4}	—	Large output signal Moderate sensitivity	Long dead time Energy dependent
Scintillation crystal	Solid†	1	10^{-7}	10	High sensitivity Rapid response Good energy resolution	Fragile Expensive
Semiconductor detector	Solid	10^{-3}	10^{-9}	1	Excellent energy response Short dead time	Requires high amplification

*Data from: "CRC Handbook of Radioactive Nuclides", Y. Wang, Ed., The Chemical Rubber Co., 1969.
†For liquid-scintillation counting a radioactive material may be mixed with a solution of an organic scintillator.

REFERENCES

"Nuclear Radiation Detection", 2nd ed., W.J. Price, McGraw-Hill Book Company, 1964.
"Reactor Physics Constants", 2nd ed., Argonne National Laboratory, ANL–5800, U.S. Atomic Energy Commission, 1963.
"National Bureau of Standards Handbook 72", U.S. Committee on Radiation Protection and Measurements, 1960.
"CRC Handbook of Radioactive Nuclides", Y.Wang, Ed., The Chemical Rubber Co., 1969.
"Theory and Practice of Scintillation Counting", J.B. Birks, Pergamon Press, 1964.
"Nuclear Engineering Handbook", H. Etherington, Ed., McGraw-Hill Book Company, 1958.
"Applied Dosimetry", K.K. Aglintsev, V.M. Kodyukov, A.F. Lyzkov, and Yu.V. Sivintsev, Iliffe Books, 1965; distributed in the United States by The Chemical Rubber Co.

Table 10–33. SOME COMMON GAMMA REFERENCE STANDARDS*

The usual method of measuring the activity of a radioactive sample with a sodium iodide counter is to compare its gamma-emission rate with a calibrated counting standard that emits rays of nearly the same energy. Absolute calibration procedures are not used because of the many factors affecting counting efficiency. A list of some of the commonly used, commercially available counting standards appears below. The principal gamma-emission and half-life are also given. It is desirable that counting standards have a relatively long half-life to reduce the frequency with which they must be replaced.

Nuclide	Principal gamma emission	Half-life	Nuclide	Principal gamma emission	Half-life
Cd^{109}	87 kev	470 days	Cs^{137}	662 kev	30 years
Co^{57}	123 kev	270 days	Mn^{54}	840 kev	310 days
Ba^{133}	357 kev	7.2 years	Na^{22}	1.28 Mev†	2.6 years
Sn^{113}	393 kev	120 days	Co^{60}	1.33 Mev	5.26 years

*Data from: "CRC Handbook of Radioactive Nuclides", Y. Wang, Ed., The Chemical Rubber Co., 1969.
†Also 540 kev β + annihilation radiation.

Table 10–34. PORTABLE RADIATION-DETECTING DEVICES*

ROBERT RADTKE

See also Section 8.5.

ABBREVIATIONS AND SYMBOLS:

c/m = counts per minute n_{th} = thermal neutron

mr/hr = milliroentgen per hour R/hr = roentgen per hour

n_f = fast neutron

Detector	Radiation detected	Ranges	Uses	Comments
Geiger–Mueller (G–M) tube	Alpha Beta X Gamma	0.04 mr/hr to 500 mr/hr	Low dose-rate surveys Area monitors Personnel radiation monitors	1. Radiation detected depends on type of G–M tube 2. Energy dependent 3. Some models saturate—do not use in high radiation fields 4. Sensitive to microwave fields 5. Ratemeter and audible pulse 6. Rapid response 7. Rugged, dependable
Ion chamber	Alpha Beta X Gamma	3 mr/hr to 10,000 R/hr	Medium and high dose-rate surveys Area monitors	1. Wide dose-rate range on a single instrument 2. Low energy dependence 3. Some models can be used in RF fields 4. Some models slow to respond
Scintillation	Alpha Beta X Gamma Neutrons	0.025 mr/hr to 200 mr/hr or to 800,000 c/m	Low-level contamination surveys	1. High sensitivity 2. Rapid response 3. Fragile 4. Audible signal and ratemeter 5. Radiation detected depends on instrument and crystal 6. Fast neutron detector where dose rate is not required
Proportional counter	Alpha Beta X Gamma Neutrons	} to 500,000 c/m to 20,000 n_{th}/cm²-sec; to 100 mrads/hr, n_f	Low-level contamination surveys Neutron survey	1. Primary use is for alpha detection or neutron surveys 2. Alpha detector can discriminate between alpha and beta-gamma 3. Neutron detector can discriminate against gamma radiation 4. Maintenance may be a problem
BF$_3$ counter	Neutrons	to 100,000 c/m	Survey	1. Rather low sensitivity 2. Bulky 3. Used with various moderators

*From: "CRC Handbook of Radioactive Nuclides", Y. Wang, Ed., The Chemical Rubber Co., 1969.

REFERENCE

For further information on rate-measuring instruments (their detectors, selection, maintenance, and calibration), see "Radiation Accidents and Emergencies", L. H. Lanzl, J. H. Pingle, and J. H. Rust, Eds., Charles C Thomas, 1965.

Table 10–35. FISSION THRESHOLD ENERGIES AND CROSS SECTIONS OF MATERIALS FOR FISSION COUNTERS*

| Isotope | Threshold energy (E_0), Mev | Cross-sections | | | Half-life, yr |
		At $2E_0$, mb	At 3 Mev, mb	At thermal energy, b	
Th^{232}	1.25	120	140	—	1.4×10^{10}
Pa^{231}	0.42	550	1,100	—	3.4×10^4
U^{233}	Thermal	—	1,900	530	1.6×10^5
U^{234}	0.26	400	1,500	—	2.5×10^5
U^{235}	Thermal	—	1,300	580	7.1×10^8
U^{236}	0.70	600	850	—	2.4×10^7
U^{238}	1.25	550	550	—	4.5×10^9
Np^{237}	0.32	850	1,500	—	2.2×10^6
Pu^{239}	Thermal	—	2,000	750	2.4×10^4

*From: "Reactor Physics Constants", 2nd ed., Argonne National Laboratory, ANL–5800, U.S. Atomic Energy Commission, July 1963.

REFERENCES

"Neutron Cross Sections", 2nd ed., D.J. Hughes and R.B. Schwartz, Brookhaven National Laboratory, BNL–325, 1958, and Supplement 1, 1960.
"Detection of Neutrons", H.H. Barschall, Springer, 1958, 45:437.
"High-Energy Neutron Detectors", R.T. Siegel, Springer, 1958, 45:487.

Table 10–36. CHARGED PARTICLE-DETECTING PHOSPHORS*

Type	Phosphor or solvent	Density, g/cm³	Relative pulse height[a]	Decay time, mμs	Wavelengths of maximum emission, Å	Remarks
Inorganic crystal	NaI(Tl)	3.67	800	250	4100	Hygroscopic
	ZnS(Ag)	4.10	800 (β), 1400 (α)	70, 900[b]	4500	Multi-crystalline powder
	ZnS(Cu)	4.10	800 (β), 1200 (α)	~1000	5200	Multi-crystalline powder
	KI(Tl)	3.13	200	<1000	4100	Phosphorescent
	CaWO₄	6.03	400	~4,000	4300	Small crystals
	CaWO₄	7.90	500	~6,000	5200	Small crystals
	CsI(Tl)[c]	4.51	350	1200	4100–5700	—
Organic solids	Anthracene ($C_{14}H_{10}$)	1.25	100	23–38	4450	—
	p-Terphenyl	1.23	40	4.5	3900; 4100	—
	trans-Stilbene	1.16	60	3–8.2	3850	—
Organic liquids	Xylene[d, e]	0.86	40	—	3250–4000[f]	—
	Xylene[d, g]	—	44	—	—	—
	Toluene[d, g]	0.87	61	<2.9	4320	—
	Toluene[h]	—	54	3.0	3820	—
Organic plastics	Polystyrene[d, i]	1.06	36	2.2	4150	—
	Polystyrene[d, j]	—	39	4.0	4450	—
	Polyvinyltoluene[d]	—	51	—	4300	—
	p-Xylene[i]	—	70	—	—	—

[a] Anthracene taken as 100 for well-collimated monoenergetic high-energy electron beam. Statistical spreads in the relative pulse height may be as large as 25% because of the intrinsic resolution of the phosphor, variations in photocathode quantum efficiencies, and variations in the optical coupling coefficient.
[b] Y. Koechlin, L. Koch, and A. Lansiart, "Etat Actual du Developpement des Compteurs a Scintillation en France", CEA–980 Centre D'Etudes (Saclay) Report, 1958.
[c] C.T. Schmidt, *IRE Trans. on Nuclear Sci.*, NS-7, 25, 1960.
[d] Primary solute is p-Terphenyl.
[e] Secondary solute is α NPO = 2-(1-naphthyl)-5-phenyloxazole.
[f] Spectral range.
[g] Secondary solute is POPOP = 1,4-di-2-(5-phenyloxazole)-benzene.
[h] Primary solute is PBD = 2-phenyl-5-(4-biphenyl)-1,3,4-oxadiazole.
[i] Primary solute is PPO = 2,5-diphenyloxazole (most soluble of primary solutes).
[j] Secondary solute is TPB = 1,1,4,4, tetraphenyl-1,3-butadiene.

*From: "Reactor Physics Constants", 2nd ed., Argonne National Laboratory, ANL–5800, U.S. Atomic Energy Commission, July 1963.

Table 10-37. THERMAL- AND RESONANCE-DETECTING FOILS AND THEIR PROPERTIES*

Isotope and reaction	Natural abundance, %	Beta radiation[a]			Gamma radiation[a]		Cross section, barn		Resonance properties		
		Half-life	Max energy, Mev	Intensity,[a] %	Energy, Mev	Intensity,[b] %	Isotope[c]	Element[c]	Resonance energy, ev	Single[d] resonance, b	Measured total resonance integral, b
Mn^{55}(n,γ)Mn^{56}	100	2.58 hr	2.85	60	0.85	70	13.3±0.2	13.2±0.2	337	—	7.7±0.5
			1.05	25							
Rh^{103}(n,γ)Rh^{104}	100	4.4 min[e]	None	—	0.077 & 0.051[f]	99	12±2	194±4	1.257	1005	656
		44 sec	2.5	98.5	0.556	1.4					
Ag^{107}(n,γ)Ag^{108}	51.35	2.3 min	1.77	98	0.63	1.0	45±4	64.8±0.4	16.5	76.6	75
									41.5	5.95	
									51.4	15	
Ag^{109}(n,γ)Ag^{110}	48.65	253 day	0.087	58	0.61	1	3.2±0.4	—	5.20	870	1160
									30.5	14.2	
									40.2	6.1	
In^{115}(n,γ)In^{116m}	95.77	54.05 min	1.0	51	1.27	75	155±10	191±3	1.457	2810	3500
			0.87	28	2.09	25			3.86	43.5	
			0.6	21					9.12	41.5	
Dy^{164}(n,γ)Dy^{165}	28.18	139.2 min	1.25	—	0.095	—	2000±200	936±20	None	—	—
			0.88		0.361						
Ir^{191}(n,γ)Ir^{192}	38.5	74.3 day	0.67	48	0.47 & 0.32[f]	48	700±200	440±20	0.654	2380	3500
			0.53	41	0.32 & 0.60[f]	41			5.36	438	
Ir^{193}(n,γ)Ir^{194}	61.5	19 hr	2.24	66	0.33	20	130±30	—	1.303	1080	1370
			1.91	15	2.05	9			40.4	36.5	
			0.98	10							
Au^{197}(n,γ)Au^{198}	100	2.7 day	0.96	98.6	0.412	95	96±10	98.2±0.5	4.906	1180	1558
									61.2	36.2	

[a] Only the most intense radiation is listed.

[b] Intensities are given per unit parent atom decay.

[c] 2,200 meter/second neutrons.

[d] Calculated from single-level resonance parameters given in "Neutron Cross Sections", 2nd ed., D.J. Hughes and R. B. Schwartz, Brookhaven National Laboratory, BNL-325, 1958, and Supplement 1, 1960.

[e] 99+% of the 4.4-minute activity decays to the 44-second activity.

[f] Cascade transition.

*From: "Reactor Physics Constants", 2nd ed., Argonne National Laboratory, ANL-5800, U.S. Atomic Energy Commission, July 1963.

Table 10–38. RESONANCE-DETECTING FOILS AND THEIR PROPERTIES*

Isotope and reaction	Natural abundance, %	Beta radiation[a]			Gamma radiation[a]		Threshold properties		
		Half-life	Max energy, Mev	Intensity, %	Energy, Mev	Intensity,[a] %	Energy, Mev	Cross section,[b] mb	Average,[c] mb
$S^{32}(n,p)P^{32}$	95.02	14.2 day	1.71	100	None	—	1.0	120	62 ± 2
$Fe^{56}(n,p)Mn^{55}$	91.68	2.58 hr	2.85	60	0.85	70	2.1	15	0.87 ± 0.4
	—	—	1.05	25	—	—	—	—	—
$F^{19}(n,2n)F^{18d}$	100	112 min	0.65^e	97	None	—	10.4	—	—
$O^{16}(n,2n)O^{15f}$	99.96	2 min	1.72^e	100	None	—	16.5	—	—
$Si^{28}(n,p)Al^{28}$	92.18	2.27 min	2.88	100	1.78	100	4	600	—
$C^{12}(n,2n)C^{11}$	98.89	20.4 min	0.98^e	99	None	—	20.2	—	—
$Al^{27}(n,p)Mg^{27}$	100	9.45 min	1.75	58	0.84	70	1.8	32	3.43
	—	—	1.59	42	1.01	30	—	—	—
$P^{31}(n,p)S^{31}$	100	2.62 sec	1.49	99.9	None	—	1.0	65	30 ± 3

[a]Only the most intense radiation is listed. Intensities are given per unit parent atom decay.
[b]Cross section determined at three times the threshold energy by linear extrapolation of measured cross section.
[c]Average is taken over a fission spectrum.
[d]Used as CaF.
[e]Positron decay.
[f]If used in the form of cellophane, there is a 20.4-min positron decay due to $C^{12}(n,2n)C^{11}$ reaction.
*From: "Reactor Physics Constants", 2nd ed., Argonne National Laboratory, ANL–5800, U.S. Atomic Energy Commission, July 1963.

Table 10–39. CHARACTERISTICS OF SOME REPRESENTATIVE ORGANIC SCINTILLATORS*

Phosphor[a]	Electrons/cm³ ($\times 10^{23}$)	Hydrogen atoms/cm³ ($\times 10^{22}$)	RPAR[b]	S.D.T.,[c] mµs	$\tau/R' (E)$,[d] mµs-Mev per photoelectron
Anthracene	4.0	4.2	1.00	35	0.035
Stilbene	4.0	4.7	0.40	6	0.015
Quaterphenyl	~3.5	~3.9	0.85	4.2	0.005
Xylene + p-Terphenyl(5) + dibiphenyloxazole(BBO) (0.015)	2.8	4.9	0.5	4.2	0.008
Toluene + p-Terphenyl(9)	2.8	4.5	0.5	4.2	0.004
Phenylcyclohexane + p-Terphenyl(5) + diphenyloxazole benzene (POPOP)	3.0	5.5	0.54	~3	0.006
Triethylbenzene + p-Terphenyl(5)	2.9	5.7	0.5	~3	0.006
Isopropylbiphenyl + p-Terphenyl(5)	3.2	4.7	0.5	—	—
Polystyrene + p-Terphenyl (2.5)	3.2	4.7	0.3	~3	0.01
Polyvinyltoluene + p-Terphenyl	2.8	4.8	0.3	~3	0.01

[a]Solute contents given in grams per liter, where designated.
[b]Relative pulse amplitude response to electrons in small scintillators, compared to anthracene.
[c]Scintillation decay time.
[d]Ratio between scintillation decay time and the number R'(E) of photoelectrons released per Mev of electron energy absorbed in the phosphor.

*From: "Reactor Physics Constants", 2nd ed., Argonne National Laboratory, ANL–5800, U.S. Atomic Energy Commission, July 1963.

REFERENCE

"Time of Flight Techniques", J.H. Neiler and W.M. Good, Interscience Publishers, 1960, p. 501.

Table 10–40. GAMMA-RAY ATTENUATION COEFFICIENTS FOR SODIUM IODIDE 0.1%-THALLIUM-ACTIVATED CRYSTALS*

Photon energy, Mev	Scattering, cm⁻¹		Photoelectric K, L, and M shells, cm⁻¹	Pair production, cm⁻¹		Total inelastic cross section, cm⁻¹	
	Coherent	Incoherent		Nucleus	Electron	0.1% Tl correction	NaI–0.1 Tl
0.01	8.3858	0.6048	448.3	—	—	0.28150	448.90
0.01268	—	0.5990	200.1	—	—	0.14535	200.70
0.015	5.1538	0.5945	140.7	—	—	—	141.29
0.01537	—	0.5912	125.5	—	—	0.51057	126.09
0.020	3.5429	0.5841	62.14	—	—	0.24583	62.72
0.030	1.9424	0.5635	18.94	—	—	0.078539	19.50
0.03323	1.80095	0.5576	13.99	—	—	—	14.55
0.03323	1.80095	0.5576	110.88	—	—	—	111.44
0.040	1.37061	0.5458	66.36	—	—	0.035084	66.91
0.050	0.88574	0.5296	36.42	—	—	0.018428	36.95
0.060	0.64993	0.5148	22.12	—		0.011024	22.635
0.080	0.39644	0.4883	9.997	—		0.004991	10.485
0.08584	—	0.4820	5.774	—	—	0.004131	6.256
0.08584	—	0.4820	5.798	—	—	0.028292	6.280
0.10	0.27265	0.4646	5.325	—	—	0.018908	5.790
0.15	0.125744	0.4190	1.6716	—	—	0.006612	2.0906
0.20	0.736885	0.3836	0.7397	—	—	0.003176	1.1233
0.30	0.338967	0.3334	0.2386	—	—	0.0012696	0.5720
0.40	0.019159	0.2995	0.10658	—	—	0.0007541	0.4061
0.50	0.0131166	0.2731	0.05775	—	—	0.0005231	0.33085
0.60	0.0085479	0.2525	0.03702	—	—	0.0004112	0.28952
0.80	0.0053056	0.2219	0.019255	—		0.0003019	0.24116
1.0	0.0026528	0.1995	0.01244	—	—	0.0002476	0.21194
1.5	0.00176852	0.16197	0.006072	0.002639	—	0.0001862	0.17070
2.0	0.00073688	0.13822	0.003851	0.009008	—	0.0001657	0.15108
3.0	—	0.10872	0.00237	0.02347	0.0000103	0.000154	0.13457
4.0	—	0.09057	0.001629	0.03676	0.0000590	0.000153	0.12902
5.0	—	0.07818	0.001185	0.04798	0.0000103	0.000155	0.12745
6.0	—	0.06903	0.001037	0.05727	0.000148	0.000159	0.12748
8.0	—	0.05650	0.000740	0.07247	0.000295	0.000166	0.13001
10.0	—	0.04809	0.000593	0.08517	0.000590	0.000178	0.13444
15.0	—	0.03555	0.000444	0.10997	0.001033	0.000202	0.14700
20.0	—	0.028543	0.000296	0.12767	0.001475	0.000223	0.15798
30.0	—	0.020755	—	0.15350	0.001918	0.000254	0.17617
40.0	—	0.016477	—	0.16974	0.002507	0.000277	0.18872
50.0	—	0.013749	—	0.18450	0.002803	0.000294	0.20105
60.0	—	0.011845	—	0.19335	0.003098	0.000307	0.20829
80.0	—	0.009323	—	0.20811	0.003540	0.000328	0.22097
100.0	—	0.007744	—	0.21845	0.003983	0.000342	0.23018

*From: "Reactor Physics Constants", 2nd ed., Argonne National Laboratory, ANL–5800, U.S. Atomic Energy Commission, July 1963.

REFERENCE

"Applied Gamma-Ray Spectrometry", G. Crouthamel, Pergamon Press, 1960.

Table 10–41.   MEASUREMENT AND TEST STANDARDS

For addresses of engineering standards organizations, see Table A–5.

Standards may be established for a variety of objectives: for the control of quantity or quality, for establishing capacity ratings, for improving interchangeable manufacture, or for the protection of users or consumers. Standards have a long history, as do many standards organizations. In fact, there are so many changes in standards that many organizations revise their lists annually. An attempt is made in the following list to give current information, but this list is in no sense exhaustive. If its currency is to be verified, the user should contact the issuing organization.

Although standards for measurement and testing are of first concern here, the list of organizations and the standards issued or promoted by them cover a much wider field, including standards for materials, processes, and performance. The fact that all standards specify numerical values that must be established by measurement is a sufficient justification for treating all types of standards in this list.

While trade practices and industrial standards set up by trade associations are only mentioned herewith, it is recognized that many national and even international standards originate from trade standards. The "engineering standards" issued by the following organizations are established by authoritative but impartial groups that have no direct interest in the manufacture, marketing, or procurement of the materials or equipment to which the standards apply. Although U.S. Government and military standards fail to meet this criterion, these standards may be fully as important to many industrial organizations as the recognized "national standards". They represent a class of their own, however, and their literature is voluminous.*

Standards set up by the national engineering societies are intended for nation-wide use and are therefore classified as national standards.

NATIONAL ENGINEERING STANDARDS

NBS: The U.S. National Bureau of Standards is a government agency within the U.S. Department of Commerce, established in 1901 and charged with the "development and maintenance of the national standards of measurement". One of its divisions is the Institute of Basic Standards. An important NBS function is the development of standard practices, codes, and specifications, but these are developed largely for Government departments and do not necessarily come into nation-wide use in engineering.

NBS regularly broadcasts frequency and time standards and maintains a Radio Standards Laboratory in Boulder, Colorado (see Table 10–44).

In keeping with its function to "provide the means and methods for making measurements consistent with the national standards of measurement", NBS conducts testing, evaluation, and calibration services, provides standard reference materials, and cooperates in the setting of international standards. NBS publications are extensive and have been the center of activity of the "National Standards Reference Data System", which undertakes to compile critically evaluated, quantitative data on substances and materials. As far as engineers in industry are concerned, NBS codes for testing and measurement are usually considered advisory only, but on basic units they are the highest authority.

ANSI: The American National Standards Institute is an independent cooperative association that promulgates and publishes national standards, reprints and sells international standards, lists certain other technical standards, and advertises the standards of various industry groups. It issues an annual catalog offering copies of some 4,000 standards. Many of these are joint standards, approved and adopted by two or more agencies, but now designated as American National Standards (formerly called "ASA Standards" and "USA Standards"). The date of each standard or of its latest revision is indicated in each case.

ASTM: The American Society for Testing and Materials is a membership society that has become the foremost U.S. source of information on the specification and testing of materials. Its membership includes individuals, corporations, libraries, and over 1,000 government agencies. It publishes annual editions of ASTM Standards covering 32 fields, as listed in the following table. The annual editions include 4,500 standards, comprising 30,000 pages and a 250-page index.

---

*The Global Engineering Documentation Services, Inc., Newport Beach, Calif., lists 250,000 documents.
See also DODISS in Sweets Microfilms, published by the McGraw-Hill Information Systems Co.

Table 10–41. MEASUREMENT AND TEST STANDARDS *(Continued)*

ASTM PART No.	Subject	Month of new issue
1	Steel-Piping Materials	April
2	Ferrous Castings	April
3	Steel Sheet, Strip, Bar, Rod, Wire; Metallic-Coated Products	April
4	Structural and Boiler Steel; Forgings; Ferro-Alloys; Filler Metal	April
5	Copper and Copper Alloys	July
6	Light Metals and Alloys	July
7	General Nonferrous Metals; Electrodeposited Coatings; Metal Powders	July
8	Magnetic Properties; Materials for Electron Tubes, Thermostats, Electrical Heating	November
9	Cement; Lime; Gypsum	November
10	Concrete and Mineral Aggregates	November
11	Bituminous Materials; Soils	April
12	Mortars; Clay and Concrete Pipe and Tile; Masonry Units; Asbestos-Cement Products	April
13	Refractories; Glass; Ceramic Materials	April
14	Thermal Insulation; Acoustical Materials; Fire Tests; Building Constructions	November
15	Paper; Packaging; Cellulose; Casein; Flexible Barrier Materials; Leather	April
16	Structural Sandwich Construction; Wood; Adhesives	July
17	Petroleum Products—Motor Fuels; Solvents; Fuel Oils; Lubricating Oils; Cutting Oils; Grease	November
18	Petroleum Products—LPG; Light Pure Hydrocarbons; Wax Petrolatum	November
19	Gaseous Fuels; Coal and Coke	November
20	Paint—Materials Specifications and Tests; Naval Stores; Aromatic Hydrocarbons	April
21	Paint—Tests for Formulated Products and Applied Coatings	April
22	Soap; Antifreezes; Wax Polishes; Halogenated Organic Solvents	July
23	Industrial Water; Atmospheric Analysis	November
24	Textiles—General Methods and Definitions	November
25	Textiles—Fibers and Products	November
26	Plastics—Specifications	July
27	Plastics—Methods of Testing	July
28	Rubber; Carbon Black; Gaskets	July
29	Electrical Insulating Materials	July
30	General Testing Methods; Quality Control; Appearance Tests; Temperature Measurement; Radiation; Spectroscopy	July
31	Metallography; Nondestructive Tests; Fatigue; Corrosion	July
32	Chemical Analysis of Metals; Metal-Bearing Ores	April
33	Index	September

NFPA: The National Fire Protection Association publishes a ten-volume set of "National Fire Codes" (7000 pages), which recommend standards and practices for fire prevention, protection, and safety.

UL: The Underwriters' Laboratories is an approval and testing agency sponsored by the American Insurance Association. UL issues test instructions and approval rules. It grants approvals of electrical and structural materials and of equipment and methods relative to fire and casualty hazards and safety. It publishes standards, specifications, and annual lists of approved materials and equipment with names of manufacturers. Located at 207 E. Ohio St., Chicago, Ill., UL also maintains testing laboratories at Chicago and Northbrook, Ill., Melville, Long Island, N.Y., and Santa Clara, Calif.

Engineering Society Standards. Most of the national engineering societies are active in promoting standards relating to their particular field of engineering. Several of the larger societies have publication programs in this field and also cooperate actively with the ANSI.* Although the list of active societies and the fields in which they prescribe standards is too long to be given here, any engineer whose work involves standardization should consult Tables A-3 and A-5 and contact the appropriate societies.

Trade-association Standards. Trade-association standards vary from simple protective agreements to large public-service operations of benefit to the technical community and to users. Because many trade standards are of great importance to engineers, their existence should not be overlooked. This is particularly true of those associations that issue "approval requirements" †

Safety Standards. This group of standard recommendations to promote safety and health is important to

*The American National Standards Institute, formerly the United States of America Standards Institute (USASI).

†For information on trade associations, see "Directory of National Trade and Professional Associations of the United States", Columbia Publishers, Washington, D.C., 1969, and "Gale Encyclopedia of Trade Associations", Gale Research Co., Book Tower, Detroit, Mich. 48226, 1968.

engineers as well as to the medical profession, industrial hygienists, toxicologists, and others.‡

INTERNATIONAL STANDARDS ORGANIZATIONS

ISO: The International Organization for Standardization is a federation of fifty-six national member bodies similar to the ANSI. It issues three annual publications: the "Memento", giving names of member organizations and lists of committees and projects; the "Catalog", listing adopted standards and draft recommendations; and the "Journal", which includes the calendar of ISO meetings. The ANSI catalog lists and offers copies of approximately 700 ISO standards.

IEC: The International Electro-technical Commission promotes about 400 international standards and publishes many of them in French and Russian as well as English. It publishes an annual "IEC Central Office Report", giving information on its technical work in the electrical field.

COPANT: The Pan-American Standards Commission issues fifty or more standards in Spanish. It coordinates the national standards programs of its member bodies in the Western hemisphere.

CEE: The International Commission on Rules for the Approval of Electrical Equipment issues specifications that are widely followed in Europe. While its membership is all European, the United States has observer status. It is somewhat similar in function to the Underwriters' Laboratories, Inc. (UL approval) in the United States.

‡For additional information consult the following handbooks published by The Chemical Rubber Co.:
"CRC Handbook of Analytical Toxicology", I. Sunshine, Ed., 1969.
"CRC Handbook of Clinical Laboratory Data", 2nd ed., W.R. Faulkner, J.W. King, and H.C. Damm, Eds., 1968.
"CRC Handbook of Laboratory Safety", 2nd ed., N.V. Steere, Ed., 1971.
"CRC Handbook of Radioactive Nuclides", Y. Wang, Ed., 1969.

Table 10–42. FIXED POINTS FOR CALIBRATION OF TEMPERATURE-MEASURING INSTRUMENTS

Substance	Phase change at standard atmosphere pressure	Temperature °C	°F	Substance	Phase change at standard atmosphere pressure	Temperature °C	°F
Helium	Melts	−270.98	−455.76	Benzophenone	Boils	305.9‡	582.6
Helium	Boils	−268.93	−452.07	Cadmium	Melts	320.9‡	609.6
Hydrogen (normal)	Boils	−252.87	−423.17	Lead	Melts	327.3‡	621.1
Oxygen	Melts	−277.	−375.6	Potassium dichromate	Melts	397.5	747.5
Nitrogen	Boils	−195.85	−320.43	Zinc	Melts	419.58‡	787.24
Oxygen	Boils	−182.96†	−297.33	Sulfur	Boils	444.6†	832.3
Isopentane	Melts	−160.	−256.	Lead chloride	Melts	501.0	933.8
Methyl cyclohexane	Melts	−126.	−194.8	Calcium nitrate	Melts	561.0	1041.8
Carbon disulfide	Melts	−112.	−169.6	Antimony	Melts	630.5‡	1166.9
Toluene	Melts	−95.	−139.	Alumium	Melts	660.1‡	1220.2
Carbon dioxide	Sublimes	−78.5‡	−109.3	Potassium chloride	Melts	770.3	1418.5
Chloroform	Melts	−63.5	−82.	Sodium chloride	Melts	801.4	1474.5
Mercury	Melts	−38.87‡	−37.97	Sodium carbonate	Melts	852.0	1565.6
Carbon tetrachloride	Melts	−22.9	−9.22	Sodium sulfate	Melts	884.7	1624.5
Water ice	Melts	0.0†	32.	Silver	Melts	961.93†	1763.47
Sodium sulfate	Melts	+32.38	90.28	Gold	Melts	1064.43†	1947.97
Acetylene dichloride	Boils	55.0	131.	Copper	Melts	1083‡	1981.4
Ethyl alcohol	Boils	78.3	172.94	Lithium silicate	Melts	1201.	2193.8
Water	Boils	100.0†	212.	Barium fluoride	Melts	1280.	2336.
Toluene	Boils	110.0	230.	Nickel	Melts	1453.	2647.4
Benzoic acid	Boils	122.36‡	252.25	Cobalt	Melts	1480.	2696.
Bromobenzene	Boils	156.6	313.88	Iron	Melts	1530.	2786.
Aniline	Boils	184.5	364.1	Palladium	Melts	1552.‡	2826.
Nitrobenzene	Boils	209.0	308.2	Platinum	Melts	1769.‡	3216.
Naphthalene	Boils	218.0‡	424.4	Rhodium	Melts	1960.	3560.
Tin	Melts	231.9‡	449.4	Iridium	Melts	2443.	4429.4
Diphenyl	Boils	254.6	490.2	Tungsten	Melts	3380.	6116.

†Primary standards.
‡Secondary standards.

Table 10–43. STANDARD AIR AND ATMOSPHERE

See also Tables 7-1 and 7-3. For other properties of air and the atmosphere, consult the Index.

PAST AND PRESENT STANDARDS

The composition and properties of air and of the atmospheric envelope surrounding the earth have been variously defined and standardized for engineering purposes. Older standards cannot be readily changed or abandoned. Various assumed standards are convenient for different purposes and locations. There are large accumulations of equipment-performance data based on "standard" test codes. Each code uses some accepted but arbitrary definition of air, often neglecting the minor constituents or even the water vapor. Great numbers of tests of materials for aeronautical or space uses have been made in atmospheres simulated according to early "standard atmosphere" patterns.

Since there is no universally accepted "standard air", the following are offered for reference and comparison. Slight differences in the assumed composition of air may be included in the variations from one standard to another.

VARIOUS STANDARDS FOR DRY AIR

Density		Absolute pressure				Temperature			Molecular	
lb/ft³	kg/m³	in. Hg (32 deg F)	psi	mm Hg (0. deg C)	MN/m² (bars/10)	deg F	deg C	K	weight	Reference
.080 722	1.293 04	29.921	14.696	760.0	.101 325	32.0	0.0	273.15	28.966	1
.076 474	1.225 0	29.921	14.696	760.0	.101 325	59.0	15.0	288.15	28.964 4	2
.075 4	1.21	29.921	14.696	760.0	.101 325	68.0	20.0	293.15	–	3
.075	1.201 4	29.921	14.696	760.0	.101 325	69.5	20.8	293.95	–	4
.070 5	1.13	29.00	14.25	738.	.098 4	85.0	29.4	302.55	–	5
.074	1.184 7	29.921	14.696	760.	.101 325	77.0	25.0	298.15	–	6

REFERENCES

1. "Tables of the Thermal Properties of Gases", J. Hilsenrath et al., NBS Circular 564, National Bureau of Standards, 1955.
2. "U.S. Standard Atmosphere, 1962", U.S. Government Printing Office, 1962; this is the standard of NASA, USAF, USWB, USN, NBS, and ICAO.
3. ASME Test Codes. The intermediate temperature of 20 deg C is used in many practical gas tables.
4. ASHRAE Standard (dry air at 69.5 deg F; saturated air at 60 deg F; air at 68 deg F, 50% relative humidity). This is approximately equal to the ISO "Standard Reference Atmosphere", 20 deg C, 65% relative humidity, 1013 mbar.
5. SAE Engine and Gas Turbine Test Codes.
6. "Room Conditions." Widely used as a typical indoor condition, i.e., dry air at 25 deg C.

Table 10–44. TIME AND FREQUENCY STANDARDS

International *standard time,* or Greenwich Mean Time (*GMT*), is scientifically known as Coordinated Universal Time (*UTC*). It is one of four values on the Universal Time scale. The basic value *UTO* is derived from mean solarrotation data. Corrected for regular periodic variations, this becomes *UT*1. Irregular variations also occur; these are reported from all parts of the world to the Bureau International de l'Heure (*BIH*) in Paris. *BIH* in turn issues the corrections to produce the astronomer's time, *UT*2. Coordinated Universal Time (*UTC*) formerly operated with a frequency offset from the atomic scale of -300×10^{-10} to agree approximately with the rotation of the earth. Occasional step adjustments in time of 0.10 second were made to compensate for unpredictable variations in the earth's rate of rotation. On January 1, 1972, the *UTC* second was made equal to the SI second; to maintain approximate agreement with *UTI*, it will be necessary to add or subtract a leap-second every 6-18 months (preferably December 31 and June 30).

International Atomic Time (*IAT*) as maintained by the *BIH* now differs from *UTC* by an integral number of seconds.

In the United States the time corrections are made and sent out by radio broadcast by the National Bureau of Standards through the operation of four radio stations—*WWVB, WWV, WWVH,* and *WWVL.* The total annual correction to maintain *UTC* in close agreement with *UT2* is approximately 200 parts in 10^{10}; the *NBS* time signals are stable to one part in 10^{12} of frequency. The *UTC (NBS)* signals are maintained within five microseconds of the time signals of the U.S. Naval Observatory, *UTC (USNO).*

Table 10–44.   TIME AND FREQUENCY STANDARDS (*Continued*)

Time signals from the Bureau's radio station WWV can be heard on the telephone. By dialing (303)499-7111, listeners can hear the accurate shortwave signals from Fort Collins, Colorado, as received at the Bureau in Boulder, Colorado. These signals are a national service provided by the U.S. Department of Commerce.

The signals include a voice announcement of Greenwich Mean Time (*GMT*) every minute, plus standard audio-frequency tones and special announcements of interest to geophysicists and navigators. The time and frequency signals are the most accurate in the U.S. available to telephone users—callers from "the lower 48" should receive time signals accurate to within 30-40 milliseconds—and are controlled ultimately by the NBS Atomic Frequency Standard in Boulder.

The eight *NBS* services are as follows: standard radio frequencies, standard audio frequencies, standard musical pitch, standard time intervals, time signals, *UT2* corrections, radio propagation forecasts, and geophysical alerts. The broadcast services of the *NBS* stations are summarized in Table A.

Table A.   RADIO-BROADCAST SERVICES OF NBS STATIONS*

Services	WWV	WWVH	WWVB	WWVL	Services	WWV	WWVH	WWVB	WWVL
Standard radio frequencies:					Standard audio frequencies:				
20 khz				×	440 hz	×	×		
60 khz			×		600 hz	×	×		
2.5 Mhz	×	×			1000 hz	×			
5 Mhz	×	×			Standard time intervals	×	×	×	
10 Mhz	×	×			Time signals	×	×	×	
15 Mhz	×	×			Time code	×		×	
20 Mhz	×				*UT2* corrections	×	×	×	
25 Mhz	×				Radio propagation forecasts	×			
					Geophysical alerts	×	×		

Note: The station locations and radiated powers are as follows:

    At Fort Collins, Colorado: *WWV*—(10 kilowatts for 5, 10, and 15 Mhz; 2.5 kilowatts for 2.5, 20, and 25 Mhz)
                       WWVB—(12 kilowatts)
                       WWVL—(1.8 kilowatts)
    At Puunene, Maui, Hawaii:*WWVH*—(2 kilowatts for 5, 10, and 15 Mhz; 1 kilowatt for 2.5 Mhz)

*Data from: "Reference Data for Radio Engineers", 5th ed., Howard W. Sams & Co., Inc., Indianapolis, Indiana, 1968.

The low frequency radio station *WWVB* (Fort Collins, Colorado) will continue to broadcast seconds pulses without offset (as before January 1, 1972). However, the designation of its time scale, *SAT* (Stepped Atomic Time), will be changed to *UTC (NBS)*, and its adjustments will be the same as for the *UTC* scale discussed above. (Previously, *SAT* incorporated step adjustments of 200 milliseconds made on the first of the month, as necessary.)

UTC (NBS) as broadcast by *WWV*, *WWVH*, and *WWVB* will be maintained within ± 0.7 second of *UT1* and with ± 1 millisecond of *UTC (BIH)*. For navigators and others who may need to know *UT1* to better than 0.7 second, a special code will be broadcast to indicate the difference between *UT1* and *UTC* with a resolution of 0.1 second.

Standard time and frequency broadcasts are also transmitted from at least fifteen other countries, with accuracy of 20 parts in 10^9 or better. Most of these stations are operating under the *UTC* system.

BASIC TIME AND FREQUENCY STANDARDS

Ephemeris time is determined by observations of the moon. In 1956 the International Committee on Weights and Measures defined the constant second in terms of ephemeris time at noon January 1, 1900, as 1/31 566 925.9747 of the tropical year. Actually, the precision indicated b6 12 significant figures cannot be justified by present experimental observations of the solar system.

Sidereal time is determined from observation of the stars; it is not perfectly uniform because of variations in the rotational speed of the earth.

Atomic time is the term usually applied to the most accurate known methods for measuring time *intervals*. Earlier standards were based on induced vibrations of tuning forks, then later of quartz crystals. Vibrations of the molecules by electrical excitation produced the "ammonia clock", in which high-frequency electronic circuits, synchronized with the ammonia-molecule vibration, controlled a standard clock with an accuracy of one or two parts in a billion. More recently the gas cell and atomic-beam devices have provided even much greater accuracy, perhaps one second in 300 years.

The cesium-beam-controlled oscillator was designated as the official frequency standard by the 12th General Conference of Weights and Measures, 1964 (see Table B).

Table B.   STANDARD FREQUENCY SOURCES*

Typical Reported Values

Characteristic	Rubidium gas cell	Cesium beam	Hydrogen maser
Stability (rms deviation, 1 hr)	1×10^{12}	8×10^{13}	3×10^{14}
Resonance events per second	10^{12}	10^{6}	10^{12}
Nominal resonance frequency, Mhz	6834.682608	9192.631770	1420.405751
Detectable drift	Yes	Yes	No

*Data from: "Reference Data for Radio Engineers", 5th ed., Howard W. Sams & Co., Inc., Indianapolis, Indiana, 1968.

REFERENCES

"NBS Frequency and Time Broadcast Services", P.P. Viezbicke, Ed., NBS Special Publication 236, National Bureau of Standards, 1971 edition.
NBS Technical News Bulletin, 53:5, May 1969.
NBS Technical News Bulletin, 55:303, December 1971.
Frequency-Time Broadcast Services, National Bureau of Standards, Boulder, Colorado 80302.

Table 10–45. ENGLISH-METRIC EQUIVALENTS FOR STANDARD TENSILE TESTS*

In view of the increasing use of the international system (SI) of units, the dimensions for standard test specimens are given in both English and metric units.

DIMENSIONS OF STANDARD ROUND TENSILE-TEST SPECIMENS

One-Half Inch Round, Two-Inch Gage Length

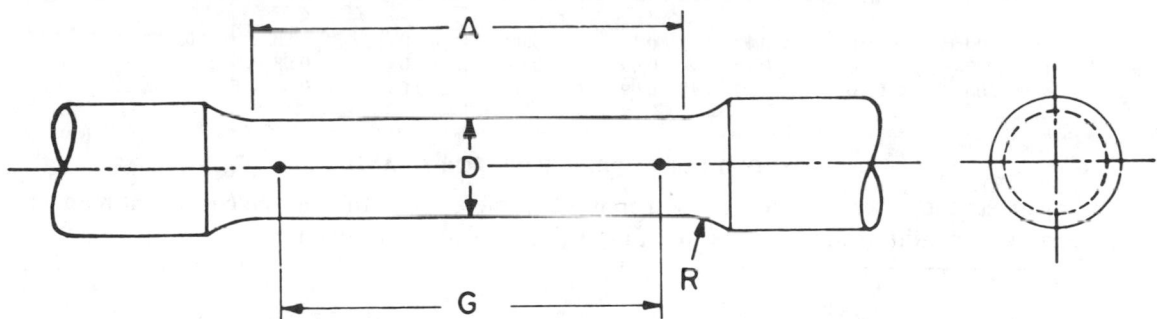

	Standard specimen		Small-size specimen proportional to standard	
	in.	mm	in.	mm
Nominal diameter	0.500	12.5	0.350	8.75
G—Gage length	2.000 ± 0.005	50.80 ± 0.13	1.400 ± 0.005	35.56 ± 0.13
D—Diameter	0.500 ± 0.010	12.70 ± 0.25	0.350 ± 0.007	8.89 ± 0.18
R—Radius of fillet, min	$^3/_8$	approx 9.5	$^1/_4$	approx 6.5
A—Length of reduced section, min	$2^1/_4$	approx 57	$1^3/_4$	approx 44.5

CONVERSION FACTORS: 1 in. = 25.4 mm (exactly)

$\qquad\qquad$ 1 lb$_f$ = 0.453 592 37 kg$_f$ (exactly)

*From: "ASTM Metric Practice Guide", National Bureau of Standards Handbook 102, U.S. Government Printing Office, 1967.

Table 10–46. BAROMETER CORRECTIONS

BAROMETER CORRECTIONS FOR TEMPERATURE

To correct the observed reading of a mercury barometer or U tube to the 32°F standard, add or subtract the following values in inches of mercury.

Actual tempera-ture of mercury column, °F	Observed reading of mercury column, in. Hg						
	20	22	24	26	28	30	32
	Add the correction						
−20	0.09	0.10	0.11	0.11	0.12	0.13	0.14
0	0.05	0.06	0.06	0.07	0.07	0.08	0.08
20	0.02	0.02	0.02	0.02	0.02	0.02	0.02
	Subtract the correction						
40	0.02	0.02	0.02	0.03	0.03	0.03	0.03
60	0.06	0.06	0.07	0.07	0.08	0.08	0.09
80	0.09	0.10	0.11	0.12	0.13	0.14	0.15
100	0.13	0.14	0.15	0.17	0.18	0.19	0.20

BAROMETER CORRECTIONS FOR ELEVATION

To correct the observed reading of a mercury barometer or U tube to the equivalent reading at a higher elevation, subtract the following values in inches of mercury for each 100-ft difference in elevation (add for lower elevation).

Mean elevation, ft	Mean atmospheric temperature, °F						
	−20	0	20	40	60	80	100
0	0.13	0.12	0.12	0.11	0.11	0.10	0.10
1,000	0.12	0.12	0.11	0.11	0.10	0.10	0.10
2,000	0.12	0.11	0.11	0.10	0.10	0.10	0.09
3,000	0.11	0.11	0.10	0.10	0.10	0.09	0.09
4,000	0.11	0.10	0.10	0.10	0.09	0.08	0.08
5,000	0.10	0.10	0.10	0.09	0.09	0.08	0.08
6,000	0.10	0.10	0.09	0.09	0.08	0.08	0.08
7,000	0.10	0.09	0.09	0.09	0.08	0.08	0.08

BAROMETER CORRECTIONS FOR GRAVITY

To correct the observed reading of a mercury barometer or U tube to the equivalent reading at standard gravity, add or subtract the following values in inches of mercury.

North latitude, deg	Elevation, ft							
	0	0	2000	2000	4000	4000	6000	6000
	Height of column, in. Hg							
	30	28	28	26	26	24	24	22
	Subtract the correction							
25	0.05	0.05	0.05	0.05	0.05	0.05	0.06	0.05
30	0.04	0.04	0.04	0.04	0.05	0.04	0.05	0.04
35	0.03	0.03	0.03	0.03	0.03	0.03	0.04	0.03
40	0.02	0.01	0.02	0.02	0.02	0.02	0.03	0.02
45	0.00	0.00	0.01	0.01	0.01	0.01	0.01	0.01
	Add the correction							
50	0.01	0.01	0.01	0.01	0.00	0.00	0.00	0.00

Table 10–47. EMERGENT STEM CORRECTION FOR LIQUID-IN-GLASS THERMOMETERS*

Accurate thermometers are calibrated with the entire stem immersed in the bath that determines the temperature of the thermometer bulb. However, for reasons of convenience, it is common practice when using a thermometer to permit its stem to extend out of the apparatus. Under these conditions both the stem and the mercury in the exposed stem are at a temperature different from that of the bulb. This introduces an error into the observed temperature. Since the coefficient of thermal expansion of glass is less than that of mercury, the observed temperature will be less than the true temperature if the bulb is hotter than the stem and greater than the true temperature, providing the thermal gradient is reversed. For exact work the magnitude of this error can only be determined by experiment. However, for most purposes it is sufficiently accurate to apply the following equation, which takes into account the difference of the thermal expansion of glass and mercury:

$$T_c = T_o + F \times L(T_o - T_m)$$

where T_c = corrected temperature

T_o = observed temperature

T_m = mean temperature of exposed stem. The mean temperature of the exposed stem may be determined by fastening the bulb of a second thermometer against the midpoint of the exposed liquid column.

L = the length of the exposed column in degrees above the surface of the substance whose temperature is being determined.

F = correction factor. For approximate work and when the liquid in the thermometer is mercury, a value for F of 0.00016 is generally used. For more accurate work with mercury-filled thermometers, values as given in the following table are used. For thermometers filled with organic liquids, it is customary to use 0.001 for the value of F.

VALUES OF F FOR VARIOUS GLASSES

$T_m°C$	Corning 0041	Corning 8800	Corning 8810	Jena 16 III	Jena 59 III
50	0.000157	0.000166	0.000156	0.000158	0.000164
150	0.000159	0.000167	0.000157	0.000158	0.000165
250	0.000163	0.000168	0.000161	0.000161	0.000170
350	0.000168	0.000173	0.000166	—	0.000177

*From: "CRC Handbook of Chemistry and Physics", 53rd ed., R.C. Weast, Ed., The Chemical Rubber Co., 1972.

Table 10–48. TECHNIQUES OF CHEMICAL ANALYSIS

For Various Types of Analytical Problems and Materials

INDEX OF TECHNIQUES

Numbers Refer to Methods Cited Below

Assay in the range of high purity: 28, 31, 32.

Elemental analysis:

Inorganic samples: 13, 14, 15, 16, 18, 20, 21, 22, 24, 34, 38, 39, 40, 46D, 47, 52C, 56.

Organic samples (for C, H, O, N, S, P, B, and halogens): 9, 11, 13, 15, 26, 34, 56.

Functional group analysis (for organic and inorganic combinations of atoms): 11, 13, 15, 16, 21, 22, 24, 34, 35, 37, 53, 56.

Gases*: 3, 8, 10, 11, 12, 28b) and c), 34, 35, 36, 38, 42, 45, 50, 52D, 54, 56.

Gross and surface examination of materials: 30, 46.

Identification and structure determination: 33, 34, 35, 36, 37, 44b, 49, 50, 51, 53, 54, 55, 56.

Metals and alloys: 4, 7, 9, 11, 13, 14, 15, 16, 24, 28a), 30, 33, 34, 38, 39, 41, 42, 43, 44, 46, 47, 49, 52BC, 56.

Molecular weight and molecular-weight distribution: 3, 12a, 13, 31, 32, 42, 49, 56, 58, 59b.

Petroleum products: 2, 3, 11, 13, 15, 16, 17, 18, 19, 24, 31, 32, 34, 35, 39, 41, 43, 45, 48, 52D, 56.

Physical properties related to composition:

Boiling point: 32.

Density: 12a, 52E.

Electric properties: 28, 29.

Freezing point: 31.

pH: 26b.

Refractive index: 45.

Thermal properties: 12bc, 33.

Vapor pressure: 59.

Trace quantities: 3, 4, 6, 20, 24, 25, 26, 34, 35, 38, 39, 40, 43, 52C, 56.

Water and sewage: 2, 4, 13, 15, 16, 17, 18, 19, 26, 28a), 29, 34, 39, 41, 52D, 53.

Table A.   METHODS OF SEPARATION

Method	Physical property serving as basis of separation	Remarks	Typical applications
1. Solvent extraction	Solubility	Includes counter-current extraction and the like; simple; rapid; applicable equally to trace and large amounts of material; can be done with very simple apparatus; use of complexing, chelating, or ion-pairing agent(s), plus adjustment of reagent concentrations and pH make this method very versatile	Organic, biochemical, and most inorganic compounds
2. Distillation analysis	Volatility	Desired fraction collected and measured; sample must not decompose thermally; azeotropic mixtures with constant composition and boiling point may be formed, prohibiting separation by ordinary methods.	Petroleum and other liquid mixtures; analysis of water in oil
3. Chromatography	Rate of migration along porous medium	Includes solution-adsorption, ion-exclusion, gel-permeation, paper, thin-layer, and gas chromatography; especially good for trace quantities; detection and estimation of separated substances by conversion to colored spots or by methods 13, 15, 16, 26ab, 29, 34, 35, 43, 45, and 52AC below	Organic, biochemical, and inorganic compounds; drugs; pesticides
4. Ion exchange		Used to remove ions from solution; to concentrate or isolate ions; to separate ions with very similar properties	Water softening; amino acids; rare-earth metals
5. Electrophoresis	Rate of migration in electric field	Used to separate and identify charged colloidal particles or macromolecular ions; mobility of species is characteristic property; position of moving boundary may be observed by discontinuity in refractive index of solution	Colloids; proteins; nucleic acids; polysaccharides
6. Electrochromatography	Rate of migration along porous medium in electric field	Combines best features of chromatography and electrophoresis; rapid preliminary evaluation of complex unknowns	Nuclear-fission products; biological tissue homogenates
7. Electroseparations	Conditions for electrodeposition	Electrodes, media, and electrolytic procedures chosen to give desired separation	Metals; alloys

*Gases can also be dissolved in a suitable liquid and determined by titrimetric or other methods.

Table 10–48. TECHNIQUES OF CHEMICAL ANALYSIS (Continued)

Table B. METHODS OF ANALYSIS

Method	Quantities measured	Remarks	Destructive	Typical applications
		GAS ANALYSIS		
8. Orsat analysis	Volume decrease after passing through series of reagents	Not positive means of identification of any component in unknown gas mixture; most useful if number of components is small and not too much alike	Yes	CH_4, C_2H_6, SO_2, CO, CO_2, H_2, O_2, H_2S, acetylenes, etc.
9. Gravimetric analysis	Weight gain of reagent	Used in organic elemental analyses	Yes	C, H, and O in organic compounds
10. Colorimetric indicators	Color formation	Detector tubes and impregnated papers commercially available; used for detecting toxic substances in air	Yes	AsH_3, $COCl_2$, HCN, Cl_2, CS_2, NH_3, CO, and other inorganic and organic compounds
11. Gasometric	Volume of gas released by reaction	Used for detection and determination of specific functional groups	Yes	Sulfamic acids; aldehydes; ketones, N in organic compounds
12. Physical measurements	a. Density or specific gravity	Standard air must be free of water and CO_2; used for mixtures where components differ appreciably in density; can be used to determine molecular weight	No	H_2 in N_2; solvent vapors in air; CO_2 in air
	b. Thermal conductivity	Suitable for monitoring to detect changes in composition	No	CO, CO_2, H_2, NH_3, SO_2, and H_2S in air, O_2, or N_2
	c. Dew point	Observe temperature to which surface exposed to sample must be lowered to just form a condensate; available in automatic form	No	Water vapor in air; also ethylene glycol and SO_2
		GRAVIMETRIC ANALYSIS		
13. Chemical precipitation	Weight of specimen	Depends on complete precipitation; convenient filtration and washing of precipitate; drying or ignition of precipitate; weighing	Yes	Metallic ions; anions; elements in alloys; cement, minerals, petroleum products
14. Electrodeposition	Weight of specimen	Deposition by electric current	Yes	Metals; alloys; Cu, Fe, Cr, and Sn
		TITRIMETRIC ANALYSIS*		
15. Visual	Color change of visual indicator	Color change must come at end-point and must not be masked by color of sample	Yes	Acid-base titrations; oxidation-reduction titrations
16. Potentiometric	Voltage	*Null-current:* end-point usually located as point of maximum potential change for specific-volume increment	Yes	Metallic ions; anions; metal ions using EDTA† titrant; Fe(II)–Ce(IV); weak acids and bases in non-aqueous media
		Constant-current: may use 1 or 2 polarized indicator electrodes; end-point usually taken as point of maximum potential; useful in cases where steady potentials are established only slowly in null-current potentiometry		Thiosulfate, dichromate, Mn, traces of As
17. pH	pH	Special case of potentiometric titrations, measuring hydrogen-ion concentration	Yes	Acids; bases; amino acids; intermediate salts of polyprotic acids, e.g., $H_2PO_4^-$
18. Amperometric	Electrical current	Potential is held constant; uses 1 polarizable electrode; end-point indicated by sharp change in direction of curve	Yes	Metallic ions; inorganic anions; certain organic compounds, e.g., styrene, *m*-aminophenol
19. Dead-stop or live-start end-point technique	Electrical current	Uses 2 polarizable electrodes; current decreases during titration until stops at end-point, or remains constant during titration and then begins to rise at end-point; apparatus extremely simple and easy to operate	Yes	Ca, Ce, Cu, U, and V ions; halogens or materials oxidizable with halogens
20. Coulometric	Quantity of electricity	End-points may be determined by spectrophotometry, potentiometry, zero-current potentiometry, amperometry, or other techniques; uses intermediate electrogenerated reactant (acid, base, oxidant, reductant, even some complexing agents) so that current and potential both remain essentially constant; extremely sensitive and precise; good for trace amounts	Yes	Various inorganic and organic substances; e.g., trinitrophenol, metal ions, Cl_2, H_2S, mercaptans, and pesticide residues
21. Conductometric	Electrical conductivity	Electrical current carried by ions in solution; ion concentration changed during titration by neutralization or precipitation reactions or by molecule formation	Yes	Acids, bases, anions, and cations

*Automatic titrators frequently use electrometric measurements for end-point determination.
†Ethylenediamine tetraacetic acid.

Table 10–48. TECHNIQUES OF CHEMICAL ANALYSIS (Continued)

Method	Quantities measured	Remarks	Destructive	Typical applications
22. High-frequency	Electrical parameters of high-frequency oscillatory circuit	Electrodes have no direct contact with solution being titrated	Yes	Metal ions; anions; organic and inorganic acids and bases; hydroxy compounds
23. Thermometric	Temperature of solution	Reagent added at constant rate; end-point indicated by abrupt change in heating or cooling rate	Yes	Metal ions titrated with EDTA†; acid-base and oxidation-reduction titrations
ELECTROMETRIC AND RELATED TECHNIQUES				
24. Voltammetry; polarography	Electrical current vs. voltage	Rapid, sensitive, and often very selective; depends on electrolysis of solutions containing oxidizable or reducible ions or molecules; applied potential to achieve $\frac{1}{2}$ of diffusion-limited current related to chemical identity of substance responsible for wave	Yes	Metal ions; anions; organic and bio-chemical compounds; elements in alloys, glass, cement, and ores
25. Chronopotentiometry	Potential vs. time	Current is held constant; curve characteristic of kind and concentration of species reacting at electrode; optimum concentration range: ~ 0.001 to $\sim 0.05 M$; at lower concentrations polarography still preferred	Yes	Mn, Pb, amines; phenols and other organic compounds; thickness of oxide and sulfide films on metals
26. Potentiometry	a. Electrical potential	Includes use of ion-selective electrodes	No	Solutions of acids and bases; bleach solutions; traces of metallic ions, such as Ag^+, Fe^{+++}, and Cu^{++}
	b. pH	Special case of potentiometry for measuring hydrogen-ion concentration	No	Weak aqueous solutions of acids and bases
27. Coulometry	Quantity of electricity	Constant potential; current decreases with time; similar to bulk electrolysis; very sensitive and accurate	Yes	Cl^-, Br^-, I^-, Fe, Cr, Sn, Ag, and Cu ions; ferrocyanide; various organic compounds
28. Conductometry	Electrical conductivity	a) Of solids, solutions, slurries, and wet solids; conductivity depends on type, mobility, and concentration of ions	No	Metals; alloys; semiconductors; water-purity determination
	Electrical conductivity	b) Of gases by absorption in liquids	Yes	SO_2, CO_2, H_2S, Cl_2
	Electrical conductivity	c) Of gases; organic halides selectively decomposed at heated anode surface to give positively charged ions	Yes	Traces of organic halides in air and other gases; Freons, methyl chloride, CCl_4
29. High-frequency techniques	Dielectric constant; complex impedance, etc.	Does not require electrode contact with sample; can be used to determine composition of binary systems; to follow chromatographic zones and distillation separations; to determine reaction-rate constants	No	Water in non-homogeneous systems; small concentrations of salt in water
30. Electrographic analysis	Location and color of spots on porous medium, e.g., paper	Paper moistened with electrolyte is sandwiched between two metal plates; electrical current causes migration of ions into paper to give colored products	No	Metals in alloys; porosity of metal coatings; elements in minerals
THERMOANALYTICAL METHODS				
31. Cryoscopy	Freezing point	Used to assay purity of high-purity materials; to calculate heat of fusion; to measure molecular weight of solute (good for weights up to $\sim 1,000$); sample must be soluble in chosen solvent at and below freezing point of latter	No	Purity in assay of benzene, phenol, Hg, styrene, vinylpyrrolidone; molecular weights of petroleum products, olefinic and acetylenic compounds, acyl chlorides, and peroxides
32. Ebullioscopy	Boiling point	Analogous to freezing-point method; care must be taken to avoid superheating	No	Organic compounds; petroleum fractions
33. Differential thermal analysis	Temperature of sample vs. temperature of inert reference substance	Includes differential-scanning calorimetry, gas-evolution analysis, etc., ambient temperature raised at controlled rate; used for qualitative detection or semiquantitative evaluation of compounds that undergo physical or thermochemical reactions, such as oxidation, dehydration, adsorption, etc.	Yes	Impurities in molten or solid salts, metals, and alloys; molecular weight and structure of organic polymer melts

†Ethylenediamine tetraacetic acid.

Table 10–48. TECHNIQUES OF CHEMICAL ANALYSIS *(Continued)*

Method	Quantities measured	Remarks	Destructive	Typical applications
		OPTICAL AND RELATED TECHNIQUES		
34. Visible and ultraviolet spectroscopy; colorimetry	Light absorption; wavelength	Speed, simplicity, and accuracy in quantitative determinations; of limited help in identification problems; sometimes preferable to infrared because stronger absorption bands permit detection of smaller quantities of unknown	No	Metallic and non-metallic elements; organic compounds; pesticides; drugs; Cl_2, SO_2, NO_2 CS_2, ozone, and Hg vapor; air pollutants
35. Infrared spectroscopy	Light absorption; wavelength	Very useful in identification and structure proof for heteroatomic compounds in gases, many non-aqueous liquids, and some solids; in quantitative determinations it is least selective and has poorest sensitivity of all spectral methods, especially when applied to mixtures; may give excellent spectra after separation	No	Organic compounds; drugs; polymers; pesticides; hydroxy compounds; carboxylic acids; hydrazides; amines
36. Microwave spectroscopy	Microwave-absorption frequency	Pure rotation spectra of many molecules lie in microwave region; limited flexibility and sensitivity, but enormous resolution; used for identification of unknowns; quantitative analysis not easy	No	Gases and vapors, e.g., H, He^+, Na, Cs, N, O, P; NH_3, HCN, H_2O, methanol
37. Raman spectroscopy	Light-emission, wavelength	Similar to infrared techniques; change in frequency of incident monochromatic light (e.g., Hg line or laser beam) due to change in rotational or vibrational energy of scattering molecules; spectrum of low intensity	No	Clear aqueous solutions; colorless, non-fluorescing solutions of inorganic substances, free of suspended solids
38. Emission spectroscopy	Light emission, wavelength	Excitation by arc or spark; specialized training necessary to achieve proper excitation and interpretation of spectra; permits simultaneous determination of ~70 elements in sample in single operation; normally only few milligrams of sample required; entire operation takes relatively short time	Yes	Metals and many non-metals, elements in alloys, minerals, organic and inorganic solids, liquids, gases that constitute less than 10% of sample
39. Flame photometry	Light emission, wavelength	Similar to emission spectroscopy, except excitation by flame and simpler technique and instrumentation are used; detects much more limited number of elements than emission spectroscopy	Yes	Alkali and alkaline earth metals in solution; Cu, Mn, Ni, B, Co, Ga, In, Ag; tetraethyl lead in gasoline; Ca, K, Mg, and Na in petroleum products
40. Atomic-absorption spectroscopy	Light absorption	Sample atomized into flame where desired element absorbs light from monochromatic source; higher sensitivity than flame photometry and applicable to more elements; only one element can be measured at a time	Yes	~65 metallic elements in liquids and dissolved solids
41. Turbidimetry	Light absorption by suspension	Used for elements that do not give good color reactions or that give precipitates difficult to separate from mother liquors; accuracy low because of variables involved	No	Elements in colorless, opaque suspensions that do not show selective absorption: e.g., Ag, Ca, Cl, S, and Sn
42. Nephelometry	Light scattering by suspension	Used for molecular-weight determination in liquids and for particle-size determination in dilute suspensions of gases and liquids	No	Macromolecules of molecular weight range ~1000 to ~5,000,000; aerosols or hydrosols in air or gas; Cl in Zr alloys
43. Fluorescence analysis	Light emission, wavelength	Including quenchofluorimetry; absorbed light (usually ultraviolet) is reemitted at longer wavelengths; often more specific and more sensitive than colorimetric methods; of greatest use in determination of concentrations too small to be easily determined by spectrophotometry or emission spectroscopy	No	Organic and biochemical substances; pigments, dyes, phosphors; drugs; petroleum products; trace amounts of elements; fluoranthene in presence of other hydrocarbons
44. Polarimetry	a. Optical activity	Angle of rotation of plane of polarization of light at some wavelength; used for quantitative determination of composition	No	Many organic compounds and a few inorganic complexes; sugar solutions; starches; resins
	b. Optical rotatory dispersion	Variation of optical activity with wavelength; used in structural and stereochemical studies	No	Organic and biochemical compounds; carbonyl, xanthate, nitrite, carbocyclic and heterocyclic aromatic compounds

Table 10–48. TECHNIQUES OF CHEMICAL ANALYSIS *(Continued)*

Method	*Quantities measured*	*Remarks*	*Destructive*	*Typical applications*
45. Refractometry	Index of refraction	Ratio of velocity of light in vacuum to that in substance; used to identify unknowns; to determine purity of substances and analyze binary mixtures quantitatively; to follow separations by distillation, chromatography, etc.; to grade optical solids such as glasses and plastics	No	Petroleum fractions; organic liquids and transparent solids; acids, salts, alcohols, soluble esters and ethers; monitoring gases for changes in composition
46. Microscopy: A. Optical	Differential absorption of light by adjacent areas of specimen	Used to determine refractive index of crystals, melting points, optical crystal properties; to examine sample for surface characteristics, homogeneity, morphology, etc.	No	Organic and inorganic compounds; drugs; minerals; metals; alloys
B. Electron	Differential scattering of electrons by adjacent areas of specimen	Magnifying power far greater than optical microscopy; one can see objects much smaller than wavelengths of light; used for study of surfaces; qualitative identification of particles and components of materials; quantitative data on particle size and size-distribution	No	Aerosols; alloys; catalysts; ceramic materials; clays; fibers; glasses; metals; pigments; polymers; plastics; resins; rubber; waxes
C. Field-ion	Differential scattering of ions by adjacent areas of specimen	Sharp tip of needle-shaped specimen viewed end-on; tip is held in vacuum at high positive voltage; trace of gas (e.g., He) admitted; when He atom touches atom of tip, it becomes positive ion and flies down particular line of force from that metal atom to fluorescent viewing screen	No	Metals; alloys
D. Electron-probe microanalysis	X-ray emission	Electrons can be sharply focused to determine microdistribution of elements in sample; used in studies of localized impurities in alloys, etc.	No	Elements from atomic number 5 and greater; alloys, semiconductor assemblies, metals in particular biological membranes or cells
47. X-ray fluorescence	Intensity and wavelength of emitted X-rays	X-rays excited by absorption of primary X-rays or high-energy electrons; used for analysis of elements in sample and to measure thickness of surface coatings	No	Elements from atomic number 9 and greater, especially rare-earth metals, Pt, Hf, Zr, Nb, Ta, W, Mo, U, in solids such as metals and alloys
48. X-ray absorption	X-ray attenuation	Used to determine heavy elements in light elements; may use polychromatic or monochromatic X-rays; can be used to measure thickness of surface coatings and in production-line control, as in gauging of sheet metal	No	Liquid or solid samples; range of elements is similar to above; S, Pb, Br, hydrocarbons, etc., in petroleum products
49. X-ray diffraction	Angle and intensity of scattered X-rays	X-rays interact with electrons and so reveal electron-density distribution in the crystal; used to determine crystal structure; identify molecules; determine effects of heat treatment; allotropic modifications; stability of phases, etc.	No	Crystalline substances; metals; alloys
50. Electron diffraction	Angle and intensity of scattered electrons	Electrons interact with electric fields of atoms and so reveal distribution of electrical potential in crystal; compared to X-ray diffraction: light atoms contribute more significantly to diffraction patterns; diffraction angles much smaller; fewer atoms needed to produce usable pattern	No	Gases; very thin specimens; surface layers of thicker specimens, e.g., corrosion and other inorganic layers on metals; pigments in paint coatings
51. Neutron diffraction	Angle and intensity of scattered neutrons	Neutrons interact primarily with atomic nuclei, revealing positions (in some respects superior to X-ray diffraction for this); neutrons also interact with electrons with unpaired spins, revealing distribution	No	Crystals

NUCLEAR AND MAGNETIC METHODS

Method	*Quantities measured*	*Remarks*	*Destructive*	*Typical applications*
52. Radiochemical techniques: A. Tracers	Radioactivity of tracer	Used to tag compound so it can be followed through chemical reactions and industrial, biological, or geological processes		Organic and biochemical compounds labeled with C^{14} or tritium

Table 10–48. TECHNIQUES OF CHEMICAL ANALYSIS (Continued)

Method	Quantities measured	Remarks	Destructive	Typical applications
B. Isotope dilution	Radioactivity of tracer	Known quantity of radionuclide isotopic with element to be determined is added to sample; some of desired element is recovered and measured, and radioactivity of accompanying tracer determined	Yes	Various metals in alloys; Th in presence of Y and trivalent rare-earth metals; organic and biochemical mixtures; geochemical and archeological samples
C. Neutron activation	Radioactivity induced by neutron irradiation	*Advantages:* sensitivity, ability to confirm purity of radionuclide measured and, in many cases, speed and simplicity; *limitations:* intense radioactivity of some samples, self-shielding, extraneous nuclear reactions	Yes	~75 trace elements; Cs, Rb in NaK alloys; Na and P in Al alloys
D. Beta-ray absorption	Beta-ray attenuation	Used to determine concentration of hydrogen and to gauge thickness of materials such as cellophane and other plastics, paper, etc.	No	H in petroleum products; H_2O in liquids
E. Gamma-ray absorption	Gamma-ray attenuation	Can be used for density determinations of liquids, powders, or slurries; used mainly for continuous in-process measurement	No	
53. Nuclear magnetic resonance (NMR)	Radio-frequency absorption in uniform magnetic field	Caused by energy transitions of nuclear spin; generally restricted to liquid (and occasionally compressed gaseous) samples; does not have great sensitivity at present for quantitative determinations; powerful tool for structure and identification problems	No	Compounds containing H^1, B^{11}, C^{13}, F^{19}, P^{31}; determination of water in liquids and solids; biochemical compounds; minerals
54. Electron-spin resonance (ESR) (paramagnetic resonance)	Microwave absorption vs. field strength	Caused by energy transitions of electron spin; quantitative estimation very useful, but requires rigorous attention to sources of error; ESR spectra obtainable on far fewer compounds than NMR; useful in solid state physics, crystallography, inorganic and organic chemistry, and biology	No	Substances containing unpaired electrons; free radicals; ions with net magnetic moment —transition series, rare-earth and actinide metals in appropriate valence states; defects in solids
55. Nuclear-quadrupole resonance (NQR)	Radio-frequency absorption in electrical field	Caused by energy transitions of nuclear quadrupole moment; NQR field not well developed at present	No	Glasses, very viscous liquids and solids; Cl, Br, I, N, B, Al, Sb, As, Cu, and others
56. Mass spectrometry	Mass numbers and ion-beam intensities	Ionized fragments of gaseous sample separated by mass in magnetic field and ion current measured; extremely sensitive; gives fragmentation pattern (unique for each compound) and molecular weight; able to distinguish isotopes	Yes	Gaseous mixtures and vapors; organic compounds; metals; semiconductors; materials in nuclear reactors; geologic-age dating
MISCELLANEOUS METHODS [a]				
57. Particle-size analysis: A. Gravitational sedimentation	Rate of settling in liquid	Suitable for particles in 2–50-micron range	Yes	Paint pigments, color pigments in plastics industry; products applied from aerosol-spray cans; flour and other dry ingredients in food industry; slip-agent particles in coated cellophane
B. Optical or electron microscopy	Counting and classifying particles by size	May be automated with electronic-scanning devices; particle-size range: >0.5 micron (optical); ~0.01 to 5 microns (electron)	No	
C. Sieve analysis	Weight remaining on screens	Usually for particle sizes >70 microns, but screens with apertures as small as 5 microns now available	Yes	
D. Special apparatus	Sedimentation, direct counting, or light scattering	Variety of apparatus available commercially		
58. Ultracentrifuge analysis	Sedimentation height vs. time	Used for molecular-weight determination from molecular weights of ~1,000 on up; determination of particle-size distributions for particles down to ~0.1 microns	Yes	Biochemical substances; proteins; polymers; high molecular-weight substances

[a]The term *micron* for 10^{-6} m is replaced by the term *micrometer,* symbol μm, in the SI system of units.

Table 10–48. TECHNIQUES OF CHEMICAL ANALYSIS (*Continued*)

Method	Quantities measured	Remarks	Destruc-tive	Typical applications
59. Vapor pressure	a. Partial vapor pressure	Pure sample of substance to be determined is sealed into bulb and immersed in boiling liquid; ratio of its vapor pressure to total vapor pressure proportional to vapor concentration	No	Liquid mixtures
	b. Vapor-pressure lowering	Used to determine weight of solute; commercial apparatus can be used for molecular weights up to 50,000 or 100,000	No	Biochemical substances; polymers; high molecular-weight substances

REFERENCES

"Modern Methods of Chemical Analysis", R.L. Pecsok and L.D. Shields, John Wiley & Sons, 1968.
"Survey of Analytical Chemistry", S. Siggia, McGraw-Hill Book Company, 1968.
"Handbook of Analytical Chemistry", L. Meites, Ed., McGraw-Hill Book Company, 1963.

Table 10–49. ACID-BASE INDICATORS, pH SCALE*

Indicator	Approximate pH range	Color change	Preparation
o-Cresolsulfonphthalein (Cresol red)	0.0–1.0	red to yel	0.1 g in 26.2 ml 0.01N NaOH + 223.8 ml water
Methyl violet	0.0–1.6	yel to bl	0.01–0.05% in water
Crystal violet	0.0–1.8	yel to bl	0.02% in water
Ethyl violet	0.0–2.4	yel to bl	0.1 g in 50 ml MeOH + 50 ml water
Malachite green	0.2–1.8	yel to bl-grn	Water
Methyl green	0.2–1.8	yel to bl	0.1% in water
2-(p-Dimethylaminophenylazo)-pyridine	0.2–1.8	yel to red	0.1% in EtOH
o-Cresolsulfonphthalein (Cresol red)	0.4–1.8	yel to red	0.1 g in 26.2 ml 0.01N NaOH + 223.8 ml water
Quinaldine red	1.0–2.2	col to red	1% in EtOH
p-(p-Dimethylaminophenylazo)-benzoic acid, Na-salt (Paramethyl red)	1.0–3.0	red to yel	EtOH
m-(p-Anilinophenylazo)benzenesulfonic acid, Na-salt (Metanil yellow)	1.2–2.4	red to yel	0.01% in water
4-Phenylazodiphenylamine	1.2–2.6	red to yel	0.01 g in 1 ml 1N HCl + 50 ml EtOH + 49 ml water
Thymolsulfonphthalein (Thymol blue)	1.2–2.8	red to yel	0.1 g in 21.5 ml 0.01N NaOH + 229.5 ml water
m-Cresolsulfonphthalein (Metacresol purple)	1.2–2.8	red to yel	0.1 g in 26.2 ml 0.01N NaOH + 223.8 ml water
p-(p-Anilinophenylazo)-benzenesulfonic acid, Na-salt (Orange IV)	1.4–2.8	red to yel	0.01% in water
4-o-Tolylazo-o-toluidine	1.4–2.8	or to yel	Water
Erythrosine, disodium salt	2.2–3.6	or to red	0.1% in water
Benzopurpurine 48	2.2–4.2	vt to red	0.1% in water
N, N-Dimethyl-p-(m-tolylazo)aniline	2.6–4.8	red to yel	0.1% in water
4,4'-bis(2-amino-1-naphthylazo)2,2'-stilbene-disulfonic acid	3.0–4.0	purp to red	0.1 g in 5.9 ml 0.05N NaOH + 94.1 ml water
Tetrabromophenolphthalein ethyl ester, K-salt	3.0–4.2	yel to bl	0.1% in EtOH
3',3'',5',5''-Tetrabromophenolsulfonphthalein (Bromophenol blue)	3.0–4.6	yel to bl	0.1 g in 14.9 ml 0.01N NaOH + 235.1 ml water
2,4-Dinitrophenol	2.8–4.0	col to yel	Saturated water solution
N, N-Dimethyl-p-phenylazoaniline (p-Dimethyl-aminoazobenzene)	2.8–4.4	red to yel	0.1 g in 90 ml EtOH + 10 ml water
Congo red	3.0–5.0	blue to red	0.1% in water
Methyl orange-Xylene cyanole solution	3.2–4.2	purp to grn	Ready solution
Methyl orange	3.2–4.4	red to yel	0.01% in water
Ethyl orange	3.4–4.8	red to yel	0.05–0.2% in water or aqueous EtOH
4-(4-Dimethylamino-1-naphthylazo)-3-methoxy-benzenesulfonic acid	3.5–4.8	vt to yel	0.1% in 60% EtOH
3',3'',5',5''-Tetrabromo-m-cresolsulfonphthalein (Bromocresol green)	3.8–5.4	yel to bl	0.1 g in 14.3 ml 0.01N NaOH + 235.7 ml water

Table 10–49. ACID-BASE INDICATORS, pH SCALE *(Continued)*

Indicator	Approximate pH range	Color change	Preparation
Resazurin	3.8–6.4	or to vt	Water
4-Phenylazo-l-naphthylamine	4.0–5.6	red to yel	0.1% in EtOH
Ethyl red	4.0–5.8	col to red	0.1 g in 50 ml MeOH + 50 ml water
2-(*p*-Dimethylaminophenylazo)-pyridine	4.4–5.6	red to yel	0.1% in EtOH
4-(*p*-Ethoxyphenylazo)-*m*-phenylenediamine monohydrochloride	4.4–5.8	or to yel	0.1% in water
Lacmoid	4.4–6.2	red to bl	0.2% in EtOH
Alizarin red S	4.6–6.0	yel to red	Dilute solution in water
Methyl red	4.8–6.0	red to yel	0.02 g in 60 ml EtOH + 40 ml water
Propyl red	4.8–6.6	red to yel	EtOH
5′,5″-Dibromo-*o*-cresolsulfonphthalein (Bromo-cresol purple)	5.2–6.8	yel to purp	0.1 g in 18.5 ml 0.01N NaOH + 231.5 ml water
3′,3″-Dichlorophenolsulfonphthalein (Chloro-phenol red)	5.2–6.8	yel to red	0.1 g in 23.6 ml 0.01N NaOH + 226.4 ml water
p-Nitrophenol	5.4–6.6	col to yel	0.1% in water
Alizarin	5.6–7.2	yel to red	0.1% in MeOH
2-(2,4-Dinitrophenylazo)-1-naphthol-3, 6-disulfonic acid, disodium salt	6.0–7.0	yel to bl	0.1% in water
3′,3″-Dibromothymolsulfonphthalein (Bromo-thymol blue)	6.0–7.6	yel to bl	0.1 g in 16 ml 0.01N NaOH + 234 ml water
6,8-Dinitro-2,4-(1H)quinazolinedione (*m*-Dinitrobenzoylene urea)	6.4–8.0	col to yel	25 g in 115 ml M NaOH + 50 ml boiling water + 0.292 g of NaCl in 100 ml water
Brilliant yellow	6.6–7.8	yel to or	1% in water
Phenolsulfonphthalein (Phenol red)	6.6–8.0	yel to red	0.1 g in 28.2 ml 0.01N NaOH + 221.8 ml water
Neutral red	6.8–8.0	red to amb	0.01 g in 50 ml EtOH + 50 ml water
m-Nitrophenol	6.8–8.6	col to yel	0.3% in water
o-Cresolsulfonphthalein (Cresol red)	7.2–8.8	yel to red	0.1 g in 26.2 ml 0.01N NaOH + 223.8 ml water
Curcumin	7.4–8.6	yel to red	EtOH
m-Cresolsulfonphthalein (Metacresol purple)	7.4–9.0	yel to purp	0.1 g in 26.2 ml 0.01N NaOH + 223.8 ml water
4,4′-bis(4-amino-l-naphthylazo)-2,2′-stilbenedi-sulfonic acid	8.0–9.0	bl to red	0.1 g in 5.9 ml 0.05N NaOH + 94.1 ml water
Thymolsulfonphthalein (Thymol blue)	8.0–9.6	yel to bl	0.1 g in 21.5 ml 0.01N NaOH + 228.5 ml water
o-Cresolphthalein	8.2–9.8	col to red	0.04% in EtOH
p-Naphtholbenzene	8.2–10.0	or to bl	1% in diluted alkali
Phenolphthalein	8.2–10.0	col to pink	0.05 g in 50 ml EtOH + 50 ml water
Ethylbis(2,4-dimethylphenyl)acetate	8.4–9.6	col to bl	Saturated solution in 50% acetone alcohol
Thymolphthalein	9.4–10.6	col to bl	0.04 g in 50 ml EtOH + 50 ml water
5-(*p*-Nitrophenylazo)salicylic acid, Na-salt (Alizarin yellow R)	10.1–12.0	yel to red	0.01% in water
Alizarin	11.0–12.4	red to purp	0.1% in MeOH
p-(2,4-Dihydroxyphenylazo)benzenesulfonic acid, Na-salt	11.4–12.6	yel to or	0.1% in water
5,5′-Indigodisulfonic acid, disodium salt	11.4–13.0	bl to yel	Water
2,4,6-Trinitrotoluene	11.5–13.0	col to or	0.1–0.5% in EtOH
1,3,5-Trinitrobenzene	12.0–14.0	col to or	0.1–0.5% in EtOH
Clayton yellow	12.2–13.2	yel to amb	0.1% in water

*From: "CRC Handbook of Chemistry and Physics", 53rd ed., R.C. Weast, Ed., The Chemical Rubber Co., 1972.

Table 10–50. FLAME AND BEAD TESTS*

FLAME COLORATIONS

Violet: Potassium compounds. Easily obscured by sodium flame. Purple-red through blue glass. Bluish-green through green glass. Rubidium and cesium compounds impart same flame as potassium compounds.

Blue: *Azure*—copper chloride. Copper bromide gives azure blue followed by green. Other copper compounds give same coloration when moistened with hydrochloric acid.
Light blue—lead, arsenic, selenium.

Green: *Emerald*—copper compounds except the halides, and when not moistened with hydrochloric acid.
Pure green—compounds of thallium and tellurium.
Yellowish—barium compounds. Some molybdenum compounds. Borates, especially when treated with sulfuric acid or when burned with alcohol.
Bluish—phosphates with sulfuric acid.
Feeble—antimony and ammonium compounds.
Whitish—zinc.

Red: *Carmine*—lithium compounds. Violet through blue glass. Invisible through green glass. Masked by barium flame.
Scarlet—strontium compounds. Violet through blue glass. Yellowish through green glass. Masked by barium flame.
Yellowish—calcium compounds. Greenish through blue glass. Green through green glass. Masked by barium flame.

Yellow: All sodium compounds. Invisible with blue glass.

BEAD TESTS

ABBREVIATIONS:

c—cold	ns—nonsaturated
h—hot	s—saturated
hc—hot or cold	ss—supersaturated

BORAX BEADS

Substance	Oxidizing flame	Reducing flame
Aluminum	Colorless (hc, ns); opaque (ss)	Colorless; opaque (s)
Antimony	Colorless; yellow or brownish (h, ss)	Gray and opaque
Barium	Colorless (ns)	—
Bismuth	Colorless; yellow or brownish (h, ss)	Gray and opaque
Cadmium	Colorless	Gray and opaque
Calcium	Colorless (ns)	—
Cerium	Red (h)	Colorless (hc)
Chromium	Green (c)	Green
Cobalt	Blue (hc)	Blue (hc)
Copper	Green (h); blue (c)	Red (c); opaque (ss); colorless (h)
Iron	Yellow or brownish red (h, ns)	Green (ss)
Lead	Colorless; yellow or brownish (h, ss)	Gray and opaque
Magnesium	Colorless (ns)	—
Manganese	Violet (hc)	Colorless (hc)
Molybdenum	Colorless	Yellow or brown (h)
Nickel	Brown; red (c)	Gray and opaque
Silicon	Colorless (hc); opaque (ss)	Colorless; opaque (s)
Silver	Colorless (ns)	Gray and opaque
Strontium	Colorless (ns)	—
Tin	Colorless (hc); opaque (ss)	Colorless; opaque (s)
Titanium	Colorless	Yellow (h); violet (c)
Tungsten	Colorless	Brown
Uranium	Yellow or brownish (h, ns)	Green
Vanadium	Colorless	Green

BEADS OF MICROCOSMIC SALT NaNH₄HPO₄

Substance	Oxidizing flame	Reducing flame
Aluminum	Colorless; opaque (s)	Colorless; not clear (ss)
Antimony	Colorless (ns)	Gray and opaque
Barium	Colorless; opaque (s)	Colorless; not clear (ss)
Bismuth	Colorless (ns)	Gray and opaque
Cadmium	Colorless (ns)	Gray and opaque
Calcium	Colorless; opaque (s)	Colorless; not clear (ss)
Cerium	Yellow or brownish red (h, s)	Colorless
Chromium	Red (h, s); green (c)	Green (c)
Cobalt	Blue (hc)	Blue (hc)
Copper	Blue (c); green (h)	Red and opaque (c)
Iron	Yellow or brown (h, s)	Colorless; yellow or brownish (h)
Lead	Colorless (ns)	Gray and opaque
Magnesium	Colorless; opaque (s)	Colorless; not clear (ss)
Manganese	Violet (hc)	Colorless
Molybdenum	Colorless; green (h)	Green (h)
Nickel	Yellow (c); red (h, s)	Yellow (c); red (h); gray and opaque
Silicon	(Swims undissolved)	(Swims undissolved)
Silver	—	Gray and opaque
Strontium	Colorless; opaque (s)	Colorless; not clear (ss)
Tin	Colorless; opaque (s)	Colorless
Titanium	Colorless (ns)	Violet (c); yellow or brownish (h)
Uranium	Green; yellow or brownish (h, s)	Green (h)
Vanadium	Yellow	Green
Zinc	Colorless (ns)	Gray and opaque

SODIUM CARBONATE BEAD

Substance	Oxidizing flame	Reducing flame
Manganese	Green	Colorless

*From: "CRC Handbook of Chemistry and Physics", 53rd ed., R.C. Weast, Ed., The Chemical Rubber Co., 1972.

Table 10–51. DIMENSIONLESS GROUPS*

J.P. Catchpole and G.D. Fulford

Dimensionless groups are frequently generated in the analysis of a complex engineering problem. The more common groups thus generated are easily recognized, while the less common ones are not. Unless the less common existing groups are recognized, an already named group could unknowingly be renamed. Table A provides a tool that may be used to avoid this occurrence, by listing the groups by the variables of which they consist. These variables—i.e., length, density, diffusivity, viscosity, etc.—are further subdivided into their exponents to which they are raised in the groups in question. Thus, Reynolds number is listed under the exponent + 1 for the variables, length, fluid velocity, and density, and the exponent − 1 for viscosity.

To illustrate the use of the tables in the analysis of a problem, the group $(kE/\eta\sigma T^3)$ might be generated in the solution of a complex heat transfer problem. From Table A the groups containing the constituent variables are checked and the groups are listed:

Thermal conductivity	(k^{+1})	F11, L7, R1
Modulus of elasticity	(E^{+1})	C1, E13, R1
Stefan-Boltzmann coefficient	(η^{-1})	R1, T6
Surface tension	(σ^{-1})	B9, C3, E8, L6, R1, W1
Temperature	(T^{-3})	R1, T6

It is immediately apparent that the only group common to all the categories listed is the Radiation number, **R1**, which is equivalent to the previously unidentified group.

The symbol assigned to a dimensionless group is usually the first two letters of its names. Several groups, however, have nonstandard symbols, particularly in the groups which are named after persons. These symbols are listed in the nomenclature.

NOMENCLATURE

a	=	annulus or clearance width, L
A	=	area, L^2
A^*	=	cooling area/unit volume, $1/L$
b	=	bearing breadth, L
B	=	groups B6, B11
c	=	specific heat, $L^2/\theta^2 T$
c_A	=	concentration, M/L^3
c_b	=	specific vapor capacity (mass/unit mass/unit pressure change), $L\theta^2/M$
$\left.\begin{array}{l} c_d \\ c_D \end{array}\right\}$	=	group D13
c_f	=	group R9
c_H	=	group H4
c_m	=	mass capacity, L^3/M
$\left.\begin{array}{l} c_p \\ c_v \end{array}\right\}$	=	specific heats at constant pressure and volume, $L^2/\theta^2 T$
c_q	=	heat capacity, $L^2/\theta^2 T$
c_Q	=	group F9
c_S	=	group S11
C	=	group C10, dimensional concentration, M/L^3
C_a	=	groups C11, C4
d	=	diameter, L
d_e	=	equivalent diameter (of particles, etc.), L
d_h	=	hydraulic diameter, L
D	=	diffusivity (molecular, unless noted otherwise), L^2/θ
D_{AB}	=	binary bulk diffusion coefficient, L^2/θ
D_{KA}	=	Knudsen diffusion coefficient, L^2/θ
e	=	voidage; porosity $(\bar{\ })$
e^*	=	surface emissivity $(\bar{\ })$
E	=	modulus of elasticity, $M/L\theta^2$
E_a	=	activation energy, L^2/θ^2
E_b	=	bulk modulus, $M/L\theta^2$
f	=	frequency, $1/\theta$, or Group F1
$f(M)$	=	group F11
F	=	force, ML/θ^2
F_b	=	force per unit length of bearing, M/θ^2
$F(M)$	=	group F6
F_R	=	resistance force in flow, ML/θ^2
g	=	acceleration due to gravity, L/θ^2
G	=	mass velocity (mass flux density; mass transfer coefficient), $M/\theta L^2$
h	=	heat transfer coefficient, $M/T\theta^3$
h_c	=	convective heat transfer coefficient, $M/T\theta^3$
H	=	energy change per unit mass (= $g \times$ head), L^2/θ^2
H'	=	fluid head, L
H_e	=	field strength, $Q/L\theta$
H_o	=	homochronicity number
$I(M)$	=	group F8
j	=	heat liberated per unit volume per unit time, $M/L\theta^3$
j_H, j_M	=	groups J2, J3

J	=	average free path/average velocity, θ, or group L6
k	=	thermal conductivity, $ML/T\theta^3$
k_c	=	mass transfer coefficient, L/θ
K	=	groups K2, K10, N5
K_1	=	group A4
K_E	=	group E4
K_F	=	group C2
K_P	=	group P13
K_Q	=	group H5
K_r	=	group E11
K_s	=	group A1
K_E	=	group E4
K_r	=	group E12
K_{rE}	=	group E13
K_s	=	group R1
K_σ	=	group C1
$K_{\sigma t}$	=	group C2
L	=	characteristic dimension (except as noted), L
L_m	=	distance from midpoint to surface, L
m_T	=	group T3
M	=	group M10
M_H	=	group H2
n	=	concentration, wt./wt. $(\bar{\ })$
n^*	=	specific mass content, mass/mass $(\bar{\ })$
n_m	=	moisture content, wt./wt. bone dry gas $(\bar{\ })$
N	=	rate of rotation, $1/\theta$, and groups M3, N4
N_{B_e}	=	groups B7, B8
N_c	=	group C5
N_{cv}	=	group C10
N_D	=	groups D7, D14
N_E	=	group E1
N_F	=	group F4
N_H	=	groups H1, H9
N_K	=	group K1
N_{KnA}	=	Knudsen number for diffusion (see Addendum)
N_l	=	group N2
N_P	=	group P7
N_{rf}	=	group R7
N_{s1}	=	group S8
N_{s2}	=	group S9
N_T	=	group N9
P	=	pressure, $M/L\theta^2$
P	=	plasticity number (see Addendum)
p_b	=	bearing pressure, $M/L\theta^2$
p_s	=	static pressure, $M/L\theta^3$
p_v	=	vapor pressure, $M/L\theta^2$
p_e	=	capillary pressure, $M/L\theta^2$
Δp_T	=	frictional pressure drop, $M/L\theta^2$
q	=	heat flux (heat flow/unit time), ML^2/θ^3
q^*	=	heat flux density (heat flux/unit area), M/θ^3
Q	=	heat liberated/unit mass, L^2/θ^2
r	=	latent heat of phase change, L^2/θ^2
r_v	=	heat of vaporization, L^2/θ^2
R	=	radius, L
R_H	=	hydraulic radius, L
R_2'	=	group R5
R_M	=	group M7
R_V	=	group V3
$\mathcal{R}$	=	gas constant, $L^2/\theta^2 T$
s	=	humid heat, $L^2/\theta^2 T$
S	=	particle area/particle volume, L^2/L^3, and group M6
$\overline{Si}$	=	group S14
t	=	temperature, T
T	=	absolute temperature, T
$\Delta t, \Delta T$	=	temperature difference, T
U^+	=	group P10
U	=	reaction rate, $M/L^3\theta$
v_s	=	velocity of surface (solid), L/θ
V	=	fluid velocity, L/θ, and group V1
V_A	=	velocity of Alfven magnetic waves, L/θ
V_f	=	volumetric flow rate, L^3/θ
V_l	=	velocity of light, L/θ
V_m	=	mass flow rate, M/θ
V_s	=	velocity of sound, L/θ
w	=	circumferential velocity, L/θ
W	=	volume of system, L^3
W^*	=	gross volume, L^3
x^+	=	entry length; distance from entrance, L
y^+	=	group P11
Z	=	group O2
α	=	thermal diffusivity (temperature conductivity), L^2/θ

Table 10–51. DIMENSIONLESS GROUPS *(Continued)*

β	= coefficient of bulk expansion, $1/T$, and group **D12**	
$\beta^{\bullet}$	= Dufour coefficient, T	
γ	= specific gravity ($^-$) and group **R3**	
$\dot{\gamma}$	= rate of shear, $1/\theta$	
Γ	= rate of change of temperature of medium, T/θ	
δ	= Soret or thermogradient coefficient, $1/T$, and group **D11**	
Δ	= difference in quantity	
ε	= height of roughness, L and group **A3**	
ε_D	= eddy mass diffusivity, L^2/θ	
ζ	= diffusion tortuosity ($^-$)	
η	= radiation coefficient (Stefan-Boltzmann coefficient), $M/T^4\theta^3$	
θ	= time, θ	
θ_r	= relaxation time, θ	
λ	= mean free path, L	
μ	= dynamic viscosity, $M/L\theta$	
μ_e	= magnetic permeability, ML/Q^2	
μ_r	= rigidity coefficient, $M/L\theta$	
ζ	= permeability, L^2	

π	= 3.1416. . . .	
Π	= power to agitator or impeller, ML^2/θ^3	
ρ	= density, M/L^3	
ρ_*	= group **P3**	
σ	= surface tension, M/θ^2 and group **S12**	
σ_c	= group **C6**	
σ_e	= electrical conductivity, $Q^2\theta/L^3M$	
σ_i	= group **T4**	
τ	= group **T8**	
τ_w	= wall shear stress, $M/L\theta^2$	
τ_y	= yield stress, $M/L\theta^2$	
φ	= group **D9**	
ψ	= groups **N8, P14, R10**	
ω	= angular velocity (of fluid, unless noted otherwise), $1/\theta$	
Ω	= mass transfer potential (concn.), M/L^3	
— (bar over)	= mean value	

N.B.: $(F) = \left(\dfrac{ML}{\theta^2}\right)$; $(H) = \left(\dfrac{ML^2}{\theta^2}\right)$

Table A.   TABLES FOR IDENTIFYING DIMENSIONLESS GROUPS

PHYSICAL PROPERTIES
General Physical Properties

Parameter	Symbol	Dimensions	Exponent	Group
Coefficient of bulk expansion	β	$1/T$	-1	**E4, 12, G2**
			$+1$	**G5, R5, K9, 6**
Density	ρ	M/L^3	-2	**M1**
			-1	**A1, B1, 10, C1, 2, 6, D3, 13, E3, 9, 10, 13, F1, 11, H5, J1, K4, K10, L7, M3, 6, N2, 3, 4, P7, 9, R10, 13, S4, 13, 15**
			$-\frac{2}{3}$	**C9, J3**
			$-\frac{1}{2}$	**E2, L11, O2, P13**
			$+\frac{1}{3}$	**J3, K5, 53**
			$+\frac{1}{2}$	**D10, G3, P10, 11, R4, W2**
			$+\frac{2}{3}$	**F2, N8, S2**
			$+1$	**A5, B1, B6, 9, C5, 7, D5, 7, 13, E4, 6, 7, 8, H6, 11, J1, 3, 4, K1, L9, P1, R5, 11, 13, S17, T1, T6, V1, W1, W3**
			$+2$	**C10, G1, 5, K9, R5, 6, T2**
Density gradient	$d\rho/dL$	M/L^4	$+1$	**R13**
Diffusivity (molecular unless noted otherwise)	D, a_m, ε_D	L^2/θ	-1	**B5, B7$^{(4)}$, D2, K4$^{(1)}$, K7$^{(2)}$, L2, 7, 10, N7, P2, P9, S4, 13**
			$-\frac{2}{3}$	**J3**
			$-\frac{1}{2}$	**T3**
			$+1$	**D12, F12$^{(1)}$, K7$^{(3)}$, L2$^{(4)}$, L9$^{(1)}$**
Diffusivity (surface)	D_s	ML/θ^2	$+1$	**S18**
Diffusion tortuosity	ζ	—	-1	**K7**
Molecular weight	M	—	-1	**D14, K9, S6**
			$+1$	**D14**
Permeability (packed bed)	k	L^2	$+\frac{1}{2}$	**L4**
Porosity (voidage)	e	—	-1	**B4**
			$-\frac{1}{2}$	**L4**
			$+1$	**K7**
Specific weight	γ	—	$\pm\frac{1}{2}$	**F2**
Surface tension	σ	M/θ^2	-3	**C2**
			-2	**C1**
			-1	**B9, C3, 13, E8, L6, R1, W1, 3**
			$-\frac{1}{2}$	**D10, G3, O2, P13, R4, W2**
			$+1$	**M9, S17, 18**

$^{(1)}$ Coefficient of potential diffusion in mass transfer.
$^{(2)}$ Knudsen diffusion coefficient.
$^{(3)}$ Binary bulk diffusion coefficient.
$^{(4)}$ Effective diffusivity $(D + \varepsilon_D)$ (molecular + eddy transfer).

PHYSICAL PROPERTIES
Electrical and Magnetic Properties

Parameter	Symbol	Dimensions	Exponent	Group
Current density	I	$Q/L^2\theta$	$+1$	**K10**
Electrical conductivity	σ_e	$Q^2\theta/L^3M$	-1	**E6**
			$+\frac{1}{2}$	**H2**
			$+1$	**L11, M3, 7**
Field strength	H_e	$Q/L\theta$	-2	**J4**
			$+1$	**H2, L11**
			$+2$	**K9, M3, 6, S6**
Magnetic permeability	μ_e	ML/Q^2	-1	**E5, J4**
			$+1$	**H2, M6, 7**
			$+\frac{3}{2}$	**L11**
			$+2$	**M3**
Voltage	E	$ML/Q\theta^2$	$+1$	**K10**

Thermal Properties

Parameter	Symbol	Dimensions	Exponent	Group
Humid heat	S	$L^2/\theta^2 T$	-1	**P15**
Latent heats of phase change	λ, r	L^2/θ^2	-1	**A3, E11, 12, 13, J1, K10, M1**
			$+1$	**B13, C9, K8, 11, N5**
Ratio of specific heats	γ	—	$\pm\frac{1}{2}$	**L5**
Specific heat	C, c	$L^2/\theta^2 T$	-1	**B2, 13, D3, 15, E1, F11, J2, K8, 11, L7, M10, N5, R3, S13**
			$-\frac{1}{3}$	**J2**
			$+\frac{1}{2}$	**F6, 7, 8**
			$+\frac{2}{3}$	**J2**
			$+1$	**A3, E4, 12, G4, J1, 4, L9, P1, 8, R3, 5, 6, 7 $^-$**
Surface emissivity	$e^{\bullet}$	—	-1	**T6**
Temperature conductivity (thermal diffusivity)	α	L^2/θ	-1	**L9, P1, 12, R5**
			$+1$	**C13, L7, 10**
Thermal conductivity	k or λ	$ML/T\theta^3$	-3	**M1**
			-2	**R6**
			-1	**B4, 12, C7, 9, 10, D4, G4, K3, L9, N6, P1, 5, 8, R5, S14**

Rheological and Elastic Behavior

Parameter	Symbol	Dimensions	Exponent	Group
Modulus of elasticity	E	$M/L\theta^2$	-1	**C5, E4, H11**
			$+1$	**C1, E13, R1**
			$+3$	**A1**
Rate of shear	$\dot{\gamma}$	$1/\theta$	$+1$	**T8**
Rigidity coefficient	μ_r	$M/L\theta$	-2	**H6**
			-1	**B3**
			-1	**E5**
Shear stress	τ	$M/L\theta^2$	$-\frac{1}{2}$	**P10**
			$+\frac{1}{2}$	**P11**
Viscosity (in all cases kinematic viscosity has been written as μ/ρ)	μ	$M/L\theta$	-2	**A1, 5, G1, 5, K1, 9, S17, T2**
			-1	**B6, C10, D5, 7, E6, 7, H8, L1, 3, 9, O1, P4, 11, R5, 6, 11, S9, 19, T1, V1**

Table 10–51. DIMENSIONLESS GROUPS (Continued)

Table A.   TABLES FOR IDENTIFYING DIMENSIONLESS GROUPS (Continued)

Rheological and Elastic Behavior

Parameter	Symbol	Dimensions	Exponent	Group
			$-\tfrac{2}{3}$	F2, K5, N8
			$-\tfrac{1}{2}$	H2
			$-\tfrac{1}{3}$	S2, 3
			$+\tfrac{1}{2}$	E2
			$+\tfrac{2}{3}$	C9, J2, 3
			$+1$	B12, C3, 13, E3, 5, M1, O2, P8, 9, S4, 8, 15, T8
			$+2$	C1
			$+4$	C2
Viscosity (surface)	μ_s	M/θ	$+1$	S19
Yield stress	τ_y	$M/L\theta^2$	$+1$	B3, H6

LENGTHS, AREAS AND VOLUMES
Characteristic Linear Dimensions
(In all cases kinematic viscosity has been written as μ/ρ)

Parameter	Symbol	Dimensions	Exponent	Group
General characteristic linear dimension	Various	L	-5	P7
			-3	F9, I.1
			-2	D9, 13, E3, F11, 17, H4, 5, N2, 3, 4, O1, R9, S8, 9, 15
			-1	B1, 10, E2, 5, 10, F1, 13, G4, H10, K6, L4, R6, 10, 15, 16, S6, 19, T5, W4
			$-\tfrac{1}{2}$	B11, D7, F14, O2
			$+\tfrac{1}{2}$	R4, T1, W2
			$+1$	B3, 4, 5, 7, 10, C7, D1, 3, 5, 11, 13, E7, 10, F1, 2, 5, G3, H2, K3, 4, 6, 10, L3, 4, 9, 11, M1, 3, 7, N6, 7, 8, O1, P1, 2, 11, R8, 10, 11, 16, S2, 7, 14, 16, 17, 18, T3, W1
			$+\tfrac{3}{2}$	D7, H7, T1
			$+2$	B9, D2, 4, E8, H6, K9, O1, P4, 5, 12, S8, 9, V1
			$+3$	A5, C10, G1.5, K1, R5
			$+4$	T2
			$+5$	R6
Dimension of agitator, impeller, etc.	Various	L	-5	P7
			-3	P9
			-2	D7, H4
			$+1$	D11, F15
			$+3$	W3
Film thickness	L_f	L	$+1$	N8
Furnace half-width	L	L	-1	T7
Larmor radius	L_L	L	$+1$	L4
Mean free path	λ	L	$+1$	K6
Particle dimension	d_e	L	-2	S15
			$+1$	F2
			$+3$	A5, G1
Pore or nozzle radius	R	L	$+1$	M12, T7
Reactor length	L	L	$+1$	B7
			-1	C3, L9
Thickness of liquid layer	L	L	$+1$	H2, L11
			$+2$	L9
			$+3$	R5
			$+4$	T2

Areas

Parameter	Symbol	Dimensions	Exponent	Group
Area	A	L^2	-1	D9, F6, 7, 8
			$+1$	B2
Area/unit volume	S, A^*	$1/L$	-1	B6
			$+1$	M11, 12

Volumes

Parameter	Symbol	Dimensions	Exponent	Group
Volume	—	L^3	$+$	A5, H9, M11

TIMES AND FREQUENCIES

Parameter	Symbol	Dimensions	Exponent	Group
Time	θ	θ	-1	D8
			$+1$	D8, 12, E3, F11, 12, H1, 10, S15, T5
Frequency	f'	$1/\theta$	$+1$	H1, 10, S16, V1

TEMPERATURES AND CONCENTRATIONS (DRIVING FORCES)
Concentrations and Related Quantities

Parameter	Symbol	Dimensions	Exponent	Group
Dimensional concentration	C, c_A	M/L^3	-1	D1, 2, R2
			$-\tfrac{1}{2}$	T3
			$+1$	B10
Dimensionless concentration—e.g., wt./wt. inert material, etc.	n	—	-1	P6
			$+1$	A4, K8
Mass capacity	C_m	L^3/M	-1	R2
Mole fraction	Y, Z	—	± 1	D14
Specific mass content, mass/unit mass	n^*	—	-1	K4
Surface concentration	Γ'	M/L^2	± 1	S18
Vapor capacity (porous body)	C_b	$L\theta^2/M$	$+1$	B13, R2

Temperatures, Temperature Differences

Parameter	Symbol	Dimensions	Exponent	Group
Temperature, temperature difference	$T, \Delta T$	T	-3	R1*, T6*
			-1	A4, 6*, B12, 13, C4*, 7, 10, D3*, 4*, 15, E1, G7, 6*, K1, 8, 9, 11, L9, N5, P5*, 12*, R14*, S6
			$-\tfrac{1}{2}$	L5
			$+\tfrac{1}{2}$	F6*, 7*, 8*
			$+1$	C4, G5, 6, J1, 4, L9, M1, P6, R5, 7, 14*
			$+3$	S14*
Rate of temperature change	—	T/θ	$+1$	P12

* Absolute temperature; others—temperature differences.

VELOCITIES, RATES, FLUXES, TRANSFER COEFFICIENTS
Velocities

Parameter	Symbol	Dimensions	Exponent	Group
Angular velocity (rate of rotation)	N	$1/\theta$	-3	P7
			-2	H4, L1
			-1	F9, R15
			$-\tfrac{1}{2}$	E2
			$+1$	S8, 11, 12, T1
			$+2$	F15, T2, W3
Fluid velocity	V	L/θ	-3	I15
			-2	C6, 11, D13, E9, 10, F1, M6, N2, 3, 4, R7, 8, 9, 10
			-1	A2, B3, C12, 14, D1, 3, F10, H7, 8, K10, L3, M3, P4, S13, 16
			$+1$	A2, B6, 7, 11, C3, 12, 14, D5, 7, E5, 7, F10, 14, H10, K2, L5, 8, M2, 4, 7, P1, 2, 3, 10, R4, 11, 15, S1, 2, 3, T5, 6, V2, W2, 4,
			$+2$	B1, 12, C5, 11, D15, E1, 11, F13, H11, W1
			$+3$	C7
Impeller or agitator circumference	U_s	L/θ	-2	P14
			-1	D9
Light	V_l	L/θ	-1	L8
Sound	V_s	L/θ	-1	M2, N1, S1
Waves	V_w	L/θ	-1	V1
			-1	P3
Velocity gradient	dV/dL	$1/\theta$	-2	R12
Velocity of Alfven waves	V_A	L/θ	-1	A2, K2, M4
			$+1$	A2, N1
			$+2$	C11
Velocity of bearing surface	v_s	L/θ	$+1$	H8, O1, S9

Flow Rates (Mass Fluxes)

Parameter	Symbol	Dimensions	Exponent	Group
Mass flow rate (mass flux)	V_m	M/θ	-1	B2, M11, T7
			$+1$	F6, 7, 8, G4, T7
Mass flux density (mass flux/unit area)	G	$M/L^2\theta$	-1	J2, 3, P15, S13
			$+1$	K4, M11

Table 10–51. DIMENSIONLESS GROUPS *(Continued)*

Table A. TABLES FOR IDENTIFYING DIMENSIONLESS GROUPS *(Continued)*

Flow Rates (Mass Fluxes)				
Parameter	Symbol	Dimen-sions	Expo-nent	Group
Mass flux/unit volume (reaction rate)	U	$M/L^3\theta$	$+\frac{1}{2}$	T3
			$+1$	D1, 2, 3, 4
Reaction rate constant	K	L/θ	-1	S5
Volumetric flow rate	V_f	L^3/θ	-1	H9
			$-\frac{1}{2}$	D11
			$+\frac{1}{2}$	S11, 12
			$+1$	D9, F9

Heat Fluxes				
Heat flux (heat flow/unit time)	q	ML^2/θ^3	$+1$	H5
Heat flux/unit area	$q^\bullet$	M/θ^3	$+1$	K3, R6
Heat liberated/unit mass	Q	L^2/θ^2 (H/M)	$+1$	D3, 4
Rate of heat liberation/unit volume (heat source power)	j	$M/L\theta^3$	$+1$	P5

Transfer Coefficients				
Heat transfer coefficient	h	$M/T\theta^3$	$+1$	B2, 4, C9, J2, N6, P15, S13
		$(H/L^2T\theta)$	$+4$	M1
Mass transfer coefficient	k_c	L/θ	$+1$	B5, J3, L9, N7, S5, 7

FORCE, HEAD, POWER, PRESSURE
Forces

Forces				
Force (resistance)	F, F_R	ML/θ^2	$-\frac{1}{3}$	S2
			$+\frac{1}{3}$	K5
			$+1$	N2, 3, 4, R9
Force/unit length	F_b	M/θ^2	$+1$	H8, O1, S9

Heads, Power

Heads, Power				
Fluid head	H	L	$-\frac{3}{4}$	S11
			$+1$	H4

Heads, Power				
Parameter	Symbol	Dimen-sions	Expo-nent	Group
Head (energy per unit mass of fluid $= gH'$)	H	L^2/θ^2	-1	P3, T4
			$-\frac{3}{4}$	S12
			$+\frac{1}{4}$	D11
			$+1$	A6, P14, T4
Power	π	ML^2/θ^3 (LF/θ)	$+1$	L1, P7

Pressures				
Pressure	P	$M/L\theta^2$	-1	F6, 7, 8, H9, S8, T8
			$+1$	B13, C6, L6, P13, R2
Pressure drop	$\Delta P, dP$	$M/L\theta$	$+1$	E9, 10, F1, H9, L3, R10
Pressure gradient	$\Delta P/L$	$M/L^2\theta^2$	$+1$	E10, F1, K1, P4, R10

CONSTANTS AND MISCELLANEOUS QUANTITIES
Gravity Acceleration

Gravity Acceleration				
Gravity acceleration	—	L/θ^2	-2	A1
			-1	B1, F13, 15, M1, S6
			$-\frac{3}{4}$	S11
			$-\frac{1}{2}$	B11, F14, P13
			$-\frac{1}{3}$	C9, S3
			$+\frac{1}{3}$	F2, N8
			$+1$	A5, B9, C2, 10, D13, E8, G1, 5, H4, R5, 6, 8, 13

Other Quantities				
Avogadro's number	N	$1/M$	$+1$	K9, S6
Boltzmann's constant	k	—	-1	K9, S6
Dufour coefficient	β	T	$+1$	A3, 4, F3
Energy of activation	E_a	L^2/θ^2	$+1$	A6
Gas constant	$\mathcal{R}$	$L^2/\theta^2 T$	-1	A6
			$-\frac{1}{2}$	L5
Shape factor	$\zeta^\bullet$	—	$+1$	V2
Soret coefficient	δ	$1/T$	$+1$	F3, P6
Stefan constant	η	$M/T^4\theta^3$	-1	R1, T6
			$+1$	S14

Table 10–51. DIMENSIONLESS GROUPS *(Continued)*

Table B. ALPHABETICAL LIST OF NAMED GROUPS

Serial No.	Name	Symbol	Definition	Significance	Field of Use	Reference
A1	Acceleration number	K_g	$E^3/\rho g^2 \mu^2 = (N_{Re}N_{Fr1})^2/(H_0)^3$	Group dependent only on physical properties	Accelerated flow	25
A2	Alfven number	N_{Al}	V_A/V (or V/V_A) [*cf.* Cowling No. Kármán No. (2), magnetic mach number]	Ratio of Alfven wave velocity/fluid velocity	Magneto-fluid dynamics	5, 6
A3	Anonymous group (1)	ε	$\beta^* c/r$ [see also Fedorov No. (2)]		Transfer processes	39
A4	Anonymous group (2)	K_1	$\beta^* \Delta n/\Delta t$; Δt = temp. diff. $[T]$; Δn = conc. diff. $[^-]$		Transfer processes	67
A5	Archimedes number	N_{Ar}	$\dfrac{gL^3\rho}{\mu^2}(\rho - \rho_o)$; ρ_o = fluid density; ρ = particle density (*cf.* N_{Ga1})	N_{Re}, gravitational force/viscous force	Fluidization, motion of liquids due to density differences	6, 13
A6	Arrhenius group	—	$E_a/\mathscr{R}T$	Activation energy/potential energy of fluid	Reaction rates	5
B1	Bagnold number	B	$3c\alpha\rho_g V^2/4d\rho_p g$, ρ_g = gas density ρ_p = particle density	Drag force/gravitational force	Saltation studies	12
B2	Bansen number	N_{Ba}	$h_r A_w/V_{mc}$; h_r = radiant heat transfer coefficient; A_w = wall area of channel; (*cf.* N_{St})	Heat transferred by radiation/thermal capacity of fluid	Radiation	1
B3	Bingham number	N_B	$\tau_y L/\mu V$ (L = channel width)	Ratio of yield stress/viscous stress	Flow of Bingham plastics	5
B4	Biot number (heat transfer)	N_{Bih}	hL_m/k (in French literature, "Biot No." $= N_{Nu}$)	Midplane thermal internal resistance/surface film resistance	Unsteady state heat transfer	5, 6, 13
B5	Biot number (mass transfer)	N_{Bim}	$k_c L/D_{int}$; L = thickness of layer, D_{int} = diffusivity at interface	Mass transfer rate at interface/mass transfer rate in interior of solid wall thickness L	Mass transfer between fluid and solid	13
B6	Blake number	B	$V\rho/[\mu(1-\epsilon)S]$	Inertial force/viscous force	Beds of particles	6
B7	Bodenstein number	N_{Bo}	$VL/D_a = N_{Pe_m}$; L = reactor length, D_a = axial diffusivity (effective) (L^2/θ)		Diffusion in reactors	6
B8	Boltzmann number	N_{Bo}	$\equiv$ Thring radiation group			1
B9	Bond number	N_{Bo}	$(\rho - \rho')L^2 g/\sigma = N_{We_1}/N_{Fr_1}$ if $\rho - \rho' \cong \rho$ (gas in liq.); ρ = drop or bubble density; ρ' = medium density	Gravitational force/surface tension force	Atomization, motion of bubbles and drops	5
B10	Bouguer number	N_{Bu}, B	$3C_D\lambda_r/4\rho DR$; C_D = wt. dust/unit bed volume (M/L^3), λ_r = mean path for radiation (L), ρ_D = dust density, R = mean particle radius. Also $N_{Bu} = kL$; L = characteristic dimension, k = absorption coefficient of medium		Radiant heat transfer to dust—gas streams	66
B11	Boussinesq number	B	$V/(2gR_h)^{1/2}$ (*cf.* N_{Fr2})	(Inertia force/gravitational force)$^{1/2}$	Wave behavior in open channels	6
B12	Brinkman number	N_{Br}	$\mu V^2/k\Delta t$; Δt = temp. diff.	Heat generation/heat transferred	Viscous flow	6
B13	Bulygin number	N_{Bu}	$r \cdot \dfrac{C_b}{C_q} \cdot \dfrac{P}{t_m - t_o}$; t_m = temp. of medium, t_o = init. temp. of body	Heat for vaporization/heat to bring liquid to boiling point	Heat transfer during evaporation	6, 13, 14
C1	Capillary number	$K\sigma$	$\mu^2 E/\rho\sigma^2 = (N/w_{e_1})^2/H_o \cdot (N_{Re})^2$	Depends only on physical properties	Action of surface tension in flowing media	25
C2	Capillarity-buoyancy number (physical properties group) (film No.)ᵃ	$K\sigma_g, K_F$	$g\mu^4/\rho\sigma^3 \cdot = \sqrt{K\sigma/Kg} = (N_{We_1})^3/(N_{Fr_1})(N_{Re})^4$	Depends only on physical properties and g	Effects of surface tension and acceleration in flowing media (two-phase flow)	21, 25
C3	Capillary number	Ca	$\mu V/\sigma = N_{We}/N_{Re}$	Viscous force/surface-tension force	Atomization, two-phase flow	6
C4	Carnot number	Ca, N_{Ca}	$(T_2 - T_1)/(T_2)$; T_1, T_2 = abs. temp. of two heat sources or sinks	Theoretical efficiency of Carnot cycle operating between T_1 and T_2		21
C5	Cauchy number	N_c	$\rho V^2/E_b = (N_{Ma})^2 = $ Hooke No.	Inertia force/compressibility force	Compressible flow	5, 25
C6	Cavitation number	σ_c	$[(p - P_v)/\rho]/(V^2/2)p$ = local static pressure (abs.); P_v = vapor pressure	Excess of local static head over vapor pressure head/velocity head	Cavitation	5
C7	Clausius number	Cl, N_{Cl}	$V^3 L\rho/k\Delta T$; ΔT = temp. diff.		Heat conduction in forced flows	25
C8	Colburn number	—	Same as Schmidt number			5
C9	Condensation number (1)	N_{Co}	$(h/k)(\mu^2/\rho^2 g)^{1/3}$	$N_{Co} = N_{Nu}\left[\dfrac{(\text{viscous force})}{(\text{gravity force})} \times \dfrac{1}{Re}\right]^{1/3}$	Condensation	5
C10	Condensation number (2)	N_{Cv}	$L^3\rho^2 gr/k\mu\Delta t$; r = latent heat of condensation		Condensation on vertical walls	5
C11	Cowling number	C	$(V_A/V)^2 \equiv$ (Alfven number)^2		Magneto-fluid dynamics	6
C12	Craya-Curtet number	C_t	$V_k/(V_d^2 - V_k^2/2)^{1/2}$; V_k = kinematic mean velocity, V_d = dynamic mean velocity		Radiant heat transfer	3, 27

ᵃ Very similar to Hu and Kintner's pH factor for drops and bubbles [*A.I.Ch.E. J.* 1, 42 (1955)].

Table 10–51. DIMENSIONLESS GROUPS *(Continued)*

Table B.   ALPHABETICAL LIST OF NAMED GROUPS *(Continued)*

Serial No.	Name	Symbol	Definition	Significance	Field of Use	Reference
C13	Crispation group	N_{Cr}	$\mu\alpha/\sigma^* L$; $\sigma^* =$ undisturbed surface tension; $L =$ layer thickness		Convection currents	55
C14	Crocco number	N_{Cr}	$V/V_{max} = \left[1 + \dfrac{2}{(\gamma-1)(N_{Ma})^2}\right]^{-1/2}$ $V_{max} =$ maximum velocity of gas expanding adiabatically	Velocity/maximum velocity	Compressible flow	5
D1	Damköhler group I	DaI	UL/V_{cA}	Chemical reaction rate/ bulk mass flow rate	Chemical reaction, momentum, and heat transfer	5
D2	Damköhler group II	DaII	UL^2/D_{cA}	Chemical reaction rate/ molecular diffusion rate	Chemical reaction, momentum, and heat transfer	5
D3	Damköhler group III	DaIII	$QUL/C_p\rho Vt$	Heat liberated/bulk transport of heat	Chemical reaction, momentum, and heat transfer	5
D4	Damköhler group IV	DaIV	QUL^2/kt	Heat liberated/conductive heat transfer	Chemical reaction, momentum, and heat transfer	5
D5	Damköhler group V	DaV	$= (N_{Re})$			5
D6	Darcy number		$4f$; see Fanning friction factor			5
D7	Dean number	N_D	$(VL\rho/\mu)(L/2R)^{1/2}$; $L =$ pipe diam.; $R =$ radius of curvature of bend	N_{Re} (centrifugal force/ inertial force)	Flow in curved channels	5, 6
D8	Deborah number	D	θ_r/θ_o; $\theta_o =$ observation time	Relaxation time/observation time	Rheology	49
D9	Delivery number	ϕ	V_f/Aw; $A =$ impeller area $= \pi d^2/4 =$ [Diameter No.]$^{-3}$ [Speed No.]$^{-1}$		Flow machines	25
D10	Deryågin number	De	$L(\rho g/2\sigma)^{1/2}$; $L =$ film thickness	Film thickness/capillary length	Coating	58
D11	Diameter group	δ	$(\pi/4)^{1/2}(2H)^{1/4} d/(V_f)^{1/2}$, $d =$ impeller diam. $=$ [pressure No.]$^{1/4} \times$ [delivery No.]$^{-1/2}$		Flow machines	25
D12	Diffusion group	β	$D\theta/L_m^2$; $D =$ diffusivity of solute through stationary solution contained in solid; cf. N_{Fom}		Mass transfer	37
D13	Drag coefficient	$C_d = C_D$	$(\rho - \rho')L_g/\rho V^2$; $\rho =$ density of object; $\rho' =$ density of medium; cf. f, ψ, N_e	Gravity force/inertial force	Free settling velocities, etc.	5
D14	Drew number	N_D	$\dfrac{Z_A(M_A - M_B) + M_B}{(Z_A - Y_{AW})(M_B - M_A)} \ln \dfrac{M_V}{M_W}$; M_A, $M_B =$ mol. wt. of components A and B; M_V, $M_W =$ mol. wt. of mixture in vapor and at wall; $Y_{AW} =$ mole fraction of A at wall; $Z_A =$ mole fraction of A in diffusing stream		Boundary layer mass transfer rates; velocity profile distortion; drag coefficients for binary system	22
D15	Dulong number	Du, N_{Du}	$V^2/C_p\Delta T =$ Eckert No.			25
E1	Eckert number	N_E	$V_\infty^2/C_p\Delta T$, $V_\infty =$ velocity of fluid far from body ($= 2$/recovery factor, q.v. $\equiv$ Dulong No.)		Compressible flow	5, 24
E2	Ekman number		$(\mu/2\rho\omega L^2)^{1/2} = (N_{Ro}/N_{Re})^{1/2}$	(Viscous force/Coriolis force)$^{1/2}$	Magneto-fluid dynamics	6
E3	Elasticity number (1)	N_{El_1}	$\theta_r\mu/\rho L^2$; $L =$ pipe radius	Elastic force/inertial force	Viscoelastic flow	5
E4	Elasticity number (2)	K_E	$\rho C_p/\beta E \equiv$ [Gay Lussac No.] $\times$ [Hooke No.] $\div$ [Dulong No.]	Depends on physical properties only	Effect of elasticity in flow processes	25
E5	Ellis number	N_{El}	$\mu_o V/2\tau_{1/2}R$; $\mu_o =$ zero shear viscosity, $\tau_{1/2} =$ shear stress at which $\mu = \mu_o/2[M/L\theta^2]$; $R =$ tube radius		Flow of non-Newtonian liquids	38
E6	Elsasser number	N_{El}	$\rho/\mu\sigma_e\mu_e \equiv N_{Re}/$[magnetic Reynolds number]		Magneto-fluid dynamics	6
E7	Entry Reynolds Number	K_E	$\chi/d_h \cdot N_{Re} = \dfrac{\chi V\rho}{\mu}$; $\chi =$ entry length	As N_{Re}	Entry or inlet processes	25
E8	Eötvös number	N_{Eo}	$(\rho - \rho')L^2 g/\sigma \equiv$ Bond No., q.v.			6
E9	Euler number (1)	N_{Eu_1}	$\Delta P_F/\rho V^2$; $\Delta P_F =$ pressure drop due to friction	Friction head/2 $\times$ velocity head	Fluid friction in conduits	5, 13, 25
E10	Euler number (2)	N_{Eu_2}	$d(-dp/dL)\rho V^2$; $d =$ pipe diam., $dp/dL =$ pressure gradient $\equiv 2 \times$ Fanning friction factor		Fluid friction in conduits	
E11	Evaporation number	K_r	V^2/r [$r =$ heat of vaporization (L^2/θ^2)]		Evaporation processes	25
E12	Evaporation number (2)	K_r	$C_p/r\beta$ (r as in E11) $\equiv$ (Gay Lussac No.) $\times$ (E11)/(Dulong No.)		Evaporation processes	25
E13	Evaporation-elasticity number	K_{rE}	$E/r\rho = K_r/$Hooke number		Evaporation processes	25
F1	Fanning friction factor	f	$d\Delta P_F/2\rho V^2 L$, $d =$ dimension of cross section; $L =$ length (cf. resistance coeff., Ne)	Shear stress at wall expressed as number of velocity heads	Fluid friction in conduits	5
F2	Fedorov number (1)	F_e, N_{Fe_1}	$d\sqrt[3]{\dfrac{4g\rho^2}{3\mu^2}\left[\dfrac{\gamma M}{\gamma g} - 1\right]}$, $d_e =$ equiv. particle diam. $\gamma M =$ sp. gr. of particles; $\gamma g =$ sp. gr. of gas (cf. N_{Ar})		Fluidized beds	6, 13

Table 10–51. DIMENSIONLESS GROUPS *(Continued)*
Table B. ALPHABETICAL LIST OF NAMED GROUPS *(Continued)*

Serial No.	Name	Symbol	Definition	Significance	Field of Use	Reference
F3	Fedorov number (2)	F_{e_2}, N_{Fe_2}	$\delta\beta^* = K_1 Pn = \varepsilon \times K_e Pn$	Mass transfer analogy of Posnov number	Transport processes	39, 67
F4	Fenske number	N_F		Number of stages in separation process		6
F5	Fineness coefficient	ψ	$L/W_D^{1/3}$; W_D = volume displacement $[L^3]$		Ship modeling	26, 43
F6 F7 } F8	Fliegner numbers	$F(M_a)$ $f(M_a)$ $I(M_a)$	$\{$ Functions of ratio of specific heats and $\{$ mach number $V_m(cT)^{1/2}/A(p_s + \rho V^2) = [\gamma M_a/(\gamma-1)^{1/2}]$ $\left[1 + \dfrac{(\gamma-1)Ma^2}{2}\right]^{1/2}$ = impulse Fliegner number; γ = ratio of specific heats, Ma = mach number, A = flow area			43
F9	Flow coefficient	C_Q	V_f/Nd, d = impeller diam.		Power required by fans, etc.	32
F10	Fluidization number		V/V_{init}, V_{init} = velocity for initial fluidization	Fluid velocity in fluidized bed/that at start of fluidization	Fluidization	59
F11	Fourier number (heat transfer)	N_{Foh}	$k\theta/\rho C_p L_m^2$		Unsteady state heat transfer	5, 13, 14, 25
F12	Fourier number (mass transfer)	N_{Fom}	$D\theta/L^2 = k_c\theta/L$ (cf. **D12**)		Unsteady state mass transfer	13
F13	Froude number (1)	N_{Fr_1}	$V^2/gL = (N_{Fr_2})^2$ (cf. Reech No., Boussinesq No., Vedernikov No.)	Inertial force/gravitational force	Wave and surface behavior	5, 13, 25
F14	Froude number (2)	N_{Fr_2}	$V/\sqrt{gL} \equiv (N_{Fr_1})^{1/2}$ (cf. Boussinesq No.)	Velocity of open channel flow/speed of very small gravity wave	Open channel flow; free surfaces	7, 50
F15	Froude No. (rotating)	Fr	DN^2/g; D = impeller diam.		Agitation	64
G1	Galileo number	N_{Ga_1}	$L^3 g\rho^2/\mu^2$ (cf. N_{Ar}, Nusselt thickness group)	$N_{Ga_1} = N_{Re} \times$ gravity force/viscous force	Circulation of viscous liquid, thermal expansion	5, 13
G2	Gay-Lussac number	Ga, N_{Ga_2}	$1/\beta\Delta T$		Thermal expansion processes	25
G3	Goucher number	N_{Go}	$R(\rho g/2\sigma)^{1/2}$; R = wall or wire radius	Gravitational force/surface tension force$^{1/2}$	Coating	24
G4	Graetz (Grätz) number	N_{Gz}	$V_m c_p/kL$	Thermal capacity fluid/convective heat transfer	Streamline flow	5, 25
G5	Grashof number	N_{Gr}	$L^3 \rho^2 g\beta\Delta t/\mu^2 \equiv N_{Ga_1}/N_{Ga_2} \equiv (N_{Re})^2/(N_{Ga_2})(N_{Fr_1})$	$N_{Gr} = N_{Re} \times$ (buoyancy force/viscous force)	Free convection	5, 13, 25
G6	Gukhman number	Gu, N_{Gu}	$(t_s - t_m)/T_s$; t_s, T_s = temp. (°C., °K.) of hot gas stream, t_m = temp. of moist surface (wet bulb temp.)	Thermodynamic criterion of evaporation under isobaric adiabatic conditions	Convective heat transfer in evaporation	6, 23, 17, 53
G7	Guldberg-Waage group	N_{Gw}	Given by equation relating volumes of reacting gases and reaction products		Chemical reaction in blast furnaces	5
H1	Hall coefficient	N_H	f_c/f (f_c = cyclotron frequency, J = av. free path/av. veloc.)		Magneto-fluid dynamics	7
H2	Hartmann number	M_H	$(\mu_e^2 H_e^2 \sigma_e L^2/\mu)^{1/2} \equiv (SR_M N_{Re})^{1/2} \equiv (N_{Re}N)^{1/2}$	Magnetically induced stress/hydrodynamic shear stress (magnetic body force/viscous force)$^{1/2}$	Magneto-fluid dynamics	5, 6
H3	Hatta number	β	$\gamma/\tanh \gamma$; $\gamma = (rCD)^{1/2}/k_c$, r = reaction rate constant $[L^3/M\theta]$ [a modified Hatta number has also been defined 55)]		Gas absorption with chemical reaction	41
H4	Head coefficient	C_H	$gH'/N^2 d^2$ (d = impeller diam.)		Flow in pumps and fans	32
H5	Heat transfer number	K_Q	$q/V^3 L^2 \rho$		Heat transfer in stream	25
H6	Hedstrom number	N_{He}	$\tau_s L^2 \rho/\mu_p^2 \equiv (N_{Re}) \times (N_{Bm})$		Flow of Bingham plastics	5
H7	Helmholtz resonator group		$(d^3/W)^{1/2}/Ma$	Proportional to frequency × residence time	Pulsating combustion	28
H8	Hersey number		$F_b/\mu v_s$ (cf. truncation number)	Load force/viscous force	Lubrication	6
H9	Hodgson number	N_H	$Wf\Delta\rho_f/\overline{V}_{f,ps}$	Time constant of system/period of pulsation	Pulsating gas flow	5
H10	Homochronous number	H_{o_1}	$V\theta/L$ (θ = time for liquid to move characteristic distance L)	Duration of process/time for liquid to move through L	Choice of time scales	13
H11	Hooke number	H_{o_2}	$\rho V^2/E \equiv$ Vauchy No., q.v.		Elasticity of flowing media	25
J1	Jakob modulus	Ja	$C_p \rho_L \Delta t/r\rho_v$ (ρ_L, ρ_v = densities of liquid and vapor; Δt = liquid superheat temperature diff.)	Maximum bubble radius/thickness of superheated film	Boiling	6
J2	J-factor (heat transfer)	j_H	$(h/c_p G)(c_p \mu/k)^{2/3} \equiv (N_{Nu})/(N_{Re})(N_{Pr})^{1/3}$		Heat, mass and momentum transfer theory	5
J3	J-factor (mass transfer)	j_M	$(k_c \rho/G)(\mu/\rho D)^{2/3} \equiv (k_c \rho/G)(N_{Sc})^{2/3}$			5
J4	Joule number	J	$2\rho C_p L \Delta t/\mu_e H_e^2 \equiv 2(N_{Re})(R_M)/(M_H)^2(N_E)$	Joule heating energy/magnetic field energy	Magneto-fluid dynamics	6
K1	Kármán number (1)	N_K	$\rho d^3(-dp/dL)/\mu^2$ (d = pipe diam., dp/dL = pressure gradient) $\equiv 2(N_{Re})^2 f^{1/2}$		Fluid friction in conduits	5

Table 10–51. DIMENSIONLESS GROUPS (Continued)
Table B.   ALPHABETICAL LIST OF NAMED GROUPS (Continued)

Serial No.	Name	Symbol	Definition	Significance	Field of Use	Reference
K2	Kármán number (2)	K	V/V_A (see Alfven No.)		Magneto-fluid dynamics	6
K3	Kirpichev number for heat transfer	Ki_q, N_{Ki_q}	$q^*L/k\Delta t$ (cf. N_{Bih}, N_{Nu})	Intensity external heat transfer/internal heat transfer intensity	Heat transfer	6, 14
K4	Kirpichev number for mass transfer	Ki_m	$GL/D\rho n^*$ (cf. N_{Pem}, N_{Bim})	Intensity external mass transfer/internal mass transfer intensity	Mass transfer	13, 14
K5	Kirpitcheff number		$(\rho F_R/\mu^2)^{1/3} = [(N_{Re})^2 cf]^{1/3}$		Flow around obstacles	34
K6	Knudsen number (1)	N_{Kn}	λ/L	Length of mean free path/characteristic dimension	Low pressure gas flow	5
K7	Knudsen number (2)	N_{KnA}	$\rho D/_{AB}D_{KA}\zeta$	Bulk diffusion/Knudsen diffusion	Gaseous diffusion in packed beds	6
K8	Kossovich number	K_o, N_{Ko}	$r_v\Delta n_m/c\Delta t$	Heat used for evaporation/heat used in raising temperature of body	Convective heat transfer during evaporation	6, 14
K9	Kronig number	Kr	$4L^2\beta\rho^2\Delta t E_S^2 N[\alpha + 2/3(p_o^2/kT)]/u^2 M$ E_S = electric field at surface, N = Avogadro's Number, α = polarization coefficient, p_o = molecular dipole moment, k = Boltzmann's constant, M = molecular weight	(N_{Re}) (electrostatic force/viscous force)	Convective heat transfer	6, 44
K10	Kutateladze number (1)	Ku	$IEL/\rho Vu'$; I = current density $[Q/L^2\theta]$, E = voltage $[ML/Q\theta^2]$, u' = enthalpy $[L^2/\theta^2]$		Electric arcs in gas streams	65
K11	Kutateladze number (2)	K	$r_v/c_p(t_o - t_w)$, (t_o, t_w = stream, wall temp.)		Combined heat and mass transfer in evaporation	17
L1	Lagrange group (1)	La_1	$\Pi/\mu L^3 N^2$; L = characteristic dimension of agitator = $N_{Re} N_p$		Agitation	6
L2	Lagrange number (2)	La_2	$(D + \varepsilon_D)/D$	Combined molecular and eddy mass transfer rate/molecular mass transfer rate	Mass transfer in turbulent systems	31
L3	Lagrange number (3)	La_3	$\Delta PR/\mu V$		Magneto-fluid dynamics	6
L4	Larmor number	R_{La}	L_L/L; (L_L = Larmor radius)		Magneto-fluid dynamics	23
L5	Laval number	La	$V\Big/\left(\dfrac{2\gamma}{\gamma+1}RT\right)^{1/2}$; γ = ratio of specific heats	Linear velocity/critical velocity of sound	Compressible flow	56
L6	Leverett function	J	$(\xi/e)^{1/2}(P\sigma/\sigma$	Characteristic dimension of surface curvature/characteristic dimension of pores	Two-phase flow in porous media	6
L7	Lewis No.	N_{Le}	$k/\rho c_p D = \alpha/D \equiv N_{Sc}/N_{Pr}$, (N.B.: Lewis number is sometimes defined as reciprocal of this quantity)		Combined heat and mass transfer	5, 25
L8	Lorentz number	N_{Lo}	V/V_1; (V_1 = velocity of light)	Fluid velocity/velocity of light	Magneto-fluid dynamics	6
L9	Luikov (Lykov) number	Lu	$k_cL/\alpha = k_cL_\rho C_p/k$	Mass diffusivity/thermal diffusivity; rate of extension of mass transfer field/rate of extension of heat transfer field	Combined heat and transfer	6, 13
L10	Lukomskii number	Lu	α/a_m; a_m = potential conductivity of mass transfer $[L^2/\theta]$		Combined heat and mass transfer	36
L11	Lundquist number	N_{Lu}	$\sigma_e H_e \mu_e^{3/2} L/\rho^{1/2} = M_H(R_M/N_{Re})^{1/2}$ (L = thickness of fluid layer)		Magneto-fluid dynamics	6
L12	Lyashchenko number	Ly	$\equiv N_{Re}^3/N_{Ar}$		Fluidization	19
L13	Lykoudis number	N_{Ly}	$(\mu_e H_e)^2\dfrac{\sigma_e}{\rho}\left[\dfrac{L}{g\beta\Delta t}\right]^{1/2} \equiv (M_H)^2/(N_{Gr})^{1/2}$		Magneto-fluid dynamics	6
M1	McAdams group		$h^4 L\mu\Delta t/k^3\rho^2 gr$	Constant for given surface orientation	Condensation	5
M2	Mach number	N_{Ma}, Ma	V/V_s; (V_s = velocity of sound in fluid) $\equiv$ $v/\sqrt{E_b/\rho}$; (E_b = bulk modulus of fluid) (cf. Sarrau number)	Linear velocity/sonic velocity	Compressible flow	5–7, 25, 50
M3	Magnetic force parameter	N	$\mu_e^2 H_e^2 \sigma_e L/\rho V$	Magnetic body force/inertia force; resistance time of fluid in field/relaxation time of lines force	Magnetic-fluid dynamics	6
M4	Magnetic mach number	M_{Ma}	V/V_a (see Alfven number)		Magneto-fluid dynamics	6
M5	Magnetic Oseen number	k	$\frac{1}{2}(1 - N_{Al}^2)R_M$	Magnetic force/inertia force	Magneto-fluid dynamics	2
M6	Magnetic pressure number	S	$\mu_e H_e^2/\rho V^2$	Magnetic pressure/2 × dynamic pressure	Magneto-fluid dynamics	5
M7	Magnetic Reynolds number	R_M	$\sigma_e\mu_e LV$ (cf. velocity number)	Mass transport diffusivity/magnetic diffusivity	Magneto-fluid dynamics	5
M8	Maievskii number		$\equiv Ma$		Compressible flow	12
M9	Marangoni number	N_{Ma}	$\dfrac{\Delta\sigma}{\Delta t}\dfrac{\Delta t}{\Delta L}L^2/\mu\alpha$; L = layer thickness		Cellular convection	46

Table 10–51. DIMENSIONLESS GROUPS *(Continued)*
Table B. ALPHABETICAL LIST OF NAMED GROUPS *(Continued)*

Serial No.	Name	Symbol	Definition	Significance	Field of Use	Reference
M10	Margoulis number	M	$\equiv N_{St}$		Forced convection	6, 7
M11	Merkel number	N_{Me}	$GA^{*}W^{*}/(V_m)$ gas	Mass of water transferred in cooling per unit humidity difference/ mass of dry gas	Cooling towers, liquid-gas contact	5
M12	Miniovich number	Mn	SR/e; R = pore radius		Drying	36
M13	Mondt number	N_{Mo}		Convective/conductive heat transfer	Heat transfer	6
N1	Naze number	N_a	$V_A/V_s \equiv (N_{Ma}.N_{A1})$	Velocity Alfven wave/ velocity of sound	Magneto-fluid dynamics	6
N2	Newton inertial force group	N_i	$F/\rho V^2 L^2$	Imposed force/inertial group	Agitation	5
N3	Newton number	N_e	$F_R/\rho V^2 L^2$; (cf. f, ψ)	Resistance force/inertia force	Friction in fluid flow	25
N4	Number of velocity heads	N	$(F/\rho L^2)/(V^2/2)$	Imposed head/velocity head	Friction in conduits	6
N5	Number for similarity of phys. and chem. changes	K	$r/C_p\Delta t$	Heat flow for phase change/superheat (supercooling) of one of the phases	Changes of phase	14
N6	Nusselt number	N_{Nu}	$hL/k = (N_{Re}N_{St})$ (cf. N_{Bi_h}, Ki_g)	Total heat transfer/conductive heat transfer	Forced convection	5, 13, 25
N7	Nusselt number for mass transfer	Nu_m, N_{Nu_m}	$k_cL/D = N_{Sh}$	Intensity of mass flux at interface/specific flux by pure molecular diffusion in layer of thickness, L	Mass transfer	13, 20
N8	Nusselt film thickness group	ψ, N_T	$L_f(\rho^2 g/\mu^2)^{1/3} = (N_{Ga})^{1/3}$; ($L_f$ = film thickness)	$= (N_{Re})^{1/3}$ (gravitational force/viscous force)$^{1/3}$	Falling films	6
O1	Ocvirk number		$(F_b//\mu V_s)(a/R)^2(D/b)^2$; ($v_s$ = shaft surface velocity; R = shaft radius; D = shaft diam.) (cf. N_S)	Load force/viscous force	Lubrication	6
O2	Ohnesorge number	Z	$\mu/(\rho L\sigma)^{1/2} = (N_{We_1})^{1/2}/(N_{Re})$	Viscous force/(inertia force × surface tension force)$^{1/2}$	Atomization	5
P1	Péclet number (heat)	Pe, N_{Pe_h}	$LV\rho C_p/k = LV/\alpha = (N_{Re}.N_{Pr})$	Bulk heat transfer/conductive heat transfer	Forced convection	5, 13, 25
P2	Péclet number (mass)	N_{Pe_m}	$LV/D = (N_{Re}.N_{Sc})$	Bulk mass transfer/diffusive mass transfer	Mass transfer	5
P3	Pipeline parameter	ρn	$V_w V_o/2H'_s$; (V_w = velocity water-hammer wave, V_o = initial velocity H'_s = static head $\propto g[L^2/\theta^2]$)	Maximum pressure rise in water hammer/? × static pressure	Water hammer	5
P4	Poiseuille number		$D^2(-dp/dL)/\mu V$ (D = pipe diam., dp/dL = pressure gradient)	= 32 for laminar flow in round pipe	Laminar fluid friction	5
P5	Pomerantsev number	P_o	$jL^2/k(t_m-t_o)$ (t_m, t_o = temp. of medium, initial temp. of body) cf. Damköhler Group IV)		Heat transfer with heat sources in medium	6, 14, 60
P6	Posnov number	Pn	$\delta\Delta t/(\Delta n_m)$ (cf. Fe_2)		Combined heat and mass transfer	13, 14, 67
P7	Power number	N_P	$\Pi/L^3\rho N^3$	Drag on (agitator impeller) or inertial force	Power consumption by agitators, fans, pumps, etc	5, 32
P8	Prandtl number	N_{Pr}	$C_p\mu/k$ = Da IV/Da III × Da V	Momentum diffusivity/ thermal diffusivity	Forced and free convection	5, 13, 25
P9	Prandtl number (mass transfer)	Pr_m	$\mu/\rho D = N_{Sc}$, v (used in Russian, German literature)	See Schmidt number		13
P10	Prandtl velocity ratio	u^+	$V/(\tau_w/\rho)^{1/2}$ (V = local fluid velocity)	Inertial force/wall shear force$^{1/2}$	Turbulence studies	5
P11	Prandtl dimensionless	y^+	$L(\rho\tau_w)^{1/2}/\mu$ (L = distance from wall, etc.)		Turbulence studies	
P12	Predvoditelev number	Pd	$\Gamma L^2/\alpha t_o = \left(\dfrac{dt}{\alpha(N_{Fo})}\right)_{max}$ where t_o = init. temp. of body, t^* = temp. of medium relative to its initial temp.	Rate of change of temp. of medium/rate of change of temp. of body	Heat transfer	13, 14, 60
P13	Pressure number (1)	K_P	$P/\{g\sigma(\rho'-\rho'')\}^{1/2}$ (ρ', ρ'' = density of liquid gas)	Absolute pressure in system (pressure jump on interface)		14
P14	Pressure number (2)		$H/\tfrac{1}{2}U_2^2$ (U_2 = circumferential velocity) $\equiv$ [diameter No.]$^{-2}$ × [Speed No.]$^{-2}$		Flow machines (turbines, pumps, etc.)	25
P15	Psychrometric ratio		h_c/Gs	Heat transfer by convection/heat transfer by mass transfer	Wet and dry bulb thermometry	5
R1	Radiation number	K_s	$kE/\eta\sigma T^3 = (N_{We_1})/(\text{Hooke No.})$ × (Stefan No.)		Radiant transfer	25
R2	Ramzin number	Ra	$\dfrac{C_bP}{C_m(\Delta\Omega)} = \dfrac{(\text{Bulgin No.})}{(\text{Kosovich No.})}$		Molar mass transfer	40
R3	Ratio of specific heats	γ	C_p/C_v (specific heats at constant pressure, volume)		Compressible flow	5
R4	Rayleigh number (1)	N_{Ra_1}	$V(\rho L/\sigma)^{1/2} = N_{We_1}$, (q.v.)	See N_{We}	Breakup of liquid jets	5
R5	Rayleigh number (2)	R'_2	$L^3\rho^2 g\beta c_p\Delta t/\mu k = L^3\rho g\beta\Delta t/\mu\alpha = (N_{Gr})\cdot(N_{Pr})$		Free convection	5, 25

Table 10–51. DIMENSIONLESS GROUPS *(Continued)*
Table B.   ALPHABETICAL LIST OF NAMED GROUPS *(Continued)*

Serial No.	Name	Symbol	Definition	Significance	Field of Use	Reference
R6	Rayleigh number (3)	Ra_3	$q^*L^5\rho^2 g\beta C_p/\mu k^2 x = (N_{Gr})(N_{Pr})$ $(N_{Nu})(L/x)$; $(L = $ pipe diam.)		Combined free and forced convection in vertical tubes	6
R7	Recovery factor	N_{rf}	$C_p(t_{aw}-t_m)/V^2$; $t_{aw} = $ attained adiabatic wall temp.; $t_m = $ temp. of moving medium (*cf.* Eckert No.)	Actual temp. recovery/ theoretical temp. recovery	Convective heat transfer in compressible flow	5
R8	Reech number		$= 1/(N_{Fr_1})$ *q.v.*		Wave and surface behavior	6
R9	Resistance coefficient (1)	C_f	$F_R/\tfrac{1}{2}\rho V^2 L^2$ (*cf.* drag coeff., Newton number, Fanning factor)		Flow resistance	25
R10	Resistance coefficient (2)	ψ	$\Delta p \cdot D_H/\tfrac{1}{2}\rho V^2 L$ ($\Delta p = $ pressure drop over length, L) (*cf.* R9)		Fluid friction in conduits	25
R11	Reynolds number	N_{Re}	$LV\rho/\mu$	Inertia force/viscous force	Dynamic similarity	5, 25
R12	Reynolds number (rotating)	Re	$L^2 N\rho/\mu$; $L = $ impeller diam.		Agitation	45
R13	Richardson number	N_{Ri}	$-(g/\rho)(d\rho/dL)/(dV/dL)_w^2$ [$L = $ height of liquid layer, $(dV/dL)_w = $ velocity gradient at wall]	Gravity force/inertial force	Stratified flow of multi-layer systems	54
R14	Romankov number	R'_o	T_D/T_{PROD}	Dry bulb temperature (abs.)/(product temperature (abs.)	Drying	6, 52
R15	Rossby number	N_{Ro}	$V/2\omega_e L \sin\Lambda$ ($\omega_e = $ angular velocity of earth's rotation [$1/\theta$]; $\Lambda = $ angle between axis of earth's rotation and direction of fluid motion [$^-$])	Inertia force/Coriolis force	Effect of earth's rotation on flow in pipes	5
R16	Roughness factor		ε/L		Fluid friction	5
S1	Sarrau number		$\equiv$ mach number, *q.v.*		Compressible flow	6
S2	Schiller number (1)		$LV(\rho^2/\mu F_R)^{1/3}$		Flow around obstacles	34
S3	Schiller number (2)	Sch	$V\left[\dfrac{3}{4}\cdot\dfrac{\rho\gamma_m}{g\mu(\gamma_M-\gamma_m)}\right]^{1/3}$; $V = $ velocity in fluidized bed; γ_m, $\gamma_M = $ specific gravity of medium and material in bed		Fluidization	66
S4	Schmidt number	N_{Sc}	$\mu/\rho D$ (*cf.* N_{Pr_m}) $(= $ Da II/Da I Da V)	Kinetic viscosity/molecular diffusivity	Diffusion in flowing	5, 25
S5	Semenov number	Sm	k_c/K; $K = $ reaction rate constant [L/θ]		Reaction kinetics	11
S6	Senftleben number	S_e	$NE_f^2[\alpha + 2/3(\rho_o^2/kT)]\cdot[1/4LM_g]$ Kronig number, *q.v.*		Convective heat transfer	6
S7	Sherwood number	N_{Sh}	$k_c L/D = Nu_m$ (also termed Taylor number)	Mass diffusivity/molecular diffusivity	Mass transfer	5
S8	Sommerfeld number (1)	N_{S_1}	$(\mu N/P_b)(D/a)^2$ ($D = $ shaft diam. (*cf.*) Ocvirk number)	Viscous force/load force	Lubrication	5
S9	Sommerfeld number (2)	N_{S_2}	$(F_b/\mu V_s)(a/R)^2$ ($V_s = $ veloc. of shaft surface; $R = $ shaft radius) ($N_{S_2} = 4/\pi N_{S_1}$)	Viscous force/load force	Lubrication	6
S10	Spalding function	Sp	$-\left(\dfrac{\delta\theta}{\delta u^+}\right)_{u^+=0}$; $\theta = (T-T_\infty)/(T_w-T_\infty)$, $T_w = $ wall temperature, $T_\infty = $ free stream temp., $u^+ = $ Prandtl velocity ratio	Dimensionless temp. gradient at wall	Convection	18, 33
S11	Specific speed	C_s	$N(V_f)^{1/2}/(gH')^{3/4}$ ($H' = $ head of liquid produced by one stage) (*cf.* speed number)		Pumps and compressors	8, 32
S12	Speed number	σ	$(4\pi)^{1/2}(V_f)^{1/2} N/(2H)^{3/4} = $ (delivery number)$^{1/2} \times$ pressure number)$^{-3/4}$ (*cf.* specific speed)		Flow machines	25
S13	Stanton number	N_{St}	$h/C_p\rho V = h/C_p G = (N_{Nu})/(N_{Re})(N_{Pr})$	Heat transferred/thermal capacity of fluid	Forced convection	5, 6, 13, 25
S14	Stefan number	St	$\eta L T^3/k$		Heat radiation	25
S15	Stokes number	St	$\mu\theta_v/\rho L^2$ ($\theta_v = $ vibration time) $\equiv (N_{S_1})^{-1}(N_{Re})^{-1}$		Particle dynamics	6
S16	Strouhal number	N_{S_1}, Sr	fL/V (*cf.* N_{Th})		Vortex streets; unsteady-state flow	6, 25
S17	Suratman number	Su	$\rho L\sigma/\mu^2 = (N_{Re})^2/(N_{We_1}) = (Z)^{-2}$		Particle dynamics	6
S18	Surface elasticity number	N_{El}	$-\dfrac{\Gamma'}{D_s}L\dfrac{(\partial\sigma)}{(\partial\Gamma')}$; $\Gamma' = $ surface concentration of surfactant in undisturbed state, $D_s = $ surface diffusivity, $L = $ film thickness		Convection cells	10
S19	Surface viscosity number	N_{Vi}	$\mu_s/\mu L$; $\mu_s = $ surface viscosity, [M/θ], $L = $ film thickness		Convection cells	10
T1	Taylor number (1)	N_{Ta_1}	$\omega_c(R_a)^{1/2}a^{3/2}\rho/\mu$; ($\omega_c = $ angular velocity of cylinder; $R_a = $ mean radius of annulus)		Stability of flow pattern in annulus with rotating cylinder	5
T2	Taylor number (2)	N_{Ta_2}	$(2\omega L^2\rho/\mu)^2[\omega = $ rate of spin $(1/\theta)$; $L = $ height of fluid layer]	α(Coriolis force/viscous force)^2	Effect of rotation on free convection	6
T3	Thiele modulus	m_T	$Q^{1/2}U^{1/2}L/k^{1/2}t^{1/2} = $ (Da IV)$^{1/2}$		Diffusion in porous catalysts	5
T4	Thoma number	σ_T	$(H_a-H_s-H_v)/H$ ($H = $ total head; $H_a = $ atm. pressure head; $H_s = $ suction head; $H_v = $ vapor pressure head)	Net positive suction head/total head	Cavitation in pumps	5
T5	Thomson number	N_{Th}	$\theta V/L$; $\theta = $ characteristic time (*cf.* N_{S_1})		Fluid flow	6
T6	Thring radiation group		$\rho C_p V/e^*\eta T^3$ (*cf.* Boltzmann number)	Bulk heat transport/heat transport by radiation	Radiation	5

Table 10–51. DIMENSIONLESS GROUPS (Continued)
Table B. ALPHABETICAL LIST OF NAMED GROUPS (Continued)

Serial No.	Name	Symbol	Definition	Significance	Field of Use	Reference
T7	Thring-Newby criterion	θ	$[(V_{m_1} + V_{mo})/V_{mo}](R/L)$; V_{mo}, V_{u_1} = mass flow rates of nozzle fluid and surrounding fluid $[M/\theta]$; R = equivalent nozzle radius; L = furnace half width		Combustion of fuels	4
T8	Truncation number	r	$\mu\gamma/P$ (cf. Hersey number)	Shear stress/normal stress	Viscous flow	6
V1	Valensi number	V	$\omega L^2 \rho/\mu$; ω = circular oscillation frequency when $\mu = 0$ $[1/\theta]$		Oscillations of drops and bubble-	57
V2	Vedernikov number	V	$\zeta^* \xi^* \overline{V}/(V_w - \overline{V}) \equiv \zeta^* \xi^*(N_{Fr_2})(\zeta^* = $ exponent of hydraulic radius in formula $[^-]$; ζ^* = shape factor of channel section; V_w = absolute velocity of disturbance wave)	Generalized Froude number	Instability of open-channel flow	6, 61
V3	Velocity number	R_v	$\equiv$ Magnetic Reynolds number, q.v.			5, 6
W1	Weber number (1)	N_{We_1}	$V^2 \rho L/\sigma = (N_{We_2})^2$	Inertia force/surface tension force	Bubble formation, etc.	2, 25
W2	Weber number (2)	N_{We_2}	$V(\rho L/\sigma)^{1/2} = (N_{We_1})^{1/2}$			
W3	Weber number (rotating)	W_e	$L^3 N^2 \rho/\sigma$; L = impeller diameter		Agitation	64
W4	Weissenberg number	N_{We}	$\omega_3 V/\omega_1 L$; $\omega_3 = \int_0^\infty s G(s)\,ds$, $\omega_1 = \int_0^\infty G(s)\,ds$, G = relaxation modulus of linear viscoelasticity, s = recoverable elastic strain	Viscoelastic force/viscous force	Viscoelastic flow	63

REFERENCES

1. Adrianov, V. N., Shorin, S. N., *AIAA J.* **1**, 1729 (1963).
2. Ahlstrom, H. G., *J. Fluid Mech.* **15**, 205 (1963).
3. Becker, H. A., Hottel, H. C., Williams, G. C., "Ninth Symposium (International) on Combustion," p. 7, Academic Press, New York, 1963.
4. Beer, J. M., Chigier, N. A., Lee, K. B., *Ibid.*, p. 892.
5. Boucher, D. F., Alves, G. E., *Chem. Eng. Progr.* **55** (9), 55 (1959).
6. *Ibid.*, **59** (8), 75 (1963).
7. British Standard 1991, "Recommendations for Letter Symbols, Signs and Abbreviations. Part 2. Chemical Engineering, Nuclear Science, and Applied Chemistry," British Standards Institution, London, 1961.
8. Brown, G. G., *et al.*, "Unit Operations," Wiley, New York, 1950.
9. Buckingham, E., *Phys. Rev.* **4**, 345 (1914).
10. Berg, J. C., Acrivos, A., *Chem. Eng. Sci.* **20**, 737 (1965).
11. Chukhanov, Z. F., *Intern. J. Heat Mass Transfer* **6**, 691 (1963).
12. Dallavalle, J. M., "Micromeritics," 2nd ed., Pitman, New York, 1948.
13. El'perin, I. T., *Inzh. Fiz. Zh. Akad. Nauk Belorussk. SSR* **4** (1), 131 (1963).
14. El'perin, I. T., *Intern. J. Heat Mass Transfer* **5**, 349 (1962).
15. Engel, F. V. A., *Z.V.D.I.* **107**, 671, 793 (1965).
16. Faller, A. J., *J. Fluid Mech.* **15**, 560 (1963).
17. Fedorov, B. I., *Inzh. Fiz. Zh. Akad. Nauk Belorussk. SSR* **7** (1), 21 (1964).
18. Gardner, G. O., Kestin, J., *Intern. J. Heat Mass Transfer* **6**, 289 (1963).
19. Gel'perin, I. T., Ainshtein, V. G., Goĭkhman, I. D., *Inzh. Fiz. Zh. Akad. Nauk Belorussk. SSR* **7** (1), 15 (1964).
20. Grassmann, P., *Chem. Ing.-Tech.* **31**, 148 (1959).
21. Grassman, P., Lemaire, L. H., *Ibid.*, **30**, 450 (1958).
22. Greene, D. F., Ph.D. Thesis, Columbia Univ., 1961 [*Dissertation Abstr.* **24** (8), 3248 (1964)].
23. Gukhman, A. A., "Introduction to the Theory of Similarity," Academic Press, New York, 1965.
24. Gutfinger, C., Tallmadge, J. A., *A.I.Ch.E. J.* **10**, 774 (1965).
25. Hahnemann, H. W., "Die Umstellung auf das internationale Einheitensystem in Mechanik und Wärmetechnik," VDI-Verlag, Düsseldorf, 1959.
26. Holt, M., "Dimensional Analysis" in "Handbook of Fluid Dynamics," V. L. Streeter, ed., McGraw-Hill, New York, 1961.
27. Hottel, H. C., Sarofim, A. F., *Intern. J. Heat Mass Transfer* **8**, 1153 (1965).
28. Hottel, H. C., Williams, G. C., Jensen, W. P., Tobey, A. C., Burrage, P. M. R., p. 923 in "Ninth Symposium (International) on Combustion," Academic Press, New York, 1963.
29. Huntley, H. E., "Dimensional Analysis," MacDonald & Co., London, 1952.
30. Johnson, S. P., "Survey of Flow Calculation Methods," p. 98, Pre-printed Papers & Program, Aeronautic & Hydraulic Divisions, A.S.M.E. Summer Meeting, June 19–21, Univ. of Calif. and Stanford Univ., 1934.
31. Kafarov, V. V., *Zh. Prikl. Khim.* **29**, 40 (1956).
32. Kay, J. M., "An Introduction to Fluid Mechanics & Heat Transfer," Cambridge Univ. Press, 1957.
33. Kestin, J., Persen, L. N., *Intern. J. Heat Mass Transfer* **5**, 143 (1962).

34. Klinkenberg, A., Mooy, H. H., *Chem. Eng. Progr.* **44**, 17 (1948).
35. Koide, K., Kubota, H., Shindo, M., *Chem. Eng.* (Japan), **28** (8), 657 (1964).
36. Lykov, A. V., Mikhaĭlov, Yu. A., "Theory of Energy & Mass Transfer," Prentice-Hall, Englewood Cliffs, N.J., 1961.
37. McCabe, W. L., Smith, J. C., "Unit Operations of Chemical Engineering," McGraw-Hill, New York, 1956.
38. Matsuhisa, S., Bird, R. B., *A.I.Ch.E. J.* **11**, 588 (1965).
39. Mikhaĭlov, Yu. A., Bornikova, R. M., *Inzh. Fiz. Zh. Akad. Nauk Belorussk. SSR* **6** (10), 45 (1963).
40. Mikhaĭlov, Yu. A., Romanina, I. V., *Ibid.*, **7** (1), 49 (1964).
41. Miyauchi, T., Nakano, K., Obata, K., Kimura, S., *Chem. Eng.* (Japan) **26** (9), 999 (1962).
42. Mkhitaryan, A. M., "Hydraulics & Fundamentals of Gas Dynamics," Israel Program for Scientific Translations, Jerusalem, 1964.
43. Mordell, D. L., Wu, J. H. T., *Can. Aeronaut. Space J.* **9** (4), 117 (1963).
44. Motulevich, V. P., Eroshenko, V. M., Petrov, Yu. P., in "Physics of Heat Exchange & Gas Dynamics," A. S. Predvoditelev, ed., Consultants Bureau, New York, 1963.
45. Nagata, S., *Chem. Eng.* (Japan) **27** (8), 592 (1962).
46. Nield, D. A., *J. Fluid Mech.* **19**, 341 (1964).
47. Potter, J. M. F., B.Sc. Thesis, Dept. of Chem. Engrg., Univ. of Birmingham, England, 1959.
48. Rayleigh, Lord, *Phil. Mag.* **48**, 321 (1899).
49. Reiner, M., *Phys. Today* **17** (1), 62 (1964).
50. Rouse, H. (ed.), "Engineering Hydraulics," Wiley, New York, 1950.
51. Rouse, H., Ince, S., "History of Hydraulics," Iowa Institute of Hydraulic Research, State University of Iowa, 1957.
52. Sazhin, B. S., *Inzh. Fiz. Zh. Akad. Nauk Belorussk. SSR* **5** (6), 13 (1962).
53. Sazhin, B. S., Miklin, Yu. A., *Ibid.*, **6** (10), 57 (1963).
54. Schlichting, H., "Boundary Layer Theory," 4th ed., McGraw-Hill, New York, 1960.
55. Scriven, L. E., Sternling, C. V., *J. Fluid Mech.* **19**, 321 (1964).
56. Sillem, H., *Z.V.D.I.* **106**, 398 (1964).
57. Szebehely, V. G., p. 771 in "Proc. 2nd U.S. Nat. Congress of Appl. Mech.," Ann Arbor, Mich., June 1954; A.S.M.E., New York, 1955.
58. Tallmadge, J. A., Labine, R. A., Wood, B. H., *Ind. Eng. Chem. Fundamentals* **4**, 400 (1965).
59. Tamarin, A. I., *Inzh. Fiz. Zh. Akad. Nauk Belorussk. SSR* **6** (7), 19 (1963).
60. Tartakovskiĭ, D. F., *Ibid.*, **7** (1), 71 (1964).
61. Vedernikov, V. V., *Compt. Rend. Acad. Sci. U.R.S.S.* **48**, 239 (1945); **52**, 207 (1946).
62. Weber, M., *Jahrb. Schiffbautechn. Ges.* **20**, 355 (1919).
63. White, J. L., *J. Appl. Polymer. Sci.* **8**, 2339 (1964).
64. Yamaguchi, I., Yabuta, S., Nagata, S., *Chem. Eng.* (Japan) **27** (8), 576 (1963).
65. Yas'ko, O. I., *Inzh.-Fiz. Zh. Akad. Nauk Belorussk. SSR* **7** (12), 112 (1964).
66. Zabrodskiĭ, S. S., "Flow & Heat Transfer in Fluidized Beds," to be published shortly by M.I.T. Press.
67. Zhuravleva, V. P., *Inzh.-Fiz. Zh. Akad. Nauk Belorussk. SSR* **6** (9), 73 (1963).

Table 10–52. DIMENSIONLESS GROUPS— SUPPLEMENTARY TABLE*

J.P. CATCHPOLE AND G.D. FULFORD

ALPHABETICAL LIST OF NEW GROUPS

Serial No.	Name	Symbol	Definition	Significance	Field of Use	Reference
A0	Absorption No.	Ab	$kc_L\sqrt{\dfrac{xL_f}{DV_f'}}$ kc_L = liquid side mass transfer coefficient; x = length of wetted surface; L_f = film thickness; V_f' = volume flow rate per wetted perimeter $[L^2/\theta]$	Dimensionless mass transfer coefficient	Gas absorption in wetted wall column	40
A1a	Advance ratio	J	V/ND V = forward speed; D = propeller diameter	Special form of Strouhal No.	Propeller studies	27
A1b	Aeroelasticity parameter	—	≡ Cauchy No., q.v.	Inertia force/compressibility force	Compressible flow	27
A4a	Anonymous group 3	ε	$Dx/V_f'L_f$ (symbols as in Absorption No.). $\varepsilon = (Ab)^2/(N_{Sh})^2$	Dimensionless diffusivity	Gas absorption in wetted wall column	40
A4b	Anonymous group 4	— $1/\alpha$ $(1/\beta)$	$\tau_w R/V_\infty\mu$; R = cylinder radius, V_∞ = velocity outside boundary layer	Frictional force/viscous force dimensionless skin friction)	Laminar boundary layer flow	12, 28, 35
B1a	Bairstow No.	—	V/V_{sw} V_{sw} = velocity of sound at wall (cf. Mach. No.)	Previously used for Mach No., now largely obsolete	—	27
B2a	Batchelor No.	—	$VL\sigma_e/V_1^2\varepsilon_e$ ε_e = electrical permittivity $[Q^2\theta^2/L^3M]$		Magnetofluid dynamics	27
B13a	Buoyancy parameter	—	$\dfrac{\Delta T}{T}\dfrac{gL}{V^2} = (N_{Gr})/(N_{Re})^2 = \dfrac{\Delta T}{T}\left(\dfrac{1}{N_{Fr_1}}\right)$	Buoyancy force/inertia force	Free convection	27
D6a	Darcy No. (2)	Da_2	VL/D'; D' = permeability coefficient of porous medium $[L^2/\theta]$	Inertia force/permeation force	Flow in porous media	19
D8a	Generalized Deborah No.	N_2	$\sqrt{I_e - I_w}\cdot\theta_n$ I_e = invariant of rate of strain tensor (sec.$^{-2}$); I_w = invariant of vorticity tensor (sec.$^{-2}$); θ_n = natural time (sec.)	Generalization of group D8	Rheology	2
D14a	Dufour No.	Du_2	$\varTheta\Theta n_{10}'/c_p$ Θ = thermodiffusion constant $= (D_T/D)/n_{10}n_{20}$ $[-]$; D_T = thermal diffusion coefficient $[L^2/\theta]$; n_{10}', $n_{20}' = n_1'/n'$, n_2'/n'; n' = total No. of molecules $= n_1' + n_2'$; n_1', n_2' = No. of molecules of components 1, 2, in binary mixture; also $Du_2 = (D_T/D)p/\rho c_p Tn_{20}$	Heat of isothermal mass transfer/enthalpy of unit mass of mixture	Thermodiffusion	22
E1a	Einstein No.	—	V/V_1 (V_1 = speed of light) (cf. Lorentz No.)	Fluid velocity/velocity of light	Magnetofluid dynamics	27
E4a	Electric field parameter	R_E	$E/V\mu_e H_e$		Magnetohydrodynamics	26
E4b	Electrical characteristic No.	El	$\rho(d\chi/dT)L^2\cdot\Delta T\cdot E_1^2/\mu^2$ E_1 = electrical field strength $[ML/Q\theta^2]$; χ = dielectric susceptibility $[Q^2\theta^2/ML^3]$		Electrical effects on transfer processes	6
E4c	Electrical Nusselt No.	N_u	$VL/D^{\bigstar}$; $D^{\bigstar} = \frac{1}{2}(D^+ + D^-)$ $[L^2/\theta]$ D^+, D^- = diffusion coefficients of ions (cf. group P2)	Convection current/diffusion current	Electrochemistry	10
E4d	Electrical Reynolds No. (1)	—	$\varepsilon_e V/Q'Lb'$ ε_e = electrical permittivity $[Q^2\theta^2/L^3M]$; Q' = space charge density $[Q/L^3]$; b' = carrier mobility $[Q\theta/M]$		Electrical effects in flow	27
E4e	Electrical Reynolds No. (2)	—	Alternate name for group E4c, q.v.			10
E10a	Modified Euler No.	Eu'	$H_L\cdot\rho_L g/V_G^2\rho_G$ H_L' = head of liquid on tray $[L]$; ρ_L, ρ_G = densities of liquid, vapor; V_G = vapor velocity based on free area, $[L/\theta]$	Friction head/velocity head	Flow of vapor across mass transfer trays	13
E13a	Expansion No.	Ex	$\left(\dfrac{gd}{V^2}\right)\left(\dfrac{\rho_L - \rho_G}{\rho_L}\right)$ d = bubble diam., V = bubble veloc., ρ_L, ρ_G = densities of liquid, gas	$1/N_{Fr_1}\times$ density ratio	Rise of bubbles	13
F12a	Fourier No. (flow)	Fo	$\mu\theta/\rho L^2$		Unsteady state flow problems	17
F12b	Frank-Kamenetskii No.	δ	$\dfrac{Q''}{k}\dfrac{E_a}{\varXi T_o^2}L^2k_o\exp(-E_a/\varXi T)$ Q'' = heat liberated per unit mass of material reacting/unit volume $[1/L\theta^2]$; k = thermal conductivity of reacting mixture $[ML/T\theta^3]$; k_o = preexponential constant in Arrhenius equation $[M/\theta]$	Dimensionless heat effect of reaction	Heat transfer in reacting systems	24
F12c	Frequency parameter	—	$\omega'L/V = 2\pi\times$ Strouhal No.; $2\pi/\omega'$ = period of motion $[\theta]$	Special form of Strouhal No. (cf. also T5)	Unsteady state flow, etc.	27
F12d	Frequency No. (2)	N_f	$\omega_r L/V$; L = packing element diameter $[L]$; V = interstitial fluid velocity $[L/\theta]$; ω_r = radial frequency (radians/sec.) $[1/\theta]$ (cf. groups H10, S16, T5)	Special form of group F12c	Flow in packed or fluidized beds	23, 29

Table 10–52. DIMENSIONLESS GROUPS—
SUPPLEMENTARY TABLE (Continued)
ALPHABETICAL LIST OF NEW GROUPS (Continued)

Serial No.	Name	Symbol	Definition	Significance	Field of Use	Reference
F12e	Frössling No. (heat transfer)	Fs_h	$(N_{Nu} - 2)/(N_{Re}^{1/2} N_{Pr}^{1/3})$	Special dimensionless heat transfer coefficient	Heat transfer to spheres in turbulent streams	8a
F12f	Frössling No. (mass transfer)	Fs_m	$(N_{Sh} - 2)/(N_{Re}^{1/2} N_{Sc}^{1/3})$	Special dimensionless mass transfer coefficient	Mass transfer to spheres in turbulent streams	8a
G2a	Geometric No.	Ge	h^*/H^*; h^* = surface area of packing element/perimeter $[L]$; H = height of packing $[L]$.	Dimensionless packed bed height	Mass transfer in packed beds	5a
G2b	Goertler parameter	—	$\dfrac{VL_b\rho}{\mu}\left(\dfrac{L_b}{R_c}\right)^{1/2}$ L_b = boundary layer momentum thickness, $[L]$; R_c = radius of curvature $[L]$	Modified Reynolds No.	Boundary layer flow on curved surfaces	27
G5a	Diffusional Grashof No.	Gr_{AB}	$L^3 \rho^2 g \beta'_A \Delta n'_A / \mu^2$ n'_A = mass fraction of species A, $[-]$; β'_A = coefficient of density change with n'_A, $[-]$	Buoyant forces × inertia forces/ (viscous forces)^2	Interphase transfer by free convection (density changes caused by concentration differences)	37
H8a	Hess No.	Je	$(KL^?/u_m)(C_o)^{n-1}$ n = order of reaction $[-]$; C_o = initial concn. $[M/L^3]$; a_m = mass transfer conductivity of reaction products $[L^2/\theta]$; K = reaction rate constant, $\left[\dfrac{1}{\theta}\dfrac{L^{3n-3}}{M^{n-1}}\right]$		Heat and mass transfer with chemical and phase changes	22
H9a	Homochronicity No.	Ho_3	$N\theta$ (N = mixer r.p.m., θ = mixing time)		Mixing, agitation	31
H11a	Hydraulic resistance group	Γc	$\Delta p_p/\rho_l g L = We_s \times$ Laplace No. $= N_{Fr_l} \times N_{Fr_l}$ Δp_p = pressure drop across liquid on tray $[M/L\theta^2]$; ρ_l = liquid density; L = depth of liquid on tray	Characterizes development of interfacial area per unit area of tray	Pressure drop in distillation columns	15
I1	Ilyushin No.	I^*	$(Vd\rho/\mu) \cdot 4\tau_D/3V^2\rho = 4\tau_D/3V^2\rho \cdot N_{Re}$; τ_D = max dynamic slip resistance $[M/L\theta^2]$		Flow of viscoplastic fluids	20a
K7a	Knudsen No. for diffusion	$N_{Kn_{A2}}$	$3eD_{AB}/4\zeta K_{oA}\bar{u}_A$ K_{oA} = Knudsen flow permeability constant; $\bar{u}_A$ = equilibrium mean molecular speed of species A	Differs from K7 by numerical constant	Gaseous diffusion in packed beds	34
K7b	Kondrat'ev No.	Kn	ΨN_{Bih}; N_{Bih} = heat transfer Biot No.; Ψ = temp. field nonuniformity parameter = $(t_s - t_a)/(t - t_a)$; t_a = temp. of surrounding medium; t_s = body surface temp.; t = body mean temp.		Heat transfer between fluid and body	21
L3a	Laplace No.	La_4	$\Delta p_p \cdot L/\sigma = \Gamma c/We_s$ Δp_p = pressure drop across liquid on tray $[M/L\theta^2]$; L = depth of liquid on tray		Interfacial behavior on distillation trays	32
L5a	Lebedev No.	Le_2	$eb_T(t_a - t_o)/c_b\rho\rho_s$ b_T = intensity of vapor expansion in capillaries of body on heating $[M/L^3T]$; t_a = temp. of surrounding medium; t_o = initial temp.; ρ_s = density of solid $[M/L^3]$	Molar expansion flux/ molar vapor transfer flux	Drying of porous materials	22
L5b	Leroux No.	—	$\equiv$ cavitation No. C6			27
L7a	Turbulent Lewis No.	Le_T	$c_p\rho\epsilon_D/k_T = l_D/l_T = \epsilon_D/\epsilon_T$ k_T = eddy thermal conductivity $[LM/T\theta^3]$; l_D, l_T = mixing lengths for mass, heat transfer $[L]$; ϵ_T = eddy thermal diffusivity $[L^2/\theta]$		Combined turbulent heat and mass transfer	11
L7b	Lewis-Semenov No.	—	$= 1/N_{Le}$; N_{Le} = group L7			27
L7c	Lock No.	—	$\rho R^4 ia'/I$ ρ = fluid density; a' = rotor lift curve slope $[L^2/M]$; i = blade chord $[L]$; R = rotor radius $[L]$; I = moment of inertia of blade about hinge $[L^4]$		Rotor blade dynamics	27
M3a	Magnetic Grashof No.	—	$4\pi\sigma_e\mu_e(\mu/\rho) \cdot N_{Gr}$ N_{Gr} = group G5		Magnetofluid dynamics	7
M4a	Magnetic No.	R_m	$\mu_e H_e(\sigma_e L/\rho V)^{1/2}$ = (magnetic force parameter)$^{1/2}$	See group M3	Magnetohydrodynamics	26
M5a	Magnetic Prandtl No.	—	$\sigma_e\mu_e\dfrac{\mu}{\rho}$ cf. Magnetic Grashof No., group M3a	Magnetic Reynolds No./Reynolds No. (properties of fluid)	Magnetofluid dynamics	27
P3a	Plasticity No.	P	$\equiv$ Bingham No.			41
P7a	Modified Power No.	N'_P	$\dfrac{\Pi}{L^5\rho N^3}\left(\dfrac{D'_e}{L'}\right)(\Delta\omega)^{-1/2}(N_bN_s)^{0.67}$ D'_e = effective agitator diameter $[L]$; L'_e = effective agitator height; $\Delta\omega$ = wall proximity factor; N_b = No. of blades on agitator; N_s = effective No. of blade edges		Agitation	36
P8a	Total Prandtl No.	Pr	$\dfrac{\epsilon_M + (\mu/\rho)}{\epsilon_T + \alpha}$ ϵ_M, ϵ_T = eddy transfer coefficient for momentum, heat $[L^2/\theta]$	Total momentum diffusivity/total thermal diffusivity	Heat transfer in combined turbulent and laminar flows	25
P8b	Turbulent Prandtl No.	Pr_T	$\epsilon_M/\epsilon_T = l/l_T$ ϵ_M, ϵ_T = eddy viscosity, eddy thermal diffusivity $[L^2/\theta]$; l, l_T = mixing lengths for momentum, heat transfer	Eddy momentum diffusivity/eddy thermal diffusivity	Heat transfer in turbulent flow	9, 25

Table 10–52. DIMENSIONLESS GROUPS— SUPPLEMENTARY TABLE (Continued)

ALPHABETICAL LIST OF NEW GROUPS (Continued)

Serial No.	Name	Symbol	Definition	Significance	Field of Use	Reference
P12a	Predvoditelev No. (mass transfer)	Pd_m	$(\Gamma_m L^2/a_m)(N_{Fo_m})$ Γ_m = rate of change of mass transfer potential of medium, (mass/unit mass)/time $[1/\theta]$; a_m = mass conductivity of material $[L^2/\theta]$; N_{Fo_m} = group **F12**	Rate of change of concn. of medium/ rate of change of concn. of body	Mass transfer	22
P15a	Pulsation No.	N_{Pu}	$fd_e\rho/G$ d_e = equiv. diam. of channel		Transfer to pulsed fluid	20
R0a	Radial frequency parameter (1)	—	$\omega_r D/V^2$, $\omega_r\alpha/V^2$ D = diffusivity or dispersion coefficient of packed bed $[L^2/\theta]$, ω_r = radial frequency (radians/sec.) $[1/\theta]$		Packed and fluidized beds	18, 33, 39
R0b	Radial frequency parameter (2)	—	$\omega_r L^2/\alpha$ ω_r as in group **R0a**; L = tube radius $[L]$		Packed and fluidized beds	29, 39
R0c	Radial frequency parameter (3)	—	$\omega_r^2 DL/V^3$ (Quantities as in groups **R0a** and **R0b**)	$\dfrac{(\text{Group R0a})^2}{(\text{Group P2})}$	Packed and fluidized beds	18, 33, 39
R0d	Radial frequency parameter (4)	—	$L(\omega_r/2D)^{1/2} \equiv [\frac{1}{2}(\text{group R0b})^{1/2}]$ (quantities as in groups **R0a** and **R0b**)	Analog of Wave No.	Packed and fluidized beds	39
R1a	Radiation parameter	Φ	$e^+\eta T_w^3 d_h/k$ e^+ = function of mean surface emissivity of walls, $[-]$; T_w = wall temp. (abs.) $[T]$		Effect of radiation on convective mass transfer	14
R6a	Reaction enthalpy No.	N_H	$(\Delta u)_A(\Delta n_A)/C_p(\Delta T)$ $(\Delta u)_A$ = enthalpy of reaction per unit mass of A produced $[L^2/\theta^2]$, n_A = mass fraction of species A $[-]$	Change in reaction energy/change in sensible energy	Interphase transfer with chemical reaction	37
R15a	Rossby No.	—	$V/\omega L$	More general form of group **R15**		27
S4a	Schmidt No. (2)	—	$\equiv$ Semenov No. (2)	No longer used		27
S4b	Schmidt No. (3)	$(Sc)_3$	$\mu\chi/\rho\sigma_e L^2$ χ as in group **E4b**	Diffusivity of vorticity/mass diffusivity of ions	Electrochemistry	10
S4c	Total Schmidt No.	Sc	$\dfrac{\varepsilon_M+(\mu/\rho)}{\varepsilon_D+D}$ ε_M = eddy viscosity $[L^2/\theta]$	Total momentum diffusivity/total mass diffusivity	Mass transfer in combined laminar and turbulent flows	25
S4d	Turbulent Schmidt No.	Sc_T	$\varepsilon_M/\varepsilon D = l/l_D$ ε_M = eddy viscosity, $[L^2/\theta]$, l, l_D = mixing lengths for momentum, mass transfer $[L]$	Eddy momentum diffusivity/eddy mass diffusivity	Mass transfer in turbulent flow	25
S5a	Semenov No. (2)	—	$\equiv 1/N_{Le}$ (see group **L7**)			27
S7a	Smoluchowski No.	—	$L/\lambda = 1/N_{Kn}$ N_{Kn} = group **K6**	See group **K6**		27
S9a	Soret No.	S_o	$\Theta n'_{20}$ (definitions as group **D14a**)	Dimensionless thermodiffusion coefficient	Coupled heat and mass transfer	22
S10a	Spalding No.	B'	$c_p\Delta T/(r_v-q_r/V_m)$; q_r = radiant heat flux $[ML^2/\theta^3]$; V_m = rate of mass transfer $[M/\theta]$	Ratio (sensible heat/ latent heat) for evaporated material	Droplet evaporation	33a
S13a	Stark No.	Sk	$\eta T^3 L/k$ L = thickness of layer $[L]$; ($\equiv$ Stefan No.)		Radiant heat transfer	1, 30
S14a	Stewart No.	—	$\mu_e^2 H_e^2\sigma_e L/V\rho$	$\dfrac{(\text{Hartmann No.})^2}{\text{Reynolds No.}}$	Magnetofluid dynamics	38
S15a	Stokes No. (2)	St_2	$1.042\, m_f g\rho(1-\rho/\rho_f)R^{*3}/\mu^2$; ρ, μ = density, viscosity of fluid; m_f, ρ_f = mass, density of float; R^* = tube radius/float radius $[-]$		Calibration of rotameters	10a
S18a	Surface tension No.	T_s	$\mu^2/h^*\sigma\rho$; h^* = surface area of packing element/perimeter $[L]$		Mass transfer in packed columns	5a
T5a	Thompson No.	N_{Th_2}	$\dfrac{-\Delta t}{\Delta L}L^2\left(\dfrac{d\sigma}{dT}\right)/\mu\alpha$ (cf. Marangoni No.)		Cellular convection	3
T7a	Thrust coefficient	T_c	$F_T/\rho V^2 d^2$ F_T = thrust force $[ML/\theta^2]$; V = forward speed $[L/\theta]$; d = tip diameter $[L]$		Propeller studies	27
T7b	Torque coefficient	Q_c	$F'/\rho V^2 d^3$ F' = propeller torque $[ML^2/\theta^2]$		Propeller studies	27
T7c	Transiency groups	K_P K_Q	$\dfrac{1}{\left(\dfrac{\partial p}{\partial L}\right)}\cdot\dfrac{\partial\left(\dfrac{\partial p}{\partial L}\right)}{\partial(Fo_f)}$ $\dfrac{1}{(N_{Re})}\cdot\dfrac{\partial(N_{Re})}{\partial(Fo_f)}$ $\partial p/\partial L$ = pressure gradient in flow direction $[M/L^2\theta^2]$; Fo_f = group **F12a**		Transient flow behavior	17
W0a	Wave No.	k	$L(\omega_r/2\alpha)^{1/2}$ ω_r = radial frequency (radians/sec.) $[1/\theta]$	Heat transfer analog group **R0d**	Cyclic heat transfer	4
W3a	Weber No. (3)	We_3	$\sigma/\rho_L g L^2$ ρ_L = liquid density $[M/L^3]$; L = depth of liquid on tray $[L]$	Surface tension force/ gravity force	Interfacial area determination in distillation equipment	32
W4a	Generalized Weissenberg No.	N_1	$\sqrt{I_e}\theta_n$ (definitions as group **D8a**)	Generalization of group **W4**	Rheology	

*Reprinted from: J.P. Catchpole and G.D. Fulford, *Ind. Eng. Chem.*, 60(3):71–78, 1968. Copyright 1968 by the American Chemical Society. Reprinted by permission of the copyright owner.

Section 11

Processes
and Control

Table 11–1. METHODS FOR BONDING DISSIMILAR METALS*

Union of	Procedure
Aluminum–beryllium	Direct dip braze
Aluminum–stainless steel	Dip braze after tin-coating stainless steel
Aluminum–stainless steel	Soldering after nickel-plating aluminum
Aluminum–stainless steel	Diffusion bonding after interface-coating application
Aluminum–stainless steel	Dip braze after Ag-plating stainless steel
Aluminum–titanium	Tungsten-arc inert-gas shielded brazing with aluminum brazing alloy
Aluminum–titanium	Soldering after plating Ni on Ti and Al
Aluminum–tungsten	Tungsten-arc inert-gas brazing with aluminum brazing alloy
Beryllium–aluminum	Direct dip braze
Beryllium–stainless steel	Vacuum braze
Columbium–molybdenum	Electron-beam welding
Columbium–stainless steel	Inert-gas braze
Copper–nickel	Diffusion bonding after tin soldering
Copper wire–nickel wire	Capacitor discharge resistance micro-welder
Copper–tungsten	Tungsten-arc inert-gas brazing with aluminum brazing alloy
Molybdenum–columbium	Electron-beam welding
Molybdenum–stainless steel	Electron-beam welding
Molybdenum–stainless steel	Vacuum braze
Molybdenum–tungsten	Electron-beam welding
Nickel–copper	Diffusion bonding after tin soldering
Nickel wire–copper wire	Capacitor discharge resistance micro-welder
Stainless steel–aluminum	Dip braze after tin-coating the stainless steel
Stainless steel–aluminum	Soldering after nickel-plating the aluminum
Stainless steel–aluminum	Diffusion bonding after interface-coating application
Stainless steel–aluminum	Dip braze after Ag-plating stainless steel
Stainless steel–beryllium	Vacuum braze
Stainless steel–columbium	Inert-gas braze
Stainless steel–molybdenum	Electron-beam welding
Stainless steel–molybdenum	Vacuum braze
Stainless steel–steel (low alloy)	Percussion stud welding
Stainless steel–titanium	Resistance-welding-machine braze
Stainless steel–tungsten	Tungsten-arc inert-gas brazing with aluminum brazing alloy
Titanium–aluminum	Tungsten-arc inert-gas shielded brazing with aluminum brazing alloy
Titanium–aluminum	Soldering after plating Ni on Ti and Al
Titanium–stainless steel	Resistance-welding-machine braze
Titanium–tungsten	Tungsten-arc inert-gas brazing with aluminum brazing alloy
Tungsten–aluminum	Tungsten-arc inert-gas brazing with aluminum brazing alloy
Tungsten–copper	Tungsten-arc inert-gas brazing with aluminum brazing alloy
Tungsten–molybdenum	Electron-beam welding
Tungsten–stainless steel	Tungsten-arc inert-gas brazing with aluminum brazing alloy
Tungsten–titanium	Tungsten-arc inert-gas brazing with aluminum brazing alloy

Ultrasonic welding, where adhesion is accomplished entirely by vibratory energy, may be used for joining thin sections of the following metals:

Gold to copper, germanium, Kovar, nickel, platinum, or silicon.
Kovar to copper, gold, nickel, or platinum.
Nickel to aluminum, copper, gold, Kovar, or molybdenum.
Platinum to aluminum, copper, gold, Kovar, nickel, or steel.
Silicon to aluminum and gold.
Steel to aluminum, copper, molybdenum, nickel, platinum, or zirconium.
Zirconium to aluminum, copper, and steel.

With the exception of germanium and silicon, these metals have also been successfully bonded to themselves by ultrasonic welding.

*Recommendations based on: "NASA Contributions to Metal Joining", NASA SP-5064, 1967.

Table 11–2. WELDING PROCESS CLASSIFICATIONS

Identifying names for welding, brazing and related processes.
Terminology of the American Welding Society.

Arc welding

Carbon arc
Shielded arc
Flux-cored arc
Gas metal arc
Gas tungsten arc
Submerged arc
Plasma arc
Stud

Resistance welding

Spot
Seam
Projection
Flash
Upset
Percussion

Gas welding

Oxyacetylene
Oxyhydrogen
Pressure gas

Solid state welding

Forge
Friction
Explosion
Diffusion
Ultrasonic

Other welding processes

Thermit
Induction
Electron beam
Electroslag
Laser beam

Brazing

Furnace, torch, induction,
resistance, dip, infrared

Soldering

Oven, torch, induction,
resistance, dip

Cutting

Oxygen, arc, laser beam

Table 11–3. CLAD OR WELDED METAL LAMINATES*

Base metal	Surface layer Name	Typical composition, percent	Bonding method	Typical usage
CLAD SHEETS OR SHAPES				
Steel	Stainless steel	18 Cr, 8 Ni	Heat and pressure[a]	Chemical pressure vessels
Aluminum	Stainless steel	18 Cr, 8 Ni	Heat and pressure[a]	Cookware, heat exchange
Aluminum	Alclad[b] (1100)	99 Al	Heat and pressure[a]	High-finish parts
Aluminum	Alclad (3003)	1.2 Mn	Heat and pressure[a]	Building siding
Aluminum	Alclad (2024)	4.4 Cu, 1.5 Mg	Heat and pressure[a]	Airplane frames
Steel	Copper	99.9 Cu	Heat and pressure[a]	Heat exchangers
Steel	Copper-nickel	70 Cu, 30 Ni	Heat and pressure[a]	Marine condensers
Steel	Monel	65 Ni, 33 Cu	Heat and pressure[a]	Marine and saline uses
Steel	Nickel-chromium	72 Ni, 16 Cr	Heat and pressure[a]	Tanks for oxidizers
Steel	Nickel	99 Ni	Heat and pressure[a]	Caustic tanks
HARD-FACING OF PARTS				
Steel, iron, alloys	Stainless steel	19 Cr, 9 Ni	Welding deposition	High-temperature wear
Steel, iron, alloys	Martensitic steel	Cr 4, Mn 1, Ni 0.5	Welding deposition	Dies
Steel, iron, alloys	Nickel-chromium steel	19 Cr, 9 Ni	Welding deposition	Exhaust valves
Steel, iron, alloys	High-speed steel	18 W, 4 Cr, 1 V	Welding deposition	High-temperature tools
Steel, iron, alloys	Austenitic manganese steel	12 Mn, 3.5 Ni	Welding deposition	Rail frogs; dipper teeth
Steel, iron, alloys	High-chromium iron	25 Cr, 6 Mn, 4 C	Welding deposition	Corrosive wear

Note: The above table lists a few typical metal coatings only. Many other combinations are available, including surface layers of various bronzes, cobalt-base alloys, and proprietary special alloys. Very hard materials such as the carbides (e.g., cemented tungsten carbide) can also be provided as surface materials or welded inserts; see Table 11-11.

[a] Includes hot rolling, explosive bonding, and friction bonding; see Table 11-4. Similar metal bonding methods, by static or dynamic pressure, are used for other purposes, such as thermostat metal elements (see ASTM B-388).

[b] The Aluminum Association has adopted *alclad* as a generic name applied to any aluminum alloy product with a coating that will electrochemically protect the core.

*Compiled from various sources.

Table 11-4. DIFFUSION WELDING AND ADHESION OF METALS

Engineering problems involving the welding of metals at temperatures well below the melting point are of two opposite kinds: (1) promotion of cold welding and its use as a fabrication process, and (2) prevention of cold welding, or adhesion, or high friction between metal surfaces in use. The following tables summarize the physical and chemical conditions for each case.

TABLE A. DIFFUSION BONDING PROCESSES FOR METAL FABRICATION*

Applications	Fabrication methods for laminar metal composites by diffusion bonding. Hot-platen press or rolls are used for flat or near-flat surfaces. Explosive bonding and friction bonding are other related diffusion-welding processes.
Metals	Most common is a noncorrosive metal, a refractory metal, or a hard metal bonded to steel. Aluminum and copper cladding are common.
Surface character	Mating surfaces must conform accurately with very smooth surfaces. Metal plating or intermediate foil is sometimes used to promote diffusion. Surfaces may be mated by pressure, heat, and friction, with simultaneous size reduction, as in lap joints or stud attachment.
Surface chemistry	Prior surface cleaning is very important; among the processes used here are scraping, heating, chemical reaction and contaminant decomposition, crushing, cleaving, and abrading. High vacuum for degassing of surfaces is favored. Friction bonding involves cleaning action.
Atmosphere	Nonoxidizing conditions are important; inert gases (helium, nitrogen, argon) may be used at 10^{-3} to 10^{-6} torr. Vacuum of 10^{-3} to 10^{-6} torr is desirable, in vacuum furnace or with vacuum envelope welded to workpiece. For high-pressure bonding the atmosphere is less important.
Pressure, duration, and dynamics	Direct and uniform pressure, timed microseconds to several hours. Mating forces are produced by clamping, by inserted ram or deadweight, or in pressure container. Pressure depends on materials and sections, but 10 000-20 000 psi is often reached. Fluid pressure contact or sand packing is used with pressures to 50 000 psi or higher. Duration is momentary in roll or impact bonding. Explosive bonding uses high detonation-velocity charge, proportioned to metal mass and yield strength and producing ballistic velocities.
Temperature	Elevated temperature, but somewhat less than halfway to melting point. Most common processing range 900-2 500°F, depending on metals and process. Hot-wall or radiation elements, friction or impact are used for heating.

*Compiled from several sources.

TABLE B. PREVENTION OF COLD WELDING OF METALS IN HIGH VACUUM*

Applications	Most acute cold welding and high-friction problems are in high-vacuum equipment or space applications; included here are other cases of mating metal surfaces in which adhesion or high friction must be avoided.
Metals	Problems occur with all metals, but tendency to cold weld is high with copper and aluminum, lower with cobalt and titanium. Coefficients of sliding friction of dry, clean metals in vacuum are very high (see Section 6.2).
Surface character	Slight surface roughness and contamination reduce cold welding under high-vacuum conditions. Grain size, orientation and hardness affect dry friction and cold welding; softer metals weld more readily.
Surface chemistry	Even normal chemical contamination on dry metal surfaces reduces the tendency to cold welding in vacuum and reduces the coefficient of sliding friction. Usual liquid or solid lubricants prevents cold welding. Slight surface oxidation on metals almost cancels dry-adhesion tendencies.
Atmosphere	Gas adsorption and atmospheric particulates function as contaminants to reduce cold welding and adhesion. Very high vacuum and degassing are required to definitely initiate cold welding. Prolonged surface exposure to normal oxygen atmosphere reduces adhesion tendencies.
Pressure, duration, and dynamics	To avoid cold welding the surface contact pressure must be kept well below that producing any deformation of metal-surface irregularities. Danger of cold welding is increased by long-period contact, even at low-contact pressure.
Temperature	Temperature rise above ambient should be avoided. Cold welding occurs more readily as temperature is increased.

*Compiled from several sources.

REFERENCES

"Composite Engineering Laminates", A.G.H. Dietz, Ed., M.I.T. Press, 1969.

"Adhesion or Cold Welding of Materials in Space Environments", (Symposium), ASTM-STP 431, American Society for Testing and Materials, 1967.

"Explosive Bonding", Battelle Memorial Institute, DMIC Memorandum 225, 1967.

"Further Studies of Diffusion Bonding", P.A. Kammer, R.E. Monroe, and D.C. Martin, *Welding Journal*, 48(3):114, 1969.

Table 11–5. SOLDERING ALLOYS

For data on hard solders, see Table 11-7.

TIN-LEAD SOLDERS

These most common solders are available in at least 15 compositions, from 2–70% tin, balance lead. High-tin solders are expensive.

Percent tin	Approximate melting range[a]		Tensile strength, kpsi	Electrical resistiv-ity,[b] Cu = 1	Corroded by	Typical uses
	Liquid, °F	Solid, °F				
2.5	580	578			Sodium hypochlorite	Seams in cans; cable sheath
5	594	518	3.4	11.7	Chlorides	Filler metal; wiping solder
15	550	440			Potassium permanganate	Plumbing
20	531	370	5.8	10.5		Radiators; tubing joints
25[c]	510	362				Torch soldering
30	491	361				Machine soldering
40	460	360			Air, will tarnish	General-purpose joining
50	420	360	6.1	9.3	Nitric acid	Sheet metal
60	374	361				Electrical
63[d]	361	361	7.5			Electronic parts
70	378	361	6.8	8.7		Coating metals

ANTIMONY SOLDERS

High strength at high temperature; harder than tin-lead solders; should not be used on metals containing zinc.

Percent tin	Percent lead	Percent antimony	Approximate melting range[a]		Tensile strength, kpsi	Electrical resistiv-ity,[b] Cu = 1	Typical uses
			Liquid, °F	Solid, °F			
20	79.	1.0	517	363	3.5		Cable sheathing; radiators
30	68.4	1.6	482	364		13.	Machine soldering
40	58.	2.0	448	365			General-purpose joining
95	0.	5.0	467	458	11.	15.	Electrical
0	95.	5.0	554	486			Metal coating and filler; batteries

[a]The number of degrees between melting and freezing is especially important for wiping and filling solders.
[b]Electrical resistivity is expressed in terms of the resistivity of copper as unity. Thermal conductivity is also roughly one-tenth that of copper. Specific heat is less than that of copper.
[c]A typical low-melting solder is 25% tin, 25% lead, and 50% bismuth (liquid at 266°F).
[d]Eutectic composition, lowest melting point for tin-lead alloys.

REFERENCES

"ASTM Special Publication 189–1956" (Symposium on Soldering), American Society for Testing and Materials.
"ASTM Specification for Solder Metal", B–32–66T, American Society for Testing and Materials, 1966.
"ASTM Standards", American Society for Testing and Materials.
"CRC Handbook of Chemistry and Physics", 53rd ed., R.C. Weast, Ed., The Chemical Rubber Co., 1972.
"Metals Handbook", Vol. 1, American Society for Metals, 1961.
"Soldering Manual", American Welding Society, 1959.
"Solders and Soldering", National Bureau of Standards Circular 492, 1950.

Table 11-6.   CONTROLLED ATMOSPHERES FOR BRAZING**

Controlled atmospheres are used to prevent oxides and scale, especially for high-quality furnace brazing.

AWS brazing atmosphere type number	Source	Maximum dew point, °F, incoming gas	Approximate composition, percent				Application		Remarks
			H_2	N_2	CO	CO_2	Filler metals	Base metals	
1	Combusted fuel gas (low hydrogen)	Room temp	.5–1	87	.5–1	11–12	BAg*, BCuP, RBCuZn*	Copper, brass*	
2	Combusted fuel gas (decarburizing)	Room temp	14–15	70–71	9–10	5–6	BCu, BAg*, RBCuZn*, BCuP	Copper†, brass*, low-carbon steel, nickel, monel, medium carbon steel§	Decarburizes
3	Combusted fuel gas, dried	−40	15–16	73–75	10–11		Same as 2	Same as 2 plus medium and high-carbon steels, monel, nickel alloys	
4	Combusted fuel gas, dried (carburizing)	−40	38–40	41–45	17–19		Same as 2	Same as 2 plus medium and high-carbon steels	Carburizes
5	Dissociated ammonia	−65	75	25			BAg*, BCuP, RBCuZn*, BCu, BNi	Same as for 1, 2, 3, 4 plus alloys containing chromium‡	
6	Cylinder hydrogen	Room temp	97–100				Same as 2	Same as 2	Decarburizes
7	Deoxygenated and dried hydrogen	−75	100				Same as 5	Same as 5 plus cobalt, chromium, tungsten alloys and carbides‡	
8	Heated volatile materials	Inorganic vapors (i.e., zinc, cadmium, lithium, volatile fluorides)					BAg	Brasses	Special purpose. May be used in conjunction with 1 thru 7 to avoid use of flux
9	Purified inert gas	Inert gas (e.g., helium, argon, etc.)					Same as 5	Same as 5 plus titanium, zirconium, hafnium	Special purpose. Parts must be *very* clean and atmosphere must be pure
10	Vacuum pumping	Vacuum					Any metal that does not vaporize	Any metal that does not vaporize	Special purpose. Elaborate equipment and procedure

*Flux required in addition to atmosphere when alloys that contain volatile components are used.
†Copper should be fully deoxidized or oxygen-free.
‡Flux must be used in addition to the atmosphere if appreciable quantities of aluminum, titanium, silicon, or beryllium are present.
§Heating time should be kept minimum to avoid objectionable decarburization.

**From: "Brazing Manual", American Welding Society, 1963.

Table 11–7. BRAZING METALS AND APPLICATIONS*

AWS–ASTM class B–	Composition, percent							Brazing range, °F
	Ag	Cu	Au	Al	Ni	Zn	Other	
Ag–1	45	15	—	—	—	16	Cd 24	1145–1400
Ag–2	35	26	—	—	—	21	Cd 18	1295–1550
Ag–4	40	30	—	—	2	28	—	1435–1650
Ag–7	56	22	—	—	—	17	Sn 5	1205–1400
Ag–18	54	40	—	—	—	5	Ni 1	1575–1775
Au–1	—	62.5	37.5	—	—	—	—	1860–2000
Au–4	—	—	82	—	18	—	—	1740–1840
Cu–1	—	99.9	—	—	—	—	—	2000–2100
Cu P–1	—	95	—	—	—	—	P 5	1450–1700
Cu P–3	5	89	—	—	—	—	P 6	1300–1500
CuZn–D	—	48	—	—	10	42	—	1720–1800
Ni–6	—	—	—	—	89	—	P 11	1700–1875
Ni–2	—	—	—	—	82.4	—	Cr 7, B 3.1, Si 4.5, Fe 3	1850–2150
AlSi 3	—	4	—	86	—	—	Si 10	1060–1120
Mg 2	—	—	—	12	—	5	Mg 83	1080–1130

Flux for brazing: Boric acid, borax, and borates are used on most metals except aluminum and magnesium. For these two metals chlorides and fluorides are used.

High-temperature service: Au-filler metals may be used for continuous service temperatures to 800°F; Ni-filler metals, to 1200°F. Other brazing metals are limited to 300–400°F, except Mg, which should not be used over 250°F.

*From: "Brazing Manual", American Welding Society, 1963.

Table 11–8. TYPICAL METAL FINISHES FOR APPEARANCE AND PROTECTION*

Material	Finish	Remarks
Aluminum alloy	Anodizing	An electrochemical-oxidation surface treatment, for improving corrosion resistance; not an electroplating process. For riveted or welded assemblies specify chromic acid anodizing. Do not anodize parts with nonaluminum inserts. Colors vary: yellow-green, gray, or black.
	"Alrok"	Chemical-dip oxide treatment. Cheap. Inferior in abrasion and corrosion resistance to the anodizing process, but applicable to assemblies of aluminum and nonaluminum materials.
Copper and zinc alloys	Bright acid dip	Immersion of parts in acid solution. Clear lacquer applied to prevent tarnish.
Brass, bronze, zinc die-casting alloys	Brass, chrome, nickel, tin	As discussed under steel.
Magnesium alloy	Dichromate treatment	Corrosion-preventive dichromate dip. Yellow color.
Stainless steel	Passivating treatment	Nitric-acid immunizing dip.

Table 11-9. TYPICAL METAL FINISHES
FOR APPEARANCE AND PROTECTION (Continued)

Material	Finish	Remarks
Steel	Cadmium	Electroplate, dull-white color, good corrosion resistance, easily scratched, good thread antiseize. Poor wear and galling resistance.
	Chromium	Electroplate, excellent corrosion resistance, and lustrous appearance. Relatively expensive. Specify hard chrome plate for exceptionally hard abrasion-resistive surface. Has low coefficient of friction. Used to some extent on nonferrous metals particularly when die-cast. Chrome-plated objects usually receive a base electroplate of copper, then nickel, followed by chromium. Used for buildup of parts that are undersized. Do not use on parts with deep recesses.
	Blueing	Immersion of cleaned and polished steel into heated saltpeter or carbonaceous material. Part then rubbed with linseed oil. Cheap. Poor corrosion resistance.
	Silver plate	Electroplate, frosted appearance, buff to brighten. Tarnishes readily. Good bearing lining. For electrical contacts, reflectors.
	Zinc plate	Dip in molten zinc (galvanizing) or electroplate of low-carbon or low-alloy steels. Low cost. Generally inferior to cadmium plate. Poor appearance. Poor wear resistance; electroplate has better adherence to base metal than hot-dip coating. For improving corrosion resistance, zinc-plated parts are given special inhibiting treatments.
	Nickel plate	Electroplate, dull white. Does not protect steel from galvanic corrosion. If plating is broken, corrosion of base metal will be hastened. Finishes in dull white, polished, or black. Do not use on parts with deep recesses.
	Black-oxide dip	Nonmetallic chemical black oxidizing treatment for steel, cast iron, and wrought iron. Inferior to electroplate. No buildup. Suitable for parts with close dimensional requirements as gears, worms, and guides. Poor abrasion resistance.
	Phosphate treatment	Nonmetallic chemical treatment for steel and iron products. Suitable for protection of internal surfaces of hollow parts. Small amount of surface buildup. Inferior to metallic electroplate. Poor abrasion resistance. Good paint base.
	Tin plate	Hot dip or electroplate. Excellent corrosion resistance, but if broken, will not protect steel from galvanic corrosion. Also used for copper, brass, and bronze parts that must be soldered after plating. Tin-plated parts can be severely worked and deformed without rupture of plating.
	Brass plate	Electroplate of copper and zinc. Applied to brass and steel parts where uniform appearance is desired. Applied to steel parts when bonding to rubber is desired.
	Copper plate	Electroplate applied before nickel or chrome plates. Also for parts to be brazed or protected against carburization. Tarnishes readily.

*Reprinted from: *Product Engineering,* 19:1, January 1948. Copyright Morgan-Grampian, Inc., 1948.

REFERENCE

For a discussion and data on surface protection methods, including metal spraying, hard facing, and surface hardening, see "Handbook of Mechanical Wear", C. Lipson and L.V. Colwell, Eds., University of Michigan Press, 1961.

Table 11–10. COATINGS FOR ALUMINUM*

Chemically deposited coatings on aluminum and aluminum alloys; for corrosion resistance, paint base, or decoration.

Table A. NINE PROCESSES

Treatment	Purpose	For use on	Operation	Finish and thickness
Zinc phosphate coating	Paint base	Wrought alloys	Power spray or dip. For light to medium coats, 1 to 3 min at 130 to 135°F.	Crystalline, 100 to 200 mg/ft².
Chromium phosphate coating	Paint base or corrosion protection	Wrought or cast alloys	Power spray, dip, brush, or spray. For light to medium coats, 20 sec to 2 min at 110 to 120°F.	Crystalline, 100 to 250 mg/ft².
Sulfuric acid anodizing	Corrosion and abrasion resistance, paint base	All alloys; uses limited on assemblies with other metals	15 to 60 min, 12 to 14 amp/ft², 18 to 20 volts, 68 to 74°F. Tank lining of plastic, rubber, lead, or brick.	Very hard, dense, clear. 0.000 2 to 0.000 8 in. thick. Withstands 250- to 1 000-hr salt spray.
Chromic acid anodizing	Corrosion resistance, paint base; also as inspection technique with dyed coatings	All alloys except those with more than 5 pct Cu	30 to 40 min, 1 to 3 amp/ft², 40 volts dc, 95°F, steel tanks and cathode, aluminum racks.	0.000 02 to 0.000 06 in. thick, 250-hr min salt spray.
Chromate conversion coating	Corrosion resistance, paint hesion, and decorative effect	All alloys	10 sec to 6 min depending on thickness, by immersion, spray, or brush, 70°F, in tanks of stainless, plastic, acid-resistant brick or chemical stoneware.	Electrically conductive, clear to yellow and brown in color, 0.000 02 in. or less thick, 150- to 2 000-hr salt spray depending on alloy composition and coating thickness.
Chemical oxidizing	Corrosion resistance, paint base	All alloys, less satisfactory on copper-bearing alloys	Basket or barrel immersion 15 to 20 min, 150 to 212°F.	May be dyed, 250-hr min salt spray.
Electropolishing	Increases smoothness and brilliance of paint of plating base	Most wrought alloys, some sand-cast and die-cast alloys	15 min, 30 to 50 amp/ft², 50 to 100 volts, less than 120°F, aluminum cathode.	35 to 85 RMS depending on treatment.
Zinc immersion	Preplate for subsequent deposition of most plating metals, improves solderability	Many alloys, modifications for others particularly regarding silicon, copper and magnesium content	30 to 60 sec, 60 to 80°F, agitated steel or rubber-lined tank.	Thin film.
Electroplating[†]	Decorative appeal and/or function	Most alloys after proper preplating		Same as on steel.

[†] See Table 3–22.

TABLE B. ELECTROPLATING

Metals	Operation
Chromium	Directly over zinc immersion coat, 65 to 70°F, 6–8 volts, 200–225 amp/ft². Transfer to bath at 120 to 125°F if copper, or copper and nickel have been applied.
Copper	Directly over zinc, or follow with copper strike, then plate in conventional copper bath.
Brass	Directly over zinc, 80 to 90°F, 2–3 volts, 3–5 amp/ft².
Nickel	Directly over zinc, or follow with copper strike, then plate in conventional nickel bath.
Cadmium	Directly over zinc, or follow with copper or nickel strike, or preferably cadmium strike, then plate in conventional cadmium bath.
Silver	Copper strike over zinc using copper cyanide bath, low pH, low temperature, 24 amp/ft² for 2 min, drop to 12 amp/ft² for 3 to 5 min; plate in silver cyanide bath, 75 to 80°F, 1 volt, 5–15 amp/ft².
Zinc	Directly over zinc immersion coating.
Tin	Directly over zinc immersion coating.
Gold	Copper strike over zinc as for silver, then plate in conventional bath.

*Reprinted by The Chemical Rubber Co. from: Iron Age, June 28, 1956, Chilton Co., Copyrighted 1956.

Table 11–11. FLAME-SPRAYED PROTECTIVE COATINGS

Metal and Non-metal Coatings by Jet Application

The appearance and life of the functional parts of a structure or a machine depend largely on the properties of their surfaces. An important part of the engineering design is, therefore, involved in the selection and processing of materials for surface qualities such as smoothness, hardness, and resistance to deterioration. Surfaces of base materials may actually be modified in molecular structure and composition, physically processed for smoothness or texture, or actually covered with a more suitable surface material by spraying, welding, cladding, plating, veneering, or coating.

The following table deals with sprayed coatings. For other data on surface-modification methods and materials see Table 11–8. Metal Finishes; Table 11–1. Bonding Dissimilar Metals; Table 3–22. Electroplating; Table 11–3. Clad or Welded Laminates; Tables 3–14 to 3–21. Paints and Coatings. For data on surface properties and surface deterioration by weathering, corrosion, wear, etc., see the index.

FLAME-SPRAYED METAL AND NON-METAL COATINGS

			Spray-deposited coating				
Material	Objective	Raw form	Typical thickness, mils	Melting point °F	Melting point °C	Deposited on	Typical applications
Metals							
Aluminum	Corrosion protection	Wire	3-6	1 220	660	Steel	Industrial atmospheres; salt spray; moisture
Aluminum	Protection; appearance	Wire; powder	2-5	1 220	660	Steel, iron	Structures, piping, tanks, piling, bridges, roofs
Aluminum	High-temperature oxidation	Wire; powder	6-15	1 220	660	Steel	Flues, exhausts, tanks, furnace parts
Aluminum	Appearance; wear; conduction	Wire; powder	1-3	1 220	660	Small objects	Metal parts, wood, cloth, reflectors, ceramics
Brass (Cu-Zn)		Wire		1 800	982	Metals, glass	Glass seals; brass castings
Bronze (Cu-Sn)		Wire				Metals, plastics, concrete	Architectural decoration, casting repairs
Bronze (Cu-Al)	Corrosion; wear resistance	Wire		1 900	1 038	Steel, bronze	Machined bronze parts for wear and good finish
Cadmium		Wire; molten	3-100	610	321	Steel	Small parts; fasteners; nuclear shielding
Cobalt (Co-Cr-W)	Hard facing	Powder		2 725	1 495	Steel	Valves, gages, molds
Copper	Protection; conduction; appearance	Wire; powder	2-10	1 980	1 082	Metals, plastics, wood	Sprayed circuits; brazing; decoration, glass seals, contacts
Bismuth (alloy)		Wire; molten	-125	520	271		Plastic molds
Lead	Corrosion protection	Wire; molten	3-20	620	327	Steel	Acid splash or fumes; acid tanks; bonding metal
Molybdenum	Hard surfacing	Wire; powder		4 750	2 621	Metals	Bonding coat; pulley, brake, and bearing surfaces
Monel (Ni-Cu)	Corrosion protection	Wire		2 400	1 316		Valves, packing glands; pump parts
Nickel	Corrosion protection	Wire; powder	15-30	2 645	1 452	Steel, iron	Paper rolls; dye vats; pump rods
Ni-Cr (80-20)	Protection, appearance	Powder	1-10			Steel, iron	Bright, high-finish; heat resistant

Table 11–11. FLAME-SPRAYED PROTECTIVE COATINGS (*Continued*)

Spray-deposited coatings

Material	Objective	Raw form	Typical thickness, mils	Melting point °F	°C	Deposited on	Typical applications
Ni-Cr-B	Hard facing	Powder	5-100	1 900	1 038	Steel	Self-fluxing, as-fused, hard, wearing surfaces; valves; glass molds
Silver	Conduction; appearance	Wire		1 760	960	Metals	Conductors; decorative effects
Solder	Bond, protection, appearance	Wire; molten		420	216	Metals	Heat-transfer fins; filling dents and seams; models
Steel		Wire		2 300	1 260	Steel	Machinable build-up on journals, housings
Steel, stainless		Wire; powder		2 600	1 427	Steel	Shafts, glands, impellers, valves, friction surfaces
Steel, superalloy		Powder		2 700	1 482	Steel	Grind-finished repairs
Tin		Molten; wire		450	232	Metals	Food-processing equipment; glass-lined tank repairs
Tungsten		Powder		6 150	3 400		
Zinc		Wire; powder	2-5	787	419	Steel	Structural steel, tanks, fasteners, window frames
Zinc		Wire; powder	8-15	787	419		Water and seawater contact; chemical solutions
Ni-Al (80-20)							
NON-METALS							
Alumina	Heat and wear resistance	Powder		3 700	2 038		High-temperature erosion protection; insulation
Cr_3C_2				3,435	1,890		High-temperature seals
Plastic	Insulation; appearance	Powder				Steel, brass, concrete	
WC				5 200	2 870	Alloy steel	
WC-Co		Powder					
Zirconia	High-temperature insulation	Powder		4 700	2 593		Thermal barriers and linings

Spray Guns.   Metal is sprayed with a heater-atomizer gun forming a high-temperature jet. The impingement force produces adhesion and metal build-up by face velocities of 300-700 fps. The metal is supplied in wire, powder, or molten form. Spray guns are designed accordingly, ranging from self-contained hand guns to large, mounted equipment under automatic control. Oxygen-fuel gas torches are combined with compressed air for atomizing. For refractory metals a d-c arc plasma jet produces temperatures of 8 000°F or more, heating an inert gas (nitrogen or argon). Complex, proprietary designs are common for nozzles, feeders, and controls.

Surface Preparation.   Surfaces must first be cleaned by washing, degreasing, solvent cleaning, or abrasive blast cleaning. Metal parts are often roughened or grooved; for heavy sprayed deposits some undercutting, shouldering, or dovetailing may be necessary. The main metal deposition may be preceded by a sprayed bonding layer, as when steel alloys are sprayed over a thin bonding layer of molybdenum. Metal work is usually preheated, for example, 50°F above room temperature for drying, or up to 200-400°F to aid bonding and to control shrinkage. Final sprayed metal coatings are often sealed for corrosion protection, surface finish, pressure tightness, or appearance. Sealers range from wax or paint to plastic or even ceramic or glass materials.

REFERENCES

"Metal Spraying Technology", 4th ed., W.E. Ballard, Charles Griffin & Company Ltd., 1963.
"Protective Coatings for Metals", 3rd ed., R.M. Burns and W.W. Bradley, ACS Monograph Series, Reinhold Publishing Corp., 1967.

Table 11–12. LIMITS OF INTERFERENCE
FOR FITTING CYLINDRICAL PARTS*

Standard limits of interference are by size range and class of service as prescribed in the American National Standard.*

Maximum and minimum interference limits are in thousandths of an inch for force and shrink fits in various diameters to 20 inches.

Nominal size range, in.		Class				
Over	To	FN1	FN2	FN3	FN4	FN5
0.04	0.12	0.05	0.2	—	0.3	0.5
		0.5	0.85	—	0.95	1.3
0.12	0.24	0.1	0.2	—	0.4	0.7
		0.6	1.0	—	1.2	1.7
0.24	0.40	0.1	0.4	—	0.6	0.8
		0.75	1.4	—	1.6	2.0
0.40	0.56	0.1	0.5	—	0.7	0.9
		0.8	1.6	—	1.8	2.3
0.56	0.71	0.2	0.5	—	0.7	1.1
		0.9	1.6	—	1.8	2.5
0.71	0.95	0.2	0.6	—	0.8	1.4
		1.1	1.9	—	2.1	3.0
0.95	1.19	0.3	0.6	0.8	1.0	1.7
		1.2	1.9	2.1	2.3	3.3
1.19	1.58	0.3	0.8	0.8	1.5	2.0
		1.3	2.4	2.4	3.1	4.0
1.58	1.97	0.4	0.8	1.2	1.8	3.0
		1.4	2.4	2.8	3.4	5.0
1.97	2.56	0.6	0.8	1.3	2.3	3.8
		1.8	2.7	3.2	4.2	6.2
2.56	3.15	0.7	1.0	1.8	2.8	4.8
		1.9	2.9	3.7	4.7	7.2
3.15	3.94	0.9	1.4	2.1	3.6	5.6
		2.4	3.7	4.4	5.9	8.4
3.94	4.73	1.1	1.6	2.6	4.6	6.6
		2.6	3.9	4.9	6.9	9.4
4.73	5.52	1.2	1.9	3.4	5.4	8.4
		2.9	4.5	6.0	8.0	11.6
5.52	6.30	1.5	2.4	3.4	5.4	10.4
		3.2	5.0	6.0	8.0	13.6
6.30	7.09	1.8	2.9	4.4	6.4	10.4
		3.5	5.5	7.0	9.0	13.6
7.09	7.88	1.8	3.2	5.2	7.2	12.2
		3.8	6.2	8.2	10.2	15.8
7.88	8.86	2.3	3.2	5.2	8.2	14.2
		4.3	6.2	8.2	11.2	17.8
8.86	9.85	2.3	4.2	6.2	10.2	14.2
		4.3	7.2	9.2	13.2	17.8
9.85	11.03	2.8	4.0	7.0	10.0	16.0
		4.9	7.2	10.2	13.2	20.0
11.03	12.41	2.8	5.0	7.0	12.0	18.0
		4.9	8.2	10.2	15.2	22.0
12.41	13.98	3.1	5.8	7.8	13.8	19.8
		5.5	9.4	11.4	17.4	24.2
13.98	15.75	3.6	5.8	9.8	15.8	22.8
		6.1	9.4	13.4	19.4	27.2
15.75	17.72	4.4	6.5	9.5	17.5	25.5
		7.0	10.6	13.6	21.6	30.5
17.72	19.69	4.4	7.5	11.5	19.5	27.5
		7.0	11.6	15.6	23.6	32.5

Class FN1: Light drive fit, thin sections, long fits, cast-iron external members.
Class FN2: Ordinary steel parts; shrink fits on tight sections; tightest fits for cast iron.
Classes FN3, 4, 5: For heavier parts, increased stress; shrink fits to avoid heavy pressing forces.

*Extracted from: American Standard Limits and Fits for Cylindrical Parts, USAS B4.1–1967, with the permission of the publisher, The American Society of Mechanical Engineers, United Engineering Center, 345 East 47th Street, New York, N.Y. 10017.

Table 11–13. RUNNING AND SLIDING FITS
FOR LUBRICATED MACHINE PARTS*

Standard limits of clearance are by size range and class of service as prescribed in the American National Standard.* Clearance limits are in thousandths of an inch.

Nominal size range, in.		Class								
Over	To	RC1	RC2	RC3	RC4	RC5	RC6	RC7	RC8	RC9
0.04	0.12	0.1	0.1	0.3	0.3	0.6	0.6	1.0	2.5	4.0
		0.45	0.55	0.8	1.1	1.4	1.8	2.6	5.1	8.1
0.12	0.24	0.15	0.15	0.4	0.4	0.8	0.8	1.2	2.8	4.5
		0.5	0.65	1.0	1.4	1.8	2.2	3.1	5.8	9.3
0.24	0.40	0.2	0.2	0.5	0.5	1.0	1.0	1.6	3.0	5.0
		0.6	0.85	1.3	1.7	2.2	2.8	3.9	6.6	10.7
0.40	0.71	0.25	0.25	0.6	0.6	1.2	1.2	2.0	3.5	6.0
		0.75	0.95	1.4	2.0	2.6	3.2	4.6	7.9	12.8
0.71	1.19	0.3	0.3	0.8	0.8	1.6	1.6	2.5	4.5	7.0
		0.95	1.2	1.8	2.4	3.2	4.0	5.7	10.0	15.5
1.19	1.97	0.4	0.4	1.0	1.0	2.0	2.0	3.0	5.0	8.0
		1.1	1.4	2.2	3.0	4.0	5.2	7.1	11.5	18.0
1.97	3.15	0.4	0.4	1.2	1.2	2.5	2.5	4.0	6.0	9.0
		1.2	1.6	2.6	3.6	4.9	6.1	8.8	13.5	20.5
3.15	4.73	0.5	0.5	1.4	1.4	3.0	3.0	5.0	7.0	10.0
		1.5	2.0	3.2	4.2	5.8	7.4	10.7	15.5	24.0
4.73	7.09	0.6	0.6	1.6	1.6	3.5	3.5	6.0	8.0	12.0
		1.8	2.3	3.6	4.8	6.7	8.5	12.5	18.0	28.0
7.09	9.85	0.6	0.6	2.0	2.0	4.0	4.0	7.0	10.0	15.0
		2.0	2.6	4.4	5.6	7.6	9.6	14.3	21.5	34.0
9.85	12.41	0.8	0.8	2.5	2.5	5.0	5.0	8.0	12.0	18.0
		2.3	2.9	4.9	6.5	9.0	11.0	16.0	25.0	38.0
12.41	15.75	1.0	1.0	3.0	3.0	6.0	6.0	10.0	14.0	22.0
		2.7	3.4	5.8	7.4	10.4	13.0	19.5	29.0	45.0
15.75	19.69	1.2	1.2	4.0	4.0	8.0	8.0	12.0	16.0	25.0
		3.0	3.8	7.2	9.0	13.0	16.0	22.0	32.0	51.0

Class RC1: Close-sliding fits; parts must assemble without perceptible play.

Class RC2: Sliding fits in which parts will move easily but are not intended to run freely.

Class RC3: Precision-running fits; cylindrical parts at slow speeds, light pressures, constant temperature.

Class RC4: Close-running fits on accurate machinery and moderate speeds and pressures.

Classes RC5, 6, 7: Medium- to free-running fits; higher speeds and pressures and/or temperature variations.

Classes RC8, 9: Loose-running fits; commercial tolerances for parts.

*Extracted from: American Standard Limits and Fits for Cylindrical Parts, USAS B4.1–1967, with the permission of the publisher, The American Society of Mechanical Engineers, United Engineering Center, 345 East 47th Street, New York, N.Y. 10017.

Table 11–14.
EXPECTED ACCURACY OF VARIOUS MACHINE PROCESSES*

Tolerance Given in Thousandths of an Inch

Quality or grade	Diameter, inches			
	0.5	1.5	2.5	3.5
		Tolerance		
Slip blocks; reference gages	.083	.12	.12	.12
High-quality gages; plug gages	.068	.083	.083	.16
Good-quality gages; gap gages	.14	.16	.20	.22
Gages; fits of extreme precision produced by lapping	.20	.30	.35	.42
Ball bearings; machine lapping; diamond or fine boring; fine grinding	.32	.45	.55	.60
Grinding; fine honing	.42	.60	.75	.82
High quality; turning; breaching; honing	.69	.95	1.2	1.4
Center lathe turning and boring; reaming; turret lathe or automatic in good condition	1.0	1.6	1.9	2.4
Worn turret lathe or automatic; horizontal or vertical boring machine	1.7	2.6	3.5	3.8
Milling; slotting; planing; metal rolling or extrusion	2.9	4.1	4.7	6.0
Drilling; rough turning and boring; precision tube drawing	4.2	6.0	8.0	9.0
Light presswork; tube drawing	7.0	10	13	15
Presswork; tube rolling	11	16	20	22
Die-casting or molding; rubber molding	17	27	35	42
Stamping (approx)	30	44	50	65
Sand casting (approx); flame cutting	46	65	80	90

*Data from: "Quality Control Handbook", J.M. Juran, Ed., McGraw-Hill Book Company, 1962.

Table 11–15.   RELATIVE VALUES OF SURFACE FINISH*

Common name for finish	rms roughness, microinches	Average peak-to-peak height, microinches	Usual tolerance specified for finished part, inches
Mirror	4	15	0.0002
Polished	8	28	0.0005
Ground	16	56	0.001
Smooth	32	118	0.002
Fine	63	220	0.003
Semifine	125	455	0.004
Medium	250	875	0.007
Semirough	500	1750	0.013
Rough	1000	3500	0.025

*From: "Mechanical Measurements", 2nd ed., T.G. Beckwith and N. Lewis Buck, Addison-Wesley Publishing Company, Inc., 1969.

Table 11–16.
STANDARD GAGES FOR WIRE, SHEET, AND TWIST DRILLS

Dimensions in Approximate Decimals of an Inch

Gage	(1) Mfrs. steel sheet	(2) USS steel sheet (old)	(3) Birmingham or Stub	(4) W & M or Roebling steel wire	(5) AWG or B & S non-ferrous wire or sheet	Numbered twist drills	Copper wire (AWG) Circular mils	Ohms/1000 ft, 77°F	Lb/1000 ft	Sheet steel Lb/sq ft
0000000		0.500		0.4900						20.00
000000		0.469		0.4615	0.580					18.75
00000		0.438		0.4305	0.516					17.50
0000		0.406	.454	0.3938	0.460		212,000	0.0500	641.0	16.25
000		0.375	.425	0.3625	0.410		168,000	0.0630	508.0	15
00		0.344	.380	0.3310	0.365		133,000	0.0795	403.0	13.75
0		0.313	.340	0.3065	0.325		106,000	0.100	319.0	12.50
1		0.281	.300	0.2830	0.289	0.2280	83,700	0.126	253.0	11.25
2		0.266	.284	0.2625	0.258	0.2210	66,400	0.159	201.0	10.625
3	.2391	0.250	.259	0.2437	0.229	0.2130	52,600	0.201	159.0	10
4	.2242	0.234	.238	0.2253	0.204	0.2090	41,700	0.253	126.0	9.375
5	.2092	0.219	.220	0.2070	0.182	0.2055	33,100	0.319	100.0	8.75
6	.1943	0.203	.203	0.1920	0.162	0.2040	26,300	0.403	79.5	8.125
7	.1793	0.188	.180	0.1770	0.144	0.2010	20,800	0.508	63.0	7.5
8	.1644	0.172	.165	0.1620	0.128	0.1990	16,500	0.641	50.0	6.875
9	.1495	0.156	.148	0.1483	0.114	0.1960	13,100	0.808	39.6	6.25
10	.1345	0.141	.134	0.1350	0.102	0.1935	10,400	1.02	31.4	5.625
11	.1196	0.125	.120	0.1205	0.0907	0.1910	8,230	1.28	24.9	5
12	.1046	0.109	.109	0.1055	0.0808	0.1890	6,530	1.62	19.8	4.375
13	.0897	0.0937	.095	0.0915	0.0720	0.1850	5,180	2.04	15.7	3.75
14	.0747	0.0781	.083	0.0800	0.0641	0.1820	4,110	2.58	12.4	3.125
15	.0673	0.0703	.072	0.0720	0.0571	0.1800	3,260	3.25	9.86	2.813
16	.0598	0.0625	.065	0.0625	0.0508	0.1770	2,580	4.09	7.82	2.5
17	.0538	0.0562	.058	0.0540	0.0453	0.1730	2,050	5.16	6.20	2.25
18	.0478	0.0500	.049	0.0475	0.0403	0.1695	1,620	6.51	4.92	2
19	.0418	0.0437	.042	0.0410	0.0359	0.1660	1,290	8.21	3.90	1.75
20	.0359	0.0375	.035	0.0348	0.0320	0.1610	1,020	10.4	3.09	1.50
21	.0329	0.0344	.032	0.0318	0.0285	0.1590	810	13.1	2.45	1.375
22	.0299	0.0312	.028	0.0286	0.0253	0.1570	642	16.5	1.94	1.25
23	.0269	0.0281	.025	0.0258	0.0226	0.1540	509	20.8	1.54	1.125
24	.0239	0.0250	.022	0.0230	0.0201	0.1520	404	26.2	1.22	1
25	.0209	0.0219	.020	0.0204	0.0179	0.1495	320	33.0	0.970	0.875
26	.0179	0.0187	.018	0.0181	0.0159	0.1470	254	41.6	0.769	0.75
27	.0164	0.0172	.016	0.0173	0.0142	0.1440	202	52.5	0.610	0.6875
28	.0149	0.0156	.014	0.0162	0.0126	0.1405	160	66.2	0.484	0.625
29	.0135	0.0141	.013	0.0150	0.0113	0.1360	127	83.4	0.384	0.5625
30	.0120	0.0125	.012	0.0140	0.0100	0.1285	101	105	0.304	0.5
31	.0105	0.0109	.010	0.0132	0.0089	0.1200	79.7	133	0.241	0.4375
32	.0097	0.0102	.009	0.0128	0.0080	0.1160	63.2	167	0.191	0.4063
33	.0090	0.0094	.008	0.0118	0.0071	0.1130	50.1	211	0.152	0.375
34	.0082	0.0086	.007	0.0104	0.0063	0.1110	39.8	266	0.120	0.3438
35	.0075	0.0078	.005	0.0095	0.0056	0.1100	31.5	335	0.0954	0.3125
36	.0067	0.0070	.004	0.0090	0.0050	0.1065	25.0	423	0.0757	0.2813
37	.0064	0.0066		0.0085	0.0045	0.1040	19.8	533	0.0600	0.2656
38	.0060	0.0062		0.0080	0.0040	0.1015	15.7	673	0.0476	0.25
39				0.0075	0.0035	0.0995	12.5	848	0.0377	
40				0.0070	0.0031	0.0980	9.9	1070	0.0200	
41				0.0066	0.0028	0.0960				
42				0.0062	0.0025	0.0935				
43				0.0060	0.0022	0.0890				
44				0.0058	.0020	0.0860				
45				0.0055	.0018	0.0820				
46				0.0052	.0016	0.0810				
47				0.0050	.0014	0.0785				
48				0.0048	.0012	0.0760				
49				0.0046	.0011	0.0730				
50				0.0044	.0010	0.0700				

Note: The present trend, especially for sheet and strip, is to quote thickness as decimal or fraction of an inch rather than gage number. ANSI Standard preferred thicknesses have been adopted. These preferred sizes for thickness of uncoated sheet, strip, and plate under 0.25 in. are as follows: .224, .220, .180, .160, .140, .125, .112, .100, .090, .080, .071, .063, .056, .050, .045, .040, .036, .032, .028, .025, .022, .020, .018, .016, .014, .012, .011, .010, .009, .008, .007, .006, .005, .004.

KEY: (1) Manufacturer's standard for hot- and cold-rolled uncoated carbon steel sheet and most alloy steel sheet.
    (2) U.S. Standard for cold-rolled steel strip and stainless and nickel alloy sheet.
    (3) Birmingham or Stub for hot-rolled carbon and alloy steel strip and tubing.
    (4) Washburn and Moen, Roebling, or U.S. Steel for steel wire.
    (5) American wire gage or Brown and Sharpe for non-ferrous wire, sheet, and strip.

Table 11-17. COMMERCIAL TWIST-DRILL SIZES

Diameter, in.	Area, sq in.	Numbered drill size (or letter size)	Fractional inch size	Tap drill for screw size	AWG (copper and nonferrous)	BWG (iron and steel)	Diameter, in.	Area, sq in.	Numbered drill size (or letter size)	Fractional inch size	Tap drill for screw size	AWG (copper and nonferrous)	BWG (iron and steel)
.0135	.000143	80			27	28	.0960	.00724	41			10	12
.0145	.000165	79			26	27	.0980	.00754	40			10	12
.0156	.000191		$\frac{1}{64}$		26	27	.0995	.00778	39			10	12
.016	.000201	78			25	27	.1015	.00809	38		5–40; $\frac{1}{8}$–40	10	12
.018	.000254	77			24	26	.1040	.00849	37		5–44	9	12
.020	.000314	76			24	25	.1065	.00891	36		6–32	9	12
.021	.000346	75			23	24	.1094	.00940		$\frac{7}{64}$		9	11
.0225	.000398	74			23	23	.1100	.00950	35			9	11
.024	.000452	73			22	23	.1110	.00968	34			9	11
.025	.000491	72			22	23	.1130	.01003	33		6–40	9	11
.026	.000531	71			21	22	.1160	.01057	32			8	11
.028	.000616	70			21	22	.1200	.01131	31			8	11
.0292	.000670	69			20	21	.1250	.01227		$\frac{1}{8}$		8	10
.031	.000755	68			20	21	.1285	.01296	30			7	10
.0313	.000765		$\frac{1}{32}$		20	21	.1360	.01453	29		8–32 or 36	7	9
.032	.000804	67			20	21	.1405	.01549	28			7	9
.033	.000855	66			19	20	.1406	.01553		$\frac{9}{64}$		7	9
.035	.000962	65			19	20	.1440	.01629	27			7	9
.036	.001018	64			18	19	.1470	.01679	26		$\frac{3}{16}$–24	6	9
.037	.001075	63			18	19	.1495	.01755	25		10 24	6	8
.038	.001134	62			18	19	.1520	.01815	24			6	8
.039	.001195	61			18	19	.1540	.01863	23			6	8
.0400	.001257	60			18	19	.1563	.01917		$\frac{5}{32}$		6	8
.0410	.001320	59			17	19	.1570	.01936	22			6	8
.0420	.001385	58			17	19	.1590	.01986	21		10–32	6	8
.0430	.001452	57			17	18	.1610	.02036	20		$\frac{3}{16}$–32	6	8
.0465	.001698	56			16	18	.1660	.02164	19			5	7
.0473	.00173		$\frac{3}{64}$	0–80	16	18	.1695	.02265	18			5	7
.0520	.00212	55			15	17	.1719	.02320		$\frac{11}{64}$		5	7
.0550	.00238	54			15	17	.1730	.02351	17			5	7
.0595	.00278	53		1–64; 1–72	14	16	.1770	.02461	16		12–24	5	7
.0625	.00307		$\frac{1}{16}$		14	16	.1800	.02545	15			5	7
.0635	.00317	52			14	16	.182	.02602	14		12–28	5	6
.0670	.00353	51			13	15	.185	.02688	13			4	6
.0700	.00385	50		2–56; 2–64	13	15	.1875	.02761		$\frac{3}{16}$		4	6
.0730	.00419	49			12	14	.189	.02806	12			4	6
.0760	.00454	48			12	14	.191	.02865	11			4	6
.0781	.00479		$\frac{5}{64}$		12	14	.1935	.02940	10			4	6
.0785	.00484	47		3–48	12	14	.196	.03017	9			4	6
.0810	.00515	46			12	14	.199	.03110	8			4	6
.0820	.00528	45		3–56	11	14	.201	.03173	7		$\frac{1}{4}$–20; 14–24	4	6
.0860	.00581	44		4–36	11	13	.2031	.03241		$\frac{13}{64}$		4	5
.0890	.00622	43		4–40	11	13	.204	.03269	6			4	5
.0935	.00687	42		4–48	10	13	.2055	.03317	5			3	5
.0938	.00690		$\frac{3}{32}$		10	13	.209	.03431	4		$\frac{1}{4}''$–24	3	5

Table 11–17. COMMERCIAL TWIST-DRILL SIZES *(Continued)*

Decimal size — Diameter, in.	Area, sq in.	Numbered drill size (or letter size)	Fractional inch size	Tap drill for screw size	AWG (copper and non-ferrous)	BWG (iron and steel)	Decimal size — Diameter, in.	Area, sq in.	Numbered drill size (or letter size)	Fractional inch size	Tap drill for screw size	AWG (copper and non-ferrous)	BWG (iron and steel)
.213	.03563	3		$\frac{1}{4}''$-28	3	5	.323	.08194	P			0	0
.2188	.03758		$\frac{7}{32}$		3	5	.3281	.08456		$\frac{21}{64}$	$\frac{1}{8}$-in. pipe	00	0
.221	.03836	2			3	4	.332	.08657	Q		$\frac{3}{8}''$-24	00	0
.228	.04083	1			3	4	.339	.09026	R			00	0
.234	.04301	A			2	4	.3438	.09281		$\frac{11}{32}$		00	00
.2344	.04314		$\frac{15}{64}$		2	4	.348	.09511	S			00	00
.238	.04449	B			2	4	.358	.1006	T			00	00
.242	.04600	C			2	3	.3594	.1014		$\frac{23}{64}$		00	00
.246	.04753	D			2	3	.368	.1064	U		$\frac{7}{16}''$-14	000	00
.2500	.04909	E	$\frac{1}{4}$		2	3	.375	.1104		$\frac{3}{8}$		000	00
.257	.05187	F		$\frac{5}{16}''$-18	2	3	.377	.1116	V			000	00
.261	.05350	G			1	2	.386	.1170	W			000	000
.2656	.05542		$\frac{17}{64}$		1	2	.3906	.1198		$\frac{25}{64}$	$\frac{7}{16}''$-20	000	000
.266	.05557	H			1	2	.397	.1238	X			000	000
.272	.05811	I		$\frac{5}{16}''$-24	1	2	.404	.1282	Y			000	000
.277	.06026	J			1	2	.4063	.1296		$\frac{13}{32}$		000	000
.281	.06202	K			1	2	.413	.1340	Z			0000	000
.2813	.06213		$\frac{9}{32}$		1	2	.4219	.1398		$\frac{27}{64}$	$\frac{1}{4}$-in. pipe; $\frac{1}{2}''$-13	0000	000
.29	.06605	L			0	1							000
.295	.06835	M			0	1	.4375	.1503		$\frac{7}{16}$		0000	0000
							.4531	.1613		$\frac{29}{64}$	$\frac{1}{2}''$-20	0000	0000
.2969	.06922		$\frac{19}{64}$		0	1							
.302	.07163	N			0	0	.4688	.1726		$\frac{15}{32}$		00000	
.3125	.07670		$\frac{5}{16}$	$\frac{3}{8}''$-16	0	0	.4844	.1843		$\frac{31}{64}$	$\frac{9}{16}''$-12	00000	
.316	.07843	O			0	0	.5000	.1963		$\frac{1}{2}$		00000	

†Equal or nearest larger.

Table 11–18. DIMENSIONAL DATA FOR STEEL PIPE*

Nominal Pipe Size, in.	Outside Diameter, in.	Schedule Number or Weight	Wall Thickness, in.	Inside Diameter, in.	Surface Area Outside, ft²/ft	Surface Area Inside, ft²/ft	Areas and Weights Cross-sectional Metal Area, in.²	Areas and Weights Cross-sectional Flow Area, in.²	Weight Pipe lb/ft
¾	1.05	40	0.113	0.824	0.275	0.216	0.333	0.533	1.131
		80	0.154	0.742	0.275	0.194	0.434	0.432	1.474
1	1.315	40	0.133	1.049	0.344	0.275	0.494	0.864	1.679
		80	0.179	0.957	0.344	0.250	0.639	0.719	2.172
1¼	1.660	40	0.140	1.38	0.434	0.361	0.668	1.496	2.273
		80	0.191	1.278	0.434	0.334	0.881	1.283	2.997
1½	1.900	40	0.145	1.61	0.497	0.421	0.799	2.036	2.718
		80	0.200	1.50	0.497	0.393	1.068	1.767	3.632
2	2.375	40	0.154	2.067	0.622	0.541	1.074	3.356	3.653
		80	0.218	1.939	0.622	0.508	1.477	2.953	5.022
2½	2.875	40	0.203	2.469	0.753	0.646	1.704	4.79	5.794
		80	0.276	2.323	0.753	0.608	2.254	4.24	7.662
3	3.5	40	0.216	3.068	0.916	0.803	2.228	7.30	7.58
		00	0.900	2.900	0.916	0.759	3.016	6.60	10.25
3½	4.0	40	0.226	3.548	1.047	0.929	2.680	9.89	9.11
		80	0.318	3.364	1.047	0.881	3.678	8.89	12.51
4	4.5	40	0.237	4.026	1.178	1.054	3.17	12.73	10.79
		80	0.337	3.826	1.178	1.002	4.41	11.50	14.99
5	5.563	10 S	0.134	5.295	1.456	1.386	2.29	22.02	7.77
		40	0.258	5.047	1.456	1.321	4.30	20.01	14.62
		80	0.375	4.813	1.456	1.260	6.11	18.19	20.78
6	6.625	10 S	0.134	6.357	1.734	1.664	2.73	31.7	9.29
		40	0.280	6.065	1.734	1.588	5.58	28.9	18.98
		80	0.432	5.761	1.734	1.508	8.40	26.1	28.58
8	8.625	10 S	0.148	8.329	2.258	2.180	3.94	54.5	13.40
		30	0.277	8.071	2.258	2.113	7.26	51.2	24.7
		80	0.500	7.625	2.258	1.996	12.76	45.7	43.4
10	10.75	10 S	0.165	10.420	2.81	2.73	5.49	85.3	18.7
		30	0.279	10.192	2.81	2.67	9.18	81.6	31.2
		Extra heavy	0.500	9.750	2.81	2.55	16.10	74.7	54.7
12	12.75	10 S	0.180	12.390	3.34	3.24	7.11	120.6	24.2
		30	0.330	12.09	3.34	3.17	12.88	111.8	43.8
		Extra heavy	0.500	11.75	3.34	3.08	19.24	108.4	65.4
14	14.0	10	0.250	13.5	3.67	3.53	10.80	143.1	36.7
		Standard	0.375	13.25	3.67	3.47	16.05	137.9	54.6
		extra heavy	0.500	13.00	3.67	3.40	21.21	132.7	72.1
16	16.0	10	0.250	15.50	4.19	4.06	12.37	188.7	42.1
		Standard	0.375	15.25	4.19	3.99	18.41	182.7	62.6
		extra heavy	0.500	15.00	4.19	3.93	24.35	176.7	82.8
18	18.0	10 S	0.188	17.621	4.71	4.61	10.52	243.9	35.8
		Standard	0.375	17.25	4.71	4.52	20.76	233.7	70.6
		extra heavy	0.500	17.00	4.71	4.45	27.49	227.0	93.5
20	20.0	10 S	0.218	19.564	5.24	5.12	13.55	300.6	46.1
		Standard	0.375	19.25	5.24	5.04	23.12	291	78.6
		extra heavy	0.500	19.00	5.24	4.97	30.6	283.5	104.1
22	22.0	10	0.250	21.50	5.76	5.63	17.1	363	58.1
		Standard	0.375	21.25	5.76	5.56	25.5	355	86.6
		extra heavy	0.500	21.00	5.76	5.50	33.8	346	114.8
24	24.0	10	0.250	23.50	6.28	6.15	18.7	434	63.4
		Standard	0.375	23.25	6.28	6.09	27.8	425	94.6
		extra heavy	0.500	23.00	6.28	6.02	36.9	415	125.5
26	26.0	Standard	0.375	25.25	6.81	6.61	30.2	501	102.6
		extra heavy	0.500	25.00	6.81	6.54	40.1	491	136.2
30	30.0	10	0.312	29.376	7.85	7.69	29.1	678	98.9
		Standard	0.375	29.250	7.85	7.66	34.9	672	118.7
		extra heavy	0.500	29.00	7.85	7.59	46.3	661	157.6
34	34.0	Standard	0.375	33.250	8.90	8.70	39.6	868	134.7
		extra heavy	0.500	33.00	8.90	8.64	52.6	855	178.9
36	36.0	Standard	0.375	35.25	9.42	9.23	42.0	976	142.7
		extra heavy	0.500	35.00	9.42	9.16	55.8	962	189.6
42	42.0	Standard	0.375	41.25	11.0	10.8	49.0	1336	166.7
		extra heavy	0.500	41.00	11.0	10.73	65.2	1320	221.6

*Reprinted with permission, from: "Design Properties of Pipe", ©1958, Chemetron Corporation.

Table 11-19. COMMERCIAL COPPER TUBING *

The following table gives dimensional data and weights of copper tubing used for automotive, plumbing, refrigeration, and heat exchanger services. For additional data see the standards handbooks of the Copper Development Association, Inc., the ASTM standards, and the "SAE Handbook."

Dimensions in this table are actual specified measurements, subject to accepted tolerances. Trade size designations are usually by actual OD, except for water and drainage tube (plumbing), which measures 1/8-in. larger OD. A 1/2-in. plumbing tube, for example, measures 5/8-in. OD, and 2-in. plumbing tube measures 2 1/8-in. OD.

KEY TO GAGE SIZES

Standard-gage wall thicknesses are listed by numerical designation (14 to 21), BWG or Stubs gage. These gage sizes are standard for tubular heat exchangers. The letter A designates SAE tubing sizes for automotive service. Letter designations K and L are the common sizes for plumbing services, soft or hard temper.

OTHER MATERIALS

These same dimensional sizes are also common for much of the commercial tubing available in aluminum, mild steel, brass, bronze, and other alloys. Tube weights in this table are based on copper at 0.323 lb/in^3. For other materials the weights should be multiplied by the following approximate factors:

aluminum	0.30	monel	0.96
mild steel	0.87	stainless steel	0.89
brass	0.95		

Size, OD		Wall Thickness			Flow Area		Metal Area, in.²	Surface Area		Weight, lb/ft
in.	mm	in.	mm	gage	in.²	mm²		Inside, ft²/ft	Outside, ft²/ft	
1/8	3.2	.030	0.76	A	0.003	1.9	0.012	0.017	0.033	0.035
3/16	4.76	.030	0.76	A	0.013	8.4	0.017	0.034	0.049	0.058
1/4	6.4	.030	0.76	A	0.028	18.1	0.021	0.050	0.066	0.080
1/4	6.4	.049	1.24	18	0.018	11.6	0.031	0.038	0.066	0.120
5/16	7.94	.032	0.81	21A	0.048	31.0	0.028	0.065	0.082	0.109
3/8	9.53	.032	0.81	21A	0.076	49.0	0.033	0.081	0.098	0.134
3/8	9.53	.049	1.24	18	0.060	38.7	0.050	0.072	0.098	0.195
1/2	12.7	.032	0.81	21A	0.149	96.1	0.047	0.114	0.131	0.182
1/2	12.7	.035	0.89	20L	0.145	93.6	0.051	0.113	0.131	0.198
1/2	12.7	.049	1.24	18K	0.127	81.9	0.069	0.105	0.131	0.269
1/2	12.7	.065	1.65	16	0.108	69.7	0.089	0.97	0.131	0.344
5/8	15.9	.035	0.89	20A	0.242	156	0.065	0.145	0.164	0.251
5/8	15.9	.040	1.02	L	0.233	150	0.074	0.143	0.164	0.285
5/8	15.9	.049	1.24	18K	0.215	139	0.089	0.138	0.164	0.344
3/4	19.1	.035	0.89	20A	0.363	234	0.079	0.178	0.196	0.305
3/4	19.1	.042	1.07	L	0.348	224	0.103	0.174	0.196	0.362
3/4	19.1	.049	1.24	18K	0.334	215	0.108	0.171	0.196	0.418
3/4	19.1	.065	1.65	16	0.302	195	0.140	0.162	0.196	0.542
3/4	19.1	.083	2.11	14	0.268	173	0.174	0.151	0.196	0.674
7/8	22.2	.045	1.14	L	0.484	312	0.117	0.206	0.229	0.455
7/8	22.2	.065	1.65	16K	0.436	281	0.165	0.195	0.229	0.641
7/8	22.2	.083	2.11	14	0.395	255	0.206	0.186	0.229	0.800
1	25.4	.065	1.65	16	0.594	383	0.181	0.228	0.262	0.740
1	25.4	.083	2.11	14	0.546	352	0.239	0.218	0.262	0.927
1 1/8	28.6	.050	1.27	L	0.825	532	0.176	0.268	0.294	0.655

*Compiled and computed.

Table 11-19. COMMERCIAL COPPER TUBING (*Continued*)

Size, OD		Wall Thickness			Flow Area		Metal Area, in.²	Surface Area		Weight, lb/ft
in.	mm	in.	mm	gage	in.²	mm²		Inside, ft²/ft	Outside, ft²/ft	lb/ft
1 1/8	28.6	.065	1.65	16K	0.778	502	0.216	0.261	0.294	0.839
1 1/4	31.8	.065	1.65	16	0.985	636	0.242	0.293	0.327	0.938
1 1/4	31.8	.083	2.11	14	0.923	596	0.304	0.284	0.327	1.18
1 3/8	34.9	.055	1.40	L	1.257	811	0.228	0.331	0.360	0.884
1 3/8	34.9	.065	1.65	16K	1.217	785	0.267	0.326	0.360	1.04
1 1/2	38.1	.065	1.65	16	1.474	951	0.294	0.359	0.393	1.14
1 1/2	38.1	.083	2.11	14	1.398	902	0.370	0.349	0.393	1.43
1 5/8	41.3	.060	1.52	L	1.779	1148	0.295	0.394	0.425	1.14
1 5/8	41.3	.072	1.83	K	1.722	1111	0.351	0.388	0.425	1.36
2	50.8	.083	2.11	14	2.642	1705	0.500	0.480	0.628	1.94
2	50.8	.109	2.76	12	2.494	1609	0.620	0.466	0.628	2.51
2 1/8	54.0	.070	1.78	L	3.095	1997	0.449	0.520	0.556	1.75
2 1/8	54.0	.083	2.11	14K	3.016	1946	0.529	0.513	0.556	2.06
2 5/8	66.7	.080	2.03	L	4.77	3078	0.645	0.645	0.687	2.48
2 5/8	66.7	.095	2.41	13K	4.66	3007	0.760	0.637	0.687	2.93
3 1/8	79.4	.090	2.29	L	6.81	4394	0.950	0.771	0.818	3.33
3 1/8	79.4	.109	2.77	12K	6.64	4284	1.034	0.761	0.818	4.00
3 5/8	92.1	.100	2.54	L	9.21	5942	1.154	0.897	0.949	4.29
3 5/8	92.1	.120	3.05	11K	9.00	5807	1.341	0.886	0.949	5.12
4 1/8	104.8	.110	2.79	L	11.92	7691	1.387	1.022	1.080	5.38
4 1/8	104.8	.134	3.40	10K	11.61	7491	1.682	1.009	1.080	6.51

Table 11-20. BLOCK AND SIGNAL-FLOW DIAGRAMS

Equivalent Representation for Basic Elements—Continuous Systems

SYMBOLS: a = input

b = output

G = transfer function

Operation	Block diagram	Signal-flow diagram	Equation
Basic element			$b = Ga$
Elements in cascade			$b = G_1 G_2 G_3 a$
Elements in parallel			$C = (G_1 + G_2)a + G_3 b -$
Feedback			$C = \dfrac{G}{1 + GH} a$

Table 11–21.   BLOCK-DIAGRAM MANIPULATIONS*

Manipulation	Original network	Equivalent network
1. Interchange of elements	$a \to K_1G_1 \to K_2G_2 \to b$	$a \to K_2G_2 \to K_1G_1 \to b$
2. Interchange of summing points	$a \to$ (sum $\uparrow b$) $\xrightarrow{a \pm b}$ (sum $\uparrow c$) $\xrightarrow{a \pm b \pm c}$	$a \to$ (sum $\uparrow c$) $\xrightarrow{a \pm c}$ (sum $\uparrow b$) $\xrightarrow{a \pm c \pm b}$
3. Rearrangement of summing points	$a \to$ (sum) $\xrightarrow{a \pm b \pm c}$, with $\xleftarrow{b \pm c}$ (sum) fed by $\uparrow b$ and $\uparrow c$	$a \to$ (sum) $\xrightarrow{a \pm b}$ (sum) $\xrightarrow{a \pm b \pm c}$, fed by $\uparrow b$ and $\uparrow c$
4. Interchange of takeoff points	$a \to K_1G_1 \xrightarrow{b} K_2G_2 \xrightarrow{c}$; takeoffs $b \leftarrow$ and $\to b$	$a \to K_1G_1 \xrightarrow{b} K_2G_2 \xrightarrow{c}$; takeoffs $b \leftarrow$ and $\to b$
5. Moving a summing point ahead of an element	$a \to K_1G_1 \xrightarrow{b}$ (sum) $\xrightarrow{d}$, $d = b \pm c$, $\uparrow c$	$a \to$ (sum) $\xrightarrow{a \pm c'} K_1G_1 \xrightarrow{d}$, $c' \uparrow$ from $\dfrac{1}{K_1G_1}$, $\xleftarrow{c}$
6. Moving a summing point beyond an element	$a \to$ (sum) $\xrightarrow{a \pm b} K_1G_1 \xrightarrow{c}$, $\uparrow b$	$a \to K_1G_1 \xrightarrow{a'}$ (sum) $\xrightarrow{c}$, $c = a' \pm b'$, $b \uparrow K_1G_1 \xrightarrow{b'}$
7. Moving a takeoff point ahead of an element	$a \to K_1G_1 \xrightarrow{b}$; takeoff $\xleftarrow{b}$	$a \to K_1G_1 \xrightarrow{b}$; $\xleftarrow{b} K_1G_1 \leftarrow$
8. Moving a takeoff point beyond an element	$a \to K_1G_1 \xrightarrow{b}$; takeoff $\xleftarrow{a}$	$a \to K_1G_1 \xrightarrow{b}$; $\xleftarrow{a} \dfrac{1}{K_1G_1} \leftarrow$
9. Moving a takeoff point ahead of a summing point	$a \to$ (sum) $\xrightarrow{c}$, $c = a \pm b$, $\xleftarrow{c}$, $\uparrow b$	$a \to$ (sum) $\xrightarrow{c}$, $c = a \pm b$; $c = a \pm b \xleftarrow{} $ (sum) $\xleftarrow{a}$, $\uparrow b$
10. Moving a takeoff point beyond a summing point	$a \to$ (sum) $\xrightarrow{c}$, $c = a \pm b$, $\xleftarrow{a}$, $\uparrow b$	$a \to$ (sum) $\xrightarrow{c}$, $c = a \pm b$; $a = c \mp b \xleftarrow{} $ (sum) $\xleftarrow{c}$, $\uparrow b$
11. Combining cascade elements	$a \to K_1G_1 \to K_2G_2 \to b$	$a \to (K_1G_1)(K_2G_2) \to b$

Table 11–21. BLOCK-DIAGRAM MANIPULATIONS (Continued)

Manipulation	Original network	Equivalent network
12. Removing an element from a forward loop	K_1G_1, K_2G_2; $d = b \pm c$	K_2G_2, $\dfrac{1}{K_2G_2}$, K_1G_1; $d = b \pm c$
13. Inserting an element in a forward loop	K_1G_1; $d = b \pm a$	$\dfrac{1}{K_2G_2}$, K_2G_2, K_1G_1, K_2G_2; $d = b \pm a$
14. Eliminating a forward loop	K_1G_1, K_2G_2; $d = b \pm c$	$K_1G_1 \pm K_2G_2$
15. Removing an element from a feedback loop	K_1G_1, K_2G_2; $c = a \pm b$	$\dfrac{1}{K_2G_2}$, K_2G_2, K_1G_1; $c' = a' \pm b'$
16. Inserting an element in a feedback loop	K_1G_1, K_2G_2; $c = a \pm b$	K_2G_2, $\dfrac{1}{K_2G_2}$, K_1G_1, K_2G_2; $c' = a' \pm b'$
17. Eliminating a feedback loop	K_1G_1, K_2G_2; $c = a \pm b$	$\dfrac{K_1G_1}{1 \mp (K_1G_1)(K_2G_2)}$
18. Special form of 17	K_1G_1; $c = a \pm b$; $b = d$	$\dfrac{K_1G_1}{1 \mp K_1G_1}$
19. Special form of 17	$a \pm b$, K_1G_1	$\dfrac{1}{1 \mp K_1G_1}$
20. Inserting a feedback loop to replace an element	K_1G_1	$a - b$, $\dfrac{K_1G_1}{1 - K_1G_1}$; $b = d$
21. Different form of 20	K_1G_1	$a - b$, $\dfrac{1}{K_1G_1} - 1$

*From: "Handbook of Automation, Computation, and Control", E.M. Grabbe, S. Ramo, and D.E. Wooldridge, Eds., Vol. 1, John Wiley & Sons, 1958, pp. 20-62 and 20-63.

Table 11–22. SIGNAL-FLOW DIAGRAMS*

Cascade

1.

$$x_3 = bx_2 = bax_1$$

Parallel

2.

$$x_2 = (a + b) x_1$$

3.

$$x_2 = x_1 + gx_2$$

$$x_2 = x_1 \left(\frac{1}{1 - g} \right)$$

4.

$$x_1 = x_0 + bx_2$$
$$x_2 = ax_1 = ax_0 + abx_2$$

$$\frac{x_3}{x_0} = \frac{a}{1 - ab}$$

Table 11–22. SIGNAL-FLOW DIAGRAMS *(Continued)*

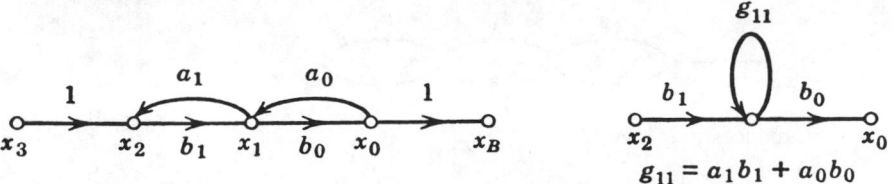

5.

$$\frac{x_B}{x_3} = \frac{b_1 b_0}{1 - g_{11}} = \frac{b_1 b_0}{1 - a_1 b_1 - a_0 b_0}$$

6.

$$\frac{x_3}{x_1} = G = \frac{g_1 g_2}{1 - g_{a1} g_{b1}}$$

$$G = \frac{x_3}{x_1} = \cfrac{g_1 g_2}{1 - \cfrac{g_{a1} g_{b1}}{1 - g_{a2} g_{b2}}}$$

7.

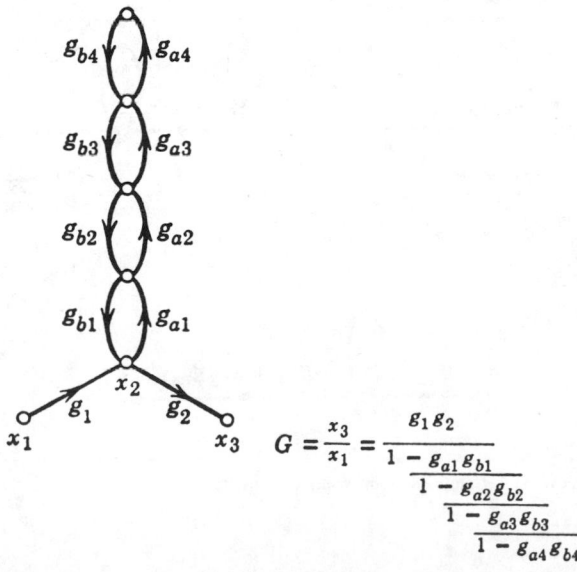

$$G = \frac{x_3}{x_1} = \cfrac{g_1 g_2}{1 - \cfrac{g_{a1} g_{b1}}{1 - \cfrac{g_{a2} g_{b2}}{1 - \cfrac{g_{a3} g_{b3}}{1 - g_{a4} g_{b4}}}}}$$

Table 11–22.   SIGNAL-FLOW DIAGRAMS *(Continued)*

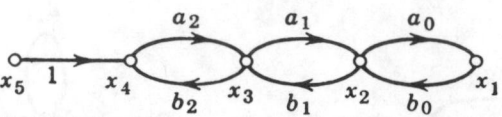

reduces to

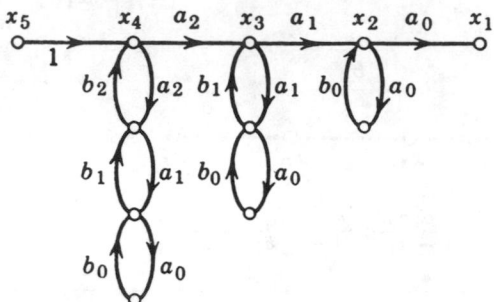

8.

which reduces to

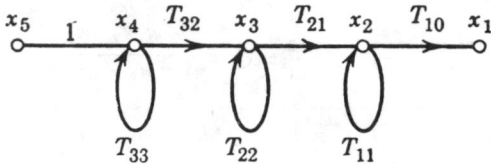

then $\dfrac{x_1}{x_5} = a_0 \left[\dfrac{1}{1 - a_0 b_0}\right] a_1 \left[\dfrac{1}{1 - \dfrac{a_1 b_1}{1 - a_0 b_0}}\right] a_2 \left[\dfrac{1}{1 - \dfrac{a_2 b_2}{1 - \dfrac{a_1 b_1}{1 - a_0 b_0}}}\right]$

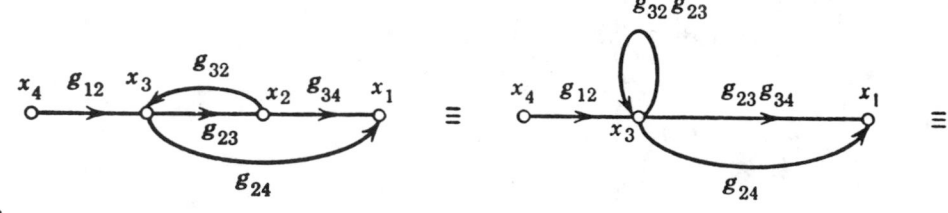

9.

*From: "Process Dynamics", D.P. Campbell, John Wiley & Sons, 1958.

Table 11–23. BLOCK AND SIGNAL-FLOW DIAGRAMS FOR SAMPLED-DATA CONTROL SYSTEMS†

NOTATIONS:

$R(s)$ = Laplace transform of input

$R^*(s)$ = Laplace transform of sampled input

$R(z)$ = z-transform of sampled input

$G(s)$ = transfer function

$G^*(s)$ = pulse transfer function

$G(z)$ = z-transform of $G(s)$

$G(z, m)$ = modified z-transform of $G(s)$

$G_1(z)G_2(z) = Z[G_1{}^*(s)G_2{}^*(s)]$
$\qquad\qquad = Z[G_1(s)] \cdot Z[G_2(s)]$

$G_1G_2(z) = Z[G_1G_2(s)]$

All samplers assumed to operate in synchronism with period T.

System block diagram	System signal-flow graph	Output transforms
		$C(s) = C^*(s)$ $C^*(s) = R^*(s)$ $C(z, m) = C(z) = R(z)$
		$C(s) = C^*(s)$ $C^*(s) = GR^*(s)$ $C(z, m) = C(z) = GR(z)$
		$C(s) = G(s)R^*(s)$ $C^*(s) = G^*(s)R^*(s)$ $C(z, m) = G(z, m)R(z)$
		$C(s) = \dfrac{G(s)R^*(s)}{1 + HG^*(s)}$ $C^*(s) = \dfrac{G^*(s)R^*(s)}{1 + HG^*(s)}$ $C(z, m) = \dfrac{G(z, m)R(z)}{1 + HG(z)}$
		$C(s) = \dfrac{R(s)G(s)[1 + HG^*(s)] - RG^*(s)H(s)G(s)}{1 + HG^*(s)}$ $C^*(s) = \dfrac{RG^*(s)}{1 + HG^*(s)}$ $C(z, m) = \dfrac{-HG(z, m)RG(z)}{1 + HG(z)} + RG(z, m)$

Table 11–23. BLOCK AND SIGNAL-FLOW DIAGRAMS FOR SAMPLED-DATA CONTROL SYSTEMS (Continued)

System block diagram	System signal-flow graph	Output transforms
		$$C(s) = \frac{G^*(s)R^*(s)}{1 + G^*(s)H^*(s)}$$ $$C^*(s) = \frac{G^*(s)R^*(s)}{1 + G^*(s)H^*(s)}$$ $$C(z, m) = \frac{G(z)R(z)}{1 + G(z)H(z)}$$
		$$C(s) = \frac{RG_1{}^*(s)G_2(s)}{1 + HG_1G_2{}^*(s)}$$ $$C^*(s) = \frac{RG_1{}^*(s)G_2{}^*(s)}{1 + HG_1G_2{}^*(s)}$$ $$C(z, m) = \frac{RG_1(z)G_2(z, m)}{1 + HG_1G_2(z)}$$
		$$C(s) = \frac{G_1{}^*(s)G_2(s)R^*(s)}{1 + G_1{}^*(s)HG_2{}^*(s)}$$ $$C^*(s) = \frac{G_1{}^*(s)G_2{}^*(s)R^*(s)}{1 + G_1{}^*(s)HG_2{}^*(s)}$$ $$C(z, m) = \frac{G_1(z)G_2(z, m)R(z)}{1 + G_1(z)HG_2(z)}$$
		$$C(s) = \frac{R^*(s)G(s)}{1 + G^*(s)H^*(s)}$$ $$C^*(s) = \frac{G^*(s)R^*(s)}{1 + G^*(s)H^*(s)}$$ $$C(z, m) = \frac{G(z, m)R(z)}{1 + G(z)H(z)}$$
		$$C(s) = \frac{G_2{}^*(s)G_3(s)G_1R^*(s)}{1 + G_2{}^*(G_1G_3H)^*}$$ $$C^*(s) = \frac{G_2{}^*(s)G_3{}^*(s)G_1R^*(s)}{1 + G_2{}^*(s)G_1G_3H^*(s)}$$ $$C(z, m) = \frac{G_2(z)G_3(z, m)G_1R(z)}{1 + G_2(z)G_1G_3H(z)}$$
		$$C(s) = GR^*(s) - \frac{G^*(s)G(s)H(s)R^*(s)}{1 + GH^*(s)}$$ $$C^*(s) = \frac{G^*(s)R^*(s)}{1 + HG^*(s)}$$ $$C(z, m) = R(z)G(z, m)\frac{G(z)GH(z, m)R(z)}{1 + HG(z)}$$

Table 11–23. BLOCK AND SIGNAL-FLOW DIAGRAMS FOR SAMPLED-DATA CONTROL SYSTEMS (*Continued*)

System block diagram	System signal-flow graph	Output transforms
(block diagram)	*(signal-flow graph)*	$C(s) = \dfrac{G_2(s)G_1 R^*(s)}{1+G_2^*(s)GH^*(s)}$ $C^*(s) = \dfrac{G_2^*(s)G_1 R^*(s)}{1+G_2^*(s)GH^*(s)}$ $C(z,m) = \dfrac{G_2(z,m)G_1 R(z)}{1+G_2(z)GH(z)}$
(block diagram)	*(signal-flow graph)*	$C(s) = \left(\dfrac{R}{1+GH_2}\right) - \dfrac{D^*\left(\dfrac{H_1}{1+GH_2}\right)\left(\dfrac{GR}{1+GH_2}\right)^*}{1+\left(\dfrac{H_1 G}{1+GH_2}\right)^* D^*}$ $C^*(s) = \dfrac{\left(\dfrac{GR}{1+GH_2}\right)^*}{1+\left(\dfrac{H_1 G}{1+GH_2}\right)^* D^*}$ $C(z,m) = \dfrac{R}{1+GH_2}(z,m) - \dfrac{D(z)\dfrac{GR}{1+GH_2}(z)\dfrac{H_1}{1+GH_2}(z,m)}{1+\left(\dfrac{H_1 G}{1+HG_2}\right)(z)D(z)}$
(block diagram)	*(signal-flow graph)*	$C(s) = \dfrac{R^*(s)\left(\dfrac{G_2 G_3 G_4}{1-G_3 G_4 G_6}\right)}{1-\left(\dfrac{G_2 G_4 G_5}{1-G_3 G_4 G_6}\right)^*}$ $C^*(s) = \dfrac{R^*(s)\left(\dfrac{G_2 G_3 G_4}{1-G_3 G_4 G_6}\right)^*}{1-\left(\dfrac{G_2 G_3 G_5}{1-G_3 G_4 G_6}\right)^*}$ $C(z,m) = \dfrac{R(z)\dfrac{G_2 G_3 G_4}{1-G_3 G_4 G_6}(z,m)}{1-\dfrac{G_2 G_3 G_5}{1-G_3 G_4 G_6}(z)}$
(block diagram)	*(signal-flow graph)*	$C(s) = \dfrac{[(RG_1)^* - (RG_3)^*]D^* G_2 G_3}{1+D^*(G_2 G_3)^*} + RG_3$ $C^*(s) = \dfrac{[(RG_1)^* - (RG_3)^*]D^*(G_2 G_3)^*}{1+D^*(G_2 G_3)^*} + (RG_3)^*$ $C(z,m) = \dfrac{[RG_1(z) - RG_3(z)]D(z)G_2 G_3(z,m)}{1+D(z)G_2 G_3(z)} + RG_3(z,m)$
(block diagram)	*(signal-flow graph)*	$C(s) = \dfrac{RG_1 G_2}{1+G_1 G_2 H_1} - \dfrac{\left(\dfrac{G_1 G_2 R}{1+G_1 G_2 H_1}\right)^*\left(\dfrac{G_2 H_2}{1+G_1 G_2 H_1}\right)}{1+\left(\dfrac{H_2 G_2}{1+G_1 G_2 H_1}\right)}$ $C^*(s) = \left(\dfrac{RG_1 G_2}{1+G_1 G_2 H_1}\right)^* - \dfrac{\left(\dfrac{G_1 G_2 R}{1+G_1 G_2 H_1}\right)^*\left(\dfrac{G_2 H_2}{1+G_1 G_2 H_1}\right)^*}{1+\left(\dfrac{H_2 G_2}{1+G_1 G_2 H_1}\right)^*}$ $C(z,m) = \dfrac{RG_1 G_2}{1+G_1 G_2 H_1}(z,m) - \dfrac{\left(\dfrac{G_1 G_2 R}{1+G_1 G_2 H_1}\right)(z)\dfrac{G_2 H_2}{1+G_1 G_2 H_1}(z,m)}{1+\dfrac{H_2 G_2}{1+G_1 G_2 H_1}(z)}$

†From: Benjamin C. Kuo, ANALYSIS AND SYNTHESIS OF SAMPLED-DATA CONTROL SYSTEMS, © 1963. Reprinted by permission of Prentice-Hall, Inc., Englewood Cliffs, New Jersey.

Table 11-24. RESPONSE OF TYPICAL SYSTEMS TO COMMON FORCING FUNCTIONS*

$$\theta_i(s) \to \boxed{G(s)} \to \theta_o(s)$$

See also Figures 6-1 to 6-5.

FORCING FUNCTION $\theta_i(t)$	$\mathcal{L}\{\theta_i(t)\}$ $\theta_i(s)$	$G(s) = \dfrac{1}{s}$ INTEGRATOR		$G(s) = \dfrac{1}{\tau s + 1}$ FIRST—ORDER LAG		$G(s) = \dfrac{1}{(\tau_1 s+1)(\tau_2 s+1)}$; $\tau_1 \neq \tau_2$ CASCADED FIRST—ORDER LAGS	
		$\theta_o(s)$	$\theta_o(t)$	$\theta_o(s)$	$\theta_o(t)$	$\theta_o(s)$	$\theta_o(t)$
UNIT IMPULSE $I(t)$ AREA = 1 $\Delta t = 0$	1	$\dfrac{1}{s}$	$U(t)$	$\dfrac{1}{\tau s + 1}$	$\dfrac{1}{\tau} e^{-t/\tau}$	$\dfrac{1}{(\tau_1 s+1)(\tau_2 s+1)}$	$\dfrac{1}{\tau_1 - \tau_2}[e^{-t/\tau_1} - e^{-t/\tau_2}]$
UNIT STEP $U(t)$	$\dfrac{1}{s}$	$\dfrac{1}{s^2}$	t SLOPE = 1	$\dfrac{1}{s(\tau s + 1)}$	$1 - e^{-t/\tau}$	$\dfrac{1}{s(\tau_1 s+1)(\tau_2 s+1)}$	$1 + \dfrac{\tau_1}{\tau_2 - \tau_1} e^{-t/\tau_1} - \dfrac{\tau_2}{\tau_2 - \tau_1} e^{-t/\tau_2}$
RAMP at SLOPE = a	$\dfrac{a}{s^2}$	$\dfrac{a}{s^3}$	at^2	$\dfrac{a}{s^2(\tau s + 1)}$	$a\tau\left(e^{-t/\tau} + \dfrac{t}{\tau} - 1\right)$ SLOPE = a	$\dfrac{a}{s^2(\tau_1 s+1)(\tau_2 s+1)}$	$a\left[\dfrac{\tau_2^2}{\tau_2-\tau_1} e^{-t/\tau_2} - \dfrac{\tau_1^2}{\tau_2-\tau_1} e^{-t/\tau_1} + t - (\tau_1 + \tau_2)\right]$ SLOPE = a
PULSE $b\{U(t) - U(t-a)\}$	$\dfrac{b(1-e^{-as})}{s}$	$\dfrac{b(1-e^{-as})}{s^2}$	$b\{t - U(t-a)(t-a)\}$	$\dfrac{b(1-e^{-as})}{s(\tau s+1)}$	$b\{1 - e^{-t/\tau} - U(t-a)\cdot(1 - e^{-\frac{t-a}{\tau}})\}$	$\dfrac{b(1-e^{-as})}{s(\tau_1 s+1)(\tau_2 s+1)}$	$b\left\{1 + \dfrac{\tau_1}{\tau_2-\tau_1} e^{-t/\tau_1} - \dfrac{\tau_2}{\tau_2-\tau_1} e^{-t/\tau_2} - u(t-a)\left[1 + \dfrac{\tau_1}{\tau_2-\tau_1} e^{-\frac{t-a}{\tau_1}} - \dfrac{\tau_2}{\tau_2-\tau_1} e^{\frac{t-a}{\tau_2}}\right]\right\}$
SINE WAVE $\sin\left(\dfrac{2\pi}{T}\right)t$ $\omega = \dfrac{2\pi}{T}$	$\dfrac{\omega}{s^2 + \omega^2}$	$\dfrac{\omega}{s(s^2+\omega^2)}$	STEADY FORCED RESP. $\frac{1}{\omega}(1 - \cos\omega t)$ SLOPE = -6DB/OCT.	$\dfrac{\omega}{(\tau s+1)(s^2+\omega^2)}$	STEADY FORCED RESP. $A \cdot \sin(\omega t + \phi)$ $A = [1 + (\omega\tau)^2]^{-1/2}$; $\phi \, \mathrm{TAN}^{-1}\omega\tau$ -6DB/OCT.	$\dfrac{\omega}{(\tau_1 s+1)(\tau_2 s+1)(s^2+\omega^2)}$	STEADY FORCED RESPONSE $A = [(1+(\tau_1\omega)^2)(1+(\tau_2\omega)^2)]^{-1/2}$ $-\phi = \mathrm{TAN}^{-1}\dfrac{\omega(\tau_1+\tau_2)}{1-\omega^2\tau_1\tau_2}$ 12 DB/OCT.

Table 11–24.   RESPONSE OF TYPICAL SYSTEMS TO COMMON FORCING FUNCTIONS (*Continued*)

FORCING FUNCTION $\theta_i(t)$	$\mathcal{L}\{\theta_i(t)\}$ $\theta_i(s)$	$G(s) = \dfrac{1}{\tau^2 s^2 + 2\zeta\tau s + 1}$; ζ=DAMPING RATIO — "THE" SECOND ORDER SYSTEM — $\theta_o(t)$	$G(s) = e^{-\alpha s}$ DEAD TIME $\theta_o(s)$	$\theta_o(t)$
UNIT IMPULSE $I(t)$ AREA =1 ∞ $\Delta t=0$	1	$\zeta<1$ $\dfrac{1}{\tau^2\omega}e^{-\zeta t/\tau}\sin\omega t$ $\zeta=1$ $\dfrac{t}{\tau^2}e^{-t/\tau}$ $\zeta>1$ $\dfrac{1}{\tau^2\omega}e^{-\zeta t/\tau}\sinh\omega t$ WHERE $\omega=\dfrac{1}{\tau}[\,|1-\zeta^2|\,]^{1/2}$	$e^{-\alpha s}$	$I(t-\alpha)$
UNIT STEP $U(t)$ 	$\dfrac{1}{s}$	$\zeta<1$ $1-\dfrac{1}{\omega\tau}e^{-\zeta t/\tau}\sin(\omega t+\phi)$ $\zeta=1$ $1-(1+t/\tau)e^{-t/\tau}$ $\zeta>1$ $1-\dfrac{\zeta t}{\omega\tau}e^{-\zeta t/\tau}\sinh(\omega t+\psi)$ $\phi=\cos^{-1}\zeta$ $\psi=\cos^{-1}\zeta$	$\dfrac{e^{-\alpha s}}{s}$	$U(t-\alpha)$
RAMP at SLOPE = a 	$\dfrac{a}{s^2}$	$\zeta<1$ $a\tau[(t/\tau-2\zeta)+\dfrac{e^{-\zeta t/\tau}}{\omega\tau}\sin(\omega t+\phi)]$ $\zeta=1$ $a\tau[(t/\tau-2)+(t/\tau+2)e^{-t/\tau}]$ $\zeta>1$ $a\tau(t/\tau-2\zeta)+\dfrac{e^{-\zeta t/\tau}}{\omega\tau}\sinh(\omega t+\psi)$ $\phi=2\cos^{-1}\zeta$ $\psi=2\cosh^{-1}\zeta$	$\dfrac{ae^{-\alpha s}}{s^2}$	$aU(t-\alpha)\ (t-\alpha)$
PULSE $b\{U(t)-U(t-a)\}$ 	$\dfrac{b(1-e^{as})}{s}$	$b[f(t)-U(t-a)f(t-a)]$ WHERE $f(t)=$ RESPONSE TO $u(t)$	$be^{-\alpha s}(1-e^{-as})$	$b\{U(t-\alpha)-U(t-a-\alpha)\}$
SINE WAVE $\sin(\frac{2\pi}{T})t$ 	$\dfrac{\omega}{s^2+\omega^2}$ $\omega=\dfrac{2\pi}{T}$	STEADY FORCED RESPONSE $A=[(1-(\tau\omega)^2)^2+(2\zeta\tau\omega)^2]^{-1/2}$; $\omega=2\pi/T$ $-\phi=\tan\dfrac{2\zeta\tau\omega}{1-(\tau\omega)^2}$ 	$\dfrac{\omega e^{-\alpha s}}{s^2+\omega^2}$	STEADY FORCED RESP $\sin\omega(t-\alpha)$

*From: "Techniques of Process Control", P.S. Buckley, John Wiley & Sons, 1964.

Table 11–25.

FREQUENCY-RESPONSE EQUATIONS FOR SOME COMMON CONTROL-SYSTEM ELEMENTS*

Description	Transfer function $G(s)$	Frequency response $G(j\omega)$	Magnitude ratio	Phase angle
1. Dead time	$\epsilon^{-T_L s}$	$\epsilon^{-j\omega T_L}$	1	$-\omega T_L$ radians
2. First-order lag	$\dfrac{1}{Ts+1}$	$\dfrac{1}{j\omega T+1}$	$\dfrac{1}{\sqrt{\omega^2 T^2+1}}$	$-\tan^{-1}(\omega T)$
3. Second-order lag	$\dfrac{1}{(Ts+1)(aTs+1)}$	$\dfrac{1}{-a\omega^2 T^2+j(1+a)\omega T+1}$	$\dfrac{1}{\sqrt{(1-a\omega^2 T^2)^2+(1+a)^2\omega^2 T^2}}$	$-\tan^{-1}\left[\dfrac{(1+a)\omega T}{1-a T^2\omega^2}\right]$
4. Quadratic (underdamped)	$\left(\dfrac{s}{\omega_n}\right)^2+\dfrac{2\zeta}{\omega_n}s+1$	$-\left(\dfrac{\omega}{\omega_n}\right)^2+j2\zeta\dfrac{\omega}{\omega_n}+1$	$\sqrt{\left(1-\dfrac{\omega^2}{\omega_n^2}\right)^2+4\zeta^2\left(\dfrac{\omega}{\omega_n}\right)^2}$	$-\tan^{-1}\left[\dfrac{2\zeta\dfrac{\omega}{\omega_n}}{1-\left(\dfrac{\omega}{\omega_n}\right)^2}\right]$
5. Ideal proportional controller	K	K	K	0
6. Ideal proportional-plus-reset controller $T_i=\dfrac{1}{r}$ r = reset rate	$K\left(1+\dfrac{1}{T_i s}\right)$ or $K\dfrac{T_i s+1}{T_i s}$	$K\left(1+\dfrac{1}{j\omega T_i}\right)$ or $K\dfrac{j\omega T_i+1}{j\omega T_i}$	$K\sqrt{1+\left(\dfrac{1}{\omega T_i}\right)^2}$	$-\tan^{-1}\left(\dfrac{1}{\omega T_i}\right)$
7. Ideal proportional-plus-rate controller	$K(1+T_d s)$	$K(1+j\omega T_d)$	$K\sqrt{1+\omega^2 T_d^2}$	$\tan^{-1}(\omega T_d)$
8. Ideal proportional-plus-reset-plus-rate controller	$K\left(1+T_d s+\dfrac{1}{T_i s}\right)$ or $K\dfrac{T_i s^2+\cdots+1}{T_i s}$	$K\left(1+j\omega T_d+\dfrac{1}{j\omega T_i}\right)$ or $K\dfrac{j\omega T_i-\omega^2 T_d T_i+1}{j\omega T_i}$	$K\dfrac{\sqrt{(\omega T_i)^2+(1-\omega^2 T_d T_i)^2}}{\omega T_i}$	$\tan^{-1}\left(\omega T_d-\dfrac{1}{\omega T_i}\right)$

*From: "Process Instruments and Controls Handbook", D.M. Considine, Ed., McGraw–Hill Book Company, 1957.

Table 11–26. ROOT LOCI*

Continuous Systems

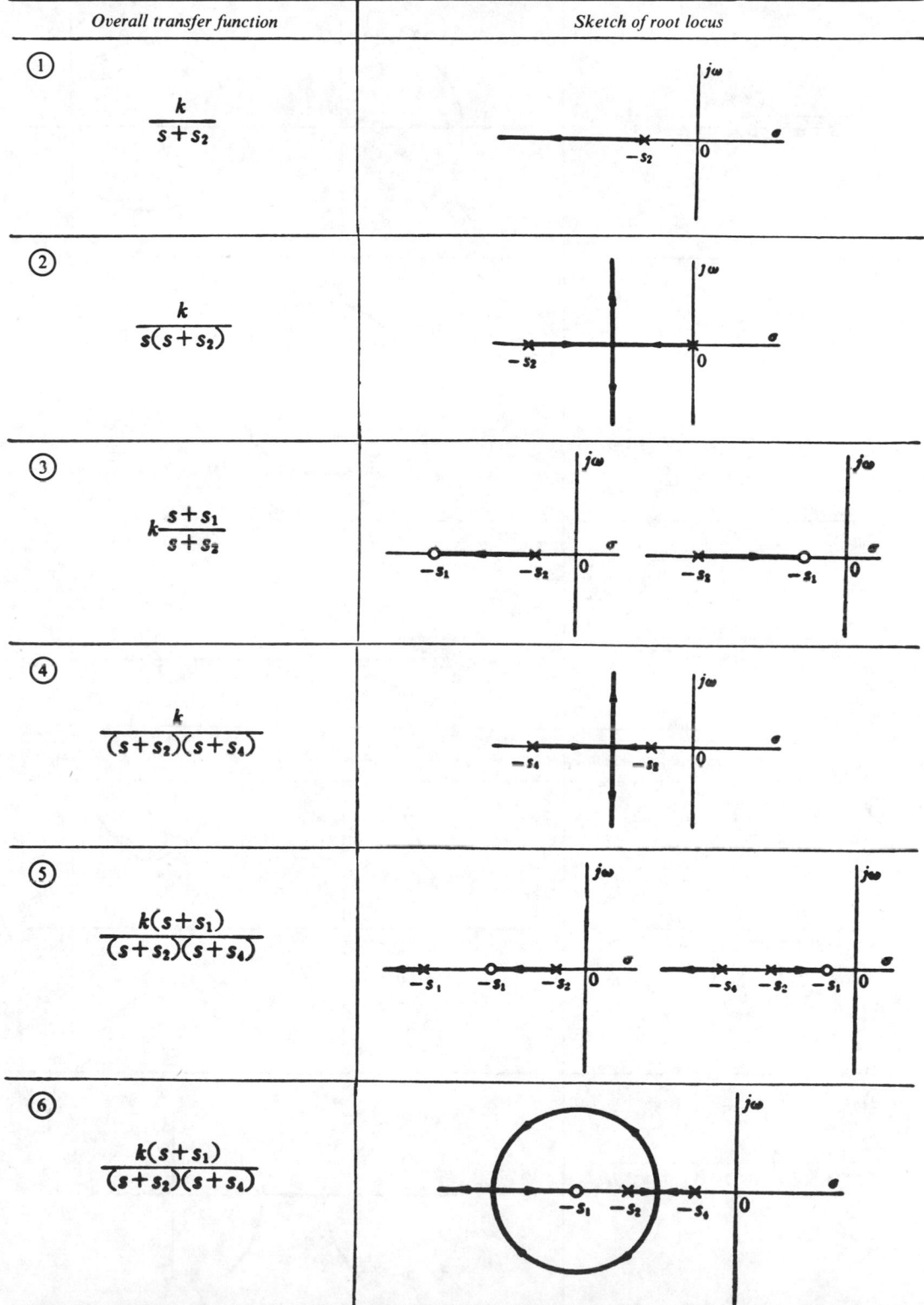

Overall transfer function	Sketch of root locus
① $\dfrac{k}{s+s_2}$	
② $\dfrac{k}{s(s+s_2)}$	
③ $k\dfrac{s+s_1}{s+s_2}$	
④ $\dfrac{k}{(s+s_2)(s+s_4)}$	
⑤ $\dfrac{k(s+s_1)}{(s+s_2)(s+s_4)}$	
⑥ $\dfrac{k(s+s_1)}{(s+s_2)(s+s_4)}$	

Table 11–26. ROOT LOCI *(Continued)*

Overall transfer function	Sketch of root locus
⑦ $$\dfrac{k(s+s_1)}{(s+\alpha+j\beta)(s+\alpha-j\beta)}$$	
⑧ $$\dfrac{k(s+s_1)}{(s+s_2)(s+s_4)}$$	
⑨ $$\dfrac{k}{(s+s_2)(s+s_4)(s+s_5)}$$	
⑩ $$\dfrac{k}{(s+s_2)(s+\alpha+j\beta)(s+\alpha-j\beta)}$$	
⑪ $$\dfrac{k}{(s+s_2)(s+\alpha+j\beta)(s+\alpha-j\beta)}$$	

Table 11–26.   ROOT LOCI *(Continued)*

Overall transfer function	Sketch of root locus
⑫ $$\dfrac{k(s+s_1)}{(s+s_2)(s+s_4)(s+s_6)}$$	
⑬ $$\dfrac{k(s+s_1)}{(s+s_2)(s+s_4)(s+s_6)}$$	
⑭ $$\dfrac{k(s+s_1)}{(s+s_2)(s+s_4)(s+s_6)}$$	
⑮ $$\dfrac{k(s+s_1)}{(s+s_2)(s+\alpha+j\beta)(s+\alpha-j\beta)}$$	
⑯ $$\dfrac{k(s+s_1)(s+s_3)}{s(s+s_2)(s+s_4)}$$	

Table 11–26. ROOT LOCI *(Continued)*

Overall transfer function	*Sketch of root locus*
(17) $\dfrac{k(s+s_1)(s+s_3)}{s(s+s_2)(s+s_4)}$	
(18) $\dfrac{k(s+s_1)(s+s_3)}{s(s+s_2)(s+s_4)}$	
(19) $\dfrac{k(s+s_1)(s+s_3)}{(s+s_2)^3}$	
(20) $\dfrac{k}{(s+s_2)(s+s_4)(s+s_6)(s+s_8)}$	
(21) $\dfrac{k}{s(s+s_2)(s+\alpha+j\beta)(s+\alpha-j\beta)}$	

Table 11-26. ROOT LOCI (Continued)

Overall transfer function	Sketch of root locus
(22) $$\frac{k}{s(s+s_2)(s+\alpha+j\beta)(s+\alpha-j\beta)}$$	
(23) $$\frac{k}{\left\{\begin{array}{c}(s+s_2)(s+s_4)(s+\alpha+j\beta)\\ \times(s+\alpha-j\beta)\end{array}\right\}}$$	
(24) $$\frac{k}{\left\{\begin{array}{c}(s+\alpha_1+j\beta)(s+\alpha-j\beta_1)\\ \times(s+\alpha_2+j\beta_2)(s+\alpha_2-j\beta_2)\end{array}\right\}}$$	
(25) $$\frac{k(s+s_1)}{\left\{\begin{array}{c}s(s+s_2)(s+\alpha+j\beta)\\ \times(s+\alpha-j\beta)\end{array}\right\}}$$	
(26) $$\frac{k(s+s_1)}{\left\{\begin{array}{c}s(s+s_2)(s+\alpha+j\beta)\\ \times(s+\alpha+j\beta)\end{array}\right\}}$$	

Table 11–26. ROOT LOCI (Continued)

Overall transfer function	Sketch of root locus
㉗ ke^{-sL}	
㉘ $\dfrac{ke^{-sL}}{s+s_2}$	

REFERENCE

"Handbook of Automation, Computation, and Control", E.M. Grabbe, S. Ramo, and D.E. Wooldridge, Eds., Vol. 1, John Wiley & Sons, 1958.

Table 11-27. HYDRAULIC COMPONENTS*

Mechanical-hydraulic network	Log magnitude characteristic	Transfer function	T_1	T_2
	G_∞, $G_0 = 0$, $G_\infty = \dfrac{b}{a}$, $\dfrac{1}{T_1}$	$\dfrac{G_\infty T_1 s}{T_1 s + 1}$	$\dfrac{a}{(a+b)K}$	$\cdots$
	G_∞, $G_0 = \dfrac{c}{a+b}$, $G_\infty = \dfrac{b+c}{a}$, $\dfrac{1}{T_1}$, $\dfrac{1}{T_2}$	$\dfrac{G_0(T_1 s + 1)}{T_2 s + 1}$	$\dfrac{b+c}{cK}$	$\dfrac{a}{(a+b)K}$
	G_∞, $G_0 = \dfrac{b}{c}$, $G_\infty = 0$, $\dfrac{1}{T_1}$	$\dfrac{G_0}{T_1 s + 1}$	$\dfrac{b+c}{cK}$	$\cdots$
	G_∞, $G_0 = \dfrac{a+b}{c}$, $G_\infty = \dfrac{a}{b+c}$, $\dfrac{1}{T_1}$, $\dfrac{1}{T_2}$	$\dfrac{G_0(T_1 s + 1)}{T_2 s + 1}$	$\dfrac{b+c}{cK}$	$\dfrac{a}{(a+b)K}$

K = velocity of piston per unit valve displacement

*From: "Handbook of Automation, Computation, and Control", E.M. Grabbe, S Ramo, and D.E. Wooldridge, Eds., Vol. 1 John Wiley & Sons, 1958, p. 23-45.

Table 11–28.   TRANSFER FUNCTION PLOTS

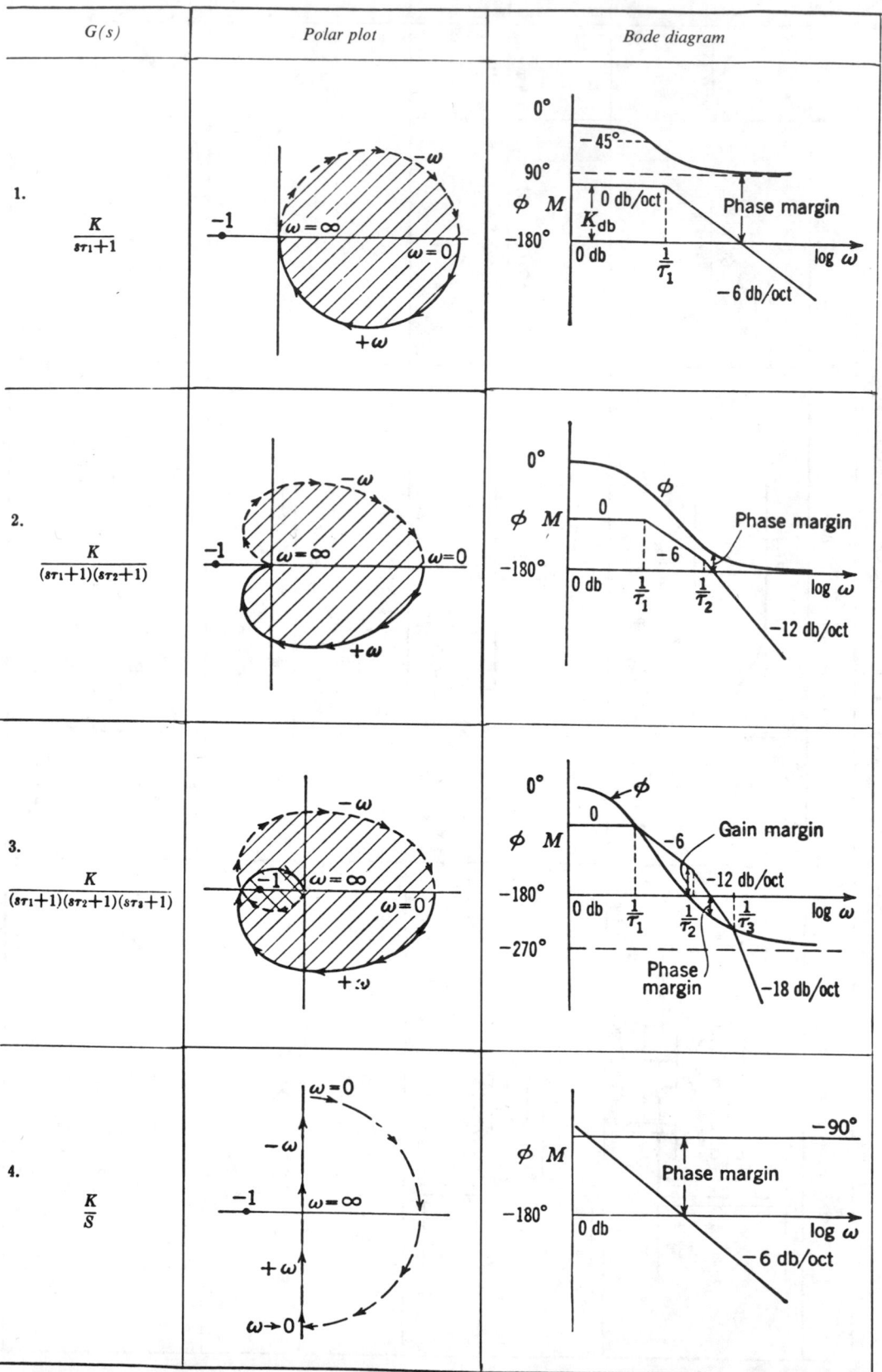

$G(s)$	Polar plot	Bode diagram
1. $\dfrac{K}{s\tau_1+1}$		
2. $\dfrac{K}{(s\tau_1+1)(s\tau_2+1)}$		
3. $\dfrac{K}{(s\tau_1+1)(s\tau_2+1)(s\tau_3+1)}$		
4. $\dfrac{K}{s}$		

FOR TYPICAL TRANSFER FUNCTION*

Nichols diagram	Root locus	Comments
		Stable; gain margin $= \infty$
		Elementary regulator; stable; gain margin $= \infty$
		Regulator with additional energy-storage component; unstable, but can be made stable by reducing gain
		Ideal integrator; stable

Table 11–28.   TRANSFER-FUNCTION PLOTS

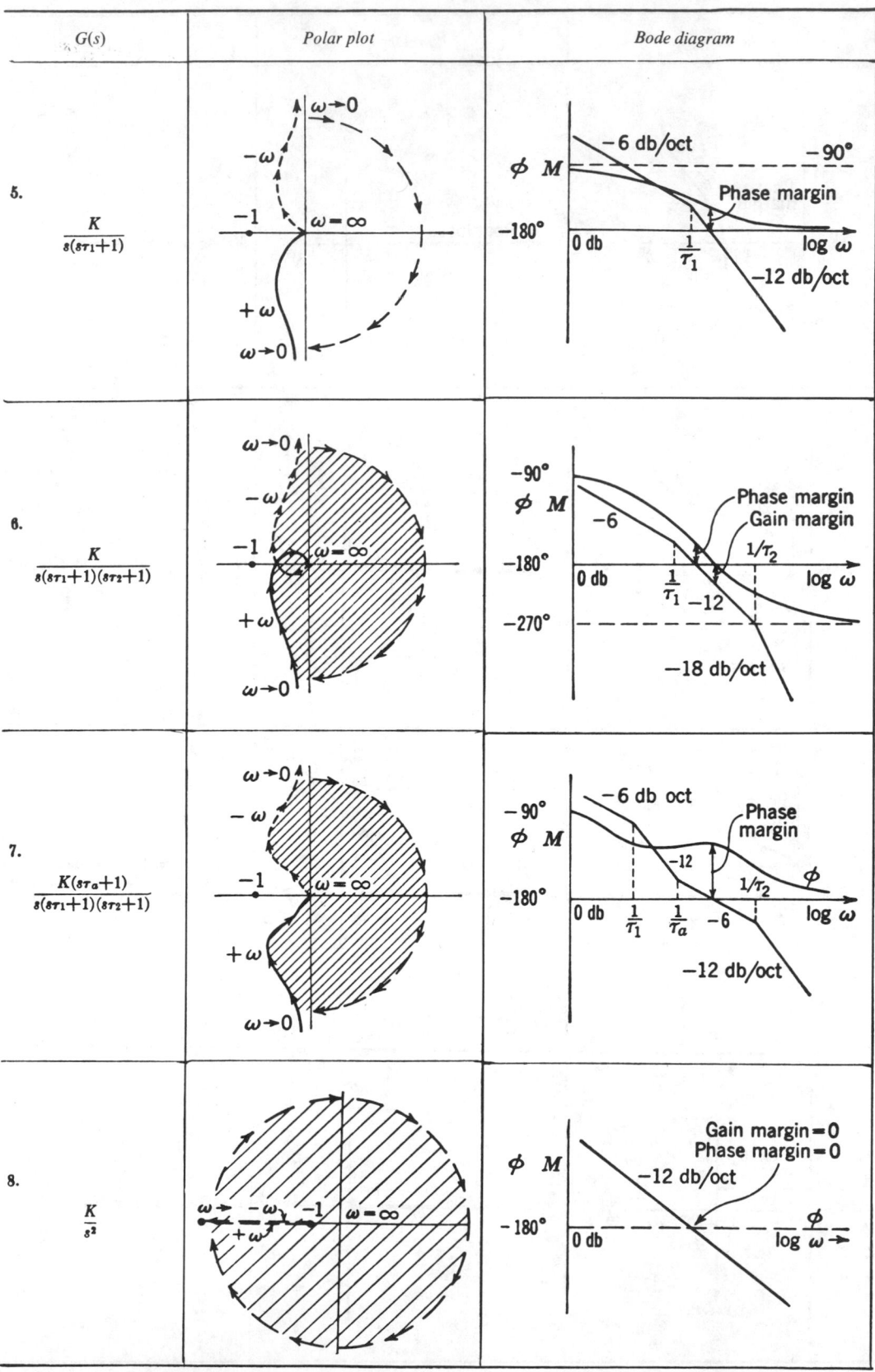

$G(s)$	Polar plot	Bode diagram
5. $\dfrac{K}{s(s\tau_1+1)}$		
6. $\dfrac{K}{s(s\tau_1+1)(s\tau_2+1)}$		
7. $\dfrac{K(s\tau_a+1)}{s(s\tau_1+1)(s\tau_2+1)}$		
8. $\dfrac{K}{s^2}$		

FOR TYPICAL TRANSFER FUNCTION *(Continued)*

Nichols diagram	Root locus	Comments
M; Phase margin; 0 db; $-180°$; $-90°$; ϕ; ω; $\omega \to \infty$	ω; R_1; $-\dfrac{1}{\tau_1}$; R_2; σ	Elementary instrument servo; inherently stable; gain margin $= \infty$
Phase margin; M; ω; 0 db; $-180°$; $-90°$; ϕ; Gain margin; $\omega \to \infty$	R_3; $\dfrac{1}{\tau_2}$; $-\dfrac{1}{\tau_1}$; R_1; R_2; ω; σ	Instrument servo with field-control motor or power servo with elementary Ward-Leonard drive; stable as shown, but may become unstable with increased gain
M; ω; Phase margin; 0 db; $-180°$; $-90°$; ϕ; $\omega \to \infty$	R_3; $-\dfrac{1}{\tau_2}$; $-\dfrac{1}{\tau_a}$; $-\dfrac{1}{\tau_1}$; R_1; R_2; ω; σ	Elementary instrument servo with phase-lead (derivative) compensator; stable
ω; M; Phase margin $= 0$; 0 db; $-270°$; $-180°$; $-90°$; ϕ; $\omega \to \infty$	Double pole; R_1; R_2; ω; σ	Inherently unstable; must be compensated

Table 11–28.   TRANSFER-FUNCTION PLOTS

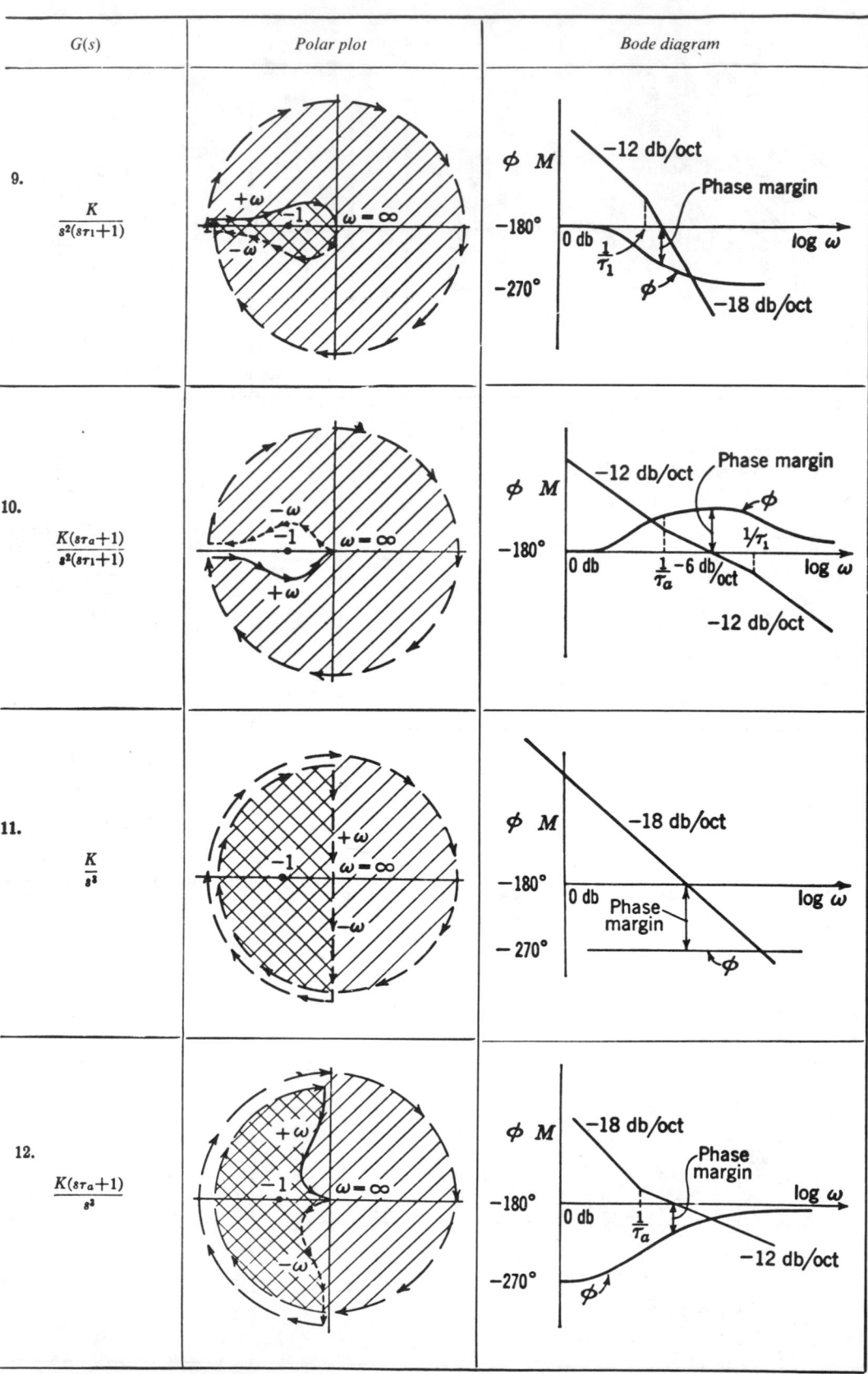

$G(s)$	Polar plot	Bode diagram
9. $\dfrac{K}{s^2(s\tau_1+1)}$		
10. $\dfrac{K(s\tau_a+1)}{s^2(s\tau_1+1)}$		
11. $\dfrac{K}{s^3}$		
12. $\dfrac{K(s\tau_a+1)}{s^3}$		

FOR TYPICAL TRANSFER FUNCTION (Continued)

Nichols diagram	Root locus	Comments
		Inherently unstable; must be compensated
		Stable for all gains
		Inherently unstable
		Inherently unstable

Table 11–28.   TRANSFER-FUNCTION PLOTS

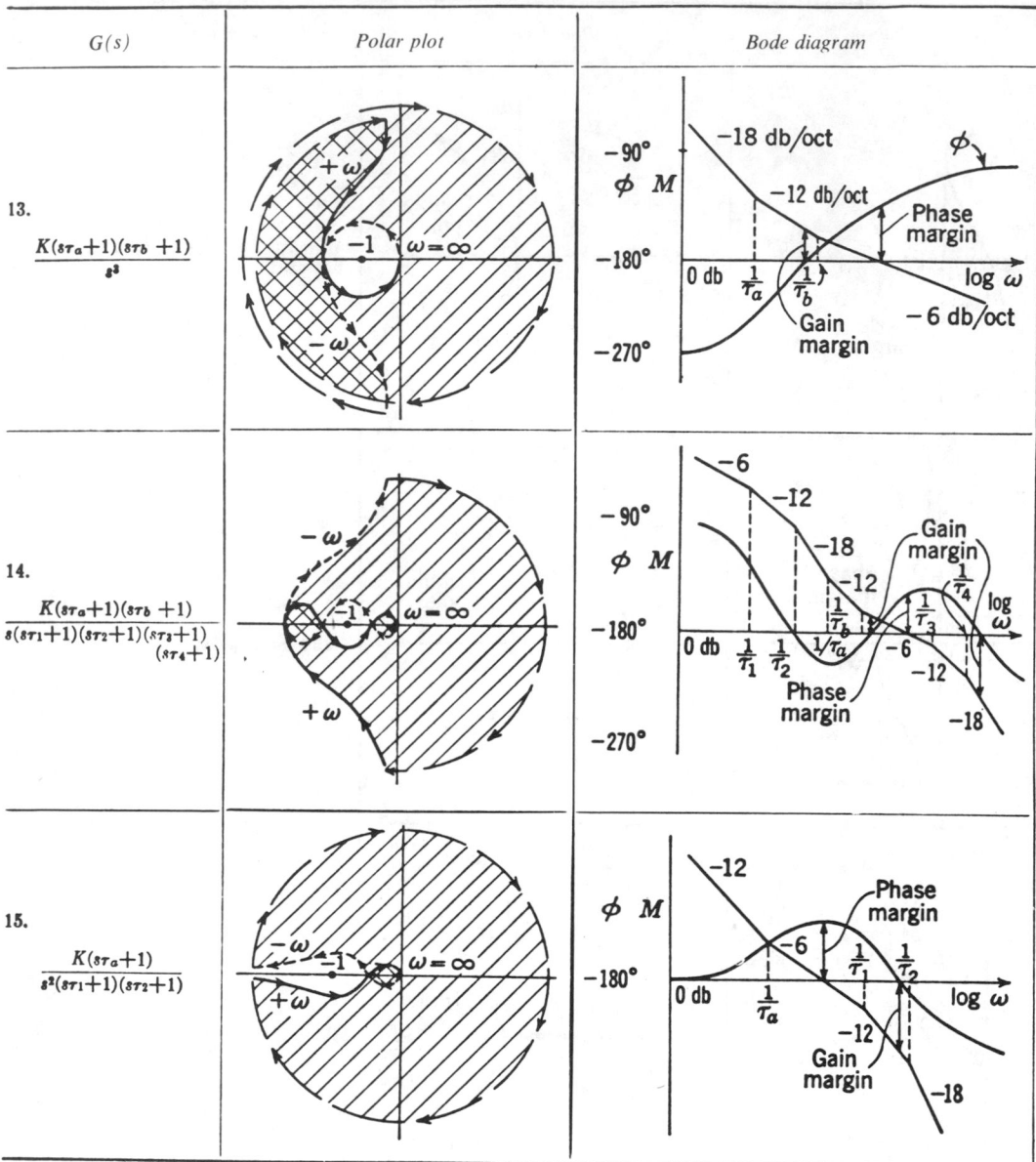

$G(s)$	Polar plot	Bode diagram
13. $\dfrac{K(s\tau_a+1)(s\tau_b+1)}{s^3}$		
14. $\dfrac{K(s\tau_a+1)(s\tau_b+1)}{s(s\tau_1+1)(s\tau_2+1)(s\tau_3+1)(s\tau_4+1)}$		
15. $\dfrac{K(s\tau_a+1)}{s^2(s\tau_1+1)(s\tau_2+1)}$		

*From: "Analysis and Design of Feedback Control Systems", 2nd ed., G.J. Thaler and R.G. Brown, McGraw-Hill Book Company, 1960.

FOR TYPICAL TRANSFER FUNCTION (*Continued*)

Nichols diagram	*Root locus*	*Comments*
		Conditionally stable; becomes unstable if gain is too low
		Conditionally stable; stable at low gain, becomes unstable as gain is raised, again becomes stable as gain is further increased, and becomes unstable for very high gains
		Conditionally stable; becomes unstable at high gain

Table 11–29. MECHANICAL COMPONENTS*

LEAD

Mechanical lead network	Log magnitude characteristic	Transfer function	T_1	T_2
(a) x_i, D_1, K_1, x_0	G_∞; $-20\ db/decade$; $\frac{1}{T_1}$; $G_\infty = 1$; $G_0 = 0$	$\dfrac{T_1 s}{T_1 s + 1}$	$\dfrac{D_1}{K_1}$	$\cdots$
(b) x_i, D_1, K_1, K_2, x_0	G_∞; $-20\ db/decade$; $\frac{1}{T_2}$; $\frac{1}{T_1}$; G_0; $G_\infty = 1$; $G_0 = \dfrac{1}{1+\dfrac{K_2}{K_1}}$	$G_0\,\dfrac{T_1 s + 1}{T_2 s + 1}$	$\dfrac{D_1}{K_1}$	$G_0 T_1$
(c) x_i, D_1, K_1, D_2, K_2, x_0	G_∞; $-20\ db/decade$; $-40\ db/decade$; $G_\infty = 1$; $G_0 = 0$	$\dfrac{T_1 T_2 s^2}{T_1 T_2 s^2 + \left[T_1 + \left(1 + \dfrac{K_2}{K_1}\right)T_2\right]s + 1}$	$\dfrac{D_1}{K_1}$	$\dfrac{D_2}{K_2}$

Table 11-29. MECHANICAL COMPONENTS (Continued)

LAG

	Mechanical lag network	Log magnitude characteristic	Transfer function	T_1	T_2
(a)		$G_0 = 1$ $G_\infty = 0$	$\dfrac{1}{T_1 s + 1}$	$\dfrac{D_1}{K_1}$	...
(b)		$G_0 = 1$ $G_\infty = \dfrac{1}{1 + D_2/D_1}$	$\dfrac{T_2 s + 1}{T_1 s + 1}$	$\dfrac{T_2}{G_\infty}$	$\dfrac{D_1}{K_1}$
(c)		$G_0 = 1$ $G_\infty = 0$	$\dfrac{1}{T_1 T_2 s^2 + [T_1 + (1 + \frac{K_2}{K_1}) T_2] s + 1}$	$\dfrac{D_1}{K_1}$	$\dfrac{D_2}{K_2}$

LAG-LEAD

	Mechanical lag-lead network	Log magnitude characteristic	Transfer function	T_1	T_2
		$G_0 = G_\infty = 1$ $G_1 = \dfrac{T_1 + T_2}{T_1 + (1 + \frac{K_2}{K_1}) T_2}$	$\dfrac{(T_1 s + 1)(T_2 s + 1)}{T_1 T_2 s^2 + [T_1 + (1 + \frac{K_2}{K_1}) T_2] s + 1}$	$\dfrac{D_1}{K_1}$	$\dfrac{D_2}{K_2}$

*From: "Handbook of Automation, Computation, and Control", E.M. Grabbe, S. Ramo, and D.E. Wooldridge, Eds., Vol. 1, John Wiley & Sons, 1958.

Table 11–30.   PNEUMATIC COMPENSATING COMPONENTS*

Approximate Relationships for High Loop Gain Controllers, $\varepsilon \ll 1$

LEAD

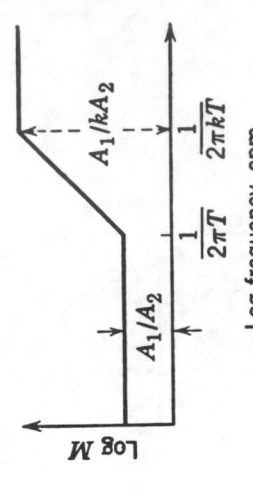

Plot for $T = T_1$

$$\frac{P_m - P_0}{P_c - P_r} = \frac{A_1}{A_2}\left[\frac{1 + T_1 s}{1 + k T_1 s}\right]$$

$$T_1 = \Re C$$

k = change in P_f for a unit change in P_m when $\Re$ is completely closed.

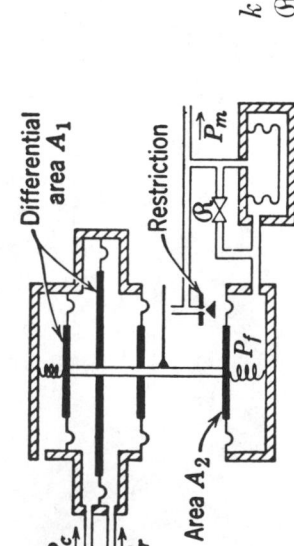

▲ = Pressure source

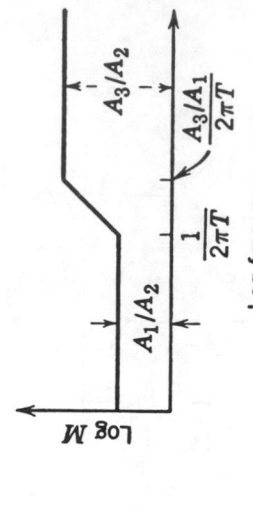

Plot for $T = T_1$

$$\frac{P_m - P_0}{P_c - P_r} = \frac{A_1}{A_2}\left[\frac{1 + (A_3/A_2)T_1 s}{1 + T_1 s}\right]$$

$$T_1 = \Re C$$

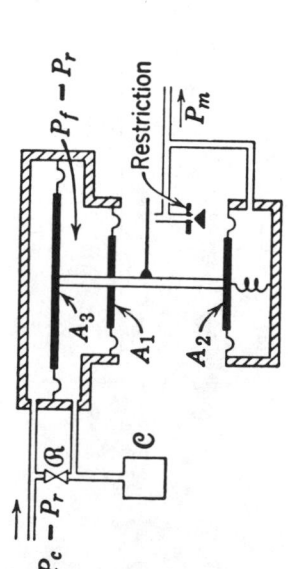

Table 11-30.   PNEUMATIC COMPENSATING COMPONENTS (*Continued*)

LAG

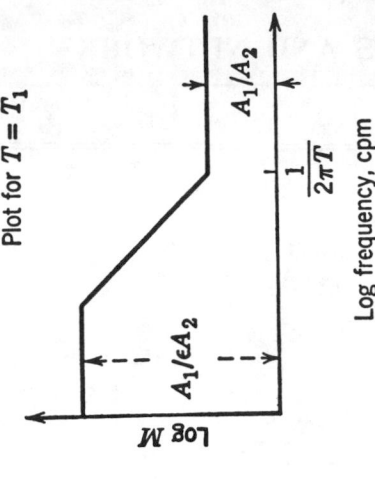

Plot for $T = T_1$

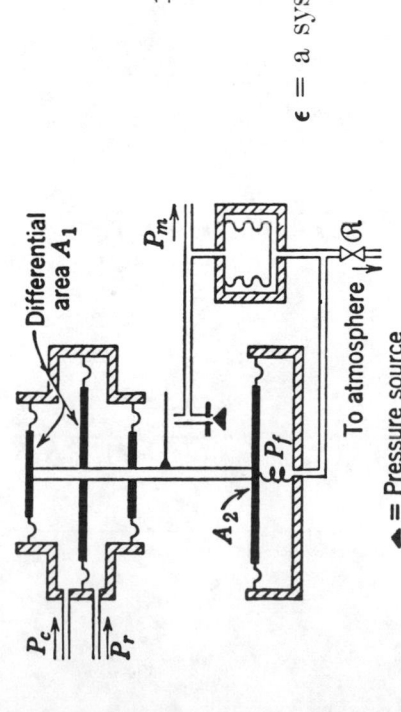

$$\frac{P_m - P_0}{P_c - P_r} = \frac{A_1}{A_2 k} \left[\frac{1 + 1/T_1 s}{1 + \epsilon/k T_1 s} \right]$$

$$T_1 = \mathcal{R}\mathcal{C}$$

ϵ = a system constant related to the loop gain.

LAG-LEAD

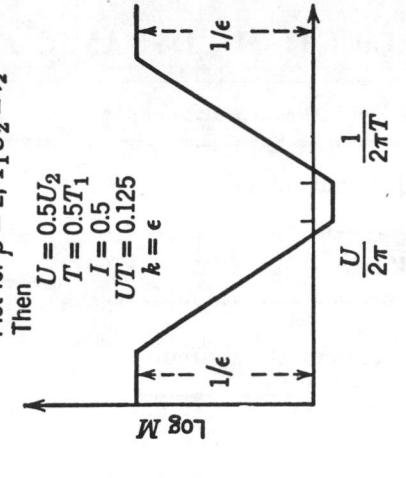

Plot for $\beta = 2$, $T_1 U_2 = \frac{1}{2}$

Then
$$U = 0.5 U_2$$
$$T = 0.5 T_1$$
$$I = 0.5$$
$$UT = 0.125$$
$$k = \epsilon$$

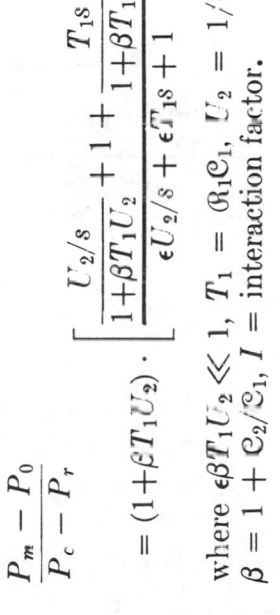

$$\frac{P_m - P_0}{P_c - P_r} =$$

$$= (1 + \epsilon T_1 U_2) \cdot \left[\frac{\dfrac{U_2/s}{1 + \beta T_1 U_2} + 1 + \dfrac{T_1 s}{1 + \beta T_1 U_2}}{\epsilon U_2/s + \epsilon T_1 s + 1} \right],$$

where $\epsilon \beta T_1 U_2 \ll 1$, $T_1 = \mathcal{R}_1 \mathcal{C}_1$, $U_2 = 1/\mathcal{R}_2 \mathcal{C}_2$, $\beta = 1 + \mathcal{C}_2/\mathcal{C}_1$, I = interaction factor.

*From: "Handbook of Automation, Computation, and Control", E.M. Grabbe, S. Ramo, and D.E. Wooldridge, Eds., Vol. 1, John Wiley & Sons, 1958, pp. 23-46 and 23-47.

Table 11–31.   DYNAMIC ELEMENTS AND NETWORKS*

Element or system	*G(s)*

1. Integrating circuit

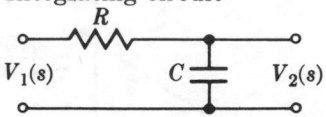

$$\frac{V_2(s)}{V_1(s)} = \frac{1}{RCS + 1}$$

2. Differentiating circuit

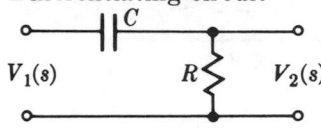

$$\frac{V_2(s)}{V_1(s)} = \frac{RCS}{RCS + 1}$$

3. Differentiating circuit

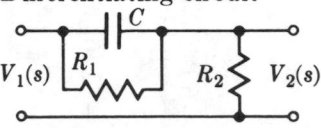

$$\frac{V_2(s)}{V_1(s)} = \frac{s + 1/R_1 C}{s + (R_1 + R_2)/R_1 R_2 C}$$

4. Lead-lag filter circuit

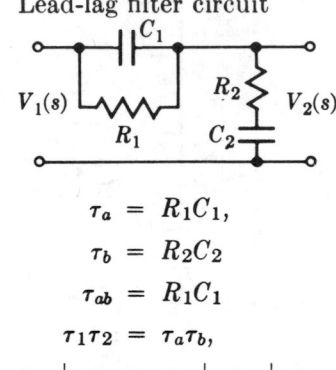

$$\tau_a = R_1 C_1,$$

$$\tau_b = R_2 C_2$$

$$\tau_{ab} = R_1 C_1$$

$$\tau_1 \tau_2 = \tau_a \tau_b,$$

$$\tau_1 + \tau_2 = \tau_a + \tau_b + \tau_{ab}$$

$$\frac{V_2(s)}{V_1(s)} = \frac{(1 + s\tau_a)(1 + s\tau_b)}{\tau_a \tau_b s^2 + (\tau_a + \tau_b + \tau_{ab})s + 1}$$

$$= \frac{(1 + s\tau_a)(1 + s\tau_b)}{(1 + s\tau_1)(1 + s\tau_2)}$$

5. dc-motor, field controlled

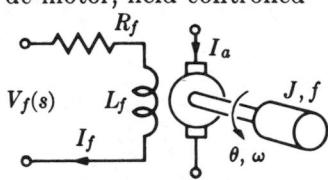

$$\frac{\theta(s)}{V_f(s)} = \frac{K_m}{s(Js + f)(L_f s + R_f)}$$

6. dc-motor, armature controlled

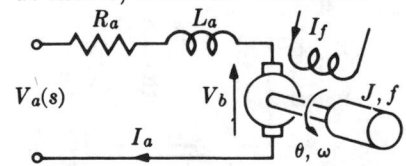

$$\frac{\theta(s)}{V_a(s)} = \frac{K_m}{s[(R_a + L_a s)(Js + f) + K_b K_m]}$$

*From: Richard C. Dorf, MODERN CONTROL SYSTEMS, 1967, Addison-Wesley, Reading, Mass.

Table 11–31. DYNAMIC ELEMENTS AND NETWORKS* *(Continued)*

Element or system	*G(s)*

7. ac-motor, two-phase control field

$$\frac{\theta(s)}{V_c(s)} = \frac{K_m}{s(\tau s + 1)}$$

$$\tau = J/(f - m)$$

m = slope of linearized torque-speed curve (normally negative)

8. Amplidyne

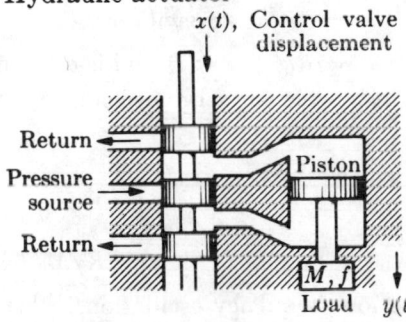

$$\frac{V_d(s)}{V_c(s)} = \frac{(K/R_c R_q)}{(s\tau_c + 1)(s\tau_q + 1)}$$

$$\tau_c = L_c/R_c, \quad \tau_q = L_q/R_q$$

For the unloaded case, $i_d \simeq 0$, $\tau_c \simeq \tau_q$, $0.05\ \text{sec} < \tau_c < 0.5\ \text{sec}$

9. Hydraulic actuator

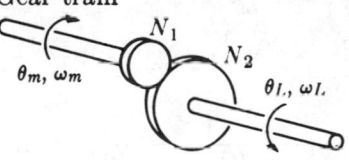

$x(t)$, Control valve displacement

Return
Pressure source
Return
Piston
M, f
Load $y(t)$

$$\frac{Y(s)}{X(s)} = \frac{K}{s(Ms + B)}$$

$$K = \frac{Ak_x}{k_P}, \quad B = \left(f + \frac{A^2}{k_P}\right)$$

$$k_x = \frac{\partial g}{\partial x}\bigg|_{x_0}, \quad k_P = \frac{\partial g}{\partial P}\bigg|_{P_0},$$

$$g = g(x, P)$$

A = area of piston

10. Gear train

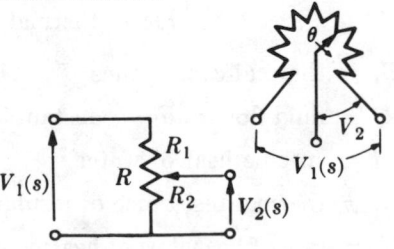

Gear ratio $= n = \dfrac{N_1}{N_2}$

$$N_2\theta_L = N_1\theta_m, \quad \theta_L = n\theta_m$$

$$\omega_L = n\omega_m$$

11. Potentiometer

$$\frac{V_2(s)}{V_1(s)} = \frac{R_2}{R} = \frac{R_2}{R_1 + R_2}$$

$$\frac{R_2}{R} = \frac{\theta}{\theta_{\max}}$$

*From: Richard C. Dorf, MODERN CONTROL SYSTEMS, 1967, Addison-Wesley, Reading, Mass.

Table 11–31. DYNAMIC ELEMENTS AND NETWORKS* *(Continued)*

Element or system	$G(s)$

12. Potentiometer error detector bridge

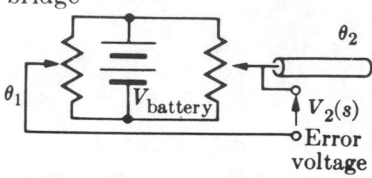

$$V_2(s) = k_s(\theta_1(s) - \theta_2(s))$$
$$V_2(s) = k_s\theta_{\text{error}}(s)$$
$$k_s = \frac{V_{\text{battery}}}{\theta_{\text{max}}}$$

13. Tachometer

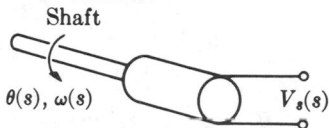

$$V_2(s) = K_t\omega(s)$$
$$= K_t s\theta(s)$$

14. dc-amplifier

$$\frac{V_2(s)}{V_1(s)} = \frac{k_a}{s\tau + 1}$$

R_0 = output resistance

C_0 = output capacitance

$\tau = R_0 C_0$, $\tau \ll 1$ and is often negligible for servomechanism amplifier

15. Accelerometer

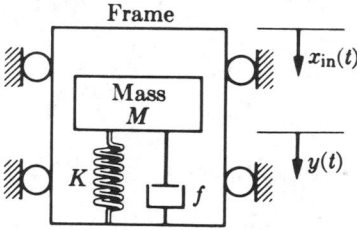

$$x_0(t) = y(t) - x_{\text{in}}(t),$$
$$\frac{X_0(s)}{X_{\text{in}}(s)} = \frac{-s^2}{s^2 + (f/M)s + K/M}$$

For low-frequency oscillations, where $\omega < \omega_n$,

$$\frac{X_0(j\omega)}{X_{\text{in}}(j\omega)} \simeq \frac{\omega^2}{K/M}$$

16. Thermal Heating System

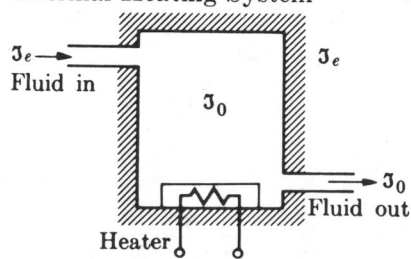

$$\frac{\Im(s)}{q(s)} = \frac{1}{C_t s + (QS + 1/R)}, \text{ where}$$

$\Im = \Im_0 - \Im_e$ = temperature difference due to thermal process

C_t = thermal capacitance

Q = fluid flow rate = constant

S = specific heat of water

R_t = thermal resistance of insulation

$q(s)$ = rate of heat flow of heating element

*From: Richard C. Dorf, MODERN CONTROL SYSTEMS, 1967, Addison-Wesley, Reading, Mass.

Engineering Organizations and Publishers

Table A-1. INFORMATION AND DATA CENTERS

Following is a partial list of centers organized for analysis and dissemination of engineering data in specific fields.

AEROSPACE RADIOISOTOPE POWER INFORMATION
CENTER (ARPIC)
Aerospace Nuclear Safety Department
Sandia Laboratory, Division 9319
Sandia Corporation
P.O. Box 5800
Albuquerque, N.Mex. 87115

ALLOY DATA CENTER
NBS Institute for Materials Research
Washington, D.C. 20234

ARGONNE NATIONAL LABORATORY
9700 South Cass Ave.
Argonne, Ill. 60439

ATOMIC AND MOLECULAR PROCESSES
INFORMATION CENTER
Oak Ridge National Laboratory
P.O. Box Y
Oak Ridge, Tenn. 37830

BROOKHAVEN NATIONAL LABORATORY
Upton, Long Island, N.Y. 11973

CHEMICAL PROPULSION INFORMATION
AGENCY (CPIA)
Applied Physics Laboratory
Johns Hopkins University
8621 Georgia Ave.
Silver Spring, Md. 20910

CHEMICAL THERMODYNAMICS DATA GROUP
NBS Institute for Basic Standards
Washington, D.C. 20234

CRYOGENICS DATA CENTER
National Bureau of Standards
Boulder, Colo. 80301

DEFENSE CERAMIC INFORMATION CENTER
Battelle Memorial Institute
Columbus Laboratories
505 King Ave.
Columbus, Ohio 43201

DEFENSE METALS INFORMATION CENTER
Battelle Memorial Institute
505 King Ave.
Columbus, Ohio 43201

ELECTRONIC PROPERTIES INFORMATION CENTER
Hughes Aircraft Center
Centinela and Teale Sts.
Culver City, Calif. 90230

ERIC CLEARINGHOUSE ON VOCATIONAL AND
TECHNICAL EDUCATION
The Ohio State University
980 Kinnear Rd.
Columbus, Ohio 43212

HIGH PRESSURE DATA CENTER
Brigham Young University
Provo, Utah 84601

ISOTOPES INFORMATION CENTER
Oak Ridge National Laboratory
P.O. Box X
Oak Ridge, Tenn. 37830

JET PROPULSION LABORATORY
California Institute of Technology
4800 Oak Grove Dr.
Pasadena, Calif. 91103

LIQUID METALS INFORMATION CENTER
Atomics International
P.O. Box 309
Canoga Park, Calif. 91304

NATIONAL OCEANOGRAPHIC DATA CENTER
Washington, D.C. 20390

NUCLEAR DESALINATION INFORMATION CENTER
Oak Ridge National Laboratory
P.O. Box Y
Oak Ridge, Tenn. 37830

NUCLEAR SAFETY INFORMATION CENTER
Oak Ridge National Laboratory
P.O. Box Y
Oak Ridge, Tenn. 37830

POWER INFORMATION CENTER
University of Pennsylvania
200 South 33rd St.
Philadelphia, Pa. 19104

RADIATION EFFECTS INFORMATION CENTER
Battelle Memorial Institute
Columbus Laboratories
505 King Ave.
Columbus, Ohio 43201

RADIATION SHIELDING INFORMATION CENTER
Oak Ridge National Laboratory
P.O. Box X
Oak Ridge, Tenn. 37830

REACTOR PHYSICS CONSTANTS CENTER
Argonne National Laboratory
9700 South Cass Ave.
Argonne, Ill. 60439

SCRIPPS INSTITUTION OF OCEANOGRAPHY
8602 La Jolla Shores Dr.
La Jolla, Calif. 92037

SIMULATED ENVIRONMENTS INFORMATION
CENTER
Technical Library
Sandia Laboratory
P.O. Box 5800
Albuquerque, N.Mex. 87115

SUPERCONDUCTIVE MATERIALS DATA CENTER
General Electric Research and Development Center
Schenectady, N.Y. 12301

THERMODYNAMIC PROPERTIES OF METALS AND
ALLOYS
Hearst Mining Building
Lawrence Radiation Laboratory
University of California
Berkeley, Calif. 94720

THERMODYNAMICS RESEARCH CENTER
Department of Chemistry
Texas A & M University
College Station, Texas 77840

THERMOPHYSICAL PROPERTIES RESEARCH CENTER
Purdue University
2595 Yeager Rd.
West Lafayette, Ind. 47906

REFERENCES

"A Directory of Information Resources in the United States: Federal Government", National Referral Center for Science and Technology, Library of Congress, June 1967.

"Directory of Special Libraries and Information Centers", 2nd ed., A.T. Kruzas, Ed., Gale Research Co., 1968.

"Research Centers Directory", 3rd ed., Gale Research Co., 1968.

"Standards and Specifications Information Sources", E.I. Struglia, Gale Research Co., 1965.

Further information also can be obtained from Information Research Center, Battelle Memorial Institute, Columbus Laboratories, 505 King Ave., Columbus, Ohio 43201; and Science Information Exchange, Smithsonian Institution, 300 Madison National Bank Building, 1730 M St., N.W., Washington, D.C. 20036.

TABLE A-2. U.S. GOVERNMENT SOURCES OF TECHNICAL DATA

Organizations, Offices, and Addresses

AEROSPACE MATERIALS INFORMATION CENTER
Air Force Materials Laboratory
AFSC Research and Technology Division
Wright-Patterson Air Force Base
Dayton, Ohio 45433

AIR FORCE SYSTEMS COMMAND
Technical Information Library
Aeronautical Systems Division
Building 12
Wright-Patterson Air Force Base
Dayton, Ohio 45433

ATOMIC ENERGY COMMISSION (AEC)
Division of Technical Information
Oak Ridge, Tennessee 37830

BUILDING RESEARCH DIVISION
National Bureau of Standards
Washington, DC 20234

BUREAU OF LABOR STATISTICS
441 G St., N.W.
Washington, DC 20212

BUREAU OF MINES
U.S. Department of the Interior
18th and C Sts., N.W.
Washington, DC 20240

BUREAU OF RECLAMATION
U.S. Department of the Interior
Office of the Chief Engineer
Denver Federal Center, Building 53
Denver, Colorado 80225

BUSINESS AND DEFENSE SERVICES
ADMINISTRATION
U.S. Department of Commerce
14th St. and Constitution Ave., N.W.
Washington, DC 20230

CENTER FOR COMPUTER SCIENCES AND
TECHNOLOGY
National Bureau of Standards
Washington, DC 20234

CIVIL AERONAUTICS BOARD
1825 Connecticut Ave., N.W.
Washington, DC 20428

COAL RESEARCH OFFICE
U.S. Department of the Interior
Washington, DC 20240

COAST GUARD
1300 E St., N.W.
Washington, DC 20226

COUNCIL ON ENVIRONMENTAL QUALITY
Executive Office of the President
622 Jackson Place, N.W.
Washington, DC 20036

DEFENSE DOCUMENTATION CENTER
Defense Supply Agency
U.S. Department of Defense
Cameron Station
Alexandria, Virginia 22314

DEPARTMENT OF AGRICULTURE
Office of Information
14th St. and Independence Ave., S.W.
Washington, DC 20250

DEPARTMENT OF HOUSING AND URBAN
DEVELOPMENT
1626 K St., N.W.
Room 102
Washington, DC 20410

DEPARTMENT OF TRANSPORTATION
400 7th St., S.W.
Washington, DC 20590

ELECTRONIC INSTRUMENTATION DIVISION
National Bureau of Standards
Washington, DC 20234

EMPLOYMENT STANDARDS ADMINISTRATION
U.S. Department of Labor
14th St. and Constitution Ave., N.W.
Washington, DC 20210

ENVIRONMENTAL DATA SERVICE
6010 Executive Blvd.
Rockville, Maryland 20852

ENVIRONMENTAL PROTECTION AGENCY
1626 K St., N.W.
Washington, DC 20460

ENVIRONMENTAL SCIENCE SERVICES
ADMINISTRATION
Washington Science Center
Rockville, Maryland 20852

FEDERAL AVIATION ADMINISTRATION
Office of Information Services
800 Independence Ave., S.W.
Washington, DC 20553

FEDERAL COMMUNICATIONS COMMISSION
Office of Chief Engineer
521 12th St., N.W.
Room 805
Washington, DC 20555

FEDERAL HIGHWAY ADMINISTRATION
U.S. Department of Transportation
Washington, DC 20591

FEDERAL POWER COMMISSION
Office of Public Information
441 G St., N.W.
Washington, DC 20426

FEDERAL RADIATION COUNCIL
1800 G St., N.W.
Washington, DC 20449

FOREST SERVICE
Division of Information and Education
14th St. and Independence Ave., S.W.
Washington, DC 20250

GEOLOGICAL SURVEY
U.S. Department of the Interior
18th and F Sts., N.W.
Washington, DC 20242

GOVERNMENT PRINTING OFFICE
North Capitol and H Sts.
Washington, DC 20401

LIBRARY OF CONGRESS
Science and Technology Division
1st St. and Independence Ave., S.E.
Washington, DC 20540

MARITIME ADMINISTRATION
Office of Research and Development
441 G St., N.W.
Washington, DC 20235

NATIONAL ACADEMY OF ENGINEERING
2101 Constitution Ave., N.W.
Washington, DC 20418

NATIONAL ACADEMY OF SCIENCES
2101 Constitution Ave., N.W.
Washington, DC 20418

NATIONAL AERONAUTICS AND SPACE
ADMINISTRATION (NASA)
Scientific and Technical Information Division
300 7th St., S.W.
Washington, DC 20546

NATIONAL BUREAU OF STANDARDS
U.S. Department of Commerce
Washington, DC 20234

NATIONAL INSTITUTES OF HEALTH
Bethesda, Maryland 20014

NATIONAL LIBRARY OF MEDICINE
8600 Rockville Pike
Bethesda, Maryland 20014

NATIONAL OCEANIC AND ATMOSPHERIC
ADMINISTRATION
U.S. Department of Commerce
Washington, DC 20235

NATIONAL REFERRAL CENTER FOR SCIENCE AND
TECHNOLOGY
Library of Congress
1st St. and Independence Ave., S.E.
Washington, DC 20540

NATIONAL SCIENCE FOUNDATION (NSF)
1800 G St., N.W.
Washington, DC 20550

NATIONAL TECHNICAL INFORMATION SERVICE
Springfield, Virginia 22151

NATIONAL WEATHER SERVICE
U.S. Department of Commerce
Washington, DC 20235

OCCUPATIONAL SAFETY AND HEALTH
ADMINISTRATION
1825 K St., N.W.
Washington, DC 20006

OFFICE OF AEROSPACE RESEARCH
1400 Wilson Blvd.
Arlington, Virginia 22209

OFFICE OF CIVIL DEFENSE – RESEARCH
The Pentagon
Washington, DC 20310

OFFICE OF THE CHIEF OF ENGINEERS
U.S. Department of the Army
Washington, DC 20315

OFFICE OF THE FEDERAL REGISTER
National Archives and Records Service
Pennsylvania Ave. and 8th St., N.W.
Washington, DC 20408

OFFICE OF NAVAL RESEARCH
Naval Research Laboratory
Washington, DC 20390

OFFICE OF SCIENCE AND TECHNOLOGY
Executive Office Bldg.
17th St. and Pennsylvania Ave., N.W.
Washington, DC 20506

OFFICE OF STANDARD REFERENCE DATA
National Bureau of Standards
Washington, DC 20234

OFFICE OF STANDARD REFERENCE MATERIALS
National Bureau of Standards
Washington, DC 20234

PATENT OFFICE
Office of Information Services
14th and E Sts., N.W.
Washington, DC 20231

PUBLIC HEALTH SERVICE
Occupational Health Program
1014 Broadway
Cincinnati, Ohio 45202

RURAL ELECTRIFICATION ADMINISTRATION
U.S. Department of Agriculture
14th St. and Independence Ave., S.W.
Washington, DC 20250

SMITHSONIAN INSTITUTION
Science Information Exchange
1730 M St., N.W.
Washington, DC 20036

SUPERINTENDENT OF DOCUMENTS
U.S. Government Printing Office
North Capitol St., between G and H Sts., N.W.
Washington, DC 20402

TECHNOLOGY TRANSFER OFFICE
Environmental Protection Agency
Washington, DC 20460

TENNESSEE VALLEY AUTHORITY (TVA)
Director of Information
New Sprankle Bldg.
Knoxville, Tennessee 37902

WATER INDUSTRIES DIVISION
U.S. Department of Commerce
14th St. and Constitution Ave., N.W.
Washington, DC 20230

"U.S. Government Organization Manual", annual publication, Office of the Federal Register, General Services Administration, Washington, DC 20408.
"Catalog of U.S. Government Publications", monthly publication, Government Printing Office.
Both publications are available from the Superintendent of Documents.

Table A-3.   ENGINEERING STANDARDS ORGANIZATIONS

ACOUSTICAL SOCIETY OF AMERICA (AcSoc)

AMERICAN ASSOCIATION OF STATE HIGHWAY OFFICIALS (AASHO)

AMERICAN ASSOCIATION OF TEXTILE CHEMISTS AND COLORISTS (AATCC)

AMERICAN CONCRETE INSTITUTE (ACI)

AMERICAN GAS ASSOCIATION (AGA)

AMERICAN GEAR MANUFACTURERS ASSOCIATION (AGMA)

AMERICAN INSTITUTE OF AERONAUTICS AND ASTRONAUTICS (AIAA)

AMERICAN INSTITUTE OF ARCHITECTS (AIA)

AMERICAN INSTITUTE OF CHEMICAL ENGINEERS (AIChE)

AMERICAN INSTITUTE OF ELECTRICAL ENGINEERS
See Institute of Electrical and Electronics Engineers

AMERICAN IRON AND STEEL INSTITUTE (AISI)

AMERICAN NATIONAL STANDARDS INSTITUTE, INC. (ANSI)
Formerly United States of America Standards Institute (USASI)—1966-1969; American Standards Association (ASA)—before 1966

AMERICAN PETROLEUM INSTITUTE (API)

AMERICAN RAILWAY ENGINEERING ASSOCIATION (AREA)

AMERICAN SOCIETY FOR METALS (ASM)

AMERICAN SOCIETY FOR QUALITY CONTROL (ASQC)

AMERICAN SOCIETY FOR TESTING AND MATERIALS (ASTM)

AMERICAN SOCIETY OF AGRICULTURAL ENGINEERS (ASAE)

AMERICAN SOCIETY OF CIVIL ENGINEERS (ASCE)

AMERICAN SOCIETY OF HEATING, REFRIGERATING AND AIR-CONDITIONING ENGINEERS (ASHRAE)

AMERICAN SOCIETY OF MECHANICAL ENGINEERS (ASME)

AMERICAN SOCIETY OF TOOL AND MANUFACTURING ENGINEERS (ASTME)

AMERICAN STANDARDS ASSOCIATION (ASA)
See American National Standards Institute, Inc. (ANSI)

AMERICAN VACUUM SOCIETY (AVS)

AMERICAN WATER WORKS ASSOCIATION (AWWA)

AMERICAN WELDING SOCIETY (AWS)

ASSOCIATION OF AMERICAN RAILROADS (AAR)

BRITISH STANDARDS INSTITUTION (BSI)

CANADA STANDARDS ASSOCIATION (CSA)

COMPRESSED GAS ASSOCIATION (CGA)

EDISON ELECTRIC INSTITUTE (EEI)

ELECTRONIC INDUSTRIES ASSOCIATION (EIA)

FEDERATION OF SOCIETIES FOR PAINT TECHNOLOGY (FSPT)

FLUID CONTROLS INSTITUTE

GRAY IRON FOUNDERS SOCIETY

HEAT EXCHANGE INSTITUTE

HYDRAULIC INSTITUTE

ILLUMINATING ENGINEERING SOCIETY (IES)

INSTITUTE OF ELECTRICAL AND ELECTRONICS ENGINEERS (IEEE)
Formerly American Institute of Electrical Engineers

INSTITUTE OF PETROLEUM (LONDON) (IP)

INSTITUTE OF TRAFFIC ENGINEERS (ITE)

INSTRUMENT SOCIETY OF AMERICA (ISA)

INTERNATIONAL COMMISSION ON RULES FOR THE APPROVAL OF ELECTRICAL EQUIPMENT (CEE)

INTERNATIONAL ELECTROTECHNICAL COMMISSION (IEC)

INTERNATIONAL ORGANIZATION FOR STANDARDIZATION (ISO)

INTERNATIONAL SPECIAL COMMITTEE ON RADIO INTERFERENCE (CISPR)

INTERSTATE COMMERCE COMMISSION (ICC)

MANUFACTURING CHEMISTS ASSOCIATION

MECHANICAL POWER TRANSMISSION ASSOCIATION (MPTA)

NATIONAL ASSOCIATION OF CORROSION ENGINEERS (NACE)

NATIONAL BOARD OF FIRE UNDERWRITERS (NBFU)

NATIONAL BUREAU OF STANDARDS (NBS)

NATIONAL COAL ASSOCIATION

NATIONAL ELECTRICAL MANUFACTURERS ASSOCIATION (NEMA)

NATIONAL FIRE PROTECTION ASSOCIATION (NFPA)

NATIONAL FLUID POWER ASSOCIATION (NFPA)

PAN AMERICAN STANDARDS COMMISSION (COPANT)

RUBBER MANUFACTURERS ASSOCIATION

SCIENTIFIC APPARATUS MAKERS ASSOCIATION

SOCIETY FOR EXPERIMENTAL STRESS ANALYSIS (SESA)

SOCIETY OF AUTOMOTIVE ENGINEERS (SAE)

TECHNICAL ASSOCIATION OF THE PULP AND PAPER INDUSTRY (TAPPI)

UNDERWRITERS' LABORATORIES (UL)

UNITED STATES OF AMERICA STANDARDS INSTITUTE (USASI)
See American National Standards Institute (ANSI)

REFERENCES

For more complete listings of government, society, and trade association groups issuing standards and codes, see "Standards and Specifications Information Sources", and "Encyclopedia of Associations", both published by Gale Research Co., Book Tower, Detroit, Mich.

Table A-4. TECHNICAL PUBLISHERS

Following are the names and addresses of technical and scientific publishers cited in footnotes and/or references in this book. The titles of frequently cited journals are also listed with their respective publishers.

ACADEMIC PRESS
111 Fifth Ave.
New York, N.Y. 10003

ACOUSTICAL SOCIETY OF AMERICA
335 East 45th St.
New York, N.Y. 10017
Journal of the Acoustical Society of America
Noise Control; now titled *Sound*

ADDISON-WESLEY PUBLISHING COMPANY
Reading, Mass. 01867

AIR-CONDITIONING AND REFRIGERATION
INSTITUTE
1815 North Fort Myer Dr.
Arlington, Va. 22209

AMERICAN CHEMICAL SOCIETY
1155 Sixteenth St., N.W.
Washington, D.C. 20036
Chemical and Engineering News
Industrial and Engineering Chemistry

AMERICAN CONCRETE INSTITUTE
P.O. Box 4754
Detroit, Mich. 48219
Proceedings of ACI

AMERICAN INSTITUTE OF AERONAUTICS AND
ASTRONAUTICS
1290 Avenue of the Americas
New York, N.Y. 10019
Astronautics & Aeronautics
Journal of Spacecraft and Rockets

AMERICAN INSTITUTE OF CHEMICAL ENGINEERS
345 East 47th St.
New York, N.Y. 10017
AIChE Journal; formerly *Transactions of AIChE*
Chemical Engineering Progress

AMERICAN INSTITUTE OF PHYSICS
335 East 45th St.
New York, N.Y. 10017
Applied Physics Letters
Journal of Applied Physics
The Physical Review

AMERICAN NATIONAL STANDARDS INSTITUTE
10 East 40th St.
New York, N.Y. 10016

AMERICAN SOCIETY OF AGRICULTURAL
ENGINEERS
420 Main St.
St. Joseph, Mich. 49085

AMERICAN SOCIETY OF HEATING, REFRIGERATING
AND AIR-CONDITIONING ENGINEERS
United Engineering Center
345 East 47th St.
New York, N.Y. 10017
ASHRAE Journal

AMERICAN SOCIETY OF LUBRICATION ENGINEERS
838 Busse Hwy.
Park Ridge, Ill. 60068
Transactions of ASLE

AMERICAN SOCIETY OF MECHANICAL ENGINEERS
United Engineering Center
345 East 47th St.
New York, N.Y. 10017
Transactions of ASME

AMERICAN SOCIETY FOR TESTING AND MATERIALS
1916 Race St.
Philadelphia, Pa. 19103

AMERICAN WELDING SOCIETY
United Engineering Center
345 East 47th St
New York, N.Y. 10017

BUTTERWORTH & CO. (PUBLISHERS) LTD.
88 Kingsway, W.C.2
London, England

CAMBRIDGE UNIVERSITY PRESS
32 East 57th St.
New York, N.Y. 10022

THE CHEMICAL RUBBER CO.
18901 Cranwood Parkway
Cleveland, Ohio 44128

CHILTON COMPANY
Chestnut & 56th Sts.
Philadelphia, Pa. 19139
Automotive Industries
The Electronic Engineer

COMPRESSED GAS ASSOCIATION
500 Fifth Ave.
New York, N.Y. 10036

CONOVER-MAST PUBLICATIONS
205 East 42nd St.
New York, N.Y. 10017
Science & Technology
Space/Aeronautics

DOVER PUBLICATIONS
180 Varick St.
New York, N.Y. 10014

ELECTROCHEMICAL SOCIETY
30 East 42nd St.
New York, N.Y. 10017
Electrochemical Society Journal
Electrochemical Technology

FARADAY PRESS
84 Fifth Ave.
New York, N.Y. 10011

GENERAL RADIO COMPANY
22 Baker Ave.
West Concord, Mass. 01781

Table A-4.   TECHNICAL PUBLISHERS (Continued)

HAYDEN PUBLISHING CO.
850 Third Ave.
New York, N.Y. 10022
Electronic Design
Microwaves

ILIFFE BOOKS LTD.
42 Russell Square
London W.C.1, England

INDUSTRIAL RESEARCH PUBLICATIONS
Beverly Shores, Ind. 46301
Electro Technology
Industrial Research

INFOSEARCH LIMITED
207 Brondesbury Park
London N.W.2, England

ILLUMINATING ENGINEERING SOCIETY
United Engineering Center
345 East 47th St.
New York, N.Y. 10017

INSTITUTE OF ELECTRICAL AND ELECTRONICS
ENGINEERS
345 East 47th St.
New York, N.Y. 10017
IEEE Transactions
Proceedings of the IEEE

INSTRUMENT SOCIETY OF AMERICA
530 William Penn Place
Pittsburgh, Pa. 15219
*Instrumentation Technology (*formerly *ISA Journal)*
ISA Transactions

INTERNATIONAL TEXTBOOK COMPANY
Scranton, Pa. 18515

INTERSCIENCE BOOKS
See John Wiley & Sons

JET PROPULSION LABORATORY
California Institute of Technology
4800 Oak Grove Dr.
Pasadena, Calif. 91103

MACMILLAN COMPANY
866 Third Ave.
New York, N.Y. 10022

McGRAW-HILL BOOK COMPANY
330 West 42nd St.
New York, N.Y. 10036

McGRAW-HILL PUBLICATIONS
330 West 42nd St.
New York, N.Y. 10036
Chemical Engineering
Electronics
Product Engineering

MIT PRESS
Cambridge, Mass. 02142

NATIONAL ACADEMY OF SCIENCES
2101 Constitution Ave.
Washington, D.C. 20037

NATIONAL FIRE PROTECTION ASSOCIATION
60 Batterymarch St.
Boston, Mass. 02110

NATIONAL FOREST PRODUCTS ASSOCIATION
1619 Massachusetts Ave., N.W.
Washington, D.C. 20036

NATIONAL RESEARCH COUNCIL
See National Academy of Sciences

OPTICAL SOCIETY OF AMERICA
1155 Sixteenth St., N.W.
Washington, D.C. 20036
Applied Optics
Journal of the Optical Society of America

PENTON PUBLISHING COMPANY
1213 West Third St.
Cleveland, Ohio 44113
Machine Design

PERGAMON PUBLISHING COMPANY
Fairview Park
Elmsford, N.Y. 10523

PLENUM PUBLISHING CORPORATION
227 West 17th St.
New York, N.Y. 10011

PRENTICE-HALL
Englewood Cliffs, N.J. 07632

PRINCETON UNIVERSITY PRESS
Princeton, N.J. 08540

REINHOLD PUBLISHING CORPORATION
See Van Nostrand Reinhold Company

HOWARD W. SAMS & CO.
4300 West 62nd St.
Indianapolis, Ind. 46268

SOCIETY OF AUTOMOTIVE ENGINEERS
485 Lexington Ave.
New York, N.Y. 10017
SAE Journal
SAE Transactions

STANFORD RESEARCH INSTITUTE
333 Ravenswood Ave.
Menlo Park, Calif. 94025

UNIVERSITY OF CALIFORNIA PRESS
2223 Fulton St.
Berkeley, Calif. 94720

UNIVERSITY OF MICHIGAN PRESS
615 East University
Ann Arbor, Mich. 48106

VAN NOSTRAND REINHOLD COMPANY
450 West 33rd St.
New York, N.Y. 10001

JOHN WILEY & SONS
605 3rd Ave.
New York, N.Y. 10016

REFERENCES

"Scientific and Technical Societies of the United States", Publication 1499, National Academy of Sciences, 1968.
"SRDS Business Publication Rates and Data", Standard Rate & Data Service; published monthly.

Table A-5. ADDRESSES OF PROFESSIONAL SOCIETIES

U.S. and Canadian National Societies in the Engineering Field

AMERICAN ASSOCIATION FOR THE ADVANCEMENT
OF SCIENCE (AAAS)
    1515 Massachusetts Ave., N.W.
    Washington, D.C. 20005

AMERICAN CHEMICAL SOCIETY (ACS)
    1155 Sixteenth St., N.W.
    Washington, D.C. 20036

AMERICAN INSTITUTE OF AERONAUTICS AND
ASTRONAUTICS (AIAA)
    1290 Sixth Ave.
    New York, N.Y. 10019

AMERICAN INSTITUTE OF CHEMICAL ENGINEERS
(AIChE)
    345 East 47th St.
    New York, N.Y. 10017

AMERICAN INSTITUTE OF INDUSTRIAL ENGINEERS,
INC. (AIIE)
    345 East 47th St.
    New York, N.Y. 10017

AMERICAN INSTITUTE OF MINING,
METALLURGICAL, AND PETROLEUM ENGINEERS,
INC. (AIME)
    345 East 47th St.
    New York, N.Y. 10017

AMERICAN INSTITUTE OF PHYSICS (AIP)
    335 E. 45th St.
    New York, N.Y. 10017

AMERICAN INSTITUTE OF PLANT ENGINEERS
(AIPE)
    1056 Delta Ave.
    Cincinnati, Ohio 45208

AMERICAN NUCLEAR SOCIETY (ANS)
    244 East Ogden Ave.
    Hinsdale, Ill. 60521

AMERICAN RAILWAY ENGINEERING ASSOCIATION
(AREA)
    59 East Van Buren St.
    Chicago, Ill. 60605

AMERICAN SOCIETY OF AGRICULTURAL
ENGINEERS (ASAE)
    2950 Niles Rd.
    St. Joseph, Mich. 49085

AMERICAN SOCIETY OF CIVIL ENGINEERS (ASCE)
    345 East 47th St.
    New York, N.Y. 10017

AMERICAN SOCIETY FOR ENGINEERING
EDUCATION (ASEE)
    2100 Pennsylvania Ave., N.W.
    Washington, D.C. 20037

AMERICAN SOCIETY OF HEATING, REFRIGERATING
AND AIR-CONDITIONING ENGINEERS, INC.
(ASHRAE)
    345 East 47th St.
    New York, N.Y. 10017

AMERICAN SOCIETY OF LUBRICATION ENGINEERS
(ASLE)
    838 Busse Highway
    Park Ridge, Ill. 60068

THE AMERICAN SOCIETY OF MECHANICAL
ENGINEERS (ASME)
    345 East 47th St.
    New York, N.Y. 10017

AMERICAN SOCIETY FOR METALS (ASM)
    Metals Park, Ohio 44073

AMERICAN SOCIETY FOR QUALITY CONTROL
(ASQC)
    161 West Wisconsin Ave.
    Milwaukee, Wis. 53203

AMERICAN SOCIETY OF SAFETY ENGINEERS (ASSE)
    5 North Wabash Ave.
    Suite 1705
    Chicago, Ill. 60602

AMERICAN SOCIETY FOR TESTING AND
MATERIALS (ASTM)
    1916 Race St.
    Philadelphia, Pa. 19103

AMERICAN WELDING SOCIETY (AWS)
    2501 N.W. 7th St.
    Miami, Florida 33125

ASSOCIATION OF IRON AND STEEL ENGINEERS
(AISE)
    1010 Empire Building
    Pittsburgh, Pa. 15222

AUDIO ENGINEERING SOCIETY, INC. (AES)
    60 East 42nd St., Room 428
    New York, N.Y. 10017

CANADIAN COUNCIL OF PROFESSIONAL
ENGINEERS (CCPE)
    116 Albert St., Suite 401
    Ottawa 4, Ontario, Canada

CONFERENCE OF STATE SANITARY ENGINEERS
(CSSE)
    P.O. Box 60630
    New Orleans, La. 70160

CONSULTING ENGINEERS COUNCIL OF THE
UNITED STATES (CEC)
    1155 15th St., N.W.
    Washington, D.C. 20005

Table A-5.   ADDRESSES OF PROFESSIONAL SOCIETIES *(Continued)*

ELECTROCHEMICAL SOCIETY, INC. (ECS)
P.O. Box 2071
Princeton, New Jersey 08540

ENGINEERING INSTITUTE OF CANADA (EIC)
2050 Mansfield St.
Montreal 2, Quebec, Canada

ENGINEERS COUNCIL FOR PROFESSIONAL
DEVELOPMENT (ECPD)
345 East 47th St.
New York, N.Y. 10017

ENGINEERS JOINT COUNCIL (EJC)
345 East 47th St.
New York, N.Y. 10017

ILLUMINATING ENGINEERING SOCIETY (IES)
345 East 47th St.
New York, N.Y. 10017

INSTITUTE OF ELECTRICAL AND ELECTRONICS
ENGINEERS (IEEE)
345 East 47th St.
New York, N.Y. 10017

INSTITUTE OF TRAFFIC ENGINEERS (ITE)
1725 DeSales St., N.W.
Washington, D.C. 20036

INSTRUMENT SOCIETY OF AMERICA (ISA)
400 Stanwix St.
Pittsburgh, Pa. 15222

NATIONAL ACADEMY OF ENGINEERING (NAE)
2101 Constitution Ave., N.W.
Washington, D.C. 20418

NATIONAL ASSOCIATION OF CORROSION
ENGINEERS (NACE)
2400 West Loop South
Houston, Texas 77027

NATIONAL ASSOCIATION OF COUNTY ENGINEERS
(NACE)
1001 Connecticut Ave., N.W.
Washington, D.C. 20036

NATIONAL INSTITUTE OF CERAMIC ENGINEERS
(NICE)
65 Ceramic Drive
Columbus, Ohio 43214

NATIONAL SOCIETY OF PROFESSIONAL ENGINEERS
(NSPE)
2029 K St., N.W.
Washington, D.C. 20006

PAN AMERICAN FEDERATION OF ENGINEERING
SOCIETIES (UPADI)
Rincón 454, esc. 414
Montevideo, Uruguay

SOCIETY FOR ADVANCEMENT OF MANAGEMENT,
INC. (SAM)
16 West 40th St.
New York, N.Y. 10018

SOCIETY OF AMERICAN MILITARY ENGINEERS
(SAME)
The Fleming Building
800 17th St., N.W.
Washington, D.C. 20006

SOCIETY OF AUTOMOTIVE ENGINEERS (SAE)
485 Lexington Ave.
New York, N.Y. 10017

SOCIETY OF DIE CASTING ENGINEERS, INC. (SDCE)
16007 West 8 Mile Rd.
Detroit, Mich. 48235

SOCIETY OF FIRE PROTECTION ENGINEERS (SFPE)
60 Batterymarch St.
Boston, Mass. 02110

SOCIETY OF MANUFACTURING ENGINEERS (SME)
20501 Ford Rd.
Dearborn, Mich. 48128

SOCIETY OF MOTION PICTURE AND TELEVISION
ENGINEERS (SMPTE)
9 East 41st St.
New York, N.Y. 10017

SOCIETY OF NAVAL ARCHITECTS AND MARINE
ENGINEERS (SNAME)
74 Trinity Place
New York, N.Y. 10006

SOCIETY OF PACKAGING AND HANDLING
ENGINEERS (SPHE)
14 East Jackson Blvd.
Chicago, Ill. 60604

SOCIETY OF PLASTICS ENGINEERS, INC. (SPE)
65 Prospect St.
Stamford, Conn. 06902

SOCIETY OF WOMEN ENGINEERS (SWE)
345 East 47th St.
New York, N.Y. 10017

REFERENCES

For a 200-page listing of societies, including full data on each, see "Directory of Engineering Societies", Engineers Joint Council (345 East 47th St., New York, N.Y. 10017), 1968, 1972.

For related organizations see "Encyclopedia of Associations", 5th ed., Gale Research Co. (Book Tower, Detroit, Mich. 48226), 1968.

Table A-6. ENGINEERING SOCIETY MEMBERSHIP
Names, Initials, and Membership

The following lists include most of the engineering societies and federations of importance in the United States and Canada. These are primarily individual-membership organizations and federations of such groups. Several international federations are listed. Government organizations and industrial trade associations are not included.

These lists are based largely on the material presented in the 1968 and 1972 editions of the "Directory of Engineering Societies", published by the Engineers Joint Council. This 200-page directory is an invaluable source of information; for each organization it lists publications, library services, permanent officers, objectives, membership qualifications, and other data.

Unlike most other professions, engineering has no single organization to voice its viewpoint or to advance its causes. The strength of the profession lies in its many specialized societies. The EJC directory is one source of full information on the organization of the engineering profession. Two other general lists that also include most of the engineering societies are "Scientific and Technical Societies of the United States"† and "Encyclopedia of Associations".‡

Addresses of many of the national engineering societies are given in Table A-5.

ENGINEERING SOCIETIES AND RELATED ORGANIZATIONS

MEMBERSHIP KEY:

- *M* = individual members
- *F* = federated societies or organized groups, affiliated or supporting
- *C* = corporate, company, or institutional members

Table A. NATIONAL AND INTERNATIONAL ORGANIZATIONS

Name	*Official initials*	*Membership*
Acoustical Society of America	AcSoc	*M* 4,700
Aerospace Electrical Society	AES	*M* 540
Air Pollution Control Association	APCA	*M* 3,385; *C* 509
Airways Engineering Society	AES	*M* 500
American Association for the Advancement of Science	AAAS	*M* 128,000
American Association of Cost Engineers	AACE	*M* 2,100
The American Association of Petroleum Geologists	AAPG	*M* 15,230
American Association of Physicists in Medicine	AAPM	*M* 600
American Automatic Control Council	AACC	*F* 8
The American Ceramic Society, Inc.	Am. Cer. Soc.	*M* 7,100
American Chemical Society	ACS	*M* 109,000; *C* 300
American Concrete Institute	ACI	*M* 13,037; *C* 450
American Congress on Surveying and Mapping	ACSM	*M* 5,500
American Crystallographic Association	ACA	*M* 1,800
American Federation of Information Processing Societies	AFIPS	*F* 12
American Foundrymen's Society	AFS	*M* 14,000
American Geological Institute	AGI	*F* 116
The American Geophysical Union	AGU	*M* 10,800
American Industrial Hygiene Association	AIHA	*M* 1,525; *F* 125
American Institute of Aeronautics and Astronautics	AIAA	*M* 33,000
The American Institute of Architects	AIA	*M* 23,000
American Institute of Chemical Engineers	AIChE	*M* 38,000
American Institute of Consulting Engineers	AICE	*M* 402
American Institute of Industrial Engineers, Inc.	AIIE	*M* 22,000
American Institute of Mining, Metallurgical, and Petroleum Engineers, Inc.	AIME	*M* 48,000
American Institute of Physics	AIP	*M* 56,000; *F* 7
American Institute of Plant Engineers	AIPE	*M* 4,700
American Iron and Steel Institute	AISI	*M* 2,825; *F* 83
American Mathematical Society	AMS	*M* 14,000

† Publication 1499. National Academy of Sciences, 1968.

‡ "Gale Encyclopedia of Trade Associations", Gale Research Co., Book Tower, Detroit, Michigan, 48226, 1968.

Table A-6. ENGINEERING SOCIETY MEMBERSHIP *(Continued)*
Table A. NATIONAL AND INTERNATIONAL ORGANIZATIONS *(Continued)*

Name	Official initials	Membership
American Nuclear Society	ANS	*M* 8,000
American Ordnance Association	AOA	*M* 42,000
American Petroleum Institute	API	*M* 8,000
The American Physical Society	APS	*M* 28,000
American Public Health Association	APHA	*M* 24,000; *F* 149
American Public Works Association	APWA	*M* 10,200; *F* 650
American Railway Engineering Association	AREA	*M* 3,328
American Society for Engineering Education	ASEE	*M* 12,000; *C* 550
American Society for Metals	ASM	*M* 38,000
American Society for Nondestructive Testing	ASNT	*M* 6,000
American Society for Quality Control, Inc.	ASQC	*M* 24,000
American Society for Testing and Materials	ASTM	*M* 19,000
American Society of Agricultural Engineers	ASAE	*M* 6,400
American Society of Body Engineers	ASBE	*M* 625
American Society of Certified Engineering Technicians	ASCET	*M* 3,200
American Society of Civil Engineers	ASCE	*M* 65,000
American Society of Heating, Refrigerating and Air-Conditioning Engineers, Inc.	ASHRAE	*M* 28,000
American Society of Lubrication Engineers	ASLE	*M* 3,200; *C* 46
The American Society of Mechanical Engineers	ASME	*M* 69,000
American Society of Naval Engineers, Inc.	ASNE	*M* 3,831
American Society of Photogrammetry	ASP	*M* 6,000
American Society of Planning Officials	ASPO	*M* 8,200
American Society of Safety Engineers	ASSE	*M* 10,000
American Society of Sanitary Engineering	ASSE	*M* 2,300
American Society of Scientific & Engineering Translators	ASSET	*M* 100
American Statistical Association	ASA	*M* 10,000
American Vacuum Society	AVS	*M* 2,437
American Water Works Association, Inc.	AWWA	*M* 22,000
American Welding Society	AWS	*M* 26,000
Association for Computing Machinery	ACM	*M* 26,000
The Association of Asphalt Paving Technologists	AAPT	*M* 650
Association of Consulting Management Engineers, Inc.	ACME	*C* 45
Association of Federal Communications Consulting Engineers	AFCCE	*M* 105
Association of Iron and Steel Engineers	AISE	*M* 12,400
Association of Senior Engineers of the Naval Ship Systems Command	ASE	*M* 650
Atomic Industrial Forum, Inc.	AIF	*M* 1,825
Audio Engineering Society, Inc.	AES	*M* 3,000; *F* 32
Building Research Institute	BRI	*M* 1,750
The Combustion Institute		*M* 2,000
Conference of Engineering Societies of Western Europe and the United States of America	EUSEC	
Conference of State Sanitary Engineers	CSSE	*M* 94
Conference of State Utility Commission Engineers		*M* 80
Construction Specifications Institute	CSI	*M* 9,700
Consulting Engineers Council of the United States	CEC	*C* 2,500
Council of Engineering and Scientific Society Executives	CESSE	*M* 102
Data Processing Management Association	DPMA	*M* 24,500
The Electrochemical Society, Inc.	ECS	*M* 4,000; *F* 132
Engineering College Magazines Associated	ECMA	*M* 55
Engineering Foundation	EF	*F* 5
Engineers' Council for Professional Development	ECPD	*F* 13
Engineers Joint Council	EJC	*F* 38
Environmental Engineering Intersociety Board, Inc.	EEIB	*M* 1,072; *F* 7
Fluid Power Society	FPS	*M* 3,800; *C* 7
Forest Products Research Society	FPRS	*M* 4,300
The Geological Society of America	GSA	*M* 8,600
Heat Transfer and Fluid Mechanics Institute	HTFMI	*M* 3,000
Highway Research Board	HRB	*M* 1,700; *C* 200
Illuminating Engineering Society	IES	*M* 11,280

Table A-6.   ENGINEERING SOCIETY MEMBERSHIP *(Continued)*
Table A.   NATIONAL AND INTERNATIONAL ORGANIZATIONS *(Continued)*

Name	Official initials	Membership
Industrial Research Institute, Inc.	IRI	*C* 212
Institute for the Certification of Engineering Technicians	ICET	*M* 28,000
The Institute of Electrical & Electronics Engineers, Inc.	IEEE	*M* 160,000
Institute of Environmental Sciences	IES	*M* 2,530
The Institute of Management Sciences	TIMS	*M* 7,000
Institute of Traffic Engineers	ITE	*M* 2,518
Instrument Society of America	ISA	*M* 20,000
Insulated Power Cable Engineers Association	IPCEA	*M* 120
International Association for the Exchange of Students for Technical Experience–United States, Inc.	IAESTE–US	*F* 115
International Astronautical Federation	IAF	*F* 50
The International Commission on Illumination–U.S. National Committee	USNC/CIE	*M* 100
International Commission on Large Dams	ICOLD	*M* and *C* 1,300; *F* 58
International Conference on Large Electric Systems	CIGRE	*M* 3,000
International Council for Scientific Management	CIOS	*F* 123; *C* 133
International Electrotechnical Commission	IEC	*F* 40
International Federation Documentation	FID	*M* and *F* 265
International Federation for Documentation	IFAC	*F* 33
International Material Management Society	IMMS	*M* 5,000
International Organization for Standardization	ISO	*F* 56
Marine Technology Society	MTS	*M* 7,000
The Mathematical Association of America	MAA	*M* 18,000
The Metallurgical Society of AIME	TMS	*M* 10,407
MTM Association for Standards and Research	MTM ASSN.	*M* 50; *F* 220
National Academy of Engineering	NAE	*M* 350
National Academy of Sciences	NAS	*M* 800
National Association of Corrosion Engineers	NACE	*M* 7,100
National Association of County Engineers	NACE	*M* 1,050
National Association of Government Engineers	NAOGE	
National Association of Power Engineers, Inc.	NAPE	*M* 12,000
National Council of Engineering Examiners	NCEE	*F* 54
National Institute of Ceramic Engineers	NICE	*M* 1,650
National Institute of Packaging, Handling and Logistic Engineers	NIPHLE	*M* 500
National Research Council	NRC	*M* 410
National Society of Professional Engineers	NSPE	*M* 70,000
The Newcomen Society in North America		*M* 17,000
Operations Research Society of America	ORSA	*M* 7,200
Optical Society of America	OSA	*M* 6,000
Radio Technical Commission for Aeronautics	RTCA	
Scientific Manpower Commission	SMC	*M* 22; *F* 11
Scientific Research Society of America	RESA	*M* 15,000
Seismological Society of America	SSA	*M* 1,236; *C* 25
Society for Advancement of Management, Inc.	SAM	*M* 5,000
Society for Applied Spectroscope	SAS	*M* 3,600
Society for Experimental Stress Analysis	SESA	*M* 2,400; *C* 70
Society for General Systems Research	SGSR	*M* 1,000; *F* 2
Society for Industrial and Applied Mathematics	SIAM	*M* 3,700
The Society for the History of Technology	SHOT	*M* 1,400
Society of Aeronautical Weight Engineers, Inc.	SAWE	*M* 1,300
Society of Aerospace Material & Process Engineers	SAMPE	*M* 1,675
Society of American Military Engineers	SAME	*M* 27,000
Society of American Value Engineers, Inc.	SAVE	*M* 2,700; *F* 15
Society of Automotive Engineers	SAE	*M* 30,000
The Society of Die Casting Engineers, Inc.	SDCE	*M* 2,234
Society of Economic Geologists	SEG	*M* 1,300
Society of Fire Protection Engineers	SFPE	*M* 1,500

Table A-6.   ENGINEERING SOCIETY MEMBERSHIP *(Continued)*

Table A.   NATIONAL AND INTERNATIONAL ORGANIZATIONS *(Continued)*

Name	Official initials	Membership
Society of Manufacturing Equipment	SME	*M* 49,000
Society of Mining Engineers of AIME	SME	*M* 17,500
Society of Motion Picture and Television Engineers	SMPTE	*M* 6,400; *F* 200
The Society of Naval Architects and Marine Engineers	SNAME	*M* 8,500
Society of Packaging and Handling Engineers	SPHE	*M* 1,694; *F* 16
Society of Petroleum Engineers of AIME	SPE of AIME	*M* 18,000
Society of Photographic Scientists and Engineers	SPSE	*M* 3,500
Society of Photo-Optical Instrumentation Engineers	SPIE	*M* 1,500
Society of Plastics Engineers, Inc.	SPE	*M* 16,000
Society of Professional Well Log Analysts, Inc.	SPWLA	*M* 875
Society of Reproduction Engineers	SRE	*M* 4,150
Society of Rheology	SOR	*M* 780
Society for Technical Communication	STC	*M* 4,000
The Society of the Sigma Xi		*M* 111,000
Society of Women Engineers	SWE	*M* 1,050
Soil Science Society of America	SSSA	*M* 3,800
The Solar Energy Society		*M* 700
Standards Engineers Society	SES	*M* 1,395
Steel Founders' Society of America	SFSA	*F* 115
The Tau Beta Pi Association, Inc.	TBπ	*M* 165,000
Technical Association of the Pulp and Paper Industry	TAPPI	*M* 12,384; *C* 560
Theta Tau	θT	*M* 18,000
Triangle Fraternity		*M* 12,500
Union of International Engineering Organizations		*F* 18
United Engineering Trustees, Inc.	UET	*F* 5
United Inventors and Scientists Corporation of America	UISA	*M* 1,065
Volunteers for International Technical Assistance, Inc.	VITA	*M* 2,600
Water Pollution Control Federation	WPCF	*M* 17,500
World Power Conference	WPC	*M* 16; *F* 84

Table A-6. ENGINEERING SOCIETY MEMBERSHIP (Continued)

Table B. STATE AND LOCAL ORGANIZATIONS

State head-quarters	Name	Initials	Membership
AL	Alabama Society of Professional Engineers		M 650
AK	Alaska Society of Professional Engineers		M 200
AZ	Arizona Council of Engineering & Scientific Association	ACE & SA	F 24
AZ	Arizona Society of Professional Engineers		M 600
AR	Arkansas Society of Professional Engineers		M 530
CA	Associated Civil Engineers and Land Surveyors of Santa Clara County	ACELSCO	M 51
CA	Bay Counties Civil Engineers and Land Surveyors Association	BCA	M 92
CA	California Council of Civil Engineers and Land Surveyors		M 508
CA	California Legislative Council of Professional Engineers	CLCPE	F 17
CA	California Society of Professional Engineers		M 3,900
CA	Consulting Engineers Association of California	CEAC	C 230
CA	County Engineers Association of California	CEAC	M 203
CA	East Bay Council on Surveying and Mapping		M 160
CA	Engineers and Architects Institute	EAI	M 215
CA	Engineers Club of Fresno		M 101
CA	The Engineers Club of San Francisco		M 1,275
CA	Los Angeles Council of Engineering Societies	LACES	F 8
CA	Los Angeles Technical Societies Council	LATSC	F 39
CA	Professional Photogrammetrists of California	PPC	F 8
CA	Sacramento Engineers Club	SEC	M 242; F 22
CA	San Francisco Bay Area Engineering Council	SFBAEC	F 22
CA	Structural Engineers Association of Central California	SEAOCC	M 140
CA	Structural Engineers Association of Northern California	SEAONC	M 560
CA	Structural Engineers Association of Southern California	SEAOSC	M 1,004
CZ	Canal Zone Society of Professional Engineers		M 80
CO	Colorado Engineering Council	CEC	F 26
CO	Colorado Scientific Society		M 360
CO	Colorado Society of Engineers	CSE	M 1,000
CO	Professional Engineers of Colorado		M 830
CO	Pueblo Engineers Society		M 142
CO	The Rocky Mountain Association of Geologists	RMAG	M 1,100
CO	Structural Engineers Association of Colorado	SEAC	M 31
CT	The Connecticut Joint Federation Inc.	CJF	M 5,534
CT	Connecticut Society of Civil Engineers	CSCE	M 1,047
CT	Connecticut Society of Professional Engineers		M 1,030
CT	The Hartford Engineers Club, Inc.	HEC	M 417
DE	Delaware Council of Engineering Societies	DCES	F 9
DE	Delaware Society of Professional Engineers		M 380
DC	District of Columbia Council of Engineering & Architectural Societies	D.C. Council	M 15,000; F 35
DC	D.C. Society of Professional Engineers		M 680
DC	Philosophical Society of Washington		M 1,195
DC	Washington Society of Engineers	WSE	M 750
FL	Florida Engineering Society		M 1,780
GA	Georgia Architectural and Engineering Society	GA&ES	M 1,100
GA	Georgia Society of Professional Engineers		M 890
HI	Big Island Engineering Association	BIEA	M 59

Table A-6.   ENGINEERING SOCIETY MEMBERSHIP *(Continued)*

Table B.   STATE AND LOCAL ORGANIZATIONS *(Continued)*

State head-quarters	Name	Initials	Membership
HI	Hawaii Society of Professional Engineers		*M* 390
ID	Idaho Society of Professional Engineers		*M* 370
IL	Illinois Society of Professional Engineers		*M* 3,500
IL	The Western Society of Engineers	WSE	*M* 1,580
IN	Fort Wayne Engineers Club	FWEC	*M* 480
IN	Indiana Society of Professional Engineers		*M* 1,180
IA	Iowa Engineering Society		*M* 680
IA	Iowa Engineering Society—Cedar Rapids Chapter		*M* 200
IA	Marshalltown Engineers Club		*M* 88
IA	North Central Iowa Engineers' Club		*M* 70
IA	South-Eastern Iowa Engineers Club		*M* 48
KS	Kansas Engineering Center		*M* 900
KS	Wichita Professional Engineering Society	WPES	*M* 198
KY	The Louisville Engineering & Scientific Societies Council	LESSC	*M* 3,500
KY	Kentucky Society of Professional Engineers		*M* 660
LA	Louisiana Engineering Society	LES	*M* 2,000
ME	Maine Association of Engineers	MAE	*M* 450
ME	Maine Society of Professional Engineers		*M* 150
MD	Engineers Joint Council of Maryland	EJCM	*F* 3
MD	The Engineering Society of Baltimore, Inc.	ESB	*M* 1,450
MD	Maryland Association of Engineers, Inc.		*M* 600
MD	Maryland Institute of Metals	MIM	*M* 28; *F* 6
MD	Maryland Society of Professional Engineers		*M* 720
MA	Boston Society of Civil Engineers	BSCE	*M* 1,110
MA	Engineering Societies of New England, Inc.	ESNE	*F* 32
MA	Engineering Society of Western Massachusetts	ESWM	*M* 305
MA	Massachusetts Society of Professional Engineers		*M* 1,380
MI	Battle Creek Engineers' Club		*M* 243
MI	The Engineering Society of Detroit	ESD	*M* 7,800
MI	Grand Rapids Engineers Club	GREC	*M* 237
MI	Michigan Engineering Society	MES	*M* 700
MI	Michigan Society of Professional Engineers		*M* 2,150
MN	Duluth Engineers Club		*M* 200
MN	The Engineers' Club of Minneapolis		*M* 950
MN	Engineers Club of Northern Minnesota	ECNM	*M* 544
MN	Engineers Society of St. Paul	ESSP	*M* 171
MN	Mesabi Range Geological Society		*M* 47
MN	Minnesota Association of Consulting Engineers	MACE	*M* 55
MN	Minnesota Federation of Engineering Societies	MFES	*F* 14
MN	Minnesota Society of Professional Engineers		*M* 1,140
MS	Mississippi Society of Professional Engineers		*M* 770
MO	Engineers Club of Kansas City		*M* 546
MO	The Engineers' Club of St. Louis		*M* 432; *F* 464; *C* 1,662
MO	Missouri Society of Professional Engineers		*M* 2,530
MT	Montana Society of Engineers		*M* 250
NB	Engineers Club of Omaha		*M* 480
NB	Professional Engineers of Nebraska		*M* 670
NV	Nevada Society of Professional Engineers		*M* 260

Table A-6.   ENGINEERING SOCIETY MEMBERSHIP *(Continued)*

Table B.   STATE AND LOCAL ORGANIZATIONS *(Continued)*

State head-quarters	Name	Initials	Membership
NH	New Hampshire Society of Professional Engineers		
NJ	Association of Scientists and Professional Engineering Personnel	ASPEP	M 1,200
NJ	International Material Management Society, New Jersey Chapter, Inc.	IMMS, NJC, INC.	M 244
NJ	Montclair Society of Engineers	MSE	M 650
NJ	New Jersey Society of Professional Engineers		M 2,630
NM	Conference of Local Environmental Health Administrators	CLEHA	M 200
NM	New Mexico Society of Professional Engineers		M 610
NY	Albany Society of Engineers	ASE	M 250
NY	Chinese Institute of Engineers, New York, Inc.	CIE	M 450
NY	The Engineering Society of Buffalo, Inc.	ESB	M 302, F 24
NY	The Municipal Engineers of the City of New York	MECNY	M 600
NY	New York State Society of Professional Engineers		M 4,220
NY	Society of Engineers of Eastern New York		M 72
NC	North Carolina Society of Engineers	NCSE	M 1,222
NC	Professional Engineers of North Carolina		M 840
NC	Raleigh Engineers Club		M 241
ND	North Dakota Society of Professional Engineers		M 250
OH	The Cleveland Engineering Society	CES	M 3,107
OH	Cleveland Technical Societies Council	CTSC	F 62
OH	The Engineering Society of Cincinnati	ESC	M 1,782
OH	The Engineers Club of Dayton		M 1,302
OH	Ohio Society of Professional Engineers		M 4,300
OH	The Technical & Scientific Societies Council of Cincinnati	TSSC	F 29
OK	Engineering Club of Oklahoma City		M 228
OK	Engineers Club of Bartlesville	ECOB	M 200
OK	Engineers' Society of Tulsa, Inc.	EST	M 415
OK	Oklahoma Society of Professional Engineers		M 1,600
OR	Consulting Engineers Council of Oregon	CECO	M 76
OR	Engineers and Architects Council of Oregon	EACO	M 14
OR	Professional Engineers of Oregon		M 960
OR	Structural Engineers Association of Oregon, Inc.	SEAO	M 98
PA	The Engineering Society of York, Pa.	ESY	M 160; F 12
PA	Engineering and Technical Societies Council of Delaware Valley, Inc.	ETSCO	F 41
PA	The Engineers' Club of Philadelphia	EC	M 1,425
PA	Engineers' Society of Western Pennsylvania	ESWP	M 1,600
PA	Erie Engineering Societies Council	EESC	M 1,300
PA	Pennsylvania Society of Professional Engineers		M 3,730
PR	Puerto Rico Society of Professional Engineers		M 160
RI	Rhode Island Society of Professional Engineers		M 240
SC	South Carolina Society of Engineers	SCSE	M 620
SC	South Carolina Society of Professional Engineers		M 490
SD	South Dakota Engineering Society		M 220
TN	The Technical Society of Knoxville		M 284
TN	Tennessee Society of Professional Engineers		M 1,140
TX	The Technical Club of Dallas, Texas		M 129
TX	Texas Society of Professional Engineers		M 5,130

Table A-6.   ENGINEERING SOCIETY MEMBERSHIP *(Continued)*

Table B.   STATE AND LOCAL ORGANIZATIONS *(Continued)*

State head-quarters	Name	Initials	Membership
UT	Utah Society of Professional Engineers		M 220
VT	Vermont Society of Engineers	VSE	M 500
VT	Vermont Society of Professional Engineers		M 130
VA	Central Virginia Engineers Club	CVEC	M 120
VA	Virginia Society of Professional Engineers		M 1,340
WA	Engineers Club		M 300
WA	Washington Society of Professional Engineers		M 780
WV	West Virginia Society of Professional Engineers		M 640
WI	Engineers and Scientists of Milwaukee, Inc.	ESM	M 1,550
WI	Wisconsin Society of Professional Engineers		M 1,370
WY	Wyoming Association of Consulting Engineers and Surveyors	WACES	M 43; C 19
WY	Wyoming Engineering Society	WES	M 350
WY	Wyoming Society of Professional Engineers		M 150

Table A-7.   NATIONAL STANDARD REFERENCE DATA SYSTEM PUBLICATIONS

U.S. Department of Commerce, National Bureau of Standards,
Office of Standard Reference Data, Washington, D.C. 20234

The NSRDS provides the quantitative data of physical science, critically evaluated and compiled for convenience. The technical scope of the NSRDS is indicated by the principal categories of data-compilation projects now active or being planned: nuclear properties, atomic and molecular properties, thermodynamic and transport properties, solid state properties, chemical kinetics, colloid and surface properties, and mechanical properties.

The NSRDS-NBS series of publications is intended primarily to include evaluated reference data and critical reviews of long-term interest to the scientific and technical community.

Publications may be obtained from the Superintendent of Documents, U.S. Government Printing Office, Washington, D.C. 20402.†

PUBLICATIONS ISSUED IN THE NSRDS SERIES

NSRDS-NBS-1, "National Standard Reference Data System Plan of Operation", E.L. Brady and M.B. Wallenstein, 1964 (15 cents).

NSRDS-NBS-2, "Thermal Properties of Aqueous Uni-univalent Electrolytes", V.B. Parker, 1965 (45 cents).

NSRDS-NBS-3, "Selected Tables of Atomic Spectra, Atomic Energy Levels and Multiplet Tables, Sect. 1, Si II, Si III, Si IV", C.E. Moore, 1965 (35 cents).

NSRDS-NBS-3, "Selected Tables of Atomic Spectra, Atomic Energy Levels, and Multiplet Tables, Sect. 2, Si I", C.E. Moore, 1967 (20 cents).

NSRDS-NBS-3, "Selected Tables of Atomic Spectra, Atomic Energy Levels, and Multiplet Tables, Sect. 3, C I, C II, C III, C IV, C V, C VI", C.E. Moore, 1970 ($1.00).

NSRDS-NBS-3, "Selected Tables of Atomic Spectra, Atomic Energy Levels, and Multiplet Tables, Sect. 4, N IV, N V, N VI, N VII", C.E. Moore, 1971 ($0.55).

NSRDS-NBS-4, "Atomic Transition Probabilities, Vol. 1, Hydrogen Through Neon", W.L. Wiese, M.W. Smith, and B.M. Glennon, 1966 ($2.50).

NSRDS-NBS-5, "The Band Spectrum of Carbon Monoxide", P.H. Krupenie, 1966 (70 cents).

NSRDS-NBS-6, "Tables of Molecular Vibrational Frequencies, Part 1", T. Shimanouchi, 1967 (40 cents).

NSRDS-NBS-7, "High Temperature Properties and Decomposition of Inorganic Salts, Part 1, Sulfates", K.H. Stern and E.L. Weise, 1966 (35 cents).

NSRDS-NBS-8, "Thermal Conductivity of Selected Materials", R.W. Powell, C.Y. Ho, and P.E. Liley, 1966 ($1.00).

NSRDS-NBS-9, "Bimolecular Gas Phase Reactions", A.F. Trotman-Dickenson and G.S. Milne, 1967 ($2.00).

NSRDS-NBS-10, "Selected Values of Electric Dipole Moments for Molecules in the Gas Phase", R.D. Nelson, Jr., D.R. Lide, Jr., and A.A. Maryott, 1967 (40 cents).

NSRDS-NBS-11, "Tables of Molecular Vibrational Frequencies, Part 2", T. Shimanouchi, 1967 (30 cents).

NSRDS-NBS-12, "Tables for the Rigid Asymmetric Rotor: Transformation Coefficients From Symmetric to Asymmetric Bases and Expectation Values of P_z^2, P_z^4, and P_z^6", R.H. Schwendeman, 1968 (60 cents).

NSRDS-NBS-13, "Hydrogenation of Ethylene on Metallic Catalysts", J. Horiuti and K. Miyahara, 1968 ($1.00).

NSRDS-NBS-14, "X-Ray Wavelengths and X-Ray Atomic Energy Levels", J.A. Bearden, 1967 (40 cents).

NSRDS-NBS-15, "Molten Salts, Vol. 1, Electrical Conductance, Density, and Viscosity Data", G. Janz, F.W. Dampier, G.R. Lakshminarayanan, P.K. Lorenz, and R.P.T. Tomkins, 1968 ($3.00).

NSRDS-NBS-16, "Thermal Conductivity of Selected Materials, Part 2", C.Y. Ho, R.W. Powell, and P.E. Liley, 1968 ($2.00).

NSRDS-NBS-17, "Tables of Molecular Vibration Frequencies, Part 3", T. Shimanouchi, 1968 (30 cents).

NSRDS-NBS-18, "Critical Analysis of the Heat-Capacity Data of the Literature and Evaluation of Thermodynamic Properties of Copper, Silver, and Gold from 0 to 300 K", G.T. Furukawa, W.G. Saba, and M.L. Reilly, 1968 (40 cents).

NSRDS-NBS-19, "Thermodynamic Properties of Ammonia as an Ideal Gas", L. Haar, 1968 (20 cents).

NSRDS-NBS-20, "Gas Phase Reaction Kinetics of Neutral Oxygen Species", H.S. Johnston, 1968 (45 cents).

NSRDS-NBS-21, "Kinetic Data on Gas Phase Unimolecular Reactions", S.W. Benson and H.E. O'Neal, 1970 ($7.00).

NSRDS-NBS-22, "Atomic Transition Probabilities, Vol. II, Sodium Through Calcium", W.L. Wiese, M.W. Smith, and B.M. Miles, 1969 ($4.50).

NSRDS-NBS-23, "Partial Grotrian Diagrams of Astrophysical Interest", C.E. Moore and P.W. Merrill, 1968 (55 cents).

NSRDS-NBS-24, "Theoretical Mean Activity Coefficients of Strong Electrolytes in Aqueous Solutions from 0 to 100°C", Walter J. Hamer, 1968 ($4.25).

Table A-7.   NATIONAL STANDARD REFERENCE
DATA SYSTEM PUBLICATIONS *(Continued)*

NSRDS-NBS-25, "Electron Impact Excitation of Atoms", B.L. Moiseiwitsch and S.J. Smith, 1968 ($2.00).

NSRDS-NBS-26, "Ionization Potentials, Appearance Potentials, and Heats of Formation of Positive Ions", J.L. Franklin, J.G. Dillard, H.M. Rosenstock, J.T. Herron, K. Draxl, and F.H. Field, 1969 ($4.00).

NSRDS-NBS-27, "Thermodynamic Properties of Argon from the Triple Point to 300 K at Pressures to 1000 Atmospheres", A.L. Gosman, R.D. McCarty, and J.G. Hust, 1969 ($1.25).

NSRDS-NBS-28, "Molten Salts, Vol. 2.1: Electrochemistry, Vol. 2.2: Surface Tension Data", G.J. Janz, Chr. G.M. Dijkhuis, G.R. Lakshminarayanan, R.P.T. Tomkins, and J. Wong, 1969.

NSRDS-NBS-29, "Photon Cross Sections, Attenuation Coefficients, and Energy Absorption Coefficients from 10 keV to 100 GeV", J.H. Hubbell, 1969 (75 cents).

NSRDS-NBS-30, "High Temperature Properties and Decomposition of Inorganic Salts, Part 2. Carbonates", K. H. Stern and E.L. Weise, 1969 (45 cents).

NSRDS-NBS-31, "Bond Dissociation Energies in Simple Molecules", B. deB. Darwent, 1970 (55 cents).

NSRDS-NBS-32, "Phase Behavior in Binary and Multicomponent Systems at Elevated Pressures: *n*-Pentane and Methane-*n*-Pentane", V.M. Berry and B.H. Sage, 1970 (70 cents).

NSRDS-NBS-33, "Electrolytic Conductance and the Conductances of the Halogen Acids in Water", W.J. Hamer and H.J. DeWane, 1970 (50 cents).

NSRDS-NBS-34, "Ionization Potentials and Ionization Limits Derived from the Analyses of Optical Spectra", C.E. Moore, 1970 (75 cents).

NSRDS-NBS-36, "Critical Micelle Concentrations of Aqueous Surfactant Systems", P. Mukerjee and K.J. Mysels, 1971 ($3.75).

NSRDS-NBS-37, "JANAF Thermochemical Tables", 2nd ed., D.R. Stull and H. Prophet, 1971 ($9.75).

NSRDS-NBS-38, "Critical Review of Ultraviolet Photoabsorption Cross Sections for Molecules of Astrophysical and Aeronomic Interest", R.D. Hudson, 1971 ($1.00).

†Cost of requested publications must accompany order. No charge is made for postage to destinations in the United States and possessions, Canada, Mexico, and certain Central and South American countries. To other countries payments for documents must cover postage; therefore, one-fourth of the price of the publication should be added for postage.

Table A-8. SOURCES OF CRITICALLY EVALUATED DATA

The following is abstracted from *Continuing Numerical Data Projects, A Survey and Analysis.* This publication was prepared under the auspices of the Office of Critical Tables, National Academy of Sciences and National Research Council. Persons requesting further information and details regarding the projects listed may obtain the publication from Printing and Publishing Office, National Academy of Sciences, 2101 Constitution Avenue, Washington, D. C., 20418. Price $5.00.

THERMODYNAMIC PROJECTS

Selected Values of Chemical Thermodynamic Properties, NBS Circular 500, F. D. Rossini, D. D. Wagman, W. H. Evans, S. Levine, and I. Jaffe, Government Printing Office, Washington, D.C. 20402, 1952, iv + 1268 pp, $7.25 (out of print). Paperback reprint issued in two parts: I. Tables; and II. References; Government Printing Office, Washington, D.C. 20402, 1961 (out of print). Revisions of C500 are being issued provisionally as NBS Technical Notes, listed below.

Selected Values of Chemical Thermodynamic Properties

Part 1: Tables for the First Twenty-three Elements in the Standard Order of Arrangement, NBS Technical Note 270-1, D. D. Wagman, W. H. Evans, I. Halow, V. B. Parker, S. M. Bailey, and R. H. Schumm, U.S. Government Printing Office, Washington, D.C. 20402, October 1965, iv + 124 pp, $0.65.

Part 2: Tables for the Elements Twenty-four through Thirty-two in the Standard Order of Arrangement, NBS Technical Note 270–2, D. D. Wagman, W. H. Evans, I. Halow, V. B. Parker, S. M. Bailey, and R. H. Schumm, U.S. Government Printing Office, Washington, D.C. 20402, May 1966, iv + 62 pp, $0.40.

Selected Values of Properties of Hydrocarbons and Related Compounds, American Petroleum Institute Research Project 44, J. B. Zwolinski, dir., Texas A&M Research Foundation, College Station, Texas. As of June 1966 there were 2391 valid data sheets, complete set 6 volumes at $0.30 per sheet.

Selected Values of Properties of Chemical Compounds, Thermodynamics Research Center, B. J. Zwolinski, dir., Texas A&M Research Foundation, College Station, Texas. As of June 1966 there were 754 valid data sheets, complete set 2 volumes at $0.30 per sheet.

JANAF (Joint Army-Navy-Air Force) Thermochemical Tables, PB 168 370, D. R. Stull, dir., Clearinghouse for Federal Scientific and Technical Information, Springfield, Virg., 1965, 945 pp, $10.00.

First Addendum, PB 168 370–1, D. R. Stull, dir., Clearinghouse for Federal Scientific and Technical Information, Springfield, Virg., 1966, vii + 197 pp, $4.00 (Supplements 18, 19, 20, and 21).

Information on further supplements is available from Dr. D. R. Stull, dir., Thermal Research Laboratory, Dow Chemical Company, Midland, Michigan.

Contributions to the Data on Theoretical Metallurgy, I–XV, published as Bureau of Mines bulletins between 1932 and 1962. Many bulletins were superceded; six are revisions and Bulletin 601 is a reprinting of four previous bulletins and out of print for some years, but still in demand as essential in present-day research. The ones currently available are listed below.

Contributions to the Data on Theoretical Metallurgy

XIII High-Temperature Heat-Content, Heat Capacity, and Entropy Data for the Elements and Inorganic Compounds, K. K. Kelley, Bulletin 584 (Bureau of Mines), U.S. Government Printing Office, Washington, D.C. 20402, 1960, 232 pp, $1.25.

XIV Entropies of Elements and Inorganic Compounds, K. K. Kelley and E. G. King, Bulletin 592 (Bureau of Mines), U.S. Government Printing Office, Washington, D.C. 20402, 1961, 149 pp, $0.75.

XV A Reprint of Bulletins 383, 384, 393, and 406 (III, IV, V, and VII), Bulletin 601, Bureau of Mines (Publications Distribution Section, 4800 Forbes Avenue), Pittsburgh, Penna., 1962, 525 pp, single copies free.

383: The Free Energies of Vaporization and Vapor Pressures of Inorganic Substances, K. K. Kelley, 1935.

384: Metal Carbonates, Correlation and Applications of Thermodynamic Data, K. K. Kelley and C. T. Anderson, 1935.

393: Heats of Fusion of Inorganic Substances, K. K. Kelley, 1936.

406: The Thermodynamic Properties of Sulphur and Its Inorganic Compounds, K. K. Kelley, 1937.

Thermodynamic Properties of Chemical Substances, V. P. Glushko et al, eds, Vol. I, 1164 pp, and Vol. II, 916 pp, USSR Academy of Sciences, Moscow, 1962, approx $16.00.

Thermodynamic Constants of Substances, Handbook in 10 parts, V. P. Glushko, ed, USSR Academy of Sciences, All-Union Institute of Scientific and Technological Information, Moscow: Part I, 1965, 146 pp, 72 kopecks ($1.75); Part II, 1966, 96 pp, 52 kopecks ($1.40). Available in the United States from Victor Kamkin, Inc., Bookstore, 1410 Columbia Rd. N.W., Washington, D.C. 20009.

Chemical Thermodynamics in Non-Ferrous Metallurgy, J. I. Gerassimov, A. N. Krestovnikov, and A. S. Shakhov, Metallurgical Publishing House, Moscow.

Vol. I: Theoretical Introduction, Thermodynamic Properties of Important Gases, Thermodynamics of Zinc and Its Important Compounds, 1960, 231 pp.

Table A-8. SOURCES OF CRITICALLY EVALUATED DATA *(Continued)*

THERMODYNAMIC PROJECTS (Continued)

Vol. II: Thermodynamics of Copper, Lead, Tin, Silver, and Their Important Compounds, 1960, 231 pp.

Vol. III: Thermodynamics of Tungsten, Molybdenum, Titanium, Zirconium, Niobium, Tantalum, and Their Important Compounds, 1963, 283 pp. English translation available as NASA-TT-F-285 or CFSTI-TT-65-50111, $6.00, from the Clearinghouse for Federal Scientific and Technical Information, Springfield, Virg. 22151.

Vol. IV: Thermodynamics of Aluminum, Antimony, Magnesium, Nickel, Bismuth, Cadmium, and Their Important Compounds, 1966, 428 pp.

Selected Values for the Thermodynamic Properties of Metals and Alloys, Ralph Hultgren, Raymond L. Orr, Philip D. Anderson, and Kenneth K. Kelley, John Wiley & Sons, Inc., New York, N.Y. 10016, 1963, xi + 963 pp, $12.50.

Thermochemistry for Steelmaking, Vol. I, J. F. Elliott and M. Gleiser, Addison-Wesley Publishing Co., Reading, Mass. 01867, 1960, viii + 296 pp, $17.50.

Thermochemistry for Steelmaking, Vol. II: Thermodynamics and Transport Properties, J. F. Elliott, M. Gleiser, and V. Ramakrishna, Addison-Wesley Publishing Co., Reading, Mass. 01867, 1963, xvi + 550 pp (pp 297–846), $25.00.

Thermodynamic Functions of Gases, F. Din, ed, Butterworth & Co. (Publishers) Ltd., 88 Kingsway, London, W.C.2, England, $12.50 per volume.

Vol. 1: Ammonia, Carbon Dioxide, Carbon Monoxide, 1956, viii + 175 pp.

Vol. 2: Air, Acetylene, Ethylene, Propane and Argon, 1956, vi + 201 pp.

Vol. 3: Methane, Nitrogen, Ethane, 1961, vi + 218 pp.

THERMOPHYSICAL PROJECTS

A Compendium of the Properties of Materials at Low Temperatures, V. J. Johnson, ed. Available from Clearinghouse for Federal Scientific and Technical Information, Springfield, Virg. 22151.

Phase I, Part I, Properties of Fluids, July 1960, 489 pp, WADD Technical Report 60-56, Part I (PB-171-618), $6.00.

Phase I, Part II, Properties of Solids, Oct. 1960, 330 pp, WADD Technical Report 60-56, Part II (PB-171-619), $4.00.

Phase I, Part III, Bibliography of References (cross-indexed), Oct. 1960, 161 pp, WADD Technical Report 60-56, Part III (PB-171-620), $3.00.

Phase II, R. B. Stewart and V. J. Johnson, eds, Dec. 1961, 501 pp, WADD Technical Report 60-56, Part IV (AD-272-769), $8.10.

Thermophysical Properties Research Center Data Book. Issued as loose-leaf sheets; sold on subscription basis at $0.10 per sheet, Thermophysical Properties Research Center, Purdue University, Research Park, 2595 Yeager Rd., West Lafayette, Ind. 47906.

Vol. I: Metallic Elements and Their Alloys.

Vol. II: Nonmetallic Elements, Compounds and Mixtures (In Liquid and Gaseous States at Normal Temperature and Pressure).

Vol. III: Nonmetallic Elements, Compounds and Mixtures (In the Solid State at Normal Temperature and Pressure).

Handbook of Thermophysical Properties of Solid Materials, A. Goldsmith, T. E. Waterman, and H. J. Hirschorn, revised edition, 1961 (8½″ × 11¼″), The Macmillan Co., 60 Fifth Ave., New York, N.Y. 10011, $90.00 per set.

Vol. I: Elements, vi + 752 pp.

Vol. II: Alloys, vi + 1,270 pp.

Vol. III: Ceramics, vi + 1,162 pp.

Vol. IV: Cermets, Intermetallics, Polymerics and Composites, vi + 798 pp.

Vol. V: Appendix (includes materials, author indexes, and a list of references), 286 pp.

PHYSICOCHEMICAL PROJECTS

Tables of Chemical Kinetics, Homogeneous Reactions, NBS Circular 510, U.S. Government Printing Office, Washington, D.C. 20402, 1951, xxiv + 732 pp (out of print).

Tables of Chemical Kinetics, Homogeneous Reactions, NBS Circular 510, Supplement 1, U.S. Government Printing Office, Washington, D.C. 20402, 1956, xiv + 422 pp (out of print).

Alphabetical Index to Tables of Chemical Kinetics, Homogeneous Reactions, NBS Circular 510, Supplement 2, U.S. Government Printing Office, Washington, D.C. 20402, 1960, iv + 37 pp, $0.35.

Tables of Chemical Kinetics, Homogeneous Reactions, Supplementary Tables, To Accompany Circular 510 and Supplements 1 and 2, NBS Monograph 34, U.S. Government Printing Office, Washington, D.C. 20402, 1961, $2.75.

Table A-8. SOURCES OF CRITICALLY EVALUATED DATA *(Continued)*

PHYSICOCHEMICAL PROJECTS (Continued)

Tables of Chemical Kinetics, Homogeneous Reactions, Supplementary Tables, NBS Monograph 34, Vol. 2, U.S. Government Printing Office, Washington, D.C. 20402, 1964, $2.00.

Phase Diagrams for Ceramists, E. M. Levin, H. F. McMurdie, and F. P. Hall, The American Ceramic Society, 4055 N. High St., Columbus, Ohio 43214, 1956, 286 pp (811 phase diagrams).

Phase Diagrams for Ceramists, Part II, E. M. Levin and H. F. McMurdie, The American Ceramic Society, 4055 N. High St., Columbus, Ohio 43214, 1959, 153 pp (462 phase diagrams).

Phase Diagrams for Ceramists, E. M. Levin, C. R. Robbins, and H. F. McMurdie (7th compilation), The American Ceramic Society, 4055 N. High St., Columbus, Ohio 43214, 1964, 601 pp (2064 phase diagrams), $18.00 (discount to members).

Phase Equilibrium Diagrams of Oxide Systems, revised and redrawn by E. F. Osborn and A. Muan, The American Ceramic Society, 4055 N. High St., Columbus, Ohio 43214, (Ten 19″ × 23″ plates for three-oxide systems containing SiO_2, four of them for oxide phases in equilibrium with metallic iron. Reproductions of these plates appear in the 1964 compilation).

Constitution of Binary Alloys, M. Hansen and K. Anderko, McGraw-Hill Book Company, New York, N.Y. 10036, 1958, xix + 1305 pp, $39.50.

Constitution of Binary Alloys, First Supplement, R. P. Elliott, McGraw-Hill Book Company, New York, N.Y. 10036, 1065, πιιπii | 877 pp, $35.00.

Stability Constants of Metal-Ion Complexes, Section I: Inorganic Ligands, compiled by Lars Gunnar Sillen; Section II: Organic Ligands, compiled by Arthur E. Martell; Special Publication No. 17, The Chemical Society, London, 1964, xviii + 754 pp, $23.00.

Solubilities of Inorganic and Metal Organic Compounds, A compilation of Solubility Data from the Periodical Literature, A. Seidell, 4th ed., W. F. Linke, American Chemical Society, 1155 16th St. N.W., Washington, D.C. 20036.

    Vol. I, 1958, iv + 1287 pp, $32.50.
    Vol. II, 1965, iv + 1914 pp, $32.50.

INDEXES

Consolidated Index of Selected Property Values: Physical Chemistry and Thermodynamics, Prepared by the Office of Critical Tables. A key to the contents of six compilations that present critically evaluated numerical property values. Publication 976, National Academy of Sciences—National Research Council Printing and Publishing Office, 2101 Constitution Ave., N.W., Washington, D.C. 20418, 1962, xxiii + 274 pp, $6.00.

Index

A

E

M

Q

properties
  miscellaneous, 120, 121, 329, 331
  superconductive, 231

S

U

V

Z